Engineering Electromagnetics

Nathan Ida

Engineering Electromagnetics

Fourth Edition

Nathan Ida
Department of Electrical Engineering
University of Akron
Akron, OH, USA

Additional material to this book can be downloaded from https://www.springer.com/us/book/9783030155568.

ISBN 978-3-030-15559-9 ISBN 978-3-030-15557-5 (eBook)
https://doi.org/10.1007/978-3-030-15557-5

This Springer imprint is published by the registered company Springer Nature Switzerland AG
The registered company address is: Gewerbestrasse 11, 6330 Cham, Switzerland

This book is lovingly dedicated to Vera, my wife and partner in life.

Preface

You can because you ought.

—Imanuel Kant

One of the main difficulties in teaching electromagnetic fields is the perception on the part of many students that electromagnetics is essentially a supportive topic. They are told that they need to study electromagnetics early in the curriculum because they will need it later to understand other topics in the electrical engineering curriculum, such as electric machines, microwaves, or communication. This, with the prevailing perception of the topic being difficult, esoteric, or boring, creates a heavy atmosphere around the subject. More often than not, this leads to self-fulfilling prophecies, and as a result, even those students who perform well do not get the full benefit of the experience such an exciting topic can impart. This is particularly sad, because electromagnetics motivates many students to pursue studies in electrical engineering. They are familiar with electromagnetic waves, electric motors, magnetic recording, radar, antennas, and many others and have been exposed to hundreds of electromagnetic devices. Yet few make the connection between these and the electromagnetics they are taught.

The answer is to study electromagnetics for what it is rather than in preparation for something that will happen in the future. The study of electromagnetic fields is not more difficult than any other topics in the electrical engineering curriculum and, in many ways, is more interesting and more applied. The number of applications is so vast that any attempt to summarize will easily fill a good-sized book. One can only guess the total share of electromagnetics to the industrial output. Huge turbo generators for generation of electricity, power transmission lines, electric motors, actuators, relays, radio, TV and microwave transmission and reception, radar, and even the mundane little magnet used to hold a paper note on the refrigerator are all electromagnetic in nature. It is indeed uncommon to find a device that works without relying on any electromagnetic principle or effect. One only has to ask oneself who is going to design these systems and what are the tools necessary to do so, and the answer to why one should study electromagnetics becomes self-evident.

This text attempts to present electromagnetics as a topic in itself with specific objectives and specific applications. The fact that it is used as a prerequisite for other subjects is merely a consequence that those other topics are based on electromagnetics. A good theoretical understanding of the electromagnetic field equations is required for electromagnetic design. The text fulfills this need by a rigorous treatment of the theoretical aspects of electromagnetics. In addition, it treats a large number of applications that the student will find interesting and useful.

The text assumes the student has the necessary background in calculus. Other required topics, including vector algebra and vector calculus, are fully covered in the text. In addition, all mathematical relations (such as integrals, derivatives, series, and others) are listed as needed in the text. In this sense, the book is fully self-contained. An effort has been made to use only quantities that have been defined previously, even if this requires, for example, change of units in mid-chapter. There will be a few exceptions to this rule, and when this happens, the reasons for doing so are also given. The reasons for this purist approach are many, but the most important is the fact that the book assumes no prior knowledge of any field quantity.

In style, the text relies on simple physical explanations, in plain language and based on known phenomena, to simplify understanding. There are many detailed examples, exercising every significant relation and topic in the book. Many of the examples rely on important applications, with particular emphasis on sensing and actuation, and contain complete step-by-step solutions and derivations as necessary. There is almost no use of acronyms. These are only used when an acronym is better known than what it represents, such as TV and FM. The presentation often relies on repetition of relations and explanations. This serves to reinforce understanding and avoids convoluted referencing to equations and text. In most cases, referencing is only done for completeness purposes, and the required equation is repeated when needed. Important or often-used relations are boxed and are always accompanied by their associated units. The notation used in the book is standard and should be familiar to students from physics and mathematics. The most important change in this respect is the use of unit vectors. Unit vectors always precede the scalar component. For example, $\mathbf{A} = \hat{\mathrm{x}} A_x + \hat{\mathrm{y}} A_y + \hat{\mathrm{z}} A_z$ is a vector with scalar components A_x in the x direction, A_y in the y direction, and A_z in the z direction. $\hat{\mathrm{x}}$, $\hat{\mathrm{y}}$, and $\hat{\mathrm{z}}$ are the corresponding unit vectors.

The structure of the book is unique in another way; most topics are discussed in two or three separate chapters. The first chapter introduces the subject and discusses the basic quantities and relations. The second chapter complements and expands on the first and introduces additional topics related to the main subject including applications. In certain cases, a third chapter discusses additional topics or a new topic related to the first two. For example, **Chapter 3** introduces the electric field and the basic relations governing it; **Chapter 4** starts with the postulates governing the electric field and continues with Gauss's law, effects of and on materials, capacitance, and other quantities associated with the electric field such as energy; **Chapter 5** then continues with analytical methods of solution of electrostatic problems. This pairing includes **Chapters 1** and **2** (vector algebra followed by vector calculus); **Chapters 3–5** (electric field, electric potential, and boundary value problems); **Chapters 8** and **9** (the static magnetic field and magnetic materials and properties); **Chapters 12** and **13** (electromagnetic waves and propagation and reflection and transmission of plane waves); and **Chapters 14–16** (theory of transmission lines, the Smith chart and transmission line circuits, and transients on transmission lines). The purpose of this grouping of chapters is twofold. First, it divides the material into more coherent, easier-to-follow, shorter units. Second, it provides intermediate breaking points at which both students and instructors can assess the situation and decide on the next steps. It also allows selection of topics without the need to skip sections within chapters. For example, whereas a chapter on time-dependent fields normally includes all material associated with Faraday's law, Maxwell's equations, and wave propagation, I have chosen to divide this material into three chapters. The first is on Faraday's law and includes all phenomena associated with induction (**Chapter 10**). The second discusses Maxwell's equations with associated topics, including the continuity equation and interface conditions (**Chapter 11**). The third discusses wave propagation as a consequence of displacement currents (**Chapter 12**). The three chapters discuss different aspects, using various approaches.

Chapters 1 and **2** discuss vector algebra and vector calculus and are rather different from the rest of the book in that the student will find no reference to electromagnetics in these chapters. This serves two purposes. First, it indicates that at this stage the student has little formal knowledge of electromagnetic field quantities but, paradoxically, he or she is aware of the properties of electromagnetic fields through knowledge acquired in other areas of physics or everyday experience. Second, it shows that the same methods and the same mathematical tools are used in other disciplines and for other applications. This approach should alleviate some of the anxiety associated with the study of electromagnetics while still acquiring all vector algebra and calculus tools needed for the study of electromagnetics. More importantly, the approach lends itself to self-study. If the student or the instructor feels that **Chapters 1** and **2** are not necessary, they may be skipped without affecting the subsequent topics.

The method of presentation of the material distinguishes between basic field relations and mathematical tools. The latter are introduced in **Chapters 1** and **2**, but wherever they are needed, they are repeated to reinforce understanding of the tools and to avoid having to refer back to **Chapter 1** or **2**. Similarly, other relations, like trigonometric functions, derivatives, and integrals, are given as needed, and as close as possible to where they are used. This should help students in reviewing material they learned previously, but do not recall or are not certain of.

A summary is provided at the end of each chapter except **Chapters 1** and **2** that are themselves summaries of vector algebra and vector calculus, respectively. The purposes of these summaries are to collect the important relations in each chapter, to reemphasize these, and to serve as a means of, first, reviewing the material and, second, referencing back to the material in the chapter. It is hoped that by doing so, the student will spend less time searching topics especially when it becomes necessary to refer back to material in a chapter that was studied earlier. For this reason, the equations listed in the summary retain the original equation numbers when appropriate. On the other hand, tables and figures are not included in the summaries but, rather, referred to when necessary. These summaries, coming as they are at the end of a chapter, also provide a retrospective view of the material, something that may help solidify understanding. However, summaries cannot replace the chapter. They are short and lack most explanations, and only the most common form of the relations is usually given. The relations in the chapters have alternative forms, some of which are useful in specific situations. The student should view these summaries simply as reminders and as an indication of the content and importance of subjects.

Each chapter includes an extensive set of problems. The problems are of two types. Some are exercises, used to ensure that the student has a chance to review the field relations and to use them in the way they were intended to be used, while the other is more involved and often based on a physical application or, in some cases, on a simplified version of a physical structure. These problems are designed to present some of the many applications in electromagnetics, in addition to their value as exercise problems. It is hoped that this will bring the student closer to the idea of design than exercise problems can.

Most chapters contain a section on applications intended to expand on material in the chapter and to expose the student to some of the myriad applications in electromagnetics or, in some cases, to physical phenomena that depend on electromagnetism. Naturally, only a very small selection of applications is given. The description is short but complete.

This textbook was written with a two-semester sequence of courses in mind but can be used equally well for a one-semester course. In a two-semester sequence, the topics in **Chapters 3–10** are expected to be covered in the first semester. If necessary, **Chapters 1** and **2** may be reviewed at the beginning of the semester, or if the students' background in vector algebra and calculus is sufficiently strong, these two chapters may be skipped or assigned for self-study. **Chapter 6** is self-standing and, depending on the instructor's preference, may or may not be covered. The second semester should ideally cover **Chapters 11–18** or at least **Chapters 11–16**. **Chapters 17** and **18** are rather extensive discussions on waveguides and antennas, respectively. They introduce new dimensions in study, building on concepts first discussed in **Chapters 11–13**, as well as new applications. These may form the basis of more advanced elective courses on these subjects.

In a one-semester course, there are two approaches that may be followed. In the first approach, **Chapters 3–5** and **7–12** are covered. This should give students a solid basis in electromagnetic fields and a short introduction to electromagnetic waves. In the second approach, selected topics from **Chapters 3–5** to **7–16** are included. It is also possible to define a program that emphasizes wave propagation by utilizing **Chapters 11–18** and excluding all topics in static electric and magnetic fields.

Chapter 6 deals with numerical methods of solution for boundary value problems in electrostatics. Although the classical methods of separation of variables and the method of images, are discussed in **Chapter 5**, they are difficult to apply with any degree of generality. The introduction of numerical methods at this stage is intended to reassure students that solutions indeed exist and that the numerical methods needed to do so are not necessarily complicated. Some methods can be introduced very early in the course of study. Finite differences and the method of moments are of this type. Finite element methods are equally simple, at least at their basic levels. These methods are introduced in **Chapter 6** and are applied to simple yet useful electrostatic configurations.

The history associated with electromagnetics is long and rich. Many of the people involved in its development had unique personalities. While information on history of science is not in itself necessary for understanding the material, I feel it has a value in its own right. It creates a more intimate association with the subject and often places things in perspective. A student can appreciate the fact that the great people in electromagnetics had to struggle with the concepts the students try to understand or that Maxwell's equations, the way we know them today, were not written by Maxwell but by Heaviside, almost 20 years after Maxwell's death. Or perhaps it is of some interest to realize that Lord Kelvin did not believe Maxwell's theory, well after it was proven experimentally by Hertz. Many will enjoy the eccentric characters of Heaviside and Tesla, or the unlikely background of Coulomb. Still others were involved in activities that had nothing to do with the sciences. Benjamin Franklin was what we might call a special envoy to England and France, and Gilbert was personal physician to Queen Victoria. All these people contributed in their own way to the development of the theory of fields, and their story is the story of electromagnetics. Historical notes are given throughout the book, primarily as footnotes.

To aid in understanding, and to facilitate some of the more complex calculations, a number of computer programs (written in MATLAB) are available for download. These are of four types. The first type is demonstrative—programs in this group are intended to display a concept such as vector addition or reflection of electromagnetic waves. Wherever appropriate in the text, the relevant program is indicated, and the student, as well as the instructor, is welcome to use these to emphasize the various concepts. An explanation file is available to explain the various inputs and outputs and the use of the programs. The second type includes simple programs used to solve a particular example or to compute values in end-of-chapter problems. Some of the programs are rather specific, but others are more general. The third type of programs includes auxiliary charts and computational tools including a full implementation of the Smith chart (**Chapter 15**). The fourth type of programs relates to **Chapter 6** exclusively. They address the use of the finite difference method, the finite element method, and the method of moments and allow the student first to duplicate the results in the various examples and then to apply these numerical techniques to other calculations including those in the problems section.

The programs are written as scripts and are mostly interactive. Needless to say, the student is more than welcome to modify these programs for use in other examples, problems, etc. To get the full benefit of these tools, the student should download the files available and read carefully the explanation files before using the various programs. All programs are available for download at https://www.springer.com/us/book/9783030155568

Also available for download are two files that may be of interest. The first describes a number of simple experiments the student or the instructor can perform to emphasize and visualize certain topics. The experiments are listed by chapter and most take very little time or effort to assemble and demonstrate. These experiments are designed to be short and simple, and to require a minimum of materials and equipment. They are qualitative experiments: no measurements are taken, and no exact relations are expected to be satisfied. The instructor may choose to use these as an introduction to a particular topic or as a means to stimulate interest. The student may view the experiments as demonstrations of possible applications. Many of the experiments can be repeated by students if they wish to do so. However, none of the experiments require laboratory facilities. The main purpose is to take electromagnetic fields off the pedestal and down to earth. I found these simple experiments particularly useful as a way of introducing a new subject. It wakes the students up, gets them to ask questions, and creates anticipation toward the subject. The simplicity of the principles involved intrigues them, and they are more inclined to look at the mathematics involved as a means rather than a goal.

The second file contains a set of review questions for each chapter. These may be used by the student or instructor to review the various subjects in each chapter.

Finally, I wish to thank those who were associated with the writing of this text in its various editions, in particular Frank Lewis, Dana Adkins, Shi Ming, and Paul Stager (all undergraduate and graduate students) who have solved some of the examples and end-of-chapter problems and provided valuable input into the writing of the first edition of this text; Professor Joao Pedro Assumpção Bastos (Federal University of Santa Catarina, Florianopolis, Brazil), who contributed a number of examples and problems; Dr. Charles Borges de Lima (Federal University of Santa Catarina, Florianopolis, Brazil), who wrote the demonstration software as well as the Smith chart program while a postdoctoral student at The University of Akron; and Prof. Guido Bassotti (National University of San Juan, Argentina), who brought to my attention a number of errors and needed modifications to the second edition. I particularly wish to thank Prof. Richard E. Denton from Dartmouth College for providing a thorough critique, raising questions as to content and method, and suggesting corrections and modifications.

I genuinely hope that the present edition improves the text both in contents and in presentation.

Akron, OH, USA
November 2018

Nathan Ida

Introduction to Electromagnetics

A Simple View of Electromagnetics

Charge is the fundamental electric quantity in nature, the same charge that will cause a spark when shuffling your shoes on a carpet and then touching a doorknob or the charge that causes your clothes to stick together. The effects that constitute electromagnetics are directly linked to charge and to the behavior of charge. Charge exists in two forms. One is the charge of the electron and is negative. The second is that of the proton and is positive (and equal in magnitude to that of the electron). Experiment has shown that like charges repel each other, whereas opposite charges attract. Electrons and protons occur in atoms, usually in pairs, and thus materials are usually charge neutral. When a material acquires excess electrons, it becomes negatively charged. Excess protons (deficiency of electrons) cause the material to be positively charged. Thus, when shuffling your shoes on a carpet, electrons are stripped off the carpet causing your body to have an excess of electrons (to become negatively charged), whereas the carpet becomes positively charged. Charges can be stationary or can move. Since charges can exert forces and can be affected by them, there is also an energy associated with charge and with its interactions. The charge itself and the way it moves define the electric, magnetic, or electromagnetic phenomena we observe. There are three possibilities that we consider which correspond to the range of phenomena that constitute electromagnetics.

(a) **Stationary charges**. A stationary charge produces an electric field intensity around itself. We shall see shortly that the electric field intensity is in fact force per unit charge and a field implies a distribution of this force in space and in materials. In a way, we can say that the charge produces a force field around itself. This gives rise to an electric energy that can be useful. Examples of this energy and the forces associated with it are the forces that cause the toner to stick to a page in a copier or dust to accumulate on display monitors, lightning, sparks in an internal combustion engine, a stunt gun, a camera flash, and many others. Since the charge is static, the phenomenon is called electrostatics (sometimes called static electricity). The electrostatic field is useful in a variety of applications including printing, copying, pollution control in power plants, painting, production of sand paper, and many others that we will discuss in **Chapters 3–5**.

(b) **Charges moving at a constant velocity**. Electrons and protons moving at constant velocity cause three effects:

1. An electric field intensity just as in **(a)**. The electric field moves with the electron.
2. An electric current due to this motion since any moving charge constitutes an electric current.
3. A magnetic field intensity, produced by the current.

We say that a moving electron produces a static magnetic field since that field is constant in time as long as currents are constant (due to the constant velocity of charges). Since both an electric and a magnetic field exist, we can also say that moving charges generate an electromagnetic field. The relative strengths of the electric and magnetic fields dictate which of the effects dominates and hence the applications associated with moving charges. The magnetic field acts on currents with a magnetic force that can be very high. Again, we may say that a current generates a force field around itself. Many of the applications commonly in use belong here. These include lighting, electric machines, heating, power distribution, etc. These are mostly DC applications but also low-frequency AC applications including power generation and distribution systems. These and other applications will be disused in **Chapters 7–9**.

(c) **Charges moving with a time-dependent velocity.** Accelerating charges produces all the effects in **(b)** plus radiation of energy. Electrons moving with a time-dependent velocity necessarily generate time-dependent effects. Energy and hence power radiate away from the time-dependent current. One can say that the electromagnetic field now propagates in space and in materials, at a specific velocity, carrying the energy associated with the electromagnetic field in a way that can best be described as a wave. The most common manifestation of this form is transmission and reception of signals to affect communication but also in applications such as microwave heating or X-ray imaging. The relations, phenomena, and applications resulting from this unique form of the electromagnetic field will be discussed in detail in **Chapters 11–18**.

Naturally, the transition between charges moving at constant velocity and charges that accelerate and decelerate is gradual and linked to alternating currents. That is, an AC current necessarily means charges move at a nonconstant velocity and, hence, any AC current will produce radiation. But, depending on the rate of acceleration of charges, the radiation effect may be dominant or may be negligible. As a rule, low-frequency currents radiate very little power and behave more like DC currents. Because of that, the domain of low-frequency currents, and in particular at power frequencies, is often called ***quasi-static***, and the fields are said to be ***quasi-static fields***, meaning that the electric and magnetic fields behave more like those of charges moving at constant velocity than those due to accelerating charges. This is fortunate since it allows the use of DC methods of analysis, such as circuit theory, to extend to low-frequency AC applications. This aspect of electromagnetic fields will be introduced in **Chapter 10** and, to an extent, in **Chapter 11** and serves as a transition to electromagnetic waves. At higher frequencies, the radiation effects dominate, the observed behavior is totally different, and the tools necessary for analysis must also change.

It is important to emphasize that the electromagnetic effects have been "discovered" experimentally and their proof is based entirely on experimental observation. All laws of electromagnetics were obtained by careful measurements, which were then cast in the forms of mathematical relations. In the learning process, we will make considerable use of the mathematical tools outlined in **Chapters 1** and **2**. It is easy to forget that the end purpose is physical design; however, every relation and every equation imply some physical quantity or property of the fields involved. It is very important to remember that however involved the mathematics may seem, electromagnetics deals with practical physical phenomena, and when studying electromagnetics, we study the effects and implications of quantities that can be measured and, more importantly, that can and are being put to practical use. There are two reasons why it is important to emphasize electromagnetics as an applied science. First, it shows that it is a useful science, and its study leads to understanding of nature and, perhaps most significantly from the engineering point of view, to understanding of the application of electromagnetics to practical and useful designs.

Units

The system of units adopted throughout this book is the Système Internationale (SI). The SI units are defined by the International Committee for Weights and Measures and include seven base units as shown in **Table 1**. The base units are as follows:

Table 1 The base SI units

Physical quantity	Unit	Symbol
Length	meter	m
Mass	kilogram	kg
Time	second	s
Electric current	ampere	a
Temperature	kelvin	K
Luminous intensity	candela	cd
Amount of substance	mole	mol

Length The **meter** (m) is the distance traveled by light in a vacuum during a time interval equal to 1/299,792,458 s.

Mass The **kilogram** (kg) is the prototype kilogram, a body made of a platinum-iridium compound and preserved in a vault in Sèvres, France.

Time The **second** (s) is the duration of 9,192,631,770 periods of the radiation corresponding to the transition between the two hyperfine levels of the ground state of the cesium-133 atom.

Electric current The **ampere** (A) is the constant current that, if maintained in two straight conductors of infinite length and of negligible circular cross section, placed 1 m apart in a vacuum, produces between the conductors a force of 2×10^{-7} newton per meter (N/m).

Temperature The **kelvin** (K) unit of thermodynamic temperature is 1/273.16 of the thermodynamic temperature of the triple point of water (the temperature and pressure at which ice, water, and water vapor are in thermodynamic equilibrium). The triple point of water is 273.16 K at a vapor pressure of 611.73 pascal (Pa) [1 Pa = 1 N/m^2].

Luminous intensity The **candela** (cd) is the luminous intensity in a given direction of a source that emits monochromatic radiation of frequency 540×10^{12} Hz and has a radiation intensity in that direction of 1/683 watts per steradian (W/sr) (see the definition of steradian below).

Amount of substance The **mole** (mol) is the amount of substance of a system that contains as many elementary entities as there are atoms in 0.012 kg of carbon-12. (The entities may be atoms, molecules, ions, electrons, or any other particles.) The accepted number of entities (i.e., molecules) is known as Avogadro's number and equals approximately 6.0221×10^{23}.

Derived Units

All other metric units in common use are derived from the base SI units. These have been defined for convenience based on some physical law, even though they can be expressed directly in the base units. For example, the unit of force is the newton [N]. This is derived from Newton's law of force as $F = ma$. The unit of mass is the kilogram, and the unit of acceleration is meters per second squared (m/s^2). Thus, the newton is in fact kilogram meters per second squared (kg$\cdot$m/s^2). Many of the common electrical units (such as the volt [V], the watt [W], and the ohm [Ω]) are derived units sanctioned by the SI standard. There are however some commonly used units that are discouraged (good examples are the watt-hour [W$\cdot$h] and the electron-volt [e$\cdot$V or eV]). We will make considerable use of derived units throughout the book, and these will be introduced and discussed as they become needed in the discussion.

Supplementary Units

The system of units includes the so-called derived nondimensional units, also termed "supplementary units." These are the unit for the plane angle, the radian [rad], and the unit for the solid angle, the steradian [sr]. The radian is defined as the planar angle at the center of a circle of radius R subtended by an arc of length R. The steradian is defined as the solid angle at the center of a sphere of radius R subtended by a section of its surface, whose area equals R^2.

Customary Units

In addition to the base SI units and derived units sanctioned by the International Committee for Weights and Measures, there are many other units, some current, some obsolete, and some nonmetric. These are usually referred to as "customary units." They include commonly used units such as the calorie [cal] or the kilowatt-hour [kW. h] and less common units (except in the United States) such as the foot, mile, gallon, psi (pounds per square inch), and many others. Some units are associated almost exclusively with particular disciplines. These units may be SI, metric (current or obsolete), or customary. These have been defined for convenience, and, as with any other unit, they represent a basic quantity that is meaningful in that discipline. For example, in astronomy, one finds the astronomical unit [AU], which is equal to the average distance between the earth and the sun (1 AU = 149,597,870.7 km). In physics, the angstrom [Å] represents atomic dimensions (1 Å= 0.1 nm). Similarly utilitarian units are the electron volt [e•V] for energy (1 e•V = 1.602×10^{-19} joule [J]), the atmosphere [atm] for pressure (1 atm = 101,325 pascal [Pa], 1 Pa = 1 N/m^2), ppm (parts per million) for chemical quantities, the sievert [sv] (1 sv = 1 J/kg) for dose equivalents in radiation exposure, and so on. Although we will stick almost exclusively to SI units, it is important to remember that should the need arise to use customary units, conversion values to and from SI units can be substituted as necessary.

Prefixes

In conjunction with units, the SI system also defines the proper prefixes that provide standard notation of very small or very large quantities. The prefixes allow one to express large and small numbers in a compact and universal fashion and are summarized in **Table 2**. Again, this is mostly a convenience, but since their use is common, it is important to use the proper notation

Table 2 Common prefixes used in conjunction with the SI system of units

Prefix	Symbol	Multiplier	Examples
atto	a	10^{-18}	as (attosecond)
femto	f	10^{-15}	fs (femtosecond)
pico	p	10^{-12}	pF (picofarad)
nano	n	10^{-9}	nH (nanohenry)
micro	μ	10^{-6}	μm (micrometer)
milli	m	10^{-3}	mm (millimeter)
centi	c	10^{-2}	cL (centiliter)
deci	d	10^{-1}	dm (decimeter)
deca	da	10^{1}	dag (decagram)
hecto	h	10^{2}	hL (hectoliter)
kilo	k	10^{3}	kg (kilogram)
mega	M	10^{6}	MHz (megahertz)
giga	G	10^{9}	GW (gigawatt)
tera	T	10^{12}	Tb (terabit)
peta	P	10^{15}	PHz (petahertz)
exa	E	10^{18}	EHz (exahertz)

to avoid mistakes and confusion. Some of the prefixes are commonly used, others are rare, and still others are used in specialized areas. Prefixes such as atto, femto, peta, and exa are rarely used, whereas prefixes such as deca, deci, and hecto are more commonly used with liquids (but see below the usage of deci in decibel).

Other Units and Measures

Units of Information

There are a few other measures that are in common use in designating specific quantities. Since digital systems use base 2, base 8, or base 16 counting and mathematics, the decimal system is not particularly convenient as a measure. Therefore, special prefixes have been devised for digital systems. The basic unit of information is the bit (a 0 or a 1). Bits are grouped into bytes, where 1 byte contains 8 bits, sometimes also called a "word." A kilobyte (kbyte or kb) is $2^{10} = 1024$ bytes or 8192 bits. Similarly a megabyte [Mb] is 2^{20} (or 1024^2) bytes or 1,048,576 bytes (or 8,388,608 bits). Although these prefixes are confusing enough, their common usage is even more confusing, as it is common to mix digital and decimal prefixes. As an example, it is common to rate a storage device or memory board as containing, say, 100 Gb. The digital prefix should mean that the device contains 2^{30} or 1024^3 bytes or approximately 107.4×10^9 bytes. Rather, the device contains 100×10^9 bytes. In digital notation, the device actually contains only 91.13 Gb.

The Decibel (dB) and Its Use

There are instances in which the use of the common prefixes is inconvenient at the very least. In particular, when a physical quantity spans a very large range of numbers, it is difficult to properly grasp the magnitude of the quantity. Often, too, a quantity only has meaning with respect to a reference value. Take, for example, a voltage amplifier. It may be a unity amplifier or may amplify by a factor of 10^6 or more with a reference at 1. Another example is the human eye, which can see in luminance (luminous intensity per unit area, measured in units of candela per unit area—cd/m^2) from about 10^{-6} to 10^6 cd/m^2. This is a vast range, and the natural reference value is the lowest luminance the eye can detect.

The use of normal scientific notation for such vast scales is inconvenient and is not particularly telling for a number of reasons. Using again the example of our eyes' response to light, it is not linear, but rather logarithmic. That is, for an object to appear twice as bright, the illuminance (luminous flux per unit area, measured in units of candela times solid angle per unit area) needs to be about ten times higher. The same applies to sound and to many other quantities. In such instances, the quantities in question are described as ratios on a logarithmic scale using the notation of decibel (dB). The basic ideas in the use of the decibel are as follows:

1. Given a quantity, divide it by the reference value for that quantity. That may be a "natural" value, such as the threshold of vision, or it may be a constant, agreed-upon value such as 1 or 10^{-6}.
2. Take the base 10 logarithm of the ratio.
3. If the quantities involved are power-related (power, power density, energy, etc.), multiply by 10:

$$p = 10 \log_{10}\left(\frac{P}{P_0}\right) \quad [\text{dB}]$$

4. If the quantities involved are field quantities (voltage, current, force, pressure, etc.), multiply by 20:

$$v = 20 \log_{10}\left(\frac{V}{V_0}\right) \quad [\mathrm{dB}]$$

In the amplifier described above, a voltage amplification of 10^6 corresponds to $20 \log_{10}(10^6/1) = 120$ dB. In the case of vision, the reference value is 10^{-6} cd/m^2. A luminance of 10^{-6} cd/m^2 is therefore 0 dB. A luminance of 10^3 cd/m^2 is $10 \log_{10}(10^3/10^{-6}) = 90$ dB. One can say that the human eye has a span of 120 dB.

When dealing with quantities of a specific range, the reference value can be selected to accommodate that range. For example, if one wishes to describe quantities that are typically in milliwatts [mW], the reference value is taken as 1 mW and power values are indicated in decibel milliwatts (dBm). Similarly, if one needs to deal with voltages in the microvolt [μV] range, the reference value is taken as 1 μV and the result is given in decibel microvolts [dBμV]. The use of a specific reference value simply places the 0 dB point at that value. As an example on the dBm scale, 0 dBm means 1 mW. On the normal scale, 0 dB means 1 W. It is therefore extremely important to indicate the scale used, or confusion may occur. There are many different scales, each clearly denoted to make sure the reference value is known.

As a final note, it should be remembered that unit analysis can facilitate understanding of the material and prevent errors in computation. It is therefore imperative that all quantities be indicated with their appropriate units.

Contents

1 Vector Algebra

The vector analysis I use may be described either as a convenient and systematic abbreviation of Cartesian analysis... In this form it is not more difficult, but easier to work than Cartesians. Of course, you have to learn it. Initially, unfamiliarity may make it difficult...

—Oliver Heaviside (1850–1925), a self-taught mathematician and electrical engineer in his introduction to Vector Analysis, originally published in 1893 (Electromagnetic Theory, Chelsea Publishing Co., N.Y., 1971, Vol. 1, p. 135).

1.1 Introduction

Vector algebra[1] is the algebra of vectors: a set of mathematical rules that allows meaningful and useful operations in the study of electromagnetics. We will define vectors and the necessary operations shortly, but, for now, it is useful to remember the following axiom which will be followed throughout this book: Nothing will be defined, no quantity or operation will be used, unless it has some utility either in explaining the observed physical quantities or otherwise simplifies the discussion of a topic. This is important because, as we increase our understanding of the subject, topics may seem to be disconnected, particularly in this and the following chapter. The discussion of vector algebra and vector calculus will be developed separately from the ideas of the electromagnetic field but for the purpose of describing the electromagnetic field. It is also implicit in this statement that by doing so, we should be able to simplify the discussion of electromagnetics and, necessarily, better understand the physical properties of fields.

Vector algebra is a set of rules that apply to vector quantities. In this sense, it is similar to the algebra we are all familiar with (which we may call scalar algebra): it has rules, the rules are defined and then followed, and the rules are self-consistent.

Because at this point we know little about electromagnetics, the examples given here will be taken from other areas: mechanics, elementary physics, and, in particular, from everyday experience. Any reference to electric or magnetic quantities will be in terms of circuit theory or generally known quantities. The principle is not to introduce quantities and relations that we do not fully understand. It sometimes comes as a surprise to find that many of the quantities involved in electromagnetics are familiar, even though we may have never thought of them in this sense. All that the rules of vector algebra do is to formalize these rather loose bits of information and define their interactions. At that point, we will be able to use them in a meaningful way to describe the behavior of fields in exact terms using a concise notation.

It is worth mentioning that vector algebra (and vector calculus, which will be discussed in the following chapter) contains a very small number of quantities and operations. For this reason, the vector notation is extremely compact. There are only two quantities required: ***scalars*** and ***vectors***. Four basic operations are required for vectors: ***addition***, ***vector scaling***, ***scalar product***, and ***vector product***.

[1] Vector analysis, of which vector algebra is a subset, was developed simultaneously and independently by Josiah Willard Gibbs (1839–1903) and Oliver Heaviside (1850–1924) around 1881, for the expressed purpose of describing electromagnetics. The notation used throughout is more or less that of Heaviside. Vector analysis did not gain immediate acceptance. It was considered to be "useless" by Lord Kelvin, and many others thought of it as "awfully difficult," as Heaviside himself mentions in his introduction to vector algebra. Nevertheless, by the end of the nineteenth century, it was in general use.

N. Ida, *Engineering Electromagnetics*, https://doi.org/10.1007/978-3-030-15557-5_1

In addition, we will define distributions of vectors and scalars in space as ***vector*** and ***scalar fields*** and will introduce the commonly used ***coordinate systems***. The discussion in this chapter starts with the definition of scalars and vectors in Cartesian coordinates. The latter is assumed to be known and is used exclusively in the first few sections, until cylindrical and spherical coordinates are defined.

1.2 Scalars and Vectors

A quantity is a ***scalar*** if it has only a magnitude at any location in space for a given time. To describe the mass of a body, all we need is the magnitude of its mass or, for a distributed mass, the distribution in space. The same applies to the altitude of a mountain or the length of a road. These are all scalar quantities and, in particular, are static scalar quantities (independent of time). In terms of quantities useful in the study of electromagnetics, we also encounter other scalars such as work, energy, time, temperature, and electric potential (voltage). Scalar sources also play an important role: The electric charge or charge distribution (for example, charge distributed in a cloud) will be seen as sources of fields. The source of a 1.5 V cell is its potential and is a scalar source.

A ***vector***, on the other hand, is described by two quantities: a magnitude and a direction in space at any point and for any given time. Therefore, vectors may be space and time dependent. Common vectors include displacement, velocity, force, and acceleration. To see that the vector definition is important, consider a weather report giving wind speeds. The speed itself is only part of the information. If you are sailing, direction of the wind is also important. For a pilot, it is extremely important to know if the wind also has a downward component (shear wind), which may affect the flight plan. Sometimes, only the magnitude may be important: The electric generating capability of a wind-driven turbine is directly proportional to the normal (perpendicular to the turbine blades) component of the wind. Other times we may only be interested in direction. For example, the news report may say: "The rocket took off straight up." Here, the direction is the important information, and although both direction and magnitude are available, for one reason or another, the liftoff speed or acceleration is not important in this statement. The unit associated with a quantity is not part of the vector notation.

The use of vectors in electromagnetics is based on two properties of the vector. One is its ability to describe both magnitude and direction. The second is its very compact form, which allows the description of quantities with great economy in notation. This economy in notation eases handling of otherwise awkward expressions but also requires familiarity with the implications of the notation. In a way, it is like shorthand. A compact notation is used, but it also requires us to know how to read it so that the information conveyed is meaningful and unambiguous.

To allow instant recognition of a vector quantity, we denote vectors by a boldface letter such as **E**, **H**, **a**, and **b**. Scalar quantities are denoted by regular letters: E, H, a, and b. In handwriting, it is difficult to make the distinction between normal and boldface lettering. A common method is to use a bar or arrow over the letter to indicate a vector. Thus, $\bar{E}, \bar{H}, \bar{a}, \bar{b}$, are also vectors. If a quantity is used only as a vector, there is no need to distinguish it from the corresponding scalar quantity. Some vector operators (which will be discussed in the following chapter) are of this type. In these instances, neither boldface nor bar notation is needed since there is no room for confusion.

1.2.1 Magnitude and Direction of Vectors: The Unit Vector and Components of a Vector

The magnitude of a vector is that scalar which is numerically equal to the length of the vector:

$$A = |\mathbf{A}| \tag{1.1}$$

The magnitude of a vector is its length and includes the units of the vector. Thus, for example, the magnitude of a velocity vector **v** is the speed v [m/s]. To define the direction of a vector **A**, we employ the idea of the unit vector. A unit vector $\hat{\mathbf{A}}$ is a vector of magnitude one (dimensionless) in the direction of **A**:

$$\boxed{\hat{\mathbf{A}} = \frac{\mathbf{A}}{|\mathbf{A}|} = \frac{\mathbf{A}}{A}} \tag{1.2}$$

Thus, $\hat{\mathbf{A}}$ can be viewed as a new dimensionless vector of unit magnitude ($|\hat{\mathbf{A}}| = 1$) in the direction of, or parallel to, the vector **A**. **Figure 1.1** shows a vector, its magnitude, and its unit vector.

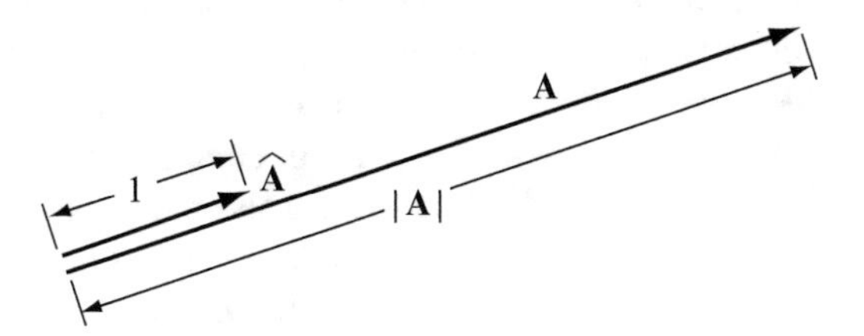

Figure 1.1 The relations between vector **A**, the unit vector $\hat{\mathbf{A}}$, and the magnitude of the vector $|\mathbf{A}|$

Vectors may have components in various directions. For example, a vehicle moving at a velocity **v** on a road that runs *SE* to *NW* has two equal velocity components, one in the *N* direction and one in the *W* direction, as shown in **Figure 1.2a**. We can write the velocity of the vehicle in terms of two velocity components as

$$\mathbf{v} = \hat{\mathbf{N}}v_N + \hat{\mathbf{W}}v_W = \hat{\mathbf{N}}v\frac{\sqrt{2}}{2} + \hat{\mathbf{W}}v\frac{\sqrt{2}}{2} \quad \left[\frac{\mathrm{m}}{\mathrm{s}}\right] \tag{1.3}$$

The two terms on the right-hand side ($\hat{\mathbf{N}}v\sqrt{2}/2$ and $\hat{\mathbf{W}}v\sqrt{2}/2$) are called the ***vector components*** of the vector. The components of the vectors can also be viewed as scalars by taking only their magnitude. These are called ***scalar components***. This definition is used extensively when standard systems of coordinates are used and the directions in space are known. In this case, the scalar components are $v\sqrt{2}/2$ in the *N* and *W* directions. To avoid confusion as to which type of component is used, we will always indicate specifically the type of component unless it is obvious which type is meant.

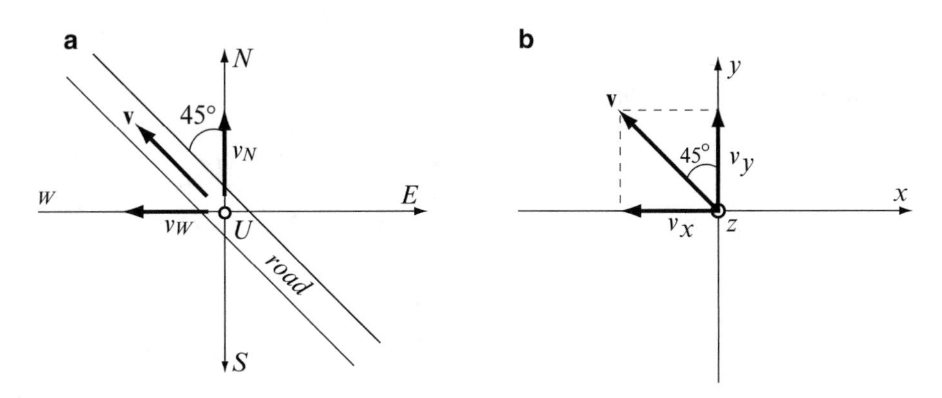

Figure 1.2 (**a**) A convenient coordinate system. (**b**) A more "standard" coordinate system

We chose here a particular system of coordinates to demonstrate that the system of coordinates is a matter of choice. The same can be accomplished by laying a standard system of coordinates, say the rectangular coordinate system over the road map shown in **Figure 1.2a**. This action transforms the road map into a standard coordinate system, and now, using **Figure 1.2b**, we can write

$$\mathbf{v} = -\hat{\mathbf{x}}v_x + \hat{\mathbf{y}}v_y = -\hat{\mathbf{x}}v\frac{\sqrt{2}}{2} + \hat{\mathbf{y}}v\frac{\sqrt{2}}{2} \quad \left[\frac{\mathrm{m}}{\mathrm{s}}\right] \tag{1.4}$$

The components of the vector are in the $-x$ and $+y$ directions. The magnitude of the vector is v, and this is written directly from the geometry in **Figure 1.2b** as

$$v = |\mathbf{v}| = |-\hat{\mathbf{x}}v_x + \hat{\mathbf{y}}v_y| = \sqrt{v_x^2 + v_y^2} \quad [\mathrm{m/s}] \tag{1.5}$$

The unit vector is in the direction of **v** and is given as

$$\hat{\mathbf{v}} = \frac{\mathbf{v}}{|\mathbf{v}|} = \frac{-\hat{\mathbf{x}}v_x + \hat{\mathbf{y}}v_y}{|-\hat{\mathbf{x}}v_x + \hat{\mathbf{y}}v_y|} = \frac{-\hat{\mathbf{x}}v_x + \hat{\mathbf{y}}v_y}{\sqrt{v_x^2 + v_y^2}} = -\hat{\mathbf{x}}\frac{\sqrt{2}}{2} + \hat{\mathbf{y}}\frac{\sqrt{2}}{2} \tag{1.6}$$

It is important to note that although the unit vector $\hat{\mathbf{v}}$ has unit magnitude, its components in the x and y directions do not. Their magnitude is $\sqrt{2}/2$. This may seem to be a minor distinction, but, in fact, it is important to realize that the vector components of a unit vector are not necessarily of unit magnitude. Note, also, that the magnitude of $\hat{\mathbf{x}}$ and $\hat{\mathbf{y}}$ is one since these are the unit vectors in the direction of the vector components of $\mathbf{v}$, namely, x and y.

In this case, both the vector and the unit vector were conveniently written in terms of a particular coordinate system. However, as a rule, any vector can be written in terms of components in other coordinate systems. An example is the one used to describe directions as N, S, W, and E. We shall discuss this separately, but from the above example, some systems are clearly more convenient than others. Also to be noted here is that a general vector in space written in the Cartesian system has three components, in the x, y, and z directions (see below). The third dimension in the above example of velocity gives the vertical component of velocity as the vehicle moves on a nonplanar surface.

In the right-handed Cartesian system (or right-handed rectangular system), we define three coordinates as shown in **Figure 1.3**. A point in the system is described as $P(x_0,y_0,z_0)$, and the general vector $\mathbf{A}$, connecting two general points $P_1(x_1,y_1,z_1)$ and $P_2(x_2,y_2,z_2)$, is given as

$$\mathbf{A}(x,y,z) = \hat{\mathbf{x}}A_x(x,y,z) + \hat{\mathbf{y}}A_y(x,y,z) + \hat{\mathbf{z}}A_z(x,y,z) \tag{1.7}$$

where the scalar components A_x, A_y, and A_z are the projections of the vector on the x, y, and z coordinates, respectively. These are

$$A_x = x_2 - x_1, \quad A_y = y_2 - y_1, \quad A_z = z_2 - z_1 \tag{1.8}$$

The length of the vector (i.e., its magnitude) is

$$A = \sqrt{A_x^2 + A_y^2 + A_z^2} = \sqrt{(x_2 - x_1)^2 + (y_2 - y_1)^2 + (z_2 - z_1)^2} \tag{1.9}$$

and the unit vector in the direction of vector $\mathbf{A}$ is

$$\hat{\mathbf{A}} = \frac{\mathbf{A}(x,y,z)}{A(x,y,z)} = \frac{\hat{\mathbf{x}}(x_2 - x_1) + \hat{\mathbf{y}}(y_2 - y_1) + \hat{\mathbf{z}}(z_2 - z_1)}{\sqrt{(x_2 - x_1)^2 + (y_2 - y_1)^2 + (z_2 - z_1)^2}} \tag{1.10}$$

or

$$\hat{\mathbf{A}} = \hat{\mathbf{x}}\frac{(x_2 - x_1)}{A} + \hat{\mathbf{y}}\frac{(y_2 - y_1)}{A} + \hat{\mathbf{z}}\frac{(z_2 - z_1)}{A} \tag{1.11}$$

We will make considerable use of the unit vector, primarily as an indicator of direction in space. Similarly, the use of components is often employed to simplify analysis.

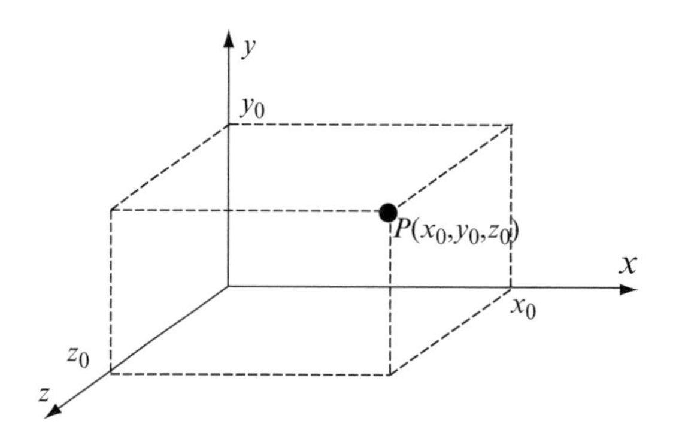

Figure 1.3 A point in the Cartesian system of coordinates

Example 1.1 A vector is given as $\mathbf{A} = -\hat{\mathbf{x}}5 - \hat{\mathbf{y}}(3x + 2) + \hat{\mathbf{z}}$. Calculate:

(a) The scalar components of the vector in the x, y, and z directions.
(b) The length of the vector.
(c) The unit vector in the direction of **A**.

Solution: The solution makes use of **Eqs. (1.8)** through **(1.11)**. In this case, the vector (and all its properties) depends on the variable x alone, although it has components in the y and z directions:

(a) The scalar components of the vector are the coefficients of the three unit vectors:

$$A_x = -5, \quad A_y = -(3x + 2), \quad A_z = 1$$

Note: The negative sign is part of the scalar component, not the unit vector.

(b) The length of the vector is given by **Eq. (1.9)**:

$$A = \sqrt{A_x^2 + A_y^2 + A_z^2} = \sqrt{(-5)^2 + (-(3x + 2))^2 + 1^2} = \sqrt{9x^2 + 12x + 30}$$

(c) The unit vector is calculated from **Eq. (1.11)**:

$$\hat{\mathbf{A}} = -\hat{\mathbf{x}}\frac{5}{\sqrt{9x^2 + 12x + 30}} - \hat{\mathbf{y}}\frac{3x + 2}{\sqrt{9x^2 + 12x + 30}} + \hat{\mathbf{z}}\frac{1}{\sqrt{9x^2 + 12x + 30}}$$

where the scalar components A_x, A_y, and A_z and the magnitude of **A** calculated in **(a)** and **(b)** were used.

Example 1.2 An aircraft takes off at a 60° angle and takeoff speed of 180 km/h in the *NE–SW* direction. Find:

(a) The velocity vector of the aircraft.
(b) Its direction in space.
(c) Its ground velocity (i.e., the velocity of the aircraft's shadow on the ground).

Solution: First, we choose a system of coordinates. In this case, *E–W*, *N–S*, and *D* (down)–*U* (up) is an appropriate choice. This choice describes the physics of the problem even though it is not the most efficient system we can use. (In the exercise that follows, the Cartesian system is used instead.) The components of velocity are calculated from the magnitude (180 km/h) of velocity and angle using projections on the ground and vertically, followed by the velocity vector and the unit vector:

(a) The aircraft velocity has two scalar components: the vertical component $v_u = 180\sin 60°$ and the ground component $v_g = 180\cos 60°$. These speeds are given in km/h. The SI units call for the second as the unit of time and the meter as the unit of distance. Thus, we convert these speeds to m/s. Since 180 km/h = 50 m/s, we get $v_g = 50\cos 60°$ and $v_u = 50\sin 60°$. The west and south components are calculated from v_g, as (see **Figure 1.4**)

$$v_w = 50\cos 60^\circ\cos 45^\circ, \quad v_s = 50\cos 60^\circ\cos 45^\circ \quad [\mathrm{m/s}]$$

The third component is v_u. Thus, the velocity vector is

$$\mathbf{v} = \hat{\mathbf{W}}\,50\cos 60^\circ\cos 45^\circ + \hat{\mathbf{S}}\,50\cos 60^\circ\sin 45^\circ + \hat{\mathbf{U}}\,50\sin 60^\circ = \hat{\mathbf{W}}\,17.678 + \hat{\mathbf{S}}\,17.678 + \hat{\mathbf{U}}\,43.3 \quad [\mathrm{m/s}]$$

(b) The direction in space is given by the unit vector

$$\hat{\mathbf{v}} = \frac{\mathbf{v}}{|\mathbf{v}|} = \frac{\hat{\mathbf{W}}\,50\cos 60^\circ\cos 45^\circ + \hat{\mathbf{S}}\,50\cos 60^\circ\sin 45^\circ + \hat{\mathbf{U}}\,50\sin 60^\circ}{\sqrt{(50\cos 60^\circ\cos 45^\circ)^2 + (50\cos 60^\circ\sin 45^\circ)^2 + (50\sin 60^\circ)^2}} = \hat{\mathbf{W}}\frac{\sqrt{2}}{4} + \hat{\mathbf{S}}\frac{\sqrt{2}}{4} + \hat{\mathbf{U}}\frac{\sqrt{3}}{2} \quad \left[\frac{\mathrm{m}}{\mathrm{s}}\right]$$

(c) Ground velocity is the velocity along the ground plane. This is calculated by setting the vertical velocity of the aircraft found in **(a)** to zero:

$$\mathbf{v} = \hat{\mathbf{W}}\,50\cos 60^{\circ}\cos 45^{\circ} + \hat{\mathbf{S}}\,50\cos 60^{\circ}\sin 45^{\circ} = \hat{\mathbf{W}}\,17.678 + \hat{\mathbf{S}}\,17.678 \qquad [\mathrm{m/s}]$$

Note: It is useful to convert the units to SI units at the outset. This way there is no confusion as to what units are used, and what the intermediate results are, at all stages of the solution.

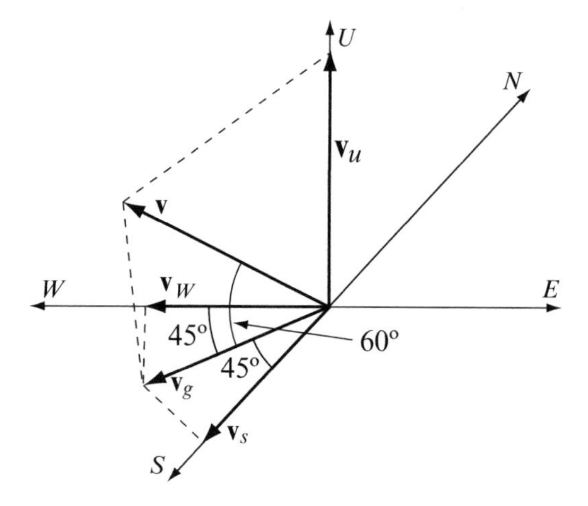

Figure 1.4 Velocity terms along the axes and on the ground

Exercise 1.1 Solve **Example 1.2** in the Cartesian system of coordinates with the positive x axis coinciding with E (east), positive y axis with N (north), and positive z axis with U (up).

Answer

(a) $\mathbf{v} = -\hat{\mathbf{x}}50\cos 60^{\circ}\cos 45^{\circ} - \hat{\mathbf{y}}50\cos 60^{\circ}\sin 45^{\circ} + \hat{\mathbf{z}}50\sin 60^{\circ} \quad [\mathrm{m/s}]$
(b) $\hat{\mathbf{v}} = -\hat{\mathbf{x}}\sqrt{2}/4 - \hat{\mathbf{y}}\sqrt{2}/4 + \hat{\mathbf{z}}\sqrt{3}/2 \quad [\mathrm{m/s}]$
(c) $\mathbf{v}_g = -\hat{\mathbf{x}}50\cos 60^{\circ}\cos 45^{\circ} - \hat{\mathbf{y}}50\cos 60^{\circ}\sin 45^{\circ} \quad [\mathrm{m/s}]$

1.2.2 Vector Addition and Subtraction **Point_Charges.m**

The first vector algebra operation that needs to be defined is vector addition. This is perhaps the most commonly performed vector operation.

The sum of two vectors results in a third vector

$$\boxed{\mathbf{A} + \mathbf{B} = \mathbf{C}} \tag{1.12}$$

To see how this operation is carried out, we use two general vectors $\mathbf{A} = \hat{\mathbf{x}}A_x + \hat{\mathbf{y}}A_y + \hat{\mathbf{z}}A_z$ and $\mathbf{B} = \hat{\mathbf{x}}B_x + \hat{\mathbf{y}}B_y + \hat{\mathbf{z}}B_z$ in Cartesian coordinates and write

$$\mathbf{C} = \mathbf{A} + \mathbf{B} = \left(\hat{\mathbf{x}}A_x + \hat{\mathbf{y}}A_y + \hat{\mathbf{z}}A_z\right) + \left(\hat{\mathbf{x}}B_x + \hat{\mathbf{y}}B_y + \hat{\mathbf{z}}B_z\right) = \left(\hat{\mathbf{x}}C_x + \hat{\mathbf{y}}C_y + \hat{\mathbf{z}}C_z\right) \tag{1.13}$$

Adding components in the same directions together gives

$$\mathbf{C} = \hat{\mathbf{x}}(A_x + B_x) + \hat{\mathbf{y}}\left(A_y + B_y\right) + \hat{\mathbf{z}}(A_z + B_z) \tag{1.14}$$

Figure 1.5 shows this process: In **Figure 1.5a**, vectors **A** and **B** are separated into their three components. **Figure 1.5b** shows that vector **C** is obtained by adding the components of **A** and **B**, which, in turn, are equivalent to translating the vector **B** (without changing its direction in space or its magnitude) so that its tail coincides with the head of vector **A**. Vector **C** is now the vector connecting the tail of vector **A** with the head of vector **B**. This sketch defines a general graphical method of calculating the sum of two vectors:

(1) Draw the first vector in the sum.
(2) Translate the second vector until the tail of the second vector coincides with the head of the first vector.
(3) Connect the tail of the first vector with the head of the second vector to obtain the sum.

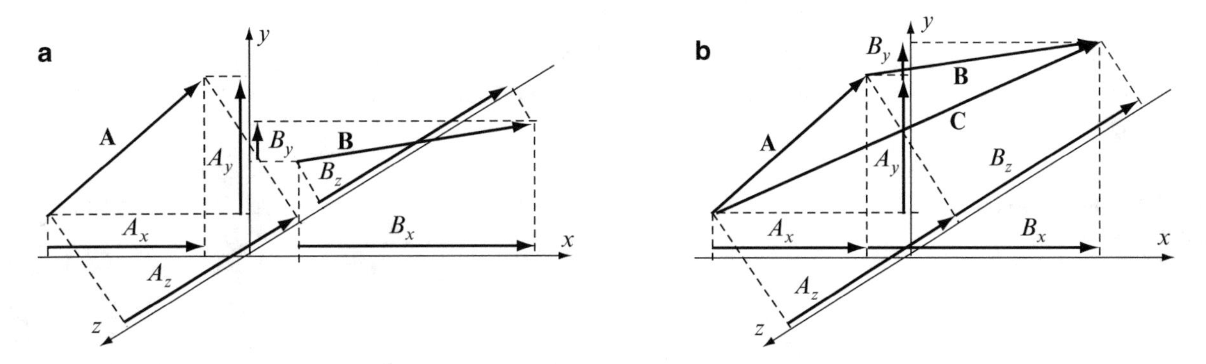

Figure 1.5 (**a**) Two vectors **A** and **B** and their x, y, and z components. (**b**) Addition of vectors **A** and **B** by adding their components

The process is shown in **Figure 1.6** in general terms. This method of calculating the sum of two vectors is sometimes called the ***head-to-tail*** method or rule. An alternative method is obtained by generating two sums **A** + **B** and **B** + **A** using the above method. The two sums are shown in **Figure 1.7a** as two separate vectors and as a single vector in **Figure 1.7b**. The result is a parallelogram with the two vectors, connected tail to tail forming two adjacent sides, and the remaining two sides are parallel lines to the vectors. This method is summarized as follows:

(1) Translate vector **B** so that its tail coincides with the tail of vector **A**.
(2) Construct the parallelogram formed by the two vectors and the two parallels to the vectors.
(3) Draw vector **C** with its tail at the tails of vectors **A** and **B** and head at the intersection of the two parallel lines (dashed lines in **Figure 1.8**).

This method is shown in **Figure 1.8** and is called the ***parallelogram rule***.

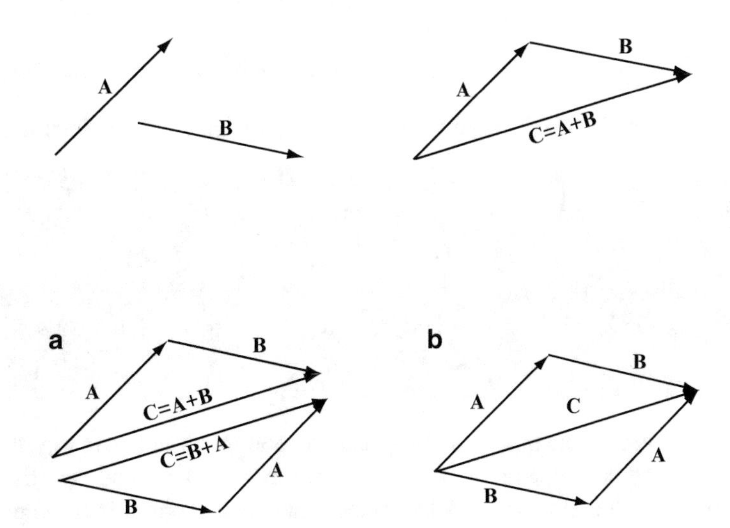

Figure 1.6 Addition of two vectors by translating vector **B** until its tail coincides with the head of vector **A**. The sum **A** + **B** is the vector connecting the tail of vector **A** with the head of vector **B**

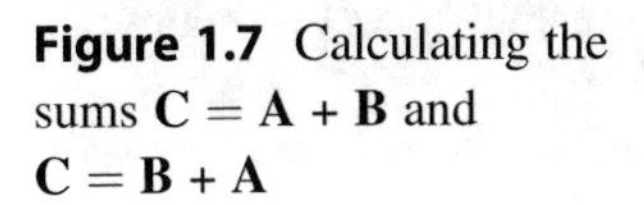

Figure 1.7 Calculating the sums **C** = **A** + **B** and **C** = **B** + **A**

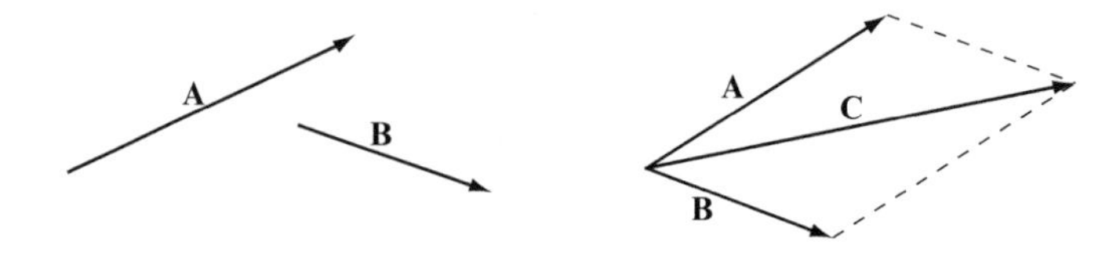

Figure 1.8 The parallelogram method. The *dashed lines* are used to show that opposite sides are equal and parallel

Vector subtraction is accomplished by noting the following:

$$\mathbf{A} - \mathbf{B} = \mathbf{A} + (-\mathbf{B}) = \hat{\mathbf{A}}A + (-\hat{\mathbf{B}})B \tag{1.15}$$

This indicates that vector subtraction is the same as the addition of a negative vector. In terms of the tail-to-head or parallelogram method, we must first reverse the direction of vector **B** and then perform summation of the two vectors. This is shown in **Figure 1.9**.

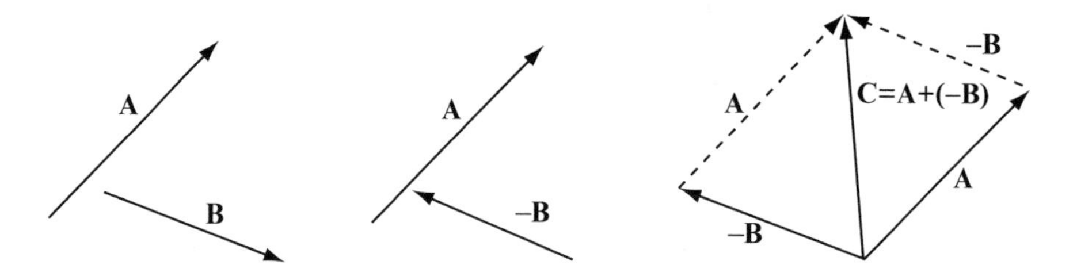

Figure 1.9 Subtraction of vector **B** from vector **A**

Summation or subtraction of more than two vectors should be viewed as a multiple-step process. For example:

$$\mathbf{A} + \mathbf{B} + \mathbf{C} = (\mathbf{A} + \mathbf{B}) + \mathbf{C} = \mathbf{D} + \mathbf{C} \tag{1.16}$$

The sum $\mathbf{D} = \mathbf{A} + \mathbf{B}$ is calculated first using the above methods and then the sum $\mathbf{D} + \mathbf{C}$ is evaluated similarly. The same applies to subtraction.

Note: Any of the two graphical methods of calculating the sum of two vectors may be used, but, in computation, it is often more convenient to separate the vectors into their components and calculate the sum of the components. This is particularly true if we also need to calculate unit vectors. The graphical methods are more useful in understanding what the sum of the vector means and to visualize the direction in space.

Vector summation and subtraction are associative and commutative processes; that is:

$$\mathbf{A} + \mathbf{B} = \mathbf{B} + \mathbf{A} \quad \text{(commutative)} \tag{1.17}$$

$$(\mathbf{A} + \mathbf{B}) + \mathbf{C} = \mathbf{A} + (\mathbf{B} + \mathbf{C}) \quad \text{(associative)} \tag{1.18}$$

The vector addition is also distributive, but we will only show this in **Section 1.2.3**.

Example 1.3 Two vectors **A** and **B** (such as the velocity vectors of two aircraft) are $\mathbf{A} = \hat{\mathbf{x}}1 + \hat{\mathbf{y}}2 + \hat{\mathbf{z}}3$ and $\mathbf{B} = \hat{\mathbf{x}}4 - \hat{\mathbf{z}}3$. Calculate:

(a) The sum of the two vectors.
(b) The difference **A** – **B** and **B** – **A** (these differences represent the relative velocities of **A** with respect to **B** and of **B** with respect to **A**).

Solution: **(a)** The vectors are placed on the system of coordinates shown in **Figure 1.10a**, and the components of **A** and **B** are found as shown. The components of vector $\mathbf{C} = \mathbf{A} + \mathbf{B}$ are now found directly from the figure. In **(b)**, we write the two expressions $\mathbf{D} = \mathbf{A} - \mathbf{B}$ and $\mathbf{E} = \mathbf{B} - \mathbf{A}$ by adding together the appropriate components:

(a) Vector **A** has scalar components of 1, 2, and 3 in the x, y, and z directions, respectively. It may therefore be viewed as connecting the origin (as a reference point) to point $P_1(1,2,3)$, as shown in **Figure 1.10a**. Vector **B** is in the x–z plane and connects the origin to point $P_2(4,0,-3)$. The vectors may be translated anywhere in space as long as their lengths and directions are not changed. Translate vector **A** such that its tail touches the head of vector **B**. This is shown in **Figure 1.10b** in terms of the components (i.e., translation of vector **A** so that its tail coincides with the head of vector **B** is the same as translating its components so that their tails coincide with the heads of the corresponding components of vector **B**). The sum **C** = **A** + **B** is the vector connecting the tail of vector **B** with the head of vector **A**. The result is (writing the projections of vector **C** onto the *x*, *y*, and *z* axes):

$$\mathbf{C} = \hat{\mathbf{x}}5 + \hat{\mathbf{y}}2$$

(b) To calculate the differences, we add the vector components of the two vectors together, observing the sign of each vector component:

$$\mathbf{D} = \mathbf{A} - \mathbf{B} = \left(\hat{\mathbf{x}}1 + \hat{\mathbf{y}}2 + \hat{\mathbf{z}}3\right) - \left(\hat{\mathbf{x}}4 - \hat{\mathbf{z}}3\right) = \hat{\mathbf{x}}(1-4) + \hat{\mathbf{y}}(2-0) + \hat{\mathbf{z}}(3-(-3))$$

or

$$\mathbf{A} - \mathbf{B} = -\hat{\mathbf{x}}3 + \hat{\mathbf{y}}2 + \hat{\mathbf{z}}6$$

$$\mathbf{E} = \mathbf{B} - \mathbf{A} = \left(\hat{\mathbf{x}}4 - \hat{\mathbf{z}}3\right) - \left(\hat{\mathbf{x}}1 + \hat{\mathbf{y}}2 + \hat{\mathbf{z}}3\right) = \hat{\mathbf{x}}(4-1) + \hat{\mathbf{y}}(0-2) + \hat{\mathbf{z}}(-3-3)$$

or

$$\mathbf{B} - \mathbf{A} = \hat{\mathbf{x}}3 - \hat{\mathbf{y}}2 - \hat{\mathbf{z}}6$$

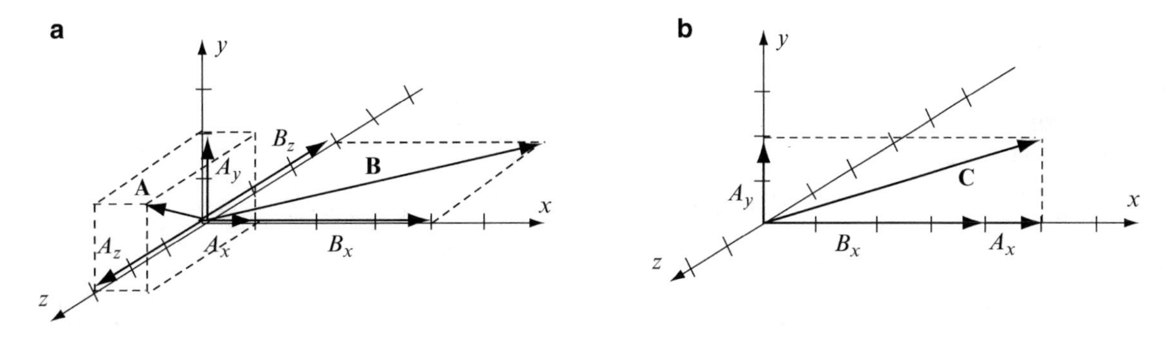

Figure 1.10 (a) Components of vectors **A** and **B**. (b) The sum **C** = **A** + **B** is obtained by summing the components of **A** and **B**

Exercise 1.2 Three vectors are given: $\mathbf{A} = \hat{\mathbf{x}}1 + \hat{\mathbf{y}}2 + \hat{\mathbf{z}}3$, $\mathbf{B} = \hat{\mathbf{x}}4 - \hat{\mathbf{y}}2 + \hat{\mathbf{z}}3$, and $\mathbf{C} = -\hat{\mathbf{x}}4$. Calculate:

(a) **A** + **B** + **C**.
(b) **A** + **B** – 2**C**.
(c) **A** – **B** – **C**.
(d) The unit vector in the direction of **A** – 2**B** + **C**.

Answer

(a) $\mathbf{A} + \mathbf{B} + \mathbf{C} = \hat{\mathbf{x}}1 + \hat{\mathbf{z}}6$.
(b) $\mathbf{A} + \mathbf{B} - 2\mathbf{C} = \hat{\mathbf{x}}13 + \hat{\mathbf{z}}6$.
(c) $\mathbf{A} - \mathbf{B} - \mathbf{C} = \hat{\mathbf{x}}1 + \hat{\mathbf{y}}4$.
(d) $-\hat{\mathbf{x}}0.8538 + \hat{\mathbf{y}}0.4657 - \hat{\mathbf{z}}0.2328$.

1.2.3 Vector Scaling

A vector can be scaled by multiplying its magnitude by a scalar value. Scaling is defined as changing the magnitude of the vector:

$$\boxed{k\mathbf{A} = k(\hat{\mathbf{A}}A) = \hat{\mathbf{A}}(kA)} \tag{1.19}$$

The term "multiplication" for vectors is not used to avoid any confusion with vector products, which we define in the following section. Scaling of a vector is equivalent to "lengthening" or "shortening" the vector without modifying its direction if k is a positive constant, as shown in **Figure 1.11a**. Increasing the velocity of an aircraft (without change in direction) from 300 to 330 km/h scales the velocity vector by a factor of $k = 1.1$. If k is negative, the resulting scaled vector has a magnitude $|k|$ times its nonscaled magnitude but also a negative direction, as shown in **Figure 1.11b**.

Vector scaling is both associative and commutative but not distributive (simply because the product of two vectors has not been defined yet); that is,

$$k\mathbf{A} = \mathbf{A}k \quad \text{(commutative)} \tag{1.20}$$

$$k(p\mathbf{A}) = (kp)\mathbf{A} \quad \text{(associative)} \tag{1.21}$$

Also,

$$k(\mathbf{A} + \mathbf{B}) = k\mathbf{A} + k\mathbf{B} \tag{1.22}$$

The latter shows that the vector sum is distributive.

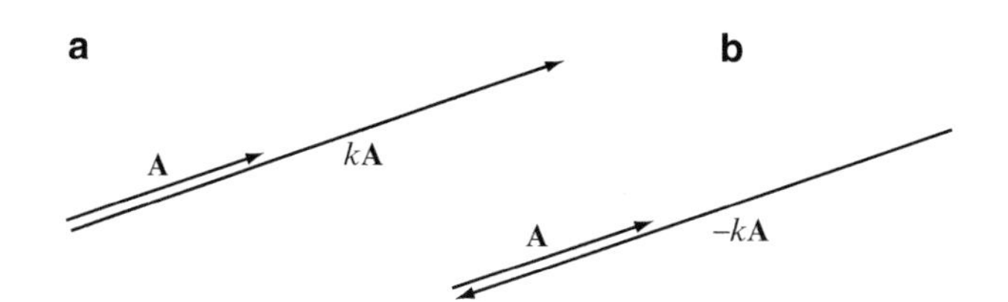

Figure 1.11 (**a**) Scaling of vector **A** by a positive scalar k. (**b**) Scaling of vector **A** by a negative scalar $-k$

1.3 Products of Vectors

The multiplication of two vectors is called a ***product***. Here, we define two types of products based on the result obtained from the product. The first type is the scalar product. This is a product of two vectors which results in a scalar. The second is a vector product of two vectors, which results in a vector. Beyond the form of the product, these have important physical and geometrical meanings, which make them some of the most useful and often encountered vector operations.

1.3.1 The Scalar Product

A ***scalar product*** of two vectors **A** and **B** is denoted as $\mathbf{A} \cdot \mathbf{B}$ and is defined as:

"the product of the magnitudes of **A** and **B** and the cosine of the smaller angle between **A** and **B**"; that is,

$$\boxed{\mathbf{A} \cdot \mathbf{B} \equiv AB\cos\phi_{AB}} \tag{1.23}$$

where the angle ϕ_{AB} is the smaller angle between **A** and **B**, as shown in **Figure 1.12**. The sign ≡ indicates that **Eq. (1.23)** is the definition of the scalar product. The result is a scalar. The scalar product is often called a ***dot product*** because of the *dot* notation used. It has a number of properties that we will exploit later:

(1) For any angle $0 \le \phi_{AB} < \pi/2$, the scalar product is positive. For angles above $\pi/2$ ($\pi/2 < \phi_{AB} \le \pi$), the scalar product is negative.

(2) The scalar product is zero for any two perpendicular vectors ($\phi_{AB} = \pi/2$).

(3) For $\phi_{AB} = 0$ (parallel vectors), the scalar product equals AB, and for $\phi_{AB} = \pi$, the product is $(-AB)$.

(4) The magnitude of the scalar product of two vectors is always smaller or equal to the product of their magnitudes ($|\mathbf{A}\cdot\mathbf{B}| \le AB$).

(5) The product can be viewed as the product of the magnitude of vector **A** and the magnitude of the projection of vector **B** on **A** or vice versa ($\mathbf{A}\cdot\mathbf{B} = A(B\cos\phi_{AB}) = B(A\cos\phi_{AB})$).

(6) The scalar product is commutative and distributive:

$$\mathbf{A}\cdot\mathbf{B} = \mathbf{B}\cdot\mathbf{A} \quad \text{(commutative)} \tag{1.24}$$

$$\mathbf{A}\cdot(\mathbf{B}+\mathbf{C}) = \mathbf{A}\cdot\mathbf{B} + \mathbf{A}\cdot\mathbf{C} \quad \text{(distributive)} \tag{1.25}$$

The scalar product can be written explicitly using two vectors **A** and **B** in Cartesian coordinates as

$$\begin{aligned}\mathbf{A}\cdot\mathbf{B} &= \left(\hat{\mathbf{x}}A_x + \hat{\mathbf{y}}A_y + \hat{\mathbf{z}}A_z\right)\cdot\left(\hat{\mathbf{x}}B_x + \hat{\mathbf{y}}B_y + \hat{\mathbf{z}}B_z\right) \\ &= \hat{\mathbf{x}}\cdot\hat{\mathbf{x}}A_xB_x + \hat{\mathbf{x}}\cdot\hat{\mathbf{y}}A_xB_y + \hat{\mathbf{x}}\cdot\hat{\mathbf{z}}A_xB_z + \hat{\mathbf{y}}\cdot\hat{\mathbf{x}}A_yB_x + \hat{\mathbf{y}}\cdot\hat{\mathbf{y}}A_yB_y + \hat{\mathbf{y}}\cdot\hat{\mathbf{z}}A_yB_z + \hat{\mathbf{z}}\cdot\hat{\mathbf{x}}A_zB_x + \hat{\mathbf{z}}\cdot\hat{\mathbf{y}}A_zB_y + \hat{\mathbf{z}}\cdot\hat{\mathbf{z}}A_zB_z\end{aligned} \tag{1.26}$$

From properties **(2)** and **(3)** and since unit vectors are of magnitude 1, we have

$$\hat{\mathbf{x}}\cdot\hat{\mathbf{y}} = \hat{\mathbf{x}}\cdot\hat{\mathbf{z}} = \hat{\mathbf{y}}\cdot\hat{\mathbf{z}} = \hat{\mathbf{y}}\cdot\hat{\mathbf{x}} = \hat{\mathbf{z}}\cdot\hat{\mathbf{x}} = \hat{\mathbf{z}}\cdot\hat{\mathbf{y}} = 0 \tag{1.27}$$

$$\hat{\mathbf{x}}\cdot\hat{\mathbf{x}} = \hat{\mathbf{y}}\cdot\hat{\mathbf{y}} = \hat{\mathbf{z}}\cdot\hat{\mathbf{z}} = 1 \tag{1.28}$$

Therefore, **Eq. (1.26)** becomes

$$\boxed{\mathbf{A}\cdot\mathbf{B} = A_xB_x + A_yB_y + A_zB_z} \tag{1.29}$$

This form affords simple evaluation of the product from the components of the vectors rather than requiring calculation of the angle between the vectors. From this, we also note that

$$\mathbf{A}\cdot\mathbf{A} = AA\cos(0) = A^2 = A_x^2 + A_y^2 + A_z^2 \tag{1.30}$$

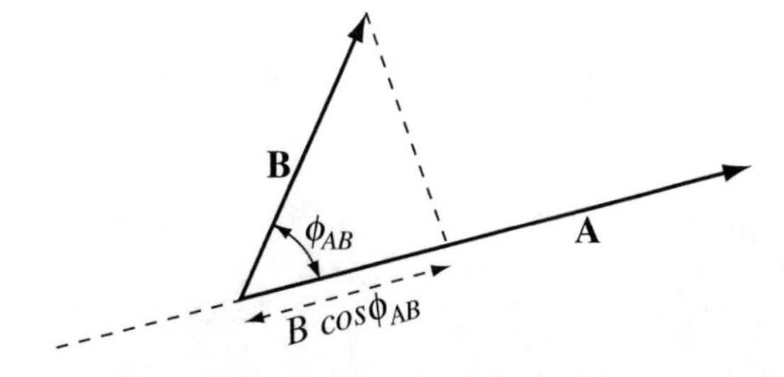

Figure 1.12 Definition of the scalar product between vectors **A** and **B**. The smaller angle between the vectors is used

Example 1.4 Calculate the projection of a general vector **A** onto another general vector **B** and the vector component of **A** in the direction of **B**.

Solution: The projection of vector **A** onto **B** is $A\cos\phi_{AB}$. This is

$$A\cos\phi_{AB} = \frac{\mathbf{A}\cdot\mathbf{B}}{B} = \frac{A_xB_x + A_yB_y + A_zB_z}{\sqrt{B_x^2 + B_y^2 + B_z^2}}.$$

To calculate the vector component of **A** in the direction of **B**, we note that the magnitude of this component is the projection calculated above, whereas the direction of the component is that of the unit vector in the direction of **B**. The latter is

$$\hat{\mathbf{B}} = \frac{\mathbf{B}}{|B|} = \frac{\hat{\mathbf{x}}B_x + \hat{\mathbf{y}}B_y + \hat{\mathbf{z}}B_z}{\sqrt{B_x^2 + B_y^2 + B_z^2}}.$$

The vector component of vector **A** in the direction of vector **B** is therefore

$$\mathbf{A}_B = \hat{\mathbf{B}}A\cos\phi_{AB} = \frac{\hat{\mathbf{x}}B_x(\mathbf{A}\cdot\mathbf{B}) + \hat{\mathbf{y}}B_y(\mathbf{A}\cdot\mathbf{B}) + \hat{\mathbf{z}}B_z(\mathbf{A}\cdot\mathbf{B})}{B_x^2 + B_y^2 + B_z^2}$$

$$= \hat{\mathbf{x}}\frac{B_x\left(A_xB_x + A_yB_y + A_zB_z\right)}{B_x^2 + B_y^2 + B_z^2} + \hat{\mathbf{y}}\frac{B_y\left(A_xB_x + A_yB_y + A_zB_z\right)}{B_x^2 + B_y^2 + B_z^2} + \hat{\mathbf{z}}\frac{B_z\left(A_xB_x + A_yB_y + A_zB_z\right)}{B_x^2 + B_y^2 + B_z^2}.$$

Example 1.5 Two vectors are given as $\mathbf{A} = \hat{\mathbf{x}} + \hat{\mathbf{y}}5 - \hat{\mathbf{z}}$ and $\mathbf{B} = -\hat{\mathbf{x}} + \hat{\mathbf{y}}5 + \hat{\mathbf{z}}$. Find the angle between the two vectors.

Solution: Using the scalar product, the cosine of the angle between the vectors is evaluated from **Eq. (1.23)** as

$$\cos\phi_{AB} = \frac{\mathbf{A}\cdot\mathbf{B}}{AB} \quad \rightarrow \quad \phi_{AB} = \cos^{-1}\left(\frac{\mathbf{A}\cdot\mathbf{B}}{AB}\right).$$

The magnitudes of **A** and **B** are

$$A = |\mathbf{A}| = \sqrt{1 + 25 + 1} = \sqrt{27} = 3\sqrt{3}, \quad B = |\mathbf{B}| = \sqrt{1 + 25 + 1} = \sqrt{27} = 3\sqrt{3}$$

The scalar product of **A** and **B** is

$$\mathbf{A}\cdot\mathbf{B} = \left(\hat{\mathbf{x}} + \hat{\mathbf{y}}5 - \hat{\mathbf{z}}\right)\cdot\left(-\hat{\mathbf{x}} + \hat{\mathbf{y}}5 + \hat{\mathbf{z}}\right) = -1 + 25 - 1 = 23.$$

Thus

$$\cos\phi_{AB} \equiv \frac{\mathbf{A}\cdot\mathbf{B}}{AB} = \frac{23}{3\sqrt{3}\times 3\sqrt{3}} = 0.85185 \quad \rightarrow \quad \phi_{AB} = \cos^{-1}(0.85185) = 31^\circ 35'.$$

Example 1.6 Application: The Cosine Formula The two vectors of the previous example are given and drawn schematically in **Figure 1.13**:

(a) Show that the distance between points P_1 and P_2 is given by

$$d = \sqrt{A^2 + B^2 - 2AB\cos\phi_{AB}} \quad [\text{m}].$$

(b) Calculate this length for the two vectors.

Figure 1.13 Diagram used to prove the cosine formula

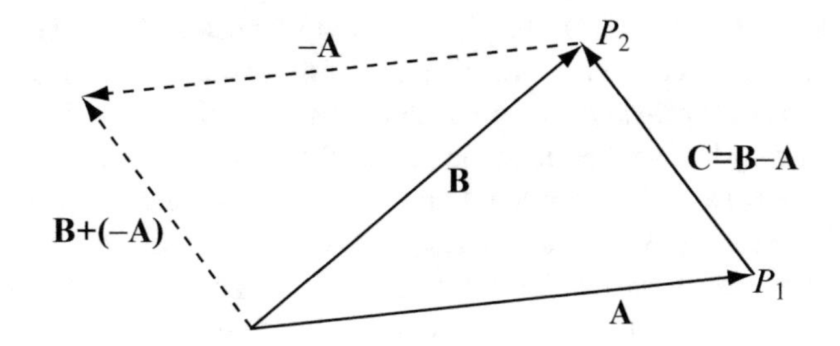

Solution: This example is recognizable as the application of the cosine formula and, in fact, may be viewed as its derivation. Assuming a third vector pointing from P_1 to P_2 as shown in **Figure 1.13**, we calculate this vector as $\mathbf{C} = \mathbf{B} - \mathbf{A}$. The scalar product $\mathbf{C} \cdot \mathbf{C}$ gives the distance C^2. This is the distance between P_1 and P_2 squared. Taking the square root gives the required result:

(a)

$$C^2 = \mathbf{C} \cdot \mathbf{C} = (\mathbf{B} - \mathbf{A}) \cdot (\mathbf{B} - \mathbf{A}) = \mathbf{B} \cdot \mathbf{B} + \mathbf{A} \cdot \mathbf{A} - 2\mathbf{B} \cdot \mathbf{A}$$

Since $\mathbf{B} \cdot \mathbf{B} = B^2$, $\mathbf{A} \cdot \mathbf{A} = A^2$, and $\mathbf{B} \cdot \mathbf{A} = \mathbf{A} \cdot \mathbf{B} = BA\cos\phi_{BA} = AB\cos\phi_{AB}$, we get

$$C^2 = A^2 + B^2 - 2AB\cos\phi_{AB} \quad \rightarrow \quad C = \sqrt{A^2 + B^2 - 2AB\cos\phi_{AB}}$$

(b) For the two vectors in **Example 1.5**

$$\mathbf{A} = \hat{\mathbf{x}} + \hat{\mathbf{y}}5 - \hat{\mathbf{z}}, \qquad \mathbf{B} = -\hat{\mathbf{x}} + \hat{\mathbf{y}}5 + \hat{\mathbf{z}}.$$

we calculated

$$A = 3\sqrt{3}, \quad B = 3\sqrt{3}, \quad \cos\phi_{AB} = 0.85185$$

The distance between P_2 and P_1 is therefore

$$d = \sqrt{A^2 + B^2 - 2AB\cos\phi_{AB}} = \sqrt{27 + 27 - 2 \times 27 \times 0.85185} = 2.828 \quad [\mathrm{m}]$$

Exercise 1.3 An airplane flies with a velocity $\mathbf{v} = \hat{\mathbf{x}}100 + \hat{\mathbf{y}}500 + \hat{\mathbf{z}}200$ [m/s]. Calculate the aircraft's velocity in the direction of the vector $\mathbf{A} = \hat{\mathbf{x}} + \hat{\mathbf{y}} + \hat{\mathbf{z}}$.

Answer $\mathbf{v}_A = \hat{\mathbf{x}}800/3 + \hat{\mathbf{y}}800/3 + \hat{\mathbf{z}}800/3$ [m/s].

1.3.2 The Vector Product

The vector product[2] of two vectors **A** and **B**, denoted as $\mathbf{A} \times \mathbf{B}$, is defined as:

> *"the vector whose magnitude is the absolute value of the product of the magnitudes of the two vectors and the sine of the smaller angle between the two vectors while the direction of the vector is perpendicular to the plane in which the two vectors lie";*

that is,

$$\boxed{\mathbf{A} \times \mathbf{B} \equiv \hat{\mathbf{n}}|AB\sin\phi_{AB}|} \tag{1.31}$$

[2] The vector product was defined by Sir William Rowan Hamilton (1805–1865) as part of his theory of quaternions around 1845. James Clerk Maxwell made use of this theory when he wrote his *Treatise on Electricity and Magnetism* in 1873, although he was critical of quaternions. Modern electromagnetics uses the Heaviside–Gibbs vector system rather than the Hamilton system.

where $\hat{\mathbf{n}}$ is the unit vector normal to the plane formed by vectors **A** and **B** and ϕ_{AB} is, again, the smaller angle between the vectors. The normal unit vector gives the direction of the product, which is obviously a vector. For this reason, it is called a ***vector product*** or a ***cross product*** because of the cross symbol used in the notation.

The unit vector may be in either direction perpendicular to the plane, and to define it uniquely, we employ the right-hand rule, as shown in **Figure 1.14**. According to this rule, if the right-hand palm is placed on the first vector in the product and rotated toward the second vector through an angle ϕ_{AB}, the extended thumb shows the correct direction of the vector product. This rule immediately indicates that moving the palm from vector **B** to vector **A** gives a direction opposite to that moving from vector **A** to **B**. Thus, we conclude that the vector product is not commutative:

$$\boxed{\mathbf{A} \times \mathbf{B} = -\mathbf{B} \times \mathbf{A} \quad \text{(noncommutative)}} \tag{1.32}$$

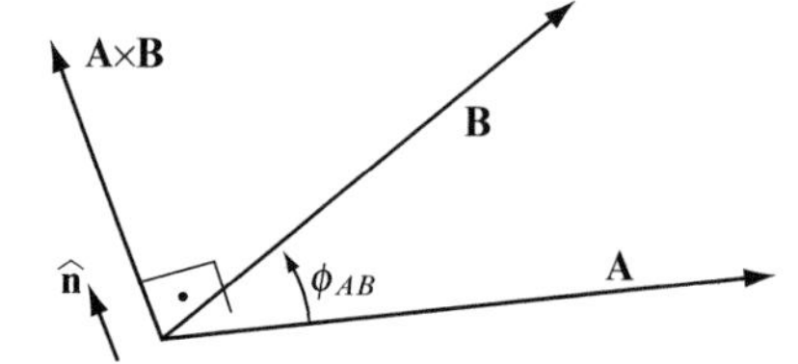

Figure 1.14 The vector product between vectors **A** and **B**

In addition to the noncommutative property of the vector product, the following properties are noted:

(1) The vector product is perpendicular to the plane of the two vectors; that is, it is perpendicular to both vectors.
(2) For two vectors which are perpendicular to each other ($\phi_{AB} = \pi/2$), the magnitude of the vector product is equal to the product of the magnitudes of the two vectors ($\sin\phi_{AB} = 1$).
(3) The vector product of two parallel vectors is zero ($\sin\phi_{AB} = 0$).
(4) The vector product of a vector with itself is zero ($\sin\phi_{AA} = 0$).
(5) The vector product is not associative (this will be discussed in the following section because it requires the definition of a triple product).
(6) The vector product is distributive:

$$\mathbf{A} \times (\mathbf{B} + \mathbf{C}) = \mathbf{A} \times \mathbf{B} + \mathbf{A} \times \mathbf{C} \tag{1.33}$$

(7) The magnitude of the vector product represents the area bounded by the parallelogram formed by the two vectors and two lines parallel to the vectors, as shown in **Figure 1.15**.

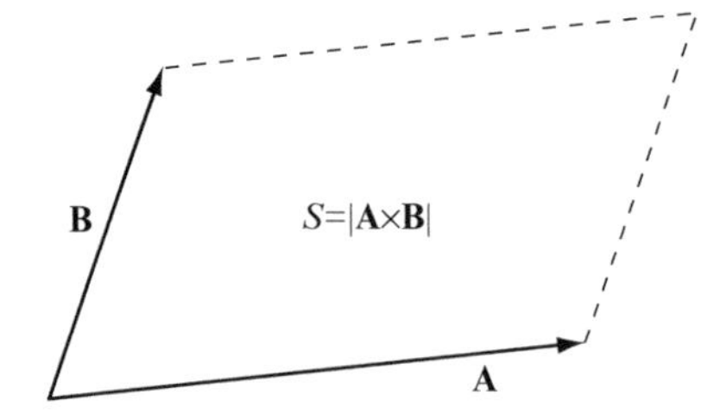

Figure 1.15 Interpretation of the magnitude of the vector product as a surface

Evaluation of the vector product is performed similarly to that for the scalar product: We write the product explicitly and expand the expression based on **Eq. (1.33)**. Using two general vectors **A** and **B** in Cartesian coordinates, we get

$$\begin{aligned}\mathbf{A} \times \mathbf{B} &= (\hat{\mathbf{x}}A_x + \hat{\mathbf{y}}A_y + \hat{\mathbf{z}}A_z) \times (\hat{\mathbf{x}}B_x + \hat{\mathbf{y}}B_y + \hat{\mathbf{z}}B_z) \\ &= (\hat{\mathbf{x}} \times \hat{\mathbf{x}})A_xB_x + (\hat{\mathbf{x}} \times \hat{\mathbf{y}})A_xB_y + (\hat{\mathbf{x}} \times \hat{\mathbf{z}})A_xB_z + (\hat{\mathbf{y}} \times \hat{\mathbf{x}})A_yB_x + (\hat{\mathbf{y}} \times \hat{\mathbf{y}})A_yB_y + (\hat{\mathbf{y}} \times \hat{\mathbf{z}})A_yB_z \\ &\quad + (\hat{\mathbf{z}} \times \hat{\mathbf{x}})A_zB_x + (\hat{\mathbf{z}} \times \hat{\mathbf{y}})A_zB_y + (\hat{\mathbf{z}} \times \hat{\mathbf{z}})A_zB_z\end{aligned} \tag{1.34}$$

Because the unit vectors $\hat{\mathbf{x}}, \hat{\mathbf{y}}, \hat{\mathbf{z}}$ are perpendicular to each other, and using the right-hand rule in **Figure 1.14**, we can write

$$\hat{\mathbf{x}} \times \hat{\mathbf{x}} = \hat{\mathbf{y}} \times \hat{\mathbf{y}} = \hat{\mathbf{z}} \times \hat{\mathbf{z}} = 0 \tag{1.35}$$

from property 4 above. Similarly, using property 2 and the right-hand rule, we can write

$$\begin{array}{lll} \hat{\mathbf{x}} \times \hat{\mathbf{y}} = \hat{\mathbf{z}}, & \hat{\mathbf{y}} \times \hat{\mathbf{z}} = \hat{\mathbf{x}}, & \hat{\mathbf{z}} \times \hat{\mathbf{x}} = \hat{\mathbf{y}}, \\ \hat{\mathbf{y}} \times \hat{\mathbf{x}} = -\hat{\mathbf{z}}, & \hat{\mathbf{z}} \times \hat{\mathbf{y}} = -\hat{\mathbf{x}}, & \hat{\mathbf{x}} \times \hat{\mathbf{z}} = -\hat{\mathbf{y}} \end{array} \tag{1.36}$$

Substitution of these products and rearranging terms gives

$$\mathbf{A} \times \mathbf{B} = \hat{\mathbf{x}}\left(A_y B_z - A_z B_y\right) + \hat{\mathbf{y}}\left(A_z B_x - A_x B_z\right) + \hat{\mathbf{z}}\left(A_x B_y - A_y B_x\right) \tag{1.37}$$

This is a rather straightforward operation, although lengthy. To avoid having to go through this process every time we use the vector product, we note that the expression in **Eq. (1.37)** has the form of the determinant of a 3 × 3 matrix:

$$\mathbf{A} \times \mathbf{B} = \begin{vmatrix} \hat{\mathbf{x}} & \hat{\mathbf{y}} & \hat{\mathbf{z}} \\ A_x & A_y & A_z \\ B_x & B_y & B_z \end{vmatrix} = \hat{\mathbf{x}}\left(A_y B_z - A_z B_y\right) + \hat{\mathbf{y}}\left(A_z B_x - A_x B_z\right) + \hat{\mathbf{z}}\left(A_x B_y - A_y B_x\right) \tag{1.38}$$

In the system of coordinates used here (right-hand Cartesian coordinates), the vector product is cyclic; that is, the products in **Eq. (1.36)** are cyclical, as shown in **Figure 1.16**. This is a simple way to generate the signs of the components of the vector product: a cross product of the unit vectors performed in the sequence shown by the arrows in **Figure 1.16a** is positive; if it is in the opposite sequence (**Figure 1.16b**), it is negative.

The vector product is used for a number of important operations. They include finding the direction of the vector product, calculation of areas, evaluation of normal unit vectors, and representation of fields.

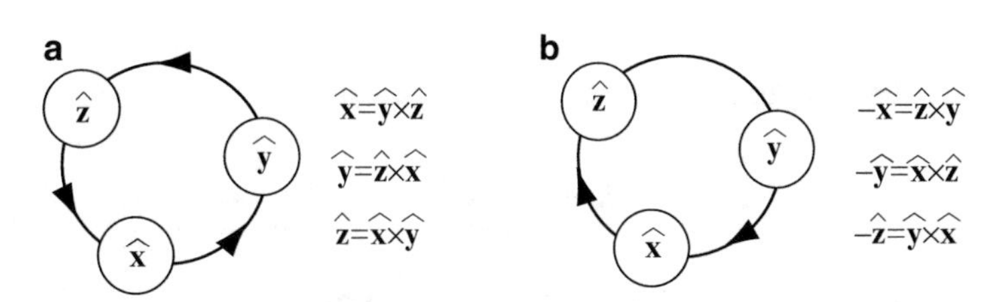

Figure 1.16 The cyclical relations between the various vector products of the unit vectors in Cartesian coordinates. (**a**) Positive sequence. (**b**) Negative sequence

Example 1.7 Application: Vector Normal to a Plane

(**a**) Find a vector normal to a plane that contains points $P_1(0,1,0)$, $P_2(1,0,1)$, and $P_3(0,0,1)$.
(**b**) Find the normal unit vector.

Solution: This is a common use for the vector product. Because the vector product of two vectors is normal to both vectors, we must first find two vectors that lie in the plane. Their vector product gives the normal vector. Calculation of the normal unit vector can be done either using the definition of the unit vector in **Eq. (1.2)** or through the use of the scalar and vector products.

Two vectors in the plane can be defined using any two pairs of points. Using P_1 and P_2, we define a vector (from P_1 to P_2) as

$$\mathbf{A} = \hat{\mathbf{x}}(x_2 - x_1) + \hat{\mathbf{y}}(y_2 - y_1) + \hat{\mathbf{z}}(z_2 - z_1) = \hat{\mathbf{x}}(1-0) + \hat{\mathbf{y}}(0-1) + \hat{\mathbf{z}}(1-0) = \hat{\mathbf{x}}1 - \hat{\mathbf{y}}1 + \hat{\mathbf{z}}1$$

Similarly for a second vector, we choose the vector between P_1 and P_3. This gives

$$\mathbf{B} = \hat{\mathbf{x}}(x_3 - x_1) + \hat{\mathbf{y}}(y_3 - y_1) + \hat{\mathbf{z}}(z_3 - z_1) = \hat{\mathbf{x}}(0-0) + \hat{\mathbf{y}}(0-1) + \hat{\mathbf{z}}(1-0) = -\hat{\mathbf{y}}1 + \hat{\mathbf{z}}1$$

The vector product, **C**, is a vector normal to both **A** and **B** and, therefore, to the plane:

$$\mathbf{A} \times \mathbf{B} = (\hat{\mathbf{x}}1 - \hat{\mathbf{y}}1 + \hat{\mathbf{z}}1) \times (-\hat{\mathbf{y}}1 + \hat{\mathbf{z}}1)$$
$$= \hat{\mathbf{x}}1 \times (-\hat{\mathbf{y}}1) + \hat{\mathbf{x}}1 \times \hat{\mathbf{z}}1 + (-\hat{\mathbf{y}}1) \times (-\hat{\mathbf{y}}1) + (-\hat{\mathbf{y}}1) \times \hat{\mathbf{z}}1 + \hat{\mathbf{z}}1 \times (-\hat{\mathbf{y}}1) + \hat{\mathbf{z}}1 \times \hat{\mathbf{z}}1$$

Using the identities in **Eqs. (1.35)** and **(1.36)**, we get

$$\mathbf{C} = \mathbf{A} \times \mathbf{B} = -\hat{\mathbf{y}}1 - \hat{\mathbf{z}}1$$

The unit vector can be found from **Eq. (1.2)** or from the definition of the vector product in **Eq. (1.31)**. We use the latter as an example of an alternative method:

$$\hat{\mathbf{n}} = \frac{\mathbf{A} \times \mathbf{B}}{|AB\sin\phi_{AB}|}$$

The angle ϕ_{AB} can be most easily calculated from the scalar product in **Eq. (1.23)** as

$$\phi_{AB} = \cos^{-1}\left(\frac{\mathbf{A} \cdot \mathbf{B}}{AB}\right)$$

To do so, we need to evaluate the scalar product and the magnitude of the vectors. These are

$$\mathbf{A} \cdot \mathbf{B} = (\hat{\mathbf{x}}1 - \hat{\mathbf{y}}1 + \hat{\mathbf{z}}1) \cdot (-\hat{\mathbf{y}}1 + \hat{\mathbf{z}}1) = 2, \quad A = \sqrt{3} \quad B = \sqrt{2}$$

Thus,

$$\phi_{AB} = \cos^{-1}\left(\frac{2}{\sqrt{6}}\right) = 35^\circ 16'$$

The unit normal vector is now

$$\hat{\mathbf{n}} = \frac{\mathbf{A} \times \mathbf{B}}{|AB\sin\phi_{AB}|} = \frac{-\hat{\mathbf{y}}1 - \hat{\mathbf{z}}1}{|\sqrt{6}\sin(35^\circ 16')|} = \frac{-\hat{\mathbf{y}}1 - \hat{\mathbf{z}}1}{1.4142} = -\hat{\mathbf{y}}0.7071 - \hat{\mathbf{z}}0.7071$$

The same result is obtained using **Eq. (1.2)**:

$$\hat{\mathbf{n}} = \frac{\mathbf{A} \times \mathbf{B}}{|\mathbf{A} \times \mathbf{B}|} = \frac{-\hat{\mathbf{y}}1 - \hat{\mathbf{z}}1}{|-\hat{\mathbf{y}}1 - \hat{\mathbf{z}}1|} = \frac{-\hat{\mathbf{y}}1 - \hat{\mathbf{z}}1}{\sqrt{2}} = -\hat{\mathbf{y}}0.7071 - \hat{\mathbf{z}}0.7071$$

Exercise 1.4 Vectors $\mathbf{A} = \hat{\mathbf{x}}1 - \hat{\mathbf{y}}2 + \hat{\mathbf{z}}3$ and $\mathbf{B} = \hat{\mathbf{x}}3 + \hat{\mathbf{y}}5 + \hat{\mathbf{z}}1$ are in a plane, not necessarily perpendicular to each other. Vector $\mathbf{C} = \hat{\mathbf{x}}17 + \hat{\mathbf{y}}8 - \hat{\mathbf{z}}11$ is perpendicular to the same plane. Show that the vector product between **C** and **A** (or between **C** and **B**) must also be in the plane of **A** and **B**.

Example 1.8 Application: Area of a Triangle Find the area of the triangle with vertices at three general points $P_1(x_1, y_1, z_1)$, $P_2(x_2, y_2, z_2)$, and $P_3(x_3, y_3, z_3)$ (**Figure 1.17a**).

Solution: In this case, the vector nature of the vector product is irrelevant, but the magnitude of the vector product in **Eq. (1.31)** is equal to the area of the parallelogram formed by the two vectors. This can be seen from the fact that the magnitude of $\mathbf{A} \times \mathbf{B}$ is $A(B\sin\phi_{AB})$. This is the area of rectangle $abb'c'$ in **Figure 1.17b**. Since triangles acc' and bdb' are identical, this is also the area of parallelogram $abdc$. Since triangles abc and cbd are identical, the area of abc is equal to half the area of *abdc*. Calculation of the area of triangle *abc* is done by calculating the magnitude of the vector product of two of the vectors forming the sides of the triangle and dividing by 2:

$$S_{abc} = \frac{|\mathbf{A} \times \mathbf{B}|}{2}$$

From **Figure 1.17a**, vectors **A** and **B** are

$$\mathbf{A} = \hat{\mathbf{x}}(x_2 - x_1) + \hat{\mathbf{y}}(y_2 - y_1) + \hat{\mathbf{z}}(z_2 - z_1),$$
$$\mathbf{B} = \hat{\mathbf{x}}(x_3 - x_1) + \hat{\mathbf{y}}(y_3 - y_1) + \hat{\mathbf{z}}(z_3 - z_1)$$

The vector product is obtained using **Eq. (1.38)**, and from this, the area of the triangle is

$$S_{abc} = \frac{|\mathbf{A} \times \mathbf{B}|}{2} = \frac{1}{2}\left\| \begin{matrix} \hat{\mathbf{x}} & \hat{\mathbf{y}} & \hat{\mathbf{z}} \\ x_2 - x_1 & y_2 - y_1 & z_2 - z_1 \\ x_3 - x_1 & y_3 - y_1 & z_3 - z_1 \end{matrix} \right\|$$

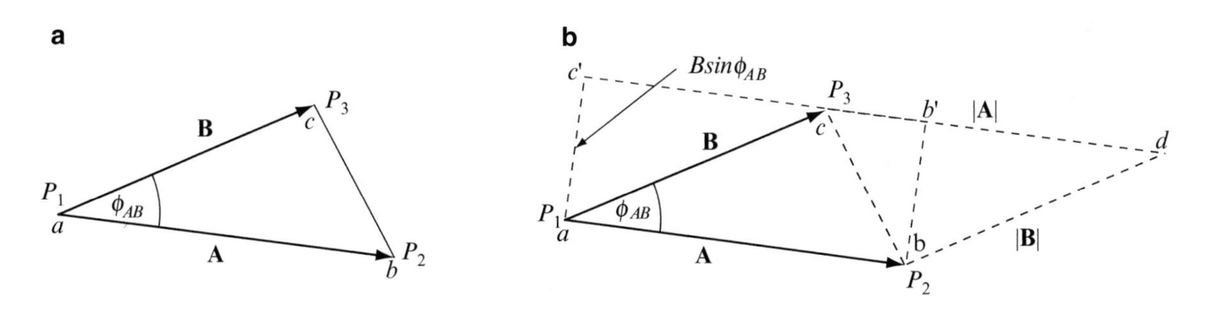

Figure 1.17 Area of a triangle. (**a**) A triangle with two of its sides shown as vectors. (**b**) The area of the triangle is half the area of the parallelogram *abdc*

Exercise 1.5 Find the area of the triangle formed by points (1,3,0), (1,2,1), and (3,5,2).

Answer $\sqrt{24}/2 = 2.4495\,\mathrm{m}^2$.

Example 1.9 Find a unit vector normal to both of the vectors

$$\mathbf{A} = \hat{\mathbf{x}}3 + \hat{\mathbf{y}}1 - \hat{\mathbf{z}}2, \quad \mathbf{B} = \hat{\mathbf{x}}1 - \hat{\mathbf{y}}5.$$

Solution: The vector products **A** × **B** or **B** × **A** result in vectors normal to both **A** and **B**, respectively:

$$\mathbf{A} \times \mathbf{B} = (\hat{\mathbf{x}}3 + \hat{\mathbf{y}}1 - \hat{\mathbf{z}}2) \times (\hat{\mathbf{x}}1 - \hat{\mathbf{y}}5) = (\hat{\mathbf{x}}3) \times (\hat{\mathbf{x}}1) + (\hat{\mathbf{x}}3) \times (-\hat{\mathbf{y}}5) + (\hat{\mathbf{y}}1) \times (\hat{\mathbf{x}}1)$$
$$+ (\hat{\mathbf{y}}1) \times (-\hat{\mathbf{y}}5) + (-\hat{\mathbf{z}}2) \times (\hat{\mathbf{x}}1) + (-\hat{\mathbf{z}}2) \times (-\hat{\mathbf{y}}5) = -\hat{\mathbf{x}}10 - \hat{\mathbf{y}}2 - \hat{\mathbf{z}}16$$

The unit vector is

$$\hat{\mathbf{n}} = \frac{-\hat{\mathbf{x}}10 - \hat{\mathbf{y}}2 - \hat{\mathbf{z}}16}{|-\hat{\mathbf{x}}10 - \hat{\mathbf{y}}2 - \hat{\mathbf{z}}16|} = \frac{-\hat{\mathbf{x}}10 - \hat{\mathbf{y}}2 - \hat{\mathbf{z}}16}{\sqrt{360}} = \frac{-\hat{\mathbf{x}}5 - \hat{\mathbf{y}} - \hat{\mathbf{z}}8}{3\sqrt{10}}$$

Using the product **B** × **A** results in a unit vector $-\hat{\mathbf{n}}$ as can be shown by application of the right-hand rule.

Example 1.10 The general equation of a plane in Cartesian coordinates is $ax + by + cz + d = 0$. The equation of the plane may be found as $f(x,y,z) = \mathbf{n} \cdot \mathbf{C} = 0$ where $\mathbf{n}$ is the normal vector to the plane and $\mathbf{C}$ is a general vector in the plane.

(a) Given three points $P_1(1,0,2)$, $P_2(3,1,-2)$, $P_3(2,3,2)$ in a plane, find the equation of the plane.
(b) Show that the equation of the plane may be written as

$$f(x,y,z) = n_x(x - x_0) + n_y(y - y_0) + n_z(z - z_0) = 0$$

where n_x, n_y, and n_z are the scalar components of the normal unit vector to the plane and (x_0,y_0,z_0) are the coordinates of a point in the plane.

Solution: The three points given define two vectors (say P_1 to P_2 and P_1 to P_3). The normal to the plane is obtained through use of the vector product. The vector **A** is then defined between a general point (x,y,z) and any of the points given. The form in **(b)** is found from **(a)**:

(a) Two vectors necessary to calculate the normal vector to the plane are
P_1 to P_2: $\mathbf{A} = \hat{\mathbf{x}}(3 - 1) + \hat{\mathbf{y}}(1 - 0) + \hat{\mathbf{z}}(-2 - 2) = \hat{\mathbf{x}}2 + \hat{\mathbf{y}}1 - \hat{\mathbf{z}}4$
P_1 to P_3: $\mathbf{B} = \hat{\mathbf{x}}(2 - 1) + \hat{\mathbf{y}}(3 - 0) + \hat{\mathbf{z}}(2 - 2) = \hat{\mathbf{x}}1 + \hat{\mathbf{y}}3$
These two vectors are in the plane. Therefore, the normal vector to the plane may be written as:

$$\mathbf{n} = \mathbf{A} \times \mathbf{B} = \left(\hat{\mathbf{x}}2 + \hat{\mathbf{y}}1 - \hat{\mathbf{z}}4\right) \times \left(\hat{\mathbf{x}}1 + \hat{\mathbf{y}}3\right) = \hat{\mathbf{x}}12 - \hat{\mathbf{y}}4 + \hat{\mathbf{z}}5$$

A general vector in the plane may be written as:

$$\mathbf{C} = \hat{\mathbf{x}}(x - 1) + \hat{\mathbf{y}}(y - 0) + \hat{\mathbf{z}}(z - 2)$$

where point P_1 was used, arbitrarily.
The equation of the plane is

$$f(x,y,z) = \mathbf{n} \cdot \mathbf{C} = \left(\hat{\mathbf{x}}12 - \hat{\mathbf{y}}4 + \hat{\mathbf{z}}5\right) \cdot \left(\hat{\mathbf{x}}(x - 1) + \hat{\mathbf{y}}(y - 0) + \hat{\mathbf{z}}(z - 2)\right) = 0$$

or

$$f(x,y,z) = 12(x - 1) - 4(y - 0) + 5(z - 2) = 12x - 4y + 5z - 22 = 0$$

(b) To show that the formula given produces the same result, we first calculate the normal unit vector:

$$\hat{\mathbf{n}} = \hat{\mathbf{x}}\frac{12}{\sqrt{185}} - \hat{\mathbf{y}}\frac{4}{\sqrt{185}} + \hat{\mathbf{z}}\frac{5}{\sqrt{185}}$$

The point (x_0,y_0,z_0) can be any point in the plane. Selecting P_3, for example, the equation of the plane is

$$f(x,y,z) = \frac{12}{\sqrt{185}}(x - 2) - \frac{4}{\sqrt{185}}(y - 3) + \frac{5}{\sqrt{185}}(z - 2) = 0$$

Multiplying both sides of the equation by $\sqrt{185}$, we get

$$f(x,y,z) = 12(x - 2) - 4(y - 3) + 5(z - 2) = 12x - 4y + 5z - 22 = 0$$

This is the same as the result obtained in **(a)**.
The formula in **(b)** is called the scalar equation of the plane whereas the form $f(x,y,z) = \mathbf{n} \cdot \mathbf{C} = 0$ is referred to as the vector equation of the plane.

1.3.3 Multiple Vector and Scalar Products

As with sums of vectors, we can define multiple products by repeatedly applying the rules of the scalar or vector product of two vectors. However, because of the particular method of defining the vector and scalar products, not all combinations of products are meaningful. For example, the result of a vector product is a vector, and therefore, it can only be obtained by scaling another vector or by a vector product with another vector. Similarly, the result of a scalar product is a scalar and cannot be used to obtain a vector. The following triple products are properly defined:

(1) The vector triple product is defined as

$$\boxed{\mathbf{A} \times (\mathbf{B} \times \mathbf{C}) = \mathbf{B}(\mathbf{A} \cdot \mathbf{C}) - \mathbf{C}(\mathbf{A} \cdot \mathbf{B})} \tag{1.39}$$

This is called a ***vector triple product*** because it involves three terms (vectors) and the result is a vector. The right-hand side can be shown to be correct by direct evaluation of the vector product (see **Exercise 1.6**). A number of properties should be noted here:

(a) The vector (double or triple) product is not associative,

$$\mathbf{A} \times (\mathbf{B} \times \mathbf{C}) \neq (\mathbf{A} \times \mathbf{B}) \times \mathbf{C}. \tag{1.40}$$

That is, the sequence in which the vector product is performed is all-important. For this reason, the brackets should always be part of the notation and should never be omitted. The product $\mathbf{A} \times \mathbf{B} \times \mathbf{C}$ is not a properly defined product.

(b) The right-hand side of **Eq. (1.39)** is often used for evaluation of the vector triple product. Because of the combination of products, this is referred to as the *BAC–CAB* rule. It provides a means of remembering the correct sequence of products for evaluation.

(c) The vector triple product can also be evaluated using the determinant rule in **Eq. (1.38)** by applying it twice. First, the product $\mathbf{B} \times \mathbf{C}$ in **Eq. (1.39)** is evaluated. This results in a vector, say $\mathbf{D}$. Then, the product $\mathbf{A} \times \mathbf{D}$ is evaluated, resulting in the vector triple product.

(2) The following are properly defined scalar triple products:

$$\mathbf{A} \cdot (\mathbf{B} \times \mathbf{C}) = \mathbf{B} \cdot (\mathbf{C} \times \mathbf{A}) = \mathbf{C} \cdot (\mathbf{A} \times \mathbf{B}) \tag{1.41}$$

This product is a ***scalar triple product*** since the result is a scalar. Note that the vectors in the product are cyclic permutations of each other. Any other order of the vectors in the triple product produces equal but negative results. Thus,

$$\mathbf{A} \cdot (\mathbf{B} \times \mathbf{C}) = -\mathbf{A} \cdot (\mathbf{C} \times \mathbf{B}) \tag{1.42}$$

because of the property of the vector product. On the other hand, from the properties of the scalar product, we have

$$\mathbf{A} \cdot (\mathbf{B} \times \mathbf{C}) = (\mathbf{B} \times \mathbf{C}) \cdot \mathbf{A} \tag{1.43}$$

We also note that since the vector product represents the area bounded by the two vectors **B** and **C**, and the scalar product is the projection of vector **A** onto the vector product, the scalar triple product represents the volume defined by the three vectors **A**, **B**, and **C** (see **Figure 1.18** and **Example 1.11**). Finally, as an aid to evaluation of the scalar triple product, we mention that this can be evaluated as the determinant of a matrix as follows:

$$\mathbf{A} \cdot (\mathbf{B} \times \mathbf{C}) = \begin{vmatrix} A_x & A_y & A_z \\ B_x & B_y & B_z \\ C_x & C_y & C_z \end{vmatrix} = A_x\left(B_yC_z - B_zC_y\right) + A_y(B_zC_x - B_xC_z) + A_z\left(B_xC_y - B_yC_x\right) \tag{1.44}$$

and, as before, can be shown to be correct by direct evaluation of the product (see **Exercise 1.7**). Note, however, that this does not imply that the scalar triple product is a determinant; the determinant is merely an aid to evaluating the product.

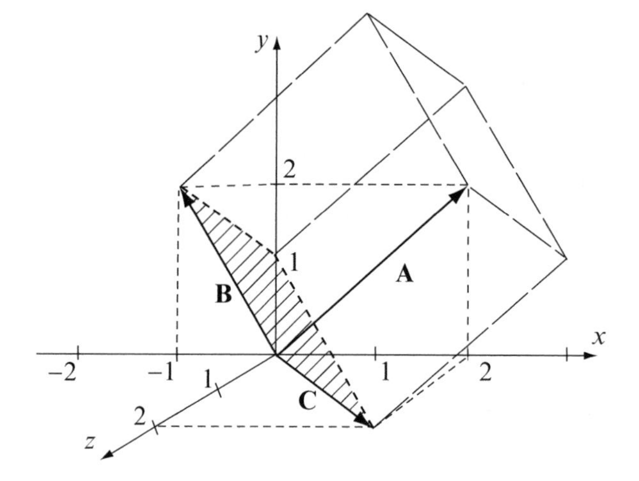

Figure 1.18 Interpretation of the scalar triple product as a volume

Important The products $\mathbf{A} \cdot (\mathbf{B} \cdot \mathbf{C})$ and $\mathbf{A} \times (\mathbf{B} \cdot \mathbf{C})$ are not defined. Can you show why?

Other products may be defined, but these are not important in electromagnetics. For example, $\mathbf{A}\times(\mathbf{B}\times(\mathbf{C} \times \mathbf{D}))$ is a properly defined vector product, but we will have no use for it in subsequent work.

Example 1.11 Application: Volume of a Parallelepiped Calculate the scalar triple product $\mathbf{A} \cdot (\mathbf{B} \times \mathbf{C})$ defined by the vectors: $\mathbf{A} = \hat{\mathbf{x}}2 + \hat{\mathbf{y}}2$, $\mathbf{B} = -\hat{\mathbf{x}}2 + \hat{\mathbf{y}}2$, and $\mathbf{C} = \hat{\mathbf{x}}1 + \hat{\mathbf{z}}2$. Show that this product represents the volume of a parallelepiped in which the tails of the three vectors form one corner of the parallelepiped.

Solution: Consider **Figure 1.18**. The three vectors form a box as shown. The magnitude of the vector product $\mathbf{D} = \mathbf{B} \times \mathbf{C}$ represents the area of the parallelogram shown cross-hatched. Considering the vector **A** and the angle it makes with the vector **D**, the scalar product $E = \mathbf{A} \cdot \mathbf{D} = AD\cos\phi_{AD}$ gives the volume of the box since $A\cos\phi_{AD}$ gives the height of the box (projection of **A** on **D**). **Equation (1.44)** is used to evaluate the volume:

$$\mathbf{A} \cdot (\mathbf{B} \times \mathbf{C}) = \begin{vmatrix} 2 & 2 & 0 \\ -2 & 2 & 0 \\ 1 & 0 & 2 \end{vmatrix} = 2(4-0) + 2(0+4) = 16 \quad [\mathrm{m}^3]$$

Example 1.12

(a) Find the vector triple product $\mathbf{B} \times (\mathbf{A} \times \mathbf{C})$ using the three vectors of the previous example.
(b) Show that the resultant vector must be in the plane formed by **A** and **C**.

Solution:

(a) The vector triple product is evaluated using the rule in **Eq. (1.39)**.
We write

$$\mathbf{D} = \mathbf{B} \times (\mathbf{A} \times \mathbf{C}) = \mathbf{A}(\mathbf{B} \cdot \mathbf{C}) - \mathbf{C}(\mathbf{B} \cdot \mathbf{A})$$

Note that this is the same relation as in **Eq. (1.39)** with vectors **A** and **B** interchanged. The vectors **A**, **B**, and **C** are

$$\mathbf{A} = \hat{\mathbf{x}}2 + \hat{\mathbf{y}}2, \quad \mathbf{B} = -\hat{\mathbf{x}}2 + \hat{\mathbf{y}}2, \quad \mathbf{C} = \hat{\mathbf{x}}1 + \hat{\mathbf{z}}2$$

The scalar products $\mathbf{B}\cdot\mathbf{C}$ and $\mathbf{B}\cdot\mathbf{A}$ are

$$\mathbf{B}\cdot\mathbf{C} = \left(-\hat{\mathbf{x}}2+\hat{\mathbf{y}}2\right)\cdot\left(\hat{\mathbf{x}}1+\hat{\mathbf{z}}2\right) = -2,$$
$$\mathbf{B}\cdot\mathbf{A} = \left(-\hat{\mathbf{x}}2+\hat{\mathbf{y}}2\right)\cdot\left(\hat{\mathbf{x}}2+\hat{\mathbf{y}}2\right) = -4+4=0$$

The vector triple product reduces to

$$\mathbf{D} = \mathbf{B}\times(\mathbf{A}\times\mathbf{C}) = \mathbf{A}(-2) - 0 = -\hat{\mathbf{x}}4 - \hat{\mathbf{y}}4$$

(b) The simplest way to show that the vector $\mathbf{D} = \mathbf{B}\times(\mathbf{A}\times\mathbf{C})$ is in the plane formed by vectors **A** and **C** is to show that the scalar triple product $\mathbf{D}\cdot(\mathbf{A}\times\mathbf{C})$ is zero. This is to say that the box formed by vectors **D**, **A**, and **C** has zero volume. This can only happen if the three vectors are in a plane. Substituting the scalar components of vectors **D**, **A**, and **C** in **Eq. (1.44)** gives

$$\mathbf{D}\cdot(\mathbf{A}\times\mathbf{C}) = \begin{vmatrix} -4 & -4 & 0 \\ 2 & 2 & 0 \\ 1 & 0 & 2 \end{vmatrix} = -16+16=0$$

Thus, vector **D** is in the plane formed by vectors **A** and **C**.

Exercise 1.6 Show that the relation in **Eq. (1.39)** is correct by direct evaluation of the vector triple product $\mathbf{A}\times(\mathbf{B}\times\mathbf{C})$ using general vectors.

$$\mathbf{A} = \hat{\mathbf{x}}A_x+\hat{\mathbf{y}}A_y+\hat{\mathbf{z}}A_z,\ \mathbf{B} = \hat{\mathbf{x}}B_x+\hat{\mathbf{y}}B_y+\hat{\mathbf{z}}B_z,\ \text{and}\ \mathbf{C} = \hat{\mathbf{x}}C_x+\hat{\mathbf{y}}C_y+\hat{\mathbf{z}}C_z.$$

Exercise 1.7 Show that the relation in **Eq. (1.44)** is correct by direct evaluation of the scalar triple product $\mathbf{A}\cdot(\mathbf{B}\times\mathbf{C})$ using general vectors.

$$\mathbf{A} = \hat{\mathbf{x}}A_x+\hat{\mathbf{y}}A_y+\hat{\mathbf{z}}A_z,\quad \mathbf{B} = \hat{\mathbf{x}}B_x+\hat{\mathbf{y}}B_y+\hat{\mathbf{z}}B_z,\ \text{and}\ \mathbf{C} = \hat{\mathbf{x}}C_x+\hat{\mathbf{y}}C_y+\hat{\mathbf{z}}C_z.$$

1.4 Definition of Fields

A field may be defined mathematically as the function of a set of variables in a given space. This rather general definition is of little use in trying to understand properties of a field from a physical point of view. Therefore, we will use a "looser" definition of a field. For the purpose of this book, ***a field is a distribution in space of any quantity: scalar, vector, time dependent, or independent of time***. The field may be defined over the whole space or a portion of space. Thus, for example, a topographical map shows the altitude of each point in a given domain; this is an "altitude field." If we can describe the wind velocity at every point in a domain, then we have defined a "velocity field." Similarly, a gravitational force field, a temperature field, a geomagnetic field and the like may be defined. Note also that although a functional dependency always exists, a field may be postulated without these dependencies being used or, for that matter, known: The "altitude field" above is obtained by measurements and is therefore experimentally found. However, at least in principle, the functional dependency exists.

Fields are fundamental to the study of electromagnetics. In this context, we will seek to understand the properties of electromagnetic fields, which, based on the definition above, are merely the distribution of the "electric and magnetic vectors." Although we do not know at this point what these are, it is easy to conceptualize the idea that if these vectors can be defined anywhere in a given space, then their distribution in that space can also be described: This process defines the field of the corresponding vector. How these fields interact with each other and with materials is what electromagnetics is all about.

1.4.1 Scalar Fields

A ***scalar field*** is a field of scalar variables; that is, if for any point in space, say (x,y,z), we know the function $f(x,y,z)$, then f is the scalar field. This may represent a temperature distribution, potential, pressure, or any other scalar function. For example,

$$f(x,y,z) = x^2 + y^2 + 5z^2 \tag{1.45}$$

is a scalar field. An example of a scalar field is shown in **Figure 1.19**. It shows a topographical map in which contour lines show various elevations. The representation in terms of contour lines (in this case, lines of constant elevation) is a simple way of representing a scalar field. If the lines were to represent pressure, a similar map may give air pressure over a continent to provide details of a meteorological report. In some cases we will find it useful to derive physical quantities from scalar fields. For example, in the field in **Eq. (1.45)**, we could calculate first-order derivatives with respect to any of the variables. If the field is, say, an altitude field, then the first-order derivative describes a slope. If we were planning to build a road, then this is extremely important information to know.

Scalar fields may be time dependent or independent of time. An example of a time-dependent scalar field could be a weather map in which temperatures vary with time or a dynamic display of the availability of parking spaces in a parking lot.

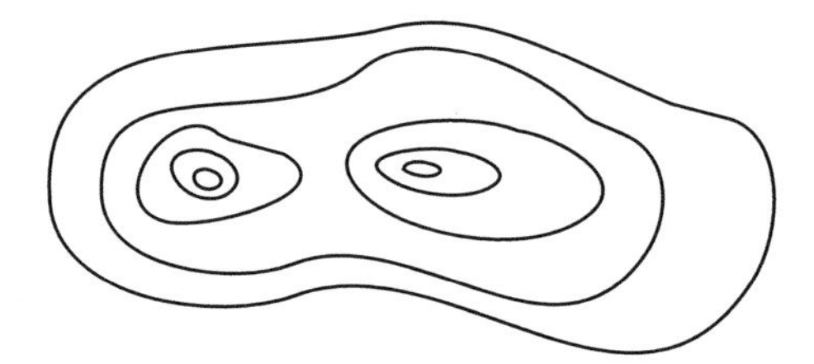

Figure 1.19 Lines of constant elevation as an example of a scalar field

1.4.2 Vector Fields

We can define a ***vector field*** as a vector function $\mathbf{F}(x,y,z,t)$. As an example,

$$\mathbf{F}(x,y,z) = \hat{\mathbf{x}}A_x(x,y,z) + \hat{\mathbf{y}}A_y(x,y,z) + \hat{\mathbf{z}}A_z(x,y,z) \tag{1.46}$$

is a static vector field. An example of a vector field is shown in **Figure 1.20**. It shows wind velocities in a hurricane. The length of the vectors indicates the magnitude of velocity and the direction gives the direction of flow.

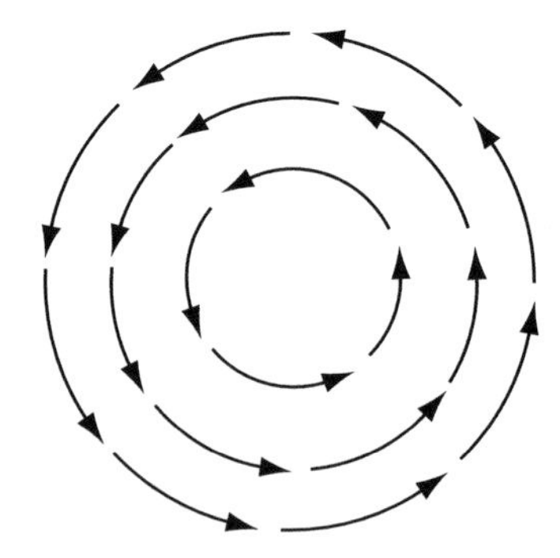

Figure 1.20 Example of a vector field: wind velocity distribution in a hurricane

A vector field may be obtained from a scalar field or a scalar field may be obtained from a vector field. As an example, if we were to use the scalar field in **Eq. (1.45)** and calculate the slopes with respect to x, y, and z, we obtain a vector field since a slope is only properly defined if both the magnitude and direction are defined. For example, when skiing on a mountain, the elevation is less important than the slope, and the slope is different in different directions. Starting at any given point, you may want to ski in the direction of maximum slope or maybe sideways on a less steep path, or you may want to follow a predetermined path, as in cross-country skiing. These may seem to be trivial notions, but they are exactly the operations that we need to perform in electromagnetic fields.

The properties of vector and scalar fields will be discussed extensively in this and the following chapter but, in particular, in the context of electromagnetic fields in the remainder of the book.

Example 1.13 Graphing Scalar Fields A scalar field is given: $\psi(x,y,z) = x^2y - 3x + 3$. Obtain a graph of the field in the range $-1 < x, y < 1$.

Solution: To obtain a graph, we substitute points (x,y,z) in the expression for the field and mark the magnitude of the field on a map. In this case, the field is in the x–y plane (it does not depend on z). There are a number of methods of representation for scalar fields. One is shown in **Figure 1.21a**. For each point (x,y), the value of the function $\psi(x,y)$ is indicated. For example, $\psi(0,0) = 3$, $\psi(0.5,0) = 1.5$, $\psi(0.5,0.5) = 1.625$, $\psi(-0.5,-0.5) = 4.375$, $\psi(0.75,0.75) = 1.172$, $\psi(-0.75,-0.75) = 4.828$, and $\psi(-1,1) = 7.0$. These points are indicated on the graph. This method is simple but does not give a complete visual picture. If this equation were to represent elevation, you would be able to show the elevation at any point, but it would be hard to see what the terrain looks like.

A second representation is shown in **Figure 1.21b**. It shows the same scalar field with a large number of points, and all points of the same magnitude are connected with a line. These are contour lines as commonly used on maps. Now, the picture is easier to read. Each contour line represents a given value $\psi = constant$ of the field.

A third method is to show the magnitude of the field above a plane for all x, y. This gives a three-dimensional picture. The individual points can then be interpolated on a grid. A representation of this kind is shown in **Figure 1.21c**. This is not unlike trying to draw an actual terrain map in which the elevation is shown.

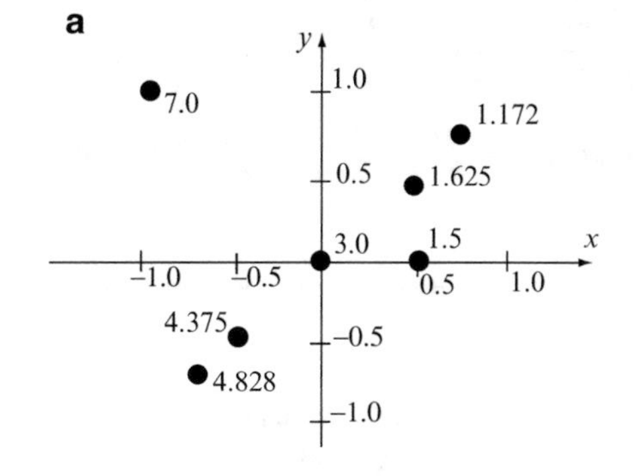

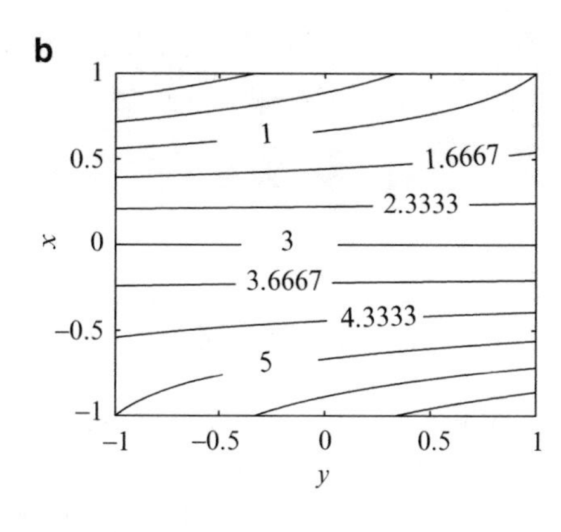

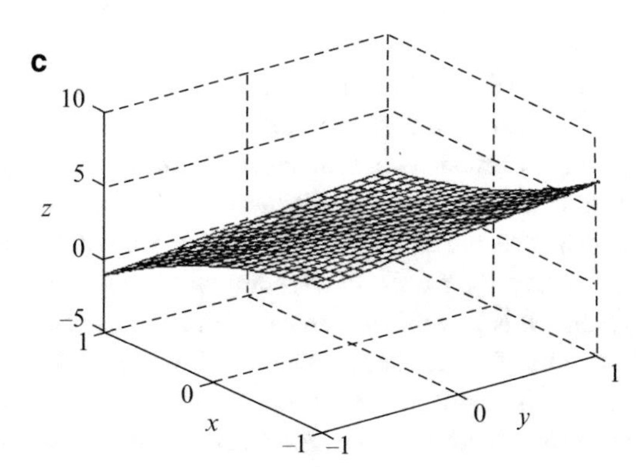

Figure 1.21 Representation of scalar fields. (**a**) Values at given coordinates. (**b**) Contours of constant value. (**c**) Values above a plane. The height at any point represents the magnitude (strength) of the scalar field

Example 1.14 Graphing Vector Fields Graph the vector field $\mathbf{A} = \hat{\mathbf{y}}x$.

Solution: For a vector field, we must show both the magnitude of the vector and its direction. Normally, this is done by locating individual points in the field and drawing an arrowed line at that point. The arrow starts at the location (point) at which the field is shown and points in the direction of the field, and the length of the arrow indicates the magnitude of the field.

In this example, the magnitude of the field is independent of the y and z directions. Thus, the magnitude is zero at $x = 0$ and increases linearly with x. The direction is in the positive y direction for $x > 0$ and in the negative y direction for $x < 0$. A simple representation of this field is shown in **Figure 1.22**.

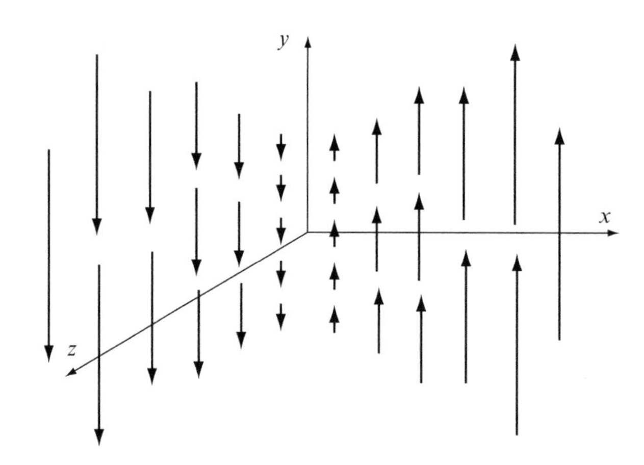

Figure 1.22 Representation of the vector field $\mathbf{A} = \hat{\mathbf{y}}x$ as *arrows*. The length of the *arrow* indicates the magnitude of the field. Only the $z = 0$ plane is shown, but this field is independent of z

Exercise 1.8 Graph the following vector fields in the range $-1 < x,\ y < 1$:

(a) $\mathbf{A} = \hat{\mathbf{x}}x + \hat{\mathbf{y}}y$.
(b) $\mathbf{A} = \hat{\mathbf{x}}x - \hat{\mathbf{y}}y$.

1.5 Systems of Coordinates

A system of coordinates is a system of reference axes used to uniquely describe the various quantities needed in the study of electromagnetics (or any other discipline). In describing scalars, vectors, products, and other quantities, it is extremely important to be able to do so in a simple, unique manner. A system of coordinates forms a unique, universally accepted and understood reference by convention; that is, we can devise many systems of coordinates, but only some of these are actually useful and only a handful have been accepted universally.

Among the various systems of coordinates, the so-called orthogonal systems are the most commonly used. These are systems in which the reference axes are normal to each other. In addition, we will only use the so-called right-hand systems. Emphasis was already given to Cartesian coordinates in the previous sections. In addition, the cylindrical coordinates system and the spherical system of coordinates will be discussed. These three systems are sufficient for our purposes.

We should mention here that as a rule, when a system of coordinates is chosen over another, it is for convenience. We know intuitively that it is easier to describe a cube in a rectangular coordinate system, whereas a spherical object must be easier to describe in a spherical system. It is possible to describe a cube in a spherical system but with considerable more difficulty. For this reason, specialized systems of coordinates have been devised. A simple example is the system used to identify location of aircraft and ships: A grid, consisting of longitude and latitude lines, has been devised, measured in degrees because they are supposed to fit the spherical surface of the globe. A rectangular grid is suitable for, say, the map of a city or a small section of a country but not as a global coordinate system. This example also indicates one of the most important aspects of working with a "convenient" system of coordinates: the need to transform from one system to another. In the above example, a ship may be sailing from point A to point B for a total of x degrees latitude. However, in practical terms, more often we need to know the distance. This means that for any longitude or latitude, we should be able to convert angles to positions in terms of distances from given points or distances between points.

There are a number of other coordinate systems designed for use in three-dimensional space. These coordinate systems have been devised and used for a variety of applications and include the bipolar, prolate spheroidal, elliptic cylindrical, and ellipsoidal systems and a handful others, in addition to the Cartesian, cylindrical, and spherical systems.

Our approach here is simple: We will define three systems we view as important and, within these systems, will present those quantities that are useful in the study of electromagnetic fields. These include length, surface, and volume as well as the required transformations from one system of coordinates to the others. The latter is an important step because it clearly indicates that the fundamental quantities we treat are independent of the system of coordinates. We can perform any operation in any system we wish and transform it to any other system if this is needed. Also, we will have to evaluate the various vector operations in the three systems of coordinates and then use these as the basis of analysis.

1.5.1 The Cartesian[3] Coordinate System

In the right-handed Cartesian system (or right-handed rectangular system), a vector **A** connecting two general points $P_1(x_1,y_1,z_1)$ and $P_2(x_2,y_2,z_2)$ is given as

$$\mathbf{A}(x,y,z) = \hat{\mathbf{x}}A_x(x,y,z) + \hat{\mathbf{y}}A_y(x,y,z) + \hat{\mathbf{z}}A_z(x,y,z) \quad (1.47)$$

where the components A_x, A_y, and A_z are the projections of the vector on the x, y, and z coordinates, respectively.

An element of length $d\mathbf{l}$, or differential of length, is a vector with scalar components dx, dy, and dz and is shown in **Figure 1.23a**:

$$\boxed{d\mathbf{l} = \hat{\mathbf{x}}dx + \hat{\mathbf{y}}dy + \hat{\mathbf{z}}dz} \quad (1.48)$$

Note: The unit vectors $\hat{\mathbf{x}}$, $\hat{\mathbf{y}}$ and $\hat{\mathbf{z}}$ are constant; they point in fixed directions in space at any point.

The elements of surface may be deduced from **Figure 1.23**. Each of the differential surfaces is parallel to one of the planes as shown in **Figure 1.23b**. Thus, we define three differential surfaces as

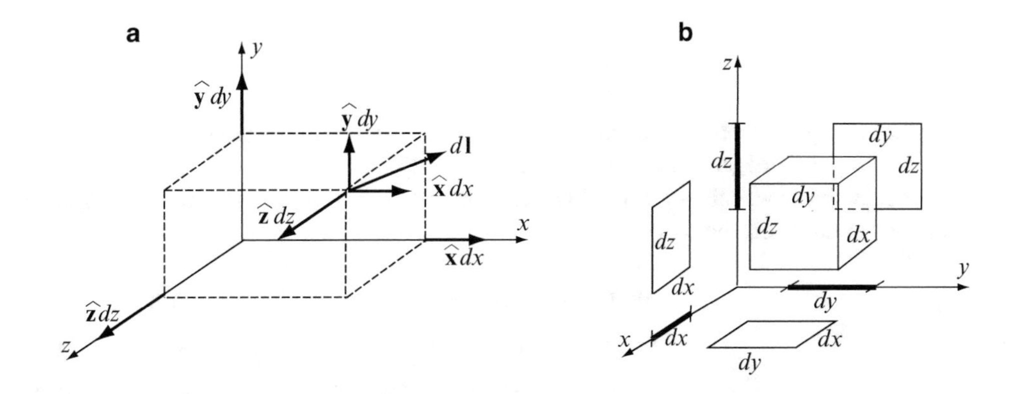

Figure 1.23 (a) The differential of length $d\mathbf{l}$ and its components in the Cartesian system of coordinates. (b) An element of volume and its surface projections on the xy, xz, and yz planes

[3] Named after René Descartes (1596–1650), French philosopher and mathematician (Cartesius is his Latinized name). The philosophical system he devised held until the Newtonian system superseded it. You may be familiar with the quote "I think, therefore I am," which Descartes coined and which was a central point in his philosophical system. The Cartesian system is named after him because he is considered to be the developer of analytical geometry. He presented the system of coordinates bearing his name in "La Geometrie," a work published in 1637.

$$\boxed{ds_x = dydz, \quad ds_y = dxdz, \quad ds_z = dxdy} \tag{1.49}$$

A differential of volume is defined as a rectangular prism with sides dx, dy, and dz, as shown in **Figure 1.23b**. The differential volume is a scalar and is written as

$$\boxed{dv = dxdydz} \tag{1.50}$$

In electromagnetics, it is often necessary to evaluate a vector function over a surface (such as integration over the surface). For this purpose, we orient the surface by defining the direction of the surface as the normal to the surface. A vector element of surface becomes a vector with magnitude equal to the element of surface [as defined in **Eq. (1.49)**] and directed in the direction of the normal unit vector to the surface. To ensure proper results, we define a positive surface vector if it points out of the volume enclosed by the surface. In a closed surface, like that shown in **Figure 1.23b**, the positive direction is easily identified. In an open surface, we must decide which side of the surface is the interior and which the exterior. **Figure 1.24** shows two surfaces. The first, in **Figure 1.24a**, is positive; the second, in **Figure 1.24b** is not defined. This, however, is often overcome from physical considerations such as location of sources. For example, we may decide that the direction pointing away from the source is positive even though the surface is not closed. In general, we define a positive direction for an open surface using the right-hand rule: If the fingers of the right hand point in the direction we traverse the boundary of the open surface, with palm facing the interior of the surface, then the thumb points in the direction of positive surface. This is shown in **Figure 1.24c** but it always depends on the direction of motion on the contour. In **Figure 1.24c**, if we were to move in the opposite direction, the surface shown would be negative. Fortunately, in practical application, it is often easy to identify the positive direction. Based on this definition, an element of surface is written as

$$d\mathbf{s} = \hat{\mathbf{n}}ds \tag{1.51}$$

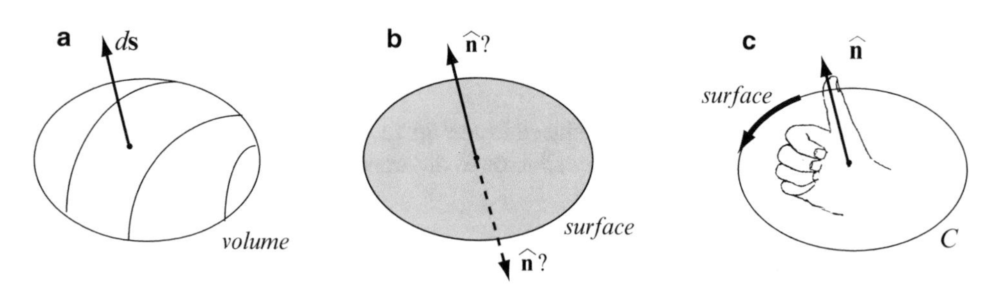

Figure 1.24 Direction of a surface. (**a**) The surface is always positive in the direction out of the volume. (**b**) In an open surface, the positive direction can be ambiguous. (**c**) The use of the right-hand rule to define direction of an open surface

$d\mathbf{s}$ should always be thought of as an element of surface ds, which is a scalar, and $\hat{\mathbf{n}}$, a normal unit vector to this surface which may be positive or negative (see **Figures 1.24** and **1.25**).

Consider the small cube shown in **Figure 1.25**. The six surfaces of the cube are parallel to the planes xy, xz, and yz. The six elemental surfaces can be written as:

$$\text{on the right face } d\mathbf{s}_x = \hat{\mathbf{x}}dydz; \text{ on the left face } d\mathbf{s}_x = -\hat{\mathbf{x}}dydz$$

$$\text{on the front face } d\mathbf{s}_z = \hat{\mathbf{z}}dxdy; \text{ on the back face } d\mathbf{s}_z = -\hat{\mathbf{z}}dxdy$$

$$\text{on the top face } d\mathbf{s}_y = \hat{\mathbf{y}}dxdz; \text{ on the bottom face } d\mathbf{s}_y = -\hat{\mathbf{y}}dxdz$$

These can be summarized as follows:

$$d\mathbf{s}_x = \pm\hat{\mathbf{x}}dydz, \quad d\mathbf{s}_y = \pm\hat{\mathbf{y}}dxdz, \quad d\mathbf{s}_z = \pm\hat{\mathbf{z}}dxdy \tag{1.52}$$

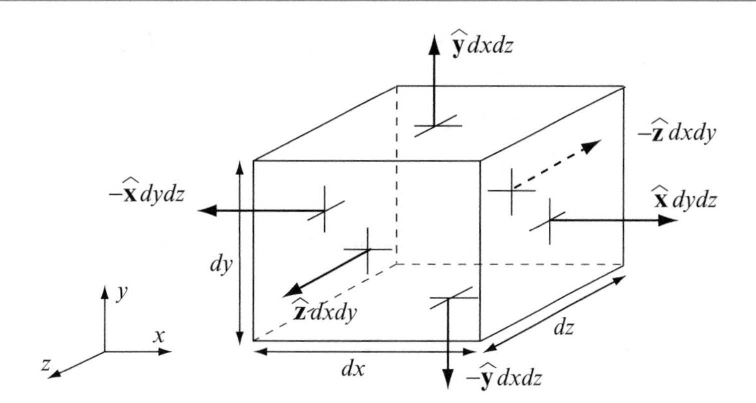

Figure 1.25 Directions of elements of surface in a cube. All surface elements are considered positive (pointing out of the volume) even though they may point in the direction of negative coordinates

Although the direction of $d\mathbf{s}$ may be in the positive or negative directions in space, the vector is considered to be positive with respect to the surface if it points out of the volume, regardless of its direction in space. Thus, all six elemental surfaces above are considered to be positive surface vectors.

Caution The elements of surface as defined above are not vector components of an area vector: they merely define the direction normal to the surface and the differential of surface, and each element should be viewed as an independent vector.

All other aspects of use of the Cartesian coordinate system including calculation of vector and scalar components, unit vectors, and the various scalar and vector operations were discussed in **Sections 1.2** through **1.4** and will not be repeated here. However, **Example 1.15** reviews some of the definitions involved.

Example 1.15 Three points are given in the Cartesian coordinate system: $P_1(2,-3,3)$, $P_2(1,1,5)$, and $P_3(3,-1,4)$.

(a) Find the three vectors: **A**, connecting P_1 to P_2; **B**, connecting P_1 to P_3; and **C**, connecting P_2 to P_3.
(b) Find the scalar component of vector **A** in the direction of vector **B**.
(c) Find the vector components of vector **B** in the direction of vector **C**.

Solution: (a) The vectors **A**, **B**, and **C** are found by calculating the components from the coordinates of the end points. For a vector connecting point (1) to point (2), the projection on each axis is the difference in the corresponding coordinates with point (2) (head) and point (1) (tail) of the vector. **(b)** The scalar component of **A** in the direction of **B** is the projection of **A** onto the unit vector $\hat{\mathbf{B}}$ and is calculated through the scalar product. **(c)** The vector components of **B** in the direction of **C** are found as in **(b)**, but now the unit vector $\hat{\mathbf{C}}$ is scaled by the scalar component $\mathbf{B}\cdot\hat{\mathbf{C}}$:

(a)

$$\mathbf{A} = \hat{\mathbf{x}}(1-2) + \hat{\mathbf{y}}(1+3) + \hat{\mathbf{z}}(5-3) = -\hat{\mathbf{x}}1 + \hat{\mathbf{y}}4 + \hat{\mathbf{z}}2$$

$$\mathbf{B} = \hat{\mathbf{x}}(3-2) + \hat{\mathbf{y}}(-1+3) + \hat{\mathbf{z}}(4-3) = \hat{\mathbf{x}}1 + \hat{\mathbf{y}}2 + \hat{\mathbf{z}}1$$

$$\mathbf{C} = \hat{\mathbf{x}}(3-1) + \hat{\mathbf{y}}(-1-1) + \hat{\mathbf{z}}(4-5) = \hat{\mathbf{x}}2 - \hat{\mathbf{y}}2 - \hat{\mathbf{z}}1$$

(b) To find the scalar component of **A** in the direction of **B**, we first calculate the unit vector $\hat{\mathbf{B}}$ and then the scalar product $\mathbf{A}\cdot\hat{\mathbf{B}}$ (see **Example 1.4**):

$$A_B = \mathbf{A}\cdot\hat{\mathbf{B}} = \frac{\mathbf{A}\cdot\mathbf{B}}{B} = \frac{(-\hat{\mathbf{x}}1 + \hat{\mathbf{y}}4 + \hat{\mathbf{z}}2)\cdot(\hat{\mathbf{x}}1 + \hat{\mathbf{y}}2 + \hat{\mathbf{z}}1)}{\sqrt{1^2+2^2+1^2}} = \frac{-1+8+2}{\sqrt{6}} = \frac{9}{\sqrt{6}}.$$

The scalar component of **A** in the direction of **B** equals $9/\sqrt{6}$.

(c) The vector component of **B** in the direction of **C** is calculated (see **Example 1.4**) as

$$\hat{\mathbf{C}}B_C = \hat{\mathbf{C}}\left(\mathbf{B}\cdot\hat{\mathbf{C}}\right) = \frac{\mathbf{C}(\mathbf{B}\cdot\mathbf{C})}{C^2} = \frac{(\hat{\mathbf{x}}2-\hat{\mathbf{y}}2-\hat{\mathbf{z}}1)\left[(\hat{\mathbf{x}}1+\hat{\mathbf{y}}2+\hat{\mathbf{z}}1)\cdot(\hat{\mathbf{x}}2-\hat{\mathbf{y}}2-\hat{\mathbf{z}}1)\right]}{C_x^2+C_y^2+C_z^2}$$

$$= \frac{(\hat{\mathbf{x}}2-\hat{\mathbf{y}}2-\hat{\mathbf{z}}1)[2-4-1]}{4+4+1} = \frac{-\hat{\mathbf{x}}2+\hat{\mathbf{y}}2+\hat{\mathbf{z}}1}{3}$$

The vector components of **B** in the direction of **C** are $-\hat{\mathbf{x}}2/3$, $\hat{\mathbf{y}}2/3$, and $\hat{\mathbf{z}}1/3$.

1.5.2 The Cylindrical Coordinate System

The need for a cylindrical coordinate system should be apparent from **Figure 1.26a**, where the cylindrical surface (such as a pipe) is located along the z axis in the Cartesian system of coordinates. To describe a point P_1 on the surface, we must give the three coordinates $P_1(x_1,y_1,z_1)$. A second point on the cylindrical surface, $P_2(x_2,y_2,z_2)$ is shown and a vector connects the two points. The vector can be immediately written as

$$\mathbf{A} = \hat{\mathbf{x}}(x_2 - x_1) + \hat{\mathbf{y}}(y_2 - y_1) + \hat{\mathbf{z}}(z_2 - z_1) \tag{1.53}$$

On the other hand, we observe that points P_1 and P_2 are at a constant distance from the z axis, equal to the radius of the cylinder. To draw this cylinder all we need is to draw a circle at a constant radius and repeat the process for each value of z. In doing so, a point at radius r is rotated an angle of 2π to describe the circle. This suggests that the process above is easiest to describe in terms of a radial distance (radius of the point), an angle of rotation, and length of the cylinder in the z direction. The result is the cylindrical system of coordinates. It is also called a circular-cylindrical system to distinguish it from the polar-cylindrical system, but we will use the short form "cylindrical" throughout this book.

The cylindrical system is shown in **Figure 1.26b**. The axes are orthogonal and the ϕ axis is positive in the counterclockwise direction when viewed from the positive z axis. Similarly, r is positive in the direction away from the z axis whereas the z axis is the same as for the Cartesian coordinate system and serves as the axis of the cylinder. The r axis extends from zero to $+\infty$, the z axis extends from $-\infty$ to $+\infty$, and the ϕ axis varies from 0 to 2π. The ϕ angle is also called the ***azimuthal angle*** and is given with reference to the x axis of a superposed Cartesian system. This reference also allows transformation between the two systems of coordinates.

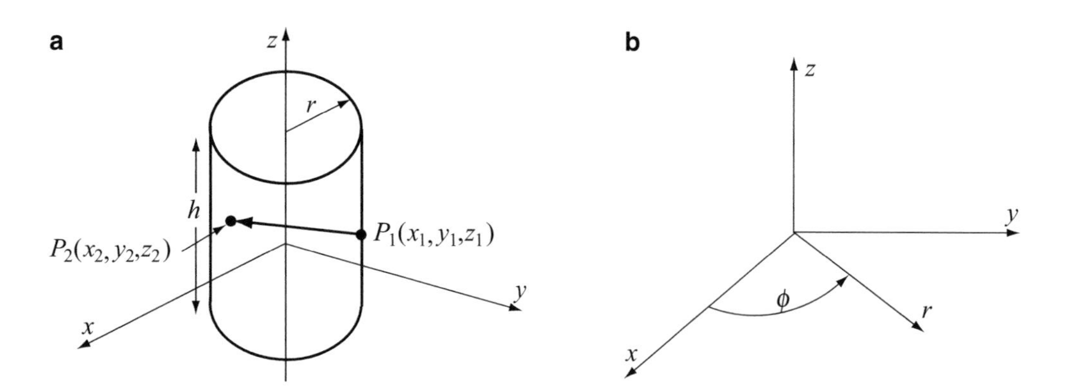

Figure 1.26 (**a**) Two points on a cylindrical surface described in terms of their Cartesian coordinates. (**b**) The cylindrical coordinate system and its relation to the Cartesian system

A general vector in the cylindrical coordinate system is given as

$$\mathbf{A} = \hat{\mathbf{r}}A_r(r,\phi,z) + \hat{\boldsymbol{\phi}}A_\phi(r,\phi,z) + \hat{\mathbf{z}}A_z(r,\phi,z) \tag{1.54}$$

All other aspects of vector algebra that we have defined are preserved. The unit vector, the magnitude of the vector, as well as vector and scalar products are evaluated in an identical fashion although the fact that one of the coordinates is an angle must be taken into account, as we shall see shortly. Since the three coordinates are orthogonal to each other, the scalar and vector products of the unit vectors are

$$\hat{\mathbf{r}}\cdot\hat{\mathbf{r}} = \hat{\boldsymbol{\phi}}\cdot\hat{\boldsymbol{\phi}} = \hat{\mathbf{z}}\cdot\hat{\mathbf{z}} = 1 \tag{1.55}$$

$$\hat{\mathbf{r}}\cdot\hat{\boldsymbol{\phi}} = \hat{\mathbf{r}}\cdot\hat{\mathbf{z}} = \hat{\boldsymbol{\phi}}\cdot\hat{\mathbf{z}} = \hat{\boldsymbol{\phi}}\cdot\hat{\mathbf{r}} = \hat{\mathbf{z}}\cdot\hat{\mathbf{r}} = \hat{\mathbf{z}}\cdot\hat{\boldsymbol{\phi}} = 0 \tag{1.56}$$

$$\hat{\mathbf{r}}\times\hat{\mathbf{r}} = \hat{\boldsymbol{\phi}}\times\hat{\boldsymbol{\phi}} = \hat{\mathbf{z}}\times\hat{\mathbf{z}} = 0 \tag{1.57}$$

$$\hat{\mathbf{r}}\times\hat{\boldsymbol{\phi}} = \hat{\mathbf{z}},\quad \hat{\boldsymbol{\phi}}\times\hat{\mathbf{z}} = \hat{\mathbf{r}},\quad \hat{\mathbf{z}}\times\hat{\mathbf{r}} = \hat{\boldsymbol{\phi}},\quad \hat{\boldsymbol{\phi}}\times\hat{\mathbf{r}} = -\hat{\mathbf{z}},\quad \hat{\mathbf{z}}\times\hat{\boldsymbol{\phi}} = -\hat{\mathbf{r}},\quad \hat{\mathbf{r}}\times\hat{\mathbf{z}} = -\hat{\boldsymbol{\phi}} \tag{1.58}$$

Next, we need to define the differentials of length, surface, and volume in the cylindrical system. This is shown in **Figure 1.27**. The differential lengths in the r and z directions are dr and dz, correspondingly. In the ϕ direction, the differential of length is an arc of length $rd\phi$, as shown in **Figure 1.27**. Thus, the differential length in cylindrical coordinates is

$$\boxed{d\mathbf{l} = \hat{\mathbf{r}}dr + \hat{\boldsymbol{\phi}}rd\phi + \hat{\mathbf{z}}dz} \tag{1.59}$$

The differentials of area are

$$\boxed{ds_r = rd\phi dz,\quad ds_\phi = drdz,\quad ds_z = rd\phi dr} \tag{1.60}$$

The differential volume is therefore

$$\boxed{dv = dr(rd\phi)dz = r\,drd\phi dz} \tag{1.61}$$

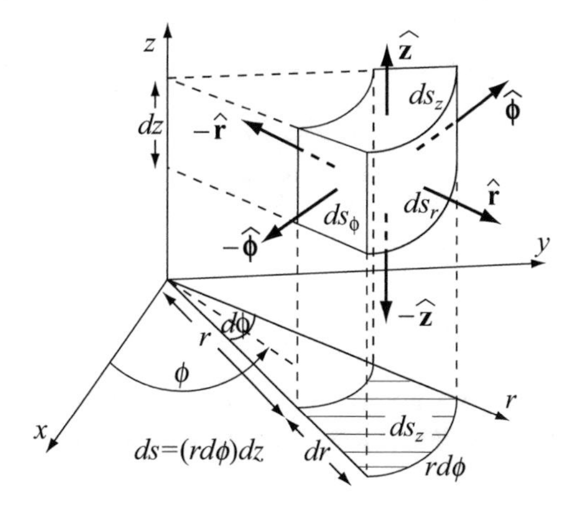

Figure 1.27 Differentials of length, surface, and volume in cylindrical coordinates

The oriented differentials of surface are defined as for the Cartesian coordinate system in terms of unit vectors normal to the surface of a cylindrical object (see **Figure 1.27**):

$$\boxed{d\mathbf{s}_r = \pm\hat{\mathbf{r}}rd\phi dz,\quad d\mathbf{s}_\phi = \pm\hat{\boldsymbol{\phi}}drdz,\quad d\mathbf{s}_z = \pm\hat{\mathbf{z}}rd\phi dr} \tag{1.62}$$

Of course, the surface vector for a general surface will vary, but it must be normal to the surface and is considered positive if it points out of the volume enclosed by the surface.

The fundamental principle that we followed in defining systems of coordinates is that the fields are independent of the system of coordinates. This also means that we can transform from one system of coordinates to another at will. All we need is to identify the transformation necessary and ensure that this transformation is unique. To find the transformation between the cylindrical and Cartesian systems, we superimpose the two systems on each other so that the z axes of the systems coincide as in **Figure 1.28a**. In the Cartesian system, the point P has coordinates (x,y,z). In the cylindrical system, the coordinates are r, ϕ, and z.

If we assume (r,ϕ,z) are known, we can transform these into the Cartesian coordinates using the relations in **Figure 1.28a**:

$$x = r\cos\phi, \quad y = r\sin\phi, \quad z = z \tag{1.63}$$

Similarly, we can write for the inverse transformation (assuming x, y, and z are known) either directly from **Figure 1.28a** or from **Eq. (1.63)**:

$$r = \sqrt{x^2 + y^2}, \quad \phi = \tan^{-1}\left(\frac{y}{x}\right), \quad z = z \tag{1.64}$$

These are the transformations for a single point. We also need to transform the unit vectors and, finally, the vectors from one system to the other. To transform the unit vectors, we use **Figure 1.28b**. First, we note that the unit vector in the z direction remains unchanged as expected. To calculate the unit vectors in the r and ϕ directions, we resolve the unit vectors $\hat{\mathbf{x}}$ and $\hat{\mathbf{y}}$ onto the r–ϕ plane by calculating their projections in the r and ϕ directions. Since all unit vectors are of unit length, they all fall on the unit circle shown. Thus,

$$\hat{\boldsymbol{\phi}} = -\hat{\mathbf{x}}\sin\phi + \hat{\mathbf{y}}\cos\phi, \quad \hat{\mathbf{r}} = \hat{\mathbf{x}}\cos\phi + \hat{\mathbf{y}}\sin\phi, \quad \hat{\mathbf{z}} = \hat{\mathbf{z}} \tag{1.65}$$

The inverse transformation is also obtained from **Figure 1.28b** by an identical process:

$$\hat{\mathbf{x}} = \hat{\mathbf{r}}\cos\phi - \hat{\boldsymbol{\phi}}\sin\phi, \quad \hat{\mathbf{y}} = \hat{\mathbf{r}}\sin\phi + \hat{\boldsymbol{\phi}}\cos\phi, \quad \hat{\mathbf{z}} = \hat{\mathbf{z}} \tag{1.66}$$

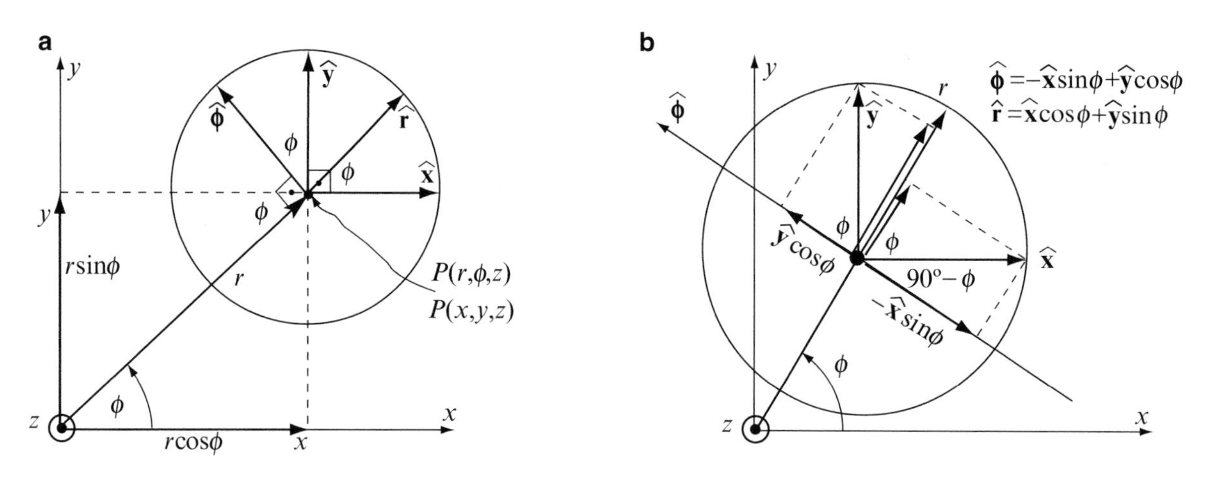

Figure 1.28 (**a**) Relation between unit vectors in the Cartesian and cylindrical systems. The circle has unit radius. (**b**) Calculation of unit vectors in cylindrical coordinates as projections of the unit vectors in Cartesian coordinates

Important Note: Unlike in Cartesian coordinates, the unit vectors $\hat{\mathbf{r}}$ and $\hat{\boldsymbol{\phi}}$ (in cylindrical coordinates) are not constant; both depend on ϕ (whereas $\hat{\mathbf{z}}$ is constant). Therefore, whenever they are used, such as in integration, this fact must be taken into account. It will often become necessary to transform the unit vectors into Cartesian coordinates using **Eq. (1.66)** to avoid this difficulty.

To obtain the transformation necessary for a vector, we use the properties of the scalar product to find the scalar components of the vector in one system of coordinates in the directions of the unit vectors of the other system. For a vector **A** given in the cylindrical system, we can write

$$A_x = \hat{\mathbf{x}} \cdot \mathbf{A} = \hat{\mathbf{x}} \cdot \left(\hat{\mathbf{r}}A_r + \hat{\boldsymbol{\phi}}A_\phi + \hat{\mathbf{z}}A_z\right) = \left(\hat{\mathbf{x}} \cdot \hat{\mathbf{r}}\right)A_r + \left(\hat{\mathbf{x}} \cdot \hat{\boldsymbol{\phi}}\right)A_\phi + \left(\hat{\mathbf{x}} \cdot \hat{\mathbf{z}}\right)A_z \tag{1.67}$$

From **Figure 1.28b**, $\hat{\mathbf{x}} \cdot \hat{\mathbf{r}} = \cos\phi$, $\hat{\mathbf{x}} \cdot \hat{\boldsymbol{\phi}} = -\sin\phi$, and $\hat{\mathbf{x}} \cdot \hat{\mathbf{z}} = 0$. Therefore,

$$A_x = A_r\cos\phi - A_\phi\sin\phi \tag{1.68}$$

Similarly, calculating the products $A_y = \hat{\mathbf{y}} \cdot \mathbf{A}$ and $A_z = \hat{\mathbf{z}} \cdot \mathbf{A}$, we get

$$A_y = A_r\sin\phi + A_\phi\cos\phi \quad \text{and} \quad A_z = A_z \tag{1.69}$$

To find the inverse transformation, we can write vector **A** in the Cartesian coordinate system and repeat the process above by finding its projections on the r, ϕ, and z axes. Alternatively, we calculate the inverse transformation by first writing **Eqs. (1.68)** and **(1.69)** as a system of equations:

$$\begin{bmatrix} A_x \\ A_y \\ A_z \end{bmatrix} = \begin{bmatrix} \cos\phi & -\sin\phi & 0 \\ \sin\phi & \cos\phi & 0 \\ 0 & 0 & 1 \end{bmatrix} \begin{bmatrix} A_r \\ A_\phi \\ A_z \end{bmatrix} \tag{1.70}$$

Calculating the inverse of this system, we get

$$\begin{bmatrix} A_r \\ A_\phi \\ A_z \end{bmatrix} = \begin{bmatrix} \cos\phi & \sin\phi & 0 \\ -\sin\phi & \cos\phi & 0 \\ 0 & 0 & 1 \end{bmatrix} \begin{bmatrix} A_x \\ A_y \\ A_z \end{bmatrix} \tag{1.71}$$

Now, for a general vector given in the cylindrical coordinate system, we can write the same vector in the Cartesian system by using the scalar components from **Eq. (1.70)** and adding the unit vectors. Similarly, if a vector in the Cartesian system must be transformed into the cylindrical system, we use **Eq. (1.71)** to evaluate its components.

Example 1.16 Two points in cylindrical coordinates are given as $P_1(r_1,\phi_1,z_1)$ and $P_2(r_2,\phi_2,z_2)$. Find the expression of the vector pointing from P_1 to P_2:

(a) In Cartesian coordinates.
(b) In cylindrical coordinates.
(c) Calculate the length of the vector.

Solution: The cylindrical coordinates are first converted into Cartesian coordinates. After the components of the vector are found, these are transformed back into cylindrical coordinates. Calculation of the length of the vector must be done in Cartesian coordinates in which each coordinate represents the same quantity (for example, length):

(a) The coordinates of points P_1 and P_2 in Cartesian coordinates are [**Eq. (1.63)**]

$$x_1 = r_1\cos\phi_1, \quad y_1 = r_1\sin\phi_1, \quad z_1 = z_1$$
$$x_2 = r_2\cos\phi_2, \quad y_2 = r_2\sin\phi_2, \quad z_2 = z_2$$

Thus, the vector **A** in Cartesian coordinates is

$$\mathbf{A} = \hat{\mathbf{x}}(x_2 - x_1) + \hat{\mathbf{y}}(y_2 - y_1) + \hat{\mathbf{z}}(z_2 - z_1) = \hat{\mathbf{x}}(r_2\cos\phi_2 - r_1\cos\phi_1) + \hat{\mathbf{y}}(r_2\sin\phi_2 - r_1\sin\phi_1) + \hat{\mathbf{z}}(z_2 - z_1)$$

(b) To find the components of the vector in cylindrical coordinates, we write from **Eq. (1.71)**

$$\begin{bmatrix} A_r \\ A_\phi \\ A_z \end{bmatrix} = \begin{bmatrix} \cos\phi & \sin\phi & 0 \\ -\sin\phi & \cos\phi & 0 \\ 0 & 0 & 1 \end{bmatrix} \begin{bmatrix} r_2\cos\phi_2 - r_1\cos\phi_1 \\ r_2\sin\phi_2 - r_1\sin\phi_1 \\ z_2 - z_1 \end{bmatrix}$$

Expanding, we get:

$$A_r = (r_2\cos\phi_2 - r_1\cos\phi_1)\cos\phi + (r_2\sin\phi_2 - r_1\sin\phi_1)\sin\phi$$

$$A_\phi = -(r_2\cos\phi_2 - r_1\cos\phi_1)\sin\phi + (r_2\sin\phi_2 - r_1\sin\phi_1)\cos\phi$$

$$A_z = z_2 - z_1$$

The vector in cylindrical coordinates is

$$\mathbf{A}(r,\phi,z) = \hat{\mathbf{r}}[(r_2\cos\phi_2 - r_1\cos\phi_1)\cos\phi + (r_2\sin\phi_2 - r_1\sin\phi_1)\sin\phi] \\ + \hat{\boldsymbol{\phi}}[-(r_2\cos\phi_2 - r_1\cos\phi_1)\sin\phi + (r_2\sin\phi_2 - r_1\sin\phi_1)\cos\phi] + \hat{\mathbf{z}}[z_2 - z_1]$$

Note that the r and ϕ components of the vector in cylindrical coordinates are not constant (they depend on the angle ϕ in addition to the coordinates of points P_1 and P_2). In Cartesian coordinates, the components only depend on the two end points. Because of this, it is often necessary to transform from cylindrical to Cartesian coordinates, especially when the magnitudes of vectors need to be evaluated.

(c) To calculate the length of the vector, we use the representation in Cartesian coordinates because it is easier to evaluate. The length of the vector is

$$|A| = \sqrt{A_x^2 + A_y^2 + A_z^2} \\ = \sqrt{(r_2\cos\phi_2 - r_1\cos\phi_1)^2 + (r_2\sin\phi_2 - r_1\sin\phi_1)^2 + (z_2 - z_1)^2} \\ = \sqrt{r_2^2 + r_1^2 - 2r_2r_1(\cos\phi_2\cos\phi_1 + \sin\phi_2\sin\phi_1) + (z_2 - z_1)^2}$$

With $\cos\phi_2\cos\phi_1 + \sin\phi_2\sin\phi_1 = \cos(\phi_2 - \phi_1)$, we get

$$|\mathbf{A}| = \sqrt{r_2^2 + r_1^2 - 2r_2r_1\cos(\phi_2 - \phi_1) + (z_2 - z_1)^2}$$

This expression may be used to calculate the magnitude of a vector when two points are given in cylindrical coordinates without first transforming into Cartesian coordinates.

Example 1.17 A vector is given in cylindrical coordinates as: $\mathbf{A} = \hat{\mathbf{r}}2 + \hat{\boldsymbol{\phi}}3 - \hat{\mathbf{z}}1$. Describe this vector in Cartesian coordinates.

Solution The vector in Cartesian coordinates is written directly from **Eqs. (1.63)**, **(1.64)**, and **(1.70)**:

$$\begin{bmatrix} A_x \\ A_y \\ Az \end{bmatrix} = \begin{bmatrix} \cos\phi & -\sin\phi & 0 \\ \sin\phi & \cos\phi & 0 \\ 0 & 0 & 1 \end{bmatrix} \begin{bmatrix} A_r \\ A_\phi \\ A_z \end{bmatrix} = \begin{bmatrix} \cos\phi & -\sin\phi & 0 \\ \sin\phi & \cos\phi & 0 \\ 0 & 0 & 1 \end{bmatrix} \begin{bmatrix} 2 \\ 3 \\ -1 \end{bmatrix}$$

$$= \begin{bmatrix} 2\cos\phi - 3\sin\phi \\ 2\sin\phi + 3\cos\phi \\ -1 \end{bmatrix} = \begin{bmatrix} \dfrac{2x}{\sqrt{x^2+y^2}} - \dfrac{3y}{\sqrt{x^2+y^2}} \\ \dfrac{2y}{\sqrt{x^2+y^2}} + \dfrac{3x}{\sqrt{x^2+y^2}} \\ -1 \end{bmatrix}$$

where the following substitutions were made using **Eq. (1.63)**:

$$\cos\phi = \frac{x}{r}, \quad \sin\phi = \frac{y}{r}, \quad r = \sqrt{x^2 + y^2}$$

Thus, vector **A** is

$$\mathbf{A} = \hat{\mathbf{x}}\left(\frac{2x}{\sqrt{x^2+y^2}} - \frac{3y}{\sqrt{x^2+y^2}}\right) + \hat{\mathbf{y}}\left(\frac{2y}{\sqrt{x^2+y^2}} + \frac{3x}{\sqrt{x^2+y^2}}\right) - \hat{\mathbf{z}}1$$

Exercise 1.9 Transform $\mathbf{A} = \hat{\mathbf{x}}2 - \hat{\mathbf{y}}5 + \hat{\mathbf{z}}3$ into cylindrical coordinates at point $(x = -2, y = 3, z = 1)$.

Answer $\mathbf{A} = -\hat{\mathbf{r}}5.27 + \hat{\boldsymbol{\phi}}1.11 + \hat{\mathbf{z}}3$

1.5.3 The Spherical Coordinate System

The spherical coordinate system is defined following the same basic ideas used for the cylindrical system. First, we observe that a point on a spherical surface is at a constant distance from the center of the sphere. To draw a sphere, we can follow the process in **Figure 1.29**. First, a point $P(x,y,z)$ is defined in the Cartesian coordinate system at a distance R from the origin as in **Figure 1.29a**. Suppose this point is rotated around the origin to form a half-circle as shown in **Figure 1.29b**. Now, we can rotate the half-circle in **Figure 1.29c** around the z axis on a 2π angle to obtain a sphere. This process indicates that a point on a sphere is best described in terms of its radial location and two angles, provided proper references can be identified. To do so, we use the Cartesian reference system in **Figure 1.30a**. The z axis serves as reference for the first angle, θ, while the x axis is used for the second angle, ϕ. The three unit vectors at point P are also shown and these form an orthogonal, right-hand system of coordinates. The range of the angle θ is between zero and π and that of ϕ between zero and 2π, whereas the range of R is between 0 and ∞. The point P is now described as $P(R,\theta,\phi)$ and a general vector is written as

$$\mathbf{A} = \hat{\mathbf{R}}\mathrm{A}_R + \hat{\boldsymbol{\theta}}A_\theta + \hat{\boldsymbol{\phi}}A_\phi \tag{1.72}$$

Since the axes are orthogonal, we have

$$\hat{\mathbf{R}}\cdot\hat{\mathbf{R}} = \hat{\boldsymbol{\theta}}\cdot\hat{\boldsymbol{\theta}} = \hat{\boldsymbol{\phi}}\cdot\hat{\boldsymbol{\phi}} = 1 \tag{1.73}$$

$$\hat{\mathbf{R}}\cdot\hat{\boldsymbol{\theta}} = \hat{\mathbf{R}}\cdot\hat{\boldsymbol{\phi}} = \hat{\boldsymbol{\theta}}\cdot\hat{\boldsymbol{\phi}} = \hat{\boldsymbol{\theta}}\cdot\hat{\mathbf{R}} = \hat{\boldsymbol{\phi}}\cdot\hat{\mathbf{R}} = \hat{\boldsymbol{\phi}}\cdot\hat{\boldsymbol{\theta}} = 0 \tag{1.74}$$

$$\hat{\mathbf{R}}\times\hat{\mathbf{R}} = \hat{\boldsymbol{\theta}}\times\hat{\boldsymbol{\theta}} = \hat{\boldsymbol{\phi}}\times\hat{\boldsymbol{\phi}} = 0 \tag{1.75}$$

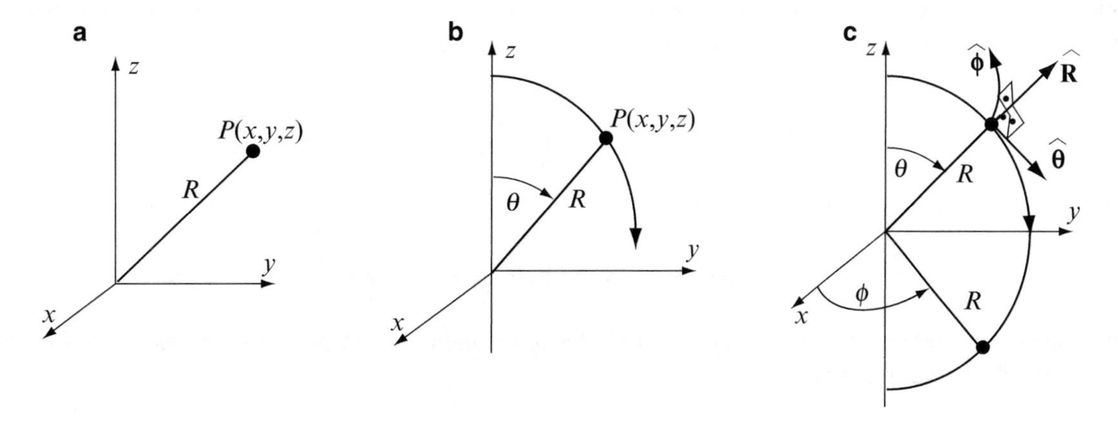

Figure 1.29 Drawing a sphere: (**a**) A point at distance R from the origin. (**b**) Keeping the distance R constant, we move away from the z axis describing an angle θ. (**c**) The half-circle is rotated at an angle ϕ. If $\phi = 2\pi$, a sphere is generated

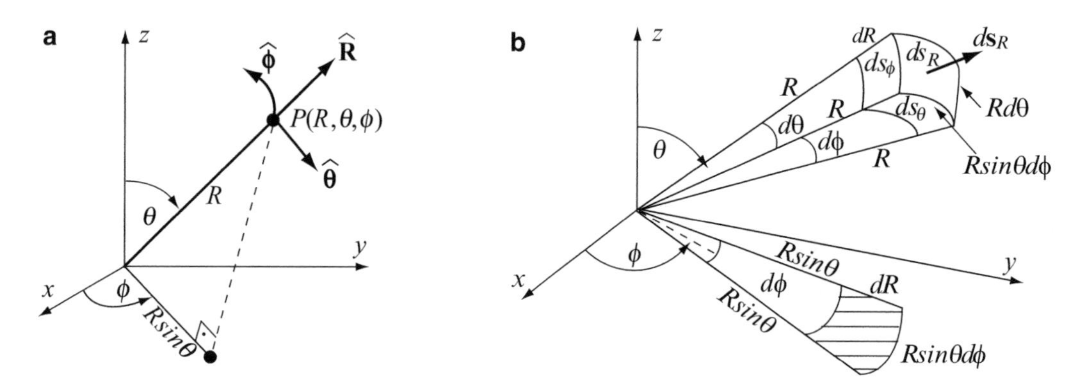

Figure 1.30 (**a**) A point in spherical coordinates and the relationship between the spherical and Cartesian coordinate systems. (**b**) Differentials of length, area, and volume in spherical coordinates

$$\hat{\mathbf{R}} \times \hat{\boldsymbol{\theta}} = \hat{\boldsymbol{\phi}}, \quad \hat{\boldsymbol{\theta}} \times \hat{\boldsymbol{\phi}} = \hat{\mathbf{R}}, \quad \hat{\boldsymbol{\phi}} \times \hat{\mathbf{R}} = \hat{\boldsymbol{\theta}}, \quad \hat{\boldsymbol{\theta}} \times \hat{\mathbf{R}} = -\hat{\boldsymbol{\phi}}, \quad \hat{\boldsymbol{\phi}} \times \hat{\boldsymbol{\theta}} = -\hat{\mathbf{R}}, \quad \hat{\mathbf{R}} \times \hat{\boldsymbol{\phi}} = -\hat{\boldsymbol{\theta}} \tag{1.76}$$

The differential length, volume, and surface are defined with the aid of **Figure 1.30b**, but, now, the lengths in both θ and ϕ directions are arc lengths. These are given as

$$\boxed{d\mathbf{l} = \hat{\mathbf{R}}dR + \hat{\boldsymbol{\theta}}Rd\theta + \hat{\boldsymbol{\phi}}R\sin\theta d\phi} \tag{1.77}$$

Differentials of surface are defined in a manner similar to the cylindrical coordinate system. These are

$$\boxed{ds_R = R^2 sin\theta d\theta d\phi, \quad ds_\theta = R\sin\theta dRd\phi, \quad ds_\phi = RdRd\theta} \tag{1.78}$$

The differential volume is

$$\boxed{dv = dR(Rd\theta)(R\sin\theta d\phi) = R^2\sin\theta dRd\theta d\phi} \tag{1.79}$$

The three basic surface vectors in the directions perpendicular to the three planes $\theta\phi$, $R\phi$, and $R\theta$ are

$$\boxed{d\mathbf{s}_R = \pm\hat{\mathbf{R}}R^2\sin\theta d\theta d\phi, \qquad d\mathbf{s}_\theta = \pm\ \hat{\boldsymbol{\theta}}R\sin\theta dRd\phi, \qquad d\mathbf{s}_\phi = \pm\ \hat{\boldsymbol{\phi}}RdRd\theta} \tag{1.80}$$

To define the coordinate transformation between spherical and Cartesian coordinates, we again use **Figure 1.30a**. The basic geometrical relations between the coordinates in the two systems are

$$R = \sqrt{x^2 + y^2 + z^2}, \quad \theta = \tan^{-1}\left(\frac{\sqrt{x^2 + y^2}}{z}\right), \quad \phi = \tan^{-1}\left(\frac{y}{x}\right) \tag{1.81}$$

Similarly, the inverse transformation is

$$x = R\sin\theta\cos\phi, \quad y = R\sin\theta\sin\phi, \quad z = R\cos\theta \tag{1.82}$$

Transformation of a general vector from spherical to Cartesian coordinates is performed by calculation of the scalar components through the scalar product:

$$A_x = \hat{\mathbf{x}} \cdot \mathbf{A} = \hat{\mathbf{x}} \cdot \left(\hat{\mathbf{R}}A_R + \hat{\boldsymbol{\theta}}A_\theta + \hat{\boldsymbol{\phi}}A_\phi\right) = \left(\hat{\mathbf{x}} \cdot \hat{\mathbf{R}}\right)A_R + \left(\hat{\mathbf{x}} \cdot \hat{\boldsymbol{\theta}}\right)A_\theta + \left(\hat{\mathbf{x}} \cdot \hat{\boldsymbol{\phi}}\right)A_\phi \tag{1.83}$$

From **Figure 1.30a**,

$$\hat{\mathbf{x}} \cdot \hat{\mathbf{R}} = \sin\theta\cos\phi, \qquad \hat{\mathbf{x}} \cdot \hat{\boldsymbol{\theta}} = \cos\theta\cos\phi \qquad \hat{\mathbf{x}} \cdot \hat{\boldsymbol{\phi}} = -\sin\phi \tag{1.84}$$

The scalar components of **A** are

$$A_x = A_R\sin\theta\cos\phi + A_\theta\cos\theta\cos\phi - A_\phi\sin\phi \tag{1.85}$$

$$A_y = A_R\sin\theta\sin\phi + A_\theta\cos\theta\sin\phi + A_\phi\cos\phi \tag{1.86}$$

$$A_z = A_R\cos\theta - A_\theta\sin\theta \tag{1.87}$$

Again making use of the matrix notation,

$$\begin{bmatrix} A_x \\ A_y \\ A_z \end{bmatrix} = \begin{bmatrix} \sin\theta\cos\phi & \cos\theta\cos\phi & -\sin\phi \\ \sin\theta\sin\phi & \cos\theta\sin\phi & \cos\phi \\ \cos\theta & -\sin\theta & 0 \end{bmatrix} \begin{bmatrix} A_R \\ A_\theta \\ A_\phi \end{bmatrix} \tag{1.88}$$

To obtain the inverse transformation, we invert the system:

$$\begin{bmatrix} A_R \\ A_\theta \\ A_\phi \end{bmatrix} = \begin{bmatrix} \sin\theta\cos\phi & \sin\theta\sin\phi & \cos\theta \\ \cos\theta\cos\phi & \cos\theta\sin\phi & -\sin\theta \\ -\sin\phi & \cos\phi & 0 \end{bmatrix} \begin{bmatrix} A_x \\ A_y \\ A_z \end{bmatrix} \tag{1.89}$$

Equation (1.89) may also be used to obtain the unit vectors $\hat{\mathbf{R}}, \hat{\boldsymbol{\theta}}$, and $\hat{\boldsymbol{\phi}}$ in terms of the unit vectors in Cartesian coordinates. These transformations are

$$\hat{\mathbf{R}} = \hat{\mathbf{x}}\sin\theta\cos\phi + \hat{\mathbf{y}}\sin\theta\sin\phi + \hat{\mathbf{z}}\cos\theta, \qquad \hat{\boldsymbol{\theta}} = \hat{\mathbf{x}}\cos\theta\cos\phi + \hat{\mathbf{y}}\cos\theta\sin\phi - \hat{\mathbf{z}}\sin\theta,$$
$$\hat{\boldsymbol{\phi}} = -\hat{\mathbf{x}}\sin\phi + \hat{\mathbf{y}}\cos\phi \tag{1.90}$$

Important Note: Clearly, $\hat{\mathbf{R}}, \hat{\boldsymbol{\theta}}$ and $\hat{\boldsymbol{\phi}}$ are not constant unit vectors in space; $\hat{\mathbf{R}}$ and $\hat{\boldsymbol{\theta}}$ depend on θ and ϕ, whereas $\hat{\boldsymbol{\phi}}$ depends on ϕ. Whenever unit vectors in spherical coordinates occur inside integrals, they must be resolved into Cartesian coordinates. The Cartesian unit vectors can then be taken outside the integral sign.

1.5.4 Transformation from Cylindrical to Spherical Coordinates

On occasion, there will also be a need to transform vectors or points from cylindrical to spherical coordinates and vice versa. We list the transformation below without details of the derivation (see **Exercise 1.11**).

The spherical coordinates R, θ, and ϕ are obtained from the cylindrical coordinates r, ϕ, and z as

$$R = \sqrt{r^2 + z^2}, \quad \theta = \tan^{-1}(r/z), \quad \phi = \phi \tag{1.91}$$

The scalar components of a vector **A** in spherical coordinates can be obtained from the scalar components of the vector in cylindrical coordinates as

$$\begin{bmatrix} A_R \\ A_\theta \\ A_\phi \end{bmatrix} = \begin{bmatrix} \sin\theta & 0 & \cos\theta \\ \cos\theta & 0 & -\sin\theta \\ 0 & 1 & 0 \end{bmatrix} \begin{bmatrix} A_r \\ A_\phi \\ A_z \end{bmatrix} \tag{1.92}$$

The cylindrical coordinates r, ϕ, and z are obtained from the spherical coordinates R, θ, and ϕ as

$$r = R\sin\theta, \quad \phi = \phi, \quad z = R\cos\theta \tag{1.93}$$

The scalar components of the vector **A** in cylindrical coordinates can be obtained from the scalar components of the vector in spherical coordinates as

$$\begin{bmatrix} A_r \\ A_\phi \\ A_z \end{bmatrix} = \begin{bmatrix} \sin\theta & \cos\theta & 0 \\ 0 & 0 & 1 \\ \cos\theta & -\sin\theta & 0 \end{bmatrix} \begin{bmatrix} A_R \\ A_\theta \\ A_\phi \end{bmatrix} \tag{1.94}$$

The scalar components of the unit vectors are obtained by replacing the scalar components in **Eqs. (1.92)** or (**1.94**) with the appropriate unit vector components (see **Exercise 1.11**).

Example 1.18 Two points are given in spherical coordinates as $P_1(R_1,\theta_1,\phi_1)$ and $P_2(R_2,\theta_2,\phi_2)$:

(a) Write the vector connecting P_1(tail) to P_2(head) in Cartesian coordinates.
(b) Calculate the length of the vector (distance between P_1 and P_2).

Solution The coordinates of P_1 and P_2 are first transformed into Cartesian coordinates, followed by evaluation of the magnitude of the vector connecting P_1 and P_2:

(a) The transformation from spherical to Cartesian coordinates [**Eq.** (**1.82**)] for points P_1 and P_2 gives

$$\begin{aligned} x_1 &= R_1\sin\theta_1\cos\phi_1, \quad & y_1 &= R_1\sin\theta_1\sin\phi_1, \quad & z_1 &= R_1\cos\theta_1 \\ x_2 &= R_2\sin\theta_2\cos\phi_2, \quad & y_2 &= R_2\sin\theta_2\sin\phi_2, \quad & z_2 &= R_2\cos\theta_2 \end{aligned}$$

The vector in Cartesian coordinates is

$$\begin{aligned} \mathbf{A} &= \hat{\mathbf{x}}(x_2 - x_1) + \hat{\mathbf{y}}(y_2 - y_1) + \hat{\mathbf{z}}(z_2 - z_1) \\ &= \hat{\mathbf{x}}(R_2\sin\theta_2\cos\phi_2 - R_1\sin\theta_1\cos\phi_1) + \hat{\mathbf{y}}(R_2\sin\theta_2\sin\phi_2 - R_1\sin\theta_1\sin\phi_1) \\ &\quad + \hat{\mathbf{z}}(R_2\cos\theta_2 - R_1\cos\theta_1) \end{aligned}$$

(b) The length of the vector in terms of spherical components can be written from **(a)**:

$$\begin{aligned} |A|^2 &= (R_2\sin\theta_2\cos\phi_2 - R_1\sin\theta_1\cos\phi_1)^2 + (R_2\sin\theta_2\sin\phi_2 - R_1\sin\theta_1\sin\phi_1)^2 + (R_2\cos\theta_2 - R_1\cos\theta_1)^2 \\ &= R_2^2\sin^2\theta_2\cos^2\phi_2 + R_1^2\sin^2\theta_1\cos^2\phi_1 - 2R_1R_2\sin\theta_1\sin\theta_2\cos\phi_1\cos\phi_2 + R_2^2\sin^2\theta_2\sin^2\phi_2 \\ &\quad + R_1^2\sin^2\theta_1\sin^2\phi_1 - 2R_1R_2\sin\theta_1\sin\theta_2\sin\phi_1\sin\phi_2 + R_2^2\cos^2\theta_2 + R_1^2\cos^2\theta_1 - 2R_1R_2\cos\theta_2\cos\theta_1 \end{aligned}$$

Rearranging terms and using the relation $\sin^2\alpha + \cos^2\alpha = 1$, we get

$$|\mathbf{A}|^2 = R_2^2 + R_1^2 - 2R_1R_2\sin\theta_1\sin\theta_2[\cos\phi_2\cos\phi_1 + \sin\phi_2\sin\phi_1] - 2R_1R_2\cos\theta_2\cos\theta_1$$

Using $\cos\alpha\cos\beta = (\cos(\alpha-\beta) + \cos(\alpha+\beta))/2$ and $\sin\alpha\sin\beta = (\cos(\alpha-\beta) - \cos(\alpha+\beta))/2$, after taking the square root of the expression, we get

$$|\mathbf{A}| = \sqrt{R_2^2 + R_1^2 - 2R_1R_2\sin\theta_1\sin\theta_2\cos(\phi_2 - \phi_1) - 2R_1R_2\cos\theta_2\cos\theta_1}$$

This is a convenient general formula for the calculation of the distance between two points in spherical coordinates, without the need to first convert the points to Cartesian coordinates.

Example 1.19 Two points are given in Cartesian coordinates as $P_1(0,0,1)$ and $P_2(2,1,3)$ and a vector **A** connects P_1(tail) to P_2(head). Find the unit vector in the direction of **A** in spherical and cylindrical coordinates.

Solution: The vector **A** connecting P_1(tail) and P_2(head) is found first, followed by the unit vector in the direction of **A**. The unit vector is then transformed into spherical and cylindrical coordinates using **Eqs. (1.89)** and **(1.71)**. The vector connecting P_1(tail) and P_2(head) is

$$\mathbf{A} = \hat{\mathbf{x}}(x_2 - x_1) + \hat{\mathbf{y}}(y_2 - y_1) + \hat{\mathbf{z}}(z_2 - z_1) = \hat{\mathbf{x}}(2-0) + \hat{\mathbf{y}}(1-0) + \hat{\mathbf{z}}(3-1) = \hat{\mathbf{x}}2 + \hat{\mathbf{y}}1 + \hat{\mathbf{z}}2$$

The unit vector in the direction of **A** is

$$\hat{\mathbf{A}} = \frac{\mathbf{A}}{A} = \frac{\hat{\mathbf{x}}2 + \hat{\mathbf{y}}1 + \hat{\mathbf{z}}2}{3} = \hat{\mathbf{x}}\frac{2}{3} + \hat{\mathbf{y}}\frac{1}{3} + \hat{\mathbf{z}}\frac{2}{3}$$

Taking this as a regular vector in Cartesian coordinates, the transformation of its components into spherical coordinates is

$$\begin{bmatrix} A_R \\ A_\theta \\ A_\phi \end{bmatrix} = \begin{bmatrix} \sin\theta\cos\phi & \sin\theta\sin\phi & \cos\theta \\ \cos\theta\cos\phi & \cos\theta\sin\phi & -\sin\theta \\ -\sin\phi & \cos\phi & 0 \end{bmatrix} \begin{bmatrix} \frac{2}{3} \\ \frac{1}{3} \\ \frac{2}{3} \end{bmatrix} = \begin{bmatrix} \frac{1}{3}(2\sin\theta\cos\phi + \sin\theta\sin\phi + 2\cos\theta) \\ \frac{1}{3}(2\cos\theta\cos\phi + \cos\theta\sin\phi - 2\sin\theta) \\ -\frac{1}{3}(2\sin\phi - \cos\phi) \end{bmatrix}$$

The unit vector in spherical coordinates is

$$\hat{\mathbf{A}} = \hat{\mathbf{R}}\frac{1}{3}(2\sin\theta\cos\phi + \sin\theta\sin\phi + 2\cos\theta) + \hat{\boldsymbol{\theta}}\frac{1}{3}(2\cos\theta\cos\phi + \cos\theta\sin\phi - 2\sin\theta) - \hat{\boldsymbol{\phi}}\frac{1}{3}(2\sin\phi - \cos\phi)$$

Although we do not show that the magnitude of the unit vector equals 1, the transformation cannot modify a vector in any way other than describing it in different coordinates. You are urged to verify the magnitude of $\hat{\mathbf{A}}$. Similarly, the transformation of the components into cylindrical coordinates is [from **Eq. (1.71)**]

$$\begin{bmatrix} A_r \\ A_\phi \\ A_z \end{bmatrix} = \begin{bmatrix} \cos\phi & \sin\phi & 0 \\ -\sin\phi & \cos\phi & 0 \\ 0 & 0 & 1 \end{bmatrix} \begin{bmatrix} \frac{2}{3} \\ \frac{1}{3} \\ \frac{2}{3} \end{bmatrix} = \begin{bmatrix} \frac{1}{3}(2\cos\phi + \sin\phi) \\ -\frac{1}{3}(2\sin\phi - \cos\phi) \\ \frac{2}{3} \end{bmatrix}$$

and the unit vector in cylindrical coordinates is

$$\hat{\mathbf{A}} = \hat{\mathbf{r}}\frac{1}{3}(2\cos\phi + \sin\phi) - \hat{\boldsymbol{\phi}}\frac{1}{3}(2\sin\phi - \cos\phi) + \hat{\mathbf{z}}\frac{2}{3}$$

Exercise 1.10 Repeat **Example 1.19**, but first transform the vector **A** into spherical and cylindrical coordinates and then divide each vector by its magnitude to find the unit vector.

Example 1.20 Given the points $P_1(2,2,-5)$ in Cartesian coordinates and $P_2(3,\pi,-2)$ in cylindrical coordinates, find:

(a) The vector connecting P_2(tail) to P_1(head) in spherical coordinates.
(b) The length of the vector in **(a)**.

Solution: In **(a)** it is easier to transform P_2 into Cartesian coordinates, write the vector in Cartesian coordinates and then transform to spherical coordinates. The Cartesian form also allows direct calculation of the length of the vector in **(b)**

(a) From **Eq. (1.63)** we have for P_2:

$$x_2 = r_2\cos\phi_2 = 3\cos\pi = -3,\ y_2 = r_2\sin\phi_2 = 3\sin\pi = 0,\ z_2 = -2.$$

P_2 in Cartesian coordinates is $(-3,0,-2)$. The vector in Cartesian coordinates is:

$$\mathbf{A} = \hat{\mathbf{x}}(x_1 - x_2) + \hat{\mathbf{y}}(y_1 - y_2) + \hat{\mathbf{z}}(z_1 - z_2) = \hat{\mathbf{x}}(2+3) + \hat{\mathbf{y}}(2-0) + \hat{\mathbf{z}}(-5+2) = \hat{\mathbf{x}}5 + \hat{\mathbf{y}}2 - \hat{\mathbf{z}}3$$

To transform the vector **A** into Spherical coordinates we use **Eq. (1.89)**:

$$\begin{bmatrix} A_R \\ A_\theta \\ A_\phi \end{bmatrix} = \begin{bmatrix} \sin\theta\cos\phi & \sin\theta\sin\phi & \cos\theta \\ \cos\theta\cos\phi & \cos\theta\sin\phi & -\sin\theta \\ -\sin\phi & \cos\phi & 0 \end{bmatrix} \begin{Bmatrix} 5 \\ 2 \\ -3 \end{Bmatrix}$$

That is

$$A_R = 5\sin\theta\cos\phi + 2\sin\theta\sin\phi - 3\cos\theta,\ A_\theta = 5\cos\theta\cos\phi + 2\cos\theta\sin\phi + 3\sin\theta,\ A_\phi = -5\sin\phi + 2\cos\phi.$$

Thus, the vector in spherical coordinates is:

$$\mathbf{A} = \hat{\mathbf{R}}(5\sin\theta\cos\phi + 2\sin\theta\sin\phi - 3\cos\theta) + \hat{\boldsymbol{\theta}}(5\cos\theta\cos\phi + 2\cos\theta\sin\phi + 3\sin\theta) + \hat{\boldsymbol{\phi}}(-5\sin\phi + 2\cos\phi)$$

(b) The length of the vector is its magnitude and is independent of the system of coordinates:

$$|\mathbf{A}| = \sqrt{A_x{}^2 + A_y{}^2 + A_z{}^2} = \sqrt{25+4+9} = 6.164$$

Exercise 1.11 Derive the transformation matrices from cylindrical to spherical and spherical to cylindrical coordinates, that is, find the coefficients in **Eqs. (1.92)** and **(1.94)** by direct application of the scalar product.

Exercise 1.12 Write the vector **A** connecting points $P_1(R_1,\theta_1,\phi_1)$ and $P_2(R_2,\theta_2,\phi_2)$, in cylindrical coordinates.

Answer

$$\begin{aligned}\mathbf{A} = {} & \hat{\mathbf{r}}[(R_2\sin\theta_2\cos\phi_2 - R_1\sin\theta_1\cos\phi_1)\cos\phi + (R_2\sin\theta_2\sin\phi_2 - R_1\sin\theta_1\sin\phi_1)\sin\phi] \\ & + \hat{\boldsymbol{\phi}}[-(R_2\sin\theta_2\cos\phi_2 - R_1\sin\theta_1\cos\phi_1)\sin\phi + (R_2\sin\theta_2\sin\phi_2 - R_1\sin\theta_1\sin\phi_1)\cos\phi] \\ & + \hat{\mathbf{z}}[R_2\cos\theta_2 - R_1\cos\theta_1]\end{aligned}$$

Exercise 1.13 Write the vector **A** connecting points $P_1(R_1,\theta_1,\phi_1)$ and $P_2(R_2,\theta_2,\phi_2)$ in spherical coordinates.

Answer

$$\mathbf{A} = \hat{\mathbf{R}}[(R_2\sin\theta_2\cos\phi_2 - R_1\sin\theta_1\cos\phi_1)\sin\theta\cos\phi + (R_2\sin\theta_2\sin\phi_2 - R_1\sin\theta_1\sin\phi_1)\sin\theta\sin\phi + (R_2\cos\theta_2 - R_1\cos\theta_1)\cos\theta]$$
$$+ \hat{\boldsymbol{\theta}}[(R_2\sin\theta_2\cos\phi_2 - R_1\sin\theta_1\cos\phi_1)\cos\theta\cos\phi + (R_2\sin\theta_2\sin\phi_2 - R_1\sin\theta_1\sin\phi_1)\cos\theta\sin\phi - (R_2\cos\theta_2 - R_1\cos\theta_1)\sin\theta]$$
$$+ \hat{\boldsymbol{\phi}}[-(R_2\sin\theta_2\cos\phi_2 - R_1\sin\theta_1\cos\phi_1)\sin\phi + (R_2\sin\theta_2\sin\phi_2 - R_1\sin\theta_1\sin\phi_1)\cos\phi]$$

1.6 Position Vectors

A ***position vector*** is defined as the vector connecting a reference point and a location or position (point) in space. The position vector always points to the position it identifies, as shown in **Figure 1.31a**. The advantage of using position vectors lies in the choice of the reference point. Normally, this point will be chosen as the origin of the system of coordinates.

A vector can always be represented by two position vectors: one pointing to its head and one to its tail as shown in **Figure 1.31b**. Vector **A** can now be written as

$$\mathbf{A} = \mathbf{P}_2 - \mathbf{P}_1 \tag{1.95}$$

This form of describing a vector will be used often because it provides easy reference to the vector **A**.

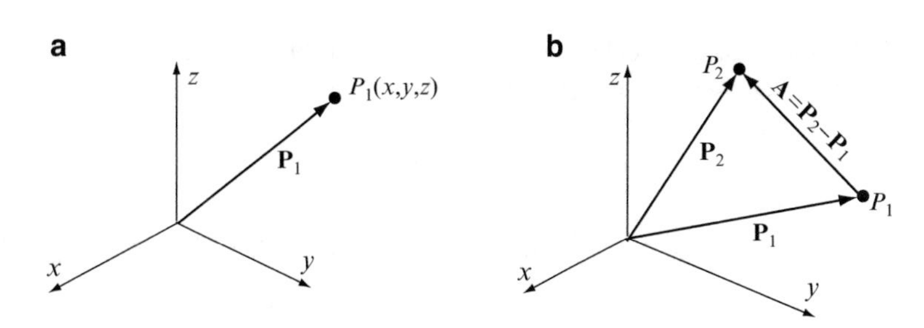

Figure 1.31 (**a**) Position vector of point P_1. (**b**) Vector **A** described in terms of the position vectors of its end points

Example 1.21 Two points are given in the Cartesian coordinate system as $P_1(1,1,3)$ and $P_2(3,1,3)$. Find the vector connecting point P_2(tail) to point P_1(head):

(a) As a regular vector.
(b) In terms of the position vectors connecting the origin of the system to points P_1 and P_2, and show that the two representations are the same.

Solution: The vector connecting points P_2 and P_1 is found as in **Example 1.19**.

The position vectors of points P_1 and P_2 are found as the vectors that connect the origin with these points. The vector **A** connecting P_2(tail) and P_1(head) is $\mathbf{A} = \mathbf{R}_1 - \mathbf{R}_2$ (see **Figure 1.32**).

(a) From **Example 1.19**, we write

$$\mathbf{A} = \hat{\mathbf{x}}(x_1 - x_2) + \hat{\mathbf{y}}(y_1 - y_2) + \hat{\mathbf{z}}(z_1 - z_2) = \hat{\mathbf{x}}(1-3) + \hat{\mathbf{y}}(1-1) + \hat{\mathbf{z}}(3-3) = -\hat{\mathbf{x}}2$$

(b) The two position vectors are

$$\mathbf{R}_1 = \hat{\mathbf{x}}(x_1 - 0) + \hat{\mathbf{y}}(y_1 - 0) + \hat{\mathbf{z}}(z_1 - 0) = \hat{\mathbf{x}}1 + \hat{\mathbf{y}}1 + \hat{\mathbf{z}}3$$
$$\mathbf{R}_2 = \hat{\mathbf{x}}(x_2 - 0) + \hat{\mathbf{y}}(y_2 - 0) + \hat{\mathbf{z}}(z_2 - 0) = \hat{\mathbf{x}}3 + \hat{\mathbf{y}}1 + \hat{\mathbf{z}}3$$

To show that $\mathbf{A} = \mathbf{R}_1 - \mathbf{R}_2$, we evaluate the expression explicitly:

$$\mathbf{A} = \mathbf{R}_1 - \mathbf{R}_2 = (\hat{\mathbf{x}}1 + \hat{\mathbf{y}}2 + \hat{\mathbf{z}}3) - (\hat{\mathbf{x}}3 + \hat{\mathbf{y}}1 + \hat{\mathbf{z}}3) = \hat{\mathbf{x}}(1-3) + \hat{\mathbf{y}}(1-1) + \hat{\mathbf{z}}(3-3) = -\hat{\mathbf{x}}2$$

and this is identical to vector **A** above.

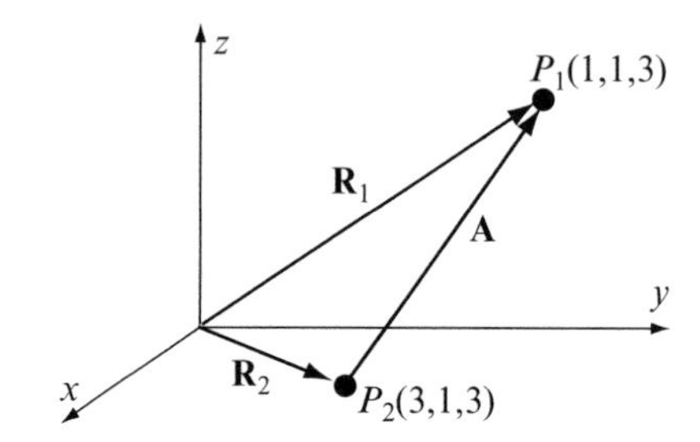

Figure 1.32 Vector **A** and its relationship with position vectors $\mathbf{R}_1$ and $\mathbf{R}_2$ and end points P_1 and P_2

Example 1.22 Two points are given in cylindrical coordinates as $P_1(1{,}30°{,}1)$ and $P_2(2{,}0°{,}2)$. Calculate the vector connecting P_1 and P_2 in terms of the position vectors of points P_1 and P_2.

Solution: First, we calculate the vectors connecting the origin with points P_1 and P_2. These are the position vectors $\mathbf{R}_1$ and $\mathbf{R}_2$. The vector connecting P_1 to P_2 is then $\mathbf{R} = \mathbf{R}_2 - \mathbf{R}_1$.

To calculate R_1 we take the tail of the vector at $P_0(0,0,0)$ and the head at $P_1(1{,}30°{,}1)$. The expression for a general vector in cylindrical coordinates was found in **Example 1.16**. With $r_0 = 0$, $\phi_0 = 0$, $z_0 = 0$,

$$\mathbf{R}_1(r,\phi,z) = \hat{\mathbf{r}}[(r_1\cos\phi_1 - 0)\cos\phi + (r_1\sin\phi_1 - 0)\sin\phi] + \hat{\boldsymbol{\phi}}[-(r_1\cos\phi_1 - 0)\sin\phi + (r_1\sin\phi_1 - 0)\cos\phi] + \hat{\mathbf{z}}[z_1 - 0]$$

or

$$\mathbf{R}_1 = \hat{\mathbf{r}}\left[\left(\sqrt{3}/2\right)\cos\phi + 0.5\sin\phi\right] + \hat{\boldsymbol{\phi}}\left[-\left(\sqrt{3}/2\right)\sin\phi + 0.5\cos\phi\right] + \hat{\mathbf{z}}1$$

Using similar steps, the position vector $\mathbf{R}_2$ is

$$\mathbf{R}_2 = \hat{\mathbf{r}}[(2\cos 0 - 0)\cos\phi + (2\sin 0 - 0)\sin\phi] + \hat{\boldsymbol{\phi}}[-(2\cos 0 - 0)\sin\phi + (2\sin 0 - 0)\cos\phi] + \hat{\mathbf{z}}[2-0] = \hat{\mathbf{r}}2\cos\phi - \hat{\boldsymbol{\phi}}2\sin\phi + \hat{\mathbf{z}}2$$

The vector $\mathbf{R}_2 - \mathbf{R}_1$ is

$$\mathbf{R} = \mathbf{R}_2 - \mathbf{R}_1 = \hat{\mathbf{r}}\left[\frac{4-\sqrt{3}}{2}\cos\phi - \frac{1}{2}\sin\phi\right] + \hat{\boldsymbol{\phi}}\left[\frac{\sqrt{3}-4}{2}\sin\phi - \frac{1}{2}\cos\phi\right] + \hat{\mathbf{z}}1$$

Exercise 1.14 Write the position vectors $\mathbf{R}_1$ and $\mathbf{R}_2$ in **Example 1.22** in Cartesian coordinates and write the vector pointing from P_1 to P_2.

Answer

$$\mathbf{R}_1 = \hat{\mathbf{x}}\frac{\sqrt{3}}{2} + \hat{\mathbf{y}}\frac{1}{2} + \hat{\mathbf{z}}1, \quad \mathbf{R}_2 = \hat{\mathbf{x}}2 + \hat{\mathbf{z}}2,$$

$$\mathbf{R}_2 - \mathbf{R}_1 = \hat{\mathbf{x}}\left(\frac{4-\sqrt{3}}{2}\right) - \hat{\mathbf{y}}\frac{1}{2} + \hat{\mathbf{z}}1$$

Problems

Vectors and Scalars

1.1 Two points $P_1(1,0,1)$ and $P_2(6,-3,0)$ are given. Calculate:

(a) The scalar components of the vector pointing from P_1 to P_2.
(b) The scalar components of the vector pointing from the origin to P_1.
(c) The magnitude of the vector pointing from P_1 to P_2.

1.2 A ship is sailing in a north–east direction at a speed of 50 km/h. The destination of the ship is 3,000 km directly east of the starting point. Note that speed is the absolute value of velocity:

(a) What is the velocity vector of the ship?
(b) How long does it take the ship to reach its destination?
(c) What is the total distance traveled from the starting point to its destination?

Addition and Subtraction of Vectors

1.3 An aircraft flies from London to New York at a speed of 800 km/h. Assume New York is straight west of London at a distance of 5,000 km. Use a Cartesian system of coordinates, centered in London, with New York in the negative x direction. At the altitude the airplane flies, there is a wind, blowing horizontally from north to south (negative y direction) at a speed of 100 km/h:

(a) What must be the direction of flight if the airplane is to arrive in New York?
(b) What is the speed in the London–New York direction?
(c) How long does it take to cover the distance from London to New York?

1.4 Vectors **A** and **B** are given: $\mathbf{A} = \hat{\mathbf{x}}5 + \hat{\mathbf{y}}3 - \hat{\mathbf{z}}1$ and $\mathbf{B} = -\hat{\mathbf{x}}3 + \hat{\mathbf{y}}5 - \hat{\mathbf{z}}2$. Calculate:

(a) $|\mathbf{A}|$.
(b) $\mathbf{A} + \mathbf{B}$.
(c) $\mathbf{A} - \mathbf{B}$.
(d) $\mathbf{B} - \mathbf{A}$.
(e) Unit vector in the direction of $\mathbf{B} - \mathbf{A}$.

Sums and Scaling of Vectors

1.5 Three vectors are given as: $\mathbf{A} = \hat{\mathbf{x}}3 + \hat{\mathbf{y}}1 + \hat{\mathbf{z}}3$, $\mathbf{B} = -\hat{\mathbf{x}}3 + \hat{\mathbf{y}}3 + \hat{\mathbf{z}}3$ and $\mathbf{C} = \hat{\mathbf{x}} - \hat{\mathbf{y}}2 + \hat{\mathbf{z}}2$:

(a) Calculate the sums $\mathbf{A} + \mathbf{B} + \mathbf{C}$, $\mathbf{A} + \mathbf{B} - \mathbf{C}$, $\mathbf{A} - \mathbf{B} - \mathbf{C}$, $\mathbf{A} - \mathbf{B} + \mathbf{C}$, $\mathbf{A} + (\mathbf{B} - \mathbf{C})$, and $(\mathbf{A} + \mathbf{B}) - \mathbf{C}$ using one of the geometric methods.
(b) Calculate the same sums using direct summation of the vectors.
(c) Comment on the two methods in terms of ease of solution and physical interpretation of results.

1.6 A geostationary satellite is located at a distance of 35,786 km above the equator. At that position it is stationary above a fixed point on the planet (i.e. it appears motionless when viewed from Earth). To force the satellite to re-enter the atmosphere a rocket is fired in the direction of motion of the satellite reducing its speed by 1,000 km/h. If the radius of the planet is 6378 km and the time for one rotation is exactly 24 h find:

(a) The velocity vectors before and after the rocket is fired.
(b) The scaling factor for the velocity produced by this action.

1.7 A particle moves with a velocity $\mathbf{v} = \hat{\mathbf{x}}300 + \hat{\mathbf{y}}50 - \hat{\mathbf{z}}100$ [m/s]. Now the velocity is reduced by a factor of 2:

(a) Calculate the direction of motion of the particle.
(b) What is the speed of the particle?

1.8 Given a quadrilateral *a, b, c, d* that has no parallel sides, show, using vector algebra, that the midpoints of its four sides, marked as *A, B, C, D* in **Figure 1.33** form the vertices of a parallelogram, that is that *AB* is parallel to *DC* and *AD* is parallel to *BC*.

Figure 1.33

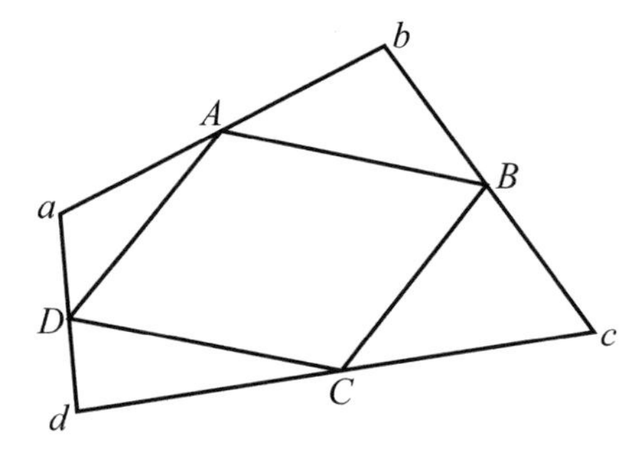

Scalar and Vector Products

1.9 Calculate the unit vector normal to the plane $3x + 4y + z = 0$.

1.10 Two vectors $\mathbf{v}_1 = \hat{\mathbf{x}}3 + \hat{\mathbf{y}}1 - \hat{\mathbf{z}}2$ [m/s] and $\mathbf{v}_2 = -\hat{\mathbf{x}}2 + \hat{\mathbf{y}}3$ [m/s] describe the velocities of two objects in space:

(a) Calculate the angle between the trajectories of the two objects.
(b) If the ground coincides with the x–y plane, calculate the ground velocities of each object.
(c) What is the angle between the ground velocities?

1.11 Find a unit vector normal to the following planes, at the given point:

(a) $z = -x - y$, at point $P(0,0,0)$.
(b) $4x - 3y + z + 5 = 0$, at point $P(0,0,-5)$.
(c) $z = ax + by$, at point $P(0,0,0)$.

1.12 A force is given as $\mathbf{F} = \hat{\mathbf{x}}a/x$ [N] where a is a given constant and x is a variable. Calculate the vector component of $\mathbf{F}$ in the direction of the vector $\mathbf{A} = \hat{\mathbf{x}}3 + \hat{\mathbf{y}}1 - \hat{\mathbf{z}}2$.

1.13 Write a formula for calculation of the area of a general triangle defined by its three vertex points: $P_1(x_1,y_1,z_1)$, $P_2(x_2,y_2,z_2)$ and $P_3(x_3,y_3,z_3)$.

1.14 Show using vector algebra that the law of sines holds in the triangle in **Figure 1.34**, where A, B, and C are the lengths of the corresponding sides; that is, show that the following is correct:

$$\frac{A}{\sin\phi_{BC}} = \frac{B}{\sin\phi_{AC}} = \frac{C}{\sin\phi_{AB}}.$$

Figure 1.34

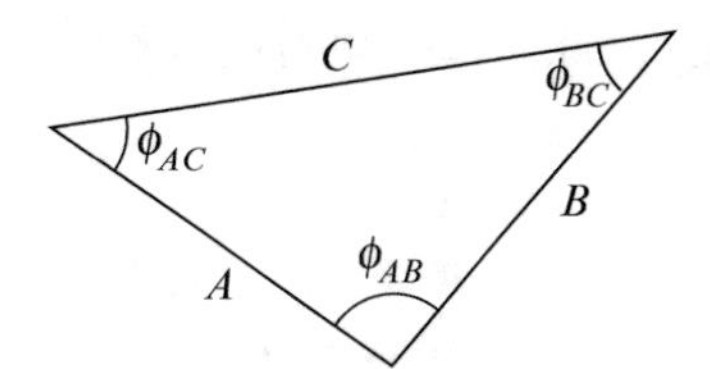

1.15 Given the general triangle in **Figure 1.34** with sides of length A, B, C, show using vector algebra that the cosines of the angles ϕ_{AB}, ϕ_{AC}, ϕ_{BC} are:

$$\cos\phi_{AB} = \frac{A^2 + B^2 - C^2}{2AB}, \quad \cos\phi_{AC} = \frac{A^2 + C^2 - B^2}{2AC}, \quad \cos\phi_{BC} = \frac{B^2 + C^2 - A^2}{2BC},$$

1.16 A vector is given as $\mathbf{A} = \hat{\mathbf{x}}3 + \hat{\mathbf{y}}1 - \hat{\mathbf{z}}2$:

(a) Find the angle between **A** and the positive z axis.
(b) Find a vector perpendicular to **A** and a unit vector in the direction of the positive z axis.

1.17 A force is given as $\mathbf{F} = \hat{\mathbf{x}}1 + \hat{\mathbf{y}}5 - \hat{\mathbf{z}}1$ [N]. Calculate the magnitude of the force in the direction of the vector $\mathbf{A} = -\hat{\mathbf{x}}3 + \hat{\mathbf{y}}2 - \hat{\mathbf{z}}2$.

1.18 Three vertices of a parallelogram are given as $P_1(7,3,1)$, $P_2(2,1,0)$, and $P_3(2,2,5)$:

(a) Find the area of a parallelogram with these vertices.
(b) Is the answer in **(a)** unique: that is, is there only one parallelogram that can be defined by these points? If not, what are the other possible solutions?

1.19 A plane through the origin of the system of coordinates is defined by two points: $P_1(1,2,-1)$ and $P_2(5,3,2)$. Find the equation of the plane.

Multiple Products

1.20 Three vectors are given as $\mathbf{A} = \hat{\mathbf{x}}3 - \hat{\mathbf{y}}x - \hat{\mathbf{z}}2x$, $\mathbf{B} = -\hat{\mathbf{x}}2x + \hat{\mathbf{y}}5$, $\mathbf{C} = \hat{\mathbf{y}}3x^2 + \hat{\mathbf{z}}y$:

(a) Calculate the product $\mathbf{A} \cdot (\mathbf{B} \times \mathbf{C})$.
(b) Calculate the product $\mathbf{B} \times (\mathbf{A} \times \mathbf{C})$.
(c) Calculate the product $(\mathbf{A} \times \mathbf{B}) \cdot \mathbf{C}$.

1.21 To define the volume of a parallelepiped, we need to define a corner of the parallelepiped and three vectors emanating from this point (see **Exercise 1.11**). Four corners of a parallelepiped are known as $P_1(0,0,0)$, $P_2(a,0,1)$, $P_3(a,2,c)$, and $P_4(1,b,1)$:

(a) Show that there are six vectors that can be defined using these nodes, but only three vectors, emanating from a node, are necessary to define a parallelepiped.
(b) Show that there are four possible parallelepipeds that can be defined using these four nodes, depending on which node is taken as the root node.
(c) Calculate the volumes of the four parallelepipeds.

1.22 Which of the following vector products yield zero and why? **A, B,** and **C** are vectors and $\mathbf{C} = \mathbf{A} \times \mathbf{B}$.

(a) $\mathbf{A} \times (\mathbf{A} \times \mathbf{B})$.
(b) $\mathbf{A} \times (\mathbf{B} \times \mathbf{B})$.
(c) $\mathbf{C} \times \mathbf{C}$.
(d) $(\mathbf{A} \times \mathbf{B}) \times \mathbf{C}$.
(e) $\mathbf{B} \times (\mathbf{A} \times \mathbf{B})$.
(f) $(\mathbf{A} \times \mathbf{A}) \times \mathbf{B}$.
(g) $\mathbf{A} \times \mathbf{C}$.
(h) $(\mathbf{C} \times \mathbf{A}) \times \mathbf{B}$.

1.23 Which of the following products are properly defined and which are not? Explain why. (a and b are scalars, **A**, **B**, and **C** are vectors.)

(a) $ab\mathbf{A} \times \mathbf{C}$.
(b) $\mathbf{A} \times \mathbf{C} \times \mathbf{B}$.
(c) $\mathbf{B} \cdot \mathbf{C} \times \mathbf{A}$.
(d) $(\mathbf{A} \times \mathbf{B}) \cdot \mathbf{A}$.
(e) $a\mathbf{B} \cdot \mathbf{C}$.
(f) $(a\mathbf{B} \times b\mathbf{A})$.

1.24 Which of the following products are meaningful? Explain. **A**, **B**, **C** are vectors and $\mathbf{C} = \mathbf{B} \times \mathbf{A}$; c is a scalar.

(a) $\mathbf{A} \cdot (\mathbf{B} \times \mathbf{C})$.
(b) $c\mathbf{A} \cdot (\mathbf{A} \times \mathbf{B})$.
(c) $\mathbf{C} \cdot \mathbf{C}$.
(d) $(\mathbf{A} \cdot \mathbf{B}) \times (\mathbf{A} \times \mathbf{B})$.
(e) $(\mathbf{A} \cdot \mathbf{B}) \cdot \mathbf{C}$.
(f) $\mathbf{A} \cdot (\mathbf{C} \times \mathbf{B})$.
(g) $\mathbf{A} \cdot (\mathbf{A} \times \mathbf{C})$.
(h) $\mathbf{A} \cdot (\mathbf{A} \times \mathbf{A})$.

1.25 Vectors $\mathbf{A} = \hat{\mathbf{x}}1 + \hat{\mathbf{y}}1 + \hat{\mathbf{z}}2$, $\mathbf{B} = \hat{\mathbf{x}}2 + \hat{\mathbf{y}}1 + \hat{\mathbf{z}}2$, and $\mathbf{C} = \hat{\mathbf{x}}1 - \hat{\mathbf{y}}2 + \hat{\mathbf{z}}3$ are given. Find the height of the parallelepiped defined by the three vectors:

(a) If **A** and **B** form the base.
(b) If **A** and **C** form the base.
(c) If **B** and **C** form the base.

Scalar and Vector Fields

1.26 A pressure field is given as $P = x(x-1)(y-2) + 1$:

(a) Sketch the scalar field in the domain $0 < x,y < 1$.
(b) Find the point(s) at which the slope of the field is zero.

1.27 Sketch the scalar fields

$$A = x + y, \quad B = x - y, \quad c = \frac{x + y}{\sqrt{x^2 + y^2}}.$$

1.28 Sketch the vector fields

$$\mathbf{A} = \hat{\mathbf{x}}y + \hat{\mathbf{y}}x, \quad \mathbf{B} = \hat{\mathbf{x}}y - \hat{\mathbf{y}}x, \quad \mathbf{C} = \frac{\hat{\mathbf{x}}x + \hat{\mathbf{y}}y}{\sqrt{x^2 + y^2}}.$$

Systems of Coordinates

1.29 Three points are given in Cartesian coordinates: $P_1(1,1,1)$, $P_2(1,1,0)$, and $P_3(0,1,1)$:

(a) Find the points P_1, P_2, and P_3 in cylindrical and spherical coordinates.
(b) Find the equation of a plane through these three points in Cartesian coordinates.
(c) Find the equation of the plane in cylindrical coordinates.
(d) Find the equation of the plane in spherical coordinates.

1.30 Write the equation of a sphere of radius a:

(a) In Cartesian coordinates.
(b) In cylindrical coordinates.
(c) In spherical coordinates.

1.31 A sphere of radius a is given. Choose any point on the sphere. Describe this point:

(a) In spherical coordinates.
(b) In Cartesian coordinates.
(c) In cylindrical coordinates.

1.32 Two points are given in cylindrical coordinates as $P_1(3,\pi/6,-1)$ and $P_2(-2,0,2)$.

(a) Write the vector connecting P_1(tail) to P_2(head) in cylindrical coordinates.
(b) Write the vector connecting P_2(tail) to P_1(head) in spherical coordinates.
(c) Find the length of the vectors in **(a)** and **(b)**.

1.33 Two points are given in spherical coordinates as $P_1(0,36°,-72°)$ and $P_2(5,-120°,45°)$.

(a) The magnitude of the vector connecting P_1(tail) to P_2(head).
(b) The unit vector in the direction of the vector in **(a)** in cylindrical coordinates.
(c) The unit vector in the direction of the vector in **(a)** in spherical coordinates.

1.34 Transform the vector $\mathbf{A} = \hat{\mathbf{x}}2 - \hat{\mathbf{y}}5 + \hat{\mathbf{z}}3$ into spherical coordinates at $(x = -2, y = 3, z = 1)$. That is, find the general transformation of the vector and then substitute the coordinates of the point to obtain the transformation at the specific point.

1.35 Vector $\mathbf{A} = \hat{\mathbf{r}}3\cos\phi - \hat{\boldsymbol{\phi}}2r^{1/2} + \hat{\mathbf{z}}r\phi$ is given:

(a) Transform the vector to Cartesian coordinates.
(b) Find the scalar components of the vector in spherical coordinates.

Position Vectors

1.36 Points $P_1(x_1,y_1,z_1)$ and $P_2(x_2,y_2,z_2)$ are given:

(a) Calculate the position vector $\mathbf{r}_1$ of point P_1.
(b) Calculate the position vector $\mathbf{r}_2$ of P_2.
(c) Calculate the vector $\mathbf{R}$ connecting P_2(tail) to P_1(head).
(d) Show that the vector $\mathbf{R}$ can be written as $\mathbf{R} = \mathbf{r}_1 - \mathbf{r}_2$.

1.37 Two points on a sphere of radius 3 are given as $P_1(3,0°,30°)$ and $P_2(3,45°,45°)$:

(a) Find the position vectors $\mathbf{P}_1$ and $\mathbf{P}_2$ pointing to P_1 and P_2 respectively.
(b) Find the vector connecting P_1(tail) to P_2(head).
(c) Using the vectors in **(a)** write the vector $\mathbf{P}_1\mathbf{P}_2$ (P_1 tail, to P_2 head) in cylindrical and Cartesian coordinates.

1.38 Given the position vectors $\mathbf{A} = \hat{\mathbf{x}}A_x + \hat{\mathbf{y}}A_y + \hat{\mathbf{z}}A_z$ and $\mathbf{B} = \hat{\mathbf{x}}B_x + \hat{\mathbf{y}}B_y + \hat{\mathbf{z}}B_z$, find the equation of the plane they form.

Vector Calculus

2

There cannot be a language more universal and more simple, more free from errors and obscurities, that is to say more worthy to express the invariable relations of natural things [than mathematics] ....
... Its chief attribute is clearness; it has no marks to express confused notions. It brings together phenomena the most diverse, and discovers the hidden analogies that unite them .... it follows the same course in the study of all phenomena; it interprets them by the same language, as if to attest the unity and simplicity of the plan of the universe ....

—Jean Baptiste Joseph Fourier (1768–1830), mathematician and physicist
Introduction to the Analytic Theory of Heat, 1822 (from the 1955 Dover edition)

2.1 Introduction

Vector calculus deals with the application of calculus operations on vectors. We will often need to evaluate integrals, derivatives, and other operations that use integrals and derivatives. The rules needed for these evaluations constitute vector calculus. In particular, line, volume, and surface integration are important, as are directional derivatives.

The relations defined here are very useful in the context of electromagnetics but, even without reference to electromagnetics, we will show that the definitions given here are simple extensions to familiar concepts and they simplify a number of important aspects of calculation.

We will discuss in particular the ideas of line, surface, and volume integration, and the general ideas of gradient, divergence, and curl, as well as the divergence and Stokes' theorems. These notions are of fundamental importance for the understanding of electromagnetic fields. As with vector algebra, the number of operations and concepts we need is rather small. These are:

Integration	Vector operators	Theorems
Line or contour integral	Gradient	The divergence theorem
Surface integral	Divergence	Stokes' theorem
Volume integral	Curl	

Vector identities

In addition, we will define the Laplacian and briefly discuss the Helmholtz theorem as a method of generalizing the definition of vector fields. These are the topics we must have as tools before we start the study of electromagnetics.

2.2 Integration of Scalar and Vector Functions

Vector functions often need to be integrated. As an example, if a force is specified, and we wish to calculate the work performed by this force, then an integration along the path of the force is required. The force is a vector and so is the path. However, the integration results in a scalar function (work). In addition, the ideas of surface and volume integrals are required for future use in evaluation of fields. The methods of setting up and evaluating these integrals will be given together with examples of their physical meaning. It should be remembered that the integration itself is identical to that performed in

N. Ida, *Engineering Electromagnetics*, https://doi.org/10.1007/978-3-030-15557-5_2

calculus. The unique nature of vector integration is in treatment of the integrand and in the physical meaning of quantities involved. Physical meaning is given to justify the definitions and to show how the various integrals will be used later. Simple applications in fluid flow, forces on bodies, and the like will be used for this purpose.

2.2.1 Line Integrals

Before defining the line integral, consider the very simple example of calculating the work performed by a force, as shown in **Figure 2.1a**. The force is assumed to be space dependent and in an arbitrary direction in the plane. To calculate the work performed by this force, it is possible to separate the force into its two components and write

$$W = \int_{x=x_1}^{x=x_2} F(x,y)\cos\alpha dx + \int_{y=y_1}^{y=y_2} F(x,y)\cos\beta dy \tag{2.1}$$

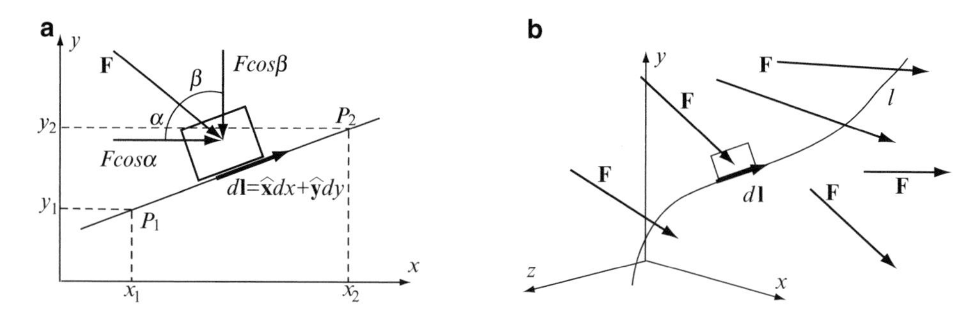

Figure 2.1 (**a**) The concept of a line integral: work performed by a force as a body moves from point P_1 to P_2. (**b**) A generalization of (**a**). Work performed in a *force field* along a general path l

An alternative and more general approach is to rewrite the force function in terms of a new parameter, say u, as $F(u)$ and calculate

$$W = \int_{u=u_1}^{u=u_2} F(u)du \tag{2.2}$$

We will return to the latter form, but, first, we note that the two integrands in **Eq. (2.1)** can be written as scalar products:

$$\mathbf{F}\cdot\hat{\mathbf{x}} = F(x,y)\cos\alpha \quad \text{and} \quad \mathbf{F}\cdot\hat{\mathbf{y}} = F(x,y)\cos\beta \tag{2.3}$$

This leads to the following form for the work:

$$W = \int_{x=x_1}^{x=x_2} \mathbf{F}\cdot\hat{\mathbf{x}}dx + \int_{y=y_1}^{y=y_2} \mathbf{F}\cdot\hat{\mathbf{y}}dy \tag{2.4}$$

We can now use the definition of $d\mathbf{l}$ in the x–y plane as $d\mathbf{l} = \hat{\mathbf{x}}dx + \hat{\mathbf{y}}dy$ and write the work as

$$W = \int_{P_1}^{P_2} \mathbf{F}\cdot d\mathbf{l} \tag{2.5}$$

where $d\mathbf{l}$ is the differential vector in Cartesian coordinates. The path of integration may be arbitrary, as shown in **Figure 2.1b**, whereas the force may be a general force distribution in space (i.e., a force field). Of course, for a general path in space, the third term in $d\mathbf{l}$ must be included ($d\mathbf{l} = \hat{\mathbf{x}}dx + \hat{\mathbf{y}}dy + \hat{\mathbf{z}}dz$).

To generalize this result even further, consider a vector field **A** as shown in **Figure 2.2a** and an arbitrary path C. The line integral of the vector **A** over the path C is written as

$$\boxed{Q = \int_C \mathbf{A}\cdot d\mathbf{l} = \int_C |\mathbf{A}||d\mathbf{l}|\cos\theta_{\mathbf{A},d\mathbf{l}}} \tag{2.6}$$

In this definition, we only employed the properties of the integral and that of the scalar product. In effect, we evaluate first the projection of the vector **A** onto the path and then proceed to integrate as for any scalar function. If the integration between two points is required, we write

$$\int_{P_1}^{P_2} \mathbf{A} \cdot d\mathbf{l} = \int_{P_1}^{P_2} |\mathbf{A}||d\mathbf{l}|\cos\theta_{\mathbf{A},d\mathbf{l}} \tag{2.7}$$

again, in complete accordance with the standard method of integration. As mentioned in the introduction, once the product under the integral sign is properly evaluated, the integration proceeds as in calculus.

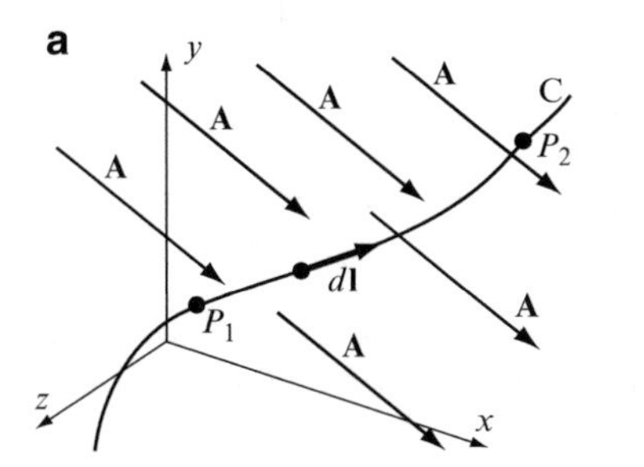

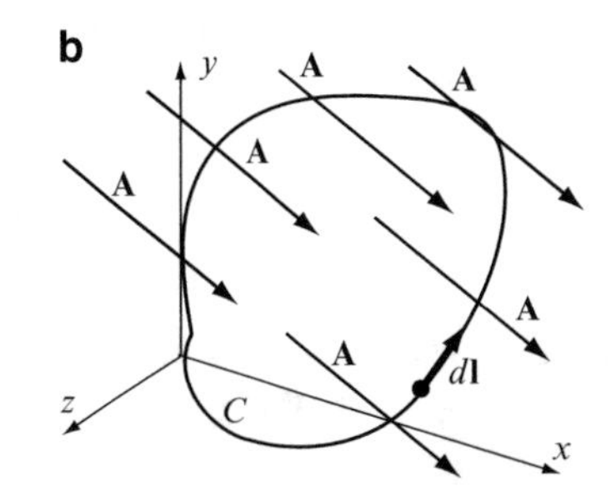

Figure 2.2 The line integral. (**a**) Open contour integration. (**b**) Closed contour integration

Extending the analogy of calculation of work, we can calculate the work required to move an object around a closed contour. In terms of **Figure 2.2b**, this means calculating the closed path integral of the vector **A**. This form of integration is important enough for us to give it a special symbol and name. It will be called a ***closed contour integral*** or a ***loop integral*** and is denoted by a small circle superimposed on the symbol for integration:

$$\oint \mathbf{A} \cdot d\mathbf{l} = \oint |\mathbf{A}||d\mathbf{l}|\cos\theta_{\mathbf{A},d\mathbf{l}} \tag{2.8}$$

The closed contour integral of **A** is also called the ***circulation of* A** around path C. The circulation of a vector around any closed path can be zero or nonzero, depending on the vector. Both types will be important in analysis of fields; therefore, we now define the following:

(1) A vector field whose circulation around any arbitrary closed path is zero is called a ***conservative field*** or a ***restoring field.*** In a force field, the line integral represents work. A conservative field in this case means that the total net work done by the field or against the field on any closed path is zero.

(2) A vector field whose circulation around an arbitrary closed path is nonzero is a ***nonconservative*** or ***nonrestoring field.*** In terms of forces, this means that moving in a closed path requires net work to be done either by the field or against the field.

Now, we return to **Eq. (2.2)**. We are free to integrate either using **Eq. (2.4)** or **Eq. (2.5)**, but which should we use? More important, are these two integrals identical? To see this, consider the following three examples.

Example 2.1 Work in a Field A vector field is given as $\mathbf{F} = \hat{\mathbf{x}}2x + \hat{\mathbf{y}}2y$.

(a) Sketch the field in space.
(b) Assume **F** is a force. What is the work done in moving from point $P_2(5,0)$ to $P_3(0,3)$ (in **Figure 2.3a**)?
(c) Does the work depend on the path taken between P_2 and P_3?

Figure 2.3

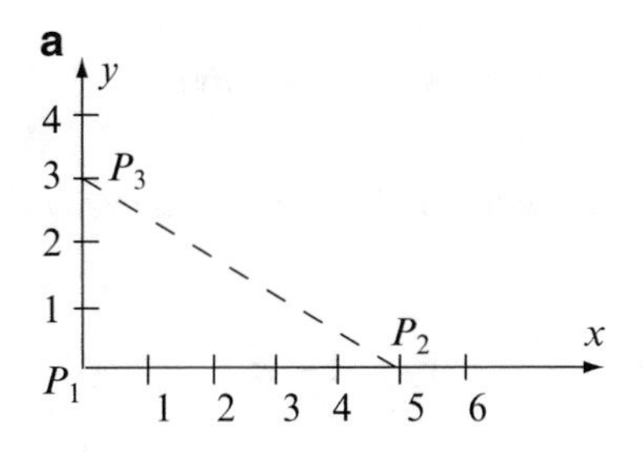

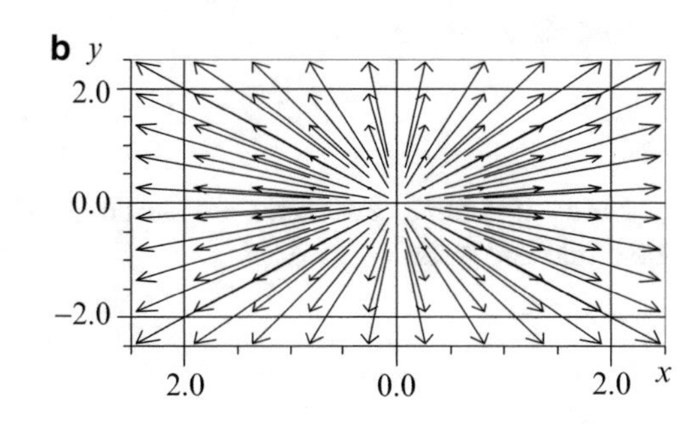

Solution: **(b)** First, we calculate the line integral of $\mathbf{F} \cdot d\mathbf{l}$ along the path between P_2 and P_3. This is a direct path. **(c)** Then, we calculate the same integral from P_2 to P_1 and from P_1 to P_3. If the two results are the same, the closed contour integral is zero.

(a) See **Figure 2.3b**. Note that the field is zero at the origin. At any point x, y, the vector has components in the x and y directions. The magnitude depends on the location of the field (thus, the different vector lengths at different locations).

(b) From P_2 to P_3, the element of path is $d\mathbf{l} = \hat{\mathbf{x}}dx + \hat{\mathbf{y}}dy$. The integration is therefore

$$\int_{P_2}^{P_3} \mathbf{F} \cdot d\mathbf{l} = \int_{P2}^{P3} (\hat{\mathbf{x}}2x + \hat{\mathbf{y}}2y) \cdot (\hat{\mathbf{x}}dx + \hat{\mathbf{y}}dy) = \int_{P_2}^{P_3} (2xdx + 2ydy) \quad [\text{J}]$$

Since each part of the integrand is a function of a single variable, x or y, we can separate the integration into integration over each variable and write

$$\int_{P_2}^{P_3} \mathbf{F} \cdot d\mathbf{l} = \int_{x=5}^{x=0} 2xdx + \int_{y=0}^{y=3} 2ydy = x^2\Big|_5^0 + y^2\Big|_0^3 = -25 + 9 = -16 \quad [\text{J}]$$

Note: This work is negative. It decreases the potential energy of the system; that is, this work is done by the field (as, for example, in sliding on a water slide, the gravitational field performs the work and the potential energy of the slider is reduced).

(c) On paths P_2 to P_1 and P_1 to P_3, we perform separate integrations. On path P_2 to P_1, $d\mathbf{l} = \hat{\mathbf{x}}dx$ and $y = 0$. The integration is

$$\int_{P_2}^{P_1} \mathbf{F} \cdot d\mathbf{l} = \int_{P_2}^{P_1} (\hat{\mathbf{x}}2x + \hat{\mathbf{y}}2y) \cdot (\hat{\mathbf{x}}dx) = \int_{P_2}^{P_1} 2xdx = \int_{x=5}^{x=0} 2xdx = x^2\Big|_5^0 = -25 \quad [\text{J}]$$

Similarly, on path P_1 to P_3, $d\mathbf{l} = \hat{\mathbf{y}}dy$ and $x = 0$. The integration is

$$\int_{P_1}^{P_3} \mathbf{F} \cdot d\mathbf{l} = \int_{P_1}^{P_3} (\hat{\mathbf{x}}2x + \hat{\mathbf{y}}2y) \cdot (\hat{\mathbf{y}}dy) = \int_{P_1}^{P_3} 2ydy = \int_{y=0}^{y=3} 2ydy = y^2\Big|_0^3 = 9 \quad [\text{J}]$$

The sum of the two paths is equal to the result obtained for the direct path. This also means that the closed contour integral will yield zero. However, the fact that the closed contour integral on a particular path is zero does not necessarily mean the given field is conservative. In other words, we cannot say that this particular field is conservative unless we can show that the closed contour integral is zero for any contour. We will discuss this important aspect of fields later in this chapter.

Example 2.2 Circulation of a Vector Field Consider a vector field $\mathbf{A} = \hat{\mathbf{x}}xy + \hat{\mathbf{y}}(3x^2 + y)$. Calculate the circulation of $\mathbf{A}$ around the circle $x^2 + y^2 = 1$.

Solution: First, we must calculate the differential of path, $d\mathbf{l}$, and then evaluate $\mathbf{A} \cdot d\mathbf{l}$. This is then integrated along the circle (closed contour) to obtain the result. This problem is most easily evaluated in cylindrical coordinates (see **Exercise 2.1**), but we will solve it in Cartesian coordinates. The integration is performed in four segments: P_1 to P_2, P_2 to P_3, P_3 to P_4, and P_4 to P_1, as shown in **Figure 2.4**.

The differential of length in the x–y plane is $d\mathbf{l} = \hat{\mathbf{x}}dx + \hat{\mathbf{y}}dy$. The scalar product $\mathbf{A} \cdot d\mathbf{l}$ is

$$\mathbf{A} \cdot d\mathbf{l} = \left(\hat{\mathbf{x}}xy + \hat{\mathbf{y}}\left(3x^2 + y\right)\right) \cdot \left(\hat{\mathbf{x}}dx + \hat{\mathbf{y}}dy\right) = xydx + \left(3x^2 + y\right)dy$$

The circulation is now

$$\oint_L \mathbf{A} \cdot d\mathbf{l} = \oint_L \left[xydx + \left(3x^2 + y\right)dy\right]$$

Before this can be evaluated, we must make sure that integration is over a single variable. To do so, we use the equation of the circle and write

$$x = \left(1 - y^2\right)^{1/2}, \quad y = \left(1 - x^2\right)^{1/2}$$

By substituting the first relation into the second term and the second into the first term under the integral, we have

$$\int_L \mathbf{A} \cdot d\mathbf{l} = \oint_L \left[x\left(1 - x^2\right)^{1/2} dx + \left(3\left(1 - y^2\right) + y\right) dy \right]$$

and each part of the integral is a function of a single variable. Now, we can separate these into four integrals:

$$\oint_L \mathbf{A} \cdot d\mathbf{l} = \int_{P_1}^{P_2} \mathbf{A} \cdot d\mathbf{l} + \int_{P_2}^{P_3} \mathbf{A} \cdot d\mathbf{l} + \int_{P_3}^{P_4} \mathbf{A} \cdot d\mathbf{l} + \int_{P_4}^{P_1} \mathbf{A} \cdot d\mathbf{l}$$

Evaluating each integral separately,

$$\int_{P_1}^{P_2} \mathbf{A} \cdot d\mathbf{l} = \int_{P_1}^{P_2} \left(x\left(1 - x^2\right)^{1/2} dx + \left(3 - 3y^2 + y\right) dy \right)$$
$$= \int_{x=1}^{x=0} x\left(1 - x^2\right)^{1/2} dx + \int_{y=0}^{y=1} \left(3 - 3y^2 + y\right) dy = -\frac{\left(1 - x^2\right)^{3/2}}{3} \Bigg|_1^0 + \left(3y + \frac{y^2}{2} - y^3\right) \Bigg|_0^1 = \frac{13}{6}$$

Note that the other integrals are similar except for the limits of integration:

$$\int_{P_2}^{P_3} \mathbf{A} \cdot d\mathbf{l} = \int_{x=0}^{x=-1} x\left(1 - x^2\right)^{1/2} dx + \int_{y=1}^{y=0} \left(3 - 3y^2 + y\right) dy = -\frac{\left(1 - x^2\right)^{3/2}}{3} \Bigg|_1^0 + \left(3y + \frac{y^2}{2} - y^3\right) \Bigg|_1^0 = -\frac{13}{6}$$

$$\int_{P_3}^{P_4} \mathbf{A} \cdot d\mathbf{l} = \int_{x=-1}^{x=0} x\left(1 - x^2\right)^{1/2} dx + \int_{y=0}^{y=-1} \left(3 - 3y^2 + y\right) dy = -\frac{\left(1 - x^2\right)^{3/2}}{3} \Bigg|_0^{-1} + \left(3y + \frac{y^2}{2} - y^3\right) \Bigg|_0^{-1} = -\frac{11}{6}$$

$$\int_{P_4}^{P_1} \mathbf{A} \cdot d\mathbf{l} = \int_{x=0}^{x=1} x\left(1 - x^2\right)^{1/2} dx + \int_{y=-1}^{y=0} \left(3 - 3y^2 + y\right) dy = -\frac{\left(1 - x^2\right)^{3/2}}{3} \Bigg|_{-1}^{0} \left(3y + \frac{y^2}{2} - y^3\right) \Bigg|_{-1}^{0} = \frac{11}{6}$$

The total circulation is the sum of the four circulations above. This gives

$$\oint_L \mathbf{A} \cdot d\mathbf{l} = 0$$

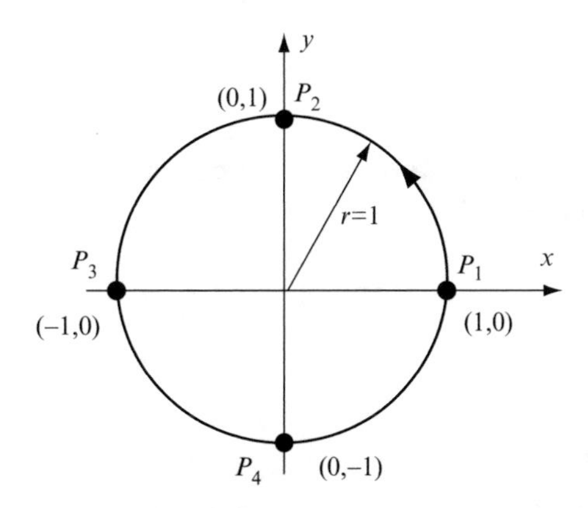

Figure 2.4 The four segments of the contour used for integration in **Example 2.2**

Exercise 2.1 Solve **Example 2.2** in cylindrical coordinates; that is, transform the vector **A** and the necessary coordinates and evaluate the integral.

Example 2.3 Line Integral: Nonconservative Field The force $\mathbf{F} = \hat{\mathbf{x}}(2x - y) + \hat{\mathbf{y}}(x + y + z) + \hat{\mathbf{z}}(2z - x)$ [N] is given. Calculate the total work required to move a body in a circle of radius 1 m, centered at the origin. The circle is in the x–y plane at $z = 0$.

Solution: To find the work, we first convert to the cylindrical system of coordinates. Also, since the circle is in the x–y plane ($z = 0$), we have

$$\mathbf{F}\big|_{z=0} = \hat{\mathbf{x}}(2x - y) + \hat{\mathbf{y}}(x + y) - \hat{\mathbf{z}}x \quad \text{and} \quad x^2 + y^2 = 1$$

Since integration is in the x–y plane, the closed contour integral is

$$\oint_L \mathbf{F} \cdot d\mathbf{l} = \oint_L (\hat{\mathbf{x}}(2x - y) + \hat{\mathbf{y}}(x + y) - \hat{\mathbf{z}}x) \cdot (\hat{\mathbf{x}}dx + \hat{\mathbf{y}}dy) = \oint_L ((2x - y)dx + (x + y)dy$$

Conversion to cylindrical coordinates gives

$$x = r\cos\phi = 1\cos\phi, \quad y = \sin\phi$$

Therefore,

$$\frac{dx}{d\phi} = -\sin\phi \quad \rightarrow \quad dx = -\sin\phi d\phi, \quad \text{and} \quad \frac{dy}{d\phi} = \cos\phi \quad \rightarrow \quad dy = \cos\phi d\phi$$

Substituting for x, y, dx, and dy, we get

$$\oint_L \mathbf{F} \cdot d\mathbf{l} = \int_{\phi=0}^{\phi=2\pi} ((2\cos\phi - \sin\phi)(-\sin\phi d\phi) + (\cos\phi + \sin\phi)\cos\phi d\phi) = \int_{\phi=0}^{\phi=2\pi} (1 - \sin\phi\cos\phi)d\phi = 2\pi \quad [\text{J}]$$

This result means that integration between zero and π and between zero and $-\pi$ gives different results. The closed contour line integral is not zero and the field is clearly nonconservative.

The function in **Example 2.1** yielded identical results using two different paths, whereas the result in **Example 2.3** yielded different results. This means that, in general, we are not free to choose the path of integration as we wish. However, if the line integral is independent of path, then the closed contour integral is zero, and we are free to choose the path any way we wish.

2.2.2 Surface Integrals

To define the surface integral, we use a simple example of water flow. Consider first water flowing through a hose of cross section s_1 as shown in **Figure 2.5a**. If the fluid has a constant mass density ρ [kg/m^3] and flows at a fixed velocity $\mathbf{v}$ [m/s], the rate of flow of the fluid (mass per unit time) is

$$w_1 = \rho s_1 v \quad [\text{kg/s}] \tag{2.9}$$

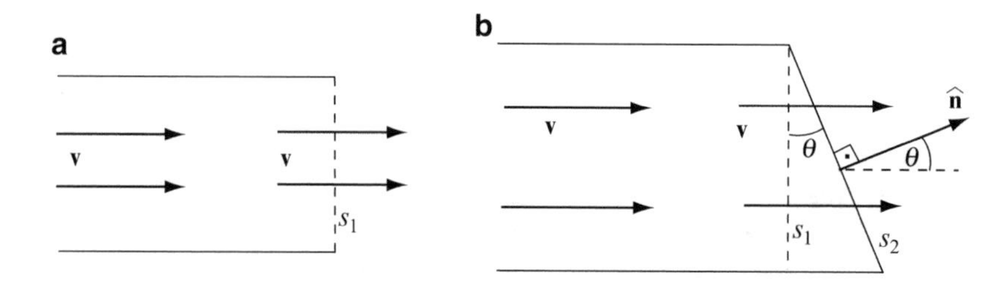

Figure 2.5 Flow through a surface. (**a**) Flow normal to surface s_1. (**b**) Flow at an angle θ to surface s_2 and the relation between the velocity vector and the normal to the surface

Now, assume that we take the same hose, but cut it at an angle as shown in **Figure 2.5b**. The cross-sectional area s_2 is larger, but the total rate of flow remains unchanged. The reason for this is that only the normal projection of the area is crossed by the fluid. In terms of area s_2, we can write

$$w_1 = \rho v s_2 \cos\theta \quad [\mathrm{kg/s}] \tag{2.10}$$

Instead of using the scalar values as in **Eqs. (2.9)** and **(2.10)**, we can use the vector nature of the velocity. Using **Figure 2.5b**, we replace the term $vs_2\cos\theta$ by $\mathbf{v} \cdot \hat{\mathbf{n}} s_2$ and write

$$w_1 = \rho \mathbf{v} \cdot \hat{\mathbf{n}} s_2 \quad [\mathrm{kg/s}] \tag{2.11}$$

where $\hat{\mathbf{n}}$ is the unit vector normal to surface s_2.

Now, consider **Figure 2.6** where we assumed that a hose allows water to flow with a velocity profile as shown. This is possible if the fluid is viscous. We will assume that the velocity across each small area Δs_i is constant and write the total rate of flow as

$$w_1 = \sum_{i=1}^{n} \rho \mathbf{v}_i \cdot \hat{\mathbf{n}} \Delta s_i \quad [\mathrm{kg/s}] \tag{2.12}$$

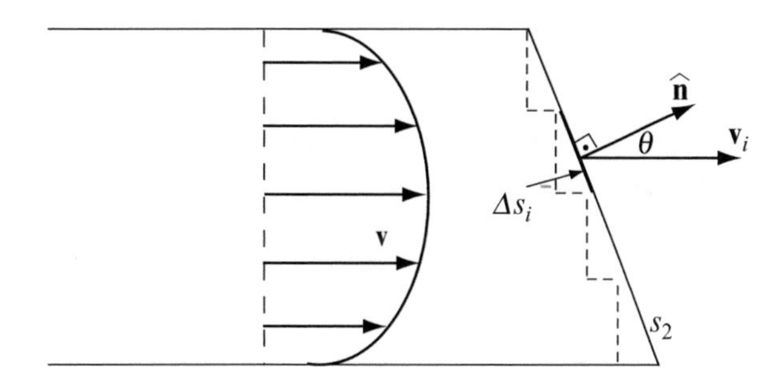

Figure 2.6 Flow with a nonuniform velocity profile

In the limit, as Δs_i tends to zero,

$$w_1 = \lim_{\Delta s \to 0} \sum_{i=1}^{\infty} \rho \mathbf{v}_i \cdot \hat{\mathbf{n}} \Delta s_i = \int_{s_2} \rho \mathbf{v} \cdot \hat{\mathbf{n}} ds_2 \quad \left[\frac{\mathrm{kg}}{\mathrm{s}}\right] \tag{2.13}$$

Thus, we obtained an expression for the rate of flow for a variable velocity fluid through an arbitrary surface, provided that the velocity profile is known, and the normal to the surface can be evaluated everywhere. For purposes of this section, we now rewrite this integral in general terms by replacing $\rho \mathbf{v}$ by a general vector **A**. This is the field. The rate of flow of the vector field **A** (if, indeed, the vector field **A** represents a flow) can now be written as

$$Q = \int_s \mathbf{A} \cdot \hat{\mathbf{n}} ds \tag{2.14}$$

This is a surface integral and, like the line integral, it results in a scalar value. However, the surface integral represents a flow-like function. In the context of electromagnetics, we call this a ***flux*** (fluxus = flow in Latin). Thus, the surface integral of a vector is the flux of this vector through the surface. The surface integral is also written as

$$\boxed{Q = \int_s \mathbf{A} \cdot d\mathbf{s}} \tag{2.15}$$

where $d\mathbf{s} = \hat{\mathbf{n}}ds$. The latter is a convenient short-form notation that avoids repeated writing of the normal unit vector, but it should be remembered that the normal unit vector indicates the direction of positive flow. For this reason, it is important that the positive direction of $\hat{\mathbf{n}}$ is always clearly indicated. This is done as follows (see also **Section 1.5.1** and **Figure 1.24**):

(1) For a closed surface, the positive direction of the unit vector is always that direction that points out of the volume (see, for example, **Figures 2.6** and **1.24a**).
(2) For open surfaces, the defining property is the contour enclosing the surface. To define a positive direction, imagine that we travel along this contour as, for example, if we were to evaluate a line integral. Consider the example in **Figure 2.7**. In this case, the direction of travel is counterclockwise along the rim of the surface. According to the right-hand rule, if the fingers are directed in the direction of travel with the palm facing the interior of the surface, the thumb points in the direction of the positive unit vector (see also **Figure 1.24c**). This simple definition removes the ambiguity in the direction of the unit vector and, as we shall see shortly, is consistent with other properties of fields.

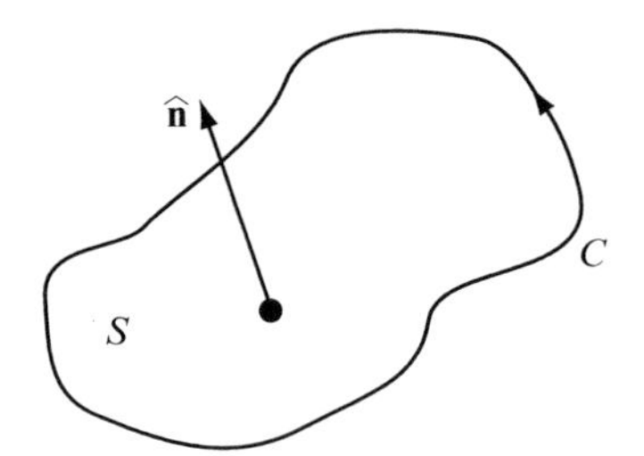

Figure 2.7 Definition of the normal to an open surface

The integration in **Eq. (2.15)** indicates the flux through a surface s. If this surface is a closed surface, we designate the integration as a ***closed surface integration***:

$$\boxed{Q = \oint_s \mathbf{A} \cdot d\mathbf{s}} \tag{2.16}$$

This is similar to the definition of closed integration over a contour. Closed surface integration gives the total or net flux through a closed surface.

Finally, we mention that since $d\mathbf{s}$ is the product of two variables, the surface integral is a double integral. The notation used in **Eq. (2.15)** or **(2.16)** is a short-form notation of this fact.

Example 2.4 Closed Surface Integral Vector $\mathbf{A} = \hat{\mathbf{x}}2xz + \hat{\mathbf{y}}2zx - \hat{\mathbf{z}}yz$ is given. Calculate the closed surface integral of the vector over the surface defined by a cube. The cube occupies the space between $0 \le x, y, z \le 1$.

Solution: First, we find the unit vector normal to each of the six sides of the cube. Then, we calculate the scalar product $\mathbf{A} \cdot \hat{\mathbf{n}}ds$, where ds is the element of surface on each side of the cube. Integrating on each side and summing up the contributions gives the net flux of $\mathbf{A}$ through the closed surface enclosing the cube.

Using **Figure 2.8**, the differentials of surface $d\mathbf{s}$ are

$$d\mathbf{s}_1 = \hat{\mathbf{x}}dydz, \quad d\mathbf{s}_2 = -\hat{\mathbf{x}}dydz$$
$$d\mathbf{s}_3 = \hat{\mathbf{z}}dxdy, \quad d\mathbf{s}_4 = -\hat{\mathbf{z}}dxdy$$
$$d\mathbf{s}_5 = \hat{\mathbf{y}}dxdz, \quad d\mathbf{s}_6 = -\hat{\mathbf{y}}dxdz$$

The surface integral is now written as

$$\oint_s \mathbf{A} \cdot d\mathbf{s} = \int_{s_1} \mathbf{A} \cdot d\mathbf{s}_1 + \int_{s_2} \mathbf{A} \cdot d\mathbf{s}_2 + \int_{s_3} \mathbf{A} \cdot d\mathbf{s}_3 + \int_{s_4} \mathbf{A} \cdot d\mathbf{s}_4 + \int_{s_5} \mathbf{A} \cdot d\mathbf{s}_5 + \int_{s_6} \mathbf{A} \cdot d\mathbf{s}_6$$

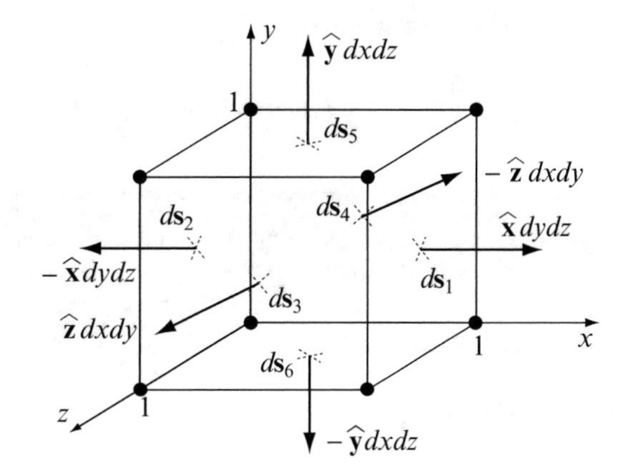

Figure 2.8 Notation used for closed surface integration in **Example 2.4**

Each term is evaluated separately. On side 1,

$$\int_{s_1} \mathbf{A} \cdot d\mathbf{s}_1 = \int_{s_1} (\hat{\mathbf{x}}2xz + \hat{\mathbf{y}}2zx - \hat{\mathbf{z}}yz) \cdot (\hat{\mathbf{x}}dydz) = \int_{s_1} 2xzdydz$$

To perform the integration, we set $x = 1$. Separating the surface integral into an integral over y and one over z, we get

$$\int_{s_1} \mathbf{A} \cdot d\mathbf{s}_1 = \int_{y=0}^{y=1} \left[\int_{z=0}^{z=1} 2z\,dz \right] dy = 2\int_{y=0}^{y=1} \left[\frac{z^2}{2}\bigg|_{z=0}^{z=1} \right] dy = \int_{y=0}^{y=1} dy = y\bigg|_{y=0}^{y=1} = 1$$

On side 2, the situation is identical, but $x = 0$ and $d\mathbf{s}_2 = -d\mathbf{s}_1$. Thus,

$$\int_{s_2} \mathbf{A} \cdot d\mathbf{s}_2 = -\int_{s_2} 2xzdydz = 0$$

On side 3, $z = 1$ and the integral is

$$\begin{aligned}\int_{s_3} \mathbf{A} \cdot d\mathbf{s}_3 &= \int_{s_3} (\hat{\mathbf{x}}2xz + \hat{\mathbf{y}}2zx - \hat{\mathbf{z}}yz) \cdot (\hat{\mathbf{z}}dxdy) \\ &= -\int_{s_3} yzdxdy = -\int_{x=0}^{x=1} \left[\int_{y=0}^{y=1} y\,dy \right] dx = -\int_{x=0}^{x=1} \left[\frac{y^2}{2}\bigg|_{y=0}^{y=1} \right] dx = -\int_{x=0}^{x=1} \frac{dx}{2} = -\frac{x}{2}\bigg|_{x=0}^{x=1} = -\frac{1}{2}\end{aligned}$$

On side 4, $z = 0$ and $d\mathbf{s}_4 = -d\mathbf{s}_3$. Therefore, the contribution of this side is zero:

$$\int_{s_4} \mathbf{A} \cdot d\mathbf{s}_4 = \int_{s_4} yzdxdy = 0$$

On side 5, $y = 1$:

$$\int_{s_5} \mathbf{A} \cdot d\mathbf{s}_5 = \int_{s_5} (\hat{\mathbf{x}}2xz + \hat{\mathbf{y}}2zx - \hat{\mathbf{z}}yz) \cdot (\hat{\mathbf{y}}dxdz) = \int_{x=0}^{1} \int_{z=0}^{1} 2zxdxdz = \frac{1}{2}$$

On side 6, $y = 0$:

$$\int_{s_6} \mathbf{A} \cdot d\mathbf{s}_6 = \int_{s_6} (\hat{\mathbf{x}}2xz + \hat{\mathbf{y}}2zx - \hat{\mathbf{z}}yz) \cdot (-\hat{\mathbf{y}}dxdz) = -\int_{x=0}^{1} \int_{z=0}^{1} 2zxdxdz = -\frac{1}{2}$$

The result is the sum of all six contributions:

$$\oint_s \mathbf{A} \cdot d\mathbf{s} = 1 + 0 - \frac{1}{2} + 0 + \frac{1}{2} - \frac{1}{2} = \frac{1}{2}$$

Example 2.5 Open Surface Integral A vector is given as $\mathbf{A} = \hat{\boldsymbol{\phi}} 5r$. Calculate the flux of the vector $\mathbf{A}$ through a surface defined by $0 \leq r \leq 1$ and $-3 \leq z \leq 3$, $\phi =$ constant. Assume the vector produces a positive flux through this surface.

Solution: The flux is the surface integral

$$\Phi = \int_s \mathbf{A} \cdot d\mathbf{s}$$

The surface s is in the $r{-}z$ plane and is therefore perpendicular to the ϕ direction, as shown in **Figure 2.9a**. Thus, the element of surface is: $d\mathbf{s}_1 = \hat{\boldsymbol{\phi}} drdz$ or $d\mathbf{s}_2 = -\hat{\boldsymbol{\phi}} drdz$. In this case, because the flux must be positive, we choose $d\mathbf{s}_1 = \hat{\boldsymbol{\phi}} drdz$. The flux is

$$\Phi = \int_{s_1} (\hat{\boldsymbol{\phi}} 5r) \cdot (\hat{\boldsymbol{\phi}} drdz) = \int_{s_1} 5rdrdz = \int_{r=0}^{r=1} \left[\int_{z=-3}^{z=+3} 5rdz \right] dr = \int_{r=0}^{r=1} 5rz \Big|_{z=-3}^{z=+3} dr = \int_{r=0}^{r=1} 30rdr = 15r^2 \Big|_{r=0}^{r=1} = 15$$

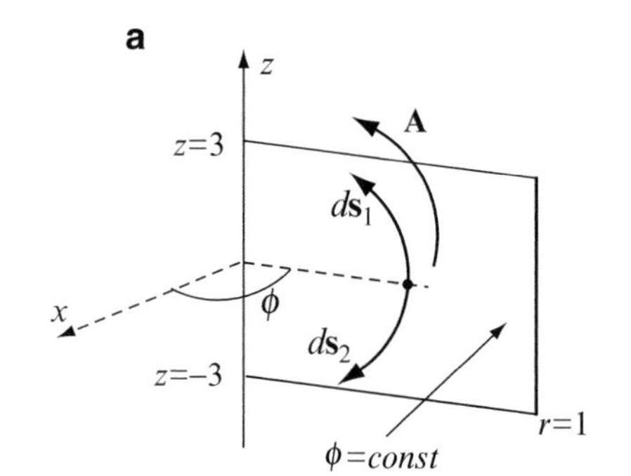

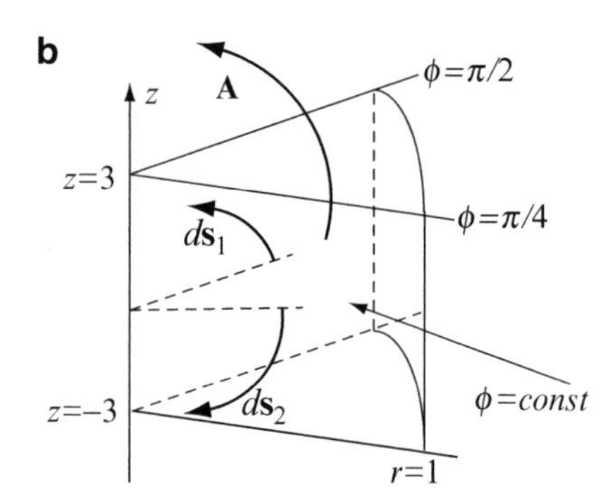

Figure 2.9 (**a**) The surface $0 \leq r \leq 1, -3 \leq z \leq 3$, $\phi =$ constant. (**b**) A wedge in cylindrical coordinates. Note that $d\mathbf{s}_1$ is in the positive ϕ direction, whereas $d\mathbf{s}_2$ is in the negative ϕ direction

Exercise 2.2 Closed Surface Integral Calculate the closed surface integral of $\mathbf{A} = \hat{\boldsymbol{\phi}} 5r$ over the surface of the wedge shown in **Figure 2.9b**.

Answer 0.

2.2.3 Volume Integrals

There are two types of volume integrals we may be required to evaluate. The first is of the form

$$W = \int_v wdv \tag{2.17}$$

where w is a scalar volume density function and dv is an element of volume. For example, if w represents the volume density distribution of stored energy (i.e., energy density), then W represents the total energy stored in volume v. Thus, the volume integral has very distinct physical meaning and will often be used in this sense. We also note that for an element of volume, such as the element in **Figure 2.10**, $dv = dxdydz$ and the volume integral is actually a triple integral (over the x, y, and z variables). The volume integral as given above is a scalar.

The second type of volume integral is a vector and is written as

$$\mathbf{P} = \int_v \mathbf{p}dv \tag{2.18}$$

This is similar to the integral in **Eq. (2.17)**, but in terms of its evaluation, it is evaluated over each component independently. The only difference between this and the scalar integral in **Eq. (2.17)** is that the unit vectors may not be constant and, therefore, they may have to be resolved into Cartesian coordinates in which the unit vectors are constant and therefore may be taken outside the integral sign (see **Sections 1.5.2**, **1.5.3**, and **Example 2.7**). In Cartesian coordinates, we may write

$$\mathbf{P} = \hat{\mathbf{x}}\int_v p_x dv + \hat{\mathbf{y}}\int_v p_y dv + \hat{\mathbf{z}}\int_v p_z dv \tag{2.19}$$

This type of vector integral is often called a regular or ordinary vector integral because it is essentially a scalar integral with the unit vectors added. It occurs in other types of calculations that do not involve volumes and volume distributions, such as in evaluating velocity from acceleration (see **Problem 2.12**).

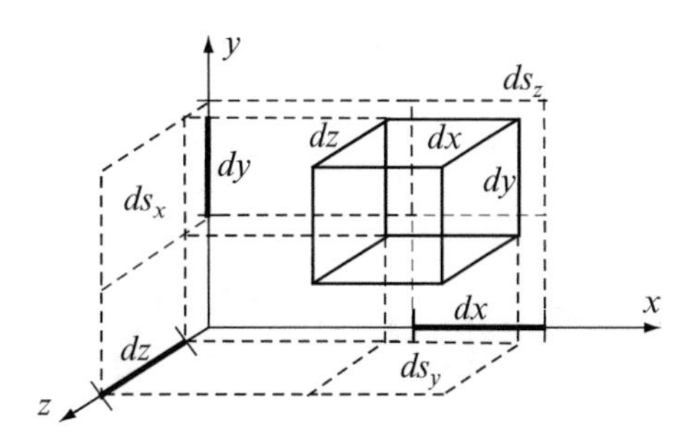

Figure 2.10 An element of volume and the corresponding projections on the axes and planes

Example 2.6 Scalar Volume Integral

(a) Calculate the volume of a section of the sphere $x^2 + y^2 + z^2 = 16$ cut by the planes $y = 0$, $z = 2$, $x = 1$, and $x = -1$.

(b) Calculate the volume of the section of the sphere cut by the planes $\theta = \pi/6$, $\theta = \pi/3$, $\phi = 0$, and $\phi = \pi/3$.

Solution: **(a)** Although, in general, the fact that integration is on a sphere may suggest the use of spherical coordinates; in this case, it is easier to evaluate the integral in Cartesian coordinates because the sphere is cut by planes parallel to the axes. The limits of integration must first be evaluated. **Figure 2.11** is used for this purpose. **(b)** Because the defining planes now are parallel to the axes in spherical coordinates, the solution is easier in spherical coordinates.

(a) The limits of integration are as follows:

(1) From the equation of the sphere, $z = \sqrt{16 - x^2 - y^2}$. From **Figure 2.11a**, the limits of integration on z are $z_1 = -\sqrt{16 - x^2 - y^2}$ and $z_2 = 2$.

(2) The limits on y are $y_1 = 0$ and $y_2 = \sqrt{16 - x^2}$ (see **Figure 2.11b**).

(3) The limits of integration on x are between $x_1 = -1$ and $x_2 = +1$ (see **Figure 2.11c**).

With the differential of volume in Cartesian coordinates, $dv = dxdydz$, we get

$$v = \int_v dv = \int_{x=-1}^{x=1} \left\{ \int_{y=0}^{y=\sqrt{16-x^2}} \left[\int_{z=-\sqrt{16-x^2-y^2}}^{z=2} dz \right] dy \right\} dx = \int_{x=-1}^{x=1} \left\{ \int_{y=0}^{y=\sqrt{16-x^2}} \left[2 + \sqrt{16-x^2-y^2} \right] dy \right\} dx$$

$$= \int_{x=-1}^{x=1} \left\{ 2y + 0.5 \left[y\sqrt{16-x^2-y^2} + (16-x^2)\sin^{-1}\left(\frac{y}{\sqrt{16-x^2}} \right) \right] \right\}_{y=0}^{y=\sqrt{16-x^2}} dx$$

$$= \int_{x=-1}^{x=1} \left\{ 2\sqrt{16-x^2} + 0.5\left[(16-x^2)\sin^{-1}(1) \right] \right\} dx = \int_{x=-1}^{x=1} 2\sqrt{16-x^2}dx + \frac{\pi}{4}\int_{x=-1}^{x=1} (16-x^2)dx$$

$$= \left[x\sqrt{16-x^2} + 16\sin^{-1}\left(\frac{x}{4} \right) \right]_{x=-1}^{x=1} + \frac{\pi}{4}\left[16x - \frac{x^3}{3} \right]_{x=-1}^{x=1} = \sqrt{15} + \sqrt{15} + 16\sin^{-1}\left(\frac{1}{4} \right) - 16\sin^{-1}\left(-\frac{1}{4} \right) + 8\pi - \frac{\pi}{6}$$

$$= 2\sqrt{15} + 32\sin^{-1}(0.25) + \pi(8 - 1/6) = 40.44$$

Thus,

$$v = 40.44 \quad [\text{m}^3]$$

(b) The limits of integration are $0 \le R \le 4$, $\pi/6 \le \theta \le \pi/3$, and $0 \le \phi \le \pi/3$. The element of volume in spherical coordinates is $dv = R^2\sin\theta dR d\theta d\phi$. The volume of the section is therefore,

$$v = \int_v dv = \int_{R=0}^{R=4} \left\{ \int_{\theta=\pi/6}^{\theta=\pi/3} \left[\int_{\phi=0}^{\phi=\pi/3} R^2\sin\theta d\phi \right] d\theta \right\} dR = \int_{R=0}^{R=4} \left\{ \int_{\theta=\pi/6}^{\theta=\pi/3} \frac{\pi}{3} R^2 \sin\theta d\theta \right\} dR$$

$$= \int_{R=0}^{R=4} \left[-\frac{\pi}{3} R^2 \cos\theta \right]_{\pi/6}^{\pi/3} dR = \int_{R=0}^{R=4} \frac{0.366\pi}{3} R^2 dR = \left[\frac{0.366\pi}{9} R^3 \right]_0^4 = \frac{64 \times 0.366\pi}{9} = 8.177 \quad [\text{m}^3]$$

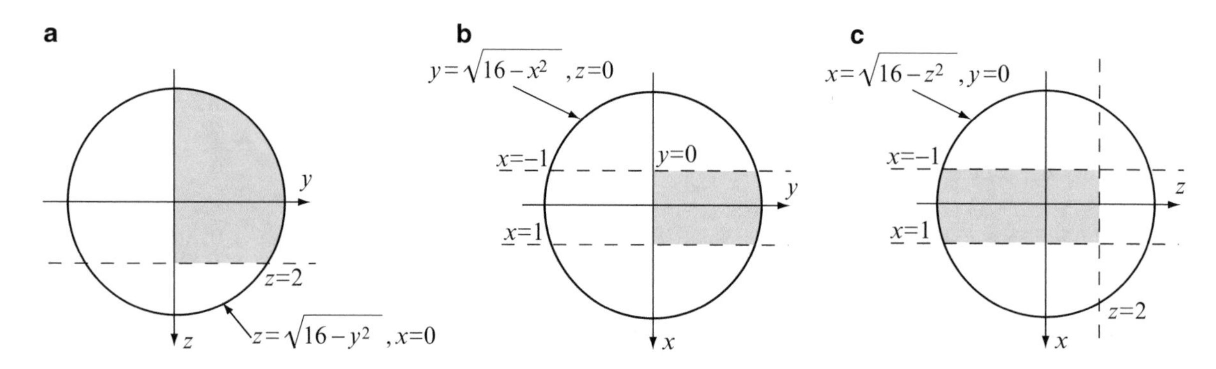

Figure 2.11 Sequence for evaluation of the volume integral in **Example 2.6**. Projections on the y–z, x–y, and x–z planes

Example 2.7 Volume Integration of a Vector Function A vector function $\mathbf{A} = \hat{\mathbf{r}}r + \hat{\mathbf{z}}3$ gives the distribution of a vector in space. This function may represent a distribution of moments or force density in volume v. Calculate the total contribution of the function in a volume defined by a cylinder of radius a and height b, centered on the z axis, above the x–y plane.

Solution: We use cylindrical coordinates to write the integral of **A** over the volume v. The vector function is integrated as follows:

$$\mathbf{F} = \int_v \mathbf{A} dv = \int_v \hat{\mathbf{r}} A_r dv + \int_v \hat{\mathbf{z}} A_z dv = \int_v \hat{\mathbf{r}} r dv + \hat{\mathbf{z}} \int_v 3dv$$

where the unit vector $\hat{\mathbf{z}}$ was taken outside the integral sign (it is constant) but the unit vector $\hat{\mathbf{r}}$ cannot be taken out of the integral since it depends on ϕ. From **Eq. (1.65)**, we write $\hat{\mathbf{r}} = \hat{\mathbf{x}}\cos\phi + \hat{\mathbf{y}}\sin\phi$ and, now, since $\hat{\mathbf{x}}$ and $\hat{\mathbf{y}}$ are constant unit vectors in Cartesian coordinates, we write (together with $dv = rdrd\phi dz$)

$$\mathbf{F} = \int_v (\hat{\mathbf{x}}\cos\phi + \hat{\mathbf{y}}\sin\phi) r dv + \hat{\mathbf{z}} \int_v 3dv = \hat{\mathbf{x}} \int_v r\cos\phi dv + \hat{\mathbf{y}} \int_v r\sin\phi dv + \hat{\mathbf{z}} \int_v 3dv$$

$$= \hat{\mathbf{x}} \int_{z=0}^{z=b} \left[\int_{\phi=0}^{\phi=2\pi} \left(\int_{r=0}^{r=a} r^2 \cos\phi dr \right) d\phi \right] dz + \hat{\mathbf{y}} \int_{z=0}^{z=b} \left[\int_{\phi=0}^{\phi=2\pi} \left(\int_{r=0}^{r=a} r^2 \sin\phi dr \right) d\phi \right] dz + \hat{\mathbf{z}} \int_{z=0}^{z=b} \left[\int_{\phi=0}^{\phi=2\pi} \left(\int_{r=0}^{r=a} 3rdr \right) d\phi \right] dz$$

$$= \hat{\mathbf{x}} \int_{z=0}^{z=b} \left[\int_{\phi=0}^{\phi=2\pi} \frac{a^3\cos\phi}{3} d\phi \right] dz + \hat{\mathbf{y}} \int_{z=0}^{z=b} \left[\int_{\phi=0}^{\phi=2\pi} \frac{a^3\sin\phi}{3} d\phi \right] dz + \hat{\mathbf{z}} \int_{z=0}^{z=b} \left[\int_{\phi=0}^{\phi=2\pi} \frac{3a^2}{2} d\phi \right] dz = \hat{\mathbf{z}} \int_{z=0}^{z=b} 3\pi a^2 dz = \hat{\mathbf{z}} 3\pi a^2 b$$

In this integration, we used the fact that $\int_0^{2\pi} \sin\phi = 0$ and $\int_0^{2\pi} \cos\phi = 0$. In summary,

$$\mathbf{F} = \hat{\mathbf{z}} 3\pi a^2 b$$

2.2.4 Symbolic Versus Numerical Integration

Integration is central to work in almost all science and engineering disciplines. Integration of vector quantities is not much different than integration of scalars except that the vector operations must be performed prior to integration and all unit vectors must be outside the integral sign, recalling that only constants with respect to the variable of integration may be taken outside the integral sign. In the case of vectors that means that the magnitude of the vector must be constant and its direction cannot vary in space. This may require transformations between systems of coordinates but once that is done, the integration itself is the same as for scalar variables.

The question is how to perform the integral itself. There are two ways of doing that. One is the so-called symbolic integration, that is, the result is in terms of variables. The second method is numerical, that is, the result is a value. Of course, a combination of the two is also possible. In general, symbolic integration is preferred because it often supplies more information on the behavior of the variables and allows one to gage if the calculations are correct. A numerical value is more concise but less desirable.

The preferred path is to perform symbolic integration and if desired substitute numerical values as the last step in the calculation. Purely numerical methods, such as integration routines on computers or calculators are sometimes convenient but they should be undertaken with great care or one may end up with a numerical value that may or may not be correct and with almost no way of verification.

2.3 Differentiation of Scalar and Vector Functions

As we might expect, in addition to the need to integrate scalar and vector expressions as described above, we also need to differentiate scalar and vector functions. The rules and implications of these operations are considered next. Three types of operations are defined: the ***gradient***, the ***divergence***, and the ***curl***. The first relates to scalar functions and the second and third to vector functions. These operations will be shown to be fundamental to understanding of vector fields.

2.3.1 The Gradient of a Scalar Function **Point_Charges.m**

The partial spatial derivatives of a scalar function $U(x,y,z)$ with nonzero first-order partial derivatives with respect to the coordinates x, y, and z are defined at a point in space as

$$\frac{\partial U}{\partial x}, \quad \frac{\partial U}{\partial y}, \quad \frac{\partial U}{\partial z} \tag{2.20}$$

Ordinary derivatives are defined in a similar manner if the function U is a function of a single variable. Obviously, the same can be done in any system of coordinates or the above can be transformed into any system of coordinates using the formulas we obtained in the previous chapter. For this reason, we start our discussion in Cartesian coordinates.

That the derivative of a scalar function describes the slope of the function is known. Also, there is no question that this is an important aspect of the function. Now, imagine that we are standing on a mountain. The slope of the mountain at any given point is not defined, unless we qualify it with something like "slope in the northeast direction" or "slope in the direction of town" or a similar statement. Also, it is decidedly different if we describe the slope up the mountain or down the mountain. If you are designing a ski path, the slope down the mountain is most important. If you were a civil engineer, designing a road, then you might be interested in the path with minimum variation in slope. Thus, an additional aspect of the derivative has entered our considerations, and this must be satisfied. For this reason, we will define a "directional derivative" which, being a vector, gives the slope (as any other derivative) but also the direction of this slope.

In particular, at any given point on a function (say, the mountain described above), there is one direction which is unique; that is, the direction of maximum slope. This direction and the slope associated with it are extremely important, and not only in electromagnetics. However, before we continue, we immediately realize that at any point, there are actually two directions which satisfy this condition. In the example of the mountain, at any point, we might go up the mountain or down the mountain. For example, flow of water at any point is in the direction of maximum slope, but it only flows down the mountain or in the direction of decrease in potential energy. On a topological map, the maximum slope is indicated by the minimum distance between two altitude lines (see **Figure 2.12**). These properties are defined by the gradient, as follows:

"*the vector which gives both the magnitude and direction of the maximum spatial rate of change of a scalar function is called the gradient of this scalar function.*"

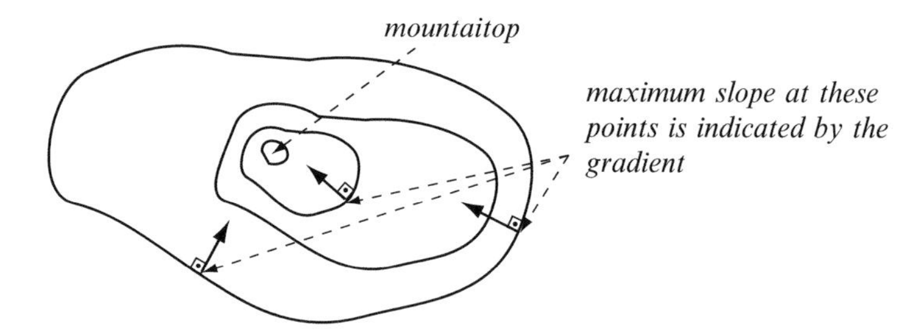

Figure 2.12 Illustration of the gradient

The rate of change is assumed to be positive in the direction of the increase in the value of the scalar function (up the mountain). Thus, returning to our example, water always flows in the direction opposite the direction of the gradient, whereas the most difficult climb on the mountain at any point is in the direction of the gradient. **Figure 2.12** shows these considerations. The gradient on the map may indicate the direction of climbing or, if this map shows atmospheric pressure, the gradient points in the direction of increased pressure. If you were to sail in the direction of the gradient in air pressure, you will always have the wind in your face.

To define the relations involved, consider **Figure 2.13**. Two surfaces are given such that the scalar function $U(x,y,z)$ (which may represent potential energy, temperature, pressure, and the like) is constant on each surface. Assuming the value of the function to be U on the lower surface and $U + dU$ on the upper surface (but still constant on each surface), then, given the scalar function $U(x,y,z)$ with partial derivatives $\partial U/\partial x$, $\partial U/\partial y$, and $\partial U/\partial z$, we can calculate the differential of U as dU by considering points $P(x,y,z)$ and $P'(x + dx,y + dy,z + dz)$ and using the total differential:

$$dU = \frac{\partial U}{\partial x}dx + \frac{\partial U}{\partial y}dy + \frac{\partial U}{\partial z}dz \tag{2.21}$$

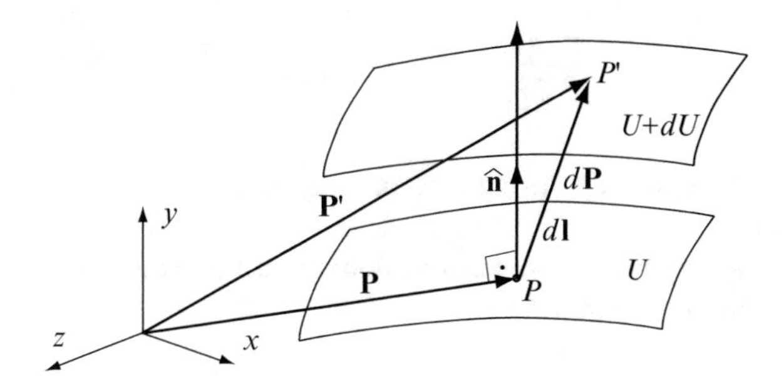

Figure 2.13 The relation between the scalar function U and its gradient

Defining the vector $d\mathbf{P} = \mathbf{P}' - \mathbf{P}$ with scalar components, dx, dy, and dz, dU can be written as the scalar product of two vectors:

$$dU = \left(\hat{\mathbf{x}}\frac{\partial U}{\partial x} + \hat{\mathbf{y}}\frac{\partial U}{\partial y} + \hat{\mathbf{z}}\frac{\partial U}{\partial z}\right) \cdot (\hat{\mathbf{x}}dx + \hat{\mathbf{y}}dy + \hat{\mathbf{z}}dz) \tag{2.22}$$

We recognize the second vector in this relation as $d\mathbf{l}$ as defined in **Eq. (1.48)** and write

$$dU = \left(\hat{\mathbf{x}}\frac{\partial U}{\partial x} + \hat{\mathbf{y}}\frac{\partial U}{\partial y} + \hat{\mathbf{z}}\frac{\partial U}{\partial z}\right) \cdot d\mathbf{l} \tag{2.23}$$

The vector in parentheses is now denoted as

$$\mathbf{D} = \hat{\mathbf{x}}\frac{\partial U}{\partial x} + \hat{\mathbf{y}}\frac{\partial U}{\partial y} + \hat{\mathbf{z}}\frac{\partial U}{\partial z} \tag{2.24}$$

Using this notation, the total differential is

$$dU = \mathbf{D} \cdot d\mathbf{l} = |\mathbf{D}||d\mathbf{l}|\cos\theta \tag{2.25}$$

From the properties of the scalar product, we know that this product is maximum when $d\mathbf{l}$ and $\mathbf{D}$ are in the same direction ($\theta = 0$). Thus, we can write the following derivative:

$$\frac{dU}{dl} = |\mathbf{D}|\cos\theta \tag{2.26}$$

This derivative depends on the direction of $d\mathbf{l}$ (in relation to $\mathbf{D}$), and, therefore, dU/dl is a directional derivative: the derivative of U in the direction of $d\mathbf{l}$. In formal terms, we can write the directional derivative in the direction $d\mathbf{l}$ in terms of the directional derivative in the normal ($\hat{\mathbf{n}}$) direction as

$$\frac{dU}{dl} = \frac{dU}{dn}\frac{dn}{dl} = \frac{dU}{dn}\cos\theta \tag{2.27}$$

where $\hat{\mathbf{n}}$ is the unit vector normal to the surface at the point at which the derivative is calculated. Thus, the maximum value of dU/dl is

$$\left.\frac{dU}{dl}\right|_{max} = \frac{dU}{dn} \tag{2.28}$$

That is, the maximum rate of change of the scalar function U is the normal derivative of the scalar function at point P. In other words, to calculate the maximum rate of change of the function, we must choose $d\mathbf{l}$ to be in the direction normal to the constant value surface. Now, returning to **Eq. (2.25)**, we get

$$\hat{\mathbf{n}}\left.\frac{dU}{dl}\right|_{max} = \hat{\mathbf{n}}\frac{dU}{dn} = \mathbf{D} = \hat{\mathbf{x}}\frac{\partial U}{\partial x} + \hat{\mathbf{y}}\frac{\partial U}{\partial y} + \hat{\mathbf{z}}\frac{\partial U}{\partial z} = \text{grad}(U) \tag{2.29}$$

This result indicates not only the meaning of the gradient but also how we can calculate it from the partial derivatives of the scalar function U.

Although the form above is correct, we normally use a special notation for the gradient of a scalar function. Again, returning to the above equation, we write

$$\text{grad}\, U = \left(\hat{\mathbf{x}}\frac{\partial}{\partial x} + \hat{\mathbf{y}}\frac{\partial}{\partial y} + \hat{\mathbf{z}}\frac{\partial}{\partial z}\right)U \tag{2.30}$$

The quantity in parentheses is a fixed operator for any scalar function we may wish to evaluate. We denote this operator in Cartesian coordinates as

$$\boxed{\nabla \equiv \left(\hat{\mathbf{x}}\frac{\partial}{\partial x} + \hat{\mathbf{y}}\frac{\partial}{\partial y} + \hat{\mathbf{z}}\frac{\partial}{\partial z}\right)} \tag{2.31}$$

This operator is called the ***nabla operator*** or the ***del operator***. We will use the latter name. The del operator is a vector operator by definition, and, therefore, it is not necessary to mark it is a vector.

Important Note: Although the del operator is a vector differential operator and we wrote it as a vector, it should be used with care since it is not a true vector (for instance, it does not have a magnitude). The reasons will become obvious later on but for now, the operator should only be used in the form given above. As an example, we have not defined (and, in fact, cannot define) the scalar or vector product between the del operator and other vectors or with itself. The extension of the considerations and notation given here to other coordinate systems should be avoided at this stage since all our discussion was in Cartesian coordinates. With this notation, the gradient of a scalar function is written as

$$\boxed{\text{grad}\, U = \nabla U = \left(\hat{\mathbf{x}}\frac{\partial}{\partial x} + \hat{\mathbf{y}}\frac{\partial}{\partial y} + \hat{\mathbf{z}}\frac{\partial}{\partial z}\right)U} \tag{2.32}$$

and is read as ***grad*** U or ***del*** U. Either form is acceptable, although the normal use in the United States is ∇U, whereas in other countries, the form grad U is often more common. From now on, we will use the notation ∇U and the pronunciation "del U" exclusively to avoid confusion.

The del operator is a mathematical operator to which, by itself, we cannot associate any geometrical meaning. It is the interaction of the del operator with other quantities that gives it geometric significance.

On the other hand, the gradient of a scalar function has a very distinct physical meaning as was shown above. The gradient has the following general properties:

(1) It operates on a scalar function and results in a vector function.
(2) The gradient is normal to a constant value surface. This can be seen from **Eq. (2.29)**. This property will be used extensively to identify the direction of vector fields.
(3) The gradient always points in the direction of maximum change in the scalar function. In terms of potential energy, the gradient shows the direction of increase in potential energy.

Example 2.8 Application: Normal to a Surface A vector normal to a surface is $\nabla f(x,y,z)$ where $f(x,y,z)$ is the equation of the surface. Consider the plane $x + \sqrt{2}y + z = 3$. Find a normal vector to this surface and the unit vector normal to the surface.

Solution: Find the gradient of the plane. This is based on the fact that the gradient is always normal to a constant value function.

We write the equation of the plane as

$$f(x,y,z) = x + \sqrt{2}y + z - 3 = 0$$

The vector normal to the plane is

$$\mathbf{A} = \nabla f(x,y,z) = \nabla\left(x + \sqrt{2}y + z - 3\right)$$
$$= \hat{\mathbf{x}}\frac{\partial}{\partial x}\left(x + \sqrt{2}y + z - 3\right) + \hat{\mathbf{y}}\frac{\partial}{\partial y}\left(x + \sqrt{2}y + z - 3\right) + \hat{\mathbf{z}}\frac{\partial}{\partial z}\left(x + \sqrt{2}y + z - 3\right) = \hat{\mathbf{x}}1 + \hat{\mathbf{y}}\sqrt{2} + \hat{\mathbf{z}}1$$

and the unit vector normal to the plane is

$$\hat{\mathbf{n}} = \frac{\mathbf{A}}{|\mathbf{A}|} = \frac{\hat{\mathbf{x}}1 + \hat{\mathbf{y}}\sqrt{2} + \hat{\mathbf{z}}1}{\sqrt{1+2+1}} = \hat{\mathbf{x}}\frac{1}{2} + \hat{\mathbf{y}}\frac{\sqrt{2}}{2} + \hat{\mathbf{z}}\frac{1}{2}$$

Note that the constant value in the equation is immaterial—it does not change the slopes in the x, y, and z directions.

Example 2.9 Application: Derivative in the Direction of a Vector Find the derivative of $xy^2 + y^2z$ at P (1,1,1) in the direction of the vector $\mathbf{A} = \hat{\mathbf{x}}3 + \hat{\mathbf{y}}4$.

Solution: The gradient of the scalar function $V = xy^2 + y^2z$ is first calculated. This gives the directional derivative in the normal direction. Then, we evaluate the gradient at point P (1,1,1) and find the projection of this vector onto the vector $\mathbf{A} = \hat{\mathbf{x}}3 + \hat{\mathbf{y}}4$ using the scalar product between the gradient and the unit vector $\hat{\mathbf{A}}$. This gives the magnitude (or scalar component) of the directional derivative and is the derivative in the required direction. The scalar function is

$$V = xy^2 + y^2z$$

The gradient of the scalar function $V(x,y,z)$ is [using **Eq. (2.32)**]

$$\nabla V = \hat{\mathbf{x}}\frac{\partial V}{\partial x} + \hat{\mathbf{y}}\frac{\partial V}{\partial y} + \hat{\mathbf{z}}\frac{\partial V}{\partial z} = \hat{\mathbf{x}}y^2 + \hat{\mathbf{y}}(2xy + 2yz) + \hat{\mathbf{z}}y^2$$

The gradient at point (1,1,1) is

$$\nabla V(1,1,1) = \hat{\mathbf{x}}1 + \hat{\mathbf{y}}4 + \hat{\mathbf{z}}1$$

The direction of $\mathbf{A}$ in space is given by the unit vector

$$\hat{\mathbf{A}} = \frac{\hat{\mathbf{x}}3 + \hat{\mathbf{y}}4}{|\hat{\mathbf{x}}3 + \hat{\mathbf{y}}4|} = \frac{\hat{\mathbf{x}}3 + \hat{\mathbf{y}}4}{5}$$

and the projection of the gradient of V onto the direction of $\mathbf{A}$ is

$$(\nabla V)\cdot\hat{\mathbf{A}} = (\hat{\mathbf{x}}1 + \hat{\mathbf{y}}4 + \hat{\mathbf{z}}1)\cdot\left(\frac{\hat{\mathbf{x}}3 + \hat{\mathbf{y}}4}{5}\right) = \frac{1}{5}(3 + 16) = \frac{19}{5}$$

This is the derivative (or, in more practical terms, the slope) of V in the direction of $\mathbf{A}$ at $P(1,1,1)$.

Example 2.10 Given two points $P(x,y,z)$ and $P'(x',y',z')$, calculate the gradient of the function $1/R(P,P')$ where R is the distance between the two points.

Solution: First, we find the scalar function that gives the distance between the two points. Then, we apply the gradient to this function. Because the coordinates (x,y,z) or (x',y',z') may be taken as the variables, the gradient with respect to each set of variables is calculated. In applications, one point may be fixed while the other varies, so there may not be a need to calculate the gradient with respect to both sets of variables.

The scalar function describing the distance between the two points can be written directly as (using (x,y,z) as variables and (x',y',z') as fixed)

$$R(x,y,z,x',y',z') = \sqrt{(x-x')^2 + (y-y')^2 + (z-z')^2} \quad \rightarrow \quad \frac{1}{R} = \left[(x-x')^2 + (y-y')^2 + (z-z')^2\right]^{-1/2}$$

To calculate the gradient, we write

$$\nabla\left(\frac{1}{R}\right) = \hat{\mathbf{x}}\frac{\partial}{\partial x}\left(\frac{1}{R}\right) + \hat{\mathbf{y}}\frac{\partial}{\partial y}\left(\frac{1}{R}\right) + \hat{\mathbf{z}}\frac{\partial}{\partial z}\left(\frac{1}{R}\right)$$

$$= \hat{\mathbf{x}}\left(-\frac{1}{2}\frac{2(x-x')}{\left[(x-x')^2 + (y-y')^2 + (z-z')^2\right]^{3/2}}\right) + \hat{\mathbf{y}}\left(-\frac{1}{2}\frac{2(y-y')}{\left[(x-x')^2 + (y-y')^2 + (z-z')^2\right]^{3/2}}\right)$$

$$+\hat{\mathbf{z}}\left(-\frac{1}{2}\frac{2(z-z')}{\left[(x-x')^2 + (y-y')^2 + (z-z')^2\right]^{3/2}}\right)$$

After simplifying,

$$\nabla\left(\frac{1}{R}\right) = -\frac{\hat{\mathbf{x}}(x-x')}{R^3} - \frac{\hat{\mathbf{y}}(y-y')}{R^3} - \frac{\hat{\mathbf{z}}(z-z')}{R^3}$$

This can also be written as

$$\nabla\left(\frac{1}{R}\right) = -\frac{1}{R^2}\left(\frac{\hat{\mathbf{x}}(x-x') + \hat{\mathbf{y}}(y-y') + \hat{\mathbf{z}}(z-z')}{R}\right) = -\frac{1}{R^2}\left(\frac{\mathbf{R}}{R}\right) = -\frac{\mathbf{R}}{R^3}$$

If we use the definition of the unit vector as $\hat{\mathbf{R}} = \mathbf{R}/R$, we get

$$\nabla\left(\frac{1}{R}\right) = -\frac{\hat{\mathbf{R}}}{R^2}$$

Of course, the following form is equivalent:

$$\nabla\left(\frac{1}{R}\right) = -\left(\frac{\hat{\mathbf{x}}(x-x') + \hat{\mathbf{y}}(y-y') + \hat{\mathbf{z}}(z-z')}{\left[(x-x')^2 + (y-y')^2 + (z-z')^2\right]^{3/2}}\right)$$

We arbitrarily calculated the derivatives with respect to the variables (x,y,z). If we wish, we can also calculate the derivatives with respect to the variables x', y', and z'. In some cases, this might become necessary. We denote the gradient so calculated as $\nabla'(1/R)$, and, by simple inspection, we get

$$\nabla'\left(\frac{1}{R}\right) = -\nabla\left(\frac{1}{R}\right) = \frac{\hat{\mathbf{R}}}{R^2}$$

since in the evaluation of the derivatives, the inner derivatives with respect to x', y', and z' are all negative.

Exercise 2.3 Given a function $f(x,y,z)$ as the distance between a point $P(x,y,z)$ and the origin $O(0,0,0)$.

(a) Determine the gradient of this function in Cartesian coordinates.
(b) What is the magnitude of the gradient?

Answer **(a)** $\nabla f = \frac{1}{f}(\hat{\mathbf{x}}x + \hat{\mathbf{y}}y + \hat{\mathbf{z}}z)$, where $f = \sqrt{x^2 + y^2 + z^2}$. **(b)** $|\nabla f| = \sqrt{\frac{1}{f^2}(x^2 + y^2 + z^2)} = 1$.

2.3.1.1 Gradient in Cylindrical Coordinates

To define the gradient in cylindrical coordinates, we can proceed in one of two ways:

(1) We may start with the definition of the total differential in **Eq. (2.21)**, rewrite it in cylindrical coordinates, and proceed in the same way we have done for the gradient in Cartesian coordinates, but using $d\mathbf{l}$ in cylindrical coordinates.
(2) Since the gradient is known in Cartesian coordinates and we have defined the proper transformation from Cartesian to cylindrical coordinates in **Section 1.5.2**, we may use this transformation to transform the gradient vector to cylindrical coordinates.

We use the second method because it will also become useful in the following sections. To do so, we write the gradient in **Eq. (2.32)** as follows:

$$\nabla U(x,y,z) = \hat{\mathbf{x}}\frac{\partial}{\partial x}U(x,y,z) + \hat{\mathbf{y}}\frac{\partial}{\partial y}U(x,y,z) + \hat{\mathbf{z}}\frac{\partial}{\partial z}U(x,y,z) \tag{2.33}$$

To transform this into cylindrical coordinates, we must write the function $U(x,y,z)$ in cylindrical coordinates as $U(r,\phi,z)$. More importantly, we must transform the unit vectors $\hat{\mathbf{x}}$, $\hat{\mathbf{y}}$, and $\hat{\mathbf{z}}$ into the unit vectors $\hat{\mathbf{r}}$, $\hat{\boldsymbol{\phi}}$, and $\hat{\mathbf{z}}$ in cylindrical coordinates and the operators $\partial/\partial x$, $\partial/\partial y$, and $\partial/\partial z$ into their counterparts in cylindrical coordinates $\partial/\partial r$, $\partial/\partial \phi$, and $\partial/\partial z$. The transformation for the unit vectors was found in **Eq. (1.66)** as follows:

$$\hat{\mathbf{x}} = \hat{\mathbf{r}}\cos\phi - \hat{\boldsymbol{\phi}}\sin\phi, \quad \hat{\mathbf{y}} = \hat{\mathbf{r}}\sin\phi + \hat{\boldsymbol{\phi}}\cos\phi, \quad \hat{\mathbf{z}} = \hat{\mathbf{z}} \tag{2.34}$$

The transformation of the partial derivatives uses the chain rule of differentiation as follows:

$$\frac{\partial}{\partial x} = \frac{\partial}{\partial r}\left(\frac{\partial r}{\partial x}\right) + \frac{\partial}{\partial \phi}\left(\frac{\partial \phi}{\partial x}\right) \quad \text{and} \quad \frac{\partial}{\partial y} = \frac{\partial}{\partial r}\left(\frac{\partial r}{\partial y}\right) + \frac{\partial}{\partial \phi}\left(\frac{\partial \phi}{\partial y}\right) \tag{2.35}$$

The derivative with respect to z remains unchanged. To evaluate the derivatives $\partial r/\partial x$, $\partial \phi/\partial x$, $\partial r/\partial y$, and $\partial \phi/\partial y$, we use the transformation for coordinates from **Eqs. (1.63)** and **(1.64)**:

$$x = r\cos\phi, \quad y = r\sin\phi, \quad z = z \tag{2.36}$$

$$r = \sqrt{x^2 + y^2}, \quad \phi = \tan^{-1}\left(\frac{y}{x}\right), \quad z = z \tag{2.37}$$

From **Eq. (2.37)**, we can write directly

$$\frac{\partial r}{\partial x} = \frac{x}{\sqrt{x^2 + y^2}} \quad \text{and} \quad \frac{\partial r}{\partial y} = \frac{y}{\sqrt{x^2 + y^2}} \tag{2.38}$$

$$\frac{\partial \phi}{\partial x} = \frac{\partial}{\partial x}\left[\tan^{-1}\left(\frac{y}{x}\right)\right] = -\frac{y}{x^2 + y^2} \quad \text{and} \quad \frac{\partial \phi}{\partial y} = \frac{\partial}{\partial y}\left[\tan^{-1}\left(\frac{y}{x}\right)\right] = \frac{x}{x^2 + y^2} \tag{2.39}$$

Substituting for x and y from **Eq. (2.36)** and using $r = \sqrt{x^2 + y^2}$ from **Eq. (2.37)**, we get

$$\frac{\partial r}{\partial x} = \cos\phi, \ \frac{\partial r}{\partial y} = \sin\phi, \ \frac{\partial \phi}{\partial x} = -\frac{\sin\phi}{r}, \ \frac{\partial \phi}{\partial y} = \frac{\cos\phi}{r} \tag{2.40}$$

Substituting these in **Eq. (2.35)**, we get

$$\frac{\partial}{\partial x} = \cos\phi\frac{\partial}{\partial r} - \frac{\sin\phi}{r}\frac{\partial}{\partial \phi} \quad \text{and} \quad \frac{\partial}{\partial y} = \sin\phi\frac{\partial}{\partial r} + \frac{\cos\phi}{r}\frac{\partial}{\partial \phi} \tag{2.41}$$

Now substituting for $\partial/\partial x$ and $\partial/\partial y$ from **Eq. (2.41)** and for $\hat{\mathbf{x}}$ and $\hat{\mathbf{y}}$ from **Eq. (2.34)** into **Eq. (2.33)** and using $U(r,\phi,z)$ for the scalar function in cylindrical coordinates, we get

$$\begin{aligned} \nabla U(r,\phi,z) = {} & \left[\hat{\mathbf{r}}\cos\phi - \hat{\boldsymbol{\phi}}\sin\phi\right]\left[\cos\phi\frac{\partial}{\partial r} - \frac{\sin\phi}{r}\frac{\partial}{\partial \phi}\right]U(r,\phi,z) \\ & + \left[\hat{\mathbf{r}}\sin\phi + \hat{\boldsymbol{\phi}}\cos\phi\right]\left[\sin\phi\frac{\partial}{\partial r} + \frac{\cos\phi}{r}\frac{\partial}{\partial \phi}\right]U(r,\phi,z) + \hat{\mathbf{z}}\frac{\partial}{\partial z}U(r,\phi,z) \end{aligned} \tag{2.42}$$

Performing the various products and using $\sin^2\phi + \cos^2\phi = 1$, we get

$$\boxed{\nabla U = \left(\hat{\mathbf{r}}\frac{\partial U(r,\phi,z)}{\partial r} + \hat{\boldsymbol{\phi}}\frac{1}{r}\frac{\partial U(r,\phi,z)}{\partial \phi} + \hat{\mathbf{z}}\frac{\partial U(r,\phi,z)}{\partial z}\right) = \left(\hat{\mathbf{r}}\frac{\partial}{\partial r} + \hat{\boldsymbol{\phi}}\frac{1}{r}\frac{\partial}{\partial \phi} + \hat{\mathbf{z}}\frac{\partial}{\partial z}\right)U(r,\phi,z)} \tag{2.43}$$

As a consequence, we can immediately write the del operator in cylindrical coordinates as

$$\boxed{\nabla \equiv \hat{\mathbf{r}}\frac{\partial}{\partial r} + \hat{\boldsymbol{\phi}}\frac{1}{r}\frac{\partial}{\partial \phi} + \hat{\mathbf{z}}\frac{\partial}{\partial z}} \tag{2.44}$$

It is important to note that the del operator in cylindrical coordinates is not the same as the del operator in Cartesian coordinates. Also to be noted is that in arriving at the definition of the gradient in cylindrical coordinates, we have not used the del operator, only the gradient in Cartesian coordinates and the transformations of coordinates and unit vectors. The process is rather tedious but is straightforward. We will use this process again in future sections but without repeating the details. The main advantage of doing so is that although we use the del operator as a symbolic description or as a notation, there is no need to perform any operations on the operator itself. We avoid these operations because the del operator is not a true vector.

Example 2.11 A scalar field is given in Cartesian coordinates as $f(x,y,z) = x + 5zy^2$. Calculate the gradient of the scalar field in cylindrical coordinates.

Solution: There are two ways to obtain the solution. One is to transform the scalar field to cylindrical coordinates and then apply the gradient to the field. The second is to calculate the gradient in Cartesian coordinates and then use the transformation matrices in **Chapter 1** to transform the gradient from Cartesian to cylindrical coordinates. We show both methods.

Method A: The coordinate transformation from cylindrical to Cartesian coordinates [**Eq. (2.36)**] is

$$x = r\cos\phi, \quad y = r\sin\phi, \quad z = z$$

Substituting these for x, y, and z in the field gives the field in cylindrical coordinates:

$$f(r,\phi,z) = r\cos\phi + 5zr^2\sin^2\phi$$

The gradient can now be calculated directly using **Eq. (2.43)**:

$$\nabla f = \left(\hat{\mathbf{r}} \frac{\partial}{\partial r} + \hat{\boldsymbol{\phi}} \frac{1}{r} \frac{\partial}{\partial \phi} + \hat{\mathbf{z}} \frac{\partial}{\partial z} \right) (r\cos\phi + 5zr^2 \sin^2\phi) = \hat{\mathbf{r}} (\cos\phi + 10zr\sin^2\phi) + \hat{\boldsymbol{\phi}} (-\sin\phi + 10zr\cos\phi\sin\phi) + \hat{\mathbf{z}} 5r^2 \sin^2\phi$$

Method B: In this method, the gradient is calculated in Cartesian coordinates and then transformed to cylindrical coordinates as a vector. The gradient in Cartesian coordinates is

$$\nabla f = \left(\hat{\mathbf{x}} \frac{\partial}{\partial x} + \hat{\mathbf{y}} \frac{\partial}{\partial y} + \hat{\mathbf{z}} \frac{\partial}{\partial z} \right) (x + 5zy^2) = \hat{\mathbf{x}} 1 + \hat{\mathbf{y}} 10zy + \hat{\mathbf{z}} 5y^2$$

Now, we use the transformation in **Eq. (1.71)** (see also **Example 1.16**):

$$\begin{bmatrix} A_r \\ A_\phi \\ A_z \end{bmatrix} = \begin{bmatrix} \cos\phi & \sin\phi & 0 \\ -\sin\phi & \cos\phi & 0 \\ 0 & 0 & 1 \end{bmatrix} \begin{bmatrix} A_x \\ A_y \\ A_z \end{bmatrix} = \begin{bmatrix} \cos\phi & \sin\phi & 0 \\ -\sin\phi & \cos\phi & 0 \\ 0 & 0 & 1 \end{bmatrix} \begin{bmatrix} 1 \\ 10zy \\ 5y^2 \end{bmatrix}$$

$$= \begin{bmatrix} \cos\phi + 10zy\sin\phi \\ -\sin\phi + 10zy\cos\phi \\ 5y^2 \end{bmatrix} = \begin{bmatrix} \cos\phi + 10zr\sin^2\phi \\ -\sin\phi + 10zr\sin\phi\cos\phi \\ 5r^2\cos^2\phi \end{bmatrix}$$

where the coordinate transformations above were again used to replace y and z. These are the scalar components of the gradient in cylindrical coordinates. If we write the vector, we get

$$\nabla f = \hat{\mathbf{r}} (\cos\phi + 10zr\sin^2\phi) + \hat{\boldsymbol{\phi}} (-\sin\phi + 10zr\sin\phi\cos\phi) + \hat{\mathbf{z}} 5r^2 \sin^2\phi$$

This is identical to the result obtained by **Method A**.

Example 2.12 Application: Slope of a Scalar Field A scalar field is given as $f(r,\phi,z) = r\phi + 3\phi z$.

(a) Calculate the slope of the scalar field in the direction of the vector $\mathbf{A} = \hat{\mathbf{r}}2 + \hat{\mathbf{z}}$.
(b) What is the slope of the field at a point $P(2,90°,1)$ in the direction of vector **A**?

Solution: The gradient of the scalar field is calculated first. This gives the derivative in the direction of maximum change in field. Find the projection of the gradient onto the direction of vector **A** using the scalar product. The direction of the slope is that of **A**. In **(b)**, the coordinates of P are substituted into the vector obtained in **(a)** to obtain the scalar component of the gradient in the direction of **A** at point P (slope).

(a) First, we calculate the gradient of the function $f(r,\phi,z)$ in cylindrical coordinates using **Eq. (2.43)**:

$$\nabla f = \left(\hat{\mathbf{r}} \frac{\partial}{\partial r} + \hat{\boldsymbol{\phi}} \frac{1}{r} \frac{\partial}{\partial \phi} + \hat{\mathbf{z}} \frac{\partial}{\partial z} \right) (r\phi + 3\phi z) = \hat{\mathbf{r}}\phi + \hat{\boldsymbol{\phi}} \left(1 + \frac{3z}{r} \right) + \hat{\mathbf{z}} 3\phi$$

Next, we need to calculate the unit vector in the direction of **A**. This is

$$\hat{\mathbf{A}} = \frac{\mathbf{A}}{A} = \frac{\hat{\mathbf{r}}2 + \hat{\mathbf{z}}}{\sqrt{2^2 + 1}} = \hat{\mathbf{r}} \frac{2}{\sqrt{5}} + \hat{\mathbf{z}} \frac{1}{\sqrt{5}}$$

The projection of ∇f in the direction of **A** is the scalar product between ∇f and $\hat{\mathbf{A}}$:

$$(\nabla f)\cdot\hat{\mathbf{A}} = \left(\hat{\mathbf{r}}\phi + \hat{\boldsymbol{\phi}}\left(1+\frac{3z}{r}\right) + \hat{\mathbf{z}}3\phi\right)\cdot\left(\hat{\mathbf{r}}\frac{2}{\sqrt{5}} + \hat{\mathbf{z}}\frac{1}{\sqrt{5}}\right) = \frac{2\phi}{\sqrt{5}} + \frac{3\phi}{\sqrt{5}} = \sqrt{5}\phi$$

This is the scalar component of the gradient in the direction of vector **A**.

(b) The gradient gives the slope of the scalar field at any point in space. To find the slope at a particular point, we substitute the coordinates of the point in the general expression of the projection of the gradient in the direction of **A**. Since the projection is independent of r and z and $\phi = \pi/2$ at P, we get

$$(\nabla f)\cdot\hat{\mathbf{A}}\,|_P = \frac{\sqrt{5}\pi}{2}$$

The slope at $P(2{,}90°{,}1)$ is $\sqrt{5}\pi/2$.

Exercise 2.4 Given the configuration of **Exercise 2.3**, calculate the gradient in cylindrical coordinates. Use the direct approach or the transformation from Cartesian to cylindrical coordinates.

Answer $\nabla f(r,\phi,z) = \frac{\mathbf{f}}{f} = \frac{\hat{\mathbf{r}}r + \hat{\mathbf{z}}z}{\sqrt{r^2+z^2}}$

2.3.1.2 Gradient in Spherical Coordinates

The gradient in spherical coordinates is defined analogously to the gradient in cylindrical coordinates; that is, we start with the gradient in Cartesian coordinates [**Eq. (2.33)**] and transform the partial derivatives, unit vectors, and variables from Cartesian to spherical coordinates. Although we will not perform all details of the derivation here (see **Exercise 2.5**), the important steps are as follows:

Step 1: We first write a general scalar function in spherical coordinates as $U(R,\theta,\phi)$.

Step 2: The unit vectors $\hat{\mathbf{x}}$, $\hat{\mathbf{y}}$, and $\hat{\mathbf{z}}$ are transformed into spherical coordinates using **Eq. (1.88)**:

$$\begin{aligned}\hat{\mathbf{x}} &= \hat{\mathbf{R}}\sin\theta\cos\phi + \hat{\boldsymbol{\theta}}\cos\theta\cos\phi - \hat{\boldsymbol{\phi}}\sin\phi,\\ \hat{\mathbf{y}} &= \hat{\mathbf{R}}\sin\theta\sin\phi + \hat{\boldsymbol{\theta}}\cos\theta\sin\phi + \hat{\boldsymbol{\phi}}\cos\phi \quad \hat{\mathbf{z}} = \hat{\mathbf{R}}\cos\theta - \hat{\boldsymbol{\theta}}\sin\theta\end{aligned} \tag{2.45}$$

Step 3: The derivatives $\partial/\partial x$, $\partial/\partial y$, and $\partial/\partial z$ are transformed into their counterparts in spherical coordinates $\partial/\partial R$, $\partial/\partial\theta$, and $\partial/\partial\phi$. To do so, we use the chain rule of differentiation, but unlike the transformation into cylindrical coordinates, now all three coordinates change and we have

$$\begin{aligned}\frac{\partial}{\partial x} &= \frac{\partial}{\partial R}\left(\frac{\partial R}{\partial x}\right) + \frac{\partial}{\partial\theta}\left(\frac{\partial\theta}{\partial x}\right) + \frac{\partial}{\partial\phi}\left(\frac{\partial\phi}{\partial x}\right),\\ \frac{\partial}{\partial y} &= \frac{\partial}{\partial R}\left(\frac{\partial R}{\partial y}\right) + \frac{\partial}{\partial\theta}\left(\frac{\partial\theta}{\partial y}\right) + \frac{\partial}{\partial\phi}\left(\frac{\partial\phi}{\partial y}\right),\\ \frac{\partial}{\partial z} &= \frac{\partial}{\partial R}\left(\frac{\partial R}{\partial z}\right) + \frac{\partial}{\partial\theta}\left(\frac{\partial\theta}{\partial z}\right) + \frac{\partial}{\partial\phi}\left(\frac{\partial\phi}{\partial z}\right)\end{aligned} \tag{2.46}$$

Step 4: Transformation of variables from Cartesian to spherical coordinates. These are listed in **Eqs. (1.82)** and **(1.81)** and are used to evaluate the partial derivatives in **Eq. (2.46)**:

$$x = R\sin\theta\cos\phi, \quad y = R\sin\theta\sin\phi, \quad z = R\cos\theta \tag{2.47}$$

$$R = \sqrt{x^2+y^2+z^2}, \quad \theta = \tan^{-1}\left(\frac{\sqrt{x^2+y^2}}{z}\right), \quad \phi = \tan^{-1}\left(\frac{y}{x}\right) \tag{2.48}$$

Although the evaluation of the various derivatives in **Eq. (2.46)** is clearly lengthier than for cylindrical coordinates, it follows identical steps, which, for the sake of brevity, we do not show. With these derivatives and substitution of these and the terms in **Eq. (2.45)** into **Eq. (2.33)**, we get the gradient in spherical coordinates as

$$\boxed{\nabla U(R,\theta,\phi) = \hat{\mathbf{R}}\frac{\partial U(R,\theta,\phi)}{\partial R} + \hat{\boldsymbol{\theta}}\frac{1}{R}\frac{\partial U(R,\theta,\phi)}{\partial \theta} + \hat{\boldsymbol{\phi}}\frac{1}{R\sin\theta}\frac{\partial U(R,\theta,\phi)}{\partial \phi}} \tag{2.49}$$

From this, we can write the del operator in spherical coordinates as

$$\boxed{\nabla \equiv \hat{\mathbf{R}}\frac{\partial}{\partial R} + \hat{\boldsymbol{\theta}}\frac{1}{R}\frac{\partial}{\partial \theta} + \hat{\boldsymbol{\phi}}\frac{1}{R\sin\theta}\frac{\partial}{\partial \phi}} \tag{2.50}$$

The del operator in spherical coordinates is different than the del operator in Cartesian or cylindrical coordinates.

Example 2.13 A sphere of radius a is given.

(a) Find the normal unit vector to the sphere at point $P(a,90°,30°)$.
(b) Find the normal unit vector at $P(a,90°,30°)$ in Cartesian coordinates.

Solution: First we write the equation of the sphere as a scalar function in spherical coordinates. Then the gradient of the scalar function is calculated. This gives the vector normal to the sphere's surface. The unit vector is found by dividing the gradient by the magnitude of the gradient. Substitution of the coordinates of P gives the unit vector at the given point.

(a) The sphere may be described in spherical coordinates as $f(R,\theta,\phi) = R$
The gradient is therefore

$$\nabla f(R,\theta,\phi) = \hat{\mathbf{R}}\frac{\partial f(R,\theta,\phi)}{\partial R} + \hat{\boldsymbol{\theta}}\frac{1}{R}\frac{\partial f(R,\theta,\phi)}{\partial \theta} + \hat{\boldsymbol{\phi}}\frac{1}{R\sin\theta}\frac{\partial f(R,\theta,\phi)}{\partial \theta} = \hat{\mathbf{R}}\frac{\partial R}{\partial R} = \hat{\mathbf{R}}$$

and the unit vector is

$$\hat{\mathbf{n}} = \nabla f_P(R,\theta,\phi) = \hat{\mathbf{R}}$$

The unit vector is independent of the location on the sphere or its radius.

(b) The unit vector $\hat{\mathbf{R}}$ is normal to the surface of the sphere and may be written in Cartesian coordinates as [see **Eq. (1.90)** or **(2.45)**]

$$\hat{\mathbf{R}} = \hat{\mathbf{x}}\sin\theta\cos\phi + \hat{\mathbf{y}}\sin\theta\sin\phi + \hat{\mathbf{z}}\cos\theta$$

And, clearly, the normal unit vector varies from point to point. At $P(a,90°,30°)$, the normal unit vector is

$$\hat{\mathbf{n}} = \hat{\mathbf{x}}\sin\left(\frac{\pi}{2}\right)\cos\left(\frac{\pi}{6}\right) + \hat{\mathbf{y}}\sin\left(\frac{\pi}{2}\right)\sin\left(\frac{\pi}{6}\right) + \hat{\mathbf{z}}\cos\left(\frac{\pi}{2}\right) = \hat{\mathbf{x}}\frac{\sqrt{3}}{2} + \hat{\mathbf{y}}\frac{1}{2}$$

Note that this is independent of the radius of the sphere.

Exercise 2.5 Derive the gradient in spherical coordinates using the steps outlined in **Eqs. (2.45)** through **(2.48)**. Verify that **Eq. (2.49)** is obtained.

Reminder The gradient is only defined for a scalar function.

2.3.2 The Divergence of a Vector Field

After defining the gradient of a scalar function, we wish now to define the spatial derivatives of a vector function. This will lead to two relations. One is the divergence of a vector, while the other is the curl of a vector. The divergence is defined first.

To understand the ideas involved, we first look at some physical quantities with which we are familiar and which lead to the definition of the divergence.

Consider first the two vector fields shown in **Figure 2.14**. In **Figure 2.14a**, the magnitude of the vector field is constant, and the direction does not vary. For example, this may represent flow of water in a channel or current in a conductor. If we draw any volume in the flow, the total net flow out of the volume is zero; that is, the total amount of water or current flowing into the volume v is equal to the total flow out of the volume.

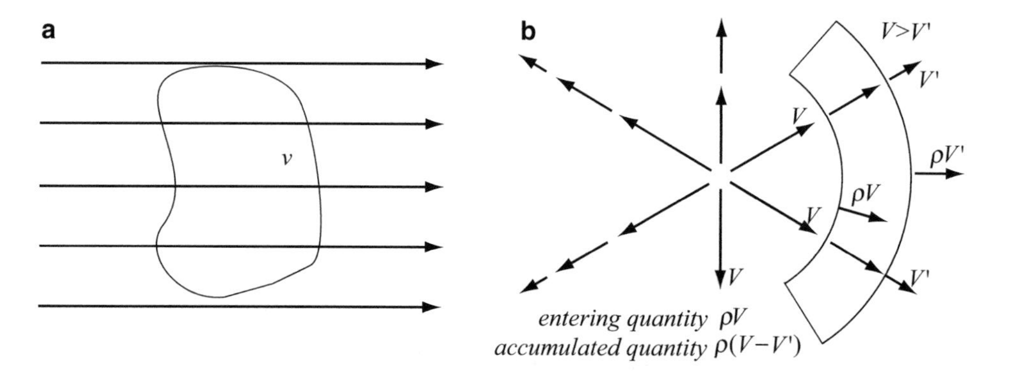

Figure 2.14 Flow through a volume. (**a**) Field is uniform and the total quantity entering volume v equals the quantity leaving the volume. (**b**) Nonuniform flow. There is an accumulation in volume v

In **Figure 2.14b**, the flow is radial from the center and the vector changes in magnitude as the flow progresses. This is indicated by the fact that the length of the vectors is reduced. A physical situation akin to this is a spherical can in which holes were made and the assembly is connected to a water hose. Water squirts in radial directions and water velocity is reduced with distance from the can. Now, if we were to draw a volume (an imaginary can), the total amount of water entering the volume is larger than the amount of water leaving the volume since water velocity changes and the amount of water is directly dependent on velocity. This fact can be stated in another way: There is a net flow of water into the volume through the surface enclosing the volume of the can where it accumulates. The latter statement is what we wish to use since it links the surface of the volume to the net flow out of the volume. In the example in **Figure 2.14b**, the net outward flow is negative. The total flux out of the volume is given by the closed surface integral of the vector **A** [see **Eq. (2.16)**]:

$$Q = \oint_s \mathbf{A} \cdot d\mathbf{s} \tag{2.51}$$

where the closed surface integral must be used since flow (into or out of the volume) occurs everywhere on the surface. Although this amount is written as a surface integral, the quantity Q clearly depends on the volume we choose. Thus, it makes sense to define the flow through the surface of a clearly defined volume such as a unit volume. If we do so, the quantity Q is the flow per unit volume. Our choice here is to do exactly that, but to define the flow through the surface, per measure of volume and then allow this volume to tend to zero. In the limit, this will give us the net outward flow at a point. Thus, we define a quantity that we will call the divergence of the vector **A** as

$$\boxed{\mathrm{Div}\,\mathbf{A} \equiv \lim_{\Delta v \to 0} \frac{\oint_s \mathbf{A} \cdot d\mathbf{s}}{\Delta v}} \tag{2.52}$$

That is,

"*the divergence of vector* **A** *is the net flux of vector* **A** *out of a small volume, through the closed surface enclosing the volume, as the volume tends to zero.*"

The meaning of the term divergence can be at least partially understood from **Figure 2.15a** where the source in **Figure 2.14** is shown again, but now we take a small volume around the source itself. Again using the analogy of water, the flow is outward only. This indicates that there is a net flow out of the volume through the closed surface. Moreover, the flow "diverges" from the point outward. We must, however, be careful with this description because divergence does not necessarily imply as clear a picture as this. The flow in **Figure 2.15b** has nonzero divergence as well even though it does not "look" divergent.

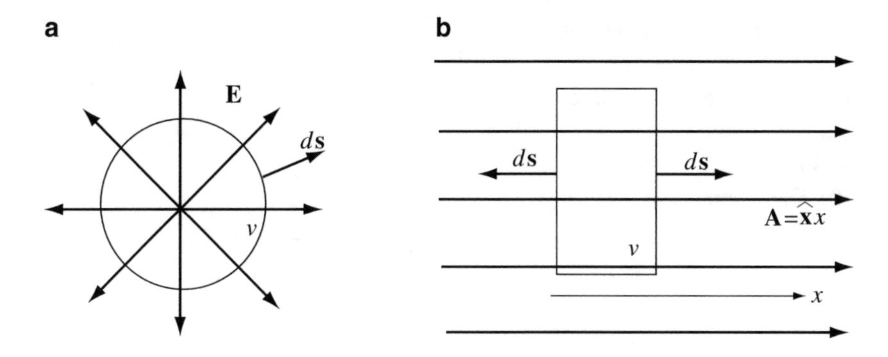

Figure 2.15 Net outward flow from a volume v. (**a**) For a radial field. (**b**) For a field varying with the coordinate x. Both fields have nonzero divergence

A second important point is that in both examples given above, nonzero divergence implies either accumulation in the volume (in this case of fluid) or flow out of the volume. In the latter case, we must conclude that if the divergence is nonzero, there must be a ***source*** of flow at the point, whereas in the former case, a negative source or ***sink*** must exist. We, therefore, have an important interpretation and use for the divergence: a measure of the (scalar) source of the vector field. From **Eqs. (2.51)** and **(2.52)**, this source is clearly a flux per unit volume. We also must emphasize here that the divergence is a point value: a differential quantity defined at a point.

2.3.2.1 Divergence in Cartesian Coordinates

The definition in **Eq. (2.52)**, while certainly physically meaningful, is very inconvenient for practical applications. It would be rather tedious to evaluate the surface integral and then let the volume tend to zero every time the divergence is needed. For this reason, we seek a simpler, more easily evaluated expression to replace the definition for practical applications. This is done by considering a general vector and a convenient but general element of volume Δv as shown in **Figure 2.16**. First, we evaluate the surface integral over the volume, then divide by the volume, and let the volume tend to zero to find the divergence at point P. To find the closed surface integral, we evaluate the open surface integration of the vector **A** over the six sides of the volume and add them. Noting the directions of the vectors $d\mathbf{s}$ on all surfaces, we can write

$$\oint_S \mathbf{A}\cdot d\mathbf{s} = \int_{s_{fr}} \mathbf{A}\cdot d\mathbf{s}_{fr} + \int_{s_{bk}} \mathbf{A}\cdot d\mathbf{s}_{bk} + \int_{s_{tp}} \mathbf{A}\cdot d\mathbf{s}_{tp} + \int_{s_{bt}} \mathbf{A}\cdot d\mathbf{s}_{bt} + \int_{s_{rt}} \mathbf{A}\cdot d\mathbf{s}_{rt} + \int_{s_{lt}} \mathbf{A}\cdot d\mathbf{s}_{lt} \tag{2.53}$$

where fr = front surface, bk = back surface, tp = top surface, bt = bottom surface, rt = right surface, and lt = left surface. Each integral is evaluated separately, and because we chose the six surfaces such that they are parallel to coordinates, their evaluation is straightforward. To do so, we will also assume the vector **A** to be constant over each surface, an assumption which is justified from the fact that these surfaces tend to zero in the limit. Since the divergence will be calculated at point $P(x,y,z)$, we take the coordinates of this point as reference at the center of the volume as shown in **Figure 2.16b**. The front surface is located at $x + \Delta x/2$, whereas the back surface is at $x - \Delta x/2$. Similarly, the top surface is at $z + \Delta z/2$ and the bottom surface at $z - \Delta z/2$, whereas the right and left surfaces are at $y + \Delta y/2$ and $y - \Delta y/2$, respectively. With these definitions in mind, we can start evaluating the six integrals. On the front surface,

$$\int_{s_{fr}} \mathbf{A}\cdot d\mathbf{s}_{fr} = \mathbf{A}_{fr}\cdot \Delta\mathbf{s}_{fr} \tag{2.54}$$

where $\mathbf{A}_{fr}$ is that component of the vector **A** perpendicular to the front surface. From the definition of the scalar product, this vector component is in the x direction, and its scalar component is equal to

$$|\mathbf{A}_{fr}| = \hat{\mathbf{x}}\cdot\mathbf{A} = A_x\left(x+\frac{\Delta x}{2}, y, z\right) \tag{2.55}$$

The latter expression requires that we evaluate the x component of **A** at a point $(x + \Delta x/2, y, z)$. To do so, it is useful to use the Taylor series expansion of $f(x + \Delta x)$ around point x:

$$f(x+\Delta x) = \sum_{k=0}^{\infty}\frac{f^{(k)}(x)}{k!}(\Delta x)^k = f(x) + \Delta x f'(x) + \frac{(\Delta x)^2}{2}f''(x) + \frac{(\Delta x)^3}{6}f'''(x) + \cdots \tag{2.56}$$

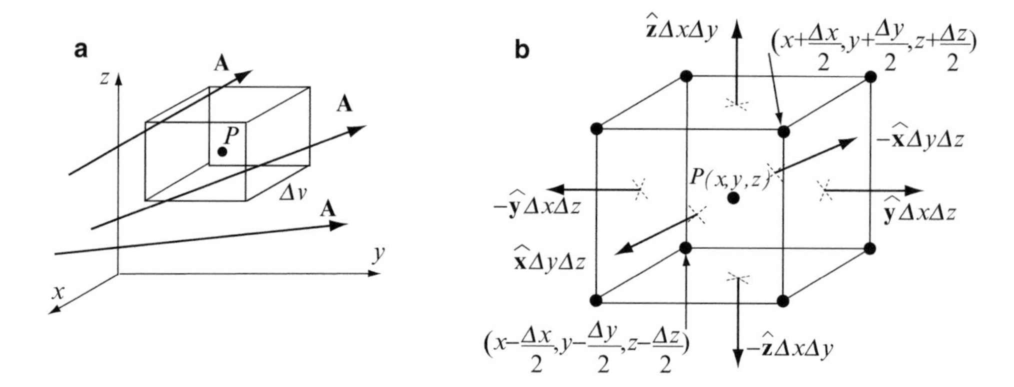

Figure 2.16 Evaluation of a closed surface integral over an element of volume. (**a**) The volume and its relation to the axes. (**b**) The elements of surface and coordinates

Anticipating truncation of the expansion after the first two terms (see **Exercise 2.6**) and replacing Δx with $\Delta x/2$, $f(x)$ with $A_x(x,y,z)$, $f(x+\Delta x)$ with $A_x(x+\Delta x/2,y,z)$, $f(x-\Delta x)$ with $A_x(x-\Delta x/2,y,z)$ and $f'(x)$ with $\partial A_x(x,y,z)/\partial x$, we get

$$A_x\left(x+\frac{\Delta x}{2},y,z\right)\approx A_x(x,y,z)+\frac{\Delta x}{2}\frac{\partial A_x(x,y,z)}{\partial x} \tag{2.57}$$

and

$$A_x\left(x-\frac{\Delta x}{2},y,z\right)\approx A_x(x,y,z)-\frac{\Delta x}{2}\frac{\partial A_x(x,y,z)}{\partial x} \tag{2.58}$$

In **Eqs. (2.57)** and **(2.58)** only the first two terms of the expansion were retained. The higher-order terms (see the general expansion in **Eq. (2.56)**) were neglected because higher powers of Δx diminish quickly.

An element of surface on the front face is

$$\Delta\mathbf{s}_{fr}=\hat{\mathbf{n}}\Delta s_{fr}=\hat{\mathbf{x}}\Delta s_{fr}=\hat{\mathbf{x}}\Delta y\Delta z \tag{2.59}$$

Substitution of this and **Eq. (2.57)** into **Eq. (2.54)** gives the surface integral as

$$\int_{s_{fr}}\mathbf{A}_{fr}\cdot d\mathbf{s}_{fr}\approx\hat{\mathbf{x}}\left(A_x(x,y,z)+\frac{\Delta x}{2}\frac{\partial A_x(x,y,z)}{\partial x}\right)\cdot\hat{\mathbf{x}}\Delta y\Delta z=\Delta y\Delta z A_x(x,y,z)+\frac{\Delta x\Delta y\Delta z}{2}\frac{\partial A_x(x,y,z)}{\partial x} \tag{2.60}$$

Since **A** has the same direction on the back surface but $d\mathbf{s}$ is in the opposite direction compared with the front surface, we get for the back surface

$$\mathbf{A}_{bk}=\hat{\mathbf{x}}\mathbf{A}_x\left(x-\frac{\Delta x}{2},y,z\right),\quad d\mathbf{s}_{bk}=-\hat{\mathbf{x}}dydz \tag{2.61}$$

With these and replacing x by $-x$ and $\Delta x/2$ by $-\Delta x/2$ in **Eq. (2.60)**, we have for the back surface

$$\int_{s_{bk}}\mathbf{A}_{bk}\cdot d\mathbf{s}_{bk}\approx-\Delta y\Delta z A_x(x,y,z)+\frac{\Delta x\Delta y\Delta z}{2}\frac{\partial A_x(x,y,z)}{\partial x} \tag{2.62}$$

Summing the terms in **Eqs. (2.60)** and **(2.62)** gives for the front and back surfaces

$$\int_{s_{fr}}\mathbf{A}\cdot d\mathbf{s}_{fr}+\int_{s_{bk}}\mathbf{A}\cdot d\mathbf{s}_{bk}\approx\Delta x\Delta y\Delta z\frac{\partial A_x(x,y,z)}{\partial x} \tag{2.63}$$

The result was obtained for the front and back surfaces, but there is nothing special about these two surfaces. In fact, if we were to rotate the volume in space such that the front and back surfaces are perpendicular to the y axis, the only difference is that the component of **A** in this expression must be taken as the y component. Although you should convince yourself that this is the case by repeating the steps in **Eqs. (2.54)** through **(2.63)** for the left and right surfaces, the following can be written directly simply because of this symmetry in calculations:

$$\int_{s_{lt}} \mathbf{A} \cdot d\mathbf{s}_{lt} + \int_{s_{rt}} \mathbf{A} \cdot d\mathbf{s}_{rt} \approx \Delta x \Delta y \Delta z \frac{\partial A_y(x,y,z)}{\partial y} \tag{2.64}$$

Similarly, for the top and bottom surfaces

$$\int_{s_{tp}} \mathbf{A} \cdot d\mathbf{s}_{tp} + \int_{s_{bt}} \mathbf{A} \cdot d\mathbf{s}_{bt} \approx \Delta x \Delta y \Delta z \frac{\partial A_z(x,y,z)}{\partial z} \tag{2.65}$$

The total closed surface integral is the sum of the surface integrals in **Eqs. (2.63), (2.64),** and **(2.65)**:

$$\oint_s \mathbf{A} \cdot d\mathbf{s} = \Delta v \frac{\partial A_x(x,y,z)}{\partial x} + \Delta v \frac{\partial A_y(x,y,z)}{\partial y} + \Delta v \frac{\partial A_z(x,y,z)}{\partial z} + (\text{higher-order terms}) \tag{2.66}$$

where the higher-order terms are those neglected in the Taylor series expansion and $\Delta v = \Delta x \Delta y \Delta z$. Now, we can return to the definition of the divergence in **Eq. (2.52)**:

$$\text{div}\,\mathbf{A} = \lim_{\Delta v \to 0} \frac{\oint_s \mathbf{A} \cdot d\mathbf{s}}{\Delta v} = \lim_{\Delta x,\, \Delta y,\, \Delta z \to 0} \frac{\oint_s \mathbf{A} \cdot d\mathbf{s}}{\Delta x \Delta y \Delta z} = \frac{\partial A_x(x,y,z)}{\partial x} + \frac{\partial A_y(x,y,z)}{\partial y} + \frac{\partial A_z(x,y,z)}{\partial z} \tag{2.67}$$

It is customary to write **Eq. (2.67)** in a short-form notation as

$$\text{div}\,\mathbf{A} = \frac{\partial A_x}{\partial x} + \frac{\partial A_y}{\partial y} + \frac{\partial A_z}{\partial z} \tag{2.68}$$

since this applies at any point in space.

The calculation of the divergence of a vector **A** is therefore very simple since all that are required are the spatial derivatives of the scalar components of the vector. The divergence is a scalar as required and may have any magnitude, including zero. The result in **Eq. (2.68)** well justifies the two pages of algebra that were needed to obtain it because now we have a simple, systematic way of evaluating the divergence. For historical reasons, the notation for divergence is $\nabla \cdot \mathbf{A}$ (read: del dot **A**).[1] The divergence of vector **A** is written as follows:

$$\boxed{\nabla \cdot \mathbf{A} = \frac{\partial A_x}{\partial x} + \frac{\partial A_y}{\partial y} + \frac{\partial A_z}{\partial z}} \tag{2.69}$$

However, it must be pointed out that a scalar product between the del operator and the vector **A** is not implied and should never be attempted. The symbolic notation $\nabla \cdot \mathbf{A}$ is just that: A notation to the right-hand side of **Eq. (2.69)**. Whenever we need to calculate the divergence of a vector **A**, the right-hand side of **Eq. (2.69)** is calculated, never a scalar product. Note also that calculation of divergence using the definition in **Eq. (2.52)** is independent of the system of coordinates. The actual evaluation of the surface integrals is obviously coordinate dependent.

Exercise 2.6 Show that the higher order terms in **Eq. (2.56)** (after the second term on the right hand side) tend to zero in the limit $\Delta x \to 0$ an therefore can be neglected.

[1] The notation used here is due to Josiah Willard Gibbs (1839–1903), who, however, never indicated or implied the notation to mean a scalar product. The implication of a scalar product between ∇ and **A** is a common error in vector calculus and for that reason alone should be avoided.

2.3.2.2 Divergence in Cylindrical and Spherical Coordinates

The divergence in cylindrical and spherical coordinates may be obtained in an analogous manner: We define a small volume with sides parallel to the required system of coordinates and evaluate **Eq. (2.52)** as we have done for the Cartesian system in **Section 2.3.2.1**. The method is rather lengthy but is straightforward (see **Exercise 2.7**). An alternative is to start with **Eq. (2.69)** and transform it into cylindrical or spherical coordinates in a manner similar to **Section 2.3.1.1**. This method is outlined next.

For cylindrical coordinates, we use **Eq. (2.41)**, which defines the transformations of the operators $\partial/\partial x$ and $\partial/\partial y$ while $\partial/\partial z$ remains unchanged. Then, from the transformations of the scalar components of a general vector from Cartesian to cylindrical coordinates given in **Eqs. (1.68)** and **(1.69)**, we get

$$A_x = A_r\cos\phi - A_\phi\sin\phi, \qquad A_y = A_r\sin\phi + A_\phi\cos\phi, \qquad A_z = A_z \tag{2.70}$$

Substitution of these and the relations in **Eq. (2.41)** into **Eq. (2.69)** gives

$$\nabla\cdot\mathbf{A}(r,\phi,z) = \left(\cos\phi\frac{\partial}{\partial r} - \frac{\sin\phi}{r}\frac{\partial}{\partial\phi}\right)(A_r\cos\phi - A_\phi\sin\phi) + \left(\cos\phi\frac{\partial}{\partial r} - \frac{\cos\phi}{r}\frac{\partial}{\partial\phi}\right)(A_r\sin\phi + A_\phi\cos\phi) + \frac{\partial A_z}{\partial z} \tag{2.71}$$

Expanding this expression and evaluating the derivatives (see **Exercise 2.8**) gives the divergence in cylindrical coordinates:

$$\boxed{\nabla\cdot\mathbf{A} = \frac{1}{r}\left(\frac{\partial(rA_r)}{\partial r}\right) + \frac{1}{r}\left(\frac{\partial A_\phi}{\partial\phi}\right) + \frac{\partial A_z}{\partial z}} \tag{2.72}$$

Similar steps may be followed to obtain the divergence in spherical coordinates. Although we do not show the steps here, the process starts again with **Eq. (2.69)**. The transformations for the operators $\partial/\partial x$, $\partial/\partial y$, and $\partial/\partial z$ from Cartesian to spherical coordinates are obtained from the expressions in **Eqs. (2.46)** through **(2.48)**, whereas the transformations of the scalar components A_x, A_y, and A_z from Cartesian to spherical coordinates are given in **Eq. (1.89)**. Substituting these into **Eq. (2.69)** and carrying out the derivatives (see **Exercise 2.9**) gives the following expression for the divergence in spherical coordinates:

$$\boxed{\nabla\cdot\mathbf{A} = \frac{1}{R^2}\frac{\partial}{\partial R}(R^2A_R) + \frac{1}{R\sin\theta}\frac{\partial}{\partial\theta}(A_\theta\sin\theta) + \frac{1}{R\sin\theta}\frac{\partial A_\phi}{\partial\phi}} \tag{2.73}$$

Reminder The notation $\nabla\cdot\mathbf{A}$ in **Eqs. (2.72)** and **(2.73)** should always be viewed as a notation only. It should never be taken as implying a scalar product.

Exercise 2.7

(a) Find the divergence in cylindrical coordinates using the method in **Section 2.3.2.1** by defining an elementary volume in cylindrical coordinates.
(b) Find the divergence in spherical coordinates using the method in **Section 2.3.2.1** by defining an elementary volume in spherical coordinates.

Exercise 2.8 Carry out the detailed operations outlined in **Section 2.3.2.2** needed to obtain **Eq. (2.72)**.

Exercise 2.9 Carry out the detailed operations outlined in **Section 2.3.2.2** needed to obtain **Eq. (2.73)**.

Example 2.14 A vector field $\mathbf{F} = \hat{\mathbf{x}}3y + \hat{\mathbf{y}}(5 - 2x) + \hat{\mathbf{z}}(z^2 - 2)$ is given. Find the divergence of **F**.

Solution: The divergence in **Eq. (2.69)** can be applied directly:

$$\nabla \cdot \mathbf{F} = \frac{\partial F_x}{\partial x} + \frac{\partial F_y}{\partial y} + \frac{\partial F_z}{\partial z} = \frac{\partial (3y)}{\partial x} + \frac{\partial (5 - 2x)}{\partial y} + \frac{\partial (z^2 - 2)}{\partial z} = 2z$$

The divergence of the vector field varies in the z direction only.

Example 2.15 Find $\nabla \cdot \mathbf{A}$ at $(R = 2, \theta = 30°, \phi = 90°)$ for the vector field

$$\mathbf{A} = \hat{\mathbf{R}}0.2R^5\phi\sin^2\theta + \hat{\boldsymbol{\theta}}0.2R^3\phi\sin^2\theta + \hat{\boldsymbol{\phi}}0.2R^3\phi\sin^2\theta.$$

Solution: We apply the divergence in spherical coordinates using **Eq. (2.73)**:

$$\begin{aligned}\nabla \cdot \mathbf{A} &= \frac{1}{R^2}\frac{\partial\left(0.2R^5\phi\sin^2\theta\right)}{\partial R} + \frac{1}{R\sin\theta}\frac{\partial\left(0.2R^3\phi\sin^3\theta\right)}{\partial \theta} + \frac{1}{R\sin\theta}\frac{\partial\left(0.2R^3\phi\sin^2\theta\right)}{\partial \phi} \\ &= R^2\phi\sin^2\theta + 0.6R^2\phi\sin\theta\cos\theta + 0.2R^2\sin\theta\end{aligned}$$

At (2,30°,90°),

$$\nabla \cdot \mathbf{A} = 4 \times \left(\frac{\pi}{2}\right) \times \left(\frac{1}{4}\right) + 0.6 \times 4 \times \left(\frac{\pi}{2}\right) \times \left(\frac{1}{2}\right) \times \left(\frac{\sqrt{3}}{2}\right) + 0.2 \times 4 \times \left(\frac{1}{2}\right) = 3.6032$$

The scalar source of the vector field **A** is equal to 3.6032 at the given point.

2.3.3 The Divergence Theorem

Consider the surface of a rectangular box whose sides are dx, dy, and dz and are parallel to the xy, xz, and yz planes, as shown in **Figure 2.17**. The surface of the lower face $PQRS$ is $dxdy$, and $d\mathbf{s}$ is in the negative z direction:

$$d\mathbf{s}_l = -\hat{\mathbf{z}}dxdy \tag{2.74}$$

The flux of $\mathbf{A} = \hat{\mathbf{x}}A_x + \hat{\mathbf{y}}A_y + \hat{\mathbf{z}}A_z$ crossing this surface is

$$d\Phi_{ls} = \mathbf{A} \cdot d\mathbf{s}_l = \hat{\mathbf{z}}A_z \cdot (-\hat{\mathbf{z}}dxdy) = -A_z dxdy \tag{2.75}$$

On the upper surface $P'Q'R'S'$, the normal to the surface is in the positive z direction and the component A_z of the vector **A** changes by an amount dA_z. Therefore, A_z on the upper face is

$$A_z + dA_z = A_z + \frac{\partial A_z}{\partial z}dz \tag{2.76}$$

The flux on the upper surface is found by multiplying by the area of the surface $dxdy$:

$$d\Phi_{us} = A_z dxdy + \frac{\partial A_z}{\partial z}dxdydz \tag{2.77}$$

The sum of the fluxes on the upper and lower surfaces gives

$$d\Phi_z = d\Phi_{ls} + d\Phi_{us} = \frac{\partial A_z}{\partial z}dv \tag{2.78}$$

Figure 2.17

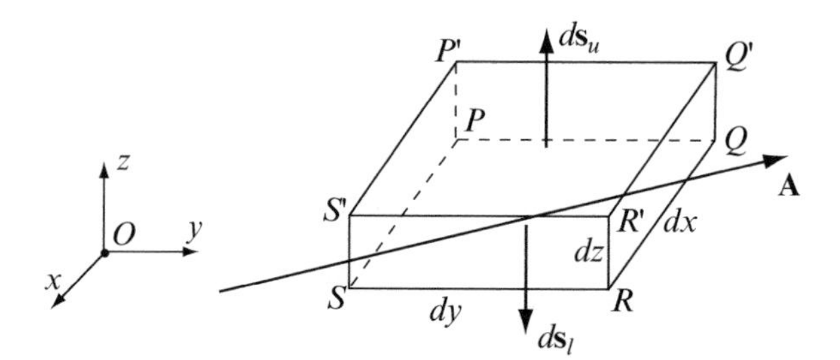

where the index z denotes that this is the total flux on the two surfaces perpendicular to the z axis and $dv = dxdydz$ is the volume of the rectangular box.

Using the same rationale on the other two pairs of parallel surfaces and summing the three contributions yields the expression

$$d\Phi = \left[\frac{\partial A_x}{\partial x} + \frac{\partial A_y}{\partial y} + \frac{\partial A_z}{\partial z}\right] dv \tag{2.79}$$

for the total flux through the box. The expression in brackets is the divergence of the vector **A**. The expression for the total flux through the small box becomes [see **Eq. (2.69)**]

$$d\Phi = (\nabla \cdot \mathbf{A}) dv \tag{2.80}$$

Now, consider an arbitrary volume v, enclosed by a surface s. Since $d\Phi$ through a differential volume is known, integration of this $d\Phi$ over the whole volume v gives the total flux passing through the volume

$$\Phi = \int_v d\Phi = \int_v (\nabla \cdot \mathbf{A}) dv \tag{2.81}$$

In **Eq. (2.51)**, the flux was evaluated by integrating over the whole surface s, which encloses the volume v. This also gives the total flux through the volume v:

$$\Phi = \oint_s \mathbf{A} \cdot d\mathbf{s} \tag{2.82}$$

Since the total flux through the volume or through the surface enclosing the volume must be the same, we can equate **Eqs. (2.81)** and **(2.82)** to get

$$\boxed{\int_v (\nabla \cdot \mathbf{A}) dv = \oint_s \mathbf{A} \cdot d\mathbf{s}} \tag{2.83}$$

This equality between the two integrals means that the flux of the vector **A** through the closed surface s is equal to the volume integral of the divergence of **A** over the volume enclosed by the surface s. We call this the ***divergence theorem***. Its most important use is the conversion of volume integrals of the divergence of a vector field into closed surface integrals. This theorem is often invoked to simplify expressions or to rewrite them in more convenient alternative forms.

Example 2.16 The vector field $\mathbf{A} = \hat{\mathbf{x}}x^2 + \hat{\mathbf{y}}y^2 + \hat{\mathbf{z}}z^2$ is given. Verify the divergence theorem for this vector over a cube 1 m on the side. Assume the cube occupies the space $0 \le x, y, z \le 1$.

Solution: First, we find the product $\mathbf{A} \cdot d\mathbf{s}$ and integrate it over the surface of the volume. Then, we integrate $\nabla \cdot \mathbf{A}$ over the whole volume of the cube of side 1 with four of its vertices at (0,0,0), (0,0,1), (0,1,0), and (1,0,0) (see **Figure 2.18**). The two results should be the same.

(a) Use the flux of **A** through the surface enclosing the volume:

$$\oint_s \mathbf{A}\cdot d\mathbf{s} = \int_{s_1} \mathbf{A}\cdot d\mathbf{s}_1 + \int_{s_2} \mathbf{A}\cdot d\mathbf{s}_2 + \int_{s_3} \mathbf{A}\cdot d\mathbf{s}_3 + \int_{s_4} \mathbf{A}\cdot d\mathbf{s}_4 + \int_{s_5} \mathbf{A}\cdot d\mathbf{s}_5 + \int_{s_6} \mathbf{A}\cdot d\mathbf{s}_6$$

where, from **Figure 2.18**

$$d\mathbf{s}_1 = \hat{\mathbf{x}}dydz, \quad d\mathbf{s}_2 = -\hat{\mathbf{x}}dydz, \quad d\mathbf{s}_3 = \hat{\mathbf{y}}dxdz$$
$$d\mathbf{s}_4 = -\hat{\mathbf{y}}dxdz, \quad d\mathbf{s}_5 = \hat{\mathbf{z}}dxdy, \quad d\mathbf{s}_6 = -\hat{\mathbf{z}}dxdy$$

Perform each surface integral separately:

(1) At $x = 1: \int_{s_1} \mathbf{A}\cdot d\mathbf{s}_1 = \int_{s_1} (\hat{\mathbf{x}}1 + \hat{\mathbf{y}}y^2 + \hat{\mathbf{z}}z^2)\cdot(\hat{\mathbf{x}}dydz) = \int_{y=0}^{y=1}\int_{z=0}^{z=1} dy\,dz = 1.$

(2) At $x = 0: \int_{s_2} \mathbf{A}\cdot d\mathbf{s}_2 = \int_{s_2} (\hat{\mathbf{y}}y^2 + \hat{\mathbf{z}}z^2)\cdot(-\hat{\mathbf{x}}dydz) = 0.$

(3) At $y = 1: \int_{s_3} \mathbf{A}\cdot d\mathbf{s}_3 = \int_{s_3} (\hat{\mathbf{x}}x^2 + \hat{\mathbf{y}}1 + \hat{\mathbf{z}}z^2)\cdot(\hat{\mathbf{y}}dxdz) = \int_{x=0}^{x=1}\int_{z=0}^{z=1} dx\,dz = 1.$

(4) At $y = 0: \int_{s_4} \mathbf{A}\cdot d\mathbf{s}_4 = \int_{s_4} (\hat{\mathbf{x}}x^2 + \hat{\mathbf{z}}z^2)\cdot(-\hat{\mathbf{y}}dxdz) = 0.$

(5) At $z = 1: \int_{s_5} \mathbf{A}\cdot d\mathbf{s}_5 = \int_{s_5} (\hat{\mathbf{x}}x^2 + \hat{\mathbf{y}}y^2 + \hat{\mathbf{z}}1)\cdot(\hat{\mathbf{z}}dxdy) = \int_{x=0}^{x=1}\int_{y=0}^{y=1} dx\,dy = 1.$

(6) At $z = 0: \int_{s_6} \mathbf{A}\cdot d\mathbf{s}_6 = \int_{s_6} (\hat{\mathbf{x}}x^2 + \hat{\mathbf{y}}y^2)\cdot(-\hat{\mathbf{z}}dxdy) = 0.$

The sum of all six integrals is 3.

(b) Use the divergence of **A** in the volume. The divergence of **A** is $\nabla\cdot\mathbf{A} = 2x + 2y + 2z$ Integration of $\nabla\cdot\mathbf{A}$ over the volume of the cube gives

$$\int_v (\nabla\cdot\mathbf{A})dv = \int_{z=0}^{z=1}\int_{y=0}^{y=1}\int_{x=0}^{x=1} (2x + 2y + 2z)dxdydz = \int_{z=0}^{z=1}\int_{y=0}^{y=1} (x^2 + 2xy + 2xz)dydz\Big|_{x=0}^{x=1}$$

$$= \int_{z=0}^{z=1}\int_{y=0}^{y=1} (1 + 2y + 2z)dydz = \int_{z=0}^{z=1} (y + y^2 + 2yz)dz\Big|_{y=0}^{y=1} = \int_{z=0}^{z=1} (1 + 1 + 2z)dz = (2z + z^2)\Big|_{z=0}^{z=1} = 3$$

Since the result in **(a)** and **(b)** are equal, the divergence theorem is verified for the given vector and volume.

Figure 2.18

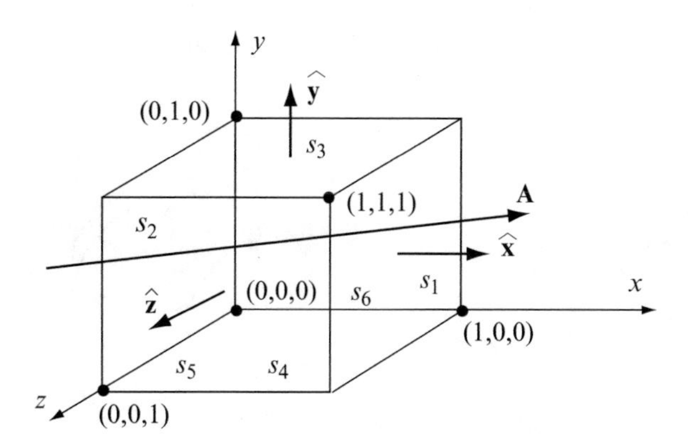

2.3.4 Circulation of a Vector and the Curl

We defined the gradient of a scalar and the divergence of a vector in the previous two sections. Both of these have physical meaning, and some applications of the two were shown in examples. In particular, the divergence of a vector was shown to be an indication of the strength of the scalar source of the vector. The question now is the following: If a vector can be generated by a scalar source (for example, a water spring is a scalar source, but it gives rise to a vector flow which has both direction and magnitude), is it also possible that a vector source gives rise to a vector field? The answer is clearly yes. Consider again the flow of a river; the flow is never uniform; it is faster toward the center of the river and slower at the banks. If you were to toss a stick into the river, perpendicular to the flow, the stick, in addition to drifting with the flow, will rotate and align itself with the direction of the flow. This rotation is caused by the variation in flow velocity: One end of the stick is dragged down the river at higher velocity than the other as shown in **Figure 2.19**. The important point here is that we cannot explain this rotation using the scalar source of the field. To explain this behavior, and others, we introduce the ***curl*** of a vector. The curl is related to circulation and spatial variations in the vector field. To define the curl, we first define the circulation of a vector. In the process, we will also try to look at the meaning of the curl and its utility.

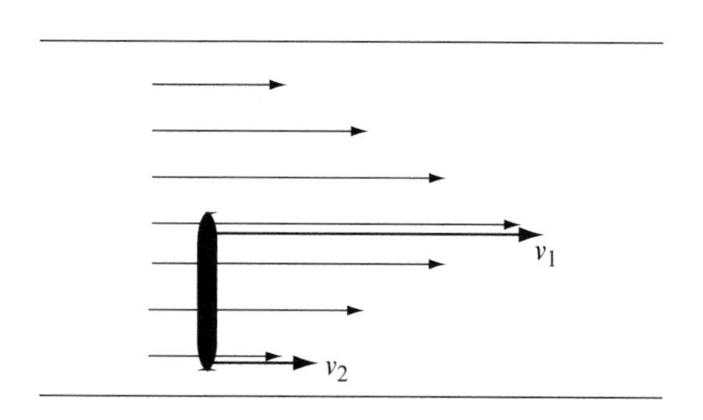

Figure 2.19 Illustration of circulation. The stick shown will rotate clockwise as it moves downstream

2.3.4.1 Circulation of a Vector Field

The closed contour integral of a vector field **A** was introduced in **Eq. (2.8)** and was called the circulation of the vector field around the contour:

$$C = \oint_L \mathbf{A} \cdot d\mathbf{l} \tag{2.84}$$

where $d\mathbf{l}$ is a differential length vector along the contour L. Why do we call this a circulation? To understand this, consider first a circular flow such as a hurricane (the wind path is circular). If **A** represents force, then the circulation represents work or energy expended. This energy increases with the circulation. If we take this as a measure for a hurricane, then measuring the circulation (if we could) would be a good measurement of the strength of the hurricane. If **A** and $d\mathbf{l}$ are parallel, as in **Figure 2.20a**, the circulation is largest. However, if **A** and $d\mathbf{l}$ are perpendicular to each other everywhere along the contour, the circulation is zero (**Figure 2.20b**). For example, an airplane, flying straight toward the eye of the hurricane, flies perpendicular to the wind and experiences no circulation. There is plenty of buffeting force but no circulation. This picture should be kept in mind since it shows that circulation as meant here does not necessarily mean geometric circulation. In other words, a vector may rotate around

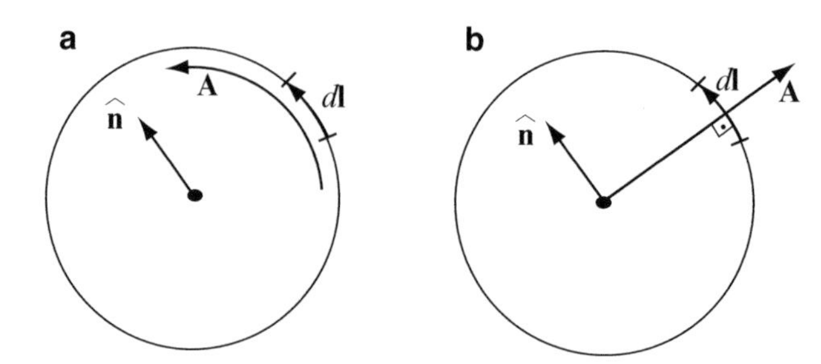

Figure 2.20 Circulation. (**a**) Maximum circulation. (**b**) Zero circulation

along a contour and its circulation may still be zero, whereas a vector that does not rotate (for example, the flow in **Figure 2.19**) may have nonzero circulation. All that circulation implies is the line integral of a vector field along a closed contour. This circulation may or may not be zero, depending on the vector field, the contour, and the relation between the two.

Although the foregoing explanation and the use of **Eq. (2.84)** as a measure of circulation are easy to understand physically, measuring the circulation in this fashion is not very useful. For one thing, **Eq. (2.84)** gives an integrated value over the contour. This tends to smooth local variations, which, in fact, may be the most important aspects of the field. Second, if we want to physically measure any quantity associated with the flow, we can only do this locally. A measuring device for wind velocity, force, etc., is a small device and the measurement may be regarded as a point measurement. Thus, we need to calculate or measure circulation in a small area. In addition to this, circulation also has a spatial meaning. In the case of a hurricane, the rotation may be regarded to be in a plane parallel to the surface of the ocean, but rotation can also be in other planes. For example, a gyroscope may rotate in any direction in space. Thus, when measuring rotation, the direction and plane of rotation are also important. These considerations lead directly to the definition of the curl. The curl is a vector measure of circulation which gives both the circulation of a vector and the direction of circulation per unit area of the field. More accurately, we define the curl using the following relation:

$$\text{curl}\,\mathbf{A} \equiv \lim_{\Delta s \to 0} \left. \frac{\hat{\mathbf{n}} \oint_L \mathbf{A} \cdot d\mathbf{l}}{\Delta s} \right|_{max} \tag{2.85}$$

"The curl of **A** *is the circulation of the vector* **A** *per unit area, as this area tends to zero and is in the direction normal to the area when the area is oriented such that the circulation is maximum."*

The curl of a vector field is, therefore, a vector field, defined at any point in space.

From the definition of contour integration, the normal to a surface enclosed by a contour is given by the right-hand rule as shown in **Figure 2.21** which also gives the direction of the curl. The definition in **Eq. (2.85)** has one drawback: It looks hopeless as far as using it to calculate the curl of a vector. To find a simpler, more systematic way of evaluating the curl, we observe that curl **A** is a vector with components in the directions of the coordinates. In the Cartesian system, for example, the vector $\mathbf{B} = \text{curl}\,\mathbf{A}$ can be written as

$$\mathbf{B} = \text{curl}\,\mathbf{A} = \hat{\mathbf{x}}\,(\text{curl}\,\mathbf{A})_x + \hat{\mathbf{y}}\,(\text{curl}\,\mathbf{A})_y + \hat{\mathbf{z}}\,(\text{curl}\,\mathbf{A})_z \tag{2.86}$$

where the indices x, y, and z indicate the corresponding scalar component of the vector. For example, $(\text{curl}\,\mathbf{A})_x$ is the scalar x component of curl **A**. This notation shows that curl **A** is the sum of three components, each a curl, one in the x direction, one in the y direction, and one in the z direction. To better understand this, consider a small general loop with projections on the x–y, y–z, and x–z planes as shown in **Figure 2.22a**. The magnitudes of the curls of the three projections are the scalar components B_x, B_y, and B_z in **Eq. (2.86)**. Calculation of these components and summation in **Eq. (2.86)** will provide the appropriate method for calculation of the curl. Now, consider an arbitrary vector **A** with scalar components A_x, A_y, and A_z. For simplicity in derivation, we assume all three components of **A** to be positive. Consider **Figure 2.22b**, which shows the projection of a small loop on the x–y plane (from **Figure 2.22a**). The circulation along the closed contour *abcda* is calculated as follows:

The projection of the vector **A** onto the x–y plane has x and y components: $\mathbf{A}_{xy} = \hat{\mathbf{x}}A_x + \hat{\mathbf{y}}A_y$. Along ab, $d\mathbf{l} = \hat{\mathbf{x}}dx$ and A_x remains constant (because Δx is very small). The circulation along this segment is

Figure 2.21 Relation between vector **A** and its curl

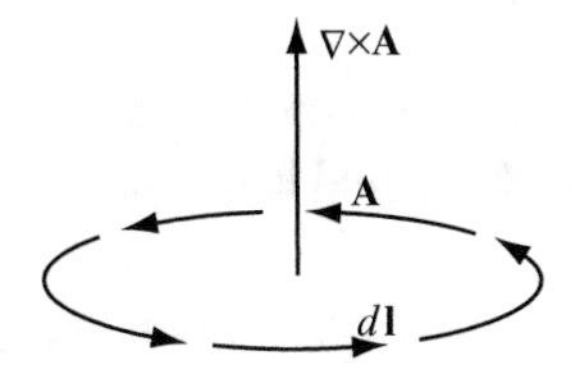

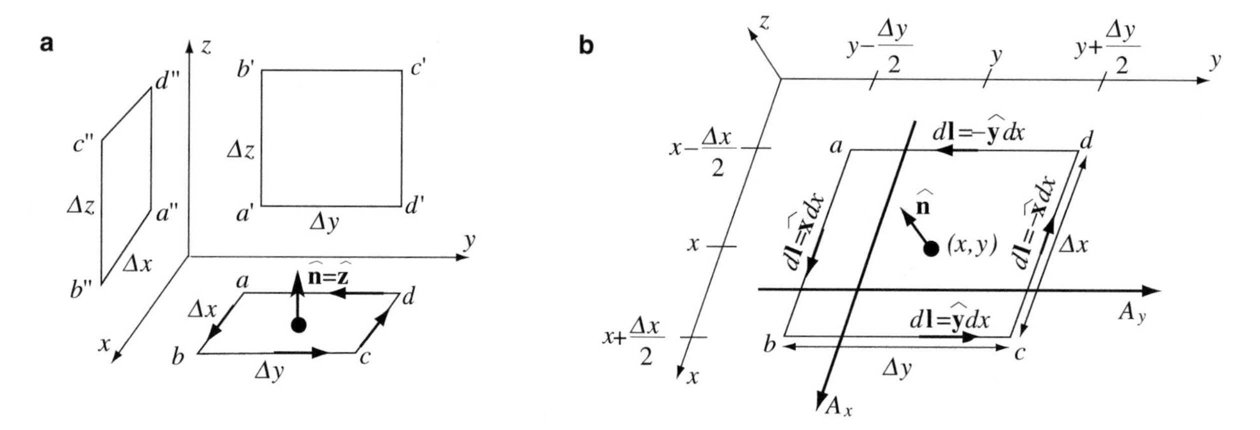

Figure 2.22 (**a**) Projections of a general loop onto the x–y, x–z, and y–z planes. (**b**) Calculation of the circulation in the x–y plane

$$\int_a^b (\hat{\mathbf{x}}A_x + \hat{\mathbf{y}}A_y) \cdot \hat{\mathbf{x}}dx = \int_a^b A_x dx = A_x\left(x, y - \frac{\Delta y}{2}, 0\right)\Delta x \approx \left(A_x(x,y,0) - \frac{\Delta y}{2}\frac{\partial A_x(x,y,0)}{\partial y}\right)\Delta x \tag{2.87}$$

The approximation in the parentheses on the right-hand side is the truncated Taylor series expansion of $A_x(x,y - \Delta y/2,0)$ around the point $P(x,y,0)$, as described in **Eqs. (2.56)** through **(2.58)**.

Along segment bc, $d\mathbf{l} = \hat{\mathbf{y}}dy$ and we assume A_y remains constant. The circulation along this segment is

$$\int_b^c (\hat{\mathbf{x}}A_x + \hat{\mathbf{y}}A_y) \cdot (\hat{\mathbf{y}}dy) = \int_b^c A_y dy = A_y\left(x + \frac{\Delta x}{2}, y, 0\right)\Delta y \approx \left(A_y(x,y,0) + \frac{\Delta x}{2}\frac{\partial A_y(x,y,0)}{\partial x}\right)\Delta y \tag{2.88}$$

Along segment cd, $d\mathbf{l} = -\hat{\mathbf{x}}dx$ and we get

$$\int_c^d (\hat{\mathbf{x}}A_x + \hat{\mathbf{y}}A_y) \cdot (-\hat{\mathbf{x}}dx) = -\int_c^d A_x dx = -A_x\left(x, y + \frac{\Delta y}{2}, 0\right)\Delta x \approx -\left(A_x(x,y,0) + \frac{\Delta y}{2}\frac{\partial A_x(x,y,0)}{\partial y}\right)\Delta x \tag{2.89}$$

Finally, along segment da, $d\mathbf{l} = -\hat{\mathbf{y}}dy$ and we get

$$\int_d^a (\hat{\mathbf{x}}A_x + \hat{\mathbf{y}}A_y) \cdot (-\hat{\mathbf{y}}dy) = -\int_d^a A_y dy = -A_y\left(x - \frac{\Delta x}{2}, y, 0\right)\Delta y \approx -\left(A_y(x,y,0) - \frac{\Delta x}{2}\frac{\partial A_y(x,y,0)}{\partial x}\right)\Delta y \tag{2.90}$$

The total circulation is the sum of the four segments calculated above:

$$\oint_{abcda} \mathbf{A} \cdot d\mathbf{l} \approx -\Delta x\Delta y\frac{\partial A_x(x,y,0)}{\partial y} + \Delta x\Delta y\frac{\partial A_y(x,y,0)}{\partial x} \tag{2.91}$$

If we now take the limit in **Eq. (2.85)** but only on the surface $\Delta x\Delta y$, we get the component of the curl perpendicular to the x–y plane. Dividing **Eq. (2.91)** by $\Delta x\Delta y$ and taking the limit $\Delta x\Delta y \to 0$ gives

$$(\text{curl}\,\mathbf{A})_z = \frac{\partial A_y}{\partial x} - \frac{\partial A_x}{\partial y} \tag{2.92}$$

As indicated above, this is the scalar component of the curl in the z direction since the normal $\hat{\mathbf{n}}$ to $\Delta x \Delta y$ is in the positive z direction.

The other two components are obtained in exactly the same manner. We give them here without repeating the process (see **Exercise 2.10**). The scalar component of the curl in the x direction is obtained by finding the total circulation around the loop $a'd'c'b'a'$ in the y–z plane in **Figure 2.22a** and then taking the limit in **Eq. (2.85)**:

$$(\text{curl}\,\mathbf{A})_x = \frac{\partial A_z}{\partial y} - \frac{\partial A_y}{\partial z} \tag{2.93}$$

Similarly, the scalar component of the curl in the y direction is found by calculating the circulation around loop $a''d''c''b''a''$ in the x–z plane in **Figure 2.22a**, and then taking the limit in **Eq. (2.85)**:

$$(\text{curl}\,\mathbf{A})_y = \frac{\partial A_x}{\partial z} - \frac{\partial A_z}{\partial x} \tag{2.94}$$

The curl of the vector **A** in Cartesian coordinates can now be written from **Eqs. (2.92)** through **(2.94)** and **Eq. (2.86)** as follows:

$$\text{curl}\,\mathbf{A} = \hat{\mathbf{x}}\left(\frac{\partial A_z}{\partial y} - \frac{\partial A_y}{\partial z}\right) + \hat{\mathbf{y}}\left(\frac{\partial A_x}{\partial z} - \frac{\partial A_z}{\partial x}\right) + \hat{\mathbf{z}}\left(\frac{\partial A_y}{\partial x} - \frac{\partial A_x}{\partial y}\right) \tag{2.95}$$

The common notation for the curl of a vector **A** is $\nabla \times \mathbf{A}$ (read: del cross **A**), and we write

$$\boxed{\nabla \times \mathbf{A} = \hat{\mathbf{x}}\left(\frac{\partial A_z}{\partial y} - \frac{\partial A_y}{\partial z}\right) + \hat{\mathbf{y}}\left(\frac{\partial A_x}{\partial z} - \frac{\partial A_z}{\partial x}\right) + \hat{\mathbf{z}}\left(\frac{\partial A_y}{\partial x} - \frac{\partial A_x}{\partial y}\right)} \tag{2.96}$$

As with the divergence, this does not imply a vector product,[2] only a notation to the operation on the right-hand side of **Eq. (2.96)**. Because of the form in **Eq. (2.96)**, the curl can be written as a determinant. The purpose in doing so is to avoid the need of remembering the expression in **Eq. (2.96)**. In this form, we write:

$$\nabla \times \mathbf{A} = \begin{vmatrix} \hat{\mathbf{x}} & \hat{\mathbf{y}} & \hat{\mathbf{z}} \\ \partial/\partial x & \partial/\partial y & \partial/\partial z \\ A_x & A_y & A_z \end{vmatrix} = \hat{\mathbf{x}}\left(\frac{\partial A_z}{\partial y} - \frac{\partial A_y}{\partial z}\right) + \hat{\mathbf{y}}\left(\frac{\partial A_x}{\partial z} - \frac{\partial A_z}{\partial x}\right) + \hat{\mathbf{z}}\left(\frac{\partial A_y}{\partial x} - \frac{\partial A_x}{\partial y}\right) \tag{2.97}$$

The latter is particularly useful as a quick way of writing the curl. Again, it should be remembered that the curl is not a determinant: only that the determinant in **Eq. (2.97)** may be used to write the expression in **Eq. (2.96)**.

The curl can also be evaluated in exactly the same manner in cylindrical and spherical coordinates. We will not do so but merely list the expressions.

In cylindrical coordinates:

$$\boxed{\nabla \times \mathbf{A} = \frac{1}{r}\begin{vmatrix} \hat{\mathbf{r}} & \hat{\boldsymbol{\phi}} r & \hat{\mathbf{z}} \\ \partial/\partial r & \partial/\partial \phi & \partial/\partial z \\ A_r & rA_\phi & A_z \end{vmatrix} = \hat{\mathbf{r}}\left(\frac{1}{r}\frac{\partial A_z}{\partial \phi} - \frac{\partial A_\phi}{\partial z}\right) + \hat{\boldsymbol{\phi}}\left(\frac{\partial A_r}{\partial z} - \frac{\partial A_z}{\partial r}\right) + \hat{\mathbf{z}}\frac{1}{r}\left(\frac{\partial (rA_\phi)}{\partial r} - \frac{\partial A_r}{\partial \phi}\right)} \tag{2.98}$$

[2] In Cartesian coordinates, the curl is equal to the cross product between the ∇ operator and the vector **A**, but this is not true in other systems of coordinates (see also footnote 1 on page 73).

In spherical coordinates:

$$\nabla \times \mathbf{A} = \frac{1}{R^2 \sin\theta}\begin{vmatrix} \hat{\mathbf{R}} & \hat{\boldsymbol{\theta}}R & \hat{\boldsymbol{\phi}}R\sin\theta \\ \partial/\partial R & \partial/\partial\theta & \partial/\partial\phi \\ A_R & RA_\theta & R\sin\theta A_\phi \end{vmatrix}$$
$$= \hat{\mathbf{R}}\frac{1}{R\sin\theta}\left(\frac{\partial(A_\phi \sin\theta)}{\partial\theta} - \frac{\partial A_\theta}{\partial\phi}\right) + \hat{\boldsymbol{\theta}}\frac{1}{R}\left(\frac{1}{\sin\theta}\frac{\partial A_R}{\partial\phi} - \frac{\partial(RA_\phi)}{\partial R}\right) + \hat{\boldsymbol{\phi}}\frac{1}{R}\left(\frac{\partial(RA_\theta)}{\partial R} - \frac{\partial A_R}{\partial\theta}\right) \tag{2.99}$$

Now that we have proper definitions of the curl and the methods of evaluating it, we must return to the physical meaning of the curl. First, we note the following properties of the curl:

(1) The curl of a vector field is a vector field.
(2) The magnitude of the curl gives the maximum circulation of the vector per unit area at a point.
(3) The direction of the curl is along the normal to the area of maximum circulation at a point.
(4) The curl has the general properties of the vector product: it is distributive but not associative

$$\nabla \times (\mathbf{A} + \mathbf{B}) = \nabla \times \mathbf{A} + \nabla \times \mathbf{B} \quad \text{but} \quad \nabla \times (\mathbf{A} \times \mathbf{B}) \neq (\nabla \times \mathbf{A}) \times \mathbf{B} \tag{2.100}$$

(5) The divergence of the curl of any vector function is identically zero:

$$\nabla \cdot (\nabla \times \mathbf{A}) \equiv 0 \tag{2.101}$$

(6) The curl of the gradient of a scalar function is also identically zero for any scalar:

$$\nabla \times (\nabla V) \equiv 0 \tag{2.102}$$

The latter two can be shown to be correct by direct evaluation of the products involved (see **Exercises 2.11** and **2.12**). These two identities play a very important role in electromagnetics and we will return to them later on in this chapter.

To summarize the discussion up to this point, you may view the curl as an indication of the rotation or circulation of the vector field calculated at any point. Zero curl indicates no rotation and the vector field can be generated by a scalar source alone. A general vector field with nonzero curl may only be generated by a scalar source (the divergence of the field) and a vector source (the curl of the field). Some vector fields may have zero divergence and nonzero curl. Thus, in this sense, the curl of a vector field is also an indication of the source of the field, but this source is a vector source. In the context of fluid flow, a curl is an indicator of nonuniform flow, whereas the divergence of the field only shows the scalar distribution of its sources. However, you should be careful with the idea of rotation. Rotation in the field does not necessarily mean that the field itself is circular; it only means that the field causes a circulation. The following examples also dwell on this and other physical points associated with the curl.

Example 2.17 Vector $\mathbf{A} = \hat{\mathbf{R}}2\cos\theta - \hat{\boldsymbol{\theta}}3R\sin\theta$ is given. Find the curl of $\mathbf{A}$.

Solution: We apply the curl in spherical coordinates using **Eq. (2.99)**. In this case, we perform the calculation for each scalar component separately:

$$(\nabla \times \mathbf{A})_R = \frac{1}{R\sin\theta}\frac{\partial(\sin\theta A_\phi)}{\partial\theta} - \frac{1}{R\sin\theta}\frac{\partial A_\theta}{\partial\phi} = \frac{1}{R\sin\theta}\frac{\partial(0)}{\partial\theta} - \frac{1}{R\sin\theta}\frac{\partial(-3R\sin\theta)}{\partial\phi} = 0$$

$$(\nabla \times \mathbf{A})_\theta = \frac{1}{R}\left(\frac{1}{\sin\theta}\frac{\partial A_R}{\partial\phi} - \frac{\partial(RA_\phi)}{\partial R}\right) = \frac{1}{R}\left(\frac{1}{\sin\theta}\frac{\partial(2R\cos\theta)}{\partial\phi} - \frac{\partial(R(0))}{\partial R}\right) = 0$$

$$(\nabla \times \mathbf{A})_\phi = \frac{1}{R}\left(\frac{\partial(RA_\theta)}{\partial R} - \frac{\partial A_R}{\partial\theta}\right) = \frac{1}{R}\frac{\partial(-3R^2\sin\theta)}{\partial R} - \frac{1}{R}\frac{\partial(2R\cos\theta)}{\partial\theta} = -6\sin\theta + 2\sin\theta = -4\sin\theta$$

Combining the components, the curl of **A** is

$$\nabla \times \mathbf{A} = -\hat{\boldsymbol{\phi}} 4\sin\theta.$$

Example 2.18 Application: Nonuniform Flow A fluid flows in a channel of width $2d$ with a velocity profile given by $\mathbf{v} = \hat{\mathbf{y}} v_0 (d - |x|)$.

(a) Calculate the curl of the velocity.

(b) How can you explain the fact that circulation of the flow is nonzero while the water itself flows in a straight line (see **Figure 2.23**)?

(c) What is the direction of the curl? What does this imply for an object floating on the water (such as a long stick)? Explain.

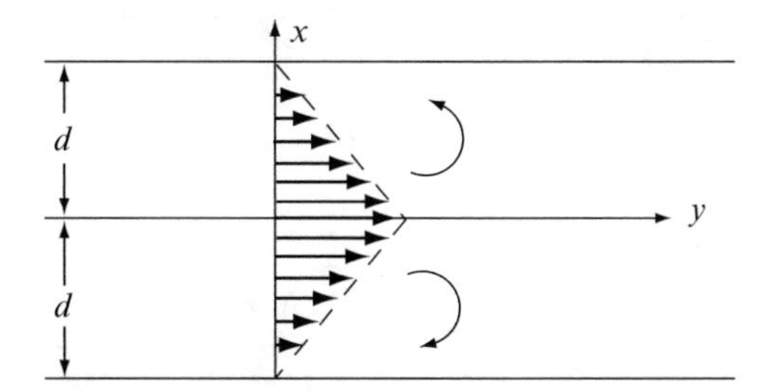

Figure 2.23 A vector field with nonzero curl. If this were a flow, a short stick placed perpendicular to the flow would rotate as shown

Solution: We calculate the curl of **v** using **Eq. (2.96)**. Even though there is only one component of the vector, this component depends on another variable. This means the curl is nonzero.

(a) Since the velocity depends on the absolute value of x, we separate the problem into three parts: one describes the solution for $x > 0$, the second for $x = 0$, the third for $x < 0$:

$$\mathbf{v} = \hat{\mathbf{y}} v_0 (d + x), \quad x < 0,$$
$$\mathbf{v} = \hat{\mathbf{y}} v_0 (d - x), \quad x > 0,$$
$$\mathbf{v} = \hat{\mathbf{y}} v_0 d, \quad x = 0.$$

For $x < 0$:

$$\nabla \times \mathbf{v} = \begin{vmatrix} \hat{\mathbf{x}} & \hat{\mathbf{y}} & \hat{\mathbf{z}} \\ \partial/\partial x & \partial/\partial y & \partial/\partial z \\ 0 & v_0(d+x) & 0 \end{vmatrix} = \hat{\mathbf{z}} \left(\frac{\partial (v_0(d+x))}{\partial x} \right) = \hat{\mathbf{z}} v_0$$

For $x > 0$:

$$\nabla \times \mathbf{v} = \begin{vmatrix} \hat{\mathbf{x}} & \hat{\mathbf{y}} & \hat{\mathbf{z}} \\ \partial/\partial x & \partial/\partial y & \partial/\partial z \\ 0 & v_0(d-x) & 0 \end{vmatrix} = \hat{\mathbf{z}} \left(\frac{\partial (v_0(d-x))}{\partial x} \right) = -\hat{\mathbf{z}} v_0$$

and for $x = 0$, $\nabla \times \mathbf{v} = 0$. Thus,

$$\nabla \times \mathbf{v} = \begin{cases} \hat{\mathbf{z}} v_0 & \text{for} \quad x < 0 \\ 0 & \text{for} \quad x = 0 \\ -\hat{\mathbf{z}} v_0 & \text{for} \quad x > 0. \end{cases}$$

(b) The curl implies neither multiple components nor a rotating vector, only that the vector varies in space. If the flow velocity were constant, the curl would be zero (the curl is zero at $x = 0$).

(c) This particular flow is unique in that the curl changes direction at $x = 0$. It is in the positive z direction for $x < 0$ and in the negative z direction for $x > 0$. Thus, if we were to place a stick anywhere in the positive part of the x axis, the stick will turn counterclockwise until it aligns itself with the flow (assuming a very thin stick). If the stick is placed in the negative part of the x axis, it will turn clockwise to align with the flow (see **Figure 2.23**). If the object is placed symmetrically about the $x = 0$ position it will not rotate, indicating zero circulation at $x = 0$.

Exercise 2.10 Following the steps in **Eqs. (2.85)** through **(2.92)**, derive the terms $(\text{curl}\, \mathbf{A})_x$ and $(\text{curl}\, \mathbf{A})_y$ as defined in **Eq. (2.86)**.

Exercise 2.11 Show by direct evaluation that $\nabla \cdot (\nabla \times \mathbf{A}) = 0$ [**Eq. (2.101)**] for any general vector **A**. Use Cartesian coordinates.

Exercise 2.12 Show by direct evaluation that $\nabla \times (\nabla V) = 0$ [**Eq. (2.102)**] for any general scalar function V. Use Cartesian coordinates.

2.3.5 Stokes'[3] Theorem

Stokes' theorem is the second theorem in vector algebra we introduce. It is in a way similar to the divergence theorem but relates to the curl of a vector. Stokes' theorem is given as

$$\boxed{\int_s (\nabla \times \mathbf{A}) \cdot d\mathbf{s} = \oint_L \mathbf{A} \cdot d\mathbf{l}} \tag{2.103}$$

It relates the integral of the curl of vector **A** over an open surface s to the closed contour integral of the vector **A** over the contour enclosing the surface s. To show that this relation is correct, we will use the relations derived from the curl and recall that curl is circulation per unit area.

Consider again the components of the curl in **Eqs. (2.92)** through **(2.94)**. These were derived for the rectangular loops in **Figure 2.22**. Now, we argue as follows: The total circulation of the vector **A** around a general loop *ABCDA* is the sum of the circulations over its projections on the x–y, x–z, and y–z planes as was shown in **Figure 2.22a**. That this is correct follows from the fact that the circulation is calculated from a scalar product. Thus, we can write the total circulation around the elementary loops of **Figure 2.22** using **Eq. (2.103)** as

$$\begin{aligned} L_{ABCDA} &= L_{abcda} + L_{a'd'c'b'a'} + L_{a''d''c''b''a''} \\ &= \left(\frac{\partial A_z}{\partial y} - \frac{\partial A_y}{\partial z}\right)\Delta y \Delta z + \left(\frac{\partial A_x}{\partial z} - \frac{\partial A_z}{\partial x}\right)\Delta x \Delta z + \left(\frac{\partial A_y}{\partial x} - \frac{\partial A_x}{\partial y}\right)\Delta x \Delta y \end{aligned} \tag{2.104}$$

[3] After Sir George Gabriel Stokes (1819–1903). Stokes was one of the great mathematical physicists of the nineteenth century. His work spanned many disciplines including propagation of waves in materials, water waves, optics, polarization of light, luminescence, and many others. The theorem bearing his name is one of the more useful relations in electromagnetics.

or

$$L_{ABCDA} = (\nabla \times \mathbf{A})_x \Delta s_x + (\nabla \times \mathbf{A})_y \Delta s_y + (\nabla \times \mathbf{A})_z \Delta s_z \tag{2.105}$$

where the indices x, y, and z indicate the scalar components of the vectors $\nabla \times \mathbf{A}$ and $\Delta\mathbf{s}$. The use of $\Delta\mathbf{s}$ in this fashion is permissible since $\Delta y \Delta z$ is perpendicular to the x coordinate and, therefore, can be written as a vector component: $\hat{\mathbf{x}}\Delta y \Delta z$, similarly for the other two projections. Thus, we can write the circulation around a loop of area Δs (assuming $\nabla \times \mathbf{A}$ is constant over Δs) as

$$\oint_L \mathbf{A} \cdot d\mathbf{l} = (\nabla \times \mathbf{A}) \cdot \Delta\mathbf{s} \tag{2.106}$$

Now, suppose that we need to calculate the circulation around a closed contour L enclosing an area s as shown in **Figure 2.24a**. To do so, we divide the area into small square loops, each of area Δs as shown in **Figure 2.24b**. As can be seen, every two neighboring contours have circulations in opposite directions on the parallel sides. This means that the circulations on each two parallel sides must cancel. The only remaining, nonzero terms in the circulations are due to the outer contour. Letting the area Δs be a differential area ds (i.e., let Δs tend to zero), the total circulation is

$$\sum_{i=1}^{\infty} \oint_{L_i} \mathbf{A} \cdot d\mathbf{l}_i = \oint_L \mathbf{A} \cdot d\mathbf{l} \tag{2.107}$$

The right-hand side of **Eq. (2.106)** becomes

$$\lim_{|\Delta s_i| \to 0} \sum_{i=1}^{\infty} (\nabla \times \mathbf{A})_i \cdot \Delta\mathbf{s}_i = \int_s (\nabla \times \mathbf{A}) \cdot d\mathbf{s} \tag{2.108}$$

Equating **Eqs. (2.107)** and **(2.108)** gives Stokes' theorem in **Eq. (2.103)**.

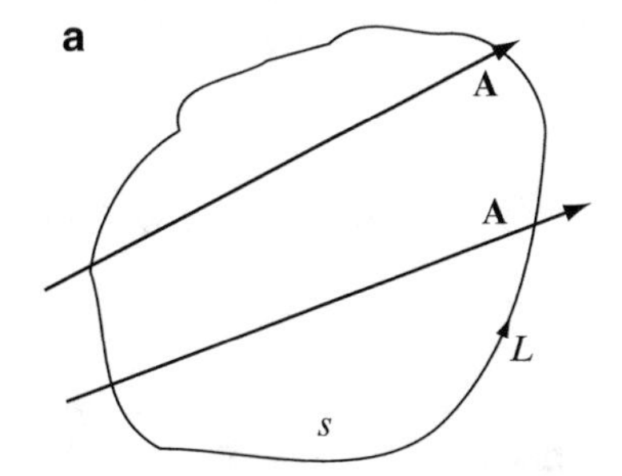

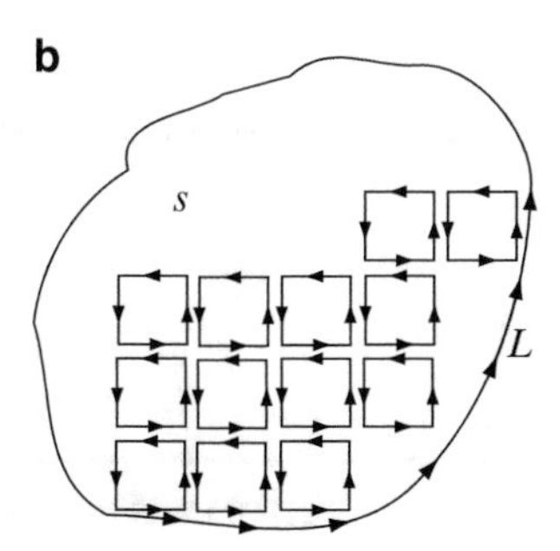

Figure 2.24 Stokes' theorem. (**a**) Vector field **A** and an open surface s. (**b**) The only components of the contour integrals on the small loops that do not cancel are along the outer contour L

Example 2.19 Verify Stokes' theorem for the vector field $\mathbf{A} = \hat{\mathbf{x}}(2x - y) - \hat{\mathbf{y}}2yz^2 - \hat{\mathbf{z}}2zy^2$ on the upper half-surface of the sphere $x^2 + y^2 + z^2 = 4$ (above the x–y plane), where the contour C is its boundary (rim of surface in the x–y plane).

Solution: To verify the theorem, we perform surface integration of the curl of **A** on the surface and closed contour integration of $\mathbf{A} \cdot d\mathbf{l}$ along C and show they are the same.

From **Eq. (2.96)**, with $A_x = (2x - y)$, $A_y = -2yz^2$, and $A_z = -2zy^2$,

$$\nabla \times \mathbf{A} = \hat{\mathbf{x}}\left(\frac{\partial(-2zy^2)}{\partial y} - \frac{\partial(-2yz^2)}{\partial z}\right) + \hat{\mathbf{y}}\left(\frac{\partial(2x-y)}{\partial z} - \frac{\partial(-2zy^2)}{\partial x}\right) + \hat{\mathbf{z}}\left(\frac{\partial(-2yz^2)}{\partial x} - \frac{\partial(2x-y)}{\partial y}\right)$$
$$= \hat{\mathbf{x}}(-4yz + 4yz) + \hat{\mathbf{y}}(0 - 0) + \hat{\mathbf{z}}(0 + 1) = \hat{\mathbf{z}}1$$

Surface Integral: We write the differential of surface on the sphere as $d\mathbf{s} = \hat{\mathbf{R}}R^2\sin\theta d\theta d\phi$:

$$\int_s (\nabla \times \mathbf{A}) \cdot d\mathbf{s} = \int_s \hat{\mathbf{z}} \cdot \hat{\mathbf{R}}R^2\sin\theta d\theta d\phi = R^2\int_{\phi=0}^{\phi=2\pi}\int_{\theta=0}^{\theta=\pi/2}\cos\theta\sin\theta d\theta d\phi$$
$$= 2\pi(2)^2\int_{\theta=0}^{\theta=\pi/2}\cos\theta\sin\theta d\theta = 8\pi\int_{\theta=0}^{\theta=\pi/2}\frac{1}{2}\sin2\theta d\theta = 4\pi\left[-\frac{\cos2\theta}{2}\right]_0^{\pi/2} = 4\pi$$

where $\hat{\mathbf{z}} \cdot \hat{\mathbf{R}} = \cos\theta$ [from **Eq. (2.45)**], $R = 2$ and $\sin\theta\cos\theta = (1/2)sin2\theta$.
Contour Integral: $d\mathbf{l} = \hat{\mathbf{x}}dx + \hat{\mathbf{y}}dy$, and using $z = 0$ on C ($x - y$ plane),

$$\oint_C \mathbf{A} \cdot d\mathbf{l} = \oint_C [\hat{\mathbf{x}}(2x - y)] \cdot [\hat{\mathbf{x}}dx + \hat{\mathbf{y}}dy] = \oint_C (2x - y)dx$$

Using cylindrical coordinates,

$$x = r\cos\phi = 2\cos\phi, \quad y = r\sin\phi = 2\sin\phi$$

and

$$\frac{dx}{d\phi} = -r\sin\phi \quad \rightarrow \quad dx = -2\sin\phi d\phi$$

Thus,

$$\oint_C \mathbf{A} \cdot d\mathbf{l} = \oint_C (2x - y)dx = \int_{\phi=0}^{2\pi}(2(2\cos\phi) - 2\sin\phi)(-2\sin\phi d\phi)$$
$$= -8\int_{\phi=0}^{2\pi}\cos\phi\sin\phi d\phi + 4\int_{\phi=0}^{2\pi}\sin^2\phi d\phi = 0 + 4\pi = 4\pi$$

and Stokes' theorem is verified since

$$\oint_C \mathbf{A} \cdot d\mathbf{l} = \int_s (\nabla \times \mathbf{A}) \cdot \hat{\mathbf{n}}\, ds = 4\pi$$

2.4 Conservative and Nonconservative Fields

A vector field is said to be conservative if the closed contour integral for any contour L in the field is zero (see also **Section 2.2.1**). It also follows from Stokes' theorem that the required condition is that the curl of the field must be zero:

$$\int_s (\nabla \times \mathbf{A}) \cdot d\mathbf{s} = \oint_L \mathbf{A} \cdot d\mathbf{l} = 0 \quad \rightarrow \quad \nabla \times \mathbf{A} = 0 \tag{2.109}$$

To see if a field is conservative, we can either show that the closed contour integral on any contour is zero or that its curl is zero. The latter is often easier to accomplish. Since the curl can be shown to be zero or nonzero in general (unlike a contour integral), the curl is the only true measure of the conservative property of the field.

Example 2.20 Two vectors $\mathbf{F}_1 = \hat{\mathbf{x}}x^2 - \hat{\mathbf{y}}z^2 - \hat{\mathbf{z}}2(zy+1)$ and $\mathbf{F}_2 = \hat{\mathbf{x}}x^2y - \hat{\mathbf{y}}z^2 - \hat{\mathbf{z}}2(zy+1)$ are given. Show that $\mathbf{F}_1$ is conservative and $\mathbf{F}_2$ is nonconservative.

Solution: To show that a vector field $\mathbf{F}$ is conservative, it is enough to show that its curl is zero. Similarly, for a vector field to be nonconservative, its curl must be nonzero.

The curls of $\mathbf{F}_1$ and $\mathbf{F}_2$ are

$$\nabla \times \mathbf{F}_1 = \hat{\mathbf{x}}\left(\frac{\partial F_{1z}}{\partial y} - \frac{\partial F_{1y}}{\partial z}\right) + \hat{\mathbf{y}}\left(\frac{\partial F_{1x}}{\partial z} - \frac{\partial F_{1z}}{\partial x}\right) + \hat{\mathbf{z}}\left(\frac{\partial F_{1y}}{\partial x} - \frac{\partial F_{1x}}{\partial y}\right) = \hat{\mathbf{x}}(-2z+2z) + \hat{\mathbf{y}}(0-0) + \hat{\mathbf{z}}(0-0) = 0$$

$$\nabla \times \mathbf{F}_2 = \hat{\mathbf{x}}(-2z+2z) + \hat{\mathbf{y}}(0-0) + \hat{\mathbf{z}}(0-x^2) = -\hat{\mathbf{z}}x^2$$

Thus, $\mathbf{F}_1$ is a conservative vector field, whereas $\mathbf{F}_2$ is clearly nonconservative.

2.5 Null Vector Identities and Classification of Vector Fields

After discussing most properties of vector fields and reviewing vector relations, we are now in a position to define broad classes of vector fields. This, again, is done in preparation of discussion of electromagnetic fields. This classification of vector fields is based on the curl and divergence of the fields and is described by the Helmholtz theorem. Before doing so, we wish to discuss here two particular vector identities because these are needed to define the Helmholtz theorem and because they are fundamental to understanding of electromagnetics. These are

$$\boxed{\nabla \times (\nabla V) \equiv 0} \tag{2.110}$$

$$\boxed{\nabla \cdot (\nabla \times \mathbf{A}) \equiv 0} \tag{2.111}$$

Both identities were mentioned in **Section 2.3.4.1** in the context of properties of the curl of a vector field and are sometimes called the ***null identities.*** These can be shown to be correct in any system of coordinates by direct evaluation and performing the prescribed operations (see **Exercises 2.11** through **2.13**). The first of these indicates that the curl of the gradient of any scalar field is identically zero. This may be written as

$$\nabla \times (\nabla V) = \nabla \times \mathbf{C} \equiv 0 \tag{2.112}$$

In other words, if a vector $\mathbf{C}$ is equal to the gradient of a scalar V, its curl is always zero. The converse is also true, if the curl of a vector field is zero, it can be written as the gradient of a scalar field:

$$\boxed{\text{If } \nabla \times \mathbf{C} = 0 \quad \rightarrow \quad \mathbf{C} = \nabla V \quad \text{or} \quad \mathbf{C} = -\nabla V} \tag{2.113}$$

Not all vector fields have zero curl, but if the curl of a vector field happens to be zero, then the above form can be used because $\nabla \times \mathbf{C}$ is zero. This type of field is called a ***curl-free field*** or an ***irrotational field***. Thus, we say that an irrotational field can always be written as the gradient of a scalar field. In the context of electromagnetics, we will use the second form in **Eq. (2.113)** by convention.

To understand the meaning of an irrotational field, consider the Stokes' theorem for the irrotational vector field $\mathbf{C}$ defined in **Eq. (2.113)**:

$$\int_s (\nabla \times \mathbf{C}) \cdot d\mathbf{s} = \oint_L \mathbf{C} \cdot d\mathbf{l} = 0 \tag{2.114}$$

This means that the closed contour integral of an irrotational field is identically zero; that is, an irrotational field is a conservative field. A simple example of this type of field is the gravitational field: if you were to drop a weight down the stairs and lift it back up the stairs to its original location, the weight would travel a closed contour. Although you may have performed strenuous work, the potential energy of the weight remains unchanged and this is independent of the path you take.

The second identity states that the divergence of the curl of any vector field is identically zero. Since the curl of a vector is a vector, we may substitute $\nabla \times \mathbf{A} = \mathbf{B}$ in **Eq. (2.111)** and write

$$\nabla \cdot (\nabla \times \mathbf{A}) = \nabla \cdot (\mathbf{B}) \equiv 0 \quad (2.115)$$

This can also be stated as follows: If the divergence of a vector field **B** is zero, this vector field can be written as the curl of another vector field **A**:

$$\boxed{\text{If} \quad \nabla \cdot \mathbf{B} = 0 \quad \rightarrow \quad \mathbf{B} = \nabla \times \mathbf{A}} \quad (2.116)$$

The vector field **B** is a special field: It has zero divergence. For this reason we call it a ***divergence-free*** or ***divergenceless field***. This type of vector field is also called ***solenoidal***.[4] We will not try to explain this term at this point; the source of the name is rooted in electromagnetic theory. We will eventually understand its meaning, but for now we simply take this as a name for divergence-free fields.

The foregoing can also be stated mathematically by using the divergence theorem:

$$\int_v (\nabla \cdot \mathbf{B}) dv = \oint_s \mathbf{B} \cdot d\mathbf{s} = 0 \quad (2.117)$$

This means that the total flux of the vector **B** through any closed surface is zero or, alternatively, that the net outward flux in any volume is zero or that the inward flux is equal to the outward flux, indicating that there are no net sources or sinks inside any arbitrary volume in the field.

Exercise 2.13 Using cylindrical coordinates, show by direct evaluation that for any scalar function ψ and vector **A**,

$$\nabla \times (\nabla \psi) = 0 \text{ and } \nabla \cdot (\nabla \times \mathbf{A}) = 0.$$

2.5.1 The Helmholtz[5] Theorem

After defining the properties of vector fields, we can now summarize these properties and draw some conclusions. In the process, we will also classify vector fields into groups, using the Helmholtz theorem which is based on the divergence and curl of the vector fields.

The Helmholtz theorem states:

"A vector field is uniquely defined (within an additive constant) by specifying its divergence and its curl."

That this must be so follows from the fact that, in general, specification of the sources of a field should be sufficient to specify the vector field. Although we could go into a mathematical proof of this theorem (which also requires imposition of conditions on the vector such as continuity of derivatives and the requirement that the vector vanishes at infinity), we will accept this theorem and look at its meaning. The Helmholtz theorem is normally written as

$$\boxed{\mathbf{B} = -\nabla U + \nabla \times \mathbf{A}} \quad (2.118)$$

[4] The term solenoid was coined by André Marie Ampère from the Greek *solen* = channel and *eidos* = form. When he built the first magnetic coil, in 1820, he gave it the name solenoid because the spiral wires in the coil reminded him of channels.

[5] Hermann Ludwig Ferdinand von Helmholtz (1821–1894). Helmholtz was one of the most prolific of the scientists of the nineteenth century. His work encompasses almost every aspect of science as well as philosophy. Perhaps his best known contribution is his statement of the law of conservation of energy. However, he is also the inventor of the ophthalmoscope—an instrument used to this day in testing eyesight. He contributed considerably to optics and physiology of vision and hearing. His work *On the Sensation of Tone* defines tone in terms of harmonics. In addition, he worked on mechanics, hydrodynamics, as well as electromagnetics. In particular, he was the person to suggest to his student Heinrich Hertz the experiments that led to the discovery of the propagation of electromagnetic waves, and by so doing laid the foundations to the age of communication.

where U is a scalar field and $\mathbf{A}$ is a vector field. That is, any vector field can be decomposed into two terms; one is the gradient of a scalar function and the other is the curl of a vector function. The vector $\mathbf{B}$ must be defined in terms of its curl and divergence. The divergence of $\mathbf{B}$ is given as

$$\nabla \cdot \mathbf{B} = \nabla \cdot (-\nabla U) + \nabla \cdot (\nabla \times \mathbf{A}) \tag{2.119}$$

The second term on the right-hand side is zero from the identity in **Eq. (2.111)**. The first term is, in general, a nonzero scalar density function and we may denote it as ρ:

$$\boxed{\nabla \cdot \mathbf{B} = \rho} \tag{2.120}$$

Because $\nabla \cdot \mathbf{B} \neq 0$, this is a nonsolenoidal field.

The curl of the vector $\mathbf{B}$ is

$$\nabla \times \mathbf{B} = \nabla \times (-\nabla U) + \nabla \times (\nabla \times \mathbf{A}) \tag{2.121}$$

Now, the first term is zero from the identity in **Eq. (2.110)**. The second term is a nonzero vector that will be denoted here as a general vector $\mathbf{J}$.

$$\boxed{\nabla \times \mathbf{B} = \mathbf{J}} \tag{2.122}$$

$\mathbf{J}$ may be regarded as the strength of the vector source. Since $\nabla \times \mathbf{B} \neq 0$ if $\mathbf{J} \neq 0$, this vector field is a rotational field.

A general field will have both nonzero curl and nonzero divergence; that is, the field is both rotational and nonsolenoidal. There are, however, fields in which the curl or the divergence or both are zero. In all, there are four types of fields that can be defined:

(1) A nonsolenoidal, rotational vector field. $\nabla \cdot \mathbf{B} = \rho$ and $\nabla \times \mathbf{B} = \mathbf{J}$. This is the most general vector field possible. The field has both a scalar and a vector source.
(2) A nonsolenoidal, irrotational vector field. $\nabla \cdot \mathbf{B} = \rho$ and $\nabla \times \mathbf{B} = 0$. The vector field has only a scalar source.
(3) A solenoidal, rotational vector field. $\nabla \cdot \mathbf{B} = 0$ and $\nabla \times \mathbf{B} = \mathbf{J}$. The vector field has only a vector source.
(4) A solenoidal, irrotational vector field. $\nabla \cdot \mathbf{B} = 0$ and $\nabla \times \mathbf{B} = 0$. The vector field has no scalar or vector sources.

The study of electromagnetics will be essentially that of defining the conditions and properties of the foregoing four types of fields. We start in **Chapter 3** with the static electric field, which is a nonsolenoidal, irrotational field (type 2 above). These properties, the curl and the divergence of the vector field, will be the basis of study of all fields.

2.5.2 Second-Order Operators

The del operator as well as the gradient, divergence, and curl are first-order operators; the result is first-order partial derivatives of the scalar or vector functions. It is possible to combine two first-order operators operating on scalar function U and vector function $\mathbf{A}$. By doing so, we obtain second-order expressions, some of which are very useful. The valid combinations are

$$\nabla \cdot (\nabla U) \quad \text{(divergence of the gradient of } U) \tag{2.123}$$

$$\nabla \times (\nabla U) \quad \text{(curl of the gradient of U)} \tag{2.124}$$

$$\nabla (\nabla \cdot \mathbf{A}) \quad \text{(gradient of the divergence of } \mathbf{A}) \tag{2.125}$$

$$\nabla \cdot (\nabla \times \mathbf{A}) \quad \text{(divergence of the curl of } \mathbf{A}) \tag{2.126}$$

$$\nabla \times (\nabla \times \mathbf{A}) \quad \text{(curl of the curl of } \mathbf{A}) \tag{2.127}$$

The relation $\nabla \cdot (\nabla U)$ [**Eq. (2.123)**] can be calculated by direct derivation using the gradient of the scalar function U. In Cartesian coordinates, the gradient is given in **Eq. (2.32)**:

$$\nabla U(x,y,z) = \hat{\mathbf{x}} \frac{\partial U(x,y,z)}{\partial x} + \hat{\mathbf{y}} \frac{\partial U(x,y,z)}{\partial y} + \hat{\mathbf{z}} \frac{\partial U(x,y,z)}{\partial z} \quad (2.128)$$

The divergence of $\nabla U\ (x,y,z)$ is now written using **Eq. (2.69)**:

$$\nabla \cdot (\nabla U) = \frac{\partial(\nabla U(x,y,z))_x}{\partial x} + \frac{\partial(\nabla U(x,y,z))_y}{\partial y} + \frac{\partial(\nabla U(x,y,z))_z}{\partial z} = \frac{\partial^2 U(x,y,z)}{\partial x^2} + \frac{\partial^2 U(x,y,z)}{\partial y^2} + \frac{\partial^2 U(x,y,z)}{\partial z^2} \quad (2.129)$$

or, in short-form notation

$$\boxed{\nabla \cdot (\nabla U) = \frac{\partial^2 U}{\partial x^2} + \frac{\partial^2 U}{\partial y^2} + \frac{\partial^2 U}{\partial z^2}} \quad (2.130)$$

From this, we can define the scalar ***Laplace operator*** (or, in short, the ***Laplacian***) as

$$\boxed{\nabla^2 = \frac{\partial^2}{\partial x^2} + \frac{\partial^2}{\partial y^2} + \frac{\partial^2}{\partial z^2}} \quad (2.131)$$

In cylindrical and spherical coordinates, we must start with the components of the vector ∇U in the corresponding system and calculate the divergence of the vector as in **Section 2.3.2** (see **Exercises 2.4, 2.5, 2.7**, **Examples 2.11** and **2.15**). The result is as follows:

In cylindrical coordinates:

$$\boxed{\nabla^2 U = \frac{\partial^2 U}{\partial r^2} + \frac{1}{r}\frac{\partial U}{\partial r} + \frac{1}{r^2}\frac{\partial^2 U}{\partial \phi^2} + \frac{\partial^2 U}{\partial z^2}} \quad (2.132)$$

In spherical coordinates:

$$\boxed{\nabla^2 U = \frac{1}{R^2}\frac{\partial}{\partial R}\left(R^2 \frac{\partial U}{\partial R}\right) + \frac{1}{R^2 \sin\theta}\frac{\partial}{\partial \theta}\left(\sin\theta \frac{\partial U}{\partial \theta}\right) + \frac{1}{R^2 \sin^2\theta}\frac{\partial^2 U}{\partial \phi^2}} \quad (2.133)$$

The expressions $\nabla \times (\nabla U)$ and $\nabla \cdot (\nabla \times \mathbf{A})$ are the null identities discussed in **Section 2.5**.

Finally, **Eqs. (2.125)** and **(2.127)** are often used together using the vector identity:

$$\boxed{\nabla^2 \mathbf{A} = \nabla(\nabla \cdot \mathbf{A}) - \nabla \times (\nabla \times \mathbf{A})} \quad (2.134)$$

where $\nabla^2 \mathbf{A}$ is called the ***vector Laplacian*** of **A**. This is written in Cartesian coordinates as

$$\nabla^2 \mathbf{A} = \hat{\mathbf{x}} \nabla^2 A_x + \hat{\mathbf{y}} \nabla^2 A_y + \hat{\mathbf{z}} \nabla^2 A_z \quad (2.135)$$

and can be obtained by direct application of the scalar Laplacian operator to each of the scalar components of the vector **A**. The scalar components of the vector Laplacian are

$$\nabla^2 A_x = \frac{\partial^2 A_x}{\partial x^2} + \frac{\partial^2 A_x}{\partial y^2} + \frac{\partial^2 A_x}{\partial z^2},$$
$$\nabla^2 A_y = \frac{\partial^2 A_y}{\partial x^2} + \frac{\partial^2 A_y}{\partial y^2} + \frac{\partial^2 A_y}{\partial z^2}, \qquad (2.136)$$
$$\nabla^2 A_z = \frac{\partial^2 A_z}{\partial x^2} + \frac{\partial^2 A_z}{\partial y^2} + \frac{\partial^2 A_z}{\partial z^2}$$

The second-order operators define second-order partial differential equations and constitute a very important area in mathematics and physics. We will use the second-order operators described here throughout this book.

The scalar and vector Laplacians as well as other vector quantities and identities in Cartesian, cylindrical, and spherical coordinate systems are listed in the appendix for easy reference.

Exercise 2.14

(a) Show that in Cartesian coordinates, the following is correct:

$$\nabla^2 \mathbf{A} = \hat{\mathbf{x}}\nabla^2 A_x + \hat{\mathbf{y}}\nabla^2 A_y + \hat{\mathbf{z}}\nabla^2 A_z.$$

(b) Show that in any other coordinate system, this relation is not correct. Use the cylindrical system as an example; that is, show that $\nabla^2 \mathbf{A} \neq \hat{\mathbf{r}}\nabla^2 A_r + \hat{\boldsymbol{\phi}}\nabla^2 A_\phi + \hat{\mathbf{z}}\nabla^2 A_z$.

2.5.3 Other Vector Identities

If U and Q are scalar functions and **A** and **B** are vector functions, all dependent on the three variables (for example, x, y, and z), we can show that

$$\nabla(UQ) = U(\nabla Q) + Q(\nabla U) \qquad (2.137)$$

$$\nabla \cdot (U\mathbf{A}) = U(\nabla \cdot \mathbf{A}) + (\nabla U) \cdot \mathbf{A} \qquad (2.138)$$

$$\nabla \cdot (\mathbf{A} \times \mathbf{B}) = -\mathbf{A} \cdot (\nabla \times \mathbf{B}) + (\nabla \times \mathbf{A}) \cdot \mathbf{B} \qquad (2.139)$$

$$\nabla \times (U\mathbf{A}) = U(\nabla \times \mathbf{A}) + (\nabla U) \times \mathbf{A} \qquad (2.140)$$

Problems

2.1 A force is described in cylindrical coordinates as $\mathbf{F} = \hat{\boldsymbol{\phi}}/r$ [N]. Find the work performed by the force along the following paths:

(a) From $P(a,0,0)$ to $P(a,b,c)$.
(b) From $P(a,0,0)$ to $P(a,b,0)$, and then from $P(a,b,0)$ to $P(a,b,c)$.

2.2 Determine whether $\int_{P_1}^{P_2} \mathbf{A} \cdot d\mathbf{l}$ between points P_1 (0,0,0) and $P_2(1,1,1)$ is path dependent for $\mathbf{A} = \hat{\mathbf{x}}y^2 + \hat{\mathbf{y}}2x + \hat{\mathbf{z}}$.

2.3 A body is moved along the path shown in **Figure 2.25** by a force $\mathbf{A} = \hat{\mathbf{x}}2 - \hat{\mathbf{y}}5$. The path between point a and point b is a parabola described by $y = 2x^2$.

(a) Calculate the work necessary to move the body from point a to point b along the parabola.
(b) Calculate the work necessary to move the body from point a to point c and then to point b.
(c) Compare the results in **(a)** and **(b)**.

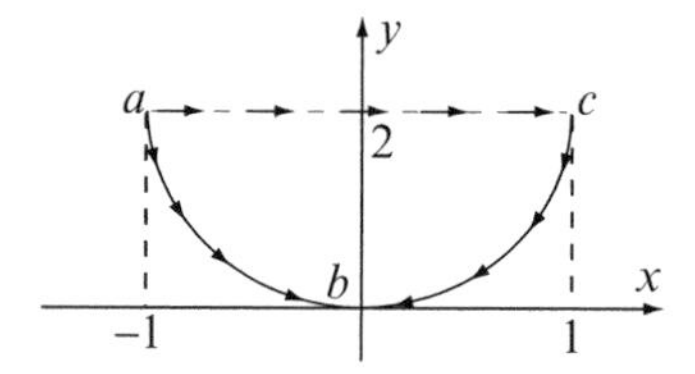

Figure 2.25

Surface Integrals (Closed and Open)

2.4 A volume is defined in cylindrical coordinates as $1 \le r \le 2$, $\pi/6 \le \phi \le \pi/3$, $1 \le z \le 2$. Calculate the flux of the vector $\mathbf{A} = \hat{\mathbf{r}}4z$ through the surface enclosing the given volume.

2.5 Given a surface $S = S_1 + S_2$ defined in spherical coordinates with S_1 defined as $0 \le R \le 1$; $\theta = \pi/6$; $0 \le \phi \le 2\pi$ and S_2 defined as $R = 1$; $0 \le \theta \le \pi/6$; $0 \le \phi \le 2\pi$. Vector $\mathbf{A} = \hat{\mathbf{R}}1 + \hat{\boldsymbol{\theta}}\theta$ is given. Find the integral of $\mathbf{A} \cdot d\mathbf{s}$ over the surface S.

2.6 Given $\mathbf{A} = \hat{\mathbf{x}}x^2 + \hat{\mathbf{y}}y^2 + \hat{\mathbf{z}}z^2$, integrate $\mathbf{A} \cdot d\mathbf{s}$ over the surface of the cube of side 1 with four of its vertices at (0,0,0), (0,0,1), (0,1,0), and (1,0,0).

2.7 The axis of a disk of radius a is in the direction of the vector $\mathbf{k} = -\hat{\mathbf{r}}2 + \hat{\mathbf{z}}3$. Vector field $\mathbf{A} = \hat{\mathbf{r}}4 + \hat{\mathbf{z}}3$ is given. Find the total flux of $\mathbf{A}$ through the disk.

2.8 A straight cylinder of radius a and height h is placed symmetrically along the z-axis so that it extends from $z = -h/2$ to $z = h/2$. Calculate the flux of the vector $\mathbf{k} = \hat{\mathbf{r}}z^2 + \hat{\boldsymbol{\phi}}3zr - \hat{\mathbf{z}}3r^2$ through the cylinder.

Volume Integrals

2.9 A mass density in space is given by $\rho(r,z) = r(r + a) + z(z + d)$ [kg/m^3] (in cylindrical coordinates).

(a) Calculate the total mass of a cylinder of length d, radius a, centered at the origin with its axis along the z axis.
(b) Calculate the total mass of a sphere of radius a centered at the origin.

2.10 A right circular cone is cut off at height h_0. The radius of the small base is a and that of the large base is b (**Figure 2.26**). The cone is filled with particles in a nonuniform distribution: $n(r,h) = 10^8 r(h - h_0)$ where h is the distance from the larger base. Find the total number of particles contained in the cone.

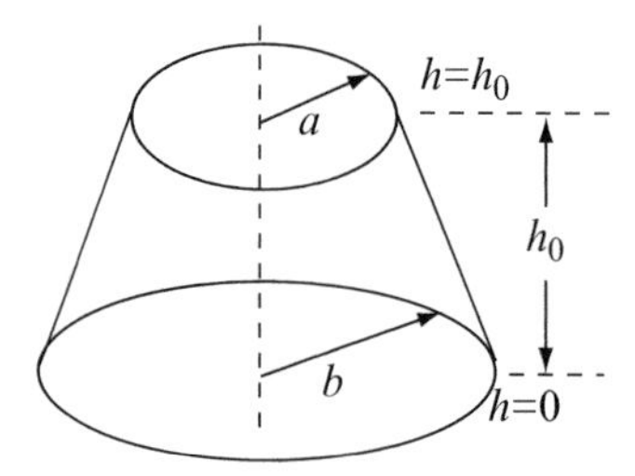

Figure 2.26

2.11 Vector field $\mathbf{f} = \hat{\mathbf{x}}2xy + \hat{\mathbf{y}}z + \hat{\mathbf{z}}y^2$ is defined as a volume force density (in N/m^3) in a sufficiently large region in space. This force acts on every particle of any body placed in the field (similar to a gravitational force).

(a) A cubic body $2 \times 2 \times 2$ m^3 in dimensions is placed in the field with its center at the origin and with its sides parallel to the system of coordinates. Calculate the total force acting on the body.
(b) The same cube as in **(a)** is placed in the first quadrant with one corner at the origin and with its sides parallel to the system of coordinates. Calculate the total force acting on the body.

Other Regular Integrals

2.12 The acceleration of a body is given as $\mathbf{a} = \hat{\mathbf{x}}(t^2 - 2t) + \hat{\mathbf{y}}3t$ [m/s^2]. Find the velocity of the body after 5 s.

2.13 Evaluate the integral $\int r^2 d\mathbf{l}$, where $r^2 = x^2 + y^2$, from the origin to the point $P(1,3)$ along the straight line connecting the origin to $P(1,3)$. $d\mathbf{l}$ is the differential vector in Cartesian coordinates.

The Gradient

2.14 Find the derivative of $xy^2 + yz$ at (1,1,2) in the direction of the vector $\hat{\mathbf{x}}2 - \hat{\mathbf{y}} + \hat{\mathbf{z}}2$.

2.15 An atmospheric pressure field is given as $P(x,y,z) = (x - 2)^2 + (y - 2)^2 + (z + 1)^2$ [N/m^2], where the x–y plane is parallel to the surface of the ocean and the z direction is vertical. Find:

(a) The magnitude and direction of the pressure gradient.
(b) The derivative of the pressure in the vertical direction.
(c) The derivative of pressure in the direction parallel to the surface, at 45° between the positive x and y axes.

2.16 The scalar field $f(r,\phi,z) = r\cos^2\phi + z\sin\phi$ is given. Calculate:

(a) The gradient of $f(r,\phi,z)$ in cylindrical coordinates.
(b) The gradient of $f(r,\phi,z)$ in Cartesian coordinates.
(c) The gradient of $f(r,\phi,z)$ in spherical coordinates.

2.17 Find the unit vector normal to the following planes:

(a) $z = -5x - 3y$.
(b) $4x - 3y + z + 5 = 0$.
(c) $z = ax + by$.

Show by explicit derivation that the result obtained is in fact normal to the plane.

2.18 Find the unit vector normal to the following surfaces at the specified points:

(a) $z = -3xy - yz$ at (0,0,0).
(b) $x = z^2 + y^2$ at (2,–1,1).
(c) $z^2 + y^2 + x^2 = 8$ at (2,–2,–2).

The Divergence

2.19 Calculate the divergence of the following vector fields:

(a) $\mathbf{A} = \hat{\mathbf{x}}x^2 + \hat{\mathbf{y}}1 - \hat{\mathbf{z}}y^2$.
(b) $\mathbf{B} = \hat{\mathbf{r}}2z^2 + \hat{\boldsymbol{\phi}}5r - \hat{\mathbf{z}}3r^2$.
(c) $\mathbf{C} = \hat{\mathbf{x}}\sqrt{x^2 + z^2} + \hat{\mathbf{y}}\sqrt{x^2 + y^2}$.

2.20 Find the divergence of $\mathbf{A} = \hat{\mathbf{x}}x^2 + \hat{\mathbf{y}}y^2 + \hat{\mathbf{z}}z^2$ at (1,–1,2).

2.21 Find the divergence of $\mathbf{A} = \hat{\mathbf{r}}2r\cos\phi - \hat{\boldsymbol{\phi}}r\sin\phi + \hat{\mathbf{z}}4z$ at $(2, 90°, 1)$.

2.22 Find the divergence of $\mathbf{A} = 0.2R^3\phi\sin^2\theta\left(\hat{\mathbf{R}} + \hat{\boldsymbol{\theta}} + \hat{\boldsymbol{\phi}}\right)$ at $(2, 30°, 90°)$.

The Divergence Theorem

2.23 Verify the divergence theorem for $\mathbf{A} = \hat{\mathbf{x}}4z - \hat{\mathbf{y}}2y^2 - \hat{\mathbf{z}}2z^2$ for the region bounded by $x^2 + y^2 = 9$ and $z = -2$, $z = 2$ by evaluating the volume and surface integrals.

2.24 A vector field is given as $\mathbf{A}(\mathbf{R}) = \mathbf{R}$, where $\mathbf{R}$ is the position vector of a point in space. Show that the divergence theorem applies to the vector $\mathbf{A}$ for a sphere of radius a.

2.25 Given $\mathbf{A} = \hat{\mathbf{x}}x^2 + \hat{\mathbf{y}}y^2 + \hat{\mathbf{z}}z^2$:

(a) Integrate $\mathbf{A} \cdot d\mathbf{s}$ over the surface of the cube of side 1 with four of its vertices at (0,0,0), (0,0,1), (0,1,0), and (1,0,0) (see **Problem 2.6**).
(b) Integrate $\nabla \cdot \mathbf{A}$ over the volume of the cube in **(a)** and show that the two results are the same.

2.26 A vector field is given at all points in space in spherical coordinates as

$$\mathbf{P} = \hat{\mathbf{R}}\frac{A}{R^2}e^{-j\alpha R}\sin^2\theta$$

where R is the radial distance from the origin, α is a constant and θ the angle between the z-axis and $\hat{\mathbf{R}}$. Calculate:
(a) The flux of the vector $\mathbf{P}$ through the surface of a volume made of two spherical concentric surfaces, centered at the origin, one of radius R_1 and one of radius $R_2 > R_1$.
(b) The divergence of $\mathbf{P}$ at a general point between the two spherical surfaces. From that calculate the flux through the volume and show that it is equal to the flux through the surface in **(a)**.

The Curl

2.27 Calculate the curl of the following three vectors:

(a) $\mathbf{A} = \hat{\mathbf{x}}x^2 + \hat{\mathbf{y}}1 - \hat{\mathbf{z}}y^2$.
(b) $\mathbf{B} = \hat{\mathbf{r}}2z^2 + \hat{\boldsymbol{\phi}}5r - \hat{\mathbf{z}}3r^2$.
(c) $\mathbf{C} = \hat{\mathbf{x}}\sqrt{x^2+z^2} + \hat{\mathbf{y}}\sqrt{x^2+y^2}$.

2.28 A fluid flows in a circular pattern with the velocity vector $\mathbf{v} = \hat{\boldsymbol{\phi}}a/r$ [m/s] where a is a constant and r [m] is the radial distance from the center of the flow.

(a) Sketch the vector field $\mathbf{v}$.
(b) Calculate the curl of the vector field.

2.29 A vector field $\mathbf{A} = \hat{\mathbf{y}}3x\cos(\omega t + 50z)$ is given (t is time, x,z are variables).

(a) What is the curl of $\mathbf{A}$?
(b) Is this a conservative field?

2.30 Two vector fields are given as: $\mathbf{A} = \hat{\mathbf{x}}y^2 + \hat{\mathbf{y}}2(x+1)yz - \hat{\mathbf{z}}(x+1)z^2$ and $\mathbf{B} = \hat{\mathbf{r}}2r\cos\phi - \hat{\boldsymbol{\phi}}4r\sin\phi + \hat{\mathbf{z}}3$.

(a) Are the vector fields $\mathbf{A}$ and $\mathbf{B}$ conservative?
(b) Calculate the of the vectors in $\mathbf{A}$ and $\mathbf{B}$.

Stokes' Theorem

2.31 Verify Stokes' theorem for $\mathbf{A} = \hat{\mathbf{r}}r + \hat{\boldsymbol{\phi}}\cos\phi + \hat{\mathbf{z}}1$ on the upper half-surface of the sphere $x^2 + y^2 + z^2 = 4$ above the xy plane. The contour bounding the surface is the rim of the half-sphere.

2.32 Vector field $\mathbf{F} = \hat{\mathbf{x}}3y + \hat{\mathbf{y}}(5-2x) + \hat{\mathbf{z}}(z^2-2)$ is given. Find:

(a) The divergence of $\mathbf{F}$.
(b) The curl of $\mathbf{F}$.
(c) The surface integral of the normal component of the curl of $\mathbf{F}$ over the open hemisphere $x^2 + y^2 + z^2 = 4$ above the x–y plane.

2.33 Vector field $\mathbf{F} = \hat{\mathbf{x}}y + \hat{\mathbf{y}}z + \hat{\mathbf{z}}x$ is given. Find the total flux of $\nabla \times \mathbf{F}$ through a triangular surface given by three points $P_1(a,0,0)$, $P_2(0,0,b)$, and $P_3(0,c,0)$.

The Helmholtz Theorem and Vector Identities

2.34 The following combinations of operators, scalars and vectors are given:

1. $\nabla \cdot (\nabla \phi)$.
2. $(\nabla \cdot \nabla)\phi$.
3. $(\nabla \times \nabla)\, \phi$.
4. $\nabla \times (\nabla \phi)$.
5. $\nabla \cdot (\nabla \times \mathbf{A})$.
6. $(\nabla \cdot \nabla) \times \mathbf{A}$.
7. $(\nabla \times \nabla) \times \mathbf{A}$.
8. $\nabla \times (\nabla \times \mathbf{A})$.
9. $\nabla \cdot (\phi\, \nabla \times \mathbf{A})$.
10. $\phi\, (\nabla \times \mathbf{A})$.
11. $\nabla\, (\nabla \times \mathbf{A})$.
12. $\nabla \times (\nabla \cdot \mathbf{A})$.

where $\mathbf{A}$ is an arbitrary vector field and ϕ an arbitrary scalar field.

(a) Which of the combinations are valid?
(b) Evaluate explicitly those that are valid (in Cartesian coordinates).

2.35 Calculate the Laplacian for the following scalar fields ((x,y,z), (r,ϕ,z) and (R,θ,ϕ) are general points in the Cartesian, cylindrical and spherical systems of coordinates respectively, a is a numerical constant):

(a) $p = (x-2)^2(y-2)^2(z+1)^2$.
(b) $p = 5r\cos\phi + 3zr^2$.
(c) $p = \dfrac{a}{R^3}\cos\theta$.

2.36 Calculate the Laplacian for the following vector fields ((x,y,z), (r,ϕ,z) and (R,θ,ϕ) are general points in the Cartesian, cylindrical and spherical systems of coordinates respectively, β is a numerical constant):

(a) $\mathbf{A} = \hat{\mathbf{x}}3y + \hat{\mathbf{y}}(5-2x) + \hat{\mathbf{z}}(z^2-2)$.
(b) $\mathbf{B} = \hat{\mathbf{r}}2r\cos\phi - \hat{\boldsymbol{\phi}}4r\sin\phi + \hat{\mathbf{z}}3$.
(c) $\mathbf{C} = \hat{\mathbf{R}}\dfrac{e^{-j\beta R}}{R^2}\sin\theta$.

2.37 Show that if $\mathbf{F}$ is a conservative field, then $\nabla^2\mathbf{F} = \nabla(\nabla \cdot \mathbf{F})$. Use cylindrical coordinates.

2.38 Given the scalar field $f(x,y,z) = 2x^2 + y$ and the vector field $\mathbf{R} = \hat{\mathbf{x}}x + \hat{\mathbf{y}}y + \hat{\mathbf{z}}z$, find:

(a) The gradient of f.
(b) The divergence of $f\mathbf{R}$.
(c) The Laplacian of f.
(d) The vector Laplacian of $\mathbf{R}$.
(e) The curl of $f\mathbf{R}$.

2.39 A vector field $\mathbf{A} = \hat{\mathbf{x}}5x + \hat{\mathbf{y}}2y + \hat{\mathbf{z}}1$ is given. What type of field is this according to the Helmholtz theorem?

2.40 A vector field $\mathbf{A} = \hat{\mathbf{R}}\phi R^2 + \hat{\boldsymbol{\theta}}R\sin\theta$ is given in spherical coordinates. What type of field is this according to the Helmholtz theorem?

2.41 The following vector fields are given:

(1) $\mathbf{A} = \hat{\mathbf{x}}x + \hat{\mathbf{y}}y$.
(2) $\mathbf{B} = \hat{\boldsymbol{\phi}}\cos\phi + \hat{\mathbf{r}}\cos\phi$.
(3) $\mathbf{C} = \hat{\mathbf{x}}y + \hat{\mathbf{z}}y$.
(4) $\mathbf{D} = \hat{\mathbf{R}}\sin\theta + \hat{\boldsymbol{\theta}}5R + \hat{\boldsymbol{\phi}}R\sin\theta$.
(5) $\mathbf{E} = \hat{\mathbf{R}}k$.

(a) Which of the fields are solenoidal?
(b) Which of the fields are irrotational?
(c) Classify these fields according to the Helmholtz theorem.

2.42 Show by direct derivation of the products that the following holds:

$$\nabla \times (\nabla \times \mathbf{A}) = \nabla(\nabla \cdot \mathbf{A}) - \nabla^2 \mathbf{A}.$$

Coulomb's Law and the Electric Field

3

I looked, and lo, a stormy wind came sweeping out of the north—a huge cloud and flashing fire, surrounded by a radiance; and in the center of it, in the center of the fire, a gleam as of amber.

—Ezekiel 1:4

3.1 Introduction

In the previous two chapters, we discussed in some detail the mathematics of electromagnetics: vector algebra and vector calculus. We are now ready to start looking into the physical phenomena of electromagnetics. It will be useful to keep this in mind: The study of electromagnetics is the study of natural phenomena. There are two reasons why it is important to emphasize electromagnetics as an applied science. First, because it is a useful science, and its study leads to understanding of nature and, perhaps most importantly from the engineering point of view, to understanding of the application of electromagnetics to practical and useful designs. Second, all aspects of electromagnetics are based on experimental observations. All laws of electromagnetics were obtained by careful measurements which were then cast in the forms of simple laws. These laws are assumed to be correct simply because there is no evidence to the contrary. This aspect of the laws of electromagnetics should not bother us too much. Although we cannot claim absolute proof to correctness of the laws, experimentation has shown that they are correct and we will view them as such. In the learning process, we will make considerable use of the mathematical tools outlined in **Chapters 1** and **2**. It is easy to forget that the end purpose is physical design; however, every relation and every equation implies some physical quantity or property of the fields involved.

As with the study of any branch of science, we must start with the basics and proceed in a logical fashion. We will start with the study of electrostatic fields. To do so, we need a few assumptions that can be verified easily by experiment. In fact, the basic assumption is the existence of positive and negative electric charges (electrons and protons). Having allowed for their existence, we can then measure forces between charges, and these forces will lead to the definition of the electric field. The electric field is, therefore, merely a manifestation of forces on charges. We may even call it an electric force field. These forces are real forces and are measurable.

The static electric field is an exceedingly useful phenomenon that permeates our lives. The number of applications and effects that rely on electrostatics is vast. From the simplest of capacitors to thunderstorms, and from sand paper deposition to laser printers and memory chips, the use of static fields is the basis either of design of the device or explanation of the effects involved.

Thus, we will try to do two things; one is to state and explain the laws. This will require a mathematical exposition of relations between forces and charges, based on experimental results. At the same time, we will discuss at least a sampling of applications of electrostatic fields.

N. Ida, *Engineering Electromagnetics*, https://doi.org/10.1007/978-3-030-15557-5_3

3.2 Charge[1] and Charge Density

The fundamental electric charge in nature is the charge of the electron. The fact that an electric charge exists was known to the ancient Greeks[2] who knew that rubbing a piece of amber with fur or cloth caused an attraction of particles such as feathers, straw, or lint. Electron is the Greek name for amber.[3] It took many years before it was understood what actually happened in this type of experiment or the amount of electric charge associated with the electron was established, but the effects of the charge were all around to be observed. For a body to contain free charge, there must be a way of removing electrons from one body and imparting them to another. The body from which electrons are removed becomes positively charged (because of excess protons) whereas the other body becomes negatively charged (excess electrons).[4] This is what happened when amber was rubbed with fur: Electrons were removed from the fur and deposited on the amber. The rubbing action supplied the energy required for electrons to be removed. Amber therefore became negatively charged and could attract small bits of material. At the same time, the fur became positively charged. We know that lightning occurs when charge accumulates beyond a certain level. This means that charges must accumulate either in a volume or over a surface. Since we know that electric charge causes forces in its vicinity, this force should also be proportional to the amount of charge. Before we can quantify the effects of electric charge, we must establish a few preliminaries. These include the charge of the electron, the unit of charge, and the definitions of point charges and distributed charges.

The unit of charge is called a ***coulomb***.[5] The charge of an electron is denoted by e and equals $e = -1.6019 \times 10^{-19}$ C. That is, one coulomb is equal to the charge of approximately 6.25×10^{18} electrons. The charge of the electron is considered to be the smallest unit of charge and all charges must be multiples of this quantity, although charge can be positive or negative. Charge may be distributed in space or may be concentrated in a small volume or a "point."

Point Charge A charge that occupies a volume in space may be considered to be a point charge for analysis purposes if this volume is small compared to the surrounding dimensions. The charges of electrons or protons are often assumed to be point charges. For practical purposes, other charges such as the charge of a sphere or any other small volume are often considered to be point charges, provided that we are far from the volume.

A charge density defines a distributed charge over a body. There are three types of charge densities:

Line Charge Density A charge distributed in a linear fashion such as along a very thin wire is given in charge per unit length. A charge density of 1 C/m means that one coulomb of charge is distributed per each meter length of the device

---

[1] The charge is a fundamental quantity of nature. It was even proposed as the basic quantity in the SI system of units, although the system as it stands now uses the ampere as the basic electric unit.

[2] Thales of Miletus (624?–546 B.C.E), one of the "seven wise men" of ancient Greece. His work was mostly in geometry (for example, the theorem concerning the right angle in a semicircle and at least four others are attributed to him). Thales is thought to be the first to record this phenomenon, although it is almost certain that it was known before him. He himself traveled and studied in many parts of the ancient world, including Egypt. Miletus, in spite of his influence on later natural philosophers, did not write any of his views and findings (or none survived). The references to him come from later writing (primarily from Aristotle who wrote Thales' record from oral records). The first written record on electricity comes from Theophrastus (371–288 B.C.E.) and dates around 300 B.C.E.

[3] Amber has a curious relation to electricity. Although merely a yellowish fossilized tree resin, it has gained some prominence in our view of electricity. The Greek name for amber is electron (ηλεκτρον) (electrum in Latin) and means "bright" or "bright one," perhaps a reference to the color of amber, a material which was held in very high esteem by the Greeks. Since amber was known to attract bodies when rubbed with fur or cloth, it eventually became a synonym to all electric phenomena, and in particular with electric charge. The actual name "electricity" was coined much later, around 1600, by William Gilbert. We should take some pride in having electricity associated with this material. It is beautiful and rare and precious. Also, it is a very good insulator.

[4] This explanation was first put forward by Benjamin Franklin (1706–1790). Franklin was, in addition to being a statesman, diplomat, and publisher, a most prolific experimenter in electricity. Whereas the legendary experiment with the kite (June 1752) led to the discovery of atmospheric electricity and the lightning rod, he is also credited with being the first to propose the so-called "one-fluid" theory of electricity. Before him, it was assumed that there are two types of electricity: vitreous (from glass or what we call positive electricity or charge) and resinous (from resin like amber, which we call negative electricity or charge). He found that these are the same except that one is excess, whereas the other a deficiency in an otherwise balanced state (this, of course, happened about 130 years before the discovery of the electron). From this, he suggested the law of conservation of charge and coined the terms "positive" and "negative," as well as the terms "charge" and "conductor." Franklin held very modern views on electricity and magnetism and his work was the turning point in electromagnetics, a turn that led directly to the modern theory we use today.

[5] After Charles Augustin de Coulomb (1736–1806). Coulomb was a colonel in the Engineering Corps of the French military who specialized in artillery, a career he abandoned shortly before the French Revolution due to health problems. The naming of the unit of charge after him indicates the importance of his work. Coulomb derived the law of force between charges, which we will investigate shortly.

(such as a wire). A more exact description of the line charge density is to say that the charge density is the charge ΔQ distributed over a length Δl as Δl approaches zero:

$$\rho_l = \lim_{\Delta l \to 0} \frac{\Delta Q}{\Delta l} \quad \left[\frac{\text{C}}{\text{m}}\right] \tag{3.1}$$

Surface Charge Density A charge distributed over a given surface such as the surface of a sphere or a sheet of paper. A surface charge density of 1 C/m^2 means that one coulomb of charge is distributed over each square meter of the surface. The surface charge density is defined mathematically as

$$\rho_s = \lim_{\Delta s \to 0} \frac{\Delta Q}{\Delta s} \quad \left[\frac{\text{C}}{\text{m}^2}\right] \tag{3.2}$$

Volume Charge Density A charge distributed over a volume such as the volume of a cloud. A volume charge density of 1 C/m^3 means that one coulomb of charge is distributed over a one meter cube of volume:

$$\rho_v = \lim_{\Delta v \to 0} \frac{\Delta Q}{\Delta v} \quad \left[\frac{\text{C}}{\text{m}^3}\right] \tag{3.3}$$

The charge densities in **Eqs. (3.1)** through **(3.3)** may be uniform or nonuniform over the dimensions given. A uniform charge density means that the charge distributed over any equal section of surface, line, or volume is the same; that is, it is independent of the variables. A nonuniform charge density occurs when the charge in different sections of the charge distribution depends on location. Note, also, that the charge densities given above as examples are rather large. Normally, much smaller charge densities are encountered in practical designs.

Example 3.1 Line, Surface, and Volume Charge Density A charge is uniformly distributed over the following three structures such that a charge of $Q = 10^{-9}$ C is distributed per unit length (1 m) of the device:

(a) A very thin wire.
(b) A conducting wire of radius $d = 10$ mm. Assume charge can only exist on the surface of the wire.
(c) A solid cylindrical, nonconducting material of radius $d = 10$ mm assuming the charge is uniformly distributed throughout the volume of the material. Calculate the charge density in each of the three structures.

Solution:

(a) Because the wire is very thin, the charge is distributed along the length of the wire (conducting or nonconducting). This is an example of line charge distribution. The line charge density is

$$\rho_l = \frac{Q}{L} = \frac{10^{-9}}{1} = 10^{-9} \quad [\text{C/m}]$$

The notation ρ_l for line charge densities will be used throughout this book. **Figure 3.1a** shows this distribution schematically. Line charge densities are typically used for thin conductors or, sometimes, as approximations of thick, charged conductors.

(b) Now, the same amount of charge is distributed per unit length of a cylindrical conductor of radius d. Since the charge on the conductor can only reside on the surface, the charge distributes itself (in this case uniformly) over the surface of the cylinder. The total external surface of a 1 m length of the cylindrical conductor is $(2\pi d)(1)$. Thus, the surface charge density on the conductor is

$$\rho_s = \frac{Q}{S} = \frac{10^{-9}}{2\pi d} = \frac{10^{-9}}{2\pi \times (0.01)} = 1.59155 \times 10^{-8} \quad [\text{C/m}^2]$$

Again, this notation (ρ_s) will be used throughout the book. **Figure 3.1b** shows a surface charge distribution. Typical situations in which surface charge density plays a role are conducting surfaces and interfaces between conducting and nonconducting materials, such as a carpet or the outer shell of a car.

(c) In this case, charges are distributed throughout the volume. The volume of a cylinder of length 1 m and radius d is $(\pi d^2 \times 1)$. The volume charge density is

$$\rho_v = \frac{Q}{v} = \frac{10^{-9}}{\pi d^2} = \frac{10^{-9}}{\pi \times (0.01)^2} = 3.183 \times 10^{-6} \quad [\mathrm{C/m^3}]$$

The notation is normally ρ_v. However, because volume charge densities are the most general, we will also use the simpler notation ρ. Unless a charge density is specifically indicated as surface or line, it is implicitly understood to be a volume charge density. **Figure 3.1c** shows a volume charge distribution. A typical example of volume charge density is charge in a cloud, space charge in a vacuum tube, or charge within the volume of a semiconductor device.

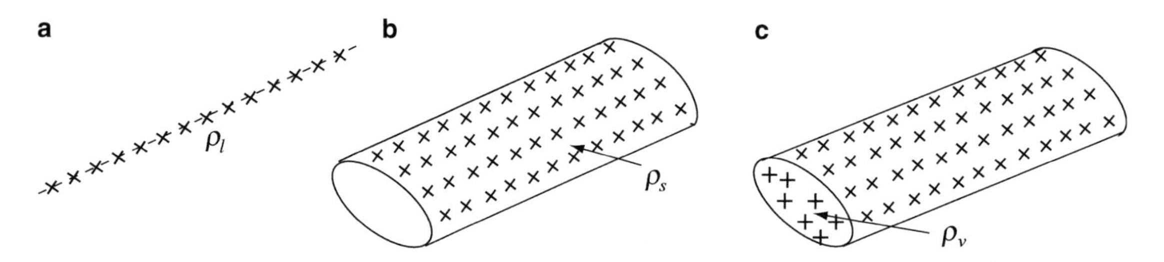

Figure 3.1 (**a**) Line charge distribution. (**b**) Surface charge distribution. (**c**) Volume charge distribution

Example 3.2 Nonuniform Charge Distribution A charge is distributed throughout a spherical volume (such as a cloud). The charge is nonuniformly distributed with a distribution function:

$$\rho_v = \rho_0 R \quad [\mathrm{C/m^3}]$$

where $\rho_0 = 10^{-7}$ and R is the distance from the center of the spherical volume. Calculate the total charge in a sphere of radius $R_0 = 10$ m.

Solution: To calculate the total charge, we integrate over the sphere of radius R_0 using the charge density ρ_v. The element of volume in spherical coordinates is $dv = R^2 \sin\theta dR d\theta d\phi$. The total charge is therefore

$$Q = \int_v \rho_v dv = \rho_0 \int_{\phi=0}^{\phi=2\pi} \int_{\theta=0}^{\theta=\pi} \int_{R=0}^{R=R_0} R\left(R^2 \sin\theta dR d\theta d\phi\right) = \pi \rho_0 R_0^4 \quad [\mathrm{C}]$$

Numerically, this gives 0.0031416 C. Note that if the charge density were uniform, we would simply multiply the charge density by the volume of the sphere.

The remainder of this chapter will discuss point charges and the interaction between them. Distributed charges will be discussed here as assemblages of point charges. In **Chapter 4**, we will view distributed charges in a somewhat different light and will develop methods suited for their treatment.

3.3 Coulomb's Law[6]

Coulomb's law is an experimental law obtained by Charles Augustin de Coulomb that defines quantitatively the force between two point charges. It states that:

[6] At the onset of the French Revolution, Charles Augustin de Coulomb started his work on electricity, following a successful military career. In or around 1785, he formulated his now famous law, which he came about in an attempt to verify previous work by Joseph Priestley (1733–1804). However Coulomb's work was much more general than Priestley's and included both attraction and repulsion forces. Coulomb was a meticulous researcher who worked on other problems in science and engineering, including friction, soil mechanics, and elasticity. To perform the necessary experiments, he used a torsion balance, which he invented a year earlier (see **Figure 3.24**). The main advantage of this balance was in its ability to measure very small forces accurately. In addition, it was perhaps the first accurate measuring device used for measurements of electric quantities.

"the force between two point charges Q_1 and Q_2 is proportional to the product of the two charges, inversely proportional to the square of the distance between the two charges, and directed along the line connecting the two charges."

The mathematical expression of Coulomb's law is

$$\boxed{\mathbf{F} = \hat{\mathbf{R}} k \frac{Q_1 Q_2}{R^2} \quad [\text{N}]} \tag{3.4}$$

where k is a proportionality factor, R is the distance between the two charges, and $\hat{\mathbf{R}}$ is a unit vector pointing from Q_1 to Q_2 if the force on Q_2 due to Q_1 is required, or from Q_2 to Q_1 if the force on Q_1 due to Q_2 is needed. We will expand on this shortly, but, for now, it is sufficient to look at the magnitude of the force. The factor k depends on the material in which the charges are located and is given as

$$k = \frac{1}{4\pi\varepsilon} \quad \left[\frac{\text{N} \cdot \text{m}^2}{\text{C}^2}\right] \tag{3.5}$$

where ε is a material constant. This material constant is called the ***permittivity*** of the material and we will define it more accurately in **Chapter 4**. For now, it is only important to understand that it has a numerical value that depends on the material in which the charges reside. The magnitude of the force between the two charges is

$$F = \frac{Q_1 Q_2}{4\pi\varepsilon R^2} \quad [\text{N}] \tag{3.6}$$

Before continuing, we should take a look at the units involved. Force is measured in ***newtons*** (N). Charges Q_1 and Q_2 are given in ***coulombs*** [C]. Thus, the unit of permittivity is given by the relation

$$\varepsilon = \frac{Q_1 Q_2}{4\pi F R^2} \quad \left[\frac{\text{C}^2}{\text{N} \cdot \text{m}^2}\right] \tag{3.7}$$

This unit is not normally employed simply because the quantity [C^2/N · m] is known as the ***farad*** [F]. In this chapter, the permittivity is taken as a given constant, so there is little reason to dwell on its meaning and the meaning of the units involved. These will become obvious in **Chapter 4**, when we talk about material properties. For now, we will simply accept the units of [F/m] or [C^2/N · m^2] for permittivity.

To avoid the need to discuss properties of materials that we have not encountered yet, we will limit our discussion here to charges in free space. The permittivity of free space is given as

$$\varepsilon_0 = 8.8541853 \times 10^{-12} \approx 8.854 \times 10^{-12} \quad [\text{F/m}] \tag{3.8}$$

The approximate value in **Eq. (3.8)** will be used throughout this book. Also, because the permittivities of free space and that of air are very close, the permittivity of free space will be used in air. The magnitude of the force between the two charges in free space becomes

$$\boxed{F = \frac{Q_1 Q_2}{4\pi\varepsilon_0 R^2} \quad [\text{N}]} \tag{3.9}$$

Equation (3.6) or **(3.9)** gives the magnitude of the force between the two charges. Because charges Q_1 and Q_2 can be positive or negative, the force can change direction. To take into account this and the fact that force is a vector quantity, consider **Figure 3.2**:

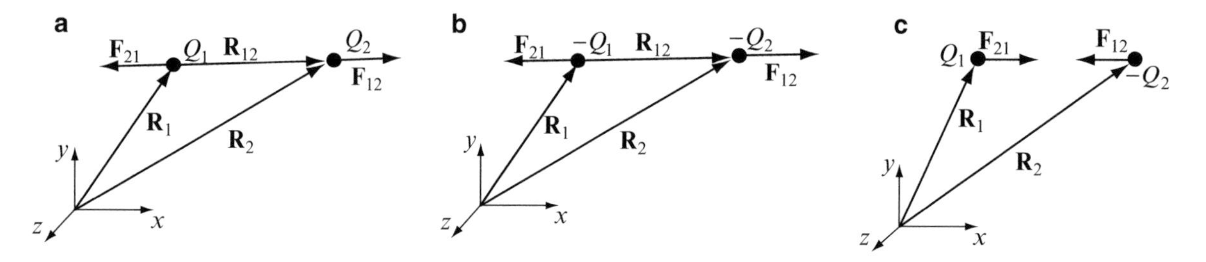

Figure 3.2 Relations between direction of forces and polarity of point charges. (**a**) Two positive charges. (**b**) Two negative charges. (**c**) Two charges of opposite polarity

(1) In **Figure 3.2a**, both charges are positive. According to the results of experimentation, the charges repel each other and the force is along the line connecting the two charges. This means that the direction of forces is away from the charges.

(2) The force on charge Q_1 due to charge Q_2 is called $\mathbf{F}_{21}$ and is equal in magnitude to the force on charge Q_2 due to charge Q_1 ($\mathbf{F}_{12}$) as expected from Newton's third law (action and reaction). The directions of these two forces are opposite each other.

(3) In **Figure 3.2b**, both charges are negative, and, again, the charges repel each other as in **Figure 3.2a**.

(4) In **Figure 3.2c**, one charge is positive, the other negative and the force is a force of attraction as shown.

To formalize these three results, we define the position vectors $\mathbf{R}_1$ and $\mathbf{R}_2$ as shown and define a vector that connects the two charges as

$$\mathbf{R}_{12} = \mathbf{R}_2 - \mathbf{R}_1 \quad \text{and} \quad \mathbf{R}_{21} = \mathbf{R}_1 - \mathbf{R}_2 \tag{3.10}$$

Now we note that the direction of the force on charge Q_1 in **Figures 3.2a** and **3.2b** is in the direction of $\mathbf{R}_{21}$, whereas the direction of the force on Q_2 is in the direction of $\mathbf{R}_{12}$. On the other hand, in **Figure 3.2c**, the forces point in opposite directions. Thus, using the vector notation, we can write the forces using the unit vectors as

$$\boxed{\mathbf{F}_{12} = \hat{\mathbf{R}}_{12}\frac{Q_1Q_2}{4\pi\varepsilon_0 R_{12}^2}, \qquad \mathbf{F}_{21} = \hat{\mathbf{R}}_{21}\frac{Q_1Q_2}{4\pi\varepsilon_0 R_{21}^2}, \quad [\mathrm{N}]} \tag{3.11}$$

Using the definition of the unit vector in the direction of $\mathbf{R}_{12}$ or $\mathbf{R}_{21}$ as

$$\hat{\mathbf{R}}_{12} = \frac{\mathbf{R}_{12}}{|\mathbf{R}_{12}|} = \frac{\mathbf{R}_2 - \mathbf{R}_1}{|\mathbf{R}_2 - \mathbf{R}_1|} \quad \text{and} \quad \hat{\mathbf{R}}_{21} = \frac{\mathbf{R}_{21}}{|\mathbf{R}_{21}|} = \frac{\mathbf{R}_1 - \mathbf{R}_2}{|\mathbf{R}_1 - \mathbf{R}_2|} \tag{3.12}$$

we can also write

$$\boxed{\mathbf{F}_{12} = \frac{Q_1Q_2\mathbf{R}_{12}}{4\pi\varepsilon_0|\mathbf{R}_{12}|^3}, \qquad \mathbf{F}_{21} = \frac{Q_1Q_2\mathbf{R}_{21}}{4\pi\varepsilon_0|\mathbf{R}_{21}|^3}, \quad [\mathrm{N}]} \tag{3.13}$$

or, in terms of position vectors,

$$\boxed{\mathbf{F}_{12} = \frac{Q_1Q_2(\mathbf{R}_2 - \mathbf{R}_1)}{4\pi\varepsilon_0|\mathbf{R}_2 - \mathbf{R}_1|^3}, \qquad \mathbf{F}_{21} = \frac{Q_1Q_2(\mathbf{R}_1 - \mathbf{R}_2)}{4\pi\varepsilon_0|\mathbf{R}_1 - \mathbf{R}_2|^3}, \quad [\mathrm{N}]} \tag{3.14}$$

We have, therefore, three alternative methods of evaluating forces between two charges. **Equation (3.9)** gives only the magnitude of the force. **Equation (3.11)** gives both magnitude and direction using unit vectors, while **Eq. (3.13)** gives the force in terms of the vector connecting the two point charges and **Eq. (3.14)** gives the force in terms of position vectors. In most cases **Eq. (3.13)** or **(3.14)** is preferable since it gives both direction and magnitude and does not require explicit calculation of the unit vector. Note that these relations also give the correct forces in the case in **Figure 3.2c**, where a negative sign is obtained from the product Q_1Q_2. This negative sign indicates that the forces are in the directions opposite to those in **Eq. (3.13)** or **(3.14)**. From **Eqs. (3.11), (3.13),** and **(3.14)**, we also note that $\mathbf{F}_{12} = -\mathbf{F}_{21}$ as required.

Example 3.3 Direction of Forces Between Point Charges Two point charges, Q_1 [C] and Q_2 [C], are located at points $P_1(1,1,0)$ and $P_2(3,2,0)$ as shown in **Figure 3.3**:

(a) Calculate the force on Q_1 and Q_2 if $Q_1 = 2 \times 10^{-9}$ C and $Q_2 = 4 \times 10^{-9}$ C.
(b) Calculate the force on Q_1 and Q_2 if $Q_1 = 2 \times 10^{-9}$ C and $Q_2 = -4 \times 10^{-9}$ C.

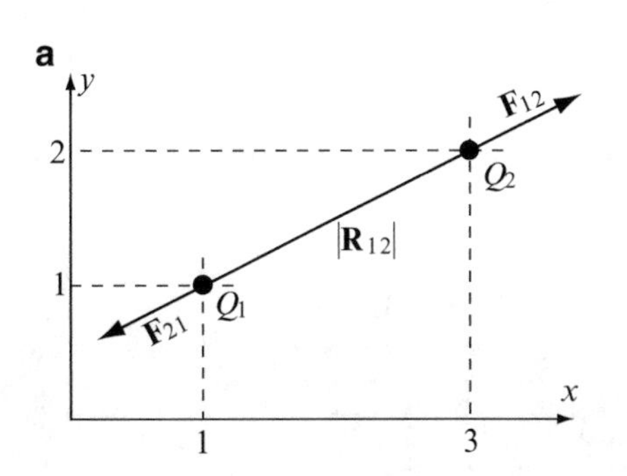

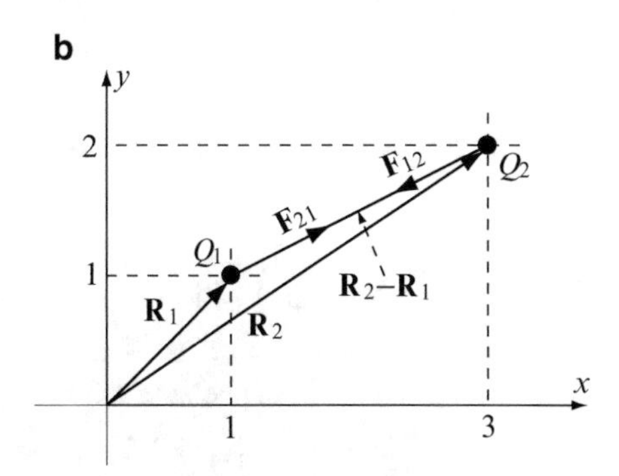

Figure 3.3 (**a**) Forces between two positive or two negative point charges. (**b**) Forces between a positive and a negative point charge

Solution: The force must be calculated as a vector using either **Eq. (3.11)**, **Eq. (3.13)**, or **Eq. (3.14)**. To demonstrate the use of the formulas, we will use **Eq. (3.11)** for **(a)** and **Eq. (3.14)** for **(b)**. In the first case, the two charges are positive and the forces are repulsion forces. In the second case, one charge is positive, the other negative, and the charges attract each other:

(a) Consider **Figure 3.3a**. The solution based on **Eq. (3.11)** is

$$\mathbf{F}_{12} = \hat{\mathbf{R}}_{12}\frac{Q_1Q_2}{4\pi\varepsilon_0 R_{12}^2}, \qquad \mathbf{F}_{21} = \hat{\mathbf{R}}_{21}\frac{Q_1Q_2}{4\pi\varepsilon_0 R_{21}^2} \quad [\mathrm{N}]$$

Calculation of the unit vector $\hat{\mathbf{R}}_{12}$ and $\hat{\mathbf{R}}_{21}$: First, we calculate the vector $\mathbf{R}_{12}$:

$$\mathbf{R}_{12} = \hat{\mathbf{x}}(x_2 - x_1) + \hat{\mathbf{y}}(y_2 - y_1) + \hat{\mathbf{z}}(z_2 - z_1) = \hat{\mathbf{x}}(3-1) + \hat{\mathbf{y}}(2-1) = \hat{\mathbf{x}}2 + \hat{\mathbf{y}}1$$

Therefore,

$$\mathbf{R}_{21} = -\mathbf{R}_{12} = -\hat{\mathbf{x}}2 - \hat{\mathbf{y}}1$$

The unit vectors are

$$\hat{\mathbf{R}}_{12} = \frac{\mathbf{R}_{12}}{|\mathbf{R}_{12}|} = \frac{\hat{\mathbf{x}}2 + \hat{\mathbf{y}}1}{\sqrt{5}}, \quad \hat{\mathbf{R}}_{21} = \frac{-\hat{\mathbf{x}}2 - \hat{\mathbf{y}}1}{\sqrt{5}}$$

Thus, the force on Q_1 (i.e., the force $\mathbf{F}_{21}$, exerted by charge Q_2 on Q_1) is

$$\mathbf{F}_{21} = \hat{\mathbf{R}}_{21}\frac{Q_1Q_2}{4\pi\varepsilon_0 \mathrm{R}_{21}^2} = \frac{-\hat{\mathbf{x}}2 - \hat{\mathbf{y}}1}{\sqrt{5}}\left(\frac{2\times10^{-9}\times4\times10^{-9}}{4\times\pi\times8.854\times10^{-12}\times5}\right) = -\hat{\mathbf{x}}12.86\times10^{-9} - \hat{\mathbf{y}}6.43\times10^{-9} \quad [\mathrm{N}]$$

From the fact that $\mathbf{F}_{12} = -\mathbf{F}_{21}$, we get the force on Q_2 as

$$\mathbf{F}_{12} = \hat{\mathbf{x}}12.86\times10^{-9} + \hat{\mathbf{y}}6.43\times10^{-9} \quad [\mathrm{N}]$$

These forces are indicated in their correct directions in **Figure 3.3a**.

(b) In this case, we use position vectors $\mathbf{R}_1$ and $\mathbf{R}_2$ and **Eq. (3.14)**. Position vectors $\mathbf{R}_1$ and $\mathbf{R}_2$ are (from **Figure 3.3b**)

$$\mathbf{R}_1 = \hat{\mathbf{x}}(x_1 - 0) + \hat{\mathbf{y}}(y_1 - 0) + \hat{\mathbf{z}}(0-0) = \hat{\mathbf{x}}1 + \hat{\mathbf{y}}1,$$
$$\mathbf{R}_2 = \hat{\mathbf{x}}(x_2 - 0) + \hat{\mathbf{y}}(y_2 - 0) + \hat{\mathbf{z}}(0-0) = \hat{\mathbf{x}}3 + \hat{\mathbf{y}}2$$

The force is now

$$\mathbf{F}_{12} = \frac{Q_1 Q_2(\hat{\mathbf{x}}3 + \hat{\mathbf{y}}2 - \hat{\mathbf{x}}1 - \hat{\mathbf{y}}1)}{4\pi\varepsilon_0|\hat{\mathbf{x}}3 + \hat{\mathbf{y}}2 - \hat{\mathbf{x}}1 - \hat{\mathbf{y}}1|^3} = \frac{\hat{\mathbf{x}}2 + \hat{\mathbf{y}}1}{5\sqrt{5}}\left(\frac{2 \times 10^{-9} \times (-4 \times 10^{-9})}{4 \times \pi \times 8.854 \times 10^{-12}}\right) = -\hat{\mathbf{x}}12.86 \times 10^{-9} - \hat{\mathbf{y}}6.43 \times 10^{-9} \quad [\text{N}]$$

and

$$\mathbf{F}_{21} = -\mathbf{F}_{12} = \hat{\mathbf{x}}12.86 \times 10^{-9} + \hat{\mathbf{y}}6.43 \times 10^{-9} \quad [\text{N}]$$

The forces now are opposite to those in part **(a)**, because Q_2 is negative. The two forces are indicated in **Figure 3.3b**.

Example 3.4 Application: Electrostatic Forces Within the Atom Consider the following model of the helium atom: The atom has two electrons and two protons; assume the electrons are stationary (which they are not) as shown in **Figure 3.4** and the two protons are located at a point. Given: mass of the electron, $m_e = 9.107 \times 10^{-31}$ kg; distance between nucleus and electron, $R_e = 0.5 \times 10^{-10}$ m; charge of electron, $e = -1.6 \times 10^{-19}$ C:

(a) Calculate the force between the two electrons and the force between each electron and the protons.
(b) Neglecting all other forces within the atom, what must the angular velocity of the electrons be to stay at the given distance from the nucleus, assuming the two electrons are always in the same relative position?

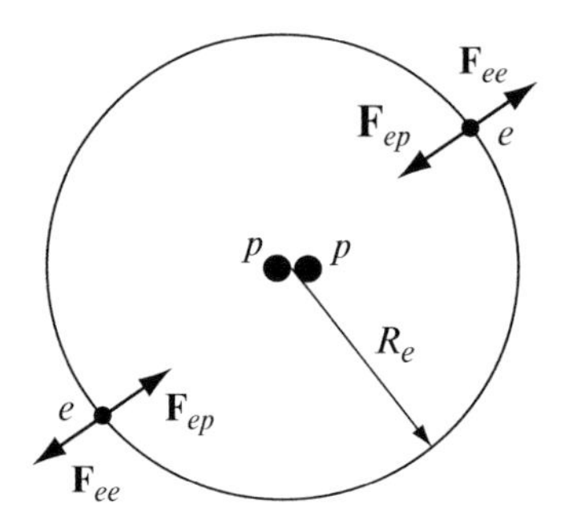

Figure 3.4 A simple model of the helium atom and the electrostatic forces involved

Solution: First, the repulsion forces between the electrons and attraction forces between electrons and protons are calculated. The net force is balanced by the centrifugal force due to orbiting of the electron. This gives the angular velocity required:

(a) The direction of the electrostatic force is radial. The two electrons repel each other, whereas the electrons are attracted to the protons. The repulsion forces between electrons are

$$\mathbf{F}_{ee} = \hat{\mathbf{R}}\frac{e^2}{4\pi\varepsilon_0(2R_e)^2} = \hat{\mathbf{R}}\frac{\left(-1.6 \times 10^{-19}\right)^2}{4 \times \pi \times 8.854 \times 10^{-12} \times \left(2 \times 0.5 \times 10^{-10}\right)^2} = \hat{\mathbf{R}}2.3 \times 10^{-8} \quad [\text{N}]$$

Attraction force between each electron and the nucleus is that between a negative electron and two positive protons. In **Figure 3.4**, this force is in the negative R direction and is indicated as F_{ep}:

$$\mathbf{F}_{ep} = -\hat{\mathbf{R}}\frac{2e^2}{4\pi\varepsilon_0(R_e)^2} = -\hat{\mathbf{R}}\frac{2 \times \left(-1.6 \times 10^{-19}\right)^2}{4 \times \pi \times 8.854 \times 10^{-12} \times \left(0.5 \times 10^{-10}\right)^2} = -\hat{\mathbf{R}}1.84 \times 10^{-7} \quad [\text{N}]$$

(b) For the electrons to remain in their orbit (based on the model used here, which does not consider atomic or gravitational forces), the net force on the electron must be balanced by the centrifugal force due to orbiting of the electron. The net electric force is an attraction force on each electron equal to $F_t = F_{ep} + F_{\text{ee}} = 1.61 \times 10^{-7}$ N.

The centrifugal force is mv^2/R, where v is the tangential speed of the electron. The centrifugal force is directed radially outward. Equating the two forces and substituting the values given yields

$$\frac{m_e v^2}{R_e} = F_t = F_{ep} + F_{ee} \quad \rightarrow \quad v = \sqrt{\frac{F_t R_e}{m_e}} = \sqrt{\frac{1.61 \times 10^{-7} \times 0.5 \times 10^{-10}}{9.107 \times 10^{-31}}} = 2.973 \times 10^6 \quad [\text{m/s}]$$

The angular velocity is found from the relation $v = \omega R$:

$$\omega = \frac{v}{R_e} = \frac{2.973 \times 10^6}{0.5 \times 10^{-10}} = 5.946 \times 10^{16} \quad [\text{rad/s}]$$

This translates to about 9.5×10^{15} orbits per second.

Note: There is much more going on in the atom than electrostatic forces, and the dimensions given here are merely assumed, but the calculation does give a flavor of the role electrostatic forces play.

3.4 The Electric Field Intensity

That there is a force acting on charges due to the presence of other charges is by now well understood. We also know how to calculate these forces for point charges. Now, we change our point of view slightly. Consider, for example, a point charge which we tie down so that it cannot be moved. Although nothing has changed in terms of the forces between the charges (remember: charges are assumed to be stationary), it is now more convenient to view the fixed charge as the source of the force acting on the second charge. Using **Figure 3.5**, the force on Q_2 due to Q_1 is

$$\mathbf{F}_{12} = \frac{\mathbf{R}_{12} Q_1 Q_2}{4\pi\varepsilon_0 |\mathbf{R}_{12}|^3} = \left(\frac{\mathbf{R}_{12} Q_1}{4\pi\varepsilon_0 |\mathbf{R}_{12}|^3} \right) Q_2 \quad [\text{N}] \tag{3.15}$$

If the two sides of the equation are divided by Q_2, we get

$$\frac{\mathbf{F}_{12}}{Q_2} = \frac{\mathbf{R}_{12} Q_1}{4\pi\varepsilon_0 |\mathbf{R}_{12}|^3} = \frac{\hat{\mathbf{R}}_{12} Q_1}{4\pi\varepsilon_0 |\mathbf{R}_{12}|^2} \quad \left[\frac{\text{N}}{\text{C}}\right] \tag{3.16}$$

Inspection of this relation reveals the following:

(1) The quantity obtained is force per unit charge [N/C].
(2) The force per unit charge varies as $1/R_{12}^2$.
(3) The right-hand side depends only on the fixed charge Q_1 and the vector $\mathbf{R}_{12}$; that is, the vector field $\mathbf{F}_{12}/Q_2$ is generated by Q_1.

Since the vector $\mathbf{R}_{12}$ is arbitrary (i.e., it simply indicates where a charge Q_2 might be located), it is defined everywhere in space. $\mathbf{F}_{12}/Q_2$ is a vector field that gives the force per unit charge anywhere in space. We call this quantity the ***electric field intensity***. In more formal fashion, the electric field intensity is defined as

$$\boxed{\mathbf{E} = \lim_{Q \to 0} \frac{\mathbf{F}}{Q} \quad \left[\frac{\text{N}}{\text{C}}\right]} \tag{3.17}$$

This indicates that the electric field intensity is a vector in the direction of the force and is proportional to the force. It gives the force per unit charge and has units of [N/C]. Another, more common unit for electric field intensity is ***volt/meter*** or [V/m]. In this chapter, we will use the [N/C] unit simply because the volt has not been defined yet. However, starting with **Chapter 4**, the [V/m] will be used exclusively.

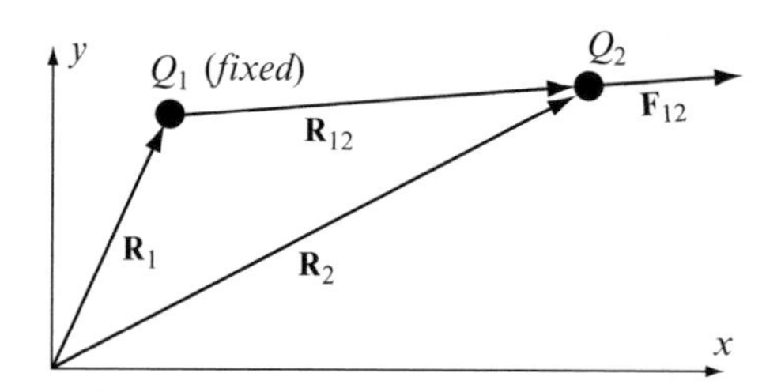

Figure 3.5 Force on charge Q_2 due to the electric field intensity produces by Q_1

Now, consider a point charge Q_1 as in **Figure 3.6a**. From **Eq. (3.13)** or **Eq. (3.11)**, the electric field intensity everywhere in space is equal to

$$\mathbf{E} = \mathbf{R}\frac{Q_1}{4\pi\varepsilon_0|\mathbf{R}|^3} \quad \text{or} \quad \mathbf{E} = \hat{\mathbf{R}}\frac{Q_1}{4\pi\varepsilon_0|\mathbf{R}|^2} \quad \left[\frac{\mathrm{N}}{\mathrm{C}}\right] \tag{3.18}$$

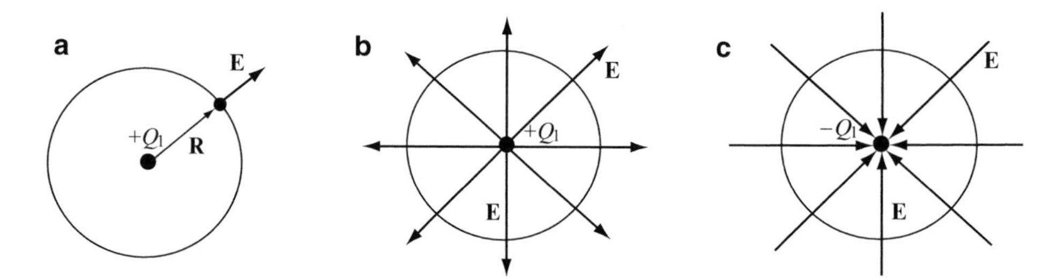

Figure 3.6 Electric field intensity due to a point charge. (**a**) Direction of the field of a positive point charge. (**b**) The electric field of a positive point charge. (**c**) The electric field of a negative point charge

A number of observations are in order here:

(1) The electric field intensity depends on the charge and the distance R from the charge.
(2) If the charge Q_1 is positive, the direction of the electric field intensity is from the point charge radially outward, since any point in space is connected with the charge along a radial line (**Figure 3.6b**).
(3) If the charge Q_1 is negative (**Figure 3.6c**), the direction of the electric field intensity is directed radially toward the charge.
(4) At equal distances from the charge, the magnitude of the electric field intensity is constant. This can be seen from the formula or simply from symmetry considerations.
(5) The electric field intensity varies as $1/R^2$ where R is the distance from the charge.

Conclusion The electric field always points away from a positive charge and toward a negative charge. The electric field can be viewed as starting at a positive charge and ending at a negative charge. This is an important point: It will provide us with a reference for electric fields and forces. Based on this simple fact, we can always determine (with or without calculations) the directions of forces and of electric fields.

If we now introduce another point charge in the electric field of the first charge, there will be a force acting on this charge. The force is given as

$$\boxed{\mathbf{F} = Q_2\mathbf{E} \quad [\mathrm{N}]} \tag{3.19}$$

This is merely an alternative form of Coulomb's law as given in **Eq. (3.13)**. However, this form is conceptually different: Q_1 is viewed as the source of the electric field intensity and this electric field intensity exerts a force on Q_2. We must hasten to say that this is only a convenience. It is equally acceptable to view Q_2 as the source of the field and the force is then exerted on Q_1. This symmetry is again due to the fact that the magnitude of the force on each charge (but not the direction) is the same.

Example 3.5 Force Exerted by an Electron An electron is located at a point in space:

(a) Calculate the electric field intensity everywhere in space. The charge of the electron is e [C].
(b) Find the force the electron exerts on a dust particle, charged with a total charge of 3.2×10^{-19} C (two protons) and located at a distance R [m] from the electron.

Solution: The electron is a negative point charge. Assuming it is located at the origin of a spherical system of coordinates, the electric field intensity is radial as in **Figure 3.6c**:

(a) The electric field intensity at a distance R from an electron is

$$\mathbf{E} = \hat{\mathbf{R}}\frac{e}{4\pi\varepsilon_0 R^2} = -\hat{\mathbf{R}}\frac{1.6 \times 10^{-19}}{4 \times \pi \times 8.854 \times 10^{-12} \times R^2} = -\hat{\mathbf{R}}\frac{1.44 \times 10^{-9}}{R^2} \quad \left[\frac{\text{N}}{\text{C}}\right]$$

(b) The force on the dust particle is calculated from **Eq. (3.19)**:

$$\mathbf{F} = Q\mathbf{E} = -\hat{\mathbf{R}}\frac{3.2 \times 10^{-19} \times 1.44 \times 10^{-9}}{R^2} = -\hat{\mathbf{R}}\frac{4.61 \times 10^{-28}}{R^2} \quad [\text{N}]$$

Although this may seem to be a small force, it can be considerable at very short distances. In the limit, as R tends to zero, the force and the electric field intensity tend to infinity.

3.4.1 Electric Fields of Point Charges **Point_Charges.m**

3.4.1.1 Superposition of Electric Fields

Before proceeding, it is well to establish the fact that superposition applies in the case of the electric field intensity. We might, of course, have suspected that it does since all relations discussed so far were linear. In our case, superposition means that the field of a charge is unaffected by the existence of other charges or by the electric fields these charges generate. For this reason, the electric field intensity at a point in space due to a number of charges is the ***vector sum*** of the electric field intensities of individual charges, each calculated in the absence of all other charges. Considering two charges Q_1 and Q_2, located at points P_1 and P_2 as shown in **Figure 3.7**, the electric field intensity at a point P_3 is calculated as

$$\mathbf{E}_3 = \hat{\mathbf{R}}_{13}\frac{Q_1}{4\pi\varepsilon_0|\mathbf{R}_{13}|^2} + \hat{\mathbf{R}}_{23}\frac{Q_2}{4\pi\varepsilon_0|\mathbf{R}_{23}|^2} \quad \left[\frac{\text{N}}{\text{C}}\right] \tag{3.20}$$

and the individual electric field intensities are those of each charge, calculated as if the other charge does not exist. These two electric field intensities have different directions in space and must be added vectorially.

The same principle applies to forces. Placing a third charge Q_3 at point P_3, the force on this charge is

$$\mathbf{F}_3 = Q_3\mathbf{E}_3 = \hat{\mathbf{R}}_{13}\frac{Q_1Q_3}{4\pi\varepsilon_0|\mathbf{R}_{13}|^2} + \hat{\mathbf{R}}_{23}\frac{Q_2Q_3}{4\pi\varepsilon_0|\mathbf{R}_{23}|^2} \quad [\text{N}] \tag{3.21}$$

as expected. Note that in addition to this force, which the electric field of charges Q_1 and Q_2 exert on charge Q_3, there is also a force between charges Q_1 and Q_2 according to Coulomb's law. This force between Q_1 and Q_2 does not affect the force on Q_3. The expressions above can now be generalized for any number of point charges: Assuming n point charges, the electric field intensity at a general point P_k is

$$\mathbf{E}_k = \sum_{\substack{i=1\\ i\neq k}}^{n}\frac{Q_i(\mathbf{R}_k - \mathbf{R}_i)}{4\pi\varepsilon_0|\mathbf{R}_k - \mathbf{R}_i|^3} = \sum_{\substack{i=1\\ i\neq k}}^{n}\hat{\mathbf{R}}_{ik}\frac{Q_i}{4\pi\varepsilon_0|\mathbf{R}_{ik}|^2} \quad \left[\frac{\text{N}}{\text{C}}\right] \tag{3.22}$$

The expression for force is

$$\mathbf{F}_k = Q_k\mathbf{E}_k = Q_k\sum_{\substack{i=1\\ i\neq k}}^{n}\frac{Q_i(\mathbf{R}_k - \mathbf{R}_i)}{4\pi\varepsilon_0|\mathbf{R}_k - \mathbf{R}_i|^3} = Q_k\sum_{\substack{i=1\\ i\neq k}}^{n}\hat{\mathbf{R}}_{ik}\frac{Q_i}{4\pi\varepsilon_0|\mathbf{R}_{ik}|^2} \quad [\text{N}] \tag{3.23}$$

When using this form of the field of multiple charges, it is important to remember that this is a summation of vectors. Each electric field intensity $\mathbf{E}_i$ or electric force $\mathbf{F}_i$ is directed in a different direction in space as indicated by its unit vector. If a total vector field is required, the components of the fields must be calculated and added to produce a single vector field.

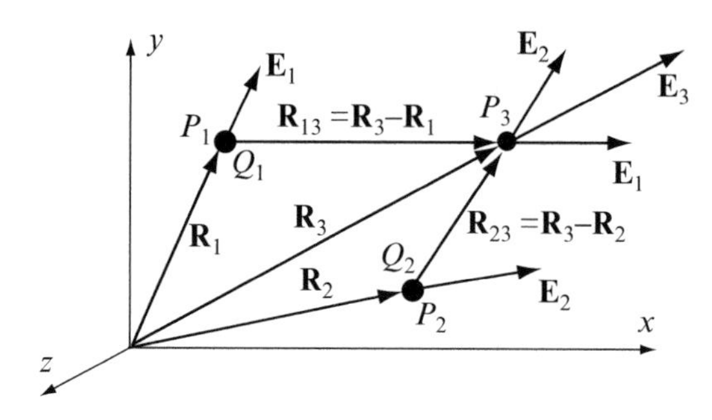

Figure 3.7 Electric field intensity at a general point in space, due to two point charges

3.4.1.2 Electric Field Lines

In an attempt to visualize the electric field, it is customary to draw the electric field intensity in terms of field lines. These are imaginary lines that show the direction of force on an infinitesimal positive point charge if it were placed in the field. The electric field intensity is everywhere tangential to field lines. Field lines can also be called force lines. Plots of field lines are quite useful in describing, qualitatively, the behavior of the electric field and of charges in the electric field. The electric field intensity of the point charge in **Figures 3.6b** and **3.6c** shows the electric field lines for a positive and a negative point charge. Similar sketches of more complicated field distributions help in understanding the field distribution in space. For example, **Figure 3.8** shows the field lines of two equal but opposite point charges. The following should be noted from this description of the field:

(1) Field lines begin at positive charges and end in negative charges. If only one type of charge exists, the lines start or end at infinity (**Figures 3.6b** and **3.6c**).
(2) Field lines show the direction of force on a positive point charge if it were placed in the field and, therefore, also show the direction of the electric field intensity. The arrows help in showing the direction of force and field.
(3) Field lines are imaginary lines; their only purpose is to visualize the electric field.

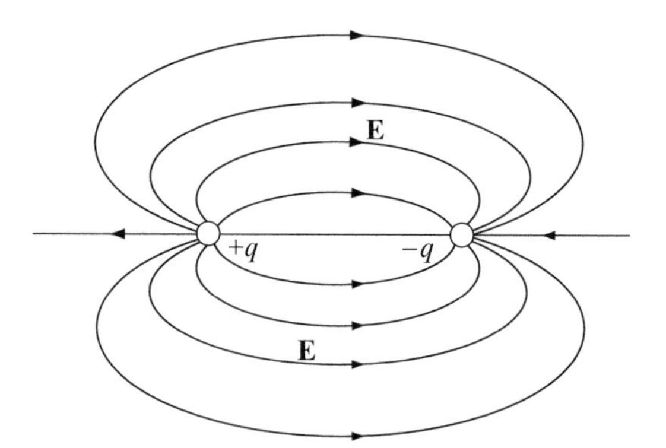

Figure 3.8 Field line representation of the electric field of two point charges

Example 3.6 Forces in a System of Charges Three equal point charges Q [C] are located as shown in **Figure 3.9a**. Each two charges are connected with a very thin string to hold them in place. The string is designed to break when a force of 0.1 N is applied.

(a) Calculate the charge Q required to break the strings, if $a = 20$ mm.
(b) What is the electric field intensity at the center of the string A–B?

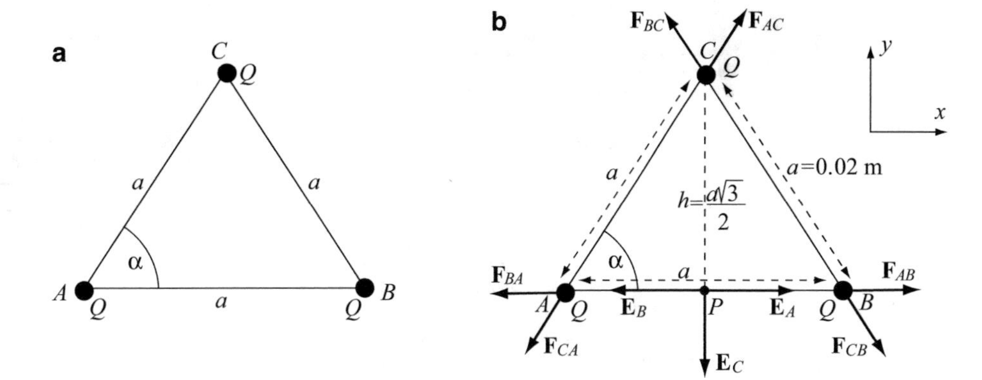

Figure 3.9 (**a**) Three point charges connected by strings. (**b**) Forces on the charges and the electric field intensity at P

Solution: Each point charge applies a force on each of the other point charges. There are, therefore, six forces, two at each vertex of the triangle. The tension on a string is due to all forces acting along the string. These are shown in **Figure 3.9b**. The electric field intensity at the center of string A–B is due to the charge at C only since charges at A and B produce electric field intensities in opposite directions at point P and they cancel each other:

(**a**) The force on any string is composed of two parts: One is the force which acts along the string and the other is the projection of the second force onto the direction of the first. For example, the total force (magnitude) on the string connecting charge A and B is $F_{BA} + F_{CA}\cos\alpha$. This force acts along the string as shown. The magnitude of any of the forces $\mathbf{F}_{AB}$, $\mathbf{F}_{BA}$, $\mathbf{F}_{AC}$, $\mathbf{F}_{CA}$, $\mathbf{F}_{BC}$, and $\mathbf{F}_{CB}$ is

$$\left|\mathbf{F}_{ij}\right| = \frac{Q_i Q_j}{4\pi\varepsilon_0 \left|\mathbf{R}_{ij}\right|^2} = \frac{Q^2}{4\pi\varepsilon_0 a^2} \quad [\text{N}]$$

The force acting on any string is therefore

$$F_t = F_{ij} + F_{ij}\cos\alpha = \frac{Q^2}{4\pi\varepsilon_0 a^2}\left(1+\cos 60^\circ\right) = \frac{3Q^2}{8\pi\varepsilon_0 a^2} \quad [\text{N}]$$

For the string to break, this force must be equal or larger than 0.1 N:

$$\frac{3Q^2}{8\pi\varepsilon_0 a^2} \geq 0.1 \quad [\text{N}] \quad \rightarrow \quad Q \geq a\sqrt{\frac{0.8\pi\varepsilon_0}{3}} = 5.45\times 10^{-8} \quad [\text{C}]$$

(**b**) The electric field intensity at point P is in the negative y direction (away from charge C), whereas the electric field intensities E_A and E_B due to charges A and B cancel as shown in **Figure 3.9b**:

$$\mathbf{E}_C = -\hat{\mathbf{y}}\frac{Q}{4\pi\varepsilon_0 h^2} = -\hat{\mathbf{y}}\frac{Q}{3\pi\varepsilon_0 a^2} \quad \left[\frac{\text{N}}{\text{C}}\right]$$

Example 3.7 Electric Field Due to a System of Charges Three point charges are arranged as shown in **Figure 3.10a**:

(**a**) Calculate the electric field intensity everywhere on the x axis.
(**b**) What are the points on the axis at which the electric field intensity is zero (other than at infinity)?

Solution: The electric field intensity is the superposition of the electric field intensities of the three charges. To find the location of zero electric field intensity, we assume a location on the axis, say $+x$, calculate the electric field intensity, and set it to zero. However, there are four distinct domains that must be considered: Two are $x > a$ and $x < -a$. The other two domains are $0 < x < a$ and $-a < x < 0$. Because the configuration of charges is symmetric, only two of these four domains need to be considered. Therefore, we solve for the domains $x > a$ and $0 < x < a$:

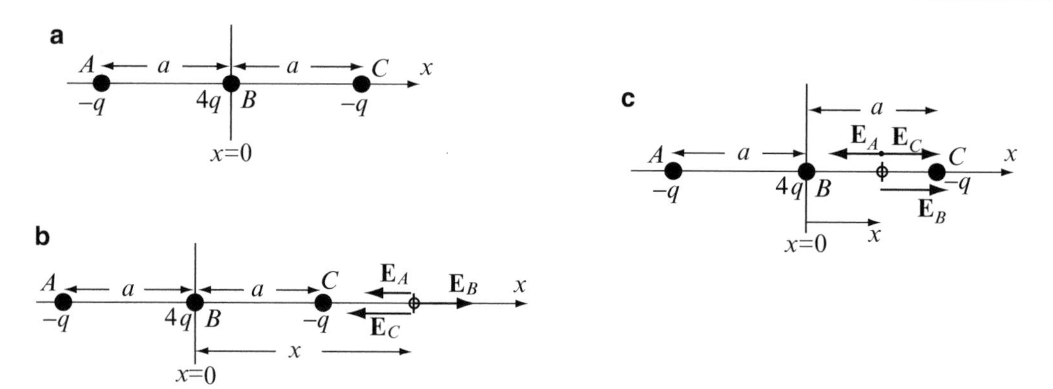

Figure 3.10 (**a**) Three point charges on a line. (**b**) The electric field intensity at $x > a$. (**c**) The electric field intensity at $x < a$

(**a**) Consider a point at a distance x from the charge $+4q$ such that $x > a$. Since the three charges are on the axis, the electric field will also be directed along the axis. The electric field intensity for any value of $x > a$ is (see **Figure 3.10b**)

$$\mathbf{E} = \hat{\mathbf{x}}\left(\frac{4q}{4\pi\varepsilon x^2} - \frac{q}{4\pi\varepsilon(x-a)^2} - \frac{q}{4\pi\varepsilon(x+a)^2}\right), \quad x > a \quad \left[\frac{\mathrm{N}}{\mathrm{C}}\right]$$

From symmetry considerations, the electric field intensity for $x < -a$ is

$$\mathbf{E} = -\hat{\mathbf{x}}\left(\frac{4q}{4\pi\varepsilon x^2} - \frac{q}{4\pi\varepsilon(x-a)^2} - \frac{q}{4\pi\varepsilon(x+a)^2}\right), \quad x < -a \quad \left[\frac{\mathrm{N}}{\mathrm{C}}\right]$$

For a point between $x = 0$ and $x = a$, the configuration is as in **Figure 3.10c**. The electric field intensity is

$$\mathbf{E} = \hat{\mathbf{x}}\left(\frac{4q}{4\pi\varepsilon x^2} + \frac{q}{4\pi\varepsilon(a-x)^2} - \frac{q}{4\pi\varepsilon(x+a)^2}\right), \quad 0 < x < a \quad \left[\frac{\mathrm{N}}{\mathrm{C}}\right]$$

and for $-a < x < 0$,

$$\mathbf{E} = -\hat{\mathbf{x}}\left(\frac{4q}{4\pi\varepsilon x^2} + \frac{q}{4\pi\varepsilon(a-x)^2} - \frac{q}{4\pi\varepsilon(x+a)^2}\right), \quad -a < x < 0 \quad \left[\frac{\mathrm{N}}{\mathrm{C}}\right]$$

(**b**) To find the points at which the electric field intensity is zero, we set the fields in (**a**) to zero and find a solution for x. For $x > a$: Taking only the magnitude of **E**, we get

$$E = \frac{4q}{4\pi\varepsilon x^2} - \frac{q}{4\pi\varepsilon(x-a)^2} - \frac{q}{4\pi\varepsilon(x+a)^2} = 0 \quad \rightarrow \quad \frac{4}{x^2} - \frac{1}{(x-a)^2} - \frac{1}{(x+a)^2} = 0$$

or

$$\frac{4}{x^2} - \frac{2(x^2+a^2)}{(x^2-a^2)^2} = 0 \quad \rightarrow \quad (x^2)^2 - 5a^2(x^2) + 2a^4 = 0$$

This is a quadratic equation in x^2: Solving for x^2 and then taking the square root to find x gives

$$x = \pm 2.13578a \quad \text{or} \quad x = \pm 0.66215a \quad [\mathrm{m}]$$

The last two solutions are not valid (because we assumed $x > a$). Thus, the first two solutions are correct and the electric field intensity is zero at $x = \pm 2.13578a$.

For $0 < x < a$: Following steps similar to the previous case:

$$\frac{4q}{4\pi\varepsilon x^2} + \frac{q}{4\pi\varepsilon(a-x)^2} - \frac{q}{4\pi\varepsilon(x+a)^2} = 0 \quad \rightarrow \quad x^4 + ax^3 - 2a^2x^2 + a^4 = 0$$

The two real solutions (the other two are imaginary) to this equation are $x = -0.671a$ [m] and $x = -1.905a$ [m]. Since neither is in the required domain (and the other two solutions are imaginary), the electric field intensity cannot be zero in the given domain. The solution in the domain $-a < x < 0$ leads to similar conclusions. Thus, the only locations at which the electric field intensity is zero are $x = \pm 2.13578a$ [m].

Note: That the electric field intensity cannot be zero between $x = 0$ and $x = a$ (or $x = 0$ and $x = -$a) can be seen from the fact that, for example, E_A must counter $E_B + E_C$ (**Figure 3.10c**), both of which are larger than E_A.

3.4.1.3 The Electric Dipole

A configuration of practical importance is that of two point charges, separated a very short distance apart as shown in **Figure 3.11a**. If, in addition, the electric field intensity at a distance $R \gg d$ is needed, the configuration is called an ***electric dipole***. The electric dipole is often encountered when dealing with electric fields of atoms and since these fields are fundamental in our attempt to understand the behavior of dielectrics, we shall now discuss the electric field intensity of the electric dipole.

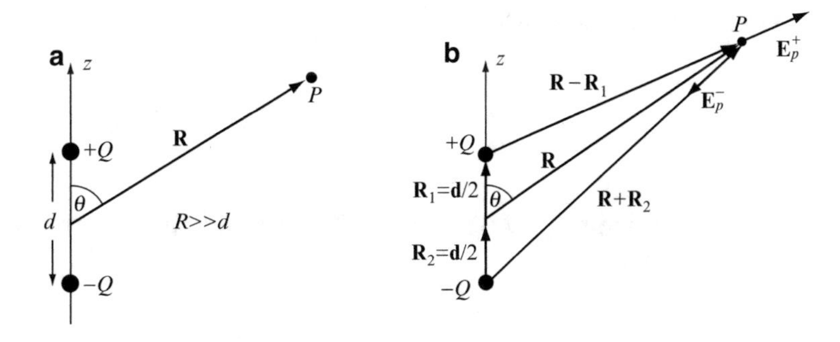

Figure 3.11 The electric dipole. (**a**) Configuration. (**b**) The electric field intensity at P

The electric field intensities of the negative and positive point charges are calculated using the position vectors indicated in **Figure 3.11b**:

$$\mathbf{E}_p^- = -\frac{Q(\mathbf{R}+\mathbf{R}_2)}{4\pi\varepsilon_0|\mathbf{R}+\mathbf{R}_2|^3} = -\frac{Q(\mathbf{R}+\mathbf{d}/2)}{4\pi\varepsilon_0|\mathbf{R}+\mathbf{d}/2|^3} \quad \left[\frac{\text{N}}{\text{C}}\right] \tag{3.24}$$

$$\mathbf{E}_p^+ = \frac{Q(\mathbf{R}-\mathbf{R}_1)}{4\pi\varepsilon_0|\mathbf{R}-\mathbf{R}_1|^3} = \frac{Q(\mathbf{R}-\mathbf{d}/2)}{4\pi\varepsilon_0|\mathbf{R}-\mathbf{d}/2|^3} \quad \left[\frac{\text{N}}{\text{C}}\right] \tag{3.25}$$

These two field intensities are in different directions in space (see **Figure 3.11b**). Their vector sum is

$$\mathbf{E}_d = \mathbf{E}_p^+ + \mathbf{E}_p^- = \frac{Q}{4\pi\varepsilon_0}\left(\frac{(\mathbf{R}-\mathbf{d}/2)}{|\mathbf{R}-\mathbf{d}/2|^3} - \frac{(\mathbf{R}+\mathbf{d}/2)}{|\mathbf{R}+\mathbf{d}/2|^3}\right) \quad \left[\frac{\text{N}}{\text{C}}\right] \tag{3.26}$$

Although **Eq. (3.26)** is exact, we can obtain an approximate, simplified form which is often more useful, by using the fact that $R \gg d$, together with the use of the binomial expansion. First we write

$$\frac{1}{|\mathbf{R}+\mathbf{d}/2|^3} = \left[\left(\mathbf{R}+\frac{\mathbf{d}}{2}\right)\cdot\left(\mathbf{R}+\frac{\mathbf{d}}{2}\right)\right]^{-3/2} = \left[R^2+\mathbf{R}\cdot\mathbf{d}+\frac{d^2}{4}\right]^{-3/2} \tag{3.27}$$

$$\frac{1}{|\mathbf{R}-\mathbf{d}/2|^3} = \left[\left(\mathbf{R}-\frac{\mathbf{d}}{2}\right)\cdot\left(\mathbf{R}-\frac{\mathbf{d}}{2}\right)\right]^{-3/2} = \left[R^2-\mathbf{R}\cdot\mathbf{d}+\frac{d^2}{4}\right]^{-3/2} \tag{3.28}$$

Taking the term $(R^2)^{-3/2}$ outside the brackets and neglecting the term $d^2/4$ (because $d \ll R$),

$$\left|\mathbf{R}+\frac{\mathbf{d}}{2}\right|^{-3} \approx R^{-3}\left[1+\frac{\mathbf{R}\cdot\mathbf{d}}{R^2}\right]^{-3/2} \quad \text{and} \quad \left|\mathbf{R}-\frac{\mathbf{d}}{2}\right|^{-3} \approx R^{-3}\left[1-\frac{\mathbf{R}\cdot\mathbf{d}}{R^2}\right]^{-3/2} \tag{3.29}$$

The binomial expansion states

$$(1+x)^n = 1+nx+\frac{n(n-1)x^2}{2!}+\frac{n(n-1)(n-2)x^3}{3!}+\cdots+\frac{n(n-1)(n-2)\cdots(n-k)x^k}{k!}+\cdots, |x|<1, n \text{ is real} \tag{3.30}$$

Using $x = \mathbf{R}\cdot\mathbf{d}/R^2$, $n = -3/2$, and neglecting all terms with orders of x larger than 1 in the expression for $|\mathbf{R} + \mathbf{d}/2|^{-3}$ (x is small because $d \ll R$), we get

$$\left|\mathbf{R}+\frac{\mathbf{d}}{2}\right|^{-3} \approx R^{-3}\left(1-\frac{3}{2}\frac{\mathbf{R}\cdot\mathbf{d}}{R^2}\right) \tag{3.31}$$

Similarly, using $x = -\mathbf{R}\cdot\mathbf{d}/\mathrm{R}^2$, $n = -3/2$, and neglecting all terms with orders of x larger than 1 in the expression for $|\mathbf{R} - \mathbf{d}/2|^{-3}$ gives

$$\left|\mathbf{R}-\frac{\mathbf{d}}{2}\right|^{-3} \approx R^{-3}\left(1+\frac{3}{2}\frac{\mathbf{R}\cdot\mathbf{d}}{R^2}\right) \tag{3.32}$$

Substituting the approximations in **Eqs. (3.31)** and **(3.32)** into **Eq. (3.26)**, we get

$$\mathbf{E}_d \approx \frac{1}{4\pi\varepsilon_0 R^3}\left[\left(\mathbf{R}-\frac{\mathbf{d}}{2}\right)\left(1+\frac{3}{2}\frac{\mathbf{R}\cdot\mathbf{d}}{R^2}\right)-\left(\mathbf{R}+\frac{\mathbf{d}}{2}\right)\left(1-\frac{3}{2}\frac{\mathbf{R}\cdot\mathbf{d}}{R^2}\right)\right] = \frac{Q}{4\pi\varepsilon_0 R^3}\left(3\frac{\mathbf{R}\cdot\mathbf{d}}{R^2}\mathbf{R}-\mathbf{d}\right) \quad \left[\frac{\mathrm{N}}{\mathrm{C}}\right] \tag{3.33}$$

The common terms in this expression are Q and $\mathbf{d}$. For convenience we define a new vector:

$$\mathbf{p} = Q\mathbf{d} \qquad [\mathrm{C}\cdot\mathrm{m}] \tag{3.34}$$

which we call the ***electric dipole moment***. With this, the electric field intensity may be written as

$$\mathbf{E}_d \approx \frac{1}{4\pi\varepsilon_0 R^3}\left(3\frac{(\mathbf{R}\cdot\mathbf{p})}{R^2}\mathbf{R}-\mathbf{p}\right) \quad \left[\frac{\mathrm{N}}{\mathrm{C}}\right] \tag{3.35}$$

In the configuration in **Figure 3.11b**, the dipole is along the z axis, and, therefore, the electric dipole moment is in the z direction ($\mathbf{p} = \hat{\mathbf{z}}p = \hat{\mathbf{z}}Qd$). If we transform this into spherical coordinates [see **Section 1.5.3**, and **Eq. (2.45)**], we get

$$\hat{\mathbf{z}} = \hat{\mathbf{R}}\cos\theta - \hat{\boldsymbol{\theta}}\sin\theta \quad \rightarrow \quad \mathbf{p} = p\left(\hat{\mathbf{R}}\cos\theta - \hat{\boldsymbol{\theta}}\sin\theta\right) \tag{3.36}$$

and

$$\mathbf{R}\cdot\mathbf{p} = \left[\hat{\mathbf{R}}R\right]\cdot\left[p\left(\hat{\mathbf{R}}\cos\theta - \hat{\boldsymbol{\theta}}\sin\theta\right)\right] = Rp\cos\theta \tag{3.37}$$

Substituting these into the expression for the electric field intensity of the dipole gives

$$\mathbf{E}_d \approx \frac{1}{4\pi\varepsilon_0 R^3}\left[3\frac{Rp\cos\theta}{R^2}\hat{\mathbf{R}}R - p\left(\hat{\mathbf{R}}\cos\theta - \hat{\boldsymbol{\theta}}\sin\theta\right)\right] = \frac{p}{4\pi\varepsilon_0 R^3}\left(\hat{\mathbf{R}}2\cos\theta + \hat{\boldsymbol{\theta}}\sin\theta\right) \quad [\mathrm{N/C}] \tag{3.38}$$

Note that the electric field intensity of a dipole varies as $1/R^3$, in contrast to that of a point charge, which varies as $1/R^2$. A plot of the electric field intensity of the dipole is shown in **Figure 3.12**. Note, in particular, that all field lines are closed through the charges and that the field distribution is symmetric about the dipole axis.

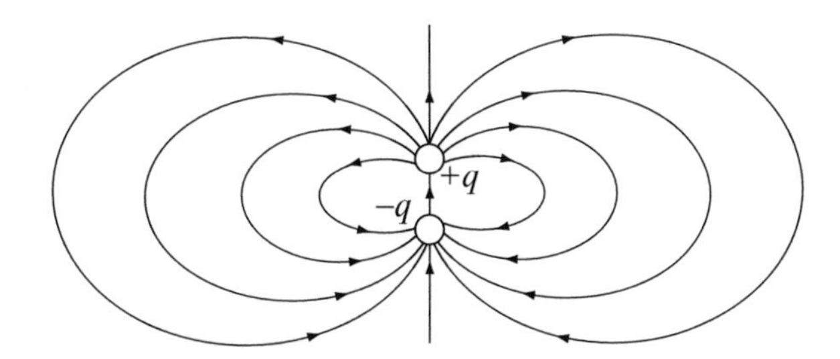

Figure 3.12 The electric field intensity of an electric dipole

Exercise 3.1 Calculate the electric field intensity of a configuration as in **Figure 3.11a** but with the positive charge twice as large as the negative charge.

Answer

$$\mathbf{E}_d \approx \frac{1}{4\pi\varepsilon_0 R^3}\left[\hat{\mathbf{R}}\left(RQ + 3Qd\cos\theta - \frac{3Qd^2\cos^2\theta}{4R}\right) + \hat{\boldsymbol{\theta}}\left(\frac{3}{2}Qd\sin\theta + \frac{3Qd^2\sin\theta\cos\theta}{4R}\right)\right] \quad \left[\frac{\mathrm{N}}{\mathrm{C}}\right]$$

3.4.2 Electric Fields of Charge Distributions

As mentioned above, point charges are only one type of charge possible. Line, surface, and volume charge distributions are also quite commonly encountered, and the question that arises naturally is how do we calculate electric fields and forces due to distributed charges. Charge distributions were defined in **Section 3.2** as charges spread over a given domain such as a volume, a surface, or a line. We may argue that all charge distributions are composed of point charges, since charges exist in multiples of the charge of electrons (or protons). However, the charge of the electron is so small that for practical purposes, we can view a charge distribution as a continuous distribution.

To treat charge distributions and the electric fields they produce, we will use the ideas of point charges and superposition; a differential point charge is defined as an elemental point charge, and the contributions of all elemental charges are summed up to produce the net effect such as the electric field intensity or force due to a charge distribution.

3.4.2.1 Line Charge Distributions

Consider **Figure 3.13** in which a charge is distributed over a line. Designating a differential length dl' at a point (x',y',z'), the total charge on this element is $\rho_l dl'$.

This charge can be viewed as an equivalent point charge $dq' = \rho_l dl'$. The electric field intensity at a point in space due to this elemental point charge is

$$d\mathbf{E} = \frac{(\mathbf{r} - \mathbf{r}')\rho_l dl'}{4\pi\varepsilon_0|\mathbf{r} - \mathbf{r}'|^3} \quad \left[\frac{\text{N}}{\text{C}}\right] \tag{3.39}$$

where $\mathbf{R} = \mathbf{r} - \mathbf{r}'$ is the vector connecting the point charge with the point at which the electric field intensity is required. This is integrated along the line of charge to obtain the field due to a segment of the line or due to the whole line

$$\boxed{\mathbf{E} = \int_{l'} \frac{(\mathbf{r} - \mathbf{r}')\rho_l dl'}{4\pi\varepsilon_0|\mathbf{r} - \mathbf{r}'|^3} \quad \left[\frac{\text{N}}{\text{C}}\right]} \tag{3.40}$$

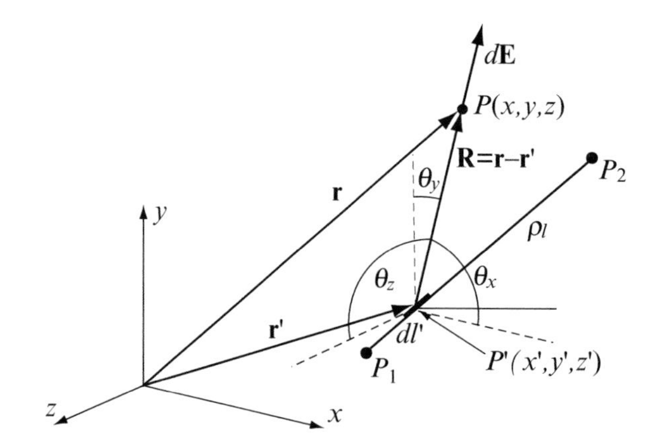

Figure 3.13 Electric field intensity due to a charged line element

The direction of the electric field intensity is in the direction of $\mathbf{r} - \mathbf{r}'$, as shown in **Figure 3.13**. Since as we integrate along the line, this direction changes, it is easier to separate the electric field intensity into its three components using the three angles indicated in the figure. To do so, we note that the angles that $d\mathbf{E}$ makes with the three coordinates are the same as those made by the vector $\mathbf{r} - \mathbf{r}'$. The latter are written from **Figure 3.13**:

$$\frac{(dE_x)}{(dE)} = \frac{(\mathbf{r} - \mathbf{r}')_x}{|\mathbf{r} - \mathbf{r}'|} = \cos\theta_x, \quad \frac{(dE_y)}{(dE)} = \frac{(\mathbf{r} - \mathbf{r}')_y}{|\mathbf{r} - \mathbf{r}'|} = \cos\theta_y, \quad \frac{(dE_z)}{(dE)} = \frac{(\mathbf{r} - \mathbf{r}')_z}{|\mathbf{r} - \mathbf{r}'|} = \cos\theta_z, \tag{3.41}$$

where $(\mathbf{r} - \mathbf{r}')_x$, $(\mathbf{r} - \mathbf{r}')_y$, and $(\mathbf{r} - \mathbf{r}')_z$ are the scalar components of the vector $\mathbf{r} - \mathbf{r}'$ in the x, y, and z directions, respectively. With $dE_x = dE\cos\theta_x$, $dE_y = dE\cos\theta_y$, $dE_z = dE\cos\theta_z$, and $|\mathbf{r} - \mathbf{r}'| = \sqrt{(x - x')^2 + (y - y')^2 + (z - z')^2}$, the electric field intensity at point $P(x,y,z)$ is

$$\begin{aligned} d\mathbf{E} &= \hat{\mathbf{x}}dE\cos\theta_x + \hat{\mathbf{y}}dE\cos\theta_y + \hat{\mathbf{z}}dE\cos\theta_z \\ &= \hat{\mathbf{x}}dE\frac{x - x'}{\sqrt{(x - x')^2 + (y - y')^2 + (z - z')^2}} + \hat{\mathbf{y}}dE\frac{y - y'}{\sqrt{(x - x')^2 + (y - y')^2 + (z - z')^2}} \\ &+ \hat{\mathbf{z}}dE\frac{z - z'}{\sqrt{(x - x')^2 + (y - y')^2 + (z - z')^2}} \quad \left[\frac{\text{N}}{\text{C}}\right] \end{aligned} \tag{3.42}$$

Now, from the relation for $d\mathbf{E}$ from **Eq. (3.39)**, we get

$$dE = |d\mathbf{E}| = \frac{\rho_l dl'}{4\pi\varepsilon_0|\mathbf{r} - \mathbf{r}'|^2} = \frac{\rho_l dl'}{4\pi\varepsilon_0\left((x-x')^2 + (y-y')^2 + (z-z')^2\right)} \quad \left[\frac{\text{N}}{\text{C}}\right] \tag{3.43}$$

Substitution of this in **Eq. (3.42)** and integration over the length of the line gives

$$\begin{aligned}\mathbf{E} = \hat{\mathbf{x}}\frac{1}{4\pi\varepsilon_0}\int_{P_1}^{P_2}\frac{\rho_l(x-x')dl'}{\left((x-x')^2+(y-y')^2+(z-z')^2\right)^{3/2}} + \hat{\mathbf{y}}\frac{1}{4\pi\varepsilon_0}\int_{P_1}^{P_2}\frac{\rho_l(y-y')dl'}{\left((x-x')^2+(y-y')^2+(z-z')^2\right)^{3/2}} \\ + \hat{\mathbf{z}}\frac{1}{4\pi\varepsilon_0}\int_{P_1}^{P_2}\frac{\rho_l(x-x')dl'}{\left((x-x')^2+(y-y')^2+(z-z')^2\right)^{3/2}} \quad \left[\frac{\text{N}}{\text{C}}\right]\end{aligned} \tag{3.44}$$

This is a rather lengthy but simple expression. It indicates that to obtain each component of the vector, we must integrate along the line. Note also that the points P_1 and P_2 are general and that the same type of result can be obtained in any system of coordinates. The charge density was left inside the integral to indicate thet it can be coordinate dependent.

Important Note: Throughout the derivation in this section, we used primed coordinates for the location of the element of charge dq. This is called the ***source point***. Unprimed coordinates are used for the location at which the field is calculated. This is called the ***field point***. This distinction between source and field points will be followed throughout the book. In practical terms, the integration required to find the field (in this case the electric field intensity) is on the primed coordinates, whereas the field point coordinates remain constant.

Example 3.8 Electric Field Intensity Due to a Charged Line Segment A thin line segment is 2 m long and charged with a uniform line charge density ρ_l [C/m]:

(a) Find the electric field intensity at a distance $a = 1$ m from the center of the segment (**Figure 3.14a**).
(b) Find the electric field intensity at a distance a from the line if the line is infinite in length.

Solution: The segment is placed in a system of coordinates as shown in **Figure 3.14b**. The cylindrical system shown is chosen because of the cylindrical nature of the segment. We set up an element of length dz' and, therefore, an equivalent point charge $dq = \rho_l dz'$ at z' along the segment and calculate the distance between the element of length dz' and point P in terms of the coordinate z' and the known distance to point P. Integrating along the line from $z' = -1$ to $z' = +1$ gives the result:

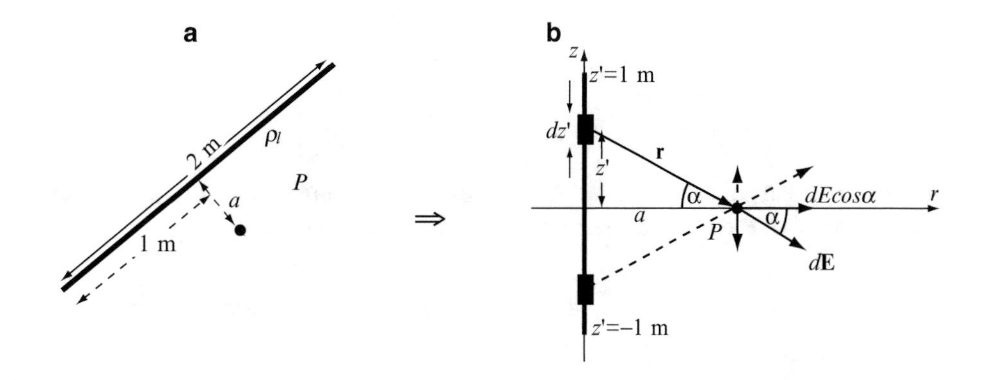

Figure 3.14 (**a**) A charged segment in space. (**b**) The same segment in a cylindrical system of coordinates

(a) The distance from a general point z' on the line and point P is $|\mathbf{r}| = \sqrt{a^2 + z'^2}$. Because of symmetry, the only field component is in the r direction (the z components from the lower and upper halves of the line cancel each other as shown in **Figure 3.14b**). Using $\cos\alpha = a/|\mathbf{r}|$,

$$\mathbf{E} = \hat{\mathbf{r}}\int_{z'=-1}^{z'=+1} \frac{\rho_l \cos\alpha dz'}{4\pi\varepsilon_0(a^2 + z'^2)} = \hat{\mathbf{r}}\int_{z'=-1}^{z'=+1} \frac{\rho_l a dz'}{4\pi\varepsilon_0(a^2 + z'^2)^{3/2}} = \hat{\mathbf{r}}\frac{\rho_l a z'}{4\pi\varepsilon_0 a^2(a^2 + z'^2)^{1/2}}\bigg|_{z'=-1}^{z'=+1}$$

$$= \hat{\mathbf{r}}\frac{\rho_l}{4\pi\varepsilon_0}\left[\frac{1}{a(a^2+1)^{1/2}} + \frac{1}{a(a^2+1)^{1/2}}\right] = \hat{\mathbf{r}}\frac{\rho_l}{2\pi\varepsilon_0 a(a^2+1)^{1/2}} \quad \left[\frac{\text{N}}{\text{C}}\right]$$

For the values given

$$\mathbf{E} = \hat{\mathbf{r}}\frac{\rho_l}{2\pi\varepsilon_0\sqrt{2}} \quad \left[\frac{\text{N}}{\text{C}}\right]$$

(b) Solution for the infinite line follows the same method except that the integration is between $z' = -\infty$ and $z' = +\infty$:

$$\mathbf{E} = \hat{\mathbf{r}}\frac{\rho_l a}{4\pi\varepsilon_0}\int_{z'=-\infty}^{z'=+\infty} \frac{dz'}{(a^2 + z'^2)^{3/2}} = \hat{\mathbf{r}}\frac{\rho_l a z'}{4\pi\varepsilon_0 a^2(a^2 + z'^2)^{1/2}}\bigg|_{z'=-\infty}^{z'=+\infty} = \hat{\mathbf{r}}\frac{\rho_l}{4\pi\varepsilon_0 a(a^2/z'^2 + 1)^{1/2}}\bigg|_{z'=-\infty}^{z'=+\infty} = \hat{\mathbf{r}}\frac{\rho_l}{2\pi\varepsilon_0 a} \quad \left[\frac{\text{N}}{\text{C}}\right]$$

Example 3.9 Electric Field Intensity Due to a Charged Half-Loop A wire is bent in the form of a half-loop of radius $a = 10$ mm and charged with a line charge density $\rho_l = 10^{-9}$ C/m. Calculate the electric field intensity at the center of the loop (point P in **Figure 3.15a**).

Solution: First, we establish an elemental point charge due to a differential arc length of the loop and calculate the electric field intensity at the center of the loop due to this elemental point charge. Because of symmetry, only a component pointing straight down may exist at the center of the loop. These aspects of calculation are shown in **Figure 3.15b**. Integration is on the angle ϕ'.

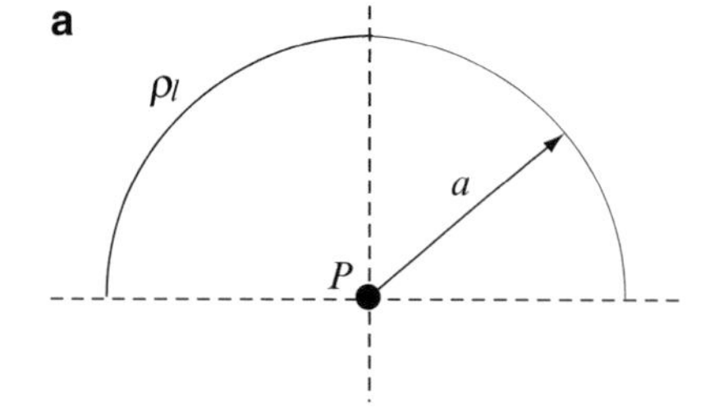

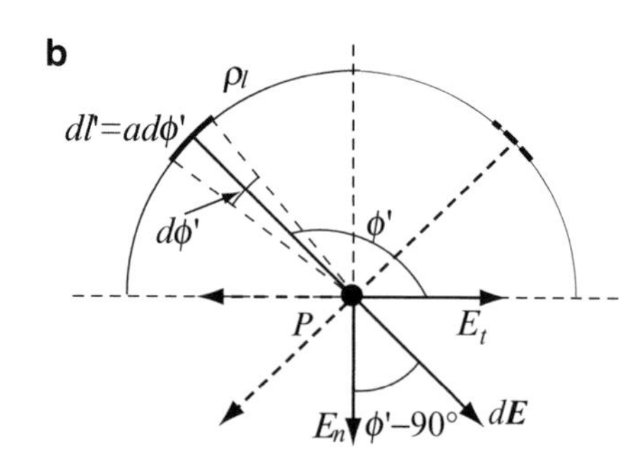

Figure 3.15 (**a**) A charged semicircular loop. (**b**) The electric field intensity due to a differential segment on the loop

The elemental point charge is $\rho_l a d\phi'$, where $a d\phi'$ is the differential arc length in cylindrical coordinates. The electric field intensity at point P due to the elemental point charge is in the r direction:

$$d\mathbf{E} = \hat{\mathbf{r}}\frac{\rho_l a\, d\phi'}{4\pi\varepsilon_0 a^2} \quad \rightarrow \quad d\mathbf{E} = \hat{\mathbf{r}}\frac{\rho_l\, d\phi'}{4\pi\varepsilon_0 a}$$

The horizontal components cancel because for each element of the arc on the left half of the loop, there is an identical element on the right half with the horizontal component in the opposite direction (see **Figure 3.15b**). The normal components sum up and we get

$$dE_n = 2dE\cos(\phi' - 90°) = 2dE\cos(-(90° - \phi')) = 2dE\sin\phi' = \frac{\rho_l \sin\phi' d\phi'}{2\pi\varepsilon_0 a}$$

The total electric field intensity is found by integrating over one-quarter of the loop (between $\phi' = 0$ and $\phi' = \pi/2$)

$$E_n = \frac{\rho_l}{2\pi\varepsilon_0 a}\int_{\phi'=0}^{\pi/2} \sin\phi' d\phi' = -\frac{\rho_l}{2\pi\varepsilon_0 a}\cos\phi'\bigg|_0^{\pi/2} = \frac{\rho_l}{2\pi\varepsilon_0 a} \quad \left[\frac{\text{N}}{\text{C}}\right]$$

For the given values, we get

$$E_n = \frac{10^{-9}}{2 \times \pi \times 8.854 \times 10^{-12} \times 0.01} = 1797.55 \quad [\text{N/C}]$$

Exercise 3.2 Calculate the electric field intensity at the center of a very thin ring of radius a [m] if a charge density ρ_l [C/m] is uniformly distributed on the ring.

Answer $\mathbf{E} = 0$.

3.4.2.2 Surface Charge Distributions

As with line distributions, surface charge densities can be uniform or nonuniform, and since charges are involved, the charge distribution generates an electric field in space. Again using the idea of a point charge, we view an element of surface as containing an elemental charge dq which is considered to be a point charge, as in **Figure 3.16**:

$$dq = \rho_s ds' \quad [\text{C}] \tag{3.45}$$

The electric field intensity due to this point charge at a distance $R = |\mathbf{r} - \mathbf{r}'|$ from the point is

$$d\mathbf{E} = \hat{\mathbf{R}}\frac{\rho_s ds'}{4\pi\varepsilon_0|\mathbf{r} - \mathbf{r}'|^2} = \hat{\mathbf{R}}\frac{\rho_s ds'(\mathbf{r} - \mathbf{r}')}{4\pi\varepsilon_0|\mathbf{r} - \mathbf{r}'|^3} \quad \left[\frac{\text{N}}{\text{C}}\right] \tag{3.46}$$

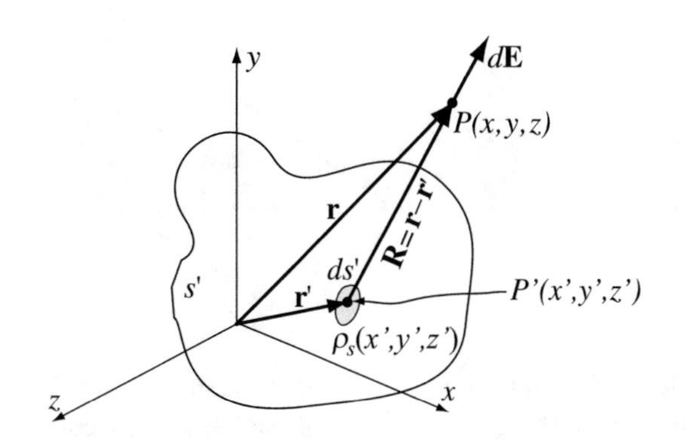

Figure 3.16 A surface charge distribution and the electric field intensity due to an element of charge on the surface. The charge density is, in general, nonuniform

Now, the electric field intensity due to a general surface distribution can be written as

$$\boxed{\mathbf{E} = \int_{s'} \frac{(\mathbf{r} - \mathbf{r}')\rho_s}{4\pi\varepsilon_0|\mathbf{r} - \mathbf{r}'|^3} ds' \quad \left[\frac{\text{N}}{\text{C}}\right]} \tag{3.47}$$

This expression is quite general, but it is of little use at this point. Before we can actually calculate the electric field intensity, it is first necessary to define the vectors $\mathbf{r}$ and $\mathbf{r}'$ in terms of the coordinates of the source point (x',y',z') and field point (x,y,z). Although it is possible to do so in general coordinates, it is much easier to understand the steps involved through examples. It is useful to view this relation as the counterpart of **Eq. (3.44)**. In this sense, the calculation of electric fields due to surface distributions is essentially the same as that for line charge distributions. The differences are in the charge density itself and in the integration.

To see the general relations involved, consider the flat surface in **Figure 3.17**. Although the figure shows a charged surface parallel to the x–y plane (at $z = z_0$), the orientation of the surface in space is not important. This choice simplifies the calculation of the electric field intensity and hence is instructive at this stage. Assuming that the surface charge density is given, an equivalent point charge is defined as shown in the figure and in **Eq. (3.45)**. The electric field intensity at point $P(x,y,z)$ due to the elemental surface charge is given in **Eq. (3.46)**. Now, before performing the integration, we separate the electric field intensity into its components. Since the separation into components in **Eq. (3.40)** was independent of the type of charge distribution we used (it only had to do with the infinitesimal point charge), we can use the same relations here. Using these together with surface integration, we get

$$\mathbf{E} = \hat{\mathbf{x}} \frac{1}{4\pi\varepsilon_0} \int_{s'} \frac{\rho_s(x - x')ds'}{|\mathbf{r} - \mathbf{r}'|^3} + \hat{\mathbf{y}} \frac{1}{4\pi\varepsilon_0} \int_{s'} \frac{\rho_s(y - y')ds'}{|\mathbf{r} - \mathbf{r}'|^3} + \hat{\mathbf{z}} \frac{1}{4\pi\varepsilon_0} \int_{s'} \frac{\rho_s(z - z')ds'}{|\mathbf{r} - \mathbf{r}'|^3} \quad \left[\frac{\text{N}}{\text{C}}\right] \tag{3.48}$$

where $|\mathbf{r} - \mathbf{r}'| = ((x - x')^2 + (y - y')^2 + (z - z')^2)^{1/2}$ and $ds' = dx'dy'$. The expression in **Eq. (3.48)** looks simple, but we still need to evaluate the surface integral. This may or may not be easy to do, depending on the surface on which the integration must be performed and on the charge density distribution. In general, for surfaces that lie in a plane, it is relatively easy to evaluate the integral. Although actual examples will be given shortly, it is worth rewriting the expression in **Eq. (3.48)** for the surface in **Figure 3.17**. Under these conditions, we replace ds' by $dx'dy'$ and the surface integral becomes a double integral over the x' and y' coordinates:

$$\begin{aligned}\mathbf{E} = \hat{\mathbf{x}} \frac{1}{4\pi\varepsilon_0} \int_{y'=y_1}^{y'=y_2} \left[\int_{x'=x_1}^{x'=x_2} \frac{\rho_s(x - x')dx'}{|\mathbf{r} - \mathbf{r}'|^3} \right] dy' + \hat{\mathbf{y}} \frac{1}{4\pi\varepsilon_0} \int_{y'=y_1}^{y'=y_2} \left[\int_{x'=x_1}^{x'=x_2} \frac{\rho_s(y - y')dx'}{|\mathbf{r} - \mathbf{r}'|^3} \right] dy' \\ + \hat{\mathbf{z}} \frac{1}{4\pi\varepsilon_0} \int_{y'=y_1}^{y'=y_2} \left[\int_{x'=x_1}^{x'=x_2} \frac{\rho_s(z - z')dx'}{|\mathbf{r} - \mathbf{r}'|^3} \right] dy' \quad \left[\frac{\text{N}}{\text{C}}\right]\end{aligned} \tag{3.49}$$

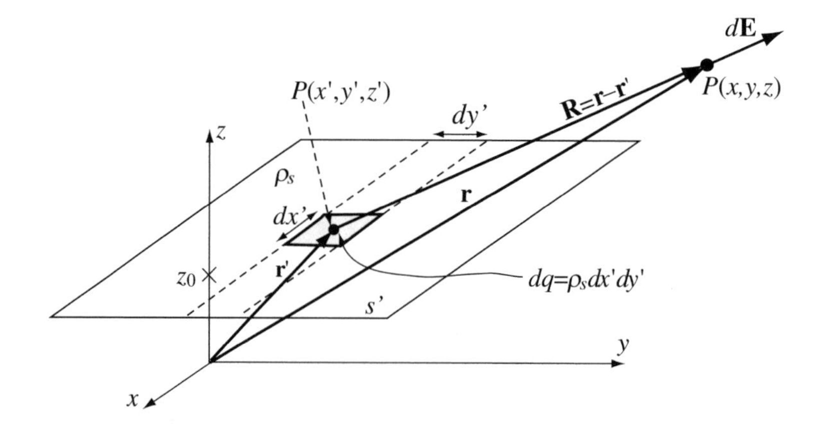

Figure 3.17 Calculation of the electric field of a surface charge distribution

It is also possible, and sometimes useful, to view the surface distribution as an assembly of infinitesimally thin strips or lines of charges as shown in **Figure 3.17**. This approach is a direct consequence of **Eq. (3.49)**. Because the inner integral for each component is identical to that for a line integral, the expression can be written as follows:

$$\mathbf{E} = \frac{1}{4\pi\varepsilon_0}\int_{y'=y_1}^{y'=y_2}\left[\hat{\mathbf{x}}\int_{x'=x_1}^{x'=x_2}\frac{\rho_s(x-x')dx'}{|\mathbf{r}-\mathbf{r}'|^3}+\hat{\mathbf{y}}\int_{x'=x_1}^{x'=x_2}\frac{\rho_s(y-y')dx'}{|\mathbf{r}-\mathbf{r}'|^3}+\hat{\mathbf{z}}\int_{x'=x_1}^{x'=x_2}\frac{\rho_s(z-z')dx'}{|\mathbf{r}-\mathbf{r}'|^3}\right]dy' \quad \left[\frac{\mathrm{N}}{\mathrm{C}}\right] \tag{3.50}$$

This may seem to be a rather minor point but it shows the use of superposition and helps understand the process of double integration because each step can be performed separately. Returning now to **Figure 3.17**, the inner integral is simply that over a line of charge with charge density $\rho_{l'} = \rho_s dy'$ (integrated over x'), whereas the outer integral sums up all lines of charge making up the surface. If the charge density is constant (uniformly distributed) it can be taken outside the integrals.

Example 3.10 Electric Field Intensity Due to Surface Charge Densities A very thin plate of size $2a \times 2b$ [m^2] is charged with a uniform charge density ρ_s [C/m^2]. Calculate the electric field intensity at the center of the plate.

Solution: To calculate the electric field intensity, we define the elemental charge on an area $dx'dy'$ at a point (x',y'). The differential components of the electric field intensity in the x and y directions at a generic point on the surface of the plate (x,y) are then calculated (see **Figure 3.18a** for the relations involved). Now, with the general relation for the electric field intensity, we substitute the required values for x and y at the center of the plate to find the electric field intensity. The distance between the two points is

$$R = \sqrt{(x-x')^2 + (y-y')^2} \quad [\mathrm{m}]$$

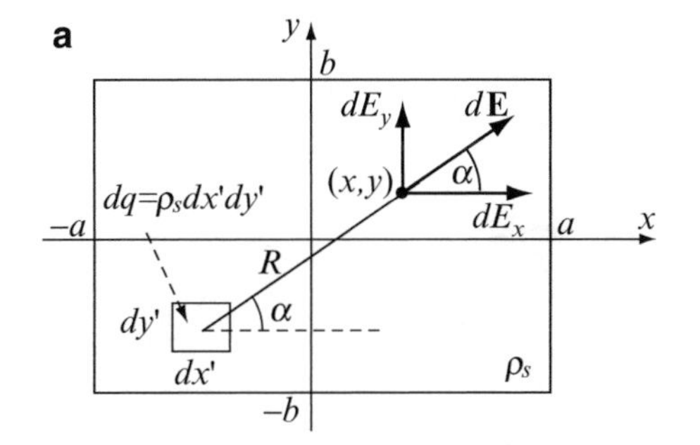

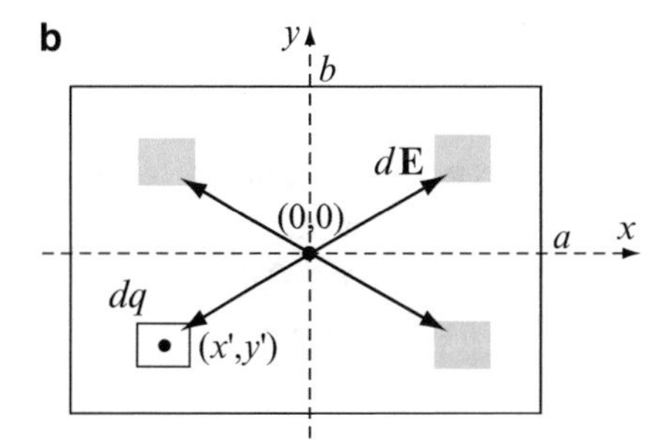

Figure 3.18 (**a**) Calculation of the electric field intensity at a general point on the plate. (**b**) The calculated electric field intensity at the center of the plate from symmetry consideration

The electric field intensity at a general point (x,y) due to an element of charge $dq = \rho_s dx'dy'$ at point (x',y') (see **Figure 3.18a**) is

$$d\mathbf{E} = \hat{\mathbf{R}}\frac{\rho_s dx'dy'}{4\pi\varepsilon_0\left[(x-x')^2+(y-y')^2\right]} \quad \left[\frac{\mathrm{N}}{\mathrm{C}}\right]$$

The electric field intensity at (x,y) is now found by first separating $d\mathbf{E}$ into its x and y components:

$$dE_x = dE\cos\alpha = \frac{\rho_s(x-x')dx'dy'}{4\pi\varepsilon_0\left[(x-x')^2+(y-y')^2\right]^{3/2}} \quad \left[\frac{\mathrm{N}}{\mathrm{C}}\right]$$

$$dE_y = dE\sin\alpha = \frac{\rho_s(y-y')dx'dy'}{4\pi\varepsilon_0\left[(x-x')^2+(y-y')^2\right]^{3/2}} \quad \left[\frac{\mathrm{N}}{\mathrm{C}}\right]$$

where $\cos\alpha = (x - x')/R$ and $\sin\alpha = (y - y')/R$. Now, we can use these relations to find the electric field intensity anywhere on the plate by integrating over x' and y'. In the case discussed here, we need to find the electric field intensity at the center of the plate. By placing the plate as shown in **Figure 3.18a**, $x = 0$, $y = 0$, and the limits of integration are between $-a$ and $+a$ in the x direction and between $-b$ and $+b$ in the y direction. Substituting these in dE_x and dE_y above and integrating over the surface gives the x and y components of the electric field intensity as

$$E_x = -\frac{\rho_s}{4\pi\varepsilon_0}\int_{y'=-b}^{y'=+b}\int_{x'=-a}^{x'=+a}\frac{x'dx'dy'}{[x'^2+y'^2]^{3/2}} = \frac{\rho_s}{4\pi\varepsilon_0}\int_{y'=-b}^{y'=+b}\frac{dy'}{\sqrt{x'^2+y'^2}}\Bigg|_{x'=-a}^{x'=+a} = \frac{\rho_s}{4\pi\varepsilon_0}\int_{y'=-b}^{y'=+b}\left(\frac{1}{\sqrt{a^2+y'^2}}-\frac{1}{\sqrt{a^2+y'^2}}\right)dy' = 0$$

$$E_y = -\frac{\rho_s}{4\pi\varepsilon_0}\int_{x'=-a}^{x'=+a}\int_{y'=-b}^{y'=+b}\frac{y'dx'dy'}{[x'^2+y'^2]^{3/2}} = \frac{\rho_s}{4\pi\varepsilon_0}\int_{x'=-a}^{x'=+a}\frac{1}{\sqrt{x'^2+y'^2}}\Bigg|_{y'=-b}^{y'=+b}dx' = 0$$

Thus, the electric field intensity at the center of the plate is zero. We could have anticipated this result from symmetry considerations. For any element of charge dq, there are three additional, symmetric elements of charge, as shown in **Figure 3.18b**. As can be seen, the electric field intensities due to the four elements cancel at the center.

Example 3.11 Force Due to Surface Charge Densities A hemisphere of radius R [m] has a uniformly distributed surface charge density ρ_s [C/m^2]. Calculate the force on a very small, positive charge q [C], placed at the center of the hemisphere (point P in **Figure 3.19**).

Solution: Because of symmetry about point P, only the vertical component of the electric field intensity (which we call the z component) is nonzero, as shown in **Figure 3.19b**. Integration is done over the surface of the hemisphere in spherical coordinates. After the electric field intensity is known, the force is found by multiplying by the charge at point P.

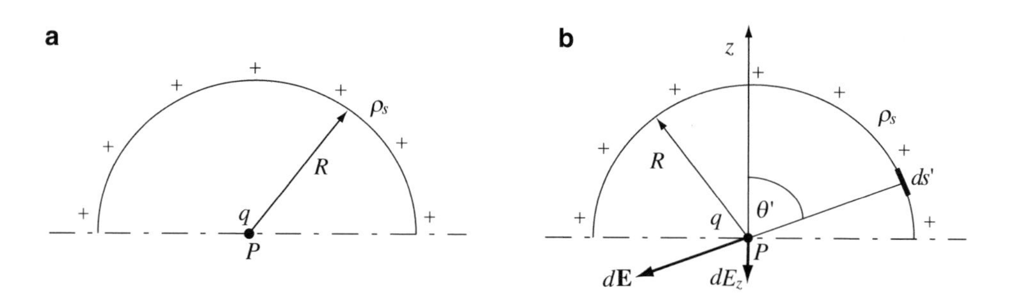

Figure 3.19 (**a**) A charged hemisphere. (**b**) The electric field intensity at point P due to a differential surface ds'

The elemental point charge on the surface of the sphere of radius R is $dQ = \rho_s R^2 \sin\theta' d\theta' d\phi'$ [C]. The electric field intensity due to this charge is

$$d\mathbf{E} = \hat{\mathbf{R}}\frac{dQ}{4\pi\varepsilon_0 R^2} = \hat{\mathbf{R}}\frac{\rho_s}{4\pi\varepsilon_0}\sin\theta' d\theta' d\phi' \quad [\text{N/C}]$$

The electric field intensity has a vertical component in the negative z direction and a horizontal component. The horizontal components due to two elements of surface which are symmetric about the vertical axis cancel each other. Thus, the only nonzero field points in the negative z direction. Its magnitude equals

$$dE_z = dE\cos\theta' = \frac{\rho_s}{4\pi\varepsilon_0}\sin\theta'\cos\theta' d\theta' d\phi' \quad [\text{N/C}]$$

The electric field intensity is found by integrating over θ' from 0 to $\pi/2$ and over ϕ' from 0 to 2π:

$$E_z = \frac{\rho_s}{4\pi\varepsilon_0}\int_{\theta'=0}^{\theta'=\pi/2}\int_{\phi'=0}^{\phi'=2\pi}\sin\theta'\cos\theta' d\theta' d\phi' = \frac{\rho_s}{4\pi\varepsilon_0}\int_{\theta'=0}^{\theta'=\pi/2} 2\pi\frac{\sin 2\theta'}{2}d\theta' = \frac{\rho_s}{4\varepsilon_0}\left(-\frac{\cos 2\theta'}{2}\right)\Big|_0^{\pi/2} = \frac{\rho_s}{4\varepsilon_0} \quad \left[\frac{\mathrm{N}}{\mathrm{C}}\right]$$

where the identity $\sin\theta'\cos\theta' = (\sin 2\theta')/2$ was used. The electric field intensity points downward (negative z direction). The force is therefore

$$\mathbf{F} = q\mathbf{E}_z = -\hat{\mathbf{z}}\frac{q\rho_s}{4\varepsilon_0} \quad [\mathrm{N}]$$

and since the point charge is positive, the force is in the same direction as **E** (a repulsion force).

3.4.2.3 Volume Charge Distributions

Treatment of volume charge distributions follows steps identical to those for surface and line charge distributions. The electric field intensity due to an element of charged volume is shown in **Figure 3.20**. The element of charge is $dq = \rho_v dv'$ and the electric field intensity due to this point charge at a distance R from the point is

$$d\mathbf{E} = \hat{\mathbf{R}}\frac{\rho_v dv'}{4\pi\varepsilon_0 R^2} \quad \left[\frac{\mathrm{N}}{\mathrm{C}}\right] \tag{3.51}$$

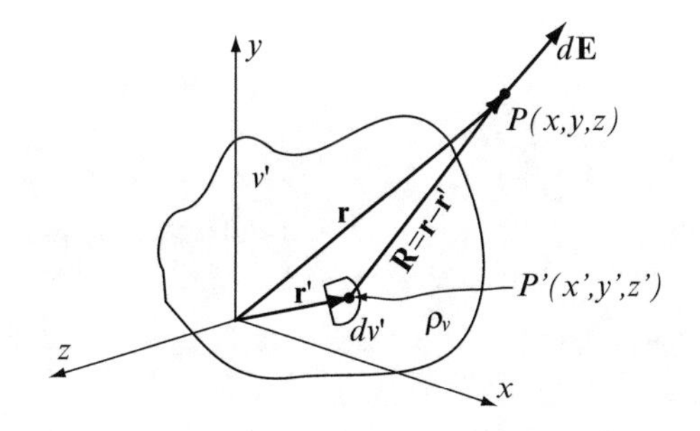

Figure 3.20 Electric field intensity due to an element of charged volume

Integrating and using the position vectors in **Figure 3.20**, we get

$$\boxed{\mathbf{E} = \int_{v'}\frac{(\mathbf{r}-\mathbf{r}')\rho_v}{4\pi\varepsilon_0|\mathbf{r}-\mathbf{r}'|^3}dv' \quad \left[\frac{\mathrm{N}}{\mathrm{C}}\right]} \tag{3.52}$$

As in the previous cases, the integration over the volume may or may not be easy to perform, but the principle of setting up the solution is straightforward. Using the same steps as for the line charge, we get for a uniform charge density ρ_v

$$\mathbf{E} = \hat{\mathbf{x}}\frac{\rho_v}{4\pi\varepsilon_0}\int_{v'}\frac{(x-x')dv'}{|\mathbf{r}-\mathbf{r}'|^3} + \hat{\mathbf{y}}\frac{\rho_v}{4\pi\varepsilon_0}\int_{v'}\frac{(y-y')dv'}{|\mathbf{r}-\mathbf{r}'|^3} + \hat{\mathbf{z}}\frac{\rho_v}{4\pi\varepsilon_0}\int_{v'}\frac{(z-z')dv'}{|\mathbf{r}-\mathbf{r}'|^3} \quad \left[\frac{\mathrm{N}}{\mathrm{C}}\right] \tag{3.53}$$

For a volume with limits as in **Figure 3.21**, we can write $dv' = dx'dy'dz'$, and the element of volume dv' is located at the source point (x',y',z'), whereas the electric field intensity is calculated at the field point (x,y,z). Also, $|\mathbf{r}-\mathbf{r}'| = ((x-x')^2 + (y-y')^2 + (z-z')^2)^{1/2}$, as was indicated earlier. Each of the three terms of **Eq. (3.53)** is integrated over the volume. This gives

$$\mathbf{E}=\hat{\mathbf{x}}\frac{\rho_v}{4\pi\varepsilon_0}\int_{z'=-z_0}^{z'=+z_0}\left\{\int_{y'=-y_0}^{y'=+y_0}\left[\int_{x'=-x_0}^{x'=+x_0}\frac{(x-x')dx'}{|\mathbf{r}-\mathbf{r}'|^3}\right]dy'\right\}dz'+\hat{\mathbf{y}}\frac{\rho_v}{4\pi\varepsilon_0}\int_{z'=-z_0}^{z'=+z_0}\left\{\int_{y'=-y_0}^{y'=+y_0}\left[\int_{x'=-x_0}^{x'=+x_0}\frac{(y-y')dx'}{|\mathbf{r}-\mathbf{r}'|^3}\right]dy'\right\}dz'$$
$$+\hat{\mathbf{z}}\frac{\rho_v}{4\pi\varepsilon_0}\int_{z'=-z_0}^{z'=+z_0}\left\{\int_{y'=-y_0}^{y'=+y_0}\left[\int_{x'=-x_0}^{x'=+x_0}\frac{(z-z')dx'}{|\mathbf{r}-\mathbf{r}'|^3}\right]dy'\right\}dz'\quad\left[\frac{\mathrm{N}}{\mathrm{C}}\right] \tag{3.54}$$

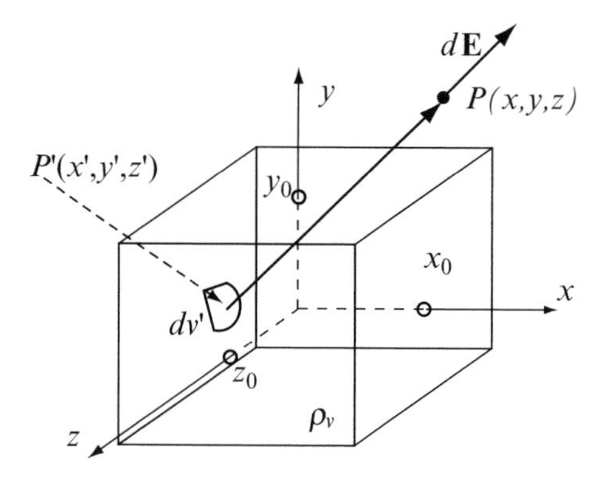

Figure 3.21 Electric field intensity due to an element of charge inside the cube

Example 3.12 Application: Electric Field Intensity In and Around a Cloud As an approximation of the electric field in and around a thundercloud, consider a thundercloud in the form of a cylinder of radius b = 1,000 m, height $2a$ = 4,000 m, and with its bottom c = 1,000 m above ground, as shown in **Figure 3.22a**. The cloud has a charge density $\rho_v = 10^{-9}$ C/m^3 uniformly distributed throughout its volume:

(a) Calculate the electric field intensity at ground level, below the center of the cloud on its axis.
(b) Calculate the electric field intensity at the bottom of the cloud, on its axis.

Note: Thunderclouds tend to be cylindrical in shape. The charge density within the cloud is not uniform and not of the same polarity. However, this example does give an idea of the quantities involved and the method of calculation of the electric field in volume charge distributions.

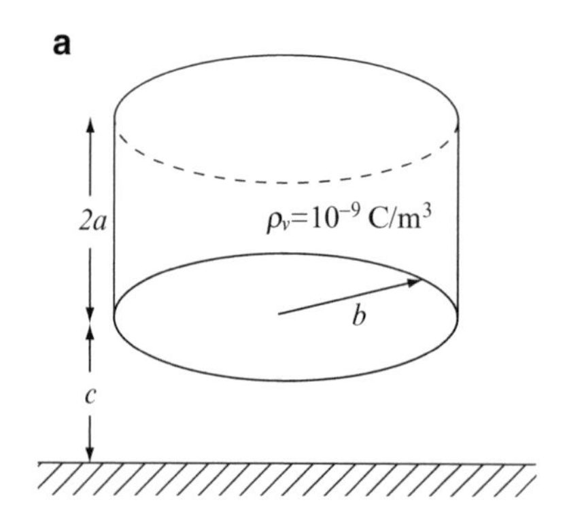

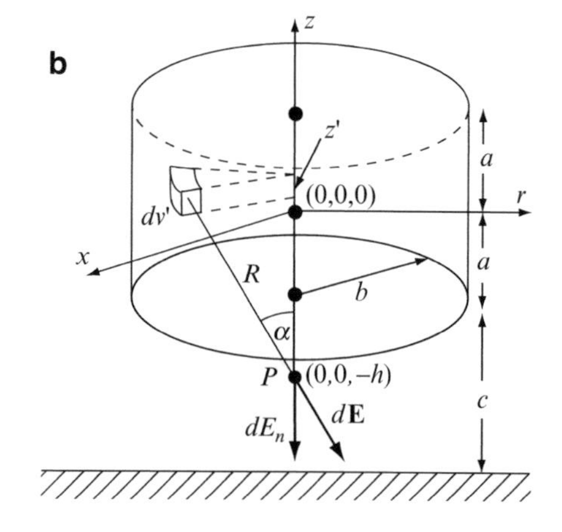

Figure 3.22 (**a**) Approximation to a thundercloud. (**b**) Calculation of the electric field intensity below the cloud

Solution: We place the cloud in a cylindrical system of coordinates as shown. The center of the cylinder is at (0,0,0). A differential of volume at a general point within the cloud (r',ϕ',z') is defined as $dv' = r'dr'd\phi'dz'$. The electric field intensity is then calculated at a point $P(0,0,-h)$ on the axis, as shown in **Figure 3.22b**. Because of symmetry, only a normal component may exist at point P. The normal component is calculated and the result integrated over the primed coordinates to obtain the total field at P:

(a) The magnitude of the electric field intensity due to the elemental charge is

$$dE = \frac{\rho_v r'dr'd\phi'dz'}{4\pi\varepsilon_0 R^2} = \frac{\rho_v r'dr'd\phi'dz'}{4\pi\varepsilon_0\left(r'^2 + (h+z')^2\right)} \quad \left[\frac{\text{N}}{\text{C}}\right]$$

where h is the distance from the center of the cloud to the field point and z' is the z coordinate of the source point, as shown in **Figure 3.22b**. The vertical distance from the element of charge to the field point is $h + z'$ and the scalar normal component of the electric field intensity is

$$dE_n = dE\cos\alpha = \frac{\rho_v r'dr'd\phi'dz'}{4\pi\varepsilon_0\left(r'^2 + (h+z')^2\right)}\frac{h+z'}{R} = \frac{\rho_v (h+z')r'dr'd\phi'dz'}{4\pi\varepsilon_0\left(r'^2 + (h+z')^2\right)^{3/2}} \quad \left[\frac{\text{N}}{\text{C}}\right]$$

Since the only component is in the negative z direction, we may also write for the total electric field intensity at P:

$$\mathbf{E} = -\hat{\mathbf{z}}\frac{\rho_v}{4\pi\varepsilon_0}\int_{z'=-a}^{z'=a}\left[\int_{r'=0}^{r'=b}\left[\int_{\phi'=0}^{\phi'=2\pi}\frac{r'(h+z')}{\left(r' + (h+z')^2\right)^{3/2}}d\phi'\right]dr'\right]dz'$$

$$= -\hat{\mathbf{z}}\frac{\rho_v}{2\varepsilon_0}\int_{z'=-a}^{z'=a}\left[\int_{r'=0}^{r'=b}\frac{r'(h+z')}{\left(r^2 + (h+z')^2\right)^{3/2}}dr'\right]dz' \quad \left[\frac{\text{N}}{\text{C}}\right]$$

where the fact that the integral over ϕ' results in 2π was used. The inner integral is evaluated as follows:

$$\int_{r'=0}^{r'=b}\frac{r'(h+z')}{\left(r' + (h+z')^2\right)^{3/2}}dr' = \left[-\frac{(h+z')}{\left(r'^2 + (h+z')^2\right)^{1/2}}\right]_{r'=0}^{r'=b} = 1 - \frac{(h+z')}{\left(b'^2 + (h+z')^2\right)^{1/2}}$$

Substituting this result in the electric field intensity, we are left with

$$\mathbf{E} = -\hat{\mathbf{z}}\frac{\rho_v}{2\varepsilon_0}\int_{z'=-a}^{z'=a}\left[1 - \frac{h+z'}{\left(b^2 + (h+z')^2\right)^{1/2}}\right]dz'$$

$$= -\hat{\mathbf{z}}\frac{\rho_v}{2\varepsilon_0}\left[z' - h\ln\left(2\sqrt{z'^2 + 2hz' + h^2 + b^2} + 2z' + 2h\right)\right.$$

$$\left.- \sqrt{b^2 + (h+z')^2} + h\ln\left(2\sqrt{z'^2 + 2hz' + h^2 + b^2} + 2z' + 2h\right)\right]_{z'=-a}^{z'=a}$$

$$= -\hat{\mathbf{z}}\frac{\rho_v}{2\varepsilon_0}\left[z' - \sqrt{b^2 + (h+z')^2}\right]_{z'=-a}^{z'=a} = -\hat{\mathbf{z}}\frac{\rho_v}{2\varepsilon_0}\left[a - \sqrt{b^2 + (h+a)^2} + a + \sqrt{b^2 + (h-a)^2}\right]$$

$$= -\hat{\mathbf{z}}\frac{\rho_v}{2\varepsilon_0}\left[2a + \sqrt{b^2 + (h-a)^2} - \sqrt{b^2 + (h+a)^2}\right] \quad \left[\frac{\text{N}}{\text{C}}\right]$$

For the given values, the electric field intensity at ground level ($h = c + a = 3{,}000$ m) is

$$\mathbf{E} = -\hat{\mathbf{z}}\frac{10^{-9}}{2 \times 8.854 \times 10^{-12}}\left[4000 + \sqrt{1000^2 + 1000^2} - \sqrt{1000^2 + 5000^2}\right] = -\hat{\mathbf{z}}17{,}799.53 \quad [\text{N/C}]$$

(b) At the bottom of the cloud, $h = a = 2{,}000$ m and we get

$$\mathbf{E} = -\hat{\mathbf{z}}\frac{\rho_v}{2\varepsilon_0}\left[2a + b - \sqrt{b^2 + 4a^2}\right] = -\hat{\mathbf{z}}\frac{10^{-9}}{2 \times 8.854 \times 10^{-12}}\left[4000 + 1000 - \sqrt{1000^2 + 4 \times 2000^2}\right] = -\hat{\mathbf{z}}49{,}519 \ [\text{N/C}]$$

Exercise 3.3 A very thin disk of radius $a = 50$ mm is charged with a uniform surface charge density $\rho_s = 10^{-7}$ C/m^2.
(a) Calculate the electric field intensity at a distance $h = 0.1$ m above the center of the disk on its axis. Assume the disk is placed in the r–ϕ plane.
(b) Calculate the electric field intensity below and above the disk at $h = 0$ (on the r–ϕ plane at the center of the disk).

Answer

(a) $\mathbf{E} = -\hat{\mathbf{z}}\frac{\rho_s}{2\varepsilon_0}\left[1 - \frac{h}{\sqrt{a^2 + h^2}}\right] = \hat{\mathbf{z}}596.2 \quad [\text{N/C}].$

(b) $\mathbf{E}_{below} = -\hat{\mathbf{z}}\frac{\rho_s}{2\varepsilon_0} = -\hat{\mathbf{z}}5{,}647 \quad [\text{N/C}] \quad \mathbf{E}_{above} = \hat{\mathbf{z}}\frac{\rho_s}{2\varepsilon_0} = \hat{\mathbf{z}}5{,}647 \quad [\text{N/C}].$

3.5 The Electric Flux Density and Electric Flux

In all relations used in the preceding sections, we made use of permittivity of materials without actually discussing what permittivity is or, for that matter, how it influences the electric field, except to say that it is material dependent. This was of relatively little concern since Coulomb's law was given as an experimental relation and any relation derived from it is therefore also experimental. In other words, dependency of the electric field intensity or force on some material property was established through experiment. In **Chapter 4,** we will discuss this again and at that point properly define permittivity and its meaning as well as its utility and applications. Simply from a convenience point of view, we observe that if we multiply the electric field intensity **E** by permittivity ε, we get a new vector:

$$\boxed{\mathbf{D} = \varepsilon\mathbf{E} \quad [\text{C/m}^2]} \tag{3.55}$$

This vector has the same direction as **E** but unlike the expression for **E**, it is independent of ε and therefore of material properties. In terms of units, the electric field intensity has units of [N/C] and permittivity has units of [C^2/N · m^2]. The units of **D** are therefore [C/m^2] as indicated **Eq. (3.55)**. Thus, because this is a density and because it relates to the electric field intensity, it is called the ***electric flux density***. The use of the term "flux" will become apparent in the following chapter. We merely want to comment here that the use of this new vector is sometimes more convenient since the electric flux density is independent of material properties [ε is not involved in calculation as shown in **Eq. (3.55)**] whereas the electric field intensity is material dependent. In terms of usage, the electric flux density is used in a manner similar to the electric field intensity. For example, the electric flux density of a point charge Q_1 at a distance R is

$$\mathbf{D} = \varepsilon_0\mathbf{E} = \frac{\varepsilon_0 Q_1\mathbf{R}}{4\pi\varepsilon_0|\mathbf{R}|^3} = \frac{Q_1\mathbf{R}}{4\pi|\mathbf{R}|^3} \quad \left[\frac{\text{C}}{\text{m}^2}\right] \tag{3.56}$$

Thus, the calculation of the electric flux density is simplified because the material is not taken into account explicitly. For example, the permittivity of water is about 80 times larger than that of free space. Therefore, the electric field intensity in water due to a given charge is 80 times smaller than the electric field intensity for the same charge in free space, but the electric flux density is the same.

On the other hand, the calculation of forces requires knowledge of permittivity. Since $\mathbf{F} = q\mathbf{E}$, the use of the electric flux density does not eliminate the need to consider permittivity. We may use either of the following expressions to calculate the force on a point charge q due to the field of point charge Q_1, and in each, permittivity appears explicitly:

$$\mathbf{F} = q\mathbf{E} = \frac{qQ_1\mathbf{R}}{4\pi\varepsilon_0|\mathbf{R}|^3} = \frac{q\mathbf{D}}{\varepsilon_0} \quad [\text{N}] \tag{3.57}$$

Once we defined the electric flux density, it is only obvious that we should also define electric flux. We note from **Eq. (3.55)** that the electric flux density is a surface density. Therefore, integrating the electric flux density over a surface provides the electric flux through that surface. We write the electric flux as:

$$\boxed{\Phi = \int_s \mathbf{D} \cdot d\mathbf{s} \quad [\text{C}]} \tag{3.58}$$

Note that the electric flux is given in units of coulombs [C] as expected.

Example 3.13 Calculation of Electric Flux Density, Electric Field Intensity, and Electric Flux in Layered Materials A point charge Q [C] is located at a point in space. The charge is surrounded by two spherical layers of materials as shown in **Figure 3.23a**. Both materials have permittivities different than free space, as indicated:

(a) Calculate the electric flux density and field intensity everywhere in space.
(b) Calculate the electric flux through a spherical surface of radius R.
(c) Plot the electric field intensity and electric flux density everywhere in space.

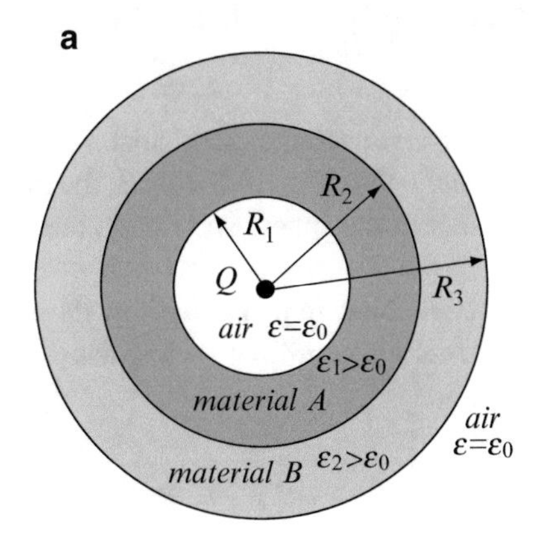

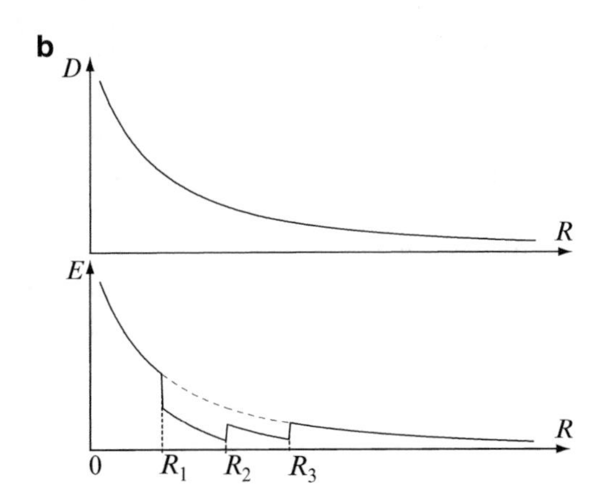

Figure 3.23 (**a**) A spherical layered configuration. (**b**) Plot of the electric flux density and electric field intensity as a function of distance from the center

Solution: After calculating the electric flux density using **Eq. (3.56)**, the electric field intensity can be calculated from **Eq. (3.55)**:

(a) The electric flux density in all materials is radially oriented and equals

$$\mathbf{D} = \hat{\mathbf{R}}\frac{Q}{4\pi R^2} \quad \left[\frac{\text{C}}{\text{m}^2}\right]$$

where R is the distance from the point charge Q. The electric field intensity in any material is

$$\mathbf{E} = \frac{\mathbf{D}}{\varepsilon} = \hat{\mathbf{R}}\frac{Q}{4\pi\varepsilon R^2} \quad \left[\frac{\mathrm{N}}{\mathrm{C}}\right]$$

where ε is the permittivity of the corresponding material. Thus, we have

$$\mathbf{E} = \frac{\mathbf{D}}{\varepsilon_0} = \hat{\mathbf{R}}\frac{Q}{4\pi\varepsilon_0 R^2} \quad \left[\frac{\mathrm{N}}{\mathrm{C}}\right], \quad 0 < R < R_1$$

$$\mathbf{E} = \frac{\mathbf{D}}{\varepsilon_1} = \hat{\mathbf{R}}\frac{Q}{4\pi\varepsilon_1 R^2} \quad \left[\frac{\mathrm{N}}{\mathrm{C}}\right], \quad R_1 < R < R_2$$

$$\mathbf{E} = \frac{\mathbf{D}}{\varepsilon_2} = \hat{\mathbf{R}}\frac{Q}{4\pi\varepsilon_2 R^2} \quad \left[\frac{\mathrm{N}}{\mathrm{C}}\right], \quad R_2 < R < R_3$$

$$\mathbf{E} = \frac{\mathbf{D}}{\varepsilon_0} = \hat{\mathbf{R}}\frac{Q}{4\pi\varepsilon_0 R^2} \quad \left[\frac{\mathrm{N}}{\mathrm{C}}\right], \quad R_3 < R < \infty$$

To be noted is the fact that the electric field intensity in one material is not affected by the presence of other materials; that is, the electric field intensity in, say, material B behaves as if the whole space were made of material B. However, we should not jump to conclusions regarding these observations since this is a unique example in that the electric field intensity in all materials is normal to the material boundaries everywhere.

(b) The electric flux density calculated in **(a)** is integrated over the spherical surface of radius R. The element of area is $d\mathbf{s} = \hat{\mathbf{R}}R^2\sin\theta d\theta d\phi$. The flux is

$$\Phi = \int_s \mathbf{D}\cdot d\mathbf{s} = \int_{\theta=0}^{\pi}\int_{\phi=0}^{2\pi} \hat{\mathbf{R}}\frac{Q}{4\pi R^2}\cdot\hat{\mathbf{R}}R^2\sin\theta d\theta d\phi = \frac{Q}{4\pi}\int_{\theta=0}^{\pi}\int_{\phi=0}^{2\pi}\sin\theta d\theta d\phi = \frac{Q}{2}\int_{\theta=0}^{\pi}\sin\theta d\theta = Q \quad [\mathrm{C}]$$

Note: the prime notation is not used here because there is no need to distinguish between source and field points.

Of course, since the flux density is constant on any spherical surface of radius R, the integration is trivial and we could just as well multiply the electric flux density by the area of the sphere and obtain the same result. Note that the flux is independent of R.

(c) **Figure 3.23b** shows the plots of the values obtained above. Note in particular how the electric field intensity varies with permittivity, while the electric flux density is independent of permittivity. Note also that the plot assumes $\varepsilon_2 < \varepsilon_1$.

3.6 Applications

Application: Coulomb's Torsional Balance In the experiments leading to development of the law bearing his name, Coulomb used a special torsional balance he devised. The balance is shown in **Figure 3.24**. It consists of a thin, insulated rod, suspended on a wire. A small conducting ball is attached to each end of the rod. One is charged while the other is used as a counterbalance weight. If a charge is now placed in the vicinity of the charged end and kept stationary, the charged end of the rod will move toward or away from the test charge. When moving, the rod twists the wire. The angle it twists is then a measure of the force between the two charges. After calibration (which could be accomplished by applying known mechanical forces), the force can be read directly (the wire acts as a spring that balances the force due to charges). This instrument was particularly useful in Coulomb's experiments because it was simple and accurate and could measure extremely small forces. To ensure accurate measurements, the balance was used in a repulsion mode and was placed in an enclosed container to minimize the effects of air movement. More modern torsional balances of this type make use of an evacuated container.

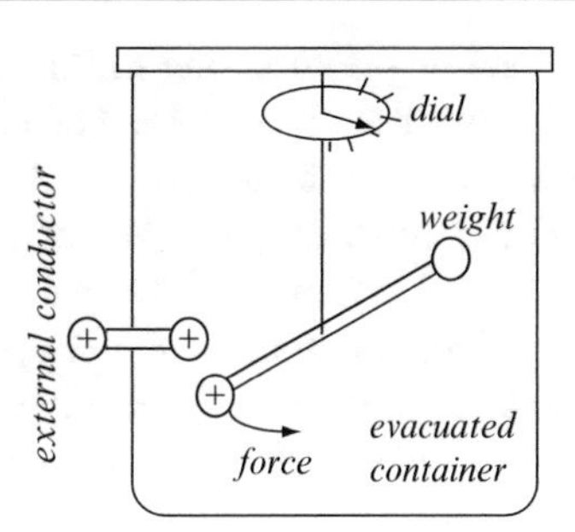

Figure 3.24 Coulomb's torsional balance

Application: Electrostatic Spraying—Spraying of Paints and Pesticides Electrostatic spraying of liquids is based on imparting a charge to the liquid to be sprayed which is forced out through a capillary opening, as shown in **Figure 3.25**. The sharp electrode generates a high electric field intensity, sufficient to charge the liquid. This is then pulled out by the effect of the anode (which is the article being sprayed). As the fluid moves out, it streams as a continuous conical jet. This jet breaks up into individual droplets because of instability in flow (depending on the charge given to the fluid) which now accelerate toward the anode, coating it. A diffuser is usually used to scatter the droplets. Spraying of pesticides with electrostatic guns can also reduce the amount of pesticides needed, since these tend to adhere to plants rather than the ground (foliage is closer to the gun than the ground). Also, because they adhere equally well on all surfaces of the plant, there is better coverage and less wash-off due to rain. Droplets can be as small as individual ions and there is minimal fogging, since all droplets are accelerated. Interestingly, the theory behind electrostatic spraying goes back to the last decade of the nineteenth century when Lord Rayleigh discussed the ideas involved. One interesting application of this principle is one type of electrostatic air cleaner designed for home use. Electrostatic spraying is useful in continuous and automatic spraying of articles and is used extensively in the automobile and appliance industries. Because of the electrostatic force exerted on the paint droplets, the paint tends to adhere better and be more uniform, with less running than other methods.

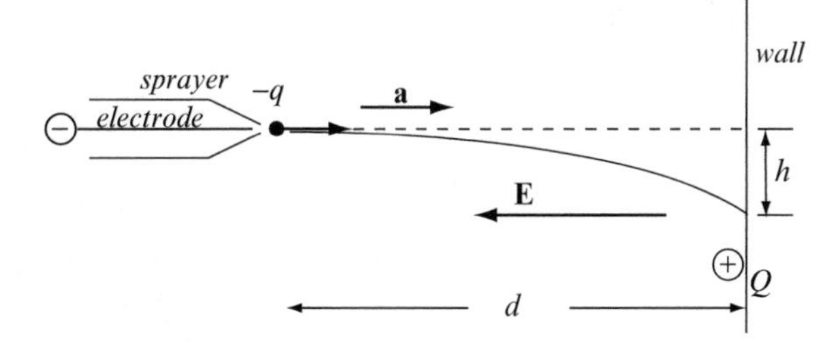

Figure 3.25 Electrostatic spraying

Application: Electrostatic Cleaners (Electrostatic Separators, Scrubbers, and Precipitators) A simple method of cleaning air is the electrostatic precipitator. It consists of a chamber through which the air is forced by means of a fan. An electric field intensity is generated inside the chamber (see **Figure 3.39**). As the air moves through a charged screen, dust (or other particles such as ash and smoke) acquire a charge (usually negative). As they move up, the charged particles are attracted to the positively charged electrode and stick to it, accumulating until they are physically removed. In small, household cleaners, this is accomplished by washing the electrode or by removing a disposable liner on the electrode. In large scrubbers in coal-fired power plants, the cleaners are installed in flue-like structures in the smoke stacks, and cleaning is usually accomplished by "shaking" the particles from the electrodes and removing them physically from a collection pit at the bottom of the stack. As might be expected, the amount of particles collected in these types of scrubbers is huge.

Application: Electrophoresis Electrophoresis is the motion of particles suspended in a fluid under the influence of electrostatic fields. Molecules in a solution have a surface charge and hence are attracted or repelled by electric fields. Suspended particles can either be naturally charged or charged by induction and will move under the influence of electric fields based on Coulomb's law. These principles are used to separate molecules based on size for analysis purposes including DNA analysis, separation and classification of proteins and in nanoparticle work.

Application: Electrostatic Deflection—Processing of Polymers by Electrostatic Extrusion An interesting extension of the electrostatic spraying mechanism described above is electrostatic spinning of fibers. A tube with capillary opening is filled with a polymer solution which, when pulled and dried, makes the fibers (such as polyesters used in synthetic fabrics). The negative electrode is introduced into the liquid material and the anode is at some distance away. Electrostatic force pulls the fluid into a continuous stream, and as it moves toward the anode it dries, forming the fiber. This fiber is deposited on the anode or spooled in its vicinity. Fibers as thin as 1 μm are possible. This method has some promise of spinning very fine, uniform fibers for a variety of applications such as specialty filters. Unlike other methods, it can be used to spin the fibers on demand at the location they are needed rather than in plants. **Figure 3.26** shows the system including the deflection mechanism used to move the fiber (because the fiber is charged, it can be deflected at will).

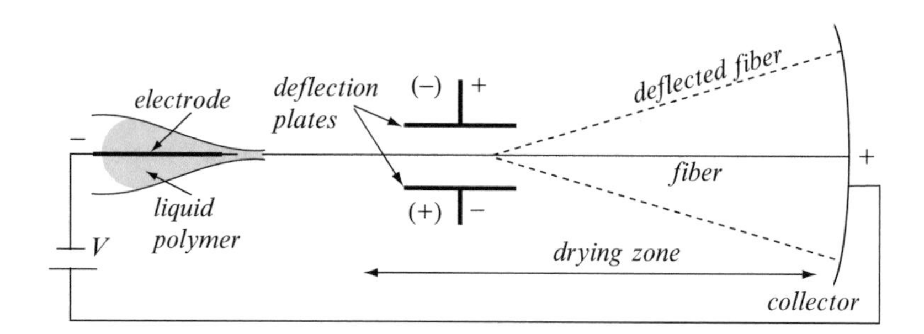

Figure 3.26 Electrostatic spinning of polymers

Application: Ion Beam Etching and Milling Ion beams as small as 0.1 μm in diameter can be focused onto a material. With the high energy densities in these highly concentrated ion beams, they can be used to etch microcircuits, for ion lithography, and to produce localized ion implantation for production of very high-density microcircuits. The ions are accelerated using large electric fields and are often focused by electrostatic lenses to a narrow point for precision machining. **Figure 3.27** shows an ion beam etching installation. Note in particular the function of the electrostatic lens shown as part of the arrangement in **Figure 3.27**. It is essentially a charged tube which squeezes the charged ions together, focusing them into a narrower beam.

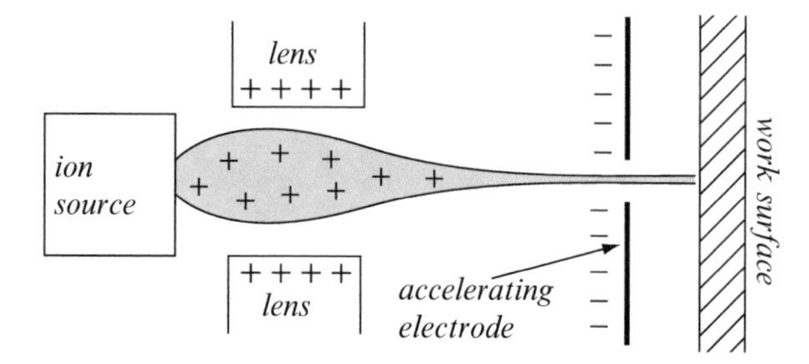

Figure 3.27 Ion beam working of materials

Application: Xerography The xerographic process as used for copying and in laser printers is a six-step electrostatic process. The method relies on, first, uniformly charging a plate or a drum in preparation for copying. This can be done in a number of ways, including discharging charge from a high-voltage source through sharp points to create a corona. The sharp point or points are moved over the surface to charge the whole surface (step 1: charging). The result is a uniformly charged surface as shown in **Figure 3.28a**. The plate or the drum is made of a photoconducting material such as selenium, usually coated on a base conducting material. A photoconducting material changes its conductivity when exposed to light. Although we have not yet discussed conductivity, conductors are understood to be materials that allow motion of charges. The conducting surface below the selenium is negatively charged, but positive charges and negative charges cannot move because selenium is normally nonconducting or poorly conducting. The material to be copied is projected onto the plate as you would project a slide on a screen (step 2: exposure). This is shown in **Figure 3.28b**. The photoconducting material becomes more conducting where light strikes (lighter areas of the image), and, therefore, some or all the charges in this area will combine with the negative charges by passing through the selenium. The effect is an electrostatic image on the plate. **Figure 3.28b** shows a black arrow on a white page. The only positive charges left are those in the area of the arrow. All other charges have diffused through the material and combined with the negative charges. In the next step, very fine, negatively charged carbon particles combined with synthetic binding materials are sprinkled onto the plate (step 3: toner). The particles will only be attracted to positively charged regions (black arrow in this case). Now, we have a layer of negative carbon particles on top of the arrow as shown in **Figure 3.28c**. Next, a positively charged page passes above the plate (or drum). The negatively charged particles are attracted to the positively charged page (step 4: transfer). To fix the image, the paper is heated to fuse the particles onto the page (step 5: fusing). After this, the drum is cleaned in preparation for the next image (step 6: cleaning).

The process described above is rather complex and its operation depends on many parameters, including reaction time of the photoconducting material (i.e., how long it takes charges to diffuse into the photoconductor), uniformity of charge, quality of the cleaning process, quality and size of particles, and many others. Nevertheless, this is one of the most useful and extensive uses of electrostatics.

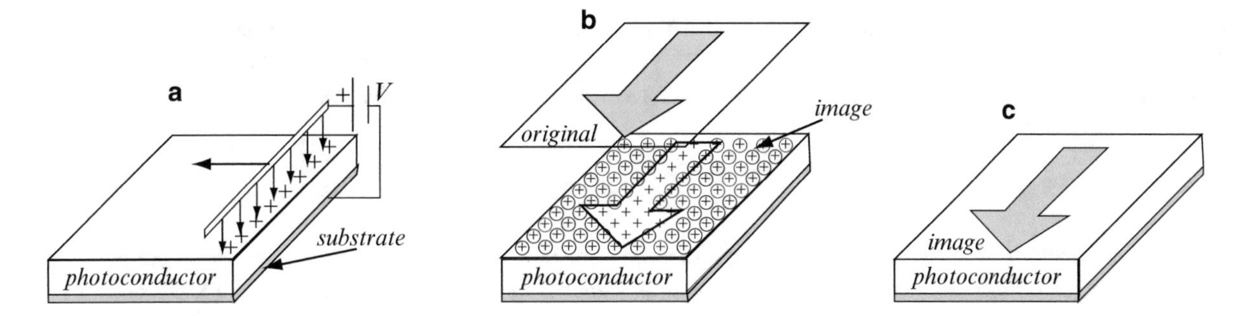

Figure 3.28 Xerography. (**a**) Charging of the photoconducting surface with positive charge. (**b**) Exposure to the original removes charges from the light portions of the original (circled charges recombine). (**c**) Negatively charged carbon particles (toner) adhere to the dark (positively charged) areas

Application: The Laser Printer The laser printer is a xerographic copier with one exception. Instead of exposing the image using a light source and lens, the image is created electronically and transferred onto the photoconducting drum by laser pulses. Essentially, for each white dot on the page, the laser beam (produced by a laser diode) is switched on, exposing this dot (or pixel), and then switched off, and moved to the next dot of the image. Typical laser printers produce an image with 600 to 1,200 dpi (dots per inch) or about 24 to 48 dots/mm. The size of a dot is of the order of 0.04 mm to 0.01 mm in diameter. A typical page requires 14.4 to 57.6 million dots and all of them must be scanned for each page. Scanning can be done optically. A laser printer that produces 10 pages per minute, scans the page in less than about 5 s. After this, the process is the same as for the copier, including the use of the same type of particles and fusing process. **Figure 3.29** shows a schematic view of the laser printer. In this scheme, the drum rotates twice: once for scanning and once for transferring the image. The fuser heats the page to fix the image. The laser printer is particularly well adapted for digital image reproduction and the resolution depends only on the number of dots. Laser printers with 48 dots/mm (1200 dpi and higher) are common, including color printers. Higher-resolution laser printers exist and are commonly used in publishing.

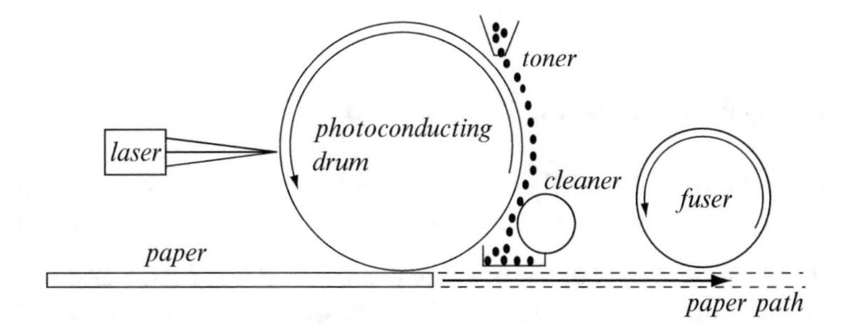

Figure 3.29 The principle of the laser printer

3.7 Summary

The starting relation in electromagnetics is ***Coulomb's law***, describing the force between two point charges. It may be written in different forms but **Eq. (3.11)** is representative:

$$\mathbf{F}_{12} = \hat{\mathbf{R}}_{12}\frac{Q_1Q_2}{4\pi\varepsilon_0 R_{12}^2}, \quad \mathbf{F}_{21} = \hat{\mathbf{R}}_{21}\frac{Q_1Q_2}{4\pi\varepsilon_0 R_{21}^2} \quad [\mathrm{N}] \tag{3.11}$$

where ε_0 is the permittivity (in this case that of free space). From Coulomb's law we defined the ***electric field intensity*** in **Eq. (3.17)** as force per unit charge. The forms in **Eq. (3.18)** are more useful:

$$\mathbf{E} = \frac{Q\mathbf{R}}{4\pi\varepsilon_0|\mathbf{R}|^3} \quad \text{or} \quad \mathbf{E} = \hat{\mathbf{R}}\frac{Q}{4\pi\varepsilon_0|\mathbf{R}|^2} \quad \left[\frac{\mathrm{N}}{\mathrm{C}}\right] \tag{3.18}$$

The relations for point charges were extended to distributed charge densities by defining elemental, differential charges based on line, surface, and volume charge distributions. Viewing the elemental charges as point charges, Coulomb's law provides a general way of calculating the electric field intensity due to distributed charges as

$$d\mathbf{E} = \frac{(\mathbf{r}-\mathbf{r}')\rho_{\Omega'}}{4\pi\varepsilon_0|\mathbf{r}-\mathbf{r}'|^3}d_{\Omega'}, \quad \mathbf{E} = \int_{\Omega'}\frac{(\mathbf{r}-\mathbf{r}')\rho_{\Omega'}}{4\pi\varepsilon_0|\mathbf{r}-\mathbf{r}'|^3}d_{\Omega'} \quad \left[\frac{\mathrm{N}}{\mathrm{C}}\right]$$

$\mathbf{r}'$ is the position vector of the elemental point charge (source point), $\mathbf{r}$ is the position vector of the field point at which the field is calculated (see **Figures 3.13, 3.16**, and **3.20**), and Ω' stands for l', s', or v' [see **Eqs. (3.40)**, **(3.47)**, and **(3.52)**].

Finally, the ***electric flux density* D** is introduced, for now simply as a means of eliminating the dependence of the electric field intensity on permittivity from calculation:

$$\mathbf{D} = \varepsilon\mathbf{E} \quad [\mathrm{C/m^2}] \tag{3.55}$$

In effect, all of this chapter deals with the electric field intensity of point charges and the forces between them—either actual point charges such as isolated point charges or elemental charges defined on line, surface, and volume distributions. Important reminders:

(1) Note the distinction between field points and source points (the latter are denoted with a (′)).
(2) Pay attention to symmetries—they can help in understanding and in solving problems.
(3) Superposition of fields and forces can be used to calculate fields of complex distributions.
(4) The electric field intensity and the force are vectors. When summing up various contribution to either of these, separation into components simplifies the task.
(5) An appropriate choice of coordinates can often simplify calculations.

Useful quantities:
Permittivity of free space: $\varepsilon_0 = 8.854853 \times 10^{-12}$ [F/m]
Charge of the electron: $e = -1.602129 \times 10^{-19}$ [C]

Problems

Point Charges, Forces and the Electric Field

3.1 Electric and Gravitational Forces. Two planets are 1,000,000 km apart (about the same as between Earth and Mars). Their mass is the same, equal to 10^{24} kg (of the same order of magnitude as the Earth):

(a) What must be the amount of free charge on each planet for the electrostatic force to equal the gravitational force? The gravitational force is $F_g = GM_1M_2/R^2$, where $G = 6.67 \times 10^{-11}$ [N · m²/kg²]. M_1 and M_2 are the masses of the planets in [kg] and R is the distance between the planets in [m]. Because of the very large distance, you may assume the planets behave like point charges, and the total charge (magnitude) on the two planets is the same.
(b) What must be the signs of the charges so that they cancel the gravitational force?

3.2 Atomic Forces. In fusion of two hydrogen nuclei, each with a charge $q = 1.6 \times 10^{-19}$ C, the nuclei must be brought together to within 10^{-20} m. Calculate the external force necessary to do so. View each nucleus as a point charge.

3.3 Repulsion Force. A very thin tube contains at its bottom a small plastic ball charged with a charge $Q = 0.1$ μC as shown in **Figure 3.30**. A second, identical ball with identical charge is inserted from above. Assume there is no friction and the tube does not affect the charges on the balls. If each ball has a mass m = 1 g and the permittivity of the tube may be assumed to be the same as that of air (free space) calculate:

(a) The distance between the balls if the tube is vertical (**Figure 3.30a**).
(b) The distance between the balls if the tube is tilted so it makes an angle $\alpha = 30°$ with the horizontal (**Figure 3.30b**).

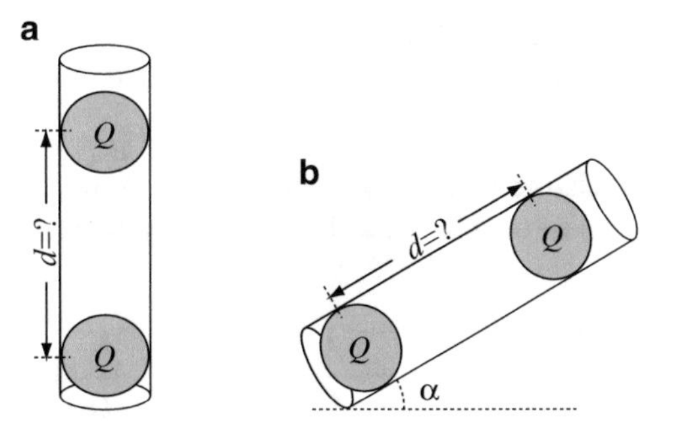

Figure 3.30

3.4 Repulsion Forces: Two point charges, each with charge $q = 10^{-9}$ C, are suspended by two strings, 0.2 m long, and connected to the same point as in **Figure 3.31**. If each charge has a mass of 10^{-4} kg, find the horizontal distance between the two charges under the assumption that α is small.

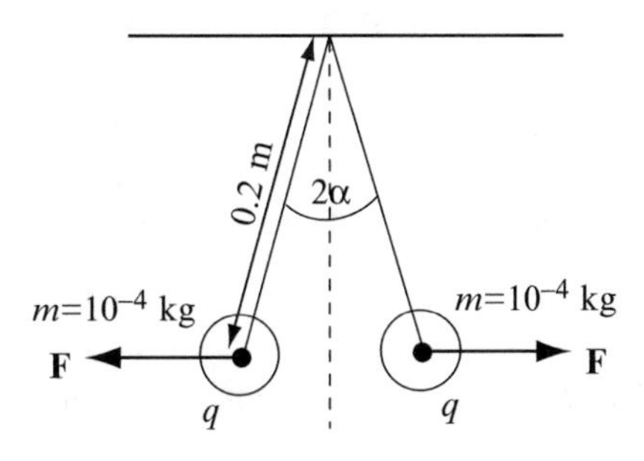

Figure 3.31

3.5 Application: The Electrometer. An electrometer is a device that can measure electric field intensities or charge. One simple implementation is shown in **Figure 3.31**. Two very small conducting balls are suspended on thin conducting wires. To apply the charge, the test charge is placed at the point of contact of the two wires. The balls repel each other since half the charge is distributed on each ball.

Suppose you need to design an instrument of this sort. Each wire is 0.2 m long and the mass of each ball is 10^{-4} kg. Assume each ball acquires half the measured charge and the balls are small enough to be considered as points:

(a) If the minimum distance measurable is 0.5 mm along the circumference of the circle the ball describes as it is deflected (arclength), what is the lowest amount of charge this instrument can measure?
(b) Can you also calculate the largest amount of charge measurable? Explain.

3.6 Force on Point Charge. Two positive point charges, each equal to Q [C], are located a distance d [m] apart. An electron, of charge e [C] and mass m [kg], is held in position on the centerline between them and at a distance x [m] from the line connecting the two charges as shown in **Figure 3.32**. The electron is now released:

(a) Calculate the acceleration the charge is experiencing (direction and magnitude). Where is the acceleration maximum?
(b) If there are no losses and no external forces, what is the path the electron describes?

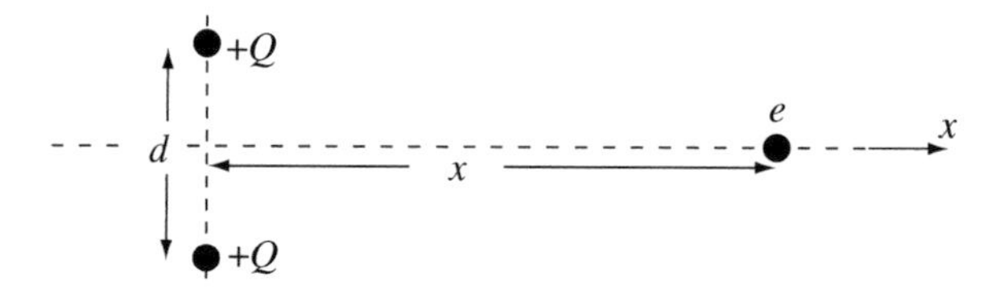

Figure 3.32

3.7 Force on Point Charge. Two point charges are located at a distance d [m] apart. One charge is $Q_1 = +q$ [C], the other is $Q_2 = +2q$ [C], as in **Figure 3.33**. A third charge, $+q$ [C], is placed somewhere on the line connecting the two charges:

(a) Assuming the third charge can only move on the line connecting Q_1 and Q_2, where will the charge move?
(b) From this stationary point, the charge is given a small vertical push (up or down) and allowed to move freely. Describe qualitatively the motion of the charge as it moves away from the two stationary charges.

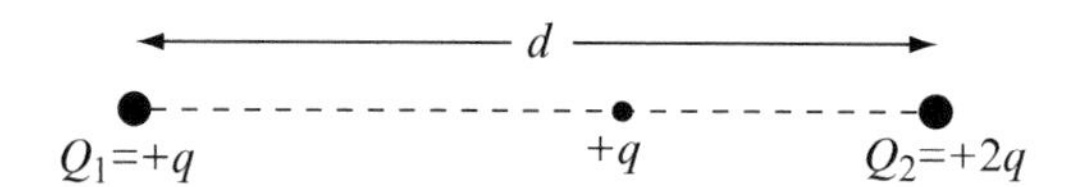

Figure 3.33

3.8 Electric Field of Point Charges. Two point charges are located as in **Figure 3.34**:

(a) Sketch the electric field intensity of this arrangement using field lines.
(b) Find a point (on the axis) where the electric field intensity is zero.

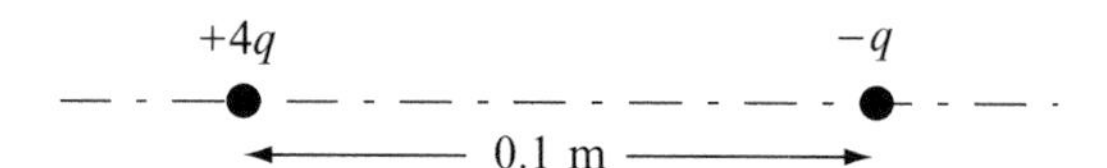

Figure 3.34

3.9 Electric Field of Point Charges. Four point charges are arranged as shown in **Figure 3.35** in free space:

(a) Find the electric field intensity at an arbitrary point in space.
(b) If $Q_1 = Q_2 = Q$ and $Q_3 = Q_4 = -Q$ show that the electric field intensity is perpendicular to the two dotted lines shown (at 45°).
(c) If $Q_1 = Q_2 = Q_3 = Q_4 = Q$ show that the electric field intensity is parallel to the two dotted lines shown (at 45°).

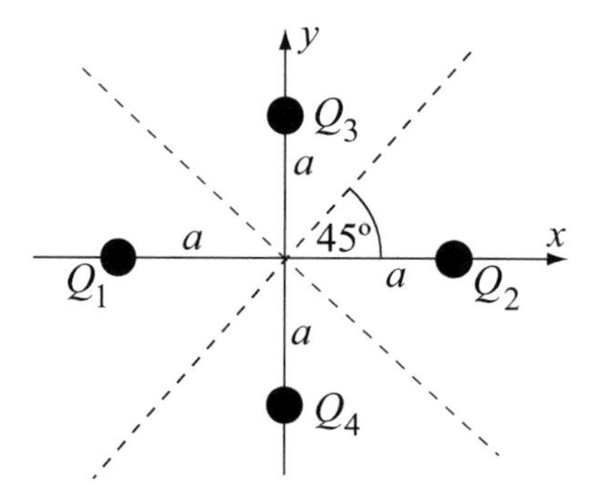

Figure 3.35

3.10 The Quadrupole. Three point charges are located as shown in **Figure 3.36**. The outer charges equal q [C] and the center charge equals $-2q$ [C]. This arrangement is called a linear quadrupole because it consists of two dipoles on a line:

(a) Find the electric field intensity at an arbitrary point P a distance R [m] from the negative charge.

(b) Use the binomial expansion to obtain an approximate solution for the electric field intensity at P for $R \gg d$. Retain the first three terms in the expansion and then neglect those terms in the approximation that can be justifiably neglected.

(c) Sketch the field lines of the quadrupole and compare with those of the dipole in **Section 3.4.1.3** (**Figure 3.12**).

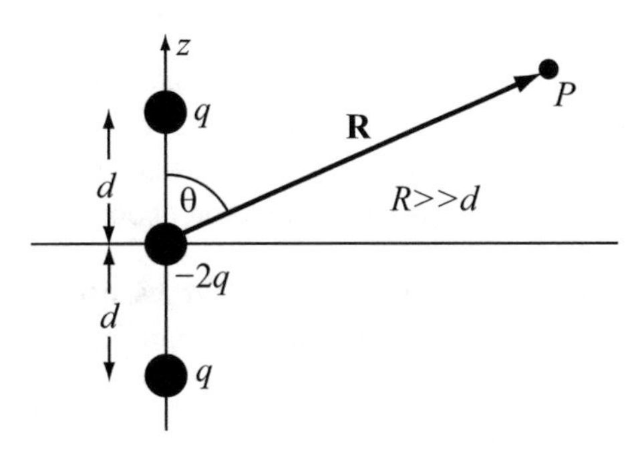

Figure 3.36

3.11 Electric Field and Forces Due to Point Charges. Two positive point charges each equal to Q [C] and two negative charges each equal to $-Q$ [C] are placed at the corners of the base of a pyramid as shown in **Figure 3.37**. The base is a square $a \times a$ m^2 in dimensions. A 5th point charge $-Q$ [C] is placed at the pinnacle of the pyramid. If all edges of the pyramid equal to a [m], calculate:

(a) The electric field intensity at the pinnacle of the pyramid.

(b) The electric field intensity at the center of the base of the pyramid.

(c) The force on the negative charge at the pinnacle.

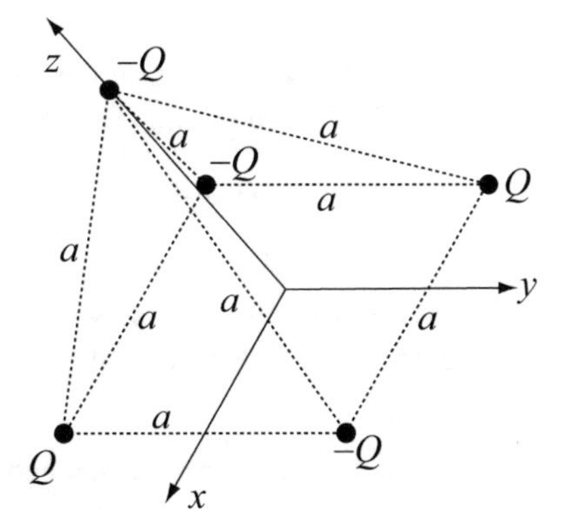

Figure 3.37

3.12 Application: Electrostatic Levitation—Repulsion Mode. The repulsion force between charges can be used to levitate bodies by ensuring the repulsion force is equal to the weight of the levitated body. Suppose four small plastic balls are placed at the corners of a rigid square, a [m] on the side, and charged each to +100 nC. Four identically charged balls are embedded into a fixed plane, with the distance between the balls also a [m]. Assume the levitation h is small compared to the length a ($h \ll a$), and the weight of the frame is negligible:

(a) If the movable frame is placed with its charges exactly above the four stationary charges and if the mass of each ball is 10^{-4} kg, calculate the elevation at which the movable frame levitates.

(b) If the movable frame is allowed to rotate but not to translate, what is the rest position of the charges? Do not calculate the position; just describe it.
(c) Suppose you push the movable frame downward slightly from the position in **(b)** and release it. What happens to the frame?
(d) Suppose you move the frame downward until the movable charges are in the same plane as the stationary charges. Disregarding mechanical questions of how this can be done, what happens to the frame now?
(e) If you continue pushing until the frame is below the plane, what will be the stationary position of the frame?

3.13 Application: Electrostatic Levitation—Attraction Mode. The situation in **Problem 3.12** is given again, but the stationary charges are positive and the movable charges are negative and of the same magnitude:

(a) Find the position at which the frame is stationary (below the plane). Assume the distance h [m] between the frames is small compared to the frame's size ($h \ll a$) and the frame is free to rotate in its plane.
(b) If the frame is pushed either downward or upward slightly, what happens?
(c) Compare with **Problem 3.12**. Which arrangement would you choose?

3.14 Force on Charge in an Electric Field. Two stationary point charges, each equal in magnitude to q [C], are located a distance d [m] apart. A third point charge is placed somewhere on the line separating the two charges and is allowed to move. The charge is free to move in any direction:

(a) If the stationary charges are positive and the moving charge is positive, it will move to the midpoint between the two stationary charges. Is this a stable position? If not, where will the charge eventually end up if its position is disturbed?
(b) What is the answer to **(a)** if the stationary charges are positive but the moving charge is negative?
(c) What is the answer to **(a)** and **(b)** if the stationary charges are negative?

3.15 Application: Accumulation of Charge. One of the common mechanisms for charge to accumulate in clouds is through friction due to raindrops. In the process, each drop acquires a positive charge, while the cloud tends to become more negative. Suppose each raindrop is 2 mm in diameter and each acquires the charge of one proton. If it rains at a rate of 10 mm/h, what is the rate of transfer of charge to earth per unit area?

3.16 Electric Field of Point Charges. Three point charges are placed at the vertices of an equilateral triangle with sides of length L [m]:

(a) Each charge is equal to q [C]. Find the points in space at which the electric field intensity is zero.
(b) Two charges are equal to q [C], the third to $2q$ [C]. Show that the electric field intensity is zero only at infinity.

Hint: To find the answers you will need to solve a transcendental equation. Use a graphical method to find the roots of the equation.

3.17 Application: Millikan's Experiment—Determination of the Charge of Electrons. A very small oil drop has a mass of 1 μg. The oil drop is placed between two plates as shown in **Figure 3.38**. A uniform electric field intensity **E** is applied between the two plates, pointing down:

(a) What is the electric field intensity that will keep the drop suspended without moving up or down if the drop contains one free electron?
(b) What is the next (lower) value of field possible and what is the charge on the drop?

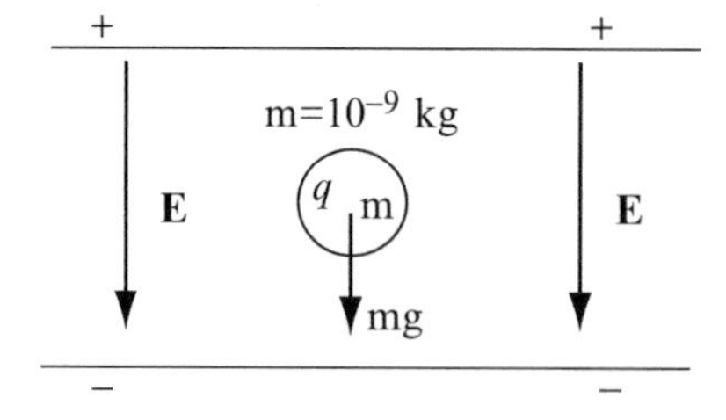

Figure 3.38

3.18 Force in a Uniform Field. A uniform electric field intensity of 1,000 N/C is directed upwards. A very small plastic ball of mass 10^{-4} kg is charged with a charge such that the ball is suspended motionless in the field:

(a) Calculate the charge on the ball assuming a point charge. Find the magnitude and sign of the charge.
(b) If the charge is doubled and the ball allowed to move starting from rest, find the acceleration of the ball.
(c) If the ball starts from rest, in a frictionless environment, find the time it takes the ball to reach a speed equal to half the speed of light under the condition given in (**b**).

3.19 Application: Electrostatic Air Cleaner. An air cleaner is made as two parallel plates with an electric field intensity of 10 kN/C between the plates as shown in **Figure 3.39**. The cleaner is installed in a vertical chimney to collect dust particles as these rise through the chimney. Assume the air and particles are pushed upwards at a velocity of 10 m/sec and that particles have been charged with an electric charge of one electron each before entering the chamber. Calculate the mass of the heaviest particle that is guaranteed not to escape the cleaner chamber.

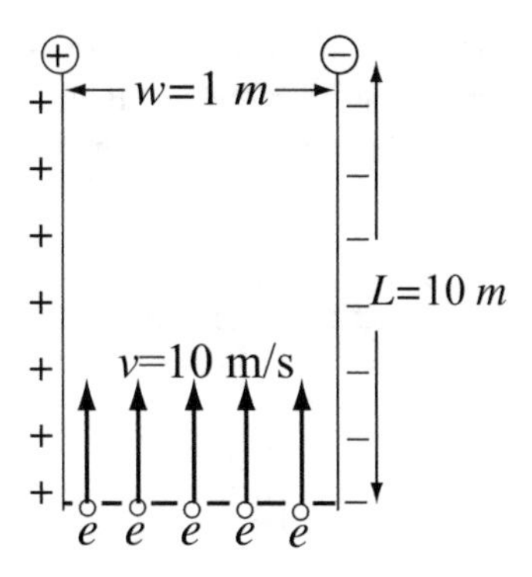

Figure 3.39

Line Charge Densities

3.20 Field Due to Line Charge Density. A short line of length L [m] is charged with a line charge density ρ_l [C/m] (see **Figure 3.40**). Calculate the electric field intensity at the four points shown in **Figure 3.40**.

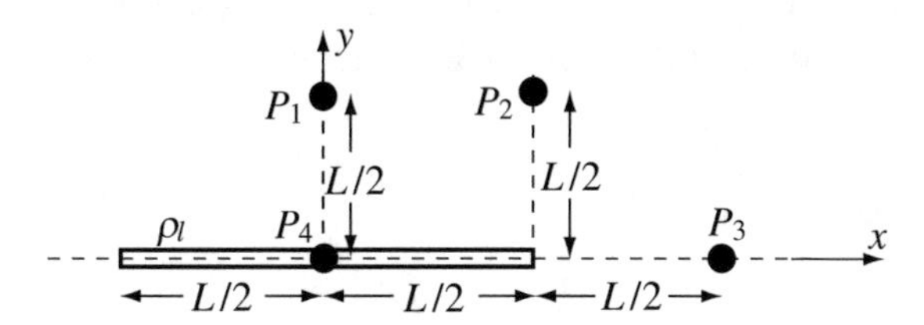

Figure 3.40

3.21 Field Due to a Charged Ring. A thin ring of radius a [m] is charged so that a uniform line charge density of ρ_l [C/m] exists on the ring:

(a) Calculate the electric field intensity at a height h [m] above the center of the ring, on its axis.
(b) Show that at very large distances ($h \gg a$), the electric field intensity is that of a point charge equal to the total charge on the ring.

3.22 Force on Line Charge Density. A charged line 1 m long is charged with a line charge density of 1 nC/m. A point charge is placed 10 mm away from its center. Calculate the force acting on the point charge if its charge equals 10 nC.

3.23 Electrostatic Forces. Two thin segments, each 1 m long, are charged, one with line charge density of 10 nC/m and the other with −10 nC/m and are placed a distance 10 mm apart, parallel to each other. Calculate the total force one segment exerts on the other.

3.24 Force Between Two Short Charged Segments. Two charged lines are colinear as shown in **Figure 3.41**. The total charge on line 1 is Q_1 [C]; the total charge on line 2 is Q_2 [C]. On each line the charge is uniformly distributed over the length of the line. Given the lengths of the lines (L_1 [m] and L_2 [m]) and the distance between them as a [m], calculate the force one line exerts on the other if both are in free space.

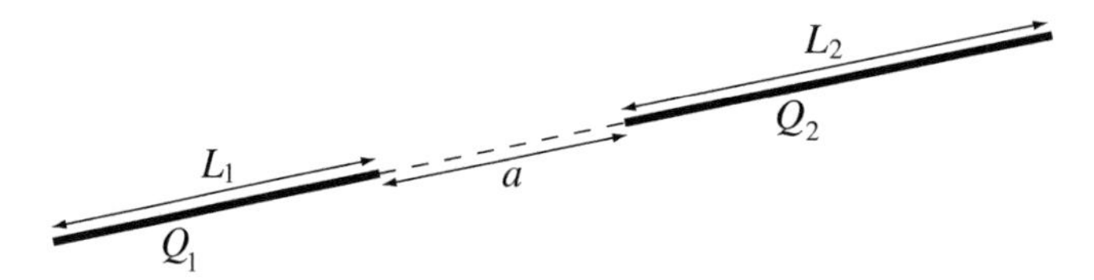

Figure 3.41

3.25 Force Between Charged Lines. A very long thin conductor is charged with a uniform line charge density ρ_l [C/m]. A second, short conductor, placed vertically on the same plane as the long conductor has length b [m] and a uniform line charge density $-\rho_l$ [C/m] (see **Figure 3.42**). Calculate the force the short line exerts on the longer line. The conductors are in free space.

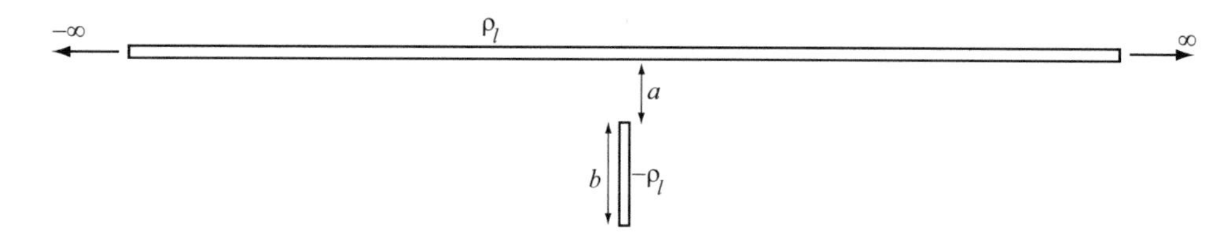

Figure 3.42

Surface Charge Densities

3.26 Electric Field Due to Surface Charge Density. A disk of radius a [m] is charged with a uniform surface charge density ρ_s [C/m^2] (see **Figure 3.43**).

(a) Calculate the electric field intensity at a distance h [m] from the center of the disk on the axis.
(b) Plot the magnitude of the electric field intensity for all values of h from $h = -5a$ to $h = 5a$ for $a = 10$ mm, $\rho_s = 2$ nC/m^2.
(c) What is the maximum electric field intensity and at what height does it occur?
(d) What is the electric field intensity in **(a)** if $a \to \infty$?

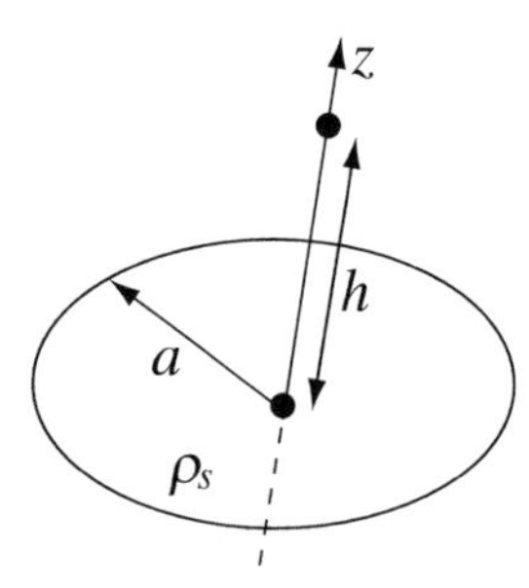

Figure 3.43

3.27 Electric Field Due to Surface Charge Density. A disk of radius a [m] is charged with a nonuniform charge density $\rho_s = \rho_0 r^2$ [C/m^2], where r [m] is the distance from the center of the disk.

(a) Calculate the electric field intensity at a distance h [m] from the center of the disk, on the axis (see **Figure 3.43**).
(b) Plot the magnitude of the electric field intensity for all values of h from $h = -5a$ to $h = 5a$ for $a = 10$ mm, $\rho_0 = 2 \times 10^{-9}$.

3.28 Electric Field of Infinite Surface. Calculate the electric field intensity in free space at a distance d [m] above an infinite plane charged with a uniform surface charge density ρ_s [C/m^2]. Show that the electric field intensity is independent of d and is perpendicular to the plane.

3.29 Charged Infinite Plate with a Hole. A thin infinite plate is charged with a uniform surface charge density ρ_s [C/m^2]. A disk of radius a [m] is removed so that a hole of radius a [m] is left at the center of the plate (no charge on the surface occupied by the hole). Calculate the electric field intensity at a height a [m] above the center of the hole in free space.

3.30 Electric Force Due to Hollow, Charged Cylindrical Surface. A very thin-walled cylindrical tube of length L [m] and radius a [m] has a surface charge density ρ_s [C/m^2] uniformly distributed as shown in **Figure 3.44**. A point charge Q [C] is placed at point P on the axis of the tube. Calculate the force on the charge Q (magnitude and direction). The tube is drawn in axial cross section.

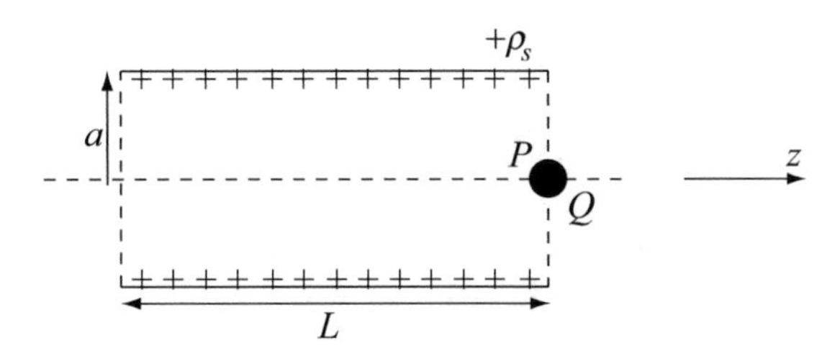

Figure 3.44

Volume Charge Densities

3.31 Field Due to Volume Charge Density. A short plastic cylinder of length L [m] and diameter $L/2$ [m] has a uniform volume charge density ρ_v [C/m^3] uniformly distributed throughout its volume.

(a) Calculate the electric field intensity at all points on the z-axis (**Figure 3.45**).
(b) Find the location of maximum field on the z-axis.

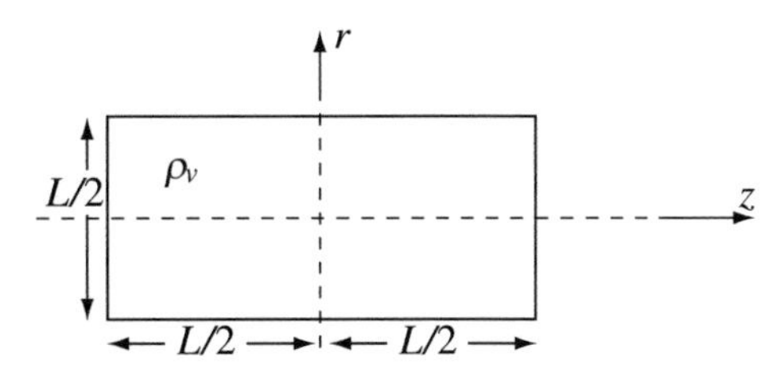

Figure 3.45

3.32 Electric Field of the Electron. The structure of the electron is not defined (i.e., its dimensions cannot be identified with certainty). Its charge, however, is considered to be a point charge. Consider the following:

(a) Suppose the electron is a point charge equal to $e = -1.602 \times 10^{-19}$ C. Calculate the electric field intensity it produces everywhere in space.
(b) Suppose now that the charge of the electron is distributed uniformly over a spherical volume of radius $R_0 = 2 \times 10^{-13}$ m. Calculate the electric field intensity everywhere in space outside the electron (that is, for $R > R_0$). Show that the result is the same as in **(a)**. **Hint:** To simplify solution, assume that the sphere is made of a stack of disks of varying radii and differential thickness dz' and calculate the electric field intensity on the axis of the disk (see, for example, **Problem 3.26**).

3.33 Electric Field of Small Volume Charge Density. A small sphere of radius a [m] has a nonuniform volume charge density given as $\rho_v = \rho_0 R(a - R)/a$ [C/m^3] where R [m] is the distance from the center of the sphere. Find the electric field intensity at a very large distance $R \gg a$. What are the assumptions you must make?

The Electric Flux Density

3.34 Electric Flux Density Due to Point Charges. Two point charges are separated a distance d [m] apart. Each charge is 10 nC and both are positive:

(a) Calculate the force between the two charges.
(b) Both charges, while still at the same distance, are immersed in distilled water ($\varepsilon = 81\varepsilon_0$). What is the force between the charges in water? Explain the difference.
(c) What can you say about the electric field intensity and electric flux density in air and water?

3.35 Electric Flux Density in Dielectrics. A point charge is located in free space. The charge is surrounded by a spherical dielectric shell, with inner diameter d_1 [m], outer diameter d_2 [m], and permittivity ε [F/m]. Calculate:

(a) The electric flux density everywhere in space.
(b) The electric field intensity everywhere in space.
(c) The total electric flux passing through the outer surface of the dielectric shell.
(d) The total flux passing through the inner surface of the dielectric.
(e) The total flux through a spherical surface of radius $R > d_2/2$ in air.
(f) What is your conclusion from the results in **(c)**, **(d)**, and **(e)**?

Gauss's Law and the Electric Potential

4

I have had my results for a long time but I do not yet know how I am to arrive at them.

—Johann Carl Friedrich Gauss (1777–1855)
(A. Arber, The Mind and the Eye, 1954)

4.1 Introduction

The fundamentals of electrostatics were given in **Chapter 3** as the definition of force through Coulomb's law and of the electric field intensity. In this chapter, we address two issues: one is to formalize some of the results obtained in the previous chapter and the second is to expand on the ideas of electric field intensity, electric flux density, and the relation between the electric field and material properties.

In the process, we will define the concepts of electric potential and energy as well as capacitance, all from simple physical concepts. Thus, our main objective is a better and deeper understanding of the electric field and, in addition, we seek methods that will allow simpler and faster solution for electrostatic field problems. These methods provide additional tools to help understand the behavior of electrostatic fields and a number of important applications.

4.2 The Electrostatic Field: Postulates

Now that we know how to calculate the electric field intensity of point charges and charge distributions, we look at it as a vector field. First, we recall Helmholtz's theorem, which states that for a vector field to be uniquely specified, both its divergence and its curl must be known. Thus, the question is: What are the curl and divergence of the electric field intensity? From a more practical point of view, the question is: What are the sources of the electric field, both scalar (divergence) and vector (curl)? If we can uniquely identify the sources, we can also calculate the field intensity. To do so, we start with the electric field intensity of a point charge. This may seem as an oversimplification, but since any electric field may be viewed as being generated by an assembly of point charges, the approach here is rather general. Consider, therefore, the electric field intensity of a positive point charge Q as shown in **Figure 4.1a**:

$$\mathbf{E} = \hat{\mathbf{R}}\frac{Q}{4\pi\varepsilon_0|\mathbf{R}|^2} \quad \left[\frac{\text{N}}{\text{C}}\right] \tag{4.1}$$

To calculate the divergence of the electric field intensity, we apply the divergence theorem to this field:

$$\int_v (\nabla\cdot\mathbf{E})dv = \oint_s \mathbf{E}\cdot d\mathbf{s} = \oint_s \hat{\mathbf{R}}\frac{Q}{4\pi\varepsilon_0|\mathbf{R}|^2}\cdot\hat{\mathbf{R}}ds = \frac{Q}{\varepsilon_0} \tag{4.2}$$

N. Ida, *Engineering Electromagnetics*, https://doi.org/10.1007/978-3-030-15557-5_4

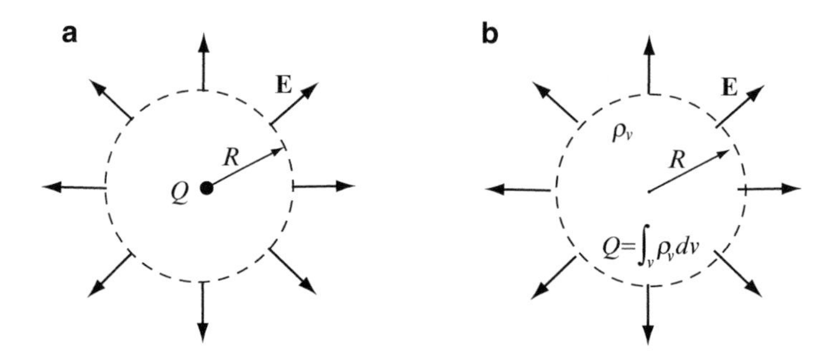

Figure 4.1 Two charge distributions that produce identical fields outside the sphere of radius R. (**a**) Electric field of a positive point charge. (**b**) Charge density in the volume. The total charge in the two cases is the same

where $d\mathbf{s}$ was taken by definition to point out of the volume. In fact, we obtained two results:

$$\int_v (\nabla \cdot \mathbf{E}) dv = \frac{Q}{\varepsilon_0} \tag{4.3}$$

and

$$\oint_s \mathbf{E} \cdot d\mathbf{s} = \frac{Q}{\varepsilon_0} \tag{4.4}$$

which are different ways of describing the same quantity.

Reverting back to the divergence theorem [**Eq. (4.2)**], we see that the closed surface integral or the volume integral is related to the total flux of **E** through the surface s enclosing the volume v. This means that the charge Q in **Eq. (4.3)** is the total charge in volume v enclosed by the surface s. Assuming the charge is distributed throughout the volume v, with a charge density ρ_v, we have in free space (see **Figure 4.1b**)

$$\int_v (\nabla \cdot \mathbf{E}) dv = \frac{1}{\varepsilon_0} \int_v \rho_v dv \tag{4.5}$$

or, equating the integrands,

$$\boxed{\nabla \cdot \mathbf{E} = \frac{\rho_v}{\varepsilon_0}} \tag{4.6}$$

Equation (4.6) applies to any charge distribution since it gives a differential or point relation. Similarly, if the divergence of an electric field is zero at a given point, the charge density at that point is zero. Note also that the result in **Eq. (4.5)** indicates that the two charge distributions in **Figures 4.1a** and **4.1b** produce identical fields at identical distance if the two surfaces enclose identical total charge.

The result in **Eq. (4.6)** [or **Eq. (4.5)**] is important in two ways: First, it defines the sources of the electric field intensity; second, it provides a means of calculating the electric field intensity. More important is the fact that we have established the divergence of the electric field intensity as one of the conditions required to define the vector field.

The second condition is the curl of the electric field intensity, which is defined through application of Stokes' theorem to the field of the point charge:

$$\int_s (\nabla \times \mathbf{E}) \cdot d\mathbf{s} = \oint_C \mathbf{E} \cdot d\mathbf{l} = \oint_C \frac{Q}{4\pi\varepsilon_0 |\mathbf{R}|^2} (\hat{\mathbf{R}} \cdot d\mathbf{l}) \tag{4.7}$$

For the case given here, any contour C that lies on the sphere of radius R produces an open, circular surface as shown in **Figure 4.2**; that is, if we were to cut a section of the sphere, the rim of the cut is the closed contour. The electric field intensity is radial, therefore perpendicular everywhere to this surface. This means that the scalar product $\hat{\mathbf{R}} \cdot d\mathbf{l} = 0$ and, therefore, the right-hand side of **Eq. (4.7)** is zero. Thus,

$$\boxed{\nabla \times \mathbf{E} = 0} \tag{4.8}$$

However, we have taken a very special example: that of a point charge and a surface on the sphere at which the electric field has constant magnitude. The question that remains to be answered is: Is this relation general? In other words, can we, in fact, state that the curl of the static electric field intensity is always zero? The answer to this is twofold. First, it is always zero for the electrostatic field. We can obviously say nothing about any time dependency that might exist since this possibility was never considered. Second, we will give more general proof to the correctness of this relation, including a very simple physical meaning. For now, we will accept this to be correct. Thus, we have at this point both the divergence [**Eq. (4.6)**] and curl [**Eq. (4.8)**] of the electric field.

According to the Helmholtz theorem **(Section 2.5.1)**, the electrostatic field is ***irrotational*** (curl-free) and ***nonsolenoidal*** (nonzero divergence); that is, the electrostatic field is generated by a scalar source alone: a charge or a charge density. These two equations are, in fact, all that is necessary to define the electrostatic vector field. For this reason, they are often taken as the basic postulates[1] for the electrostatic field. From these, we can derive all previous relations. The postulates were derived here for two purposes. One is the simple fact that they describe the fundamental properties of the static electric field as a vector and in very compact form. The second is to show how the properties of vectors can be used to derive useful field relations. This process, while not as intuitive as, say, Coulomb's law, is extremely useful in field relations and we will make considerable use of both relations in **Eqs. (4.6)** and **(4.8)**.

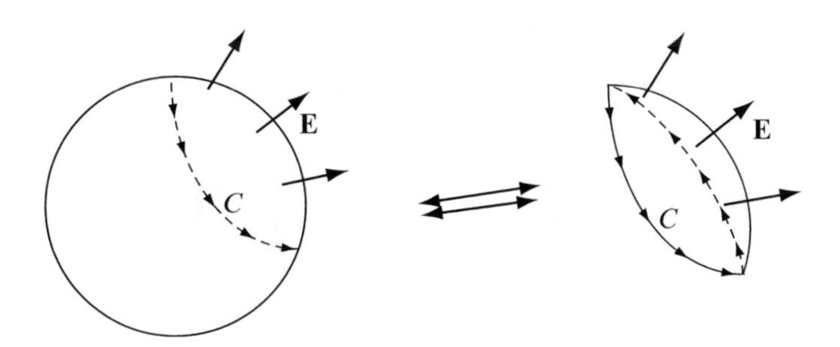

Figure 4.2 A closed contour on the surface of a sphere

To derive the differential relations in **Eqs. (4.6)** and **(4.8)**, we made use of the divergence and Stokes' theorems, both of which are integral relations. If we now return to **Eqs. (4.4)** and **(4.7)**, we can also write directly

$$\boxed{\oint_s \mathbf{E} \cdot d\mathbf{s} = \frac{Q}{\varepsilon_0}, \quad \oint_C \mathbf{E} \cdot d\mathbf{l} = 0} \tag{4.9}$$

The relations in **Eq. (4.9)** are the integral equations from which the differential forms in **Eqs. (4.6)** and **(4.8)** were obtained. They are, therefore, the integral form of the postulates.

[1] Postulates are axioms that have not been disproved. As mentioned before, the postulates, or fundamental relations, are experimentally evaluated quantities, and since experiments have not shown anything to the contrary, we can use them as postulates. In the very unlikely event that one or more of our postulates turn out to be wrong, we would have then to "adjust" all derived relations. There is very little danger of that happening, and, until it does, we can safely use the current relations.

To summarize, the two required postulates for the electrostatic field are

$$\begin{array}{|llr|}\hline \nabla\cdot\mathbf{E}=\dfrac{\rho_v}{\varepsilon_0}, & \nabla\times\mathbf{E}=0 & \text{in differential form} \\ \text{or} & & \\ \oint_s \mathbf{E}\cdot d\mathbf{s}=\dfrac{Q}{\varepsilon_0}, & \oint_C \mathbf{E}\cdot d\mathbf{l}=0 & \text{in integral form} \\ \hline\end{array} \tag{4.10}$$

Any electrostatic field must satisfy these conditions.

Example 4.1 The Electrostatic Field The following vector fields are given in free space:
(1) $\mathbf{A}_1=\hat{\mathbf{x}}x$, (2) $\mathbf{A}_2=\hat{\mathbf{x}}5$, (3) $\mathbf{A}_3=\mathbf{A}_1+\mathbf{A}_2=\hat{\mathbf{x}}(x+5)$

(a) Are these electrostatic fields?
(b) For those fields that are electrostatic, calculate the equivalent charge density that generates these fields in free space.

Solution: For a vector field to be an electrostatic field, it must satisfy the two postulates in **Eq. (4.10)**; that is, its curl must be zero and its divergence must be nonzero within the charge distribution, if any:

(a) The divergence of each of the three vector fields is

$$\nabla\cdot\mathbf{A}_1=\frac{\partial A_x}{\partial x}+\frac{\partial A_y}{\partial y}+\frac{\partial A_z}{\partial z}=\frac{d(x)}{dx}=1,\quad \nabla\cdot\mathbf{A}_2=0,\quad \nabla\cdot\mathbf{A}_3=1$$

The first condition, namely, that the divergence be nonzero (or zero if there are no charge densities), is satisfied. The curl of $\mathbf{A}$ is

$$\nabla\times\mathbf{A}_1=\hat{\mathbf{x}}\left(\frac{\partial A_z}{\partial y}-\frac{\partial A_y}{\partial z}\right)+\hat{\mathbf{y}}\left(\frac{\partial A_x}{\partial z}-\frac{\partial A_z}{\partial x}\right)+\hat{\mathbf{z}}\left(\frac{\partial A_y}{\partial x}-\frac{\partial A_x}{\partial y}\right)=\hat{\mathbf{x}}(0-0)+\hat{\mathbf{y}}\left(\frac{\partial(x)}{\partial z}-0\right)+\hat{\mathbf{z}}\left(0-\frac{\partial(x)}{\partial y}\right)=0$$

$$\nabla\times\mathbf{A}_2=0,\quad \nabla\times\mathbf{A}_3=0$$

Thus, since both postulates are satisfied, the vector fields $\mathbf{A}_1$, $\mathbf{A}_2$, and $\mathbf{A}_3$ represent electrostatic fields. Note in particular that vector field $\mathbf{A}_2$ has zero divergence. This means that within the domain considered, there are no charge densities and the field is generated by charges outside the solution domain. How this is done will be shown later in this chapter. In this respect, it is worth recalling that the Helmholtz theorem defines the vector field to within an additive constant: the divergence or the curl is not changed by adding a constant vector to $\mathbf{A}_1$, $\mathbf{A}_2$, or $\mathbf{A}_3$.

(b) The charge density everywhere is calculated from the first postulate [**Eq. (4.6)**]:

$$\nabla\cdot\mathbf{A}_1=\frac{\rho_1}{\varepsilon_0}=1\ \rightarrow\ \rho_1=\varepsilon_0=8.854\times10^{-12}\quad \left[\text{C/m}^3\right]$$

Similarly,

$$\rho_2=0,\quad \rho_3=\varepsilon_0=8.854\times10^{-12}\quad [\text{C/m}^3]$$

These volume charge densities are uniform throughout space.

Exercise 4.1 Is the vector field $\mathbf{A}=\hat{\boldsymbol{\phi}}r^2-\hat{\mathbf{z}}5z$ an electrostatic field? Explain.

Answer No, because its curl is nonzero: $\nabla\cdot\mathbf{A}=-5$, $\nabla\times\mathbf{A}=\hat{\mathbf{z}}3r$.

4.3 Gauss's Law[2]

After calculating the electric fields of point and distributed charges, where we had to perform rather intricate integrations, we should ask ourselves the following question: Is there a way to calculate the electric field and forces in the electric field in a simpler manner? The answer is, at least partially, yes and we will proceed now to define both the relations required and the conditions under which this is possible. To do so, we start again with the electric field intensity of a point charge [**Eq. (4.1)**]. The electric field intensity of a point charge at any radial distance R is constant in magnitude and points radially outward (or radially inward for a negative charge). If we know the electric field intensity, we can also calculate the total electric flux passing through the surface using the first relation in **Eq. (4.9)**. This is done by integrating over the surface of the sphere of radius R:

$$\oint_s \mathbf{E} \cdot d\mathbf{s} = \oint_s \left(\hat{\mathbf{R}} \frac{Q}{4\pi\varepsilon_0 |\mathbf{R}|^2} \right) \cdot (\hat{\mathbf{R}} ds) = \frac{Q}{\varepsilon_0} \tag{4.11}$$

That is, the surface integral of **E** over the sphere of radius R, on which the electric field intensity is constant in magnitude, equals the total charge enclosed by the sphere. Alternatively, we may use **Eq. (3.55)** to write the electric flux density as $\mathbf{D} = \varepsilon_0 \mathbf{E}$ and substitute this in **Eq. (4.11)**. This leads to a relation called ***Gauss's law***, which may be written in one of two forms:

$$\boxed{\oint_s \mathbf{D} \cdot d\mathbf{s} = Q \quad [\mathrm{C}]} \tag{4.12}$$

$$\boxed{\oint_s \mathbf{E} \cdot d\mathbf{s} = \frac{Q}{\varepsilon_0}} \tag{4.13}$$

where Q is the total charge enclosed by the surface s. This law is very important in electrostatics, but we will meet it many more times when discussing other types of fields. Either the form in **Eq. (4.12)** or **Eq. (4.13)** may be used, depending on application and on convenience. **Equation (4.12)** is particularly useful because it is independent of material properties. We note here a number of properties of Gauss's law:

(1) The charge Q is the total charge enclosed by the surface s. In this case, s is the surface of the sphere of radius R. Here, we used a point charge, but any other charge distribution can be thought of as an assembly of point charges. Thus, Gauss's law will be useful for other types of distributions.

(2) The left-hand side of **Eq. (4.12)** is the total electric flux passing through the surface s. Since the unit of flux density **D** is $\mathrm{C/m^2}$, the unit of electric flux is the ***coulomb*** [C].

(3) Gauss's law states that the total electric flux through a closed surface is equal to the *charge enclosed by this surface*. It is extremely important to remember this relationship; it implies that any charge outside the surface s does not contribute to the flux through the surface.

(4) Gauss's law is an alternate form of Coulomb's law. It does not introduce any new quantities. It simply restates what we already know about the electric field intensity or the electric flux density in a more general form.

(5) Gauss's law may be used either to calculate the equivalent charges from known electric fields or electric fields due to known charges.

---

[2] Johann Carl Friedrich Gauss (1777–1855) is considered to be one of the greatest mathematicians that ever lived. He also contributed to mechanics and optics, as well as electricity and magnetism, in which connection he is noted here. Gauss was an unusual man of science. Most of his discoveries in mathematics date to the time he was between 14 and 17 years old. For example, he developed the least squares method in 1794 at age 17. He was a modest man, reluctant to take credit. Much of his work was only published after his death, some almost 100 years after he died. Gauss is also remembered for practical inventions in electromagnetics, most notably an electromagnetic telegraph and a magnetometer (instrument for measurement of magnetic fields). Because of the latter, a unit of magnetic field (albeit in the cgs system of units) was named after him.

(6) Note that if the electric field intensity were to be negative, the charge would be negative.
(7) The scalar product used in the integrand in Gauss's law means that the vector notation is lost in the calculation process. If the vector forms are needed (for example, when calculating the electric field intensity), these must be restored based on the charge distribution or other physical considerations.

4.3.1 Applications of Gauss's Law

There are two basic forms in which Gauss's law may be used:

(1) Calculation of the electric field intensity or electric flux density from known charge configurations.
(2) Calculation of the equivalent charge in a volume, provided the electric field intensity (or electric flux density) is known everywhere in space, but, in particular, at the location Gauss's law is applied.
The evaluation of electric field intensities and equivalent charge densities is discussed next and in examples.

4.3.1.1 Calculation of the Electric Field Intensity

To use Gauss's law for the calculation of the electric field intensity **E**, there is a slight difficulty: the unknown value in **Eq. (4.13)** is inside the integral and, in addition, the scalar product between the electric field intensity **E** and the vector *d***s** must be evaluated. This certainly cannot be done in general configurations. However, under two special conditions, the evaluation of the electric field intensity is very simple. These are as follows:

(1) The angle between the electric field intensity and the surface s is constant anywhere on the surface. This allows the evaluation of the scalar product and, therefore, of the integral. In particular, if the electric field intensity is in the direction of *d***s** (perpendicular to the surface), the scalar product is equal to the product of the magnitudes of **E** and *d***s**. If **E** is in the direction opposite *d***s**, the scalar product is the negative of the product of the magnitudes of **E** and *d***s**. If the electric field intensity is parallel to the surface, the scalar product $\mathbf{E} \cdot d\mathbf{s}$ is zero (see **Figures 4.3a** and **4.3b**).
(2) The electric field intensity is constant in magnitude over the surface s. This means that the electric field intensity can be taken out of the integral after evaluating the scalar product in **Eq. (4.13)**.

A surface that satisfies these conditions is called a ***Gaussian surface***. With the above conditions, we can write on the Gaussian surface

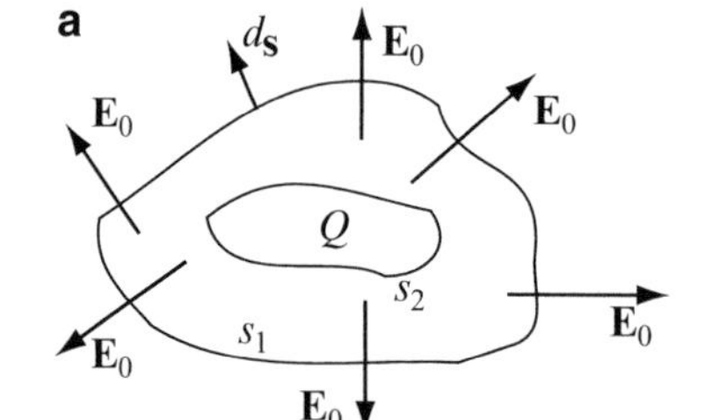

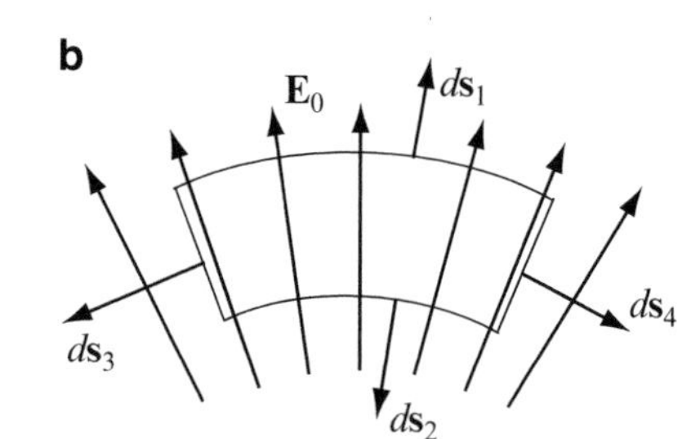

Figure 4.3 Conditions necessary for the application of Gauss's law:
(**a**) $\mathbf{E}_0 \cdot d\mathbf{s} = E_0 ds$.
(**b**) $\mathbf{E}_0 \cdot d\mathbf{s}_1 = E_0 ds_1$,
$\mathbf{E}_0 \cdot d\mathbf{s}_2 = -E_0 ds_2$,
$\mathbf{E}_0 \cdot d\mathbf{s}_3 = \mathbf{E}_0 \cdot d\mathbf{s}_4 = 0$

$$\oint_s \mathbf{E} \cdot d\mathbf{s} = E\oint_s ds = \frac{Q}{\varepsilon_0} \quad \text{or} \quad E = \frac{Q}{\varepsilon_0 \oint_s ds} \quad \left[\frac{\mathrm{N}}{\mathrm{C}}\right] \tag{4.14}$$

Similarly, for the electric flux density **D**, we can write

$$D = \frac{Q}{\oint_s ds} \quad \left[\frac{\mathrm{C}}{\mathrm{m}^2}\right] \tag{4.15}$$

Note that the vector notation has been lost since the scalar product was evaluated beforehand. However, from knowledge of location and signs of charges, we can easily restore the direction of the fields, as will be shown shortly.

Before continuing, we must note that although the simplifications afforded by the assumptions used here restrict its application to a small set of simple configurations, the idea of Gauss's law is completely general. Also, the charge in **Eqs. (4.14)** and **(4.15)** may be replaced by a line, surface, or volume charge distribution by using one of the relations

$$Q = \int_{l'} \rho_l dl' \quad Q = \int_{s'} \rho_s ds' \quad Q = \int_{v'} \rho_v dv' \quad [\text{C}] \tag{4.16}$$

as long as the basic assumptions of a constant, perpendicular (or parallel) field are preserved. Also, note that the surface in **Eq. (4.16)** is not necessarily the same as in **Eq. (4.14)** or **(4.15)** although it can be. The surface s', contour l', and volume v' indicate the domains in which charge densities exist. The integrals in **Eq. (4.16)** are over the sources enclosed by the Gaussian surface, whereas the surface integrals in **Eqs. (4.14)** and **(4.15)** are over the Gaussian surface itself.

For the conditions of a Gaussian surface to be satisfied, the charges or charge distributions must be highly symmetric, as we shall see in examples. In fact, there are only four general classes of problems that can be solved under these conditions:

(1) Point charges.
(2) Symmetrical spherical charge distributions and charged spherical layers.
(3) Infinite lines of charges or symmetrically charged cylindrical objects.
(4) Infinite charged surfaces or infinite charged layers.

These classes of problems (see examples that follow) are rather limited, but their usefulness can be extended through the use of superposition of solutions and by using Gauss's law on geometries which are approximately the same as one of the classes above. For example, a square cross-sectional object may be approximated as an assembly of thinner, round objects, whereas a finite length line may be solved as if it were infinite in extent, provided the line is "long."

Before demonstrating the use of Gauss's law, we wish to outline the method in terms of the basic steps required for solution:

(1) Inspect the charges or charge distributions. Careful study of symmetries and charge distributions can provide hints on how to approach the problem or, in fact, if the problem is solvable using Gauss's law.
(2) Sketch the electric field (direction and magnitude) to see if a surface can be identified on which the electric field intensity is both constant in magnitude and directed perpendicular to the surface, or a surface to which the electric field intensity is parallel. In some cases, sections of a surface may be found on which the electric field intensity is perpendicular and constant, whereas on other sections, the electric field intensity is parallel to the surface. For example, for a point charge, the surface will be spherical, whereas for a straight line, a cylindrical surface will be required. The Gaussian surface must pass through the location at which the field is evaluated and must be a closed surface.
(3) If a Gaussian surface cannot be found, try to separate the charge distribution into two or more charge configurations (without modifying the problem) for each of which a Gaussian surface can be obtained. The electric field intensity of the configuration will then be the superposition of the individual fields.
Calculate the electric fields on the Gaussian surface or surfaces.

Example 4.2 Electric Field of Spherical Charge Distributions Two very thin, spherical, conducting shells are charged as in **Figure 4.4a**. The inner shell has a total charge $-Q$, whereas the outer shell has a total charge $+Q$. Charges are distributed uniformly on the surfaces of the corresponding shells:

(a) Calculate the electric field intensity everywhere in space.
(b) A point charge of magnitude $+Q$ is now inserted at the center of the two shells as shown in **Figure 4.4b**. Calculate the electric field intensity everywhere in space.
(c) Plot the two solutions.

Solution: Because the electric field intensity for this symmetric charge distribution is everywhere radial and constant at constant distances from the center, the Gaussian surfaces required are spheres of various radii. Here, we calculate the field in

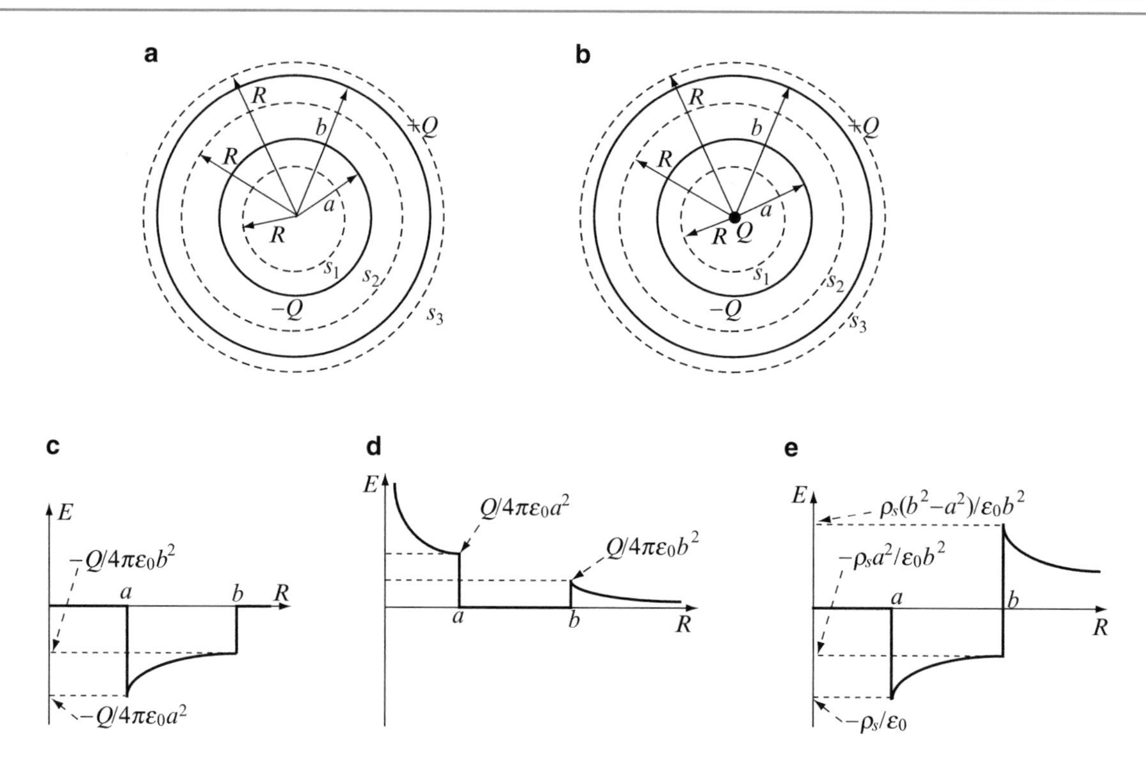

Figure 4.4 (**a**) Two uniformly charged spherical shells. (**b**) A point charge surrounded by two uniformly charged spherical shells. (**c**) Plot of the electric field intensity in (**a**). (**d**) Plot of the electric field intensity in (**b**). (**e**) Plot of the electric field intensity in **Exercise 4.2**

three sections: inside the inner shell, between the two shells, and outside the outer shell. The three surfaces are shown as dotted lines in **Figures 4.4a** and **4.4b**.

(a) Draw a Gaussian surface inside the inner shell in **Figure 4.4a**. Since this shell encloses no charge, the electric field intensity is zero:

$$\mathbf{E} = 0, \quad 0 \leq R < a$$

The field between the two shells is calculated on s_2. Because $\mathbf{E}$ is perpendicular to this surface and the Gaussian surface encloses a total charge $-Q$, we can write

$$E4\pi R^2 = \frac{-Q}{\varepsilon_0} \quad \rightarrow \quad E = \frac{-Q}{4\pi\varepsilon_0 R^2} \quad \text{or} \quad \mathbf{E} = -\hat{\mathbf{R}}\frac{Q}{4\pi\varepsilon_0 R^2} \quad \left[\frac{\text{N}}{\text{C}}\right], \quad a < R < b$$

where the fact that the field is known to be radial was used to write the vector form of the field. Note also that this solution is the same as if the charge were located at the center of the sphere as a point charge.

Outside the outer shell, the Gaussian surface is s_3. This encloses both charged shells symmetrically. The total charge enclosed by s_3 is zero. Thus, the electric field intensity outside the outer shell is zero:

$$\mathbf{E} = 0, \quad b < R < \infty$$

(b) The solution in part **(a)** can be used here as well, but the charges enclosed by the various Gaussian surfaces are different. Inside surface s_1, there is a point charge $+Q$. Gauss's law gives

$$E4\pi R^2 = \frac{Q}{\varepsilon_0} \quad \rightarrow \quad E = \frac{Q}{4\pi\varepsilon_0 R^2} \quad \text{or} \quad \mathbf{E} = \hat{\mathbf{R}}\frac{Q}{4\pi\varepsilon_0 R^2} \quad \left[\frac{\text{N}}{\text{C}}\right], \quad 0 < R < a$$

Between the two shells, surface s_2 now encloses zero charge (a point charge +Q and a surface charge totaling –Q). Thus, the electric field intensity between the two shells is zero:

$$\mathbf{E} = 0, \quad a < R < b$$

Surface s_3 encloses a net positive charge +Q. Thus, the electric field intensity outside the outer shell is

$$\mathbf{E} = \hat{\mathbf{R}}\frac{Q}{4\pi\varepsilon_0 R^2} \quad \left[\frac{\mathrm{N}}{\mathrm{C}}\right], \quad b < R < \infty$$

(c) The two solutions are drawn in **Figures 4.4c** and **4.4d**. Note the jump in the electric field intensity at the shells themselves. This is due to the charges on the shells. Also, in **(b)**, the electric field intensity at $R = 0$ tends to $+\infty$, because of the existence of a point charge at this location.

Exercise 4.2 Consider the geometry in **Figure 4.4a**. Assume a uniform surface charge density $-\rho_s$ [C/m^2] is placed on the inner shell and a uniform surface charge density $+\rho_s$ [C/m^2] is placed on the outer shell. Calculate the electric field intensity everywhere in space. Plot the solution.

Answer

$$\mathbf{E} = 0, \ 0 \le R < a, \qquad \mathbf{E} = -\hat{\mathbf{R}}\frac{a^2\rho_s}{\varepsilon_0 R^2} \quad \left[\frac{\mathrm{N}}{\mathrm{C}}\right], \ a < R < b,$$

$$\mathbf{E} = \hat{\mathbf{R}}\frac{(b^2 - a^2)\rho_s}{\varepsilon_0 R^2} \quad \left[\frac{\mathrm{N}}{\mathrm{C}}\right], \ b < R < \infty$$

See **Figure 4.4e** for a plot of the solution.

Example 4.3 Application: Electric Field of Surface Layers The electric field of flat layers is very important, especially in design of capacitors and capacitor-like devices. Here, we show how Gauss's law can be used to calculate the electric field intensity of simple charged layers.

A charge density ρ_s [C/m^2] is uniformly distributed over a very large surface (such as a large, thin aluminum foil), as in **Figure 4.5**. Calculate the electric field intensity everywhere in space.

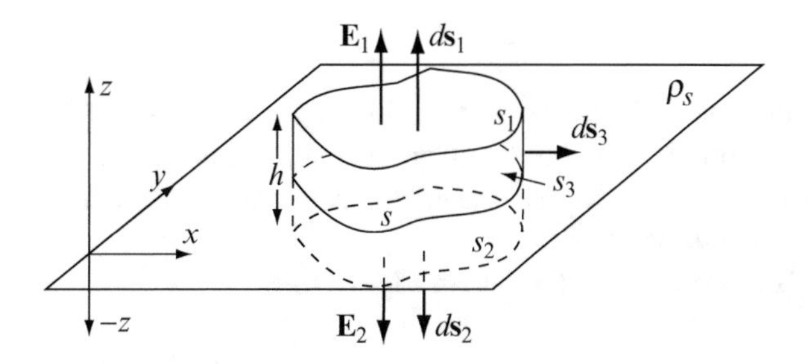

Figure 4.5 A large (infinite) charged layer and the Gaussian surface required for calculation of the electric field intensity

Solution: From symmetry considerations, the electric field must be perpendicular to the layer and constant in magnitude on any surface parallel to the layer. The electric field points in the negative z direction below the layer and in the positive z direction above the layer. The appropriate Gaussian surface is shown in **Figure 4.5**. Note that $d\mathbf{s}$ is positive in the direction out of the volume. Assuming $s_1 = s_2 = s$, the Gaussian surface cuts a section of the charged layer equal to s. The charge enclosed by the surface is, therefore, $\rho_s s$.

The scalar product $\mathbf{E} \cdot d\mathbf{s}$ is nonzero on s_1 and s_2 but is zero on the lateral surface ($\mathbf{E} \cdot d\mathbf{s}_3 = 0$). Therefore,

$$\oint_s \mathbf{E} \cdot d\mathbf{s} = \int_{s_1} \mathbf{E}_1 \cdot d\mathbf{s}_1 + \int_{s_2} \mathbf{E}_2 \cdot d\mathbf{s}_2 = E_1 s_1 + E_2 s_2 = 2Es$$

where $E_1 = E_2 = E$ from symmetry considerations. Thus,

$$2Es = \frac{\rho_s s}{\varepsilon_0} \quad \rightarrow \quad E = \frac{\rho_s s}{2s\varepsilon_0} = \frac{\rho_s}{2\varepsilon_0} \quad \left[\frac{\text{N}}{\text{C}}\right]$$

The electric field intensity is in the positive z direction above the layer and in the negative z direction below the layer (recall that the electric field intensity always points away from positive charges). The solution is

$$\mathbf{E} = \hat{\mathbf{z}}\frac{\rho_s}{2\varepsilon_0} \quad \left[\frac{\text{N}}{\text{C}}\right] \quad (z > 0) \quad \text{and} \quad \mathbf{E} = -\hat{\mathbf{z}}\frac{\rho_s}{2\varepsilon_0} \quad \left[\frac{\text{N}}{\text{C}}\right] \quad (z < 0)$$

Example 4.4 Application: Electric Field of a DC Overhead Transmission Line—Gauss's Law and Superposition Overhead transmission lines generate an electric field everywhere in space. The electric field intensity must be kept low enough for safety reasons. The following example gives an idea of how this electric field behaves and introduces the idea of superposition of solutions using Gauss's law.

Consider a very long (infinite) line, located at a distance $d = 10$ m above ground and charged with a uniform, line charge density $\rho_l = 10^{-7}$ C/m as shown in **Figure 4.6a**. Neglect the ground's influence:

(a) Calculate the electric field intensity everywhere in space.
(b) What is the magnitude of the electric field intensity at ground level, directly below the line?
(c) A second, identical line is now placed at a distance $a = 2$ m below the first line as shown in **Figure 4.7a** and is charged with a line charge density$-\rho_l$ [C/m]. Calculate the electric field intensity at a general point in space as well as at ground level directly below the lines. Compare with **(a)** and **(b)**.

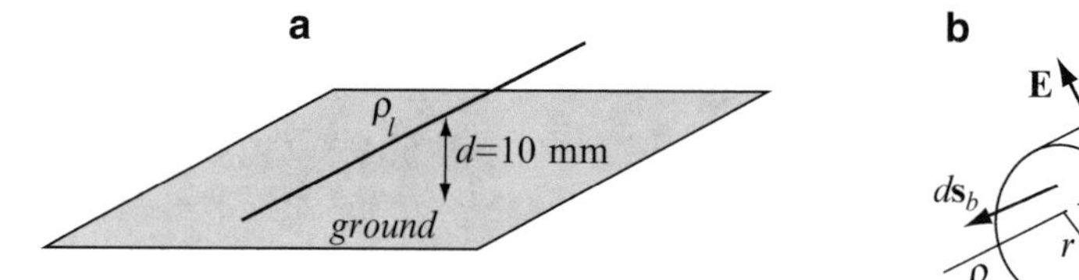

Figure 4.6 (**a**) A single line above ground. (**b**) The Gaussian surface for a long line of charge. The surface is a cylinder of radius r, arbitrary length L, and coaxial with the line

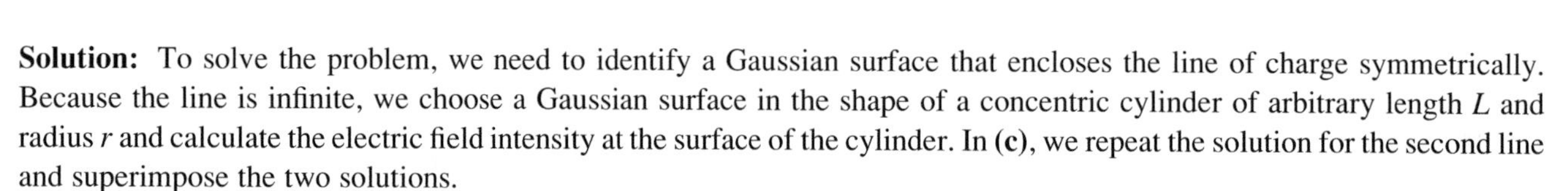

Solution: To solve the problem, we need to identify a Gaussian surface that encloses the line of charge symmetrically. Because the line is infinite, we choose a Gaussian surface in the shape of a concentric cylinder of arbitrary length L and radius r and calculate the electric field intensity at the surface of the cylinder. In **(c)**, we repeat the solution for the second line and superimpose the two solutions.

(a) From Gauss's law applied to the line in **Figure 4.6a**,

$$\oint_s \mathbf{E} \cdot d\mathbf{s} = \frac{\rho_l L}{\varepsilon_0}$$

where L is an arbitrary length of the line of charge enclosed by the Gaussian surface. The Gaussian surface is shown in **Figure 4.6b**. The electric field intensity at a distance r from the line is therefore

$$\oint_s \mathbf{E} \cdot d\mathbf{s} = \int_{s_a} E ds_a = 2\pi r L E = \frac{\rho_l L}{\varepsilon_0} \quad \rightarrow \quad E = \frac{\rho_l}{2\pi\varepsilon_0 r} \quad \left[\frac{\text{N}}{\text{C}}\right]$$

In vector notation (the electric field is in the positive r direction),

$$\mathbf{E} = \hat{\mathbf{r}}\frac{\rho_l}{2\pi\varepsilon_0 r} \quad \left[\frac{\text{N}}{\text{C}}\right]$$

Note that in this case, $d\mathbf{s}_a = \hat{\mathbf{r}}ds_a$ and $\mathbf{E} = \hat{\mathbf{r}}E$. Also, the scalar products $\mathbf{E}\cdot d\mathbf{s}_b$ and $\mathbf{E}\cdot d\mathbf{s}_c$ are zero since $\mathbf{E}$ is perpendicular to $d\mathbf{s}_b$ and $d\mathbf{s}_c$. Therefore, the only nonzero contribution to the scalar product is $\mathbf{E}\cdot d\mathbf{s}_a$ and this equals Eds_a. Because $\mathbf{E}$ is constant in magnitude on the surface, it is taken outside the integral.

(b) At ground level ($r = 10$ m, directly below the line), the field points downward and its magnitude is

$$E = \frac{\rho_l}{2\pi r\varepsilon_0} = \frac{\rho_l}{2\pi d\varepsilon_0} = \frac{10^{-7}}{2\times\pi\times 10\times 8.854\times 10^{-12}} = 179.75 \quad [\text{N/C}]$$

(c) Now, we have two lines, each at a different location in space. Therefore, we will use a Cartesian coordinate system and place the negatively charged line at ($x_2 = 0$, $y_2 = d - a = 8$ m) and the positively charged line at ($x_1 = 0$, $y_1 = d = 10$ m). Ground level is at $y = 0$. We use the result in **(a)** and write (see **Figure 4.7b**)

$$r_1 = \sqrt{x^2 + (y - y_1)^2}, \quad r_2 = \sqrt{x^2 + (y - y_2)^2} \quad [\text{m}]$$

The electric field intensity at $P(x,y)$ due to each line is

$$\mathbf{E}^+ = \hat{\mathbf{r}}_1\frac{\rho_l}{2\pi\varepsilon_0\sqrt{x^2 + (y - y_1)^2}}, \quad \mathbf{E}^- = \hat{\mathbf{r}}_2\frac{-\rho_l}{2\pi\varepsilon_0\sqrt{x^2 + (y - y_2)^2}} \quad \left[\frac{\text{N}}{\text{C}}\right]$$

The total electric field intensity at $P(x,y)$ is the superposition of the two fields. To calculate the vector sum of the two electric field intensities at $P(x,y)$, we separate the two fields into their components. From **Figure 4.7b**, the components of the two electric fields are

$$\mathbf{E}^+ = \hat{\mathbf{x}}E^+\cos\theta_1 + \hat{\mathbf{y}}E^+\sin\theta_1, \quad \mathbf{E}^- = -\hat{\mathbf{x}}E^-\cos\theta_2 - \hat{\mathbf{y}}E^-\sin\theta_2 \quad [\text{N/C}]$$

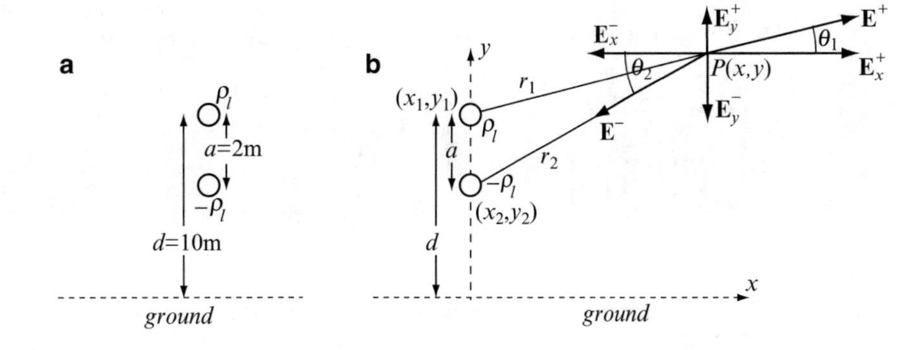

Figure 4.7 (**a**) A transmission line. (**b**) The electric field intensities produced by the two lines at $P(x,y)$

Also from **Figure 4.7b**,

$$\cos\theta_1 = \frac{x}{r_1} = \frac{x}{\sqrt{x^2 + (y - y_1)^2}}, \quad \sin\theta_1 = \frac{y - y_1}{r_1} = \frac{y - y_1}{\sqrt{x^2 + (y - y_1)^2}}$$

$$\cos\theta_2 = \frac{x}{r_2} = \frac{x}{\sqrt{x^2 + (y - y_2)^2}}, \quad \sin\theta_2 = \frac{y - y_2}{r_2} = \frac{y - y_2}{\sqrt{x^2 + (y - y_2)^2}}$$

Substituting these into $\mathbf{E}^+$ and $\mathbf{E}^-$ and summing the components gives the electric field intensity at $P(x,y)$

$$\mathbf{E} = \hat{\mathbf{x}}\frac{\rho_l x}{2\pi\varepsilon_0}\left[\frac{1}{x^2 + (y - y_1)^2} - \frac{1}{x^2 + (y - y_2)^2}\right] + \hat{\mathbf{y}}\frac{\rho_l}{2\pi\varepsilon_0}\left[\frac{y - y_1}{x^2 + (y - y_1)^2} - \frac{y - y_2}{x^2 + (y - y_2)^2}\right] \quad \left[\frac{\text{N}}{\text{C}}\right]$$

At ground level, immediately below the two lines ($x = 0$, $y = 0$) and writing $y_1 = d$ and $y_2 = d - a$, the magnitude of the electric field intensity is

$$E(0,0) = \frac{\rho_l}{2\pi\varepsilon_0}\left[\frac{1}{y_2} - \frac{1}{y_1}\right] = \frac{\rho_l}{2\pi\varepsilon_0}\frac{a}{d(d - a)} = \frac{10^{-7}}{2 \times \pi \times 8.854 \times 10^{-12}} \times \frac{2}{10 \times (10 - 2)} = 44.94 \quad [\text{N/C}]$$

This electric field intensity is much lower than that of a single line. It can be reduced further by decreasing the distance between the two lines.

4.3.1.2 Calculation of Equivalent Charges

The second application of Gauss's law is for the calculation of charge distributions from known electric field intensities. This relies on Gauss's law in the following form:

$$Q = \varepsilon_0 \oint_s \mathbf{E} \cdot d\mathbf{s} = \oint_s \mathbf{D} \cdot d\mathbf{s} \quad \left[\frac{\text{C}}{\text{m}^2}\right] \tag{4.17}$$

Thus, as long as the electric field intensity is known, the equivalent charge can be evaluated. We note, however, that only the equivalent charge is evaluated, not the exact distribution. To see that this is the case, consider again **Figure 4.1**. The point charge Q in **Figure 4.1a** and the volume charge density ρ_v in **Figure 4.1b** generate an identical electric field intensity $\mathbf{E}$ at a distance R if the total charge in **Figure 4.1b** is the same as that in **Figure 4.1a**. Thus, even though the two charge distributions are quite different, the equivalent charges calculated from Gauss's law are the same and equal to Q. This fact is not very surprising because of the integral relation used, but in practical terms, it also means that, in general, it is not possible to distinguish between the fields of point and distributed charges from knowledge of the electric field intensity alone. Sometimes, there may be additional information on the charge which allows us to distinguish between point charges and other distributions and their corresponding electric field intensities. The following two examples show applications of Gauss's law to calculation of equivalent charges using **Eq. (4.17)**.

Example 4.5 Application: Electric Field Intensity and Charge Density In and Around a Thundercloud A spherical cloud of charged particles of radius $R_0 = 1$ km produces a known electric field intensity inside the cloud ($R \leq R_0$), given as $\mathbf{E} = \hat{\mathbf{R}}R^2$ [N/C]. Calculate:

(a) The total charge in the cloud.
(b) The charge density in the cloud.

Solution: The total charge in the cloud is found from Gauss's law from the fact that outside the charge distribution, the electric field intensity only depends on the total charge enclosed by the Gaussian surface, not on its distribution. From this, both the charge distribution and the electric field intensity may be found:

(a) Enclosing the sphere by a Gaussian surface of radius R_0, we write from Gauss's law,

$$Q = 4\pi R_0^2 \varepsilon_0 E = 4\pi R_0^2 \varepsilon_0 R_0^2 = 4\pi \varepsilon_0 R_0^4 \quad [\mathrm{C}]$$

Numerically we get

$$Q = 4\pi \times 8.854 \times 10^{-12} \times 1000^4 = 111.26 \quad [\mathrm{C}].$$

(b) One method of calculating the charge density is to assume a radius-dependent charge density of general form, integrate over the volume of the sphere, and equate the result to that found in **(a)**. A simpler method is to calculate the divergence of the electric field intensity in the charge distribution. We show both.

Method A Assume a charge distribution of the form $\rho_v(R)$ [C/m^3] and calculate the total charge in the sphere:

$$\int_v \rho_v(R)dv = \int_{R=0}^{R_0}\int_{\theta=0}^{\pi}\int_{\phi=0}^{2\pi} \rho_v(R)R^2 \sin\theta dR d\theta d\phi \qquad [\mathrm{C}]$$

Although the charge distribution is not known, we can still integrate over θ and ϕ. This gives

$$\int_v \rho_v(R)dv = 4\pi \int_{R=0}^{R_0} \rho_v(R)R^2 dR = 4\pi R_0^4 \varepsilon_0 \qquad [\mathrm{C}]$$

where the right-hand side is the result in **(a)**. Clearly, for this to be satisfied, we must have

$$\rho_v(R) = 4\varepsilon_0 R \qquad [\mathrm{C/m^3}]$$

Method B The divergence of the electric field intensity [**Eq. (4.6)**] in spherical coordinates gives

$$\nabla \cdot \mathbf{E} = \frac{1}{R^2}\frac{\partial (R^2 E_R)}{\partial R} = \frac{1}{R^2}\frac{\partial (R^4)}{\partial R} = 4R = \frac{\rho_v}{\varepsilon_0} \quad \rightarrow \quad \rho_v = 4\varepsilon_0 R \;\; [\mathrm{C/m^3}]$$

which, of course is the same as in method **A**.

Example 4.6 Application: Generation of a Uniform Electric Field Intensity It is necessary to generate a uniform electric field intensity of magnitude $E_0 = 10^5$ N/C between two very large parallel plates, separated a distance $d = 10$ mm as shown in **Figure 4.8a**. The material between the plates is free space and the electric field intensity outside the plates is zero. Calculate the required charge densities on the surface of each plate.

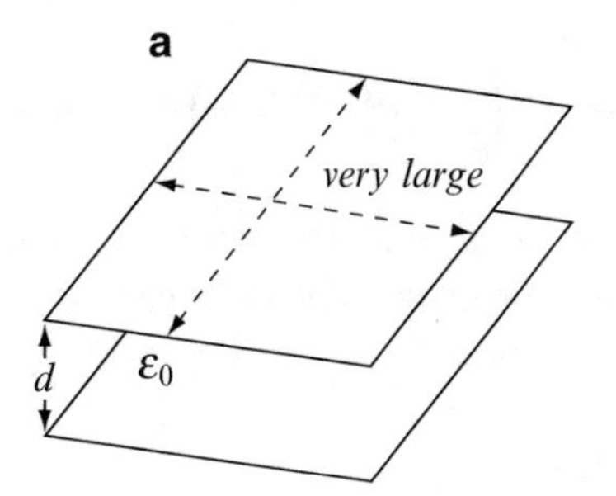

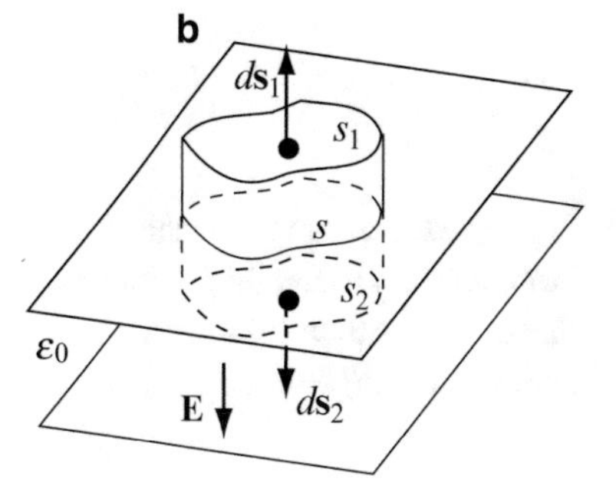

Figure 4.8 (**a**) Two very large charged plates. (**b**) The Gaussian surface necessary to calculate the electric charge density on the upper plate

Solution: A Gaussian surface is drawn in the form of an arbitrarily shaped volume symmetric about one of the plates, as shown in **Figure 4.8b**. This volume has surfaces either parallel to the plate or perpendicular to it to facilitate calculation of the scalar product in Gauss's law. Also, we assume the electric field points from the upper to the lower plate. From Gauss's law, we have

$$\oint_s \mathbf{E} \cdot d\mathbf{s} = E\int_{s2} ds_2 = Es_2 = \frac{Q}{\varepsilon_0}$$

The only charge is on the surface s in the form of a uniform charge density ρ_s [C/m^2]. The total charge on surface s is $\rho_s s$ [C]. Since $s = s_1 = s_2$, we get

$$Es_2 = \frac{\rho_s s_2}{\varepsilon_0} \quad \rightarrow \quad \rho_s = \varepsilon_0 E = 8.854 \times 10^{-12} \times 10^5 = 8.854 \times 10^{-7} \quad [\mathrm{C/m^2}]$$

The charge density on the lower plate must be identical and negative (since $\mathbf{E}$ and $d\mathbf{s}$ are in opposite directions and $\mathbf{E}$ points from upper to lower plate). Thus, the solution is

$\rho_s = 8.854 \times 10^{-7}$ $[\mathrm{C/m^2}]$ on the upper plate

and

$\rho_s = -8.854 \times 10^{-7}$ $[\mathrm{C/m^2}]$ on the lower plate

4.4 The Electric Potential

Up to this point, we tried to calculate the electric field intensity directly from charges and charge densities. This, of course, is a natural choice for calculation since charges are the sources of electric fields. In practice, however, it is sometimes useful to consider indirect methods of calculation. One general method of calculating fields is through the use of potentials. In particular, the scalar electric potential will be defined here from fundamental ideas of force and work related to the electric field. In subsequent sections, we will also view the scalar electric potential from the vector relations point of view and, at that point, will formalize the calculation of electric fields from the scalar potential. In many cases, this will be easier to do than direct calculation of fields because of the scalar nature of the potential.

To define the electric potential, we must first define work and must decide what constitutes positive and negative work. Positive work is defined as work performed ***against the system***, whereas negative work is that performed ***by the system***. Thus, moving a positive charge in an electric field produced by a positive charge, against the field, is considered to be positive work because by doing so, we increase the potential energy of the system. On the other hand, if the charge is allowed to move due to forces in the electric field, the charge will eventually come to rest in a position of minimum potential energy. In the process, the system has lost potential energy, which produces negative work. With these concepts in mind, the differential of work in moving a positive charge at constant velocity against the field is

$$dW = -\mathbf{F} \cdot d\mathbf{l} = -Q\mathbf{E} \cdot d\mathbf{l} \ \ [\mathrm{J}] \tag{4.18}$$

As we move the charge at constant velocity along any path in the field, the work performed is the integral of dW along the path. Referring to **Figure 4.9**, the increase in potential energy as the charge Q is moved between points a and b is

$$W = -Q\int_a^b \mathbf{E} \cdot d\mathbf{l} = W_b - W_a \quad [\mathrm{J}] \ \text{or} \ [\mathrm{N \cdot m}] \tag{4.19}$$

This expression gives the difference in potential energy between the two points in much the same way as raising a weight from height a to height b. Since the work performed depends on the charge Q, it is more convenient to define work per unit of charge rather than total work. To do so, we divide both sides of **Eq. (4.19)** by Q. This quantity is the ***scalar electric potential or potential difference***:

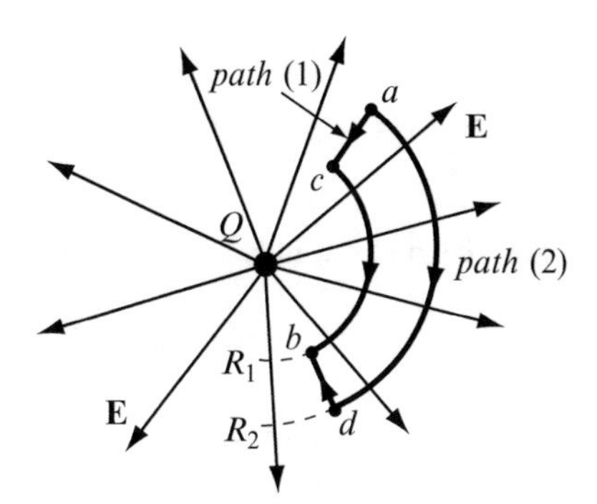

Figure 4.9 Calculation of potential difference. In moving from point a to point b, we choose convenient paths to facilitate the evaluation of the scalar product $\mathbf{E} \cdot d\mathbf{l}$

$$V_{ba} = \frac{W}{Q} = -\int_a^b \mathbf{E} \cdot d\mathbf{l} = \frac{W_b - W_a}{Q} = V_b - V_a \quad \left[\frac{\mathrm{J}}{\mathrm{C}}\right] \text{ or } \left[\frac{\mathrm{N} \cdot \mathrm{m}}{\mathrm{C}}\right] \tag{4.20}$$

Note the convention V_{ba}; it indicates that this is the potential difference between point b (end point in moving the potential) and point a (starting point). Thus, ***potential*** is only properly defined as a difference in the potential of two points. We can view the potential at point a as a reference potential. In particular, if the potential at point a happens to be zero, then the potential at point b can be viewed as the absolute potential. This will be discussed further in the following section.

Before continuing we note that the units of potential are [J/C] or [N · m/C]. This unit is designated as the ***volt***.[3] Until now, the electric field intensity was given in units of [N/C] to indicate its definition as force per unit charge. Now, we replace this with [V/m] which is a more convenient and more commonly used unit of electric field intensity and also shows the relation between electric field intensity and potential.

To summarize:

Units of potential:

$$[\mathrm{V}] = [\mathrm{volt}] = \left[\frac{\mathrm{joule}}{\mathrm{coulomb}}\right] = \left[\frac{\mathrm{newton} \times \mathrm{meter}}{\mathrm{coulomb}}\right]$$

Units of electric field intensity:

$$\left[\frac{\mathrm{V}}{\mathrm{m}}\right] = \left[\frac{\mathrm{volt}}{\mathrm{meter}}\right] = \left[\frac{\mathrm{newton} \times \mathrm{meter}}{\mathrm{coulomb} \times \mathrm{meter}}\right] = \left[\frac{\mathrm{newton}}{\mathrm{coulomb}}\right] = \left[\frac{\mathrm{N}}{\mathrm{C}}\right]$$

4.4.1 Electric Potential Due to Point Charges

For a point charge Q located at the origin of the spherical system of coordinates, the electric field intensity anywhere in space is given by **Eq. (4.1)**. The resulting electric field is shown in **Figure 4.9**. Now, consider two points a and b at distances R_2 and R_1, respectively, from the origin. To calculate the potential, we choose *path* (1) in **Figure 4.9** and integrate from point a to point c and then from point c to point b. We get

[3] After Count Alessandro Giuseppe Antonio Anastasio Volta (1745–1827). Volta is best known for his work on the electric battery. His invention, the Volta pile, was the first practical battery and is the predecessor to the common modern batteries. However, he also did considerable work in electrostatics, including the invention of one of the simplest electrostatic generators. This was known as the electroforo perpetuo (perpetual electric machine) because it could retain its charge for long periods and could be used to charge other devices repeatedly without need for recharging the device itself. Volta had a long and distinguished career in the sciences, including work on what was at the time called "animal electricity." After Luigi Aloisio Galvani (1737–1798) showed that a dead frog's muscles responded to touching with metals, Volta concluded that a potential is present. This experiment led, indirectly, to the invention of the battery, but also to a long and bitter controversy between the two men on the nature of the phenomenon. Galvani believed that the observation indicated a new type of electricity he called "animal electricity," whereas Volta believed and later proved it to be one and the same as that obtained from batteries.

$$V_{ba} = V_b - V_a = (V_c - V_a) + (V_b - V_c) = -\int_a^c \mathbf{E} \cdot d\mathbf{l}_1 - \int_c^b \mathbf{E} \cdot d\mathbf{l}_2 \quad [\text{V}] \tag{4.21}$$

In this case, calculation is best carried out in spherical coordinates. With $d\mathbf{l} = \hat{\mathbf{R}}dR + \hat{\boldsymbol{\theta}}Rd\theta + \hat{\boldsymbol{\phi}}R\sin\theta d\phi$, $d\mathbf{l}_1$ and $d\mathbf{l}_2$ are

$$d\mathbf{l}_1 = \hat{\mathbf{R}}dR, \qquad d\mathbf{l}_2 = \hat{\boldsymbol{\theta}}Rd\theta \tag{4.22}$$

Before performing the integration in **Eq. (4.21)**, we must evaluate the integrand. In the first integral in **Eq. (4.21)**, **E** and $d\mathbf{l}_1$ are in the same direction; therefore, $\mathbf{E} \cdot d\mathbf{l}_1 = QdR/4\pi\varepsilon_0|\mathbf{R}^2|$. In the second integral, **E** is in the positive R direction and $d\mathbf{l}_2$ is in the positive θ direction. Thus, $\mathbf{E} \cdot d\mathbf{l}_2 = 0$ since **E** and $d\mathbf{l}_2$ are perpendicular to each other. Substituting these into **Eq. (4.21)**, and noting that $R_c = R_b$ (points b and c are at same distance from the origin) and $|\mathbf{R}| = R$, we get

$$V_{ba} = -\int_{R_a}^{R_b} \frac{Q}{4\pi\varepsilon_0 R^2} dR = \left(\frac{Q}{4\pi\varepsilon_0 R_b} - \frac{Q}{4\pi\varepsilon_0 R_a}\right) = \frac{Q}{4\pi\varepsilon_0}\left(\frac{1}{R_b} - \frac{1}{R_a}\right) = V_b - V_a \quad [\text{V}] \tag{4.23}$$

First, we should note that since $R_b < R_a$, this potential is positive. In other words, the closer to the point charge we get, the larger V_b becomes. In fact, at $R_b \to 0$, the potential V_b tends to infinity.

So far, we dealt strictly with potential differences, although in **Eq. (4.23)**, V_b and V_a are the absolute potentials at points b and a, respectively. In general, we cannot calculate these potentials independently. However, in the case of a point charge and other finite-size charge distributions, it is possible to calculate the potential difference between point b and point a, with point a at ∞. In this case, $V_a = 0$ and we get the absolute potential at point b as

$$V_b = -\int_{\infty}^{R_b} \frac{Q}{4\pi\varepsilon_0 R^2} dR = \frac{Q}{4\pi\varepsilon_0 R_b} \quad [\text{V}] \tag{4.24}$$

The potential at point b is now the work per unit charge required to bring a unit charge (test charge) from infinity to point b. Therefore, although we call this an absolute potential, it is a potential difference between point b and the reference point at infinity.

Calculation of absolute potentials is often possible and sometimes preferred because once absolute potentials are known, potential differences are easy to calculate. This is also compatible with our normal practice of assigning a reference value (such as ground potential in a circuit) and measuring all potentials with respect to the reference value. However, there will be instances where the zero potential reference at ∞ is either impossible to satisfy or inconvenient to apply. For example, an infinitely long line or infinite plane extends to infinity and zero potential at infinity would imply zero potential on the line or plane. Now, the potential at infinity cannot be assumed to be zero. But the reference point does not have to be at infinity and does not have to be zero. For example, if you measure potential (voltage) in an electric circuit, the measurement is with respect to the "ground" or "chassis" of the instrument. We usually assume ground is at zero potential but not necessarily so. The chassis itself may be at a certain potential above ground level and planet Earth may be at yet another potential level with respect to the stars. Other reference values and other locations in space can then be chosen for this purpose, as we shall see in examples.

All relations given so far were defined with the charge at the origin. This is obviously only a convenient choice. If the charge is located at any other point in space, the use of position vectors allows a more general expression. Referring to **Figure 4.10**, we can write for the charge Q,

$$V_b = \frac{Q}{4\pi\varepsilon_0|\mathbf{R}_b - \mathbf{R}'|} \quad \text{and} \quad V_a = \frac{Q}{4\pi\varepsilon_0|\mathbf{R}_a - \mathbf{R}'|} \quad [\text{V}] \tag{4.25}$$

and the potential difference between the two points is

$$V_b - V_a = \frac{Q}{4\pi\varepsilon_0}\left(\frac{1}{|\mathbf{R}_b - \mathbf{R}'|} - \frac{1}{|\mathbf{R}_a - \mathbf{R}'|}\right) \quad [\text{V}] \tag{4.26}$$

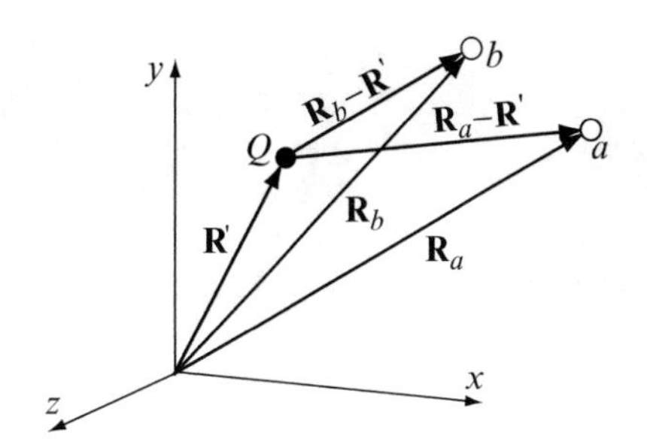

Figure 4.10 Use of position vectors to calculate the potentials at points a and b

Perhaps the most remarkable aspect of the results obtained above is that the potential difference is independent of the path we take to evaluate it. Two paths are shown in **Figure 4.9**. On either path shown, the scalar product between **E** and $d\mathbf{l}$ can be separated into a component along the electric field intensity and one perpendicular to the electric field intensity and only the component along the electric field intensity contributes to the scalar product. Thus, movement in the radial direction changes the potential (contributes to the integral), whereas movement of the test charge perpendicular to the radial direction (along any circular path of constant radius) does not change the potential. The latter aspect of the field has two consequences:

(1) The potential only depends on the radial distances between points a and b and the point charge, regardless of path taken. In a more general sense, the potential difference is independent of path for any electrostatic field, regardless of how it may have been generated.
(2) The potential on any spherical surface enclosing a point charge (centered at the charge) is constant. These spheres are surfaces of constant potential. We note that potential is constant on lines or surfaces perpendicular to the electric field intensity. In other configurations, the lines of constant potential may not be spherical, but they are always perpendicular to the electric field intensity.

Now, for the first advantage in using the potential: its scalar nature. Consider first a number of point charges as shown in **Figure 4.11**. The potential at a point $P(\mathbf{R})$ in space is a scalar summation of the potentials of the individual charges. Extending the result of the single point charge, we can write

$$V(\mathbf{R}) = \frac{Q_1}{4\pi\varepsilon_0|\mathbf{R}-\mathbf{R}_1|} + \frac{Q_2}{4\pi\varepsilon_0|\mathbf{R}-\mathbf{R}_2|} + \frac{Q_3}{4\pi\varepsilon_0|\mathbf{R}-\mathbf{R}_3|} + \cdots + \frac{Q_n}{4\pi\varepsilon_0|\mathbf{R}-\mathbf{R}_n|} \quad [\mathrm{V}] \tag{4.27}$$

or, in a more compact form,

$$\boxed{V(\mathbf{R}) = \frac{1}{4\pi\varepsilon_0}\sum_{i=1}^{n}\frac{Q_i}{|\mathbf{R}-\mathbf{R}_i|} \quad [\mathrm{V}]} \tag{4.28}$$

where the distance $|\mathbf{R}-\mathbf{R}_i|$ is simply the distance between the ith charge and point P as shown in **Figure 4.11**. The potential of each point charge is calculated as if all other point charges do not exist and the principle of superposition is used to find the total potential.

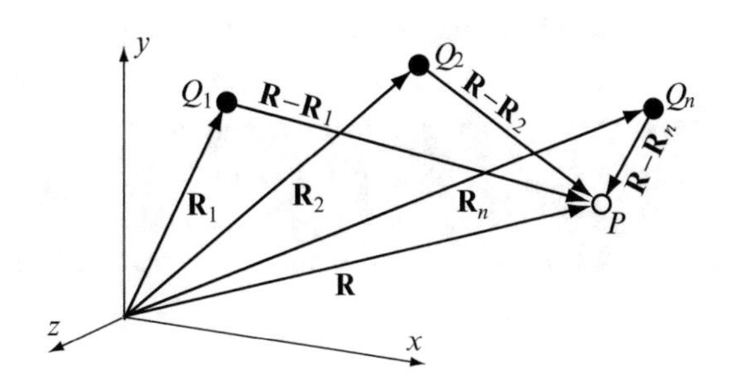

Figure 4.11 Calculation of potential at point P due to n point charges

Example 4.7 Potential of Point Charges Three point charges are arranged as shown in **Figure 4.12**. Calculate the potential at a general point $P(x,y)$ in the plane of the charges.

Solution: The potential due to each point charge is calculated separately and summed up to obtain the total potential due to the three charges.

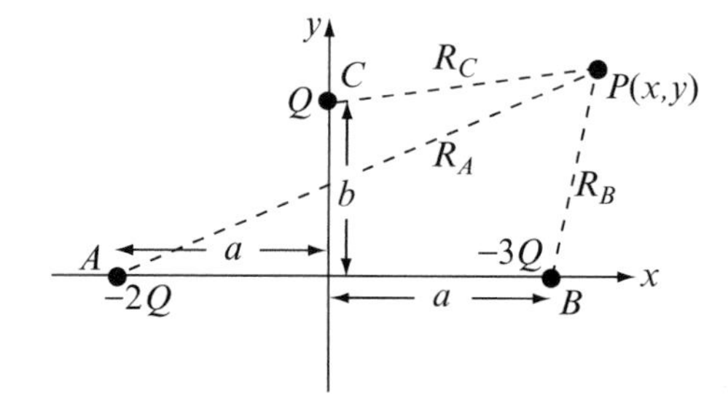

Figure 4.12 Calculation of potential at point $P(x,y)$ due to the three charges shown

The potential at a distance R from a point charge is

$$V = \frac{Q}{4\pi\varepsilon_0 R} \quad [\text{V}]$$

where R is the distance between the point charge and the location at which the potential is calculated. From **Figure 4.12**, we have

$$R_A = \sqrt{(x+a)^2 + y^2}, \quad R_B = \sqrt{(x-a)^2 + y^2}, \quad R_c = \sqrt{x^2 + (y-b)^2} \quad [\text{m}]$$

The potentials are

$$V_{PA} = \frac{-2Q}{4\pi\varepsilon_0\sqrt{(x+a)^2 + y^2}}, \quad V_{PB} = \frac{-3Q}{4\pi\varepsilon_0\sqrt{(x-a)^2 + y^2}}, \quad V_{PC} = \frac{Q}{4\pi\varepsilon_0\sqrt{x^2 + (y-b^2)}} \quad [\text{V}]$$

The potential at P is

$$V_P = V_{PA} + V_{PB} + V_{PC} = \frac{Q}{4\pi\varepsilon_0}\left[-\frac{2}{\sqrt{(x+a)^2 + y^2}} - \frac{3}{\sqrt{(x-a)^2 + y^2}} + \frac{1}{\sqrt{x^2 + (y-b)^2}}\right] \quad [\text{V}]$$

This simple calculation indicates the ease with which the potential due to any number of point charges may be calculated. Note also that in the final result, the term in the square brackets is a geometrical term and does not include charge or material properties. This will be exploited in **Chapter 6** to find numerical solutions to a number of much more complex problems.

Exercise 4.3 Three point charges, each equal to $Q = 10^{-7}$ C, are located on the x axis at $x = 0$, $x = 1$ m, and $x = 2$ m, as shown in **Figure 4.13**. Calculate the potential at $(x = 0, y = 1)$.

Answer 1,936 V.

Figure 4.13 Calculation of potential at point $P(0,1)$ due to the three charges shown

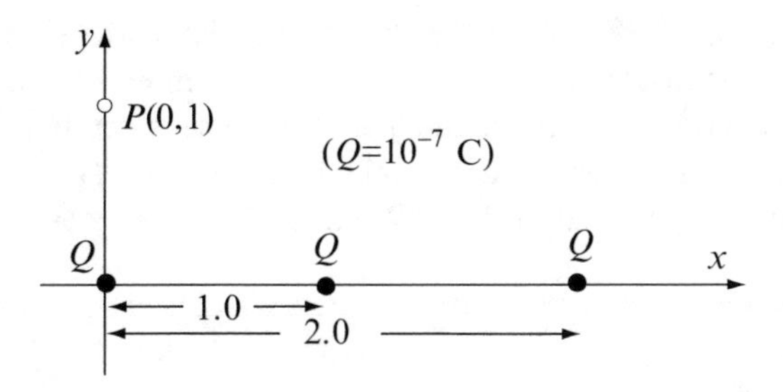

4.4.2 Electric Potential Due to Distributed Charges

The extension of the results of the previous section to distributed charges is based on the fact that any distributed charge can be viewed as an assembly of small (actually infinitesimal) point charges. This aspect and its relation to calculation of electric potential are shown in **Figure 4.14** for line, surface, and volume charge distributions. In all three cases, a small point charge is defined and the electric potential calculated as

$$\Delta V(\mathbf{R}) = \frac{1}{4\pi\varepsilon_0}\frac{\Delta Q_i}{\left|\mathbf{R}-\mathbf{R}_i'\right|} \quad [\mathrm{V}] \tag{4.29}$$

where ΔQ_i is the charge of element i of the distribution and can be any of those given in **Figure 4.14**. Performing summation over all elements of the distribution **(Figure 4.14a)** and letting $\Delta \to 0$ yields (for line charge distributions)

$$V(\mathbf{R}) = \lim_{\substack{\Delta l' \to 0 \\ (n\to\infty)}} \frac{1}{4\pi\varepsilon_0}\sum_{i=1}^{n}\frac{\rho_{l'}(\mathbf{R}_i')\Delta l'}{|\mathbf{R}-\mathbf{R}_i'|} = \frac{1}{4\pi\varepsilon_0}\int_{l'}\frac{\rho_{l'}(R')dl'}{|\mathbf{R}-\mathbf{R}'|} \quad [\mathrm{V}] \tag{4.30}$$

where the prime indicates integration on the charged segment. For a surface charge distribution **(Figure 4.14b)**,

$$V(\mathbf{R}) = \frac{1}{4\pi\varepsilon_0}\int_{s'}\frac{\rho_{s'}(R')\,ds'}{|\mathbf{R}-\mathbf{R}'|} \quad [\mathrm{V}] \tag{4.31}$$

and for a volume distribution **(Figure 4.14c)**:

$$V(\mathbf{R}) = \frac{1}{4\pi\varepsilon_0}\int_{v'}\frac{\rho_{v'}(R')\,dv'}{|\mathbf{R}-\mathbf{R}'|} \quad [\mathrm{V}] \tag{4.32}$$

Thus, the net effect is replacement of the elemental charge in **Eq. (4.29)** by the infinitesimal charge and replacement of the summation by integration. The charge densities were assumed to be location dependent and, therefore, left inside the integral sign. If charge density is uniform, it should be taken out of the integral. Also, note that the integration is performed over the primed coordinates; that is, only on that part of space over which the charge density is nonzero.

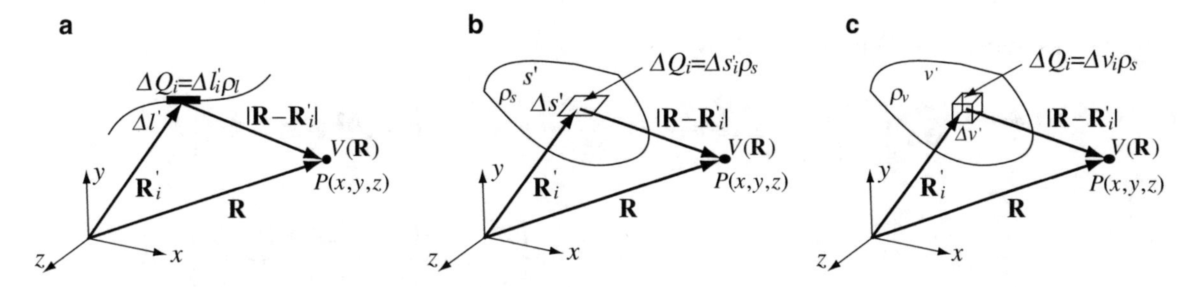

Figure 4.14 Calculation of potential due to **(a)** line charge distribution, **(b)** surface charge distribution, **(c)** volume charge distribution

In **Eqs. (4.30)** through **(4.32)**, we have used the potential of an infinitesimal (point) charge to calculate the potential of a charge distribution. The main premise is that the scalar summation (integration) of the potentials of point charges can be performed relatively easily. This is certainly true in some cases but not always. Sometimes, in particular when highly symmetric charge distributions are involved, it is simpler to calculate the electric field intensity first, using Gauss's law, and then use the definition of potential in **Eq. (4.20)** to evaluate the potential. The following examples explore both methods.

Example 4.8 A short segment of length $L = 2$ m is charged with a uniform line charge density $\rho_l = 10^{-7}$ C/m:

(a) Calculate the potential at a general point in space.
(b) What is the potential at a radial distance $d = 0.1$ m from the center of the segment?

Solution: The geometry is shown in **Figure 4.15** where the segment was placed in a cylindrical system of coordinates. Clearly, Gauss's law cannot be used to calculate **E.** Thus, an element of length dz' is chosen at location z', with charge $dq = \rho_l dz'$. The distance between this point and the general point $P(r,\phi,z)$ is calculated and the potential found by integration along the segment:

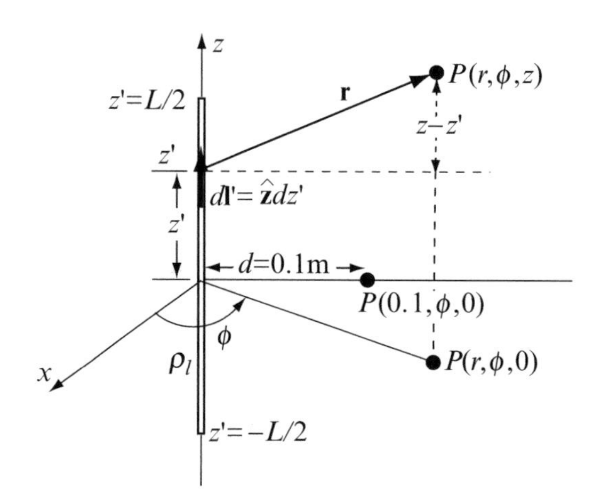

Figure 4.15 Calculation of potential at $P(r,\phi,z)$ due to a short, charged line segment

(a) Although, in general, the distance between point z' and P depends on all three coordinates, in this case it is independent of the angle ϕ. The distance $|\mathbf{r}|$ is

$$|\mathbf{r}| = \sqrt{r^2 + (z - z')^2} \quad [\text{m}]$$

The potential due to the element of charge (with respect to zero potential at infinity) is [see **Eq. (4.29)**]

$$dV = \frac{\rho_l dz'}{4\pi\varepsilon_0|\mathbf{r}|} = \frac{\rho_l dz'}{4\pi\varepsilon_0\sqrt{r^2 + (z - z')^2}} \quad [\text{V}]$$

Integrating this from $z' = -L/2$ to $z' = +L/2$ gives

$$V(r,\phi,z) = \frac{\rho_l}{4\pi\varepsilon_0}\int_{z'=-L/2}^{z'=L/2}\frac{dz'}{\sqrt{r^2 + (z - z')^2}} = \frac{\rho_l}{4\pi\varepsilon_0}\ln\left(2\sqrt{r^2 + (z - z')^2} + 2z' - 2z\right)\Bigg|_{-L/2}^{L/2}$$

$$= \frac{\rho_l}{4\pi\varepsilon_0}\ln\left(\frac{2\sqrt{r^2 + (z - L/2)^2} + L - 2z}{2\sqrt{r^2 + (z + L/2)^2} - L - 2z}\right) \quad [\text{V}]$$

For $L = 2$ m, $\rho_l = 10^{-7}$ C/m, and $\varepsilon_0 = 8.854 \times 10^{-12}$ F/m, the potential at a general point is

$$V(r,\phi,z) = \frac{10^{-7}}{4 \times \pi \times 8.854 \times 10^{-12}} \ln\left(\frac{2\sqrt{r^2 + (z-1)^2} + 2 - 2z}{2\sqrt{r^2 + (z+1)^2} - 2 - 2z}\right) = 898.8 \ln\left(\frac{\sqrt{r^2 + (z-1)^2} + 1 - z}{\sqrt{r^2 + (z+1)^2} - 1 - z}\right) \quad [\mathrm{V}]$$

(b) At a radial distance $d = 0.1$ m (point $P(0.1,\phi,0)$) the potential is

$$V(0.1,\phi,0) = 898.8 \ln\left(\frac{\sqrt{0.1^2 + (-1)^2} + 1}{\sqrt{0.1^2 + (1)^2} - 1}\right) = 5{,}389.6 \quad [\mathrm{V}]$$

Example 4.9 Potential Due to a Charge Distribution A very long, conducting cylindrical shell (a tube) of radius a [m] is coated with a dielectric layer between $r = a$ and $r = b$ and connected to zero potential as shown in **Figure 4.16a**. A uniform volume charge density $\rho_v = \rho_0$ [C/m^3] is distributed throughout the dielectric layer:

(a) Calculate the potential everywhere inside the charge layer.
(b) Calculate the potential outside the charge layer.
(c) Plot the potential in space.

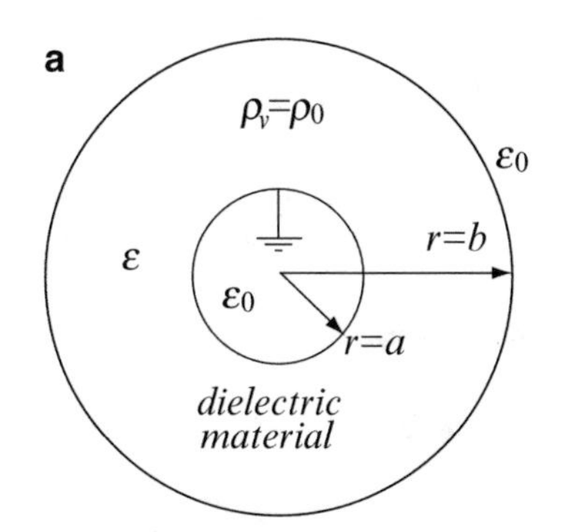

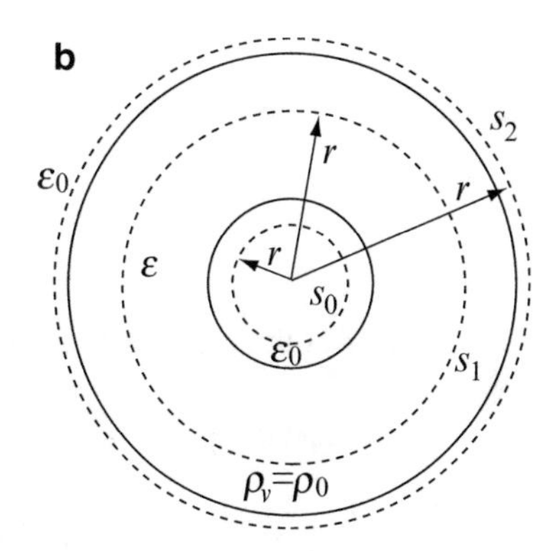

Figure 4.16 Calculation of potential in a cylindrical geometry shown in cross section. (**a**) Configuration. (**b**) Gaussian surfaces needed to calculate the electric field intensity

Solution: It is simpler to calculate the electric field intensity first, using Gauss's law, because the structure is very long. After the electric field intensity is calculated everywhere, the potential is evaluated by integration, using **Eq. (4.20)**. The Gaussian surfaces needed for the various regions are shown in **Figure 4.16b**.
(a) Electric field intensity inside the inner cylinder: $r < a$.
$\mathbf{E} = 0$ (no charge inside any Gaussian surface drawn in this region).

Between the two cylinders: $a < r < b$.
The Gaussian surface is a cylinder of radius r ($r > a$) and arbitrary length L:

$$\int_{s_1} \mathbf{E} \cdot d\mathbf{s}_1 = \frac{1}{\varepsilon}\int_{v_1'} \rho_v dv_1'$$

where v_1' is that part of the volume enclosed by the Gaussian surface s_1 that contains charge density. For a cylindrical Gaussian surface at r ($a < r < b$), the volume that contains charge is a hollow cylinder of length L, inner radius a, and outer radius r. Because charge density is uniform, the integral on the right-hand side is

$$\int_{v_1'} \rho_v dv_1' = \rho_0\left(\pi r^2 - \pi a^2\right)L \quad [\mathrm{C}]$$

Substituting in Gauss's law, and since both $d\mathbf{s}$ and $\mathbf{E}$ point radially out,

$$E2\pi rL = \frac{\rho_0(\pi r^2 - \pi a^2)L}{\varepsilon} \quad \rightarrow \quad \mathbf{E} = \hat{\mathbf{r}}\frac{\rho_0(r^2 - a^2)}{2r\varepsilon} \quad \left[\frac{\mathrm{V}}{\mathrm{m}}\right]$$

Note: In the dielectric, we use ε not ε_0. The potential inside the charge layer may be calculated from the definition of potential in **Eq. (4.20)**:

$$V_{r0,a} = V_{r_0} - V_a = V_{r_0} = -\int_a^{r_0} \mathbf{E}\cdot d\mathbf{l} = -\int_a^{r_0} \hat{\mathbf{r}}\frac{\rho_0(r^2 - a^2)}{2r\varepsilon}\cdot d\mathbf{l} \quad [\mathrm{V}]$$

The integration is along the radial direction from a to an arbitrary point $a < r_0 < b$ and $d\mathbf{l} = \hat{\mathbf{r}}dr$:

$$V_{r_0} = -\int_a^{r_0} \hat{\mathbf{r}}\frac{\rho_0(r^2 - a^2)}{2r\varepsilon}\cdot\hat{\mathbf{r}}dr = -\frac{\rho_0}{2\varepsilon}\int_a^{r_0}\frac{(r^2 - a^2)}{r}dr = -\frac{\rho_0}{2\varepsilon}\left[\frac{r^2}{2} - a^2\ln r\right]_a^{r_0} = -\frac{\rho_0}{4\varepsilon}\left[r_0^2 - a^2 - 2a^2\ln\frac{r_0}{a}\right] \quad [\mathrm{V}]$$

Note: The constant of integration is zero since the reference voltage $V_a = 0$. Thus, at $r_0 = b$, the potential is

$$V_b = -\frac{\rho}{4\varepsilon}\left[b^2 - a^2 - 2a^2\ln\frac{b}{a}\right] \quad [\mathrm{V}]$$

(b) For the domain outside the outer cylinder, $b < r < \infty$, again using a Gaussian surface at s_2, we get

$$\mathbf{E} = \hat{\mathbf{r}}\frac{\rho_0(b^2 - a^2)}{2r\varepsilon_0} \quad \left[\frac{\mathrm{V}}{\mathrm{m}}\right]$$

This is the same form as for the domain inside the dielectric, but the total charge is contained between $r = a$ and $r = b$.

To calculate the potential at a point r_1 outside, we integrate as in the previous case from $r = b$ to $r = r_1$:

$$V_{r_1,b} = V_{r_1} - V_b = -\int_b^{r_1}\frac{\rho_0(b^2 - a^2)}{2r\varepsilon_0}dr = -\frac{\rho_0(b^2 - a^2)}{2\varepsilon_0}\ln\frac{r_1}{b} \quad \rightarrow \quad V_{r_1} = V_b - \frac{\rho_0(b^2 - a^2)}{2\varepsilon_0}\ln\frac{r_1}{b} \quad [\mathrm{V}]$$

In this case, the reference point is at $r = b$ and V_b was calculated above. Substituting this, we get

$$V_{r_1,b} = -\frac{\rho_0}{4\varepsilon}\left[b^2 - a^2 - 2a^2\ln\frac{b}{a}\right] - \frac{\rho_0(b^2 - a^2)}{2\varepsilon_0}\ln\frac{r_1}{b} \quad [\mathrm{V}]$$

(c) See **Figure 4.17** for a sketch of the potential. Note that the potential is continuous everywhere. The potential at $r = b = 0.2$ m is -113.91 V.

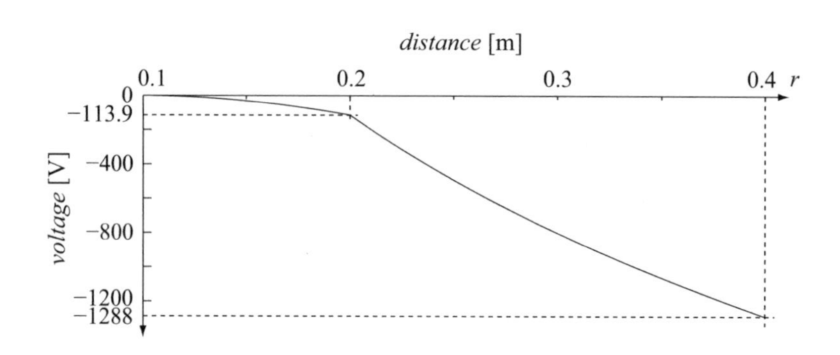

Figure 4.17 The potential everywhere in space due to the configuration in **Figure 4.16a**

Example 4.10 Two very long, thick cylindrical conducting lines, each of radius a [m], are separated a distance d [m]. A charge density ρ_s [C/m^2] is distributed uniformly on the surface of one line and $-\rho_s$ [C/m^2] on the surface of the second, as shown in **Figure 4.18a**:

(a) Calculate the potential difference between the two lines.
(b) Can you tell what the absolute potential of each line is?

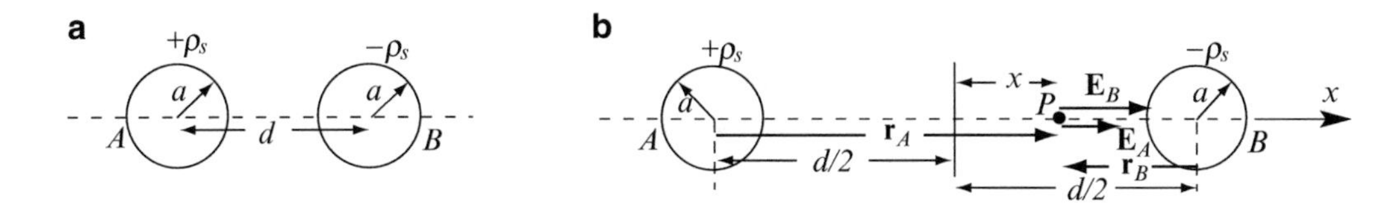

Figure 4.18 Two thick lines with uniform surface charge density. (**a**) Geometry and dimensions. (**b**) Calculation of the electric field intensity at point P

Solution: The electric field intensity everywhere in space may be calculated from Gauss's law. To calculate the potential of the lines, it is necessary that we find a reference point at which the potential is known. The solution at infinity cannot be used as a reference potential because the lines themselves extend to infinity, but the point midway between the two lines is at a constant potential anywhere along the line. In fact, the potential is zero although the actual value is not important. Another possibility is to assume one line is at a constant (perhaps zero) potential and calculate the potential of the other relative to the first:

(a) Suppose we choose a general point $P(x,0)$ and calculate the electric field intensity at this point using **Figure 4.18b**. The electric field intensity due to line A is found by constructing a Gaussian surface around conductor A in the form of a cylinder of radius r_A and length L. From Gauss's law, we get

$$E_A 2\pi r_A L = \frac{\rho_s 2\pi a L}{\varepsilon_0} \quad \rightarrow \quad E_A = \frac{\rho_s a}{\varepsilon_0 r_A} \quad \left[\frac{\text{V}}{\text{m}}\right]$$

The solution for conductor B is identical except that the electric field intensity is negative (pointing toward the conductor):

$$E_B 2\pi r_B L = -\frac{\rho_s 2\pi a L}{\varepsilon_0} \quad \rightarrow \quad E_B = -\frac{\rho_s a}{\varepsilon_0 r_B} \quad \left[\frac{\text{V}}{\text{m}}\right]$$

where $\rho_s 2\pi a L$ is the total charge on a cylinder of radius a and length L. In vector notation,

$$\mathbf{E}_A = \hat{\mathbf{r}}_A \frac{\rho_s a}{\varepsilon_0 r_A}, \quad \mathbf{E}_B = \hat{\mathbf{r}}_B \frac{\rho_s a}{\varepsilon_0 r_B} \quad \left[\frac{\text{V}}{\text{m}}\right]$$

From geometrical considerations alone, the potential midway between the two conductors must be zero whereas the electric field intensity on the line connecting their centers is horizontal. However, we will assume the potential midway is some fixed value V_0 for generality. Using **Figure 4.18b** and taking point P at a general location between the cylinders, we can write

$$r_A = \frac{d}{2} + x, \quad r_B = \frac{d}{2} - x \quad [\text{m}]$$

The electric field intensity is in the positive x direction and equals

$$\mathbf{E}(x,0) = \hat{\mathbf{x}} \left(\frac{\rho_s a}{\varepsilon_0 r_A} + \frac{\rho_s a}{\varepsilon_0 r_B} \right) = \hat{\mathbf{x}} \frac{\rho_s a}{\varepsilon_0} \left(\frac{2}{d + 2x} + \frac{2}{d - 2x} \right) \quad \left[\frac{\text{V}}{\text{m}}\right]$$

To find the potential on line B, we integrate in the direction of the electric field from $x = 0$ to $x = (d/2) - a$. In this case, $d\mathbf{l} = \hat{\mathbf{x}}dx$:

$$V_{B,0} = V_B - V_0 = -\int_{x=0}^{d/2-a} \mathbf{E} \cdot d\mathbf{l} = -\int_{x=0}^{d/2-a} \hat{\mathbf{x}} \frac{\rho_s a}{\varepsilon_0}\left(\frac{2}{d+2x} + \frac{2}{d-2x}\right) \cdot \hat{\mathbf{x}}dx$$

$$= -\frac{2\rho_s a}{\varepsilon_0}\int_{x=0}^{d/2-a}\left(\frac{1}{d+2x} + \frac{1}{d-2x}\right) = -\frac{\rho_s a}{\varepsilon_0}\ln\frac{d+2x}{d-2x}\bigg|_0^{d/2-a} = -\frac{\rho_s a}{\varepsilon_0}\ln\frac{d-a}{a} \quad \text{[V]}$$

To obtain the potential on line A, we integrate against the electric field from $x = 0$ to $x = -(d/2) + a$, again with $d\mathbf{l} = \hat{\mathbf{x}}dx$. The result is a positive potential difference $V_{A,0}$ of identical magnitude:

$$V_{A,0} = V_A - V_0 = -\int_{x=0}^{-d/2+A} \mathbf{E} \cdot d\mathbf{l} = -\frac{2\rho_s a}{\varepsilon_0}\int_{x=0}^{-d/2+a}\left(\frac{1}{d+2x} + \frac{1}{d-2x}\right) = \frac{\rho_s a}{\varepsilon_0}\ln\frac{d-a}{a} \quad \text{[V]}$$

The potential difference between lines A and B is

$$V_{AB} = V_{A,0} - V_{B,0} = V_A - V_B = \frac{\rho_s a}{\varepsilon_0}\ln\frac{d-a}{d} - \left(-\frac{\rho_s a}{\varepsilon_0}\ln\frac{d-a}{d}\right) = \frac{2\rho_s a}{\varepsilon_0}\ln\frac{d-a}{a} \quad \text{[V]}$$

Note that V_0 cancels, meaning that the reference potential is irrelevant.

(b) In general, we cannot calculate the absolute potential of the lines. However, in this case, the lines are charged with equal and opposite charge densities. This means that the potential midway between them is zero and the potentials calculated above are the absolute potentials. If, on the other hand, the potential at the center between the lines is raised by, say, V_0, then the potentials V_A and V_B are also raised by the same value. Under this condition, only the potential difference between the lines is known.

Exercise 4.4 Two parallel plates are very large (infinite), separated a distance d [m], in air, and charged with equal but opposite charge density ρ_s [C/m^2]. Calculate the potential difference between the two plates.

Answer $V = \dfrac{\rho_s d}{\varepsilon_0}$ [V].

Exercise 4.5 Two parallel plates are very large (infinite), separated a distance d [m] and connected to a potential difference V. Calculate the electric field intensity between the two plates.

Answer $E = \dfrac{V}{d} \quad \left[\dfrac{\text{V}}{\text{m}}\right]$, pointing from the positive to the negative plate.

4.4.3 Calculation of Electric Field Intensity from Potential

It was stated at the beginning of the previous section that one of the reasons to define the electric potential is its scalar nature, therefore simplifying calculations. It is indeed easier to evaluate the electric potential than it is to evaluate the electric field intensity (at least in most cases described up to this point; we also saw examples in which it is easier to calculate the electric field intensity). However, we have not shown how the electric field intensity or forces may be calculated from the potential. Unless we can do so, the potential cannot be an alternative way of calculating the electric field intensity or electric flux density. This aspect of the calculation will be shown next.

Calculation of the electric field intensity from potential is based on the curl-free nature of the electric field intensity [**Eq. (4.8)** or **Eq. (4.10)**]. In **Section 2.5**, we discussed the identity $\nabla \times (\nabla V) = 0$ and stated that if the curl of a vector field is zero, then based on this identity, the vector field may be written as the gradient of the scalar field V [see **Eq. (2.113)**]. For the electric field intensity and the electric potential, this may be stated as follows:

$$\text{Because } \nabla \times \mathbf{E} = 0, \quad \mathbf{E} = -\nabla V \tag{4.33}$$

The negative sign indicates that the electric field intensity points from high to low potential. For example, in Cartesian coordinates,

$$\boxed{\mathbf{E} = -\nabla V = -\hat{\mathbf{x}}\frac{\partial V}{\partial x} - \hat{\mathbf{y}}\frac{\partial V}{\partial y} - \hat{\mathbf{z}}\frac{\partial V}{\partial z} \quad \left[\frac{\text{V}}{\text{m}}\right]} \tag{4.34}$$

Now, we have a method of calculating the electric field intensity from the electric potential. The method is as follows:

(1) Calculate the electric potential at a general point in the space of interest.
(2) Calculate the gradient of potential in a convenient system of coordinates and take its negative.
(3) If necessary, substitute numerical values of the coordinates to obtain the potential at a specific point.

The negative of the components of the gradient are therefore the components of the electric field intensity, and **Eq. (4.33)** may be viewed as just another way of stating the curl-free postulate. Also, from the properties of the gradient, we conclude that the electric field intensity is perpendicular to a constant potential surface such as the surface of a conductor. **Equation (4.34)** calculates the electric field intensity in Cartesian coordinates. The gradient in cylindrical or spherical coordinates may be used when appropriate.

Example 4.11 Calculation of Electric Field Intensity for Point Charges Calculate the electric field intensity due to the three charges in **Figure 4.12**.

Solution: Although the electric field intensity of point charges can be calculated easily using Gauss's law, each point charge generates a field in a different direction in space. After the electric field intensity due to each point charge is calculated, we must sum the vectors to get the electric field intensity. For a few charges, this is acceptable, but for a large number of charges, it is extremely lengthy and tedious. In such cases, it is easier to calculate the potential at a general point in space and, using the gradient of the potential [**Eq. (4.34)**], to obtain the electric field intensity from the general expression of potential.

The expression for potential at a general point (x,y) was obtained in **Example 4.7**:

$$V_p = \frac{Q}{4\pi\varepsilon_0}\left[-\frac{2}{\sqrt{(x+a)^2+y^2}} - \frac{3}{\sqrt{(x-a)^2+y^2}} + \frac{1}{\sqrt{x^2+(y-b)^2}}\right] \quad [\text{V}]$$

The electric field intensity is written using **Eq. (4.34)**. Substituting V_p for V, we get

$$\begin{aligned}\mathbf{E}(x,y) = -\nabla V_p(x,y) = &-\hat{\mathbf{x}}\frac{Q}{4\pi\varepsilon_0}\frac{\partial}{\partial x}\left[-\frac{2}{\sqrt{(x+a)^2+y^2}} - \frac{3}{\sqrt{(x-a)^2+y^2}} + \frac{1}{\sqrt{x^2+(y-b)^2}}\right] \\ &-\hat{\mathbf{y}}\frac{Q}{4\pi\varepsilon_0}\frac{\partial}{\partial y}\left[-\frac{2}{\sqrt{(x+a)^2+y^2}} - \frac{3}{\sqrt{(x-a)^2+y^2}} + \frac{1}{\sqrt{x^2+(y-b)^2}}\right] \quad \left[\frac{\text{V}}{\text{m}}\right]\end{aligned}$$

where the derivative with respect to z is zero. Performing the differentiation gives

$$\mathbf{E}(x,y) = -\hat{\mathbf{x}}\frac{Q}{4\pi\varepsilon_0}\left[\frac{2(x+a)}{\left((x+a)^2+y^2\right)^{3/2}} + \frac{3(x-a)}{\left((x-a)^2+y^2\right)^{3/2}} - \frac{x}{\left(x^2+(y-b)^2\right)^{3/2}}\right]$$
$$-\hat{\mathbf{y}}\frac{Q}{4\pi\varepsilon_0}\left[\frac{2y}{\left((x+a)^2+y^2\right)^{3/2}} + \frac{3y}{\left((x-a)^2+y^2\right)^{3/2}} - \frac{(y-b)}{\left(x^2+(y-b)^2\right)^{3/2}}\right] \quad \left[\frac{\mathrm{V}}{\mathrm{m}}\right]$$

Although this may seem lengthy, it is rather straightforward, and in many practical cases, it is simpler than the direct calculation of the electric field intensity.

Exercise 4.6 Calculate the electric field intensity in **Figure 4.13**, at point $P(0,1,0)$.

Hint First obtain an expression for potential in terms of general coordinates at point $P(x,y,z)$.
Answer $\mathbf{E}(0,1,0) = -\hat{\mathbf{x}}478.54 + \hat{\mathbf{y}}1296.92 \quad [\mathrm{V/m}]$.

Example 4.12 Application: The Dipole
Calculate the potential of a dipole. Assume a positive charge Q [C] and a negative charge $-Q$ [C] are separated by a distance d [m] ($d \ll R$; see **Section 3.4.1.3**). Calculate the electric field intensity from the potential. Compare with the result in **Eq. (3.38)**.

Solution: Consider **Figure 4.19**. We assume $R \gg d$ and use the same notation as in **Section 3.4.1.3**. The potential at point P is then calculated as for any two point charges. From the general expression for the potential, we can calculate the electric field intensity using the gradient: $\mathbf{E} = -\nabla V$.

$$V_P = \frac{Q}{4\pi\varepsilon_0}\left[\frac{1}{|\mathbf{R}_1|} - \frac{1}{|\mathbf{R}_2|}\right] = \frac{Q}{4\pi\varepsilon_0}\left[\frac{1}{|\mathbf{R}-\mathbf{d}/2|} - \frac{1}{|\mathbf{R}+\mathbf{d}/2|}\right] \quad [\mathrm{V}]$$

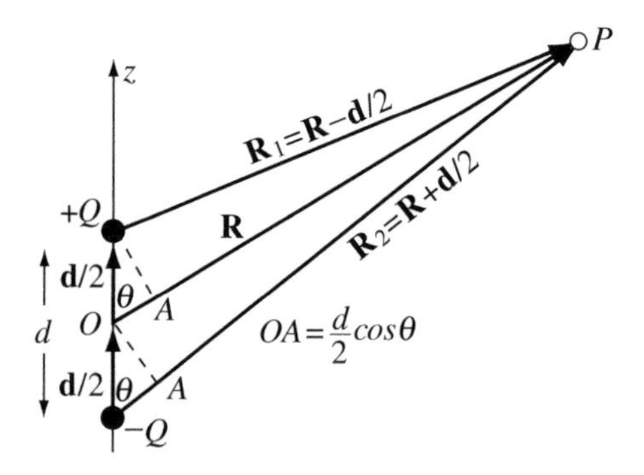

Figure 4.19 The electric dipole and calculation of potential at point P due to the dipole

The magnitudes of $\mathbf{R}_1$ and $\mathbf{R}_2$ can be evaluated by observing that if $\mathbf{R}$ makes an angle θ with the axis on which the charges are located, then for $R \gg d$, both $\mathbf{R}_1$ and $\mathbf{R}_2$ make approximately the same angle with the axis. With the approximation $OA \approx (d/2)\cos\theta$ (see **Figure 4.19**), we can write

$$\left|\mathbf{R}-\frac{\mathbf{d}}{2}\right|^{-1} \approx \left[R-\frac{d}{2}\cos\theta\right]^{-1} = R^{-1}\left[1+\frac{d}{2R}\cos\theta\right]$$

$$\left|\mathbf{R}+\frac{\mathbf{d}}{2}\right|^{-1} \approx \left[R+\frac{d}{2}\cos\theta\right]^{-1} = R^{-1}\left[1-\frac{d}{2R}\cos\theta\right]$$

where the truncated binomial expansion $(1 + x)^n \approx 1 + nx$ with $x = \pm(d/2R)\cos\theta$ and $n = -1$ was used [see **Eq. (3.30)**]. Substituting these in the expression for the potential, we get

$$V_P \approx \frac{Q}{4\pi\varepsilon_0}\left[\frac{[1+(d/2R)\cos\theta]}{R} - \frac{[1-(d/2R)\cos\theta]}{R}\right] = \frac{Qd\cos\theta}{4\pi\varepsilon_0 R^2} \quad [\mathrm{V}]$$

To calculate the electric field intensity, we use the gradient in spherical coordinates and write

$$\mathbf{E}_P = -\nabla V_p = -\left[\hat{\mathbf{R}}\frac{\partial V_p(R,\theta)}{\partial R} + \hat{\boldsymbol{\theta}}\frac{1}{R}\frac{\partial V_p(R,\theta)}{\partial \theta}\right] \quad \left[\frac{\mathrm{V}}{\mathrm{m}}\right]$$

From the definition of the dipole moment in **Eq. (3.34)**, $Qd = p$ and $Qd\cos\theta = \mathbf{p}\cdot\hat{\mathbf{R}}$. The potential and electric field intensity of the dipole can also be written as

$$V_P \approx \frac{\mathbf{p}\cdot\hat{\mathbf{R}}}{4\pi\varepsilon_0 R^2} \quad [\mathrm{V}] \quad \text{and} \quad \mathbf{E}_P = \frac{p}{4\pi\varepsilon_0 R^3}\left[\hat{\mathbf{R}}2\cos\theta + \hat{\boldsymbol{\theta}}\sin\theta\right] \quad [\mathrm{V/m}]$$

The electric field intensity obtained here is the same electric field intensity we obtained in **Eq. (3.38)**.

4.5 Materials in the Electric Field

Until now most discussion of electric fields was in free space although we have used other materials by simply indicating that they have a different permittivity ε. This was deliberately done to keep the discussion simple because we wished to avoid any need to deal with the effect that either materials have on the electric field or the electric field has on materials. In the engineering environment, it is often necessary to evaluate fields and potentials inside and in the presence of various materials. Therefore, we discuss here materials and material properties in conjunction with the electric field because of their importance in practical design. This will open a whole new range of applications and devices.

There are, in effect, two types of materials that will be discussed here. The first type is conductors. The second is dielectric materials or nonconductors (also called insulators). Because at this stage we cannot take into account any motion of charges, we can only discuss perfect dielectrics and perfect conductors.

4.5.1 Conductors

In electrostatics, charges do not move. This certainly does not mean that charges cannot move. They can and they do, but, if they do, the relations developed so far cannot be used because of our basic assumptions. These assumptions will be modified later to take into account the motion of charges, but for now, we must ask ourselves what is really meant by a conductor in the electrostatic environment. The answer is surprisingly simple even though the consequences are far-reaching: a conductor is a material that allows free movement of charge within its volume. In other words, if a charge is introduced into a conductor, it *can* move freely until something prevents it from moving. This something may be an electric field or the surface of the conductor. Whereas charges are free to move, we will only treat them *after* they have settled in their final distribution. Thus, the movement of charges is merely a mechanism to reach the steady state. After charges have reached their final state, the conductor has no effect on the charges. The fact that a conductor allows charges to move is now irrelevant since charges do not move any more. This means that for electrostatic purposes, the ability of a conductor to conduct is the important fact, not its conductivity (i.e., not how "good" or how "poor" a conductor is). For this reason, we will use the term conductor in a very unique way. At no time will the actual (numerical) conductivity of a material play any role in electrostatics. Conductors in the electrostatic field are said to be ***perfect conductors***.

The forgoing arguments are exemplified in **Figure 4.20**. In **Figure 4.20a**, body A has a certain amount of charge distributed in some fashion. Body B is neutral (has no charge). This is an electrostatic situation and we can calculate the electric field or potential if necessary. In **Figure 4.20b**, the two bodies are connected with a conductor. Now, charges can move freely from body A to body B (charges will move because this will allow them a lower potential energy, as we shall see shortly). The situation in this figure is not an electrostatic situation, and while charges are moving, we cannot calculate the electric field or potential. In **Figure 4.20c**, a steady state has been reached and charges cease moving. Note that charges may actually exist on the conductor itself, but because they are not moving, this is again a static situation. In **Figure 4.20d**, the conductor is removed to isolate the two bodies. Each body is now charged and an electrostatic condition again exists.

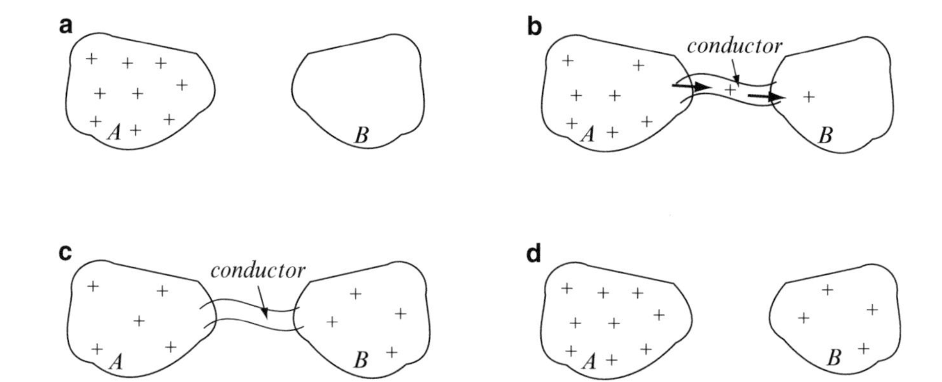

Figure 4.20 The charging process. (**a**) Body A is charged, B is not. (**b**) Connecting the two bodies causes flow of charge from A to B. (**c**) Steady state is reached when motion of charges ceases. (**d**) Disconnecting the two bodies leaves both charged

Another property of conductors is the ease with which free electrons can be separated. This means that if a force is applied on these electrons, they will separate and will move to the surface of the conductor. This is shown in **Figure 4.21**. An electric field is produced between two plates by applying an external source to the plates. The presence of the electric field separates the negative and positive charges in the conductor and the charges move to the surface. Because these charges were not introduced externally (i.e., the conductor has not acquired additional charge), they are called ***induced charges***. When the force that has induced the charges is removed, the charges return to their neutral state.

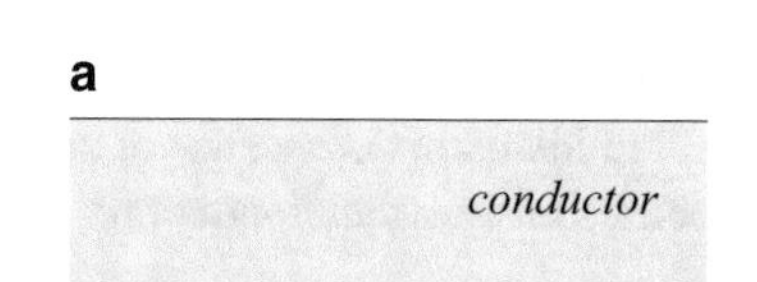

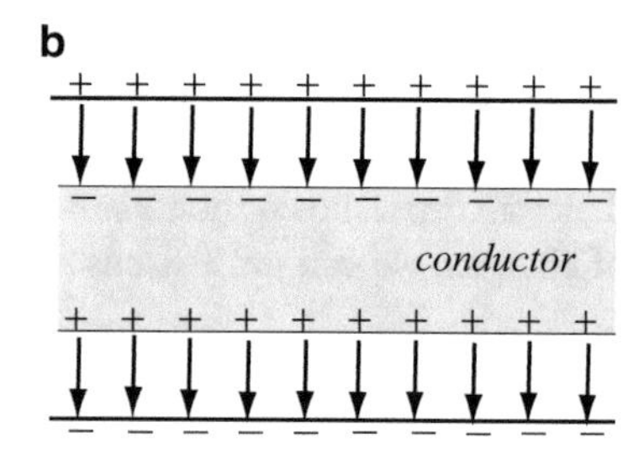

Figure 4.21 Charging of a conductor by induction. (**a**) Conductor in neutral state. (**b**) Separation of charges by an external electric field

Properties of conductors in the electrostatic field are as follows:

(**1**) Charge introduced into a conductor moves to the surface of the conductor. No charge exists in the interior.
(**2**) Charges distribute themselves on the surface of the conductor to produce zero electric field intensity in the interior of the conductor, regardless of the shape of the conductor or the external charges that may exist. This property is one of the most useful properties in application of conductors in electrostatics.
(**3**) A single point charge may exist anywhere in the volume of a conductor because there are no forces acting on it. In practice, this cannot happen unless the charge is a single electron. Any other charge will redistribute itself on the surface of the conductor.

(4) The volume charge density inside a conductor is zero. Surface charges may exist on conducting surfaces.
(5) The electric field intensity on the surface of the conductor (external to the surface) must be perpendicular to the surface at any point on the surface. Because of this, the potential on the surface of any conductor is constant.

The latter property is explained using **Figure 4.22** as follows. If a charge exists on the surface of a conductor and it produces both a normal and a tangential electric field, the normal component "pushes" any other charge against the surface. Since the charge cannot escape the conductor, this component of the electric field cannot move the charge. The tangential component will generate a force (with other charges in the vicinity), which will force the charges to move along the surface. This motion can only stop if there is no force to sustain it (remember, charges can move freely in the conductor). The charges in a conductor will come to rest in a configuration which cancels the tangential (parallel to the surface) electric field intensity component. The only remaining force (and therefore electric field) is perpendicular to the surface.

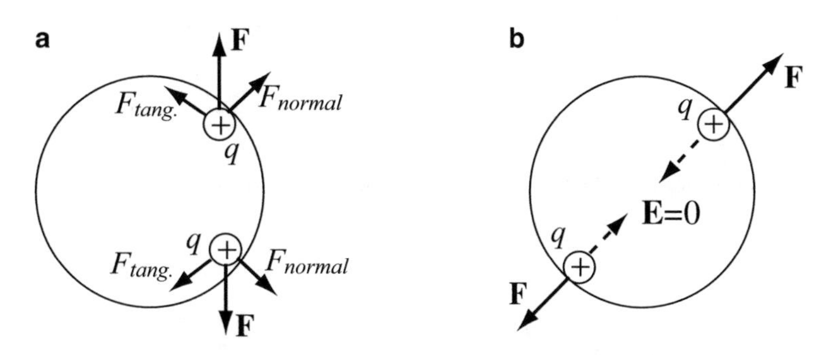

Figure 4.22 The process by which tangential components of the electric field intensity at the surface of a conductor vanish. (**a**) Tangential components force the charges to move. (**b**) Charges stop moving when the tangential field components cancel

4.5.1.1 Electric Field at the Surface of a Conductor

We mentioned that the tangential component of the electric field intensity at the surface of a conductor must be zero simply from the definition of the conductor. What can we say about the normal component? To consider this, we use **Figure 4.23**. A charge density is distributed on the surface of an arbitrary conductor. As required, we assume that only a normal component of the electric field intensity exists and the electric field intensity inside the conductor must be zero. Because the assumed charge density is positive, the electric field intensity points away from the surface. Using Gauss's law, we can write

$$E_n = \frac{\rho_s}{\varepsilon_0} \quad \left[\frac{\mathrm{V}}{\mathrm{m}}\right] \tag{4.35}$$

Thus, the conditions at the surface of a conductor are given as

$$E_t = 0, \quad E_n = \frac{\rho_s}{\varepsilon_0} \quad \left[\frac{\mathrm{V}}{\mathrm{m}}\right], \quad \text{or} \quad D_n = \rho_s \quad \left[\mathrm{C/m^2}\right] \tag{4.36}$$

The relation in **Eq. (4.36)**[4] not only gives the electric flux density on the surface but also allows the calculation of the charge density on the surface of the conductor when the conductor is inserted into an electric field. We will make considerable use of this property in examples and in many instances in future chapters.

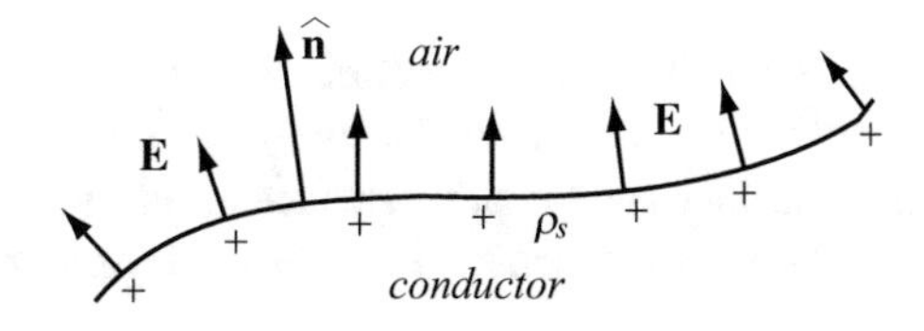

Figure 4.23 Charge density and electric field intensity at the surface of a conductor

[4] These properties, namely, that the electric field intensity at the surface of a conductor is normal to the surface and directly proportional to the surface charge density, were discovered by Coulomb in the course of his investigations on charged bodies.

Example 4.13 Surface Charge Density
The electric field intensity at the surface of a conductor is known to be perpendicular, pointing towards the surface, equal to 100 V/m, and the conductor is in free space:

(a) What is the charge density on the surface of the conductor?
(b) What are the electric field intensity and the electric potential inside the conductor if the potential on the surface of the conductor is $V_s = 10$ V?

Solution: The charge density at the surface of a conductor is found using **Eq. (4.35)** or **(4.36)**. The electric field intensity in any conductor under static conditions is zero; the potential is constant and equal to the surface potential.

Thus,

(a)

$$\rho_s = -D_n = -\varepsilon_0 E_n = -8.854 \times 10^{-12} \times 100 = -8.854 \times 10^{-10} \quad [\mathrm{C/m^2}]$$

The charge density is negative since the electric field intensity points towards the surface.

(b)

$$\mathbf{E} = 0 \quad [\mathrm{V/m}], \quad V = V_s = 10 \quad [\mathrm{V}]$$

Example 4.14 Conductor in an Electrostatic Field
A point charge Q [C] is surrounded by a hollow conducting sphere as shown in **Figure 4.24a**. Calculate:

(a) The electric field intensity and electric potential everywhere in space. Plot the electric field intensity and the potential.
(b) The surface charge densities on the inner and outer surfaces of the conducting shell.

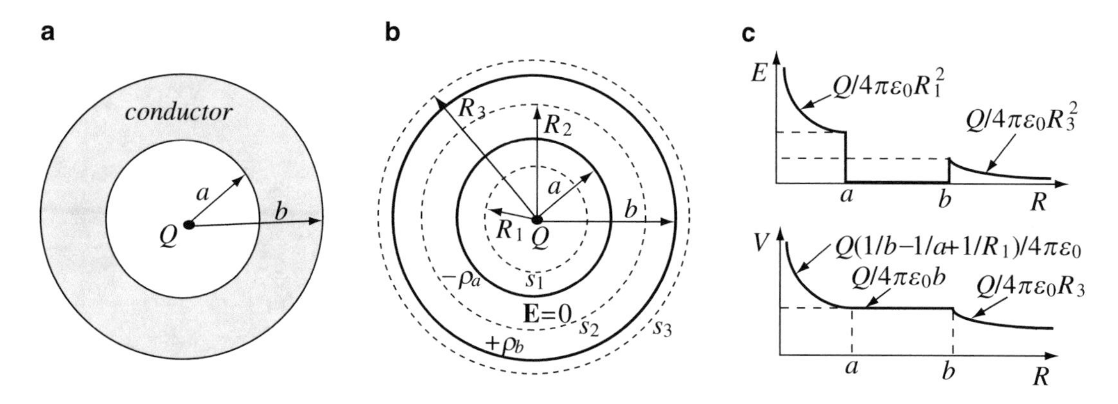

Figure 4.24 Electric field due to a point charge in the presence of a conductor. (**a**) Point charge surrounded by a hollow conducting sphere. (**b**) The Gaussian surfaces used to calculate the electric field intensity. (**c**) Plot of the electric field intensity and potential

Solution: The electric field intensity inside the conducting sphere is zero, but everywhere else, the field of the point charge is not disturbed by the conductor. This means that the inner surface of the spherical shell has a total charge equal to $-Q$ uniformly distributed on the surface and the outer surface acquires a total positive charge Q. These can be found from direct application of Gauss's law. After the electric field intensity is found everywhere, the electric potential is found by integrating from ∞ to the required point:

(a) A Gaussian surface at $R < a$ is used to find the electric field intensity inside the shell and one at $R > b$ to find the electric field intensity outside the sphere: these are shown in **Figure 4.24b**. Since the electric field of a point charge is radial (and, therefore, perpendicular to any spherical surface), we have from Gauss's law for R_1 and R_3,

$$\int_s \mathbf{E}\cdot d\mathbf{s} = E4\pi R_1^2 = \frac{Q}{\varepsilon_0} \quad \rightarrow \quad E = \frac{Q}{4\pi R_1^2\varepsilon_0} \quad \left[\frac{\text{V}}{\text{m}}\right], \quad 0 \le R_2 < a$$

$$\mathbf{E} = 0, \quad a \le R_2 \le b$$

$$\int_s \mathbf{E}\cdot d\mathbf{s} = E4\pi R_3^2 = \frac{Q}{\varepsilon_0} \quad \rightarrow \quad E = \frac{Q}{4\pi R_3^2\varepsilon_0} \quad \left[\frac{\text{V}}{\text{m}}\right], \quad R_3 > b$$

The electric potential is found by integrating from ∞ to the required point. Outside the sphere, $R = R_3 > b$:

$$V_{R_3} = -\int_\infty^{R_3} \mathbf{E}\cdot d\mathbf{l} = -\int_\infty^{R_3} \hat{\mathbf{R}}E\cdot\hat{\mathbf{R}}dR = -\int_\infty^{R_3} \frac{QdR}{4\pi\varepsilon_0 R^2} = \frac{Q}{4\pi\varepsilon_0 R_3} \quad [\text{V}]$$

At the outer surface of the conductor, the potential is found by setting $R_3 = b$:

$$V_b = \frac{Q}{4\pi\varepsilon_0 b} \quad [\text{V}]$$

To find the potential inside the conductor, we continue integrating from $R = b$ to $R = R_2$ over E. Using the same relation as for V_{R_2} we get

$$V_{R_2} = -\int_\infty^{R_2} \mathbf{E}\cdot d\mathbf{l} = -\int_\infty^{b} \frac{QdR}{4\pi\varepsilon_0 R^2} - \int_b^{R_2} 0dR = \frac{Q}{4\pi\varepsilon_0 b} \quad [\text{V}]$$

Thus, the electric potential remains constant in the conductor. We know this to be true from the fact that a perfect conductor is everywhere at constant potential. The potential inside the sphere ($0 < R_1 < a$) is

$$V_{R_1} = -\int_\infty^{R_1} \mathbf{E}\cdot d\mathbf{l} = -\int_\infty^{b} \frac{QdR}{4\pi\varepsilon_0 R^2} - \int_b^{a} 0dR - \int_a^{R_1} \frac{QdR}{4\pi\varepsilon_0 R^2} = \frac{Q}{4\pi\varepsilon_0}\left[\frac{1}{b} - \frac{1}{a} + \frac{1}{R_1}\right] \quad [\text{V}]$$

The potential at $R_1 = a$ equals that at b and the potential at the location of the point charge ($R_1 = 0$) is infinite as required. A plot of the electric field intensity and electric potential is shown in **Figure 4.24c**. Note that the integration is done in segments because the expression for the electric field intensity in each segment is different.

(b) To calculate the charge densities on the inner surface, we use the fact that $E = 0$ inside the conductor. We draw a Gaussian surface at $R = a^+$ (immediately inside the conductor so that the surface is included). For this Gaussian surface to produce zero field intensity, the total charge enclosed must be zero:

$$\int_s \mathbf{E}\cdot d\mathbf{s} = E4\pi R_2^2 = \frac{Q}{\varepsilon_0} + \frac{\rho_{sa}4\pi a^2}{\varepsilon_0} = 0 \quad \rightarrow \quad \rho_{sa} = -\frac{Q}{4\pi a^2} \quad \left[\frac{\text{C}}{\text{m}^2}\right]$$

The surface charge density on the outer surface is calculated from the fact that the total charge in the system is Q [C] (no net charge in the conductor). Therefore, the total charge on the outer surface equals the total charge on the inner surface. This gives

$$Q_b = -Q_a = Q \quad \rightarrow \quad \rho_{sb} = \frac{Q}{4\pi b^2} \quad \left[\frac{\text{C}}{\text{m}^2}\right]$$

4.5.2 Dielectric Materials

Unlike conductors, ***dielectrics*** are materials in which charges are not free to move. This is an ideal situation since, even though electrons may be bound in molecules, there is always some movement of electrons. However, it will be useful here to neglect such effects and assume dielectrics are "ideal" in the same way we assumed conductors to be "perfect" conductors. Later, when the need to include conduction of free electrons arises, we will have to reevaluate this approach and allow partial conduction in dielectrics.

4.5.3 Polarization and the Polarization Vector

A perfect dielectric, therefore, is a material which has bound charges but no free charges. What happens to these charges if the dielectric is placed in an electric field? In a simplistic model, we may view each atom or each molecule as consisting of positive and negative charges in a charge-neutral configuration. Therefore, the net external electric field produced by the atoms of the material is zero (although the fields inside the atom are certainly not). For example, if we view electrons as spinning around the atom in circular orbits, the outer layer of the atom is negatively charged whereas the nucleus is positively charged, as shown in **Figure 4.25a**. The external electric field is zero, as can be seen from Gauss's law by enclosing the atom by a spherical Gaussian surface. If we now apply an external electric field such as by an external point charge or, more easily, by charged surfaces (**Figure 4.25b**), the external field will exert a force on both the positive and negative charges. This will tend to displace the charges, that is, to polarize the atoms. The net effect is that each molecule becomes an electric dipole, as shown in **Figure 4.25c**. As expected, weak external electric fields will cause weak electric dipole moments (the equivalent distance between the displaced charges is small), whereas strong electric fields cause strong dipole moments. The equivalent dipole and the equivalent dipole moment drawn in **Figure 4.25c** constitute a model that is consistent with the measured macroscopic effect of polarized materials.

Figure 4.25 Polarization of charges in a dielectric. (**a**) A charge-neutral atom. (**b**) The electron cloud is shifted slightly due to the external electric field. (**c**) The net effect is an equivalent dipole

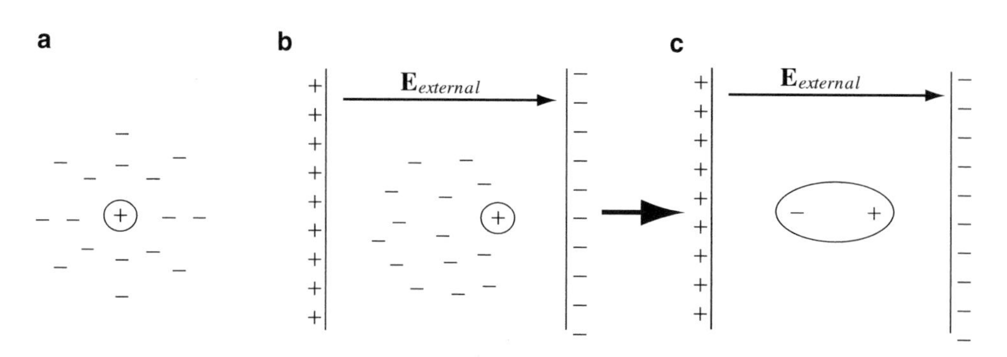

From **Figure 4.25c** and from Coulomb's law, it is clear that the dipole aligns itself with the external electric field. Now, consider a large number of molecules, all aligned in the direction of the externally applied electric field, as shown in **Figure 4.26**. Two effects must now be taken into account. First, the total dipole moment that the material exhibits depends on the number of molecules and on the strength of individual molecular dipole moments. Second, each dipole moment, viewed as two point charges, generates an electric field pointing from the positive to the negative charge. This field opposes the external field, indicating that the net field inside the dielectric is lower than the external field. A total dipole moment per unit volume can be defined by summing the dipole moments of individual molecules:

$$\mathbf{P} = \lim_{\Delta v \to 0} \frac{1}{\Delta v} \sum_{i=1}^{N} \mathbf{p}_i \quad \left[\frac{\mathrm{C}}{\mathrm{m}^2}\right] \tag{4.37}$$

This vector is called a ***polarization vector*** and, based on the preceding discussion, is dependent on the external electric field intensity. While a useful definition, the polarization vector as given above is not really calculable in general. For one, the dipole moment of individual molecules is not known and is not easily measurable. In addition, not all molecules are polarized in an identical fashion. For example, we expect molecules near the surface to have different dipole moments than molecules inside the material. The net effect of the electric field due to polarization of the molecules is to reduce the electric field intensity inside the material.

Figure 4.26 Polarization of charges in a dielectric by means of an external field. The short arrows show the direction of the fields of dipoles

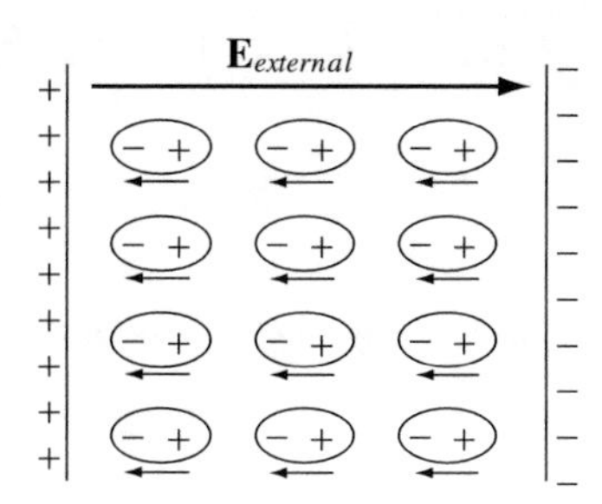

For an element of volume dv', we can use the idea of the dipole and calculate the potential at a distance R (see **Example 4.12**) from dv' as

$$dV = \frac{\mathbf{P} \cdot \hat{\mathbf{R}}}{4\pi\varepsilon_0 |\mathbf{R}|^2} dv' \quad [\mathrm{V}] \tag{4.38}$$

In this expression, $\mathbf{P}dv'$ is the equivalent dipole moment of the volume dv' since $\mathbf{P}$ was defined as the dipole moment per unit volume in **Eq. (4.37)**. To obtain the potential due to the whole volume, we integrate over the volume of the charge distribution to obtain

$$V = \int_{v'} dV = \int_{v'} \frac{\mathbf{P} \cdot \hat{\mathbf{R}}}{4\pi\varepsilon_0 |\mathbf{R}|^2} dv' \quad [\mathrm{V}] \tag{4.39}$$

It is well worth comparing this expression with that for the potential of a charge density ρ_v in any arbitrary volume. This was obtained in **Eq. (4.32)** as

$$V = \frac{1}{4\pi\varepsilon_0} \int_{v'} \frac{\rho_v dv'}{|\mathbf{R}|} \quad [\mathrm{V}] \tag{4.40}$$

Comparing the two, we note that **Eq. (4.39)** can be written in the form of **Eq. (4.40)** if the equivalent charge density in the volume v' is written as

$$\rho_{pv} = \frac{\mathbf{P} \cdot \hat{\mathbf{R}}}{|\mathbf{R}|} \quad \left[\frac{\mathrm{C}}{\mathrm{m}^3}\right] \tag{4.41}$$

Thus, we conclude that due to the polarization vector, there is an equivalent polarization volume charge density throughout the volume v'. It is useful to separate the charge density into a volume and a surface charge density because these can be very different from each other. To do so, we rewrite **Eq. (4.39)** as

$$V = \frac{1}{4\pi\varepsilon_0} \int_{v'} \mathbf{P} \cdot \frac{\hat{\mathbf{R}}}{|\mathbf{R}|^2} dv' = -\frac{1}{4\pi\varepsilon_0} \int_{v'} \mathbf{P} \cdot \nabla\left(\frac{1}{R}\right) dv' = \frac{1}{4\pi\varepsilon_0} \int_{v'} \mathbf{P} \cdot \nabla'\left(\frac{1}{R}\right) dv' \quad [\mathrm{V}] \tag{4.42}$$

The later step needs some explanation. We obtained in **Example 2.10**

$$\nabla'\left(\frac{1}{R}\right) = -\nabla\left(\frac{1}{R}\right) = \frac{\hat{\mathbf{R}}}{R^2} \tag{4.43}$$

where (in Cartesian coordinates) $R = ((x - x')^2 + (y - y')^2 + (z - z')^2)^{1/2}$. The integration is on the volume v' and therefore must be carried out in the primed coordinates.

The relation in **Eq. (4.43)** was used in **Eq. (4.42)**.

Now, we use the following vector identity [see **Eq. (2.138)**]:

$$\nabla \cdot U\mathbf{A} = U\nabla \cdot \mathbf{A} + \mathbf{A} \cdot \nabla U \tag{4.44}$$

where U is any scalar function and $\mathbf{A}$ any vector function. In this case, we can write the integrand in **Eq. (4.42)** using $U = 1/R$ and $\mathbf{A} = \mathbf{P}$:

$$\mathbf{P} \cdot \nabla'\left(\frac{1}{R}\right) = \nabla' \cdot \frac{1}{R}\mathbf{P} - \frac{1}{R}(\nabla' \cdot \mathbf{P}) \tag{4.45}$$

Substituting this into **Eq. (4.42)** gives

$$V = \frac{1}{4\pi\varepsilon_0}\int_{v'} \nabla' \cdot \frac{1}{R}\mathbf{P}dv' - \frac{1}{4\pi\varepsilon_0}\int_{v'}\frac{1}{R}(\nabla' \cdot \mathbf{P})dv' \quad [\text{V}] \tag{4.46}$$

The first integral can be converted into a surface integral using the divergence theorem:

$$\int_{v'} \nabla' \cdot \frac{1}{R}\mathbf{P}dv' = \int_{s'}\frac{1}{R}\mathbf{P} \cdot d\mathbf{s}' = \int_{s'}\frac{1}{R}(\mathbf{P} \cdot \hat{\mathbf{n}})ds' \tag{4.47}$$

and, finally, substituting this for the first integral in **Eq. (4.46)**, we get

$$V = \frac{1}{4\pi\varepsilon_0}\int_{s'}\frac{1}{R}(\mathbf{P} \cdot \hat{\mathbf{n}})ds' + \frac{1}{4\pi\varepsilon_0}\int_{v'}\frac{-(\nabla' \cdot \mathbf{P})}{R}dv' \quad [\text{V}] \tag{4.48}$$

Comparing the two terms with the general expressions for the potential due to surface charge densities and volume charge densities in **Eqs. (4.31)** and **(4.32)**, we can rewrite the potential as that due to an equivalent polarization surface charge density ρ_{ps} [C/m^2] and an equivalent polarization volume charge density ρ_{pv} [C/m^3] as

$$V = \frac{1}{4\pi\varepsilon_0}\int_{s'}\frac{\rho_{ps}}{R}ds' + \frac{1}{4\pi\varepsilon_0}\int_{v'}\frac{\rho_{pv}}{R}dv' \quad [\text{V}] \tag{4.49}$$

where the surface and volume charge densities are

$$\boxed{\rho_{ps} = \mathbf{P} \cdot \hat{\mathbf{n}}' \quad [\text{C/m}^2] \quad \text{and} \quad \rho_{pv} = -\nabla' \cdot \mathbf{P} \quad [\text{C/m}^3]} \tag{4.50}$$

In general, we can drop the primed notation since this relation only applies in the volume in which polarization exists (and, therefore, there is no specific need to identify the volume), but it is left intact here to signify the fact that polarization can only take place inside the material.

When molecules of dielectrics align with the external field, there are two possible situations:

(1) The dielectric is uniform in its properties. All molecules have identical dipole moments, and because of this, all internal polarization charges cancel each other with the exception of charges on the surface of the material, as shown in **Figure 4.27a**. This can be explained simply by proximity of the positive and negative charges (see **Figure 4.26**). The net effect is an induced (polarization) surface charge density due to the effect of the external electric field. The total field now is the combined external field and the internal field due to induced surface charges as shown.

(2) The dielectric is not uniform in its properties. Molecules at different locations are polarized differently. Although some of the polarization charges cancel each other, some do not, and the net effect is a polarization surface charge density on the surface and a polarization volume charge density in the interior of the material, as shown in **Figure 4.27b**. Again, the net effect is that of a combined electric field intensity due the external and internal fields.

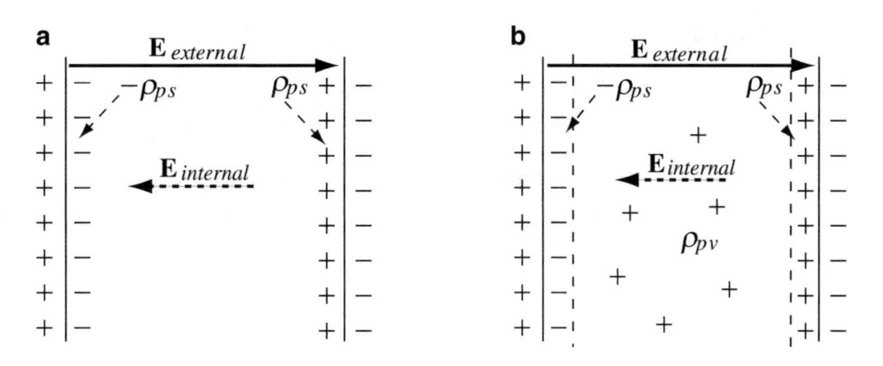

Figure 4.27 (**a**) Surface polarization charge density in a uniformly polarized dielectric. (**b**) Surface and volume polarization charge densities in a nonuniformly polarized dielectric

4.5.4 Electric Flux Density and Permittivity

After discussing polarization, we still have a basic problem: although qualitatively we know what the polarization vector is and that surface and volume charge densities may be induced due to the external electric field, these quantities are not yet calculable in quantitative terms. However, if we were to assume that both surface and volume charge densities in the dielectric are known, we could then proceed to calculate the total electric field intensity as well as the potential anywhere in space. To do so, we take a more general view and look at a volume that has both induced charges due to polarization of dielectrics and source, or free, charges due to external, applied sources. Polarization charges are represented by a volume charge density ρ_{pv} [C/m^3]. Source charges are also represented by a volume charge density and are denoted as ρ_v [C/m^3]. At any point in the volume, we have a total charge density ρ [C/m^3]:

$$\boxed{\rho = \rho_v + \rho_{pv} \quad [\text{C/m}^3]} \tag{4.51}$$

We now invoke the first postulate of the electric field from **Eq. (4.6)**. This applies to any electric field and therefore we can write inside the volume of the dielectric:

$$\nabla \cdot \mathbf{E} = \frac{\rho}{\varepsilon_0} = \frac{\rho_v + \rho_{pv}}{\varepsilon_0} = \frac{\rho_v}{\varepsilon_0} + \frac{\rho_{pv}}{\varepsilon_0} \tag{4.52}$$

The first term on the right-hand side is the divergence of the electric field intensity, in the absence of polarization charges. This term is the easier to evaluate and measure because it is generated by external means rather than by internal, atomic level effects. Thus, it is sensible to rewrite the expression in terms of this charge density by multiplying both sides by ε_0:

$$\nabla \cdot \varepsilon_0 \mathbf{E} - \rho_{pv} = \rho_v \qquad [\text{C/m}^3] \tag{4.53}$$

The second term, ρ_{pv}, can be written as the negative divergence of **P** [from **Eq. (4.50)**] and the relation now becomes

$$\nabla \cdot \varepsilon_0 \mathbf{E} + \nabla \cdot \mathbf{P} = \rho_v \qquad [\text{C/m}^3] \tag{4.54}$$

From the first postulate **[Eq. (4.6)]**, we know that the free charge density is related to the electric field intensity as (see also **Section 3.5**)

$$\nabla \cdot \mathbf{E} = \frac{\rho_v}{\varepsilon_0} \quad \text{or} \quad \nabla \cdot \mathbf{D} = \rho_v \tag{4.55}$$

Taking the second form in **Eq. (4.55)**, we conclude that the following must hold:

$$\nabla \cdot \varepsilon_0 \mathbf{E} + \nabla \cdot \mathbf{P} = \nabla \cdot \mathbf{D} \tag{4.56}$$

or, alternatively, that

$$\boxed{\mathbf{D} = \varepsilon_0 \mathbf{E} + \mathbf{P} \quad [\mathrm{C/m^2}]} \tag{4.57}$$

This relation has three important implications. First, it gives a useful relation by which we can calculate the polarization vector in dielectrics from measurable quantities, for if we measure the electric field intensity **E** and electric flux density **D**, the polarization vector is immediately available. Second, since the polarization vector adds to the flux density in the absence of polarization ($\varepsilon_0\mathbf{E}$), it can be viewed as a means of decreasing the electric field intensity in the dielectric for a given external flux density. This property is the basis of many applications and devices, as we shall see shortly. The third implication from **Eq. (4.57)** is that the units of polarization are [C/m^2].

Also, because polarization depends on the external electric field intensity, we can write the polarization vector in terms of a new parameter χ_e, which is called ***electric susceptibility***:

$$\mathbf{P} = \varepsilon_0 \chi_e \mathbf{E} \quad [\mathrm{C/m^2}] \tag{4.58}$$

Electric susceptibility indicates how susceptible the dielectric is to polarization. It is, however, a difficult parameter to measure. For this reason, it is seldom used. The purpose of using it here is as an intermediate term which we will do away with shortly. Substituting this back into **Eq. (4.57)**, we get

$$\mathbf{D} = \varepsilon_0 \mathbf{E} + \mathbf{P} = \varepsilon_0 \mathbf{E} + \varepsilon_0 \chi_e \mathbf{E} = \varepsilon_0 (1 + \chi_e) \mathbf{E} \quad [\mathrm{C/m^2}] \tag{4.59}$$

Now we have a direct relation between **D** and **E** for any material. Writing $\mathbf{D} = \varepsilon\mathbf{E}$, we define the permittivity ε of any material as

$$\varepsilon = \varepsilon_0 (1 + \chi_e) \qquad [\mathrm{F/m}] \tag{4.60}$$

The term in parentheses on the right-hand side of **Eq. (4.60)** is that part of permittivity that relates to the dielectric itself as opposed to the permittivity of free space; therefore, we write

$$\varepsilon_r = (1 + \chi_e) \quad \to \quad \varepsilon = \varepsilon_0 \varepsilon_r \qquad [\mathrm{F/m}] \tag{4.61}$$

and, as a consequence,

$$\boxed{\mathbf{D} = \varepsilon \mathbf{E} = \varepsilon_0 \varepsilon_r \mathbf{E} \quad [\mathrm{C/m^2}]} \tag{4.62}$$

where ε_r is called the ***dielectric constant*** or ***relative permittivity*** of the dielectric. Although we could also use the relation $\varepsilon_r = 1 + \chi_e$, we will not do so for the very simple reason that ε_r can be measured directly without the need to measure χ_e. The relative permittivity of free space is 1, which also implies that in free space $\chi_e = 0$ (or $\mathbf{P} = 0$). Both ε_r and χ_e are dimensionless parameters.

Note the units in **Eq. (4.60)**. Permittivity can be written as $\varepsilon = |\mathbf{D}|/|\mathbf{E}|$, hence the unit of permittivity is [(C/m^2)/(V/m)] = [C/V · m]. The [C/V] is known as the Farad, hence the unit of permittivity is [F/m].

The advantage of using the relation in **Eq. (4.62)** is obvious: a simple relation exists between the electric field intensity and the electric flux density, a relation that is based on the experimental evaluation of permittivity. This approach avoids most of the difficulties involved in the evaluation of polarization. In a way, we have bundled all polarization effects into one relatively easy term to measure. Although this is proper to do, we must not forget that permittivity of a material is a measure of its polarization and that polarization must sometimes be evaluated. Permittivities of materials can be measured directly and these are available in tables. **Table 4.1** shows relative permittivities of a few materials.

Table 4.1 Relative permittivities of some dielectrics

Material	ε_r	Material	ε_r	Material	ε_r
Quartz	3.8–5	Paper	3.0	Silica	3.8
GaAs*	13	Bakelite	5.0	Quartz	3.8
Nylon	3.1	Glass	6.0 (4–7)	Snow	3.8
Paraffin	3.2	Mica	6.0	Soil (dry)	2.8
Perspex	2.6	Water (distilled)	81	Wood (dry)	1.5–4
Polystyrene foam	1.05	Polyethylene	2.2	Silicon	11.8
Teflon	2.0	PVC***	6.1	Ethyl alcohol	25
$BaTiO_3$**	10,000	Germanium	16	Amber	2.7
Air	1.0006	Glycerin	50	Plexiglas	3.4
Rubber	3.0	Nylon	3.5	Aluminum oxide	8.8

*Gallium Arsenide, **Barium Titanate, ***PolyVinyl Chloride

Example 4.15 Polarization Vector in a Coaxial Cable Two conducting concentric, cylindrical shells are separated by a dielectric of permittivity ε [F/m]. This is the common form of coaxial cables where the dielectric is the insulating material between the two conductors. A total charge $+Q$ [C] is distributed uniformly per unit length of the inner shell (see **Figure 4.28a**). Calculate:

(a) The magnitude and direction of the polarization vector inside the dielectric.
(b) The polarization volume charge density in the dielectric.
(c) The polarization surface charge densities on the surfaces of the dielectric.

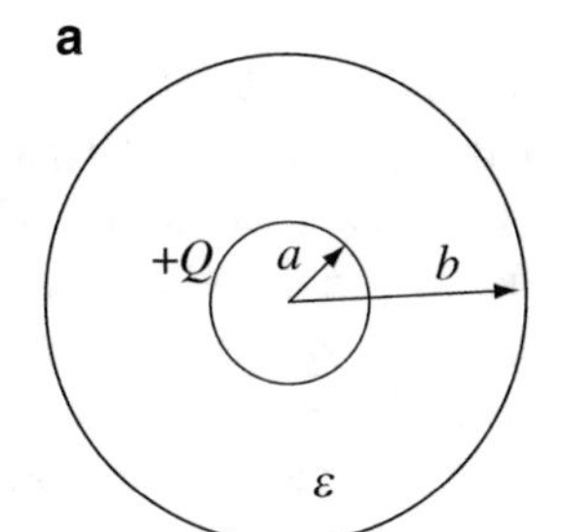

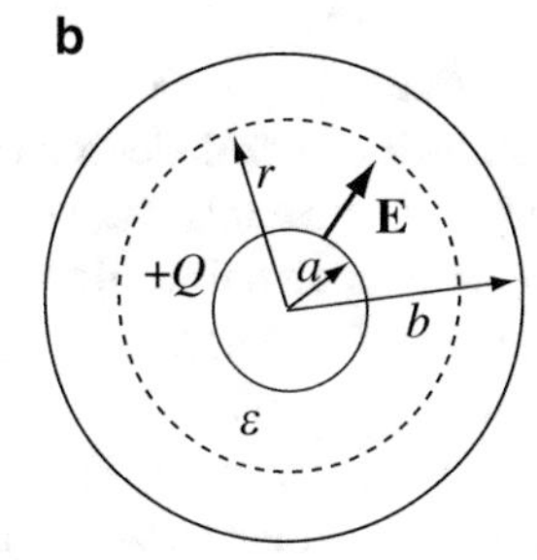

Figure 4.28 (**a**) Coaxial cable made of two cylindrical conductors separated by a dielectric. (**b**) A cylindrical Gaussian surface of radius r and length L (into the page) used to calculate the field

Solution:

(a) Here, we use the relations among the electric field intensity, electric flux density, and permittivity **[Eq. (4.57)]** to find the polarization **P**. The relations in **Eq. (4.50)** are used to calculate charge densities in **(b)** and **(c)** from the polarization vector **P**.

$$\mathbf{D} = \varepsilon_0\mathbf{E} + \mathbf{P} \quad \text{or} \quad \mathbf{P} = \mathbf{D} - \varepsilon_0\mathbf{E} \qquad [\mathrm{C/m^2}]$$

To calculate the electric field intensity **E** for $a < r < b$, we use Gauss's law. Using a cylindrical surface of length L and radius r as shown in **Figure 4.28b**:

$$2\pi rLD = QL \;\rightarrow\; D = \frac{Q}{2\pi r}\left[\frac{\mathrm{C}}{\mathrm{m^2}}\right] \;\rightarrow\; E = \frac{Q}{2\pi\varepsilon r}\left[\frac{\mathrm{V}}{\mathrm{m}}\right]$$

Substituting **D** and **E** in the above relation gives

$$\mathbf{P} = \mathbf{D} - \varepsilon_0 \mathbf{E} = \hat{\mathbf{r}} \left[\frac{Q}{2\pi r} - \varepsilon_0 \frac{Q}{2\pi \varepsilon r} \right] = \hat{\mathbf{r}} \frac{Q}{2\pi r} \left[1 - \frac{\varepsilon_0}{\varepsilon} \right] \quad \left[\frac{\mathrm{C}}{\mathrm{m}^2} \right]$$

(b) The polarization vector **P** has only an r component. From **Eq. (4.50)**, the polarization volume charge density is

$$\rho_{pv} = -\nabla \cdot \mathbf{P} = -\frac{1}{r} \frac{\partial}{\partial r}(r P_r) = -\frac{1}{r} \frac{\partial}{\partial r} \left(\frac{rQ}{2\pi r} \left[1 - \frac{\varepsilon_0}{\varepsilon} \right] \right) = 0.$$

(c) The polarization surface charge densities are also calculated from **Eq. (4.50)**. On the inner surface, the direction of the normal is in the negative r direction, whereas on the outer surface, it is in the positive r direction. Thus,

$$\rho_{psa} = \mathbf{P} \cdot \hat{\mathbf{n}} = \hat{\mathbf{r}} \frac{Q}{2\pi a} \left[1 - \frac{\varepsilon_0}{\varepsilon} \right] \cdot (-\hat{\mathbf{r}}) = -\frac{Q}{2\pi a} \left[1 - \frac{\varepsilon_0}{\varepsilon} \right] \quad \left[\frac{\mathrm{C}}{\mathrm{m}^2} \right]$$

$$\rho_{psb} = \mathbf{P} \cdot \hat{\mathbf{n}} = \hat{\mathbf{r}} \frac{Q}{2\pi b} \left[1 - \frac{\varepsilon_0}{\varepsilon} \right] \cdot (\hat{\mathbf{r}}) = \frac{Q}{2\pi b} \left[1 - \frac{\varepsilon_0}{\varepsilon} \right] \quad \left[\frac{\mathrm{C}}{\mathrm{m}^2} \right]$$

Important

(1) ρ_{psa} and ρ_{psb} are charge densities on the surface of the dielectric (not on the conducting shells).
(2) If the dielectric is removed (replaced with free space), the polarization vector and the surface and volume charge densities become zero since now $\varepsilon = \varepsilon_0$.

4.5.4.1 Linearity, Homogeneity, and Isotropy

Now that the electric properties of materials have been defined through electric permittivity (and, indirectly, through polarization), it is appropriate to classify these properties. Specifically, we define three important properties: linearity, homogeneity, and isotropy of materials.

A material is ***linear*** in a particular property (like permittivity) if the property does not change when the fields change. For example, if permittivity of a dielectric is independent of the applied field, the material is said to be ***linear in permittivity***. It is very important to note that this is not the same as linear variation: if the permittivity of the material were to vary linearly with the applied field, the material would be a nonlinear material. Similarly, we may discuss a linear distribution of some sort or the linear behavior of a device. These should not be confused with the linearity property of materials involved.

A ***homogeneous*** material is a material whose physical properties do not vary from point to point in the material. The permittivity of a homogeneous dielectric material is the same at any point in the material.

An ***isotropic*** material is one whose properties are independent of direction in space.

Materials may be linear or nonlinear, homogeneous or nonhomogeneous, isotropic or anisotropic. Linearity, homogeneity, and anisotropy are associated with each of the material properties. A material may be linear in one property but nonlinear in another. The same consideration applies to homogeneity and isotropy. For example, the expression in **Eq. (4.62)** was obtained by implicitly assuming that the dielectric was linear (i.e., its properties were independent of the applied electric field) and isotropic (for which **E** and **D** are in the same direction).

A linear, homogeneous, isotropic material is called a ***simple material***. Although there are some very important engineering materials which do not behave in this fashion, we will not discuss them here.

4.5.5 Dielectric Strength

Charges in a dielectric are bound and cannot move freely. In the presence of an external electric field, the forces exerted on molecules polarize them and generate an internal electric field intensity. It is reasonable to assume that if the external electric field is increased, there will be a point at which the force the electric field exerts on electrons is sufficiently high to remove electrons from molecules. At this stage, these electrons become free electrons and move under the influence of the external field. In effect, the material has become conducting. The maximum electric field intensity at which this *does not occur* is called the ***dielectric strength*** of the material, denoted as K [V/m]. Beyond this electric field intensity, the material becomes conducting and therefore useless as a dielectric. The material is said to have undergone ***dielectric breakdown***. We are rather familiar with this effect. Lightning is a breakdown effect in air. It occurs when the electric field intensity produced by charges exceeds a certain value (approximately 3,000 V/mm or 3 million volts per meter). In the case of lightning, breakdown means "shorting" of the sources of charge, causing a rapid discharge. In other cases, such as in electronic circuits, breakdown can occur at relatively low potential differences, destroying the circuit.

From **Table 4.2**, we note that dielectrics have a dielectric strength larger than air. In other words, if a large potential exists between two locations in space and we wish to reduce the chance of breakdown (sparking), a dielectric inserted at that location would probably do the trick.

Table 4.2 Dielectric strength of some common dielectrics

Material	Dielectric strength K [V/m]	Material	Dielectric strength K [V/m]
Air	3×10^6	Glass	3×10^7
Mineral oil	1.5×10^7	Quartz	4×10^7
Wax paper	5×10^7	Polystyrene	1×10^7
Porcelain	2×10^7	Mica	2×10^8
Alumina	1.4×10^7	Teflon	1×10^8
Distilled water	6×10^7	Rubber	2.5×10^7
Diamond	2×10^9	Vacuum	1×10^{18}

Example 4.16 Application: Dielectric Strength and Breakdown in Semiconductor Devices

Integrated circuits owe much of their success to miniaturization. This allows increased device density by closer spacings of conducting films and other circuit elements on the semiconductor layer. Conducting layers are separated from each other by dielectric layers, normally in the form of oxides of the semiconductors.

One limit to miniaturization is the possibility of breakdown of the insulating oxide layer due to application of operating voltages or accumulation of electrostatic charges while the device is not operating. Consider the following example.

Two conducting strips in an integrated circuit are separated a distance d. The layer between the strips is silicon oxide SiO_2, which has dielectric strength of 30,000 V/mm:

(a) Assuming the electric field intensity is uniform between the two strips, calculate the smallest separation possible between the strips at an operating voltage of 5 V.

(b) Suppose in a commercially available device, the smallest separation is 0.2 μm. What is the maximum potential that the device can withstand without damage?

Solution: In **(a)**, because the electric field intensity is uniform, we assume the strips form two parallel plates, which may be viewed as large compared to the distance between them, and calculate the distance d for breakdown. In **(b)**, the potential that will produce breakdown is calculated based on the relation in **(a)**:

(a) From the relation between potential and electric field intensity between two parallel plates in **Exercise 4.5**,

$$E = \frac{V}{d} \quad \rightarrow \quad d = \frac{V}{E} = \frac{5}{3 \times 10^7} = 1.67 \times 10^{-7} \quad [\text{m}]$$

Note: The smallest separation possible is 0.167 μm. In practice, a larger separation will almost always be required to prevent damage by electrostatic discharge (due to accumulation of static charges in storage or while handling the device).

(b) Assuming the same conditions as in part **(a)**, the largest potential difference allowable is

$$V = Ed = 3 \times 10^7 \times 0.2 \times 10^{-6} = 6 \quad [\mathrm{V}]$$

This may seem sufficiently large to allow safe operation. Recall, however, that as you walk on a carpet, you can acquire sufficient charge to create a spark as you touch a door knob. The potential acquired in this fashion can easily exceed a few kilovolts. If you were to touch a semiconductor device instead, the electrostatic discharge (ESD) would destroy the device. Thus, sensitive devices such as processors and memory chips are shipped with their pins short circuited (by embedding them in a conducting foam) or protected by conducting plastic packaging to avoid damage. Some devices have built-in active circuitry that limits the potential on any pin to a safe level. Of course, there are other limits to miniaturization, and the calculation here is by no means exact, but this example shows one difficulty that must be addressed in design and packaging of semiconductor devices.

Example 4.17 Application: Diagnosing Ignition Systems The fact that dielectrics reduce the electric field intensity can also work against us. One example is the internal combustion engine. The spark plug is designed to ignite the gasoline mixture. For this, the electric field intensity must be large enough to cause breakdown (create a spark). If the electric field intensity is reduced because of faulty connections or other reasons, it may happen that a spark is still obtained in air but not in the presence of gasoline vapor since gasoline has a high permittivity and a high dielectric strength. The same effect occurs when spark plugs are wet or contaminated with oil.

Consider that the electric system in a car has deteriorated and produces an ignition voltage of 10 kV. This is connected across a spark plug (see **Figure 4.29a**) in air. The electrodes of the spark plug are separated 0.8 mm (typical spark gaps are between 0.6 and 1.5 mm). As part of a test, the spark is checked, and since 10 kV can produce a spark about 3 mm long (in air), the system seems to be in good order:

(a) The spark plug is now inserted into the cylinder. Suppose the gasoline–air mixture **(Figure 4.29b)** has a dielectric strength of 15 kV/mm. Will the engine run? If not, what must be the minimum voltage for the engine to operate?
(b) As an emergency measure, the gap on the spark plug can be reduced to produce a spark. What must be the maximum gap to produce a spark with an ignition voltage of 10 kV? Assume the electric field intensity between the electrodes of the spark plug is uniform.

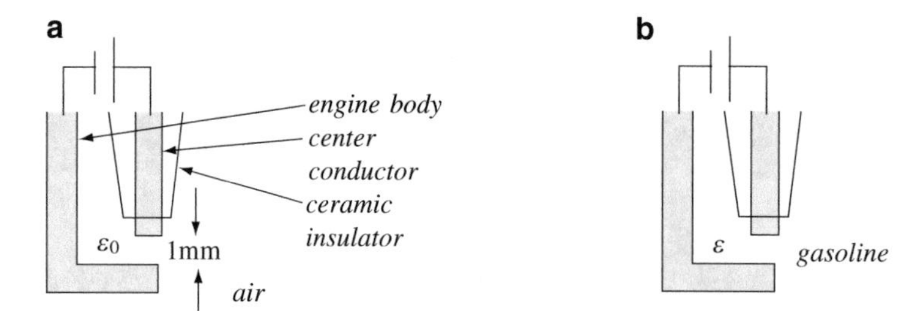

Figure 4.29 A spark plug used in an internal combustion engine (partial view). The spark is generated when the electric field intensity in the gap exceeds the dielectric strength of the air–fuel mixture. (**a**) Structure. (**b**) Air-gasoline mixture in the gap changes the permittivity

Solution:

(a) No: the minimum required voltage for breakdown in the gasoline mixture is $0.8 \times 15 = 12$ kV.
(b) The electric field intensity is the ratio between potential and distance, where, because the gap is small, we assume the electric field intensity between the electrodes is the same as between two parallel plates (see **Example 4.16** or **Exercise 4.5**):

$$E = \frac{V}{d} \quad \rightarrow \quad d = \frac{V}{E} = \frac{10 \times 10^3}{15 \times 10^3} = 0.667 \qquad [\mathrm{mm}]$$

The gap must be smaller than 0.667 mm. This result is an approximate solution but it gives a rough idea of the design requirements for this device.

4.6 Interface Conditions

Up to this point, we assumed that the electric field exists in space which is either a vacuum (characterized by ε_0) or a dielectric (characterized by permittivity $\varepsilon = \varepsilon_0\varepsilon_r$). In practical applications, it often becomes necessary to use combinations of materials. The interface between any two materials is a physical discontinuity in material properties. The question that arises is: What happens to the electric field intensity and the electric flux density at such an interface? That, in general, the interface influences the field we know from the fact that inside conductors, the electric field intensity is zero, whereas on the surface of the conductor, only a normal component of the field exists. This indicates that the electric field intensity is discontinuous at the conductor's interface, at least for the normal component. Moreover, this change is abrupt: the electric field intensity is finite in free space and is zero in the conductor. What are the consequences of this sudden change and how do we define the change? What are the conditions at interfaces between dielectrics or between dielectrics and conductors? These questions are answered by the interface conditions.

The principle used in defining interface conditions is to apply the two postulates in **Eq. (4.9)** at the interface. The conditions on the tangential components of the fields are derived from the closed contour integral of $\mathbf{E} \cdot d\mathbf{l}$, whereas the conditions on the normal component are derived from Gauss's law.

In electrostatics, there are two useful interfaces:

(1) Interface between two dielectrics including that between a dielectric and free space.
(2) Interface between dielectrics and conductors, including that between a conductor and free space.

We start with a dielectric–dielectric interface.

4.6.1 Interface Conditions Between Two Dielectrics

Figure 4.30a shows two materials with different permittivities: ε_1 in material 1 and ε_2 in material 2. We assume that a uniformly distributed positive static electric charge density exists on the interface between the two materials as a surface charge density ρ_s [C/m^2]. We start with the second postulate:

$$\oint_C \mathbf{E} \cdot d\mathbf{l} = 0 \tag{4.63}$$

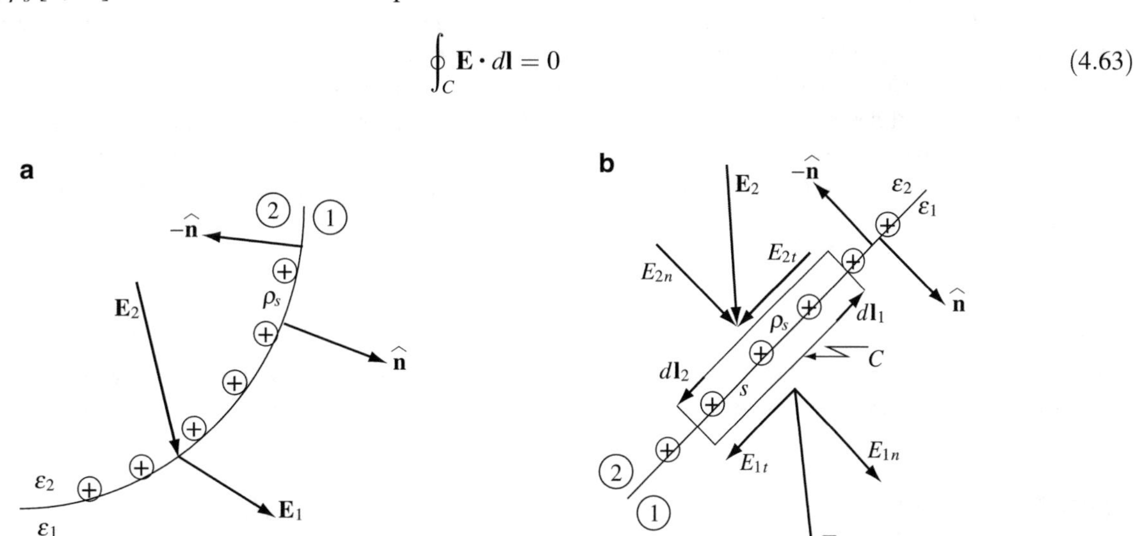

Figure 4.30 (**a**) Refraction of the electric field at an interface between two different materials. (**b**) A contour enclosing a small section of the interface used to evaluate the tangential components of the electric field intensity. $\hat{\mathbf{n}}$ is positive pointing into medium (1)

To evaluate this over a closed loop, we define the small loop shown in **Figure 4.30b**. This loop has two sides parallel and infinitely close to the interface. Although this is a very unique loop, it does not detract from the generality of the relation since the contour in **Eq. (4.63)** is arbitrary. Also, because the length of $d\mathbf{l}$ along the interface is very small, $\mathbf{E}_{1t}$ and $\mathbf{E}_{2t}$ are considered constant on the respective sides of the interface. Neglecting the contribution to the circulation of the smaller sides (the sides perpendicular to the interface tend to zero in length because of our choice of the loop), we obtain

$$\oint_C \mathbf{E} \cdot d\mathbf{l} = \int_{C_1} \mathbf{E}_1 \cdot d\mathbf{l}_1 + \int_{C_2} \mathbf{E}_2 \cdot d\mathbf{l}_2 = 0 \quad (4.64)$$

where C_1 and C_2 are the parts of the contour in materials 1 and 2, respectively. We note that

$$\mathbf{E}_1 \cdot d\mathbf{l}_1 = (\mathbf{E}_{1t} + \mathbf{E}_{1n}) \cdot d\mathbf{l}_1 = -E_{1t} dl_1 \quad \text{and} \quad \mathbf{E}_2 \cdot d\mathbf{l}_2 = E_{2t} dl_2 \quad (4.65)$$

where the negative sign comes from the fact that $\mathbf{E}_{1t}$ and $d\mathbf{l}_1$ are in opposite directions. Also, E_{1n} is normal to the contour and therefore produces zero scalar product with $d\mathbf{l}_1$, similarly for $\mathbf{E}_2$. For the rectangular loop, we assume the length of the parallel paths to be $\Delta l_1 = \Delta l_2 = \Delta l$ and write

$$-E_{1t} \int_{\Delta l_1} dl + E_{2t} \int_{\Delta l_2} dl = 0 \quad \rightarrow \quad -E_{1t} \Delta l + E_{2t} \Delta l = 0 \quad (4.66)$$

or

$$\boxed{E_{1t} = E_{2t}} \quad (4.67)$$

The interface condition for the electric flux density $\mathbf{D}$ can be written by substituting $E_{1t} = D_{1t}/\varepsilon_1$ and $E_{2t} = D_{2t}/\varepsilon_2$:

$$\boxed{\frac{\mathrm{D}_{1t}}{\varepsilon_1} = \frac{\mathrm{D}_{2t}}{\varepsilon_2}} \quad (4.68)$$

The tangential components of the electric field intensity remain unchanged (are continuous) across the interface, but the tangential components of the electric flux density are discontinuous across the same interface. The discontinuity is due to the change in permittivity. The electric field intensity or the electric flux density may have two tangential components on the interface, depending on the orientation of the interface and the direction of the electric field intensity. The relations above indicate that each tangential component remains unchanged across the interface.

We turn now to the first postulate in **Eq. (4.9)** (Gauss's law) and apply it at the interface. To do so, we define an infinitesimal volume at the interface as shown in **Figure 4.31**. Since the volume is infinitesimal (i.e., both its end surfaces s_1 and s_2 are small, and the height of the cylinder tends to zero), we may view the interface as a flat surface of large extent and the normal component of the electric field intensity as constant on the parts of the Gaussian surface parallel to the interface. Gauss's law gives

$$\oint_s \mathbf{D} \cdot d\mathbf{s} = Q \quad [\mathrm{C}] \quad (4.69)$$

where Q is the total charge enclosed in the volume. This charge is located entirely on the interface between the two materials on the surface s within the cylinder and equals $\rho_s s$.

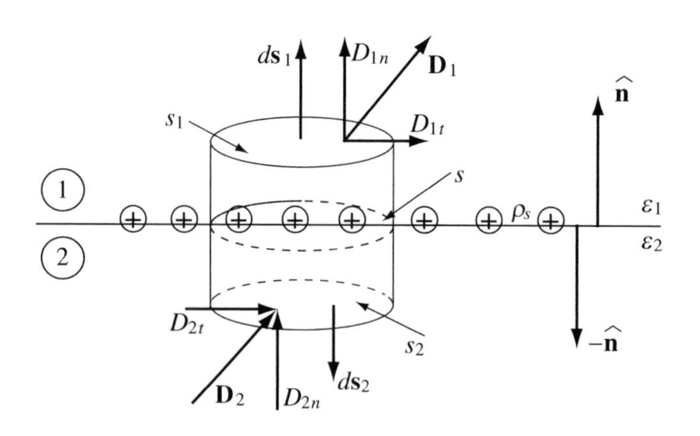

Figure 4.31 Gaussian surface at the interface used to evaluate the normal components of the electric flux density. Note that $\hat{\mathbf{n}}$ is positive pointing into medium (1)

The flux of the vector **D** is divided into two parts; because the flux on the lateral surface of the cylinder is zero, we have for an infinitesimally short cylinder

$$\oint_s \mathbf{D} \cdot d\mathbf{s} = \int_{s1} \mathbf{D}_1 \cdot d\mathbf{s}_1 + \int_{s2} \mathbf{D}_2 \cdot d\mathbf{s}_2 \tag{4.70}$$

where s_1 and s_2 are the surfaces of the bases of the cylinder. From **Figure 4.31**,

$$\mathbf{D}_1 \cdot d\mathbf{s} = (\mathbf{D}_{1n} + \mathbf{D}_{1t}) \cdot d\mathbf{s}_1 = D_{1n} ds_1 \quad \text{and} \quad \mathbf{D}_2 \cdot d\mathbf{s}_2 = -D_{2n} ds_2 \tag{4.71}$$

where $\mathbf{D}_{1n}$ and $d\mathbf{s}_1$ are in the same direction (point out of the volume), whereas $\mathbf{D}_{2n}$ and $d\mathbf{s}_2$ are in opposite directions. We get, using **Eq. (4.69)**,

$$\mathrm{D}_{1\mathrm{n}} \int_{s1} ds_1 - D_{2n} \int_{s2} ds_2 = \rho_s s \quad [\mathrm{C}] \tag{4.72}$$

Since $s_1 = s_2 = s$, we obtain

$$\boxed{D_{1n} - D_{2n} = \rho_s \quad [\mathrm{C/m^2}]} \quad \text{or} \quad \boxed{\varepsilon_1 E_{1n} - \varepsilon_2 E_{2n} = \rho_s \quad [\mathrm{C/m^2}]} \tag{4.73}$$

Thus, the change in the normal component of the electric flux density at the surface is equal to the surface charge density at the interface between the two materials.

From **Figure 4.31** and from **Eq. (4.71)**, we note that $D_{1n} = \hat{\mathbf{n}} \cdot \mathbf{D}_1$ and $D_{2n} = -\hat{\mathbf{n}} \cdot \mathbf{D}_2$ where $\hat{\mathbf{n}}$ is the normal unit vector pointing into material (1). Using this notation, we can write **Eq. (4.73)** as

$$\boxed{\hat{\mathbf{n}} \cdot (\mathbf{D}_1 - \mathbf{D}_2) = \rho_s \quad [\mathrm{C/m^2}]} \quad \text{or} \quad \boxed{\hat{\mathbf{n}} \cdot (\varepsilon_1 \mathbf{E}_1 - \varepsilon_2 \mathbf{E}_2) = \rho_s \quad [\mathrm{C/m^2}]} \tag{4.74}$$

This form is more general than the form in **Eq. (4.73)**, but in practical use, **Eq. (4.73)** is sufficient since **D** (or **E**) only has one normal component.

In the particularly common case, when there are no static charges on the interface ($\rho_s = 0$), we get

$$E_{1t} = E_{2t} \text{ and } \varepsilon_1 E_{1n} = \varepsilon_2 E_{2n} \quad \text{if } \rho_s = 0 \tag{4.75}$$

Using **Figure 4.32** and the above relations, we can write the following expressions for $\rho_s = 0$:

$$\tan\theta_1 = \frac{E_{1t}}{E_{1n}} \quad \text{and} \quad \tan\theta_2 = \frac{E_{2t}}{E_{2n}} \tag{4.76}$$

$$\frac{\tan\theta_1}{\tan\theta_2} = \frac{E_{2n}}{E_{1n}} = \frac{D_{2n}/\varepsilon_2}{D_{1n}/\varepsilon_1} = \frac{\varepsilon_1}{\varepsilon_2} \tag{4.77}$$

The larger the change in material properties, the larger the angular change between $\mathbf{E}_1$ and $\mathbf{E}_2$. However, we must point out that the variation in ε between most dielectric materials is rather small. As an example, the maximum ratio between permittivities of air and mica (these two materials are frequently used in insulating electric devices) is no higher than 6. The angular change in the direction of the electric field intensity at the interface is called ***refraction*** and is similar to the refraction of light rays passing between two materials with different indices of refraction. In fact, the optical index of refraction is related to the relative permittivity of materials, as we shall see in **Chapter 13**.

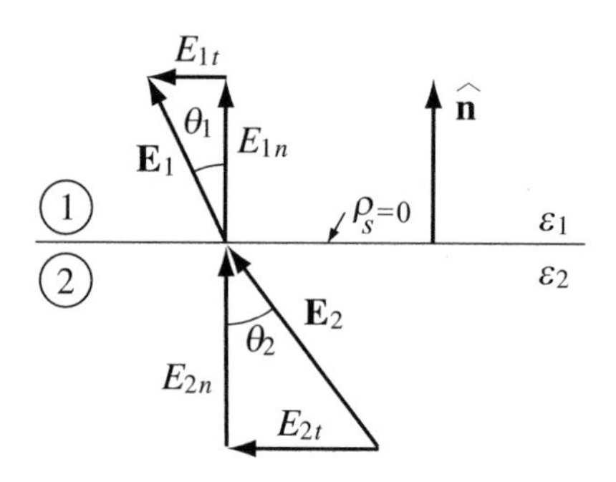

Figure 4.32 Relations between the components of the electric field intensity at the interface between two different dielectrics. Note that the electric field intensity may have two tangential components on the interface

4.6.2 Interface Conditions Between Dielectrics and Conductors

The interface conditions between a dielectric and a conductor can be found using the interface conditions between two dielectrics and requiring that the electric field intensity in one of them be zero as required for a conductor. Thus, assuming that material 2 in **Figures 4.30** and **4.31** is a conductor ($\mathbf{E}_2 = 0$), we get from the interface conditions in **Eqs. (4.67)** and **(4.69)**

$$\boxed{E_{1t} = E_{2t} = 0, \quad D_{1t} = D_{2t} = 0} \tag{4.78}$$

that is, the tangential components of the electric field intensity and electric flux density must be zero on both sides of the interface. If there is an electric field, the electric field intensity must be normal to the surface of a conductor, a result we already got from physical considerations of forces on charges.

Similarly, from **Eq. (4.73)** for the normal components of the electric field intensity and electric flux density, we can write

$$\boxed{E_{1n} = \frac{\rho_s}{\varepsilon_1} \quad \left[\frac{\mathrm{V}}{\mathrm{m}}\right] \quad \text{and} \quad D_{1n} = \rho_s \quad \left[\mathrm{C/m^2}\right]} \tag{4.79}$$

It should be noted that these relations give the correct surface charge density for the conditions in **Figure 4.30a**; that is, the surface charge density is positive because the electric flux density points away from the conductor's surface (material 2). If it were to point toward the conductor's surface, the surface charge density would be negative. If the surface charge density on the conductor is zero, the electric field intensity at the surface must also be zero. This is the same as saying that if a conductor is placed in an electric field, there must be a charge density induced on the surface. More than that, the magnitude of the surface charge density must be equal to the magnitude of the electric flux density $|\mathbf{D}|$, which has only a normal component at the conductor's surface. Unlike the interface between dielectrics on which a charge density may or may not exist, the interface between a conductor and dielectric will always have a surface charge density when an electric field exists.

The various interface conditions are summarized in **Table 4.3**.

Table 4.3 Interface conditions for the electric field intensity and electric flux density

Type of interface	Tangential components of **E**	Tangential components of **D**	Normal components of **E**	Normal components of **D**
Dielectric–dielectric with surface charge density	$E_{1t} = E_{2t}$	$\frac{D_{1t}}{\varepsilon_1} = \frac{D_{2t}}{\varepsilon_2}$	$\varepsilon_1 E_{1n} - \varepsilon_2 E_{2n} = \rho_s$ or $\hat{\mathbf{n}} \cdot (\varepsilon_1 \mathbf{E}_1 - \varepsilon_2 \mathbf{E}_2) = \rho_s$	$D_{1n} - D_{2n} = \rho_s$ or $\hat{\mathbf{n}} \cdot (\mathbf{D}_1 - \mathbf{D}_2) = \rho_s$
Dielectric–dielectric no surface charge density	$E_{1t} = E_{2t}$	$\frac{D_{1t}}{\varepsilon_1} = \frac{D_{2t}}{\varepsilon_2}$	$\varepsilon_1 E_{1n} = \varepsilon_2 E_{2n}$	$D_{1n} = D_{2n}$
Dielectric–conductor	$E_{1t} = 0, E_{2t} = 0$	$D_{1t} = 0, D_{2t} = 0$	$E_{1n} = \frac{\rho_s}{\varepsilon_1}, E_{2n} = 0$	$D_{1n} = \rho_s, D_{2n} = 0$

Note: If any of the dielectrics is free space, the permittivity for the corresponding dielectric becomes ε_0. The surface charge density is taken as positive. This means that the net electric field intensity points away from the surface charge, if this charge exists.

Example 4.18 A uniform electric field intensity of magnitude 1,000 V/m is measured underwater and points at an angle $\theta_2 = 30°$ at the interface between water and air, as shown in **Figure 4.33**. If it is known that no surface charges can exist and the relative permittivity of water is 80, calculate the electric field intensity in air (direction and magnitude).

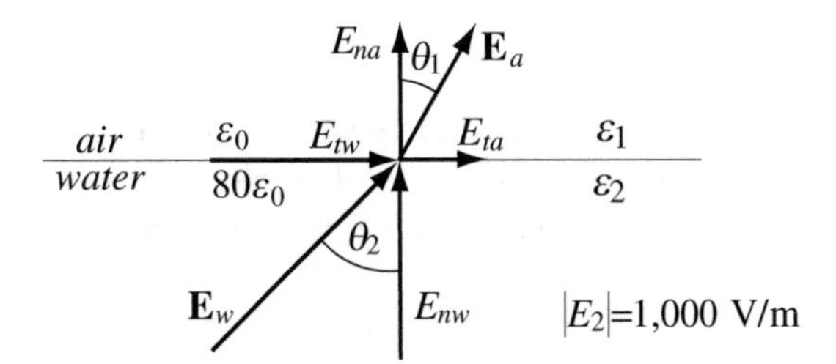

Figure 4.33 The electric field intensity at the surface between water and air

Solution: The electric field intensity in air is found by separating the electric field intensity in water into its tangential and normal components and then imposing the continuity in the tangential component and discontinuity in the normal component **[Eq. (4.75)]**. The angle between the electric field intensity and the normal is given in **Eq. (4.76)** (taking water as material 2).

The tangential and normal components of the electric field intensity in water are

$$E_{tw} = E\sin\theta_2 = 1{,}000\sin 30° = 500 \quad [\text{V/m}],$$

$$E_{nw} = E\cos\theta_2 = 1{,}000\cos 30° = 866 \quad [\text{V/m}]$$

In air,

$$E_{ta} = E_{tw} = 500 \quad [\text{V/m}]$$

$$\varepsilon_0 E_{na} = 80E_{nw} \quad \rightarrow \quad E_{na} = \frac{80\varepsilon_0}{\varepsilon_0}E_{nw} = 80 \times 866 = 69{,}280 \quad [\text{V/m}]$$

The normal electric field intensity is much larger in air than in water and the total electric field intensity is therefore also larger. The magnitude of the total electric field intensity in air is

$$E_a = \sqrt{E_{ta}^2 + E_{tn}^2} = \sqrt{500^2 + 69{,}280^2} = 69{,}281 \quad [\text{V/m}]$$

From **Eq. (4.76)**,

$$\tan\theta_1 = \frac{E_{1t}}{E_{1n}} = \frac{500}{69{,}281} = 7.21 \times 10^{-3} \quad \rightarrow \quad \theta_1 = 0.413°$$

The electric field intensity in air points at 0.413° from the normal.

Example 4.19 Interface Conditions Between Dielectrics with a Surface Charge Density on the Interface An interface between two dielectrics coincides with the $x = 0$ plane as shown in **Figure 4.34**. The electric field intensity in medium (1) is $\mathbf{E}_1 = \hat{\mathbf{x}}100 + \hat{\mathbf{y}}300 - \hat{\mathbf{z}}50$ V/m. A surface charge density $\rho_s = 5 \times 10^{-10}$ C/m^2 exists on the interface:

(a) Calculate the electric field intensity in medium (2).
(b) Calculate the electric flux density in medium (2).

Solution: The electric field intensity has two tangential components—the x and z components—and these remain unchanged. The normal component is the z component:

(a) The tangential electric field intensity may be written in terms of its components by writing $E_{2y} = E_{1y} = 300$ [V/m] and $E_{2z} = E_{1z} = -50$ [V/m]
The tangential component of the electric field intensity in medium (2) is

$$\mathbf{E}_{2t} = \mathbf{E}_{1t} = \hat{\mathbf{y}}300 - \hat{\mathbf{z}}50 \quad [\text{V/m}]$$

The normal component is calculated from **Eq. (4.73)** or **Eq. (4.74)**. The latter is selected to demonstrate its use:

$$\hat{\mathbf{n}} \cdot (\varepsilon_1 \mathbf{E}_1 - \varepsilon_2 \mathbf{E}_2) = \rho_s \quad [\text{C/m}^2]$$

where the normal points into medium (1), that is, $\hat{\mathbf{n}} = -\hat{\mathbf{x}}$. That is,

$$\hat{\mathbf{n}} \cdot (\varepsilon_0 \mathbf{E}_1 - \varepsilon_2 \mathbf{E}_2) = -\hat{\mathbf{x}} \cdot (\varepsilon_0(\hat{\mathbf{x}}100 + \hat{\mathbf{y}}300 - \hat{\mathbf{z}}50) - \varepsilon_2 \mathbf{E}_2) = 5 \times 10^{-10} \quad [\text{C/m}^2]$$

Performing the scalar product and writing formally $\mathbf{E}_2 = \hat{\mathbf{x}}E_{2x} + \hat{\mathbf{y}}E_{2y} + \hat{\mathbf{z}}E_{2z}$, we have

$$-100\varepsilon_0 + \varepsilon_2 E_{2x} = 5 \times 10^{-10} \quad [\text{C/m}^2] \quad \rightarrow \quad E_{2x} = \frac{5 \times 10^{-10} + 100\varepsilon_0}{4\varepsilon_0} \quad \left[\frac{\text{V}}{\text{m}}\right]$$

Since $\varepsilon_0 = 8.854 \times 10^{-12}$ F/m, we get

$$E_{2x} = \frac{5 \times 10^{-10} + 100\varepsilon_0}{4\varepsilon_0} = \frac{5 \times 10^{-10} + 100 \times 8.854 \times 10^{-12}}{4 \times 8.854 \times 10^{-12}} = 39.12 \quad [\text{V/m}]$$

Thus, the electric field intensity in medium (2) is

$$\mathbf{E}_2 = \hat{\mathbf{x}}39.12 + \hat{\mathbf{y}}300 - \hat{\mathbf{z}}50 \quad [\text{V/m}]$$

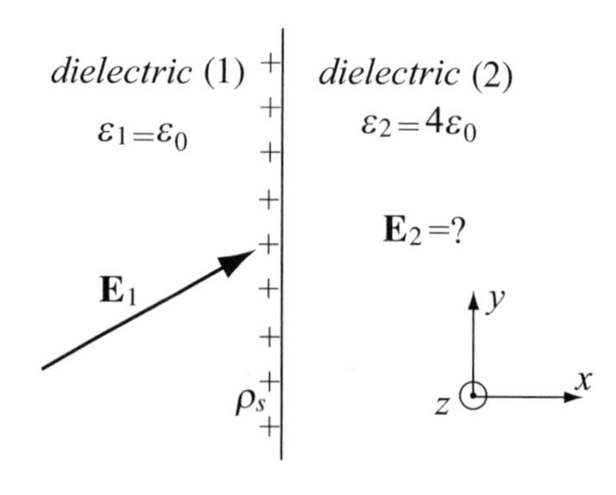

Figure 4.34 Interface between two dielectrics with a surface charge density on the interface. The electric field intensity in medium (1) is given

(b) One can repeat the process in **(a)** for the electric flux density, but since the electric field intensity in medium (2) is known, it is sufficient to write

$$\mathbf{D}_2 = \varepsilon_2\mathbf{E}_2 = 4\varepsilon_0(\hat{\mathbf{x}}39.12 + \hat{\mathbf{y}}300 - \hat{\mathbf{z}}50) = 4 \times 8.854 \times 10^{-12}(\hat{\mathbf{x}}39.12 + \hat{\mathbf{y}}300 - \hat{\mathbf{z}}50)$$
$$= \hat{\mathbf{x}}1.385 \times 10^{-9} + \hat{\mathbf{y}}1.062 \times 10^{-8} - \hat{\mathbf{z}}1.771 \times 10^{-9} \quad [\mathrm{C/m^2}]$$

Example 4.20 Interface Conditions Between Dielectrics and Conductors An electric field intensity with magnitude $E_0 = 1{,}000$ V/m is perpendicular to the surface of a perfect conductor as shown in **Figure 4.35a**:

(a) Calculate the surface charge density at the conductor–air interface.
(b) Suppose the conductor is coated with a dielectric with relative permittivity of 2.5 as shown in **Figure 4.35b**. The electric field intensity at the air–dielectric interface is the same as in **(a)**. Does this coating change the conductor's surface charge density?

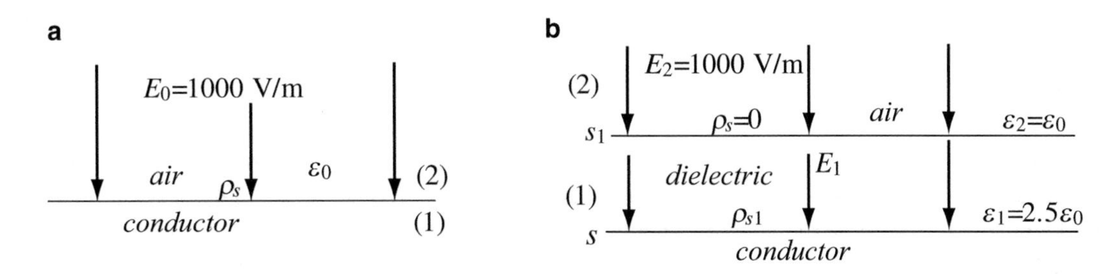

Figure 4.35 Electric field intensity at the surface of a conductor. (**a**) A conductor–air interface. (**b**) A Conductor–dielectric interface and an air–dielectric interface

Solution: The electric field intensity in the conductor is zero. The result is a surface charge density as in **Eq. (4.79)**. Note that the surface charge density must be negative to produce an opposing field in the volume occupied by the conductor. We will assume material (1) is the conductor and material (2) is air so that the relations in **Table 4.3** may be used directly. In this case, the electric field intensity points toward the conductor's surface; therefore, the surface charge density is negative.

(a) The magnitude of the charge density is

$$E_{0n} = \frac{\rho_s}{\varepsilon_0} \quad \rightarrow \quad |\rho_s| = \varepsilon_0 E_{0n} = 8.854 \times 10^{-12} \times 1000 = 8.854 \times 10^{-9} \quad [\mathrm{C/m^2}]$$

The charge density is negative and equals -8.854×10^{-9} C/m^2.

(b) At the interface between air and dielectric (no surface charge density at this interface) and assuming air is material (2) and the dielectric is material (1), we get

$$\varepsilon_2 E_{2n} = \varepsilon_1 E_{1n} = \varepsilon_0 E_{0n} \quad \rightarrow \quad E_{1n} = \frac{\varepsilon_0 E_{0n}}{\varepsilon_1} = \frac{8.854 \times 10^{-12} \times 1000}{2.5 \times 8.854 \times 10^{-12}} = 400 \quad [\mathrm{V/m}]$$

And, at the dielectric–conductor interface, the magnitude of the electric charge density is

$$E_{1n} = \frac{\rho_{s1}}{\varepsilon_1} \quad \rightarrow \quad |\rho_{s1}| = \varepsilon_1 E_{1n} = 2.5 \times 8.854 \times 10^{-12} \times 400 = 8.854 \times 10^{-9} \quad [\mathrm{C/m^2}]$$

Again, the charge density must be negative: $\rho_{s1} = -8.854 \times 10^{-9}$ C/m^2. Thus, the charge density at the interface does not change with the introduction of the dielectric layer.

Exercise 4.7 Assume a charge density ρ_{s1} [C/m^2] at the interface between dielectric and air in **Figure 4.35b**. What must this charge density be for the charge density at the conductor–dielectric interface to be zero?

Answer $\rho_{s1} = -8.854 \times 10^{-9}$ [C/m^2].

4.7 Capacitance

Whereas capacitance is a familiar concept, primarily as the property of capacitors in circuits, the field concept of capacitance is somewhat different and more general. Capacitance is defined as the ratio between charge and potential; that is, it indicates the amount of charge a body can store for a given, applied potential. Using this definition, any body has a certain capacitance, but, as we shall see, even though capacitance relates charge and potential, the capacitance itself is independent of charge or potential; it is a function of body dimensions and material properties. Any device that has capacitance may be called a ***capacitor***. The general definition of capacitance is

$$\boxed{C = \frac{Q}{V} \quad [\mathrm{F}]} \tag{4.80}$$

This definition applies to any configuration of bodies and potentials. In practical terms, however, potential is only properly defined as potential difference (such as between a body and infinity or between two bodies). Also, for a body to be at a given constant potential, it must be a conductor. Therefore, it is convenient to define capacitance between two conducting bodies or capacitance of one conducting body with respect to another. This is shown schematically in **Figure 4.36**. In **Figure 4.36a**, two general bodies, such as two conductors, are shown. A potential difference $V_{BA} = V_B - V_A$ is connected between the two, shown as a voltage source (i.e., a cell or a battery) which supplies the charge. Because of this source, there is a charge $-Q$ on A and $+Q$ on B. The capacitance can now be defined as the magnitude of the charge on one body (since the other is equal in magnitude) divided by the magnitude of the potential difference between the two bodies:

$$C = \frac{|Q|}{|V_B - V_A|} \quad [\mathrm{F}] \tag{4.81}$$

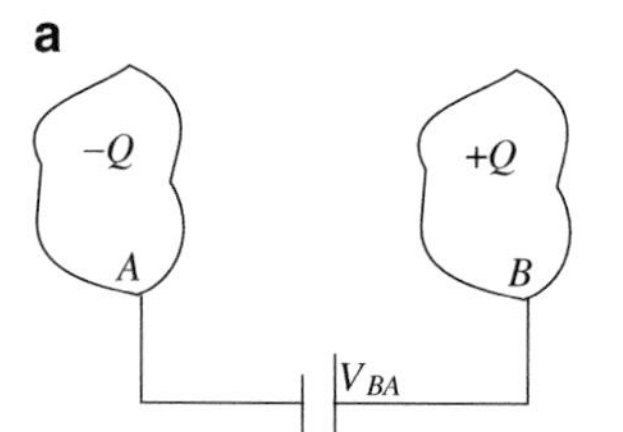

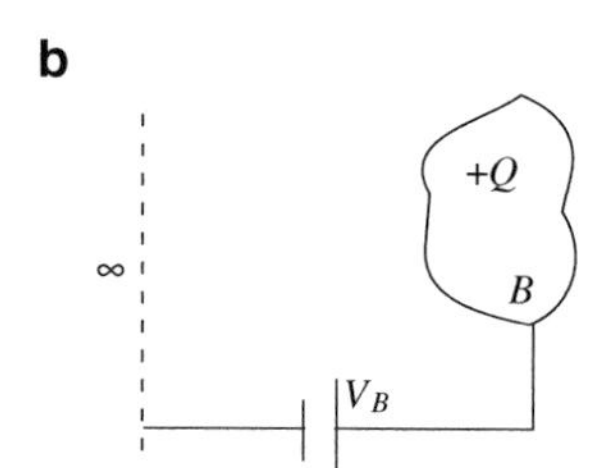

Figure 4.36 Definition of capacitance. (**a**) Capacitance between two (conducting) bodies. (**b**) Capacitance of a single (conducting) body

Capacitance is always positive. Using this definition, the capacitance of a single conductor can be viewed as the capacitance of this conductor with respect to a reference at infinity. This is shown schematically in **Figure 4.36b**, where we have connected a battery between B and "infinity." Since the reference potential can often be taken to be zero at infinity (see however **Section 4.7.2**), we obtain from the above result:

$$C = \frac{|Q|}{|V_B - V_\infty|} = \frac{|Q|}{|V_B - 0|} = \frac{|Q|}{|V_B|} \quad [\mathrm{F}] \tag{4.82}$$

The unit of capacitance is the ***farad***[5] or [F] (1 F = 1 C/V). The farad happens to be a very large unit and therefore it is common to encounter capacitors and capacitances in units of ***microfarad*** (denoted [μF]; 1 μF = 10^{-6} F) ***nanofarad*** (denoted [nF]; 1 nF = 10^{-9} F) and ***picofarad*** (denoted [pF]; 1 pF = 10^{-12} F).

Calculation of capacitance can be performed either using the above formula or directly from the electric field through Gauss's law and the definition of potential:

$$C = \frac{|Q|}{|V_B - V_A|} = \frac{\left|\oint_s \varepsilon \mathbf{E} \cdot d\mathbf{s}\right|}{\left|-\int_A^B \mathbf{E} \cdot d\mathbf{l}\right|} \quad [\mathrm{F}] \tag{4.83}$$

where the surface integral is over the surface enclosing the charge and the line integral is along any contour between the two bodies forming the capacitor or from infinity to a single-body capacitor.

Calculation of capacitance is accomplished using the following steps:

A. If the applied potential is known or may be assumed:

(a) For a two-conductor capacitor:

(1) Apply an arbitrary potential difference between the two conductors.
(2) Calculate the charge on one of the conductors. The second conductor has equal charge but opposite in sign.
(3) Find the ratio between charge on one conductor and the potential difference between the conductors. Take the absolute value to obtain capacitance.

(b) For a single-conductor capacitor:

(1) Apply a potential on the conductor (with zero reference potential at infinity for any finite size conducting body).
(2) Calculate the total charge on the conductor using either Gauss's law or direct integration.
(3) Find the capacitance by dividing the charge by the potential.

B. If the applied charge or charge density is known or may be assumed:

(a) For a two-conductor capacitor:

(1) Apply an arbitrary charge or charge density on the two conductors. The total charge on one conductor must be equal in magnitude and opposite in sign to the total charge on the second conductor.
(2) Calculate the potential difference between the two conductors. This may require the calculation of the electric field intensity first.
(3) Find the ratio between charge on one conductor and the potential difference between the conductors. Take the absolute value to obtain capacitance.

(b) For a single-conductor capacitor:

(1) Apply a charge or charge density on the given conductor. Usually a positive charge is assumed.
(2) Calculate the potential on the conductor with reference to infinity.
(3) Find the capacitance by dividing the charge by the assumed potential.

Example 4.21 Capacitance of the Globe Calculate the capacitance of the Earth assuming it is a conducting sphere of radius R_0 = 6,400 km.

---

[5] After Michael Faraday (1791–1867), who probably had more influence on the development of electricity and magnetism than anyone else. Son of a blacksmith, he started as a bookbinder's apprentice and became a foremost experimentalist and discoverer of many phenomena in electromagnetics. His work has laid the foundation of the unified theory espoused a few years after his death by James Clerk Maxwell. The naming of the farad after him is in recognition to his very many contributions, which included important experiments in electrostatics. We will meet him again in **Chapter 10**.

Solution: The Earth is conducting (although it is not a perfect conductor, we will assume that it approximates this condition). The capacitance is calculated by assuming a charge Q, uniformly distributed on its surface. From this, we calculate the potential on the surface with reference to zero at infinity. The ratio between charge and voltage is the capacitance of the planet.

Assuming a charge Q is distributed uniformly over the surface, the electric field intensity at any location outside the sphere ($R > R_0$) is found by creating a spherical, Gaussian surface, concentric with the sphere:

$$4\pi R^2 E = \frac{Q}{\varepsilon_0} \quad \rightarrow \quad E = \frac{Q}{\varepsilon_0 4\pi R^2} \quad \left[\frac{\mathrm{V}}{\mathrm{m}}\right]$$

The electric field intensity is radial, but for the purpose of this example, it is not important to calculate its vector form. This electric field intensity is the same as that of a point charge of magnitude Q. The electric potential at any point a distance R from the point charge is therefore

$$V(R) = \frac{Q}{4\pi\varepsilon_0 R} \quad [\mathrm{V}]$$

At $R = R_0$,

$$V(R_0) = \frac{Q}{4\pi\varepsilon_0 R_0} \quad \rightarrow \quad Q = 4\pi\varepsilon_0 R_0 V(R_0) \quad [\mathrm{V}]$$

From the relation $C = Q/V$, the capacitance of the Earth is

$$C = \frac{Q}{V(R_0) - 0} = 4\pi\varepsilon_0 R_0 \quad [\mathrm{F}]$$

Numerically, this gives

$$C = 4 \times \pi \times 8.854 \times 10^{-12} \times 6400 \times 1000 = 7.12 \times 10^{-4} \quad [\mathrm{F}]$$

The capacitance of the Earth is only about 712 μF. This figure gives a good indication just how large a unit the farad is.

4.7.1 The Parallel Plate Capacitor

Point_Charges.m

The parallel plate capacitor is one of the most common devices in electronic circuits. It consists of two conducting plates and a dielectric separating the two. The conducting plates may, in fact, be conducting layers deposited on two sides of a dielectric or may be thin foils. The dielectric may be a solid such as mica, paper, glass, or mylar, a liquid such as oil, an oxide formed on the conductor itself or at the interface between the metal and a dielectric, or a gas such as air. The most important point is that the separating material must be a dielectric. Each type of capacitor finds uses in different applications. For example, oil capacitors are often used in power devices such as starting capacitors in electric machines, whereas electrolytic capacitors are used in electronic circuits whenever large capacitances together with small physical dimensions are required. The principle of the parallel plate capacitor is shown in **Figure 4.37**.

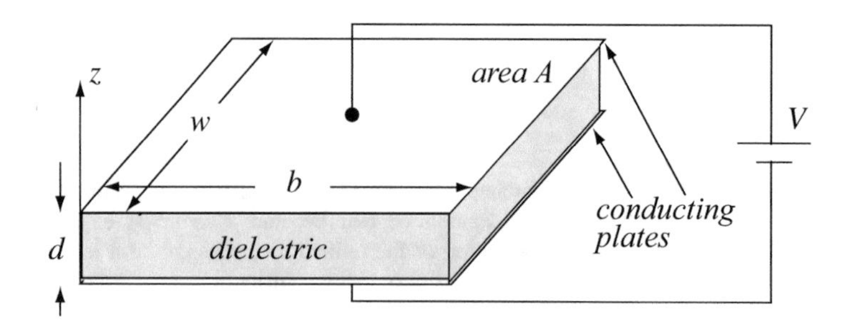

Figure 4.37 The parallel plate capacitor

To calculate the capacitance of the parallel plate capacitor, we use an approximate model. There are two seemingly contradictory assumptions involved in this model:

(1) The plates are finite in size.
(2) The electric field intensity between the plates is the same as if the plates were infinite in size.

In other words, we will use the electric field of infinite, charged plates because this provides a very simple way of determining the electric field intensity even though capacitors are physically limited in size (and often very small). These approximations can be easily justified if the distance d between the plates is much smaller than the other two dimensions. We say that in the capacitor model, the fringing effects of the electric field at the edges are neglected. Under these conditions, the charge density on each plate is uniformly distributed and equal to

$$\rho_s = \frac{Q}{A} \quad \left[\frac{\mathrm{C}}{\mathrm{m}^2}\right] \quad \text{on the upper plate,} \quad \rho_s = -\frac{Q}{A} \quad \left[\frac{\mathrm{C}}{\mathrm{m}^2}\right] \quad \text{on the lower plate} \tag{4.84}$$

Using now the results of **Example 4.3** for infinite, charged plates, the electric field intensity between the plates in **Figure 4.37** is

$$\mathbf{E} = -\hat{\mathbf{z}}\frac{\rho_s}{\varepsilon} = -\hat{\mathbf{z}}\frac{Q}{\varepsilon A} \quad \left[\frac{\mathrm{V}}{\mathrm{m}}\right] \tag{4.85}$$

To calculate the capacitance, we must evaluate the potential difference between the two plates. From the definition of potential,

$$\mathbf{V} = -\int_{z=0}^{z=d} \mathbf{E} \cdot d\mathbf{l} = -\int_{z=0}^{z=d} \left(-\hat{\mathbf{z}}\frac{Q}{\varepsilon A}\right) \cdot \hat{\mathbf{z}}dz = \frac{Qd}{\varepsilon A} \quad [\mathrm{V}] \tag{4.86}$$

and, therefore, the capacitance is

$$C = \frac{Q}{V} = \frac{Q}{(Qd/\varepsilon A)} = \frac{\varepsilon A}{d} \quad [\mathrm{F}] \tag{4.87}$$

This result points out the fundamental properties of capacitance:

(1) Capacitance is independent of sources; charge or potential is not part of the formula for capacitance.
(2) Capacitance is a property of the geometry; it depends only on the physical dimensions of the geometry and the material properties between the conductors.
(3) Calculation of capacitance assumes the existence of charges or charge densities to allow calculation of fields and potentials, but these charges are arbitrarily chosen and cancel out in the final result.

Example 4.22 Application: High-Voltage Oil Capacitors A high-voltage parallel plate capacitor is made of two aluminum foils, $w = 30$ mm wide and $b = 2$ m long (**Figure 4.37**). Between them there is a paper layer so that the foils are insulated from each other. The paper is impregnated with oil to give it a relative permittivity $\varepsilon_r = 2.5$. The oil used has a dielectric strength $V_b = 20{,}000$ V/mm. The paper is $d = 0.1$ mm thick. For practical purposes of space, the foil paper assembly is rolled as on a spool and placed in a protective shell. Assume the electric field intensity is uniform between the foils and calculate:

(a) The capacitance of the device.
(b) The maximum voltage rating of the capacitor (maximum potential difference allowed).
(c) In the production process, a tear in the paper causes a small section of the capacitor to lack the dielectric although the distance between the two foils is maintained. What is the voltage rating now?

Solution: The capacitance is calculated using **Eq. (4.87)**, whereas the voltage rating is the thickness of the dielectric multiplied by the breakdown voltage.

(a) The capacitance is

$$C = \frac{\varepsilon A}{d} = \frac{\varepsilon_0 \varepsilon_r A}{d} = \frac{\varepsilon_0 \varepsilon_r wb}{d} = \frac{8.854 \times 10^{-12} \times 2.5 \times 0.03 \times 2}{0.0001} = 1.328 \times 10^{-8} \quad [\text{F}]$$

This is 13.28 nF.

(b) The maximum voltage allowed is 0.1 mm $\times$ 20,000 V/mm $=$ 2,000 V. The capacitor's ratings are

$$C = 13.28 \quad [\text{nF}], \quad V_{max} = 2{,}000 \quad [\text{V}]$$

(c) In the area of the tear, the breakdown voltage is that of air which is 3,000 V/mm. Thus, the maximum voltage allowed is now only 300 V, but the capacitance has not changed because the tear is small. The capacitor's ratings are now

$$C = 13.28 \quad [\text{nF}], \; V_{max} = 300 \quad [\text{V}]$$

Flaws in the dielectric as well as evaporation, leaks, contamination, and the like can affect the ratings of capacitors and reduce their performance or even render them useless.

4.7.2 Capacitance of Infinite Structures

It is sometimes required to calculate the capacitance of structures which are very large or infinite in size. A simple example of a structure of this sort is an overhead power line. Another is the infinite plate used above to define the parallel plate capacitor. From the above definitions, it is clear that if the physical dimensions of the capacitor are very large, the capacitance is also very large; if the dimensions are infinite, the capacitance is infinite. Thus, the total capacitance of an infinite structure is not a very useful term. However, capacitance is a property of the device; it is quite important to be able to associate capacitance with structures like transmission lines. To do so, we define capacitance per unit length as shown in **Figure 4.38**.

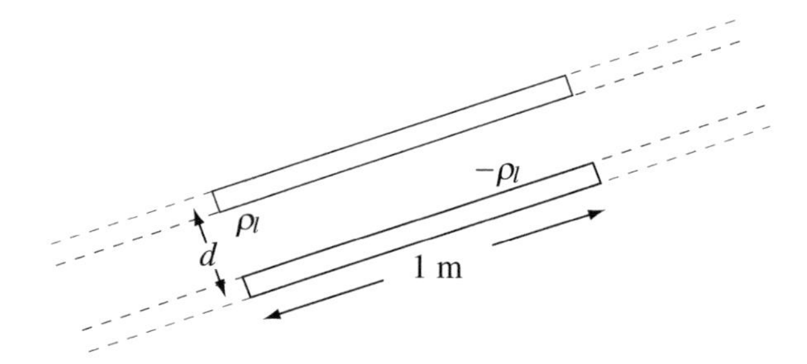

Figure 4.38 Principle of calculation of capacitance per unit length of a device

In this method, we assume that all properties of the fields remain unaltered. The only change is that the charge is distributed per unit length of the device. Now, we calculate the capacitance of the transmission line of unit length, but assuming it has the properties of an infinitely long line; that is, the charge is uniformly distributed over the unit length and there are no "end effects." These concepts are explained in the following two examples.

Example 4.23 Application: Capacitance of Cables A long coaxial cable is made with an internal conductor of radius $a = 2$ mm and an external conductor of radius $b = 6$ mm. The design calls for three layers of insulation between the two conductors. The inner layer is 1 mm thick and is made of rubber ($\varepsilon_r = 4.0$), the next layer is a plastic ($\varepsilon_r = 9$), 1 mm thick, and the third layer is a foam ($\varepsilon_r = 1.5$), 2 mm thick. (See **Figure 4.39**.) Calculate the capacitance per unit length of the cable.

Figure 4.39 Configuration for **Example 4.23**

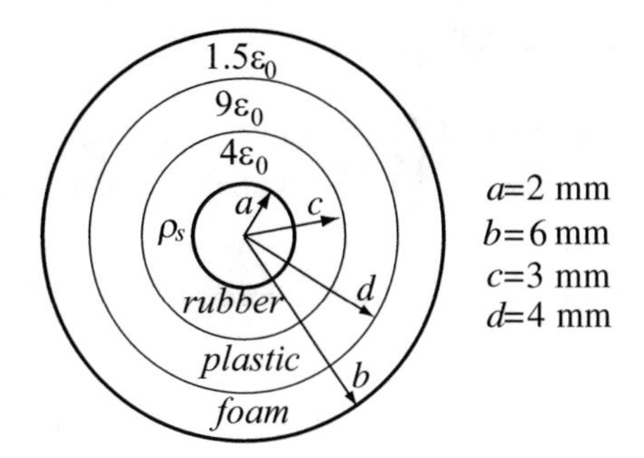

Solution: To calculate the capacitance per unit length, we assume a charge per unit area of the inner conductor of ρ_s [C/m^2], take a unit length of the cable, and calculate the electric field intensity, followed by potential, and divide charge per unit length by potential difference to find the capacitance per unit length.

If a charge density ρ_s is placed on the inner conductor, the electric flux density between the conductors at a point $a < r < b$ can be calculated using a Gaussian surface in the form of a cylinder of radius r and length L, concentric with the inner conductor (see **Section 4.3.1**):

$$\int_s \mathbf{D} \cdot d\mathbf{s} = q \quad \rightarrow \quad D2\pi rL = \rho_s 2\pi aL$$

The electric flux density is in the r direction and equals

$$\mathbf{D} = \hat{\mathbf{r}}\frac{\rho_s a}{r} \quad \left[\frac{\mathrm{C}}{\mathrm{m}^2}\right]$$

The electric field intensity in the three layers is
In rubber ($2 < r < 3$ mm), $\varepsilon_r = 4$:

$$\mathbf{E}_{rubber} = \frac{\mathbf{D}}{4\varepsilon_0} = \hat{\mathbf{r}}\frac{\rho_s a}{4\varepsilon_0 r} \quad \left[\frac{\mathrm{V}}{\mathrm{m}}\right]$$

In plastic ($3 < r < 4$ mm), $\varepsilon_r = 9$:

$$\mathbf{E}_{plastic} = \frac{\mathbf{D}}{9\varepsilon_0} = \hat{\mathbf{r}}\frac{\rho_s a}{9\varepsilon_0 r} \quad \left[\frac{\mathrm{V}}{\mathrm{m}}\right]$$

In foam ($4 < r < 6$ mm), $\varepsilon_r = 1.5$:

$$\mathbf{E}_{foam} = \frac{\mathbf{D}}{1.5\varepsilon_0} = \hat{\mathbf{r}}\frac{\rho_s a}{1.5\varepsilon_0 r} \quad \left[\frac{\mathrm{V}}{\mathrm{m}}\right]$$

The potential difference between the inner and outer shells (integrating from outer to inner shell, against the field) is

$$V_{ab} = -\int_b^a \mathbf{E} \cdot d\mathbf{l} = -\int_b^{b-0.002} E_{foam} dr - \int_{b-0.002}^{b-0.003} E_{plastic} dr - \int_{b-0.003}^{a} E_{rubber} dr$$

$$= -\frac{\rho_s a}{\varepsilon_0}\left(\frac{1}{1.5}\ln\frac{b-0.002}{b} + \frac{1}{9}\ln\frac{b-0.003}{b-0.002} + \frac{1}{4}\ln\frac{a}{b-0.003}\right) \quad [\mathrm{V}]$$

With $a = 0.002$ m and $b = 0.006$ m, we get

$$V_{ab} = -\frac{\rho_s \times 0.002}{\varepsilon_0}\left(\frac{1}{1.5}\ln\frac{0.004}{0.006} + \frac{1}{9}\ln\frac{0.003}{0.004} + \frac{1}{4}\ln\frac{0.002}{0.003}\right) = 8.073 \times 10^{-4}\frac{\rho_s}{\varepsilon_0} \quad [\mathrm{V}]$$

The total charge per unit length of the inner conductor ($L = 1$ m) is $2\pi a\rho_s$. Dividing the charge per unit length by the potential difference gives

$$C = \frac{Q}{V_{ab}} = \frac{2\pi a\rho_s}{8.073 \times 10^{-4}\rho_s/\varepsilon_0} = \frac{2\pi a\varepsilon_0}{8.073 \times 10^{-4}} = \frac{2 \times \pi \times 0.002 \times 8.854 \times 10^{-12}}{8.073 \times 10^{-4}} = 1.378 \times 10^{-10} \quad [\mathrm{F/m}]$$

or 137.8 pF/m.

4.7.3 Connection of Capacitors

In many applications, capacitors must be connected in series or in parallel. Similarly, capacitances of systems of conductors can be viewed as being composed of two or more capacitances, connected in one mode or another, or in a combination of series and parallel connections.

To obtain the formula for two capacitors in parallel, consider two parallel plate capacitors as shown in **Figure 4.40a**. For generality, the two capacitors are shown as different in physical shape and capacitance. Since both capacitors are connected to the same source, $V_1 = V_2 = V$. The charge is Q_1 on C_1 and Q_2 on C_2. The total charge supplied by the source is therefore $C = C_1 + C_2$.

The capacitance the source sees, by definition, is

$$C_{total} = \frac{Q_{total}}{V_{total}} = \frac{Q_1 + Q_2}{V} = \frac{Q_1}{V} + \frac{Q_2}{V} \quad [\mathrm{F}] \tag{4.88}$$

Thus, we conclude

$$C_{total} = C_1 + C_2 \quad [\mathrm{F}] \tag{4.89}$$

Extending the argument above to n capacitors in parallel, we get

$$\boxed{C_{total} = C_1 + C_2 + \ldots + C_n = \sum_{i=1}^{n} C_i \quad \text{for } n \text{ parallel capacitors}} \tag{4.90}$$

The connection of capacitors in series is treated similarly. Using the same capacitors, connected as in **Figure 4.40b**, we can write

$$Q_1 = Q_2 = \mathrm{Q} \quad [\mathrm{C}] \quad \text{and} \quad V_1 + V_2 = V \quad [\mathrm{V}] \tag{4.91}$$

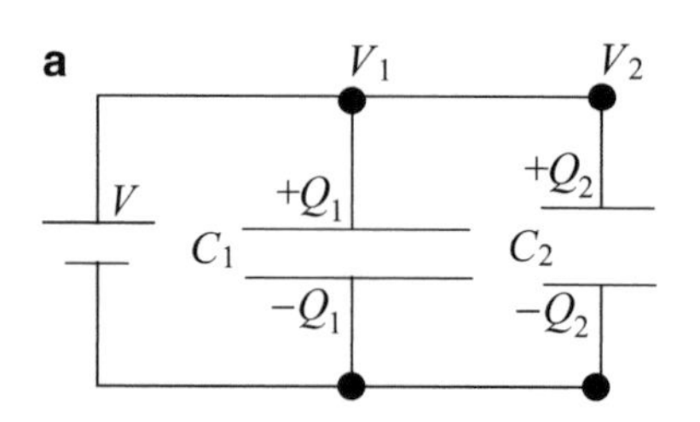

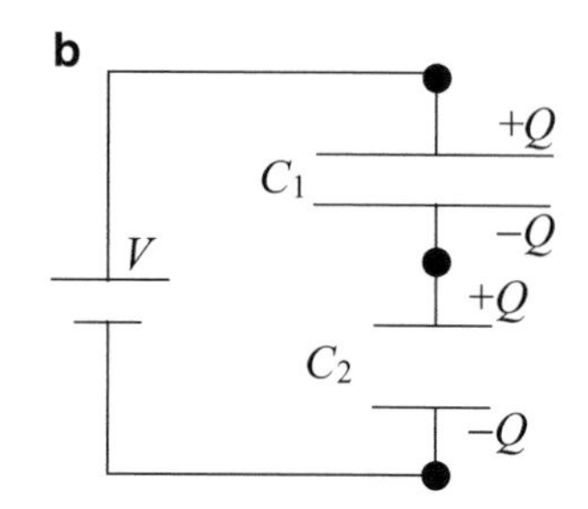

Figure 4.40 Connection of capacitors in parallel and in series. (**a**) Parallel connection of capacitors to a source. (**b**) Series connection of capacitors to a source

Note that the charge on each of the capacitors is the same in spite of the fact that the capacitors are different in size. This is because the lower plate of C_1 and the upper plate of C_2 are connected together and can be viewed as one conductor (see also **Figure 4.21**). As such, the charge that can exist on the two plates must be equal in magnitude and opposite in sign as shown. Thus, we can write

$$V_1 = \frac{Q_1}{C_1},\ V_2 = \frac{Q_2}{C_2},\ V_{total} = \frac{Q}{C_{total}} \quad \to \quad \frac{Q_1}{C_1} + \frac{Q_2}{C_2} = \frac{Q}{C_1} + \frac{Q}{C_2} = \frac{Q}{C_{total}} \tag{4.92}$$

or

$$\frac{1}{C_{total}} = \frac{1}{C_1} + \frac{1}{C_2} \tag{4.93}$$

Extending this to any number of capacitors,

$$\boxed{\frac{1}{C_{total}} = \frac{1}{C_1} + \frac{1}{C_2} + \ldots + \frac{1}{C_n} = \sum_{i=1}^{n} \frac{1}{C_i} \quad \text{for } n \text{ series capacitors}} \tag{4.94}$$

Example 4.24 Application: Switched Capacitor DC to DC Converter A common principle used in low-power DC-to-DC converters is shown in **Figure 4.41**. The method is called switched capacitor DC-to-DC converter or charge pump DC-to-DC converter and is used in many low-power applications to convert one DC voltage level to another. Consider the following example: Two capacitors, each $C = 1{,}000\ \mu\text{F}$, are connected across a 12 V source as shown in **Figure 4.41a**. Each capacitor acquires a voltage $V = 12$ V and a charge Q [C]. Now, the two capacitors are disconnected and reconnected across the load as shown in **Figure 4.41b**. The voltage across the load is 24 V. If the capacitors are charged quickly and discharged slowly, the voltage across the load will be close to 24 V. The switching between the circuit in **Figure 4.41a** and that in **Figure 4.41b** is done by a suitable switching circuit. The waveform across the load is shown in **Figure 4.41c**, where the average load voltage V_{av} is also indicated. This waveform is normally filtered to obtain a more constant voltage. Calculate:

(a) The total capacitance during charging and discharging periods.
(b) The total charge transferred to the load, assuming each capacitor is fully discharged when connected across the load.

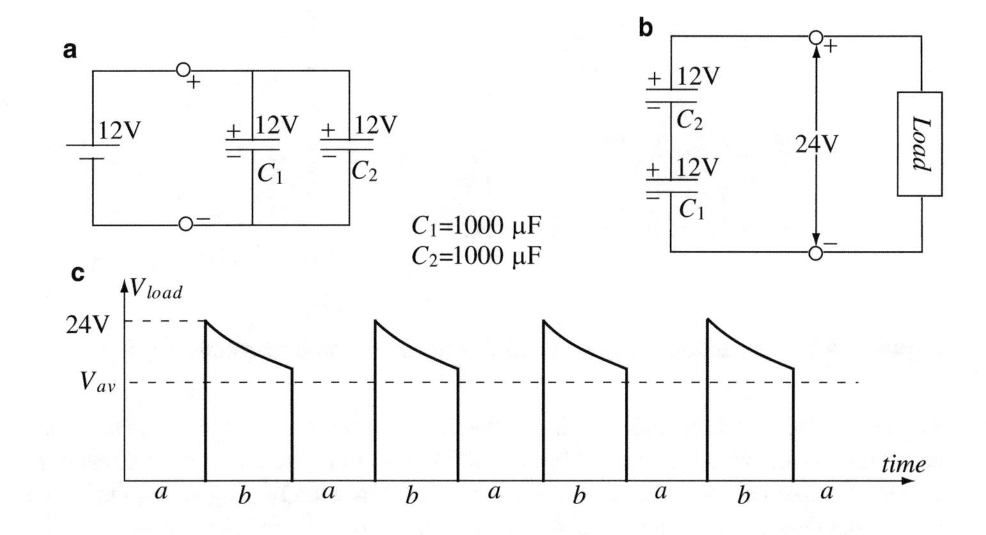

Figure 4.41 Principle of a switched capacitor DC-to-DC converter. (**a**) Equal capacitors C_1 and C_2 are charged to 12 V each. (**b**) The capacitors are disconnected and reconnected in series to deliver 24 V to the load. (**c**) Approximate load waveform. Charging occurs during periods a, discharging during periods b

Solution:

(**a**) During charging, the total capacitance is $C_t = 2C = 2{,}000$ μF, and during discharging, the total capacitance is $C_t = C/2 = 500$ μF. In the first case, the capacitors are connected in parallel and in the second in series.
In parallel:

$$C_t = C_1 + C_2 = 2C = 2000 \qquad [\mu\text{F}]$$

In series:

$$\frac{1}{C_t} = \frac{1}{C_1} + \frac{1}{C_2} = \frac{1}{C} + \frac{1}{C} = \frac{2}{C} \quad \rightarrow \quad C_t = \frac{C}{2} = 500 \qquad [\mu\text{F}]$$

(**b**) The charge acquired by each capacitor connected in parallel is

$$Q = CV = 1000 \times 10^{-6} \times 12 = 1.2 \times 10^{-2} \qquad [\text{C}]$$

When connecting the two capacitors in series, the charge for the total capacitance is that of one capacitor. Thus, the charge transferred is 1.2×10^{-2} C.

Example 4.25 Application: The Capacitive Fuel Gauge A simple capacitive fuel gauge can be built as two concentric cylinders forming a capacitor. The fuel is allowed to enter the gap between the two cylinders. The capacitance of the cylinders depends on the fuel level. This device works very well with nonconducting fuels such as gasoline, oil, and kerosene but also with distilled water and other dielectric fluids. **Figure 4.42a** shows a possible arrangement. The device is simple, is accurate, has no moving parts, and can serve as an anti-sloshing chamber that avoids high fluctuations in fuel level indication due to motion.

Assume two copper cylinders form a coaxial capacitor. The inner cylinder is $2a = 10$ mm in diameter and the outer is $2b = 20$ mm. The cylinders (and the tank) are $d = 500$ mm high. Relative permittivity of the fluid is $\varepsilon_r = 15$. Neglect fringe effects at the ends of the cylinder and find:

(**a**) A general relation between fluid level and capacitance of the device.
(**b**) Plot the capacitance versus fluid level. This is the gauge's calibration curve. Is the reading of this fuel gauge linear?

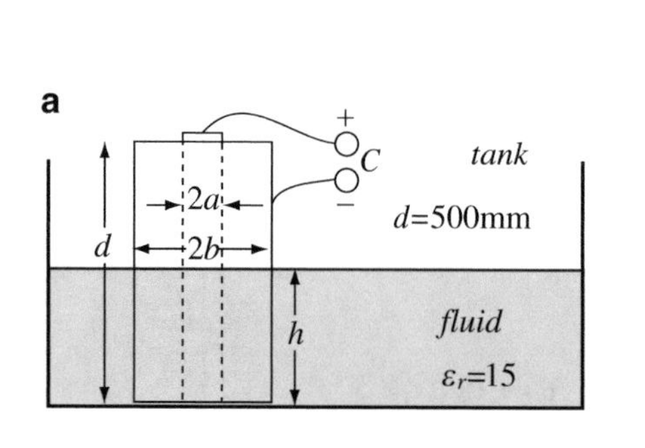

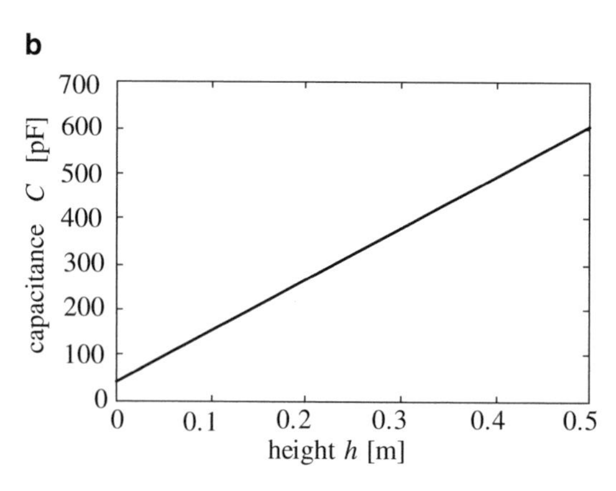

Figure 4.42 A capacitive fuel gauge. (**a**) Arrangement and dimensions. (**b**) Calibration curve

Solution: The part of the cylinders inside the fluid has larger permittivity than that above the fluid. Each part is viewed as a capacitor and the two are connected in parallel (the voltage across each part is the same). Assume the fluid level is h [m]. The various dimensions and configuration are shown in **Figure 4.42a**. Neglecting the fringe effects at the ends of the capacitor means that Gauss's law can be applied as if the capacitor were infinitely long.

(a) The potential on the immersed capacitor is V_{ab}. Due to this potential, there will be a charge Q distributed on the immersed part of the inner cylinder and $-Q$ on the immersed part of the outer cylinder. The electric field intensity between the two cylinders is found by Gauss's law by drawing a cylinder at $a < r < b$. The height of the cylinder is h:

$$E2\pi rh = \frac{Q}{\varepsilon} \quad \rightarrow \quad E = \frac{Q}{2\pi rh\varepsilon} \quad \rightarrow \quad \mathbf{E} = \hat{\mathbf{r}}\frac{Q}{2\pi rh\varepsilon} \quad \left[\frac{\text{V}}{\text{m}}\right]$$

where the direction of the electric field intensity is in the positive r direction (from the given polarity of V_{ab}). To calculate the capacitance, we integrate the electric field intensity along a path between the two cylinders to find the potential difference between the cylinders:

$$V_{ab} = -\int_b^a \mathbf{E} \cdot d\mathbf{r} = -\int_b^a \frac{Qdr}{2\pi rh\varepsilon} = \frac{Q}{2\pi h\varepsilon}\ln\frac{b}{a} \quad [\text{V}]$$

The capacitance of the immersed part of the capacitor is

$$C_{im.} = \frac{Q}{V_{ab}} = \frac{2\pi h\varepsilon}{\ln(b/a)} \quad [\text{F}]$$

The part of the capacitor above the liquid is evaluated in an identical fashion, but the length of the capacitor is $d - h$ and the dielectric constant is ε_0. Substituting these values in the above expressions, we obtain

$$C_{ab.} = \frac{2\pi(d-h)\varepsilon_0}{\ln(b/a)} \quad [\text{F}]$$

The total capacitance of the two cylinders is the sum of these two capacitances:

$$C_t = C_{im.} + C_{ab.} = \frac{2\pi}{\ln(b/a)}(h\varepsilon + (d-h)\varepsilon_0) = \frac{2\pi\varepsilon_0}{\ln(b/a)}(h\varepsilon_r + d - h) \quad [\text{F}]$$

The relation between fluid level and capacitance is

$$C(h) = \frac{2\pi\varepsilon_0}{\ln(b/a)}(h\varepsilon_r + d - h) \quad [\text{F}]$$

(b) The calibration curve for the fuel gauge is shown in **Figure 4.42b**. The curve is linear but does not pass through zero because when the tank is empty, the capacitance is that of an air-filled capacitor.

Example 4.26 Application: Multilayer Capacitors A parallel plate capacitor is made of two plates, 100 mm by 100 mm in size, separated a distance of 1 mm. The design calls for three layers of insulation between the two conductors. The first layer is polystyrene foam ($\varepsilon_r = 1.05$), the next layer is a mica ($\varepsilon_r = 6.0$), and the third layer is paper ($\varepsilon_r = 3.0$). The thickness of the first layer is 0.2 mm, the second is 0.6 mm, and the third is 0.2 mm (**Figure 4.43**). Calculate the capacitance of the device.

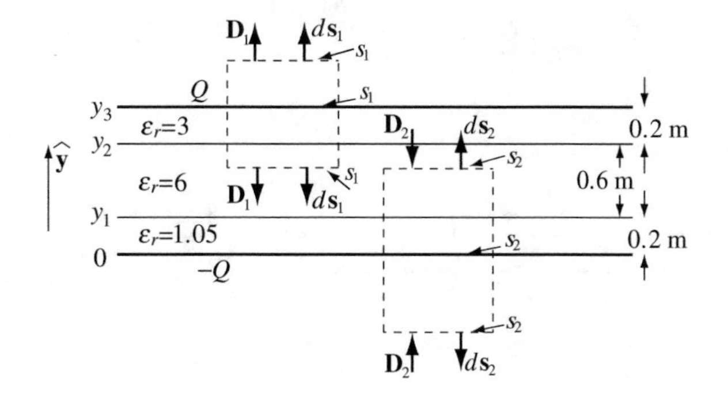

Figure 4.43 Parallel plate capacitor with a three-layer dielectric between the plates, shown in cross section. The Gaussian surfaces used to calculate the electric flux density between the plates are also shown

Solution: The electric field intensity between two parallel plates is calculated by assuming a positive charge on one plate and negative on the other. The fact that materials are placed between the plates only changes the magnitude of the field. In this case, we calculate the electric flux density anywhere between the two plates using Gauss's law for flux density because the flux density is independent of permittivity. From this, the electric field intensity in the various materials is calculated and then the electric field intensity is integrated to calculate the potential difference between the plates. Division of the assumed charge by the potential difference gives the capacitance.

From Gauss's law, the electric flux density due to the upper plate carrying a charge Q is (see **Figure 4.43**)

$$\oint_s \mathbf{D} \cdot d\mathbf{s} = \rho_s s \rightarrow \quad 2\int_{s_1} \mathbf{D}_1 \cdot d\mathbf{s}_1 = \rho_s s_1 \rightarrow \quad D_1 = \frac{\rho_s}{2} \quad \left[\frac{\mathrm{C}}{\mathrm{m}^2}\right]$$

where ρ_s [C/m^2] is the charge density on the upper plate. An identical calculation produces the field due to the lower plate, that is

$$D_2 = \frac{\rho_s}{2} \quad \left[\frac{\mathrm{C}}{\mathrm{m}^2}\right]$$

At any given location between the plates, D_1 and D_2 are equal in magnitude and both point in the negative y direction as can be seen from **Figure 4.43**. The total flux density is the sum of D_1 and D_2. The charge density on the upper plate equals the total charge on the plate, Q, divided by the area of the plate, s_{cap}. Similarly for the lower plate. The electric flux density outside the capacitor and on the lateral sides of the Gaussian surface is zero. Thus,

$$\mathbf{D} = -\hat{\mathbf{y}}\frac{Q}{s_{cap}} \quad \left[\frac{\mathrm{C}}{\mathrm{m}^2}\right]$$

Now, we can calculate the electric field intensity everywhere by dividing $\mathbf{D}$ by ε. The electric field intensities in the three materials are

$$\mathbf{E}_{foam} = -\hat{\mathbf{y}}\frac{Q}{1.05\varepsilon_0 s_{cap}}, \qquad \mathbf{E}_{mica} = -\hat{\mathbf{y}}\frac{Q}{6\varepsilon_0 s_{cap}}, \qquad \mathbf{E}_{paper} = -\hat{\mathbf{y}}\frac{Q}{3\varepsilon_0 s_{cap}} \qquad \left[\frac{\mathrm{V}}{\mathrm{m}}\right]$$

The potential across the capacitor is ($\mathbf{E}$ and $d\mathbf{l}$ are in opposite directions and $dl = dy$):

$$V = -\int_0^{y_3} E \cdot d\mathbf{l} = -\int_0^{y_1} \left(-E_{foam}dy\right) - \int_{y_1}^{y_2} \left(-E_{mica}dy\right) - \int_{y_2}^{y_3} \left(-E_{paper}dy\right) \qquad [\mathrm{V}]$$

Substituting the various values gives

$$V = \frac{Q}{\varepsilon_0 s_{cap}}\left[\int_0^{y_1}\frac{dy}{1.05} + \int_{y_1}^{y_2}\frac{dy}{6} + \int_{y_2}^{y_3}\frac{dy}{3}\right] = \frac{Q}{\varepsilon_0 s_{cap}}\left[\frac{y_1}{1.05} + \frac{y_2 - y_1}{6} + \frac{y_3 - y_2}{3}\right] \quad [\mathrm{V}]$$

The capacitance is

$$C = \frac{Q}{V} = \frac{\varepsilon_0 s_{cap}}{\frac{y_1}{1.05} + \frac{y_2 - y_1}{6} + \frac{y_3 - y_2}{3}} = \frac{8.854 \times 10^{-12} \times 0.01}{\frac{2 \times 10^{-4}}{1.05} + \frac{6 \times 10^{-4}}{6} + \frac{2 \times 10^{-4}}{3}} = 247.9 \times 10^{-12} \quad [\mathrm{F}]$$

Exercise 4.8 Application: Two-Layer Capacitor A parallel plate capacitor of unit area and separation d [m] is filled with two insulating layers. One layer is d_1 [m] thick and has permittivity $\varepsilon = 2\varepsilon_0$ [F/m], and the second is d_2 [m] thick ($d = d_1 + d_2$) with permittivity $\varepsilon = 3\varepsilon_0$ [F/m]. What is the capacitance of the capacitor?

Answer $C = \dfrac{6\varepsilon_0}{2d_2 + 3d_1}$ [F]

Exercise 4.9 The capacitor in **Exercise 4.8** is connected across a voltage V. Calculate the voltage across each dielectric layer and the electric field intensity in each dielectric layer. Use the idea of capacitors in series to simplify the solution.

Answer

$$V_1 = \frac{3d_1 V}{3d_1 + 2d_2}, \quad V_2 = \frac{2d_2 V}{2d_2 + 3d_1} \quad [\mathrm{V}], \quad E_1 = \frac{3V}{3d_1 + 2d_2}, \quad E_2 = \frac{2V}{2d_2 + 3d_1} \quad \left[\frac{\mathrm{V}}{\mathrm{m}}\right]$$

4.8 Energy in the Electrostatic Field: Point and Distributed Charges

The electrostatic field is a force field acting on electric charges. Thus, the ideas of work and potential energy that we used to define electric potential were natural extensions to the basic postulates (or of Coulomb's law). In this sense, we have already defined energy. The purpose of this section is to formalize the relations energy obeys in fields generated by point charges and charge distributions and, in the process, to derive yet another tool for analysis of electrostatic fields. The use of energy in calculation of fields has the same advantage as the use of potential: it is a scalar function and, therefore, simpler to evaluate. We should note here that the only type of energy that can be considered in the context of electrostatic fields is electric potential energy. This implies that we can calculate potential energy in a system of charges from the amount of work done by the electric field or against the electric field.

As could be suspected at this stage, our definition of energy is based on potential and charge. We recall that potential was defined in **Eq. (4.20)** as work per unit charge. More exactly, the potential difference between two points was defined as the amount of work required to move the test charge Q between the two points:

$$V_{ba} = \frac{W}{Q} = -\int_a^b \mathbf{E} \cdot d\mathbf{l} \quad [\mathrm{V}] \tag{4.95}$$

where the negative sign indicates that for work to be positive, it must be done against the existing electric field intensity **E**. In other words, work is only done if a field exists. Similarly, only an existing field can perform work on a charge. More important in this context is the definition of potential as work per unit charge. We can, therefore, view the potential energy of a system as the product of potential and charge:

$$W = QV_{ba} \quad [\mathrm{J}] \tag{4.96}$$

It is important to understand what is meant by "system" and where this energy is stored. From previous discussions, we understand that the energy must be stored in the electric field; that is, the energy is associated with the field not with the charge. The "system" therefore means the domain in which the field exists. If this field is confined to the region between the plates of a capacitor, then this region is the system. If the electric field exists throughout space (such as in the case of a point charge in an infinite medium), the system encompasses the whole of space. We will actually use terms such as "energy in a system of charges" or "energy due to charge densities," but it is understood to mean energy in the electric field due to charges or charge densities. Nevertheless, in many cases, it will be more convenienet to calculate energy in terms of capacitance and charge or in terms of charge and potential but here again, the energy is associated with the electric field even though we may not actually use the field explicitly. We strat the discussion in terms of charge and potential based on Eq. (**4.96**), then discuss energy in terms of capacitance followed by energy due to distributed charge densities and only then turn our attention to calculation of energy in terms of the electric field intensity.

How do we calculate potential energy in a system of charges? There are two related methods. One is based on **Eq. (4.96)** and we look into it first. The second involves the use of the electric field intensity directly.

Consider **Figure 4.44**. Initially, there are no charges anywhere in space and the electric field intensity everywhere in space is zero. Now, a point charge Q_1 is brought from infinity to point P_1. Since the existing field is zero, the amount of work performed in bringing the charge from infinity to point P_1 is zero (**Figure 4.44a**). This may seem unusual at first, but not if we remember the definition: work must be done against an existing field. This is so even though Q_1 produces an electric field of its own. You may wish to think of this in terms of gravitation: if there were only one star in the universe, this star could "move" anywhere in space without being affected by any gravitational field. The fact that it has a gravitational field of its own does not change this. Denoting the energy due to charge Q_1 as W_1, we write

$$W_1 = 0 \tag{4.97}$$

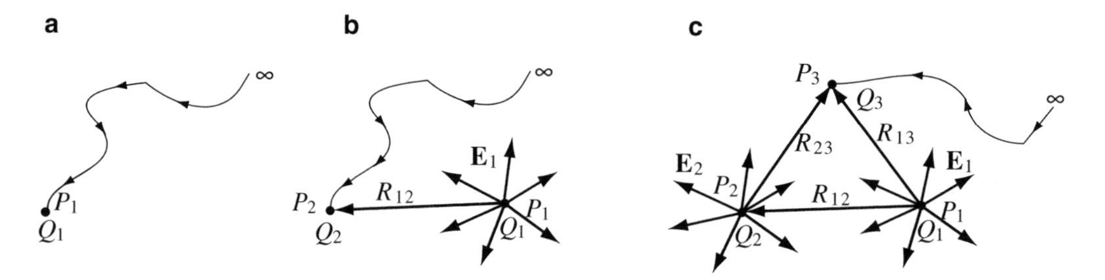

Figure 4.44 Potential energy in a system of charges. (**a**) Q_1 adds zero potential energy because the existing electric field intensity in space is zero. (**b**) Q_2 is brought against the field of Q_1. (**c**) Q_3 is brought against the fields of Q_1 and Q_2

Now, consider another point charge Q_2 being brought from infinity to point P_2 (**Figure 4.44b**). To do so, we must consider the existing electric field intensity, which is that due to charge Q_1. This electric field intensity is

$$\mathbf{E} = \hat{\mathbf{R}}\frac{Q_1}{4\pi\varepsilon R^2} \quad \left[\frac{\text{V}}{\text{m}}\right] \tag{4.98}$$

where we used the general permittivity ε to indicate that this and the following results apply in general dielectrics as well as in free space. The potential at point P_2 is

$$V_2 = -\int_{\infty}^{P_2} \mathbf{E} \cdot d\mathbf{l} = -\int_{\infty}^{P_2} \hat{\mathbf{R}}\frac{Q_1}{4\pi\varepsilon R^2} \cdot \hat{\mathbf{R}}dR = \frac{Q_1}{4\pi\varepsilon R_{12}} \quad [\text{V}] \tag{4.99}$$

where R_{12} is the distance between points P_1 and P_2. Thus, the potential energy due to charge Q_2 is

$$W_2 = Q_2V_2 = Q_2\frac{Q_1}{4\pi\varepsilon R_{12}} \quad [\text{J}] \tag{4.100}$$

A third charge Q_3 is now brought from infinity to point P_3 (**Figure 4.44c**). This charge is brought against the combined fields of charge Q_1 and Q_2. Using superposition, we can write directly

$$V_3 = -\int_{\infty}^{P_3} \mathbf{E}_1 \cdot d\mathbf{l} - \int_{\infty}^{P_3} \mathbf{E}_2 \cdot d\mathbf{l} = \frac{Q_1}{4\pi\varepsilon R_{13}} + \frac{Q_2}{4\pi\varepsilon R_{23}} \quad [\text{V}] \tag{4.101}$$

where R_{13} is the distance between P_1 and P_3, and R_{23} is the distance between points P_2 and P_3. The energy added to the system is

$$W_3 = Q_3 V_3 = Q_3 \frac{Q_1}{4\pi\varepsilon R_{13}} + Q_3 \frac{Q_2}{4\pi\varepsilon R_{23}} \quad [\mathrm{J}] \tag{4.102}$$

The total potential energy in the system of charges is therefore the sum of the three terms:

$$W = W_1 + W_2 + W_3 = 0 + Q_2 \frac{Q_1}{4\pi\varepsilon R_{12}} + Q_3 \frac{Q_1}{4\pi\varepsilon R_{13}} + Q_3 \frac{Q_2}{4\pi\varepsilon R_{23}} \quad [\mathrm{J}] \tag{4.103}$$

This energy was obtained by arbitrarily bringing charge Q_1 first, followed by charge Q_2, and then by charge Q_3. The order is obviously irrelevant: the energy depends only on the charges and the potentials at the respective location, to which they are brought. Thus, if we started with charge Q_3, followed by Q_2, and then Q_1, we would obtain exactly the same expressions except that the indices 3 and 1 are interchanged. Thus, we can write directly

$$W = W_3 + W_2 + W_1 = 0 + Q_2 \frac{Q_3}{4\pi\varepsilon R_{32}} + Q_1 \frac{Q_3}{4\pi\varepsilon R_{31}} + Q_1 \frac{Q_2}{4\pi\varepsilon R_{21}} \quad [\mathrm{J}] \tag{4.104}$$

We now sum the two energies in **Eqs. (4.103)** and **(4.104)** for the sole purpose of obtaining an expression for energy which does not show this apparent dependency on order of charges. Adding **Eqs. (4.103)** and **(4.104)** and collecting terms gives

$$2W = Q_1 \left(\frac{Q_3}{4\pi\varepsilon R_{31}} + \frac{Q_2}{4\pi\varepsilon R_{21}} \right) + Q_2 \left(\frac{Q_1}{4\pi\varepsilon R_{12}} + \frac{Q_3}{4\pi\varepsilon R_{32}} \right) + Q_3 \left(\frac{Q_1}{4\pi\varepsilon R_{13}} + \frac{Q_2}{4\pi\varepsilon R_{23}} \right) \quad [\mathrm{J}] \tag{4.105}$$

The terms in parentheses are the potentials at the corresponding nodes due to the charges at the other nodes:

$$V_1 = \frac{Q_3}{4\pi\varepsilon R_{31}} + \frac{Q_2}{4\pi\varepsilon R_{21}}, \quad V_2 = \frac{Q_1}{4\pi\varepsilon R_{12}} + \frac{Q_3}{4\pi\varepsilon R_{32}}, \quad V_3 = \frac{Q_1}{4\pi\varepsilon R_{13}} + \frac{Q_2}{4\pi\varepsilon R_{23}} \quad [\mathrm{V}] \tag{4.106}$$

and, therefore,

$$\boxed{W = \frac{1}{2}(Q_1 V_1 + Q_2 V_2 + Q_3 V_3) \quad [\mathrm{J}]} \tag{4.107}$$

This expression is symmetric and shows that the potential energy is calculable from potentials and charges. Note that the potential at the location of any point charge is calculated using all charges except the charge at the location at which the potential is calculated. The expression in **Eq. (4.107)** can be generalized to any number of charges by repeating the above process. Thus, for N charges, located at points P_1 through P_N, the energy of the system is

$$\boxed{W = \frac{1}{2}\sum_{i=1}^{N} Q_i V_i \quad [\mathrm{J}]} \tag{4.108}$$

The expression in **Eq. (4.108)** also shows that although the ***joule*** [J] is the correct unit for energy, we can just as easily use a unit which is [J] = coulomb × volt or [C·V]. While there is no compelling reason to do so, except to show that the joule, which is normally associated with mechanical or thermal energy, is properly defined in terms of electric units, the unit is rather large. A unit which is sometimes convenient, and used extensively in physics, is the ***electron volt*** [eV]. This is the energy of a charge equal to one electron at a location with potential 1 V. Thus, $1\ \mathrm{eV} = 1.6 \times 10^{-19}\ \mathrm{C} \times 1\ \mathrm{V} = 1.6 \times 10^{-19}\ \mathrm{J}$. This is a very small unit, but it finds applications in, for example, measuring energy of accelerated particles and energy bands in semiconductors. In the first case, energies of the order of trillion eV are possible, while the second is of the order of a few eV.

From the definition of capacitance ($C = Q/V$ or $Q = CV$ or $V = Q/C$), two alternate forms of **Eq. (4.108)** may be written:

$$W = \frac{1}{2}\sum_{i=1}^{N} C_i V_i^2 = \frac{1}{2}\sum_{i=1}^{N} \frac{Q_i^2}{C_i} \quad [\mathrm{J}] \tag{4.109}$$

In this form, the energy in the system is associated with the capacitances of the different parts of the system and either the charge or potential differences on the various capacitances. The latter form is sometimes easier to apply than **Eq. (4.108)**, especially when capacitance is easy to identify and calculate. From this, the energy stored in a capacitor may be written as

$$\boxed{W = \frac{CV^2}{2} = \frac{QV}{2} = \frac{Q^2}{2C} \quad [\mathrm{J}]} \tag{4.110}$$

Next, we look at the energy due to a continuous distribution of charges. To do so, we envision a general distribution of charges in a generic volume as shown in **Figure 4.45**. To calculate the energy, we designate a small portion of volume as $\Delta v_i'$, located at a point $P_i(x',y',z')$. The total charge in this elemental volume is ΔQ_i. The potential at point P_i is V_i. Thus, we can write for the volume, which now is viewed as being made of small point charges, as

$$W = \frac{1}{2}\sum_{i=1}^{N} \Delta Q_i V_i = \frac{1}{2}\sum_{i=1}^{N} \rho_v(x',y',z')\Delta v_i' V_i(x',y',z') \quad [\mathrm{J}] \tag{4.111}$$

where the dependency on the point $P_i(x',y',z')$ is written explicitly. Now, we let $\Delta v_i' = \Delta v'$ (all volume elements are equal in size) and let $\Delta v'$ tend to zero (so that N tends to infinity) and write

$$\boxed{W = \lim_{\Delta v' \to 0}\left[\frac{1}{2}\sum_{i=1}^{\infty} \rho_v(x',y',z')\Delta v' V_i(x',y',z')\right] = \frac{1}{2}\int_{v'} \rho_v V dv' \quad [\mathrm{J}]} \tag{4.112}$$

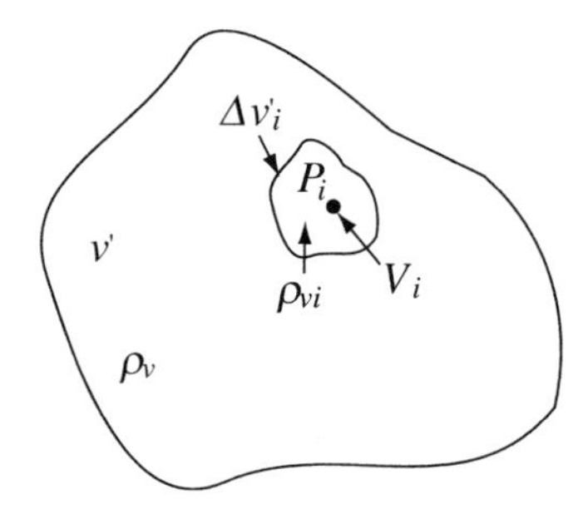

Figure 4.45 Calculation of the potential energy in a continuous distribution of charges

Note that the integration must be performed over the domain in which the charge density exists. Also, neither voltage nor charge density need be constant in the volume. If either is, it can be taken out of the integral.

Since charges can be distributed over surfaces or lines, the same basic principle applies to these charges. By replacing the volume integration with surface or line integration, we can write the energy due to these charge densities as

$$\boxed{W = \frac{1}{2}\int_{s'} \rho_s V ds' \quad [\mathrm{J}]} \tag{4.113}$$

And

$$\boxed{W = \frac{1}{2}\int_{l'} \rho_l V dl' \quad [\mathrm{J}]} \tag{4.114}$$

In these relations V is the potential at the location of the element of charge (dl', ds', or dv').

Example 4.27 Energy in a System of Point Charges Four point charges are located in a rectangular pattern as shown in **Figure 4.46**. Calculate the total energy stored in the system of charges.

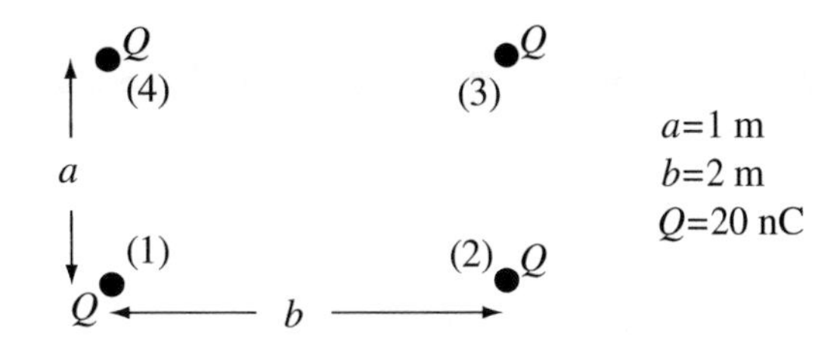

Figure 4.46 A system of four point charges

Solution: To calculate the energy in the system, we use **Eq. (4.108)**, but we must first calculate the potential at the location of each of the four charges due to the other three. In this particular case, the potential at each of the four points is the same since all four charges are identical and symmetrically placed.

For example, if we calculate the potential at point (3), the distance between charges (4) and (3) is b, between (2) and (3) is a, and between (1) and (3) is $(a^2 + b^2)^{1/2}$. Thus,

$$V_3 = \frac{Q}{4\pi\varepsilon_0 a} + \frac{Q}{4\pi\varepsilon_0 b} + \frac{Q}{4\pi\varepsilon_0\sqrt{a^2+b^2}} = \frac{Q}{4\pi\varepsilon_0}\left[\frac{1}{a} + \frac{1}{b} + \frac{1}{\sqrt{a^2+b^2}}\right] \quad [\mathrm{V}]$$

In this case, $V_1 = V_2 = V_3 = V_4$. The energy stored in the system of charges is

$$W = \frac{1}{2}\sum_{i=1}^{4} Q_i V_i = \frac{1}{2}(Q_1V_1 + Q_2V_2 + Q_3V_3 + Q_4V_4) = 2QV_3 \quad [\mathrm{J}]$$

Substituting the value for V_3, we get

$$W = \frac{Q^2}{2\pi\varepsilon_0}\left[\frac{1}{a} + \frac{1}{b} + \frac{1}{\sqrt{a^2+b^2}}\right] = \frac{(20\times 10^{-9})^2}{2\times\pi\times 8.854\times 10^{-12}}\left[1 + \frac{1}{2} + \frac{1}{\sqrt{5}}\right] = 1.4\times 10^{-5} \quad [\mathrm{J}]$$

The energy is very small because the charges are small and relatively far apart.

Exercise 4.10 Two electrons are separated a distance of 10^{-10} m (typical distances in atoms). Calculate the energy in the system made up of the two electrons.

Answer $W = 2.3 \times 10^{-18}$ J $= 14.4$ eV.

Example 4.28 Energy in a Cloud A cloud in the form of a sphere of radius a has a uniform volume charge density ρ_v. Calculate the total energy associated with the cloud. Use $a = 1$ km, $\rho_v = 10$ nC/m^3. Note: the dimensions and charge density may approximate those found in a thundercloud. This energy, or part of it, is released during lightning.

Solution: To calculate the total energy stored in the charge distribution, we use **Eq. (4.112)**. From symmetry considerations, the potential at a distance R from the center of the sphere is constant. Thus, we calculate the potential at an arbitrary radius R inside the charge distribution and multiply this by the element of charge dQ contained in a spherical shell of radius R and thickness dR. The resulting expression is then integrated between $R = 0$ and $R = a$.

The electric field intensity at a point R inside the sphere is calculated using a Gaussian surface in the form of a sphere of radius $R < a$. Because the field is radial, we get

$$\int_{s_R} \mathbf{E} \cdot d\mathbf{s} = \int_{v_R} \frac{\rho_v dv}{\varepsilon_0} \quad \rightarrow \quad E4\pi R^2 = \frac{\rho_v}{\varepsilon_0}\frac{4\pi R^3}{3} \quad \text{or} \quad \mathbf{E} = \hat{\mathbf{R}}\frac{\rho_v R}{3\varepsilon_0} \quad \left[\frac{\text{V}}{\text{m}}\right], \quad R < a$$

The electric field intensity outside the sphere at a point $R > a$ is calculated as

$$\int_{s_a} \mathbf{E} \cdot d\mathbf{s} = \int_{v_a} \frac{\rho_v dv}{\varepsilon_0} \quad \rightarrow \quad E4\pi R^2 = \frac{\rho_v}{\varepsilon_0}\frac{4\pi a^3}{3} \quad \text{or} \quad \mathbf{E} = \hat{\mathbf{R}}\frac{\rho_v a^3}{3\varepsilon_0 R^2} \quad \left[\frac{\text{V}}{\text{m}}\right], \quad R > a$$

To calculate the potential at a point R_0 inside the sphere, we integrate the electric field intensity from infinity to that point, with $d\mathbf{l} = \hat{\mathbf{R}}dR$:

$$V_{R_0} = -\int_{\infty}^{a} \hat{\mathbf{R}}\frac{\rho_v a^3}{3\varepsilon_0 R^2} \cdot \hat{\mathbf{R}}dR - \int_{a}^{R_0} \hat{\mathbf{R}}\frac{\rho_v R}{3\varepsilon_0} \cdot \hat{\mathbf{R}}dR = \frac{\rho_v a^3}{3\varepsilon_0 R}\bigg|_{\infty}^{a} - \frac{\rho_v R^2}{6\varepsilon_0}\bigg|_{a}^{R_0} = \frac{\rho_v a^2}{2\varepsilon_0} - \frac{\rho_v R_0^2}{6\varepsilon_0} \quad [\text{V}]$$

Now, we define a spherical shell of charge of thickness dR_0 at $R = R_0$. The shell contains a charge:

$$dQ = \rho_v dv' = \rho_v 4\pi R_0^2 dR$$

Since the potential at the location of this shell is V_{R_0}, these two quantities are now substituted into **Eq. (4.112)**:

$$W = \frac{1}{2}\int_{v'} \rho_v V dv' = \frac{1}{2}\int_{R_0=0}^{a} \rho_v 4\pi R_0^2 \left[\frac{\rho_v a^2}{2\varepsilon_0} - \frac{\rho_v R_0^2}{6\varepsilon_0}\right] dR_0 = \frac{\rho_v^2 \pi}{\varepsilon_0}\left[\frac{R_0^3 a^2}{3} - \frac{R_0^5}{15}\right]_{R_0=0}^{R_0=a} = \frac{4\pi a^5 \rho_v^2}{15\varepsilon_0} \quad [\text{J}]$$

that is,

$$W = \frac{4\pi a^5 \rho_v^2}{15\varepsilon_0} = \frac{4 \times \pi \times 1000^5 \times \left(10 \times 10^{-9}\right)^2}{15 \times 8.854 \times 10^{-12}} = 9.46 \times 10^9 \quad [\text{J}]$$

This is almost 10 billion joules! The high energy associated with lightning explains some of its dramatic effects.

Note: Part of this energy is stored in the volume of the cloud, part of it in the space surrounding the cloud (see **Problem 4.50**).

4.8.1 Energy in the Electrostatic Field: Field Variables

The second method of calculating the energy in the electrostatic field is in terms of the electric field intensity itself. Although this is not physically different than the previous method, it results in a different form and is more convenient when the electric field intensity or the electric flux density is known as opposed to potential. In addition, this method shows the fact that energy is associated with the electric field rather than charges. We start with **Eq. (4.112)** since this is the most general expression for energy we have so far.

From **Eqs. (4.6)** and **(4.62)**, we can write for the charge density ρ_v,

$$\rho_v = \nabla \cdot \mathbf{D} = \nabla \cdot (\varepsilon \mathbf{E}) \quad [\text{C/m}^3] \tag{4.115}$$

Thus, the energy in **Eq. (4.112)** is

$$W = \frac{1}{2}\int_{v'} (\nabla \cdot \mathbf{D}) V \, dv' \quad [\text{J}] \tag{4.116}$$

To simplify this expression, we make use of the following vector identity:

$$\nabla \cdot (V\mathbf{D}) = V(\nabla \cdot \mathbf{D}) + \mathbf{D} \cdot (\nabla V) \tag{4.117}$$

which applies to any vector **D** and any scalar V [see **Eq. (2.138)**]. Substitution of the term $V(\nabla \cdot \mathbf{D})$ from this expression into **Eq. (4.116)** gives

$$W = \frac{1}{2}\int_{v'} (\nabla \cdot (V\mathbf{D}) - \mathbf{D} \cdot (\nabla V))dv' = \frac{1}{2}\int_{v'} \nabla \cdot (V\mathbf{D})dv' - \frac{1}{2}\int_{v'} \mathbf{D} \cdot (\nabla V)dv' \quad [\mathrm{J}] \tag{4.118}$$

The first integral can now be transformed into a closed surface integral using the divergence theorem:

$$\frac{1}{2}\int_{v'} \nabla \cdot (V\mathbf{D})dv' = \frac{1}{2}\oint_{s'} V\mathbf{D} \cdot d\mathbf{s}' \tag{4.119}$$

and **Eq. (4.118)** becomes

$$W = \frac{1}{2}\oint_{s'} V\mathbf{D} \cdot d\mathbf{s}' - \frac{1}{2}\int_{v'} \mathbf{D} \cdot (\nabla V)dv' \tag{4.120}$$

The closed surface integral, calculated over any surface that encloses the volume v' is zero based on the following argument: taking a large sphere as the volume, its surface varies with radius as R^2. At the same time, VD varies at least as R^{-3} (for point charges; for dipoles, the variation is R^{-5}). Thus, as R varies, the integrand varies at least as R^{-1}. As the surface increases, the integrand decreases. For a very large surface (R tends to ∞), this integral diminishes to zero. Now, v' in **Eq. (4.120)** is replaced by v, which, in general, is the whole of space. Also, we can replace the gradient of V by the electric field intensity using the relation $\mathbf{E} = -\nabla V$ and write

$$W = -\frac{1}{2}\int_{v} \mathbf{D} \cdot (\nabla V)dv = \frac{1}{2}\oint_{v} \mathbf{D} \cdot \mathbf{E}dv \quad [\mathrm{J}] \tag{4.121}$$

Finally, we can also write

$$\boxed{W = \frac{1}{2}\int_{v} \mathbf{D} \cdot \mathbf{E}dv = \frac{1}{2}\int_{v} \varepsilon\mathbf{E} \cdot \mathbf{E}dv = \frac{1}{2}\int_{v} \varepsilon E^2 dv \quad [\mathrm{J}]} \tag{4.122}$$

where the relations, $\mathbf{D} = \varepsilon\mathbf{E}$ and $\mathbf{E} \cdot \mathbf{E} = E^2$, were used. Note, however, that the volume v now does not necessarily contain charges; therefore, the prime is not used—that is, this energy is calculated from the electric field intensity in space rather than from charge and potential. The advantage of viewing energy in this fashion is that it indicates that energy is distributed in the volume in which the electric field exists and, therefore, can be viewed as being associated with the field. This is particularly important if the volume in which **E** exists has no charges, because it still stores energy.

Any of these relations can be used for the calculation of energy in the electrostatic field. We should choose the relation that is most convenient for our purposes and this depends on the application. Finally, we note that the integrand in any of these relations is given in terms of energy per unit volume. Thus, we can define the integrand as the ***electrostatic energy density*** as

$$\boxed{w_E = \frac{1}{2}\mathbf{D} \cdot \mathbf{E} = \frac{1}{2}\varepsilon E^2 \quad \left[\frac{\mathrm{J}}{\mathrm{m}^3}\right]} \tag{4.123}$$

Alternative forms of the energy density are

$$w_E = \frac{D^2}{2\varepsilon} = \frac{\mathbf{D}\cdot(-\nabla V)}{2} \quad \left[\frac{\text{J}}{\text{m}^3}\right] \tag{4.124}$$

and, from **Eq. (4.112)**, we can also write

$$w_E = \frac{\rho_v V}{2} \quad \left[\frac{\text{J}}{\text{m}^3}\right] \tag{4.125}$$

Example 4.29 Energy Due to Surface Charge Distribution A long, thick cylindrical conductor of radius $a = 20$ mm, charged with a surface charge density $\rho_s = 0.01$ C/m^2, is located in free space:

(a) Calculate the energy stored per unit length of the conductor within a radius $b = 100$ m from the conductor.
(b) What is the energy per unit length stored inside the conductor?

Solution: First, we calculate the electric field intensity due to a long conductor. The energy per unit length is then the total energy stored in the space outside the conductor per unit length of the conductor. From the electric field intensity we calculate the energy density and then integrate this energy density over the cylindrical surface extending from the conductor to $b = 100$ m:

(a) Using Gauss's law, the electric field intensity at a distance r from the cylinder is

$$\oint_s \mathbf{E}\cdot d\mathbf{s} = \frac{q}{\varepsilon_0} \quad \to \quad E2\pi rL = \frac{\rho_s 2\pi aL}{\varepsilon_0}$$

where the integration is over the area of a Gaussian surface in the form of a cylinder of radius r and length L. The electric field intensity is in the r direction and equals

$$\mathbf{E} = \hat{\mathbf{r}}\frac{\rho_s a}{\varepsilon_0 r} \quad \left[\frac{\text{V}}{\text{m}}\right]$$

The energy density in space is therefore

$$w = \frac{\varepsilon_0 E^2}{2} = \frac{\rho_s^2 a^2}{2\varepsilon_0 r^2} \quad \left[\frac{\text{J}}{\text{m}^3}\right]$$

To find the total energy stored in the space outside the conductor, we define a shell of thickness dr and length 1 m and write the volume of this shell at r as $dv = 2\pi r dr$. The total energy stored becomes

$$W = \int_v w dv = \int_{r=a}^{r=b} \frac{\pi\rho_s^2 a^2}{\varepsilon_0 r} dr = \frac{\pi\rho_s^2 a^2}{\varepsilon_0}\ln\frac{b}{a} \quad \left[\frac{\text{J}}{\text{m}}\right]$$

For $a = 0.02$ m, $b = 100$ m, $\rho_s = 10^{-2}$ C/m^2, we get

$$W = \frac{\pi \times 10^{-4} \times (0.02)^2}{8.854 \times 10^{-12}}\ln\frac{100}{0.02} = 1.21 \times 10^5 \quad [\text{J/m}].$$

(b) The electrostatic energy stored in the conductor itself is zero because the electric field intensity inside the conductor is zero.

Example 4.30 Energy Due to a Point Charge A point charge is placed at the origin of the spherical system of coordinates. Calculate the total energy stored:

(a) In the space between $R = a$ [m] and $R = b$ [m] $(b > a)$.
(b) Between $R = b$ [m] and infinity.
(c) In the whole space due to the field of the point charge. Explain the meaning of this result.

Solution: We calculate the electric field intensity and then the energy density at a general element of volume dv. Integration of this energy density gives the stored energy in any section of space.

(a) The electric field intensity of a point charge at a distance R from the charge is

$$\mathbf{E} = \hat{\mathbf{R}}\frac{Q}{4\pi\varepsilon_0 R^2} \quad \left[\frac{\text{V}}{\text{m}}\right]$$

Since we wish to calculate energy in spherical volumes, the most convenient element of volume is a spherical shell at radius R and thickness dR. The volume of this differential shell is

$$dv = 4\pi R^2 dR \quad [\text{m}^3]$$

The energy density at any distance R from the point charge is

$$w_E = \frac{1}{2}\varepsilon_0 E^2 = \frac{Q^2}{32\pi^2\varepsilon_0 R^4} \quad \left[\frac{\text{J}}{\text{m}^3}\right].$$

(b) The total energy stored in the volume between $R = a$ and $R = b$ is

$$W = \int_{R=a}^{R=b} w_E dv = \int_{R=a}^{R=b} \frac{Q^2 4\pi R^2 dR}{32\pi^2\varepsilon_0 R^4} = \frac{Q^2}{8\pi\varepsilon_0}\int_{R=a}^{R=b}\frac{dR}{R^2} = \left[\frac{Q^2}{8\pi\varepsilon_0}\left(-\frac{1}{R}\right)\right]_a^b = \frac{Q^2}{8\pi\varepsilon_0}\left(\frac{1}{a} - \frac{1}{b}\right) \quad [\text{J}]$$

The energy stored between $R = b$ and infinity is

$$W = \int_{R=b}^{R=\infty} w_E dv = \frac{Q^2}{8\pi\varepsilon_0}\int_{R=b}^{R=\infty}\frac{dR}{R^2} = \frac{Q^2}{8\pi\varepsilon_0}\left(-\frac{1}{R}\right)\Big|_b^\infty = \frac{Q^2}{8\pi\varepsilon_0 b} \quad [\text{J}].$$

(c) Using the result of part **(b)**, the total energy in the system is

$$W = \int_{R=0}^{R=\infty} w_E dv = \frac{Q^2}{8\pi\varepsilon_0}\int_{R=0}^{R=\infty}\frac{dR}{R^2} \quad \to \quad \infty \quad [\text{J}]$$

This may seem as a surprising result. However, it indicates that the energy required to generate or assemble this charge is infinite; that is, it shows that a point charge cannot be created. This should be contrasted with the form in **Eq. (4.97)**, where we found that moving a single charge in space does not add any energy to the system. This, however, does not mean the system has zero energy. In fact, if we were to bring a point charge of any magnitude to the location where there is already another point charge, the energy required would be infinite. This result has many implications. One is the fact that the energy required to fuse two point charges of identical polarity is infinite. For example, the fusion of two atoms requires very large energies. If the nuclei were true point charges, fusion would require infinite energy.

Exercise 4.11 Repeat **Example 4.28** using **Eq. (4.122)** and show that the result in **Example 4.28** is obtained.

4.8.2 Forces in the Electrostatic Field: The Principle of Virtual Work

Forces on charges were discussed in **Chapter 3**. Here, we wish to present yet another method of calculating forces, from a different point of view. Instead of starting with charge or charge density and using the definition of the electric field intensity, we go the other way around: since potential energy is due to work done and work is related to force through displacement, we should be able to calculate forces from energy considerations. This is the same as asking the following question: if the potential energy of a system is known, what are the forces in the system? The method we discuss here is rather general. First, we identify the body or bodies on which the existing force must act. This may be a point charge or a charged body. If these charges are free to move, they will; if not, the force will act on the body to which these charges are attached. Thus, for example, we can talk about forces on a conducting body if charges are constrained from leaving the conductor. The actual calculation of force follows the principle of measuring forces: the charge or body is allowed to move and the force required to constrain the body from moving is calculated. For this reason, the method is called the principle of ***virtual work*** or ***virtual displacement***.

Consider, first, a point charge in an electric field, isolated from any other sources. If the force on the point charge is $\mathbf{F}$, then if the charge were to move a very short distance $d\mathbf{l}$, the change in work involved would be

$$dW = \mathbf{F}\cdot d\mathbf{l} \quad [\mathrm{J}]. \tag{4.126}$$

Now, the question is: Who performs this work? Is it performed *against* the field or *by* the field? The answers lie in the measuring process: any work done by the field reduces the potential energy of the system (negative work). Thus, the work $\mathbf{F}\cdot d\mathbf{l}$ must be equal to the change in potential energy of the system. Suppose the system performs this work (negative work):

$$\mathbf{F}\cdot d\mathbf{l} = \left(\mathbf{F}\cdot \hat{d\mathbf{l}}\right)dl = -dW_e \quad [\mathrm{J}] \tag{4.127}$$

From vector calculus (see **Section 2.3.1**), we can write dW_e in terms of the gradient of W_e as

$$dW_e = (\nabla W_e)\cdot d\mathbf{l} \quad [\mathrm{J}] \tag{4.128}$$

Substituting this in **Eq. (4.127)**, we get

$$\mathbf{F}\cdot d\mathbf{l} = -(\nabla W_e)\cdot d\mathbf{l} \quad [\mathrm{J}] \tag{4.129}$$

or

$$\mathbf{F} = -\nabla W_e \quad [\mathrm{N}] \tag{4.130}$$

In other words, the force on the isolated charge is equal to the negative of the gradient of potential energy. The larger the energy gradient (change in energy with position in space), the larger the force acting on the charge.

Since the force is now known everywhere in space, we can also calculate the electric field intensity or the electric flux density in terms of energy as a simple extension to the calculation of force as

$$\mathbf{E} = \frac{\mathbf{F}}{Q} = -\frac{\nabla W_e}{Q} \quad \left[\frac{\mathrm{V}}{\mathrm{m}}\right] \tag{4.131}$$

Note, however, that in this case, we assumed that the charge is isolated: the fact that the charge was allowed to "move" did not change the amount of charge in the system; it remained constant. This meant that the energy needed to move the charge had to be supplied by the system itself and, therefore, the negative sign in **Eq. (4.130)**. A similar situation is shown in **Figure 4.47**, where a parallel plate capacitor is connected to a potential difference. If we wish to calculate the force acting on a plate, we again allow the plate to "move" a virtual distance dl. However, now the displacement causes a change in the charge of the capacitor since the potential is constant whereas the capacitance changes. If charge is removed from the plates (plates move apart and, therefore, the capacitance is reduced), the system performs work and the potential energy stored in the capacitor is reduced. If the plates are forced closer together, the work is done against the system, capacitance increases as does charge on the plates, and the potential energy is increased. In the latter case, the change in potential energy in the system is positive and the sign in **Eq. (4.130)** is positive.

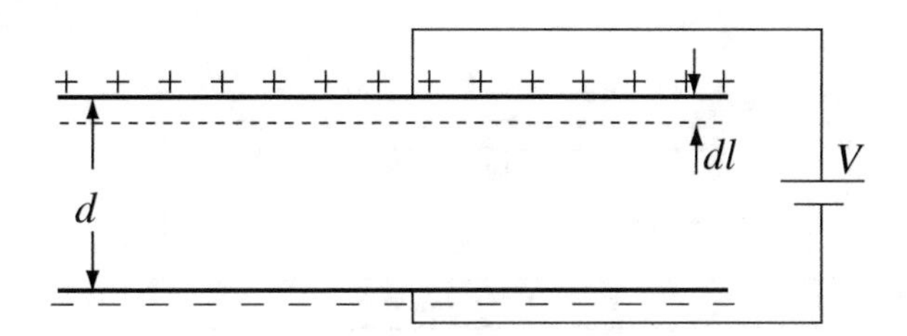

Figure 4.47 The method of virtual displacement

Example 4.31 Coulomb's Law from Energy Considerations Two point charges, Q_1 [C] and Q_2 [C], are placed a distance R [m] apart. Using the general expression for two point charges, show that the force obtained from energy considerations is equal to that obtained from Coulomb's law.

Solution: The energy in the system may be calculated using **Eq. (4.108)**. Then, the force between the charges is found from the negative of the gradient of energy, using **Eq. (4.130)**. The potentials V_1 due to Q_2 at the location of Q_1 and V_2 due to Q_1 at the location of Q_2 are

$$V_1 = \frac{Q_2}{4\pi\varepsilon_0 R}, \quad V_2 = \frac{Q_1}{4\pi\varepsilon_0 R} \quad [\text{V}]$$

The energy in the system of two charges [from **Eq. (4.108)**] is

$$W = \frac{1}{2}(Q_1 V_1 + Q_2 V_2) = \frac{Q_1 Q_2}{4\pi\varepsilon_0 R} \quad [\text{J}]$$

Assuming the charges are placed on an arbitrary axis as shown in **Figure 4.48**, the negative of the gradient of W in the direction of $\mathbf{R}$ is

$$\mathbf{F} = -\nabla W = -\hat{\mathbf{R}}\frac{\partial}{\partial R}\left(\frac{Q_1 Q_2}{4\pi\varepsilon_0 R}\right) = \hat{\mathbf{R}}\left(\frac{Q_1 Q_2}{4\pi\varepsilon_0 R^2}\right) \quad [\text{N}]$$

Thus, the force is in the positive R direction if both Q_1 and Q_2 have the same sign as required. This is the same as the force obtained using Coulomb's law.

Figure 4.48 Two point charges used to calculate force from energy considerations

Example 4.32 Application: The Electrostatic Speaker Electrostatic speakers use the idea of the capacitor, but the force between the two parallel plates, rather than capacitance, is exploited. In a capacitor, the two plates are charged with opposite charges; therefore, the plates attract each other. The force depends on the potential between the plates since the amount of charge depends on the potential ($Q = CV$). The moving plate is mounted on springs to create a restoring force (See **Figure 4.49a**). Now, the movement of the plate is directly proportional to the voltage applied to the plate. Although high voltages are normally required, electrostatic speakers are highly linear devices and, therefore, reproduce sound very well. Electrostatic headphones also exist. As an example, we will calculate here the pressure generated between the plates of a parallel plate capacitor to show the principle of the electrostatic speaker.

A capacitor with very large plates separated a distance d as shown in **Figure 4.49b** is given. The material between the plates is air. A potential V is connected across the plates. Assume the distance between the plates is small. Dimensions and data are $d = 1$ mm and $V = 1{,}000$ V:

(a) Calculate the pressure the plates exert on the air between the plates.
(b) What is the pressure if the air is replaced with a gas with relative permittivity $\varepsilon_r = 8$?

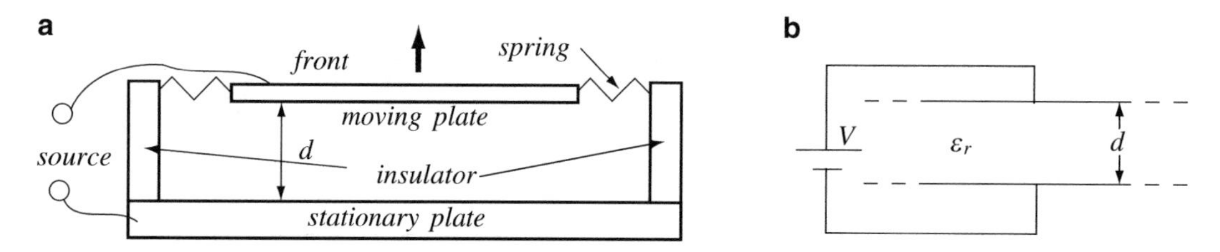

Figure 4.49 (**a**) Principle of the electrostatic speaker. (**b**) Capacitor model of the electrostatic speaker

Solution: We calculate the force per unit area of the plates (pressure) by calculating the energy stored per unit area of the plates of the capacitor. The force is then calculated from the gradient of energy. The lower plate is assumed to be stationary.

(**a**) The electric field intensity between the plates is $E = V/d$ (see **Example 4.16**).
The energy density in the space between the plates is

$$w = \frac{1}{2}\varepsilon_0 E^2 = \frac{1}{2}\varepsilon_0 \left(\frac{V}{d}\right)^2 \quad \left[\frac{\text{J}}{\text{m}^3}\right]$$

Now, suppose the plates are moved an infinitesimal distance dy apart. The change in energy due to this movement is

$$dW = wdv = \frac{1}{2}\varepsilon_0 \left(\frac{V}{d}\right)^2 (1^2)dy \quad [\text{J}]$$

where 1^2 signifies the unit area of the plates. The total (normal) force on a 1 m^2 area of the plates is

$$F = \frac{dW}{dy} = \frac{1}{2}\varepsilon_0 \left(\frac{V}{d}\right)^2 \times 1^2 \quad [\text{N}]$$

Dividing this by the area of the plates gives

$$p = \frac{1}{2}\varepsilon_0 \left(\frac{V}{d}\right)^2 \quad \left[\frac{\text{N}}{\text{m}^2}\right]$$

The pressure is

$$p = \frac{8.854 \times 10^{-12}}{2} \left(\frac{1{,}000}{0.001}\right)^2 = 4.427 \quad [\text{N/m}^2]$$

This is a relatively low pressure but it will produce a sound pressure level of 107 dB. For a loudspeaker to be practical, the size of the plates is usually much smaller than 1 m^2. Therefore, only low-power speakers are really practical. Also, the voltage on the plates cannot be increased much beyond about 1,000 V/mm to avoid breakdown unless a different material is used as a dielectric.

(**b**) If the material is changed to one of relative permittivity ε_r, the pressure is increased by this a factor ε_r to

$$p = \frac{dW}{dy} = \frac{1}{2}\varepsilon_0\varepsilon_r \left(\frac{V}{d}\right)^2 = p = \frac{8 \times 8.84 \times 10^{-12}}{2} \left(\frac{1{,}000}{0.001}\right)^2 = 35.4 \quad [\text{N/m}^2]$$

In this case, the pressure is increased by a factor of 8. Unless the dielectric is a gas, the device cannot be used as a speaker. However, similar devices are used to generate ultrasonic waves in materials. The virtual displacement method assumes the potential, and hence the energy density between the plates remains constant. A real, physical motion of the plates changes both and in that case we must also take into account energy supplied or absorbed by external sources.

4.9 Applications

Application: Shielding in the Electrostatic Field—Faraday's Cage; the Car as a Partially Shielded Compartment In case of lightning, you are probably safer in a car than in any other shelter, except a shelter equipped with a lightning rod. This is because the electric field intensity inside a conductor is zero. A lightning strike to the car should not cause a direct injury. (Sensitive electric equipment might be damaged because of large currents that, in turn, produce magnetic fields, and in extreme cases, the fuel tank might explode, but you most probably will not be electrocuted. There is a cheerful thought.) The car is a partial electrostatic shield (Faraday's cage). A better shield would be a car without windows, but that may be difficult to drive. Vans with few, small windows, usually make better electrostatic shields because they approximate a continuous metallic surface more closely. Other safe locations are low ditches and uniform, dense, low bushes. Anything that sticks out such as trees, umbrellas, a raised golf club, or walking in an exposed, flat area causes a larger electric field intensity in the vicinity of the object and, therefore, a natural point for a discharge. Thus, in a severe thunderstorm, it is safer to sit down than to stand up and safer yet to crouch close to the ground (crouching is safer than lying down) or in a ditch (if you do not drown in the runoff, you should be all right).

The same principles can be used to protect sensitive microcircuits that can be easily damaged by electrostatic discharge. For individual circuits, it is common to embed the pins of the device in a conducting foam. Boards and other sensitive pieces of equipment are placed inside conducting plastic bags, as shown in **Figure 4.50**. Since the electric field intensity inside a conductor is zero, there is no danger of damage until the device is removed. At that point, special precautions are taken to avoid damage. Some of these precautions include conducting wrist straps for workers, conducting mats to stand on, and control of humidity in the assembly area to prevent accumulation of charge. Certain types of clothing (primarily synthetic materials) are taboo in these assembly rooms.

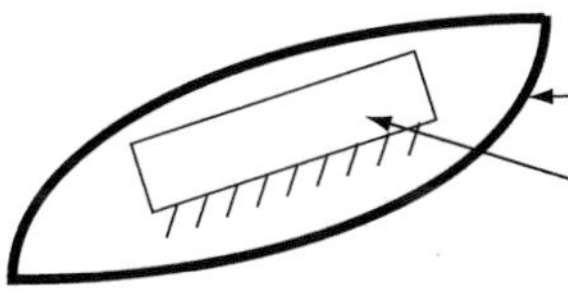

Figure 4.50 Use of the principle of Faraday's cage to protect sensitive electronic equipment before installation

Application: Repair of Energized (Live) Power Lines In servicing electric power lines, it is not always feasible to disconnect power to the lines because of the disruption this may cause. To solve this problem, many repairs are done under live conditions. In repairing high-voltage power lines (above 150 kV) under power, an insulated boom lifts the repairman who wears a stainless steel mesh suit that covers his whole body except parts of his face. At some distance (about 1 m) from the energized cable, he touches the cable with a metallic wand. This equalizes the potential of the lineman and the cable, allowing repair without danger. An alternative is to use a helicopter with the lineman sitting on a cradle below the helicopter. Both helicopter and lineman are at line potential and both lineman and pilot wear the protective suit. The principle is that of Faraday's cage. As the wand touches, there will be an arc and current will flow, which charges the suit. Repairs are routinely done using these methods on power lines up to 765 kV.

Birds rely on a slightly different approach: since their body is very small, they can sit on a power line without ill effects; the potential difference on such a small body is minimal. This is not the case for very large birds or for any bird on very high-voltage lines; there the potential difference can be large enough either to kill the bird or to render its perch very uncomfortable (think about it: have you ever seen a bird on a high-voltage line or a very large bird on any electric line?).

Application: Nondestructive Testing and Evaluation of Materials: Detection of Cracks in Glass and Ceramics by Electrostatic Attraction of Powders The attraction of charged particles has many applications. One of the more useful is in nondestructive testing of materials. A method of testing for cracks in dielectrics is shown in **Figure 4.51**. The dielectric (such as glass) is placed in an electric field produced by a capacitor. The electric field is disturbed by any change in material properties such as cracks, inclusions in the glass, etc., creating field variations around discontinuities. By sprinkling a dielectric powder on the test material, the particles will be attracted to the cracks. These particles are usually treated with fluorescent agents so that they may be viewed under special lighting (black light). The same basic ideas are used for diagnosis in other high-voltage apparatus. For example, distributor caps in cars often show dust paths where discharges occur or where cracks in the cap are present. These dust paths are a simple method of diagnosing ignition problems. Similarly, dust streaks on a high-voltage insulator may indicate cracks and the need for replacement.

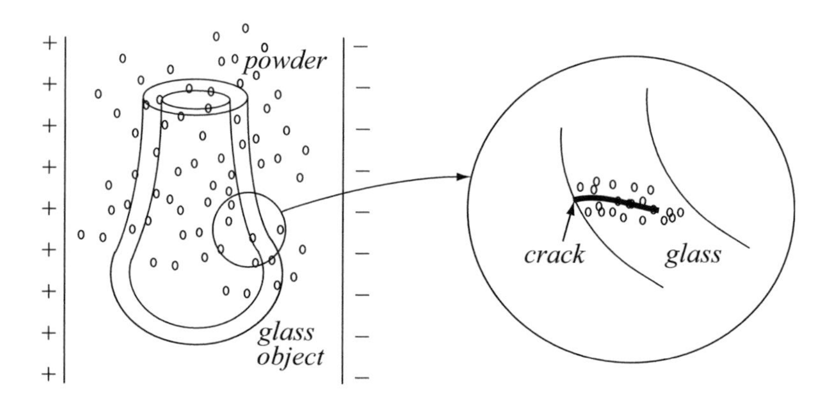

Figure 4.51 Nondestructive testing of glass. The electric field variations around cracks attract charged powders to indicate the location and shape of the crack

Application: Electrostatic Production of Sandpaper An industrial sandpaper deposition method is shown in **Figure 4.52**. It consists of a very large capacitor (two conducting plates). On one (negative side), a nonconducting belt moves and carries the "sand" (actually aluminum oxide particles of different sizes, depending on the grit required). Close to the upper positive plate, the paper or cloth is backed with glue facing down. As the paper moves, particles which have acquired negative charges are attracted to the positive plate and intercept the glued paper. The velocity of movement and size of particles that can be deposited depend on the potential difference. The speed of movement of the glued paper defines the amount of particles deposited. The process is followed by drying and cutting. This type of production has many advantages, not the least is uniformity of product, speed, and the fact that particles align themselves with the sharp point out, making a better product. Also, unless the potential is too low, full coverage of particles is obtained.

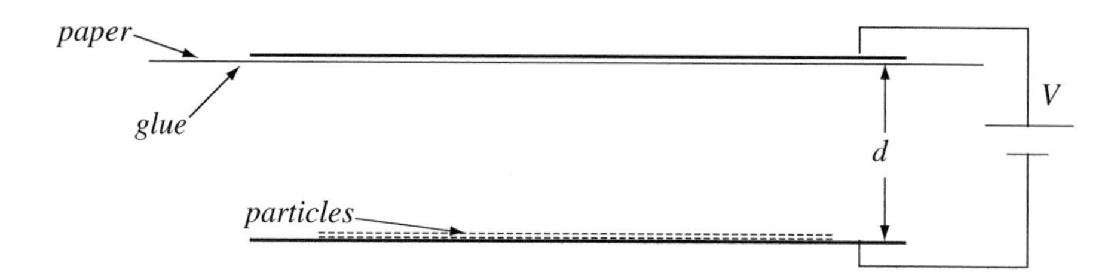

Figure 4.52 Electrostatic production of sandpaper. Negative charged particles are attracted to the positive plate, intercepting the glued paper

Application: Lightning Lightning is the discharge of a charged region (such as a cloud) in a very short period of time. Cloud-to-earth or earth-to-cloud lightning is most familiar because it is a most impressive phenomenon. However, all discharges can be considered in the same genre. Natural lightning occurs when a thundercloud acquires charges both from motion of the cloud and, in particular, from rising ice crystals and falling hailstones. Friction and collisions strip electrons from ice particles, causing the cloud to be primarily negatively charged. The ground under the cloud and to some distance away becomes positively charged by induction. Some areas of the cloud may become positively charged by induction as well. The mechanism of initiating a discharge is not well understood, but once the potential has exceeded the dielectric strength, a discharge will occur as air becomes essentially conductive. Either a cloud-to-cloud, cloud-to-earth, earth-to-cloud or discharge within a cloud is possible. A negatively charged cloud will cause a cloud-to-ground flow of electrons, thus causing a cloud-to-ground discharge. Positively charged clouds can also initiate a cloud-to-ground discharge. Ground-to-cloud discharge has a different branching characteristic and can be positive or negative. A cloud-to-ground discharge starts at the lower part of the thundercloud, perhaps by weak sparks caused by the large negative charge and small, scattered positive charge within the cloud. This then causes the main discharge through what is called the air channel. Normally, the downward negative leader and upward positive leader meet, causing a shock wave due to expansion of hot gases (the thunderclap) and spreading the stroke down and up in the form of a flash. The return stroke, that is, the main, bright discharge, is the visible part of the event and may repeat a number of times using the same air channel. This gives some flashes a pulsating nature. Two return strokes are quite common, but as many as 20 have been observed. The channel is usually quite thin, but the currents are extremely high—over 50,000 A. Temperatures within the channel can reach about 30,000 °C. The power associated with a single lightning strike is very high: over 100 million watts per meter length of the lightning channel. For a 1 km lightning channel (a relatively short strike since lightning can be as long as 10 km), this is about 10^{11} W (100 GW). However, the duration is extremely short—only a few microseconds for peak currents, about

100 ms when multiple strokes occur and a trailing edge of longer duration (perhaps 500–600 ms of low, continuous current, as low as 100–200 A). The latter part of the strike is, in fact, the most dangerous because it lasts longer and can ignite materials. This trailing current is responsible for most lightning-initiated forest fires. The total duration is about 0.8 s, and discharges of the order of 100 C of charge are common. The energy associated with lightning is converted into light, sound, heat, and high-frequency waves (which you can detect using your radio, especially on the AM dial). Voltages present before lightning occurs are of the order of 10^8 to 10^9 V.

Because lightning strikes are rather short, the danger is not as high as might be expected. Many people hit by lightning live to tell their story. The initial current, which tends to flow mostly on the skin, can cause the person to stop breathing, but resuscitation will almost always restore breathing or it may resume on its own. Damage to trees and aircraft is also relatively minor, again because of the tendency to flow over the surface.

Lightning is a very difficult phenomenon to observe and quantify, and much more so for natural lightning. For this reason, many measurements are done with triggered lightning, which is normally done by firing a small rocket, with a trailing wire into a thundercloud, thus creating a favorable path for the return strike.

Lightning occurs mostly in hot regions and seems to be associated with rain. There is even evidence that it is a factor in spawning of tornadoes. However, it can occur in winter and some of the most spectacular lightning occurs during eruptions of volcanoes where flying ash generates the required charge. Lightning seems to occur on other planets as well; lightning flashes were detected on Jupiter by the Voyager 1 spacecraft in 1979.

Application: Volta's Glass Pistol and the Internal Combustion Engine Alessandro Volta used an "electrostatic pistol" to identify "bad air"—essentially methane, a by-product gas emitted by rotting organic material in marshes. A glass pistol—essentially a small container—accumulated the gas and then was corked. A spark plug in the pistol was then connected to a source to produce a spark which ignited the gases. **Figure 4.53** shows this curious device and its operation. He used this device to identify bad air or what was believed at the time (wrongly) to cause malaria (literally mal-bad, aria-air in Italian). He suggested this idea be used as a sort of an electric gun, which, at the time, was not practical. However, the idea leads directly to the internal combustion engine where a gas mixture is ignited by a spark.

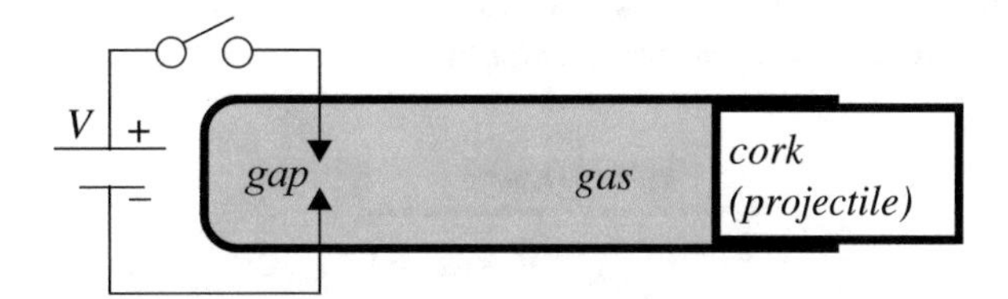

Figure 4.53 Volta's glass pistol. This idea was later incorporated in the ignition system of internal combustion engines

Application: Electrostatic Paper Holder—Electrostatic Clamping, Example of Surface Charge One application of electrostatic force is in holding or clamping paper to the a surface such as that of a plotter. The method is shown in **Figure 4.54**. The surface is a conducting (nonmetallic) surface, and after the paper is placed on the surface, it is charged with a surface charge by alternate positive/negative wires. Each wire can be viewed as a source of electric field which polarizes a section of the paper. The net effect is attraction between the paper and the surface. A few hundred volts difference between neighboring conductors is sufficient to hold a page down for various purposes such as plotting or drawing, posting on bulletin boards and the like. The method can also be used to hold semiconductor wafers for processing and in similar applications.

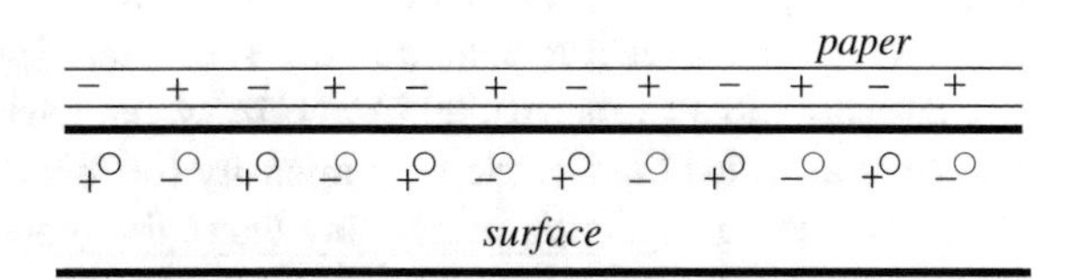

Figure 4.54 Principle of electrostatic clamping as used to hold down paper for plotting

Application: Ion Propulsion An interesting and potentially useful electrostatic device is the ion engine. The idea is no more than action and reaction. In this type of engine, ions at very high speeds are emitted from an ion thruster. The force acting on the ion acts also on the thruster as a reaction force. If a very dense stream of ions at high speeds can be generated, enough thrust to move the engine forward can be obtained. The method is proposed for interstellar travel since it requires small amounts of fuel (as compared to chemical jets). The thrusts generated in this method are relatively small but useful in low-drag environments such as space. It is also clean and simple, allowing long-range travel. The power required for acceleration of ions can be derived from solar cells or a small nuclear generator on board the space vehicle. It is believed that in deep-space missions, these boosters should be able to propel higher payloads at higher speed primarily because the fuel quantities required are small. Speeds of the order of 150,000 km/h should be possible in free space. One type of ion thruster is shown in **Figure 4.55**. Ion thrusters are not yet used for propulsion but are commonly used as correction thrusters in satellites.

The propellant (such as cesium or lithium) is ionized in an ionization chamber, usually as positive ions. These are then accelerated toward a negative electrode and pass through the electrode. At this point, the ions are neutralized by injecting electrons into the stream at the same velocity as the ions. This neutralization is necessary to retain the neutral charge of the engine; otherwise, emitted charged particles will be drawn back to the engine (for positive ions, the engine and, therefore, the vehicle become increasingly negative as particles are emitted).

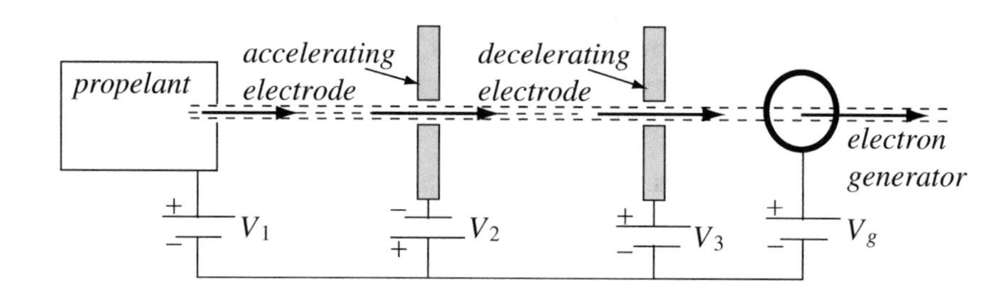

Figure 4.55 Principle of the ion thruster

4.10 Summary

Chapter 4 continues the discussion on electrostatic fields. We start with the ***postulates***—the definition of the divergence and curl of **E** followed by their integral representation:

$$\overbrace{\nabla\cdot\mathbf{E}=\frac{\rho_v}{\varepsilon_0},\quad \nabla\times\mathbf{E}=0}^{\text{differential form}}\quad \text{or}\quad \overbrace{\oint_s \mathbf{E}\cdot d\mathbf{s}=\frac{Q}{\varepsilon_0},\quad \oint_l \mathbf{E}\cdot d\mathbf{l}=0}^{\text{integral form}} \tag{4.10}$$

The first of these (divergence equation) is Gauss's law. It is particularly useful in its integral form for evaluation of the electric field intensity or electric flux density as follows:

$$\oint_s \mathbf{E}\cdot d\mathbf{s}=\frac{Q_{enc.}}{\varepsilon_0}\quad (4.13)\quad \text{or}\quad \oint_s \mathbf{D}\cdot d\mathbf{s}=Q_{enc.} \tag{4.12}$$

where $Q_{enc.}$ is the total charge enclosed by the surface s.

Gauss's law is universally applicable for calculation of fields from charge distributions or of the equivalent charge from fields. For practical use we require the electric field intensity **E** (or electric flux density **D**) to be constant in magnitude and either perpendicular or tangential to a surface called ***Gaussian surface,*** which encloses a total charge $Q_{enc.}$ symmetrically and passes through the point at which the electric field intensity (or flux density) is calculated. This allows us to take the quantities **D** or **E** outside the integral sign after performing the scalar product and evaluate them. Note:

(1) Only highly symmetric charge configurations can be treated analytically.
(2) The magnitude but not the direction of the field is found.
(3) When Gauss' law is used for calculation of charges from known electric fields, only the total equivalent charge enclosed by the Gaussian surface is calculable, not its actual distribution.

The ***electric potential*** is defined as the difference in potential energy per unit charge. This establishes the potential difference between two arbitrary points, b, a, due to an electric field intensity **E** as

$$V_{ba} = -\int_a^b \mathbf{E}\cdot d\mathbf{l} = V_b - V_a \quad \left[\frac{\text{J}}{\text{C}}\right] \text{ or } \left[\frac{\text{N}\cdot\text{m}}{\text{C}}\right] \text{ or } [\text{V}] \tag{4.20}$$

The absolute potential at a point is the potential difference between that point and a zero reference point, usually, but not always, at infinity. A system of n point charges at position vectors $\mathbf{R}_i$ produces a potential at a position vector **R** as

$$V(\mathbf{R}) = \frac{1}{4\pi\varepsilon_0}\sum_{i=1}^{n}\frac{Q_i}{|\mathbf{R}-\mathbf{R}_i|} \quad [\text{V}] \tag{4.28}$$

By defining differential point charges on line, surface, and volume charge distributions (see **Figure 4.14**), **Eq. (4.28)** extends to potentials of any charge distribution [see **Eqs. (4.30)** through **(4.32)**]:

$$V(\mathbf{R}) = \frac{1}{4\pi\varepsilon_0}\int_{\Omega'}\frac{\rho_{\Omega'}}{|\mathbf{R}-\mathbf{R}'|}d\Omega' \quad [\text{V}]\text{, where } \Omega' \text{ stands for } l',\ s',\text{ or } v'.$$

The electric field intensity may be evaluated from the potential as

$$\mathbf{E} = -\nabla V \quad [\text{V/m}] \tag{4.34}$$

This method requires calculation of V in general coordinates but is particularly useful because calculation of potential, being a scalar function, is often easier than direct calculation of **E**.

Materials: Perfect conductors and dielectrics

In **perfect conductors**:

(1) The electric field intensity is zero inside the medium.
(2) The electric field intensity outside the conductor is normal at every point of the surface.
(3) A surface charge density must exist equal in magnitude to $\rho_s = \varepsilon E_n$. The sign of the charge is defined by the direction of the normal electric field intensity E_n.

Dielectrics are insulating (nonconducting) media. We restrict ourselves here to perfect dielectrics. ***Polarization*** in dielectrics is due to the effect of the external electric field on the atoms of the medium and manifests itself through the existence of surface and volume polarization charge densities

$$\rho_{ps} = \mathbf{P}\cdot\hat{\mathbf{n}}' \quad [\text{C/m}^2] \quad \text{and} \quad \rho_{pv} = -\nabla'\cdot\mathbf{P} \quad [\text{C/m}^3] \tag{4.50}$$

A ***polarization vector*** **P** may be postulated either as a vector sum of the polarization fields of individual atoms [**Eq. (4.37)**] or from the macroscopic effect on the external electric flux density: $\mathbf{D} = \varepsilon_0\mathbf{E} + \mathbf{P} = \varepsilon_0\varepsilon_r\mathbf{E}$ $[\text{C/m}^2]$ [**Eqs. (4.57)** and **(4.62)**].

The ***relative permittivity*** ε_r [nondimensional] incorporates the effects of polarization and is a property of the dielectric. It is the ratio between the permittivity of the material and that of free space.

Dielectric strength is the maximum electric field intensity, beyond which the dielectric breaks down—essentially conducts current. Each dielectric has a given, measurable dielectric strength. The dielectric strength in air is approximately 3,000 V/mm.

Interface conditions describe the relations between the electric field intensity on the two sides of the interface between different materials (see **Figures 4.30** and **4.31**). See **Table 4.3** for a summary.

$$E_{1t} = E_{2t} \text{ or } \frac{D_{1t}}{\varepsilon_1} = \frac{D_{2t}}{\varepsilon_2} \left[\frac{\text{V}}{\text{m}}\right] \text{ and } D_{1n} - D_{2n} = \rho_s \text{ or } \varepsilon_1 E_{1n} - \varepsilon_2 E_{2n} = \rho_s \ [\text{C/m}^2] \quad (4.67), (4.68), \text{and } (4.73)$$

The surface charge density ρ_s is nonzero on conductors but may also exist on dielectrics.

A more general representation of the interface conditions for the normal components of the field is

$$\hat{\mathbf{n}} \cdot (\mathbf{D}_1 - \mathbf{D}_2) = \rho_s \quad \text{or} \quad \hat{\mathbf{n}} \cdot (\varepsilon_1 \mathbf{E}_1 - \varepsilon_2 \mathbf{E}_2) = \rho_s \qquad [\text{C/m}^2] \tag{4.74}$$

Capacitance is the property of a body to store charge when connected to a potential. As the ratio between charge and potential it applies to conducting bodies (on which potential is constant):

$$C = \frac{Q}{V} \quad [\text{F}] \tag{4.80}$$

(1) The capacitance of an isolated body is the charge on that body divided by its potential and may be viewed as the capacitance between the body and infinity.
(2) The capacitance between two bodies is the magnitude of the charge on one body divided by the potential difference between them.
(3) Capacitance is calculated by either assuming a known charge or charge density on the bodies and calculating the resulting potential difference or assuming a known potential difference and calculating the charge.
(4) In either case capacitance is independent of charge or potential—it only depends on physical dimensions and on permittivity.

For parallel plate capacitors with plates of area A, separation d, and permittivity ε between the plates:

$$C = \frac{\varepsilon A}{d} \quad [\text{F}] \tag{4.87}$$

The capacitance of N capacitors connected in series or in parallel is calculated as follows:
Series

$$\frac{1}{C_{total}} = \sum_{i=1}^{N} \frac{1}{C_i} \quad \left[\frac{1}{\text{F}}\right] \tag{4.94}$$

Parallel

$$C_{total} = \sum_{i=1}^{N} C_i \quad [\text{F}] \tag{4.90}$$

Energy in the electrostatic field is defined from potential since the latter is potential energy per unit charge. Given a system of N point charges, the energy in the system is

$$W = \frac{1}{2}\sum_{i=1}^{N} Q_i V_i \quad [\text{J}] \tag{4.108}$$

where V_i is the potential at the location of charge Q_i due to all charges in the system except Q_i. When the capacitance of a system can be identified, the energy may be calculated as follows:

$$W = \frac{CV^2}{2} = \frac{QV}{2} = \frac{Q^2}{2C} \quad [\text{J}] \tag{4.110}$$

By defining differential line, surface and volume charge distributions, the definition of energy can be extended to charge distributions as [see **Eqs. (4.112)** through **(4.114)**]:

$$W = \frac{1}{2}\int_{\Omega'} \rho_{\Omega'} V d\Omega' \quad [\text{J}] \ (\Omega' \text{ stands for } l',\ s',\ \text{or } v')$$

The energy may be written in terms of the electric field intensity as well:

$$W = \frac{1}{2}\int_v \mathbf{D}\cdot\mathbf{E}dv = \frac{1}{2}\int_v \varepsilon E^2 dv = \frac{1}{2}\int_v \frac{D^2}{\varepsilon}dv \quad [\mathrm{J}] \tag{4.122}$$

Integration is over the space in which the electric field intensity is nonzero.

The integrands in **Eq. (4.122)** are the energy densities. These may be written as follows:

$$w = \frac{\mathbf{D}\cdot\mathbf{E}}{2} = \frac{\varepsilon E^2}{2} = \frac{D^2}{2\varepsilon} = \frac{\rho_v V}{2} \quad \left[\frac{\mathrm{J}}{\mathrm{m}^3}\right] \qquad (4.123), (4.124), \text{and } (4.125)$$

Since energy may be viewed as integrated force over distance it is also possible to calculate the force in terms of energy, provided energy is expressed as a function of space variables:

$$\mathbf{F} = -\nabla W_e \quad [\mathrm{N}]. \tag{4.130}$$

Problems

Postulates

4.1 Charge Density Required to Produce a Field. A layer of charge fills the space between $x = -a$ [m] and $x = +a$ [m]. The layer is charged with a nonuniform volume charge density $\rho(x)$ [C/m^3]. The electric field intensity everywhere inside the charge distribution is given by $\mathbf{E}(x) = \hat{\mathbf{x}}Kx^3$ [N/C], where K is a given constant. Assume the layer is in free space:

(a) Calculate the volume charge density $\rho(x)$.
(b) What can you say about the charge density for $x > 0$ and for $x < 0$?
(c) How does the charge density vary in the y and z directions?

4.2 Charge Density Required to Produce a Given Field. A volume charge density is distributed throughout the free space. The electric field intensity at a distance R from the origin is given as $\mathbf{E} = \hat{\mathbf{R}}kR^2$ [V/m] in free space, where k is a constant.

(a) Calculate the volume charge density that produces this field.
(b) Show that the given field is a valid electric field.

4.3 Electrostatic Fields. Which of the following vector fields are electrostatic fields? Justify your answer.

(a) $\mathbf{A} = \hat{\mathbf{x}}yz + \hat{\mathbf{y}}2x$.
(b) $\mathbf{A} = \hat{\mathbf{R}}bR^3 + \hat{\boldsymbol{\theta}}c$ where b and c are constants.
(c) $\mathbf{A} = \hat{\boldsymbol{\phi}}r$.
(d) $\mathbf{A} = \hat{\mathbf{R}}\left(10^{-8}/R^3\right)\cos\theta + \hat{\boldsymbol{\theta}}\left(10^{-8}/2R^3\right)\sin\theta$.

Gauss's Law: Calculation of Electric Field Intensity from Charge Distributions

4.4 Electric Field Due to Spherical Distribution of Charge. A spherical region $0 \le R \le 2$ mm contains a uniform volume charge density of 100 nC/m^3, whereas another region, 4 mm $\le R \le$ 6 mm, contains a uniform charge density of –200 nC/m^3. If the charge density is zero elsewhere, find the electric field intensity for (assume $\varepsilon = \varepsilon_0$):

(a) $R \le 2$ mm.
(b) 2 mm $\le R \le$ 4 mm.
(c) 4 mm $\le R \le$ 6 mm.
(d) $R \ge 6$ mm.

4.5 Application of Superposition. A dielectric sphere of radius $3a$ [m] has two spherical hollows, one of radius a [m] and one of radius $2a$ [m] as shown in **Figure 4.56**. A total charge Q [C] is distributed uniformly throughout the solid part of the volume. Assume the sphere is made of a material with permittivity ε_0 [F/m]. Calculate the electric field intensity at point P.

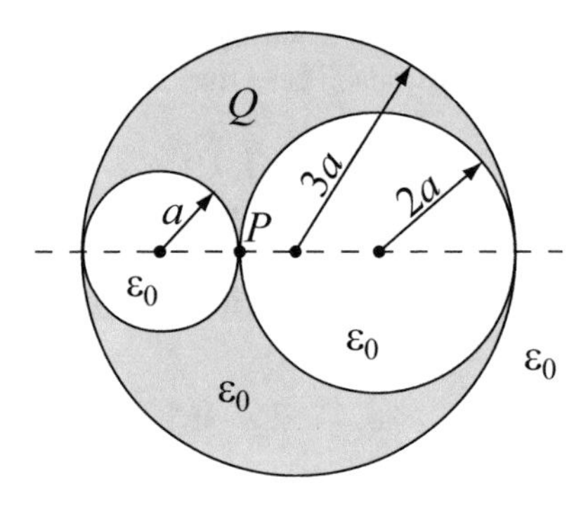

Figure 4.56

4.6 Fields of Point and Distributed Charges. A point charge Q [C] is placed at the origin of the system of coordinates. A concentric spherical surface with radius $R = a$ [m] with a uniform surface charge density ρ_s [C/m^2] in free space surrounds the point charge:

(a) Find **E** everywhere.
(b) Find the required surface charge density ρ_s in terms of Q such that the electric field intensity is zero for $R > a$.

4.7 Electric Field Due to Planar Charge Density. An infinite plane, charged with a uniform surface charge density ρ_0 [C/m^2], is immersed in a dielectric of permittivity ε [F/m]. Calculate the electric field intensity:

(a) Everywhere.
(b) Everywhere if the dielectric is removed.

4.8 Superposition of Fields of Planar Charges. A charge distribution consists of parallel, infinite plates, charged with equal but opposite surface charge densities ρ_s [C/m^2] separated a distance $2a$ [m] apart, two additional infinite plates separated a distance $2b$ [m] apart and charged in a similar fashion, and a uniformly distributed volume charge density ρ_v [C/m^3] between the two inner plates (**Figure 4.57**). Calculate the electric field intensity everywhere. Assume permittivity of free space everywhere.

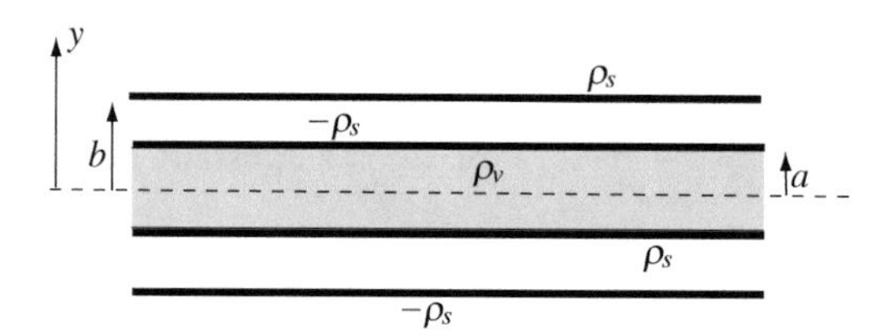

Figure 4.57

4.9 Superposition of Fields of Planar Charges. Two very large (infinite) charged planes intersect at 90° as shown in **Figure 4.58**. The four quadrants are each made of a different material as shown. The planes are charged uniformly with positive surface charge density $+\rho_s$ [C/m^2]:

(a) Find the electric field intensity everywhere.
(b) A positive point charge $+q$ [C] is placed at a general point in space $P(x,y)$, but not on the planes. Calculate the force (magnitude and direction) on this charge.
(c) Assuming $\varepsilon_3 > \varepsilon_2 > \varepsilon_1 > \varepsilon_0$, where is the force largest?

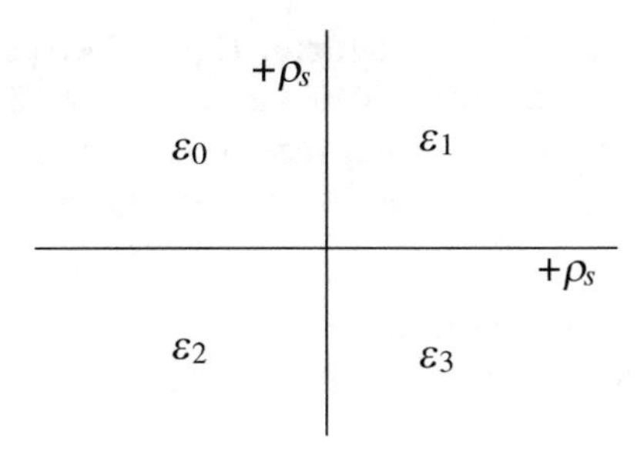

Figure 4.58

4.10 Electric Field Due to a Layer of Charge. A volume charge density is distributed inside an infinite section of the space as shown in **Figure 4.59**. The distribution is uniform in the y and z directions. The volume charge density depends on x as $\rho_v = \rho_0|x|$ [C/m^3]. Calculate the electric field intensity everywhere.

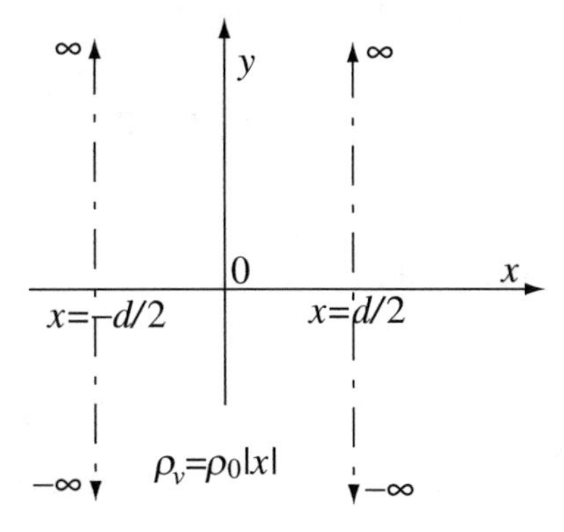

Figure 4.59

4.11 Superposition of Fields Due to Line and Surface Charge Densities. Given a very large sheet with uniform surface charge density ρ_s [C/m^2] located at $z = z_0$ and a long line with uniform line charge density ρ_l [C/m] placed at $z = 0$, on the x axis in free space:

(a) Find the electric field intensity everywhere.
(b) Find the electric flux density at point (0,0,1).
(c) Find the electric field intensity if the whole geometry is immersed in oil ($\varepsilon = 4\varepsilon_0$).

4.12 Application: Electric Field Intensity in a Coaxial Line. Two very long cylindrical shells are arranged as shown in **Figure 4.60**. The shells are uniformly charged with equal and opposite charge densities. The space between the two shells is filled with a dielectric with permittivity ε [F/m]:

(a) Find the electric field intensity everywhere.
(b) Sketch the electric field intensity everywhere.
(c) A line of charge is now introduced on the center axis to produce a zero field outside the outer shell. Calculate the required line charge density.

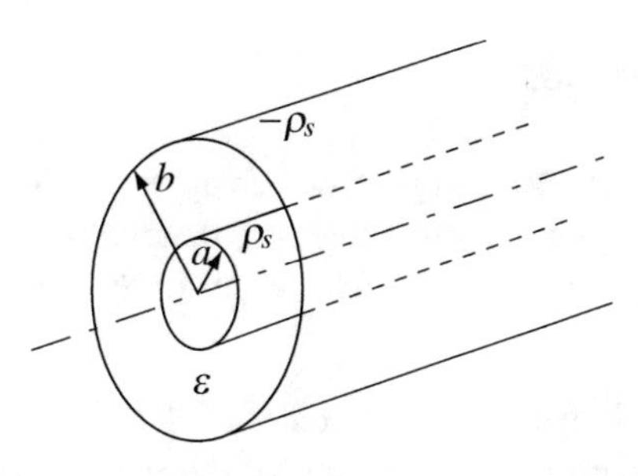

Figure 4.60

4.13 Application: Electric Field intensity in a Two-Conductor Cable. Two parallel cylinders, each 10 mm in radius, are placed in free space and each contains a uniformly distributed charge density of 0.5 μC per meter length. The axes of the cylinders are separated by 4 m. For orientation purposes, place the two conductors on the x–z plane with their axes along the z direction, with the z-axis midway between the conductors and the y axis pointing up. What is the magnitude of the electric field intensity:

(a) At a point P_1 midway between the two cylinders?
(b) At a point P_2 midway between the two cylinders and two meters below them?

Gauss's Law: Calculation of Equivalent Charge from the Electric Field Intensity

4.14 Charge Distribution Required to Produce a Field. A flat dielectric with permittivity ε [F/m] and thickness d [m] is charged with an unknown volume charge density (**Figure 4.61**). The dielectric may be assumed to be very large (essentially an infinite slab). The electric field intensity is known inside the dielectric and is given as

$$\mathbf{E} = \hat{\mathbf{y}}\frac{y^2}{2} \quad \left[\frac{\mathrm{N}}{\mathrm{C}}\right] \quad \text{for} \quad \frac{d}{2} > y > 0, \qquad \mathbf{E} = -\hat{\mathbf{y}}\frac{y^2}{2} \quad \left[\frac{\mathrm{N}}{\mathrm{C}}\right] \quad \text{for} \quad -\frac{d}{2} < y < 0$$

(a) What is the volume charge density inside the dielectric?
(b) What is the electric field intensity outside the dielectric?

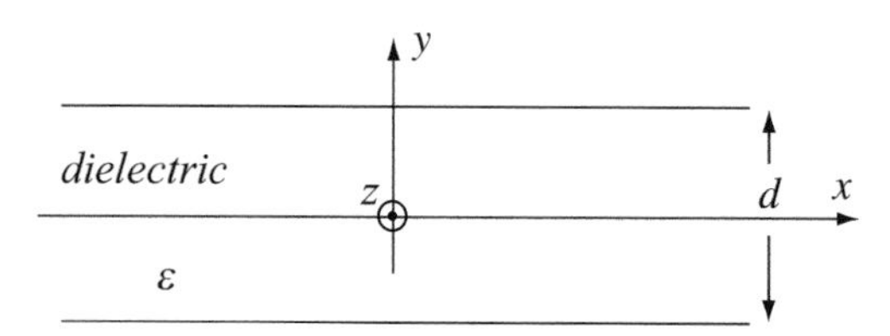

Figure 4.61

4.15 Volume Charge Density Needed to Produce a Field. The electric field intensity is given inside a sphere of radius $R \leq b$ [m] as $\mathbf{E} = \hat{\mathbf{R}}4R^2$ [N/C]. The sphere has permittivity ε [F/m]. Find:

(a) The volume charge density ρ_v at $R = b/4$ [m].
(b) The total electric flux leaving the spherical surface of radius $R = b/2$ [m].

4.16 Maximum Allowable Charge Density on Power Lines. A DC power transmission line is made of two conductors, each 20 mm in diameter and separated a distance 6 m apart. The maximum electric field intensity allowed anywhere in space is 3×10^6 N/C. If one line is positively charged and one is negatively charged, calculate:

(a) The maximum surface charge density allowable on the conductors.
(b) Suppose the conductors are covered with a 3 mm thick rubber insulation which has a relative permittivity of 3.0. The maximum electric field intensity allowed in rubber is 2.5×10^7 N/C. What is now the maximum surface charge density allowable on the conductors?

Potential: Point and Distributed Charges

4.17 Potential Due to a System of Point Charges. Eight equal charges $q = 3$ nC are placed at the vertices of a cube in a vacuum. Place the cube in a Cartesian system of coordinates so that the faces are parallel to the x–y, x–z, and y–z planes, with one vertex at the origin. The cube is in the positive octant of the system of coordinates. The side of the cube is $a = 0.5$ m. Find:

(a) The potential and the electric field intensity at the center of the cube.
(b) The potential and electric field intensity at the center of the face parallel to the x–y plane (at $x = a/2$, $y = a/2$, $z = a$).

4.18 Potential Due to Point Charge and Spherical Charge Distribution. A point charge Q [C] is surrounded by a uniform spherical charge distribution of radius a [m], having a uniform volume charge density ρ_v [C/m^3]:

(a) Calculate the potential everywhere.
(b) Plot the potential with respect to position.

4.19 Application: Potential and Field Between Charged Plates. The two plates in **Figure 4.62** are very large (infinite). The dielectric between the plates is air:

(a) Calculate the electric field intensity between and outside the plates.
(b) If the potential on the left plate is zero, calculate and plot the potential everywhere.
(c) Assume that the potential difference between the plates is increased by V_0. This potential is connected with its positive side to the right plate with the zero potential reference kept as in **(b)**. How does this affect the potential outside and between the plates?

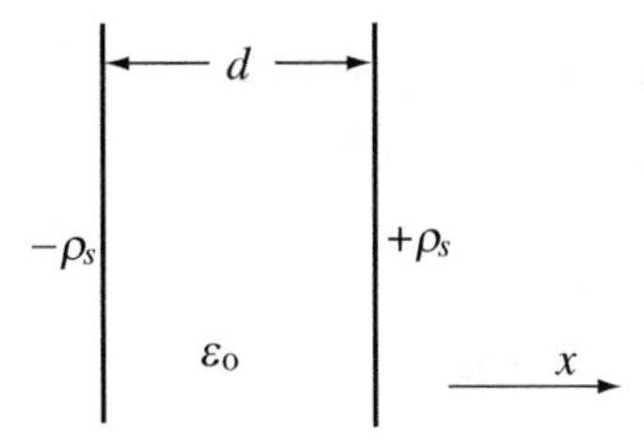

Figure 4.62

4.20 Application: Charge Density in Coaxial Cable Connected to a Battery. A very long, solid cylindrical conductor of radius b [m] is placed at the center of an equally long, thin, cylindrical shell of radius a [m] as shown in **Figure 4.63** to form a coaxial line and a voltage source is connected as shown. Calculate:

(a) Charge density per unit area on the conductor.
(b) Charge density per unit area on the outer shell.

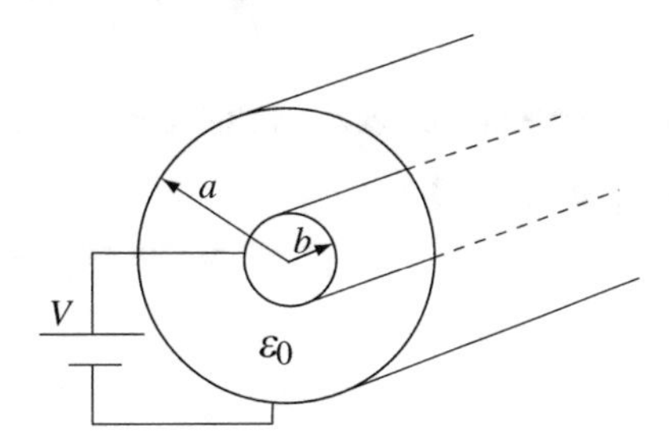

Figure 4.63

4.21 Application: Potential Due to a Charged Disk. A thin disk of radius a [m] carries a uniform surface charge density ρ_0 [C/m^2]. Find the potential at a distance d [m], on the axis of the disk (directly above its center) in free space.

4.22 Field and Potential Due to Charged Dielectric. Two very large, conducting, parallel plates separated a distance $2a$ [m] contain a uniform volume charge density ρ_v [C/m^3] between them. The plates are very thin and are both at zero potential. Permittivity between the plates is ε [F/m], outside the plates ε_0 [F/m]. Calculate:

(a) The electric field intensity between and outside the plates. Sketch.
(b) The potential everywhere between the two and outside the plates. Sketch.

Electric Field from Potential

4.23 The Potential and Field of a System of Point Charges $2N + 1$ point charges are distributed on the x-axis as shown in **Figure 4.64**. The distance between each two point charges is a [m] and the charges are in free space. Calculate:

(a) The potential at a general point $P(x,y)$.
(b) The electric field intensity at $P(x,y)$.

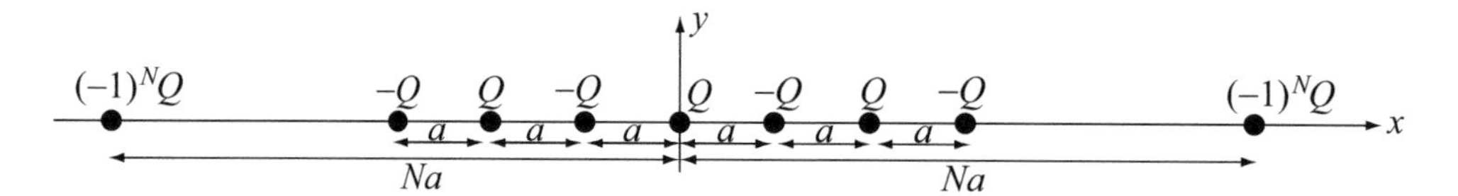

Figure 4.64

4.24 Charge Density and Electric Field Intensity Due to Potential. The electric potential inside a dielectric sphere of radius b [m] and permittivity ε [F/m] is given as $V = aR^2$ [V] where a is a constant. The sphere is located in free space. Find:

(a) The volume charge density ρ_v inside the dielectric sphere.
(b) The electric field intensity **E**, inside and outside the sphere.

4.25 Electric Field Intensity Due to Potential. A potential is given at a general point (x,y,z) in free space as $V(x,y,z) = 0.5\,[1/(x-2) + 1/(y-1) - 2/(z+2)]$ [V].

(a) Calculate the electric field intensity at any point in space.
(b) Show that the result in **(a)** is a valid electric field intensity.
(c) Calculate the volume charge density that produces this field.

4.26 The Quadrupole. The potential of a system of point charges is known at a general point (R,θ,ϕ) at a large distance from the charges as

$$V(R,\theta) = C\frac{3\cos^2\theta - 1}{R^3} \quad [\text{V}]$$

where R is the distance from the origin, θ the angle between the vector **R** and the reference z axis and C is a constant.

(a) Calculate the electric field intensity at that point.
(b) Explain why it is not possible to calculate the point charges that produced this field.

Conductors in the Electric Field

4.27 Conductor in a Uniform Electric Field. Two infinite plates are separated a distance $2b$ [m] in free space. A conducting slab of thickness $2c$ [m] ($c < b$) is inserted between the plates as shown in **Figure 4.65**. The plates are connected to a potential V [V] with the lower plate at zero (reference) potential:

(a) Calculate the potential everywhere between the two plates before inserting the conductor.
(b) Calculate the potential everywhere after inserting the conductor.
(c) Does the vertical position of the conductor between the plates affect the electric field intensity? Explain.

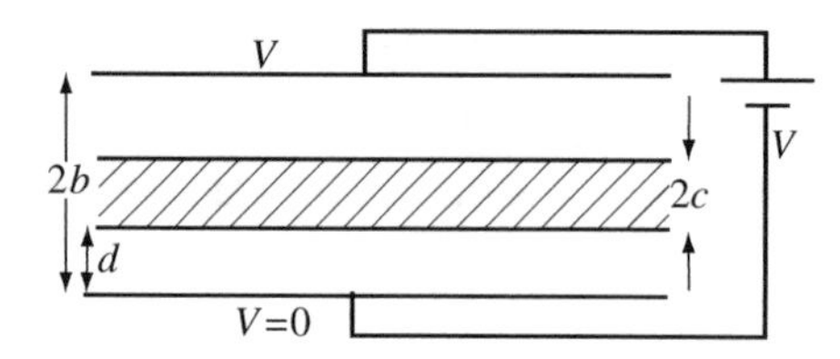

Figure 4.65

4.28 Potential in a Conductor. A point charge Q [C] is placed at the center of a conducting sphere of radius R [m], in a small hollow cavity so that it does not touch the conductor. Calculate the potential at a point $r = R/2$ [m].

4.29 Layered Structures. The device in **Figure 4.66** is made as follows: A spherical conductor of radius a [m] is surrounded by a dielectric shell of outer radius b [m] and permittivity ε_1 [F/m]. This shell is surrounded by another spherical shell of outer radius c [m] and permittivity ε_2 [F/m]. On top of this, there is a conducting shell with outer radius d [m]. A potential V_0 [V] is connected between the outer and inner conductors as shown in **Figure 4.66.** Calculate the charge density on the inner and outer conductors. Where is this charge density distributed?

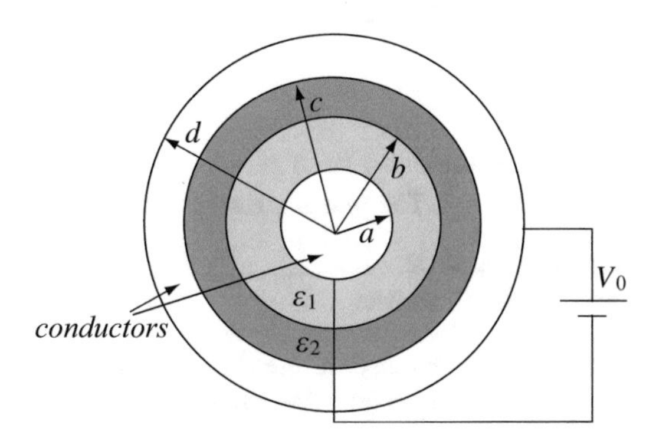

Figure 4.66

4.30 Electric Field and Potential in Spherical Shells and Conductors. Two geometries are given in **Figures 4.67a** and **4.67b**. In **Figure 4.67a**, the two shells are spherical, with a surface charge density ρ_{sa} [C/m^2] on the inner shell and $-\rho_{sb}$ [C/m^2] on the outer shell. The shells and the point charge at the center are in free space. In **Figure 4.67b**, the space between $R = a$ [m] and $R = b$ [m] is filled by a conductor. In both cases a negative point charge $-Q$ [C] is located at the origin:

(a) Calculate the electric field intensity and potential everywhere in space for both geometries.
(b) Show the difference between the two geometries by two simple plots, one for the electric field intensity and one for the potential, for each geometry.
(c) What is the relation between charge densities ρ_{sa} and ρ_{sb} that will make the solution for the two geometries identical?

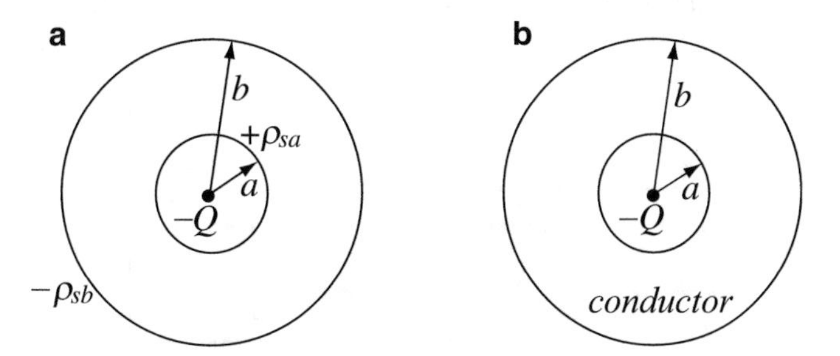

Figure 4.67

Polarization

4.31 Polarization in Dielectrics. Two long, conducting concentric cylindrical shells are separated by two dielectric layers of permittivity ε_1 [F/m] and ε_2 [F/m] as shown in **Figure 4.68**. A line of charge with line charge density ρ_l [C/m] is placed at the center of the inner shell. Calculate the magnitude and direction of the polarization vector inside the two dielectrics.

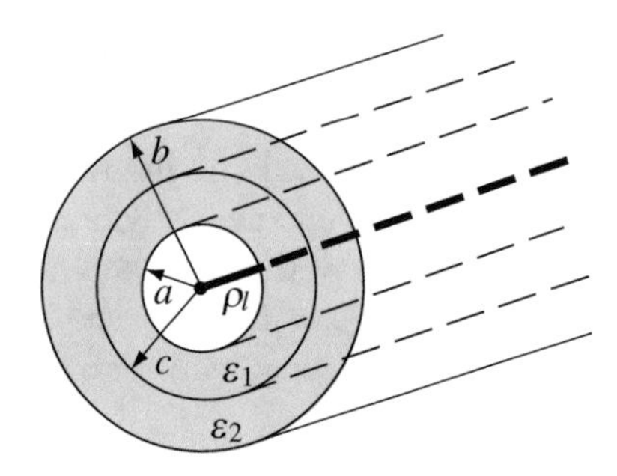

Figure 4.68

4.32 Polarization in Dielectrics. The space between two large parallel plates, separated a distance $d = 0.1$ mm, is filled with a dielectric with relative permittivity of 4. The plates are connected to a 12 V battery:

(a) What is the electric field intensity in the dielectric?
(b) Calculate the polarization vector in the dielectric.

4.33 Electric Flux Density in Polarized Medium. A uniform line charge density of $\rho_l = 4$ μC/m is located along the z axis in free space. A planar surface charge density $\rho_s = 20$ μC/m^2 is placed at $x = 1$, parallel to the y–z plane:

(a) Find the electric flux density at a general point $P(x,y,z)$.
(b) Find the electric flux density at point $P(1,0,0)$.
(c) What is the polarization vector at a general point $P(x,y,z)$, if space has a dielectric constant $\varepsilon = 4\varepsilon_0$?

Dielectric Strength

4.34 Application: Multilayer Devices. Two parallel plates are connected to a constant electrostatic potential as shown in **Figure 4.69**. Half the space between the plates is filled with air, one-quarter with a conductor and the fourth quarter with a dielectric with relative permittivity of 80:

(a) What is the maximum electric field intensity in the system? Where does this occur?
(b) If the breakdown potential in air is 3,000 V/mm and in the dielectric it is 20,000 V/mm, what is the maximum voltage V allowed across the device before breakdown occurs either in air or in the dielectric?

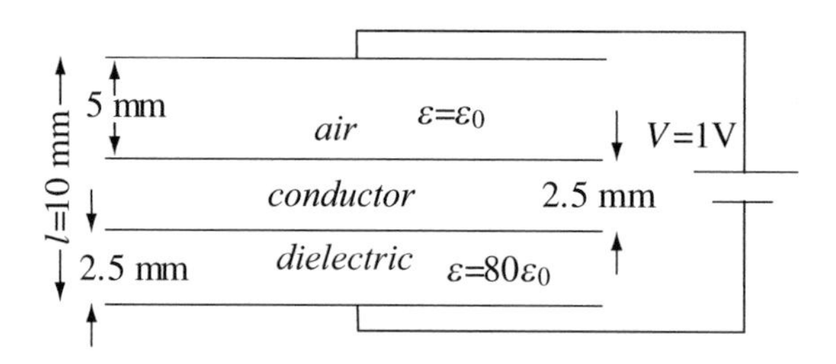

Figure 4.69

4.35 Application: Maximum Charge Density and Maximum Potential on a Sphere. A conducting sphere of radius a [m] is located in air. If the electric field intensity in air cannot exceed 3,000 V/mm:

(a) What is the maximum surface charge density the sphere can accumulate?
(b) What is the maximum potential that can be connected to the sphere without causing breakdown?
(c) What is the maximum possible surface charge density on the surface of planet Earth if its radius is 6370 km and the charge may be assumed to be uniformly distributed on its surface.

4.36 Breakdown in Power Lines. A two-conductor power transmission system operates at a voltage V. The radius of each line is 20 mm and the lines are separated a distance of 1 m (between centers):

(a) Calculate the maximum voltage difference between the two conductors, V, allowed without causing breakdown. Assume the two lines are charged with equal but opposite sign charge per unit length.
(b) If the two lines have an insulation layer made of rubber of thickness 2 mm which has a breakdown voltage much higher than the breakdown voltage in air, what is the estimated maximum voltage allowed now assuming the permittivity of the insulation is close to that of air?

Interface Conditions

4.37 Interface Conditions for D. Two dielectric regions are defined by the normal $\hat{\mathbf{n}} = \hat{\mathbf{x}}$ to the interface pointing into region 1. The interface is charge-free. If $\varepsilon_r = 1$, and $\mathbf{E} = \hat{\mathbf{x}}5 + \hat{\mathbf{y}}3$ [V/m] in region 1, find the expression for **D** in region 2 where $\varepsilon_r = 2$.

4.38 Interface Conditions on Dielectric Sphere. The electric field intensity inside a dielectric sphere of radius $a = 100$ mm and relative permittivity $\varepsilon_r = 12$ is given as $\mathbf{E} = \hat{\mathbf{R}}10^4R^2$ [V/m]. The electric field is generated by an appropriate volume charge density in the dielectric.

(a) Calculate the electric field intensity outside the sphere at its surface.
(b) Calculate the surface charge density required on the surface of the sphere that will cancel the electric field intensity outside the sphere.

4.39 Interface Conditions in Layered Structures. Two flat dielectrics of thickness d [m], both having dielectric constant ε_1 [F/m], are placed on the two sides of a conducting sheet of thickness t [m]. The electric field intensity in free space to the left of the materials is given and is normal to the surface of the dielectric, as shown in **Figure 4.70**:

(a) Calculate the electric field intensity everywhere.
(b) What is the potential difference between the left and right surfaces (the potential difference on the three layers)?

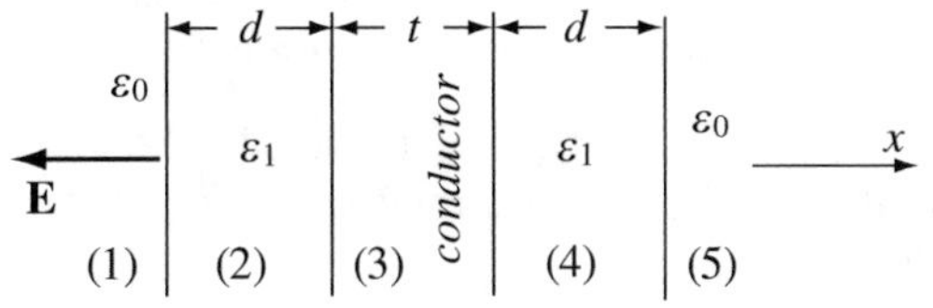

d t d
ε0 ε1 conductor ε1 ε0
E x
(1) (2) (3) (4) (5)

Figure 4.70

4.40 Interface Conditions with Charge Density on the Interface. The electric field intensity in air, in fair weather, is 100 V/m. Suppose the electric field intensity at the surface of the Earth in a flat, desert area points downward and there is a charge density on the surface of -5×10^{-10} C/m^2. Assume air has properties of free space and the Earth is a dielectric with relative permittivity of 2:

(a) Calculate the electric field intensity immediately below the surface of the Earth.
(b) What is the electric field intensity below the surface if the charge density on the surface is zero?

4.41 Interface Conditions in Layered Dielectrics. Two large planar layers of dielectric materials are bonded together. An electric field intensity in free space exists to the left of the layers. The layers are each of thickness d [m] (see **Figure 4.71**):

(a) Calculate the electric field intensity everywhere.
(b) Calculate the potential difference between two points: one at $x = +a$ [m], the second at $x = -a$ [m].
(c) Show that the angle of the electric field intensity in the space to the right of the two layers is $\theta°$.

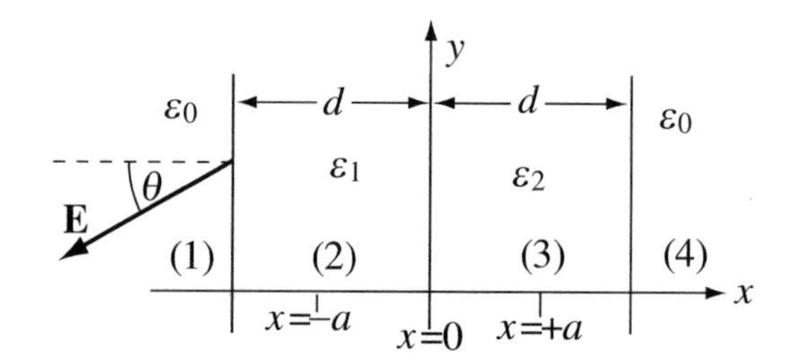

Figure 4.71

4.42 Interface Conditions for the Polarization Vector. Given the interface conditions for the electric field intensity and the electric flux density in **Table 4.3**:

(a) Find the interface conditions required for the polarization vector at the interface between two dielectrics of permittivity ε_1 [F/m] and ε_2 [F/m]. First, assume no free charges on the interface and then allow for existence of free charges on the interface.

(b) Determine the ratio of the normal components and the ratio of the tangential components of the polarization vector at the interface between two perfect dielectrics with permittivities ε_1 and ε_2 in terms of the relative permittivity of the two perfect dielectrics. Assume there are no charges at the interfaces.

Capacitance

4.43 Application: Parallel Plate Capacitor. A parallel plate capacitor of unit area and separation d [m] is filled with thickness d_1 [m] of a material with $\varepsilon = 20\varepsilon_0$ [F/m] and thickness d_2 [m] with a material with $\varepsilon = 3\varepsilon_0$ [F/m]. If $d_1 + d_2 = d$ [m], what is the capacitance of the device?

4.44 Application: Capacitance per Unit Length of Coaxial Cable. Calculate the capacitance per unit length of the coaxial cable in **Figure 4.63**.

4.45 Application: Spherical Capacitor. Two concentric, conducting spherical shells with radii $a = 49$ mm and $b = 50$ mm are separated by a dielectric with relative permittivity $\varepsilon_r = 3.3$. Calculate the capacitance of this device.

4.46 Layered Spherical Capacitor. Suppose it is required to double the capacitance of the spherical capacitor of **Problem 4.45** by introducing a layer of dielectric material with relative permittivity $\varepsilon_r = 15$, as shown in **Figure 4.72**. Calculate the thickness of the layer, x:

(a) If the layer is placed on the inner shell (**Figure 4.72a**).

(b) If it is placed below the outer shell (**Figure 4.72b**).

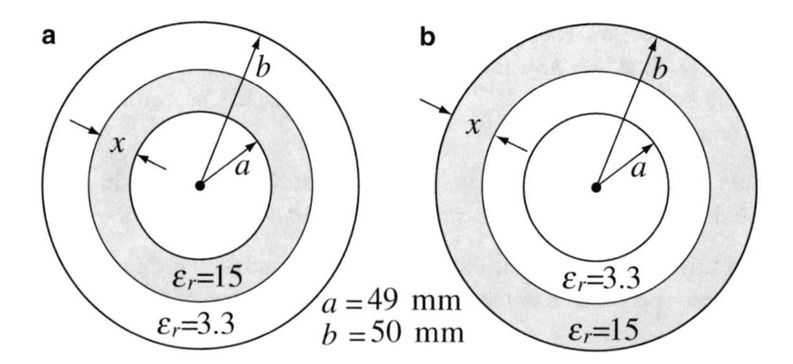

Figure 4.72

4.47 Layered Spherical Capacitor. Calculate the capacitance of the device in **Figure 4.66**.

4.48 Application: Capacitance per Unit Length of Coaxial Cable. A coaxial cable is made as shown in **Figure 4.73**. The inner conductor has radius a [m], the outer conductor has radius b [m]. To separate the two conductors 4 wedges made of a dielectric with permittivity $4\varepsilon_0$ [F/m] run the length of the cable as shown. Each wedge occupies 30 degrees of the cross section (i.e. 1/12 of the volume between the two conductors). Calculate the capacitance per unit length of the cable.

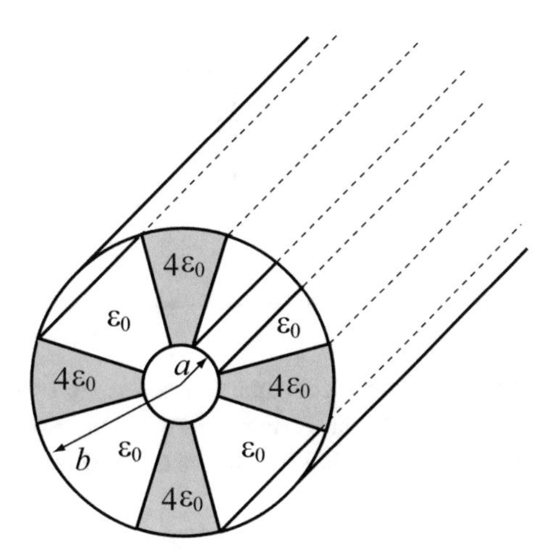

Figure 4.73

Energy in the Electric Field

4.49 Potential Energy of a System of Point Charges. Three point charges are arranged as shown in **Figure 4.74**. Calculate the potential energy in the system.

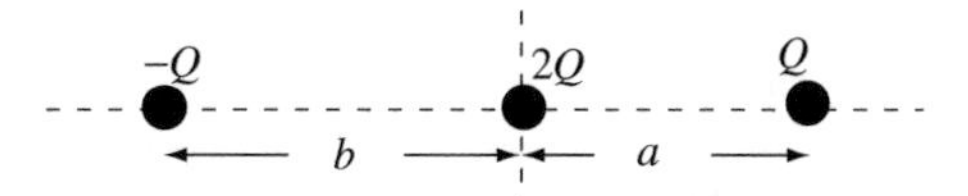

Figure 4.74

4.50 Potential Energy of Distributed Charges. A charged sphere of radius a [m] is charged with a nonuniform volume charge density $\rho_v = \rho_0 R(R - a)$ [C/m^3], is located in free space, and has permittivity ε [F/m]. Calculate:

(a) The energy stored in the volume of the sphere.
(b) The energy stored in the space surrounding the sphere.
(c) The total energy associated with the sphere.

4.51 Energy Stored in a Capacitor. Assume a solid conducting sphere of radius a [m] is connected to a potential V [V] with respect to infinity. Calculate the total energy stored in the electric field produced by the sphere. Assume the charge on the surface of the sphere is uniformly distributed.

4.52 Application: Energy and Charge Stored in Parallel Plate Capacitor. A parallel plate capacitor is connected to a voltage source as shown in **Figure 4.75a**. The dielectric is free space. A flat perfect conductor of thickness a [m] is inserted as shown in **Figure 4.75b**:

(a) What is the change in energy per unit area of the capacitor due to insertion of the conductor (i.e., what is the change in energy for a 1 m^2 section of the capacitor)?

(b) If the capacitor has surface area of 100 mm × 100 mm, $d = 1$ mm, $a = 0.2$ mm and is connected to a 12 V source, calculate the total change in charge on the surface of the capacitor as the conductor is inserted assuming the voltage remains unchanged.

(c) Repeat **(a)** and **(b)** if the conductor is replaced with a dielectric of the same dimensions and a relative permittivity of 4.

(d) Describe how this device can be used to detect the conductor (for example, it may be used as a proximity sensor).

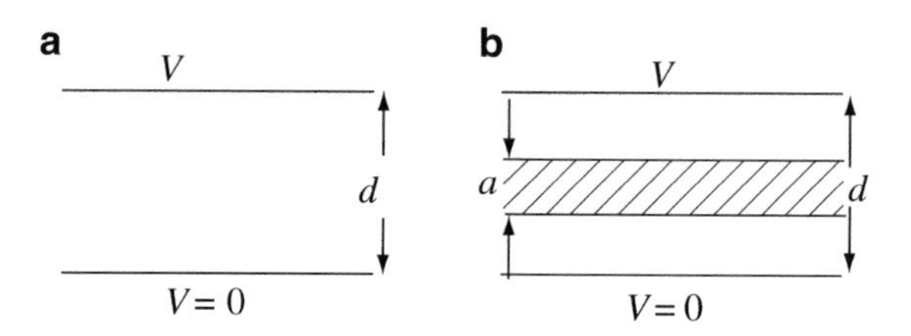

Figure 4.75

4.53 Energy Stored in a Spherical Capacitor. A spherical capacitor is formed by two concentric, conducting shells with a dielectric between them (**Figure 4.76**). What is the change in stored energy if the dielectric is removed? Assume a known total charge + Q [C] on the inner shell and $-Q$ [C] on the outer shell, and these charges remain constant as the dielectric is removed.

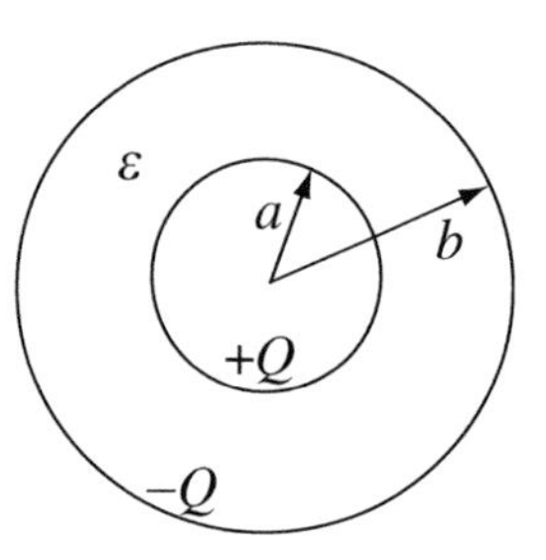

Figure 4.76

4.54 Energy Density. A point charge of magnitude q [C] is located at the center of two concentric, thin spherical shells as shown in **Figure 4.77**. The inner shell carries a surface charge density ρ_0 [C/m^2] and the outer shell a surface charge density $-\rho_0$ [C/m^2]:

(a) Calculate the energy density everywhere in space. Plot the energy density.

(b) How much energy is stored in the volume between the two shells?

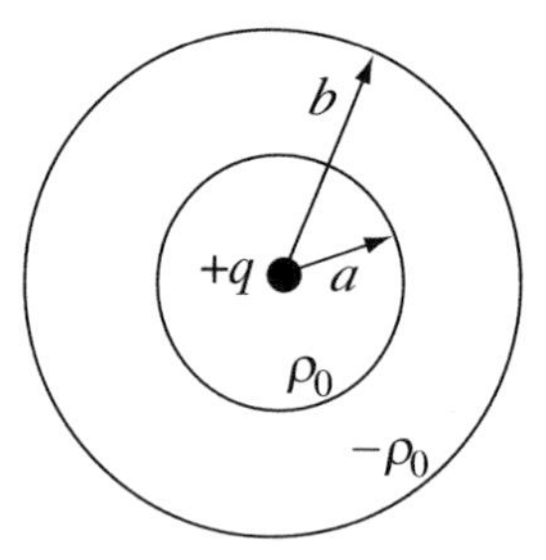

Figure 4.77

4.55 Energy per Unit Length. A line of charge with line charge density ρ_l [C/m] is surrounded by a dielectric layer (filling the space between a and b) as shown in **Figure 4.78**. The permittivity of the layer is ε [F/m]. Calculate the energy per unit length stored in the dielectric layer.

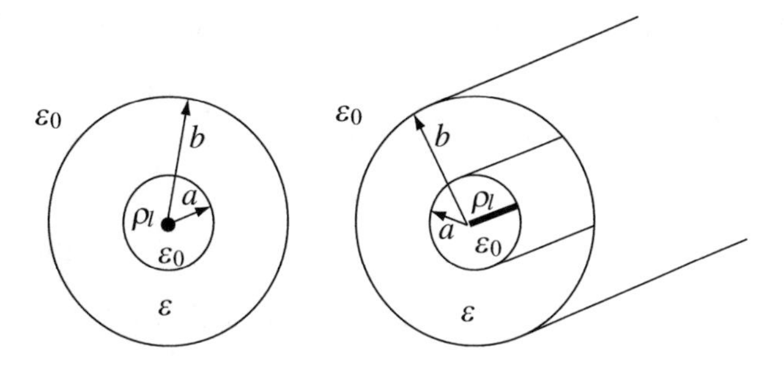

Figure 4.78

Forces

4.56 Application: Pressure in the Electrostatic Field. A parallel plate capacitor with very large plates, separated a distance d [m] apart, is given. The space between the plates is filled with a dielectric with relative permittivity of 2. A potential V [V] is connected across the plates. Assume the distance between the plates is small. Calculate the pressure the plates exert on the material between the plates.

4.57 Application: Force in the Electrostatic Field. A square conducting plate, b [m] on the side weighs k [kg], is located over a very large conducting plate, with a dielectric of thickness d [m] and permittivity ε [F/m] between the plates. The plate is connected to a potential V [V] as shown in **Figure 4.79** and a string connects it to a weight P [kg]. Neglecting fringing of the electric field, calculate the potential required so that the plate does not move (the weight is balanced by the electrostatic force). The acceleration of gravity is g [m/s^2].

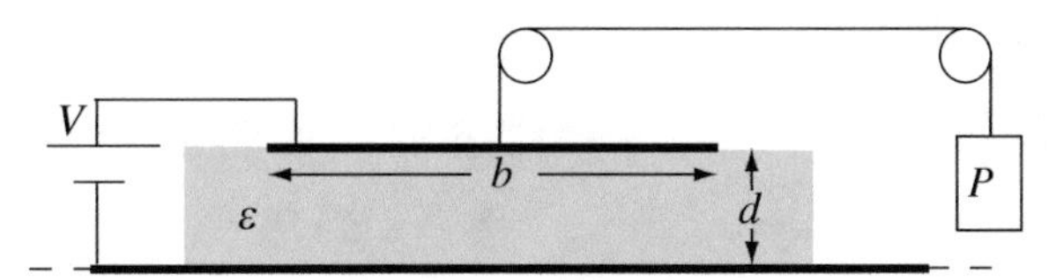

Figure 4.79

4.58 Application: Force in the Electrostatic Field. Two plates, b [m] wide, are separated a distance d [m] apart and overlap a distance c [m] as shown in **Figure 4.80**. Between the plates there is a dielectric with relative permittivity ε_r. Assume the plates are very close to each other and you can neglect any fringing. A potential V [V] is connected across the plates. Calculate the horizontal force acting on the plates. What is the direction of this force?

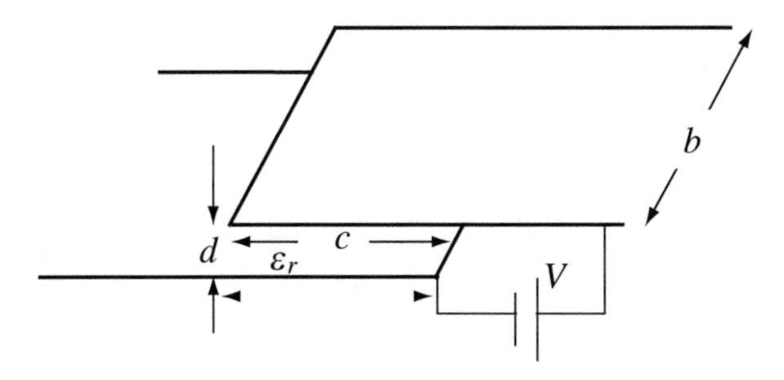

Figure 4.80

4.59 Application: Force on Dielectric in a Capacitor. A parallel plate capacitor is given as shown in **Figure 4.81**. The dielectric of permittivity ε [F/m] occupies a section of length x [m] and width b [m] between the plates. The rest of the space between the plates is air with a permittivity ε_0 [F/m]. With the dimensions given, calculate the force on the dielectric if a potential difference V_0 [V] exists between the plates. **Hint:** The force tends to pull the dielectric into the capacitor. If the dielectric were free to move, it would fill the space between the plates.

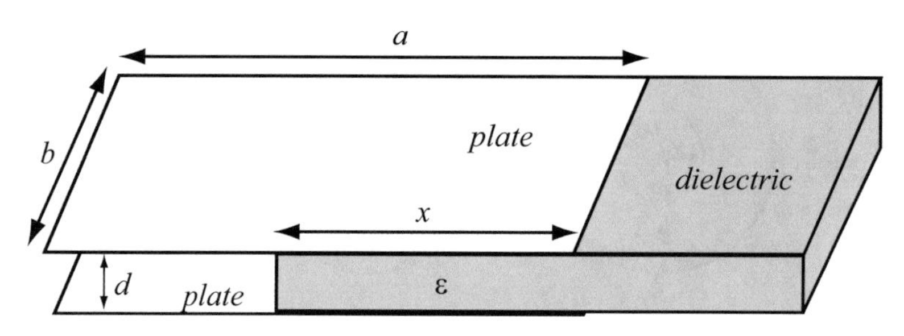

Figure 4.81

4.60 Inflation of a Balloon by Electrostatic Forces. In the early time of space communication, inflatable balloons were used as reflectors to evaluate propagation properties in space. But the use of gas to inflate large space structures is problematic because of leaks, diffusion across the structure skin, and finite supply. A possible alternative is to use charge and the forces it can generate to keep the structure "inflated" without the need for gas. Since charge can be generated as needed using harvested energy through solar cells, the structure can maintain its shape indefinitely. Consider a balloon of radius a [m], charged with a surface charge density ρ_0 [C/m^2] in free space:

(a) Calculate the pressure inside the balloon.
(b) What is the required charge density for a balloon of radius 10 m if the pressure required is 100 pascals (Pa). (1 pascal $=$ 1 N/m^2)?
(c) If the maximum electric field intensity that can be generated is 10^9 V/m, what is the maximum pressure achievable using the dimensions in **(b)**?

Boundary Value Problems: Analytic Methods of Solution

5

If there is no other use discover'd of Electricity, this, however, is something considerable, that it may help to make a vain man humble.

—Benjamin Franklin (1706–1790)
statesman, scientist, inventor

5.1 Introduction

The relations and methods introduced in **Chapters 3** and **4** dealt primarily with point and distributed charges and the electric fields they produce. If the charges were known, the electric field intensity and potential could be determined. However, many practical situations exist in which the charges are either unknown or are distributed in a complex fashion. The use of the simple formulas for the calculation of fields in these geometries is not always possible. In still other geometries, we have no knowledge of charges but only of fields and potentials. For example, in an overhead transmission line, we may know the potential but not the charge on the line. How can we then calculate the electric field intensity everywhere in space? Similarly, when designing an electric instrument, such as an electrostatic filter, the engineer is not going to calculate "how much charge must be present on the electrode." This information, while important in itself, is not normally a design parameter simply because we do not usually use "charge supply sources" and we are ill equipped to measure charge or charge density. The more common problem in design would be to calculate the required potential on the electrodes of the device to produce the needed effect. This information is important because with it, the power supply required can be designed. Although the principles in **Chapters 3** and **4** and the formulas developed for calculation of fields, potentials, and energy are applicable to these types of problems as well, the difficulty is in applying them.

In this chapter, we look at yet another technique for finding the electric fields and potential differences in a given geometry. The method is that of solution to boundary value problems; that is, we ask ourselves: Given the boundary conditions of an electrostatic problem, what is the potential (as well as electric field intensity and charge distributions) throughout the geometry? In the case of an overhead transmission line, the potentials on the line and perhaps at ground level are known. These are boundary conditions of the geometry. Based on these conditions and the postulates we derived earlier, can we calculate the potential everywhere? The answer is yes, but, as we will see, the actual calculation can be rather involved, depending on the geometry and configuration of potentials. The approach here is straightforward and includes four steps:

(1) Rewrite the electrostatic field postulates in terms of the potential as a second-order partial differential equation.
(2) Find a general solution to this partial differential equation using any suitable method of solution.
(3) Satisfy the general solution in Step (2) for the particular boundary conditions that define the problem to obtain a particular solution.
(4) Calculate any other quantity, such as the electric field intensity or charge density, from the potential.

It is important to mention here that some seemingly simple problems cannot be solved using the methods in the previous chapters and in this, not because the solution does not exist but because the general solutions cannot be satisfied everywhere in the solution domain and on the boundaries of the problem. These types of problems are deferred to the following chapter, in which we discuss numerical methods of solution.

N. Ida, *Engineering Electromagnetics*, https://doi.org/10.1007/978-3-030-15557-5_5

5.2 Poisson's[1] Equation for the Electrostatic Field

The required postulates for the electrostatic field in linear, homogeneous, isotropic dielectric materials are

$$\nabla \times \mathbf{E} = 0, \qquad \nabla \cdot \mathbf{E} = \frac{\rho_v}{\varepsilon} \tag{5.1}$$

We recall the fact that the curl of **E** is zero (the electrostatic field is a conservative field) allowed us to define the electric field intensity in terms of the electric scalar potential as

$$\mathbf{E} = -\nabla V \qquad [\mathrm{V/m}] \tag{5.2}$$

This is now substituted into the second postulate in **Eq. (5.1)**:

$$\nabla \cdot \mathbf{E} = -\nabla \cdot (\nabla V) = \frac{\rho_v}{\varepsilon} \tag{5.3}$$

We recall from **Chapter 2** [**Eq. (2.130)**] that $\nabla \cdot (\nabla V) = \nabla^2 V$. Substitution of this into **Eq. (5.3)** gives

$$\nabla^2 V = -\frac{\rho_v}{\varepsilon} \tag{5.4}$$

It is worth looking at this equation in expanded form. For example, in Cartesian coordinates,

$$\boxed{\frac{\partial^2 V}{\partial x^2} + \frac{\partial^2 V}{\partial y^2} + \frac{\partial^2 V}{\partial z^2} = -\frac{\rho_v}{\varepsilon}} \tag{5.5}$$

This is called Poisson's equation. It is a linear, scalar partial differential equation and, most importantly, it is equivalent to solving for the electric field intensity **E** using the basic postulates. Any technique that solves this equation is an appropriate method for calculation of electrostatic fields. Since the left-hand side of **Eq. (5.4)** is the divergence of the gradient of V [i.e., $\nabla \cdot (\nabla V)$], we can immediately obtain Poisson's equation in cylindrical and spherical coordinates. In fact, we have already written the left-hand side of **Eq. (5.4)** in the three systems of coordinates in **Eqs. (2.130), (2.132),** and **(2.133)**. Poisson's equation in cylindrical coordinates is

$$\boxed{\frac{1}{r}\frac{\partial}{\partial r}\left(r\frac{\partial V}{\partial r}\right) + \frac{1}{r^2}\frac{\partial^2 V}{\partial \phi^2} + \frac{\partial^2 V}{\partial z^2} = -\frac{\rho_v}{\varepsilon}} \tag{5.6}$$

and in spherical coordinates

$$\boxed{\frac{1}{R^2}\frac{\partial}{\partial R}\left(R^2\frac{\partial V}{\partial R}\right) + \frac{1}{R^2 \sin\theta}\frac{\partial}{\partial \theta}\left(\sin\theta\frac{\partial V}{\partial \theta}\right) + \frac{1}{R^2 \sin^2\theta}\frac{\partial^2 V}{\partial \phi^2} = -\frac{\rho_v}{\varepsilon}} \tag{5.7}$$

[1] Siméon Denis Poisson (1781–1840) was an applied mathematician. Most of his work was in application of mathematics to physics and, in particular, to electrostatics and magnetism. A contemporary of Lagrange, Laplace, and Fourier, he was also an astronomer and is known for his work on probability theory and definite integrals, in addition to electromagnetics. He also contributed to the understanding of Fourier series, the theory of heat, and mechanics.

5.3 Laplace's[2] Equation for the Electrostatic Field

Under charge-free conditions, the right-hand side in **Eq. (5.5)** is zero. In Cartesian coordinates,

$$\boxed{\frac{\partial^2 V}{\partial x^2}+\frac{\partial^2 V}{\partial y^2}+\frac{\partial^2 V}{\partial z^2}=0} \tag{5.8}$$

This is Laplace's equation, and it is obvious that whenever there are no charge densities in the domain in which a solution is sought, we will use Laplace's equation. Laplace's equation in cylindrical and spherical coordinates is as in **Eqs. (5.6)** and **(5.7)** but with zero on the right-hand side. Laplace's equation applies to a number of other physical problems in addition to electrostatic fields. These include the gravitational field in the absence of mass, the pressure field in the absence of sources, and the temperature field in the absence of heat sources in the solution domain. We will meet both Laplace's and Poisson's equations in the following chapters.

To summarize, the main reasons to try to solve Laplace's or Poisson's equations rather than using the field equations in **Eq. (5.1)** are three:

(1) Both Laplace's and Poisson's equations as used here are scalar equations.
(2) Design parameters are often in terms of potentials on structures rather than charges and fields.
(3) We can take advantage of mathematical techniques that have been developed for solution of second-order partial differential equations.

5.4 Solution Methods

As a general rule, methods of solution for second-order partial differential equations are appropriate here. The first method that comes to mind is the method of separation of variables. However, for Poisson's equation, it is much more difficult to obtain a solution using separation of variables. The method we use is essentially a comparative method: that is, we do not actually solve the partial differential equation but rather obtain a solution by superposition of known solutions, each of which satisfies Poisson's equation. The method is known as the method of images. Both the separation of variables method and the method of images are quite powerful although, as we shall see shortly, limited in scope. In this chapter, we discuss the following three methods:

(1) Direct integration for Laplace's and Poisson's equations.
(2) Method of images for problems described by Poisson's equation.
(3) Separation of variables for problems described by Laplace's equation.

The first of these can only be used in one-dimensional problems: applications in which the potential varies in one dimension in space. The method of images is a rather interesting and quite general method of solution that also has applications beyond the calculation of electrostatic fields.

It is also worth mentioning that any of the solutions we obtained in **Chapter 4**, such as by using Gauss's law, are, necessarily, solutions to Poisson's or Laplace's equations. In particular, the expressions in **Eqs. (4.30)** through **(4.32)** are solutions of Poisson's equation.

5.4.1 Uniqueness of Solution

Any of the methods mentioned in the previous section and many others can be used to solve electrostatic problems provided that the boundary conditions of the problem can be satisfied. This aspect of solution will become evident shortly, but, before

[2] Pierre Simon Laplace (1749–1827) was a distinguished mathematician and astronomer. Whereas we are aware of his work on the Laplace transform and calculation of the determinant by expansion by minors, Laplace also contributed to the understanding of gravitation, motion of planets, theory of probability, and others. His career was long and varied. In 1799, he was appointed minister of the interior by Napoleon (apparently he was not very successful because this episode only lasted about 6 weeks). After that, he was made a count and, later, a marquis.

we do so, it is worthwhile stating that whatever method we use for solution of either Laplace's or Poisson's equations, we must obtain the same solution. In other words, the solution is unique and independent of method.

Uniqueness of solution of Laplace's and Poisson's equations is guaranteed by the ***uniqueness theorem***. The theorem states that if two solutions to either of the equations are obtained, then the solutions must be the same. In formal terms, we must prove the theorem for each equation or at least for Poisson's equation (since Laplace's equation is a special case of Poisson's equation). Instead, we will take this as given and will only resort to substituting a given solution into the equations to show that it satisfies the equation (see **Problems 5.1** and **5.2**).

5.4.2 Solution by Direct Integration

In principle, to solve a differential equation, we need to "integrate" the equation. This, of course, is not possible in general for a second-order partial differential equation. If, however, the potential only varies in one dimension in space, the partial derivatives can be replaced by ordinary derivatives and the equation can be integrated directly. Consider the following one-dimensional Poisson equation in Cartesian coordinates:

$$\frac{d^2V(x)}{dx^2} = -\frac{\rho_v(x)}{\varepsilon} \tag{5.9}$$

Because the potential varies only with x, ordinary derivatives were used to replace the partial derivatives in **Eq. (5.5)**. Integrating both sides of the equation once, we get

$$\frac{dV(x)}{dx} = \int\left[-\frac{\rho_v(x)}{\varepsilon}\right]dx + a \tag{5.10}$$

Integrating again, we get

$$V(x) = \int\left[\int\left[-\frac{\rho_v(x)}{\varepsilon}\right]dx\right]dx + ax + b \quad [\mathrm{V}] \tag{5.11}$$

The notation used here, that of using the indefinite integral and also adding the constants, is unusual, but it indicates the process and allows for space-dependent charge densities. In practice, we start with **Eq. (5.9)** with known charge density on the right-hand side and integrate once. Only then do we proceed to evaluate **Eq. (5.11)**. The two constants of integration a and b must be determined to obtain a particular solution, and this, of course, is obtained from known values of V; the known values are the boundary conditions of the problem.

If instead of solving Poisson's equation, we solve Laplace's equation (i.e., a solution domain in which there are no charge densities), we simply remove the contribution of charge densities and obtain

$$V(x) = ax + b \tag{5.12}$$

and the constants of integration are again evaluated from the boundary conditions of the problem.

Example 5.1 Potential and Fields in a Capacitor The plates of a capacitor are separated a distance d [m] and grounded as in **Figure 5.1**. The dielectric between the plates has permittivity ε [F/m], and a volume charge density $\rho_v(x) = \rho_0 x(x-d)$ [C/m^3] is distributed throughout the volume of the dielectric. Find the potential and electric field intensity everywhere between the plates of the capacitor.

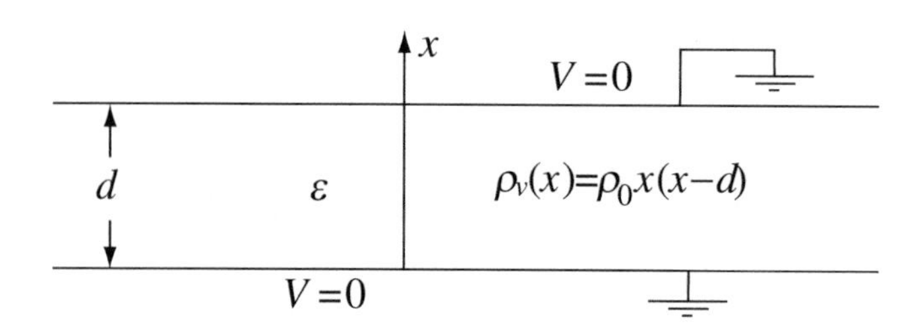

Figure 5.1 A parallel plate capacitor with grounded plates and a charge density between the plates

Solution: Since the charge density only depends on x, this is a one-dimensional problem. Direct integration of Poisson's equation provides the potential. Calculation of the gradient of the potential gives the electric field intensity.

Substitution of the charge density into Poisson's equation gives

$$\frac{d^2V}{dx^2} = -\frac{\rho_0 x(x-d)}{\varepsilon}$$

with boundary conditions $V(x = 0) = 0$ and $V(x = d) = 0$. Integrating twice the general form of the solution is obtained as

$$V(x) = -\frac{\rho_0 x^4}{12\varepsilon} + \frac{\rho_0 x^3 d}{6\varepsilon} + C_1 x + C_2 \quad [\mathrm{V}]$$

With the given boundary conditions, we get

$$V(x=0) = 0 = C_2 \quad \rightarrow \quad C_2 = 0$$

$$V(d) = 0 = -\frac{\rho_0 d^4}{12\varepsilon} + \frac{\rho_0 d^3 d}{6\varepsilon} + C_1 d + 0 \quad \rightarrow \quad C_1 = -\frac{\rho_0 d^3}{12\varepsilon}$$

Thus, the solution to this boundary value problem is

$$V(x) = -\frac{\rho_0 x^4}{12\varepsilon} + \frac{\rho_0 x^3 d}{6\varepsilon} - \frac{\rho_0 x d^3}{12\varepsilon} \quad [\mathrm{V}]$$

This solution satisfies the boundary conditions and the potential is everywhere negative (between the plates).

The electric field intensity is calculated from the gradient of the potential:

$$\mathbf{E}(x) = -\nabla V(x) = -\hat{\mathbf{x}}\frac{dV(x)}{dx} = \hat{\mathbf{x}}\left(\frac{\rho_0 x^3}{3\varepsilon} - \frac{\rho_0 x^2 d}{2\varepsilon} + \frac{\rho_0 d^3}{12\varepsilon}\right) \quad \left[\frac{\mathrm{V}}{\mathrm{m}}\right]$$

Note that the electric field intensity is in the negative x direction for any value $x > d/2$, is zero for $x = d/2$, and is in the positive x direction for any value $x < d/2$.

Check Verify this solution by substituting it back into Poisson's equation.

Exercise 5.1 What is the electric field intensity and the electric potential outside the plates in **Example 5.1**? Hint: You must use Gauss's law.

Answer

$$\mathbf{E} = \hat{\mathbf{x}}\frac{\rho_0 d^3}{12\varepsilon_0} \left[\frac{\mathrm{V}}{\mathrm{m}}\right], \quad V = -\frac{\rho_0 d^3 x}{12\varepsilon_0} \quad [\mathrm{V}], \quad x < 0,$$

$$\mathbf{E} = -\hat{\mathbf{x}}\frac{\rho_0 d^3}{12\varepsilon_0} \left[\frac{\mathrm{V}}{\mathrm{m}}\right], \quad V = \frac{\rho_0 d^3 (x-d)}{12\varepsilon_0} \quad [\mathrm{V}], \quad x > d.$$

Example 5.2 Application: The Stud Sensor A stud sensor is made as a simple capacitor with its two plates located on a plane. The principle is shown in **Figure 5.2a**. The two plates are connected to a simple balanced bridge. The bridge is unbalanced if any material is placed below the plates. When using the device to detect wood studs in a wall, the bridge is balanced to zero in the absence of studs. When a stud is detected, the device turns on a light or sound. A simpler configuration is solved in the following example, assuming the plates are semi-infinite in extent.

Two semi-infinite conducting plates are placed on the r–z plane as shown in **Figure 5.2b**. The plates are very close to each other at $r = 0$ but do not touch. The right plate is connected to a potential V_0 and the left to a zero potential. The plates are very thin:

(a) Calculate the electric potential everywhere.
(b) Calculate the electric field intensity everywhere.

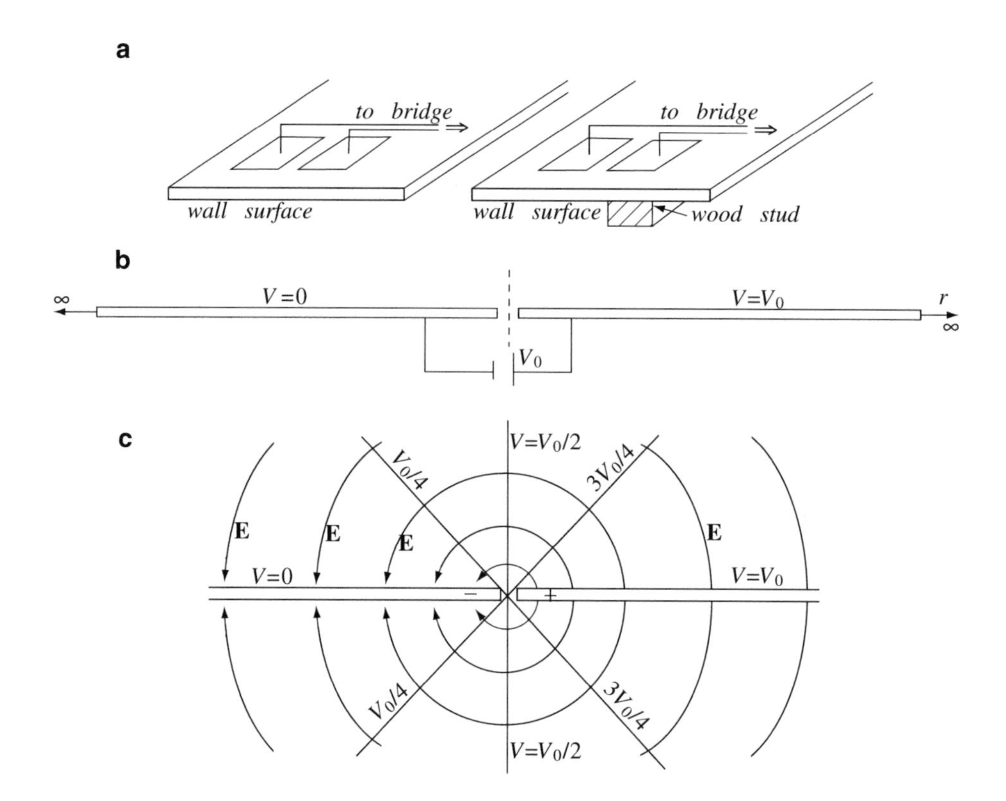

Figure 5.2 The stud sensor. (**a**) Balanced and unbalanced conditions. (**b**) Configuration for solution. (**c**) Plot of electric field intensity and potential

Solution: If we use cylindrical coordinates, both plates are along the r axis but are separated a distance π in the ϕ direction. In this case, we must solve Laplace's equation in cylindrical coordinates (no charges between the plates) and only the ϕ component is nonzero. **Figure 5.2b** shows the arrangement:

(a) From **Eq. (5.6)**, because the potential is constant in the r and z directions, we get

$$\frac{1}{r^2}\frac{\partial^2 V}{\partial \phi^2} = 0 \quad \rightarrow \quad \frac{d^2 V}{d\phi^2} = 0$$

The solution to this equation is obtained by directly integrating twice:

$$V(\phi) = a\phi + b \quad [\mathrm{V}]$$

From the boundary conditions of the problem, we get:
At $\phi = 0$, $V = V_0$: this gives $b = V_0$.

At $\phi = \pi$, $V = 0$ and we get

$$V(\pi) = a\pi + V_0 = 0 \quad \rightarrow \quad a = -\frac{V_0}{\pi}$$

The solution is therefore

$$V(\phi) = -\frac{V_0}{\pi}\phi + V_0 \quad [\mathrm{V}]$$

The solution is linear with the angle ϕ. For example, at $\pi/2$, the potential equals $V_0/2$.

(b) The electric field intensity is calculated as

$$\mathbf{E} = -\nabla V = -\hat{\boldsymbol{\phi}}\frac{1}{r}\frac{\partial V}{\partial \phi} = -\hat{\boldsymbol{\phi}}\frac{1}{r}\frac{d}{d\phi}\left[-\frac{V_0}{\pi}\phi + V_0\right] = \hat{\boldsymbol{\phi}}\frac{V_0}{\pi r} \quad \left[\frac{\mathrm{V}}{\mathrm{m}}\right]$$

The electric field intensity is circular (in the positive ϕ direction above the plates) and depends on r. As we proceed from $r = 0$ (where the plates are closest to each other), the electric field intensity decreases. A plot of the electric field intensity and potential is shown in **Figure 5.2c**. The solution below the plates is similar, but the electric field intensity is in the opposite direction.

Example 5.3 Potential Distribution in a Coaxial Line Two coaxial conductors are shown in **Figure 5.3.** The conductors are very long. A voltage V_0 [V] is connected across the conductors as shown:

(a) Assuming only the potential difference V_0 is known, find the potential distribution everywhere between the conductors.

(b) What is the electric field intensity between the conductors?

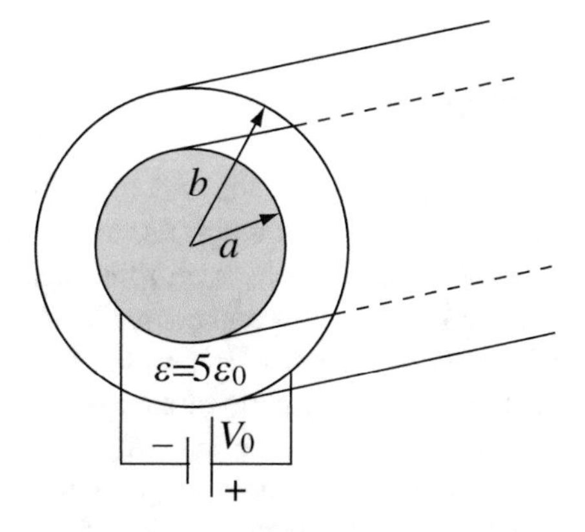

Figure 5.3 Coaxial line connected to a source. The potential and electric field intensity distributions between the conductors is sought

Solution: The geometry is cylindrical and there are no sources in the solution domain enclosed by the two cylinders. Thus, we will use Laplace's equation in cylindrical coordinates. Since the only possible variation in voltage is in the r direction, a one-dimensional Laplace's equation describes the solution:

(a) The equation to solve is

$$\frac{1}{r}\frac{d}{dr}\left(r\frac{dV}{dr}\right) = 0 \quad \rightarrow \quad \frac{d}{dr}\left(r\frac{dV}{dr}\right) = 0$$

The latter form is allowable, since in this geometry, $r \neq 0$. Integrating once, we get

$$r\frac{dV}{dr} = C_1 \quad \rightarrow \quad \frac{dV}{dr} = \frac{C_1}{r}$$

Integrating this again gives

$$V(r) = C_1 \ln r + C_2 \qquad [\mathrm{V}]$$

The constants of integration are evaluated from the boundary conditions.
These are $V(r = a) = 0$ and $V(r = b) = V_0$:

$$V(a) = C_1 \ln a + C_2 = 0 \quad \text{and} \quad V(b) = C_1 \ln b + C_2 = V_0 \qquad [\mathrm{V}]$$

Solving for C_1 and C_2 gives

$$C_1 = \frac{V_0}{\ln(b/a)}, \quad C_2 = -\frac{V_0 \ln a}{\ln(b/a)}$$

and the solution for potential everywhere between the conductors is

$$V(r) = \frac{V_0}{\ln(b/a)}(\ln r - \ln a) \qquad [\mathrm{V}].$$

(b) The electric field intensity is found from the gradient of the potential. Since the potential only depends on r, the gradient in cylindrical coordinates has an r component only:

$$\mathbf{E}(r) = -\nabla V(r) = -\hat{\mathbf{r}}\frac{\partial}{\partial r}\left[\frac{V_0}{\ln(b/a)}(\ln r - \ln a)\right] = -\hat{\mathbf{r}}\frac{V_0}{r\ln(b/a)} \qquad \left[\frac{\mathrm{V}}{\mathrm{m}}\right]$$

Note that the solution to this problem can also be obtained using Gauss's law (see **Section 4.3).**

5.4.3 The Method of Images

The method of images is based on the principle that a field produced by a source or system of sources in part of the space can be produced by other sets of sources. Although the uniqueness theorem specifies that any solution obtained using Laplace's or Poisson's equations is unique, which, in turn, means that a given source produces a unique solution, it is possible to produce two identical fields by different source configurations, provided this field is not the total field. By this we mean that different combinations of sources can produce the same solution in a given part of space although they will have completely different solutions in other parts of space. Thus, the method of images solves for the fields in a given problem by replacing it with an equivalent problem, which is easier to solve and which has the same solution in the required domain. In particular, problems that involve point and distributed charges and conducting bodies can often be replaced by equivalent point charges or charge distributions for which the solution is known or can be obtained readily. As an illustration, recall that in using Gauss's law, we could replace a spherical charge distribution with an equivalent point charge if all we required was the electric field intensity outside the charge distribution. This provided a correct solution only outside the charge distribution.

The question is, how can we find the equivalent charges or charge distributions? The basic idea is that a charge (point charge or charge distribution) reflects in a conductor much the same as light reflects in a mirror. Under these conditions, the conductor is replaced by the image charge and the field due to the two charges is calculated. The method is usually associated with perfect conductors and we will limit ourselves to images due to perfect conductors, but this is not a necessary condition. Image methods can be used with conductors or dielectrics.

In more exact terms, the method of images applies whenever a system of charges can be identified and an equal potential surface produced by these charges exists. As an example, charge Q in **Figure 5.4a** represents any point or distributed charge above a conducting surface. The method of images requires that we remove the conductor (which is at a constant potential, that we assume for now to be zero) and replace it by the image of the charge Q, as shown in **Figure 5.4b**. By doing so, the surface of the conductor is replaced by an identical constant potential surface and the solution above this surface is the same as in the original problem in **Figure 5.4a**. Note, however, that below the plane (where the conductor was located), the solution is incorrect: The electric field intensity in a perfect conductor is zero, whereas the electric field intensity for the system of charges in **Figure 5.4b** is not. This minor difficulty is resolved in a very simple manner: we obtain the solution of **Figure 5.4b** and use it only above the plane whereas the solution below the plane is known ($\mathbf{E} = 0$).

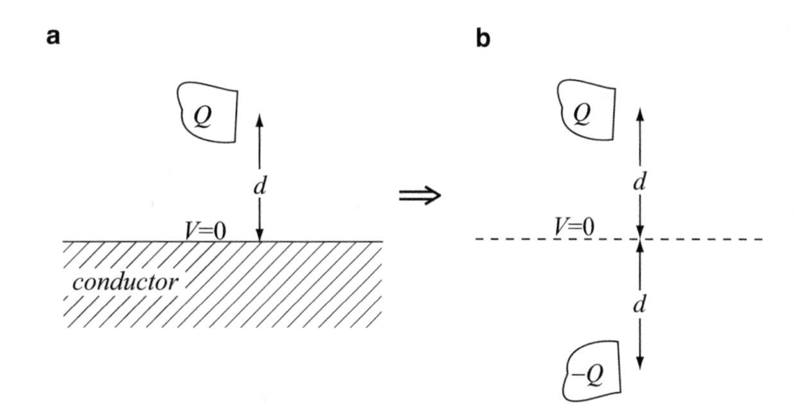

Figure 5.4 The general principle of the method of images. (**a**) The original problem to be solved. (**b**) The equivalent or image problem. The two configurations have the same solution above the conductor

A useful analogy can be obtained by viewing the conducting surface as a mirror. When looking into a mirror, we see an image looking in the opposite direction. If we were to look at an angle into the mirror, so would the image be at an angle, looking at us. We also note that the image itself does not exist: there is nothing behind the mirror. Therefore, only the "solution" in front of the mirror is valid. You may think of it in this sense: A light source in front of the mirror reflects off the mirror. Any object in front of the mirror is lit by two sources: one is the original source, the second is the reflected light from the mirror. If we were to remove the mirror, the reflected light would disappear. If we now place an identical source at the location of the image in the mirror, the light distribution in the area in front of the mirror would be the same as if the mirror were still there. Thus, the image is a convenient artifice. We also note that the name "method of images"[3] comes from the analogy with optical images. From this simple analogy, we can write the following general properties for the method of images in planar geometries (see **Figure 5.4**):

(1) The image is the negative of the source. The magnitude of the image is the same as the source.
(2) The constant potential surface is assumed to be infinite (or very large)
(3) The geometry is reflected in the constant potential surface as in a mirror.
(4) The image and source are at the same distance from the mirror.
(5) Multiple sources produce multiple images, again like reflection in a mirror.
(6) Single or multiple point or distributed charges in front of multiple mirrors also produce multiple images.

Important Rules **(1)**, **(2)** and **(4)** only apply to planar surfaces. The rest apply in curved surfaces as well.

5.4.3.1 Point and Line Charges Point_Charges.m

The simplest application of the method of images is that of a single point charge or a single line of charge over a conducting plane. These are discussed now, as they serve to demonstrate the method. Using superposition of solutions, these are then extended to multiple charges or lines and multiple conducting surfaces. Consider first a point charge Q over a conducting surface, as shown in **Figure 5.5a**. The conductor is assumed to be at zero potential, but this condition is not necessary since it

[3] The method of images is attributed to Lord Kelvin [Sir William Thomson (1824–1907)], who introduced it in 1848 for solution of electric problems. It is, however, a general method that applies equally well to magnetic and electromagnetic problems, as well as to other problems described by Poisson's equation.

is only a reference value. In other words, if the conducting surface is not at zero potential, we may assume it is and, after solving the image problem, add the solution due to the potential on the conductor (a visual explanation of this process in curved geometries can be seen in **Figure 5.25**). The conductor is now removed and an image charge, equal in magnitude but negative, is located at a distance d below the equipotential surface (which is at the location of the surface of the conductor), as shown in **Figure 5.5b**. The two charges produce a field distribution as shown in **Figure 5.5c**. We note that the potential midway between two opposite charges is zero, as was our assumption. Thus, removing the conductor and replacing it with the image has not changed this condition. We also conclude immediately that the solution above the surface of the conductor is the solution in the upper part of **Figure 5.5c**.

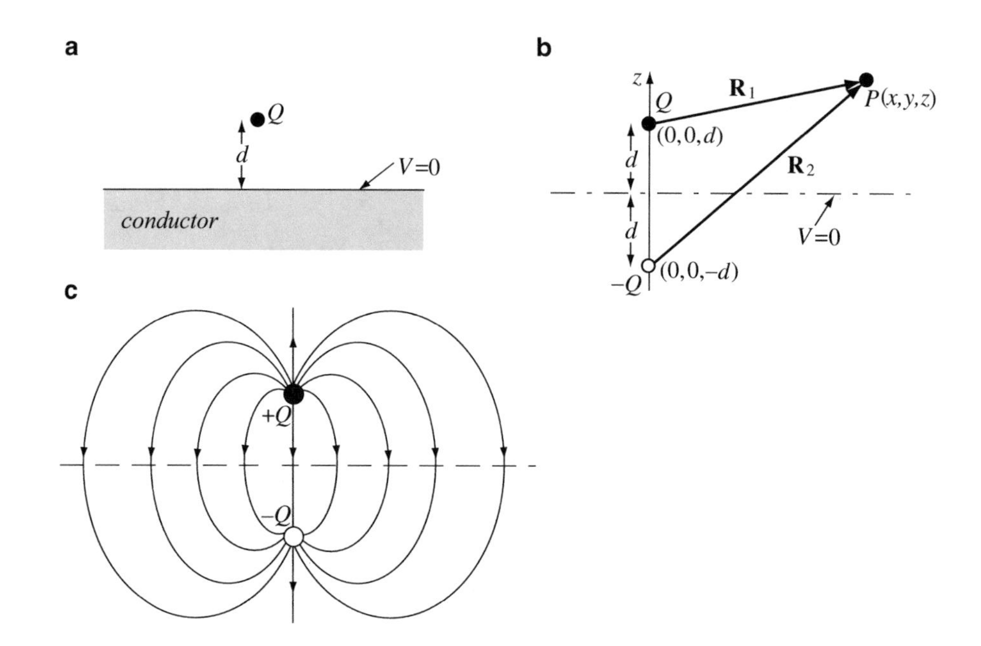

Figure 5.5 (**a**) Point charge over a conducting surface. (**b**) The conducting surface is removed and an image charge equal but opposite in sign is introduced. (**c**) The electric field intensity for the configuration in (**b**)

To find the solution, all we need to do is find the electric field intensity, or the potential, everywhere in space due to two point charges. If the potential is calculated, then the electric field intensity must be calculated from the potential through the use of the gradient. If, on the other hand, the electric field intensity is evaluated, the potential is calculated by integration. Both methods will be demonstrated here:

(1) Calculation of Potential. The potential at any point in space $P(x,y,z)$ due to two point charges at points $P_1(0,0,d)$ and $P_2(0,0,-d)$ is (see **Figure 5.5b**)

$$V(x,y,z) = \frac{Q}{4\pi\varepsilon_0 R_1} - \frac{Q}{4\pi\varepsilon_0 R_2} = \frac{Q}{4\pi\varepsilon_0}\left[\frac{1}{\left[x^2+y^2+(z-d)^2\right]^{1/2}} - \frac{1}{\left[x^2+y^2+(z+d)^2\right]^{1/2}}\right] \quad (\mathrm{V}) \tag{5.13}$$

Note that at $z = 0$, the potential is zero for any point in the x–y plane, as we have assumed. Now, we can calculate the electric field intensity as the gradient of the electric potential:

$$\begin{aligned}\mathbf{E}(x,y,z) &= -\nabla V(x,y,z) = -\hat{\mathbf{x}}\frac{\partial V(x,y,z)}{\partial x} - \hat{\mathbf{y}}\frac{\partial V(x,y,z)}{\partial y} - \hat{\mathbf{z}}\frac{\partial V(x,y,z)}{\partial z} \\ &= \frac{Q}{4\pi\varepsilon_0}\left[\frac{\hat{\mathbf{x}}x+\hat{\mathbf{y}}y+\hat{\mathbf{z}}(z-d)}{\left[x^2+y^2+(z-d)^2\right]^{3/2}} - \frac{\hat{\mathbf{x}}x+\hat{\mathbf{y}}y+\hat{\mathbf{z}}(z+d)}{\left[x^2+y^2+(z+d)^2\right]^{3/2}}\right] \quad \left[\frac{\mathrm{V}}{\mathrm{m}}\right]\end{aligned} \tag{5.14}$$

Setting $z = 0$, we obtain the electric field intensity at the surface of the conductor:

$$\mathbf{E}(x,y,0) = -\hat{\mathbf{z}}\frac{2Qd}{4\pi\varepsilon_0\left[x^2 + y^2 + d^2\right]^{3/2}} \quad \left[\frac{\text{V}}{\text{m}}\right] \tag{5.15}$$

At the surface of the conductor, the electric field intensity is in the negative z direction and its magnitude depends on the location in the plane as well as on the height d. The electric field intensity above the conducting surface is that given in **Eq. (5.14)**, whereas in the conductor, the solution is zero, resulting in the solution in **Figure 5.6**. The charge density on the surface of the conductor is also indicated. This charge density is a direct consequence of the fact that the electric field must "end" in a negative charge, indicating that the actual solution is due to the point charge Q and the equivalent induced charges on the surface of the conductor. The induced charge on the surface of the conductor can be easily calculated from the interface conditions at the surface since the electric field intensity only has a normal component at the conducting surface. From the interface conditions between air and the conductor, we get [see **Eq. (4.79)**]

$$D_t = 0, \quad D_n = \rho_s \quad \to \quad \varepsilon_0 E_n = \rho_s \quad [\text{C/m}^2] \tag{5.16}$$

or using the electric field intensity in **Eq. (5.15)**,

$$\rho_s(x,y) = -\frac{2Qd}{4\pi\left[x^2 + y^2 + d^2\right]^{3/2}} \quad \left[\frac{\text{C}}{\text{m}^2}\right] \tag{5.17}$$

The charge density is negative and distributed on the surface according to location. The maximum charge density is located at ($x = 0$, $y = 0$), directly below the charge Q, as is evident from **Eq. (5.17)**. Also, the total charge on the infinite plane $z = 0$ is equal to $-Q$, as can be shown by direct integration (see **Exercise 5.2**). The charge density distribution is shown in **Figure 5.7a** in the x direction ($y = 0$). The charge distribution on the plane is in the shape of a bell, with its highest point exactly below the point charge, as shown in **Figure 5.7b**.

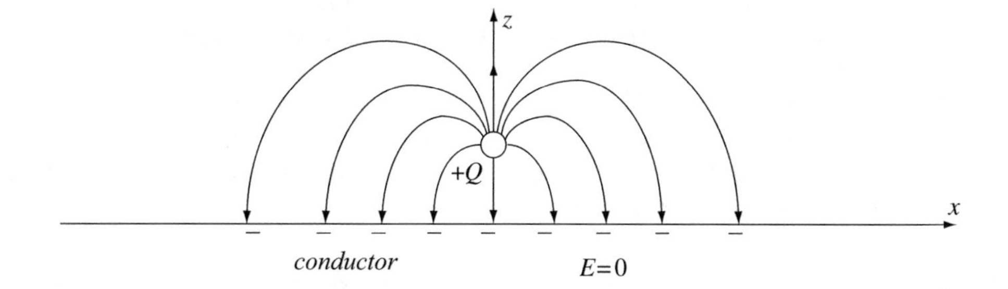

Figure 5.6 Complete solution to the point charge over a conducting plane shown in **Figure 5.5a**

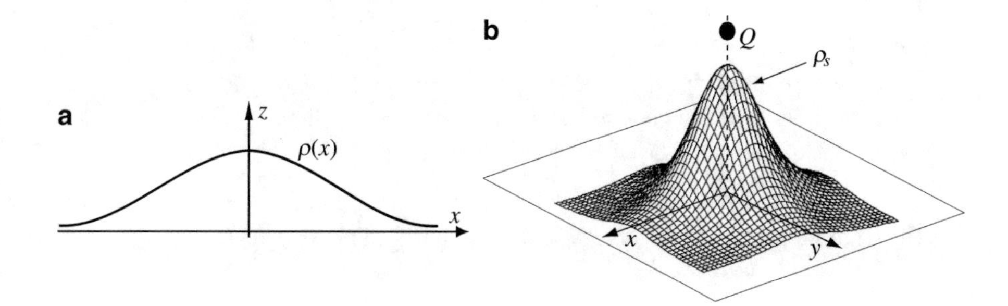

Figure 5.7 (**a**) Induced charge density (magnitude) on the surface of the conductor in **Figure 5.5a**. (**b**) Induced charge density shown as magnitude over the x–y plane

(2) Calculation of the Electric Field Intensity. As mentioned earlier, we can start by calculating the electric field intensity due to the two point charges. Starting with **Figure 5.8**, and noting the two vectors $\mathbf{R}_1$ and $\mathbf{R}_2$, we write

$$\mathbf{E}(x,y,z) = \frac{Q\mathbf{R}_1}{4\pi\varepsilon_0 R_1^3} + \frac{(-Q)\mathbf{R}_2}{4\pi\varepsilon_0 R_2^3} = \frac{Q}{4\pi\varepsilon_0}\left[\frac{\hat{\mathbf{x}}x + \hat{\mathbf{y}}y + \hat{\mathbf{z}}(z-d)}{\left[x^2+y^2+(z-d)^2\right]^{3/2}} - \frac{\hat{\mathbf{x}}x + \hat{\mathbf{y}}y + \hat{\mathbf{z}}(z+d)}{\left[x^2+y^2+(z+d)^2\right]^{3/2}}\right] \quad \left[\frac{\text{V}}{\text{m}}\right] \tag{5.18}$$

which, not surprisingly, is the same result as in **Eq. (5.14)**. To calculate the potential, we use **Eq. (5.18)** and write

$$V = -\int \mathbf{E}(x,y,z)\cdot d\mathbf{l} = -\int \frac{Q\mathbf{R}_1}{4\pi\varepsilon_0 R_1^3}\cdot d\mathbf{R} + \int \frac{Q\mathbf{R}_2}{4\pi\varepsilon_0 R_2^3}\cdot d\mathbf{R} = \frac{Q}{4\pi\varepsilon_0 R_1} - \frac{Q}{4\pi\varepsilon_0 R_2} \quad [\text{V}] \tag{5.19}$$

where the integration is carried out from ∞ to point $P(x,y,z)$. This result is exactly the potential in **Eq. (5.13)**, again as expected.

Now, we can summarize the basic steps involved in the solution by the method of images in planar geometries:

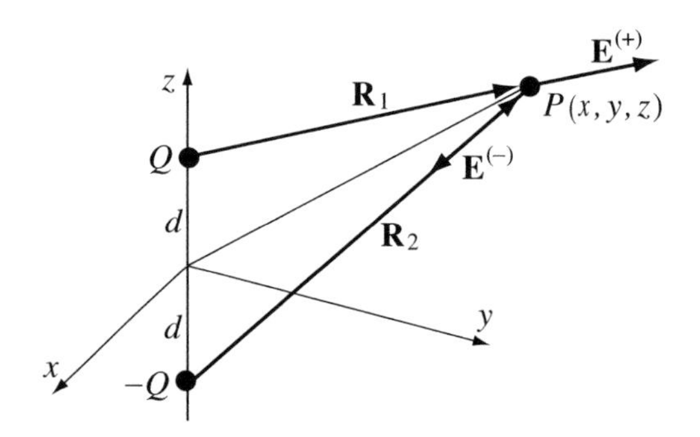

Figure 5.8 The configuration in **Figure 5.5b** as used to calculate the electric field intensity

(1) Find a surface on which the potential is constant. This potential will normally be on the surface of a conductor but does not have to be zero.

(2) Locate the image charge (or charges) equal to the source charge, opposite in sign and at equal distance on the other side of the surface. The conductor itself is removed since its effects have been replaced by the image charge(s).

(3) Find either the electric field intensity (if there are few charges) or the electric potential (if the number of charges is large). In either case the electric field intensity or the potential must be calculated at a general point in space.

(4) Calculate other quantities if required, based on the basic formulas of the electric field intensity or the electric potential as given here and in **Chapters 3** and **4**.

Example 5.4 Application: Electric Field Intensity at Ground Level Due to Charged Clouds If the charge distribution in a cloud can be estimated, the electric field intensity can be calculated everywhere, assuming the ground to be a conductor. As a simple example of this type of problem, we consider here a spherical charge distribution as an approximation to a cloud.

A spherical cloud has a uniform negative volume charge density $-\rho_v$ [C/m^3] and is located above ground, as shown in **Figure 5.9a**. The dimensions and charge density are shown in the figure. Assume the conducting ground is at zero potential and does not disturb the charge distribution within the cloud.

(a) Calculate the electric field intensity everywhere at ground level.

(b) What is the largest electric field intensity at ground level? Where does it occur?

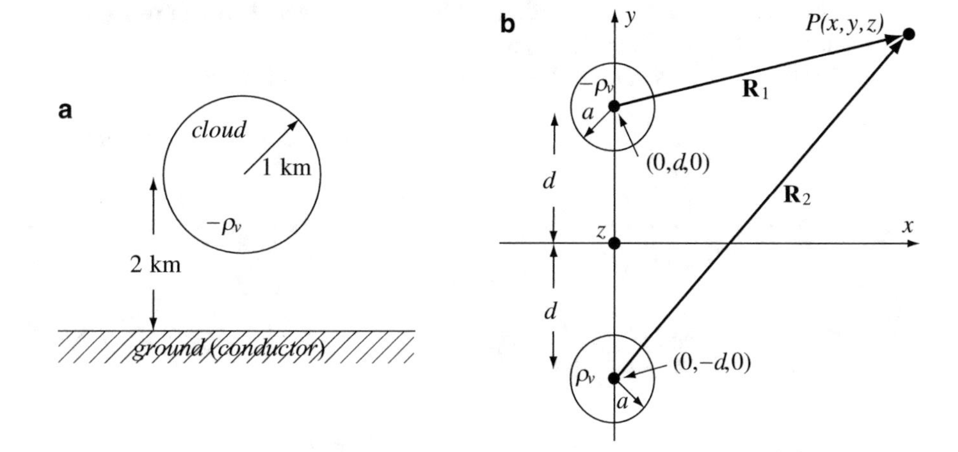

Figure 5.9 (**a**) A charged cloud over a conducting ground. (**b**) The image representation of the configuration in (**a**)

Solution: To calculate the electric field intensity, we use the method of images. The ground is removed and an identical sphere, with positive charge, is placed a distance $d = 2$ km below the location of the ground, as shown in **Figure 5.9b**. The potential is calculated outside the spheres as if the spheres were point charges, placed at their center (a result we obtained from Gauss's law). After calculating the potential at a general point $P(x,y,z)$; we find the negative gradient of the potential to calculate the electric field intensity. Setting $y = 0$ gives the electric field intensity on the ground:

(a) The electric potential at a distance R_1 from the upper sphere and R_2 from the lower sphere is

$$V(P) = -\frac{Q}{4\pi\varepsilon_0 R_1} + \frac{Q}{4\pi\varepsilon_0 R_2} \quad [\mathrm{V}]$$

The total charge Q in each sphere is the volume of the sphere multiplied by the charge density

$$Q = \frac{4\pi a^3 \rho_v}{3} \quad [\mathrm{C}]$$

The distances R_1 and R_2 are $R_1 = (x^2 + (y - d)^2 + z^2)^{1/2}$ and $R_2 = (x^2 + (y + d)^2 + z^2)^{1/2}$. The electric potential at $P(x,y,z)$ is

$$V(x,y,z) = -\frac{a^3\rho_v}{3\varepsilon_0\left(x^2 + (y-d)^2 + z^2\right)^{1/2}} + \frac{a^3\rho_v}{3\varepsilon_0\left(x^2 + (y+d)^2 + z^2\right)^{1/2}} \quad [\mathrm{V}]$$

Now, we calculate the electric field intensity at point P as $\mathbf{E} = -\nabla V$:

$$\mathbf{E}(x,y,z) = -\hat{\mathbf{x}}\frac{\partial V(x,y,z)}{\partial x} - \hat{\mathbf{y}}\frac{\partial V(x,y,z)}{\partial y} - \hat{\mathbf{z}}\frac{\partial V(x,y,z)}{\partial z}$$

$$= \frac{a^3\rho_v}{3\varepsilon_0}\left[-\frac{\hat{\mathbf{x}}x + \hat{\mathbf{y}}(y-d) + \hat{\mathbf{z}}z}{\left(x^2 + (y-d)^2 + z^2\right)^{3/2}} + \frac{\hat{\mathbf{x}}x + \hat{\mathbf{y}}(y+d) + \hat{\mathbf{z}}z}{\left(x^2 + (y+d)^2 + z^2\right)^{3/2}}\right] \quad \left[\frac{\mathrm{V}}{\mathrm{m}}\right]$$

To find the electric field intensity on the ground plane, we set $y = 0$:

$$\mathbf{E}(x,0,z) = \hat{\mathbf{y}}\frac{2a^3\rho_v}{3\varepsilon_0}\left[\frac{d}{\left(x^2 + d^2 + z^2\right)^{3/2}}\right] \quad \left[\frac{\mathrm{V}}{\mathrm{m}}\right].$$

(b) From the general expression of the electric field intensity on the ground plane, the intensity is largest at $z = x = 0$ because the denominator is then smallest. The maximum electric field intensity occurs at (0,0,0), exactly under the center of the cloud:

$$\mathbf{E}_{max} = \mathbf{E}(0,0,0) = \hat{\mathbf{y}}\frac{2a^3\rho_v}{3\varepsilon_0 d^2} \quad \left[\frac{\mathrm{V}}{\mathrm{m}}\right]$$

This is the same as the electric field intensity midway between two point charges, each of magnitude $4\pi a^3\rho_v/3\varepsilon_0$, separated a distance $2d$ apart and of opposite signs.

For the values given here, the maximum electric field intensity at ground level is

$$\mathbf{E}(0,0,0) = \hat{\mathbf{y}}1.885 \times 10^{13}\rho_v \quad [\mathrm{V/m}].$$

Exercise 5.2 In **Example 5.4**:

(a) Calculate the charge density on the ground.

(b) Find the total charge on the ground and show it is equal to the total charge in the cloud but of opposite sign.

Answer **(a)** $\rho_s = \dfrac{2a^3\rho_v}{3}\left[\dfrac{d}{\left(x^2+d^2+z^2\right)^{3/2}}\right] \quad \left[\dfrac{\mathrm{C}}{\mathrm{m}^2}\right]$, **(b)** $Q = \dfrac{4\pi a^3\rho_v}{3} \quad [\mathrm{C}]$.

Exercise 5.3 In **Example 5.4**, Calculate the electric field intensity at the center of the cloud if the volume charge density is –0.5 μC/m^3.

Answer

$$\mathbf{E} = \hat{\mathbf{y}}1.176 \times 10^6 \quad \mathrm{V/m}$$

Example 5.5 Charge in Front of a Right-Angle Conductor A point charge q [C] is located in front of an infinitely long, right-angle conductor shown in **Figure 5.10a** in cross section. Assuming the conductor is at zero potential calculate:

(a) The electric field intensity everywhere in the plane of the charge ($z = 0$).

(b) The charge density induced on the surfaces of the conductor in the same plane as in part **(a)**.

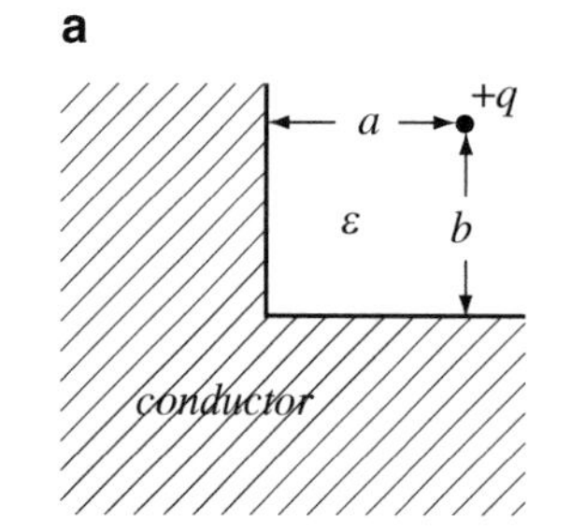

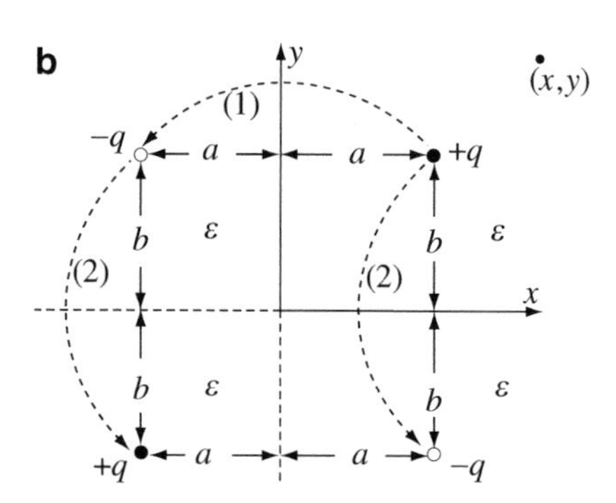

Figure 5.10 (**a**) Point charge in front of a conducting corner. (**b**) Replacement of conducting surfaces by equivalent image charges. The arrowed lines show the sequence of generating the image charges

Solution: First, it is necessary to find the system of images that guarantees zero potential on the planes corresponding to the conductor surface after the conductor is removed. This is shown in **Figure 5.10b**. From this, the potential at a general point in space is calculated, and then the electric field intensity is calculated as $\mathbf{E} = -\nabla V$. The charge density is calculated using the expression $\rho_s = \varepsilon E_n$ at the surface (i.e., at $x = 0$ for the vertical surface or $y = 0$ for the horizontal surface):

(a) The system of charges is found by assuming two perpendicular planes, coinciding with the $x = 0$ and $y = 0$ planes. The image charges shown in **Figure 5.10b** are found by reflecting the charge first about the $x = 0$ plane. This gives an image charge at ($-a,b$). Then, these two charges are reflected about the $y = 0$ plane to obtain the charges shown in **Figure 5.10b**.

The potential at (x,y) due to the four charges is

$$V(x,y) = \frac{q}{4\pi\varepsilon}\left[\frac{1}{\left((x-a)^2+(y-b)^2\right)^{1/2}} - \frac{1}{\left((x-a)^2+(y+b)^2\right)^{1/2}} + \frac{1}{\left((x+a)^2+(y+b)^2\right)^{1/2}} - \frac{1}{\left((x+a)^2+(y-b)^2\right)^{1/2}}\right] \quad [\mathrm{V}]$$

From this, we calculate $\mathbf{E}$ as

$$\mathbf{E}(x,y) = -\hat{\mathbf{x}}\frac{\partial V(x,y)}{\partial x} - \hat{\mathbf{y}}\frac{\partial V(x,y)}{\partial y}$$

$$= \hat{\mathbf{x}}\frac{q}{4\pi\varepsilon}\left[\frac{x-a}{\left((x-a)^2+(y-b)^2\right)^{3/2}} - \frac{x-a}{\left((x-a)^2+(y+b)^2\right)^{3/2}} + \frac{x+a}{\left((x+a)^2+(y+b)^2\right)^{3/2}} - \frac{x+a}{\left((x+a)^2+(y-b)^2\right)^{3/2}}\right]$$

$$+\hat{\mathbf{y}}\frac{q}{4\pi\varepsilon}\left[\frac{y-b}{\left((x-a)^2+(y-b)^2\right)^{3/2}} - \frac{y+b}{\left((x-a)^2+(y+b)^2\right)^{3/2}} + \frac{y+b}{\left((x+a)^2+(y+b)^2\right)^{3/2}} - \frac{y-b}{\left((x+a)^2+(y-b)^2\right)^{3/2}}\right] \quad \left[\frac{\mathrm{V}}{\mathrm{m}}\right].$$

(b) To calculate the charge density on the vertical surface, we set $x = 0$ in the expression of the electric field intensity. This gives $\mathbf{E}(0,y)$. Collecting terms, we get

$$\mathbf{E}(0,y) = -\hat{\mathbf{x}}\frac{qa}{2\pi\varepsilon}\left[\frac{1}{\left(a^2+(y-b)^2\right)^{3/2}} - \frac{1}{\left(a^2+(y+b)^2\right)^{3/2}}\right] \quad \left[\frac{\mathrm{V}}{\mathrm{m}}\right]$$

Note that the y component of the field cancels. $E(0,y)$ is the normal component to the vertical surface. Therefore, the charge density on this surface is

$$\rho_s(0,y) = \varepsilon E(0,y) = -\frac{qa}{2\pi}\left[\frac{1}{\left(a^2+(y-b)^2\right)^{3/2}} - \frac{1}{\left(a^2+(y+b)^2\right)^{3/2}}\right] \quad \left[\frac{\mathrm{C}}{\mathrm{m}^2}\right]$$

The charge density is zero at $y = 0$ and at $y = \infty$. It is maximum at $y = b$ and is negative everywhere since the first term is always larger than the second term in the square brackets.

The charge density on the horizontal surface is calculated in exactly the same way, but setting $y = 0$ in the general expression for $\mathbf{E}$. After simplification, we get

$$\mathbf{E}(x,0) = -\hat{\mathbf{y}}\frac{qb}{2\pi\varepsilon}\left[\frac{1}{\left((x-a)^2+b^2\right)^{3/2}} - \frac{1}{\left((x+a)^2+b^2\right)^{3/2}}\right] \quad \left[\frac{\mathrm{V}}{\mathrm{m}}\right]$$

The electric field intensity on the horizontal surface is normal to the surface. The charge density on this surface is

$$\rho_s(x,0) = -\frac{qb}{2\pi}\left[\frac{1}{\left((x-a)^2+b^2\right)^{3/2}} - \frac{1}{\left((x+a)^2+b^2\right)^{3/2}}\right] \quad \left[\frac{\mathrm{C}}{\mathrm{m}^2}\right]$$

The charge density is zero at $x = 0$ and at $x = \infty$ and is maximum at $x = a$. A plot of the charge distribution is shown in **Figure 5.11a** and for the electric field intensity in **Figure 5.11b**.

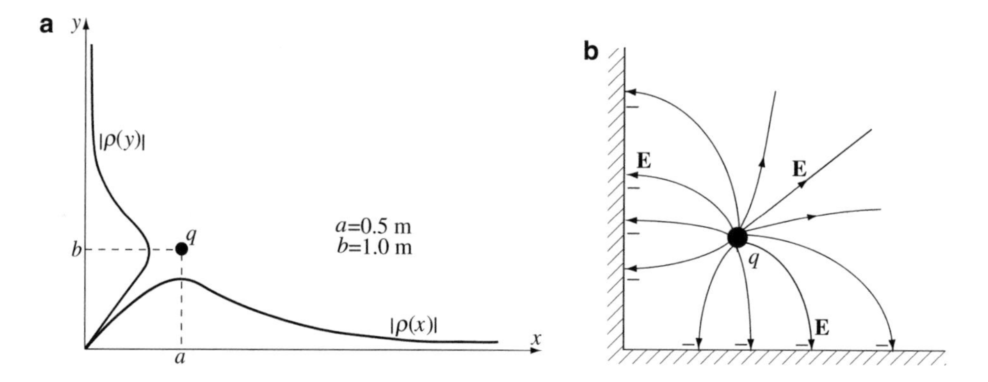

Figure 5.11 (**a**) Plot of the charge distribution on the surfaces of the conductor in **Figure 5.10a**. (**b**) Plot of the electric field intensity in **Figure 5.10a**

Example 5.6 Two Surfaces at Any Angle **Point_Charges.m**

Two conducting planes intersect at an angle smaller than 180°. Both surfaces are at zero potential. A point charge q [C] is placed midway between the two planes at a distance a [m] from the intersection, as shown in **Figure 5.12a**:

(**a**) Show the image charges.
(**b**) How many charges are needed for an angle that is an integer divisor of 180°?
(**c**) How many charges are needed for an angle that is a non-integer divisor of 180°?
(**d**) What happens if the charge is not midway between the two planes?

Solution: An image charge is created for any charge reflecting in a conducting surface. Multiple surfaces generate multiple charges. If the system of images is finite and unique, the problem can be solved. If not, the problem cannot be solved using the method of images:

(**a**) To find the images, we first extend both surfaces to infinite planes and reflect the charge in each surface in turn. In the first step, charge q is reflected about the upper and lower surfaces, generating image charges $q_1' = -q$ and $q_1'' = -q$ as shown in **Figure 5.12b**. Now, charge q_1' is reflected by the upper surface, generating charge $q_2' = +q$, and charge q_1'' is reflected about the lower surface, generating image charge $q_2'' = +q$. Note that the two planes are extended to show that the lines connecting the charge with its image always intersect the plane, or its extension, at right angles. The sequence now is repeated indefinitely: charge $+q_2'$ is reflected about the upper surface to generate image charge $q_3' = -q$ and charge $+q_2''$ reflects about the lower surface to generate image charge $q_3'' = -q$, and so on. Note that all image charges as well as the original charge are located on a circle of radius a. Careful drawing of the charges should convince you of that. The example in **Figure 5.12b** was done for a 60° angle between the plates. Therefore, the last image charge is q_3''. For other angles, the number of charges would be different as discussed in parts (**b**), (**c**), and (**d**) of this example.

(**b**) In part (**a**), we used a 60° angle between the plates. Reflection of the charge generated a sequence of image charges until we reached $q_3' = -q$ and $q_3'' = -q$. These two images fell on the same point, and since they are both of the same sign, this completes the generation of images since continuing the reflection process does not generate new images. The number of image charges is 5. The divisor for 60° is $n = 180/60 = 3$. Thus, the number of image charges is $2n - 1 = 5$. For any integer divisor of 180°, n, there are $2n - 1$ images, for a total of $2n$ charges necessary to produce zero potential on planes A and B.

(**c**) If the angle is not an integer divisor of 180°, the image charges cannot satisfy the zero potential condition on the two planes. One of two situations can happen: either the number of charges is infinite (including charges of opposite sign falling on top of each other) and therefore the sequence of generating the charges cannot be completed or the image charges eventually fall on the plates or their extensions. In either case, the zero potential condition cannot be satisfied. As an example, consider $n = 1.5$. In this case, the angle between the plates is 120°. The first two images are the reflection about plates A and B in **Figure 5.12c** and are shown as q_1' and q_1''. These images are negative and lie on the extension of the planes themselves, and therefore, the potential on each of the two planes varies with position on the planes, contradicting the requirement of constant potential. Therefore, the problem cannot be solved using the method of images.

(**d**) If the angle is an integer divisor of 180°, the answer is as in (**b**), but the charges are nonuniformly spaced on the circle (see **Exercise 5.4**). If it is a non-integer divisor, the answer is as in (**c**).

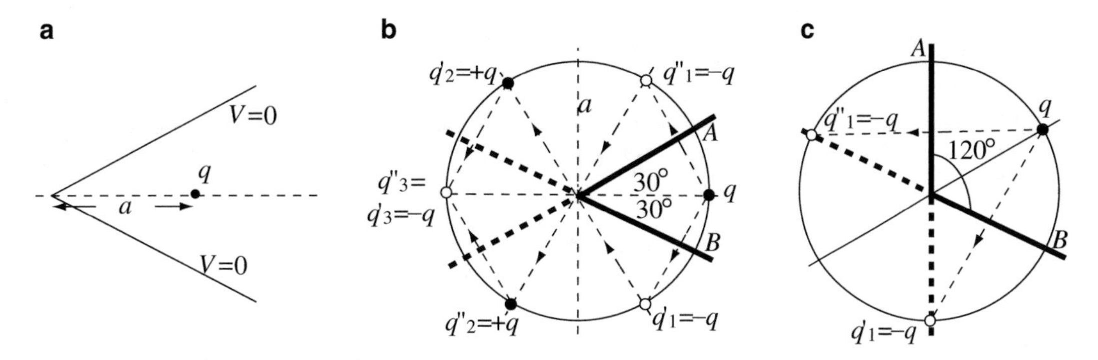

Figure 5.12 (**a**) Point charge in front of two intersecting conducting planes. (**b**) System of images equivalent to (**a**). (**c**) System of images for a 120° angle between the planes

Exercise 5.4 Find the location and number of image charges if the charge in **Figure 5.12a** is placed at a distance d [m] from the upper plate and a distance b [m] from the lower plate while the angle between the plates is 30°.

Answer 11 image charges. See **Figure 5.13** for location of the charges.

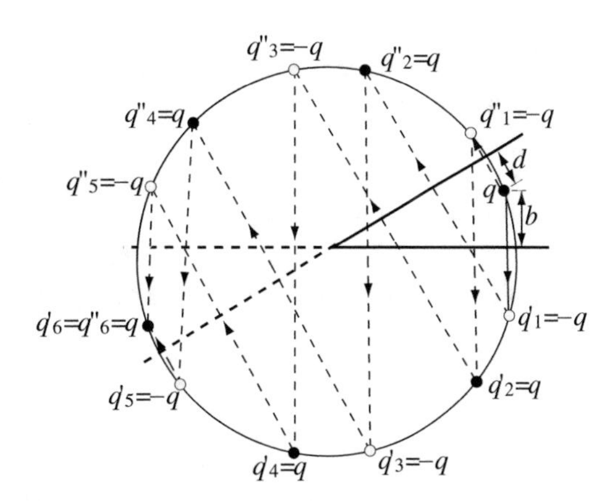

Figure 5.13 Location of image charges for **Exercise 5.4**

Exercise 5.5

(**a**) Find the electric potential at a general point in the x–y plane for a charge placed as in **Figure 5.14** with the angle between the plates equal to 60°. Assume the intersection between the plates occurs at $x = 0, y = 0$, the upper plate coincides with the y axis, and the charge q [C] is on the bisector (at 30° from the y axis), at a distance a [m] from the intersection of the plates.

(**b**) Find the electric field intensity on the vertical plate in the x–y plane.

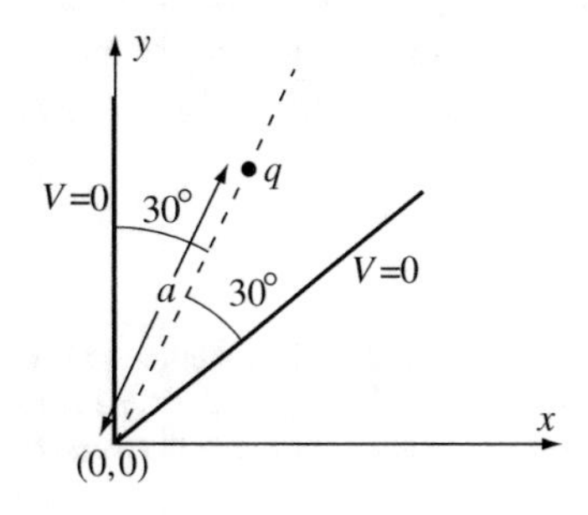

Figure 5.14 Configuration for **Exercise 5.5**

Answer

$$V(x,y) = \frac{q}{4\pi\varepsilon_0}\left[\frac{1}{\sqrt{(x-a/2)^2+\left(y-a\sqrt{3}/2\right)^2}} - \frac{1}{\sqrt{(x+a/2)^2+\left(y-a\sqrt{3}/2\right)^2}} - \frac{1}{\sqrt{(x-a)^2+y^2}} + \frac{1}{\sqrt{(x+a)^2+y^2}}\right.$$
$$\left. + \frac{1}{\sqrt{(x-a/2)^2+\left(y+a\sqrt{3}/2\right)^2}} - \frac{1}{\sqrt{(x+a/2)^2+\left(y+a\sqrt{3}/2\right)^2}}\right] \quad [\mathrm{V}]$$

$$\mathbf{E}(0,y) = \hat{\mathbf{x}}\frac{qa}{4\pi\varepsilon_0}\left[\frac{2}{(a^2+y^2)^{3/2}} - \frac{1}{\left(a^2/4+\left(y-a\sqrt{3}/2\right)^2\right)^{3/2}} - \frac{1}{\left(a^2/4+\left(y+a\sqrt{3}/2\right)^2\right)^{3/2}}\right] \quad \left[\frac{\mathrm{V}}{\mathrm{m}}\right].$$

5.4.3.2 Charged Line over a Conducting Plane

A problem similar to that of a point charge over a plane is that of a charged line over a plane. In fact, if the charged conductor is long, we may be able to assume it is infinite and the problem becomes a two-dimensional problem (the electric field intensity or potential does not vary along the length of the conductor). A physical application of this type is a cable or a pair of wires such as a power line over a conducting ground or a conducting strip over a conducting plane in a printed circuit board.

The basic configurations we consider are shown in **Figure 5.15**. The thin line is analyzed here, whereas the thick and finite-length lines are shown in examples. For ease of solution, we will again assume zero potential on the ground (a fixed potential can always be added as a reference potential). The problem described here was once proposed as a solution to telephone communication. In the 1830s, when all telephone communication required copper wires, it was proposed to use a single wire for each telephone line and use the ground as a common, return conductor.[4] The idea was eventually abandoned, primarily because of "noise" in the system (the sources of which will be understood in later chapters) and the difficulty of making good ground connections, but was used for a while as a practical system.

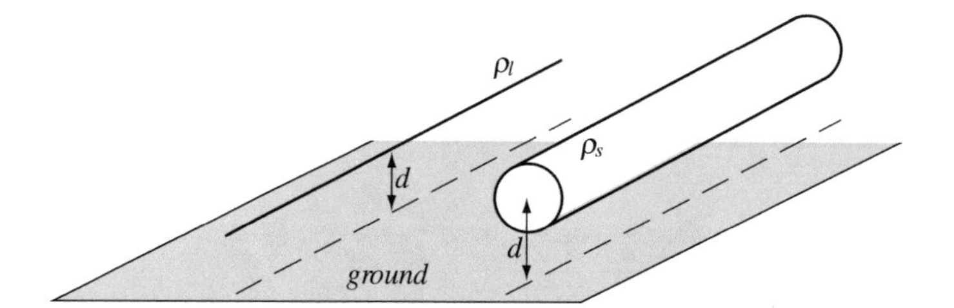

Figure 5.15 Thin and thick charged lines over a conducting plane

As with the point charge in **Figure 5.5**, the conducting plane is removed and replaced with a constant potential plane and an image, equal in magnitude, shape, and dimensions and opposite in sign, placed at a distance d below the constant potential surface, as shown in **Figure 5.16a**. Now, the solution is obtained as the solution due to two line charges, one positive and one negative, at a distance $2d$ apart. In this case, because the lines are long, it is more convenient to calculate the electric field intensity directly since we can apply Gauss's law to each line separately and add the solutions together using superposition. With the dimensions in **Figure 5.16a**, we calculate the electric field intensity at point $P(x,y)$ due to the upper line as if the lower (negative) line did not exist. The electric field intensity is constant on a circle of radius r_1 and in the direction of r_1. Using Gauss's law we can write for the electric field intensity at a distance r_1 from the positive line,

[4] The idea of using the ground as the return conductor is sometimes attributed to Joseph Henry who is known to have used a system of this type in the early 1830s to communicate from his home to his laboratory on the Campus of Albany Academy in Albany, NY. During the same period, and apparently preceding Henry, Carl August Steinheil produced the same type of telegraph and installed working devices in Munich, Germany.

$$\int_s \mathbf{E} \cdot d\mathbf{s} = E2\pi r_1 L = \frac{\rho_l L}{\varepsilon_0} \tag{5.20}$$

where L is an arbitrary length of the line (see **Chapter 4**), $\rho_l L$ is the total charge on a length L of the line, and $2\pi r_1 L$ is the cylindrical surface of the cylinder of length L and radius r_1. Thus, the electric field intensity (considering the direction $\hat{\mathbf{r}}_1$) is

$$\mathbf{E}^+ = \hat{\mathbf{r}}_1 \frac{\rho_l}{2\pi r_1 \varepsilon_0} \quad \left[\frac{\mathrm{V}}{\mathrm{m}}\right] \tag{5.21}$$

Similarly, for the lower line of charge, the direction is in the $-\hat{\mathbf{r}}_2$ direction:

$$\mathbf{E}^- = -\hat{\mathbf{r}}_2 \frac{\rho_l}{2\pi r_2 \varepsilon_0} \quad \left[\frac{\mathrm{V}}{\mathrm{m}}\right] \tag{5.22}$$

The total electric field intensity is the sum of the two electric fields:

$$\mathbf{E} = \mathbf{E}^+ + \mathbf{E}^- = \frac{\rho_l}{2\pi\varepsilon_0}\left(\frac{\hat{\mathbf{r}}_1}{r_1} - \frac{\hat{\mathbf{r}}_2}{r_2}\right) \quad \left[\frac{\mathrm{V}}{\mathrm{m}}\right] \tag{5.23}$$

A more explicit expression is obtained by substituting the vectors $\mathbf{r}_1 = \hat{\mathbf{x}}x + \hat{\mathbf{y}}(y-d)$ and $\mathbf{r}_2 = \hat{\mathbf{x}}x + \hat{\mathbf{y}}(y+d)$. Writing the magnitudes of the vectors as $r_1 = \sqrt{x^2 + (y-d)^2}$ and $r_2 = \sqrt{x^2 + (y+d)^2}$ and the unit vectors as $\hat{\mathbf{r}}_1 = \mathbf{r}_1/r_1$ and $\hat{\mathbf{r}}_2 = \mathbf{r}_2/r_2$, the electric field intensity in **Eq. (5.23)** becomes

$$\mathbf{E} = \frac{\rho_l}{2\pi\varepsilon_0}\left(\frac{\hat{\mathbf{x}}x + \hat{\mathbf{y}}(y-d)}{x^2 + (y-d)^2} - \frac{\hat{\mathbf{x}}x + \hat{\mathbf{y}}(y+d)}{x^2 + (y+d)^2}\right) \quad \left[\frac{\mathrm{V}}{\mathrm{m}}\right] \tag{5.24}$$

To check that this solution is correct, we note that the electric field intensity diminishes with distance as expected. Also, at $y = 0$, we get

$$\mathbf{E} = \frac{\rho_l}{2\pi\varepsilon_0}\left(\frac{\hat{\mathbf{x}}x + \hat{\mathbf{y}}(-d)}{x^2 + d^2} - \frac{\hat{\mathbf{x}}x + \hat{\mathbf{y}}d}{x^2 + d^2}\right) = -\hat{\mathbf{y}}\frac{\rho_l d}{\pi\varepsilon_0}\frac{1}{(x^2 + d^2)} \quad \left[\frac{\mathrm{V}}{\mathrm{m}}\right] \tag{5.25}$$

Thus, on the conducting surface, the electric field is perpendicular, as required.

We can now easily calculate the electric potential everywhere in space and the charge density on the surface of the conductor as for the point charge over a plane. The charge density on the surface of the conductor is again negative (because the source charge, i.e., the line over the plane in **Figure 5.16a**, is positive). The charge density is

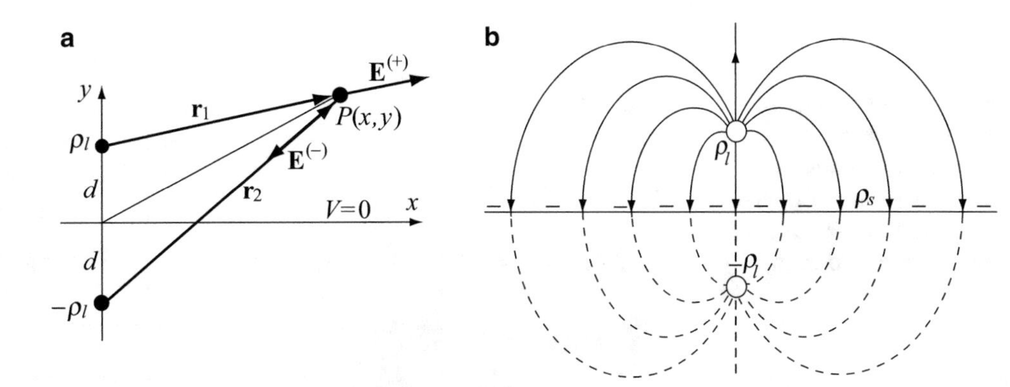

Figure 5.16 (**a**) The charged line and its image with the conducting plane removed. (**b**) The electric field intensity due to a charged line over a conducting plane

$$\rho_s = \varepsilon_0 E_n = -\frac{\rho_l d}{\pi} \frac{1}{\left(x^2 + d^2\right)} \quad \left[\frac{\text{C}}{\text{m}^2}\right] \tag{5.26}$$

where the fact that at the surface there is only a normal component of the electric field intensity in the negative y direction was used. Note that the charge density only varies with the x dimension. This is a consequence of the fact that the line is assumed to be infinitely long. We must also remember that ρ_l is a line charge density [C/m] and ρ_s is a charge per unit area [C/m^2]. If we integrate the surface charge density on the conducting surface from $x = -\infty$ to $x = +\infty$, the result is $-\rho_l$; that is,

$$-\int_{x=-\infty}^{x=+\infty} \frac{\rho_l d}{\pi} \frac{1}{\left(x^2 + d^2\right)} dx = -\frac{\rho_l}{\pi} \tan^{-1}\frac{x}{d}\bigg|_{-\infty}^{+\infty} = \frac{\rho_l}{\pi}\left(\frac{\pi}{2} - \left(-\frac{\pi}{2}\right)\right) = -\rho_l \quad \left[\frac{\text{C}}{\text{m}}\right] \tag{5.27}$$

The negative sign indicates that the charge per unit length of the conducting surface is the opposite of the charge on the line. The solution now is as shown in **Figure 5.16b**, where the electric field distribution above the plane is that obtained in **Eq. (5.24)** and the surface charge density is that in **Eq. (5.26)**. The solution below the surface (in the conductor) is zero, as required, since all electric field lines that start on the line of charge end on the surface of the conductor. Note the similarity of this solution to that in **Figure 5.5c**. However, the two solutions are not the same; the surface charge distributions on the conductor are completely different.

Example 5.7 Overhead Transmission Lines Two thin overhead wires carry line charge densities as shown. Find the electric field intensity at point P (magnitude and direction) in **Figure 5.17a**.

Solution: The method of images may be used to remove the conducting surface and replace its effect with image lines. Now, there are four charged lines as shown in **Figure 5.17b**. Calculate the x and y components of the electric field intensity at P due to the four lines.

The electric field intensities of the four lines are

$$\mathbf{E}_1 = \frac{\hat{\mathbf{r}}_1 \rho_l}{2\pi r_1 \varepsilon_0}, \quad \mathbf{E}_2 = -\frac{\hat{\mathbf{r}}_2 \rho_l}{2\pi r_2 \varepsilon_0}, \quad \mathbf{E}_3 = \frac{\hat{\mathbf{r}}_3 \rho_l}{2\pi r_3 \varepsilon_0}, \quad \mathbf{E}_4 = -\frac{\hat{\mathbf{r}}_4 \rho_l}{2\pi r_4 \varepsilon_0} \quad \left[\frac{\text{V}}{\text{m}}\right]$$

where

$$r_1 = \sqrt{c^2 + (a-b)^2}, \quad r_2 = c - a, \quad r_3 = \sqrt{(c-a)^2 + 4a^2}, \quad r_4 = \sqrt{(a+b)^2 + c^2} \quad [\text{m}]$$

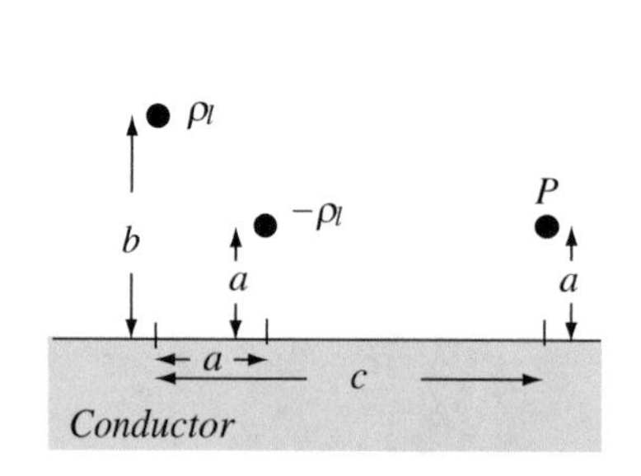

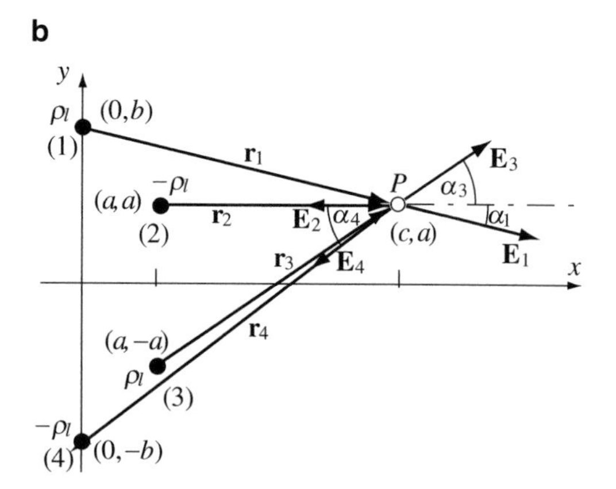

Figure 5.17 (**a**) Two charged lines above a conducting plane. (**b**) The system of images necessary to calculate the electric field in (**a**)

Also

$$E_{1x} = E_1 \cos\alpha_1 = E_1 \frac{c}{r_1}, \qquad E_{1y} = E_1 \sin\alpha_1 = E_1 \frac{b-a}{r_1} \quad \left[\frac{\text{V}}{\text{m}}\right]$$

$$E_{2x} = E_2, \quad E_{2y} = 0$$

$$E_{3x} = E_3 \cos\alpha_3 = E_3 \frac{c-a}{r_3}, \qquad E_{3y} = E_3 \sin\alpha_3 = E_3 \frac{2a}{r_3} \quad \left[\frac{\text{V}}{\text{m}}\right]$$

$$E_{4x} = E_4 \cos\alpha_4 = E_4 \frac{c}{r_4}, \qquad E_{4y} = E_4 \sin\alpha_4 = E_4 \frac{b+a}{r_4} \quad \left[\frac{\text{V}}{\text{m}}\right]$$

The x and y components are

$$E_x = E_{1x} + E_{3x} - E_{2x} - E_{4x}, \quad E_y = -E_{1y} + E_{3y} - E_{4y}$$

Substituting the values of r_i in the expressions for the electric field intensities and separating them into their x and y components, we get

$$E_x = \frac{\rho_l}{2\pi\varepsilon_0}\left[\frac{c}{c^2 + (a-b)^2} + \frac{c-a}{(c-a)^2 + 4a^2} - \frac{1}{c-a} - \frac{c}{(a+b)^2 + c^2}\right] \quad \left[\frac{\text{V}}{\text{m}}\right]$$

$$E_y = \frac{\rho_l}{2\pi\varepsilon_0}\left[-\frac{b-a}{c^2 + (a-b)^2} + \frac{2a}{(c-a)^2 + 4a^2} - \frac{b+a}{(a+b)^2 + c^2}\right] \quad \left[\frac{\text{V}}{\text{m}}\right]$$

The electric field intensity may now be written as $\mathbf{E} = \hat{\mathbf{x}}E_x + \hat{\mathbf{y}}E_y$ [V/m].

Example 5.8 Application: Thick Cylinder Above a Conducting Surface A very long cylinder (you may assume it is infinitely long) of radius a [m] made of a perfect dielectric has a charge density uniformly distributed on its surface equal to ρ_s [C/m^2]. The cylinder is located at a height d [m] above a very thick conductor, as shown in **Figure 5.18a**. Calculate the electric field intensity everywhere in space assuming the presence of the conductor does not disturb the charges on the surface of the cylinder. Assume permittivity of the dielectric is ε_0 [F/m].

Solution: The conductor is replaced by an image cylinder with equal and negative surface charge density. The electric field intensity of each cylinder is calculated separately and the results superimposed to obtain the solution.

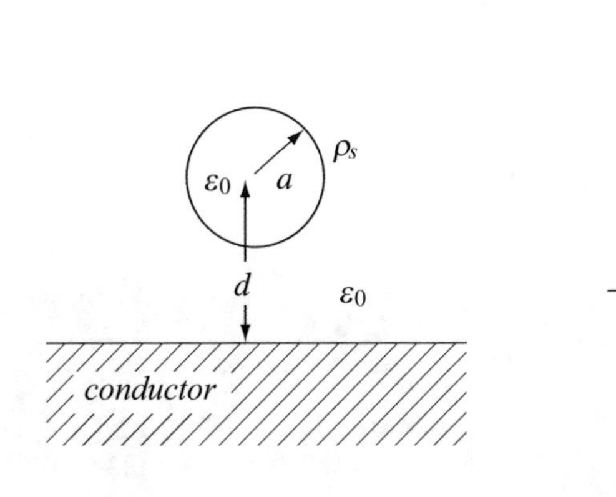

Figure 5.18 (**a**) A charged dielectric cylinder over a conducting surface. (**b**) The cylinder in (**a**) and its image with the conducting surface removed

Refer to **Figure 5.18b** for the images and coordinates. The electric fields are separated into the x and y components as shown. The electric field intensities are

$$\mathbf{E}_1 = \hat{\mathbf{r}}_1 \frac{a\rho_s}{\varepsilon_0 r_1}, \quad \mathbf{E}_2 = -\hat{\mathbf{r}}_2 \frac{a\rho_s}{\varepsilon_0 r_2} \quad \left[\frac{\mathrm{V}}{\mathrm{m}}\right]$$

where a Gaussian surface for each cylinder was used. The x and y components of the electric field intensity are

$$E_{1x} = \frac{a\rho_s}{\varepsilon_0 r_1}\cos\theta_1 = \frac{a\rho_s x}{\varepsilon_0 r_1^2}, \quad E_{2x} = -\frac{a\rho_s}{\varepsilon_0 r_2}\cos\theta_2 = -\frac{a\rho_s x}{\varepsilon_0 r_2^2} \quad \left[\frac{\mathrm{V}}{\mathrm{m}}\right]$$

$$E_{1y} = \frac{a\rho_s}{\varepsilon_0 r_1}\sin\theta_1 = \frac{a\rho_s (y-d)}{\varepsilon_0 r_1^2}, \quad E_{2y} = -\frac{a\rho_s}{\varepsilon_0 r_2}\sin\theta_2 = -\frac{a\rho_s (y+d)}{\varepsilon_0 r_2^2} \quad \left[\frac{\mathrm{V}}{\mathrm{m}}\right]$$

where r_1 and r_2 are

$$r_1 = \sqrt{x^2 + (y-d)^2}, \quad r_2 = \sqrt{x^2 + (y+d)^2} \quad [\mathrm{m}]$$

The solution anywhere outside the cylinders is

$$E_x = E_{1x} + E_{2x} = \frac{a\rho_s x}{\varepsilon_0}\left[\frac{1}{r_1^2} - \frac{1}{r_2^2}\right], \quad E_y = E_{1y} + E_{2y} = \frac{a\rho_s}{\varepsilon_0}\left[\frac{y-d}{r_1^2} - \frac{y+d}{r_2^2}\right] \quad \left[\frac{\mathrm{V}}{\mathrm{m}}\right]$$

Therefore:

$$\mathbf{E} = \hat{\mathbf{x}}E_x + \hat{\mathbf{y}}E_y = \hat{\mathbf{x}}\frac{a\rho_s x}{\varepsilon_0}\left[\frac{1}{x^2+(y-d)^2} - \frac{1}{x^2+(y+d)^2}\right] + \hat{\mathbf{y}}\frac{a\rho_s}{\varepsilon_0}\left[\frac{y-d}{x^2+(y-d)^2} - \frac{y+d}{x^2+(y+d)^2}\right] \quad \left[\frac{\mathrm{V}}{\mathrm{m}}\right]$$

This solution is valid everywhere above ground ($y > 0$) except inside the upper cylinder. Inside the upper cylinder, the field due to itself is zero (because the charge is only distributed on the surface). The solution inside the upper cylinder is

$$E_x = E_{2x} = -\frac{1}{r_2^2}\frac{a\rho_s x}{\varepsilon_0}, \quad E_y = E_{2y} = -\frac{a\rho_s}{\varepsilon_0}\frac{(y+d)}{r_2^2} \quad \left[\frac{\mathrm{V}}{\mathrm{m}}\right]$$

Or:

$$\mathbf{E} = -\hat{\mathbf{x}}\frac{a\rho_s x}{\varepsilon_0\left(x^2+(y+d)^2\right)} - \hat{\mathbf{y}}\frac{a\rho_s}{\varepsilon_0}\frac{y+d}{\left(x^2+(y+d)^2\right)} \quad \left[\frac{\mathrm{V}}{\mathrm{m}}\right].$$

Note: The electric field intensity below ground ($y < 0$) is zero. Also, the electric field intensity inside the cylinders could only be calculated because their permittivity is the same as free space.

Example 5.9 Short Line Segment over a Conducting Plane A line of charge with uniform charge density ρ_l [C/m] and length d [m] is placed at a height h [m] above a thick conducting plane, as shown in **Figure 5.19a**:

(a) Calculate the electric potential everywhere in space.
(b) Calculate the electric field intensity at the surface of the conducting plane, below the center of the wire.

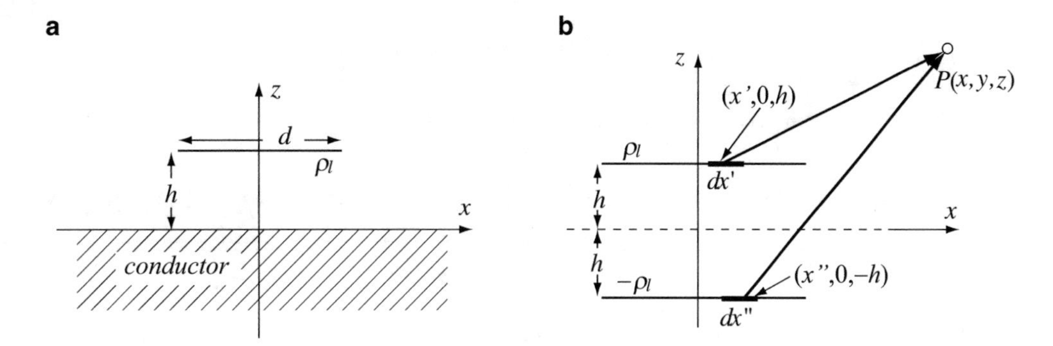

Figure 5.19 (**a**) A short, charged line segment over a conducting plane. (**b**) The line segment and its image with the conducting plane removed

Solution: The conductor is replaced by an image of the line segment as in **Figure 5.19b**. The line segments are placed on the x–z plane. The solution now is that of two short wires with opposite charge densities as shown in **Figure 5.19b**. Using $dl' = dx'$ for the upper wire and $dl'' = dx''$ for the lower wire, a point $(x',0,h)$ on the upper wire and a point $(x'',0,-h)$ on the lower wire, the potential at a general point (x,y,z) is

(**a**) Due to the upper wire:

$$V_P^+ = \frac{1}{4\pi\varepsilon_0}\int_{x'=-d/2}^{x'=d/2}\frac{\rho_l dx'}{\left[(x-x')^2+(y-0)^2+(z-h)^2\right]^{1/2}} \quad [\text{V}]$$

Due to the lower wire:

$$V_P^- = -\frac{1}{4\pi\varepsilon}\int_{x''=-d/2}^{x''=d/2}\frac{\rho_l dx''}{\left[(x-x'')^2+(y-0)^2+(z+h)^2\right]^{1/2}} \quad [\text{V}]$$

The total potential is

$$V_P = V_P^+ + V_P^- = \frac{\rho_l}{4\pi\varepsilon_0}\left[\int_{x'=-d/2}^{x'=d/2}\frac{dx'}{\left[(x-x')^2+y^2+(z-h)^2\right]^{1/2}} - \int_{x''=-d/2}^{x''=d/2}\frac{dx''}{\left[(x-x'')^2+y^2+(z+h)^2\right]^{1/2}}\right]$$

$$= \frac{\rho_l}{4\pi\varepsilon_0}\ln\left[2\sqrt{(x-x')^2+y^2+(z-h)^2}+2x'-2x\right]_{x'=-d/2}^{x'=d/2} - \frac{\rho_l}{4\pi\varepsilon_0}\ln\left[2\sqrt{(x-x'')^2+y^2+(z+h)^2}+2x''-2x\right]_{x''=-d/2}^{x''=d/2}$$

$$= \frac{\rho_l}{4\pi\varepsilon_0}\ln\frac{\left[2\sqrt{(x-d/2)^2+y^2+(z-h)^2}+d-2x\right]}{\left[2\sqrt{(x+d/2)^2+y^2+(z-h)^2}-d-2x\right]} + \frac{\rho_l}{4\pi\varepsilon_0}\ln\frac{\left[2\sqrt{(x+d/2)^2+y^2+(z+h)^2}-d-2x\right]}{\left[2\sqrt{(x-d/2)^2+y^2+(z+h)^2}+d-2x\right]} \quad [\text{V}]$$

(**b**) The electric field intensity is required at $x = 0$, $y = 0$, $z = 0$. At this point, the electric field intensity points in the negative z direction; therefore, there is no need to calculate the x and y components of **E**. The z component of the electric field intensity at a general point in space is

$$E_z(x,y,z) = -\frac{\partial V_P(x,y,z)}{\partial z} = -\frac{\rho_l\left[2(z-h)\left((x-d/2)^2+y^2+(z-h)^2\right)^{-1/2}\right]}{4\pi\varepsilon_0\left[2\sqrt{(x-d/2)^2+y^2+(z-h)^2}+d-2x\right]}$$

$$+\frac{\rho_l\left[2(z-h)\left((x+d/2)^2+y^2+(z-h)^2\right)^{-1/2}\right]}{4\pi\varepsilon_0\left[2\sqrt{(x+d/2)^2+y^2+(z-h)^2}-d-2x\right]}-\frac{\rho_l\left[2(z+h)\left((x+d/2)^2+y^2+(z+h)^2\right)^{-1/2}\right]}{4\pi\varepsilon_0\left[2\sqrt{(x+d/2)^2+y^2+(z+h)^2}-d-2x\right]}$$

$$+\frac{\rho_l\left[2(z+h)\left((x-d/2)^2+y^2+(z+h)^2\right)^{-1/2}\right]}{4\pi\varepsilon_0\left[2\sqrt{(x-d/2)^2+y^2+(z+h)^2}+d-2x\right]}\quad\left[\frac{\text{V}}{\text{m}}\right]$$

At $x = 0$, $y = 0$, $z = 0$, the electric field intensity becomes

$$\mathbf{E}(0,0,0) = -\hat{\mathbf{z}}\,\frac{\rho_l h}{\pi\varepsilon_0}\left[\frac{\left(d^2/4+h^2\right)^{-1/2}}{2\sqrt{d^2/4+h^2}-d}-\frac{\left(d^2/4+h^2\right)^{-1/2}}{2\sqrt{d^2/4+h^2}+d}\right]\quad\left[\frac{\text{V}}{\text{m}}\right].$$

5.4.3.3 Charges Between Parallel Planes

In **Section 5.4.3.1** we saw that if there are multiple conducting planes in which charges reflect, the situation is similar to two or more mirrors, which, of course, produce multiple reflections. Similarly, if we have multiple charges, and perhaps multiple surfaces, the number of image charges may be rather high. However, the method of treating the charges and their reflections is systematic.

The method we used so far was essentially one of finding the reflections of the charges due to the conducting surfaces. Once all image charges were established, a general solution in the space between the conducting surfaces was found.

An alternative way to look at the same process is as follows: Since all conducting surfaces must be at constant potentials, assume all are at zero potential (any other potential can be added later as a reference potential). Removing the conductors, we must now add charges that will ensure that the potential on all surfaces which previously coincided with conducting surfaces is zero. Any number of charges, positive or negative and of any magnitude, can be added, provided the zero potential condition is satisfied. The location of the charges can also be adjusted if necessary. Note also that this particular way of looking at the problem does not require that conductors be flat and, in fact, they can be curved, as we will see shortly.

Example 5.10 Point Charge Between Parallel Plates A point charge q [C] is placed midway between two very large, grounded, conducting surfaces, as shown in **Figure 5.20a**. Find the potential at P.

Solution: First, we must find the image charges that will produce a zero potential on the two plates when the plates are removed. This is done by first reflecting the charge q about the left plate to produce charge $q_1' = -q$. This produces zero potential on the left plate but not on the right plate (see **Figure 5.20b**). The charge q is then reflected by the right plate to produce image charge $q_1'' = -q$, and charge $-q_1'$ is reflected by the right plate to produce image charge $q_2' = +q$. Now, the potential is zero on the right plate, but the potential on the left plate is disturbed (see **Figure 5.20c**). The next step is to reestablish zero potential on the left plate. This is done by reflecting both charges on the right-hand side of the plates by the left plate to produce image charges $q_2'' = +q$ and $q_3' = -q$ (**Figure 5.20d**). Since there are three charges on each side of the left plane, the potential is now zero on the left plate but not on the right plate. This process repeats indefinitely, but

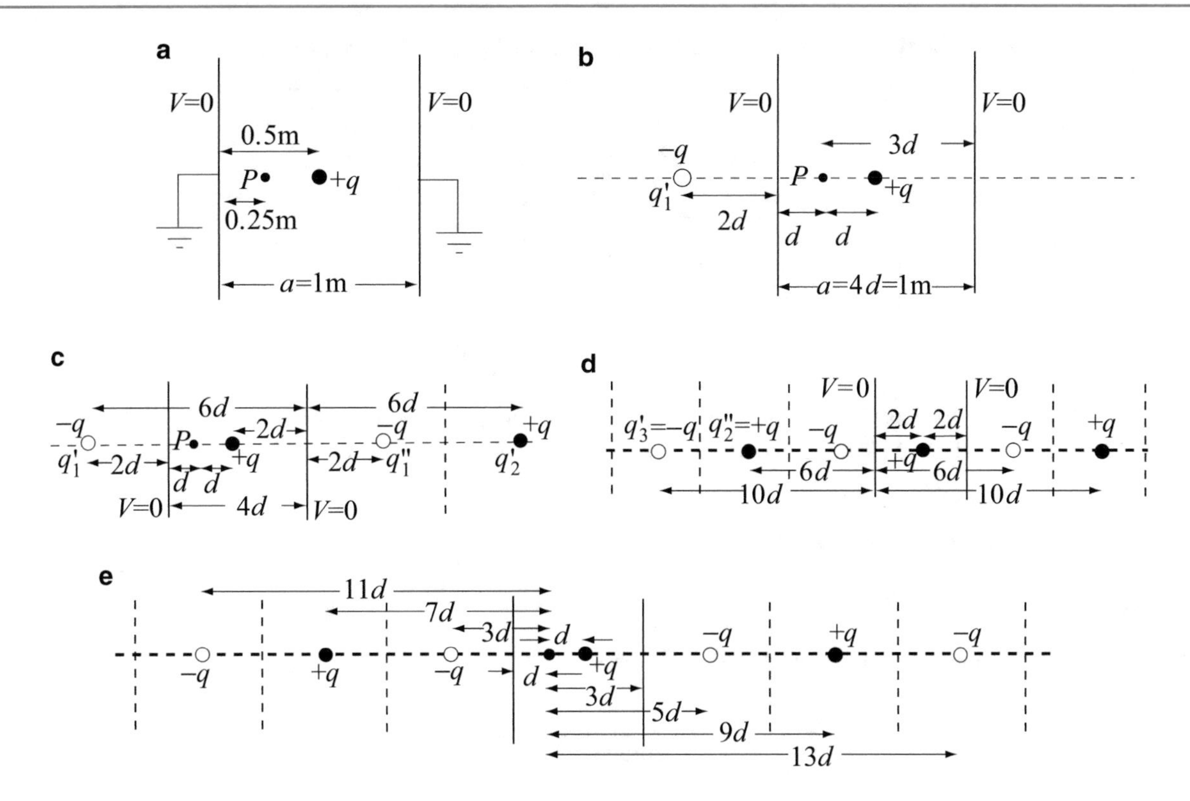

Figure 5.20 (**a**) Point charge between two infinite planes. (**b**) An image charge q_1' about the left plane ensures zero potential on the left plane. (**c**) Reflection of q and its image q_1' results in zero potential on the right plane. (**d**) Reflection of q_1'' and q_2' about the left plane restores zero potential on the left plane. (**e**) "Final" system of charges required to replace the conducting planes

each step takes us further away from the plates, reducing the disturbance we introduce. The net effect is shown in **Figure 5.20e**: There is an infinite number of charges, each placed at the center between two imaginary plates indicated by the dotted lines. The charges alternate in sign in both directions starting from the original charge and are all equal in magnitude.

To calculate the electric potential at point P, we use the dimensions in **Figure 5.20e** with $d = 0.25$ m and $a = 4d = 1$ m:

$$V_P = \frac{q}{4\pi\varepsilon_0}\left[\frac{1}{d} - \frac{1}{3d} - \frac{1}{5d} + \frac{1}{7d} + \frac{1}{9d} - \frac{1}{11d} - \frac{1}{13d} + \frac{1}{15d} + \frac{1}{17d} - \frac{1}{19d} - \frac{1}{21d} + \frac{1}{23d} + \frac{1}{25d} - \frac{1}{27d} - \frac{1}{29d} + \cdots\right. \quad [\mathrm{V}]$$

To evaluate this expression, we rewrite it as follows:

$$V_P = \frac{q}{4\pi\varepsilon_0 d}\left[1 - \sum_{i=1}^{N}\frac{1}{8i-5} - \sum_{i=1}^{N}\frac{1}{8i-3} + \sum_{i=1}^{N}\frac{1}{8i-1} + \sum_{i=1}^{N}\frac{1}{8i+1}\right] \quad [\mathrm{V}]$$

where $N = \infty$ for an exact solution. The latter can be computed for any finite number of terms. The accuracy of the result depends on how many charges we use. The following shows the results for $N = 1, 2$:

$$V_P = \frac{q}{4\pi\varepsilon_0 d}0.7206 = \frac{q}{4\times\pi\times 8.854\times 10^{-12}\times 0.25}0.7206 = 2.591\times 10^{10}q \quad [\mathrm{V}],\quad N=1$$

$$V_P = \frac{q}{4\pi\varepsilon_0 d}0.6783 = \frac{q}{4\times\pi\times 8.854\times 10^{-12}\times 0.25}0.6783 = 2.438\times 10^{10}q \quad [\mathrm{V}],\quad N=2$$

The following table shows the solution for a few additional values of N, up to $N = 10{,}000$.

N	V [V]	*Relative error* [%]
1	$2.591 \times 10^{10}q$	
2	$2.438 \times 10^{10}q$	–5.9%
10	$2.287 \times 10^{10}q$	–6.2%
100	$2.248 \times 10^{10}q$	–1.7%
1,000	$2.244 \times 10^{10}q$	–0.18%
10,000	$2.244 \times 10^{10}q$	0

After about $N = 1{,}000$, the solution changes very little and, therefore, may be regarded as an "exact" solution. Note that using $N = 10$ means a total of 43 charges. The result with $N = 10$ is less than 1.92% below the exact solution. This value is calculated as

$$\frac{V_P(N = 10{,}000) - V_P(N = 10)}{V_P(N = 10{,}000)} \times 100\% = \frac{2.244 \times 10^{10}q - 2.287 \times 10^{10}q}{2.244 \times 10^{10}q} \times 100\% = -1.92\%$$

The relative error in the third column is calculated as follows:

$$e = \frac{V_P(\text{current}) - V_P(\text{previous})}{V_P(\text{previous})} \times 100\%$$

where "current" refers to the current row (say for $N = 100$) and "previous" to the previous row (say for $N = 10$). For the values given ($d = 0.25$ m, $\varepsilon_0 = 8.854 \times 10^{-12}$ F/m), the potential at P is $2.244 \times 10^{10}q$ [V].

Note: The results above were obtained for a specific value of d. It should be remembered that the method outlined here can be used to calculate the potential at any point P between the plates (and only between the plates). The electric field intensity can be similarly calculated.

Example 5.11 Point Charge in a Conducting Box Point_Charges.m

A point charge q [C] is placed at the center of a grounded, conducting box as shown in **Figure 5.21a**. Find the system of charges required to calculate the potential inside the box.

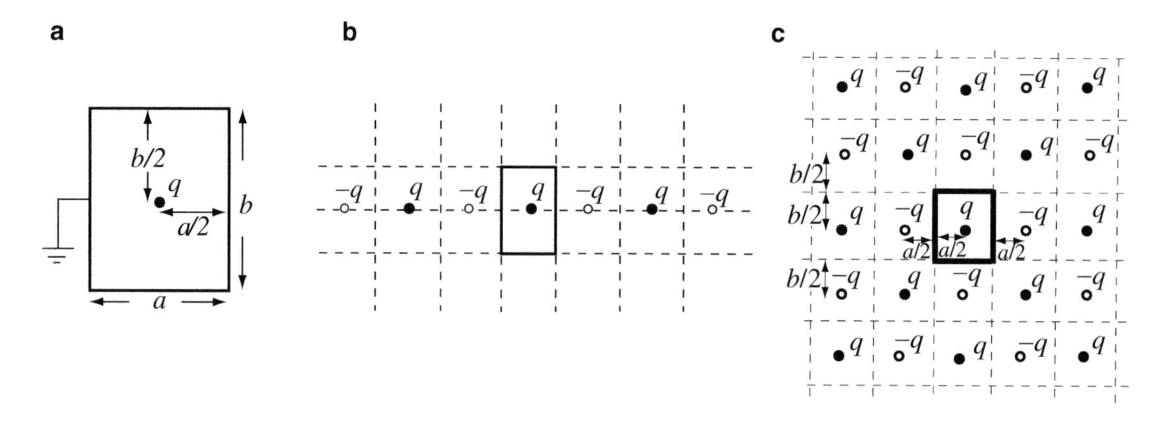

Figure 5.21 (**a**) A point charge at the center of a conducting box. (**b**) Reflection of charges about the vertical planes. (**c**) Reflection of the charges obtained in (**b**) about the horizontal planes

Solution: The box in **Figure 5.21a** may be viewed as the intersection of two sets of parallel planes, one horizontal and one vertical. We obtained the system of charges for the vertical plates in **Figure 5.20e**. Now, we can view this system of charges as being placed at the center of the horizontal plates as shown in **Figure 5.21b**. These charges are all reflected by the two horizontal surfaces with the net result given in **Figure 5.21c**. Again, the number of charges is infinite, their signs alternate, and each is inside an imaginary box identical to the original box.

Once a satisfactory number of image charges have been identified and properly placed (a number which depends on the accuracy required), the electric field intensity or potential everywhere within the box can be calculated, although hand calculation is not usually feasible. It should also be remembered that only the solution inside the original box is valid.

5.4.3.4 Images in Curved Geometries

The solution of Poisson's equation in curved geometries for a number of important physical applications can also be carried out using the method of images. An example in which this might be important is in the calculation of the electric field intensity and voltage distribution in an underground cable, as shown in **Figure 5.22**. The outer conducting shield is a surface at constant (zero) potential, and the voltage distribution inside the cable depends on the charges or potentials on the two charged conductors. This application is rather difficult to analyze by other methods, but it is relatively simple with the method of images:

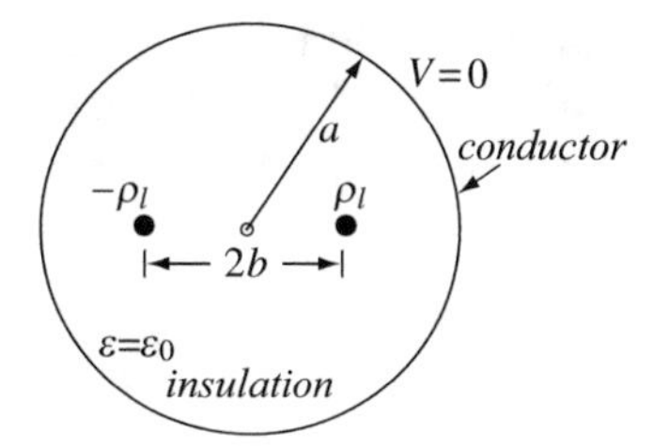

Figure 5.22 Shielded underground cable: example for use of the method of images

(1) Line Charge Outside a Conducting Cylinder. Before addressing the problem in **Figure 5.22**, we will look at a simpler configuration: that of a thin line of charge in the vicinity of a cylindrical conductor.

First, we invoke the optical equivalent: A line source (such as a thin filament) reflected by a cylindrical mirror produces an image similar to that obtained by a flat mirror, but the distance between the image and the mirror is different than between the source and the surface (see **Figure 5.23a**). In fact, we know that the image must be at a location on the r axis, at a distance $0 < b < a$, simply from the optical experience. As the source approaches the conducting surface, so does the image and vice versa, but the image is always in the range given.

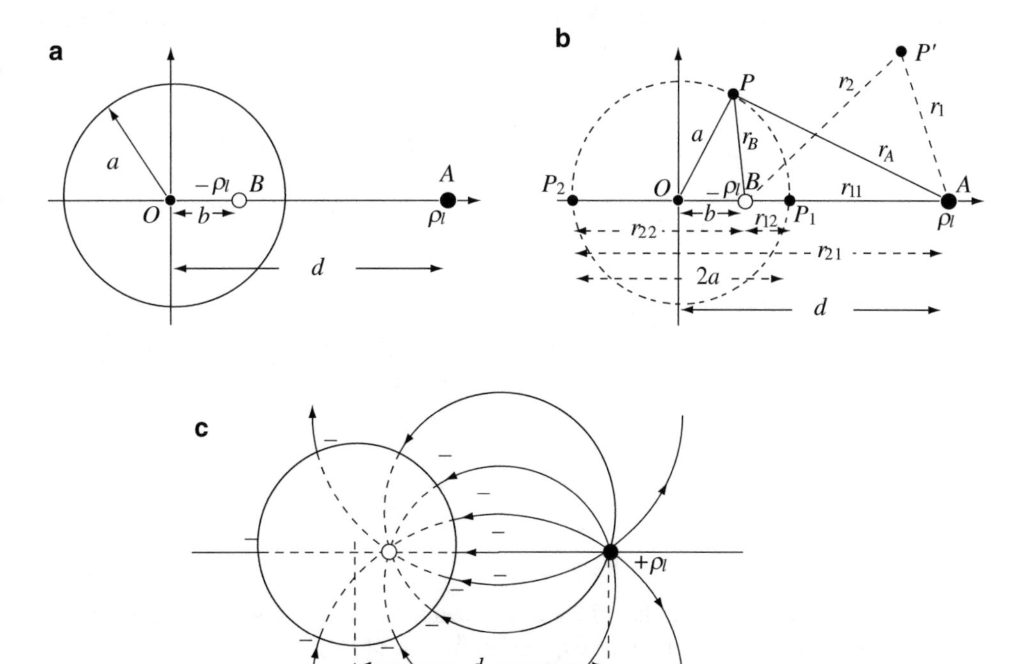

Figure 5.23 (**a**) A charged line and its image required to produce constant potential on a parallel cylindrical conductor. (**b**) Calculation of location of the image line and potential on the cylinder. (**c**) The electric field intensity outside the cylinder

To solve the problem, we must place the image at a position b such that the potential at the location of the surface (dashed circle in **Figure 5.23b**) is constant. Here, we specifically require a constant potential rather than a zero potential; the latter cannot be satisfied on a cylindrical surface with only two equal line charges but the former can. The magnitude of the image is assumed to be equal to the source but negative ($-\rho_l$ [C/m]).

We wish now to calculate the potential at a general point P' in space. Once we have the general expression, we use this to calculate the potential at the location of the cylinder using the electric field intensity of the charged line at a distance r from the line. This was calculated in **Section 3.4.2** (see **Example 3.8)** and is equal to

$$\mathbf{E} = \hat{\mathbf{r}} \frac{\rho_l}{2\pi\varepsilon_0 r} \quad \left[\frac{\text{V}}{\text{m}}\right] \tag{5.28}$$

However, we are interested in the potential. To calculate the potential difference between two points r_1 and r_0, we integrate the electric field intensities due to the two lines of charge using the definition

$$V^+ = -\int_{r_0}^{r_1} \mathbf{E} \cdot d\mathbf{l} = -\int_{r_0}^{r_1} \frac{\hat{r}' \rho_l}{2\pi\varepsilon_0 r} \cdot d\mathbf{r} = -\frac{\rho_l}{2\pi\varepsilon_0} \ln\frac{r_1}{r_0} \quad [\text{V}] \tag{5.29}$$

This calculation gives the voltage at a distance r_1 from the positive line, with reference (zero potential) at a distance r_0 from the positive line. r_0 is, of course, arbitrary as long as it is not zero or infinite.

Similarly, the potential at a distance r_2 from the negative line of charge is

$$V^- = -\int_{r_0}^{r^2} \frac{(-\rho_l)}{2\pi\varepsilon_0 r} = \frac{\rho_l}{2\pi\varepsilon_0} \ln\frac{r_2}{r_0} \quad [\text{V}] \tag{5.30}$$

where we have chosen the same reference point for both potentials. The total potential at point P' is

$$V_{P'} = V^+ + V^- = \frac{\rho_l}{2\pi\varepsilon_0}\left(\ln\frac{r_2}{r_0} - \ln\frac{r_1}{r_0}\right) = \frac{\rho_l}{2\pi\varepsilon_0} \ln\frac{r_2}{r_1} \quad [\text{V}] \tag{5.31}$$

Note that the potential at P' is nowhere zero except for $r_2 = r_1$. This can be anywhere on the plane midway between the two lines of charge.

The potential on the cylinder cannot, in fact, be calculated unless we know the exact location of the line charges, which we do not know yet. However, we know that whatever the potential on this surface, it must be constant. To calculate the location of the image line of charge, we use this property and equate the two potentials at points P_1 and P_2. Using the expression above, with r_1 and r_2 as shown in **Figure 5.23b**, we get

$$V_{P1} = \frac{\rho_l}{2\pi\varepsilon_0} \ln\frac{r_{12}}{r_{11}} = \frac{\rho_l}{2\pi\varepsilon_0} \ln\frac{a-b}{d-a} \quad [\text{V}] \tag{5.32}$$

$$V_{P2} = \frac{\rho_l}{2\pi\varepsilon_0} \ln\frac{r_{22}}{r_{21}} = \frac{\rho_l}{2\pi\varepsilon_0} \ln\frac{a+b}{a+d} \quad [\text{V}] \tag{5.33}$$

Equating the two gives

$$V_{P1} = V_{P2} = \frac{\rho_l}{2\pi\varepsilon_0} \ln\frac{a-b}{d-a} = \frac{\rho_l}{2\pi\varepsilon_0} \ln\frac{a+b}{a+d} \quad \rightarrow \quad \frac{a-b}{d-a} = \frac{a+b}{a+d} \tag{5.34}$$

or

$$\boxed{b = \frac{a^2}{d} \quad [\text{m}]} \tag{5.35}$$

Now that we know the location of the image charge, we can calculate the actual potential on the surface of the cylinder by substituting this result into the expression for V_{P1} or V_{P2}:

$$V_{P1} = \frac{\rho_l}{2\pi\varepsilon_0}\ln\frac{a-b}{d-a} = \frac{\rho_l}{2\pi\varepsilon_0}\ln\frac{da-a^2}{d(d-a)} = \frac{\rho_l}{2\pi\varepsilon_0}\ln\frac{a}{d} \quad [\mathrm{V}] \tag{5.36}$$

We must also show that this holds true for any point on the cylinder, not only at points P_1 and P_2. To do so, we calculate the potential at point P, which is a general point on the cylinder. This potential is

$$V_P = \frac{\rho_l}{2\pi\varepsilon_0}\ln\frac{r_B}{r_A} \quad [\mathrm{V}] \tag{5.37}$$

To evaluate this potential, we note that r_B/r_A must be a constant. Also, from the geometry in **Figure 5.23a**, triangles OBP and OPA must be similar triangles to ensure that the circular surface is an equal potential surface. To satisfy this condition we select b so that the angles OPB and OAP are equal. Similarity of the triangles gives

$$\frac{b}{a} = \frac{a}{d} = \frac{r_B}{r_A} \tag{5.38}$$

Thus, the potential on the surface of the cylinder is

$$\boxed{V_P = \frac{\rho_l}{2\pi\varepsilon_0}\ln\frac{a}{d} \quad [\mathrm{V}]} \tag{5.39}$$

Our assumption was correct and the potential is the same everywhere on the cylinder.

Now we can also justify the assumption that the location of the image line must be between $0 < b < a$ since as d approaches a (the source line approaches the surface of the cylinder), b approaches a. Similarly, for d approaching infinity, b approaches zero [**Eq. (5.35)**].

The configuration in **Figure 5.23b** is that of a line of charge ρ_l [C/m] and its image $-\rho_l$ [C/m]. The potential and electric field intensity due to this configuration were calculated in **Section 5.4.3.2** and that solution applies here as well, at any location outside the conducting cylinder. Inside the cylinder the electric field intensity is zero and the potential is constant and given by **Eq. (5.39)**. A plot of the electric field is shown in **Figure 5.23c**. Note that the charge density on that part of the surface closer to the positive line of charge is higher than on the opposite side, indicating a higher normal electric field intensity on that section of the conducting cylinder. The dotted lines inside the cylinder indicate the solution obtained from the two charged lines and the fact that only the solid parts of the lines exist in the physical configuration which, in this case, was that of a line of charge outside the conducting cylinder. The dotted lines are the solution to the complementary problem: that of a line of (negative) charge inside a conducting shell.

It is also worth noting again the fact that we could not reasonably assume zero potential on the surface of the cylindrical conductor. From hindsight, this is clear: two equal, opposite charged lines only produce zero potential on the plane midway between the two.

Example 5.12 Application: Underground Cable Find the electric field intensity inside the cable shown in **Figure 5.22**. The two internal lines are charged with charge densities ρ_l [C/m] and $-\rho_l$ [C/m] and are symmetric about the center of the shield. The shield is conducting and has radius a [m]. Assume the lines are very long, thin, and separated a distance $2b$ [m] apart ($b < a$).

Solution: This is the application with which we started the section of image charges in curved geometries, and we solve it here by superposition of solutions using **Eqs. (5.35)** and **(5.28)**. Taking first one of the conductors as a line of charge, we find the image line of charge outside the surface of the shield. Then, repeating the process for the second line of charge, we obtain a system of four charged lines, two positive and two negative, which produce a solution identical to that of the two lines and shield. The solution is only valid inside the shielded cable.

Taking first the positive line of charge, we obtain a situation similar to that in **Figure 5.23a**, except that the signs of the two lines of charge are reversed. From **Eq. (5.35)**, the location of the image line of charge outside the cylindrical shield is

$$b = \frac{a^2}{d} \quad \rightarrow \quad d = \frac{a^2}{b} \quad [\text{m}]$$

The image line of charge is negative and situated at a distance a^2/b from the center of the cable. Taking now the negative line of charge inside the shield, the image due to this line is positive and situated at the same distance as the image to the positive line but to the left of the cable. The four charged lines are shown in **Figure 5.24**.

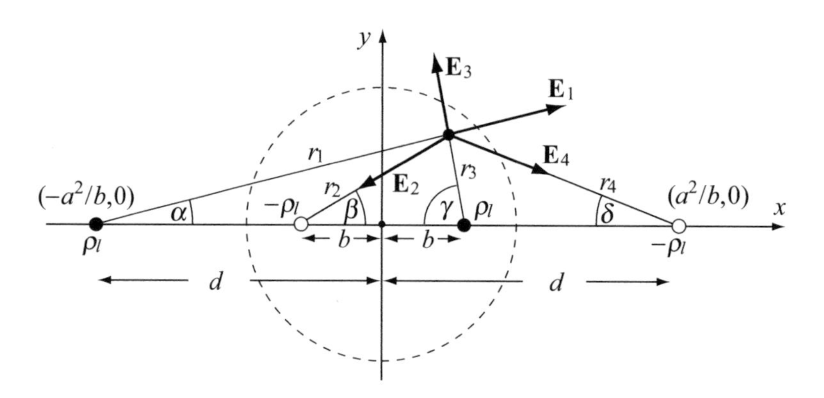

Figure 5.24 Complete system of lines of charge needed for analysis of the shielded cable in **Figure 5.22**. Dimensions and distances to a general point inside the shield are also shown

To find the electric field intensity inside the cable, we use Gauss's law. Writing the field intensity in terms of components and using the formula for the electric field intensity of a charged line, we get (see **Figure 5.24**)

$$\mathbf{E} = \hat{\mathbf{x}}\left(\frac{\rho_l\cos\alpha}{2\pi\varepsilon_0 r_1} - \frac{\rho_l\cos\beta}{2\pi\varepsilon_0 r_2} - \frac{\rho_l\cos\gamma}{2\pi\varepsilon_0 r_3} + \frac{\rho_l\cos\delta}{2\pi\varepsilon_0 r_4}\right) + \hat{\mathbf{y}}\left(\frac{\rho_l\sin\alpha}{2\pi\varepsilon_0 r_1} - \frac{\rho_l\sin\beta}{2\pi\varepsilon_0 r_2} + \frac{\rho_l\sin\gamma}{2\pi\varepsilon_0 r_3} - \frac{\rho_l\sin\delta}{2\pi\varepsilon_0 r_4}\right) \quad \left[\frac{\text{V}}{\text{m}}\right]$$

Writing r_1, r_2, r_3, r_4, $\sin\alpha$, $\cos\alpha$, $\sin\beta$, $\cos\beta$, $\sin\gamma$, $\cos\gamma$, $\sin\delta$, and $\cos\delta$ in terms of the coordinates of a general point (x,y) and the coordinates of the four charged lines, we get from **Figure 5.24**

$$r_1 = \sqrt{(x + a^2/b)^2 + y^2}, \quad r_2 = \sqrt{(x + b)^2 + y^2}, \quad r_3 = \sqrt{(x - b)^2 + y^2}, \quad r_4 = \sqrt{(x - a^2/b)^2 + y^2}$$

$$\cos\alpha = \frac{x + a^2/b}{\sqrt{(x + a^2/b)^2 + y^2}}, \quad \cos\beta = \frac{x + b}{\sqrt{(x + b)^2 + y^2}}, \quad \cos\gamma = \frac{x - b}{\sqrt{(x - b)^2 + y^2}}, \quad \cos\delta = \frac{x - a^2/b}{\sqrt{(x - a^2/b)^2 + y^2}}$$

$$\sin\alpha = \frac{y}{\sqrt{(x + a^2/b)^2 + y^2}}, \quad \sin\beta = \frac{y}{\sqrt{(x + b)^2 + y^2}}, \quad \sin\gamma = \frac{y}{\sqrt{(x - b)^2 + y^2}}, \quad \sin\delta = \frac{y}{\sqrt{(x - a^2/b)^2 + y^2}}$$

Substituting these into the expression for the electric field intensity above gives

$$\mathbf{E}(x,y) = \hat{\mathbf{x}}\frac{\rho_l}{2\pi\varepsilon_0}\left(\frac{x + a^2/b}{(x + a^2/b)^2 + y^2} - \frac{x + b}{(x + b)^2 + y^2} - \frac{x - b}{(x - b)^2 + y^2} + \frac{x - a^2/b}{(x - a^2/b)^2 + y^2}\right)$$
$$+ \hat{\mathbf{y}}\frac{\rho_l}{2\pi\varepsilon_0}\left(\frac{y}{(x + a^2/b)^2 + y^2} - \frac{y}{(x + b)^2 + y^2} + \frac{y}{(x - b)^2 + y^2} - \frac{y}{(x - a^2/b)^2 + y^2}\right) \quad \left[\frac{\text{V}}{\text{m}}\right]$$

The electric field intensity obtained here is that of four line charges in the absence of the shield. This solution is only valid inside the shield. Outside the shield, the electric field intensity is zero. The reason for this is that the charged lines produce induced charges on the shield and the combination of the fields of the line charges and those of the induced charges on the shield produce zero electric field intensity outside the shield.

(2) Point Charge Outside a Conducting Sphere. Another problem that can be solved using the method of images is that of a point charge outside a conducting sphere. We start first by looking at a point charge outside a grounded sphere (sphere at zero potential) and then extend the method to spheres at given potentials or with a given charge density on their surface. The sphere is assumed to be conducting.

The geometry and the charge are shown in **Figure 5.25a**. The charge q' is an image charge and is shown in **Figure 5.25b**, but we do not know its magnitude and location; we will calculate both. From experience with the cylindrical geometry of the previous section, we know it must be negative if q is positive and must lie somewhere between $b = 0$ and $b = a$. The means of calculation is the potential on the surface of the sphere. Calculating the potential at two points like P_1 and P_2 in **Figure 5.25b** gives the necessary relations. The potential at P_1 is

$$V_1 = \frac{1}{4\pi\varepsilon_0}\left(\frac{q}{d-a} + \frac{q'}{a-b}\right) = 0 \tag{5.40}$$

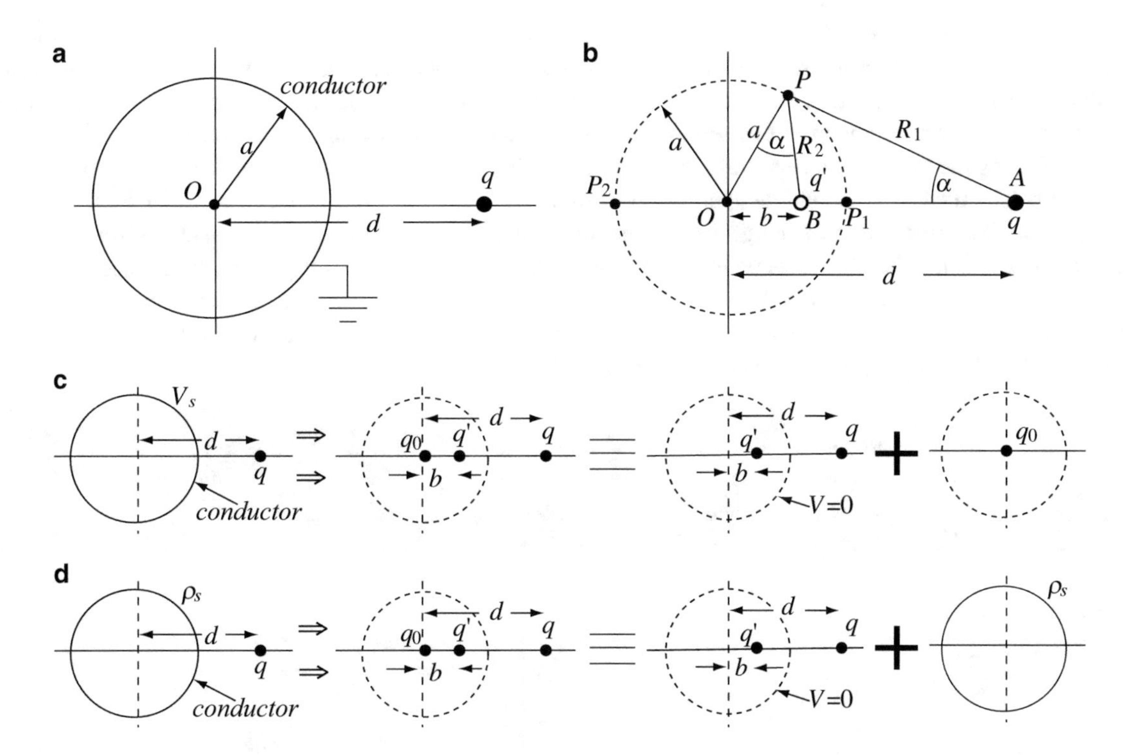

Figure 5.25 (**a**) Point charge outside a conducting sphere held at zero potential. (**b**) Assumed magnitude and location of the image point charge. (**c**) Application of superposition to take into account a constant potential on the surface. (**d**) Application of superposition to take into account surface charge density

Note that the potential everywhere on the sphere can be zero since the magnitude of the image charge is assumed to be different than that of q. The potential at P_2 similarly is

$$V_2 = \frac{1}{4\pi\varepsilon_0}\left(\frac{q}{d+a} + \frac{q'}{a+b}\right) = 0 \tag{5.41}$$

Solving these two equations for q and b gives

$$\boxed{b = \frac{a^2}{d} \quad [\mathrm{m}], \quad q' = -\frac{qa}{d} \quad [\mathrm{C}]} \tag{5.42}$$

The potentials V_1 and V_2 are zero, but is the potential zero everywhere else on the sphere? To show that it is, and therefore our solution is correct, we calculate the potential at a general point P on the sphere.

Using the law of cosines, we calculate first the distances R_1 and R_2 from the charge and its image to point P:

$$R_1 = \left(d^2 + a^2 + 2dacos\alpha\right)^{1/2}, \quad R_2 = \left(b^2 + a^2 + 2bacos\alpha\right)^{1/2} \tag{5.43}$$

where the equality of angles BPO and OAP was used. The potential at P is therefore

$$V_P = \frac{1}{4\pi\varepsilon_0}\left(\frac{q}{R_1} + \frac{q'}{R_2}\right) = \frac{1}{4\pi\varepsilon_0}\left(\frac{q}{\left(d^2 + a^2 + 2adcos\alpha\right)^{1/2}} - \frac{qa}{d\left(\left(a^2/d\right)^2 + a^2 + 2\left(a^2/d\right)acos\alpha\right)^{1/2}}\right) = 0 \tag{5.44}$$

where b and q' from **Eq. (5.42)** were substituted.

Now that we have established the location, magnitude, and sign of the image charge, the electric field intensity and potential may be calculated anywhere in space. This solution is, again, only valid outside the conducting sphere. The electric field intensity inside the conducting sphere is zero and its potential is constant and equals the potential on the surface.

A charged sphere can be analyzed in a similar manner. We start by first assuming the potential on the sphere is zero. Then, after finding the image charge as above, we add a point charge at the center to account for the charge on the surface of the sphere. The method is merely an application of superposition as shown in **Figure 5.25c** or **Figure 5.25d**. From Gauss's law, we know that the point charge at the center of the sphere must be equal to the total charge on the sphere. For a charge density ρ_s [C/m^2], the equivalent point charge is $q_0 = 4\pi a^2\rho_s$ [C]. If, on the other hand, the surface charge density is not specified, but, instead, the potential V_s on the surface of the sphere is known, then again from Gauss's law, the charge q_0 must be that point charge at the center of the sphere that will produce the given potential at its surface. From the definition of potential of a point charge, it must equal $q_0 = 4\pi\varepsilon_0 aV_s$ [C].

Example 5.13 Point Charge Inside a Conducting, Hollow Sphere A point charge Q [C] is placed in a spherical cavity in a conductor as shown in **Figure 5.26**. Calculate the electric field intensity inside the cavity and the charge density at point $P(0,0,0)$. Assume the conductor is grounded (zero potential).

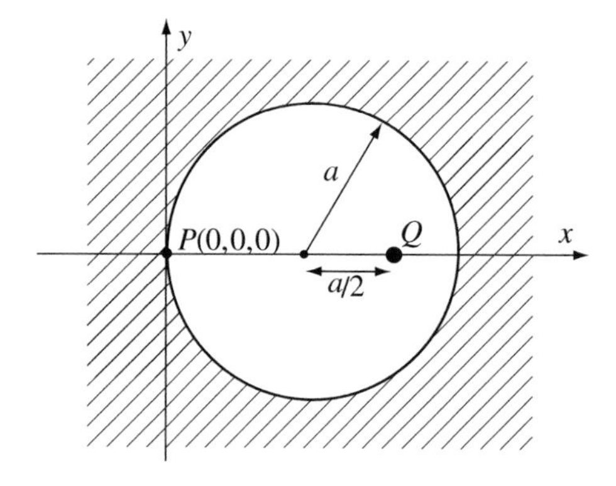

Figure 5.26 Point charge inside a hollow spherical cavity with conducting walls

Solution: Consider first the complementary problem, that of a charge outside a conducting sphere, as shown in **Figure 5.27a**. This is the same configuration considered in **Figure 5.25**, in which q' was the image charge due to the external charge q. For this configuration, the charge and its location were found in **Eq. (5.42)**:

$$q' = -\frac{qa}{d} \quad [\text{V}], \quad b = \frac{a^2}{d} \quad [\text{m}]$$

where $b = a^2/d = a/2$. For the given values,

$$b = \frac{a^2}{d} = \frac{a}{2} \rightarrow d = 2a \quad [\text{m}]$$

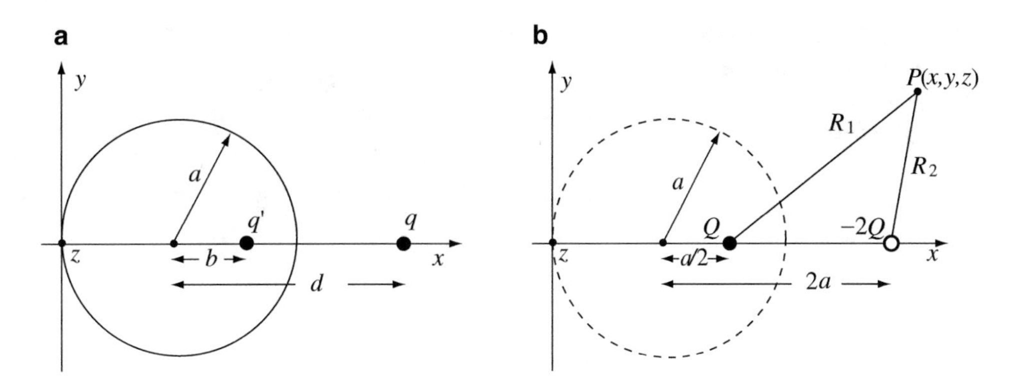

Figure 5.27 (**a**) The complementary problem to that in **Figure 5.26**: a point charge q outside a conducting sphere and its image q'. (**b**) The configuration in **Figure 5.26** with its image charge and conductor removed

However, in the configuration in **Figure 5.27b**, the charge q' is given ($q' = Q$). Therefore, we view q' as the actual charge and calculate q as its image outside the conducting sphere. Comparing **Figure 5.27a** with **Figure 5.25b**, we write

$$Q = q' = -\frac{qa}{d} \quad \rightarrow \quad q = -\frac{Qd}{a} = -2Q \quad [\mathrm{C}]$$

The configuration is now shown in **Figure 5.27b**. The solution due to this system of charges inside the sphere is identical in form to the solution inside the spherical cavity in **Figure 5.26**. The potential at a general point $P(x,y,z)$ anywhere in space is

$$V(x,y,z) = \frac{Q}{4\pi\varepsilon_0}\left[\frac{1}{R_1} - \frac{2}{R_2}\right] = \frac{Q}{4\pi\varepsilon_0}\left[\frac{1}{\left[(x-3a/2)^2 + y^2 + z^2\right]^{1/2}} - \frac{2}{\left[(x-3a)^2 + y^2 + z^2\right]^{1/2}}\right] \quad [\mathrm{V}]$$

The electric field intensity anywhere in space is calculated through the gradient. After collecting terms,

$$\mathbf{E}(x,y,z) = -\nabla V(x,y,z) = -\hat{\mathbf{x}}\frac{\partial V(x,y,z)}{\partial x} - \hat{\mathbf{y}}\frac{\partial V(x,y,z)}{\partial y} - \hat{\mathbf{z}}\frac{\partial V(x,y,z)}{\partial z}$$

$$= \frac{Q}{4\pi\varepsilon_0}\left(\frac{\hat{\mathbf{x}}(x-3a/2) + \hat{\mathbf{y}}y + \hat{\mathbf{z}}z}{\left[(x-3a/2)^2 + y^2 + z^2\right]^{3/2}} - \frac{\hat{\mathbf{x}}2(x-3a) + \hat{\mathbf{y}}2y + \hat{\mathbf{z}}2z}{\left[(x-3a)^2 + y^2 + z^2\right]^{3/2}}\right) \quad \left[\frac{\mathrm{V}}{\mathrm{m}}\right]$$

The electric charge density at $x = 0$, $y = 0$, $z = 0$ is found from the fact that at the surface of a conductor, the relation between the normal component of the electric field intensity and charge density is

$$|E_n(x,y,z)| = \frac{|\rho_s|}{\varepsilon_0} \quad \rightarrow \quad |\rho_s| = \varepsilon_0|E_n(x,y,z)| \quad [\mathrm{C/m^2}]$$

at any point (x,y,z) on the internal surface of the cavity. Although the calculation gives the magnitude of the charge density, we know the charge density must be negative since the charge that induces it is the point charge Q inside the sphere, which is positive. At (0,0,0), the normal component of the electric field intensity is in the negative x direction. Taking the x component of **E**, substituting $x = 0$, $y = 0$, $z = 0$ and multiplying by ε_0 gives the charge density at point $P(0,0,0)$:

$$\rho_s(0,0,0) = -\frac{Q}{18\pi a^2} \quad \left[\frac{\mathrm{C}}{\mathrm{m}^2}\right]$$

The solution to this problem inside the spherical cavity is shown in **Figure 5.28**. The dashed lines are the solution outside the sphere, which, although a correct solution to the complementary problem (that of a charge outside a conducting sphere), is not valid here.

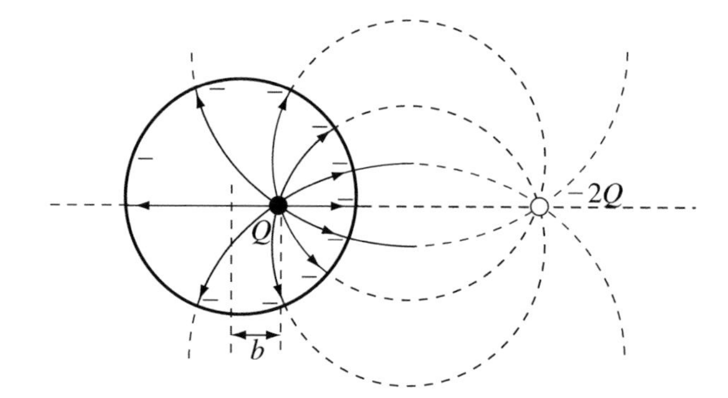

Figure 5.28 The electric field intensity inside the spherical cavity of **Figure 5.26**

Exercise 5.6 A point charge of 10 nC is placed outside a conducting sphere of radius 1 m and at a distance 0.1 m from the surface of the sphere. Calculate the potential midway between the surface of the sphere and the charge on the line connecting the charge and the center of the sphere.

Answer 1,218.13 V.

(3) **Cylindrical Distributions: Calculation of Capacitance**. Most of the applications discussed in the previous subsection dealt with point sources or thin line charge distributions. The method of images can be easily applied to other geometries. One particularly important class of applications is the calculation of capacitance per unit length of long or infinite conductors in the presence of conducting surfaces. The following example shows how this can be done.

Example 5.14 Application: Capacitance of Overhead Line A cylindrical wire of radius $a = 10$ mm runs parallel to the ground at a height $h = 10$ m. Calculate the capacitance per unit length between the wire and ground. This problem is important in determining capacitance per unit length of transmission lines and, therefore, coupling between wires and conducting surfaces such as cables inside instruments.

Solution: To calculate the capacitance per unit length, we assume that a line charge density ρ_l [C/m] exists on the conductor and calculate the potential difference between the charged line and its image, thus removing the ground. However, the potential between the two lines is twice the potential between line and ground. Thus, the capacitance between the conductor and its image will be half the capacitance between line and ground.

The geometry is shown in **Figure 5.29a**. The potential at ground level is zero, whereas the potential of the line is some constant value, yet to be determined. The first step in the solution is to replace the ground by an image of the line, as shown in **Figure 5.29b**. Since the image line has opposite sign charge, the potential at ground level remains zero. Now, we remove the conducting surfaces of both cylindrical wires and replace them by line charges, so that the potential on the upper line remains constant and positive, while the potential on the lower line remains constant and negative. The sequence is as follows: Starting with the upper conductor, we view the lower line charge as an external line charge to the upper line. Accordingly, the image inside the upper line is ρ_l and is placed a distance $b = a^2/d$ from the center of the line as shown in **Figure 5.29c**. Viewing the charge on the upper line as external to the lower line, we obtain the situation shown in **Figure 5.29d**. Combining **Figure 5.29c** and **Figure 5.29d**, the two conductors are replaced by the two line charges shown in **Figure 5.29e**. These two line charges produce constant positive potential on the conductor, constant negative potential on its image (lower conductor), and zero potential midway between them (at ground level).

The two equivalent lines of charge are at a distance $b = a^2/d$ from the center of each of the conductors. Thus, the two line charges are a distance $d{-}a^2/d$ or $2h{-}2b$ apart. Rewriting d in terms of b gives $d = a^2/b$. In terms of the known distance between ground and conductor h, we get

$$2h = b + d = b + \frac{a^2}{b} \quad \rightarrow \quad b^2 - 2hb + a^2 = 0$$

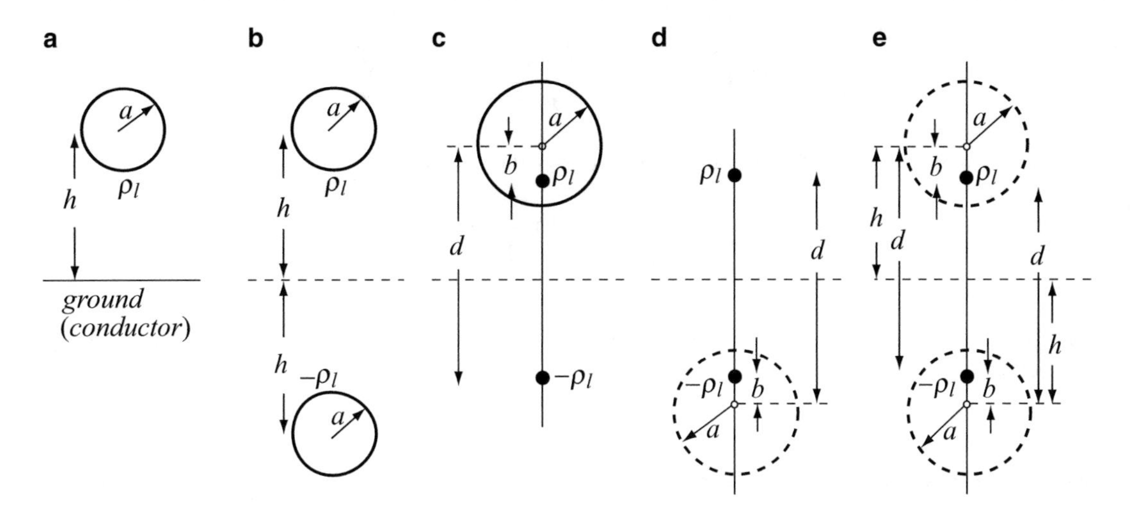

Figure 5.29 (**a**) Thick overhead conductor above ground. (**b**) The conductor and its image. (**c**) Replacement of the image by a thin line of charge that maintains the potential at the locations of the image conductor. (**d**) Replacement of the upper conductor by an equivalent thin charged line. (**e**) The equivalent configuration used for calculation

The two possible solutions for b are

$$b_1 = h + \sqrt{h^2 - a^2} \text{ and } b_2 = h - \sqrt{h^2 - a^2} \quad [\text{m}]$$

In this example, $h = 10$ m and $a = 0.01$ m. Thus, b_1 is ignored because it gives the location of the charge outside the conductor. With $b = b_2$, the distance d is (from **Figure 5.29e**)

$$d = 2h - b = h + \sqrt{h^2 - a^2} \quad [\text{m}]$$

From **Eq. (5.39)**, the potential on the upper conductor is

$$V_{uc} = -\frac{\rho_l}{2\pi\varepsilon_0}\ln\frac{a}{d} = -\frac{\rho_l}{2\pi\varepsilon_0}\ln\frac{a}{h + \sqrt{h^2 - a^2}} \quad [\text{V}]$$

This potential is positive because $a < d$. On the lower conductor, the charge is negative. Therefore, again from **Eq. (5.39)**, the potential on the lower conductor is

$$V_{lc} = \frac{\rho_l}{2\pi\varepsilon_0}\ln\frac{a}{d} = \frac{\rho_l}{2\pi\varepsilon_0}\ln\frac{a}{h + \sqrt{h^2 - a^2}} \quad [\text{V}]$$

The potential difference between the two conductors is the potential on the upper conductor minus that on the lower conductor:

$$V_t = V_{uc} - V_{lc} = -\frac{\rho_l}{2\pi\varepsilon_0}\ln\frac{a}{h + \sqrt{h^2 - a^2}} - \frac{\rho_l}{2\pi\varepsilon_0}\ln\frac{a}{h + \sqrt{h^2 - a^2}} = \frac{\rho_l}{\pi\varepsilon_0}\ln\frac{h + \sqrt{h^2 - a^2}}{a} \quad [\text{V}]$$

The total charge per unit length of the cable is ρ_l [C] (by assumption). Thus, the capacitance per unit length of the cable is

$$C = \frac{\rho_l}{V_t} = \frac{\rho_l}{\frac{\rho_l}{\pi\varepsilon_0}\ln\frac{h + \sqrt{h^2 - a^2}}{a}} = \frac{\pi\varepsilon_0}{\ln\left(\frac{h + \sqrt{h^2 - a^2}}{a}\right)} \quad \left[\frac{\text{F}}{\text{m}}\right]$$

This is the capacitance per unit length of two cylindrical conductors, separated a distance $2h$ apart. We are interested here in the capacitance between the upper conductor and ground. Because the potential difference between the upper conductor and ground is half the potential calculated above whereas the charge per unit length remains the same, the capacitance of the conductor-ground arrangement is twice that calculated above:

$$C = \frac{2\rho_l}{V_t} = \frac{2\pi\varepsilon_0}{\ln\left[\left(h + \sqrt{h^2 - a^2}\right)/a\right]} \quad \left[\frac{\text{F}}{\text{m}}\right]$$

For the values given above ($a = 0.01$ m, $h = 10$ m), the capacitance per unit length is

$$C = \frac{2\pi\varepsilon_0}{\ln\left[\left(h + \sqrt{h^2 - a^2}\right)/a\right]} = \frac{2 \times \pi \times 8.854 \times 10^{-12}}{\ln\left[\left(10 + \sqrt{10^2 - 0.01^2}\right)/0.01\right]} = 7.32 \quad \text{pF/m}$$

Note: In this solution, we implicitly assumed that the charge distribution on one wire is not affected by the charge on the ground (or the second wire). This is a good assumption if $h \gg a$.

5.4.4 Separation of Variables: Solution to Laplace's Equation

Another method of solving Laplace's equation is the general method of separation of variables. It is a common method of solution for second-order partial differential equations and should be familiar from calculus. We repeat here the basic steps because this will give us a better insight into the physical solution. The method provides general solutions to Laplace's equation. In fact, this generality will turn out to be somewhat of a disappointment because although a general solution is rather easy to obtain, a particular solution for a particular problem is not. Nevertheless, the method is one of the most useful methods for the solution of electrostatic problems as well as others. Separation of variables will only be shown in Cartesian and cylindrical coordinates. Separation of variables in spherical coordinates requires additional mathematical tools and since its utility is limited will not be discussed here.

5.4.4.1 Separation of Variables in Cartesian Coordinates **Electric_Potential.m**

To outline the method, consider first Laplace's equation in Cartesian coordinates:

$$\frac{\partial^2 V}{\partial x^2} + \frac{\partial^2 V}{\partial y^2} + \frac{\partial^2 V}{\partial z^2} = 0 \tag{5.45}$$

where V indicates any scalar function including the electric scalar potential (voltage). Assuming linear behavior, the solution $V(x,y,z)$ can be written as the product of three separate solutions:

$$V(x,y,z) = X(x)Y(y)Z(z) \tag{5.46}$$

where $X(x)$ is only dependent on the x variable, $Y(y)$ on the y variable, and $Z(z)$ on the z variable. This dependence on a single variable will allow the separation of Laplace's equation into three scalar equations, each dependent on a single variable. Substitution of the general solution into **Eq. (5.45)** gives

$$Y(y)Z(z)\frac{d^2X(x)}{dx^2} + X(x)Z(z)\frac{d^2Y(y)}{dy^2} + X(x)Y(y)\frac{d^2Z(z)}{dz^2} = 0 \tag{5.47}$$

where the partial derivatives were replaced by ordinary derivatives because of the dependency on a single variable. Dividing both sides by $V(x,y,z)$ gives

$$\frac{1}{X(x)}\frac{d^2X(x)}{dx^2} + \frac{1}{Y(y)}\frac{d^2Y(y)}{dy^2} + \frac{1}{Z(z)}\frac{d^2Z(z)}{dz^2} = 0 \tag{5.48}$$

In this form, each term depends on a single variable and, therefore, can be separated. For the separation to be valid, each term must be equal to a constant to be determined. The equation can be written as

$$\frac{1}{X(x)}\frac{d^2X(x)}{dx^2} = -k_x^2 \tag{5.49}$$

$$\frac{1}{Y(y)}\frac{d^2Y(y)}{dy^2} = -k_y^2 \tag{5.50}$$

$$\frac{1}{Z(z)}\frac{d^2Z(z)}{dz^2} = -k_z^2 \tag{5.51}$$

where the three constants must satisfy

$$\boxed{k_x^2 + k_y^2 + k_z^2 = 0} \tag{5.52}$$

and the negative sign in **Eqs. (5.49)**, through **(5.51)** as well as the use of the squared constants are arbitrary; that is, any combination of constants can be chosen, provided **Eq. (5.52)** is satisfied. This choice will be convenient later when we evaluate the constants for specific applications. We also note that only two of the constants can be chosen independently and the third is defined by **Eq. (5.52)**. The following three differential equations are obtained from **Eqs. (5.49)** through **(5.51)**:

$$\frac{d^2X(x)}{dx^2} + k_x^2X(x) = 0 \tag{5.53}$$

$$\frac{d^2Y(y)}{dy^2} + k_y^2Y(y) = 0 \tag{5.54}$$

$$\frac{d^2Z(z)}{dz^2} + k_z^2Z(z) = 0 \tag{5.55}$$

Now, the three equations are completely independent and can be solved separately. Further, these three equations have standard solutions which we can now use. The first two equations are identical in form and, therefore, have the same form of solution. Assuming that k_x^2 and k_y^2 are positive, the general solutions for $X(x)$ and $Y(y)$ are

$$\boxed{X(x) = A_1\sin(k_xx) + A_2\cos(k_xx) = B_1e^{jk_xx} + B_2e^{-jk_xx}} \tag{5.56}$$

$$\boxed{Y(y) = A_3\sin(k_yy) + A_4\cos(k_yy) = B_3e^{jk_yy} + B_4e^{-jk_yy}} \tag{5.57}$$

To find the solution for the third term, we first note from **Eq. (5.52)** that $k_z^2 = -(k_x^2 + k_y^2)$. Therefore, because k_z^2 is negative, **Eq. (5.55)** has the following standard solution:

$$\boxed{Z(z) = A_5\sinh(|k_z|z) + A_6\cosh(|k_z|z) = B_5e^{|k_z|z} + B_6e^{-|k_z|z}} \tag{5.58}$$

The constants A_i or B_i in **Eqs. (5.56)** through **(5.58)** must be evaluated to obtain a particular solution. These are evaluated from the boundary conditions of the problem. Either the harmonic form of solution or the exponential form may be used, depending on which form is most appropriate or most convenient. Thus, the general solution for the three-dimensional Laplace equation is written as

$$\boxed{\begin{aligned} V(x,y,z) &= X(x)Y(y)Z(z) \\ &= [A_1\sin(k_xx) + A_2\cos(k_xx)][A_3\sin(k_yy) + A_4\cos(k_yy)][A_5\sinh(|k_z|z) + A_6\cosh(|k_z|z)] \end{aligned}} \tag{5.59}$$

or

$$V(x,y,z) = \left[B_1 e^{jk_x x} + B_2 e^{-jk_x x}\right]\left[B_3 e^{jk_y y} + B_4 e^{-jk_y y}\right]\left[B_5 e^{|k_z|z} + B_6 e^{-|k_z|z}\right] \quad (5.60)$$

First, we note that any problem described in terms of Laplace's equation will have a solution in terms of spatial sine, cosine, hyperbolic sine, and hyperbolic cosine functions. Second, any of the constants A_i or B_i can be zero, depending on the boundary condition of the problem. Finally, although the solution appears in terms of nine unknowns ($A_1, A_2, A_3, A_4, A_5, A_6, k_x, k_y, k_z$), only six unknowns are independent. This can be verified in general, but instead of doing so, we will show this in **Example 5.16**.

The solution in **Eq. (5.59)** or **(5.60)** is a solution in three dimensions. If the potential does not vary in one of the dimensions, the problem becomes two-dimensional and the general solution is the product of two variables. In two dimensions, for example, in the x–z plane, $\partial V(x,y,z)/\partial y = 0$ and $k_y = 0$. Thus, the solutions are essentially identical with the middle term (the term $Y(y)$) removed and with $k_z^2 = -k_x^2$. The general solution is

$$V(x,z) = X(x)Z(z) = [A_1 \sin(k_x x) + A_2 \cos(k_x x)][A_5 \sinh(|k_x|z) + A_6 \cosh(|k_x|z)] \quad (5.61)$$

$$V(x,z) = \left[B_1 e^{jk_x x} + B_2 e^{-jk_x x}\right]\left[B_5 e^{|k_x|z} + B_6 e^{-|k_x|z}\right] \quad (5.62)$$

Of course, if the solution is in the x–y plane rather than in the x–z plane, z is replaced by y in the equations above without any other change being necessary. We also note that if any of the constants k_x or k_y are zero (k_z can only be zero if k_x and k_y are zero in three dimensions or if one of the constants is zero in two dimensions), the corresponding equation has a simple linear solution. If, for example, $k_x = 0$, the solution for $X(x)$ is $X(x) = A_1 x + A_2$, as can be shown by direct integration of **Eq. (5.53)** after setting $k_x = 0$.

Example 5.15 Potential Inside a Two-Dimensional Box Two very large, parallel conducting surfaces, separated a distance a and held at zero potential, are placed above but insulated from a third conducting surface that is held at a constant potential V_0 [V] as shown in **Figure 5.30a**. Assuming the potential at infinity is zero, calculate the potential everywhere in the channel defined by the three surfaces.

Solution: The geometry described here is a two-dimensional problem since the potential does not vary in the z direction. Thus, we can view the geometry as an infinitely large conducting sheet folded at $x = 0$ and $x = a$ [m]. (See **Figure 5.30a**). The solution therefore occurs in the cross section. The solution is obtained by use of **Eqs. (5.61)** and **(5.62)**, but with the y variable replacing the z variable.

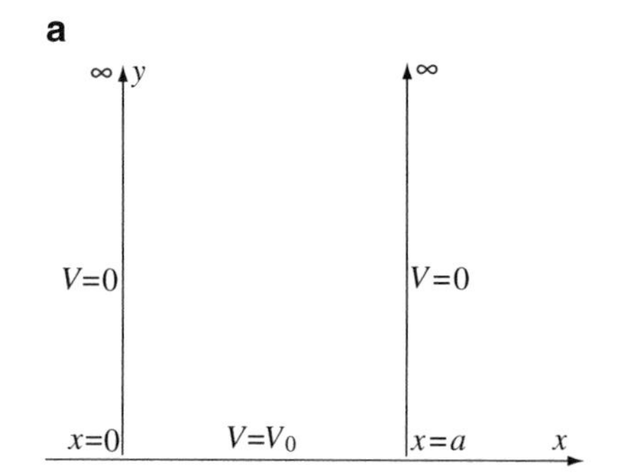

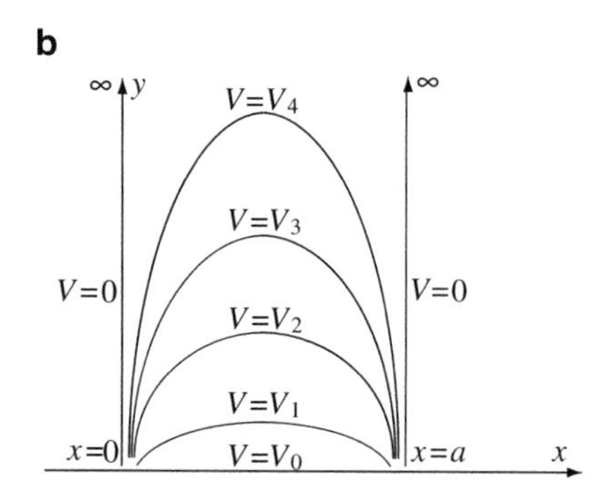

Figure 5.30 (a) Infinite box defined by $x = 0$, $x = a$, $y = 0$, $y \to \infty$. (b) Potential distribution inside the box

The general solution is

$$V(x,y) = (A_1 \sin kx + A_2 \cos kx)\left(B_1 e^{|k|y} + B_2 e^{-|k|y}\right) \quad \text{[V]}$$

where $k_y^2 = -k_x^2 = -k^2$ was used. For the solution in the y direction, we used the exponential form. This is because we anticipate using values of y that tend to infinity. For such values, exponential forms are more convenient than the hyperbolic forms. To satisfy the boundary conditions, we write

(1) At $x = 0, \rightarrow V(0,y) = A_2(B_1e^{|k|y} + B_2e^{-|k|y}) = 0 \rightarrow A_2 = 0.$

(2) At $x = a, \rightarrow V(a,y) = (A_1\sin ka)(B_1e^{|k|y} + B_2e^{-|k|y}) = 0 \rightarrow A_1\sin ka = 0$
This gives

$$ka = m\pi \quad \rightarrow \quad k = \frac{m\pi}{a} \quad \left[\frac{\text{rad}}{\text{m}}\right]$$

where m is any integer, including zero. We will, however, exclude $m = 0$ from the solution because it leads to $k = 0$ and a linear solution of the form $Ax + B$.

Similarly, the negative values of m need not be considered because negative m will only change the sign of the solution. The general solution at this stage is

$$V(x,y) = A_1\sin\left(\frac{m\pi}{a}x\right)\left(B_1e^{m\pi y/a} + B_2e^{-m\pi y/a}\right) \quad [\text{V}]$$

(3) At $y = \infty, \quad \rightarrow \quad V(x,\infty) = A_1\sin\left(\frac{m\pi x}{a}\right)\left(B_1e^{m\pi\infty/a}\right) = 0 \quad \rightarrow \quad B_1 = 0$

The solution at this stage is

$$V(x,y) = C\sin\left(\frac{m\pi x}{a}\right)e^{-m\pi y/a} \quad [\text{V}]$$

where $C = A_1B_2$.

(4) At $y = 0, V(x,0) = V_0$: To satisfy this condition, we cannot simply substitute $y = 0$ in the general solution. If we did, the solution would be sinusoidal in the x direction and no constant C can satisfy the boundary condition. However, the solution may also be written as a superposition of solutions of the above form. We write

$$V(x,y) = \sum_{m=1}^{\infty} C_m\sin\left(\frac{m\pi x}{a}\right)e^{-m\pi y/a} \quad [\text{V}]$$

Now, we substitute $y = 0$:

$$V(x,0) = V_0 = \sum_{m=1}^{\infty} C_m\sin\left(\frac{m\pi x}{a}\right) \quad [\text{V}]$$

The latter form is a Fourier sine series which, in effect, approximates the pulse $V(x,0) = V_0, 0 \leq x \leq a$, by an infinite series. In this sense, C_m are the amplitudes of the coefficients of the series. To obtain C_m, we multiply both sides by $\sin(p\pi x/a)$, where p is an integer, and integrate both sides from *zero* to a. This is a general technique we will use again and is due to Fourier himself:

$$\int_{x=0}^{a} V_0\sin\left(\frac{p\pi x}{a}\right)dx = \int_{x=0}^{a}\sum_{m=1}^{\infty} C_m\sin\left(\frac{m\pi x}{a}\right)\sin\left(\frac{p\pi x}{a}\right)dx = \sum_{m=1}^{\infty}\int_{x=0}^{a} C_m\sin\left(\frac{m\pi x}{a}\right)\sin\left(\frac{p\pi x}{a}\right)dx$$

where the integration and the sum were interchanged. Each side of the relation is integrated separately. The left-hand side gives

$$\int_{x=0}^{a} V_0\sin\left(\frac{p\pi x}{a}\right)dx = \begin{cases} \dfrac{2aV_0}{p\pi} & \text{for } p \text{ odd} \\ 0 & \text{for } p \text{ even} \end{cases}$$

For the right-hand side, we integrate each integral in the sum. For any value of m, we get

$$\int_{x=0}^{a} C_m \sin\left(\frac{m\pi x}{a}\right)\sin\left(\frac{p\pi x}{a}\right)dx = \begin{cases} \dfrac{C_m a}{2} & \text{for } p = m \\ 0 & \text{for } p \neq m \end{cases}$$

To satisfy both conditions above, m must be odd and $p = m$. Any other value yields zero. Equating the left and right hand results we obtain:

$$C_m = \frac{4V_0}{m\pi}, \quad m = 1, 3, 5, \ldots.$$

If we substitute this in the general solution, we obtain the solution inside the box:

$$V(x,y) = \frac{4V_0}{\pi} \sum_{m=1,3,5,\ldots}^{\infty} \frac{1}{m}\sin\left(\frac{m\pi x}{a}\right) e^{-m\pi y/a} \quad [\text{V}]$$

The potential distribution in the box is shown in **Figure 5.30b**.

Example 5.16 Potential in a Box A hollow rectangular box with conducting surfaces is given in **Figure 5.31**. Five of the walls are at zero potential, whereas the wall at $z = c$ [m] is insulated from the others and connected to a constant potential $V = V_0$ [V]. Calculate the potential everywhere inside the box.

Figure 5.31

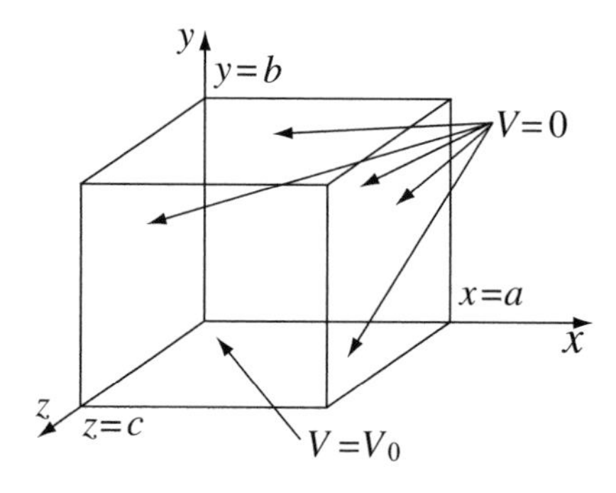

Solution: The general solution is given in **Eq. (5.59)**:

$$V(x,y,z) = [A_1\sin(k_x x) + A_2\cos(k_x x)]\left[A_3\sin(k_y y) + A_4\cos(k_y y)\right][A_5\sinh(|k_z|z) + A_6\cosh(|k_z|z)] \quad [\text{V}]$$

$$\text{at } x = 0, \quad V(0,y,z) = 0 \quad \text{at } x = a, \quad V(a,y,z) = 0$$

$$\text{at } y = 0, \quad V(x,0,z) = 0 \quad \text{at } y = b, \quad V(x,b,z) = 0$$

$$\text{at } z = 0, \quad V(z,y,0) = 0 \quad \text{at } z = c, \quad V(x,y,c) = V_0$$

We start by satisfying the boundary conditions at $x = 0$, $x = a$, $y = 0$, $y = b$, and $z = 0$:

(1) At $x = 0$:

$$V(0,y,z) = [A_2]\left[A_3\sin(k_y y) + A_4\cos(k_y y)\right][A_5\sinh(|k_z|z) + A_6\cosh(|k_z|z)] = 0 \rightarrow A_2 = 0$$

(2) At $x = a$:

$$V(a,y,z) = A_1\sin(k_x a)\left[A_3\sin(k_y y) + A_4\cos(k_y y)\right]\left[A_5\sinh(|k_z|z) + A_6\cosh(|k_z|z)\right] = 0$$

This condition can be satisfied if

$$\sin(k_x a) = 0 \;\rightarrow\; k_x a = m\pi \;\rightarrow\; k_x = \frac{m\pi}{a}\;\left[\frac{\text{rad}}{\text{m}}\right], \quad m = 1, 2, 3, \ldots$$

(3) At $y = 0$:

$$V(x,0,z) = A_1\sin\left(\frac{m\pi x}{a}\right)[A_4][A_5\sinh(|k_z|z) + A_6\cosh(|k_z|z)] = 0 \;\rightarrow\; A_4 = 0$$

(4) At $y = b$:

$$V(x,b,z) = A_1\sin\left(\frac{m\pi x}{a}\right)A_3\sin(k_y b)\,[A_5\sinh(|k_z|z) + A_6\cosh(|k_z|z)] = 0$$

This condition can be satisfied if

$$\sin(k_y b) = 0 \;\rightarrow\; k_y b = n\pi \;\rightarrow\; k_y = \frac{n\pi}{b}\;\left[\frac{\text{rad}}{\text{m}}\right], \quad n = 1, 2, 3, \ldots$$

(5) At $z = 0$:

$$V(x,y,0) = A_1\sin\left(\frac{m\pi}{a}x\right)A_3\sin\left(\frac{n\pi}{b}x\right)[A_6] = 0 \;\rightarrow\; A_6 = 0$$

Note: Both m and n can be zero, but this leads to a linear solution (see **Example 5.15**). Also, m and n can be negative, but since negative values merely change the sign of the solution, we exclude these values. Now, since $k_z^2 = -(k_x^2 + k_y^2)$ [see **Eq. (5.52)**] and combining the constants A_1, A_3, and A_5 into a single constant, we can write the general solution at this stage as

$$V(x,y,z) = C\sin\left(\frac{m\pi}{a}x\right)\sin\left(\frac{n\pi}{b}y\right)\sinh\left(\sqrt{k_x^2 + k_y^2}z\right) = C\sin\left(\frac{m\pi}{a}x\right)\sin\left(\frac{n\pi}{b}y\right)\sinh\left(\pi z\sqrt{\frac{m^2}{a^2} + \frac{n^2}{b^2}}\right) \quad [\text{V}]$$

First, we note that there is only one condition to be satisfied: the sixth and final boundary condition at $z = c$. Before using this condition, we note that any superposition of solutions is also a solution to the original equation. The following sum may be taken as a general solution (see discussion in **Example 5.15**):

$$V(x,y,z) = \sum_{m=1}^{\infty}\sum_{n=1}^{\infty} C_{mn}\sin\left(\frac{m\pi}{a}x\right)\sin\left(\frac{n\pi}{b}y\right)\sinh\left(\pi z\sqrt{\frac{m^2}{a^2} + \frac{n^2}{b^2}}\right) \quad [\text{V}]$$

(6) The sixth boundary condition is substituted into this sum in order to evaluate the constant C_{mn}. For $z = c$, $V(x,y,c) = V_0$, and using the short form notation for $|k_z|$

$$k_z = \pi\sqrt{\frac{m^2}{a^2} + \frac{n^2}{b^2}}\quad\left[\frac{\text{rad}}{\text{m}}\right]$$

we get the general solution as

$$V(x,y,c) = V_0 = \sum_{m=1}^{\infty}\sum_{n=1}^{\infty} C_{mn}\sin\left(\frac{m\pi x}{a}\right)\sin\left(\frac{n\pi y}{b}\right)\sinh(k_z c) \qquad [\text{V}]$$

As in **Example 5.15**, this is a Fourier series, but now it is a series in two variables and applies anywhere on the plane $z = c$ $(0 \le x \le a, 0 \le y \le b)$. To find the constant C_{mn}, we perform two operations. First, we multiply both sides by the term sin $(p\pi x/a)$ $\sin(q\pi y/b)$, where p and q are integers (see **Example 5.15**). This gives

$$V_0\sin\left(\frac{p\pi x}{a}\right)\sin\left(\frac{q\pi y}{b}\right) = \sum_{m=1}^{\infty}\sum_{n=1}^{\infty} C_{mn}\sin\left(\frac{m\pi x}{a}\right)\sin\left(\frac{n\pi y}{b}\right)\sin\left(\frac{p\pi x}{a}\right)\sin\left(\frac{q\pi y}{b}\right)\sinh(k_z c)$$

In the second step, both sides of the expression are integrated on the plane $z = c$ in the limits $0 \le x \le a$, $0 \le y \le b$. To simplify the evaluation, we perform the integration on the left-hand side and right-hand side separately. For the left-hand side,

$$\int_{x=0}^{a}\int_{y=0}^{b} V_0\sin\left(\frac{p\pi x}{a}\right)\sin\left(\frac{q\pi y}{b}\right)dxdy = V_0\left[\int_{x=0}^{a}\sin\left(\frac{p\pi x}{a}\right)dx\right]\left[\int_{y=0}^{b}\sin\left(\frac{q\pi y}{b}\right)dy\right]$$

$$= \begin{cases} V_0\left[\dfrac{2a}{p\pi}\right]\left[\dfrac{2b}{q\pi}\right] = \dfrac{4abV_0}{pq\pi^2} & \text{for } p, q, \text{ odd} \\ 0 & \text{for } p, q, \text{ even} \end{cases}$$

For the right-hand side, interchanging the summation and integration and evaluating each of the integrals in the sum separately yields

$$\int_{x=0}^{a}\int_{y=0}^{b} C_{mn}\left[\sin\left(\frac{m\pi x}{a}\right)\sin\left(\frac{p\pi x}{a}\right)\right]\left[\sin\left(\frac{n\pi y}{b}\right)\sin\left(\frac{q\pi y}{b}\right)\right]\sinh(k_z c)dx\,dy$$

$$= C_{mn}\sinh(k_z c)\left[\int_{x=0}^{a}\sin\left(\frac{m\pi x}{a}\right)\sin\left(\frac{p\pi x}{a}\right)dx\right]\left[\int_{y=0}^{b}\sin\left(\frac{n\pi y}{b}\right)\sin\left(\frac{q\pi y}{b}\right)dy\right]$$

$$= C_{mn}\sinh(k_z c)\left[\int_{x=0}^{a}\frac{1}{2}\left[\cos\left(\frac{(m-p)\pi x}{a}\right) - \cos\left(\frac{(m+p)\pi x}{a}\right)\right]dx\right]$$

$$\times\left[\int_{y=0}^{b}\frac{1}{2}\left[\cos\left(\frac{(n-q)\pi y}{b}\right) - \cos\left(\frac{(n+q)\pi y}{b}\right)\right]dy\right] = \begin{cases} \dfrac{C_{mn}ab\sinh(k_z c)}{4} & \text{for } \quad m = p, n = q \\ 0 & m \ne p, n \ne q \end{cases}$$

Now, satisfying both the conditions $m = p$, $n = q$, and m and n, odd, we get

$$C_{mn} = \frac{16V_0}{mn\pi^2\sinh(k_z c)}$$

Substituting this back into the general solution, we get by equating the left- and right-hand sides

$$V(x,y,z) = \frac{16V_0}{\pi^2}\sum_{m=1,3,5,\ldots}^{\infty}\;\sum_{n=1,3,5,\ldots}^{\infty}\frac{\sin\left(\frac{m\pi}{a}x\right)\sin\left(\frac{n\pi}{b}y\right)\sinh\left(\pi z\sqrt{\left(\frac{m}{a}\right)^2 + \left(\frac{n}{b}\right)^2}\right)}{mn\sinh\left(\pi c\sqrt{\left(\frac{m}{a}\right)^2 + \left(\frac{n}{b}\right)^2}\right)} \qquad [\text{V}]$$

Not a simple result but doable. In particular, note the similarities between this and the previous example: the difference is that here we used a Fourier series in two variables to approximate the solution.

Finally, it is worth reiterating that to obtain the solution, we used six boundary conditions. This means that only six independent constants needed to be evaluated. In this case, these were A_2, k_x, A_4, k_y, A_6, and C_{mn}.

5.4.4.2 Separation of Variables in Cylindrical Coordinates

An essentially identical process to that in the previous section can be followed in cylindrical coordinates. We can start with the general Laplace equation in cylindrical coordinates and proceed in exactly the same manner by assuming a general solution and separating the equation into three independent equations. However, the equations themselves are expected to be different and, in fact, they are more difficult to solve as they require, in addition to harmonic functions, knowledge of Bessel functions. For this reason, we will not pursue here the solution of the general Laplace equation in cylindrical coordinates. Rather, we solve a simplified form of Laplace's equation in two dimensions, which only requires harmonic functions. In spite of this simplification, the process is illustrative of the general method and has important practical application in solution of real problems.

The type of physical configuration we look at is any physical problem in which the potential does not vary with the z dimension. In other words, the partial derivative with respect to z is zero. With this condition, Laplace's equation in cylindrical coordinates becomes

$$\frac{1}{r}\frac{\partial}{\partial r}\left(r\frac{\partial V}{\partial r}\right)+\frac{1}{r^2}\frac{\partial^2 V}{\partial \phi^2}=0 \tag{5.63}$$

This equation represents physical applications in which the third dimension is large (or perhaps infinite) and the potential does not vary in this dimension. Following the steps of the previous section, we write a general solution as the product of two independent solutions:

$$V(r,\phi)=R(r)\Phi(\phi) \tag{5.64}$$

and, as before, the implicit assumption is that the solutions are linear. Substitution of the general solution in **Eq. (5.63)** gives

$$\frac{\Phi(\phi)}{r}\frac{\partial}{\partial r}\left(r\frac{\partial R(r)}{\partial r}\right)+\frac{R(r)}{r^2}\frac{\partial^2\Phi(\phi)}{\partial \phi^2}=0 \tag{5.65}$$

The equation is now divided by $V(r,\phi)$:

$$\frac{1}{rR(r)}\frac{1}{\partial r}\frac{\partial}{\partial r}\left(r\frac{\partial R(r)}{\partial r}\right)+\frac{1}{r^2\Phi(\phi)}\frac{\partial^2\Phi(\phi)}{\partial \phi^2}=0 \tag{5.66}$$

This equation is still not separable since both terms depend on r. To eliminate this dependence, we multiply both sides of the equation by r^2 and obtain

$$\frac{r}{R(r)}\frac{\partial}{\partial r}\left(r\frac{\partial R(r)}{\partial r}\right)+\frac{1}{\Phi(\phi)}\frac{\partial^2\Phi(\phi)}{\partial \phi^2}=0 \tag{5.67}$$

Now the first term depends only on r whereas the second term depends only on ϕ. Choosing a constant k_r^2 for the first term and k_ϕ^2 for the second term, we can write

$$\frac{r}{R(r)}\frac{d}{dr}\left(r\frac{dR(r)}{dr}\right)+k_r^2=0 \tag{5.68}$$

$$\frac{1}{\Phi(\phi)}\frac{d^2\Phi(\phi)}{d\phi^2}+k_\phi^2=0 \tag{5.69}$$

and

$$\boxed{k_r^2 + k_\phi^2 = 0} \tag{5.70}$$

The choice here is to use a positive constant for k_ϕ^2 since, then, the solution is identical in form to that in **Eq. (5.56)**. The solution for $\Phi(\phi)$ is

$$\boxed{\Phi(\phi) = A_1 \sin(k\phi) + A_2 \cos(k\phi) = B_1 e^{jk\phi} + B_2 e^{-jk\phi}} \tag{5.71}$$

where the term k was used as $k^2 = k_\phi^2 = -k_r^2$. The equation for $R(r)$ can now be written as

$$r^2 \frac{d^2R(r)}{dr^2} + r\frac{dR(r)}{dr} - k^2 R(r) = 0 \tag{5.72}$$

Also, because the solution for $\Phi(\phi)$ is periodic, k must be an integer. The solution of **Eq. (5.72)** is

$$R(r) = B_3 r^k + B_4 r^{-k} \tag{5.73}$$

and the solution for $V(r,\phi)$ is

$$\boxed{V(r,\phi) = R(r)\Phi(\phi) = [A_1 \sin(k\phi) + A_2 \cos(k\phi)]\left[B_3 r^k + B_4 r^{-k}\right]} \tag{5.74}$$

or

$$\boxed{V(r,\phi) = R(r)\Phi(\phi) = \left[B_1 e^{jk\phi} + B_2 e^{-jk\phi}\right]\left[B_3 r^k + B_4 r^{-k}\right]} \tag{5.75}$$

Again, we obtained a general solution for any problem that can be described in r–ϕ coordinates. As previously, the constants are found by satisfying the equation for known boundary conditions. We also note that we must calculate a total of four independent constants (i.e., k and three constant coefficients).

If $k = 0$, **Eqs. (5.68)** and **(5.69)** become one-dimensional Laplace equations and each can be integrated directly as was done in **Section 5.4.2**. In this case, the solutions are (see **Section 5.4.2**)

$$\boxed{V(r) = A \ln r + B} \quad \boxed{V(\phi) = C\phi + D} \tag{5.76}$$

Example 5.17 Application: The Electrostatic Precipitator An electrostatic precipitator is a device designed to collect both positively and negatively charged ash in the stack of a power plant. The stack is high and is built as shown in **Figure 5.32a**. The central conductor is in the form of a mesh to allow particles to move around. With the dimensions and potentials as given, calculate the potential distribution everywhere in the stack.

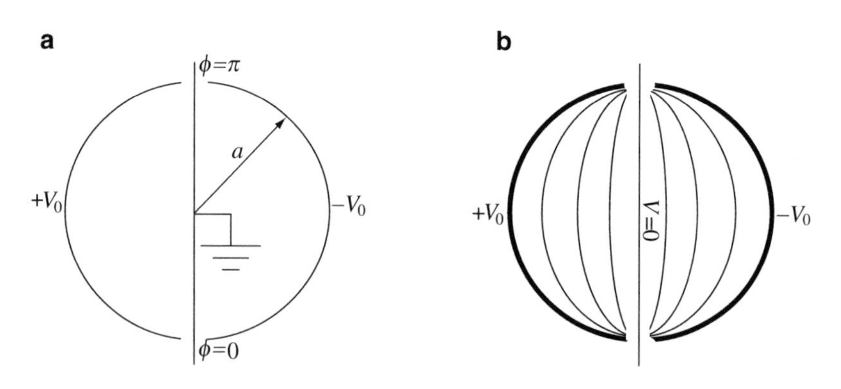

Figure 5.32 (**a**) Cross section of an electrostatic precipitator. (**b**) Potential distribution inside the precipitator

Solution: Because the stack is circular, a cylindrical system of coordinates is used. If the system is placed with the z axis along the axis of the stack, the potential can only vary in the r and ϕ directions. Because the stack is high, we may assume that any edge effects at the bottom and top of the stack can be neglected. The equation describing the potential is **Eq. (5.63)** and the general solution is given in **Eq. (5.74)**:

$$V(r,\phi) = [A_1\sin(k\phi) + A_2\cos(k\phi)]\left[B_3r^k + B_4r^{-k}\right]$$

The boundary conditions are

$$V(a,\phi) = -V_0 \quad \text{for} \quad 0 < \phi < \pi \text{ and } V(a,\phi) = V_0 \quad \text{for} \quad \pi < \phi < 2\pi$$

Because the potential must be finite everywhere in the stack, the term containing r^{-k} must be zero (i.e., $B_4 = 0$); otherwise, as we approach the center of the stack, the potential would approach infinity. The solution therefore becomes

$$V(r,\phi) = B_3r^k[A_1\sin(k\phi) + A_2\cos(k\phi)] \qquad [\text{V}]$$

Now, we note that the solution $V(r,\phi)$ must be an odd function of ϕ. This can be seen from the fact that the boundary condition is negative for $0 < \phi < \pi$, $2\pi < \phi < 3\pi$, etc., and positive for $\pi < \phi < 2\pi$, $3\pi < \phi < 4\pi$, etc. Therefore, the cosine term must also vanish everywhere ($A_2 = 0$). The solution is

$$V(r,\phi) = Cr^k\sin(k\phi) \qquad [\text{V}]$$

This solution, while correct, is not sufficient to satisfy the problem and its boundary conditions. To ensure that the problem is satisfied everywhere, we take an infinite sum of all possible solutions of the above form:

$$V(r,\phi) = \sum_{k=1}^{\infty} C_k r^k\sin(k\phi) \qquad [\text{V}]$$

To evaluate the constants C_k, we use the method in **Example 5.15**. First, we note that at $r = a$, the conditions are

$$V(a,\phi) = -V_0 = \sum_{k=1}^{\infty} C_k a^k\sin(k\phi) \quad \text{for} \quad 0 < \phi < \pi$$

$$V(a,\phi) = V_0 = \sum_{k=1}^{\infty} C_k a^k\sin(k\phi) \quad \text{for} \quad \pi < \phi < 2\pi$$

Now, we follow the process discussed in **Example 5.15** and multiply both sides by $\sin(p\pi\phi/\pi) = \sin(p\phi)$ and integrate between $\phi = 0$ and $\phi = \pi$. Taking the first of these expressions, we get

$$-\int_{\phi=0}^{\pi} V_0\sin(p\phi)d\phi = \sum_{k=1}^{\infty}\int_{\phi=0}^{\pi} C_k a^k\sin(k\phi)\sin(p\phi)d\phi$$

From the left-hand side,

$$-\int_{\phi=0}^{\pi} V_0\sin(p\phi)d\phi = \begin{cases} -\dfrac{2V_0}{p} & \text{for } p \text{ odd} \\ 0 & \text{for } p \text{ even} \end{cases}$$

From the right-hand side,

$$-\int_{\phi=0}^{\pi} C_k a^k\sin(k\phi)\sin(p\phi)d\phi = \begin{cases} \dfrac{\pi C_k a^k}{2} & \text{for } p = k \\ 0 & \text{for } p \neq k \end{cases}$$

Thus, k must be odd and $p = k$. The constant C_k is therefore

$$C_k = -\frac{4V_0}{k\pi a^k}$$

Substituting this in the general solution, we get

$$V(r,\phi) = \sum_{k=1,3,5,\ldots}^{\infty} \frac{-4V_0}{k\pi a^k} r^k \sin(k\phi) \quad [\text{V}], \quad 0 < \phi < 2\pi$$

A plot of the solution is shown in **Figure 5.32b**.

Exercise 5.7 The configuration in **Example 5.17** is given again. Using the method shown, calculate and plot the potential outside the stack. **Hint**: Outside the stack the term containing r^k must be zero so that the potential decreases to zero at infinity, but the term containing r^{-k} must be retained.

Answer

$$V(r,\phi) = \sum_{k=1,3,5,\ldots}^{\infty} \frac{-4V_0 a^k}{k\pi r^k} \sin(k\phi) \quad [\text{V}], \quad 0 < \phi < 2\pi$$

5.5 Summary

Chapter 5 continues discussion of methods of solution for the electrostatic field. The methods are called collectively boundary value problems because they essentially solve Poisson's (or Laplace's) equation under given boundary conditions.

Poisson's equation for the electric potential is the basis of the methods that follow:

$$\nabla^2 V = -\frac{\rho_v}{\varepsilon}, \quad \text{in Cartesian coordinates}: \quad \frac{\partial^2 V}{\partial x^2} + \frac{\partial^2 V}{\partial y^2} + \frac{\partial^2 V}{\partial z^2} = -\frac{\rho_v}{\varepsilon} \qquad \textbf{(5.4)} \text{ and } \textbf{(5.5)}$$

If the right-hand side is zero, we obtain ***Laplace's equation***. The equation can be written in cylindrical and spherical coordinates as well [**Eqs. (5.6)** and **(5.7)**]. Solution provides the potential and, if necessary, the electric field intensity. ***Direct integration*** of either Poisson's or Laplace's equations is possible if the potential only depends on one variable (one-dimensional problem). If direct integration can be used, the process is as follows:

(1) Integrate both sides of the equation twice to obtain the potential as a function of the variable.
(2) Substitute known values of the potential (boundary conditions) to obtain the constants of integration.
(3) Substitute the constants into the function in **(1)** to obtain the particular solution for V.

Method of Images The method solves for fields and potentials due to charges (point charges or charge distributions) in the presence of conductor(s) by removing the conductor(s) and replacing them with image charges that maintain the potential on the surface(s) unchanged. Then the solution outside the conductor(s) is due to the original charges and the image charges.

Flat Conducting Surfaces

(1) Image charges are equal in magnitude to the original charges, opposite in signs, and placed at the same distance from the surface, in the space occupied by the conductor. The conductor is removed.
(2) The number of image charges depends on the number of conducting surfaces.
(3) For two parallel conducting plates with point charges between them, the number of image charges is infinite.
(4) For conducting surfaces at an angle α, if $n = 180°/\alpha$ is an integer, the number of image charges is $2n - 1$. If n is not an integer, the method cannot be used (see **Example 5.6**).

Line, surface, or volume charge distributions in the presence of conductors behave the same way as point charges since they are assemblies of point charges. The images are then lines, surfaces, or volumes of the same geometric shape and size, mirrored about the conducting surface(s).

Cylindrical Surfaces Lines of charge parallel to conducting cylindrical surfaces reflect as follows:

(1) The image line of charge is parallel and equal in magnitude to the original line of charge. The sign of the image line of charge is opposite (**Figure 5.23**).
(2) A line of charge at a distance d outside a cylindrical conductor of radius $a < d$ produces an image charge at a distance b from the center of the conductor such that

$$b = \frac{a^2}{d} \quad [\mathrm{m}]. \tag{5.35}$$

Point Charges and Conducting Spherical Surfaces A point charge q at a distance d from the center of a conducting sphere of radius $a < d$ creates an image with magnitude q' inside the sphere at a distance b from its center (**Figure 5.27**):

$$\boxed{b = \frac{a^2}{d} \quad [\mathrm{m}], \quad q' = -\frac{qa}{d} \quad [\mathrm{C}]} \tag{5.42}$$

Notes:

(1) In all cases, the images do not actually exist—they are artificially postulated for calculation purposes. They replace the effect of conductors.
(2) The fields found are only valid outside conductors.
(3) The potential of the conducting surface is assumed to be constant and, except for cylindrical surfaces, to be zero.
(4) The effect of any other potential, charge, or charge distribution must be added to the solution found by the method of images.

Separation of Variables Laplace's equation is a second-order partial differential equation and can be solved using the method of separation of variables in any system of coordinates. The general solution is the following.

In Cartesian coordinates (3 dimensions):

$$V(x,y,z) = [A_1\sin(k_x x) + A_2\cos(k_x x)][A_3\sin(k_y y) + A_4\cos(k_y y)][A_5\sinh(|k_z|z) + A_6\cosh(|k_z|z)] \quad [\mathrm{V}] \tag{5.59}$$

or

$$V(x,y,z) = \left[B_1 e^{jk_x x} + B_2 e^{-jk_x x}\right]\left[B_3 e^{jk_y y} + B_4 e^{-jk_y y}\right]\left[B_5 e^{|k_z|z} + B_6 e^{-|k_z|z}\right] \quad [\mathrm{V}] \tag{5.60}$$

where

$$k_x^2 + k_y^2 + k_z^2 = 0 \tag{5.52}$$

The constants A_i and B_i are evaluated from known boundary conditions (see **Example 5.15**).

In Cartesian coordinates (2 dimensions in x–z), with $k_z^2 = -k_x^2$:

$$V(x,y) = [A_1\sin(k_x x) + A_2\cos(k_x x)][A_5\sinh(|k_x|z) + A_6\cosh(|k_x|z)] \quad [\mathrm{V}] \tag{5.61}$$

or

$$V(x,z) = \left[B_1 e^{jk_x x} + B_2 e^{-jk_x x}\right]\left[B_5 e^{j|k_x|z} + B_6 e^{-j|k_x|z}\right] \quad [\mathrm{V}] \tag{5.62}$$

In 2-dimensional cylindrical coordinates (variation in the r–ϕ plane, no variation of the field in the z direction):

$$V(r,\phi) = [A_1\sin(k\phi) + A_2\cos(k\phi)][B_3 r^k + B_4 r^{-k}] \quad [\mathrm{V}] \tag{5.74}$$

or

$$V(r,\phi) = [A_1 e^{jk\phi} + A_2 e^{-jk\phi}][B_3 r^k + B_4 r^{-k}] \quad [\text{V}] \quad \textbf{(5.75)}, \quad k^2 = k_\phi^{\,2} = -k_r^{\,2}$$

The main difficulty in using these general solutions is the evaluation of the constants A_i and B_i from the boundary conditions. The task is highly dependent on geometry and the given boundary conditions.

Problems

Laplace's and Poisson's Equations

5.1 Solution to Laplace's Equation. The solution of Laplace's equation is known to be $V = f(x)g(y) = fg$, where f and g are scalar functions, f is a linear function of x alone, and g is a linear function of y alone. Which of the following are also solutions to Laplace's equation?

(a) $V = 2fg$.
(b) $V = fg + x + yg$.
(c) $V = 2f + cg$, where c is a constant.
(d) $V = f^2 g$.

5.2 Solution to Laplace's Equation. The potential produced by a point charge Q [C] at a distance R [m] in free space is given as $V(R) = Q/4\pi\varepsilon_0 R$ [V].

(a) Show that this potential satisfies Laplace's Equation at any R except $R = 0$.
(b) What happens at $R = 0$?

Direct Integration

5.3 Application: Potential Distribution in a Capacitor. A parallel plate capacitor with distance between plates of 2 mm contains between its plates a dielectric with permittivity $\varepsilon = 4\varepsilon_0$ [F/m]. The potential difference between the plates is 5 V. Suppose the plate at $x = 0$ is at zero potential, the plate at $x = 0.002$ m is at 5 V. Find $V(x)$ everywhere assuming there are no edge effects.

5.4 Application: Uniform Space Charge Density in a Capacitor. The parallel plate capacitor in **Figure 5.33** contains a dielectric with permittivity ε_0 [F/m] and a constant charge density $\rho_v = \rho_0$ [C/m^3] distributed uniformly throughout the dielectric. Assume the plates are infinite and calculate the potential everywhere between the plates.

5.5 Nonuniform Space Charge in a Capacitor. A parallel plate capacitor with plates separated a distance $d = 2$ mm is connected to a 100 V source as in **Figure 5.33**. The space between the plates is air, but there is also a charge distribution between the plates given as $\rho_v = 10^{-6}x(x - d)$ [C/m^3] where x is the distance from the zero voltage plate. Assume the plates are large and calculate the potential everywhere between the plates.

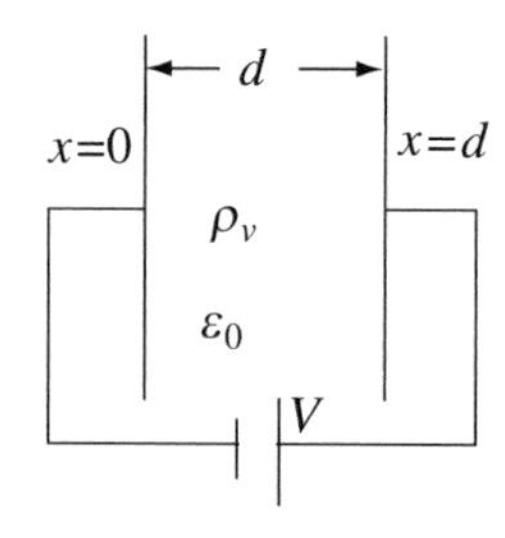

Figure 5.33

5.6 Application: Potential and Field in Coaxial Cables. Coaxial cables used for cable TV also distribute power to amplifiers and other devices on the lines. Suppose a standard coaxial cable with an inner conductor made of a solid wire 0.5 mm in diameter and an outer, thin conducting shell 8 mm in diameter is used. A DC voltage of 64 V is connected with the positive pole connected to the inner conductor. Calculate:

(a) The potential distribution everywhere in the coaxial cable.
(b) The electric field intensity everywhere in the coaxial cable.
(c) Plot the potential and the electric field intensity as a function of distance from the center of the cable.

5.7 Potential Due to Large Planes. Two conducting planes at 45° are connected to potentials as shown in **Figure 5.34**. The two planes do not meet at the origin (a small gap exists between them). If a potential V_0 is given on the horizontal plane, with reference to the inclined plane, calculate:

(a) The potential everywhere between the plates in the 1st octant.
(b) The potential everywhere in the rest of the space.
Hint: use cylindrical coordinates

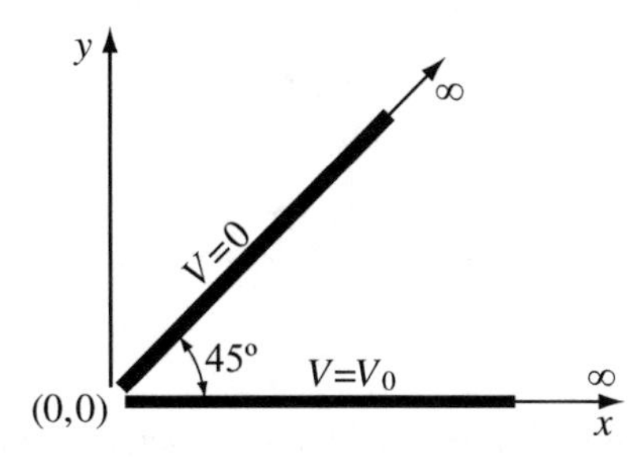

Figure 5.34

5.8 Application: Potential and Field in Spherical Capacitor. A spherical capacitor is made of two concentric shells. The inner shell is 10 mm in diameter. The outer shell is 10.1 mm in diameter. A potential $V_0 = 50$ V is connected across the two shells so that the positive pole of the battery is connected to the inner shell. The space between the shells is filled with a dielectric with a relative permittivity of 2.25. Calculate:

(a) The potential everywhere between the shells. Does the potential depend on permittivity? Explain.
(b) The electric field intensity everywhere between the shells.
(c) Plot the potential and electric field intensity as a function of distance from the inner shell.

5.9 Spherical Charge Distribution The potential inside a spherical charge distribution centered at the origin is given as $V(R) = cR^2$ [V] where c is a constant. The permittivity in the charge distribution is ε [F/m]. Calculate the volume charge density that produces this potential.

Method of Images: Point and Line Charges in Planar Configurations

5.10 Point Charge Above a Conducting Plane. A 5 nC point charge lies 2 m above a conducting plane. Find the surface charge density:

(a) Directly below the point charge, on the conducting plane.
(b) 1 m from the point in **(a)** (sideway).

5.11 Point Charges Above a Conducting Plane. Two point charges are located above a conducting plane as shown in **Figure 5.35**. Calculate the electric field intensity:

(a) In the space above the plane.
(b) At the surface of the conductor.

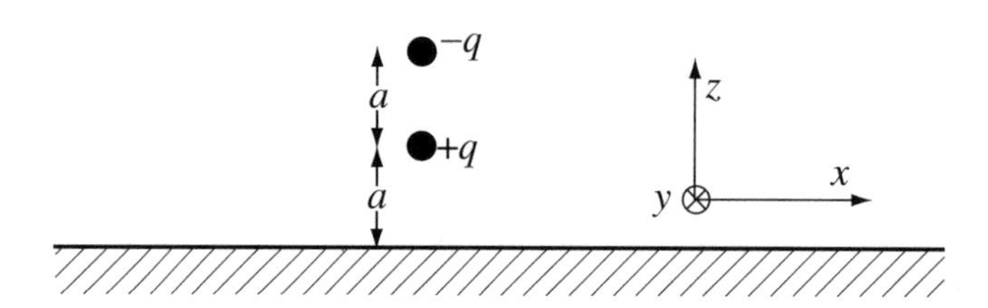

Figure 5.35

5.12 Charged Line Above a Conducting Plane. A very long charged line is located above a perfectly conducting plane in free space as shown in **Figure 5.36**. The charge density on the line is ρ_l [C/m]:

(a) Find the surface charge density on the conducting plane.
(b) Show that the total charge per unit depth of the conducting surface equals $-\rho_l$ [C/m].

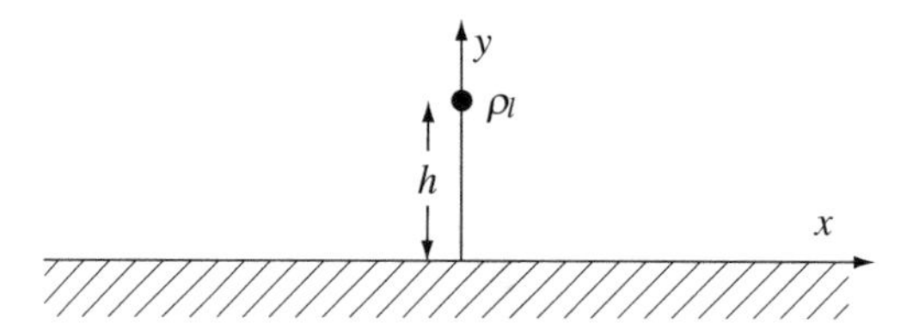

Figure 5.36

5.13 Charged Line and Multiple Conducting Planes. An infinitely long line charged with a line charge density ρ_l [C/m] is located at a distance d [m] from a very large conductor, as in **Figure 5.37**:

(a) Calculate the electric field intensity and the potential on the dotted line shown.
(b) A second conducting surface is now placed to fill the space to the right of the dotted line. What are now the electric field intensity and potential on this line? Compare with the results in **(a)**.

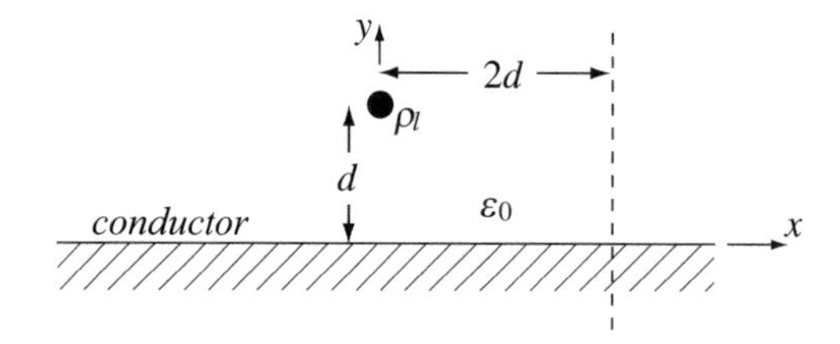

Figure 5.37

Method of Images: Multiple Planes

5.14 Point Charge Between Conducting Planes. A point charge Q [C] is located midway between two conducting plates in free space. The plates intersect at 45° to each other:

(a) Show that the electric field intensity is perpendicular to the plates everywhere.
(b) Calculate the charge density induced on the plates if the radial distance from the intersection of the plates to the point charge is d [m].

5.15 Point Charge Between Intersecting Conducting Planes. Two conducting planes intersect at 30°, as in **Figure 5.38**. Both surfaces are at zero potential. A point charge q [C] is placed midway between the two planes at point A, as shown.

(a) Draw the system of image charges and calculate the electric field intensity everywhere between the planes, assuming free space.

(b) Show by means of a drawing what happens if the charge is moved from the center toward the upper plane so that it is at a distance b [m] from the upper plane and a distance c [m] from the lower plane while the radial distance is maintained (point B).

(c) Show by means of a drawing what happens if the angle between the planes is not an integer divisor of 180°. Use an angle of 75° as an example.

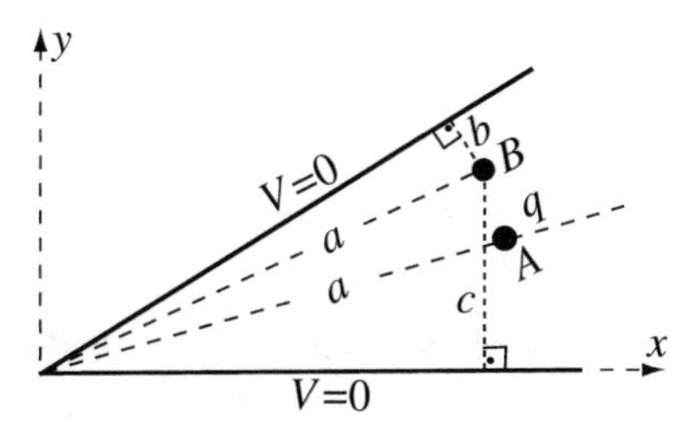

Figure 5.38

5.16 Point Charge Between Parallel Conducting Planes. A point charge is placed between two infinite planes in free space, as in **Figure 5.39**:

(a) Show the location and magnitude of the first few image charges.

(b) Place a system of coordinates so that the upper plate is at $x = 0$ and the point charge is at ($x = 3d/4$, $y = 0$), and calculate the potential at the middle point between the plates ($x = d/2$, $y = 0$) using the first six image charges (plus the original point charge).

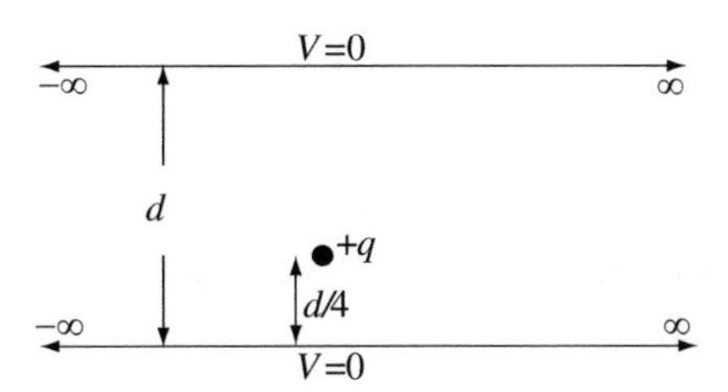

Figure 5.39

5.17 Point Charge Between Parallel Conducting Planes. A point charge is placed between two infinite planes in free space, as in **Figure 5.39**:

(a) Place a system of coordinates so that the upper plate is at $x = 0$ and the point charge is at $(3d/4,0)$. Find an expression for the potential at any point between the two plates as a sum on N image charges.

(b) Optional: Write a computer program or a script that will evaluate the sum in **(a)** for $N = 7$, $N = 100$, $N = 1{,}000$, and $N = 100{,}000$, at $x = 0.05$ m, $y = 2.0$ m using $q = 1 \times 10^{-12}$ C, $d = 0.5$ m. Compare the results and provide an indication of how many image charges are necessary for a maximum relative error of 1%.

5.18 Point Charge Between Parallel Conducting Planes. Consider a point charge embedded in a dielectric material as shown in **Figure 5.40**. The dielectric, of permittivity ε_1 [F/m], is bound by two parallel plates, each held at a constant potential V_0 [V]:

(a) Can you use the method of images to find the field inside the dielectric. How?

(b) If so, find the first four image charges and their location and calculate the potential at any point between the plates. Place a system of coordinates so that the upper plate is at $x = 0$ and the point charge is at $(d_2,0)$.

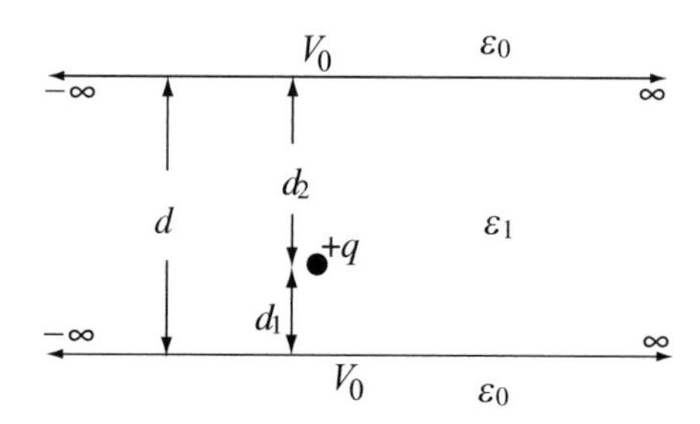

Figure 5.40

5.19 Point Charge in Infinite Closed Channel. Figure 5.41 shows a hollow cavity in a conducting medium. The cavity is infinite in the dimension perpendicular to the plane shown:

(a) Show the location of the image charges necessary to calculate the electric field intensity inside the cavity.
(b) Use the nearest four images and the charge q [C] to calculate an approximate value for the electric field intensity at the center of the box, in the plane shown. Place a system of coordinates so that the lower left corner of the cavity is at (0,0).

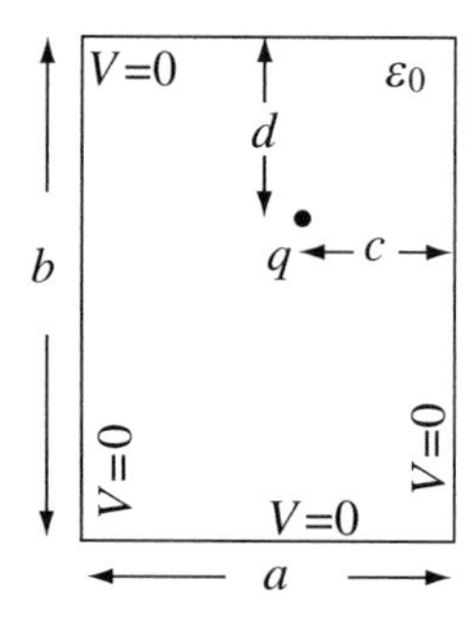

Figure 5.41

5.20 Point Charge in Infinite Closed Channel. The cavity in **Figure 5.41** is given. The cavity is infinite in the dimension perpendicular to the plane shown:

(a) Show the image charges locations.
(b) Place a system of coordinates so that the lower left corner of the cavity is at (0,0) and find a general expression for the potential at a general point within the cavity.
(c) Evaluate the expression in **(b)** using the closest eight image charges (plus the original charge) at the center of the cavity.
(d) Optional: Write a computer program or a script that will evaluate the sum in **(b)** at $x = 1.0$ m, $y = 2.0$ m, for the closest $N = 9$, $N = 100$, $N = 1{,}000$, and $N = 100{,}000$ charges. Use $q = 1 \times 10^{-9}$ C, $a = 5$m, $b = 10$ m, $c = 2.5$ m, $d = 2.5$ m. Compare the results and provide an indication on how many image charges are necessary for a maximum incremental error of 1%.

Method of Images in Curved Geometries

5.21 Application: Cable in a Tunnel. A cable runs through a mine shaft suspended from the ceiling as shown in **Figure 5.42**. The cable is thin and carries a charge of 10 nC per meter. The shaft is 10 m in diameter. Calculate the electric field intensity at the center of the shaft (magnitude and direction). Assume soil is conducting, and the air in the shaft has permittivity of free space.

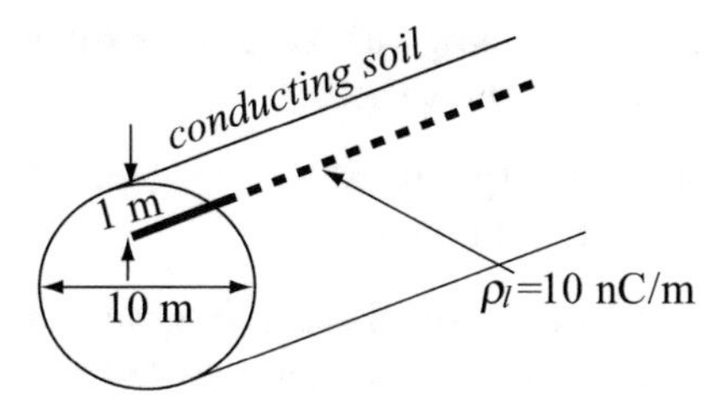

Figure 5.42

5.22 Application: Two Charged Wires Next to a Conducting Cylinder. Two wires, one charged with a positive line charge density and one with a negative line charge density, run parallel to a thick conducting pipe in free space as shown in **Figure 5.43**. Calculate the electric field intensity everywhere.

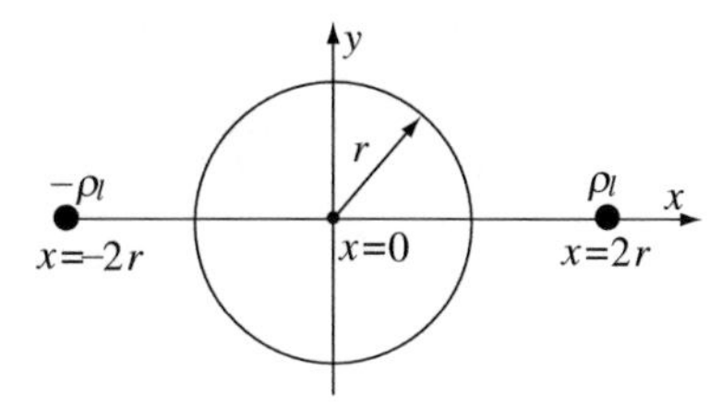

Figure 5.43

5.23 Point Charge in a Conducting Shell. A point charge $+q$ [C] is located inside a very thin conducting spherical shell halfway between the center and the shell shown in **Figure 5.44**. Assume the conducting shell is at zero potential:

(a) Calculate the electric field intensity at the center of the sphere assuming free space inside the shell.
(b) What is the field outside the sphere?
(c) Plot the electric field intensity inside the sphere by plotting the field lines (approximate plot is sufficient).

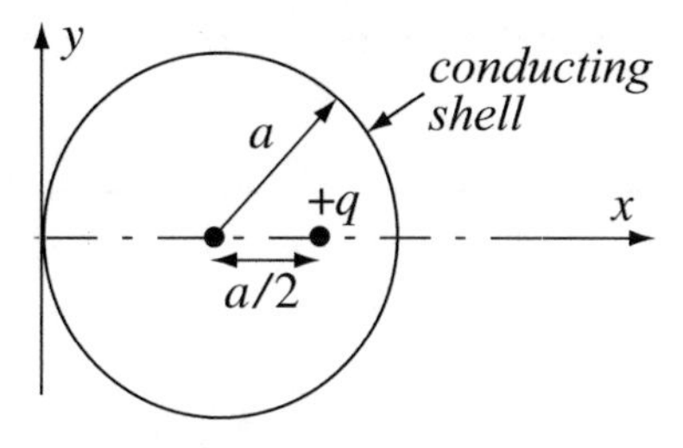

Figure 5.44

5.24 Point Charge Outside a Conducting Sphere. For the conducting sphere and point charge shown in **Figure 5.45**, find:

(a) The image charge (magnitude and location).
(b) The electric field intensity on the surface of the sphere.
(c) The induced charge density on the surface of the sphere.
(d) Sketch the electric field intensity outside the sphere.

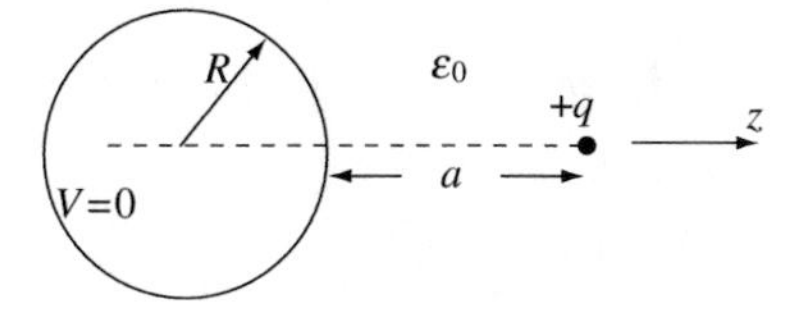

Figure 5.45

5.25 Point Charge Inside Hollow, Charged Conducting Sphere. Solve **Problem 5.23** if the shell is at a constant potential V_0 [V]. Sketch the electric field intensity outside the shell.

5.26 Point Charge Outside Charged Conducting Sphere. Solve **Problem 5.24** if the shell has a positive surface charge density ρ_s [C/m^2]. The potential on the shell need not be zero.

5.27 Two Point Charges Outside Grounded Conducting Sphere. A grounded (zero potential) conducting spherical shell of radius a [m] and two point charges are located in free space as shown in **Figure 5.46**:

(a) Calculate the electric field intensity at a general point outside the sphere in thgw y–z plane if $c > a$ and $d > a$.
(b) What is the electric field intensity outside the sphere if $c, d \to \infty$? Explain.
(c) What is the electric field intensity at a general point inside the sphere if $c < a$ and $d < a$.
(d) What is the electric field intensity at a general point inside the sphere if $c, d \to 0$? Explain.

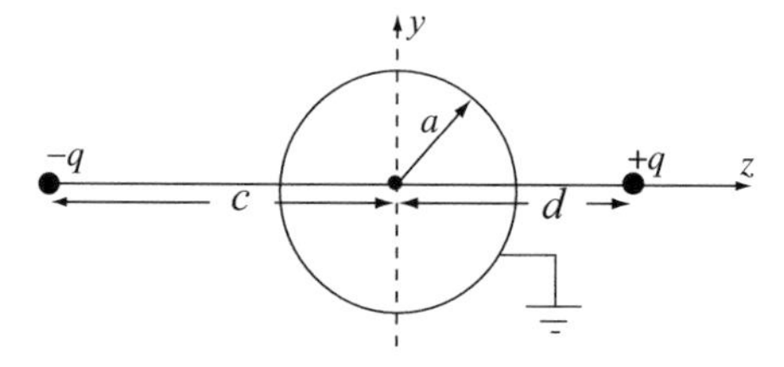

Figure 5.46

Separation of Variables in Planar Geometries

5.28 Potential in Infinite Channel. The geometry in **Figure 5.47** is made of a semi-infinite channel bounded by the planes $x = 0$, $x = \infty$, $y = 0$, $y = b$ [m]. Three sides are held at zero potential and the fourth is isolated from the others (between $y = 0$ and $y = b$) and held at potential V_0 [V]. Find the potential within the channel.

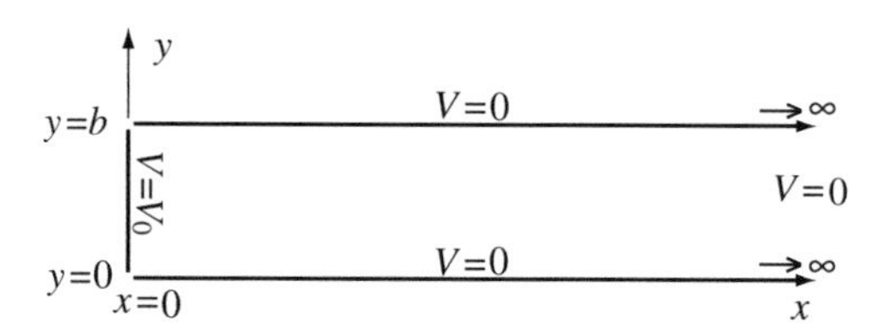

Figure 5.47

5.29 Potential in Infinite Channel. The infinite channel shown in **Figure 5.48** is made of three conducting walls connected to a zero potential, and a fourth is an isolated wall (top cover) connected to a potential V_u. The cover and sides are close to each other but are not touching. The material in the channel is free space:

(a) Calculate the potential everywhere in the channel if $V_u = 10$ V (DC).
(b) Plot the potential as a sequence of constant potential lines inside the box.

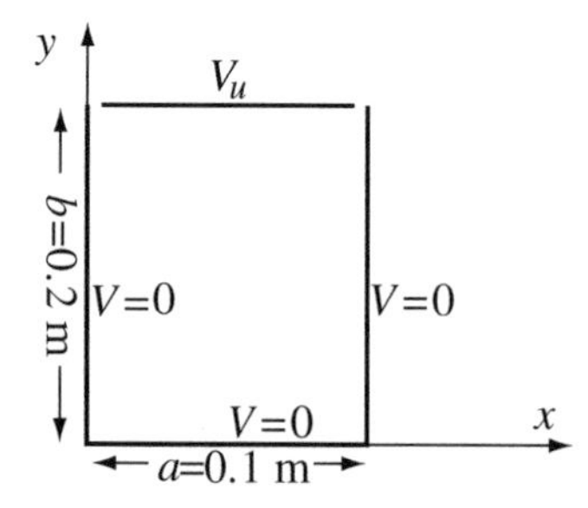

Figure 5.48

5.30 Potential in Infinite Channel. Figure 5.48 shows the cross section through an infinitely long channel. Three sides are connected to ground potential, and the top is insulated from the rest of the structure and connected to a potential V_u as shown. The channel is filled with a material with permittivity ε [F/m]. Calculate the potential everywhere inside the cross section if $V_u = 10\sin(\pi x/a)$ [V].

5.31 Potential in Infinite Channel. A two-dimensional geometry consists of four infinitely long plates, shown in cross section in **Figure 5.49**. For the given plate potentials, calculate the electric potential at a general point between the plates.

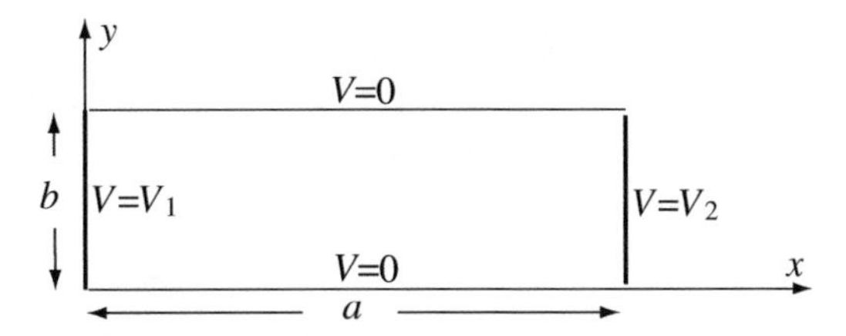

Figure 5.49

5.32 Potential in a Box with Conducting Walls. A three-dimensional cubic box, with dimensions $a \times a \times a$ [m^3], is grounded on four sides. The two sides at $z = a$ [m] and at $x = a$ [m] are connected to a constant potential as shown in **Figure 5.50**. Calculate the potential everywhere inside the box.

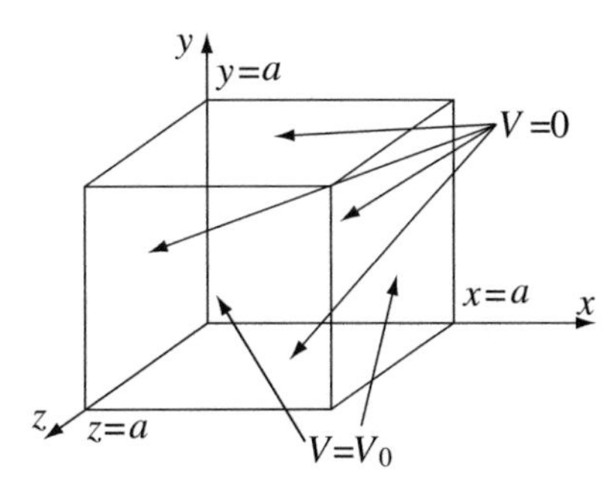

Figure 5.50

Separation of Variables in Cylindrical Geometries

5.33 Application: Electrostatic Precipitator. An electrostatic precipitator is made by lining the interior of a smokestack with two half-shells as shown in cross section in **Figure 5.51**. Assume the stack is very long and a potential difference of 100 kV is connected as shown. Ground is at zero potential so that the left half-shell is at – 50 kV and the right half-shell at +50 kV. Calculate:

(a) The potential inside and outside the stack.
(b) The electric field intensity inside and outside the stack.

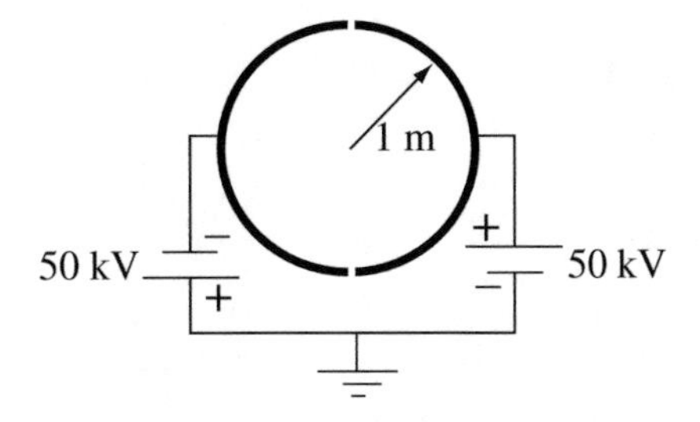

Figure 5.51

5.34 Application: Electrostatic Precipitator. The designers of the precipitator in **Problem 5.34** found that they cannot make two half-shells as needed, but they can make the precipitator of four quarter-shells and connect them as shown in **Figure 5.52**. Calculate:

(a) The potential inside and outside the stack.
(b) The electric field intensity inside and outside the stack.

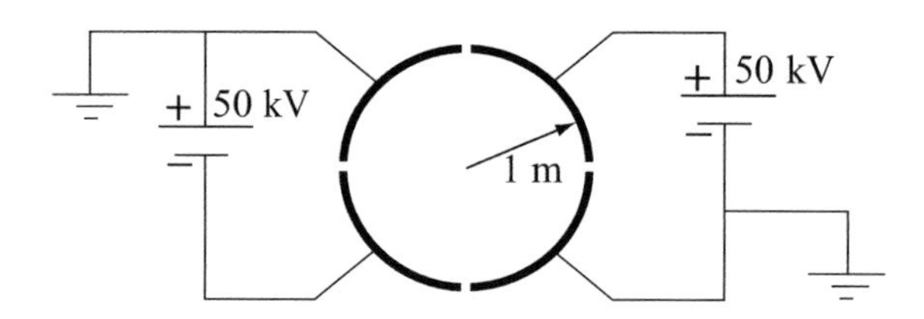

Figure 5.52

Boundary Value Problems: Numerical (Approximate) Methods

6

Although this may seem a paradox, all exact science is dominated by the idea of approximation

Bertrand Russell, (1872–1970),
philosopher, mathematician, historian

6.1 Introduction

In **Chapters 3** through **5**, we discussed analytic methods of solution for electrostatic problems. The most outstanding feature of these methods was that the solution was exact and in the form of a mathematical relation. On the other hand, only certain classes of problems could be solved. In the case of the method of images, the question was one of finding the correct system of images, a requirement that meant, almost always, that a constant potential surface exists or can be stipulated. If this conducting surface is very complex, the condition of constant potential on the surface may not be as easy to satisfy. Similarly, separation of variable, while certainly valid in general, is often difficult to apply because of the need to satisfy complex boundary conditions. If the boundaries of the problem are not parallel to coordinates, it is next to impossible to find the constants required for solution. Even for the simple planar geometries discussed in **Chapter 5**, the solution required considerable skill. Furthermore, precious little was said about solution of Poisson's equation. For another example of the difficulty in analytic solutions, consider **Figure 6.1**, which shows a simple parallel plate capacitor. Suppose we need to calculate its capacitance. One method we used before is to assume that the plates are very close to each other, neglect fringing, and use the formula for the parallel plate capacitor. In **Figure 6.1a**, the distance between the plates is very small, and the capacitance may be approximated as $C = \varepsilon A/d$ [F]. **Figure 6.1b** shows the same two plates, but now the plates are much farther apart, perhaps because our design requires that this capacitor withstand high voltages. Here, we cannot neglect fringing, and, therefore, the use of the formula for parallel plate capacitors is incorrect. How can we solve this problem? It seems that none of the methods of the previous chapters applies here. Yet, this type of problem is quite common. Other examples are the voltage and fields of power lines in the presence of conducting objects (such as transformers, towers, buildings, etc.), field distributions between nonparallel surfaces, and many others. Although, in some cases, assumptions can be made to simplify the problem, this is not always possible and we are faced with the need to solve problems for which analytic methods cannot be used.

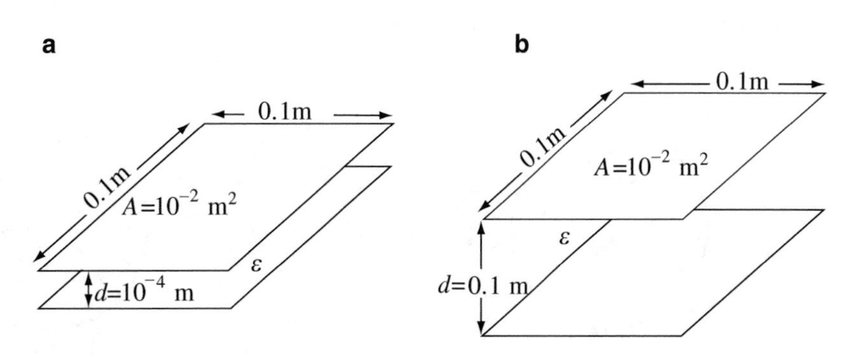

Figure 6.1 (**a**) Parallel plate capacitor for which fringing fields can be neglected. (**b**) d is large and fringing cannot be neglected

There are a number of numerical methods that can be used when analytic methods fail. These methods, some of which are very old and others quite new, are all based on availability of computers to solve the field equations. Some of the more

N. Ida, *Engineering Electromagnetics*, https://doi.org/10.1007/978-3-030-15557-5_6

common numerical techniques will be described next. We will first describe the general idea of a numerical method and then discuss in detail three representative methods for numerical solution of electrostatic problems.

Before we plunge into numerical methods of solutions, it is well to reiterate that, in a way, the only reason for doing so is our inability to solve the problem analytically; that is, the goal should always be an analytic, closed-form solution. Even when numerical solutions are pursued, and they are pursued quite often, analytic methods are still important as a means of checking numerical solutions. It should also be noted that the methods described here can be used beyond electrostatics although the details of implementation are necessarily different.

6.1.1 A Note on Scripts and Computer Programs

Numerical methods require computers, and computers require computer programs or scripts. Computer programs are written in a computer language and require compilation prior to execution. Scripts are written in a scripting language and do not require compilation. The examples solved in this chapter as well as many of the problems at the end of the chapter use Matlab scripts for solutions. The scripts required to solve the various examples and the data files listed in the text are available from https://www.springer.com/us/book/9783030155568 and can be downloaded when required. The site also lists the various input and output files referred to in the text as well as an information file with explanations. It is recommended that these scripts and data files be downloaded before any attempt at understanding the examples given in this chapter is undertaken. The site contains other scripts as well as computer programs that may be useful. These were written in the simplest possible way with a minimum of constructs. To keep things simple and easily understandable, the scripts were specifically written for the examples given. Because of that, they are not general. However, they can serve as the basis for more general scripts (or programs) and, in particular, may be adapted for solution of the end-of-chapter problems. Most of the checks usually found in programs (such as limits on arrays, correctness of values, and the like) have been taken out to keep them short. The user interface is primitive and minimalistic. Thus, the user should be careful with the data or incorrect results will be obtained. To aid in this task, the input data for the given example is also included with each script and a free-format simple interrogative input is included in each script. Scripts are referred in the text by name.

6.2 The General Idea of Numerical Solutions

What is then a numerical method? Quite simply, any method that solves a class of problems, based on discretization of the continuum and approximation of the solution variables in some systematic and, preferably, simple way. An analogy is in order here: Suppose you need to make a soccer ball out of leather. A perfect ball made out of a single piece of leather is impossible to make; yet, a perfect ball is what we want—that is the analytic solution. Instead, we might proceed to form two half-spheres by, say, pressing and stretching the leather in the same way shoe tips are made. This is one possible approximation, but it is not a simple approximation because it requires complicated machinery. Instead, we could choose small, simple patches of some defined shape, all planar, and stitched together. One example is the triangular pattern. Another is a hexagonal patch. A third (the way soccer balls are actually made) is a combination of pentagonal (black) and hexagonal (white) patches. The ball has 32 patches sewn together, to form a 32-faceted volume that approximates a sphere. Any of these approximations is valid, although, in this case, the pentagonal/hexagonal approximation is used by convention. The approximation process is shown in **Figure 6.2**.

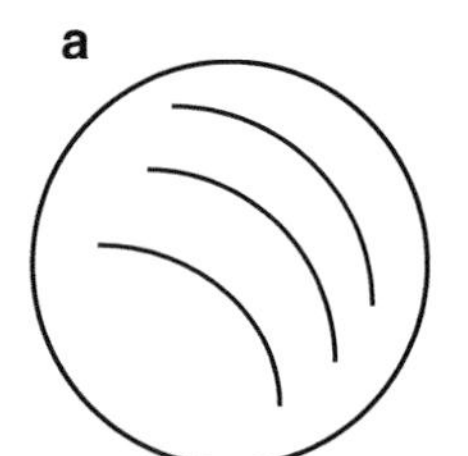

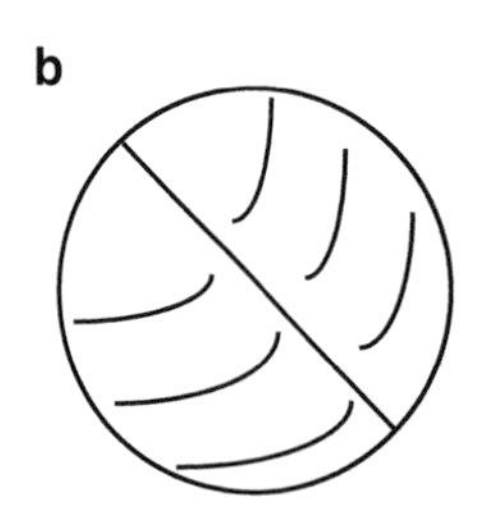

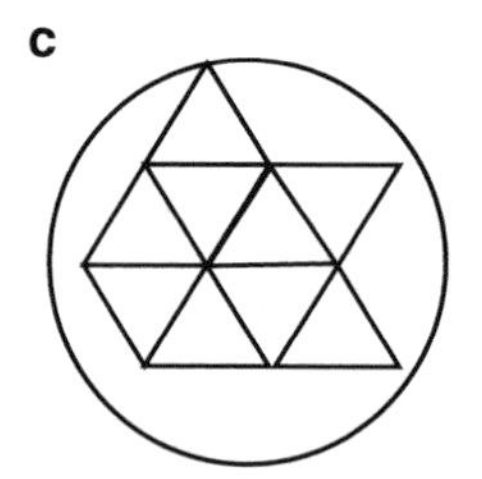

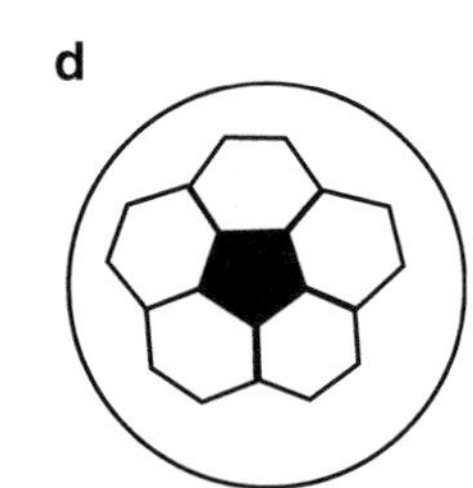

Figure 6.2 The approximation process. (**a**) A perfect ball. (**b**) A ball made of two half-spheres. (**c**) A ball made of triangular patches. (**d**) A soccer ball made of hexagonal and (white) pentagonal (black) patches

The above process is the essence of any approximation method. The whole process is based on the premise that if the approximation is not good enough (such as using 32 patches for a soccer ball), we can increase the number of patches while decreasing their size as much as we wish. In the limit, the number of patches tends to infinity, with their size tending to zero; that is, the patches are reduced to points and the approximation becomes "exact."

A similar example is shown in **Figure 6.3a**. A general surface contains a nonuniform charge density and we need to calculate the total charge or the electric field intensity at a point in space. We know how to solve this problem analytically: All we have to do is integrate the electric field intensity of charged differential surface elements over the surface. In practice, however, unless the surface representation is simple, we cannot perform the integration. An approximate solution may be found by dividing the surface into any number of small subdomains and assuming each subdomain has a constant but different charge density, depending on the location of the subdomain (**Figure 6.3b**). If the subdomains are small, and we are free to make them as small as we wish, the total charge of each subdomain may be taken as a point charge at the center of the subdomain and the problem is solved as if we had as many point charges as we have subdomains, with the point charges located at the center of the subdomains. Now, of course, the problem is simple, except for the fact that we have many, perhaps many thousands of points to evaluate, hence the need for a computer, even for the simple integration described here.

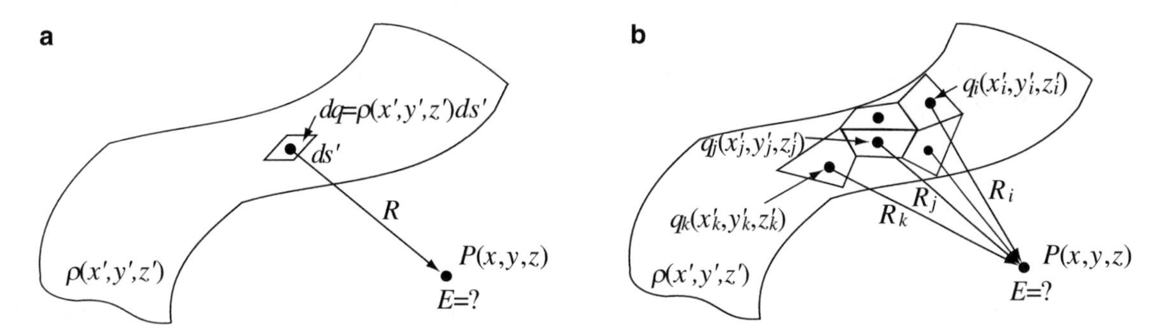

Figure 6.3 (**a**) Nonuniform charge distribution on a complex surface. (**b**) Division of the surface into patches, each with a different but constant charge density

6.3 The Finite Difference Method: Solution to the Laplace and Poisson Equations

The finite difference method is the oldest of the numerical methods; its origins have been traced to Gauss. However, as we shall see shortly, it only really came into widespread use with the advent of the computer because, whereas the calculations necessary are very simple, there is a large number of calculations that need to be performed. The finite difference method consists of replacing the partial derivatives in the partial differential equation describing the physical process by an algebraic approximation based on simple relations between the values of the function we need to evaluate. Since these relations are in the form of differences between values of the function, and these values are at small but finite distances from each other, the method is called a ***finite difference method***.

We will present the finite difference method for Laplace's and Poisson's equations based on the definition of the derivative. This is simple and intuitive, and most importantly, it shows the physical interpretation of the approximations involved. The formal definition of the finite difference method is based on truncated Taylor series. These have the advantage that they allow the definition of approximations to any derivative and also, by truncating the series at different locations, different approximation errors may be allowed. However, for the type of equations we need to solve here (and, in fact, throughout this book), only first- and second-order derivatives are needed. Therefore, we first define the general approximation for first-order derivatives. From this approximation, we then derive an approximation to second-order derivatives and then apply these to the solution of Laplace's and Poisson's equations.

6.3.1 The Finite Difference Approximation: First-Order Derivative

The approximation to a first-order derivative can be found in the definition of the derivative itself. Consider **Figure 6.4a**, where a general function of a single variable is shown. The function is shown as a continuous function, but it may also be a sequence of points such as may be obtained from measurements. The derivative $f'(x) = df/dx$ at a point x_i is the tangent to the curve at this point (line f'). An approximation to the derivative can be found by taking two points, say on both sides of the point x_i, and passing a straight line through them. If the two points are chosen to be equally spaced about the point x_i, as in **Figure 6.4a** (line f_i'), the following expression for the slope of the line is obtained:

$$\frac{df(x_i)}{dx} \approx \frac{f(x_i + \Delta x) - f(x_i - \Delta x)}{2\Delta x} \tag{6.1}$$

Because the derivative is evaluated using two symmetric points around x_i, this expression is called a ***central difference formula*** and is immediately recognized as the definition of the derivative if we allow Δx to tend to zero. This is a valid approximation to the derivative, but it is not the only possible approximation. Another valid approximation (slope) at point x_i is (line f_2')

$$f_2' = \frac{df(x_i)}{dx} \approx \frac{f(x_i + \Delta x) - f(x_i)}{\Delta x} \tag{6.2}$$

This is called a ***forward difference formula*** because it uses the point x_i and the point ahead of it ($x_i + \Delta x$). A third approximation is that of line f_3':

$$f_3' = \frac{df(x_i)}{dx} \approx \frac{f(x_i) - f(x_i - \Delta x)}{\Delta x} \tag{6.3}$$

This is called a ***backward difference formula*** because it uses the point x_i and the point behind it ($x_i - \Delta x$). The three approximations are not identical, and, therefore, one may be a better approximation than the other, but all three are valid because in the limit, as Δx approaches zero, all three lead to the correct slope of the function at point x_i.

6.3.2 The Finite Difference Approximation: Second-Order Derivative

An approximation to the second-order derivative may be obtained from the first-order derivatives using the method of the previous section. Suppose that we used the formula in **Eq. (6.3)** to calculate the first-order derivative at all points of the function in **Figure 6.4a**. The result is the function in **Figure 6.4b**. This function is drawn as a continuous function to indicate that the number of points can be as large as necessary, Δx can be as small as necessary, and, therefore, the function describing the derivative can be as close to a continuous function as is practical. Now, we start with the function in **Figure 6.4b** and calculate the first-order derivative at a point x_i using **Eq. (6.2)**. This gives

$$f''(x_i) = \frac{df'(x_i)}{dx} \approx \frac{f'(x_i + \Delta x) - f'(x_i)}{\Delta x} \tag{6.4}$$

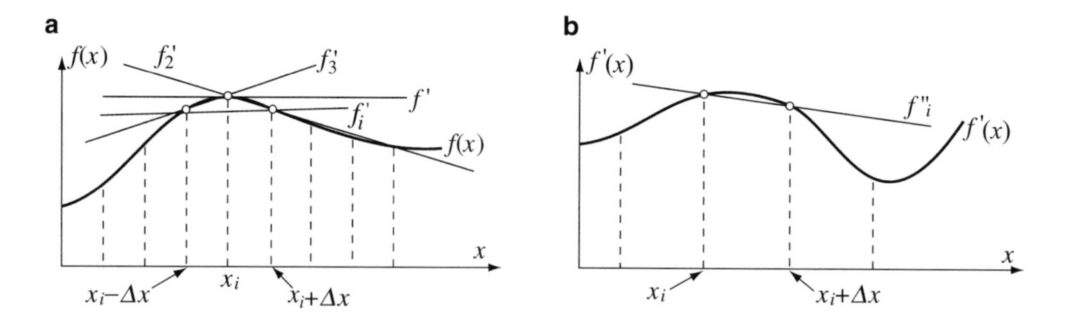

Figure 6.4 (**a**) Exact first derivative f' and approximations to the first derivative at point x_i. (**b**) Approximation to the second derivative at point x_i

From **Figure 6.4a** the derivatives $f'(x_i + \Delta x)$ and $f'(x_i)$ are given in terms of the original function [using **Eq. (6.3)**] as

$$f'(x_i + \Delta x) = \frac{f(x_i + \Delta x) - f(x_i)}{\Delta x} \quad \text{and} \quad f'(x_i) = \frac{f(x_i) - f(x_i - \Delta x)}{\Delta x} \tag{6.5}$$

Substitution of these two approximations into **Eq. (6.4)** gives

$$\frac{d^2 f(x_i)}{dx^2} = \frac{df'(x_i)}{dx} \approx \frac{1}{\Delta x}\left[\frac{f(x_i + \Delta x) - f(x_i)}{\Delta x} - \frac{f(x_i) - f(x_i - \Delta x)}{\Delta x}\right] \tag{6.6}$$

Collecting terms, we get the approximation for the second-order derivative:

$$\frac{d^2 f(x_i)}{dx^2} \approx \frac{f(x_i + \Delta x) - 2f(x_i) + f(x_i - \Delta x)}{(\Delta x)^2} \tag{6.7}$$

To derive this equation, we used **Eqs. (6.2)** and **(6.3)**, but other possibilities exist (see **Exercise 6.1**).

Before continuing, we rewrite **Eq. (6.7)** in short-form notation by denoting $x_i + \Delta x$ as x_{i+1}, $x_i - \Delta x$ as x_{i-1}, and Δx as h:

$$\boxed{\frac{d^2 f_i}{dx^2} \approx \frac{f_{i+1} - 2f_i + f_{i-1}}{h^2}} \tag{6.8}$$

If the function f were a function of the y or z variables, the approximation in each case would be

$$\frac{d^2 f(y)}{dy^2} \approx \frac{f_{j+1} - 2f_j + f_{j-1}}{(\Delta y)^2} \quad \text{and} \quad \frac{d^2 f(z)}{dz^2} \approx \frac{f_{k+1} - 2f_k + f_{k-1}}{(\Delta z)^2} \tag{6.9}$$

The last step in our approximation is to assume that the approximation to the second-order ordinary derivative above is also a good approximation to the second-order partial derivative.

The above functions were each dependent on one variable alone. To define the second-order derivative of a function of two variables, we assume that the function $f(x,y)$ is defined over a surface S (**Figure 6.5**) and its values are known everywhere on the surface and therefore at the nodes of the mesh created by the discretization. For simplicity, we will also assume that $\Delta x = \Delta y = h$. Now, we can write the following using **Figure 6.5** and **Eq. (6.8)**:

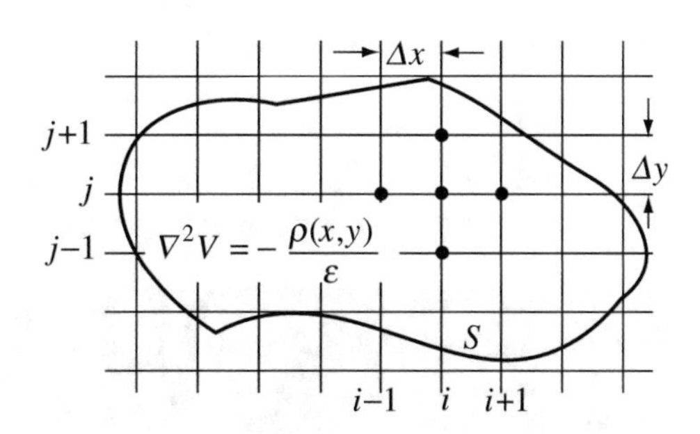

Figure 6.5 A surface divided into a grid (or mesh) over which the function $f(x,y)$ is approximated using finite differences

$$\frac{\partial^2 f(x,y)}{\partial x^2} \approx \frac{f_{i-1,j} - 2f_{i,j} + f_{i+1,j}}{h^2} \quad \text{and} \quad \frac{\partial^2 f(x,y)}{\partial y^2} \approx \frac{f_{i,j-1} - 2f_{i,j} + f_{i,j+1}}{h^2} \tag{6.10}$$

Combining the two, we get an expression for the two-dimensional Laplacian:

$$\frac{\partial^2 f(x,y)}{\partial x^2} + \frac{\partial^2 f(x,y)}{\partial y^2} \approx \frac{f_{i-1,j} - 2f_{i,j} + f_{i+1,j}}{h^2} + \frac{f_{i,j-1} - 2f_{i,j} + f_{i,j+1}}{h^2} \tag{6.11}$$

or, combining terms,

$$\frac{\partial^2 f(x,y)}{\partial x^2} + \frac{\partial^2 f(x,y)}{\partial y^2} \approx \frac{f_{i-1,j} + f_{i+1,j} + f_{i,j-1} + f_{i,j+1} - 4f_{i,j}}{h^2} \tag{6.12}$$

The Laplacian in three dimensions is obtained similarly. All we need to do is add the partial derivative with respect to z. The form of the derivative is identical to that in **Eq. (6.12)**, but now there are three indices, one in each dimension. Assuming that $\Delta x = \Delta y = \Delta z = h$ gives (see **Exercises 6.2** and **6.3**)

$$\frac{\partial^2 f(x,y,z)}{\partial x^2} + \frac{\partial^2 f(x,y,z)}{\partial y^2} + \frac{\partial^2 f(x,y,z)}{\partial z^2} \approx \frac{f_{i-1,j,k} + f_{i+1,j,k} + f_{i,j-1,k} + f_{i,j+1,k} + f_{i,j,k-1} + f_{i,j,k+1} - 6f_{i,j,k}}{h^2} \tag{6.13}$$

The approximation in three dimensions is just as simple as in two dimensions. However, because three-dimensional problems are more difficult to visualize and the bookkeeping tasks on the various indices are more complex, most (but not all) of the problems solved here are two-dimensional applications.

We are now in a position to use these approximations to find solutions to electrostatic field problems. First, we define the approximation in terms of the scalar potential V for Laplace's equation in two dimensions. In terms of the electric potential, this is [from **Eq. (6.12)**]

$$\frac{\partial^2 V(x,y)}{\partial x^2} + \frac{\partial^2 V(x,y)}{\partial y^2} \approx \frac{V_{i-1,j} + V_{i+1,j} + V_{i,j-1} + V_{i,j+1} - 4V_{i,j}}{h^2} = 0 \tag{6.14}$$

or, multiplying both sides by $h^2 = (\Delta x)^2$, the approximation for the potential at point (i,j) is

$$V_{i,j} = \frac{V_{i-1,j} + V_{i+1,j} + V_{i,j-1} + V_{i,j+1}}{4} \quad [\mathrm{V}] \tag{6.15}$$

This approximation is independent of h. The approximation to Poisson's equation is [see **Eq. (5.5)**]

$$\frac{V_{i-1,j} + V_{i+1,j} + V_{i,j-1} + V_{i,j+1} - 4V_{i,j}}{h^2} = -\frac{\rho(x,y)}{\varepsilon(x,y)} \quad [\mathrm{V}] \tag{6.16}$$

where $\rho(x,y)$ is the surface charge density (if any). From this, the approximation for the potential at point (i,j) is

$$V_{i,j} = \frac{V_{i-1,j} + V_{i+1,j} + V_{i,j-1} + V_{i,j+1}}{4} + \frac{h^2 \rho_{i,j}}{4\varepsilon_{i,j}} \quad [\mathrm{V}] \tag{6.17}$$

Exercise 6.1 Find an approximation to the second-order derivative as follows:

(a) Use **Eq. (6.1)** alone, noting that the derivative is approximated using two points symmetric about point x_i.
(b) Use **Eq. (6.2)** alone, noting that the derivative at point x_i is approximated using points x_i and x_{i+1}.
(c) Why are these formulas less attractive than the one obtained in **Eq. (6.7)**?

Answer

(a) $\dfrac{d^2 f(x_i)}{dx^2} \approx \dfrac{f_{i+2} - 2f_i + f_{i-2}}{(2\Delta x)^2}$. **(b)** $\dfrac{d^2 f(x_i)}{dx^2} \approx \dfrac{f_{i+2} - 2f_{i+1} + f_i}{(\Delta x)^2}$.

(c) The approximation in **(a)** requires five points (even though $f_i(x_i + \Delta x)$ and $f_i(x_i - \Delta x)$ are not explicitly used), whereas the approximation in **(b)** is not symmetric about the calculation point. Both of these approximations, although valid, lead to reduced accuracy.

6.3.3 Implementation

Now that we have the approximations for Poisson's and Laplace's equations, we need to define the solution procedure. To do so, we use the example in **Figure 6.6**. It consists of a box, with three sides connected to zero potential and the fourth set to a constant potential $V_0 = 10$ V. The third dimension of the box (perpendicular to the page) is infinite, making this a two-dimensional problem. This particular problem is described by Laplace's equation; therefore, we will use the approximation in **Eq. (6.14)** or **(6.15)**, depending on the method we adopt for solution. This particular problem has an analytic solution, obtained by separation of variables (see **Section 5.4.4.1** and **Problem 5.29**), so that the finite difference solution may be verified.

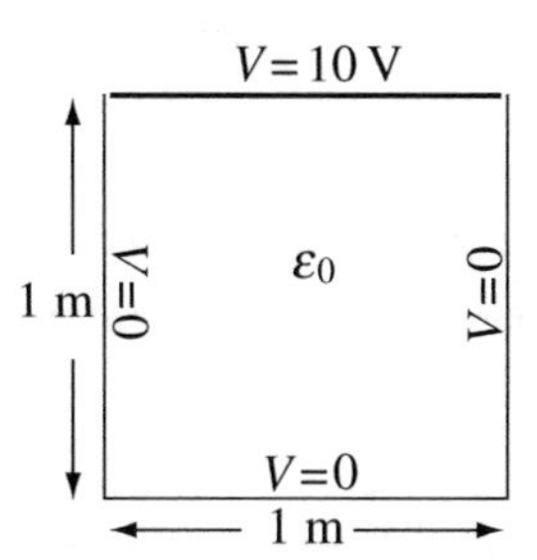

Figure 6.6 A simple geometry used to demonstrate finite difference solutions for electrostatic fields (Laplace's equation)

We have now two ways to solve the problem: The first is called an ***implicit method*** in which all potential values inside the box are taken as unknown values at the nodes created by discretization of the geometry (see Figures **6.5** and **6.7**). Using the approximation in **Eq. (6.14)**, we write an equation at each internal node, relating the unknown potential at the node with the unknown potentials at the neighboring nodes. This results in a system of equations in N unknowns which is solved to obtain the unknown potentials. The second method is an ***explicit method***. In this method, all potential values are assumed to be known. Since we do not know the potentials at the internal nodes, we guess their value and then proceed to calculate new, updated values at the nodes, based on the known values at the nodes around them. To do so, we assume that the potential at node (i,j) is not known but all other nodes are known and use **Eq. (6.15)**. This process results in an updated value at node (i,j). The process is repeated iteratively until the solution at all nodes does not change or changes very little. The resulting potentials are the "correct" potentials at the nodes.

In general, explicit methods are preferred in finite difference problems. Their main advantage is in that there is no need to create, store, and solve a large system of equations. Although explicit methods cannot always be applied, they are applicable to all electrostatic problems we will encounter as well as many others. Both methods are shown here, starting with the implicit solution.

6.3.3.1 Implicit Solution

Implicit finite difference solution to an electrostatic problem is performed in six basic steps:

Step 1 First, we define a solution domain. This is the domain in which the potential must be obtained and is shown in **Figure 6.6** for a particular example. In this case, the solution domain is clearly defined by the actual, conducting boundaries, but in other applications, the boundaries must be decided upon based on the physical configuration. An example of this need to define boundaries is given in **Example 6.2**.

Step 2 Next, we define the boundary conditions of the problem. Remember that without boundary conditions, we cannot obtain a particular solution to a differential equation. This is true in analytic solutions and in numerical solutions. The boundary conditions are also shown in **Figure 6.6** as the known potentials on the surfaces of the box. This type of boundary condition is known as a ***Dirichlet boundary condition***.

Step 3 The solution domain is divided into a grid such that $\Delta x = \Delta y$. The intersections of the grid lines define the nodes of the solution domain at which the potentials will be calculated. The nodes are marked with the appropriate i,j indices as shown in **Figure 6.7a** for one choice of the grid. To simplify the discussion, we only divided the box into a 5×5 grid, with a total of 36 nodes. Of these, 16 are internal to the boundary and 20 are on the boundary. The 20 potentials on the boundary are known, whereas the 16 potentials inside the solution domain are not known and must be evaluated.

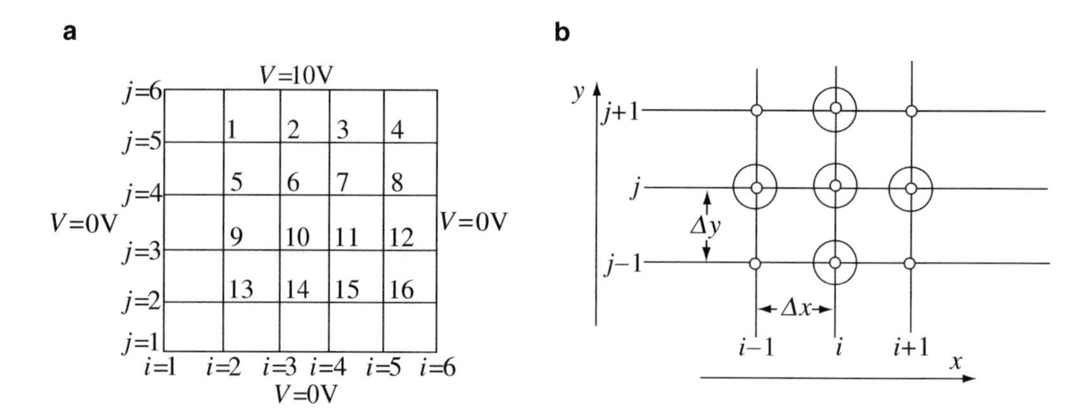

Figure 6.7 (**a**) Finite difference grid and node numbering for internal nodes. (**b**) Notation used for finite difference approximation. The five nodes marked are those used to approximate the Laplacian at point (i,j)

Step 4 The approximation in **Eq. (6.14)** is applied at each of the internal nodes. The approximation uses five points, including the node at which the potential is evaluated as shown in **Figure 6.7b**. The resulting "stencil" is applied at each node of the grid. This results in 16 equations in 16 unknowns as follows.

At a general node, for which the stencil does not include a border node such as node 11, we get: At node 11: $i = 4, j = 3$:

$$-4V_{4,3} + V_{3,3} + V_{5,3} + V_{4,2} + V_{4,4} = 0 \quad \rightarrow \quad 4V_{11} - V_{10} - V_{12} - V_{15} - V_7 = 0$$

where the final expression is written in terms of node numbers rather than the indices i, j. Similar expressions are obtained for nodes 6, 7, and 10.

At a node near the boundary for which the stencil includes a boundary node, we get: At node 2: $i = 3, j = 5$:

$$-4V_{3,5} + V_{2,5} + V_{4,5} + V_{3,4} + V_{3,6} = 0 \quad \rightarrow \quad 4V_2 - V_1 - V_3 - V_6 - 10 = 0$$

where the boundary value for the node at index $i = 3, j = 6$ was substituted as a boundary condition. Similar expressions are obtained for nodes 3, 5, 8, 9, 12, 14, and 15.

At a corner node such as node 4, two of the points in the stencil are on the boundary. The two values at the boundary nodes are substituted in the appropriate index locations: At node 4: $i = 5, j = 5$:

$$-4V_{5,5} + V_{4,5} + V_{6,5} + V_{5,4} + V_{5,6} = 0 \quad \rightarrow \quad 4V_4 - V_3 - 0 - V_8 - 10 = 0$$

Similar expressions are obtained for nodes 1, 13, and 16. After applying the approximation, we obtain a system of 16 equations in 16 unknowns. We write these relations with the free value on the right-hand side. For example, the equations for nodes 11, 2, and 4 become

$$4V_{11} - V_{10} - V_{12} - V_{15} - V_7 = 0,$$

$$4V_2 - V_1 - V_3 - V_6 = 10,$$

$$4V_4 - V_3 - V_8 = 10$$

Performing identical operations for all other interior nodes, and writing the resulting 16 equations into a matrix form, we get

$$\begin{bmatrix} 4 & -1 & 0 & 0 & -1 & 0 & 0 & 0 & 0 & 0 & 0 & 0 & 0 & 0 & 0 & 0 \\ -1 & 4 & -1 & 0 & 0 & -1 & 0 & 0 & 0 & 0 & 0 & 0 & 0 & 0 & 0 & 0 \\ 0 & -1 & 4 & -1 & 0 & 0 & -1 & 0 & 0 & 0 & 0 & 0 & 0 & 0 & 0 & 0 \\ 0 & 0 & -1 & 4 & 0 & 0 & 0 & -1 & 0 & 0 & 0 & 0 & 0 & 0 & 0 & 0 \\ -1 & 0 & 0 & 0 & 4 & -1 & 0 & 0 & -1 & 0 & 0 & 0 & 0 & 0 & 0 & 0 \\ 0 & -1 & 0 & 0 & -1 & 4 & -1 & 0 & 0 & -1 & 0 & 0 & 0 & 0 & 0 & 0 \\ 0 & 0 & -1 & 0 & 0 & -1 & 4 & -1 & 0 & 0 & -1 & 0 & 0 & 0 & 0 & 0 \\ 0 & 0 & 0 & -1 & 0 & 0 & -1 & 4 & 0 & 0 & 0 & -1 & 0 & 0 & 0 & 0 \\ 0 & 0 & 0 & 0 & -1 & 0 & 0 & -1 & 4 & 0 & 0 & 0 & -1 & 0 & 0 & 0 \\ 0 & 0 & 0 & 0 & 0 & -1 & 0 & 0 & -1 & 4 & -1 & 0 & 0 & -1 & 0 & 0 \\ 0 & 0 & 0 & 0 & 0 & 0 & -1 & 0 & 0 & -1 & 4 & -1 & 0 & 0 & -1 & 0 \\ 0 & 0 & 0 & 0 & 0 & 0 & 0 & -1 & 0 & 0 & -1 & 4 & -1 & 0 & 0 & -1 \\ 0 & 0 & 0 & 0 & 0 & 0 & 0 & 0 & -1 & 0 & 0 & 0 & 4 & -1 & 0 & 0 \\ 0 & 0 & 0 & 0 & 0 & 0 & 0 & 0 & 0 & -1 & 0 & 0 & -1 & 4 & -1 & 0 \\ 0 & 0 & 0 & 0 & 0 & 0 & 0 & 0 & 0 & 0 & -1 & 0 & 0 & -1 & 4 & -1 \\ 0 & 0 & 0 & 0 & 0 & 0 & 0 & 0 & 0 & 0 & 0 & -1 & 0 & 0 & -1 & 4 \end{bmatrix} \begin{bmatrix} V_1 \\ V_2 \\ V_3 \\ V_4 \\ V_5 \\ V_6 \\ V_7 \\ V_8 \\ V_9 \\ V_{10} \\ V_{11} \\ V_{12} \\ V_{13} \\ V_{14} \\ V_{15} \\ V_{16} \end{bmatrix} = \begin{bmatrix} 10 \\ 10 \\ 10 \\ 10 \\ 0 \\ 0 \\ 0 \\ 0 \\ 0 \\ 0 \\ 0 \\ 0 \\ 0 \\ 0 \\ 0 \\ 0 \end{bmatrix} \quad (6.18)$$

This matrix is unique to finite difference methods. The diagonal in each row is 4 (for three-dimensional applications it is 6 and for one-dimensional applications it is 2). The off-diagonal terms are -1 at the locations of the outer points of the finite difference stencil and zero elsewhere. The location of the nonzero, off-diagonal terms depends on the numbering sequence adopted. The right-hand side contains all free terms. Because of this very simple structure, the matrix can be built directly from the finite difference grid without the need to construct the equations explicitly.

Step 5 The potentials are obtained by solving the system of equations in **Eq. (6.18)**. Any method of solution may be employed. For example, we may use the Gaussian elimination method, Gauss–Seidel method, or any other solution method applicable to the solution of linear systems of equations. Computational software tools such as Matlab, Mathematica, Maple may also be used.

Script **fdmimp.m** assembles and solves the system of equations using Gaussian elimination. The script is rather general in that it can be applied to other finite difference problems but is simple enough to be understood. The potentials V_1 through V_{16} obtained from the script are shown in **Figure 6.8**, in the same sequence as the nodes are numbered in **Figure 6.7a**, together with the boundary values.

Figure 6.8 Potentials at the interior and boundary nodes in **Figure 6.7a**. The solution at node #5 is emphasized

10.0	10.0	10.0	10.0	10.0	10.0
0.0	4.5455	5.9470	5.9470	4.5455	0.0
0.0	**2.2348**	3.2955	3.2995	2.2348	0.0
0.0	1.0985	1.7045	1.7045	1.0985	0.0
0.0	0.4545	0.7197	0.7197	0.4545	0.0
0.0	0.0	0.0	0.0	0.0	0.0

Step 6 The final step in the solution of a numerical method may be called a "data processing" step and includes any additional calculations such as calculation of electric fields from potentials, display of data, and interpretation of results. One way to present the results is given in **Figure 6.8**, where the potentials at the nodal points are shown as a list. **Figure 6.9a** shows a different form of the solution by showing equal potential lines in a plot. This type of representation is useful in design because it shows where the potential is high and where the gradients are high (distance between lines is small). **Figure 6.9b** shows the same result in a three-dimensional plot. Various types of plots, including these shown here, are easily obtained with commercially available graphing software.[1] In the present problem, the density is highest at and around the upper left and upper right corners, where the potential changes from 10 V to 0 V in a very short distance. In other cases, the

[1] The plots in **Figures 6.9a** and **6.9b** were drawn with Matlab, using the data in **Figure 6.8**.

solution might mean the capacitance of a device, breakdown voltage in a gap, or the electric field intensity in the solution domain or at a point. The latter is calculated by employing the finite difference method as an approximation to the derivative and calculating the terms of the gradient in the potential as we shall see in **Example 6.1**.

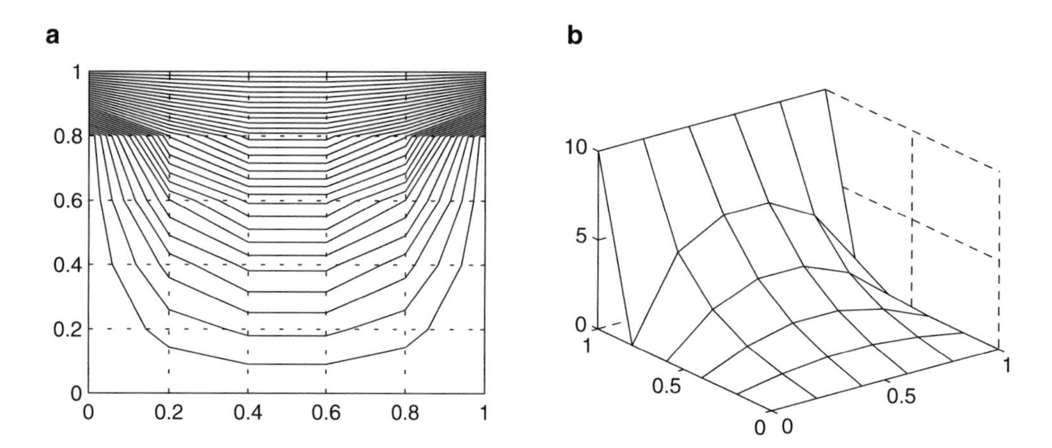

Figure 6.9 (**a**) Representation of results as a contour plot of constant potentials at increments of 0.3226 V in a cross-section of **Figure 6.6**. (**b**) Representation of the results as a three-dimensional plot. Height indicates potential

6.3.3.2 Explicit Solution

In an explicit solution, we start with **Eq. (6.15)** and assume that all potentials are known, including at interior nodes, at any step of the solution. Steps 1, 2, 3, and 6 are identical to those for the implicit solution. The differences between the implicit and explicit methods are in Steps 4 and 5.

Step 4: Approximation Before applying the approximation in **Eq. (6.15)**, all interior potentials are set to zero for lack of a better choice. This is the guess required to start the solution. If we have any basis for a better guess, it should be used, but the initial guess is not terribly important. Any reasonable guess will lead to a correct solution.

Evaluation of the potential at node (i,j) consists of calculating the average of the four potentials above, below, to its left, and to its right as indicated in **Eq. (6.15)**. The only additional constraint is that any potential that is updated is then used for the evaluation of subsequent potentials.

As examples, the potentials at points 1, 2, 3, 4, 5, and 6 (in **Figure 6.7a**) are calculated as

$$V_1 = \frac{10+0+0+0}{4} = 2.5\ \text{V}, \qquad V_2 = \frac{2.5+10+0+0}{4} = 3.125\ \text{V}$$

$$V_3 = \frac{3.125+10+0+0}{4} = 3.28125\ \text{V}, \qquad V_4 = \frac{3.28125+10+0+0}{4} = 3.3203125\ \text{V}$$

$$V_5 = \frac{0+2.5+0+0}{4} = 0.625\ \text{V}, \qquad V_6 = \frac{0.625+3.125+0+0}{4} = 0.9375\,\text{V}$$

Note that in the approximation, the latest, most up-to-date values of points is used.

Step 5: Solution The solution proceeds similarly with all other nodes. At each step, a nodal value is calculated as an average of the previously known nodes. After 16 steps, one iteration through the mesh is completed and a new solution is obtained. This solution is shown in **Figure 6.10a**. Comparing this solution with that obtained for the implicit solution in **Figure 6.8** (which is exact within the round-off errors of computation) or with the analytic solution, this is obviously not the "correct" solution. Thus, we repeat the process, again starting with node 1, assuming that the value at node 1 is unknown and evaluating it in terms of the neighboring nodes using the solution in **Figure 6.10a** as the new guess. After repeating the solution three more times, we obtain the solution in **Figure 6.10b**. This solution is closer but still not "correct." To obtain an "accurate" solution, we repeat this process until the change in solution is lower than a predetermined tolerance value. To see how the solution progresses, consider **Figure 6.11**, which shows the solution at point 5 as a function of the number of iterations employed. After about 15 iterations, the solution is within 0.1% of the exact solution, whereas after 20 iterations, it is within 0.01% of the exact solution. The solution, therefore, converges to the exact solution (in this plot, $V_5 = 2.2348$ V) as the number of iterations increases, and provided the number of iterations is sufficiently large, an accurate solution will be obtained.

a

10.0	10.0	10.0	10.0	10.0	10.0
0.0	2.5	3.125	3.2813	3.3203	0.0
0.0	**0.625**	0.9375	1.0547	1.0938	0.0
0.0	0.1563	0.2734	0.3320	0.3564	0.0
0.0	0.0391	0.0781	0.1025	0.1147	0.0
0.0	0.0	0.0	0.0	0.0	0.0

b

10.0	10.0	10.0	10.0	10.0	10.0
0.0	4.1418	5.4189	5.5217	4.3353	0.0
0.0	**1.7464**	2.6459	2.7655	1.9699	0.0
0.0	0.7227	1.1993	1.2890	0.8892	0.0
0.0	0.2691	0.4702	0.5142	0.3508	0.0
0.0	0.0	0.0	0.0	0.0	0.0

Figure 6.10 Solution to geometry in **Figure 6.6**. (**a**) After one iteration. (**b**) After four iterations. Node 5 is emphasized for reference and comparison with the solution in **Figure 6.8**

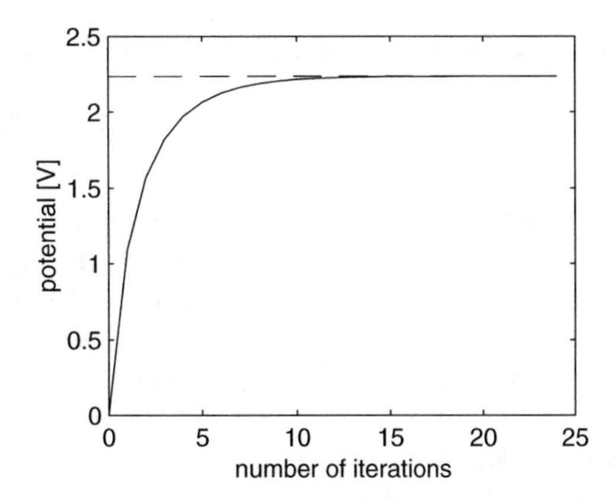

Figure 6.11 Solution at node 5 (**Figure 6.7a**) as a function of number of iterations

The only remaining question is how to stop the iteration procedure so that an acceptable solution is obtained. We could obviously choose one node and follow the solution at this node. Since, in general, the correct solution is not known, all we can do is look at the change in solution, and when the change in the solution between subsequent iterations is lower than a given error e, we view the solution is acceptable.

This can be written as

$$\left|V_i^n - V_i^{n-1}\right| \leq e \tag{6.19}$$

This is very simple but has the disadvantage that we infer the correctness of the solution throughout the solution domain based on a single node. If the solution also converges at all other nodes at the same rate (a fact we normally do not know and have no reason to assume), this approach is very good. More appropriate would be a measure of convergence that takes into account the changes at all nodes. A possible error criterion is an average error per node:

$$\frac{1}{N}\sum_{i=1}^{N}\left|V_i^n - V_i^{n-1}\right| \leq e \tag{6.20}$$

The process described here is an iterative process and is explicit in that the solution is known at each iteration of the process. Explicit procedures are preferred over implicit methods, particularly when very large grids are required and the implicit procedure would be prohibitively slow.

The results shown here were obtained with the script **fdmexp.m**. The results for the last (24th) iteration with an error tolerance $e = 10^{-5}$ are shown in **Figure 6.12**. Note that the results are very close to those from the implicit solution in **Figure 6.8**.

Figure 6.12 Two-dimensional finite difference results. Explicit solution for a 5 × 5 mesh with an error tolerance of 5×10^{-5}. The output for the last (24th) iteration is listed with node 5 emphasized

10.0	10.0	10.0	10.0	10.0	10.0
0.0	4.5454	5.9469	5.9469	4.5454	0.0
0.0	**2.2347**	3.2953	3.2953	2.2348	0.0
0.0	1.0984	1.7044	1.7045	1.0984	0.0
0.0	0.4545	0.7196	0.7197	0.4545	0.0
0.0	0.0	0.0	0.0	0.0	0.0

6.3.4 Solution to Poisson's Equation

So far, we discussed only the solution to Laplace's equation. The extension of the above results to Poisson's equation is rather simple. The difference is merely in the approximation which was given in **Eq. (6.17)**. In the implicit approach, we normally write the approximation as

$$4V_{i,j} - V_{i-1,j} - V_{i+1,j} - V_{i,j-1} - V_{i,j+1} = h^2\frac{\rho_{i,j}}{\varepsilon_{i,j}} \tag{6.21}$$

In the explicit solution, the form in **Eq. (6.17)** is used directly:

$$V_{i,j} = \frac{V_{i-1,j} + V_{i+1,j} + V_{i,j-1} + V_{i,j+1}}{4} + \frac{h^2\rho_{ij}}{4\varepsilon_{ij}} \tag{6.22}$$

Note, however, that these approximations are fundamentally different than those for Laplace's equation in that now the approximation depends on the distance between the nodes ($\Delta x = \Delta y = h$). We will have to take this into account and choose a small enough distance to provide an accurate solution. This may seem to be a difficult task because, after all, how do we know what value to choose? In practice, we will chose some value and solve the problem, and if we want to make sure the choice is good, we may choose a second, smaller value (such as half the previous value of h) and repeat the solution. If the solution does not change, or changes very little, the choice was acceptable. If not, we may repeat this process until we are satisfied with the choice. Note, also, that the extra term in Poisson's equation applies only at those nodes at which there is a charge density. At any other node, we use the approximation to Laplace's equation.

Exercise 6.2 Find the finite difference approximation for the three-dimensional Laplace's equation for implicit and explicit solutions.

Answer

$$6V_{i,j,k} - V_{i-1,j,k} - V_{i+1,j,k} - V_{i,j-1,k} - V_{i,j+1,k} - V_{i,j,k-1} - V_{i,j,k+1} = 0 \quad \text{(implicit)}$$
$$V_{i,j,k} = \frac{V_{i-1,j,k} - V_{i+1,j,k} - V_{i,j-1,k} - V_{i,j+1,k} - V_{i,j,k-1} - V_{i,j,k+1}}{6} \quad \text{(explicit)}.$$

Exercise 6.3 Find the finite difference approximation for the three-dimensional Poisson's equation for implicit and explicit solutions.

Answer

$$6V_{i,j,k} - V_{i-1,j,k} - V_{i+1,j,k} - V_{i,j-1,k} - V_{i,j+1,k} - V_{i,j,k-1} - V_{i,j,k+1} = \frac{h^2\rho_{i,j,k}}{\varepsilon_{i,j,k}} \quad \text{(implicit)}$$
$$V_{i,j,k} = \frac{V_{i-1,j,k} - V_{i+1,j,k} - V_{i,j-1,k} - V_{i,j+1,k} - V_{i,j,k-1} - V_{i,j,k+1}}{6} + \frac{h^2\rho_{i,j,k}}{6\varepsilon_{i,j,k}} \quad \text{(explicit)}.$$

Example 6.1 Solution to the One-Dimensional Poisson's Equation Consider a parallel plate capacitor as in **Figure 6.13a**. The capacitor is connected to a potential difference of 100 V, but, in addition, the dielectric between the plates contains a volume charge density which is a function of the distance between the plates given as $\rho_v = -2.5 \times 10^{-3}(x^2 - 0.1x)$ [C/m^3]. Assume the y and z dimensions of the capacitor are very large so that edge effects may be neglected and the material between the plates to be dielectric with relative permittivity of 4:

(a) Find the potential distribution everywhere between the two plates.
(b) Find the electric field intensity everywhere between the plates.

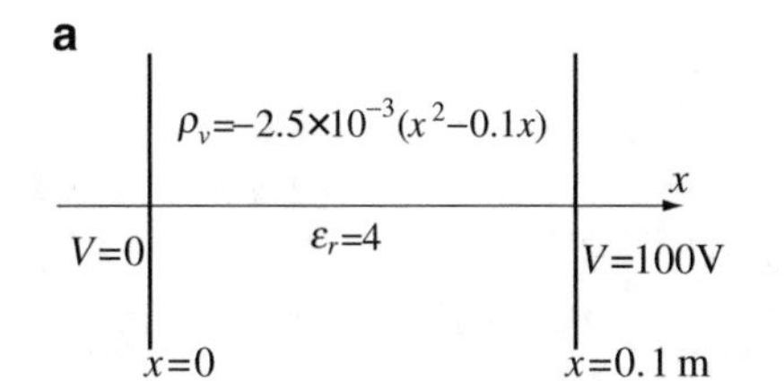

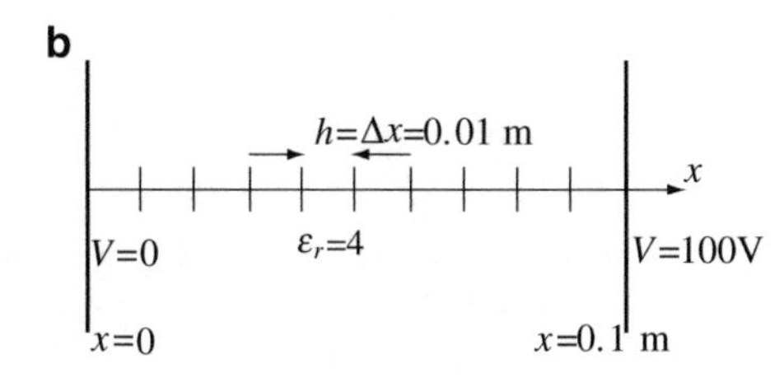

Figure 6.13 (**a**) Parallel plate capacitor with charge density between the plates connected to a potential difference. (**b**) Finite difference discretization

Solution:

(a) Because the potential only varies with the x dimension, the problem may be solved using the one-dimensional Poisson's equation. The capacitor is replaced with an equivalent geometry shown in **Figure 6.13b**. The distance between the plates is divided into $k = 10$ subdomains, resulting in nine unknown potential values in addition to the two known potentials on the boundaries. Although this problem has an analytic solution and the problem may, in fact, be solved by hand, we will use a script because, in general, hand calculation is too tedious.

The potential distribution in the capacitor as obtained using script **fdm1d.m** is shown in **Figure 6.14a**. The analytic solution for this problem was obtained in **Section 5.4.2** [**Eq. (5.11)**; see also **Example 5.1**].

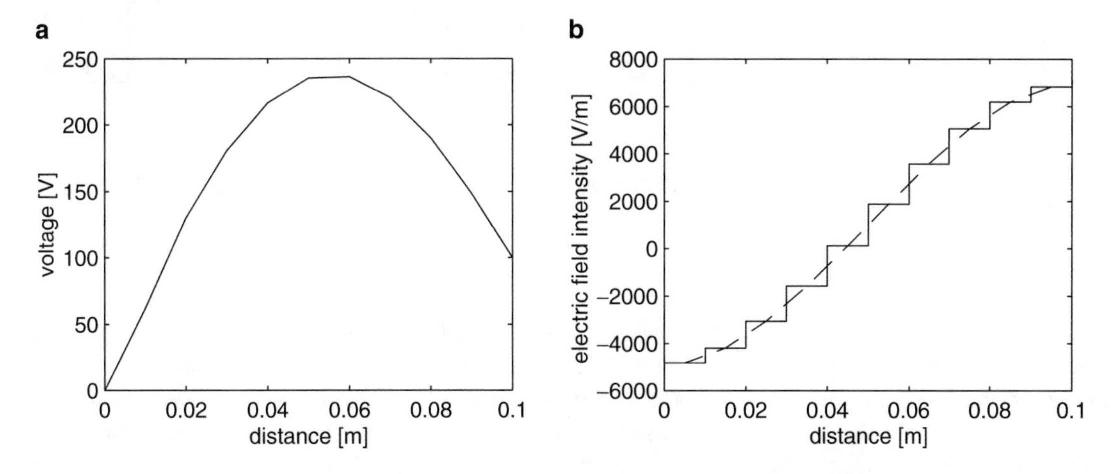

Figure 6.14 Solution for the problem in **Figure 6.13a**. (**a**) Solution for potential. (**b**) Solution for electric field intensity. Negative sign indicates the field is in the negative x direction

(b) The electric field intensity is calculated from the potentials at the various points using **Eqs. (6.1), (6.2)**, or **(6.3)**. By definition, the electric field intensity is

$$\mathbf{E} = -\nabla V = -\hat{\mathbf{x}}\frac{dV(x)}{dx} \quad \left[\frac{\mathrm{V}}{\mathrm{m}}\right]$$

After the potential values at all points in the mesh are known, the electric field intensity is written using **Figure 6.13b** and **Eq. (6.2)** as

$$\mathbf{E}_i = -\hat{\mathbf{x}}\left(\frac{V_{i+1} - V_i}{\Delta x}\right) \quad \left[\frac{\mathrm{V}}{\mathrm{m}}\right]$$

This electric field intensity is calculated for each space between two potential values and is assumed to be constant between the potentials.

Script **fdm1d.m** solves both for the potential and electric field intensity. A plot of the electric field intensity is shown in **Figure 6.14b**. Note that it is a staircase plot because the gradient of the potential remains constant between each two potential points. The accuracy of these results may be easily improved by using a larger number of points.

Example 6.2 Two-Dimensional Poisson's Equation: Solution Using Irregular Boundaries Consider the geometry in **Figure 6.15a**. The upper plate is a half-cylinder and forms, together with the lower plate, a capacitor. The potential difference between the upper and lower plates is $V = 24$ V and the plates are separated a distance of 20 mm at the edges. Assume free space throughout and large (infinite) dimension perpendicular to the page:

(a) Calculate the potential distribution between the plates.
(b) A volume charge density of 10^{-6} C/m^3 is added, uniformly distributed in the space between the plates. Calculate the potential distribution everywhere and compare with **(a)**.

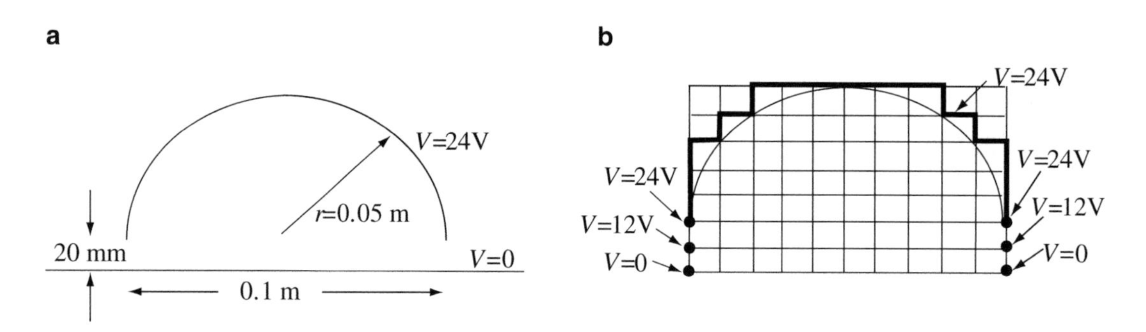

Figure 6.15 (**a**) Capacitor made of a cylindrical plate over a flat plate (shown in cross section). (**b**) Rectangular grid over the geometry in (**a**)

Solution: Although the geometry here is two-dimensional, the upper surface is curved and it would seem that a rectangular, uniform grid is not suitable for analysis of this problem. To analyze the problem, we first fit a uniform grid over the geometry as shown in **Figure 6.15b**. Then, we modify the surface of the upper plate such that it follows points on the grid by moving the surface to the nearest grid point, as shown in **Figure 6.15b** (thick line). This may seem as taking too much liberty with the geometry, and, in fact, for the grid shown, it is. However, in the limit, as we increase the number of points, the approximate and exact surfaces become closer. To simplify analysis, we will assume that all points above the upper surface are at the potential of the upper surface. The latter assumption is equivalent to making the upper conductor fill the space above the surface up to the top boundary. This does not entail any approximation on the potential between the two surfaces.

Now, we have a uniform grid in which all points which are not between the two surfaces are boundary nodes. At the point where the two surfaces are closest to each other, we must also specify boundary conditions. Since we do not know the conditions at these locations, we simply assume that the potential varies linearly and specify the value at the nodes as half the potential difference between the plates as shown in **Figure 6.15b**.

The potential is found using script **fdm2d.m**, using the iterative method for Poisson's equation. In this case, it is most convenient to plot equal potential lines because the surface is curved. The plots in **Figure 6.16** were obtained using the grid in **Figure 6.15b**. Note that in **Figure 6.16b**, the potential between the plates can be higher than the boundary potentials because of the charge density that exists between the plates.

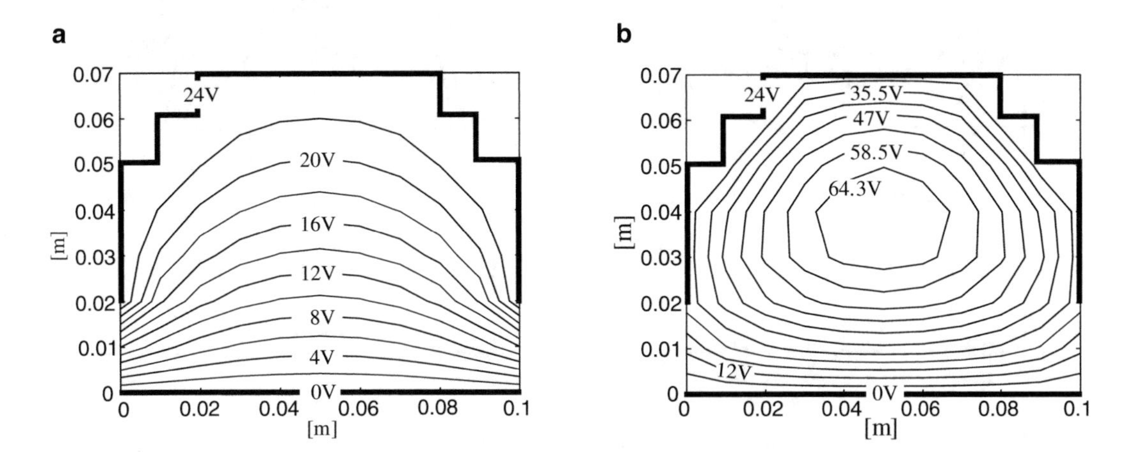

Figure 6.16 Solution for the capacitor in **Figure 6.15a** Equal potential lines are shown. (**a**) Potential distribution without space charge. (**b**) Distribution with charge density of 10^{-6} C/m^3 between the plates

Example 6.3 Three-Dimensional Laplace's Equation The three-dimensional box in **Figure 6.17a** is given. The upper plate is connected to a 10 V potential and the other five plates are grounded (0 V). Find the potential and electric field intensity everywhere inside the box.

Figure 6.17 (**a**) Conducting box with given boundary potentials. (**b**) 5 × 5 × 5 finite difference grid over the box

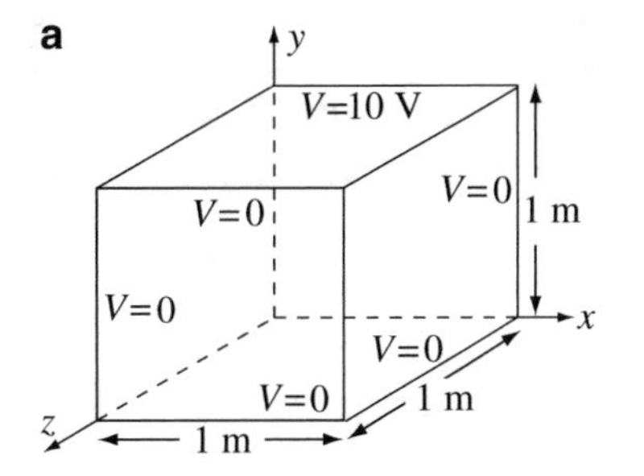

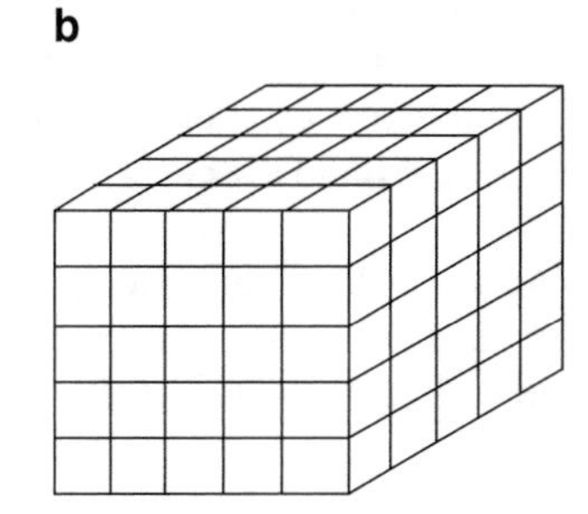

Solution: The solution of three-dimensional problems is usually much more complicated than two-dimensional applications. However, the finite difference method uses a rectangular grid and the bookkeeping is not too difficult. The approximation here is that for Laplace's equation given in **Exercise 6.2**.

A three-dimensional grid, with five divisions in each direction, is defined over the mesh as shown in **Figure 6.17b**. The mesh has a total of 64 internal nodes which must be evaluated. The solution follows almost identical steps as for the two-dimensional solution in **Section 6.3.3**.

One difficulty encountered in three-dimensional applications is in the display of data. Usually, a few cross sections are cut and the potential drawn on these planes. Two contour plots are shown in **Figures 6.18a** and **6.18b**. The first is for a cross section cut vertically at $x = 0.6$, parallel to the y–z plane. The second is a horizontal cut through at $z = 0.6$, parallel to the x–y plane. Three-dimensional plots may also be drawn as shown in **Figure 6.19**. The latter is drawn for a 10 × 10 × 10 grid.

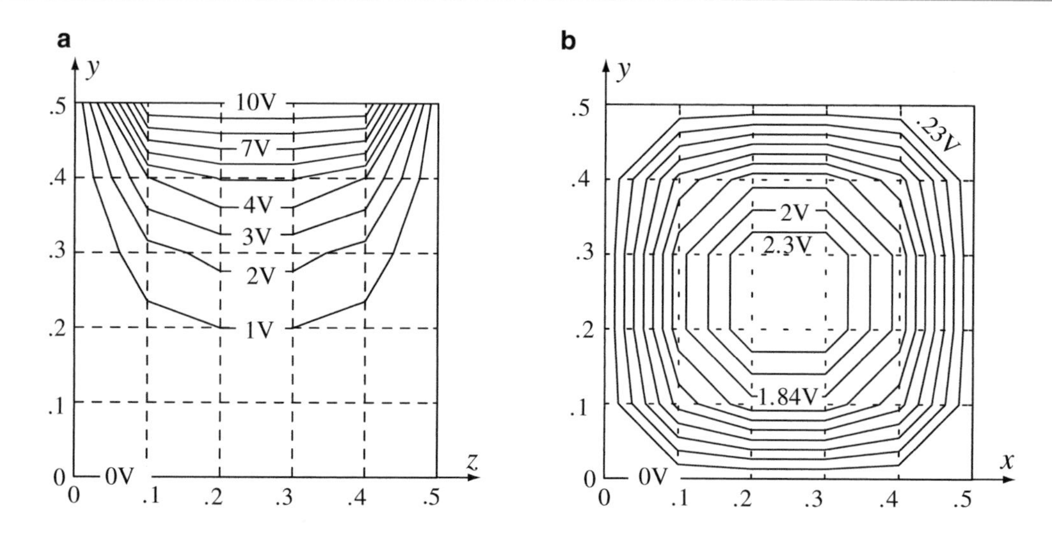

Figure 6.18 (**a**) Contour plot on a cross section cut vertically at $x = 0.6$. (**b**) Contour plot on a cross section cut horizontally at $z = 0.6$. Both plots are for the $5 \times 5 \times 5$ grid

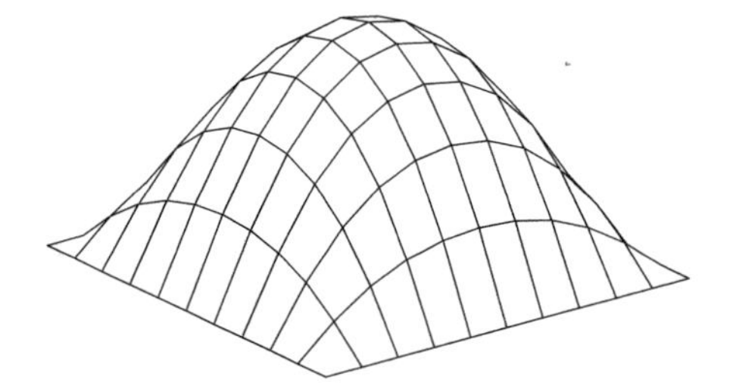

Figure 6.19 Three-dimensional plot of the potential distribution in the box ($10 \times 10 \times 10$ mesh, horizontal cut at $z = 0.5$ m)

To calculate the electric field intensity, we use the potential difference between every two neighboring nodes, as was done in **Example 6.1**. However, now there are three directions in space. Application of **Eq. (6.2)** in the x, y, and z directions provides the electric field intensities on the edges of the cells of the grid. The method is shown in **Figure 6.20** and a list of the electric field intensities at a few mesh locations is given in **Table 6.1**. Note that different components are evaluated at different locations on the grid. It is also possible to calculate the average of the electric field intensity in each direction and associate these averages with the center of the cell in **Figure 6.20**. These calculations were performed using script **fdm3d.m**.

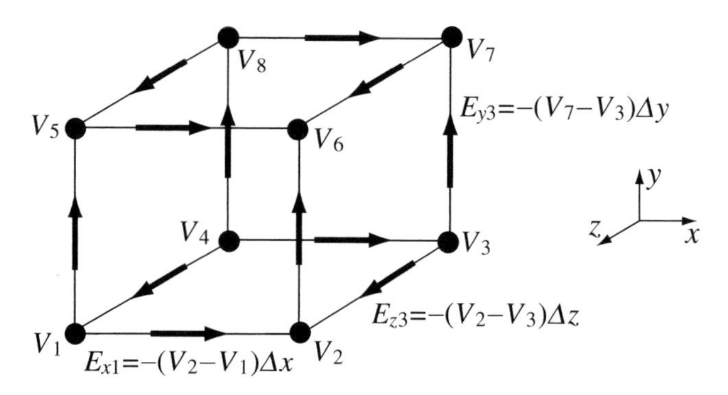

Figure 6.20 Calculation of electric field intensities from potentials, using **Eq. (6.2)**

Table 6.1 Electric field intensity at a few locations in the $10 \times 10 \times 10$ grid

x	y	z	E_x	E_y	E_z
0.3 m	0.2 m	0.2 m	0.4235 V/m	x	x
0.6 m	0.5 m	0.8 m	x	0.0 V/m	x
0.6 m	0.6 m	0.9 m	x	x	24.652 V/m
0.7 m	0.2 m	0.2 m	−0.4235 V/m	x	x
0.2 m	0.7 m	0.2 m	x	−0.4235 V/m	x
0.2 m	0.2 m	0.3 m	x	x	1.422 V/m

x indicates that the corresponding component is not calculated at this location

6.4 The Method of Moments: An Intuitive Approach

Consider, again, the calculation of the electric potential from known charge density distributions. Suppose we know the charge distribution in a given section of the space Ω. Then, from **Eq. (4.32)**, the potential a distance R from the charge distribution is

$$V(x,y,z) = \int_{\Omega'} \frac{1}{4\pi\varepsilon R} \rho_{\Omega'}(x',y',z')d\Omega' \quad [\text{V}] \tag{6.23}$$

where Ω' denotes the domain in which the charge distribution is known and primed coordinates are used to distinguish between the source and observation points (at which the potential is calculated). The domain Ω' can be a surface s', a length l', or a volume v'. In all cases, the formula for the calculation of the potential has exactly the same form. In even more general terms, the equation may be written as

$$V(x,y,z) = \int_{\Omega'} K(x,y,z,x',y',z')\rho_{\Omega}'(x',y',z')d\Omega' \quad [\text{V}] \tag{6.24}$$

where K is a geometric function (called the kernel of the integral) relating the measured quantity V and the source ρ. In this notation, the method can be applied to any integral of this form, whatever the measured quantity and source.

The question now is the following: Suppose the charge density ρ is not known but, instead, the potential is known. Can we also use this relation to calculate the charge density distribution? The answer is yes, but only under certain conditions. To see how this can be done in general, we use the example in **Figure 6.21a**. It consists of a very thin charged conducting surface with a general surface charge density $\rho_s(x',y',z')$ [C/m^2]. The charge density is not known, but the potential on the conducting surface is known and equal to V_0 [V]. The goal is to calculate the charge density everywhere on the surface.

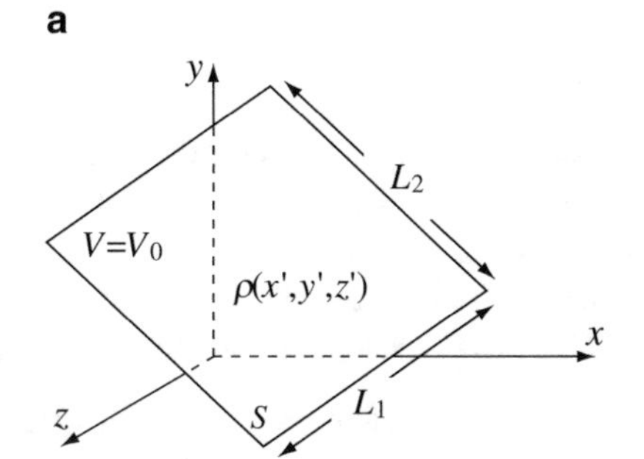

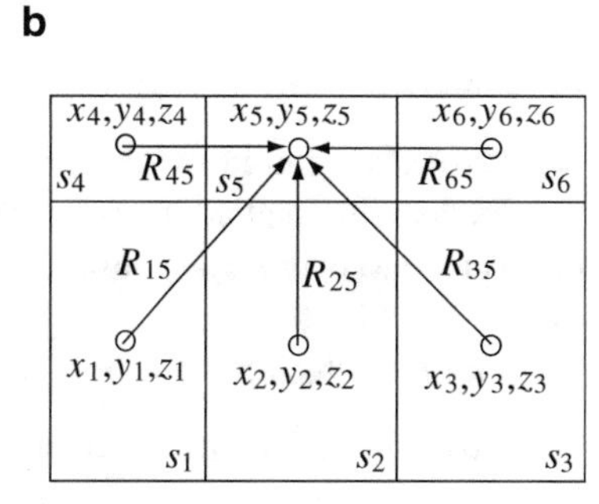

Figure 6.21 (**a**) A charged surface with known potential and unknown surface charge density distribution. (**b**) Division of the charged surface in (**a**) into six subdomains. The charge density in each subdomain is unknown but assumed constant within each subdomain

The first hint at what we must do is from **Eq. (6.23)**. For the charge density to be calculable, it must be taken outside the integral sign. This, in turn, means that the charge density must be constant on the surface. This condition cannot be met in general; therefore, we resort to dividing the surface into a large number of subsurfaces and assume that for each subsurface, the charge density is constant, but it may vary from subsurface to subsurface. An example of the subdivision of the surface into subdomains is shown in **Figure 6.21b**, where six subdomains are used. The premise behind this approach is that since we are free to divide the surface into as many subdomains as we wish, we can make this approximation as good as needed by simply increasing the number of subdomains. In the limit, each subdomain is infinitesimally small and we are back to the expression in **Eq. (6.23)**. To formulate the method we divide the surface into N subdomains and assume a uniform charge density ρ_{si} [C/m^2] on subdomain s_i'. The potential at an arbitrary point (x_j,y_j,z_j) on subdomain j is

$$V\left(x_j,y_j,z_j\right) = \sum_{i=1}^{N} \int_{s_i'} \frac{\rho_{si}}{4\pi\varepsilon R_{ij}(x_i',y_i',z_i',x_j,y_j,z_j)} ds_i' = \sum_{i=1}^{N} \frac{\rho_{si}}{4\pi\varepsilon} \int_{s_i'} \frac{1}{R_{ij}(x_i',y_i',z_i',x_j,y_j,z_j)} ds_i' \quad [\text{V}] \tag{6.25}$$

where R_{ij} is the distance between the location of ds_i' and the point at which the potential is calculated. Next, we assume that the total charge on subdomain i, which is equal to $q_{si} = \rho_{si}s_i$, is located at the center of subdomain i. The potential on subdomain j is constant and may be written as:

$$V_j = \frac{1}{4\pi\varepsilon}\sum_{\substack{i=1\\ i\neq j}}^{N} \frac{\rho_{si}s_i}{R_{ij}(x_i', y_i', z_i', x_j, y_j, z_j)} + V_{\rho_{sj}} \quad [\text{V}] \tag{6.26}$$

The distance R_{ij} is

$$R_{ij}(x_i', y_i', z_i', x_j, y_j, z_j) = \sqrt{\left(x_j - x_i'\right)^2 + \left(y_j - y_i'\right)^2 + \left(z_j - z_i'\right)^2} \tag{6.27}$$

The first term in **Eq. (6.26)** is the potential at the center of subdomain j, due to $N - 1$ point charges located at the centers of all subdomains except subdomain j. The second term is the potential at the center of subdomain j due to the charge density on subdomain j. This term must be calculated separately since if we use the formula for point charges we obtain an infinite potential ($R_{ij} = 0$ for $j = i$).

The computation sequence for the potential at $j = 5$ is shown schematically in **Figure 6.21b**. The relationship between subdomains i and j ($i \neq j$) is shown in **Figure 6.22a**.

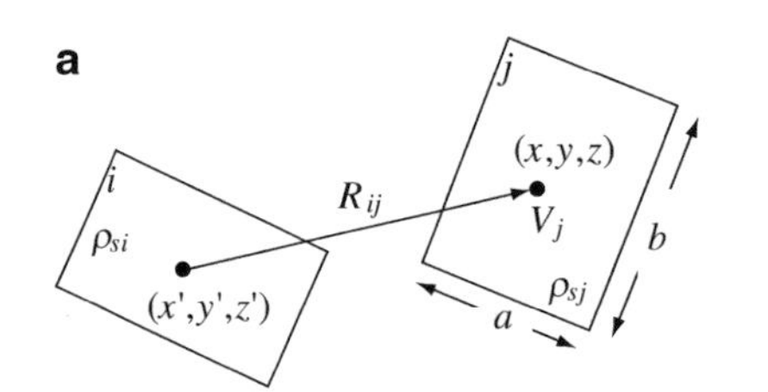

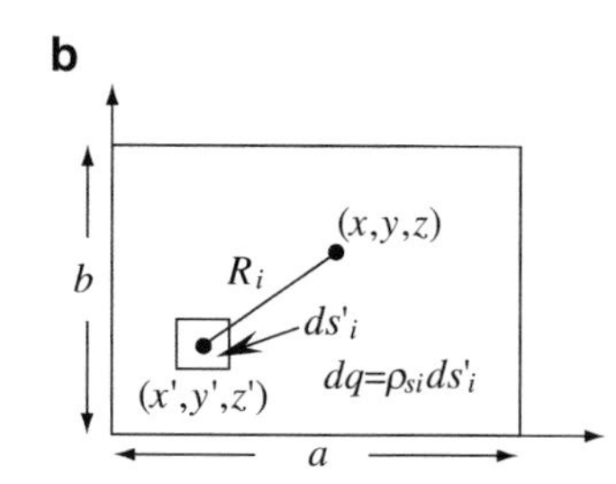

Figure 6.22 (**a**) Relation between potential on subdomain j and charge density on subdomain i. (**b**) Calculation of the potential at the center of subdomain i due to its own charge density

A number of points must be indicated here:

(1) The expression in the integral in **Eq. (6.25)** is a purely geometrical expression. It only has to do with the surface of subdomain i and the distance between this subdomain and the point at which the potential is calculated.
(2) There are N unknown values of ρ_s, one on each subdomain.
(3) To be able to evaluate N unknown values of the surface charge density, we must evaluate the potential at N distinct points at which the potential is known. This provides N equations for the N unknown charge densities.

To use this method for practical calculation, we must define a systematic way of dividing the surface into subdomains and of calculating the potential V_j at N points. These decisions are made as follows:

(1) The surface is divided into N subdomains. The subdomains do not need to be equal in size but equal-sized subdomains simplify the solution and allow simple computer implementation.
(2) The number of subdomains can be as large as we wish. The larger the number of subdomains, the more accurate the solution, but also the lengthier the computation.
(3) The potential is evaluated at the center of each of the subdomains and set to the known value at that point.

In effect, we have replaced the charged surface by equivalent point charges, each equal to $\rho_i s_i$, and placed these point charges at the center of the corresponding subdomain. It is now a simple matter of calculating the potentials and setting these to the known values at the centers of the subdomains. The general expressions for the potentials are

$$V_j = V_{\rho_{sj}} + \sum_{i=1, i\neq j}^{N} \rho_{si}\frac{s_i}{4\pi\varepsilon R_{ij}} \quad [\text{V}], \quad j = 1, 2, \ldots, N \tag{6.28}$$

To obtain a form amenable to computation, we rewrite **Eq. (6.28)** as

$$V_j = K_{jj}\rho_{sj} + \sum_{\substack{i=1\\ i\neq j}}^{N} \rho_{si}K_{ij} \quad [\text{V}], \quad j = 1, 2, \ldots, N \tag{6.29}$$

where the geometric function[2] is

$$K_{ij} = \frac{s_i}{4\pi\varepsilon\sqrt{\left(x_j - x_i'\right)^2 + \left(y_j - y_i'\right)^2 + \left(z_j - z_i'\right)^2}}, \quad i,j = 1,2,\ldots,N, \quad i \neq j \tag{6.30}$$

Note that the terms for which $i = j$ cannot be calculated using this expression because if we place a charge at a point, the potential at that point is infinite. The potential at point j due to the charge density on the subsurface j is calculated by direct integration. To do so, we assume a flat surface, of dimensions a and b, and with charge density ρ_s = constant as in **Figure 6.22b**. The potential at the center of the plate [point $(x_i,y_i,z_i) = (x_j,y_j,z_j)$] is found by direct integration (see **Exercise 6.4**) as

$$V\left(x_j, y_j, z_j\right) = \frac{\rho_{sj}}{4\pi\varepsilon}\left(2a_j \ln\frac{b_j + \sqrt{a_j^2 + b_j^2}}{a_j} + 2b_j \ln\frac{a_j + \sqrt{a_j^2 + b_j^2}}{b_j}\right) \quad [\text{V}] \tag{6.31}$$

Therefore, the geometric function for $i = j$ is

$$K_{jj} = \frac{1}{4\pi\varepsilon}\left(2a_j \ln\frac{b_j + \sqrt{a_j^2 + b_j^2}}{a_j} + 2b_j \ln\frac{a_j + \sqrt{a_j^2 + b_j^2}}{b_j}\right) \tag{6.32}$$

With this, we can use **Eq. (6.29)** for all values of i,j.

The computation sequence starts by dividing the surface in **Figure 6.21a** into N equal subdomains ($N = 6$ in this case), as shown in **Figure 6.21b**. For convenience, we will assume equal-sized rectangles with constant charge density on each rectangle. The centers of the subdomains are points (x_j,y_j,z_j). Using **Eq. (6.29)**, in expanded form, the potential at each of the six points is

$$\begin{aligned}
\rho_{s1}K_{1,1} + \rho_{s2}K_{1,2} + \rho_{s3}K_{1,3} + \cdots + \rho_{s(N-1)}K_{1,N-1} + \rho_{sN}K_{1,N} &= V_1 \\
\rho_{s1}K_{2,1} + \rho_{s2}K_{2,2} + \rho_{s3}K_{2,3} + \cdots + \rho_{s(N-1)}K_{2,N-1} + \rho_{sN}K_{2,N} &= V_2 \\
&\vdots \\
\rho_{s1}K_{N-1,1} + \rho_{s2}K_{N-1,2} + \rho_{s3}K_{N-1,3} + \cdots + \rho_{s(N-1)}K_{N-1,N-1} + \rho_{sN}K_{N-1,N} &= V_{N-1} \\
\rho_{s1}K_{N,1} + \rho_{s2}K_{N,2} + \rho_{s3}K_{N,3} + \cdots + \rho_{s(N-1)}K_{N,N-1} + \rho_{sN}K_{N,N} &= V_N
\end{aligned} \tag{6.33}$$

Rewriting this as a matrix system, we get

$$\begin{bmatrix} K_{1,1} & K_{1,2} & K_{1,3} & \cdots & K_{1,N-1} & K_{1,N} \\ K_{2,1} & K_{2,2} & K_{2,3} & \cdots & K_{2,N-1} & K_{2,N} \\ \vdots & \vdots & \vdots & \vdots & \vdots & \vdots \\ K_{N-1,1} & K_{N-1,2} & K_{N-1,3} & \cdots & K_{N-1,N-1} & K_{N-1,N} \\ K_{N,1} & K_{N,2} & K_{N,3} & \cdots & K_{N,N-1} & K_{N,N} \end{bmatrix} \begin{bmatrix} \rho_{s1} \\ \rho_{s2} \\ \vdots \\ \rho_{s(N-1)} \\ \rho_{sN} \end{bmatrix} = \begin{bmatrix} V_1 \\ V_2 \\ \vdots \\ V_{N-1} \\ V_N \end{bmatrix} \tag{6.34}$$

Note that $K_{ij} = K_{ji}$; that is, the system of equations is symmetric. Recall as well that V_1 through V_N must be known values.

This system can be solved in any convenient manner. For a small system, hand calculation or the use of a programmable calculator may be sufficient. For larger systems, a computer will almost always have to be used. Once the charge density has been calculated everywhere on the conducting surface, it may be used in many ways. For example, we may want to calculate the capacitance of the plate or the electric field intensity anywhere in space. Some of the possible uses of the method are shown in the examples that follow. The method is rather general and applies to other types of fields as well. Also, we may use the same method to calculate the charge density on surfaces, inside volumes, or on thin or thick lines. The only fundamental requirement of the method is to find a simple relation between charge and potential in this case, or between the source and the fields in the general case.

[2] The kernel K_{ij} is called here a "geometric" function. In fact, it is the potential at point j due to a unit point charge at point i. This function is also called the Green's function for the geometry. The resulting Green's function method of solution for Poisson's equation is an important method of solution for both analytic and numerical methods because once the solution for a unit source is obtained, the solution for any source is relatively easy to obtain. However, we will call the kernel a geometric function because we do not pursue the idea of a Green's function in any detail.

Exercise 6.4 Derive **Eq. (6.31)** for a rectangular surface, charged with a constant surface charge density ρ_s [C/m^2]. Assume dimensions of the plate are a [m] and b [m], define a differential area $ds = dx'dy'$, and calculate the potential at the center of the plate by direct integration over the surface of the plate. Use **Figure 6.22b**, assume the plate is in the x–y plane and the potential is calculated at (0,0).

Example 6.4 Application: Charge and Capacitance of an Electrode The flat electrode in **Figure 6.23a** is given:

(a) If the potential on the plate is 1 V, what is the total charge on the plate?
(b) Find the capacitance of the plate.
(c) Find the electric field intensity at a height of 2 m above the center of the plate.

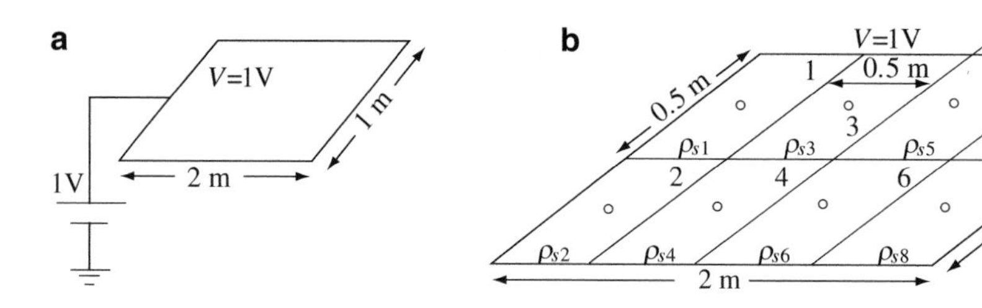

Figure 6.23 (**a**) Charged rectangular electrode held at a constant potential of 1 V. (**b**) Division of the electrode into eight equal subdomains, each with constant unknown charge density

Solution: For convenience, we place the plate on the x–y plane with the center at the origin and divide the plate into eight equal subsurfaces as shown in **Figure 6.23b**. This subdivision may not be sufficient to obtain an accurate solution, but, for now, this discretization is used so that we may give the actual matrix entries. Also, with a matrix of this size, hand calculation is possible. The script **mom1.m** is used to obtain more accurate solutions:

(a) To calculate the unknown charge densities, the locations of the centers of the subdomains are first identified. With the sides of the plate parallel to the axes, the centers of the subdomains are P_1 (−0.75,0.25), P_2 (−0.75,−0.25), P_3 (−0.25, 0.25), P_4 (−0.25,−0.25), P_5 (0.25,0.25), P_6 (0.25,−0.25), P_7 (0.75,0.25), and P_8 (0.75,−0.25). The permittivity of free space is 8.854×10^{-12} F/m and the size of each subdomain is $a_i = 0.5$ m and $b_i = 0.5$ m, and, therefore, $s_i = 0.25$ m^2. The plate is at 1 V potential with respect to infinity. With these, we can now go back to **Eq. (6.30)** and calculate K_{ij} and to **Eq. (6.32)** to calculate K_{jj}. The resulting matrix is

$$10^{10} \times \begin{bmatrix} 1.59 & .450 & .450 & .318 & .225 & .201 & .150 & .142 \\ .450 & 1.59 & .318 & .450 & .201 & .225 & .142 & .150 \\ .450 & .318 & 1.59 & .450 & .450 & .318 & .225 & .201 \\ .318 & .450 & .450 & 1.59 & .318 & .450 & .201 & .225 \\ .225 & .201 & .450 & .318 & 1.59 & .450 & .450 & .318 \\ .201 & .225 & .318 & .450 & .450 & 1.59 & .318 & .450 \\ .150 & .142 & .225 & .201 & .450 & .318 & 1.59 & .450 \\ .142 & .150 & .201 & .225 & .318 & .450 & .450 & 1.59 \end{bmatrix} \begin{bmatrix} \rho_{s1} \\ \rho_{s2} \\ \rho_{s3} \\ \rho_{s4} \\ \rho_{s5} \\ \rho_{s6} \\ \rho_{s7} \\ \rho_{s8} \end{bmatrix} = \begin{bmatrix} 1 \\ 1 \\ 1 \\ 1 \\ 1 \\ 1 \\ 1 \\ 1 \end{bmatrix}$$

Using the script **mom1.m** or solving the matrix using a programmable calculator gives the charge densities as follows:

$$\rho_{s1} = \rho_{s2} = \rho_{s7} = \rho_{s8} = 0.31572 \times 10^{-10} \quad [\text{C/m}^2]$$

and

$$\rho_{s3} = \rho_{s4} = \rho_{s5} = \rho_{s6} = 0.22241 \times 10^{-10} \quad [\text{C/m}^2]$$

The total charge on the plate is the sum of the charges of the individual domains:

$$Q = \sum_{i=1}^{8} \rho_{si} s_i = 4 \times 0.25 \times 0.31572 \times 10^{-10} + 4 \times 0.25 \times 0.22241 \times 10^{-10}$$
$$= 0.538 \times 10^{-10} \quad [\text{C}].$$

(b) To find the capacitance of the plate, we divide the total charge by the potential difference between the plate and the reference point at infinity:

$$C = \frac{Q}{V} = \frac{0.538 \times 10^{-10}}{1} = 53.8 \quad [\text{pF}]$$

Because we only used eight subdomains to describe a 2 m^2 plate, we should expect some error. To check the relative error, the number of subdomains is increased and the capacitance calculated using the script **mom1.m**. The results are as follows:

Number of subdomains	4 × 2	8 × 4	16 × 8	20 × 10	24 × 12
Capacitance	53.8 pF	56.5 pF	57.9 pF	58.2 pF	58.4 pF

The result can be improved as much as we wish by increasing the number of subdomains. Unfortunately, this has its limits: The computer can be easily overwhelmed if we get carried away with the number of subdomains. Note that for a division of 24 × 12, the script solves a 288 × 288 system of equations with a full matrix. If we were to increase the division to, say, 50 × 50, a 2,500 × 2,500 system must be solved.

(c) To calculate the electric field intensity at the center and above the plate, we note that the subdomains are symmetric about the center of the plate. This means that each two symmetric charges cancel each other's horizontal component of the electric field intensity. Using **Figure 6.24**, the electric field intensity at point P may be written as follows:

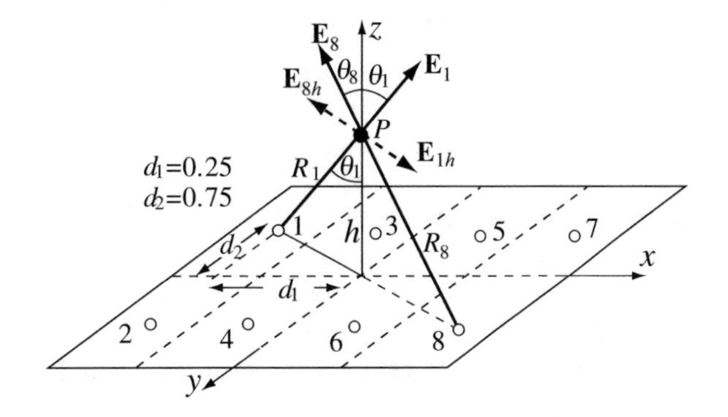

Figure 6.24 Calculation of the electric field intensity above the center of the plate in **Figure 6.23a**, showing the electric field intensities due to points 1 and 8

Due to point 1: The total charge at point (1) is $Q_1 = 0.25 \times 0.31572 \times 10^{-10} = 7.893 \times 10^{-12}$ C. The vertical component of the electric field intensity due to point (1) is

$$E_1 = \frac{Q_1}{4\pi\varepsilon_0 R_1^2}\cos\theta_1 = \frac{Q_1 h}{4\pi\varepsilon_0 R_1^2 R_1} = \frac{Q_1 h}{4\pi\varepsilon_0 \left(d_1^2 + d_2^2 + h^2\right)^{3/2}}$$
$$= \frac{7.893 \times 10^{-12} \times 2}{4 \times \pi \times 8.854 \times 10^{-12} \times \left(0.25^2 + 0.75^2 + 2^2\right)^{3/2}} = 0.0143 \quad [\text{V/m}]$$

Each of the four corner subdomains (1), (2), (7), and (8) produces identical fields. Similarly, due to point (5), the total charge at point (5) is $Q_5 = 0.25 \times 0.22241 \times 10^{-10} = 5.56 \times 10^{-12}$ C. The electric field intensity at point P due to this charge is

$$E_5 = \frac{Q_5}{4\pi\varepsilon_0 R_5^2}\cos\theta_5 = \frac{5.56 \times 10^{-12} \times 2}{4 \times \pi \times 8.854 \times 10^{-12} \times \left(0.25^2 + 0.25^2 + 2^2\right)^{3/2}} = 0.0119 \quad [\text{V/m}]$$

The fields due to points (3), (4), (5), and (6) are identical. Thus, the total field is in the positive z direction (positive charges) and equal to

$$\mathbf{E}_{total} = \hat{\mathbf{z}}4(0.0119 + 0.0143) = \hat{\mathbf{z}}0.105 \quad [\text{V/m}]$$

Script **mom1.m** solves for the charge and capacitance for a rectangular plate of any dimensions and with any number of subdomains. The Script may be easily adapted for other applications.

Example 6.5 Capacitance Between Two Conducting Strips The two strips in **Figure 6.25a** represent two general conductors. The strips are 50 mm wide and 1 m long and are separated a distance of 100 mm vertically. Calculate the capacitance between the two strips if:

(a) The two strips are parallel to each other ($\theta = 0°$).
(b) The strips are perpendicular to each other ($\theta = 90°$).

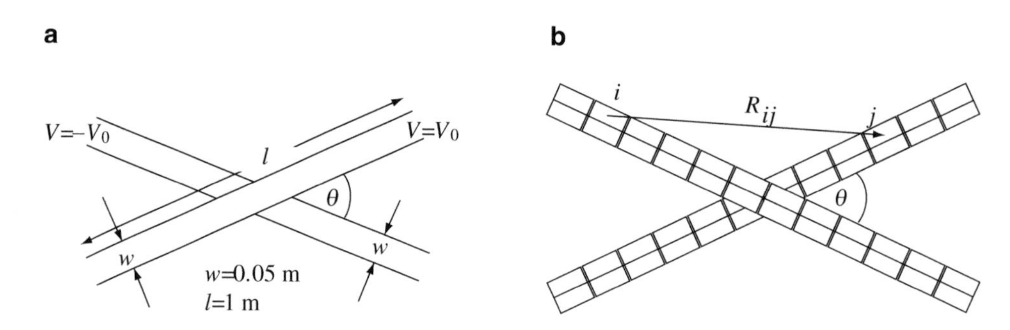

Figure 6.25 (**a**) Two conducting strips separated a distance d (vertically) and intersecting at an angle θ. (**b**) Division of the two strips into 24 subdomains each

Solution: Although the two strips form a parallel plate capacitor, the capacitance cannot be calculated using the parallel plate capacitor formula because of the very complicated electric fields which exist throughout space. Instead, we use the method of moments. To calculate capacitance, we assume a potential +1 V on the top plate and −1 V on the lower plate, if the plates are identical in dimensions. The magnitude of the potential is immaterial but it must have a numerical value (see **Example 6.6** on how this potential is selected if the plates are not equal in dimensions). The potential difference between the two conductors is 2 V.

Each plate is divided into an arbitrary number of subdomains which, in general, can be of arbitrary size. The division is shown schematically in **Figure 6.25b**, where each strip is shown subdivided into 24 equal subdomains. The number of domains that should be used depends on the accuracy required. Applying the division and using the dimensions in **Figure 6.25a**, with a total of 24 subdomains on each strip, the capacitance between the two strips is found to be 12.62 pF, using script **mom1.m**. This discretization is rather coarse. The capacitance for a number of discretization levels is shown in **Table 6.2**.

Table 6.2 Capacitance for parallel and crossed strips at various discretization levels

Discretization (each strip)	6×1	12×2	18×3
Capacitance–strips parallel to each other	12.59 pF	12.62 pF	12.74 pF
Capacitance–strips perpendicular to each other	9.75 pF	9.78 pF	10.1 pF

Example 6.6 Application: Inter-gate Capacitances in CMOS Devices In a CMOS (complementary metal oxide semiconductor) integrated circuit, the gates of two FETs (field effect transistors) are deposited on the surface of a silicon chip as shown schematically in **Figure 6.26a**. One gate is 10 × 10 μm in size, the second is 20 × 20 μm, and the two are separated by 2 μm. The capacitance between the gates is important because it defines coupling between the gates; that is, when one gate is charged, the other acquires a charge too, because of the capacitance between them. If this capacitance is large, the second gate may be activated when it should not be. It is required to find the capacitance between the two gates. Assume the two plates are in free space and calculate the capacitance between the plates. In reality, one side of the plates is on silicon, which has a relative permittivity of about 12, but, for simplicity, we will assume the plates to be in free space.

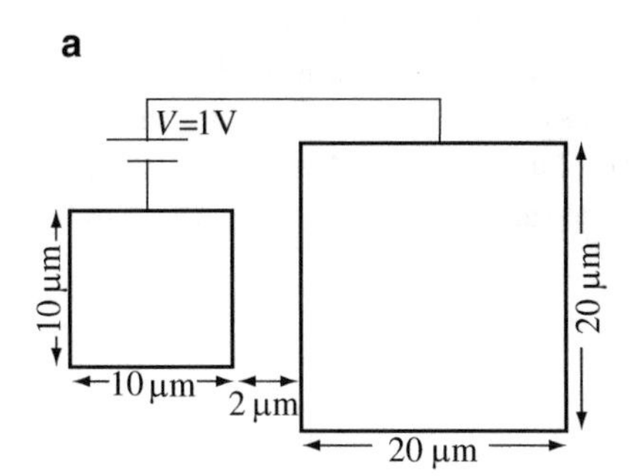

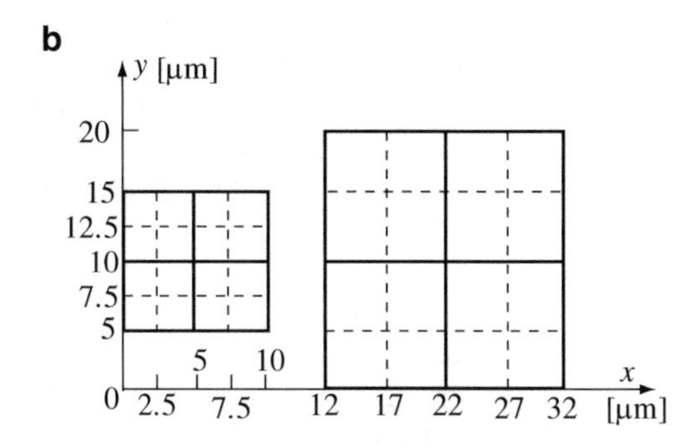

Figure 6.26 (**a**) A model for the gates of two field-effect transistors. (**b**) Coordinates and dimensions for evaluation of capacitance between the two gates in (**a**)

Solution: The solution here is similar to that in **Example 6.5**, but now we have two unequal plates. Therefore, we divide each plate into a number of subdomains and assume a potential difference of $V_0 = 1$ V between the two plates. Since the plates are unequal in size, it is not possible to assume that the plates are at equal and opposite potentials. Therefore, we assume one plate to be at potential V_A and the second at potential V_B, where $V_A + V_B = 1$ V. This leaves us with one unknown potential. The necessary additional relation comes from the fact that the total charge on the smaller plate and the total charge on the larger plate are equal in magnitude; that is, the potential must be such as to produce equal and opposite total charges on the two plates.

Next, we need to divide the plates into subdomains. We use different numbers of divisions on the two plates but, on each plate, the subdomains are kept equal in size. To simplify the evaluation, we use the system of coordinates shown in **Figure 6.26b**. Each plate is divided into four equal subdomains for illustration purposes (solid lines). Other divisions may be used (such as the dashed lines, see **Table 6.3**). The location of the centers is evaluated from the coordinates given and number of domains on each plate. Now, we again employ **Eqs. (6.30)** and **(6.32)** to evaluate the terms K_{ij} and K_{jj} for $i = 1$ to 8 and $j = 1$ to 8 (for the discretization shown in **Figure 6.26b**). The resulting matrix is 8 by 8 in size. To solve for the charge densities of the subdomains, we must first know the potential on each plate. Because the two plates are unequal in size, we do not know the exact values of the potentials on the plates that guarantee equal charge distribution on the plates. The actual potentials required for the charges on two plates to be equal in magnitude are computed iteratively: We assume an initial zero potential on one plate and 1 V on the second. In each iteration, the potential on both plates is reduced and the charges are evaluated and checked to see if the charges on the two plates are equal in magnitude. If they are, the solution is complete. If not, the iteration process is continued. The results for a number of discretizations obtained with script **mom1.m** are shown in **Table 6.3**.

Note: In **Table 6.3**, the first discretization in the first row is for the smaller plate, the second for the larger plate.

Table 6.3 Capacitance for different discretization levels

Discretization	2 × 2, 2 × 2	2 × 2, 4 × 4	2 × 2, 8 × 8	4 × 4, 4 × 4	4 × 4, 8 × 8
Capacitance [F]	0.3276×10^{-15}	0.3446×10^{-15}	0.3541×10^{-15}	0.3710×10^{-15}	0.3829×10^{-15}

Example 6.7 Method of Moments: Hand Computation In general, the system of equations produced by the method of moments is difficult to solve by hand and, if it is large, almost impossible. However, for a relatively small number of subdomains, hand computation is possible, especially if symmetry conditions are used to reduce computation. To show how hand computation may be used, **Example 6.6** is used here again, but, first, we look into the calculation of capacitance of a single plate.

Solution:

(a) Capacitance of the small plate.

Consider the smaller plate in **Figure 6.26a**. To calculate its capacitance, we divide the plate into 16 equal subdomains as shown in **Figure 6.27a**. At first glance, it would seem that we must solve a 16×16 system of equations. This is not the case. Inspection of the geometry shows that the four corner subdomains must have equal charge and, therefore, equal charge densities because of symmetry. Similarly, subdomains 2, 3, 5, 9, 8, 12, 14, and 15 have the same charge densities (but different than all other subdomains). Subdomains 6, 7, 10, and 11 also have the same charge density (and again different than all other subdomains). Thus, in effect, there are only three distinct charge densities that need to be calculated and only three equations in three unknowns need to be assembled and solved.

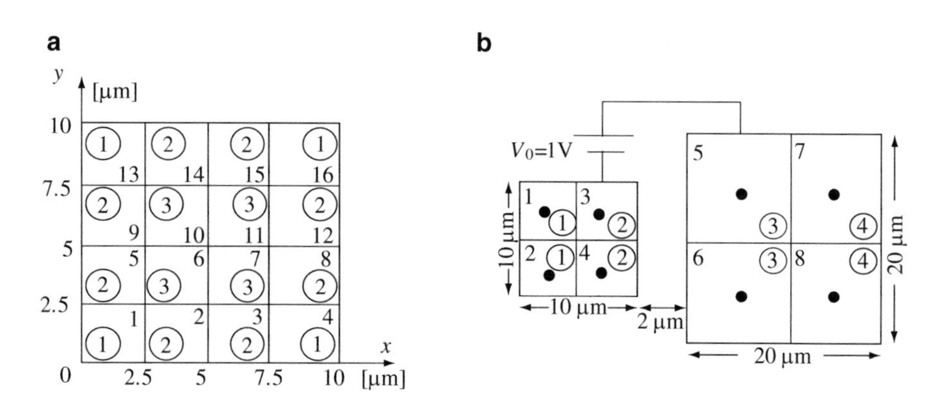

Figure 6.27 (**a**) Division of the smaller plate in **Figure 6.26a** into 16 subdomains. (**b**) Division of the geometry in **Figure 6.26a** for hand computation. Circled numbers indicate equal charge density

The solution proceeds by writing three distinct equations: one for subdomain 1, one for subdomain 2, and the third for subdomain 6. The three equations are

$$V_1 = \frac{1}{4\pi\varepsilon_0}\left\{\rho_1\left[K'_{1,1} + \frac{2s}{R_{1,4}} + \frac{s}{R_{1,16}}\right] + \rho_2 s\left[\frac{2}{R_{1,2}} + \frac{2}{R_{1,3}} + \frac{2}{R_{1,8}} + \frac{2}{R_{1,12}}\right] + \rho_3 s\left[\frac{1}{R_{1,6}} + \frac{2}{R_{1,7}} + \frac{1}{R_{1,11}}\right]\right\} \quad [\mathrm{V}]$$

$$V_2 = \frac{1}{4\pi\varepsilon_0}\left\{\rho_1 s\left[\frac{1}{R_{1,2}} + \frac{1}{R_{2,13}} + \frac{1}{R_{2,4}} + \frac{1}{R_{2,16}}\right] + \rho_2\left[K'_{2,2} + \frac{s}{R_{2,3}} + \frac{s}{R_{2,5}} + \frac{2s}{R_{2,8}} + \frac{s}{R_{2,12}} + \frac{s}{R_{2,14}} + \frac{s}{R_{2,15}}\right]\right.$$

$$\left. + \rho_3 s\left[\frac{1}{R_{2,6}} + \frac{1}{R_{2,7}} + \frac{1}{R_{2,10}} + \frac{1}{R_{2,11}}\right]\right\} \quad [\mathrm{V}]$$

$$V_6 = \frac{1}{4\pi\varepsilon_0}\left\{\rho_1 s\left[\frac{1}{R_{1,6}} + \frac{2}{R_{4,6}} + \frac{1}{R_{6,16}}\right] + \rho_2 s\left[\frac{2}{R_{2,6}} + \frac{2}{R_{3,6}} + \frac{2}{R_{6,8}} + \frac{2}{R_{6,12}}\right] + \rho_3\left[K'_{6,6} + \frac{2s}{R_{6,7}} + \frac{s}{R_{6,11}}\right]\right\} \quad [\mathrm{V}]$$

where the term K_{jj}' is the term in brackets in **Eq. (6.32)**. The various constants are as follows:

$$V_1 = V_2 = V_6 = 1 \quad [\mathrm{V}]$$
$$a = 2.5 \times 10^{-6}\ \mathrm{m},\ \ b = 2.5 \times 10^{-6}\ \mathrm{m}$$
$$s = 2.5 \times 10^{-6} \times 2.5 \times 10^{-6} = 6.25 \times 10^{-12}\ \mathrm{m}^2$$
$$R_{1,2} = R_{2,3} = R_{2,6} = R_{6,7} = 2.5 \times 10^{-6}\,\mathrm{m}$$
$$R_{1,3} = R_{2,4} = R_{2,10} = R_{6,8} = 5.0 \times 10^{-6}\,\mathrm{m}$$
$$R_{1,4} = R_{2,14} = 7.5 \times 10^{-6}\,\mathrm{m}$$
$$R_{1,16} = 7.5\sqrt{2} \times 10^{-6}\ \mathrm{m}$$
$$R_{1,8} = R_{2,13} = R_{2,15} = \sqrt{(7.5)^2 + (2.5)^2} \times 10^{-6} = \sqrt{62.5} \times 10^{-6}\,\mathrm{m}$$
$$R_{1,12} = R_{2,16} = \sqrt{(7.5)^2 + (5)^2} \times 10^{-6} = \sqrt{81.25} \times 10^{-6}\,\mathrm{m}$$
$$R_{1,7} = R_{2,8} = R_{2,11} = R_{4,6} = R_{6,12} = \sqrt{(5)^2 + (2.5)^2} \times 10^{-6} = \sqrt{31.25} \times 10^{-6}\,\mathrm{m}$$
$$R_{1,11} = R_{6,16} = R_{2,12} = 5\sqrt{2} \times 10^{-6}\ \mathrm{m}$$
$$R_{2,5} = R_{4,7} = R_{2,7} = R_{6,11} = R_{1,6} = R_{3,6} = 2.5\sqrt{2} \times 10^{-6}\,\mathrm{m}$$
$$K_{1,1}' = K_{2,2}' = K_{6,6}' = 10 \times 10^{-6} \ln\left(1 + \sqrt{2}\right)$$

Denoting ρ_1 as the charge density on any of the corner subdomains (1, 4, 13, and 16); ρ_2 the charge density on any of subdomains 2, 3, 5, 9, 8, 12, 14, or 15; and ρ_3 the charge density on any of subdomains 6, 7, 10, or 11, and substituting the values above in the three equations above gives

$$1 = \frac{1}{4 \times \pi \times 8.854 \times 10^{-6}} \left\{ \rho_1 \left[10\ \ln\left(1 + \sqrt{2}\right) + \frac{12.5}{7.5} + \frac{6.25}{7.5\sqrt{2}} \right] \right.$$
$$\left. + 6.25 \times \rho_2 \frac{2}{2.5} + \frac{2}{5} + \frac{2}{\sqrt{62.5}} + \frac{2}{\sqrt{81.25}} \right] + 6.25 \times \rho_3 \left[\frac{1}{2.5\sqrt{2}} + \frac{2}{\sqrt{31.25}} + \frac{1}{5\sqrt{2}} \right] \right\} \quad [\mathrm{V}]$$

$$1 = \frac{1}{4 \times \pi \times 8.854 \times 10^{-6}} \left\{ 6.25 \times \rho_1 \left[\frac{1}{2.5} + \frac{1}{\sqrt{62.5}} + \frac{1}{5} + \frac{1}{\sqrt{81.25}} \right] + \rho_2 \left[10\ \ln\left(1 + \sqrt{2}\right) + \frac{6.25}{2.5} + \frac{6.25}{2.5\sqrt{2}} \right. \right.$$
$$\left. \left. + \frac{12.5}{\sqrt{31.25}} + \frac{6.25}{5\sqrt{2}} + \frac{6.25}{7.5} + \frac{6.25}{\sqrt{62.5}} \right] + 6.25 \times \rho_3 \left[\frac{1}{2.5} + \frac{1}{2.5\sqrt{2}} + \frac{1}{5} + \frac{1}{\sqrt{31.25}} \right] \right\} \quad [\mathrm{V}]$$

$$1 = \frac{1}{4 \times \pi \times 8.854 \times 10^{-6}} \left\{ 6.25 \times \rho_1 \left[\frac{1}{2.5\sqrt{2}} + \frac{1}{\sqrt{31.25}} + \frac{1}{5\sqrt{2}} \right] \right.$$
$$\left. + 6.25 \times \rho_2 \left[\frac{2}{2.5} + \frac{2}{2.5\sqrt{2}} + \frac{2}{5.0} + \frac{2}{\sqrt{31.25}} \right] + \rho_3 \left[10\ \ln\left(1 + \sqrt{2}\right) + \frac{12.5}{2.5} + \frac{6.25}{2.5\sqrt{2}} \right] \right\} \quad [\mathrm{V}]$$

Simplifying these expressions gives

$$1 = 9.949 \times 10^4 \rho_1\ +\ 9.408 \times 10^4 \rho_2\ +\ 4.393 \times 10^4 \rho_3 \quad [\mathrm{V}]$$
$$1\ =\ 4.704 \times 10^4 \rho_1\ +\ 1.602 \times 10^5 \rho_2\ +\ 5.964 \times 10^4 \rho_3 \quad [\mathrm{V}]$$
$$1\ =\ 4.390 \times 10^4 \rho_1\ +\ 1.193 \times 10^5 \rho_2\ +\ 1.400 \times 10^5 \rho_3 \quad [\mathrm{V}]$$

Solution of this system of equations gives

$$\rho_1 = 5.4755 \times 10^{-6}\ \mathrm{C/m^2},$$
$$\rho_2 = 3.8295 \times 10^{-6}\ \mathrm{C/m^2},$$
$$\rho_3 = 2.1613 \times 10^{-6}\ \mathrm{C/m^2}$$

To calculate the capacitance of the plate, we first calculate the total charge. The latter is

$$Q = 4 \times 6.25 \times 10^{-12} \times 5.4755 \times 10^{-6} + 8 \times 6.25 \times 10^{-12} \times 3.8295 \times 10^{-6} + 4 \times 6.25 \times 10^{-12} \times 2.1613 \times 10^{-6} = 3.824 \times 10^{-16} \quad [\mathrm{C}]$$

The capacitance is

$$C = \frac{Q}{V} = \frac{3.824 \times 10^{-16}}{1} = 3.824 \times 10^{-16} \quad [\mathrm{F}].$$

(b) Capacitance between plates.

As a second example of hand computation, consider, again, **Figure 6.26a**. We wish to calculate the capacitance between the two plates. Because of symmetry about a horizontal line passing through the center of the geometry (see **Figure 6.27b**), there are only four different charge densities as shown. The equations for the four distinct charge densities (written for subdomains 1, 3, 5, and 7) are

$$V_1 = \frac{\rho_1}{4\pi\varepsilon_0}\left[K'_{11} + \frac{s_1}{R_{12}}\right] + \frac{\rho_2 s_1}{4\pi\varepsilon_0}\left[\frac{1}{R_{13}} + \frac{1}{R_{14}}\right] + \frac{\rho_3 s_2}{4\pi\varepsilon_0}\left[\frac{1}{R_{15}} + \frac{1}{R_{16}}\right] + \frac{\rho_4 s_2}{4\pi\varepsilon_0}\left[\frac{1}{R_{17}} + \frac{1}{R_{18}}\right] \quad [\mathrm{V}]$$

$$V_1 = \frac{\rho_1 s_1}{4\pi\varepsilon_0}\left[\frac{1}{R_{13}} + \frac{1}{R_{23}}\right] + \frac{\rho_2}{4\pi\varepsilon_0}\left[K'_{33} + \frac{s_1}{R_{34}}\right] + \frac{\rho_3 s_2}{4\pi\varepsilon_0}\left[\frac{1}{R_{35}} + \frac{1}{R_{36}}\right] + \frac{\rho_4 s_2}{4\pi\varepsilon_0}\left[\frac{1}{R_{37}} + \frac{1}{R_{38}}\right] \quad [\mathrm{V}]$$

$$V_2 = \frac{\rho_1 s_1}{4\pi\varepsilon_0}\left[\frac{1}{R_{15}} + \frac{1}{R_{25}}\right] + \frac{\rho_2 s_1}{4\pi\varepsilon_0}\left[\frac{1}{R_{35}} + \frac{1}{R_{45}}\right] + \frac{\rho_3}{4\pi\varepsilon_0}\left[K'_{55} + \frac{s_2}{R_{56}}\right] + \frac{\rho_4 s_2}{4\pi\varepsilon_0}\left[\frac{1}{R_{57}} + \frac{1}{R_{58}}\right] \quad [\mathrm{V}]$$

$$V_2 = \frac{\rho_1 s_1}{4\pi\varepsilon_0}\left[\frac{1}{R_{17}} + \frac{1}{R_{27}}\right] + \frac{\rho_2 s_1}{4\pi\varepsilon_0}\left[\frac{1}{R_{37}} + \frac{1}{R_{47}}\right] + \frac{\rho_3 s_2}{4\pi\varepsilon_0}\left[\frac{1}{R_{57}} + \frac{1}{R_{67}}\right] + \frac{\rho_4}{4\pi\varepsilon_0}\left[K'_{77} + \frac{s_2}{R_{78}}\right] \quad [\mathrm{V}]$$

In these, K'_{jj} are defined as in the previous example and the following are used:

On the small plate: $a = b = 5 \times 10^{-6}$ m, $s_1 = 5 \times 10^{-6} \times 5 \times 10^{-6} = 25 \times 10^{-12}\ \mathrm{m^2}$

On the large plate: $a = b = 10 \times 10^{-6}$ m, $s_2 = 10 \times 10^{-6} \times 10 \times 10^{-6} = 100 \times 10^{-12}\ \mathrm{m^2}$

$$R_{1,2} = R_{1,3} = R_{3,4} = 5 \times 10^{-6}\ \mathrm{m}, \quad R_{14} = R_{23} = 5\sqrt{2} \times 10^{-6}\ \mathrm{m}$$

$$R_{1,5} = \sqrt{(17-2.5)^2 + (5-2.5)^2} \times 10^{-6} = \sqrt{216.5} \times 10^{-6}\ \mathrm{m}$$

$$R_{1,6} = R_{2,5} = \sqrt{(17-2.5)^2 + (-5-2.5)^2} \times 10^{-6} = \sqrt{266.5} \times 10^{-6}\ \mathrm{m}$$

$$R_{1,7} = \sqrt{(27-2.5)^2 + (5-2.5)^2} \times 10^{-6} = \sqrt{606.5} \times 10^{-6}\ \mathrm{m}$$

$$R_{1,8} = R_{2,7} = \sqrt{(27-2.5)^2 + (-5-2.5)^2} \times 10^{-6} = \sqrt{656.5} \times 10^{-6}\ \mathrm{m}$$

$$R_{3,5} = \sqrt{(17-7.5)^2 + (2.5-5)^2} \times 10^{-6} = \sqrt{96.5} \times 10^{-6}\ \mathrm{m}$$

$$R_{3,6} = R_{4,5} = \sqrt{(17-7.5)^2 + (2.5+5)^2} \times 10^{-6} = \sqrt{146.5} \times 10^{-6}\ \mathrm{m}$$

$$R_{3,7} = \sqrt{(27-7.5)^2 + (5-2.5)^2} \times 10^{-6} = \sqrt{386.5} \times 10^{-6}\ \mathrm{m}$$

$$R_{3,8} = R_{4,7} = \sqrt{(27-7.5)^2 + (-5-2.5)^2} \times 10^{-6} = \sqrt{436.5} \times 10^{-6}\ \mathrm{m}$$

$$R_{5,6} = \mathrm{R}_{5,7} = R_{7,8} = 10 \times 10^{-6}\ \mathrm{m}, \quad R_{5,8} = R_{6,7} = 10\sqrt{2} \times 10^{-6}\ \mathrm{m}$$

In this case, $K'_{1,1} = K'_{3,3}$ and $K'_{5,5} = K'_{7,7}$. From **Eq. (6.32)**, we get

$$K'_{1,1} = K'_{3,3} = 20 \times 10^{-6} \ln\left(1+\sqrt{2}\right), \quad K'_{5,5} = K'_{7,7} = 40 \times 10^{-6} \ln\left(1+\sqrt{2}\right)$$

Substituting these values into the four equations, we get

$$V_1 = \frac{1}{4 \times \pi \times 8.854 \times 10^{-6}} \left\{ \rho_1 \left[20 \ln\left(1+\sqrt{2}\right) + \frac{25}{5} \right] + 25 \times \rho_2 \left[\frac{1}{5} + \frac{1}{5\sqrt{2}} \right] \right.$$

$$\left. + 100 \times \rho_3 \left[\frac{1}{\sqrt{216.5}} + \frac{1}{\sqrt{266.5}} \right] + 100 \times \rho_4 \left[\frac{1}{\sqrt{606.5}} + \frac{1}{\sqrt{656.5}} \right] \right\} \quad [\mathrm{V}]$$

$$V_1 = \frac{1}{4 \times \pi \times 8.854 \times 10^{-6}} \left\{ 25 \times \rho_1 \left[\frac{1}{5} + \frac{1}{5\sqrt{2}} \right] + \rho_2 \left[20 \ln\left(1+\sqrt{2}\right) + \frac{25}{5} \right] \right.$$

$$\left. + 100 \times \rho_3 \left[\frac{1}{\sqrt{96.5}} + \frac{1}{\sqrt{146.5}} \right] + 100 \times \rho_4 \left[\frac{1}{\sqrt{386.5}} + \frac{1}{\sqrt{436.5}} \right] \right\} \quad [\mathrm{V}]$$

$$V_2 = \frac{1}{4 \times \pi \times 8.854 \times 10^{-6}} \left\{ 100 \times \rho_1 \left[\frac{1}{\sqrt{216.5}} + \frac{1}{\sqrt{266.5}} \right] + 25 \times \rho_2 \left[\frac{1}{\sqrt{96.5}} + \frac{1}{\sqrt{146.5}} \right] \right.$$

$$\left. + \rho_3 \left[40 \ln\left(1+\sqrt{2}\right) + \frac{100}{10} \right] + 100 \times \rho_4 \left[\frac{1}{10} + \frac{1}{10\sqrt{2}} \right] \right\} \quad [\mathrm{V}]$$

$$V_2 = \frac{1}{4 \times \pi \times 8.854 \times 10^{-6}} \left\{ 100 \times \rho_1 \left[\frac{1}{\sqrt{606.5}} + \frac{1}{\sqrt{656.5}} \right] + 25 \times \rho_2 \left[\frac{1}{\sqrt{386.5}} + \frac{1}{\sqrt{436.5}} \right] \right\}$$

$$+ 100 \times \rho_3 \left[\frac{1}{10} + \frac{1}{10\sqrt{2}} \right] + \rho_4 \left[40 \ln\left(1+\sqrt{2}\right) + \frac{100}{10} \right] \Bigg\} \quad [\mathrm{V}]$$

Since the potentials V_1 and V_2 are not known but the potential difference is known, we write

$$V_1 - V_2 = V_0 = 1.0 \text{ V} \quad \rightarrow \quad V_1 = 1.0 + V_2 \quad [\mathrm{V}]$$

The fifth equation necessary is the equality of charge on the two plates:

$$2\rho_1 s_1 + 2\rho_2 s_2 = -(2\rho_3 s_3 + 2\rho_4 s_4)$$

The negative sign on the right-hand side indicates the fact that the charge on plate (2) must be negative if the charge on plate (1) is positive. Thus, we assume potential V_2 to be an unknown to be determined and write the following five equations:

$$1.0 + V_2 = 2.034 \times 10^5 \rho_1 + 7.672 \times 10^4 \rho_2 + 1.161 \times 10^5 \rho_3 + 7.157 \times 10^4 \rho_4$$
$$1.0 + V_2 = 7.672 \times 10^4 \rho_1 + 2.034 \times 10^5 \rho_2 + 1.658 \times 10^5 \rho_3 + 8.874 \times 10^4 \rho_4$$
$$V_2 = 2.904 \times 10^4 \rho_1 + 4.148 \times 10^4 \rho_2 + 4.067 \times 10^5 \rho_3 + 1.534 \times 10^5 \rho_4$$
$$V_2 = 1.789 \times 10^4 \rho_1 + 2.218 \times 10^4 \rho_2 + 1.534 \times 10^5 \rho_3 + 4.067 \times 10^5 \rho_4$$
$$50\rho_1 + 50\rho_2 + 200\rho_3 + 200\rho_4 = 0$$

Solution of these five equations in five unknowns gives the charge densities ρ_1, ρ_2, ρ_3, and ρ_4 and the potential V_2:

$$\rho_1 = 3.0345 \times 10^{-6} \text{ C/m}^2$$
$$\rho_2 = 3.5176 \times 10^{-6} \text{ C/m}^2$$
$$\rho_3 = -1.0194 \times 10^{-6} \text{ C/m}^2$$
$$\rho_4 = -6.1860 \times 10^{-7} \text{ C/m}^2$$
$$V_2 = -0.27569 \text{ V}.$$

Note that the potential V_2 does not equal V_1 in magnitude ($V_1 = 1.0 + V_2 = 0.72431$ V) as required. To calculate the capacitance, we calculate the total charge density on either plate. On the small plate

$$Q_1 = 2 \times 3.0345 \times 10^{-6} \times 25 \times 10^{-12} + 2 \times 3.5176 \times 10^{-6} \times 25 \times 10^{-12} = 3.276 \times 10^{-16} \text{ C}$$

On the larger plate

$$Q_2 = -2 \times 1.0194 \times 10^{-6} \times 100 \times 10^{-12} - 2 \times 6.1860 \times 10^{-7} \times 100 \times 10^{-12} = -3.276 \times 10^{-16} \text{ C}$$

The two charges are equal in magnitude (within the computation error) and opposite in sign as required. Taking the charge on the smaller (or larger) plate and dividing by the potential difference between the plates gives the capacitance:

$$C = \frac{Q}{V} = \frac{3.276 \times 10^{-16}}{1} = 3.276 \times 10^{-16} \text{ F}$$

Note that the capacitance calculated here is virtually identical to that calculated in **Example 6.6** (first column in **Table 6.3**).

Exercise 6.5 Find the equations required for hand calculation of the charge densities in **Example 6.4**, using the subdomains in **Figure 6.23b**. Find the capacitance and compare to that found in **Example 6.4**.

Answer

$$\frac{\rho_1}{4\pi\varepsilon_0}\left[K'_{1,1} + \frac{s_2}{R_{1,2}} + \frac{s_7}{R_{1,7}} + \frac{s_8}{R_{1,8}}\right] + \frac{\rho_3}{4\pi\varepsilon_0}\left[\frac{s_3}{R_{1,3}} + \frac{s_4}{R_{1,4}} + \frac{s_5}{R_{1,5}} + \frac{s_6}{R_{1,6}}\right] = 1$$

$$\frac{\rho_1}{4\pi\varepsilon_0}\left[\frac{s_1}{R_{1,3}} + \frac{s_2}{R_{2,3}} + \frac{s_7}{R_{3,7}} + \frac{s_8}{R_{3,8}}\right] + \frac{\rho_3}{4\pi\varepsilon_0}\left[K'_{3,3} + \frac{s_4}{R_{3,4}} + \frac{s_5}{R_{3,5}} + \frac{s_6}{R_{3,6}}\right] = 1$$

where

$$s_1 = s_2 = s_3 = s_4 = s_5 = s_6 = s_7 = s_8 = 0.25 \text{ m}^2$$
$$K'_{1,1} = K'_{3,3} = 1.76275$$
$$\rho_1 = \rho_2 = \rho_7 = \rho_8, \qquad \rho_3 = \rho_4 = \rho_5 = \rho_6$$
$$R_{1,2} = R_{1,3} = R_{3,4} = R_{3,5} = 0.5 \text{ m}, \quad R_{1,4} = R_{2,3} = R_{3,6} = 0.5\sqrt{2} \text{ m},$$
$$R_{3,7} = R_{1,5} = 1 \text{ m}, \quad R_{3,8} = R_{1,6} = \sqrt{1.25} \text{ m},$$
$$R_{1,7} = 1.5 \text{ m}, \quad R_{1,8} = \sqrt{2.5} \text{ m}$$

$C = 53.81$ pF, virtually the same as in **Example 6.4** (first column).

6.5 The Finite Element Method: Introduction

The finite element method only dates to the early 1950s, but it has evolved into a highly sophisticated and useful method for the solution of a very large number of engineering problems, in all disciplines. The method is different than both the finite difference method and the method of moments. It consists of the division of the solution domain into subdomains, called ***finite elements***, in the form of spatial subdomains of finite length, area, or volume, as shown in **Figure 6.28** for surface elements. Each element is defined by a number of edges, which define its space. The intersection of every two edges defines a ***node***. The assemblage of elements forms a ***mesh*** which must follow certain rules. The first and foremost of these is that elements must be of finite size. Second, the elements must be compatible. The latter property is explained in **Figures 6.29a** and **6.29b**.

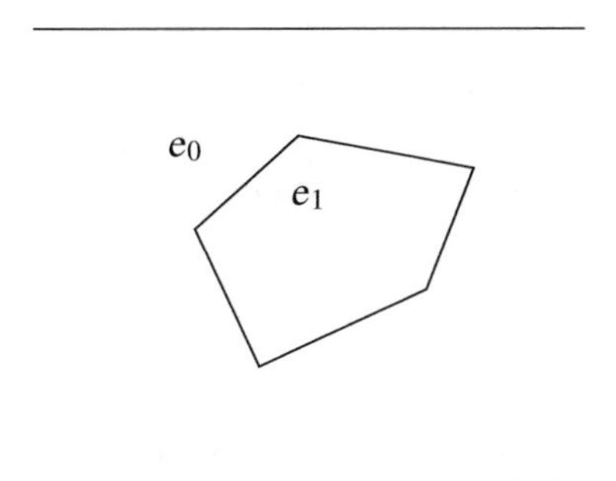

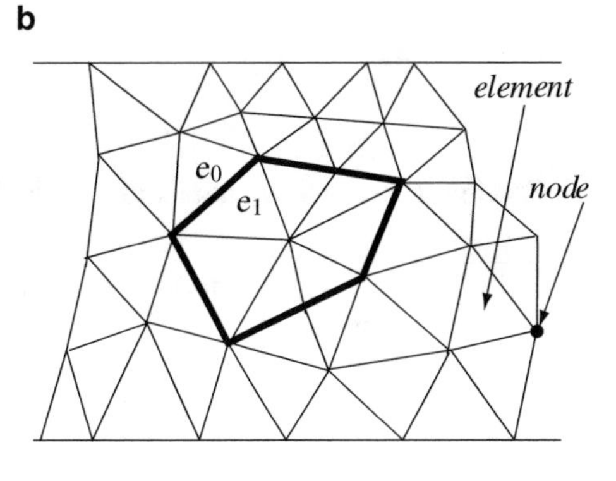

Figure 6.28 (**a**) A simple geometry with two materials to be discretized into finite elements. (**b**) A finite element discretization (partial) of the geometry in (**a**) using triangular finite elements

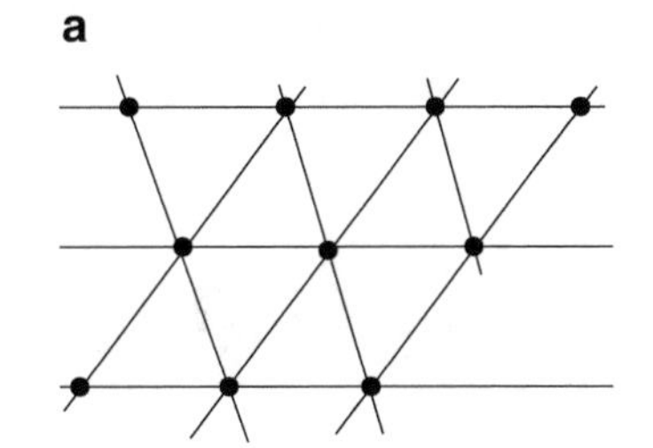

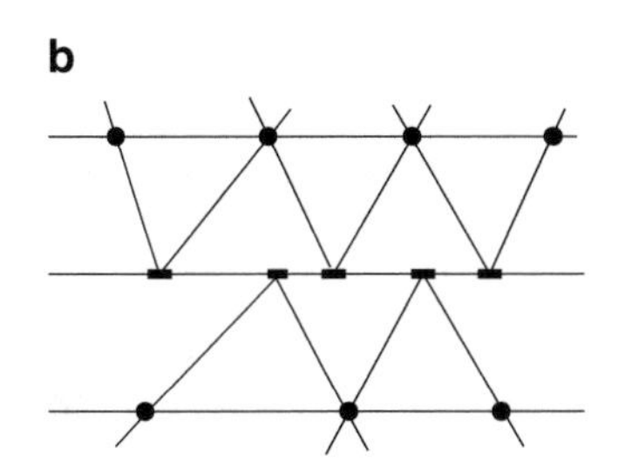

Figure 6.29 (**a**) A compatible mesh. Only triangle vertices form nodes. (**b**) A noncompatible mesh. Some of the nodes are located along edges

Finite elements can be of different sizes such that in regions where we anticipate larger variations in fields, the number of elements and their sizes can be changed to obtain higher element densities. In regions of lower fields, the number of elements can be lower. Each element can contain a different material; therefore, the interface between two different materials must coincide with element boundaries, as shown in **Figure 6.28b**.

The finite element method is somewhat more involved than the finite difference method and the method of moments, both in terms of the approximation it uses and in geometrical aspects of setting up the problem. Like the method of moments, it results in a system of equations which must be solved.

We introduce the method by first discussing the concept of a finite element and the approximation involved, followed by the implementation of the finite element method for Poisson's equation. Then, we define the general procedure for solution and apply this to a number of examples.

6.5.1 The Finite Element

Definition of a finite element is the first step in any finite element analysis. There are many ways by which a finite element may be defined as there are different types of elements and methods of analysis. We will restrict ourselves, however, to one element, in two dimensions: the triangular element. The element should be viewed as a volume with unit depth. Therefore, whenever we discuss the area of the element, a volume is in fact implied. The finite element chosen also defines the approximation we can use. The only possible approximation on a simple triangle is a linear approximation in both spatial variables.

6.5.1.1 The Triangular Element

The triangle in **Figure 6.30** has three nodes and a finite area. The approximation within the element is assumed to be linear and of the form

$$V(x,y) = \alpha_1 + \alpha_2 x + \alpha_3 y \quad [\mathrm{V}] \tag{6.35}$$

where the coefficients α_1, α_2, and α_3 remain to be determined and $V(x,y)$ is the solution defined within the finite element, including at its nodes and on its edges. In this discussion, the solution is the electric potential, but it may represent other quantities. To determine the coefficients, we use **Figure 6.30** and write for the three nodes of the element:

$$V_i = \alpha_1 + \alpha_2 x_i + \alpha_3 y_i \quad [\mathrm{V}] \tag{6.36}$$

$$V_j = \alpha_1 + \alpha_2 x_j + \alpha_3 y_j \quad [\text{V}] \tag{6.37}$$

$$V_k = \alpha_1 + \alpha_2 x_k + \alpha_3 y_k \quad [\text{V}] \tag{6.38}$$

Solving these three equations for the three unknowns α_1, α_2, and α_3 gives

$$\alpha_1 = \frac{1}{2\Delta}\left[\left(x_j y_k - x_k y_j\right)V_i + (x_k y_i - x_i y_k)V_j + \left(x_i y_j - x_j y_i\right)V_k\right] \tag{6.39}$$

$$\alpha_2 = \frac{1}{2\Delta}\left[\left(y_j - y_k\right)V_i + (y_k - y_i)V_j + \left(y_i - y_j\right)V_k\right] \tag{6.40}$$

$$\alpha_3 = \frac{1}{2\Delta}\left[(x_k - x_j)V_i + (x_i - x_k)V_j + \left(x_j - x_i\right)V_k\right] \tag{6.41}$$

where Δ is the area of the triangle and is given by

$$2\Delta = \begin{vmatrix} 1 & x_i & y_i \\ 1 & x_j & y_j \\ 1 & x_k & y_k \end{vmatrix} = \left(x_j y_k - x_k y_j\right) + (x_k y_i - x_i y_k) + \left(x_i y_j - x_j y_i\right) \tag{6.42}$$

These are now substituted into the approximation in **Eq. (6.35)** to obtain

$$\boxed{V(x,y) = N_i V_i + N_j V_j + N_k V_k \qquad [\text{V}]} \tag{6.43}$$

where the functions N_i, N_j, and N_k are

$$N_i = \frac{1}{2\Delta}\left[\left(x_j y_k - x_k y_j\right) + \left(y_j - y_k\right)x + \left(x_k - x_j\right)y\right] \tag{6.44}$$

$$N_j = \frac{1}{2\Delta}\left[\left(x_k y_i - x_i y_k\right) + \left(y_k - y_i\right)x + \left(x_i - x_k\right)y\right] \tag{6.45}$$

$$N_k = \frac{1}{2\Delta}\left[\left(x_i y_j - x_j y_i\right) + \left(y_i - y_j\right)x + \left(x_j - x_i\right)y\right] \tag{6.46}$$

These three functions are called the ***shape functions*** for the element because they only depend on the dimensions or "shape" of the element. The shape functions have the following properties:

(1) $N_i = 1$ at node i and $N_i = 0$ at all other nodes, $N_j = 1$ at node j and $N_j = 0$ at all other nodes, and $N_k = 1$ at node k and $N_k = 0$ at all other nodes.

(2) The sum of the shape functions is equal to 1 at any point within the element and on its boundaries:

$$N_i + N_j + N_k = \sum_{l=i,j,k} N_l = 1 \tag{6.47}$$

This property can be verified directly from **Eqs. (6.44)** through **(6.46)** for any triangle (see **Exercise 6.6**). The above properties are general and hold for any finite element, regardless of shape, dimensionality, and method of defining the finite element.

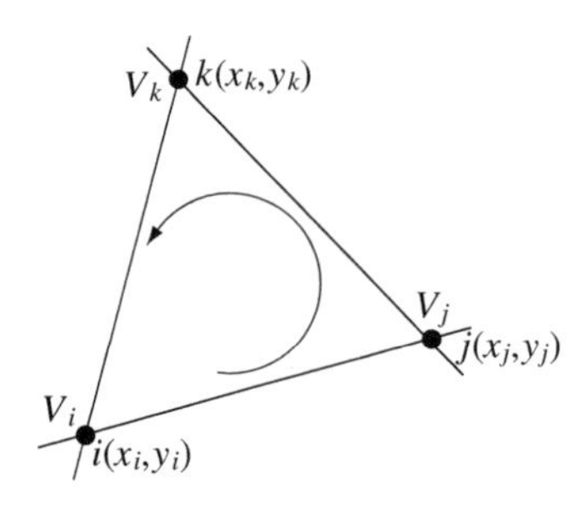

Figure 6.30 The three nodes of a general triangular element. Nodes must be ordered in a counterclockwise sequence

The derivatives of the shape functions with respect to x and y are evaluated directly from **Eqs. (6.44)** through **(6.46)**:

$$\frac{\partial N_i}{\partial x}=\frac{1}{2\Delta}\left(y_j-y_k\right),\qquad \frac{\partial N_i}{\partial y}=\frac{1}{2\Delta}\left(x_k-x_j\right) \tag{6.48}$$

$$\frac{\partial N_j}{\partial x}=\frac{1}{2\Delta}\left(y_k-y_i\right),\qquad \frac{\partial N_j}{\partial y}=\frac{1}{2\Delta}\left(x_i-x_k\right) \tag{6.49}$$

$$\frac{\partial N_k}{\partial x}=\frac{1}{2\Delta}\left(y_i-y_j\right),\qquad \frac{\partial N_k}{\partial y}=\frac{1}{2\Delta}\left(x_j-x_i\right) \tag{6.50}$$

These provide the approximation to the derivatives of the potential as

$$\boxed{\frac{\partial V(x,y)}{\partial x}=\frac{\partial N_i}{\partial x}V_i+\frac{\partial N_j}{\partial x}V_j+\frac{\partial N_k}{\partial x}V_k} \tag{6.51}$$

$$\boxed{\frac{\partial V(x,y)}{\partial y}=\frac{\partial N_i}{\partial y}V_i+\frac{\partial N_j}{\partial y}V_j+\frac{\partial N_k}{\partial y}V_k} \tag{6.52}$$

Equation (6.43) is the approximation of the function everywhere in the finite element and **Eqs. (6.51)** and **(6.52)** are its derivatives. These are at the heart of the finite element method. Before proceeding, we wish to write the shape functions, their derivatives, and the approximations for potential and their derivatives in a more compact form to avoid tedious operations later. We therefore denote the following:

$$p_i=x_jy_k-x_ky_j,\qquad p_j=x_ky_i-x_iy_k,\qquad p_k=x_iy_j-x_jy_i \tag{6.53}$$

$$q_i=y_j-y_k,\qquad q_j=y_k-y_i,\qquad q_k=y_i-y_j \tag{6.54}$$

$$r_i=x_k-x_j,\qquad r_j=x_i-x_k,\qquad r_k=x_j-x_i \tag{6.55}$$

The shape functions and their derivatives may be written as

$$N_l=\frac{1}{2\Delta}(p_l+q_lx+r_ly)\quad\text{and}\quad\frac{\partial N_l}{\partial x}=\frac{q_l}{2\Delta},\quad\frac{\partial N_l}{\partial y}=\frac{r_l}{2\Delta},\quad l=i,j,k \tag{6.56}$$

With these, the approximations for the potential and its derivatives are

$$\boxed{V(x,y)=\sum_{l=i,j,k}\frac{1}{2\Delta}(p_l+q_lx+r_ly)V_l=\sum_{l=i,j,k}N_lV_l\quad[\mathrm{V}]} \tag{6.57}$$

$$\boxed{\frac{\partial V(x,y)}{\partial x}=\sum_{l=i,j,k}\frac{q_l}{2\Delta}V_l,\qquad\frac{\partial V(x,y)}{\partial y}=\sum_{l=i,j,k}\frac{r_l}{2\Delta}V_l} \tag{6.58}$$

We will use these in the next section to define the finite element solution process.

After the potential is evaluated everywhere (as shown in the following sections), the electric field intensity may be evaluated from the potential, thus providing a complete solution to any electrostatic problem. The electric field intensity is found from the relation $\mathbf{E}=-\nabla V$ as

$$\mathbf{E}=\hat{\mathbf{x}}E_x+\hat{\mathbf{y}}E_y=-\hat{\mathbf{x}}\frac{\partial V}{\partial x}-\hat{\mathbf{y}}\frac{\partial V}{\partial y}\quad\left[\frac{\mathrm{V}}{\mathrm{m}}\right] \tag{6.59}$$

Using **Eq. (6.58)**, we get

$$E_x = -\sum_{l=i,j,k} \frac{q_l}{2\Delta} V_l \quad \text{and} \quad E_y = -\sum_{l=i,j,k} \frac{r_l}{2\Delta} V_l \quad \left[\frac{\text{V}}{\text{m}}\right] \tag{6.60}$$

These components are independent of x and y; they are constant in an element, as can be expected since the potential varies linearly within the element. Now, the nature of the approximation is even more apparent. The analytic solution is normally continuous and the electric field varies smoothly in the solution domain (except at interfaces between different materials). On the other hand, the numerical solution is discontinuous.

Exercise 6.6

(a) Show that $N_i = 1$ at node $i(x_i,y_i)$ and zero at nodes $j(x_j,y_j)$ and $k(x_k,y_k)$ by setting $x = x_i$ and $y = y_i$, then setting $x = x_j$ and $y = y_j$, and then setting $x = x_k$ and $y = y_k$ in **Eq. (6.43)**.

(b) Show that the sum of the shape functions anywhere within the finite element equals 1 by summing **Eqs. (6.44)** through **(6.46)**.

6.5.2 Implementation of the Finite Element Method

6.5.2.1 The Field Equations

So far, we obtained a general approximation for the potential in a finite element. The approximation should satisfy Poisson's and Laplace's equations simply because of the uniqueness theorem. A quick inspection reveals that if we substitute the approximation in **Eq. (6.57)** into Laplace's or Poisson's equations, the left-hand side is identically zero since the approximation is first-order in x and y, whereas the equations require second-order derivatives. There are two approaches we can take to extricate ourselves from this difficulty. One, and the most obvious, is to use higher-order approximations in x and y. The second is to modify the equations we solve so that only first-order derivatives are required. Surprisingly perhaps, we take the second route but with a twist: We look for a new function, which contains first-order derivatives only and whose solution provides a correct solution to the original equation (Laplace's or Poisson's equation in this case).

How do we come up with the appropriate function and how do we guarantee that it provides the correct solution to the original problem? The answer to the first question is that there are systematic methods to obtain the appropriate function. As for the second, we can prove mathematically that the function provides the correct solution. However, we will do neither. Instead, we will rely on the time-honored method of "guessing" the function, finding the solution, and then verifying that it is a correct solution. Fortunately, this need only be done once since the function then applies to all electrostatic problems. In electrostatic applications as well as others, an appropriate function is the potential energy in the solution domain. The method consists of minimization of an energy function instead of solving directly the physical equation (in this case $\nabla^2 V = 0$ or $\nabla^2 V = -\rho/\varepsilon$). For a given domain, extending over a volume v, the energy function is

$$F(E) = \int_v \left[\frac{\varepsilon}{2}E^2 - \rho V\right] dv \quad [\text{J}] \tag{6.61}$$

where the first part is the energy stored in the electric field and the second is the energy in the sources that may exist in the solution domain. However, our problem is in two dimensions. Thus, we assume the solution domain to be made of a volume of unit depth in the third dimension (for example, z) and of area s in the plane. In this case, the element of volume is $dv = 1ds$ and the integration is on the surface of the solution domain:

$$F(E) = \int_s \left[\frac{\varepsilon}{2}E^2 - \rho V\right] ds \quad [\text{J}] \tag{6.62}$$

Now that we have established the energy in the system, we use this as the function to minimize. First, we must rewrite **Eq. (6.62)** in terms of the potential rather than the electric field intensity. (In Laplace's equation, there are no sources ($\rho = 0$) and the only potential energy is due to the electric field intensity in the solution domain). Because $E^2 = \mathbf{E} \cdot \mathbf{E}$, the energy function in **Eq. (6.62)** is [see **Eq. (6.59)**]

$$F(V) = \int_s \left[\frac{\varepsilon}{2} \left\{ \left(\frac{\partial V}{\partial x} \right)^2 + \left(\frac{\partial V}{\partial y} \right)^2 \right\} - \rho V \right] ds \tag{6.63}$$

From here on, we will discuss only Poisson's equation and the energy function associated with it. Laplace's equation, its energy function, and all other aspects of finite element implementation are obtained by simply setting the volume charge density ρ to zero. Note that **Eq. (6.63)** contains only first-order derivatives as required, and the first-order approximation for the potential can be used.

6.5.2.2 Discretization

The second step in a finite element implementation is discretization of the solution domain into any number of finite elements that may be required to properly model the physical domain. After the physical domain is properly defined, the interior of the domain is divided into elements. This simply means that the coordinates of all nodes in the solution domain are defined and the appropriate nodes are associated with each element. Thus, for each element in the solution domain, we obtain:

(1) Coordinates of the three nodes of element $N: x_i, y_i, x_j, y_j, x_k, y_k$.
(2) Node numbers associated with element $N: i, j, k$.
(3) The material properties of all elements. In this case, this means associating with each element a value for permittivity.
(4) On those elements that have charge density, the charge density must be identified with the element.

These steps constitute the definition of a finite element mesh. In addition, we must identify those nodes that happen to be on the boundary of the mesh and the potential values prescribed on these nodes.

6.5.2.3 Minimization

In the minimization process, we perform the following:

(1) The energy function is evaluated for each element in turn.
(2) The derivative of the energy function with respect to each unknown in the element is calculated and set to zero.

The idea of discretization is to form the solution domain by an assemblage of finite elements. The energy function for the solution domain is then the sum of the energy functions of individual finite elements:

$$F(V) = \sum_{m=1}^{M} F_m \tag{6.64}$$

where M is the total number of elements in the solution domain. Assuming that the mesh has a total of N nodes, the energy function must be minimized with respect to each unknown in the solution domain. For the nth node potential V_n,

$$\boxed{\frac{\partial F(V)}{\partial V_n} = \sum_{m=1}^{M} \frac{\partial F_m}{\partial V_n} = 0} \tag{6.65}$$

This means that to minimize the energy function over the whole solution domain, we can minimize the energy function over each individual element separately and then add the results to obtain minimization over the global domain. Thus, the important point in establishing a numerical procedure of calculation is to obtain the generic term $\partial F_m / \partial V_n$ in the sum of **Eq. (6.65)**. This is done next.

Substituting the approximation for $\partial V / \partial x$ and $\partial V / \partial y$ from **Eq. (6.58)** into **Eq. (6.63)** and separating the integral over element m into two parts gives

$$F_m = \frac{\varepsilon}{2} \int_s \left[\left(\sum_{l=i,j,k} \frac{1}{2\Delta} q_l V_l \right)^2 + \left(\sum_{l=i,j,k} \frac{1}{2\Delta} r_l V_l \right)^2 \right] ds - \rho \int_s \sum_{l=i,j,k} N_l V_l ds \tag{6.66}$$

In this relation, we also assumed ε and ρ are constant within element m and, therefore, were taken outside the integral signs. The first integral is due to the electric field intensity and will be evaluated first. Denoting the first integral in **Eq. (6.66)** as F_e and noting that neither q_l nor r_l depends on the coordinates x or y, the surface integral simply means multiplying the integrand by the surface of element i, which equals Δ [see **Eq. (6.42)**]. Thus, we can write

$$F_e = \frac{\varepsilon}{2}\int_s \left[\left(\sum_{l=i,j,k} \frac{1}{2\Delta} q_l V_l\right)^2 + \left(\sum_{l=i,j,k} \frac{1}{2\Delta} r_l V_l\right)^2\right] ds = \frac{\varepsilon\Delta}{2}\left[\left(\sum_{l=i,j,k} \frac{1}{2\Delta} q_l V_l\right)^2 + \left(\sum_{l=i,j,k} \frac{1}{2\Delta} r_l V_l\right)^2\right] \quad (6.67)$$

Expanding the sums

$$F_e = \frac{\varepsilon\Delta}{2}\left[\frac{1}{(2\Delta)^2}\left(q_i V_i + q_j V_j + q_k V_k\right)^2 + \frac{1}{(2\Delta)^2}\left(r_i V_i + r_j V_j + r_k V_k\right)^2\right] \quad (6.68)$$

Now, we are ready to calculate the second integral in **Eq. (6.66)**, which we denote as F_s:

$$F_s = \rho \int_s \sum_{l=i,j,k} N_l V_l ds \quad (6.69)$$

This is much more difficult to evaluate because the approximation for V depends both on x and on y [see **Eq. (6.57)**]. Instead of performing the integration, we note that the integral in **Eq. (6.69)** is proportional to the charge density ρ and potential V in the element as $F_s \propto \rho V \Delta$. An approximation to this quantity may be obtained assuming the voltage in the element to be the average between the three nodal potentials. Using this approximation, we get

$$F_s = \frac{\rho\Delta(V_1 + V_2 + V_3)}{3} \quad (6.70)$$

where the charge density ρ and the average potential are assumed to be at the centroid (center of gravity) of the element. Again, our reason for accepting this approximation is that as the size of elements decreases, the approximation becomes more accurate.

Now, the general expression of the energy function in a general finite element becomes

$$F_m = F_e - F_S = \frac{\varepsilon}{8\Delta}\left[\left(q_i V_i + q_j V_j + q_k V_k\right)^2 + \left(r_i V_i + r_j V_j + r_k V_k\right)^2\right] - \frac{\rho\Delta(V_1 + V_2 + V_3)}{3} \quad (6.71)$$

The minimization is completed by taking the derivative of the energy function with respect to each unknown potential value in element m:

$$\frac{\partial F_m}{\partial V_i} = \frac{\varepsilon}{4\Delta}\left[\left(q_i^2 V_i + q_i q_j V_j + q_i q_k V_k\right) + \left(r_i^2 V_i + r_i r_j V_j + r_i r_k V_k\right)\right] - \frac{\rho\Delta}{3} \quad (6.72)$$

$$\frac{\partial F_m}{\partial V_j} = \frac{\varepsilon}{4\Delta}\left[\left(q_j q_i V_i + q_j^2 V_j + q_j q_k V_k\right) + \left(r_j r_i V_i + r_j^2 V_j + r_j r_k V_k\right)\right] - \frac{\rho\Delta}{3} \quad (6.73)$$

$$\frac{\partial F_m}{\partial V_k} = \frac{\varepsilon}{4\Delta}\left[\left(q_k q_i V_i + q_k q_j V_j + q_k^2 V_k\right) + \left(r_k r_i V_i + r_k r_j V_j + r_k^2 V_k\right)\right] - \frac{\rho\Delta}{3} \quad (6.74)$$

Rewriting these three equations as a matrix gives

$$\boxed{\begin{bmatrix} \partial F_m/\partial V_i \\ \partial F_m/\partial V_j \\ \partial F_m/\partial V_k \end{bmatrix} = \frac{\varepsilon}{4\Delta}\begin{bmatrix} q_i q_i + r_i r_i & q_i q_j + r_i r_j & q_i q_k + r_i r_k \\ q_j q_i + r_j r_i & q_j q_j + r_j r_j & q_j q_k + r_j r_k \\ q_k q_i + r_k r_i & q_k q_j + r_k r_j & q_k q_k + r_k r_k \end{bmatrix}\begin{bmatrix} V_i \\ V_j \\ V_k \end{bmatrix} - \frac{\rho\Delta}{3}\begin{bmatrix} 1 \\ 1 \\ 1 \end{bmatrix}} \quad (6.75)$$

This is called an ***elemental matrix*** or an elemental contribution. We note the following:

(1) The matrix is symmetric ($a_{ij} = a_{ji}$). This is typical of most finite element formulations.
(2) For each element in the solution domain, we get a similar matrix. The differences are in the area Δ of the element and individual coefficients, but the form remains the same.
(3) The approximation for Laplace's equation is obtained by setting $\rho = 0$ in **Eq. (6.75)**.

6.5.2.4 Assembly of the Elemental Matrices

The next step is assembly of the elemental matrices into a ***global system*** of equations. The assembly is required because of the approach we took: that of minimizing the energy function in each element separately. The assembly is then the summation of all individual elemental matrices into the global matrix. To perform the summation, we create a large system of equations whose size is equal to the number of unknowns in the finite element discretized domain. To outline the method for matrix assembly, consider **Figure 6.31a**, where a small finite element mesh with four triangular elements and six nodes is shown. Note the element numbering (circled) and node numbering. The numbering of elements and nodes is arbitrary but must be sequential.

Figure 6.31 (**a**) A simple finite element mesh used to demonstrate matrix assembly. (**b**) The elemental matrix for element 2 in (**a**)

a

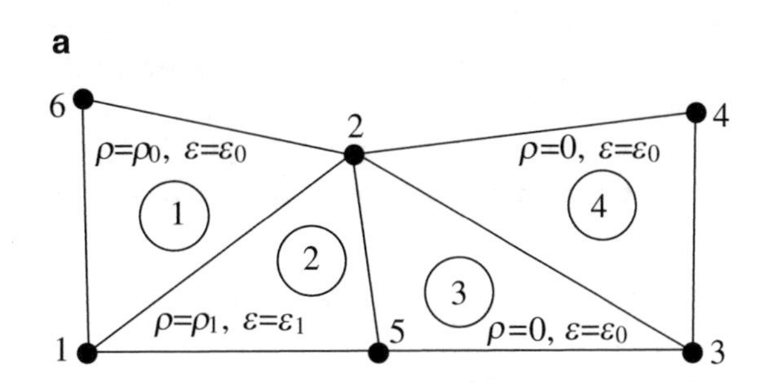

b

	i=1	j=5	k=2
i=1	s_{11}	s_{12}	s_{13}
j=5	s_{21}	s_{22}	s_{23}
k=2	s_{31}	s_{32}	s_{33}

Figure 6.31b shows the form of the elemental matrix for element 2. The row and column numbers correspond to the node numbers of the corresponding element. The row and column numbers define the actual location in the global matrix. The coefficients of the elemental matrix in **Figure 6.31b** are entered in the global matrix as follows:
coefficient $s_{1,1}$ is added to coefficient $S_{1,1}$ in the global matrix
coefficient $s_{1,2}$ is added to coefficient $S_{1,5}$ in the global matrix
coefficient $s_{1,3}$ is added to coefficient $S_{1,2}$ in the global matrix
coefficient $s_{2,1}$ is added to coefficient $S_{5,1}$ in the global matrix
coefficient $s_{2,2}$ is added to coefficient $S_{5,5}$ in the global matrix
coefficient $s_{2,3}$ is added to coefficient $S_{5,2}$ in the global matrix
coefficient $s_{3,1}$ is added to coefficient $S_{2,1}$ in the global matrix
coefficient $s_{3,2}$ is added to coefficient $S_{2,5}$ in the global matrix
coefficient $s_{3,3}$ is added to coefficient $S_{2,2}$ in the global matrix

This process is repeated for all elements in the solution domain. In terms of the algorithm used for assembly, the normal practice is to generate an elemental matrix and immediately assemble it in the global matrix, before the next elemental matrix is assembled. We also note that an identical process applies to the right-hand side vector as well (elemental vectors q_i are assembled in global vector Q). After assembly, a global system of equations is obtained as

$$[S]\{V\} - \{Q\} = 0 \tag{6.76}$$

This system has six equations in six unknowns. Assembly of all four elements into the global system results in the following:

$$\begin{bmatrix} s_{11}^1 + s_{11}^2 & s_{12}^1 + s_{13}^2 & 0 & 0 & s_{12}^2 & s_{13}^1 \\ s_{31}^2 + s_{21}^1 & s_{22}^1 + s_{33}^2 + s_{33}^3 + s_{33}^4 & s_{32}^3 + s_{31}^4 & s_{32}^4 & s_{32}^2 + s_{31}^3 & s_{23}^1 \\ 0 & s_{23}^3 + s_{13}^4 & s_{22}^3 + s_{11}^4 & s_{12}^4 & s_{21}^3 & 0 \\ 0 & s_{23}^4 & s_{21}^4 & s_{22}^4 & 0 & 0 \\ s_{21}^2 & s_{23}^2 + s_{13}^3 & s_{12}^3 & 0 & s_{22}^2 + s_{11}^3 & 0 \\ s_{31}^1 & s_{32}^1 & 0 & 0 & 0 & s_{33}^1 \end{bmatrix} \begin{bmatrix} V_1 \\ V_2 \\ V_3 \\ V_4 \\ V_5 \\ V_6 \end{bmatrix} - \begin{bmatrix} (\rho_0\Delta_1)/3 + (\rho_1\Delta_2)/3 \\ (\rho_0\Delta_1)/3 + (\rho_1\Delta_2)/3 \\ 0 \\ 0 \\ (\rho_1\Delta_2)/3 \\ (\rho_0\Delta_1)/3 \end{bmatrix} = 0 \tag{6.77}$$

where the superscript indicates the element number from which the corresponding contribution was assembled.

6.5.2.5 Application of Boundary Conditions

As with any boundary value problem, we must apply boundary conditions to obtain a particular solution to the problem. In most cases, some of the nodes on the boundary of the problem are specified. These are then introduced into the matrix and the matrix is then reduced accordingly. For example, suppose the potentials at node 3 and node 6 are known and equal to V_0, their value is inserted in the matrix and the matrix reduced to a 4 × 4 matrix as follows:

$$\begin{bmatrix} s_{11}^1+s_{11}^2 & s_{12}^1+s_{13}^2 & 0 & s_{12}^2 \\ s_{31}^2+s_{21}^1 & s_{22}^1+s_{33}^2+s_{33}^3+s_{33}^4 & s_{32}^4 & s_{32}^2+s_{31}^3 \\ 0 & s_{23}^4 & s_{22}^4 & 0 \\ s_{21}^2 & s_{23}^2+s_{13}^3 & 0 & s_{22}^2+s_{11}^3 \end{bmatrix} \begin{bmatrix} V_1 \\ V_2 \\ V_4 \\ V_5 \end{bmatrix} - \begin{bmatrix} (\rho_0\Delta_1)/3+(\rho_1\Delta_2)/3-s_{13}^1V_0 \\ (\rho_0\Delta_1)/3+(\rho_1\Delta_2)/3-\left(s_{32}^3+s_{31}^4\right)V_0-s_{23}^1V_0 \\ -s_{21}^4V_0 \\ (\rho_1\Delta_2)/3-s_{12}^3V_0 \end{bmatrix} = 0 \tag{6.78}$$

This process is natural when only a few equations are involved or if the solution must be performed by hand. In finite element calculations, it is rather time-consuming, especially when the matrix is large. In addition, it has a big disadvantage in that the locations of the unknowns are in the "wrong" places. For example, unknown potential V_5 is in location No. 4 in the solution vector. Every time we reduce the size of the matrix, we must keep track of the location of the unknowns. Again, in hand computation, this is a minor problem, but in a computer program or a script, it is all too easy to lose track of the unknowns.

An alternative method is to leave the matrix unchanged but to force the unknowns V_3 and V_6 to be equal to the value V_0. This is done as follows:

(1) Replace the diagonal in the appropriate rows (row 3 and row 6) with a very large number P. This can be any convenient value, say $P = 10^{20}$.
(2) Replace the right-hand side by the known value V_0 multiplied by P (the right-hand side is now PV_0).

In effect, rows 3 and 6 have only a diagonal term and a right-hand side term because all other values in the row are negligible in comparison to P. For example, for row No. 3, we have

$$PV_3 - PV_0 = 0 \quad \rightarrow \quad V_3 = V_0 \tag{6.79}$$

This method, although only approximate, is much more compatible with a computational method and is often used to impose boundary conditions in finite element applications.

So far, we discussed one type of boundary condition: specified potentials on the boundary. This type of boundary condition is called a ***Dirichlet boundary condition***. Another type of boundary condition is often useful in finite element calculations. It is called a ***Neumann boundary condition*** and occurs whenever the normal component of the electric field intensity in electrostatic applications is zero on a boundary. As an example of the types of boundary conditions applicable, consider the parallel plate capacitor shown in **Figure 6.32**. If the solution domain is the interior of the capacitor, then only the potentials on the two plates are required. These potentials are known and, therefore, are Dirichlet boundary conditions. Now, suppose we cut the capacitor vertically in the middle on the line $A - A'$ and wish to solve for the potential on the right-hand side of the geometry. Line $A - A'$ becomes a boundary. This is permissible because we know the potential on the right-hand side is the same as that on the left-hand side of the boundary $A - A'$. The x component of the electric field intensity (normal to line $A - A'$) is zero. Thus, the boundary condition on line $A - A'$ is a Neumann boundary condition. Fortunately, in finite element applications, Neumann boundary conditions do not need to be specified. We simply leave the values on $A - A'$ unspecified. The Neumann boundary condition can be utilized every time we have a symmetry in the potentials in the problem. It is usually quite easy to identify symmetries. However, we must make sure that the normal component of the electric field intensity on any symmetry line is zero.

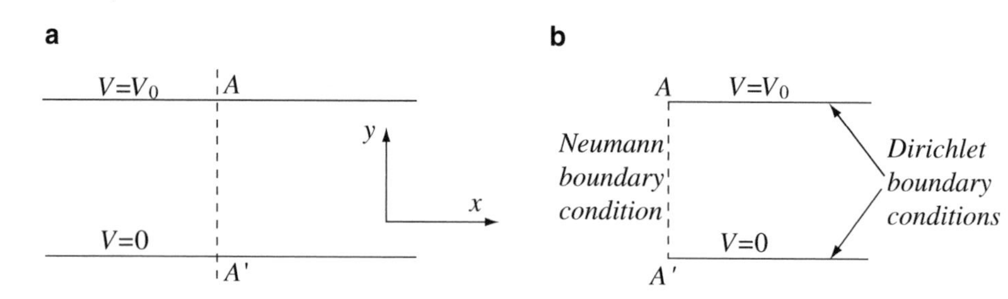

Figure 6.32 Boundary conditions in finite element applications. (**a**) Line $A - A' =$ is a geometric and potential symmetry line. (**b**) The upper and lower plates are Dirichlet boundary conditions. Line $A - A'$ is a Neumann boundary condition and is left unspecified

6.5.2.6 Solution

After applying boundary conditions, the system of equations is solved for the unknown values of V, as we shall see in examples. The solution of the system of equations can be performed using any applicable method. Gaussian elimination is often used for this purpose but other methods can be used.

Example 6.8 Consider the problem shown in **Figure 6.33a**. It consists of an infinitely long enclosed, air-filled channel, with grounded sides ($V = 0$). The top is insulated from the sides and connected to a potential $V_0 = 100$ V. Calculate the potential distribution everywhere in the channel.

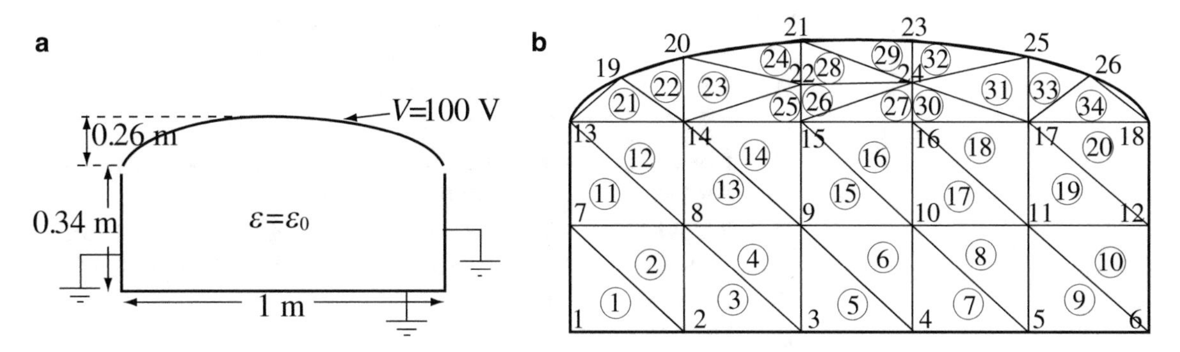

Figure 6.33 (**a**) Infinite channel (shown in cross section) with given boundary conditions. (**b**) Finite element mesh for the geometry in (**a**). Circled numbers are element numbers; the rest are node numbers

Solution: The steps involved in the solution are as follows:

(a) A finite element mesh must be defined. The mesh in **Figure 6.33b** is one choice which takes into account the shape of the geometry and uses a relatively small number of elements. A total of 34 elements and 26 nodes are used. Sixteen of the nodes are on the boundary and, therefore, are specified as Dirichlet boundary conditions. Nodes 1 through 6, 7, 12, 13, and 18 are set to zero, whereas nodes 19, 20, 21, 23, 25, and 26 are set to 100 V.

(b) The input data for the finite element script is defined from the mesh. This consists of the following:

(1) General mesh data. These include the number of nodes, number of elements, number of boundary conditions, and number of different materials in the geometry. The first record in **Figure 6.34** is the general mesh data for this example. The data, as used by the script, are listed in **dat2.**

(2) Element and material data. Each element is defined by three nodes and contains a single material. The element data consist of 34 records, one for each element, in the sequence they are numbered. Each line lists the three nodes of the element and a material index. Records 2 through 35 in **Figure 6.34** are the element data. The material index for all elements is 1 indicating that only one material is present (free space). Note also that the nodes of each element are numbered in a counterclockwise sequence.

(3) Node data. To calculate areas of elements, the coordinates of each node are needed. These are listed in sequence (coordinates of node 1, followed by node 2, and so on) as the next 26 records in the mesh data in **Figure 6.34**.

(4) Boundary and boundary condition data. The first line of the mesh data indicates how many boundary conditions are specified. In this case, there are two: one is zero, specified on the bottom and sides, and the second is 100 V, specified on the top. The boundary condition data are specified by entering the boundary condition value first, followed by the node numbers on which the value is specified. Thus, on line 62 (**Figure 6.34**), the boundary condition 100.0 is specified. Line 63 gives 20 node numbers. The first six (19, 20, 21, 23, 25, and 26) are the nodes of the top boundary. The rest are zeros and are disregarded by the script. By entering a fixed number of boundary nodes, the data reading is simplified. Similarly, line 64 shows zero, followed by 20 node numbers on line 65. The first 10 are nodes on the bottom and side boundaries; the rest are disregarded.

Figure 6.34 Input mesh data for **Example 6.8.**
* These lines were added for explanation purposes: they are not part of the data file. The first column is a sequential line number

```
* General mesh data:
1    26 34 2 1
* Element data
2    1 2 7 1
3    2 8 7 1
4    2 3 8 1
5    3 9 8 1
6    3 4 9 1
7    4 10 9 1
8    4 5 10 1
9    5 11 10 1
10   5 6 11 1
11   6 12 11 1
12   7 8 13 1
13   8 14 13 1
14   8 9 14 1
15   9 15 14 1
16   9 10 15 1
17   10 16 15 1
18   10 11 16 1
19   11 17 16 1
20   11 12 17 1
21   12 18 17 1
22   13 14 19 1
23   14 20 19 1
24   14 22 20 1
25   22 21 20 1
26   14 15 22 1
27   15 24 22 1
28   15 16 24 1
29   22 24 21 1
30   24 23 21 1
31   16 17 24 1
32   24 17 25 1
33   24 25 23 1
34   17 26 25 1
35   17 18 26 1
* Node coordinates
36   0.0 0.0
37   0.2 0.0
38   0.4 0.0
39   0.6 0.0
40   0.8 0.0
41   1.0 0.0
42   0.0 0.16
43   0.2 0.16
44   0.4 0.16
45   0.6 0.16
46   0.8 0.16
47   1.0 0.16
48   0.0 0.34
49   0.2 0.34
50   0.4 0.34
51   0.6 0.34
52   0.8 0.34
53   1.0 0.34
54   0.05 0.46
55   0.2 0.55
56   0.4 0.6
57   0.4 0.46
58   0.6 0.6
59   0.6 0.46
60   0.8 0.55
61   0.95 0.46
* Boundary conditions
62   100.0
63   19 20 21 23 25 26 0 0
     0 0 0 0 0 0 0 0 0 0 0 0
64   0.0
65   1 2 3 4 5 6 7 12 13 18
     0 0 0 0 0 0 0 0 0 0
* Material properties
66   1.0 0.0
```

(5) **Material data**. For each material in the mesh, the relative permittivity and the volume charge density are entered. In this case, the only material in the mesh is air and, therefore, line 66 lists the relative permittivity as 1.0 and the volume charge density as 0.0.

The input data are quite extensive although rather simple. In most finite element applications, these data are generated by special programs called mesh generators. The user only has to properly define the geometry of the problem and its properties. However, in this example and the examples that follow, the data may be entered by hand; this method is more tedious but does not require a mesh generator.

(c) **Solution**. Now that the mesh data are available (and, hopefully, correct), the finite element script is run to solve for the potentials at the nodes of the solution domain. Before doing so, the input data in **dat2** must be copied into a file named **dat1** which is used as the default input file for the finite element script. The script used here is called **fem1.m**. With the input data in **Figure 6.34**, the script produces the node potentials and electric field intensities and places these into two files. One is **out1** and lists the numerical output produced by the script. The second is **out2** and contains identical results as **out1** but with comments. The potentials are calculated at the nodes and vary linearly within the element according to **Eq. (6.57)**. The electric field intensity as calculated here is averaged throughout each element and is associated with the centroid of the element. The listing produced by the script shows the x component, y component, and magnitude of the electric field intensity at the centroid of each element in the mesh.

(d) **Postprocessing of data**. The node potentials are rarely the only results needed. For example, we may wish to know the potential distribution in the solution domain. One way to see this is to show the contour plot (plot of constant potentials) as in **Figure 6.35**. Other types of results may be the energy per unit length stored in the geometry, potentials at points other than nodes, and many others. These may be performed on the data obtained from the program using any suitable software tool.

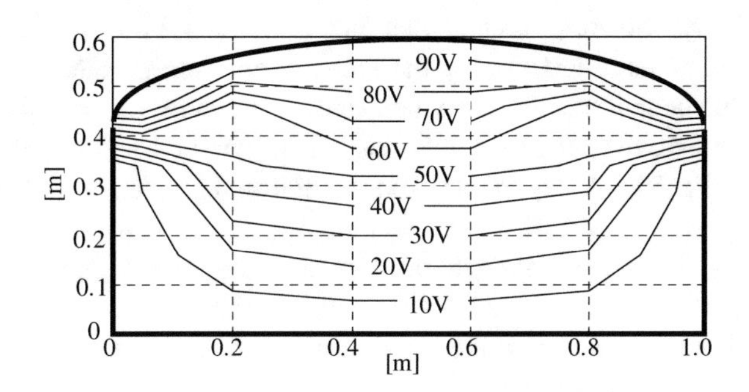

Figure 6.35 Plot of constant potential contours in the cross section of the channel in **Figure 6.33a**

Example 6.9 Application: Flaw in the Dielectric of a Capacitor A parallel plate capacitor is given. The insulation between the plates is mica with relative permittivity $\varepsilon_r = 4$. Because of problems in production, there is a fault in the mica in the form of a rectangular vein, as shown in **Figure 6.36**. The vain may be considered to be air:

(a) Calculate the potential everywhere inside the capacitor if a potential $V = 100$ V is connected across the plates.
(b) Calculate the electric field intensity at the center of the fault for the conditions in **(a)**.
(c) What is the maximum potential difference allowable with and without the fault if the dielectric strength in mica is 20×10^6 V/m, whereas in air it is 3×10^6 V/m?

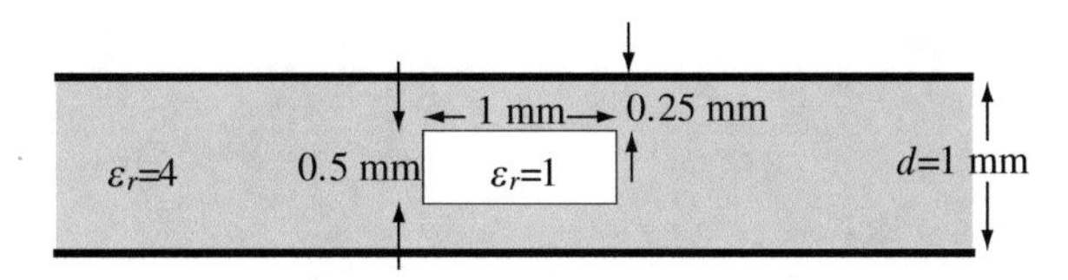

Figure 6.36 A mica insulated, parallel plate capacitor with a small flaw in the dielectric

Solution: The first step in the solution is to define a mesh. To do so, we note the following:

(1) The geometry in **Figure 6.36** is symmetric about a vertical line through the center of the flaw. Therefore, only half the geometry need be analyzed. We take the right-hand side.
(2) The capacitor is very large in relation to the distance d. In practice, we must analyze a finite-size geometry. Because the flaw is rather small, we expect any variation in the field to be around the flaw. Therefore, the right-hand side boundary is taken, arbitrarily, at 4 mm from the center of the flaw. The mesh with dimensions and boundary conditions is shown in **Figure 6.37**. The left side boundary is a symmetry line and, therefore, is a Neumann boundary. Nothing on this boundary need be specified. The top and bottom boundaries are Dirichlet boundaries since the potentials are known. The right boundary is an artificial boundary. It may be specified in one of two ways:

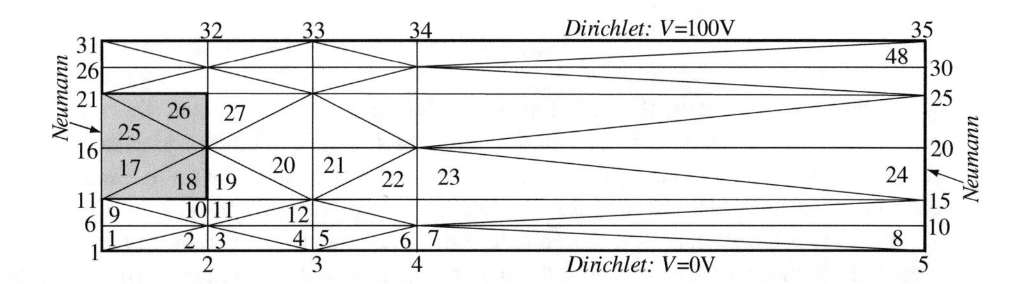

Figure 6.37 Finite element mesh and boundary conditions for the capacitor in **Figure 6.36**

(i) The potential at the boundary is the same as in an infinite capacitor (far enough from the flaw). Therefore, it is linearly distributed on the boundary. The potentials at nodes 10, 15, 20, 25, and 30 are 12.5 V, 25 V, 50 V, 75 V, and 87.5 V, respectively. These are Dirichlet boundary conditions. The total number of Dirichlet boundary conditions is seven: one on the top surface, one on the bottom, and five on the right-hand boundary.

(ii) Alternatively, we may assume that the right boundary is also a Neumann boundary condition and leave it unspecified. This is justified by the fact that the normal electric field intensity on this boundary is zero, again, because the boundary is far from the defect. We choose this method here because it simplifies data input. With this, there are only two boundary conditions: one on the bottom boundary, one on the top. The top boundary is at 100 V (on nodes 31, 32, 33, 34, and 35) and the bottom boundary is at zero (on nodes 1, 2, 3, 4, and 5).

(a) The input data for the mesh are generated as in **Example 6.8**. and are listed in **dat3**. The solution is produced by first copying **dat3** into **dat1** (the input file specified in script **fem1.m**) and then running the script. The script produces output files **out1** and **out2**, as was described in **Example 6.8**. The output data (with comments) are listed in file **out3** (this is a copy of **out2**).
A contour plot of the potential is shown in **Figure 6.38a**.

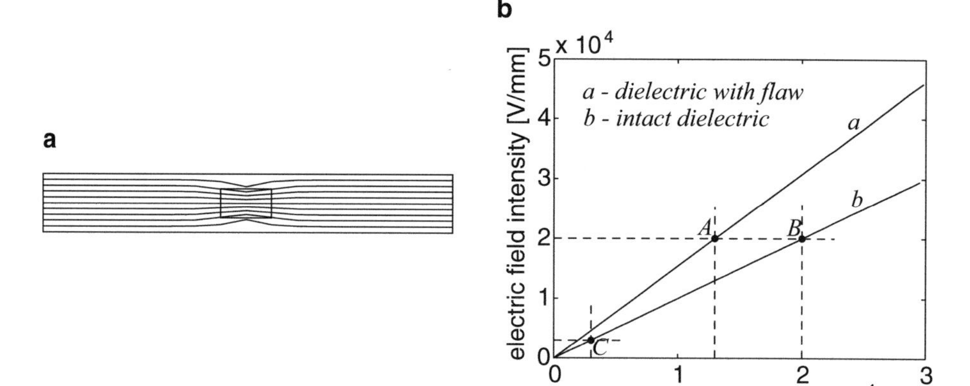

Figure 6.38 (**a**) Contours of constant potential in the capacitor in **Figure 6.36**. (**b**) Magnitude of the electric field intensity in the flaw as a function of applied potential

(b) The electric field intensity can only be calculated at the center of elements. There are four elements in the fault: elements 17, 18, 25, and 26. Of these, elements 17 and 25 are closest to the center of the fault. The electric field intensities at the center of these two elements are (from **out3**)

$$\mathbf{E}_{17} = \hat{\mathbf{x}}7.629 \times 10^{-5} - \hat{\mathbf{y}}1.539 \times 10^2, \quad \mathbf{E}_{25} = \hat{\mathbf{x}}7.629 \times 10^{-5} - \hat{\mathbf{y}}1.539 \times 10^2 \quad [\text{V/mm}]$$

Taking the average between these two fields (which just happen to be the same) gives the approximate electric field intensity at the center of the flaw as

$$\mathbf{E}_{center} = \hat{\mathbf{x}}7.629 \times 10^{-5} - \hat{\mathbf{y}}1.539 \times 10^2 \quad [\text{V/mm}]$$

Note that the field has a small x component and a dominant y component as expected.

(c) The maximum potential difference allowable without the flaw is calculated analytically as

$$V_{max} = Ed = 20 \times 10^6 \times 0.001 = 20{,}000 \quad [\text{V}]$$

When the flaw is present, the potential is not uniformly distributed and we must calculate the potential difference numerically. However, there is a slight difficulty here: We are in effect trying to find the boundary conditions that will provide the maximum electric field intensity allowable in the flaw. The finite element method requires known boundary conditions to calculate the field intensity. The way we approach this problem is to start with a known potential difference and increment the potential on the boundary until the electric field intensity in the flaw (air) equals the breakdown electric field intensity. The potential difference thus obtained is the maximum allowable potential difference. In other words, we run the finite element script with known trial potentials and choose that potential which provides the required result. **Figure 6.38b** shows a plot of the magnitude of the electric field intensity in element 25 obtained for potential differences starting at 0 and ending at 30,000 V, in increments of 500 V. For each boundary potential value, the boundary conditions must be modified accordingly. From this figure, the maximum potential difference allowable is 13,000 V (point A). This is significantly lower than the 20,000 V (point B) allowed with an intact dielectric, but still much higher than about 3,000 V (point C) for an air-filled dielectric; that is, damage to the dielectric of the capacitor reduces its breakdown voltage significantly.

Example 6.10 Electric Fields Near a DC Busbar A busbar used in the distribution of electric power in a distribution box is at potential 220 V above ground. The busbar is a rectangular conductor as shown in **Figure 6.39a**. The ground and the busbar may be considered to be perfect conductors:

(a) Find the electric potential everywhere in space.
(b) Find the location and magnitude of the maximum electric field intensity.

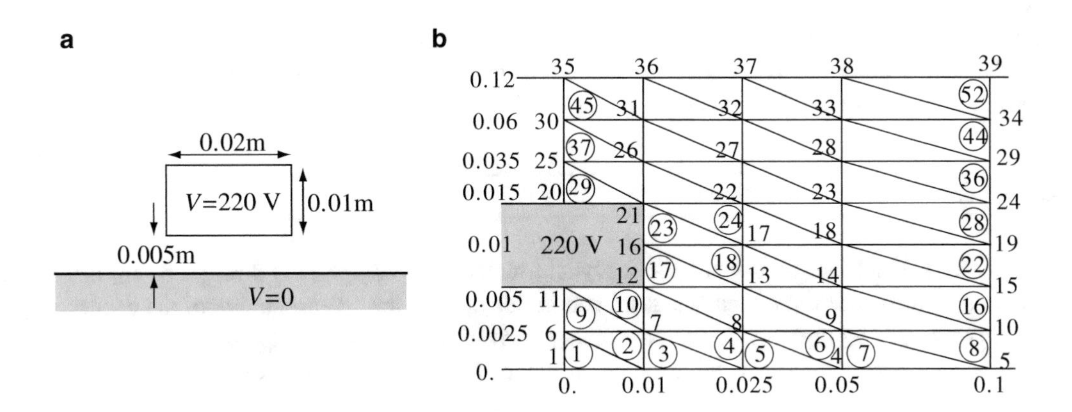

Figure 6.39 (**a**) A busbar at 220 V over a conducting ground. (**b**) Placement of artificial boundary and symmetry line for the geometry in (**a**). A mesh is also shown

Solution: We again start by defining the geometry and the boundary conditions. Using a symmetry line vertically through the center of the bar, we eliminate half the geometry. The cutting line becomes a Neumann boundary. Next, we must place artificial boundaries at some distance from the bar. By placing these boundaries at a reasonable distance from the source, the solution can be accurate while the mesh required is reasonably small. In this case, the boundaries are placed at 0.1 m from the symmetry line and 0.12 m from the ground plane, as shown in **Figure 6.39b** and are set to potential zero. These boundaries are chosen assuming their influence on the solution is minimal. If this turns out to be wrong, they will have to be taken further away. A total of 52 elements and 39 nodes are used. The left boundary is left unspecified with the exception of the bar which is held at 100 V.

(a) The mesh input data are again generated as in **Example 6.8** and are listed in **dat5.** This data file is now copied into **dat1** (default input file to the script) and script **fem1.m** is executed. The output appears in **out1** and **out2. out1** is used to plot the results and **out2**, which contains comments, is copied onto **out5**, which is also available so that the results may be inspected. A contour plot of the potential is shown in **Figure 6.40**. Note that the results have been reflected about the symmetry line to show the potential everywhere.

Figure 6.40 Contour plot of the solution for **Example 6.10**

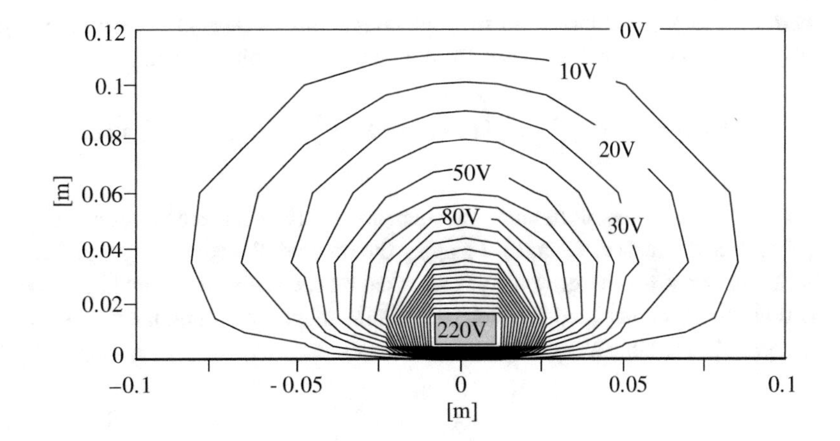

(b) From the results in **out5**, the magnitude of the electric field intensity is highest in element No. 11 and equals 4.496×10^4 V/m. This value is calculated at the centroid of element No. 11 and corresponds to a distance of about 1.7 mm below and about 10 mm to the right of the corner of the busbar. That the maximum electric field intensity should be around the corner is expected, but the exact location and magnitude depends on the dimensions and on the mesh used.

The three numerical methods discussed in this chapter for the solution of electrostatic problems were presented in their simplest possible forms for the purpose of demonstrating numerical techniques. They can be used in much more complex situations and each of them is available in many software packages both commercially and as open source software, in varying degrees of complexity and with different levels of user interfaces. But these are not the only numerical methods available. Many more have been developed for computation and design of electrical devices. Numerical methods have also been developed for time-dependent electric fields as well as magnetic and electromagnetic fields. It is safe to say that if an electromagnetic problem, of any complexity, needs to be solved a numerical technique has been developed for that purpose.

6.6 Summary

Numerical methods of solution are used when the analytical methods in **Chapters 3** through **5** fail, usually because the geometry is too complicated. The three methods described in this chapter are representative of the concepts involved.

The ***finite difference method*** replaces partial or ordinary derivatives with simple approximations. Given points x_i on a line, at distance $\Delta x = h$ from each other, with unknown values $f(x_i) = f_i$, the approximation to first- and second-order derivatives may be written as

$$\frac{df(x_i)}{dx} \approx \frac{f(x_i + \Delta x) - f(x_i - \Delta x)}{2\Delta x} = \frac{f_{i+1} - f_{i-1}}{2h} \tag{6.1}$$

$$\frac{d^2 f_i}{dx^2} \approx \frac{f_{i+1} - 2f_i + f_{i-1}}{h^2} \tag{6.8}$$

Similar expressions are written for derivatives with respect to y and z if needed. To use these approximations, the space in which a solution is sought is divided into a grid with points i,j, generated by parallel lines, separated distances $\Delta x = \Delta y = h$, in the two directions in space forming a two-dimensional grid (**Figure 6.5**). An unknown potential $V_{i,j}$, is assumed at each point of the grid using the approximation in **Eq. (6.8)**

$$V_{i,j} = \frac{V_{i-1,j} + V_{i+1,j} + V_{i,j-1} + V_{i,j+1}}{4} + h^2 \frac{\rho_{i,j}}{4\varepsilon_{i,j}} \tag{6.17}$$

This equation is repeated at each internal point of the grid. Boundary values are incorporated in the approximation when any of the values of $V_{i,j}$ corresponds to a boundary point. Solution of the system of equations provides the unknown values over the grid. Other quantities such as electric fields, forces, and energy can then be calculated from the potential. The extension to three dimensions is straightforward (see **Exercise 6.3**). The charge density at points of the grid can vary from point to point as can the permittivity.

The ***method of moments*** solves for the equivalent sources that produce a known potential. Given a charge distribution $\rho_{\Omega'}(x',y',z')$, the potential at a distance R from a differential point charge is

$$V(x,y,z) = \int_{\Omega'} \frac{1}{4\pi\varepsilon R} \rho_{\Omega'}(x',y',z') d\Omega' = \int_{\Omega'} K(x,y,z,x',y',z') \rho_{\Omega'}(x',y',z') d\Omega' \tag{6.23), (6.24}$$

$K(x,y,z,x',y',z')$ is a geometric function that depends on dimensions and permittivity and Ω' is the space in which the charge density is distributed (surface, volume). To apply the method, the space (say a plate) is divided into any number of subspaces (subsurfaces), an unknown charge density is assumed on each subsurface, and the potential on the surface must be known. The geometric function K is calculated for each pair of subsurfaces assuming the charge is concentrated at its center. Assuming the surface has been divided into N rectangular subsurfaces, each with an area $s_i = a_i \times b_i$, the known potential on subsurface j is

$$V_j = K_{jj}\rho_{sj} + \sum_{\substack{i=1 \\ i \neq j}}^{N} \rho_{si} K_{ij} \quad [\text{V}], \quad j = 1, 2, \ldots, N \tag{6.29}$$

where

$$K_{ij}=\frac{s_i}{4\pi\varepsilon\sqrt{(x_j-x_i')^2+\left(y_j-y_i'\right)^2+(z_j-z_i')^2}},\quad i,j=1,2,\ldots,N,\quad i\neq j \tag{6.30}$$

$$K_{jj}=\frac{1}{4\pi\varepsilon}\left(2a_j\ \ln\frac{b_j+\sqrt{a_j^2+b_j^2}}{a_j}+2b_j\ \ln\frac{a_j+\sqrt{a_j^2+b_j^2}}{b_j}\right) \tag{6.32}$$

Equation (6.29) is written for the potential on each subdomain to obtain N equations in N unknown values ρ_{sj}. Solution of this system provides the charge density on each subdomain from which we can then calculate potentials and fields anywhere in space using the methods of **Chapters 3** and **4**.

Notes:

(1) The larger the number of subdomains, the more accurate the solution.
(2) Subdomains can be of different sizes or all equal in size.
(3) The potential must be known on each subdomain.
(4) When solving for the charge densities on plates of capacitors, the total charge on one plate must equal in magnitude to the total charge on the second. The potential of each plate must be adjusted to satisfy this condition (see **Example 6.7**).

The ***finite element method*** assumes the space (line, surface, volume) is divided into finite-size sections or elements and a potential distribution (constant, linear, quadratic, etc.) is assumed within each element based on the nodes (vertices) of the element. The potentials at these vertices are the unknowns we seek. The approximation for each element is then used to generate a system of equations for the unknown potentials. Given a triangular element with vertices (x_i,y_i), the potential within the element is

$$V(x,y)=N_iV_i+N_jV_j+N_kV_k \tag{6.43}$$

where N_i, N_j, N_k are called shape functions and V_i, V_j, V_k are the unknown potentials at the nodes of the element. For a triangular element of area Δ and node coordinates (x_i,y_i), (x_j,y_j), and (x_k,y_k):

$$N_i=\frac{1}{2\Delta}\left[(x_jy_k-x_ky_j)+(y_j-y_k)x+(x_k-x_j)y\right] \tag{6.44}$$

$$N_j=\frac{1}{2\Delta}\left[(x_ky_i-x_iy_k)+(y_k-y_i)x+(x_i-x_k)y\right] \tag{6.45}$$

$$N_k=\frac{1}{2\Delta}\left[(x_iy_j-x_jy_i)+(y_i-y_j)x+(x_j-x_i)y\right] \tag{6.46}$$

To solve an electrostatic problem, we write the energy in a volume as

$$F(E)=\int_v\left(\frac{1}{2}\varepsilon E^2-\rho V\right)dv \tag{6.61}$$

For the two-dimensional problems discussed in this chapter, assuming unit thickness for the geometry ($dv=1ds$) and using $\mathbf{E}=-\nabla V$:

$$F(V)=\int_s\left(\frac{\varepsilon}{2}\left\{\left(\frac{\partial V}{\partial y}\right)^2+\left(\frac{\partial V}{\partial y}\right)^2\right\}-\rho V\right)ds \tag{6.63}$$

The first derivatives of $V(x,y)$ are calculated from **Eq. (6.43)** as follows:

$$\frac{\partial V(x,y)}{\partial x}=\frac{\partial N_i}{\partial x}V_i+\frac{\partial N_j}{\partial x}V_j+\frac{\partial N_k}{\partial x}V_k \tag{6.51}$$

$$\frac{\partial V(x,y)}{\partial y} = \frac{\partial N_i}{\partial y}V_i + \frac{\partial N_j}{\partial y}V_j + \frac{\partial N_k}{\partial y}V_k \tag{6.52}$$

These and $V(x,y)$ from **Eq. (6.43)** are now substituted in **Eq. (6.63)**, resulting in the energy function in terms of the unknown potentials in the element and the shape functions of the element.

This energy is minimized with respect to each unknown value in each element to produce a solution

$$\frac{\partial F(V)}{\partial V_n} = \sum_{m=1}^{M} \frac{\partial F_m}{\partial V_n} = 0 \tag{6.65}$$

M is the number of elements in the assembly (called a mesh) and N the number of nodes. **Equation (6.65)** produces N equations in N unknowns which, when boundary conditions are applied and the system solved, produces the potentials at the nodes of the finite element mesh.

The solution consists of the following:

(1) Definition of a finite element (triangle in two dimensions, tetrahedron in three dimensions, or any other defined shape that divides the space).
(2) Approximation of the potential over the element.
(3) The equation to solve is an energy related function [see **Eq. (6.63)** for example].
(4) Minimization of the energy function over the space, done one element at a time.
(5) Application of boundary conditions and solution for potentials.
(6) Electric fields can then be calculated from potentials if necessary (see **Example 6.10**).

Problems

Finite Differences

6.1 One-Dimensional Geometry. A parallel plate capacitor is shown in **Figure 6.41**. The parallel plates may be viewed as infinite in extent. With $d = 1$ m and free space between the plates:

(a) Calculate the potential distribution everywhere within the capacitor using the finite difference method with four equal divisions.
(b) Repeat the solution in **(a)** with eight equal division.
(c) Calculate the potential distribution using direct integration. Show by comparison with **(a)** and **(b)** that the division does not matter in this case. Why?

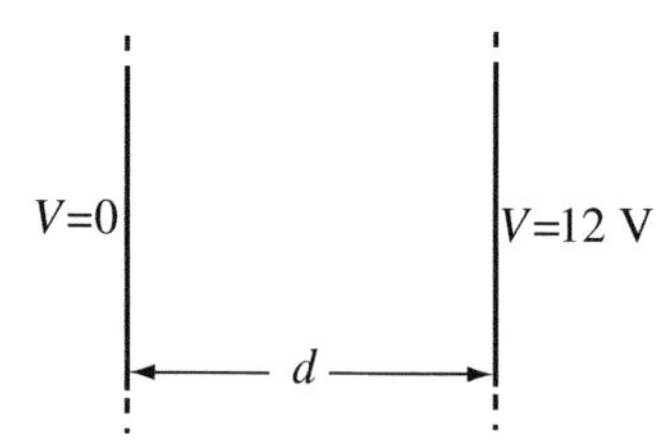

Figure 6.41

6.2 Application: Capacitor with Space Charge Between Plates. The parallel plate capacitor in **Figure 6.41** is given. In addition to the data in **Problem 6.1**, there is also a charge density everywhere inside the material such that $\rho/\varepsilon = 1$ C/F·m^2, where ρ is the charge density [C/m^3] and ε is the permittivity [F/m] of the dielectric between the plates:

(a) Find the potential distribution everywhere within the capacitor using four equal divisions.
(b) Repeat the solution in **(a)** with eight equal divisions. Show that the division chosen is important. Which division gives a better result? Compare with the analytical solution.

6.3 Capacitor with Space Charge. Consider the parallel plate capacitor in **Figure 6.42**. A uniform charge density $\rho_0 = 10^{-6}$ C/m^3 exists everywhere inside the capacitor and both plates are grounded. The permittivity of the material inside the capacitor equals $4\varepsilon_0$ [F/m]. Assume the plates are very large:

(a) Calculate the potential and electric field intensity everywhere inside the capacitor using a one-dimensional finite difference method.

(b) Find the analytic solution by direct integration and compare the numerical and analytic solutions.

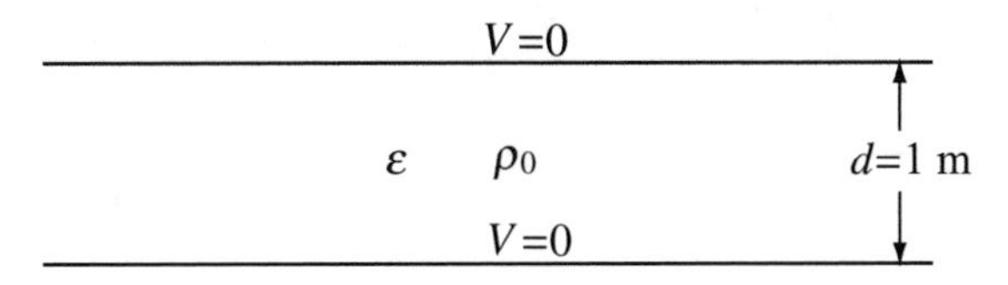

Figure 6.42

6.4 Infinite Channel. Solve for the electric potential in the geometry in **Figure 6.43**. This is a two-dimensional problem in the form of an infinite channel shown in cross section. The top conducting surface is physically separated from the side plates and is held at a potential 10 V whereas the side and lower boundaries are grounded (held at zero potential). Given: $\varepsilon = \varepsilon_0$ [F/m], $\rho_v = 0$. Use an explicit method of solution.

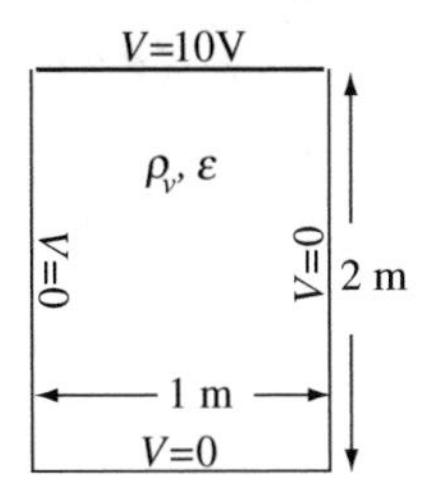

Figure 6.43

6.5 Infinite Channel with Interior Charge. Solve for the electric potential in the geometry given in **Figure 6.43**. The outer boundaries of the channel are grounded except for the top surface which is at 10 V and a uniformly distributed volume charge distribution exists inside the channel as shown. Given: $\rho_v = 10^{-6}$ C/m^3, $\varepsilon = 80\varepsilon_0$ [F/m].

6.6 Infinite Channel with Interior Charge. Solve for the electric potential in the geometry in **Figure 6.44**. This is a two-dimensional problem similar to that in **Problem 6.4**, with the outer boundaries grounded and a charge distribution exists inside part of the channel as shown. Given: $\rho_v = 10^{-6}$ C/m^3, $\varepsilon_1 = 80\varepsilon_0$ [F/m].

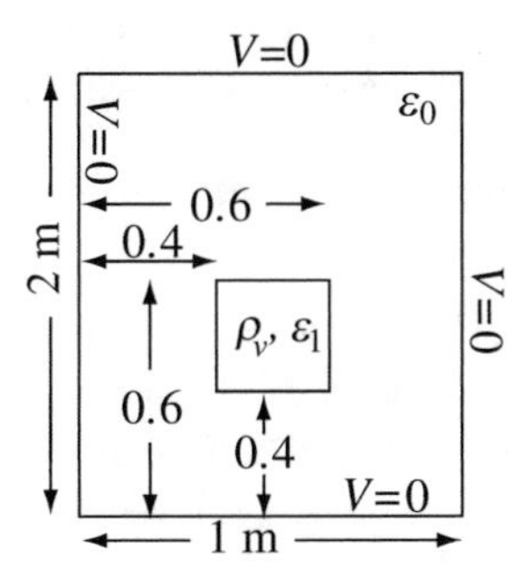

Figure 6.44

Method of Moments

6.7 Application: Capacitance of Small Plates. Two rectangular plates are used as a variable capacitor by rotating one plate about the axis *A–A* while the other plate is fixed. In the two extremes, the plates are either parallel to each other or flat on a plane as shown in **Figures 6.45a** and **6.45b**. The plates are 50 mm by 20 mm in size. When parallel to each other, they are 2 mm apart, as shown in **Figure 6.45b**. When on a plane, they are separated 2 mm while the edges remain parallel. Calculate the range of the capacitor's capacitance in free space.

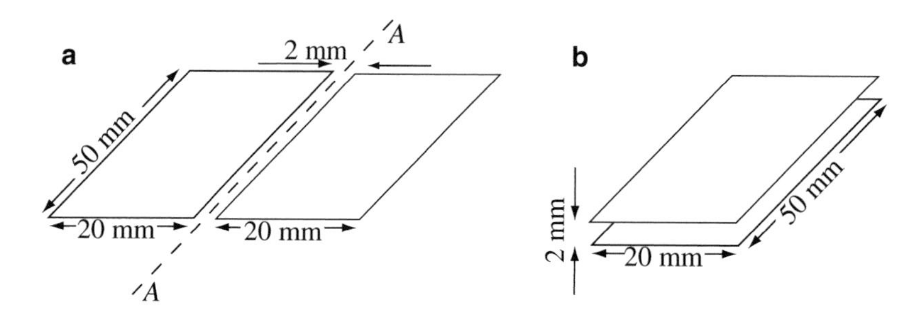

Figure 6.45

6.8 Application: Small Capacitor. A parallel plate capacitor is made from two plates, as shown in **Figure 6.46a**. The material between the plates is free space:

(a) Calculate the capacitance of the capacitor.

(b) Now the capacitor is cut in two as shown in **Figure 6.46b**. Calculate the capacitance of one of the two smaller capacitors thus created. Is the sum of the capacitance of the two halves in **Figure 6.46b** equal to the capacitance of the whole capacitor in **Figure 6.46a**? If not, why not?

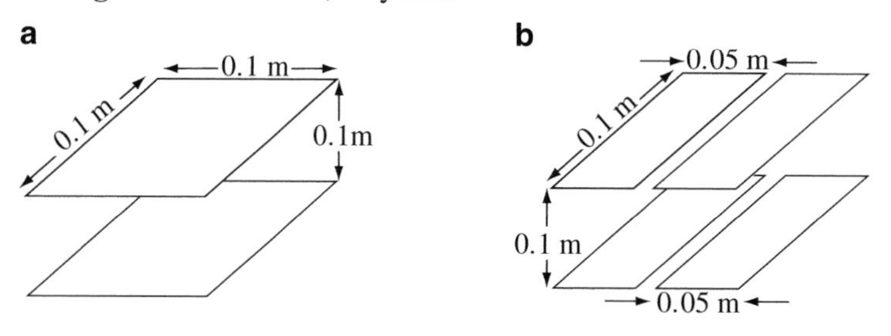

Figure 6.46

6.9 Application: Small Capacitor. Consider **Figure 6.46a**:

(a) Show by a series of calculations with increasing number of subdomains that as the number of subdomains increases, the capacitance of the small capacitor convergences to a constant value. What is this value?

(b) Plot the charge density on the upper plate on a line parallel to one of the sides of the plate and that crosses through the center of the plate (or very close to it). Comment on the shape of this plot and its meaning.

6.10 Application: Capacitance of a Washer. Calculate the capacitance of a thin, flat washer of internal radius $c = 10$ mm and external radius $d = 60$ mm.

6.11 Application: Coupled-Charge Devices (CCD). In a memory array or a CCD (coupled-charge device), capacitive devices are arranged in a two-dimensional array. A 3×3 section is shown in **Figure 6.47**. The purpose of these capacitors is to store charge for a relatively short period of time. Each plate is 2 μm by 2 μm and the separation between two plates is 0.5 μm. As part of the analysis of the device, it is required to calculate the capacitances as follows (assume the plates are in free space):

(a) Between each two nearest plates (plates A and B).

(b) Between each plate and the plate on the diagonal (plates A and C).

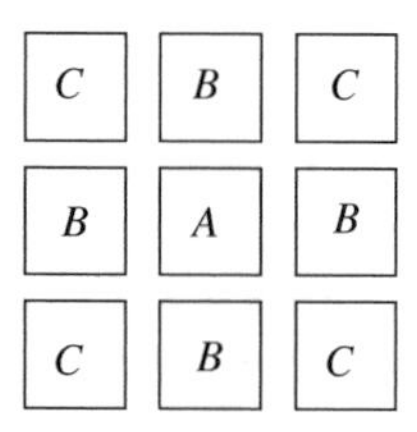

Figure 6.47

6.12 Capacitance of Perpendicular Plates. Two plates with given dimensions form a capacitor as shown in **Figure 6.48**. The plates are perpendicular to each other. With the dimensions given and assuming the plates are in air (free space):

(a) Calculate the capacitance between the plates using the method of moments and hand computation. Use a small, reasonable number of subdomains.

(b) If the potential difference between the plates is 1 V, what is the potential and the electric field intensity at point P? Use the charge densities obtained in **(a)** to calculate the potential at P.

(c) Write a program (or use script **mom1.m**) to calculate a sequence of results, each with increasing number of subdomains until the change in solution is less than 5%. What is the minimum number of subdomains needed if all subdomains must be of the same size?

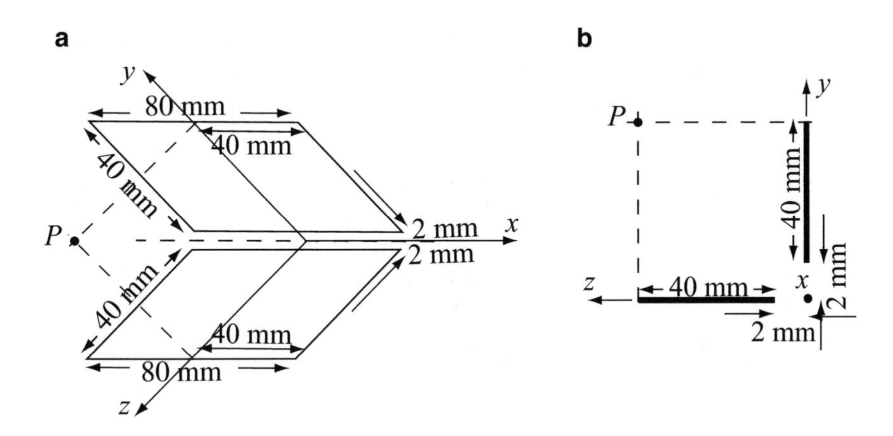

Figure 6.48

6.13 Capacitance Between Two Unequal Plates. The two plates in **Figure 6.49** are given. Calculate the capacitance between the plates assuming they are in free space.

(a) Use hand computation with 4 equal elements on each of the plates.

(b) Use script **mom1.m** and a sequence of divisions of the plates to obtain a more accurate solution. Use the same number of divisions on each plate and increase the number of divisions until the capacitance does not change by more than 1%. What is the capacitance and how many divisions on each plate are needed?

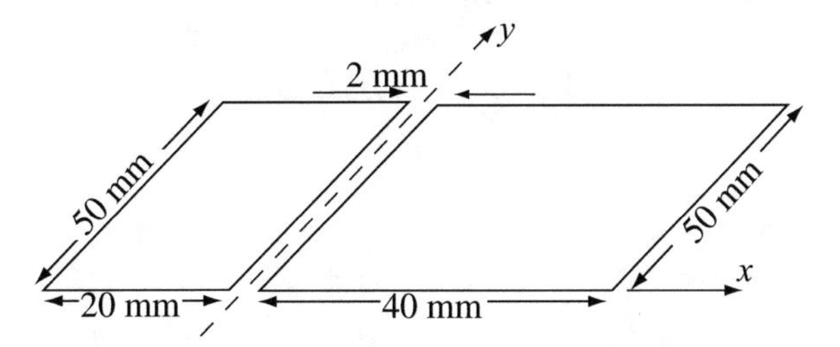

Figure 6.49

Finite Elements

6.14 One-Dimensional Finite Elements. A one-dimensional element may be defined in a manner similar to that in **Section 6.5.1.1** by assuming a section of a line of finite length, two points at its ends with coordinates (x_1) and (x_2), and unknown values ϕ_1 and ϕ_2 as shown in **Figure 6.50**:

(a) For this element, calculate the shape functions necessary to define the element by assuming a linear variation inside the element of the form $\phi(x) = a + bx$.

(b) With the shape functions in **(a)**, find an expression for the function $\phi(x)$ inside the element in terms of the shape functions.

(c) Show that the magnitude of N_1 [shape function at node (1)] is 1 at node (1) and zero at node (2).

(d) Show that the magnitude of N_2 [shape function at node (2)] is 1 at node (2) and zero at node (1).

(e) Show that the sum of the shape functions at any point $x_1 \leq x \leq x_2$ equals 1.

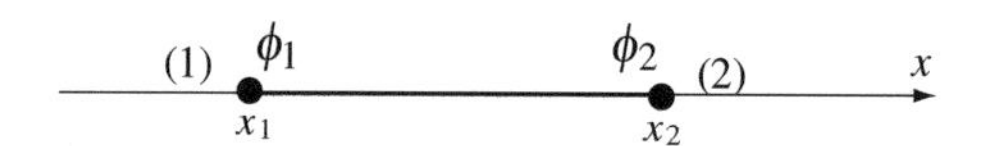

Figure 6.50

6.15 Application: Field and Potential in Parallel Plate Capacitor. Consider the parallel plate capacitor with a dielectric between the plates shown in **Figure 6.51**. The dielectric has relative permittivity of 6 and a volume charge distribution as shown: Assume plates are very large:

(a) Set up a finite element solution and solve for the potential, using the shape functions in **Problem 6.14**. Use hand computation and as many elements as necessary.
(b) From the potential distribution, calculate the electric field intensity in the capacitor. Sketch the solutions.
(c) Find the analytic solution for potential by direct integration and compare with the results in **(a)**.

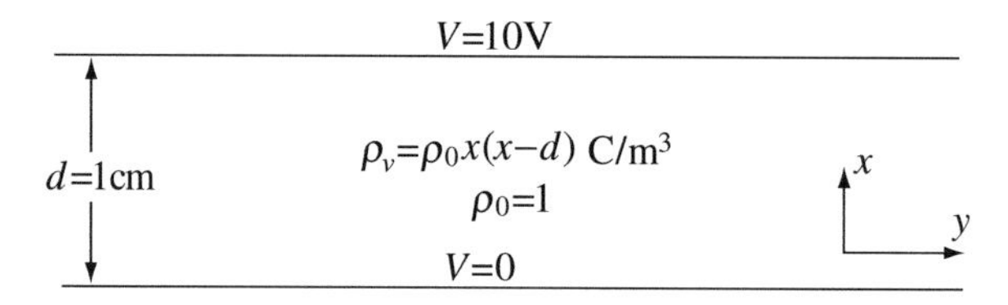

Figure 6.51

6.16 Application: Field and Potential in Parallel Plate Capacitor. Consider **Figure 6.52**. Using the triangular shape functions defined in **Section 6.5.1** and the implementation in **Section 6.5.2**:

(a) Calculate the potential distribution inside the capacitor. Assume plates are very large.
(b) From the potential distribution, calculate the electric field intensity in the capacitor. Sketch the solutions.
(c) Find the analytic solution and compare with **(a)** and **(b)**.

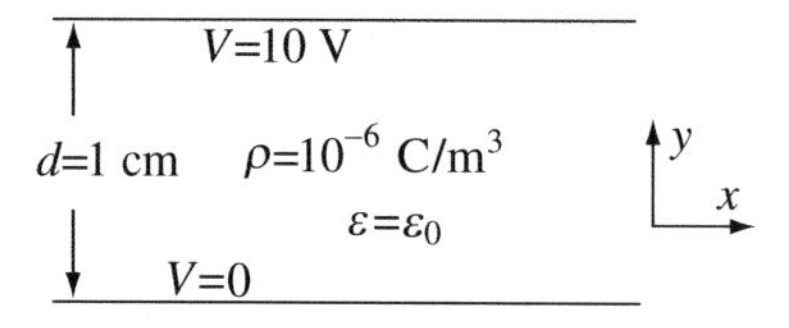

Figure 6.52

6.17 Quadrilateral Finite Elements. **Figure 6.53** shows a two-dimensional, four-node element (quadrilateral element). Assume an approximation of the form $\phi(x,y) = a + bx + cy + dxy$:

(a) Define the shape functions and their derivatives for a particular element with nodes at $P_1(0,0)$, $P_2(1,0)$, $P_3(1,1)$, and $P_4(0,1)$.
(b) Define the shape functions and their derivatives for the general element in **Figure 6.53**.
(c) Discuss the method and comment on its extension to more complex finite elements.

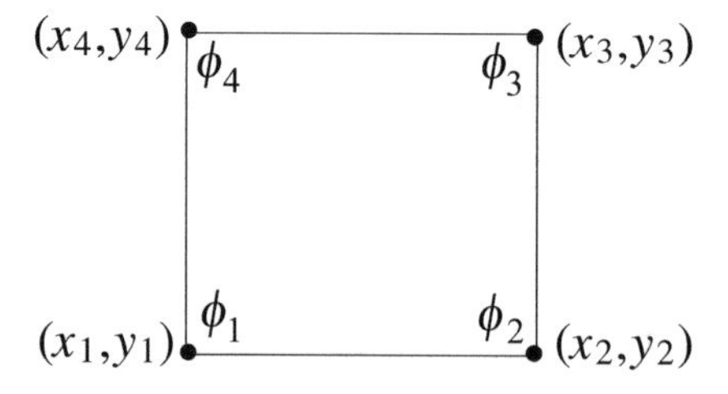

Figure 6.53

6.18 Application: Breakdown in Microcircuits. One of the challenges of microcircuits is the need for an ever-decreasing size of components. This means that components must be closer to each other and, therefore, the danger of breakdown between components and, in particular, between lines leading to them. To see the difficulty involved, consider the following: In a microcircuit, the smallest width conducting lines are 0.5 μm. Two such lines run side by side as shown in **Figure 6.54**. Assume the material below the lines is silicon with a relative permittivity of 12, and above and between the lines, it is free space. Breakdown in air occurs at 3×10^6 V/m and in silicon at 3×10^7 V/m. Assume the silicon layer is very thick:

(a) If the minimum distance between lines is $d = 0.5$ μm, what is the maximum potential difference allowable between the two lines? This is usually taken as the maximum source voltage for the circuit.

(b) If the circuit must operate on a maximum potential difference of 5 V, what is the minimum distance d [μm] allowed between the lines?

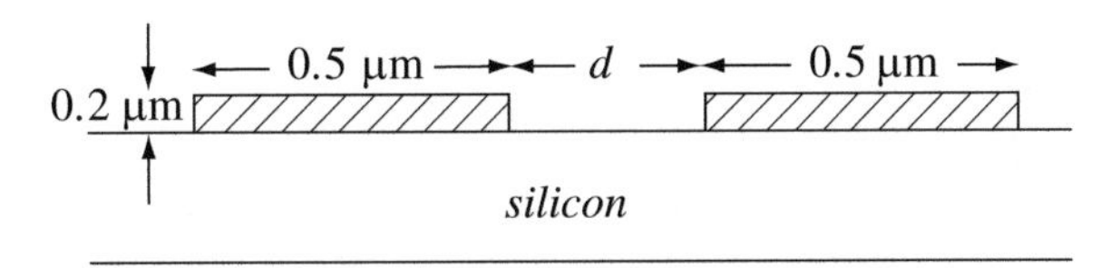

Figure 6.54

6.19 Application: Breakdown in Printed Circuits. The width of two strips on opposite sides of a printed circuit board is 1 mm and their thickness is 0.1 mm as shown in **Figure 6.55a**. The material is 0.5 mm thick, made of fiberglass with relative permittivity of 3.5. Breakdown voltage in fiberglass occurs at 30 kV/mm:

(a) What is the maximum potential difference allowable between the two strips?

(b) Suppose the printed circuit board has a flaw so that the material between the strips is missing, as shown in **Figure 6.55b**. What is now the maximum electric potential allowable?

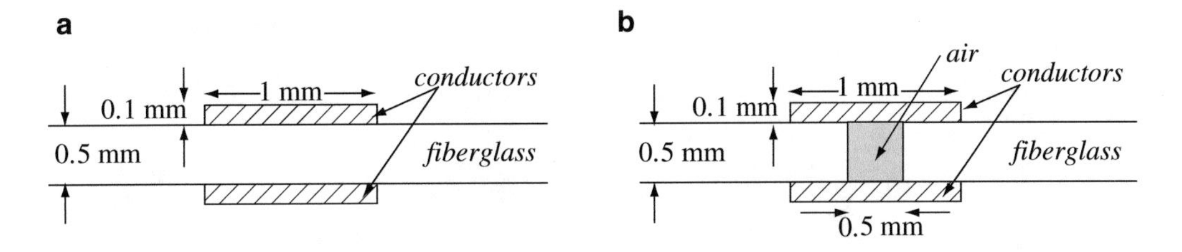

Figure 6.55

6.20 Application: Potential in Three-Phase Underground Cable. A three-phase power line operates at 240 V peak and is enclosed in a conducting shield which is at zero potential. The space between conductors and shield is filled with a material with relative permittivity 2.0:

(a) Assume the lines are very thin (**Figure 6.56a**) and calculate the potential distribution everywhere in the cable. Plot constant potential lines. The conditions are $r = 0.05$ m, $d = 0.01$ m, $V_1 = 240$ V, $V_2 = 240\cos(120°)$ [V], and $V_3 = 240\cos(240°)$ [V].

(b) The actual lines are each 10 mm in diameter (**Figure 6.56b**). Calculate the potential distribution everywhere in the cable. The conditions are $r = 0.05$ m, $d = 0.01$ m, $V_1 = 240$ V, $V_2 = 240\cos(120°)$ [V], and $V_3 = 240\cos(240°)$ [V]. **Hint**: The conductor surfaces become boundary conditions with known potentials. No need to discretize the interior of the conductors.

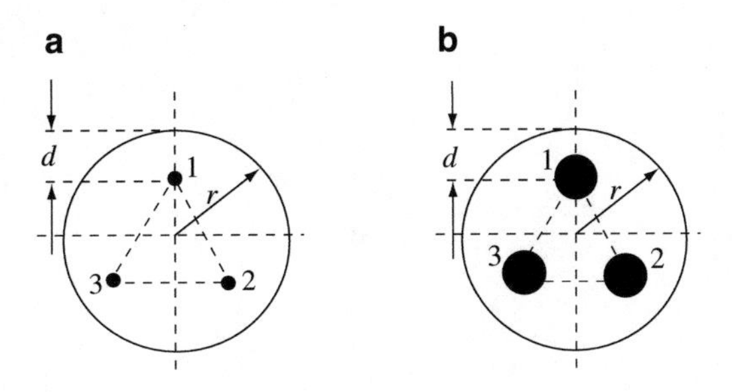

Figure 6.56

The Steady Electric Current

7

I believe there are 15,747,724,136,275,002,577,605,653,961,181,555,468,044,717,914, 527, 116,709, 366,231,425,076,185,631,031, 296 protons in the universe and the same number of electrons.

—Sir Arthur Eddington (1882–1944), physicist
The Philosophy of Physical Science (Cambridge, 1939)

7.1 Introduction

In the previous four chapters, we discussed the electrostatic field and its applications. The only reference to time was in the fact that whenever charges are placed in a conductor, they move until a steady state charge distribution is obtained. However, no attempt was made to characterize their motion or any effects motion might have on fields.

The present chapter discusses the motion of charges in space and in materials. More specifically, we look into what happens when charges are allowed to move from one point to another at given velocities or when charges have intrinsic velocities (i.e., they drift or are forced to move). As we shall see shortly, this type of time dependency is rather simple, but, for the first time, we discuss the possibility of charges moving. The complete discussion of time-dependent electric fields is not addressed until we reach **Chapters 10** and **11** because we do not yet have all the necessary tools to do so. The only aspect of time dependency allowed here is the motion of charges at a constant velocity. This will lead to the notion of ***steady current***, which, in practical terms, means a current that is constant in time. Also, as might be expected, because now we discuss both voltage and current, the various aspects of circuit theory such as Ohm's, Kirchhoff's and Joule's laws, and their relation to the electric field are established.

The mechanism of motion of charges allows the definition of the various properties of current, whereas the type of motion itself defines the type of current that may exist. Also, if current exists in materials, different materials can be expected to "conduct" it differently. We know from experience that some materials are "good" conductors and others are "poor" conductors. We will therefore define conductivity of materials, a property that, like permittivity, is characteristic for given materials.

As was our approach before, we start with the known properties and modify these to take into account the additional properties that need to be defined. If the postulates need to be modified, then a new set of postulates are put forth without, in any way, changing what we already have; that is, nothing we do here will modify the properties of the electrostatic fields as defined in **Chapters 3** and **4**. If a postulate is modified, then it applies under the conditions given here and the "old" postulates apply under the assumptions given for the electrostatic field. We will make no attempt at this stage to generalize the postulates or to "unify" the theory of electric fields simply because we do not yet have all the properties of the electric field to do so. This unification will occur in **Chapter 11**.

Current and current density are the main themes of this chapter. By current it is meant any motion of electric charges. In a metal, current is generated by the motion of electrons between atoms. This is called a ***conduction current***. Positive and negative ions in a gas also generate a current as they move. Similarly, electrons emitted from the cathode of a vacuum tube or released from a photovoltaic tube as well as electrons and holes (positive charges) in semiconductors also generate currents. A charged speck of dust or a charged drop of water generates a current as they move through the atmosphere. This type of current is called ***convection current***. We will distinguish between conduction and convection currents because they differ in behavior.

N. Ida, *Engineering Electromagnetics*, https://doi.org/10.1007/978-3-030-15557-5_7

Conduction currents, which are generated by motion of electrons between atoms, obey Ohm's law, whereas convection currents, which are generated by the motion of free charges in nonconducting materials (vacuum or gases), do not obey Ohm's law. Based on this definition, currents in semiconductor materials are conduction currents, whereas currents in the ionosphere are convection currents.

7.2 Conservation of Charge

Conservation of charge is a fundamental law of nature; charge cannot be created or destroyed (see **Example 4.30** where it is shown that assembly of a point charge requires infinite energy). The total amount of charge in the universe is fixed over time. Although there are physical processes that seem to generate charge, they always generate pairs of charges such that the total amount remains constant. For example, neutrons can sometimes turn into a proton and an electron, but no net charge has been created in the process. The reason conservation of charge is important for our purposes is that as a consequence, if charge moves from one point to another, we have a way of accounting for all charges at all times. Conservation of charge leads to important relations, such as Kirchhoff's laws, but also governs such mundane processes as the charging of a capacitor. Later in our studies, we will also apply this idea to show that electromagnetic waves can propagate in space.

7.3 Conductors, Dielectrics, and Lossy Dielectrics

7.3.1 Moving Charges in an Electric Field

For a charge to move, there must be a force acting on it. This force can be mechanical (such as motion of charged particles in a cloud due to wind), electrical (due to an electric field), or thermal (such as thermionic emission of electrons from hot surfaces), or any other means of releasing charges from atoms (for example, by collisions). In these processes, charges in the form of electrons, ions, or protons are involved. Motion of charges constitutes a flux, or current of charges, or, in short, current. Different mechanisms of motion of charges produce different types of currents. Two types, the conduction and convection currents, will be discussed here. Although the principles involved are similar, the laws currents obey are different, depending on the type of current. This notion of "types of currents" is perhaps new since we are normally used to currents in circuits which are conduction currents. To understand how currents are generated and what laws they obey, it is useful to look at the two types of currents separately. The main difference is that in convection currents, the electric field is not necessary for a current to exist, whereas conduction currents can only be generated by an electric field.

7.3.2 Convection Current and Convection Current Density

Consider the possibility of charged particles being moved from point to point by the wind. These charges have a velocity, but there is no electric field that acts on them. Similarly, charged particles (electrons and protons), emitted by the Sun as part of the solar wind in the form of a rarefied plasma, speed toward the Earth in the absence of an externally applied electric field. Whereas the charges themselves generate an electric field as any charge does and they may interact with each other, this interaction is not the reason they move toward the Earth. The reason is the initial velocity imparted to them by atomic forces within the Sun. Suppose we wished to measure or calculate the current generated by a source that emits charges. First, we must define what we mean by current. Each charged particle generates a small current. The total current is that emitted from the source. Thus, a current is only properly defined if we also define the area through which it flows; the current through a large area (perpendicular to the direction of motion of charges) such as s_2 in **Figure 7.1a** is larger than the current through a small area such as s_1 because more charges pass through s_2 per unit time. Thus, we define an element of volume Δv with cross-sectional area Δs and length Δl as shown in **Figure 7.1b**. Since the volume is very small, we can safely assume that all moving charges move at an average velocity v_z in the direction perpendicular to Δs. A certain number of charges will be present at all times in the volume. Assuming there are N charges per unit volume, the total number of charges in Δv is $N\Delta v$. Each charge is equal to q. The total charge present in volume Δv is

$$\Delta Q = Nq\Delta v \quad [\mathrm{C}] \tag{7.1}$$

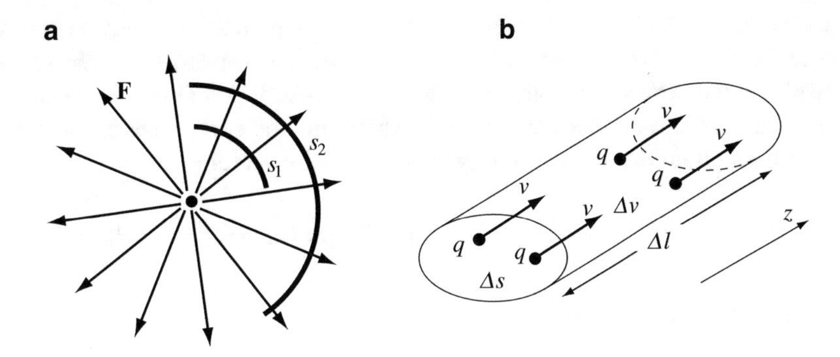

Figure 7.1 (**a**) Flux (current) through a surface. The flux through s_1 is smaller than through s_2. (**b**) An element of volume through which charged particles move

The element of volume Δv can also be expressed as $\Delta v = \Delta s v_z \Delta t$, where $v_z \Delta t$ is the length Δl of the element of volume; that is, it defines the length in which a particle of velocity v_z is found during a time Δt. With these the amount of charge in the element of volume Δv is

$$\Delta Q = Nq(\Delta s)(v_z \Delta t) \quad [\text{C}] \tag{7.2}$$

We now divide both sides of the expression by Δt. This gives the amount of charge ΔQ per time Δt or the time rate of change of charge in the given volume:

$$I = \frac{\Delta Q}{\Delta t} = Nqv_z \Delta s \qquad [\text{C/s}] \tag{7.3}$$

This quantity is the current I through the element of surface Δs. We have now an initial definition of current: "Current is the time rate of change of charge." The term "initial definition" refers to the fact that we will redefine the current as $I = -dQ/dt$. For now, the initial definition in **Eq. (7.3)** is quite sufficient. We note the following:

(1) Current depends on the number of charges per unit volume. If the source generates very few charges, the current will obviously be very small since, then, N is small.
(2) Current is directly proportional to the component of the velocity of the charged particles, v_z (perpendicular to the cross section Δs). Faster moving charges generate a larger current.
(3) The current also depends on the charge of the particles. Most often, the charge is that of an electron.
(4) The current depends on the surface. The larger the cross-sectional area Δs, the larger the current.

The first thing to notice is that the unit of current as given in **Eq. (7.3)** is the coulomb/second. This is what we call the ***ampere***,[1] denoted as [A]. In other words, the SI unit of charge is actually ampere · second [A•s]. We normally use the derived unit coulomb [C] for convenience.

Next we note that whereas the current depends on Δs, the quantity Nqv_z only depends on the charges themselves and the source that produces them. The current per unit area or ***current density*** is defined as

$$J = Nqv_z \quad [\text{A/m}^2] \tag{7.4}$$

Because the current density is independent of the cross section through which it flows, it is a more fundamental quantity than current and, in the context of electromagnetics, it is often a more useful quantity. Only when the current is confined to a known cross section (current in a wire or in a channel), it becomes convenient to use the current itself. Of course, in circuits, where conductors are well defined, current is more useful than current density.

[1] After André-Marie Ampère (1775–1836), French physicist and professor of mathematics. One of the most intense researchers ever, he is responsible more than anyone else, for connecting electricity and magnetism. About one week after Oersted discovered the magnetic field of a current in 1820, Ampère explained the observed phenomenon and then developed the force relations between current-carrying conductors. By doing so, he laid the foundations of a new form of science which he called electrodynamics. We call this electromagnetism. Ampère was active for many years working on a variety of subjects, including the integration of partial differential equations. Early in his life, he suffered the trauma of seeing his father guillotined during the Reign of Terror following the French Revolution, but he overcame this within a year to become one of the most notable scientists of the period. When James Clerk Maxwell came to unify the theory of electromagnetics, he acknowledged many of his predecessors' work but mostly Ampère's and Faraday's work. The unit of current and the law bearing his name attest to the importance of Ampère's contributions.

We assumed, arbitrarily, that the velocity of the charged particles has only a component in the z direction. This, of course, is not necessarily so; charges can flow in any direction. However, only that component of the charge velocity that is perpendicular to the surface Δs produces a current that crosses Δs. Thus, we may assume that velocity is a general vector $\mathbf{v}$, and the surface Δs has a normal unit vector $\hat{\mathbf{n}}$. Therefore, the component of the velocity vector perpendicular to s is $\hat{\mathbf{n}} \cdot \mathbf{v}$. The current through Δs due to an arbitrary charge velocity vector $\mathbf{v}$ is

$$I = \frac{\Delta Q}{\Delta t} = Nq\mathbf{v} \cdot \hat{\mathbf{n}}\, \Delta s \quad [\mathrm{A}] \tag{7.5}$$

and the current density is a vector:

$$\mathbf{J} = Nq\mathbf{v} \quad [\mathrm{A/m^2}] \tag{7.6}$$

The quantity $Nq = \rho_v$ gives the charge per unit volume (charge density in the volume of interest). The current density is, therefore, the product of volume charge density and velocity of charges:

$$\boxed{\mathbf{J} = \rho_v \mathbf{v} \quad [\mathrm{A/m^2}]} \tag{7.7}$$

Now, we are in a position to generalize these results. Since from **Eq. (7.5)** we can write $\Delta I = \mathbf{J} \cdot \hat{\mathbf{n}}\, \Delta s$, the current density can be defined as a limit:

$$\mathbf{J} = \hat{\mathbf{n}} \lim_{\Delta s \to 0} \frac{\Delta I}{\Delta s} \quad \left[\frac{\mathrm{A}}{\mathrm{m^2}}\right] \tag{7.8}$$

This definition has the advantage that the current density is now defined at a point. From the current density, the current through any surface can be calculated by integration over the surface:

$$\boxed{I = \int_s \mathbf{J} \cdot d\mathbf{s} \quad [\mathrm{A}]} \tag{7.9}$$

This relation clearly indicates that I is the flux of $\mathbf{J}$ through an area s.

Example 7.1 Convection Current Due to Rain

(a) Calculate the current density due to raindrops if the charge on each droplet is equal to that of 1,000 electrons, free-fall velocity near the surface of the Earth is 20 m/s, and the number of drops per second per meter squared of the surface is 10,000.

(b) What is the total current flowing between cloud and Earth if the cloud extends over an area of 10 km^2 in the absence of any wind?

(c) What happens if a horizontal wind at 30 km/h blows?

Solution: First, we must calculate the number of charges per unit volume. In this case, a convenient unit volume is 1 m^3 and the unit time is one second. From this, we can calculate the current density using **Eq. (7.4)**. The total current is then evaluated from **Eq. (7.9)**. The current density for a horizontal wind has both horizontal and vertical components.

(a) If 10,000 drops hit the surface at 20 m/s, these charges travel a distance of 1 m in 1/20 s. Thus, the total number of charged drops in the unit volume must be (10,000 drops/m^2/s)/(20 m/s) = 500 drops/m^3. In other words, if we could isolate the volume and monitor the charges, it would completely empty of charges in 1/20 s. The magnitude of the current density is therefore

$$J = Nqv_z = 500 \times 1000 \times 1.6 \times 10^{-19} \times 20 = 1.6 \times 10^{-12} \quad [\mathrm{A/m^2}]$$

Note the scalar notation and the use of the magnitude of the electron's charge. The velocity is vertical, which was taken here to be the z direction.

(b) The total current is this current density integrated over the area over which it rains. In this case, this is a simple multiplication of the current density by the surface area since the flow of charges is uniform:

$$I = JS = 1.6 \times 10^{-12} \times 10 \times (1{,}000)^2 = 1.6 \times 10^{-5} \quad [\mathrm{A}].$$

(c) Starting with the current density in part **(a)**, the velocity component in the vertical direction has not changed. However, now there is an additional horizontal velocity of 30 km/h = 8.34 m/s. Thus, the horizontal component of current density is equal to the ratio between horizontal (v_x) and vertical (v_z) velocities multiplied by the vertical current density:

$$J_x = J\frac{v_x}{v_z} = 1.6 \times 10^{-12} \times \frac{8.34}{20} = 6.672 \times 10^{-13} \quad [\mathrm{A/m^2}]$$

The vertical velocity has not changed and the current passing from cloud to Earth remains the same since the number of charges per unit time remains constant. Only the horizontal location at which charges are deposited has changed.

Note: The current due to rain is real and is an important aspect of the charge balance in the atmosphere. In general, the current is not constant, and rain can cause accumulation of charges. We cannot talk about resistance of air to the flow of this current since the current does not flow through air because of an electric field. The current flows because of the gravitational force on the raindrops. Thus, convection current does not satisfy the basic circuit assumptions implicit in Ohm's law.

Example 7.2 Application: Current in a Photovoltaic Tube A photovoltaic tube is made of a cathode that emits electrons when illuminated by light and an anode that collects them. These electrons are emitted at an initial velocity and are accelerated toward the anode, a distance $d = 20$ mm from the cathode, by a potential difference between anode and cathode of 1,000 V (**Figure 7.2**). The electrons are emitted uniformly over the surface of the cathode, which is a disk of radius $a = 10$ mm, with an initial velocity $v = 100$ m/s and a current of 0.01 A is measured in the circuit. Assume that charge accumulating in the space between the anode and cathode does not affect the motion of charges:

(a) Calculate the charge density next to the cathode.
(b) What is the charge density next to the anode?

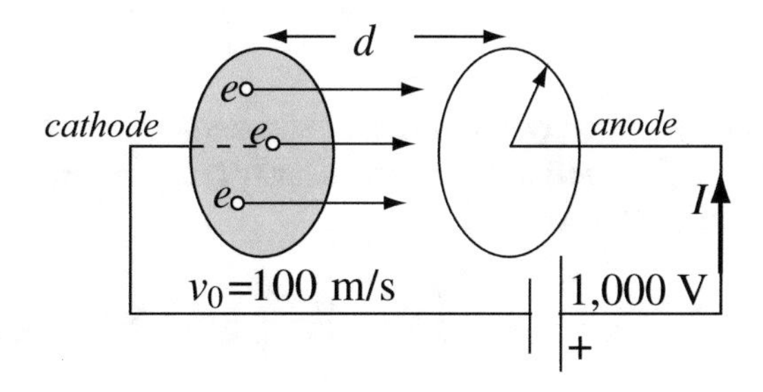

Figure 7.2 A photovoltaic tube and the external current it produces in a circuit

Solution: The current density is calculated from the total current which remains constant through the tube. From the current density, we calculate the volume charge density at the cathode using **Eq. (7.7)**. To calculate the charge density at the anode, we must first calculate the velocity of electrons at the anode and then use **Eq. (7.7)** again.

(a) The current density is

$$J = \frac{I}{S} = \frac{0.01}{\pi a^2} = \frac{0.01}{\pi \times (0.01)^2} = 31.83 \quad [\mathrm{A/m^2}].$$

From **Eq. (7.7)**, the charge density next to the cathode is

$$\rho_v = \frac{J}{v} = \frac{31.83}{100} = 0.3183 \quad [\text{C/m}^3]$$

This charge density is equivalent to 1.989×10^{18} electrons/m^3.

(b) To calculate the velocity at the anode, we must first calculate the acceleration of the charges. This is calculated from the electrostatic force, $\mathbf{F} = Q\mathbf{E}$. In this case, the electric field intensity is the potential difference divided by the distance between the anode and cathode:

$$E = \frac{V}{d} = \frac{1{,}000}{0.02} = 50{,}000 \quad [\text{V/m}]$$

The magnitude of the force is

$$F = QE = 1.6 \times 10^{-19} \times 50{,}000 = 8.0 \times 10^{-15} \quad [\text{N}]$$

The acceleration is calculated from $\mathbf{F} = m\mathbf{a}$, where m is the mass of the electron:

$$a = \frac{F}{m} = \frac{8.0 \times 10^{-15}}{9.109 \times 10^{-31}} = 8.782 \times 10^{15} \quad [\text{m/s}^2]$$

To calculate the final velocity, we use the equations of motion:

$$v_t = v_0 + at = 100 + 8.782 \times 10^{15} t \quad [\text{m/s}]$$

$$d = v_0 t + \frac{at^2}{2} = 100t + \frac{8.782 \times 10^{15} t^2}{2} = 0.02 \quad [\text{m}]$$

where v_t is the final velocity of charges. Solving first for t from the second equation and substituting in the first, we get a velocity of

$$v_t = 18{,}742{,}465 \quad [\text{m/s}]$$

This is a rather high velocity, but it is only about 6% of the speed of light.

Now, using **Eq. (7.7)** and since the current density remains constant (otherwise there will be continuous accumulation of charge in the tube), we get

$$\rho_v = \frac{J}{v} = \frac{31.83}{18{,}742{,}465} = 1.698 \times 10^{-6} \quad [\text{C/m}^3]$$

Since the velocity has increased and the current remains the same, the charge density has decreased considerably. This is typical of vacuum tubes in which the electron density is highest at the cathode.

Note: Here, we assumed that the space charge between anode and cathode has no effect. This is only true for very high anode voltages (the condition is called temperature-limited operation). At normal operating voltages, the flow of charges through the tube is space-charge limited and the space charge dominates the flow. The charge current relation through the tube is then described by the Langmuir–Child law and this differs from the flow described in this example.

7.3.3 Conduction Current and Conduction Current Density

Conduction current occurs in materials which allow the exchange of electrons. The way this current is normally understood is that free (or valence) electrons in the outer shells of atoms move randomly from atom to atom. In the process, the material itself remains neutral. This process occurs in materials that we call conductors. It is very pronounced in metals but can occur in almost any material, to one degree or another. Now, if an electric field is applied on a conductor, the electric field will exert a force $\mathbf{F} = q\mathbf{E}$ [N] on the free electrons and the free electrons will be accelerated by the field. In their motion, electrons encounter atoms, collide, and are slowed down. The end result is that in spite of the acceleration due to the electric field, the electrons travel at a fixed average velocity, called ***drift velocity***. The number of collisions determines this velocity, and the larger the number of collisions, the lower the velocity or, in simpler terms, the higher the resistance to motion of electrons and, therefore, to current. To obtain a quantitative expression for current in a conductor due to an externally applied electric field, we use **Figure 7.3**. Because the drift velocity is proportional to the applied electric field intensity, we can write

$$\mathbf{v} \propto \mathbf{E} \tag{7.10}$$

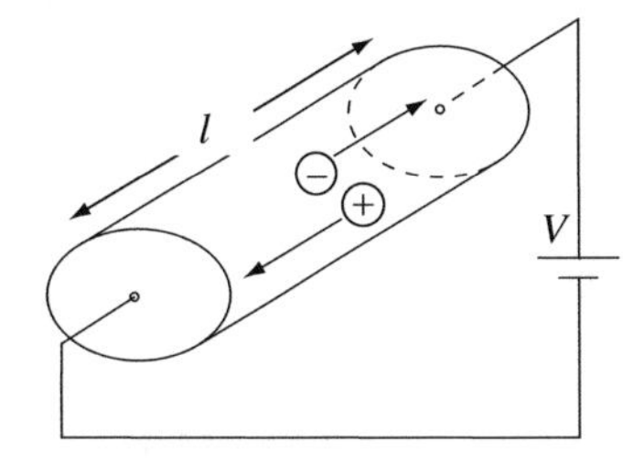

Figure 7.3 Model of conduction current in a material

The number of available charges (either electrons or electrons and protons) depends on the material itself and the current density is proportional to charge velocity. Thus we can also write using **Eq. (7.6)**:

$$\mathbf{J} \propto \mathbf{v} \tag{7.11}$$

The proportionality factor is a material property that can be determined experimentally for any material. Since **J** is proportional to velocity and the velocity is also proportional to the electric field intensity, the relation between current density and the applied electric field intensity in a conductor may be written as

$$\boxed{\mathbf{J} = \sigma\mathbf{E} \qquad [\mathrm{A/m^2}]} \tag{7.12}$$

where σ is called the ***conductivity*** of the material. The units of conductivity are determined from **Eq. (7.12)** by writing $\sigma = J/E$. This gives units of [A/m^2]/[V/m] = [A/(V·m)]. This unit is designated as the ***siemens per meter*** [S/m] and the ***siemens***[2] [S] is equal to the ampere/volt [A/V]. This unit is the reciprocal of the ohm ([S] = [1/Ω]).

Conductivity of materials depends on many parameters, including material composition, atomic structure, material working, and temperature. For this reason, conductivity is determined experimentally for each material at given conditions. Conductivity of materials is usually given at a known temperature (normally at 20° C). Some important engineering materials and their conductivities are listed in **Tables 7.1** and **7.2**. The distinction between conductors and insulators is based on the relative magnitude of conductivity. **Table 7.2** lists some special materials, including biological materials.

[2] After Werner von Siemens (1816–1892), inventor and industrialist (founder of the Siemens Company). He is best known for the invention of the self-excited dynamo in 1867 which, at the time, was the first dynamo that did not require permanent magnets for its operation and became one of the most important electromagnetic devices. His brother, Sir William Siemens, was also an inventor, but his work was mostly in mechanics. A third Siemens was Sir William's nephew, who was one of the contenders for the invention of the light bulb.

Table 7.1 Conductivities of some conductors and insulators

Conductors		Insulators	
Material	Conductivity [S/m]	Material	Conductivity [S/m]
Silver (Ag)	6.1×10^{7}	Distilled water	1.0×10^{-4}
Copper (Cu)	5.7×10^{7}	Ferrite	1.0×10^{-3}
Gold (Au)	4.1×10^{7}	Water	1.0×10^{-3}
Aluminum (Al)	3.5×10^{7}	Bakelite	1.0×10^{-9}
Tungsten (W)	1.8×10^{7}	Glass	1.0×10^{-12}
Brass	1.1×10^{7}	Rubber	1.0×10^{-13}
Iron (Fe)	1.0×10^{7}	Mica	1.0×10^{-15}
Nichrome alloy	1.0×10^{6}	Quartz	1.0×10^{-17}
Mercury (Hg)	1.0×10^{6}	Diamond	1.0×10^{-16}
Graphite	1.0×10^{6}	Wood	1.0×10^{-8}
Carbon (C)	3.0×10^{5}	Polystyrene	1.0×10^{-16}

Table 7.2 Conductivities of semiconductors, biological materials, water, soil, and concrete

Semiconductors		Biological and other materials	
Material	Conductivity [S/m]	Material	Conductivity [S/m]
Gallium antimony (GaSb)	0.25	Blood	0.7
Indium phosphide (InP)	1.25	Body tissue	1.0×10^{-2}
Indium arsenide (InAs)	0.34	Skin (dry)	2.0×10^{-4}
Cadmium telluride (CdTe)	1×10^{-12}	Bone	1.0×10^{-2}
Aluminum arsenide (AlAs)	0.1	Fat	0.04
Aluminum phosphide (AlP)	1,000	Seawater	4.0
Germanium (Ge)	2.2	Distilled water	1.0×10^{-4}
Silicon (pure) (Si)	4.35×10^{-6}	Soil	1.0×10^{-2}
Gallium arsenide (GaAs)	$< 1 \times 10^{-8}$	Concrete	0.06

Equation (7.12) is known as Ohm's law. This looks different than Ohm's law as normally used in circuits but only because it is written in its point form (see **Section 7.4**). Several properties of current are evident from this relation:

(1) Conduction current density is in the direction of the electric field intensity. Note that this is in spite of the fact that electrons flow in the direction opposite that of the electric field. You will recall (from circuit theory) that this is by convention: Positive current is assumed to flow from the positive pole of the source. In fact, electrons flow in the opposite direction; therefore, the current is normally a flow of negative charges, flowing in the direction opposite that of the electric field intensity.

(2) Conductivity is a property of the material. Each material is characterized by a conductivity which may be very high or very low.

(3) Ohm's law in **Eq. (7.12)** is a material constitutive relation: It defines the relation between current density and electric field intensity. We will make considerable use of this relation in the future. However, it should be remembered at all times that this relation is only correct for conduction current densities.

Note the lower conductivities of semiconductors. This conductivity can be controlled over wide ranges by inclusion of other materials in the semiconductor (a process called doping, which is fundamental to production of semiconductor devices).

Example 7.3 Velocity of Free Electrons (Drift Velocity) Copper has a free electron density of about 8×10^{28} electrons/m^3. A copper wire, 3 mm in diameter, carries a steady current of 50 A. Calculate:

(a) The velocity of electrons (drift velocity) in the wire.
(b) The maximum electron velocity in copper if the maximum current density allowable is 10^9 A/m^2.

Solution: The current density is calculated first from current and dimensions of the wire. The velocity is then calculated from **Eq. (7.6)**.

(a) The current density in the conductor is

$$J = \frac{I}{S} = \frac{50}{\pi r^2} = \frac{50}{\pi \times \left(1.5 \times 10^{-3}\right)^2} = 7.073553 \times 10^6 \quad \left[\mathrm{A/m^2}\right].$$

The velocity of the electrons is [from **Eq. (7.6)**]

$$v = \frac{J}{Nq} = \frac{7.073553 \times 10^6}{1.6 \times 10^{-19} \times 8 \times 10^{28}} = 5.5262 \times 10^{-4} \quad [\mathrm{m/s}].$$

This is a mere 0.55 mm/s or about 2 m/h. Not very fast.

(b) Assuming the maximum current density is 10^9 A/m^2, and the charge density remains constant (which it does), we get

$$v = \frac{J}{Nq} = \frac{1 \times 10^9}{1.6 \times 10^{-19} \times 8 \times 10^{28}} = 7.8125 \times 10^{-2} \quad [\mathrm{m/s}].$$

This is still very slow: only about 281 m/h.

Note: This velocity is exceedingly small. The current occurs not from electrons moving through the wires but, rather, from electrons moving from one atom to the next in very short jumps, but all of them jumping at almost the same time. Thus, when an electron jumps from atom to atom at one end of the wire, another electron jumps from atom to atom at the other end. To us, it appears as if an electron has moved through the wire when, in fact, an electron may never travel the length of the wire. Because of these simultaneous jumps, the electrons drift in a sort of "musical chairs" motion, and a signal can propagate through the wire at speeds close to the speed of light.

7.4 Ohm's Law

Ohm's law[3] as defined in **Eq. (7.12)** can be easily cast into the more familiar form known from circuit theory. The latter is known as

$$V = IR \quad [\mathrm{V}] \tag{7.13}$$

where R is the resistance (say, of a length of wire), I the current through the wire, and V the voltage across the wire. The relation between the point form of Ohm's law and the circuit theory form is shown in **Figure 7.4**. In **Figure 7.4a**, a segment

[3] Named after Georg Simon Ohm (1787–1854), who experimented with conductors and determined the relation between voltage and current in conductors. This relation is Ohm's law, which he proposed in 1827. In addition, the unit of resistance is named after him. Ohm had a hard time with his now famous law; initially, it was dismissed as wrong. It took many years for the scientific community to accept it. Acceptance eventually came from the Royal Society in England and eventually spread to his own country, Germany.

of conductor of length l is connected to a source. This source produces a uniform electric field inside the material and therefore a uniform current density **J**. The relation between the electric field intensity and current density is given in **Eq. (7.12)**. **Figure 7.4b** shows the equivalent circuit of the same conductor. All that is needed now is to calculate the current in the conductor and the voltage across it. Using the definition of current in **Eq. (7.9)** as the flux of current density gives

$$I = \int_s \mathbf{J} \cdot d\mathbf{s} = \int_s \sigma \mathbf{E} \cdot d\mathbf{s} \quad [\text{A}] \tag{7.14}$$

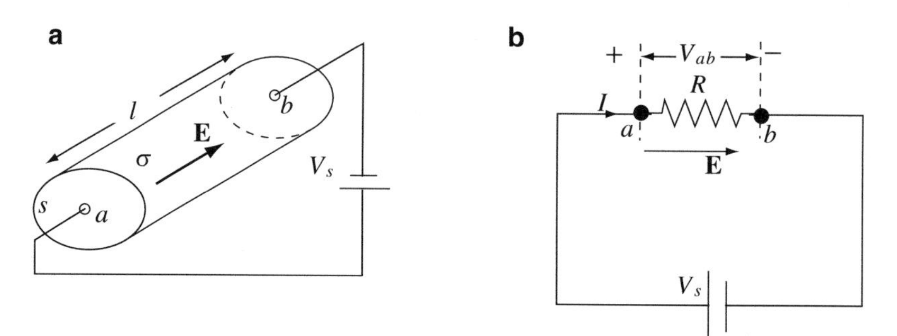

Figure 7.4 (**a**) Field representation of Ohm's law. (**b**) Equivalent circuit representation of Ohm's law

The voltage across the conductor is calculated using the definition of potential and potential difference as

$$V_{ab} = -\int_b^a \mathbf{E} \cdot d\mathbf{l} \quad [\text{V}] \tag{7.15}$$

Substitution of these two relations in **Eq. (7.13)** gives

$$-\int_b^a \mathbf{E} \cdot d\mathbf{l} = R \int_s \sigma \mathbf{E} \cdot d\mathbf{s} \quad \rightarrow \quad R = \left(-\int_b^a \mathbf{E} \cdot d\mathbf{l}\right) \Big/ \left(\int_s \sigma \mathbf{E} \cdot d\mathbf{s}\right) \tag{7.16}$$

The latter relation gives the generalized resistance in any conductor. Note that the electric field intensity **E** and conductivity σ can vary along the conductor. The proper (positive) sign is obtained if the line integral is performed against the electric field intensity (V_{ab} is positive if I is positive). Alternatively, taking the absolute values of the two integrals, we get

$$\boxed{R = \frac{V_{ab}}{I} = \left| \frac{-\int_b^a \mathbf{E} \cdot d\mathbf{l}}{\int_s \sigma \mathbf{E} \cdot d\mathbf{s}} \right| \quad [\Omega]} \tag{7.17}$$

In all these relations, it is important to remember that the surface integral is over that area through which the current flows and it may be an open surface integral (such as the cross section of a conductor) or a closed surface if the current flows throughout the surface of the conductor. In the particular case of constant conductivity and uniform electric field intensity (the assumptions associated with circuit theory), the integrals in **Eq. (7.17)** can be evaluated in general terms; that is,

$$R = \frac{V_{ab}}{I} = \frac{El}{\sigma ES} = \frac{l}{\sigma S} \quad [\Omega] \tag{7.18}$$

This relation for resistance is well known as the resistance of a uniform wire of length l with constant conductivity σ and cross-sectional area S. The quantity $1/\sigma$ is the resistivity of the wire (often denoted ρ and given in $\Omega{\cdot}\text{m}$).We will retain the form in **Eq. (7.17)** as a more general form and will not use resistivity, to avoid confusion.

Example 7.4 Application: Resistance and Current Density in a Fuse The fuse, in its various forms, is one of the most important, and often unappreciated, components in circuits and devices. In general terms, fuses rely on melting (or fusing) of a piece of conductor when the current in the fuse exceeds a given value.

A fuse is made of copper in the shape shown in **Figure 7.5**. The two large sections are intended to clamp the fuse, whereas the narrow section is the actual fuse. Copper can carry a safe current density of 10^8 A/m^2. Above this current, copper will melt (because of heat generated by the current). The thickness of the copper sheet is $t = 0.1$ mm:

(a) Design the width c of the fusing section so that it will break at currents above $I = 20$ A. This design is typical of automotive fuses (although pure copper is almost never used because of oxidation problems).
(b) Calculate the total resistance of the fuse for the dimensions given in **Figure 7.5** and the width found in **(a)**.

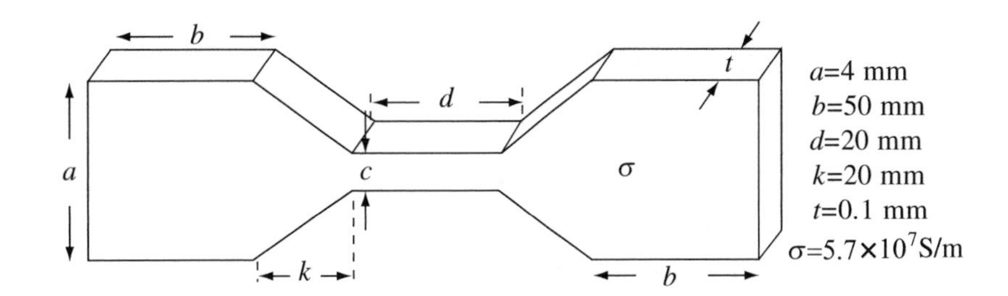

Figure 7.5 A simple fuse. The middle narrow section melts when current exceeds rated value

Solution: Here, we need to calculate the dimension c shown in **Figure 7.5**. The total resistance is that of the narrow strip, the two flat flanges, and the two trapezoidal sections connecting the two:

(a) The maximum current through the fuse is 20 A. With the maximum given current density, we have

$$I = JS = 20 = 1 \times 10^8 \times 0.0001 \times c \quad \rightarrow \quad c = 0.002 \quad [\text{m}]$$

The fuse strip must be 2 mm wide.

(b) We will calculate the resistance of each section separately and then sum them as series elements. The resistance of the middle narrow section ($L_1 = d = 0.02$ m, $S_1 = 0.0001 \times 0.002$ m^2 in **Figure 7.6a**) is

$$R_1 = \frac{L_1}{\sigma S_1} = \frac{0.02}{5.7 \times 10^7 \times 0.0001 \times 0.002} = 1.75439 \times 10^{-3} \quad [\Omega]$$

The resistance of one of the two rectangular end sections ($L_2 = b = 0.05$ m, $S_2 = 0.0001 \times 0.004$ m^2 in **Figure 7.6b**) is

$$R_2 = \frac{L_2}{\sigma S_2} = \frac{0.05}{5.7 \times 10^7 \times 0.0001 \times 0.004} = 2.19298 \times 10^{-3} \quad [\Omega]$$

To calculate the resistance of one of the two trapezoidal sections connecting the two ends with the middle section, we use **Figure 7.6c**. The section at position x has a cross-sectional area S and length dx and, therefore, a resistance dR_3. The length y_1 is

$$y_1 = \frac{0.002 - 0.001}{0.02} x = 0.05x \qquad [\text{m}]$$

Thus, the cross-sectional area of the conductor at position x is

$$S(x) = 2 \times (0.002 - y_1) \times 0.0001 = (0.004 - 0.1x) \times 0.0001 \qquad [\text{m}^2]$$

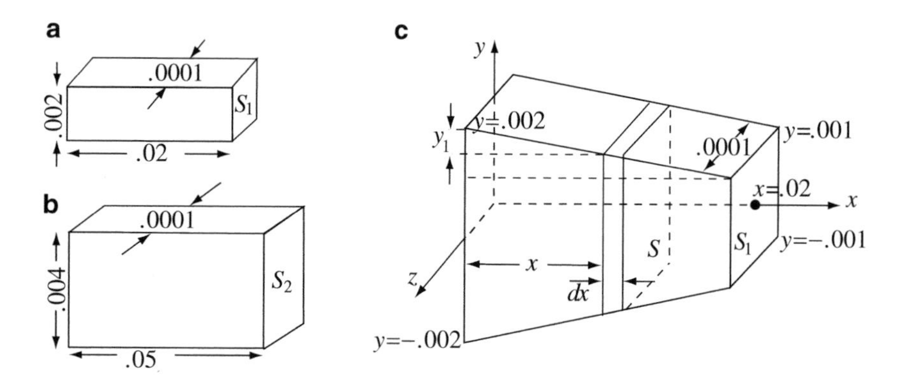

Figure 7.6 Sections of a fuse. (**a**) Fusing section. (**b**) Clamping section. (**c**) Transition section showing the method of calculation of resistance

The resistance of the differential section is

$$dR_3 = \frac{dx}{\sigma S(x)} = \frac{dx}{5.7 \times 10^7 \times 0.0001 \times (0.004 - 0.1x)} \quad [\Omega]$$

Integrating this over the length from $x = 0$ to $x = 0.02$, we get

$$R_3 = \frac{1}{5.7 \times 10^7 \times 0.0001} \int_{x=0}^{x=0.02} \frac{dx}{(0.004 - 0.1x)} = -1.7544 \times 10^{-4} \times \ln(0.004 - 0.1x)\Big|_{x=0}^{x=0.02}$$

$$= -1.7544 \times 10^{-4} \times (-6.2146 + 5.52146) = 1.216 \times 10^{-4} \quad [\Omega].$$

The total resistance is

$$R = R_1 + 2R_2 + 2R_3 = 1.7544 \times 10^{-3} + 4.38596 \times 10^{-3} + 2.432 \times 10^{-4} = 6.3836 \times 10^{-3} \quad [\Omega].$$

Exercise 7.1 Because of shortage of copper, it was decided to make the fuse in **Example 7.4** out of iron (not a very good idea because of corrosion). If the same dimensions, including the width calculated in **Example 7.4a**, are used and the current density at which the fuse melts is half the current density in copper, calculate:

(**a**) The current rating of the fuse.
(**b**) The resistance of the fuse.

Answer (**a**) $I = 10$ A, (**b**) $R = 0.03638\ \Omega$.

Example 7.5 Application: Resistive Position Sensor; Resistances in Series and Parallel A resistive position sensor is built of three identical flat bars of graphite, $L = 0.15$ m long, $w = 0.02$ m wide, and $d = 0.01$ m thick. One bar slides between two stationary bars as shown in **Figure 7.7a**. The resistance measured between the two ends is a measure of the position x. The conductivity of graphite is $\sigma = 10^4$ S/m:

(**a**) Calculate the resistance of the sensor when fully closed and when fully extended.
(**b**) Plot the calibration curve of the sensor.

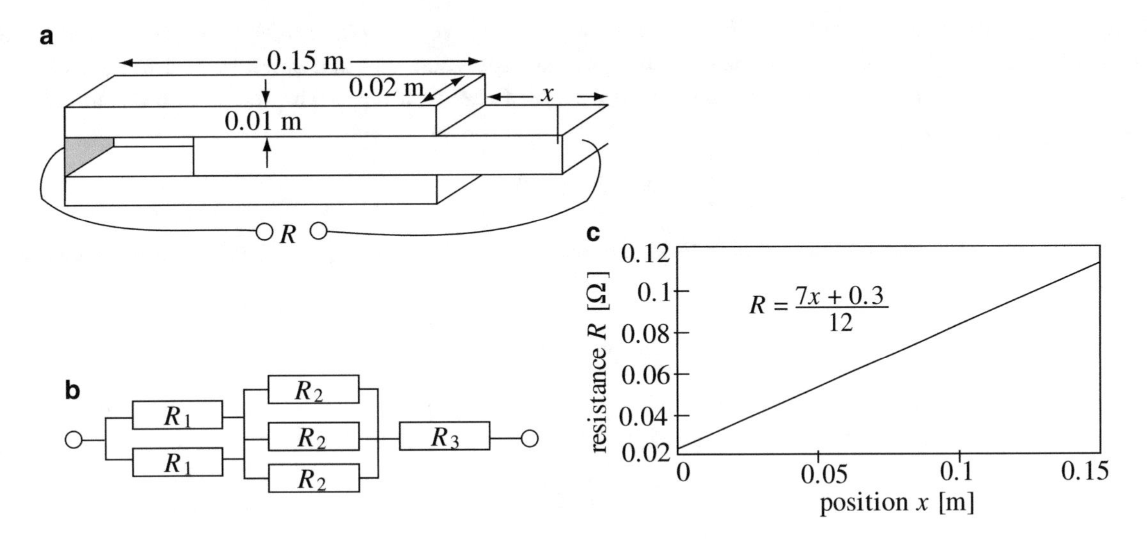

Figure 7.7 (**a**) Configuration of a resistive position sensor. (**b**) Equivalent circuit of the sensor in (**a**). (**c**) Calibration curve for the sensor

Solution: The sensor can be viewed as being made of three sections: the part of the moving bar (of length x) that sticks out, the section where the three bars overlap (of length $0.15 - x$), and the section of the two stationary bars to the left of the overlapping area (also of length x). Each has a resistance which varies with x and the total resistance is the sum of the three resistances.

(**a**) An equivalent circuit in terms of resistances is shown in **Figure 7.7b**. The resistances are as follows:

$$R_1 = \frac{x}{\sigma S}, \quad R_2 = \frac{L - x}{\sigma S}, \quad R_3 = \frac{x}{\sigma S} \quad [\Omega]$$

where $S = wd$ is the cross-sectional area of each bar. The two resistances marked as R_1 are in parallel, as are the three resistances marked as R_2. Thus, the total resistance of the device is

$$R = \frac{R_1}{2} + \frac{R_2}{3} + R_3 = \frac{x}{2\sigma wd} + \frac{L - x}{3\sigma wd} + \frac{x}{\sigma wd} = \frac{7x + 2L}{6\sigma wd} \quad [\Omega]$$

Maximum resistance occurs when $x = 0.15$ m (sensor fully extended):

$$R_{max} = \frac{7 \times 0.15 + 0.3}{6 \times 10^4 \times 0.02 \times 0.01} = \frac{7 \times 0.15 + 0.3}{6 \times 10^4 \times 0.02 \times 0.01} = 0.1125 \quad [\Omega]$$

Minimum resistance occurs with the sensor closed ($x = 0$):

$$R_{min} = \frac{0.3}{6 \times 10^4 \times 0.02 \times 0.01} = 0.025 \quad [\Omega].$$

(**b**) A plot of the resistance versus displacement x is shown in **Figure 7.7c**. Note that the plot is linear and its range is sufficiently large for meaningful measurements of position. In effect this device is a linear potentiometer.

7.5 Power Dissipation and Joule's Law

Resistance to flow of current means that energy must be dissipated. More exactly, when charges move (due to the force exerted by the electric field), electrons collide with atoms in the material, as was discussed at the introduction to this chapter. This collision is accompanied by the loss of some of the energy of the electron. This energy must be absorbed by the atom, increasing its thermal energy. The heat generated can, if sufficiently high, melt the material. In an incandescent light bulb, the current flowing through the filament heats it up to a sufficiently high temperature to emit light. Further heating (such as caused by increasing the voltage across the bulb) may result in melting the filament and destroying the bulb.

Although we already have an expression for the power dissipated in a resistor from circuit theory, it is useful to start from the basic definition of power since we seek here a more general expression, one that is not subject to the assumptions of circuit theory. To do so, we recall that power is the rate of change of energy or the scalar product of force and velocity. The latter is more convenient here:

$$P = \mathbf{F} \cdot \mathbf{v} \quad [\mathrm{W}] \tag{7.19}$$

The unit of power is the ***watt*** [W] or [N·m/s]. From the force on a single point charge $\mathbf{F} = q\mathbf{E}$, we get the power due to this charge as

$$p = q\mathbf{E} \cdot \mathbf{v} \quad [\mathrm{W}] \tag{7.20}$$

If we now take an element of volume dv and wish to calculate the total power in this volume, we argue that there are N charges per unit volume and, therefore, Ndv charges in the element of volume dv. Thus,

$$dP = Npdv = Nq\mathbf{E} \cdot \mathbf{v}dv \quad [\mathrm{W}] \tag{7.21}$$

Now, from **Eq. (7.6)**, $Nq\mathbf{v}$ is the current density $\mathbf{J}$ and we get

$$dP = \mathbf{E} \cdot \mathbf{J}dv \quad [\mathrm{W}] \tag{7.22}$$

If the power in a given volume is required, this expression is integrated over the volume:

$$\boxed{P = \int_v \mathbf{E} \cdot \mathbf{J}dv \quad [\mathrm{W}]} \tag{7.23}$$

where $\mathbf{E}$ is the electric field intensity inside the volume. This is Joule's law in integral form. The law gives the total power dissipated in a volume in which a current density exists. We can also define a volume power density from **Eq. (7.22)** as

$$\boxed{\frac{dP}{dv} = \mathbf{E} \cdot \mathbf{J} \quad [\mathrm{W/m^3}]} \tag{7.24}$$

This is the point form of Joule's law, which gives the dissipated power density at any point in space. Using the constitutive relation $\mathbf{J} = \sigma\mathbf{E}$, we can also write

$$\boxed{p_d = \frac{dP}{dv} = \mathbf{E} \cdot \mathbf{J} = \sigma\mathbf{E} \cdot \mathbf{E} = \frac{J^2}{\sigma} = \sigma E^2 \quad [\mathrm{W/m^3}]} \tag{7.25}$$

We can now link between the point form and the more common form ($P = VI$, or $P = I^2R$) using **Eq. (7.23)**. First, we write $dv = dlds$. With this, the integral over the volume in **Eq. (7.23)** becomes a double integral, one over the surface s and one over length l:

$$P = \int_v \mathbf{E} \cdot \mathbf{J}dv = \left(\int_l \mathbf{E} \cdot d\mathbf{l}\right)\left(\int_s \mathbf{J} \cdot d\mathbf{s}\right) = VI \quad [\mathrm{W}] \tag{7.26}$$

The separation of the two integrals in **Eq. (7.26)** is allowable because $\mathbf{E}$ is independent of s, and $\mathbf{J}$ is independent of l, and the separation above merely represents pulling a constant in front of an integral sign.

Example 7.6 Application: Power Dissipated in a Fuse—Power Density A fuse is made of copper in the shape shown in **Figure 7.5**. The fuse is designed to carry a current of up to 20 A. Assume that the fuse "blows" at 20 A and that the fusing current density in copper is 10^8 A/m^2:

(a) What is the electric field intensity in the narrow section of the fuse?
(b) What is the total power dissipated in the narrow section at maximum current?

Solution: We have already calculated the current densities as well as resistances of the various sections of the fuse in **Example 7.4**. The easiest method to calculate the power dissipated in the fuse is to use **Eq. (7.26)** and view the fuse as a circuit element. However, we will use **Eq. (7.25)** to calculate the power density in the fuse. The maximum power density occurs in the narrow section as required for fusing:

(a) In the narrow section, $\sigma = 5.7 \times 10^7$ S/m and $J = 10^8$ A/m^2. Thus, the electric field intensity E is

$$E = \frac{J}{\sigma} = \frac{10^8}{5.7 \times 10^7} = 1.754 \quad \left[\frac{\text{V}}{\text{m}}\right]$$

This is a small electric field intensity, as we would expect from a good conductor.

(b) Using **Eq. (7.25)**, we can write the power density in copper as

$$p_d = \frac{J^2}{\sigma} = \frac{10^{16}}{5.7 \times 10^7} = 1.754 \times 10^8 \quad \left[\frac{\text{W}}{\text{m}^3}\right]$$

This seems to be a large power density. It is, but the amount of copper in a meter cube is also very large.

The total power dissipated in the narrow section is the power density multiplied by the volume of the section. From the dimensions in **Figure 7.5**, we get

$$P = p_d v = p_d tcd = 1.754 \times 10^8 \times 0.0001 \times 0.002 \times 0.02 = 0.7018 \quad [\text{W}]$$

where $t = 0.1$ mm is the thickness of the fuse, $c = 2$ mm is the width, and $d = 20$ mm is the length of the fusing strip as calculated in **Example 7.4**. The total power dissipated in the fuse is relatively small although not negligible. Fuses do get hot, sometimes very hot.

Note: There is additional power dissipated in the wider sections of the fuse, but they do not cause fusing and, therefore, were neglected in this calculation.

Example 7.7 Application: Power Density in Semiconductors A diode is made as shown in **Figure 7.8**. The junction between the two electrodes is $d = 0.1$ mm long and has a cross-sectional area $S = 1$ mm^2. The diode is made of silicon which can safely dissipate up to 0.5 W/mm^3. If the forward bias potential on the junction is 0.7 V, calculate:

(a) The maximum current the diode can carry without damage.
(b) The conductivity of the silicon at the current value calculated in **(a)**.

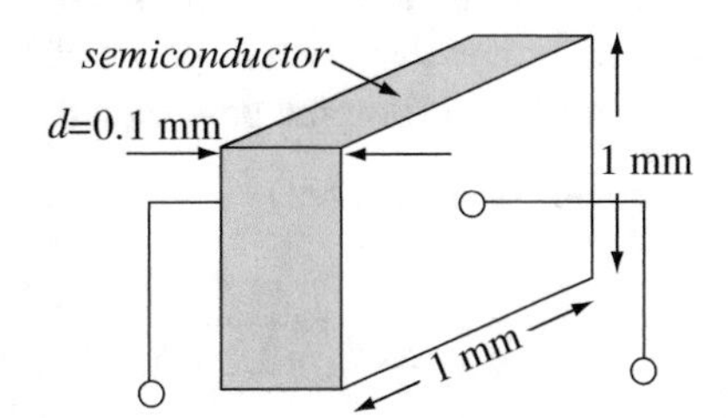

Figure 7.8 Simplified model of a semiconductor diode

Solution: The current is calculated from the maximum power dissipated assuming it is uniform throughout the junction. Conductivity is calculated from the current density and power using Joule's law in **Eq. (7.25)**.

(a) Total maximum power dissipated in the silicon is

$$P = p_d v = p_d S d = 0.5 \times 1 \times 0.1 = 0.05 \quad [\mathrm{W}]$$

The maximum current in the diode is

$$I = \frac{P}{V} = \frac{0.05}{0.7} = 0.0714 \quad [\mathrm{A}].$$

(b) From Joule's law [see **Eq. (7.25)**]

$$p_d = \frac{J^2}{\sigma} = \frac{I^2}{S^2 \sigma} \quad \rightarrow \quad \sigma = \frac{I^2}{S^2 p_d} \quad \left[\frac{\mathrm{S}}{\mathrm{m}}\right]$$

With the values given ($p_d = 0.5$ W/mm^3 $= 5 \times 10^8$ W/m^3, $S = 1$ mm^2 $= 10^{-6}$ m^2), the conductivity is

$$\sigma = \frac{0.0714^2}{10^{-12} \times 5 \times 10^8} = 10.2 \quad [\mathrm{S/m}].$$

Note: the conductivity calculated here is an effective conductivity at the given operating point. The relation between current in and voltage on the diode is nonlinear hence conductivity depends on the operating point.

Example 7.8 Application: Power Dissipated in a Lightning Strike—Power Density A lightning strike generates relatively large power during short periods of time. This translates in high power being dissipated in the form of heat and sound as well as electromagnetic energy. The latter is usually detectable in radios in the form of noise. This example gives an idea of the magnitudes involved.

A lightning strike lasts 250 μs and carries a current of $I = 25{,}000$ A in a channel $r = 3$ mm in radius and $l = 2$ km long. Assume that the conductivity in the channel is $\sigma = 2.2 \times 10^5$ S/m (typical in plasmas), that all energy is dissipated within the channel and calculate:

(a) The total power dissipated in the lightning strike.
(b) The power density in the channel in W/m^3.
(c) The electric field intensity in the channel.
(d) The total energy dissipated (energy is power integrated over time).
(e) The length of time the energy dissipated in a single lightning strike could power the United States if this energy could be harnessed. The total average power needed to power the entire United States is 4.2×10^{12} W.

Solution: The resistance of the channel is calculated as for any cylindrical conductor. From this, the total power is calculated using Joule's law. The power density and power dissipated are calculated from the total power. The electric field intensity in the channel is calculated from the definition of current density.

(a) First, we calculate the resistance of the channel:

$$R = \frac{l}{\sigma S} = \frac{l}{\sigma \pi r^2} = \frac{2{,}000}{2.2 \times 10^5 \times \pi \times (0.003)^2} = 321.53 \qquad [\Omega]$$

The power dissipated in the channel is

$$P = VI = I^2 R = \left(25{,}000\right)^2 \times 321.53 = 2 \times 10^{11} \quad [\mathrm{W}].$$

(b) The power density in the channel is constant and is the total power divided by the volume of the channel:

$$p_d = \frac{P}{v} = \frac{P}{\pi r^2 l} = \frac{2 \times 10^{11}}{\pi \times (0.003)^2 \times 2{,}000} = 3.537 \times 10^{12} \quad [\mathrm{W/m^3}].$$

(c) The electric field intensity in the channel is

$$E = \frac{J}{\sigma} = \frac{I}{\sigma \pi r^2} = \frac{25{,}000}{2.2 \times 10^5 \times \pi \times (0.003)^2} = 4019 \quad [\mathrm{V/m}].$$

Note that this electric field intensity is much lower than the electric field intensity required for breakdown (about 3×10^6 V/m). The breakdown electric field intensity (required to exceed the dielectric strength) only exists before breakdown occurs. When the current flows, the electric field intensity is much smaller.

(d) The energy dissipated is the power multiplied by the time length of the strike:

$$W = Pt = 2 \times 10^{11} \times 250 \times 10^{-6} = 5.0 \times 10^7 \quad [\mathrm{J}].$$

(e) The consumption is 4.2×10^{12} J/s. This means that a lightning strike of this magnitude could power the United States for 13.6 μs ($t = 5.7 \times 10^7 / 4.2 \times 10^{12} = 13.57 \times 10^{-6}$ s).

7.6 The Continuity Equation and Kirchhoff's Current Law

The idea of conservation of charge was discussed briefly at the beginning of this chapter. We can accept this idea based on experience and experiment. A more important engineering question is: What are the consequences of this law? To answer this, we must first put the law into an explicit form called the ***continuity equation.***

Consider **Figure 7.9a**, which shows an isolated volume, charged with a charge density ρ_v. No charge leaves or enters the volume and, therefore, the charge is conserved. This is a trivial example of conservation of charge. Now, consider **Figure 7.9b** where we connect the volume through a wire and allow the charge to flow through the wire to some other body (not shown). Because charges flow, there is a current I in the wire. At the same time, the charge in volume v diminishes. The charge that flows out of the volume over a time dt (i.e., the charge that flows through the wire) is dQ. We now define a quantity called the time rate of decrease of charge in the volume v as $-dQ/dt$. From these considerations, we conclude that the rate of decrease in the charge in volume v must equal the current out of the volume:

$$\boxed{I = -\frac{dQ}{dt} \quad [\mathrm{A}]} \tag{7.27}$$

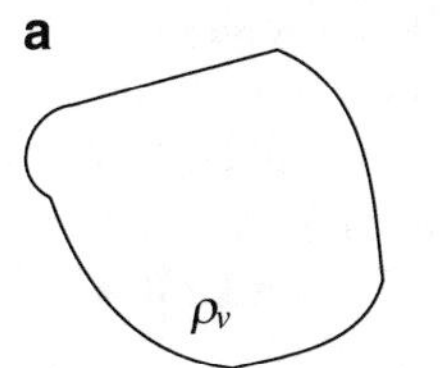

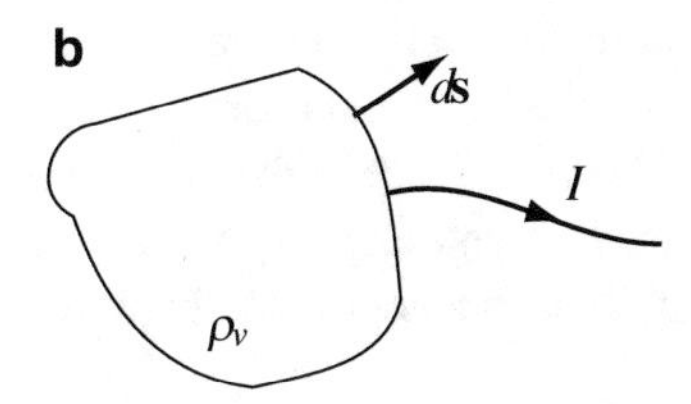

Figure 7.9 (**a**) Isolated volume charge. (**b**) The rate of decrease of charge in volume v is the current out of the volume

This, in fact, is the basic definition of current. The current is considered to be positive because it flows through the surface of the volume v in the direction of $d\mathbf{s}$. This definition of the current is similar to our initial definition of current in **Section 7.3.2** except for the negative sign.

The total charge in volume v is

$$Q = \int_v \rho_v dv \quad [\mathrm{C}] \tag{7.28}$$

Substituting this in **Eq. (7.27)**, we get

$$I = -\frac{d}{dt}\int_v \rho_v dv = \int_v \left(-\frac{\partial \rho_v}{\partial t}\right) dv \qquad [\text{A}] \tag{7.29}$$

Note that moving the derivative into the integral requires the use of the partial derivative since the charge density may also be a function of space. From **Eq. (7.9)**, we write the current I flowing out of the volume v in terms of the current density $\mathbf{J}$ that flows through the surface enclosing the volume v:

$$\oint_s \mathbf{J} \cdot d\mathbf{s} = \int_v \left(-\frac{\partial \rho_v}{\partial t}\right) dv \tag{7.30}$$

The surface s is a closed surface (encloses the volume v). Thus, we can apply the divergence theorem to the left-hand side to convert the closed surface integral to a volume integral. This will allow us to equate the integrands and get a simpler expression. **Equation (7.30)** now becomes

$$\oint_s \mathbf{J} \cdot d\mathbf{s} = \int_v \nabla \cdot \mathbf{J}\, dv = \int_v \left(-\frac{\partial \rho_v}{\partial t}\right) dv \tag{7.31}$$

Since both integrals are over the same volume, we get

$$\boxed{\nabla \cdot \mathrm{J} = -\frac{\partial \rho_v}{\partial t}} \tag{7.32}$$

This is the general form of the continuity equation. It states that

"*the divergence of the current density is equal to the negative rate of change of volume charge density anywhere in space.*"

This expression holds at any point in space and is not limited to conductors.

In the particular case of steady currents, the charge density ρ_v does not vary with time. The rate of change of charge with time is zero and, therefore, the charge depleted from volume v must be replenished constantly to maintain a steady current. Under this condition, we can write the continuity equation as

$$\boxed{\nabla \cdot \mathbf{J} = 0 \quad \text{if}\, \rho_v = \text{constant}} \tag{7.33}$$

Equation (7.32) indicates that a time-dependent charge density causes nonzero divergence and therefore a nonsteady (time-dependent) current density and electric field intensity. Although we normally associate current density with conductors, we must insist that the relation in **Eq. (7.32)** does not require conductivity, only that a time rate of change of charge exists. Later, starting with **Chapter 11**, we will place considerable importance on this relation because it will allow us to stipulate a current density in nonconducting materials based solely on the rate of change of charge density.

Equation (7.33) means that a steady current must flow in closed circuits. It cannot end in a point (such as a point charge) because, then, the divergence at that point would not be zero, invalidating the requirement of steady current. This also means that the total current entering any volume must equal the total current leaving this volume. A direct consequence of the divergence-free requirement of steady currents is Kirchhoff's current law.

7.6.1 Kirchhoff's Current Law

Kirchhoff's current law (KCL) for steady currents is obtained from **Eq. (7.30)** by setting the time rate of change of charge density to zero. We get the following expression:

$$\boxed{\oint_s \mathbf{J} \cdot d\mathbf{s} = 0} \tag{7.34}$$

which states that for any volume, the total net current flowing through the surface of the volume is zero. **Figure 7.10** shows this concept. **Figure 7.10a** shows a number of general channels (or conductors) leading currents in and out of the volume v through the surface s. The total net current through the surface is zero, as indicated from **Eq. (7.34)**, provided the currents are steady currents. **Figure 7.10b** shows a single current distribution and an arbitrary volume v through which the current flows. Again, the net steady current flowing into the volume is zero. This must be so since, otherwise, there must be an accumulation of charges or depletion of charges in the volume v.

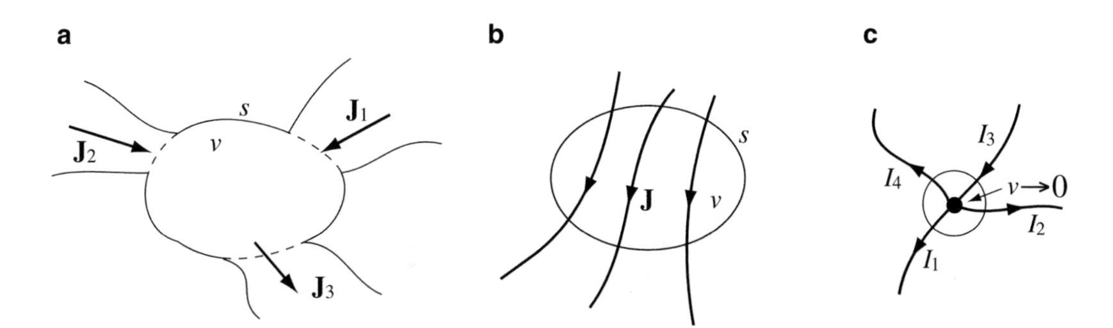

Figure 7.10 Kirchhoff's current law. (**a**) For steady currents, the total net flow out of any volume is zero. (**b**) The steady current through a volume is zero. (**c**) The circuit equivalent of Kirchhoff's law

Kirchhoff's law in circuit theory is a simplified form of this relation. It assumes a number of conductors, each very thin, converging to a node as shown in **Figure 7.10c**. In this case, we write the discrete form of **Eq. (7.34)** as

$$\boxed{\sum_{i=1}^{n} I_i = 0} \tag{7.35}$$

To obtain this relation, we must assume infinitely thin conductors and that the volume in **Eq. (7.34)** tends to zero. By doing so, we simplify the calculation but also restrict ourselves to thin wires and uniform currents in each wire.

Perhaps we should ask ourselves what happens with Kirchhoff's current law if currents are not steady? In the context of circuit analysis, it makes no difference since the volume of a junction is assumed to be zero and there can be no accumulation of charges in zero volume. Thus, Kirchhoff's current law as given in **Eq. (7.35)** holds true for nonsteady currents as well. However, in the more general case of **Eq. (7.30)**, the sum of all currents entering a volume is not zero if currents are not steady. There can be accumulation or depletion of charges. Therefore, Kirchhoff's law, as used in circuits, is not valid and we must use **Eq. (7.30)**, which is a more general form than Kirchhoff's current law.

Example 7.9 Divergence of Current Density A current source is connected to a resistive grid as shown in **Figure 7.11a**. What is the total net current flowing into the volume shown?

Figure 7.11 (**a**) A resistive grid and an arbitrary volume enclosing part of the grid. (**b**) Calculation of the net current in volume v

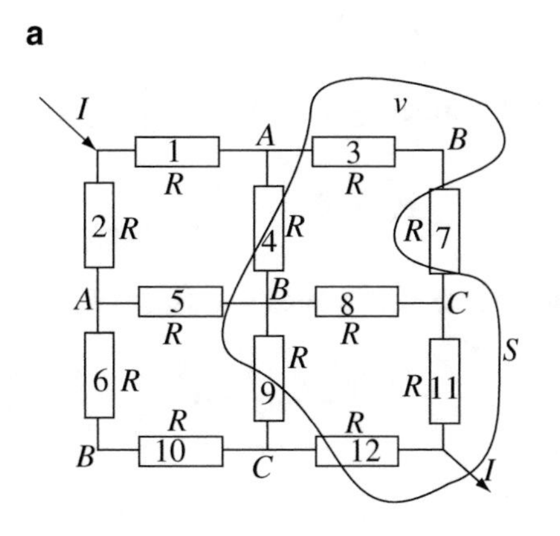

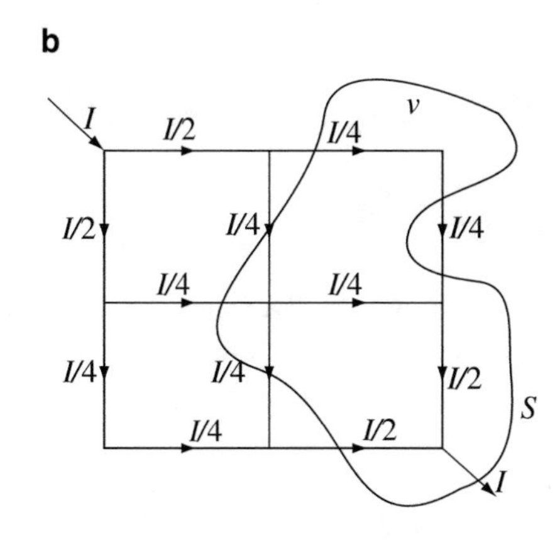

Solution: To calculate the total net current, we could either invoke the divergence-free condition for steady currents which states that the total net current in any volume is zero, or we could calculate the current entering and leaving the volume and show that it is zero. The latter approach is shown here.

First, the currents in the various resistors are calculated. We note that points A are at the same potential, points B are at the same potential, and points C are at the same potential. Connecting the corresponding points (all A points together, all B points together, and all C points together) gives the currents in resistors 1, 2, 11, and 12 as $I/2$ and in all other resistors as $I/4$. The currents entering and leaving the volume are shown in **Figure 7.11b**. The total net current is the sum of the currents in the eight branches the volume cuts:

$$I_t = \frac{I}{4} + \frac{I}{4} + \frac{I}{4} + \frac{I}{4} + \frac{I}{2} - I - \frac{I}{4} - \frac{I}{4} = 0$$

Example 7.10 Divergence of Current Density and Relaxation: The Field Approach In **Chapter 4**, we stated that charges in conductors under static conditions are distributed on the surface of the conductor in such a way that the potential energy in the system is minimum. We also said that if charges are placed in the interior of a conductor, they will move to the surface. This motion of charges constitutes a current. Since the current ceases once the charges are static on the surface, the current is a transient current. To see how this happens, what the time constants involved are, and as an example of nonzero divergence of current density, consider the following example:

A chunk of silicon made in the form of a sphere of radius 100 mm is given. The conductivity of silicon is 4×10^{-4} S/m, its relative permittivity is 12, and both are constant. Suppose that by some means a uniform volume charge density $\rho_0 = 10^{-6}$ C/m^3 is placed in the interior of the sphere at $t = 0$. Calculate:

(a) The current produced by the charges as they move to the surface.
(b) The time constant of the charge decay in the silicon.
(c) The divergence of the current density during the transient.

Solution: At time $t = 0$, the charges start moving toward the surface. The charge density must satisfy the continuity equation at all times. In addition, the current in the conducting material must satisfy Ohm's law. From these two relations, we can calculate the time-dependent charge density in the material. The current in the material is calculated from the current density and its divergence is calculated from the continuity equation:

(a) From the continuity equation [**Eq. (7.32)**]

$$\nabla \cdot \mathbf{J} = -\frac{\partial \rho_v}{\partial t}$$

From Ohm's law [**Eq. (7.12)**]

$$\mathbf{J} = \sigma \mathbf{E} \qquad [\text{A/m}^2]$$

Taking the divergence on both sides and combining the two equations gives

$$\sigma(\nabla \cdot \mathbf{E}) = -\frac{\partial \rho_v}{\partial t}$$

From the divergence postulate for the electric field intensity in **Eq. (4.6)**, $\nabla \cdot \mathbf{E} = \rho_v/\varepsilon$. Therefore,

$$\frac{\partial \rho_v}{\partial t} + \frac{\sigma}{\varepsilon}\rho_v = 0$$

The solution to this differential equation is

$$\rho_v(t) = \rho_v(t=0)e^{-t\sigma/\varepsilon} \quad [\text{C/m}^3]$$

The charge density at $t = 0$ is $\rho_0 = 10^{-6}$ C/m^3, $\varepsilon = 12\varepsilon_0 = 1.06 \times 10^{-10}$ F/m, and $\sigma = 4 \times 10^{-4}$ S/m. With these, the charge density at any time is

$$\rho_v(t) = 10^{-6}e^{-4\times10^{-4}t/1.06\times10^{-10}} = 10^{-6}e^{-t/2.65\times10^{-7}} \quad [\text{C/m}^3]$$

To calculate the current density, we argue as follows: at a radius R, the total current crossing the surface defined by the sphere of radius R ($R < d$) is

$$I(R,t) = -\frac{dQ_R}{dt} = -\frac{4\pi R^3}{3}\frac{d}{dt}(\rho_v(t)) = 15.8R^3e^{-t/2.65\times10^{-7}} \quad [\text{A}]$$

Note that the current depends on the location and increases with the radius. Therefore, it is not constant in space or time.

(b) The time constant of the charge decay is ε/σ. This time constant depends on material alone and is called the ***relaxation time***. A long time constant (poor conductors) means charges take longer to "relax" or to reach the surface. A short time (good conductors) means the charges quickly reach their static state. In the present case,

$$\tau = \frac{\varepsilon}{\sigma} = 2.65 \times 10^{-7} \quad [\text{s}]$$

This is a rather long relaxation time. In good conductors, the relaxation time is much shorter (see **Exercise 7.3**).

(c) The divergence of the current density is

$$\nabla \cdot \mathbf{J} = -\frac{\partial \rho_v}{\partial t} = -\frac{d}{dt}\left(\rho_0 e^{-t\sigma/\varepsilon}\right) = \frac{\rho_0\sigma}{\varepsilon}e^{-t\sigma/\varepsilon} = 3.77e^{-t/26.52\times10^{-6}}$$

The divergence of the current density is clearly not zero, but decays with time.

Exercise 7.2

(a) Draw an arbitrary closed contour on **Figure 7.11** and show that it yields zero net flow of current.

(b) Suppose a capacitor is inserted into the circuit instead of any of the resistors and the current is switched on. Draw a closed contour for which the total net current is not necessarily zero.

Answer **(b)** Any contour passing between the plates of the capacitor.

Exercise 7.3 Calculate the relaxation time in the following materials:

(a) Copper: $\sigma = 5.7 \times 10^7$ [S/m], $\varepsilon = \varepsilon_0$ [F/m].

(b) Quartz: $\sigma = 1.0 \times 10^{-17}$ [S/m], $\varepsilon = 80\varepsilon_0$ [F/m].

(c) A superconducting material ($\sigma \to \infty$).

Answer

(a) $\tau = 1.553 \times 10^{-19}$ s.

(b) $\tau = 7.083 \times 10^6$ s (almost 82 days).

(c) $\tau \to 0$.

7.7 Current Density as a Field

In this chapter, we introduced a new vector field: the current density **J**. Since the current density is directly related to the electric field intensity [through **Eq. (7.12)**], it may seem that the current density need not be discussed separately as a vector field. However, we must recall that the postulates used in **Chapters 3** and **4** for the electrostatic field did not allow movement of charges. The postulates $\nabla \times \mathbf{E} = 0$ and $\nabla \cdot \mathbf{E} = \rho/\varepsilon$ were only correct for stationary charges. Under the conditions given here, namely, those required for steady currents, we must use the properties of the current density as indicated above. To completely specify the current density, we must provide an expression for its divergence and for its curl (this is required by the Helmholtz theorem, which states that a vector field is uniquely specified if its divergence and curl are known; see **Chapter 2, Section 2.5.1**). The divergence of the current density for steady currents is zero. The curl of **J** can be obtained from Ohm's law in **Eq. (7.12)** and the curl of the electric field intensity **E** as given in **Eq. (4.8)**. These give

$$\nabla \times \mathbf{E} = 0 \quad \rightarrow \quad \nabla \times \frac{\mathbf{J}}{\sigma} = 0 \tag{7.36}$$

We can integrate **Eq. (7.36)** over a surface s. However, we know from **Chapter 4** [**Eq. (4.9)**] that the integral form of **Eq. (7.36)** is

$$\oint_C \mathbf{E} \cdot d\mathbf{l} = 0 \tag{7.37}$$

Substituting $\mathbf{E} = \mathbf{J}/\sigma$ gives the integral form of **Eq. (7.36)** as

$$\boxed{\oint_C \frac{\mathbf{J} \cdot d\mathbf{l}}{\sigma} = 0} \tag{7.38}$$

That is, the steady current density is conservative (closed contour integral of **J** is zero). The current density is an ***irrotational, solenoidal*** field.

To be noted here is that the divergence of the current density is different than that for the static electric field intensity. In **Section 4.3** we saw that because the divergence of the electric field intensity at any point is nonzero (and may be positive or negative), the electric charge can exist in isolation.

On the other hand, the divergence of the steady current density is identically zero. This means that there is no ***source*** of the steady electric current density in the same way as for the electric field. In fact, it means that the lines of current or ***stream lines*** (similar to ***lines of electric field***) must close on themselves. Only then can the divergence be zero everywhere. **Figure 7.12** shows this. As long as all currents close on themselves, any volume we choose will have a net zero current flowing into the volume, indicating zero divergence. It is useful to have this picture in mind for any divergence-free field since in every case that a field is divergence-free, it will be represented by some form of closed line (although the line represents different quantities for different fields).

To summarize, the postulates of the steady current are as follows:

In differential form,

$$\boxed{\nabla \cdot \mathbf{J} = 0 \quad \text{and} \quad \nabla \times \frac{\mathbf{J}}{\sigma} = 0} \tag{7.39}$$

In integral form,

$$\boxed{\oint_s \mathbf{J} \cdot d\mathbf{s} = 0 \quad \text{and} \quad \oint_c \frac{\mathbf{J} \cdot d\mathbf{l}}{\sigma} = 0} \tag{7.40}$$

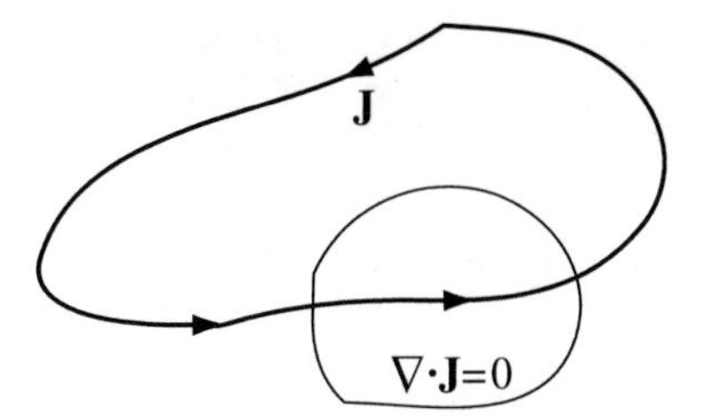

Figure 7.12 Implications of the divergence-free nature of steady current density. Lines of current density must close on themselves, and the net flow of current into an arbitrary volume is zero

Example 7.11 A conducting cylindrical rod of radius $a = 10$ mm has a conductivity which varies with the radial distance from the center of the conductor as: $\sigma = 10^7(1 - r^2)$ [S/m]. The conductor is 20 m long and its ends are connected to a voltage source of 1 V:

(a) Calculate the total current passing through any cross section of the conductor.
(b) A 1 m long section of the conductor is inspected. What is the total net current flowing into this section?

Solution: The current is found from the voltage and resistance, which must be calculated first. Since the current density is divergence-free, the total current entering any volume equals the current leaving the volume. Thus, the net current entering the volume in **(b)** is zero.

(a) First, we need to calculate the resistance of the cylinder. To do so, we use a section in the form of a hollow cylinder of radius r and thickness dr, as in **Figure 7.13**. The resistance of this cylinder when connections are made to its ends is

$$dR = \frac{l}{\sigma ds} = \frac{20}{2\pi r \times 10^7(1 - r^2)dr} \quad [\Omega]$$

where $ds = 2\pi r dr$ is the cross-sectional area of the elemental hollow cylinder in **Figure 7.13**. The resistance cannot be calculated directly because, in effect, each subsequent elemental cylinder connects in parallel to the one shown.

However, if we use the admittance, we can write

$$dY = \frac{1}{dR} = \frac{\sigma ds}{l} = \frac{10^7(1 - r^2)2\pi r dr}{20} = 10^6\pi\left(r - r^3\right)dr \quad [1/\Omega]$$

and the total admittance is

$$Y = 10^6\pi\int_{r=0}^{r=0.01}\left(r - r^3\right)dr = 10^6\pi\left[\frac{r^2}{2} - \frac{r^4}{4}\right]_0^{0.01} \approx 50\pi \quad [1/\Omega]$$

The resistance of the cylinder is $0.02/\pi$ [Ω]. The current in the conductor is

$$I = \frac{V}{R} = \frac{1 \times \pi}{0.02} = 157 \quad [\text{A}].$$

(b) This current exists in any cross section of the conductor. Therefore, the net current entering any cross section equals that leaving it. The total net current flowing into the section is zero.

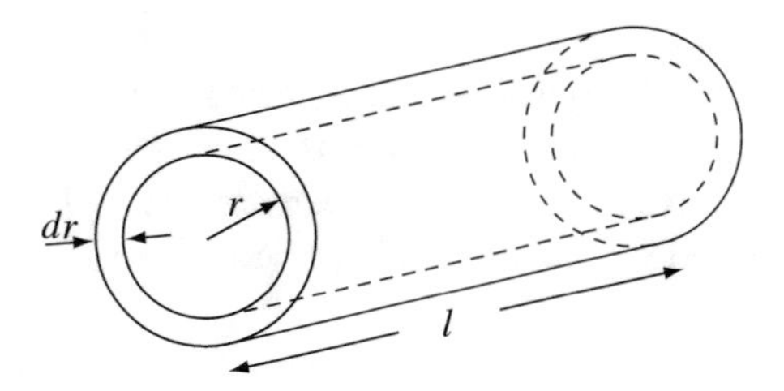

Figure 7.13 Method of calculation of the resistance in **Example 7.11**

Exercise 7.4 A conductor carries a current density given as $\mathbf{J} = \hat{\mathbf{x}}xy^2 + \hat{\mathbf{y}}5$ [A/m^2]. Is this a steady current?

Answer No, since the divergence of **J** is nonzero, $\nabla \cdot \mathbf{J} = y^2 \neq 0$.

7.7.1 Sources of Steady Currents

A steady current implies a constant flow of charge from a source. How can a source be built that will do exactly that? Perhaps the most obvious source of charge is a charged capacitor. Connection of a charged capacitor to a circuit allows the flow of charge through the circuit. Because the amount of charge available in a capacitor is finite and the voltage across the capacitor if $V = Q/C$, the current diminishes with time until the capacitor has been discharged. A very simple example is shown in **Figure 7.14a**. From circuit theory you will recall that the voltage across the capacitor and hence the current in the resistor is exponential with a time constant RC (**Figure 7.14b**). This is obviously a time-dependent rather than a steady current.

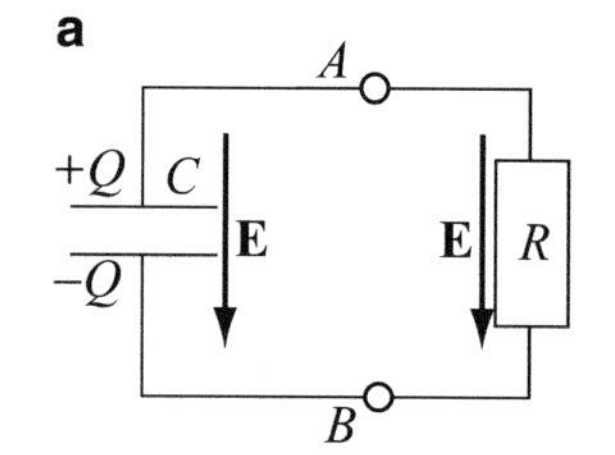

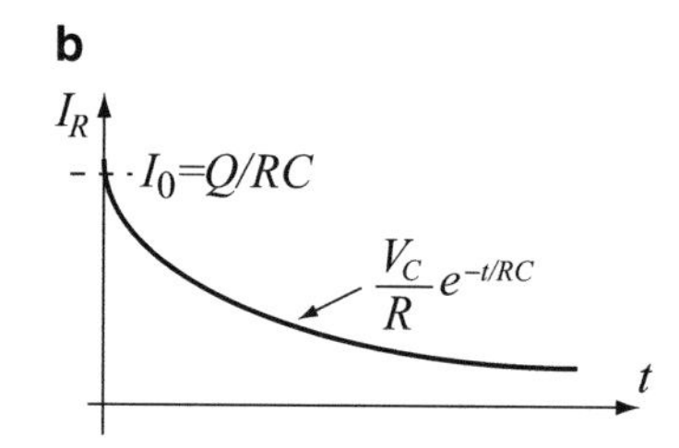

Figure 7.14 A time-dependent source. (**a**) A charged capacitor connected to a resistor. (**b**) Current through the resistor. Note the direction of the electric field intensity in the capacitor and in the resistor

What is then a source of steady current? Any mechanism such as an electrostatic generator that can produce charges at a given rate can serve as a steady source. As long as the generator operates, there will be a steady supply of charge. If we do not remove more charge than can be replenished in a given time, the current will remain constant. To maintain this constant current, we must expend energy (for example, in driving the generator). This mechanical energy is converted into electric energy which, in turn, is dissipated in the load.

Another common source is chemical. In a battery, the separation of charges occurs between each electrode and the electrolyte inside the battery. The potential difference between the electrodes is maintained by a constant charge on each electrode. As charge is removed (in the form of current), the potential difference remains constant, which, in turn, means that the charge on the electrodes remains constant. The current in the circuit in **Figure 7.15a** is therefore a steady current. Thus, a battery is essentially a capacitor, in which charges are generated (always in pairs) by chemical action. When the chemicals are spent, replenishment of charges cannot occur and the battery is "dead."

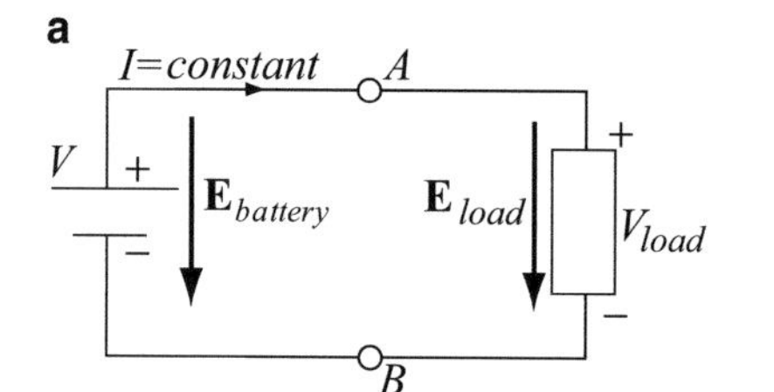

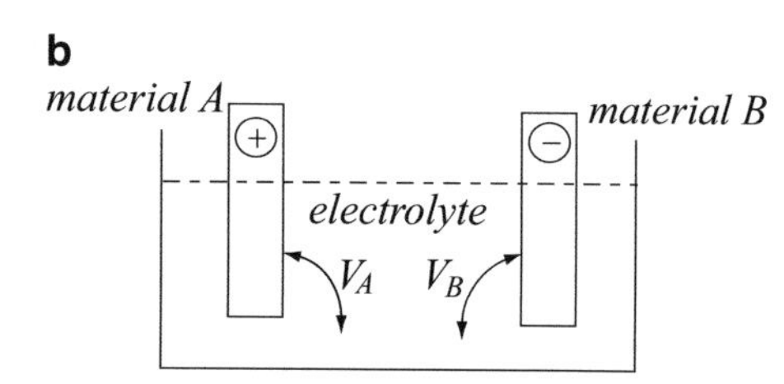

Figure 7.15 Source of a steady current. (**a**) A battery supplies a constant current through a load. (**b**) The principle of the electrolytic cell. If $V_A > V_B$, $V_{out} = V_A - V_B$ and polarity is as shown

In a dry cell, the electrolyte is a gel; a wet battery contains a liquid such as sulfuric acid. The output potential of the battery is called an ***electromotive force*** (**emf**) and is considered to be constant. The use of electromotive force indicates it as a source, although it is not a force but a potential difference. We will use the name electromotive force to indicate any potential source and not only in batteries. The name voltage will be used for potential drop and emf will be used for sources.

The electromotive force of a battery depends on the contact potentials between each electrode and the electrolyte. The two potentials shown in **Figure 7.15b** must be different; otherwise, the electromotive force would be zero. The larger the difference between V_A and V_B, the larger the electromotive force of the cell. For example, in a regular dry cell, the electromotive force is 1.5 V, whereas in a wet cell (such as used in car batteries), the electromotive force of a cell is 2 V. For a 12 V battery, we must connect six such cells in series.[4]

7.7.2 Kirchhoff's Voltage Law

To define Kirchhoff's voltage law (KVL) in general terms, consider **Figure 7.16**. It consists of a battery and a load. A current density **J** flows in the load. Because of the polarity of the source, the internal electric field intensity **E** and the external electric field intensities are in opposite directions. The current density in the circuit is $\mathbf{J} = \sigma\mathbf{E}$, where σ is the conductivity of the load material. From **Eq. (7.37)**, we can write

$$\oint_c \mathbf{E} \cdot d\mathbf{l} = \int_{L_b} \mathbf{E}_b \cdot d\mathbf{l} + \int_{L_l} \mathbf{E}_l \cdot d\mathbf{l} = 0 \tag{7.41}$$

where L_b is the path inside the battery and L_l is the path in the load. Since $E_l = -E_b$ and, $E_l = J/\sigma$ we can write

$$V_b = \int_{L_l} \frac{\mathbf{J}}{\sigma} \cdot d\mathbf{l} \quad [\mathrm{V}] \tag{7.42}$$

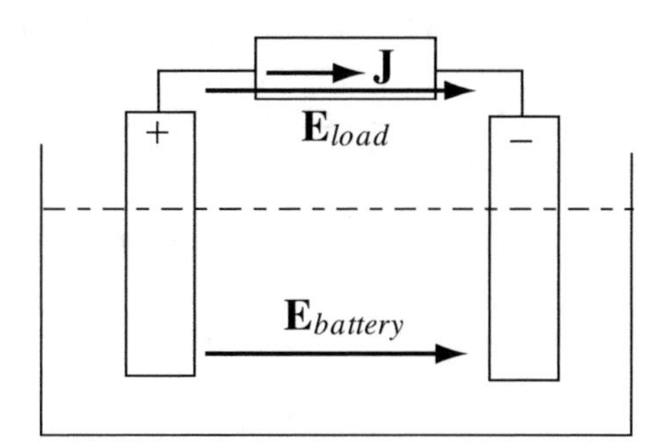

Figure 7.16 Relation between the internal and external electric field intensities in a battery

The left-hand side of **Eq. (7.42)** is the electromotive force of the battery, whereas the right-hand side is the voltage drop across the load, made of all voltage drops in the circuit, including possibly a voltage drop within the battery due to internal resistance. We can, in fact, say that the battery is an ideal source and the right-hand side is the closed contour integral of the electric field intensity (all voltage drops in the circuit). Thus,

$$\boxed{V_b = \oint_c \frac{\mathbf{J}}{\sigma} \cdot d\mathbf{l} \quad [\mathrm{V}]} \tag{7.43}$$

[4] The invention of the dry and wet cell is due to Alessandro Volta. In 1800, he announced his invention, which consisted of disks of silver and disks of zinc, separated by cloth impregnated with a saline or acidic solution. This became known as a Volta pile. He also experimented with wet copper and zinc cells, a combination used to this day in many dry cells although in many cells, the copper has been replaced with carbon. For this purpose he used cups with saline solution in which he dipped zinc and copper plates to form cells which, when connected in series, produced the desired electromotive force. To this arrangement of cups and electrodes he gave the name "crown of cups." Based on these principles, many other cells were developed, the difference being mostly in the materials used and construction.

This is Kirchhoff's voltage law. If we write the current as $I = JS$ (where S is the cross section of the load resistor), we get

$$V_b = \oint_c \frac{I}{\sigma S} dl \quad [\mathrm{V}] \tag{7.44}$$

The left-hand side is the total electromotive force in the circuit. The right-hand side is immediately recognized as the resistance multiplied by current [see, for example, **Eq. (7.18)**], that is, the sum of all potential drops. Thus, in the context of circuits, the relation in **Eq. (7.44)** can be written as

$$\sum V_b = \sum RI \tag{7.45}$$

The use of summation instead of integration simply means that a circuit always contains a finite, well-defined number of voltage drops and electromotive forces (emfs).

7.8 Interface Conditions for Current Density

When a current flows through an interface between two materials with different conductivities, such as when two different materials are soldered together or butted against each other, there will be an effect on the current at the interface. This effect, as in the case of the electric field intensity at the interface between two dielectrics, is obtained from the application of the postulates on the interface. The conditions on the tangential and normal components are obtained, which can then be used to define the behavior of the current density (and therefore current) on both sides of the interface and at the interface itself. In calculating the interface conditions, we follow steps similar to those in **Section 4.6**.

However, because the steady current density is directly related to the electric field intensity, the interface conditions for the electric current density may be defined based on those for the electric field intensity. From **Chapter 4**, we recall that the interface conditions for the electric field intensity are

$$E_{1t} = E_{2t} \tag{7.46}$$

$$D_{1n} - D_{2n} = \rho_s \quad \text{or} \quad \varepsilon_1 E_{1n} - \varepsilon_2 E_{2n} = \rho_s \tag{7.47}$$

where we assumed that the relations between the electric field intensities are as shown in **Figure 7.17a**. Suppose now that a current density flows across the interface as shown in **Figure 7.17b**. The materials on both sides of the interface are conducting with conductivities σ_1 and σ_2. They also have a permittivity as indicated.

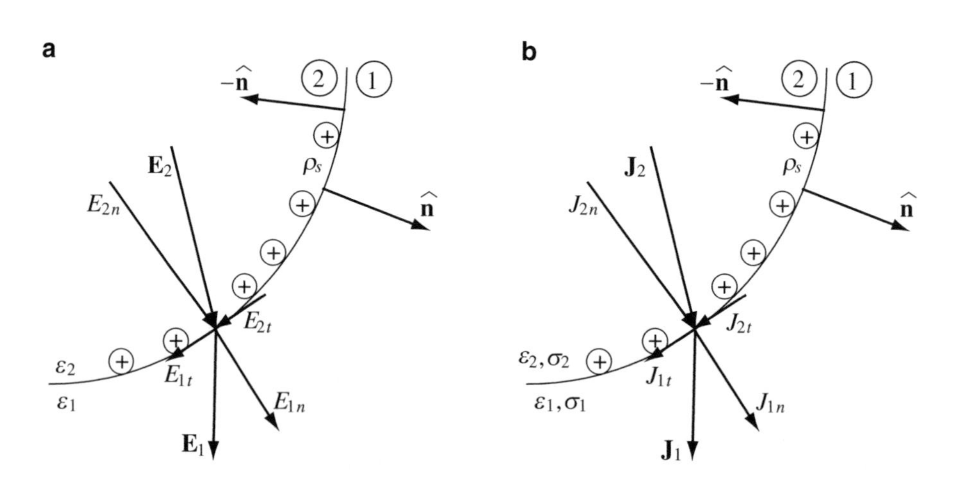

Figure 7.17 (**a**) Relations between the electric field intensities on two sides of an interface. (**b**) Relations between the current densities on two sides of an interface

Now, because $\mathbf{J} = \sigma\mathbf{E}$, we can write for the tangential components of the current density in **Eq. (7.46)**:

$$J_{1t} = \sigma_1 E_{1t}, \quad J_{2t} = \sigma_2 E_{2t} \quad \rightarrow \quad \frac{J_{1t}}{\sigma_1} = \frac{J_{2t}}{\sigma_2} \tag{7.48}$$

The current density may have two tangential components on the interface. **Equation (7.48)** applies to each component. Because the divergence of the current density must be zero ($\nabla \cdot \mathbf{J} = 0$), the total net current flowing through any section of the interface must be zero (Kirchhoff's current law). This requires that

$$J_{1n} = J_{2n} \tag{7.49}$$

The interface conditions for current density are summarized as

$$\boxed{\frac{J_{1t}}{\sigma_1} = \frac{J_{2t}}{\sigma_2} \quad \text{and} \quad J_{1n} = J_{2n}} \tag{7.50}$$

In addition, from **Eq. (7.47)**, we can write

$$\varepsilon_1 \frac{J_{1n}}{\sigma_1} - \varepsilon_2 \frac{J_{2n}}{\sigma_2} = \rho_s \quad \left[\frac{\text{C}}{\text{m}^2}\right] \tag{7.51}$$

The latter relation indicates that whenever the two conductors on the two sides of the interface have different conductivities and/or different permittivities, there is a surface charge density generated at the interface. Only if both $\sigma_1 = \sigma_2$ and $\varepsilon_1 = \varepsilon_2$ or if $\varepsilon_1/\sigma_1 = \varepsilon_2/\sigma_2$ will the surface charge density be zero.

Example 7.12 Example of Current at an Interface Consider a very large (infinite) conductor made of two materials as shown in **Figure 7.18**. The current density in conductor No. 1 is in the positive y direction and is equal to J [A/m^2]. Calculate the magnitude and direction of the electric field intensity in conductor No. 2 at the interface.

Figure 7.18

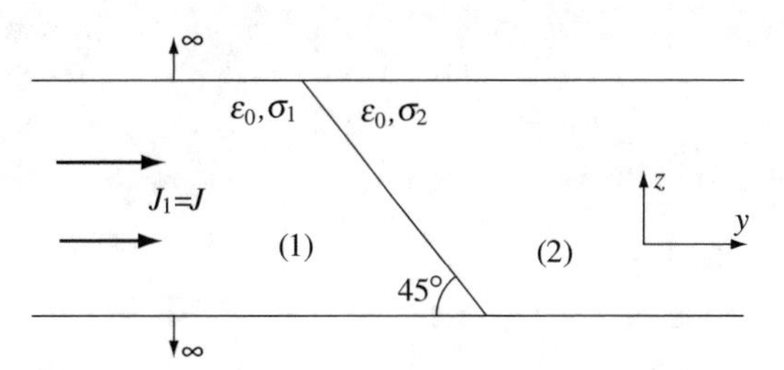

Solution: The current density has a normal and a tangential component to the interface. To find the current density in material (2), we impose the continuity condition on the two components (see **Figure 7.18**).

The tangential and normal components of the current density at the interface in material (1) are

$$J_{1n} = \frac{\sqrt{2}}{2}J, \quad J_{1t} = \frac{\sqrt{2}}{2}J \quad \left[\frac{\text{A}}{\text{m}^2}\right]$$

From **Eq. (7.48)**, we have for the tangential components

$$\frac{J_{1t}}{\sigma_1} = \frac{J_{2t}}{\sigma_2} \quad \rightarrow \quad J_{2t} = \frac{\sigma_2\sqrt{2}}{2\sigma_1}J \quad \left[\frac{\text{A}}{\text{m}^2}\right]$$

From the continuity of the normal components, we get

$$J_{2n} = J_{1n} = \frac{\sqrt{2}}{2}J \quad \left[\frac{\mathrm{A}}{\mathrm{m}^2}\right]$$

The components of the electric field intensity in material (2) at the interface are

$$E_{2n} = \frac{J_{2n}}{\sigma_2} = \frac{\sqrt{2}}{2\sigma_2}J, \quad E_{2t} = \frac{J_{2t}}{\sigma_2} = \frac{\sqrt{2}}{2\sigma_1}J \quad \left[\frac{\mathrm{V}}{\mathrm{m}}\right]$$

Taking the components in the y direction, we get

$$E_{2y} = E_{2n}\sin 45^\circ + E_{2t}\cos 45^\circ = \frac{J}{2}\left(\frac{1}{\sigma_2} + \frac{1}{\sigma_1}\right) \quad \left[\frac{\mathrm{V}}{\mathrm{m}}\right]$$

The electric field intensity in the z direction is

$$E_{2z} = E_{2n}\cos 45^\circ - E_{2t}\sin 45^\circ = \frac{J}{2}\left(\frac{1}{\sigma_2} - \frac{1}{\sigma_1}\right) \quad \left[\frac{\mathrm{V}}{\mathrm{m}}\right]$$

The electric field intensity in material (2) at the interface is

$$\mathbf{E}_2 = \hat{\mathbf{y}}\frac{J}{2}\left(\frac{1}{\sigma_2} + \frac{1}{\sigma_1}\right) + \hat{\mathbf{z}}\frac{J}{2}\left(\frac{1}{\sigma_2} - \frac{1}{\sigma_1}\right) \quad \left[\frac{\mathrm{V}}{\mathrm{m}}\right]$$

Note that if $\sigma_1 = \sigma_2$, the electric field intensity at the interface in material (2) has only a y component as expected.

Example 7.13 Application: Connection of Aluminum and Copper Wires The electrical code specifies that copper and aluminum wires should not be connected together in wiring systems unless special connectors are used. One reason is that aluminum tends to create a high-resistivity (oxide) layer at the contact. This layer can easily overheat under high loads and cause fires. Consider the following example.

In a wiring system in a house, a copper wire and aluminum wire are connected as shown in **Figure 7.19**. A thin layer of oxide has formed between the two conductors. The layer is $d = 0.0025$ mm thick and has the same diameter as the wires (2 mm). Aluminum oxide has very low conductivity, but because of contamination, the actual conductivity of the oxide layer is 10^3 S/m. Conductivity of aluminum is 3×10^7 S/m and that of copper is 5.7×10^7 S/m. A current of 12 A passes through the circuit. Calculate:

(a) The power density in aluminum, copper, and oxide layer. Compare.
(b) The total power dissipated in the aluminum oxide layer.

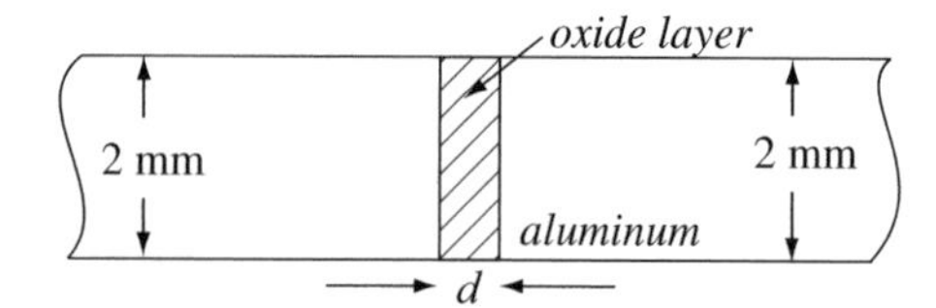

Figure 7.19 Connection of an aluminum wire and copper wire and the oxide layer between them

Solution: The current density in each segment is the same since the cross-sectional area of the wires and oxide layers is the same. From current density and conductivity, we calculate the power density in each segment.

(a) The current density in each segment is

$$J = \frac{I}{S} = \frac{I}{\pi r^2} = \frac{12}{\pi \times (0.001)^2} = 3.82 \times 10^6 \quad [\mathrm{A/m^2}]$$

From the conductivities of copper, aluminum oxide, and aluminum, the power densities are calculated from Joule's law **(Eq. (7.25))** as

$$p_d(\mathrm{cu}) = \frac{J^2}{\sigma_{cu}} = \frac{(3.82 \times 10^6)^2}{5.7 \times 10^7} = 2.56 \times 10^5 \quad [\mathrm{W/m^3}]$$

$$p_d(\mathrm{al}) = \frac{J^2}{\sigma_{al}} = \frac{(3.82 \times 10^6)^2}{3.5 \times 10^7} = 4.17 \times 10^5 \quad [\mathrm{W/m^3}]$$

$$p_d(\mathrm{oxide}) = \frac{J^2}{\sigma_{oxide}} = \frac{(3.82 \times 10^6)^2}{10^3} = 1.46 \times 10^{10} \quad [\mathrm{W/m^3}]$$

The dissipated power density in the oxide layer is much higher than in copper or aluminum.

(b) The power dissipated in the oxide layer is the power density multiplied by the volume of the layer:

$$P = p_d \times v = p_d \times \pi \times r^2 \times d = 1.46 \times 10^{10} \times \pi \times 10^{-6} \times 2.5 \times 10^{-6} = 0.115 \quad [\mathrm{W}]$$

This may seem small, but it is dissipated in a layer which is only 2.5 μm thick. An equivalent layer of copper only dissipates 2×10^{-6} W and an equivalent aluminum layer only dissipates 3.275×10^{-6} W. Clearly the concentrated power in the oxide layer can cause local heating and, therefore, the danger of fire.

Example 7.14 Interface Conditions in Poor Conductors Perfect conductors are defined by their conductivities and perfect dielectrics are completely defined by their permittivity. There are, however, materials that are in between: They are neither perfect dielectrics nor perfect conductors. Materials with relatively high conductivity may be characterized as poor conductors, whereas low-conductivity materials may be viewed as dielectrics with finite conductivity. These materials are also called lossy dielectrics. The following example shows that a surface charge density may be generated at the interface between two poor conductors (or lossy dielectrics) when current flows across the interface.

Two cylindrical, poorly conducting materials with cross-sectional area s [m^2] and with conductivities σ_1 and σ_2 [S/m] are connected to a current source as shown in **Figure 7.20**. Calculate the surface charge density on the interface between the two conductors. Assume the permittivities of the two conducting materials are equal to ε_0 [F/m].

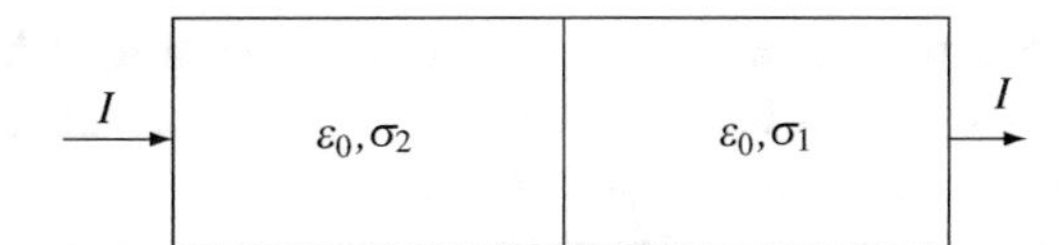

Figure 7.20 Interface between two poorly conducting materials

Solution: The electric field intensity is first calculated from the current density, which is the same in the two materials. Then, the electric flux density in each material is calculated and the continuity of the normal component of the electric flux density is used to calculate the charge density on the interface.

The current densities in the two conductors are

$$J_1 = J_2 = \frac{I}{s} \quad \left[\frac{\text{A}}{\text{m}^2}\right]$$

The electric field intensities are

$$E_1 = \frac{J_1}{\sigma_1} = \frac{I}{s\sigma_1}, \qquad E_2 = \frac{J_2}{\sigma_2} = \frac{I}{s\sigma_2} \quad \left[\frac{\text{V}}{\text{m}}\right]$$

The continuity condition on the electric flux density (which is normal to the interface) is

$$D_{1n} - D_{2n} = \rho_s \quad \rightarrow \quad \rho_s = \varepsilon_0(E_1 - E_2) \quad [\text{C/m}^2]$$

With the electric field intensities calculated, the surface charge density on the interface is

$$\rho_s = \frac{\varepsilon_0 I}{s}\left(\frac{1}{\sigma_1} - \frac{1}{\sigma_2}\right) \quad \left[\frac{\text{C}}{\text{m}^2}\right]$$

Note that if $\sigma_1 = \sigma_2$, the surface charge density is zero. This charge density does not affect the current (because the current is steady and $\nabla \cdot \mathbf{J} = 0$), but it affects the electric field intensity (see **Chapter 4**.)

7.9 Applications

Application: Resistors One of the simplest and common methods of producing resistors for electronics equipment is to use a relatively low-conductivity powder. The resistivity of the material is then controlled by the amount of powder in the resistor. In common resistors, carbon powder is added to a binder material and shaped to the right size. These resistors are reliable and inexpensive, but they tend to be noisy at high-resistance values. One of the reasons for using carbon for resistors is its very small temperature coefficient. Carbon film resistors are made as a film of carbon deposited onto a ceramic substrate. Resistance is controlled by cutting groves into the surface (mechanically or by lasers). These resistors are less noisy and more accurate, but have lower power dissipation. The same method is used for variable resistors (potentiometers). The resistance is varied by moving a sliding wire on the surface of the carbon layer.

Wire resistors are wound in various forms. The idea is to use a length of wire to obtain the required resistance. Their main advantage is in very accurate resistance (low-tolerance resistors) and high-power dissipation capabilities. A common method of construction, especially in surface mount resistors, is to use a metal film and control the resistance by changing the thickness and/or composition of the metal film.

Application: Potential Drop Method of Nondestructive Testing of Materials One method of nondestructive testing of conducting materials is based on the change in resistance of an article when the properties of the material change. For example, if a conducting material cracks, its resistance increases. In the extreme, if the conduction path is interrupted (cut), the resistance increases to infinity. In the method in **Figure 7.21a**, the resistance is proportional to the depth of the crack. By keeping the distance between the two probes constant and comparing the resistance with that of an intact article, it is possible to detect a variety of material changes including cracks, corrosion, thinning, etc.

In practical applications, two probes are kept at a fixed location and connected to a source as shown in **Figure 7.21b**. Two sensing probes then measure the potential drop between two other points. The same method can be used for prospecting for minerals and oil. In this case, a number of rods are inserted deep into the ground and the resistances between rods are evaluated. When compared with resistances in similar ground layers, it is possible to detect variations which may indicate the existence of deposits in the ground. **Figure 7.21c** shows this concept.

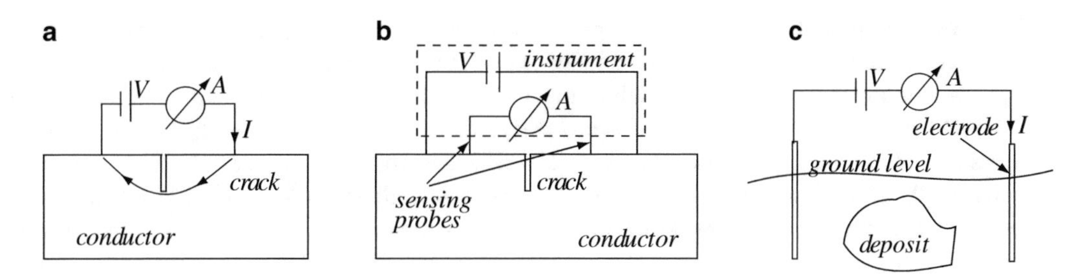

Figure 7.21 (**a**) Potential drop method of testing and prospecting. (**b**) The four-probe method of testing and prospecting. (**c**) Potential drop method of prospecting for minerals

Application: Cathodic Protection of Metal Structures A common method of protecting buried pipes, tanks and other structures from corrosion is based on the observation that corrosion is initiated because of the potential difference between conducting materials and the slightly acidic soil or between different locations on the same metal. Similar effects occur in any salt solution. The metal and corrosion products serve as anode (metal) and cathode (corrosion products such as rust). In essence, a voltaic cell is created between the conductor and its corrosion products, causing electrolysis and electrochemical corrosion. Metal ions are transferred to the electrolyte (anodic reaction), leaving behind electrons which then move through the metal to non-corroded areas where they react with water or oxygen (cathodic reaction). Thus, an electrolytic cell has been formed. A cathodic protection method is based on the fact that the potential difference between anodic and cathodic locations can be neutralized. What is needed is an opposing current to flow from any source to the protected metal through the surrounding electrolyte. This is accomplished by burying electrodes in the ground and connecting a DC source between the electrodes and the conductor being protected. If no current passes through the cell, no corrosion occurs. Because the negative terminal is connected to the buried object, it is called cathodic protection. **Figure 7.22** shows the basic principle. The anode is buried in the ground, in the vicinity of the buried object, to produce a flow of current to the object. If this current is properly adjusted, all electrolytic currents will cease, as will corrosion.

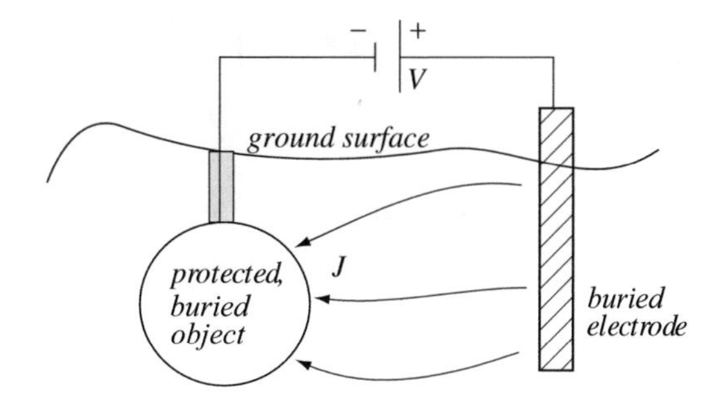

Figure 7.22 Cathodic protection of buried conducting objects

Application: Electrical Welding—An Example of the Use of Joule's Law In electrical welding, a current passes through the two metals to be welded. Because of the large current density and finite conductivity of the materials, the temperature at the weld site is raised to the point of locally melting the materials and fusing them together. In general electrical welding, a third material is used. This metal, also called an electrode, is melted to form the weld as shown in **Figure 7.23a**. In spot welding, there is no third material involved: The two materials are fused together. The latter is particularly attractive for automatic welding of sheet stock because it is very fast, can be handled by robots, and is very efficient with thin materials. This method (**Figure 7.23b**) is used almost exclusively in welding of car body components, home appliances, and the like. Currents in excess of 100 A are used, and, considering that the spot weld is usually no more than 6 mm in diameter and of the order of 1–2 mm in thickness, the power density involved is considerable.

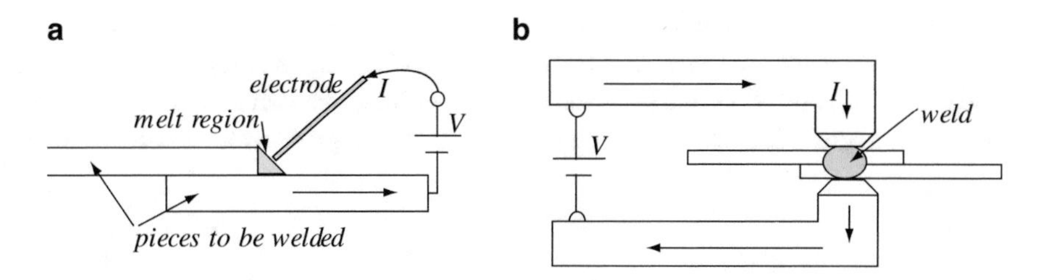

Figure 7.23 Electric welding. (**a**) The use of a third material or "electrode welding." (**b**) Spot welding

Application: The Strain Gauge—Measurement of Strain, Force, and Deformation of Materials Elongation of a wire can be used as a measure of strain in the wire. If the elongation is reversible, this may be used as a strain or force sensor. This is the basic principle of a strain gauge. Although in deforming a solid, both its conductivity and dimensions change, we will explain the operation of the strain gauge entirely on dimensional changes, keeping conductivity constant. Consider a length of wire L with cross-sectional area A and conductivity σ. The resistance of the wire is $L/\sigma A$. Now, suppose the wire is stretched to twice its length. The volume of the wire has not changed and neither has its conductivity. Thus, since the length is $2L$, its cross-sectional area must be $A/2$. The resistance of the stretched wire is now $(2L)/(\sigma A/2) = 4L/\sigma A$, which is four times the resistance of the original wire. It is therefore an easy matter to use the resistance of the wire as a measure of force, pressure, strain, or any related quantity such as acceleration, deflection, weight, etc. In practical applications the elongation is limited to a few percents of L.

Strain gauges are made of conducting materials such as nickel or nickel–copper alloys (constantan is commonly used because it has a low-temperature coefficient) or semiconductors. The material is normally deposited on a substrate layer in the form of a thin film and then etched to form a long, continuous wire as shown in **Figure 7.24**. Now, any stretching in the direction of sensitivity of the gauge will produce a change in resistance and allow measurement of force.

An example of the application of strain gauges in measuring the flexing of a bridge deck is shown in **Figure 7.25** together with a possible bridge measurement method. The voltmeter now indicates the deflection of the bridge deck and can also be used to automatically measure weight on the bridge as well as permanent deformation such as sagging of the deck.

The strain gauge is a simple, accurate, and durable sensor which is used extensively in measurements of forces and strains. Multiple sensors can be used to measure multiple axis motion and forces. Similarly, multiple sensors in two or three axes can be produced on a single substrate, providing an extremely versatile sensing device.

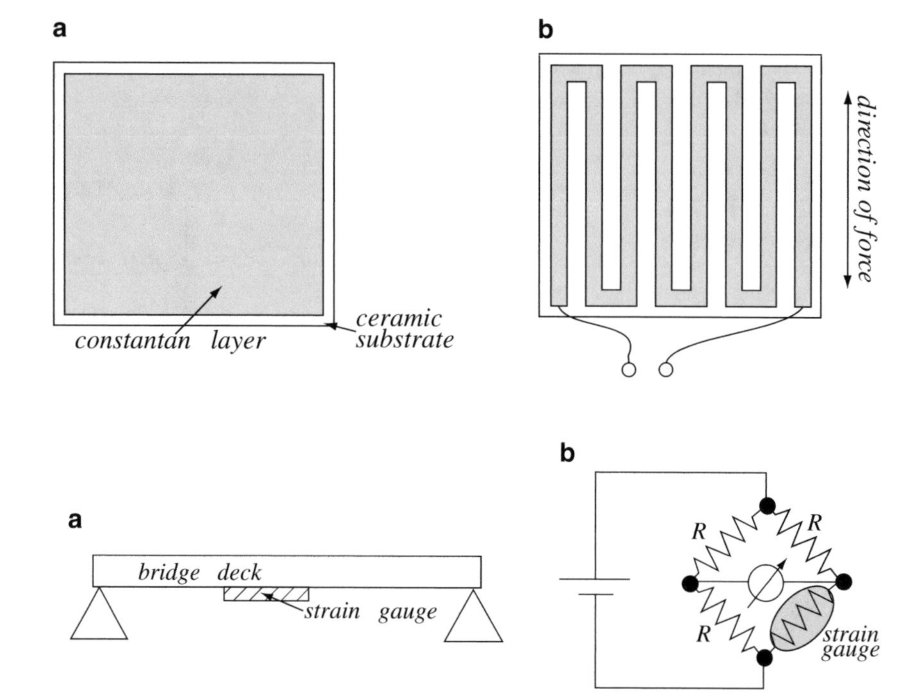

Figure 7.24 Construction of a strain gauge. (**a**) Thin-film deposition. (**b**) Etching of the gauge

Figure 7.25 Use of the strain gauge. (**a**) Measurement of strain in a bridge deck. (**b**) The strain is measured in a bridge circuit

Application: Electrocardiography One of the most useful cardiac diagnosis methods is the electrocardiograph (ECG). Electrocardiography is the graphic recording of the electrical potentials produced by cardiac tissue. Electrical pulses produced by the heart are conducted throughout the body. Weak currents generated by these pulses produce potentials between any two locations in the body. Monitoring of these potentials is used to determine the state of the heart, arteries, and ventricles. Measurements are made by placing electrodes at different locations on the body and measuring the potential differences between them. For example, potentials between the extremities of the body (legs, hands, torso) can be measured. The standard leads are to left arm (*LA*), right arm (*RA*), and left leg (*LL*). These measure potential differences between (*LA*) and (*RA*), (*LL*) and (*RA*), and (*LL*) and (*LA*). Other locations can and often are measured, in particular on the chest and around the heart. **Figure 7.26** shows the standard electrocardiograph connections. The diagnosis involves inspection of the signals obtained for any abnormality that may be an indication of cardiac problems. The cardiogram is a pulsed signal since the heart does not produce steady currents. However, the methods involved are based on the principles described in this chapter.

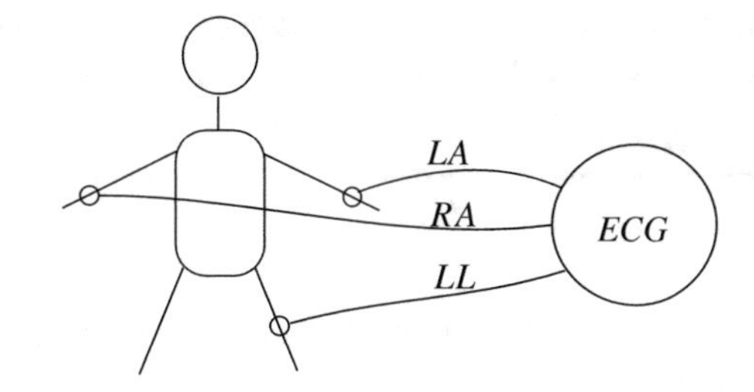

Figure 7.26 The electrocardiogram. Standard connections to the body are shown

7.10 Summary

The relation between electromagnetics and circuits is the main theme in this chapter. Common relations such as Ohm's, Joule's, and Kirchoff's laws are shown to be simplified forms of more general electromagnetics relations. We start with the ***current density*** derived from motion of charge. Given N point charges per unit volume, each q [C] [Nq is the charge density in the volume ($\rho_v = Nq$)] moving at a velocity $\mathbf{v}$ produce a current density $\mathbf{J}$

$$\mathbf{J} = Nq\mathbf{v} = \rho_v\mathbf{v} \quad [\mathrm{A/m^2}] \tag{7.6) and (7.7}$$

This current density exists in any medium—conducting or not. Currents flowing in nonconductors are called ***convection currents***. The current through any surface s is

$$I = \int_s \mathbf{J} \cdot d\mathbf{s} \quad [\mathrm{A}] \tag{7.9}$$

Conduction currents are related to the electric field intensity $\mathbf{E}$ [V/m] and conductivity σ [S/m] as

$$\mathbf{J} = \sigma\mathbf{E} \quad [\mathrm{A/m^2}] \tag{7.12}$$

This is known as ***Ohm's law***. The more conventional form of Ohm's law can be obtained from **Eq. (7.12)**. One form convenient for calculation of resistance in conductors is

$$R = \frac{V_{ab}}{I} = \left| \frac{-\int_b^a \mathbf{E} \cdot d\mathbf{l}}{\int_s \sigma\mathbf{E} \cdot d\mathbf{s}} \right| \quad [\Omega] \tag{7.17}$$

Joule's law is obtained from the electric field intensity and current density as follows:

$$P = \int_v \mathbf{E} \cdot \mathbf{J} dv \quad [\mathrm{W}] \tag{7.23}$$

The integrand is the ***power density***

$$p_d = \frac{dP}{dv} = \mathbf{E} \cdot \mathbf{J} = \sigma E^2 = \frac{J^2}{\sigma} \quad \left[\frac{\mathrm{W}}{\mathrm{m^3}}\right] \tag{7.25}$$

From the definition of current as the time rate of decrease in charge ($I = -dQ/dt$), we obtain the very important ***equation of continuity***

$$\nabla \cdot \mathbf{J} = -\frac{\partial \rho_v}{\partial t} \tag{7.32}$$

The divergence of current density is zero for ***steady currents*** but nonzero for ***time-dependent currents***.

The current density **J** is a vector. Therefore, the postulates governing it in conductors for steady currents are

$$\overbrace{\nabla \cdot \mathbf{J} = 0 \quad \text{and} \quad \nabla \times \frac{\mathbf{J}}{\sigma} = 0}^{\textit{differential form}} \quad \text{or} \quad \overbrace{\oint_s \mathbf{J} \cdot d\mathbf{s} = 0 \quad \text{and} \quad \oint_l \frac{\mathbf{J} \cdot d\mathbf{l}}{\sigma} = 0}^{\textit{integral form}} \qquad (7.39) \text{ and } (7.40)$$

The ***interface conditions*** for the normal and tangential components of current density are derived from the electric field intensity [see **Eqs. (7.46), (7.47), (7.12),** and **Figure 7.18**]:

$$\frac{J_{1t}}{\sigma_1} = \frac{J_{2t}}{\sigma_2} \quad \text{and} \quad J_{1n} = J_{2n} \tag{7.50}$$

In addition, in poor (non-perfect) conductors, there will be a surface charge density at the interface between two media with different conductivities generated due to flow of current from medium (2) into medium (1) (see **Figure 7.17** and **Example 7.14**):

$$\varepsilon_1 \frac{J_{1n}}{\sigma_1} - \varepsilon_2 \frac{J_{2n}}{\sigma_2} = \rho_s \quad [\mathrm{C/m^2}] \tag{7.51}$$

Problems

Convection and Conduction Current

7.1 Application: Current and Charge in an Electric Vehicle Battery. An electric vehicle is powered by a Lithium-Ion battery rated at 40 kWh at a voltage of 360 V:

(a) How much charge can the battery supply?

(b) What must be the area of a parallel plate capacitor, with plates separated a distance $d = 0.01$ mm and relative permittivity of 4 for the dielectric between the plates to supply the same charge as the battery in (**a**)?

(c) An alternative to batteries are supercapacitors. Most supercapacitors are rated at 2.7 V. If commercial supercapacitors rated at 3,000 F, 2.7 V are used, how many capacitors are needed to replace the battery? How are the supercapacitors connected?

7.2 Application: Convection Current. The solar wind is a stream of mostly protons emitted by the Sun as part of its normal reaction. The particles move at about 500 km/s and, at the Earth's orbit, the particle density is about 10^6 protons per m^3. Most of these protons are absorbed in the upper atmosphere and never reach the surface of the Earth:

(a) Taking the radius of the Earth as 6,400 km and the charge of the proton as 1.6×10^{-19} C, calculate the current intercepting the globe if these charges were not absorbed in the atmosphere.

(b) During a solar flare, the solar wind becomes much more intense. The velocity increases to 1,000 km/s (typical) and the number of particles increases to about 10^7 protons/m^3. Calculate the current intercepting the globe during the solar flare.

7.3 Drift Velocity and Electric Field Intensity in Conductors. A copper conductor with conductivity $\sigma = 5.7 \times 10^7$ S/m carries a current of 10 A. The conductor has a diameter of 2 mm. Calculate:

(a) The electric field intensity inside the conductor.

(b) The average drift velocity of electrons in the conductor assuming a free electron density of 8.48×10^{28} electrons/m^3.

7.4 Application: Velocity of Electrons in Conductors. A High Voltage DC (HVDC) power line between a power generating station and the city it supplies is 1,200 km long, is 60 mm in diameter and carries a current of 2,500 A.

The line is made of aluminum with a free electron density of 6×10^{28} electrons/m^3. How long would it take an electron to traverse the length of the line?

7.5 Electric Field Due to a Charged Beam. A cylindrical electron beam consists of a uniform volume charge density moving at a constant axial velocity $v_0 = 5 \times 10^6$ m/s. The total current carried by the beam is $I_0 = 5$ mA and the beam has a diameter $d = 1$ mm. Assuming a uniform distribution of charges within the beam's volume, calculate the electric field intensity inside and outside the electron beam.

Conductivity and Resistance

7.6 Carrier Velocity and Conductivity. A power device is made of silicon and carries a current of 100 A. The cross-section of the device is 10 mm $\times$ 10 mm and it is 1 mm thick. The potential drop on the device is 0.6 V, measured across the thickness of the device:

(a) What is the conductivity of the silicon used?
(b) If the velocity of the carriers is 2,000 m/s, how many free charges (electrons) must be present in the device at all times?

7.7 Lossy Coaxial Cable. In a coaxial cable of length L, the inner conductor is of radius a [m] and the outer conductor of radius b [m], and these are separated by a lossy dielectric with permittivity ε [F/m] and conductivity σ [S/m]. A DC source, V [V], is connected with the positive side to the outer conductor of the cable. Neglecting edge effects determine:

(a) The electric field intensity in the dielectric.
(b) The current density in the dielectric.
(c) The total current flowing from outer to inner conductor.
(d) The resistance between the outer and inner conductors.

7.8 Application: Resistance in Layered Conductors. A cylindrical conductor, 1 m long, is made of layered materials as shown in **Figure 7.27**. The inner layer is copper, the intermediate layer is brass, and the outer layer is a thin coating of nichrome, used for appearances' sake. Use conductivity data from **Table 7.1**:

(a) Calculate the resistance of the conductor before coating with nichrome, as measured between the two ends of the cylinder.
(b) What is the percentage change in resistance of the conductor due to the nichrome coating?

Figure 7.27

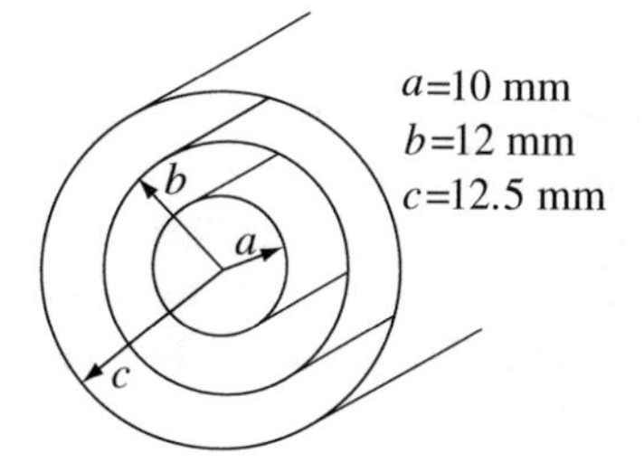

7.9 Application: Temperature Dependency of Resistance. Conductivity of copper varies with temperature as $\sigma = \sigma_0/[1 + \alpha(T - T_0)]$, where $\alpha = 0.0039/°C$ for copper, T is the actual temperature and T_0 is the temperature at which σ_0 is known (usually 20°C). A cylindrical copper conductor of radius 10mm and length 0.3 m passes through a wall as shown in **Figure 7.28**. The temperature inside is constant at 25°C and the temperature outside varies from 30°C in summer and -10°C in winter. Conductivity of copper is $\sigma_0 = 5.7 \times 10^7$ S/m at $T_0 = 20$°C. The temperature distribution along the section of the bar that is inside the wall is linear:

(a) What is the resistance of the conductor bar section embedded in the wall in summer?
(b) What is the resistance of the conductor bar section embedded in the wall in winter?
(c) How can this conductor be used to measure the temperature difference between outside and inside?

Figure 7.28

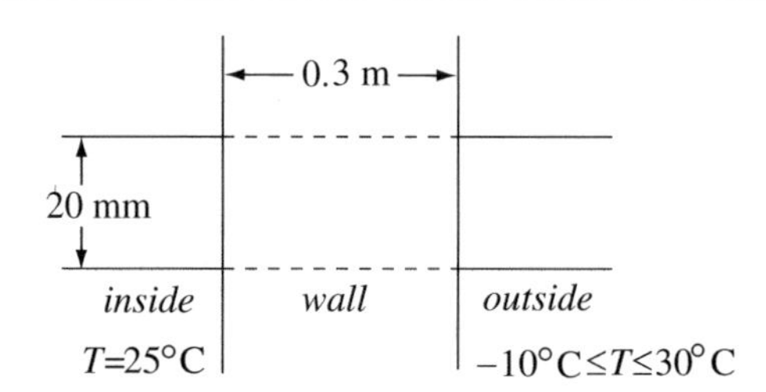

7.10 Application: Temperature Sensor. It is required to design a device to measure the temperature of an internal combustion engine. The minimum temperature expected is −30°C and maximum temperature is that of boiling water. A copper wire is wound in the form of a coil and is enclosed in a watertight enclosure which is immersed in the coolant of the engine. If the copper wire is 10 m long, and 0.1 mm in diameter, calculate the minimum and maximum resistance of the sensor provided its conductivity is given by $\sigma = \sigma_0/[1 + \alpha(T - T_0)]$, where $\alpha = 0.0039$/°C, $\sigma_0 = 5.7 \times 10^7$ S/m is the conductivity at $T_0 = 20$°C. Assume the dimensions of the wire do not change with temperature.

7.11 Application: Measurement of Conductivity. A very simple (and not very accurate) method of measuring conductivity of a solid material is to make a cylinder of the material and place it between two cylinders of known dimensions and conductivity, as shown in **Figure 7.29**. A current source, with constant current I_0 [A], is connected to the three cylinders. A voltmeter is connected to two very thin disks made of silver. The voltage read is directly proportional to the resistance of the cylinder, and since the dimensions are known, the conductivity of the material can be related to the potential:

(a) Calculate the conductivity of the material as a function of dimensions and potential. Use $I_0 = 1$ A, $a = 10$ mm, $b = 100$ mm.

(b) What are the implicit assumptions in this measurement?

Figure 7.29

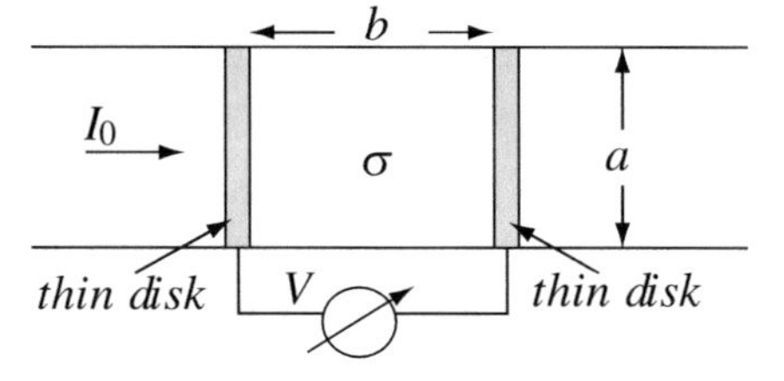

7.12 Application: Measurement of Conductivity. A method of measuring conductivity of fluids is to construct a cylinder and drill a hole at the center of the cylinder. Both ends are plugged with cylindrical plugs of radius a [m], known length and of the same conductivity as the cylinder (**Figure 7.30a**). Assuming the conductivity of the cylinder is σ_0 [S/m] and dimensions are as shown, calculate the resistance of the device. Now, one plug is removed, the hollow space is filled with the fluid and the plug replaced as shown in **Figure 7.30b**. If the resistance of the device is now lower by 10%, what is the conductivity of the fluid?

Figure 7.30

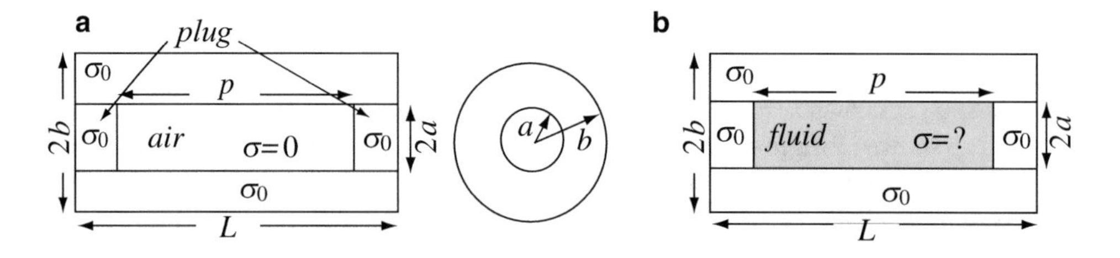

7.13 Potential Difference and Resistance. Three cylinders are attached as shown in **Figure 7.31**. A current I [A] passes through the cylinders. If σ_1, σ_2, and σ_3 [S/m] are the conductivities of the cylinders, calculate the potential difference on the three cylinders. Assume current densities are uniform in each cylinder.

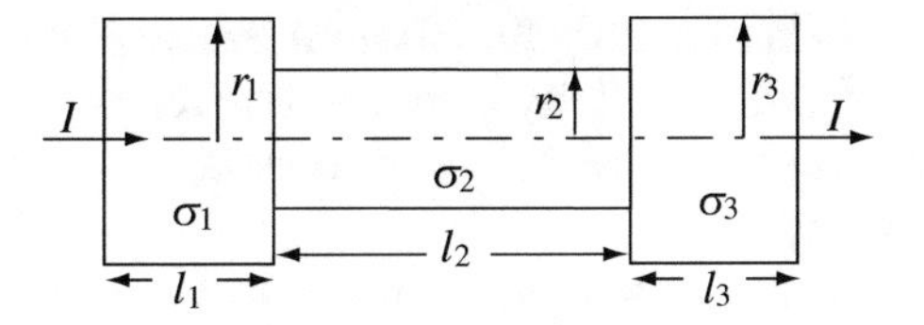

Figure 7.31

7.14 Nonhomogeneous Conductivity. A block of material with dimensions as shown in **Figure 7.32** has conductivity $\sigma = \sigma_0 + xd(d - x)$ [S/m], where $x = 0$ at the left end of the block. A current I [A] passes through the block. σ_0 is a constant. Calculate:

(a) The resistance of the block.
(b) The potential difference between the two ends of the block using Ohm's law.
(c) The potential difference between the two ends of the block using the electric field intensity in the conductor. Compare with **(b).**

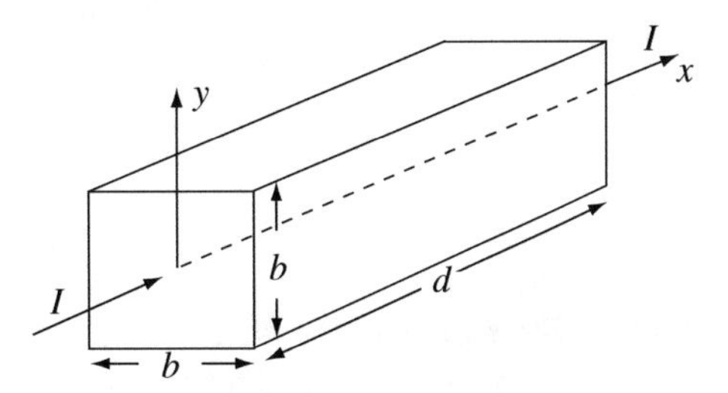

Figure 7.32

7.15 Potential Difference and Resistance. A conducting sphere of radius a [m], with conductivity σ [S/m], is cut into two hemispheres. Then, each hemisphere is trimmed so that it has a flat surface on top and the two sections are connected as shown in **Figure 7.33**. A uniform current density **J** [A/m^2] crosses the upper flat surface as shown. Calculate the potential difference between A and B.

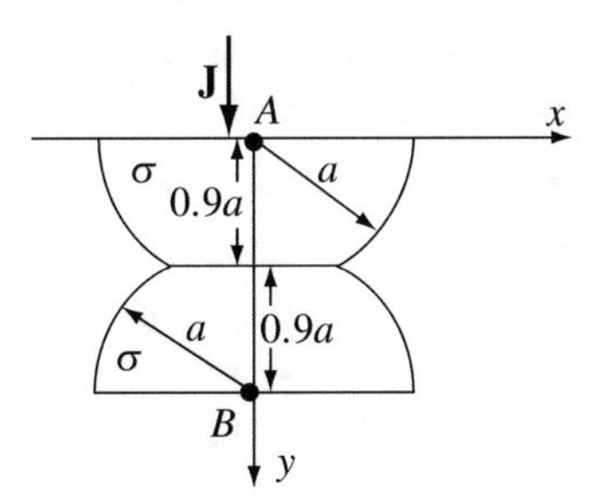

Figure 7.33

7.16 Application: Ground Resistance of Electrodes. A ground electrode is built in the form of a hemisphere, of radius $a = 0.1$ m, buried in the ground, flush with the surface as shown in **Figure 7.34**. Calculate the contact resistance (resistance between the wire and surrounding ground). You may assume the ground is a uniform medium semi-infinite in extent, with conductivity $\sigma = 0.1$ S/m. **Hint:** Calculate the resistance between the hemisphere and a concentric conducting surface at infinity.
Note: This is an important calculation in design of lightning and power fault protection systems.

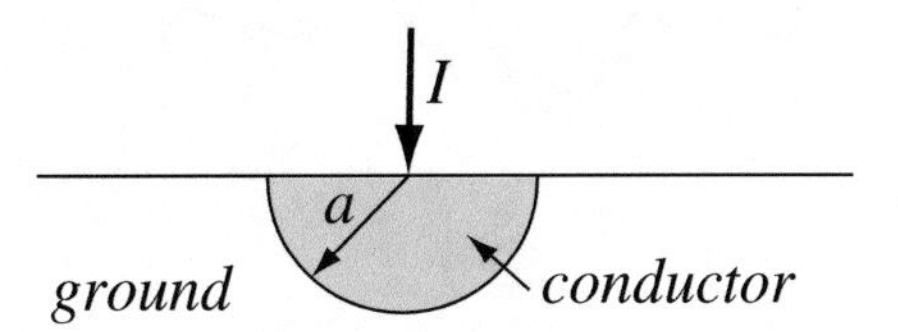

Figure 7.34

7.17 Application: Measurement of Soil Conductivity for Mineral Prospecting. The measurement of soil conductivity (or its reciprocal, resistivity) is an important activity in prospecting for minerals. Usually two or more electrodes are inserted in the ground and the current measured for a known voltage. The conductivity of the ground can then be correlated with the presence of minerals.

Two conducting electrodes, each made in the form of a half sphere, one of radius a [m], the second of radius b [m], are buried in the ground flush with the surface (**Figure 7.35**). The horizontal distance between the centers of the electrodes is large so that one electrode does not influence the other (i.e., the current around one electrode is not distorted by the proximity of the other). If a battery of current V_0 [V] is connected as shown and a current I_0 [A] is measured, calculate the conductivity of the soil. You may assume that the ground is semi-infinite. Assume the conductivity of the electrodes is much higher than that of the soil and hence their resistance can be neglected.

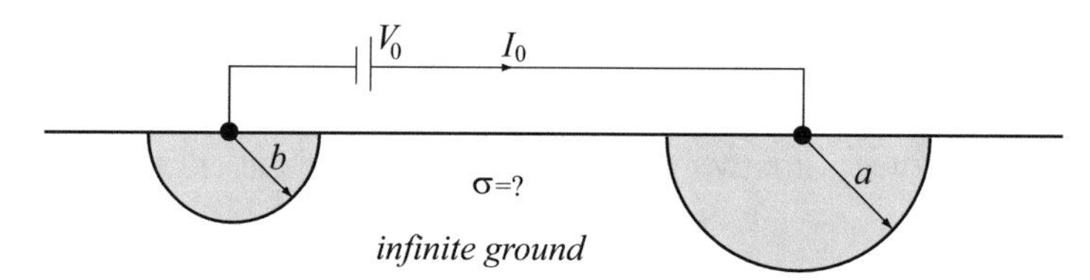

Figure 7.35

Power Dissipation and Joule's Law

7.18 Application: Series and Parallel Loads. The early electric distribution for lighting (Edison's system) was in series. All electric bulbs were connected in series and the source voltage was raised to the required value. Today's systems are almost always in parallel. Suppose you wanted to calculate which system is more economical in terms of power loss for street lighting. A street is 1 km long and a light bulb is placed every 50 m (total of 20 bulbs), each bulb rated at 240 V, 500 W:

(a) What is the minimum wire thickness required for the parallel and for the series connections (copper, $\sigma = 5.7 \times 10^7$ S/m) if the current density cannot exceed 10^6 A/m^2 and the wires must be of the same thickness along the street?

(b) The wire calculated in **(a)** is used. Calculate: (1) The total weight of copper used for the series and for the parallel systems (copper weighs approximately 8 tons/m^3). (2) Total power loss in the copper wires for the two systems. (3) Which system is more economical in terms of amount of copper needed and in terms of power loss?

(c) What are your conclusions from the calculations above?

7.19 Application: Resistance and Power in Faulty Cables. An underwater cable is made as a coaxial cable with internal conductor of diameter $a = 2$ mm and a thin external conductor of diameter $b = 20$ mm (see **Figure 7.36**). The cable is used for communication between two islands and also supplies power to devices such as amplifiers connected to the cable. The cable is 1.2 km long, and operates at 48 V. Because of a leak, seawater entered the cable, contaminating the insulating material between the two conductors. As a result the conductivity of the insulation has been reduced to 10^{-4} S/m:

(a) What is the resistance seen by a generator connected to the cable at one end?

(b) How much power is dissipated in the contaminated insulation inside the cable?

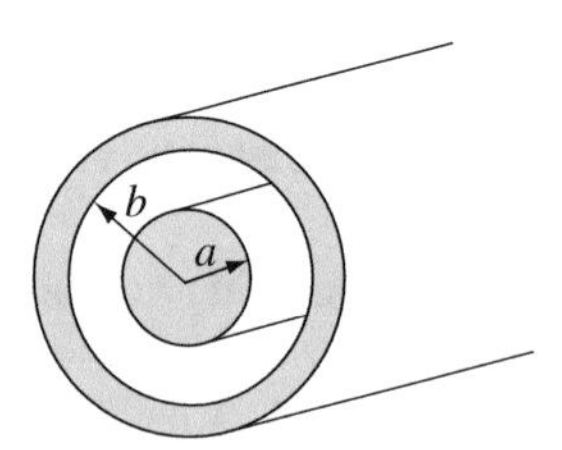

Figure 7.36

7.20 Application: Bimetal Conductor. A conductor is made of two materials as shown in **Figure 7.37**. This is typical of bimetal devices. The device is connected to a voltage source:

(a) Calculate the current density in each material.

(b) Calculate the total power dissipated in each material.

(c) Assuming that each material has the same coefficient of expansion and the same power dissipating capability, which way will the bimetal device bend? Explain.

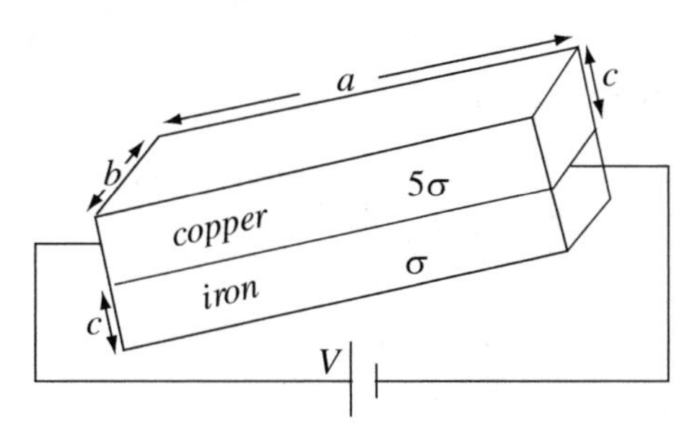

Figure 7.37

7.21 Series and Parallel Resistances. An antenna element is made of a hollow aluminum tube, $L = 800$ mm long, with inner diameter $a = 10$ mm and outer diameter $b = 12$ mm. For operational reasons, it becomes necessary to lengthen the antenna to 1,200 mm. To do so, it is proposed to use the same tube but add a cylindrical solid graphite coupler that fits in the tube with a shoulder to support the extension as shown in **Figure 7.38**. Conductivity of aluminum is $\sigma_a = 3.6 \times 10^7$ S/m and that of graphite is $\sigma_g = 10^4$ S/m. The coupler's dimensions are $c = 120$ mm, $d = 40$ mm and each of the sections that fit inside the tube is 40 mm long. The extension tube is $e = 360$ mm long:

(a) Calculate the change in resistance of the antenna element (resistance is measured between the two ends of the structure) due to the change in the structure.

(b) Suppose that because of cost the graphite coupler is replaced with an identical aluminum coupler. What is now the change in resistance?

Note: The resistance of antennas is an important parameter in antenna efficiency.

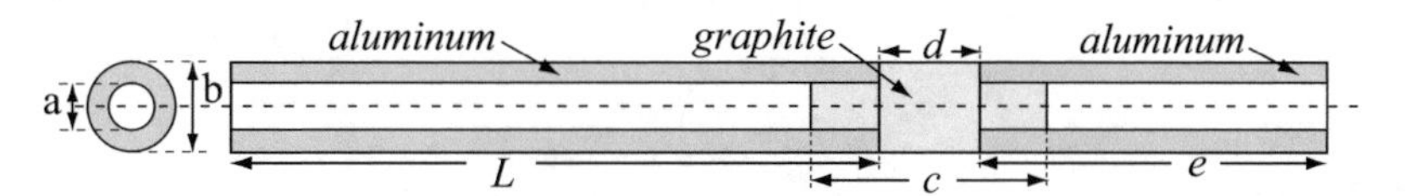

Figure 7.38

7.22 Series and Parallel Resistances. Consider again **Problem 7.21**. After fitting the coupler and the extension, it became evident that the structure is not sufficiently strong. To improve its strength, a graphite tube is added over the graphite coupler as shown in **Figure 7.39**. The graphite tube is $k = 200$ mm long and has inner diameter equal to the outer diameter of the tube ($b = 12$ mm) and outer diameter $h = 16$ mm. The tube is centered over the coupler. Given the dimensions and material properties, calculate:

(a) The total resistance of the antenna.

(b) The total resistance of the antenna if both the coupler and the external tube are made of aluminum.

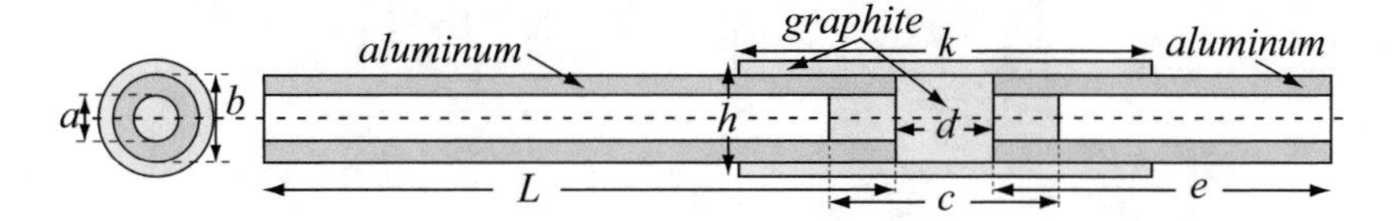

Figure 7.39

Continuity and Circuit Laws

7.23 Application: Conservation of Charge. A small ion thruster is made in the form of a box with a round opening through which the ions are ejected (**Figure 7.40**). Suppose that by external means (not shown) the ions are ejected at a velocity of 10^6 m/s, the ions are all positive with charge equal to that of a proton, and the ion density in the ion beam is 10^{18} ions/m^3. The opening is 100 mm^2 in area and the beam may be considered to be cylindrical. Calculate:

(a) The current in the beam.
(b) Discuss the behavior of this current over time.

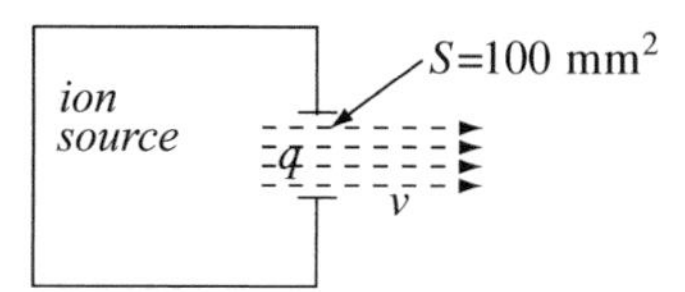

Figure 7.40

7.24 Conservation of Charge in a Circuit. A capacitor of capacitance $C = 1{,}000$ μF is connected in series with a switch and a 12 V battery. The resistance of the wires is 10 Ω, and the wire used in the circuit is 1 mm in diameter. The switch is switched on at $t = 0$:

(a) Show that if you take a volume that includes one plate of the capacitor (see **Figure 7.41**), Kirchhoff's current law as defined in circuits is not satisfied.
(b) Calculate the rate of change of charge on one plate of the capacitor in **Figure 7.41**.
(c) Show that the current cannot be a steady current.

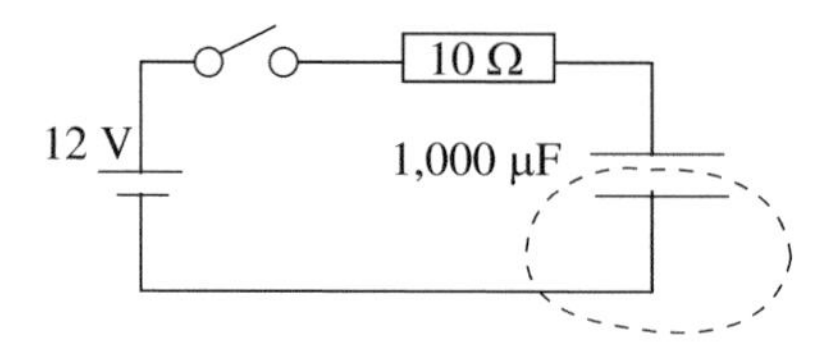

Figure 7.41

Interface Conditions

7.25 Surface Charge Produced by Current. Two cylindrical, lossy dielectrics of radius b [m] but of different conductivities are connected to a potential V [V]. Dimensions and properties are shown in **Figure 7.42**. Two perfectly conducting cylinders are connected to the two ends for the purpose of connecting to the source.

(a) Calculate the total charge at the interface between each two materials (indicated by arrows in **Figure 7.42**).
(b) Repeat **(a)** with the polarity of the battery reversed.

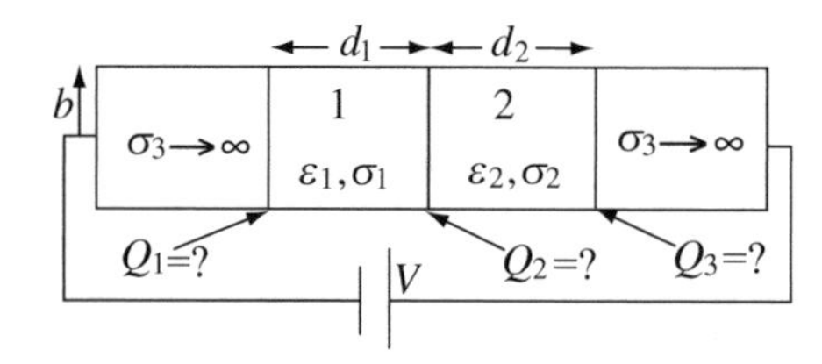

Figure 7.42

7.26 Application: Electric Field Intensity and Power Density in Conductors. A copper and an aluminum conductor are connected as shown in **Figure 7.43**. Both have a diameter $d = 10$ mm. A 10 A current flows through the two conductor as shown. Properties are $\sigma_{copper} = 5.7 \times 10^7$ S/m and $\sigma_{aluminum} = 3.6 \times 10^7$ S/m whereas permittivity of both materials is ε_0. Calculate:

(a) The electric field intensity in copper and in aluminum.
(b) The dissipated power density in copper and in aluminum.
(c) The charge density on the interface between copper and aluminum.

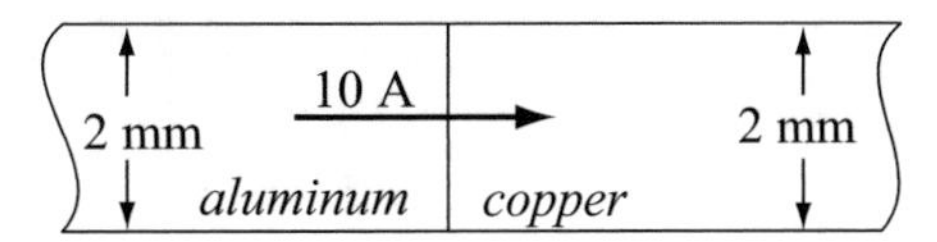

Figure 7.43

7.27 Application: Spot Welding of Sheet Stock. In an electric spot welding machine, the electrodes are circular with diameter $d = 10$ mm. Two sheets of steel are welded together (**Figure 7.44**). If the power density (in steel) required for welding is 10^8 W/m^3, calculate:

(a) The potential difference needed on the steel sheets and the current in the electrodes.
(b) The dissipated power density in the electrodes.
(Assume the weld area equals the cross-sectional area of the electrodes and all current flows in the weld area.)

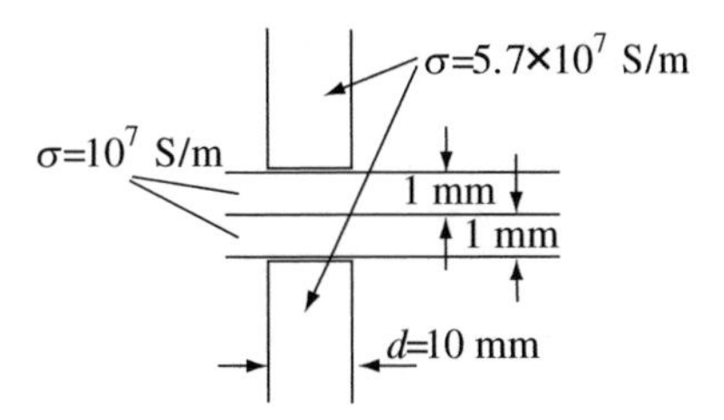

Figure 7.44

The Static Magnetic Field

8

"There's a South Pole," said Christopher Robin, "and I expect there's an East Pole and a West Pole, though people don't like talking about them."

Winnie-The-Pooh

8.1 Introduction

After discussing the static electric field and steady currents, we are now ready to take another significant step in the study of electromagnetics, the study of the static magnetic field. But what exactly is a magnetic field? This question will be answered gradually, but, for a simple description, we may say it is a new type of force field in the same sense that the electric field is a force field. Take, for example, a magnet. It attracts or repels other magnets and generates a "magnetic field" around itself. The permanent magnet generates a static (time-independent) magnetic field. A direct current can also generate a static magnetic field. How do we know that? As with many other aspects of electromagnetics, we know by experiment.

The various properties of the magnetic field will be discussed primarily from the point of view of currents. The importance of this for engineering design is twofold; first, it indicates that a static magnetic field can be generated to suit design purposes. Second, it provides a link between the electric field and the magnetic field. Thus, we can view the electric field as a source of current, which, in turn, is the source for the magnetic field and, therefore, at least a partial explanation for the term "electromagnetics." The reason for the qualification is that the opposite does not happen in the case of static fields: a static magnetic field does not generate an electric field. We will see in **Chapters 10** and **11** that a time-dependent magnetic field does generate a time-dependent electric field, and at that point the link between the electric and magnetic field will be complete.

For many years, electric and magnetic fields were thought to be separate phenomena even though both were known since antiquity. It was not until 1819 when Hans Christian Oersted[1] found that the needle of a compass moved in the presence of a current-carrying wire that a link between the two fields was found. He concluded that the only way this can happen is if the current generates a magnetic field around the wire. He used a compass to "map" the behavior of the magnetic field around the wire. Following his initial discovery, André-Marie Ampère[2] quickly established the correct relation between current and the magnetic field in what is now known as Ampère's law. With all that, as late as the beginning of the twentieth century, it was still common to use separate units for electric and magnetic quantities.

[1] Hans Christian Oersted (1777–1851), Danish scientist and professor of physics. He tried for many years to establish the link between electricity and magnetism, a link that was suspected to exist by him and many other scientists of the same period. He finally managed to do so in his now famous experiment of 1819 in which he showed that a current in a wire affects a magnetic needle (compass needle). He disclosed his experiments, all made in the presence of distinguished witnesses, in 1820. Oersted was very careful to ensure that what he saw was, in fact, a magnetic phenomenon by repeating the experiments many times and with various "needles," in addition to the magnetic needle (to show that the effect does not exist in conducting materials such as copper or insulating materials such as glass—only in magnetized materials). Intervening materials between the wire and needle were also tested. As was the custom of the day, his work was written and communicated in Latin in a pamphlet titled: "Experimenta circa efficaciam, conflictus eletrici in acum magneticam."

[2] See the footnote on page 337.

N. Ida, *Engineering Electromagnetics*, https://doi.org/10.1007/978-3-030-15557-5_8

In this chapter, we discuss the relationships between the steady electric current and the static magnetic field. These will be in the form of basic postulates which, as in the case of the static electric field, are experimental in nature. Two important relations, the Biot–Savart and Ampère's laws, will allow us to calculate the magnetic field due to electric currents.

8.2 The Magnetic Field, Magnetic Field Intensity, and Magnetic Flux Density Electric_Current.m

What, then, is the magnetic field? We might start to answer this question by playing with two magnets. The first effect to notice is that there is a force between the two magnets; the magnets either attract or repel each other as shown in **Figure 8.1**. Since attraction happens at a distance, each magnet must have a domain in which it attracts the other magnet. This is exactly what we called a field. That the field is a vector field we can establish using a compass as a measuring device. The direction of the force is established by the direction of the compass needle in space. Placing the compass at as many positions as we wish, a complete map of the vector field is established. This simple measurement establishes the following:

(1) A field exists throughout space.
(2) The field is stronger closer to the magnet.
(3) The two ends of the magnet behave differently; one attracts the north pole of the compass and is labeled the south pole; the other attracts the south pole of the compass and is labeled the north pole of the magnet (**Figure 8.1c**). This arbitrary identification is convenient because of its relation with the Earth's magnetic field.
(4) By placing the compass at different locations in space, we can map the magnetic field. One field line is shown in **Figure 8.1d**.

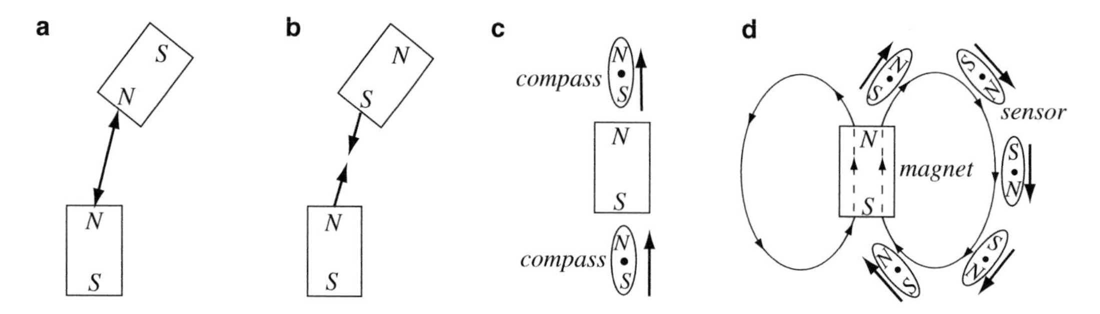

Figure 8.1 The permanent magnet. (**a**) Two permanent magnets repel each other. (**b**) Two permanent magnets attract each other. (**c**) Identification of the poles of a magnet. (**d**) Mapping the magnetic field of a permanent magnet using a compass

This simple experiment is all that is necessary to establish the existence of the magnetic field of a magnet. What we need to do now is to show that the same occurs due to steady electric currents and, more importantly, to find the exact relationship between the electric current and the magnetic field. For now, we will say that the magnetic field is a ***field of force*** acting on a magnet; the only qualification is that the magnet must be small (for example, the needle of a compass) to render the measurement valid.

The first of these, showing that a force exists due to the magnetic field produced by a current, is easily performed using the above idea but for a steady current. This is, in fact, what Oersted did in his historic experiment. Consider a straight wire, carrying a steady (direct) current I as shown in **Figure 8.2a**. Positioning the compass at various locations in space, we note that a force exists on the compass. Since this force cannot be an electrostatic force (no charges on the compass) and it certainly cannot be gravitational (since disconnecting the current will cancel the force), we must conclude that the force is similar to that between two magnets. Thus, the current in the wire has generated a magnetic field with properties identical to those of an equivalent magnet.

Experiment shows that the direction of the compass now is tangential to any circle centered at the current and depends on its direction. In **Figure 8.2b**, the direction is counterclockwise, whereas in **Figure 8.2c**, it is clockwise. The direction of the compass (arrow pointing from S to N) is taken as the direction of the magnetic field. From this experiment, we establish two very general and important properties:

(1) The direction of the magnetic field due to a current is defined by the right-hand rule shown in **Figure 8.2d**. The rule states that *"if the thumb of the right-hand shows the direction of current, the curled fingers show the direction of the magnetic field."*

(2) The lines of magnetic field (i.e., the lines to which the compass is tangential) are always closed lines. We will return to this property shortly, after we have a better definition of the magnetic field.

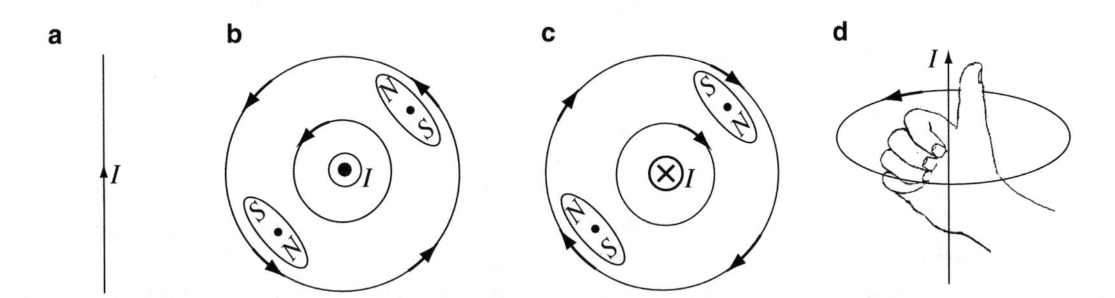

Figure 8.2 Oersted's experiment. (**a**) Current in a wire. (**b**) Direction of the compass needle for a current out of the page. (**c**) Direction of the compass for a current into the page. (**d**) The right-hand rule establishes the relation between the direction of current and the resulting magnetic field

The above experiment is lacking in one respect: Nothing has been established about the strength of the force exerted on the compass. We can intuitively say that the further away from the current we are, the lower the force, but exact values of force, or its relationship with the current, cannot be established from this experiment.

Since a magnetic field is generated by a steady current but not by a static charge, we conclude that the magnetic field and, therefore, the force in the magnetic field are related to the motion of charges (current) or, more specifically, to the velocity of charges. We recall that in the case of a static electric charge, a force acted on the charge as $\mathbf{F} = q\mathbf{E}$. Similarly, there is a force acting on a moving charge due to the magnetic field. As with the electrostatic force, we can establish the force in a magnetic field by performing a series of experiments from which both the magnitude and direction of the force are found. These experiments lead to the following relation between the magnetic force $\mathbf{F}_m$, electric charge q, charge velocity $\mathbf{v}$, and a new quantity $\mathbf{B}$:

$$\boxed{\mathbf{F}_\mathrm{m} = q\mathbf{v} \times \mathbf{B} \quad [\mathrm{N}]} \tag{8.1}$$

where $\mathbf{B}$ is called the ***magnetic flux density***. This relation is known as the Lorentz[3] force equation for a moving charge in the magnetic field, whereas $\mathbf{F}_e = q\mathbf{E}$ is the Coulomb force for a charge q in the electric field. For a general field, which includes both electric and magnetic components, the force on a charge is the sum of the two force vectors and may be written as

$$\boxed{\mathbf{F}_{total} = \mathbf{F}_e + \mathbf{F}_m = q\mathbf{E} + q\mathbf{v} \times \mathbf{B} \quad [\mathrm{N}]} \tag{8.2}$$

This relation is known as the Lorentz–Coulomb force equation (sometimes only as the Lorentz force equation). Note that a stationary charge ($\mathbf{v} = 0$) only experiences an electric force, whereas a moving charge experiences both an electric and a magnetic force.

The relation $q\mathbf{v} \times \mathbf{B}$ may be viewed as the defining relation for the magnetic flux density $\mathbf{B}$ since all other quantities are known or measurable. This is similar to using the relation $\mathbf{F} = q\mathbf{E}$ to define the electric field intensity as $\mathbf{E} = \mathbf{F}/q$ (force per unit charge). Although we will define the magnetic flux density in a different way shortly, the discussion here is useful in that it shows the physical meaning of the magnetic flux density.

[3] Hendrik Antoon Lorentz (1853–1928). Dutch physicist who is best known for his work on the effects of magnetism on radiation. For this he received the Nobel Prize in 1902. In his attempt to explain electricity, magnetism, and light, he arrived at the Lorentz transformation, which helped Albert Einstein in formulating his theory of relativity.

In terms of units, the magnetic flux density has units of (newton/coulomb)/ (meter/second) or newton/(ampere • meter). This unit is called the ***tesla*** [T] or weber/meter^2 [Wb/m^2]. The weber is equal to newton.meter/ampere. The common SI unit is the tesla [T], whereas the [Wb/m^2] is used sometimes to emphasize that **B** is a flux density. Both units are appropriate and commonly used.

In addition to the magnetic flux density, we define the magnetic field intensity **H** using the constitutive relation

$$\mathbf{B} = \mu \mathbf{H} \quad [\mathrm{T}] \tag{8.3}$$

where μ is called the ***magnetic permeability*** (or, in short, permeability) of materials. In free space, the relation is

$$\mathbf{B} = \mu_0 \mathbf{H} \quad [\mathrm{T}] \tag{8.4}$$

where μ_0 is the ***permeability of free space***. The magnitude of μ_0 is $4\pi \times 10^{-7}$ and its unit is the henry/meter or [H/m]. This unit will become obvious as we discuss the magnetic field intensity **H** and the magnetic flux density **B**. The relations in **Eqs. (8.3)** and **(8.4)** are fundamental and we will discuss them at some length in the following chapter. At this point, it is sufficient to indicate that both can be shown to be correct by performing whatever experiments are necessary to do so. This approach will allow us to introduce the important laws of Biot and Savart and Ampère.

Equation (8.1) may be viewed as defining the magnetic flux density **B**, but this definition is not convenient for calculation because it involves the vector product of **v** and **B** rather than **B** directly and is defined on individual moving charges. We will, therefore, seek a more convenient mathematical relation, one that will allow calculation of the magnetic flux density directly from the current I. Also, the magnetic flux density is material dependent (through μ). For this reason, we calculate **H** rather than **B**, at least until we had a chance to discuss behavior of materials in the magnetic field, which we will do in the following chapter. For these reasons, most of the discussion in this chapter is in terms of **H** rather than **B**. This does not create much difficulty since $\mathbf{B} = \mu\mathbf{H}$ can be used to calculate **B**. We will return to using **B** in latter parts of this chapter and in the next chapter because the force in the magnetic field depends on **B** (and, by consequence, on μ) and because force is fundamental in magnetic applications.

There are, in fact, two relations that accomplish the task of calculating the magnetic field intensity: the Biot–Savart and Ampère's laws. We start by describing the Biot–Savart law since Ampère's law may be derived from the Biot–Savart law.

8.3 The Biot–Savart Law

The law due to Biot and Savart[4] was deduced from a series of experiments on the effects of currents on each other and on permanent magnets (such as the compass needle Oersted used). The law gives a relation for the magnetic field intensity **H** at a point in space due to a current I. It states that the element of magnetic field intensity $d\mathbf{H}$ at a point $P(x,y,z)$ due to an element of current Idl' located at a point $P'(x',y',z')$ is proportional to the current element, the angle ψ between $d\mathbf{l}'$ and the position vector $\mathbf{R} = \mathbf{r} - \mathbf{r}'$, and inversely proportional to the distance R squared.

These quantities are shown in **Figure 8.3**. The magnitude of $d\mathbf{H}$ is

$$\boxed{dH(x,y,z) = \frac{I(x',y',z')dl'(x',y',z')\sin\psi}{4\pi R^2} \quad \left[\frac{\mathrm{A}}{\mathrm{m}}\right]} \tag{8.5}$$

where the coordinates (x,y,z) indicate a general point at which the field is calculated (field point), while coordinates (x',y',z') indicate the location of the current element (source point). Thus, I and dl' only depend on the primed coordinates. The unit of H is the ampere/meter [A/m], as can be seen from **Eq. (8.5)**.

[4] Jean-Baptiste Biot (1774–1862) was a professor of physics and astronomy. Among other topics, he worked on optics and the theory of light. Félix Savart (1791–1841) was a physician by training but later became professor of physics. This law is attributed to Biot and Savart because they were the first to present quantitative results for the force of a current on a magnet (in fact quantifying Oersted's discovery). The experiments mentioned here are described in detail in James Clerk Maxwell's *Treatise on Electricity and Magnetism* (Part IV, Chapter II), as Ampère's experiments. The law was presented in 1820, about a year after the landmark experiment by Oersted established the link between current and the magnetic field.

To find the total magnetic field intensity at point $P(x,y,z)$, the above quantity is integrated along the current. For the segment in **Figure 8.3**, we get

$$\boxed{H(x,y,z) = \int_a^b \frac{I(x',y',z')dl'(x',y',z')\sin\psi}{4\pi R^2} \quad \left[\frac{\text{A}}{\text{m}}\right]} \tag{8.6}$$

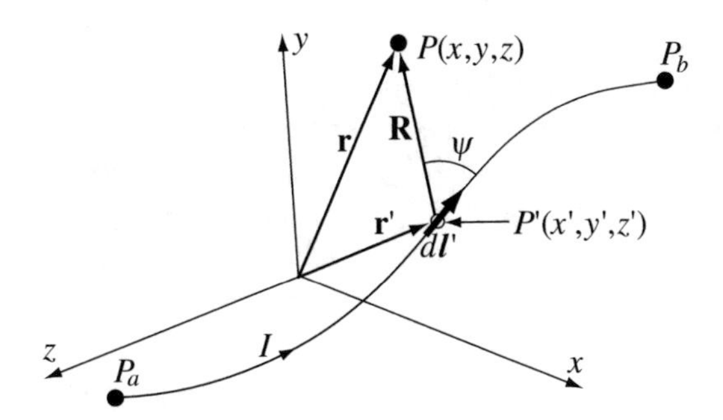

Figure 8.3 Derivation of the Biot–Savart law for a current element

Since the magnetic field intensity **H** is a vector, it is more appropriate to calculate the vector **H** rather than its magnitude. From the right-hand rule described above, we note that the magnetic field intensity **H** is perpendicular to I (or $d\mathbf{l}'$) as well as to the plane formed by **r** and $\mathbf{r}'$. Thus, it must be described as a cross product between $d\mathbf{l}'$ and $\mathbf{r} - \mathbf{r}'$. Using the cross product $d\mathbf{l} \times (\mathbf{r} - \mathbf{r}') = \hat{\mathbf{n}}\, dlR\sin\psi$ in **Eq. (8.6)** gives

$$\mathbf{H}(x,y,z) = \frac{1}{4\pi}\int_a^b \frac{Id\mathbf{l}' \times (\mathbf{r} - \mathbf{r}')}{|\mathbf{r} - \mathbf{r}'|^3} \quad \left[\frac{\text{A}}{\text{m}}\right] \tag{8.7}$$

Also, the current I is constant in most cases we will treat and may be taken outside the integral. Alternatively, using the definition of the unit vector, $\hat{\mathbf{R}} = (\mathbf{r} - \mathbf{r}')/|\mathbf{r} - \mathbf{r}'|$,

$$\boxed{\mathbf{H}(x,y,z) = \frac{1}{4\pi}\int_a^b \frac{Id\mathbf{l}' \times \hat{\mathbf{R}}}{|\mathbf{R}|^2} \quad \left[\frac{\text{A}}{\text{m}}\right]} \tag{8.8}$$

In either case, the direction of **H** is perpendicular to the vectors $\mathbf{R} = \mathbf{r} - \mathbf{r}'$ and $d\mathbf{l}'$, as required. If we use the definition of the magnetic flux density $\mathbf{B} = \mu\mathbf{H}$, the Biot–Savart law becomes

$$\mathbf{B}(x,y,z) = \frac{\mu}{4\pi}\int_a^b \frac{Id\mathbf{l}' \times \hat{\mathbf{R}}}{|\mathbf{R}|^2} \quad [\text{T}] \tag{8.9}$$

This is then the basic law we need to define and calculate the magnetic field intensity or magnetic flux density, given a current or current distribution. We must, however, note the following:

(1) The current element is assumed to be infinitely thin. A conductor of this type is called a ***filament***.

(2) The shape of the contour (i.e., the shape of the filament or wire) is not important except for the evaluation of the line integral.

(3) Strictly speaking, we must always have a closed contour along which we integrate; otherwise, there can be no current. However, it is permissible to calculate the contribution to the field due to a segment of the contour, assuming the current closes somehow.

(4) The space in which the current flows and in which the field is calculated is assumed to be of the same material and homogeneous.

The application of the law for practical calculations is straightforward. One aspect that must be followed strictly is the fact that integration is always along the current (i.e., in primed coordinates such as (x',y',z') or (r',ϕ',z')), whereas the field is calculated at a fixed point such as (x,y,z) or (r,ϕ,z). The two systems of coordinates should not be confused.

The Biot–Savart law is rather general and can be used for any current configuration, including distributed currents and current densities. Before we discuss these, it is useful to review a few examples of calculation of fields due to filaments (thin conducting wires).

Example 8.1 Field Intensity Due to a Short, Straight Segment The thin, finite-length wire in **Figure 8.4a** carries a current $I = 1$ A:

(a) Calculate the magnetic field intensity at point A shown in **Figure 8.4a**.
(b) Calculate the magnetic field intensity at point B shown in **Figure 8.4a**.
(c) What is the magnetic field intensity at a distance $h = 1$ m from the wire if the wire is infinitely long?

Solution: Because the wire produces a circular field, the problem is best solved in cylindrical coordinates (r,ϕ,z). An element of length $d\mathbf{l}' = \hat{\mathbf{z}}dz'$ is identified at point $(0,0,z')$ in **Figure 8.4b** or **8.4c**. The magnetic field intensity at point $P(r,\phi,z)$ is calculated using the Biot–Savart law. The solution is independent of the ϕ coordinate, simply from symmetry considerations:

(a) At point A, the coordinates are $(r = 1,\ z = 0)$. The magnetic field intensity at point A due to the current in element $d\mathbf{l}' = \hat{\mathbf{z}}dz'$ is

$$d\mathbf{H}(r,z) = \frac{I\hat{\mathbf{z}}dz' \times \hat{\mathbf{R}}}{4\pi|\mathbf{R}|^2} \quad \left[\frac{\mathrm{A}}{\mathrm{m}}\right]$$

The vector $\mathbf{R}$ (see **Figure 8.4b**) and the unit vector $\hat{\mathbf{R}}$ are, respectively,

$$\mathbf{R} = \mathbf{r} - \mathbf{r}' = \hat{\mathbf{r}}h - \hat{\mathbf{z}}z', \quad \hat{\mathbf{R}} = \frac{\hat{\mathbf{r}}h - \hat{\mathbf{z}}z'}{\sqrt{h^2 + z'^2}}$$

Thus, $d\mathbf{H}$ is

$$d\mathbf{H}(r,z) = \frac{I\hat{\mathbf{z}}dz' \times (\hat{\mathbf{r}}h - \hat{\mathbf{z}}z')}{4\pi\left(h^2 + z'^2\right)\left(\sqrt{h^2 + z'^2}\right)} = \hat{\boldsymbol{\phi}}\frac{Ih\,dz'}{4\pi\left(h^2 + z'^2\right)^{3/2}} \quad \left[\frac{\mathrm{A}}{\mathrm{m}}\right]$$

The total magnetic field intensity at A is found setting $h = 1$ m and integrating from $z' = -1$ m to $z' = +1$ m:

$$\mathbf{H}_A = \hat{\boldsymbol{\phi}}\frac{I}{4\pi}\int_{z'=-1}^{z'=1}\frac{dz'}{\left(1 + z'^2\right)^{3/2}} = \hat{\boldsymbol{\phi}}\frac{I}{2\pi}\int_{z'=0}^{z'=1}\frac{dz'}{\left(1 + z'^2\right)^{3/2}} = \hat{\boldsymbol{\phi}}\frac{I}{2\pi}\frac{z'}{\sqrt{z'^2 + 1}}\bigg|_0^1 = \hat{\boldsymbol{\phi}}\frac{I}{2\sqrt{2}\pi} \quad \left[\frac{\mathrm{A}}{\mathrm{m}}\right]$$

The magnetic field intensity is in the ϕ direction, as indicated by the right hand rule.

(b) To calculate the magnetic field intensity at point B, it is convenient to shift the r axis so that point B is on the axis. This allows the use of the previous result, with a change in the limits of integration from $(-1,+1)$ to $(0,2)$, as shown in **Figure 8.4c**:

$$\mathbf{H}_B = \hat{\boldsymbol{\phi}}\frac{I}{4\pi}\int_{z'=0}^{z'=2}\frac{dz'}{\left(1 + z'^2\right)^{3/2}} = \hat{\boldsymbol{\phi}}\frac{I}{4\pi}\frac{z'}{\sqrt{z'^2 + 1}}\bigg|_0^2 = \hat{\boldsymbol{\phi}}\frac{I}{2\sqrt{5}\pi} \quad \left[\frac{\mathrm{A}}{\mathrm{m}}\right]$$

Note that the magnetic field intensity at point B is lower than at point A.

(c) For an infinitely long wire, we use the result in **(a)** but integrate between $-\infty$ and $+\infty$. The magnetic field intensity at a distance h from the wire is

$$\mathbf{H}_h = \hat{\boldsymbol{\phi}}\frac{I}{4\pi}\int_{z'=-\infty}^{z'=\infty}\frac{h\,dz'}{\left(h^2+z'^2\right)^{3/2}} = \hat{\boldsymbol{\phi}}\frac{I}{2\pi}\frac{z'}{h\sqrt{h^2+z'^2}}\bigg|_0^{\infty} = \hat{\boldsymbol{\phi}}\frac{I}{2\pi h}\quad\left[\frac{\text{A}}{\text{m}}\right]$$

For $I = 1$ A and $h = 1$ m,

$$\mathbf{H} = \hat{\boldsymbol{\phi}}\frac{1}{2\pi}\quad\left[\frac{\text{A}}{\text{m}}\right].$$

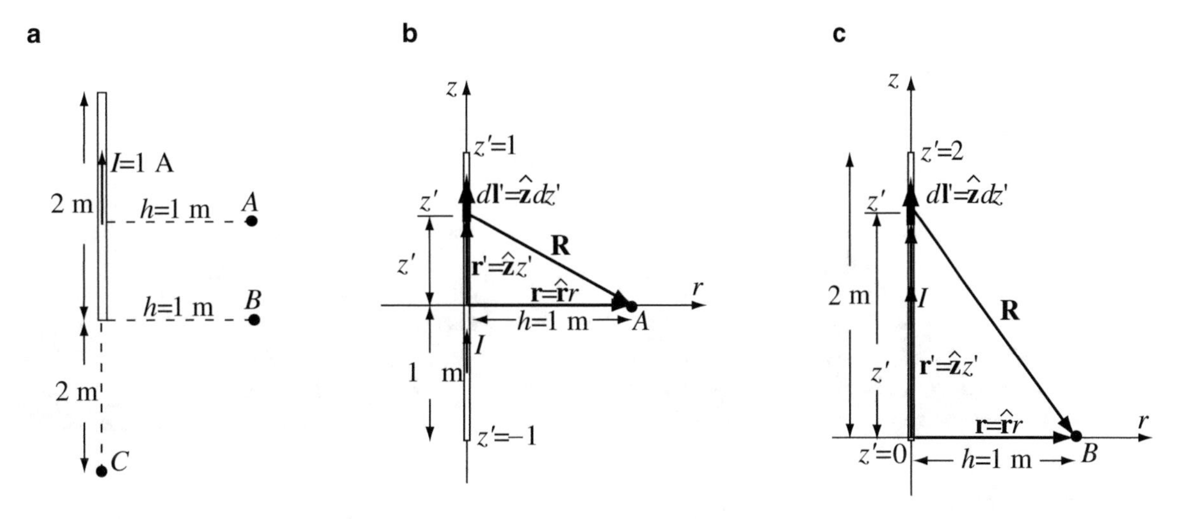

Figure 8.4 Magnetic field intensity due to a short, straight segment carrying current I. (**a**) Geometry and dimensions. (**b**) Calculation of **H** at point A. (**c**) Calculation of **H** at point B

Exercise 8.1 For the geometry in **Example 8.1a**:

(a) Calculate the magnetic field intensity at $h = 0$ in the middle of the segment in **Figure 8.4a**.
(b) Calculate the magnetic field intensity at any point along the axis of the wire, but outside the wire. Use, for example, point C in **Figure 8.4a**.

Answer
(a) 0. **(b)** 0.

Example 8.2 Magnetic Field Intensity and Magnetic Flux Density Due to a Half-Loop A current I [A] flows in the circuit shown in **Figure 8.5**. Calculate the magnetic flux density and the magnetic field intensity at the center of the half-loop assuming the circuit is in free space.

Figure 8.5 Calculation of the magnetic flux density at the center of a semicircular current loop

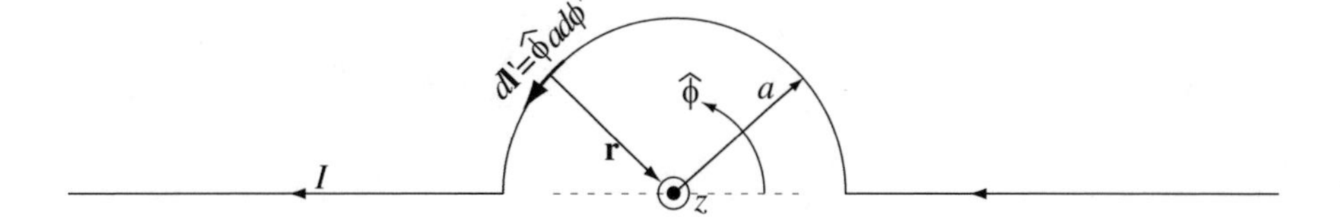

Solution: The Biot–Savart law is used to integrate around the semicircular loop. The vector **r** is in the negative r direction, and the magnetic flux density is in the direction perpendicular to the loop (out of the page for any point inside the loop). The contribution of the straight wire is zero since for any point on the straight wire, $\sin\phi$ in **Eq. (8.5)** is zero (see **Exercise 8.1**).

The two vectors necessary are $d\mathbf{l}'$ (in the direction of current) and **r** (pointing to the center of the loop). Taking the current to flow in the positive ϕ direction, the two vectors are

$$\mathbf{r} = -\hat{\mathbf{r}}a, \quad d\mathbf{l}' = \hat{\boldsymbol{\phi}}ad\phi' \quad \rightarrow \quad d\mathbf{l}' \times \mathbf{r} = \hat{\boldsymbol{\phi}}ad\phi' \times (-\hat{\mathbf{r}}a) = \hat{\mathbf{z}}a^2d\phi'$$

Now the Biot–Savart law gives

$$\mathbf{B} = \mu_0\mathbf{H} = \hat{\mathbf{z}}\frac{\mu_0 I}{4\pi}\int_0^{\pi}\frac{a^2d\phi'}{|\mathbf{r}|^3} = \hat{\mathbf{z}}\frac{\mu_0 I}{4\pi}\int_0^{\pi}\frac{d\phi'}{a} = \hat{\mathbf{z}}\frac{\mu_0 I}{4a} \quad [\mathrm{T}]$$

Thus,

$$\mathbf{B} = \hat{\mathbf{z}}\frac{\mu_0 I}{4a} \quad [\mathrm{T}] \quad \text{and} \quad \mathbf{H} = \hat{\mathbf{z}}\frac{I}{4a} \quad \left[\frac{\mathrm{A}}{\mathrm{m}}\right]$$

Example 8.3 Magnetic Field Intensity of a Circular Loop

(a) Calculate the magnetic field intensity **H** at point P in **Figure 8.6** generated by the current I [A] in the loop. Point P is at a height h [m] along the axis of the loop.
(b) Calculate the magnetic field intensity at the center of the loop (point O).

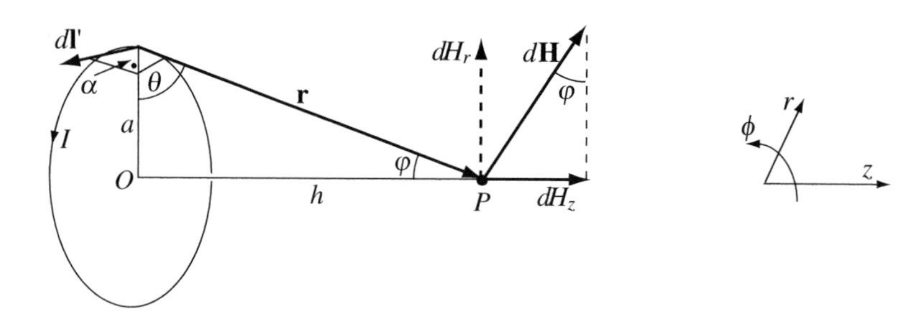

Figure 8.6 Calculation of the magnetic field intensity at height h above a current-carrying loop

Solution: The solution is most easily carried out in cylindrical coordinates. From **Figure 8.6**, the element of length is $d\mathbf{l}' = \hat{\boldsymbol{\phi}}ad\phi'$ and is directed in the positive ϕ direction. This produces a magnetic field intensity perpendicular to **r**. This magnetic field intensity has an axial component and a component perpendicular to the axis. The latter cancels because of symmetry of the current: An element $d\mathbf{l}'$ diametrically opposed to the element shown produces an identical field but the component normal to the axis is opposite in direction. Since the only nonzero field intensity is in the z direction, the calculations may be carried out in scalar components.

(a) The magnetic field intensity due to current in an element of the loop of length dl' is

$$d\mathbf{H} = \frac{Id\mathbf{l}' \times \mathbf{r}}{4\pi\mathbf{r}^3} \quad \rightarrow \quad dH = \frac{Idl'}{4\pi r^2}\sin\alpha = \frac{Idl'}{4\pi r^2} \quad \left[\frac{\mathrm{A}}{\mathrm{m}}\right]$$

where α is the angle between $d\mathbf{l}'$ and **r** and $dl' = ad\phi'$. In this case, $\alpha = 90°$. The component dH_z is

$$dH_z = \frac{Idl'}{4\pi r^2}\sin\varphi \quad [\mathrm{A/m}]$$

The angle φ is constant for any position on the loop and we have:

$$\sin\varphi = \frac{a}{\sqrt{a^2 + h^2}}$$

Thus,

$$dH_z = \frac{Idl'}{4\pi(a^2 + h^2)} \frac{a}{\sqrt{a^2 + h^2}} = \frac{Ia\,dl'}{4\pi(a^2 + h^2)^{3/2}} \quad \left[\frac{\text{A}}{\text{m}}\right]$$

Substituting for dl' and integrating over the loop, we get

$$H_z = \frac{Ia^2}{4\pi(a^2 + h^2)^{3/2}} \int_{\phi'=0}^{\phi'=2\pi} d\phi' = \frac{Ia^2}{2(a^2 + h^2)^{3/2}} \quad \left[\frac{\text{A}}{\text{m}}\right]$$

or, in a more formal notation,

$$\mathbf{H}(0{,}0{,}h) = \hat{\mathbf{z}} \frac{Ia^2}{2(a^2 + h^2)^{3/2}} \quad \left[\frac{\text{A}}{\text{m}}\right].$$

(b) At the center of the loop, $h = 0$ [m]. Thus,

$$\mathbf{H}(0{,}0{,}0) = \hat{\mathbf{z}} \frac{I}{2a} \quad [\text{A/m}]$$

Comparison with **Example 8.2** shows this result to be twice the intensity of the half-loop.

Example 8.4 Magnetic Field Intensity Due to a Rectangular Loop: Superposition of Fields A rectangular loop carries a current I [A] as shown in **Figure 8.7a**. Calculate the magnetic field intensity at the center of the loop.

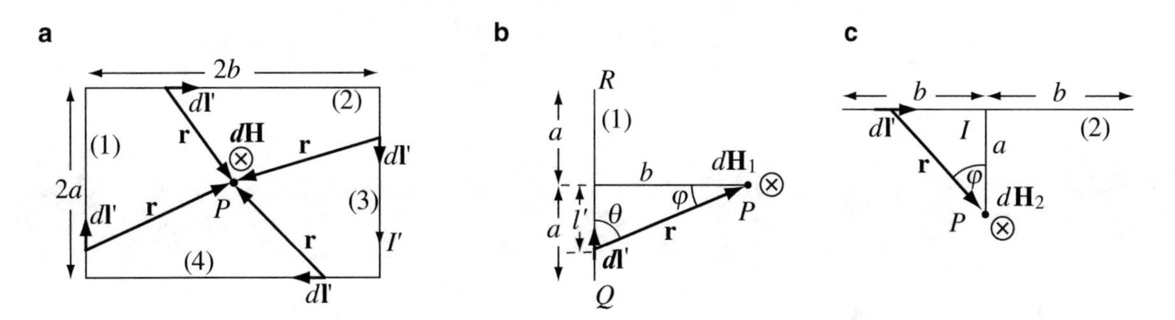

Figure 8.7 (**a**) Calculation of the magnetic field intensity at the center of a rectangular loop. (**b**) Calculation of the magnetic field intensity due to the left segment in (**a**). (**c**) Calculation of the magnetic field intensity due to the top segment in (**a**)

Solution: To calculate the magnetic field intensity, we use the Biot–Savart law for each of the segments forming the loop, following a slightly different method than in **Example 8.1**. **Figures 8.7b** and **8.7c** show how the field of each segment is calculated separately. Superposition of the four fields completes the solution.

Starting with the left, vertical segment, which we denote as segment (1) (see **Figure 8.7b**), and using scalar components, we get at the center of the loop (see also **Example 8.1** for the magnetic field intensity of a short, straight segment)

$$dH_1 = \frac{I|d\mathbf{l}' \times \hat{\mathbf{r}}|}{4\pi r^2} = \frac{Idl'}{4\pi r^2}\sin\theta = \frac{Idl'}{4\pi r^2}\cos\varphi \quad \left[\frac{\text{A}}{\text{m}}\right]$$

where $|\hat{\mathbf{r}}| = 1$ was used. Instead of using the method in **Example 8.1**, the variables are changed to be a function of φ as follows. From **Figure 8.7b**,

$$\tan\varphi = \frac{l'}{b} \quad \rightarrow \quad l' = b\tan\varphi \quad \rightarrow \quad \frac{dl'}{d\varphi} = b\sec^2\varphi \quad \rightarrow \quad dl' = b\sec^2\varphi d\varphi$$

In addition, $\cos\varphi = b/r$, which gives $r = b/\cos\varphi$. Substituting dl' and r in the expression for dH and noting that the integration is symmetric about the centerline, we can write

$$H_1 = \frac{I}{4\pi b}\int_{-\varphi_1}^{\varphi_1} \cos\varphi d\varphi = \frac{I\sin\varphi}{4\pi b}\bigg|_{-\varphi_1}^{\varphi_1} = \frac{I\sin\varphi_1}{2\pi b} \quad \left[\frac{\text{A}}{\text{m}}\right]$$

where angles $+\varphi_1$ and $-\varphi_1$ are the limiting angles corresponding to points Q and R in **Figure 8.7b**. From **Figure 8.7b**, we can write $\sin\varphi_1$ as

$$\sin\varphi_1 = \frac{a}{\sqrt{a^2 + b^2}}$$

Substituting this in the solution gives

$$H_1 = \frac{Ia}{2\pi b\sqrt{a^2 + b^2}} \quad \left[\frac{\text{A}}{\text{m}}\right]$$

The magnetic field intensity of segment (3) is identical in magnitude and direction. The magnetic field intensity of segment (2) is calculated by analogy since the only difference between segments (1) and (2) is that a and b are interchanged (see **Figure 8.7c**). Thus, the magnetic field intensity at the center of the loop due to segment (2) and therefore also due to segment (4) is

$$H_2 = H_4 = \frac{Ib}{2\pi a\sqrt{a^2 + b^2}} \quad \left[\frac{\text{A}}{\text{m}}\right]$$

The total field is $H_t = 2H_2 + 2H_1$, or

$$H_t = \frac{I}{\pi\sqrt{a^2 + b^2}}\left(\frac{a}{b} + \frac{b}{a}\right) = \frac{I}{\pi ab}\sqrt{a^2 + b^2} \quad \left[\frac{\text{A}}{\text{m}}\right] \quad \text{(directed into the page)}$$

Exercise 8.2 Consider again **Example 8.4** and calculate the magnetic field intensity at one of the corners of the loop in **Figure 8.7a**.

Answer $H = \frac{I}{8\pi ab}\sqrt{a^2 + b^2} \quad \left[\frac{\text{A}}{\text{m}}\right]$ (directed into the page).

8.3.1 Applications of the Biot–Savart Law to Distributed Currents

The fact that the Biot–Savart law is written as a line integral does not mean that current distributions (say, current in a thin sheet of metal or current in a volume) cannot be treated as well. In fact, all that is required is to view the current distribution as an assembly of thin wires and calculate the field as a superposition of the fields due to each thin wire using **Eq. (8.8)** or **(8.9)**. In practice, each thin wire is a wire of differential thickness, and the contribution is found by integration rather than by summation.

There are two specific current configurations which are particularly important and will be treated here. One is a planar distribution of currents shown in **Figure 8.8a** and the second is a volume distribution shown in **Figure 8.8b**. These represent thin flat conductors and thick conductors, respectively. Consider first a very thin, flat conductor of length $2L$ and width $2d$ as shown in **Figure 8.9a**. A total current I flows in the conductor. Note that the conductor must be part of a larger closed circuit for the current to flow, but we wish to calculate the magnetic field intensity due to this section alone. If we "cut" out of the sheet, a wire of width dx' and the same thickness as the sheet, located at x', we obtain the element shown in **Figure 8.9b**. This is exactly the same configuration as in **Figure 8.3** for a straight, thin wire. The current in this differential wire is

$$dI = \left(\frac{I}{2d}\right)dx' \quad [\text{A}] \tag{8.10}$$

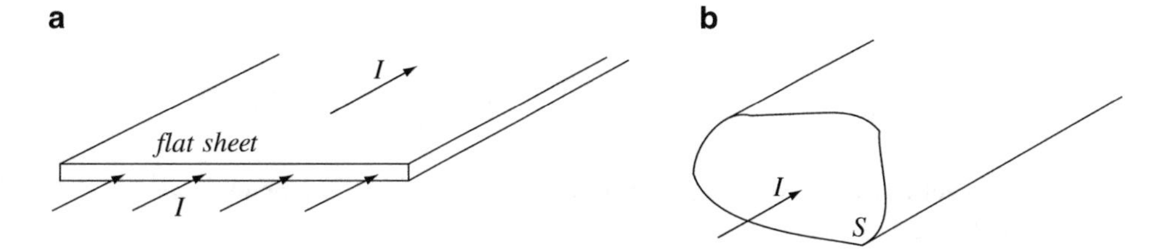

Figure 8.8 (**a**) Surface current distribution. (**b**) Volume current distribution

and the total contribution due to this differential wire is found using **Eq. (8.8)**:

$$d\mathbf{H}(x,y,z) = \left[\int_{z'=-L}^{z'=+L} \frac{I\,dx'}{2d}\frac{d\mathbf{l}'\times\hat{\mathbf{R}}}{4\pi|\mathbf{r}-\mathbf{r}'|^2}\right] \quad \left[\frac{\text{A}}{\text{m}}\right] \tag{8.11}$$

where integration is on $d\mathbf{l}$ and $\mathbf{r}$ is the vector connecting $d\mathbf{l}$ and $P(x,y,z)$ (see **Figure 8.3**). To obtain the total field intensity, we integrate over the width of the current sheet in **Figure 8.9a**. We get

$$\mathbf{H}(x,y,z) = \int_{x'=-d}^{x'=+d}\left[\int_{z'=-L}^{z'=+L} \frac{I}{2d}\frac{d\mathbf{l}'\times\hat{\mathbf{R}}}{4\pi|\mathbf{r}-\mathbf{r}'|^2}\right]dx' \quad \left[\frac{\text{A}}{\text{m}}\right] \tag{8.12}$$

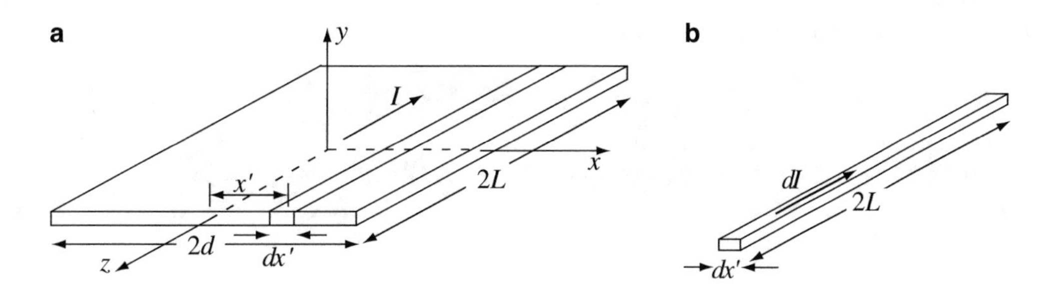

Figure 8.9 (**a**) Flat sheet with a current distribution along its width. (**b**) An element of current used to apply the Biot–Savart law

Now, we must define the various vectors, $d\mathbf{l}'$, $\mathbf{r}$, $\mathbf{r}'$, $\mathbf{R}$, and perform the integration. An example of how this may be done is shown in **Example 8.5**. In **Eq. (8.12)**, the term $I/2d$ has units of current per unit length [A/m]. This is a surface current density and is a convenient term when dealing with thin sheets of current since, in this case, the thickness of the current sheet is small and fixed.

If the current is distributed throughout a conductor of finite volume, we proceed in the same fashion: Define a differential wire and integrate the contributions of the individual filaments over the cross section of the conductor. To see how this is done, consider the conductor in **Figure 8.10a**. The current I is assumed to be uniformly distributed throughout the cross-sectional area of the conductor. Thus, referring to **Figure 8.10b**, the geometry is that of a thin wire of cross-sectional area $dx'dy'$, carrying a current dI and located at a general point (x',y') in the conductor, shown separately in **Figure 8.10b**:

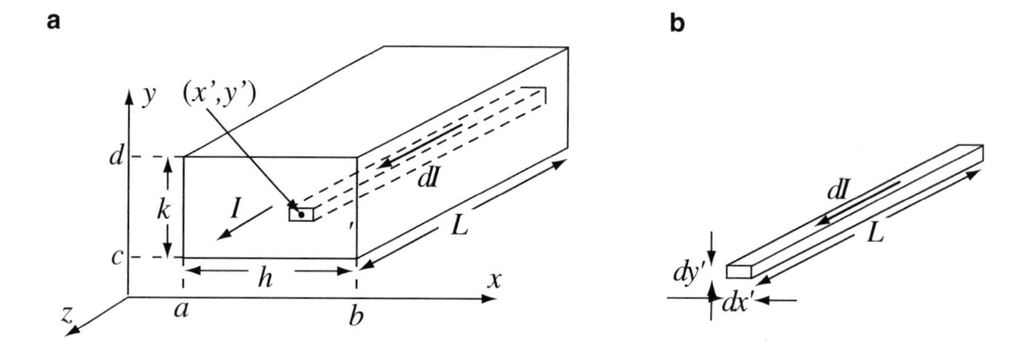

Figure 8.10 (**a**) A thick conductor with a current distributed throughout its cross section. (**b**) An element of current in the form of a wire of differential cross-sectional area

$$dI = \frac{I}{kh} dx'dy' \quad [\text{A}] \tag{8.13}$$

where the quantity I/kh is the current density in the wire [A/m^2]. Now, we substitute **Eq. (8.13)** into **Eq. (8.8)** to find the contribution of the differential wire:

$$d\mathbf{H}(x,y,z) = \int_{z'=0}^{z'=L} \frac{I}{kh} dx'dy' \frac{d\mathbf{l}' \times \hat{\mathbf{R}}}{4\pi|\mathbf{r} - \mathbf{r}'|^2} \quad \left[\frac{\text{A}}{\text{m}}\right] \tag{8.14}$$

where, based on **Figure 8.10**, we substitute $d\mathbf{l}' = \hat{\mathbf{z}}dz'$ (integrating in the direction of the current). To find the total magnetic field intensity due to the conductor, we integrate over the cross-sectional area of the conductor to account for the total current:

$$\mathbf{H}(x,y,z) = \int_{y'=c}^{y'=d} \left[\int_{x'=a}^{x'=b} \left[\int_{z'=0}^{z'=L} \frac{I}{kh} \frac{\hat{\mathbf{z}} \times \hat{\mathbf{R}}}{4\pi|\mathbf{r} - \mathbf{r}'|^2} dz' \right] dx' \right] dy' \quad \left[\frac{\text{A}}{\text{m}}\right] \tag{8.15}$$

where $d\mathbf{l}$, $\mathbf{r}$, $\mathbf{r}'$, and $\mathbf{R}$ are defined in **Figure 8.3**. These expressions may look intimidating, but they are nothing more than the integration of the magnetic field intensity of a differential area wire over the cross-sectional area of a thick conductor **[Eq. (8.15)]** or width of a flat conductor **[Eq. (8.12)]**.

Example 8.5 Magnetic Field Intensity Due to a Long Thin Sheet: Application as a Ground Plane A thin sheet of conducting material, $b = 1$ m wide, and very long carries a current $I = 100$ A. Calculate the magnetic field intensity at a distance $h = 0.1$ m above the center of the conductor (see **Figure 8.11a**).

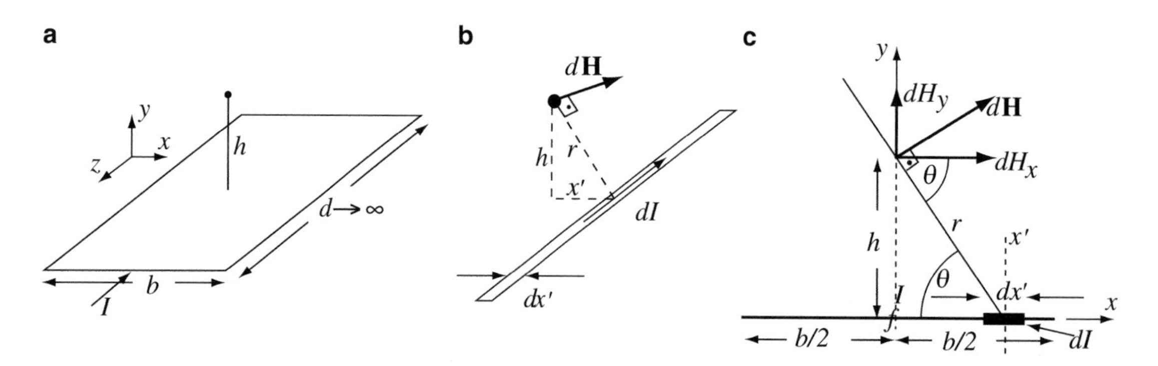

Figure 8.11 Magnetic field intensity due to a conducting ground plane. (**a**) Configuration. (**b**) Calculation of the magnetic field intensity of a differential strip. (**c**) Calculation of the field of the ground plane as an assembly of infinitesimal strips

Solution: The magnetic field intensity may be evaluated by viewing the sheet as an assembly of thin conductors, each dx' [m] wide, infinitely long, and carrying a current $(I/b)dx'$. The magnetic field intensity for a thin segment was found in **Example 8.1** by direct application of the Biot–Savart law. The field of the infinitesimally thin conductor is integrated over the width of the sheet to find the magnetic field intensity due to the sheet.

The current dI in an infinitesimally thin section of the sheet (cut lengthwise as in **Figure 8.11b)** is

$$dI = \frac{I}{b}dx' = 100dx' \quad [\mathrm{A}]$$

The magnetic field intensity at a distance r from a long line carrying a current dI (see **Example 8.1c)** is

$$d\mathbf{H} = \hat{\boldsymbol{\phi}}\frac{dI}{2\pi r} \quad \left[\frac{\mathrm{A}}{\mathrm{m}}\right]$$

Now we place this infinitesimal segment at a general location x' on the sheet, as shown in **Figure 8.11c**. The distance r is the distance to the point at which we wish to calculate the magnetic field intensity. Because as x' changes, the angle θ and, therefore, the direction of the field intensity change, it is easier to calculate the vertical and horizontal components separately and then integrate each. From symmetry alone, we know that the vertical component must cancel because symmetric segments have equal and opposite vertical components. The horizontal component is

$$dH_x = \frac{dI}{2\pi r}\cos(90-\theta) = \frac{dI}{2\pi r}\sin\theta = \frac{h\,dI}{2\pi r^2} \quad \left[\frac{\mathrm{A}}{\mathrm{m}}\right]$$

where $\sin\theta = h/r$ was used. Substituting $r = (x'^2 + h^2)^{1/2}$ and rearranging terms, we get

$$dH_x = \frac{h\,dI}{2\pi(x'^2 + h^2)} \quad \left[\frac{\mathrm{A}}{\mathrm{m}}\right]$$

Substituting $dI = 100dx'$ and $h = 0.1$ [m] and integrating between $x' = -b/2 = -0.5$ m and $x' = b/2 = 0.5$ m gives

$$H_x = \frac{10}{2\pi}\int_{-0.5}^{0.5}\frac{dx'}{(x'^2 + 0.1^2)} = \frac{10}{2\pi}\left[\frac{1}{0.1}\tan^{-1}\frac{x'}{0.1}\right]_{-0.5}^{0.5} = \frac{50 \times 2.7468}{\pi} = 43.72 \quad [\mathrm{A/m}]$$

The resulting magnetic field intensity is parallel to the sheet (in the x direction in **Figure 8.11c)**.

8.4 Ampère's Law

Ampère's law,[5] also called the Ampère circuital law, states that

"the circulation of **H** *around a closed path C is equal to the current enclosed by the path."*

That is,

$$\boxed{\oint_C \mathbf{H}\cdot d\mathbf{l} = I_{enclosed} \quad [\mathrm{A}]} \tag{8.16}$$

[5] Ampère's law is named after André-Marie Ampère (see footnote on page 337). Ampère derived this law from study of the solenoidal (circular) nature of the magnetic field of straight wires.

The circulation is defined by the line integral around a closed contour of the scalar product $\mathbf{H} \cdot d\mathbf{l}$. Thus, in fact, only the component of **H** tangential to the contour of integration is included in the calculation.

Why do we need another law that, as stated, may be derived from the Biot–Savart law (even though we show no proof)? There are a couple of reasons. First, Ampère's law is much easier to apply to some problems. In particular, highly symmetric current configurations are easily evaluated using Ampère's law, whereas they may be more complex using the Biot–Savart law. It should be remembered, however, that Ampère's law is universally applicable. Second, in the next section, we will show that Ampère's law is, in fact, one of the postulates of the magnetic field. This will become even more important when we discuss time-dependent fields in **Chapter 11**. In a practical sense, Ampère's law is another tool which we can use whenever it makes sense to do so.

When applying Ampère's law, we must remember that under normal circumstances, the unknown quantity is **H**. Since **H** is inside the integral sign, we must find a closed contour, enclosing the current I such that the component of **H** tangential to the contour is constant along the contour. Under these conditions, $\mathbf{H} \cdot d\mathbf{l} = H_{tan}\, dl$ and H_{tan} can be taken outside the integral sign. Evaluation of the integral is now possible. Thus, the requirements here are that **H** be tangential to a contour and constant along the contour. These conditions are satisfied for highly symmetric current configurations. These include the following:

(1) Current in an infinite (or as an approximation, in a very long) filament.
(2) Current or current density in an infinite (or very long) solid or hollow cylindrical conductor. Normally, the current is uniformly distributed in the conductor, but any radially symmetric current distribution is allowed.
(3) Infinite sheet of current (or very large, flat current sheet).
(4) Multiple conductors in a symmetric configuration.
(5) Nonsymmetric current distributions that are superpositions of symmetric current distributions as described in (1) through (3).

It should also be noted that in the use of Ampère's law, the vector notation of the magnetic field intensity is lost–we can only calculate its magnitude. The vector notation can be restores through the use of the right hand rule.

The key requirement for the application of Ampère's law is symmetry of current or current density distribution. The law applies to any current configurations, but it is not generally possible to find a contour over which the tangential magnetic field intensity is constant and, therefore, evaluate the field. The following examples show various important aspects of application of Ampère's law.

Example 8.6 Application: Field Intensity Due to a Single, Thin Wire–Magnetic Field of Overhead Transmission Lines Calculate the magnetic field intensity due to a long filamentary conductor carrying a current I at a distance h from the wire. The conductor is very long (infinite). Compare this result with the result in **Example 8.1c**.

Solution: Since the magnetic field intensity of the wire is circular, a circle of radius h (in the $r-\phi$ plane) may be used as the contour in Ampère's law. Taking the current to be in the z direction for convenience (see **Figure 8.2b** or **Figure 8.4b**), the magnetic field intensity is in the ϕ direction. Taking $d\mathbf{l} = \hat{\boldsymbol{\phi}} h d\phi$ we get from Ampère's law

$$\oint \mathbf{H} \cdot d\mathbf{l} = \oint \hat{\boldsymbol{\phi}} H \cdot \hat{\boldsymbol{\phi}} h d\phi = Hh \int_0^{2\pi} d\phi = 2\pi H h = I \quad \rightarrow \quad H = \frac{I}{2\pi h} \quad \left[\frac{\text{A}}{\text{m}}\right]$$

This result is identical to the result obtained in **Example 8.1c** but is much easier to obtain.

Example 8.7 Application: Field Intensity Due to a Long, Thick Conductor—The Transmission Line A wire of radius $a = 10$ mm carries a current $I = 400$ A. The wire may be assumed to be infinite in length.

(a) Calculate the magnetic field intensity everywhere.
(b) Plot the magnetic field intensity as a function of distance from the center of the conductor.

Solution: The solution in **Example 8.6** is the magnetic field intensity at any distance outside a wire carrying a current I. We also recall that only the current enclosed by the contour contributes to the magnetic field intensity. Thus, outside the conductor, at a distance $r > a$, the magnetic field intensity is given in **Example 8.6**. Inside the wire ($r \le a$), we still use the same relation, but the current is only that enclosed by the contour.

(a) For $0 < r \le a$: The current density in the wire is the total current divided by the cross-sectional area of the wire (see **Figure 8.12a**):

$$J = \frac{I}{\pi a^2} \quad \left[\frac{\mathrm{A}}{\mathrm{m}^2}\right]$$

The current enclosed by a contour of radius r equals the current density multiplied by the cross-sectional area enclosed by the contour:

$$I_r = JS_r = \frac{I\pi r^2}{\pi a^2} = I\frac{r^2}{a^2} \quad [\mathrm{A}]$$

The magnetic field intensity at a distance r from the center of the wire ($0 < r \le a$) is

$$H = \frac{I_r}{2\pi r} = \frac{Ir}{2\pi a^2} = \frac{400r}{2 \times \pi \times (0.01)^2} = 6.366 \times 10^5 r \quad [\mathrm{A/m}]$$

For $r > a$ the result of **Example 8.6** gives

$$H = \frac{I_r}{2\pi r} = \frac{I}{2\pi r} = \frac{400}{2 \times \pi \times r} = \frac{63.66}{r} \quad \left[\frac{\mathrm{A}}{\mathrm{m}}\right].$$

To restore the vector form of H, we note that the magnetic field intensity is circular, following the right-hand rule. If the current is in the z direction, $\mathbf{H}$ is in the ϕ direction. That is,

$$\mathbf{H} = \hat{\boldsymbol{\phi}}\frac{63.66}{r} \quad \left[\frac{\mathrm{A}}{\mathrm{m}}\right].$$

(b) The plot of H versus distance is shown in **Figure 8.12b**. Note that the magnetic field intensity is zero at the center of the wire and rises linearly up to the wire surface where the magnetic field intensity is 6366 A/m. Then, it decays as $1/r$ until it is zero at infinity. The plot in **Figure 8.12b** gives only the magnitude of **H**.

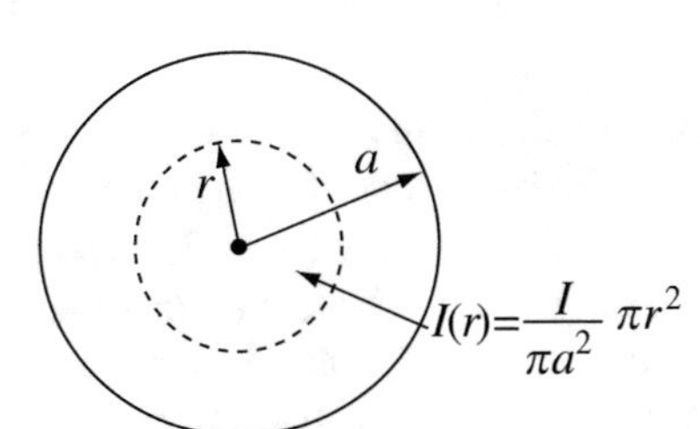

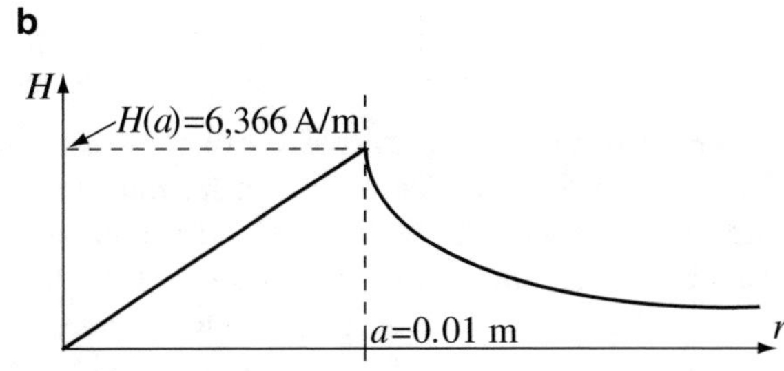

Figure 8.12 Magnetic field intensity due to a thick conductor carrying a uniformly distributed current. (**a**) Method of calculation of **H** inside the conductor. (**b**) Plot of H versus radial distance from the center of the conductor

Example 8.8 Application: Field Intensity Due to an Infinite Sheet—Application as a Ground Plane for Lightning Protection When it becomes necessary to protect devices from overvoltages such as produced by lightning, ground planes are often employed. Underneath power substations and in computer rooms, it is common to use a grid of conductors (as an approximation to a continuous, conducting plane). The example that follows shows how the magnetic field intensity due to a large ground plane is calculated.

A very large sheet of conducting material may be assumed to be infinite and carries a surface current density $J_l = 10$ A/m as shown in **Figure 8.13a**. Calculate the field intensity everywhere in space.

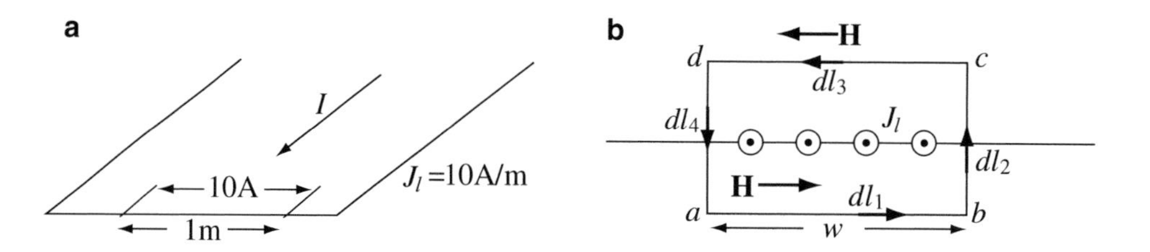

Figure 8.13 Magnetic field intensity of a large (infinite) ground plane. (**a**) Geometry and dimensions. (**b**) The contour used for calculation using Ampère's law

Solution: To calculate the field, we draw a contour on which the magnetic field intensity is either perpendicular or parallel to the contour. From symmetry considerations, the magnetic field intensity at any two points at equal distance from the plane must be the same. The proper contour is a rectangular loop of arbitrary size, as shown in **Figure 8.13b**. The dimensions *bc* and *da* are not important since the same amount of current is enclosed by the contour for any dimension *bc* and *da*.

We assume the magnetic field intensity on both sides of the sheet is in the direction of integration (right-hand rule) and evaluate the integral:

$$\oint \mathbf{H} \cdot d\mathbf{l} = I = J_l w$$

where J_l is the current density on the surface of the conductor and has units of ampere/meter [A/m]. Performing the integration in segments gives

$$\oint \mathbf{H} \cdot d\mathbf{l} = \int_a^b \mathbf{H}_1 \cdot d\mathbf{l}_1 + \int_b^c \mathbf{H}_2 \cdot d\mathbf{l}_2 + \int_c^d \mathbf{H}_3 \cdot d\mathbf{l}_3 + \int_d^a \mathbf{H}_4 \cdot d\mathbf{l}_4$$

The second and fourth integrals are zero because **H** and $d\mathbf{l}$ are perpendicular to each other on these segments. The first and third integrals are evaluated observing that $|H_1| = |H_3| = |\mathrm{H}|$ and, from the symmetry consideration above, are constant along the segments *ab* and *cd*. Also, the distances *ab* and *cd* are equal to *w*. Thus,

$$\oint \mathbf{H} \cdot d\mathbf{l} = H\int_a^b dl_1 + H\int_c^d dl_3 = 2Hw = J_l w \quad \rightarrow \quad H = \frac{J_l}{2} = 5 \quad [\mathrm{A/m}]$$

The magnetic field intensity is constant in space, equals half the magnitude of the current density, and is in opposite directions on the two sides of the plate, following the right-hand rule. The magnetic field intensity is parallel to the surface of the sheet. The result above is exact for an infinite sheet but may be used as an approximation for large sheets or even for smaller sheets if the field very close to the sheet is required. Note also that the infinite sheet geometry may be viewed as being made of single, thin wires, very closely spaced, with *N* wires per unit width of the sheet, and each carrying a current *I*/*N* as shown in **Figure 8.13b**. If *N* tends to infinity, the continuous distribution in **Figure 8.13a** is obtained.

Example 8.9 Application: Magnetic Field Intensity in a Toroidal Coil A torus with the dimensions shown in **Figure 8.14a** is wound with $N = 100$ turns of wire, uniformly wound around the torus. The coil thus formed carries a current $I = 1$ A. Calculate the magnetic flux density everywhere in space. The torus has a rectangular cross section and the permeability of the core is μ_0 [H/m].

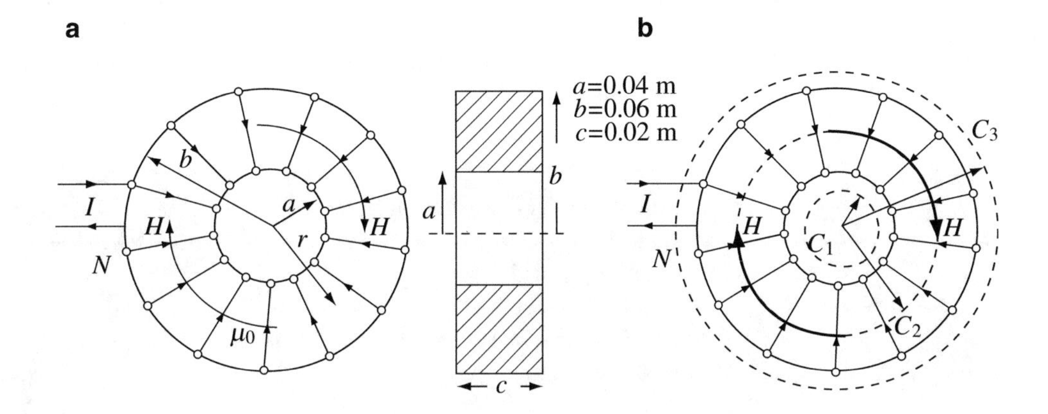

Figure 8.14 A toroidal coil. (**a**) Dimensions and construction. (**b**) Contours used for calculation of the magnetic flux density

Solution: To use Ampère's law, we need to find a contour on which the magnetic field intensity is constant in magnitude and, preferably, in the direction of the contour. **Figure 8.14b** shows that any contour which is concentric with the torus' axis is symmetric about all conductors and, therefore, can be used in conjunction with Ampère's law. From the right-hand rule, applied to individual wires, the direction of the flux density is parallel to the contours, as shown in **Figure 8.14b**.

For $r < a$: Contour C_1 does not enclose any current. Therefore, the magnetic field intensity and magnetic flux density are zero. For $a < r < b$: Contour C_2 encloses a current equal to NI. From Ampère's law,

$$\oint_{C_2} \mathbf{H} \cdot d\mathbf{l} = H \oint dl_2 = H 2\pi r = NI \quad \rightarrow \quad H = \frac{NI}{2\pi r} = \frac{100}{2\pi r} = \frac{15.915}{r} \quad \left[\frac{\text{A}}{\text{m}}\right]$$

The magnetic field intensity varies within the torus. It is higher toward the inner surface and lower toward the outer surface. Sometimes, it is convenient to approximate the magnetic field intensity in the torus as an average between the outer and inner fields and assume that this average field intensity exists everywhere within the torus. Calculating the magnetic field intensity at $r = a$ and $r = b$ and taking the average between the two values gives

$$H_{av} = \frac{NI}{4\pi}\left(\frac{1}{a} + \frac{1}{b}\right) = \frac{100}{4\pi}\left(\frac{1}{0.04} + \frac{1}{0.06}\right) = 331.57 \quad [\text{A/m}]$$

This approximation is quite good if a and b are large compared to the radial thickness of the torus $(b - a)$. The magnetic flux density is found from the relation $\mathbf{B} = \mu_0 \mathbf{H}$:

$$B = \mu_0 H = \frac{\mu_0 NI}{2\pi r} = \frac{4 \times \pi \times 10^{-7} \times 100}{2 \times \pi \times r} = \frac{2 \times 10^{-5}}{r} \quad [\text{T}]$$

and the average magnetic flux density is

$$B_{av} = \frac{\mu_0 NI}{4\pi}\left(\frac{1}{a} + \frac{1}{b}\right) = \frac{4 \times \pi \times 10^{-7} \times 100}{4 \times \pi}\left(\frac{1}{0.04} + \frac{1}{0.06}\right) = 4.167 \times 10^{-4} \quad [\text{T}]$$

For $r > b$: Contour C_3 encloses N currents into the page and N currents out of the page. Thus, the total net current enclosed is zero and the magnetic field intensity and magnetic flux density are zero.

Note that the magnetic flux density does not depend on the width c of the torus.

Notes:

(1) The magnetic field is contained entirely within the torus. This property is used in many applications, including high-quality transformers and other coils which require containment of the field. Because of the field containment, the magnetic field generated in a torus does not affect or interfere with other devices.
(2) The effect of having 100 turns is to multiply the current by 100. In other words, from Ampère's law, the total current enclosed in the contour is $100I$. If this current is due to 10 turns carrying 10 A each, or 10,000 turns carrying 0.01 A each, the result is the same. For this reason, we will often use the term ***ampere • turns*** [A • t] to indicate the total current in a device.
(3) This problem, although easily solved using Ampère's law, is difficult to solve using the Biot–Savart law; integration along all current paths is very tedious.

Example 8.10 Application: Magnetic Field Inside a Long Solenoid A solenoid is a long coil of wires, wound on a circular (sometimes rectangular) form. The turns of the solenoid are most often tightly wound and normally in a single layer. Short, multilayer solenoids are called coils.

A very long cylindrical solenoid is wound with a turn density of 1,000 turns per meter length of the solenoid. The current in the turns of the solenoid is $I = 1$ A. Calculate the magnetic field intensity and magnetic flux density everywhere in space. The current and dimensions are shown in **Figure 8.15a** (in axial cross section).

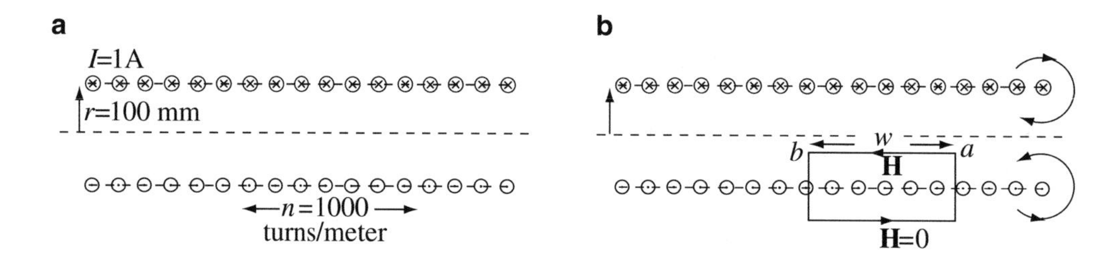

Figure 8.15 A long solenoid. (**a**) Dimensions and properties (axial cross section shown). (**b**) Contour used for calculation

Solution: Because the coil is cylindrical, the cross section shown in **Figure 8.15a** is identical through any axial cut that includes the axis of the coil. The configuration looks as if we had two planar current sheets, with currents in opposite directions. From the right-hand rule, we note that the fields of the two opposing current layers are in opposite directions outside the solenoid but are in the same direction inside the solenoid. Thus, the field outside the solenoid must be zero.

A contour is chosen as shown in **Figure 8.15b**. The width of the contour is w and is arbitrary. The total current enclosed by the contour is nwI. Since the field outside the solenoid is zero and inside it must be axial (from the right-hand rule and symmetry considerations), Ampère's law gives

$$\oint_{C_2} \mathbf{H} \cdot d\mathbf{l} = H \int_a^b dl = Hw = nwI \quad \rightarrow \quad H = nI \quad [\text{A/m}]$$

Thus, the magnetic field intensity inside the solenoid is

$$H = nI = 1{,}000 \quad [\text{A/m}]$$

The magnetic flux density is found by multiplying the magnetic field intensity by the permeability of free space:

$$B = \mu_0 H = \mu_0 nI = 4 \times \pi \times 10^{-7} \times 1{,}000 = 1.257 \times 10^{-3} \quad [\text{T}]$$

Like the torus in **Example 8.9**, the field of a long solenoid is zero outside the solenoid. Inside the solenoid, the field is constant. Practical solenoids are finite in length and, for these, the result obtained here is only an approximation. Note also that an infinite solenoid may be viewed as a torus with infinite radius (see **Exercise 8.3**). Note, again, that the same result would be obtained if, for example, the turn density were 100 turns/m while the current were 10 A. What is important for the solution is the term nI, which is the number of *ampere • turns* per unit length of the solenoid. For this reason, a solenoid may also be made of a single, bent sheet, carrying a given current.

Exercise 8.3 Show that the magnetic field intensity of a circular cross-sectional torus of average radius a [m] equals the field of an infinite solenoid as the average radius of the torus tends to infinity.

8.5 Magnetic Flux Density and Magnetic Flux

We started this chapter with the Lorentz force on a moving charge and showed that this force is proportional to the magnetic flux density **B** [see **Eq. (8.1)**]. The relation between the magnetic field intensity **H** and the magnetic flux density **B** for general materials was also given as $\mathbf{B} = \mu\mathbf{H}$ in **Eq. (8.3)**. The permeability μ is material related. We will discuss this in more detail in the following chapter. For now it is sufficient to say that every material has a given and measurable permeability μ and we may assume that permeability is known, even though it may not be a constant value.

The units of permeability can now be defined in terms of **B** and **H**. The SI unit for **B** is the *tesla*[6] [T], whereas that for **H** is the *ampere/meter* [A/m]. Thus, permeability has units of *tesla • meter/ampere* or *weber • meter/(meter^2 • ampere)*. In the latter form, the *weber/ampere* is called a *henry* [H]. Therefore, the units of permeability are *henry/meter* [H/m]:

$$\mu \quad \rightarrow \quad \left[\frac{\text{tesla}\cdot\text{meter}}{\text{ampere}}\right] = \left[\frac{\text{weber}\cdot\text{meter}}{\text{meter}^2\cdot\text{ampere}}\right] = \left[\frac{\text{weber}}{\text{ampere}}\frac{1}{\text{meter}}\right] = \left[\text{henry}\frac{1}{\text{meter}}\right] = \left[\frac{\text{H}}{\text{m}}\right]$$

The magnetic flux density is an area density vector. In the case of the electric field, we also had a flux density (**D**), and in the case of steady currents, the current density (**J**) was also an area density vector. In both cases, we integrated the density function over the area to obtain the total flux through an area. For example, the surface integral of **J** was the total current through the surface. Because **B** is a flux per unit area, we can write

$$\boxed{\Phi = \int_s \mathbf{B}\cdot d\mathbf{s} \quad [\text{Wb}]} \tag{8.17}$$

The quantity Φ is called the ***magnetic flux*** and has units of *tesla • meter*^{2} [**T** • m^2] or weber [Wb]. Whereas the tesla is usually used for **B,** the weber is more convenient for the flux. The concept of magnetic flux is shown in **Figure 8.16**. The analogy between the magnetic flux to any real flow is exactly that—an analogy. Nothing flows in the real sense. On the other hand, this analogy allows the use of analogous quantities. In flow of a fluid, we can define lines of flow or streamlines. These lines show the direction of flow. Similarly, a magnetic flux line is a line along which the magnetic flux density **B** is directed; that is, the magnetic flux density is tangential to flux lines everywhere in the magnetic field. The direction of **B** must be shown specifically by an arrow if required (see **Figure 8.16b**). Note also that flux lines do not imply that the flux density is constant. The magnitude of the flux density is indicated by the line density. At any point, the distance between adjacent flux lines indicates the magnitude of the flux density **B**. The concept of flux lines or field lines is artificial. It serves to conceptualize the magnetic field just as it serves to conceptualize the flow in a river.

[6] After Nicola Tesla (1856–1943). Tesla is well known for his invention of AC electric machines and development of the multiphase system of AC power. This happened in the early 1880s, followed by a number of other important engineering designs, including the installation of the first, large-scale AC power generation and distribution system at Niagara Falls (approx. 12 MW), in 1895. His induction motor became a staple of AC power systems, but his patents (about 500 in all) include transformers, generators, and systems for transmission of power. Tesla also had designs of grandiose scale, including "worldwide aerial transmission of power," high-frequency, high-voltage generators and many others. Tesla was always somewhat of a "magician" and liked to keep his audience guessing, but later in his life, he became more of an eccentric and a recluse. Although some of his designs were more dreams than engineering, he is considered by many as the greatest inventor in electrical engineering. His statement regarding his student pay that there are "too many days after the first of the month" is even more familiar than his inventions, even to those that never heard it.

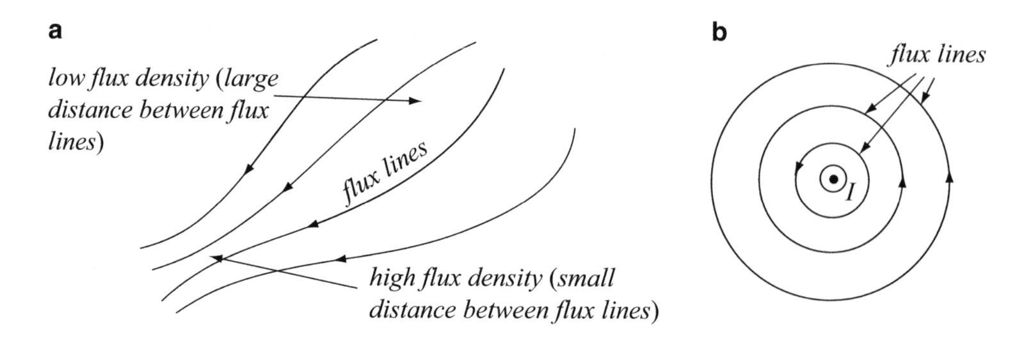

Figure 8.16 The concept of magnetic flux. (**a**) Flux in space. (**b**) Flux lines around a filamentary current

Example 8.11 Application: Flux Through a Loop A square loop is placed near a current-carrying wire as shown in **Figure 8.17a**. The loop and the wire are in the r–z plane in free space:

(**a**) Calculate the flux through this loop.

(**b**) The loop is now turned around its vertical axis by 90° (**Figure 8.17b**) so that the loop is now perpendicular to the r–z plane and symmetric about the wire. Calculate the flux through the loop.

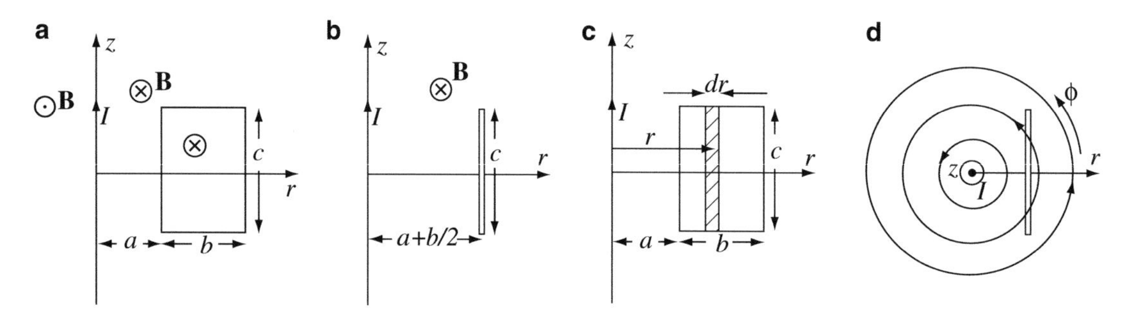

Figure 8.17 (**a**) Loop placed near a current-carrying wire. Loop and wire are in a plane. (**b**) Loop perpendicular to the plane in (**a**). (**c**) Method of calculating the flux in the loop. (**d**) Total flux in (**b**) is zero

Solution: The flux is calculated using **Eq. (8.17)**, whereas the flux density **B** is calculated using Ampère's law in **Eq. (8.16)**. The magnetic field intensity of a wire was calculated in **Example 8.6**, but we will calculate it here anew.

(**a**) To calculate the magnetic flux density, we define a contour of radius r and use cylindrical coordinates with the wire coinciding with the z axis. For a current in the positive z direction, the magnetic field intensity is in the positive ϕ direction (right-hand rule, see **Figure 8.17d**). Thus, Ampère's law gives

$$\oint \mathbf{B}\cdot d\mathbf{l} = \oint \hat{\boldsymbol{\phi}} B \cdot \hat{\boldsymbol{\phi}} r d\phi = Br\int_{\phi=0}^{2\pi} d\phi = 2\pi r B = \mu_0 I$$

or

$$B = \frac{\mu_0 I}{2\pi r} \quad \text{or} \quad \mathbf{B} = \hat{\boldsymbol{\phi}}\frac{\mu_0 I}{2\pi r} \quad [\text{T}]$$

To calculate the flux, we define an element of area $ds = cdr$ as shown in **Figure 8.17c** and integrate this from $r = a$ to $r = a + b$:

$$\Phi = \int_s \hat{\mathbf{\phi}} \frac{\mu_0 I}{2\pi r} \cdot \hat{\mathbf{\phi}} ds = \frac{\mu_0 I c}{2\pi} \int_a^{a+b} \frac{dr}{r} = \frac{\mu_0 I c}{2\pi} \ln_r \bigg|_{r=a}^{r=a+b} = \frac{\mu_0 I_c}{2\pi} \ln \frac{a+b}{a} \quad [\text{Wb}].$$

(b) The loop is now perpendicular to the r–z plane. Although direct integration as in **(a)** can be performed, the total (net) flux through the loop must be zero. This can be seen from the fact that any flux that enters the loop equals the flux that exits it, as shown in **Figure 8.17d**. Note, however, that if the loop were not symmetric about the current, this argument would not be correct.

8.6 Postulates of the Static Magnetic Field

From the relations obtained so far, it is now possible to define a minimum set of required relations for the static magnetic field: the postulates. To do so, the curl and divergence of the field must be specified (Helmholtz's theorem). However, instead of trying to define these directly, we will use the relations already obtained. One relation is Ampère's law in **Eq. (8.16)**. We can rewrite the enclosed current in **Eq. (8.16)** in terms of the current density **J** using **Figure 8.18.**

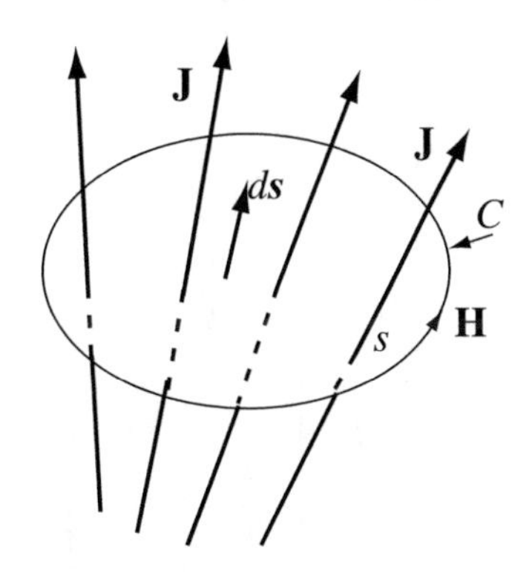

Figure 8.18 Ampère's law in terms of current density

The current enclosed by a closed contour C is the integral of the normal component of **J** over the surface bounded by the contour C. Thus, using the scalar product to calculate the normal component, we can write

$$\oint_C \mathbf{H} \cdot d\mathbf{l} = \int_s \mathbf{J} \cdot d\mathbf{s} \tag{8.18}$$

Using Stokes' theorem, we convert the closed contour integral to a surface integral as

$$\oint_C \mathbf{H} \cdot d\mathbf{l} = \int_s (\nabla \times \mathbf{H}) \cdot d\mathbf{s} = \int_s \mathbf{J} \cdot d\mathbf{s} \tag{8.19}$$

Equating the integrands of the two surface integrals gives the required curl condition:

$$\boxed{\nabla \times \mathbf{H} = \mathbf{J}} \tag{8.20}$$

The second relation we must identify is the divergence of the magnetic field intensity or that of the magnetic flux density. Before doing so, we will give two examples which will serve as an introduction to the general result. First, consider **Figure 8.16b**, in which a few flux lines of a line current are shown. If we draw any volume, anywhere in space, the total number of flux lines entering the volume must be equal to the total number of flux lines leaving the volume. If it were otherwise, some flux lines would either terminate in the volume or start in the volume. The conclusion is that the total net flux entering any volume must be zero; that is, there cannot be a source of flux (or a sink) inside the volume v. In other words, regardless of the volume, we choose, there cannot be a single magnetic pole (north or south) inside the volume although there can be pairs of poles. Thus, the conclusion is that the total flux through a closed surface (enclosing a volume v) must be zero regardless of the shape or size of the surface:

$$\boxed{\Phi = \oint_s \mathbf{B} \cdot d\mathbf{s} = 0} \tag{8.21}$$

The second example that is useful in this regard is that of a permanent magnet. Consider a number of small bar magnets, two of which are shown in **Figure 8.19a**, together with a representation of their magnetic fields using flux lines. North and south poles are shown on each magnet. For simplicity, we will assume that the magnets are identical in all respects. Now, we perform the following simple experiment: Place the south pole of magnet A on the north pole of magnet B. The result is shown in **Figure 8.19b**. It is clearly a single magnet and the N and S poles shown in the middle do not show a field of their own. In fact, these poles seem to have disappeared. If we now separate the two magnets again, each magnet returns to its original state. The two bar magnets are again as shown in **Figure 8.19a**; each has two poles and each produces a magnetic field identical to that of the composite magnet except, perhaps, to the strength of the field. We can repeat this process as many times as we wish with magnets as small as possible **(Figure 8.19c)**. The same effect occurs: Each magnet has two poles regardless of size or how many smaller magnets make it. In the limit, the long magnet is made of differential length magnets, and since each of these also has two poles, we must conclude that magnetic poles can only exist in pairs. Because of this, the flux lines must close through the magnet as shown in **Figure 8.19d**. While the discussion above is not a proof, it is an experiment-supported conclusion and we will have to be satisfied with this.[7]

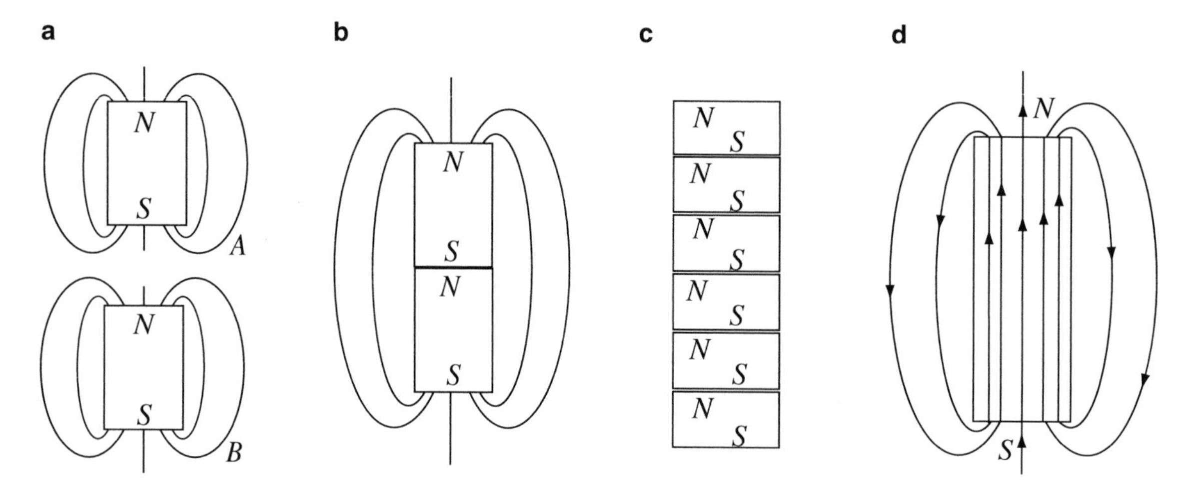

Figure 8.19 Experiment that shows magnetic poles must exist in pairs. (**a**) Two permanent magnets and their fields. (**b**) Connecting the two magnets as shown results in a single magnet. (**c**) A permanent magnet as a stack of elementary magnets, each with two poles. (**d**) The magnetic field of a permanent magnet must close through the magnet

Applying the divergence theorem to **Eq. (8.21)** gives

$$\oint_s \mathbf{B} \cdot d\mathbf{s} = \int_v (\nabla \cdot \mathbf{B}) dv = 0 \quad \rightarrow \quad \nabla \cdot \mathbf{B} = 0 \tag{8.22}$$

Equations (8.20) and **(8.22)** define the curl of **H** and the divergence of **B** and, through the constitutive relation $\mathbf{B} = \mu\mathbf{H}$, the curl of **B** and the divergence of **H**. We choose these two equations as the postulates of the magnetostatic field.

$$\boxed{\nabla \times \mathbf{H} = \mathbf{J}, \quad \nabla \cdot \mathbf{B} = 0} \tag{8.23}$$

[7] The search for monopoles (single magnetic poles) is still continuing in basic physics research. No evidence of their existence has ever been found either on macroscopic or microscopic levels.

These relations were obtained from **Eq. (8.16)** or **(8.19)** and **Eq. (8.21)**. The latter two equations are therefore the integral form of the postulates:

$$\oint_C \mathbf{H} \cdot d\mathbf{l} = I_{enclosed}, \quad \oint_s \mathbf{B} \cdot d\mathbf{s} = 0 \tag{8.24}$$

The first equation in each set is Ampère's law, whereas the second equation simply represents the fact that no single magnetic pole may exist. From these equations, we note the following:

(1) The magnetic field is nonconservative; the closed contour integral of the magnetic field intensity is nonzero.
(2) The magnetic field is rotational; the curl of the magnetic field intensity is nonzero.
(3) The magnetic field is solenoidal; the divergence of the magnetic flux density is zero.
(4) The magnetic flux is conserved; the total net flux through any closed surface is zero.

8.7 Potential Functions

According to the previous section, the static magnetic field is completely defined by the curl and divergence given in **Eq. (8.23)** as required by the Helmholtz theorem. However, the theorem does not imply that the forms given above are the only possible forms and, more importantly, it does not give any clue as to which relations are easier to use. In fact, we have already seen that Ampère's law is easier to use in some cases and the Biot–Savart law is more practical in others. We also recall from electrostatics that the use of the electric scalar potential (voltage) was one function that allowed simplification of the solutions of many otherwise complicated problems. The question now is: Are there any scalar or vector functions that can be used in conjunction with the magnetic field to effect the same results, that is, to simplify solution?

In fact, there are two functions that can be used. One is a vector function and can be used in general to describe the magnetic field. The second is a scalar function which may be used under certain conditions. These two functions are described next.

8.7.1 The Magnetic Vector Potential

Magnetostatics.m

Since the divergence of the magnetic flux density is zero **[Eq. (8.22)]**, we can invoke the vector identity $\nabla \cdot (\nabla \times \mathbf{A}) = 0$ (see **Section 2.5**) and write:

$$\mathbf{B} = \nabla \times \mathbf{A} \tag{8.25}$$

This is justified because when we substitute it back into the above vector identity we get $\nabla \cdot (\nabla \times \mathbf{A}) = \nabla \cdot \mathbf{B} = 0$. This can always be done for a divergence-free (solenoidal) field **B**. The vector **A** is called the ***magnetic vector potential***[8] and is defined through **Eq. (8.25)**. It is important to note the following:

(1) The magnetic vector potential is defined based on the divergence-free condition of **B**.
(2) The definition of **A** is based entirely on the mathematical properties of the vector **B**, not on its physical characteristics. In this sense, **A** is viewed as an auxiliary function rather than a fundamental field quantity. Nevertheless, the magnetic vector potential is an important function with considerable utility. We will make considerable use of the magnetic vector potential here and in subsequent chapters.

[8] The magnetic vector potential was considered by James Clerk Maxwell to be a fundamental quantity from which the magnetic flux density was derived. He called it the "Electrokinetic Momentum." In fact, Maxwell used the scalar and vector potentials to define fields. The fields as we use them today were introduced as fundamental quantities by Oliver Heaviside and Heinrich Hertz. Oliver Heaviside, in particular, had some harsh words about potential functions. He considered the magnetic vector potential an "absurdity" and "Maxwell's monster," which should be "murdered." Harsh words, but then Heaviside had harsh words for many people and subjects. Although we use Heaviside's form of electromagnetics in using field variables, we also make considerable use of potential functions. The guiding rule for us is simplicity and convenience.

(3) Since the magnetic vector potential is a vector function, both its curl and divergence must be specified. The curl is defined in **Eq. (8.25)**, but we have said nothing about its divergence. At this stage, we will assume that the divergence of **A** is zero ($\nabla \cdot \mathbf{A} = 0$) and delay the discussion of this quantity.

(4) The magnetic vector potential does not have a simple physical meaning in the sense that it is not a measurable physical quantity like **B** or **H**. (We will try to give it some physical interpretation shortly.) It may seem a bit unsettling to define a physical quantity based on the mathematical properties of another function and then use this secondary function to evaluate physical properties of the magnetic field. In fact, there is nothing unusual about this process. You can view the definition of the magnetic vector potential as a transformation. As long as the inverse transformation is unique, there is nothing wrong in **A** not having a readily defined physical meaning. We can use the magnetic vector potential in any way that is consistent with the properties of a vector field and the rules of vector algebra. If we then transform back to the magnetic flux density using **Eq. (8.25)**, all results thus obtained are correct.

(5) Because the magnetic vector potential relates to the magnetic flux density through the curl, the magnetic vector potential **A** is at right angles to the magnetic flux density **B**.

(6) The unit of **A** is the Wb/m, as can be seen from **Eq. (8.25)**.

Now, it remains to be seen that the use of the magnetic vector potential does have some advantage in the calculation of the magnetic field. To show this, we start with the Biot–Savart law for the magnetic flux density [**Eq. (8.9)**]:

$$\mathbf{B} = \nabla \times \mathbf{A} = \frac{\mu I}{4\pi}\int_a^b \frac{d\mathbf{l}' \times (\mathbf{r} - \mathbf{r}')}{|\mathbf{r} - \mathbf{r}'|^3} \quad [\mathrm{T}] \tag{8.26}$$

where a and b are two general points on the current contour. Note that we prefer to use $\mathbf{r} - \mathbf{r}'$ instead of $\mathbf{R}$ [as in **Eq. (8.9)**] to preserve the distinction between the source (primed) and field (unprimed) points. For evaluation, we will try to get the right-hand side into the form $\nabla \times \mathbf{F}$ such that we can then write $\mathbf{A} = \mathbf{F}$.

First, we note the following relation:

$$\frac{\mathbf{r} - \mathbf{r}'}{|\mathbf{r} - \mathbf{r}'|^3} = -\nabla\frac{1}{|\mathbf{r} - \mathbf{r}'|} \tag{8.27}$$

This relation, which is not immediately obvious, was derived in **Example 2.10**, but it may be shown to be correct by direct derivation. Substituting this in **Eq. (8.26)**,

$$\mathbf{B} = \nabla \times \mathbf{A} = -\frac{\mu_0 I}{4\pi}\int_a^b d\mathbf{l}' \times \nabla\left(\frac{1}{|\mathbf{r} - \mathbf{r}'|}\right) \quad [\mathrm{T}] \tag{8.28}$$

To transform this into the form required, we use the vector identity $\nabla \times (\varphi\mathbf{f}) = \varphi(\nabla \times \mathbf{f}) + (\nabla\varphi) \times \mathbf{f}$ (where $\mathbf{f}$ is any vector function and φ any scalar function). In our case, $\mathbf{f} = d\mathbf{l}'$ and $\varphi = 1/|\mathbf{r} - \mathbf{r}'|$, and we get

$$\nabla \times \left(\frac{d\mathbf{l}'}{|\mathbf{r} - \mathbf{r}'|}\right) = \frac{1}{|\mathbf{r} - \mathbf{r}'|}(\nabla \times d\mathbf{l}') + \left(\nabla\frac{1}{|\mathbf{r} - \mathbf{r}'|}\right) \times d\mathbf{l}' \tag{8.29}$$

The gradient is taken with respect to the general (unprimed) coordinates. Thus, the first term on the right-hand side must vanish since $\nabla \times d\mathbf{l}' = 0$. We therefore have

$$\nabla \times \left(\frac{d\mathbf{l}'}{|\mathbf{r} - \mathbf{r}'|}\right) = \left(\nabla\frac{1}{|\mathbf{r} - \mathbf{r}'|}\right) \times d\mathbf{l}' \tag{8.30}$$

or since $\mathbf{A} \times \mathbf{B} = -\mathbf{B} \times \mathbf{A}$ for any two vector fields **A** and **B**, we get

$$d\mathbf{l}' \times \left(\nabla\frac{1}{|\mathbf{r} - \mathbf{r}'|}\right) = -\nabla \times \left(\frac{d\mathbf{l}'}{|\mathbf{r} - \mathbf{r}'|}\right) \tag{8.31}$$

Substituting this result for the integrand in **Eq. (8.28)** gives

$$\mathbf{B} = \nabla \times \mathbf{A} = \frac{\mu_0 I}{4\pi} \int_a^b \nabla \times \left(\frac{d\mathbf{l}'}{|\mathbf{r} - \mathbf{r}'|} \right) \quad [\mathrm{T}] \tag{8.32}$$

The integration is along the current and, therefore, on the primed coordinates; the curl inside the integral is on the unprimed coordinates. This means that the curl and the integral operators can be interchanged (the curl is independent of the primed coordinates). Also, since $\mu_0 I/4\pi$ is a constant, it can be moved inside the curl operator:

$$\mathbf{B} = \nabla \times \mathbf{A} = \nabla \times \left[\frac{\mu_0 I}{4\pi} \int_a^b \frac{d\mathbf{l}'}{|\mathbf{r} - \mathbf{r}'|} \right] \quad [\mathrm{T}] \tag{8.33}$$

From this,

$$\boxed{\mathbf{A} = \frac{\mu_0 I}{4\pi} \int_a^b \left(\frac{d\mathbf{l}'}{|\mathbf{r} - \mathbf{r}'|} \right) \quad \left[\frac{\mathrm{Wb}}{\mathrm{m}} \right]} \tag{8.34}$$

This is the Biot–Savart law in terms of the magnetic vector potential.

In comparison with the Biot–Savart law in **Eq. (8.26)**, this is simpler to evaluate since there is no need to evaluate the vector product in **Eq. (8.26)**. In addition, the magnetic vector potential **A** is in the direction of $d\mathbf{l}'$, which, by definition, is taken in the direction of flow of current. Thus, the magnetic vector potential is everywhere in the direction of the current (and perpendicular to **B**).

After the magnetic vector potential is calculated, it may be used directly to evaluate other quantities or it may be used to calculate the magnetic flux density using **Eq. (8.33)**.

The premise behind the magnetic vector potential is that it can serve in lieu of the magnetic flux density or the magnetic field intensity; that is, it is up to us which function to choose for calculation. The ultimate choice should be based on ease of use, but as long as we can evaluate the various expressions either function yields the correct results. This is important since otherwise we must qualify the definition of **A.** To see that solution in terms of the magnetic vector potential is equivalent to solution in terms of **B** or **H**, we now define Ampère's law and magnetic flux in terms of the magnetic vector potential. We start with Ampère's law in **Eq. (8.20)**. In free space, μ_0 is constant and we get

$$\nabla \times \mathbf{H} = \mathbf{J} \quad \rightarrow \quad \nabla \times \mathbf{B} = \mu_0 \mathbf{J} \tag{8.35}$$

Substituting the definition of the magnetic vector potential $\mathbf{B} = \nabla \times \mathbf{A}$ gives

$$\nabla \times (\nabla \times \mathbf{A}) = \mu_0 \mathbf{J} \tag{8.36}$$

The left-hand side can be evaluated directly by applying the curl twice. Instead, we can exploit the vector equality $\nabla \times (\nabla \times \mathbf{A}) = \nabla(\nabla \cdot \mathbf{A}) - \nabla^2 \mathbf{A}$ [see **Eq. (2.134)**]. Thus, by direct substitution into Ampère's law, we get

$$\nabla(\nabla \cdot \mathbf{A}) - \nabla^2 \mathbf{A} = \mu_0 \mathbf{J} \tag{8.37}$$

Now, we must decide on the divergence of **A** which, until now, has not been defined. Taking $\nabla \cdot \mathbf{A} = 0$ is one choice mentioned earlier. This condition is called ***Coulomb's gauge***. As long as we do not violate the field equations (i.e., as long as we do not modify the properties of the magnetic field), the choice of the divergence of **A** is arbitrary. Later, when discussing the time-dependent field, we will revisit this issue, but for static fields, the choice above is the best choice. Thus, we get

$$\boxed{\nabla^2 \mathbf{A} = -\mu_0 \mathbf{J}} \tag{8.38}$$

This is a vector Poisson equation. The term $\nabla^2 \mathbf{A}$ is called the vector Laplacian and was discussed in **Chapter 2**. This equation can be solved for the vector field **A** in a manner similar to that for the scalar Poisson equation (for the electric scalar potential) discussed in **Chapters 5** and **6**. If we write the vectors **A** and **J** explicitly (in Cartesian coordinates in this case), each has three vector components:

$$\nabla^2(\hat{\mathbf{x}}A_x + \hat{\mathbf{y}}A_y + \hat{\mathbf{z}}A_z) = -\mu_0(\hat{\mathbf{x}}J_x + \hat{\mathbf{y}}J_y + \hat{\mathbf{z}}J_z) \quad (8.39)$$

Equating vector components on both sides of the equation, we obtain three separate, scalar equations, one for each scalar component of **A**:

$$\boxed{\nabla^2 A_x = -\mu_0 J_x, \quad \nabla^2 A_y = -\mu_0 J_y, \quad \nabla^2 A_z = -\mu_0 J_z} \quad (8.40)$$

Each of these equations is a one-dimensional scalar Poisson equation and can be solved separately for the x, y, and z components of **A**. The question is, how do we obtain general solutions for these equations? One way to do so is to go back to the solution of the scalar Poisson equation for the electric field discussed in **Chapter 5**. Poisson's equation for the electric scalar potential was given in **Eq. (5.4)**:

$$\nabla^2 \mathrm{V} = -\frac{\rho_v}{\varepsilon_0} \quad (8.41)$$

The general solution for this equation is [see **Eq. (4.32)** and **Section 5.4**]

$$V(\mathbf{r}) = \frac{1}{4\pi\varepsilon_0}\int_v \frac{\rho_v(r')dv'}{|\mathbf{r}-\mathbf{r}'|} \quad [\mathrm{V}] \quad (8.42)$$

Poisson's equations in **Eq. (8.40)** are similar in form to **Eq. (8.41)**. If we replace ρ_v/ε_0 by $\mu_0 J$, the equations are identical in form and, therefore, have the same form of solutions. Thus, we can write a solution for each component as

$$A_x(\mathbf{r}) = \frac{\mu_0}{4\pi}\int_{v'} \frac{J_x(r')dv'}{|\mathbf{r}-\mathbf{r}'|} \quad \left[\frac{\mathrm{Wb}}{\mathrm{m}}\right]$$

$$A_y(\mathbf{r}) = \frac{\mu_0}{4\pi}\int_{v'} \frac{J_y(r')dv'}{|\mathbf{r}-\mathbf{r}'|} \quad \left[\frac{\mathrm{Wb}}{\mathrm{m}}\right] \quad (8.43)$$

$$A_z(\mathbf{r}) = \frac{\mu_0}{4\pi}\int_{v'} \frac{J_z(r')dv'}{|\mathbf{r}-\mathbf{r}'|} \quad \left[\frac{\mathrm{Wb}}{\mathrm{m}}\right]$$

These can be combined in a vector solution by writing

$$\mathbf{A}(\mathbf{r}) = \frac{\mu_0}{4\pi}\int_{v'} \frac{\mathbf{J}(\mathbf{r}')dv'}{|\mathbf{r}-\mathbf{r}'|} \quad \left[\frac{\mathrm{Wb}}{\mathrm{m}}\right] \quad (8.44)$$

Now, consider a conductor of length $d\mathbf{l}'$ and cross-sectional area ds' as shown in **Figure 8.20.** Writing $dv' = d\mathbf{l}' \cdot \hat{\mathbf{n}}\,ds'$, substituting this relation into **Eq. (8.44)**, and separating the volume integral into a surface and line integral, we get for any point outside the conductor

$$\mathbf{A}(\mathbf{r}) = \frac{\mu_0}{4\pi}\int_L \left[\int_s \frac{\mathbf{J}(\mathbf{r}')\cdot\hat{\mathbf{n}}\,ds'}{|\mathbf{r}-\mathbf{r}'|}\right] d\mathbf{l}' = \frac{\mu_0 I}{4\pi}\int_L \frac{d\mathbf{l}'}{|\mathbf{r}-\mathbf{r}'|} \quad \left[\frac{\mathrm{Wb}}{\mathrm{m}}\right] \quad (8.45)$$

where the integration of the current density over the cross-sectional area is independent of the distance of the volume element and the point at which the field is calculated.

This result is the same as the magnetic vector potential in **Eq. (8.34)**, which, of course, it should be. It indicates that the solution using Ampère's law or the Biot–Savart law is the same regardless of our starting point.

Figure 8.20 A conductor of length $d\mathbf{l}'$ and cross-sectional area ds'

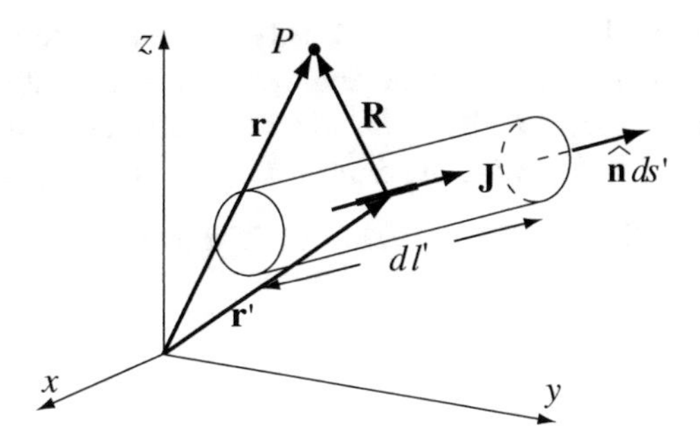

Finally, we look into the calculation of flux using **Eq. (8.17)**. Substituting $\mathbf{B} = \nabla \times \mathbf{A}$, we get

$$\Phi = \int_s \mathbf{B} \cdot d\mathbf{s} = \int (\nabla \times \mathbf{A}) \cdot d\mathbf{s} \quad [\text{Wb}] \tag{8.46}$$

Using Stokes' theorem, we can convert this to a closed contour integral. The final result is

$$\boxed{\Phi = \oint_C \mathbf{A} \cdot d\mathbf{l} \quad [\text{Wb}]} \tag{8.47}$$

where C is the contour bounding the surface s. This relation shows that the total flux through a surface s is equal to the line integral of the magnetic vector potential along the contour of the surface. In other words, we do not need to calculate the magnetic flux density $\mathbf{B}$ if we wish to evaluate the total flux. This relation also gives a physical meaning to the magnetic vector potential as a measure of flux.

Example 8.12 Magnetic Vector Potential Due to a Short, Straight Segment A thin, finite-length wire as shown in **Figure 8.21a** carries a current $I = 1$ A. Calculate:

(a) The magnetic vector potential at point P_1 shown in **Figure 8.21a**.
(b) The magnetic vector potential at point P_2 shown in **Figure 8.21b**.
(c) The magnetic vector potential at a general point $P(r,\phi,z)$ in **Figure 8.21c**.

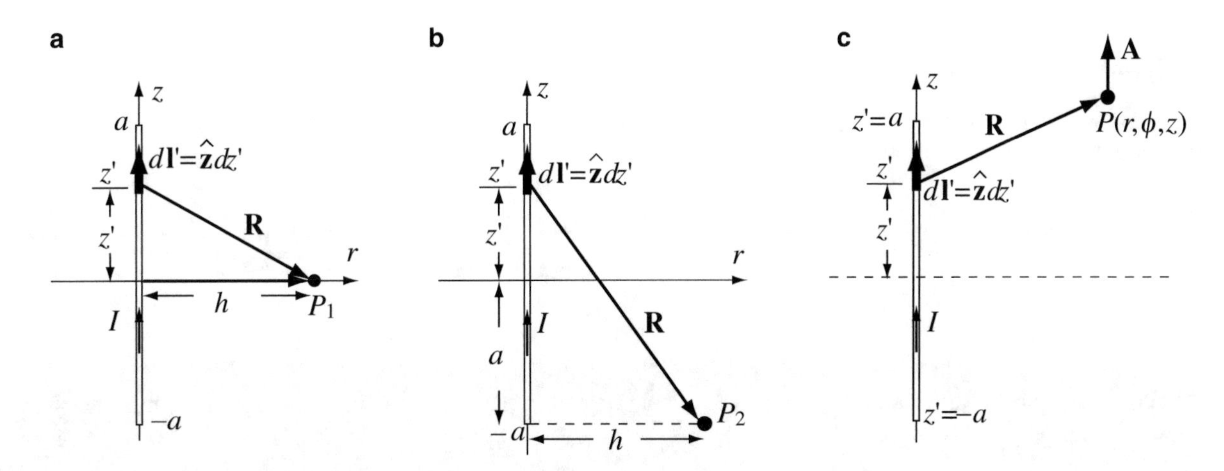

Figure 8.21 Magnetic vector potential of a short segment. (**a**) Calculation of **A** at P_1. (**b**) Calculation of **A** at P_2. (**c**) Calculation of **A** at a general point $P(r,\phi,z)$

Solution: We solved for the magnetic flux density and magnetic field intensity for the same segment in **Example 8.1**. The approach here is similar: An element of length $d\mathbf{l}' = \hat{\mathbf{z}}dz'$ is identified at point $P(0,0,z')$. The magnetic vector potential at point $P(r,\phi,z)$ is calculated using the Biot–Savart law in terms of the magnetic vector potential given in **Eq. (8.34)**. Since the magnetic vector potential is always in the direction of current, only a z component exists.

(a) At point P_1, the coordinates are $(r = h,\ z = 0)$. The magnetic vector potential at point P_1 due to the current in element $d\mathbf{l}' = \hat{\mathbf{z}}dz'$ is

$$d\mathbf{A}(r,z) = \hat{\mathbf{z}}\frac{\mu_0 I\,dz'}{4\pi|\mathbf{R}|} \quad \left[\frac{\text{Wb}}{\text{m}}\right]$$

The vector $\mathbf{R}$ and its magnitude are

$$\mathbf{R} = \hat{\mathbf{r}}h - \hat{\mathbf{z}}z' \quad \rightarrow \quad R = \sqrt{h^2 + z'^2} \quad [\text{m}]$$

The magnetic vector potential is found by integrating over the length of the segment from z$' = -a$ to z$' = +a$:

$$\mathbf{A}(h,0) = \hat{\mathbf{z}}\frac{\mu_0 I}{4\pi}\int_{z'=-a}^{z'=a}\frac{dz'}{\sqrt{h^2+z'^2}} = \hat{\mathbf{z}}\frac{\mu_0 I}{4\pi}\ln\left(z' + \sqrt{h^2+z'^2}\right)\Big|_{z'=-a}^{z'=a} = \hat{\mathbf{z}}\frac{\mu_0 I}{4\pi}\ln\frac{a+\sqrt{h^2+a^2}}{\sqrt{h^2+a^2}-a} \quad \left[\frac{\text{Wb}}{\text{m}}\right].$$

(b) To calculate the magnetic vector potential at point P_2, we use **Figure 8.21b**. $d\mathbf{l}'$ remains the same as in **(a)**, but the vector $\mathbf{R}$ and its magnitude become

$$\mathbf{R} = \hat{\mathbf{r}}h - \hat{\mathbf{z}}(z'+a) \quad \rightarrow \quad R = \sqrt{h^2 + (z'+a)^2} \quad [\text{m}]$$

Substituting this in the expression for $d\mathbf{A}$ and integrating from $z' = -a$ to $z' = a$,

$$\mathbf{A}(h,-a) = \hat{\mathbf{z}}\frac{\mu_0 I}{4\pi}\int_{z'=-a}^{z'=a}\frac{dz'}{\sqrt{z'^2 + 2az' + (h^2+a^2)}}$$

$$= \hat{\mathbf{z}}\frac{\mu_0 I}{4\pi}\ln\left(2\sqrt{z'^2 + 2az' + (h^2+a^2)} + 2z' + 2a\right)\Big|_{z'=-a}^{z'=a} = \hat{\mathbf{z}}\frac{\mu_0 I}{4\pi}\ln\frac{\sqrt{4a^2+h^2}+2a}{h} \quad \left[\frac{\text{Wb}}{\text{m}}\right].$$

(c) As in the previous two calculations, we place the element of length $d\mathbf{l}'$ at a point z' along the element of current and calculate the length of the vector $\mathbf{R}$ (see **Figure 8.21c**):

$$\mathbf{R} = \hat{\mathbf{r}}r - \hat{\mathbf{z}}(z - z') \quad \rightarrow \quad R = \sqrt{r^2 + (z+z')^2} \quad [\text{m}]$$

The magnetic vector potential at $P(r,\phi,z)$ is

$$\mathbf{A}(r,\phi,z) = \hat{\mathbf{z}}\frac{\mu_0 I}{4\pi}\int_{z'=-a}^{z'=a}\frac{dz'}{\sqrt{z'^2 - 2zz' + (r^2+z^2)}}$$

$$= \hat{\mathbf{z}}\frac{\mu_0 I}{4\pi}\ln\left(2\sqrt{z'^2 - 2zz' + (r^2+z^2)} + 2z' - 2z\right)\Big|_{z=-a}^{z=a} = \hat{\mathbf{z}}\frac{\mu_0 I}{4\pi}\ln\frac{\sqrt{a^2 - 2za + (r^2+z^2)} + a - z}{\sqrt{a^2 + 2za + (r^2+z^2)} - a - z} \quad \left[\frac{\text{Wb}}{\text{m}}\right].$$

Exercise 8.4 Using **Example 8.12**, calculate the magnetic flux density at point P_1 using the definition of the magnetic vector potential in **Eq. (8.25)** and show that the result in **Example 8.1a** is obtained for $h = a = 1$ m if we write $\mathbf{H} = \mathbf{B}/\mu_0$.

Example 8.13 Flux Due to a Square Loop A square loop carries a current $I = 1$ A. The length of each side of the loop is $2a = 1$ m. A smaller loop with dimensions $2b = a = 0.5$ m is placed inside the larger loop as shown in **Figure 8.22a**. Calculate the total flux produced by the outer loop that passes through the inner loop.

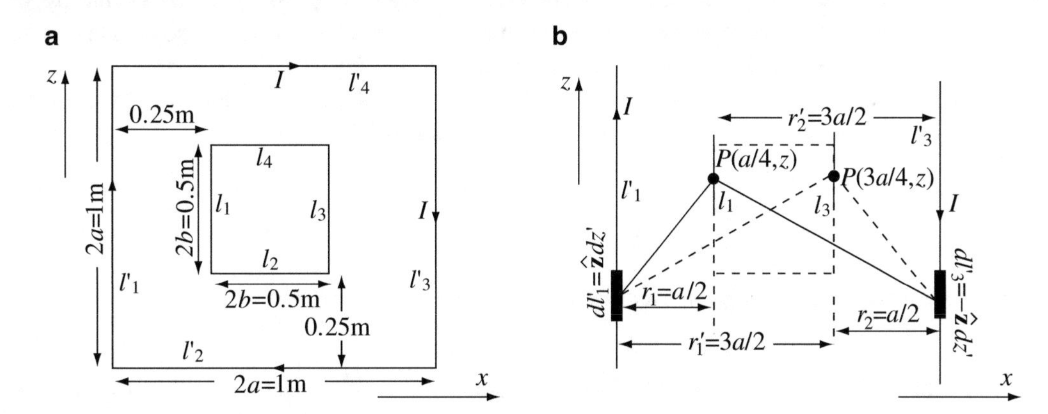

Figure 8.22 Calculation of flux through the inner loop, generated by the outer loop. (**a**) Geometry and dimensions. (**b**) Calculation of the magnetic vector potential due to segments l_1' and l_3' on l_1 and due to segments l_1' and l_3' on l_3

Solution: In general, to calculate the magnetic flux, we need to calculate the magnetic flux density and then integrate over the area through which the flux flows. An alternative method is to calculate the magnetic vector potential along the boundaries of the surface and integrate this potential to obtain the flux using **Eq. (8.47)**. We show the latter method here. The magnetic vector potential due to the outer loop at the boundaries of the inner loop is first calculated and then integrated over the boundaries of the inner loop.

The magnetic vector potential due to a z-directed current-carrying segment of length $2a$, at a distance r from the segment, was calculated in **Example 8.12c**.

$$\mathbf{A}(r,\phi,z) = \hat{\mathbf{z}}\frac{\mu_0 I}{4\pi}\ln\frac{\sqrt{a^2 - 2za + (r^2 + z^2)} + a - z}{\sqrt{a^2 + 2za + (r^2 + z^2)} - a - z} = \hat{\mathbf{z}}\frac{\mu_0 I}{4\pi}\left[\sinh^{-1}\left(\frac{a-z}{r}\right) + \sinh^{-1}\left(\frac{z+a}{r}\right)\right] \quad \left[\frac{\text{Wb}}{\text{m}}\right]$$

where the relation $\ln(x + (x^2 + 1)^{1/2}) = \sinh^{-1}x$ was used to write the second form of $\mathbf{A}$. Also, because $\mathbf{A}$ does not depend on ϕ, this dependence will be dropped for the remainder of calculations in this example.

We will use this expression as follows: Because the line integral in **Eq. (8.47)** requires only the component of the magnetic vector potential in the direction of integration, and the boundaries of the two loops are parallel to each other, we calculate the magnetic vector potential due to segments l_1' and l_3' at segments l_1 and l_3 since segments l_2' and l_4' generate components of the magnetic vector potential perpendicular to l_1 and l_3. Similarly, the magnetic vector potential on segments l_2 and l_4 is due to l_2' and l_4'. Using **Figure 8.22b**, the magnetic vector potential at segment l_1 due to segment l_1' is

$$\mathbf{A}(r_1, z) = \hat{\mathbf{z}}\frac{\mu_0 I}{4\pi}\left(\sinh^{-1}\frac{2(a-z)}{a} + \sinh^{-1}\frac{2(z+a)}{a}\right) \quad \left[\frac{\text{Wb}}{\text{m}}\right]$$

The magnetic vector potential at segment l_1 due to segment l_3' is

$$\mathbf{A}(r_2', z) = -\hat{\mathbf{z}}\frac{\mu_0 I}{4\pi}\left(\sinh^{-1}\frac{2(a-z)}{3a} + \sinh^{-1}\frac{2(z+a)}{3a}\right) \quad \left[\frac{\text{Wb}}{\text{m}}\right]$$

Note that this magnetic vector potential is in the opposite direction to that due to segment l_1 because the currents are in opposite directions. The total magnetic vector potential at any location on segment l_1 is

$$\mathbf{A}_{l1} = \hat{\mathbf{z}}\frac{\mu_0 I}{4\pi}\left(\sinh^{-1}\frac{2(a-z)}{a} + \sinh^{-1}\frac{2(z+a)}{a} - \sinh^{-1}\frac{2(a-z)}{3a} - \sinh^{-1}\frac{2(z+a)}{3a}\right) \quad \left[\frac{\text{Wb}}{\text{m}}\right]$$

Similarly, interchanging between r_1 and r_2 in **Figure 8.22b**, the magnetic vector potential on segment l_3 is identical in magnitude but in the opposite direction. Instead of calculating the magnetic vector potential on the upper and lower segments, we note from symmetry considerations that the magnitudes of the magnetic vector potentials on these segments must be the same as for segments l_1 and l_3 even though the direction of **A** is horizontal.

The total flux is calculated by integrating the magnetic vector potential along the boundaries of the inner loop. Going along the current, all four segments contribute equal and positive values to the integral:

$$\Phi = 4\int_{z=a/2}^{z=3a/2}\mathbf{A}_{l1}\cdot d\mathbf{l}_1 = \frac{\mu_0 I}{\pi}\int_{z=a/2}^{z=3a/2}\left(\sinh^{-1}\frac{2(a-z)}{a} + \sinh^{-1}\frac{2(z+a)}{a} - \sinh^{-1}\frac{2(a-z)}{3a} - \sinh^{-1}\frac{2(z+a)}{3a}\right)dz \quad [\text{Wb}]$$

Performing the integration of the four terms above gives

$$\Phi = \frac{\mu_0 I}{\pi}\left[-(a-z)\sinh^{-1}\frac{2(a-z)}{a}\right] + \sqrt{(a-z)^2+\frac{a^2}{4}} + (a+z)\sinh^{-1}\frac{2(a+z)}{a} - \sqrt{(a+z)^2+\frac{a^2}{4}} + (a-z)\sinh^{-1}\frac{2(a-z)}{3a}$$

$$\left. - \sqrt{(a-z)^2+\frac{9a^2}{4}} - (a+z)\sinh^{-1}\frac{2(a+z)}{3a} + \sqrt{(a+z)^2+\frac{9a^2}{4}}\;\right]_{z=a/2}^{z=3a/2}$$

$$= \frac{\mu_0 I}{\pi}\left(\frac{a}{2}\sinh^{-1}(-1) + \frac{a}{\sqrt{2}} + \frac{5a}{2}\sinh^{-1}5 - \frac{\sqrt{26}a}{2} - \frac{a}{2}\sinh^{-1}\left(-\frac{1}{3}\right) - \frac{\sqrt{10}a}{2} - \frac{5a}{2}\sinh^{-1}\left(\frac{5}{3}\right) + \frac{\sqrt{34}a^2}{2}\right)$$

$$-\frac{\mu_0 I}{\pi}\left(-\frac{a}{2}\sinh^{-1}1 + \frac{a}{\sqrt{2}} + \frac{3a}{2}\sinh^{-1}3 - \frac{\sqrt{10}a}{2} + \frac{a}{2}\sinh^{-1}\frac{1}{3} - \frac{\sqrt{10}a}{2} - \frac{3a}{2}\sinh^{-1}1 + \frac{\sqrt{18}a}{2}\right)$$

$$= \frac{\mu_0 I}{\pi}\left(\frac{\sqrt{10}a}{2} - \frac{\sqrt{26}a}{2} + \frac{\sqrt{34}a}{2} - \frac{\sqrt{18}a}{2} + \frac{5a}{2}\left[\sin^{-1}5 - \sinh^{-1}\frac{5}{3}\right] + \frac{3a}{2}\left[\sinh^{-1}1 - \sinh^{-1}3\right]\right) \quad [\text{Wb}]$$

For the values given ($a = 0.5$ m, $I = 1$ A, $\mu_0 = 4\pi \times 10^{-7}$), the total flux is

$$\Phi = 2\times10^{-7}\left(\frac{\sqrt{10}}{2} - \frac{\sqrt{26}}{2} + \frac{\sqrt{34}}{2} - \frac{\sqrt{18}}{2} + \frac{5}{2}\left[\sin^{-1}5 - \sinh^{-1}\frac{5}{3}\right] + \frac{3}{2}\left[\sinh^{-1}1 - \sinh^{-1}3\right]\right) = 1.984\times10^{-7} \quad [\text{Wb}]$$

Note: The flux through the inner loop may also be calculated by integrating the flux density produced by the outer loop over the area of the inner loop using **Eq. (8.17)**. However, the method shown here is much simpler for the given configuration. This is not always the case. For example, the magnetic vector potential for infinitely long current carrying conductors is infinite. Calculation of flux due to infinitely long currents is best done using Ampère's law and integration of the flux density over the area of interest.

8.7.2 The Magnetic Scalar Potential Magnetostatics.m

The condition for a scalar potential to be defined is that the vector field must be curl-free; that is, the field must be a conservative field. Any vector field **F** that satisfies the curl-free condition $\nabla \times \mathbf{F} = 0$ may be described as the gradient of a scalar function φ by

$$\mathbf{F} = -\nabla \varphi. \tag{8.48}$$

The basis of this statement is that now we can substitute this back into the curl and obtain $\nabla \times \mathbf{F} = \nabla \times (-\nabla \varphi) = 0$, based on one of the vector identities defined in **Chapter 2 [Eq. (2.110)]**. Whenever this is possible, it has the very distinct advantage of allowing calculation in terms of a scalar function rather than in terms of a vector function. Note that it would have been appropriate to define $\mathbf{F} = \nabla \varphi$ instead of $\mathbf{F} = -\nabla \varphi$. The negative sign is introduced by convention, as was done for the electric scalar potential in **Section 4.4.3**.

Inspection of **Eq. (8.20)** shows that the magnetic field intensity is not curl-free in general and, therefore, we cannot describe it in terms of a scalar function. There are, however, a number of important applications in magnetics in which a magnetic field exists, but there are no current densities involved. The most obvious are those involving permanent magnets. In this case, $\mathbf{J} = 0$ and we can write

$$\nabla \times \mathbf{H} = 0 \quad \rightarrow \quad \mathbf{H} = -\nabla \psi \quad [\mathrm{A/m}] \tag{8.49}$$

where ψ is called the ***magnetic scalar potential***. This potential has properties similar to those of the electric potential since it was defined in exactly the same way. Thus, the closed contour integral of $\mathbf{H} \cdot d\mathbf{l}$ is now zero (the magnetic field intensity is conservative because $\mathbf{J} = 0$). Also, we may define the magnetic scalar potential difference as

$$\psi_{ba} = \psi_b - \psi_a = -\int_a^b \mathbf{H} \cdot d\mathbf{l} \quad [\mathrm{A}] \tag{8.50}$$

where ψ_a may be viewed as a reference magnetic scalar potential. The unit of the magnetic scalar potential is the ampere [A].

The magnetic scalar potential satisfies the Laplace equation exactly like the electric scalar potential:

$$\boxed{\nabla^2 \psi = 0} \tag{8.51}$$

Thus, whenever a magnetic problem can be described such that there are no current sources in the region of interest, the properties of the electric scalar potential can be invoked for the magnetic scalar potential.

The following example shows the use of the magnetic scalar potential in a simple geometry.

Example 8.14 The Magnetic Scalar Potential Due to a Magnet A very large magnet is made as shown in **Figure 8.23**, with the two poles separated a distance, $d = 0.2$ m. Because of the size of the magnet, we may assume the field to be perpendicular to the surface of the poles. Suppose the magnetic flux density between the poles is known to be 0.5 T:

(a) Calculate the magnetic scalar potential everywhere between the poles.
(b) Calculate the magnetic scalar potential difference between the two poles.
(c) Suppose we now move the poles apart to twice the distance. What is the magnetic flux density if the magnetic scalar potential remains the same?

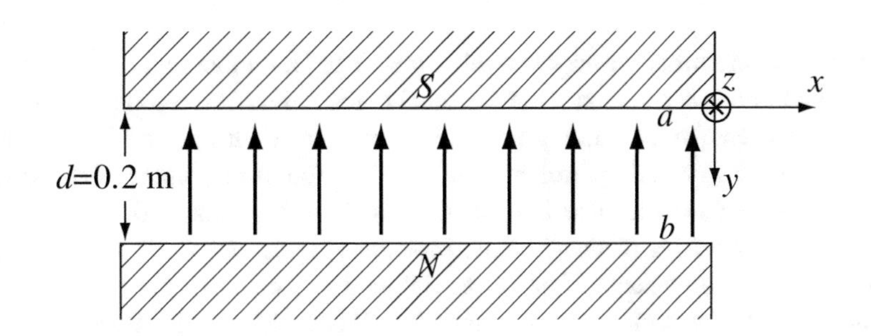

Figure 8.23 Gap between the poles of a large magnet and the magnetic flux density between them

Solution: This problem is solved by letting the upper surface be at a reference magnetic scalar potential (zero value) and then integrating against the magnetic flux density from the upper surface toward the lower surface using **Eq. (8.50)**. In **(c)**, we assume that since the magnetic potential difference remains the same, the magnetic flux density must diminish in the same way the electric field diminishes when the plates of a capacitor are moved apart while keeping the potential difference constant:

(a) Using **Eq. (8.50)** and assuming the magnetic field intensity is in the negative y direction as shown in **Figure 8.23**, we integrate from the upper pole toward the lower pole between $y = 0$ and a general point y between the poles:

$$\psi_{ya} = \psi_y - \psi_a = -\int_0^y (-\hat{\mathbf{y}}H) \cdot (\hat{\mathbf{y}}dy) = \int_0^y \frac{B}{\mu_0} dy = \frac{By}{\mu_0} \quad [\text{A}]$$

The magnetic scalar potential varies linearly with the distance from the upper pole.

(b) The magnetic scalar potential difference between the two poles is

$$\psi_{ba} = \psi_b - \psi_a = -\int_{y=a}^{b} (-\hat{\mathbf{y}}H) \cdot (\hat{\mathbf{y}}dy) = \int_{y=0}^{d} \frac{B}{\mu_0} dy = \frac{Bd}{\mu_0} = \frac{0.5 \times 0.2}{4\pi \times 10^{-7}} = 7.96 \times 10^4 \quad [\text{A}]$$

Note: The positive direction for potential is against the field. Therefore, the magnetic scalar potential at b is higher than at a.

(c) The scalar potential difference remains the same and, because the poles are large, the magnetic flux density remains uniform between the poles. Using the result in **(b)**, we get

$$\psi_{ba} = \frac{Bd}{\mu_0} = 7.96 \times 10^4 \quad \rightarrow \quad B = \frac{\psi_{ba}\mu_0}{2d} = \frac{7.96 \times 10^4 \times 4\pi \times 10^{-7}}{0.4} = 0.25 \quad [\text{T}].$$

8.8 Applications

Application: Magnetic Prospecting—Geomagnetism It is well known that a magnetic compass will not function properly in certain environments. For example, near volcanoes, where basalt rocks are present, there is a distortion in the terrestrial magnetic field due to concentration of iron in basalt. Similar occurrences can be observed in the presence of large deposits of iron ore, especially in the presence of magnetite. These observations are the basis of a specialized form of prospecting called magnetic prospecting or magnetic surveying. Since most rocks contain small amounts of magnetic materials (mostly magnetite, Fe_3O_4, and hematite Fe_2O_3), measurement of variations in the geomagnetic field can detect deposits or lack of deposits in the crust. Measurements can be simple, such as direct measurement and recording of the surface field variations, or may require specialized techniques, such as the use of special magnetometers (a magnetometer is a sensitive instrument for measurement of magnetic fields). In some cases, the existence of ore may be detected directly. In other applications, specific materials are found in rock layers which exhibit detectable magnetic properties. For example, gold is often found in igneous rocks, which may be traced with a magnetometer. Diamonds are often found in kimberlite veins which also have specific magnetic anomalies. Similarly, the lack of any anomaly may indicate very deep base sedimentary rocks which are essential for accumulation of oil. Magnetic surveying of this type, which may be called passive magnetic surveying, is often done from aircraft, especially when large areas need to be surveyed. The method can also be used in archaeological and geological research. There are also active methods of prospecting which are often used.

Application: The Helmholtz and Maxwell Coils The Helmholz coil or Helmholtz pair is a simple configuration of two loops or two short coils separated a distance equal to the radius of the coils as shown in **Figure 8.24a**. The two loops or coils carry identical currents in the same direction. The uniqueness of this coil is that it produces a uniform magnetic field in the volume between the two coils with better uniformity around the center of the system. The coil has many applications in measurements. One particular use is to cancel the terrestrial magnetic field to produce a region with essentially zero magnetic field for sensitive measurements. In some applications three mutually orthogonal Helmholtz pairs are used to ensure that all three components of the terrestrial field cancel. Other applications rely on the coil to produce a well defined exactly known field. A variation of the Helmholtz coil is the Maxwell coil shown in **Figure 8.24b**. In this configuration there are three coils with very specific dimensions for the purpose of improving the uniformity of the field in the volume between the outer coils. In the Maxwell coil, the central coil is of radius R whereas the two smaller coil are of radius $R_1 = R\sqrt{4/7}$ and the distance between the central coil to each of the smaller coils is $d/2 = R\sqrt{3/7}$. The currents in the three coils are equal and in the same direction as shown.

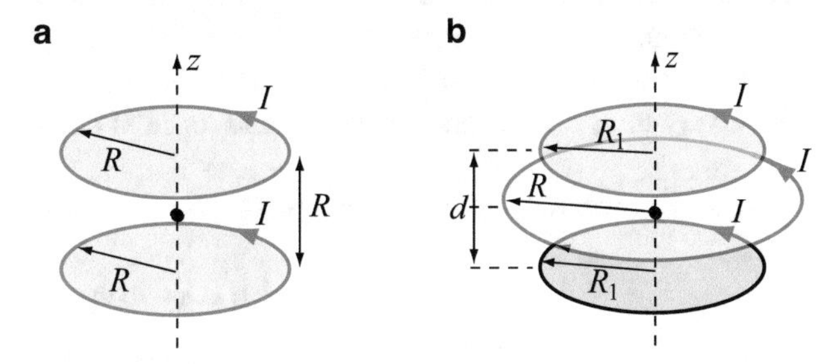

Figure 8.24 (**a**) The Helmholtz coil. (**b**) The Maxwell coil. Both are shown as loops but in practice each loop may be replaced with a short n-turn coil

Application: High-Field Coils It is sometimes necessary to produce very high magnetic fields. One application is in magnetic resonance imaging (MRI) devices. In these applications, a magnetic flux density of between 0.5 and 7 T (or more) is required. To produce fields of this magnitude, a solenoid, about 1 m long and at least 0.5 m in diameter is made and a current, sufficiently high to produce the field, is passed through the coil. The currents are extremely high and the losses in the coils cannot be dissipated under normal conditions. Thus, it becomes necessary to cryogenically cool the solenoids through use of liquid nitrogen (77°K) or even at superconducting temperatures through use of liquid helium (4.2°K). Usually the higher temperatures are preferred because most metals used for coils become extremely brittle at superconducting temperatures. Only a few alloys can operate at these very low temperatures (niobium–titanium and lead-gold are the most commonly used).

Application: High-Voltage Overhead Transmission Lines A pair of overhead transmission lines generates a magnetic field intensity in space. A growing concern in society is the effect these magnetic (and electric) fields have on living organisms. In particular, some studies have implicated high magnetic fields due to overhead transmission lines with higher than average cancer rates for those exposed. Most of these studies suggest, without being conclusive, that low-frequency fields, such as those produced by AC distribution lines, are at fault. As power requirements grow, there is an increasing need to use higher voltages and higher currents on transmission lines. Power distributions above 1 MV (million volts) already exist (for example, the trans-Siberian line in Russia uses 1.2 MV). What are the magnetic field intensities we can expect at ground levels? Typically, the distance between lines is a few meters to prevent the lines from touching during storms, and the lines may be as high as 20–25 m. This produces magnetic field intensities at ground level that may exceed 5 A/m.

8.9 Summary

The relation between currents and the magnetic field was explored, primarily through the Biot–Savart and Ampère's laws and served as an introduction to magnetostatics. The starting point was the magnetic field intensity **H** [A/m] due to a filament segment carrying current I [A] and was calculated using the ***Biot–Savart law*** (see **Figure 8.3**)

$$\mathbf{H}(x,y,z) = \frac{1}{4\pi}\int_a^b \frac{Id\mathbf{l}' \times \hat{\mathbf{R}}}{|\mathbf{R}|^2} \quad \left[\frac{\mathrm{A}}{\mathrm{m}}\right] \tag{8.8}$$

The Biot–Savart law calculates the magnetic field intensity **H** due to filamentary currents but can be used in thick conductors by stipulating a filament with differential cross section and integration over all such filaments. The magnetic flux density **B** [T] is related to **H** as $\mathbf{B} = \mu\mathbf{H}$ where μ is the permeability of the medium.

Ampère's law is the circulation of the magnetic field intensity around a closed contour:

$$\oint_c \mathbf{H} \cdot d\mathbf{l} = I_{enclosed} \quad [\mathrm{A}] \tag{8.16}$$

For this to be useful in calculation of **H** (or **B**), we require that **H** be constant and either parallel or perpendicular to $d\mathbf{l}$ along the path of integration. That is, we must find a contour, enclosing the current I, on which the scalar product in the integrand can be evaluated a priori so that H can be taken outside the integral. This relation is particularly useful for calculation of fields of very long conductors, solenoids, and toroidal coils.

Magnetic Flux Currents produce magnetic fields and magnetic fields produce flux, Φ

$$\Phi = \int_s \mathbf{B} \cdot d\mathbf{s} \quad [\mathrm{Wb}] \tag{8.17}$$

Postulates The relations above lead to the postulates of the magnetostatic field, specifying its curl and divergence:

$$\overbrace{\nabla \times \mathbf{H} = \mathbf{J} \quad \text{and} \quad \nabla \cdot \mathbf{B} = 0}^{\textit{differential form}} \quad \text{or} \quad \overbrace{\oint_c \mathbf{H} \cdot d\mathbf{l} = I_{enc.} \quad \text{and} \quad \oint_s \mathbf{B} \cdot d\mathbf{s} = 0}^{\textit{integral form}} \qquad \text{(8.23) and (8.24)}$$

Magnetic Vector Potential From $\nabla \cdot \mathbf{B} = 0$ we can define a magnetic vector potential as follows:
Since $\nabla \cdot \mathbf{B} = 0$, we have (see **Section 2.5** and the Helmholtz theorem):

$$\mathbf{B} = \nabla \times \mathbf{A} \tag{8.25}$$

Substitution of this into **Eqs. (8.8)** and **(8.17)** leads to the Biot–Savart law and the flux in terms of the magnetic vector potential. These are often easier to calculate:

$$\mathbf{A} = \frac{\mu I}{4\pi}\int_a^b \frac{d\mathbf{l}'}{|\mathbf{R}|} \quad \left[\frac{\mathrm{Wb}}{\mathrm{m}}\right] \quad \text{and} \quad \Phi = \oint_c \mathbf{A} \cdot d\mathbf{l} \quad [\mathrm{Wb}] \qquad \text{(8.34) and (8.47)}$$

A magnetic scalar potential ψ is defined if $\mathbf{J} = 0$ in **Eq. (8.23)** based on the Helmholtz theorem:

$$\textit{If } \nabla \times \mathbf{H} = 0 \quad \rightarrow \quad \mathbf{H} = -\nabla\psi \tag{8.49}$$

ψ is used in the same fashion as the electric scalar potential V, but its units are the ampere.

Reminder Permeability of free space is $\mu_0 = 4\pi \times 10^{-7}$ [H/m]

Problems

Note: In the problems that follow, all materials and spaces have permeability $\mu_0 = 4\pi \times 10^{-7}$ [H/m] unless otherwise specified.

The Biot–Savart Law

8.1 Magnetic Flux Density Due to Filamentary Currents. A current I [A] flows in a conductor shaped as an equilateral triangle **(Figure 8.25)**. Calculate the magnetic flux density:

(a) At the center of gravity of the triangle (where the three normals to the sides meet).
(b) At a vertex of the triangle.

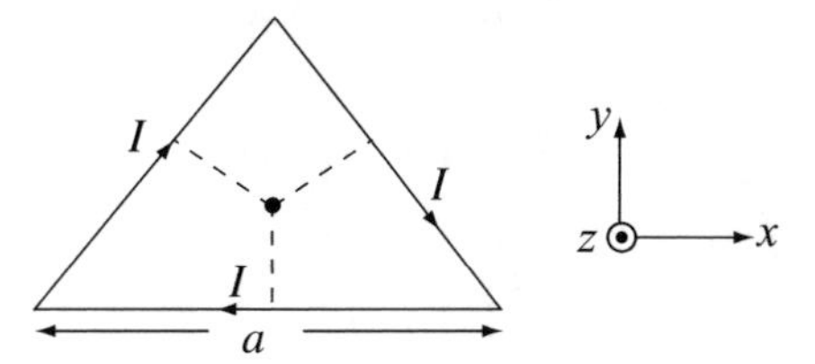

Figure 8.25

8.2 Magnetic Flux Density of Semi-infinite Segments. An infinitely long wire carrying a current I [A] is bent as shown in **Figure 8.26**. Find the magnetic flux density at points P_1 and P_2. (P_2 is at the center of the horizontal wire, at a distance d [m] from the bend.)

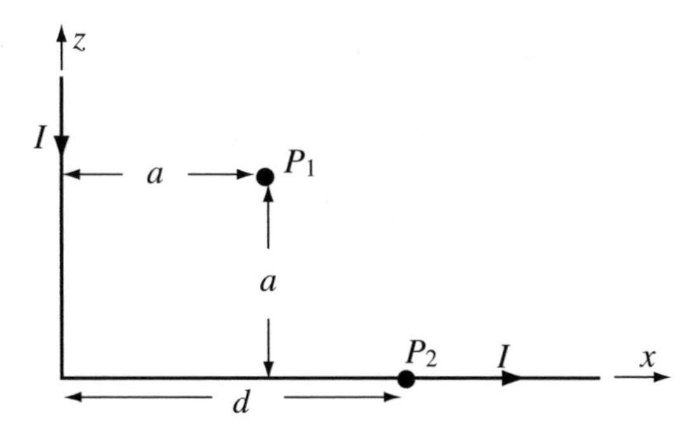

Figure 8.26

8.3 Magnetic Flux Density of Circular and Hexahedral Loops. In an application it is required to produce a flux density B_0 [T] at the center of a wire loop of radius R [m], carrying a current I_L [A]. To simplify the laying of the wire, the loop is replaced with a hexagon circumscribed in the loop as shown in **Figure 8.27**. The hexagonal wire carries a current I_H [A]. Calculate the ratio I_H/I_L so that the hexagonal wire produces the same flux density B_0 [T] at P as the circular loop.

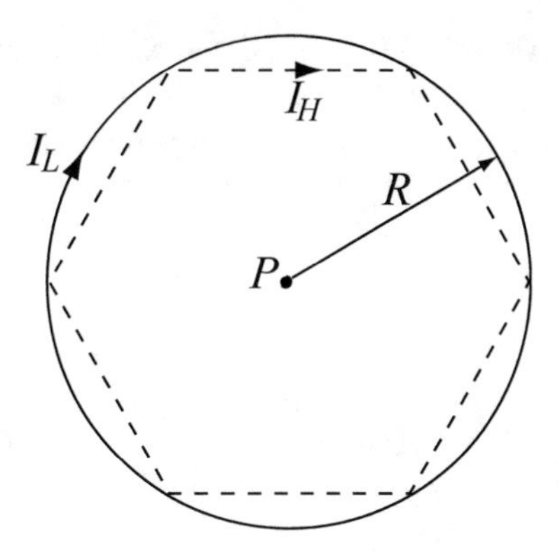

Figure 8.27

8.4 Magnetic Flux Density of Polygonal Loop. Show that the magnetic flux density at the center of a loop of wire carrying a current I [A] and shaped like a regular plane polygon of $2n$ sides, the distance between parallel sides being $2a$ [m], is $B = (\mu_0 nI/\pi a)\sin(180°/2n)$ [T].

8.5 Application: Magnetic Field of Moving Charges. A thin insulating disk of outer radius b has a uniform charge density ρ_s [C/m^2] distributed over the surface between $r = a$ [m] and $r = b$ [m]. The charges are bound and cannot move from their location. If the disk is rotating at an angular velocity ω, calculate the magnetic field intensity at the center of the disk (**Figure 8.28**). **Note**. The first experiment to show the equivalency between moving charges and magnetic field was produced by Henry A. Rowland (1849–1901) in 1875. He showed that a charge placed on a disk (rotating at 3,660 rpm) produced a magnetic field exactly like that of a closed-loop current.

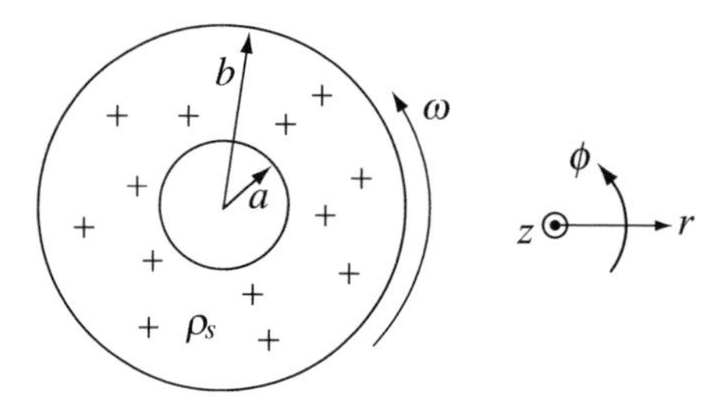

Figure 8.28

8.6 Magnetic Flux Density of a Loop. The rectangular wire loop in **Figure 8.29** carries a current I [A] as shown.

(a) Calculate the magnetic flux density (magnitude and direction) at a point h [m] high above the center of the loop (see **Figure 8.29**).

(b) Show that as the ratio b/a or a/b tends to zero, the flux density in **(a)** tends to zero.

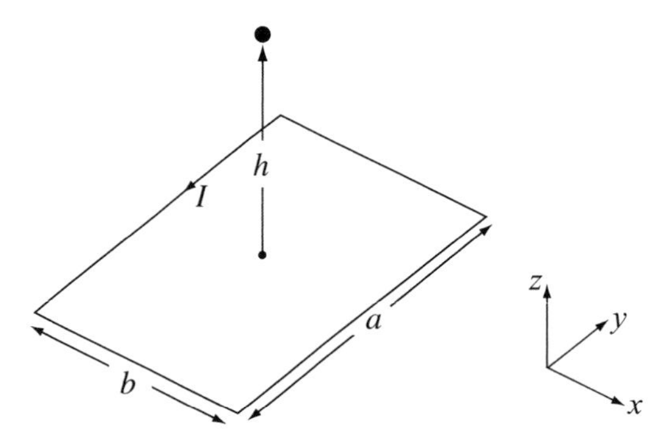

Figure 8.29

8.7 Application: Magnetic Flux Density of a Spiral Coil. A wire is bent to form a flat spiral coil and carries a current I [A]. Calculate the flux density at a point h [m] high above the center of the spiral (**Figure 8.30**). The coil radius is a [m]. The spiral has N uniformly distributed turns. Assume each turn is a perfect circle.

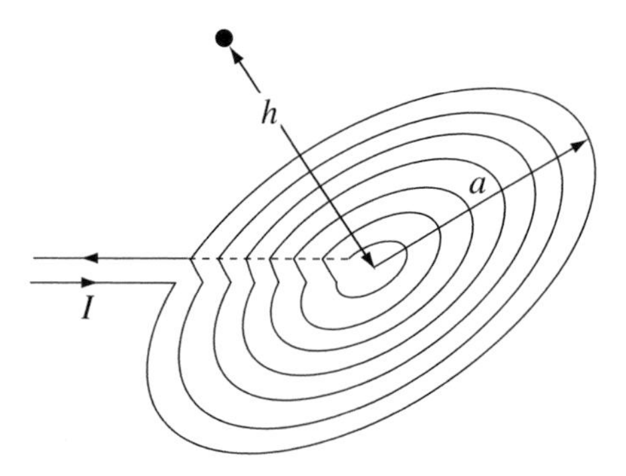

Figure 8.30

8.8 Application: Magnetic Fields of Planar Structures. A semiconductor material is made as shown in **Figure 8.31**. The long ends of the piece are connected to a voltage source. This causes electrons to move toward the positive connection and holes toward the negative connection. Holes and electrons move at the same velocity v [m/s] and have the same charge magnitude q [C] each. The hole and electron densities are equal to ρ holes/m^3 and ρ electrons/m^3, respectively:

(a) Set up the integrals needed to calculate the magnetic flux density at a height h [m] above the center of the piece (see **Figure 8.31**). Do not evaluate the integrals.

(b) Optional: Evaluate the integrals in **(a)** for the following data: $a = 0.02$ m, $b = 0.05$ m, $d = 0.5$ m, $h = 0.1$ m, $\rho = 10^{20}$ (holes or electrons/m^3), $v = 1$ m/s, $q = 1.6 \times 10^{-19}$ C. **Note**. The integration is very tedious. You may wish to perform the integration numerically.

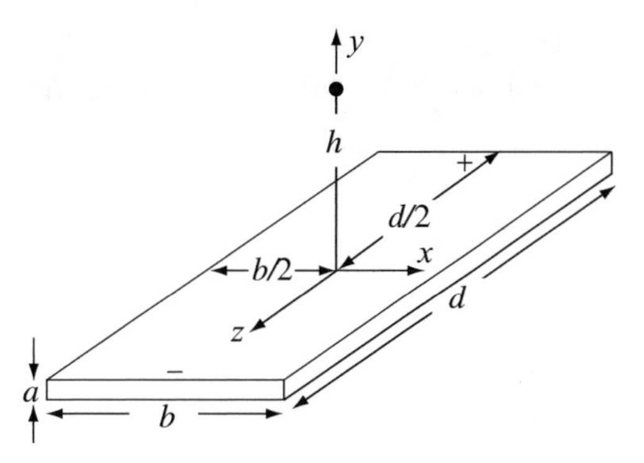

Figure 8.31

Ampère's Law

8.9 Magnetic Flux Density of Thick Conductors. Two, infinitely long cylindrical conductors of radius a [m] carry a current I [A] each. One conductor coincides with the y axis, whereas the second is parallel to the z axis, in the x–z plane, at a distance d [m] from the origin as shown in **Figure 8.32**. The currents are in the positive y direction in the first conductor and in the positive z direction in the second conductor. Calculate the magnetic flux density (magnitude and direction) at all points on the z axis.

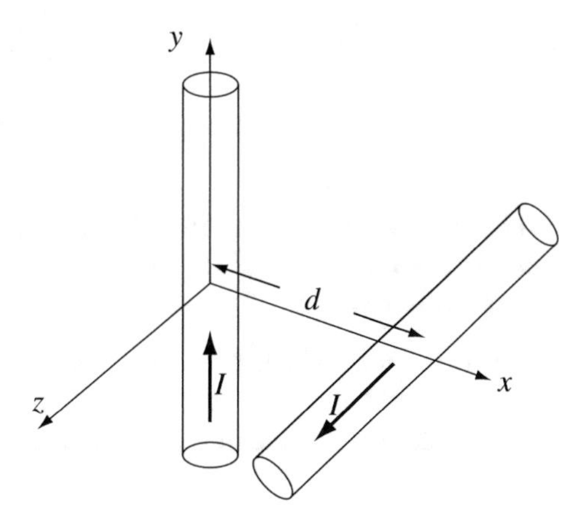

Figure 8.32

8.10 Magnetic Flux Density of Thick Conductors. Two infinitely long cylinders, one of radius a [m] and one of radius b [m], are separated as shown in **Figure 8.33**. One cylinder is directed parallel to the z axis, in the x–z plane, the other parallel to the y axis, in the x–y plane. A current density J [A/m^2] flows in each, directed in the positive y and z directions as shown. Calculate the magnetic flux density at a general point $P(x,y,z)$ outside the cylinders.

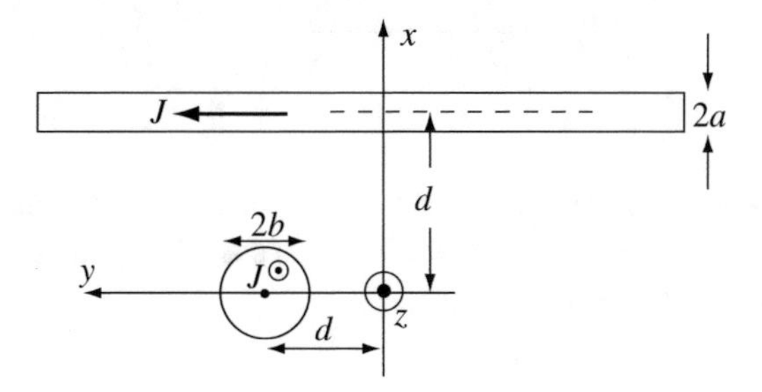

Figure 8.33

8.11 Application: Magnetic Flux Density Due to Power Lines. A high-voltage transmission line operates at 750 kV and 2,000 A maximum. The towers used to support the lines are $h = 20$ m high and the lines are separated a distance $2a = 6$ m **(Figure 8.34)**. Calculate:

(a) The magnetic flux density anywhere at ground level.
(b) The magnetic flux density (magnitude and direction) at ground level, midway between the two wires.

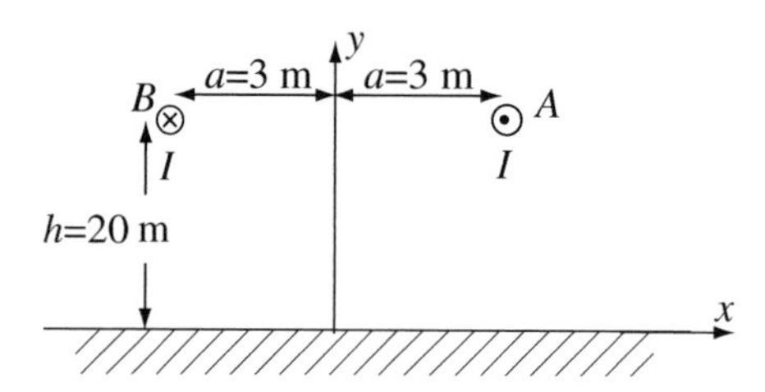

Figure 8.34

Ampère's Law, Superposition

8.12 Magnetic Flux Density Due to Current Distribution Two long copper strips on a printed circuit board are shown in cross section. The strips are long and their thickness t is negligible. A current flows in strip (1) into the page and returns (out of the page) in strip (2) as shown in **Figure 8.35**. The current is uniformly distributed in the cross-section of each strip. Calculate the magnetic flux density at point P (direction and magnitude). P is midway between the edges of the strips.

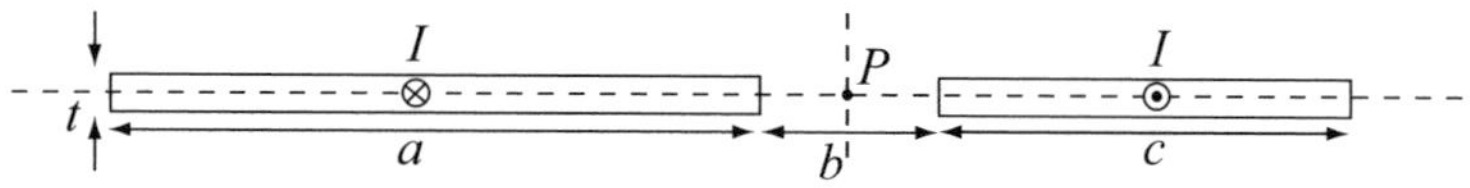

Figure 8.35

8.13 Magnetic Flux Density Due to Volume Distribution of Currents. A very long, L-shaped conductor with dimensions shown in cross section in **Figure 8.36** carries a total current I [A]. The current is in the positive z direction (out of the paper) and is uniformly distributed in the cross-section:

(a) Set up the integrals needed to calculate the magnetic flux density (magnitude and direction) at point $P(a,a)$.
(b) **Optional**: Evaluate the integrals in **(a)** for the following data: $a = 0.05$ m, $b = 0.025$ m, $I = 100$ A. **Note**. The integration is very tedious. You may wish to perform the integration numerically.

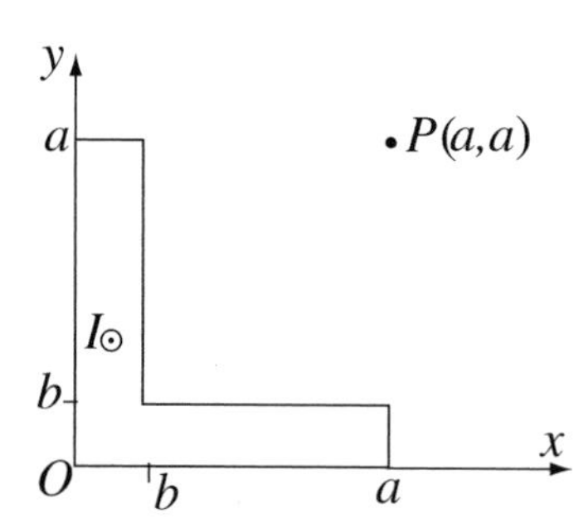

Figure 8.36

8.14 Magnetic Field of a Current Sheet. A thin layer of wires forms an infinite sheet of current. There are N wires per meter and each wire carries a current I [A] as shown in **Figure 8.37**. Calculate the magnitude and direction of the magnetic flux density everywhere in space.

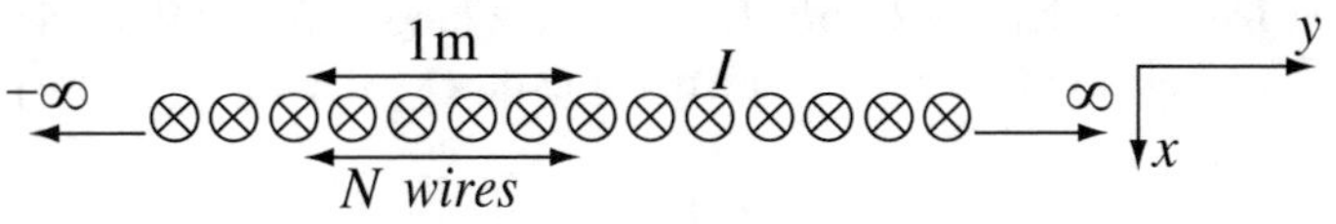

Figure 8.37

8.15 Application: Magnetic Field of Thin and Thick Conductors. An infinitely long thin wire is placed at the center of a hollow, infinitely long cylindrical conductor as shown in **Figure 8.38**. The conductor carries a current density J [A/m^2] and the wire carries a current I [A]. The direction of J is out of the page. Find the magnetic flux density for the regions $0 < r < \mathrm{r}_1, r_1 < r < r_2, r > r_2$ for the following conditions: **(a)** $I = 0$. **(b)** I in the direction of J. **(c)** I in the direction opposing J.

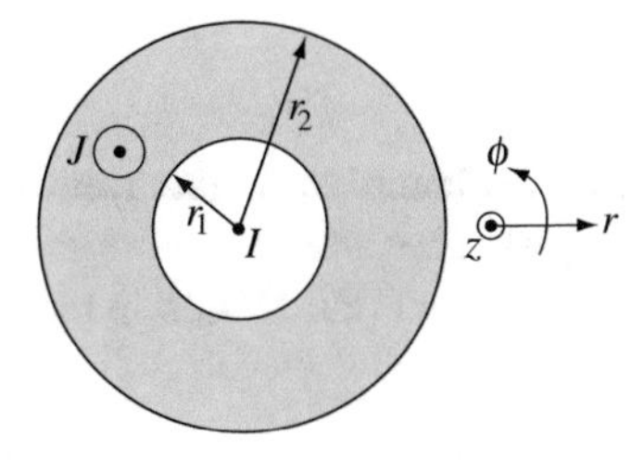

Figure 8.38

8.16 Magnetic Flux Density Due to Rotating Surface Charge Density. A long conducting cylinder of radius b [m] contains a uniformly distributed surface charge density ρ_s [C/m^2] on its outer surface. The cylinder spins at an angular velocity ω [rad/s] around its axis **(Figure 8.39)**. Calculate the magnetic field intensity everywhere.

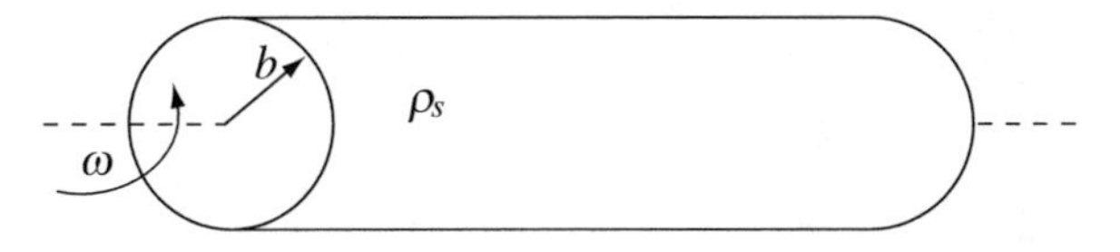

Figure 8.39

8.17 Superposition of Magnetic Flux Densities Due to Thick Conductors. A hollow, infinitely long cylindrical conductor has an outer radius b [m] and an inner radius a [m]. An offset solid cylinder of radius c [m] is located inside the large cylinder as shown in **Figure 8.40**. The two cylinders are parallel and their centers are offset by a distance d [m]. Assuming that the current density in each cylinder is J [A/m^2] and is uniform, calculate the magnetic flux density at point A (outside the larger cylinder). The current in the larger cylinder is out of the page, and in the smaller cylinder, it is into the page.

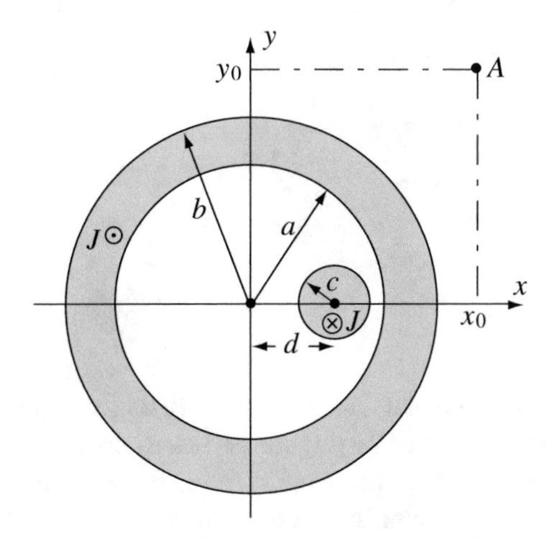

Figure 8.40

8.18 Application: Magnetic Field in a Coaxial Cable. Given a very long (infinite) thin wire with current I [A] and a very long (infinite) cylindrical tube of thickness $r_2 - r_1$ with uniform current density J_1 [A/m^2] as shown in **Figure 8.41**, find:

(a) **B** for $0 < r < r_1$, **(b)** **B** for $r_2 < r < \infty$.

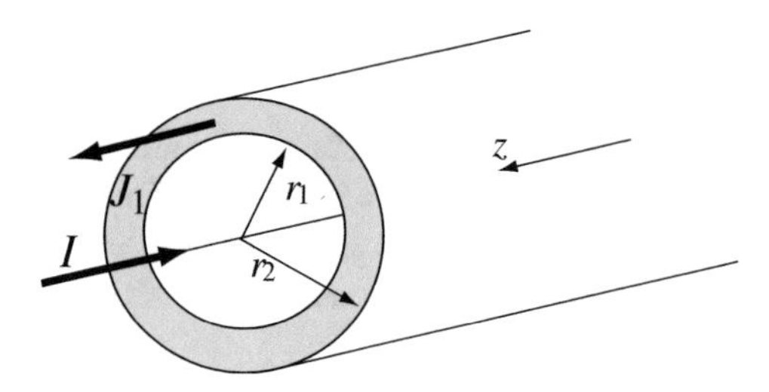

Figure 8.41

8.19 Magnetic Field in Coaxial Solenoids. A long solenoid (a single layer of very thin wire turns) of radius r_1 [m] is placed inside a second long solenoid of radius r_2 [m]. The currents in the solenoids are equal but in opposite directions, as shown in **Figure 8.42a** in axial cross section and in Figure **8.42b** in radial cross-section, and each solenoid has n turns per meter length:

(a) Calculate the magnetic flux density for $0 < r < r_1$, $r_1 < r < r_2$ and $r > r_2$.
(b) Repeat **(a)**, if the current in the inner solenoid is reversed.

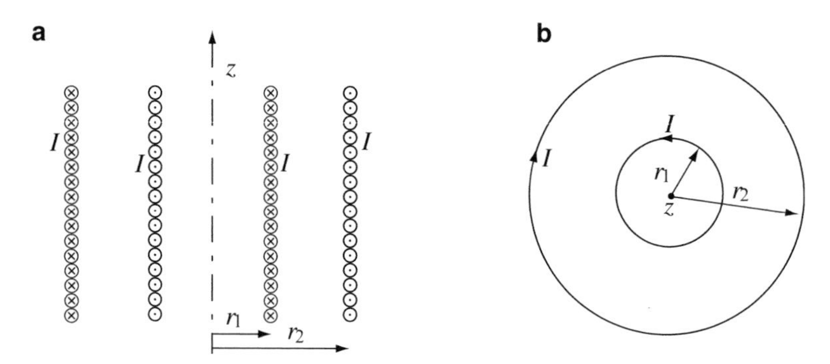

Figure 8.42

8.20 Superposition of Solutions in Solid Conductors. A very long (infinite) tubular conductor has radius b [m] and an offset hole of radius c as shown in **Figure 8.43**. The center of the hole is offset a distance d [m] from the center of the conductor. If the current density in the conductor flows out of the paper, is uniform in the conductor's cross-section (shaded area) and equals J [A/m^2], calculate the magnetic field intensity at A.

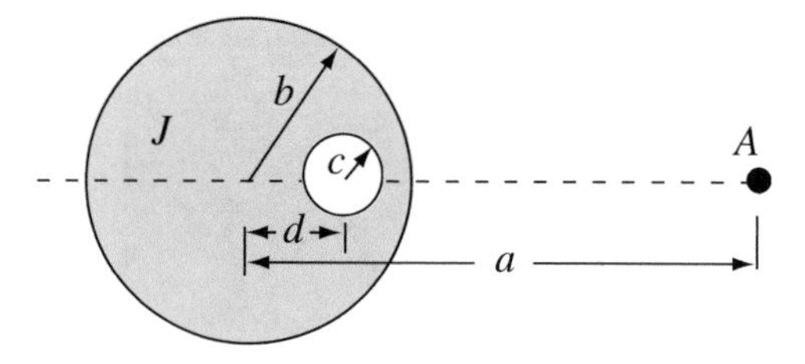

Figure 8.43

8.21 Application: Magnetic Flux Density Due to a Cable. Three very long cylindrical, solid conductors are arranged on a line as shown. The center conductor has radius $2a$ [m] and carries a current I [A] flowing into the page. The two outer conductors have radius a [m] and each carries a current $I/2$ [A] out of the page. They form a cable so that a current I [A] flows in one direction in the center conductor and returns (in the opposite direction) in the outer two conductors (**Figure 8.44**). Find the magnetic flux density (magnitude and direction) at points A, B, C and D.

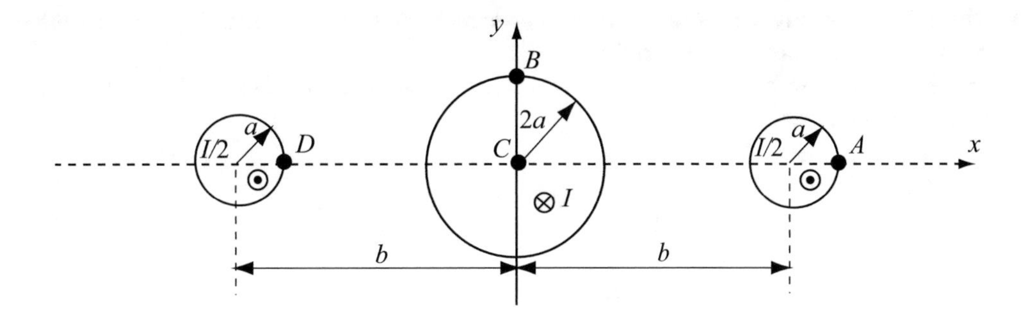

Figure 8.44

Biot–Savart Law, Magnetic Vector Potential

8.22 Magnetic Vector Potential of Current Segment. A very long (but not infinite) straight wire carries a current I [A] and is located in air:

(a) Calculate the magnetic vector potential at a distance a [m] from the center of the wire. Assume for simplicity that the wire extends from $-L$ [m] to $+L$ [m] and that $L \gg a$.
(b) From the result in **(a)**, what must be the magnetic vector potential for an infinitely long current-carrying wire?

8.23 Magnetic Vector Potential of Loops. A circular loop and a square loop are given as shown in **Figure 8.45**:

(a) Find the ratio between the magnetic flux densities at the center of the two loops.
(b) What is the magnetic vector potential at the center of each loop? Hint. Use Cartesian rather than cylindrical coordinates for the integration for both loops.

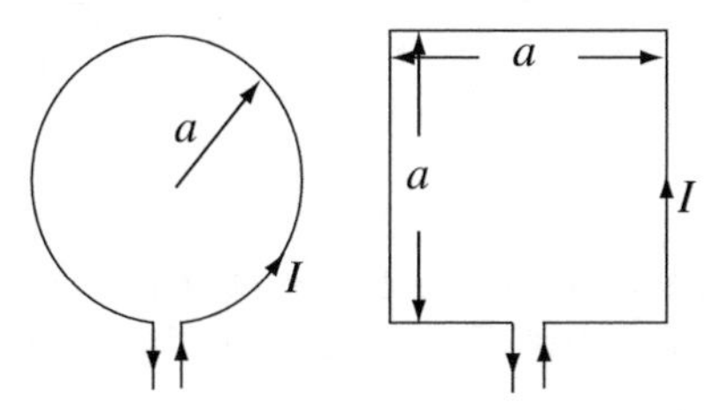

Figure 8.45

8.24 Application: Magnetic Flux. A straight wire carries a current $I = 1$ A. A square loop is placed flat in the plane of the wire as shown in **Figure 8.46**. Calculate:

(a) The flux, using the magnetic flux density.
(b) The flux, using the magnetic vector potential. (**Hint**. Use the result of **Problem 8.22**. Compare the results in **(a)** and **(b)**.)

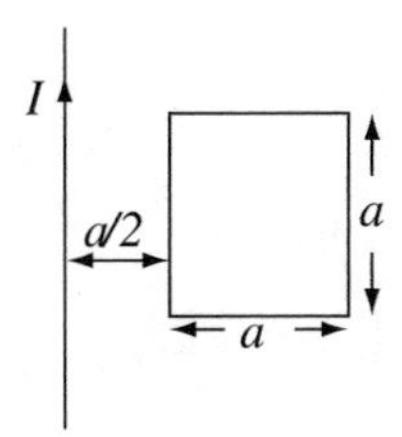

Figure 8.46

8.25 Magnetic Flux. A straight, infinitely long wire carries a current $I = 10$ A and passes either above or next to a loop of radius $a = 0.1$ m as shown in **Figure 8.47**:

(a) If the wire passes as in **Figure 8.47a**, show from symmetry considerations that the total flux through the loop is zero, regardless of how close the loop and the wire are.

(b) If the loop is moved sideways as in **Figure 8.47b**, calculate the total flux through the loop. Assume the loop and wire are in the same plane.

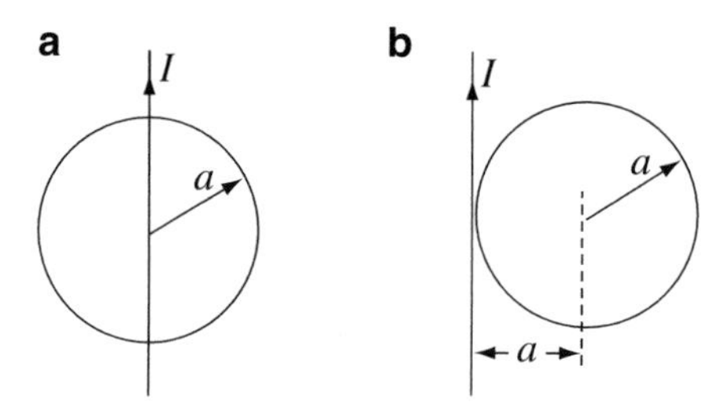

Figure 8.47

Magnetic Scalar Potential

8.26 Magnetic Scalar Potential Between the Poles of a Permanent Magnet. The magnetic flux density between the poles of a large magnet is 0.1 T and is uniform everywhere between the poles in **Figure 8.48**. Calculate the magnetic scalar potential difference between the poles.

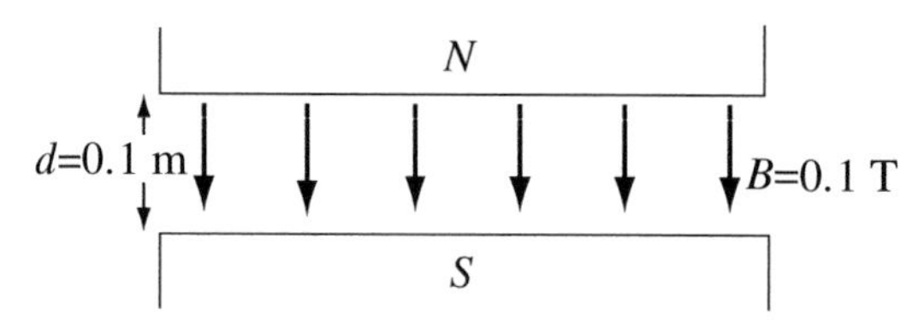

Figure 8.48

8.27 Magnetic Scalar Potential Between the Poles of a Permanent Magnet. The magnetic scalar potential difference between the poles of a magnet is 100 A (**Figure 8.49**). Calculate the magnetic field intensity if the space between the poles is air and the magnetic field intensity is uniform.

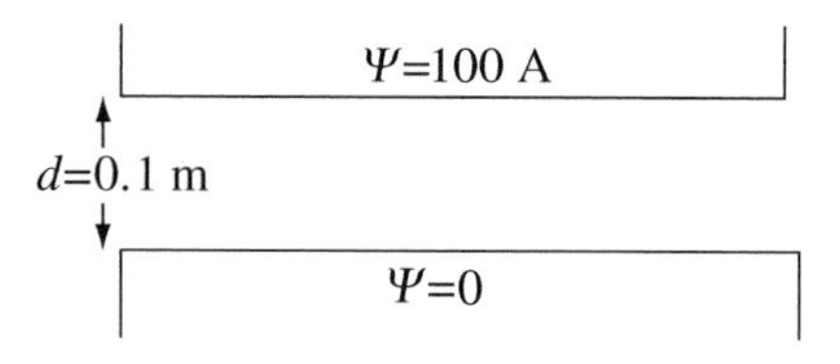

Figure 8.49

Magnetic Materials and Properties

9

Now sing my muse, fir 'tis a weighty cause.
Explain the Magnet, why it strongly draws,
And brings rough iron to its fond embrace.
This men admire; for they have often seen
Small rings of iron, six, or eight, or ten,
Compose a subtile chain, no tye between;
But, held by this, they seem to hang in air,
One to another sticks and wantons there;
So great the Loadstone's force, so strong to bear!.

—Titus Lucretius Carus (94–50 BCE), *De Rerum Natura* (On the nature of things), T. Creech, Translation, London 1714.

9.1 Introduction

The above reference to lodestone is interesting in that it is over 2000 years old. The property of the magnetic field to attract or generate a force is universally known and is used in practical devices, probably more than most of us realize. How many applications of the permanent magnet do you recall? Did you know, for example, that many electric motors use permanent magnets or that the ignition in cars is commonly controlled by a permanent magnet and a Hall-element switch? It is therefore quite useful to identify the properties of the permanent magnet since sooner or later you will encounter it in design. Thus follows the study of magnetic properties of materials. Many materials exhibit magnetic properties, some quite surprising. The permanent magnet is only one of them. Iron, nickel, or chromium oxides on magnetic data storage devices store information in the form of magnetic field variations. Solid nickel contracts when placed in a magnetic field, whereas strong magnetic fields cause atoms to tilt about their spin axes, a phenomenon that leads directly to magnetic resonance imaging (MRI). Naturally occurring materials, such as magnetite (Fe_3O_4), are found in bacteria and in brains of many animals which use this material as a biological compass for navigation in the geomagnetic field.

Many of the magnetic properties of materials were known from antiquity, although not necessarily understood. The magnet was known at least from the times of Thales of Miletus (sixth century BCE) who spoke of the lodestone as a matter of fact (also called loadstone or leading stone, from the fact that the stone leads the mariner at sea). The name magnet was given after Magnesia, a region in Macedonia where, it is held by many, the lodestone was first found. According to a legend related by Pliny the Elder,[1] the magnet was discovered when a herdsman, apparently wearing iron-studded shoes, and using an iron-pointed staff, took his herd to pasture and got stuck on a hillside on Mount Ida (in Crete). Interesting, if unlikely story. What is certain is that magnets were known for ages. Some evidence points as far as 2700 BCE and definite knowledge of the

[1] Caius Plinius Secundus. Pliny died during the eruption of Mount Vesuvius in 79 CE, the same eruption that buried Pompeii. In his book on Natural History, published around 77 CE, he mentions both the legend of discovery of the lodestone as well as an interesting story about magnetic suspension. He goes on to describe the various types of lodestone found at the time and locations where these are found. One particularly interesting is the haematite (blood-stone, from its red color). This is essentially rust: Fe_2O_3 which is only very slightly magnetic. The common lodestone is made of magnetite (Fe_3O_4 and tends to be black). His description of the stone as the most marvelous thing there is or "lifelike" is poetic, but many attributes such as curative powers are totally absurd. According to Pliny, other common names for the stone were "live iron," "Heraclion" (a reference to either the city of Heraclea or Hercules, referring to its power over iron), and "Sideratis" (iron earth).

N. Ida, *Engineering Electromagnetics*, https://doi.org/10.1007/978-3-030-15557-5_9

lodestone traces back to 600 BCE, to the time of Thales of Miletus.[2] There are even some intriguing ancient legends of structures made of lodestone for the purpose of suspending iron statues.[3] The lodestone was also considered to have medicinal values and features in the most notable medical books of antiquity and of the Middle Ages. Magic powers were also to be found and some strange concepts of magnetic properties were prevalent. In the absence of understanding of magnetic properties, it is not surprising that magic should be associated with such a remarkable material. But perhaps the most beautiful reference is in the French name for magnet—aimant, which means loving or affectionate, or the Chinese name "loving stone" alluding to the attraction between magnets and between magnets and iron.

Systematic study of magnetism started with William Gilbert[4] who, in 1600, wrote the first serious account on magnetism. His book, *De Magnete* (originally written in Latin), is considered by many to be the beginning of electromagnetics. You may wish to read at least parts of it (in translation, if you prefer) as it gives a special flavor of ancient notions and modern experiment and, for a 400-year-old book, is surprisingly modern and readable.

Magnets are not the main topic of this chapter. However, the first true magnetic device was the permanent magnet in the form of the compass. It was known in China at least as early as 1000 CE and in Europe at least as early as 1200 CE. However, like much else about magnets, its origins are shrouded in mystery.[5] It is then not surprising that the magnet was, for a long time, the subject of inquiry. In our study, we discuss magnets because they generate forces and, therefore, have energy and work associated with them, properties we can show by simple experiment. This serves as a natural introduction to the relations between magnetic fields and force and energy and leads to the derivation of useful relations for the calculation of forces, work, and energy in the magnetic field.

In addition to magnetic properties of materials, this chapter introduces some of the most important and most useful aspects of static magnetic fields. The behavior of magnetic fields at the interface between two materials will be discussed in the form of interface conditions. The interface conditions are useful in understanding how the magnetic field behaves, say, at the contact region between a magnet and a piece of iron or in the gap between the stator and rotor of an electric motor.

Considerable space is devoted to magnetization and properties of permanent magnets both in terms of design and in terms of their general magnetic properties. Following this, we discuss inductance, energy, forces, and torque in the magnetic field as well as the concept of magnetic circuits. The latter, in particular, is a common design tool, especially in power devices.

[2] Thales of Miletus is believed to be the earliest to mention the lodestone, although he probably learned of it during his travels in Egypt (see footnote 2 on page 98). Thales is said to have believed (according to Aristotle) that the magnet has soul since it attracted (had "sympathy" to) iron.

[3] Pliny writes that an architect by the name of Timochares by order from King Ptolemy II, Phyladelphus, began to put a vaulted roof of lodestone on the temple of Arsinoe (Ptolemy's wife) in Alexandria, so that her statue, made of iron, would be suspended in air. Another mention by Ruffinus says that in the temple of Serapis (in Memphis, Egypt), there was an iron chariot suspended by lodestones. When the stones were removed, the chariot fell and was smashed to pieces. Beda (the Venerable) says that a statue of Bellerophon's horse (Pegasus) framed of iron was placed between lodestones, with wings expanded, floating in air. These are legends, but that the idea of suspending iron by lodestone was mentioned almost 2000 years ago is remarkable in itself. Magnetic levitation seems to be not all that new, at least not in concept.

[4] William Gilbert (1540–1603) is most often mentioned as physician to Queen Elisabeth I (even though this occurred only during the last 2 years of his life), but his education included Mathematics and Physics. His main activity prior to the publication of *De Magnete* was in medicine, but he also experimented in chemistry and, certainly, in magnetics. His book became to be appreciated as one of the first to describe systematic experiment without regard to mysticism, opinions, and perpetuation of unfounded information and was the culmination of 18 years of careful (and expensive) experimental work. It is fascinating to read how he disclaims opinions of ancients and contemporaries about magnetic properties attributed to planets or constellations or that rubbing of a magnet with garlic will cause it to lose its properties. These seemingly trivial experiments were necessary to dispute prevalent notions of the time, including that magnet will not attract iron in the presence of diamonds or that it will lose its attraction if rubbed by onions, garlic, or goat's blood. For example, in Book III, Chapter XIII, he describes an experimant in which he surrounded a lodestone by 75 "excellent" diamonds in the presence of witnesses and could observe no effect on the magnetic field or that rubbing a magnetic needle with a diamond does not affect it (one belief held was that the diamond will cause the needle to reverse its action). Gilbert died in the plague of 1603.

[5] The magnetic compass is believed to have been invented in China and perhaps later, separately, in Europe. There are also some stories about Marco Polo having brought it from China, but this cannot be true, as mention of the compass as a known instrument in Europe can be traced as far back as 1180 CE. Compasses were known to be made in Italy before 1300 CE (at Amalfi, near Naples) whereas Marco Polo returned from China in 1295. According to some accounts, the compass was known in China as early as 1100 BCE. One account gives details of a chariot, on which a figurine with outstretched arms points to the south. The figurine was pivoted and had in it lodestones to act as a compass. The time: 2637 BCE in China.

9.2 Magnetic Properties of Materials

One question that was alluded to in the previous chapter is the following: What, if any, is the effect of a magnetic field on materials? We asked a similar question about the electric field and, as a result, obtained a definition of polarization in materials and a simple definition of electric permittivity. Similar results will be obtained here for the magnetic field. The basis of discussion is to consider the atomic structure of the various materials and the effect a magnetic field can have on atoms. The effects are fascinating and lead to surprising and highly useful applications, as we will see shortly. The effects of magnetic fields on materials are varied, with some materials exhibiting little or no effect and others exhibiting very pronounced effects to the extent that we often call these materials "magnetic materials."

Consider again a permanent magnet. That it is permanently "magnetized" is the reason why it will attract another magnet or a piece of iron. Also, by experiment, we know that if we place a piece of iron on the pole of a magnet, the iron becomes "magnetized" and acquires the properties of the permanent magnet. These properties are lost after removal of the magnet. A piece of hard steel will behave similarly but will retain the magnetic property for some time. You may convince yourself of this by placing a paper clip on a magnet. When removing the clip, it loses its magnetic properties. The blade of a screwdriver, which is made of steel, becomes permanently magnetized after being placed on a magnet. On the other hand, other materials such as copper, aluminum, or plastics do not seem to be affected by the presence of the magnet.

Now for the big questions: What is magnetization? How is it imparted to materials? Why are some materials affected while others are not? We will try to answer these questions and others in the following sections, but first we must introduce some tools and, in particular, the idea of the magnetic dipole.

9.2.1 The Magnetic Dipole

The magnetic dipole is essentially a small loop, carrying a current. It is a useful concept both in itself and as a means of explaining the behavior of magnetic materials. To define the dipole and its properties, consider first a filamentary loop of any size as shown in **Figure 9.1a**. A current I flows in the loop and we wish to calculate the magnetic flux density at an arbitrary point in space $P(R,\theta,\phi)$ (in spherical coordinates). This is the field, or observation point. To calculate the field, we define an element of current of length dl' at point $P'(R',\theta',\phi')$, carrying a current I. The magnetic flux density can be calculated directly from **Eq. (8.9)** in terms of the magnetic flux density or from **Eq. (8.34)** in terms of the magnetic vector potential. The latter is easier to apply and is therefore used here. The magnetic vector potential at point P due to the loop in **Figure 9.1a** is

$$\mathbf{A} = \frac{\mu_0 I}{4\pi}\oint_{C'}\frac{d\mathbf{l}'}{|\mathbf{R}-\mathbf{R}'|} \quad \left[\frac{\text{Wb}}{\text{m}}\right] \tag{9.1}$$

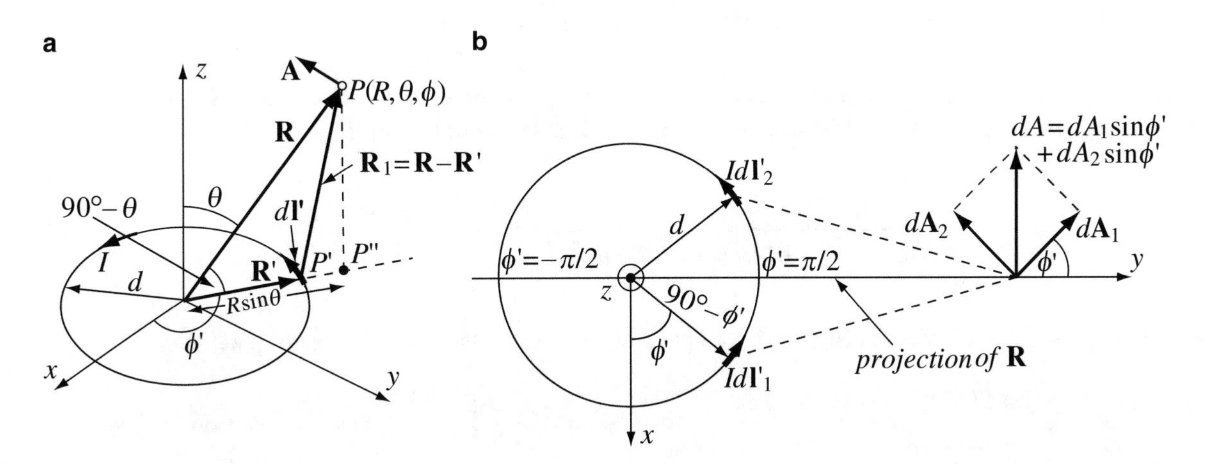

Figure 9.1 The magnetic dipole. (**a**) A loop carrying a current I and the dimensions necessary to calculate the magnetic vector potential at $P(R,\theta,\phi)$. (**b**) The magnetic vector potential at $P(R,\theta,\phi)$ is in the ϕ direction

where the magnitude of $\mathbf{R}'$ is equal to the radius of the loop and the vector $\mathbf{R}_1 = \mathbf{R} - \mathbf{R}'$ is the vector connecting the source element $d\mathbf{l}'$ to the field point P. The element of length dl' along the loop (in spherical coordinates at $\theta = \pi/2$) is [see **Eq. (1.77)**]:

$$d\mathbf{l}' = \hat{\boldsymbol{\phi}} d\, d\phi' \tag{9.2}$$

and, as expected, the integration is independent of R' or θ'. This current element produces a magnetic vector potential in the ϕ direction anywhere in space. To understand the contribution of the current element to the magnetic vector potential, consider **Figure 9.1b**, which is a top view of **Figure 9.1a**. First, we note that the magnetic vector potential at point P is everywhere parallel to the current in the loop. Second, the contribution at this point is due to two symmetric elements of current $d\mathbf{l}_1'$ and $d\mathbf{l}_2'$. The tangential components add up and the normal components (those in the direction radial to the loop) cancel everywhere in space. Thus, we can write for the magnetic vector potential at point P:

$$\mathbf{A} = \hat{\boldsymbol{\phi}} \frac{\mu_0 I d}{2\pi} \int_{-\pi/2}^{\pi/2} \frac{\sin\phi' d\phi'}{|\mathbf{R} - \mathbf{R}'|} \quad \left[\frac{\text{Wb}}{\text{m}}\right] \tag{9.3}$$

The distance $|\mathbf{R} - \mathbf{R}'|$ from the current element to the point P is

$$|\mathbf{R} - \mathbf{R}'| = \sqrt{(\mathbf{R} - \mathbf{R}') \cdot (\mathbf{R} - \mathbf{R}')} = \sqrt{\mathbf{R} \cdot \mathbf{R} + \mathbf{R}' \cdot \mathbf{R}' - 2\mathbf{R} \cdot \mathbf{R}'} \tag{9.4}$$

Since $\mathbf{R} \cdot \mathbf{R} = R^2$, $\mathbf{R}' \cdot \mathbf{R}' = d^2$, and $\mathbf{R} \cdot \mathbf{R}' = Rd\cos(90 - \theta)\sin\phi'$, we have at P:

$$\mathbf{A} = \hat{\boldsymbol{\phi}} \frac{\mu_0 I d}{2\pi} \int_{-\pi/2}^{\pi/2} \frac{\sin\phi' d\phi'}{\left(R^2 + d^2 - 2Rd\cos(90 - \theta)\sin\phi'\right)^{1/2}} \quad \left[\frac{\text{Wb}}{\text{m}}\right] \tag{9.5}$$

This is an exact result. To simplify the integration, we calculate the magnetic vector potential at large distances from the loop $R \gg d$. In this case, d^2 in the above relation is negligible (but not the term involving $2Rd$, which can be much larger than d^2). With this, and using $\cos(90 - \theta) = \sin\theta$, we can write

$$\frac{1}{\left(R^2 + d^2 - 2Rd\cos(90 - \theta)\sin\phi'\right)^{1/2}} \approx \frac{1}{\left(R^2 - 2Rd\sin\theta\sin\phi'\right)^{1/2}} = \frac{\left(1 - 2(d/R)\sin\theta\sin\phi'\right)^{-1/2}}{R} \tag{9.6}$$

Using the binomial expansion approximation $(1 + x)^p \approx 1 + px$ (where all but the first two terms of the binomial expansion are neglected), with $x = -2d\sin\theta\sin\phi'/R$ and $p = -1/2$, **Eq. (9.6)** becomes

$$\frac{1}{\left(R^2 + d^2 - 2Rd\cos(90 - \theta)\sin\phi'\right)^{1/2}} \approx \frac{1 + (d/R)\sin\theta\sin\phi'}{R} \tag{9.7}$$

Substituting this approximation in **Eq. (9.5)**, we get an expression for the magnetic vector potential:

$$\begin{aligned}\mathbf{A} &\approx \hat{\boldsymbol{\phi}} \frac{\mu_0 I d}{2\pi} \int_{\phi'=-\pi/2}^{\phi'=\pi/2} \frac{1 + (d/R)\sin\theta\sin\phi'}{R} \sin\phi' d\phi' = \hat{\boldsymbol{\phi}} \frac{\mu_0 I d}{2\pi R} \int_{\phi'=-\pi/2}^{\phi'=\pi/2} \left(\sin\phi' + \frac{d}{R}\sin\theta\sin^2\phi'\right) d\phi' \\ &= \hat{\boldsymbol{\phi}} \frac{\mu_0 I d}{2\pi R} \left(\Big[-\cos\phi'\Big]_{\phi'=-\pi/2}^{\phi'=\pi/2} + \left[\frac{\phi' d\sin\theta}{2R} - \frac{d\sin\theta\sin 2\phi'}{4R}\right]_{\phi'=-\pi/2}^{\phi'=\pi/2} \right) = \hat{\boldsymbol{\phi}} \frac{\mu_0 I d^2 \sin\theta}{4R^2} \quad \left[\frac{\text{Wb}}{\text{m}}\right]\end{aligned} \tag{9.8}$$

The magnetic flux density is now obtained using the curl of **A**:

$$\mathbf{B} = \nabla \times \mathbf{A} = \hat{\mathbf{R}}\frac{1}{R\sin\theta}\left[\frac{\partial}{\partial\theta}\left(\frac{\mu_0 I d^2 \sin^2\theta}{4R^2}\right)\right] - \hat{\boldsymbol{\theta}}\frac{1}{R}\left[\frac{\partial}{\partial R}\left(\frac{\mu_0 I d^2 \sin\theta}{4R}\right)\right] \quad [\mathrm{T}] \tag{9.9}$$

In this case, **A** only has a ϕ component. Therefore, the curl of **A** has an R and a θ component. Evaluating the derivatives in **Eq. (9.9)** gives

$$\boxed{\mathbf{B} \approx \frac{\mu_0 I d^2}{4R^3}\left(\hat{\mathbf{R}}2\cos\theta + \hat{\boldsymbol{\theta}}\sin\theta\right) \quad [\mathrm{T}]} \tag{9.10}$$

This was a rather tedious calculation and, in the end, we only got an approximation to the solution. However, this result is rather accurate under the following two equivalent conditions:

(1) The field point is at very large distances from the loop (R large).
(2) The loop radius, d, is very small.

Both conditions are satisfied by the requirement that $R \gg d$.

This result is important because it allows the calculation of the field due to electrons orbiting around the nucleus of the atom. These will be viewed as small loops or magnetic dipoles, which produce a magnetic field outside the atom. This, in turn, will provide a model that explains the magnetic field of permanent magnets. We will also use the discussion leading to the magnetic dipole in analysis of loop antennas in **Chapter 18**.

An interesting comparison can be made at this point with the electric dipole discussed in **Section 3.4.1.3** [see **Eq. (3.38)**]. The electric field intensity of the electric dipole and the magnetic flux density of the magnetic dipole are [after multiplying numerator and denominator in **Eq. (9.10)** by π]

$$\mathbf{E} \approx \frac{p}{4\pi\varepsilon_0 R^3}\left(\hat{\mathbf{R}}2\cos\theta + \hat{\boldsymbol{\theta}}\sin\theta\right) \quad [\mathrm{V/m}] \tag{9.11}$$

$$\mathbf{B} \approx \frac{I\pi d^2}{4\pi(1/\mu_0)R^3}\left(\hat{\mathbf{R}}2\cos\theta + \hat{\boldsymbol{\theta}}\sin\theta\right) \quad [\mathrm{T}] \tag{9.12}$$

where, in the electric dipole, we defined p [C · m] as the magnitude of the electric dipole moment, whereas the dipole moment, which is a vector, was defined as $\mathbf{p} = q\mathbf{d}$, where q [C] is the charge of the dipole and $\mathbf{d}$ [m] is the vector connecting the two charges as shown in **Figure 9.2a**. Using the two expressions and the analogy between the two, we can now define a magnetic dipole moment. Its magnitude equals $I\pi d^2$ and the direction, following the convention in **Figure 9.2b**, is in the z direction. In this case, $\mathbf{m} = \hat{\mathbf{z}}I\pi d^2$ [A·m^2], but, in general, for an arbitrary orientation of the loop, the direction of the magnetic dipole moment is in the direction perpendicular to the loop. Thus, we write, in general,

$$\mathbf{m} = \hat{\mathbf{n}}I\pi d^2 = \hat{\mathbf{n}}IS \quad [\mathrm{A \cdot m^2}] \tag{9.13}$$

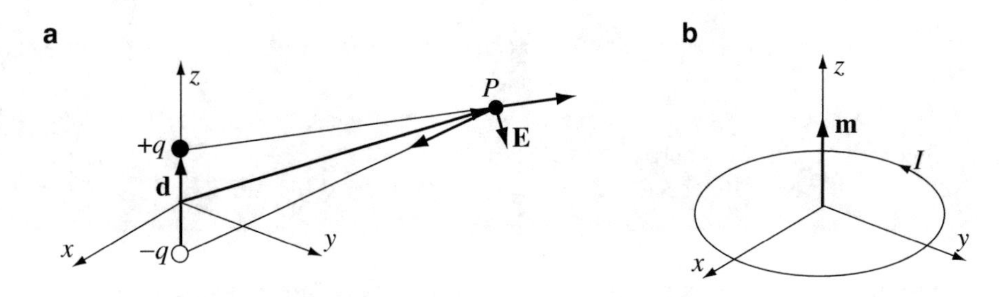

Figure 9.2 (**a**) The electric dipole. (**b**) The magnetic dipole

where S is the area of the planar loop. To simplify notation, we will assume that $\hat{\mathbf{n}} = \hat{\mathbf{z}}$ for the purposes of this section. The magnetic flux density in **Eq. (9.10)** can be written now in terms of the magnetic dipole moment **m** as

$$\mathbf{B} \approx \frac{\mu_0 m}{4\pi R^3}\left(\hat{\mathbf{R}}2\cos\theta + \hat{\mathbf{\theta}}\sin\theta\right) \quad [\mathrm{T}] \tag{9.14}$$

Note that the direction of the magnetic flux density depends on the location in space whereas the direction of **m** is always perpendicular to the loop. Using the magnetic dipole moment **m** and the definition of the cross product, we can write the following from **Eq. (9.8)**:

$$\mathbf{A} \approx \hat{\mathbf{\phi}}\frac{\mu_0 I d^2 \sin\theta}{4R^2} = \frac{\mu_0\left(\hat{\mathbf{\phi}} m \sin\theta\right)}{4\pi R^2} \quad \left[\frac{\mathrm{Wb}}{\mathrm{m}}\right] \tag{9.15}$$

Since $\hat{\mathbf{\phi}} m\sin\theta = \mathbf{m} \times \hat{\mathbf{R}}$, the magnetic vector potential of a small loop at a large distance R can be written in terms of the magnetic dipole moment as

$$\mathbf{A} \approx \frac{\mu_0 \mathbf{m} \times \hat{\mathbf{R}}}{4\pi R^2} \quad \left[\frac{\mathrm{Wb}}{\mathrm{m}}\right] \tag{9.16}$$

Example 9.1 Dipole Moment and Equivalent Loop Current of an Atom The magnetic dipole moment of a hydrogen atom is given as $m = 9 \times 10^{-24}\ \mathrm{A \cdot m^2}$. Assuming that this moment is produced by a small, circular, atomic level current with radius equal to the orbit radius of the electron ($d = 5 \times 10^{-11}$ m), calculate the equivalent current produced by the atom.

Solution: The magnetic dipole moment is calculated from **Eq. (9.13)** assuming the orbiting electron to be a dipole. This assumption is fully justified for the dimensions of the atom:

$$|\mathbf{m}| = I\pi d^2 \quad [\mathrm{A \cdot m^2}]$$

The equivalent current of the electron is

$$I = \frac{m}{\pi d^2} = \frac{9 \times 10^{-24}}{\pi \times \left(5 \times 10^{-11}\right)^2} = 1.15 \times 10^{-3} \quad [\mathrm{A}]$$

This current of 1.15 mA is a very large current for a single electron.

Example 9.2 Dipole Moment of a Square Loop A square loop with dimension $a = 1$ m on the side carries a current $I = 0.1$ A.

(a) Calculate the exact magnetic flux density of the loop at a height h above the center of the loop. What is the magnetic flux density at large distances?
(b) Calculate the magnetic flux density at the same location using the dipole approximation; that is, calculate the field of a dipole of identical current and area.
(c) Compare the two results at $h = 1$ m, 10 m, 100 m, and 1,000 m.

Solution: The exact flux density is calculated using the Biot–Savart law. Then, we compare this field with the field of a dipole. Consider **Figure 9.3a**. The magnetic flux density at point P equals four times the perpendicular component of the magnetic flux density of a single segment of length 1 m. The horizontal components of the four segments cancel because each two opposite segments produce fields in opposite horizontal directions. **Figure 9.3b** shows how **B** is calculated.

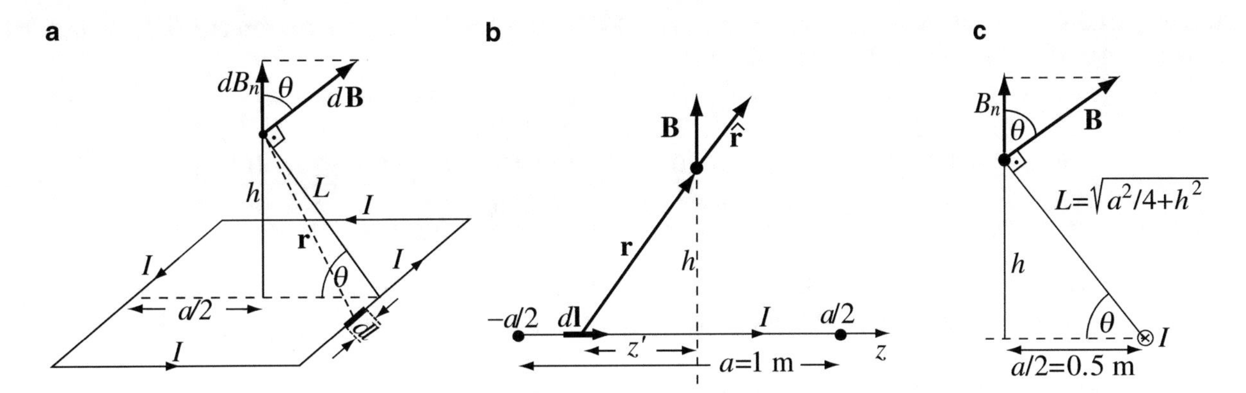

Figure 9.3 Magnetic flux density at height h [m] above the center of a square loop. (**a**) Configuration and the magnetic flux density due to an element of current. (**b**) Calculation of the flux density of one segment of the loop. (**c**) Components of **B** at height h

(**a**) From **Figures 9.3a** and **9.3b** and assuming the current segment is in the z direction in a cylindrical system of coordinates, the magnetic flux density **B** is [from **Eq. (8.9)**], setting $d\mathbf{l} = \hat{\mathbf{z}}dz'$

$$\mathbf{B} = \frac{\mu I}{4\pi}\int_{z'=-a/2}^{z'=a/2}\frac{\hat{\mathbf{z}}dz'\times(\hat{\mathbf{r}}L-\hat{\mathbf{z}}z')}{|\mathbf{r}|^3} = \hat{\mathbf{\phi}}\frac{\mu_0 I}{4\pi}\int_{z'=-a/2}^{z'=a/2}\frac{Ldz'}{\left(L^2+z'^2\right)^{3/2}} = \hat{\mathbf{\phi}}\frac{\mu_0 I}{4\pi}\frac{z'}{L\sqrt{L^2+z'^2}}\Bigg|_{z'=-a/2}^{z'=a/2} = \hat{\mathbf{\phi}}\frac{\mu_0 aI}{4\pi L\sqrt{a^2/4+L^2}} \quad [\mathrm{T}]$$

This flux density points in the ϕ direction (with respect to the current segment) or at an angle θ with the normal as shown in **Figure 9.3c**. Since we are only interested in the normal component, we write

$$B_n = B\cos\theta = \frac{\mu_0 aI\cos\theta}{4\pi L\sqrt{a^2/4+L^2}} \quad [\mathrm{T}]$$

From **Figures 9.3a** and **9.3c**

$$L = \sqrt{a^2/4+h^2}, \quad \cos\theta = \frac{a}{2L} = \frac{a}{2\sqrt{a^2/4+h^2}}$$

The total magnetic flux density due to all four segments is four times B_n. Substituting L and $\cos\theta$,

$$B_t = 4B_n = \frac{\mu_0 a^2 I}{2\pi\left(a^2/4+h^2\right)^{3/2}} \quad [\mathrm{T}]$$

At very large distances, $h^2 \gg a^2/4$ and we get

$$B_t \approx \frac{\mu_0 a^2 I}{2\pi h^3} \quad [\mathrm{T}].$$

(**b**) If we assume the loop is a dipole, the magnitude of its magnetic moment is $|\mathbf{m}| = IS = Ia^2$, where $S = a^2$ is the area of the loop. The direction of the dipole is in the direction of the axis (z direction), but here we only use the magnitude. From **Eq. (9.14)**, and setting $\theta = 0$ for the vertical component, we get

$$B_d \approx \frac{\mu_0 m}{2\pi R^3} = \frac{\mu_0 a^2 I}{2\pi h^3} \quad [\mathrm{T}].$$

(c) The magnetic flux density at 1 m, 10 m, 100 m, and 1,000 m calculated with the exact formula (B_t) and the dipole approximation (B_d) are shown in the table below:

	$h = 1$ m	$h = 10$ m	$h = 100$ m	$h = 1{,}000$ m
B_t	1.431×10^{-8} [T]	1.99×10^{-11} [T]	1.999×10^{-14} [T]	2×10^{-17} [T]
B_d	2×10^{-8} [T]	2×10^{-11} [T]	2×10^{-14} [T]	2×10^{-17} [T]

Note that at 10 m, the difference between the two results is less than 1%. Therefore, at large distances, the dipole approximation is a good approximation to the exact flux density of the square loop.

Exercise 9.1 A rectangular loop with sides $a = 10$ mm and $b = 20$ mm is placed on the x–y plane, in free space, centered at the origin and carries a current $I = 0.5$ A.

(a) Calculate the magnetic field intensity of the loop at a general point P under the conditions used for the dipole; that is, $R \gg b$, where R is the distance from the center of the loop to point P.
(b) Calculate the magnetic flux density at any point on the x–y plane for $R \gg b$.

Answer

(a) $\mathbf{H} \approx \dfrac{7.958 \times 10^{-6}}{R^3}\left(\hat{\mathbf{R}}2\cos\theta + \hat{\boldsymbol{\theta}}\sin\theta\right)$ [A/m]. **(b)** $\mathbf{B} \approx \hat{\boldsymbol{\theta}}\dfrac{1 \times 10^{-11}}{R^3}$ [T].

9.2.2 Magnetization: A Model of Magnetic Properties of Materials

The simplest atomic model of a material is that of a positive nucleus and a negative cloud of electrons, each orbiting around the nucleus and each spinning about its own axis. A similar model was used in **Chapter 4** to discuss polarization of dielectrics, but, there, we neglected motion of electrons. Here, we look at a slightly modified model since we are only interested in the magnetic properties of the material. Electrons orbiting around the atom generate an equivalent current. For modeling purposes, we will assume that the electron orbits in a planar circular path and, therefore, is equivalent to a magnetic dipole as discussed in the previous section. The magnetic dipole moment of an electron is $\mathbf{m} = \hat{\mathbf{n}} I\pi d^2 = \hat{\mathbf{n}} IS$ where $\hat{\mathbf{n}}$ is the direction normal to the plane in which the electron orbits. This model of the atom is shown in **Figure 9.4a**. The equivalent model is that of a loop of radius d, carrying a current I shown in **Figure 9.4b**. This model is appropriate both for the orbiting electrons and for the electron spin in that both produce an internal magnetic field. An identical model can be obtained by replacing the loop with a very small magnet, as shown in **Figure 9.4c**, if the magnetic flux density produced by this elementary magnet is the same as that of the loop. Here, we prefer the current loop model, but the magnet model is also useful. In general, electrons in various atoms rotate in different planes and they spin around randomly oriented axes. Therefore, we can view the magnetic dipole moments of atoms in a volume as being randomly oriented, unless there is some mechanism which will force them to align in certain directions. Although we cannot actually calculate the magnetic dipole moment of any particular atom (we can certainly estimate it, as was shown in **Example 9.1**) or find its direction in space, it is reasonable to say that if there are N atoms in a volume, the total magnetic moment of the volume is the vector sum of all individual dipoles in the volume. However, since the volume dipole density can vary from point to point (depending on the local distribution of dipole moments), it is more useful to calculate this quantity in an infinitesimal volume and thus obtain a point measure of the magnetic moment. This quantity is called ***magnetization*** and is defined as

$$\mathbf{M} = \lim_{\Delta v \to 0} \frac{1}{\Delta v} \sum_{i=1}^{N} \mathbf{m}_i \quad \left[\frac{\text{A}}{\text{m}}\right] \tag{9.17}$$

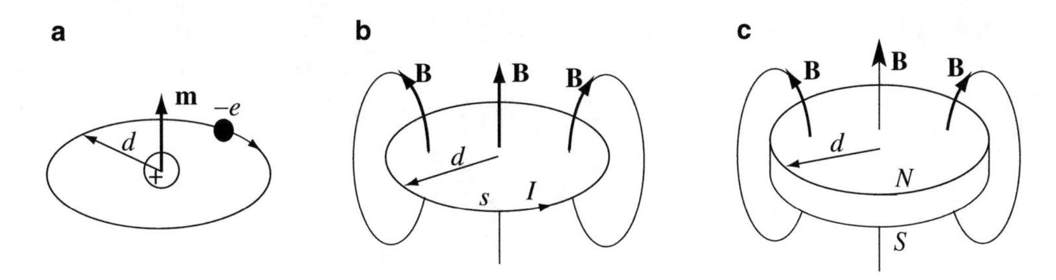

Figure 9.4 Magnetic dipole model of the atom. (**a**) Orbiting electron and its magnetic dipole moment. (**b**) The magnetic flux density due to the magnetic dipole. (**c**) Equivalent permanent magnet model

Note that since the magnetic dipole moment has units of ampere • meter^2 [A • m^2], the magnetization, which is a volume density of dipole moments, has units of ampere/meter [A/m]. The magnetization of a material can be (and often is) zero. Random orientation of dipole moments produces a zero sum vector and, therefore, zero magnetization.

What are the conditions under which nonzero magnetization can exist? Before we answer this question, consider what happens if all (or most) dipole moments are aligned in a given direction in space. The aligned dipoles have nonzero magnetization and therefore produce a net magnetic flux density **B**, as shown schematically in **Figure 9.5b**. In effect, we can view this dipole distribution as if it were a magnet of some known strength and shape, composed of a large number of elementary magnets as shown in **Figure 9.5c**. Thus, we conclude that a permanent magnet is any material in which the magnetic dipoles are aligned in a preferred direction and stay that way. Based on this simple model, a "stronger" permanent magnet is one in which more of the dipoles are aligned in a preferred direction. Why a material should align in a preferred direction and why it should stay that way will be discussed in the following section. The important question now is: given the magnetization of a material everywhere in its volume, what is the magnetic flux density produced by the magnetization? The reason why we pursue this path is that it provides a relation between magnetization and current density and, therefore, a very simple, physical explanation to the concept of magnetism. In this sense, the permanent magnet becomes merely an equivalent current density distribution and there is nothing mysterious about a current distribution. The model also allows evaluation of fields due to permanent magnets. To do so, we must associate the effect of the field with material volume. This is easily done from the definition of magnetization. Given a magnetization **M**, everywhere in a volume **[Eq. (9.17)]**, the magnetic dipole moment due to an element of volume dv' can be written as

$$d\mathbf{m} = \mathbf{M}dv' \quad \left[\mathrm{A}\cdot\mathrm{m}^2\right] \tag{9.18}$$

where v' indicates only that part of the volume which is magnetized. Substituting this in **Eq. (9.16)**, the contribution to **A** due to $d\mathbf{m}$ is

$$d\mathbf{A} \approx \frac{\mu_0\, d\mathbf{m}\times\hat{\mathbf{R}}}{4\pi R^2} = \frac{\mu_0 \mathbf{M}\times\hat{\mathbf{R}}}{4\pi R^2}dv' = \frac{\mu_0 \mathbf{M}\times\mathbf{R}}{4\pi R^3}dv' \quad \left[\frac{\mathrm{Wb}}{\mathrm{m}}\right] \tag{9.19}$$

where $\hat{\mathbf{R}} = \mathbf{R}/R$ was used. Thus, if a volume v' is magnetized, the magnetic vector potential at a point in space is calculated using **Figure 9.6** as

$$\mathbf{A} = \frac{\mu_0}{4\pi}\int_{v'}\frac{\mathbf{M}\times\mathbf{R}}{R^3}dv' \quad \left[\frac{\mathrm{Wb}}{\mathrm{m}}\right] \tag{9.20}$$

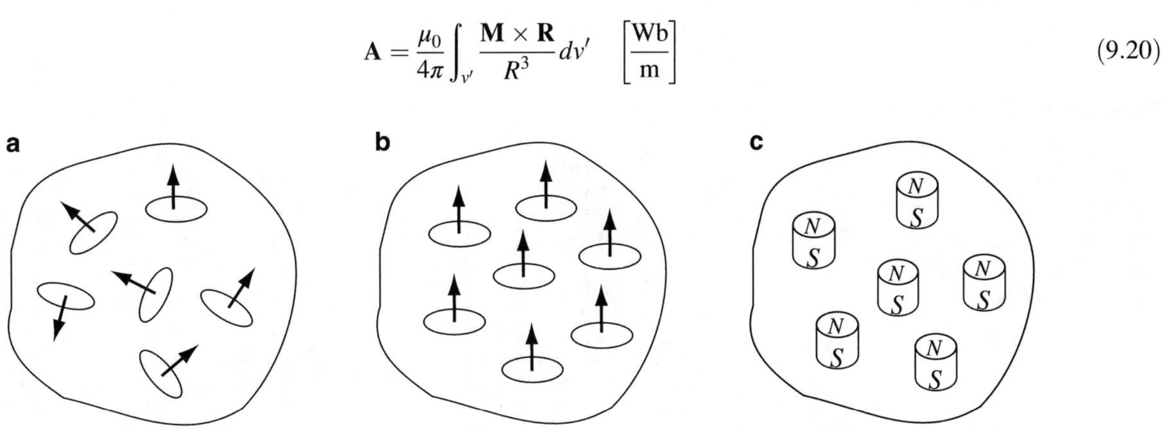

Figure 9.5 Model for magnetization in materials. (**a**) Randomly oriented dipoles. (**b**) Orientation of dipoles produces magnetization. (**c**) Permanent magnet model of magnetization

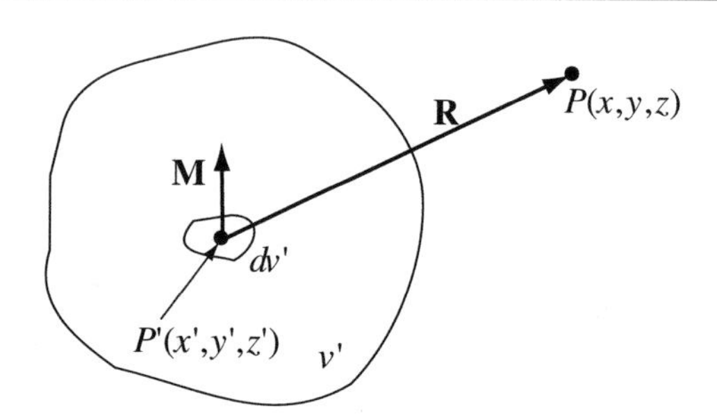

Figure 9.6 Calculation of the magnetic vector potential at point $P(x,y,z)$ due to magnetization **M**

This result can be viewed as the magnetic vector potential due to a volume of infinitesimal magnets, each with magnetization $d\mathbf{m} = \mathbf{M}dv'$. After the magnetic vector potential is known for a given volume magnetization, the magnetic flux density can be calculated directly from $\mathbf{B} = \nabla \times \mathbf{A}$.

The relation in **Eq. (9.20)** is not very useful for practical calculations, other than perhaps the fact that it indicates the direction of the magnetic vector potential and gives a general idea as to the magnitude of **A**. The magnetization is not often known, with the exception of permanent magnets. For the purpose of calculation of the fields in permanent magnets, we seek a model in terms of equivalent currents instead of magnetization based on the model shown in **Figure 9.5b**. There are two ways to do so. One is essentially mathematical and separates the integral in **Eq. (9.20)** into a surface and a volume integral. We will not pursue it here. The method we adopt views the magnetized volume as being made of very small permanent magnets, each equivalent to an oriented dipole. Thus, consider a very large magnetized volume as shown in **Figure 9.7a**. All magnetic dipoles point upward, and each is viewed as a small loop with magnetic dipole moment **m**. Now, consider one stack of dipoles as shown in **Figure 9.7b**. This stack is assumed to be very long and with uniform magnetization (all dipole moments aligned in the same direction and uniformly distributed), with n dipoles per unit length of the stack. Each dipole is of radius d and carries a current I. By analogy, this stack can be viewed as a solenoid of radius d, with n turns per unit length and carrying a current I. From the result in **Example 8.10**, the magnetic flux density anywhere inside the solenoid is

$$\mathbf{B}_m = \hat{\mathbf{z}}\mu_0 nI = \hat{\mathbf{z}}\mu_0 \frac{N}{L} I \quad [\mathrm{T}] \tag{9.21}$$

where $n = N/L$ is the number of turns per unit length, written here for a finite length L, in which there are a total of N loops or dipoles. The index m indicates that this field is produced by the magnetization as opposed to applied current densities. Multiplying both numerator and denominator by the area of the dipole (the cross-sectional area of the equivalent solenoid) gives

$$\mathbf{B}_m = \hat{\mathbf{z}}\mu_0 \frac{N(\pi d^2)I}{(\pi d^2)L} \quad [\mathrm{T}] \tag{9.22}$$

The quantity $\pi d^2 I$ is the magnitude of the dipole moment as given in **Eq. (9.13)**, whereas $\hat{\mathbf{z}}\pi d^2 I$ is the magnetic dipole moment **m**. The denominator is the volume of the solenoid of length L. Since the dipoles are uniformly distributed in the stack, we can write

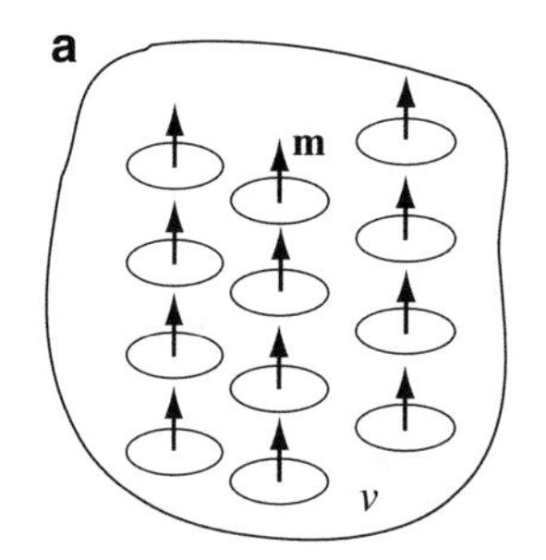

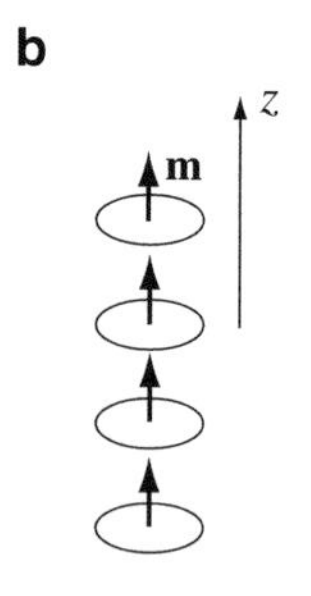

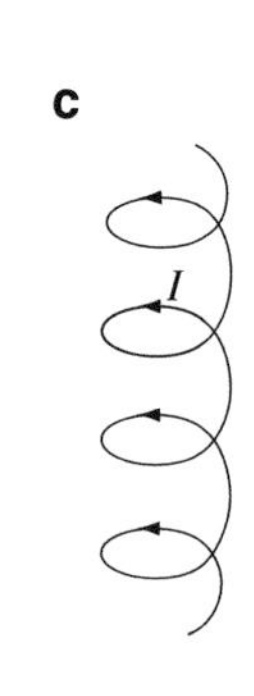

Figure 9.7 (**a**) Magnetized material viewed as distributed dipoles. (**b**) A vertical stack of dipoles within the material. (**c**) Solenoid equivalent of (**b**)

$$\mathbf{B}_m = \mu_0 \frac{N\mathbf{m}}{v} = \mu_0 \mathbf{M} \quad [\text{T}] \tag{9.23}$$

This is a somewhat surprising result. Because $\mathbf{B} = \mu\mathbf{H}$, it follows that the magnetic field intensity **H** inside a permanent magnet is equal to **M**. Perhaps this should have come as no surprise since we already mentioned that the units of magnetization are [A/m] and these are also the units of the magnetic field intensity. For uniform magnetization, we get

$$\mathbf{H}_m = \mathbf{M} \quad [\text{A/m}] \tag{9.24}$$

Applying Ampère's law ($\nabla \times \mathbf{H} = \mathbf{J}$) to **Eq. (9.24)** gives

$$\nabla \times \mathbf{H}_m = \nabla \times \mathbf{M} = \mathbf{J}_m \quad [\text{A/m}^2] \tag{9.25}$$

where $\mathbf{J}_m$ is the equivalent current density that generates the magnetization. Since this current occurs in the volume of the magnetized material, we call it a ***magnetization volume current density***. This definition of the magnetization volume current density is completely general, but, in our very simple example, the magnetization **M** is constant (we assumed it to be so). Thus, it immediately follows that for the conditions given in **Figure 9.7**, the magnetization volume current density $\mathbf{J}_m$ must be zero. If, however, **M** were a function of space, $\nabla \times \mathbf{M}$ would be nonzero and there would be a nonzero volume magnetization current density.

If **M** is constant, there is no magnetization volume current density, but there is an equivalent current on the surface of the magnetized material. To see how this current is produced, consider a long, cylindrical, uniformly magnetized material such as the rod magnet in **Figure 9.8a**. **Figure 9.8b** shows a slice through the magnet, with the dipoles associated with this slice. Since the dipoles are uniformly distributed, all internal dipoles cancel each other (currents in opposite directions). The only net current is that of the outer dipoles, shown in **Figure 9.8c**. Thus, the permanent magnet reduces to a solenoid with a thin sheet of current on the outer layer. This current is due to dipoles, and if we use again the idea of a stack of these currents as in **Figure 9.7b**, we can write the current per unit length of the solenoid as a surface current density $\mathbf{J}_{ms}$. This current density has units [A/m]; that is, it indicates the amount of current per unit length of the solenoid and is proportional to magnetization. Also, because of the equivalency between the solenoid and magnet, we must assume that both produce identical fields. The magnetic field of the solenoid is in the axial direction (in the direction of **M**). For **M** to produce a field in this direction, the equivalent current density $\mathbf{J}_{ms}$ must be perpendicular to the magnetization as shown in **Figure 9.9a**. This condition can be satisfied if $\mathbf{J}_{ms}$ is in the direction $\mathbf{M} \times \hat{\mathbf{n}}$ on the curved surface of the cylinder. To calculate the magnitude of the current density $\mathbf{J}_{ms}$, we assume a solenoid formed by a current sheet with current density $\mathbf{J}_{ms}$ [A/m] as shown in **Figure 9.9a**. The magnetic flux density inside the solenoid (it is zero outside) is calculated using Ampère's law as shown in **Figure 9.9b**. This gives

$$BL = \mu_0 J_{ms} L \quad \rightarrow \quad B = \mu_0 J_{ms} \quad [\text{T}] \tag{9.26}$$

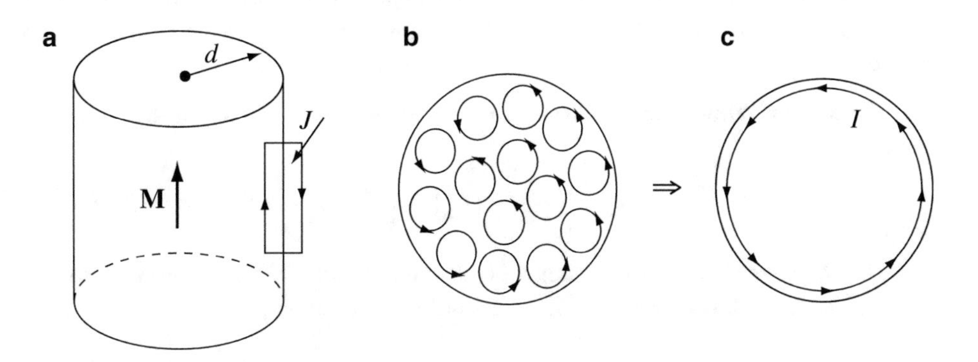

Figure 9.8 (**a**) A permanent magnet. (**b**) A slice through the magnet showing the dipoles. For constant magnetization, all internal currents cancel. (**c**) The net effect is a surface current density

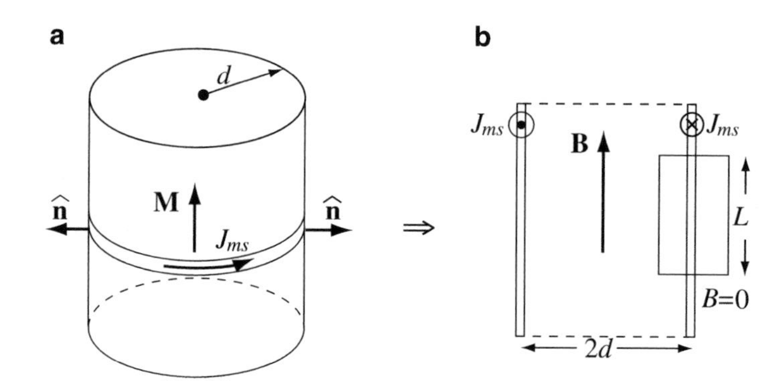

Figure 9.9 Magnetization and equivalent surface current density. (**a**) Direction of surface current density for uniform magnetization. (**b**) The solenoid equivalent model for the magnet in (**a**)

Comparing this with **Eq. (9.23)**, we see that the magnitude of $\mathbf{J}_{ms}$ must be equal to the magnitude of **M**. Thus, we can write

$$\mathbf{J}_{ms} = \mathbf{M} \times \hat{\mathbf{n}} \quad [\mathrm{A/m}] \tag{9.27}$$

The current density in **Eq. (9.27)** is called a ***magnetization surface current density***. Thus, the magnetization **M** in any material is equivalent to two current densities: a magnetization volume current density which appears whenever the magnetization within the material is nonuniform, and a magnetization surface current density which exists in uniformly or nonuniformly magnetized materials. The two current densities are

$$\mathbf{J}_m = \nabla \times \mathbf{M} \quad [\mathrm{A/m^2}] \quad \text{and} \quad \mathbf{J}_{ms} = \mathbf{M} \times \hat{\mathbf{n}} \quad [\mathrm{A/m}] \tag{9.28}$$

In other words, a magnetized material can always be modeled by a surface and a volume current density. However, we must insist at this point on the fact that these currents do not exist in the sense that we cannot measure them. They are real currents on the atomic level (see **Figure 9.5**), but, normally, we will view them as fictitious currents that allow calculation in lieu of magnetization. Only the magnetization is considered to be real and measurable. To understand this and the use of equivalent currents for calculation of magnetic fields due to permanent magnets, consider the following two examples.

Example 9.3 Application: Magnetic Flux Density of a Long, Uniformly Magnetized Magnet A very long cylindrical magnet has constant magnetization everywhere inside the magnet equal to $M = 5{,}000$ A/m directed along the axis. The diameter of the magnet is 40 mm.

(a) Calculate the magnetic flux density due to this magnet everywhere in space.
(b) Design an equivalent solenoid (made of wire turns rather than a current sheet) that produces an identical magnetic flux density everywhere.

Solution: Since **M** is constant, there are no equivalent volume current densities ($\nabla \times \mathbf{M} = 0$), but there is a surface current density according to **Eq. (9.27)**. Assuming the magnetization to be in the z direction (arbitrarily), the current density is in the ϕ direction.

(a) The magnetization is perpendicular to the cross-sectional surface of the cylinder (**M** is directed along the axis of the cylinder).

$$|\mathbf{J}_{ms}| = |\mathbf{M} \times \hat{\mathbf{n}}| = M = 5{,}000 \quad [\mathrm{A/m}]$$

The equivalent current-sheet solenoid is shown in **Figure 9.10b**. Using Ampère's law, the total current per unit length is 5,000 A/m and, from **Eq. (9.26)**, we get inside the solenoid,

$$B = \mu_0 J_{ms} 1 = \mu_0 I = 4\pi \times 10^{-7} \times 5{,}000 = 0.00628 \quad [\mathrm{T}]$$

The magnetic flux density outside the solenoid is zero.

Note: Permeability of most magnets is close to μ_0, although this information is not strictly required here since the equivalent solenoid is always taken to be in free space. Also, the diameter of the magnet is immaterial as long as the magnetization is constant.

(b) The equivalent solenoid in **Figure 9.10b** in the form of a current sheet is not practical. In most cases, a solenoid will be made of a single layer of wires, wound tightly. However, the total current per meter length must remain the same (5,000 A/m) or, more appropriately, 5,000 A • t/m (ampere • turns/meter). Any choice of number of turns is correct in principle. For example, we may choose 1,000 turns per meter. Each turn is 1 mm in diameter and must then carry 5 A. This is a little high for a 1 mm diameter wire but not unrealistic (current density in the 1 mm copper wire is 6.4×10^6 A/m^2 which is acceptable for copper wires). The equivalent solenoid is shown in **Figure 9.10c**.

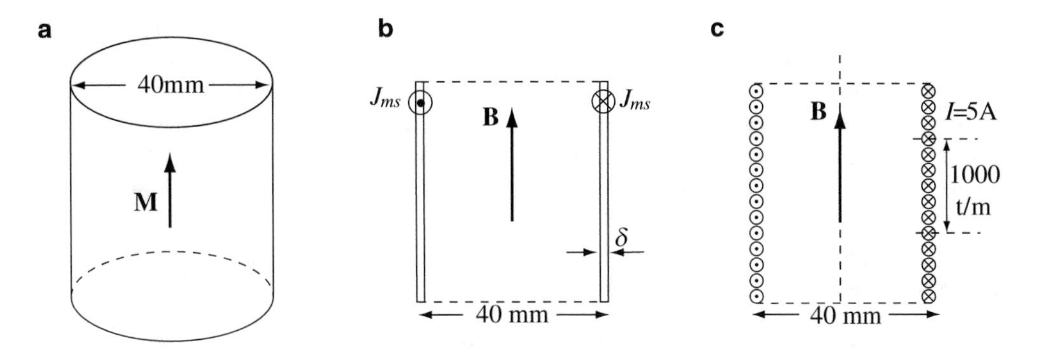

Figure 9.10 (**a**) Permanent magnet with constant magnetization **M**. (**b**) Equivalent surface current model. (**c**) Solenoid implementation of the model in (**b**)

Example 9.4 Application: Magnetic Flux Density and Equivalent Current Densities: Design of an Electromagnet A very long (infinite) cylindrical magnet of radius $a = 20$ mm has magnetization $\mathbf{M}(r) = \hat{\mathbf{z}}500,000$ $(a - r)$ A/m where r is the distance from the center of the magnet (**Figure 9.11a**).

(a) Calculate the magnetic flux density inside and outside the magnet.

(b) Design an equivalent solenoid, or combination of solenoids that will produce the same magnetic flux density everywhere.

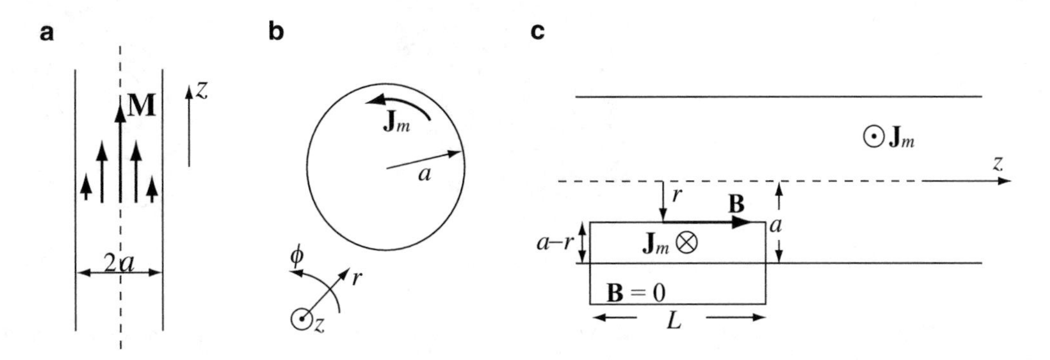

Figure 9.11 (**a**) Long magnet with magnetization **M**(r). (**b**) Equivalent volume magnetization current density. (**c**) Use of Ampère's law to calculate the magnetic flux density inside the model

Solution: We first calculate the equivalent current densities: from the current densities, the magnetic flux density everywhere in space may then be calculated, and the solenoids designed to produce the same magnetic flux density. In this particular case, the magnetization is maximum at the center of the magnet and goes down linearly until it becomes zero at $r = a$.

(a) From the curl in cylindrical coordinates, the volume magnetization current density is

$$\mathbf{J}_m = \nabla \times \mathbf{M} = -\hat{\boldsymbol{\phi}}\frac{d}{dr}(500{,}000(a - r)) = \hat{\boldsymbol{\phi}}500{,}000 \quad [\text{A/m}^2]$$

On the surface (at $r = a$),

$$\mathbf{J}_{ms} = \mathbf{M} \times \hat{\mathbf{n}} = \mathbf{M} \times \hat{\mathbf{r}} = \hat{\mathbf{z}}500{,}000(a - r) \times \hat{\mathbf{r}} = \hat{\boldsymbol{\phi}}500{,}000(a - a) = 0 \quad [\text{A/m}]$$

Because the magnetization is zero at the surface, the equivalent surface current density is also zero. This particular magnet is therefore described in terms of a volume magnetization current density alone, as shown in **Figure 9.11b**. Note that although this current density is uniform, it produces a linear magnetization, which is highest on the axis of the magnet and zero at its outer surface, as required.

Since the current density is circular, at any distance r from the center it forms a solenoid (see **Figure 9.12b** for a schematic view of the magnetization current density in the volume of the magnet) and the resulting magnetic flux density outside the magnet is zero. To calculate the magnetic flux density inside the magnet, we use Ampère's law on the contour shown in **Figure 9.11c**. The current enclosed by the contour is due to the area between r and a; hence the total current is $I = J_m L(a - r)$. The magnetic flux density outside the magnet is zero; hence the magnetic flux density at a distance r from the center of the magnet is

$$BL = \mu_0(a - r)LJ_m \quad \rightarrow \quad B = \mu_0(a - r)500{,}000 = 0.6283(a - r) \quad [\text{T}]$$

This flux density is in the positive z direction, is zero at $r = a = 0.02$ m, is equal to 0.01257 [T] at $r = 0$, and is zero outside the solenoid. The flux density increases linearly from the surface of the magnet to its center.

(b) An equivalent solenoid can be designed based on the final result. However, we cannot hope to construct a single, simple solenoid as in the previous example because the magnetic flux density inside the solenoid is r dependent; that is, because a single-layer solenoid produces a constant magnetic flux density in its interior, it cannot be used to model a varying magnetic flux density. One solution is to use a number of solenoids, one inside the other so that as we progress from the surface inward, the flux density increases. In this approach, thin layers of conductors are used (each layer is a solenoid) as shown schematically in **Figure 9.12**.

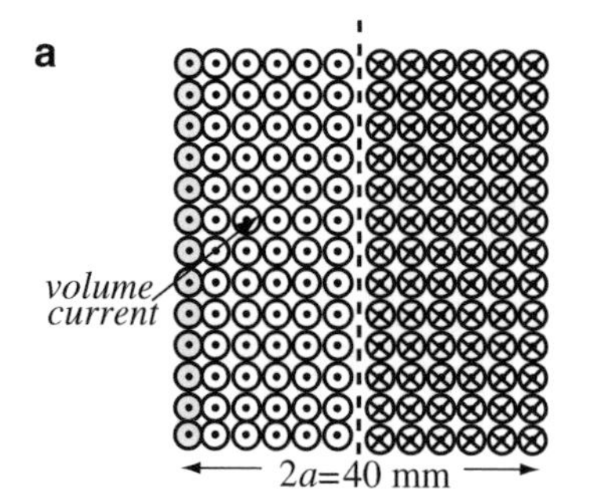

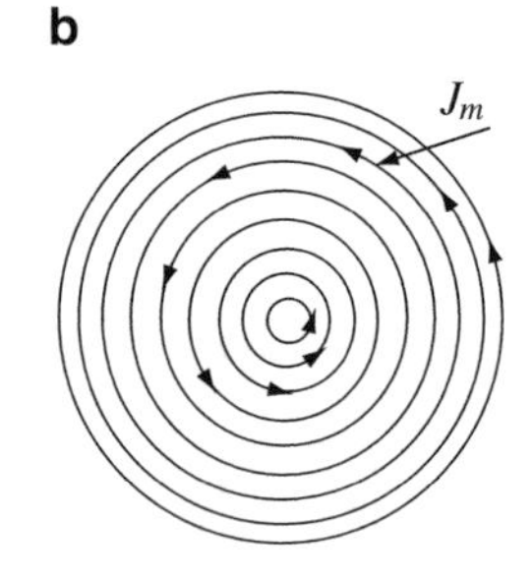

Figure 9.12 Design of a coil that models the magnet in **Figure 9.11a**. (**a**) The structure of the windings and direction of current. (**b**) Top view of **Figure 9.12a**

Now, we must choose a wire diameter that will carry enough current so that the current density is equal to J_m [A/m]. We argue as follows: if we were to wind a tight coil with an axial cross section of 1 m^2, the total current flowing through half the cross section would be 500,000 A. If the wires are 1 mm in diameter, the coil will contain 500,000 turns, and, therefore, each turn will carry 1 A. This is a relatively low current for this wire diameter. If this is our choice, there will be 20 layers in the coil, each layer with 1,000 turns/meter. We could choose a 0.5 mm diameter wire. This will double the number of turns and reduce the current in each wire by a factor of 2, to 0.5A. Now there will be 40 layers representing the volume of the magnet, with 2,000 turns/meter in each layer. Either choice is valid, although the second is closer to a continuous current distribution. It should be noted that this solution is only an approximation, as the magnetic flux density does not vary continuously but rather in steps; each additional solenoid increases the flux density by 1/40th of the maximum flux density (for the 0.5 mm wire). In practical terms, the solution in **Figure 9.12** means that a multilayer coil has maximum field at the center which decreases toward the outer layer.

Note: We neglected here some important engineering considerations for the sake of simplicity. One is the question of power dissipation in the solenoids. For such a massive coil, this may be significant and, in practice, may require cooling. The second is economical: each of the designs mentioned above has implications on cost of wires, weight, power requirements, power supply design, and so on.

The main point in these examples is that magnets are not different than solenoids and can always be replaced by equivalent current densities. This has distinct advantages in that no special treatment is needed when we incorporate magnets in design. It also shows that permanent magnets are electromagnetic devices: they rely on their operation on currents, albeit currents we cannot measure. From an application point of view, a magnetic field can be produced either with a permanent magnet or an electromagnet (coil). The choice often depends on requirements, cost, and convenience, as well as mechanical properties.

Now, we must return to the question posed at the beginning of this section: why should magnetic dipoles align themselves in a specific direction in space? As with other effects, there must be a force exerted on them to align. This force is supplied by an external magnetic field. Although we have not yet discussed the question of forces in the magnetic field in any detail, we know from experiment that a magnet will align itself with another magnet if allowed to do so. Thus, following the model of small magnets, each produced by a dipole, dipoles will align with an externally produced magnetic field as shown in **Figure 9.13**.

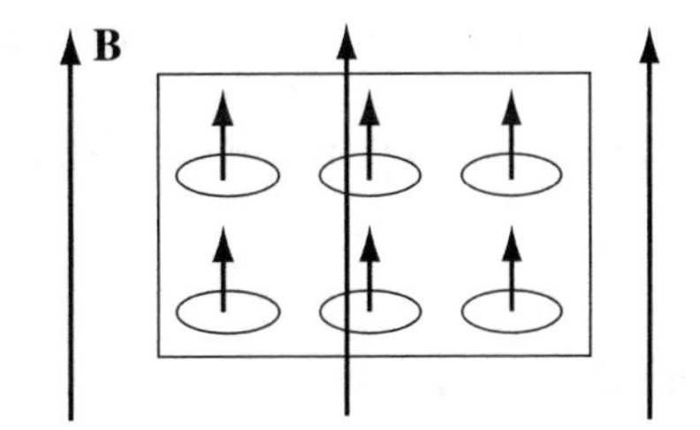

Figure 9.13 Alignment of dipole moments in an externally applied magnetic field

From this figure, it is clear that the external magnetic flux density $\mathbf{B}_e$ and the magnetic flux density produced by the magnetized material $\mathbf{B}_m$ are in the same direction and therefore add

$$\mathbf{B}_t = \mathbf{B}_e + \mathbf{B}_m \quad [\mathrm{T}] \tag{9.29}$$

where $\mathbf{B}_m$ is given in **Eq. (9.23)**. With this, the total magnetic flux density in the magnetized material can be written as

$$\boxed{\mathbf{B}_t = \mu_0\mathbf{H}_e + \mu_0\mathbf{M} = \mu_0(\mathbf{H}_e + \mathbf{M}) \quad [\mathrm{T}]} \tag{9.30}$$

Now, we argue that since magnetization is generated by the external field $\mathbf{H}_e$, then $\mathbf{M}$ must be proportional to $\mathbf{H}_e$. The proportionality factor is called ***magnetic susceptibility*** and indicates how susceptible the material is to magnetization. Magnetic susceptibility is a fundamental property of materials and is denoted by χ_m. The relation between the external magnetic field intensity and magnetization is written as

$$\mathbf{M} = \chi_m\mathbf{H}_e \quad [\mathrm{A/m}] \tag{9.31}$$

Substituting this back into **Eq. (9.30)** gives

$$\mathbf{B}_t = \mu_0(\mathbf{H}_e + \chi_m\mathbf{H}_e) = \mu_0(1 + \chi_m)\mathbf{H}_e \quad [\mathrm{T}] \tag{9.32}$$

Now, since the relation $\mathbf{B}_t = \mu\mathbf{H}_e$ holds in general, we can write

$$\mu = \mu_0(1 + \chi_m) \quad [\mathrm{H/m}] \tag{9.33}$$

The term μ is called the ***magnetic permeability*** of the material. Because it depends on susceptibility, it differs from material to material. In free space, $\chi_m = 0$ and, therefore, $\mu = \mu_0$. In other materials, permeability can be larger or smaller than in free space, depending on susceptibility of the material. The theoretical range of magnetic susceptibilities is between –1 and infinity, with an equivalent permeability from zero to infinity. Most materials range in permeability between μ_0 and about $10^6\ \mu_0$. The quantity $1 + \chi_m$ is called the ***relative permeability*** of the material since it gives the ratio between permeability μ of the given material and permeability of free space:

$$\boxed{\mu = \mu_0\mu_r \quad [\mathrm{H/m}]} \tag{9.34}$$

This form is particularly convenient because μ_0 is a small value ($\mu_0 = 4\pi \times 10^{-7}$ H/m). Thus, the relative permeability simply indicates how much larger or smaller permeability is in relation to permeability of free space and, therefore, is a dimensionless quantity. The unit of permeability is the henry/meter [H/m], or using **Eq. (9.32)**, the tesla/(ampere/meter) or (weber/meter^2)/(ampere/meter) [Wb/A · m]. The quantity Wb/A is called a ***henry*** [H].[6] These units were also discussed in **Chapter 8**.

The most remarkable aspect of the definition of relative permeability is that it avoids the need to deal with magnetization and susceptibility which, in general, are difficult to use and, instead, the effects are lumped in a single, experimentally measurable quantity, μ. This is similar to the way we treated polarization in the electric field where we defined the permittivity to take into account polarization. With the exception of permanent magnets, most of our work will be in terms of the external magnetic field intensity **H**, permeability μ, and the magnetic flux density **B**.

9.2.3 Behavior of Magnetic Materials

In the previous section, we discussed magnetization and, more importantly, its relation with the magnetic field intensity **H**, the magnetic flux density **B**, and the equivalent current density **J** (both surface and volume current densities). Now, we wish to discuss the basic magnetic properties of materials from a macroscopic point of view; that is, we look at those parameters which are important for engineering design.

One question has been left open in the previous discussion: what is the source of the magnetization and, perhaps more significant, why should one material be capable of magnetization (for example, iron) while another material seems to be unaffected by external magnetic fields? The answer to this lies in the magnetic susceptibility or, alternatively, in the magnetic permeability of materials. Some materials are highly susceptible to magnetization while others are not. Based on our previous approach, we will view magnetic susceptibility as a fundamental, experimentally obtained property. However, it is much easier to discuss magnetic properties in terms of the relative permeability of materials. Thus, materials with high relative permeability exhibit high magnetization when placed in a magnetic field, whereas materials with low permeability

[6] After Joseph Henry (1797–1878), Professor of Mathematics and Natural Philosophy at Albany and later at Princeton. His major contributions to electricity and magnetism are the development of the electromagnet and the study of induced currents. He was the first to show the potential uses of the magnetic force. When he developed his powerful electromagnets (the Smithsonian Institution still has one of his early electromagnets, capable of lifting up to 1.5 tons), it became evident that magnetism was of practical importance and led directly to the development of electric motors and other electromagnetic power devices. These investigations were made in parallel to those of Michael Faraday. Henry became the first secretary of the Smithsonian Institution in 1846 and served in this capacity until his death. Others of his interests led to the demonstration of telegraphy, a primitive form of which he used to communicate between his office and home. He was an enthusiastic meteorologist and is credited with influencing the establishment of the U.S. Weather Bureau.

do not. In practical terms, we define three basic types of materials. These are diamagnetic, paramagnetic, and ferromagnetic materials. All materials fall under one of these groups, except vacuum, which is considered to be nonmagnetic.

9.2.3.1 Diamagnetic and Paramagnetic Materials

Diamagnetic materials are materials with relative permeabilities smaller than 1 ($\mu_r < 1$). This class includes important materials such as mercury, gold, silver, copper, lead, silicon, and water. The relative permeability of most diamagnetic materials varies between 0.9999 and 0.99999 (susceptibility varies between -10^{-5} and -10^{-4}), and for most applications, they may be assumed to be nonmagnetic (i.e., $\mu_r \sim 1$). The relative permeabilities of some diamagnetic materials are listed in **Table 9.1**.

One notable exception to the small negative susceptibility of most diamagnetic materials are superconducting materials.

Table 9.1 Relative permeabilities for some diamagnetic and paramagnetic materials

Diamagnetic materials		Paramagnetic materials	
Material	Relative permeability	Material	Relative permeability
Silver	0.999974	Air	1.00000036
Water	0.9999991	Aluminum	1.000021
Copper	0.999991	Palladium	1.0008
Mercury	0.999968	Platinum	1.00029
Lead	0.999983	Tungsten	1.000068
Gold	0.999998	Magnesium	1.00000693
Graphite (Carbon)	0.999956	Manganese	1.000125
Hydrogen	0.999999998	Oxygen	1.0000019

These materials are purely diamagnetic ($\chi_m = -1$), and, from **Eq. (9.33)**, the magnetic permeability is equal to zero ($\mu_r = 0$). Inside these materials, the magnetic flux density is zero ($\mathbf{B} = 0$) for any external magnetic field intensity.

An interesting aspect of diamagnetism is the fact that the magnetic flux density inside the diamagnetic material is lower than the external magnetic field. If we were to place a piece of diamagnetic material over a permanent magnet, the magnet will repel the diamagnetic material, as shown in **Figure 9.14**. This can be easily explained from the equivalent magnets representation in **Figure 9.14b**. Since the magnet and the equivalent magnetic field (due to magnetization of the diamagnetic material) oppose each other, the diamagnetic material is repelled from the magnetic field in the same way that two magnets repel each other when their magnetic flux densities oppose each other. However, this force is extremely small for all diamagnetic materials, except superconductors, in which it is very large. This repulsion is the reason why a permanent magnet floats above a superconductor.

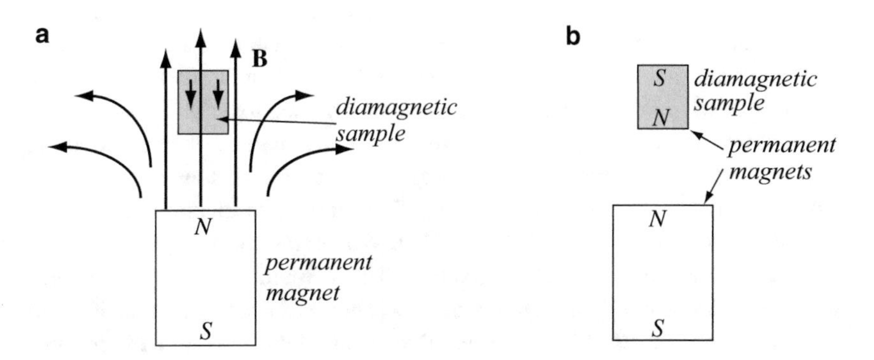

Figure 9.14 (**a**) Repulsion of a diamagnetic sample by a magnetic field due to opposite internal magnetization. (**b**) A permanent magnet model of (**a**)

The behavior of diamagnetic materials can be explained based on the atomic model of materials. One model of diamagnetic materials has the magnetic moments due to orbiting electrons and electron spins cancel each other under normal conditions. Under the influence of an external field, the field due to orbiting electrons is slightly smaller than that of spins, causing a net magnetic field which opposes the external field and, thus, the lower permeability of diamagnetic materials. In fact, all materials possess this property; an external magnetic field induces a magnetization in the material which opposes the external field. However, in other types of materials, this property is obscured by larger magnetic effects, as we shall see shortly.

Paramagnetic materials are materials in which the relative permeability is slightly larger than 1. In these materials, the orbital and spin moments do not cancel and atoms have a net magnetic moment in the absence of an external magnetic field. However, since moments are oriented randomly, the net external field observed is either zero or very close to zero. Unlike diamagnetic materials, in which the net magnetic moments of atoms are induced by an external field, the magnetic moments in paramagnetic materials always exist. In the presence of an external magnetic field, these moments tend to align with the external field and increase the total field.

Relative permeability of paramagnetic materials ranges between about 1.0000001 and 1.001 (magnetic susceptibility varies between 10^{-7} and 10^{-3} and is always positive). Some common materials such as aluminum, palladium, tungsten, and air are paramagnetic. An interesting and surprising consequence is that a piece of aluminum is attracted to a magnet. However, you will need a very strong magnet to "feel" this attraction since the relative permeability of aluminum is only 1.000021. A few other materials which exhibit paramagnetism are listed in **Table 9.1**.

9.2.3.2 Ferromagnetic Materials

By far the most useful magnetic materials are the so-called ***ferromagnetic materials***. These derive their name from iron (ferrum) as the most common of the ferromagnetic materials. The relative permeability of ferromagnetic materials is much larger than 1 and can be in the thousands or higher. Some typical ferromagnetic materials are iron, cobalt, and nickel. Other materials and their relative permeabilities are given in **Table 9.2**.

Table 9.2 Relative permeabilities for some ferromagnetic materials

Material	μ_r	Material	μ_r
Cobalt	250	Permalloy (78.5% Ni)*	100,000
Nickel	600	Fe_3O_4 (Magnetite)	100
Iron	6,000	Ferrites	5,000
Supermalloy (5% Mo, 79% Ni)*	10^7	Mumetal (75% Ni, 5% Cu, 2% Cr)*	100,000
Steel (0.9%C)	100	Permendur	5,000
Silicon iron (4% Si)	7,000		

*Iron completes the composition to 100%. These compositions are approximate and other percentages exist commercially

Ferromagnetic materials tend to magnetize in the direction of the magnetic field and some of them retain this magnetization after the external magnetic field has been removed. When they do so, and the magnetization is permanently retained, the material becomes a permanent magnet. An additional important property of ferromagnetic materials is the dependence of magnetization on the level of the external field. Thus, magnetization in ferromagnetic materials is a nonlinear process.

The large magnetization of ferromagnetic materials cannot be explained in terms of the simple model of electron spins and orbits since these exist in all materials and most materials are not ferromagnetic. The model for ferromagnetic materials is a modified model which has been proven experimentally to be correct. Unlike other materials, the individual electron spins, instead of being randomly oriented, are oriented together in domains; that is, a number of spins in a small volume of the material are aligned in the same direction in the absence of an external magnetic field, held together by atomic coupling forces. This small volume is called a ***magnetic domain***.[7] A domain is of the order of 0.001 mm to 1.0 mm in width and has a volume between 10^{-9} mm^3 and 1 mm^3. The number of electron spins in a domain varies with the domain size, but the average is of the order of 10^{16}. Thus, although domains are small, they are very large on the atomic level. Their existence has been measured and even photographed. The domain structure of ferromagnetic materials is shown schematically in **Figure 9.15a**. Individual domains may be aligned in any direction in space. The transition region between domains is called a ***domain wall***. When an external magnetic field is applied, the magnetic domains that are aligned in the direction of

[7] Pierre Weiss (1865–1940), French physicist and perhaps the first to initiate the study of atomic level magnetic fields, theorized in 1907 that coupling forces in domains must exist, and these forces hold the spins together. The existence of domains as well as loss of magnetization beyond a certain temperature (Curie point) were also part of Weiss' work. The domain walls are sometimes called Weiss walls or Bloch walls (after Felix Bloch, Swiss physicist and recipient of the 1952 Nobel Prize in Physics).

the field remain aligned, but the domain walls move, causing them to grow at the expense of neighboring domains (**Figure 9.15b**). As the external field increases further, most of the domains will align in the direction of the magnetic field (**Figure 9.15c**). The external magnetic field intensity required to do so is relatively small whereas the resulting internal magnetic flux density is large.

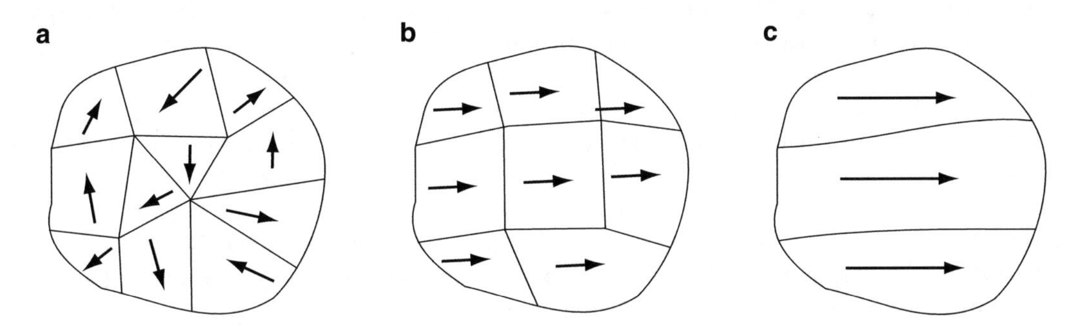

Figure 9.15 Domain model of magnetization in ferromagnetic materials. (**a**) Random orientation of domains. (**b**) External field causes aligned domains to grow. (**c**) Aligned domains occupy all or most of the volume

This aspect of magnetization is best explained using the so-called magnetization curve. The magnetization curve is an experimental plot of the internal magnetic flux density against an applied external magnetic field intensity. A typical curve for iron is shown in **Figure 9.16a**. To understand this behavior, consider first a sample of non-magnetized material, which has no internal magnetization (domains are randomly oriented) and therefore is represented by point O in **Figure 9.16a**. As the external field increases, the magnetic flux density in the sample increases along the curve shown. This is because domains start aligning with the external field. As the external magnetic field increases further, there are fewer domains left to align and, therefore, the slope of the curve decreases. At some point, all domains will be aligned with the external magnetic filed. Any increase in the external magnetic field will only increase the internal field by the increase in the external magnetic field: there is no increase in magnetization. This point is shown as point M_1 and is called a ***saturation point***. The whole curve segment between O and M_1 is called an ***initial magnetization curve***.

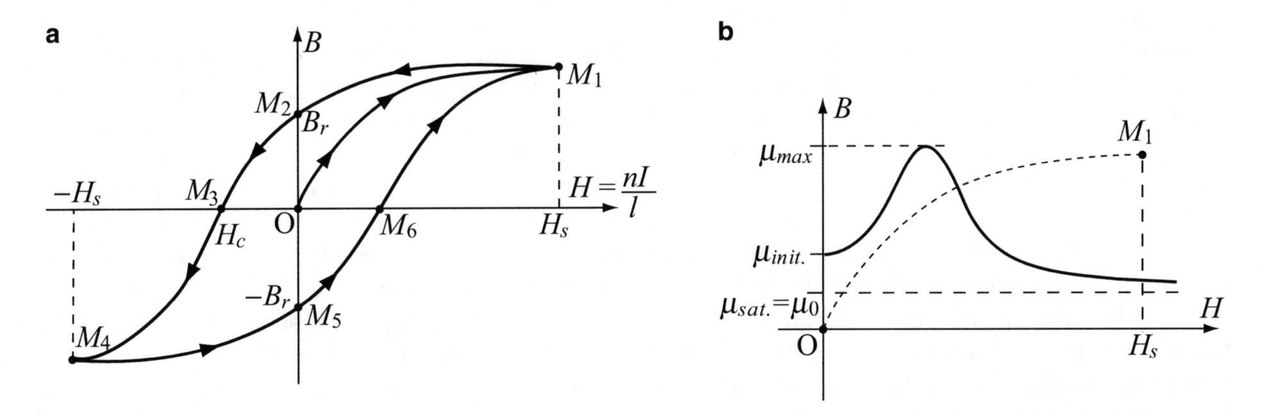

Figure 9.16 (**a**) Magnetization (**B**(**H**)) curve for iron. (**b**) Relative permeability along the initial magnetization curve

If we now decrease the external field, the magnetization curve does not follow the initial magnetization curve. Rather, the internal magnetic flux density decreases slower due to a lag in the realignment of the domains. In other words, domains tend to "retain" their magnetization. This lag in alignment gives the magnetization curve its common name: ***hysteresis curve*** ("to lag" in Greek). Reducing the field further, we eventually reach point M_2. At this point, the external magnetic field intensity is zero, but the internal magnetic flux density is B_r. B_r is called the ***remnant*** or ***residual magnetic flux density*** since it is "left over" after the external magnetic field intensity has been removed. Thus, at this point, we have a piece of material which has an internal magnetic field in the absence of an external magnetic field.

To reduce the magnetic flux density to zero, we must apply a magnetic field intensity in the opposite direction. This process is called demagnetization. If the negative external magnetic field intensity is increased to H_c, the magnetic flux density is reduced to zero. This value of **H** is called the ***coercive field intensity*** (sometimes also called coercive force). Further increasing the demagnetizing field causes a negative flux density; that is, after domains have relaxed to a random pattern at H_c, they now align in the opposite direction in the sample. Eventually, all domains will be aligned in the opposite direction and a saturation point has again been reached at point M_4. Note that the magnitudes of **H** and **B** at M_1 and M_4 are the same since the domain structure is the same except for directions.

Reducing the demagnetizing field reduces the flux density to $-B_r$, which, again, is a remnant flux density and is equal in magnitude to B_r, at point M_2. To reduce this remnant field to zero, we must increase the magnetic field intensity. At point M_6, the magnetic flux density has been reduced to zero (this is the coercive field intensity). A further increase in the magnetizing field intensity will eventually bring us to point M_1, which is the saturation point we reached before. Now, if we continue the cycle of increasing and decreasing the magnetic field intensity, we follow the outer loop in **Figure 9.16a**. The only way we can get on the initial magnetization curve again is by completely demagnetizing the material first or starting with a material that has not been subjected to magnetization. This indicates that the material has a "history": the state of the material depends on what happened to it previously.

From the magnetization curve (also called **B**−**H** or **B**(**H**) curve), we note the following important properties:

(1) The important values, saturation point H_s, remnant flux density B_r and coercive field intensity H_c, are material dependent. Each material saturates at different field levels and has different H_c and B_r values. These values are shown in **Table 9.3** for a sample of important magnetic materials.

(2) Since the magnetization curve gives the relation between **B** and **H** (**B** = μ**H**), the slope of the curve at any point gives of the permeability of the ferromagnetic material at that field level. This slope depends on the location on the curve. Thus, permeability of ferromagnetic materials is a nonlinear function of the magnetic field intensity **H**. We usually write this as

$$\mathbf{B}=\mu(H)\mathbf{H} \quad [\mathrm{T}] \tag{9.35}$$

Also to be noted is that permeability is negative anywhere between point M_2 and M_3 because for a negative field intensity H, the magnetic flux density is positive. Similarly, between points M_5 and M_6, the magnetic field intensity is positive and the magnetic flux density is negative. Permeability is positive in the first and third quadrants of the curve. Permeability is high along the initial magnetization curve except when approaching saturation. At saturation, the relative permeability approaches 1. The permeability curve corresponding to the initial magnetization curve is shown in **Figure 9.16b**. The most important consequence of this behavior of materials is that an increase in the magnetic field intensity (which normally increases the magnetic flux density in ferromagnetic materials) may cause saturation and, therefore, a much lower increase in the magnetic flux density at higher field levels. For this reason, the magnetic flux density in magnetic devices is normally not allowed to reach saturation.

(3) The region between M_2 and M_3 is characterized by an internal magnetic flux density without an externally applied magnetic field intensity or in the presence of a demagnetizing external field. This region is where permanent magnets operate. An ideal permanent magnet will operate at point M_2, whereas most magnets operate somewhere between points M_2 and M_3.

(4) At any point on the curve, the behavior depends on the magnetization prior to reaching that point (history). The only exception is at very low magnetization levels. At the beginning of the initial magnetization curve, the external field is very low and the magnetization is reversible; that is, the magnetization does not follow a hysteresis loop but rather reverses itself along the initial magnetization curve.

(5) The surface area of the magnetization curve represents energy: the energy needed to move the domain walls and align domains. This energy is lost in the process. Thus, when magnetization is done with a periodic field (AC field), each cycle of the field traces the loop once, and during each cycle, there is a loss of energy per unit volume of the material. Although we do not calculate this energy, it is qualitatively obvious that the narrower the loop, the lower the energy loss.

(6) Narrow, tall loops represent materials with low loss per cycle, high remnant magnetic flux density, and low coercive field intensity. The low coercive field, in particular, indicates that the material can be easily magnetized and demagnetized. This is useful in applications where it is necessary to magnetize and demagnetize the material repeatedly and quickly such as in electric motors, transformers, relays, and the like. These types of materials are called soft magnetic materials and are used in machines and transformers and other alternating current devices. Some typical materials are listed in **Table 9.3**. Note that the coercive field intensity of soft magnetic materials is low: typically less than 50 A/m.

Table 9.3 Properties of soft magnetic materials

Material	Relative permeability (max.) μ_r	Coercive field intensity H_c [A/m]	Remnant flux density B_r [T]	Saturation flux density B_s [T]
Iron (0.2% impure)	9,000	80	0.77	2.15
Pure iron (0.05% impure)	2×10^5	4	1.3	2.15
Silicon iron (3% Si)	55,000	8	0.95	2.0
Permalloy	10^6	4	0.6	1.08
Supermalloy (5% Mo, 79% Ni)	10^7	0.16	0.5	0.79
Permendur	5,000	160	1.4	2.45
Nickel	600	60	0.4	0.6

(7) Broad, low loops represent materials with lower remnant magnetic flux density but higher coercive field intensity. The main advantage of these materials is that they are "hard" to demagnetize once magnetized. In other words, they require a larger reverse field to reduce the remnant magnetization to zero. They are, therefore, the main candidates for production of permanent magnets. For this purpose, we actually would prefer both high remnant flux density and high coercive field intensity. Some useful hard magnetic materials and their properties are shown in **Table 9.4**. Note the very large coercive fields in some hard magnetic materials in contrast to those of soft magnetic materials.

Table 9.4 Properties of hard magnetic materials

Material	μ_r	H_c [kA/m]	B_r [T]	Curie temp. [°C]
Alnico (Aluminum–Nickel–Cobalt)	3–5	60	1.25	850
Ferrite (Barium–Iron)	1.1	240	0.38	600
Sm–Co (Sammarium–Cobalt)	1.05	700	0.9	700
Ne–Fe–B (Neodymium–Iron–Boron)	1.05	800	1.15	300

Magnetic properties of all materials are temperature dependent to a certain degree. In particular, each material has a temperature beyond which it loses its magnetization, called the ***Curie temperature.*** At this temperature, ferromagnetic materials change their magnetic behavior to that of paramagnetic materials. As a consequence, it is not possible to magnetize a ferromagnetic material above the Curie temperature. For iron, this value is approximately 770°C. The same happens to permanent magnets. If the permanent magnet is heated above its Curie temperature, it loses its magnetization and becomes just a piece of (paramagnetic) material. It can normally be magnetized again, but the magnetization process depends on the material and may involve more than inserting the material in a magnetic field.

9.2.3.3 Other Magnetic Materials

Although all materials are either diamagnetic, paramagnetic, or ferromagnetic, there are subclasses of materials which are sufficiently different and important in engineering to be considered as separate from the three general groups. Two of these, with important engineering applications, are the ***ferrimagnetic*** and ***superparamagnetic*** materials.

Ferrimagnetic materials, better known as ferrites, are based on ferromagnetic particles, formed and compressed together with bonding agents to form solids. The magnetic moment in ferrites is weaker than in the base ferromagnetic material from which ferrites are made (mostly iron compounds mixed with other oxides). Ferrites have low conductivity and, being made of small particles (powders), can be made into almost any shape required. Most ferrites are used at high frequencies where their low conductivity is an advantage (lower losses). For example, the antenna core of most portable radios is made of a ferrite, as are high-frequency transformers in switching power supplies.

Superparamagnetic materials are made of small ferromagnetic particles, suspended in nonconducting substrates such as resins and plastics. The most common applications are magnetic recording tapes, disks, magnetic strips and other media which contain iron or chromium particles, suspended in a solidified solution. Each particle is a separate magnetic entity and can be magnetized or demagnetized separately or together with its neighbors. The state of the particles is a measure of the magnetization applied and, therefore, can be used to record, store and retrieve data. The uses of this type of materials are vast: from magnetic tapes (audio, video, magnetic strips on credit cards) to recording media (disks). With various media, very high-density recording is possible.

Another interesting type of material in this group is the magnetic fluid. These are similar to particles used on tapes but are suspended in a liquid. The liquid may be water, oil, or a solvent (such as kerosene). Magnetic fluids are useful in the detection of magnetic fields, testing for cracks in magnetic materials, and a score of other applications in which the fluid state is more convenient than the solid. One such application, used experimentally, is the treatment of localized tumors. These are injected with a magnetic fluid made of a biocompatible fluid and superparamagnetic nanoparticles and then the whole body or the specific organ is exposed to a microwave field (similar to a microwave oven but of much lower intensity). The magnetic material absorbs much more energy than the surrounding tissue, locally heating the tumor and destroying it with minimum damage to surrounding tissue. Magnetic fluids are also used in specialty loudspeakers where they also conduct heat by flowing through heat exchangers. An important application of magnetic fluids is in magnetic seals in conjunction with magnetic bearings.

9.3 Magnetic Interface Conditions

In this section we discuss the interface conditions for the static magnetic field following steps similar to those in **Section 4.6** for the electric field. Consider two materials with permeabilities μ_1 and μ_2, respectively, as in **Figure 9.17**. Because we are interested in the interface itself, we must also assume that a current may exist at the interface between the two materials. This assumption does not imply that this current exists in general; only that it may exist under certain conditions, as implied from Ampère's law ($\nabla \times \mathbf{H} = \mathbf{J}$). Thus, we assume that, in general, a current density exists on the interface between the two materials and calculate the general conditions at the interface. From these conditions, we will also be able to tell when a current density exists and calculate it.

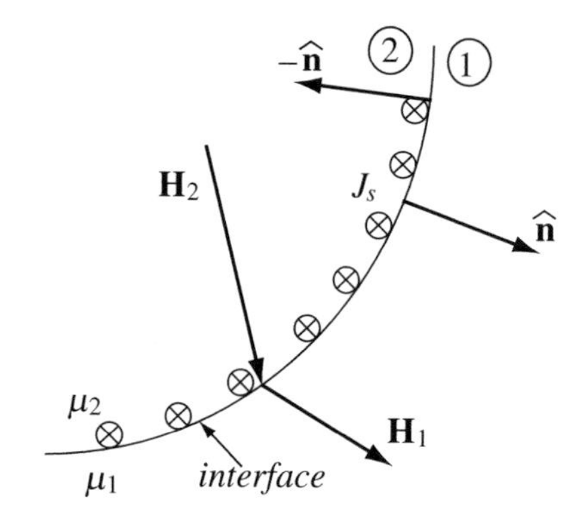

Figure 9.17 Interface between two general magnetic materials, with an assumed surface current density

9.3.1 Interface Conditions for the Tangential and Normal Components of the Magnetic Field Intensity H

From the constitutive relation $\mathbf{B} = \mu\mathbf{H}$, we know that the magnetic field undergoes changes at the interface between two materials. We assume arbitrarily directed magnetic field intensities and magnetic flux densities in both materials and calculate the required relations between the fields at the interface. For convenience, this is done by separating the fields into the components tangential and normal to the interface and applying the magnetic field postulates to each component separately. To define the interface conditions for the magnetic field intensity **H**, consider **Figure 9.18a** which also indicates a possible surface current density. We first apply Ampère's law on the contour *abcda*:

$$\oint_{abcda} \mathbf{H} \cdot d\mathbf{l} = I_{enclosed} \quad [\mathrm{A}] \tag{9.36}$$

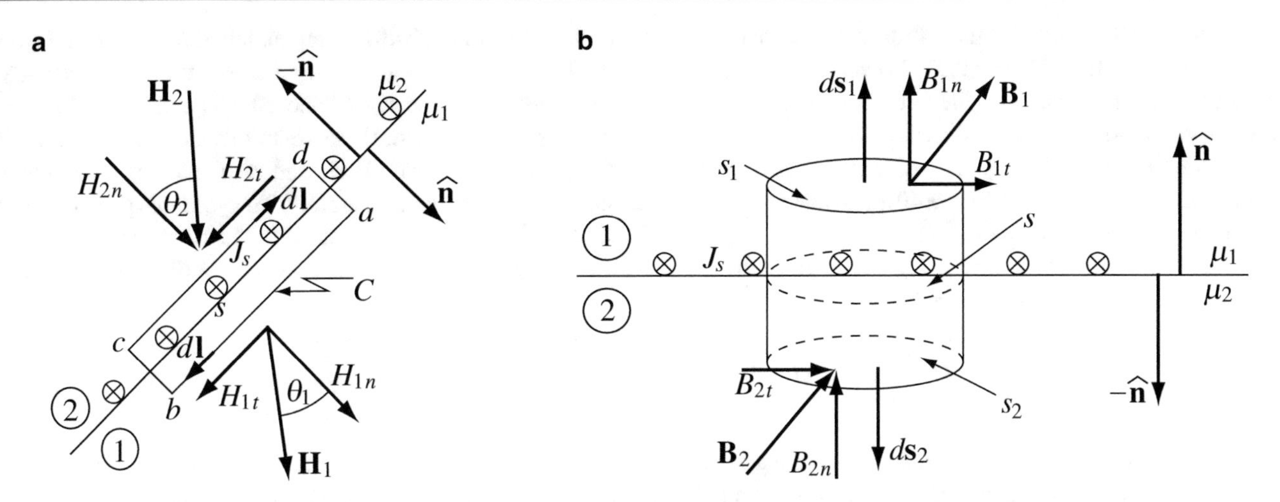

Figure 9.18 (**a**) Calculation of the relation between tangential components of the magnetic field intensity. (**b**) Calculation of the relation between normal components of the magnetic flux density

Allowing the distances bc and da to tend to zero, the total contribution due to this part of the contour is zero. Only the integration along ab and cd contributes to the left-hand side of **Eq. (9.36)**. Also, the scalar product $\mathbf{H} \cdot d\mathbf{l}$ means that only the tangential components are used. In other words, $\mathbf{H} \cdot d\mathbf{l} = H_t dl$. Denoting the surface current density by J_s [A/m] (see **Figure 9.18a**), we get

$$\int_{ab} H_{1t}\ dl_1 - \int_{cd} H_{2t}\ dl_2 = \int_{ab} J_s\ dl \tag{9.37}$$

In this expression, the vector notation was dropped since the fields are collinear with the path (H_{2t} and dl_2 are in opposite directions, hence the negative sign). The current density is perpendicular to the loop $abcda$, and since it can only depend on the width of the loop, it is integrated over the path dl. Integrating over the two segments ab and cd assuming the distances $ab = cd$ tend to zero, we get

$$\boxed{H_{1t} - H_{2t} = J_s \quad [\text{A/m}]} \tag{9.38}$$

This is the first condition at the interface; the discontinuity of the tangential component of the magnetic field intensity is equal to the surface current density on the interface, if such a current density exists. Note also that this current density has units of A/m indicating that it is limited to the surface and does not extend inside either medium. If there is no current on the surface, the tangential components of **H** are equal and **H** is said to be continuous at the interface:

$$H_{1t} = H_{2t} \quad \text{if} \quad J_s = 0 \tag{9.39}$$

From $\mathbf{B} = \mu\mathbf{H}$, we can also write for the general relation in **Eq. (9.38)**:

$$\boxed{\frac{B_{1t}}{\mu_1} - \frac{B_{2t}}{\mu_2} = J_s \quad \left[\frac{\text{A}}{\text{m}}\right]} \tag{9.40}$$

or

$$\frac{B_{1t}}{\mu_1} = \frac{B_{2t}}{\mu_2} \quad \text{if} \quad J_s = 0 \tag{9.41}$$

Equations (9.38) and **(9.40)** give the basic relations between the tangential components of **H** and **B** on both sides of the interface. In particular, the tangential components of **B** are discontinuous regardless of any current density at the interface. The discontinuity in **B** is related to permeabilities of the two materials.

It must be noted here that the interface current density J_s in **Eq. (9.38)** or in **Eq. (9.40)** is perpendicular to H_{1t} and to H_{2t} as can also be seen from **Figure 9.17**. However, **Eq. (9.38)** [or **Eq. (9.40)**] only provides the magnitude of the current density. It is essential to recall that the direction of the current density is given by the right-hand rule (see **Section 8.2** and **Figure 8.2c**). As long as the direction of the current density is properly established, these equations can be used. This is rather simple when the magnetic field intensities have only one tangential component. If $\mathbf{H}_1$ and/or $\mathbf{H}_2$ have two tangential components, **Eq. (9.38)** [or **Eq. (9.40)**] must be applied to each tangential component separately. To avoid this difficulty and to establish a more general relation, we note from **Figure 9.17** that the tangential components of the fields may be written as $\mathbf{H}_{1t} = \hat{\mathbf{n}} \times \mathbf{H}_1$ and $\mathbf{H}_{2t} = -\hat{\mathbf{n}} \times \mathbf{H}_2$ where $\hat{\mathbf{n}}$ is the normal unit vector pointing into material (1). With these relations, **Eqs. (9.38)** and **(9.40)** may be written as

$$\boxed{\hat{\mathbf{n}} \times (\mathbf{H}_1 - \mathbf{H}_2) = \mathbf{J}_s \quad [\text{A/m}]} \quad \text{or} \quad \boxed{\hat{\mathbf{n}} \times \left(\frac{\mathbf{B}_1}{\mu_1} - \frac{\mathbf{B}_2}{\mu_2}\right) = \mathbf{J}_s \quad [\text{A/m}]} \tag{9.42}$$

This form of the interface conditions guarantees correct magnitude and direction for the components of $\mathbf{J}_s$ without resorting to the right-hand rule. **Equation (9.42)** is a more general form of **Eqs. (9.38)** and **(9.40)** and should be used in all instances except, perhaps, when the fields have only one tangential component.

For the normal components of the field, we use **Eq. (8.21)** and calculate the relations between the normal components of **B** on both sides of the interface by calculating the total normal flux through the interface. Because this requires calculation of the magnetic flux Φ over a closed surface, we define a cylindrical volume as shown in **Figure 9.18b**, where the normal components of **B** are perpendicular to the bases of the cylinder. Thus, B_{n1} is in the direction of ds_1, and B_{n2} is in the direction opposite ds_2. Allowing the volume of the cylinder to tend to zero (i.e., the volume of the cylinder encloses only the interface), we have

$$\oint_s \mathbf{B} \cdot d\mathbf{s} = 0 \quad \rightarrow \quad \int_{s_1} B_{1n} ds_1 - \int_{s_2} B_{2n} ds_2 = 0 \tag{9.43}$$

With $s_1 = s_2$, the boundary condition becomes

$$\boxed{B_{1n} = B_{2n}} \tag{9.44}$$

Again, with the use of the constitutive relation $\mathbf{B} = \mu\mathbf{H}$, we can write the interface conditions for the normal component of **H** as

$$\boxed{\mu_1 H_{1n} = \mu_2 H_{2n}} \tag{9.45}$$

The normal component of the magnetic flux density is continuous across an interface, but the normal component of the magnetic field intensity is not. The discontinuity in H_n is again related to the permeability of the two materials. The interface conditions are summarized in **Table 9.5**.

Table 9.5 Interface conditions for the static magnetic field

General conditions at an interface	Conditions at an interface without surface current density
$\hat{\mathbf{n}} \times (\mathbf{H}_1 - \mathbf{H}_2) = \mathbf{J}_s \quad [\text{A/m}]$ or: $H_{1t} - H_{2t} = J_s^* \quad [\text{A/m}]$ $B_{1n} = B_{2n}$	$\hat{\mathbf{n}} \times (\mathbf{H}_1 - \mathbf{H}_2) = 0$ or: $H_{1t} = H_{2t}$ $B_{1t} = B_{2t}$
$\hat{\mathbf{n}} \times \left(\frac{\mathbf{B}_1}{\mu_1} - \frac{\mathbf{B}_2}{\mu_2}\right) = \mathbf{J}_s \quad [\text{A/m}]$	$\hat{\mathbf{n}} \times \left(\frac{\mathbf{B}_1}{\mu_1} - \frac{\mathbf{B}_2}{\mu_2}\right) = 0$
or	or
$\frac{B_{1t}}{\mu_1} - \frac{B_{2t}}{\mu_2} = J_s^* \quad [\text{A/m}]$	$\frac{B_{1t}}{\mu_1} = \frac{B_{2t}}{\mu_2}$
$B_{1n} = B_{2n}$	$B_{1n} = B_{2n}$

*This form requires the use of the right-hand rule to establish the vector relation between the tangential components of the fields and the current density

Now that the general interface conditions have been defined, it is appropriate to dwell a bit on the surface current density. Perhaps the most important question is: When do current densities exist and when not? First, since current cannot flow on the surface of nonconducting materials, at least one of the materials must be a conductor. Also, whenever the conductivities of materials are finite, currents will flow inside conductors rather than on the surface. Thus, the only practical case in which we must assume true surface current densities is at the interfaces between perfect conductors and other materials. In most applications related to steady currents (magnetostatics), the interface conditions in the second column of **Table 9.5** are used. The only exception is a situation in which the currents are constrained to flow on a surface such as a highly conducting coating on a poorly conducting base material. Whenever current density on the surface can be neglected, the tangential component of the magnetic field intensity and the normal component of the magnetic flux density are continuous across the interface.

The concept of surface current is somewhat difficult to understand since our concept of current is that of volume current densities; that is, in circuit theory, we always assumed that a conductor has finite thickness even though the thickness of the conductor was not normally used in calculation. This, of course, is still correct in the sense that only for perfect conductors ($\sigma \to \infty$) is the current limited to the surface. Since this condition is not satisfied in most practical applications and the conductivity of materials is finite, a true surface current doesn't usually occur. We defer discussion of the surface current density for materials other than perfect conductors until we discuss time-dependent fields. **Example 9.6** shows why a surface current density must exist on the surface of a superconductor.

As a direct consequence of the above interface conditions, the magnetic field (either **H** or **B**) is refracted at the interface between two materials with different permeabilities. Assuming there are no current densities on the surface, we can write for continuity of the tangential components of **H** and normal components of **B**:

$$H_{1t} = H_{2t} \quad \text{and} \quad B_{1n} = B_{2n} \tag{9.46}$$

From **Figure 9.18a**

$$\tan\theta_1 = \frac{H_{1t}}{H_{1n}} \quad \text{and} \quad \tan\theta_2 = \frac{H_{2t}}{H_{2n}} \tag{9.47}$$

and

$$\frac{\tan\theta_1}{\tan\theta_2} = \frac{H_{2n}}{H_{1n}} = \frac{B_{2n}/\mu_2}{B_{1n}/\mu_1} = \frac{\mu_1}{\mu_2} \tag{9.48}$$

where the relation $\mathbf{H} = \mathbf{B}/\mu$ and $B_{1n} = B_{2n}$ were used

Thus, if $\mu_1 > \mu_2$, $\theta_1 > \theta_2$, and if $\mu_1 < \mu_2$, $\theta_1 < \theta_2$. The refraction can be rather high because the ratio between permeabilities of different materials can be high. This means that a magnetic field almost parallel to an interface in one material may be almost perpendicular in the other material.

Example 9.5 Refraction at the Interface of a Ferromagnetic Material and Air The magnetic field intensity in an iron piece is directed at 85° to the normal to the surface as shown in **Figure 9.19**.

(a) Calculate the direction of the magnetic field intensity in air.
(b) If the magnetic flux density in iron is 1 T, what is the magnetic flux density in air?

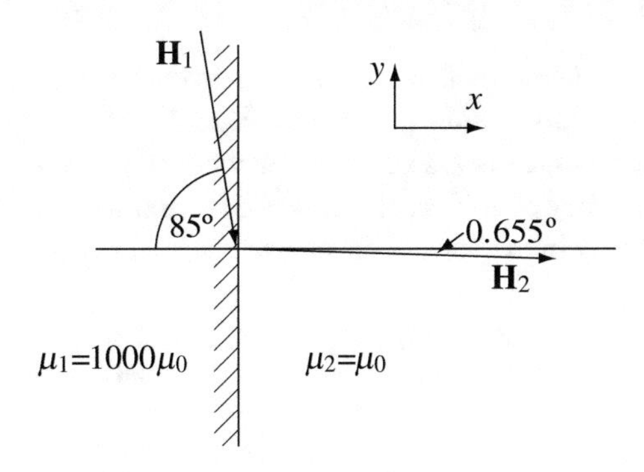

Figure 9.19 The magnetic field intensity relations at the interface between a high-permeability material and free space

Solution:

(a) The solution may be obtained by calculating the tangential and normal components of the magnetic field intensity and then applying the continuity relations for the tangential and normal components. Alternatively, we can use **Eq. (9.48)**. The latter gives

$$\tan\theta_2 = \frac{\tan 85^\circ}{1000} \quad \rightarrow \quad \theta_2 = 0.655^\circ$$

Thus, the magnetic field intensity in air is almost perpendicular to the surface. The opposite is also true: a magnetic field intensity at almost any angle (except 90°) in air will refract to an almost parallel field in the ferromagnetic material.

(b) Taking iron to be material (1) and air to be material (2), the interface conditions are

$$\frac{B_{1t}}{\mu_1} = \frac{B_{2t}}{\mu_2} \quad \rightarrow \quad B_{2t} = \frac{B_{1t}\mu_2}{\mu_1} \quad [\mathrm{T}]$$

$$B_{2n} = B_{1n}$$

The tangential and normal components of the magnetic flux density in iron are calculated from the magnitude of the flux density and the angle between the flux density B_1 and the normal:

$$B_{1n} = B_1\cos 85^\circ \quad B_{1t} = B_1\sin 85^\circ \quad [\mathrm{T}]$$

The components of the flux density in air are

$$B_{2t} = \frac{B_1\mu_2\sin 85^\circ}{\mu_1} = \frac{B_1\mu_0\sin 85^\circ}{1000\mu_0} = \frac{1\times 0.9962}{1000} = 9.962\times 10^{-4} \quad [\mathrm{T}]$$

$$B_{2n} = B_1\cos 85^\circ = 8.716\times 10^{-2} \quad [\mathrm{T}]$$

The magnitude of the flux density is

$$B_2 = \sqrt{B_{2t}^2 + B_{2n}^2} = 8.716\times 10^{-2} \quad [\mathrm{T}]$$

This flux density is at an angle of 0.655° to the normal as in **Figure 9.19**.

Note: If a system of coordinates is given, the vector form can be written. For the system shown, we get

$$\mathbf{B}_{air} = \hat{\mathbf{x}}B_{2n} - \hat{\mathbf{y}}B_{2t} = \hat{\mathbf{x}}8.716\times 10^{-2} - \hat{\mathbf{y}}9.962\times 10^{-4} \quad [\mathrm{T}]$$

Example 9.6 Current Density at the Surface of a Perfect Conductor A magnetic flux density of 1 T exists parallel to the surface of a large, flat superconductor. Calculate the surface current density produced by this magnetic flux density.

Solution: The magnetic field intensity in the superconducting material is zero; that is, both tangential and normal components are zero (see discussion in **Section 9.2.3.1**). The only way a tangential component of the magnetic field intensity can exist at the surface is if a current density exists on the surface of the superconductor.

From the general boundary conditions in **Table 9.5**, denoting air as medium (1) and the superconductor as medium (2), and then setting the flux density in the superconductor to $B_{2t} = 0$ we get

$$\frac{B_{1t}}{\mu_1} - \frac{B_{2t}}{\mu_2} = J_s \quad \rightarrow \quad J_s = \frac{B_{1t}}{\mu_0} = \frac{1}{4\pi\times 10^{-7}} = 796 \quad [\mathrm{kA/m}]$$

The direction of the surface current density J_s is such that it cancels the external magnetic flux density inside the superconductor; that is, the current density on the surface of the perfect conductor may be viewed as the source of an additional field which cancels the external field in the superconductor. Note also that J_s is perpendicular to H_{1t} and H_{2t} as required [see **Eq. (9.42)**].

9.4 Inductance and Inductors

By now, we should have enough tools to calculate the magnetic field intensity due to currents in conductors either using Ampère's law or, if the configuration is complex, using the Biot–Savart law. From **Eq. (8.17)** or **(8.47)** we can also calculate the total magnetic flux passing through any surface. We will use these concepts to define a new and very useful quantity: the inductance.

First, consider a conducting loop C_1 in free space. This may be a bent wire or one turn of a solenoid. If a current I_1 flows in the loop, a magnetic flux density $\mathbf{B}_1$ is produced. This flux density exists everywhere in space and varies from point to point (see **Figure 9.20**). We discuss the field due to loops because they are easier to evaluate and understand and have specific applications in engineering. However, the discussion is completely general and the results apply to any current configuration, since any current must be part of a closed loop.

Now, consider a second conducting loop C_2 at some other physical location in space also shown in **Figure 9.20**. Since loop C_1 produces the magnetic flux density, all flux lines pass through loop C_1, but only some of the flux produced by loop C_1 passes through loop C_2. Assuming the flux density **B** is known (calculated using the Biot–Savart law), we can calculate the magnetic flux passing through each loop. To calculate the various fluxes we use the following notation:

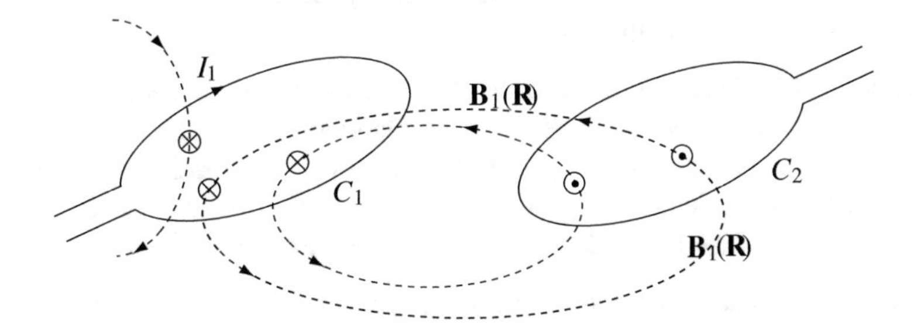

Figure 9.20 Loop (1) produces a magnetic flux, part of which passes through loop (2)

$\mathbf{B}_1$ = flux density produced by loop C_1
I_1 = current in loop C_1
Φ_{11} = magnetic flux, produced by loop C_1 that passes through loop C_1
Φ_{12} = magnetic flux, produced by loop C_1 that passes through loop C_2

In the discussion that follows, it is assumed that the medium in which the loops are placed is linear; that is, the permeability of the medium is independent of the currents in either loop.

The total flux in loop C_1 is

$$\Phi_{11} = \int_{s_1} \mathbf{B}_1 \cdot d\mathbf{s}_1 \quad [\text{Wb}] \tag{9.49}$$

Similarly, the flux through loop C_2 is

$$\Phi_{12} = \int_{s_2} \mathbf{B}_1 \cdot d\mathbf{s}_2 \quad [\text{Wb}] \tag{9.50}$$

The flux density $\mathbf{B}_1$ is not known unless the actual dimensions of C_1 and the value of I_1 are known. However, in very general terms, we recall that in the Biot–Savart law, if the current I is a constant, it may be taken outside the integral sign [see **Eq. (8.9)**]. Therefore, $\mathbf{B}_1$ is directly proportional to the current I_1:

$$\mathbf{B}_1 = I_1 \left[\frac{\mu_0}{4\pi} \oint_{C_1} \frac{d\mathbf{l}' \times \hat{\mathbf{R}}}{|\mathbf{R}|^2} \right] \quad [\text{T}] \tag{9.51}$$

where $\mathbf{R} = \mathbf{r}_1 - \mathbf{r}'$, $\mathbf{r}_1$ is the position vector of the point at which $\mathbf{B}_1$ is calculated, and $\mathbf{r}'$ is the position vector of the current element $d\mathbf{l}'$. Suppose we use this equation and calculate the magnetic flux density everywhere in space due to the loop and then substitute it in **Eqs. (9.49)** and **(9.50)** and perform the integration over s_1 and s_2. The result is the magnetic fluxes Φ_{11} and Φ_{12}, which are directly proportional to I_1:

$$\Phi_{11} = L_{11} I_1 \quad [\text{Wb}] \tag{9.52}$$

$$\Phi_{12} = L_{12} I_1 \quad [\text{Wb}] \tag{9.53}$$

The terms L_{11} and L_{12} are constants, independent of current; only the geometry of the configuration plays a role in the values of L_{11} and L_{12} as can be seen from the term in brackets in **Eq. (9.51).** L_{11} is defined as

$$L_{11} = \frac{\Phi_{11}}{I_1} \quad \left[\frac{\text{Wb}}{\text{A}} \; or \; \text{H}\right] \tag{9.54}$$

This is called the ***self-inductance*** of loop C_1. It is a self-inductance because all terms involved in its calculation relate only to itself. Loop C_2 has no effect on the self-inductance of loop C_1. Similarly,

$$L_{12} = \frac{\Phi_{12}}{I_1} \quad [\text{H}] \tag{9.55}$$

L_{12} is called the ***mutual inductance*** between loop C_1 and loop C_2. It is a mutual inductance because the flux Φ_{12} depends both on loop C_1 and on loop C_2, as is evident from **Eq. (9.50)**.

Before proceeding we note the following:

(1) Inductance (self or mutual) has units of [Wb/A]. This unit is designated as the henry [H].
(2) Inductance depends only on the geometrical configuration of the circuits and is independent of current. Although the current produces the flux, once we divide by the current, the result contains only geometrical terms (and permeability).
(3) Any device that has inductance may be called an ***inductor***. In practice, an inductor is usually a coil or a solenoid designed specifically for its inductance and often used as a circuit element.

The above results can be generalized in two ways: One is to ask ourselves what happens if instead of a single turn, we have two or more turns either in C_1 or C_2 or both. This is shown in **Figure 9.21**, where N_1 loops are bundled together to form a circuit C_1 and N_2 loops are bundled together to form a circuit C_2. We argue as follows: if the flux due to a single loop carrying a current I is equal to Φ, the flux due to N loops carrying the same current I is $N\Phi$ from the principle of superposition, provided that the same flux passes through all loops. Thus, the total flux in C_1 is $N_1\Phi_{11}$, and the total flux passing through C_2 is $N_2\Phi_{12}$. The equations for self- and mutual inductance can now be written as

$$\boxed{L_{11} = \frac{N_1\Phi_{11}}{I_1} \quad \text{and} \quad L_{12} = \frac{N_2\Phi_{12}}{I_1} \quad [\text{H}]} \tag{9.56}$$

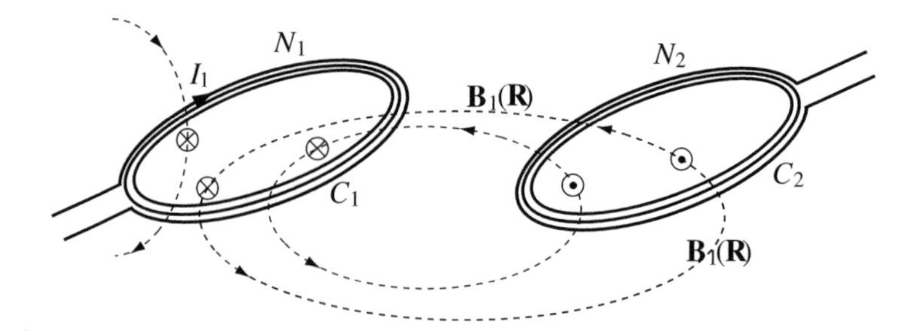

Figure 9.21 Inductance of multiple loop circuits. Circuit (1) produces a magnetic flux, part of which passes through circuit (2)

The term $N_1\Phi_{11}$ is the total flux that links circuit C_1 with itself or the ***flux linkage*** of C_1 with itself. The term $N_2\Phi_{12}$ is the total flux that links circuit C_1 with circuit C_2 or the flux linkage between circuit C_1 and C_2. These terms are denoted as

$$\Lambda_{11} = N_1\phi_1 \quad \text{and} \quad \Lambda_{12} = N_2\Phi_{12} \quad [\text{weber} \cdot \text{turns}] \tag{9.57}$$

The unit of flux linkage is the weber • turns [Wb • t]. With this notation, the self- and mutual inductances are defined as

$$\boxed{L_{11} = \frac{\Lambda_{11}}{I_1} \quad \text{and} \quad L_{12} = \frac{\Lambda_{12}}{I_1} \quad [\text{H}]} \tag{9.58}$$

This definition is more general and includes in it the definitions in **Eqs. (9.54)** and **(9.55)**.

The second generalization we wish to make is to assume that circuit C_2 also carries a current I_2 as shown in **Figure 9.22** and see how this affects inductance. To do so, we use superposition as follows: if the indices 1 and 2 in **Figure 9.21** are interchanged, we get an identical situation, except that now the self-inductance of circuit C_2 is calculated as L_{22} and the mutual inductance between circuit C_2 and circuit 1 is calculated as L_{21}. Thus

$$\boxed{L_{22} = \frac{N_2\Phi_{22}}{I_2} = \frac{\Lambda_{22}}{I_2} \quad \text{and} \quad L_{21} = \frac{N_1\Phi_{21}}{I_2} = \frac{\Lambda_{21}}{I_2} \quad [\text{H}]} \tag{9.59}$$

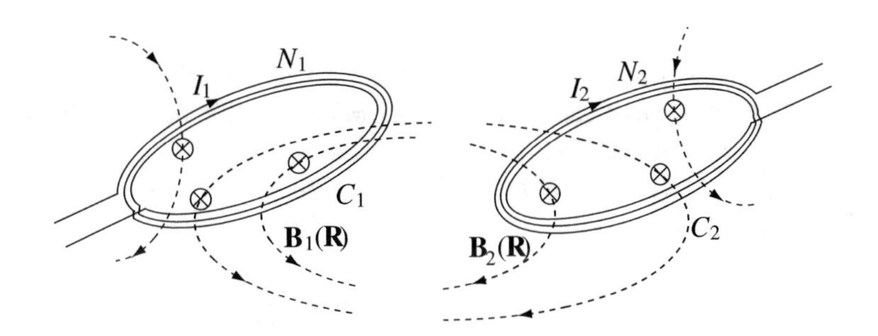

Figure 9.22 Both circuits (1) and (2) carry currents and produce flux

where the following definitions were used:

$\mathbf{B}_2$ = flux density produced by loop C_2
I_2 = current in loop C_2
Φ_{22} = magnetic flux, produced by circuit C_2 that passes through circuit C_2
Φ_{21} = magnetic flux, produced by circuit C_2 that passes through circuit C_1
Λ_{22} = magnetic flux linkage between circuit C_2 and itself
Λ_{21} = magnetic flux linkage between circuit C_2 and circuit C_1

We will see in examples that $L_{12} = L_{21}$. Although we do not prove it (it requires calculation of energy), this is a general relation and will be used in subsequent derivations.

Calculation of inductance involves the following steps:

(1) For each current-carrying conductor, calculate the flux density in space. If the conductor for which you need to calculate self-inductance does not carry a current, assume an arbitrary current. The calculation may be done using the Biot–Savart or Ampère's law.
(2) Calculate the flux passing through any circuit as required. For self-inductance, only the flux in the circuit producing the flux is required.
(3) Calculate the total flux linkage linking the circuit with itself (for self-inductance) or with other circuits (for mutual inductance).
(4) Divide the appropriate flux linkage by the current that produced it to obtain the self- or mutual inductance.

It is important to remember that the relations for inductance as given here only apply for a linear medium; that is, it is assumed that the inductance is independent of current.

Example 9.7 Application: Self-inductance of a Toroidal Coil A torus with rectangular cross section and dimensions as shown in **Figure 9.23a** is wound with a coil. The coil consists of $N = 1{,}000$ turns of very fine wire wound in a uniform single layer on the surface of the torus. Permeability of the torus is μ_0 [H/m]. Use $b = 50$ mm, $c = 20$ mm, $d = 70$ mm. Calculate the self-inductance of the coil.

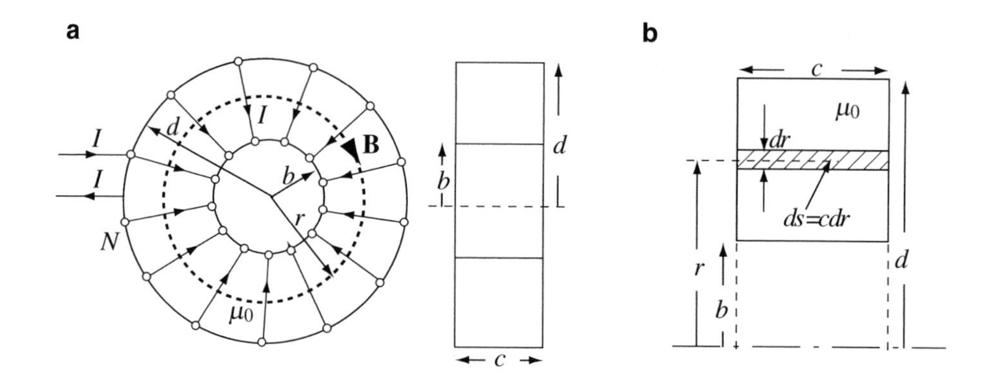

Figure 9.23 Calculation of inductance in a toroidal coil. (**a**) Dimensions. (**b**) Calculation of flux

Solution: The flux density in the torus is calculated first using Ampère's law. The flux density outside the torus is zero (see **Example 8.9**). From this, we calculate the total flux in the torus. Multiplying by the number of turns to find the flux linkage and dividing by the current in the turns gives the self-inductance.

From Ampère's law, a contour at radius $b < r < d$ inside the torus encloses N turns, each carrying a current I. The magnetic flux density is therefore

$$B = \frac{\mu_0 NI}{2\pi r} \quad [\mathrm{T}]$$

The flux density is not constant but varies with the radius r. To calculate the flux, we must integrate the flux density over the cross-sectional area of the torus. To do so, we take an elementary area $ds = cdr$ and, therefore, an element of flux $d\Phi = Bd\mathbf{s}$ as shown in **Figure 9.23b** and write

$$\Phi = \int_s Bds = \int_{r=b}^{r=d} \frac{\mu_0 NI}{2\pi r} cdr = \frac{\mu_0 NIc}{2\pi} \int_{r=b}^{r=d} \frac{dr}{r} = \frac{\mu_0 NIc}{2\pi} \ln\frac{d}{b} \quad [\mathrm{Wb}]$$

This flux passes through all windings of the coil. Therefore, the total flux linkage is

$$\Lambda = N\Phi = \frac{\mu_0 N^2 Ic}{2\pi} \ln\frac{d}{b} \quad [\mathrm{Wb \cdot t}]$$

and the self-inductance is

$$L = \frac{\Lambda}{I} = \frac{N\Phi}{I} = \frac{\mu_0 N^2 c}{2\pi} \ln\frac{d}{b} \quad [\mathrm{H}]$$

The self-inductance of the toroidal coil (not the torus) is

$$L_{11} = \frac{\mu_0 N^2 c}{2\pi} \ln\frac{d}{b} = \frac{4 \times \pi \times 10^{-7} \times 1000^2 \times 0.02}{2 \times \pi} \ln\frac{0.07}{0.05} = 1.346 \times 10^{-3} \quad [\mathrm{H}].$$

Exercise 9.2 Assume the torus in **Example 9.7** is made of iron with relative permeability of μ_1=1,000μ_0 [H/m] and the coil consists of 10 turns. What is the self-inductance of the coil if the flux is entirely contained within the volume of the torus?

Answer 13.46 μH.

Example 9.8 Application: Mutual Inductance Between a Wire and a Toroidal Coil—Core Memory The geometry shown in this example was used in the past as a memory device. It consisted of a small magnetic torus (usually made of ferrite) and a wire passing through the torus (see **Figure 9.24**). A current in the wire was used to magnetize the magnetic torus, which then retained this magnetization until an opposite current was used to erase it. Magnetic memories before the advent of semiconductor memories were almost exclusively used and were made of very large matrices of miniature toroids, about 1 mm in diameter. Each torus was a single memory bit. The main advantage of this type of memory was in its nonvolatility: memory could be retained after the computer was switched off. In addition to the magnetization (or write) wire, there was an erase wire and a sensing (read) wire which allowed the computer to read the memory without erasing it. Thus, each torus had three wires passing through it. The following example calculates the mutual inductance between the magnetizing wire and a coil on the torus.

A torus with mean radius r_0 [m] and a cross-sectional area as shown in **Figure 9.24b** is given. A coil with N turns is wound uniformly around the torus. A straight, long wire carrying a current I [A] passes through the center of the torus.

(a) Calculate the mutual inductance between the wire and coil if the torus is air filled. Assume $r_0 \gg b$.
(b) In an attempt to increase the mutual inductance, the torus is made of ferrite with relative permeability of 5,000. Calculate the mutual inductance between the wire and coil.

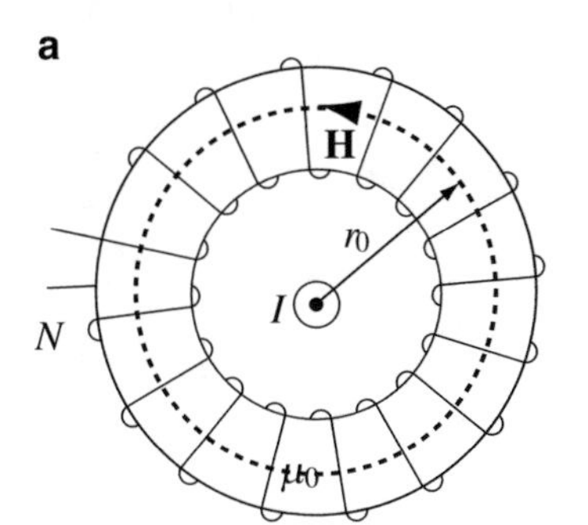

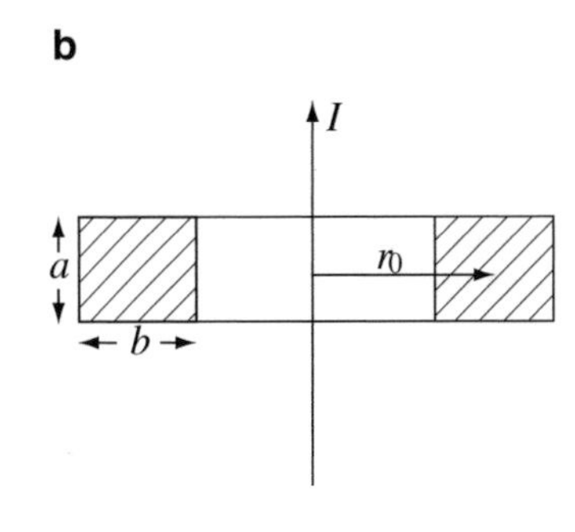

Figure 9.24 (**a**) A current-carrying wire passing through a toroidal coil. (**b**) Cross-sectional view

Solution: The wire generates a flux density in the circumferential direction. The magnetic field intensity of the wire is calculated as if the torus did not exist, since the magnetic field intensity is independent of permeability. From this, the magnetic flux density is calculated by multiplying H by the permeability μ of the torus. Because $r_0 \gg b$, we can use the flux density at r_0 as an average flux density, which is approximately uniform throughout the cross section. The total flux is the product of flux density and the cross-sectional area of the torus. The flux linkage with the N turns of the coil is found by multiplying the flux by N. Division by the current in the wire gives the mutual inductance.

(a) Using Ampère's law and a contour of radius r_0 around the wire, the field intensity is calculated as

$$H2\pi r_0 = I \quad \rightarrow \quad H = \frac{I}{2\pi r_0} \quad \left[\frac{\text{A}}{\text{m}}\right]$$

Because $r_0 \gg b$ the flux density in the torus is approximately uniform and equals $B = \mu_0 H$. The total flux then becomes

$$\Phi = Bab = \mu_0 Hab = \frac{\mu_0 abI}{2\pi r_0} \quad [\text{Wb}]$$

This flux links all N turns of the coil. The total flux linkage is

$$\Lambda_{12} = N\Phi = \frac{\mu_0 abNI}{2\pi r_0} \quad [\text{Wb}\cdot\text{t}]$$

and the mutual inductance is

$$L_{12} = \frac{\Lambda_{12}}{I_1} = \frac{\mu_0 abN}{2\pi r_0} \quad [\text{H}]$$

The mutual inductances are equal: $L_{21} = L_{12}$. This fact can be used to our advantage. For example, in the above example, it is relatively easy to calculate the mutual inductance L_{12}. The mutual inductance L_{21} is much more difficult to define since, now, we must calculate the magnetic flux density due to the torus inside the loop created by the wire and its return at infinity. Because the two inductances are the same, we can choose to calculate that which is easiest.

(b) If the torus is made of ferrite, the only change is in the permeability of ferrite. The permeability is 5,000 times larger and so is the mutual inductance:

$$L_{12} = \frac{\Lambda_{12}}{I_1} = \frac{5{,}000\mu_0 abN}{2\pi r_0} \quad [\text{H}].$$

Example 9.9 Self- and Mutual Inductances in Multiple Coils Three coils are wound on a toroidal core with properties and dimensions as shown in **Figure 9.25**. Assume $a \gg (b - a)$ and that all three coils are uniformly wound around the torus (one on top of the other) and calculate:

(a) The self-inductances of each of the three coils.
(b) The mutual inductances between coils (1) and (2), between coils (2) and (3), and between coils (1) and (3).

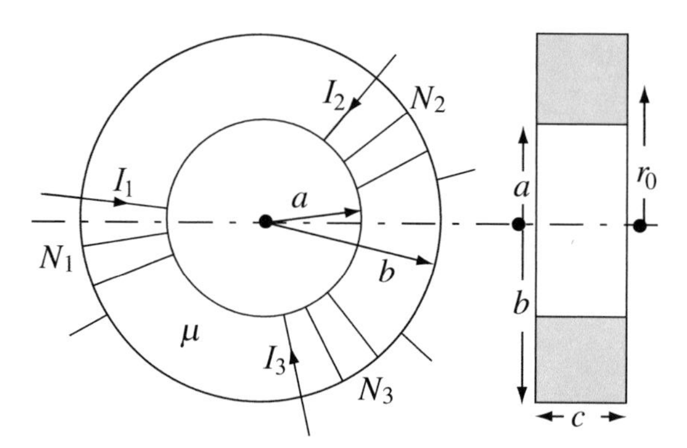

Figure 9.25 Three coils wound on a common torus. Each coil is uniformly wound around the core

Solution: To calculate the self-inductance, we assume each coil, separately, carries an arbitrary current, calculate the flux, then the flux linkage, and then divide by current in the coil to find the self-inductance of the coil. The mutual inductance is calculated similarly by taking one coil at a time, assuming a current through the coil, calculating the flux in the core, and then calculating the flux that links each pair of coils. Division by the current generating the flux gives the mutual inductance.

(a) We start with coil (1). Assuming a current I_1, the flux density in the coil is (see **Example 9.8**):

$$B_1 = \frac{\mu N_1 I_1}{2\pi r_0} \quad [\text{T}]$$

where $r_0 = (a + b)/2$ is the average radius (since $a \gg (b - a)$).

The flux generated by coil (1) is

$$\Phi_1 = B_1 S = \frac{\mu N_1 I_1 (b-a)c}{\pi(b+a)} \quad [\text{Wb}]$$

Because, in a torus, all flux is contained within the core of the torus, this flux links all N_1 turns of coil (1). Thus, the flux linkage is

$$\Lambda_{11} = N_1 \Phi_1 = N_1 B_1 S = \frac{\mu N_1^2 I_1 (b-a)c}{\pi(b+a)} \quad [\text{Wb} \cdot \text{t}]$$

The self-inductance of coil (1) is

$$L_{11} = \frac{\Lambda_{11}}{I_1} = \frac{\mu N_1^2 (b-a)c}{\pi(b+a)} \quad [\text{H}]$$

Repeating the process for each of the remaining two coils, we get

$$L_{22} = \frac{\Lambda_{22}}{I_2} = \frac{\mu N_2^2 (b-a)c}{\pi(b+a)}, \quad L_{33} = \frac{\Lambda_{33}}{I_3} = \frac{\mu N_3^2 (b-a)c}{\pi(b+a)} \quad [\text{H}]$$

Note: The self-inductance may also be written as

$$L_{ii} = \frac{\mu N_i^2 S}{l} \quad [\text{H}]$$

where N_i is the number of turns in coil i, S is the cross-sectional area of the core, and l is the average length of the core.

(b) To calculate the mutual inductance, say, between coils (1) and (3), we first calculate the magnetic flux density and magnetic flux Φ_1 produced by coil (1) as in **(a)**. All of this flux links with the N_3 turns of coil (3). Thus, the flux linkage between coils (1) and (3) is

$$\Lambda_{13} = \Phi_1 N_3 = \frac{\mu N_1 N_3 I_1 (b-a)c}{2\pi r_0} = \frac{\mu N_1 N_3 I_1 (b-a)c}{\pi(b+a)} \quad [\text{Wb} \cdot \text{t}]$$

The mutual inductance between coil (1) and coil (3) is

$$L_{13} = \frac{\Lambda_{13}}{I_1} = \frac{\mu N_1 N_3 (b-a)c}{\pi(b+a)} \quad [\text{H}]$$

Before continuing, we note that if we were to calculate the flux due to coil (3) rather than coil (1), we would obtain

$$\Phi_3 = B_3 S = \frac{\mu N_3 I_3 (b-a)c}{\pi(b+a)} \quad [\text{Wb}]$$

The flux linkage between coils (3) and (1) is

$$\Lambda_{31} = \Phi_3 N_1 = \frac{\mu N_3 N_1 I_3 (b-a)c}{\pi(b+a)} \quad [\text{Wb} \cdot \text{t}]$$

and the mutual inductance between coil (3) and coil (1) is

$$L_{31} = \frac{\Lambda_{31}}{I_3} = \frac{\mu N_3 N_1 (b-a)c}{\pi(b+a)} \quad [\text{H}]$$

Clearly, $L_{13} = L_{31}$. Repeating the process, we obtain for the other two pairs of mutual inductances:

$$L_{12} = L_{21} = \frac{\mu N_1 N_2 (b-a)c}{\pi(b+a)}, \quad L_{23} = L_{32} = \frac{\mu N_2 N_3 (b-a)c}{\pi(b+a)} \quad [\text{H}]$$

As for the self-inductance, the mutual inductance may also be written in general for the mutual inductance between coils (i) and (j) on the same core:

$$L_{ij} = \frac{\mu N_i N_j S}{l} \quad [\text{H}].$$

where N_i is the number of turns in coil i, N_j is the number of turns in coil j, S is the cross-sectional area of the core, and l is the average length of the core.

9.4.1 Inductance per Unit Length

The self- and mutual inductances defined in **Eqs. (9.58)** and **(9.59)** are the total self- or mutual inductances of the corresponding loops because the flux linkage was calculated based on the total flux through the loops. However, if the loops are very large in size as is the case of long cables, the flux linkage becomes very large (or infinite) and, therefore, the total inductance is very large (or infinite). In such cases it is more useful to define the ***inductance per unit length*** of the structure. All results obtained so far apply except that we calculate the flux linkage of a 1 m section of the structure instead of the total flux linkage. Naturally, the unit of inductance now becomes the [H/m]. **Examples 9.10** and **9.11** and **Exercise 9.3** discuss some of the details involved in the calculation of inductance per unit length.

Example 9.10 Application: Self-inductance of a Long Solenoid—Inductance per Unit Length Find the self-inductance per unit length of a long solenoid with $n = 100$ turns per unit length and a diameter $d = 50$ mm.

Solution: We assume a current I in the turns and use Ampère's law to calculate the magnetic flux density inside the solenoid. Since the flux density is constant in the solenoid, the total flux is obtained by multiplying the magnetic flux density by the cross-sectional area and then the flux linkage is found by multiplying the total flux by the number of turns per unit length. Division of the flux linkage per unit length by the current gives the inductance per unit length. Note that only inductance per unit length of a solenoid has practical meaning. The total inductance is very large (infinite).

To calculate the flux density, we use Ampère's law as in **Example 8.10**, using contour a as shown in **Figure 9.26**. The length of the contour is arbitrary. The flux density outside the solenoid is zero (as was shown in **Example 8.10**). Thus,

$$B = \frac{\mu_0 n I L}{L} = \mu_0 n I \quad [\text{T}]$$

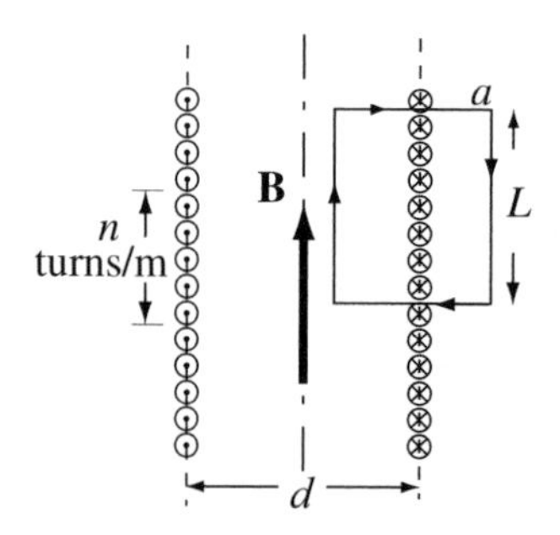

Figure 9.26 Calculation of the flux density in a solenoid

This is the same result we obtained in **Example 8.10**. The total flux is

$$\Phi = BS = \mu_0 n I \pi \frac{d^2}{4} \quad [\text{Wb}]$$

where d is the diameter of the solenoid. The flux linkage per unit length of the solenoid is

$$\Lambda = \Phi n = \mu_0 n^2 I \pi \frac{d^2}{4} \quad \left[\frac{\text{Wb} \cdot \text{t}}{\text{m}}\right]$$

since the flux passes through all turns of the solenoid. The self-inductance per unit length is

$$L_{11} = \frac{\Lambda}{I} = \mu_0 n^2 \pi \frac{d^2}{4} \quad \left[\frac{\text{H}}{\text{m}}\right]$$

Thus, whereas the flux density is independent of the diameter of the solenoid, its inductance depends on the diameter. For the values given, the inductance per unit length is

$$L_{11} = 4 \times \pi \times 10^{-7} \times 100^2 \times \pi \times \frac{0.05^2}{4} = 2.47 \times 10^{-5} \quad [\text{H/m}]$$

or in more standard notation $L_{11} = 24.7$ μH/m.

9.4.2 External and Internal Inductance

So far we have defined and calculated the self- and mutual inductance of a number of configurations, including the inductance per unit length of a long solenoid. In all of these configurations we assumed the current-carrying conductors to be thin and that the flux used for computation of the flux linkage and inductance was external to the conductors—the flux inside the conductors, if any, was neglected. The inductance calculated from the flux linkage external to the conductors is also called the ***external inductance***. If the current-carrying conductors are of finite dimensions, part of the flux produced will actually exist inside the conductors and produce a flux linkage of its own. As a result, part of the inductance is due to the flux linkage interior to the conductors. This inductance is called the ***internal inductance***. The total inductance is the sum of the external and internal inductances. In many instances, the internal inductance is small compared to the external inductance and is therefore neglected. For this reason, what is normally called inductance refers more often to external inductance. **Example 9.11** and **Exercise 9.3** explore the concepts of internal, external, and total inductance per unit length.

Example 9.11 Application: Inductance per Unit Length of Coaxial Cables A coaxial cable is made of an inner solid conductor of radius b [m] and an outer thin, flexible conductor of radius a [m] and negligible thickness, separated by a dielectric. Assume all materials have permeability μ_0 [H/m]. Calculate the self-inductance of the coaxial cable per unit length of the cable.

Solution: We assume a current I flows into the outer conductor and out of the inner conductor as shown in **Figure 9.27a** and that the current is uniformly distributed in the inner conductor (uniform current density). For ease of calculation, we divide the conductor into three domains: One is the inner conductor ($0 < r < b$). The second is the domain between the two conductors ($b \leq r \leq a$). The third is the exterior of the outer conductor ($r > a$). The flux in each of the domains is calculated first. Then, we calculate the flux linkage and, finally, divide the total flux linkage by the current to obtain the inductance. Unlike calculation of inductance in solenoids, where the flux linkage is clearly defined, here we will have to define the flux linkage for a differential element of current and calculate the total flux linkage by integration.

(1) Inside the inner conductor ($0 < r < b$). We draw a contour at radius $0 < r < b$. The total current enclosed by the contour is the area enclosed by the contour (**Figure 9.27b)** multiplied by the current density in the conductor. The current density in the inner conductor is

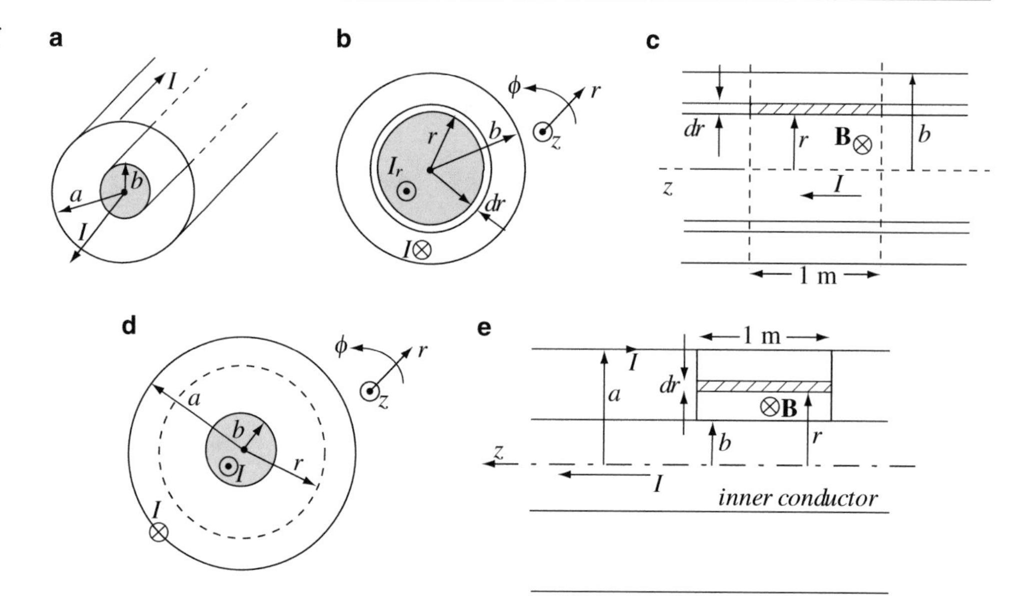

Figure 9.27 Calculation of internal and external self-inductance in a coaxial cable. (**a**) Geometry and dimensions. (**b**) Calculation of flux density in the inner conductor. (**c**) Calculation of flux linkage in the inner conductor. (**d**) Calculation of flux density between conductors. (**e**) Calculation of flux linkage between conductors

$$J_i = \frac{I}{\pi b^2} \quad \left[\frac{\text{A}}{\text{m}^2}\right]$$

The total current enclosed by the contour of radius $r < b$ is

$$I_r = \pi r^2 J_i = \frac{\pi r^2 I}{\pi b^2} \quad [\text{A}]$$

The magnetic flux density at radius r is found by applying Ampère's law around the contour of radius r:

$$2\pi r B_r = \mu_0 I_r = \frac{\mu_0 r^2 I}{b^2} \quad \rightarrow \quad B_r = \frac{\mu_0 r I}{2\pi b^2} \quad [\text{T}], \quad (0 < r < b)$$

Now, suppose we define an element of area in the form of a ring of radius r, thickness dr, and width 1 m as shown in **Figures 9.27b** and **9.27c**. The flux in the ring (i.e., the flux that passes through the area of the ring $ds = 1 \times dr$) perpendicular to the area is

$$d\Phi_1 = \mathbf{B}_r \cdot d\mathbf{s} = \frac{\mu_0 I r}{2\pi b^2} \times 1 \times dr = \frac{\mu_0 I r dr}{2\pi b^2} \quad [\text{Wb}]$$

This flux is produced by the current enclosed by the ring (current outside the ring produces zero flux density at the location of the ring). Thus, the flux linking the ring equals the flux $d\Phi_1$ multiplied by the ratio of the current enclosed and the total current in the conductor (**Figure 9.27b**):

$$d\Lambda_1 = d\Phi_1 \frac{I_r}{I} = d\Phi_1 \frac{\pi r^2}{\pi b^2} = \frac{\mu_0 I r^3 dr}{2\pi b^4} \quad [\text{Wb} \cdot \text{t}]$$

The total flux linkage is found by integrating over r from $r = 0$ to $r = b$:

$$\Lambda_1 = \int_{r=0}^{r=b} \frac{\mu_0 I r^3 dr}{2\pi b^4} = \left. \frac{\mu_0 I r^4}{8\pi b^4} \right|_{r=0}^{r=b} = \frac{\mu_0 I}{8\pi} \quad [\text{Wb} \cdot \text{t}]$$

This flux linkage is independent of the radius of the conductor. Because it is entirely due to flux linkages within the conductor's volume this is an internal flux linkage. The internal inductance due to the inner conductor is therefore

$$L_{int} = \frac{\mu_0}{8\pi} \quad \left[\frac{\text{H}}{\text{m}}\right].$$

(2) The second domain is between the two conductors $b \le r \le a$. On the contour shown in **Figure 9.27d**, using Ampère's law (and the right-hand rule),

$$B2\pi r = \mu_0 I \quad \rightarrow \quad B = \frac{\mu_0 I}{2\pi r} \quad \text{or} \quad \mathbf{B} = \hat{\boldsymbol{\phi}} \frac{\mu_0 I}{2\pi r} \quad [\text{T}]$$

Using **Figure 9.27e**, the element of area perpendicular to the flux density is $1dr$ and the flux per unit length through the area shown is

$$\Phi = \int_{r=b}^{r=a} \mathbf{B} \cdot d\mathbf{s} = \int_{r=b}^{r=a} \frac{\mu_0 I}{2\pi r} 1dr = \frac{\mu_0 I}{2\pi} \ln \frac{a}{b} \quad [\text{Wb}]$$

The flux linkage equals Φ ($N = 1$) and the inductance per unit length is

$$L_{ext} = L_{11} = \frac{\Lambda}{I} = \frac{\Phi}{I} = \frac{\mu_0}{2\pi} \ln \frac{a}{b} \quad \left[\frac{\text{H}}{\text{m}}\right]$$

Note, again, that the current I is irrelevant. It is required for formal computation but cancels the final result and inductance depends only on physical dimensions and permeability. This inductance is due to flux linkages external to the conductors and, therefore, is the external inductance per unit length of the cable.

For $r > a$, the flux density is zero (the total net current enclosed by any contour of radius $r > a$ is zero). Thus, there is no contribution to inductance due to this domain.

The total inductance per unit length of the coaxial cable is the sum of the internal and external inductances:

$$L = L_{int} + L_{11} = L_{int} + L_{ext} = \frac{\mu_0}{8\pi} + \frac{\mu_0}{2\pi} \ln \frac{a}{b} \quad \left[\frac{\text{H}}{\text{m}}\right].$$

Exercise 9.3 A coaxial cable as used in TV antennas has an inner conductor which is 0.5 mm in diameter and an outer conductor 5 mm in diameter. The outer conductor is a very thin shell.

(a) Calculate the total inductance per unit length of the cable.
(b) What are the internal and external inductances per unit length?
(c) What is the total inductance per unit length if you assume the inner conductor to be a very thin shell with outer diameter 0.5 mm instead of a solid conductor?

Answer

(a) $L = 5.1 \times 10^{-7}$ [H/m].
(b) External inductance: $L_{ext} = 4.6 \times 10^{-7}$ H/m. Internal inductance: $L_{int} = 0.5 \times 10^{-7}$ H/m.
(c) $L = 4.6 \times 10^{-7}$ H/m.

From these examples, it is apparent that self- and mutual inductances exist even in cases when these are not self-evident. In particular, the inductance of straight wires and infinitely long structures requires us to "look" for a generalized loop. However, if there is a current in a conductor, this conductor must be part of a closed circuit, perhaps closing at infinity, and a loop always exists. For infinite structures, the loop is somewhat artificial since we only calculate the inductance per unit length.

We defined inductance as the ratio between flux linkage and current, but the basic question of why we do so remains; that is, what does inductance do that cannot be done from field relations directly? Part of the answer lies in the fact that inductance is geometry dependent rather than current dependent. Thus, the calculation of flux, for example, can be carried out by first calculating the inductance and then the flux linkage from **Eqs. (9.56)** through **(9.59)**. The second part of the answer will be given in detail in the following section, where we discuss energy. However, at this point, you might wish to recall that in electrostatics, we defined capacitance as the ratio between charge and potential. Inductance has equivalent use in magnetics as capacitance has in electrostatics, including storage of energy.

9.5 Energy Stored in the Magnetic Field

The normal process of defining energy in a field is to start with the force in the field, calculate work performed by the field (negative work) or against the field (positive work), that is, start with $dW = \mathbf{F} \cdot d\mathbf{l}$, and then find the energy density associated with the field. We used this approach in the calculation of electrostatic energy (see **Section 4.8**). In the magnetic field, this approach cannot be followed. The reason is that a proper definition of energy in the magnetic field must start with the time-dependent field. Since we have not discussed this aspect of fields yet (and will not do so until **Chapter 10**), we must either accept the equations for energy as given, define them by analogy to the relations in electrostatics, or not define them at all at this stage. A compromise seems to be appropriate here: we will use known relations from circuit theory to find the energy stored in the magnetic field of an inductor. Later, in **Chapter 12**, we will devote considerable time to power and energy in the general electromagnetic field (time dependent as well as time harmonic). The relations we define here for energy will then become simplifications of the more general energy relations.

Consider the *RL* circuit in **Figure 9.28**. From Kirchhoff's circuital law, the voltage across the *RL* circuit is

$$V = V_R + V_L = RI + L\frac{dI}{dt} \quad [\text{V}] \tag{9.60}$$

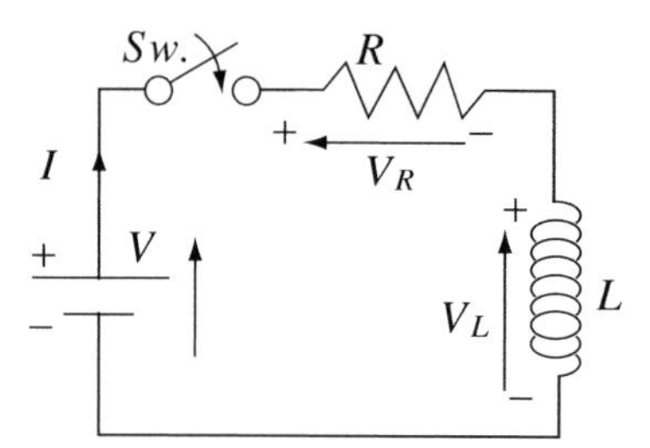

Figure 9.28 *RL* circuit used to define energy stored in the inductor *L*

Here, we are only interested in the second term of **Eq. (9.60)** (the inductive term). The instantaneous power in the inductor is

$$P_L = V_L I = LI\frac{dI}{dt} \quad [\text{W}] \tag{9.61}$$

and the magnetic energy stored in the inductor is the integrated power over time:

$$\boxed{W_m = \int_0^t P_L dt = \int_0^I LI\frac{dI}{dt}dt = \int_0^I LIdI = \frac{1}{2}LI^2 \quad [\text{J}]} \tag{9.62}$$

It is useful to compare this to the electrostatic energy stored in a capacitor, which is given as $CV^2/2$, where C is capacitance and V the voltage on the capacitor. This comparison shows that the role of C is taken by L, whereas that of V is taken by I. It also indicates that if storage of energy in the capacitor is characterized by the potential across the plates, storage of energy in an inductor is characterized by the current in the inductor. Any change in this current changes the stored energy in the inductor.

This relation for magnetic energy is convenient for the calculation of energy stored in an inductor, such as the one shown in **Figure 9.28**. However, in other configurations, this is not so. Consider, for example, the two coils in **Figure 9.22**. Each circuit can be viewed as an inductor with self-inductances L_{11} and L_{22}. The energy stored in each of these inductors is properly given by **Eq. (9.62)**. However, there are also two mutual inductances L_{12} and L_{21}. How do we calculate the energy stored in these mutual inductances and what is the total energy stored in a system? Furthermore, suppose we have N coupled circuits. Can we find a simple way of calculating the energy in the system? To find some answers to these questions, we start with **Eq. (9.62)** but rewrite it in terms of the magnetic flux to obtain a more general statement of energy in the magnetic field.

Consider, first, the two circuits in **Figure 9.22**. Each circuit produces a flux, some of which links with the other circuit. These are as follows:

Φ_{11} = Flux that links C_1 alone (produced by C_1 and passing through C_1)
Φ_{22} = Flux that links C_2 alone (produced by C_2 and passing through C_2)
Φ_{12} = Flux that links C_1 and C_2 and is produced by C_1 (dashed lines)
Φ_{21} = Flux that links C_2 and C_1 and is produced by C_2 (dashed lines)

Now, we can calculate the total flux in circuit C_1 as the flux produced by the circuit itself (Φ_{11}) and that portion of the flux produced by C_2 that links with C_1 (i.e., Φ_{21}). Depending on the directions of currents in C_1 and C_2, these fluxes may oppose each other or may be in the same direction (the fluxes in **Figure 9.22** are shown as opposing each other, but inverting the current in one of the circuits will make both fluxes in the same direction). Thus, we can write for the total flux in each circuit:

Flux in circuit C_1:

$$\Phi_1 = \Phi_{11} \pm \Phi_{21} \quad [\text{Wb}] \tag{9.63}$$

Flux in circuit C_2:

$$\Phi_2 = \Phi_{22} \pm \Phi_{12} \quad [\text{Wb}] \tag{9.64}$$

The flux linkage in each circuit is the flux through the circuit multiplied by the number of loops in the circuit:

$$\Lambda_1 = N_1\Phi_1 = N_1\Phi_{11} \pm N_1\Phi_{21} \quad [\text{Wb} \cdot \text{t}] \tag{9.65}$$

$$\Lambda_2 = N_2\Phi_2 = N_2\Phi_{22} \pm N_2\Phi_{12} \quad [\text{Wb} \cdot \text{t}] \tag{9.66}$$

We can now define the total inductance of each circuit by dividing the flux linkage by its current. These inductances are denoted L_1 and L_2 and include both the self-inductance and the mutual inductance:

$$L_1 = \frac{\Lambda_1}{I_1} = \frac{N_1\Phi_1}{I_1} = \frac{N_1\Phi_{11}}{I_1} \pm \frac{N_1\Phi_{21}}{I_1} \quad [\text{H}] \tag{9.67}$$

$$L_2 = \frac{\Lambda_2}{I_2} = \frac{N_2\Phi_2}{I_2} = \frac{N_2\Phi_{22}}{I_2} \pm \frac{N_2\Phi_{12}}{I_2} \quad [\text{H}] \tag{9.68}$$

Using the definition of energy in **Eq. (9.62)**, we can write for the total energy in the two circuits:

$$W_m = \frac{1}{2}LI^2 = \frac{1}{2}L_1I_1^2 + \frac{1}{2}L_2I_2^2 \quad [\text{J}] \tag{9.69}$$

Substituting **Eqs. (9.67)** and **(9.68)** into this relation gives

$$W_m = \frac{1}{2}\left(\frac{N_1\Phi_{11}}{I_1} \pm \frac{N_1\Phi_{21}}{I_1}\right)I_1^2 + \frac{1}{2}\left(\frac{N_2\Phi_{22}}{I_2} \pm \frac{N_2\Phi_{12}}{I_2}\right)I_2^2 \quad [\text{J}] \tag{9.70}$$

Thus, we get a relation in terms of the total fluxes Φ_1 and Φ_2:

$$W_m = \frac{1}{2}N_1(\Phi_{11} \pm \Phi_{21})I_1 + \frac{1}{2}N_2(\Phi_{22} \pm \Phi_{12})\ I_2 = \frac{1}{2}N_1\Phi_1I_1 + \frac{1}{2}N_2\Phi_2I_2 \quad [\text{J}] \tag{9.71}$$

This can be written as a general expression for any number of circuits n:

$$\boxed{W_m = \frac{1}{2}\sum_{i=1}^{n} N_i \Phi_i I_i \quad [\text{J}]} \tag{9.72}$$

where Φ_i is the total flux, that is, the flux in circuit i due to all circuits in the system.

From the definition of inductance, we can also write, in general [see, for example, **Eq. (9.56)**]:

$$LI = N\Phi \tag{9.73}$$

Thus, we get

$$N_1\Phi_{11} = L_{11}I_1, \quad N_1\Phi_{21} = L_{21}I_2, \quad N_2\Phi_{22} = L_{22}I_2, \quad N_2\Phi_{12} = L_{12}I_1 \tag{9.74}$$

Substituting these in **Eq. (9.71)** gives

$$W_m = \frac{1}{2}L_{11}I_1^2 \pm \frac{1}{2}L_{21}I_1I_2 + \frac{1}{2}L_{22}I_2^2 \pm \frac{1}{2}L_{12}I_2I_1 \quad [\text{J}] \tag{9.75}$$

Collecting terms, using the relation $L_{12} = L_{21}$, and because both mutual fluxes are positive, gives

$$W_m = \frac{1}{2}L_{11}I_1^2 + \frac{1}{2}L_{22}I_2^2 \pm L_{12}I_1I_2 \quad [\text{J}] \tag{9.76}$$

We can again generalize this relation for any number of inductors, each with its own inductance and mutual inductances between each two inductors by properly defining the sign of the energy term due to mutual inductance. To do so, we define mutual inductance L_{ij} to be positive if the current in L_i produces a flux which adds to the flux in the loop L_j with which it couples. Otherwise, it is negative. With this assumption, the energy in a system of n inductors is

$$\boxed{W_m = \frac{1}{2}\sum_{i=1}^{n}\sum_{j=1}^{n}\left(L_{ij}I_iI_j\right) \quad [\text{J}]} \tag{9.77}$$

Note that in this expression, if a current, say I_i produces a flux opposing the flux produced by a current I_j, any term I_iI_j, $i \neq j$, in the expression is negative, whereas terms I_iI_j, $i = j$, are always positive.

A simple consequence of **Eq. (9.76)** is a formula for the calculation of inductances in series. If L_1 and L_2 are connected in series, then $I_2 = I_1$ and we can write

$$W_m = \frac{1}{2}L_{11}I_1^2 + \frac{1}{2}L_{22}I_1^2 \pm L_{12}I_1^2 = \frac{L_{eq}I_1^2}{2} \quad [\text{J}] \quad \rightarrow \quad L_{eq} = L_{11} + L_{22} \pm 2L_{12} \quad [\text{H}] \tag{9.78}$$

This expression can be easily extended to any number of series inductances (see **Example 9.13**). The sign of the mutual inductance term is determined as above, based on the direction of fluxes produced by the currents in each inductor.

Example 9.12 Application: Superconducting Magnetic Energy Storage (SMES) One method of storing large amounts of energy is to use a superconducting coil. The total energy stored in the coil is given by **Eq. (9.62)**. To store considerable amounts of energy, the coil must have a large inductance and negligible resistance (therefore the need for superconducting coils). Proposed systems include underground coils that can be used during peak power consumption. To do so, special switches connect the coil to the grid whenever necessary.

A proposed superconducting storage ring is made as a toroidal coil. The cross-sectional radius is $d = 1$ m and the average radius of the torus is $r = 1$ km. The torus is wound with $N = 150{,}000$ turns and can carry a current $I = 100{,}000$ A. The geometry is shown in **Figure 9.29**.

(a) Calculate the magnetic flux density in the torus.
(b) What is the total amount of energy stored in this torus?
(c) A city requires 100 MW of power. How long can a storage ring of this type power the city in case of a blackout in power generation? Assume there are no losses in the conversion of energy from stored DC energy to AC energy required by standard grids.

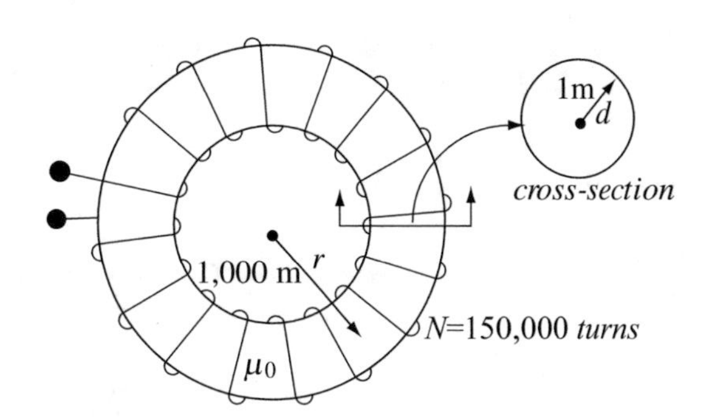

Figure 9.29 A large superconducting ring for energy storage

Solution: The flux density in a torus has been calculated in **Example 9.7** by assuming it is uniform in the torus. This assumption is based on the fact that the radius of the ring (1,000 m) is large compared to the radius of the cross section (1 m). The inductance of the torus is calculated from the flux density by multiplying the flux density by the cross-sectional area S and the number of turns to obtain the flux linkage and then dividing by the current in the turns of the torus.

(a) For the dimensions, properties, and current given here, the flux density in the torus is (see **Example 9.7**):

$$B = \frac{\mu_0 NI}{2\pi r} = \frac{4 \times \pi \times 10^{-7} \times 150{,}000 \times 100{,}000}{2 \times \pi \times 1{,}000} = 3 \quad [\mathrm{T}].$$

(b) To calculate the total energy stored in the inductor, we calculate first the inductance and then use **Eq. (9.62)** or **Eq. (9.77)** with $n = 1$. The total flux in the torus is

$$\Phi = BS = \frac{\pi d^2 \mu_0 NI}{2\pi r} = \frac{d^2 \mu_0 NI}{2r} \quad [\mathrm{Wb}]$$

The total flux linkage is $N\Phi$:

$$\Lambda = N\Phi = \frac{d^2 \mu_0 N^2 I}{2r} \quad [\mathrm{Wb \cdot t}]$$

The self-inductance of the storage ring is

$$L = \frac{\Lambda}{I} \quad \to \quad L = \frac{d^2 \mu_0 N^2}{2r} \quad [\mathrm{H}]$$

The stored energy in the ring is

$$W = \frac{LI^2}{2} = \frac{d^2 \mu_0 N^2 I^2}{4r} = \frac{1^2 \times 4 \times \pi \times 10^{-7} \times 2.25 \times 10^{10} \times 10^{10}}{4 \times 1{,}000} = 7.07 \times 10^{10} \quad [\mathrm{J}].$$

(c) Since energy is power integrated over time and the city requires 100 MW, the energy needs of the city may be met for

$$t = \frac{W}{P} = \frac{7.07 \times 10^{10}}{1 \times 10^8} = 707 \quad [\mathrm{s}]$$

This is approximately 11 min, 47 s.

Example 9.13 Connection of Coils in Series An inductance of 150 μH is required in an electric circuit. However, because of a shortage of this type of inductor, an engineer decides to use three inductors, each 50 μH, in series as shown in **Figure 9.30a**. The inductors, when connected, have a mutual inductance $L_{12} = L_{23} = 10$ μH and $L_{13} = 5$ μH. The three inductors are identical and identical currents produce identical flux in each inductor.

(a) Calculate the inductance of the circuit and the total stored energy for a current $I = 1$ A.
(b) What is the maximum inductance that can be obtained with the three inductors?
(c) What is the minimum inductance possible with the three inductors?
(d) Is it possible to obtain an inductance of 150 μH using the three inductors given? How?

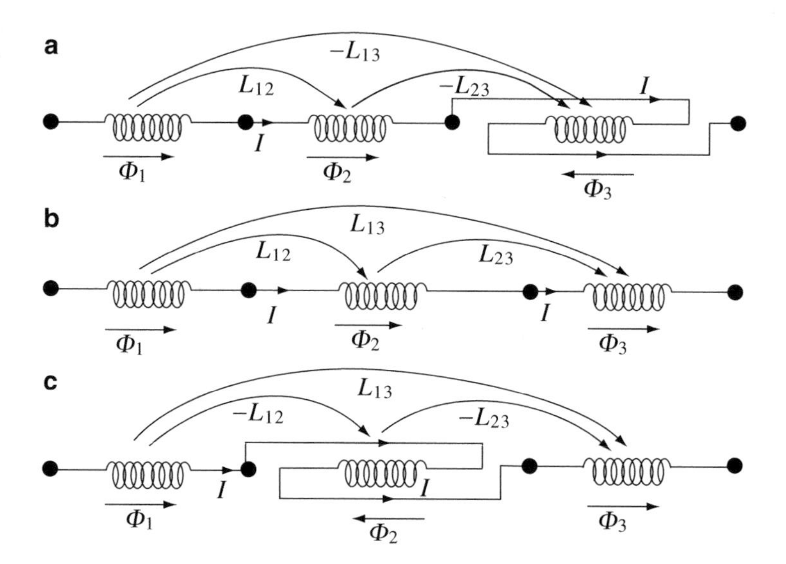

Figure 9.30 Connection of inductors in series. (**a**) Φ_3 opposes the fluxes Φ_1 and Φ_2. (**b**) All three fluxes are in the same direction. (**c**) Φ_2 opposes Φ_1 and Φ_3

Solution: (**a**) Although the inductors are identical, L_3 produces a flux which opposes the flux in coils (1) and (2). Therefore, the mutual inductance terms between coil (3) and coils (1) and (2) are negative. The mutual inductance between coils (1) and (2) is positive because their fluxes are in the same direction. **(b)** For the inductance to be maximum, all mutual inductance terms must be positive [see **Eq. (9.78)**]. Similarly, in **(c)**, as many of the mutual inductance terms as possible must be negative for minimum inductance.

(a) Because L_{13}, L_{23}, L_{31}, and L_{32} are negative, the total inductance is

$$L_{eq} = L_{11} + L_{22} + L_{33} + 2L_{12} - 2L_{23} - 2L_{13} = 150 + 20 - 20 - 10 = 140 \quad [\mu\text{H}]$$

The energy stored is

$$W_t = \frac{L_{11}I^2}{2} + \frac{L_{22}I^2}{2} + \frac{L_{33}I^2}{2} + \frac{2L_{12}I^2}{2} - \frac{2L_{23}I^2}{2} - \frac{2L_{13}I^2}{2} = \frac{L_{eq}I^2}{2} = 70 \quad [\mu\text{J}]$$

The total energy stored in the three 50 μH inductors is 70 μJ.

(b) Maximum inductance is obtained by flipping the connections on L_3 (**Figure 9.30b**). The result is

$$L_{max} = L_{11} + L_{22} + L_{33} + 2L_{12} + 2L_{23} + 2L_{13} = 200 \quad [\mu\text{H}]$$

(c) To obtain minimum inductance, the connections must be such that as many of the largest mutual inductance terms are negative. We cannot make them all negative because when we flip another coil, those terms connected to this coil that were initially negative become positive. To minimize inductance, we flip the coil (or coils) that produces the largest negative mutual inductances. In this case, because L_{12} and L_{23} are larger than L_{13}, flipping coil (2) will make L_{12} and L_{23} negative, minimizing the total inductance. Choosing to reverse coil (2) (see **Figure 9.30c**), we get

$$L_{min} = L_{11} + L_{22} + L_{33} - 2L_{12} - 2L_{23} + 2L_{13} = 120 \quad [\mu\text{H}].$$

(d) The only way three 50 μH coils can produce a series inductance of 150 μH is if their mutual inductances are zero. This can be done if the fluxes of the three coils are noninteracting; that is, if the flux in one coil does not couple into any other coil. This means the coils are shielded or their fluxes are enclosed. For example, if each coil is made in the form of a torus, the flux of each coil is contained in the torus and there will be no flux linkage between the coils.

9.5.1 Magnetostatic Energy in Terms of Fields

In the previous section, we found the energy in a system of inductors to be a simple summation process. This, while useful, has the distinct disadvantage that it only allows the calculation of energy in a system of inductors. What do we do if, for example, we need to calculate the energy from known field quantities such as **B** or **H**? A good example is the calculation of energy in the field of a permanent magnet. In this case, the inductance is not useful and, yet, the energy stored in the magnetic field must often be calculated. Thus, it is important that we generalize the expressions above further, to include all physical situations, not only discrete inductors.

This generalization starts with **Eq. (9.72)**. However, now we argue as follows: If any system of currents is given, these can always be divided into any number of loops or current segments. For example, a solenoid can be viewed as an infinite number of loops, a straight, solid conductor can be viewed as an infinite number of filamentary currents, and a general conductor can be viewed as a collection of current segments of some defined shape. Thus, we may view the above sum as an infinite sum over single loops or current segments (i.e., $N_i = 1$). With this assumption, Φ_i is the flux that links with the elementary segment or loop. Since the sum is infinite, we will replace the sum by an integral, but before we do so, recall that the flux can be calculated by integrating the flux density over the surface through which the flux flows.

Now, consider **Figure 9.31**. A thick, closed conductor of arbitrary shape carries a current I. The cross section of the conductor is shown separately and is denoted by s' to indicate that the current in the conductor flows through this area. We wish to calculate the magnetic energy stored in the field produced by this conductor. The method is as follows: Take an elementary conductor with cross-sectional area $\Delta s'$. Now, we can calculate the flux passing through the loop formed by this elementary closed conductor as

$$\Phi_i = \int_{s_i} \mathbf{B} \cdot \hat{\mathbf{n}}\, ds_i \quad [\text{Wb}] \tag{9.79}$$

where the surface enclosed by the elementary closed conductor is shaded in **Figure 9.31** and we assume that the flux density **B** can be evaluated using the Biot–Savart law. Note that the area s_i is the area of the loop, not that of the cross section of the conductor! For each elementary closed conductor of this type, we can write a similar equation. The current in the elementary conductor shown is ΔI_i

$$\Delta I_i = \frac{I}{s'} \Delta s_i' \quad [\text{A}] \tag{9.80}$$

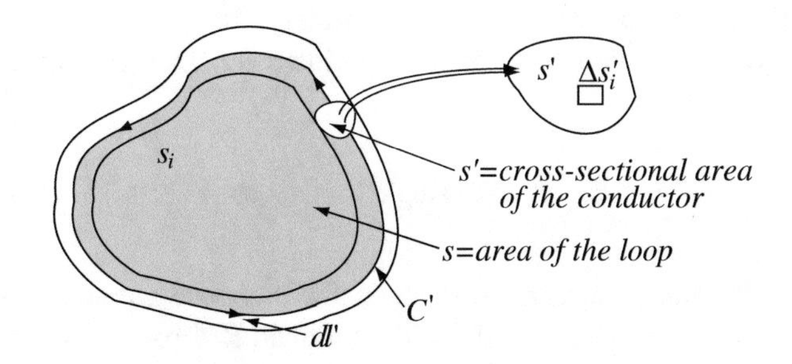

Figure 9.31 A general closed contour in a conductor of cross-sectional area s' used to calculate energy stored in the magnetic field

Now, we can write the total energy stored in the magnetic field using **Eq. (9.72)** as

$$W_m = \frac{1}{2} \sum_{i=1}^{n} \frac{I}{s'} s_i' \int_{s_i} \mathbf{B} \cdot \hat{\mathbf{n}} ds_i \quad [\text{J}] \tag{9.81}$$

where I/s' is the current density J in the conductor. The magnetic flux density can be written in terms of the magnetic vector potential $\mathbf{A}$ as $\mathbf{B} = \nabla \times \mathbf{A}$:

$$W_m = \frac{1}{2} \sum_{i=1}^{n} J\Delta s_i' \int_{s_i} (\nabla \times \mathbf{A}) \cdot \hat{\mathbf{n}} ds_i \quad [\mathrm{J}] \tag{9.82}$$

Using Stokes' theorem to convert the surface integral to a closed contour integral around the surface, substituting this in **Eq. (9.82)**, and noting that **J** is in the direction of $d\mathbf{l}'$ (as defined in **Chapter 8**), we can write for the energy

$$W_m = \frac{1}{2}\sum_{i=1}^{n} \left[J\Delta s_i' \oint_{C'} \mathbf{A} \cdot d\mathbf{l}' \right] = \frac{1}{2}\sum_{i=1}^{n} \left[\Delta s_i' \oint_{C'} \mathbf{A} \cdot J d\mathbf{l}' \right] \quad [\mathrm{J}] \tag{9.83}$$

As we allow $\Delta s_i'$ to tend to zero, the sum becomes an integral:

$$\boxed{W_m = \frac{1}{2}\int_{s'} \left[\oint_{C'} \mathbf{A} \cdot \mathbf{J} dl' \right] ds' = \frac{1}{2}\int_{v'} \mathbf{A} \cdot \mathbf{J} dv' \quad [\mathrm{J}]} \tag{9.84}$$

where s' is the cross-sectional area of the conductor and C' is the circumference of the loop. Thus, the above integration is over the volume of the conductor, v'. This must be so because outside the conductor **J** is zero and the contribution to this integral is zero. Note, also, that we started with the flux Φ, which exists both outside and inside the conductor itself. However, the use of Stokes' theorem allowed us to convert this surface integral to a contour integral and this contour is entirely within the volume of the conductor.

This relation is important because it allows the calculation of energy from the source: from the current density and the resulting magnetic vector potential. Also, recalling the Biot–Savart law, the magnetic vector potential is directly proportional to the current I (or current density J) and, therefore, energy is related to I^2 or J^2. We know this to be true from Joule's law in **Section 7.5**. Also, this relation shows that energy is volume related.

The energy in **Eq. (9.84)** seems to be stored in the conducting volume itself since it is obtained by integration over the volume of the conductor. However, we may argue that energy can also be associated with the field **B** or **H**, regardless of location of the current that produces the fields. The same can be said about the field of a permanent magnet. Otherwise, a magnet will have no energy associated with it (no current). To show that this is the case, we can proceed in two directions. From Ampère's law we can replace **J** in **Eq. (9.84)** by $\nabla \times \mathbf{H}$ and obtain a relation for **B**. It is, however, easier to start with the general energy relation in **Eq. (9.72)** and the definition of flux in **Eq. (9.79)**.

Consider a simple current-carrying conductor such as the wire shown in **Figure 9.32**. The conductor can be thin or thick. We wish to calculate the energy associated with an element of volume due to a conductor. **Equation (9.72)** is now

$$W_m = \frac{1}{2}\sum_{i=1}^{1} \Phi_i I_i = \frac{1}{2}\Phi_1 I_1 \quad [\mathrm{J}] \tag{9.85}$$

where we assumed $N = 1$. This does not diminish from the generality of the derivation. If there are multiple currents, the energy due to each current can be calculated separately. The flux generated by the current can now be calculated by first using Ampère's law to calculate the magnetic field intensity: The surface s over which the flux is evaluated is arbitrary and can be as small as necessary:

$$I = \oint_C \mathbf{H} \cdot d\mathbf{l} \quad \text{for single loops} \quad \text{or :} \quad NI = \oint_C \mathbf{H} \cdot d\mathbf{l} \quad \text{for multiple loops} \tag{9.86}$$

The contour C is again arbitrary. However, we will choose a convenient contour such that **H** and $d\mathbf{l}$ are in the same direction to simplify the evaluation of the scalar product. Referring to **Figure 9.32**, we created a volume, in the form of a general tube, with cross section s in which the flux density is constant so that the flux Φ is independent of the location in the cross section along the tube. With these considerations in mind, we can write from **Eq. (9.85)**:

Figure 9.32 An arbitrary volume in the form of a general ring around the current I

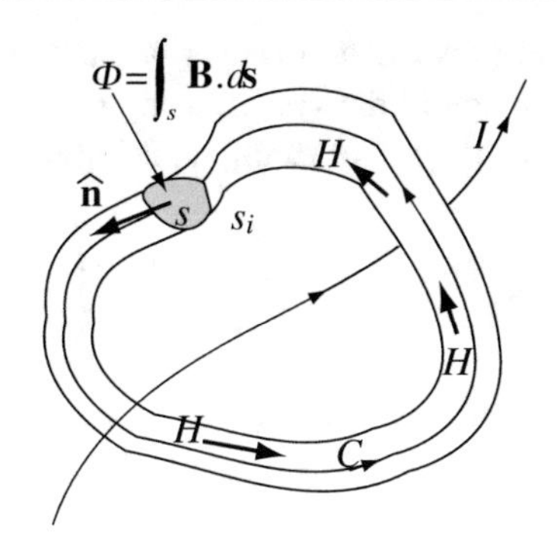

$$W_m = \frac{1}{2}\int_s \mathbf{B}\cdot\hat{\mathbf{n}}ds \oint_C \mathbf{H}\cdot d\mathbf{l} = \frac{1}{2}\oint_C \left(\int_S \mathbf{B}\cdot\hat{\mathbf{n}}ds\right)\mathbf{H}\cdot d\mathbf{l} \quad [\mathrm{J}] \tag{9.87}$$

Note that the surface integral representing flux was inserted into the line integral because the flux is assumed to be constant along the contour C. Since $\mathbf{B}\cdot\hat{\mathbf{n}}$ is the scalar component of $\mathbf{B}$ in the direction normal to s (see **Figure 9.32**), we can write

$$W_m = \frac{1}{2}\oint_C \left(\int_s \mathbf{B}\cdot\hat{\mathbf{n}}ds\right)\mathbf{H}\cdot d\mathbf{l} = \frac{1}{2}\oint_C \left(\int_s \mathbf{B}\cdot\mathbf{H}ds\right)dl = \frac{1}{2}\int_v \mathbf{B}\cdot\mathbf{H}dv \quad [\mathrm{J}] \tag{9.88}$$

This magnetic energy represents the total energy enclosed in the (arbitrary) volume of the tube in **Figure 9.32**. This tube is entirely outside any current or current density distribution and, therefore, its energy represents energy in space (or any other material that may be present in the volume). Because of this, the integrand of **Eq. (9.88)** may be viewed as a volume energy density in space. Using the relation $\mathbf{B} = \mu\mathbf{H}$, we can rewrite this relation in a number of useful forms:

$$\boxed{W_m = \frac{1}{2}\int_v \mathbf{B}\cdot\mathbf{H}dv = \frac{1}{2}\int_v \mu\mathbf{H}\cdot\mathbf{H}dv = \frac{1}{2}\int_v \mathbf{B}\cdot\frac{\mathbf{B}}{\mu}dv \quad [\mathrm{J}]} \tag{9.89}$$

Any of these relations is appropriate to use depending on which is more convenient. Note, also, that the integration may be done over any volume, including the whole of space. The result is the energy stored in the chosen volume. If the energy is over the whole space, it represents the total energy in the system. If permeability μ is a constant, then $\mathbf{B}$ and $\mathbf{H}$ are in the same direction and we can write the scalar products as $\mathbf{B}\cdot\mathbf{H} = BH$, $\mathbf{H}\cdot\mathbf{H} = H^2$, and $\mathbf{B}\cdot\mathbf{B} = B^2$. Viewing the integrand in **Eq. (9.89)** as an energy density, w_m, we can write

$$\boxed{w_m = \frac{\mathbf{B}\cdot\mathbf{H}}{2} = \frac{BH}{2} = \frac{\mu H^2}{2} = \frac{B^2}{2\mu} \quad \left[\frac{\mathrm{J}}{\mathrm{m}^3}\right]} \tag{9.90}$$

Similarly, the integrand in **Eq. (9.84)** is an energy density in terms of the magnetic vector potential $\mathbf{A}$ and current density $\mathbf{J}$ and we can write

$$\boxed{w_m = \frac{\mathbf{A}\cdot\mathbf{J}}{2} \quad \left[\frac{\mathrm{J}}{\mathrm{m}^3}\right]} \tag{9.91}$$

In practical use it is common to first calculate the magnetic energy density and then integrate over the volume of a device, or over some part of space to calculate the total magnetic energy stored in a given volume.

We conclude this section by returning to the initial result in **Eq. (9.62)**. If the energy stored in the inductor is known, it can be used to calculate the inductance as

$$W_m = \frac{1}{2}LI^2 \quad \rightarrow \quad L = \frac{2W_m}{I^2} \quad [\mathrm{H}] \tag{9.92}$$

Since it is sometimes easier to calculate the energy due to a current-carrying conductor or a system of conductors, it is often easier to calculate the energy in the system using any of the above equations [such as **Eq. (9.89)**] and calculate the inductance of the system from energy. For example, using the general result for energy 1

$$L = \frac{1}{I^2}\int_v \mathbf{B}\cdot\mathbf{H}dv \quad [\mathrm{H}] \tag{9.93}$$

However, we should be careful since **Eq. (9.93)** gives the total inductance and it is not always easy to differentiate between self- and mutual inductances. This relation should only be used if only a self-inductance exists or if the total inductance of the system is needed.

Example 9.14 Stored Magnetic Energy An infinitely long solenoid with radius a [m] and n turns per unit length is given. The turns carry a current I [A]. A long piece of iron, of radius b [m], is placed in the solenoid, as shown in **Figure 9.33**. The relative permeability of iron is μ_r and that of free space is 1. Calculate the total work per unit length of the solenoid necessary to pull the iron completely out of the solenoid. Assume iron does not saturate and the **B**(**H**) curve of iron is linear. Does this work increase or decrease the potential energy of the system? Explain.

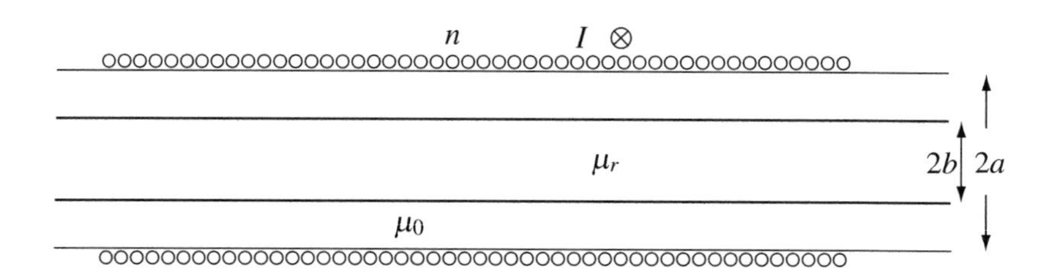

Figure 9.33 Long solenoid with an iron core

Solution: The energy per unit length of the solenoid with the iron and, separately, with air is calculated. The difference between the two energies is the work per unit length necessary to remove the iron.

Applying Ampère's law inside the iron:

$$B = \mu_0\mu_r nI \quad [\mathrm{T}]$$

In air, outside the iron:

$$B = \mu_0 nI \quad [\mathrm{T}]$$

The energy per unit length with the iron is

$$W_m = \int_v \frac{B^2}{2\mu}dv = \frac{B^2\pi b^2}{2\mu} + \frac{B^2\pi(a^2-b^2)}{2\mu_0} = \frac{\mu n^2I^2\pi b^2}{2} + \frac{\mu_0 n^2I^2\pi(a^2-b^2)}{2} \quad \left[\frac{\mathrm{J}}{\mathrm{m}}\right]$$

After removing the iron, the flux density is that in free space and the energy in the solenoid is

$$W_0 = \int_v \frac{B^2}{2\mu_0}dv = \frac{B^2\pi a^2}{2\mu_0} = \frac{\mu_0 n^2I^2\pi a^2}{2} \quad \left[\frac{\mathrm{J}}{\mathrm{m}}\right]$$

The work per unit length required to remove the iron is the difference between the final energy (after removing the iron) and initial energy (with iron in):

$$W = W_0 - W_m = \frac{\mu_0 n^2I^2\pi a^2}{2} - \frac{\mu_0 n^2I^2\pi b^2}{2} - \frac{\mu_0 n^2I^2\pi(a^2-b^2)}{2} = -\frac{n^2I^2\pi b^2(\mu-\mu_0)}{2} \quad \left[\frac{\mathrm{J}}{\mathrm{m}}\right]$$

This work is negative ($\mu > \mu_0$); that is, removing the iron decreases the potential energy in the system.

Example 9.15 Total Inductance Two solenoids are placed one inside the other (see **Figure 9.34**). Each carries a current I [A] as shown and has n turns per unit length. Calculate the total inductance per unit length of the system using the energy method.

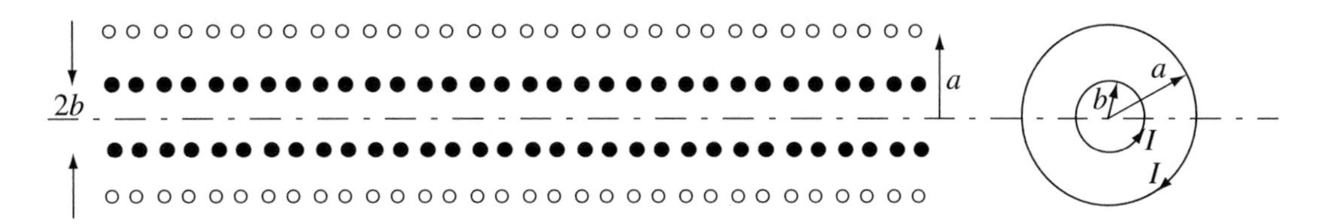

Figure 9.34 Two solenoids, one inside the other, shown in axial cut and in cross section

Solution: We first calculate the flux density everywhere in space due to each solenoid separately. Inside the smaller solenoid, the flux density is zero since the flux of the two solenoids is equal and opposite to each other. In the area between the smaller and larger solenoids, the flux density due to the outer solenoid is nonzero whereas outside the outer solenoid the flux density is again zero.

The flux density between the two solenoids ($b < r < a$) is

$$B = \mu_0 n I \qquad [\mathrm{T}]$$

Because this is constant throughout the area, the stored magnetic energy per unit length of the solenoids is

$$W = \frac{B^2}{2\mu_0}\pi\left(a^2 - b^2\right) = \frac{\mu_0 n^2 I^2 \pi\left(a^2 - b^2\right)}{2} \quad \left[\frac{\mathrm{J}}{\mathrm{m}}\right]$$

This must also be equal to:

$$W = L\frac{I^2}{2} \quad \left[\frac{\mathrm{J}}{\mathrm{m}}\right] \quad \rightarrow \quad L = \frac{2W}{I^2} \quad \left[\frac{\mathrm{H}}{\mathrm{m}}\right]$$

The total inductance per unit length is

$$L_t = \mu_0 n^2 \pi\left(a^2 - b^2\right) \quad [\mathrm{H/m}]$$

The same result may be obtained by evaluating L_{11}, L_{22}, L_{12}, and L_{21}, but the energy calculation is simpler (see **Exercise 9.4**).

Exercise 9.4 In **Example 9.15**, evaluate the self-inductances per unit length (L_{11}, L_{22}) of the two solenoids and the mutual inductances between them (L_{12}, L_{21}) and show that the total inductance is the same as that found in **Example 9.15**.

Answer

$$L_{11} = \mu_0 n^2 \pi b^2, \quad L_{22} = \mu_0 n^2 \pi a^2, \quad L_{12} = L_{21} = \mu_0 n^2 \pi b^2 \quad [\mathrm{H/m}]$$

$$L_t = L_{11} + L_{22} - L_{12} - L_{21} = \mu_0 n^2 \pi\left(a^2 - b^2\right) \quad [\mathrm{H/m}]$$

9.6 Magnetic Circuits **Magnetostatics.m**

A useful and relatively simple tool in design of magnetic devices is the idea of the magnetic circuit. It is based on an analogy between the fundamental equations and properties of the static magnetic field and the static electric field. This analogy is based on the following relations:

Electric field Magnetic field

$$\mathbf{E} = \frac{1}{\sigma}\mathbf{J} \quad \left[\frac{\mathrm{V}}{\mathrm{m}}\right] \qquad \mathbf{H} = \frac{1}{\mu}\mathbf{B} \quad \left[\frac{\mathrm{A}}{\mathrm{m}}\right] \tag{9.94}$$

$$V = \oint_C \mathbf{E}\cdot d\mathbf{l} \quad [\mathrm{V}] \qquad NI = \oint_C \mathbf{H}\cdot d\mathbf{l} \quad [\mathrm{A}\cdot\mathrm{t}] \tag{9.95}$$

$$I = \int_s \mathbf{J}\cdot d\mathbf{s} \quad [\mathrm{A}] \qquad \Phi = \int_s \mathbf{B}\cdot d\mathbf{s} \quad [\mathrm{Wb}] \tag{9.96}$$

The "current" in the magnetic circuit is represented by the magnetic flux, whereas "voltage" is represented by the term NI. The latter, defined in Ampère's law [for example, **Eq. (9.86)**], is often called the ***magnetomotive force***. The voltage V is an ***electromotive force*** supplied by a source such as a battery. We must note, however, a number of points:

(1) The magnetic circuit is only an analogy: flux is not a "magnetic current" and the magnetomotive force is not a "magnetic voltage."
(2) This analogy between circuits only applies if the conditions for a circuit are satisfied. In particular, an electric circuit requires that current flows in closed circuits in conductors. Similarly, the flux must be contained in closed "magnetic paths." These concepts must be carefully defined and understood before we can properly use the analogy between electric and magnetic circuits.
(3) The magnetomotive force in a magnetic circuit is supplied either by a coil with N turns and current I or by an equivalent permanent magnet, as shown in **Figure 9.35**.

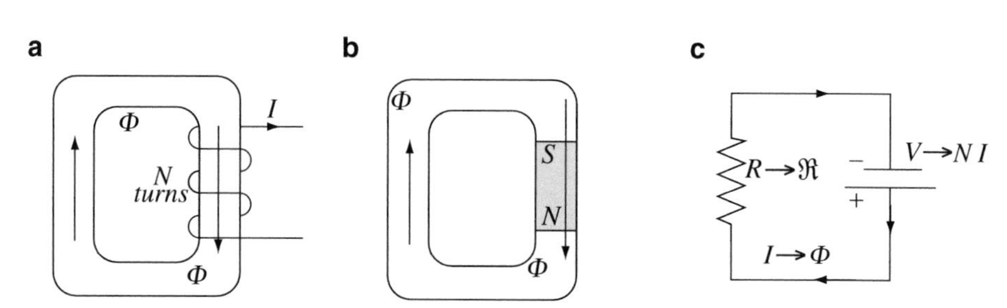

Figure 9.35 (**a**) Magnetomotive force in a magnetic circuit generated by a coil with N turns and current I. (**b**) A magnetomotive force generated by a permanent magnet. (**c**) The equivalent circuit for (**a**) or (**b**)

In the equivalent circuit in **Figure 9.35c**, the meaning of the equivalent resistance $\Re$ has not yet been defined. To do so, we use the simple magnetic circuit in **Figure 9.36**. The torus is used here because it satisfies the basic condition of a magnetic circuit; namely, all flux is contained within the magnetic circuit (magnetic core). The magnetic field intensity for the torus was calculated in **Example 9.7**. Its magnitude is

$$H = \frac{NI}{2\pi r_a} \quad \left[\frac{\mathrm{A}}{\mathrm{m}}\right] \tag{9.97}$$

where r_a is the average radius of the torus. The field intensity is uniform in the cross-sectional area of the torus (approximately) and, therefore, the flux is given by

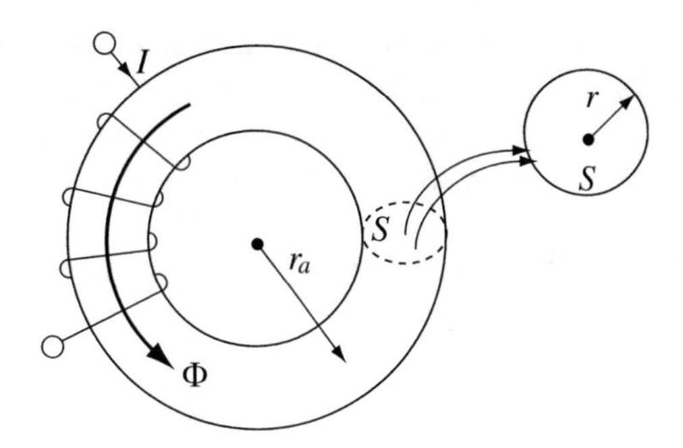

Figure 9.36 A closed magnetic circuit used to define reluctance

$$\Phi = \int_s \mathbf{B} \cdot d\mathbf{s} = \int_s \mu \mathbf{H} \cdot d\mathbf{s} = \mu HS = \mu \frac{NI}{2\pi r} S \quad [\text{Wb}] \tag{9.98}$$

where μ is the permeability of the core material and S is the cross-sectional area of the magnetic circuit. The magnetomotive force (NI) and the flux (Φ) are known. From the equivalent circuit in **Figure 9.35c**, the equivalent "resistance" in the circuit is

$$\Re = \frac{NI}{\Phi} = \frac{NI}{\mu \frac{NI}{2\pi r_a} S} = \frac{2\pi r_a}{\mu S} = \frac{l}{\mu S} \quad \left[\frac{1}{\text{H}}\right] \tag{9.99}$$

where $l = 2\pi r_a$ is the average length of the magnetic path. $\Re$ is called ***magnetic reluctance*** (often shortened to ***reluctance***) and is analogous to resistance in an electric circuit. Note that in this sense, μ is viewed as a "magnetic conductivity" and $1/\mu$ as a "magnetic resistivity." The term $1/\mu$ is also called ***reluctivity*** of the magnetic material in analogy to the *resistivity* of conductors.

The calculation of reluctance of a circuit is straightforward: It is equal to the length of the magnetic circuit divided by the cross-sectional area and magnetic permeability. It depends on the physical size of the device and on its permeability.

The use of a torus in the above derivation was arbitrary: Any closed magnetic circuit would do. In fact, we may consider a magnetic circuit made of a number of materials with perhaps many branches. As long as all flux is contained within the magnetic circuit, the device can be analyzed using this method. In particular, consider the magnetic circuit in **Figure 9.37,** which is made of two sections, one with permeability μ_1 and the other μ_2.

The reluctance of each section is calculated from **Eq. (9.99)**. In this case, there are two reluctances, connected in series:

$$\Re_1 = \frac{l_1}{\mu_1 S}, \quad \Re_2 = \frac{l_2}{\mu_2 S} \quad \left[\frac{1}{\text{H}}\right] \tag{9.100}$$

The flux in the magnetic circuit is

$$\Phi = \frac{NI}{\Re} = \frac{NI}{\Re_1 + \Re_2} = \frac{NI}{l_1/\mu_1 S + l_2/\mu_2 S} \quad [\text{Wb}] \tag{9.101}$$

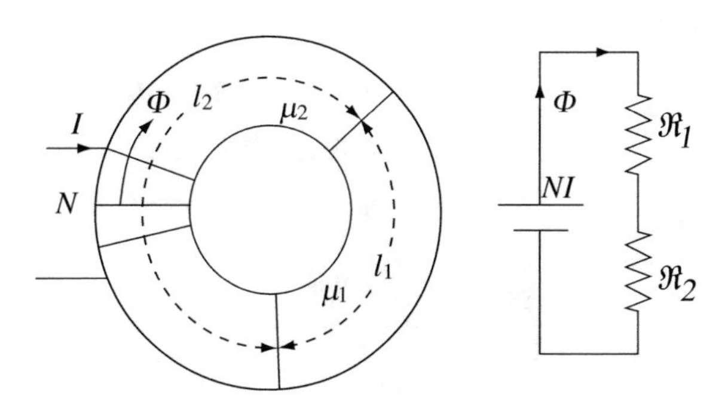

Figure 9.37 Magnetic circuit made of two magnetic materials and equivalent circuit

Note also that any material can be included in the circuit, as long as the conditions of the circuit are satisfied. If, however, a very high reluctance material is included (low μ), such as free space, the length of this material must be kept to a minimum; otherwise the flux will spread out and the magnetic circuit is not a true circuit any more. In other words, when large air gaps are included in the magnetic circuit, the analysis of the magnetic circuit using this method may not be valid or a large error may be introduced.

The expression in **Eq. (9.101)** may be generalized for any number of magnetomotive forces and reluctances in a closed circuit as

$$\boxed{\Phi = \frac{\sum_{i=1}^{n} N_i I_i}{\sum_{j=1}^{k} \Re_j} \quad [\text{Wb}]} \tag{9.102}$$

Example 9.16 Magnetic Circuit with a Gap The magnetic circuit in **Figure 9.38** is given. Calculate the magnetic field intensity H in the gap.

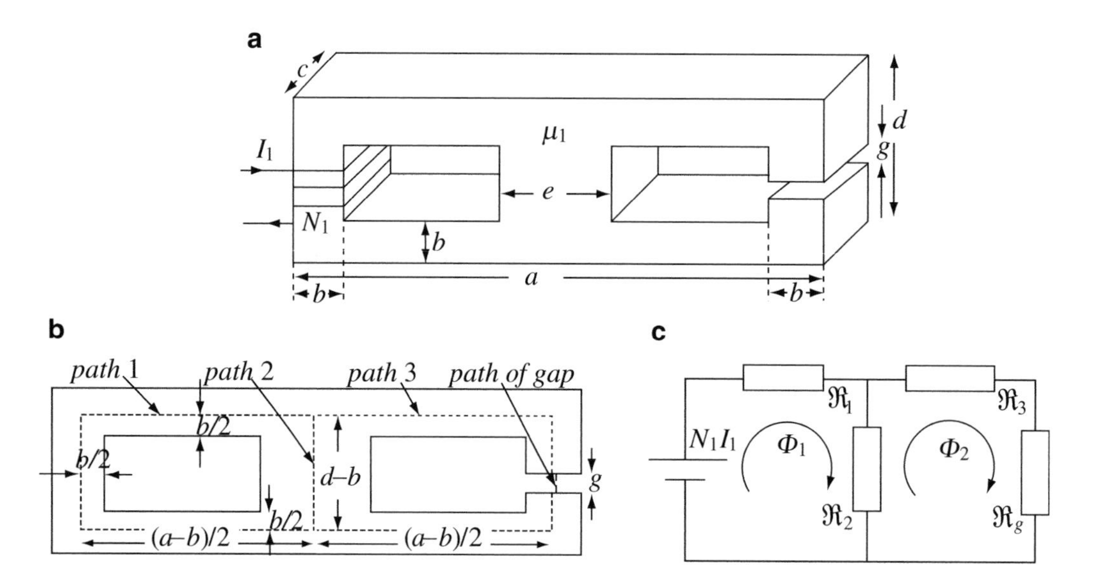

Figure 9.38 A magnetic circuit. (**a**) Dimensions. (**b**) Definition of average magnetic paths. (**c**) Equivalent circuit

Solution: The equivalent circuit with a source equal to $N_1 I_1$ and reluctances in each path is shown in **Figure 9.38c**. The three reluctances are calculated using the average paths shown in **Figure 9.38b**. Note that the cross-sectional area everywhere is bc except in the central leg, where it is ec. Since the flux in the gap is needed, it is best to use two loops as shown in **Figure 9.38c**, but other methods can be used.

The two loop equations are

$$N_1 I_1 = \Phi_1(\Re_1 + \Re_2) - \Phi_2 \Re_2$$

$$\Phi_2(\Re_2 + \Re_3 + \Re_g) - \Phi_1 \Re_2 = 0$$

The fluxes are calculated as

$$\Phi_1 = \frac{N_1 I_1 (\Re_2 + \Re_3 + \Re_g)}{(\Re_2 + \Re_3 + \Re_g)(\Re_1 + \Re_2) - \Re_2^2}, \quad \Phi_2 = \frac{N_1 I_1 \Re_2}{(\Re_2 + \Re_3 + \Re_g)(\Re_1 + \Re_2) - \Re_2^2} \quad [\text{Wb}]$$

For the leg in which the gap is located, the flux density is

$$B_2 = B_g = \frac{\Phi_2}{bc} = \frac{N_1 I_1 \mathfrak{R}_2}{bc\left\{(\mathfrak{R}_2 + \mathfrak{R}_3 + \mathfrak{R}_g)(\mathfrak{R}_1 + \mathfrak{R}_2) - \mathfrak{R}_2^2\right\}} \quad [\mathrm{T}]$$

and the magnetic field intensity in the gap is

$$H_g = \frac{B_g}{\mu_0} = \frac{N_1 I_1 \mathfrak{R}_2}{\mu_0 bc\left\{(\mathfrak{R}_2 + \mathfrak{R}_3 + \mathfrak{R}_g)(\mathfrak{R}_1 + \mathfrak{R}_2) - \mathfrak{R}_2^2\right\}} \quad \left[\frac{\mathrm{A}}{\mathrm{m}}\right]$$

The reluctances needed are calculated using the average path lengths in **Figure 9.38b**, as follows:

$$\mathfrak{R}_1 = \frac{a + d - 2b}{\mu_1 bc}, \quad \mathfrak{R}_2 = \frac{d - b}{\mu_1 ec}, \quad \mathfrak{R}_3 = \frac{a + d - 2b - g}{\mu_1 bc}, \quad \mathfrak{R}_g = \frac{g}{\mu_0 bc} \quad \left[\frac{1}{\mathrm{H}}\right]$$

Thus, the magnetic field intensity in the gap is

$$H_g = \frac{N_1 I_1 \dfrac{d - b}{\mu_1 ec}}{\mu_0 bc\left\{\left(\dfrac{d - b}{\mu_1 ec} + \dfrac{a + d - 2b - g}{\mu_1 bc} + \dfrac{g}{\mu_0 bc}\right)\left(\dfrac{a + d - 2b}{\mu_1 bc} + \dfrac{d - b}{\mu_1 ec}\right) - \left(\dfrac{d - b}{\mu_1 ec}\right)^2\right\}} \quad \left[\frac{\mathrm{A}}{\mathrm{m}}\right]$$

Writing $\mu_1 = \mu_0 \mu_{r1}$ and simplifying the expression gives

$$H_g = \frac{\mu_{r1} N_1 I_1 (d - b) b}{[b(d - b) + e(a + d - 2b - g) + \mu_{r1} eg][e(a + d - 2b) + b(d - b)] - b^2 (d - b)^2} \quad \left[\frac{\mathrm{A}}{\mathrm{m}}\right]$$

Example 9.17 Application: Use of Permanent Magnets in Magnetic Circuits A cylindrical magnet of length a [m], radius b [m], and uniform magnetization **M** [A/m] is inserted in a magnetic path as shown in **Figure 9.39**. The cross section of the magnetic path is constant and equal to that of the magnet. Two small gaps, each of length d [m], are also present. Other dimensions and properties are given in the figure. Assume permeabilities of the path are infinite, except in gaps and magnet, where the permeability is μ_0 [H/m]. Calculate the magnetic field intensity H in each of the gaps.

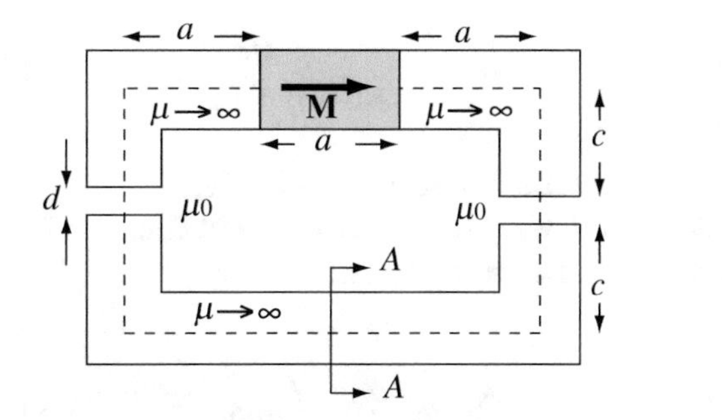

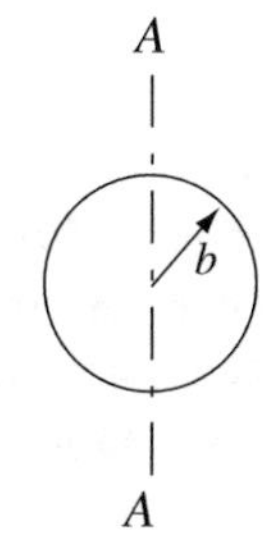

Figure 9.39 Use of a permanent magnet to generate the flux in a magnetic circuit

Solution: To calculate the flux, it is first necessary to find the equivalent magnetomotive force that produces the magnetization in the magnet as an equivalent solenoid problem. The flux in the magnetic circuit is calculated assuming there is no flux leakage in the gaps. The equivalent circuit includes the two gap reluctances as well as the reluctance of the magnet itself. The flux and flux density in the gaps are now found from the equivalent circuit.

For a uniformly magnetized magnet, the equivalent current density (current per unit length of the magnet) is [see **Eq. (9.27)**]:

$$\mathbf{M} \times \hat{\mathbf{n}} = \mathbf{J} = \hat{\boldsymbol{\phi}} M \quad [\text{A/m}]$$

This current density is on the surface of the magnet, is directed circumferentially, and produces a flux density in the same direction as **M**. The total equivalent current producing the flux density is equal to Ma. Thus, the magnetomotive force is equal to Ma. The reluctance of the path includes the two gaps and the length of the magnet since, now, the magnet has been replaced by a solenoid, the volume occupied by the magnet having permeability of free space:

$$\mathfrak{R} = \frac{2d + a}{\mu_0 \pi b^2} \quad \left[\frac{1}{\text{H}}\right]$$

Because the reluctance in iron is zero ($\mu \to \infty$), the flux in the circuit is

$$\Phi = \frac{Ma}{\mathfrak{R}} = \frac{\mu_0 \pi b^2 Ma}{2d + a} \quad [\text{Wb}]$$

Dividing by the cross-sectional area πb^2 gives the magnetic flux density and further dividing by permeability gives the magnetic field intensity:

$$B = \frac{\Phi}{S} = \frac{\mu_0 Ma}{2d + a} \quad [\text{T}] \quad \to \quad H = \frac{B}{\mu_0} = \frac{Ma}{2d + a} \quad \left[\frac{\text{A}}{\text{m}}\right]$$

Note: We assumed that the magnet can be replaced by a solenoid and that by doing so, the permeability of the magnet equals that of free space. In practice, this approximation is good for long magnets and for magnets in a closed magnetic path, as is the case here. The relation for B shows that in effect the gap has increased by the length of the magnet (because the magnet has low permeability). Nevertheless, the flux is contained within the magnet and thus the magnetic circuit method is valid. Good magnets have a relative permeability between 1 and 3. If the actual permeability of the permanent magnet is known, it should be used in the design.

Exercise 9.5 Calculate the magnetic field intensity in the gaps of **Example 9.17** for finite permeability of iron. Assume the relative permeability of iron is μ_r.

Answer

$$H = \frac{\mu_r Ma}{\mu_r(2d + a) + (5a + 4c)} \quad \left[\frac{\text{A}}{\text{m}}\right]$$

It is useful to generalize these results for more complex circuits, including circuits with multiple loops. To do so, we write Kirchhoff's laws for magnetic circuits based on the discussion in **Chapter 7**. By analogy between currents and fluxes and between voltages and magnetomotive forces, we can write directly:

Kirchhoff's current law

$$\sum_i \Phi_i = 0 \quad \left(\text{from} \quad \sum_i I_i = 0\right) \tag{9.103}$$

Kirchhoff's voltage law

$$\sum_i N_i I_i = \sum_j \mathfrak{R}_j \Phi_j \quad \left(\text{from} \sum_i V_i = \sum_j R_j I_j\right) \tag{9.104}$$

In addition, the connection of reluctances in series and parallel follows the same rules as the connection of resistances in an electric circuit.

In summary, the solution to a magnetic problem using magnetic circuits follows the following steps:

(1) Determine if the flux generated in the circuit is contained within the magnetic circuit. Look in particular if permeabilities of materials in the magnetic paths are high, since flux tends to follow high-permeability paths (i.e., low-reluctivity paths). Large air gaps and low-permeability materials will tend to allow flux to "leak" out of the circuit, invalidating the assumptions of a circuit.
(2) Determine the average path lengths for each material or section of material and find the reluctances of each material using **Eq. (9.99)**.
(3) Locate and calculate the magnetomotive forces in the circuit using **Eq. (9.95)** These are either the ampere-turns of coils or the equivalent ampere-turns of permanent magnets.
(4) Draw an equivalent circuit in terms of voltages (magnetomotive forces), currents (fluxes), and resistances (reluctances).
(5) Use **Eqs. (9.103)** and **(9.104)** to find the unknowns. These are usually the fluxes in various parts of the magnetic circuit. From flux, other magnetic circuit parameters can be obtained.

Once an equivalent circuit has been obtained, you can use any analysis tool you wish. For complex magnetic circuits, you may even wish to use a DC circuit analysis computer program.

9.7 Forces in the Magnetic Field

We started **Chapter 8** with a short discussion on force in the magnetic field. The initial purpose was only to show that the magnetic field is related to force in a way similar to the relation of the electric field to force. Now, it is time to revisit force, quantify it, and see what the differences between forces in the magnetic and electric field are.

The magnetic field exerts a force on a moving charge which is directly proportional to the velocity and magnitude of the charge. Measuring both the force in a magnetic field in the absence of an electric field and the force in an electric field in the absence of a magnetic field, we obtain a force relation which is both general and distinguishes between electric and magnetic forces. The relation is called the Lorentz force equation (also called the Coulomb–Lorentz equation) and is written as

$$\boxed{\mathbf{F}_{total} = \mathbf{F}_e + \mathbf{F}_m = q\mathbf{E} + q\mathbf{v} \times \mathbf{B} \quad [\mathrm{N}]} \tag{9.105}$$

where $\mathbf{F}_e$ is the force due to the electric field and $\mathbf{F}_m$ is the force due to the magnetic field. If the electric field intensity is zero, only a magnetic force is present. The Lorentz force equation for the magnetic field gives the magnetic force as

$$\mathbf{F}_m = q\mathbf{v} \times \mathbf{B} \quad [\mathrm{N}] \tag{9.106}$$

The importance of the Lorentz force equation is that it is fundamental in electromagnetics; that is, it serves the same purpose as any of the postulates we have introduced for the electric or magnetic fields. When we come to the point of unifying the theory of electromagnetics, **Eq. (9.105)** will become one of the seven basic relations needed to completely define the theory.

We must note here the following:

(1) From a dimensional point of view, the term $\mathbf{v} \times \mathbf{B}$ must have units of electric field intensity [V/m]. This means that, in fact, the force can be viewed as an electric force, proportional to the magnetic flux density. More on this connection between the electric and magnetic field will be said in **Chapter 10**.
(2) The magnetic force term can only exist if the velocity of charges is nonzero. If all charges are stationary, only an electric force exists.
(3) The magnetic force is directed perpendicular to both the direction of motion (velocity) and the direction of the magnetic flux density **B**. This is a consequence of the vector product between **v** and **B**.
(4) A velocity perpendicular to the magnetic field will cause the charge or charged body to move in a circular motion, with the plane of the circle perpendicular to the direction of the magnetic field (**Figure 9.40a**).

(5) A charge moving in the direction of the magnetic flux density experiences no force (**Figure 9.40b**) because the vector product of two parallel vectors is zero.

(6) A charge with velocity components perpendicular and in the direction of the field experiences a helical motion, with the normal component causing the circular motion and the tangential component causing the translation of the path of motion (**Figure 9.40c**).

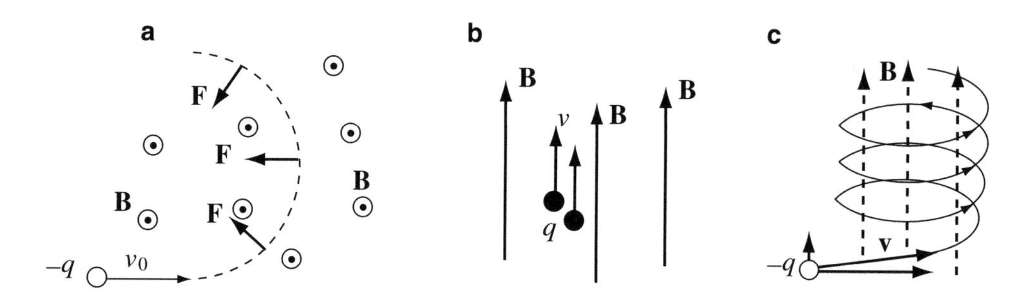

Figure 9.40 (**a**) A negative charge moving perpendicular to the magnetic field experiences a radial force causing circular motion. (**b**) Motion parallel to the field generates no force. (**c**) Motion with components perpendicular and parallel to the field causes the charge to move in a helical path

Example 9.18 Path of Electrons in a Magnetic Field An electron is injected into a uniform magnetic field at right angle to the flux density. Mass of the electron is 9×10^{-31} kg.

(a) If the velocity of the electron is constant at $v = 100{,}000$ m/s and the magnetic flux density is $\mathbf{B} = \hat{\mathbf{z}}\,0.5\,\mathrm{T}$, calculate the path of the electron. Assume velocity is perpendicular to **B**.

(b) Describe qualitatively the path of the electron if, in addition to the magnetic flux density in (**a**), there is also an electric field intensity of magnitude $E = 10^5$ V/m, in the direction opposite **B**.

Solution: The magnetic and electric forces on a charged particle are defined by the Lorentz force equation. The magnetic force is perpendicular to the magnetic field and direction of motion (velocity) and the electric force is in the direction of the electric field.

(a) The magnetic force is given as

$$\mathbf{F}_m = -q\mathbf{v} \times \mathbf{B} \quad \rightarrow \quad F = |q\mathbf{v} \times \mathbf{B}| = 1.6 \times 10^{-19} \times 10^5 \times 0.5 = 0.8 \times 10^{-14} \quad [\mathrm{N}]$$

The direction of the force is perpendicular to the plane formed by **v** and **B** (r direction) and hence is perpendicular to both the flux density **B** and velocity **v** as shown in **Figure 9.41a**. Because of this, the electron will move in a circle, counterclockwise (see also **Figure 9.40a**). The radius of the circle is defined by the centrifugal force on the electron and the latter depends on its mass. The radius is calculated by equating the magnitudes of the two forces:

$$\frac{mv^2}{r} = F_m \quad \rightarrow \quad r = \frac{mv^2}{F_m} = \frac{9.1 \times 10^{-31} \times 10^{10}}{0.8 \times 10^{-14}} = 1.138 \times 10^{-6} \quad [\mathrm{m}]$$

The radius of motion of the electron is 1.138 μm.

(b) If an electric field intensity **E** exists in the direction opposite that of the magnetic field, the electron also experiences a force due to the electric field. The magnitude of the electric force is

$$|F_e| = qE = 1.6 \times 10^{-19} \times 10^5 = 1.6 \times 10^{-14} \quad [\mathrm{N}]$$

This force is in the direction of **B** because the charge is negative. The combined effect of the magnetic and electric field causes the electron to move in an upward, counterclockwise spiral path, as shown in **Figure 9.41b**. The radius is defined by **B**, the pitch by **E**. The radius is as above (1.138 μm). The electric force is constant, causing the electron to accelerate in the z direction. The acceleration at low speeds is $a_z = F_e/m = 1.6 \times 10^{-14}/9.1 \times 10^{-31} = 1.758 \times 10^{16}$ m/s^2, but as the speed increases, the electron becomes relativistic and the rest mass cannot be used. The electron will eventually reach a speed close to the speed of light, provided the electric field can be maintained over the path of the electron.

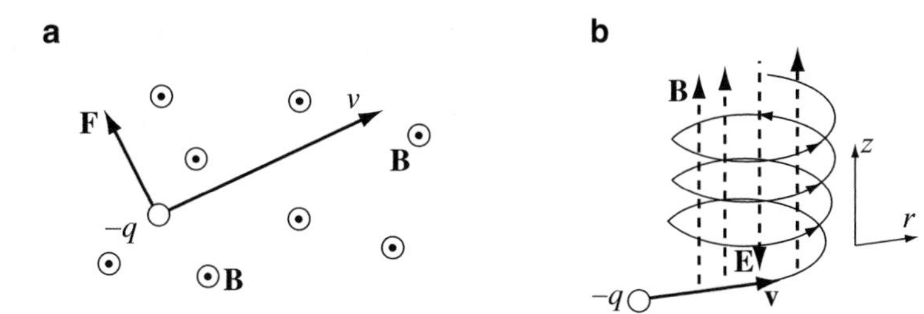

Figure 9.41 (**a**) Force on a moving electron in the presence of a magnetic field. (**b**) Path of an electron in the presence of an electric, and a magnetic field

The magnetic force given in **Eq. (9.106)** is not particularly convenient to apply since it relies on forces on individual charges. It is usually not possible to account for individual charges (say electrons) and we certainly do not know the velocities of individual charged particles. Instead, it is more convenient to develop relations for forces on currents or current densities in the magnetic field. This is not done only for convenience's sake: In practical applications, the forces on current-carrying conductors are extremely important and there is a real benefit to be gained from knowledge of these forces.

To find appropriate relations, we recall the definition of current density in terms of charges in a conductor. Consider **Figure 9.42a**, where a volume $dv' = ds'dl'$ is given. The volume contains N charges per unit volume. The number of charges in the differential volume is Ndv', all moving at an average velocity in the direction of the conductor in response to an externally applied electric field intensity $\mathbf{v} \times \mathbf{B}$. The force on the differential volume can, therefore, be written as

$$d\mathbf{F}_m = Ndv'q\mathbf{v} \times \mathbf{B} \quad [\mathrm{N}] \tag{9.107}$$

where q is the charge of an individual particle (such as an electron). The use of $d\mathbf{F}_m$ indicates the force due to the volume dv'. Following the discussion in **Section 7.3.2 [Eq. (7.6)]**, the current density in this segment of conductor is given as $\mathbf{J} = Nq\mathbf{v}$ [A/m^2]. Thus, the force in **Eq. (9.107)** can be written as

$$d\mathbf{F}_m = \mathbf{J} \times \mathbf{B}dv' \quad [\mathrm{N}] \tag{9.108}$$

where the product $\mathbf{J} \times \mathbf{B}$ may be viewed as a volumetric force density $\mathbf{f}$:

$$\mathbf{f} = \mathbf{J} \times \mathbf{B} \quad [\mathrm{N/m^3}] \tag{9.109}$$

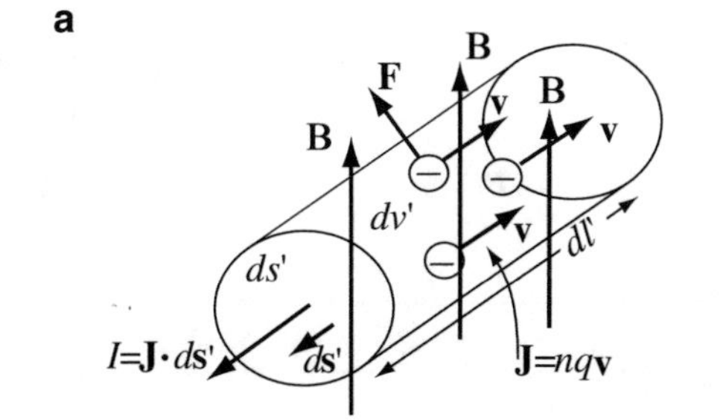

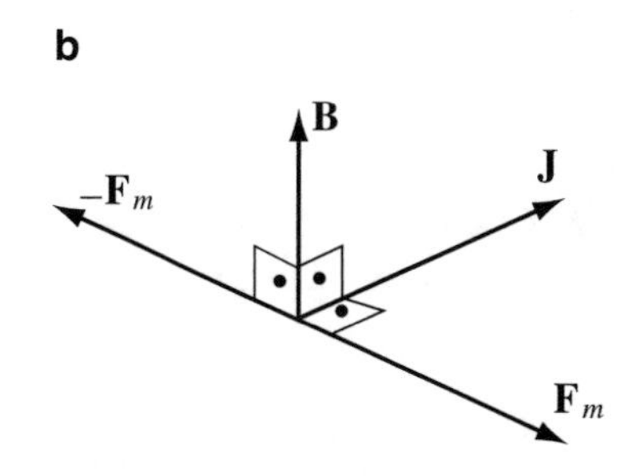

Figure 9.42 Forces on currents in the magnetic field. (**a**) Force on a differential volume carrying current density **J**. (**b**) Direction of force on a current-carrying conductor in the magnetic field

The direction of force is given by the right-hand rule and is shown in **Figure 9.42b**. For a given volume, the total force is an integration of this relation over the volume:

$$\boxed{\mathbf{F}_m = \int_{v'} \mathbf{J} \times \mathbf{B}dv' \quad [\mathrm{N}]} \tag{9.110}$$

This is a common statement of the magnetic (Lorentz) force and, in many ways, is a more useful relation since it gives the force on distributed currents. From this, we can now find the force on a current-carrying conductor such as a thin wire or a current loop (closed circuit).

The force on a current-carrying conductor can be written from **Eq. (9.110)** and **Figure 9.42a** by noting that the total current through the conductor is equal to the current density multiplied by the cross-sectional area of the conductor:

$$I = \mathbf{J} \cdot d\mathbf{s}' \quad [\mathrm{A}] \quad \text{or} \quad I d\mathbf{l}' = \mathbf{J} dv' \quad [\mathrm{A \cdot m}] \tag{9.111}$$

Substituting this in **Eq. (9.108)** we obtain the force on an element of current as[8]

$$\boxed{d\mathbf{F}_m = I d\mathbf{l}' \times \mathbf{B} \quad [\mathrm{N}]} \tag{9.112}$$

where the direction of $d\mathbf{l}'$ is in the direction of the current I in the conductor. For a conducting segment of length L and carrying a current I, the total force is

$$\mathbf{F}_m = \oint_L I d\mathbf{l}' \times \mathbf{B} \quad [\mathrm{N}] \tag{9.113}$$

Similarly, if the force on a closed circuit is needed, a closed contour integration is required:

$$\mathbf{F}_m = \oint_C I d\mathbf{l}' \times \mathbf{B} \quad [\mathrm{N}] \tag{9.114}$$

The fundamental relation for force in the present approach is that of **Eq. (9.112)** or **Eq. (9.108)**. It gives the force on an element of current in a magnetic field. Now, we can also calculate the force between two current elements by noting that the magnetic flux density can be generated by an element of a second current. Referring to **Figure 9.43**, an element of current $d\mathbf{l}_1'$ produces a magnetic flux density $d\mathbf{B}_{12}$ at the location of a second element of current $d\mathbf{l}_2'$. This second element will experience a force $d\mathbf{F}_{12}$ caused by the action of the field of current segment $d\mathbf{l}_1'$ on current segment $d\mathbf{l}_2'$. The opposite is also true: current element $d\mathbf{l}_2'$ produces a magnetic flux density at the location of $d\mathbf{l}_1'$ and this produces a force $d\mathbf{F}_{21}$ on element $d\mathbf{l}_1'$ due to the field of $d\mathbf{l}_2'$. Let's take this in three steps:

(1) The magnetic flux density due to current element $d\mathbf{l}_1'$ at a distance R_{12} in space is given by the Biot–Savart law as

$$d\mathbf{B}_{12} = \frac{\mu_0 I_1}{4\pi} \frac{d\mathbf{l}_1' \times \hat{\mathbf{R}}_{12}}{R_{12}^2} \quad [\mathrm{T}] \tag{9.115}$$

where $\mathbf{R}_{12}$ is the vector connecting $d\mathbf{l}_1'$ and $d\mathbf{l}_2'$, as shown in **Figure 9.43**.

(2) Calculate the force acting on $d\mathbf{l}_2'$ using **Eq. (9.112)**:

$$d\mathbf{F}_{12} = I_2 d\mathbf{l}_2' \times d\mathbf{B}_{12} = \frac{\mu_0 I_1 I_2}{4\pi} d\mathbf{l}_2' \times \left[\frac{d\mathbf{l}_1' \times \hat{\mathbf{R}}_{12}}{R_{12}^2}\right] \quad [\mathrm{N}]. \tag{9.116}$$

(3) Calculate the total force on any segment of C_2 (P_3 to P_4) due to a segment of the circuit C_1 (P_1 to P_2) by integrating along C_2 and then along C_1:

$$\boxed{\mathbf{F}_{12} = \frac{\mu_0 I_1 I_2}{4\pi} \int_{P_3}^{P_4} \int_{P_1}^{P_2} \frac{d\mathbf{l}_1' \times \left(d\mathbf{l}_2' \times \hat{\mathbf{R}}_{12}\right)}{R_{12}^2} \quad [\mathrm{N}]} \tag{9.117}$$

To find this relation, we assumed that a circuit exists, but only part of each circuit contributes to the force. This is useful

[8] This force is called the Laplace force although we derived it from the Lorentz force. The distinction between the two forces is in the fact that the Laplace force is a macroscopic force on the ensemble of moving charges whereas the Lorentz force is the microscopic force on a moving charged particle.

whenever we have current segments or when a force per unit length is required. If, instead, we need to calculate the force due to an entire closed circuit on a second closed circuit, the integration must be carried out as closed contour integrals over each circuit. Using the notation of **Figure 9.43**, we can write:

$$\mathbf{F}_{12} = \frac{\mu_0 I_1 I_2}{4\pi} \oint_{C_2} \oint_{C_1} \frac{d\mathbf{l}_2' \times \left(d\mathbf{l}_1' \times \hat{\mathbf{R}}_{12}\right)}{R_{12}^2} \quad [\mathrm{N}] \tag{9.118}$$

From Newton's third law we can also write $\mathbf{F}_{21} = -\mathbf{F}_{12}$.

Equation (9.118) is known as the Ampère force law. This force was obtained for two thin current-carrying conductors (the thin current assumption is required for Biot–Savart's law to apply). However, superposition of filamentary currents can be used to calculate the forces between thick conductors. Note that the cross products must be taken properly (the vector product in parentheses is calculated first) and the direction of the vector $\mathbf{R}_{12}$ is from $d\mathbf{l}_1'$ or $d\mathbf{l}_2'$. Note also that I_1 and I_2 are always positive numbers, whereas $d\mathbf{l}_1'$ or $d\mathbf{l}_2'$ can be either positive (if taken in the direction of the respective current) or negative (if taken in the direction opposite that of the respective current). Taking the contours as positive in the directions of the currents ensures the correct direction for the force.

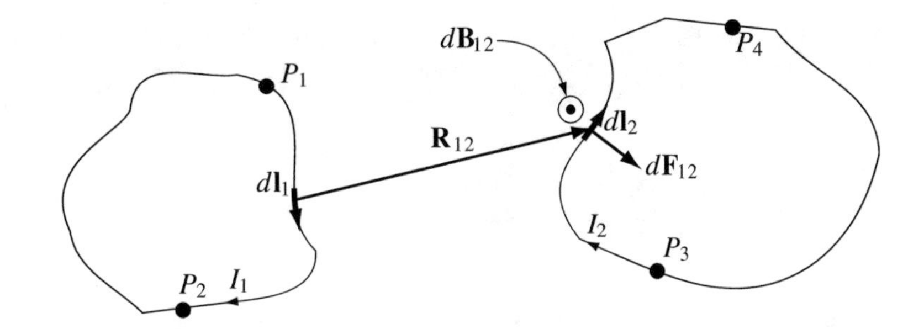

Figure 9.43 Relation between a current element in loop (1) and the force it exerts on a current element in loop (2)

Example 9.19 Application: Force Exerted by the Geomagnetic Field on Power Distribution Lines A 500 km long DC power distribution line carries a current of 1,000 A.

(a) Assuming that the perpendicular component of the terrestrial magnetic flux density is 50 μT (typical), calculate the total force exerted on one conductor of the distribution line.

(b) Find the direction of the force for the two-conductor line carrying current as above.

Solution: The force may be calculated using **Eq. (9.113)** directly since the magnetic field can be assumed to be constant everywhere on the line.

(a) The force is

$$\mathbf{F}_m = \int_C I d\mathbf{l}' \times \mathbf{B} \quad [\mathrm{N}]$$

The magnitude of the force is calculated as

$$F_m = IB \int_C dl' = lIB \quad [\mathrm{N}]$$

where l is the length of the line. Note the use of the open integration. The force for a segment of the circuit is calculated, but the implicit assumption that the circuit closes must be made for the current to flow. The total force on a conductor is

$$F_m = 500{,}000 \times 1{,}000 \times 0.00005 = 25{,}000 \quad [\mathrm{N}]$$

This is a significant force, but it is only 0.05 N per meter length of the cable. This is negligible compared to other forces, including the weight of the cable, wind and snow loads, tension due to temperature variations, and forces between conductors due to their currents (see **Example 9.20**).

(b) Using the right-hand rule for the force, the forces on each conductor are such that the conductors are attracted toward each other, as shown in **Figure 9.44**.

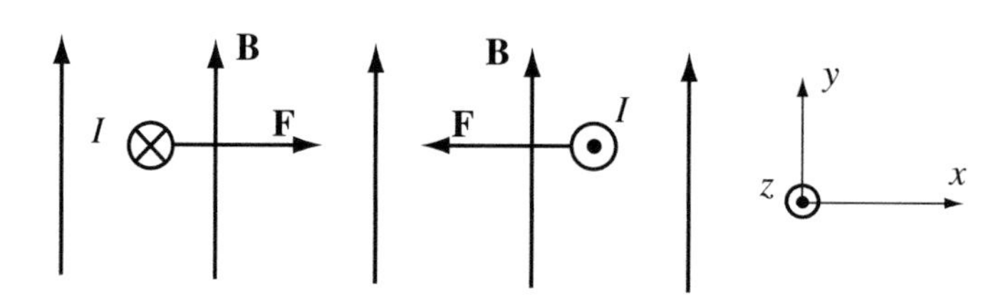

Figure 9.44 Forces on overhead current-carrying conductors due to the terrestrial magnetic field

Example 9.20 Application: Force Between Two Overhead Transmission Lines Carrying Direct Currents Two currents cause a force on each other. In overhead transmission lines, the currents may be quite high. Because of this force, when the current is switched on under load or a short circuit exists on the line, the cables may swing violently from side to side. The distance between the cables must be such that under the most severe swing, either due to faults on the line or wind, the cables do not touch.

A long power transmission line carries a current $I = 1{,}200$ A. The distance between two towers is $l = 100$ m and the distance between the two conductors is $d = 3$ m. Calculate the total force between the two cables between each two towers. What is the direction of this force?

Solution: The force may be calculated using Ampère's force law or, alternatively, using **Eq. (9.112)** directly. The latter is possible because the force is constant along the line, and because it is much simpler, we use it here. The configuration is shown in **Figure 9.45**.

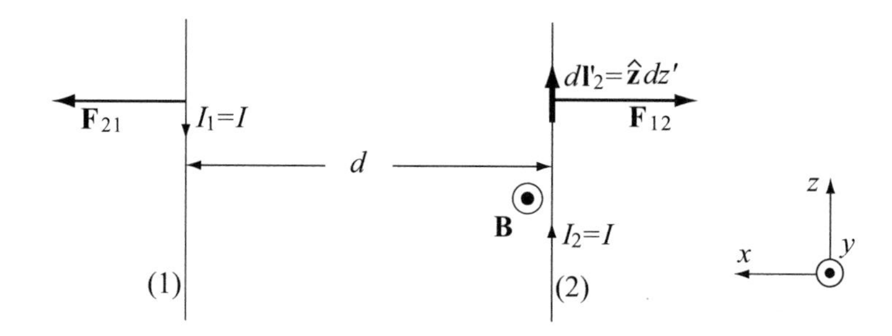

Figure 9.45 Forces on power lines due to the currents in the lines

The magnetic flux density at a distance d from the current I_1 in **Figure 9.45** at the location of the second conductor [current I_2] is (see **Example 9.8**):

$$\mathbf{B} = \hat{\mathbf{y}}\frac{\mu_0 I}{2\pi d} \quad [\mathrm{T}]$$

From **Eq. (9.112)** and taking $d\mathbf{l}_2 = \hat{\mathbf{z}}dz$ we get

$$d\mathbf{F}_m = I\,d\mathbf{l}_2' \times \mathbf{B} = I\hat{\mathbf{z}}dz' \times \hat{\mathbf{y}}\frac{\mu_0 I}{2\pi d} = -\hat{\mathbf{x}}\frac{\mu_0 I^2 dz'}{2\pi d} \quad [\mathrm{N}]$$

This is the force exerted on segment (2) by segment (1). Since the force is independent of y, the force on a section of length $l = 100$ m is simply 100 times larger:

$$\mathbf{F}_{12} = -\hat{\mathbf{x}}\frac{\mu_0 I^2 L}{2\pi d} = -\hat{\mathbf{x}}\frac{4\times\pi\times10^{-7}\times1200^2\times100}{2\times\pi\times3} = -\hat{\mathbf{x}}9.6 \quad [\mathrm{N}]$$

The force on the left cable (1) is to the left, and on the right cable ($\mathbf{F}_{12} = -\mathbf{F}_{21}$) is to the right and equal in magnitude. This force tends to separate the wires.

Example 9.21 A very long, thin wire carrying a current I_0 [A] and a rectangular loop carrying a current I_1 [A] are placed on a plane as shown in **Figure 9.46a**. The wire carries a current $I_0 = 5$ A and the loop carries a current $I_1 = 2$ A. Calculate the total force on the wire due to the loop.

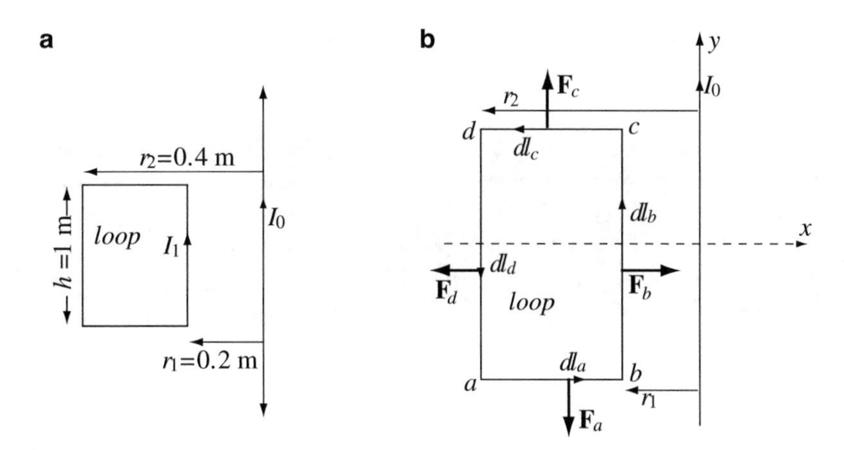

Figure 9.46 Forces between a loop and a current-carrying wire

Solution: There are two ways this problem may be solved. One is to use the general expression in **Eq. (9.118)**. The force due to the loop is evaluated by integrating over the two closed paths: around the loop and the wire from $-\infty$ to $+\infty$. A simpler way, one we follow here, is to calculate the magnetic flux density due to the infinite wire at the location of the loop and then calculate the force the wire exerts on each of the segments of the loop using **Eq. (9.112)** or **Eq. (9.114)**. Since the force on the wire is then the negative of the force on the loop, the force on the wire is immediately available.

The magnetic flux density at any location in space due to the infinitely long wire is given as $\mu_0 I_0/2\pi r$. In this case, the direction of the magnetic flux density is in the positive z direction to the left of the wire, based on the right hand rule:

$$\mathbf{B}_0 = \hat{\mathbf{z}}\frac{\mu_0 I}{2\pi r} \quad [\text{T}] \quad \text{for} \quad x < 0$$

With the magnetic flux densities above, the force on the loop (denoted as $\mathbf{F}_l$) is [see **Eq. (9.114)**]

$$\mathbf{F}_l = \oint_{C_1} I_1 d\mathbf{l} \times \mathbf{B}_0 = \int_a^b I_1 d\mathbf{l}_a \times \mathbf{B}_0 + \int_b^c I_1 d\mathbf{l}_b \times \mathbf{B}_0 + \int_c^d I_1 d\mathbf{l}_c \times \mathbf{B}_0 + \int_b^a I_1 d\mathbf{l}_d \times \ \mathbf{B}_0 \quad [\text{N}]$$

The four components of the force on the loop are shown in **Figure 9.46b**. The first thing to note is that $\mathbf{F}_a$ and $\mathbf{F}_c$ cancel each other. Second, the forces on opposite sides of the loop are in opposite directions, but $\mathbf{F}_b > \mathbf{F}_d$. Third, only parallel currents exert forces on each other. Thus, the total force on the loop, exerted by the wire, is

$$\mathbf{F}_l = \mathbf{F}_b + \mathbf{F}_d = \int_b^c I_1 d\mathbf{l}_b \times \mathbf{B}_0 + \int_d^a I_1 d\mathbf{l}_d \times \mathbf{B}_0 \quad [\text{N}]$$

Now, we must evaluate the magnetic flux density at the locations of the segments of the loop. The directions of the various elements of length are in the direction of I_1 and are shown in **Figure 9.46b**. With these, the force on the loops is

$$\mathbf{F}_l = \int_b^c I_1(\hat{\mathbf{y}}dy) \times \left(\hat{\mathbf{z}}\frac{\mu_0 I_0}{2\pi r_1}\right) + \int_a^d I_1(-\hat{\mathbf{y}}dy) \times \left(\hat{\mathbf{z}}\frac{\mu_0 I_0}{2\pi r_2}\right) \quad [\text{N}]$$

Evaluating the two integrals gives

$$\mathbf{F}_l = \hat{\mathbf{x}}\frac{\mu_0 I_0 I_1 h}{2\pi r_1} - \hat{\mathbf{x}}\frac{\mu_0 I_0 I_1 h}{2\pi r_2} \quad [\text{N}]$$

With the values given, the total force is

$$\mathbf{F}_l = \hat{\mathbf{x}}\frac{\mu_0 I_0 I_1 h}{2\pi}\left[\frac{1}{r_1} - \frac{1}{r_2}\right] = \hat{\mathbf{x}}\frac{4 \times \pi \times 10^{-7} \times 5 \times 2 \times 1}{2 \times \pi}\left[\frac{1}{0.1} - \frac{1}{0.2}\right] = \hat{\mathbf{x}}10^{-5} \quad [\text{N}]$$

The force on the wire($\mathbf{F}_w = -\mathbf{F}_l$) is $-\hat{\mathbf{x}}10^{-5}$ [N].

Exercise 9.6 Application: Experimental Definition of the ampere One method of defining the ampere is as follows: A 1 m length of thin wire is placed parallel to and a distance 1 m from a very long straight wire as shown in **Figure 9.47**. Each wire carries a current I. The force between the segment and the infinite wire is measured. If the current in both wires is 1 A, what is the measured force between the segment and the long wire? (The standard for the ampere is a modified form of this arrangement which uses coils rather than straight wires to increase the measured force and, therefore, reduce the error involved in the measurement of very small forces).

Answer 2×10^{-7} [N].

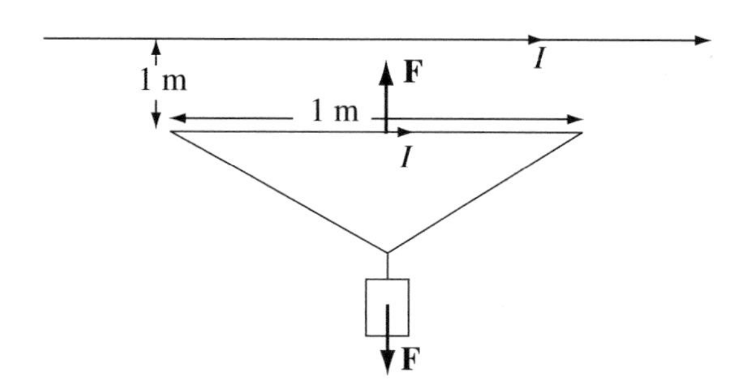

Figure 9.47 Method of defining the ampere in terms of measured forces

9.7.1 Principle of Virtual Work: Energy in a Gap

The calculation of force using Ampère's force law in **Eq. (9.118)** is particularly suited for simple configurations such as segments of wires or long conductors. There is, however, a large number of applications in which this method is next to impossible to apply. For example, the force between two magnets cannot be calculated using this method. Another method, which is sometimes easier to apply, is the method of ***virtual work*** or ***virtual displacement.*** We discussed the principle in **Section 4.8.2**. It relies on the basic method of measuring force; when measuring force (as, for example, in weighing), we allow the force to move the body on which the force operates a small distance such that a mechanism or sensor may be activated. This distance is then a measure of the force. If a force **F** exists between two sections of a magnetic path as shown in **Figure 9.48a**, energy is also associated with the system. If we now allow the two sections to move a distance dl, the energy in the gap will be reduced by an amount $\mathbf{F} \cdot d\mathbf{l}$. If allowed, the two pieces will move closer together reducing the gap volume (**Figure 9.48b**). Thus, assuming the potential magnetic energy of the system to be W before the motion, the change in the potential energy of the system is

$$\mathbf{F} \cdot d\mathbf{l} = -dW \quad [\text{N}] \tag{9.119}$$

Since the total derivative dW can be written in terms of the gradient (see **Sections 2.3.1** and **4.8.2**) as

$$dW = (\nabla W) \cdot d\mathbf{l} \tag{9.120}$$

Figure 9.48 Method of virtual work. (**a**) Force between two magnets. (**b**) Calculation of force through virtual work

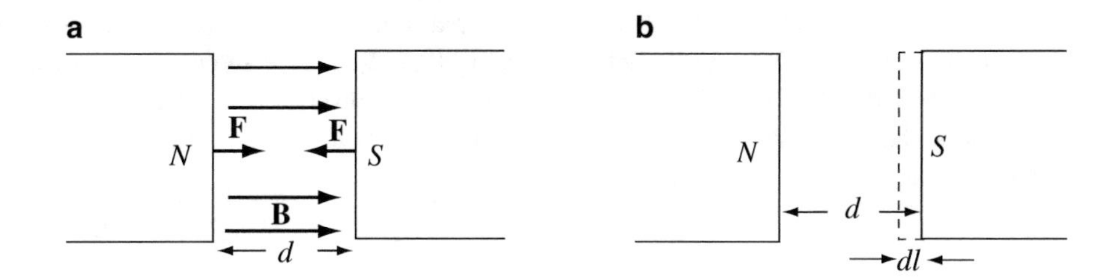

The force is

$$\boxed{\mathbf{F} = -\nabla W \quad [\mathrm{N}]} \tag{9.121}$$

Thus, the magnetic force is the negative of the gradient in potential energy. In expanded form, we can write in Cartesian coordinates

$$\mathbf{F} = \hat{\mathbf{x}}F_x + \hat{\mathbf{y}}F_y + \hat{\mathbf{z}}F_z = -\hat{\mathbf{x}}\frac{\partial W}{\partial x} - \hat{\mathbf{y}}\frac{\partial W}{\partial y} - \hat{\mathbf{z}}\frac{\partial W}{\partial z} \quad [\mathrm{N}] \tag{9.122}$$

As expected, the force may have three components in space provided that energy changes as the body moves in a particular direction. If it does not, the force in that particular direction is zero.

Note: It is implicit that the virtual displacement $d\mathbf{l}$ does not change the energy density or total energy in the gap since in reality, nothing moves.

Example 9.22 Application: Forces on Iron Pieces in a Uniform Magnetic Field—The Solenoid Valve An infinitely long solenoid of inner radius b [m] has n turns per unit length and carries a current I [A]. Two very long iron bars, each of radius b [m], are placed in the solenoid and are separated by a very small gap of length d [m], as shown in **Figure 9.49**. Permeability of iron is μ [H/m] and that of free space is μ_0 [H/m]. Calculate the force between the two iron bars. What is the direction of this force?

Figure 9.49

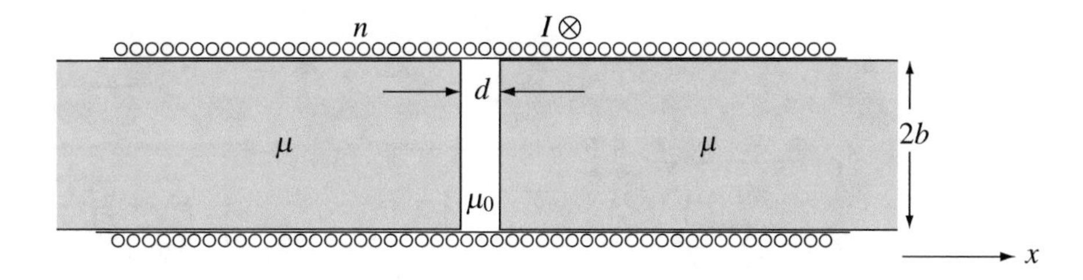

Solution: The magnetic field intensity in the solenoid is calculated, and from it, the magnetic flux density in the gap. Now, allowing a virtual displacement of the iron bars, we calculate the change in energy due to this displacement, from which we obtain the force.

The magnetic flux density is in the negative x direction with magnitude

$$B = \mu H = \mu n I \quad [\mathrm{T}]$$

Since the gap is small, we may assume that the same flux density exists in the gap and that the flux density is constant throughout the gap. The total energy in the gap is [see **Eq. (9.89)**]

$$W_m = \frac{B^2}{2\mu_0}v_{gap} = \frac{B^2\pi b^2 d}{2\mu_0} = \frac{\mu^2 n^2 I^2 \pi b^2 d}{2\mu_0} \quad [\mathrm{J}]$$

To calculate the force, we allow the gap to decrease or increase by dl. The change in energy in the gap is $dW = Wdl$. The total force on the bars is then calculated using **Eq. (9.122)**. This is the same as to calculate the change in energy due to the change in volume caused by the displacement dl. We assume dl is positive and, therefore, increases the potential energy in the system; that is, dW as well as dl are positive. If the right bar moves a distance dl to the right, the force due to this positive displacement is

$$\mathbf{F} = -\hat{\mathbf{x}}\frac{dW_m}{dl} = -\hat{\mathbf{x}}\frac{\mu^2 n^2 I^2 \pi b^2}{2\mu_0} \quad [\text{N}]$$

We could also assume dl to be negative (the right bar moves a distance dl to the left). This reduces the gap, and, therefore, both dW and dl are negative. The resulting force is again in the negative x direction. In other words, it does not matter what direction we assume for dl as long as the change in energy dW is calculated accordingly. This force acts on the right bar and tends to close the gap because this reduces the energy in the system. The force on the left bar is in the positive x direction (from Newton's law) and it also tends to close the gap. If allowed, the two bars would move toward each other.

Example 9.23 Application: Force in a Magnetic Circuit—The Airless Sprayer An airless sprayer is a device used to spray fluids such as paints or pesticides without the use of compressed air. The device uses a simple piston action to push the fluid through an orifice at high velocity. The pump used in most sprayers is a hinged electromagnet with a gap, as shown in **Figure 9.50a**. When the current is switched on, the gap closes, moving the piston. When the current is switched off, the piston returns to its original position. This motion is sufficient to pump fluid in and to expel it at high velocity. Commercial sprayers use an AC source, causing the piston to move 100 or 120 times per second, depending on the frequency of the electrical grid. The same idea is used in small air pumps such as those used in fish tanks. The most characteristic aspect of these devices is relatively high noise levels and heat produced in the magnetic core. This type of device is also used extensively to open and close valves and in other types of linear actuators.

A pump used to spray paint is made as shown in **Figure 9.50b**. The coil contains $N = 5{,}000$ turns and carries a current $I = 0.1$ A. Permeability of iron is $\mu = 1{,}000\mu_0$ [H/m] and the gaps are $d = 5$ mm long. Assume all flux is contained in the gaps (no leakage of flux). Other dimensions are given in the figure.

(a) Calculate the force exerted by the moving piece on the piston.
(b) What is the force if the gaps are reduced to 1 mm?

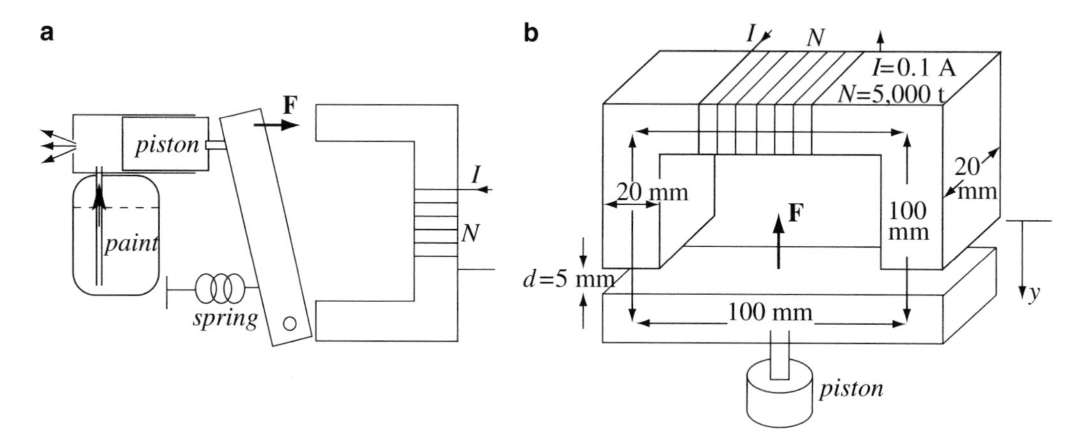

Figure 9.50 (**a**) A hinged electromagnet used in an airless sprayer. (**b**) A modified form of the electromagnet in (**a**) used for calculation of forces

Solution: In this case, there are two gaps. The energy in the two gaps is calculated and then we assume the lower piece moves up a distance dl, thereby reducing the gaps and reducing the potential energy stored in the gaps. Thus, both dl and dW are negative.

(a) The magnetic flux density in the gap is calculated using the magnetic circuit shown in **Figure 9.50b**. The reluctances of the iron and gaps are

$$\Re_i = \frac{l_{iron}}{\mu_{iron}S} = \frac{0.39}{1{,}000 \times 4 \times \pi \times 10^{-7} \times 0.02^2} = 7.759 \times 10^5 \quad [1/\mathrm{H}]$$

$$\Re_g = \frac{l_{gap}}{\mu_0 S} = \frac{0.01}{4 \times \pi \times 10^{-7} \times 0.02^2} = 1.99 \times 10^7 \quad [1/\mathrm{H}]$$

where S is the cross-sectional area of the magnetic circuit. The flux density in iron and, therefore, in the gap is

$$B_g = \frac{\Phi_g}{S} = \frac{NI}{S(\Re_i + \Re_g)} = \frac{5{,}000 \times 0.1}{0.02^2 \times (7.759 \times 10^5 + 1.99 \times 10^7)} = 0.06 \quad [\mathrm{T}]$$

Since the flux density is constant in the gaps, the total energy stored in the two gaps is

$$W_m = \frac{B^2}{\mu_0} v_{gap} = \frac{B^2 S d}{\mu_0} \quad \rightarrow \quad dW = \frac{B^2 S\, dl}{\mu_0} \quad [\mathrm{J}]$$

For a length $-dl$ (which reduces the gap), dW is also negative. From **Eq. (9.122)**, we get

$$\mathbf{F} = -\hat{\mathbf{y}} \frac{-dW_m}{-dl} = -\hat{\mathbf{y}} \frac{B^2 S}{\mu_0} = -\hat{\mathbf{y}} \frac{0.06^2 \times 0.02^2}{4 \times \pi \times 10^{-7}} = -\hat{\mathbf{y}} 1.146 \quad [\mathrm{N}].$$

(b) The reluctance in iron remains the same, but the reluctance in the gap has decreased by a factor of 5:

$$\Re_g = \frac{l_{gap}}{\mu_0 S} = \frac{0.002}{4 \times \pi \times 10^{-7} \times 0.02^2} = 3.979 \times 10^6 \quad [1/\mathrm{H}]$$

The flux density is therefore

$$B_g = \frac{NI}{S(\Re_i + \Re_g)} = \frac{5{,}000 \times 0.1}{0.02^2 \times (7.759 \times 10^5 + 3.979 \times 10^6)} = 0.263 \quad [\mathrm{T}]$$

The force is

$$\mathbf{F} = -\hat{\mathbf{y}} \frac{B^2 S}{\mu_0} = -\hat{\mathbf{y}} \frac{0.263^2 \times 0.02^2}{4 \times \pi \times 10^{-7}} = -\hat{\mathbf{y}} 22.02 \quad [\mathrm{N}]$$

The force increases as the gap decreases because the flux density is higher. In practical designs of this type, the gap is kept at a minimum and the motion of the moving piece is also small to keep the force more or less constant. If, for example, the moving piece were to move 4 mm the gap would change from 5 mm to 1 mm and the force would vary from 1.146 N to 22.02 N.

9.8 Torque

Torque is the product of force **F** and length of arm d about a pivot (or an axis), as shown in **Figure 9.51**. In our case, the force is a magnetic force. Torque is an important aspect of all rotating machinery. It is responsible for the rotation of the magnetic needle in a compass, as well as rotation of the rotor in an electric motor.

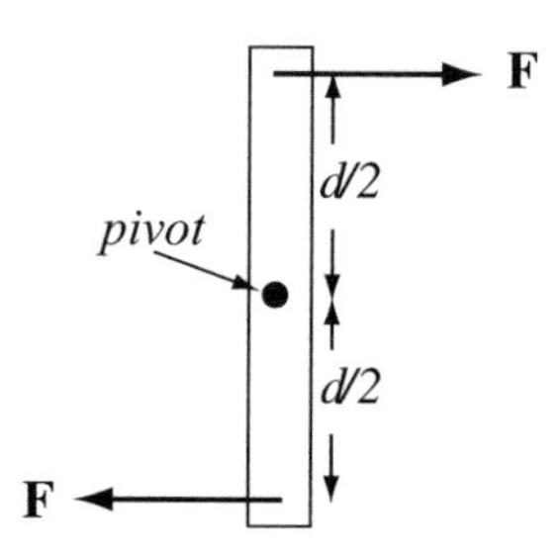

Figure 9.51 Relation between force and torque

To define torque, consider a small, square loop, w by d meters in size, placed in a uniform magnetic flux density **B**, as shown in **Figure 9.52a**. The loop is constrained to rotate about an axis. A current flows in the loop as shown. Using **Eq. (9.113)**, we can calculate the forces on each of the four sides of the loop. On side ab, the current is in the positive z direction, and **B** is in the positive x direction:

$$\mathbf{F}_{ab} = \int_a^b I\,d\mathbf{l}' \times \mathbf{B} = \int_a^b I(\hat{\mathbf{z}}dl') \times \hat{\mathbf{x}}B = \hat{\mathbf{y}}wIB \quad [\mathrm{N}] \tag{9.123}$$

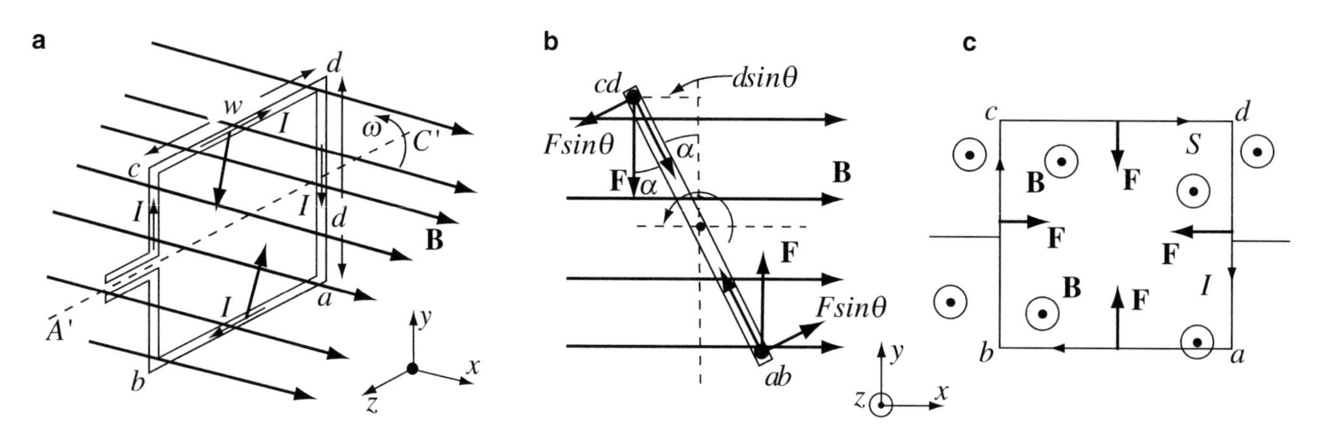

Figure 9.52 Forces and torque on a loop in a magnetic field. (**a**) Configuration and dimensions. (**b**) Plane of the loop at an angle α to the normal. (**c**) Force components in the plane of the loop do not contribute to torque

On side cd, the current is in the opposite direction and therefore

$$\mathbf{F}_{cd} = -\hat{\mathbf{y}}wIB \quad [\mathrm{N}] \tag{9.124}$$

The forces on sides ad and bc are in the direction of the axis (z axis; see **Figure 9.52c**) and, therefore, cannot contribute to torque. Thus, only the forces on ab and cd contribute to torque. The magnitude of torque is therefore

$$T = 2F\frac{d\sin\alpha}{2} \quad [\mathrm{N \cdot m}] \tag{9.125}$$

where $(d/2)sin\alpha$ is the armlength on which the force operates and F is the magnitude of $\mathbf{F}_{cd}$ (or $\mathbf{F}_{ab}$) (**Figure 9.52b**). The factor of 2 is due to contribution of the two forces to torque. Thus, the torque is

$$T = wBId\sin\alpha \quad [\mathrm{N \cdot m}] \tag{9.126}$$

This is the general expression for the magnitude of torque. We can also write this in a different way by noting the following:

(a) The product $wId = m$ is the magnitude of the magnetic dipole moment of the loop. In vector notation we write $\mathbf{m} = \hat{\mathbf{n}} wId$ where $\hat{\mathbf{n}}$ is the normal to the loop.

(b) The term $wBId\sin\alpha = mB\sin\alpha$ can be written as the magnitude of the vector product of the vectors **m** and **B**, since torque is perpendicular to **B** and to **m**.

Thus,

$$\mathbf{T} = \mathbf{m} \times \mathbf{B} \quad [\mathrm{N \cdot m}] \tag{9.127}$$

This expression has the advantage that the direction of torque becomes evident. Because torque is perpendicular to both **m** and **B**, it must be in the direction of the axis (**m** is perpendicular to the surface of the loop; see also **Section 9.2.1**).

Example 9.24 Application: Torque on a Square Coil and the Principle of Electric Motors A square coil consists of $N = 100$ turns, tightly packed together. The coil is placed in a uniform magnetic flux density $B = 0.2$ T as shown in **Figure 9.53**. The coil is $a = 200$ mm on the side and carries a current $I = 10$ A. Calculate the maximum torque on the coil.

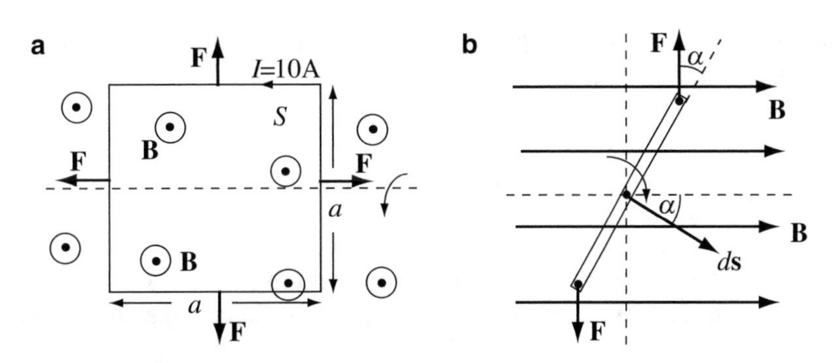

Figure 9.53 Square coil in a uniform magnetic field. (**a**) Top view. (**b**) Side view along the axis of rotation

Solution: The torque of the square loop is given in **Eq. (9.126)**. Since all 100 turns of the coil are close together, we may assume they are at the same angle with respect to the field and, therefore, the torque of the coil is N times larger.

With $w = d = a$, the torque is

$$T = Na^2IB\sin\alpha = 100 \times 0.2^2 \times 10 \times 0.2\sin\alpha = 8\sin\alpha \quad [\mathrm{N \cdot m}]$$

The maximum torque is 8 N·m and occurs when the plane of the coil (loop) is aligned with the magnetic flux density ($\alpha = 90°$). This torque is responsible for rotation in electric motors. The use of coils, rather than single loops, allows a significant increase in torque. In practical motors of this type, when the coil has rotated one half-turn, the direction of the current is reversed so that the coil continues to rotate.

Example 9.25 Torque on an Electron The electron, as it rotates around its axis, produces a current as was shown in **Example 9.1**. This current is in effect a small loop. Suppose an electron rotates around the hydrogen atom in a circle of radius $r = 5 \times 10^{-11}$ m and produces the equivalent of 1 mA current. Calculate the magnitude and direction (with respect to the axis of rotation) of the torque if the electron is placed in the field of a magnetic resonance imaging magnet of 3 T. Assume the magnetic flux density and the axis of rotation of the electron are at an arbitrary angle as shown in **Figure 9.54**.

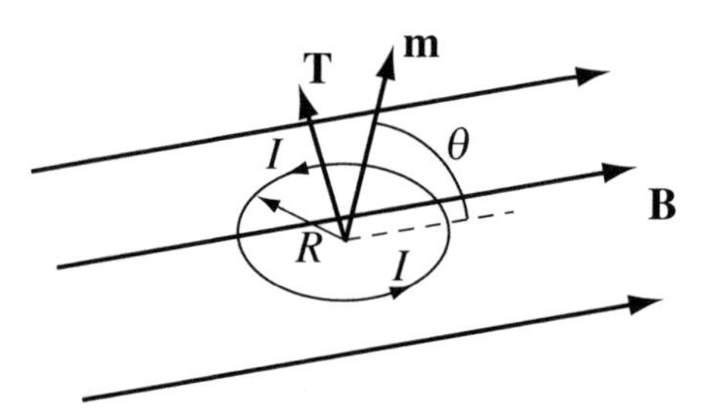

Figure 9.54 An electron spinning in the field of a magnetic resonance imaging device

Solution: After calculating the magnetic dipole moment of the electron, we use **Eq. (9.127)** to calculate the torque.

The magnitude of the magnetic moment of the loop is

$$m = \pi r^2 I = \pi (5 \times 10^{-11})^2 \times 10^{-3} = 7.854 \times 10^{-24} \quad [\text{A} \cdot \text{m}^2]$$

First, we note that the torque is in the plane of the loop, perpendicular to the plane formed by **B** and the axis of the loop. The magnitude of the torque is

$$T = |\mathbf{m} \times \mathbf{B}| = mB\sin\theta = 7.854 \times 10^{-24} \times 3 \sin\theta = 2.356 \times 10^{-23} \sin\theta \quad [\text{N} \cdot \text{m}]$$

The torque is maximum when the axis of rotation of the electron and the magnetic flux density are perpendicular to each other, and zero when they are parallel. The torque is always in the plane of the loop created by the rotating electron, and perpendicular to the flux density.

9.9 Applications

Application: Flipping of Magnetization in Lava as an Indication of Flipping of Earth's Magnetic Field An interesting manifestation of magnetization and the Curie temperature can be found in lava layers that have solidified while under the influence of the terrestrial magnetic field. There is evidence that the magnetic field of the Earth flips direction every 30,000 to 50,000 years or so. Lava deposits on the bottom of oceans, in particular in the Atlantic Ocean near Iceland and in the south Pacific, show a pattern of very long strips, alternately magnetized north to south and south to north. This gave rise to the theory that the magnetic field of the Earth reverses regularly. The explanation of the regular magnetization features on the seabed is that molten rock, as it surged from the Atlantic ridge, has solidified in the presence of the geomagnetic field. As the flowing lava cooled below the Curie temperature, the magnetic particles in the rock gained the preferred direction of the field and remained this way after solidification, permanently recording the magnetic field at the time. Subsequent lava flow pushed these strips further from the ridge, forming new strips, again forming a record of the geomagnetic field. This is an interesting aspect of magnetization because it indicates how permanent a magnetic recording can be. The magnetic record remained intact over millions of years and, in fact, this magnetization should remain unchanged until the rock melts again. The fact that the geomagnetic field can reverse does not have a ready explanation, especially since such reversals seem to be relatively quick and most likely accompanied by severe disturbances to the balance of life on the planet. The phenomenon of magnetized strips was discovered when sensitive magnetic field measuring devices (magnetometers) were used to detect submarines by detecting the disturbance their iron hull causes in the surrounding geomagnetic field.

In old manuals for production of magnetic needles for compasses, the recipe calls for the iron needle to be red hot and held in the north–south direction. In this position, the needle is dipped into water. In this process, the iron becomes hard steel and the domains remain aligned in the direction north–south (lengthwise in the needle). Thus, a magnetic needle is produced in the same way that magnetization in the Earth's crust has been produced on the ocean floor.

Application: Ferrite Cores for Coils and Transformers We will only discuss transformers in **Chapter 10**, but it is easy to see that if we need a very complicated shape for a core, it is difficult to prepare these shapes from solid materials such as iron. On the other hand, ferrite materials are made of powdered ferromagnetic materials, mixed together, molded into the desired shape, and then sintered. In principle, any ferromagnetic material can be made into a ferrite. Iron oxides form the basis of most ferrites (for example, Fe_3O_4) mixed with bivalent compounds such as Nickel oxide (NiO), Barium oxide (BaO), Manganese oxide (MnO), or others of the same type. The final product can have almost any shape and size and the properties of the solid material depend both on the materials used and the processing of the materials. Because of the sintering process used to bind the compounds together, the final material is very hard, is brittle, and can only be worked by grinding. Conductivity is low (normally below 10^{-5} S/m), but some materials have a conductivity of the order of 1 S/m. Ferrite cores are used extensively in coils and transformers, particularly where operation at high frequencies is required. As a ferromagnetic material, ferrites have a relatively low coercive field intensity (usually below 100 A/m), relative permeability between 10 and 10,000, remnant flux density below about 0.5 T, and a rather square magnetization curve.

Application: Magnetic Recording; Magnetic Tapes, Strips, and Drives A very useful application of magnetic materials is in recording of signals. Although the properties of various magnetic recording media vary in composition and quality, the principles are the same. In its simplest form, we may think of a magnetic medium as containing a layer of magnetic particles (not unlike very small permanent magnets) on a substrate and a means of orienting these particles by external fields. A magnetic recording medium, either tape (or strip) or disk, is made by one of two methods: one consists of coating a base material with ferromagnetic particles in a binding material. Each particle is independently suspended and its domains can be oriented by an external magnetic field. The second method is deposition of a ferromagnetic alloy in the form of a thin layer on a nonferromagnetic base material such as aluminum. The materials used are various iron and cobalt alloys. This method is normally used for magnetic disk drives.

In particulate media, the particles are diluted in a binder to about 20–50% volume and coated onto a substrate film made of polyethylene or on rigid aluminum disks. The binder contains substances that protect the film and the recording head from physical damage. The particles may be any number of available ferromagnetic materials, including cobalt, nickel, iron, their oxides, and others. Most tapes use iron oxides and in particular Fe_2O_3. A simple recording system is shown in **Figure 9.55**. The recording head is a closed magnetic circuit in which a very thin gap (as small as 1 μm) has been cut. The flux produced by the coil passes through the gap and part of it leaks around the gap. This leakage field is proportional to the current in the coil. As it interacts with the particles on the tape, it magnetizes them at levels that are proportional to the signal being recorded. The magnetization in the particles is the recorded signal. When required, the signal can be read by a process that will be described in **Chapter 10**. Although magnetic tapes were used in the past most current applications include magnetic strips on credit cards and access cards and, in particular, magnetic hard drives. New advanced methods of recording focus on higher densities for ever increasing hard drives capacities. A common method of recording data on high-density drives is the magneto-optic method, which is another application of the Curie temperature. In this application, a laser beam is used to heat the recording medium beyond the Curie temperature, at the point at which we wish to record information. Then, the recording magnetic field is applied, and as the point cools below the Curie temperature, the information magnetization is retained. Because the laser beam can be very small in diameter, the heating is local and fast, as is the cooling and the required magnetic field is low and higher recording densities are possible.

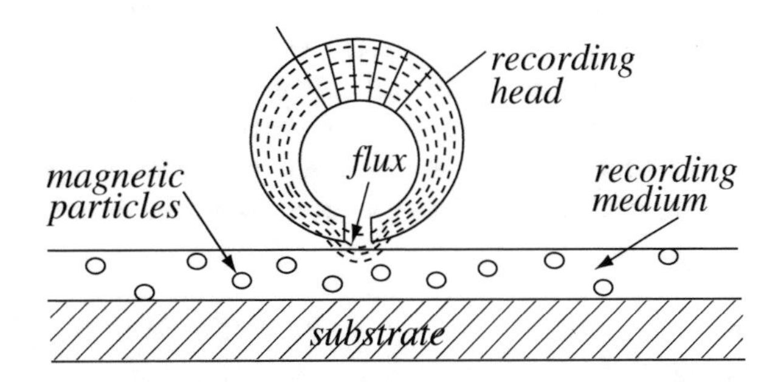

Figure 9.55 A simple recording mechanism showing a recording head and magnetic tape

Application: Magnetic Particle Testing A common method of nondestructive testing of ferromagnetic products is the magnetic particle testing. The idea is simplicity itself. Any ferromagnetic material when subjected to an external magnetic field is magnetized to a certain extent. If the material is uniform, the magnetic field is confined to the interior of the ferromagnetic material. If, however, there is a discontinuity in the material such as cracks, some of the magnetic flux "leaks" out. This field can then interact with external agents such as magnetic particles. If we now sprinkle ferromagnetic particles on the object, these will be attracted to the location of the flaw clearly indicating it. To many, this process is simply known as "magnafluxing" (Magnaflux is a trademark, not a method). Magnetic particles can be as simple as iron filings but can be much finer, with antioxidants and wetting materials added. Some processes call for dry powders, and others for wet solutions. Magnetic particles are sometimes dyed to facilitate detection; some are fluorescent to be observed under special (black) lighting. There are many particle sizes used in industry ranging from 1 to 25 μm in size for wet methods and from 100 to 1,000 μm for dry methods, depending on application. Magnetic particles are normally spherical or elongated and are mostly ferromagnetic oxides. Pure ferromagnetic materials can also be used, but these are not stable. Oxides, on the other hand, are stable and can be stored and used without their properties changing in the process.

Application: Magnetite as a Guiding Mechanisms Used by Bacteria and Animals An intriguing use of magnetic materials by microorganisms and animals is in guidance in the terrestrial magnetic field. Magnetite is a naturally occurring substance and traces of the material have been found in some organisms as well as birds and other animals, including humans. There is a strong indication that many animals are, or were in the past, using this magnetic material as a natural compass. Pigeons almost certainly use terrestrial magnetism as part of their guiding mechanism. Migrating birds also use this compass, in addition to other methods. Although this sense is not developed in humans, the occurrence of magnetite in our brains suggests that at some time in our past, we may have had at least a partial ability to navigate using the terrestrial magnetic field. However, there is no definite proof of this and the magnetite in our brains may well be leftover particles with no specific function since their occurrence is at very low densities.

Some microbes use magnetite to orient themselves along the magnetic field. In these bacteria, called ***magnetotactic bacteria***, a string of single-domain magnetite (Fe_3O_4) particles (easily visible under microscopes) forms a compass that keeps the bacteria oriented along the geomagnetic or any other local field. The magnetic moment of the cells is fixed and the bacteria have no control over it. In other animals and insects, there is experimental evidence of active use of the magnetic field, indicating that these animals have magnetic sensors. Experimental data have been collected on pigeons whereby small magnets attached to their heads have shown to disorient them. Bees seem to use the magnetic field, as do amphibians such as salamanders. Some experiments have even focused on training animals and insects to detect low-level magnetic fields and local anomalies in the terrestrial magnetic field.

Application: Magnetic Shielding—Passive Magnetic Shielding of MRI Equipment It is often required to shield instruments from the effects of naturally or artificially produced magnetic fields. For example, we might want to measure very low magnetic fields at levels much below the geomagnetic field. In other cases, we might need to calibrate a magnetic device, and the effects of external fields must be eliminated. In still other cases, the very large magnetic fields produced by electromagnets must be contained so they do not interfere with instruments in their vicinity. Examples of this type of work abound. In a clinical magnetic resonance imaging (MRI) device, the fields produced by the magnet are of the order of 1 to 3 T. Yet, within a few meters, we need to operate computers, testing equipment, and place operators who might not appreciate being exposed to high magnetic fields continuously. The threat of such high magnetic fields to equipment is real. For example, the magnetic field required to erase a computer disk is about 0.008 T. Clearly, any disk brought close to a magnetic resonance imaging device will be affected. Other examples are the need to measure the magnetic field produced by the brain. This field is orders of magnitude smaller than the geomagnetic field. The solution to these problems is magnetic shielding. In its simplest form, magnetic shielding provides a low-reluctance path which the magnetic field takes, thus steering the field away from the shielded area. An example is shown in **Figure 9.56**. The coil at the center of the shielded structure produces a magnetic field. By placing an iron box around the coil, most of the field is contained in the iron because

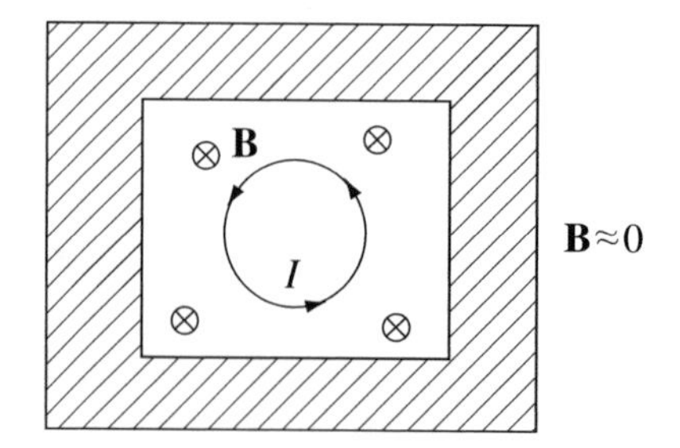

Figure 9.56 A simple magnetic shield used to contain the field inside the structure. The same structure will shield the interior from external magnetic fields

it has a high relative permeability and, therefore, low reluctivity. This method is equivalent to shorting an electric circuit to prevent current from flowing into a certain part of the circuit. Magnetic shields can be rather big structures. For example, in an MRI installation, the shield may be inserted in the walls of the room. For a room $4 \times 4 \times 2.5\ \text{m}^3$ (not a very large room) with an iron shield, 50 mm thick in all four walls, ceiling, and floor, the total volume of iron is $3.6\ \text{m}^3$. At a weight of approximately 7.8 tons/m^3, the total weight of the shield is 28 tons. This is not a trifling weight, if nothing else, then for design of foundations for the building. In the above example, shielding is used to contain the magnetic field in a given area. If the source of the magnetic field is outside the shield, the magnetic field is excluded from the area inside the shield.

9.10 Summary

The current chapter applies the Biot–Savart and Ampère's laws and introduces most of the concepts in magnetostatics including inductance, energy, and interface conditions. We start with the magnetic flux density and the dipole moment of a small loop of radius d carrying current I, at large distances ($R \gg d$) (see **Figure 9.1**).

$$\mathbf{B} \approx \frac{\mu_0 m}{4\pi R^3}\left(\hat{\mathbf{R}}2\cos\theta + \hat{\boldsymbol{\theta}}\sin\theta\right) \quad [\text{T}] \tag{9.14}$$

$\mathbf{m} = \hat{\mathbf{n}} I\pi d^2$ [A · m^2] is the ***magnetic dipole moment*** ($\hat{\mathbf{n}}$ is the normal to the loop, $m = |\mathbf{m}|$).

The dipole moment leads to the definition of ***magnetization*** and ***magnetization current density***. The magnetization $\mathbf{M}$ is due to a ***magnetization volume current density*** $\mathbf{J}_m$ and a ***magnetization surface current density*** $\mathbf{J}_{ms}$:

$$\mathbf{J}_m = \nabla \times \mathbf{M} \quad [\text{A/m}^2], \quad \mathbf{J}_{ms} = \mathbf{M} \times \hat{\mathbf{n}} \quad [\text{A/m}] \tag{9.28}$$

The magnetization manifests itself in the permeability μ of the material:

$$\mathbf{B} = \mu_0 \mathbf{H}_e + \mu_0 \mathbf{M} = \mu_0 \mu_r \mathbf{H}_e \quad [\text{T}] \tag{9.30–9.34}$$

where $\mu = \mu_0 \mu_r$ [H/m] is the magnetic permeability of the material, μ_r [dimensionless] is its relative permeability, and $\mathbf{H}_e$ is the magnetic field intensity external to the medium. The higher the magnetization, the higher the permeability of the medium.

Magnetic Materials—Properties

Diamagnetic materials are materials with relative permeability slightly smaller than 1.

Paramagnetic materials have relative permeability slightly higher than 1.

Ferromagnetic materials are characterized by very high permeability ($\mu_r \gg 1$).

Ferromagnetic materials exhibit ***hysteresis***—a nonlinear effect due to magnetic domains whereby the relation between the magnetic flux density and magnetic field intensity under AC conditions follows a closed path **(Figure 9.16).**

Hysteresis is responsible for losses but also for the existence of permanent magnets—magnetized materials that retain their magnetization.

Soft magnetic materials are those materials that can be easily demagnetized.

Hard magnetic materials are "hard" to demagnetize and are used for production of permanent magnets.

Interface conditions for the magnetic field define the behavior at interfaces. The magnetic interface conditions between two materials are [see **Eqs. (9.38)** through **(9.45)** and **Table 9.5**]:

$$\hat{\mathbf{n}} \times (\mathbf{H}_{1t} - \mathbf{H}_{2t}) = \mathbf{J}_s, \quad \hat{\mathbf{n}} \times \left(\frac{\mathbf{B}_{1t}}{\mu_1} - \frac{\mathbf{B}_{2t}}{\mu_2}\right) = \mathbf{J}_s \quad \text{and} \quad B_{1n} = B_{2n}, \quad \mu_1 H_{1n} = \mu_2 H_{2n}$$

$\hat{\mathbf{n}}$ points into material (1) and a surface current density $\mathbf{J}_s$ [A/m] may exist at the interface between conductors and nonconductors (see **Figure 9.17**).

Inductance is the ratio of flux linkage and the current that produces it. It is independent of current and only depends on physical dimensions and permeability. Given two circuits made of N_1 and N_2 loops and carrying currents I_1 and I_2, respectively, we define ***self-inductances*** L_{11}, L_{22} and ***mutual inductances*** L_{12}, L_{21} as [see **Eqs. (9.56)** through **(9.59)**]:

$$L_{11}=\frac{N_1\Phi_{11}}{I_1}=\frac{\Lambda_{11}}{I_1},\quad L_{22}=\frac{N_2\Phi_{22}}{I_2}=\frac{\Lambda_{22}}{I_1},\quad L_{12}=\frac{N_2\Phi_{12}}{I_1}=\frac{\Lambda_{12}}{I_1},\quad L_{21}=\frac{N_1\Phi_{21}}{I_2}=\frac{\Lambda_{21}}{I_1}\quad [\mathrm{H}]$$

Φ_{11} is the flux produced by circuit (1) linking all turns of circuit (1), Φ_{12} is the flux produced by circuit (1) linking all turns of circuit (2), and so on. $\Lambda = N\Phi$ is called ***flux linkage***.

Inductance entails assumption of a current in a circuit, calculation of the flux density, calculation of flux, and flux linkage followed by division by the current that generated the flux. The inductance is independent of the assumed current. In infinite structures a more useful relation is self- and mutual inductance per unit length of the device.

External inductance—inductance due to flux outside conductors
Internal inductance—inductance due to flux within the conductor's volume.

Energy stored in the magnetic field is closely related to inductance even where inductors cannot be clearly identified. The basic definition starts with the energy stored in an inductor L due to passage of current I:

$$W_m=\frac{LI^2}{2}\quad [\mathrm{J}] \tag{9.62}$$

In a system of n loops or coils, each with N_i turns and current I_i, the magnetic energy is

$$W_m=\frac{1}{2}\sum_{i=1}^{n}N_i\Phi_i I_i\quad [\mathrm{J}] \tag{9.72}$$

where Φ_i is the total flux in loop i due to all current-carrying loops in the system. Alternatively it may be written in terms of inductance as

$$W_m=\frac{1}{2}\sum_{i=1}^{n}\sum_{j=1}^{n}L_{ij}I_iI_j\quad [\mathrm{J}] \tag{9.77}$$

where L_{ij} is the self-inductance ($i = j$) or mutual inductance ($i \neq j$) between circuits i,j, and the current I_i, I_j are the currents in the circuits. If the currents I_i, I_j produce fluxes in the same direction, L_{ij} is considered positive, if not, negative.

A more general approach is in terms of fields. This is particularly useful when inductances cannot be identified and calculated such as in space. The magnetic energy can be calculated from various field quantities as follows:

$$W_m=\frac{1}{2}\int_{v'}\mathbf{A}\cdot\mathbf{J}dv'=\frac{1}{2}\int_{v}\mathbf{B}\cdot\mathbf{H}dv=\frac{1}{2}\int_{v}\mu H^2dv=\frac{1}{2}\int_{v}\frac{B^2}{\mu}dv\quad [\mathrm{J}] \tag{9.89, 9.90}$$

v' here is the volume in which the current density $\mathbf{J}$ exists whereas v is the volume in which the magnetic field is nonzero. The first of these is useful where $\mathbf{J}$ is easily identified such as in conducting media. In space, where currents may not exist, calculation in terms of the magnetic flux density and field intensity is used.

The integrand in the energy expressions is the energy density:

$$w_m=\frac{\mathbf{A}\cdot\mathbf{J}}{2}=\frac{\mathbf{B}\cdot\mathbf{H}}{2}=\frac{BH}{2}=\frac{\mu H^2}{2}=\frac{B^2}{2\mu}\quad \left[\frac{\mathrm{J}}{\mathrm{m}^3}\right] \tag{9.90, 9.91}$$

Magnetic circuits are based on the equivalence between currents in closed electric circuits and flux in closed magnetic paths. The requirements are for permeability to be high and any gaps in the circuits to be as small as possible to avoid flux leakage around the gaps. Under these conditions the flux in any closed path within the magnetic circuit is

$$\Phi=\frac{\sum_{i=1}^{n}N_iI_i}{\sum_{j=1}^{k}\mathfrak{R}_j}\quad [\mathrm{Wb}] \tag{9.102}$$

With

$$\Re_j = \frac{l_j}{\mu_j S_j} \quad \left[\frac{1}{\mathrm{H}}\right] \tag{9.99}$$

where N_i is the number of turns in coil i, I_i its current, and $\Re_j$ is the magnetic reluctance of segment j of the magnetic path with l_j the length of the segment, μ_j its permeability, and S_j its cross-sectional area (see **Figure 9.35**). Magnetic circuits allow simple calculation of fluxes and magnetic fields in devices that satisfy the basic requirements of a magnetic circuit.

Forces in the magnetic field are defined based on the Lorentz force equation ($\mathbf{F}_m = q\mathbf{v} \times \mathbf{B}$) in **Eq. (9.106)**, which governs forces on moving charges. Since moving charges constitute currents, **Eq. (9.106)** can be developed into more useful relations. For currents in a volume we have

$$\mathbf{F}_m = \int_{v'} \mathbf{J} \times \mathbf{B} dv' \quad [\mathrm{N}] \tag{9.110}$$

where $\mathbf{J} \times \mathbf{B}$ is a volume force density [N/m^3]. In a thin wire carrying current I in a magnetic field:

$$\mathbf{F}_m = \int_{L'} I d\mathbf{l}' \times \mathbf{B} \quad [\mathrm{N}] \tag{9.113}$$

In all cases, the flux density **B** is due to sources other than I (I cannot cause a force on itself).

Equation (9.113) is the basis of Ampère's force law, which defines a force between any two current-carrying wire segments that are part of closed circuits (see **Figure 9.43**):

$$\mathbf{F}_{12} = \frac{\mu_0 I_1 I_2}{4\pi} \int_{p_3}^{p_4} \int_{p_1}^{p_2} \frac{d\mathbf{l}_2' \times \left(d\mathbf{l}_1' \times \hat{\mathbf{R}}_{12}\right)}{R_{12}^2} \quad [\mathrm{N}] \tag{9.117}$$

Here one segment extends between p_1 and p_2, the second between p_3 and p_4, and the vector $\mathbf{R}_{12}$ connects $d\mathbf{l}_1'$ and $d\mathbf{l}_2'$. Both **Eqs. (9.113)** and **(9.117)** can be extended for closed loops [see **Eqs. (9.114)** and **(9.118)**].

Principle of Virtual Work Given the magnetic energy in a system, a force, such as between two faces of a gap, may be calculated by allowing a virtual displacement of one of the surfaces. The change in energy is related to force (**Section 9.7.1**):

$$\mathbf{F} = -\nabla W \quad [\mathrm{N}] \tag{9.121}$$

If there is a force on a system, there may also be a torque. ***Torque*** is simply the force multiplied by armlength. In the case of a loop with magnetic dipole moment **m**, the torque may be written as

$$\mathbf{T} = \mathbf{m} \times \mathbf{B} \quad [\mathrm{N \cdot m}] \tag{9.127}$$

Problems

Magnetic Dipoles and Magnetization

9.1 Application: Magnetic Dipole. A small circular loop of radius a [m] and a small, square loop, $a \times a$ [m^2] are given.

- **(a)** Calculate the magnetic dipole moment of the circular loop.
- **(b)** Justify based on physical considerations the following statement: "At large distances, the magnetic dipole of the square loop is the same as that of a circular loop of identical area."
- **(c)** Calculate the magnetic flux density at a general point in space due to each loop for the conditions in **(b)**.

9.2 Magnetic Dipole. A loop of radius a [m] carries a current I [A]. A second, smaller loop of radius b [m] is placed at point P_1 (on the axis of the first loop, at height h [m] above the loop), at point P_2 (at a horizontal distance h [m], on the plane of the first loop) or at point P_3 so that the planes of the two loops are parallel in all three cases (**Figure 9.57**). In all three cases, $h \gg b$, $a > b$ and $h \gg a$. Calculate the total flux through the small loop produced by the large loop (in free space) when:

(a) The small loop is at point P_1.
(b) The small loop is at point P_2.
(c) The small loop is at point P_3.

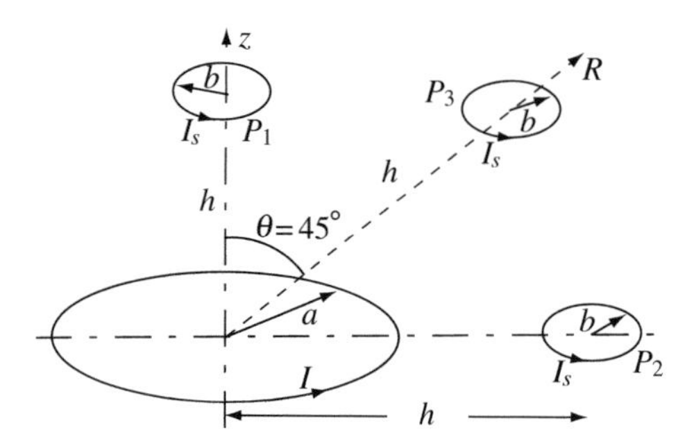

Figure 9.57

9.3 Magnetization. A very long cylindrical magnet has constant magnetization **M** [A/m], directed as in **Figure 9.58**. A solenoid made of thin wires is wound tightly around the magnet, with n turns per unit length of the solenoid. What must be the current (magnitude and direction) in the coil to cancel the magnetization of the magnet?

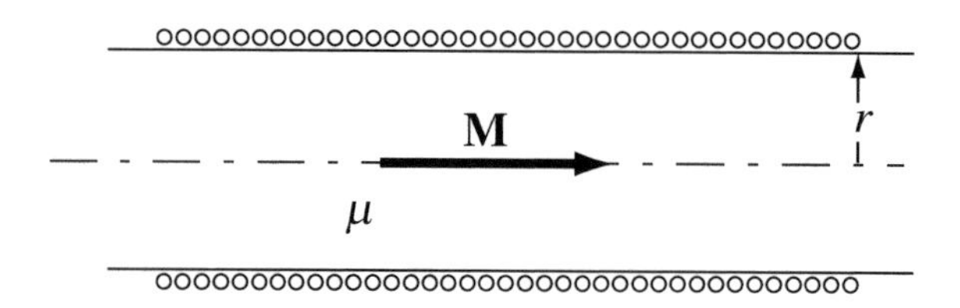

Figure 9.58

9.4 Application: Magnetization and Magnetic Field in a Permanent Magnet. A very long cylindrical magnet has magnetization **M** [A/m] as shown in **Figure 9.58**. The magnet is made of a material with permittivity μ [H/m]. Calculate the magnetic field intensity inside the magnet. Disregard the solenoid shown.

Magnetic Interface Conditions

9.5 Magnetic Interface Conditions. A magnetic field intensity **H** [A/m] is given at the interface between materials (1) and (2) as shown in **Figure 9.59**. Calculate the direction and magnitude of the magnetic flux density in materials (2) and (3). Assume there are no surface currents on the interfaces.

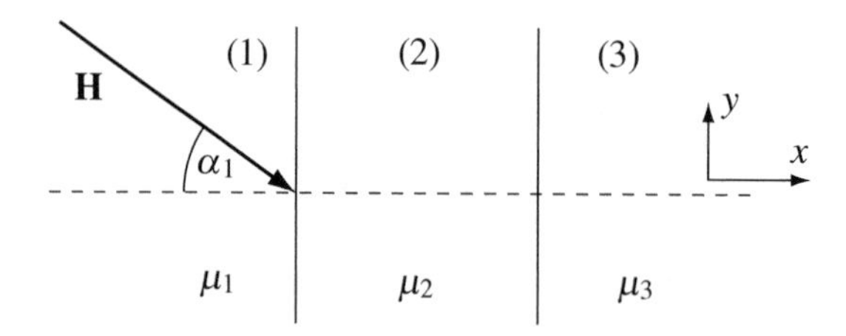

Figure 9.59

9.6 Interface Conditions for Ferromagnetic Media. The relative permeability of a large, flat piece of iron is 100. If the magnetic field intensity in the iron must be at 60° to the surface, what must be the direction of the magnetic field intensity at the surface of the iron piece in air? Assume there are no currents on the interfaces.

9.7 Interface Conditions and Flux Density. A two-layer magnetic sheet is made as shown in **Figure 9.60**. Each sheet is d [m] thick. Permeabilities are $\mu_1 = \mu_0, \mu_2 = 200\mu_0, \mu_3 = 50\mu_0$, and $\mu_4 = \mu_0$ [H/m]. A magnetic flux density in material (1) is given at 30° to the normal and of magnitude $B = 0.01$ T. Calculate the magnetic flux density (magnitude and direction) in materials (2), (3) and (4).

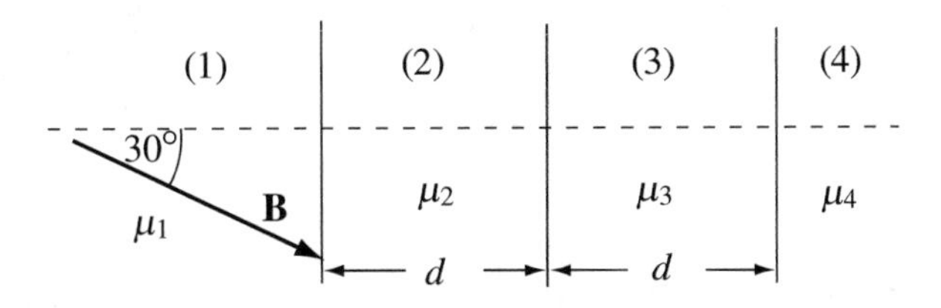

Figure 9.60

Inductance

9.8 Application: Self- and Mutual Inductances of Coils. A coil is wound uniformly in the form of a torus (see **Figure 9.61**). A long solenoid, of radius $a < b$ [m] and n turns per unit length, is inserted in the central hole of the torus as shown. Calculate:

(a) The self-inductance of the toroidal coil.
(b) The self-inductance per unit length of the solenoid.
(c) The mutual inductance between the toroidal coil and the solenoid.

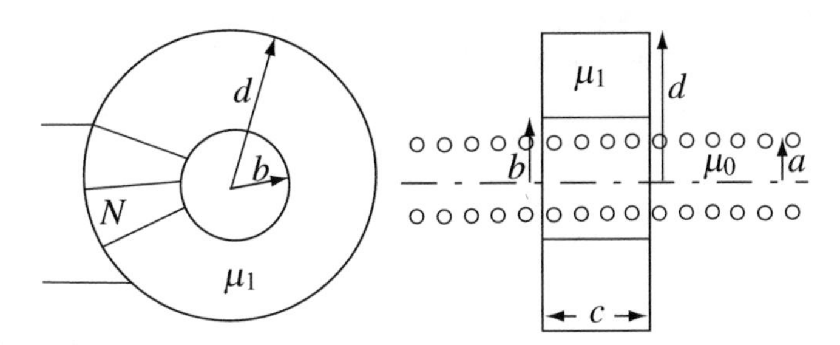

Figure 9.61

9.9 Application: Self- and Mutual Inductances of Coils. Three coils are wound on two toroidal cores as follows: Coil 1 is wound uniformly with N_1 turns on one of the cores. This is shown in two views in **Figure 9.62a** including the dimensions of the core. Coil 2 is wound uniformly with N_2 turns on a second identical core. Now both cores are placed side by side as shown in **Figure 9.62b**. Coil 3 is wound uniformly with N_3 turns on the two cores (on top of coils 1 and 2) as shown in **Figure 9.62b** in cross-section. The permeability of the cores is μ [H/m]. Calculate:

(a) The self inductance of each of the coils
(b) The mutual inductances between each pair of coils
(c) The total inductance of each of the coils if the fluxes produced by the three coils are in the same direction.

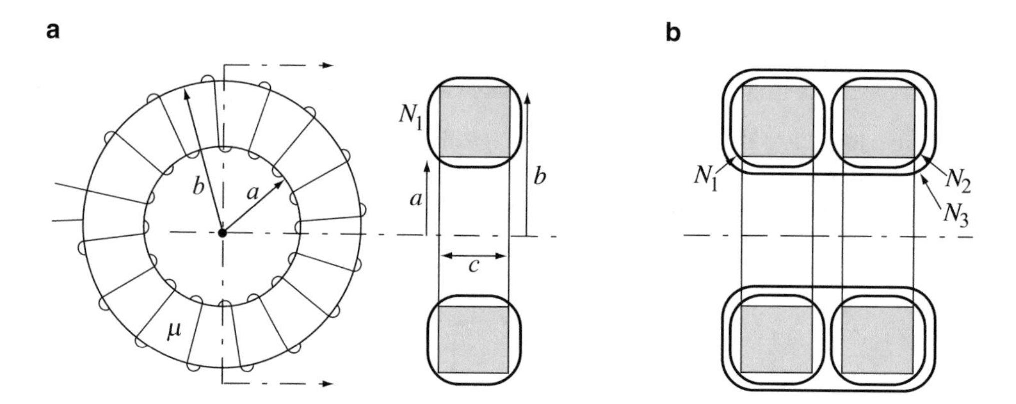

Figure 9.62

9.10 Application: Self-inductance of Long Solenoid. A long solenoid is wound on a hollow cylinder made of iron. Dimensions are given in **Figure 9.63**. Permeability of iron is $\mu_1 > \mu_0$ and that of free space is μ_0 [H/m]. The solenoid carries a current I [A] and has n turns per unit length. Calculate the self-inductance per unit length of the solenoid.

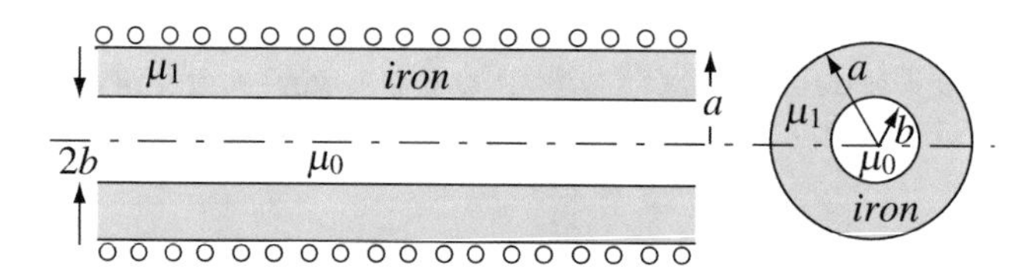

Figure 9.63

9.11 Inductance of a Double Cylinder. A thin sheet of copper is bent into an infinitely long double cylinder as shown in cross section and three-dimensional view in **Figure 9.64**. Calculate the inductance per unit length of the double cylinder. For purposes of calculation, assume $d \to 0$.

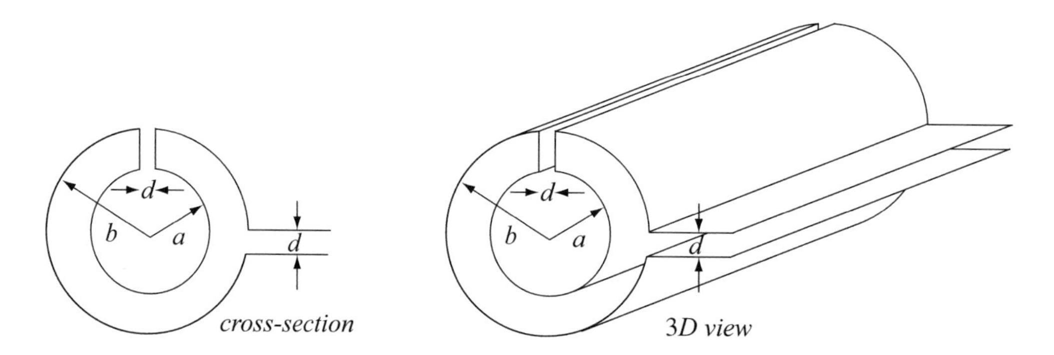

Figure 9.64

9.12 Mutual Inductance Between Straight Wire and Loop. A straight, long wire is placed in a plane in free space. A rectangular loop is also in the same plane as shown in **Figure 9.65a**.

(a) Calculate the mutual inductance between wire and loop.

(b) Now the wire is moved so that it is above the center of the loop without actually touching it as shown in **Figure 9.65b**. What is now the mutual inductance between wire and loop?

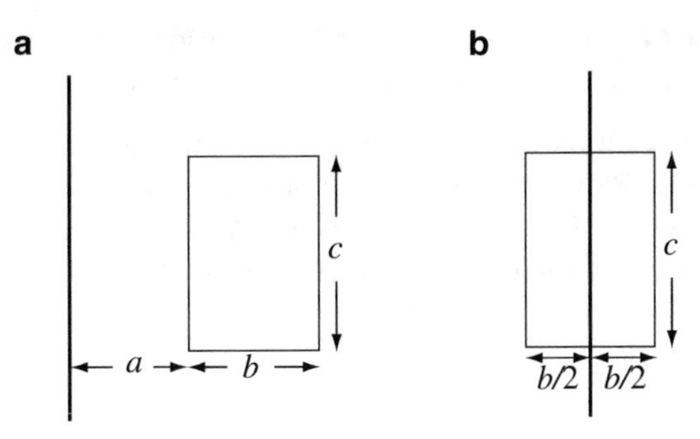

Figure 9.65

9.13 Mutual Inductance Between Wire and Solenoid. A very long wire (infinite for practical purposes) is located at a distance d [m] from the center of an infinite solenoid in free space (**Figure 9.66**). The solenoid has a radius b [m] ($b < d$) and n turns per unit length. The solenoid and the wire are at 90° to each other, as shown. Calculate the mutual inductance between solenoid and wire.

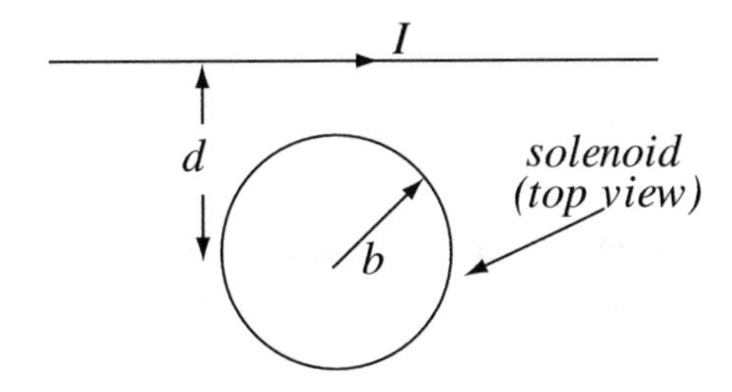

Figure 9.66

9.14 Mutual Inductance Between Coil and Solenoid. A very long (infinite) solenoid with N turns per unit length and radius a [m] is given. A coil with two turns and radius b [m] ($b > a$) is located (in free space) as shown in **Figure 9.67**. The coil and solenoid do not touch.

(a) Calculate the mutual inductance between the coil and solenoid.
(b) Suppose the coil is now rotated so that it makes a 45° angle with the axis of the solenoid. What is now the mutual inductance between the coil and solenoid? (The solenoid is entirely enclosed within the coil.)

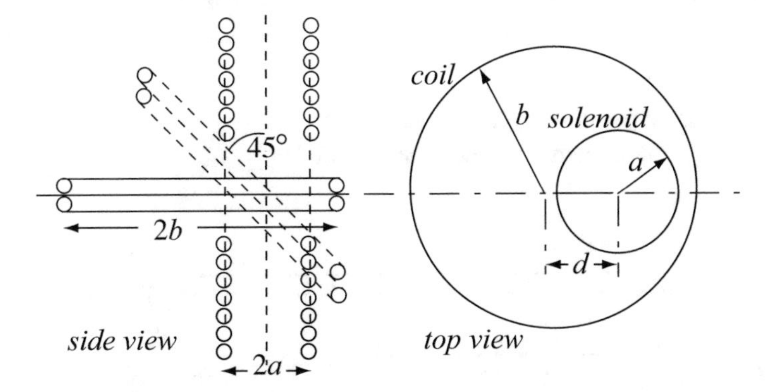

Figure 9.67

9.15 Application: Mutual Inductance Between Loops. The geometry and conditions in **Problem 9.2** are given. Calculate the mutual inductances L_{ab} and L_{ba} between the large loop and the small loop if:

(a) The small loop is at point P_1 in **Figure 9.57** and parallel to the large loop. Write the necessary assumptions.
(b) The small loop is at point P_2 in **Figure 9.57** and parallel to the large loop. Write the necessary assumptions.
(c) T small loop is at point P_3 in **Figure 9.57** and parallel to the large loop. Write the necessary assumptions.
(d) Based on the calculation in **(a)** through **(c)**, does the relation $L_{ab} = L_{ba}$ hold? Explain.

9.16 Application: Inductors in Series and Parallel. Three inductors, each of inductance L [H] are connected as shown in **Figure 9.68**. Assume first that the inductors experience no mutual inductance.

(a) Calculate the total inductance for the series connection in **Figure 9.68a**.
(b) Calculate the total inductance for the parallel connection in **Figure 9.68b**.
(c) Calculate the total inductance for the connection in **Figure 9.68c**.
(d) Repeat (**a**) through (**c**) if between each two inductors there is a mutual inductance equal to M [H]. Assume the fluxes in all three coils are in the same direction.

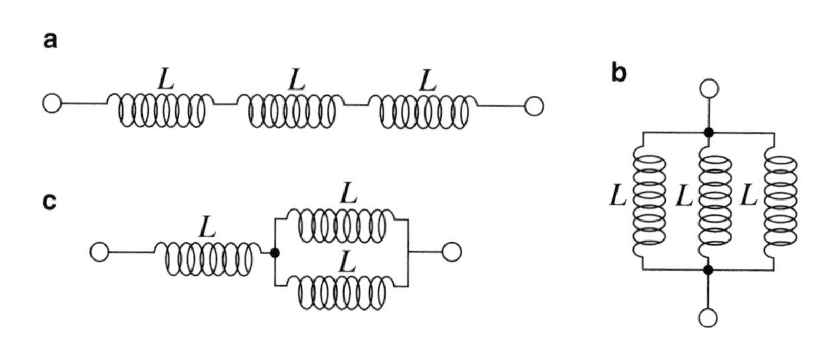

Figure 9.68

9.17 Application: Inductance per Unit Length of Coaxial Cables. A coaxial cable is made of an inner cylindrical shell of radius b [m] and an outer, concentric cylindrical shell of radius a [m], separated by a dielectric with permeability μ_0 [H/m]. Calculate the inductance per unit length of the coaxial cable. Assume the shells are very thin.

9.18 Application: Internal and External Inductances. A coaxial cable is made of a solid inner conductor of radius b [m] and a concentric, outer solid conductor of inner radius a [m] and outer radius c [m]. The conductors are separated by a material of permeability μ_0 [H/m]. Assume any current in the conductors is uniformly distributed throughout the conductor's cross-sectional area and the permeability of the conductors is μ_0 [H/m]. Calculate:

(a) The internal inductance per unit length due to the inner conductor.
(b) The internal inductance per unit length due to the outer conductor.
(c) The external inductance per unit length of the cable.
(d) The total inductance per unit length of the cable.

9.19 Application: Inductance Per Unit Length of Cables. Two two-wire cables carry a current I_1 [A] and I_2 [A] as shown in **Figure 9.69**. Both cables can be assumed to be very long (infinite) and both are placed flat on a plane in free space. Calculate:

(a) The external self-inductance per unit length of each cable.
(b) The mutual inductance per unit length between the two cables.

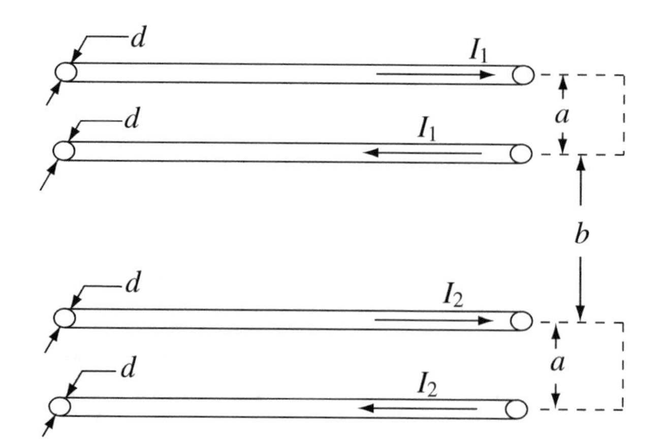

Figure 9.69

9.20 Application: Self- and Mutual Inductance in Printed Circuit Boards. The configuration shown in **Figure 9.70** represents part of a printed circuit board (PCB). The two outer conductors represent the power supply traces and may be assumed to be long, whereas the loop in the center is part of the functional circuit and is small. The outer traces are of width $2t$. The permeability of the PCB is μ_0 [H/m] and the dimensions are as follows: $a = 25$ mm, $b = 20$ mm, $c = 4$ mm, $d = 6$ mm, and $t = 1$ mm. Calculate:

(a) The external self-inductance per unit length of the outer circuit.
(b) The total self-inductance per unit length of the outer circuit.
(c) The mutual inductance between the loop and the outer circuit. Neglect the thickness of the traces of the small loop for the purpose of this calculation.

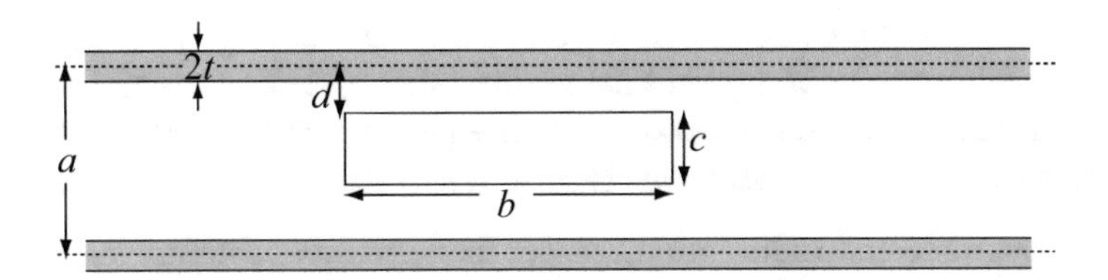

Figure 9.70

Energy

9.21 Stored Energy in a System of Inductors. An infinite solenoid with n_1 turns per unit length and radius a [m] carries a current I [A]. A second infinite solenoid, with n_2 turns per unit length and radius b [m], is placed over the first as shown in **Figure 9.71a**. The two solenoids do not actually touch. **Figure 9.71b** shows the relation of the two solenoids from an axial view. The second solenoid also carries a current I [A]. Calculate the change in energy per unit length in the system (made of the two solenoids) when:
(a) The current in the inner solenoid is reversed.
(b) The current in the outer solenoid is switched off.
(c) The current in the inner solenoid is switched off.

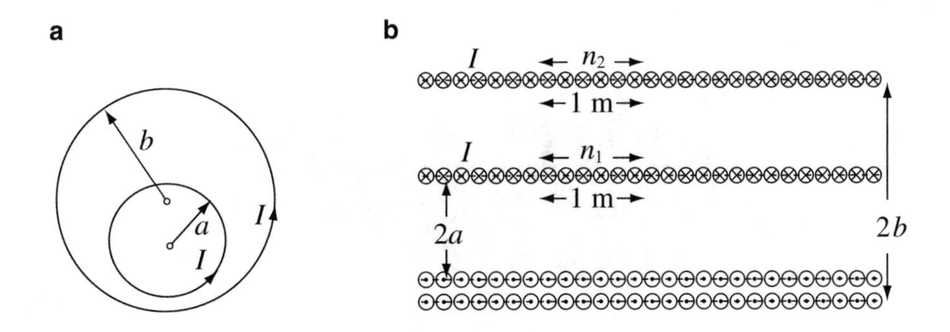

Figure 9.71

9.22 Change in Stored Energy Due to Change in Material properties. A toroidal coil of inner radius r_1 [m], outer radius r_2 [m], and square cross section is shown in **Figure 9.72**. The material of the torus has permeability $\mu_1 > \mu_0$, the coil has N turns and carries a current I [A]. Calculate the change in the stored magnetic energy if the material inside the torus is removed (the core of the torus is in effect replaced with free space).

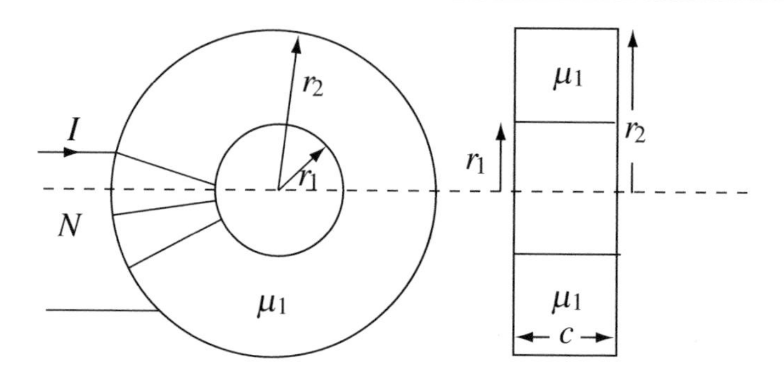

Figure 9.72

9.23 Stored Energy in a Gap. A toroidal coil of inner radius r_1 [m], outer radius r_2 [m], and square cross section is given as shown in **Figure 9.72**. The material of the torus has permeability $\mu_1 > \mu_0$, the coil has N turns and carries a current I [A]. Calculate the change in the stored magnetic energy if a gap of length l_g [m] is cut in the torus. Assume the gap does not create flux leakage and the average radius is $(r_2 + r_1)/2$ [m].

9.24 Stored Energy in Series-Connected Inductors. Consider the torus and configuration given in **Figure 9.25**. For the given values and with $N_3 = 2N_2 = 2N_1$ turns calculate:

(a) The total energy stored in the torus, if all three fluxes produced by the coils are in the same direction.

(b) Now, all three coils are connected in series to a current I [A] ($I_1 = I_2 = I_3 = I$). It is required that the torus store the minimum possible energy. Show how the coils must be connected and calculate the energy stored and the equivalent inductance under this condition.

(c) The configuration now is as in **(b)**, but it is required that the torus store maximum energy. Show how the connections must be made to accomplish this and calculate the stored energy and the equivalent inductance.

9.25 Application: Energy Stored in a Magnet. A cylindrical permanent magnet of radius a [m] with constant magnetization $\mathbf{M} = \hat{\mathbf{z}}M_0$ [A/m] and infinite in length is given. Calculate the energy stored per unit volume of the magnet. Assume permeability of the magnet is μ_0 [H/m].

9.26 Work Necessary to Modify a Magnetic Core. A long (infinite) solenoid of radius a [m] has n turns per unit length. The turns carry a current I [A]. An equally long piece of iron also of radius a [m] is placed in the solenoid, as shown in **Figure 9.73**. The relative permeability of iron is μ_r and that of free space is 1. Calculate the total work per unit length of the solenoid necessary to pull the iron completely out of the solenoid. Does this work increase or decrease the potential energy of the system? Explain.

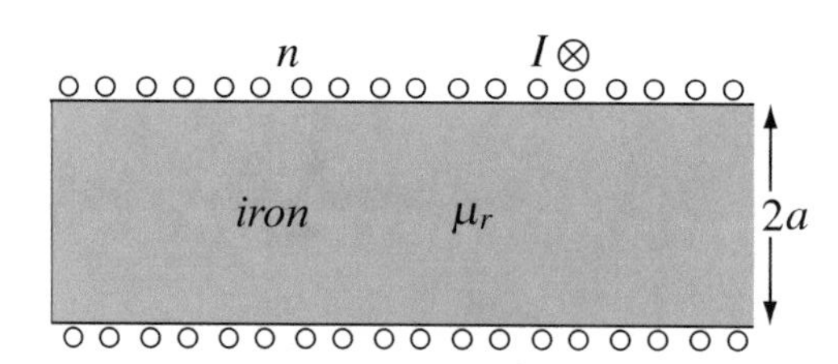

Figure 9.73

9.27 Change in Energy Due to Change in Mutual Inductance. A square loop b [m] by b [m] is located outside a very long solenoid with a total of N turns. Both the solenoid and the loop carry a current I [A] as shown in **Figure 9.74a**. Assume that the solenoid is d [m] long but that its field is identical to an infinite solenoid of the same radius and that there is no mutual inductance between loop and solenoid in **Figure 9.74a**. Assume free space.

(a) Calculate the change in energy in the system if the loop is placed inside the solenoid ($b < a\sqrt{2}$) while the loop is perpendicular to the axis of the solenoid (**Figure 9.74b**).

(b) Calculate the change in energy in the system if the loop is placed over the solenoid ($b > a$) while the loop is perpendicular to the axis of the solenoid (**Figure 9.74c**).

(c) What are the answers to **(a)** and **(b)** if the current in the square loop is reversed?

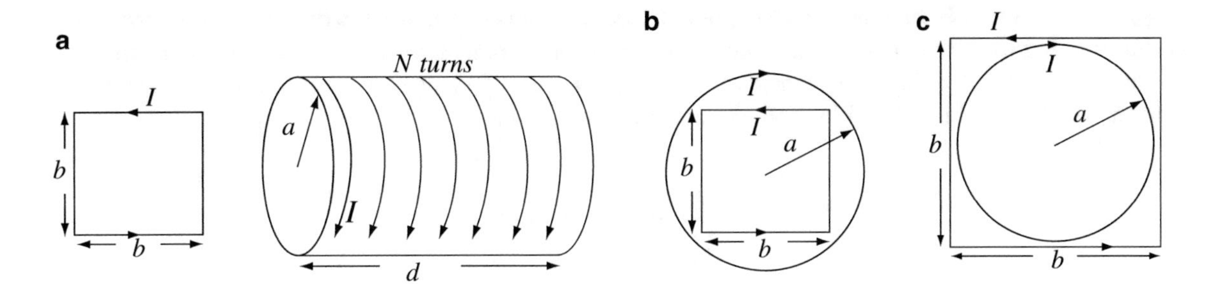

Figure 9.74

Magnetic Circuits

9.28 Application: Flux in a Toroidal Magnetic Circuit. A straight, long wire passes at the center of a torus as shown in **Figure 9.75** and carries a current I [A]. The torus has permeability μ [H/m].

(a) Calculate the total flux in the torus due to the current in the wire.
(b) What is the flux if a gap of length l_g [m] is cut in the torus (dotted lines)? Assume $(d - b) \gg b$.

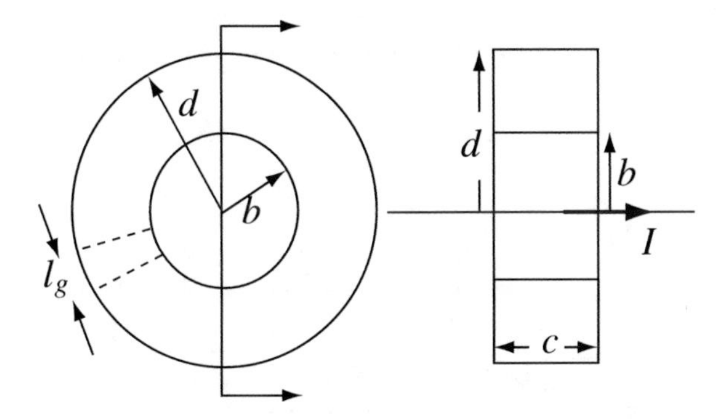

Figure 9.75

9.29 Application: Magnetic Circuits. Calculation of Flux and Field Intensity. A magnetic circuit is given in **Figure 9.76.** The magnetic circuit has two gaps as shown. A single turn of wire carrying a current I [A] is placed in the center of one of the gaps. The loop is b by c in dimensions. Assume $e > b$, that all flux is contained within the magnetic circuit and that the magnetic path length is the average length of the corresponding section. Use the following: $a = 10$ cm, $b = 1$ cm, $c = 2$ cm, $d = 5$ cm, $e = 4$ cm, $g = 0.1$ cm $\mu_1 = 200\mu_0$ [H/m], $I = 0.4$ A. Calculate:

(a) The flux in gap (2) if the loop is placed in gap (1).
(b) The minimum and maximum magnetic field intensity (H) in the circuit if the loop is placed in gap (1). Where do these occur?

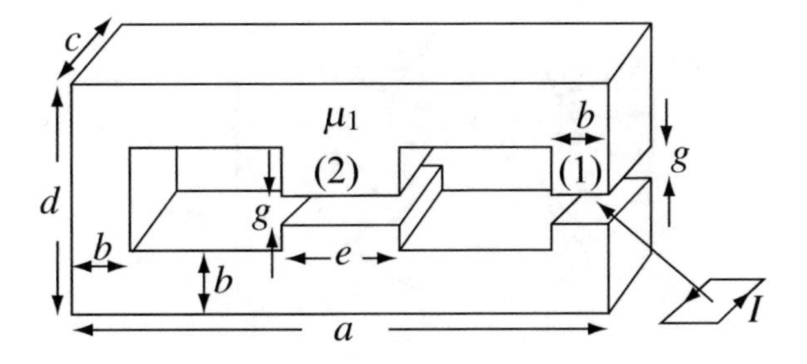

Figure 9.76

9.30 Application: Mutual Inductance in a Magnetic Circuit. A torus with average radius r_0 [m] and cross-sectional area as shown in **Figure 9.77** is given. A coil with N turns is wound around the torus. A small gap of length l_g [m] is cut in the torus and a loop of radius a [m] is inserted in the gap, such that the loop is centered in the gap. Calculate the mutual inductance between the loop and the coil. Assume $r_0 \gg b$, $a < b$.

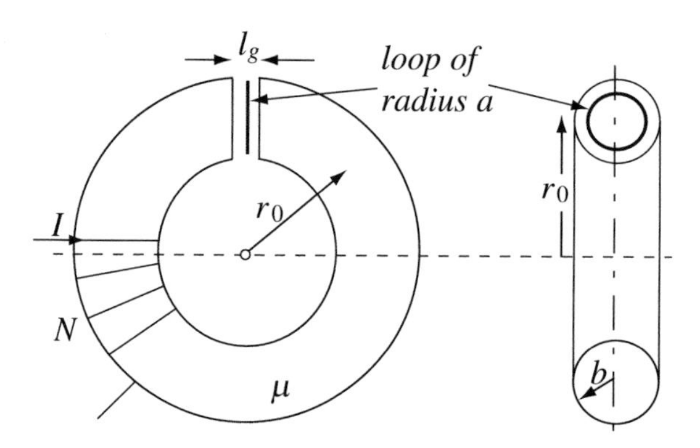

Figure 9.77

9.31 Magnetic Circuit with Different Materials. The magnetic circuit in **Figure 9.78** is given. The two halves are made of different materials with different permeabilities. The length of the gap is l_g [m]. The cross-sectional area of the core is the same everywhere. Calculate the magnetic field intensity in the gap. Assume $a \gg l_g$ and $d \gg l_g$.

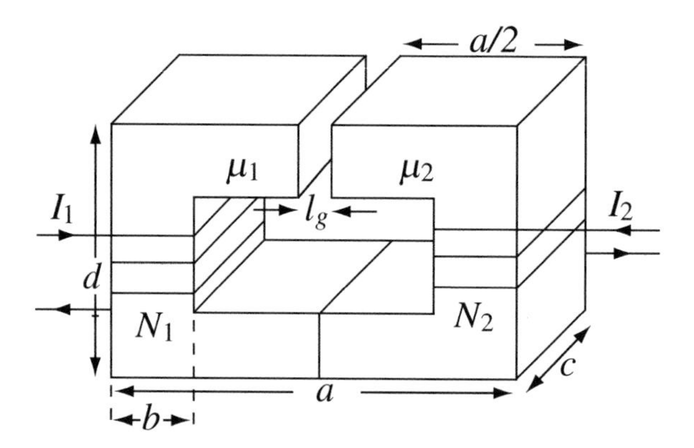

Figure 9.78

9.32 Magnetic Circuit with Different Materials. A torus is made of two types of materials with four small gaps as shown in **Figure 9.79**. Assume the gaps have properties of free space. The coil has N turns, uniformly wound around the torus and carries a current I [A]. Calculate the flux density in the gaps assuming there is no flux leakage at the edges of the gaps.

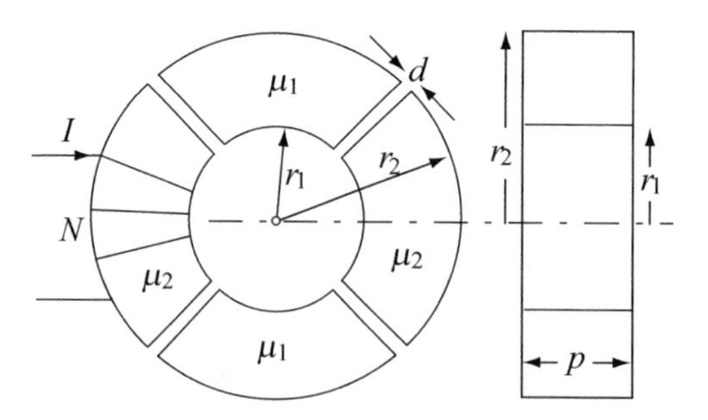

Figure 9.79

9.33 Application: Stored Energy and Magnetic Flux Density in a Magnetic Circuit. The magnetic circuit in **Figure 9.80** is given. The two halves are made of different materials with different permeabilities. The length of each gap is l_g [m]. The cross-sectional area of the core is constant. Take the magnetic path as the average length of the core:

(a) If I_1 and I_2 are known, calculate the total energy stored in the magnetic field.
(b) If I_1 is known, calculate I_2 such that the magnetic field intensity in the gap is zero. Show its direction.

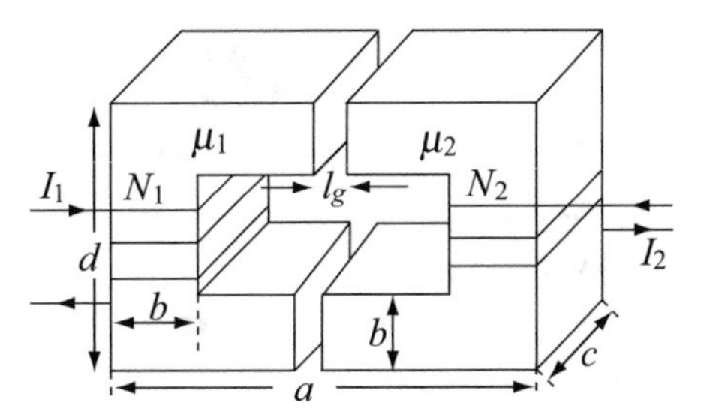

Figure 9.80

9.34 Stored Energy in a Magnetic Device. The core shown in **Figure 9.81** is made of two pieces. Two gaps exist between the two pieces as shown. The cross-sectional area of the two pieces is the same and is constant. The relative permeability of the upper and lower piece tends to infinity. Currents, number of turns, and dimensions are as shown. Calculate the total magnetic energy stored in the magnetic field of this device.

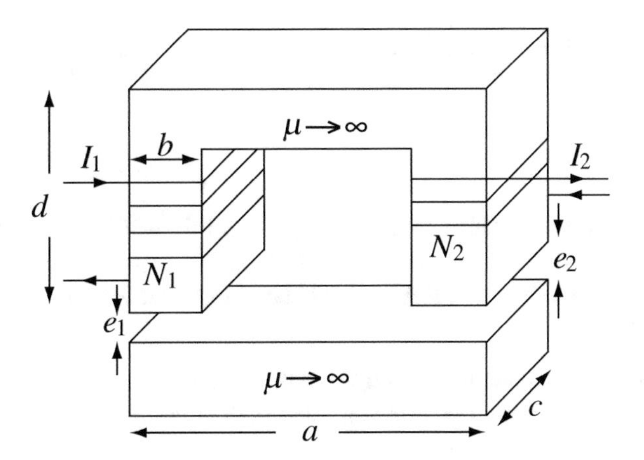

Figure 9.81

9.35 Mutual Inductance. An iron core is made as shown in **Figure 9.82**. A coil (L_1) with N_1 turns is wound on the right leg of the core and a coil (L_2) with N_2 turns is wound on the left leg of the core. Calculate the mutual inductance between the two coils.

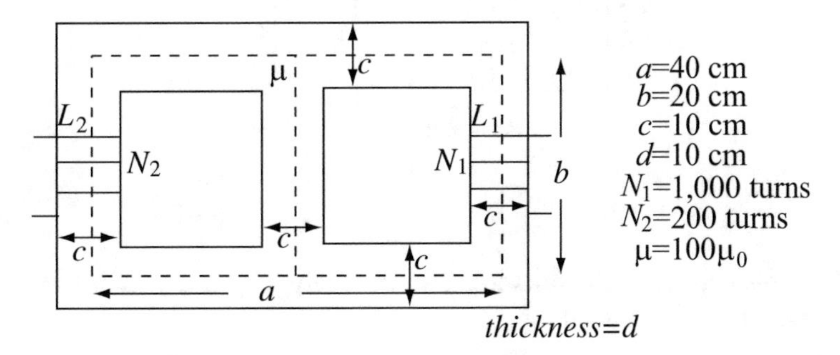

Figure 9.82

Forces

9.36 Application: Force on a Current-Carrying Conductor. A thin bar 1 m long carries a current $I = 10$ A. What is the maximum force that the terrestrial magnetic field exerts on the bar? Assume the magnetic field of the planet is parallel to the surface of the earth and equal to $B = 50$ μT. Show the required orientation of the bar so that the force is maximum.

9.37 Application: Force on a Loop. A current I_1 [A] flows in a thin conducting wire in the positive z direction. A rectangular loop carries a current I_2 [A] as shown in **Figure 9.83a**. Dimensions are as shown and the loop and wire are on a plane. In **Figure 9.83b** assume the wire and loop are on a plane but do not touch. Assuming free space, find:

(a) The force on the loop in **Figure 9.83a** (magnitude and direction).
(b) The force on the wire in **Figure 9.83a** (magnitude and direction).
(c) The force on the wire as a function of the distance a in **Figure 9.83b** for $0 < a < b$.
(d) What is the force in **(c)** if $a = b/2$?
(e) What is the force if $a \to 0$?

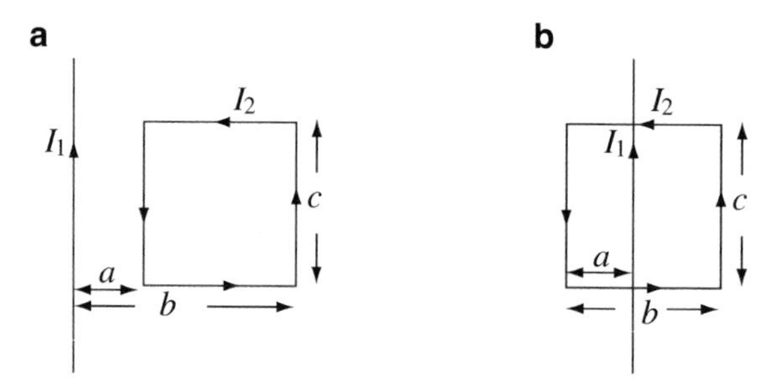

Figure 9.83

9.38 Forces Between Current-Carrying Conductors. A very long, thin wire passes between two rectangular loops as shown in **Figure 9.84**. The wire carries a current $I_0 = 3$ A, loop (1) carries a current $I_1 = 4$ A, and loop (2) carries a current $I_2 = 2$ A, in the directions shown in **Figure 9.84**. The loops and the wire are on the y–z plane with the loops symmetric about the y-axis.

(a) Calculate the force on each loop due to the current in the wire.
(b) Calculate the total force on the wire due to the currents in the two loops.

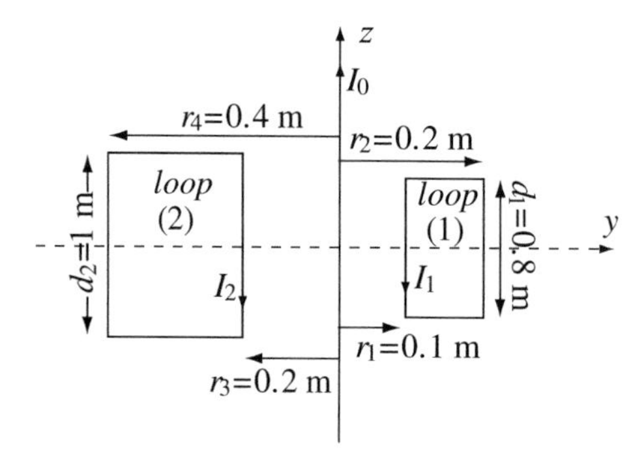

Figure 9.84

9.39 Forces on Thick, Current-Carrying Conductors. Two infinitely long conducting bars (shown in cross section in **Figure 9.85**) carry a current I [A] as shown.

(a) Write an expression for the force per unit length between the two conductors (do not evaluate the integrals).
(b) Is this an attraction or repulsion force? Explain.

(c) **Optional:** Integrate the expression in **(a)** for $a = 0.05$ m, $b = 0.2$ m, $d = 0.2$ m, $c = 0.1$ m, $I = 1{,}000$ A, and $\mu = \mu_0$ [H/m] everywhere.

Note: The integral is very tedious. You may want to write a script or computer program to accomplish the integration.

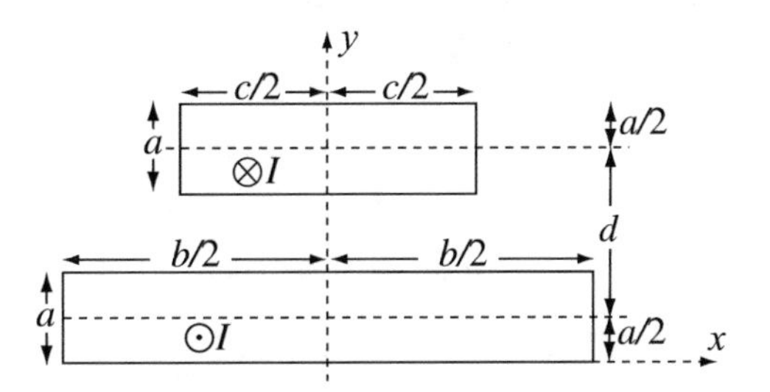

Figure 9.85

9.40 Force Between Loops. Two square loops in free space are parallel to each other and carry currents as shown in **Figure 9.86**.

(a) Show, without calculations, that the two loops attract each other.
(b) Write the expression for the total force on one of the loops.
(c) **Optional:** Evaluate the force in **(b)** for $a = d = 1$ m, $b = 2$ m, and $I_1 = I_2 = 1$ A.

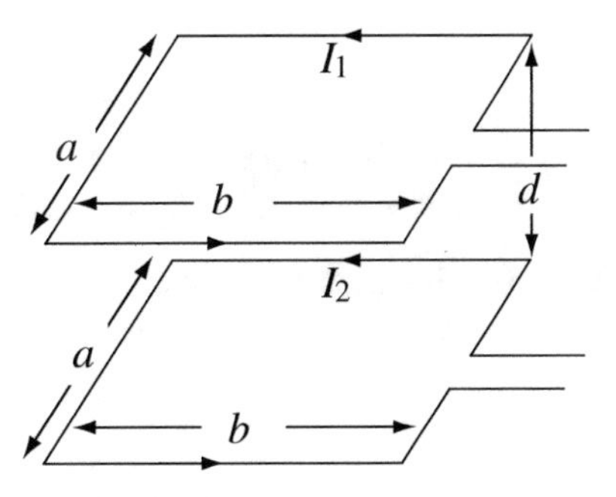

Figure 9.86

9.41 Application: Forces in a Magnetic Circuit. A lifting electromagnet designed to lift steel bars is built as shown in **Figure 9.87**. To protect the poles, they are coated with a thin polymer of thickness t so that the minimum gap between the magnet and the bar being lifted equals t. Assuming the permeability of both the electromagnet core and the steel bar equal $200\mu_0$ [H/m], calculate:

(a) The weight per unit current in the coil that the device can hold.
(b) What is the current needed to hold a 1 ton bar.

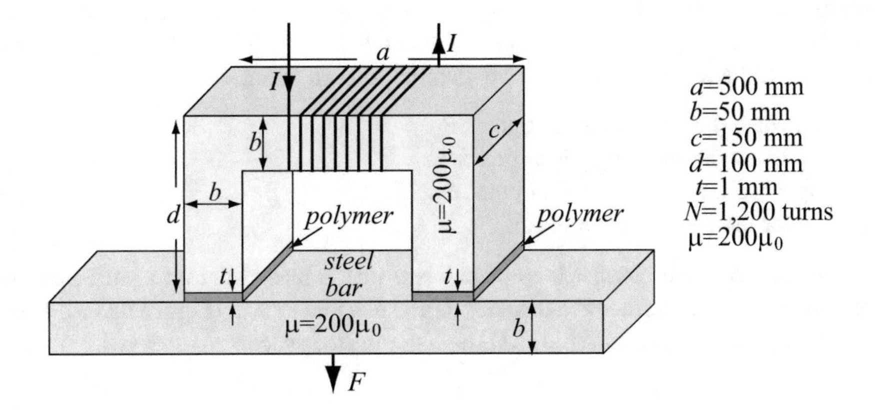

Figure 9.87

9.42 Application: Forces on Magnetic Poles of a Gap. In **Figure 9.88**, calculate the force that exists between the two faces of the gap. The magnetic path has a permeability equal to μ [H/m] and the core is d [m] thick. a, b, and g are the average path lengths of the corresponding magnetic path sections.

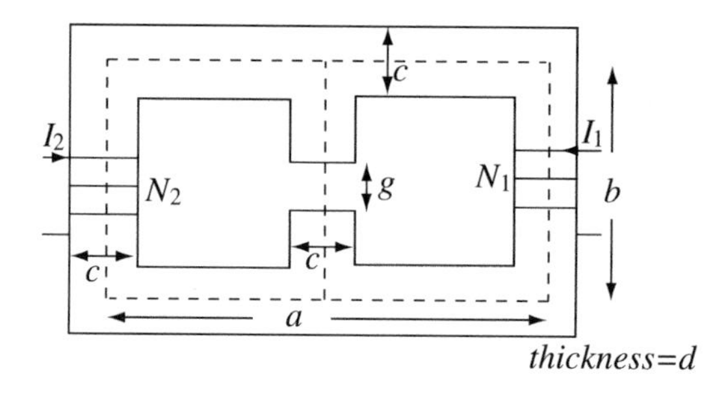

Figure 9.88

Torque

9.43 Application: Torque on a Current-Carrying Loop. Calculate:

(a) The torque on the loop in **Figure 9.83a**. Assume the loop and wire are on a plane in free space.

(b) The torque on the loop in **Figure 9.83b**. Assume the loop and wire are on a plane in free space, are very close to each other but do not touch.

9.44 Torque on a Current-Carrying Bar. A thin bar of length $a = 1$ m carries a current I and is placed in a uniform magnetic field. The magnetic flux density is $\hat{\mathbf{B}} = \hat{\mathbf{z}}B_0$ where $B_0 = 0.4$ T and the current in the bar is 0.1 A. The bar is pivoted at its center and connected to the current as shown in **Figure 9.89a**.

(a) Calculate the torque on the bar. Show the direction of forces.

(b) Suppose the source is connected as in **Figure 9.89b**. What is now the torque on the bar?

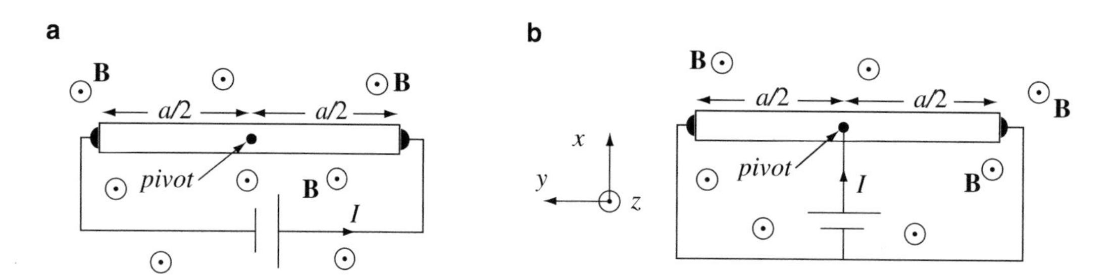

Figure 9.89

9.45 Application: Torque on Magnetic Dipoles. A loop of radius a [m] carries a current I [A]. A second, small loop of radius b [m] carries a current I_s [A] and is placed at point P_1, P_2, or P_3 as shown in **Figure 9.57**. In all cases, $h \gg b$, $a \gg b$, $h \gg a$, and the axes of the loops are parallel to each other. Calculate the torque:

(a) On the small loop when the small loop is at point P_1.

(b) On the small loop when the small loop is at point P_2.

(c) On the small loop when the small loop is at point P_3.

(d) On the large loop for the configuration in **(c)**.

9.46 Torque on Small Loops. A small circular loop of radius a [m] is placed flat on a surface. A second, square loop, $a \times a$ [m^2], is placed on the same surface at a distance d [m] such that $d \gg a$ (**Figure 9.90a**). The circular loop carries a current I_c [A] and the square loop a current I_s [A], both are in free space.

(a) Find the magnetic dipole moments of the circular and square loops in **Figure 9.90a**.

(b) Now, the square loop is rotated around its axis as in **Figure 9.90b** so that the side *de* is up without changing the directions of the currents. Calculate the torque on the round loop.

(c) Assuming the conditions in **(b)**, the number of turns in the round loop is increased to N. How does this affect the torque on the square loop?

(d) Assuming the conditions in **(c)**, the current in the round loop is changed to flow in the direction opposite that shown in **Figure 9.90b** and the number of turns in the square loop is also increased to N. Calculate the torque on the round loop.

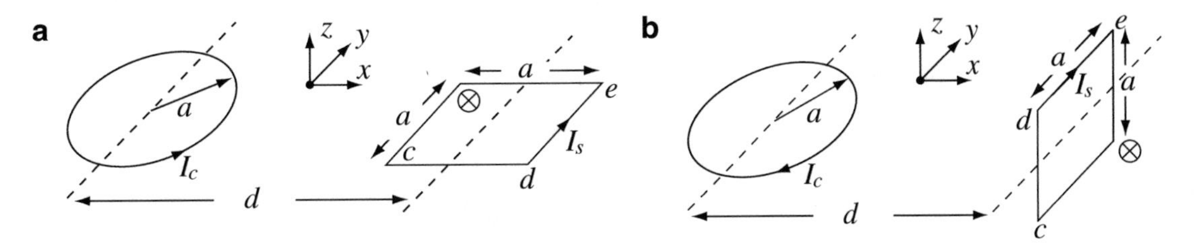

Figure 9.90

Faraday's Law and Induction

10

Whether Ampere's beautiful theory were adopted, or any other, or whatever reservation were mentally made, still it appeared very extraordinary, that as every electric current was accompanied by a corresponding intensity of magnetic action at right angles to the current, good conductors of electricity, when placed within the sphere of this action, should not have any current induced through them, or some sensible effect produced equivalent in force to such a current.

—Michael Faraday (1791–1867), article 3 in the first series of "Experimental Researches in Electricity," Nov. 24, 1831

10.1 Introduction

In the previous chapters, we found it useful to treat the electric and magnetic phenomena separately. In **Chapters 4** through **7**, we treated electrostatic fields by relying on the two postulates:

$$(\nabla \times \mathbf{E} = 0 \quad \text{and} \quad \nabla \cdot \mathbf{D} = \rho_v) \quad \text{or} \quad \left(\oint_C \mathbf{E} \cdot d\mathbf{l} = 0 \quad \text{and} \quad \oint_s \mathbf{D} \cdot d\mathbf{s} = Q \right)$$

In **Chapters 8** and **9**, some of the basic magnetic phenomena were introduced, now relying on the following two postulates:

$$(\nabla \times \mathbf{H} = \mathbf{J} \quad \text{and} \quad \nabla \cdot \mathbf{B} = 0) \quad \text{or} \quad \left(\oint_C \mathbf{H} \cdot d\mathbf{l} = I \quad \text{and} \quad \oint_s \mathbf{B} \cdot d\mathbf{s} = 0 \right)$$

These postulates allowed us to treat a large number of applications and gain insight into the behavior of the electric and magnetic fields governed by the postulates.

Two aspects of this approach should be apparent by now:

(1) The discussion of electrostatic fields was independent of that for magnetostatic fields. Even though we saw in **Chapter 7** that a static electric field can cause a flow of charge in a conductor and, therefore, a current, the electric field does not depend on this current and any of its consequences; that is, the static electric field is uniquely defined from the charge distribution in the system. The current thus generated does, in turn, generate a static magnetic field as defined by Ampere's law. However, this magnetic field, while it may coexist with the electric field, does not affect it.

(2) The discussion was limited to static applications. We have not specifically stated that the postulates of the magnetostatic field are not valid under time-varying conditions, but because the magnetostatic field relied on steady electric currents, time-dependent phenomena could not be included.

The question is: What happens if the fields are time varying? Or, perhaps, a better statement would be: How do we need to modify the fundamental postulates to treat time-varying electric and magnetic fields?

This chapter discusses the question of time dependency in some detail. We will see that the electric and magnetic fields under time-varying conditions are interdependent. A time-varying electric field generates a time-varying magnetic field and vice versa. Thus, the time-dependent magnetic flux density **B** (and therefore **H**) is dependent on the electric field intensity **E** (and therefore on **D**). The above postulates will be modified to account for this dependency. Since now we must treat both the electric

N. Ida, *Engineering Electromagnetics*, https://doi.org/10.1007/978-3-030-15557-5_10

and magnetic fields as coupled vectors, the postulates for both fields must be included. A total of four relations are required to specify the electromagnetic field under time-dependent conditions: the curl of the electric field intensity **E**, the divergence of the electric flux density **D**, the curl of the magnetic field intensity **H**, and the divergence of the magnetic flux density **B**.

Because of the dependency between the electric and magnetic fields, we should use the term electromagnetic field when dealing with time-dependent fields. This term indicates that the two fields cannot be treated separately. The remaining part of this books deals with the electromagnetic field. In some cases, we will find it easier or more useful to treat the electric field or the magnetic field alone, but whichever field we choose to emphasize, it should be remembered that the other field can always be derived if necessary and, more importantly, it always exists.

After introducing Faraday's law as the first basic law governing time-dependent fields, we discuss some applications of Faraday's law. These include traditional power devices like transformers, motors, and generators as well as more recently developed devices like linear motors and levitating mechanisms. Other applications are acceleration of particles, electromagnetic testing of materials, electromagnetic ore prospecting, heating and melting of materials, magnetic braking, and many others.

The main difference at this stage, between the time-dependent and time-independent fields, is in the process of induction: A time-dependent magnetic field produces an electric field and that, in turn, induces currents in conducting materials. This induction is fundamental to the operation of many very important devices, including transformers and generators. There are two mechanisms of induction: one is due to change in the magnetic flux; the other due to motion in the magnetic field. Both are discussed since they can exist simultaneously in the same device and because both are important in design.

10.2 Faraday's Law

The coupling between the electric and magnetic fields is based on an experimental relation known as Faraday's law. This law was formulated in 1831 by Michael Faraday following a series of experiments. Faraday[1] observed that if he moved a closed loop in the magnetic field of a magnet or if he moved the magnet while the loop remained stationary, a current flowed in the loop. This current was not due to external sources but rather was induced in the loop by the change in the magnetic flux. He also found that the current was proportional to the rate of change of flux.

The current observed by Faraday is due to an induced voltage in the loop. This voltage is called an ***electromotive force*** or ***emf***. The electromotive force produced in this experiment can be written as

$$\text{emf} = -\frac{d\Phi}{dt} \quad [\text{V}] \tag{10.1}$$

where Φ is the flux through the loop. The physical situation is shown in **Figure 10.1**. In general, there may be more than one loop in the same location; therefore, a more general relation for the electromotive force is

$$\text{emf} = -N\frac{d\Phi}{dt} \quad [\text{V}] \tag{10.2}$$

where N is the number of loops. This type of emf is also called a ***transformer action emf*** for reasons we will see shortly.

To find the relation between the electric field intensity and the magnetic flux density, we use the definitions of flux and of electromotive force and substitute these in Faraday's law.

In **Chapter 7** (see, for example, **Section 7.7.1**), we discussed the idea of an electromotive force as the source of steady currents in conductors and in conjunction with Kirchhoff's laws. There, we found that the electromotive force is the closed contour integral of the nonconservative electric field intensity (the closed contour integral of a conservative electric field intensity is zero):

[1] Michael Faraday (1791–1867) had little schooling, mostly as self-education. At age 22, he became assistant to Sir Humphry Davy (a well-known chemist and member of the Royal Institution) who helped his education. In 1821, Faraday demonstrated the operation of an electric motor. In 1825 he became director of the Royal Institution Laboratory and, there, in 1831 he discovered what is now known as Faraday's law. His experiment showed that the motion of a magnet near a wire loop produces an electromotive force, and therefore a current, in the closed loop. Faraday experimented in other areas as well. These included dielectrics, electrolysis, and polarization of light. The farad unit of capacitance is named after him in honor of his many and varied achievements. Faraday worked long and hard and is considered the ultimate experimentalist. His work was not all fun. In 1862, he became very ill from inhaling mercury vapors (he was using mercury for contacts) and it took him almost 2 years to recover. Faraday, more than anyone else, laid the foundation for the electromagnetic theory later developed by James Clerk Maxwell into what is today known as Maxwell's theory, and this is summarized by Maxwell's equations. Faraday's law is one of the four equations. Later, when Maxwell unified the electromagnetic theory, he gave credit to many, but mostly to Faraday.

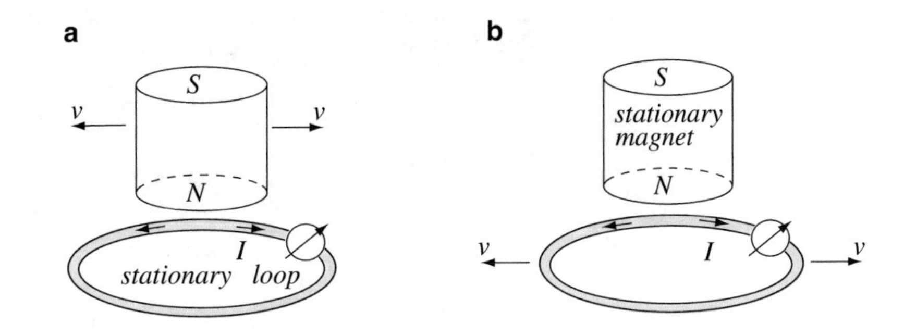

Figure 10.1 Faraday's law: (**a**) Movement of a permanent magnet in the presence of a loop generates an electromotive force and a current in the loop. (**b**) Movement of the loop in the presence of the stationary magnet generates an identical current

$$\text{emf} = \oint_C \mathbf{E} \cdot d\mathbf{l} \quad [\text{V}] \tag{10.3}$$

The magnetic flux used in **Eq. (10.1)** is calculated from the magnetic flux density as

$$\Phi = \int_s \mathbf{B} \cdot d\mathbf{s} \quad [\text{Wb}] \tag{10.4}$$

Substituting the relations for flux and emf in **Eq. (10.1)** gives

$$\oint_C \mathbf{E} \cdot d\mathbf{l} = -\frac{d\Phi}{dt} = -\frac{d}{dt}\int_s \mathbf{B} \cdot d\mathbf{s} \quad [\text{V}] \tag{10.5}$$

Now, applying Stokes' theorem to the left-hand side of **Eq. (10.5)**,

$$\oint_C \mathbf{E} \cdot d\mathbf{l} = \int_s (\nabla \times \mathbf{E}) \cdot d\mathbf{s} = -\frac{d}{dt}\int_s \mathbf{B} \cdot d\mathbf{s} \quad [\text{V}] \tag{10.6}$$

where s is the surface bounded by the closed contour C. Because the integral on the right-hand side is independent of time, we can perform the differentiation with respect to time inside the integral

$$\int_s (\nabla \times \mathbf{E}) \cdot d\mathbf{s} = -\int_s \frac{\partial \mathbf{B}}{\partial t} \cdot d\mathbf{s} \qquad [\text{V}] \tag{10.7}$$

and, therefore, for a loop with a constant surface (the surface is not time dependent), the integrands must be equal:

$$\boxed{\nabla \times \mathbf{E} = -\frac{\partial \mathbf{B}}{\partial t}} \tag{10.8}$$

This is the differential statement of Faraday's law and it clearly indicates the relation between the electric and magnetic fields. The electric field is nonconservative (the curl of the electric field intensity is not zero). For this reason, the electric field intensity cannot be defined as the gradient of a scalar potential and the definition of electrostatic potential cannot be used here. **Equation (10.8)** also indicates that Faraday's law is general and the loop is not necessary for the induced emf to exist. This aspect of the relation will be used later to define and evaluate induced currents in conducting volumes.

The relation in **Eq. (10.8)** is one of the required postulates for the time-dependent electric field intensity **E**. The remaining postulates, namely, the divergence of **D**, the curl of **H**, and the divergence of **B**, remain unchanged. Therefore, the required postulates for the time-dependent electromagnetic field are as follows.

Differential form	Integral form
$\nabla \times \mathbf{E} = -\dfrac{\partial \mathbf{B}}{\partial t}$	$\oint_C \mathbf{E} \cdot d\mathbf{l} = -N\dfrac{d\Phi}{dt}$ [V]
$\nabla \times \mathbf{H} = \mathbf{J}$ [A/m^2]	$\oint_C \mathbf{H} \cdot d\mathbf{l} = I$ [A]
$\nabla \cdot \mathbf{B} = \mathbf{0}$	$\oint_s \mathbf{B} \cdot d\mathbf{s} = 0$
$\nabla \cdot \mathbf{D} = \rho$ [C/m^3]	$\oint_s \mathbf{D} \cdot d\mathbf{s} = Q$ [C]

The only difference between these relations and those for the static electric and static magnetic fields is in the first equation, but this seemingly small difference has far-reaching implications, as we shall see shortly. For now, we merely observe that Faraday's law couples between the electric and magnetic fields. From now on, we will discuss the electric field and magnetic field as coupled fields.

10.3 Lenz's Law[2]

In **Eq. (10.1)**, the negative sign indicates that if the magnetic flux linking the loop increases, the induced emf produces a current whose flux opposes the increase in the flux linking the loop and if the flux decreases, the emf produces a current whose flux augments the flux linking the loop. To see this, consider first the loop in **Figure 10.2a**. Using the right-hand rule for the flux generated by current I_1 in loop (1), an increase in the current increases the flux through loops (1) and (2). Loop (1) is then the source of the flux. An emf is generated in loop (2), but, because loop (2) is open, there is no current in the loop and therefore no induced flux due to loop (2). Now, consider the situation in **Figure 10.2b** where, again using the right-hand rule, an increase in the flux due to loop (1) produces a negative emf in loop (2). This corresponds to Faraday's law. This induced emf, shown in **Figure 10.2c**, can be viewed as the source of a flux that is in the opposite direction to the flux in **Figure 10.2a**. Thus, the induced current due to the induced emf produces a flux that opposes the flux that generated the emf. This relation between induced electromotive force and the magnetic flux linkage is defined by Lenz's law, which states:

"*the direction of the electromotive force is such that the flux generated by the induced current opposes the change in flux.*"

As a direct consequence of Lenz's law, the flux linking a circuit tends to maintain its value (magnitude and direction), resisting, as it were, any change. Although Faraday's law deals with electromotive force, and an electromotive force may be generated in an open or closed circuit, the statement of Lenz's law requires a current and, therefore, a closed circuit.

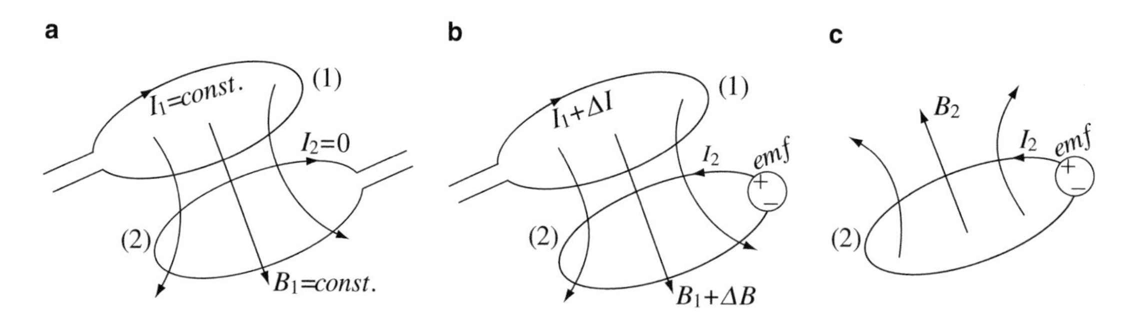

Figure 10.2 (**a**) Flux generated by a constant current in a loop. (**b**) Electromotive force induced in loop (2) due to loop (1). (**c**) Flux in loop (2) due to induced emf opposes the inducing flux. The emf is shown at an instant in time to indicate the direction of the induced current but it can be a DC or a time dependent emf depending on how it is generated

[2] Emil Khristianovich Lenz (1804–1865) was a Russian scientist with wide ranging interests. He deduced this law in 1833, but he was involved in many activities in addition to his research in electromagnetics and electrothermal and electrochemical applications. He also worked on measurement methods and developed preliminary laws similar to Kirchhoff's laws before Kirchhoff did so and contributed to measurements of salinity in the seas as well as contributions to understanding of weather, galvanic cells, and electric machines. (He is also known by his German name as Heinrich Friedrich Emil Lenz).

10.4 Motional Electromotive Force: The DC Generator

The force on a charge moving in a constant magnetic field was calculated in **Chapter 9 [Eq. (9.106)]** as

$$\mathbf{F} = q\mathbf{v} \times \mathbf{B} \quad [\text{N}] \tag{10.9}$$

Because the electric field intensity is defined as force per unit charge, we can write

$$\mathbf{E} = \frac{\mathbf{F}}{q} = \mathbf{v} \times \mathbf{B} \quad [\text{V/m}] \tag{10.10}$$

The term $\mathbf{v} \times \mathbf{B}$ is therefore an electric field intensity, generated by motion of charges. If we substitute this in **Eq. (10.3)**, we get

$$\text{emf} = \oint_C \mathbf{E} \cdot d\mathbf{l} = \oint_C (\mathbf{v} \times \mathbf{B}) \cdot d\mathbf{l} \quad [\text{V}] \tag{10.11}$$

where C is the contour on which the electromotive force is desired.

To see how these relations apply, we consider here a classical example of induced emf: that of a sliding bar on two parallel rails in a magnetic field as shown in **Figure 10.3a**. The rails are separated a distance d and are shorted together on one side. The magnetic flux density points out of the page. The bar and the shorted rails form a loop *abcd*. If we move the bar to the right at a velocity $\mathbf{v}$, there will be an induced emf in the bar because the loop increases in area and therefore encloses a larger, changing flux. We wish now to calculate this emf for a bar of length d, flux density $\mathbf{B}$, and velocity $\mathbf{v}$, because this calculation leads directly to the idea of the DC generator. There are two ways to calculate the emf in this case. One is from the motional effect; the second is from the change in flux through the loop. The two methods are equivalent, but the reasoning is quite different. We look at both solution methods and show that they are equivalent:

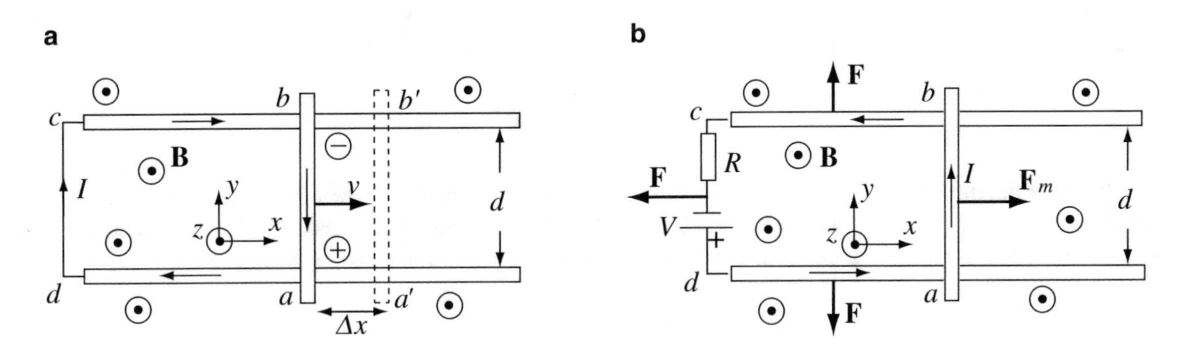

Figure 10.3 (**a**) A bar sliding on rails at velocity $\mathbf{v}$, perpendicular to the magnetic field. (**b**) Force on a conducting bar which carries current perpendicular to the magnetic field

(a) Calculation of emf from motional considerations. As the bar moves at a velocity $\mathbf{v}$, the electrons in the bar move with it at the same velocity. This motion creates a force on the electrons and, therefore, an electric field intensity $\mathbf{v} \times \mathbf{B}$ in the bar. The induced emf is, therefore,

$$\text{emf} = \int_a^b (\mathbf{v} \times \mathbf{B}) \cdot d\mathbf{l} = Bvd \quad [\text{V}]. \tag{10.12}$$

Note that we have made use of **Eq. (10.11)** but only that part of the loop *abcd* that moves contributes to the emf hence the closed contour integral in **Eq. (10.11)** reduces to integration along the moving bar in **Eq. (10.12)**.

(b) Calculation of emf from change in flux. The total flux enclosed by the loop is

$$\Phi = \int_s \mathbf{B} \cdot d\mathbf{s} = BS \quad [\text{Wb}] \tag{10.13}$$

where S is the area of the loop *abcd* and $\mathbf{B}$ is constant throughout the loop.

Now, if the bar moves at a velocity **v**, the location of the bar after a time Δt will be at a distance Δx from the previous point (dashed bar in **Figure 10.3a**). The flux through the loop is now

$$\Phi' = \int_s \mathbf{B} \cdot d\mathbf{s} = BS' = B(S + \Delta S) = B(S + \Delta xd) \quad [\mathrm{Wb}] \tag{10.14}$$

Therefore, the change in flux due to this motion is

$$\Delta\Phi = \Phi - \Phi' = -Bd\Delta x \quad [\mathrm{Wb}] \tag{10.15}$$

Using the magnitude of velocity as $v = \Delta x/\Delta t$ and taking the limit as $\Delta t \rightarrow 0$, we get

$$d\Phi = -Bd(v\,dt) \quad [\mathrm{Wb}] \tag{10.16}$$

and the emf is found from the definition in **Eq. (10.1)**:

$$\mathrm{emf} = -\frac{d\Phi}{dt} = Bvd \quad [\mathrm{V}] \tag{10.17}$$

This is the same result as previously obtained, but there is a distinction between the two results. In the first case, the emf is associated with the moving bar itself. In the second, it is associated with the whole loop. **Figure 10.3a** also shows the polarity of the bar and the direction of current in the loop caused by the generated emf.

We can view this device as a linear generator since application of a mechanical force produces an emf. How about the opposite: Does the connection of a source produce motion of the bar and, therefore, produce a motor effect? The answer is yes. To see this, consider **Figure 10.3b**, which shows the bar and rail discussed above, but now a source V and a resistance (representing the internal resistance of the source and that of the rails and bar) are connected as shown. We wish to calculate the force, if any, on the bar. The current in the rail and bar is $I = V/R$ and flows as shown. The force on a current-carrying conductor in a magnetic field is given by the following relation [see **Eq. (9.114)**]:

$$\mathbf{F} = I\oint_C d\mathbf{l}' \times \mathbf{B} \quad [\mathrm{N}] \tag{10.18}$$

where $d\mathbf{l}'$ is taken in the direction of the current. Each segment of the loop will experience a force. The forces are shown in **Figure 10.3b** and each is equal to BIl where l is the length of the corresponding segment. These forces tend to expand the loop, but, since only the bar is free to move, it will move to the right. The force moving the bar is

$$\mathbf{F} = \hat{\mathbf{x}}BId = \hat{\mathbf{x}}B\frac{V}{R}d \quad [\mathrm{N}] \tag{10.19}$$

However, as the loop expands, the length of the rail increases and the resistance increases. This, in turn, decreases the force with the expansion of the loop. This is in addition to the effect of the emf induced due to the motion of the bar (also called ***back emf***) which generates a force that tends to decrease the applied force on the bar (see **Example 10.2**).

The relations above deal primarily with constant motion in a DC magnetic field and hence the emf generated is a DC emf. However, if the magnetic field is time dependent, or if motion is not constant (for example, if a conductor moves back and forth in a DC magnetic field), the emf will also be time dependent. That is, motion action can produce either DC or AC emfs.

Example 10.1 Induced emf Due to Motion: Motional Electromotive Force A straight, thin conducting bar of length $L = 1$ m moves in a constant magnetic field at a speed $v = 20$ m/s. The magnitude of the magnetic flux density is $B_0 = 0.5$ T. Calculate the induced emf in the bar under the following conditions:

(a) The bar moves perpendicular to the magnetic field as shown in **Figure 10.4a**. Calculate the emf and show its polarity.
(b) The bar moves at 30° to the x axis as shown in **Figure 10.4b**. Calculate the emf and show its polarity.

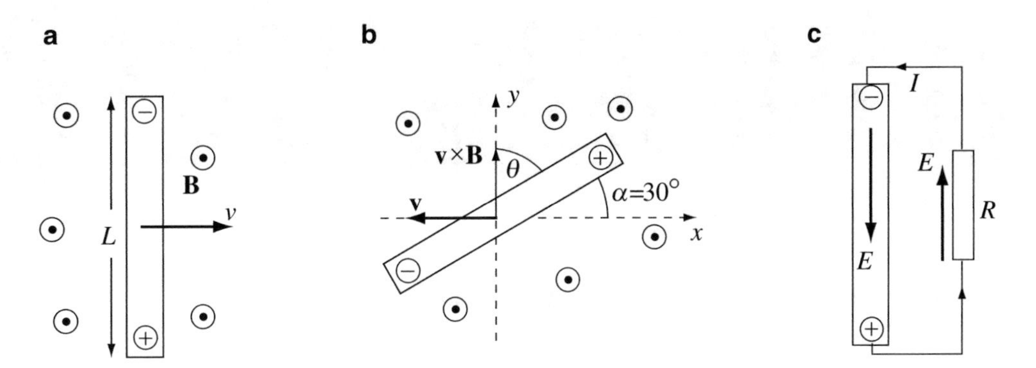

Figure 10.4 The emf in a moving conductor. (**a**) Conductor moves perpendicular to the magnetic field and velocity is perpendicular to the conductor. (**b**) Conductor and velocity vector at an angle α. (**c**) Equivalent circuit of the bar in (**a**) as a DC generator, showing polarity of the emf

Solution: The solution is based on calculation of the electric field intensity at the location of the bar, $\mathbf{E} = \mathbf{v} \times \mathbf{B}$. We can either create a system of coordinates to use the vector notation or, since the induced emf is required in the bar, use the bar as a reference axis and calculate all quantities with respect to this axis:

(**a**) The cross-product $\mathbf{v} \times \mathbf{B}$ is found to be in the direction of the bar, pointing downward. This is the electric field intensity in the bar: $\mathbf{E} = \mathbf{v} \times \mathbf{B}$. The emf is therefore

$$\text{emf} = \int_l \mathbf{E} \cdot d\mathbf{l} = \int_l (\mathbf{v} \times \mathbf{B}) \cdot d\mathbf{l} = vB_0L = 20 \times 0.5 \times 1 = 10 \quad [\text{V}].$$

The potential difference (emf) between the ends of the bar is 10 V. The polarity is as shown in **Figure 10.4c**. The polarity can be deduced from the direction of the field $\mathbf{E} = \mathbf{v} \times \mathbf{B}$ or from the fact that the magnetic field exerts a force on the electrons which moves them upward. This bar can now be used as a source in a circuit, and in this sense, the motion in the field makes the bar a generator.

(**b**) The solution in this case is the same except that the cross-product $\mathbf{v} \times \mathbf{B}$ and the bar are at $\theta = 90 - \alpha = 60°$ to each other. The induced emf is found in two steps. First, we find the vector product $\mathbf{E} = \mathbf{v} \times \mathbf{B}$. This gives the electric field intensity in the direction perpendicular to $\mathbf{v}$ and $\mathbf{B}$. The direction of this field is straight up (y-direction):

$$\mathbf{E} = \mathbf{v} \times \mathbf{B} = -\hat{\mathbf{x}}v \times \hat{\mathbf{z}}B = \hat{\mathbf{y}}vB \qquad [\text{V/m}]$$

However, we are only interested in that component of the electric field intensity that is parallel to the bar. Thus, the second step is to calculate the projection of $\mathbf{E}$ on the bar. The component parallel to the bar is

$$E_b = vB\cos(90 - \alpha) = vB\sin\alpha \quad [\text{V/m}]$$

and the induced electromotive force is now

$$\text{emf} = \int_l \mathbf{E} \cdot d\mathbf{l} = vB_0L\sin\alpha = 20 \times 0.5 \times 1 \times 0.5 = 5 \quad [\text{V}]$$

The smaller the angle α, the lower the induced emf. Polarity of the emf is opposite that in (**a**) and is shown in **Figure 10.4b**. Note that $\mathbf{E} = \mathbf{v} \times \mathbf{B}$ points from the negative to the positive ends in the bar. This may seem strange at first, but we must remember that this is the field on a positive charge (electrons move in the direction opposite the electric field). The external field (such as in a circuit) will point from positive to negative as required (see **Figure 10.4c**).

Exercise 10.1 What is the induced emf and its polarity if:

(a) The bar in **Figure 10.4a** moves straight up at a velocity **v**.
(b) The bar in **Figure 10.4b** moves straight up at a velocity **v**.

Answer

(a) Zero. **(b)** 10cos30° = 8.66 V. Polarity is shown in **Figure 10.4b**.

Example 10.2 Application: The Linear Generator A DC generator can be built as shown in **Figure 10.5a**. The bar, load, and rails form a closed circuit. The two rails are separated a distance $d = 1.5$ m and the load resistance is $R = 10\ \Omega$. Assume the rails and bar are perfectly conducting and the bar moves to the right at a constant velocity $v = 10$ m/s. The magnetic flux density is constant, directed upward, and equal to $B = 0.1$ T. Calculate:

(a) The electromotive force produced by this generator.
(b) The force required to move the bar at constant velocity.
(c) The mechanical power required for generation.

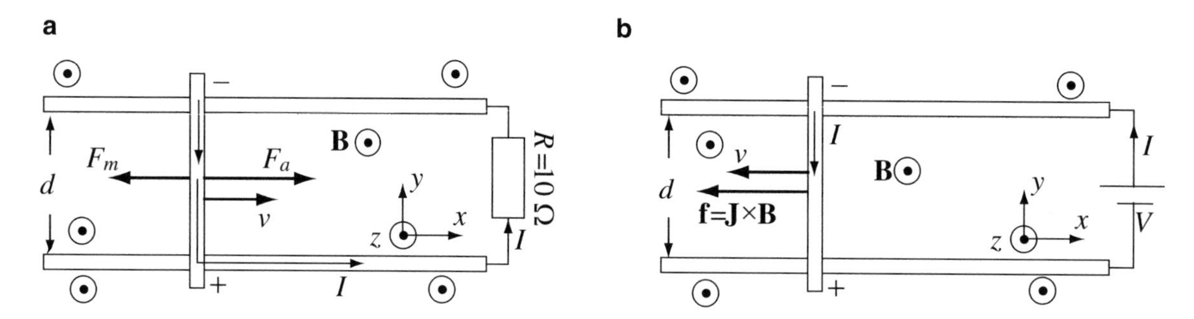

Figure 10.5 (**a**) Principle of a linear generator. (**b**) Principle of a linear motor

Solution: A force is applied to the bar shown in **Figure 10.5a**, the bar moves, and this generates an emf on the bar. The generator creates a force $\mathbf{F}_m$ which opposes the applied force $\mathbf{F}_a$ (otherwise, the bar would accelerate). The mechanical power is calculated from velocity and force, which, in turn, is calculated from the current in the circuit and the magnetic flux density:

(a) To calculate the emf on the bar, we must calculate the rate of change of flux with time. As the bar moves to the right at a velocity v, after a time dt, the bar is at a new location, a distance $dx = vdt$ to the right of the original location, changing the loop area by $d(vdt)$. Since the flux density is constant, the change in flux is

$$d\Phi = Bd(vdt) \quad [\text{Wb}]$$

The magnitude of the emf is

$$|\text{emf}| = \frac{d\Phi}{dt} = Bvd = 0.1 \times 1.5 \times 10 = 1.5 \quad [\text{V}].$$

The polarity of the emf is shown in **Figure 10.5a**. As the bar moves to the right, the term $\mathbf{v} \times \mathbf{B}$ points down (negative y direction). This is the direction of force on a positive charge and, therefore, the direction of current.

(b) The existence of a current in the circuit gives rise to a force $\mathbf{F}_m$. This force can be calculated from **Eq. (10.18)**:

$$\mathbf{F}_m = I\int_0^d d\mathbf{l}' \times \mathbf{B} = I\int_0^d -\hat{\mathbf{y}}dl' \times \hat{\mathbf{z}}B = -\hat{\mathbf{x}}IBd \quad [\text{N}].$$

This force is directed in the direction opposing $\mathbf{F}_a$, as shown. The induced current I is emf/R = vBd/R. To maintain movement at constant velocity, the two forces must balance:

$$F_m = F_a = IBd = vB^2d^2/R = 10 \times (0.1)^2 \times 1.5^2/10 = 0.0225 \quad [\text{N}].$$

(c) The mechanical power P is given by $P = \mathbf{F} \cdot \mathbf{v}$, giving the power of the applied force $\mathbf{F}_a$ as

$$P = F_a v = v^2B^2d^2/R = 10^2 \times (0.1)^2 \times 1.5^2/10 = 0.225 \quad [\text{W}].$$

This mechanical power is equal to the power dissipated through Joule's losses which, using the current $I = vBd/R$, is

$$RI^2 = v^2B^2d^2/R = 0.225 \quad [\text{W}].$$

Exercise 10.2 Suppose the rails in **Figure 10.5a** have a resistance per unit length r [Ω/m] each. The bar is perfectly conducting. Calculate the current in the resistor R at a time t if the bar starts at $t = 0$ at $x = 0$ ($x = 0$ is at the location of R) and moves at constant speed v in the negative x direction.

Answer $I(t) = \dfrac{Bvd}{R + 2rvt}$ [A]

Example 10.3 Application: The Linear Motor A linear DC motor is obtained by replacing the resistance R in **Figure 10.5a** with a source V with polarity such that the current flows in the same direction as the current in the generator (**Figure 10.5b**).

The two rails are separated a distance $d = 1.5$ m and a battery with potential $V = 12$ V is connected at one end of the rails. Assume the rails are perfectly conducting, but the bar has a resistance $R = 1$ Ω and the bar is free to move. The magnetic flux density is constant, directed upward, and equal to $B = 0.1$ T:

(a) Calculate the force acting on the moving bar if the bar is held stationary.
(b) If the bar is allowed to move, find the maximum velocity of the bar.
(c) Find the equation of motion of the bar assuming the bar has mass m. Explain.

Solution: If the bar is held stationary, there is a DC current flowing through the circuit equal to V/R. The magnetic field exerts a force density on this current. This force is to the left (in the negative x direction according to the right-hand rule). If the bar is allowed to move, it will accelerate and its velocity will increase with time. At the same time, since the bar moves in the magnetic field, there will be an emf induced in the bar due to motion. This emf produces a current which opposes the source current. The equation of motion is found from the relation $\mathbf{F} = m\mathbf{a}$:

(a) If the bar is stationary, the current in the bar is

$$I = \frac{V}{R} = 12 \quad [\text{A}]$$

This current flows in the negative y direction as shown. The force, which is in the negative x direction, is

$$\mathbf{F}_m = I\int_0^d d\mathbf{l}' \times \mathbf{B} = I\int_0^d -\hat{\mathbf{y}}dl' \times \hat{\mathbf{z}}B = -\hat{\mathbf{x}}BId = -\hat{\mathbf{x}}0.1 \times 12 \times 1.5 = -\hat{\mathbf{x}}1.8 \quad [\text{N}].$$

(b) If the bar is allowed to move, it will move in the negative x direction under the force in part (a). An induced emf will exist in the bar which, according to **Eq. (10.12)**, is

$$\text{emf} = vBd \quad [\text{V}]$$

We do not know the velocity v, but we can still write

$$V - \text{emf} - IR = 0 \quad \rightarrow \quad I = \frac{V - \text{emf}}{R} \quad [\text{A}]$$

Substituting this into the force equation, the magnitude of $\mathbf{F}_m$ is given as

$$F_m = IBd = \frac{V - \text{emf}}{R}Bd = \frac{V - vBd}{R}Bd \quad [\text{N}]$$

This force is maximum at the start of motion when $v = 0$. As the bar moves, an emf, equal to vBd, is induced in the circuit, reducing the current in the bar. In other words, we start with a maximum force (starting force) given in (**a**), and as the bar moves, the force is reduced. In this case, there is no friction; therefore, the force must decrease to zero for the velocity to be constant. This gives

$$V - vBd = 0 \quad \rightarrow \quad v = \frac{V}{Bd} = \frac{12}{0.1 \times 1.5} = 80 \quad [\text{m/s}].$$

(c) The balance of forces (at any time before the acceleration diminishes to zero) requires that

$$ma = m\frac{dv}{dt} = F_m \quad [\text{N}]$$

The differential equation describing the motion is

$$m\frac{dv}{dt} = \frac{V - vBd}{R}Bd$$

In a practical linear motor, the bar, or, more often, a number of conductors, will move back and forth at a constant or varying speed, and we will also have to deal with friction forces as well as losses in the motor itself. However, this example serves to describe the principle.

10.5 Induced emf Due to Transformer Action

In describing Faraday's law, we mentioned that the emf produced by a changing flux is called a transformer action emf. The reason for this name is to distinguish it from the motional emf discussed in the previous section and because it is commonly encountered in transformers. The distinction here is between a moving conductor in the field causing a change in flux (motional emf) and a stationary loop in a changing magnetic field (transformer action emf). If the emf is connected in a closed circuit, there will also be a current in the loop and this current will produce a flux density according to Lenz's law. In the previous section, we saw that the motional emf can be calculated either from the forces applied on the moving electrons in the conductors or as a change in flux. However, whenever we discuss stationary circuits, we must use the transformer action relation in **Eq. (10.2)**.

Any closed circuit placed in a time-varying magnetic field experiences a current due to the induced emf. If, for example, you were to place a circuit, such as an amplifier or a computer board, in a time-dependent magnetic field, an emf will be induced in any closed circuit on the board producing a current. This type of induced emf can sometimes be detected as an annoying hum in audio amplifiers due to induced currents from power lines.

Example 10.4 Application: The AC Generator A uniform magnetic field is generated by a time-varying source as shown in **Figure 10.6a**. A stationary square loop is placed such that its plane is perpendicular to the magnetic flux density **B**. Assume the magnetic flux density is sinusoidal and given as $B = B_0 \sin\omega t$ [T] with $B_0 = 0.1$ T, $\omega = 100\pi$ rad/s, $a = 0.1$ m, $b = 0.1$ m:

(a) Calculate the induced emf in the loop.
(b) What is the induced emf if the loop is made of $N = 100$ turns, placed in essentially the same location?
(c) What is the solution in **(a)** and **(b)** if the loop is at an angle α to the field, as shown in **Figure 10.6c**?

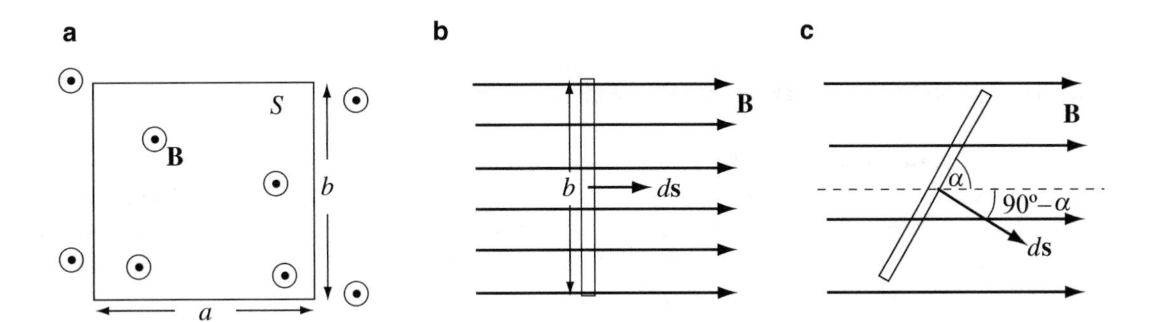

Figure 10.6 **(a)** A square loop with its plane perpendicular to a uniform, time-dependent magnetic flux density. **(b)** Side view of the figure in **(a)**. **(c)** A square loop with its plane at an angle α to a uniform, time-dependent magnetic flux density

Solution: The transformer action emf is found using **Eqs. (10.1)** and **(10.7)**. The magnetic flux density is uniform; therefore, the flux is immediately available. The emf of N identical loops in the same field is N times larger than the emf of a single loop [see **Eq. (10.2)**]. In **(c)**, the emf also depends on the angle the loop makes with the field because the flux depends on the angle:

(a) From **Eq. (10.7)**, the emf in the loop is

$$\text{emf} = -\int_s \frac{\partial \mathbf{B}}{\partial t} \cdot d\mathbf{s} = -\frac{d}{dt}\int_s \mathbf{B} \cdot d\mathbf{s} = -\frac{dB}{dt} S = -S\omega B_0 \cos\omega t \quad [\text{V}]$$

where we used the fact that for parallel vectors **B** and d**s**, $\mathbf{B} \cdot d\mathbf{s} = Bds$ (**Figure 10.6b**). With $S = ab$, the induced emf in the loop is

$$\text{emf} = -\omega ab B_0 \cos\omega t = -0.314\cos(314t) \quad [\text{V}]$$

Thus, the peak induced emf is $\pm$ 0.314 V.

(b) If there are $N = 100$ identical loops linking the same flux, the induced emf is N times larger. This can be understood from the fact that turns are connected in series:

$$\text{emf} = -\omega abNB_0 \cos\omega t = -31.4\cos(314t) \quad [\text{V}].$$

(c) The solution now is similar, but the scalar product between **B** and d**s** is $\mathbf{B} \cdot d\mathbf{s} = Bds\cos(90 - \alpha)$ as shown in **Figure 10.6c**:

$$\text{emf} = -\omega B_0 \cos\omega t \cos(90 - \alpha)\int_s ds = -\omega ab B_0 \sin\alpha \cos\omega t = -0.314\sin\alpha\cos 314t \quad [\text{V}]$$

or, for N loops,

$$\text{emf} = -\omega abNB_0\sin\alpha\cos\omega t = -31.4\sin\alpha\cos 314t \quad [\text{V}].$$

The induced emf is proportional to the orientation of the loop with respect to the flux density. To maximize the induced electromotive force, the loop must be perpendicular to the flux. Similarly, a loop parallel to the field ($\alpha = 0$) has zero induced emf since no flux passes through the loop. These aspects are important in design of motors, generators, and other devices.

Although this is a simple example, it indicates that the induced emf is directly proportional to frequency, number of turns, the magnetic flux density, and the area of the loop. Any one of these parameters may be used to optimize a device. As an example, if a particularly small device operating at a given frequency is needed, the flux density might be increased to decrease the area of the loop. If frequency can be increased, the same effect can be achieved without the need to increase the flux density. For example, most electric machines used in aircraft operate at 400 Hz as opposed to the more common 50 or 60 Hz devices in industry. This decreases their size and weight for the same design parameters.

10.6 Combined Motional and Transformer Action Electromotive Force

The electromotive forces in **Eqs. (10.2)** and **(10.11)** were obtained from two different situations. The first is generated in a stationary circuit by a change in flux. The second is generated due to motion in a magnetic field. The distinction indicates the source of the electromotive force and points to possible uses of the two electromotive forces. Both electromotive forces can exist together in a single circuit, as we will see shortly in an example. Therefore, the total induced electromotive force in a circuit is the sum due to motion and transformer action:

$$\text{emf} = \oint_C (\mathbf{v} \times \mathbf{B}) \cdot d\mathbf{l}' - N\frac{d\Phi}{dt} \quad [\text{V}] \tag{10.20}$$

or, in a more consistent form,

$$\text{emf} = \oint_C (\mathbf{v} \times \mathbf{B}) \cdot d\mathbf{l}' - N\int_s \frac{\partial \mathbf{B}}{\partial t} \cdot d\mathbf{s}' \quad [\text{V}] \tag{10.21}$$

This is a more general expression of Faraday's law in that both effects (the transformer and the motional action emfs) can exist together. The surface integration in **Eq. (10.21)** must be done over the whole area in which the change in flux takes place (usually area of a loop), whereas the line integral is around the contour (usually a conducting loop or many conducting loops) in which the emf is induced.

It should be noted again that for a transformer action emf to exist, the magnetic flux density itself must be time dependent, whereas a motion action emf can be generated by motion in a time-dependent or DC magnetic flux density.

10.6.1 The Alternating Current Generator

Consider the situation in **Figure 10.7**. The loop rotates around the axis in a constant, uniform magnetic field at angular velocity ω [rad/s]. At any given time, the normal to the loop surface makes an angle α with the magnetic field. Since the magnetic flux density is constant in time, there is no emf due to the change of flux (transformer action emf), but since the loop is rotating, and each section of the loop moves at some velocity $\mathbf{v}$, there is an induced emf due to this motion. We calculate the induced emf in the four sections of the loop indicated in **Figure 10.7a**:

$$\text{emf} = \text{emf}_{ab} + \text{emf}_{bc} + \text{emf}_{cd} + \text{emf}_{da} \quad [\text{V}] \tag{10.22}$$

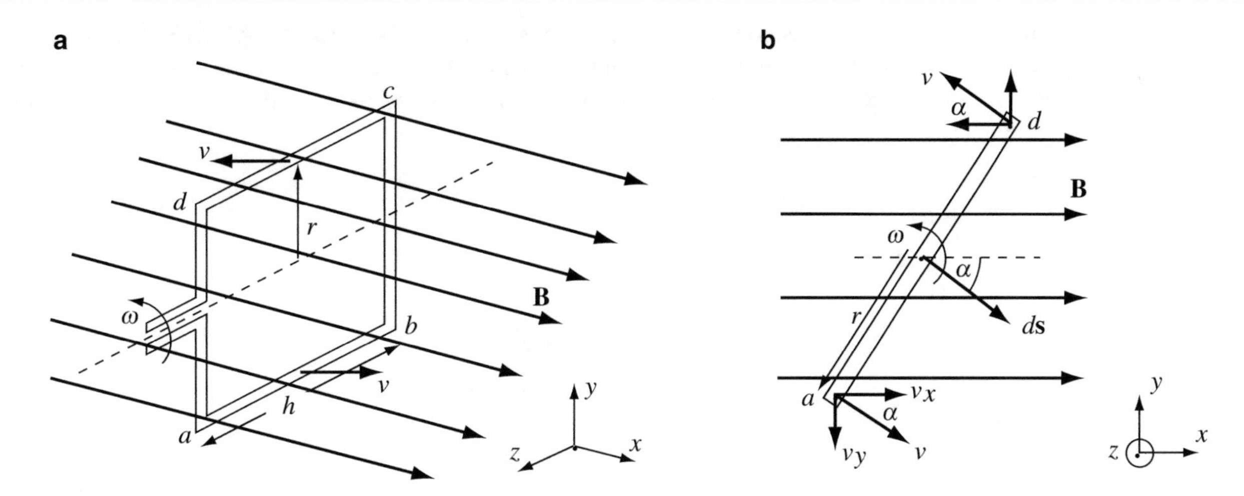

Figure 10.7 The alternating current generator. (**a**) A loop rotating in a magnetic field at angular frequency ω. (**b**) The relation between the loop and magnetic field at a given instant in time

The emf in section *bc* and *da* is zero because $\mathbf{dl}$ and $\mathbf{v} \times \mathbf{B}$ are perpendicular to each other throughout the length of these two segments as can be seen in **Figure 10.7b** ($\mathbf{dl}$ is along the segments, whereas $\mathbf{v} \times \mathbf{B}$ is perpendicular to the segments, as can be verified using the right-hand rule). However, $\mathbf{v} \times \mathbf{B}$ and $\mathbf{dl}$ are in the same direction on segment *ab* and in opposite directions on segment *cd*, producing a nonzero emf. The velocity of segments *ab* and *cd* (in the system of coordinates shown in **Figure 10.7b**) is

$$\mathbf{v}_{ab} = r\omega(\hat{\mathbf{x}}\cos\alpha - \hat{\mathbf{y}}\sin\alpha) \quad \text{and} \quad \mathbf{v}_{cd} = r\omega(-\hat{\mathbf{x}}\cos\alpha + \hat{\mathbf{y}}\sin\alpha)[\text{m/s}] \tag{10.23}$$

The vector products $\mathbf{v} \times \mathbf{B}$ on segments *ab* and *cd* are

$$(\mathbf{v} \times \mathbf{B})_{ab} = r\omega(\hat{\mathbf{x}}\cos\alpha - \hat{\mathbf{y}}\sin\alpha) \times \hat{\mathbf{x}}B_0 = \hat{\mathbf{z}}B_0 r\omega\sin\alpha \tag{10.24}$$

$$(\mathbf{v} \times \mathbf{B})_{cd} = r\omega(-\hat{\mathbf{x}}\cos\alpha + \hat{\mathbf{y}}\sin\alpha) \times \hat{\mathbf{x}}B_0 = -\hat{\mathbf{z}}B_0 r\omega\sin\alpha \tag{10.25}$$

Performing the product $(\mathbf{v} \times \mathbf{B}) \cdot \mathbf{dl}$ and integrating along segment *ab* and *cd* gives the total emf in the loop:

$$\text{emf} = B_0 2rh\omega\sin\alpha = B_0 S\omega\sin\alpha \quad [\text{V}] \tag{10.26}$$

where $S = 2rh$ is the area of the loop. N identical loops rotating together produce an emf that is N times larger:

$$\text{emf} = B_0 N 2rh\omega\sin\alpha = B_0 NS\omega\sin\alpha \quad [\text{V}] \tag{10.27}$$

Also, the angle α after a time t (starting with $\alpha = 0$ at $t = 0$) is ωt. Thus the emf of the generator is

$$\text{emf} = NB_0 S\omega\sin\omega t \quad [\text{V}] \tag{10.28}$$

This device is clearly an AC generator. The emf is directly proportional to the angular velocity, the magnetic flux density, and the area of the loop. The generator can be designed as a trade-off between the various parameters. If the frequency must be constant, the loop must be rotated at a fixed angular velocity. Most generators operate on this or a very similar principle. As an example, the constant flux density may be generated by a permanent magnet or by an electromagnet and a DC source. We also note that it is actually easier to generate an AC emf than it is to generate a DC emf. In fact, in most cases, a DC generator is an AC generator with a means of rectifying (or converting) the AC emf into a DC emf. This can be done through use of diodes or through commutators. For example, a car alternator is a three-phase AC machine supplying DC through three diodes. A commutator can be used to disconnect the loop and reverse the connection every time the output goes through zero. This is done by connecting the loop through sliding connectors on the axis of the generator. A DC generator based on commutation of the connection to a rotating loop is shown in **Figure 10.8** together with the generator waveform. The only differences between the simplified forms of the generator discussed here and practical generators are in the way the magnetic fields are generated, the magnetic paths, the arrangement of loops, and the details of mechanical construction.

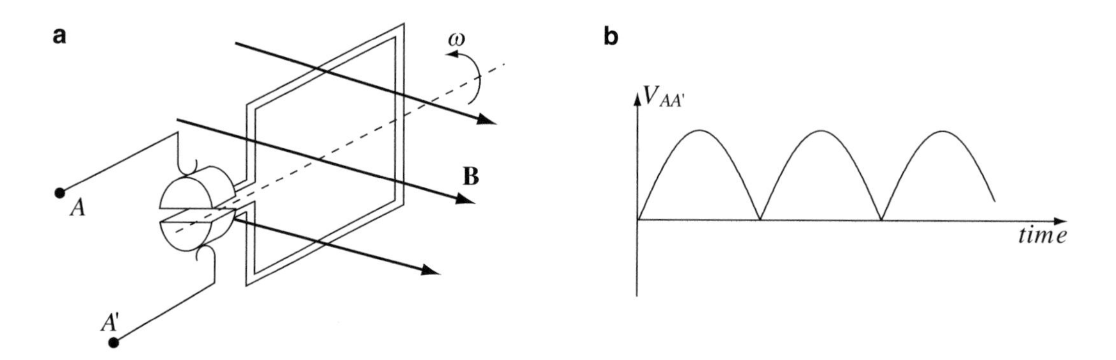

Figure 10.8 (**a**) A DC generator with commutating contacts. (**b**) The resulting output waveform

A more general situation is shown in **Figure 10.9**, where a loop rotates at an angular velocity ω_1 in a uniform, time-varying magnetic field given as $B = B_0 \sin\omega_2 t$.

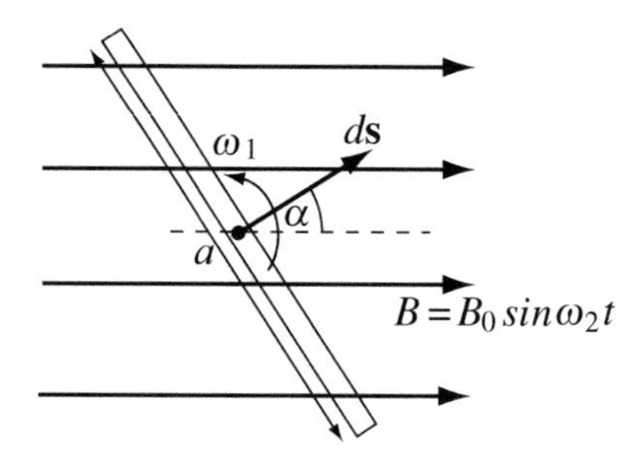

Figure 10.9 A loop rotating inside an AC magnetic field. The loop is b [m] wide (into the page)

The emf is now a superposition of induced emfs due to change of flux and due to motion of the loop. We treat each of these separately:

(a) emf due to change of flux (transformer action emf). Consider **Figure 10.9** where the loop is shown at an arbitrary fixed angle to the time-dependent magnetic flux density. The emf is

$$\text{emf}_t = -N\frac{d\Phi}{dt} = -N(S\cos\alpha)\frac{dB}{dt} = -\omega_2 SNB_0 \cos\omega_2 t \cos\alpha \quad [\text{V}] \tag{10.29}$$

where $S = ab$ is the area of the loop and N is the number of loops ($N = 1$ in **Figure 10.9**).

(b) emf due to motion. The emf due to motion is given by **Eq. (10.27)** except that now the flux density is time dependent. Therefore:

$$\text{emf}_m = \omega_1 SNB_0 \sin\omega_2 t \sin\alpha \quad [\text{V}] \tag{10.30}$$

where $S = ab$ is the area of the loop and N is the number of loops.

The total emf is

$$\text{emf} = \text{emf}_t + \text{emf}_m = -\omega_2 SNB_0 \cos\omega_2 t \cos\alpha + \omega_1 SNB_0 \sin\omega_2 t \sin\alpha \quad [\text{V}] \tag{10.31}$$

If we start with $\alpha = 0$ at $t = 0$, then $\alpha = \omega_1 t$. Thus

$$\begin{aligned}\text{emf} &= -\omega_2 SNB_0 \cos\omega_2 t \cos\omega_1 t + \omega_1 SNB_0 \sin\omega_2 t \sin\omega_1 t \\ &= -SNB_0[\omega_2 \cos\omega_2 t \cos\omega_1 t - \omega_1 \sin\omega_2 t \sin\omega_1 t] \quad [\text{V}]\end{aligned} \tag{10.32}$$

If $\omega_1 = \omega_2 = \omega$, the expression can be further simplified:

$$\text{emf} = -\omega SNB_0 (\cos^2\omega t - \sin^2\omega t) = -\omega SNB_0 \cos 2\omega t \quad [\text{V}] \tag{10.33}$$

Thus, the device in **Figure 10.9** constitutes an AC generator with AC field excitation. Note however that this generator is different than the generator described in **Eq. (10.28)**. In particular, the frequency of the electromotive force now depends on both the frequency of rotation and on the frequency of the magnetic flux, whereas in **Eq. (10.28)**, the frequency depended only on the frequency of rotation. The generator in **Eq. (10.28)** is usually preferred because it provides a constant frequency. If we can regulate the rotation of the mechanical system (steam generator, water turbine, or diesel engine used to drive the loops), a constant-frequency generator is obtained. Because the amplitude also depends on frequency, it is only possible to obtain a constant amplitude if the frequency is kept constant. The magnetic field (also called excitation field) can be produced by DC sources such as a battery or permanent magnets. In large machines, such as turbogenerators or hydrogenerators, AC generators are used to generate the power required for excitation. The output from these machines is then rectified to provide DC excitation to the generators.

Example 10.5 A simple AC generator is made by inserting a loop of radius $d = 50$ mm inside a long solenoid of radius $b = 60$ mm. The number of turns per unit length of the solenoid is $n = 1{,}000$ turns/m and these carry a DC current $I = 1$ A. The loop is connected to the outside and is provided with an axis to rotate, as shown in **Figure 10.10**:

(a) If the loop rotates at 3,000 rpm, calculate the emf in the loop.
(b) If the loop is made of copper wire, and the wire has a diameter $D = 1$ mm, calculate the maximum current the loop can supply (shorted output). Neglect resistance of wires leading to the loop. The conductivity of copper is 5.7×10^7 S/m.
(c) How much energy must be expended in rotating the loop in one hour if the terminals of the loop are shorted?

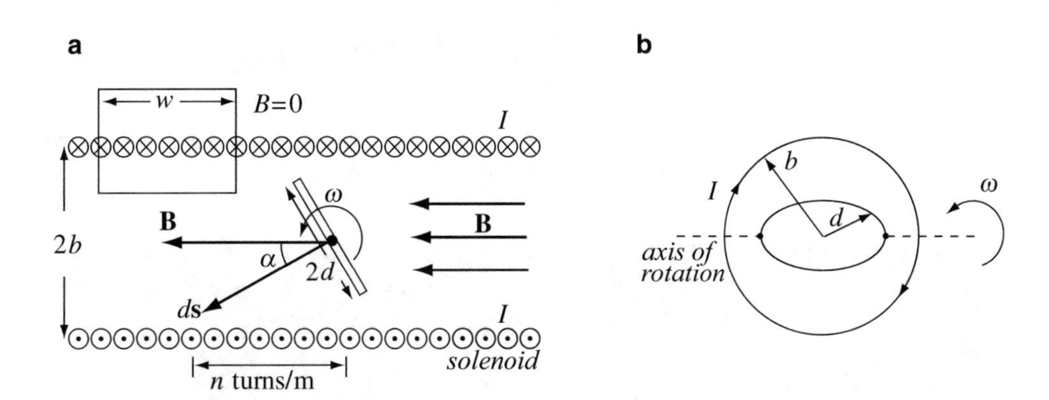

Figure 10.10 A loop rotating inside a long solenoid. (**a**) Axial cross section. (**b**) Side view

Solution: The flux density inside the solenoid is constant and may be calculated using Ampère's law. The electromotive force is calculated using the notation in **Figure 10.7b** and the transformer action emf in **Eq. (10.1)**. In **(b)**, the current is limited only by the internal resistance of the loop. In **(c)**, the power dissipated multiplied by time gives the required energy:

(a) The magnetic flux density in the solenoid is calculated using a contour as shown in **Figure 10.10a**. The total current enclosed by the contour is wnI and the magnetic flux density is zero outside the solenoid:

$$Bw = \mu_0 nIw \quad \rightarrow \quad B = \mu_0 nI = 4 \times \pi \times 10^{-7} \times 1{,}000 \times 1 = 0.0004\pi \quad [\text{T}]$$

The direction of the flux density is found from the right-hand rule and is shown in **Figure 10.10a**. The magnetic flux through the loop is

$$\Phi = \int_s \mathbf{B} \cdot d\mathbf{s} = \int_s B\cos\alpha\, ds = BS\cos\alpha \quad [\text{Wb}]$$

where α is the angle between **B** and $d\mathbf{s}$ as the loop rotates. Assuming zero-phase angle (i.e., $\alpha = 0$ at $t = 0$), the angle α after a time t is ωt and we get

$$\Phi = BS\cos\omega t \quad [\text{Wb}]$$

The emf is therefore

$$|\text{emf}| = \left| -\frac{d\Phi}{dt} \right| = \omega BS\sin\omega t = \left(2\times\pi\times\frac{3000}{60}\right)\times\left(\pi\times 0.05^2\right)\times(0.0004\times\pi)\times\sin\left(2\times\pi\times\frac{3000}{60}t\right)$$
$$= 3.1\times 10^{-3}\sin 314t \quad [\text{V}]$$

Note: The emf was calculated using the transformer action approach. The motional action approach gives the same result, but because the loop is circular, it is much more difficult to calculate.

(b) For a copper wire of radius r, made into a loop of radius d, the resistance is

$$R = \frac{l}{\sigma S} = \frac{2\pi d}{\sigma\pi r^2} = \frac{2d}{\sigma(D/2)^2} = \frac{2\times 0.05}{5.7\times 10^7\times 0.0005^2} = 0.007 \quad [\Omega]$$

The maximum current occurs for a shorted loop at $\omega t = \pi k/2$, $k = 1, 3, 5, \ldots$ and equals

$$I = \frac{\text{emf}}{R} = \frac{3.1\times 10^{-3}}{0.007} = 0.443 \quad [\text{A}]$$

Thus, the device described here is an AC generator that can supply a peak current of 0.443 A at peak voltage (emf) of 3.1 mV, operates at 50 Hz, and has an internal resistance of 0.007 Ω.

(c) To calculate power, and therefore energy, we use the root mean square value of the current. The energy expended in 1 h is

$$W = \frac{I^2 R}{2}t = \frac{0.443^2\times 0.007}{2}\times 3{,}600 = 2.47 \quad [\text{J}]$$

This is 2.47 W • s or approximately 0.687 mW • h. This energy is the work required to rotate the loop against the magnetic forces on the loop (there are no other losses in this system).

Exercise 10.3 Use the dimensions and data in **Example 10.5**. Assume the loop is replaced with a very short coil with $N = 50$ turns, with the same diameter as the loop. The large solenoid is now rotated around the short coil at 3,600 rpm. Calculate:

(a) The emf in the short coil.
(b) The frequency and internal resistance of the generator.
(c) The peak power the generator supplies with shorted terminals.

Answer

(a) $\text{emf} = 0.186\sin 120\pi t$ [V]. **(b)** 60 Hz, 0.35 Ω. **(c)** 98.8 mW.

Example 10.6 Application: The AC Generator An AC generator is made as shown in **Figure 10.11**. The coil contains $N = 500$ turns and is supplied with a sinusoidal current of amplitude $I = 10$ A and frequency 60 Hz. The relative permeability of iron is $\mu_r = 1{,}000$. The loop is 10 mm × 10 mm and rotates at 3,600 rpm. Assume the flux density in the gap is uniform and perpendicular to the iron surfaces and the $B(H)$ curve is linear:

(a) Calculate the emf of the loop.
(b) What is the waveform of the emf?
(c) Suppose you need to generate a 10 V output (peak) using this device. How many turns are required in the rotating coil?

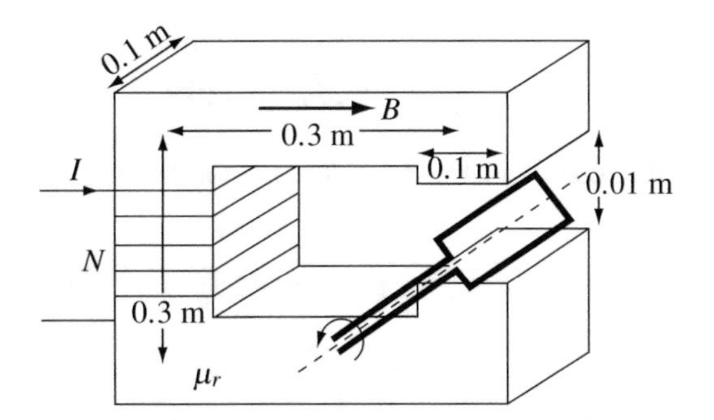

Figure 10.11 A simple AC generator

Solution: Because the coil rotates in an AC field, there are two components of the induced emf: one due to rotation and may be viewed as a motional emf. The second is due to the transformer effect and occurs even if the loop does not rotate. The emf in the coil is the sum of these two emfs. We could use the general expression in **Eq. (10.32)**, but at this stage, it is best to calculate each emf separately:

(a) (1) emf due to change of flux. The transformer action emf is given in **Eq. (10.29)**. However, we must first calculate the magnetic flux density in the gap of the magnetic structure. The latter is calculated using magnetic circuits assuming all flux is contained within the gap. The flux density in the gap is

$$B = \frac{\Phi}{S} = \frac{NI}{S(\Re_m + \Re_g)} = \frac{NI}{\left(\frac{l_m}{\mu_0\mu_r} + \frac{l_g}{\mu_0}\right)} = \frac{500 \times 10\sin 120\pi t}{\left(\frac{1.19}{4 \times \pi \times 10^{-7} \times 1{,}000} + \frac{0.01}{4 \times \pi \times 10^{-7}}\right)} = 0.5615\sin 120\pi t \quad [\mathrm{T}]$$

where $\Re_m$ is the reluctance of the magnetic path in iron, $\Re_g$ the reluctance in the gap, l_m is the length of the magnetic path in iron, l_g is the length of the gap, and μ_r is the relative permeability of iron. The emf in the loop is calculated from **Eq. (10.29)** using $\omega_2 = 2\pi \times 60 = 120\pi$ rad/s and $\omega_1 = 2\pi \times 3600/60 = 2\pi \times 60 = 120\pi$ rad/s, but the number of turns in the loop is $N_{loop} = 1$:

$$\begin{aligned} \mathrm{emf}_t &= -\omega_2 S N_{loop} B_0 \cos\omega_2 t \cos\alpha \\ &= -2 \times \pi \times 60 \times 0.01 \times 0.01 \times 1 \times 0.5615 \cos(2 \times \pi \times 60t) \cos\left(2 \times \pi \times \frac{3{,}600}{60} t\right) = -0.0212 \cos^2 120\pi t \quad [\mathrm{V}] \end{aligned}$$

where the notation in **Figure 10.9** was used and N_{loop} was used to distinguish the number of turns in the rotating loop from those in the magnetic circuit.

(2) emf due to motion of the loop in the magnetic field. This is given in **Eq. (10.30)** where $\omega_1 = \omega_2 = 120\pi$ rad/s, $\alpha = \omega_1 t = 120\pi$ rad, and the number of turns in the rotating loop is again $N_{loop} = 1$:

$$\mathrm{emf}_m = \omega_1 S N_{loop} B_0 \sin\omega_2 t \sin\alpha = 120 \times \pi \times 0.01 \times 0.01 \times 0.5615 \sin^2 120\pi t = 0.0212 \sin^2 120\pi t \quad [\mathrm{V}]$$

The total emf is the sum of the two emfs and is also given in **Eq. (10.33)**:

$$\text{emf} = \text{emf}_t + \text{emf}_m = -0.0212(\cos^2 120\pi t - \sin^2 120\pi t) = -0.0212 \cos 240\pi t \quad [\text{V}].$$

(b) The waveform is cosinusoidal but at a frequency twice the frequency of the field, or 120 Hz. In this case the frequency of the magnetic field and that of rotation happen to be the same. If they are not, then **Eq. (10.32)** must be used instead.

(c) Since the amplitude of the emf is directly proportional to the number of turns and the emf above was generated in a single turn, we can write the emf per turn as

$$\text{emf}_0 = 0.0212 \qquad [\text{V/turn}]$$

A 10 V peak (20 V peak-to-peak) output requires

$$\frac{10}{\text{emf}_0} = \frac{10}{0.0212} = 471.7 \quad [\text{turns}]$$

The rotating coil should contain 472 turns (the number of turns is usually given in integer numbers).

Exercise 10.4 In **Example 10.6**,

(a) calculate the emf if the coil is supplied by a DC source.
(b) How many turns are required for a 10 V peak output?

Answer

(a) emf $= -0.0212\cos 120\pi t$ [V]. **(b)** 472 [turns].

Example 10.7 A coil is wound on a torus as shown in **Figures 10.12a** and **10.12b**. The coil consists of N turns, carries a current $I = I_0 \sin\omega t$ [A], and has an average radius r [m] ($r \gg b$) as shown:

(a) Calculate the induced emf in the coil. What is the meaning of this emf?
(b) Show that the induced emf is proportional to the inductance of the coil.
(c) Suppose the permeability of the torus is very large. What is the induced emf in the coil?

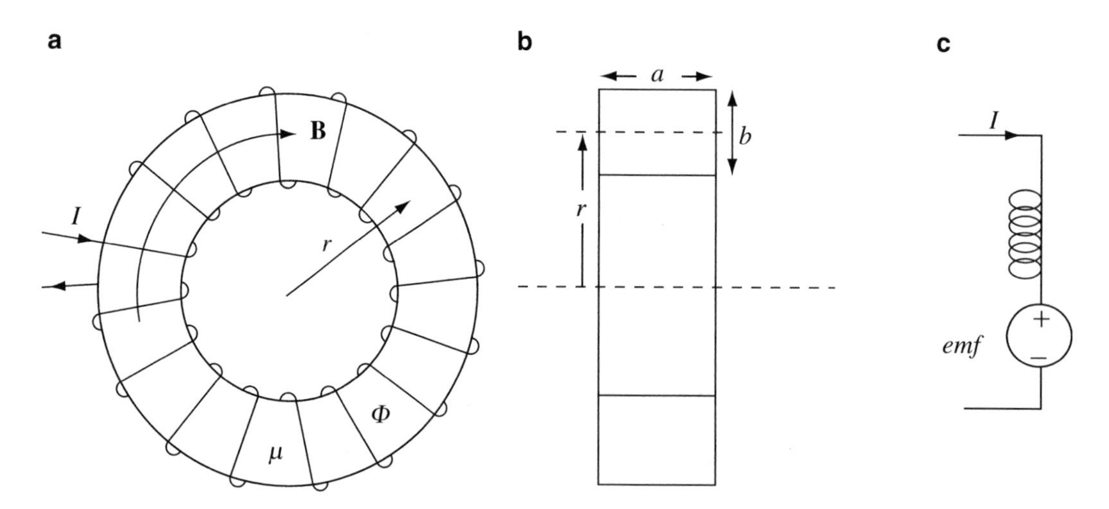

Figure 10.12 (**a**) A wound torus. (**b**) Cross section showing dimensions. (**c**) Equivalent circuit of the inductor

Solution: The induced emf in the coil is due to the change in flux (transformer action) in the coil itself. That is, because the coil is fed with an AC source, it produces an AC flux which in return induces an emf in itself. After calculating the flux in the core of the torus, the emf is calculated from Faraday's law. In **(c)**, the induced emf is also very large because the inductance of the coil is directly proportional to permeability:

(a) Since $r \gg b$, we can use the flux density at r as the average, uniform flux density throughout the core. The flux density is found from Ampere's law (see **Example 8.9** or **9.10**):

$$2\pi r B = \mu N I \quad \rightarrow \quad B = \frac{\mu N I_0 \sin\omega t}{2\pi r} \quad [\mathrm{T}]$$

and the direction of the flux density is as shown in **Figure 10.12a**. The flux in the core is

$$\Phi = BS = \frac{\mu N a b I_0 \sin\omega t}{2\pi r} \quad [\mathrm{Wb}]$$

The induced emf in the coil due to change of flux through the core is

$$\mathrm{emf} = -N\frac{d\Phi}{dt} = -\frac{\mu N^2 a b \omega I_0 \cos\omega t}{2\pi r} \quad [\mathrm{V}]$$

The induced emf is shown in **Figure 10.12c** as a source in series with the ideal coil. This source opposes the current as required by Lenz's law and is the AC voltage measured on the coil. Without this emf (such as if the coil is driven by a DC source), the voltage on the coil would be zero except for any voltage drop that might exist because of the resistance of the coil.

(b) The inductance of any device is the flux linkage divided by current. In this case, the self-inductance of the coil is

$$L = \frac{N\Phi}{I} = \frac{\mu N^2 a b}{2\pi r} \quad [\mathrm{H}]$$

Thus, the emf in the coil is

$$\mathrm{emf} = -L\frac{dI}{dt} = -\frac{\mu N^2 a b \omega I_0 \cos\omega t}{2\pi r} \quad [\mathrm{V}]$$

This expression is a direct result of Lenz's law and is extremely important in AC analysis of circuits. We will use it in the following section.

(c) If the permeability is large, so is the inductance of the coil. Therefore, the emf is also very large. If permeability tends to infinity, so does the induced emf. The same effect can be obtained with finite permeability by increasing the magnitude of the time derivative of the current (in this case, by increasing the frequency). This effect is responsible for large pulses that occur when the current in inductive circuits is changed quickly (such as when connecting or disconnecting a circuit). In electronic circuits, it is often required to protect devices, such as output power stages in amplifiers, motor and relay drivers, and the like, from being damaged due to inductive pulses.

Exercise 10.5 A small coil has a self-inductance of 10 μH. A sinusoidal current of amplitude 0.1 A and frequency 1 kHz passes through the coil.

(a) Calculate the emf measured on the coil.
(b) What is the emf if frequency changes to 100 kHz?
(c) Calculate the emf at 1 kHz for a 10 mH coil.

Answer

(a) $-6.283 \times 10^{-3}\cos 2{,}000\pi t$ [V]. **(b)** $-0.6283\cos 200{,}000\pi t$ [V]. **(c)** $-6.283\cos 2{,}000\pi t$ [V].

10.7 The Transformer

The transformer is a device designed to transform voltages and currents (and, therefore, impedances). It is an AC device and operates on the principles of Faraday's law. The transformer consists of two or more coils and a magnetic path that links the coils. There is a variety of transformers with different types of paths, but they all operate on the same principles. Power transformers are designed primarily for voltage transformation and operate at relatively high currents. The magnetic path is made of a ferromagnetic material like iron to produce a low-reluctance magnetic path (see **Figure 10.13**). Typically, the iron core of the transformer is laminated to reduce induction of currents in the core which contribute to losses. Impedance-matching transformers are normally designed for low-power applications. There are other types of transformers, some with iron cores, some without a core (air-core transformers), and still others with ferrite cores. There are also transformers which do not look like transformers but act as such. For a device to be considered a transformer, it must have two or more coils, coupled together by a common flux, whatever the physical construction of the device.

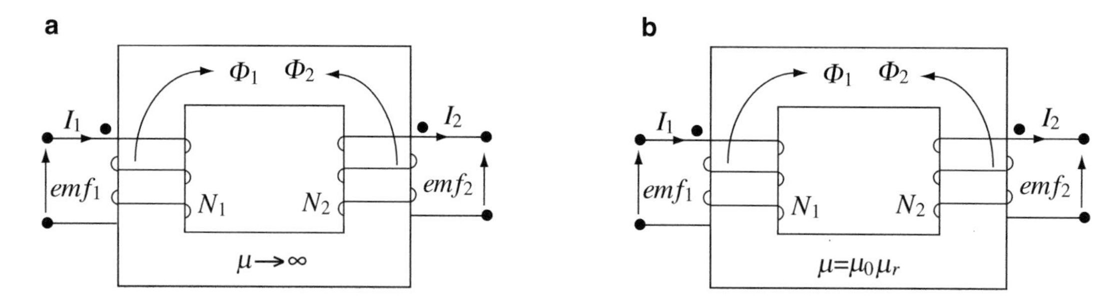

Figure 10.13 The transformer. emf_2 and I_2 are induced quantities. (**a**) The ideal transformer has a core with infinite permeability. (**b**) Core with finite permeability

10.7.1 The Ideal Transformer

An ideal transformer is one in which all flux links the coils of the transformer (i.e., flux does not leak out of the magnetic path). This implies that the permeability of the magnetic path is high (ideally, it should be infinite) and the path is closed. In addition, we assume there are no losses in the transformer.

The flux in a magnetic circuit was calculated in **Eq. (9.102)**:

$$\Phi = \frac{\sum_{i=1}^{n} N_i I_i}{\sum_{j=1}^{k} \Re_j} \quad [\mathrm{Wb}] \tag{10.34}$$

where N_i is the number of turns of coil i, I_i is the current in this coil, and $\Re_j$ is the magnetic reluctance of the jth segment of the path. The reluctance of the magnetic path was given in **Eq. (9.99)**:

$$\Re = \frac{l_m}{\mu S} \quad \left[\frac{1}{\mathrm{H}}\right] \tag{10.35}$$

where l_m is the length of the magnetic path, μ its permeability, and S the cross-sectional area of the path. Assuming that there are no losses in the core or in the coils of the transformer in **Figure 10.13**, the flux in **Eq. (10.34)** becomes

$$\Phi = \frac{N_1 I_1 - N_2 I_2}{\Re} \quad [\mathrm{Wb}] \tag{10.36}$$

where the negative sign in front of $N_2 I_2$ is due to the fact that I_2 is an induced current and Lenz's law stipulates that the induced current must produce a flux opposing the flux that produces it. Because the core is made of iron with high permeability, we may assume $\mu \to \infty$ and, therefore, $\Re \to 0$ and we can write

$$N_1 I_1 - N_2 I_2 = \Phi \Re \approx 0 \tag{10.37}$$

This, however, is only an approximation. In many cases, this approximation is very good and gives results that are very close to the exact values. From this relation, we get

$$N_1 I_1 = N_2 I_2 \tag{10.38}$$

The total flux in the core is the same in both coils and the emfs across the two coils are

$$V_1 = \text{emf}_1 = -N_1 \frac{d\Phi}{dt}, \quad V_2 = \text{emf}_2 = -N_2 \frac{d\Phi}{dt} \quad [\text{V}] \tag{10.39}$$

From these, the voltage ratio between primary and secondary is

$$\boxed{\frac{\text{emf}_1}{\text{emf}_2} = \frac{V_1}{V_2} = \frac{I_2}{I_1} = \frac{N_1}{N_2} = a \qquad [\text{dimensionless}]} \tag{10.40}$$

where $a = N_1/N_2$ denotes the turn ratio, also called the ***transformer ratio***.

Note: The emf in each coil is opposite in sign to the applied voltage on the coil, if any, so that Kirchhoff's voltage law gives zero in the primary and secondary circuits. This is the same as saying that the total flux in an ideal transformer is zero. In a real transformer, Kirchhoff's voltage law in each circuit results in a small voltage difference ($V_1 > \text{emf}_1, V_2 < \text{emf}_2$). The difference is due to losses in the transformer and the transformer ratio changes accordingly.

While most transformers are designed either to transform currents or voltages, they also change the impedance of the circuit. The impedance of the primary circuit is given by the ratio of emf_1 and I_1:

$$Z_1 = \frac{\text{emf}_1}{I_1} = \frac{a\text{emf}_2}{I_2/a} = a^2 Z_2 \quad [\Omega] \tag{10.41}$$

or, if Z_L is a load impedance,

$$\boxed{\frac{Z_1}{Z_L} = a^2} \tag{10.42}$$

The impedance Z_1 is, in fact, the effective load impedance seen by the source. Impedance matching is sometimes the primary function of the transformer. However, regardless of the function, the impedance seen from the primary or the effective impedance in the primary circuit depends on the turn ratio squared and the impedance of the secondary.

Although an actual transformer includes losses due to resistance of the conductors, induced currents in the core, and currents needed to magnetize the core (as well as capacitive losses), the main approximation used to define an ideal transformer was the assumption that the permeability is infinite and, therefore, that the reluctance of the magnetic path is zero. In practical applications, this is never the case, and in some transformers, like air-core transformers, the above approximations cannot be used at all. In many transformers, the losses are relatively small (sometimes less than 1%) and the above approximations are quite good. However, in low-power transformers, losses may be high relative to the total power capacity of the transformer.

10.7.2 The Real Transformer: Finite Permeability

The transformer in **Figure 10.13b** consists of a core with relatively high reluctance (low permeability); therefore, the approximation of infinite permeability cannot be used. If we wish to calculate the ratio between primary and secondary, we must calculate the flux in the magnetic circuit. We first assume that all flux links both coils of the transformer (no flux leakage), but permeability is finite. The net flux in the magnetic circuit linking both coils in **Figure 10.13b** is

$$\Phi = \frac{\mu S}{l}(N_1 I_1 - N_2 I_2) \quad [\text{Wb}] \tag{10.43}$$

Now, using Faraday's law, the emf in each coil is calculated as

$$\mathrm{emf}_1 = N_1 \frac{d}{dt}\left(\frac{\mu S}{l}(N_1 I_1 - N_2 I_2)\right), \quad \mathrm{emf}_2 = N_2 \frac{d}{dt}\left(\frac{\mu S}{l}(N_1 I_1 - N_2 I_2)\right) \quad [\mathrm{V}] \tag{10.44}$$

Rearranging the terms, we get

$$\boxed{\mathrm{emf}_1 = \frac{\mu S}{l} N_1^2 \frac{dI_1}{dt} - \frac{\mu S}{l} N_1 N_2 \frac{dI_2}{dt} \quad [\mathrm{V}]} \tag{10.45}$$

$$\boxed{\mathrm{emf}_2 = \frac{\mu S}{l} N_1 N_2 \frac{dI_1}{dt} - \frac{\mu S}{l} N_2^2 \frac{dI_2}{dt} \quad [\mathrm{V}]} \tag{10.46}$$

In **Example 9.9**, we calculated the self-inductances and mutual inductances of three coils on a closed magnetic core of finite permeability μ. Using those results for the first two coils, we have

$$L_{11} = \frac{\mu S}{l} N_1^2, \quad L_{12} = L_{21} = \frac{\mu S}{l} N_1 N_2, \quad L_{22} = \frac{\mu S}{l} N_2^2 \quad [\mathrm{H}] \tag{10.47}$$

Using these relations, **Eqs. (10.45)** and **(10.46)** become

$$\boxed{\mathrm{emf}_1 = L_{11} \frac{dI_1}{dt} - L_{12} \frac{dI_2}{dt} \quad [\mathrm{V}]} \tag{10.48}$$

$$\boxed{\mathrm{emf}_2 = L_{21} \frac{dI_1}{dt} - L_{22} \frac{dI_2}{dt} \quad [\mathrm{V}]} \tag{10.49}$$

Here, we have used the fact that the current in the secondary is due to induction; that is, it only exists if the current in the primary exists. According to Lenz's law, the flux produced by this current is always in opposition to the flux due to the primary. Therefore, the flux in the core is small (it is zero for an ideal transformer and for a nonideal transformer with zero losses).

To more easily identify the emfs induced in various coils, the so-called dot convention is used. A dot is placed on the terminal of the coil which, when a current flows into the dot, produces a flux in the direction of the net flux in the core. In the case of transformers, this means that when the current increases on a dotted terminal, all dotted terminals experience an increase in emf. A current flowing into a dot produces a positive emf and a current flowing away from a dot produces a negative emf. In **Figure 10.13b**, I_1 flows into the dot and I_2 flows away from the dot. The emfs in **Eqs. (10.48)** and **(10.49)** that are associated with I_1 are positive whereas those associated with I_2 are negative.

The induced emfs in **Eqs. (10.48)** and **(10.49)** may also be understood in terms of impedances. In particular, in the frequency domain, d/dt is replaced with $j\omega$ and the emfs in **Eqs. (10.48)** and **(10.49)** become

$$\mathrm{emf}_1 = j\omega L_{11} I_1 - j\omega L_{12} I_2 \quad [\mathrm{V}] \tag{10.50}$$

$$\mathrm{emf}_2 = j\omega L_{21} I_1 - j\omega L_{22} I_2 \quad [\mathrm{V}] \tag{10.51}$$

where I_1 and I_2 as well as emf_1 and emf_2 are now phasors. Note also that the emf and current in each coil are 90° out of phase.

Although the relations in **Eqs. (10.48)** and **(10.49)** may look very different than those for the ideal transformer, they are, in fact, very similar. In particular, because all flux is contained within the core and there are no losses, the ratio between the voltage (emf) in the primary and secondary remains the same as for the ideal transformer in **Eq. (10.40)**. Therefore, any transformer in which all flux is contained within the core and which has no losses behaves as an ideal transformer regardless of the permeability of the core. This can be seen most easily from **Eq. (10.44)**.

10.7.3 The Real Transformer: Finite Permeability and Flux Leakage

In the previous two sections, we assumed that all flux produced by a coil is contained within the core. This is not always the case, as we have seen in **Section 9.4**. There are conditions under which some of the flux closes outside the core or there is no core to begin with. Consider two coils in air. In this case, we do not know how much of the flux connects the two coils, but we can assume that a fraction of the flux produced by one coil links the second coil. Suppose this fraction is k. Because of that, the mutual inductances also change by this fraction; that is, if all flux links both coils, the inductances in **Eq. (10.47)** are obtained. If only a fraction k links coil 1 and coil 2, we get

$$L_{11} = \frac{\mu S}{l}N_1^2, \quad L_{12} = L_{21} = k\frac{\mu S}{l}N_1N_2, \quad L_{22} = \frac{\mu S}{l}N_2^2 \quad [\text{H}] \tag{10.52}$$

We note that

$$L_{11}L_{22} = \frac{\mu S}{l}N_1^2\frac{\mu S}{l}N_2^2 \quad \rightarrow \quad \sqrt{L_{11}L_{22}} = \frac{\mu S}{l}N_1N_2 \tag{10.53}$$

or

$$L_{12} = L_{21} = k\sqrt{L_{11}L_{22}} \quad [\text{H}] \tag{10.54}$$

The constant k is called a ***coupling coefficient*** and indicates how much of the flux produced by one coil links the other. For the ideal transformer in **Section 10.7.1** and for the transformer in **Section 10.7.2**, the coupling coefficient is equal to 1 (all flux links both coils). In air-core transformers, the coupling coefficient is almost always smaller than 1. If the coupling coefficient is known, the emfs in each coil can be calculated from the self-inductances of the two coils and the coupling coefficient. For the transformer in **Figure 10.13b**, but now assuming that $k \leq 1$, the emfs are

$$\boxed{\text{emf}_1 = L_{11}\frac{dI_1}{dt} - k\sqrt{L_{11}L_{22}}\frac{dI_2}{dt} \quad [\text{V}]} \tag{10.55}$$

$$\boxed{\text{emf}_2 = k\sqrt{L_{11}L_{22}}\frac{dI_1}{dt} - L_{22}\frac{dI_2}{dt} \quad [\text{V}]} \tag{10.56}$$

Thus, the smaller the coupling coefficient, the closer the emf is to that of a simple coil, and the closer k is to 1, the closer the behavior is to that of an ideal transformer.

Example 10.8 Application: The Toroidal Transformer A widely used transformer in a variety of products including high-quality audio and test equipment is built around a toroidal core, as in **Figure 10.14**. The transformer shown is designed to supply 48 V at 2 A for the output stage of an amplifier. The toroidal core has a cross-sectional area of $S = 400$ mm^2, a mean magnetic path of $l = 200$ mm, and the relative permeability of the core is very large (infinite). The primary operates at 240 V, 60 Hz. If the primary coil must have 800 turns to generate the required flux in the core, calculate:

(a) The number of turns in the secondary and current in the primary.
(b) Show that the result in **(a)** remains unchanged if the permeability of the core is finite, as long as all flux remains contained within the core.

Figure 10.14 A toroidal power transformer

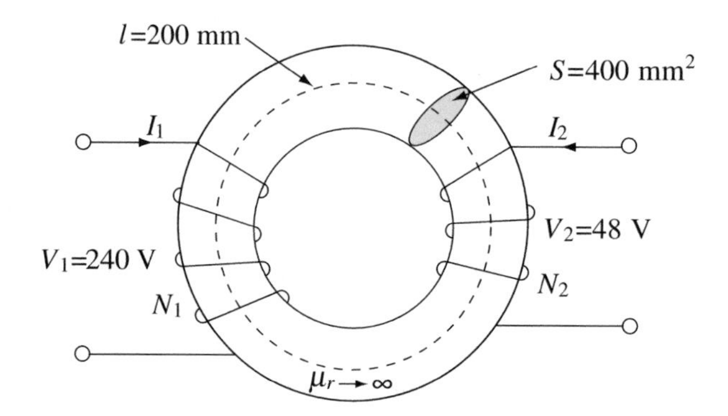

Solution: In **(a)**, we may use the expressions for the ideal transformer since the reluctivity of the magnetic path is zero. In **(b)** we use the expressions in **Eqs. (10.45)** and **(10.46)** and calculate the ratio between the emfs:

(a) From **Eq. (10.40)**

$$\frac{V_1}{V_2}=\frac{N_1}{N_2} \quad \rightarrow \quad N_2=\frac{N_1V_2}{V_1}=\frac{800\times 48}{240}=160 \quad [\text{turns}]$$

where V_1 is the voltage on the primary and V_2 is the voltage on the secondary. The current in the primary is also calculated from **Eq. (10.40)**:

$$\frac{I_1}{I_2}=\frac{N_2}{N_1} \quad \rightarrow \quad I_1=\frac{N_2I_2}{N_1}=\frac{160\times 2}{800}=0.4 \quad [\text{A}]$$

(b) The ratio between emf_1 and emf_2 from **Eqs. (10.45)** and **(10.46)** is

$$\frac{\text{emf}_1}{\text{emf}_2}=\frac{\frac{\mu S}{l}N_1^2\frac{dI_1}{dt}-\frac{\mu S}{l}N_1N_2\frac{dI_2}{dt}}{\frac{\mu S}{l}N_1N_2\frac{dI_1}{dt}-\frac{\mu S}{l}N_2^2\frac{dI_2}{dt}}=\frac{N_1\left(N_1\frac{dI_1}{dt}-N_2\frac{dI_2}{dt}\right)}{N_2\left(N_1\frac{dI_1}{dt}-N_2\frac{dI_2}{dt}\right)}=\frac{N_1}{N_2}$$

This is the same as in **(a)**; therefore, the ratio remains unchanged for any value of μ. However, if μ is low, the flux will tend to leak, invalidating the assumptions used to obtain **Eqs. (10.45)** and **(10.46)**. If this happens, **Eqs. (10.55)** and **(10.56)** must be used. In practical design, if μ is large (but finite), it is safe to use the assumptions for the ideal transformer.

Notes:

(1) Because $N_1I_1=N_2I_2$ and since I_2 produces a flux which opposes that due to I_1, the net flux in the core is zero. In reality, there will be a small flux due to losses.

(2) Because this is an ideal transformer, the dimensions of the core and frequency of the source are not important: They do not figure in the calculation. In practical transformers, the dimensions define the maximum flux density allowable without the core reaching saturation.

Toroidal transformers are favored for audio and test equipment applications because they are inherently low-leakage transformers, even at low core permeability. Typically, they offer the shortest magnetic path and, therefore, the lowest reluctance in addition to being economical in both winding and core materials. However, since the winding of coils is usually done after the core is assembled (made of stacked-up laminations or of strips of the lamination material wound in the form of a torus), it is complicated and requires special winding equipment. Toroidal transformers are also very useful in switching and high-frequency applications.

Example 10.9 A high-frequency transformer is made in the form of two coils on a nonmagnetic form ($\mu = \mu_0$), as in **Figure 10.15**. The self-inductance of coil (1) is 10 μH and of coil (2) 20 μH. The current in the primary is $I_0\sin\omega t$ where $f = 1$ MHz, $I_0 = 0.1$ A, and the secondary is open. The coupling coefficient between the two coils is 0.2. Calculate:

(a) The voltage (emf) required in the primary to sustain the given current
(b) The voltage (emf) in the secondary

Figure 10.15 A high-frequency transformer; $k \leq 1$

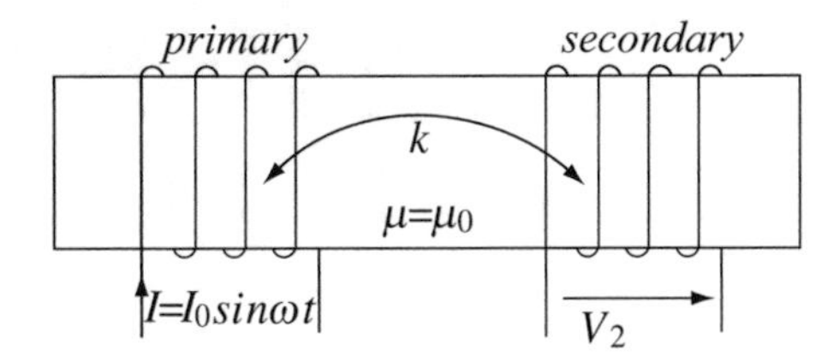

Solution: This transformer must be treated as a real transformer, using **Eqs. (10.55)** and **(10.56)**. In addition, because the secondary coil is open, $I_2 = 0$:
(a) From **Eq. (10.55)**, the induced emf in the primary, induced by itself, is

$$\text{emf}_1 = L_{11}\frac{dI_1}{dt} = L_{11}\frac{d(I_0\sin\omega t)}{dt} = L_{11}\omega I_0\cos\omega t$$
$$= 10 \times 10^{-6} \times 2 \times \pi \times 10^6 \times 0.1\cos(2 \times \pi \times 10^6 t) = 6.283\cos(2\pi \times 10^6 t) \quad [\text{V}]$$

(b) From **Eq. (10.56)**, the emf induced in the secondary by the primary is

$$\text{emf}_2 = k\sqrt{L_{11}L_{22}}\frac{dI_1}{dt} = k\sqrt{L_{11}L_{22}}\omega I_0\cos\omega t$$

$$= 0.2 \times \sqrt{10 \times 10^{-6} \times 20 \times 10^{-6}} \times 2 \times \pi \times 10^6 \times 0.1\cos(2 \times \pi \times 10^6 t) = 1.78\cos(2\pi \times 10^6 t) \quad [\text{V}]$$

This emf is the open circuit voltage on the secondary coil.

Example 10.10 Application: The Current Transformer Although **Eqs. (10.55)** and **(10.56)** define the general transformer, the current transformer is unique in that its primary coil is connected in series with the circuit in which it operates. In many cases, the primary coil is part of the circuit and the transformer core surrounds it. Three examples of current transformers are shown in **Figure 10.16**. The first, in **Figure 10.16a**, is a simple transformer that, in principle, can also be used as a voltage transformer. What makes it unique is the low number of turns in the coils, especially in the primary coil. This is necessary since it is connected in series with the circuit and it should have low impedance. The transformer in **Figure 10.16b** is similar except that the primary is a single turn passing through the core. In this case, the turn ratio a is $1/N_2$. This particular arrangement is often used because it does not require connections into the circuit and is particularly useful for measuring purposes (i.e., $aI_1 = I_2$). If appropriate, or necessary, the primary may be made of two or more turns by passing the wire through the core two or more times. A common measuring device based on this principle is shown in **Figure 10.16c**. This is a clamping ampere meter. It is essentially a current transformer without the primary coil. The secondary coil is connected to a measuring device such as a digital meter or bridge. The core is split and hinged such that it can be opened and closed around the wire in which we wish to measure the current. With N_2 turns in the secondary coil and since the primary in this case has a single turn, the measured current in the primary is equal to $1/a$ times the current in the secondary, which is either measured directly or is deduced from the voltage on the secondary. The advantage of this device is that it measures current without the need to cut the circuit, but, as you might expect, it is only accurate at relatively high currents.

(continued)

Example 10.10 Application: The Current Transformer (continued)

It is required to design a current transformer that will continuously measure a sinusoidal current supplied to an installation by placing the transformer over one of the wires leading to the installation. The peak current expected is 100 A at 60 Hz. A toroidal core, made of iron with average radius $a = 30$ mm, is available. The cross section of the torus is circular, with a radius $b = 10$ mm. Relative permeability of the iron is $\mu_r = 200$. The torus is inserted over the wire as shown in **Figure 10.17** and the secondary coil is connected to a voltmeter. The voltmeter can measure between $V = 0$ and $V = 1$ V (peak):

(a) Calculate the number of turns in the secondary of the current transformer for full-scale reading at 100 A.
(b) Suppose you do not wish to use an iron core for the solenoid because of induced currents in the iron. Can you use an air-core torus? If so, what is the number of turns required if the torus is made of plastic, with the same dimensions as before and for the same reading?

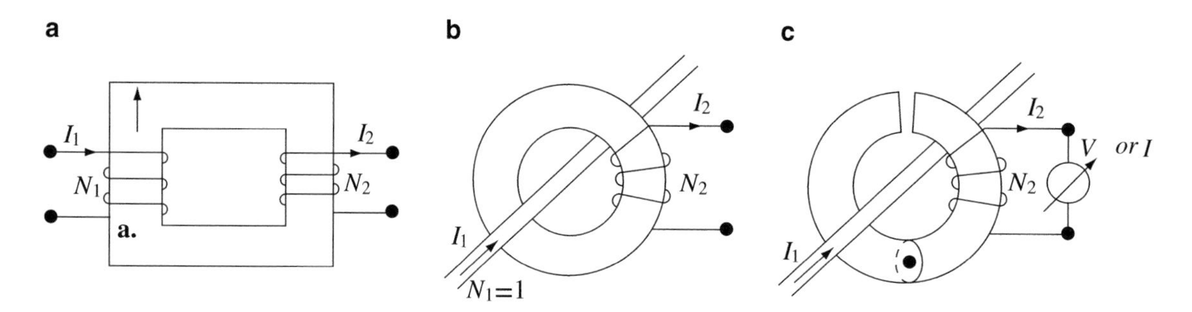

Figure 10.16 (**a**) A current transformer. (**b**) A single-turn transformer. (**c**) A clamping ampere meter

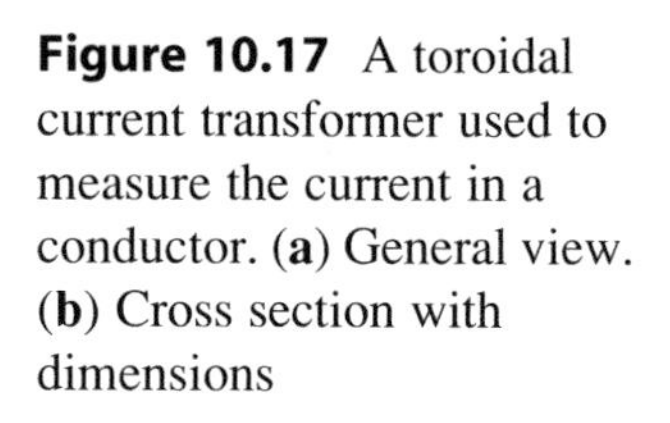
Figure 10.17 A toroidal current transformer used to measure the current in a conductor. (**a**) General view. (**b**) Cross section with dimensions

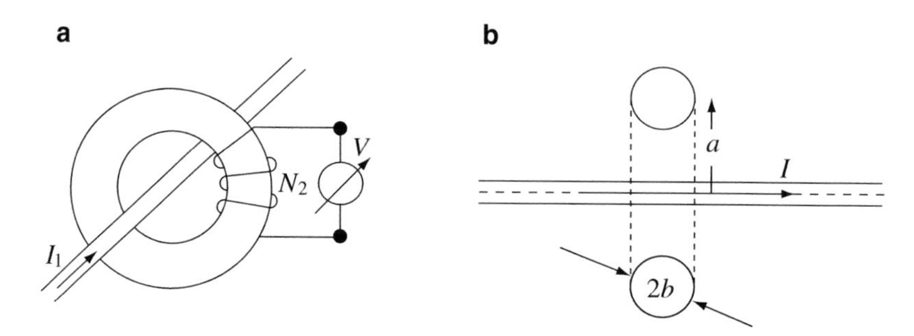

Solution: The magnetic flux density inside the torus is calculated as for any infinitely long wire carrying a current I. The flux in the torus is then calculated, and from the flux, the emf is calculated using Faraday's law for a single turn. The number of turns is the ratio between the full-scale reading and the emf of a single turn:

(a) The magnetic flux density at a distance a from the wire at full-scale current is (see **Example 8.11**)

$$B = \frac{\mu_0\mu_r I\sin\omega t}{2\pi a} = \frac{4\times\pi\times10^{-7}\times200\times100\sin120\pi t}{2\times\pi\times0.03} = 0.133\sin(120\pi t) \quad [\mathrm{T}]$$

Assuming this to be uniform in the cross-sectional area of the torus, the flux in the core is the flux density multiplied by the cross-sectional area:

$$\Phi = BS = B\pi b^2 = 0.133\times\pi\times0.01^2\sin(120\pi t) = 4.178\times10^{-5}\sin(120\pi t) \quad [\mathrm{Wb}]$$

The induced emf in a loop is

$$\mathrm{emf}_0 = -\frac{d\Phi}{dt} = -\omega B\pi b^2 = -2\times\pi\times60\times4.189\times10^{-5}\cos(120\pi t) = -0.01575\cos(120\pi t) \quad [\mathrm{V}]$$

This is the emf per turn. The peak emf per turn is 0.01575V. Thus, the number of turns required in the secondary coil is

$$N_2 = \frac{V_2}{emf_0} = \frac{1}{0.01575} = 63.49 \quad [\text{turns}] \quad \rightarrow \quad N_2 = 63.5 \quad [\text{turns}].$$

(b) As long as the torus is centered with the wire and the turns on the torus are uniformly wound around its circumference, an air-filled toroidal coil may be used just as well. However, the magnetic flux density and magnetic flux in the torus are 200 times smaller since, now, $\mu_r = 1$. The emf per turn will also be 200 times smaller and the number of required turns is 200 times larger or 12,700 turns. The iron core is therefore a better solution. Lamination of the core can reduce losses and heating in the core to a minimum.

Note: The current transformer discussed here is an ideal transformer because we assumed there are no losses and all flux in the core remains contained, in spite of the relatively low permeability of the core.

10.8 Eddy Currents

Up to this point, we assumed that an induced emf (induced voltage) can be generated in a loop, or any conducting wire, regardless of shape. If the loop is closed in a circuit, the induced emf produces an induced current. However, Faraday's law as written in **Eq. (10.1)** does not require the existence of a physical loop: Induction of flux, and therefore electromotive force, exists even if an actual loop is not obvious. To see this, consider a time-dependent, uniform magnetic flux density as in **Figure 10.18a**. If we place a loop in this flux density, an induced emf is generated in the loop as expected. Now, consider the situation in **Figure 10.18b**, where a solid cylindrical conductor is placed in the changing magnetic flux density. We do not have a loop per se, but we can view the cylinder as being composed of thin short-circuited loops, as in **Figure 10.18c**. Each one of these loops will have an emf that produces an induced current in it. Now, considering again the cylinder in **Figure 10.18b**, it is obvious that the magnetic flux density induces currents in the volume of the cylinder. These currents are also called eddy currents, or Foucault currents[3]. Normally, we view eddy currents as undesirable because they dissipate power in the volume of materials and therefore generate heat (losses) in the material. This is certainly the case in transformers and in some electric motors. However, as we will see in examples that follow, there are important applications of eddy currents including levitation, heating and melting of materials, nondestructive testing for material integrity, and induction motors, where eddy currents are essential.

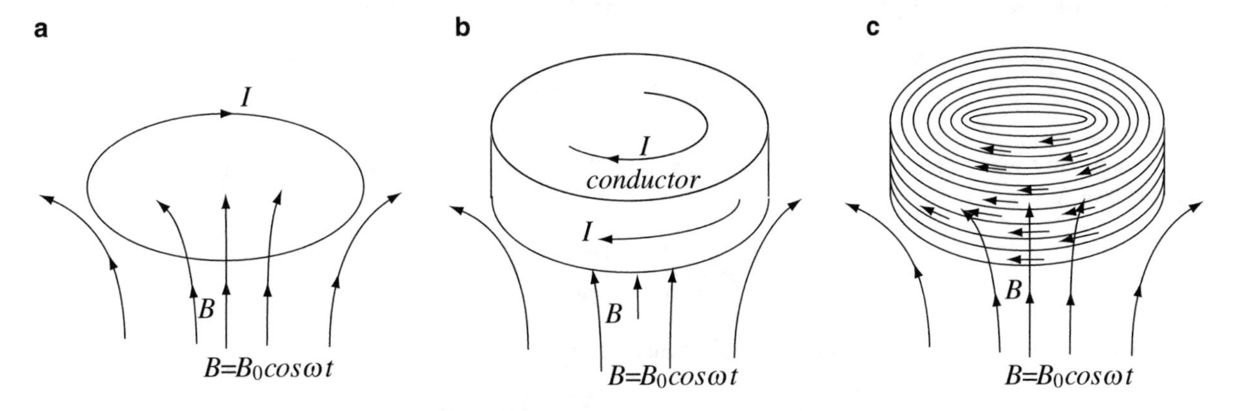

Figure 10.18 (**a**) A time-dependent flux density generates an induced current in a loop. (**b**) A time-dependent magnetic flux density generates induced currents in a conducting volume. (**c**) The conducting volume is seen as being made of short-circuited conducting loops similar to the loop in (**a**)

[3] *Induced currents* is the generic name associated with currents in the bulk of conducting materials. The term *eddy currents* is the common name used to distinguish induced currents occurring in the bulk of materials with induced currents in thin wire loops. The name *Foucault currents* is commonly used in France and is named after Jean Bernard Léon Foucault (1819–1868) as a tribute to his extensive contribution to many areas of science, most notably to optics and electromagnetics. Foucault is best remembered for his pendulum, which measured, for the first time, the rotation of the Earth (1851), but he also invented the gyroscope (1852), a method of photographing stars (1845) and showed that heat has wave properties. Many other techniques, including the modern method of making mirrors, are due to him.

Example 10.11 Application: Losses in Conducting Materials A thin circular disk of radius $d = 100$ mm and thickness $c = 1$ mm is placed in a uniform, AC magnetic field as in **Figure 10.19a**. The magnetic flux density varies as $B = B_0 \sin\omega t$ [T] and is directed perpendicular to the disk. The conductivity of the disk is $\sigma = 10^7$ S/m, $f = 50$ Hz, and the amplitude of the magnetic flux density is 0.2 T:

(a) Calculate the instantaneous power dissipated in the disk due to induced (eddy) currents. Assume the magnetic flux density is not modified by the induced currents and the field remains constant throughout the disk.
(b) What is the peak power loss at 100 Hz, under the same assumptions?

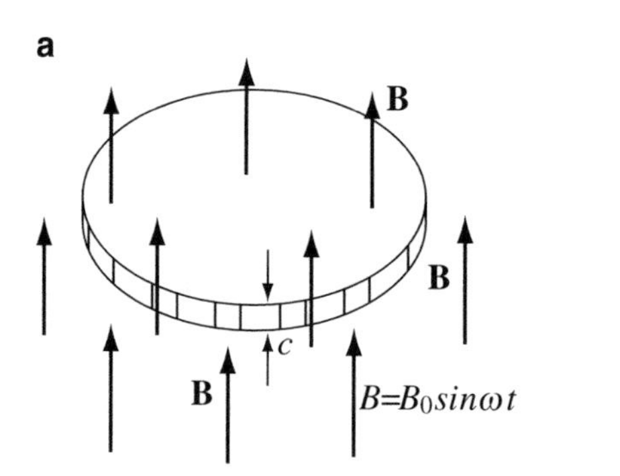

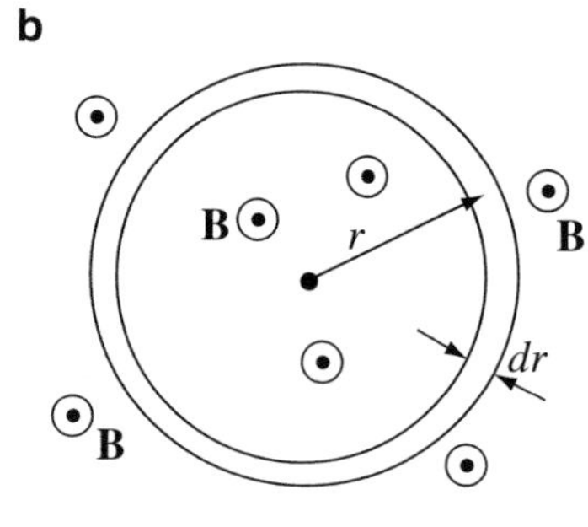

Figure 10.19 Induced currents due to change in flux. (**a**) Geometry. (**b**) A ring of radius r and differential width used to calculate the flux

Solution: To calculate the eddy current, an infinitesimal ring is "cut" out of the disk and viewed as a loop. Now, we can calculate the emf induced in this ring and its resistance. From these, we obtain the power dissipated in the infinitesimal ring. To find the total power dissipated in the disk, we integrate the power over all rings that make up the disk:

(a) Consider **Figure 10.19b**. The total flux enclosed in the ring of radius r is

$$\Phi = \pi r^2 B = \pi r^2 B_0 \sin\omega t \quad [\text{Wb}]$$

The emf (neglecting the sign since only the power is needed, not the direction of current) is

$$\left|\frac{d\Phi}{dt}\right| = \omega\pi r^2 B_0 \cos\omega t \quad [\text{V}]$$

To calculate power, we need the resistance of the ring. This is calculated for a ring of length $2\pi r$ and cross-sectional area equal to cdr as

$$R = \frac{l}{\sigma S} = \frac{2\pi r}{\sigma c dr} \quad [\Omega]$$

The instantaneous power dissipated in this infinitesimal ring is

$$dP(t) = \frac{V^2}{R} = \frac{\text{emf}^2}{R} = \frac{(\omega\pi r^2 B_0 \cos\omega t)^2}{2\pi r/\sigma c dr} = \frac{\omega^2 \pi r^3 B_0^2 \sigma c(\cos^2\omega t)dr}{2} \quad [\text{W}]$$

Since the disk is made of an infinite number of rings varying in radius from zero to d, we integrate this expression over r and get

$$P(t) = \int_{r=0}^{r=d} \frac{\omega^2 \pi r^3 B_0^2 \sigma c(\cos^2\omega t)dr}{2} = \frac{\omega^2 \pi d^4 B_0^2 \sigma c(\cos^2\omega t)}{8} \quad [\text{W}]$$

For the values given above, this power is

$$P(t) = \frac{(2 \times \pi \times 50)^2 \times \pi \times 0.1^4 \times 0.2^2 \times 10^7 \times 10^{-3} \times \cos^2(2 \times \pi \times 50t)}{8} = 1550.3 \cos^2(314.16t) \quad [\text{W}]$$

The peak power dissipated is 1,550.3 W.

(b) Since the power dissipated is proportional to the square of the frequency and all other parameters remain unchanged, the peak power dissipated at 100 Hz is four times larger or 6,201.2 W. This power is very large considering the small volume involved. The result is quick heating of the material or even melting. This method of heating metals is commonly used in both melting (induction melting) and heat treatment of conducting materials. Perhaps the most common method of surface hardening (such as on bearing surfaces and rotating shafts) is the use of induction heating coils to locally heat the surface that needs to be hardened followed by quenching in oil. Because coils can be made to fit rather awkward surfaces, the method is versatile, and because heating is quick, it is fast and efficient.

In practice, the magnetic flux density does change in the material (we shall see in **Chapter 12** why and how) and the power dissipated is smaller than that found here. Also, because of the change in the magnetic flux density in the material, more power is dissipated on the surface of the conductor than in its interior. This property is often used to produce localized surface heating such as in hardening of surfaces of rotating shafts.

Example 10.12 Induced Currents Due to Change in Flux Consider the thin conducting ring in **Figure 10.20a**. The flux density $\mathbf{B} = \mathbf{B}_0$ [T] is constant and uniform throughout its cross section. At a given time $t = 0$, the flux density $\mathbf{B}$ starts to increase as

$$\mathbf{B}(t) = \mathbf{B}_0(1 + kt) \qquad [\text{T}]$$

where k is a constant. Calculate the induced current in the ring. Assume the ring is thin and the induced currents do not affect the magnetic field. Numerical values are: cross-sectional area of the ring $s = 1$ mm^2, $r = 10$ mm, $\sigma = 10^7$ S/m, $B_0 = 1$ T, $k = 60$ T/s.

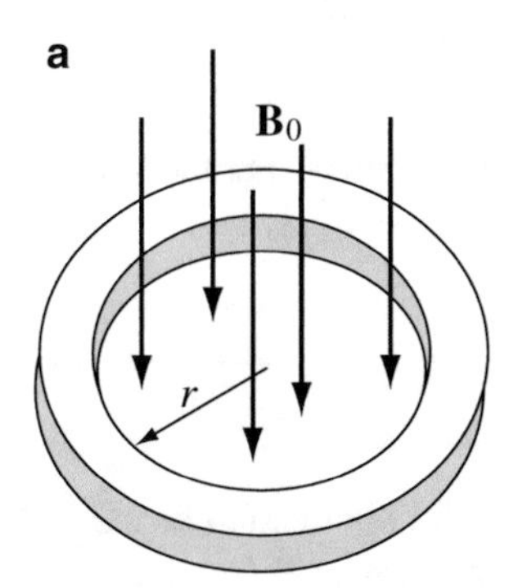

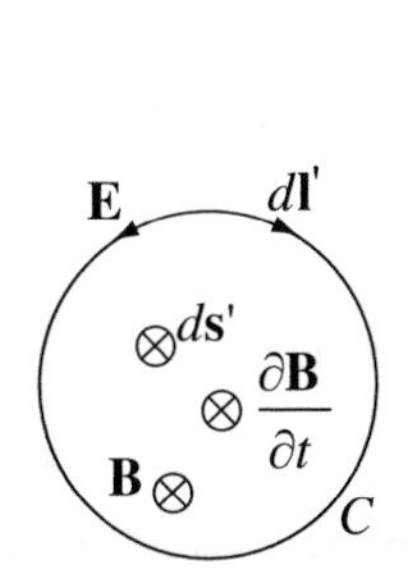

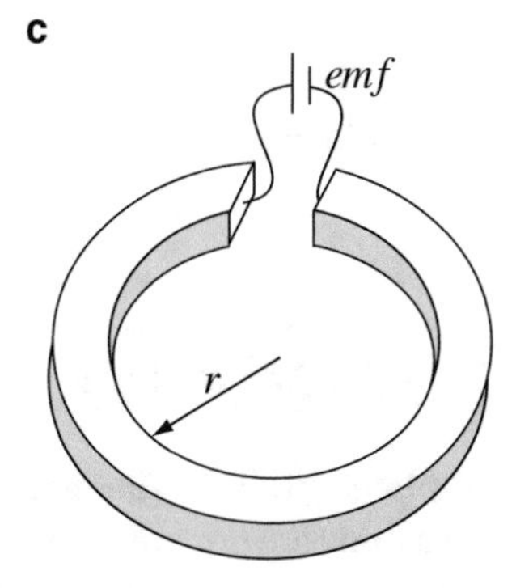

Figure 10.20 Induced currents due to change in flux. (**a**) A conducting ring in a magnetic field. (**b**) Relation between magnetic and electric fields. (**c**) Equivalent circuit showing the induced emf in the ring

Solution: There are two methods to solve this problem: (1) The increase in the magnetic flux density causes an induced electric field intensity in the closed loop, which may be calculated using **Eq. (10.5)**. This electric field generates a current density in the material of the loop equal to $\sigma\mathbf{E}$. Assuming the current density is uniform in the conductor, the current is found by multiplying the current density by the cross-sectional area of the conducting ring. (2) The increase in the magnetic flux density induces an emf in the ring. This emf produces a current equal to the emf divided by the resistance of the loop. We show both methods.

Method (1) The change in flux density produces an electric field intensity $\mathbf{E}$ in the direction shown in **Figure 10.20b**, assuming that the flux density increases as indicated. The induced electric field intensity, induced current density, and induced current can now be calculated using **Eq. (10.5)**:

$$\oint_C \mathbf{E} \cdot d\mathbf{l}' = \int_{s'} -\frac{\partial \mathbf{B}}{\partial t} \cdot d\mathbf{s}'$$

where s' is the surface defined by the circular ring and C is the circumference of the ring. Noting that $\mathbf{B}$ depends only on time and that the pairs of vectors $\mathbf{E}$, $d\mathbf{l}'$, and $\partial\mathbf{B}/\partial t$, $d\mathbf{s}'$ are collinear gives:

$$E2\pi r = -\pi r^2 \frac{\partial B}{\partial t} \quad \rightarrow \quad |E| = \frac{r}{2}\frac{\partial}{\partial t}(B_0 + B_0 kt) = \frac{krB_0}{2} \quad \left[\frac{\text{V}}{\text{m}}\right]$$

From this, the current density $J = \sigma E$ and current are

$$J = \frac{\sigma k B_0 r}{2} \quad \left[\frac{\text{A}}{\text{m}^2}\right] \quad \text{and} \quad I = \frac{\sigma k B_0 s r}{2} \quad [\text{A}]$$

where s is the cross-sectional area of the ring.
Method (2). The emf in the closed loop equals

$$|\text{emf}| = \left| \int_{s'} \frac{\partial \mathbf{B}}{\partial t} \cdot d\mathbf{s}' \right| = \pi r^2 B_0 k \qquad [\text{V}]$$

The emf can be viewed as a voltage source in the ring, as shown in **Figure 10.20c**. Viewing the loop as a circuit, the emf is

$$\text{emf} = RI \quad [\text{V}]$$

where R is the resistance of the loop and I the current in the loop. The resistance of the ring is

$$R = \frac{l}{\sigma s} = \frac{2\pi r}{\sigma s} \quad [\Omega]$$

Combining the last three relations, we get

$$\pi r^2 B_0 k = \frac{2\pi r}{\sigma s} I \quad \rightarrow \quad I = \frac{\sigma k B_0 s r}{2} \quad [\text{A}]$$

This is identical to the result obtained in method (1). With the given numerical values, the current in the loop is 3 A.

10.9 Applications

Application: The Magnetic Brake An interesting and very useful application of induced currents is the magnetic brake. To outline the principle involved, consider **Figure 10.21a**. An electromagnet generates a flux density $\mathbf{B}$ in the gap. This field is assumed to be constant. A pendulum-like flat piece, made of a conducting material, is placed such that it can move into the gap. If the current in the electromagnet, I, is zero, the oscillation of the pendulum is not affected by the structure. If there is a current in the coil, the movement of the conducting plate into the magnetic field (**Figure 10.21b)** generates induced currents in the plate itself due to the motion of the conductor in the magnetic field. The flux of the induced currents is such that it opposes the field $\mathbf{B}$. According to Lenz's law, the induced currents tend to maintain this condition by opposing the flux. **Figure 10.21b** gives the direction of the fields. The electric field intensity due to the induced currents is given as $\mathbf{E} = \mathbf{v} \times \mathbf{B}$ and we get

$$\mathbf{J} = \sigma \mathbf{E} = \sigma \mathbf{v} \times \mathbf{B} \quad [\text{A/m}^2] \tag{10.57}$$

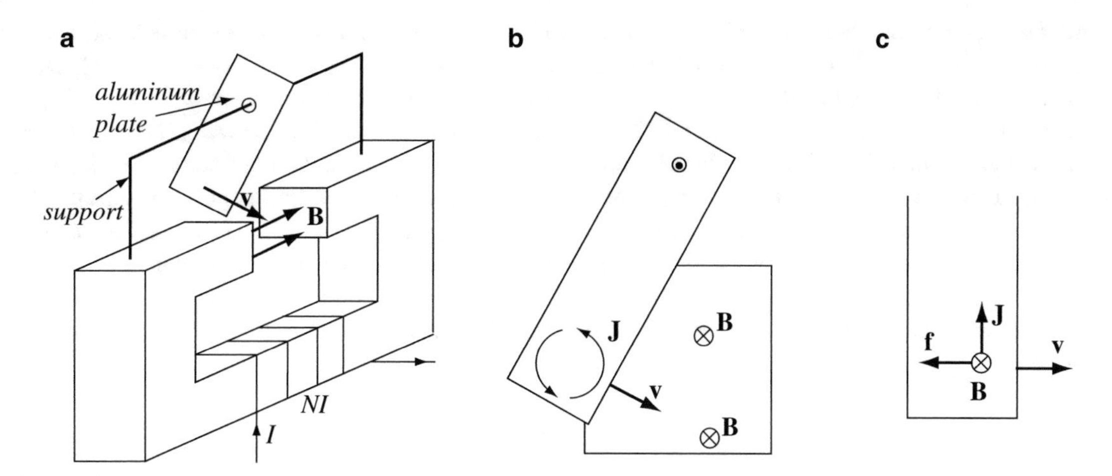

Figure 10.21 (**a**) The magnetic brake. (**b**) Direction of fields in the magnetic brake. (**c**) Direction of induced currents in the plate

The velocity at which the plate penetrates into the gap is responsible for the magnitude of the induced currents. The relation between the current density **J**, magnetic flux density **B**, and force density **f** at a point is shown in **Figure 10.21c**.

Using **Eq. (9.109)** we get the volumetric force density **f**:

$$\mathbf{f} = \mathbf{J} \times \mathbf{B} = \sigma(\mathbf{v} \times \mathbf{B}) \times \mathbf{B} \quad [\mathrm{N/m^3}] \tag{10.58}$$

If all vectors are mutually orthogonal, as is the case in this example, the total force is

$$F = \sigma v B^2 V_0 l \quad [\mathrm{N}] \tag{10.59}$$

and its direction, given by the cross-product $\mathbf{J} \times \mathbf{B}$, opposes the direction of **v**. This has the effect of damping the movement of the plate into the gap. If the conductivity σ of the plate were infinite, the plate would be repelled from the gap. In reality, σ is finite and the plate decelerates as the power due to induced currents is dissipated in the plate. The plate penetrates into the gap, slows down and, eventually, reaches a state of static equilibrium at the lowest point of its oscillation. This principle is used extensively on locomotives and trucks. Conducting disks are installed on the axles of the vehicle and electromagnets are placed around them such that the disks move in the gap of the electromagnets, as in **Figure 10.22**.

Figure 10.22 A practical magnetic brake. Braking takes place by the interaction of the electromagnet and eddy currents in the disk

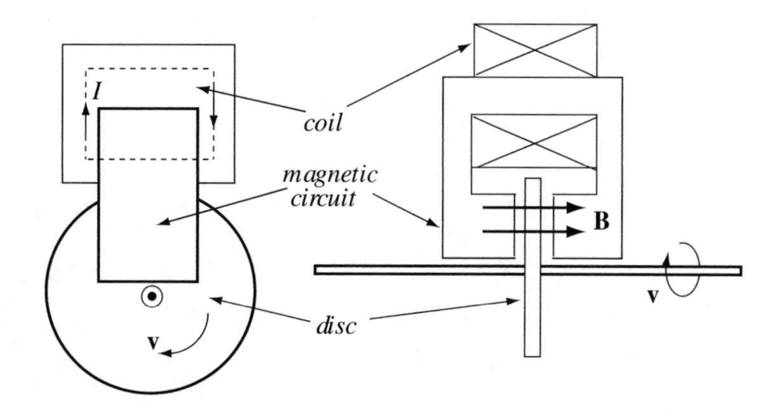

When the mechanical brakes are applied, a current is also applied to the electromagnet and the braking effects of the mechanical and magnetic brakes are added together. We note, however, that the braking effect assumes a velocity **v**. For this reason, electromagnetic brakes cannot be used to completely stop a vehicle, only to slow it down. The magnetic brake is therefore more appropriately called a magnetic retarder or damper. Electric brakes have many advantages. First, they brake better at high speeds and are natural antilocking brakes, since locking of the wheels will immediately release the brakes. Similarly, dragging and binding are not possible because they are noncontact devices. On the other hand, they dissipate large amounts of energy, need considerable electric power, and must be supplemented by mechanical brakes.

Application: The Linear Variable Differential Transformer (LVDT). A useful position sensor is made as a transformer with a single primary coil and two identical secondary coils wound on a short core as shown in **Figure 10.23**. The Secondary coils are connected in a differential mode so that when the coils are centered on the core, the output is zero (**Figure 10.23a**). If now the core moves to the left, the emf in coil (2) decreases because the coupling between the primary and coil (2) diminishes while that in coil (1) remains unchanged. Similarly, if the core moves to the right, the emf in coil (1) decreases. The device is linear within short distances of motion of the core and serves as a sensitive position sensor. The sensor is bidirectional, the direction of motion being indicated by the emfs in coils (1) and (2). The LVDT is used in many applications because of its simplicity, sensitivity, accuracy and ruggedness. These include accurate position sensing of machine tools, steering, and industrial controls. LVDTs operate at frequencies ranging from 50 Hz to about 20 kHz and have ranges from a few mm to over 25 cm. They are enclosed and sealed (**Figure 10.23b**) and can be used to sense force, acceleration and any other quantity related to position and displacement.

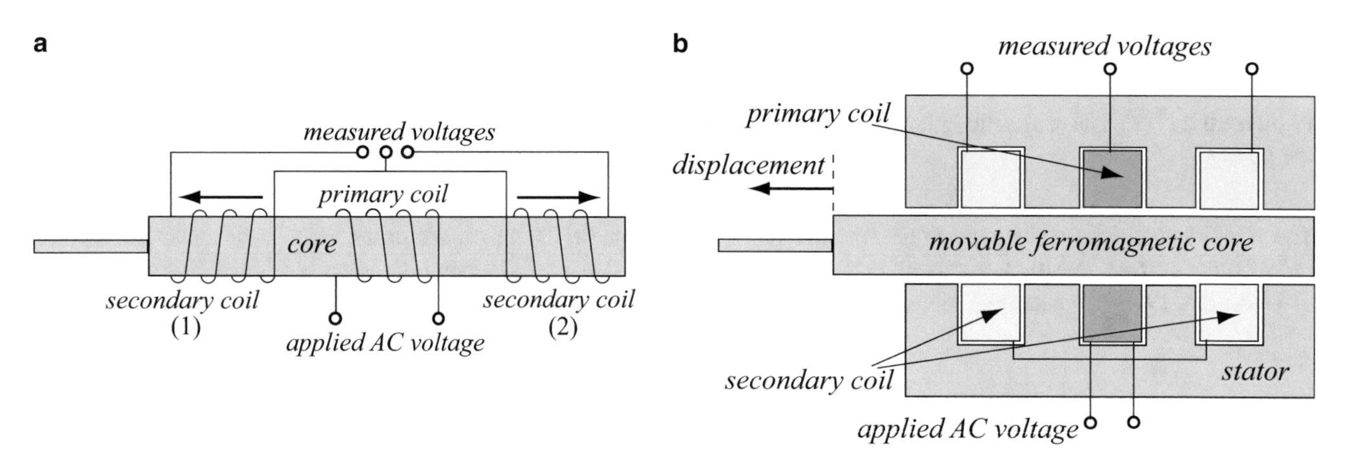

Figure 10.23 The Linear Variable Differential Transformer (LVDT) sensor. (**a**) Principle. (**b**) construction

Application: The Acyclic[4] (Homopolar) Generator—Faraday's Disk The idea of moving a bar in a magnetic field is fundamental to all generators in one way or another, as was amply shown in the previous sections. One particularly simple method is to rotate the bar in a magnetic field rather than translate it. By doing so, the motion is greatly simplified. The basic idea is shown in **Figure 10.24**. It consists of a bar, pivoted at one end and rotated in the magnetic field. Two connections are made: one at the pivot (axis) and one at the moving end. An emf is generated in the bar which is proportional to the speed of motion. In this case, the output is proportional to the frequency of rotation, but it remains DC. A more common implementation

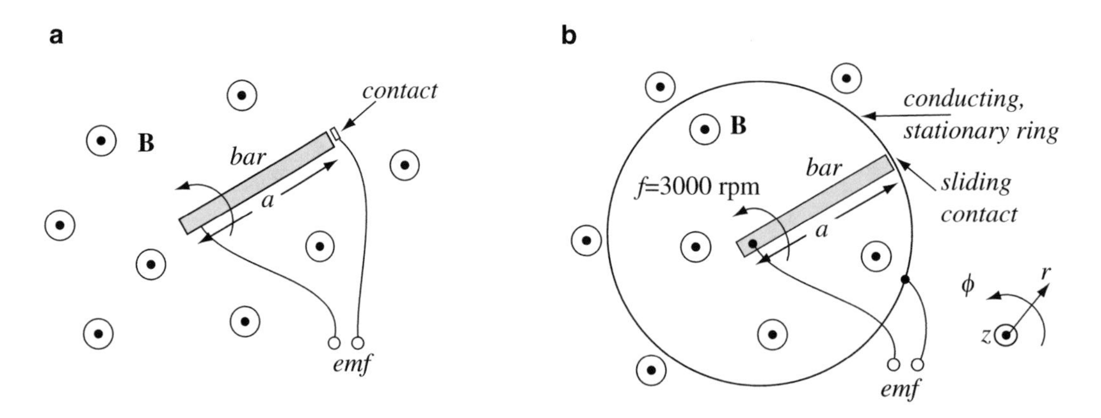

Figure 10.24 The principle of the acyclic generator. (**a**) A rotating bar perpendicular to a magnetic field. (**b**) The bar in (**a**) with a more practical sliding ring connection

[4] An acyclic or homopolar generator is a machine in which the emf induced in the moving conductors maintains the same polarity with respect to the conductors as the conductors move.

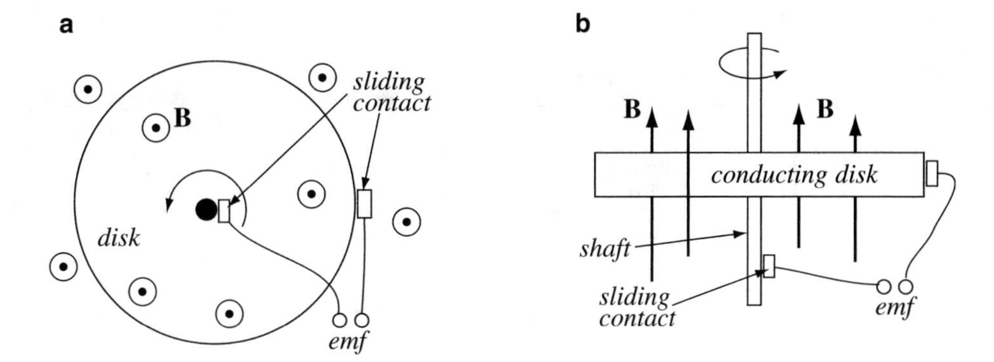

Figure 10.25 A practical acyclic generator. (**a**) A conducting disk rotates in a magnetic field. Connections are made through sliding contacts. (**b**) Side view of the generator

of the same idea, one that simplifies the connections, is a disk on a shaft rotating in the magnetic field, as shown in **Figure 10.25**. The rotating disk acts the same as the rotating bar, but in practical terms, the connections are easier to make and the device is balanced. This method of generation is one of only a small number of methods that allow direct generation of DC power.

Application: The Acyclic (Homopolar) Motor Consider now the opposite problem: The contacts on the disk or the pivoted bar of the previous application are connected to an external source as shown in **Figure 10.26**. The disk now rotates as a motor (see **Problem 10.14**). The homopolar motor is particularly suitable for applications where low-voltage, high-current sources are available or for applications that require high torque.

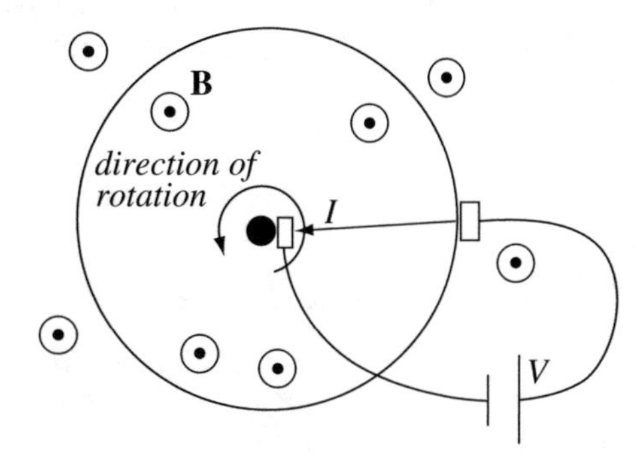

Figure 10.26 The acyclic motor

Application: The Watt-Hour Meter A useful device based on the interaction of induced eddy currents in a conductor is the common watt-hour meter found in many homes. The meter is built of an aluminum disk on a spindle. The disk is placed between the poles of a magnetic yoke, as shown in **Figure 10.27**. Three coils are wound on the yoke. The upper, center coil is called a voltage coil since the current in the coil and, therefore, the field it generates depends on the line voltage. The two lower coils are connected in series with the load. These coils generate a field that is proportional to the current in the load. Both the current and voltage coils generate eddy currents in the conducting disk. However, either one, by itself, produces no torque in the disk. When both are present, the interaction of the current and voltage fields produces a torque, proportional to the product of current and voltage. The speed of the disk is therefore proportional to power. In addition, the meter employs permanent magnets as retarders or braking devices whose braking force is proportional to the speed of the disk. This is required to avoid acceleration of the disk and, therefore, inaccurate measurements. The shaft of the disk is geared and turns indicator dials that register the energy consumption. Although there are more modern instruments that use direct measurement of current and voltage (as well as phase angle) and display power consumption digitally, the common watt-hour meter is still used extensively because it is an accurate, reliable, and inexpensive instrument. The rotating disk in watt-hour meters is normally set such that it can be seen through a window in front of the instrument, giving a simple indication of its operation.

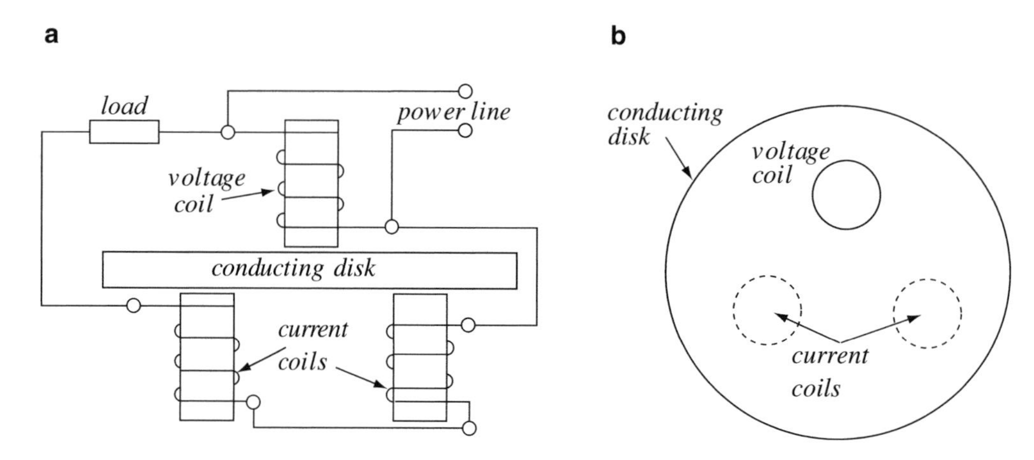

Figure 10.27 The watt-hour meter. (**a**) Side view. (**b**) Top view

Application: The Rail Launcher One application of the magnetic force exerted on currents is the rail launcher or rail gun. In its simplest form, it consists of two rails with a conducting projectile that shorts the rails as shown in **Figure 10.28a**. The current in the rails generates a flux density between the rails and the interaction of the current in the projectile, and this field will force the projectile out. This method has been used to demonstrate the possibility of firing projectiles at velocities much higher than those possible with explosives for both peaceful and military applications. One possible application that has been proposed is to use this device to fire satellites into orbit, since the initial velocities that can be achieved are high enough to permit such applications. This means that escape velocities are possible or that much greater damage can be done when the gun is used for military purposes. The rail gun also finds applications in the acceleration of very small masses in particle research.

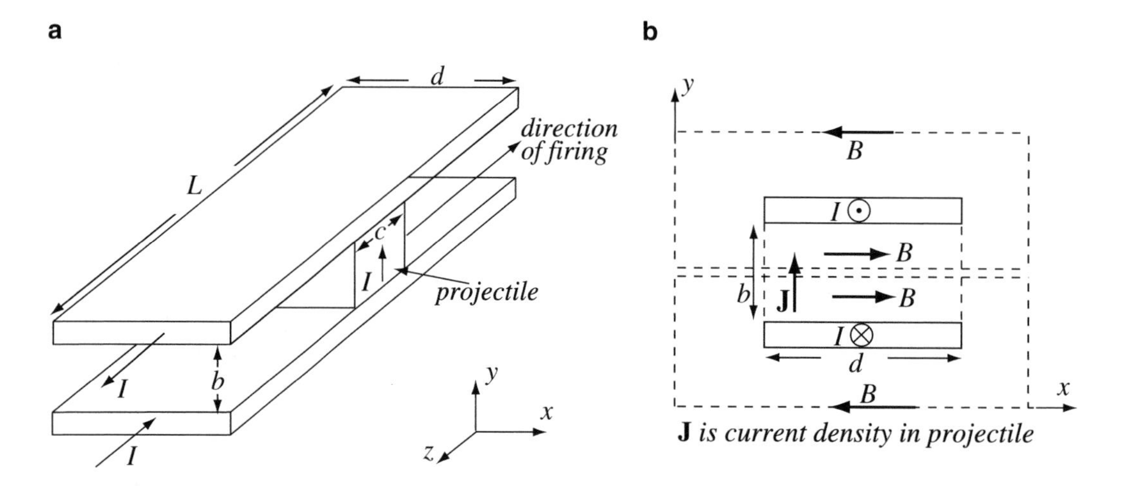

Figure 10.28 The rail launcher. (**a**) Configuration. (**b**) The magnetic flux density between the rails and its relation to current density in the projectile

Application: Eddy Current Testing of Materials One of the most common methods of testing conducting materials for flaws is the eddy current method. It consists essentially of a coil, connected to a constant current or constant voltage AC source. A current flows in the coil and produces a magnetic field in the vicinity of the coil as shown in **Figure 10.29a**. If the conditions in the space around the coil do not change, the coil inductance, and therefore its AC impedance, remains constant. Now, suppose we bring the coil near a conducting material as in **Figure 10.29b**. There are now induced currents in the conducting material due to the induced emf. This causes the impedance of the coil to decrease (i.e., more power must be provided by the source). The current in the coil changes (for constant voltage supply) or the voltage on the coil changes (for constant current supply). This establishes the reference reading (current, voltage, or impedance). If, however, there is a flaw in the material, such as a crack or inclusion, the induced currents in the conducting material change and so does the impedance of the coil. Monitoring the coil impedance (measuring the current for constant voltage sources or the voltage for constant current sources) gives a direct reading of the condition of the material. Any variation from the constant reading obtained with the "sound" material is an indication of some change in the material, either material condition (cracks, inclusions, corrosion) or material properties (changes in conductivity or permeability of the material). This configuration is shown in **Figure 10.29c**. A test and the voltage on the coil (for constant current supply) is shown in **Figure 10.30**. This method of testing is commonly employed for testing of conductors, including airframes in aircraft, aluminum skins on wings, and other critical parts, and in testing of tubing in air-conditioning units and power plants. In testing tubes, two coils are inserted inside the tubes and the two coils connected in series (**Figure 10.31a**) to provide a differential output as in **Figure 10.31b**. As long as both coils are in the vicinity of sound material, the output is zero. If one coil approaches a defect, such as a hole or crack, it will show a different impedance than the other coil and the output will be nonzero, as shown in **Figure 10.31b**.

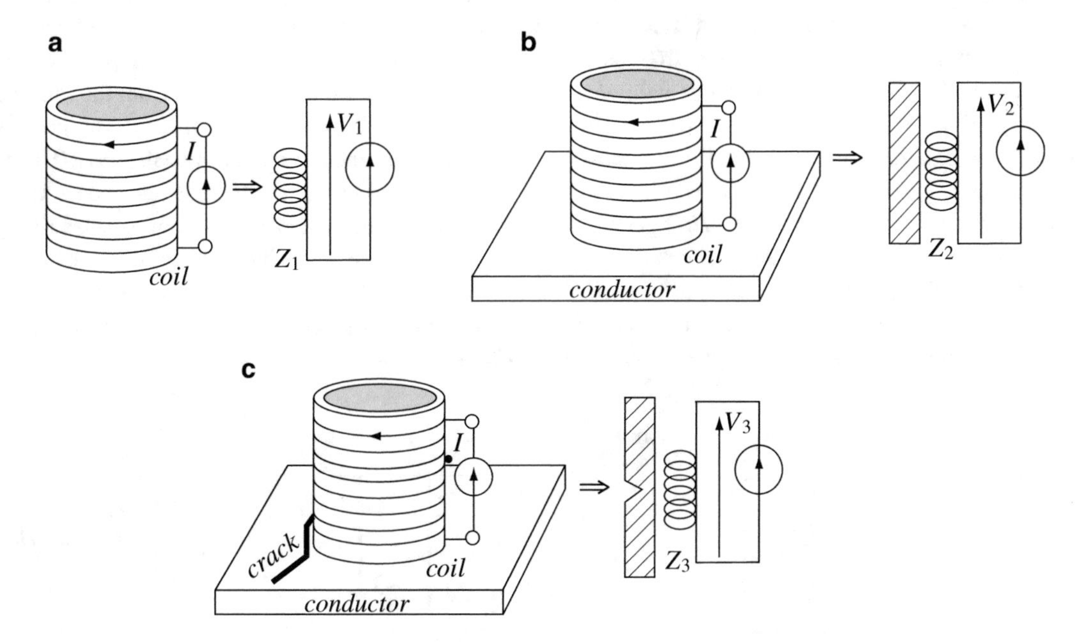

Figure 10.29 Eddy current testing of materials. (**a**) The coil and its equivalent circuit. (**b**) Testing of an intact material produces an output V_2. (**c**) A flaw produces a different output V_3

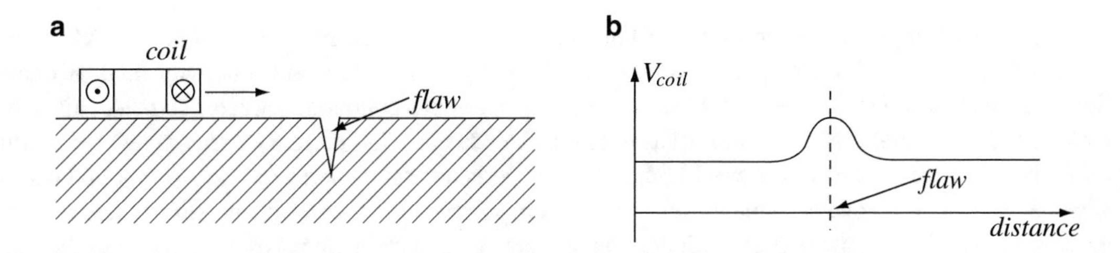

Figure 10.30 Eddy current testing for surface flaws. (**a**) A simple coil passes over the flaw. (**b**) The output signal due to a flaw

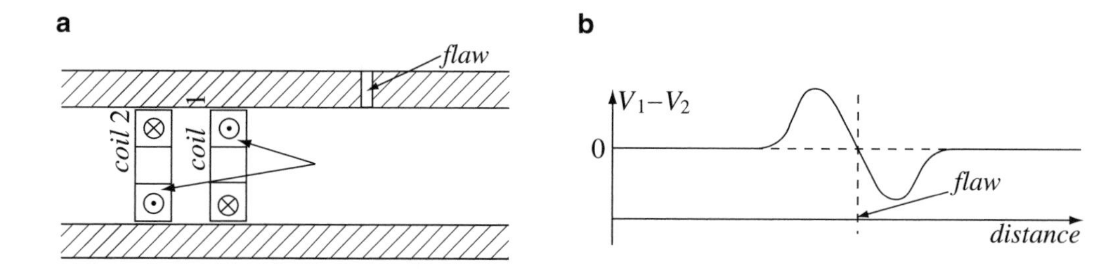

Figure 10.31 Eddy current testing of tubulars. (**a**) Two coils connected in opposition pass through the tube. (**b**) The output signal due to a flaw

Application: The Magnetohydrodynamic (MHD) DC Generator One method of generating electricity is the magnetohydrodynamic method. The principle is that of a moving bar in a magnetic field, not unlike that used in **Example 10.1**. However, instead of the moving bar, a conducting fluid moves between two conducting electrodes, as shown in **Figures 10.32a** and **10.32b**. Two magnets generate a very high magnetic field in a channel between them, called a ***magnetohydrodynamic*** (MHD) ***channel***. Two conducting electrodes are placed at right angles to the field, insulated from the magnets and each other. A conducting fluid is now forced through the channel. The magnetic field and velocity product $\mathbf{v} \times \mathbf{B}$ are as shown in the figure which indicates that the left plate becomes negative whereas the right plate is positive. The conducting fluid can be any fluid. Most experimental generators use highly ionized exhaust gases from the burning of fuels. A very simple *MHD* generator is made of a channel as above, connected to the exhaust of a jet engine. To increase conductivity of the gases, these are seeded with conducting ions such as alkali metal vapors. In coal-fired *MHD* generators, potassium carbonate is used as seed to increase conductivity. The most attractive feature of these generators is the fact that they are stationary and produce DC directly. *MHD* generators have low efficiency (below 15%), but because the system can be fully contained and operating in closed circuits, it has considerable promise, especially for solar power generation. In spite of its simplicity, the engineering challenges of *MHD* are quite difficult to handle. One is the need for very large magnetic fields which can only be obtained at superconducting temperatures. The other is the very high speeds and pressures, as well as temperatures in the channel. See **Problem 10.24** for a sample calculation.

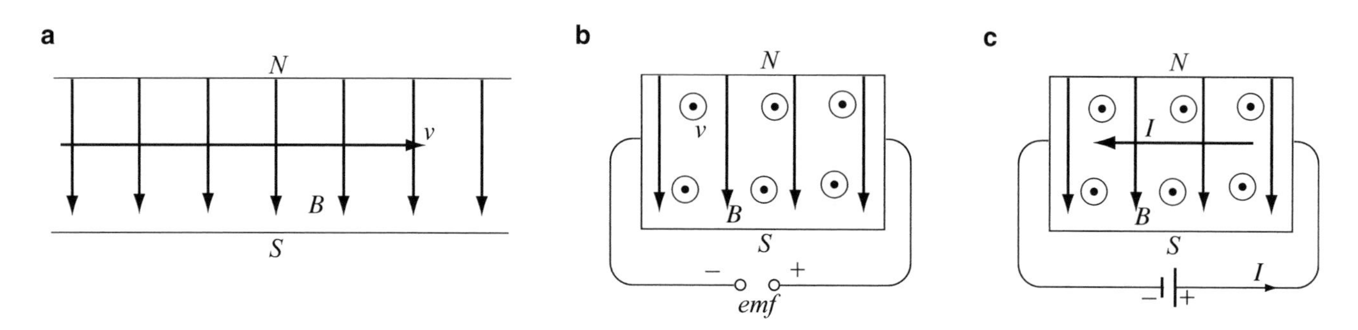

Figure 10.32 The magnetohydrodynamic generator. (**a**) Relation between fluid velocity and magnetic field. (**b**) Cross-sectional view showing the emf. (**c**) The magnetohydrodynamic pump. The pumped fluid moves out of the page

Application: Magnetohydrodynamic Pump for Molten Metal The force in the rail gun in **Figure 10.28** can be used in a somewhat different way. Assume that the space between the rails is filled with a high-conductivity fluid. A current passing through the fluid will generate a force on the fluid which is then pushed out in a continuous stream. To make this into a practical pump, the magnetic field is generated across a channel as in **Figure 10.32a** and a DC current passes through the conducting fluid at right angles to the magnetic field as in **Figure 10.32c**. This device is a ***magnetohydrodynamic pump*** and can be used for pumping molten metals and, in particular, molten sodium as used in some nuclear power plants. Since these metals have high conductivities, it is a very efficient method in particular since it acts as a pump without moving parts. In the movement of corrosive materials, or molten metals at high temperatures, this is an overriding requirement. The MHD pump has also been proposed for propulsion of submarines since seawater is conducting (its conductivity is 4 S/m). The main advantage would be reduced noise (one of the main means of detection of submarines is the noise they generate by their propellers, a noise which travels quite far in water). However, the currents required are large and the efficiency of the system is very low, mainly because of the very low conductivity of seawater. To increase efficiency, it would be necessary to use very large magnetic fields which, in turn, require cryogenically cooled magnets (superconducting magnets) with all the associated cooling equipment, weight, and energy.

10.10 Summary

Following the study of electrostatics and magnetostatics, we now look into time-dependent phenomena, starting with ***Faraday's law*** of induction. Faraday's law was originally observed as an induced voltage (or ***electromotive force*** (***emf***)) in a loop due to motion of a magnet in its vicinity. For a single loop or for N loops in the same location, it takes the forms:

$$emf = -\frac{d\Phi}{dt} \quad [\mathrm{V}] \quad or \quad emf = -N\frac{d\Phi}{dt} \quad [\mathrm{V}] \tag{10.1) or (10.2}$$

This observation modifies the first postulate of the electric field (curl equation):

$$\nabla \times \mathbf{E} = -\frac{\partial \mathbf{B}}{\partial t} \quad or \quad \oint_C \mathbf{E} \cdot d\mathbf{l} = -\frac{\partial \Phi}{\partial t} \tag{10.8) or (10.5}$$

Lenz's law accompanies Faraday's law and gives meaning to the negative sign. It states: "The direction of the *emf* is such that the flux generated by the induced current opposes the change in flux."

An emf may be viewed as being generated by motion or by inherent time dependency of the field.

Motion action emf is produced by motion of a conductor in a magnetic field:

$$emf = \int_a^b (\mathbf{v} \times \mathbf{B}) \cdot d\mathbf{l} \quad [\mathrm{V}] \tag{10.12}$$

where a conductor extending from a to b moves at a velocity $\mathbf{v}$ in a magnetic flux density $\mathbf{B}$.

Transformer action emf requires that the magnetic flux density be time dependent:

$$emf = \oint_C \mathbf{E} \cdot d\mathbf{l} = \int_s (\nabla \times \mathbf{E}) \cdot d\mathbf{s} = -\frac{\partial}{\partial t}\int_s \mathbf{B} \cdot d\mathbf{s} \quad [\mathrm{V}] \tag{10.6}$$

The ***transformer*** is a device that relies on its operation on induced emfs. In an ideal transformer, there are no losses and the magnetic path has low reluctance. For a two-coil closed path transformer with path reluctance $\Re$ (**Figure 10.13b**), the flux along the path is

$$\Phi = \frac{N_1 I_1 - N_2 I_2}{\Re} \tag{10.36}$$

The terminal voltages, currents, and impedances of the primary (1) and secondary (2) coils are related by the transformer ratio a

$$\frac{emf_1}{emf_2} = \frac{V_1}{V_2} = \frac{I_2}{I_1} = \frac{N_1}{N_2} = a \quad and \quad \frac{Z_1}{Z_2} = a^2 \tag{10.40) and (10.42}$$

In the real transformer, the reluctance is not necessarily very low but we still assume a closed magnetic path. The emfs in the primary (1) and secondary (2) are now given in terms of self and mutual inductances of the two coils (see **Figure 10.13b**):

$$emf_1 = L_{11}\frac{dI_1}{dt} - L_{12}\frac{dI_2}{dt} \quad [\mathrm{V}] \tag{10.48}$$

$$emf_2 = L_{21}\frac{dI_1}{dt} - L_{22}\frac{dI_2}{dt} \quad [\mathrm{V}] \tag{10.49}$$

If the magnetic path is not closed, the coupling between the coils is weaker and we define a coupling coefficient $0 < k < 1$. Since $L_{12}=L_{21}$ we write

$$emf_1 = L_{11}\frac{dI_1}{dt} - k\sqrt{L_{11}L_{22}}\frac{dI_2}{dt} \quad [\mathrm{V}] \tag{10.55}$$

$$emf_2 = k\sqrt{L_{11}L_{22}}\frac{dI_1}{dt} - L_{22}\frac{dI_2}{dt} \quad [\mathrm{V}] \tag{10.56}$$

Problems

Motional emf

10.1 Motional emf. Two trains approach each other at velocities $v_1 = 100$ km/h and $v_2 = 120$ km/h as shown in **Figure 10.33**. Assume the trains' axles are good conductors, are separated a distance $d = 2$ m and each rail has a resistance $r = 0.1$ Ω/km. The trains are $P = 10$ km apart. The vertical component of the terrestrial magnetic flux density is $B_0 = 0.05$ mT. Calculate:

(a) The current in the rails produced by the motion of the trains in the terrestrial magnetic flux density at the instant shown in **Figure 10.33**.
(b) The current in the rails as a function of time taking $t = 0$ at the instance shown in **Figure 10.33**.

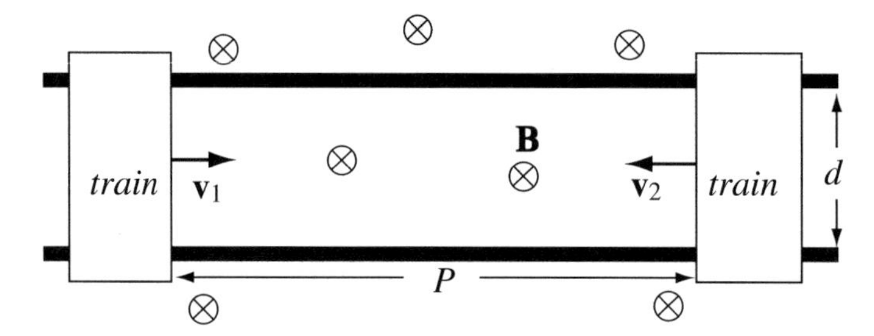

Figure 10.33

10.2 Application: Motional emf as a Motion Detector Mechanism. Suppose you want to know if a train is moving on a rail but it is too far to see. Assume the rails are a distance d [m] apart and are insulated from each other and from the ground. To check for motion of the train, you measure the potential difference between the two rails:

(a) Calculate the velocity of the train assuming that the terrestrial magnetic flux density is equal to B_0 [T] and is perpendicular to the surface of the Earth, pointing downward. The potential difference measured is V [V] with its positive pole on the top rail in **Figure 10.34**.
(b) In which direction is the train moving?
(c) If the resistance per meter length of a rail equals r [Ω/m], calculate the distance P from the measurement point to the train. How can this distance be found by a simple measurement at the point where the observer is located?
(d) Can these measurements be used to avoid collision of trains?

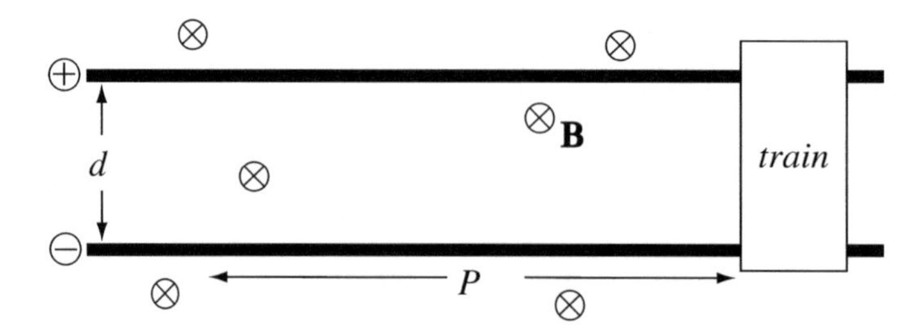

Figure 10.34

Induced emf

10.3 Forces Due to Induction. Two very long, parallel rails are placed in a uniform, time-dependent magnetic field. The resistance of the rails is negligible. On the left side, the rails are shorted with a wire of resistance R [Ω]. The wire is fixed and not allowed to move. On the right, a bar with zero resistance is placed on the rails (**Figure 10.35**) so that it is free to move. The magnetic flux density is given as $\mathbf{B} = -\hat{\mathbf{z}}B_0\cos(\omega t)$ [T]:

(a) Calculate the magnitude and direction of the force acting on this bar in the instance shown in **Figure 10.35**.
(b) What is the force at $t = 0$?
(c) Suppose now that the bar is free to move. Describe qualitatively the motion of the bar.

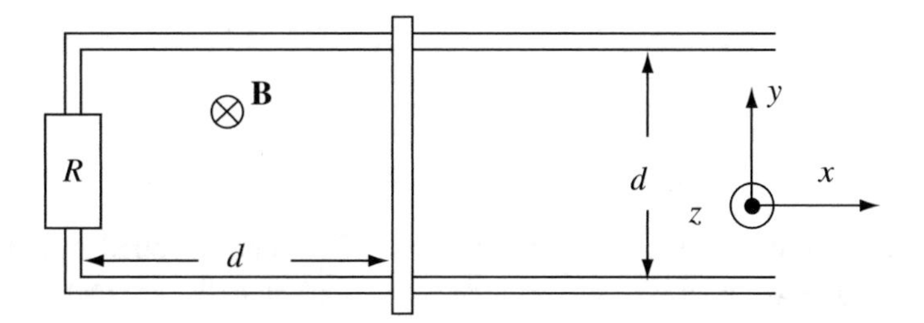

Figure 10.35

10.4 Motional emf. An infinitely long, straight wire of radius r [m] carries a uniform current density J [A/m^2]. The direction of the current is as shown in **Figure 10.36a**. Another infinite, thin wire parallel to the first wire moves to the right at a velocity v_0 [m/s]. The wires are shown in cross-section.

(a) Calculate the induced voltage (emf) per unit length in the moving wire, when the moving wire is at a distance d [m] from the thick wire. Show polarity on the visible end.
(b) Now the current density in the thick wire is removed and a current I [A] equal to the current in the thick wire in **Figure 10.36a** is applied to the thin conductor as shown in **Figure 10.36b**. The direction of the current is the same as that of the current density in **Figure 10.36a**. The thick conductor moves to the left while the thin wire is stationary. Calculate the voltage (emf) per unit length in the moving wire when the moving wire is at a distance d [m] from the thin wire. Show polarity on the visible end.

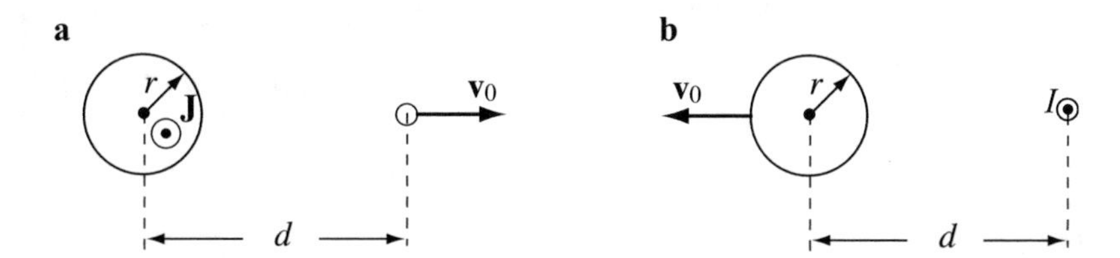

Figure 10.36

10.5 Induced emf in a Strip. An infinitely long strip of conducting material of width d [m] moves in a uniform, sinusoidal magnetic flux density $B = B_0\cos\omega t$ [T] at a constant velocity v [m/s]. The field makes a 60° angle with the direction of movement of the strip. Calculate the induced emf between the two sliding (stationary) contacts on the opposite sides (a, b) of the strip (**Figure 10.37**). Show polarity.

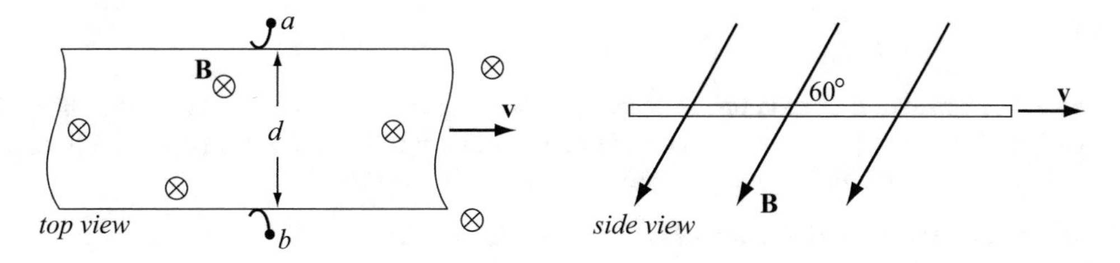

Figure 10.37

10.6 Motion and Transformer Action emf in Loops. A square loop and a round loop move at a constant velocity v_0 [m/s] in the y direction ($\mathbf{v} = \hat{\mathbf{y}}v_0$) along the axis (dashed line), as shown in **Figure 10.38**. The magnetic field intensity is given as $\mathbf{H} = \hat{\mathbf{z}}H_0\cos(\omega t)$ [A/m]. Calculate the ratio between the induced voltages (emf) in the two loops.

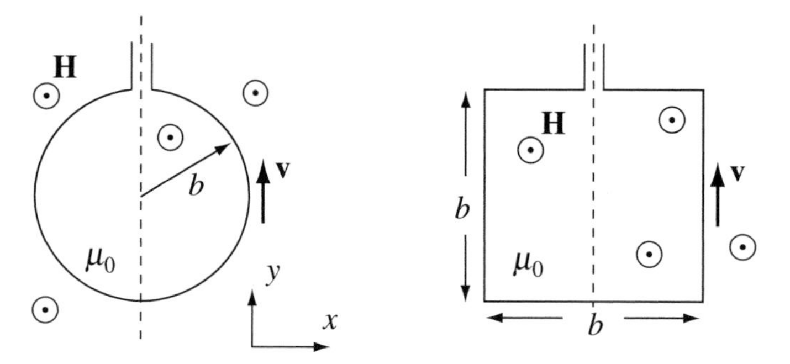

Figure 10.38

10.7 Application: Motion and Transformer Action emf in Loops. Consider the round and square loops in **Figure 10.38**. The magnetic field intensity is given as $\mathbf{H} = \hat{\mathbf{z}}H_0\cos(\omega_0 t)$ [A/m]. Assume the vertical velocity shown is zero ($\mathbf{v} = 0$).

(a) Calculate the emf in the square loop if the loop rotates around the axis shown (dashed line) at ω_1 radians per second.

(b) Calculate the emf in the round loop if the loop rotates around the axis shown (dashed line) at ω_0 radians per second.

10.8 Application: Power Line Current Sensor. A sensor designed to detect and estimate the current in a power line remotely is proposed as follows: A small coil of area S is placed at ground level at the center between the two conductors of an AC power line (see **Figure 10.39**). The line carries a current I_0 [A] (rms) at frequency f [Hz] and is sinusoidal. The surface of the coil is parallel to the ground. If the height of the lines and their separation is known, find a relation between the current in the line and the emf produced in the coil. Assume the coil has area S [m^2] and N turns. All materials involved have permeability of free space.

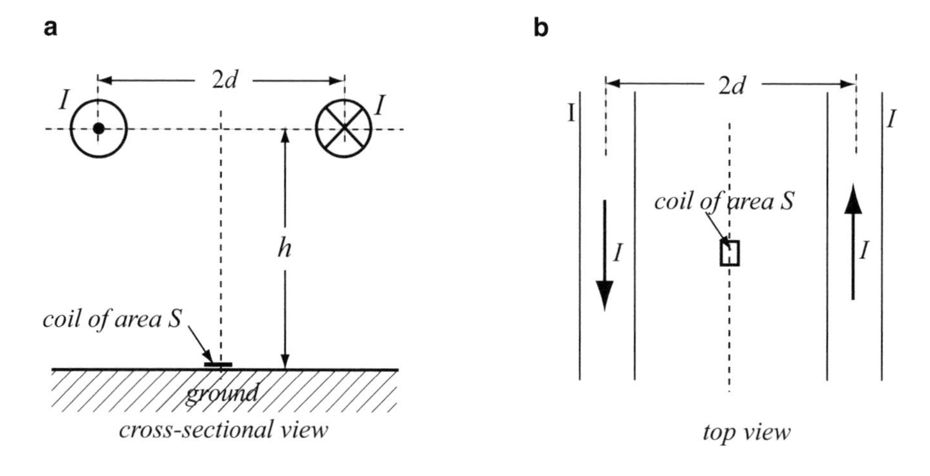

Figure 10.39

Generator emf

10.9 Application: emf in Moving Conductor. A very long wire carries a current I [A] as shown in **Figure 10.40**. A conducting bar of length b [m] moves to the right at a velocity v [m/s] so that its upper end is kept at a constant distance a [m] from the wire. The bar and wire are on a plane in free space.

(a) Calculate the emf generated by the bar if $I = I_0$ [A] is a DC current.

(b) Show the polarity of the emf and explain how this emf can be measured.

(c) Repeat (a) if the current is $I = I_0 \sin\omega t$ [A].
(d) Repeat (a) if the velocity is $v = v_0 \sin\omega t$ [m/s].

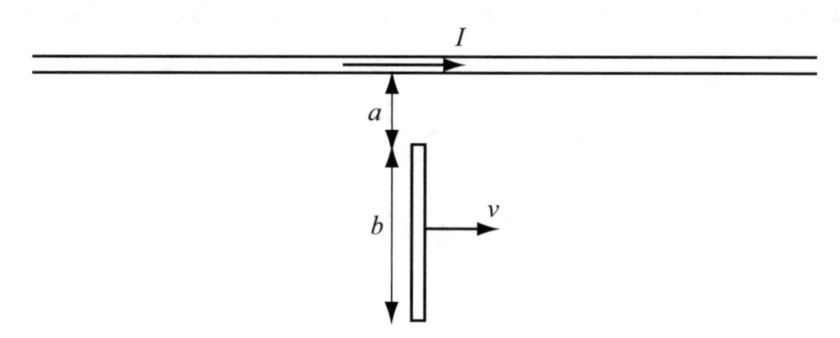

Figure 10.40

10.10 Application: emf in a Generator. A piece of wire is bent and rotated in a uniform, magnetic field **B** [T] as shown in **Figure 10.41.** The frequency of rotation is f (rotations per second). Calculate the induced emf between the terminals a and b. Show the polarity of the induced voltage (emf).

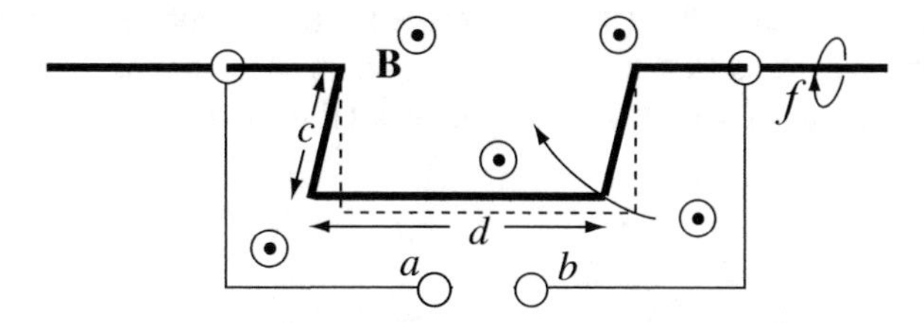

Figure 10.41

10.11 emf in Rotating Bar. A conducting bar of length l [m] is pivoted at one end and rotates at an angular velocity ω [rad/s] in a magnetic field **B** [T]. Find the induced emf on the bar if it rotates perpendicular to the flux density in the counterclockwise direction for:

(a) $\mathbf{B} = \hat{\mathbf{z}}B_0$.
(b) $\mathbf{B} = \hat{\mathbf{z}}B_0 e^{-r}, \quad 0 \leq r \leq l.$

10.12 Homopolar **Generator.** A spoked, conducting wheel rotates in an AC magnetic field as shown in **Figure 10.42.** The frequency of rotation is ω [rad/s]. The magnetic field intensity is $H = 10^6 \cos(\omega t)$ [A/m]. A conducting hoop is provided on which the outer wire slides freely:

(a) Calculate the induced voltage (emf) between the center of the wheel and its outer perimeter.
(b) How does the solution change if the wheel is a solid conducting wheel?

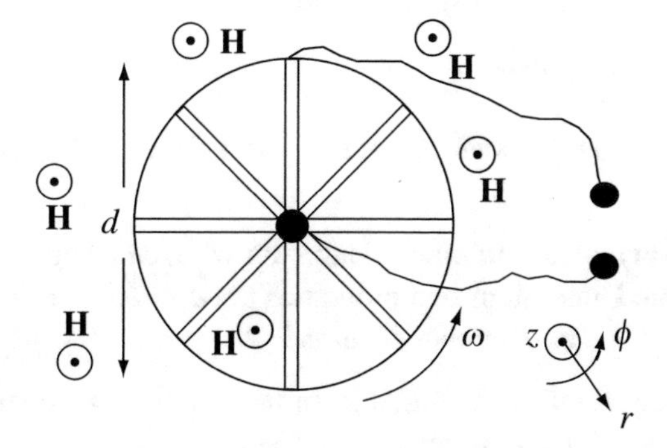

Figure 10.42

10.13 Application: Homopolar Generator. A disk of radius 100 mm rotates in a uniform magnetic flux density at 6,000 rpm as shown in **Figure 10.43**. The magnetic flux density is sinusoidal, with amplitude 0.1 T and frequency 400 Hz. A wire is connected to the center of the disk and one is sliding on its edge, making good contact. Calculate the voltage (motional emf) between points A and B.

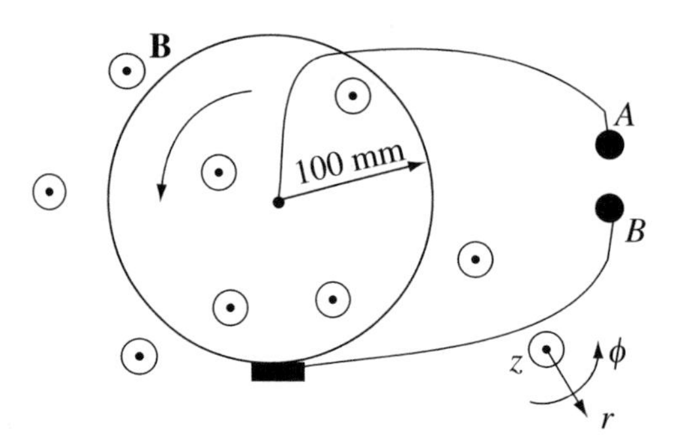

Figure 10.43

10.14 Torque in Acyclic Motor. A disk of radius 100 mm is placed in a constant, uniform magnetic flux density of 0.2 T. A wire is connected to the center of the disk, and one is sliding on its edge, making good contact. The disk connected to a DC source that supplies a current of 100 A through the disk **(Figure 10.44)**. Calculate the torque generated by the device.

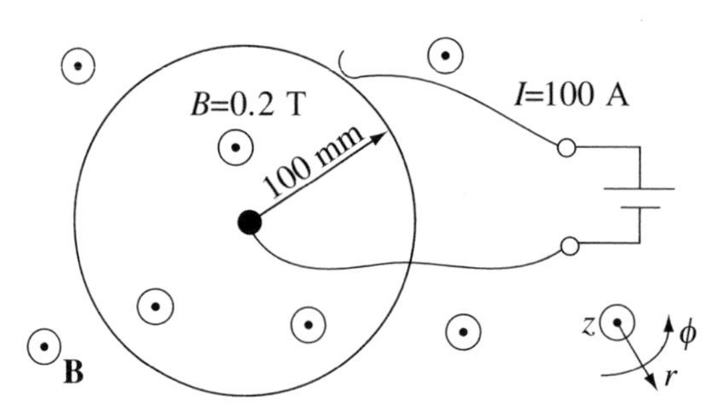

Figure 10.44

Transformers

10.15 Ideal Transformer. An ideal transformer has a primary to secondary turn ratio of 10. If the transformer's primary coil is connected to 110 V and transfers 150 W, calculate:

(a) The voltage and current in the secondary.
(b) The current in the primary.
(c) The impedance of the primary and secondary coils. What is the ratio between the impedance in the primary and secondary?

10.16 Application: Current Transformer. An infinitely long, thin wire, carrying a current I, is located at the center of a torus as shown in **Figure 10.45**. Dimensions and properties are $a = 20$ mm, $b = 30$ mm, $d = 10$ mm, frequency is 100 Hz, $I_0 = 1$ A, $\mu_0 = 4\pi \times 10^{-7}$ H/m, number of turns is 200.

(a) Calculate the induced voltage (emf) in the coil if its turns are uniformly distributed around the torus. The torus is made of a ferromagnetic material with relative permittivity of 100.
(b) Repeat **(a)** if the coil is wound around a toroidal form so that the relative permeability in the torus is the same as that of air.

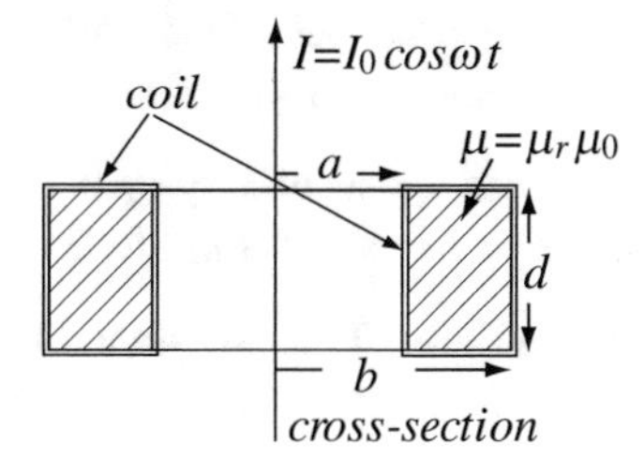

Figure 10.45

10.17 Application: High-Power Ideal Transformer. A transformer is used in a hydroelectric power plant to step up the output from a 720 MW generator. The generator operates at 18 kV, 60 Hz. The output voltage required for transmission is 750 kV:

(a) If an ideal transformer is used, calculate the turn ratio and input and output currents at full load.
(b) Ideal transformers do not exist. Suppose the transformer has 1% losses (typical for these transformers). Calculate the required turn ratio to maintain an output voltage of 750 kV at full load. Assume all power loss occurs in the iron core of the transformer.
(c) Calculate the output current and the flux in the magnetic core for rated input power with the turn ratio in **(b)**. Use a relative permeability of 500, length of the magnetic path of 4 m, and cross-sectional area of 0.2 m^2. Assume that the number of turns in the primary is given as $N_1 = 20$.

10.18 Application: Design of a Current Transformer. You have an AC digital voltmeter with its basic range 199.9 mV rms and wish to monitor the input power to a house by measuring the input current. The maximum current expected is 150 A (rms) at 60 Hz. You choose to build a simple current transformer by cutting an iron ring (so that the ring can be slipped over the wire) and winding a coil on the ring. The emf in this coil is your reading. The ring is of average radius of 20 mm, a cross-sectional area of 100 mm^2, and relative permeability of 200:

(a) How many turns are required to produce full-scale reading for maximum input?
(b) Suppose you wish now to use the same device to measure currents up to 500 A (rms). How many turns are required for this purpose?

10.19 Application: Transformer with Multiple Windings. The magnetic circuit in **Figure 10.46** is given. Coil 1 has $N_1 =$ 100 turns and carries a current $I_1 = I_0\cos(\omega t)$ and coil 2 has $N_2 = 50$ turns and carries a current $I_2 = I_0\cos(\omega t)$ where $I_0 = 1$ A, $\omega = 314$ rad/s. The currents in the two coils are such that their magnetic flux densities are in the directions shown in the figure. Coil 3 has $N_3 = 1{,}000$ turns. Dimensions are average lengths and are given in the figure. The cross-sectional area of the core is 100 cm^2. Relative permeability of the core is 100:

(a) Calculate V_1 and V_2 (the voltages across the terminals of coils 1 and 2).
(b) Calculate the induced voltage in coil 3.

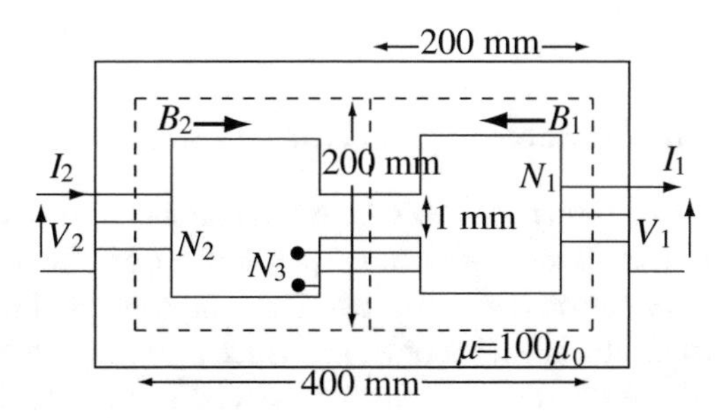

Figure 10.46

10.20 Application: Loosely Coupled Transformer. Two coils, each with inductance 10 mH, are placed next to each other to form a transformer. It was determined that the coefficient of coupling between the coils is 0.1 (i.e., 10% of the flux produced by each coil passes through the second coil). One coil, designated as the primary, is connected to a sinusoidal current with amplitude 1 A and frequency of 100 kHz. Calculate:

(a) The emf in the primary and secondary with the secondary open circuited.
(b) The emf in the primary and secondary if a current of amplitude 0.05 A is drawn from the secondary.

10.21 Induced Current in a Loop: A Nonconventional Transformer. Determine the induced voltage (emf) $v(t)$ in the rectangular loop shown next to a very long wire carrying a time-varying current $I(t) = I_0 \sin\omega t$ [A] (**Figure 10.47**).

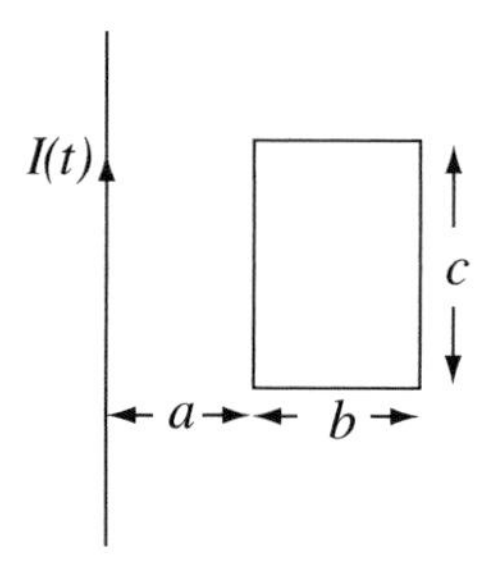

Figure 10.47

10.22 Application: Power Line Power Scavenging for Sensing. As part of a smart grid, one proposes to place sensors on power lines to sense a variety of parameters such as temperature, current, corrosion, vibrations, and others and transmit these parameters wirelessly to a central location. To do so, the sensors are placed on the power line itself by attaching them to one conductor of the power line. To power the sensors and the wireless transmitter, one can use the magnetic field produced by the line by designing an appropriate transformer. A solution is to place a toroidal coil around the conductor as shown in **Figure 10.48**. Suppose a power line carries a sinusoidal current $I = 500$ A (rms) at a frequency $f = 60$ Hz. The toroidal core has an inner radius $a = 30$ mm, outer radius $b = 50$ mm, and thickness $c = 20$ mm. Its relative permeability is 100:

(a) Calculate the number of turns needed on the torus to produce an RMS voltage of 6 V.
(b) What is the maximum theoretical power available for use in the sensors?

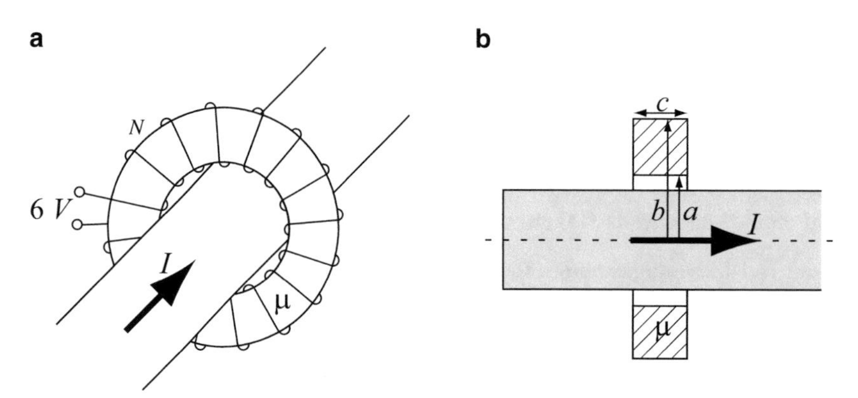

Figure 10.48 Structure of power line power scavenging. (**a**) General view, (**b**) axial cross section

10.23 Application: Ground Fault Circuit Interrupt (GFCI). An important safety device is the GFCI (also called a residual current device (RCD)). It is intended to disconnect electrical power if current flows outside of the intended circuit, usually to ground, such as in the case when a person is electrocuted. The schematic in **Figure 10.49** shows the concept. The two conductors supplying power to an electrical socket or an appliance pass through the center of a toroidal coil. Normally the currents in the two conductors are the same and the net induced emf due to the two conductors cancel each other, producing a net zero output in the current sensor. If there is a fault and current flows to ground, say a current I_g, the return wire will carry a smaller current and the current sensor produces an output proportional to the ground current I_g. If that current exceeds a set value (typically 4 −30 mA), the induced emf causes the circuit to disconnect. These devices are common in many locations and are required by code in any location in close proximity to water (bathrooms, kitchens, etc.).

Consider the GFCI shown schematically in **Figure 10.49**. The device is designed to operate in a 50 Hz installation and trip when the voltage on the toroidal coil is $V_{out} = 100\ \mu\text{V}$ rms. For a toroidal coil with average diameter $a = 30$ mm and a cross-sectional diameter of $b = 10$ mm:

(a) Calculate the number of turns needed if an air-filled toroidal coil is used and the device must trip at a ground current of 6 mA.

(b) Calculate the number of turns needed to trip at a ground current of 6 mA if a ferromagnetic torus with relative permeability of 2,000 is used as a core for the coil.

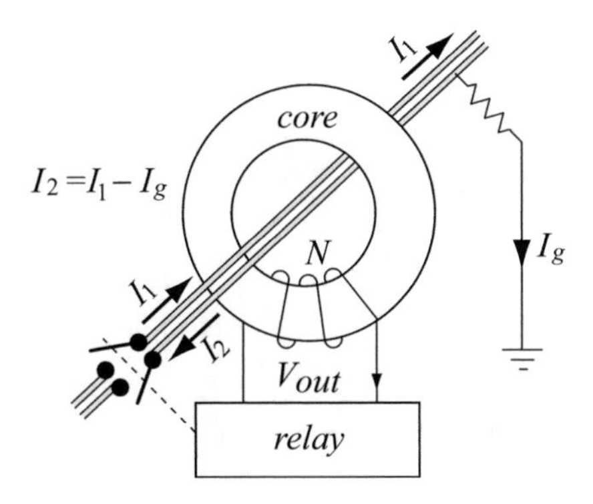

Figure 10.49 Principle of a GFCI sensor

10.24 Application: The MHD Generator. A small magnetohydrodynamic (MHD) generator is proposed as part of the exhaust system of a jet engine. The MHD channel is 200 mm by 200 mm in cross section and is 1 m long. A magnetic flux density of 1 T is generated between two surfaces as shown in **Figure 10.50**. The exhaust moves at 200 m/s and is seeded to produce a conductivity of 50 S/m in the channel. Calculate:

(a) The output voltage (emf) of the generator. Show polarity.

(b) The internal resistance of the generator.

(c) Maximum power output to a load under maximum power conditions.

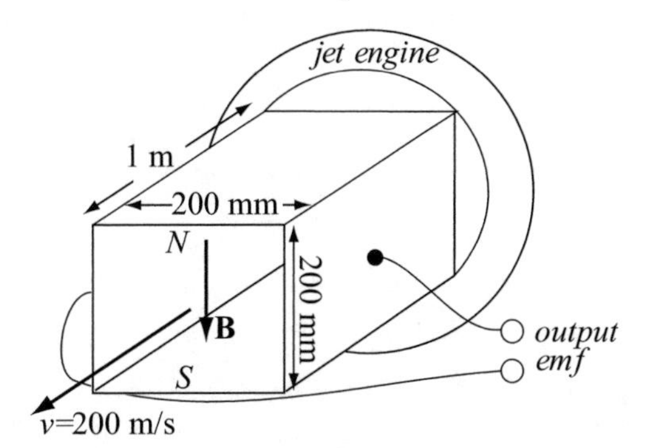

Figure 10.50

Maxwell's Equations 11

From a long view of the history of mankind—seen from say, ten thousand years from now—there can be little doubt that the most significant event of the 19th century will be judged as Maxwell's discovery of the laws of electrodynamics. The American civil war will pale into provincial insignificance in comparison with this important scientific event of the same decade.

Richard P. Feynman (1918–1988)
Physicist, Nobel laureate

11.1 Introduction: The Electromagnetic Field

Most students in engineering and science have heard the term "Maxwell's equations," and some may also have heard that Maxwell's equations "describe all electromagnetic phenomena." However, it is not always clear what exactly do we mean by these equations. How are they any different from what we have studied in the previous chapters? We recall discussing static and time-dependent fields and, in the process, discussed many applications. To do so, we used the definitions of the curl and divergence of the electric and magnetic fields: what we called "the postulates." Are Maxwell's equations different? Do they add anything to the previously described phenomena? Perhaps the best question to ask is the following: Is there any other electromagnetic phenomenon that was not discussed in the previous chapters because the definitions we used were not sufficient to do so? If so, do Maxwell's equations define these yet unknown properties of the electromagnetic field? The answer to the latter is emphatically yes.

In fact, we do not need to go far to find applications which could not be treated using, for example, Faraday's law. The most obvious is transmission of power as, for example, in communication. All applications related to transmission of power (radar, communication, radio, etc.) were conspicuously missing in the previous chapters, but there is an even more important (and related) aspect of the electromagnetic field which was not mentioned until now. Take, for example, induction of voltage in a loop. Faraday's law gives an accurate statement of how the induction occurs and the magnitude of the induced emf. Now let's say that two loops are located a short distance from each other and one loop induces an electromotive force in the second. If we were to separate the loops a very long distance from each other and measure the induced voltage in the second loop, the magnitude will be very small. The question is, however, this: Is there any lag in time between switching on the current in the first loop and detection of the induced voltage in the second loop because of the distance between the loops? Faraday's law says nothing about that and neither do any of the postulates used previously. Intuitively, we know there must be a time lag since nothing can occur instantaneously. In this regard, consider the following: On January 22, 2003, NASA received the last transmission from the *Pioneer* 10 space probe. At that time, *Pioneer* 10 was 5 weeks shy of its 31st year of space flight and was over 12.2 billion km from the Earth.[1] At that distance, the transmission took approximately 11 h 18 min

[1] *Pioneer* 10 was launched on March 2, 1972, with the intention of moving out of the solar system. Designed for a mission of 21 months, it was the first spacecraft to pass through the asteroid belt and out of the solar system. The spacecraft carries a unique plaque that identifies the Earth and its inhabitants as the designers of the craft. *Pioneer* 10 is still flying, heading for the red star Aldebaran in the constellation Taurus where it is expected to arrive in about two million years. As of January 1, 2019, the spacecraft was 18.2 billion miles away but its power source is too weak to transmit over this distance.

N. Ida, *Engineering Electromagnetics*, https://doi.org/10.1007/978-3-030-15557-5_11

to reach the Earth. This is hardly instantaneous. In fact, if we divide distance by time, we find that the information has traveled through space at the speed of light. It should then be obvious that something must be missing since we cannot account for this time lag when using Faraday's or Ampère's laws.

Maxwell's equations are, in fact, the four postulates we introduced in the previous chapter with a modification to Ampère's law to account for finite speed of propagation of power. This modification is rather simple and we usually refer to it as displacement current or displacement current density and is simply a statement of conservation of charge. However, in spite of its simplicity, it is far reaching and, as we shall see shortly, crucial to all applications involving transmission of signals or power. In a way, the remainder of this book is dedicated to the discussion of this aspect of electromagnetic fields.

To return now to the applications discussed in the previous chapters, we should ask ourselves the following question: If the postulates used need to be modified, do we also need to go back and discuss all that we have done and perhaps modify all previous relations? The answer, fortunately, is no. In all applications until now, there was no need for these modifications even though the solutions obtained were often only approximations. However, these were very good approximations and there would be very little to be gained by including the ideas introduced in this chapter into the previous results, as we shall see. For example, two coils, near each other, experience a force. The force is not instantaneous, but because the distance between the coils is small, the time lag is so small as to be justifiably neglected.

11.2 Maxwell's Equations

When, in 1873, James Clerk Maxwell[2] wrote his now famous *Treatise on Electricity and Magnetism*, he wrote in the preface to the book that his purpose was essentially that of explaining Faraday's ideas (published in *Experimental Researches in Electricity* in 1839) into a mathematical and, therefore, more universal form. He makes it amply clear that his treatise is a sort of summary or unification of the knowledge in electrical and magnetic fields as put forward by others, including those who preceded Faraday (Ampère, Gauss, Coulomb, and others). We might add that the notation we use today to write Maxwell's equations was introduced by Oliver Heaviside[3] almost 20 years after Maxwell's theory appeared. If you were to read Maxwell's book, you might not recognize the equations written in the previous chapters or in this. What then is Maxwell's unique contribution? Why do we normally refer to the electromagnetic field equations as "Maxwell's equations"? Surely, it is more than simply because he summarized what others have done.

His main contribution is in proposing the inclusion of displacement currents[4] in Ampère's law. This seemingly minor change in the field equations as known before his time was, in fact, a fundamental change in the theory of electromagnetics. Maxwell's ideas, which were often expressed in mechanical terms, were not immediately accepted since they implied a number of aspects of the electric and magnetic fields that had no proof at the time. Maxwell himself had no experimental

[2] James Clerk Maxwell (1831–1879), Scottish scientist, trained as a mathematician. Between 1856 and 1860 he lectured in Aberdeen and in 1860 became Professor of Natural Philosophy and Astronomy at King's College, London, until 1865. After that, he resigned and busied himself in writing, including on the *Treatise on Electricity and Magnetism*, published in 1873. Maxwell's work was not limited to electricity and magnetism. He wrote on the theory of gases, on heat, and on such topics as light, color, color blindness, the rings of Saturn, and others. The three-color combination (red, green, and blue) used to this day in defining color processes such as television and monitor screens was invented by Maxwell in 1855. Although a modest man, he knew the value of his work and was proud of it. The publication of the *Treatise on Electricity and Magnetism* was a turning point in electromagnetics. It was for the first time since Oersted's discovery of the link between electricity and magnetism that this link extended to the generation and propagation of waves. This was shown experimentally by Hertz 15 years later and opened the way to the invention of radio and the communication era.

[3] Oliver Heaviside (1850–1925). Oliver Heaviside was by all accounts the "enfant terrible" of electromagnetics. A brilliant man with a natural gift for mathematical analysis, he was the incarnation of antiestablishment. Heaviside dropped out of school at age 16 and at age 18 started working for the Anglo-Danish cable company. He worked for about 6 years during which he taught himself the theories of electricity and magnetism and, apparently, applied mathematics. After that, Heaviside set out to understand and explain Maxwell's theory, and in the process, he derived the modern form of Maxwell's equations (at about the same time Hertz did). He was one of the first to use phasors and did much to propagate the use of vector analysis. Heaviside may well be considered the developer of operational calculus and of the theory of transmission lines. Much of his work remains uncredited, a process which started in his lifetime. He was reclusive and abrasive, qualities that did not win him friends. Constant attacks on the "mathematicians of Cambridge" caused much friction, as did the fact that his papers were difficult to understand. For example, he insisted that potentials have no value. In his words, these were "Maxwell's monsters" and should be "murdered." Similarly, he dismissed the theory of relativity. With all his failings, quite a bit of what we study today in electromagnetics and circuits must be credited to Heaviside.

[4] The term displacement and displacement current were coined by Maxwell from the analogies he used. To Maxwell, flux lines were analogous to lines of flow in an incompressible fluid. Using this analogy, he called the quantity **D**, the electric displacement and therefore the current density produced by its time derivative, the displacement current density.

proof for the existence of displacement currents, but it is obvious from reading his book that he considered both displacement currents and the implications of their existence as fact. Experimental proof of the existence of electromagnetic waves at frequencies well below those of light came only in 1888, when the young Heinrich Hertz[5] showed through his famous experiments that an electromagnetic disturbance travels through air and can be received at a distance. This was almost 10 years after Maxwell died and 15 years after he wrote the Treatise. Some of the most important implications of displacement currents and of Maxwell's equations in general are as follows:

(1) Interdependence of the electric and magnetic fields.
(2) The existence of electromagnetic waves.
(3) Finite speed of propagation of electromagnetic waves.
(4) Propagation in free space is at the speed of light, and light itself is an electromagnetic wave.

11.2.1 Maxwell's Equations in Differential Form

To understand the importance of these concepts and Maxwell's contribution to the theory of electromagnetics, it is well worth looking at the electromagnetic field equations as they existed before Maxwell's introduction of displacement currents (below on the left) and after Maxwell's modification (below on the right):

Field equations before Maxwell's modification		Field equations after Maxwell's modification	
$\nabla \times \mathbf{E} = -\dfrac{\partial \mathbf{B}}{\partial t}$	(11.1)	$\nabla \times \mathbf{E} = -\dfrac{\partial \mathbf{B}}{\partial t}$	(11.5)
$\nabla \times \mathbf{H} = \mathbf{J} \quad [\text{A/m}^2]$	(11.2)	$\nabla \times \mathbf{H} = \mathbf{J} + \dfrac{\partial \mathbf{D}}{\partial t} \quad \left[\dfrac{\text{A}}{\text{m}^2}\right]$	(11.6)
$\nabla \cdot \mathbf{D} = \rho_v \quad [\text{C/m}^3]$	(11.3)	$\nabla \cdot \mathbf{D} = \rho_v \quad [\text{C/m}^3]$	(11.7)
$\nabla \cdot \mathbf{B} = 0$	(11.4)	$\nabla \cdot \mathbf{B} = 0$	(11.8)

Comparison between the two sets reveals that the only difference is in the last term of **Eq. (11.6)**. This term is a current density and is called the ***displacement current density***.

Perhaps the easiest way to show that the pre-Maxwell equations are not, in general, adequate is to show that they are not consistent with the continuity equation or that conservation of charge is not satisfied unless the displacement current density in **Eq. (11.6)** is introduced. To do so, we take the divergence on both sides of **Eq. (11.2)**:

$$\nabla \cdot (\nabla \times \mathbf{H}) = \nabla \cdot \mathbf{J} \tag{11.9}$$

On the left-hand side, the divergence of the curl of a vector is identically zero. Thus, we get

$$\nabla \cdot \mathbf{J} = 0 \tag{11.10}$$

On the other hand, if we do the same with **Eq. (11.6)**, we get

$$\nabla \cdot (\nabla \times \mathbf{H}) = \nabla \cdot \mathbf{J} + \nabla \cdot \left(\frac{\partial \mathbf{D}}{\partial t}\right) \tag{11.11}$$

[5] Heinrich Rudolph Hertz (1857–1894). Hertz was trained as an engineer but had considerable interest in other areas, including mathematics and languages. At the suggestion of Herman von Helmholtz, he undertook a series of experiments which, in 1888, led to verification of Maxwell's theory. This included proof of propagation of waves which he showed by receiving the disturbance produced by a spark with a receiver which was essentially a loop with a gap. In the process, he measured the speed of propagation and found it to be that of light. The conclusion that light itself was an electromagnetic wave was a logical extension from these experiments. At the age of 31, Hertz succeeded where others failed. In his experiments, he showed many of the properties of waves, including reflection, polarization, refraction, periodicity, resonance, and even the use of parabolic antennas. It is all the more tragic that he died only 6 years later at the age of 37. The notation we use in this book is due to Hertz (and Heaviside) rather than due to Maxwell. Unlike Maxwell, who emphasized the use of potentials, Hertz and Heaviside emphasized the use of the electric and magnetic fields.

Again, the left-hand side is identically zero and we can write

$$\nabla \cdot \mathbf{J} = -\frac{\partial}{\partial t}(\nabla \cdot \mathbf{D}) \tag{11.12}$$

where the time derivative and the divergence were interchanged. Substituting $\nabla \cdot \mathbf{D}$ from **Eq. (11.7)** gives

$$\nabla \cdot \mathbf{J} = -\frac{\partial \rho_v}{\partial t} \tag{11.13}$$

which is exactly the continuity equation [see **Eq. (7.32)**]. Thus, introduction of the displacement current density in **Eq. (11.2)** is equivalent to enforcing the law of conservation of charge as was discussed in **Section 7.6**. That this should be so is intuitively understood: If the law of conservation of charge is correct and if the field equations must obey this law, then the law must be incorporated into the equations. However, there is one question that we have alluded to in the introduction. If, indeed, the displacement current density must be included, how is it possible that all the results obtained in previous chapters were considered to be correct while explicitly neglecting the displacement current term? An even more important question is: What are the implications of the new results that the inclusion of displacement current suggests? The answer to the first question is, in fact, implicit in the continuity equation itself. If we assume that the time derivative of current density is zero, that is, that the charge density is constant with time, the continuity equation states that $\nabla \cdot \mathbf{J} = 0$. This corresponds to the pre-Maxwell equations. In other words, as long as we deal with steady (DC) currents or with zero currents (static charges), all relations we have developed in previous chapters are correct (and, in fact, exact). The answer to the second question is much more involved and will be answered gradually in the following chapters. At this point we simply note that as long as the displacement current is small [low value of the time derivative in **Eq. (11.6)** or **(11.13)** or, alternatively, low frequencies], the displacement current density may be neglected. A more quantitative explanation will follow in **Chapter 12**. We now give a very simple example that indicates the importance of displacement currents for understanding some of the simplest aspects of electromagnetics.

Example 11.1 Displacement Current in a Capacitor Consider the capacitor in **Figure 11.1a**. The capacitor is connected to an AC source to form a closed circuit. Calculate the displacement current in the capacitor.

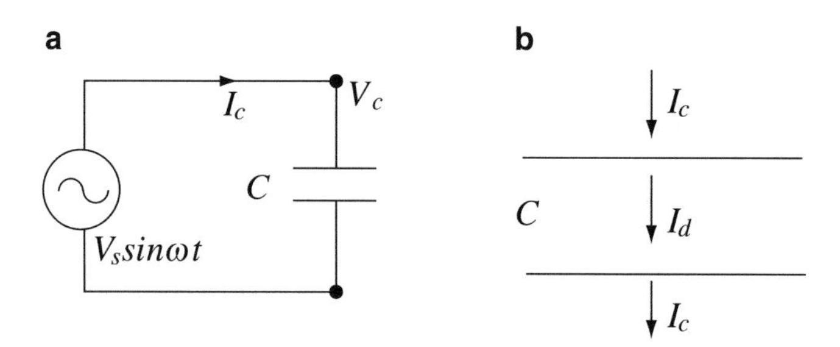

Figure 11.1 Displacement current in a capacitor. (**a**) Capacitor connected to an AC generator. (**b**) Relation between the conduction and displacement current

Solution: A current flowing in the circuit can be physically measured using an AC amperemeter. The current may be calculated using circuit concepts, and since current in a closed circuit is the same everywhere in the circuit, this must also be equal to the displacement current in the capacitor. A second method is to calculate the electric flux density **D** in the capacitor and then calculate $\partial \mathbf{D}/\partial t$ in the capacitor to obtain the displacement current density.

Method (1) The current in the circuit in **Figure 11.1a** is

$$I = C\frac{dV_c}{dt} = CV_s\omega\cos\omega t \quad [\text{A}]$$

Because this is the current at any point in the circuit, and a closed circuit must have the same current everywhere, this current must also exist inside the capacitor. There is no other way. We are therefore forced to allow for the existence of a current through the dielectric in the capacitor even though a dielectric cannot support a conduction current (after all, it is an insulator). Therefore, the displacement current in the dielectric must be equal to the conduction current in the circuit (**Figure 11.1b**); that is,

$$I_d = CV_s\omega\cos\omega t \quad [\mathrm{A}]$$

Method (2) An alternative method is to calculate the electric flux density **D** in the capacitor and the displacement current from its time derivative. For a parallel plate capacitor, the electric field intensity between the plates is uniform and its magnitude is

$$E = \frac{V_c}{d} = \frac{V_s\sin\omega t}{d} \quad \left[\frac{\mathrm{V}}{\mathrm{m}}\right]$$

where V_c is the potential across the capacitor's plates and d the distance between the plates. Using this relation and $\mathbf{D} = \varepsilon\mathbf{E}$, the displacement current density is calculated using **Eq. (11.6)**:

$$J_d = \frac{\partial D}{\partial t} = \varepsilon\frac{\partial E}{\partial t} = \varepsilon\frac{\partial}{\partial t}\left(\frac{V_s\sin\omega t}{d}\right) = \frac{\varepsilon}{d}V_s\omega\cos\omega t \quad [\mathrm{A/m^2}]$$

To find the total displacement current, the current density must be integrated over the surface area of the plates of the capacitor. Denoting this area as S, we get

$$I_d = \int_{s'} \mathbf{J}_d \cdot d\mathbf{s}' = \frac{\varepsilon V_s\omega\cos\omega t}{d}\int_{s'} ds' = \frac{\varepsilon S V_s\omega\cos\omega t}{d} = CV_s\omega\cos\omega t \quad [\mathrm{A}]$$

where $C = \varepsilon S/d$ is the capacitance of a parallel plate capacitor of plate area S and distance d between the plates. This result is identical to that in Method (1).

Here, we have an example in which the displacement current is actually necessary to account for the behavior of the circuit. In its absence, we must assume that current cannot flow through the capacitor. In circuit theory, displacement currents are not normally used or assumed. To account for their effects, the common explanation is that the plates of the capacitor are alternately charged with positive and negative charges.

Example 11.2 A slab of perfect dielectric material ($\varepsilon_r = 2$) is placed in a microwave oven. The oven produces an electric field (as well as a magnetic field). Assume that the electric field intensity is uniform in the slab and sinusoidal in form and that it is perpendicular to the surface of the slab. The microwave oven operates at a frequency of 2.45 GHz (1 GHz $= 10^9$ Hz) and produces an electric field intensity with amplitude 500 V/m inside the dielectric:

(a) Calculate the displacement current density in the dielectric.
(b) Is there a displacement current in air? If so, calculate it.

Solution: Calculate the electric flux density D in the dielectric and calculate its derivative with respect to time to get the displacement current density. The displacement current only requires that an electric field exists and is always in the direction of the electric field intensity. Therefore, any material in which there is an electric field will support a displacement current, including free space:

(a) The electric field intensity and electric flux density in the dielectric are

$$E = E_0\sin\omega t \quad \rightarrow \quad D = \varepsilon_r\varepsilon_0 E_0\sin\omega t \quad [\mathrm{V/m}]$$

The displacement current density in the dielectric is

$$J_d = \frac{\partial D}{\partial t} = \varepsilon_r\varepsilon_0 E_0\omega\cos\omega t \quad [\mathrm{A/m^2}]$$

For the values given, this current density is

$$J_d = 2\times 8.854\times 10^{-12}\times 500\times 2\times \pi\times 2.45\times 10^9\times \cos\left(2\times \pi\times 2.45\times 10^9\right)t = 136.3\cos\left(4.9\times 10^9\pi\right)t \quad [\mathrm{A/m^2}]$$

with a peak current density of 136.3 $\mathrm{A/m^2}$.

(b) In air, the electric flux density D is the same as in the dielectric ($D_{1n} = D_{2n}$, although the electric field intensity is twice as high in air). Since D is the same, so must the displacement current density be the same. The displacement current density in air is

$$J_d = 136.3\cos\left(4.9\times 10^9\pi\right)t \quad [\mathrm{A/m^2}].$$

11.2.2 Maxwell's Equations in Integral Form

Equations (11.5) through **(11.8)** are the differential or point form of Maxwell's equations. As such, they describe the fields in space and, as we shall see shortly, lead to partial differential equations. This form, however, is not always convenient. For the calculation of fields and field-related quantities, an integral expression is often more convenient and, sometimes, more descriptive of the phenomenon involved. It is therefore useful to obtain the integral forms of Maxwell's equations. This is done by integrating the two curl equations over an arbitrary open surface and the two divergence equations over an arbitrary volume. We start with **Eq. (11.6)** since the integral form of the remaining equations was obtained in **Chapters 4**, **8**, and **10**. Integrating over an arbitrary open surface s, we get

$$\int_s (\nabla\times\mathbf{H})\cdot d\mathbf{s} = \int_s \mathbf{J}\cdot d\mathbf{s} + \int_s \frac{\partial\mathbf{D}}{\partial t}\cdot d\mathbf{s} \quad [\mathrm{A}] \tag{11.14}$$

The expression on the left-hand side is converted to a contour integral using Stokes' theorem. Ampère's law now becomes

$$\int_s (\nabla\times\mathbf{H})\cdot d\mathbf{s} = \oint_C \mathbf{H}\cdot d\mathbf{l} = I_c + \int_s \frac{\partial\mathbf{D}}{\partial t}\cdot d\mathbf{s} = I_c + I_d \quad [\mathrm{A}] \tag{11.15}$$

The term on the left-hand side is the circulation of the magnetic field intensity, whereas on the right-hand side, the first term is the conduction current (all currents except displacement currents, including induced currents) and the second is the displacement current. The integral form of Maxwell's equations is therefore

$$\oint_C \mathbf{E}\cdot d\mathbf{l} = -\frac{d\Phi}{dt} \quad \text{(Faraday's law)} \tag{11.16}$$

$$\oint_C \mathbf{H}\cdot d\mathbf{l} = I_c + \int_s \frac{\partial\mathbf{D}}{\partial t}\cdot d\mathbf{s} \quad \text{(Ampère's law)} \tag{11.17}$$

$$\oint_s \mathbf{D}\cdot d\mathbf{s} = Q \quad \text{(Gauss's law)} \tag{11.18}$$

$$\oint_s \mathbf{B}\cdot d\mathbf{s} = 0 \quad \text{(no monopoles)} \tag{11.19}$$

The first of these is Faraday's law as derived in **Chapter 10**. The second is Ampère's law. We first introduced this law in **Chapter 8** for the magnetic fields of steady currents. The law as given here is sometimes called the modified Ampère's law to distinguish it from the pre-Maxwell form defined in **Chapter 8**. The third relation is Gauss's law, which was discussed in **Chapter 4**. The fourth indicates the divergence-free condition of the magnetic flux density which was discussed at length in **Chapter 8** as indicating the fact that the magnetic field is always generated by a pair of poles (i.e., no single magnetic poles exist).

In practical applications, we may be required to solve for any or all of the variables in Maxwell's equations. It is well worth pausing here to discuss these equations. In particular, we ask ourselves if, indeed, these equations are all that we need to solve an electromagnetic problem.

First, we note that the equations [either in differential form in **Eqs. (11.5)** through **(11.8)** or in integral form in **Eqs. (11.16)** through **(11.19)**], contain four vector variables **E**, **D**, **B**, and **H** and two sources: **J** (or I) and ρ_v (or Q). The first is a vector source, whereas the second a scalar source. Each vector variable has three components in space, and, therefore, we actually have 12 unknown values for the 12 components of the fields. Since the first two equations are vector equations, they are equivalent to six scalar equations. The last two equations [**Eqs. (11.18)** and **(11.19)**] are scalar equations. Thus, we have 8 scalar equations in 12 unknowns. Clearly, some additional relations must be added in order to solve the equations. Before we add any relations, we must also ascertain if the four Maxwell's equations are independent. If they are not, additional relations might be required.

Recall the way that Maxwell's equations were derived. They were based on the definition of the curl and divergence. At no point did we require that the equations be independent. In fact, the last two equations in each set can be derived from the first two with the aid of the continuity equation. To see that this is the case, consider **Eq. (11.6)**. If we take the divergence on both sides of the equation, we get

$$\nabla \cdot (\nabla \times \mathbf{H}) = \nabla \cdot \mathbf{J} + \nabla \cdot \frac{\partial \mathbf{D}}{\partial t} \tag{11.20}$$

The left-hand side is zero (the divergence of the curl of any vector is identically zero). If we interchange the time derivative and the gradient in the last term on the right-hand side, we get

$$0 = \nabla \cdot \mathbf{J} + \frac{\partial}{\partial t}(\nabla \cdot \mathbf{D}) \tag{11.21}$$

Now, from the continuity equation [**Eq. (11.13)**]

$$\nabla \cdot \mathbf{J} = -\frac{\partial \rho_v}{\partial t} \quad \rightarrow \quad 0 = -\frac{\partial \rho_v}{\partial t} + \frac{\partial}{\partial t}(\nabla \cdot \mathbf{D}) \tag{11.22}$$

and, finally,

$$\nabla \cdot \mathbf{D} = \rho_v \tag{11.23}$$

This is exactly **Eq. (11.7)**. A similar calculation shows that **Eq. (11.8)** can be derived from **Eq. (11.5)**.

This dependency of the equations means that, in fact, we only have two independent vector equations (the two curl equations) in four vector unknowns. Thus, only 6 scalar equations are available for solution and the number of unknowns is 12. We therefore need two more independent vector equations to solve the system. These equations are the two constitutive relations $\mathbf{B} = \mu\mathbf{H}$ and $\mathbf{D} = \varepsilon\mathbf{E}$. That this must be so can also be seen from the fact that Maxwell's equations as written in **Eqs. (11.5)** through **(11.8)** or **(11.16)** through **(11.19)** do not refer to material properties at all. On the other hand, we know that fields are very much dependent on materials. This dependency is expressed by the constitutive relations.

In addition, the Lorentz force equation, first introduced in **Chapter 8**, should be considered as part of a complete set of equations required for the solution of an electromagnetic field problem. The complete set of equations is summarized in **Table 11.1**. Thus, a total of seven equations are normally considered to constitute a complete set. An additional constitutive relation was defined in **Eq. (7.12)** as $\mathbf{J} = \sigma\mathbf{E}$. This is not included in **Table 11.1** because it is limited to conducting regions and because it will be generalized in **Chapter 12**. In time-dependent applications of electromagnetics, only the first two Maxwell's equations are independent and need to be used for the solution, together with the constitutive relations. The force equation is included in the complete set because it cannot be derived from Maxwell's equations. For this reason, the above complete set is sometimes called the Maxwell–Lorentz equations.

Finally, because the material constitutive relations are an integral part of the electromagnetic field equations, it is worth reiterating the fact that material properties may be, in general, linear or nonlinear, isotropic or anisotropic, homogeneous or nonhomogeneous. These properties were first defined in **Section 4.5.4.1** and are repeated below as a reminder:

Table 11.1 Summary of the electromagnetic field equations in differential and integral forms

Maxwell's equations	Differential form	Integral form
Faraday's law	$\nabla\times\mathbf{E}=-\dfrac{\partial\mathbf{B}}{\partial t}$ (11.24)	$\oint_C \mathbf{E}\cdot d\mathbf{l}=-\dfrac{d\Phi}{dt}$ [V] (11.28)
Ampère's law	$\nabla\times\mathbf{H}=\mathbf{J}+\dfrac{\partial\mathbf{D}}{\partial t}$ [A/m²] (11.25)	$\oint_C \mathbf{H}\cdot d\mathbf{l}=\int_s\left(\mathbf{J}+\dfrac{\partial\mathbf{D}}{\partial t}\right)\cdot d\mathbf{s}$ [A] (11.29)
Gauss's law	$\nabla\cdot\mathbf{D}=\rho_v$ [C/m³] (11.26)	$\oint_s \mathbf{D}\cdot d\mathbf{s}=Q$ [C] (11.30)
No monopoles	$\nabla\cdot\mathbf{B}=0$ (11.27)	$\oint_s \mathbf{B}\cdot d\mathbf{s}=0$ (11.31)
Constitutive relations	$\mathbf{B}=\mu\mathbf{H}$ [T] (11.32) $\mathbf{D}=\varepsilon\mathbf{E}$ [C/m²] (11.33)	
The Lorentz force equation	$\mathbf{F}=q(\mathbf{E}+\mathbf{v}\times\mathbf{B})$ [N] (11.34)	

Linear: Linearity in material properties (μ, ε, σ) means these properties do not change as the fields vary.
Homogeneous: Material properties do not depend on position: The material properties do not vary from point to point in space.
Isotropic: Material properties are independent of direction in space.

As an example, we may speak of nonlinear electromagnetic field equations if the equations are used in nonlinear media. In the following chapters, we will deal with linear, isotropic, homogeneous media (simple media) exclusively.

Example 11.3 Show that if $\mathbf{J}=0$ in **Eq. (11.29)** and $Q=0$ in **Eq. (11.30)**, the two divergence equations in **Eqs. (11.30)** and **(11.31)** can be obtained from **Eqs. (11.28)** and **(11.29)** without the need to invoke the continuity equation.

Solution: By taking the divergence of **Eqs. (11.24)** and **(11.25)**, we obtain the divergence equations in differential form. Integration on both sides of the result gives the answer:

$$\nabla\cdot(\nabla\times\mathbf{E})=-\nabla\cdot\frac{\partial\mathbf{B}}{\partial t}=-\frac{\partial(\nabla\cdot\mathbf{B})}{\partial t}$$

where the time derivative and the divergence were interchanged since these operations are mutually exclusive. Because the divergence of the curl of any vector field is identically zero, it follows that

$$\nabla\cdot\mathbf{B}=0$$

Taking the volume integral of this relation and using the divergence theorem, we get

$$\int_v(\nabla\cdot\mathbf{B})dv=\oint_s\mathbf{B}\cdot d\mathbf{s}=0\quad\rightarrow\quad\oint_s\mathbf{B}\cdot d\mathbf{s}=0$$

This is **Eq. (11.31)**.
Similarly, starting with **Eq. (11.25)**, setting $\mathbf{J}=0$, and taking the divergence on both sides, we get

$$\nabla\cdot(\nabla\times\mathbf{H})=0=\nabla\cdot\frac{\partial\mathbf{D}}{\partial t}=\frac{\partial(\nabla\cdot\mathbf{D})}{\partial t}\quad\rightarrow\quad\nabla\cdot\mathbf{D}=0$$

Again taking the volume integral and using the divergence theorem gives

$$\int_v(\nabla\cdot\mathbf{D})dv=\oint_s\mathbf{D}\cdot d\mathbf{s}=0\quad\rightarrow\quad\oint_s\mathbf{D}\cdot d\mathbf{s}=0$$

The latter is identical to **Eq. (11.30)** for $Q=0$.

11.3 Time-Dependent Potential Functions

The concept of potential functions was introduced in **Section 8.7** where the magnetic scalar and vector potentials were discussed. Also, the electric scalar potential was discussed at length in **Chapter 4**. The utility of these potential functions in the calculation of the electric and magnetic fields was shown and this utility was the justification of deriving the potentials in the first place. Here, we revisit the idea of scalar and vector potentials, but now the potentials are time-dependent although their purpose is still the same: to allow alternative, often simpler calculation of field quantities. We will also show that the potentials, and in particular the magnetic vector potential, lead to second-order partial differential equations representation of Maxwell's equations, which can then be solved by standard differential equation methods.

We recall that for a vector function to be represented by a scalar function alone, the vector function must be curl-free; that is,

$$\text{If}\ \ \nabla\times\mathbf{F}=0\quad\rightarrow\quad\mathbf{F}=-\nabla\Omega \tag{11.35}$$

where Ω is a scalar function and is called a scalar potential in the context of electromagnetics. Similarly, for a vector field to be represented by an auxiliary vector function, the vector field must be divergence-free:

$$\text{If}\ \ \nabla\cdot\mathbf{F}=0\quad\rightarrow\quad\mathbf{F}=\nabla\times\mathbf{W} \tag{11.36}$$

The vector function **W** is now a vector potential.

Vector and scalar potentials may be used even if the vector field is neither curl-free nor divergence-free. To do so, we invoke the Helmholtz theorem and write the vector field as the sum of an irrotational term and a solenoidal term (see **Section 2.5.1**):

$$\mathbf{G}=-\nabla U+\nabla\times\mathbf{C} \tag{11.37}$$

where U is a scalar potential and **C** a vector potential. The first term (the gradient of U) is irrotational since taking its curl yields zero. The second term is solenoidal since taking its divergence yields zero. Thus, the general process of defining scalar and vector potentials for vector fields is as follows:

(1) If the vector field is curl-free (irrotational), a scalar potential may be defined which completely describes the vector field.
(2) If the vector field is divergence-free (solenoidal), a vector potential may be defined which completely describes the vector field.
(3) For a general vector field, both a scalar and a vector potential are required to describe the vector field. The gradient of the scalar potential is used to describe the irrotational part of the field, whereas the vector potential is used to describe the solenoidal part of the field.

The potentials we define need not have any physical meaning, although they often do. Their definition is based on the vector properties of the fields and may be viewed as transformations. As such, as long as the transformation is unique and is properly defined, the potentials are valid. We will discuss here the electric scalar potential and the magnetic vector potential; these are needed for our discussion of electromagnetic fields. There are, however, other potential functions that may be defined. We will only touch on some of these as examples.

11.3.1 Scalar Potentials

Regarding Maxwell's equations above, there are two scalar potentials that may be defined: the electric scalar potential and the magnetic scalar potential. However, inspection of Maxwell's equations shows that the equations are not, in general, curl-free. Therefore, scalar potentials alone cannot be used to solve for general electric and magnetic fields. There are, however, two situations in which scalar potentials may be used:

(1) If the time derivative of the magnetic flux density in Faraday's law **[Eq. (11.24)]** is zero, the electric field intensity is curl-free, and the electric potential may be used in lieu of the electric field intensity:

$$\mathbf{E}=-\nabla V\quad\text{if}\quad\nabla\times\mathbf{E}=0 \tag{11.38}$$

(2) If the displacement current density and the conduction current density in Ampère's law **[Eq. (11.25)]** are zero, the magnetic field intensity is curl-free, and the magnetic scalar potential may be used in lieu of the magnetic field intensity:

$$\mathbf{H} = -\nabla\psi \quad \text{if} \quad \nabla\times\mathbf{H} = 0 \tag{11.39}$$

The electric scalar potential V and the magnetic scalar potential ψ can only be used by themselves for static electric and magnetic fields because only under static conditions can the electric and magnetic field intensity be curl-free. However, they are also useful in general electromagnetic fields in combination with vector potentials, as we shall see later.

11.3.2 The Magnetic Vector Potential

The magnetic vector potential was defined in **Section 8.7.1** as

$$\mathbf{B} = \nabla\times\mathbf{A} \quad \text{because} \quad \nabla\cdot\mathbf{B} = 0 \tag{11.40}$$

This also applies to time-dependent magnetic fields because the definition is based entirely on the divergence-free condition of the magnetic flux density. The magnetic vector potential may be used in many ways, one of which was given in **Chapter 8**, where the Biot–Savart law in terms of the magnetic vector potential was used. Here, we will use the function to represent Maxwell's equations and, in the process, show that it may be used for calculation of field quantities. To do so, we substitute the definition into Maxwell's first and second equations:

$$\nabla\times\mathbf{E} = -\frac{\partial(\nabla\times\mathbf{A})}{\partial t} \tag{11.41}$$

$$\nabla\times\left(\frac{1}{\mu}\nabla\times\mathbf{A}\right) = \mathbf{J} + \frac{\partial\mathbf{D}}{\partial t} \tag{11.42}$$

From **Eq. (11.41)** by interchanging the time derivative with the ∇ operator, we write

$$\nabla\times\mathbf{E} = \nabla\times\left(-\frac{\partial\mathbf{A}}{\partial t}\right) \quad \rightarrow \quad \nabla\times\left(\mathbf{E} + \frac{\partial\mathbf{A}}{\partial t}\right) = 0 \tag{11.43}$$

Now the term in the parentheses is curl-free and it may be written as the gradient of the electric scalar potential:

$$\nabla\times\left(\mathbf{E} + \frac{\partial\mathbf{A}}{\partial t}\right) = 0 \quad \rightarrow \quad \left(\mathbf{E} + \frac{\partial\mathbf{A}}{\partial t}\right) = -\nabla V \tag{11.44}$$

Rearranging this expression gives the general electric field intensity as

$$\boxed{\mathbf{E} = -\frac{\partial\mathbf{A}}{\partial t} - \nabla V \qquad \left[\frac{\text{V}}{\text{m}}\right]} \tag{11.45}$$

This form is exactly that required by the Helmholtz theorem **[Eq. (11.37)]**. This, when substituted back into Faraday's law, gives the same relation as in **Eq. (11.43)** because $\nabla\times(\nabla V) = 0$. Also, the expression gives the correct result for the static electric field for which $\mathbf{E} = -\nabla V$. Now, we substitute the electric field intensity from **Eq. (11.45)** into **Eq. (11.42)**:

$$\nabla\times\left(\frac{1}{\mu}\nabla\times\mathbf{A}\right) = \mathbf{J} + \frac{\partial}{\partial t}\left[\varepsilon\left(-\frac{\partial\mathbf{A}}{\partial t} - \nabla V\right)\right] \tag{11.46}$$

For simplicity, we will assume that the material in which this relation is defined is linear, isotropic and homogeneous such that the permeability μ and permittivity ε are independent of position. This gives

$$\nabla \times (\nabla \times \mathbf{A}) = \mu \mathbf{J} - \mu\varepsilon \frac{\partial^2 \mathbf{A}}{\partial t^2} - \mu\varepsilon \frac{\partial}{\partial t}(\nabla V) \tag{11.47}$$

The left-hand side of **Eq. (11.47)** can be expanded using the identity $\nabla \times (\nabla \times \mathbf{A}) = \nabla(\nabla \cdot \mathbf{A}) - \nabla^2 \mathbf{A}$ [see **Eq. (2.134)**]. Substituting this and rearranging terms, we get

$$\nabla(\nabla \cdot \mathbf{A}) - \nabla^2 \mathbf{A} = \mu \mathbf{J} - \mu\varepsilon \frac{\partial^2 \mathbf{A}}{\partial t^2} - \nabla\left(\mu\varepsilon \frac{\partial V}{\partial t}\right) \tag{11.48}$$

Inspection of **Eq. (11.48)** as well as the process leading to it reveals that Maxwell's equations were now replaced by a single equation in terms of the magnetic vector potential and the electric scalar potential. In fact, since the magnetic vector potential has three components, the equation is equivalent to three scalar equations. However, we need one more relation to take into account the electric scalar potential V. One possibility is to assume that V is zero. The second is to assume it is independent of time and the third is to assume it is constant in space. None of these is a general property of the electric field, and, therefore, we cannot assume these in general, although one of these may be valid occasionally depending on the application. To resolve this difficulty, we use the fact that the divergence of **A** has not yet been defined. Since a vector is only uniquely defined if both its curl and its divergence are specified, we are free to choose the divergence of the magnetic vector potential. The second relation needed is therefore the divergence of the magnetic vector potential. From **Eq. (11.48)**, we note that if we choose

$$\boxed{\nabla \cdot \mathbf{A} = -\mu\varepsilon \frac{\partial V}{\partial t}} \tag{11.49}$$

the first and the last terms in **Eq. (11.48)** disappear and we get

$$\boxed{-\nabla^2 \mathbf{A} = \mu \mathbf{J} - \mu\varepsilon \frac{\partial^2 \mathbf{A}}{\partial t^2}} \tag{11.50}$$

and this equation is now the equivalent form of Maxwell's equations; that is, instead of solving Maxwell's equations for the magnetic field intensity and the electric field intensity, we can solve for the magnetic vector potential and then obtain the magnetic and electric field intensities as well as flux densities from the magnetic vector potential.

The relation in **Eq. (11.49)** is called the ***Lorenz condition*** or the ***Lorenz gauge***. There are three questions associated with **Eqs. (11.49)** and **(11.50)**. First, is the Lorenz gauge the only possible choice? Second, how do we know that this choice is correct? Third, why should we use the magnetic vector potential in the first place since we can obtain a second-order equation in terms of **H** or **E**, as will be shown shortly? The answer to the first question is no. There are other choices that may be used, but this particular choice eliminates the scalar potential in the equation and therefore simplifies the equation which, in turn, also should simplify its solution. A commonly used gauge, particularly in static applications, is $\nabla \cdot \mathbf{A} = 0$, which is called the ***Coulomb's gauge*** and was introduced in **Chapter 8 [Eqs. (8.37)** and **(8.38)]**. The answer to the second question is that this choice is "consistent with the field equations." The latter statement means that Lorenz's condition is consistent with the principle of conservation of charge. The answer to the third question is twofold: First, it allows representation in terms of a single field variable **A**, instead of the need for **E** and **H**. Second, and perhaps more important, the magnetic vector potential is sometimes more convenient to use than the electric field intensity **E** or the magnetic field intensity **H**. While it is not the purpose here to prove this, it should be noticed that the magnetic vector potential is everywhere in the direction of the current density **J**. This means that if the current density has a single component in space, the magnetic vector potential also has a single component. On the other hand, the magnetic field intensity has two components (perpendicular to the current). Without actually solving the equations, it is intuitively understood that solving for a single component of a field in space should be easier than solving for two components.

11.3.3 Other Potential Functions

By now, it should be understood that a potential function can be defined based on the properties of the original field, for the purpose of replacing the field with an equivalent but perhaps simpler representation. Other potential functions may be defined in addition to the magnetic vector potential and the electric scalar potential discussed here. In **Section 11.3.1**, we defined the magnetic scalar potential in current-free regions. A current vector potential for steady currents in conducting media may be defined in a manner similar to the magnetic vector potential using the condition $\nabla \cdot \mathbf{J} = 0$ (see **Exercise 11.1**). Other potentials are the Hertz, Lorentz, and Whittaker potentials. However, because these potentials are not required for the development of the concepts presented in this book, we do not pursue these here (but see **Exercise 11.1**, **Problems 11.21** through **11.24**, and **12.5**).

Example 11.4 Vector and Scalar Potentials in Conducting Media It is required to define the electromagnetic field equations for low frequencies in a highly conductive material in terms of the magnetic vector potential **A**. The material may be assumed to be linear, homogeneous, and isotropic in all its material properties. Show that by using Coulomb's gauge ($\nabla \cdot \mathbf{A} = 0$), the field equations reduce to:

$$\nabla^2 \mathbf{A} = -\mu \mathbf{J}$$

Solution: The field equations are manipulated as in **Eqs. (11.41)** through **(11.48)**. This results in

$$\nabla(\nabla \cdot \mathbf{A}) - \nabla^2 \mathbf{A} = \mu \mathbf{J} - \mu\varepsilon \frac{\partial^2 \mathbf{A}}{\partial t^2} - \nabla\left(\mu\varepsilon \frac{\partial V}{\partial t}\right)$$

To simplify this equation, we must find a way of eliminating the first term on the left-hand side and the last term on the right-hand side. The first is obtained by substituting Coulomb's gauge:

$$-\nabla^2 \mathbf{A} = \mu \mathbf{J} - \mu\varepsilon \frac{\partial^2 \mathbf{A}}{\partial t^2} - \nabla\left(\mu\varepsilon \frac{\partial V}{\partial t}\right)$$

Now, we note that in a highly conducting material, the electric field intensity **E** is very low (the electric field intensity is zero in a perfectly conducting medium). From **Eq. (11.45)**, we conclude that this can happen only if both $\partial \mathbf{A}/\partial t$ and ∇V are small. If we rewrite the field equation as

$$-\nabla^2 \mathbf{A} = \mu \mathbf{J} + \mu\varepsilon \frac{\partial}{\partial t}\left(-\frac{\partial \mathbf{A}}{\partial t} - \nabla V\right)$$

the term in parentheses is the electric field intensity [see **Eq. (11.45)**] and because both ∇V and $\partial \mathbf{A}/\partial t$ are small, the entire second term on the right-hand side may be neglected, leading to the result:

$$\nabla^2 \mathbf{A} = -\mu \mathbf{J}$$

Exercise 11.1 Vector and Scalar Potential Functions A vector potential **T** is defined in a conducting material with steady current ($\nabla \cdot \mathbf{J} = 0$) as

$$\nabla \times \mathbf{T} = \mathbf{J}, \quad \nabla \cdot \mathbf{T} = \sigma\mu \frac{\partial \psi}{\partial t}, \quad \text{and} \quad \mathbf{H} = \mathbf{T} - \nabla \psi$$

where ψ is a scalar function:

(a) If displacement currents are neglected and the only currents are induced currents, show from Maxwell's equations that

$$\nabla^2 \mathbf{T} = \sigma\mu \frac{\partial \mathbf{T}}{\partial t}$$

(b) What would you call the potentials **T** and ψ?

Answer **(b)** **T** is a current vector potential (also called an electric vector potential) and ψ is a magnetic scalar potential.

11.4 Interface Conditions for the Electromagnetic Field

We now return to a question we asked earlier: What happens to the electromagnetic field at the interface between two different materials? In the application of Maxwell's equations to various problems, we often encounter interfaces between different materials, with different material properties. The constitutive relations for the electric and magnetic fields indicate that fields are different in different materials. This, of course, is not new: The electrostatic field and the magnetostatic field were shown to behave differently in different materials. The interface conditions for the general, time-dependent electromagnetic fields as described by Maxwell's equations are essentially those we used for the static electric and magnetic fields together with any added condition that time dependency and the addition of the displacement current density in Ampère's law might add. In fact, we find that the modifications necessary are minor.

To define interface conditions, we must apply Maxwell's equations for the general electromagnetic field at the interface. From **Eqs. (11.1)** through **(11.8)**, **Table 11.1**, and the discussion in **Section 11.2**, it is clear that the only equations that have changed due to the time-dependent nature of the electromagnetic field are Faraday's law (discussed in **Chapter 10**) and Ampère's law (through the addition of displacement current). To see what needs to be done, consider **Table 11.2**, which lists the electrostatic and magnetostatic field equations together with the interface conditions we obtained in **Sections 4.6** and **9.3**, as well as Maxwell's equations. This table clearly indicates that the interface conditions on the normal components of the magnetic flux density and electric flux density did not change since the equations that defined them did not change. However, the fields D_{1n}, D_{2n}, B_{1n}, and B_{2n} as well as the charge density ρ_s are now time-dependent quantities. The interface conditions for the tangential components of the electric and magnetic field intensities need to be derived anew because the equations have been modified. This need is indicated by question marks in **Table 11.2**. This is our next task.

Table 11.2 Summary of the electromagnetic field equations and interface conditions

Static field equations	Interface conditions for static fields	Maxwell's equations in integral form	Interface conditions for time-dependent fields
$\oint_C \mathbf{E}\cdot d\mathbf{l} = 0$	$E_{1t} = E_{2t}$	$\oint_C \mathbf{E}\cdot d\mathbf{l} = -\dfrac{d\Phi}{dt}$ [V]	?
$\oint_C \mathbf{E}\cdot d\mathbf{l} = I_{enc.}$ [A]	$\hat{\mathbf{n}}\times(\mathbf{H}_1-\mathbf{H}_2)=\mathbf{J}_s$ [A/m] or $H_{1t}-H_{2t}=J_s^*$ [A/m]	$\oint_C \mathbf{H}\cdot d\mathbf{l} = \oint_s\left(\mathbf{J}+\dfrac{\partial\mathbf{D}}{\partial t}\right)\cdot d\mathbf{s}$ [A]	?
$\oint_s \mathbf{D}\cdot d\mathbf{s} = Q$ [C]	$\hat{\mathbf{n}}\cdot(\mathbf{D}_1-\mathbf{D}_2)=\rho_s$ [C/m^2] or $D_{1n}-D_{2n}=\rho_s$ [C/m^2]	$\oint_s \mathbf{D}\cdot d\mathbf{s} = Q$ [C]	$\hat{\mathbf{n}}\cdot(\mathbf{D}_1-\mathbf{D}_2)=\rho_s$ [C/m^2] or $D_{1n}-D_{2n}=\rho_s$ [C/m^2]
$\oint_s \mathbf{B}\cdot d\mathbf{s} = 0$	$B_{1n}=B_{2n}$	$\oint_s \mathbf{B}\cdot d\mathbf{s} = 0$	$B_{1n}=B_{2n}$

*This form requires the use of the right-hand rule to establish the vector relation between the tangential fields and the current density

Interface conditions are derived as follows:

(1) Two general materials are assumed to be in contact, forming an interface. By an interface, we mean an infinitely thin boundary, with no properties of its own. On each side of the interface, the materials have uniform properties and are linear. The interface may contain surface current densities and surface charge densities.

(2) Maxwell's equations in integral form are applied to the interface. From these, we find the conditions on the electric and magnetic fields on both sides of the interface since the fields must satisfy Maxwell's equations everywhere.

(3) Conditions at the interface are given in terms of components of the field at the interface. The tangential and normal components at the interface are the required interface conditions.

(4) The interface condition for the tangential components of **E** is derived from **Eq. (11.28)**.

(5) The interface condition for the tangential components of **H** is derived from **Eq. (11.29)**.

11.4.1 Interface Conditions for the Electric Field

To derive the interface conditions, we assume two different materials with properties as shown in **Figure 11.2a**. We assume a time-dependent surface charge density ρ_s at the interface between the two different materials and apply **Eq. (11.28)**. Because **Eq. (11.28)** is a closed contour integral, we choose a closed contour that includes the interface alone, with two sides parallel to the interface and two sides perpendicular to the interface. The integration is in the direction shown. The closed contour is formed by the segments *ab*, *bc*, *cd*, and *da*:

$$\oint_{abcda} \mathbf{E} \cdot d\mathbf{l} = -\frac{d\Phi_{abcda}}{dt} = 0 \tag{11.51}$$

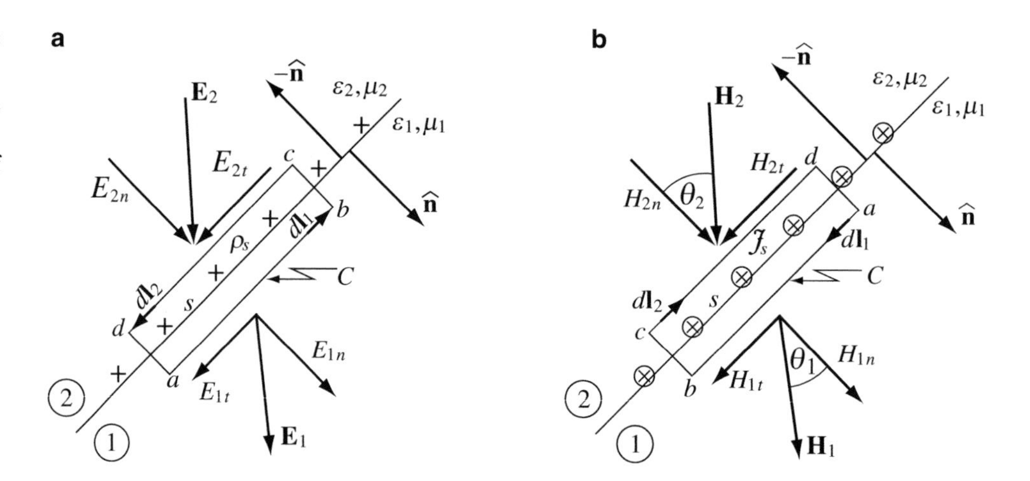

Figure 11.2 (**a**) Conditions for the tangential components of **E** at a general interface. (**b**) Conditions for the tangential components of **H** at a general interface

We note the following:

(1) The total flux Φ enclosed in the loop *abcda* is zero because, in the limit, the area enclosed by the loop is zero. This follows from the requirement that the contour only enclose the interface itself.

(2) The distances *bc* and *da* also tend to zero. Thus, the contribution of these two segments of the contour to the closed contour integration must be zero.

(3) The electric field in material (2) is in the direction of integration (in the direction of $d\mathbf{l}$); in material (1), the electric field intensity is in the direction opposite the direction of integration.

From these considerations we can now write **Eq. (11.51)** as

$$\oint_{abcda} \mathbf{E} \cdot d\mathbf{l} = \int_{ab} \mathbf{E}_1 \cdot d\mathbf{l}_1 + \int_{cd} \mathbf{E}_2 \cdot d\mathbf{l}_2 = 0 \tag{11.52}$$

Since the tangential component of E_2 is in the direction $d\mathbf{l}_2$ and the tangential component of E_1 is in the negative $d\mathbf{l}_1$ direction, we get

$$-\int_{ab} E_{1t} dl_1 + \int_{cd} E_{2t} dl_2 = 0 \tag{11.53}$$

By choosing the distance $ab = cd$, we get

$$\boxed{E_{1t} = E_{2t} \quad \text{or} \quad \frac{D_{1t}}{\varepsilon_1} = \frac{D_{2t}}{\varepsilon_2}} \tag{11.54}$$

where $D_{1t} = \varepsilon_1 E_{1t}$ and $D_{2t} = \varepsilon_2 E_{2t}$.

The following should be noted from these relations:

(1) The tangential component of the electric field intensity is continuous across an interface between two general materials, regardless of charge densities on the surface.
(2) The tangential component of the electric flux density is discontinuous across the interface. The discontinuity is equal to the ratio between the permittivities of the materials.
(3) Interface conditions for the electric field intensity are independent of the magnetic field.
(4) The interface conditions for the time-dependent electric field are identical to those for the static electric field as given in **Chapter 4**. This is a consequence of Faraday's law: The connection between the electric and the magnetic fields is through the flux and the total flux on the interface is zero (the interface has "zero" area).

11.4.2 Interface Conditions for the Magnetic Field

An almost identical sequence follows for the evaluation of the interface conditions for the magnetic field. The same interface between two general materials as in **Figure 11.2a** is used except that now, because we use Ampère's law, there are no charge densities on the interface but, rather, a current density. The conditions for application of Ampère's law are shown in **Figure 11.2b**. The following conditions are used:

(1) The contour *abcda* encloses the interface alone; that is, *bc* and *da* tend to zero and the area of the contour tends to zero.
(2) The tangential component of the magnetic field intensity in material (1) is in the same direction as the direction of integration, whereas in material (2), it is in the opposite direction.
(3) The total current enclosed by the contour is equal to the (surface) current density on the surface multiplied by the length *ab* (or *cd*, since these may be taken to be equal).

From these, we can write

$$\oint_c \mathbf{H} \cdot d\mathbf{l} = \int_{abcda} \mathbf{H} \cdot d\mathbf{l} = \int_{ab} H_{1t} dl_1 - \int_{cd} H_{2t} dl_2 = \int_{ab} J_s dl \qquad (11.55)$$

Note that in **Eq. (11.29)**, the integration for current is a surface integration because, in general, **J** is a current distributed over a volume. However, here, the current is distributed over a surface; therefore, the integration is on the line *ab*. The closed contour integral of $\mathbf{H} \cdot d\mathbf{l}$ always equals the current enclosed by the contour. Note also that the contribution to the line integral due to displacement current densities is zero. This can be best understood from the fact that as we approach the interface, the area enclosed by the contour *abcda* tends to zero, and, therefore, the surface integral of $\partial \mathbf{D}/\partial t$ tends to zero. Choosing $ab = cd$, we get

$$\boxed{H_{1t} - H_{2t} = J_s \quad \text{and} \quad \frac{B_{1t}}{\mu_1} - \frac{B_{2t}}{\mu_2} = J_s \qquad \left[\frac{\text{A}}{\text{m}}\right]} \qquad (11.56)$$

where $\mathbf{B} = \mu\mathbf{H}$ was used to obtain the second relation from the first.

The interface conditions for the magnetic field may be summarized as follows:

(1) The tangential component of the magnetic field intensity is discontinuous in the presence of surface current densities. The discontinuity is equal to the surface current density. In the absence of surface current densities, the tangential component of the magnetic field intensity is continuous across the interface.
(2) The tangential component of the magnetic flux density is discontinuous.

The interface condition in **Eq. (11.56)** assumes the magnetic field intensity only has one tangential component or that each tangential component is treated separately. We have faced the same issue in **Section 9.3.1**. To establish a more general relation, we note from **Figure 11.2** that the tangential components of the magnetic field intensity may be written as $\mathbf{H}_{1t} = \hat{\mathbf{n}} \times \mathbf{H}_1$ [A/m] and $\mathbf{H}_{2t} = -\hat{\mathbf{n}} \times \mathbf{H}_2$ [A/m] where $\hat{\mathbf{n}}$ points into material (1) [see **Eq. (9.42)**]. With these observations, **Eq. (11.56)** is written as

$$\boxed{\hat{\mathbf{n}} \times (\mathbf{H}_1 - \mathbf{H}_2) = \mathbf{J}_s \qquad [\text{A/m}]} \quad \text{or} \quad \boxed{\hat{\mathbf{n}} \times \left(\frac{\mathbf{B}_1}{\mu_1} - \frac{\mathbf{B}_2}{\mu_2}\right) = \mathbf{J}_s \qquad [\text{A/m}]} \qquad (11.57)$$

As was discussed in **Section 9.3.1**, this form guarantees the correct magnitude and direction of the fields without resorting to the right-hand rule. **Equation (11.57)** should be used in all instances, although, when the magnetic field intensity has only one tangential component, **Eq. (11.56)** is equally suitable. These interface conditions are identical to the conditions we obtained for the magnetostatic field in **Chapter 9**. The conditions obtained for general materials are summarized in **Table 11.3**. There are a total of eight interface conditions, although only the four relations in the first and third row of **Table 11.3** were obtained from Maxwell's equations directly. The other four were obtained from the constitutive relations. But, do we need all these relations or, more importantly, are all these relations independent relations?

In **Section 11.2.2**, we mentioned that Gauss's law can be derived from Ampère's law and the equation of continuity and the zero divergence condition for the magnetic flux density can be derived from Faraday's law and the equation of continuity. Therefore, the last two interface conditions (those derived from the divergence equations) are not independent conditions. This means that to specify the continuity of the tangential electric field intensity and the continuity of the normal magnetic flux density is equivalent: The two can be derived from the same equations. Clearly, there is no need to specify both, and if we do, this may lead to overspecification. Similarly, the conditions for the tangential component of the magnetic field intensity and the normal component of the electric flux density are equivalent and only one should be specified.

Important Note: The electric and magnetic fields are mutually dependent on each other only in time-dependent cases. This also applies to interface conditions. The static electric and magnetic fields are independent of each other and we are therefore free to specify any and all boundary conditions.

Table 11.3 Electromagnetic interface conditions for general materials

	Electric field	Magnetic field	
Tangential components	$E_{1t} = E_{2t}$	$\hat{\mathbf{n}} \times (\mathbf{H}_1 - \mathbf{H}_2) = \mathbf{J}_s$ or $H_{1t} - H_{2t} = J_s^*$	[A/m] [A/m]
	$\dfrac{D_{1t}}{\varepsilon_1} = \dfrac{D_{2t}}{\varepsilon_2}$	$\hat{\mathbf{n}} \times \left(\dfrac{\mathbf{B}_1}{\mu_1} - \dfrac{\mathbf{B}_2}{\mu_2}\right) = \mathbf{J}_s$ or $\dfrac{B_{1t}}{\mu_1} - \dfrac{B_{2t}}{\mu_2} = J_s^*$	[A/m] [A/m]
Normal components	$\hat{\mathbf{n}} \cdot (\mathbf{D}_1 - \mathbf{D}_2) = \rho_s$ [C/m^2] or $D_{1n} - D_{2n} = \rho_s$ [C/m^2]	$B_{1n} = B_{2n}$	
	$\hat{\mathbf{n}} \cdot (\varepsilon_1 \mathbf{E}_1 - \varepsilon_2 \mathbf{E}_2) = \rho_s$ [C/m^2] or $\varepsilon_1 E_{1n} - \varepsilon_2 E_{2n} = \rho_s$ [C/m^2]	$\mu_1 H_{1n} = \mu_2 H_{2n}$	

* This form requires the use of the right-hand rule to establish the vector relation between the tangential componenets and the current density

The interface conditions as discussed here are for two general materials. Because of this, we had to include both surface charge densities and surface current densities in the conditions. A surface charge density may exist on the surface of a dielectric due to polarization or due to physical charges being placed or generated on the surface (for example, by friction). Another possible source of charges at an interface is due to flow of current across the interface between lossy dielectrics (see **Section 7.8**). Surface charges may also exist at the interface between perfect conductors and dielectrics. Surface current densities may exist at the surface of conductors and in particular perfect conductors. However, in many practical applications, we do not need to worry about charge or current densities at the interface. In particular, two types of interfaces are unique and often useful:

(1) Interfaces between perfect dielectrics (lossless dielectrics).
(2) Interfaces between a perfect dielectric and a perfect conductor.

In the first of these, there are neither current densities nor charge densities at the interface. The interface conditions therefore reduce to those in **Table 11.4**.

Table 11.4 Summary of interface conditions between two perfect dielectrics

	Electric field		Magnetic field	
Tangential components	$E_{1t} = E_{2t}$	$D_{1t}/\varepsilon_1 = D_{2t}/\varepsilon_2$	$H_{1t} = H_{2t}$	$B_{1t}/\mu_1 = B_{2t}/\mu_2$
Normal components	$D_{1n} = D_{2n}$	$\varepsilon_1 E_{1n} = \varepsilon_2 E_{2n}$	$B_{1n} = B_{2n}$	$\mu_1 H_{1n} = \mu_2 H_{2n}$

The second type of interface discussed here is that between a perfect dielectric and a perfect conductor. In this case, the overriding condition is that of the conductor, that is, that all fields in the perfect conductor must be zero. Assuming material (2) is the perfect conductor, E_{2t}, H_{2t}, D_{2n}, and B_{2n} are zero. The interface conditions are given in **Table 11.5**.

Table 11.5 Summary of interface conditions between a perfect dielectric and a perfect conductor

	Electric field		Magnetic field	
Tangential components	$E_{1t} = E_{2t} = 0$	$D_{1t} = D_{2t} = 0$	$H_{1t} = J_s^*$	$B_{1t} = \mu_1 J_s^*$
Normal components	$D_{1n} = \rho_s$	$E_{1n} = \rho_s/\varepsilon_1$	$B_{1n} = 0$	$H_{1n} = 0$

*The directions of J_s and H_{1t} or B_{1t} are related through the right-hand rule

Example 11.5 Interface Conditions for Electromagnetic Fields A magnetic field intensity exists in material (1) in **Figure 11.3a** as $\mathbf{H}_1(x,y,z,t) = (\hat{\mathbf{x}} + \hat{\mathbf{y}}2 - \hat{\mathbf{z}}3)\cos 377t$ [A/m]. Material (1) has a relative permeability of 100. The interface between material (1) and free space [material (2)] is on the x–y plane, at $z = 0$ and there are no currents on the interface. Calculate the magnetic field intensity **H** and the magnetic flux density **B** in material (2).

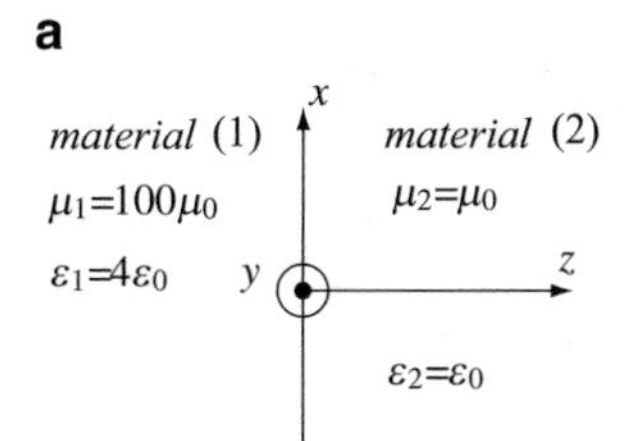

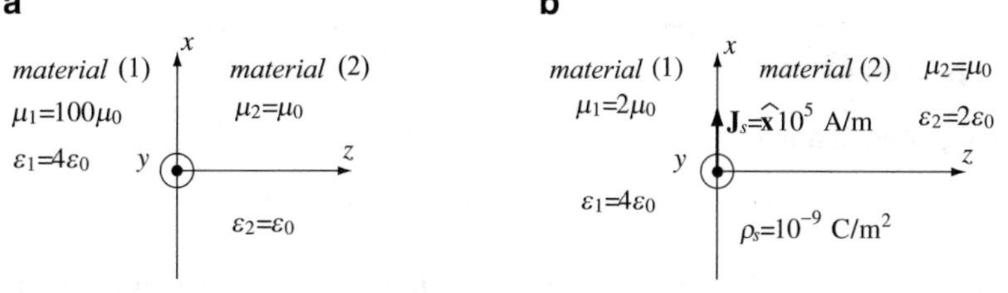

Figure 11.3 (**a**) Interface between two materials. (**b**) Interface between two general materials with a current density on the interface

Solution: The magnetic field intensity has a tangential component and a normal component at the interface. In vector components, these are

$$\mathbf{H}_{1t}(x,y,z,t) = (\hat{\mathbf{x}} + \hat{\mathbf{y}}2)\cos 377t, \quad \mathbf{H}_{1n}(x,y,z,t) = -\hat{\mathbf{z}}3\cos 377t \quad [\text{A/m}]$$

The tangential components of the magnetic field intensity are continuous across the interface (no current density on the interface):

$$\mathbf{H}_{2t} = \mathbf{H}_{1t} = (\hat{\mathbf{x}} + \hat{\mathbf{y}}2)\cos 377t \quad [\text{A/m}]$$

The normal component of the magnetic flux density is continuous across the interface:

$$\mu_1 \mathbf{H}_{1n} = \mu_2 \mathbf{H}_{2n} \rightarrow \mathbf{H}_{2n} = \frac{\mu_1}{\mu_2}\mathbf{H}_{1n} = \frac{\mu_{r1}\mu_0}{\mu_0}\mathbf{H}_{1n} = \mu_{r1}\mathbf{H}_{1n} = -\hat{\mathbf{z}}300\cos 377t \quad [\text{T}]$$

Thus, the magnetic field intensity in material (2) is

$$\mathbf{H}_2 = \mathbf{H}_{2t} + \mathbf{H}_{2n} = (\hat{\mathbf{x}} + \hat{\mathbf{y}}2 - \hat{\mathbf{z}}300)\cos 377t \quad [\text{A/m}]$$

The magnetic flux density in material (2) is

$$\mathbf{B}_2 = \mu_0 \mathbf{H}_2 = (\hat{\mathbf{x}} + \hat{\mathbf{y}}2 - \hat{\mathbf{z}}300)\mu_0 \cos 377t \quad [\text{T}]$$

Exercise 11.2 The configuration in **Example 11.5** is given. The electric field intensity in material (1) is $\mathbf{E}_1(x,y,z,t) = k(\hat{\mathbf{x}} + \hat{\mathbf{y}}2 - \hat{\mathbf{z}}3)\cos 377t$ [V/m], where k is a constant. Calculate the electric field intensity and electric flux density in material (2). Assume there are no charges on the interface.

Answer

$$\mathbf{E}_2(x,y,z,t) = k(\hat{\mathbf{x}} + \hat{\mathbf{y}}2 - \hat{\mathbf{z}}12)\cos 377t \quad [\mathrm{V/m}],$$
$$\mathbf{D}_2(x,y,z,t) = k(\hat{\mathbf{x}} + \hat{\mathbf{y}}2 - \hat{\mathbf{z}}3)\varepsilon_0\cos 377t \quad [\mathrm{C/m^2}]$$

Example 11.6 Interface Conditions for the Static Electric and Magnetic Fields An interface between two general materials contains both a current density given as $\mathbf{J}_s = \hat{\mathbf{x}}10^5$ A/m and a uniform surface charge density given as $\rho_s = 10^{-9}$ C/m^2. The static magnetic field intensity and static electric field intensity in material (1) in **Figure 11.3b** are

$$\mathbf{H}_1 = \hat{\mathbf{x}}10^5 + \hat{\mathbf{y}}10^5 - \hat{\mathbf{z}}10^5 \quad [\mathrm{A/m}], \qquad \mathbf{E}_1 = \hat{\mathbf{x}}100 + \hat{\mathbf{y}}20 - \hat{\mathbf{z}}100 \quad [\mathrm{V/m}]$$

For the material properties given in **Figure 11.3b**, ($\mu_1 = 2\mu_0$, $\mu_2 = \mu_0$ [H/m], $\varepsilon_1 = 4\varepsilon_0$, and $\varepsilon_2 = 2\varepsilon_0$ [F/m]), find:

(a) The electric field intensity in material (2).
(b) The magnetic flux density in material (2).

Note: Static electric and magnetic fields are independent of each other.

Solution: Since both current densities and charge densities exist on the interface, we must use the general interface conditions in **Table 11.2**:

(a) The tangential and normal vector components of **E** in material (1) are

$$\mathbf{E}_{1t} = \hat{\mathbf{x}}100 + \hat{\mathbf{y}}20, \quad \mathbf{E}_{1n} = -\hat{\mathbf{z}}100 \quad [\mathrm{V/m}]$$

The tangential component of the electric field intensity is continuous across the interface:

$$\mathbf{E}_{2t} = \mathbf{E}_{1t} = \hat{\mathbf{x}}100 + \hat{\mathbf{y}}20 \quad [\mathrm{V/m}]$$

The normal component of the electric field intensity is discontinuous across the interface:

$$\varepsilon_1 E_{1n} - \varepsilon_2 E_{2n} = \rho_s \quad \rightarrow \quad E_{2n} = \frac{\varepsilon_1 E_{1n} - \rho_s}{\varepsilon_2} \quad [\mathrm{V/m}]$$

where we assume E_{1n} points away from the interface and E_{2n} points toward the interface. This gives

$$E_{2n} = \frac{4\varepsilon_0(-100) - 10^{-9}}{2\varepsilon_0} = -200 - \frac{10^{-9}}{2 \times 8.854 \times 10^{-12}} = -256.47 \quad [\mathrm{V/m}]$$

Thus, the electric field intensity in material (2) is

$$\mathbf{E}_2 = \mathbf{E}_{2t} + \mathbf{E}_{2n} = \hat{\mathbf{x}}100 + \hat{\mathbf{y}}20 - \hat{\mathbf{z}}256.47 \quad [\mathrm{V/m}]$$

(b) First, we write the magnetic flux density ($\mathbf{B} = \mu\mathbf{H}$)

$$\mathbf{B}_1 = \hat{\mathbf{x}}2\mu_0 \times 10^5 + \hat{\mathbf{y}}2\mu_0 \times 10^5 - \hat{\mathbf{z}}2\mu_0 \times 10^5 \quad [\mathrm{T}]$$

For convenience we separate the magnetic flux density into its tangential and normal components as follows:

$$\mathbf{B}_{1t} = \mu_1\mathbf{H}_{1t} = \hat{\mathbf{x}}2\mu_0 \times 10^5 + \hat{\mathbf{y}}2\mu_0 \times 10^5, \quad \mathbf{B}_{1n} = \mu_1\mathbf{H}_{1n} = -\hat{\mathbf{z}}2\mu_0 \times 10^5 \quad [\mathrm{T}]$$

The tangential component of the magnetic flux density is discontinuous across the interface (**Table 11.3**):

$$\hat{\mathbf{n}} \times \left(\frac{\mathbf{B}_1}{\mu_1} - \frac{\mathbf{B}_2}{\mu_2} \right) = \mathbf{J}_s \quad \rightarrow \quad \hat{\mathbf{n}} \times \mathbf{B}_2 = \mu_2 \left(\frac{\hat{\mathbf{n}} \times \mathbf{B}_1}{\mu_1} - \mathbf{J}_s \right) \quad [\mathrm{T}]$$

Since the normal must point into medium (1), we write $\hat{\mathbf{n}} = -\hat{\mathbf{z}}$ and get

$$-\hat{\mathbf{z}} \times \mathbf{B}_2 = \mu_2 \left[\frac{-\hat{\mathbf{z}} \times \mathbf{B}_1}{\mu_1} - \mathbf{J}_s \right] \quad [\mathrm{T}]$$

or

$$\mathbf{B}_{2t} = \mu_2 \left(\frac{\hat{\mathbf{z}} \times \mathbf{B}_1}{\mu_1} + \mathbf{J}_s \right) \quad [\mathrm{T}]$$

Substituting for $\mathbf{B}_1$ and $\mathbf{J}_s$,

$$\mathbf{B}_{2t} = \mu_0 \left(\frac{\hat{\mathbf{z}} \times (\hat{\mathbf{x}} 2\mu_0 \times 10^5 + \hat{\mathbf{y}} 2\mu_0 \times 10^5 - \hat{\mathbf{z}} 2\mu_0 \times 10^5)}{2\mu_0} + \hat{\mathbf{x}} \times 10^5 \right) = \hat{\mathbf{y}} 10^5 \mu_0 \quad [\mathrm{T}]$$

The normal component of $\mathbf{B}$ is continuous across the interface:

$$\mathbf{B}_{2n} = \mathbf{B}_{1n} = -\hat{\mathbf{z}} 2 \times 10^5 \mu_0 \quad [\mathrm{T}]$$

Thus, the magnetic flux density in material (2) is

$$\mathbf{B}_2 = \hat{\mathbf{y}} 10^5 \mu_0 - \hat{\mathbf{z}} 2 \times 10^5 \mu_0 \quad [\mathrm{T}]$$

11.5 Particular Forms of Maxwell's Equations

Maxwell's equations as given in **Section 11.2.2** are general and apply to all electromagnetic situations and for any type of time dependency. In this sense, whenever there is a need to solve an electromagnetic problem, we can start with **Eqs. (11.24)** through **(11.27)** or, if integral representation is more convenient, with **Eqs. (11.28)** through **(11.31)**. However, more often than not, there is no need to resort to the general system. For example, we might need to solve the equations at low frequencies, in which case the displacement currents might be negligible or do not exist. In still other situations the current densities or charge densities in the system, or both, are negligible. Some of these representations are particularly useful, and, therefore, we discuss these here, before we apply them to particular electromagnetic problems in the following chapters. In particular, the time-harmonic representation of the equations is often useful.

11.5.1 Time-Harmonic Representation

In a time-harmonic field, the time dependency is sinusoidal. This is a form often encountered in engineering and, as is well known from circuit theory, offers distinct advantages in analysis. As with circuits, the time-harmonic form can be used for almost any waveform through the use of Fourier series. This approach implies linearity in relations. The same is true in electromagnetics. Much of the remaining material in this book will be based on the time-harmonic representation of the electromagnetic field equations. For this reason, we present now Maxwell's equations in time-harmonic form.

There are two basic differences between time-dependent and time-harmonic forms:

(1) Field variables as well as sources are phasors.
(2) The time derivative operator d/dt is replaced by $j\omega$.

Before discussing these any further, it is useful to review the concept of phasors, particularly because we will use phasors in conjunction with vector variables.

The phasor notation is a method of representing complex numbers. Consider a complex number $b = u_0 + jv_0$, where $j = \sqrt{-1}$. The complex number b can be represented in a plane, called a ***complex plane***, as in **Figure 11.4a**. The real part of b is u_0, and it is the projection of b on the real axis, whereas the imaginary part of a is v_o and represents its projection on the imaginary axis.

Instead of writing b in the above form, we can also write b in terms of a magnitude and an angle. The phasor notation is based on the latter form and arises from Euler's equation:

$$be^{j\varphi} = b\cos(\varphi) + jb\sin(\varphi) \tag{11.58}$$

That is, if the radius of a circle of magnitude b makes an angle φ with the real axis, its projections on the real and imaginary axes are $b\cos(\varphi)$ and $b\sin(\varphi)$, respectively. These concepts are shown in **Figure 11.4b**. The phase angle φ can be general and we will assume that it has the form $\varphi = \omega t + \theta$, where ω is an angular frequency, t is time, and θ is some fixed phase angle. Substituting this in **Eq. (11.58)** gives

$$be^{j\varphi} = be^{j(\omega t+\theta)} = be^{j\omega t}e^{j\theta} = b\cos(\omega t + \theta) + jb\sin(\omega t + \theta) \tag{11.59}$$

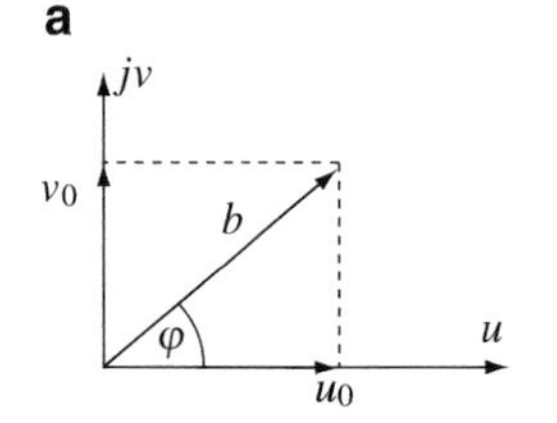

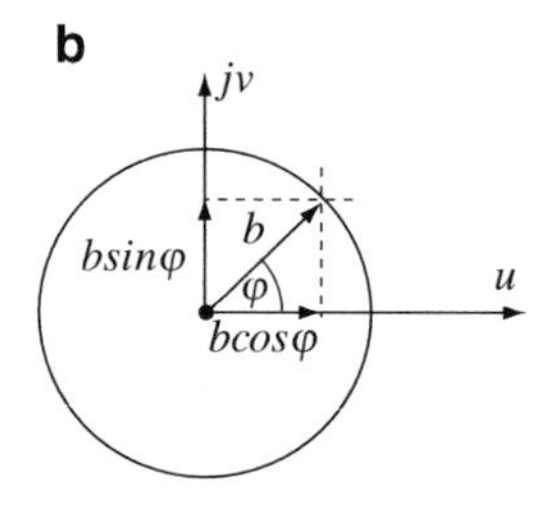

Figure 11.4 (a) Representation of a complex number b. (**b**) Harmonic representation of a general complex number of magnitude b

A sinusoidal function of the type often used in fields is

$$f_1(x,y,z,t) = A_0(x,y,z)\cos(\omega t + \theta) \quad \text{or} \quad f_2(x,y,z,t) = A_0(x,y,z)\sin(\omega t + \theta) \tag{11.60}$$

The phasor notation now allows us to write

$$A_0(x,y,z)\cos(\omega t + \theta) = \mathrm{Re}\{A_0(x,y,z)e^{j\omega t}e^{j\theta}\} \tag{11.61}$$

$$A_0(x,y,z)\sin(\omega t + \theta) = \mathrm{Im}\{A_0(x,y,z)e^{j\omega t}e^{j\theta}\} \tag{11.62}$$

where A_0 is real and independent of time, Re{ } means the real part of the function, and Im{ } means the imaginary part of the function. Finally, we define the ***phasor*** as that part of the function which does not contain time; that is,

$$A_p(x,y,z) = A_0(x,y,z)e^{j\theta} \tag{11.63}$$

This is sometimes written as an amplitude and phase as

$$A_p(x,y,z) = A_0(x,y,z)\angle\theta \tag{11.64}$$

Summarizing, the phasor can be written in three different forms:

$$A_p(x,y,z) = A_0(x,y,z)e^{j\theta} = A_0(x,y,z)\angle\theta = A_0(x,y,z)\cos\theta + jA_0(x,y,z)\sin\theta \tag{11.65}$$

The first form is called the exponential form, the second is the polar form, and the third is the rectangular form. Most of our work in this book will be carried out in the exponential form. On occasion, we will use the polar form, particularly for presentation of results because it is a more compact method. The rectangular form is also convenient in some cases because of its explicit representation in complex variables. Using the exponential form of the phasor, the time domain form, $A(x,y,z,t)$, can be written as

$$A(x,y,z,t) = \mathrm{Re}\{A_0(x,y,z)e^{j\theta}e^{j\omega t}\} \qquad (11.66)$$

Now, the reason for the use of phasors is apparent: It allows representation of fields in terms of a magnitude (A_0) and a phase angle (θ) without explicitly considering time, since the phasor does not contain the term $e^{j\omega t}$. When we need to convert phasors to time, the term $e^{j\omega t}$ is included as in **Eq. (11.66)**.

In the above discussion, we assumed A to be a scalar function. However, the definition of the phasor, because it has to do with any complex number, applies equally well to vectors. All we need to do is replace the scalar A by a vector $\mathbf{A}$. Similarly, the amplitude A_0 now becomes a vector $\mathbf{A}_0$. The phasor form of $\mathbf{A}$ is $\mathbf{A}_p$. Thus, given a time-dependent vector $\mathbf{A}(x,y,z,t)$, the phasor form of $\mathbf{A}(x,y,z,t)$ is $\mathbf{A}_p(x,y,z)$, and given the phasor form $\mathbf{A}_p$, the time-dependent vector is $\mathbf{A}(x,y,z,t) = \mathrm{Re}\{\mathbf{A}_p(x,y,z)e^{j\omega t}\}$.

One of the most distinct advantages in working with phasors is the ease with which time derivatives are performed. The time derivative of a general vector $\mathbf{A}(x,y,z,t)$ is

$$\frac{d}{dt}(\mathbf{A}(x,y,z,t)) = \mathrm{Re}\{j\omega\mathbf{A}_p(x,y,z)e^{j\omega t}\} \qquad (11.67)$$

In the practical use of phasors, we do not keep the term $e^{j\omega t}$, but it is understood to exist. Neither do we denote the phasor in any other way. In this section, the phasor was denoted with a subscript p. In later use we will drop this notation because it will normally be understood from the context if we are using phasors or not.

Example 11.7 Phasors A time-dependent electric field intensity is given as $\mathbf{E} = \hat{\mathbf{x}}(10\pi + j20\pi)\cos(10^6t - 120y)$ [V/m]. Write the electric field intensity as a phasor using the following:

(a) The rectangular notation.
(b) The polar representation.
(c) The exponential representation.

Solution:

(a) First, we must write the electric field intensity as follows:

$$\begin{aligned}\mathbf{E} &= \hat{\mathbf{x}}(10\pi\cos(10^6t - 120y) + j20\pi\cos(10^6t - 120y)) \\ &= \hat{\mathbf{x}}(10\pi\cos(10^6t - 120y) + 20\pi\cos(10^6t - 120y + \pi/2)) \quad [\mathrm{V/m}]\end{aligned}$$

Each term can be written in rectangular form noting its amplitude and phase. Comparison with **Eq. (11.59)** shows that $\omega t = 10^6t$, $\theta_1 = -120y$, and $\theta_2 = -120y + \pi/2$. Removing the term $e^{j\omega t}$, the phasor form becomes [see **Eq. (11.65)**]

$$\begin{aligned}\mathbf{E} = \hat{\mathbf{x}}(&10\pi\cos(-120y) + j10\pi\sin(-120y) + 20\pi\cos(-120y + \pi/2) \\ &+ j20\pi\sin(-120y + \pi/2)) \quad [\mathrm{V/m}]\end{aligned}$$

Or, writing $\cos(-120y) = \cos(120y)$, $\sin(-120y) = -\sin(120y)$, $\cos(-120y + \pi/2) = \sin(120y)$, and $\sin(-120y + \pi/2) = \cos(120y)$, we can simplify the expression

$$\begin{aligned}\mathbf{E} &= \hat{\mathbf{x}}[10\pi\cos 120y - j10\pi\sin 120y + 20\pi\sin 120y + j20\pi\cos 120y] \\ &= \hat{\mathbf{x}}[(10\pi + 20\pi)\cos 120y - j(10\pi - 20\pi)\sin 120y] \quad [\mathrm{V/m}].\end{aligned}$$

(b) In polar representation, we get

$$\mathbf{E} = \hat{\mathbf{x}}10\pi\angle{-120y} + \hat{\mathbf{x}}20\pi\angle{-120y + \pi/2} \quad [\mathrm{V/m}].$$

(c) In exponential form,

$$\mathbf{E} = \hat{\mathbf{x}}(10\pi e^{-j120y} + 20\pi e^{-j120y}e^{j\pi/2}) \quad [\text{V/m}].$$

Example 11.8 Phasors The magnetic field intensity in a material is given as a phasor:

$$\mathbf{H} = [\hat{\mathbf{x}}(100 + j50) + \hat{\mathbf{y}}50 + \hat{\mathbf{z}}100]e^{j60}e^{-j3x} \quad [\text{A/m}]$$

(a) Write the magnetic field intensity in rectangular form.
(b) What is the time-dependent magnetic field intensity **H** in the material?

Solution:

(a) First, we write the magnetic field intensity as

$$\mathbf{H} = \hat{\mathbf{x}}(100e^{j60}e^{-j3x} + 50e^{j60}e^{-j3x}e^{j\pi/2}) + \hat{\mathbf{y}}50e^{j60}e^{-j3x} + \hat{\mathbf{z}}100e^{j60}e^{-j3x} \quad [\text{A/m}]$$

The magnetic field intensity in rectangular form is

$$\begin{aligned}\mathbf{H} = {} & \hat{\mathbf{x}}[100\cos(60 - 3x) + j100\sin(60 - 3x) + 50\cos(60 - 3x + \pi/2) + j50\sin(60 - 3x + \pi/2)] \\ & + \hat{\mathbf{y}}\left[50\cos(60 - 3x) + j50\sin(60 - 3x)\right] + \hat{\mathbf{z}}\left[100\cos(60 - 3x) + j100\sin(60 - 3x)\right] \quad \left[\text{A/m}\right].\end{aligned}$$

(b) The time-dependent field is written as

$$\begin{aligned}\mathbf{H}(t) = \text{Re}\left\{\mathbf{H}e^{j\omega t}\right\} = {} & \hat{\mathbf{x}}\left[100\cos(\omega t + 60 - 3x) + 50\cos(\omega t + 60 - 3x + \pi/2)\right] \\ & + \hat{\mathbf{y}}50\cos(\omega t + 60 - 3x) + \hat{\mathbf{z}}100\cos(\omega t + 60 - 3x) \qquad \left[\text{A/m}\right].\end{aligned}$$

11.5.2 Maxwell's Equations: The Time-Harmonic Form

With the notation given above, we can now write Maxwell's equations in terms of phasors. Implicit in this development is linearity of material properties. Assuming that all vector and scalar quantities are phasors, we simply replace d/dt or $\partial/\partial t$ by $j\omega$ in **Eqs. (11.24)**, **(11.25)**, **(11.28)** and **(11.29)**. The time-harmonic differential and integral forms of Maxwell's equations together with the constitutive relations and the Lorentz force are summarized in **Table 11.6**.

Table 11.6 Summary of the time-harmonic electromagnetic field equations

	Differential form		Integral form	
Maxwell's equations	$\nabla \times \mathbf{E} = -j\omega\mathbf{B}$	(11.68)	$\oint_C \mathbf{E} \cdot d\mathbf{l} = -j\omega \int_s \mathbf{B} \cdot d\mathbf{s}$ [V]	(11.72)
	$\nabla \times \mathbf{H} = \mathbf{J} + j\omega\mathbf{D}$	(11.69)	$\oint_C \mathbf{H} \cdot d\mathbf{l} = \int_s (\mathbf{J} + j\omega\mathbf{D}) \cdot d\mathbf{s}$ [A]	(11.73)
	$\nabla \cdot \mathbf{D} = \rho_v$	(11.70)	$\oint_s \mathbf{D} \cdot d\mathbf{s} = Q$ [C]	(11.74)
	$\nabla \cdot \mathbf{B} = 0$	(11.71)	$\oint_s \mathbf{B} \cdot d\mathbf{s} = 0$	(11.75)
Constitutive relations			$\mathbf{B} = \mu\mathbf{H}$ [T]	(11.76)
			$\mathbf{D} = \varepsilon\mathbf{E}$ [C/m^2]	(11.77)
The Lorentz force equation			$\mathbf{F} = q(\mathbf{E} + \mathbf{v} \times \mathbf{B})$ [N]	(11.78)

Note that the constitutive relations and the Lorentz force equations have not changed although all vector quantities are now phasors. Of course, velocity is still a real number. ε and μ remain unaffected by the phasor notation. The charge Q or the charge density ρ_v may, in some cases, be time dependent, in which case they also become phasors.

Another important point to be noted here is that if displacement currents in **Eq. (11.69)** or **(11.73)** are neglected, the pre-Maxwell system of equations is obtained, but the fields are now time-harmonic fields. This system of equations, which is characterized by slow varying fields (and hence the neglection of displacement currents), is called the ***quasi-static field equations***. The term quasi-static means that the equations are static-like in the sense that the equations satisfy Laplace's or Poisson's equations. One of the advantages of this form is that it extends many of the properties as well as the methods used for static fields to low frequency time-dependent fields.

Finally, if we set all time derivatives to zero, the purely static equations as encountered in electrostatics and magnetostatics are obtained. As was said previously, under this condition, the electric field in **Eqs. (11.68)** and **(11.70)** and the magnetic field in **Eqs. (11.69)** and **(11.71)** (or their integral counterparts) are decoupled, and there is no need to discuss them as a system of equations.

11.5.3 Source-Free Equations

The general forms of Maxwell's equations can sometimes be simplified if the sources do not need to be taken into account. Under these conditions, the current density **J**, the charge density ρ_v, or both are removed from the equations and a much simpler form of the equations is obtained. This is true in the time-dependent or phasor forms of the equations. The time-dependent and time-harmonic source-free Maxwell's equations are summarized in **Tables 11.7** and **11.8**.

Table 11.7 The source-free time-dependent Maxwell's equations

	Differential	Integral
Faraday's law	$\nabla \times \mathbf{E} = -\dfrac{\partial \mathbf{B}}{\partial t}$	$\oint_C \mathbf{E} \cdot d\mathbf{l} = -\dfrac{d\Phi}{dt}$ [V]
Ampere's law	$\nabla \times \mathbf{H} = \dfrac{\partial \mathbf{D}}{\partial t} \quad \left[\dfrac{\mathrm{A}}{\mathrm{m}^2}\right]$	$\oint_C \mathbf{H} \cdot d\mathbf{l} = \int_s \dfrac{\partial \mathbf{D}}{\partial t} \cdot d\mathbf{s}$ [A]
Gauss's law	$\nabla \cdot \mathbf{D} = 0$	$\oint_s \mathbf{D} \cdot d\mathbf{s} = 0$
No monopoles	$\nabla \cdot \mathbf{B} = 0$	$\oint_s \mathbf{B} \cdot d\mathbf{s} = 0$

Table 11.8 The source-free time-harmonic Maxwell's equations

Faraday's law	$\nabla \times \mathbf{E} = -j\omega\mathbf{B}$	$\oint_C \mathbf{E} \cdot d\mathbf{l} = -j\omega \oint_s \mathbf{B} \cdot d\mathbf{s}$ [V]
Ampere's law	$\nabla \times \mathbf{H} = j\omega\mathbf{D}$ [A/m^2]	$\oint_C \mathbf{H} \cdot d\mathbf{l} = \int_s j\omega\mathbf{D} \cdot d\mathbf{s}$ [A]
Gauss's law	$\nabla \cdot \mathbf{D} = 0$	$\oint_s \mathbf{D} \cdot d\mathbf{s} = 0$
No monopoles	$\nabla \cdot \mathbf{B} = 0$	$\oint_s \mathbf{B} \cdot d\mathbf{s} = 0$

That these equations are simpler than those given in **Eqs. (11.24)** through **(11.27)** or **(11.28)** through **(11.31)** is obvious. For example, the divergence of **D** (or **E**) is zero, which makes the electric field solenoidal. The fact that we do not need to treat sources makes the solution of field problems much simpler, provided that the conditions under which these equations apply are satisfied.

We will not expand on this here except to point out that treatment of fields under source-free conditions is quite common. You may wish to think about it in this fashion: If you are interested in evaluating the field distribution in a volume such as a room due to say, terrestrial magnetism, or the transmission from a distant TV station, there is little choice but to solve the problem in the absence of sources. In both of these cases, we have no knowledge of the sources strengths or their location. The fields, however, are real. We can measure them at various locations, we can find their distribution in space, and we can calculate a number of other properties related to the fields.

11.6 Summary

The main topic in this chapter is the introduction of displacement current density in Ampère's law and its consequences. The final result is Maxwell's equations, which include the postulates in the previous chapters but also the modification due to displacement currents. The displacement current density modifies Ampère's law by adding the term $\mathbf{J}_d = \partial\mathbf{D}/\partial t$ [A/m^2] as follows:

$$\nabla \times \mathbf{H} = \mathbf{J} + \frac{\partial \mathbf{D}}{\partial t} \quad \left[\frac{\text{A}}{\text{m}^2}\right] \tag{11.6}$$

This, together with Faraday's and Gauss's laws, forms what are call ***Maxwell's equations*** given below in differential form (left) and integral form (right):

$$\nabla \times \mathbf{E} = -\frac{\partial \mathbf{B}}{\partial t} \quad (11.24) \quad \text{or :} \quad \oint_C \mathbf{E} \cdot d\mathbf{l} = -\frac{\partial \Phi}{\partial t} \quad [\text{V}] \tag{11.28}$$

$$\nabla \times \mathbf{H} = \mathbf{J} + \frac{\partial \mathbf{D}}{\partial t} \quad \left[\frac{\text{A}}{\text{m}^2}\right] \quad (11.25) \quad \text{or :} \quad \oint_C \mathbf{H} \cdot d\mathbf{l} = \int_s \left(\mathbf{J} + \frac{\partial \mathbf{D}}{\partial t}\right) \cdot d\mathbf{s} \quad [\text{A}] \tag{11.29}$$

$$\nabla \cdot \mathbf{D} = \rho_v \quad \left[\text{C/m}^3\right] \quad (11.26) \quad \text{or :} \quad \oint_s \mathbf{D} \cdot d\mathbf{s} = Q \quad [\text{C}] \tag{11.30}$$

$$\nabla \cdot \mathbf{B} = 0 \quad (11.27) \quad \text{or :} \quad \oint_s \mathbf{B} \cdot d\mathbf{s} = 0 \tag{11.31}$$

The material constitutive relations $\mathbf{D} = \varepsilon\mathbf{E}$ and $\mathbf{B} = \mu\mathbf{H}$ and the Lorentz force $\mathbf{F} = q(\mathbf{E} + \mathbf{v} \times \mathbf{B})$ are part of the general system of equations called the Maxwell–Lorentz equations (see **Table 11.1**). The third constitutive relation, $\mathbf{J} = \sigma\mathbf{E}$, applies in conducting media.

Time-dependent potentials are defined based on the properties of the curl and divergence of fields:

$$\mathbf{E} = -\nabla V, \quad \text{if} \quad \nabla \times \mathbf{E} = 0 \tag{11.38}$$

V is the electric scalar potential (voltage).

$$\mathbf{H} = -\nabla \psi, \quad \text{if} \quad \nabla \times \mathbf{H} = 0 \tag{11.39}$$

ψ is the magnetic scalar potential.

$$\mathbf{B} = \nabla \times \mathbf{A}, \quad \text{because} \quad \nabla \cdot \mathbf{B} = 0 \tag{11.40}$$

A is the magnetic vector potential.

The time-dependent electric field intensity, based on Ampère's law **[Eq. (11.24)]**, is

$$\mathbf{E} = -\frac{\partial \mathbf{A}}{\partial t} - \nabla V \quad \left[\frac{\text{V}}{\text{m}}\right] \tag{11.45}$$

Gauges define the divergence of vector potentials (in this case the magnetic vector potential).
$\nabla \cdot \mathbf{A} = 0$ for static fields (Coulomb's gauge) and

$$\nabla \cdot \mathbf{A} = -\mu\varepsilon \frac{\partial V}{\partial t} \tag{11.49}$$

for time-dependent fields (Lorenz's gauge).

Interface conditions for time-dependent fields are identical to those for static fields as discussed in **Chapters 4** and **9**. These are summarized in **Tables 11.2** through **11.5** (see **Figure 11.2** for reference):

$$E_{1t} = E_{2t}, \quad \frac{D_{1t}}{\varepsilon_1} = \frac{D_{2t}}{\varepsilon_2} \quad \text{and} \quad D_{1n} - D_{2n} = \rho_s, \quad \varepsilon_1 E_{1n} - \varepsilon_2 E_{2n} = \rho_s$$

$$\hat{\mathbf{n}} \times (\mathbf{H}_1 - \mathbf{H}_2) = \mathbf{J}_s, \quad \hat{\mathbf{n}} \times \left(\frac{\mathbf{B}_1}{\mu_1} - \frac{\mathbf{B}_2}{\mu_2}\right) = \mathbf{J}_s \quad \text{and} \quad B_{1n} = B_{2n}, \quad \mu_1 H_{1n} = \mu_2 H_{2n}$$

Electromagnetic fields are often represented in terms of phasors. ***Phasor representation*** of any function A (scalar or vector) is as follows:

$$A_p(x,y,z) = A_0(x,y,z)e^{j\theta} = A_0(x,y,z)\angle\theta = A_0(x,y,z)\cos\theta + jA_0(x,y,z)\sin\theta \tag{11.65}$$

Transformation into the time domain is as follows:

$$A(x,y,z,t) = \text{Re}\left\{A_0(x,y,z)e^{j\theta}e^{j\omega t}\right\} \tag{11.66}$$

$$\frac{d}{dt}(\mathbf{A}(x,y,z,t)) = \text{Re}\left\{j\omega\mathbf{A}_p(x,y,z)e^{j\omega t}\right\} \tag{11.67}$$

Time-harmonic field equations play an important role in electromagnetics. Maxwell's equations in the frequency domain (see **Table 11.6**) are:

$$\nabla \times \mathbf{E} = -j\omega\mathbf{B} \quad (11.68) \quad \text{or:} \quad \oint_C \mathbf{E} \cdot d\mathbf{l} = -j\omega \int_s \mathbf{B} \cdot d\mathbf{s} \quad [\text{V}] \tag{11.72}$$

$$\nabla \times \mathbf{H} = \mathbf{J} + j\omega\mathbf{D} \quad \left[\text{A/m}^2\right] \quad (11.69) \quad \text{or:} \quad \oint_C \mathbf{H} \cdot d\mathbf{l} = \int_s (\mathbf{J} + j\omega\mathbf{D}) \cdot d\mathbf{s} \quad [\text{A}] \tag{11.73}$$

$$\nabla \cdot \mathbf{D} = \rho_v \quad \left[\text{C/m}^3\right] \quad (11.70) \quad \text{or:} \quad \oint_s \mathbf{D} \cdot d\mathbf{s} = Q \quad [\text{C}] \tag{11.74}$$

$$\nabla \cdot \mathbf{B} = 0 \quad (11.71) \quad \text{or:} \quad \oint_s \mathbf{B} \cdot d\mathbf{s} = 0 \tag{11.75}$$

where **E, H, D, B**, and **J** are vector phasors and Q and ρ_v are scalar phasors. Note however that we do not mark these in any particular way—it is understood from the context when these quantities must be phasors.

Source-free equations are obtained by setting $\mathbf{J} = 0$, $\rho_v = 0$ in either the time or frequency domain equations. These are summarized in **Tables 11.7** and **11.8**.

Problems

Maxwell's Equations, Displacement Current, and Continuity

11.1 Displacement Current Density. A magnetic flux density, $\mathbf{B} = \hat{\mathbf{y}}(0.1\cos 5z)\cos 100t$ [T] exists in a linear, isotropic, homogeneous material characterized by ε and μ. Find the displacement current density in the material if there are no source charges or current densities in the material.

11.2 Displacement Current in Spherical Capacitor. Determine the displacement current I_d [A] which flows between two concentric, conducting spherical shells of radii a and b [m] where $b > a$ in free space with a voltage difference $V_0\sin\omega t$ [V] applied between the spheres.

11.3 Application: Displacement Current in Cylindrical Capacitor. A voltage source $V_0\sin\omega t$ [V] is connected between two concentric conductive cylinders of radii $r = a$ and $r = b$ [m], where $b > a$, with length L [m]. $\varepsilon = \varepsilon_r\varepsilon_0$ [F/m], $\mu = \mu_0$ [H/m], and $\sigma = 0$ for $a < r < b$. Neglect any end effects and find:

(a) The displacement current density at any point $a < r < b$.
(b) The total displacement current I_d flowing between the two cylinders.

11.4 Conservation of Charge and Displacement Current. Show that the displacement current density in Maxwell's second equation (Ampère's law) is a direct consequence of the law of conservation of charge.

11.5 Application: Displacement and Conduction Current Densities in Lossy Capacitor. A lossy dielectric is located between two parallel plates which are connected to an AC source (**Figure 11.5**). Material properties of the dielectric are $\varepsilon = 9\varepsilon_0$ [F/m], $\mu = \mu_0$ [H/m], and $\sigma = 4$ S/m. The source is given as $V = 1\cos\omega t$ [V]. Calculate the frequency at which the magnitude of the displacement current density is equal to the magnitude of the conduction current density. Assume all material properties are independent of frequency.

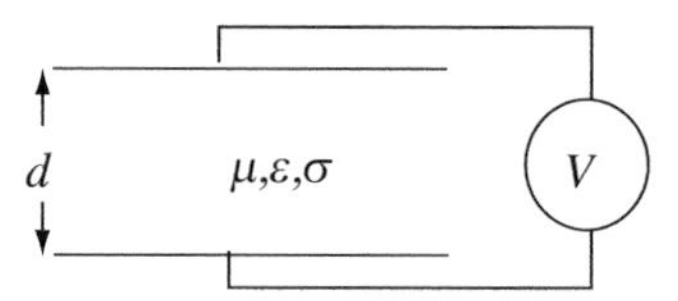

Figure 11.5

11.6 Application: Displacement and Conduction Current Densities. A capacitor is made of two parallel plates with a dielectric between them. The area of each plate is A [m^2] and the dielectric is d [m] thick. The dielectric is lossy with a permittivity ε [F/m] and conductivity σ [S/m]. If the capacitor is connected to a sinusoidal AC source of amplitude V [V] and angular frequency ω [rad/s]:

(a) Find the ratio between the amplitudes of the conduction current density and displacement current density.
(b) Show that the conduction and displacement current densities are 90° out of phase.

11.7 Application: Lossy Capacitor. The capacitor in **Problem 11.6** is connected to a 12 V DC source and charged for a long period of time. Now the source is disconnected. Find the time constant of discharge of the capacitor.

11.8. Application: Displacement Current. An AC generator operating at a frequency of 1 GHz is connected with a wire to a small conducting sphere of radius $a = 10$ mm at some distance away (see **Figure 11.6**). If the sphere is in free space, calculate the current in the wire. Neglect any effect the ground may have. The generator generates a sinusoidal voltage of amplitude 100 V.

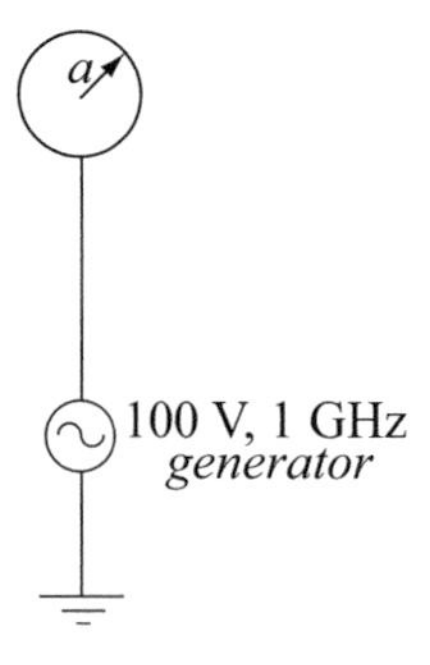

Figure 11.6

11.9 Application: Displacement current. Two conducting spheres, each of radius $a = 0.2$ m are connected to an AC source of amplitude 120 V and frequency of 100 kHz as shown in **Figure 11.7**, separated a distance $d = 10$ m apart. Assume the voltage source is sinusoidal ($V(t) = 120\sin 2\pi ft$) and free space everywhere. Calculate the current supplied by the source. **Hint:** think about the spheres as forming a capacitor and assume charge on the spheres at any instant in time is uniformly distributed on their surfaces.

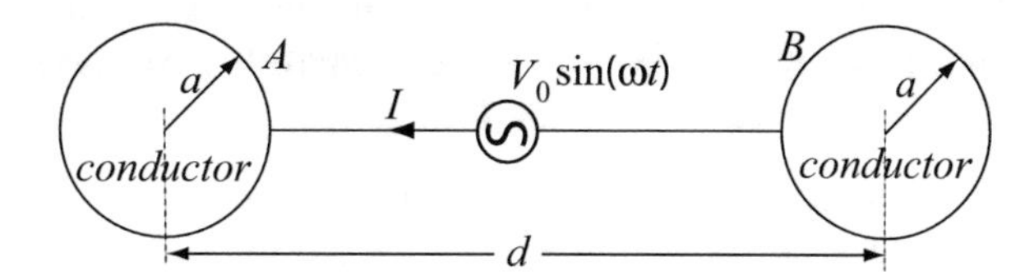

Figure 11.7

Maxwell's Equations

11.10 Maxwell's Equations. An electric field intensity and a magnetic field intensity are given as follows:

$$\mathbf{E} = \hat{\mathbf{y}} 120\pi\cos\left(10^6\pi t + \beta z\right) \quad [\mathrm{V/m}]$$
$$\mathbf{H} = \hat{\mathbf{x}} H_0\cos\left(10^6\pi t + \beta z\right) \quad [\mathrm{A/m}]$$

(a) What values of H_0 and β are required for the two fields to satisfy Maxwell's equations in free space?

(b) What values of H_0 and β are required for the two fields to satisfy Maxwell's equations in a perfect dielectric with relative permittivity 4 and relative permeability 1? Assume there are no charge densities in the dielectric.

11.11 Dependency in Maxwell's Equations. Show that **Eq. (11.8)** ($\nabla \cdot \mathbf{B} = 0$) can be derived from **Eq. (11.5)** and, therefore, is not an independent equation.

11.12 Dependency in Maxwell's Equations. Show that **Eq. (11.7)** ($\nabla \cdot \mathbf{D} = \rho_v$) can be derived from **Eq. (11.6)** with the use of the continuity equation [**Eq. (11.13)**] and, therefore, is not an independent equation.

11.13 The Lorenz Condition (Gauge). Show that the Lorenz condition in **Eq. (11.49)** leads to the continuity equation. **Hint**: Use the expression for electric potential due to a general volume charge distribution and the expression for the magnetic vector potential due to a general current density in a volume.

11.14 Application: Displacement Current in a Conductor. A copper wire with conductivity $\sigma = 5.7 \times 10^7$ S/m carries a sinusoidal current of amplitude $I_0 = 0.1$ A and frequency $f = 100$ MHz. Assume the current is uniformly distributed within the conductor's cross-section. Calculate the amplitude of the displacement current in the wire.

11.15 Maxwell's Equations in Cylindrical Coordinates. Write Maxwell's equations explicitly in cylindrical coordinates by expanding the expressions in **Eqs. (11.24)** through **(11.27)**.

11.16 Maxwell's Equations. A time-dependent magnetic field is given as $\mathbf{B} = \hat{\mathbf{x}} 20 e^{j\left(10^4 t + 10^{-4} z\right)}$ [T] in a material with properties $\varepsilon_r = 9$ and $\mu_r = 1$. Assume there are no sources in the material. Using Maxwell's equations:

(a) Calculate the electric field intensity in the material.

(b) Calculate the electric flux density and the magnetic field intensity in the material.

11.17 Maxwell's Equations. A time-dependent electric field intensity is given as $\mathbf{E} = \hat{\mathbf{x}} 10\pi\cos\left(10^6 t - 50z\right)$ [V/m]. The field exists in a material with properties $\varepsilon_r = 4$ and $\mu_r = 1$. Given that $\mathbf{J} = 0$ and $\rho_v = 0$, calculate the magnetic field intensity and magnetic flux density in the material.

Potential Functions

11.18 Current Density as a Primary Variable in Maxwell's Equations. Given Maxwell's equations in a linear, isotropic, homogeneous medium. Assume that there are no source current densities and no charge densities anywhere in the solution space. An induced current density $\mathbf{J}_e$ [A/m^2] exists in conducting materials. Assume the whole space is conducting, with a very low conductivity, σ [S/m]. Rewrite Maxwell's equations in terms of the current density $\mathbf{J}_e = \sigma\mathbf{E}$. In other words, assume you need to solve for $\mathbf{J}_e$ directly.

11.19 Magnetic Scalar Potential. Write an equation, equivalent to Maxwell's equations in terms of a magnetic scalar potential in a linear, isotropic, homogeneous medium. State the conditions under which this can be done:

(a) Show that Maxwell's equations reduce to a second-order partial differential equation in the scalar potential. What are the assumptions necessary for this equation to be correct?
(b) What can you say about the relation between the electric and magnetic field intensities under the given conditions?

11.20 Magnetic Vector Potential. Given: Maxwell's equations and the vector $\mathbf{B} = \nabla \times \mathbf{A}$, in a linear, isotropic, homogeneous medium. Assume that $\mathbf{E} = 0$ for static fields:

(a) By neglecting the displacement currents, show that Maxwell's equations reduce to a second-order partial differential equation in **A** alone.
(b) What is the electric field intensity?
(c) Show that by using the Coulomb's gauge, the equation in **(a)** becomes a simple Poisson equation.

11.21 An Electric Vector Potential. A vector potential may be derived as $\nabla \times \mathbf{F} = -\mathbf{D}$ where **D** is the electric flux density:

(a) What must be the static magnetic field intensity (other than $\mathbf{H} = 0$ or $\mathbf{H} = \mathbf{C}$, where **C** is a constant vector) if we know that in the static case, $\nabla \times \mathbf{H} = 0$?
(b) Find a representation of Maxwell's equations in terms of the vector potential **F** in a current-free region (i.e., a region without source currents).
(c) What might the divergence of **F** be for the representation in **(b)** to be useful? Explain.

11.22 Modified Vector Potential. A modified vector potential may be defined as $\mathbf{F} = \mathbf{A} + \nabla\psi$, where **A** is the magnetic vector potential as defined in **Eq. (11.40)** and ψ is any scalar function:

(a) Show that this is a correct definition of the vector potential.
(b) Find an expression of Maxwell's equations in terms of **F** alone.
(c) How would you name the two potentials **F** and ψ?

11.23 The Hertz Potential. In a linear, isotropic, homogeneous medium devoid of sources, one can derive the fields from a single potential called the Hertz potential, $\boldsymbol{\pi}$, as follows:

$$\mathbf{A} = j\omega\mu\varepsilon\boldsymbol{\pi}, \quad V = -\nabla \cdot \boldsymbol{\pi}$$

where **A** is the magnetic vector potential and V the electric scalar potential.
Find the expressions for the electric and magnetic field intensities to show that they are dependent on $\boldsymbol{\pi}$ alone.

11.24 The Use of a Gauge. In a linear, isotropic, homogeneous medium devoid of sources, one can define the magnetic Hertz potential $\boldsymbol{\pi}_m$ as

$$\mathbf{E} = -j\omega\mu\nabla \times \boldsymbol{\pi}_m \quad [\text{V/m}]$$

Show that one can write Maxwell's equations in the frequency domain in terms of $\boldsymbol{\pi}_m$ alone provided a proper gauge is defined. What is that gauge?

Interface Conditions for General Fields

11.25 Application: Displacement Current Density in a Dielectric. A time-dependent electric field intensity is applied on a dielectric as shown in **Figure 11.8**. The electric field intensity in free space is given as $\mathbf{E} = \hat{\mathbf{z}}E_0\cos\omega t$ [V/m]. The relative permittivity of the material is $\varepsilon_r = 25$. For $E_0 = 100$ V/m and $\omega = 10^9$ rad/s, calculate the peak displacement current density in the dielectric (there are no surface charges at the interface between air and material).

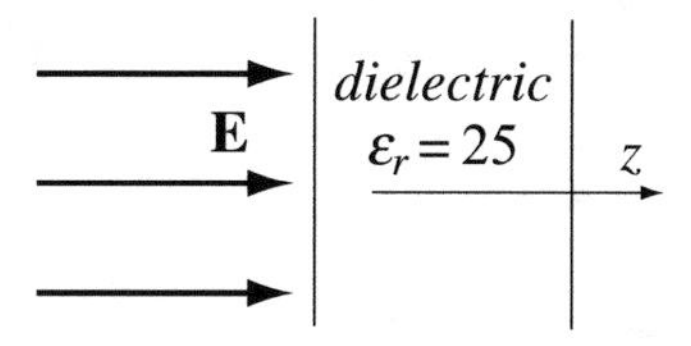

Figure 11.8

11.26 Interface Conditions for General Materials. Two dielectrics meet at an interface (see **Figure 11.9**) at $x = 0$. A sinusoidal electric field intensity of peak value 80 V/m and frequency 1 kHz exists in dielectric (1) directed as shown. For $x < 0$, $\varepsilon = 2\varepsilon_0$ [F/m], and $\mu = \mu_0$ [H/m]. For $x > 0$, $\varepsilon = 3\varepsilon_0$, and $\mu = 2\mu_0$. If the electric field intensity vector is incident at 30° from the normal, find the magnitudes of the electric field intensity and the electric flux density on each side of the interface if:

(a) There is no charge density on the interface.
(b) A charge density of magnitude 0.6 nC/m^2 exists on the interface and the electric field intensity in medium (1) is the same as in **(a)**. The charge density varies with time exactly the same as the electric field intensity.

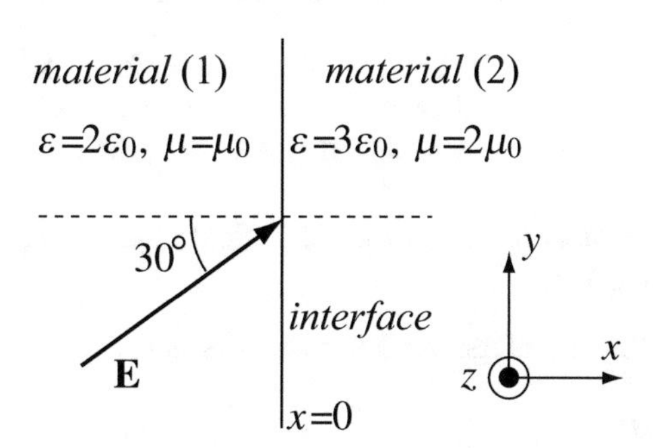

Figure 11.9

11.27 Calculation of Fields Across Interfaces. A region, denoted as region (1), occupies the space $x < 0$ and has relative permeability $\mu_{r1} = 6$. The magnetic field intensity in region (1) is $\mathbf{H}_1 = \hat{\mathbf{x}}4 + \hat{\mathbf{y}} - \hat{\mathbf{z}}2$ [A/m]. Region (2) is defined as $x > 0$ with $\mu_{r2} = 5.0$. No current exists at the interface. Find **B** in region (2).

11.28 Interface Conditions for Permeable Materials. An interface between free space and a perfectly permeable material exists. In free space (1), $\mu = \mu_0$ [H/m], $\varepsilon = \varepsilon_0$ [F/m], and $\sigma = 0$. In the permeable material (2), $\mu = \infty$, $\sigma = 0$, and $\varepsilon = \varepsilon_0$. Define the interface conditions at the interface between the two materials.

11.29 Surface Current Density at Interfaces. Two materials meet at an interface as shown in **Figure 11.10**. Material (1) has relative permeability 4 and material (2) has relative permeability 50. The interface is at $z = 0$. The magnetic flux density in material (1) is given as $\mathbf{B} = \hat{\mathbf{x}}0.01 + \hat{\mathbf{y}}0.02 + \hat{\mathbf{z}}0.015$ [T]. In material (2), it is known that all tangential components of **H** are zero. It is assumed that material (2) is conducting.

(a) Calculate the surface current density that must exist on the interface for this condition to be satisfied.
(b) Calculate the magnetic flux density in material (2).

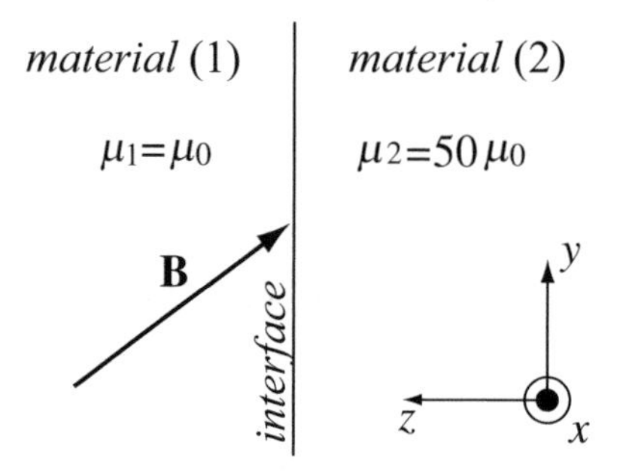

Figure 11.10

11.30 Surface Current Density at an Interface. The interface between a conducting medium and free space coincides with the x–y plane. The conducting medium occupies the space $z > 0$ and has permeability μ_0 [H/m]. The magnetic field intensity in free space is given as $\mathbf{H} = \hat{\mathbf{x}}10^5 + \hat{\mathbf{y}}2 \times 10^5 + \hat{\mathbf{z}}10^4$ [A/m].

(a) The magnetic field intensity in the conducting medium.

(b) The surface current density that must exist on the interface required to cancel the tangential components of the magnetic field intensity in the conductor.

Time-Harmonic Equations/Phasors

11.31 Vector Operations on Phasors. Two complex vectors are given as $\mathbf{A} = \mathbf{a} + j\mathbf{b}$ and $\mathbf{B} = \mathbf{c} + j\mathbf{d}$, where **a**, **b**, **c**, and **d** are real vectors. Calculate (* indicates complex conjugate) and compare the following products:

$$\mathbf{A}\cdot\mathbf{A} \quad \mathbf{A}\cdot\mathbf{A}^* \quad \mathbf{A}\cdot\mathbf{B} \quad \mathbf{A}\cdot\mathbf{B}^* \quad \mathbf{A}^*\cdot\mathbf{B}$$
$$\mathbf{A}\times\mathbf{A} \quad \mathbf{A}\times\mathbf{A}^* \quad \mathbf{A}\times\mathbf{B} \quad \mathbf{A}\times\mathbf{B}^* \quad \mathbf{A}^*\times\mathbf{B}$$
$$\mathbf{B}\times\mathbf{A} \quad \mathbf{B}\times\mathbf{A}^* \quad \mathbf{B}^*\times\mathbf{A}$$

11.32 Conversion of Phasors to the Time Domain. A magnetic field intensity is given as $\mathbf{H} = \hat{\mathbf{y}}5e^{-j\beta z}$ [A/m]. Write the time-dependent magnetic field intensity.

11.33 Conversion to Phasors. The following magnetic field intensity is given in a domain $0 \le x \le a$, $0 \le y \le b$:

$$H(x,y,z,t) = H_0 \sin\frac{m\pi x}{a}\cos\frac{n\pi y}{b}\cos(\omega t - kz) \quad [\mathrm{A/m}]$$

where x, y, and z are the space variables, m and n are integers, and k is a constant. Find the rectangular, polar, and exponential phasor representations of the field.

11.34 Conversion to Phasors. An electric field intensity is given as

$$E(z,t) = E_1\cos(\omega t - kz + \psi) + E_2\cos(\omega t + kz + \psi) \quad [\mathrm{V/m}]$$

Write the phasor form of E in polar and exponential forms.

11.35 Conversion of Phasors to the Time Domain. A phasor is given as

$$E(x,z) = E_0 e^{-j\beta_0(x\sin\theta_i + z\cos\theta_i)} \quad [\mathrm{V/m}]$$

where x and z are variables and β_0 and θ_i are constants. Find the time-dependent form of the field E.

11.36 Conversion to Phasors. The electric field intensity in a domain is given as

$$E_x(z,t) = E_0\cos(\omega t - kz + \phi) \quad [\mathrm{V/m}]$$

Find:

(a) The phasor representation of the field in exponential form.
(b) The first-order time derivative of the phasor.

11.37 Time-Harmonic Fields. The electric field intensity

$$\mathbf{E} = \hat{\mathbf{x}}10\pi\cos\left(10^6 t - 50z\right) + \hat{\mathbf{y}}10\pi\cos\left(10^6 t - 50z\right) \quad [\text{V/m}]$$

is given in a linear, isotropic, homogeneous medium of permeability μ_0 [H/m] and permittivity ε_0 [F/m]. Write the magnetic field intensity and the magnetic flux density:

(a) In terms of the time-dependent electric field intensity.
(b) In terms of the time-harmonic electric field intensity.

11.38 Time-Harmonic Fields. The magnetic field intensity in air is given as $\mathbf{H} = \left(\hat{\mathbf{x}}H_x + \hat{\mathbf{y}}H_y + \hat{\mathbf{z}}H_z\right)e^{j\beta z}e^{j\phi}$ [A/m], where H_x, H_y and H_z are complex numbers given as $H_x = h_x + jg_x$, $H_y = h_y + jg_y$ and $H_z = h_z + jg_z$:

(a) What is the time-dependent magnetic field intensity **H** in air?
(b) Write the magnetic field intensity in terms of amplitude and phase.

11.39 Conversion of Phasors. Two vector fields are given in phasor form as

$$\mathbf{E}_1 = \hat{\mathbf{x}}(20 + j20)e^{j0.3\pi z} + \hat{\mathbf{y}}(10 - j20)e^{j0.3\pi z}, \quad \mathbf{E}_2 = -\hat{\mathbf{x}}(20 - j10)e^{j0.3\pi z} + \hat{\mathbf{y}}(20 + j20)e^{j0.3\pi z} \quad [\text{V/m}]$$

Calculate:

(a) The time domain representation of the two fields.
(b) The sum $\mathbf{E}_1 + \mathbf{E}_2$ in phasor form and in the time domain.
(c) The difference $\mathbf{E}_1 - \mathbf{E}_2$ in phasor form and in the time domain.
(d) The vector product of the two fields in the time domain and in phasor form.
(e) The scalar product of the two fields in the time domain and in phasor form.

Electromagnetic Waves and Propagation 12

As to the properties of electromagnetic radiation, I need first of all to come up with a little more than just words to the idea that a changing magnetic field makes an electric field, a changing electric field makes a magnetic field and that this pumping cycle produces an electromagnetic wave...

J. Robert Oppenheimer (1904–1967), physicist
from *The Flying Trapeze*, Oxford University Press, 1964

12.1 Introduction

After summarizing Maxwell's equations in **Chapter 11**, we are now ready to discuss their implications. In particular, we will deal directly or indirectly with the displacement current term in Maxwell's equations.

We have alluded to the fact that displacement currents are responsible for the wave or propagating nature in the field equations. Although we have some understanding of what a wave is, this understanding probably does not extend to electromagnetic waves. This will be our first task: to understand what electromagnetic waves are, why they must exist, and, later, to define the properties of the waves. The applications resulting from this new view of electromagnetics are vast and exciting, and we will have the opportunity to discuss some of them here and in the remainder of this book.

12.2 The Wave

What is a wave? Why is it important? What does it add to the physics of electromagnetics that was not present in the time-dependent field as discussed in **Chapter 10**? We will try to answer the first two questions directly whereas the third will become self-evident as a result.

Consider, first, an example: An earthquake, centered 100 km from a city, causes some damage in the city. For this damage to occur, there must be a mechanism by which energy generated at the center of the earthquake is propagated. The earthquake produces stresses in the Earth's crust and these stresses are relieved by propagating the energy and dissipating it over large areas of the Earth's surface, causing Earth movement at relatively large distances from the source. The same effect is felt when pounding with a hammer on a piece of wood or other material. The material flexes under the pressure of the hammer and the effect can be felt at a distance from the source.

A few observations on the above scenario are useful here:

(1) As the distance from the center increases, the magnitude of Earth movement is reduced (attenuated).
(2) Recording of the earthquake on a seismograph looks as in **Figure 12.1a**, showing a repetitive movement of the crust, at some frequency or range of frequencies.
(3) The tremor is felt at different locations at different times; propagation of the earthquake is at finite (relatively low) speeds.

The earthquake is an elastic wave because it requires that the intervening material sustains stress. In other words, if a location within the range of the earthquake could be isolated from the crust with a vacuum gap, air, or any other flexible material (such as rubber), the earthquake would not affect that location. For example, you would not be able to feel an

N. Ida, *Engineering Electromagnetics*, https://doi.org/10.1007/978-3-030-15557-5_12

earthquake while flying over an affected area. This is because air can sustain very little stress. We refer to this effect as attenuation: The elastic wave is attenuated rapidly in air. On the other hand, earthquakes propagate very well in water.

Another example of a wave is the motion of a string, fixed at one end to a post and the other end free to move. The free end moves up and down at a constant rate as shown in **Figure 12.1b**. The amplitude at the free end depends on how far the end moves, and the frequency of the wave depends on the rate of motion. In the process, the string oscillates, and energy is transferred through the motion. For example, if you were to hold your hand in the path of the string, you would feel the motion as the string hits your hand. The amplitude becomes smaller closer to the fixed end. At the fixed end, the amplitude is constrained to zero.

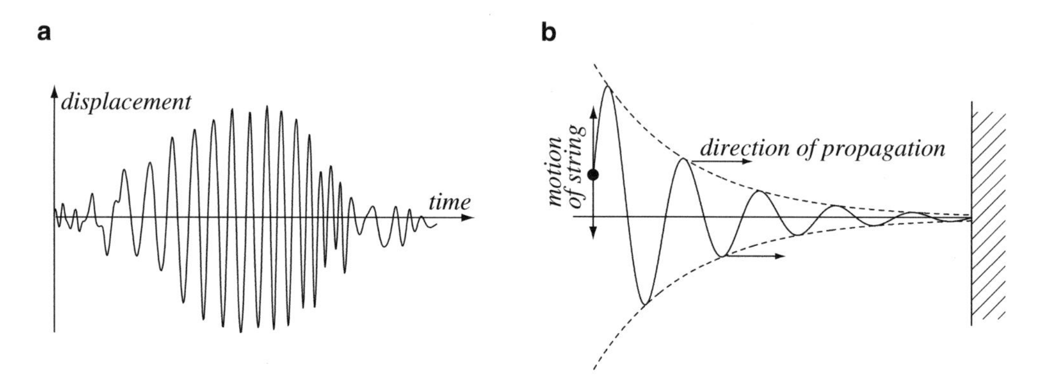

Figure 12.1 Two types of waves. (**a**) Seismograph recording of the wave nature of an earthquake. (**b**) Wave motion generated by moving the free end of a string up and down

Now, we can define the wave and its properties based on the above observations:

A wave is a disturbance in the surrounding medium with the following properties:

(1) The disturbance occurs in space and must be time dependent. Hitting a material with a hammer produces a wave; the action is a disturbance in space and is time dependent. Other examples are the operation of a loudspeaker, turning on the light, moving a paddle in water, plucking a string, etc.

(2) The disturbance can be a single event (hitting a nail with a hammer), repetitive (a paddle moving in water to generate waves), or time harmonic (for example, sinusoidal motion of the loudspeaker diaphragm).

A wave propagates in the medium with the following properties:

(1) The disturbance, which we will now call a wave, propagates in the medium or across media, at finite speeds. A sound wave propagates in water at a speed of about 1,500 m/s, whereas in air, it propagates at about 340 m/s. Light propagates in air at 3×10^8 m/s. Earthquakes propagate at speeds between about 2,000 and over 8,000 m/s, depending on the composition of the Earth's crust and location in the crust. In seawater, the propagation is at about 1,500 m/s, again varying with depth. Tsunamis (tidal waves), generated by earthquakes, propagate at speeds between about 250 km/h and over 1,000 km/h depending on the depth of the ocean.

(2) Waves propagate with attenuation. After being generated at the source, the wave propagates outward from the source and, in the process, loses its "strength." Sound becomes fainter the further we are from the source, whereas the wake of a boat becomes weaker at larger distances. The reduction in amplitude of a wave may be either because of losses (such as friction or absorption) in the material or simply because the power in the wave is spread in a continually increasing volume as the wave propagates. Attenuation due to losses is very much material dependent, as is the speed of propagation. For example, waves in water and waves in oil are attenuated differently. Sound propagates farther in cold, dense air and to shorter distances in warm air. Light is attenuated more by particulate matter in air and is attenuated very rapidly by most solids. There are large variations in attenuation in various materials. For this reason, we assume waves propagate in all materials, the differences between them being characterized by the attenuation. For example, light is assumed to propagate in solids but with such high attenuation that a thin layer of the solid seems to block light. However, a thin enough layer would be transparent.

(3) A wave transports energy. We may say that loud noises "hurt" or are "painful." The only way this can happen is if energy is transferred by the wave to our ear drums. Similarly, a very loud sound, such as a sonic boom or an explosion, may shatter windows: energy is coupled from the source through the sound waves to the window.

(4) Propagation of the wave is directional: waves propagate away from a boat and away from a loudspeaker. Electromagnetic waves transmitted from a satellite propagate from satellite to Earth.
(5) Waves can be reflected, transmitted, refracted, and diffracted. Reflection, transmission, refraction, and diffraction of light are well-known examples to these properties of electromagnetic waves.
(6) There are different types of waves with different properties. In the example of earthquakes, the P (primary or compressional) wave is generated by alternate compression and stretching of material. The S (shear or transverse) wave is generated by the motion of particles perpendicular to the direction of propagation. There are also surface waves, and each type of wave travels at different speeds. The same general principles apply to electromagnetic waves, including the existence of surface waves.

To see how wave properties manifest themselves, consider the motion of a tight string. When you pluck the string of a guitar, it moves from side to side, generating a sound wave. There are, in fact, two effects here. One is the motion of the string itself, which is a wave motion. The second is the change in air pressure generated by the string's motion, which are the sound waves we hear. The sound (frequency) depends on the size of the string (both length and thickness) and the amplitude depends on the displacement. The wave produced in the string in **Figure 12.2**, in the form of displacement of the string, y, is described by the following equation:

$$\frac{\partial^2 y}{\partial t^2} = \frac{Tg}{w}\frac{\partial^2 y}{\partial x^2} = v^2\frac{\partial^2 y}{\partial x^2} \tag{12.1}$$

where T is the tension in the string [N], g is the gravity acceleration [m/s^2], and w is weight per unit length of the string [N/m]. The term Tg/w has units of [m^2/s^2] and is, therefore, a velocity squared. This is the velocity of propagation of the wave in the string. This equation is a scalar wave equation and its solution should be familiar from physics. The important point is that it defines the form of a wave equation; the function (displacement in this case) is both time dependent and space dependent. There are other terms that may exist (such as a source term or a loss term), but the two terms above are essential. The field so represented is a wave and has all the properties described above. **Equation (12.1)** is normally written in more convenient forms as

$$\frac{1}{v^2}\frac{\partial^2 y}{\partial t^2} = \frac{\partial^2 y}{\partial x^2} \quad \text{or} \quad \frac{\partial^2 y}{\partial x^2} - \frac{1}{v^2}\frac{\partial^2 y}{\partial t^2} = 0 \tag{12.2}$$

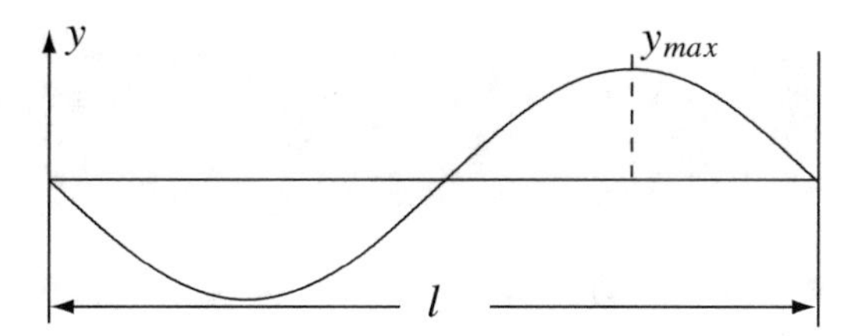

Figure 12.2 Wave motion of a tight string of length l

What about the solution to this equation? This can be obtained in a number of ways. One is by separation of variables. The second is to introduce two new independent variables $\xi = x - vt$ and $\eta = x + vt$, substitute these into the wave equation, and perform the derivatives. Then, by integration on the two new variables, we obtain a general solution of the form[1]

$$y(x,t) = g(x - vt) + f(x + vt) \tag{12.3}$$

where $g(x,t)$ and $f(x,t)$ are arbitrary functions, which describe the shape of the wave. These may be the displacements of the string at any given time and location. For example, $\sin(x - vt)$ and $\cos(x - vt)$ may be appropriate functions. We can get a better feel for what the solution means by taking a very long string (such as a wire between two posts).We will assume here that the string is infinite. Now, we create a disturbance such as plucking the string at a time $t = 0$. This gives the initial condition $y(x,0) = g(x) + f(x)$. Consider, for example, the disturbance shown in **Figure 12.3b**, created by moving the string as shown. If we let go of the string, the disturbance moves in both directions at a velocity v. After a time t_1, the disturbances

[1] This solution is known as the D'Alembert's solution of the wave equation.

have moved to the right and left a distance vt_1 as shown. The disturbances propagating in the positive and negative x directions propagate away from the source and are called forward-propagating waves. For a vector field, such as the electric or magnetic field, an equation equivalent to **Eq. (12.1)** is the vector wave equation:

$$\frac{1}{v^2}\frac{\partial^2 \mathbf{A}}{\partial t^2} = \frac{\partial^2 \mathbf{A}}{\partial z^2} \quad \text{or} \quad \frac{\partial^2 \mathbf{A}}{\partial z^2} - \frac{1}{v^2}\frac{\partial^2 \mathbf{A}}{\partial t^2} = 0 \tag{12.4}$$

where **A** stands for any of the field vectors (**E**, **H**, etc.), v is the speed of propagation of the wave, and, in this case, the wave is assumed to propagate in the z direction.

The electromagnetic wave equation will be solved in phasors, in the frequency domain rather than in the time domain. The propagation of the wave is real nonetheless. Speed of propagation, amplitudes, and all other aspects of the wave are similar to the above simple problem, although the displacement of the string will be replaced by the amplitude of the electric or magnetic fields and the propagation will be in space since the field is defined in space. On the other hand, we will talk about propagation in a certain direction in space exactly like the propagation along the string and a material-related velocity.

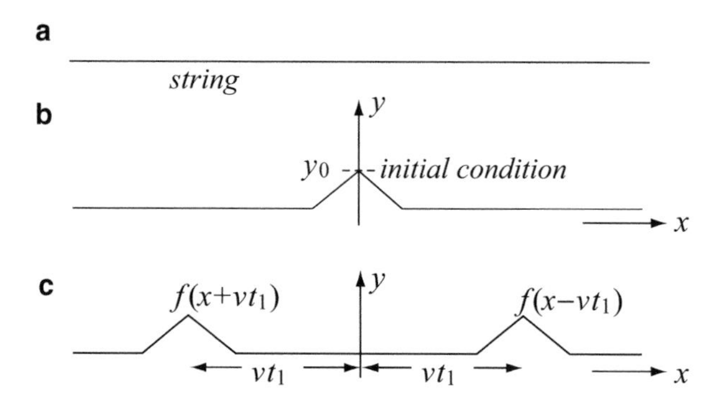

Figure 12.3 Propagation of a disturbance in a tight string. (**a**) String before the disturbance occurs. (**b**) The disturbance is introduced at $x = 0$, $t = 0$. (**c**) The disturbance propagates in the positive and negative x directions at speed v

Example 12.1 Show that the solution $y(x,t) = (1/2)[(x - vt)^2 + (x + vt)^2]$ is a solution to the scalar wave equation in **Eq. (12.1)**.

Solution: Substitution of the given solution in **Eq. (12.1)** and performing the required derivatives should result in an equality. Starting with the wave equation:

$$\frac{\partial^2 y(x,t)}{\partial t^2} = v^2 \frac{\partial^2 y(x,t)}{\partial x^2}$$

Performing the left-hand time derivatives, we get

$$\frac{\partial^2}{\partial t^2}\left(\frac{1}{2}(x - vt)^2 + \frac{1}{2}(x + vt)^2\right) = 2v^2$$

On the right-hand side, we have

$$v^2 \frac{\partial^2}{\partial x^2}\left(\frac{1}{2}(x - vt)^2 + \frac{1}{2}(x + vt)^2\right) = 2v^2$$

Because the two sides are identical, the solution satisfies the wave equation.

Example 12.2 An electric field intensity is given in free space as $\mathbf{E} = \hat{\mathbf{x}}100\cos\left(10^6 t - 10^6\sqrt{\mu_0\varepsilon_0}z\right)$ [V/m] where μ_0 [H/m] and ε_0 [F/m] are the permeability and permittivity of free space, respectively. **E** is the solution of a vector wave equation.

(a) Calculate the amplitude, frequency, and speed of propagation of the wave.
(b) Show that this solution is of the same form as that in **Eq. (12.3)**.
(c) What is the direction of propagation?

Solution: The given electric field intensity has a single component in the x direction, but the component is independent of x. Therefore, if it is a solution to the wave equation, it must be a solution to an equation of the form of **Eq. (12.4)** because the solution is a vector. To calculate the wave properties, we substitute the electric field intensity in **Eq. (12.4)**.

(a) The relevant wave equation in this case is

$$\frac{\partial^2 \mathbf{E}}{\partial z^2} - \frac{1}{v^2}\frac{\partial^2 \mathbf{E}}{\partial t^2} = 0$$

Substituting the electric field intensity and performing the derivatives:

$$\hat{\mathbf{x}}\frac{\partial^2\left[100\cos\left(10^6 t - 10^6\sqrt{\mu_0\varepsilon_0 z}\right)\right]}{\partial z^2} - \hat{\mathbf{x}}\frac{1}{v^2}\frac{\partial^2\left[100\cos\left(10^6 t - 10^6\sqrt{\mu_0\varepsilon_0 z}\right)\right]}{\partial t^2}$$
$$= -\hat{\mathbf{x}}100\times\left(10^6\right)^2\mu_0\varepsilon_0\cos\left(10^6 t - 10^6\sqrt{\mu_0\varepsilon_0 z}\right) + \hat{\mathbf{x}}\frac{1}{v^2}100\times\left(10^6\right)^2\cos\left(10^6 t - 10^6\sqrt{\mu_0\varepsilon_0 z}\right) = 0$$

After dividing both sides by $100\times\left(10^6\right)^2\cos\left(10^6 t - 10^6\sqrt{\mu_0\varepsilon_0 z}\right)$, we get

$$\mu_0\varepsilon_0 = \frac{1}{v^2} \quad \rightarrow \quad v = \frac{1}{\sqrt{\mu_0\varepsilon_0}}$$

Thus, the speed of propagation of the wave must be

$$v = \frac{1}{\sqrt{\mu_0\varepsilon_0}} = c = 3\times 10^8 \quad [\mathrm{m/s}]$$

From the electric field intensity itself, we can write the amplitude as $E_0 = 100$ V/m and frequency as $f = 10^6/2\pi$ [Hz] by simply writing the solution as

$$\mathbf{E} = \hat{\mathbf{x}}100\cos\left(10^6 t - 10^6\sqrt{\mu_0\varepsilon_0 z}\right) = \hat{\mathbf{x}}E_0\cos(\omega t + \varphi) \quad [\mathrm{V/m}].$$

(b) The electric field intensity may be written as follows:

$$\mathbf{E} = \hat{\mathbf{x}}100\cos\left(10^6 t - 10^6\sqrt{\mu_0\varepsilon_0 z}\right) = \hat{\mathbf{x}}100\cos\left[-\omega\sqrt{\mu_0\varepsilon_0}\left(z - \frac{t}{\sqrt{\mu_0\varepsilon_0}}\right)\right] = \hat{\mathbf{x}}g(z - vt) \quad [\mathrm{V/m}]$$

This is clearly of the same form as **Eq. (12.3)**, except that now the solution is a vector and only the first part in **Eq. (12.3)** is present (the second part is zero).

(c) The direction of propagation of the wave can be determined from comparison of the electric field intensity (the solution) with the general solution in **Eq. (12.3)**. In this case, the direction of propagation is in the z direction. Note, in particular, that the electric field intensity is directed in the x direction but propagates in the z direction.

Exercise 12.1

(a) Show that the function $y(x,t) = \cos(x - wt) + \cos(x + wt)$ is a solution to the scalar wave equation.
(b) What must be the speed of propagation of the wave, in this case?

Answer (**b**) $v = w$ [m/s].

12.3 The Electromagnetic Wave Equation and Its Solution

Based on the introduction of the displacement currents in Ampère's law, Maxwell predicted the existence of propagating waves, a prediction that was verified experimentally in 1888 by Heinrich Hertz. This prediction was based on the nature of the equations one obtains by using Maxwell's equations. We will show here that Maxwell's equations result, in general, in wave equations. These can be written in a number of useful forms, each useful under certain conditions. The solutions to the electromagnetic wave equations lead to a number of useful definitions, including phase velocity, wave impedance, and others.

Two types of equations will be discussed. One is the source-free wave equation, also called a ***homogeneous wave equation***. The second is a complete equation, including source terms, and is called a ***nonhomogeneous wave equation***. We first use the equations in the time domain, but most of our work here and in the following chapters will be in terms of phasors and the time-harmonic wave equation. It should also be remembered that homogeneity here relates to the form of the equation and should not be confused with material homogeneity, which merely states that material properties are independent of position.

12.3.1 The Time-Dependent Wave Equation

How do we know that Maxwell's equations in fact represent wave equations? If they do, how do we show that is the case? A hint to what needs to be done is the form in **Eq. (12.1)**; we need to rewrite Maxwell's equations in this form. To do so, we must obtain a second-order equation in time and space, in terms of a single variable. In fact, we have already done so in **Chapter 11**. There, we wrote Maxwell's equations in terms of the magnetic vector potential [see **Eq. (11.50)**] as

$$\frac{\partial^2 \mathbf{A}}{\partial x^2} + \frac{\partial^2 \mathbf{A}}{\partial y^2} + \frac{\partial^2 \mathbf{A}}{\partial z^2} = -\mu \mathbf{J} + \mu\varepsilon \frac{\partial^2 \mathbf{A}}{\partial t^2} \tag{12.5}$$

This equation is of the same form as **Eq. (12.1)** except that now the magnetic vector potential varies in all three spatial directions and, in addition, a source term (current density) is included. This is a nonhomogeneous wave equation and is much more general than **Eq. (12.1)**. If the source does not exist, we obtain the source-free or homogeneous wave equation for the magnetic vector potential:

$$\frac{\partial^2 \mathbf{A}}{\partial x^2} + \frac{\partial^2 \mathbf{A}}{\partial y^2} + \frac{\partial^2 \mathbf{A}}{\partial z^2} = \mu\varepsilon \frac{\partial^2 \mathbf{A}}{\partial t^2} \tag{12.6}$$

This particular form of the equations is only one possible form. Other potential functions or the various field variables themselves may be used to obtain similar wave equations. The principle is to substitute the corresponding variable into Maxwell's equations and manipulate the equation until the resulting equation is in terms of a single variable.

Example 12.3 Obtain a wave equation in terms of the electric scalar potential, V.

Solution: We start with Faraday's law (Maxwell's first equation) and use the Lorenz condition given in **Eq. (11.49)**. From Faraday's law, we have

$$\nabla \times \mathbf{E} = -\frac{\partial \mathbf{B}}{\partial t} = -\frac{\partial(\nabla \times \mathbf{A})}{\partial t} = -\nabla \times \frac{\partial \mathbf{A}}{\partial t} \quad \rightarrow \quad \nabla \times \left(\mathbf{E} + \frac{\partial \mathbf{A}}{\partial t}\right) = 0 \quad \rightarrow \quad \mathbf{E} + \frac{\partial \mathbf{A}}{\partial t} = -\nabla V$$

The electric field intensity is [see also **Eqs. (11.41)** through **(11.45)**]

$$\mathbf{E} = -\frac{\partial \mathbf{A}}{\partial t} - \nabla V \quad \left[\frac{\mathrm{V}}{\mathrm{m}}\right]$$

This relation is substituted in Maxwell's third equation

$$\nabla \cdot \mathbf{D} = \rho_v \quad \rightarrow \quad \nabla \cdot \mathbf{D} = \varepsilon \nabla \cdot \mathbf{E} = \varepsilon \nabla \cdot \left(-\frac{\partial \mathbf{A}}{\partial t} - \nabla V\right) = \rho_v$$

Expanding the terms in parentheses and dividing by ε on both sides of the equation, we obtain

$$-\nabla \cdot \frac{\partial \mathbf{A}}{\partial t} - \nabla \cdot (\nabla V) = \frac{\rho_v}{\varepsilon}$$

Now, we multiply both sides by -1, interchange between the divergence and time derivative and recall that $\nabla \cdot (\nabla V) = \nabla^2 V$ [vector identity in **Eq. (2.130)**]:

$$\frac{\partial}{\partial t}(\nabla \cdot \mathbf{A}) + \nabla^2 V = -\frac{\rho_v}{\varepsilon}$$

The Lorenz condition in **Eq. (11.49)** [$\nabla \cdot \mathbf{A} = -\mu\varepsilon(\partial V/\partial t)$] is now used to eliminate the magnetic vector potential. Substituting this for $\nabla \cdot \mathbf{A}$ and rearranging terms gives

$$\nabla^2 V + \frac{\partial}{\partial t}\left(-\mu\varepsilon \frac{\partial V}{\partial t}\right) = \nabla^2 V - \mu\varepsilon \frac{\partial^2 V}{\partial t^2} = -\frac{\rho_v}{\varepsilon}$$

If $\rho_v = 0$, we obtain the homogeneous wave equation in terms of the electric scalar potential V. The nonhomogeneous and homogeneous wave equations are

$$\nabla^2 V - \mu\varepsilon \frac{\partial^2 V}{\partial t^2} = -\frac{\rho_v}{\varepsilon} \quad \text{(nonhomogeneous)}$$

$$\nabla^2 V - \mu\varepsilon \frac{\partial^2 V}{\partial t^2} = 0 \quad \text{(homogeneous)}.$$

Example 12.4 Obtain a source-free wave equation for the magnetic field intensity, **H**.

Solution: To obtain a wave equation in terms of **H**, we start with Ampère's law in **Eq. (11.6)**

$$\nabla \times \mathbf{H} = \mathbf{J} + \frac{\partial \mathbf{D}}{\partial t} = \mathbf{J} + \frac{\partial(\varepsilon \mathbf{E})}{\partial t}$$

where we substituted $\mathbf{D} = \varepsilon \mathbf{E}$. Now, we seek to substitute for **E**, in order to eliminate it. To do so, we use Faraday's law:

$$\nabla \times \mathbf{E} = -\frac{\partial \mathbf{B}}{\partial t} = -\frac{\partial \mu \mathbf{H}}{\partial t} = -\mu \frac{\partial \mathbf{H}}{\partial t}$$

where the constitutive relation $\mathbf{B} = \mu\mathbf{H}$ was used to write Faraday's law in terms of **H** rather than **B**. To be able to substitute this relation into Ampère's law, we first take the curl on both sides of Ampère's law:

$$\nabla \times (\nabla \times \mathbf{H}) = \nabla \times \mathbf{J} + \nabla \times \varepsilon \frac{\partial \mathbf{E}}{\partial t} = \nabla \times \mathbf{J} + \varepsilon \frac{\partial}{\partial t}(\nabla \times \mathbf{E})$$

Two conditions are implicit here: that the curl and time derivatives are independent of each other and that permittivity is constant in space. The first is always correct. The second is an assumption and does not have to hold in all situations. As long as the materials are homogeneous, we should have no difficulty with this assumption. The left-hand side can now be written as $\nabla \times (\nabla \times \mathbf{H}) = -\nabla^2\mathbf{H} + \nabla(\nabla \cdot \mathbf{H})$. The term $\nabla \times \mathbf{E}$ from Faraday's law is now substituted into Ampère's law

$$-\nabla^2\mathbf{H} + \nabla(\nabla \cdot \mathbf{H}) = \nabla \times \mathbf{J} - \varepsilon \frac{\partial}{\partial t}\left(\mu \frac{\partial \mathbf{H}}{\partial t}\right)$$

From Maxwell's fourth equation **[Eq. (11.8)]**, assuming μ is also constant in space (material homogeneity condition), $\nabla \cdot \mathbf{H} = \mathbf{0}$. Thus, we get

$$\nabla^2\mathbf{H} - \mu\varepsilon \frac{\partial^2 \mathbf{H}}{\partial t^2} = -\nabla \times \mathbf{J} \quad \text{(nonhomogeneous)}$$

$$\nabla^2\mathbf{H} - \mu\varepsilon \frac{\partial^2 \mathbf{H}}{\partial t^2} = 0 \quad \text{(homogeneous)}$$

Note that this wave equation is identical in form to the wave equation for the electric scalar potential in **Example 12.3**. Also, note that if we need the homogeneous wave equation only, it is best to start with the source-free Maxwell's equations.

Exercise 12.2 Obtain the homogeneous wave equation in terms of the electric flux density **D**.

Answer $\nabla^2\mathbf{D} - \mu\varepsilon \dfrac{\partial^2 \mathbf{D}}{\partial t^2} = 0.$

12.3.2 Time-Harmonic Wave Equations

The time-harmonic wave equation is obtained either by starting with the time-harmonic Maxwell's equations and following steps similar to those in the previous section, or with the time-dependent equation and then transforming the resulting time-dependent wave equations to time-harmonic wave equations.

If we choose the latter approach, we simply replace $\partial/\partial t$ by $j\omega$. For example, the time-dependent wave equation in **Eq. (12.6)** can be written in the time-harmonic form as

$$\frac{\partial^2 \mathbf{A}}{\partial x^2} + \frac{\partial^2 \mathbf{A}}{\partial y^2} + \frac{\partial^2 \mathbf{A}}{\partial z^2} = \mu\varepsilon(j\omega)^2\mathbf{A} = -\omega^2\mu\varepsilon\mathbf{A} \tag{12.7}$$

However, in doing so, we also implicitly changed the variable **A** from a real variable to a phasor even though the same notation is used. The term $e^{j\omega t}$ is implicit in **A**.

We will show next how to obtain a wave equation in terms of the electric field intensity **E**, starting from the time-harmonic Maxwell equations, and how to obtain the wave equation for **H** by transforming the time-dependent wave equation in **Example 12.4**.

To obtain the time-harmonic wave equation in terms of the electric field intensity **E**, we start with Maxwell's equations in time-harmonic form [see **Eqs. (11.68)** through **(11.71)**], but written in terms of **E** and **H**. Assuming linear, isotropic, homogeneous materials, these are

$$\nabla \times \mathbf{E} = -j\omega\mathbf{B} = -j\omega\mu\mathbf{H} \tag{12.8}$$

$$\nabla \cdot \mathbf{H} = 0 \tag{12.9}$$

$$\nabla \times \mathbf{H} = \mathbf{J} + j\omega\mathbf{D} = \mathbf{J} + j\omega\varepsilon\mathbf{E} \tag{12.10}$$

$$\nabla \cdot \varepsilon\mathbf{E} = \rho_v \tag{12.11}$$

We start by taking the curl on both sides of Faraday's law [**Eq. (12.8)**]:

$$\nabla \times (\nabla \times \mathbf{E}) = -j\omega\mu(\nabla \times \mathbf{H}) \tag{12.12}$$

Substituting for $\nabla \times \mathbf{H}$ from Ampère's law **[Eq. (12.9)]**

$$\nabla \times (\nabla \times \mathbf{E}) = -j\omega\mu(\mathbf{J} + j\omega\varepsilon\mathbf{E}) \tag{12.13}$$

Again using the identity $\nabla \times (\nabla \times \mathbf{E}) = -\nabla^2\mathbf{E} + \nabla(\nabla \cdot \mathbf{E})$

$$-\nabla^2\mathbf{E} + \nabla(\nabla \cdot \mathbf{E}) = -j\omega\mu(\mathbf{J} + j\omega\varepsilon\mathbf{E}) \tag{12.14}$$

The divergence of **E** is given in **Eq. (12.11)**. Separating the current density into source and induced current densities (i.e., $\mathbf{J} = \mathbf{J}_0 + \mathbf{J}_e = \mathbf{J}_0 + \sigma\mathbf{E}$ [A/m^2], where $\mathbf{J}_0$ [A/m^2] is an applied source current density and $\mathbf{J}_0 = \sigma\mathbf{E}$ [A/m^2] is an induced current density), substituting, and rearranging terms gives

$$\nabla^2\mathbf{E} = \nabla\left(\frac{\rho_v}{\varepsilon}\right) + j\omega\mu\mathbf{J}_0 + j\omega\mu(\sigma\mathbf{E} + j\omega\varepsilon\mathbf{E}) \tag{12.15}$$

There are three sources of the electric fields: One is due to charge distribution in space in the form of the gradient of the charge density [first term on the right-hand side in **Eq. (12.15)**]. The second is due to applied current densities [second term on the right-hand side of **Eq. (12.15)**]. These are external, applied sources. In addition, the time derivative of **B** generates induced current densities as required from Faraday's law. These current densities are represented by the term $\sigma\mathbf{E}$ and are not externally applied. The last term on the right-hand side is due to displacement current densities. The source-free wave equation is obtained if the external sources ρ_v and $\mathbf{J}_s$ are eliminated:

$$\boxed{\nabla^2\mathbf{E} = j\omega\mu(\sigma\mathbf{E} + j\omega\varepsilon\mathbf{E})} \tag{12.16}$$

If, in addition, the losses are zero ($\sigma = 0$), the source-free, lossless wave equation is obtained:

$$\nabla^2\mathbf{E} = j\omega\mu(j\omega\varepsilon)\mathbf{E} \tag{12.17}$$

Multiplying the terms on the right-hand side and rearranging gives

$$\boxed{\nabla^2\mathbf{E} + \omega^2\mu\varepsilon\mathbf{E} = 0} \tag{12.18}$$

This equation is a source-free wave equation in lossless media. It is a commonly used form of the wave equation and forms the basis of the remaining chapters of this book. **Equation (12.18)** is called the ***Helmholtz equation*** for the electric field intensity in lossless media. In lossy media, we use **Eq. (12.16)**.

As mentioned earlier, the time-harmonic wave equation may be obtained from the time-dependent wave equation through the phasor transformation. As an example, consider the source-free wave equation in terms of **H**, obtained in **Example 12.4**. If we replace $\partial/\partial t$ by $j\omega$, we get the time-harmonic, source-free, lossless wave equation for **H**:

$$\boxed{\nabla^2\mathbf{H} + \omega^2\mu\varepsilon\mathbf{H} = 0} \tag{12.19}$$

This equation is identical in form to **Eq. (12.18)** (it is a Helmholtz equation in terms of **H**) and, therefore, must also have an identical form of solution. Both **Eqs. (12.18)** and **(12.19)** are extremely important in electromagnetics, as we shall see shortly.

Exercise 12.3 Find the nonhomogeneous, time-harmonic wave equation in terms of the magnetic flux density **B** in lossless media.

Answer $\nabla^2\mathbf{B} + \omega^2\mu\varepsilon\mathbf{B} + \mu(\nabla \times \mathbf{J}) = 0$.

Exercise 12.4 Find the time-harmonic, source-free, wave equation in terms of the electric scalar potential in lossless media.

Answer $\nabla^2 V + \omega^2\mu\varepsilon V = 0$.

12.3.3 Solution of the Wave Equation

Now that various forms of the electromagnetic wave equation have been obtained, it is time to solve them. First, we must decide which wave equation to solve and under what conditions. In principle, it does not matter if we solve one wave equation or another, but, in practice, it is important to solve for the electric and magnetic fields in the domain of interest rather than, say, for the electric scalar potential, since these will be more useful in subsequent discussions. Therefore, we will solve first for the electric field intensity **E** and the magnetic field intensity **H** in lossless media and in the absence of sources. The starting point is **Eq. (12.18)** or **(12.19)**. To observe the behavior of fields and define the important aspects of propagation, we use a one-dimensional wave equation; that is, we assume that the electric field intensity **E** or the magnetic field intensity **H** has a single component in space. The conditions under which we solve the equations are:

(1) Fields are time harmonic.
(2) The electric field intensity is directed in the x direction but varies in the z direction; that is, the field is perpendicular to the direction of propagation.
(3) The medium in which the wave propagates is lossless ($\sigma = 0$).
(4) The wave equation is source free ($\mathbf{J}_0 = 0$, $\rho_v = 0$).

This set of assumptions seems to be rather restrictive. In fact, it is not. Although the direction in space is fixed, we are free to choose this direction and we can repeat the solution with a field in any other direction in space. Also, and perhaps more importantly, many of the above assumptions are actually satisfied, at least partially in practice. For example, if the electric field intensity at the antenna of a receiver is needed, there is no need to take into account the actual current at the transmitting antenna: only the equivalent field in space. Similarly, propagation in general media, although not identical to propagation in lossless media, is quite similar in many cases. The benefit of this approach is in keeping the solution simple while still capturing all important properties of the wave. The alternative is a more general solution but one that is hopelessly complicated.

In fact, the conditions stated in this section specify what is called a uniform plane wave.

12.3.4 Solution for Uniform Plane Waves

A ***uniform plane wave*** is a wave (i.e., a solution to the wave equation) in which the electric and magnetic field intensities are directed in fixed directions in space and are *constant in magnitude and phase on planes perpendicular to the direction of propagation.*

Clearly, for a field to be constant in amplitude and phase on infinite planes, the source must also be infinite in extent. In this sense, a plane wave cannot be generated in practice. However, many practical situations can approximate plane waves to such an extent that plane waves are actually more common and more useful than one might think. For example, suppose a satellite in geosynchronous orbit transmits a TV program to Earth. The satellite is at a distance of approximately 36,000 km. For all practical purposes, it looks to us as a point source and the transmission will be at an approximately constant amplitude and phase on a small section of the surface of a sphere of radius 36,000 km that covers the area of the planet seen from the satellite (in reality, satellite communication is in a fairly narrow beam covering only a small section of the sphere, but on this section, the above conditions apply). This is as good an approximation to a plane wave as one can wish. More important, the receiving antenna is of such small size compared with the distances involved that it sees a plane wave locally. Thus, analysis of the wave as a plane wave is fully justified, even if the distances involved were smaller. You may want to think in the same way about a radio transmitter on the other side of town or a TV station 50 km away. Thus, the use of plane waves is rather useful, quite general, and will restrict our solutions very little while allowing simplification in both discussion and calculation.

12.3.5 The One-Dimensional Wave Equation in Free-Space and Perfect Dielectrics

With the assumptions in **Section 12.3.3**, the electric field intensity is

$$\mathbf{E} = \hat{\mathbf{x}}E_x(z) \quad [\mathrm{V/m}] \tag{12.20}$$

where **E** is a phasor (i.e., $e^{j\omega t}$ is implied). These assumptions imply the following conditions:

$$E_y = E_z = 0 \quad \text{and} \quad \frac{\partial E_*}{\partial x} = \frac{\partial E_*}{\partial y} = 0 \tag{12.21}$$

where $*$ denotes any component of **E**. Substitution of these into **Eq. (12.18)** results in

$$\frac{d^2E_x}{dz^2} + \omega^2\mu\varepsilon E_x = 0 \tag{12.22}$$

where the partial derivative was replaced with the ordinary derivative because of the field dependence on z alone. Also, since the electric field is directed in a fixed direction in space, a scalar equation is sufficient. We denote:

$$\boxed{k = \sqrt{\omega^2\mu\varepsilon} = \omega\sqrt{\mu\varepsilon} \quad [\mathrm{rad/m}]} \tag{12.23}$$

Equation (12.22) is identical in form to **Eq. (12.1)** (except of course, it is written here in the time-harmonic form); therefore, it has the same type of solution. All we need is to find the functions f and g in the general solution in **Eq. (12.3)**. In this case, since **Eq. (12.22)** describes simple harmonic motion, it has a solution

$$E_x(z) = E_0^+e^{-jkz} + E_0^-e^{jkz} \quad [\mathrm{V/m}] \tag{12.24}$$

where E_0^+ and E_0^- are constants to be determined from the boundary conditions of the problem. The notations (+) and (−) indicate that the first term is a propagating wave in the positive z direction called a ***forward-propagating wave*** and the second a propagating wave in the negative z direction called a ***backward-propagating wave***, as in **Figure 12.4a** (horizontal arrows indicate the direction of propagation; the electric field intensity components are vertical). The amplitudes E_0^+ and E_0^- are real (but they may, in general, be complex) and are arbitrary. This solution can be verified by direct substitution into **Eq. (12.22)**.

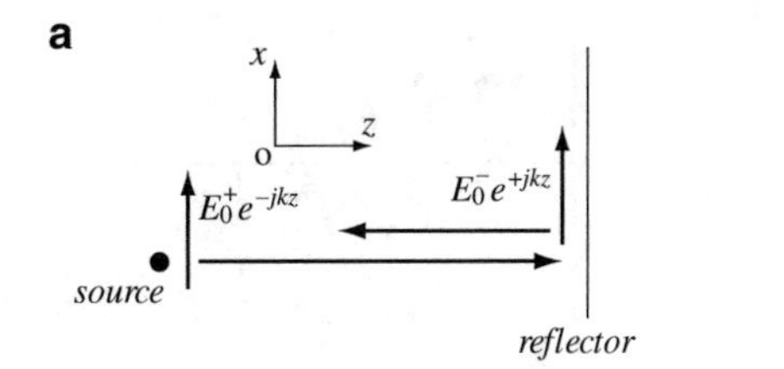

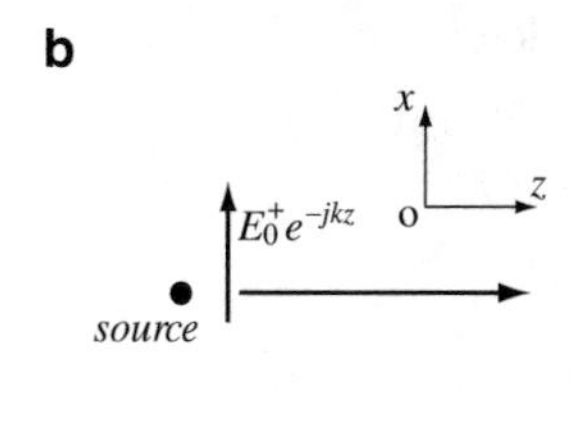

Figure 12.4 (**a**) Forward- and backward-propagating waves in bounded space. (**b**) Forward-propagating wave in unbounded space (the horizontal arrows show the direction of propagation)

Using the phasor transformation, we can write the solution in the time domain as

$$\boxed{E_x(z,t) = \text{Re}\{E_x(z)e^{j\omega t}\} = E_0^+\cos(\omega t - kz + \phi) + E_0^-\cos(\omega t + kz + \phi) \quad [\text{V/m}]} \tag{12.25}$$

where the initial (arbitrary) phase angle ϕ was added for completeness. Note that this solution is of the same form as the solution in **Eq. (12.3)**.

If the wave propagates in boundless space, only an outward wave exists and E_0^- is zero (all power propagates away from the source and there can be no backward-propagating waves). If the forward-propagating wave is reflected without losses (i.e., for an electric field, this means reflection by a perfect conductor; for ripples in the lake, it means a rigid shore), the amplitudes of the two waves are equal. **Figure 12.4b** shows schematically a forward-propagating wave without reflection. Assuming only a forward-propagating wave, the solution is

$$\boxed{E_x(z) = E_0^+e^{-jkz}e^{j\phi} \quad \text{or} \quad E_x(z,t) = E_0^+\cos(\omega t - kz + \phi) \quad [\text{V/m}]} \tag{12.26}$$

Examining these expressions, it becomes apparent that what changes with time is the phase of the wave. In other words, the phase of the wave "travels" at a certain velocity. To see what this velocity is, we use **Figure 12.5** and follow a fixed point on the wave, for which the phase of the field is $\omega t - kz + \phi =$ constant:

$$z = \frac{\omega t}{k} + \frac{\phi}{k} - \text{constant} \tag{12.27}$$

The speed of propagation of the phase is

$$v_p = \frac{dz}{dt} = \frac{\omega}{k} = \frac{1}{\sqrt{\mu\varepsilon}} \quad \left[\frac{\text{m}}{\text{s}}\right] \tag{12.28}$$

where $k = \omega\sqrt{\mu\varepsilon}$ was used [see **Eq. (12.23)**]. v_p is called the ***phase velocity*** of the wave. If you need a better feel for this velocity, think of a surfer catching a wave. The surfer rides the wave at a fixed point on the wave itself but moves forward at a given velocity. The surfer's velocity is equal to the phase velocity in the case of ocean waves. (The surfer is not moved forward by the wave; rather, the surfer slides down the wave. If it were not for this sliding, only bobbing up and down would occur as the phase of the wave moves forward.) In unbounded, lossless space, the phase velocity and the velocity of the wave or the velocity of transport of energy in the wave are the same. This is not always the case, as we will clearly see in **Chapter 17**. In this chapter, until we start discussing propagation of waves in bounded media, the terms phase velocity and speed or velocity of propagation can be used interchangeably since they happen to be the same. In general, however, they are different. The speed of propagation is a real speed, the speed at which energy propagates (or in the case of an ocean wave, the surfer's speed). The phase velocity is not a real speed in the sense that nothing material moves at that speed; only an imaginary point on the wave moves at this velocity. Because the phase velocity does not relate to physical motion, it can be smaller or larger than the speed of light and, as mentioned, may be different than the velocity of transport of energy. The phase velocity of electromagnetic waves is material dependent. In particular, in free space,

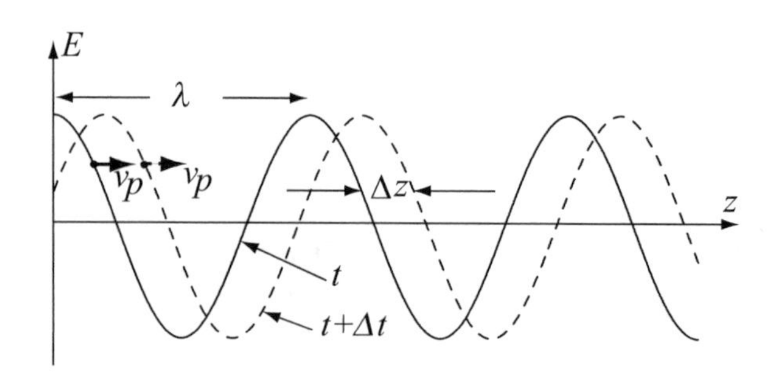

Figure 12.5 Definition of wavelength and calculation of phase velocity

$$v_p = \frac{1}{\sqrt{\mu_0 \varepsilon_0}} = \frac{1}{\sqrt{4 \times \pi \times 10^{-7} \times 8.8541853 \times 10^{-12}}} = 2.997925 \times 10^8 \approx 3 \times 10^8 = c \quad [\text{m/s}] \tag{12.29}$$

The phase velocity of electromagnetic waves in free space equals the speed of light. Perhaps this should have been suspected since light is an electromagnetic wave.

The phase velocity in most materials is lower than c since $\mu_r, \varepsilon_r \geq 1$. In fact, the phase velocity in good conductors can be a small fraction of the speed of light.

As the wave propagates, the distance between two successive crests of the wave depends both on the frequency of the wave and its phase velocity. We define the ***wavelength*** λ (in meters) as that distance a wave front (a front of constant phase) travels in one cycle:

$$\lambda = \frac{v_p}{f} = \frac{2\pi v_p}{\omega} = \frac{2\pi}{\omega\sqrt{\mu\varepsilon}} = \frac{2\pi}{k} \quad [\text{m}] \tag{12.30}$$

The wavelength for the surfer is the distance between two successive crests of ocean waves. This distance is relatively long (perhaps 50 to 100 m). In the electromagnetic case, the wavelength can be very short or very long, depending on frequency and phase velocity. For example, the wavelength in free space for a wave at 50 Hz is 6,000 km. At 30 GHz (a frequency used to communicate with satellites), the wavelength is 10 mm. From the definition of the wavelength in **Eq. (12.30)**, we can write k as

$$k = \frac{2\pi}{\lambda} \quad \left[\frac{\text{rad}}{\text{m}}\right] \tag{12.31}$$

k is called the ***wave number***. If the wavelength in free space is given, then k is called the ***free-space wave number***.

Example 12.5 Propagation of Electromagnetic Waves in Water An electromagnetic wave propagates downward from an aircraft and into water. The wave is at a frequency of 10 GHz. Assume that at this frequency fresh water has a relative permittivity of 24 (no losses) and neglect any effect the interface between air and water may have:

(a) Calculate phase velocity, wavelength, and wave number in air.
(b) Calculate phase velocity, wavelength, and wave number in water.
(c) Write the electric field intensities in air and water. Assume the electric and magnetic fields are parallel to the surface of the water, with known but different amplitudes in air and water.

Solution: The phase velocity is calculated from the permeability and permittivity of air and water using **Eq. (12.28)** or **(12.29)**. **Equation (12.26)** is then used to write the electric field intensity since only a forward-propagating wave is assumed to exist. The amplitude of the wave in air and water is generally different. Here, we simply assume E_a is the amplitude in air and E_w is the amplitude in water to indicate this difference. The actual relation between the two amplitudes will be discussed in **Chapter 13**.

(a) The permeability and permittivity in air are μ_0 [H/m] and ε_0 [F/m]. The phase velocity in air is therefore

$$v_{pa} = \frac{1}{\sqrt{\mu_0 \varepsilon_0}} = \frac{1}{\sqrt{4 \times \pi \times 10^{-7} \times 8.854 \times 10^{-12}}} = 2.998 \times 10^8 = c \quad [\text{m/s}]$$

The wavelength and wave number in air are

$$\lambda_a = \frac{v_{pa}}{f} = \frac{2.998 \times 10^8}{10^{10}} = 0.03 \quad [\text{m}], \quad k_a = \frac{2\pi}{\lambda_a} = \frac{2\pi}{0.03} = 209.44 \quad [\text{rad/m}].$$

(b) In water, $\mu = \mu_0$ [H/m] and $\varepsilon = 24\varepsilon_0$ [F/m]. The phase velocity, wavelength, and wave numbers are

$$v_{pw} = \frac{1}{\sqrt{\mu_0\varepsilon_r\varepsilon_0}} = \frac{1}{\sqrt{\varepsilon_r}\sqrt{\mu_0\varepsilon_0}} = \frac{1}{\sqrt{24\times 4\times\pi\times 10^{-7}\times 8.854\times 10^{-12}}} = \frac{2.998\times 10^8}{\sqrt{24}} = 6.12\times 10^7 \quad [\text{m/s}],$$

$$\lambda_w = \frac{v_{pw}}{f} = \frac{6.12\times 10^7}{10^{10}} = 0.00612 \quad [\text{m}],$$

$$k_w = \frac{2\pi}{\lambda_w} = \frac{2\pi}{0.00612} = 1{,}026.66 \quad [\text{rad/m}]$$

Note that the phase velocity in water is lower by a factor of $\sqrt{\varepsilon_r} = 4.899$, the wavelength is shorter by a factor of 4.899, and the wave number is 4.899 times larger.

(c) Using **Eq. (12.26)**, we can write the electric fields in water and air. We assume that the normal direction is z and the electric field intensity is in the x or y direction (arbitrarily, but parallel to the surface of the water). The fields in air are

$$E_{air}(z) = E_a e^{-jk_a z} = E_a e^{-j209.44z} \quad \text{or} \quad E_{air}(z,t) = E_a\cos\left(2\pi\times 10^{10}t - 209.44z\right) \quad [\text{V/m}]$$

where we assumed zero initial phase angle and real amplitude. The fields in water are

$$E_{water}(z) = E_w e^{-jk_w z} = E_w e^{-j1{,}026.66z} \quad \text{or} \quad E_{water}(z,t) = E_w\cos\left(2\pi\times 10^{10}t - 1{,}026.66z\right) \quad [\text{V/m}]$$

The differences are in the amplitude and wave number. Since the wave number multiplied by distance z is a phase, the phase of the wave changes much faster in water than in air.

Example 12.6 Suppose a permanent space station is built on Mars and the station communicates regularly with Earth. The distance between Earth and Mars is approximately 100 million km. Calculate the delay between transmission and reception of a signal sent from Mars and received on Earth.

Solution: The intervening medium is free space and, therefore, the speed of propagation is c. The time required for a transmission to reach Earth is

$$t = \frac{d}{v_p} = \frac{100\times 10^6\times 10^3}{3\times 10^8} = 3.33\times 10^2 = 333 \quad [\text{s}]$$

This is a delay of over 5.5 min. In other words, if a response is required, it cannot be had before 11 min have passed. This also means that timing of any radio-controlled equipment must take into account this delay. Try to imagine the difficulties in communication with distant stars. The nearest star is 3–4 light years away. Any two-way communication will take 6–8 years (if, of course, such vast distances can be covered at all); you better make every word count!

So far, we only discussed the electric field intensity **E**. Maxwell's equations tell us that a magnetic field intensity **H** exists whenever an electric field intensity **E** exists. Thus, for a complete discussion of the electromagnetic wave, we must discuss the magnetic field as well. Rather than repeating the process above, we simply substitute the electric field intensity we obtained in Maxwell's first equation **[Eq. (12.8)]** to obtain the magnetic field intensity. The equation in terms of components is

$$\hat{\mathbf{x}}\left(\frac{\partial E_z}{\partial y} - \frac{\partial E_y}{\partial z}\right) + \hat{\mathbf{y}}\left(\frac{\partial E_x}{\partial z} - \frac{\partial E_z}{\partial x}\right) + \hat{\mathbf{z}}\left(\frac{\partial E_y}{\partial x} - \frac{\partial E_x}{\partial y}\right) = -j\omega\mu\left(\hat{\mathbf{x}}H_x + \hat{\mathbf{y}}H_y + \hat{\mathbf{z}}H_z\right) \qquad (12.32)$$

From the assumption that **E** has only an x component, which varies only in z, only the term $\partial E_x/\partial z$ exists. This means that $H_x = H_z = 0$, and **Eq. (12.32)** becomes a scalar equation:

$$\frac{dE_x}{dz} = -j\omega\mu H_y \tag{12.33}$$

or writing this for the forward-propagating wave, for H_y,

$$H_y^+(z) = \frac{j}{\omega\mu}\frac{dE_x^+}{dz} \quad \left[\frac{\text{A}}{\text{m}}\right] \tag{12.34}$$

Calculating the derivative of E_x^+ with respect to z from **Eq. (12.26)**, we get

$$\frac{dE_x^+}{dz} = \frac{d}{dz}\left(E_0^+ e^{-jkz}\right) = -jk\left(E_0^+ e^{-jkz}\right) = -jkE_x^+(z) \tag{12.35}$$

Substituting this result in **Eq. (12.34)** gives

$$H_y^+(z) = \frac{k}{\omega\mu}E_x^+(z) \quad \left[\frac{\text{A}}{\text{m}}\right] \tag{12.36}$$

As was mentioned earlier, the reference field is **E** (an arbitrary choice used in electromagnetics as a convention). Thus, we define the ratio between $E_x(z)$ and $H_y(z)$ as

$$\eta = \frac{E_x^+(z)}{H_y^+(z)} = \frac{\omega\mu}{k} = \sqrt{\frac{\mu}{\varepsilon}} \quad [\Omega] \tag{12.37}$$

This quantity is an impedance because the electric field intensity is given in [V/m] and the magnetic field intensity is given in [A/m]. The quantity η is called the ***intrinsic impedance*** or ***wave impedance*** of the material since it is only dependent on material properties, as the right-hand side of **Eq. (12.37)** shows. The intrinsic impedance of free space is

$$\eta_0 = \sqrt{\frac{\mu_0}{\varepsilon_0}} = \sqrt{\frac{4\pi \times 10^{-7}}{8.854 \times 10^{-12}}} = 376.7343\ \Omega \tag{12.38}$$

For application purposes, we use 377 Ω as an approximate value (or sometimes, $120\pi = 376.99\ \Omega$) for the intrinsic impedance of free space:

$$\boxed{\eta_0 \cong 377\ \Omega} \tag{12.39}$$

The equation for H_y^+ can now be written from **Eq. (12.37)** as

$$\boxed{H_y^+(z) = \frac{1}{\eta}E_x^+(z) \quad [\text{A/m}]} \tag{12.40}$$

Note that **H** and **E** propagate in the same direction and are orthogonal to each other and to the direction of propagation. This property makes **E** and **H** ***transverse electromagnetic*** (TEM) waves. The relation between the electric and magnetic field intensities in space is shown in **Figure 12.6**. This is a very special relation: for an electric field intensity in the positive x direction, the magnetic field intensity must be in the positive y direction (for the wave to propagate in the positive z direction), as the above results indicate. This aspect of the propagation will be defined later in this chapter in terms of the vector product between the electric and magnetic field intensities.

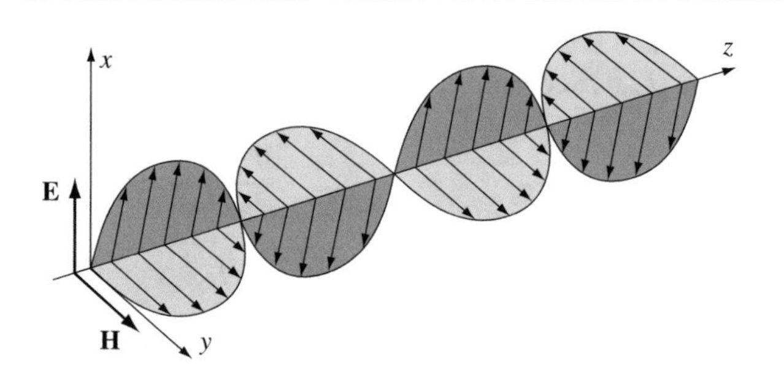

Figure 12.6 The relation between the electric and magnetic field intensities in a plane wave

The above discussion was restricted to a single component of the electric and magnetic field intensities. However, the same can be done with any other component of the electric or magnetic field and any other direction of propagation. The only real restriction on the above properties was the use of the lossless wave equation. This will be relaxed later in this chapter when we discuss propagation of waves in materials.

The properties defined above are important properties of electromagnetic waves. We defined them for time-harmonic uniform plane waves, and, therefore, they are only meaningful for time-harmonic fields. Wavelength and wave number can only properly be defined for time-harmonic fields. On the other hand, phase velocity and intrinsic impedance can be defined in terms of material properties alone and therefore do not depend on the time-harmonic form of the equations.

Example 12.7 An AM radio station transmits at 1 MHz. At some distance from the antenna, the amplitude of the electric field intensity is 10 V/m. The wave propagates from the station outward uniformly in all directions and the electric field intensity is everywhere perpendicular to the direction of propagation. Assume air has the properties of free space:

(a) Find the magnetic field intensity of the wave.
(b) Write the electric and magnetic field intensities in the time domain.
(c) During very heavy rain, the effective relative permittivity of air changes from $\varepsilon_r = 1.0$ to $\varepsilon_r = 1.5$. Calculate the change in phase velocity, intrinsic impedance, and the magnetic field intensity, assuming the amplitude of the electric field intensity remains the same.

Solution: Because the transmission is uniform, propagation is on a spherical surface. At large distances from the source, the spherical surface may be viewed as a plane and, therefore, the transmission may be approximated as a plane wave. **(a)** Assuming that the vertical direction coincides with the z direction, we may write the electric field intensity as z directed and propagating in the x (or, if we wish, in the y) direction, parallel to the surface of the Earth. The magnetic field intensity is then found such that propagation is, indeed, in this direction and away from the station. **(b)** Changes in permittivity affect the intrinsic impedance and therefore the ratio between the electric and magnetic field intensities.

(a) The electric field intensity only varies with x and the wave only propagates outward. Therefore, the electric field intensity has the form

$$\mathbf{E}(x) = \hat{\mathbf{z}} E_0 e^{-jk_0 x} \quad [\mathrm{V/m}]$$

where k_0 is the wave number in free space. To find the magnetic field intensity, we use Faraday's law in Cartesian coordinates:

$$\hat{\mathbf{x}}\left(\frac{\partial E_z}{\partial y} - \frac{\partial E_y}{\partial z}\right) + \hat{\mathbf{y}}\left(\frac{\partial E_x}{\partial z} - \frac{\partial E_z}{\partial x}\right) + \hat{\mathbf{z}}\left(\frac{\partial E_y}{\partial x} - \frac{\partial E_x}{\partial y}\right) = -j\omega\mu_0\left(\hat{\mathbf{x}}H_x + \hat{\mathbf{y}}H_y + \hat{\mathbf{z}}H_z\right)$$

Since **E** only has a z component and it may only vary with x, only the term $\partial E_z/\partial x$ exists on the left-hand side. On the right-hand side therefore, only the y component may exist. Under these conditions we have

$$\hat{\mathbf{y}}\frac{dE_z}{dx} = \hat{\mathbf{y}}j\omega\mu_0 H_y$$

The derivative of E_z with respect to x is

$$\frac{dE_z}{dx} = \frac{d}{dx}\left[E_0 e^{-jk_0x}\right] = -jk_0E_0e^{-jk_0x}$$

where we assumed the amplitude is independent of x (plane wave). The magnetic field intensity is

$$H_y = \frac{1}{j\omega\mu_0}\frac{dE_z}{dx} = -\frac{k_0}{\omega\mu_0}E_0e^{-jk_0x} \quad [\text{A/m}]$$

In vector form,

$$\mathbf{H}(x) = -\hat{\mathbf{y}}\frac{k_0}{\omega\mu_0}E_0e^{-jk_0x} = -\hat{\mathbf{y}}\frac{E_0}{\eta_0}e^{-jk_0x} \quad [\text{A/m}].$$

With the given data:

$$k_0 = \omega\sqrt{\mu_0\varepsilon_0} = \frac{2\pi f}{c} = \frac{2\times\pi\times 10^6}{3\times 10^8} = 0.021 \quad [\text{rad/m}]$$

and

$$\mathbf{H}(x) = -\hat{\mathbf{y}}\frac{10}{377}e^{-j0.021x} = -\hat{\mathbf{y}}0.0265\,e^{-j0.021x} \quad [\text{A/m}]$$

(b) Since we have no other information, we must assume that the initial phase angle is zero. Also, the amplitude of the electric field intensity is known and is a real value. The electric and magnetic field intensities in the time domain are

$$\mathbf{E}(x,t) = \hat{\mathbf{z}}\text{Re}\left[E_0e^{-jk_0x}e^{j\omega t}\right] = \hat{\mathbf{z}}E_0\cos(\omega t - k_0x) = \hat{\mathbf{z}}10\cos\left(2\pi\times 10^6t - k_0x\right) \quad [\text{V/m}]$$

$$\mathbf{H}(x,t) = \hat{\mathbf{y}}\text{Re}\left[\frac{-k_0}{\omega\mu_0}E_0e^{-jk_0x}e^{j\omega t}\right] = -\hat{\mathbf{y}}\frac{k_0}{\omega\mu_0}E_0\cos(\omega t - k_0x) = -\hat{\mathbf{y}}\frac{E_0}{\eta_0}\cos(\omega t - k_0x) \quad [\text{A/m}]$$

With $\eta_0 = 377\ \Omega$, we get

$$\mathbf{H}(x,t) = -\hat{\mathbf{y}}0.0265\cos\left(2\pi\times 10^6t - 0.021x\right) \quad [\text{A/m}].$$

(c) The phase velocity and intrinsic impedance in air (free space) are $v_p = 3\times 10^8$ m/s and $\eta_0 = 377\ \Omega$. The phase velocity in heavy rain is reduced by a factor of $\sqrt{\varepsilon_r} = \sqrt{1.5} = 1.2247$. Thus, in heavy rain the phase velocity is $v_p = 2.4495\times 10^8$ m/s. The intrinsic impedance is given in **Eq. (12.37)**:

$$\eta = \sqrt{\frac{\mu}{\varepsilon}} = \sqrt{\frac{\mu_r}{\varepsilon_r}}\sqrt{\frac{\mu_0}{\varepsilon_0}} = \eta_0\sqrt{\frac{\mu_r}{\varepsilon_r}} = \frac{377}{\sqrt{1.5}} = 307.8 \quad [\Omega]$$

Note also, that k increases by the same factor. Thus, the electric and magnetic field intensities become

$$\mathbf{E}(x,t) = \hat{\mathbf{z}}10\cos\left(2\pi\times 10^6t - 0.021\sqrt{1.5}x\right) \quad [\text{V/m}]$$

$$\mathbf{H}(x,t) = -\hat{\mathbf{y}}0.0325\cos\left(2\pi\times 10^6t - 0.021\sqrt{1.5}x\right) \quad [\text{A/m}]$$

The wave number (as well as the wavelength) has changed. The wave number increases, whereas the wavelength decreases. Because the amplitude of the electric field intensity remains the same, the amplitude of the magnetic field intensity has increased by a factor of $\sqrt{1.5}$.

Note: We arbitrarily assigned x as the direction of propagation. Similar results can be obtained by rotating the system of coordinates so that the direction of propagation coincides with any other axis, or one may use the spherical system of coordinates as we will do in **Chapter 18** to define propagation from antennas.

Example 12.8 A radar installation transmits a wave whose magnetic field intensity is

$$\mathbf{H} = \hat{\mathbf{x}}H_0\cos(\omega t - k_0 z) \quad [\mathrm{A/m}]$$

where $H_0 = 25$ A/m and $f = 30$ GHz. Propagation is in free space and z is the vertical direction. Assume plane waves and lossless propagation. Calculate:

(a) The wave number for the wave.
(b) The electric field intensity of the wave in phasor form.

Solution: The free-space wave number is calculated from the intrinsic impedance of free space which is known. With the intrinsic impedance, we can calculate the magnetic field intensity, using Faraday's law in Cartesian coordinates:

(a) From **Eq. (12.37)**, we write

$$\eta_0 = \frac{\omega\mu_0}{k_0} \quad \rightarrow \quad k_0 = \frac{\omega\mu_0}{\eta_0} = \frac{2\times\pi\times 3\times 10^{10}\times 4\times\pi\times 10^{-7}}{377} = 628.3 \quad [\mathrm{rad/m}]$$

(b) The magnitude of the electric field intensity can be written directly from **Eq. (12.40)**:

$$|\mathbf{E}| = \eta_0|\mathbf{H}| = 377\times 25 = 9425 \quad [\mathrm{V/m}]$$

However, to find the direction of the electric field intensity, we must use Ampère's law in lossless media and in the frequency domain, written here in component form:

$$\hat{\mathbf{x}}\left(\frac{\partial H_z}{\partial y} - \frac{\partial H_y}{\partial z}\right) + \hat{\mathbf{y}}\left(\frac{\partial H_x}{\partial z} - \frac{\partial H_z}{\partial x}\right) + \hat{\mathbf{z}}\left(\frac{\partial H_y}{\partial x} - \frac{\partial H_x}{\partial y}\right) = j\omega\varepsilon_0\left(\hat{\mathbf{x}}E_x + \hat{\mathbf{y}}E_y + \hat{\mathbf{z}}E_z\right)$$

Because **H** has only a component in the x direction and only varies with z, only the derivative $\partial H_x/\partial z$ is nonzero on the left-hand side. This term is in the y direction; therefore, the right-hand side can only have a y-directed component:

$$\hat{\mathbf{y}}\frac{dH_x}{dz} = \hat{\mathbf{y}}j\omega\varepsilon_0 E_y$$

To calculate the derivative of **H** with respect to z, we write **H** in phasor form:

$$\mathbf{H} = \hat{\mathbf{x}}H_0 e^{-jk_0 z} \quad [\mathrm{A/m}]$$

The electric field intensity is therefore

$$\mathbf{E} = \hat{\mathbf{y}}\frac{1}{j\omega\varepsilon_0}\frac{\partial H_x}{\partial z} = \hat{\mathbf{y}}\frac{-jk_0}{j\omega\varepsilon_0}H_0 e^{-jk_0 z} = -\hat{\mathbf{y}}\frac{k_0}{\omega\varepsilon_0}H_0 e^{-jk_0 z} = -\hat{\mathbf{y}}\frac{628.3}{2\times\pi\times 3\times 10^{10}\times 8.854\times 10^{-12}}\times 25e^{-j628.3z}$$
$$= -\hat{\mathbf{y}}9{,}425e^{-j628.3z} \quad [\mathrm{V/m}]$$

The electric field intensity is in the negative y direction. Note that this relation also implies the following:

$$\eta_0 = \frac{k_0}{\omega\varepsilon_0} = \frac{\omega\mu_0}{k_0} \quad [\Omega]$$

That this must be correct can be shown by direct multiplication (see **Exercise 12.5**)

Exercise 12.5 Show that the following two forms of the intrinsic impedance are identical:

$$\eta = \frac{\omega\mu}{k} \quad \text{and} \quad \eta = \frac{k}{\omega\varepsilon} \quad [\Omega]$$

Now that we know how to write the electric field intensity from the magnetic field intensity and vice versa, it is well to return to the beginning of the section and discuss the idea of a plane wave and its generation a bit more. Suppose that we built an infinite sheet of current with a total line current density J_s [A/m]. **Figure 12.7** shows how this might be accomplished, at least in principle, by the use of a stack of infinite wires. If the current is an AC current, then a wave is generated which propagates away from the sheet of current. The electric and magnetic field intensities are shown in the figure (the latter is calculated from Ampère's law, as in **Example 8.8**). From Ampère's law, the magnetic field intensity (and therefore the electric field intensity) is constant on any plane parallel to the sheet, and since the sheet is perpendicular to the direction of propagation (negative or positive z directions), this constitutes a true plane wave. Of course, because we cannot physically build the current sheet, we cannot obtain a true plane wave in any physical application.

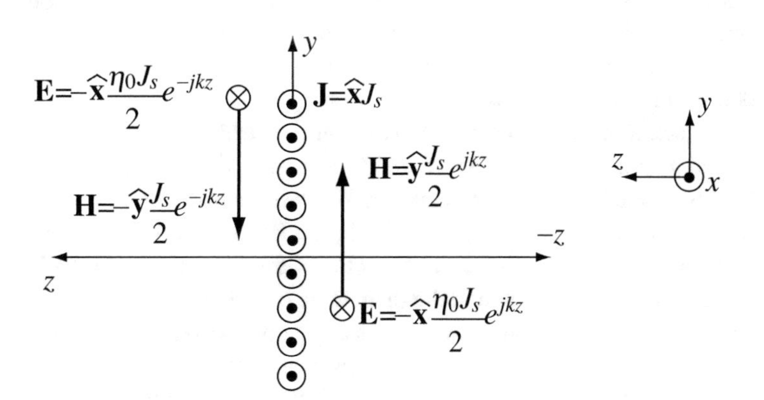

Figure 12.7 Generation of a true uniform plane wave by an infinite current sheet. The plane wave propagates away from the sheet

12.4 The Electromagnetic Spectrum

The previous section alluded to the fact that a low-frequency wave has a long wavelength and a high-frequency wave has a short wavelength, based on **Eq. (12.30)**. This fact is quite important in applications of electromagnetic waves. By analogy, we know that low-frequency sound waves propagate to larger distances. Whales use these very low frequencies to communicate. Similarly, a foghorn on a ship produces low-frequency notes. Dolphins, on the other hand, use high-frequency sounds to locate prey in water, as do bats in air. Thus, different portions of the spectrum of sound are used and are useful for different applications. The same applies to the electromagnetic spectrum. You may know, for example, that FM radio transmission in the US is between 87.5 MHz and 108 MHz or that most satellite communication occurs between 1000 MHz (1 GHz) and 30,000 MHz (30 GHz). Similarly your cell phone may be operating in the 1.9 GHz band (1850 to 1990 MHz) or in the 800 MHz band (824 to 894 MHz). Although we cannot explain at this stage why the various application use (or perhaps require) various frequencies, it must be that some frequency ranges are more appropriate or better suited for specific applications.

The electromagnetic spectrum is divided into bands, based either on frequencies or the equivalent wavelength in free space. These bands are, to a large extent, arbitrary and are designated for identification purposes. A simplified graphical representation of the electromagnetic spectrum is shown in **Figure 12.8**. The following should be noted:

(1) The spectrum of electromagnetic waves is between zero and ∞. Although we may not be able to use waves above certain frequencies, they can be used and do exist.

(2) Infrared, visible light, ultraviolet rays, X-rays, γ-rays, cosmic rays, etc., are electromagnetic waves.

(3) The narrower bands below the infrared band are arbitrarily divided by wavelength and designated names. Each band is one decade in wavelength.

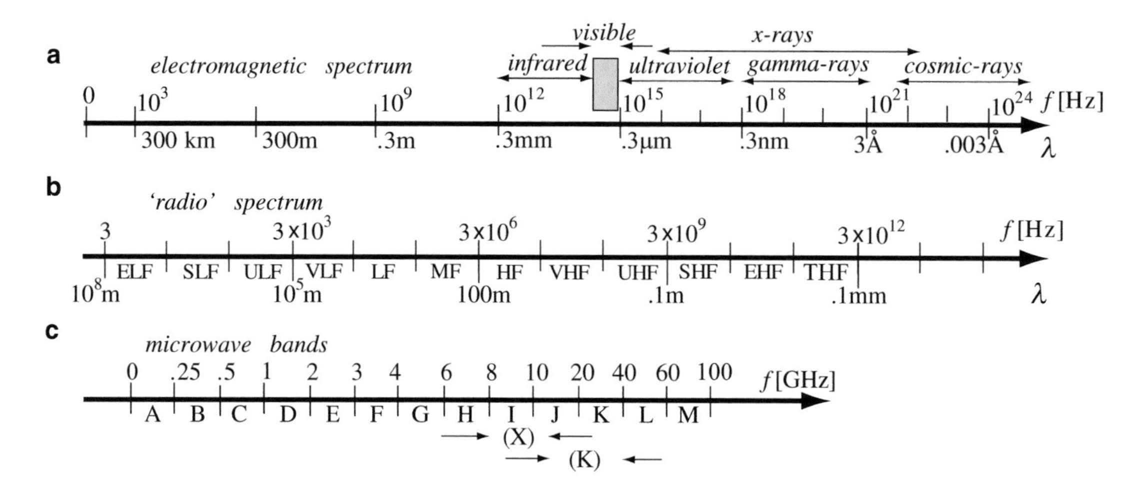

Figure 12.8 The electromagnetic spectrum

Most of our work will have to do with the spectrum below the infrared region since much of the work on light is treated in optics. However, the relations we develop (Snell's laws, reflection, transmission, and refraction of waves) apply equally well to higher frequencies. In fact, we will see that some of the relations in optics are simplifications of the electromagnetic relations obtained at lower frequencies.

Use of the electromagnetic spectrum is based on the needs of the various applications and do not follow any particular, designated band. For example, RADAR (**R**adio **D**etection **A**nd **R**anging), as used for aircraft detection, guidance, and weather, operates in the SHF and EHF domains. Radar can also be used to detect objects buried underground. Typical frequencies for this application of radar can be as low as 20 MHz, in the HF band. Similarly, communication with submarines can be done at frequencies below 100 Hz in the SLF band. Sometimes, the frequency used is allocated by convention. For example, one of the frequency ranges allowed for amateur radio (ham radio) is 3.5 to 4 MHz. This is by convention. Other frequency bands may be used as well (and some are), but these have been decided upon so that one group of users does not interfere with another. After all, we may not wish to mix, say, military use of the FM band with FM radio stations or citizens band radio with air traffic control. Other frequencies such as those for radar or communication with submarines are dictated by the application. In radar equipment, the higher the frequency, the higher the resolution. In communication with satellites, the size of the antennas is dictated by frequency (the higher the frequency, the smaller the antenna). It is therefore of some advantage to use higher frequencies. In communication with submarines, the main effect is that of penetration of waves in water. Low frequencies penetrate well, whereas high frequencies do not. Similarly, microwave ovens operate mostly at 2,450 MHz because at that frequency water absorbs electromagnetic energy and can be heated.

The spectrum may be further subdivided for specific purposes. For example, the VHF band may be divided by frequency allocations to TV channels. Again, this is by convention. Similarly, the microwave region is often divided in bands, each designated with a letter as shown in **Figure 12.8c**. In this definition, microwave ovens operate in the E band and police radar detectors operate in the I (previously known as the X band; microwave bands shown in brackets are old designations shown here for comparison) or K band. With this designation, you can at least get the satisfaction of knowing in which band the radar detector works if you get caught speeding.

12.5 The Poynting Theorem and Electromagnetic Power

One of the most important characteristics of waves is their ability to transport energy and the power associated with the process. Without this ability, many of the most important applications of electromagnetics could not be realized. To examine power and energy relations in the electromagnetic wave, it is convenient to look first at the general time-dependent expression for the rate of energy transfer that includes time rate of change in stored magnetic and stored electric energy and dissipated power. As always, the starting point must be with Maxwell's equations.

Before formalizing the expressions for energy transfer, first consider Ampère's law **[Eq. (11.25)]**:

$$\nabla \times \mathbf{H} = \mathbf{J} + \frac{\partial \mathbf{D}}{\partial t} \quad \left[\frac{\mathrm{A}}{\mathrm{m}^2}\right] \tag{12.41}$$

where $\mathbf{J}$ includes all possible current densities as follows:

$$\mathbf{J} = \mathbf{J}_0 + \mathbf{J}_e = \mathbf{J}_0 + \sigma\mathbf{E} \quad \left[\mathrm{A/m}^2\right] \tag{12.42}$$

where $\mathbf{J}_0$ indicates source current densities and $\mathbf{J}_e$ indicates induced current densities in conducting media. Now, suppose we take the scalar product of the electric field intensity $\mathbf{E}$ with **Eq. (12.41)**:

$$\mathbf{E} \cdot (\nabla \times \mathbf{H}) = \mathbf{E} \cdot \mathbf{J} + \mathbf{E} \cdot \frac{\partial \mathbf{D}}{\partial t} \tag{12.43}$$

In **Chapter 7 [Eq. (7.24)]**, we defined Joule's law as

$$\frac{dP}{dv} = \mathbf{E} \cdot \mathbf{J} \quad \left[\frac{\mathrm{W}}{\mathrm{m}^3}\right] \tag{12.44}$$

Thus, the first term on the right-hand side in **Eq. (12.43)** is the volume power density due to current densities. Although we have not yet discussed the meaning of the second term, it is also a volume power density. Both terms on the right-hand side of **Eq. (12.43)** depend only on the electric field intensity $\mathbf{E}$. Therefore, these are electric power density terms. We could now integrate **Eq. (12.43)** over a volume to calculate the total electric power in the volume. However, a more useful relation is obtained by proceeding with the following vector identity **[Eq. (2.139)]**:

$$\nabla \cdot (\mathbf{E} \times \mathbf{H}) = \mathbf{H} \cdot (\nabla \times \mathbf{E}) - \mathbf{E} \cdot (\nabla \times \mathbf{H}) \tag{12.45}$$

The second term on the right-hand side is **Eq. (12.43)** and, therefore, all three terms in **Eq. (12.45)** represent power densities. The first term on the right-hand side results from taking the scalar product of the magnetic field intensity $\mathbf{H}$ and Faraday's law **[Eq. (11.24)]**:

$$\mathbf{H} \cdot (\nabla \times \mathbf{E}) = -\mathbf{H} \cdot \left(\frac{\partial \mathbf{B}}{\partial t}\right) \tag{12.46}$$

According to **Eq. (12.45)**, **Eq. (12.46)** represents the magnetic power density. Using the vector identity in **Eq. (12.45)** and the two relations in **Eqs. (12.43)** and **(12.46)**, we get

$$\nabla \cdot (\mathbf{E} \times \mathbf{H}) = -\mathbf{H} \cdot \frac{\partial \mathbf{B}}{\partial t} - \mathbf{E} \cdot \left(\mathbf{J} + \frac{\partial \mathbf{D}}{\partial t}\right) = -\mathbf{H} \cdot \frac{\partial \mathbf{B}}{\partial t} - \mathbf{E} \cdot \frac{\partial \mathbf{D}}{\partial t} - \mathbf{E} \cdot \mathbf{J} \quad \left[\frac{\mathrm{W}}{\mathrm{m}^3}\right] \tag{12.47}$$

Assuming that we consider the power relations in a volume v, bounded by an area s, the total power in the volume is obtained by integrating over the volume:

$$\int_v \nabla \cdot (\mathbf{E} \times \mathbf{H}) dv = -\int_v \left(\mathbf{H} \cdot \frac{\partial \mathbf{B}}{\partial t} + \mathbf{E} \cdot \frac{\partial \mathbf{D}}{\partial t}\right) dv - \int_v \mathbf{E} \cdot \mathbf{J}\, dv \quad [\mathrm{W}] \tag{12.48}$$

The left-hand side is transformed from a volume integral to a closed surface integral using the divergence theorem. We also use the following identities:

$$\mathbf{E}\cdot\frac{\partial\mathbf{D}}{\partial t}=\frac{\partial}{\partial t}\left(\frac{\mathbf{E}\cdot\mathbf{D}}{2}\right),\quad \mathbf{H}\cdot\frac{\partial\mathbf{B}}{\partial t}=\frac{\partial}{\partial t}\left(\frac{\mathbf{H}\cdot\mathbf{B}}{2}\right) \tag{12.49}$$

With these, **Eq. (12.48)** becomes

$$\oint_s(\mathbf{E}\times\mathbf{H})\cdot d\mathbf{s}=-\frac{\partial}{\partial t}\int_v\left(\frac{\mathbf{H}\cdot\mathbf{B}}{2}+\frac{\mathbf{E}\cdot\mathbf{D}}{2}\right)dv-\int_v\mathbf{E}\cdot\mathbf{J}dv\quad[\mathrm{W}] \tag{12.50}$$

or, performing the scalar products

$$\mathbf{H}\cdot\mathbf{B}=\mu\mathbf{H}\cdot\mathbf{H}=\mu H^2,\quad \mathbf{E}\cdot\mathbf{D}=\varepsilon\mathbf{E}\cdot\mathbf{E}=\varepsilon E^2 \tag{12.51}$$

we get

$$\boxed{\oint_s(\mathbf{E}\times\mathbf{H})\cdot d\mathbf{s}=-\frac{\partial}{\partial t}\int_v\left(\frac{\mu H^2}{2}+\frac{\varepsilon E^2}{2}\right)dv-\int_v\mathbf{E}\cdot\mathbf{J}dv\quad[\mathrm{W}]} \tag{12.52}$$

The left-hand side of **Eq. (12.52)** represents the total outward flow of power through the area s bounding the volume v or, alternately, the energy per unit time crossing the surface s. If this flow is inwards, it is a negative flow; if outward, it is positive (because $d\mathbf{s}$ is always positive pointing out of the volume). The expression $\mathbf{E}\times\mathbf{H}$ has units of [V/m] × [A/m] = [W/m^2] and is therefore a surface power density. This power density is called the ***Poynting[2] vector*** $\mathcal{P}$:

$$\boxed{\mathcal{P}=\mathbf{E}\times\mathbf{H}\quad[\mathrm{W/m^2}]} \tag{12.53}$$

The advantage of this expression is that it also indicates the direction of power flow, information that is important for wave propagation calculations. Thus, power flows in the direction perpendicular to both **E** and **H**, according to the right-hand rule we used for the vector product (see **Section 1.3.2**). This will often be used to define or identify the direction of propagation of a wave. If the electric field intensity **E** is known, the magnetic field intensity **H** can always be calculated from the appropriate Maxwell's equation. Then, the direction of propagation of the wave can be found from the vector product of the two.

The first term on the right-hand side of **Eq. (12.52)** represents the time rate of decrease in the potential or stored energy in the system. It has two components: One is the time rate of change of the stored electric energy and the other is time rate of change of the stored magnetic energy.

The second term is due to any sources that may exist in the volume. There are two possibilities that must be considered here. One is that the current density is a source current density such as produced by a battery or a generator inside volume v. The second is a current density provided by external sources (outside the volume v). To understand this, suppose a battery is connected to a resistor as shown in **Figure 12.9**. Note that the current densities in the resistor and in the battery are in the same direction as you would expect from a closed circuit. However, the internal electric field intensity in the battery is opposite the electric field intensity in the resistor. This, again, was discussed in **Chapter 7** and it clearly indicates the difference between a source and a dissipative or load term. When we introduce the term $\mathbf{E}\cdot\mathbf{J}$ in the Poynting theorem, the term will be negative if we introduce the load and positive if we introduce the source. Thus, we distinguish between two situations:

(1) No Sources in the Volume v: In this case, all sources are external to the volume, but there may be induced current densities inside the volume. All power in the volume must come from outside sources, whereas in any conducting material, the term **J** can be written as $\mathbf{J}=\sigma\mathbf{E}$. Because all power comes from outside the volume, this case is also called the ***receiver case***. The Poynting theorem now reads

[2] John Henry Poynting (1852–1914) published in 1884 what are now known as the Poynting theorem and the Poynting vector in a paper titled "On the transfer of energy in the electromagnetic field." Poynting also performed extensive experiments aimed to determine the gravitational constant and wrote on radiation and radiation pressure. Although the Poynting vector *is* a pointing vector, remember that pointing is not the same as Poynting.

$$\oint_s (\mathbf{E} \times \mathbf{H}) \cdot d\mathbf{s} = -\frac{\partial}{\partial t}\int_v \left(\frac{\mu H^2}{2} + \frac{\varepsilon E^2}{2}\right) dv - \sigma \int_v E^2 dv \quad [\mathrm{W}] \tag{12.54}$$

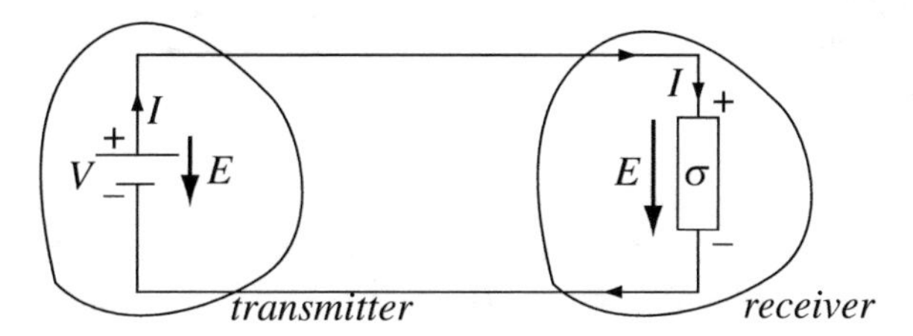

Figure 12.9 Use of a source and load to distinguish between transmitted and received power

Note that both terms on the right-hand side are negative. This means that power flows into the volume v. For the term $(\mathbf{E} \times \mathbf{H}) \cdot d\mathbf{s}$ to be negative, the term $\mathbf{E} \times \mathbf{H}$ must be opposite $d\mathbf{s}$, or into the volume. This situation is shown in **Figure 12.10a**. The receiver shown receives energy from outside its volume. This energy is partly dissipated (in resistive elements) and partly stored in the form of electric and magnetic energy (in capacitors and inductors). Another example is shown in **Figure 12.10b**, which shows food cooking in a microwave oven. The terms of **Eq. (12.54)** now are the rate of decrease in stored electric and magnetic energies in the volume occupied by the food. The third term is that part of the energy flowing into the volume that is converted to heat and does the cooking. Note that the stored energy cannot cook the food: if there are no losses in the food, there will be no energy dissipated and no cooking can take place.

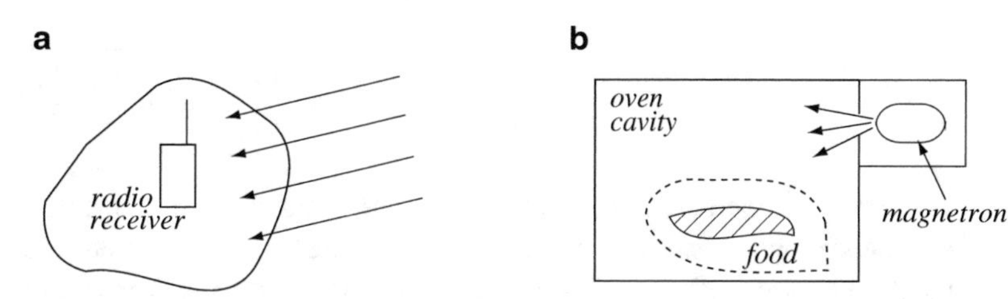

Figure 12.10 (**a**) The receiver case: power enters the volume from outside. (**b**) Example of a receiver case: power enters the food from outside its volume

(2) **Sources in the Volume v**: The situation discussed here is shown in **Figure 12.11**. A source is located in the volume, but there are no losses in the same volume. Since the electric field intensity $\mathbf{E}$ and current density $\mathbf{J}$ are in opposite directions in the source (see **Figure 12.9**), the product $\mathbf{E} \cdot \mathbf{J}$ is negative. Thus, **Eq. (12.52)** now becomes

$$\oint_s (\mathbf{E} \times \mathbf{H}) \cdot d\mathbf{s} = -\frac{\partial}{\partial t}\int_v \left(\frac{\mu H^2}{2} + \frac{\varepsilon E^2}{2}\right) dv + \int_v EJdv \quad [\mathrm{W}] \tag{12.55}$$

As expected, the flow of power is out of the volume (away from the source); therefore, the term $(\mathbf{E} \times \mathbf{H}) \cdot d\mathbf{s}$ is positive. This is also called the ***transmitter case***. Two examples are shown in **Figure 12.11**. The first shows a transmitting antenna such as from a mobile telephone. Power is transmitted out. The second shows the microwave oven again, but the volume v now encloses the magnetron (microwave source tube). The power required for cooking is generated in this volume but must be transferred out to do the cooking. No dissipation should occur in the magnetron (otherwise it will itself be "cooked"). In practice, there is quite a bit of power dissipated in the magnetron, and because of this, it must be cooled.

Of course, both energy sources and dissipative terms may exist in the same volume. In this case, the dissipative term in **Eq. (12.54)** and the source term in **Eq. (12.55)** are present in the equation as two distinct terms. From a practical point

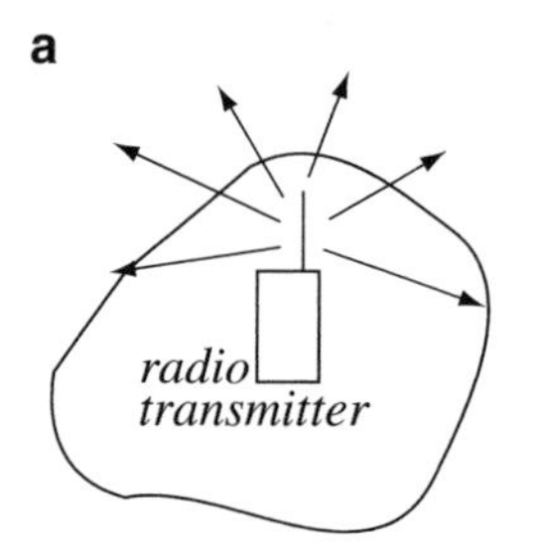

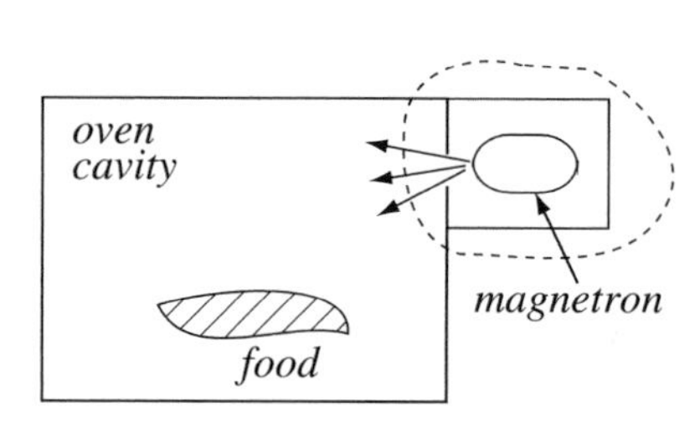

Figure 12.11 (**a**) The general transmitter case: power generated in the volume exits through its surface. (**b**) Power generated in the magnetron is transmitted out into the microwave oven cavity

of view, dissipative terms may represent losses in generators such as the energy dissipated in the magnetron of the microwave oven. However, for our purposes, it is best to keep the two terms separate.

We chose to work with the Poynting vector using the time-dependent Maxwell's equations. If the electric and magnetic fields do not depend on time, we set the time derivative to zero, in which case only a dissipative term or a source term may exist. There cannot be a rate of change in stored electric or magnetic energy, as we have seen in **Chapter 7**. On the other hand, the direction of the Poynting vector is still valid: it indicates the direction of flow of power (from the source or into a dissipative volume) as we shall see in **Example 12.10**. Also, for electrostatic or magnetostatic applications, there is no dissipation and the Poynting vector shows zero for the very simple reason that the electrostatic field is not accompanied by a magnetic field and the magnetostatic field is not accompanied by an electric field. Thus, the Poynting theorem describes all power relations in a system whether they are electrostatic, magnetostatic, or time dependent. Because the vector product between the electric field intensity and the magnetic field intensity is taken, these two quantities must be related (i.e., they must be generated by the same sources); otherwise the results obtained will have no meaning.

The expressions in **Eqs. (12.52)** through **(12.55)** are instantaneous quantities. For practical purposes, a time-averaged quantity is sometimes more useful. For a periodic time variation of fields, this can be obtained by averaging over a time T (usually a cycle of the field), giving the time-averaged Poynting vector:

$$\boxed{\mathcal{P}_{av} = \frac{1}{T}\int_0^T \mathcal{P}(t)dt \quad \left[\frac{\text{W}}{\text{m}^2}\right]} \tag{12.56}$$

The time-averaged Poynting vector is a time-averaged power density. To calculate the total power, either instantaneous or time averaged, the Poynting vector must be integrated over the surface through which the power crosses. This usually means a closed surface enclosing a volume, but not always. The instantaneous power is given as:

$$P(t) = \oint_s \mathcal{P}(t)\cdot d\mathbf{s} = \oint_s \left(\mathbf{E}(t)\times\mathbf{H}(t)\right)\cdot d\mathbf{s} \quad [\text{W}] \tag{12.57}$$

whereas the time-averaged power through a closed surface s is

$$P_{av} = \oint_s \mathcal{P}_{av}\cdot d\mathbf{s} \quad [\text{W}] \tag{12.58}$$

There are many cases in which the surface s is an open surface. For example, power may be entering or leaving a volume through a "window." This simply means that the power density on the surface enclosing the volume is zero except on the window and, therefore, the closed surface integration reduces to integration over the window. Because the Poynting theorem in **Eq. (12.52)** is defined over a closed surface (since it requires the power stored and dissipated in a volume, which is enclosed by a surface s), the relations in **Eqs. (12.57)** and **(12.58)** are written as closed surface integrals.

The important properties of the Poynting theorem and the Poynting vector are as follows:

(1) The Poynting theorem gives the power relations of the fields in any volume.
(2) The Poynting vector is the power density on the surface of a volume.
(3) The Poynting vector gives the direction of propagation of electromagnetic power.
(4) The Poynting theorem gives the net flow of power out of a given volume through its enclosing surface.

Example 12.9 Consider a plane wave with an electric field intensity $\mathbf{E} = -\hat{\mathbf{y}}E_0\cos(\omega t - kz)$ [V/m], where $E_0 = 1{,}000$ V/m and $f = 300$ MHz. Propagation is in free space:

(a) What is the direction of propagation of the wave?
(b) Calculate the instantaneous and time-averaged power densities in the wave.
(c) Calculate the total instantaneous and time-averaged power carried by the wave.
(d) Suppose a receiving dish antenna is 1 m in diameter. How much power intersects the receiving antenna if the surface of the dish is perpendicular to the direction of propagation of the wave?

Solution: From our discussion on plane waves, the direction of propagation must be in the z direction. However, we can show this from the Poynting vector. For this, we first calculate the magnetic field intensity using Faraday's law. Power density, total power, etc., are all calculated from the Poynting vector:

(a) The magnetic field intensity is found from **Eq. (12.32)** by noting that the electric field intensity has only a y component and varies only with z:

$$-\hat{\mathbf{x}}\frac{dE_y}{dz} = \hat{\mathbf{x}}j\omega\mu H_x$$

Using the phasor form of **E**

$$\mathbf{E} = -\hat{\mathbf{y}}E_0 e^{-jkz} \quad [\text{V/m}]$$

we get

$$\mathbf{H} = \hat{\mathbf{x}}\frac{k}{\omega\mu_0}E_0 e^{-jkz} = \hat{\mathbf{x}}\frac{1}{\eta_0}E_0 e^{-jkz} \quad [\text{A/m}]$$

or, in the time domain,

$$\mathbf{H} = \hat{\mathbf{x}}\frac{1}{\eta_0}E_0\cos(\omega t - kz) \quad [\text{A/m}]$$

The Poynting vector is

$$\boldsymbol{\mathcal{P}} = \mathbf{E}\times\mathbf{H} = \left[-\hat{\mathbf{y}}E_0\cos(\omega t - kz)\right]\times\left[\hat{\mathbf{x}}\frac{1}{\eta_0}E_0\cos(\omega t - kz)\right] = \hat{\mathbf{z}}\frac{E_0^2}{\eta_0}\cos^2(\omega t - kz) \quad \left[\text{W/m}^2\right]$$

The direction of flow of power is the z direction. This is also the direction of propagation of the wave.

(b) The instantaneous power density emitted by the antenna is that given in **(a)**:

$$\boldsymbol{\mathcal{P}}(z,t) = \mathbf{E}(z,t)\times\mathbf{H}(z,t) = \hat{\mathbf{z}}\frac{E_0^2}{\eta_0}\cos^2(\omega t - kz) \quad \left[\text{W/m}^2\right]$$

The time-averaged power density is found by integrating the instantaneous power density over one cycle of the wave ($T = 1/f = 2\pi/\omega$):

$$\boldsymbol{\mathcal{P}}_{av}(z) = \hat{\mathbf{z}}\frac{1}{T}\int_0^T\frac{E_0^2}{\eta_0}\cos^2(\omega t - kz)dt = \hat{\mathbf{z}}\frac{1}{T}\frac{E_0^2}{\eta_0}\int_0^T\left[\frac{1}{2}+\frac{1}{2}\cos 2(\omega t - kz)\right]dt$$

$$= \hat{\mathbf{z}}\frac{1}{T}\frac{E_0^2}{\eta_0}\int_0^T\frac{dt}{2} + \hat{\mathbf{z}}\frac{1}{T}\frac{E_0^2}{\eta_0}\int_0^T\left[\frac{1}{2}\cos 2(\omega t - kz)\right]dt \quad \left[\frac{\text{W}}{\text{m}^2}\right]$$

The second integral is zero and the first equals $T/2$. The time-averaged power density is therefore

$$\mathcal{P}_{av}(z) = \hat{\mathbf{z}}\frac{E_0^2}{2\eta_0} = \hat{\mathbf{z}}\frac{1{,}000^2}{2\times 377} = \hat{\mathbf{z}}1{,}326.26 \quad \left[\mathrm{W/m^2}\right]$$

(c) The power density is uniform throughout space and does not depend on location (except for phase, which varies in the z direction). Thus, both the total instantaneous and time-averaged power are infinite. This is true of any plane wave.

(d) The power intersecting the antenna equals the power density multiplied by the surface of the antenna. Thus, for a dish of diameter $d = 1$ m, the instantaneous power is

$$P(t) = \left|\mathcal{P}(z,t)\right|S = \frac{E_0^2\pi d^2}{4\eta_0}\cos^2(\omega t - kz) = \frac{1{,}000^2\times\pi\times 1}{4\times 377}\cos^2\left(6\pi\times 10^8 t - kz\right)$$
$$= 2{,}083.28\cos^2\left(6\pi\times 10^8 t - kz\right) \quad [\mathrm{W}]$$

The time-averaged power is

$$P_{av} = |\mathcal{P}_{av}|S = \frac{E_0^2\pi d^2}{8\eta_0} = 1{,}326.26\times\pi\times(0.5)^2 = 1{,}041.64 \quad [\mathrm{W}]$$

Example 12.10 A cylindrical conductor of radius a [m] is made of a material with conductivity σ [S/m] and carries a direct current I [A]. Calculate the power loss for a segment of the conductor L [m] long. The conductor is shown in **Figure 12.12**.

Solution: This problem can be solved most easily using the methods of **Chapter 7**. In particular, the resistance of the conductor may be calculated directly followed by calculation of losses using Joule's law. Instead, we will use the Poynting theorem (using the receiver case) to calculate the losses and in the process gain some insight into the loss process. We use **Eq. (12.54)** and evaluate the left-hand side directly. In this case, the time derivatives in **Eq. (12.54)** are zero. The Poynting theorem is therefore

$$\oint_s (\mathbf{E}\times\mathbf{H})\cdot d\mathbf{s} = -\sigma\int_v E^2\, dv \quad [\mathrm{W}]$$

We could calculate the right-hand side, which is exactly Joule's law. Instead, we evaluate the left-hand side. The current density in the conductor is uniform and equal to $\mathbf{J} = \hat{\mathbf{z}}I/S$, where $S = \pi a^2$. In the conductor, $\mathbf{J} = \sigma\mathbf{E}$ and is directed in the z direction. Therefore

$$\mathbf{E} = \hat{\mathbf{z}}\frac{I}{\sigma\pi a^2} \quad \left[\frac{\mathrm{V}}{\mathrm{m}}\right]$$

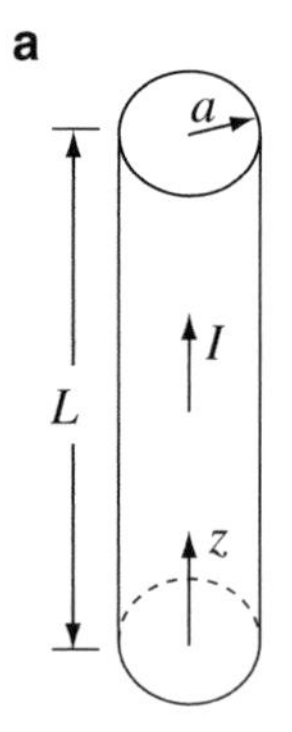

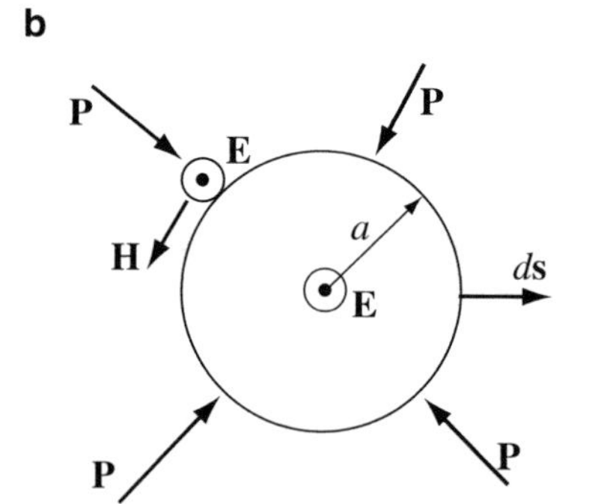

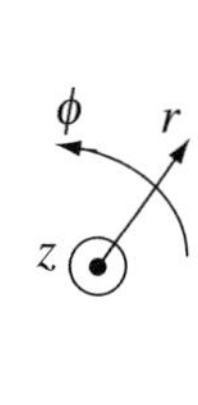

Figure 12.12 (**a**) A segment of a current-carrying conductor. (**b**) Cross section of the conductor viewed from the top of **Figure 12.12a**. The electric and magnetic field intensities and the Poynting vector are shown

The magnetic field intensity **H** at the surface of the conductor may be calculated from Ampère's law. Taking a contour around the conductor at $r = a$ and using the right-hand rule, we have (see **Example 8.7**)

$$\oint_C \mathbf{H}\cdot \mathrm{d}\mathbf{l} = I \quad \rightarrow \quad \mathbf{H} = \hat{\boldsymbol{\phi}}\frac{I}{2\pi a} \quad \left[\frac{\mathrm{A}}{\mathrm{m}}\right]$$

Note that we chose the contour at $r = a$ so that all the conducting material is enclosed by this contour. The power density at the surface of the conductor is

$$\boldsymbol{\mathcal{P}} = \mathbf{E}\times\mathbf{H} = \hat{\mathbf{z}}\frac{I}{\sigma\pi a^2}\times\hat{\boldsymbol{\phi}}\frac{I}{2\pi a} = -\hat{\mathbf{r}}\frac{I^2}{2\sigma\pi^2 a^3} \quad \left[\frac{\mathrm{W}}{\mathrm{m}^2}\right]$$

This is the power density entering the conductor through its outer surface. Note that for a given current, the dissipated power density is inversely proportional to conductivity.

The most interesting aspect of this calculation is that this power is directed into the conductor, which means that it must be dissipated in the conductor. The total power is the integral of this power density over the entire surface of the conductor. In the case of the cylinder of length L, the surface is made of the cylindrical surface and the two bases. Thus, we evaluate the total power P from **Eq. (12.54)**. In this case, all time derivatives are zero so there is no change in the stored electric or magnetic energy:

$$P = \oint_s \boldsymbol{\mathcal{P}}\cdot d\mathbf{s} = \int_{cyl}\boldsymbol{\mathcal{P}}\cdot d\mathbf{s}_{cyl} + \int_{lb}\boldsymbol{\mathcal{P}}\cdot d\mathbf{s}_{lb} + \int_{ub}\boldsymbol{\mathcal{P}}\cdot d\mathbf{s}_{ub} \quad [\mathrm{W}]$$

where cyl = cylindrical surface, lb = lower base, and ub = upper base. The three surface vectors are $d\mathbf{s}_{cyl} = \hat{\mathbf{r}}a\,d\phi dz$, $d\mathbf{s}_{lb} = -\hat{\mathbf{z}}ds$, and $d\mathbf{s}_{ub} = \hat{\mathbf{z}}ds$. The last two integrals vanish (because $d\mathbf{s}_{lb}$ and $d\mathbf{s}_{ub}$ are perpendicular to $\hat{\mathbf{r}}$, the scalar products $\boldsymbol{\mathcal{P}}\cdot d\mathbf{s}_{ub}$ and $\boldsymbol{\mathcal{P}}\cdot d\mathbf{s}_{lb}$ are zero), and we can write

$$P = \int_{cyl}\boldsymbol{\mathcal{P}}\cdot d\mathbf{s}_{cyl} = -\int_{cyl}\left(\hat{\mathbf{r}}\frac{I^2}{2\sigma\pi^2 a^3}\right)\cdot\hat{\mathbf{r}}a\,d\phi\,dz = -\int_{\phi=0}^{\phi=2\pi}\left[\int_{z=0}^{z=L}\frac{I^2 a\ dz}{2\sigma\pi^2 a^3}\right]d\phi = -\frac{I^2 L}{\sigma\pi a^2} \quad [\mathrm{W}]$$

This power is negative indicating losses. Also, from the calculation of resistance in **Chapter 7**, we know the resistance of a conductor of length L is $L/\sigma S$, where $S = \pi a^2$. Thus, not surprisingly, this result is the same as that obtained from Joule's law with the exception of the negative sign:

$$P = I^2 R = \frac{I^2 L}{\sigma\pi a^2} \quad [\mathrm{W}]$$

As expected from **Eq. (12.54)**, this is the negative of the power obtained using the Poynting vector. It demonstrates that the dissipated power in the conductor (due to resistance) can be obtained through the power penetrating into the conductor by the electromagnetic fields. By use of the Poynting vector, the power dissipated in the conductor does not enter the conductor through its connections but through the electric and magnetic fields generated by the power source (i.e., a battery) and penetrates through the outer surface of the conductor. Thus, unlike the common view of current flowing through the conductor and encountering resistance, the use of the Poynting vector indicates that power is propagated by the electric and magnetic fields. In effect, the conductor is not necessary for the propagation of power but is used to guide the power where it is needed. Dissipation is a consequence of the conductors not being ideal.

Example 12.11 Time-Averaged Power Density in Sinusoidal Fields Consider an electric field intensity and a magnetic field intensity generated by a time-harmonic source as $\mathbf{E} = \mathbf{E}_p e^{j\omega t}$ [V/m] and $\mathbf{H} = \mathbf{H}_p e^{j\omega t}$ [A/m], where **E** and **H** are phasors. $\mathbf{E}_p$ and $\mathbf{H}_p$ are complex, given as $\mathbf{E}_p = \mathbf{E}_r + j\mathbf{E}_i$ [V/m] and $\mathbf{H}_p = \mathbf{H}_r + j\mathbf{H}_i$ [A/m]:

(a) Calculate the time-averaged Poynting vector.
(b) Using the properties of complex numbers, show that the time-averaged Poynting vector may be written as

$$\mathcal{P}_{av} = \frac{1}{2}\text{Re}\{\mathbf{E} \times \mathbf{H}^*\} \quad [\text{W/m}^2]$$

where $*$ indicates the complex conjugate.

Solution: We first write the time-dependent form of the electric and magnetic field intensities and then use these to calculate the time-dependent Poynting vector using **Eq. (12.53)**. The time-averaged Poynting vector is then obtained using **Eq. (12.56)**, where T is the time of one cycle $(T = 2\pi/\omega)$:

(a) The time domain form of the electric field intensity is written from the definition of phasors as

$$\mathbf{E}(t) = \text{Re}\{(\mathbf{E}_r + j\mathbf{E}_i)e^{j\omega t}\} = \text{Re}\{(\mathbf{E}_r + j\mathbf{E}_i)(\cos\omega t + j\sin\omega t)\} = \mathbf{E}_r\cos\omega t - \mathbf{E}_i\sin\omega t \quad [\text{V/m}]$$

and similarly for the magnetic field intensity

$$\mathbf{H}(t) = \mathbf{H}_r\cos\omega t - \mathbf{H}_i\sin\omega t \quad [\text{A/m}]$$

Now, the time-dependent Poynting vector may be written as

$$\begin{aligned}\mathcal{P}(t) &= \mathbf{E}(t) \times \mathbf{H}(t) = (\mathbf{E}_r\cos\omega t - \mathbf{E}_i\sin\omega t) \times (\mathbf{H}_r\cos\omega t - \mathbf{H}_i\sin\omega t) \\ &= \mathbf{E}_r \times \mathbf{H}_r\cos^2\omega t + \mathbf{E}_i \times \mathbf{H}_i\sin^2\omega t - \mathbf{E}_i \times \mathbf{H}_r\sin\omega t\cos\omega t - \mathbf{E}_r \times \mathbf{H}_i\sin\omega t\cos\omega t \quad [\text{W/m}^2]\end{aligned}$$

The time-averaged Poynting vector is calculated from **Eq. (12.56)**:

$$\begin{aligned}\mathcal{P}_{av} &= \frac{1}{T}\int_0^{t=T} \mathbf{E}(t) \times \mathbf{H}(t)dt = \mathbf{E}_r \times \mathbf{H}_r\frac{\omega}{2\pi}\int_0^{t=2\pi/\omega}\cos^2\omega t dt + \mathbf{E}_i \times \mathbf{H}_i\frac{\omega}{2\pi}\int_0^{t=2\pi/\omega}\sin^2\omega t dt \\ &- (\mathbf{E}_i \times \mathbf{H}_r + \mathbf{E}_r \times \mathbf{H}_i)\frac{\omega}{2\pi}\int_0^{t=2\pi/\omega}\sin\omega t\cos\omega t dt \quad [\text{W/m}^2]\end{aligned}$$

where $T = 2\pi/\omega$ was used. For clarity, we integrate each term separately:

$$\mathbf{E}_r \times \mathbf{H}_r\frac{\omega}{2\pi}\int_0^{t=2\pi/\omega}\cos^2\omega t dt = \mathbf{E}_r \times \mathbf{H}_r\frac{\omega}{2\pi}\left[\frac{t}{2} + \frac{\sin 2\omega t}{4\omega}\right]_{t=0}^{t=2\pi/\omega} = \frac{\mathbf{E}_r \times \mathbf{H}_r}{2}$$

$$\mathbf{E}_i \times \mathbf{H}_i\frac{\omega}{2\pi}\int_0^{t=2\pi/\omega}\sin^2\omega t dt = \mathbf{E}_i \times \mathbf{H}_i\frac{\omega}{2\pi}\left[\frac{t}{2} - \frac{\sin 2\omega t}{4\omega}\right]_{t=0}^{t=2\pi/\omega} = \frac{\mathbf{E}_i \times \mathbf{H}_i}{2}$$

$$(\mathbf{E}_i \times \mathbf{H}_r + \mathbf{E}_r \times \mathbf{H}_i)\frac{\omega}{2\pi}\int_0^{t=2\pi/\omega}\sin\omega t\cos\omega t dt = (\mathbf{E}_i \times \mathbf{H}_r + \mathbf{E}_r \times \mathbf{H}_i)\frac{\omega}{2\pi}\left[\frac{\sin^2\omega t}{2\omega}\right]_{t=0}^{t=2\pi/\omega} = 0$$

Therefore, the time-averaged Poynting vector is

$$\mathcal{P}_{av} = \frac{\mathbf{E}_r \times \mathbf{H}_r + \mathbf{E}_i \times \mathbf{H}_i}{2} \quad [\text{W/m}^2].$$

(b) Starting with the phasor description of the vectors **E** and **H**, we write

$$\mathbf{E}\times\mathbf{H}^* = (\mathbf{E}_r + j\mathbf{E}_i)e^{j\omega t}\times(\mathbf{H}_r - j\mathbf{H}_i)e^{-j\omega t} = \mathbf{E}_r\times\mathbf{H}_r + \mathbf{E}_i\times\mathbf{H}_i + j(\mathbf{E}_i\times\mathbf{H}_r - \mathbf{E}_r\times\mathbf{H}_i)$$

Comparing this with the result in **(a)**, we can write the time-averaged Poynting vector as

$$\mathcal{P}_{av} = \frac{1}{2}\mathrm{Re}\{\mathbf{E}\times\mathbf{H}^*\} \quad \left[\mathrm{W/m^2}\right].$$

12.6 The Complex Poynting Vector

As pointed out earlier, most electromagnetic relations encountered here, including most applications, are handled in the frequency domain, assuming sinusoidal excitation. Thus, it often becomes necessary to define the Poynting vector in the frequency domain. This definition also shows the relation between real and reactive power and is closely related to time-averaged power.

In **Example 12.11**, we calculated the time-averaged Poynting vector in a general field under sinusoidal conditions as

$$\boxed{\mathcal{P}_{av} = \frac{1}{2}\mathrm{Re}\{\mathbf{E}\times\mathbf{H}^*\} \quad \left[\mathrm{W/m^2}\right].} \tag{12.59}$$

where $*$ indicates the complex conjugate form. Since the fields used to find this relation were completely general phasors, this relation applies for any sinusoidal field. Comparing this to **Eq. (12.53)**, we are led to define a complex Poynting vector as

$$\boxed{\mathcal{P}_c = \mathbf{E}\times\mathbf{H}^* = \mathbf{E}^*\times\mathbf{H} \quad \left[\mathrm{W/m^2}\right].} \tag{12.60}$$

where **E** and **H** are phasors. Clearly, the value of the complex Poynting vector is the ease with which the time-averaged power density and, therefore, time-averaged power are evaluated.

A formal derivation of the complex Poynting vector starts with Maxwell's first two equations in the frequency domain [**Eqs. (11.68)** and **(11.69)**]:

$$\nabla\times\mathbf{E} = -j\omega\mu\mathbf{H} \tag{12.61}$$

$$\nabla\times\mathbf{H} = \mathbf{J} + j\omega\varepsilon\mathbf{E} \tag{12.62}$$

The conjugates of **Eqs. (12.61)** and **(12.62)** are

$$\nabla\times\mathbf{E}^* = j\omega\mu\mathbf{H}^* \tag{12.63}$$

$$\nabla\times\mathbf{H}^* = \mathbf{J}^* - j\omega\varepsilon\mathbf{E}^* \tag{12.64}$$

The current density $\mathbf{J}^*$ includes source and induced current densities [see **Eq. (12.42)**]

$$\mathbf{J}^* = \mathbf{J}_0^* + \mathbf{J}_e^* = \mathbf{J}_0^* + \sigma\mathbf{E}^* \tag{12.65}$$

First we write the scalar product between $\mathbf{H}^*$ and **Eq. (12.61)** as

$$\mathbf{H}^*\cdot(\nabla\times\mathbf{E}) = -j\omega\mu\mathbf{H}\cdot\mathbf{H}^* \tag{12.66}$$

Next we write the scalar product between **E** and **Eq. (12.64)** as

$$\mathbf{E}\cdot(\nabla\times\mathbf{H}^*) = (\mathbf{J}^* - j\omega\varepsilon\mathbf{E}^*)\cdot\mathbf{E} \tag{12.67}$$

Equations (12.66) and **(12.67)** may be combined using the following vector identity [see **Eq. (2.139)**]:

$$\mathbf{H}^* \cdot (\nabla \times \mathbf{E}) - \mathbf{E} \cdot (\nabla \times \mathbf{H}^*) = \nabla \cdot (\mathbf{E} \times \mathbf{H}^*) \tag{12.68}$$

Substituting for $\mathbf{H}^* \cdot (\nabla \times \mathbf{E})$ from **Eq. (12.66)** and for $\mathbf{E} \cdot (\nabla \times \mathbf{H}^*)$ from **Eq. (12.67)** and rearranging terms gives

$$\nabla \cdot (\mathbf{E} \times \mathbf{H}^*) = j\omega(\varepsilon \mathbf{E} \cdot \mathbf{E}^* - \mu \mathbf{H} \cdot \mathbf{H}^*) - \mathbf{E} \cdot \mathbf{J}^* \tag{12.69}$$

The first two terms on the right-hand side represent the electric and magnetic power densities. The third term represents the input and dissipated power densities.

Using the ideas of the transmitter and receiver cases discussed in the previous section, the term $\mathbf{E} \cdot \mathbf{J}^*$ is replaced by $\sigma \mathbf{E} \cdot \mathbf{E}^*$ for the receiver case and by $-\mathbf{E} \cdot \mathbf{J}_0^*$ for the transmitter case, as was done earlier. To write this in terms of power rather than power density, we integrate **Eq. (12.69)** over an arbitrary volume v:

$$\int_v \nabla \cdot (\mathbf{E} \times \mathbf{H}^*) dv = j\omega \int_v (\varepsilon \mathbf{E} \cdot \mathbf{E}^* - \mu \mathbf{H} \cdot \mathbf{H}^*) dv - \int_v \mathbf{E} \cdot \mathbf{J}_0^* dv - \int_v \sigma \mathbf{E} \cdot \mathbf{E}^* dv \quad [\mathrm{W}] \tag{12.70}$$

Using the divergence theorem on the left-hand side, we get

$$\oint_s (\mathbf{E} \times \mathbf{H}^*) \cdot d\mathbf{s} = j\omega \int_v (\varepsilon \mathbf{E} \cdot \mathbf{E}^* - \mu \mathbf{H} \cdot \mathbf{H}^*) dv - \int_v \mathbf{E} \cdot \mathbf{J}_0^* dv - \int_v \sigma \mathbf{E} \cdot \mathbf{E}^* dv \quad [\mathrm{W}] \tag{12.71}$$

where $\mathbf{J}_0$ indicates a source current density. The left-hand side is the complex power flow through the surface s enclosing the volume v. The first term on the right-hand side is the reactive power in the volume, the second term on the right-hand side is the complex source power (either positive or negative depending on the location of the source), and the last term is the dissipated power in the volume if dissipation occurs (in conducting media).

Equation (12.71) is the complex Poynting theorem. As mentioned at the beginning of this section when using the complex Poynting vector, it is for the purpose of calculating time-averaged quantities. It is therefore more useful to write this relation as two terms as follows [using the notation in **Eq. (12.59)**]:

$$\frac{1}{2}\mathrm{Re}\left\{\oint_s \boldsymbol{\mathcal{P}}_c \cdot d\mathbf{s}\right\} = -\frac{1}{2}\int_v \mathbf{E} \cdot \mathbf{J}_0^* dv - \frac{1}{2}\int_v \sigma \mathbf{E} \cdot \mathbf{E}^* dv \quad [\mathrm{W}] \tag{12.72}$$

$$\frac{1}{2}\mathrm{Im}\left\{\oint_s \boldsymbol{\mathcal{P}}_c \cdot d\mathbf{s}\right\} = \omega \int_v \left(\frac{\varepsilon \mathbf{E} \cdot \mathbf{E}^*}{2} - \frac{\mu \mathbf{H} \cdot \mathbf{H}^*}{2}\right) dv \quad [\mathrm{W}] \tag{12.73}$$

Equation (12.72) gives the real power balance in the volume. The left-hand side is the net outward flow of time averaged power through the surface enclosing the volume. The first term on the right-hand side is the net source power (in this case, the source is outside the volume hence the negative sign) and the last term is the dissipated power in the volume. $\mathbf{E} \cdot \mathbf{J}_0^*$ is positive for the receiver case and negative for the transmitter case as was discussed in **Section 12.5**. Therefore, the second term on the right-hand side of **Eq. (12.72)** is negative for the receiver case and positive for the transmitter case.

Usually, in the transmitter case, we will assume there are no losses in the volume, whereas in the receiver case, there are no sources in the volume. If this is so, the corresponding terms are deleted from **Eqs. (12.72)**.

Equation (12.73) is the balance of reactive power. It shows the rate of flow of reactive power across the surface. From the result in **Example 12.11** and **Exercise 12.6**, we can write the stored, time-averaged magnetic and electric energy densities as

$$w_{m(av)} = \frac{1}{4}\mu \mathbf{H} \cdot \mathbf{H}^*, \quad w_{e(av)} = \frac{1}{4}\varepsilon \mathbf{E} \cdot \mathbf{E}^* \quad \left[\frac{\mathrm{J}}{\mathrm{m}^3}\right] \tag{12.74}$$

To emphasize the time-averaged power densities, **Eq. (12.73)** may be written as

$$\frac{1}{2}\mathrm{Im}\left\{\oint_s \boldsymbol{\mathcal{P}}_c \cdot d\mathbf{s}\right\} = 2\omega \int_v \left(\frac{\varepsilon \mathbf{E} \cdot \mathbf{E}^*}{4} - \frac{\mu \mathbf{H} \cdot \mathbf{H}^*}{4}\right) dv \quad [\mathrm{W}] \tag{12.75}$$

Example 12.12 Consider again the magnetic and electric fields obtained in **Example 12.8**, but now these are given in the frequency domain as

$$\mathbf{E} = -\hat{\mathbf{y}}\eta_0 H_0 e^{-jkz} \quad \left[\frac{\mathrm{V}}{\mathrm{m}}\right] \quad \text{and} \quad \mathbf{H} = \hat{\mathbf{x}} H_0 e^{-jkz} \quad \left[\frac{\mathrm{A}}{\mathrm{m}}\right]$$

where $H_0 = 25$ A/m and frequency is 30 GHz. Propagation is in free space:

(a) Calculate the time-averaged power density in the wave.
(b) Write the stored electric and magnetic energy densities separately.

Solution: The time-averaged Poynting vector is calculated using **Eq. (12.59)** and the stored electric and magnetic energy densities are given in **Eq. (12.74)**:

(a) First, we need to calculate the complex conjugate of **H**. To do so, we note that H_0 is real and the complex conjugate of e^{-jkz} is e^{+jkz}. Thus,

$$\mathbf{H}^* = \hat{\mathbf{x}} H_0 e^{jkz} \quad [\mathrm{A/m}]$$

The time-averaged Poynting vector is

$$\mathcal{P}_{av} = \frac{1}{2}\mathrm{Re}(\mathbf{E} \times \mathbf{H}^*) = \frac{1}{2}\mathrm{Re}\left(-\hat{\mathbf{y}}\eta_0 H_0 e^{-jkz} \times \hat{\mathbf{x}} H_0 e^{jkz}\right) = \hat{\mathbf{z}}\frac{\eta_0 H_0^2}{2} \quad \left[\frac{\mathrm{W}}{\mathrm{m}^2}\right]$$

Thus, power is transferred in the positive z direction. The time-averaged power density is

$$\mathcal{P}_{av} = \hat{\mathbf{z}}\frac{\eta_0 H_0^2}{2} = \hat{\mathbf{z}}\frac{377 \times 25^2}{2} = \hat{\mathbf{z}}117.81 \times 10^3 \quad [\mathrm{W/m^2}].$$

(b) The time-averaged stored electric and magnetic energy densities are

$$w_m = \frac{1}{4}\mu_0 \mathbf{H} \cdot \mathbf{H}^* = \frac{1}{4}\mu_0\left(\hat{\mathbf{x}} H_0 e^{-jkz}\right) \cdot \left(\hat{\mathbf{x}} H_0 e^{jkz}\right) = \frac{\mu_0 H_0^2}{4} = \frac{4 \times \pi \times 10^{-7} \times 25^2}{4} = 1.963 \times 10^{-4} \quad [\mathrm{J/m^3}]$$

$$w_e = \frac{1}{4}\varepsilon_0 \mathbf{E} \cdot \mathbf{E}^* = \frac{\varepsilon_0}{4}\left(-\hat{\mathbf{y}}\eta_0 H_0 e^{-jkz}\right) \cdot \left(-\hat{\mathbf{y}}\eta_0 H_0 e^{jkz}\right) = \frac{\eta_0^2 \varepsilon_0 H_0^2}{4} = \frac{8.854 \times 10^{-12} \times 377^2 \times 25^2}{4} = 1.963 \times 10^{-4} \quad [\mathrm{J/m^3}]$$

Note that the stored electric and magnetic energy densities are rather small and are equal in magnitude.

Exercise 12.6 From the results in **Example 12.11**, show that the time-averaged stored electric and magnetic energy densities for time-harmonic fields are

$$w_{m(av)} = \frac{1}{4}\mu \mathbf{H} \cdot \mathbf{H}^*, \quad w_{e(av)} = \frac{1}{4}\varepsilon \mathbf{E} \cdot \mathbf{E}^* \quad \left[\frac{\mathrm{J}}{\mathrm{m}^3}\right].$$

Exercise 12.7 Given a plane electromagnetic wave with electric field intensity

$$\mathbf{E} = \hat{\mathbf{x}} E_0 e^{jkz} \quad [\mathrm{V/m}]$$

show that the time-averaged power density at any point in space may be written as

$$\mathcal{P}_{av} = \frac{E_0^2}{2\eta} \quad \left[\frac{\mathrm{W}}{\mathrm{m}^2}\right].$$

12.7 Propagation of Plane Waves in Materials waves.m

That waves are affected by the material in which they propagate has been shown in **Section 12.3.5**, where propagation in lossless dielectrics, including free space, was discussed. The phase velocity, wavelength, wave number, and intrinsic impedance are material dependent. We also know from day-to-day experience that different materials affect waves differently. For example, when you pass under a bridge or through a tunnel, your radio ceases to receive. Propagation in water is vastly different than propagation in free space. If you ever listened to a shortwave radio, you experienced much better reception during the night than during the day. All these are due to effects of materials or environmental conditions on waves.

This aspect of propagation of waves is discussed next because it is extremely important both to understanding of propagation and to applications of electromagnetic waves. Based on the propagation properties of waves, we can choose the appropriate frequencies, type of wave, power, and other parameters needed for design.

In the process, we define the important parameters of propagating waves which, in addition to those defined in **Section 12.3**, describe an electromagnetic wave. These parameters include the propagation, phase, and attenuation constants, as well as the skin depth and the complex permittivity. These parameters will then be used for the remainder of the book to describe the behavior of waves in a number of important configurations, including transmission lines, waveguides, and antennas.

12.7.1 Propagation of Plane Waves in Lossy Dielectrics

A lossy dielectric is a material which, in addition to polarization of charges, conducts free charges to some extent. In simple terms, it is a poor insulator, whereas a perfect dielectric is a perfect insulator. For our purpose, a lossy dielectric is characterized by its permittivity and conductivity. Thus, we may assume that in addition to displacement currents, there are also conduction currents in the dielectric. The assumption that there are no sources in the solution domain is still valid.

The source-free wave equation with losses was written in **Eq. (12.16)** for the electric field intensity as

$$\nabla^2\mathbf{E} = j\omega\mu(\sigma + j\omega\varepsilon)\mathbf{E} \tag{12.76}$$

Compare this with the lossless equation ($\sigma = 0$)

$$\nabla^2\mathbf{E} = j\omega\mu(j\omega\varepsilon)\mathbf{E} \tag{12.77}$$

Note that the two equations are of exactly the same form if the term $\sigma + j\omega\varepsilon$ in **Eq. (12.76)** is replaced with a complex term $j\omega\varepsilon_c$; that is,

$$j\omega\varepsilon_c = \sigma + j\omega\varepsilon \tag{12.78}$$

The term ε_c can be written as

$$\varepsilon_c = \frac{\sigma + j\omega\varepsilon}{j\omega} = \varepsilon - j\frac{\sigma}{\omega} = \varepsilon\left[1 - j\frac{\sigma}{\omega\varepsilon}\right] = \varepsilon + j\varepsilon'' \quad [\mathrm{F/m}] \tag{12.79}$$

This is called the ***complex permittivity*** and, in general, replaces the permittivity ε in the field equations. The imaginary part of the complex permittivity is associated with losses. Now, the term lossless dielectric becomes obvious: these are dielectrics in which $\sigma = 0$ and ε_c is real and equal to ε. The definition of complex permittivity is not merely a mathematical nicety: it is an accurate model of material behavior. The real and imaginary parts of the complex permittivity are measurable.

The ratio between the imaginary and real parts of the complex permittivity is called the ***loss tangent***[3] of the material and is a common measure of how lossy materials are:

$$\tan\theta_{loss} = \frac{\sigma}{\omega\varepsilon} = \frac{\varepsilon''}{\varepsilon'} \quad [\text{dimensionless}] \tag{12.80}$$

Since the loss tangent may be viewed as the ratio between induced and displacement current densities we will use it to define approximation limits to the complex permittivity. A very low conductivity means that the permittivity is real, whereas a high conductivity means that the imaginary part of the complex permittivity dominates and the real part may be neglected.

To obtain a solution to the wave equation in lossy media, we will rely on the solution we already obtained for the lossless equation. Since the two are identical in form if the permittivity in the lossless equation is replaced with the complex permittivity, we can write the wave equation in lossy dielectrics as

$$\nabla^2\mathbf{E} = j\omega\mu(j\omega\varepsilon_c)\mathbf{E} = j\omega\mu\left(j\omega\varepsilon\left[1 - j\frac{\sigma}{\omega\varepsilon}\right]\right)\mathbf{E} \tag{12.81}$$

or, writing this in the form of the Helmholtz equation in **Eq. (12.18)**,

$$\nabla^2\mathbf{E} - j\omega\mu\left(j\omega\varepsilon\left[1 - j\frac{\sigma}{\omega\varepsilon}\right]\right)\mathbf{E} = 0 \tag{12.82}$$

Comparing this with the source-free (Helmholtz) equation and denoting

$$\boxed{\gamma = j\omega\sqrt{\mu\varepsilon}\sqrt{\left[1 - j\frac{\sigma}{\omega\varepsilon}\right]}} \tag{12.83}$$

Equation (12.82) can be written as

$$\boxed{\nabla^2\mathbf{E} - \gamma^2\mathbf{E} = 0} \tag{12.84}$$

The quantity γ is called the ***propagation constant*** and is, in general, a complex number. The propagation constant can also be written directly from **Eq. (12.76)** by comparison with **Eq. (12.84)** as

$$\gamma = \sqrt{j\omega\mu(\sigma + j\omega\varepsilon)} \tag{12.85}$$

Equation (12.84) for lossy materials is similar to **Eq. (12.18)** for lossless materials. We can put them in exactly the same form if we write

$$\gamma = jk_c \tag{12.86}$$

where

$$k_c = \omega\sqrt{\mu\varepsilon}\sqrt{\left[1 - j\frac{\sigma}{\omega\varepsilon}\right]} \quad \left[\frac{\text{rad}}{\text{m}}\right] \tag{12.87}$$

The importance of this is that now we can use all the relations obtained for the lossless propagation of waves by replacing the term jk in **Eqs. (12.24)** and **(12.26)** by γ.

The general solution for propagation in a lossy dielectric has the same two wave components as in **Eq. (12.24)**: one traveling in the positive z direction, the other in the negative z direction

$$E_x(z) = E_0^+ e^{-\gamma z} + E_0^- e^{+\gamma z} \quad [\text{V/m}] \tag{12.88}$$

---

[3] A more accurate description of complex permittivity includes also polarization losses which are due to friction between molecules in the dielectric. These losses add to the real part in **Eq. (12.78)** and therefore to the loss tangent. Our view here is that polarization losses add to conductivity, making σ a total effective conductivity. We will use the loss tangent in **Sections 12.7.2** and **12.7.3** for the sole purpose of defining limits of approximation for the complex permittivity.

Similarly, assuming only an outgoing wave, we have from **Eq. (12.26)**

$$E_x(z) = E_0^+ e^{-\gamma z} \quad [\mathrm{V/m}] \tag{12.89}$$

Since the propagation constant is a complex number (see **Exercise 12.8**), it can also be written as

$$\boxed{\gamma = \alpha + j\beta} \tag{12.90}$$

This gives for the general solution

$$E_x(z) = E_0^+ e^{-\alpha z} e^{-j\beta z} + E_0^- e^{+\alpha z} e^{+j\beta z} \quad [\mathrm{V/m}] \tag{12.91}$$

Similarly, in the case of forward propagation only

$$E_x(z) = E_0^+ e^{-(\alpha + j\beta)z} = E_0^+ e^{-\alpha z} e^{-j\beta z} \quad [\mathrm{V/m}] \tag{12.92}$$

The general solution in the time domain may be written as

$$E_x(z,t) = E_0^+ e^{-\alpha z} \cos(\omega t - \beta z) + E_0^- e^{+\alpha z} \cos(\omega t + \beta z) \quad [\mathrm{V/m}] \tag{12.93}$$

For a wave propagating in the positive z direction only, this reduces to the first term of **Eq. (12.93)**:

$$E_x(z,t) = E_0^+ e^{-\alpha z} \cos(\omega t - \beta z) \quad [\mathrm{V/m}] \tag{12.94}$$

In this form, the propagating wave has the same form as **Eq. (12.26)** where β has replaced k and the exponential term $e^{-\alpha z}$ multiplies the amplitude. This is therefore a wave, propagating in the positive z direction, with a phase velocity v_p and with an exponentially decaying amplitude. Thus, unlike the lossless case, in which the amplitude remained constant as the wave propagated, this time the amplitude changes as the wave propagates (**Figure 12.13**). Much more will be said about this decay in amplitude in future sections and chapters. Perhaps the most important general comment is that the decay can be quite rapid and that it depends on conductivity. If $\sigma = 0$, $\alpha = 0$, $e^{-\alpha z} = 1$, and the amplitude does not decay as the wave propagates.

For now, we simply note that α causes an attenuation of the amplitude of the wave and is called the ***attenuation constant***. The attenuation constant α is measured in nepers/meter [Np/m]. The ***neper*** is a dimensionless constant and defines the fraction of the attenuation the wave undergoes in 1 m. Attenuation of 1 neper/meter [Np/m] reduces the wave amplitude to $1/e$ as it propagates a distance of 1 m. Therefore, it is equivalent to 8.69 dB/m ($20 \log_{10} e = 8.69$), that is, 1 Np/m = 8.69 dB/m.

The imaginary part, β, only affects the phase of the wave and is called the ***phase constant***. The phase constant for lossless materials is identical to the wave number k as defined in **Eq. (12.31)**. However, we will use k as notation for wave number and use the phase constant β for all media, including lossless dielectrics.

A propagating wave in a lossy material is shown schematically in **Figure 12.13**. As the wave propagates in space, its amplitude is reduced exponentially. All aspects of propagation presented in the previous section remain the same except for replacing k by β and including the exponential decay in the amplitude.

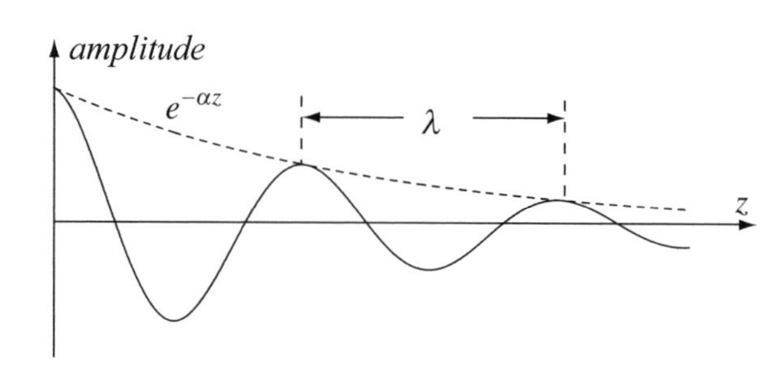

Figure 12.13 Propagation of a wave in a lossy material showing exponential attenuation

The attenuation and phase constants for a general, lossy material are found by separating the real and imaginary parts of γ in **Eq. (12.83)**. These are (see **Exercise 12.8**)

$$\boxed{\alpha = \omega \sqrt{\frac{\mu\varepsilon}{2}\left[\sqrt{1 + \left(\frac{\sigma}{\omega\varepsilon}\right)^2} - 1\right]} \quad \left[\frac{\mathrm{Np}}{\mathrm{m}}\right]} \tag{12.95}$$

$$\beta = \omega\sqrt{\frac{\mu\varepsilon}{2}\left[\sqrt{1+\left(\frac{\sigma}{\omega\varepsilon}\right)^2}+1\right]} \quad \left[\frac{\text{rad}}{\text{m}}\right] \tag{12.96}$$

The other parameters required for description of the wave in general lossy media are the phase velocity, wavelength, and intrinsic impedance. The phase velocity and wavelength are now

$$v_p = \frac{\omega}{\beta} = \frac{1}{\sqrt{\frac{\mu\varepsilon}{2}\left[\sqrt{1+\left(\frac{\sigma}{\omega\varepsilon}\right)^2}+1\right]}} \quad \left[\frac{\text{m}}{\text{s}}\right] \tag{12.97}$$

$$\lambda = \frac{2\pi}{\beta} = \frac{2\pi}{\omega\sqrt{\frac{\mu\varepsilon}{2}\left[\sqrt{1+\left(\frac{\sigma}{\omega\varepsilon}\right)^2}+1\right]}} \quad [\text{m}] \tag{12.98}$$

Thus, both phase velocity and wavelength are smaller in lossy dielectrics, depending on conductivity. For lossless materials ($\sigma = 0$), **Eqs. (12.97)** and **(12.98)** reduce to those for lossless materials given in **Eqs. (12.28)** and **(12.30)**. To find the intrinsic impedance, we return to **Eqs. (12.35)** and **(12.36)**. The magnetic field intensity can be written from **Eq. (12.35)** as

$$\frac{dE_x^+}{dz} = \frac{d}{dz}\left(E_0^+ e^{-\gamma z}\right) = -\gamma E_x^+(z) \tag{12.99}$$

Substituting this in **Eq. (12.33)** gives

$$-\gamma E_x^+(z) = -j\omega\mu H_y \tag{12.100}$$

The intrinsic impedance is now written as

$$\eta = \frac{E_x^+(z)}{H_y^+(z)} = \frac{j\omega\mu}{\gamma} \quad [\Omega] \tag{12.101}$$

In the case considered here, the intrinsic impedance becomes

$$\eta = \frac{j\omega\mu}{\gamma} = \frac{j\omega\mu}{\sqrt{j\omega\mu(\sigma + j\omega\varepsilon)}} = \sqrt{\frac{j\omega\mu}{\sigma + j\omega\varepsilon}} \quad [\Omega] \tag{12.102}$$

The intrinsic impedance (also called the wave impedance) is now a complex number. It has both a resistive and a reactive part. In practical terms, this means that **E** and **H** are out of time phase in all but lossless materials and are out of phase in space for all materials (i.e., for plane waves they are perpendicular to each other).

Some important general observations are appropriate here:

(1) The phase velocity in lossy dielectrics is lower than in perfect dielectrics. This can be seen from **Eq. (12.97)** since β for lossy materials is larger than k for perfect dielectrics. The larger the losses, the lower the phase velocity (see **Example 12.13**).

(2) The intrinsic impedance (wave impedance) in lossy dielectrics is complex, indicating a phase difference between the electric and magnetic field intensity in the same way as the phase difference between voltage and current in a circuit which contains reactive components. The magnitude of the intrinsic impedance is lower in conductive media. The higher the conductivity (losses), the lower the magnitude of the impedance.

(3) The electric and magnetic field intensities remain perpendicular to each other and to the direction of propagation regardless of losses. This is a property of the uniform plane waves we assumed.

(4) Attenuation of the wave in lossy media is exponential. This means that in materials with high conductivity, the attenuation is rapid. These materials will be called high-loss materials. Low-loss materials are materials with low conductivity.

Example 12.13 The electric field intensity of a plane electromagnetic wave is given as $\mathbf{E}(z) = \hat{\mathbf{x}}8\cos(10^6\pi t)$ [V/m] at a point $P(x = 0, y = 0, z = 0)$. The magnetic field intensity is in the positive y direction and the wave propagates in a material with properties $\varepsilon = \varepsilon_0$ [F/m], $\mu = \mu_0$ [H/m], and $\sigma = 1.5 \times 10^{-5}$ S/m:

(a) Calculate the magnetic field intensity at a distance of 1 km from point P in the direction of propagation.
(b) How much faster does the wave propagate if $\sigma = 0$?

Solution: To evaluate the magnetic field intensity at point P, all we need is the intrinsic impedance in **Eq. (12.101)**. However, to evaluate the wave at a distance $z = 1{,}000$ m, we must also evaluate the attenuation and phase constants in **Eqs. (12.95)** and **(12.96)**.

(a) The intrinsic impedance of the material is

$$\eta = \sqrt{\frac{j\omega\mu_0}{\sigma + j\omega\varepsilon_0}} = \sqrt{\frac{j10^6 \times \pi \times 4 \times \pi \times 10^{-7}}{1.5 \times 10^{-5} + j10^6 \times \pi \times 8.854 \times 10^{-12}}} = 342.69 + j86.5 \quad [\Omega]$$

The attenuation and propagation constants are

$$\alpha = \omega\sqrt{\frac{\mu_0\varepsilon_0}{2}\left[\sqrt{1 + \left(\frac{\sigma}{\omega\varepsilon_0}\right)^2} - 1\right]}$$

$$= 10^6 \times \pi\sqrt{\frac{4 \times \pi \times 10^{-7} \times 8.854 \times 10^{-12}}{2}\left[\sqrt{1 + \left(\frac{1.5 \times 10^{-5}}{10^6 \times \pi \times 8.854 \times 10^{-12}}\right)^2} - 1\right]} = 0.00273 \quad [\text{Np/m}]$$

$$\beta = \omega\sqrt{\frac{\mu_0\varepsilon_0}{2}\left[\sqrt{1 + \left(\frac{\sigma}{\omega\varepsilon_0}\right)^2} + 1\right]}$$

$$= 10^6 \times \pi\sqrt{\frac{4 \times \pi \times 10^{-7} \times 8.854 \times 10^{-12}}{2}\left[\sqrt{1 + \left(\frac{1.5 \times 10^{-5}}{10^6 \times \pi \times 8.854 \times 10^{-12}}\right)^2} + 1\right]} = 0.0108 \quad [\text{rad/m}]$$

In addition, the amplitude of the magnetic field intensity is

$$H(z) = \frac{E(z)}{\eta} \quad \left[\frac{\text{A}}{\text{m}}\right]$$

These are now combined together to write the magnetic field intensity. We will write it in the time domain. At point $P(0,0,0)$, the magnetic field intensity is

$$\mathbf{H}(z = 0) = \hat{\mathbf{y}}\frac{8}{\eta}\cos(10^6\pi t) \quad [\text{A/m}]$$

At a distance z in the direction of propagation, the magnetic field intensity is

$$\mathbf{H}(z) = \hat{\mathbf{y}}\frac{8}{\eta}e^{-\alpha z}\cos\left(10^6\pi t - \beta z\right) \quad [\text{A/m}]$$

For the values given, this is

$$\mathbf{H}(z = 1{,}000) = \hat{\mathbf{y}}\frac{8}{199.1\sqrt{2.777 + j1.5}}e^{-2.73}\cos\left(10^6\pi t - 10.8\right)$$
$$= \hat{\mathbf{y}}\,(0.0014 - j0.0004)e^{-2.73}\cos\left(10^6\pi t - 10.8\right) \quad [\text{A/m}]$$

The attenuation reduces the amplitude of the electric and magnetic fields to $e^{-2.73} = 0.0652$ of their amplitude at $z = 0$ in 1 km (a factor of over 15). This attenuation is rather high, indicating a rather lossy dielectric.

(b) The phase velocity of the wave in free space (lossless) is $c = 3 \times 10^8$ m/s. In the lossy dielectric, the phase velocity is given in **Eq. (12.97)**:

$$v_p = \frac{\omega}{\beta} = \frac{10^6 \times \pi}{0.0108} = 2.91 \times 10^8 \quad [\text{m/s}]$$

Thus, the wave propagates about 3% faster in free space than in the lossy dielectric.

Exercise 12.8 Show that **Eqs. (12.95)** and **(12.96)** are the real and imaginary parts of the propagation constant given in **Eq. (12.83)**.

Exercise 12.9 In the situation given in **Example 12.12**, assume that in addition to the properties given, there is also a small attenuation of 0.01 Np/m due to charged particles in air:

(a) Calculate the time-averaged power density in air at any location z.
(b) What are the stored magnetic and electric energy densities at the same location?

Answer

(a) $\mathcal{P}_{av} = \hat{\mathbf{z}}117.81e^{-0.02z} \quad [\text{W/m}^2]$.

(b) $w_m = 1.963 \times 10^{-4}e^{-0.02z}, \quad w_e = 1.963 \times 10^{-4}e^{-0.02z} \quad [\text{J/m}^3]$.

12.7.2 Propagation of Plane Waves in Low-Loss Dielectrics

We define low-loss materials as those materials in which the loss tangent is small: $\sigma/\omega\varepsilon \ll 1$ (or, equivalently, that the imaginary part of the complex permittivity in **Eq. (12.79)** is small compared to the real part). This relation also indicates that a material may be considered to be low loss in a given frequency range, whereas in another frequency range, this assumption may not hold. For example, conductivity of seawater is 4 S/m and its relative permittivity is 80 (at low frequencies). At 1 GHz, $\sigma/\omega\varepsilon = 0.899$. At 100 GHz, $\sigma/\omega\varepsilon = 0.00899 \ll 1$. At 1 MHz, $\sigma/\omega\varepsilon = 899 \gg 1$. Thus, the classification of materials changes, depending on frequencies, but at 100 GHz, seawater (based on the properties given here) is clearly low loss and at lower frequencies, it is a high-loss dielectric. In practice, the permittivity of the material also changes with frequency, changing the range in which a material may be considered to be a low-loss material.

All properties of the wave propagating in low-loss dielectrics remain the same as for any lossy material. But the above condition for low-loss materials simplifies some of these relations, allowing easier application and better understanding of behavior of waves propagating in these materials. The propagation constant now can be approximated using the binomial expansion (because $\sigma/\omega\varepsilon < 1$) as

$$\gamma = j\omega\sqrt{\mu\varepsilon}\sqrt{\left(1 - \frac{j\sigma}{\omega\varepsilon}\right)} \approx j\omega\sqrt{\mu\varepsilon}\left(1 - \frac{j\sigma}{2\omega\varepsilon} + \frac{1}{8}\left(\frac{\sigma}{\omega\varepsilon}\right)^2 + \frac{j}{16}\left(\frac{\sigma}{\omega\varepsilon}\right)^3 - \frac{5}{128}\left(\frac{\sigma}{\omega\varepsilon}\right)^4 + \ldots\right) \tag{12.103}$$

Deciding, somewhat arbitrarily, to neglect all but the first three terms in the expansion, the attenuation constant is approximated by the second (real) term in **Eq. (12.103)**:

$$\boxed{\alpha \approx \frac{\sigma}{2}\sqrt{\frac{\mu}{\varepsilon}} = \frac{\sigma}{2}\eta_n \quad \left[\frac{\text{Np}}{\text{m}}\right]} \tag{12.104}$$

where η_n is the no-loss intrinsic impedance (i.e., the intrinsic impedance of a material with the same μ and ε but in which $\sigma = 0$), and the phase constant is

$$\boxed{\beta \approx \omega\sqrt{\mu\varepsilon}\left(1 + \frac{1}{8}\left(\frac{\sigma}{\omega\varepsilon}\right)^2\right) \quad \left[\frac{\text{rad}}{\text{m}}\right]} \tag{12.105}$$

In very low-loss cases, the second term in **Eq. (12.105)** may also be neglected and the phase constant may often be approximated as

$$\beta \approx \omega\sqrt{\mu\varepsilon} \quad [\text{rad/m}] \tag{12.106}$$

The phase constant and, therefore, phase velocity and wavelength for low-loss dielectrics are essentially unchanged from those for the lossless dielectric because the second term in **Eq. (12.105)** is small, but the attenuation constant can be quite significant. Thus, the phase velocity and wavelength are

$$v_p \approx \frac{1}{\sqrt{\mu\varepsilon}\left(1 + \frac{1}{8}\left(\frac{\sigma}{\omega\varepsilon}\right)^2\right)} \quad \left[\frac{\text{m}}{\text{s}}\right], \quad \lambda = \frac{2\pi}{\beta} = \frac{v_p}{f} = \frac{1}{f\sqrt{\mu\varepsilon}\left(1 + \frac{1}{8}\left(\frac{\sigma}{\omega\varepsilon}\right)^2\right)} \quad [\text{m}] \tag{12.107}$$

for general low-loss materials and

$$v_p \approx \frac{1}{\sqrt{\mu\varepsilon}} \quad \left[\frac{\text{m}}{\text{s}}\right], \quad \lambda \approx \frac{2\pi}{\beta} = \frac{v_p}{f} \quad [\text{m}] \tag{12.108}$$

for very low-loss materials. The intrinsic impedance in low-loss dielectrics is still a complex number. Substituting the value of γ from **Eq. (12.103)** in **Eq. (12.101)**, and using the expansion again, η can be approximated as

$$\eta = \frac{j\omega\mu}{\gamma} = \sqrt{\frac{\mu}{\varepsilon}}\frac{1}{\sqrt{\left(1 - \frac{j\sigma}{\omega\varepsilon}\right)}} \approx \sqrt{\frac{\mu}{\varepsilon}}\left(1 + \frac{j\sigma}{2\omega\varepsilon}\right) = \eta_n\left(1 + \frac{j\sigma}{2\omega\varepsilon}\right) \quad [\Omega] \tag{12.109}$$

where η_n is the no-loss intrinsic impedance for the same material. The reactive part of the intrinsic impedance is quite small since $\sigma/\omega\varepsilon \ll 1$. Thus, for many practical applications, the intrinsic impedance of the lossless material may be used with little error.

Example 12.14 A satellite in geosynchronous orbit (36,000 km above the equator) is used for communication at 30 GHz. The atmosphere is 15 km thick. Assume free space above the atmosphere and plane wave propagation. Properties of the atmosphere are $\varepsilon = 1.05\varepsilon_0$ [F/m], $\mu = \mu_0$ [F/m], and $\sigma = 10^{-6}$ S/m:

(a) Calculate the phase velocity in the atmosphere and in free space.
(b) Calculate the attenuation and phase constants in the atmosphere and in free space.
(c) Calculate the propagation constant in the atmosphere (air). Compare with the propagation constant in free space.
(d) Compare the intrinsic impedance in free space and in the atmosphere.
(e) If the minimum electric field intensity required for reception is 10 mV/m, what must be the minimum amplitude of the electric field intensity at the transmitter? Assume the satellite does not amplify the signal but only reflects it, and both transmitter and receiver are on Earth.

Solution: First, we check that the low-loss equations apply to see if they can be used to calculate the phase velocity, attenuation, and phase constants. The attenuation constant is used to calculate the field after it propagates twice through the atmosphere (up and down). Only the atmosphere need be considered because there are no losses in free space:

(a) The low-loss condition is

$$\frac{\sigma}{\omega\varepsilon}=\frac{10^{-6}}{2\times\pi\times 3\times 10^{10}\times 8.854\times 1.05\times 10^{-12}}=5.7\times 10^{-7}\ll 1$$

The low-loss approximation applies here. The phase velocity in free space is $c = 3 \times 10^8$ m/s. In the atmosphere it is equal to $c/\sqrt{1.05} = 2.928 \times 10^8$ m/s. This is only about 2.4% change in the phase velocity.

(b) The attenuation and phase constants are given in **Eqs. (12.104)** and **(12.106)**:

$$\alpha\approx\frac{\sigma}{2}\sqrt{\frac{\mu_0}{\varepsilon_0\varepsilon_r}}=\frac{\sigma\eta_0}{2\sqrt{\varepsilon_r}}=\frac{10^{-6}\times 377}{2\times\sqrt{1.05}}=1.84\times 10^{-4}\quad[\mathrm{Np/m}]$$

$$\beta\approx\omega\sqrt{\mu\varepsilon}=\omega\sqrt{\mu_0\varepsilon_0\varepsilon_r}=\omega\frac{\sqrt{\varepsilon}}{c}=2\times\pi\times 3\times 10^{10}\frac{\sqrt{1.05}}{3\times 10^8}=643.83\quad[\mathrm{rad/m}]$$

Note: The more accurate expression in **Eq. (12.105)** may be used but, because the loss is very low, an identical result is obtained. For most applications, unless $\sigma/\omega\varepsilon$ is close to 1, **Eq. (12.106)** should be used rather than **Eq. (12.105)**.
The attenuation constant in free space is zero. The phase constant in free space is

$$\beta_0=\omega\sqrt{\mu_0\varepsilon_0}=\frac{\omega}{c}=628.32\quad\left[\frac{\mathrm{rad}}{\mathrm{m}}\right].$$

(c) The propagation constant in the atmosphere is $\gamma = \alpha + j\beta = 1.84 \times 10^{-4} + j643.83$, where α and β are those given in **(b)**. In free space, the propagation constant is $\gamma_0 = j\beta_0 = j628.32$.

(d) The intrinsic impedance in free space is 377 Ω. In the atmosphere, it is calculated using **Eq. (12.109)**, where $\eta_n = \eta_0/\sqrt{1.05}$

$$\eta\approx\eta_n\left(1+\frac{j\sigma}{2\omega\varepsilon}\right)=\frac{377}{\sqrt{1.05}}\left(1+\frac{j10^{-6}}{2\times 2\times\pi\times 3\times 10^{10}\times 1.05\times 8.854\times 10^{-12}}\right)=367.9+j1.07\times 10^{-4}\quad[\Omega]$$

The intrinsic impedance has a small imaginary part. If we neglect this, the only change in the intrinsic impedance is due to change in permittivity.

(e) Because only the amplitude is required and this is only attenuated in the atmosphere, we write

$$E=10\times 10^{-3}=E_0e^{-\alpha d}=E_0e^{-1.84\times 10^{-4}\times 30{,}000}=0.004E_0\quad\rightarrow\quad E_0=2.5\quad[\mathrm{V/m}]$$

Note: This is an ideal example: there are losses in free space as well, but these are usually smaller than in the atmosphere. Because attenuation occurs mostly in the atmosphere, satellite communication requires relatively little power. If space were entirely lossless (which it is not because of existence of charged particles), communication with satellites would require about the same power levels as communication on Earth at distances of about 30 km (twice the assumed thickness of the atmosphere). In reality, the atmosphere is much thicker, but its density and losses diminish with altitude. Also, the transmission spreads over a relatively large area. Therefore, the required output from satellites is larger than that calculated here, but satellites with power outputs of between 100 W and 200 W are common.

12.7.3 Propagation of Plane Waves in Conductors

In highly conductive materials, the losses are high and we can assume that $\sigma \gg \omega\varepsilon$ (i.e., conduction currents dominate). Under this condition, the complex propagation constant can be approximated from **Eq. (12.83)** as

$$\gamma \approx j\omega\sqrt{\mu\varepsilon}\sqrt{-\frac{j\sigma}{\omega\varepsilon}} = \sqrt{\frac{j\omega\mu\varepsilon\sigma}{\varepsilon}} = (1+j)\sqrt{\frac{\omega\mu\sigma}{2}} \tag{12.110}$$

by neglecting 1 compared to $j\sigma/\omega\varepsilon$ in **Eq. (12.83)** and using $\sqrt{\mathrm{j}} = (1+\mathrm{j})/\sqrt{2}$. From this, we get

$$\alpha = \sqrt{\frac{\omega\mu\sigma}{2}} = \sqrt{\pi f\mu\sigma} \quad \left[\frac{\mathrm{Np}}{\mathrm{m}}\right], \quad \beta = \sqrt{\frac{\omega\mu\sigma}{2}} = \sqrt{\pi f\mu\sigma} \quad \left[\frac{\mathrm{rad}}{\mathrm{m}}\right] \tag{12.111}$$

The attenuation and phase constants are equal in magnitude and are very large. The wave is attenuated rapidly to the point where propagation in conducting media can only exist within short distances. The propagating wave can now be written as

$$E_x(z) = E_0^+ e^{-z/\delta} e^{-jz/\delta} \quad [\mathrm{V/m}] \tag{12.112}$$

where the term

$$\delta = \sqrt{\frac{2}{\omega\mu\sigma}} = \sqrt{\frac{1}{\pi f\mu\sigma}} = \frac{1}{\alpha} \quad [\mathrm{m}] \tag{12.113}$$

is known as the ***skin depth*** or ***depth of penetration*** of the wave. It is defined as that distance in which the amplitude of a plane wave is attenuated to $1/e$ of its original amplitude. The skin depth in conductors is small. In the microwave range, it can be of the order of a few microns (depending on material properties and frequency). Because waves at these high frequencies penetrate very little in conductors, it is quite common to use the perfect conductor approximation for conducting materials.

The phase velocity in good conductors is [from **Eqs. (12.97)** and **(12.111)**]

$$v_p = \frac{\omega}{\beta} = \omega\delta = \sqrt{\frac{2\omega}{\mu\sigma}} \quad \left[\frac{\mathrm{m}}{\mathrm{s}}\right] \tag{12.114}$$

and is obviously small compared to the phase velocity in dielectrics or free space, because δ is small.

The wavelength also changes drastically compared to free space or lossless dielectrics. It is very short and given by

$$\lambda = \frac{2\pi}{\beta} = 2\pi\delta \quad [\mathrm{m}] \tag{12.115}$$

The intrinsic impedance is [using **Eq. (12.102)**]

$$\eta \approx \frac{j\omega\mu}{\gamma} = \frac{j\omega\mu}{(1+j)\sqrt{\omega\mu\sigma/2}} = (1+j)\sqrt{\frac{\omega\mu}{2\sigma}} = (1+j)\frac{1}{\sigma\delta} = (1+j)\frac{\omega\mu\delta}{2} \quad [\Omega] \tag{12.116}$$

where $j/(1+j) = (1+j)/2$ was used. The phase angle of the intrinsic impedance is, therefore, 45°. This is characteristic of good conductors for which the magnetic field intensity lags behind the electric field intensity by 45°. The intrinsic impedance of conductors can be very low and is much lower than the intrinsic impedance of free space. For example, the intrinsic impedance in copper at 1 GHz is $(1+j) \times 8.3 \times 10^{-3}$ Ω compared to 377 Ω in free space.

Example 12.15 Application: Shielded Enclosures As an engineer you are asked to design a shielded room so that high-frequency waves cannot penetrate into the room. The shield is made of aluminum, in the form of a box. Assume that waves are plane waves and the lowest frequency at which the shielded room should satisfy the requirements is 1 MHz. The shield must reduce the amplitude of the electric field intensity inside the box by a factor of 10^6 compared to the amplitude outside. The conductivity of aluminum is 3.7×10^7 S/m. Assume that the electric field intensity at the outer surfaces of the conductor is the same as immediately below the surfaces and calculate the thickness required. (This is not true in conductors because of reflection of waves at the surface, a subject we will treat in the next chapter, but for the purpose of this problem, we will assume no reflection at the surface of the conductor). The calculation here will give a "worst-case solution" since reflection of waves reduces the field in the conducting layer and increases shielding effectiveness.

(a) Calculate the minimum thickness of the walls of the shielded room to satisfy the design criterion.
(b) Suppose you ran out of aluminum and needed to use iron instead of aluminum. If the conductivity of iron is 1×10^7 S/m and its relative permeability is 100, what is the thickness of the shield?

Solution: The amplitude of the electric field intensity is E_0 on one side of the conducting sheet and needs to be $10^{-6}E_0$ on the other side, as shown in **Figure 12.14**. The thickness necessary is calculated from the attenuation constant and the electric field intensities.

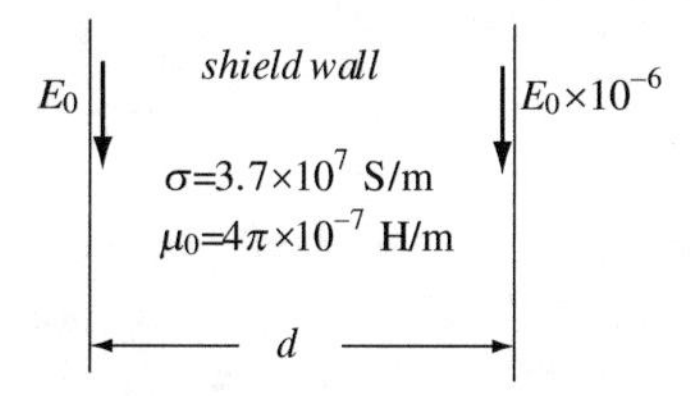

Figure 12.14 The relation between the electric field intensities at the two surfaces of a conducting shield wall

(a) The attenuation constant is given in **Eq. (12.111)**:

$$\alpha = \sqrt{\pi f \mu_0 \sigma} = \sqrt{\pi \times 10^6 \times 4 \times \pi \times 10^{-7} \times 3.7 \times 10^7} = 1.21 \times 10^4 \quad [\text{Np/m}]$$

The electric field intensity at the inner and outer surfaces inside the conducting sheet is related as

$$10^{-6}E_0 = E_0 e^{-\alpha d}$$

or taking the logarithm on both sides

$$-6\ln 10 = -\alpha d \quad \rightarrow \quad d = \frac{6\ln 10}{\alpha} = \frac{6 \times \ln 10}{1.21 \times 10^4} = 0.001142 \quad [\text{m}]$$

The aluminum box should be 1.142 mm thick.

(b) The conductivity of iron is lower but its permeability is higher. The attenuation constant in iron at the same frequency is

$$\alpha = \sqrt{\pi f 100 \mu_0 \sigma} = \sqrt{\pi \times 10^6 \times 100 \times 4 \times \pi \times 10^{-7} \times 10^7} = 6.28 \times 10^4 \quad [\text{Np/m}]$$

and the thickness d is

$$d = \frac{6 \ln 10}{\alpha} = \frac{6 \times \ln 10}{6.28 \times 10^4} = 0.00022 \quad [\mathrm{m}]$$

Iron is a better shielding material because the iron shield is over 5 times thinner, and, even though iron is about 2.5 times heavier than aluminum, the total weight is lower. Higher-permeability materials may be used to obtain an even more effective shield. On the other hand, high conductivity, nonmagnetic materials like copper and aluminum are commonly used for shields at high frequencies because they are easier to work with and have other properties that make them attractive for design. This calculation shows why your radio will not work inside a reinforced concrete garage or a cell phone call may interrupted in an elevator.

Example 12.16 Application: Communication with Submarines The most severe restriction to communication with submerged submarines is the high loss exhibited by seawater. The following example shows this difficulty.

Suppose we wish to communicate with submarines using a conventional communication system at 1 MHz (in the AM radio range). Properties of seawater are $\varepsilon = 78\varepsilon_0$ [F/m], $\mu = \mu_0$ [H/m], and $\sigma = 4$ S/m. Assume that a magnetic field intensity of 10,000 A/m can be generated at the surface of the ocean and that the receiver in the submarine can receive magnetic fields as low as 1 μA/m:

(a) Calculate the maximum range for this communication system.
(b) Suppose the frequency is lowered to 100 Hz. What is the range now?
(c) Suppose the antenna on the submarine must be one half wavelength in length. Calculate the required antenna lengths in **(a)** and in **(b)**.

Solution: First, we must check if seawater should be treated as a conductor or as a lossy dielectric. Then we need to calculate the attenuation constant and the range for communication as in **Example 12.15**. The same applies to the lower frequency in **(b)**, but we need to check again if seawater can be treated as a conductor at the lower frequency:

(a) To see if seawater is a conductor or not at 1 MHz, we write

$$\frac{\sigma}{\omega\varepsilon} = \frac{4}{2 \times \pi \times 10^6 \times 78 \times 8.854 \times 10^{-12}} = 921 \gg 1$$

This is a high-loss medium and the attenuation constant is

$$\alpha = \sqrt{\pi f \mu_0 \sigma} = \sqrt{\pi \times 10^6 \times 4 \times \pi \times 10^{-7} \times 4} = 3.974 \quad [\mathrm{Np/m}]$$

The maximum range is calculated from the following relation (see **Example 12.15**):

$$1 \times 10^{-6} = H_0 e^{-\alpha d} = 10{,}000 e^{-\alpha d} \quad \rightarrow \quad 1 \times 10^{-10} = e^{-\alpha d} \quad \rightarrow \quad d = \frac{10 \ln 10}{3.974} = 5.794 \quad [\mathrm{m}]$$

The range for communication is less than 6 m. This perhaps is not surprising since the skin depth in seawater at 1 MHz is only about 0.25 m.

(b) At 100 Hz, we have

$$\frac{\sigma}{\omega\varepsilon} = \frac{4}{2 \times \pi \times 100 \times 10^6 \times 78 \times 8.854 \times 10^{-12}} = 9.21 \times 10^6 \gg 1$$

and this is certainly a conductor. The attenuation constant is

$$\alpha = \sqrt{\pi f \mu_0 \sigma} = \sqrt{\pi \times 100 \times 4 \times \pi \times 10^{-7} \times 4} = 0.03974 \quad [\mathrm{Np/m}]$$

Using the formula for distance in **(a)**, we get

$$d = \frac{10\ln 10}{0.03974} = 579.4 \quad [\text{m}]$$

This is not a very long range but is feasible for communication underwater.

Note: The attenuation in seawater is very high and increases with frequency as this example shows. For this reason, communication in seawater is very difficult, and if done, it must be done at very low frequencies.

(c) The wavelength is calculated as

$$\lambda = \frac{2\pi}{\beta} = \frac{2\pi}{\alpha} \quad [\text{m}]$$

The wavelength (underwater) at 100 Hz is 158 m and at 1 MHz it is 1.58 m. A half-wavelength antenna at 100 Hz must be 79 m long and at 1 MHz it must be 0.79 m. Note that the wavelength underwater is significantly shorter than in free space.

Low-frequency trailing antennas for submarines are used for communication at very low frequencies (usually below 150 Hz). These antennas may be hundreds of meters long.

Example 12.17 Application: Current Distribution and AC Resistance of Conductors One important consequence of the skin depth is the fact that currents in conductors decay exponentially from the surface inward. Thus, in AC systems, the current-carrying capacity is reduced since more of the current flows on the surface, whereas the current density allowable is fixed. This also means that the AC resistance of a conductor is larger than its DC resistance.

A conductor with a 4 mm diameter carries a total current $I = 100$ A. The conductor is made of copper with a conductivity of 5.7×10^7 S/m. Calculate:

(a) DC resistance per meter length and current density in the wire.

(b) AC resistance per meter length and maximum current in the wire. Use a frequency of 60 Hz and assume exponential decay of the current density from the surface to the center of the wire with maximum current density allowable being the same as in the DC case.

Solution: The electric field intensity in the conductor decays exponentially from its value on the surface. Since $\mathbf{J} = \sigma\mathbf{E}$, the current density $\mathbf{J}$ also decays exponentially. Thus, to calculate the DC and AC resistances, we assume in each case a known current density at the surface of the conductor. In the DC case, this current density is uniform throughout the conductor's cross section. In the AC case, it is largest at the surface. Therefore, for a given maximum current density, the total current in the DC case is larger and its DC resistance is lower:

(a) The DC resistance of the conductor was calculated in **Chapter 7 [Eq. (7.18)]** as

$$R_{dc} = \frac{L}{\sigma S} = \frac{1}{5.7 \times 10^7 \times \pi \times 0.002^2} = 0.001396 \quad [\Omega/\text{m}]$$

The current density in the conductor is the total current divided by its cross-sectional area:

$$J = \frac{I}{S} = \frac{100}{\pi \times 0.002^2} = 7.958 \times 10^6 \quad \left[\text{A/m}^2\right].$$

(b) To calculate the AC resistance, we assume this current density at the surface. Then, we calculate the current density everywhere in the conductor and integrate the current density to obtain the total current. The same voltage that produced the current in **(a)** must produce the current in **(b)**. Thus, the resistance is calculated from the total current and voltage. The attenuation constant is

$$\alpha = \sqrt{\pi f \mu \sigma} = \sqrt{\pi \times 60 \times 4 \times \pi \times 10^{-7} \times 5.7 \times 10^7} = 116.2 \quad [\text{Np/m}]$$

The skin depth in the conductor is

$$\delta = \frac{1}{\alpha} = 0.0086 \quad [\text{m}]$$

The current density in the conductor can now be written as

$$J(r) = J_0 e^{-\alpha(a-r)} \quad [\text{A/m}^2]$$

where J_0 is the current density at the surface, a is the radius of the conductor, and r is the distance from the center of the conductor at which the current density is calculated. The total current in the conductor is now calculated using **Figure 12.15**. The current in a ring of radius r and thickness dr is

$$dI_{ac} = 2\pi r dr J(r) = 2\pi r J_0 e^{-\alpha(a-r)} dr \quad [\text{A}]$$

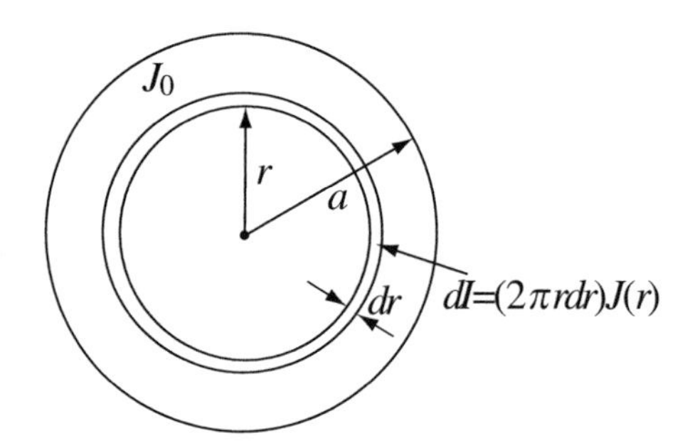

Figure 12.15 Calculation of AC current in a conductor

Integrating over the radius of the conductor

$$I_{ac} = 2\pi J_0 e^{-\alpha a} \int_{r=0}^{r=a} r e^{\alpha r} dr = 2\pi J_0 e^{-\alpha a} \frac{e^{\alpha r}}{\alpha^2} (\alpha r - 1) \Big|_0^a$$

this gives

$$I_{ac} = 2\pi J_0 e^{-\alpha a} \left[\frac{e^{\alpha a}}{\alpha^2} (\alpha a - 1) + \frac{1}{\alpha^2} \right]$$

$$= 2 \times \pi \times 7.958 \times 10^6 e^{-116.2 \times 0.002} \left[\frac{e^{116.2 \times 0.002}}{116.2^2} (116.2 \times 0.002 - 1) + \frac{1}{116.2^2} \right] = 92.7 \quad [\text{A}]$$

The voltage that produced the current in **(a)** is $V = I_{dc} R_{dc}$. If we connect an AC voltage with RMS value equal to the DC voltage in **(a)** to a 1 m length of the conductor, the AC resistance is

$$R_{ac} = \frac{V}{I_{ac}} = \frac{I_{dc} R_{dc}}{I_{ac}} = \frac{0.00139 \times 100}{92.7} = 0.0015 \quad [\Omega/\text{m}]$$

The AC resistance is about 8% higher than the DC resistance, causing more losses on the line.

This example shows a very important aspect of power transmission; DC systems are more efficient in transferring power. For the same allowable current density in a conductor, they transfer more power than AC systems. However, because of questions of transformation of voltages, AC power is usually used for transmission. When power must be transmitted over long distances, high-voltage direct current (HVDC) systems are more economical in spite of the need for converters and inverters on the ends of the power transmission line. HVDC distribution systems are in use throughout the world, especially for transmission from remote power stations. Voltages in excess of 1 million volts may be used for this purpose.

12.7.4 The Speed of Propagation of Waves and Dispersion

In **Section 12.3.5**, we defined the phase velocity and wave number of a plane wave propagating in free space. These definitions were extended to perfect dielectrics, lossy dielectrics, and conductors in **Sections 12.7.1** through **12.7.3**. However, there are some difficulties with the definition of phase velocity which we have not addressed and which will be discussed here. First, the phase velocity, defined as $v_p = 1/\sqrt{\mu\varepsilon}$ [see **Eq. (12.29)**], cannot be used in general. For example, we had to modify v_p in lossy dielectrics and conductors as shown in **Eqs. (12.97)** and **(12.107)**, respectively. Clearly, the definition in **Eq. (12.29)** is only valid in free space and, by extension, in lossless dielectrics.

Similarly, the wave number $k = \omega\sqrt{\mu\varepsilon}$ in **Eq. (12.23)** or the phase constant β in **Eq. (12.96)** is frequency dependent. The definition of phase velocity as used up to now, of course, only applied to monochromatic waves (waves at a single-valued frequency). Can we still use the idea of a phase velocity for nonmonochromatic waves, that is for waves that contain a range of frequencies? The answer is clearly no, except for propagation in lossless unbounded media. To characterize the speed of propagation of a nonmonochromatic wave, we will introduce here the idea of group velocity. Finally, in this regard, we might ask the following question: Are the phase velocity or group velocity also the speed at which energy propagates? Again, the answer, in general, is no, and we will have to define a new velocity: the velocity of energy transport to answer this question.

The meaning of phase velocity is usually taken as the speed with which the phase of a wave propagates in space. It was alluded to in **Section 12.3.5** that this is not the speed of any real quantity propagating and therefore can be, and often is, larger than the speed of light.

Fortunately, there are many cases in which the many velocities of electromagnetic waves (or light) yield the same results and there is no need to worry about different velocities. In particular, in monochromatic plane waves, propagating in free space in an unbounded domain, the phase velocity, group velocity, and velocity of energy transport are the same. Similarly, in lossless dielectrics as well as in very low-loss dielectrics, either the phase velocity is independent of frequency or may be approximated as frequency independent, simplifying analysis.

In addition to phase velocity, group velocity, and the velocity of energy transport, there are other definitions of wave velocity in electromagnetics, each with its own assumptions and uses.[4]

12.7.4.1 Group Velocity

Group velocity is the velocity of a wave packet consisting of a narrow range or band of frequencies. An example akin to this is a frequency-modulated (FM) wave as used in FM radio transmission. In this type of wave, a carrier wave at an angular frequency ω_0 is modulated by another wave of angular frequency $\Delta\omega \ll \omega_0$. The angular frequency of the wave will vary between $\omega_0 - \Delta\omega$ and $\omega_0 + \Delta\omega$. Clearly, we cannot now talk about the phase velocity of the wave because phase velocity is only defined for a single frequency.

To define the group velocity of a packet of waves, we will consider here the case of amplitude modulation (AM). Consider two waves with the same amplitude and propagating in the same direction, but the two waves are at slightly different frequencies. One wave is at an angular frequency $\omega_1 = \omega_0 - \Delta\omega$ and the other at $\omega_2 = \omega_0 + \Delta\omega$. The amplitudes of the waves are E and the waves propagate in the z direction in a lossless medium. The phase constants of each of the waves are written from the definition $\beta = \omega\sqrt{\mu\varepsilon}$ [rad/m]. Therefore, $\beta_1 = \beta_0 - \Delta\beta$ and $\beta_2 = \beta_0 + \Delta\beta$. With these, the waves are

$$E_1(z,t) = E\cos(\omega_1 t - \beta_1 z) = E\cos\big((\omega_0 - \Delta\omega)t - (\beta_0 - \Delta\beta)z\big) \quad [\text{V/m}] \tag{12.117}$$

$$E_2(z,t) = E\cos(\omega_2 t - \beta_2 z) = E\cos\big((\omega_0 + \Delta\omega)t - (\beta_0 + \Delta\beta)z\big) \quad [\text{V/m}] \tag{12.118}$$

[4] R. L. Smith, "The Velocities of Light," American Journal of Physics, Vol. 38, No. 8, Aug. 1970, pp. 978–983.

The sum of these two waves gives the total wave:

$$\begin{aligned} E(z,t) &= E_1(z,t) + E_2(z,t) \\ &= E\cos((\omega_0 + \Delta\omega)t - (\beta_0 + \Delta\beta)z) + E\cos((\omega_0 - \Delta\omega)t - (\beta_0 - \Delta\beta)z) \\ &= E\cos(\omega_0 + \Delta\omega)t\cos(\beta_0 + \Delta\beta)z + E\sin(\omega_0 + \Delta\omega)t\sin(\beta_0 + \Delta\beta)z \\ &\quad + E\cos(\omega_0 - \Delta\omega)t\cos(\beta_0 - \Delta\beta)z + E\sin(\omega_0 - \Delta\omega)t\sin(\beta_0 - \Delta\beta)z \\ &= [2E\cos(\Delta\omega t - \Delta\beta z)]\cos(\omega_0 t - \beta_0 z) \quad [\mathrm{V/m}] \end{aligned} \tag{12.119}$$

This is a wave with amplitude equal to the sum of the amplitudes of the individual waves and a fundamental or carrier frequency ω_0. The amplitude of the wave varies cosinusoidally with frequency $\Delta\omega$ as can be seen in **Figure 12.16**. The carrier travels at a velocity v_p, which is calculated [see **Eq. (12.28)**] as follows.

By assuming a constant point on the carrier, the phase velocity of the single frequency carrier is

$$\omega_0 t - \beta_0 z = const. \quad \rightarrow \quad \frac{d(\omega_0 t - \beta_0 z)}{dt} = \omega_0 - \beta_0 \frac{dz}{dt} = 0 \quad \rightarrow \quad v_p = \frac{dz}{dt} = \frac{\omega_0}{\beta_0} \quad \left[\frac{\mathrm{m}}{\mathrm{s}}\right] \tag{12.120}$$

This much we have seen for a monochromatic plane wave.

The modulation, or the envelope, also travels but at a different velocity. Performing the same operation for the modulation, we write

$$\Delta\omega t - \Delta\beta z = \text{const.} \quad \rightarrow \quad \frac{d(\Delta\omega t - \beta z)}{dt} = \Delta\omega - \Delta\beta\frac{dz}{dt} = 0 \quad \rightarrow \quad v_g = \frac{dz}{dt} = \frac{\Delta\omega}{\Delta\beta} = \frac{1}{\Delta\beta/\Delta\omega} \quad \left[\frac{\mathrm{m}}{\mathrm{s}}\right] \tag{12.121}$$

In the limit, as $\Delta\beta$ tends to zero, we write

$$\boxed{v_g = \lim_{\Delta\beta\to 0}\frac{1}{\Delta\beta/\Delta\omega} = \frac{1}{d\beta/d\omega} \quad \left[\frac{\mathrm{m}}{\mathrm{s}}\right]} \tag{12.122}$$

This velocity is called the ***group velocity*** or the velocity of the wave packet with a narrow frequency width ($\Delta\omega \ll \omega_0$). The latter is actually an informative name since the modulation, or the group, is traveling at this velocity, which can be very different than the phase velocity. The definition given here does not apply to wide-band signals.

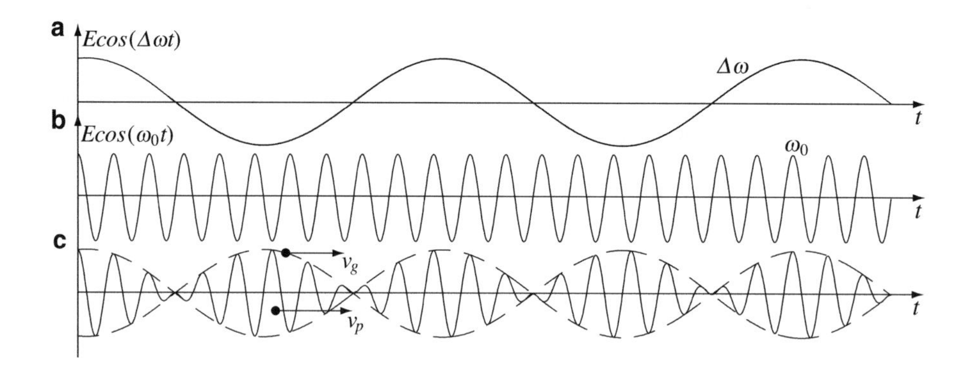

Figure 12.16 Amplitude modulation. (**a**) The modulating signal. (**b**) The high-frequency carrier. (**c**) The amplitude-modulated carrier. Signals shown at $z = 0$

12.7.4.2 Velocity of Energy Transport

The velocity of energy transport is based on the fact that when calculating the Poynting vector of a wave, we, in fact, calculate the rate of change of energy in a given volume. The energy density in the volume is given in [J/m^3], whereas the Poynting vector is given in [W/m^2]. The ratio between the Poynting vector and the energy density is the velocity of energy transport. In a lossless medium, the velocity of energy transport is defined as the ratio between the time-averaged propagated power density (time-averaged Poynting vector $\mathcal{P}_{av}$) and the time-averaged stored energy density:

$$\mathbf{v}_e = \mathcal{P}_{av}/w_{av} \quad [\text{m/s}] \tag{12.123}$$

This definition of velocity is convenient because it gives the velocity at which energy is transported and is always lower or equal to the speed of light in the medium in which the wave propagates. In addition, the direction of propagation of energy is also immediately available. This definition will become handy in cases where the phase velocity becomes larger than the speed of light, a situation often encountered in the propagation of waves in the presence of conducting bodies (but not inside the conductors). Because the velocity of energy transport is defined for lossless media, it is always equal to the phase velocity in lossless, unbounded space. It is, however, different when waves propagate in bounded space, as we shall see in the next chapter.

12.7.4.3 Dispersion

Now that we have the group and phase velocities, it is relatively easy to understand what dispersion means. To see this, consider again the two waves in **Eqs. (12.117)** and **(12.118)**, but suppose that each propagates a distance z_0, separately, in a lossy dielectric. The phase velocity of each wave is given in **Eq. (12.97)** and is different for each wave. The two waves arrive at their target at different times and with different phase angles. If the two waves carry information, then this information will arrive distorted. A simple example is transmission of music from a radio station to your radio receiver. Each frequency in the signal will propagate at different speeds and arrive with different phases. The signal you will hear will have components which are delayed and shifted in phase. If the phase velocity is frequency dependent, we say that the signal disperses and, therefore, that the medium through which the wave propagates is dispersive. We call any material in which the phase velocity depends on frequency a ***dispersive medium***. Fortunately, not all materials are dispersive. We saw that in free space, the phase constant is linearly dependent on frequency ($\beta = \omega\sqrt{\mu\varepsilon}$). Therefore, the phase velocity $v_p = \omega/\beta = 1/\sqrt{\mu\varepsilon}$ is independent of frequency. Free-space and perfect dielectrics are nondispersive media. In nondispersive media, all waves propagate at the same speed and, therefore, the group velocity and the phase velocity are equal: $v_g = v_p$. In other materials, the permittivity and, sometimes, the permeability are frequency dependent; therefore, these materials are dispersive.

In most dispersive materials, the phase velocity decreases as frequency increases. This dispersion is called ***normal dispersion***. In some other materials, the phase velocity increases with frequency. This is called ***anomalous dispersion***.

A dispersion relation is the relation between β and ω shown in **Figure 12.17**. **Figure 12.17a** shows a number of nondispersive materials. The relation between β and ω is linear. Therefore, when taking the derivative in **Eq. (12.122)**, the result is a straight line whose slope is the phase velocity in the medium. **Figure 12.17b** shows a nonlinear dispersion relation. The group velocity is the tangent to this curve at any point on the curve, indicating that the group velocity is frequency dependent. The line asymptotically tangential to the curve is the lossless phase velocity (since, as frequency approaches infinity, $\sigma/\omega\varepsilon$ approaches zero and the material becomes lossless).

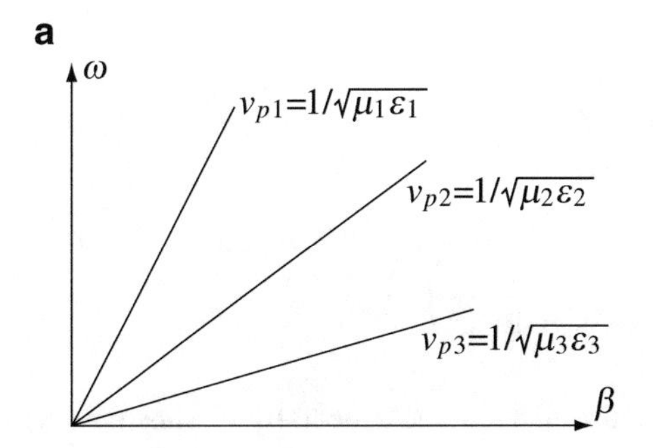

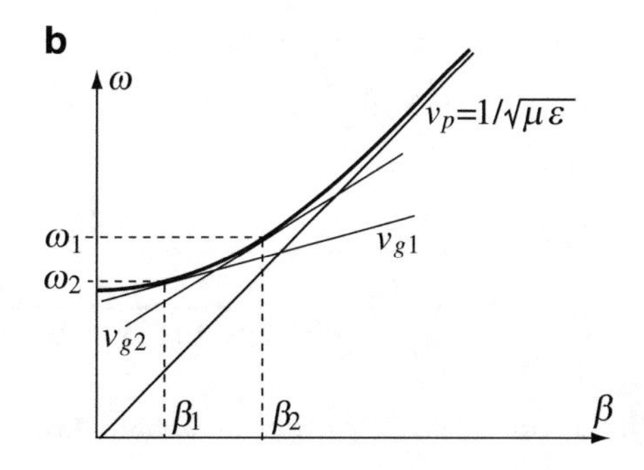

Figure 12.17 Nondispersive and dispersive media. (**a**) A nondispersive medium has a linear relation between frequency and phase constant. (**b**) A dispersive medium has a nonlinear relation between frequency and phase constant

Example 12.18 Application: Dispersion in the Atmosphere A TV station operates with a carrier signal of 96 MHz. The video signal, which is modulated on the carrier signal, is 6 MHz in width, making the frequency of the wave vary between 93 and 99 MHz. The waves generated by the station may be viewed as plane waves at large distances. Assume the station transmits in a lossy atmosphere with permeability μ_0 [H/m], permittivity $1.05\varepsilon_0$ [F/m], and conductivity 10^{-3} S/m:

(a) Calculate the phase and group velocities of the wave.
(b) Show that if we can assume the medium to be a very low-loss material, phase velocity and group velocity are essentially the same.

Solution: Before doing anything else, we must decide which approximations, if any, may be used for calculations by evaluating $\sigma/\omega\varepsilon$. Based on this, we choose the appropriate formulas for phase velocity, phase constant, etc. The group velocity is then calculated from the phase constant using **Eq. (12.122)**.

In this case,

$$\frac{\sigma}{\omega\varepsilon}=\frac{10^{-3}}{2\times\pi\times 96\times 10^{6}\times 1.05\times 8.854\times 10^{-12}}=0.178<1.0$$

However, it would be inappropriate to assume that this is a very low-loss dielectric and, therefore, we must use **Eq. (12.105)** for the phase constant and **Eq. (12.107)** for phase velocity.

(a) The phase velocity is frequency dependent, as can be seen from **Eq. (12.107)**. Thus, we calculate the minimum (at the lowest frequency transmitted by the station) and maximum phase velocities (at the highest frequency transmitted):
At 93 MHz:

$$v_{pmin}\approx\frac{1}{\sqrt{\mu_0\varepsilon_r\varepsilon_0}\left(1+\frac{1}{8}\left(\frac{\sigma}{\omega_{min}\varepsilon}\right)^2\right)}=\frac{c}{\sqrt{\varepsilon_r}\left(1+\frac{1}{8}\left(\frac{\sigma}{\omega_{min}\varepsilon}\right)^2\right)}$$

$$=\frac{3\times 10^{8}}{\sqrt{1.05}\left(1+\frac{1}{8}\left(\frac{10^{-3}}{2\times\pi\times 93\times 10^{6}\times 1.05\times 8.854\times 10^{-12}}\right)^2\right)}=2.9154\times 10^{8}\quad[\mathrm{m/s}]$$

At 99 MHz:

$$v_{p\,max}\approx\frac{c}{\sqrt{\varepsilon_r}\left(1+\frac{1}{8}\left(\frac{\sigma}{\omega_{max}\varepsilon}\right)^2\right)}$$

$$=\frac{3\times 10^{8}}{\sqrt{1.05}\left(1+\frac{1}{8}\left(\frac{10^{-3}}{2\times\pi\times 99\times 10^{6}\times 1.05\times 8.854\times 10^{-12}}\right)^2\right)}=2.9168\times 10^{8}\quad[\mathrm{m/s}]$$

Although the velocities only differ by about 0.06%, this can cause distortions, especially if transmission is over long distances.

To calculate the group velocity, we use **Eq. (12.122)** with the phase constant given in **Eq. (12.105)**. From this, we evaluate

$$\frac{d\beta}{d\omega}=\frac{d}{d\omega}\left[\omega\sqrt{\mu\varepsilon}\left(1+\frac{1}{8}\left(\frac{\sigma}{\omega\varepsilon}\right)^2\right)\right]=\sqrt{\mu\varepsilon}\left(1-\frac{1}{8}\left(\frac{\sigma}{\omega\varepsilon}\right)^2\right)$$

The group velocity is then given from **Eq. (12.122)** as

$$v_g = \frac{1}{d\beta/d\omega} = \frac{1}{\sqrt{\mu_0\varepsilon_r\varepsilon_0}\left(1-\frac{1}{8}\left(\frac{\sigma}{\omega\varepsilon}\right)^2\right)} = \frac{c}{\sqrt{\varepsilon_r}\left(1-\frac{1}{8}\left(\frac{\sigma}{\omega\varepsilon}\right)^2\right)}$$

$$= \frac{3\times 10^8}{\sqrt{1.05}\left(1-\frac{1}{8}\left(\frac{10^{-3}}{2\times\pi\times 96\times 10^6\times 1.05\times 8.854\times 10^{-12}}\right)^2\right)} = 2.939384\times 10^8 \quad [\mathrm{m/s}]$$

Note that the group velocity is calculated at the center (carrier) frequency and is different than the phase velocity at that same frequency which is 2.9161 $\times$ 10^8 m/s. However, the differences are small and for this reason, we usually cannot sense distortions in normal radio transmissions due to dispersion.

(b) In a very low-loss material, $\sigma/\omega\varepsilon$ can be neglected and the phase velocity becomes that of the lossless medium. Under this condition,

$$v_p = v_g = \frac{1}{\sqrt{\mu\varepsilon_r}} = \frac{c}{\sqrt{\varepsilon_r}} = \frac{3\times 10^8}{\sqrt{1.05}} = 2.9277\times 10^8 \quad [\mathrm{m/s}].$$

12.8 Polarization of Plane Waves

The electric (or magnetic) field intensity of a uniform plane wave has a direction in space. This direction may either be constant or may change as the wave propagates. Because the direction of the fields is important we define what is called polarization of waves. The polarization of a plane wave is

"the figure traced by the tip of the electric field intensity vector as a function of time, at a fixed point in space."

Polarization is defined for the electric field intensity only, since the magnetic field intensity is obtainable from the electric field intensity by use of Maxwell's equations [(**Eq. (12.32)**] and there is no need to define its polarization separately. The reason why polarization of plane waves is important is because the propagation properties of the wave are affected by the polarization. A simple example is the direction of an antenna. If the electric field intensity is, say, horizontally directed in space (polarized in this direction), then a receiving antenna must also be directed horizontally if the electric field intensity (and therefore the current density) in the antenna is to be maximized. Polarization of plane waves can be intentional or a result of propagation through materials.

We will treat linear polarization first as the simplest polarization and then extend linear polarization to include the important cases of circular and elliptical polarization.

The polarization of a plane wave is determined as follows:

(a) Write the electric field intensity in the time domain.
(b) An observation point in space is chosen so that the wave propagates straight toward the observer at that point.
(c) The direction of the electric field intensity **E** is followed as time changes. The tip of the vector traces some pattern in the plane perpendicular to the direction of propagation. This trace is the polarization of the wave. If the figure traced by the tip of the electric field intensity is a straight line, we call this ***linear polarization***. In general, the trace is an ellipse and the polarization is said to be ***elliptical polarization***. A special case of the elliptical polarization is ***circular polarization***.
(d) In addition, the vector may seem to rotate clockwise or counterclockwise as time changes. If the vector rotates in a clockwise direction, the vector is said to be ***left-hand polarized***. If the wave rotates in a counterclockwise direction, it is said to be ***right-hand polarized***. A wave may be, for example, ***left-hand elliptically polarized*** (or, in short, ***left elliptically polarized***) or ***right-hand elliptically polarized*** (in short, ***right elliptically polarized***). The sense of rotation of the wave is determined by the simple use of the right-hand rule: If the fingers of the right-hand curl in the direction of rotation of the electric field, the thumb must show in the direction of propagation for a right-hand sense and in the direction opposite the direction of propagation for a left-hand sense.

These definitions and their physical meaning are described next.

12.8.1 Linear Polarization

The electric field intensity

$$\mathbf{E} = \hat{\mathbf{y}}E_y(z) = \hat{\mathbf{y}}E_y e^{-\gamma z} \quad [\text{V/m}] \tag{12.124}$$

of a wave propagating in the z direction and directed in the y direction is linearly polarized in the y direction. The electric field intensity varies in the y direction (its amplitude may be constant or decaying, but, as it propagates, the phase changes). This linearly polarized wave is shown in **Figure 12.18a** as viewed by an observer, placed on the z axis looking onto the xy plane where the electric field is shown as it changes with time. To see how this occurs, we write the electric field intensity in **Eq. (12.124)** in the time domain:

$$\mathbf{E}(z,t) = \text{Re}\{\hat{\mathbf{y}}E_y e^{-\gamma z}e^{j\omega t}\} = \text{Re}\{\hat{\mathbf{y}}E_y e^{-\alpha z - j\beta z}e^{j\omega t}\} = \hat{\mathbf{y}}E_y e^{-\alpha z}\cos(\omega t - \beta z) \quad [\text{V/m}] \tag{12.125}$$

For any constant value of z, the direction of the vector remains in the y direction as the amplitude changes between a negative value $-E_y e^{-\alpha z}$ to a positive value $E_y e^{-\alpha z}$.

The electric field intensity

$$\mathbf{E}(z,t) = \hat{\mathbf{x}}E_x e^{-\alpha z}\cos(\omega t - \beta z) + \hat{\mathbf{y}}E_y e^{-\alpha z}\cos(\omega t - \beta z) \quad [\text{V/m}] \tag{12.126}$$

has two components: one in the x direction, one in the y direction. **Figure 12.18b** shows that each component describes a line on the corresponding axis while its amplitude is attenuated. The resultant wave is in a direction that depends on the amplitudes E_x and E_y. As an example, if E_x and E_y are equal, the electric field is linearly polarized at 45° to the x axis. In general, the superposition of two fields, or two components of the same field, each with linear polarization, produces a linearly polarized wave at an angle $\tan^{-1}(E_y/E_x)$, if the two electric fields are in time phase (or $\tan^{-1}(-E_y/E_x)$, if the two are 180° out of time phase).

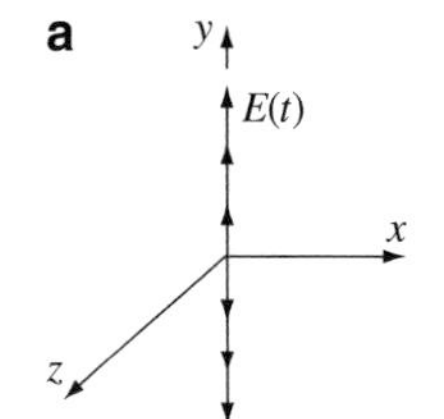

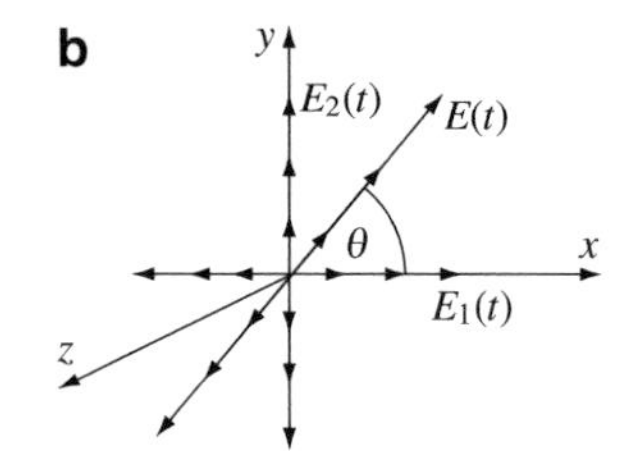

Figure 12.18 (**a**) Time variation of a wave, linearly polarized in the y direction traveling in the z direction. (**b**) General linear polarization obtained by superposition of two linearly polarized waves which are in time phase

12.8.2 Elliptical and Circular Polarization

Suppose an electromagnetic wave has an electric field intensity with two components in space, say one in the x direction and one in the y direction. The two components are out of time phase with the y component leading the x component by an angle φ. The electric field intensity in the time domain is

$$\mathbf{E}(z,t) = \hat{\mathbf{x}}E_1\cos(\omega t - \beta z) + \hat{\mathbf{y}}E_2\cos(\omega t - \beta z + \varphi) \quad [\text{V/m}] \tag{12.127}$$

This is simply the superposition of the two components of the electric field. To define the polarization and sense of rotation of the wave, we use the above four steps:

Step a: Write the electric field intensity in the time domain **[Eq. (12.127)]**.

Step b: Fix a value for z. In most cases, $z = 0$ is most convenient, for which we have

$$\mathbf{E}(z=0,t) = \hat{\mathbf{x}}E_1\cos(\omega t) + \hat{\mathbf{y}}E_2\cos(\omega t + \varphi) \quad [\text{V/m}] \tag{12.128}$$

Now, the values defining the behavior of the equation are the amplitudes E_1 and E_2 and the phase angle φ. Although these may have any values in general, we distinguish two important values for E_1 and E_2, namely, $E_1 \neq E_2$ and $E_1 = E_2$, and two values for φ, $\varphi = -\pi/2$ and $\varphi = +\pi/2$. This defines four distinct cases as follows:

Case 1: $E_1 \neq E_2$ and $\varphi = -\pi/2$. The electric field intensity now is

$$\mathbf{E}(z=0,t) = \hat{\mathbf{x}}E_1\cos(\omega t) + \hat{\mathbf{y}}E_2\cos\left(\omega t - \frac{\pi}{2}\right) = \hat{\mathbf{x}}E_1\cos(\omega t) + \hat{\mathbf{y}}E_2\sin(\omega t) \quad [\text{V/m}] \tag{12.129}$$

Two relations may be extracted from this equation. If we denote $E_x = E_1\cos\omega t$ and $E_y = E_2\sin\omega t$, we can write

$$\cos(\omega t) = \frac{E_x(0,t)}{E_1}, \quad \sin(\omega t) = \frac{E_x(0,t)}{E_2} \tag{12.130}$$

and

$$\cos^2(\omega t) + \sin^2(\omega t) = 1 \quad \rightarrow \quad \frac{E_x^2(0,t)}{E_1^2} + \frac{E_y^2(0,t)}{E_2^2} = 1 \tag{12.131}$$

Step c: **Equation (12.131)** is the equation of an ellipse; that is, as time changes, the tip of the electric field intensity **E** (whose components are $E_x(0,t)$ and $E_y(0,t)$) describes an ellipse on the $x-y$ plane. The wave is therefore elliptically polarized, as shown in **Figure 12.19a**.

Step d: To find the sense of rotation of the field vector, we use the following simple method. Using the time domain field expression as written after step **b** [**Eq. (12.129)** in this case], two values of ωt are substituted and the vector **E** evaluated. From these, it is possible to determine the rotation. For example, we may choose convenient values $\omega t = 0$ and $\omega t = \pi/2$. These give

$$\mathbf{E}(z=0, \omega t = 0) = \hat{\mathbf{x}}E_1, \quad \mathbf{E}(z=0, \omega t = \pi/2) = \hat{\mathbf{y}}E_2 \quad [\text{V/m}] \tag{12.132}$$

As ωt changed from 0 to $\pi/2$, the electric field intensity vector moved counterclockwise from being in the positive x direction to the positive y direction (see **Figure 12.19b**). Using the right-hand rule, the wave is found to be a right elliptically polarized wave. This method is quite simple but it is important to choose values of ωt that are convenient for calculation and that give unambiguous answers. If, for example, we found that at the second value of ωt the vector was in the negative x direction, it would have been impossible to determine if the rotation is clockwise or counterclockwise.

Case 2: $E_1 \neq E_2$ and $\varphi = \pi/2$. Again, we follow the above four steps. The electric field intensity in the time domain (**step a**) is given in **Eq. (12.128)**. Setting $z = 0$ (**step b**) and $\varphi = +\pi/2$ gives the electric field intensity as

$$\mathbf{E}(z=0,t) = \hat{\mathbf{x}}E_1\cos(\omega t) + \hat{\mathbf{y}}E_2\cos\left(\omega t + \frac{\pi}{2}\right) = \hat{\mathbf{x}}E_1\cos(\omega t) - \hat{\mathbf{y}}E_2\sin(\omega t) \quad [\text{V/m}] \tag{12.133}$$

Equation (12.131) remains the same and, therefore, the wave is still elliptically polarized. To find the sense of rotation, we again choose two values of ωt; $\omega t = 0$ and $\omega t = \pi/2$, and substitute in the expression for **E** in **Eq. (12.133)**. These give

$$\mathbf{E}(z=0, \omega t = 0) = \hat{\mathbf{x}}E_1, \quad \mathbf{E}(z=0, \omega t = \pi/2) = -\hat{\mathbf{y}}E_2 \quad [\text{V/m}] \tag{12.134}$$

As ωt changes from zero to $\pi/2$, the direction of the vector **E** changes from $+x$ to $-y$. This can only happen if the rotation of the vector **E** is clockwise (see **Figure 12.19c**). From the right-hand rule, this is then a left elliptically polarized wave.

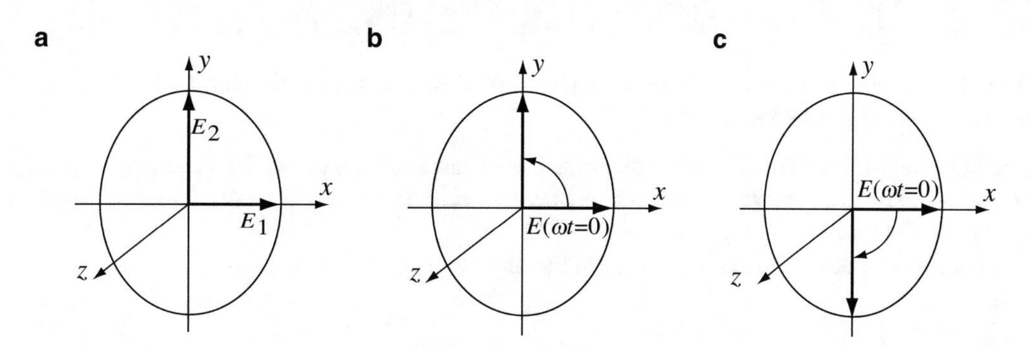

Figure 12.19 (**a**) An elliptically polarized wave. (**b**) The electric field rotates counterclockwise: the wave is right elliptically polarized. (**c**) The electric field rotates clockwise: the wave is left elliptically polarized

Case 3: $E_1 = E_2$ and $\varphi = -\pi/2$. This is clearly similar to case 1 above. If we set $E_1 = E_2 = E_0$ in **Eq. (12.131)**, we get

$$E_x^2 + E_y^2 = E_0^2 \tag{12.135}$$

This is the equation of a circle; therefore, the polarization is circular. Since **Eq. (12.129)** remains unchanged, except for the fact that $E_1 = E_2$, the wave is right circularly polarized. (see **Figure 12.19b** but with $E_1 = E_2$).
Clearly then, circular polarization is a special case of elliptical polarization.

Case 4: $E_1 = E_2$ and $\varphi = +\pi/2$. From the discussion in **case 3** and **case 2**, this is a left circularly polarized wave (see **Figure 12.19c** but with $E_1 = E_2$).
It is also worth noting the following:

(1) If $\varphi = 0$, the wave in **Eq. (12.127)** becomes

$$\mathbf{E}(z,t) = \hat{\mathbf{x}}E_1\cos(\omega t - \beta z) + \hat{\mathbf{y}}E_2\cos(\omega t - \beta z) \tag{12.136}$$

This is identical in form to **Eq. (12.126)** and the wave is, therefore, a ***linearly polarized*** wave with polarization at an angle $\tan^{-1}(E_2/E_1)$ with respect to the x axis.

(2) If $\varphi = 180°$, the wave in **Eq. (12.127)** is

$$\mathbf{E}(z,t) = \hat{\mathbf{x}}E_1\cos(\omega t - \beta z) + \hat{\mathbf{y}}E_2\cos(\omega t - \beta z + \pi) = \hat{\mathbf{x}}E_1\cos(\omega t - \beta z) - \hat{\mathbf{y}}E_2\cos(\omega t - \beta z) \quad [\text{V/m}] \tag{12.137}$$

This is again a linearly polarized wave at an angle $\tan^{-1}(-E_2/E_1)$ with respect to the x axis.

(3) From (2) and (1), it is clear that linear polarization is a special case of the general elliptical polarization.

(4) From **Eqs. (12.127)** and **(12.131)**, we can also conclude that an elliptically polarized wave can always be written as the superposition of two linearly polarized waves. Two superposed linearly polarized waves produce an elliptically polarized wave if the amplitudes of the two waves are different and if there is a phase difference between the two. The phase difference also defines the sense of rotation of the wave. If the phase difference between the two linearly polarized waves is zero, the result is a linearly polarized wave as a special case of elliptical polarization. If the two linearly polarized waves have equal amplitudes but there is also a phase difference between the two, the superposed wave is a circularly polarized wave.

The lead or lag in **Eqs. (12.129)** and **(12.133)** were set to 90° because the results lead to simple expressions. However, the phase difference between two waves can be arbitrary. If this is the case, the waves are elliptically polarized (or circularly polarized if the amplitudes are the same). The only difference is that the ellipse the vectors **E** describe is rotated in space (its axes do not coincide with the x and y axes).

Example 12.19 The following waves are given:

(a) $\mathbf{E}(z,t) = -\hat{\mathbf{y}}25e^{-0.001z}\cos(10^3 t - 1000z) \quad [\text{V/m}]$.

(b) $\mathbf{H}(z) = -\hat{\mathbf{x}}H_0 e^{-j\beta z} + \hat{\mathbf{y}}2H_0 e^{-j\beta z} \quad [\text{A/m}]$.

(c) $\mathbf{H}(z) = -\hat{\mathbf{y}}H_0 e^{-j\beta z} + j\hat{\mathbf{x}}H_1 e^{-j\beta z} \quad [\text{A/m}]$.

Find the polarization in each case.

Solution: The polarization is obtained by systematic application of the four steps in **Section 12.8.2**. However, in **(b)** and **(c)**, the electric field intensities must be found first.

(a) The electric field intensity is directed in the y direction for all values of t and z. As the wave propagates, its amplitude is attenuated and its phase changes. Thus, the wave is linearly polarized in the y direction. Another way to look at it is to formally apply the four-step method given in **Section 12.8.2**.
Step a is not necessary because the field is given in the time domain.
Step b: We set $z = 0$:

$$\mathbf{E}(z=0,t) = -\hat{\mathbf{y}}25\cos(10^3 t) \quad [\text{V/m}]$$

Step c: As time changes, the vector **E** may be either in the positive y or negative y direction. The field is linearly polarized in the y direction.

Step d: For a linearly polarized wave, there can be no rotation. This can be seen from the fact that for any two values of t the vector remains on the y axis.

(b) The magnetic field intensity has two components: one in the y direction and one in the x direction with amplitude half that of the y component. The two components are in phase; therefore, the polarization is linear, but for proper characterization, we must first find the electric field intensity. This is found from Ampère's law:

$$\nabla \times \mathbf{H} = \hat{\mathbf{x}}\left(\frac{\partial H_z}{\partial y} - \frac{\partial H_y}{\partial z}\right) + \hat{\mathbf{y}}\left(\frac{\partial H_x}{\partial z} - \frac{\partial H_z}{\partial x}\right) + \hat{\mathbf{z}}\left(\frac{\partial H_y}{\partial x} - \frac{\partial H_x}{\partial y}\right) = j\omega\varepsilon\mathbf{E}$$

With H_z =0, $\partial H_y/\partial x$ =0, and $\partial H_x/\partial y$ =0 and calculating the derivatives $\partial H_x/\partial z$ and $\partial H_y/\partial z$, we get

$$j\omega\varepsilon\mathbf{E} = \hat{\mathbf{x}}\, j\beta 2H_0 e^{-j\beta z} + \hat{\mathbf{y}}\, j\beta H_0 e^{-j\beta z}$$

Dividing both sides by $j\omega\varepsilon$ and setting $\eta = \beta/\omega\varepsilon$ gives the expression for the electric field intensity:

$$\mathbf{E} = \hat{\mathbf{x}}\eta 2H_0 e^{-j\beta z} + \hat{\mathbf{y}}\eta H_0 e^{-j\beta z} \quad [\mathrm{V/m}]$$

Now, we apply steps **(a)** through **(d)** to find the polarization and sense of rotation (if any).

Step a: The vector **E** is written in the time domain:

$$\mathbf{E}(z,t) = \mathrm{Re}\left\{\hat{\mathbf{x}}\eta 2H_0 e^{-j\beta z} e^{-j\beta t} + \hat{\mathbf{y}}\eta H_0 e^{-j\beta t}\right\} = \hat{\mathbf{x}}\eta 2H_0 \cos(\omega t - \beta z) + \hat{\mathbf{y}}\eta H_0 \cos(\omega t - \beta z) \quad [\mathrm{V/m}]$$

Step b: We set $z = 0$:

$$\mathbf{E}(z=0,t) = \hat{\mathbf{x}}\eta 2H_0 \cos(\omega t) + \hat{\mathbf{y}}\eta H_0 \cos(\omega t) \quad [\mathrm{V/m}]$$

Step c: At t =0, the vector **E** has components $2\eta H_0$ in the positive x direction and ηH_0 in the positive y direction. This ratio remains constant as t changes. Thus, **E** is linearly polarized at an angle equal to $\tan^{-1}(Hy/Hx) = \tan^{-1}(1/2) = 26°34'$ with respect to the positive x axis (see **Figure 12.20a**).

(c) First, we find the electric field intensity. Applying Ampère's law as in **(b)**, setting $H_z = 0$, $\partial H_y/\partial x = \partial H_x/\partial y = 0$, and calculating the derivatives $\partial H_x/\partial z$ and $\partial H_y/\partial z$ gives

$$j\omega\varepsilon\mathbf{E} = -\hat{\mathbf{x}} j\beta H_0 e^{-j\beta z} + \hat{\mathbf{y}} j\beta H_1 e^{-j\beta z}$$

Now, we divide by $j\omega\varepsilon$ and set $\eta = \beta/\omega\varepsilon$:

$$\mathbf{E} = -\hat{\mathbf{x}}\eta H_0 e^{-j\beta z} - \hat{\mathbf{y}} j\eta H_1 e^{-j\beta z} \quad [\mathrm{V/m}]$$

Step a: The electric field intensity in the time domain is

$$\mathbf{E}(z,t) = \mathrm{Re}\left\{-\hat{\mathbf{x}}\eta H_0 e^{-j\beta z} e^{-j\omega t} - \hat{\mathbf{y}} j\eta H_1 e^{-j\beta z} e^{-j\omega t}\right\} = -\hat{\mathbf{x}}\eta H_0 \cos(\omega t - \beta z) - \hat{\mathbf{y}}\eta H_1 \cos(\omega t - \beta z + \pi/2) \quad [\mathrm{V/m}]$$

where $j = e^{j\pi/2}$ was used.

Step b: Setting $z = 0$ gives

$$\mathbf{E}(z=0,t) = -\hat{\mathbf{x}}\eta H_0 \cos(\omega t) - \hat{\mathbf{y}}\eta H_1 \cos(\omega t + \pi/2) = -\hat{\mathbf{x}}\eta H_0 \cos(\omega t) - \hat{\mathbf{y}}\eta H_1 \sin(\omega t) \quad [\mathrm{V/m}]$$

This is clearly an elliptically polarized wave [see **Eqs. (12.129)** through **(12.131)**] since $H_0 \neq H_1$ and, as t changes, the vector **E** describes an ellipse (**step c**).

Step d: The rotation of **E** is found by setting $\omega t = 0$ and $\omega t = \pi/2$. These give

$$\mathbf{E}(z=0, \omega t = 0) = -\hat{\mathbf{x}}\eta H_0, \quad \mathbf{E}(z=0, \omega t = \pi/2) = -\hat{\mathbf{y}}\eta H_1 \quad [\mathrm{V/m}]$$

This indicates rotation in the counterclockwise direction. The wave is ***right elliptically polarized*** (see **Figure 12.20b**).

Note: Because polarization is defined on the electric field intensity, we had to first find the electric field intensity from the magnetic field intensity in **(b)** or **(c)**. It is also possible to find the polarization of the magnetic field intensity since the magnetic field intensity in a plane wave is always perpendicular to the electric field intensity and rotates with the electric field intensity (if there is any rotation). See **Exercise 12.10**.

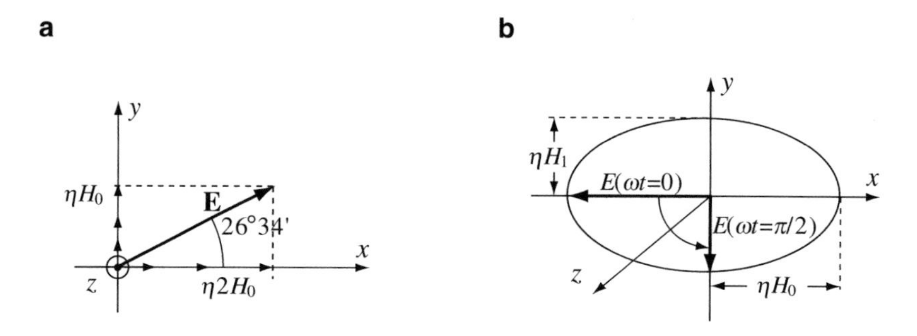

Figure 12.20 Identification of polarization. (**a**) The wave in **Example 12.19b** shown in the second quadrant. (**b**) The wave in **Example 12.19c** and its sense of rotation (assuming $H_0 > H_1$)

Exercise 12.10 In **Example 12.19**, find the polarization of the magnetic field intensity.

Answer (**a**) **H** is linearly polarized in the x direction. (**b**) **H** is linearly polarized at $116°34'$ with respect to the positive x axis. (**c**) **H** is right elliptically polarized.

12.9 Applications

Application: Communication with Spacecraft One of the most challenging communication problems is that with distant spacecraft. Although communication with objects in space is relatively simple and in many ways seems to be easier than communication on Earth, it has its own challenges. One of these is the vast distances involved. When communicating from Earth stations to spacecraft, large power and special antennas can be used, but when communicating from spacecraft to Earth, both power and antenna size are very limited. As spacecraft push further in space, the available power decreases because of reduced solar intensity, yet the requirements for range (and therefore for power) increase. A uniquely interesting example is offered by the Voyager spacecraft. Voyager 2 was launched on August 20, 1977, followed by Voyager 1 on September 15, 1977. In August 2018, Voyager 1 was at a distance of 19 billion km from Earth traveling in interstellar space, which it entered in August 2012. It is expected to cease transmission by 2025 when its RTG (radioisotope thermoelectric generator) output will not be sufficient to power the spacecraft (The RTG uses Plutonium 238 which has a half lifetime of 87.74 years). Communication with the spacecraft at such distances takes over 19 h, 43 m, each way. More interesting is the fact that the spacecraft transmitter is a mere 23 W (at launch), perhaps much less because of deterioration of its plutonium-powered power sources after over 40 years of continuous operation. Compare this small power with the 50 kW of some AM stations or the 5 MW some radar equipment use or to that to some portable radios (such as citizens band (CB) radios) that use between 3 and 5 W of power for a range of a few kilometers. From this vast distance the waves travel, it is also clear that the attenuation in space is rather small but is not zero. Both the time delay and the attenuation of waves will be a big problem in any long-range mission to the stars, when undertaken. In fact, because of these limitations, any deep-space exploration will have to be autonomous, with the spacecraft traveling, perhaps for generations, and returning with information that we will never get but future generations will. Each of the Voyager spacecraft carries a golden record with recordings and images representing life on earth. Should it encounter any civilization in the outer space, they should be able to identify the source of the spacecraft. See also the introduction to **Chapter 11** for a short description of the Pioneer 10 spacecraft.

Application: Range of TV and Radio Transmission TV and radio stations are regulated as to frequency, maximum power allowable, type of polarization, and other aspects of their operation. However, the range of a station depends, among other things, on attenuation in air. This, in turn, depends on a host of environmental conditions, including amount of moisture in air, pressure, pollutants, surface conditions, and others. Because of all these, the range, particularly of TV stations, is rather short. Also, some types of transmissions, such as microwaves, travel in a line-of-sight manner. Any obstruction such as hills, buildings, etc., prevents reception at the obstructed site. The range of a transmission system is a rather complex problem which must take into account antennas, attenuation, environmental conditions, and background noise, among others. Because of attenuation, the range of a TV station transmitting 50 kW of power is no more than about 100 km and even this range may be too large for good reception. AM transmission occurs at much lower frequencies (540 kHz–1.6 MHz). At these frequencies the attenuation is generally lower and the range is longer. On the other hand, FM transmission is in the VHF range (88–106 MHz) and therefore has a range similar to that of TV transmission. Typically, the attenuation constant in air (below about 3 GHz) is about 0.01 dB/km. Although we cannot calculate the range of transmission accurately, we can get a pretty good idea for the range using known attenuation constants and assuming an isotropic antenna (antenna which transmits uniformly in all directions). This then gives a worst-case maximum range which can be improved through use of more directional transmission, better antennas, etc.

Application: Superconducting Power Transmission The idea of skin depth[5] has been proposed for an unusual application: transmission of power in superconducting cables. Assuming that superconductivity will always require refrigeration, one proposed system is to use hollow conductors in which the refrigerant is passed keeping the temperature of the cables low enough for superconductivity to be maintained. Because at superconducting temperatures the conductivity is very high and the depth at which current exists is small, only the outer surface and a small depth below it will conduct current. A proposed method is to immerse the conductor in the form of a thin film of superconducting material coated on a metallic tube, in liquid helium or, when high-temperature superconductors are available, in liquid nitrogen. A possible method of using superconductors is demonstrated by the simplified superconducting power conductor in **Figure 12.21** and could carry vast amounts of power with little or no losses. This is particularly attractive for power distributions where losses in conductors account for some 3–7% of all losses in power generation and distribution.

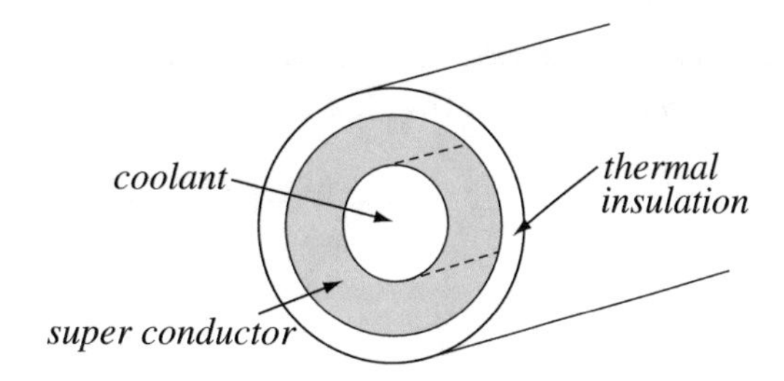

Figure 12.21 Simplified structure of a superconducting cable

Application: Optical Fiber Magnetometer The use of optical fibers for communication is well established, but optical fibers have many other applications. One useful application is in the measurement of very low magnetic fields. The principle is quite simple and is based on two fundamental properties. One is the magnetostrictive properties of some materials and the second is the change in phase of a wave as the length of path it travels changes. The method is shown in **Figure 12.22**. Two very long optical fibers of identical length are connected to the same laser source. One fiber is coated with a magnetostrictive material such as nickel. When no magnetic fields are present, the paths of light are identical and the outputs of the two fibers are in phase. Phase comparison between the two fibers shows a zero output. If both fibers are placed in a magnetic field, the magnetostrictive fiber changes its length by contraction. The contraction is rather small, and the total change in length is given as

[5] Skin depth per se does not apply to superconducting materials. There is, however, an equivalent relation that governs depth of current penetration in superconductors usually written as $\delta = \sqrt{\Lambda/4\pi}$, where Λ is known as the London order parameter. This relation is called the London relation after Heintz London (1907–1970) and his older brother Fritz London (1900–1954), who, among other important contributions to superconductivity, studied AC losses in superconductors.

$$\frac{\Delta l}{l} = cB_0^2$$

where c is the magnetostrictive material constant given in units of $1/\mathrm{T}^2$ (for example, $c = 10^{-4} 1/\mathrm{T}^2$ for nickel). Even though the effect is rather small, a very small change in length of the fiber will change the phase considerably. Since the propagation is at optical frequencies, the wavelength is of the order of a few hundred nanometers. For example, if a He–Ne (helium–neon) laser emitting at 633 nm (red) is used, a change in length of 10 nm changes the phase by $20\pi/633 = 0.031\pi$ or 5.7°. The length of the fibers, l, can be as long as we wish since the fibers can be placed on spools or in any convenient configuration. Sensitivities below 10^{-9} T are obtainable with a device which is a relatively simple, rugged, passive device.

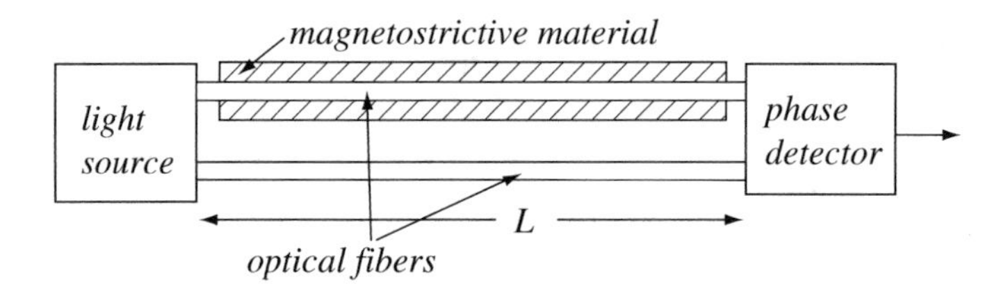

Figure 12.22 The principle of a magnetostrictive optical fiber magnetometer

12.10 Summary

The fundamentals of wave propagation and the behavior of waves in various media are the subjects of the present chapter. We start with the source-free electromagnetic wave equation in general, lossy **[Eq. (12.16)]**, and lossless **[Eq. (12.18)]** media (see **Examples 12.3** and **12.4** and **Exercises 12.2** through **12.4**):

$$\nabla^2 \mathbf{E} = j\omega\mu(\sigma\mathbf{E} + j\omega\varepsilon\mathbf{E}) \quad \text{and} \quad \nabla^2\mathbf{E} + \omega^2\mu\varepsilon\mathbf{E} = 0 \qquad (12.16) \text{ and } (12.18)$$

Wave equations identical in form may be written for **H**, **B**, **D**, **A**, or V and may also be written in the time domain.

Uniform plane waves are waves in which the amplitude and phase are constant at any point on any plane perpendicular to the direction of propagation of the wave. The form we assume for the purpose of discussion is $\mathbf{E} = \hat{\mathbf{x}}E(z)$.

Solution of the lossless wave equation **[Eq. (12.18)]** for plane waves in lossless media is

$$E_x(z) = E_0^+ e^{-jkz} + E_0^- e^{jkz} \quad [\mathrm{V/m}] \qquad (12.24)$$

$$k = \omega\sqrt{\mu\varepsilon} \quad [\mathrm{rad/m}] \qquad (12.23)$$

or, in the time domain,

$$E_x(z,t) = \mathrm{Re}\{E_x(z)e^{j\omega t}\} = E_0^+\cos(\omega t - kz) + E_0^-\cos(\omega t + kz) \quad [\mathrm{V/m}] \qquad (12.25)$$

An arbitrary phase angle ϕ may also be added to either solution (due to, for example, the complex nature of E_0^+ or E_0^-). The first term is a forward-propagating wave (in the positive z direction), the second a backward-propagating wave (negative z direction in this case).

Properties of the Wave

$$\text{Phase velocity}: \quad v_p = \frac{1}{\sqrt{\mu\varepsilon}} \quad \left[\frac{\mathrm{m}}{\mathrm{s}}\right] \qquad (12.28)$$

$$\text{In free space}: \quad v_p \approx 3 \times 10^8 \quad [\mathrm{m/s}] \qquad (12.29)$$

$$\text{Wavelength}: \quad \lambda = \frac{v_p}{f} = \quad [\mathrm{m}] \qquad (12.30)$$

$$\text{Wave number}: \quad k = \frac{2\pi}{\lambda} \quad \left[\frac{\mathrm{rad}}{\mathrm{m}}\right] \qquad (12.31)$$

$$\text{Intrinsic impedance}: \quad \eta = \sqrt{\frac{\mu}{\varepsilon}} \quad [\Omega] \tag{12.37}$$

$$\text{In free space}: \quad \eta_0 \approx 377 \quad [\Omega] \tag{12.39}$$

Poynting Theorem, Poynting Vector, Power, and Power Density The ***Poynting vector*** gives the magnitude and direction of propagation of the instantaneous power density:

$$\mathcal{P} = \mathbf{E} \times \mathbf{H} \quad \left[\text{W/m}^2\right] \tag{12.53}$$

The ***Poynting theorem*** gives the net power entering or leaving a volume v, enclosed by area s:

$$\mathcal{P}(t) = \oint_s (\mathbf{E} \times \mathbf{H}) \cdot d\mathbf{s} = -\frac{\partial}{\partial t}\int_v \left(\frac{\mu H^2}{2} + \frac{\varepsilon E^2}{2}\right) dv - \int_v \mathbf{E} \cdot \mathbf{J} dv \quad [\text{W}] \tag{12.52}$$

The first term on the right-hand side is the time rate of change of stored energy, the second is power due to source and induced currents. A net negative power (power flow into the volume) is called the ***receiver case***. Net positive power (out of the volume) is called the ***transmitter case***.

Time-averaged power density can be calculated from instantaneous power density or from the complex Poynting vector:

$$\mathcal{P}_{av} = \frac{1}{T}\int_0^T \mathcal{P}(t)\, dt \quad \left[\frac{\text{W}}{\text{m}^2}\right] \tag{12.56}$$

or

$$\mathcal{P}_{av} = \frac{1}{2}\text{Re}\{\mathbf{E} \times \mathbf{H}^*\} \quad \left[\frac{\text{W}}{\text{m}^2}\right] \tag{12.59}$$

where $T = 1/f = 2\,\pi/\omega$ and $*$ indicates the complex conjugate. The complex Poynting vector is $\mathcal{P}_c = \mathbf{E} \times \mathbf{H}^*$ [W/m^2]. The complex Poynting theorem may be written as

$$\oint_s (\mathbf{E} \times \mathbf{H}^*) \cdot d\mathbf{s} = j\omega \int_v (\varepsilon \mathbf{E} \cdot \mathbf{E}^* - \mu \mathbf{H} \cdot \mathbf{H}^*)\, dv - \int_v \mathbf{E} \cdot \mathbf{J}_0^* dv - \int_v \sigma \mathbf{E} \cdot \mathbf{E}^* dv \quad [\text{W}] \tag{12.71}$$

where $\mathbf{E} \cdot \mathbf{J}_0^*$ may be negative or positive depending on the source of $\mathbf{J}_0^*$. The last term represents ohmic losses. Time-averaged power is usually calculated using **Eq. (12.71)**.

Time-averaged energy densities (electric and magnetic) are

$$w_{e(av)} = \frac{\varepsilon \mathbf{E} \cdot \mathbf{E}^*}{4}, \quad w_{m(av)} = \frac{\mu \mathbf{H} \cdot \mathbf{H}^*}{4} \quad \left[\frac{\text{J}}{\text{m}^3}\right] \tag{12.74}$$

Propagation of Plane Waves in General Media Given properties (ε,μ,σ) the wave equation is written in terms of the complex permittivity ε_c as

$$\nabla^2 \mathbf{E} = j\omega\mu(j\omega\varepsilon_c)\mathbf{E} \tag{12.81}$$

where

$$\varepsilon_c = \varepsilon\left[1 - j\frac{\sigma}{\omega\varepsilon}\right] \quad \left[\frac{\text{F}}{\text{m}}\right] \tag{12.79}$$

The wave equation to solve is

$$\nabla^2 \mathbf{E} - \gamma^2 \mathbf{E} = 0 \tag{12.84}$$

where

$$\gamma = \alpha + j\beta = j\omega\sqrt{\mu\varepsilon}\sqrt{1 - j\frac{\sigma}{\omega\varepsilon}} \tag{12.83}$$

and

$$\alpha = \omega\sqrt{\frac{\mu\varepsilon}{2}\left[\sqrt{1 + \left(\frac{\sigma}{\omega\varepsilon}\right)^2} - 1\right]} \quad \left[\frac{\text{Np}}{\text{m}}\right] \tag{12.95}$$

$$\beta = \omega\sqrt{\frac{\mu\varepsilon}{2}\left[\sqrt{1 + \left(\frac{\sigma}{\omega\varepsilon}\right)^2} + 1\right]} \quad \left[\frac{\text{rad}}{\text{m}}\right] \tag{12.96}$$

$\gamma = \alpha + j\beta$ is the ***propagation constant***, α is the ***attenuation constant***, and β the ***phase constant***. Phase velocity and wavelength are also dependent on conductivity [see **Eqs. (12.97)** and **(12.98)**].

The ***intrinsic impedance*** is now complex:

$$\eta = \frac{j\omega\mu}{\gamma} = \sqrt{\frac{j\omega\mu}{\sigma + j\omega\varepsilon}} \quad [\Omega] \tag{12.102}$$

The main effect that is different than propagation in lossless media is attenuation of the waves. The solution for attenuated plane waves includes an attenuation factor:

$$E_x(z) = E_0^+ e^{-\alpha z} e^{-j\beta z} + E_0^- e^{\alpha z} e^{j\beta z} \quad [\text{V/m}] \tag{12.91}$$

or, in the time domain,

$$E_x(z,t) = E_0^+ e^{-\alpha z}\cos(\omega t - \beta z) + E_0^- e^{\alpha z}\cos(\omega t + \beta z) \quad [\text{V/m}] \tag{12.93}$$

Low-Loss Dielectrics $\sigma/\omega\varepsilon \ll 1$. Approximations are defined as follows:

$$\alpha \approx \frac{\sigma}{2}\sqrt{\frac{\mu}{\varepsilon}} \quad \left[\frac{\text{Np}}{\text{m}}\right] \tag{12.104}$$

$$\beta \approx \omega\sqrt{\mu\varepsilon}\left(1 + \frac{1}{8}\left(\frac{\sigma}{\omega\varepsilon}\right)^2\right) \quad \left[\frac{\text{rad}}{\text{m}}\right] \tag{12.105}$$

$$\eta \approx \sqrt{\frac{\mu}{\varepsilon}}\left(1 + \frac{j\sigma}{2\omega\varepsilon}\right) \quad [\Omega] \tag{12.109}$$

High-Loss Materials $\sigma/\omega\varepsilon \gg 1$. Approximations are:

$$\alpha \approx \sqrt{\pi f\mu\sigma} \quad [\text{Np/m}], \quad \beta \approx \sqrt{\pi f\mu\sigma} \quad [\text{rad/m}] \tag{12.111}$$

$$\eta \approx (1+j)\sqrt{\frac{\omega\mu}{2\sigma}} = (1+j)\frac{1}{\sigma\delta} = (1+j)\frac{\omega\mu\delta}{2} \quad [\Omega] \tag{12.116}$$

Skin depth is the depth at which the amplitude of the wave reduces to $1/e$ of its value:

$$\delta \approx \frac{1}{\sqrt{\pi f\mu\sigma}} = \frac{1}{\alpha} \quad [\text{m}] \tag{12.113}$$

Group velocity is the velocity of a packet of waves in a narrow range of frequencies. It is different from phase velocity except in perfect dielectrics

$$v_g = \frac{1}{d\beta/d\omega} \quad \left[\frac{\text{m}}{\text{s}}\right] \tag{12.122}$$

Dispersion is the frequency dependence of the phase velocity which causes waves of different frequencies to travel at different velocities. Perfect dielectrics are dispersionless.

Polarization of plane waves is the path described by the tip of the electric field intensity as it propagates in space toward the observer:

(1) Linear polarization—the tip of the electric field intensity describes a straight line.
(2) Circular polarization—the tip of the electric field intensity describes a circle.
(3) Elliptical polarization—the tip of the electric field intensity describes an ellipse.
(4) Rotation: circularly and elliptically polarized waves can rotate clockwise or counterclockwise as they propagate. Counterclockwise rotation is said to be ***right elliptically*** (or ***circularly***) ***polarized*** because it follows the right-hand rule—the thumb is in the direction of propagation of the wave and the curled fingers show the direction of rotation of the electric field intensity. ***Left circularly*** (or ***elliptically***) ***polarized*** waves rotate clockwise as they propagate toward the observer.

Problems

The Time-Dependent Wave Equation

12.1 The Wave Equation. Starting with the general time-dependent Maxwell's equations in a linear, isotropic, homogeneous medium, write a wave equation in terms of the electric field intensity.

(a) Show that if you neglect displacement currents, the equation is not a wave equation.
(b) Write the source-free wave equation from the general equation you obtained.

12.2 Source-Free Wave Equation. Obtain the source-free time-dependent wave equation for the magnetic flux density in a linear, isotropic, homogeneous medium.

The Time-Harmonic Wave Equation

12.3 Time-Harmonic Wave Equation. Using the source-free Maxwell's equations, show that a Helmholtz equation can be obtained in terms of the magnetic vector potential. Use the definition $\mathbf{B} = \nabla \times \mathbf{A}$ and a simple medium (linear, isotropic, homogeneous material). Justify the choice of the divergence of **A**.

12.4 The Helmholtz Equation for D. Using Maxwell's equations, find the Helmholtz equation for the electric flux density **D** in a linear, isotropic, homogeneous material.

12.5 The Electric Hertz Potential. In a linear, isotropic, homogeneous medium, in the absence of sources, the Hertz vector potential $\mathbf{\Pi}_e$ may be defined such that $\mathbf{H} = j\omega\varepsilon\nabla \times \mathbf{\Pi}_e$:

(a) Express the electric field intensity in terms of $\mathbf{\Pi}_e$.
(b) Show that the Hertz potential satisfies a homogeneous Helmholtz equation provided a correct gauge is chosen. What is this appropriate gauge?

Solution for Uniform Plane Waves

12.6 Plane Wave. The electric field intensity of a plane wave has an amplitude of 120 V/m, is directed in the negative z direction, has a frequency of 76 MHz and propagates in the positive y direction in free space (properties of free space are $\varepsilon = \varepsilon_0$ [F/m], $\mu = \mu_0$ [H/m].)

(a) Write the expression of the electric field intensity in phasor and time-dependent forms.
(b) Write the expression for the magnetic field intensity in phasor and time-dependent forms.

12.7 Plane Wave. A wave propagates in free space and its electric field intensity is $\mathbf{E} = \hat{\mathbf{x}}100e^{-j220z} + \hat{\mathbf{x}}100e^{j220z}$ [V/m]:

(a) Show that **E** satisfies the source-free wave equation.
(b) What are the wave's phase velocity and frequency?

The Poynting Vector

12.8 Use of the Poynting Vector. A simple and common use of the Poynting vector is identification of direction of propagation of a wave or the direction of fields in space. Consider the magnetic field intensity of a plane electromagnetic wave propagating in free space:

$$\mathbf{H} = \hat{\mathbf{x}}H_0e^{j\beta y} + \hat{\mathbf{z}}H_1e^{j\beta y} \quad [\text{A/m}]$$

(a) Calculate the electric field intensity using the properties of the Poynting vector.
(b) Show that the result in **(a)** is correct by substituting the magnetic field intensity into Maxwell's equations and evaluating the electric field intensity through Maxwell's equations.

12.9 Application: Power Relations in a Microwave Oven. The peak electric field intensity at the bottom of a microwave oven is equal to 2,500 V/m. Assuming that this is uniform over the area of the oven which is equal to 400 cm^2, calculate the peak power the oven can deliver. Permeability and permittivity are those of free space.

12.10 Application: Heating Food in a Microwave Oven. A frozen pizza is marketed to be heated in a microwave oven. As an engineer you are asked to write heating instructions, specifically how long should it be heated to reach a proper temperature. The average residential microwave oven is 50 cm wide, 40 cm deep, and 30 cm high. The oven has a low and a high heating level. At low, it produces a time averaged power of 500 W, whereas at high, the power is 1000 W. We will assume the pizza is placed flat on the bottom of the oven and any power coupled into the pizza enters from above. The pizza is 25 cm in diameter, 1.5 cm thick, and is 75% water by volume. The microwave oven heats the water in the pizza. The frozen pizza is at –20 °C and must be heated to 75 °C. Heat capacity of water is 4.1885 J/(g • K) and the latent heat (of melting ice) is 334 J/g (heat capacity is the energy needed to raise the temperature of one gram of substance (water in this case) by 1 degree Kelvin, and latent heat is the energy required to melt a gram of ice at 0 °C to water at 0 °C). Assume that the heat transfer from the electromagnetic waves to the pizza is 80% efficient and calculate:

(a) The time it takes to heat the pizza on the low setting of the oven.
(b) The time it takes to heat the pizza on the high setting of the oven.
(c) The cost in electricity to heat the pizza if a kW • h costs $0.16.

12.11 Application: Power Dissipation in Cylindrical Conductor. A cylindrical conductor of radius R and infinite length carries a current of amplitude I [A] and frequency f [Hz]. The conductivity of the conductor is σ [S/m] (**Figure 12.23**). Calculate the time-averaged dissipated power per unit length in the conductor, neglecting displacement currents in the conductor. Assume the current is uniformly distributed throughout the cross section.

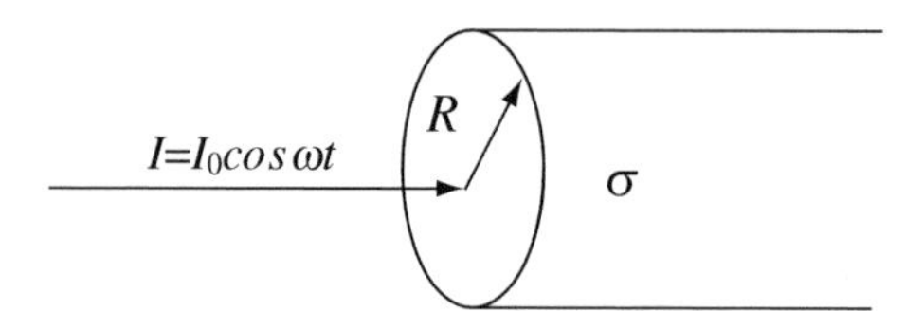

Figure 12.23

12.12 Application: Power Radiated by an Antenna. An antenna produces an electric field intensity in free space as follows:

$$\mathbf{E} = \hat{\mathbf{\theta}}\frac{12\pi}{R}e^{-j2\pi R}\sin\theta \quad \left[\frac{\text{V}}{\text{m}}\right]$$

where R is the radial distance from the antenna and the field is described in a spherical coordinate system. The field of the antenna behaves as a plane wave in the spherical system of coordinates. Calculate:

(a) The magnetic field intensity of the antenna.
(b) The time-averaged power density at a distance R from the antenna.
(c) The total radiated power of the antenna.

12.13 Application: Electromagnetic Radiation Safety. The allowable time-averaged microwave power density exposure in industry in the United States is 10 mW/cm^2. As a means of understanding the thermal effects of this radiation level (nonthermal effects are not as well defined and are still being debated), it is useful to compare this radiation level with thermal radiation from the Sun. The Sun's radiation on Earth is about 1,400 W/m^2 (time averaged). To compare the fields associated with the two types of radiation, view these two power densities as the result of a Poynting vector. Calculate:

(a) The electric and magnetic field intensity due to the Sun's radiation on Earth.
(b) The maximum electric and magnetic field intensities allowed by the standard. Compare with that due to the Sun's radiation.

12.14 Poynting Vector and Theorem in a Capacitor. A parallel plate capacitor is made of two parallel disks as shown in **Figure 12.24**. The capacitor is connected to a sinusoidal source, $V(t) = V_0 \sin\omega t$. The dielectric between the plates is free sapce:

(a) Calculate the complex Poynting vector in the capacitor.
(b) Show that the time-averaged power in the capacitor is zero.
(c) Show that reactive power propagates in the negative r direction.
(d) How do the results in **(b)** and **(c)** square with the circuit theory behavior of capacitors?
Hint: start with the calculation of the electric and magnetic field intensities inside the capacitor.

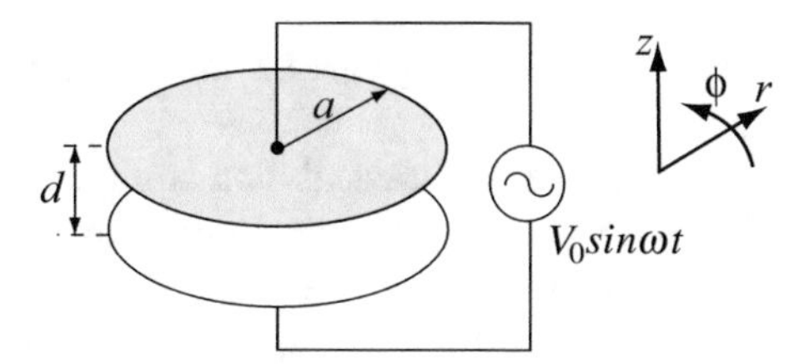

Figure 12.24

12.15 Stored Energy. In a region of space where there are no currents, the time-averaged Poynting vector equals 120 W/m^2. Assume that this power density is uniform on the surface of a sphere of radius $a = 0.1$ m, pointing outwards. Calculate the total stored energy in the sphere. The frequency is 1 GHz and the sphere has properties of free space.

Propagation in Lossless, Low-Loss, and Lossy Dielectrics

12.16 Energy Density in Dielectrics. A plane wave of given frequency propagates in a perfect dielectric:

(a) Calculate the time-averaged stored electric energy density and show that it is equal to the magnetic volume energy density.
(b) Suppose now the dielectric is lossy: calculate the time-averaged-stored electric and magnetic energy densities. Are they still the same?

12.17 Application: Propagation in Lossy Media. Seawater has a conductivity of 4 S/m. Its permittivity depends on frequency; relative permittivity at 100 Hz is 80, at 100 MHz it is 32, and at 10 GHz it is 24:

(a) How can seawater be characterized in terms of its loss at these three frequencies? That is, is seawater a low-loss, high-loss, or a general lossy medium for which no approximations can be made?
(b) Calculate the intrinsic impedance at the three frequencies.
(c) What can you conclude from these calculations for the propagation properties of seawater?

12.18 Generation of a Plane Wave in a Lossy Dielectric. A very thin conducting layer carries a surface current density J_0 [A/m] as shown in **Figure 12.25**. The frequency is f and the current is directed in the positive z direction. The layer is immersed in seawater, which has permittivity ε, permeability μ_0, and conductivity σ. If the layer is at $x = 0$, calculate the electric and magnetic fields at $x = x_0$. Given: $J_0 = 1$ A/m, $f = 100$ MHz, $\sigma = 4$ S/m, $\varepsilon_0 = 8.854 \times 10^{-12}$ F/m, $\varepsilon_r = 80$, $\mu_0 = 4\pi \times 10^{-7}$ H/m, $x_0 = 1$ m.

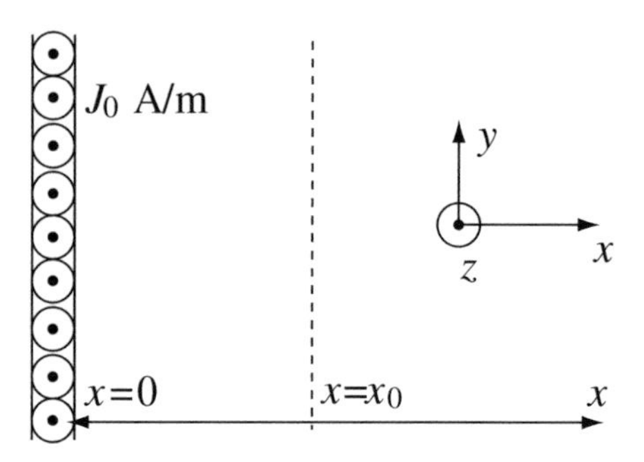

Figure 12.25

12.19 Application: Communication in the Atmosphere. A parabolic antenna of radius $b = 1$ m transmits at a frequency $f = 300$ MHz to a receiving antenna a distance $d = 200$ km away. The receiving antenna is also of radius b and the transmission is parallel to the ground, in the atmosphere, as in **Figure 12.26a**. Assume that the wave propagates as a plane wave and the beam remains constant in diameter (same diameter as the antennas). Use the following properties: $\varepsilon = \varepsilon_0$ [F/m], $\mu = \mu_0$ [H/m], $\sigma = 2 \times 10^{-7}$ S/m.

(a) Calculate the time-averaged power the transmitting antenna must supply if the receiving antenna must receive a magnetic field intensity of magnitude 1 mA/m.

(b) In an attempt to reduce the power required, the transmission is directed to a satellite which contains a perfect reflector, as shown in **Figure 12.26b**. The waves propagate through the atmosphere, into free space to the satellite and back to the receiving antenna. If the satellite is at a height $h = 36{,}000$ km and the waves propagate in the atmosphere for a distance $q = 20$ km in each direction, what is the power needed in this case for the same reception condition as in **(a)**? Compare the result with that in **(a)**.

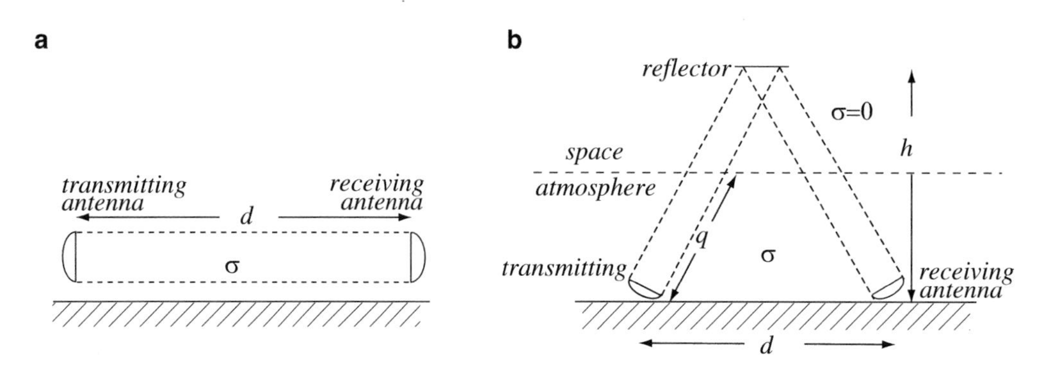

Figure 12.26

12.20 Power Relations in the Atmosphere. A plane wave propagates in the atmosphere. Properties of the atmosphere are ε_0 [F/m], μ_0 [H/m], and $\sigma = 10^{-6}$ S/m. The electric field intensity has an amplitude $E_0 = 100$ V/m in the x direction and $E_0 = 100$ V/m in the z direction. The frequency of the wave is $f = 100$ MHz. Find:

(a) The direction of propagation.

(b) The instantaneous power per unit area in the direction of propagation.

(c) The magnitude of the magnetic field intensity after the wave has propagated a distance $d = 10$ km.

12.21 Application: Radar Detection and Ranging of Aircraft. A radar antenna transmits 50 kW at 10 GHz. Assume transmission is in a narrow beam, 1 m^2 in area, and that within the beam, waves are plane waves. The wave is reflected from an aircraft but only 1% of the power propagates back in the direction of the transmitting antenna. If the airplane is at a distance of 100 km, calculate the total power received by the antenna assuming the reflected power propagates in the same narrow beam as the transmitted power. Assume permittivity and permeability of free space and conductivity of 10^{-7} S/m.

12.22 Application: Fiber Optics Communication. Two optical fibers are used for communication. One is made of glass with properties $\mu = \mu_0$ [H/m] and $\varepsilon = 1.75\varepsilon_0$ [F/m] and has an attenuation of 2 dB/km. The second is made of plastic with properties $\mu = \mu_0$ and $\varepsilon = 2.5\varepsilon_0$ and attenuation of 10 dB/km. Suppose both are used to transmit signals over a length of 10 km. The input to each fiber is a laser, operating at a free-space wavelength of 800 nm and input power of 0.1 W. Calculate:

(a) The power available at the end of each fiber.
(b) The wavelength, intrinsic (wave) impedance, and phase velocity in each fiber.
(c) The phase difference between the two fibers at their ends.

12.23 Application: Wave Properties and Remote Sensing in the Atmosphere. A plane wave of frequency *f* propagates in free space and encounters a large volume of heavy rain. The permittivity of air increases by 8% due to the rain:

(a) Calculate the intrinsic impedance, phase constant, and phase velocity in rain and the percentage change in wavelength.
(b) Compare the properties calculated in **(a)** with those in free space. Can any or all of these be used to monitor atmospheric conditions (such as weather prediction)? Explain.

12.24 Application: Attenuation in the Atmosphere. Measurements with satellites show that the average solar radiation (solar constant) in space is approximately 1,400 W/m^2. The total radiation reaching the surface of the Earth on a summer day is approximately 1,100 W/m^2. 50% of this radiation is in the visible range. To get some insight into the radiation process, assume an atmosphere which is 15 km thick. From this calculate the average attenuation constant in the atmosphere over the visible range assuming it is constant throughout the range and that attenuation is the only loss process.

Propagation in High-Loss Dielectrics and Conductors

12.25 Intrinsic Impedance in Copper. Copper has the following properties: $\mu = \mu_0$ [H/m], $\varepsilon = \varepsilon_0$ [F/m], and $\sigma = 5.7 \times 10^7$ S/m. Calculate the intrinsic impedance of copper at 100 MHz:

(a) Using the exact formula for general lossy materials.
(b) Assuming a high-loss material. Compare with **(a)** and with the intrinsic impedance of free space.

12.26 Application: Communication with Trapped Miners. A serious problem with mine accidents is the difficulty of communicating with trapped miners. The main problem is the lossy nature of soils. Properties of soil depend on composition and moisture but may be approximated as $\varepsilon_r = 3.2$ and $\sigma = 0.02$ S/m. Suppose that one wishes to communicate with miners trapped underground by generating a wave at the surface with an electric field intensity of amplitude 400 V/m and a frequency of 300 MHz.

(a) If the minimum electric field intensity required at the receiver is 10^{-6} V/m, what is the maximum depth of communication possible?
(b) Suppose one tries the same at a frequency 10 times higher. What is now the maximum depth?
(c) Suppose one tries the same at a frequency 10 times lower. What is now the maximum depth?
(d) What is your conclusion from these calculations regarding the possibility of wireless communication with trapped miners?

Note: There are other effects that limit communication underground, some of which will be discussed in the following chapter including reflections and effects on antennas but these are neglected here.

12.27 Skin Depth and Penetration in Lossy Media. Two plane waves propagate in two materials as shown in **Figure 12.27**:

(a) What is the ratio between the distances the waves travel in each material before the electric and magnetic field intensities are attenuated to 1% of their amplitude at the surface? Assume that the waves enter the materials without losses or reflections.
(b) What is the ratio between the phase velocities?

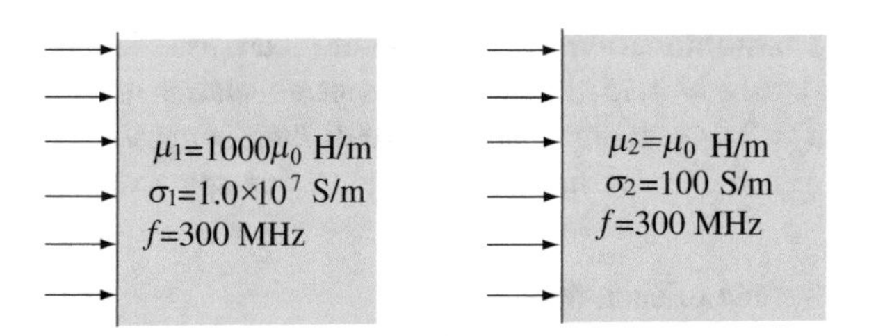

Figure 12.27

12.28 Application: Measurement of Conductivity in Lossy Materials. In an attempt to measure conductivity of a material, a plane wave is applied to one surface of a slab and measured at the other surface. Suppose the electric field intensity just below the left surface of the material is measured as E_0 [V/m]. The electric field intensity at the right surface (again, just below the surface) is $0.1E_0$ [V/m]. Permeability and permittivity are those of free space, the material is known to have high conductivity, and the slab is $d = 10$ mm thick. The measurements are performed at 400 Hz. Calculate the conductivity of the material and its attenuation constant.

12.29 Application: **Underwater Communication**. Suppose a submarine could generate a plane wave and use it to communicate with another submarine in seawater. If the ratio between the amplitude at the receiver and that at the transmitter must be 10^{-12} or higher, what is the maximum range of communication at:

(a) 10 MHz.
(b) 100 Hz.

Assume relative permittivity in both cases is 72 and conductivity of seawater is 4 S/m.

12.30 Application: Skin Depth in Conductors. Calculate the skin depth for the following conditions:

(a) Copper: $f = 10$ GHz, $\mu = \mu_0$ [H/m], $\sigma = 5.7 \times 10^7$ S/m.
(b) Mercury: $f = 10$ GHz, $\mu = \mu_0$ [H/m], $\sigma = 1 \times 10^6$ S/m.

12.31 Wave Impedance in Conductors. Calculate the intrinsic (wave) impedance of copper and iron at 60 Hz and 10 GHz. The conductivity of copper is 5.7×10^7 S/m and that of iron is 1×10^7 S/m. The permeability of copper is μ_0 [H/m] and that of iron is $1{,}000\mu_0$ [H/m] at 60 Hz and $2.3\mu_0$ [H/m] at 10 GHz:

(a) Compare these with the wave impedance in free space.
(b) What can you conclude from these calculations for the propagation properties of conductors in general and ferromagnetic conductors in particular?

12.32 Classification of Lossy Materials. Three materials are considered for their propagation properties: ethanol, concrete and seawater. In ethanol, $\varepsilon/\varepsilon_0 = 24$ and $\sigma = 1.3\text{x}10^{-7}$ S/m, in concrete, $\varepsilon/\varepsilon_0 = 4.5$ and $\sigma = 0.012$ S/m and in seawater, $\varepsilon/\varepsilon_0 = 78$ and $\sigma = 4$ S/m. How do you classify these materials for propagation purposes at 1 MHz and 2.4 GHz assuming properties do not change in the given range of frequencies? Explain.

12.33 Application: **Skin Depth and Communication in Seawater**. How deep does an electromagnetic wave transmitted by a radar operating at 3 GHz propagate in seawater before its amplitude is reduced to 10^{-6} of its amplitude just below the surface? Use the following properties: $\sigma = 4$ S/m, $\varepsilon = 76\varepsilon_0$ [F/m] (at 3 GHz), and $\mu = \mu_0$ [H/m]. How good is radar for detection of submarines? Explain.

12.34 Penetration of Light in Copper. Since light is an electromagnetic wave and electromagnetic waves penetrate in any material except perfect conductors, calculate the depth of penetration of light into a sheet of copper. The conductivity of copper is 5.7×10^7 S/m, and its permeability is that of free space. Assume the frequency of light (in mid-spectrum) is 5×10^{14} Hz.

12.35 Properties of Seawater at Different Frequencies. The relative permittivity of seawater changes with frequency in a rather complex fashion. A simple approximation is the following: the relative permittivity stays constant at 78 up to 1 GHz. Then it goes down linearly to 5 at 100 GHz and stays constant at a value of 5 above 100 GHz. Its conductivity is 4 S/m and may be assumed frequency independent for the purpose of this problem. Calculate:

(a) The approximate range of frequencies over which seawater may be assumed to be a high loss medium, assuming permittivity remains constant.
(b) The approximate range of frequencies over which seawater may be assumed to be a low loss medium.

12.36 AC Current Distribution in a Conductor. A cable made of iron, with properties as shown in **Figure 12.28**, carries a current at 100 Hz. If the current density allowed (maximum) is 100 A/mm^2, find the current density at the center of the conductor ($\sigma = 1 \times 10^7$ S/m, $\mu = 20\mu_0$ [H/m], $r = 0.1$ m).

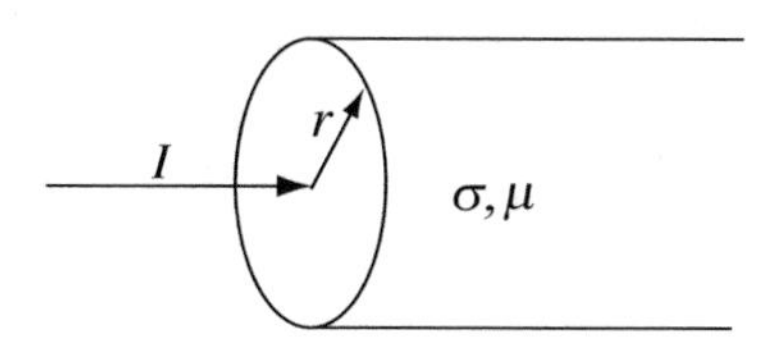

Figure 12.28

12.37 Application: Skin Depth and Design of Cables for AC Power Distribution. A thick wire (cable) is made of steel as shown in **Figure 12.29a**. If the maximum current density allowed at any point in the material is 10 A/mm^2:

(a) What is the total current the cable can carry at 60 Hz?
(b) To improve the current-carrying capability, a cable is made of 10 thinner wires (see **Figure 12.29b**) such that the total cross-sectional area is equal to that of the cable in **(a)**. What is the total current the new cable can carry for the same maximum current density, frequency, and material properties?
(c) Compare the results in **(a)** and **(b)** with the DC current the wire can carry at.

Note: Assume that the solution for plane waves applies here even though the surface is curved. This is, in general, applicable if the skin depth is small compared to the radius.

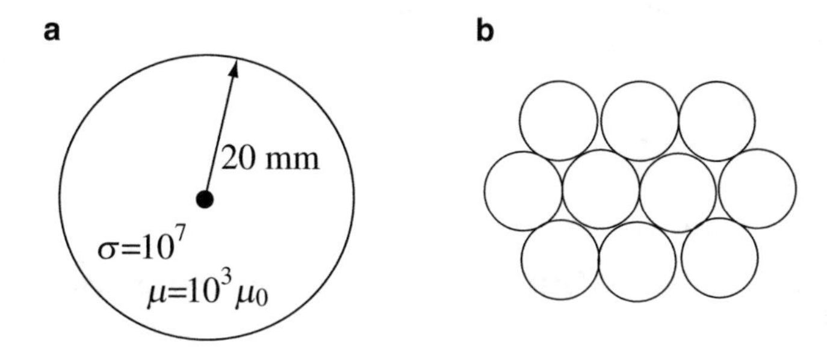

Figure 12.29

12.38 Application: Electromagnetic Shielding. A shielded room is designed to reduce the amplitude of high-frequency waves that penetrate into the room. The shield is made of a nonconducting, nonmagnetic material with a thin coating of conducting material on the outer surface. The shield must reduce the electric field intensity by a factor of 10^6 compared to the field outside at a frequency of 10 MHz. The conducting layer may be made of aluminum, copper, mu-metal, or a conducting polymer. Conductivities, permeabilities, cost, and mass of the three materials are given in the table below.

	Cu	Al	Mu-metal	Polymer
Conductivity [S/m]	5.7×10^7	3.6×10^7	0.5×10^7	0.001
Permeability [H/m]	μ_0	μ_0	$10^5\mu_0$	μ_0
Mass [kg/m^3]	8,960	2,700	7,800	1,200
Cost [unit/kg]	1	1.5	100	0.01

(a) Calculate the minimum thickness required for each of the four materials.
(b) Which material should we choose if the overall important parameter is: 1. Cost. 2. Mass. 3. Volume of conducting material. Use a unit area of the wall for comparison.

Dispersion and Group Velocity

12.39 Dispersion and Group Velocity. Under certain conditions, a wave propagates with the following phase constant:

$$\beta = \omega\sqrt{\mu\varepsilon}\sqrt{1 - \frac{\omega_c^2}{\omega^2}} \quad \left[\frac{\text{rad}}{\text{m}}\right]$$

where ω [rad/s] is the angular frequency of the wave, and ω_c [rad/s] is a fixed angular frequency. Because this gives the relation between β and ω, it is a dispersion relation:

(a) Plot the dispersion relation for the wave: use $\omega_c = 10^7$, $10^7 \le \omega \le 2 \times 10^7$ [rad/s].
(b) What are the phase and group velocities?
(c) What happens at $\omega = \omega_c$? Explain.

12.40 Dispersion and Group Velocity. The dispersion relation for a wave is given as

$$\beta = \sqrt{\omega^2\mu\varepsilon - \frac{\pi^2}{a}} \quad \left[\frac{\text{rad}}{\text{m}}\right]$$

(a) Plot the dispersion relation. ***a*** is a constant and $\pi^2/\text{a} < \omega^2\mu\varepsilon$.
(b) Find the group and phase velocities.

12.41 Phase and Group Velocities. Show that the following relation between group and phase velocity exists:

$$v_g = v_p + \beta\frac{dv_p}{d\beta} \quad \left[\frac{\text{m}}{\text{s}}\right]$$

12.42 Phase and Group Velocities in Lossy Dielectrics. A plane wave with electric field intensity $\mathbf{E} = \hat{\mathbf{x}}100e^{-j2z}$ [V/m] propagates in rubber, which has properties $\mu = \mu_0$ [H/m], $\varepsilon = 4\varepsilon_0$ [F/m], and $\sigma = 0.001$ S/m and may be considered to be a very low-loss dielectric. Calculate:

(a) The phase velocity in the material.
(b) The group velocity in the material.
(c) The energy transport velocity.
(d) What is the conclusion from the results in **(a)**–**(c)**?

12.43 Group Velocity and Dispersion in Low-Loss Media. A general low-loss medium is given in which the term $\sigma/\omega\varepsilon$ is small but not negligible:

(a) Calculate the group velocity.
(b) Plot the group velocity as a function of frequency.
(c) Is the medium dispersive and, if so, is the dispersion normal or anomalous? Explain.

Polarization of Plane Waves

12.44 Polarization of Plane Waves. The magnetic field intensity of a plane wave is given as $\mathbf{H}(x) = \hat{\mathbf{y}}10e^{-j\beta x}$ [A/m]. What is the polarization of this wave?

12.45 Polarization of Plane Waves. The magnetic field intensity of a plane wave is given as $\mathbf{H}(x) = \hat{\mathbf{y}}10e^{-j\beta x} + \hat{\mathbf{z}}15e^{-j\beta x}$ [A/m]. What is the polarization of this wave?

12.46 Polarization of Plane Waves. The electric field intensity of a plane wave is given as $\mathbf{E}(x) = \hat{\mathbf{y}}10e^{-0.1x}\left(e^{-j\beta x} + e^{j\beta x}\right) + \hat{\mathbf{y}}10e^{-0.1x}\left(e^{-j\beta x} - e^{j\beta x}\right)$ [V/m]. What is the polarization of this wave?

12.47 Polarization of Plane Waves. The electric field intensity of a plane wave is given as $\mathbf{E}(x,t) = \hat{\mathbf{y}}100\cos(\omega t - \beta x) + \hat{\mathbf{z}}200\cos(\omega t - \beta x - \pi/2)$ [V/m]. What is the polarization of this wave?

12.48 Polarization of Superposed Plane Waves. Two plane waves propagate in the same direction. Both waves are at the same frequency and have equal amplitudes. Wave A is polarized linearly in the x direction, and wave B is polarized in the direction of $\hat{\mathbf{x}} + \hat{\mathbf{y}}$. In addition, wave B lags behind wave A by a small angle θ. What is the polarization of the sum of the two waves?

12.49 Polarization of Plane Waves. The magnetic field intensity of a plane wave is given as $\mathbf{H}(x,t) = \hat{\mathbf{y}}100\cos(\omega t - \beta x) + \hat{\mathbf{z}}200\cos(\omega t - \beta x + \pi/2)$ [A/m]. What is the polarization of this wave?

Reflection and Transmission of Plane Waves

13

God runs electromagnetics on Monday, Wednesday and Friday by the wave theory and the devil runs it on Tuesday, Thursday and Saturday by the Quantum theory.

Sir William Bragg (1862–1942),
physicist, Nobel laureate, 1915, on Electromagnetics

13.1 Introduction

The propagation of waves in free space and in materials was discussed at some length in **Chapter 12**. In this chapter, we discuss properties of waves as they propagate through different materials and changes in their amplitudes, directions and phases as they propagate through the interfaces between materials. This aspect of the propagation of waves is fundamental and many of the properties of waves are defined by materials and their interfaces. As an example, waves are reflected from conducting and dielectric surfaces giving rise to so-called standing waves. The various properties depend on the materials involved, the direction of propagation, and the polarization of the waves. To keep the discussion simple and within the context of plane waves, we will look at a number of simple interface conditions. These include perpendicular and oblique incidence on conducting and dielectric interfaces, conditions often encountered in applications.

The results we obtain here are useful in a variety of applications, ranging from radar operation to fiber optics. For example, reflection from interfaces can be used to detect water levels in underground aquifers as well as oil deposits. The same principle can be used to measure and monitor the thickness of materials on a production line or flaws in plastics. The design of radomes for radar and communication equipment requires that no reflections at interfaces exist, whereas radar evading (stealth) aircraft absorb all incoming wave energy or reflect it in directions other than the radar antenna. All these can be accomplished by the proper choice of materials and conditions at the interfaces between materials.

The basic principle involved in describing the behavior of a wave at the interface between materials is to write the waves on both sides of the interface and to match the components of the electric and magnetic fields at the interface. In general, this means applying the interface conditions on the fields at the interface. The general conditions are shown in **Figure 13.1**. The wave in material (1) propagates at an angle to the normal to the interface. This angle is called the ***incidence angle***, θ_i. The tangential components of the electric field intensity are continuous at the interface, as we have seen in **Chapter 11**. The normal components are discontinuous. We assume that there are no charge densities or current densities at the interface (except for conducting interfaces). The wave is partly transmitted into material (2) and partly reflected at an angle θ_r called the ***reflection angle***. We will also show that the incidence and reflection angle are equal. The ***transmission angle*** (or ***refraction angle***) θ_t is different than the incidence angle θ_i, as one would expect from two different materials. Note however that the discussion of waves at interfaces goes beyond simple interface conditions. Although interface conditions as discussed in **Chapter 11** must be satisfied, the propagation properties of the wave must also be taken into account. This means that we must consider such properties as polarization of the wave and its speed of propagation.

N. Ida, *Engineering Electromagnetics*, https://doi.org/10.1007/978-3-030-15557-5_13

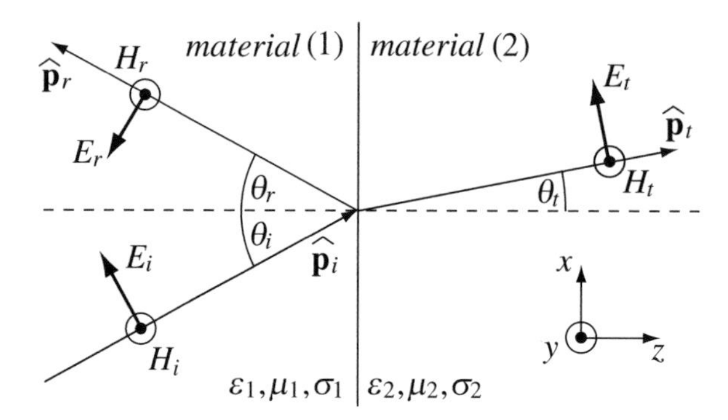

Figure 13.1 The general relationships between fields at an interface between two materials. Directions of the reflected fields are assumed

As we discuss the behavior of waves at interfaces, it is useful to recall the behavior of light waves. We expect similar behavior, including reflection, transmission, and refraction of waves at the interface. We start with a general, lossy dielectric interface and then proceed to discuss lossless and low-loss dielectrics and conductors. Normal incidence of the wave on the interface is considered first, followed by oblique incidence with polarized waves at conducting and dielectric interfaces.

13.2 Reflection and Transmission at a General Dielectric Interface: Normal Incidence

Figure 13.2 shows an incident wave $\mathbf{E}_i$ propagating in a dielectric with permittivity ε_1, conductivity σ_1, and permeability μ_1. The wave encounters the interface between material (1) and material (2), which is also a lossy dielectric with permittivity ε_2, conductivity σ_2, and permeability μ_2. Based on the definition of angle of incidence in the previous section, the incident wave hits the interface at a zero-degree angle (perpendicular to the interface) and is reflected at the same angle. The transmitted wave also propagates perpendicular to the interface.

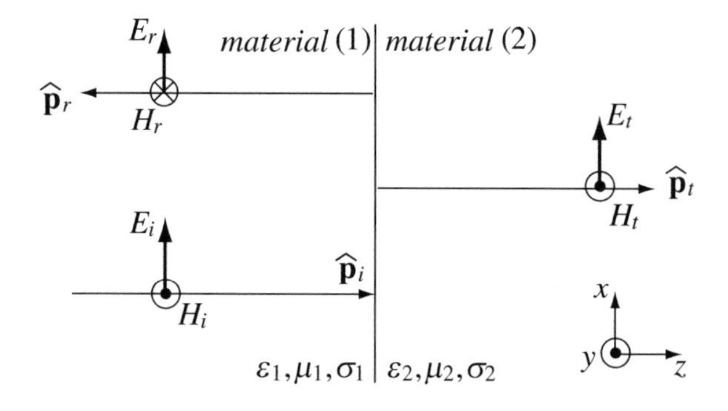

Figure 13.2 Reflection and transmission at a general interface: normal incidence. The directions of the reflected field vectors are assumed

To simplify the discussion, we assume a wave propagating in the positive z direction, as in **Eq. (12.89)**. The incident electric field intensity is

$$\mathbf{E}_i = \hat{\mathbf{x}} E_{i1} e^{-\gamma_1 z} \quad [\mathrm{V/m}] \tag{13.1}$$

where E_{i1} is the amplitude of the incident wave and the propagation constant is

$$\gamma_1 = \alpha_1 + j\beta_1 \tag{13.2}$$

The wave reflected from the interface propagates backward in material (1) and is assumed to be

$$\mathbf{E}_r = \hat{\mathbf{x}} E_{r1} e^{+\gamma_1 z} \quad [\mathrm{V/m}] \tag{13.3}$$

The backward propagation is indicated by the positive sign in the exponent (the wave propagates in the negative z direction). The amplitude E_{r1} is, in general, different than E_{i1} and may be zero (no reflection at the interface). The reflected electric field intensity is assumed, arbitrarily, to be in the same direction as the electric field intensity of the incident wave. The actual direction will be found by imposing the interface conditions at the interface. The total wave in material (1) has two components: one propagating in the positive z direction and one in the negative z direction:

$$\mathbf{E}_1 = \hat{\mathbf{x}} E_{i1} e^{-\gamma_1 z} + \hat{\mathbf{x}} E_{r1} e^{+\gamma_1 z} \quad [\mathrm{V/m}] \tag{13.4}$$

The magnetic field intensity also has two components and can be calculated from **Eq. (12.40)** or directly from Faraday's law as

$$\mathbf{H}_1 = \hat{\mathbf{y}} \frac{E_{i1}}{\eta_1} e^{-\gamma_1 z} - \hat{\mathbf{y}} \frac{E_{r1}}{\eta_1} e^{+\gamma_1 z} \quad \left[\frac{\mathrm{V}}{\mathrm{m}}\right] \tag{13.5}$$

Note that the backward propagating magnetic field intensity is negative if the backward electric field intensity is positive. This sign can be found directly from Maxwell's first equation (Faraday's law), but, here, we simply point out that the sign must be negative since the propagation of power due to this term is in the negative z direction (direction of the Poynting vector). In practice, we may find that E_{r1} itself is negative, but, regardless of the sign of E_{r1}, H_{r1} must be such that the Poynting vector $\mathbf{E}_r \times \mathbf{H}_r$ points away from the interface. Thus, from the Poynting vector, we have

$$\mathcal{P}_1^+ = \left(\hat{\mathbf{x}} E_{i1} e^{-\gamma_1 z}\right) \times \left(\hat{\mathbf{y}} \frac{E_{i1}}{\eta_1} e^{-\gamma_1 z}\right) = \hat{\mathbf{x}} \times \hat{\mathbf{y}} \frac{E_{i1}^2}{\eta_1} e^{-2\gamma_1 z} = \hat{\mathbf{z}} \frac{E_{i1}^2}{\eta_1} e^{-2\gamma_1 z} \quad \left[\frac{\mathrm{W}}{\mathrm{m}^2}\right] \tag{13.6}$$

which propagates in the positive z direction. Similarly, the Poynting vector for the backward propagating wave is

$$\mathcal{P}_1^- = \left(\hat{\mathbf{x}} E_{r1} e^{\gamma_1 z}\right) \times \left(-\hat{\mathbf{y}} \frac{E_{r1}}{\eta_1} e^{\gamma_1 z}\right) = -\hat{\mathbf{x}} \times \hat{\mathbf{y}} \frac{E_{r1}^2}{\eta_1} e^{2\gamma_1 z} = -\hat{\mathbf{z}} \frac{E_{r1}^2}{\eta_1} e^{2\gamma_1 z} \quad \left[\frac{\mathrm{W}}{\mathrm{m}^2}\right] \tag{13.7}$$

In these relations, the intrinsic impedance η_1 and the propagation constant γ_1 are completely general and given as [see **Eq. (12.102)** and **Eq. (12.85)**]

$$\eta_1 = \sqrt{\frac{j\omega\mu_1}{\sigma_1 + j\omega\varepsilon_1}} \quad [\Omega], \quad \gamma_1 = \sqrt{j\omega\mu_1(\sigma_1 + j\omega\varepsilon_1)} \tag{13.8}$$

To define the transmitted waves, we may assume that **E** and **H** in material (2) are of the same form as in material (1) except that in this case, the backward propagating wave in material (2) cannot exist. This is because material (2) extends to infinity and there is no mechanism for a reflected wave in material (2) to exist. Thus, we get

$$\mathbf{E}_t = \mathbf{E}_2 = \hat{\mathbf{x}} E_2 e^{-\gamma_2 z} \quad [\mathrm{V/m}] \tag{13.9}$$

$$\mathbf{H}_t = \mathbf{H}_2 = \hat{\mathbf{y}} \frac{E_2}{\eta_2} e^{-\gamma_2 z} \quad \left[\frac{\mathrm{A}}{\mathrm{m}}\right] \tag{13.10}$$

and, as required, this wave propagates in the positive z direction. The intrinsic impedance and the propagation constant in material (2) are

$$\eta_2 = \sqrt{\frac{j\omega\mu_2}{\sigma_2 + j\omega\varepsilon_2}} \quad [\Omega], \quad \gamma_2 = \sqrt{j\omega\mu_2(\sigma_2 + j\omega\varepsilon_2)} \tag{13.11}$$

Now that we wrote the fields in each material, we can define a reflection coefficient as the ratio between the amplitudes of the reflected and incident waves:

$$\boxed{\Gamma = \frac{E_{r1}}{E_{i1}} \quad [\text{dimensionless}]} \tag{13.12}$$

The reflection coefficient is a dimensionless quantity which gives the fraction of the incident wave amplitude reflected back from the interface. It can vary from zero (no reflection) to 1 (total reflection) and can be either positive or negative. Since both amplitudes are, in general, complex numbers, the reflection coefficient may also be a complex number.

Similarly, a transmission coefficient is defined as the ratio between the amplitudes of the transmitted and incident waves:

$$\boxed{T = \frac{E_t}{E_{i1}} \quad [\text{dimensionless}]} \tag{13.13}$$

The transmission coefficient is also a dimensionless quantity and gives the fraction of the incident wave amplitude transmitted across the interface. The transmission coefficient is, in general, complex and can vary in magnitude between zero and 2, as we shall see shortly.

Using the definition of the reflection coefficient, the reflected wave can be written in terms of the incident wave as

$$\mathbf{E}_r = \hat{\mathbf{x}}E_{r1}e^{+\gamma_1 z} = \hat{\mathbf{x}}\Gamma E_{i1}e^{+\gamma_1 z} \quad [\text{V/m}] \tag{13.14}$$

Using the transmission coefficient T, we can write

$$\mathbf{E}_2 = \hat{\mathbf{x}}E_2 e^{-\gamma_2 z} = \hat{\mathbf{x}}TE_{i1}e^{-\gamma_2 z} \quad [\text{V/m}] \tag{13.15}$$

Similarly, the incident, reflected, and transmitted magnetic field intensities are

$$\mathbf{H}_i = \hat{\mathbf{y}}\frac{E_{i1}}{\eta_1}e^{-\gamma_1 z} \quad \left[\frac{\text{A}}{\text{m}}\right] \tag{13.16}$$

$$\mathbf{H}_r = -\hat{\mathbf{y}}\Gamma\frac{E_{i1}}{\eta_1}e^{+\gamma_1 z} \quad \left[\frac{\text{A}}{\text{m}}\right] \tag{13.17}$$

$$\mathbf{H}_2 = \hat{\mathbf{y}}T\frac{E_{i1}}{\eta_2}e^{-\gamma_2 z} \quad \left[\frac{\text{A}}{\text{m}}\right] \tag{13.18}$$

The transmission and reflection coefficients can now be evaluated from the relations at the interface. To do so, we place the interface at $z = 0$, write the total electric and magnetic fields on each side of the interface, and equate the tangential components (which are continuous across the interface):

$$\mathbf{E}_i + \mathbf{E}_r = \mathbf{E}_2 \quad \rightarrow \quad E_{i1} + \Gamma E_{i1} = TE_{i1} \quad [\text{V/m}] \tag{13.19}$$

Similarly, from the continuity of the tangential components of the magnetic field intensity,

$$\mathbf{H}_i + \mathbf{H}_r = \mathbf{H}_2 \quad \rightarrow \quad \frac{E_{i1}}{\eta_1} - \frac{\Gamma E_{i1}}{\eta_1} = \frac{TE_{i1}}{\eta_2} \quad \left[\frac{\text{A}}{\text{m}}\right] \tag{13.20}$$

From **Eqs. (13.19)** and **(13.20)**, we can write

$$\boxed{1 + \Gamma = T} \tag{13.21}$$

$$\frac{1}{\eta_1} - \frac{\Gamma}{\eta_1} = \frac{T}{\eta_2} \tag{13.22}$$

or, solving for Γ and, separately, for T

$$\boxed{\Gamma = \frac{\eta_2 - \eta_1}{\eta_1 + \eta_2} \quad [\text{dimensionless}]} \tag{13.23}$$

and

$$\boxed{T = \frac{2\eta_2}{\eta_1 + \eta_2} \quad [\text{dimensionless}]} \tag{13.24}$$

Γ can be negative (depending on the relative values of the intrinsic impedances), but T is always positive. Both coefficients are ratios of impedances and, therefore, are dimensionless.

The reflection and transmission coefficients are important properties in wave propagation. We will meet them in many diverse situations. It is, therefore, important to familiarize ourselves with their general properties. It is often possible to draw simple conclusions on behavior of electromagnetic waves from calculation or estimation of these coefficients. As a simple example, if you were to try to design a stealth aircraft, the immediate conclusion from **Eqs. (13.23)** and **(13.24)** is that the reflected wave should be as small as possible since the lower the reflection, the smaller the amplitude of the wave returning from the aircraft. Thus, the requirement for an aircraft "invisible to radar" is that the reflection coefficient is zero, which, in turn, means that $\eta_2 = \eta_1$. This conclusion was drawn without actually calculating the coefficient or, for that matter, without even commenting on the practicability of the solution. Similarly, when using a microwave oven to heat a material, the lower the reflection coefficient, the more energy couples into the material and the more efficient the heating process. Another way to look at it is in the generator-load context. A perfectly matched load causes no reflection of energy into the generator. All energy on the line is transferred from the line to the load (zero reflection coefficient). An unmatched load means that part of the energy is transferred into the load and part of it is reflected back into the generator. On the other hand, there are situations in which we may wish to maximize reflections. One example is in the design of reflector antennas (most "dish" antennas are parabolic reflectors; the antenna itself is a small "feed" at the focal point of the parabola). In this case, the transmission coefficient should be as close as possible to zero so that power does not penetrate into the reflector and the reflection coefficient as close as possible to -1 [see **Eq. (13.21)**].

The following properties of the reflection and transmission coefficients will be useful in this and future sections:

(1) Both coefficients are in general complex numbers.
(2) If the conductivities of both materials are zero, both Γ and T are real numbers. This happens whenever the materials on the two sides of the interface are perfect dielectrics.
(3) If material (2) is a perfect conductor, its intrinsic impedance is zero (i.e., $\sigma_2 \to \infty$ and, therefore, $\eta_2 \to 0$). This means that the reflection coefficient is -1 and the transmission coefficient is zero. The wave does not penetrate into a perfect conductor and the reflected wave is in the opposite direction in space; that is, the reflected wave is 180° out of phase in space with the incident wave.
(4) Both constants are frequency dependent in most cases, as can be seen from **Eqs. (13.8)** and **(13.11)**. In this sense, the constants are not really constant. In other situations, including those of perfect dielectrics and perfect conductors, the constants may be regarded as true constants since the intrinsic impedance of perfect dielectrics is real and independent of frequency, and for perfect conductors it is zero.

The main utility of the reflection and transmission coefficients at this point is to describe the transmitted and reflected waves in terms of the incident wave. This approach is sensible since the incident wave is normally known, whereas the reflection and transmission coefficients are only dependent on materials and possibly on frequency.

Returning now to the fields in material (1), the total electric field intensity can be expressed as the sum of the incident and reflected waves as in **Eq. (13.4)**:

$$\mathbf{E}_1(z) = \mathbf{E}_i(z) + \mathbf{E}_r(z) = \hat{\mathbf{x}}E_{i1}(e^{-\gamma_1 z} + \Gamma e^{+\gamma_1 z}) \quad [\text{V/m}] \tag{13.25}$$

By adding and subtracting the term $\Gamma e^{-\gamma_1 z}$ in the brackets on the right-hand side, we can write

$$\mathbf{E}_1(z) = \hat{\mathbf{x}}E_{i1}((1+\Gamma)e^{-\gamma_1 z} + \Gamma(e^{+\gamma_1 z} - e^{-\gamma_1 z})) \quad [\text{V/m}] \tag{13.26}$$

or using the identity $e^{+\gamma_1 z} - e^{-\gamma_1 z} = 2\sinh(\gamma_1 z) = -j2\sin(j\gamma_1 z)$ and the fact that $1 + \Gamma = T$,

$$\boxed{\mathbf{E}_1(z) = \hat{\mathbf{x}}E_{i1}(Te^{-\gamma_1 z} - \Gamma j2\sin(j\gamma_1 z)) \quad [\text{V/m}]} \tag{13.27}$$

Similarly, the magnetic field intensity in material (1) is the sum of the incident and reflected fields:

$$\mathbf{H}_1(z) = \hat{\mathbf{y}}\frac{E_{i1}}{\eta_1}(e^{-\gamma_1 z} - \Gamma e^{+\gamma_1 z}) \quad [\text{A/m}] \tag{13.28}$$

Again adding and subtracting $\Gamma e^{-\gamma_1 z}$ in the brackets, we get

$$\mathbf{H}_1(z) = \hat{\mathbf{y}}\frac{E_{i1}}{\eta_1}((1+\Gamma)e^{-\gamma_1 z} - \Gamma(e^{+\gamma_1 z} + e^{-\gamma_1 z})) \quad [\text{A/m}] \tag{13.29}$$

With $T = 1 + \Gamma$ and $e^{+\gamma_1 z} + e^{-\gamma_1 z} = 2\cosh(\gamma_1 z) = 2\cos(j\gamma_1 z)$, this gives

$$\boxed{\mathbf{H}_1(z) = \hat{\mathbf{y}}\frac{E_{i1}}{\eta_1}(Te^{-\gamma_1 z} - \Gamma 2\cos(j\gamma_1 z)) \quad [\text{A/m}]} \tag{13.30}$$

The fields in material (2) are

$$\boxed{\mathbf{E}_2(z) = \hat{\mathbf{x}}TE_{i1}e^{-\gamma_2 z} \quad [\text{V/m}]} \tag{13.31}$$

and

$$\boxed{\mathbf{H}_2(z) = \hat{\mathbf{y}}T\frac{E_{i1}}{\eta_2}e^{-\gamma_2 z} \quad \left[\frac{\text{A}}{\text{m}}\right]} \tag{13.32}$$

Equations (13.27) and **(13.30)** are useful because they indicate that the sum of the incident and reflected waves consists of a wave component propagating in the positive z direction and a non-propagating wave component. The latter [second term in **Eq. (13.27)** or **Eq. (13.30)**] is called a ***standing wave***. We also observe that if the transmission coefficient is zero, only a standing wave exists, whereas if the reflection coefficient is zero, there are no standing waves. Standing waves will be discussed separately in conjunction with reflection from conducting surfaces, but we point out here that standing waves exist any time there is reflection of a wave, and it is caused by interference between the forward and backward propagating waves.

The above discussion is quite general and assumes nothing about material properties. In practice, there are a number of combinations of materials which are important. For example, the interfaces between free space (air) and conductors or dielectrics are commonly encountered. Some of the more important material interfaces are described in the following sections.

Example 13.1 Application: Optical Fiber Connectors Two optical fibers are connected through a connector to form an interface as shown in **Figure 13.3**. In optical fibers, the attenuation is indicated in dB/km. Fiber (1) is rated as 1 dB/km (a good fiber) and the second as 10 dB/km. The source of light is in free space (not shown), at a wavelength of 700 nm (a red laser or light emitting diode). Assume both fibers are low-loss dielectrics (which in practice they are; conductivity of glass is about 10^{-12} S/m) and that propagation is from fiber (1) into fiber (2). Calculate:

(a) The reflection and transmission coefficients at the interface.
(b) The amplitude of the electric and magnetic field intensities at $d = 10$ km from the interface, in material (2), assuming the amplitude of the incident electric field intensity in material (1) is known at the interface as E_{i1}.
(c) Show that power is conserved across the interface, that is, that the transmitted time-averaged power density must equal the incident time-averaged power density minus the reflected time-averaged power density.

Figure 13.3 Conditions at the interface between two optical fibers

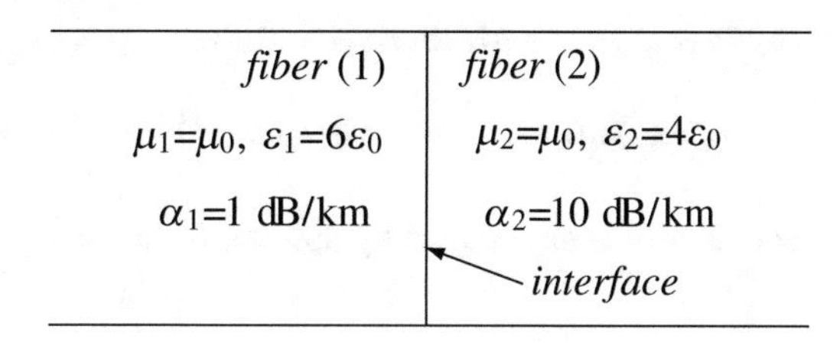

Solution: From the given attenuation, we calculate the attenuation constant, and from **Eqs. (13.8)** and **(13.11)**, we calculate the intrinsic impedance of fibers (1) and (2). Then, the reflection and transmission coefficients are calculated from **Eqs. (13.23)** and **(13.24)**. The electric and magnetic field intensities in material (2) are given in **Eqs. (13.31)** and **(13.32)**.

The attenuation and phase constants for the two fibers are calculated first. Since the attenuation constant is given in dB/km and 1 neper/meter = 8.69 dB/m (see **Section 12.7.1**), the attenuation constants are

$$\alpha_1 = \frac{1}{1000 \times 8.69} = 1.15 \times 10^{-4}, \quad \alpha_2 = \frac{10}{1000 \times 8.69} = 1.15 \times 10^{-3} \quad [\text{Np/m}]$$

The phase constant is calculated from the relation $\beta = 2\pi/\lambda$ (low-loss dielectric), where λ is the wavelength in each fiber. The wavelength is given in free space as 700 nm. In the fibers, the speed of propagation is reduced by a factor of $\sqrt{\varepsilon_r}$ and the wavelength is also reduced by this factor since $\lambda = v_p/f$. Thus, the phase constants in the two fibers are

$$\beta_1 = \frac{2\pi \times \sqrt{6}}{700 \times 10^{-9}} = 21.986 \times 10^6, \quad \beta_2 = \frac{2\pi \times \sqrt{4}}{700 \times 10^{-9}} = 17.952 \times 10^6 \quad [\text{rad/m}]$$

To calculate the intrinsic impedance, we write [see **Eq. (13.8)**]:

$$\eta = \sqrt{\frac{j\omega\mu}{\sigma + j\omega\varepsilon}} = \frac{j\omega\mu}{\sqrt{j\omega\mu(\sigma + j\omega\varepsilon)}} = \frac{j\omega\mu}{\gamma} = \frac{j\omega\mu}{\alpha + j\beta} \approx \frac{\omega\mu}{\beta} = \sqrt{\frac{\mu}{\varepsilon}} \quad [\Omega]$$

where the fact that the phase constant (β_1 or β_2) is much larger numerically than the attenuation constant (α_1 or α_2) was used. The intrinsic impedances in the two fibers are

$$\eta_1 = \sqrt{\frac{\mu_0}{6\varepsilon_0}} = \frac{377}{\sqrt{6}} = 153.91, \quad \eta_2 = \sqrt{\frac{\mu_0}{4\varepsilon_0}} = \frac{377}{2} = 188.5 \quad [\Omega]$$

The transmission and reflection coefficients are calculated in terms of the intrinsic impedances. From **Eqs. (13.23)** and **(13.24)**

$$\Gamma = \frac{\eta_2 - \eta_1}{\eta_1 + \eta_2} = \frac{188.5 - 153.91}{188.5 - 153.91} = 0.101$$

$$T = \frac{2\eta_2}{\eta_1 + \eta_2} = \frac{188.95}{188.95 + 153.91} = 1.101$$

Note that since the reflection coefficient is positive, the transmission coefficient must be larger than 1, as can be seen from **Eq. (13.21)**.

The incident electric field intensity in fiber (1) is given as E_{i1}. First, we calculate the electric field intensity across the interface and then, using the attenuation in material (2), calculate the amplitude of the electric field intensity at a distance of 10 km. The amplitude of the electric field intensity in material (2) is

$$E_2(0) = TE_{i1} = 1.101E_{i1} \quad [\text{V/m}]$$

where (0) indicates that this is at the interface. At $d = 10$ km, the amplitude is

$$E_2(d) = E_2(0)e^{-\alpha_2 d} = 1.101E_{i1}e^{-1.15\times10^{-3}\times10^4} = 1.115 \times 10^{-5}E_{i1} \quad [\text{V/m}]$$

The electric field intensity has been reduced by about a factor of 10^5 in 10 km. The magnetic field intensity at the same location is

$$H_2 = \frac{E_2}{\eta_2} = \frac{1.115 \times 10^{-5}}{188.5}E_{i1} = 5.915 \times 10^{-8}E_{i1} \quad [\text{A/m}]$$

The time-averaged power across the interface may be written from the Poynting theorem (see **Example 12.9**):

$$\frac{E_{i1}^2}{2\eta_1} - \frac{E_r^2}{2\eta_1} = \frac{E_t^2}{2\eta_2} \quad \rightarrow \quad \frac{E_{i1}^2}{2\eta_1} - \frac{(\Gamma E_{i1})^2}{2\eta_1} = \frac{(TE_{i1})^2}{2\eta_2}$$

or

$$\frac{1}{\eta_1} - \frac{\Gamma^2}{\eta_1} = \frac{T^2}{\eta_2} \quad \rightarrow \quad \eta_2 - \eta_2\Gamma^2 = \eta_1 T^2$$

Using the reflection and transmission coefficients and the intrinsic impedances in **(b)** we get

$$188.5 - 188.5 \times 0.101^2 = 153.91 \times 1.101^2 \quad \rightarrow \quad 186.57 = 186.57$$

The equality shows that power density is conserved across the interface.

Exercise 13.1 In **Example 13.1**, assume the propagation is from material (2) into material (1) and calculate:

(a) The reflection and transmission coefficients at the interface.
(b) The amplitude of the electric and magnetic field intensities at a distance 1 km from the interface in material (1) assuming the amplitude of the incident electric field intensity in material (2) at the interface equals E_2. Use the same properties and assumptions as in **Example 13.1**.

Answer **(a)** $\Gamma = -0.101$, $T = 0.899$. **(b)** $E_1(1{,}000 \text{ m}) = 0.8E_2$ [V/m], $H_1(1{,}000 \text{ m}) = 5.21 \times 10^{-3}E_2$ [A/m].

Exercise 13.2 In **Example 13.1**, calculate the exact complex reflection and transmission coefficients.

Answer $\Gamma = 0.101 + j2.9114 \times 10^{-11}$, $\Gamma = 1.101 + j2.9114 \times 10^{-11}$.

13.2.1 Reflection and Transmission at an Air-Lossy Dielectric Interface: Normal Incidence

In many practical applications, material (1) is free space; that is, an incident wave propagates in free space and encounters a lossy dielectric. A situation of this type is in a wave transmitted in free space (say from a radar or communication antenna) and encountering a concrete wall or a body of water. The difference with respect to the previous section is that η_1 becomes η_0 and in the expressions for $\mathbf{E}_i$ and $\mathbf{E}_r$, γ_1 is replaced by $j\beta_0$. This situation is shown in **Figure 13.2** for $\sigma_1 = 0$, $\varepsilon_1 = \varepsilon_0$, and $\mu_1 = \mu_0$. Thus, if material (1) is free space and material (2) is a lossy dielectric, the reflection and transmission coefficients are

$$\Gamma = \frac{\eta_2 - \eta_0}{\eta_0 + \eta_2}, \quad T = \frac{2\eta_2}{\eta_0 + \eta_2} \quad \text{[dimensionless]} \tag{13.33}$$

The electric and magnetic field intensities are [from **Eqs. (13.27)**, **(13.30)**, **(13.31)**, and **(13.32)**]

$$\mathbf{E}_1(z) = \hat{\mathbf{x}} E_{i1}\left(Te^{-j\beta_0 z} + j2\Gamma \sin(\beta_0 z)\right) \quad [\text{V/m}] \tag{13.34}$$

$$\mathbf{H}_1(z) = \hat{\mathbf{y}}\frac{E_{i1}}{\eta_0}\left(Te^{-j\beta_0 z} - \Gamma 2\cos(\beta_0 z)\right) \quad [\text{A/m}] \tag{13.35}$$

$$\mathbf{E}_2(z) = \hat{\mathbf{x}} E_{i1} T e^{-j\gamma_2 z} \quad [\text{V/m}] \tag{13.36}$$

$$\mathbf{H}_2(z) = \hat{\mathbf{y}}\frac{E_{i1}}{\eta_2} T e^{-j\gamma_2 z} \quad \left[\frac{\text{A}}{\text{m}}\right] \tag{13.37}$$

The phase constant and intrinsic impedance in material (1) are

$$\beta_0 = \omega\sqrt{\mu_0\varepsilon_0} \quad [\text{rad/m}], \quad \eta_0 = \sqrt{\frac{\mu_0}{\varepsilon_0}} \quad [\Omega] \tag{13.38}$$

and γ_2 and η_2 remain as in **Eq. (13.11)**. Note that even though the expressions for $\mathbf{E}_2$ and $\mathbf{H}_2$ are the same as in **Eqs. (13.31)** and **(13.32)**, the transmission coefficient and the propagation constant are different, and, therefore, the fields in material (2) are also different.

These expressions can be easily adapted to any other lossless dielectric in place of material (1). To do so, the permittivity and permeability of free space in **Eq. (13.38)** are replaced by those of the dielectric.

Example 13.2 Application: Radar Sensing of the Environment In an attempt to map the thickness of the polar ice caps, a special airplane is equipped with a downward-looking radar operating at 10 GHz. The idea is to fly over at a known height and measure the time it takes for narrow pulses to reach the bottom of the ice and return to the receiver. From the knowledge of speed of propagation in air and in ice, it is possible to calculate the thickness of the ice. The properties of ice at 10 GHz are $\mu = \mu_0$ [H/m], $\varepsilon = 3.5\varepsilon_0$ [F/m], $\sigma = 10^{-6}$ S/m and those of free space are $\mu = \mu_0$ [H/m] and $\varepsilon = \varepsilon_0$ [F/m] $\sigma = 0$. The antenna transmits a uniform beam 1 m in diameter and the time-averaged power is 1 kW. Assume plane waves and that the beam remains of constant diameter:

(a) If the ice is 10 km deep at a measurement point and the surface below the ice is perfectly reflecting, calculate the amplitude of the electric field intensity that reaches back to the aircraft antenna.
(b) Is this measurement feasible?

Solution: To calculate the electric field intensity below the surface of the ice, we need the transmission coefficient from air to ice. To calculate the field at the aircraft, we also need the transmission coefficient from ice into air (for the returning wave). The electric field intensity at the antenna is calculated from the Poynting vector. This electric field intensity is then allowed to propagate through air, transmit into the ice, propagate to the bottom and back to the surface, transmit through the surface, and propagate back to the airplane. The propagation path and coefficients are shown in **Figure 13.4**:

(a) Before calculating the electric field intensity below the ice surface, we calculate the electric field intensity at the antenna. The power density at the antenna is 1 kW/S, where S is the area of the beam:

$$\mathcal{P}_{av} = \frac{P}{S} = \frac{1{,}000}{\pi \times (0.5)^2} = 1{,}273.2 \quad [\text{W/m}^2]$$

From the discussion in **Chapter 12** (see **Example 12.9**), the amplitude of the time-averaged Poynting vector is

$$\mathcal{P}_{av} = \frac{E^2}{2\eta_0} = 1{,}273.2 \quad [\text{W/m}^2]$$

In free space, $\eta_0 = 377\ \Omega$ and, therefore, the amplitude of the electric field intensity is

$$E = \sqrt{2\eta_0 \mathcal{P}_{av}} = \sqrt{2 \times 377 \times 1{,}273.2} = 979.8 \quad [\text{V/m}]$$

This is E_0 in **Figure 13.4**. The electric field propagates in air without attenuation. Thus, $E_1 = E_0 = 979.8$ V/m in air, at the surface of the ice.

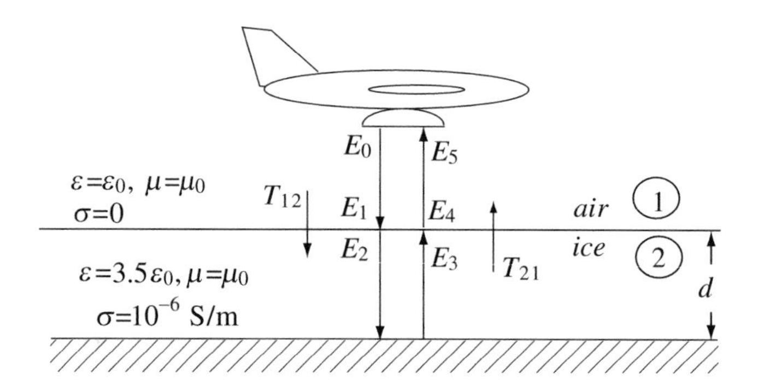

Figure 13.4 Remote sensing. Measurement of thickness of the polar ice caps

To calculate the amplitude of the electric field intensity below the ice, E_2, we need the transmission coefficient between air and ice. Because ice is a low-loss dielectric, we use the formula for intrinsic impedance for low-loss dielectrics given in **Eq. (12.109)**:

$$\eta = \eta_n\left(1 + \frac{j\sigma}{2\omega\varepsilon}\right) = \frac{\eta_0}{\sqrt{\varepsilon_r}}\left(1 + \frac{j10^{-6}}{2 \times 2 \times \pi \times 10^{10} \times 8.854 \times 3.5 \times 10^{-12}}\right) = \frac{377}{\sqrt{3.5}}\left(1 + j2.57 \times 10^{-7}\right) \approx 201.51 \quad [\Omega]$$

where η_n is the no-loss intrinsic impedance of ice. Thus, the transmission coefficient from air to ice is

$$T_{12} = \frac{2\eta}{\eta + \eta_0} = \frac{2 \times 201.51}{201.51 + 377} = 0.6967$$

The electric field intensity below the surface of the ice is therefore

$$E_2 = T_{12}E_1 = 0.6967 \times 979.8 = 682.63 \quad [\text{V/m}].$$

The electric field intensity in ice propagates with attenuation until it reaches the bottom. Then, it is reflected and propagated with the same attenuation until it reaches the surface again. From the relations for low-loss dielectrics **[Eq. (12.104)]**, the attenuation constant is

$$\alpha \approx \frac{\sigma}{2}\sqrt{\frac{\mu}{\varepsilon}} = \frac{\sigma}{2}\sqrt{\frac{\mu_0}{3.5\varepsilon_0}} = \frac{10^{-6} \times 377}{2 \times \sqrt{3.5}} = 1.0076 \times 10^{-4} \quad [\text{Np/m}]$$

The total distance traveled is 20 km. The electric field intensity below the surface of the ice after it has propagated to the bottom and back is

$$E_3 = E_2 e^{-2\alpha d} = 682.63 \times e^{-2 \times 1.0076 \times 10^{-4} \times 10^4} = 91 \quad [\text{V/m}]$$

This is now transmitted across the interface with the transmission coefficient between ice and air, which is

$$T_{21} = \frac{2\eta_0}{\eta + \eta_0} = \frac{2 \times 377}{201.51 + 377} = 1.303$$

The electric field intensity in air, just above the surface of the ice, is

$$E_4 = E_3 T_{21} = 91 \times 1.303 = 118.6 \quad [\text{V/m}]$$

This wave propagates in air and reaches the antenna without further change. The electric field intensity returning to the antenna is 118.6 V/m.

(b) The measurement is feasible because the attenuation is rather low. This type of measurement is also used to measure thickness of snow.

Note: there are direct reflections off the surface of the ice as well as multiple reflections within the ice layer. These were neglected in this example.

13.2.2 Reflection and Transmission at an Air-Lossless Dielectric Interface: Normal Incidence

A common combination of materials at an interface is free space and a lossless dielectric. An example is a radome over a radar antenna (a radome is a dielectric cover designed to protect the antenna and, at the same time, should be transparent to the waves). Using **Figure 13.2** again, material (1) is characterized by $\varepsilon_1 = \varepsilon_0$, $\mu_1 = \mu_0$, and $\sigma_1 = 0$ and material (2) is characterized by ε_2, μ_2, and $\sigma_2 = 0$. The transmission and reflection coefficients can be written in terms of the relative permeability and relative permittivity of material (2) alone:

$$\Gamma = \frac{\eta_2 - \eta_0}{\eta_0 + \eta_2} = \frac{\sqrt{\mu_{r2}\mu_0/\varepsilon_{r2}\varepsilon_0} - \sqrt{\mu_0/\varepsilon_0}}{\sqrt{\mu_{r2}\mu_0/\varepsilon_{r2}\varepsilon_0} + \sqrt{\mu_0/\varepsilon_0}} = \frac{\sqrt{\mu_{r2}/\varepsilon_{r2}} - 1}{\sqrt{\mu_{r2}/\varepsilon_{r2}} + 1} = \frac{\sqrt{\mu_{r2}} - \sqrt{\varepsilon_{r2}}}{\sqrt{\mu_{r2}} + \sqrt{\varepsilon_{r2}}} \tag{13.39}$$

$$T = \frac{2\sqrt{\mu_{r2}/\varepsilon_{r2}}}{\sqrt{\mu_{r2}/\varepsilon_{r2}} + 1} = \frac{2\sqrt{\mu_{r2}}}{\sqrt{\mu_{r2}} + \sqrt{\varepsilon_{r2}}} \tag{13.40}$$

Both Γ and T are real numbers. The electric and magnetic field intensities are

$$\mathbf{E}_1(z) = \hat{\mathbf{x}} E_{i1}\left(Te^{-j\beta_0 z} + j2\Gamma \sin(\beta_0 z)\right) \quad [\text{V/m}] \tag{13.41}$$

$$\mathbf{H}_1(z) = \hat{\mathbf{y}} \frac{E_{i1}}{\eta_0}\left(Te^{-j\beta_0 z} - 2\Gamma \cos(\beta_0 z)\right) \quad [\text{A/m}] \tag{13.42}$$

$$\mathbf{E}_2(z) = \hat{\mathbf{x}} E_{i1} T e^{-j\beta_2 z} \quad [\text{V/m}] \tag{13.43}$$

$$\mathbf{H}_2(z) = \hat{\mathbf{y}}\,\frac{E_{i1}}{\eta_2} T e^{-j\beta_2 z} \quad \left[\frac{\mathrm{A}}{\mathrm{m}}\right] \tag{13.44}$$

where

$$\beta_0 = \omega\sqrt{\mu_0\varepsilon_0}, \quad \beta_2 = \omega\sqrt{\mu_2\varepsilon_2} \quad [\mathrm{rad/m}], \quad \eta_0 = \sqrt{\frac{\mu_0}{\varepsilon_0}}, \quad \eta_2 = \sqrt{\frac{\mu_2}{\varepsilon_2}} \quad [\Omega] \tag{13.45}$$

If material (1) is not free space but a general lossless dielectric, the permittivity and permeability of this material are substituted for those of free space.

We note from these expressions that if the ratio $\mu_{r2}/\varepsilon_{r2}$ is equal to that of free space (i.e., $\mu_{r2}/\varepsilon_{r2} = 1$), there is no reflection at the interface and the whole wave is transmitted across the boundary. This is the case of perfect impedance matching since both materials have the same intrinsic impedance. While this situation is certainly not common in dielectrics, some materials, such as ferrites, can be made to closely resemble this condition by adjusting their permeability.

Similarly, if we are interested in large reflections, the permittivity of material (2) must be made as large as possible. Some materials have relatively high permittivity and are, therefore, reflective. One such material is water ($\varepsilon_r = 81$ at low frequencies). The permittivity of materials is also frequency dependent. For example, the permittivity of water decreases with frequency until, in the optical range, it is only about $1.75\varepsilon_0$.

Example 13.3 Application: Transparent Materials Suppose you are required to design a material which is completely transparent to electromagnetic waves at 1 GHz. For the purpose of this example, assume the material is very thick so that you can consider only the interface between it and free space. You are free to choose any relative permittivity between 2 and 9 and any permeability (it is possible to change the permeability of a dielectric by adding to it ferromagnetic particles). Assume the material remains lossless:

(a) Find the combinations of material properties that will accomplish this design requirement.
(b) What happens at a different frequency, say 2 GHz?

Solution: For an interface to be transparent, the intrinsic impedance on both sides of the interface must be the same:

(a) Equating the intrinsic impedance in free space and a general material, we get

$$\frac{\mu_0}{\varepsilon_0} = \frac{\mu_2}{\varepsilon_2} = \frac{\mu_0}{\varepsilon_0}\frac{\mu_{r2}}{\varepsilon_{r2}} \quad \rightarrow \quad \frac{\mu_{r2}}{\varepsilon_{r2}} = 1$$

Thus, for $\varepsilon_{r2} = 2, \mu_{r2} = 2$, and for $\varepsilon_{r2} = 9, \mu_{r2} = 9$. The range of possible relative permabilities is between 2 and 9.

(b) The reflection coefficient in lossless materials is independent of frequency. The material is transparent at all frequencies. In lossy materials, this is not true.

13.2.3 Reflection and Transmission at an Air-Conductor Interface: Normal Incidence

We consider next the propagation of a wave from free space (or any lossless dielectric), impinging on a high-conductivity conductor, as in **Figure 13.5**. This condition is often encountered either accidentally, such as a wave impinging on a metal structure, or purposely, in reflecting a wave from a parabolic reflector or off the body of an airplane. For perfect conductors, the transmission coefficient must be zero and we can write directly from **Eq. (13.21)**

$$T = 0 \quad \text{and} \quad \Gamma = -1 \tag{13.46}$$

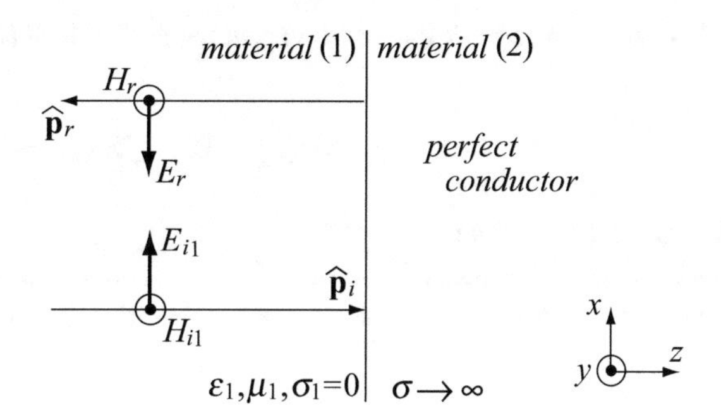

Figure 13.5 Reflection and transmission at a dielectric–conductor interface: normal incidence. The directions of the waves are assumed

Substituting these in **Eqs. (13.27)** and **(13.30)**, and noting that in medium (1) $\gamma = j\beta_1$, the electric and magnetic field intensities to the left of the conductor in **Figure 13.5** are

$$\mathbf{E}_1(z) = -\hat{\mathbf{x}}j2E_{i1}\sin(\beta_1 z) \quad [\text{V/m}] \tag{13.47}$$

$$\mathbf{H}_1(z) = \hat{\mathbf{y}}2\frac{E_{i1}}{\eta_1}\cos(\beta_1 z) \quad [\text{A/m}] \tag{13.48}$$

The more general notation β_1 rather than β_0 was used here to indicate that the same results apply to any lossless dielectric. The wave formed by these fields is unique in that it does not propagate; that is, as z changes, the amplitudes change, but there is no propagation in space. This is a standing wave. The principle of a standing wave is shown in **Figure 13.6**, where the amplitude of the wave changes with time, but the location of the peaks and zeros (nodes) is constant in space. To see that this is the case, it is most convenient to calculate the time-averaged Poynting vector for the wave:

$$\mathcal{P}_{av1} = \frac{1}{2}\text{Re}\{\mathbf{E}_1(z) \times \mathbf{H}_1^*(z)\} = -\hat{\mathbf{z}}\text{Re}\left\{j\frac{2}{\eta_1}|E_{i1}|^2\sin\beta_1 z\cos\beta_1 z\right\} = 0 \tag{13.49}$$

Because $\mathcal{P}_{av}$ is purely imaginary in material (1), no real power is transferred in the direction of propagation of the incident

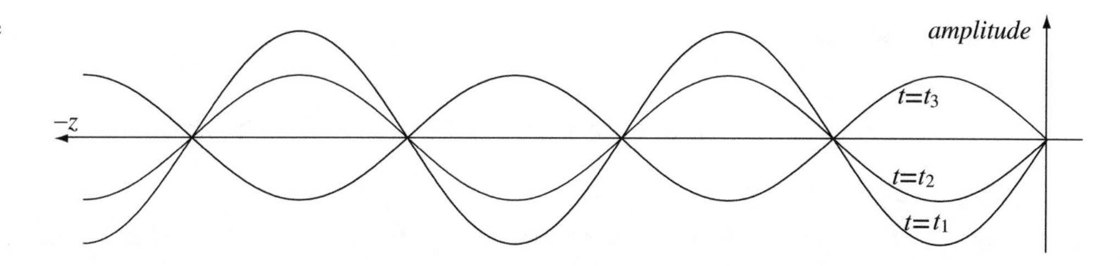

Figure 13.6 Standing wave due to reflection at a conducting interface. Total reflection (complete standing wave) without attenuation is shown

wave and, therefore, into the conductor. This, of course, is exactly what the reflection and transmission coefficients in **Eq. (13.46)** show, but it also means that no power is propagated anywhere to the left of the interface either. The energy in the system is "standing" or, we may say, it propagates back and forth with the same net effect.

To better understand the importance and behavior of standing waves, we first write the waves in the time domain and also calculate the parameters characterizing the standing waves. We start with the electric field intensity in **Eq. (13.47)**. In the time domain, this becomes

$$\begin{aligned}\mathbf{E}_1(z,t) &= \text{Re}\{\mathbf{E}_1(z)e^{j\omega t}\} = \text{Re}\{-\hat{\mathbf{x}}j2E_{i1}\sin(\beta_{1z})e^{j\omega t}\} = \text{Re}\{\hat{\mathbf{x}}2E_{i1}\sin(\beta_1 z)e^{j\omega t}e^{-j\pi/2}\}\\ &= \hat{\mathbf{x}}2E_{i1}\sin(\beta_1 z)\cos\left(\omega t - \frac{\pi}{2}\right) = -\hat{\mathbf{x}}2E_{i1}\sin(\beta_1 z)\sin\omega t \quad [\text{V/m}]\end{aligned} \tag{13.50}$$

where $-j = e^{-j\pi/2}$ was used. Now, we can analyze this wave by inspection.

(1) First, we note that the amplitude of the wave varies from zero to a maximum of $2E_{i1}$, depending on the position z:

$$(E_1)_{min} = 0, \quad (E_1)_{max} = 2E_{i1} \tag{13.51}$$

(2) The wave varies as sin(ωt) in time.

(3) The wave varies as $\sin(\beta_1 z)$ in space. If we take the interface as the reference point (i.e., $z = 0$), the wave amplitude is zero for any value that makes $\beta_1 z$ a multiple of π. For z negative (i.e., to the left of the conducting surface in **Figure 13.5**), we have

$$E_1 = 0 \quad \text{for} \quad \beta_1 z = -n\pi, \quad n = 0,1,2, \ldots \tag{13.52}$$

β_1 may be written in terms of the wavelength as $\beta_1 = 2\pi/\lambda_1$. With this, **Eq. (13.52)** becomes

$$E_1 = 0 \quad \text{at} \quad z = -\frac{n\lambda_1}{2}, \quad n = 0,1,2, \ldots \tag{13.53}$$

The amplitude of the wave is zero at $z = 0$, $z = -\lambda_1/2$, $z = -\lambda_1$, $z = -3\lambda_1/2$, etc. These points (also called nodes) are shown in **Figure 13.7a**. If there is only a standing wave, we call this a ***complete standing wave***. An incomplete standing wave means that in addition to the standing wave, there is also a propagating wave (such as when a wave reflects off a dielectric). In a complete standing wave, the amplitude varies between zero and twice the amplitude of the incident wave, whereas in an incomplete standing wave, the amplitude varies between a minimum and a maximum value, which depend on the amplitudes of the incident and reflected waves. The magnitude of the ratio between the maximum and minimum amplitude of the standing wave is called the ***standing wave ratio*** (SWR) and will be discussed at length in **Chapter 14**. **Equation (13.50)** is also called the ***standing wave pattern*** for the electric field intensity (a similar standing wave pattern may be obtained for the magnetic field intensity). Either pattern may be plotted as in **Figure 13.6**.

(4) The standing wave is maximum (positive or negative) for any value of $\beta_{1z} = -(n\pi + \pi/2)$:

$$E_1 = 2E_{i1} \quad \text{for} \quad \beta_1 z = -\left(n\pi + \frac{\pi}{2}\right), \quad n = 0,1,2, \ldots \tag{13.54}$$

or using the wavelength

$$E_1 = 2E_{i1} \quad \text{at} \quad z = -\frac{n\lambda_1}{2} - \frac{\lambda_1}{4}, \quad n = 0,1,2, \ldots \tag{13.55}$$

Thus, the maxima of the standing wave are at $z = -\lambda_1/4$, $z = -3\lambda_1/4$, $z = -5\lambda_1/4$, etc., as shown in **Figure 13.7a**.

(5) Because of the sinusoidal behavior of the wave in space, we can place another conducting surface at any node of the standing wave without affecting the wave behavior. For example, we could place a conducting surface at $z = -\lambda_1/2$, $-\lambda_1$, etc., as shown by dashed lines in **Figure 13.7a**. We will take up this aspect of propagation again in **Chapter 17**. At this point, we simply mention that introducing the plate does not alter the field between the plate and the conducting surface at $z = 0$.

(6) The above analysis was based on the electric field intensity in **Eq. (13.47)**. A similar analysis for the magnetic field intensity in **Eq. (13.48)** may be performed. The magnetic field intensity in the time domain is

$$\mathbf{H}_1(z,t) = \hat{\mathbf{y}}2\frac{E_{i1}}{\eta_1}\cos(\beta_1 z)\cos(\omega t) \quad [\text{A/m}] \tag{13.56}$$

Comparison of **Eqs. (13.56)** and **(13.50)** shows that the electric and magnetic fields are in time quadrature as well as shifted in space by 90° (one-quarter wavelength). Therefore, the magnetic field intensity is maximum wherever the electric field intensity is minimum and vice versa. The relations are therefore

$$H_1 = 0 \quad \text{at} \quad z = -\frac{n\lambda_1}{2} - \frac{\lambda_1}{4}, \quad n = 0,1,2, \ldots \tag{13.57}$$

$$H_1 = 2\frac{E_{i1}}{\eta_1} \quad \text{at} \quad z = -\frac{n\lambda_1}{2}, \quad n = 0,1,2, \ldots \tag{13.58}$$

These relations are shown in **Figure 13.7b**.

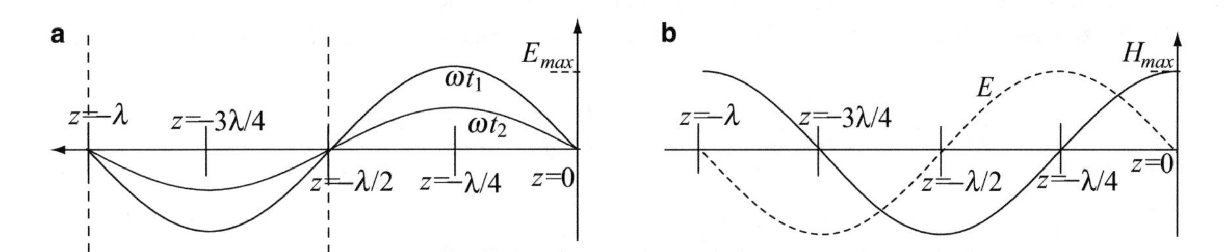

Figure 13.7 Standing waves. (**a**) Location of minima and maxima for the electric field intensity for two values of ωt. (**b**) Standing wave for the magnetic field intensity and its relation to the standing wave for the electric field intensity

In the above discussion we assumed that the conductor is a perfect conductor. What happens if the conductor is not a perfect conductor? Intuitively, we can say that there should be a transmitted wave, but this should be small. In other words, only a small fraction of the incident wave should propagate into the conductor. The reflection coefficient is large and close to -1. To see how the waves behave at the interface between air and a good conductor, we make use of the propagation properties of good conductors as defined in **Section 12.7.3**, for which $\sigma/\omega\varepsilon \gg 1$. For a good but not perfect conductor we obtained in **Eqs. (12.111)** and **(12.113)**

$$\alpha = \sqrt{\pi f \mu \sigma} \;\; [\text{Np/m}], \quad \beta = \sqrt{\pi f \mu \sigma} \;\; [\text{rad/m}], \quad \delta = \frac{1}{\sqrt{\pi f \mu \sigma}} \;\; [\text{m}] \tag{13.59}$$

where α is the attenuation constant, β the phase constant, and δ the skin depth in the conductor.

Consider now the general electric and magnetic fields in material (1) as given in **Eqs. (13.27)** and **(13.30)**:

$$\mathbf{E}_1(z) = \hat{\mathbf{x}}E_{i1}\left(Te^{-\gamma_1 z} - j2\Gamma\sin(j\gamma_1 z)\right) \;\; [\text{V/m}], \quad \mathbf{H}_1(z) = \hat{\mathbf{y}}\frac{E_{i1}}{\eta_0}\left(Te^{-\gamma_1 z} - 2\Gamma\cos(j\gamma_1 z)\right) \;\; [\text{A/m}] \tag{13.60}$$

and in material (2) as given in **Eqs. (13.31)** and **(13.32)**:

$$\mathbf{E}_2(z) = \hat{\mathbf{x}}TE_{i1}e^{-\gamma_2 z} \;\; [\text{V/m}], \quad \mathbf{H}_2(z) = \hat{\mathbf{y}}T\frac{E_{i1}}{\eta_2}e^{-\gamma_2 z} \;\; \left[\frac{\text{A}}{\text{m}}\right] \tag{13.61}$$

In this particular case, material (1) is air (or a perfect dielectric) and we can write $\gamma_1 = j\beta_1$. For material (2), $\gamma_2 = \alpha_2 + j\beta_2$, where α and β are given in **Eq. (13.59)**.

The reflection and transmission coefficients in **Eqs. (13.23)** and **(13.24)** still apply, but because the intrinsic impedance of a good conductor is rather small, we write the reflection and transmission coefficients in terms of the skin depth in the conductor. The intrinsic impedance in a good conductor is $\eta_2 = (1 + j)/\sigma_2\delta_2$ [from **Eq. (12.116)**]. Substituting this in **Eqs. (13.23)** and **(13.24)**, we obtain

$$\Gamma = \frac{1 + j - \sigma_2\delta_2\eta_0}{1 + j + \sigma_2\delta_2\eta_0}, \quad T = \frac{2(1 + j)}{\sigma_2\delta_2\eta_0 + (1 + j)} \tag{13.62}$$

These expressions are both complex, and according to the discussion on propagation in conductors [**Eq. (12.116)**], these are only approximations since the intrinsic impedance itself is an approximation. It is interesting to note that if the skin depth decreases, the term $\sigma_2\delta_2\eta_0$ becomes smaller, until, in the limit, it can be neglected with respect to $1 + j$. In this limit, the reflection coefficient approaches +1 and the transmission coefficient approaches 2. This may seem a contradiction at first since this would imply that there is transmission into the conductor when, in fact, we stated in **Chapter 12** that the lower the skin depth, the lower the penetration into the conductor. However, it is worth restating that the whole idea of skin depth is based on $\sigma/\omega\varepsilon \gg 1$. For a given material, the skin depth can be decreased by increasing the frequency but this also reduces the magnitude of $\sigma/\omega\varepsilon$. Therefore, at some frequency, the conductor ceases to behave as a conductor. This happens when displacement currents dominate, and under these conditions, the skin depth has no meaning. **Equations (13.59)** and **(13.62)** are only valid if the condition for a conductor is satisfied; that is, only if $\sigma/\omega\varepsilon \gg 1$.

Using these expressions in **Eq. (13.60)**, we obtain the electric and magnetic field intensities in material (1) as

$$\mathbf{E}_1(z) = \hat{\mathbf{x}}E_{i1}\left(\frac{2(i+j)}{\sigma_2\delta_2\eta_0 + (i+j)}e^{-j\beta_1 z} + \frac{1+j-\sigma_2\delta_2\eta_0}{1+j+\sigma_2\delta_2\eta_0}(j2\sin(\beta_1 z))\right) \quad \left[\frac{\text{V}}{\text{m}}\right] \tag{13.63}$$

$$\mathbf{H}_1(z) = \hat{\mathbf{y}}\frac{E_{i1}}{\eta_0}\left(\frac{2(i+j)}{\sigma_2\delta_2\eta_0 + (i+j)}e^{-j\beta_1 z} - \frac{1+j-\sigma_2\delta_2\eta_0}{1+j+\sigma_2\delta_2\eta_0}(2\cos(\beta_1 z))\right) \quad \left[\frac{\text{A}}{\text{m}}\right] \tag{13.64}$$

where $\gamma_1 = j\beta_1$ was used. The fields in material (2) are

$$\mathbf{E}_2 = \hat{\mathbf{x}}TE_{i1}e^{-\gamma_2 z} = \hat{\mathbf{x}}E_{i1}\frac{2(1+j)}{\sigma_2\delta_2\eta_0 + (1+j)}e^{-\gamma_2 z} \quad \left[\frac{\text{V}}{\text{m}}\right] \tag{13.65}$$

$$\mathbf{H}_2 = \hat{\mathbf{y}}T\frac{E_{i1}}{(1+j)(1/\sigma_2\delta_2)}e^{-\gamma_2 z} = \hat{\mathbf{y}}E_{i1}\frac{2\sigma_2\delta_2}{\sigma_2\delta_2\eta_0 + (1+j)}e^{-\gamma_2 z} \quad \left[\frac{\text{A}}{\text{m}}\right] \tag{13.66}$$

The fields in material (2) are rather small and decay fast, as expected for a good conductor, but they are not zero.

The expressions in **Eqs. (13.63)** through **(13.66)** are not different from those in **Eqs. (13.25)** through **(13.32)**, except for the fact that the standing wave component is large, as it should be for a good conductor. A useful calculation at this point is the power flow into the conductor. This is found by calculating the Poynting vector for the transmitted wave:

$$\begin{aligned}\mathbf{E}\times\mathbf{H}^* &= \hat{\mathbf{z}}(TE_{i1}e^{-\gamma_2 z})\left(\frac{TE_{i1}e^{-\gamma_2 z}}{\eta_2}\right)^* = \hat{\mathbf{z}}\left(TE_{i1}e^{-(\alpha_2+j\beta_2)z}\right)\left(\frac{\sigma_2\delta_2 TE_{i1}e^{-(\alpha_2+j\beta_2)z}}{1+j}\right)^* \\ &= \hat{\mathbf{z}}E_{i1}^2\sigma_2\delta_2 T\left(\frac{T}{1+j}\right)^* e^{-2\alpha_2 z} = \hat{\mathbf{z}}4E_{i1}^2\sigma_2\delta_2\left(\frac{1+j}{(\sigma_2\delta_2\eta_0+1)^2+1}\right)e^{-2\alpha_2 z} \quad \left[\frac{\text{W}}{\text{m}^2}\right]\end{aligned} \tag{13.67}$$

where $e^{-\gamma z}(e^{-\gamma z})^* = e^{-(\alpha+j\beta)z}e^{-(\alpha-j\beta)z} = e^{-2\alpha z}$ was used. Assuming that $|\eta_2| \ll |\eta_0|$ (this is a good approximation in a good conductor), the total power flow into the conductor is minimal. The real part of the Poynting vector represents the dissipated power in the conductors and this decays as $e^{-2\alpha z}$, indicating very rapid attenuation. For a perfect conductor $\eta_2 = 0$ $(T = 0)$ and the power density in the conductor is zero, as required. This aspect of power propagation shows why a microwave oven, while heating lossy substances such as food, will not dissipate much power in its own walls, which are made of steel (see **Example 13.6**).

Example 13.4 Application: Reflectometry It is required to measure the distance from an antenna to a reflecting surface (such as a wall, planet, etc.). A wave is transmitted to the wall and a zero (minimum reception) in the standing wave pattern is recorded using a second antenna at a distance d_1 [m] from the sending antenna, as shown in **Figure 13.8**. The frequency of the wave is f_1 [Hz]. Now, the frequency is decreased until the receiving antenna reads a maximum in the electric field at the same location. If the frequency for the maximum reading is f_2 [Hz], calculate the distance between the transmitting antenna to the conducting surface. The values are $f_1 = 100$ MHz, $f_2 = 99.9$ MHz, and $d_1 = 10$ m. Use the properties of free space without attenuation.

Figure 13.8 Reflectometry: measurement of distance to a target by identifying the nodes in the standing wave pattern

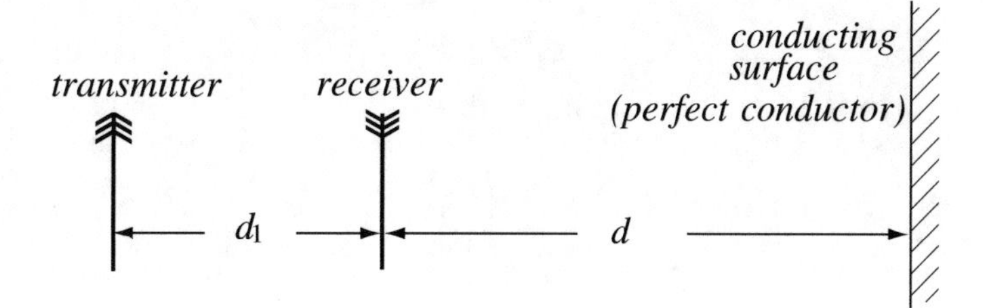

Solution: To calculate the distance between the receiving antenna and wall, we first calculate the number of minima in the standing wave pattern. The distance is obtained from the wavelength and the number of half-wavelengths in d.

For frequency f_1, assuming there are n minima between the receiving antenna and wall,

$$d = n\frac{\lambda_1}{2} = n\frac{c}{2f_1} \quad \text{since} \quad \lambda_1 f_1 = c$$

As the frequency decreases, the wavelength increases and the distance between minima increases. The maxima (as well as minima) move to the left until, at frequency f_2, the first maximum to the right of the receiving antenna moves to the location of the receiving antenna. There are now $n - 1$ minima in the standing wave pattern between the wall and the receiver (because the minimum at the receiver has now moved to the left) plus the distance between a minimum and a maximum. Thus, the distance between the wall and the receiver is

$$d = (n-1)\frac{\lambda_2}{2} + \frac{\lambda_2}{4} = (n-1)\frac{c}{2f_2} + \frac{c}{4f_2} \quad \text{since} \quad \lambda_2 f_2 = c$$

Equating the two expressions for d,

$$n\frac{c}{2f_1} = (n-1)\frac{c}{2f_2} + \frac{c}{4f_2}$$

From this,

$$n = \frac{f_1}{2(f_1 - f_2)}$$

For the values given,

$$n = \frac{100 \times 10^6}{2 \times \left(100 \times 10^6 - 99.9 \times 10^6\right)} = \frac{100}{2 \times (0.1)} = 500$$

The distance d is therefore

$$d = 500\frac{3 \times 10^8}{2 \times 100 \times 10^6} = 750 \text{ m}$$

and adding the 10 m between the two antennas, we get 760 m.

Note: This type of measurement is quite sensitive and is routinely performed in many applications such as measurement of thickness of conductors or even measurement of coatings such as paint. In practical applications, the standing wave pattern is not complete and, therefore, only minima and maxima are detected (not zeros of the pattern). In most cases, minima are easier to detect than maxima because they are sharper. At still higher frequencies, the distance between minima and maxima is very small and very sensitive measurements are possible even over short distances.

Example 13.5 A plane wave propagates in free space and encounters a perfect conductor. The frequency of the wave is 1 GHz and the amplitude of its electric field intensity is 1,500 V/m. The electric field intensity is directed in the x direction and the wave propagates in the z direction, as shown in **Figure 13.5**. Calculate:

(a) The electric and magnetic field intensities everywhere to the left of the conductor's surface.
(b) The power relations everywhere to the left of the interface.

Solution: The conducting interface reflects the wave and, since there is no attenuation in free space, the amplitude of the reflected wave equals that of the incident wave but opposite in sign. The electric field intensity to the left of the interface is given by **Eq. (13.50)**.

(a) The instantaneous electric field intensity is

$$\mathbf{E}_1(z,t) = -\hat{\mathbf{x}}2E_{i1}\sin(\beta_1 z)\sin(\omega t) \quad [\mathrm{V/m}]$$

The phase constant at the given frequency in free space is

$$\beta_1 = \omega\sqrt{\mu_0\varepsilon_0} = \frac{\omega}{c} = \frac{2\pi \times 10^9}{3 \times 10^8} = \frac{20\pi}{3} \quad \left[\frac{\mathrm{rad}}{\mathrm{m}}\right]$$

Thus, the electric field intensity is

$$\mathbf{E}_1(z,t) = -\hat{\mathbf{x}}3{,}000\sin\left(\frac{20\pi z}{3}\right)\sin\left(2\pi \times 10^9 t\right) \quad [\mathrm{V/m}]$$

From **Eq. (13.56)**, the instantaneous magnetic field intensity is

$$\mathbf{H}_1(z,t) = \hat{\mathbf{y}}\,\frac{3{,}000}{377}\cos\left(\frac{20\pi z}{3}\right)\cos\left(2\pi \times 10^9 t\right) \quad [\mathrm{A/m}]$$

The electric and magnetic fields are in time quadrature (90° phase difference) and shifted in space by $\lambda/4$, as well as being orthogonal in their direction in space. The electric field intensity is zero at

$$\frac{20\pi z}{3} = -n\pi \quad \rightarrow \quad z = -\frac{3n}{20}, \quad n = 0,1,2,3,\ldots$$

that is, the electric field intensity is zero at $z = 0$, -0.15 m, -0.3 m, -0.45 m, etc.
The magnetic field intensity is zero at

$$\frac{20\pi z}{3} = -\frac{n\pi}{2}, \quad z = -\frac{3n}{40}, \quad n = 1,3,5,7,\ldots$$

that is, the nodes of the magnetic field intensity are at $z = -0.075$ m, -0.225 m, -0.375 m, -0.525 m, etc.
Note that the distance between two nodes of either **E** or **H** is one-half wavelength and the distance between a node of the electric field intensity and the nearest node of the magnetic field intensity is one-quarter wavelength.
The electric and magnetic fields also vary with time, but the nodes remain fixed in space, thus, again, the meaning of standing waves.

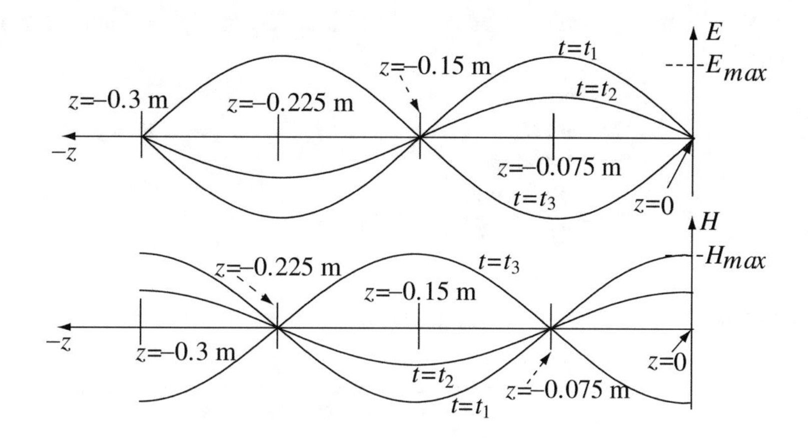

Figure 13.9 Electric and magnetic field intensities to the left of an air–conductor interface for different times. Note the relative location of the nodes

The solution for the electric and magnetic field intensities is shown in **Figure 13.9**, where the field at three time instances is shown. Note the locations of the zeros and the variation in time.

(b) We look first at the instantaneous power density in space using the instantaneous Poynting vector is

$$\mathcal{P}_1(z,t) = \mathbf{E}_1(z,t) \times \mathbf{H}_1(z,t) = -\hat{\mathbf{z}}\frac{4}{377}|E_{i1}|^2 \sin\left(\frac{20\pi z}{3}\right)\cos\left(\frac{20\pi z}{3}\right)$$
$$\times \sin\left(2\pi \times 10^9 t\right)\cos\left(2\pi \times 10^9 t\right) \quad [\mathrm{W/m^2}]$$

This exists anywhere to the left of the conducting interface. The time-averaged power density may be calculated by integrating the instantaneous power density over one cycle of the wave, but it is easier to evaluate it from the time-harmonic forms of the electric and magnetic fields given in **Eqs. (13.47)** and **(13.48)**:

$$\mathbf{E}_1(z) = -\hat{\mathbf{x}} j2E_{i1}\sin(\beta_1 z) \quad [\mathrm{V/m}], \quad \mathbf{H}_1(z) = \hat{\mathbf{y}}2\frac{E_{i1}}{\eta 1}\cos(\beta_1 z) \quad [\mathrm{A/m}]$$

The time-averaged power density is therefore

$$\mathcal{P}_{av1}(z) = \frac{1}{2}\mathrm{Re}\{\mathbf{E}(z) \times \mathbf{H}^*(z)\} = \frac{1}{2}\mathrm{Re}\left\{-\hat{\mathbf{x}} j2E_{i1}\sin(\beta_1 z) \times \hat{\mathbf{y}}2\frac{E_{i1}}{\eta 1}\cos(\beta_1 z)\right\} = 0$$

that is, there is no propagation of real power.

Example 13.6 Application: Microwave Cooking—Why the Oven Itself Does Not Get Hot? The microwave oven is a closed cavity and, therefore, the waves that exist inside are not plane waves. However, to approximate the conditions in the walls of the oven, we take an equivalent sheet of metal, equal in thickness and area to that of the oven, and expose it to the same power density it would be subject to in the oven. We assume here that the total area of the walls is 1 m^2 (a medium-sized domestic oven), operating at 2.45 GHz and the walls are made of steel.

A very large sheet of steel is illuminated by a plane wave at 2.45 GHz. The steel is 1 mm thick, has conductivity of 10^7 S/m, and has a relative permeability of 200. The time-averaged power density impinging on the sheet is 1,000 $\mathrm{W/m^2}$ and the wave impinges on the walls perpendicularly. Calculate the power dissipated in the steel per unit area.

Solution: Steel is a good conductor; therefore, we expect little penetration. From the given data, we first calculate the incident electric and magnetic field intensities at the interface. From material data, we calculate the skin depth δ and then calculate the transmission coefficient in **Eq. (13.62)** or use **Eqs. (13.65)** and **(13.66)** directly to calculate the electric and magnetic field intensities just below the surface of the walls. Using the Poynting vector with these fields will give the power density at the walls. Since all transmitted power is dissipated in the walls, the product of the power density in the wall and the area of the walls gives the total power dissipated in the walls.

The amplitude of the incident electric field intensity E_{i1} is calculated from the time-averaged power density (see **Example 12.9**):

$$\mathcal{P}_{av} = \frac{E_{i1}^2}{2\eta_0} \quad \rightarrow \quad E_{i1} = \sqrt{2\eta_0 \mathcal{P}_{av}} = \sqrt{2 \times 377 \times 1{,}000} = 868 \quad [\text{V/m}],$$

$$H_{i1} = \frac{E_{i1}}{\eta_0} = \frac{868}{377} = 2.3 \quad [\text{A/m}]$$

The skin depth is

$$\delta = \frac{1}{\sqrt{\pi f \mu \sigma}} = \frac{1}{\sqrt{\pi \times 2.45 \times 10^9 \times 200 \times 4 \times \pi \times 10^{-7} \times 10^7}} = 2.27 \times 10^{-7} \quad [\text{m}]$$

The skin depth is only 0.227 μm. Therefore, we can assume the sheet to be infinitely thick for practical purposes. The electric field intensity immediately inside the iron sheet [from **Eq. (13.65)**, with $z = 0$] is

$$E_2 = \frac{2(1+j)E_{i1}}{\eta_0 \sigma_2 \delta_2 + (1+j)} = \frac{2(1+j)868}{377 \times 10^7 \times 2.27 \times 10^{-7} + (1+j)} \approx \frac{2(1+j)868}{855.8} = 2.028(1+j) \quad [\text{V/m}]$$

where j in the denominator was neglected in comparison to the large real part. The magnetic field intensity is calculated (using $\eta_2 = (1 + j)/\sigma_2\delta_2$) as

$$H_2 = \frac{E_2}{\eta_2} = \frac{2.028(1+j)}{(1+j)/\sigma_2\delta_2} = 2.028\sigma_2\delta_2 = 2.028 \times 10^{-7} \times 2.27 \times 10^{-7} = 4.6 \quad [\text{A/m}]$$

The time-averaged Poynting vector gives the power per unit area entering the steel sheet:

$$\mathcal{P}_{av} = \left|\frac{1}{2}\text{Re}\{\mathbf{E}_2 \times \mathbf{H}_2^*\}\right| = \frac{1}{2}\text{Re}\{2.028 \times (1+j) \times 4.6\} = 4.66 \quad [\text{W/m}^2]$$

For a 1 m square of the conductor only 4.66 W enters the steel sheet. This is dissipated in the conductor. Thus, less than 0.5% of the power generated by the oven enters its walls and contributes to heating. Things are more complicated than this in a real oven but this example gives a sense of the quantities involved.

Exercise 13.3 Derive **Eq. (13.67)** from **Eqs. (13.65)** and **(13.66)**.

Exercise 13.4 A plane wave propagates from free space into copper. Calculate the time-averaged power density entering the copper and show that the amount of power per unit area of the interface, going into the copper, is very small. Assume frequency is 100 MHz and the magnitude of the incident electric field intensity at the surface, in air, is 100 V/m. The conductivity of copper is 5.7×10^7 S/m.

Answer The time-averaged power density into copper is 3.7×10^{-4} W/m^2. The time-averaged incident power density in air is 13.26 W/m^2.

13.3 Reflection and Transmission at an Interface: Oblique Incidence on a Perfect Conductor waves.m

A plane wave, obliquely incident on an interface between two materials, undergoes changes similar to those for normal incidence. Part of the wave is transmitted and part of it is reflected. In some cases, there is either only transmission or only reflection, depending on the values of the reflection and transmission coefficients. In oblique incidence, the continuity of the field components is taken into account as interface conditions at the interface between the two materials. The behavior of the wave at the interface depends on the polarization of the wave. To assist us in describing the waves at the interface, a ***plane of incidence*** is defined as the plane described by the direction of propagation of the incident wave (i.e., the Poynting vector) and the normal to the surface at the interface, as shown in **Figure 13.10a**. The directions of the electric and magnetic field intensities for uniform plane waves are normal to the direction of propagation. If the electric field intensity is parallel to the plane of incidence, we refer to this as ***parallel polarization*** (sometimes called H polarization) and is shown in **Figure 13.10a**. ***Perpendicular polarization*** refers to the case of the electric field intensity perpendicular to the plane of incidence (sometimes called E polarization), shown in **Figure 13.10b**. The general case of arbitrary polarization can be treated by separating the electric field intensity into its normal and parallel components as a combination of parallel and perpendicular polarizations.

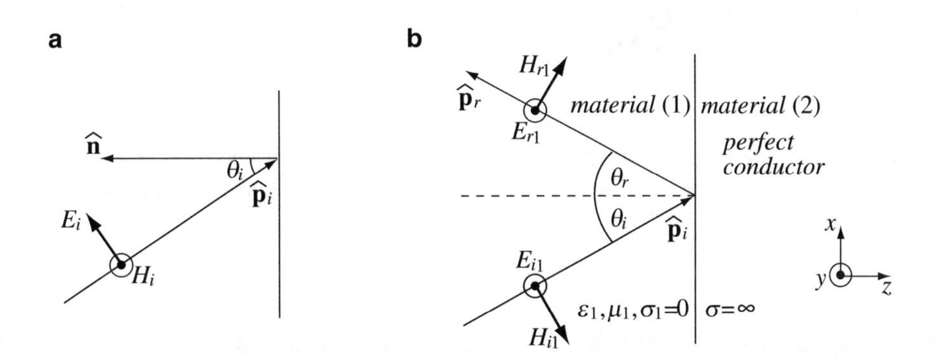

Figure 13.10 (**a**) Definition of the plane of incidence. The incident electric field intensity is parallel to the plane of incidence. (**b**) Oblique incidence at a dielectric–conductor interface. The electric field is polarized perpendicular to the plane of incidence. The directions of the reflected fields are assumed

We treat the problem of oblique incidence by looking at the propagation of waves obliquely incident at an interface for parallel and perpendicular polarization separately for dielectrics and conductors. The case of incidence on conducting interfaces is given first, followed by incidence on lossless dielectrics. The more general case of oblique incidence on lossy dielectrics is not treated here because it results in nonuniform plane waves and this is a subject for more advanced study. However, the principles used here are directly applicable to any interface.

13.3.1 Oblique Incidence on a Perfectly Conducting Interface: Perpendicular Polarization

In this configuration, the electric field intensity is perpendicular to the plane of incidence, as shown in **Figure 13.10b**. Incidence is from a dielectric with permittivity ε_1, permeability μ_1 and zero conductivity on a perfect conductor surface. The direction of propagation can be written directly from **Figure 13.10b**. For the incident wave, the unit vector in the direction of propagation is

$$\hat{\mathbf{p}}_i = \hat{\mathbf{x}}\sin\theta_i + \hat{\mathbf{z}}\cos\theta_i \tag{13.68}$$

For the reflected wave, the direction of propagation is

$$\hat{\mathbf{p}}_r = \hat{\mathbf{x}}\sin\theta_r - \hat{\mathbf{z}}\cos\theta_r \tag{13.69}$$

where the unit vectors $\hat{\mathbf{p}}_i$ and $\hat{\mathbf{p}}_r$ refer to the direction of the Poynting vector for the incident and reflected waves, respectively.

Inspection of the electric field intensity in **Eq. (13.1)** shows that the variable z in the exponent is a distance the wave propagates from some reference point, whereas β is the phase constant of the wave. The product βz is the phase of the wave after it has propagated a distance z from some reference point. Since the phase constant β_1 is given for the wave propagating in the $\hat{\mathbf{p}}_i$ direction, we can view the wave in **Figure 13.10b** as two components, one propagating in the positive x direction with phase constant β_{1x} and one in the positive z direction with phase constant $\beta_1 z$ (see **Figure 13.11a**), where

$$\beta_{1x} = \beta_1 \sin\theta_i, \quad \beta_{1z} = \beta_1 \cos\theta_i \quad [\text{rad/m}] \tag{13.70}$$

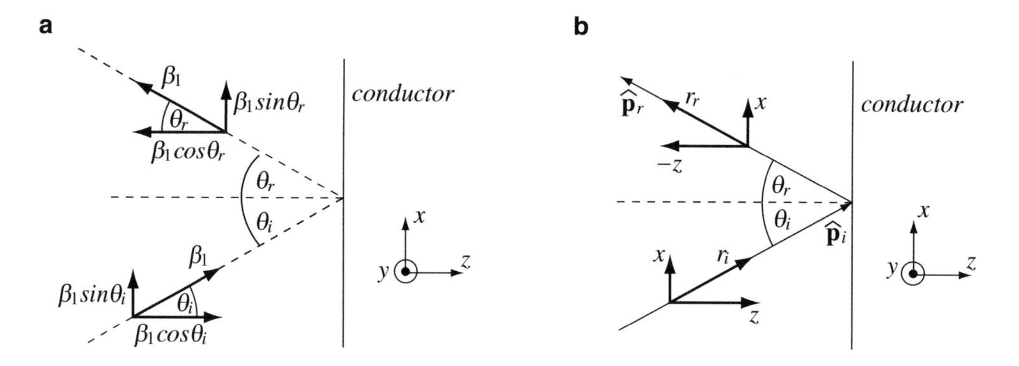

Figure 13.11 Direction of propagation and phase constants at an air-conductor interface. (**a**) Relation between phase constants. (**b**) Distances traveled by the incident and reflected waves

When the wave propagates a distance r_i along $\hat{\mathbf{p}}_i$ (see **Figure 13.11b**), the vertical component of the wave propagates a distance x with phase constant $\beta_{1x,}$ and the horizontal component propagates a distance z with a phase constant β_{1z}. The phase of the incident wave is therefore

$$\beta_1 r_i = \beta_1 x \sin\theta_i + \beta_1 z \cos\theta_i \quad [\text{rad}] \tag{13.71}$$

Similarly, for the backward propagating wave, the components of β are as in **Eq. (13.70)**, but now the wave travels a distance x in the vertical direction and a distance $-z$ in the horizontal direction and the angle is θ_r. The phase of the reflected wave is

$$\beta_1 r_r = \beta_1 x \sin\theta_r - \beta_1 z \cos\theta_r \quad [\text{rad}] \tag{13.72}$$

Using the directions of the incident electric and magnetic field intensities shown in **Figure 13.10b**, and the phase of the incident wave from **Eq. (13.71)**, the incident electric field intensity is in the positive y direction and the magnetic field intensity has both a negative x and a positive z component. These are

$$\mathbf{E}_i(x,z) = \hat{\mathbf{y}} E_{i1} e^{-j\beta_1 (x\sin\theta_i + z\cos\theta_i)} \quad [\text{V/m}] \tag{13.73}$$

$$\mathbf{H}_i(x,z) = \frac{E_{i1}}{\eta_1}(-\hat{\mathbf{x}}\cos\theta_i + \hat{\mathbf{z}}\sin\theta_i) e^{-j\beta_1 (x\sin\theta_i + z\cos\theta_i)} \quad [\text{A/m}] \tag{13.74}$$

The reflected fields are obtained using **Eq. (13.72)** for the phase of the wave:

$$\mathbf{E}_r(x,z) = \hat{\mathbf{y}} E_{r1} e^{-j\beta_1 (x\sin\theta_r - z\cos\theta_r)} \quad [\text{V/m}] \tag{13.75}$$

$$\mathbf{H}_r(x,y) = \frac{E_{r1}}{\eta_1}(\hat{\mathbf{x}}\cos\theta_r + \hat{\mathbf{z}}\sin\theta_r) e^{-j\beta_1 (x\sin\theta_r - z\cos\theta_r)} \quad [\text{A/m}] \tag{13.76}$$

In a perfect conductor, there is no transmitted wave; therefore, at the interface ($z = 0$), the total tangential electric field intensity (in this case the electric field intensity has only a tangential component) must be zero:

$$\mathbf{E}_i(x,0) + \mathbf{E}_r(x,0) = \hat{\mathbf{y}}\left[E_{i1} e^{-j\beta_1 \sin\theta_i} + E_{r1} e^{-j\beta_1 x \sin\theta_r}\right] = 0 \tag{13.77}$$

For this to be satisfied, the following must hold:

$$E_{r1} = -E_{i1} \quad \text{and} \quad \theta_r = \theta_i \tag{13.78}$$

that is, the sum of the amplitudes must be zero whereas the phases at the interface remain constant ($\beta_1 x \sin\theta_i = \beta_1 x \sin\theta_r$). The latter, relation, namely, $\theta_r = \theta_i$ is called Snell's law of reflection. The amplitudes of the reflected and incident waves are the same in absolute values and the angle of incidence and reflection are the same. The reflected electric and magnetic field intensities can be written in terms of the incident electric field intensity as

$$\mathbf{E}_r(x,z) = -\hat{\mathbf{y}}E_{i1}e^{-j\beta_1(x\sin\theta_i - z\cos\theta_i)} \quad [\text{V/m}] \tag{13.79}$$

$$\mathbf{H}_r(x,z) = -\frac{E_{i1}}{\eta_1}(\hat{\mathbf{x}}\cos\theta_i + \hat{\mathbf{z}}\sin\theta_i)e^{-j\beta_1(x\sin\theta_i - z\cos\theta_i)} \quad [\text{A/m}] \tag{13.80}$$

The reflection coefficient in this case is equal to -1. This indicates that the reflected electric and magnetic field intensity that we assumed in **Figure 13.10b** are in the directions opposite those shown. The total electric and magnetic field intensities are (after rearranging terms)

$$\mathbf{E}_1(x,z) = \hat{\mathbf{y}}E_{i1}\left[e^{-j\beta_1 z\cos\theta_i} - e^{j\beta_1 z\cos\theta_i}\right]e^{-j\beta_1 x\sin\theta_i} = -\hat{\mathbf{y}}\,j2E_{i1}\sin(\beta_1 z\cos\theta_i)e^{-j\beta_1 x\sin\theta_i} \quad [\text{V/m}] \tag{13.81}$$

$$\mathbf{H}_1(x,z) = -2\frac{E_{i1}}{\eta_1}[\hat{\mathbf{x}}\cos\theta_i\cos(\beta_1 z\cos\theta_i) + \hat{\mathbf{z}}\,j\sin\theta_i\sin(\beta_1 z\cos\theta_i)]e^{-j\beta_1 x\sin\theta_i} \quad [\text{A/m}] \tag{13.82}$$

To obtain these relations, **Eqs. (13.73)** and **(13.75)** were summed together and **Eqs. (13.74)** and **(13.76)** were summed together, after substituting the relations in **Eq. (13.78)**. In addition, the relations $e^{-j\beta_1 z\cos\theta_i} - e^{j\beta_1 z\cos\theta_i} = -j2\sin(\beta_1 z\cos\theta_i)$ and $e^{-j\beta_1 z\cos\theta_i} + e^{j\beta_1 z\cos\theta_i} = 2\cos(\beta_1 z\cos\theta_i)$ were used to simplify the expressions. The term $\beta_{1x} = \beta_1\sin\theta_i$ may be viewed as the modified phase constant in the x direction due to the presence of the conductor; that is, the conducting surface causes the wave to propagate parallel to the surface with a phase constant $\beta_{1x} = \beta_1\sin\theta_i$. As a consequence, the phase velocity of the wave propagating parallel to the conducting surface is different than the phase velocity in material (1) since $\beta = \omega/v_p$. The phase velocity in the x direction is now $v_{px} = \omega/\beta_{1x} = \omega/(\beta_1\sin\theta_i) = v_p/\sin\theta_i$. A similar discussion shows that the phase velocity in the z direction is $v_{pz} = v_p/\cos\theta_i$. In fact, if material (1) is free space, the phase velocities in the x and y directions are greater than the speed of light. This, of course, is admissible and does not imply that energy propagates faster than the speed of light (see **Section 12.7.4**).

To see how the wave propagates, it is useful to write the time-averaged Poynting vector:

$$\begin{aligned}\mathcal{P}_{av} &= \frac{1}{2}\text{Re}\{\mathbf{E}_1(x,z)\times\mathbf{H}_1^*(x,z)\} = \frac{1}{2}\text{Re}\Big\{\left(-\hat{\mathbf{y}}j2E_{i1}\sin(\beta_1 z\cos\theta_i)e^{-j\beta_1(x\sin\theta_i)}\right)\\ &\quad\times\left(-2\frac{E_{i1}}{\eta_1}[\hat{\mathbf{x}}\cos\theta_i\cos(\beta_1 z\cos\theta_i) - \hat{\mathbf{z}}\,j\sin\theta_i\sin(\beta_1 z\cos\theta_i)]e^{+j\beta_1(x\sin\theta_i)}\right)\Big\}\\ &= \text{Re}\Big\{(\hat{\mathbf{y}}\times\hat{\mathbf{x}})j2\frac{E_{i1}^2}{\eta_1}\sin(\beta_1 z\cos\theta_i)\cos(\beta_1 z\cos\theta_i)\cos\theta_i\\ &\quad+(-\hat{\mathbf{y}}\times\hat{\mathbf{z}})2j^2\frac{E_{i1}^2}{\eta 1}\sin(\beta_1 z\cos\theta_i)\sin(\beta_1 z\cos\theta_i)\sin\theta_i\Big\} \quad [\text{W/m}^2]\end{aligned} \tag{13.83}$$

where the conjugate of $\mathbf{H}_1(x,z) = (-\hat{\mathbf{x}}\mathbf{H}_{1x} - \hat{\mathbf{z}}\,j\mathbf{H}_{1z})e^{-j\beta_1 x\sin\theta_i}$ is $\mathbf{H}_1^*(x,z) = (-\hat{\mathbf{x}}\mathbf{H}_{1x} + \hat{\mathbf{z}}\,j\mathbf{H}_{1z})e^{+j\beta_1 x\cos\theta_i}$. This can be simplified by writing $\hat{\mathbf{y}}\times\hat{\mathbf{x}} = -\hat{\mathbf{z}}$, $\hat{\mathbf{y}}\times\hat{\mathbf{z}} = \hat{\mathbf{x}}$, $\sin(\beta_1 z\cos\theta_i)\cos(\beta_1 z\cos\theta_i) = (1/2)\sin(2\beta_1 z\cos\theta_i)$ and $j^2 = -1$:

$$\begin{aligned}\mathcal{P}_{av} &= \text{Re}\left\{-\hat{\mathbf{z}}j\frac{E_{i1}^2}{\eta_1}\sin(2\beta_1 z\cos\theta_i)\cos\theta_i + \hat{\mathbf{x}}\frac{2E_{i1}^2}{\eta_1}\sin^2(\beta_1 z\cos\theta_i)\sin\theta_i\right\}\\ &= \hat{\mathbf{x}}\frac{2E_{i1}^2}{\eta_1}\sin^2(\beta_1 z\cos\theta_i)\sin\theta_i \quad [\text{W/m}^2]\end{aligned} \tag{13.84}$$

Note that the wave in the z direction is a standing wave since the power density in this direction is purely imaginary, but the wave in the x direction is propagating. There are a number of important properties associated with these results, properties which we will use extensively in **Chapter 17**. We note the following:

(1) For any angle $0 < \theta_i < \pi/2$, there is a propagating term and a standing wave term. The standing wave becomes smaller as θ_i increases, whereas the propagating term becomes smaller as θ_i decreases.
(2) For perpendicular incidence ($\theta_i = 0$), the propagating term is zero and the wave is a pure standing wave. The time-averaged power density is zero.
(3) The wave propagates parallel to the surface of the conductor (in this case, in the positive x direction) for any angle $0 < \theta_i < \pi/2$, as can be seen from **Eq. (13.84)**. The conducting surface has the net effect of guiding the waves parallel to its surface. In **Chapter 17**, we will call these guided waves and will use the results of this and the following section to define the properties of guided waves.
(4) The amplitude of the propagating wave depends on the z-coordinate. This means that the amplitude is not constant on the plane perpendicular to the direction of propagation and hence this is not a uniform plane wave. Note in particular that whereas the incident and reflected waves are uniform plane waves, their sum is not.

Example 13.7 Application: Reflection of Waves by the Ionosphere A perpendicularly polarized plane wave propagates in air and impinges on the ionosphere as shown in **Figure 13.12a**. The amplitude of the electric field intensity is 100 V/m, its frequency is 3 GHz, and the angle of incidence is 30°. Assume air has properties of free space and the ionosphere is a perfect conductor at the frequency of the wave.

(a) Calculate the total electric and magnetic field intensity in air.
(b) Calculate the time-averaged power density in air.

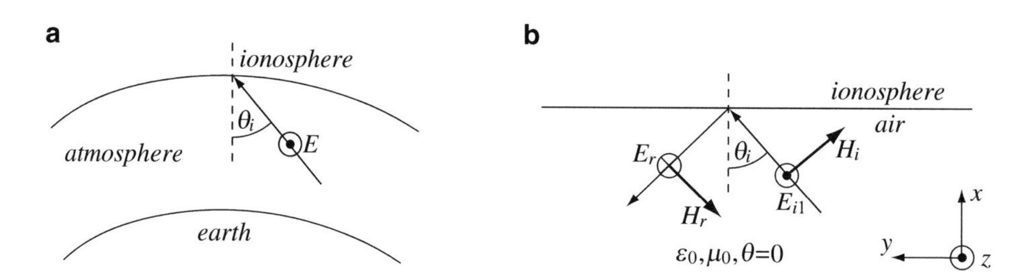

Figure 13.12 (a) A wave impinging on the ionosphere which is assumed to be a perfect conductor. (b) The system of coordinates used for solution

Solution: The electric and magnetic field intensities in the dielectric are given in **Eqs. (13.81)** and **(13.82)** and the time-averaged power density in **Eq. (13.84)**, but before calculating the fields, we must define a system of coordinates which, of course, is arbitrary. One possible system is shown in **Figure 13.12b**. Also, whereas the electric field intensity is perpendicular to the plane of incidence, it can be either in the negative or positive z direction. We choose the latter.

(a) The incident electric field intensity and magnetic field intensity are [see **Eqs. (13.73)** and **(13.74)**]

$$\mathbf{E}_i(x,y) = \hat{\mathbf{z}}E_{i1}e^{-j\beta_1(y\sin\theta_i + x\cos\theta_i)} \quad [\text{V/m}]$$

$$\mathbf{H}_i(x,y) = \frac{E_{i1}}{\eta_1}(-\hat{\mathbf{y}}\cos\theta_i + \hat{\mathbf{x}}\sin\theta_i)e^{-j\beta_1(y\sin\theta_i + x\cos\theta_i)} \quad [\text{A/m}]$$

From these, because the tangential components of E_i and E_r are in opposite directions (see **Figure 13.12b**), $\Gamma = -1$, $E_{r1} = -E_{i1}$, and the reflected fields are [see **Eqs. (13.79)** and **(13.80)**]

$$\mathbf{E}_r(x,y) = -\hat{\mathbf{z}}E_{i1}e^{-j\beta_1(y\sin\theta_i - x\cos\theta_i)} \quad [\text{V/m}]$$

$$\mathbf{H}_r(x,y) = -\frac{E_{i1}}{\eta_1}(\hat{\mathbf{y}}\cos\theta_i + \hat{\mathbf{x}}\sin\theta_i)e^{-j\beta_1(y\sin\theta_i - x\cos\theta_i)} \quad [\text{A/m}]$$

and the total fields in air are

$$\mathbf{E}_1(x,y) = -\hat{\mathbf{z}}j2E_{i1}\sin(\beta_1 x\cos\theta_i)e^{-j\beta_1(y\sin\theta_i)} \quad [\text{V/m}]$$

$$\mathbf{H}_1(x,y) = -2\frac{E_{i1}}{\eta_1}[\hat{\mathbf{y}}\cos\theta_i\cos(\beta_1 x\cos\theta_i) + \hat{\mathbf{x}}j\sin\theta\sin(\beta_1 x\cos\theta_i)]e^{-j\beta_1(y\sin\theta_i)} \quad [\text{A/m}]$$

With the values given ($\beta_1 = \beta_0 = 2\pi f/c = 2\pi \times 3 \times 10^9/3 \times 10^8 = 20\pi$ [rad/m], $\eta_1 = \eta_0 = 377\ \Omega$, $E_{i1} = 100$ V/m), the fields are

$$\mathbf{E}_1(x,y) = -\hat{\mathbf{z}}j200\sin(17.32\pi x)e^{-j10\pi y} \quad [\text{V/m}]$$

$$\mathbf{H}_1(x,y) = -0.265[\hat{\mathbf{y}}1.732\cos(17.32\pi x) + \hat{\mathbf{x}}j\sin(17.32\pi x)]e^{-j10\pi y} \quad [\text{A/m}].$$

(b) The time-averaged power density is as in **Eq. (13.84)** except that the propagation is in the y direction:

$$\mathcal{P}_{av} = \frac{1}{2}\text{Re}\{\mathbf{E}\times\mathbf{H}^*\} = \hat{\mathbf{y}}\frac{2E_{i1}^2}{\eta_1}\sin^2(\beta_1 x\cos\theta_i)\sin\theta_i \quad [\text{W/m}^2]$$

With the given values, we get

$$\mathcal{P}_{av} = \hat{\mathbf{y}}\,26.53\sin^2(17.32\pi x) \quad [\text{W/m}^2]$$

Exercise 13.5 For the oblique incidence described in this section (perpendicular polarization), calculate for the incident wave shown in **Figure 13.10b**:

(a) The phase velocities of the wave in the x direction (along the conductor's surface), z direction (perpendicular to the conductor's surface), and $\hat{\mathbf{p}}_i$ direction (along the direction of propagation of the incident wave).
(b) What are the three velocities if $\theta_i \to 0$?

Answer **(a)** $v_x = c/\sin\theta_i$, $v_z = c/\cos\theta_i$, $v_{pi} = c$ [m/s]. **(b)** $v_x \to \infty$, $v_z \to c$, $v_{pi} = c$ [m/s].

13.3.2 Oblique Incidence on a Perfectly Conducting Interface: Parallel Polarization

The discussion of the previous section applies here as well, but the components of the fields are different. The electric field intensity now lies in the incidence plane and, therefore, will have x and z components, whereas the magnetic field intensity is perpendicular to the plane of incidence (y direction in **Figure 13.13**). The direction of propagation of the incident and reflected waves is identical to those in **Eqs. (13.68)** and **(13.69)**. Also, the phases of the incident and reflected waves are as in **Eqs. (13.71)** and **(13.72)**. However, the incident electric field intensity now has two components; one is in the positive x direction, the second in the negative z direction. Note that once we choose the direction of **E**, the direction of **H** must be such that the cross product between **E** and **H** is in the direction of propagation. From these considerations and from **Figure 13.13**, the incident electric and magnetic field intensities are

$$\mathbf{E}_i(x,z) = E_{i1}(\hat{\mathbf{x}}\cos\theta_i - \hat{\mathbf{z}}\sin\theta_i)e^{-j\beta_1(x\sin\theta_i + z\cos\theta_i)} \quad [\text{V/m}] \tag{13.85}$$

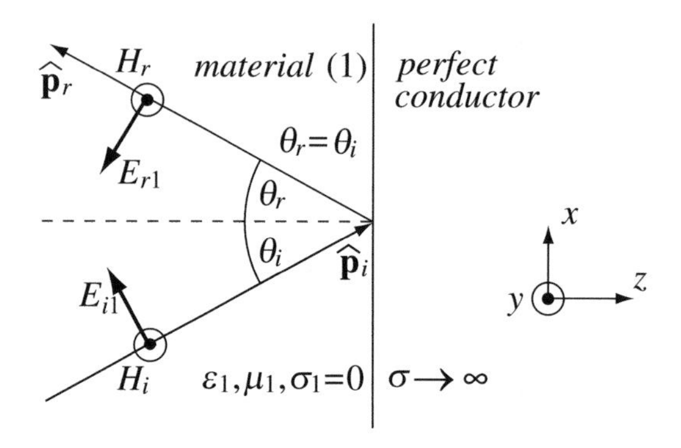

Figure 13.13 Incident and reflected waves for oblique incidence at a dielectric–conductor interface: parallel polarization

$$\mathbf{H}_i(x,z) = \hat{\mathbf{y}}\,\frac{E_{i1}}{\eta_1}e^{-j\beta_1(x\sin\theta_i + z\cos\theta_i)} \quad [\mathrm{A/m}] \tag{13.86}$$

Similarly, the reflected electric field intensity has components in the negative x and negative z directions, whereas the reflected magnetic field intensity is in the positive y direction. The reflected fields are assumed to be

$$\mathbf{E}_r(x,z) = E_{r1}(-\hat{\mathbf{x}}\cos\theta_i - \hat{\mathbf{z}}\sin\theta_i)e^{-j\beta_1(x\sin\theta_i - z\cos\theta_i)} \quad [\mathrm{V/m}] \tag{13.87}$$

$$\mathbf{H}_r(x,z) = \hat{\mathbf{y}}\,\frac{E_{r1}}{\eta_1}e^{-j\beta_1(x\sin\theta_i - z\cos\theta_i)} \quad [\mathrm{A/m}] \tag{13.88}$$

where, again, we used the relation $\theta_r = \theta_i$.

At the conducting interface, the tangential components of the electric field intensity $\mathbf{E}(x,0) = \mathbf{E}_i(x,0) + \mathbf{E}_r(x,0)$ must be zero. Setting $z = 0$ in **Eqs. (13.85)** and **(13.87)** and summing the tangential (x) components of $\mathbf{E}$, we get

$$E_{i1}(\hat{\mathbf{x}}\cos\theta_i)e^{-j\beta_1(x\sin\theta_i)} + E_{r1}(-\hat{\mathbf{x}}\cos\theta_i)e^{-j\beta_1(x\sin\theta_i)} = 0 \quad \to \quad E_{r1} = E_{i1} \tag{13.89}$$

Substituting this in **Eqs. (13.87)** and **(13.88)**, and summing the incident and reflected waves in **Eqs. (13.85)** and **(13.87)** for the total electric field intensity and **Eqs. (13.86)** and **(13.88)** for the total magnetic field intensity, we get the total fields in material (1) as the sum of the incident and reflected fields:

$$\mathbf{E}_1(x,z) = -2E_{i1}[\hat{\mathbf{x}}j\cos\theta_i\sin(\beta_1 z\cos\theta_i) + \hat{\mathbf{z}}\sin\theta_i\cos(\beta_1 z\cos\theta_i)]e^{-j\beta_1 x\sin\theta_i} \quad [\mathrm{V/m}] \tag{13.90}$$

$$\mathbf{H}_1(x,z) = \hat{\mathbf{y}}2\frac{E_{i1}}{\eta_1}\cos(\beta_1 z\cos\theta_i)e^{-j\beta_1 x\sin\theta_i} \quad [\mathrm{A/m}] \tag{13.91}$$

As with perpendicular polarization, the wave propagating in the z direction consists of E_{1x} and H_{1y} and these are out of phase. Therefore, we have a standing wave, oscillating exactly as for the perpendicular polarization. The wave in the x direction is a propagating wave (see **Exercise 13.6**). Again, as for perpendicular polarization, the propagating wave is not a uniform plane wave since its amplitude depends on z (see **Exercise 13.6**).

Exercise 13.6

(a) Show that **Eq. (13.90)** is the sum of **Eqs. (13.85)** and **(13.87),** and **Eq. (13.91)** is the sum of **Eqs. (13.86)** and **(13.88)**.

(b) Calculate the time-averaged Poynting vector for the total fields and show that waves propagate in the positive x direction and waves in the z direction are standing waves.

Answer **(b)** $\mathcal{P}_{av}(x,z) = \hat{\mathbf{x}}\,\dfrac{2E_{i1}^2}{\eta_1}\cos^2(\beta_1 z\cos\theta_i)\sin\theta_i \quad [\mathrm{W/m^2}]$.

13.4 Oblique Incidence on Dielectric Interfaces

Based on the previous two sections, oblique incidence on a dielectric should be quite similar: an incident wave gives rise to a reflected wave, both propagating in the same material. However, unlike incidence on perfect conductors in the previous two sections, there is also a wave propagating in material (2) in a fashion similar to that of **Section 13.2**, but, since the incident wave is at an angle θ_i to the normal, we expect the wave in material (2) to also propagate at an angle to the normal. These considerations are shown in **Figure 13.14**. The reflection angle θ_r and transmission angle θ_t both depend on the incident angle θ_i. We have already shown in **Section 13.3.1** that $\theta_r = \theta_i$ [see **Eq. (13.78)**] and will use this relation (Snell's law of reflection) from now on without comment. To be able to describe all wave properties in terms of the incident wave alone, we must also define a relation between the transmission angle θ_t and the incidence angle θ_i. Also, we must expect that the reflection and transmission coefficient should be different than those obtained for normal incidence. We start with perpendicularly polarized waves.

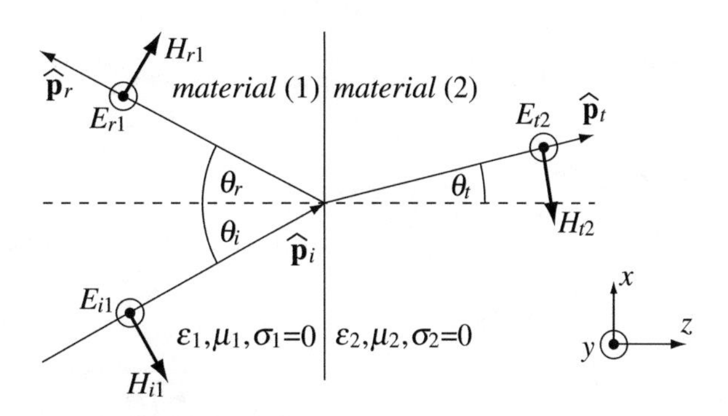

Figure 13.14 Incident, reflected, and transmitted waves for oblique incidence at a dielectric–dielectric interface. Perpendicular polarization

13.4.1 Oblique Incidence on a Dielectric Interface: Perpendicular Polarization

To define the conditions for the reflection and transmitted waves, we use **Figure 13.14**, write the electric and magnetic field intensities on both sides and apply the boundary conditions on the interface for the tangential components of the electric field intensity. The incident electric and magnetic field intensities are the same as **Eqs. (13.73)** and **(13.74)**:

$$\mathbf{E}_i(x,z) = \hat{\mathbf{y}} E_{i1} e^{-j\beta_1(x\sin\theta_i + z\cos\theta_i)} \quad [\text{V/m}] \tag{13.92}$$

$$\mathbf{H}_i(x,z) = \frac{E_{i1}}{\eta_1}(-\hat{\mathbf{x}}\cos\theta_i + \hat{\mathbf{z}}\sin\theta_i) e^{-j\beta_1(x\sin\theta_i + z\cos\theta_i)} \quad [\text{A/m}] \tag{13.93}$$

where, again, the direction of propagation of the incident wave is given by $\hat{\mathbf{p}}_i = \hat{\mathbf{x}}\sin\theta_i + \hat{\mathbf{z}}\cos\theta_i$. The reflected electric and magnetic field intensities are the same as in **Eqs. (13.75)** and **(13.76)**:

$$\mathbf{E}_r(x,z) = \hat{\mathbf{y}} E_{r1} e^{-j\beta_1(x\sin\theta_i - z\cos\theta_i)} \quad [\text{V/m}] \tag{13.94}$$

$$\mathbf{H}_r(x,z) = \frac{E_{r1}}{\eta_1}(\hat{\mathbf{x}}\cos\theta_i + \hat{\mathbf{z}}\sin\theta_i) e^{-j\beta_1(x\sin\theta_i - z\cos\theta_i)} \quad [\text{A/m}] \tag{13.95}$$

Similarly, the transmitted electric and magnetic field intensities have the same form as the incident wave but with different amplitudes and propagate at a different angle (see **Figure 13.14**):

$$\mathbf{E}_t(x,z) = \hat{\mathbf{y}} E_{t2} e^{-j\beta_2(x\sin\theta_t + z\cos\theta_t)} \quad [\text{V/m}] \tag{13.96}$$

$$\mathbf{H}_t(x,z) = \frac{E_{t2}}{\eta_2}(-\hat{\mathbf{x}}\cos\theta_t + \hat{\mathbf{z}}\sin\theta_t) e^{-j\beta_2(x\sin\theta_t + z\cos\theta_t)} \quad [\text{A/m}] \tag{13.97}$$

To determine the transmission and reflection coefficients, the tangential components of the electric field intensity and those of the magnetic field intensity on both sides of the interface (i.e., at $z = 0$) are equated. From **Figure 13.14** and **Eqs. (13.92)** through **(13.97)**, and taking only the tangential components (y component for **E** and x component for **H**) at $z = 0$, we have

$$(E_{i1} + E_{r1})e^{-j\beta_1 x\sin\theta_i} = E_{t2}e^{-j\beta_2 x\sin\theta_t} \quad \text{and} \quad \left(-\frac{E_{i1}}{\eta_1} + \frac{E_{r1}}{\eta_1}\right)\cos\theta_i e^{-j\beta_1 x\sin\theta_i} = -\frac{E_{t2}}{\eta_2}\cos\theta_t e^{-j\beta_2 x\sin\theta_t} \tag{13.98}$$

There are three relations that must be satisfied:

$$e^{-j\beta_1 x\sin\theta_i} = e^{-j\beta_2 x\sin\theta_t} \quad \text{or} \quad \beta_1\sin\theta_i = \beta_2\sin\theta_t \tag{13.99}$$

and

$$E_{i1} + E_{r1} = E_{t2} \quad \text{and} \quad -\frac{E_{i1}}{\eta_1}\cos\theta_i + \frac{E_{r1}}{\eta_1}\cos\theta_i = -\frac{E_{t2}}{\eta_2}\cos\theta_t \tag{13.100}$$

From **Eq. (13.99)**, we get

$$\omega\sqrt{\varepsilon_1\mu_1}\sin\theta_i = \omega\sqrt{\varepsilon_2\mu_2}\sin\theta_t \tag{13.101}$$

or

$$\boxed{\sin\theta_t = \frac{\sqrt{\varepsilon_1\mu_1}}{\sqrt{\varepsilon_2\mu_2}}\sin\theta_i} \tag{13.102}$$

This relation between the incident and refraction angle is Snell's law of refraction. Since $\varepsilon_1 = \varepsilon_0\varepsilon_{r1}$, $\varepsilon_2 = \varepsilon_0\varepsilon_{r2}$, $\mu_1 = \mu_0\mu_{r1}$ and $\mu_2 = \mu_0\mu_{r2}$ (where ε_{r1}, ε_{r2}, μ_{r1}, μ_{r2} are the relative permittivities and relative permeabilities of the two media), and since the phase velocities in medium (1) and (2) are $v_{p1} = 1/\sqrt{\varepsilon_1\mu_1}$ and $v_{p2} = 1/\sqrt{\varepsilon_2\mu_2}$ respectively, we can also write Snell's law of refraction as

$$\boxed{\frac{\sin\theta_t}{\sin\theta_i} = \frac{n_1}{n_2} = \frac{v_{p2}}{v_{p1}}} \tag{13.103}$$

where $n_1 = \sqrt{\varepsilon_{r1}\mu_{r1}}$ is the (optical) index of refraction in medium (1) and $n_2 = \sqrt{\varepsilon_{r2}\mu_{r2}}$ is the (optical) index of refraction in medium (2). Snell's law of refraction in **Eq. (13.102)** or **Eq. (13.103)** is the same for perpendicular and parallel polarization.

Now returning to **Eq. (13.100)**, the solution of the two relations for E_{r1} and E_{t2} gives

$$E_{r1} = E_{i1}\frac{\eta_2\cos\theta_i - \eta_1\cos\theta_t}{\eta_2\cos\theta_i + \eta_1\cos\theta_t}, \qquad E_{t2} = E_{i1}\frac{2\eta_2\cos\theta_i}{\eta_2\cos\theta_i + \eta_1\cos\theta_t} \quad \left[\frac{\text{V}}{\text{m}}\right] \tag{13.104}$$

Because E_{r1} and E_{i1} are in the same direction, the reflection coefficient may be written as $\Gamma_\perp = E_{r1}/E_{i1}$ and the transmission coefficient as $T_\perp = E_{t1}/E_{i1}$:

$$\boxed{\Gamma_\perp = \frac{E_{r1}}{E_{i1}} = \frac{\eta_2\cos\theta_i - \eta_1\cos\theta_t}{\eta_2\cos\theta_i + \eta_1\cos\theta_t} \quad [\text{dimensionless}]} \tag{13.105}$$

$$\boxed{T_\perp = \frac{E_{t2}}{E_{i1}} = \frac{2\eta_2\cos\theta_i}{\eta_2\cos\theta_i + \eta_1\cos\theta_t} \quad [\text{dimensionless}]} \tag{13.106}$$

The notation $\perp$ indicates these are the reflection and transmission coefficients for perpendicular polarization, because, as we shall see, the coefficients for parallel polarization differ.

Now, the total fields in each material can be written directly. In material (1), the fields are the sum of the incident and reflected waves [from **Eqs. (13.92)** and **(13.94)** for E_1 and from **Eqs. (13.93)** and **(13.95)** for H_1]:

$$\mathbf{E}_1(x,z) = \hat{\mathbf{y}}E_{i1}\left[e^{-j\beta_1 z\cos\theta_i} + \Gamma_\perp e^{j\beta_1 z\cos\theta_i}\right]e^{-j\beta_1 x\sin\theta_i} \quad [\text{V/m}] \tag{13.107}$$

$$\begin{aligned}\mathbf{H}_1(x,z) &= \hat{\mathbf{x}}\frac{E_{i1}\cos\theta_i}{\eta_1}\left[\Gamma_\perp e^{j\beta_1 z\cos\theta_i} - e^{-j\beta_1 z\cos\theta_i}\right]e^{-j\beta_1 x\sin\theta_i} \\ &+ \hat{\mathbf{z}}\frac{E_{i1}\sin\theta_i}{\eta_1}\left[e^{-j\beta_1 z\cos\theta_i} + \Gamma_\perp e^{j\beta_1 z\cos\theta_i}\right]e^{-j\beta_1 x\sin\theta_i} \quad [\text{A/m}]\end{aligned} \tag{13.108}$$

In medium (2), where the only wave is the transmitted wave, **Eqs. (13.96)** and **(13.97)** describe the wave. Using the transmission coefficient, we can write

$$\mathbf{E}_t(x,z) = \hat{\mathbf{y}}T_\perp E_{i1}e^{-j\beta_2(x\sin\theta_t + z\cos\theta_t)} \quad [\text{V/m}] \tag{13.109}$$

$$\mathbf{H}_t(x,z) = \frac{T_\perp E_{i1}}{\eta_2}(-\hat{\mathbf{x}}\cos\theta_t + \hat{\mathbf{z}}\sin\theta_t)^{-j\beta_2(x\sin\theta_t + z\cos\theta_t)} \quad [\text{A/m}] \tag{13.110}$$

In all these relations, we could also use **Eq. (13.102)** to write the refraction angle θ_t in terms of the incident angle θ_i. However, this would complicate the expressions considerably.

The electric field intensity $\mathbf{E}_1$ is in the y direction, but $\mathbf{H}_1$ has components in the x and z directions. As was the case for conducting interfaces, we have a propagating wave in the x direction and a standing wave in the z direction (see **Exercise 13.7**).

Exercise 13.7 Show that the wave in material (1) in **Figure 13.14** propagates in the positive x direction, whereas in the z direction, there are both a standing wave and a propagating wave, by calculating the time-averaged Poynting vector in material (1).

Exercise 13.8 Show that the wave in material (2) in **Figure 13.14** propagates in the x and z directions and that there are no standing waves in this region, by calculating the Poynting vector.

Exercise 13.9 Show that for perpendicular polarization the relation $1 + \Gamma_\perp = T_\perp$ holds.

Example 13.8 A perpendicularly polarized plane wave impinges on a very thick sheet of plastic at an angle $\theta_i = 30^\circ$ from free space, as shown in **Figure 13.14**. The relative permittivity of the plastic is $\varepsilon_r = 4$ and its relative permeability is $\mu_r = 1$. If the amplitude of the incident electric field intensity is $E_{i1} = 100$ V/m, calculate the time-averaged power density transmitted into the plastic.

Solution: First, we express the refraction angle and transmission coefficient in terms of the incident angle θ_i. Then, using **Eqs. (13.109)** and **(13.110)**, we calculate the time-averaged power density in material (2), which should have both x and z components.

The reflection and refraction angles are evaluated from the following relations:

$$\theta_r = \theta_i, \qquad \sin\theta_t = \sin\theta_i \frac{n_1}{n_2} = \sin\theta_i \sqrt{\frac{\varepsilon_1}{\varepsilon_2}} = \sin\theta_i \sqrt{\frac{1}{\varepsilon_{r2}}} = 0.5\sqrt{\frac{1}{4}} = 0.25$$

The transmission coefficient for perpendicular polarization **[Eq. (13.106)]** with $\mu_1 = \mu_2 = \mu_0$ is

$$T_\perp = \frac{2\eta_2\cos\theta_i}{\eta_2\cos\theta_i + \eta_1\cos\theta_t} = \frac{2\sqrt{\mu_0/\varepsilon_2}\cos\theta_i}{\sqrt{\mu_0/\varepsilon_2}\cos\theta_i + \sqrt{\mu_0/\varepsilon_0}\cos\theta_t} = \frac{2\cos\theta_i}{\cos\theta_i + \sqrt{\varepsilon_{r2}}\cos\theta_t} = \frac{2\cos\theta_i}{\cos\theta_i + 2\cos\theta_t}$$

Rewriting $\cos\theta_t$ in terms of $\sin\theta_t$ as $\cos\theta_t = \sqrt{1-\sin^2\theta_t} = \sqrt{1-\sin^2\theta_i/\varepsilon_{r2}}$, and then substituting the values for θ_i and $\sin\theta_t$, we get

$$T_\perp = \frac{2\cos\theta_i}{\cos\theta_i + 2\sqrt{1-\sin^2\theta_i/\varepsilon_{r2}}} = \frac{2\cos\theta_i}{\cos\theta_i + \sqrt{4-\sin^2\theta_i}} = \frac{1.732}{0.866 + \sqrt{4-0.25}} = 0.618$$

Now, the transmitted electric and magnetic field intensities may be calculated using **Eqs. (13.109)** and **(13.110)**. However, we are interested in the time-averaged Poynting vector:
Poynting vector:

$$\mathcal{P}_{av} = \frac{1}{2}\text{Re}\{\mathbf{E}\times\mathbf{H}^*\} = \hat{\mathbf{y}}\,\frac{T_\perp E_{i1}}{2}e^{-j\beta_2(x\sin\theta_t + z\cos\theta_t)} \times \frac{T_\perp E_{i1}}{\eta_2}(-\hat{\mathbf{x}}\cos\theta_t + \hat{\mathbf{z}}\sin\theta_t)e^{j\beta_2(x\sin\theta_t + z\cos\theta_t)}$$

$$= \frac{T_\perp^2 E_{i1}^2}{2\eta_2}\hat{\mathbf{y}} \times (-\hat{\mathbf{x}}\cos\theta_t + \hat{\mathbf{z}}\sin\theta_t) = \frac{T_\perp^2 E_{i1}^2}{2\eta_2}(\hat{\mathbf{z}}\cos\theta_t + \hat{\mathbf{x}}\sin\theta_t) \quad \left[\text{W/m}^2\right]$$

where $\eta_2 = \eta_0/2 = 377/2\ \Omega$. With $\sin\theta_t = 0.25$, $\cos\theta_t = \sqrt{1-\sin^2\theta_t} = \sqrt{1-(0.25)^2} = 0.968$, the time-averaged power density is

$$\mathcal{P}_{av} = \frac{0.618^2 \times 100^2}{377}(\hat{\mathbf{z}}0.968 + \hat{\mathbf{x}}0.25) = \hat{\mathbf{x}}\,2.533 + \hat{\mathbf{z}}\,9.8 \quad \left[\text{W/m}^2\right]$$

13.4.2 Oblique Incidence on a Dielectric Interface: Parallel Polarization

The situation considered here is shown in **Figure 13.15a**. The incident electric field intensity is parallel to the plane of incidence and the magnetic field intensity is perpendicular in the y direction so that the incident wave propagates toward the interface. The directions of the reflected fields in **Figure 13.15a** are assumed. The correct directions are found from the interface conditions that must be satisfied.

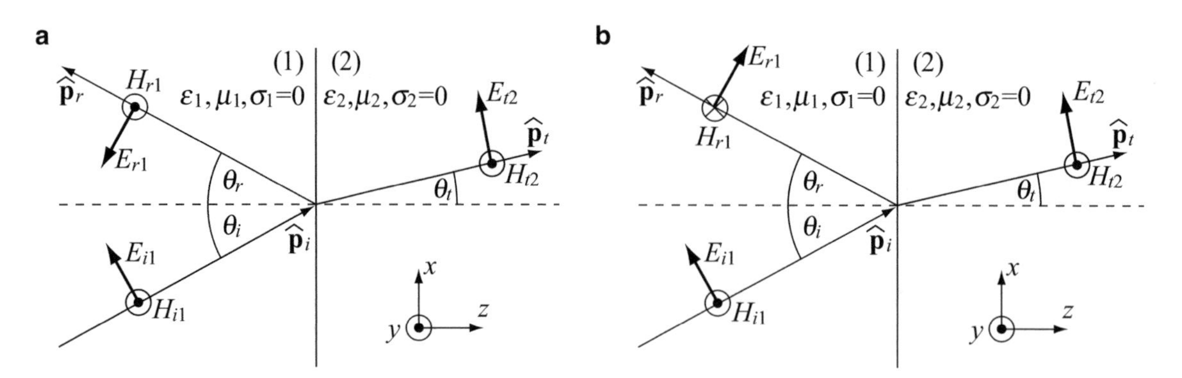

Figure 13.15 (**a**) Assumed incident, reflected, and transmitted waves for oblique incidence at a dielectric–dielectric interface. The electric field is polarized parallel to the plane of incidence. (**b**) Correct, calculated waves (for $\Gamma_\parallel$ positive)

The incident and reflected electric and magnetic field intensities for the configuration in **Figure 13.15a** are as given in **Eqs. (13.85)** through **(13.88)**:

$$\mathbf{E}_i(x,z) = E_{i1}(\hat{\mathbf{x}}\cos\theta_i - \hat{\mathbf{z}}\sin\theta_i)e^{-j\beta_1(x\sin\theta_i + z\cos\theta_i)} \quad [\text{V/m}] \tag{13.111}$$

$$\mathbf{H}_i(x,z) = \hat{\mathbf{y}}\frac{E_{i1}}{\eta_1}e^{-j\beta_1(x\sin\theta_i + z\cos\theta_i)} \quad [\text{A/m}] \tag{13.112}$$

$$\mathbf{E}_r(x,z) = E_{r1}(-\hat{\mathbf{x}}\cos\theta_i - \hat{\mathbf{z}}\sin\theta_i)e^{-j\beta_1(x\sin\theta_i - z\cos\theta_i)} \quad [\text{V/m}] \tag{13.113}$$

$$\mathbf{H}_r(x,z) = \hat{\mathbf{y}}\frac{E_{r1}}{\eta_1}e^{-j\beta_1(x\sin\theta_i - z\cos\theta_i)} \quad [\text{A/m}] \tag{13.114}$$

The transmitted wave into material (2) can be written directly from **Figure 13.15a**:

$$\mathbf{E}_t(x,z) = E_{t1}(\hat{\mathbf{x}}\cos\theta_t - \hat{\mathbf{z}}\sin\theta_t)e^{-j\beta_2(x\sin\theta_t + z\cos\theta_t)} \quad [\text{V/m}] \tag{13.115}$$

$$\mathbf{H}_t(x,z) = \hat{\mathbf{y}}\frac{E_{t2}}{\eta_2}e^{-j\beta_2(x\sin\theta_t + z\cos\theta_t)} \quad [\text{A/m}] \tag{13.116}$$

At the interface between the two media (at $z = 0$), the continuity conditions on the tangential components of the electric and magnetic field intensities are

$$E_{i1}\cos\theta_i - E_{r1}\cos\theta_i = E_{t2}\cos\theta_t \quad \text{and} \quad \frac{E_{i1}}{\eta_1} + \frac{E_{r1}}{\eta_1} = \frac{E_{t2}}{\eta_2} \tag{13.117}$$

Solving for E_{r1} and E_{t2}, we get

$$E_{r1} = E_{i1}\frac{\eta_1\cos\theta_i - \eta_2\cos\theta_t}{\eta_2\cos\theta_t + \eta_1\cos\theta_i}, \qquad E_{t2} = E_{i1}\frac{2\eta_2\cos\theta_i}{\eta_2\cos\theta_t + \eta_1\cos\theta_i} \tag{13.118}$$

To define the reflection coefficient for parallel polarization, we note from **Figure 13.15a** and from **Eq. (13.117)** that E_{i1} and E_{r1} are in opposite directions. Therefore, the reflection coefficient for parallel polarization is defined as

$$\boxed{\Gamma_{\|} = -\frac{E_{r1}}{E_{i1}} = \frac{\eta_2\cos\theta_t - \eta_1\cos\theta_i}{\eta_2\cos\theta_t + \eta_1\cos\theta_i} \quad [\text{dimensionless}]} \tag{13.119}$$

On the other hand, E_{t2} and E_{i1} are in the same direction and therefore the transmission coefficient is

$$\boxed{T_{\|} = \frac{E_{t2}}{E_{i1}} = \frac{2\eta_2\cos\theta_i}{\eta_2\cos\theta_t + \eta_1\cos\theta_i} \quad [\text{dimensionless}]} \tag{13.120}$$

The total fields in medium (1) are calculated by summing the incident and reflected waves. With the use of the reflection coefficient (i.e., using $E_{r1} = -\Gamma_{\|}E_{i1}$), these become

$$\mathbf{E}_1(x,z) = \hat{\mathbf{x}}E_{i1}\cos\theta_i\left(\Gamma_{\|}e^{j\beta_1 z\cos\theta_i} + e^{-j\beta_1 z\cos\theta_i}\right)e^{-j\beta_1 x\sin\theta_i} + \hat{\mathbf{z}}E_{i1}\sin\theta_i\left(\Gamma_{\|}e^{j\beta_1 z\cos\theta_i} - e^{-j\beta_1 z\cos\theta_i}\right)e^{-j\beta_1 x\sin\theta_i} \quad [\text{V/m}] \tag{13.121}$$

$$\mathbf{H}_1(x,z) = -\hat{\mathbf{y}}\frac{E_{i1}}{\eta_1}\left(\Gamma_{\|}e^{j\beta_1 z\cos\theta_i} - e^{-j\beta_1 z\cos\theta_i}\right)e^{-j\beta_1 x\sin\theta_i} \quad [\text{A/m}] \tag{13.122}$$

Using $E_{t2} = T_{\|}E_{i1}$ in **Eqs. (13.115)** and **(13.116)**, we get the fields in medium (2):

$$\mathbf{E}_t(x,z) = T_{\|}E_{i1}(\hat{\mathbf{x}}\cos\theta_t - \hat{\mathbf{z}}\sin\theta_t)e^{-j\beta_2(x\sin\theta_t + z\cos\theta_t)} \quad [\text{V/m}] \tag{13.123}$$

$$\mathbf{H}_t(x,z) = \hat{\mathbf{y}}\frac{T_{\|}E_{i1}}{\eta_2}e^{-j\beta_2(x\sin\theta_t + z\cos\theta_t)} \quad [\text{A/m}] \tag{13.124}$$

Because of the relation between the incident and reflected electric fields, the direction of the reflected electric field intensity in **Figure 13.15a** must be reversed if $\Gamma_{\|}$ is positive. The correct electric and magnetic field intensities are shown in **Figure 13.15b** for $\Gamma_{\|}$ positive.

Example 13.9 A plane wave is generated underwater and propagates toward the surface at an angle α as shown in **Figure 13.16a**. Assume lossless conditions (distilled water) with relative permittivity $\varepsilon_r = 25$, relative permeability $\mu_r = 1$, and a wavelength of 1 m (in water). The electric field intensity is in the plane as shown and has an amplitude of 1 V/m. Calculate:

(a) The electric and magnetic field intensities in air for an incidence angle $\alpha = 5^\circ$.
(b) The incident instantaneous and time-averaged power densities just below the water surface.
(c) The transmitted instantaneous and time-averaged power densities in air immediately above the water's surface.

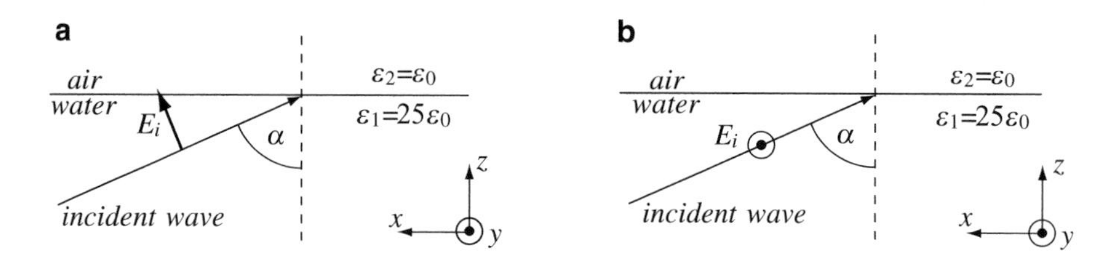

Figure 13.16 A wave propagating from water into air. (**a**) Parallel polarization. (**b**) Perpendicular polarization

Solution: The electric field intensity is parallel to the plane of incidence as shown in **Figure 13.16a**. The polarization is therefore parallel. At the interface, we use the transmission coefficient to find the electric and magnetic field intensities in air, using the relations for parallel polarization.

(a) To define the electric and magnetic fields, we need to calculate the intrinsic impedances in air and water, the transmission coefficient, and the transmission angle θ_t. We know the incident angle: $\theta_i = \alpha = 5^\circ$. The intrinsic impedances are

$$\eta_2 = \eta_0 = 377\ \Omega \text{ in air}, \quad \eta_1 = \frac{\eta_0}{\sqrt{\varepsilon_r}} = \frac{377}{\sqrt{25}} = 75.4\ \Omega \text{ in water}$$

The relation between the incidence and refraction angle is given in **Eq. (13.102)** for any polarization:

$$\sin\theta_t = \frac{n_{water}}{n_{air}}\sin\alpha = \frac{\sqrt{25}}{1}\sin\alpha = 5\sin\alpha$$

where the refraction angle is larger than the incident angle as expected for propagation from a high- to a low-permittivity dielectric. For the given incidence angle, the transmission angle is $\theta_t = \sin^{-1}(5\sin\alpha) = 25.83^\circ$ (or $25^\circ 50'$). The transmission coefficient is

$$T_{\|} = \frac{2\eta_2\cos\alpha}{\eta_2\cos\theta_t + \eta_1\cos\alpha} = \frac{2\eta_0\cos\alpha}{\eta_0\sqrt{1-25\sin^2\alpha} + (\eta_0/5)\cos\alpha}$$

$$= \frac{10\cos\alpha}{5\sqrt{1-25\sin^2\alpha} + \cos\alpha} = \frac{10\cos 5^\circ}{5\sqrt{1-25\sin^2 5^\circ} + \cos 5^\circ} = 1.812$$

where the relations $\cos\theta_t = \sqrt{1-\sin^2\theta_t}$ and $\eta_1 = \eta_0/5$ were used.

The electric and magnetic field intensities generated underwater are known and are of the same general form as in **Eqs. (13.111)** and **(13.112)**, although directions of individual components are different. Substituting the intrinsic impedance of water ($\eta_0/5$), using the coordinate system in **Figure 13.16a** and noting that the phase constant in water is $\beta_1 = 2\pi/\lambda = 2\pi$, the incident electric and magnetic field intensities are

$$\mathbf{E}_i(x,z) = 1(\hat{\mathbf{x}}\cos\alpha + \hat{\mathbf{z}}\sin\alpha)e^{-j2\pi(-x\sin\alpha + z\cos\alpha)} \quad [\mathrm{V/m}]$$

$$\mathbf{H}_i(x,z) = \hat{\mathbf{y}}\frac{5}{\eta_0}e^{-j2\pi(-x\sin\alpha + z\cos\alpha)} \quad [\mathrm{A/m}]$$

The transmitted electric and magnetic fields have the same form as the incident fields, but the propagation is at an angle θ_t and the intrinsic impedance is that of free space [see **Eqs. (13.123)** and **(13.124)**]:

$$\mathbf{E}_t(x,z) = E_i T_{\|}(\hat{\mathbf{x}}\cos\theta_t + \hat{\mathbf{z}}\sin\theta_t)e^{-j\beta_2(-x\sin\theta_t + z\cos\theta_t)} \quad [\mathrm{V/m}]$$

$$\mathbf{H}_t(x,z) = \hat{\mathbf{y}}\frac{E_i T_{\|}}{\eta_0}e^{-j\beta_2(-x\sin\theta_t + z\cos\theta_t)} \quad [\mathrm{A/m}]$$

The phase constant in air is $\beta_2 = 2\pi/(\lambda\sqrt{25}) = 2\pi/5 = 0.4\pi$. With $\sin\theta_t = 5\sin\alpha$, $T_{\|} = 1.812$, $E_i = 1$ V/m, $\cos\theta_t = \sqrt{1-25\sin^2\alpha}$, we get

$$\mathbf{E}_t(x,z) = 1.812\left(\hat{\mathbf{x}}\sqrt{1-25\sin^2\alpha} + \hat{\mathbf{z}}5\sin\alpha\right)e^{-j0.4\pi\left(-x5\sin\alpha + z\sqrt{1-25\sin^2\alpha}\right)} \quad [\mathrm{V/m}]$$

$$\mathbf{H}_t(x,z) = \hat{\mathbf{y}}\frac{1.812}{377}e^{-j0.4\pi\left(-x5\sin\alpha + z\sqrt{1-25\sin^2\alpha}\right)} \quad [\mathrm{A/m}].$$

(b) To calculate the incident, instantaneous Poynting vector, we must first write the time-dependent fields:

$$\mathbf{E}_i(x,z,t) = \mathrm{Re}\{\mathbf{E}_i(x,z)e^{j\omega t}\} = (\hat{\mathbf{x}}\cos\alpha + \hat{\mathbf{z}}\sin\alpha)\cos\left(\omega t - 2\pi(-x\sin\alpha + z\cos\alpha)\right) \quad [\mathrm{V/m}]$$

$$\mathbf{H}_i(x,z,t) = \mathrm{Re}\{\mathbf{H}_i(x,z)e^{j\omega t}\} = \hat{\mathbf{y}}\frac{5}{\eta_0}\cos\left(\omega t - 2\pi(-x\sin\alpha + z\cos\alpha)\right) \quad [\mathrm{A/m}]$$

The instantaneous power density at the surface is given by the Poynting vector at $z = 0$:

$$\begin{aligned}
\mathcal{P}_i(x,0,t) &= \mathbf{E}_i(x,0,t) \times \mathbf{H}_i(x,0,t) \\
&= (\hat{\mathbf{x}}\cos\alpha + \hat{\mathbf{z}}\sin\alpha)\cos(\omega t + \beta_1 x\sin\alpha) \times \hat{\mathbf{y}}\frac{5}{\eta_0}\cos(\omega t + 2\pi x\sin\alpha) \\
&= \hat{\mathbf{z}}\frac{5\cos\alpha}{\eta_0}\cos^2(\omega t + 2\pi x\sin\alpha) - \hat{\mathbf{x}}\frac{5\sin\alpha}{\eta_0}\cos^2(\omega t + 2\pi x\sin\alpha) \\
&= (\hat{\mathbf{z}}1.32 \times 10^{-2} - \hat{\mathbf{x}}1.16 \times 10^{-3})\cos^2(\omega t + 2\pi x\sin\alpha) \quad [\mathrm{W/m^2}]
\end{aligned}$$

Note that the phase of the instantaneous power varies with x as the wave propagates in this direction. It also varies with z, but because we calculated the power density at the surface ($z = 0$), this variation is not shown.

The time-averaged power density may be calculated by integration over one cycle of the wave or using the complex Poynting vector and the fields in **(a)**. The latter is simpler:

$$\mathcal{P}_{iav}(x,0) = \frac{1}{2}\text{Re}\{\mathbf{E}_i(x,0) \times \mathbf{H}_i^*(x,0)\} = \frac{1}{2}\text{Re}\left\{(\hat{\mathbf{x}}\cos\alpha + \hat{\mathbf{z}}\sin\alpha)e^{-j2\pi(-x\sin\alpha)} \times \hat{\mathbf{y}}\frac{5}{\eta_0}e^{j2\pi(-x\sin\alpha)}\right\}$$

$$= \hat{\mathbf{z}}\frac{5\cos\alpha}{2\eta_0} - \hat{\mathbf{x}}\frac{5\sin\alpha}{2\eta_0} = \hat{\mathbf{z}}0.66 \times 10^{-2} - \hat{\mathbf{x}}0.58 \times 10^{-3} \quad [\text{W/m}^2].$$

(c) To calculate the transmitted power (instantaneous or time averaged) at the surface, we use the same steps as in **(b)**, but now we must use the equations for the transmitted wave given in **(a)**. The time-dependent forms (at $z = 0$) are

$$\mathbf{E}_t(x,0,t) = 1.812\left(\hat{\mathbf{x}}\sqrt{1 - 25\sin^2\alpha} + \hat{\mathbf{z}}5\sin\alpha\right)\cos(\omega t - 0.4\pi(-5x\sin\alpha)) \quad [\text{V/m}]$$

$$\mathbf{H}_t(x,0,t) = \hat{\mathbf{y}}\frac{1.812}{\eta_0}\cos(\omega t - 0.4\pi(-5x\sin\alpha)) \quad [\text{A/m}]$$

The transmitted instantaneous power density is

$$\mathcal{P}_t(x,0,t) = \mathbf{E}_t(x,0,t) \times \mathbf{H}_t(x,0,t)$$

$$= 1.812\left(\hat{\mathbf{x}}\sqrt{1 - 25 - \sin^2\alpha} + \hat{\mathbf{z}}5\sin\alpha\right)\cos(\omega t - 0.4\pi(-5x\sin\alpha)) \times \hat{\mathbf{y}}\frac{1.812}{\eta_0}\cos(\omega t - 0.4\pi(-5x\sin\alpha))$$

$$= \hat{\mathbf{z}}\frac{1.812^2 \times \sqrt{1 - 25\sin^2\alpha}}{\eta_0}\cos^2(\omega t + 2\pi x\sin\alpha) - \hat{\mathbf{x}}\frac{1.812^2 \times 5\sin\alpha}{\eta_0}\cos^2(\omega t + 2\pi x\sin\alpha)$$

$$= \left(\hat{\mathbf{z}}7.84 \times 10^{-3} - \hat{\mathbf{x}}3.79 \times 10^{-3}\right)\cos^2(\omega t + 2\pi x\sin\alpha) \quad [\text{W/m}^2]$$

The transmitted time-averaged power density (calculated at $z = 0$) is

$$\mathcal{P}_{tav}(x,0) = \frac{1}{2}\text{Re}\{\mathbf{E}_t(x,0) \times \mathbf{H}_t^*(x,0)\}$$

$$= \frac{1}{2}\text{Re}\left\{1.812\left(\hat{\mathbf{x}}\sqrt{1 - 25\sin^2\alpha} + \hat{\mathbf{z}}5\sin\alpha\right)e^{-j0.4\pi(-x9\sin x)} \times \hat{\mathbf{y}}\frac{1.812}{\eta_0}e^{j0.4\pi(-x9\sin x)}\right\}$$

$$= \frac{1.812}{2}\left(\hat{\mathbf{x}}\sqrt{1 - 25\sin^2\alpha} + \hat{\mathbf{z}}5\sin\alpha\right) \times \hat{\mathbf{y}}\frac{1.812}{\eta_0} = \frac{1.812^2}{2 \times \eta_0}\left(\hat{\mathbf{z}}\sqrt{1 - 25\sin^2\alpha} - \hat{\mathbf{x}}5\sin\alpha\right)$$

$$= \hat{\mathbf{z}}3.92 \times 10^{-3} - \hat{\mathbf{x}}1.895 \times 10^{-3} \quad [\text{W/m}^2]$$

Exercise 13.10 Show that the following relations hold:

$$1 + \Gamma_{\|} = T_{\|}\left(\frac{\cos\theta_t}{\cos\theta_i}\right), \quad 0 \le \theta_i \le \pi/2$$

$$|\Gamma_{\|}|^2 < |\Gamma_{\perp}|^2, 0 < \theta_i < \pi/2$$

$$\Gamma_{\|} = \Gamma_{\perp} = \Gamma, \theta_i = 0$$

Exercise 13.11 Repeat **Example 13.9** for the configuration in **Figure 13.16b**.

$$\mathbf{E}_t(x,z) = \hat{\mathbf{y}}1.694e^{-j0.4\pi\left(-x5\sin\alpha+z\sqrt{1-25\sin^2\alpha}\right)} \quad [\mathrm{V/m}]$$

$$\mathbf{H}_t(x,z) = \frac{1.67}{\eta_0}\left(-\hat{\mathbf{x}}\sqrt{1-25\sin^2\alpha}+\hat{\mathbf{z}}5\sin\alpha\right)e^{-j0.4\pi\left(-x5\sin\alpha+z\sqrt{1-25\sin^2\alpha}\right)} \quad [\mathrm{A/m}]$$

$$\mathcal{P}_i(x,0,t) = \left(-\hat{\mathbf{x}}1.156\times10^{-3}+\hat{\mathbf{z}}1.32\times10^{-2}\right)\cos^2(\omega t+2\pi x\sin\alpha) \quad [\mathrm{W/m^2}]$$

$$\mathcal{P}_{iav}(x,0) = -\hat{\mathbf{x}}0.578\times10^{-3}+\hat{\mathbf{z}}0.66\times10^{-2} \quad [\mathrm{W/m^2}]$$

$$\mathcal{P}_t(x,0,t) = \left(\hat{\mathbf{z}}6.85\times10^{-3}-\hat{\mathbf{x}}3.317\times10^{-3}\right)\cos^2(\omega t+2\pi x\sin\alpha) \quad [\mathrm{W/m^2}]$$

$$\mathcal{P}_{tav}(x,0) = \hat{\mathbf{z}}3.425\times10^{-3}-\hat{\mathbf{x}}1.6585\times10^{-3} \quad [\mathrm{W/m^2}].$$

13.4.3 Brewster's Angle

The reflection and transmission coefficients we obtained in the previous sections depended on the incidence and transmission angles and on the intrinsic impedances of the two materials. A closer inspection of the reflection coefficients and their behavior is now in order. The reflection coefficients for perpendicular and parallel polarizations [from **Eqs. (13.105)** and **(13.119)**] are

$$\Gamma_\perp = \frac{\eta_2\cos\theta_i-\eta_1\cos\theta_t}{\eta_2\cos\theta_i+\eta_1\cos\theta_t}, \qquad \Gamma_\parallel = \frac{\eta_2\cos\theta_t-\eta_1\cos\theta_i}{\eta_2\cos\theta_t+\eta_1\cos\theta_i} \tag{13.125}$$

Either reflection coefficient is zero if the numerator is zero; that is, $\Gamma_\perp = 0$ if $\eta_2\cos\theta_i = \eta_1\cos\theta_t$ and $\Gamma_\parallel = 0$ if $\eta_2\cos\theta_t = \eta_1\cos\theta_i$. The angle at which either condition is satisfied is called the ***Brewster angle***. We will now explore these possibilities starting with the reflection coefficient for parallel polarization in lossless dielectrics.

13.4.3.1 Brewster's Angle for Parallel Polarization

For parallel polarization, the reflection coefficient is zero if [from **Eq. (13.125)**]

$$\eta_2\cos\theta_t = \eta_1\cos\theta_i \tag{13.126}$$

To find the angle θ_i at which this is satisfied, we rewrite sin θ_t in terms of θ_i. From **Eq. (13.102)**, we have

$$\sin\theta_t = \sqrt{\frac{\mu_1\varepsilon_1}{\mu_2\varepsilon_2}}\sin\theta_i \tag{13.127}$$

Using $\cos\theta_\mathrm{i} = \sqrt{1-\sin^2\theta_i}$ and $\cos\theta_t = \sqrt{1-\sin^2\theta_t}$, we can write **Eq. (13.126)** as

$$\eta_2\sqrt{1-\frac{\mu_1\varepsilon_1}{\mu_2\varepsilon_2}\sin^2\theta_i} = \eta_1\sqrt{1-\sin^2\theta_i} \tag{13.128}$$

Now, using $\eta_1 = \sqrt{\mu_1/\varepsilon_1}$ and $\eta_2 = \sqrt{\mu_2/\varepsilon_2}$, squaring **Eq. (13.128)**, separating $\sin^2\theta_i$, and then taking the square root, we get

$$\sin\theta_b = \sqrt{\frac{\varepsilon_2(\mu_2\varepsilon_1 - \mu_1\varepsilon_2)}{\mu_1(\varepsilon_1{}^2 - \varepsilon_2{}^2)}} \tag{13.129}$$

The index b indicates θ_b as the Brewster's angle. This may also be written as

$$\theta_b = \sin^{-1}\sqrt{\frac{\varepsilon_2(\mu_2\varepsilon_1 - \mu_1\varepsilon_2)}{\mu_1(\varepsilon_1{}^2 - \varepsilon_2{}^2)}} \tag{13.130}$$

Thus, for any two materials except two materials of identical permittivity ($\varepsilon_1 = \varepsilon_2$), there is a specific angle at which there is no reflected wave. In the particular but very common case in which both materials have the permeability of free space ($\mu_1 = \mu_2 = \mu_0$), the expression for Brewster's angle is greatly simplified:

$$\boxed{\sin\theta_b = \sqrt{\frac{\varepsilon_2}{\varepsilon_1 + \varepsilon_2}} \quad \text{or} \quad \theta_b = \sin^{-1}\sqrt{\frac{\varepsilon_2}{\varepsilon_1 + \varepsilon_2}} \quad \text{if} \quad \mu_1 = \mu_2} \tag{13.131}$$

The importance of Brewster's angle is twofold. First, it shows that by proper choice of the angle of incidence, the reflection from a lossless material for a parallel polarized wave can be canceled. Second, if a wave has an electric field intensity which has components parallel and perpendicular to the plane of incidence and if the wave impinges on a material interface at the Brewster's angle, the reflection of the parallel polarized component is canceled but not that of the perpendicularly polarized component. The reflected wave consists of the perpendicularly polarized component of the wave alone. Thus, for any general wave (polarized or unpolarized), the reflected wave at the Brewster angle of incidence is linearly polarized perpendicular to the plane of incidence. For this reason, Brewster's angle is also called a ***polarizing angle***. Because the parallel polarized wave at the Brewster angle is not reflected, it follows that it must be transmitted across the interface. Thus, the Brewster angle may also be called the angle of total transmission.

13.4.3.2 Brewster's Angle for Perpendicular Polarization

An angle of no reflection may also be defined for perpendicular polarization by starting with $\Gamma_\perp$ in **Eq. (13.125)**. The condition for zero reflection is now

$$\eta_2\cos\theta_i = \eta_1\cos\theta_t \tag{13.132}$$

Following steps identical to those for parallel polarization, the Brewster angle is

$$\boxed{\theta_b = \sin^{-1}\sqrt{\frac{\mu_2(\varepsilon_1\mu_2 - \varepsilon_2\mu_1)}{\varepsilon_1(\mu_2^2 - \mu_1^2)}}} \tag{13.133}$$

However, for two dielectrics with identical permeabilities, the condition for no reflection cannot be satisfied. (When $\mu_1 = \mu_2$, the denominator in **Eq. (13.133)** is zero.) If μ_1 and μ_2 are not the same, the condition can be satisfied and a Brewster angle exists. For materials with identical permittivities but different permeabilities, the Brewster angle for perpendicular polarization is

$$\boxed{\theta_b = \sin^{-1}\sqrt{\frac{\mu_2}{\mu_2 + \mu_1}}} \quad \textit{if} \quad \varepsilon_1 = \varepsilon_2 \tag{13.134}$$

While both **Eqs. (13.133)** and **(13.134)** describe correct relationships, most materials do not fall in this category; that is, very few dielectrics have different permeabilities and almost none have different permeabilities and the same permittivity. For this reason, the Brewster angle is most often associated with parallel polarization rather than perpendicular polarization.

Example 13.10 A plane wave is generated underwater ($\varepsilon = 25\varepsilon_0$ [F/m], $\mu = \mu_0$ [H/m]). The wave is parallel polarized, propagates in water, and reflects and/or transmits through the interface between water and air as shown in **Figure 13.17**. Calculate the angle α for which there is no reflection at the interface

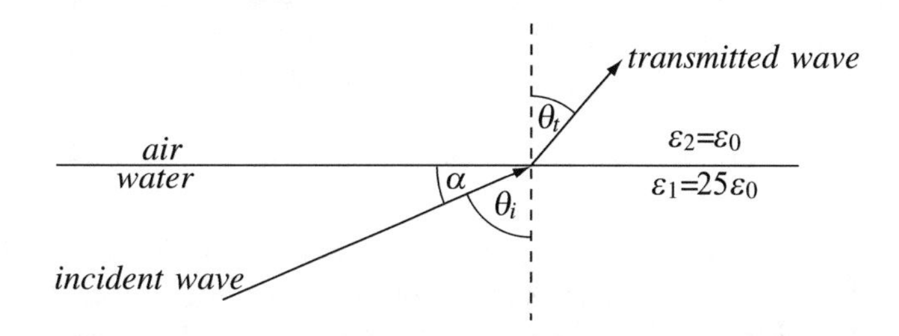

Figure 13.17 A wave incident on the water surface from below

Solution: The incident angle is $\theta_i = 90 - \alpha$. The angle of no reflection is the Brewster angle. In this case, both dielectrics (air and water) have the same permeability; therefore, the Brewster angle is given in **Eq. (13.131)**.

The Brewster angle for a parallel polarized wave propagating from water (ε_1) into air (ε_2) is

$$\theta_b = \sin^{-1}\sqrt{\frac{\varepsilon_2}{\varepsilon_1 + \varepsilon_2}} = \sin^{-1}\sqrt{\frac{\varepsilon_0}{25\varepsilon_0 + \varepsilon_0}} = \sin^{-1}\sqrt{\frac{1}{25+1}} = 11.31^\circ$$

The angle α at which there is no reflection is

$$\alpha = 90 - \theta_b = 78.69^\circ$$

13.4.4 Total Reflection

If the wave propagates across an interface such that the angle of refraction is larger than the angle of incidence, an increase in the angle of incidence leads to an angle at which the transmitted wave propagates at 90° to the normal (see **Figure 13.18**). This angle is called a ***critical angle***. Any increase in the angle of incidence results in total reflection of the incident wave since what would have been the transmitted wave in material (2) is transmitted into material (1). This condition occurs in lossless dielectrics if $\varepsilon_1 > \varepsilon_2$. For example, waves incident on the surface of water from below satisfy this condition. If we view the water surface at a low angle to the normal (i.e., almost perpendicular to the surface of the water), we can see from under the surface. At high angles of incidence, the surface looks as a mirror because of total reflection. The phenomenon exists in either perpendicular or parallel polarization. In terms of the reflection coefficient, total reflection occurs when the reflection coefficient is equal to unity. Substitution of this angle ($\theta_t = 90°$) into the relations for the reflection coefficient in **Eqs. (13.105)** and **(13.119)** gives

$$\Gamma_\perp = 1, \Gamma_{||} = -1, \quad \text{at} \quad \theta_t = 90^\circ \tag{13.135}$$

To define the critical angle, we again use Snell's law in **Eq. (13.127)**:

$$\sin\theta_t = \sqrt{\frac{\mu_1\varepsilon_1}{\mu_2\varepsilon_2}}\sin\theta_i \tag{13.136}$$

Substituting $\theta_t = 90°$ gives the critical angle:

$$\boxed{\sin\theta_c = \sqrt{\frac{\mu_2\varepsilon_2}{\mu_1\varepsilon_1}} \quad \text{for} \quad \mu_2\varepsilon_2 \le \mu_1\varepsilon_1} \tag{13.137}$$

The condition $\mu_2\varepsilon_2 \le \mu_1\varepsilon_1$ is also necessary otherwise, $\sin\theta_c$ would be larger than 1.

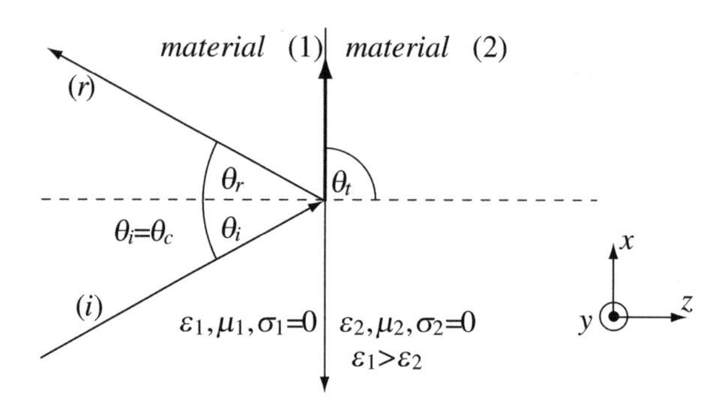

Figure 13.18 Total reflection at the interface between two dielectrics occurs when $\theta_t \geq 90°$

Now, suppose we increase the angle of incidence above θ_c. This leads to $\sin\theta_t > 1$; that is, $\sin\theta_t = 1$ for $\theta_i = \theta_c$. Because $\theta_i = \theta_c < 90°$, an increase in θ_i increases the right-hand side of **Eq. (13.136)** above 1. When substituting this condition in the reflection coefficients in **Eqs. (13.105)** and **(13.119)**, it leads to complex values for the reflection coefficients. The magnitude of the reflection coefficients remains equal to 1, but they are no longer real values.

Therefore, total reflection occurs for $\theta_i \geq \theta_c$

$$\boxed{\theta_i \geq \sin^{-1}\sqrt{\frac{\mu_2\varepsilon_2}{\mu_1\varepsilon_1}} \quad \text{for} \quad \mu_2\varepsilon_2 \leq \mu_1\varepsilon_1} \tag{13.138}$$

In dielectric media for which the permeability is equal in both materials, the condition for total reflection is

$$\boxed{\theta_i \geq \sin^{-1}\sqrt{\frac{\varepsilon_2}{\varepsilon_1}} \quad \text{for} \quad \varepsilon_2 \leq \varepsilon_1,\ \mu_1 = \mu_2} \tag{13.139}$$

The relations in **Eqs. (13.138)** and **(13.139)** are independent of polarization.

Example 13.11 Application: Propagation Within Dielectric Layers A wave propagates inside a glass pane of thickness d [m] at angle of incidence θ_i. Material properties are shown in **Figure 13.19**. What must be the minimum angle of incidence θ_i so that the wave inside the pane does not escape (total reflection at the interior surfaces)? Assume that the wave has entered the glass pane from an edge (not shown). Use the following properties: $\varepsilon_0 = 8.854 \times 10^{-12}$ F/m, $\mu_0 = 4\pi \times 10^{-7}$ H/m, $\varepsilon_2 = 3\varepsilon_0$ [F/m], and $d = 10$ mm.

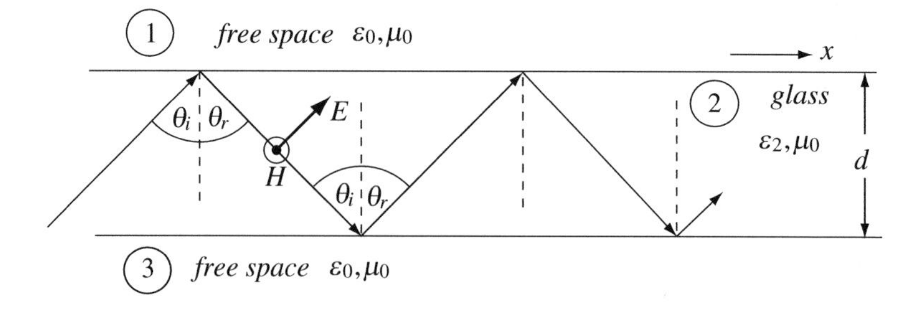

Figure 13.19 Total internal reflection at the interface between glass and air: the glass is said to guide the wave

Solution: For total internal reflection, there are two conditions that must be satisfied: (1) the material outside must be less dense (lower permittivity) than inside and (2) the angle must be above the critical angle. This is calculated from **Eq. (13.136)**:

$$\frac{\sin\theta_t}{\sin\theta_i} = \sqrt{\frac{\varepsilon_2}{\varepsilon_0}}$$

since for total reflection $\theta_t = \pi/2$, we have

$$\sin\theta_c = \sqrt{\frac{\varepsilon_0}{\varepsilon_2}} = \sqrt{\frac{1}{3}} = 0.57753 \quad \text{or} \quad \theta_c = 35°16'$$

For any angle $\theta_i \geq \theta_c$, the wave is totally reflected from either interface. We say that the wave is now guided by the glass pane. This is the basic principle of wave guiding in optical fibers. The requirements are quite simple: a lossless (or low loss) material with dielectric constant higher than the surrounding medium and a means of coupling electromagnetic waves into the material. Much more will be said about this in **Chapter 17**, but see also the following example.

Example 13.12 Application: Integrated Optical Waveguides In integrated optical devices, it is often necessary to propagate waves in particular materials or to "guide" them from one point to another. The idea of channeling a wave is not different in principle than that of guiding a sound wave between two points by means of a pipe. The main requirement here is that the wave should not penetrate through the boundaries of the dielectric. This is easily accomplished by means of a dielectric layer, provided that the materials with which the layer interfaces have lower permittivities.

In the optical device shown in **Figure 13.20**, it is required that light does not escape through either interface. The device is made of a low-permittivity layer (silicon dioxide) on which a high-permittivity layer is grown or deposited (guiding layer, made of silicon nitride), and on top of this, there is a cladding of a low-permittivity layer. In practical devices, the substrate is silicon, followed by a silicon dioxide layer to produce a low-permittivity material. The guiding layer may be silicon or some other high-permittivity material such as silicon nitride. In this case, there is no cladding, but in practice, there will be some cladding to protect the guiding layer. The light is incident from the left at an angle θ as shown in **Figure 13.20**.

(a) What must be the angle θ for light not to escape through the upper surface?
(b) What must be the angle θ for light not to escape through the lower surface?
(c) What must be the angle θ for light not to escape the guiding layer?

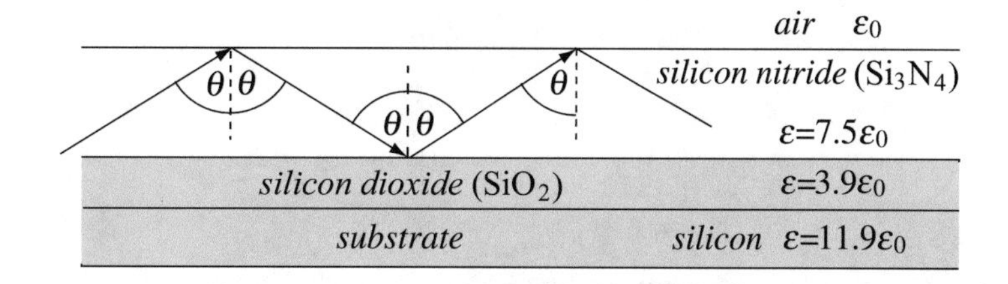

Figure 13.20 An integrated optical. The device relies on total internal reflection at angles above the to confine the wave

Solution: The critical angles for the upper and lower surface are different. Thus, the condition for each surface is different. For light not to escape either surface, the incidence angle must be larger than the larger of the two angles.

(a) The critical angle for a wave propagating from the guiding layer into free space at the upper surface is

$$\sin\theta_{us} = \sqrt{\frac{\varepsilon_0}{7.5\varepsilon_0}} = 0.365, \quad \theta_{us} = 21.42°$$

For angles equal to or larger than 21.42°, there is total reflection on the upper surface and light cannot escape into air.

(b) The critical angle at the lower surface (for a wave propagating from the guiding layer into the substrate) is

$$\sin\theta_{ls} = \sqrt{\frac{3.9\varepsilon_0}{7.5\varepsilon_0}} = 0.721, \quad \theta_{ls} = 46.15°$$

For any angle smaller than this, there will be transmission through the lower surface.

(c) Since for angles above 21.42°, there is no transmission through the upper surface but there is transmission through the lower surface up to 46.15°, the angle of incidence must be above 46.15°.

Note: This device is an optical waveguide similar to optical fibers. One important aspect of waveguides has been shown here: that of confining the wave between surfaces which do not allow transmission (total internal reflection). In these types of devices, any transmission through interfaces means losses. It is therefore important to ensure that total internal reflection occurs at all allowable angles of incidence.

13.5 Reflection and Transmission for Layered Materials at Normal Incidence

waves.m

At multiple interfaces, such as in dielectric slabs and layered dielectrics, we expect both reflection and transmission at each interface. The methods used in **Sections 13.2.1** through **13.2.3** are difficult to apply directly to layered media. It is much easier to calculate the general fields on each side of each layer and then apply the interface conditions at each interface. From these conditions, the fields on each side of each interface are calculated, completely specifying the problem. To outline the method, we treat here a lossy dielectric slab between two general, lossy dielectrics as shown in **Figure 13.21**, where the electric and magnetic field intensities are also shown and are chosen so that power propagates to the right, incident from material (1). The field intensities in each medium are as follows.

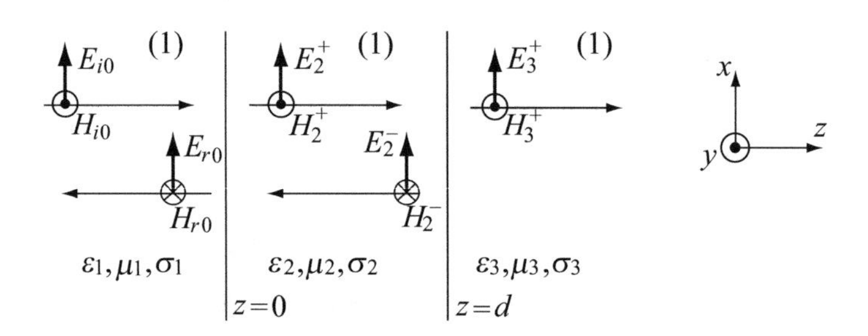

Figure 13.21 Reflection and transmission for a lossy dielectric slab: normal incidence

In material (1) the propagation constant is $\gamma_1 = \alpha_1 + j\beta_1$, and we can write

$$\mathbf{E}_1 = \hat{\mathbf{x}}\left[E_{i0}e^{-\gamma_1 z} + E_{r0}e^{\gamma_1 z}\right] \quad [\text{V/m}] \tag{13.140}$$

$$\mathbf{H}_1 = \hat{\mathbf{y}}\left[\frac{E_{i0}}{\eta_1}e^{-\gamma_1 z} - \frac{E_{r0}}{\eta_1}e^{\gamma_1 z}\right] \quad \left[\frac{\text{A}}{\text{m}}\right] \tag{13.141}$$

In material (2), $\gamma_2 = \alpha_2 + j\beta_2$ and we can write

$$\mathbf{E}_2 = \hat{\mathbf{x}}\left[E_2^+ e^{-\gamma_2 z} + E_2^- e^{\gamma_2 z}\right] \quad [\text{V/m}] \tag{13.142}$$

$$\mathbf{H}_2 = \hat{\mathbf{y}}\left[\frac{E_2^+}{\eta_2}e^{-\gamma_2 z} - \frac{E_2^-}{\eta_2}e^{\gamma_2 z}\right] \quad \left[\frac{\text{A}}{\text{m}}\right] \tag{13.143}$$

In material (3), $\gamma_3 = \alpha_3 + j\beta_3$ and there is only forward propagation:

$$\mathbf{E}_3 = \hat{\mathbf{x}}E_3^+ e^{-\gamma_3 z} \quad [\text{V/m}] \tag{13.144}$$

$$\mathbf{H}_3 = \hat{\mathbf{y}}\eta_3 E_3^+ e^{-\gamma_3 z} \quad [\text{A/m}] \tag{13.145}$$

Because both **E** and **H** are tangential to the various interfaces, the interface conditions are as follows:
At $z = 0$,

$$E_1(0) = E_2(0) \quad \rightarrow \quad E_{i0} + E_{r0} = E_2^+ + E_2^- \tag{13.146}$$

$$H_1(0) = H_2(0) \quad \rightarrow \quad \frac{E_{i0}}{\eta_1} - \frac{E_{r0}}{\eta_1} = \frac{E_2^+}{\eta_2} - \frac{E_2^-}{\eta_2} \tag{13.147}$$

At $z = d$,

$$E_2(d) = E_3(d) \quad \rightarrow \quad E_2^+ e^{-\gamma_2 d} + E_2^- e^{\gamma_2 d} = E_3^+ e^{-\gamma_3 d} \tag{13.148}$$

$$H_2(d) = H_3(d) \quad \rightarrow \quad \frac{E_2^+}{\eta_2} e^{-\gamma_2 d} - \frac{E_2^-}{\eta_2} e^{\gamma_2 d} = \frac{E_3^+}{\eta_3} e^{-\gamma_3 d} \tag{13.149}$$

For convenience, we write **Eqs. (13.146)** through **(13.149)** as a matrix:

$$\begin{bmatrix} -1 & 1 & 1 & 0 \\ \dfrac{1}{\eta_1} & \dfrac{1}{\eta_2} & -\dfrac{1}{\eta_2} & 0 \\ 0 & e^{-\gamma_2 d} & e^{-\gamma_2 d} & -e^{-\gamma_3 d} \\ 0 & \dfrac{e^{-\gamma_2 d}}{\eta_2} & -\dfrac{e^{-\gamma_2 d}}{\eta_2} & -\dfrac{e^{-\gamma_3 d}}{\eta_3} \end{bmatrix} \begin{Bmatrix} E_{r0} \\ E_2^+ \\ E_2^- \\ E_3^+ \end{Bmatrix} = \begin{Bmatrix} E_{i0} \\ E_{i0}/\eta_1 \\ 0 \\ 0 \end{Bmatrix} \tag{13.150}$$

The system in **Eq. (13.150)** may be extended to any number of layers by following the method above (see **Problem 13.42**). This system of equations may be solved numerically once the various constants ($\eta_1, \eta_2, \eta_3, \gamma_1, \gamma_2, \gamma_3$, and d) and the incident electric field intensity E_{i0} are specified. It may also be solved in general terms to obtain the general forms of E_{r0}, E_2^+, E_2^-, and E_3^+, and then, by substitution in **Eqs. (13.140)** through **(13.145)**, the total fields in the various media. A slab reflection coefficient as well as a slab transmission coefficient may also be calculated. These calculations are discussed next.

By direct solution of **Eq. (13.150)**, we obtain the following relations:

$$E_{r0} = \frac{\Gamma_{12} + \Gamma_{23} e^{-2\gamma_2 d}}{1 + \Gamma_{12}\Gamma_{23} e^{-2\gamma_2 d}} E_{i0} \quad \left[\frac{\text{V}}{\text{m}}\right] \tag{13.151}$$

$$E_2^- = \frac{T_{12}\Gamma_{23} e^{-2\gamma_2 d}}{1 + \Gamma_{12}\Gamma_{23} e^{-2\gamma_2 d}} E_{i0} \quad \left[\frac{\text{V}}{\text{m}}\right] \tag{13.152}$$

$$E_2^+ = \frac{T_{12}}{1 + \Gamma_{12}\Gamma_{23} e^{-2\gamma_2 d}} E_{i0} \quad \left[\frac{\text{V}}{\text{m}}\right] \tag{13.153}$$

$$E_3^+ = \frac{T_{12} T_{23} e^{-\gamma_2 d} e^{\gamma_3 d}}{1 + \Gamma_{12}\Gamma_{23} e^{-2\gamma_2 d}} E_{i0} \quad \left[\frac{\text{V}}{\text{m}}\right] \tag{13.154}$$

where Γ_{12} and T_{12} are the reflection and transmission coefficients at the interface between materials (1) and (2) and Γ_{23} and T_{23} are the reflection and transmission coefficients at the interface between materials (2) and (3). These are given as

$$\Gamma_{12} = \frac{\eta_2 - \eta_1}{\eta_2 + \eta_1}, \quad \Gamma_{23} = \frac{\eta_3 - \eta_2}{\eta_3 + \eta_2}, \quad T_{12} = \frac{2\eta_2}{\eta_2 + \eta_1}, \quad T_{23} = \frac{2\eta_3}{\eta_3 + \eta_2} \tag{13.155}$$

In turn, the intrinsic impedances are

$$\eta_1 = \sqrt{\frac{j\omega\mu_1}{\sigma_1 + j\omega\varepsilon_1}}, \quad \eta_2 = \sqrt{\frac{j\omega\mu_2}{\sigma_2 + j\omega\varepsilon_2}}, \quad \eta_3 = \sqrt{\frac{j\omega\mu_3}{\sigma_3 + j\omega\varepsilon_3}} \quad [\Omega] \tag{13.156}$$

From the definition of the reflection coefficient as the ratio between the reflected and incident waves, we define the ***slab reflection coefficient*** from **Eq. (13.151)** as

$$\Gamma_{slab} = \frac{E_{r0}}{E_{i0}} = \frac{\Gamma_{12} + \Gamma_{23} e^{-2\gamma_2 d}}{1 + \Gamma_{12}\Gamma_{23} e^{-2\gamma_2 d}} \quad [\text{dimensionless}] \tag{13.157}$$

Similarly, the ***slab transmission coefficient*** [from **Eq. (13.154)**] is

$$T_{slab} = \frac{E_3^+}{E_{i0}} = \frac{T_{12} T_{23} e^{-\gamma_2 d} e^{\gamma_3 d}}{1 + \Gamma_{12}\Gamma_{23} e^{-2\gamma_2 d}} \quad [\text{dimensionless}] \tag{13.158}$$

The slab reflection and transmission coefficients indicate the degree of transparency of the slab. A low transmission coefficient indicates a material opaque to propagation of electromagnetic waves, whereas a high transmission coefficient indicates a more transparent material. However, it should be noted that the transmission coefficient may be small even if the reflection coefficient is small since the lossy dielectric slab attenuates the field in addition to internal reflections in the slab.

The results in **Eqs. (13.151)** through **(13.158)** were obtained assuming general lossy dielectrics. Lossless dielectrics as well as perfect conductors may be treated by simply replacing the appropriate properties. For example, a common application is a lossless dielectric in free space. Under these conditions, $\gamma_1 = \gamma_3 = j\beta_0$, $\gamma_2 = j\beta_2$, $-\Gamma_{23} = \Gamma_{12}$, $T_{12} = 1 + \Gamma_{12}$, $T_{23} = 1 - \Gamma_{12}$ where $\beta_0 = \omega\sqrt{\mu_0 \varepsilon_0}$ is the phase constant in free space and $\beta_2 = \omega\sqrt{\mu_2 \varepsilon_2}$ the phase constant in the dielectric slab. With these the slab reflection and transmission coefficients are (see also **Example 13.13)**

$$\Gamma_{slab} = \frac{\Gamma_{12}\left(1 - e^{-j2\beta_2 d}\right)}{1 - \Gamma_{12}^2 e^{-j2\beta_2 d}}, \qquad T_{slab} = \left[\frac{\left(1 - \Gamma_{12}^2\right) e^{-j2\beta_2 d} e^{j2\beta_0 d}}{1 - \Gamma_{12}^2 e^{-j2\beta_2 d}}\right] \tag{13.159}$$

These properties may be designed for minimum or maximum transparency of the slab either through specification of the thickness of the layer or through specification of its permittivity. Two particular methods of reducing reflections are widely used as follows.

Half-Wavelength Impedance Matching Section Consider again **Figure 13.21** but with all three dielectrics assumed to be lossless ($\sigma_1 = \sigma_2 = \sigma_3 = 0$). If $\varepsilon_{r1} = \varepsilon_{r3}$, such as for a lossless dielectric layer in free space, and assuming the thickness of the layer to be $d = \lambda/2$, we get, from **Eq. (13.159)**,

$$\Gamma_{slab} = \frac{\Gamma_{12}(1 - e^{-j2\pi})}{1 - \Gamma_{12}^2 e^{-j2\pi}} = 0 \tag{13.160}$$

where $\beta_2 d = (2\pi/\lambda)(\lambda/2) = \pi$ and $e^{-j2\pi} = \cos 2\pi - j\sin 2\pi = 1$ were used.

Thus, a $\lambda/2$ layer guarantees no reflection into medium (1). One common application of this method is in radomes designed to both protect equipment such as antennas and to allow transmission without reflections through the radome.

Quarter-Wavelength Impedance Matching Section If $\varepsilon_{r1} \neq \varepsilon_{r3}$ and $d = \lambda/4$, **Eq. (13.157)** becomes

$$\Gamma_{slab} = \frac{\Gamma_{12} + \Gamma_{23} e^{-2\beta_2 d}}{1 + \Gamma_{12}\Gamma_{23} e^{-2\beta_2 d}} = \frac{\Gamma_{12} + \Gamma_{23} e^{-j\pi}}{1 + \Gamma_{12}\Gamma_{23} e^{j\pi}} = \frac{\Gamma_{12} - \Gamma_{23}}{1 - \Gamma_{12}\Gamma_{23}} \tag{13.161}$$

where $\beta_2 d = (2\pi/\lambda)(\lambda/4) = \pi/2$ and $e^{-j2\pi/2} = \cos\pi - j\sin\pi = -1$ were used.

For this to be zero, we must have

$$\Gamma_{12} = \Gamma_{23} \quad \rightarrow \quad \eta_2 = \sqrt{\eta_1 \eta_3} \tag{13.162}$$

or, if $\mu_1 = \mu_2 = \mu_3 = \mu_0$ [H/m],

$$\varepsilon_2 = \sqrt{\varepsilon_1 \varepsilon_3} \quad [\mathrm{F/m}] \tag{13.163}$$

This method is widely used to reduce reflections in optical devices such as lenses. Proper choice of the coating's permittivity guarantees reduction in reflections.

Another practical application is a lossy or lossless dielectric slab backed by a perfect conductor (see **Figure 13.22**). In this case, the reflection coefficient at the slab-conductor interface is $\Gamma_{23} = -1$, and again we can obtain simple expressions by substituting this in **Eqs. (13.157)** and **(13.158)** (see **Example 13.14**). Finally, it should be noted that if the slab itself is a perfect conductor, then $\Gamma_{12} = -1$, $\Gamma_{23} = +1$ and substitution of these in **Eqs. (13.157)** and **(13.158)** results in $\Gamma_{slab} = -1$, $T_{slab} = 0$ as expected.

Example 13.13 Application: Antenna Radomes A radome is a protective dielectric cover placed over antennas to protect them from the environment. These can be a "dielectric window" in the skin of an airplane to allow its radar to transmit while still maintaining the required smooth surface or may be a dome over a large antenna on the ground or on a ship. In either case, one of the main requirements is that the radome be transparent to transmitted and received waves at the frequency or frequencies at which the antenna operates:

(a) What must be the relative permittivity of a lossless radome material 0.05 m thick for the slab reflection coefficient to be zero at 1 GHz? The permeability of the material is $\mu = \mu_0$ [H/m].

(b) Suppose you cannot find the material required in **(a)** but have plenty of Perspex, which has relative permittivity of 6 and may be assumed to be lossless. Calculate the required thickness to avoid any reflection by the radome at 1 GHz.

Solution: **(a)** From the slab reflection coefficient in **Eq. (13.159)**, the condition for zero reflection is

$$\Gamma_{slab} = \frac{\Gamma_{12}\left(1 - e^{-j2\beta_2 d}\right)}{1 - \Gamma_{12}^2 e^{-j\beta_2 d}} = 0$$

This reduces to

$$1 - e^{-j2\beta_2 d} = 0$$

since, for any dielectric, $\Gamma_{12} \neq 0$. Expanding this, we have

$$1 - e^{-j2\beta_2 d} = 1 - \cos(2\beta_2 d) + j\sin(2\beta_2 d) \quad \rightarrow \quad \cos(2\beta_2 d) - j\sin(2\beta_2 d) = 1$$

Since the right-hand side is real, we have

$$\cos(2\beta_2 d) = 1 \quad \rightarrow \quad 2\beta_2 d = 2n\pi \quad \rightarrow \quad \beta_2 d = n\pi, \quad n = 0,1,2\ldots$$

Although any value of n will do, n cannot be zero; otherwise the thickness must be zero. For the given thickness and taking $n = 1$, we get

$$\beta_2 = \frac{\pi}{d} = \frac{\pi}{0.05} = 62.832 \quad [\mathrm{rad/m}]$$

For lossless dielectrics, the phase constant is $\beta = \omega\sqrt{\mu\varepsilon}$ [rad/m] and we have

$$\beta_2 = \omega\sqrt{\mu\varepsilon} \quad \rightarrow \quad \varepsilon = \frac{\beta_2^2}{\omega^2\mu} \quad \left[\frac{\text{F}}{\text{m}}\right]$$

Thus,

$$\varepsilon = \frac{3947.84}{4\times\pi^2\times 10^{18}\times 4\times\pi\times 10^{-7}} = 7.958\times 10^{-11} \quad \rightarrow \quad \varepsilon_r = \frac{7.958\times 10^{-11}}{8.854\times 10^{-12}} = 8.988.$$

(b) Now, we go in reverse. We start with known permeability, permittivity, and frequency. This gives the phase constant in the radome. From this, we calculate

$$\beta_2 = \omega\sqrt{\mu_0 6\varepsilon_0} = \omega\sqrt{\mu_0\varepsilon_0}\sqrt{6} = \frac{2\times\pi\times 10^9\times\sqrt{6}}{3\times 10^8} = 51.3 \quad [\text{rad/m}]$$

The thickness is (again using the thinnest possible radome: $n = 1$)

$$d = \frac{\pi}{\beta_2} = \frac{\pi}{51.3} = 0.06124 \quad [\text{m}]$$

Note that the reflection coefficient is frequency dependent. This means that at other frequencies the reflection coefficient is not zero (the radome is not completely transparent).

Example 13.14 Application: Measurement of Dielectric Constant of Dielectric Coatings A simple method for measurement of the permittivity of dielectric coatings such as paints is shown in **Figure 13.22**. The measurement consists of a source producing a wave that impinges on the dielectric. The reflection coefficient is then measured. The frequency of the source is varied until the reflection coefficient is either maximum or minimum. The dielectric constant is then calculated from the formula for the slab reflection coefficient.

A perfect dielectric 10 mm thick has permeability of free space and is placed next to a perfect conductor as shown in **Figure 13.22**. A wave impinges on the dielectric and the reflected wave is measured. As the frequency is varied, it is found that the reflected wave's magnitude is maximum at $f = 10$ GHz. Calculate the dielectric constant of the dielectric.

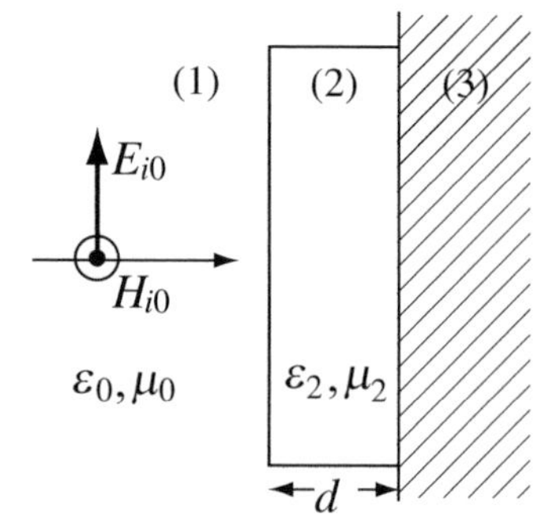

Figure 13.22 A conductor-backed dielectric

Solution: Using **Eq. (13.157)** for the reflection coefficient (with $\gamma_2 = j\beta_2$) we get

$$\Gamma_{slab} = \frac{\Gamma_{12} + \Gamma_{23}e^{-j2\beta_2 d}}{1 + \Gamma_{12}\Gamma_{23e^{-j2\beta_2 d}}}$$

The reflection coefficient at the dielectric–conductor interface is

$$\Gamma_{23} = \frac{\eta_3 - \eta_2}{\eta_3 + \eta_2} = \frac{0 - \eta_2}{0 + \eta_2} = -1$$

Thus,

$$\Gamma_{slab} = \frac{\Gamma_{12} - e^{-j2\beta_2 d}}{1 - \Gamma_{12} e^{-j2\beta_2 d}} = -1$$

The reason the reflection coefficient should be -1 is that Γ_{12} itself is also negative for any dielectric since $\varepsilon_2 > \varepsilon_0$. We have

$$\Gamma_{12} - e^{-j2\beta_2 d} = -1 + \Gamma_{12} e^{-j2\beta_2 d} \quad \rightarrow \quad (\Gamma_{12} + 1) = (\Gamma_{12} + 1) e^{-j2\beta_2 d}$$

or

$$e^{-j2\beta_2 d} = 1 \quad \rightarrow \quad \cos 2\beta_2 d = 1 \quad \rightarrow \quad 2\beta_2 d = 2n\pi \quad \rightarrow \quad \beta_2 \frac{n\pi}{d}, \quad n = 1, 2, 3, \ldots$$

Thus, for the first maximum, $n = 1$,

$$\beta_2 = \omega\sqrt{\mu_0 \varepsilon_0 \varepsilon_r} = \frac{\omega}{c}\sqrt{\varepsilon_r} = \frac{\pi}{d} \quad \rightarrow \quad \sqrt{\varepsilon_r} = \frac{\pi c}{\omega d}$$

The relative permittivity is therefore

$$\varepsilon_r = \left(\frac{\pi c}{\omega d}\right)^2 = \left(\frac{\pi \times 3 \times 10^8}{2 \times \pi \times 10^{10} \times 0.01}\right)^2 = 2.25.$$

13.6 Applications

Application: Microwave Cooking We mentioned microwave cooking earlier. Its utility comes from the fact that at certain frequencies, water absorbs energy from electromagnetic waves. The water itself has a relatively low conductivity but it has dielectric losses which are high at certain frequencies. A particularly useful frequency is 2.45 GHz, which is universally used in residential microwave ovens. Although some foods are, in fact, lossy dielectrics in the true sense (animal tissue, for example, has a relatively large conductivity because of salinity), the main function of microwave ovens is to act on water present in the substance being cooked. A moist substance will cook well but a dry substance will not. Although microwave ovens of moderate size (around 1 kW or less) are common in the kitchen, they also find considerable utility in industrial applications where quick drying of wet substances is needed. Examples are drying of grain before storage and shipment, drying and curing of polymers, large-scale cooking, dielectric welding of plastics, and many others.

Application: Freeze-Drying One application of wave propagation in lossy dielectrics is freeze-drying of foods. It consists of freezing the substance to be dried and then placing it in a vacuum chamber. At that point, heat is applied to evaporate the ice (by sublimation) and thus extract the moisture from the substance. This is different than cooking the substance at high temperature in that the water in the material is evaporated and extracted without boiling and, therefore, with minimum damage to the substance itself. The result is a dehydrated substance (food, plasma, etc.) without damage to tissue. The structure remains essentially unaltered, as do color and texture. Rehydration restores it to the original condition. The idea of using microwaves for this purpose is almost natural, especially since heat conduction through vacuum is very poor, as is the conduction through dried-up tissue. With a microwave source, the waves are absorbed in the ice itself, and if the energy is high enough, ice is evaporated as required.

Application: Radomes and Dielectric Windows It is often necessary to transmit electromagnetic waves from one area into another through a physical barrier. One example was mentioned earlier: the radome. Whenever an antenna must be physically separated from the environment, its energy can only be transmitted and received through this barrier. For example, in an airplane, the radar antenna must be located within the body of the airplane for aerodynamic purposes. A dielectric window is then provided to allow transmission and reception. On ships, the antenna must be protected from the environment by a cover. Another example is the magnetron (a microwave generator). The waves generated in a vacuum chamber must be coupled to the outside such as into a microwave oven with as little reflected energy as possible. These covers and windows operate like radomes (see **Figure 13.23a**) and are designed to transfer energy without any reflection (see **Example 13.13**). A radome may be designed in two ways. One is to choose permittivities and permeabilities such that the intrinsic impedance of the radome material equals that of the surrounding domain (air or free space). For this purpose, the material must be a lossless dielectric with material properties such that $\mu/\varepsilon = \mu_0/\varepsilon_0$. This method has the distinct advantage that the design is independent of frequency. The second method is to choose a very low loss dielectric and design its thickness such that there are no reflections at the required frequency. In practical applications, it is often required to switch frequencies, and the design of the radome must be such that it is transparent at all required frequencies, a design which is often difficult to achieve.

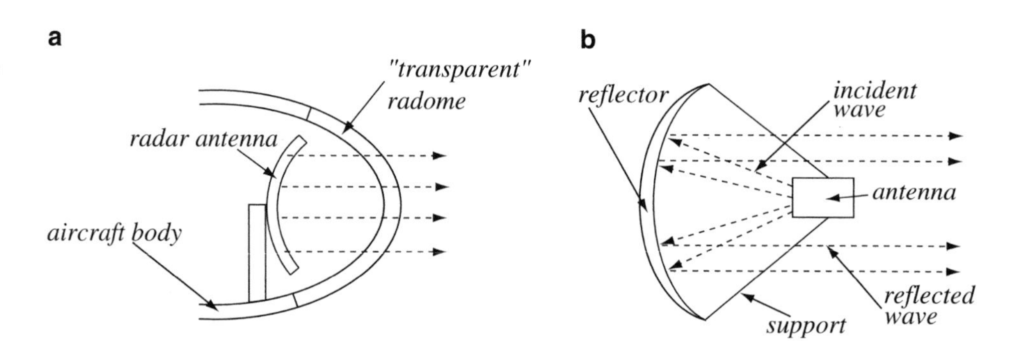

Figure 13.23 (**a**) An aircraft antenna radome. (**b**) The "dish" or parabolic is made of a small feed and a large reflector

Application: Microwave Reflectors Many antennas rely on reflectors to direct the beam in specific directions. A typical parabolic (dish) antenna is shown in **Figure 13.23b**. It consists of a parabolic dish very much like the surface in a car's headlights. The antenna itself (also called a feed) is a small horn located in the focal point of the reflector (similar to the bulb in the headlight). The feed radiates toward the reflector and the reflector then reflects the beam into the direction required. These antennas are highly useful because they transmit energy in narrow beams in the required direction. They are common in satellites and other communication systems. Reflectors may also be used in a passive form: their purpose is to reflect waves or perhaps to change their direction. For example, the actual antenna, which may include a reflector, may be placed at the bottom of a transmission tower, whereas a reflector may be placed on top of the tower to reflect the waves. The antenna transmits upward and the reflector changes the direction of propagation horizontally. This has the advantage that the antenna may be serviced on the ground, and the reflector may be made much lighter than the antenna itself. One particularly interesting aspect of reflectors dates to the early time of satellites. Back then, it was seriously considered using reflectors in space to bounce transmissions. Experiments with aluminized balloons in space were performed and these, while not considered for permanent service, provided the first experimental data for satellite communication. Other ideas included the use of the moon's surface as a reflector. Needless to say, these had minimal success. However, the reflection off surfaces is the common method of radar and microwave mapping of the Earth and the planets. The common thread in these applications is the need for good reflectivity. In antenna reflectors, this is obtained by use of highly conductive, polished materials, whereas in mapping surfaces, the natural reflectivity of surfaces (or any other reflecting materials or regions such as clouds) is used. The variations in reflectivity of materials and surfaces are then used for remote sensing and monitoring.

Application: Scattering of Waves The reflection of waves by any material, including perfect dielectrics, is due to the fact that the reflection coefficient is almost always nonzero. For example, the atmosphere has a permittivity different than free space and the permittivity differs from place to place depending on atmospheric pressure and weather conditions. Similarly, any substance in the atmosphere such as a dust cloud, an airplane, rain or snow, or a pressure front will have permittivities that differ from that of air. These variations may be detrimental to communication in that some energy is reflected in various directions (scattered) rather than serving a useful purpose, whereas in some cases, this scattering is quite useful. Because of scattering, many of the effects mentioned above can be detected by measuring the reflectivity of the materials or conditions present. This is extremely important in weather prediction and remote sensing of the environment. Other applications include communication such as the tropospheric scattering method shown in **Figure 13.24a**. In this method, the transmitter sends a rather narrow beam upward into the troposphere. The waves are scattered and some of the scattered waves are then reflected back into the receiver. With this method there is no need for a reflector to reflect the waves back into the receiver; use is made of the natural reflections that occur in the troposphere. Another simple use of scattering is in microwave testing of lossless dielectrics, shown in **Figure 13.24b**. A microwave beam illuminates the test sample. Some of the waves propagate through the material and some are reflected back into the transmitter. However, none will be coupled to the upper receiver. If, however, there are inclusions, defects, etc., in the material, these will scatter waves in many directions, some of which will be received in the upper receiver. This reception is then an indication of the defects or foreign materials in the test sample.

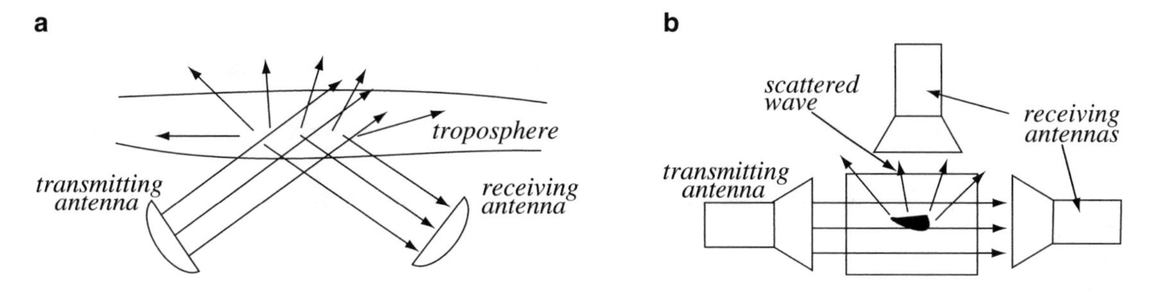

Figure 13.24 (**a**) Tropospheric scattering method of communication. (**b**) Microwave method of testing dielectrics for defects and inclusions

Application: Stealth Aircraft There are two methods of avoiding detection by radar. One is to ensure that the reflection coefficient of the aircraft, as a whole, is as nearly as possible close to zero. If there is no energy reflected back from the aircraft, there is no energy reaching the antenna of the radar and the aircraft is "transparent" to the incoming wave. To do so, the aircraft is coated with materials which have the same intrinsic impedance as air but which also absorb (dissipate) energy. The latter is required because if it were not for this, the wave would propagate through the coating and reflect off the metallic surfaces of the aircraft. Materials appropriate for this purpose are those for which $\mu/\varepsilon = \mu_0/\varepsilon_0$. The required ratio is usually obtained by varying the permeability by adding ferromagnetic powders. In addition, these materials must have some loss to attenuate the wave. Alternatively, a number of layers of different materials are used so that the general reflection coefficient is zero or nearly zero. The absorption of energy must be over a wide enough spectrum to avoid detection by shifting frequencies of the radar system (shifting frequencies is the simplest way to detect "undetectable" aircraft). Absorbing paints and coatings such as rubber and polymers exist that will absorb certain frequencies or range of frequencies. In most cases, radar-absorbing materials are used only where necessary (such as engine intakes, wing tips and edges, etc.) to reduce rather than eliminate the aircraft radar visibility.

A second method of avoiding detection is to reflect the incoming waves but to deflect these in directions away from the radar antenna. In this method, no energy is absorbed, but little is reflected back to the antenna. Aircraft of this type will have sharp angles, as shown in **Figure 13.25**. The sharper the corners, the less energy will be reflected. Note, however, that the flat surfaces employed are quite visible if viewed from a steep enough angle. In the example in **Figure 13.25**, the aircraft is visible from underneath or even from above, but these are not normal angles of observation. Typically a radar installation will try to detect aircraft at low angles, possibly from the front or side. For these angles, the bottom flat surface in **Figure 13.25** is not detectable.

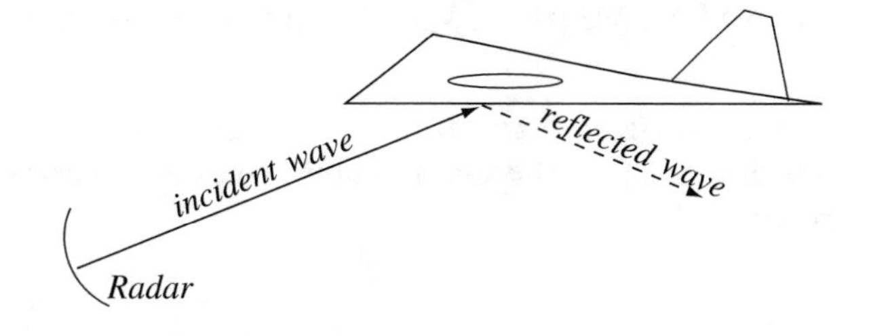

Figure 13.25 A "stealth aircraft" based on sharp corners and flat surfaces that reflect waves away from the radar antenna

Wave-absorbing materials have many applications that are not related to the military. For instance, in evaluation of antennas, it is important to avoid reflections from structures around the antenna so that a proper evaluation can take place. These structures are routinely coated with absorbing materials, usually in the form of narrow foam pyramids, impregnated with conductive materials to attenuate the incoming waves. These form so-called anechoic chambers because they do not reflect waves (no echoes).

13.7 Summary

This chapter takes up the issues of transmission, reflection, and refraction of plane waves at the interface between two different media. The dominant quantities are the reflection and transmission coefficients at interfaces between media.

Definitions

Plane of incidence: the plane formed by the direction of propagation of the incident wave and the normal to the interface (**Figures 13.1** and **13.10**).

Incidence angle: the angle between the direction of propagation of the incident wave and the normal to the interface (**Figure 13.1**).

Reflection angle: the angle between the direction of propagation of the reflected wave and the normal to the interface (**Figure 13.1**).

Transmission angle: the angle between the direction of propagation of the transmitted wave and the normal to the interface (**Figure 13.1**).

Perpendicular (normal) incidence: the wave impinges on an interface perpendicularly (**Figure 13.2**).

Oblique incidence: the wave impinges on an interface at an angle (**Figures 13.10** and **13.1**).

Perpendicular polarization: the electric field intensity is perpendicular to the plane of incidence (see **Figure 13.14**).

Parallel polarization: the electric field intensity is parallel to the plane of incidence (see **Figure 13.15**).

Perpendicular Incidence on general media. For a wave propagating from medium (1) into medium (2), the reflection and transmission coefficients are (see **Figure 13.2**)

$$\Gamma = \frac{E_{r1}}{E_{i1}} = \frac{\eta_2 - \eta_1}{\eta_2 + \eta_1} \quad (13.23) \qquad T = \frac{E_t}{E_{i1}} = \frac{2\eta_2}{\eta_2 + \eta_1} \quad (13.24) \qquad 1 + \Gamma = T \quad (13.21)$$

where η_1 and η_2 are given in **Eqs. (13.8)** and **(13.11)**. η_1 and η_2 and therefore Γ and T can be complex.

The total fields (sum of incident and reflected waves) in medium (1) with E_{i1} known (**Figure 13.2**) are

$$\mathbf{E}_1(z) = \hat{\mathbf{x}}E_{i1}(Te^{-\gamma_1 z} - j\Gamma 2\sin(j\gamma_1 z)) \quad [\text{V/m}] \quad (13.27) \qquad \mathbf{H}_1(z) = \hat{\mathbf{y}}\,\tfrac{E_{i1}}{\eta_1}(Te^{-\gamma_1 z} - \Gamma 2\cos(j\gamma_1 z)) \quad [\text{A/m}] \quad (13.30)$$

where $\gamma_1 = \sqrt{j\omega\mu_1(\sigma_1 + j\omega\varepsilon_1)}$ [see **Eq. (13.8)**]. The total fields in medium (2) are

$$\mathbf{E}_2(z) = \hat{\mathbf{x}}TE_{i1}e^{-\gamma_2 z} \quad [\text{V/m}] \quad (13.31) \qquad \mathbf{H}_2(z) = \hat{\mathbf{y}}T\frac{E_{i1}}{\eta_2}e^{-\gamma_2 z} \quad \left[\frac{\text{A}}{\text{m}}\right] \quad (13.32)$$

where $\gamma_2 = \sqrt{j\omega\mu_2(\sigma_2 + j\omega\varepsilon_2)}$ [see **Eq. (13.11)**].

At the interface between a perfect dielectric and a perfect conductor: $\Gamma = -1$, $T = 0$ and $\gamma_1 = j\beta_1$. Only ***standing waves***, that is, waves that oscillate but do not propagate, can exist in the dielectric:

$$\mathbf{E}_1(z) = \hat{\mathbf{x}}j2E_{i1}\sin(\beta_1 z) \quad [\text{V/m}] \quad (13.47) \qquad \mathbf{H}_1(z) = \hat{\mathbf{y}}2\frac{E_{i1}}{\eta_1}\cos(\beta_1 z) \quad [\text{A/m}] \quad (13.48)$$

Nodes of the standing wave (zero electric field intensity, maximum magnetic field intensity) are at $z = -n\lambda_1/2$, with λ_1 the wavelength in the dielectric ($z = 0$ is assumed at the conducting interface). Maxima in **E** or minima in **H** are $\lambda_1/4$ on either side of the minima in **E**.

Perpendicular Polarization, Oblique Incidence on a Conductor (Figure 13.10b)

$$\Gamma = -1,\ T = 0 \quad \to \quad E_{r1} = -E_{i1} \quad \text{and} \quad \theta_r = \theta_i \tag{13.78}$$

Total fields in the dielectric [medium (1)]:

$$\mathbf{E}_1(x,z) = -\hat{\mathbf{y}}\, j2E_{i1}\sin(\beta_1 z\cos\theta_i)e^{-j\beta_1 x\sin\theta_i} \quad [\text{V/m}] \tag{13.81}$$

$$\mathbf{H}_1(x,z) = -2\frac{E_{i1}}{\eta_1}[\hat{\mathbf{x}}\cos\theta_i\cos(\beta_1 z\cos\theta_i) + \hat{\mathbf{z}}\, j\sin\theta_i\sin(\beta_1 z\cos\theta_i)]e^{-j\beta_1 x\sin\theta_i} \quad [\text{A/m}] \tag{13.82}$$

Parallel Polarization, Oblique Incidence on a Conductor (Figure 13.13) Reflections and transmission coefficients are the same as for perpendicular polarization. The total fields are

$$\mathbf{E}_1(x,z) = -2E_{i1}[\hat{\mathbf{x}}\, j\cos\theta_i\sin(\beta_1 z\cos\theta_i) + \hat{\mathbf{z}}\sin\theta_i\cos(\beta_1 z\cos\theta_i)]e^{-j\beta_1 x\sin\theta_i} \quad [\text{V/m}] \tag{13.90}$$

$$\mathbf{H}_1(x,z) = \hat{\mathbf{y}}2\frac{E_{i1}}{\eta_1}\cos(\beta_1 z\cos\theta_i)e^{-j\beta_1 x\sin\theta_i} \quad [\text{A/m}] \tag{13.91}$$

Conclusions for Both Types of Polarizations

(1) The wave propagates parallel to the conducting surface.
(2) Only standing waves exist perpendicular to the surface.
(3) Power is guided along the conducing surface.

Oblique Incidence on Dielectrics (Figure 13.14) Snell's law of reflection (lossless dielectrics):

$$\theta_r = \theta_i \quad (13.78) \qquad \frac{\sin\theta_t}{\sin\theta_i} = \frac{\sqrt{\varepsilon_1\mu_1}}{\sqrt{\varepsilon_2\mu_2}} = \frac{n_1}{n_2} = \frac{v_{p2}}{v_{p1}} \qquad (13.102) \text{ and } (13.103)$$

where $n_i = \sqrt{\varepsilon_{ri}\mu_{ri}}$ is the *index of refraction* of medium i and ε_{ri}, μ_{ri} are the relative permittivity and relative permeability of the medium.

Perpendicular Polarization, Oblique Incidence on a Dielectric:
Reflection and transmission coefficients:

$$\Gamma_\perp = \frac{E_{r1}}{E_{i1}} = \frac{\eta_2\cos\theta_i - \eta_1\cos\theta_t}{\eta_2\cos\theta_i + \eta_1\cos\theta_t} \ [\text{dimensionless}] \quad (13.105) \qquad T_\perp = \frac{E_{t2}}{E_{i1}} = \frac{2\eta_2\cos\theta_i}{\eta_2\cos\theta_i + \eta_1\cos\theta_t} \ [\text{dimensionless}] \quad (13.106)$$

and $1 + \Gamma_\perp = T_\perp$

The electric and magnetic field intensities in both media are given in **Eqs. (13.107)** through **(13.110)**.

Parallel polarization, Oblique Incidence on a Dielectric Reflection and transmission coefficients

$$\Gamma_\| = -\frac{E_{r1}}{E_{i1}} = \frac{\eta_2\cos\theta_t - \eta_1\cos\theta_i}{\eta_2\cos\theta_t + \eta_1\cos\theta_i} \ [\text{dimensionless}] \quad (13.119) \qquad T_\| = \frac{E_{t2}}{E_{i1}} = \frac{2\eta_2\cos\theta_i}{\eta_2\cos\theta_t + \eta_1\cos\theta_i} \ [\text{dimensionless}] \quad (13.120)$$

and $1 + \Gamma_\| = T_\|\left(\frac{\cos\theta_t}{\cos\theta_i}\right)$ (see **Exercise 13.10**).

The electric and magnetic field intensities in both media are given in **Eqs. (13.121)** through **(13.124)**.

Brewster's angle is the angle of no reflection (also called polarizing angle) for waves propagating from medium (1) into medium (2). For parallel polarization, provided $\varepsilon_1 \neq \varepsilon_2$,

$$\theta_b = \sin^{-1}\sqrt{\frac{\varepsilon_2}{\mu_1}\frac{(\mu_2\varepsilon_1 - \mu_1\varepsilon_2)}{\varepsilon_1^2 - \varepsilon_2^2}} \quad (13.130) \qquad \text{or} \qquad \theta_b = \sin^{-1}\sqrt{\frac{\varepsilon_2}{\varepsilon_1 + \varepsilon_2}} \quad \text{if} \quad \mu_1 = \mu_2 \quad (13.131)$$

For perpendicular polarization, provided $\mu_1 \neq \mu_2$

$$\theta_b = \sin^{-1}\sqrt{\frac{\mu_2}{\varepsilon_1}\frac{(\mu_2\varepsilon_1 - \mu_1\varepsilon_2)}{\mu_2^2 - \mu_1^2}} \quad (13.133) \qquad \text{or} \qquad \theta_b = \sin^{-1}\sqrt{\frac{\mu_2}{\mu_2 + \mu_1}} \quad \text{if} \quad \varepsilon_1 = \varepsilon_2 \quad (13.134)$$

Critical Angle and ***Total Reflection*** A wave propagating in medium (1) is reflected back into medium (1) without transmission if

$$\theta_i \geq \sin^{-1}\sqrt{\frac{\mu_2\varepsilon_2}{\mu_1\varepsilon_1}}, \quad \text{for} \quad \mu_2\varepsilon_2 \leq \mu_1\varepsilon_1 \quad (13.138)$$

That is, total reflection can only occur when propagating from a higher to a lower permittivity dielectric (most dielectrics have the permeability of free space) at and above the critical angle.

Reflection from Layered Structures, Normal Incidence The slab reflection and transmission coefficients (lossy slab of thickness d between lossy dielectrics) (see **Figure 13.21**) are

$$\Gamma_{slab} = \frac{E_{r0}}{E_{i0}} = \frac{\Gamma_{12} + \Gamma_{23}e^{-2\gamma_2 d}}{1 + \Gamma_{12}\Gamma_{23}e^{-2\gamma_2 d}} \quad (13.157) \qquad T_{slab} = \frac{E_3^-}{E_{i0}} = \frac{T_{12}T_{23}e^{-2\gamma_2 d}e^{\gamma_3 d}}{1 + \Gamma_{12}\Gamma_{23}e^{-2\gamma_2 d}} \quad (13.158)$$

where Γ_{ij} and T_{ij} are the reflection and transmission coefficients at the interface as the wave propagates from material i into material j.

Notes:

(1) The reflection and transmission coefficients are defined only for the electric field intensity based on the continuity of the tangential components.

(2) The reflected/transmitted magnetic field intensity components are calculated from the electric field intensity by dividing by the appropriate intrinsic impedance.

Problems

Reflection and Transmission at a General Dielectric Interface: Normal Incidence

13.1 Reflection and Transmission at Air-Lossy Dielectric Interface. A plane wave impinges perpendicularly on a half-space made of a low loss dielectric. Calculate the reflected and transmitted electric field intensities. Use **Figure 13.26** for reference. Assume the material to the left has properties of free space and to the right has the properties: $\sigma_2 = 10^{-9}$ S/m, $\varepsilon_2 = 12\varepsilon_0$ [F/m] , $\mu_2 = \mu_0$ [H/m]), and the frequency is 100 MHz.

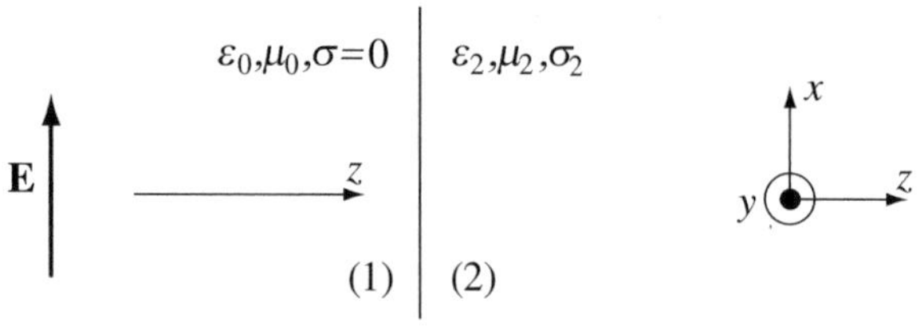

Figure 13.26

13.2 Incident and Reflected Waves at a Lossless Dielectric Interface. A plane wave is given as $E = E_0 e^{-j\beta z}$ [V/m] and propagates in free space. The wave hits a dielectric wall ($\varepsilon = 2\varepsilon_0$ [F/m]) at normal incidence. With $E_0 = 10$ V/m, $\mu_0 = 4\pi \times 10^{-7}$ H/m, $\varepsilon_0 = 8.854 \times 10^{-12}$ F/m, $f = 1$ GHz, calculate:

(a) The peak electric field intensity, left of the wall.
(b) The peak magnetic field intensity, left of the wall.

13.3 Incident and Reflected Waves at a Lossy Dielectric Interface. The configuration in **Figure 13.26** is given. A wave propagates in the direction perpendicular to the interface between free space and a general lossy material (z direction) and has an electric field intensity directed as shown. Calculate the ratio between the maximum and minimum electric field amplitudes in material 1.

13.4 Application: Transmission of Power into Solar Cells. Consider the question of generating electricity with silicon solar cells. The relative permittivity of silicon at optical wavelengths is 1.75 and it may be considered to be lossless. Assume uniform plane waves, perpendicular incidence, and that 30% of the power entering the cells is converted into electric power. The Sun power density at the location of the cells is 1,120 W/m^2.

(a) Calculate the power per unit area of the cell it can generate and its overall efficiency.
(b) Suppose a new type of material is designed which has properties identical to those of silicon except that its permittivity equals that of free space. How much larger is the power that solar cells made of this material can generate and its efficiency?

13.5 Application: Power Transmitted into Glass at Normal Incidence. A laser beam is incident on a glass surface from free space. The beam is narrow, 0.1 mm in diameter, with a power density in the beam of 0.1 W/m^2. Assume normal incidence on the surface and plane wave behavior. Glass is lossless and has a relative permittivity of 1.8 at the frequency used:

(a) Calculate the amplitude of the incident electric and magnetic field intensities in space and the transmitted electric and magnetic field intensities in the glass.
(b) Calculate the total power transmitted into the glass.

13.6 Application: Transmission Through Optical Fibers. In the optical fiber in **Figure 13.27** calculate the power available at the detector. Assume plane wave behavior. The source transmits a uniform power density of 10 W/m^2 across the cross-section of the fiber as shown. Air is considered lossless. The optical fiber is $L = 10$ km long, is $a = 1$ mm in diameter, has permeability of free space and a permittivity equal to $1.8\varepsilon_0$ [F/m]. The losses in the fiber are 5 dB/km.

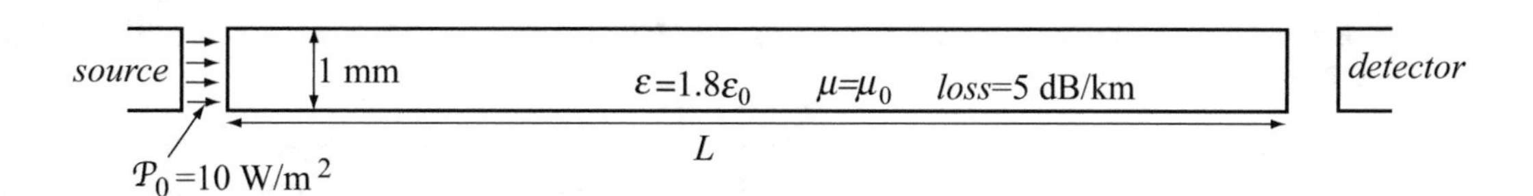

Figure 13.27

13.7 Application: Optical Fiber Strain Sensor. Consider the optical fiber system in **Figure 13.27** but assume, for simplicity that the fiber is lossless and its length is $L = 0.4$ m. To use it as a sensor, the ends of the sensor are fixed to the object in which strain is measured (for example, the deck of a bridge). Strain is defined as the elongation of the sensor per unit length of the sensor ($\Delta L/L$). A strain applied to the optical fiber causes it to elongate and that changes the phase of the detected signal because light propagates a longer distance. If the detector can distinguish a change of 1° in the phase of the signal, calculate the resolution of the sensor, that is, the smallest strain measurable. The wavelength in the fiber is 850 nm.

13.8 Conservation of Power Across an Interface. Show that time-averaged power is conserved across an interface between two media for:

(a) Lossless media, perpendicular incidence.
(b) Lossy media, perpendicular incidence.

13.9 Application: The Sun at the Beach or: Why Do We Get Sunburns? The Sun impinges on the ground at 1,300 W/m^2 (time-averaged power density). If the properties of the skin are known as $\sigma = 0.01$ S/m, $\mu = \mu_0$ [H/m] and $\varepsilon = 24\varepsilon_0$ [F/m], calculate the amount of power dissipated in the skin of a person. Assume the area exposed is 1 m^2, the Sun radiates at an average frequency of 5×10^{14} Hz and is perpendicular to the surface of the skin.

13.10 Application: Radiation Exposure. One of the main concerns in exposure to microwave radiation is heating effects in the body. The US radiation safety code specifies that the total amount of radiation should not exceed 10 mW/cm^2 of skin for 6 hours. Suppose an average person is exposed to this radiation at a frequency of 10 GHz. The effective area of the skin is 1.5 m^2, and the body properties are $\sigma = 0.01$ S/m, $\mu = \mu_0$ [H/m], and $\varepsilon = 24\varepsilon_0$ [F/m], at the given frequency. Calculate the total power absorbed by the body and the total energy absorbed during maximum exposure.

13.11 Application: Transmission of Power Through an Interface. A plane wave with an electric field intensity equal to E_0 [V/m] propagates from free space into a lossy dielectric with properties ε_1, μ_1, σ_1, and thickness 1 m. The direction of propagation is perpendicular to the surface of the material. Calculate the time-averaged power dissipated in the lossy dielectric per unit area of the material. Given: $\varepsilon_1 = 2\varepsilon_0$ [F/m], $\mu_1 = 50\mu_0$ [H/m], $\sigma_1 = 10$ S/m, $f = 100$ MHz, $E_0 = 100$ V/m.

Reflection and Transmission at a Dielectric Conductor Interface: Normal Incidence

13.12 Application: Standing Waves and Reflectometry. An antenna generates an electric field intensity directed upward. The amplitude of the wave is $E_0 = 100$ V/m, at a wavelength of 12 m:

(a) Calculate the location of the antenna in relation to a perfectly conducting wall such that a standing wave is generated with three positive maxima in the electric field intensity between the wall and antenna, and the antenna is at the location of the fourth positive peak. Assume propagation in free space.
(b) If propagation occurs in a low-loss medium, $\varepsilon_1 = 4\varepsilon_0$ [F/m], $\mu_1 = \mu_0$ [H/m], $\sigma_1 = 10^{-5}$ S/m, calculate the amplitude of the electric field intensity at the location of the first positive maximum to the right of the antenna.

13.13 Application: Reflection of Waves from Conducting Surfaces. A wave propagates in free space and impinges perpendicularly on a perfectly conducting surface. Show that the ratio between the electric field intensity and the magnetic field intensity anywhere to the left of the conducting surface is purely imaginary or that the electric and magnetic field intensities are out of phase. **Hint:** Use the relations $e^{j\beta z} - e^{-j\beta z} = j2\sin\beta z$ and $e^{j\beta z} + e^{-j\beta z} = 2\cos\beta z$.

13.14 Application: Surface Current Generated by Incident Waves. A wave impinges perpendicularly on a perfectly conducting surface. The amplitude of the incident electric field intensity is 10 V/m and the wave propagates in free space. For orientation purposes, assume the wave propagates in the positive z direction and the electric field intensity is directed in the negative y direction:

(a) Calculate the surface current density (A/m) produced by the incident field.
(b) Show that the total field in free space is the sum of the incident field and the field produced by the surface current density.

Oblique incidence on a Conducting Interface: Perpendicular Polarization

13.15 Interface Conditions at a Conductor Interface. A uniform plane wave impinges on a good conductor at an arbitrary angle. The wave is polarized perpendicular to the plane of incidence.

(a) What are the interface conditions that exist at the interface between the conductor and air?
(b) What happens to the wave inside the conductor (i.e., describe the relations for phase velocity, depth of penetration, intrinsic impedance, and propagation constant)?

13.16 Oblique Incidence on a Conducting Surface: Perpendicular Polarization. A perpendicularly polarized plane wave impinges on a flat metallic interface at an angle of incidence α. The incident electric field intensity is in the positive x direction, has amplitude 100 V/m, frequency 100 GHz, and propagates in free space. Assume the incident magnetic field intensity has components in the positive y and negative z directions and that the interface coincides with the x-y plane. Calculate:

(a) The incident magnetic field intensity.
(b) The reflected electric and magnetic field intensities.
(c) The surface current density as a function of the incidence angle α on the surface of the conductor. Plot its magnitude and show for what values of the incidence angle the current density is maximum and for what values it is minimum.

Oblique Incidence on a Conducting Interface, Parallel Polarization

13.17 Oblique Incidence on a Conductor: Parallel Polarization. A uniform plane wave impinges on a good conductor at an arbitrary angle. The wave is polarized parallel to the plane of incidence.

(a) What are the interface conditions that exist at the interface between the conductor and air?
(b) Compare the results obtained here with those in **Problem 13.15**.

13.18 Oblique Incidence on a Conductor: Parallel Polarization. A parallel polarized plane wave impinges on a flat metallic reflector at an angle of incidence α. The incident magnetic field intensity is in the positive x direction, has amplitude 100 A/m, and propagates in free space at a frequency of 100 GHz. Assume the incident electric field intensity has components in the negative y and positive z directions and the interface is on the x–y plane.

(a) Calculate the incident electric field intensity.
(b) Calculate the reflected electric and reflected magnetic field intensities.
(c) Calculate the surface current density as a function of the incidence angle α on the surface of the conductor. Plot and show for what values of the incidence angle the current density is maximum and for what values it is minimum.
(d) Compare the results obtained here with those in **Problem 13.16**.

13.19 Standing Waves for Oblique Incidence on a Conductor. A plane wave is polarized parallel to the plane of incidence, its magnetic field intensity is directed in the positive x direction, and it has an amplitude of 15 A/m. Assume the incident electric field intensity has components in the negative y and positive z directions. The phase constant of the wave is 200 rad/m. The wave impinges on a conducting surface on the x–y plane at 30°.

(a) Calculate the standing wave pattern.
(b) Find the location and amplitude of the standing wave peaks.
(c) Calculate the total time-averaged power density in space. Show that real power propagates parallel to the surface.

13.20 Application: Propagation of Waves in the Presence of a Conducting Surface. A plane wave is parallel polarized and impinges on the surface of a perfect conductor at an angle. For a given amplitude and frequency and assuming the wave propagates in free space before reaching the conductor:

(a) Determine the phase velocity in the direction parallel to the surface of the conductor (in which real power propagates) as a function of the angle of incidence.
(b) What is the phase velocity if the incident wave is parallel to the surface of the conductor?
(c) Compare the results in **(a)** and **(b)** with the phase velocity in free space in the absence of the conductor.

13.21 Application: Surface Currents Induced by an Obliquely Incident Wave. A plane wave impinges on a perfectly conducting surface at 30° to the normal. The amplitude of the incident electric field intensity is 10 V/m and the wave propagates in free space. Assume the surface is in the x–y plane and calculate:

(a) The surface current density (A/m) produced by the field if the polarization is perpendicular and the incident electric field intensity is in the positive y direction.
(b) The surface current density for parallel polarization if the incident magnetic field intensity is in the positive y direction.

Parallel and Perpendicular Polarization in Dielectrics

13.22 Conservation of Power Across an Interface. Show that time-averaged power is conserved across an interface between two media for:

(a) Lossless media, incidence at an angle, perpendicular polarization.
(b) Lossless media, incidence at an angle, parallel polarization.
Hint: Recall that the transmission and reflection coefficients are defined for the tangential components of the electric field intensity.

13.23 Oblique Incidence on a Dielectric: Perpendicular Polarization. A perpendicularly polarized plane wave impinges on a perfect dielectric from free space. The electric field intensity is in the positive x direction, has amplitude E_{i1}, and the incident wave propagates so that it makes an angle θ_i to the normal. Assume the magnetic field intensity has components in the positive y and negative z directions and the interface is on the x–y plane. The properties of the dielectric are μ [H/m] and ε [F/m].

(a) Calculate the time-averaged power density in air.
(b) Calculate the time-averaged power density in the dielectric.
(c) What is the most fundamental difference between the two power densities calculated above?

13.24 Oblique Incidence on a Dielectric. A uniform plane wave is incident at an angle on an interface between two perfect dielectrics incoming from dielectric (1). The interface coincides with the y–z plane and the dielectrics have properties $\mu_2 = \mu_1 = \mu_0$ [H/m], $\varepsilon_2 = 3\varepsilon_0$ [F/m] and $\varepsilon_1 = 2\varepsilon_0$ [F/m]. The scalar components of the incident electric field intensity are $E_{ix} = 10$ V/m and $E_{iy} = 5$ V/m:

(a) Find the angle of incidence and the transmission angle.
(b) Identify the polarization of the wave in relation to the given geometry.
(c) Calculate the reflection and transmission coefficients.
(d) From **(a)** and **(c)**, find the scalar components of the reflected and transmitted waves.

13.25 Application: Phase Shift of Transmitted and Reflected Waves. A plane wave at given amplitude and frequency propagates in free space, is polarized perpendicular to the plane of incidence, and impinges on the surface of a high-loss dielectric at an angle θ:

(a) Find the phase shift of the wave at the interface between air and the lossy dielectric; that is, find the phase shift of the transmission coefficient.
(b) Is there also a phase shift in the reflected wave? If so, calculate this phase shift.

13.26 Phase Velocity and its Dependence on Incidence Angle. A plane wave is parallel polarized and impinges at 30° to the normal, on the surface of a perfect dielectric with relative permittivity ε_r and relative permeability μ_r. For a given amplitude and frequency, and assuming the wave propagates in free space before hitting the dielectric:

(a) Determine the phase velocity in the direction parallel to the surface of the dielectric (in which real power propagates) as a function of the angle of incidence.
(b) What is the phase velocity if the incident wave is parallel to the surface of the dielectric?
(c) Compare the results in **(a)** and **(b)** with the phase velocity in free space in the absence of the dielectric.

13.27 Application: Reflection Coefficient and its Dependency on Angle of Incidence. Calculate the reflection coefficient for a planar surface of Teflon versus incidence angle when the electric field intensity remains tangential to the surface and when the electric field intensity has both a normal and a tangential component. Properties of Teflon are $\varepsilon = 2.1\varepsilon_0$ [F/m] and $\mu = \mu_0$ [H/m]. Plot the reflection coefficients for an incidence angle between zero and $\pi/2$.

13.28 Application: Grazing and Normal Incidence: Reflection and Transmission Coefficients. The reflection and transmission coefficients for perpendicular polarization in **Eqs. (13.105)** and **(13.106)** and for parallel polarization in **Eqs. (13.119)** and **(13.120)**, apply for any angle of incidence. Using the general expressions find:

(a) The reflection and transmission coefficients for grazing incidence ($\theta_i = 90°$) for perpendicular and parallel polarization. Explain.

(b) The reflection and transmission coefficients for normal incidence ($\theta_i = 0°$) and show that these are the same for either polarization. Explain.

13.29 Application: Measurement of Thickness of Dielectrics. To measure the thickness of a dielectric material (or if thickness is known its dielectric constant), a collimated wave is sent at an angle θ_1 (**Figure 13.28**). By receiving the reflection from the first surface and from the second surface, the distances d_1 and d_2 can be directly related to the thickness or dielectric constant of the material. Find the relation needed to measure d. Assume parallel polarization, all material properties are known, dielectric and free space are lossless, and the beam is a narrow beam. Given ε_0 [F/m] and μ_0 [H/m] for free space, $\varepsilon_1 = 4\varepsilon_0$ [F/m], $\mu_1 = \mu_0$ [H/m] for the dielectric, d_1 and d_2 are known, and $d_3 = 10$ mm.

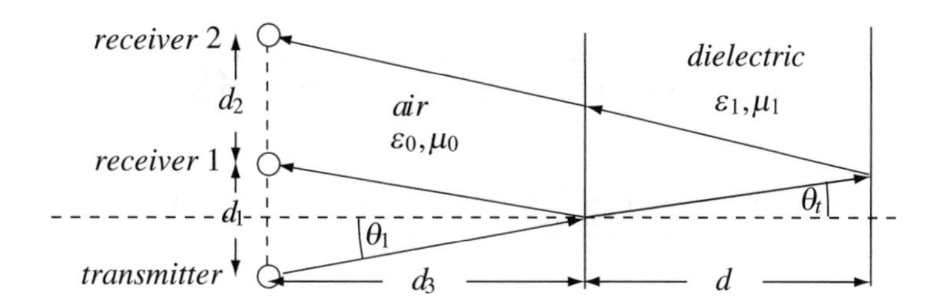

Figure 13.28

Brewster's Angle

13.30 Brewster's Angle in Dielectrics. Calculate the Brewster angle for the following dielectric interfaces for a wave propagating from material (2) into material (1):

(a) Distilled water (1) and air (2): $\varepsilon_1 = 24\varepsilon_0$ [F/m], $\varepsilon_2 = \varepsilon_0$ [F/m], $\mu_2 = \mu_1 = \mu_0$ [H/m].
(b) Plexiglas (1) and air (2): $\varepsilon_{r1} = 4$, $\varepsilon_{r2} = 1$, $\mu_2 = \mu_1 = \mu_0$ [H/m].
(c) Teflon (1) and air (2): $\varepsilon_{r1} = 2.1$, $\varepsilon_{r2} = 1$, $\mu_2 = \mu_1 = \mu_0$ [H/m].

13.31 Calculation of Permittivity from the Brewster Angle. A plane electromagnetic wave is incident on the surface of a dielectric at 62° from air (free space). Calculate the permittivity of the dielectric if at this angle there is no reflection from the surface. Assume parallel polarization of the wave.

Total Reflection

13.32 Critical Angles in Dielectrics. What are the critical angles for the following dielectric interfaces? The wave propagates from material (1) into material (2) and all materials have permeability of free space:

(a) Distilled water (1) and air (2): $\varepsilon_1 = 24\varepsilon_0$ [F/m], $\varepsilon_2 = \varepsilon_0$ [F/m].
(b) Plexiglas (1) and glass (2), $\varepsilon_{r1} = 4.0$, $\varepsilon_{r2} = 1.75$.
(c) Teflon (1) and air (2), $\varepsilon_{r1} = 2.1$, $\varepsilon_{r2} = 1$.

13.33 Application: Use of Critical Angle to Measure Permittivity. A plane electromagnetic wave is incident on the surface of a dielectric at 36° from within the dielectric, at the interface between the dielectric and free space. Calculate the relative permittivity of the dielectric if at this angle there is total reflection from the surface. Assume the dielectric has permeability of free space.

13.34 Critical Angle in Dielectric. A plane wave with parallel polarization is incident on the interface between a perfect dielectric and free space, at 28° from within the dielectric. The dielectric has permeability of free space and relative permittivity $\varepsilon_r = 4$. Calculate:

(a) The reflection and transmission coefficients at the interface.
(b) The critical angle.

13.35 Application: Design of Sheathing for Optical Fibers. An optical fiber is made of glass with a thin plastic coating as shown in **Figure 13.29**. Both materials are transparent at the frequencies of interest. Relative permittivity of glass is 1.85 and the relative permittivity of the plastic coating may be chosen as 1.95 or 1.75. The optical fiber operates in free space:

(a) Which coating is a better choice and why?
(b) Calculate the critical angle for propagation only inside the glass layer, based on your answer in (**a**).
(c) Suppose propagation is also allowed in the coating. What is now the critical angle if the coating selected in (**a**) is used?
(d) What is the critical angle if the second coating (the one not selected in (**a**)) is used?

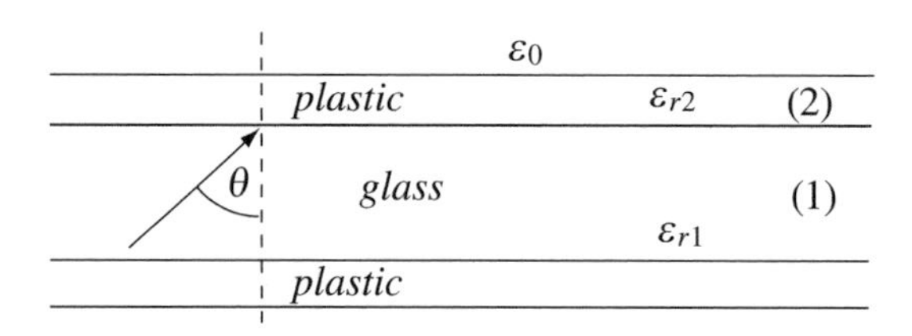

Figure 13.29

Reflection and Transmission for Lossy and Lossless Dielectric Slabs at Normal Incidence

13.36 Application: Propagation Through Lossless Slab. A lossless dielectric layer of thickness d [m] and material constants ε_2 [F/m], μ_2 [H/m], and $\sigma_2 = 0$ is given. The dielectric is in free space (**Figure 13.30**). Assume a plane wave at frequency f impinges on the dielectric (perpendicular to the surface). Calculate the total electric field intensity in the slab (medium (2)), to its left (medium (1)) and to its right (medium (3)) and the slab reflection and transmission coefficients.

13.37 Application: Propagation Through Lossless Dielectric Slab. A dielectric layer of thickness d [m] and material constants ε_2 [F/m], μ_2 [H/m], σ_2 [S/m] is given as shown in **Figure 13.30**. The dielectric is lossless, in free space. Assume a plane wave impinges on the dielectric (perpendicular to the surface) from the left. Given: $\mu_2 = \mu_0$ [H/m], $\varepsilon_2 = 4\varepsilon_0$ [F/m], $\sigma_2 = 0, f = 1$ GHz, $d = 0.01$ m.

(a) Calculate the intrinsic impedances in material (1), (2), and (3).
(b) Assuming the incident electric field intensity in medium (1) is known, find expressions for the electric and magnetic field intensities in each material.
(c) Evaluate the electric and magnetic field intensities in material (1) at the interface, in material (3) at the interface, and in material (2) at $d/2$. Assume the amplitude of the incident electric field intensity is 1 [V/m].

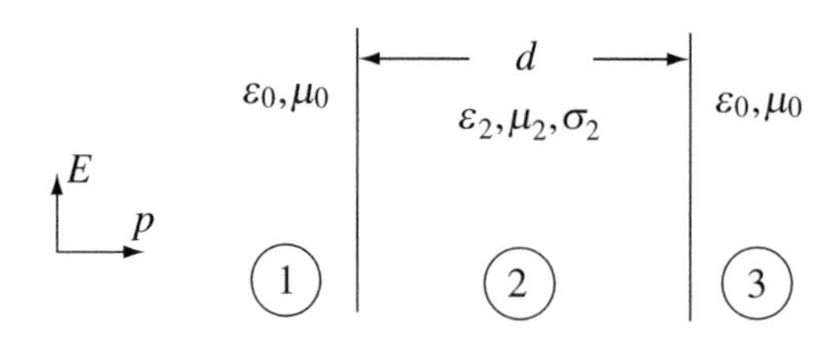

Figure 13.30

13.38 Conditions for Transparency of Dielectrics. Given a dielectric slab in free space, with material properties $\mu = \mu_0$ [H/m], $\varepsilon = 4\varepsilon_0$ [F/m], $\sigma = 0$, what must be the thickness of the slab so that there is no reflection from the material at 1 GHz (i.e., the slab reflection coefficient at the surface is zero)? Is this at all possible with the material properties given?

13.39 Application: Design of Radomes. A radome is placed in front of a radar antenna to protect the antenna from the elements. Material properties of the radome are known and the radome is a perfect dielectric. Calculate the minimum thickness of the radome so that it is transparent for waves propagating from the antenna and to the antenna. The radome properties are $\mu = \mu_0$ [H/m], $\varepsilon = 4\varepsilon_0$ [F/m], $\sigma = 0, f = 10$ GHz. Use properties of free space for air.

13.40 Application: Design of a Dielectric Window. In a microwave oven it is necessary to place a transparent dielectric window made of a quartz sheet between the magnetron (microwave power generator) and the cavity of the oven, so that the magnetron operates under vacuum. If the oven operates at 2.45 GHz, what must be the thickness of the quartz sheet? Use $\mu = \mu_0$ [H/m], $\sigma = 0$ and $\varepsilon = 3.8\varepsilon_0$ [F/m] for quartz and μ_0, ε_0 everywhere else.

13.41 Propagation Through a Lossy Dielectric Slab. Solve **Problem 13.37** but now the dielectric also has a conductivity $\sigma_2 = 0.001$ S/m.

13.42 Propagation Through a Two-Layer Slab. A two-layer dielectric slab in free space is given as shown in **Figure 13.31**. An incident wave propagates from material (1) and impinges on the first dielectric interface at $z = 0$. The wave has an electric field intensity of magnitude $E_0 = 1$ V/m directed in the x direction and is at frequency 150 MHz. Hint: Set up the general equations in each section of space and match the fields at the interfaces. Set up a system of equations based on these relations to solve for the forward and backward components in each section and solve for the numerical values of the fields. Find:

(a) The electric and magnetic field intensities everywhere.
(b) The reflection and transmission coefficient of the composite slab.

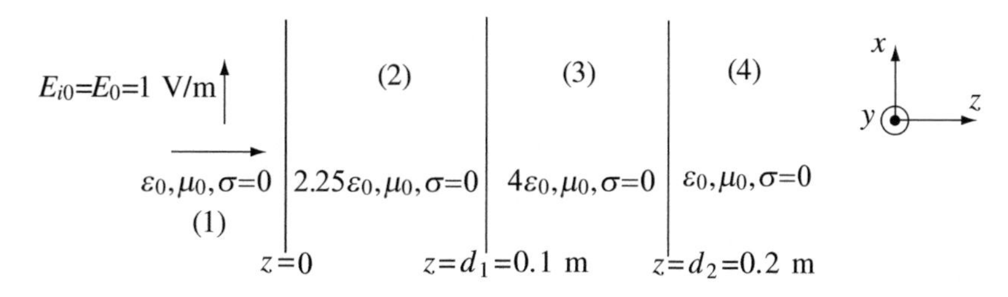

Figure 13.31

Reflection and Transmission for a Dielectric Slab Backed by a Perfect Conductor: Normal Incidence

13.43 Reflection from a Conductor-Backed Slab. A perfect dielectric of permittivity ε_2 [F/m], permeability μ_2 [H/m], and thickness d [m] is backed by a perfect conductor as shown in **Figure 13.32**. A plane wave is incident from the left as shown:

(a) Assuming that the incident wave is known, calculate the reflection coefficient at the interface between free space and dielectric.
(b) Calculate the required thickness d for the reflection to be maximum.
(c) Show that the conductor backed slab cannot be made transparent for any thickness d.

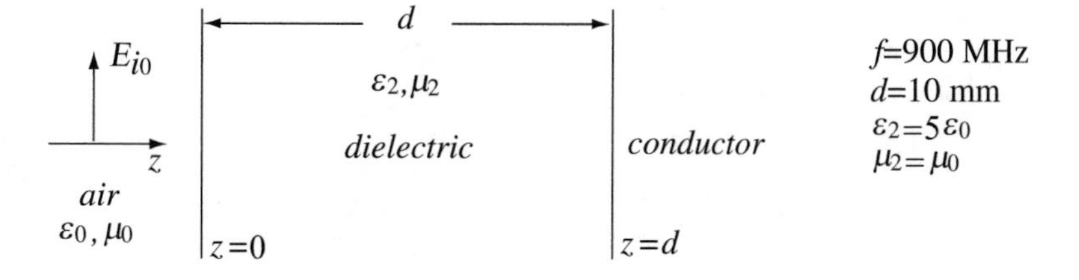

Figure 13.32

Theory of Transmission Lines

14

O tell me, when along the line
From my full heart the message flows,
What currents are induced in thine?
One click from thee will end my woes.

—James C. Maxwell (1831–1879),
mathematician, physicist
Valentine from a telegraph clerk to a telegraph clerk,

14.1 Introduction

Hopefully, by now you have a good understanding of waves propagating in space and in materials, including reflection and transmission at interfaces. Although not mentioned often enough, there were a number of assumptions implicit in this type of propagation. The most important was the fact that only plane waves were treated. In most cases, we also assumed the waves only propagate forward from the source, although reflections from interfaces cause waves to also propagate backward toward the source and these were treated in **Chapter 13**. Whereas the existence of interfaces complicates treatment, it also allows for applications such as radar to be feasible. If we were to summarize the previous two chapters in a few words, we would say that all wave phenomena were treated in essentially infinite space; that is, plane waves were not restricted in space except for the occasional interface.

There are, however, many applications in which this type of propagation is either impractical, not feasible, or inefficient. For example, consider the following situation: A spacecraft is flying at a distance of two million kilometers from Earth toward a distant planet. How can we communicate with the spacecraft? It makes no sense to use plane waves for this purpose even if true plane waves could be generated. A more practical approach would be a narrow beam, perhaps not much larger than the spacecraft, tracking the vehicle. Doing so reduces the power requirements and minimizes interference with other systems.

In Earth-bound systems, there is a third approach: connect a pair of conductors between two points and transmit the information over the two conductors. We call a connection, of this sort a ***transmission line***. Although at first glance this approach seems like a very simple circuit, it is far from it. We have not yet defined the properties of the transmission line proposed here, but the following simple analysis of the line, an analysis that does not require knowledge of line properties, should point out the special properties of this type of line.

Consider the following example: A power transmission line connects a power station with a load at a very large distance. Neglect losses on the line. There is an additional assumption implicit in treating this problem as a simple circuit, that of instantaneous propagation. In other words, we assume that any change in the load appears instantaneously at the generator regardless of the length of the line. If we were to short the load, the generator will see a short circuit at the same instant.

We know that this is not true; all propagation of energy takes time. Even in DC circuits, we often take into account time constants because of the capacitive and inductive terms in the circuits which influence the transient characteristics of the circuits. What happens if we look at a wave propagating on the same line? For a power line, the frequency is

N. Ida, *Engineering Electromagnetics*, https://doi.org/10.1007/978-3-030-15557-5_14

60 Hz. Therefore, assuming propagation at the speed of light (propagation on lines is much slower than this in most cases), the wavelength of the wave is 5,000 km. Imagine now that you could measure the voltage on the line at any location and at the generator and compare the two at any given time. The two voltages will be different. For example, using **Figure 14.1**, the voltage at a distance of 1,250 km (a quarter wavelength) is zero when the generator is at its peak. One-quarter of a cycle later (1/240 s), the generator voltage goes to zero and the voltage at point A is now maximum. There is clearly a delay of 1/240 s between the two locations—this is the time required for the voltage wave to propagate from the generator to point A.

Figure 14.1 A very long power transmission line and conditions on the line at different times and locations

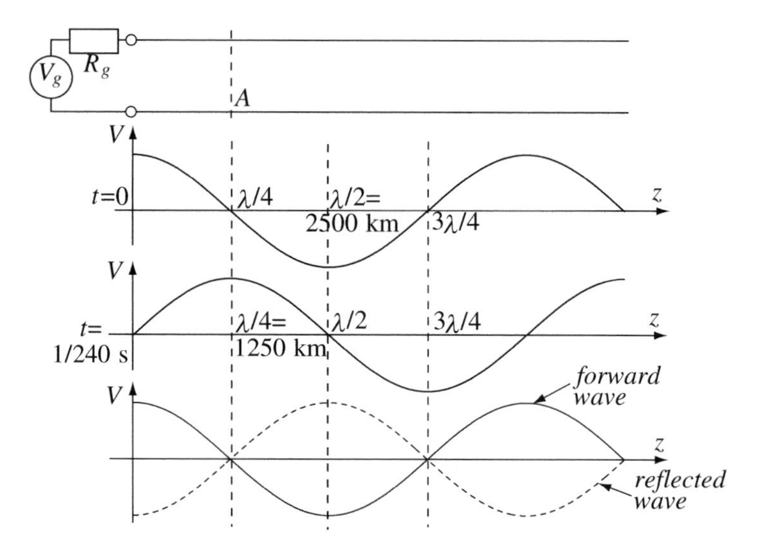

Now suppose we short the line at some point. In circuit theory, this is a disaster: the current will immediately rise to dangerous levels and destroy the circuit unless proper protection (such as a fuse) is available. However, for the above finite-speed line, suppose we short the circuit at exactly one-quarter wavelength. A short means that the energy propagated on the line cannot reach the load. Since the line is ideal, it cannot dissipate energy either. Thus, the energy must propagate back on the line toward the generator. The short is a disturbance, and by the time it reaches back to the generator, another quarter cycle has passed and now the generator is at negative maximum voltage. The voltage of the disturbance propagating back from the load is at positive maximum voltage. The total voltage at the generator is now zero. This does not look like such a disastrous event. Certainly, the generator will not be destroyed. Of course, if the short occurs at any other location, the result would be different. Also, we only looked at the short at a given instant in time. However, the point here is to show that the behavior of a transmission line is different than a circuit, because the assumptions we use are different.

Although this "analysis" leaves out more than it includes and certainly does not take into account all effects on the line; it indicates that propagation on lines is not the same as flow of power in simple circuits.

Consider another example with less "bang" to it but with similarly important consequences. Suppose you designed a circuit consisting of two sensors connected to an AND gate. The two sensors produce pulses (a) and (b) as shown in **Figure 14.2b**. A normal circuit theory approach would give a "1" if both inputs are "1." The result in **Figure 14.2b** (c) is expected. Now, assume the inputs of the same gate are connected to the same sensors but one is a distance a and the second a distance b from the inputs as shown in **Figure 14.2a**. Because the speed of propagation v_p of the pulses on lines is finite, the pulse on line A reaches the gate after a time a/v_p. The second reaches it at a time b/v_p. Thus, if the pulses are narrow, the two pulses reach the gate at different times and the output is "wrong" **[Figure 14.2b (d)]**. The longer line has a longer delay. The design is correct as far as circuits are concerned but may not operate properly because of delays on the lines. For these circuits to operate properly, both lines must be of the same length, such as strips of the same length on a printed circuit board, or the pulse on the shorter line must be delayed so that the two pulses reach the gate at the same time. This aspect of propagation is extremely important in high-speed computers. As the speed of computation increases, the limitation of propagation on physical lines, even within a board or a single chip, becomes more important and, in the end, imposes upper limits on computation speed.

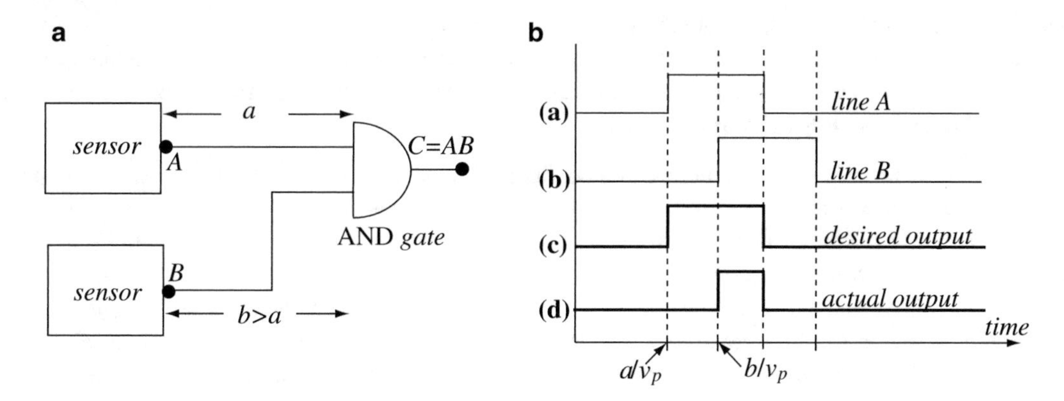

Figure 14.2 (**a**) A logic circuit with two input lines of different lengths. (**b**) Desired and actual output for a set of inputs

The two examples given suggest that transmission lines are not simple circuits, and for proper design, simple properties such as finite speed of propagation must be taken into account. The approach here is to view all lines as transmission lines, define their properties, and then see what the connection between transmission lines and circuits is. We will find that circuits are essentially transmission lines in which the distances are so short as to allow us to neglect the finite speed of propagation at the operating frequency. This, however, can only be correct at relatively low frequencies. In the case of the above power transmission line, a few kilometers is a very short distance because the wavelength is 5,000 km. On the other hand, at 10 GHz, the wavelength is only 3 cm. Any circuit connection longer than a few centimeters will be a "long" connection and propagation effects cannot be neglected.

Thus, you may view the theory of transmission lines as a more general approach to treatment of transfer of energy on lines. Many of the methods used here will be familiar from circuit theory, and in most cases, the results will be in terms of voltages and currents on the line. The connection between voltages and currents and electric and magnetic fields also affords a different point of view of the field variables.

Example 14.1 Two memory boards are connected to the processor of a computer. Suppose we wish to add two values stored in the two memory banks. One memory bank is 0.1 m from the processor and the second is 0.15 m away. The processor can add the two values in 1 ns (a 1 GHz computer). Propagation on the transmission lines connecting the processor to memory banks is at $0.2c$ [m/s] (typical of copper lines). What is the minimum time needed for computation?

Solution: The total time of computation is the time needed for propagation (rounded to the nearest cycle since all computation is done in cycles) plus the time needed by the processor. In this case, computation can only start after the signal on the longer line reaches the processor. A delay of $0.05/0.2c$ must also be introduced on the shorter line so that both signals from memory reach the processor at the same time.

The delay due to propagation is

$$\Delta t = 0.15/0.2c = 0.15/\left(0.2 \times 3 \times 10^8\right) = 2.5 \times 10^{-9} \quad [\text{s}]$$

Thus, the time required for computation is four cycles or 4 ns instead of 1 ns. In effect, the computer has been slowed down by a factor of 4. There are many ways of dealing with this problem, including shorter lines, cache memory on the processor chip, propagation while the processor performs other tasks, and other methods of scheduling.

14.2 The Transmission Line

What then is a transmission line? Well, it is no more than a physical connection between two locations through two conductors. We must indicate at the outset that any transmission of energy through conducting or nonconducting media may be considered a transmission line. Also, any guiding of energy by physical structures may be included in this general definition. However, we will restrict our discussion here to conducting lines with the following properties:

(1) The transmission line is made of two conductors in any configuration.
(2) The electric and magnetic field intensities on the line are perpendicular to each other and perpendicular to the direction of propagation of power. This type of propagation was defined in **Chapter 12** as transverse electromagnetic (TEM) propagation and has all the properties of plane waves.

Examples of lines that we may consider are parallel conducting wires such as the two-wire power cable used to power your toaster or the overhead power transmission line made of thick cables and suspended from towers. Similarly, a twisted pair of wires as used in some telephone lines is of this type. These three transmission lines are shown in **Figures 14.3a–14.3c**. Another common type of transmission line is the coaxial transmission line shown in **Figure 14.3d** (also discussed in **Chapter 9**). It is made of two coaxial conductors: an inner, thin, solid conductor and an outer hollow cylindrical conductor. The latter is usually stranded to allow flexibility and the two conductors are insulated with some dielectric material. Dimensions of coaxial cables and their properties vary, but good examples of often used coaxial cables are antenna cables for televisions, input cables for oscilloscopes, or input leads in audio equipment. A third type of transmission line which we will concern ourselves with is the parallel strip line shown in **Figure 14.3e**. This line may be made of two strips, very close to each other, such as strips on printed circuit boards or of two parallel plates.

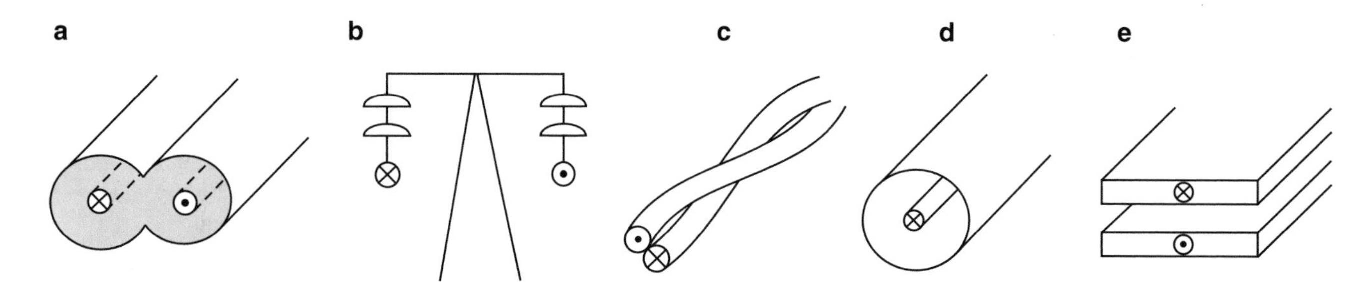

Figure 14.3 (**a**) Simple two-lead cable. (**b**) Overhead power line. (**c**) Twisted pair. (**d**) The coaxial transmission line. (**e**) Parallel plate transmission line (strip line)

Although each line has its own properties and parameters, our discussion will be general and will encompass all lines that satisfy the above requirements. In doing so, we first discuss infinite lines, followed by finite, load terminated lines. The lossless (ideal) line is discussed first since it is the simplest, followed by lossy or attenuating lines. In terms of sources connected to the line, we start with steady state AC sources but will also discuss transients and the effect of line parameters on propagation of these transients.

14.3 Transmission Line Parameters

A transmission line has three types of parameters:

(1) Dimensional parameters: These include length, dimensions of each conductor (thickness, width, diameter, etc.), spacing between lines, thickness of insulation, and the like. These parameters define the physical configuration of the line but also play a role in defining its electrical properties.
(2) Material parameters: The line is made of conductors and insulators. The electrical properties of these materials are their conductivities, permittivities, and permeabilities. These obviously affect the way a line performs its task.
(3) Electric or circuit parameters: These are the resistance, capacitance, inductance, and conductance per unit length of the line. Although we could calculate these parameters for the whole line (lumped parameters), we will have little use for lumped parameters. The reason for this was hinted at in the introduction: The voltage and current vary along the line, making the use of lumped parameters useless. Instead, we will use distributed parameters. The four line parameters are:

R: Series resistance of the line in ohms per unit length [Ω/m].
L: Series inductance of the line in henrys per unit length [H/m].
C: Shunt capacitance of the line in farads per unit length [F/m].
G: Shunt conductance of the line in siemens per unit length [S/m].

Before we discuss the properties of transmission lines, it is important to be able to define the various line parameters. These are evaluated from known electromagnetic relations and we, in fact, have performed these tasks in previous chapters. However, we will repeat the steps involved in these calculations here to review the principles involved. To do so, we consider as an example, the parallel plate transmission line in **Figure 14.3e**. The line is very long but we will evaluate the parameters for a length $l = 1$ m. The procedure given here is rather general and applies to many transmission lines although the details of evaluation of the expressions for different lines vary.

14.3.1 Calculation of Line Parameters

14.3.1.1 Resistance per Unit Length

Any transmission line, made of conducting materials, has a finite resistance because of the finite conductivity of the material. Because the current in the transmission line is time dependent, and especially at high frequencies, the current is dominated by the skin depth as was discussed in **Chapter 12**. This is shown schematically in **Figure 14.4a**. Only a small depth of the conductor contains current (at a depth of 5 skin depths, the current density is less than 0.7% the current density at the surface). At the frequencies normally used in transmission lines and in good conductors, this depth is often only a few micrometers. Therefore, we will call this current a surface current and the current density it produces, a surface current density. The resistance we need to evaluate is, therefore, a surface resistance in the sense that only a small volume close to the surface contributes to this resistance. This resistance is the series resistance we need. To see how the series resistance R_s can be calculated for the parallel plate transmission line in **Figure 14.3e**, we use **Figure 14.4b**. The current in the lower conductor flows in the positive z direction. We will also assume initially that the thickness t tends to infinity, calculate the surface resistance per unit length of the lower plate, and then multiply the value of this resistance by 2 to obtain the total series resistance per unit length of the line.

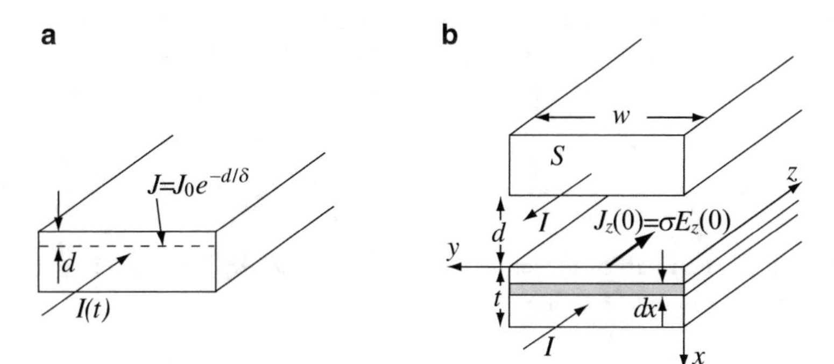

Figure 14.4 (**a**) The AC distribution of current density in a transmission line decays exponentially from the surface. (**b**) Calculation of the series resistance in a parallel plate transmission line based on the AC current distribution

For a current density to exist in the conductor, there must be an electric field intensity E_z inside the conductor (in the direction of flow of current) for any conductor except a perfect conductor. The electric field intensity inside the conductor decays exponentially with depth; that is, a wave propagating into the conductor produces an electric field:

$$E_z(x) = E_z(0)e^{-\alpha x}e^{-j\beta x} = E_z(0)e^{-x/\delta}e^{-jx/\delta} = E_z(0)e^{-(1+j)x/\delta} \quad [\text{V/m}] \tag{14.1}$$

where $E_z(0)$ is the electric field intensity at the surface ($x = 0$) of the conductor. The attenuation and phase constants for the given conductor are α and β, respectively, and the skin depth is δ. For a good conductor, the constants α, β, and δ are [see **Eqs. (12.111)** and **(12.113)**]

$$\alpha = \sqrt{\pi f \mu_c \sigma_c} \ [\text{Np/m}], \quad \beta = \sqrt{\pi f \mu_c \sigma_c} \ [\text{rad/m}], \quad \delta = \frac{1}{\alpha} = \frac{1}{\sqrt{\pi f \mu_c \sigma_c}} \ [\text{m}] \tag{14.2}$$

In these relations, σ_c and μ_c indicate conductivity and permeability, respectively, of the conducting material to distinguish them from the material properties of the dielectric between the conductors.

The current density in the conductor is

$$J_z(x) = \sigma_c E_z(x) = \sigma_c E_z(0)e^{-(1+j)x/\delta} \quad \left[\text{A/m}^2\right] \tag{14.3}$$

To calculate the total current in the lower conductor, we note that the current density only varies with depth. Thus, an element of current $dI = J(x)wdx$ **(Figure 14.4b)** is defined. This is now integrated over the thickness of the conductor, which we take to be infinitely thick:

$$I = w\int_{x=0}^{\infty} J_z(x)dx = w\sigma_c E_z(0)\int_{x=0}^{\infty} e^{-(1+j)x/\delta}dx = \frac{w\sigma_c \delta E_z(0)}{1+j} \quad [\text{A}] \tag{14.4}$$

Note that the thickness t can be taken to be infinite since the current density after about 10 skin depths is so small as to contribute almost nothing to the total current. This approximation is permissible for any conductor which is thick compared to the skin depth.

The impedance of a section of length $l = 1$ m (the line is directed in the z direction) of the conductor is, by definition,

$$Z = \frac{V}{I} = \frac{1}{I}\int_{z=0}^{l} \mathbf{E}\cdot d\mathbf{l} = \int_{z=0}^{l}\left(\frac{E_z}{I}\right)dz \quad [\Omega] \tag{14.5}$$

where E_z is the electric field intensity in the direction of the current. The impedance per unit length is the ratio between the tangential component of the electric field intensity [V/m] at the surface of the conductor and the surface current [A], shown in parentheses in **Eq. (14.5)**:

$$Z = \frac{E_z(0)}{I} = \frac{E_z(0)(1+j)}{w\delta\sigma_c E_z(0)} = \frac{(1+j)}{w\delta\sigma_c} = \frac{1}{w}\left(\frac{1}{\delta\sigma_c} + \frac{j}{\delta\sigma_c}\right) \quad \left[\frac{\Omega}{\text{m}}\right] \tag{14.6}$$

where I from **Eq. (14.4)** was used. The expression in parentheses contains only quantities related to material properties, which are independent of dimensions and have units of [Ω]. The real part of this relation is the surface resistance of the conductor and is independent of dimensions—it is a property of the conductor at the given frequency:

$$\boxed{R_s = \frac{1}{\sigma_c \delta} \quad [\Omega]} \tag{14.7}$$

Multiplying the real part of the impedance in **Eq. (14.6)** by 2 to take into account the upper conductor, we obtain the resistance per unit length of the line:

$$\boxed{R = \frac{2}{w\sigma_c\delta} = \frac{2}{w}\sqrt{\frac{\pi f \mu_c}{\sigma_c}} \quad \left[\frac{\Omega}{\text{m}}\right]} \tag{14.8}$$

The imaginary part of the surface impedance is due to the inductive nature of the conductor. We can write the surface inductance L_s and the inductance per unit length L as

$$L_s = \frac{1}{\sigma_c\delta\omega} \quad [\text{H}], \quad X = \frac{j}{w\sigma_c\delta} = j\omega L \quad \rightarrow \quad L = \frac{1}{w\sigma_c\delta\omega} \quad \left[\frac{\text{H}}{\text{m}}\right] \tag{14.9}$$

The inductance per unit length L obtained here is an internal series inductance per unit length of the lower conductor and *should not* be confused with the inductance per unit length of the transmission line. The latter is the external inductance which we will calculate shortly. The internal surface inductance is quite small, especially at very high frequencies and in good conductors. For this reason, we normally neglect this term in the analysis of transmission lines.

14.3.1.2 Inductance per Unit Length

The inductance per unit length of any transmission line can be obtained by calculating the magnetic flux density due to an assumed current in the line, calculating the total flux linkage with the line per unit length, and then dividing by the current to obtain the inductance. This method was described in detail in **Section 9.4**. To calculate the inductance per unit length, we use again the geometry in **Figure 14.4b**. First, we calculate the magnetic flux density from Ampère's law. Because $w \gg d$, the magnetic flux density between the plates can be assumed to be uniform and parallel to the plates (**Figure 14.5a**). A contour is drawn around one of the conductors, as shown in **Figure 14.5a**. The magnetic field intensity outside the plates is zero since a contour enclosing both conductors encloses a zero net current. Thus, only the path section contained between the conductors (a – d) contributes to the flux density. From Ampere's law,

$$I = \oint \mathbf{H} \cdot d\mathbf{l} = Hw \quad \rightarrow \quad H = \frac{I}{w} \quad \rightarrow \quad B = \frac{\mu I}{w} \quad [\mathrm{T}] \tag{14.10}$$

where $\mathbf{B} = \mu\mathbf{H}$ was used and μ is the permeability of the dielectric material between the conducting plates.

Now, we need to calculate the total flux linkage. Since there is only one closed circuit (out of which we only calculate the flux for a 1 m section), the total flux contained between the two conductors is the flux linkage of the segment. This flux is calculated by integrating the flux density over the shaded area in **Figure 14.5b**, which shows the transmission line from a side view. Because the flux density is uniform, this gives

$$\Phi = \Lambda = BS = B(d \times 1) = \frac{\mu I d}{w} \quad [\mathrm{Wb}] \tag{14.11}$$

Dividing by the current I gives the external inductance per unit length of the transmission line:

$$\boxed{L = \frac{\Lambda}{I} = \frac{\mu d}{w} \quad \left[\frac{\mathrm{H}}{\mathrm{m}}\right]} \tag{14.12}$$

In general, we assume that the internal inductance given in **Eq. (14.9)** is small compared with the external inductance in **Eq. (14.12)** and, therefore, the internal inductance is usually neglected.

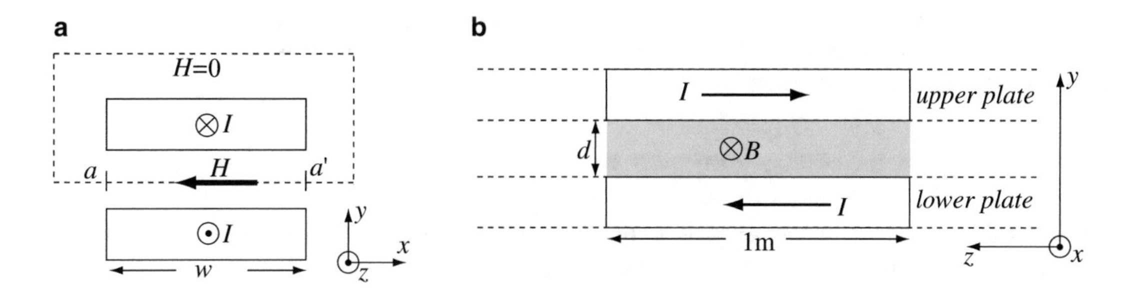

Figure 14.5 Calculation of inductance per unit length of the parallel plate transmission line. (**a**) Calculation of magnetic field intensity between the plates. (**b**) Calculation of flux per unit length

14.3.1.3 Capacitance per Unit Length

Capacitance of any system of two conductors is calculated by assuming a given charge or charge density on one conductor, equal and opposite charge on the second conductor, and then calculating the potential difference between the two conductors. From the calculated potential difference and charge, the capacitance is calculated as $C = Q/V$. This method of computation was discussed in detail in **Section 4.7.2**. We assume a total charge Q is uniformly distributed on the inner surface of the upper conductor and a total charge $-Q$ is uniformly distributed on the inner surface of the lower conductor. This forms a capacitor, with two plates, each of length 1 m and width w (**Figure 14.6**). Thus, the surface charge densities are $Q/(\omega \times 1)$ [C/m^2] on the upper conductor and $-Q/(\omega \times 1)$ [C/m^2] on the lower conductor. Assuming no fringing ($w \gg d$), the use of Gauss's law gives

$$E = \frac{Q}{w\varepsilon} \quad \left[\frac{\mathrm{V}}{\mathrm{m}}\right] \tag{14.13}$$

where the Gaussian surface is shown in **Figure 14.6** and ε is the permittivity of the material between the conducting plates. The charge only exists on the inner surfaces of the conductors and we have also taken into account the fact that the electric field intensity is zero outside the plates. The potential difference between the plates is

$$|V| = \int_0^d E dl = \frac{Qd}{w\varepsilon} \quad [\mathrm{V}] \tag{14.14}$$

The capacitance per unit length is therefore

$$\boxed{C = \frac{Q}{V} = \frac{w\varepsilon}{d} \quad \left[\frac{\mathrm{F}}{\mathrm{m}}\right]} \tag{14.15}$$

The same result may be obtained from the formula for parallel plate capacitors, but the method given here is general.

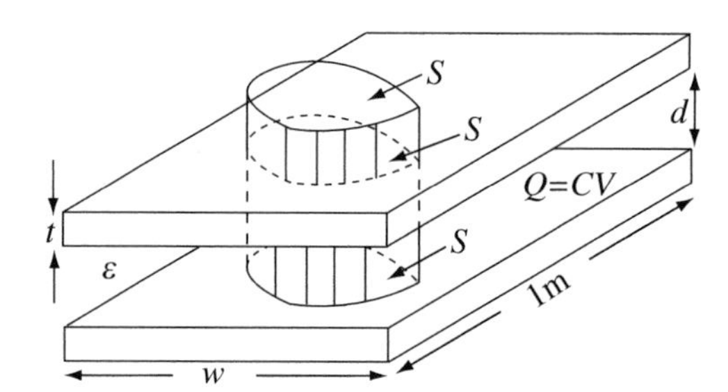

Figure 14.6 Calculation of the electric field intensity between the plates of the transmission line using Gauss's law with an assumed surface charge density on the upper plate equal to Q/w [C/m^2]

14.3.1.4 Conductance per Unit Length

Conductance is calculated in a manner similar to that for calculation of resistance in **Chapter 7**. We assume the material between the plates has a uniform conductivity σ and apply a known, arbitrary potential difference between the plates which generates a uniform electric field intensity $E = V/d$ between the plates. This gives rise to a current density $\mathbf{J} = \sigma\mathbf{E}$. The total current is then calculated by integrating $\mathbf{J}$ over the area of the plates ($w \times 1$). From Ohm's law, we can now calculate the resistance and its reciprocal is the conductance G. To outline the method, consider **Figure 14.7**. The potential V produces an electric field intensity and a current density:

$$E = \frac{V}{d} \quad and \quad J = \sigma E \quad \rightarrow \quad J = \sigma\frac{V}{d} \quad \left[\frac{\mathrm{A}}{\mathrm{m}^2}\right] \tag{14.16}$$

This current density flows from the upper plate to the lower plate, and is uniform between the plates. Thus, the total current is this current density multiplied by the area of the plate:

$$I = JS = \sigma\frac{V}{d}w \quad [\mathrm{A}] \tag{14.17}$$

Assuming the unknown resistance between the plates is R, we can write from Ohm's law

$$V = IR = R\sigma\frac{V}{d}w \quad \rightarrow \quad R = \frac{d}{w\sigma} \quad [\Omega] \tag{14.18}$$

The conductance per unit length is therefore

$$G = \frac{1}{R} = \frac{w\sigma}{d} \quad \left[\frac{\mathrm{S}}{\mathrm{m}}\right] \tag{14.19}$$

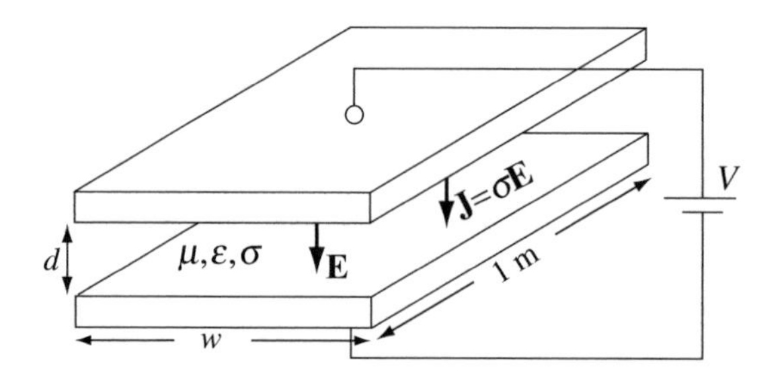

Figure 14.7 Calculation of conductance per unit length by calculating the resistance between upper and lower plates

The methods outlined in **Sections 14.3.1.1** through **14.3.1.4** are completely general and apply equally well to other types of transmission lines. **Table 14.1** shows the line parameters for the coaxial transmission line, the two-wire transmission line and, the parallel plate transmission line evaluated above.

Table 14.1 Transmission line parameters for three common transmission lines

Two-wire line (**Figure 14.3a**).	Coaxial line (**Figure 14.3d**).	Parallel plate line (**Figure 14.3e**).
a = radius of conductor, d = distance between centers of conductors.	a = radius of inner conductor, b = inner radius of outer conductor.	w = width of plates, d = distance between plates.
$R = \dfrac{1}{\pi a \delta \sigma_c} \quad \left[\dfrac{\Omega}{\text{m}}\right]$	$R = \dfrac{1}{2\pi \delta \sigma_c}\left[\dfrac{1}{a}+\dfrac{1}{b}\right] \quad \left[\dfrac{\Omega}{\text{m}}\right]$	$R = \dfrac{2}{w \delta \sigma_c} \quad \left[\dfrac{\Omega}{\text{m}}\right]$
$L = \dfrac{\mu}{\pi}\cosh^{-1}\dfrac{d}{2a} \quad \left[\dfrac{\text{H}}{\text{m}}\right]$	$L = \dfrac{\mu}{2\pi}\ln\dfrac{b}{a} \quad \left[\dfrac{\text{H}}{\text{m}}\right]$	$L = \dfrac{\mu d}{w} \quad \left[\dfrac{\text{H}}{\text{m}}\right]$
$G = \dfrac{\pi \sigma}{\cosh^{-1}(d/2a)} \quad \left[\dfrac{\text{S}}{\text{m}}\right]$	$G = \dfrac{2\pi \sigma}{\ln(b/a)} \quad \left[\dfrac{\text{S}}{\text{m}}\right]$	$G = \dfrac{\sigma w}{d} \quad \left[\dfrac{\text{S}}{\text{m}}\right]$
$C = \dfrac{\pi \varepsilon}{\cosh^{-1}(d/2a)} \quad \left[\dfrac{\text{F}}{\text{m}}\right]$	$C = \dfrac{2\pi \varepsilon}{\ln(b/a)} \quad \left[\dfrac{\text{F}}{\text{m}}\right]$	$C = \dfrac{w \varepsilon}{d} \quad \left[\dfrac{\text{F}}{\text{m}}\right]$

Note: If $(d/2a)^2 \gg 1$, $\cosh^{-1}(d/2a) \approx \ln(d/a)$. For widely separated, two-wire, thin lines, this approximation can be used to simplify the expressions. σ_c and μ_c are the conductivity and permeability of the conductor, respectively. σ, μ, and ε are the properties of the dielectric between the conductors.

In summary, the transmission line parameters are evaluated as any other lumped circuit parameters for a line of unit length.

Example 14.2 A two-wire transmission line is made of two bare, round wires and operates at 400 Hz. The conductors are made of copper and placed in air. The two wires need to pass through a very thick wall made of alumina and into a hot oven to connect to a temperature sensor. Because copper does not withstand high temperatures very well, the section inside the wall is made of tungsten but with identical dimensions to the outside wire. The geometry is shown in **Figure 14.8**. Calculate the line parameters for the segment inside the wall and the line outside the wall. Assume properties of air are the same as of vacuum.

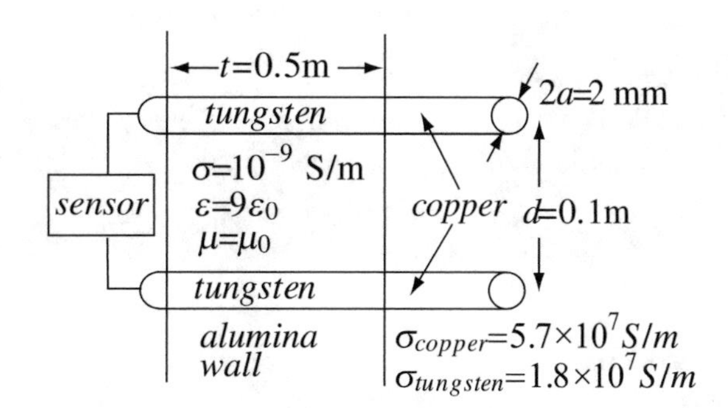

Figure 14.8 Two transmission line segments with different properties, connected in series

Calculate the line parameters for the segment inside the wall and the line outside the wall.

Solution: The line parameters of both sections are given in column 1 of **Table 14.1**, but they have different values.

(1) Copper line in air:

$$R = \frac{1}{\pi a \delta \sigma_c} = \frac{\sqrt{\pi f \mu_0 \sigma_c}}{\pi a \sigma_c} = \frac{1}{a}\sqrt{\frac{f \mu_0}{\pi \sigma_c}} = \frac{1}{0.001}\sqrt{\frac{400 \times 4 \times \pi \times 10^{-7}}{\pi \times 5.7 \times 10^7}} = 1.675\pi \times 10^{-3} \quad [\Omega/\text{m}]$$

Where $\delta = 1/\sqrt{\pi f \mu_0 \sigma_c}$, μ_0 is the permeability and σ_c is the conductivity of copper.

$$L = \frac{\mu}{\pi}\cosh^{-1}\frac{d}{2a} \approx \frac{\mu_0}{\pi}\ln\frac{d}{a} = \frac{4\times\pi\times10^{-7}}{\pi}\ln\frac{0.1}{0.001} = 1.842 \quad [\mu\text{H/m}]$$

where the approximation $\cosh^{-1}(d/2a) \approx \ln(d/a)$ was used (since $(d/2a)^2 = 50^2 \gg 1$). The same approximation is used for the calculation of capacitance and conductance per unit length.

$$G = \frac{\pi\sigma}{\cosh^{-1}(d/2a)} \approx \frac{\pi\sigma}{\ln(d/a)} = 0$$

The conductance in air is zero because conductivity of air is zero:

$$C = \frac{\pi\varepsilon}{\cosh^{-1}(d/2a)} \approx \frac{\pi\varepsilon_0}{\ln(d/a)} = \frac{\pi\times8.854\times10^{-12}}{\ln 100} = 6.04 \quad [\text{pF/m}]$$

(2) Tungsten line in alumina

$$R = \frac{1}{a}\sqrt{\frac{f\mu_0}{\pi\sigma_t}} = \frac{1}{0.001}\sqrt{\frac{400\times4\times\pi\times10^{-7}}{\pi\times1.8\times10^7}} = 2.98\times10^{-3} \quad [\Omega/\text{m}]$$

$$L \approx \frac{\mu_0}{\pi}\ln\frac{d}{a} = 1.842 \quad [\mu\text{H/m}]$$

$$G \approx \frac{\pi\sigma_a}{\ln(d/a)} = \frac{\pi\times10^{-9}}{\ln 100} = 6.822\times10^{-10} \quad [\text{S/m}]$$

$$C \approx \frac{\pi 9\varepsilon_0}{\ln(d/a)} = \frac{\pi\times9\times8.854\times10^{-12}}{\ln 100} = 54.36 \quad [\text{pF/m}]$$

Thus, while the inductance per unit length remains the same (because permeability of alumina and that of air are the same), all other parameters are different. We, therefore, expect these two line segments to have different properties, including different speeds of propagation.

Exercise 14.1

(a) Calculate the line parameters for a coaxial line with inner radius $a = 1$ mm, outer radius $b = 4$ mm, and material parameters as in **Figure 14.9**. The line operates at 60 Hz.
(b) Does it matter how thick the outer conductor is?

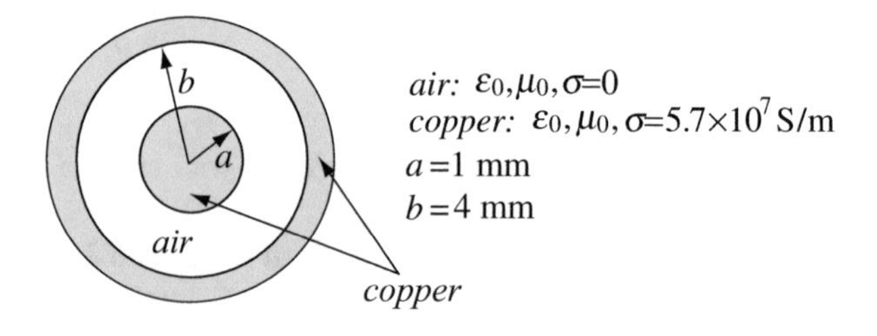

Figure 14.9 Cross section and properties of a coaxial transmission line

Answer **(a)** $R = 0.41$ mΩ/m, $L = 0.277$ μH/m, $G = 0$, $C = 40.1$ pF/m. **(b)** Yes, Calculation of resistance per unit length assumes conductors are thick compared to the skin depth.

14.4 The Transmission Line Equations

As discussed above, the lumped parameter approach to transmission lines is not feasible. Instead, we define the transmission line equations using a distributed parameter approach. The transmission line is viewed as being made of a large number of short segments, each of length Δl as shown in **Figure 14.10** which also shows the parameters of one segment. In this notation, $R\Delta l$ is the resistance of the line of length Δl, $L\Delta l$ is the inductance, $C\Delta l$ is the capacitance, and $G\Delta l$ is the conductance, where R, L, C, and G are given per unit length. The total series impedance of the line segment is therefore

$$Z = R\Delta l + j\omega L\Delta l \quad [\Omega] \tag{14.20}$$

and the parallel line admittance is

$$Y = G\Delta l + j\omega C\Delta l \quad [1/\Omega] \tag{14.21}$$

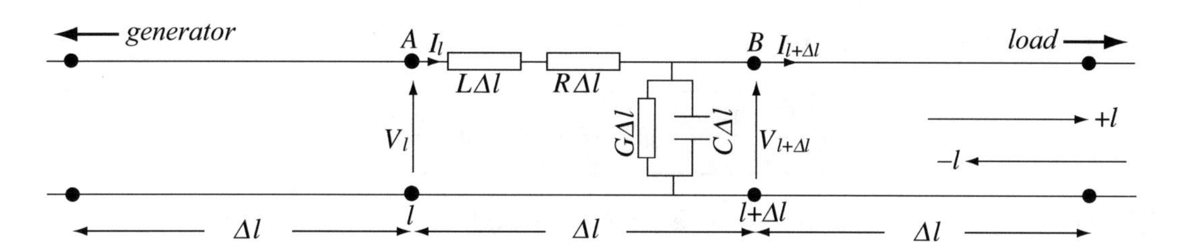

Figure 14.10 A transmission line viewed as a distributed parameter circuit built of segments of arbitrary but small length Δl. One segment is shown in detail. Note the general direction l. Later, we will replace this with a specific coordinate

These parameters can now be used to build a transmission line of any length, as shown in **Figure 14.10**. The Δ notation was used to indicate that the segment of line used is arbitrary but must be small compared to wavelength. The circuit equations are written using Kirchhoff's laws for one of the segments to obtain the transmission line equations, assuming that both current and voltage are phasors, that is, we derive the equations in the frequency domain, assuming time-harmonic excitation. The voltage across the line segment of length Δl can be written in terms of the voltages at points A and B and the current in the segment. With the notation in **Figure 14.10**, we have,

$$V(l + \Delta l) - V(l) = -I(l)[R\Delta l + j\omega L\Delta l] \quad [\text{V}] \tag{14.22}$$

Dividing both sides by Δl

$$\frac{V(l + \Delta l) - V(l)}{\Delta l} = -I(l)[R + j\omega L] \tag{14.23}$$

The term on the left-hand side becomes the derivative of V with respect to l if we let Δl tend to zero. Thus, since Δl is arbitrarily small, we may write

$$\boxed{\frac{dV(l)}{dl} = -I(l)[R + j\omega L]} \tag{14.24}$$

This relation holds at any point on the line. Similarly, the current in the segment can be written in terms of the current at points A and B and the voltage at point B as

$$I(l + \Delta l) - I(l) = -V(l + \Delta l)[G\Delta l + j\omega C\Delta l] \quad [\text{A}] \tag{14.25}$$

Following steps identical to **Eqs. (14.23)** and **(14.24)**, we get

$$\frac{dI(l)}{dl} = -V(l + \Delta l)[G + j\omega C] \tag{14.26}$$

To obtain an equation of the same form as for the voltage in **Eq. (14.24)**, we expand the term $V(l + \Delta l)$ in a Taylor series about l as $V(l + \Delta l) = V(l) + (dV(l)/dl)\Delta l/1! + (d^2V(l)/dl^2)(\Delta l)^2/2! + \ldots$ Neglecting all terms that contain Δl gives an approximation $V(l + \Delta l) \approx V(l)$. Substitution of this in **Eq. (14.26)** gives

$$\boxed{\frac{dI(l)}{dl} = -V(l)[G + j\omega C]} \tag{14.27}$$

The transmission line equations are the current and voltage relations in **Eqs. (14.24)** and **(14.27)**. These are two coupled first-order differential equations. Before attempting to solve for current and voltage, we can eliminate one of the variables and obtain separate equations for $V(l)$ and $I(l)$. To do so, we substitute $I(l)$ from **Eq. (14.24)** into **Eq. (14.27)** and $V(l)$ from **Eq. (14.27)** into **Eq. (14.24)**. From **Eq. (14.24)**,

$$I(l) = -\frac{dV(l)}{dl}\frac{1}{[R + j\omega L]} \quad [\text{A}] \tag{14.28}$$

Substitution of this into **Eq. (14.27)** gives

$$\boxed{\frac{d^2V(l)}{dl^2} - V(l)[G + j\omega C][R + j\omega L] = 0} \tag{14.29}$$

Similarly, substituting $V(l)$ from **Eq. (14.27)** into **Eq. (14.24)**, we get

$$\boxed{\frac{d^2I(l)}{dl^2} - I(l)[G + j\omega C][R + j\omega L] = 0} \tag{14.30}$$

These two equations are wave equations of the same form as given in **Eq. (12.84)** for the electric field intensity **E** (see **Section 12.7.1**). In fact, we can rewrite **Eqs. (14.29)** and **(14.30)** as

$$\frac{d^2V}{dl^2} - \gamma^2 V = 0 \tag{14.31}$$

and

$$\frac{d^2I}{dl^2} - \gamma^2 I = 0 \tag{14.32}$$

where

$$\boxed{\gamma = \alpha + j\beta = \sqrt{(R + j\omega L)(G + j\omega C)}} \tag{14.33}$$

The first of these is the wave equation for the voltage on the line and the second is the wave equation for current in the line. Therefore, γ is the propagation constant in analogy with the definition of the propagation constant in **Chapter 12 [Eq. (12.83)]**. This is fortunate because we can now use the solutions obtained in **Chapter 12** for plane waves. In fact, all we have to do is replace the electric field intensity in **Eq. (12.88)** by the voltage $V(l)$, the magnetic field intensity by $I(l)$, and the constant of propagation γ, by the term in **Eq. (14.33)**.

The propagation constant in **Eq. (14.33)** is complex. α is the attenuation constant along the line and β is the phase constant. The attenuation constant is given in nepers/m and the phase constant in radians/meter.

Based on the form of these equations and the similarity to the equations for plane waves **[Eq. (12.82)]**, we can now solve them by simply performing the above substitutions and using the solutions for plane waves. Thus, for the general transmission line described here, the solution for voltage and current can be written with the aid of **Eq. (12.88)** as

$$\boxed{V(l) = V^+e^{-\gamma l} + V^-e^{\gamma l} \quad [\text{V}]} \tag{14.34}$$

$$\boxed{I(l) = I^+ e^{-\gamma l} + I^- e^{\gamma l} \quad [\text{A}]} \tag{14.35}$$

Direct substitution of these solutions into **Eqs. (14.31)** and **(14.32)** shows they are correct. The solution to these equations has two parts: one propagating in the positive l direction, the other in the negative l direction, along the line, exactly as for plane waves. V^+ and V^- are the amplitudes of the voltage waves propagating in the positive and negative l directions, respectively. For the current solution, I^+ and I^- are the respective amplitudes of the current waves. The amplitudes of the forward and backward propagating waves, V^+ and V^-, can be calculated from the terminal voltages on the transmission line as we shall see shortly.

It is interesting to note here that whereas plane waves were a convenient simplification for wave propagation, their use in the transmission lines we discuss here is exact; that is, the waves in these transmission lines behave exactly as plane waves.

So far, we have defined one characteristic quantity of the line: the propagation constant in **Eq. (14.33)**. Now that we obtained the voltages and currents on the line, we can define the second characteristic quantity of any transmission line: the characteristic line impedance.

The ***characteristic line impedance*** Z_0 of a transmission line is defined as the ratio between the forward-propagating voltage amplitude and the forward-propagating current amplitude:

$$\boxed{Z_0 = \frac{V^+}{I^+} \quad [\Omega]} \tag{14.36}$$

To evaluate the characteristic impedance in terms of the line parameters (since these are known and independent of line current), we substitute the general solution from **Eqs. (14.34)** and **(14.35)** into the transmission line relation in **Eqs. (14.24)** and **(14.27)**. Starting with **Eq. (14.24)**, we get

$$\frac{d\left(V^+ e^{-\gamma l} + V^- e^{\gamma l}\right)}{dl} = -\left(I^+ e^{-\gamma l} + I^- e^{\gamma l}\right)[R + j\omega L] \tag{14.37}$$

or, after evaluating the derivatives,

$$-\gamma V^+ e^{-\gamma l} + \gamma V^- e^{\gamma l} = -\left(I^+ e^{-\gamma l} + I^- e^{\gamma l}\right)[R + j\omega L] \tag{14.38}$$

Similarly, using **Eq. (14.27)**, we get

$$-\gamma I^+ e^{-\gamma l} + \gamma I^- e^{\gamma l} = -\left(V^+ e^{-\gamma l} + V^- e^{\gamma l}\right)[G + j\omega C] \tag{14.39}$$

Now, suppose, first, that only a forward-propagating wave exists by setting $V^- = 0, I^- = 0$ in **Eqs. (14.38)** and **(14.39)**, We get

$$\boxed{-\gamma V^+ e^{-\gamma l} = -I^+ e^{-\gamma l}[R + j\omega L] \quad \text{and} \quad -\gamma I^+ e^{-\gamma l} = -V^+ e^{-\gamma l}[G + j\omega C]} \tag{14.40}$$

Thus, the characteristic impedance can be written as

$$Z_0 = \frac{V^+}{I^+} = \frac{R + j\omega L}{\gamma} = \frac{\gamma}{G + j\omega C} \quad [\Omega] \tag{14.41}$$

The first form is obtained from the first expression in **Eq. (14.40)** and the second from the second expression. Also, by substituting for γ from **Eq. (14.33)**, we obtain

$$\boxed{Z_0 = \sqrt{\frac{R + j\omega L}{G + j\omega C}} \quad [\Omega]} \tag{14.42}$$

Now suppose that only a backward-propagating wave exists. By setting $V^+ = 0$, $I^+ = 0$ in **Eqs. (14.38)** and **(14.39)**, we get

$$\gamma V^- e^{\gamma l} = -I^- e^{\gamma l}[R + j\omega L] \quad \text{and} \quad \gamma I^- e^{\gamma l} = -V^- e^{\gamma l}[G + j\omega C] \tag{14.43}$$

Dividing each of these two equations by I^-, we can write

$$\frac{V^-}{I^-} = -\frac{R + j\omega L}{\gamma} = -\frac{\gamma}{G + j\omega C} = -Z_0 \tag{14.44}$$

We can summarize these results as follows:

$$Z_0 = \frac{V^+}{I^+} = -\frac{V^-}{I^-} = \frac{R + j\omega L}{\gamma} = \frac{\gamma}{G + j\omega C} = \sqrt{\frac{R + j\omega L}{G + j\omega C}} \quad [\Omega] \tag{14.45}$$

The characteristic impedance Z_0 is independent of location on the line and only depends on line parameters. Thus, the name characteristic impedance. The characteristic impedance is, in general, a complex value. However, whereas all other line parameters are given in per meter units, the characteristic impedance is a line property, independent of length. In other words, for any given line, if we were to measure the characteristic impedance, the above value would be obtained for any length of line and at any location on the line.

Using **Eq. (14.45)**, the line current given in **Eq. (14.35)** can be written as

$$I(l) = \frac{V^+}{Z_0} e^{-\gamma l} - \frac{V^-}{Z_0} e^{\gamma l} \quad [\text{A}] \tag{14.46}$$

Finally, we also mention that the wavelength and phase velocity for any propagating voltage or current wave on a lossless line ($\alpha = 0$ in **Eq. (14.33)**) are given as

$$\boxed{\lambda = \frac{2\pi}{\beta} \quad [\text{m}]} \quad \boxed{v_p = \frac{\omega}{\beta} \quad \left[\frac{\text{m}}{\text{s}}\right]} \tag{14.47}$$

The quantity βl has units of radians. It is called the ***electrical length*** of the line and may be considered an additional line parameter.

The discussion in this section assumed time-harmonic quantities. This was done on purpose, since phasor calculations are usually simpler to perform and the final result is also simpler. More important, this choice allowed us to use the results already obtained for transverse electromagnetic wave propagation. In turn, this choice shows that propagation along transmission lines is similar to propagation in free space and other materials, as long as the basic assumptions of transverse electromagnetic waves are satisfied. Both plane waves in materials and waves in transmission lines satisfy these conditions. Thus, we can expect that other parameters such as reflection and transmission of energy as well as the reflection and transmission coefficients should be similar. We will discuss these topics separately.

Instead of using the time-harmonic forms for voltage and current, we could start with the time-dependent voltage and current to obtain the time-dependent transmission line equations following essentially identical steps as above. Since we have the frequency domain representations in **Eqs. (14.24)**, **(14.27)**, **(14.29)** and **(14.30)**, we could obtain the time dependent wave equations by simply transforming these into the time domain using the tranformations $d[f(t)]/dt \Leftrightarrow j\omega[F(\omega)]$ and $d^2[f(t)]/dt^2 \Leftrightarrow (j\omega)^2[F(\omega)]$. This entails no more than replacing $j\omega$ by d/dt, $(j\omega)^2 = -\omega^2$ by d^2/dt^2 and $V(l)$ and $I(l)$ by $V(l,t)$ and $I(l,t)$ (see **Exercises 14.2** and **14.3**).

To simplify matters we restrict ourselves to the time-harmonic representation and do not pursue the time dependent transmission line equations any further. However, we will touch on some aspects of time-dependent behavior of transmission lines in **Chapter 16** when we discuss transients on transmission lines.

Exercise 14.2 Obtain the time dependent representation of the transmission line equation starting with the frequency domain representation in **Eqs. (14.24)** and **(14.27)**.

Answer $\frac{dV(l,t)}{dl} = -I(l,t)R - L\frac{dI(l,t)}{dt}, \quad \frac{dI(l,t)}{dl} = -V(l,t)G - C\frac{dV(l,t)}{dt}$

Exercise 14.3 Obtain the time dependent wave equations starting with the frequency domain representation in **Eqs. (14.29)** and **(14.30)**.

Answer

$$\frac{d^2V(l,t)}{dl^2} - LC\frac{d^2V(l,t)}{dt^2} - (LG + RC)\frac{dV(l,t)}{dt} - RGV(l,t) = 0$$

$$\frac{d^2I(l,t)}{dl^2} - LC\frac{d^2I(l,t)}{dt^2} - (LG + RC)\frac{dI(l,t)}{dt} - RGI(l,t) = 0$$

14.5 Types of Transmission Lines

The transmission line equations in **Section 14.4** were obtained for a completely general transmission line. As can be seen, the equations are rather involved. The propagation constant as well as the line impedance are complex and are not always easy to evaluate. Both a phase constant and an attenuation constant exist; therefore, we can expect the waves along the line to decay due to attenuation as well as change their phases. The fact that both a forward- and backward-propagating wave exists indicates that the line may be finite in length whereby the backward-propagating wave is due to a reflection from the load, a connection on the line, or any other discontinuity that may exist.

For practical applications, we distinguish between a number of special types of transmission lines in addition to the above general lossy line. These are the ***lossless transmission line***, the ***long transmission line***, the ***low resistance transmission line*** and the ***distortionless transmission line***. The wave characteristics on these lines are simplified because of the assumptions associated with them but, more importantly, they represent useful, practical lines. These are described next.

14.5.1 The Lossless Transmission Line

A lossless transmission line is a line for which both the series resistance and the shunt conductance are zero ($R = 0, G = 0$). In practice, this implies that the line is made of perfect conducting materials and perfect dielectrics. Although no practical line satisfies these conditions exactly, many lines satisfy them approximately. The implications of these conditions are that the attenuation constant is zero, the propagation constant is purely imaginary, and the characteristic impedance of the line is real.

If we substitute $R = 0$ and $G = 0$ in the propagation constant in **Eq. (14.33)**, we get

$$\gamma = j\beta = j\omega\sqrt{LC} \tag{14.48}$$

Similarly, the characteristic impedance of the line [from **Eq. (14.42)**] is real and equal to

$$\boxed{Z_0 = \sqrt{\frac{L}{C}} \quad [\Omega]} \tag{14.49}$$

A number of propagation parameters can now be easily evaluated. The phase and attenuation constants are found from the propagation constant:

$$\boxed{\beta = \omega\sqrt{LC} \quad [\text{rad/m}], \quad \alpha = 0} \tag{14.50}$$

The wavelength is defined as

$$\boxed{\lambda = \frac{2\pi}{\beta} = \frac{2\pi}{\omega\sqrt{LC}} \quad [\mathrm{m}]} \tag{14.51}$$

and the speed of propagation of the wave along the line (phase velocity) is

$$\boxed{v_p = \frac{\omega}{\beta} = \frac{1}{\sqrt{LC}} \quad \left[\frac{\mathrm{m}}{\mathrm{s}}\right]} \tag{14.52}$$

Because the dielectric is lossless, the phase velocity may also be written as

$$v_p = \frac{1}{\sqrt{\mu\varepsilon}} \quad \left[\frac{\mathrm{m}}{\mathrm{s}}\right] \tag{14.53}$$

From this, the following relation is obtained:

$$\boxed{\mu\varepsilon = LC} \tag{14.54}$$

In particular, the phase constant and the phase velocity only depend on the inductance and capacitance per unit length. The voltage or current waves propagate along the line without attenuation at a speed dictated by the inductance and capacitance per unit length of the line.

Example 14.3 Application: Antenna Down-Cables A common transmission line is the antenna cable used for rooftop TV antennas. The cable is made of two wires separated by a thin dielectric in the form of a flat cable. The characteristic impedance of these lines is 300 Ω. If the conductors are made of copper, separated by air (free space), and are 1 mm thick, calculate:

(a) The required distance between the two wires to produce a 300 Ω impedance.
(b) Calculate the phase velocity and the phase constant when receiving VHF channel 3 (63 MHz) and UHF channel 69 (803 MHz).

Solution: The characteristic impedance is given in **Eq. (14.49)**. From this and the relations for L and C in **Table 14.1**, column 1, we calculate the required distance.

(a) The intrinsic impedance is

$$Z_0 = \sqrt{\frac{L}{C}} = \sqrt{\frac{\frac{\mu}{\pi}\cosh^{-1}(d/2a)}{\pi\varepsilon/\cosh^{-1}(d/2a)}} = \frac{1}{\pi}\sqrt{\frac{\mu}{\varepsilon}}\left(\cosh^{-1}\frac{d}{2a}\right) = 300 \quad [\Omega]$$

For $\mu = \mu_0$ and $\varepsilon = \varepsilon_0$ and with $a = 0.0005$ m

$$\cosh^{-1}\frac{d}{2a} = \frac{300\pi}{\sqrt{\mu_0/\varepsilon_0}} = \frac{300\pi}{376.99} = 2.5 \quad \rightarrow \quad d = 2a\cosh 2.5 = 0.00613 \quad [\mathrm{m}]$$

The distance between the wires should be 6.13 mm.

(b) To calculate the phase velocity and phase constant, we need the capacitance and inductance per unit length. However, since $LC = \mu_0\varepsilon_0$, the phase velocity must be that of free space, regardless of frequency ($v_p = c$). The phase constant depends on frequency. From **Eq. (14.50)**

$$\beta = \frac{\omega}{c} \quad \rightarrow \quad \beta_{63\,\mathrm{MHz}} = \frac{2 \times \pi \times 63 \times 10^6}{3 \times 10^8} = 1.32 \quad [\mathrm{rad/m}]$$

$$\beta_{803\,\mathrm{MHz}} = \frac{2 \times \pi \times 803 \times 10^6}{3 \times 10^8} = 16.82 \quad [\mathrm{rad/m}].$$

Example 14.4 Application: CAble TeleVision (CATV) Cables A cable TV coaxial cable is designed with a characteristic impedance of 75 Ω. The inner conductor is 0.5 mm thick and the internal diameter of the outer conductor is 8 mm.

(a) Calculate the dielectric constant required for the material between the conductors to produce this impedance. Assume permeability of free space.
(b) Calculate the phase velocity on the line.

Solution:

(a) The characteristic impedance is given in **Eq. (14.49)** and the capacitance and inductance per unit length in column 2 in **Table 14.1**.

$$Z_0 = \sqrt{\frac{L}{C}} = \sqrt{\frac{\mu_0 (\ln(b/a))^2}{4\pi^2 \varepsilon}} = \frac{\ln(b/a)}{2\pi} \sqrt{\frac{\mu_0}{\varepsilon}} = 75 \quad [\Omega]$$

Solving for ε, with all other values known

$$\varepsilon = \left(\frac{\ln(b/a)}{2\pi}\right)^2 \frac{\mu_0}{75^2} = \left(\frac{\ln \dfrac{0.004}{0.00025}}{2 \times \pi}\right)^2 \frac{4 \times \pi \times 10^{-7}}{75^2} = 43.5 \times 10^{-12} \quad [\text{F/m}]$$

or in terms of relative permittivity $\varepsilon_r = \varepsilon/\varepsilon_0 = 43.5 \times 10^{-12}/8.854 \times 10^{-12} = 4.92$. This relative permittivity may be attained with some plastics, although, in actual design, it is just as likely to choose the dielectric first and work with the other parameters around it to obtain the required impedance.

(b) To calculate the phase velocity, we use the permeability and permittivity of the material:

$$v_p = \frac{1}{\sqrt{\mu_0 \varepsilon}} = \frac{1}{\sqrt{4 \times \pi \times 10^{-7} \times 43.5 \times 10^{-12}}} = 1.353 \times 10^8 \quad [\text{m/s}]$$

Thus, the speed of propagation in the cable is 2.22 (actually $\sqrt{4.92}$) times slower than in free space, or in the same cable but with air as the dielectric.

14.5.2 The Long Transmission Line

A long transmission line is a line that for practical purposes may be considered to be infinite. The infinite transmission line is characterized by transmission without backward-propagating waves since, as we have seen in **Chapter 13**, a backward-propagating wave can only exist if the incident wave is reflected from a discontinuity in the wave's path. The long line may be lossy or lossless. For a lossy line, the voltage and current waves are found from **Eqs. (14.34)** and **(14.46)** by removing the backward-propagating wave:

$$V(l) = V^+ e^{-\gamma l} \quad [\text{V}] \quad \text{and} \quad I(l) = I^+ e^{-\gamma l} = \frac{V^+}{Z_0} e^{-\gamma l} \quad [\text{A}] \qquad (14.55)$$

The propagation constant γ is given in **Eq. (14.33)** and the characteristic impedance of the line is given in **Eq. (14.45)**. If the long line is lossless, the voltage and current waves are

$$V(l) = V^+ e^{-j\beta l} \quad \text{and} \quad I(l) = \frac{V^+}{Z_0} e^{-j\beta l} \qquad (14.56)$$

The phase constant is given in **Eq. (14.50)** and the characteristic impedance in **Eq. (14.49)**.

The infinite transmission line cannot be realized physically, but it will prove to be a convenient approximation for very long lines or for short lines before the forward wave has reached the load.

Example 14.5 Application: Propagation and Attenuation in Coaxial Cables The line in **Example 14.4** is used to connect a cable TV distribution center to a TV 20 km away. Assume that the material between the conductors has an attenuation of 1 dB/km, which may be considered a low-loss line.

The permeability of the dielectric in the line is μ_0 [H/m] and its permittivity is $4.92\varepsilon_0$ [F/m]. The frequency is 80 MHz (approximately the middle frequency of VHF channel 5).

(a) Calculate the propagation constant of the wave.
(b) Write the voltage and current everywhere on the line. Assume the voltage at the generator is 1 V and there are no reflections of waves anywhere on the line.
(c) If a TV requires a signal of at least 100 mV to receive properly, what must be the signal amplitude at the generator?

Solution: We calculate the phase constant from the relations for low-loss dielectrics given in **Section 12.7.2**. The attenuation constant is calculated directly from the attenuation given.

(a) The attenuation is 1 dB/km. Since we require the attenuation constant in nepers/m and one Np/m equals 8.69 dB/m, the attenuation constant is

$$\alpha = \frac{1}{8.69 \times 1000} = 1.15 \times 10^{-4} \quad [\text{Np/m}]$$

The phase velocity of a low-loss dielectric is approximately the same as that in the lossless dielectric. Thus, the phase constant is approximately

$$\beta \approx \omega\sqrt{\mu\varepsilon} = \omega\sqrt{\mu_0\varepsilon_0}\sqrt{\varepsilon_r} = \frac{\omega}{c}\sqrt{\varepsilon_r} = \frac{2 \times \pi \times 80 \times 10^6 \times \sqrt{4.92}}{3 \times 10^8} = 3.716 \quad [\text{rad/m}]$$

These give the propagation constant as

$$\gamma = \alpha + j\beta = 1.15 \times 10^{-4} + j3.716.$$

(b) Assuming zero phase at the generator and the generator is taken as the origin ($l = 0$), we write for the voltage

$$V(z) = 1e^{-\gamma z} = e^{-1.15\times10^{-4}z}e^{-j3.716z} \quad [\text{V}]$$

The current is

$$I(z) = \frac{V(z)}{Z_0} = \frac{1}{75}e^{-1.15\times10^{-4}z}e^{-j3.716z} \quad [\text{A}].$$

(c) The distance between the generator and the TV is 20,000 m. The required signal at the TV is 100 mV. Thus, we write

$$\begin{aligned} V(l = 20,000\ m) &= V(l = 0)e^{-1.15\times10^{-4}\times20,000}e^{-j3.71\times20,000} \\ &= V(l = 0) \times 0.1e^{-j74,200} \quad [\text{V}] \end{aligned}$$

The magnitude of the voltage at the generator is, therefore, V(l = 0) = 1 V. This is an extremely low-loss line. Practical lines have much higher losses (see the following exercise). The very large change in phase (3.71 rad/m) means that in practice the change in phase over large distances as in this example is not very useful.

Exercise 14.4 Suppose the line in **Example 14.5** has an attenuation of 10 dB/km.

(a) What is the required voltage amplitude at the generator to produce a signal of 10 mV at the TV a distance of 10 km away.
(b) Would you characterize this line as a low-loss or a high-loss line?

Answer **(a)** 987.16 V. **(b)** High loss.

14.5.3 The Distortionless Transmission Line

The propagation constant and characteristic impedance for general lossy lines were obtained in **Eqs. (14.33)** and **(14.42)**, respectively. These are rather complicated expressions and are frequency dependent. Whenever transmission lines are used for propagation of a single frequency wave (monochromatic wave), the fact that the line impedance and propagation constant are frequency dependent is less important, but when a wave has a range of frequencies, such as in the communication of information, each frequency component will be attenuated differently, the phase of each component will propagate at different speeds, and each component will see a different line impedance. This inevitably leads to distortion of the wave (see **Sections 12.7.4.1** through **12.7.4.3**).

The question is: How can we design a general lossy line so that the attenuation constant, phase velocity, and characteristic impedance of the line are independent of frequency? If we can do that, we would obtain a distortionless transmission line. To do so, we note that if $R/L = G/C$, the propagation constant in **Eq. (14.33)** becomes

$$\gamma = j\omega\sqrt{LC}\sqrt{1+\frac{R}{j\omega L}}\sqrt{1+\frac{R}{j\omega L}} = j\omega\sqrt{LC}\left[1+\frac{R}{j\omega L}\right] = j\omega\sqrt{LC}+R\sqrt{\frac{C}{L}} \tag{14.57}$$

From this, the attenuation and phase constants are

$$\boxed{\alpha = R\sqrt{\frac{C}{L}} \quad \left[\frac{\text{Np}}{\text{m}}\right], \quad \beta = \omega\sqrt{LC} \quad [\text{rad/m}]} \tag{14.58}$$

and, therefore, the phase velocity is

$$\boxed{v_p = \frac{\omega}{\beta} = \frac{1}{\sqrt{LC}} \quad \left[\frac{\text{m}}{\text{s}}\right]} \tag{14.59}$$

Thus, the first two conditions (i.e., that the attenuation constant and phase velocity are independent of frequency) are satisfied. What about the characteristic impedance? If we substitute the condition $R/L = G/C$ in **Eq. (14.42)**, we get

$$\boxed{Z_0 = \sqrt{\frac{R+j\omega L}{RC/L+j\omega C}} = \sqrt{\frac{L}{C}} \quad [\Omega]} \tag{14.60}$$

The characteristic impedance is also constant and the above requirements are satisfied. Thus, for a line to be distortionless, the line parameters must be designed so that

$$\boxed{\frac{R}{L} = \frac{G}{C}} \tag{14.61}$$

With this condition, the distortionless transmission line[1] has the same phase constant and characteristic impedance as the lossless line but a nonzero, constant attenuation.

Example 14.6 A line is made of two parallel conductors embedded in a low-loss dielectric, as shown in **Figure 14.11**. Material properties and dimensions are given in the figure. The design calls for a distortionless transmission line because the line is intended for communication of information. Assume the frequency used is 100 MHz, and that the dielectric extends far from the conductors.

(a) Calculate the required distance d between the wires to produce a distortionless line at the given frequency.
(b) What are the characteristic impedance of the line and its attenuation constant?
(c) If, reduction of at most 40 dB is allowed before an amplifier is required, calculate the distance between each two amplifiers on the line.

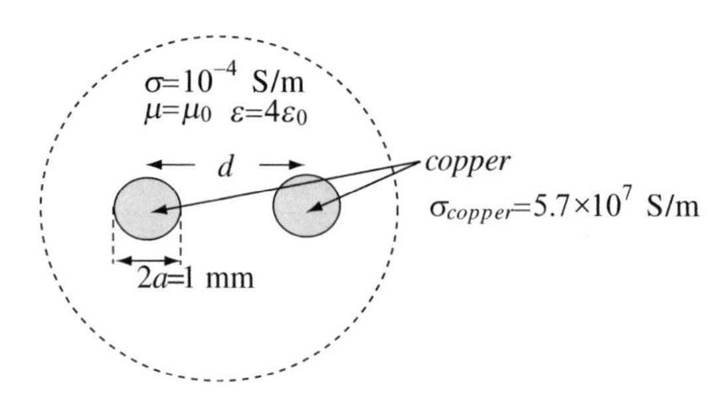

Figure 14.11 A two-wire transmission line. The distance d is designed so that the line is distortionless

Solution: From the distortionless line requirement,

$$\frac{R}{L} = \frac{G}{C}$$

(a) Substituting the parameters of the two-conductor line from **Table 14.1** (column 1),

$$\frac{1/\pi a\delta\sigma_c}{\frac{\mu}{\pi}\cosh^{-1}(d/2a)} = \frac{\frac{\pi\sigma}{\cosh^{-1}(d/2a)}}{\frac{\pi\varepsilon}{\cosh^{-1}(d/2a)}} \quad \rightarrow \quad \frac{1}{a\mu_0\delta\sigma_c\cosh^{-1}(d/2a)} = \frac{\sigma}{\varepsilon}$$

where $\delta = 1/\sqrt{\pi f \mu_c \sigma_c}$ and the index c indicates conductor material properties:

$$\cosh^{-1}\frac{d}{2a} = \frac{\varepsilon}{a\sigma}\sqrt{\frac{\pi f}{\mu_0\sigma_c}} = \frac{4\times 8.854\times 10^{-12}}{0.0005\times 10^{-4}}\sqrt{\frac{\pi\times 10^8}{5.7\times 10^7\times 4\times\pi\times 10^{-7}}} = 1.483$$

Thus,

$$\frac{d}{2a} = \cosh(1.483) = 2.317 \quad \rightarrow \quad d = 2\times 2.317\times 0.0005 = 2.317\times 10^{-3} \quad [\mathrm{m}]$$

The two wires must be separated by 2.317 mm to produce a distortionless line.

[1] The formula for distortionless lines is due to Oliver Heaviside. It was devised in 1897 as a solution to distortions on long (intercontinental) telephone lines. Following this, telephone lines were routinely "loaded" with additional series inductance at regular intervals to adjust their parameters so that distortionless lines are obtained. This practice is now rare. Lines are produced with parameters that guarantee they are distortionless.

(b) To calculate the characteristic impedance and the attenuation constant, the line parameters are needed. These are obtained from **Table 14.1** and, with the above dimensions, are

$$R = 1.675 \quad [\Omega/\text{m}], \quad L = 0.592 \quad [\mu\text{H/m}],$$
$$G = 2.12 \times 10^{-4} \quad [\text{S/m}], \quad C = 75 \quad [\text{pF/m}]$$

It is worth verifying that these parameters indeed make a distortionless line:

$$\frac{R}{L} = \frac{1.675}{0.592 \times 10^{-6}} = 2.83 \times 10^{6}, \quad \frac{G}{C} = \frac{2.12 \times 10^{-4}}{75 \times 10^{-12}} = 2.83 \times 10^{6}$$

Therefore, the conditions for distortionless operation are satisfied. Now, the characteristic impedance and attenuation constant are

$$Z_0 = \sqrt{\frac{L}{C}} = \sqrt{\frac{0.592 \times 10^{-6}}{75 \times 10^{-12}}} = 88.84 \quad [\Omega]$$

$$\alpha = R\sqrt{\frac{L}{C}} = 1.675\sqrt{\frac{75 \times 10^{-12}}{0.592 \times 10^{-6}}} = 0.0188 \quad [\text{Np/m}].$$

(c) The attenuation in the line is 0.0188 Np/m. 1 Np/m = 8.69 dB/m and, therefore, the attenuation is 0.1634 dB/m. For a total of 40 dB the distance is

$$d = \frac{40}{0.1634} = 244.84 \quad [\text{m}]$$

An amplifier is required every 245 m or so. This means the line is too lossy. Note, also, that this type of line is not normally used at high frequencies; coaxial lines are more common.

Note: The resistance per unit length is frequency dependent. Therefore, the line can only be distortionless in a narrow band of frequencies around 100 MHz.

14.5.4 The Low-Resistance Transmission Line

It was mentioned before that a transmission line is made of two conductors in a given configuration. In a line of this type, it is often possible to assume that the conductivity of the conductor is so high as to have negligible resistance. In other words, the propagation on the transmission line is not affected by the conductor itself. The conductors are required only to guide the waves, but all propagation parameters are affected by the properties of the dielectric alone. Substituting $R = 0$ in **Eqs. (14.33)** and **(14.42)**, we get

$$\gamma = j\omega\sqrt{LC}\sqrt{1 + \frac{G}{j\omega C}} \tag{14.62}$$

$$Z_0 = \sqrt{\frac{j\omega L}{G + j\omega C}} \quad [\Omega] \tag{14.63}$$

Since the conductor's effect can be neglected, we can view this as a transverse electromagnetic wave propagating in a lossy dielectric material with properties ε, μ, and σ as if the conductors were not there.

For a general lossy dielectric, we obtained the propagation constant in **Eq. (12.83)** as

$$\gamma = j\omega\sqrt{\mu\varepsilon}\sqrt{\left[1 + \frac{\sigma}{j\omega\varepsilon}\right]} \tag{14.64}$$

The propagation constants in the transmission line and in the general dielectric are of exactly the same form. Direct comparison between **Eqs. (14.64)** and **(14.62)** gives the following two relations:

$$\boxed{LC = \mu\varepsilon \quad \text{and} \quad \frac{\sigma}{\varepsilon} = \frac{G}{C}} \tag{14.65}$$

where σ is the conductivity of the dielectric between the conductors. These two relations are important for two reasons:

(1) They hold for lossless and lossy transmission lines even if the series resistance is not zero. This can be easily verified for the three transmission lines listed in **Table 14.1**.
(2) The relations provide one of the simplest methods of evaluating the parameters of the line. If, for example, C is known, L and G can be evaluated directly. This is useful because in many cases, one of the line parameters is easier to evaluate than the other two. In such cases, these two relations provide a simple means of finding the line parameters.

Note also that if, in addition, $G = 0$, the line becomes lossless.

Example 14.7 Application: Superconducting Power Lines A transmission line designed for power transmission at 60 Hz is made with superconducting cables. The two conductors that make the transmission line are separated by 3 m. The size of the wires is not known, but their inductance per unit length is known to be 0.5 μH/m. The permittivity and permeability are those of free space and the conductivity of air is 10^{-7} S/m. Calculate:

(a) The attenuation constant on the line.
(b) The characteristic impedance of the line.

Solution: The attenuation constant, which is entirely due to losses in air, is calculated as the real part of **Eq. (14.62)**, after the capacitance per unit length is calculated from **Eq. (14.65)**. The characteristic impedance is given in **Eq. (14.63)**. Because this is a superconducting transmission line, the series resistance of the line is zero.

(a) The capacitance per unit length **[Eq. (14.65)]** is

$$C = \frac{\mu_0\varepsilon_0}{L} = \frac{1}{c^2 L} = \frac{1}{9 \times 10^{16} \times 0.5 \times 10^{-6}} = 22.22 \times 10^{-12} \quad [\text{F/m}]$$

The conductance is also calculated from **Eq. (14.73)**

$$G = \frac{\sigma C}{\varepsilon_0} = \frac{10^{-7} \times 22.22 \times 10^{-12}}{8.854 \times 10^{-12}} = 2.51 \times 10^{-7} \quad [\text{S/m}]$$

Substituting these, we get the propagation constant:

$$\gamma = j\omega\sqrt{LC}\sqrt{1 + \frac{G}{j\omega C}} = j \times 2 \times \pi \times 60 \times \sqrt{0.5 \times 10^{-6} \times 22.22 \times 10^{-12}} \times \sqrt{1 + \frac{2.51 \times 10^{-7}}{j \times 2 \times \pi \times 60 \times 22.22 \times 10^{-12}}}$$

$$= j1.257 \times 10^{-6}\sqrt{1 - j29.96} = (4.783 + j4.946) \times 10^{-6}$$

The attenuation constant is very small and equal to 4.783×10^{-6} Np/m or 4.16×10^{-5} dB/m. Since the attenuation is frequency dependent, this value would change at other frequencies.

(b) The characteristic impedance of the line is

$$Z_0 = \sqrt{\frac{j\omega L}{G + j\omega C}} = \sqrt{\frac{j \times 2 \times \pi \times 60 \times 0.5 \times 10^{-6}}{2.51 \times 10^{-7} + j \times 2 \times \pi \times 60 \times 22.22 \times 10^{-12}}} = 19.6926 + j19.0463 \approx 19.7(1 + j) \quad [\Omega].$$

Note: Although we treated this problem as a wave problem, and, in fact, waves do exist at power frequencies, they are much less important than effects such as the induction of eddy currents and flux leakage. In spite of the fact that at low frequencies, distributed parameters are not normally necessary, their use is correct.

14.6 The Field Approach to Transmission Lines

The discussion in the previous sections was in terms of general line parameters and therefore applies to any transmission line. Unlike previous chapters, the primary variables were the voltage on and current in the line. This choice is natural if we view the line as a distributed parameter circuit. It is, however, possible to arrive at exactly the same results from a field point of view. In this case, the primary variables are the electric and magnetic field intensities and the discussion is much the same as that for plane waves. One advantage of using field variables is that these are vectors and, therefore, the direction of propagation at any point is always available and indicates the direction in which power is transferred. A second advantage is that fields can be used when currents and voltages cannot. This is the case for transmission lines made of a single conductor or of dielectrics. These will be called waveguides and will be treated in **Chapter 17**. To demonstrate this approach, we look now at the wave characteristics on a parallel plate transmission line.

Suppose the transmission line shown in **Figure 14.12** is given. The line is very long and $w \gg d$. The material between the plates is a general dielectric. At a given instant in time, the potential between the two plates and the currents in the plates are as shown. For the given condition, the electric field intensity points from the upper plate to the lower plate (x direction) and the magnetic field intensity is parallel to the plates, pointing in the y direction. Because of our assumption that $w \gg d$, we may assume that the electric field intensity is everywhere perpendicular to the plates (no fringing at the edges) and the magnetic field intensity is everywhere parallel to the plates.

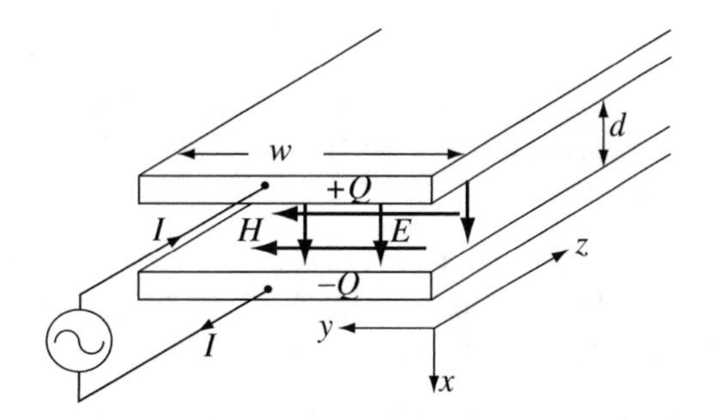

Figure 14.12 Relations between current and charge on the conductor and the electric and magnetic field intensities in the dielectric of a parallel plate transmission line

We know that the two fields are a solution to the source-free wave equation since there are no sources in this domain and propagation takes place; that is, the fields obey the general Maxwell equations. Also, because the transmission line is infinite in extent, in the z direction, there can only be a forward-propagating wave. Without knowing what the electric field intensity amplitude is, we can write, in general terms:

$$\mathbf{E} = \hat{\mathbf{x}} E_0 e^{-\gamma z} \quad [\mathrm{V/m}] \tag{14.66}$$

where we have replaced the generic coordinate l with z. The magnetic field intensity is perpendicular to the electric field intensity and, using the intrinsic impedance of the dielectric, we can write

$$\mathbf{H} = \hat{\mathbf{y}} \frac{E_0}{\eta} e^{-\gamma z} \quad \left[\frac{\mathrm{A}}{\mathrm{m}}\right] \tag{14.67}$$

where η is the intrinsic impedance of the dielectric between the plates. The wave is a transverse electromagnetic wave (**E** and **H** are perpendicular to each other and to the direction of propagation). The direction of propagation of the wave is in the positive z direction, as shown by the Poynting vector:

$$\mathcal{P}(z) = \mathbf{E} \times \mathbf{H} = \hat{\mathbf{x}} E_0 e^{-\gamma z} \times \hat{\mathbf{y}} \frac{E_0}{\eta} e^{-\gamma z} = \hat{\mathbf{z}} \frac{E_0^2}{\eta} e^{-2\gamma z} \quad \left[\frac{\mathrm{W}}{\mathrm{m}^2}\right] \tag{14.68}$$

The electric field intensity E_0 was arbitrarily chosen, but, in practice, its sources are the charge distribution on the conducting surfaces and the current density in the conducting plates. The voltage between the two plates can be written as

$$V = \int_{l_1} \mathbf{E} \cdot d\mathbf{l}_1 \quad [\mathrm{V}] \tag{14.69}$$

and the current in one of the plates (upper) as

$$I = \oint_{l_2} \mathbf{H} \cdot d\mathbf{l}_2 \quad [\mathrm{A}] \tag{14.70}$$

where the path l_1 and contour l_2 are shown in **Figure 14.13b**. These are the line voltage and current and may be substituted in **Eqs. (14.34)** and **(14.35)** to obtain the transmission line voltage and current in terms of the electric and magnetic field intensities. This approach will be used, indirectly, in **Chapter 17**, but we will not pursue it here.

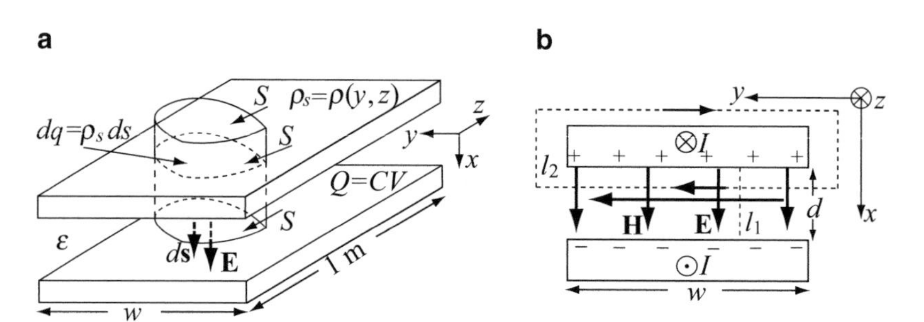

Figure 14.13 (a) Calculation of charge density using Gauss's law. (b) Calculation of current density using Ampere's law

We can now calculate the charge density and the current density in the conductors that will produce the required electric and magnetic fields from **Eqs. (14.69)** and **(14.70)**. This calculation is not absolutely necessary for the discussion here, but it emphasizes two important points:

(1) The sources of the fields produced by the transmission line are the charges on and currents in the line.
(2) The charge and current distributions must be of a form that produces these fields; not all charge and current distributions will produce a propagating wave in the transmission line.

Suppose that a charge distribution exists on the upper and lower plates as shown in **Figure 14.13a**. To calculate the electric field intensity, we use Gauss's law. A small volume, with two surfaces parallel to the upper plate, is defined as shown in **Figure 14.13a**. The electric field intensity outside the plates is zero (as for parallel plate capacitors) and the electric field intensity between the plates is given by **Eq. (14.66)**. Taking a surface S as shown, we get from Gauss's law

$$\int_s \mathbf{E} \cdot d\mathbf{s} = \int_s (\hat{\mathbf{x}} E_0 e^{-\gamma z}) \cdot (\hat{\mathbf{x}} ds) = \frac{1}{\varepsilon} \int_s \rho ds \tag{14.71}$$

or

$$\rho(y, z) = \varepsilon E_0 e^{-\gamma z} \quad [\mathrm{C/m^2}] \tag{14.72}$$

Thus, the charge density is uniform in the y direction (independent of y) but varies along the line. This variation is better seen if the charge density is written in the time domain as

$$\rho(y, z, t) = \mathrm{Re}\left\{\varepsilon E_0 e^{-(\alpha + j\beta) z} e^{j\omega t}\right\} = \varepsilon E_0 e^{-\alpha z} \cos(\omega t - \beta z) \quad [\mathrm{C/m^2}] \tag{14.73}$$

In other words, the charge distribution must be cosinusoidal in the z direction. The attenuation constant produces a decaying charge density magnitude with distance. If propagation is without attenuation, then $\alpha = 0$ and there is no decay in amplitude of the electric field intensity. The charge density distribution on the lower plate is the same as on the upper plate but opposite in sign.

The current density in the line is calculated from Ampere's law. Using the upper plate again and assuming some current density in the plate, we can enclose this current density with an arbitrary contour as shown in **Figure 14.13b**. The magnetic field intensity outside the plates is zero and between the plates is given by **Eq. (14.67)**. In our case, **H** is in the positive y direction, as is $d\mathbf{l}$. Thus, the current density is in the positive z direction (**H** and **J** are always perpendicular to each other). Since the current is uniform in the y direction in this case, we can write $I(y,z) = wJ(y,z)$ and, performing the integration in **Eq. (14.70)** with the field in **Eq. (14.67)**, we get

$$\frac{E_0}{\eta} e^{-\gamma z} = J(y, z) \quad [\mathrm{A/m}] \tag{14.74}$$

This gives the magnitude of the current density in the upper plate. This current must be in the positive z direction to produce a magnetic field intensity in the positive y direction (based on our notation in **Figure 14.12**); therefore,

$$\mathbf{J}(y, z) = \hat{\mathbf{z}} \frac{E_0}{\eta} e^{-\gamma z} \quad [\mathrm{A/m}] \tag{14.75}$$

The current density in the lower plate is the same in magnitude but in the negative z direction. The variation of current density along the line is also cosinusoidal, as for the charge density.

Note that the same results could be obtained from the boundary conditions for a perfect conductor as discussed in **Section 11.4.2**. If we do so, the required conditions at the surface of the conductors are given in **Table 11.5**.

Example 14.8 Parallel Plate Transmission Line Consider the parallel plate transmission line shown in **Figure 14.14**. The distance between the plates is very small compared to the width of the line ($w \gg d$); the plates are perfectly conducting and separated by free space. A voltage is applied to one end of the line: $V = V_0 \cos \omega t$, where $V_0 = 12$ V and $\omega = 3 \times 10^9$ rad/s. Calculate:

(a) The surface charge density on the plates. Calculate the minimum and maximum charge density.
(b) The surface current density on the plates.
(c) The time-averaged power propagated in the line if all power is contained within the cross-sectional area of the line (i.e., no fields exist outside the line).

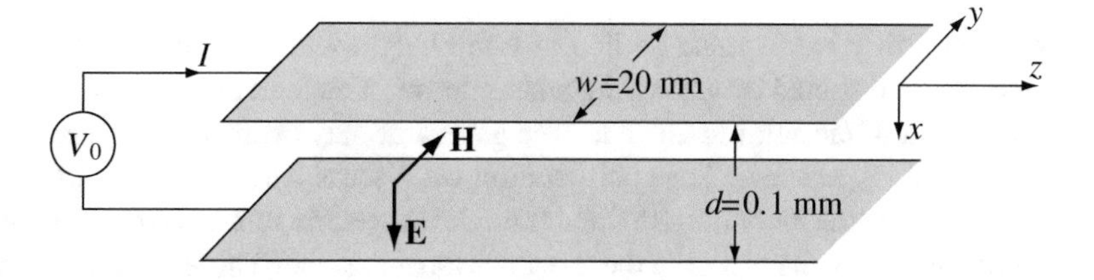

Figure 14.14 A parallel plate transmission line and the electric and magnetic field intensities between the plates

Solution: Since there is no fringing of the fields, the electric field intensity anywhere on the line equals the potential divided by the separation d as can be seen from **Figure 14.13b** and **Eq. (14.69)**, exactly like in a parallel plate capacitor. The line is lossless and long; therefore, the propagation constant is $\gamma = j\beta_0$.

(a) At the generator, the electric and magnetic field intensities are:

$$E_0 = \frac{V_0}{d} \cos\omega t \quad \left[\frac{\mathrm{V}}{\mathrm{m}}\right] \quad \text{and} \quad H_0 = \frac{E_0}{\eta_0} = \frac{V_0}{\eta_0 d} \cos\omega t \quad \left[\frac{\mathrm{A}}{\mathrm{m}}\right]$$

If we assume the electric field to be in the positive x direction as in **Figure 14.14**, the magnetic field must be in the positive y direction for the wave to propagate in the positive z direction. Taking this convention, the electric and magnetic field intensity vector phasors (at $z = 0$) are

$$\mathbf{E} = \hat{\mathbf{x}}\frac{V_0}{d} \quad \left[\frac{\mathrm{V}}{\mathrm{m}}\right], \quad \mathbf{H} = \hat{\mathbf{y}}\frac{V_0}{\eta_0 d} \quad \left[\frac{\mathrm{A}}{\mathrm{m}}\right]$$

These fields propagate in the positive z direction. At a distance z from the generator, the fields are

$$\mathbf{E}(z) = \hat{\mathbf{x}}\frac{V_0}{d}e^{-j\beta_0 z} \quad \left[\frac{\mathrm{V}}{\mathrm{m}}\right], \quad \mathbf{H}(z) = \hat{\mathbf{y}}\frac{V_0}{\eta_0 d}e^{-j\beta_0 z} \quad \left[\frac{\mathrm{A}}{\mathrm{m}}\right]$$

From the electric field intensity, the charge density on the line is

$$\rho(z) = \varepsilon_0 E_0 e^{-\gamma z} = \varepsilon_0 \frac{V_0}{d}e^{-j\beta_0 z} = 8.854 \times 10^{-12} \times \frac{12}{0.0001}e^{-j3\times 10^9 z/3\times 10^8} = 1.062 \times 10^{-6}e^{-j10z} \quad [\mathrm{C/m^2}].$$

(b) The surface current density is

$$\mathbf{J} = \hat{\mathbf{z}}\frac{V_0}{\eta_0 d}e^{-j\beta_0 z} = \hat{\mathbf{z}}\frac{12}{377 \times 0.0001}e^{-j10z} = \hat{\mathbf{z}}318.3e^{-j10z} \quad [\mathrm{A/m}].$$

(c) The time-averaged power density may be calculated anywhere on the line. However, because the line is lossless, it is best to calculate this at the generator. The time-averaged power density is

$$\mathcal{P}_{av}(z) = \hat{\mathbf{z}}\frac{E_0^2}{2\eta_0} = \hat{\mathbf{z}}\frac{V_0^2}{2\eta_0 d^2} = \hat{\mathbf{z}}1.9098 \times 10^7 \quad [\mathrm{W/m^2}]$$

and the total power is the power density multiplied by the cross-sectional area of the line. The latter is $S = wd$. Thus,

$$P = \mathcal{P}_{av}S = 1.9098 \times 10^7 \times 0.02 \times 0.0001 = 38.196 \quad [\mathrm{W}]$$

The same result can be obtained by multiplying the time-averaged current by time-averaged voltage. This gives:

$$P = \frac{V_0 I_0}{2} = \frac{V_0 J w}{2} = \frac{12 \times 318.3 \times 0.02}{2} = 38.196 \quad [\mathrm{W}]$$

14.7 Finite Transmission Lines

By a finite transmitting line is meant a line of finite length with a generator at one end and a load at the other. Both the generator and the load should be viewed in generic terms: The load may actually be a short circuit, an open circuit, another transmission line, or a transmitting antenna. The generator may be an actual source, the output of another transmission line, or, perhaps, a receiving antenna. The configuration we discuss here is shown in **Figure 14.15** for a line of length d.

Until now, we discussed only infinite lines or made no specific reference to the length of the line. Now, we have to discuss the distance on the line with respect to the fixed points of the line: These are the locations of the load and the generator. We could use the generator or the load for this purpose, but it is common to use the load as a reference point. This choice is partly arbitrary, partly based on convenience, and mostly on convention. At any rate, the only important point here is to be consistent and not flip between points of reference.

Refering to **Figure 14.15**, if we move the reference point to the load, that is, if $l = 0$ at the load, the generator is then located at $l = -d$. To avoid confusion, we define a new variable for position on the line as z. Defining $z = 0$ at the load, the generator is at $z = -d$. In other words, we replace l by $-z$ for all points between the generator and load.

The general wave solutions for voltage and current on a line were written in **Eqs. (14.34)** and **(14.35)** with reference to the generator as

$$V(l) = V^+e^{-\gamma l} + V^-e^{\gamma l} \quad [\mathrm{V}] \quad \text{and} \quad I(l) = I^+e^{-\gamma l} + I^-e^{\gamma l} \quad [\mathrm{A}] \qquad (14.76)$$

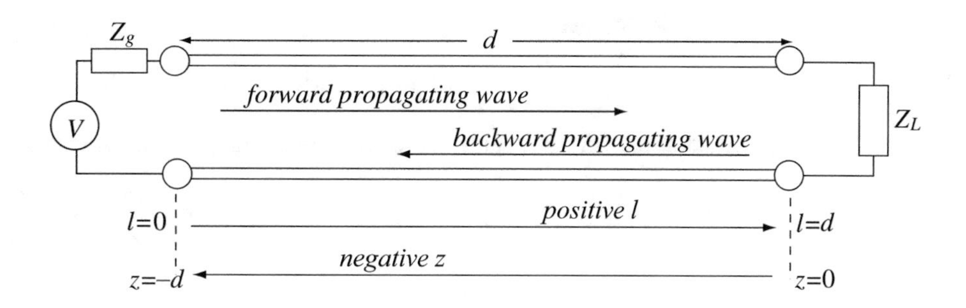

Figure 14.15 A finite transmission line with the reference shifted to the load

The wave solution on the terminated (finite) transmission line with reference to the load (replacing l with $-z$ and $-l$ with z) becomes

$$V(z) = V^+ e^{\gamma z} + V^- e^{-\gamma z} \quad [\mathrm{V}] \quad \text{and} \quad I(z) = I^+ e^{\gamma z} + I^- e^{-\gamma z} \quad [\mathrm{A}] \tag{14.77}$$

From **Eq. (14.46)** the current can also be written in terms of the voltage and characteristic impedance. The voltage and current on the line become:

$$\boxed{V(z) = V^+ e^{\gamma z} + V^- e^{-\gamma z} \quad [\mathrm{V}]} \quad \text{and} \quad \boxed{I(z) = \frac{V^+}{Z_0} e^{\gamma z} - \frac{V^-}{Z_0} e^{-\gamma z} \quad [\mathrm{A}]} \tag{14.78}$$

The first term in each relation is still the forward-propagating wave (towards the load) and the second is the backward-propagating wave (towards the generator).

Example 14.9 The Amplitudes of the Forward and Backward Propagating Waves The amplitudes V^+ and V^- have been used so far assuming they are known. They can be calculated from the terminal voltages on the transmission line.

(a) Given the lossless line in **Figure 14.16**, calculate the amplitudes of the forward and backward propagating waves if the load voltage is given as $V_L = 50$ V.

(b) Given the lossless line in **Figure 14.16**, calculate the amplitudes of the forward and backward propagating waves if the line input voltage and currents are given as $V_i = 50$ V and $I_i = 1$A.

Figure 14.16

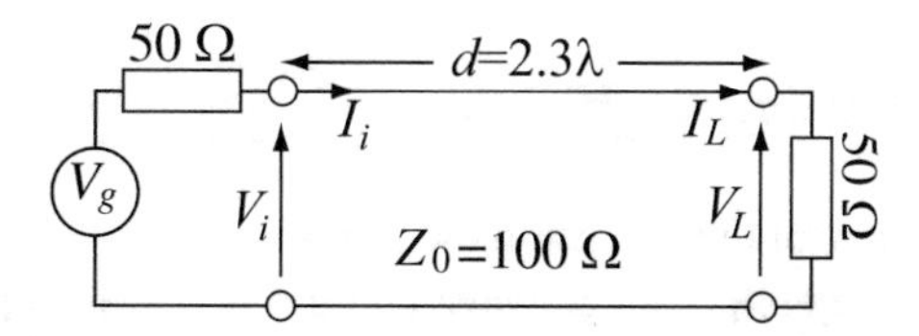

Solution: We use the general relations in **Eq. (14.78)**.

(a) In **Figure 14.16**, the load voltage is given. Since our reference point is at the load, we substitute $z = 0$ in **Eq. (14.78)** and get

$$V_L = V^+ + V^- \quad [\mathrm{V}] \quad \text{and} \quad I_L = \frac{V^+}{Z_0} - \frac{V^-}{Z_0} \quad [\mathrm{A}]$$

For the data given

$$V_L = V^+ + V^- = 50 \quad [\mathrm{V}] \quad \text{and} \quad I_L = \frac{V^+}{100} - \frac{V^-}{100} = \frac{V_L}{Z_L} = \frac{50}{50} = 1 \quad [\mathrm{A}] \quad \rightarrow \quad V^+ - V^- = 100 \quad [\mathrm{V}]$$

Solving for V^+ and V^- we get

$$V^+ = 75 \quad [\text{V}], \quad V^- = -25 \quad [\text{V}].$$

(b) In this case the terminal voltage is given at the line input. Since the input is at a distance d from the load, we write

$$V_i = V^+ e^{j\beta d} + V^- e^{-j\beta d} = 50 \quad [\text{V}] \quad \text{and} \quad I_i = \frac{V^+}{Z_0} e^{j\beta d} - \frac{V^-}{Z_0} e^{-j\beta d} = 1 \quad [\text{A}]$$

For the given data (with $\beta = 2\pi/\lambda$)

$$V^+ e^{j2\pi\times 2.3} + V^- e^{-j2\pi\times 2.3} = 50 \quad \text{and} \quad V^+ e^{j2\pi\times 2.3} - V^- e^{-j2\pi\times 2.3} = 100 \quad [\text{V}]$$

Solving for V^+ and V^- we get

$$V^+ = 75e^{-j4.6\pi} \quad \text{and} \quad V^- = -25e^{-j4.6\pi} \quad [\text{V}]$$

or

$$V^+ = 75\angle 108^\circ \text{ and } V^- = 25\angle -72^\circ \ [\text{V}]$$

Exercise 14.5

(a) Calculate the line input voltage and current (V_i, I_i) in the line in **Figure 14.16** if the load voltage is given as $V_L = 50$ V.

(b) Calculate the load voltage and current (V_L, I_L) in **Figure 14.16** if the line input voltage and currents are given as $V_i = 50$ V and $I_i = 1$ A.

Answer

(a) $V_i = -15.45 + j95.1 \quad [\text{V}], \quad I_i = -0.31 + j0.475 \quad [\text{A}]$
(b) $V_L = -15.45 - j95.1 \quad [\text{V}], \quad I_L = -0.31 - j1.9 \quad [\text{A}]$

14.7.1 The Load Reflection Coefficient waves.m

First, we recall the definition of the characteristic impedance Z_0. This was defined for an infinite transmission line as the ratio between the forward-propagating voltage wave and the forward-propagating current wave. Thus, for any line, the characteristic impedance is

$$Z_0 = \frac{V^+ e^{\gamma z}}{I^+ e^{\gamma z}} = \frac{V^+}{I^+} = -\frac{V^-}{I^-} \quad [\Omega] \tag{14.79}$$

as was shown in **Eq. (14.45)**. This impedance is characteristic of the line and has nothing to do with generator or load. Similarly, the propagation constant γ is independent of load or generator, as are the parameters R, L, G, and C.

Since the load is very important for our analysis and since it is one of the few variables an engineer has any control over, it is only natural that we should wish to analyze the transmission line behavior in terms of the load impedance and the line variables. Thus, we first write the load impedance:

$$Z_L = \frac{V_L}{I_L} \quad [\Omega] \tag{14.80}$$

where V_L and I_L are the total load voltage and total load current. By total voltage and current is meant the sum of forward and backward voltages and currents, respectively.

The load is located at $z = 0$. In terms of the current and voltage of the line, this becomes

$$Z_L = \frac{V(0)}{I(0)} = \frac{V^+ + V^-}{I^+ + I^-} = \frac{V^+ + V^-}{V^+/Z_0 - V^-/Z_0} = Z_0 \frac{V^+ + V^-}{V^+ - V^-} \quad [\Omega] \tag{14.81}$$

Note that if only forward-propagating waves exist ($V^- = 0$), the load impedance must be equal to the characteristic impedance of the line. This condition defines matching between load and line. Matching in transmission lines only requires that the load and line impedances be equal, unlike circuits where matching usually means maximum transfer of power (conjugate matching). Under matched conditions ($Z_L = Z_0$), there are no backward propagating waves.

On the other hand, if $Z_L \neq Z_0$, there will be both forward-propagating and backward-propagating waves. At the load ($z = 0$) we can calculate the backward propagating wave amplitude V^- from **Eq. (14.81)** as

$$V^- = V^+ \frac{Z_L - Z_0}{Z_L + Z_0} \quad [\mathrm{V}] \tag{14.82}$$

The backward-propagating wave is due to the reflection of the forward-propagating wave at the load. Thus, we define the ***load reflection coefficient*** as

$$\boxed{\Gamma_L = \frac{V^-}{V^+} = \frac{Z_L - Z_0}{Z_L + Z_0} \quad [\text{dimensionless}]} \tag{14.83}$$

It is important to remember that this is the reflection coefficient at the load only. At other locations on the line, the reflection coefficient is, in general, different and we should never confuse the load reflection coefficient with any other reflection coefficient that may be convenient to define. The load reflection coefficient will always be denoted with a subscript L as in **Eq. (14.83)**. Note also that in general, the load reflection coefficient is a complex number since it is the ratio of the complex amplitudes V^- and V^+. Thus, we can also write the reflection coefficient as

$$\boxed{\Gamma_L = |\Gamma_L| e^{j\theta_\Gamma}} \tag{14.84}$$

where θ_Γ is the phase angle of the load reflection coefficient. This form will become handy later in our study.

Example 14.10 Application: Mismatched Antenna and Line A transmission line used to connect a transmitter to its antenna has characteristic impedance $Z_0 = 50\ \Omega$. The antenna, with impedance $Z_L = 50 + j50$, is connected as a load to the line. Calculate the load reflection coefficient.

Solution: Using **Eq. (14.83)**

$$\Gamma_L = \frac{Z_L - Z_0}{Z_L + Z_0} = \frac{50 + j50 - 50}{50 + j50 + 50} = \frac{j50}{100 + j50} = \frac{1 + j2}{5}$$

or in terms of magnitude and phase

$$\Gamma_L = \frac{1}{5} + j\frac{2}{5} = \frac{1}{\sqrt{5}} e^{j0.352\pi}$$

This mismatch is not very healthy for the transmitter because of the backward propagating waves (and, therefore, power) returning to the generator and, therefore, should be avoided.

Example 14.11 A long power transmission line supplies 1,500 MW at 750 kV rms to a matched load (i.e., the load impedance equals the line impedance).

(a) Suppose the load is disconnected. What is the reflection coefficient at the load?

(b) Because of a fault on the line, the load changes from the matched condition to $Z_L = 200 + j100\ \Omega$. What is the reflection coefficient at the load now?

Solution: The load impedance is calculated from the load power and the reflection coefficient is then calculated from **Eq. (14.83)**.

(a) The load impedance under matched conditions is

$$P = \frac{V_L^2}{Z_L} \quad \rightarrow \quad Z_L = \frac{V_L^2}{P} = \frac{(750{,}000)^2}{1.5 \times 10^9} = 375 \quad [\Omega]$$

The characteristic line impedance is $Z_0 = 375\ \Omega$.

If the load is disconnected, the load impedance becomes infinite ($Z_L \gg Z_0$) and the load reflection coefficient is

$$\Gamma_L = \frac{Z_L - Z_0}{Z_L + Z_0} \approx \frac{Z_L}{Z_L} = +1$$

(b) The load reflection coefficient is

$$\Gamma_L = \frac{Z_L - Z_0}{Z_L + Z_0} = \frac{200 + j100 - 375}{200 + j100 + 375} = \frac{-175 + j100}{575 + j100} = -0.26605 + j0.22018 = 0.345e^{j0.78\pi}$$

The magnitude of the reflection coefficient is $|\Gamma_L| = 0.345$ and the phase angle of the load reflection coefficient is $\theta_L = 0.78\pi$.

14.7.2 Line Impedance and the Generalized Reflection Coefficient

After calculating the characteristic impedance and the reflection coefficient at the load, we can now tackle the question of the impedance at any other point on the line. This is an important question because it will allow us to connect the line to, say, a generator, ensuring that the line is matched to the generator, or to connect one line to another. These are questions of practical engineering importance. The simple example in **Figure 14.17** shows the concepts involved. A loudspeaker is to be connected to a power amplifier through a transmission line. We know that for optimal operation, the output of the amplifier must be matched to the load. At the amplifier, the load consists of the speaker and the line and the amplifier must be matched to the line. We defer the question of matching until the next chapter, but for any attempt at matching, we must be able to calculate the input impedance of the line.

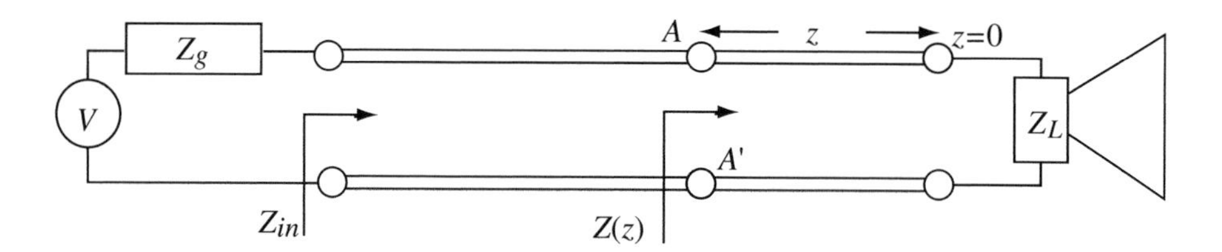

Figure 14.17 Distinction between load, input, and line impedances

This input impedance, which, in general, is different than the characteristic impedance of the line, must in some way depend on the load impedance. That this must be so should be obvious from our experience: suppose the above amplifier is matched to the line for the given load. If we now change the load, say by shorting the speaker, the system is not matched any more. In fact, by shorting the load, we may well have damaged the amplifier. It is, therefore, important to be able to calculate the line impedance for any load condition. Before continuing, we distinguish between two terms associated with impedance of the line. These are as follows:

Input line impedance is the impedance at the input or generator side of the line. In the above example, this impedance is the impedance of the line at the end, which is connected to the source (amplifier in this case). This impedance is denoted as Z_{in}.

Line impedance is the impedance at any point on the line. The distinction between the two terms is shown in **Figure 14.17**. The line impedance is denoted as $Z(z)$. The distinction is not terribly important since if we were to cut the line at the points A–A', the line impedance would then become the input impedance. We will, however, distinguish between the two terms wherever appropriate.

To calculate the line impedance, we need to calculate the total voltage and total current at any point on the line and divide the voltage by current. Using **Figure 14.18a** as a guide, the voltage and current at point z on the line are (from **Eq. 14.78**)

$$V(z) = V^+e^{\gamma z} + V^-e^{-\gamma z} \quad [\text{V}], \quad I(z) = \frac{V^+}{Z_0}e^{\gamma z} - \frac{V^-}{Z_0}e^{-\gamma z} \quad [\text{A}] \tag{14.85}$$

We can divide $V(z)$ by $I(z)$ to obtain $Z(z)$, but this would not be very helpful now because the result would be in terms of both the forward and backward waves. Instead, we use the load reflection coefficient in **Eq. (14.83)** to write

$$\boxed{V(z) = V^+\left(e^{\gamma z} + \Gamma_L e^{-\gamma z}\right) \quad [\text{V}]} \quad \text{and} \quad \boxed{I(z) = \frac{V^+}{Z_0}\left(e^{\gamma z} - \Gamma_L e^{-\gamma z}\right) \quad [\text{A}]} \tag{14.86}$$

The line impedance at point z is

$$\boxed{Z(z) = \frac{V(z)}{I(z)} = Z_0\frac{\left(e^{\gamma z} + \Gamma_L e^{-\gamma z}\right)}{\left(e^{\gamma z} - \Gamma_L e^{-\gamma z}\right)} \quad [\Omega]} \tag{14.87}$$

This expression is quite useful because it requires only knowledge of the reflection coefficient at the load, the characteristic impedance of the line, and the value of z (distance from the load). We will make considerable use of this expression here and in the following chapter.

Another way to look at the expression in **Eq. (14.87)** is to use the definition of the reflection coefficient in **Eq. (14.83)** and substitute it in **Eq. (14.87)**. Doing so and rearranging terms gives

$$Z(z) = Z_0\frac{\left((Z_L + Z_0)e^{\gamma z} + (Z_L - Z_0)e^{-\gamma z}\right)}{\left((Z_L + Z_0)e^{\gamma z} - (Z_L - Z_0)e^{-\gamma z}\right)} = Z_0\frac{Z_L\left(e^{\gamma z} + e^{-\gamma z}\right) + Z_0\left(e^{\gamma z} - e^{-\gamma z}\right)}{Z_0\left(e^{\gamma z} + e^{-\gamma z}\right) + Z_L\left(e^{\gamma z} - e^{-\gamma z}\right)} \quad [\Omega] \tag{14.88}$$

Now, we can use the identities $(e^{\gamma z} + e^{-\gamma z})/2 = \cosh\gamma z$ and $(e^{\gamma z} - e^{-\gamma z})/2 = \sinh\gamma z$ and write

$$\boxed{Z(z) = Z_0\frac{Z_L\cosh\gamma z + Z_0\sinh\gamma z}{Z_0\cosh\gamma z + Z_L\sinh\gamma z} = Z_0\frac{Z_L + Z_0\tanh\gamma z}{Z_0 + Z_L\tanh\gamma z} \quad [\Omega]} \tag{14.89}$$

where the relation $\tanh\gamma z = \sinh\gamma z/\cosh\gamma z$ was used. With these relations, we can now calculate the line impedance at any location, including at the input of the line.

Now, we can argue as follows: If the line impedance at a point on the line is equal to $Z(z)$, then cutting the line at this point and replacing the cut section by an equivalent load equal to $Z(z)$ should not change the conditions on the line to the left of the cut. This is shown in **Figure 14.18a**. The equivalent line in **Figure 14.18b** can be viewed as a new line with load impedance $Z(z)$. There is no reason we cannot calculate the reflection coefficient at this point on the line using **Eq. (14.83)** with $Z(z)$ instead of Z_L. Using **Eqs. (14.83)** and **(14.86)**, we get for the reflection coefficient at point z on the line

$$\Gamma(z) = \frac{V^-(z)}{V^+(z)} = \frac{V^+\Gamma_L e^{-\gamma z}}{V^+e^{\gamma z}} = \frac{\Gamma_L e^{-\gamma z}}{e^{\gamma z}} = \Gamma_L e^{-2\gamma z} \tag{14.90}$$

or using the form in **Eq. (14.84)** and also the relation $\gamma = \alpha + j\beta$,

$$\boxed{\Gamma(z) = \Gamma_L e^{-2\gamma z} = \Gamma_L e^{-2\alpha z - j2\beta z} = \left|\Gamma_L\right| e^{-2\alpha z} e^{j\theta_\Gamma} e^{-j2\beta z}} \tag{14.91}$$

The reflection coefficient $\Gamma(z)$ is called the ***generalized reflection coefficient*** to distinguish it from the load reflection coefficient. The generalized reflection coefficient on a general, lossy line can be viewed as having an amplitude $|\Gamma_L|$ at the load, which decays exponentially (for a lossy line) as we move toward the generator, and a phase which varies linearly with z and is equal to

$$\boxed{\phi_{\Gamma(z)} = \theta_\Gamma - 2\beta z \quad [\text{rad}]} \tag{14.92}$$

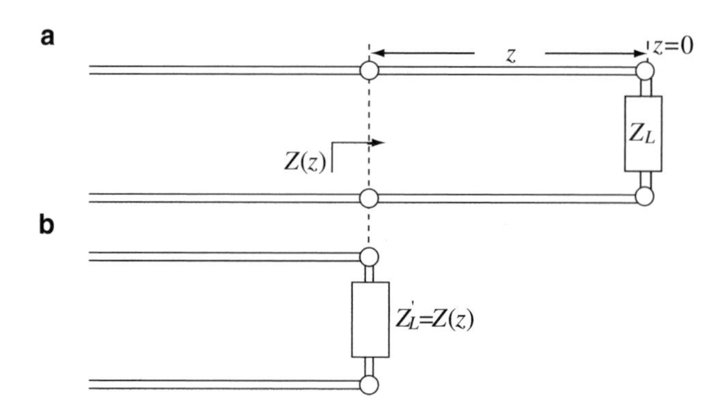

Figure 14.18 Method of calculation of the generalized reflection coefficient. (**a**) The impedance on the line at a general point, viewed as a new load to the line to the *left* of the point. (**b**) The new line and load

Although these relations are rather general, we will, for the most part, use lossless transmission lines. This simply means that $\alpha = 0$ and $\gamma = j\beta$, but doing so will simplify analysis considerably.

Example 14.12 A transmission line has propagation constant $\gamma = 0.01 + j0.05$, characteristic impedance $Z_0 = 50\ \Omega$, and a load $Z_L = 50 + j50\ \Omega$ is connected at one end. Calculate:

(a) The impedance on the line at the load.
(b) The impedance at a distance of 10 m from the load.
(c) Plot the line impedance as a function of distance from load.

Solution: The impedance on the line may be calculated from **Eq. (14.87)** or from **Eq. (14.89)**. The latter is usually more convenient.

(a) To find the impedance at the load, we set $z = 0$,

$$Z(z=0) = Z_0 \frac{Z_L + Z_0 \tanh(0)}{Z_0 + Z_L \tanh(0)} = Z_L = 50 + j50 \quad [\Omega]$$

This, of course, could have been guessed, but the calculation shows that the line impedance formula applies anywhere on the line.

(b) The impedance at a distance $z = 10$ m on the line may also be calculated using **Eq. (14.89)** by setting $z = 10$ m. However, we will use **Eq. (14.87)** to demonstrate its use. To do so, we first calculate the reflection coefficient at the load (see **Example 14.10**):

$$\Gamma_L = \frac{Z_L - Z_0}{Z_L + Z_0} = \frac{50 + j50 - 50}{50 + j50 + 50} = \frac{j}{2 + j1} = \frac{1 + j2}{5} = \frac{1}{\sqrt{5}} e^{j0.352\pi}$$

The line impedance at $z = 10$ m is

$$Z(z) = Z_0 \frac{\left(e^{\gamma z} + \frac{1}{\sqrt{5}} e^{j0.352\pi} e^{-\gamma z}\right)}{\left(e^{\gamma z} - \frac{1}{\sqrt{5}} e^{j0.352\pi} e^{-\gamma z}\right)} = 50 \frac{\left(\sqrt{5} e^{(0.1 + j0.5)} + e^{(-0.1 + j(0.352\pi - 0.5))}\right)}{\left(\sqrt{5} e^{(0.1 + j0.5)} - e^{(-0.1 + j(0.352\pi - 0.5))}\right)} = 50 \frac{2.91 + j1.7}{1.4249 + j0.6695}$$

$$= 106.68 + j9.53 \quad [\Omega]$$

where all angles were measured in radians.

(c) The line impedance (magnitude) as a function of distance from the load is shown in **Figure 14.19** for the first 200 m. Because of attenuation, the load's effect diminishes with distance.

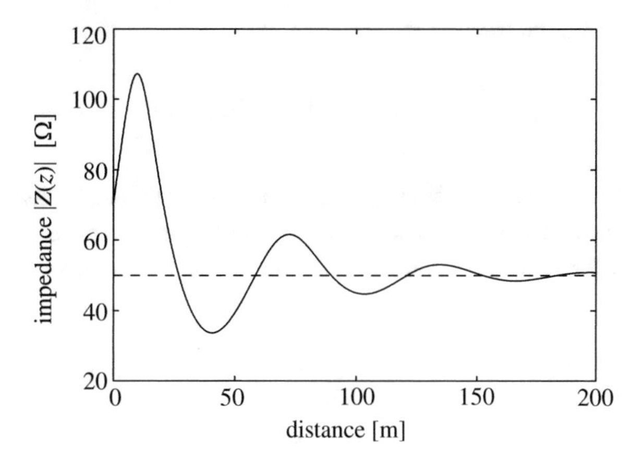

Figure 14.19 Impedance [Ω] along a lossy transmission line

14.7.3 The Lossless, Terminated Transmission Line

In all of the above relations, we assumed a general lossy transmission line in which the propagation constant is a general complex number. There was no reason to do otherwise since we could always replace γ by $\alpha + j\beta$ to obtain the expressions in terms of the attenuation and phase constants α and β for any condition. However, the expression in **Eq. (14.89)** requires the use of hyperbolic sine, cosine, and tangent functions. If the line is lossless, then $\alpha = 0$ and $\gamma = j\beta$. Under these conditions, the voltage and current on the line [setting $\gamma = j\beta$ in **Eq. (14.86)**] are

$$V(z) = V^{+}\left(e^{j\beta z} + \Gamma_L e^{-j\beta z}\right) \quad [\text{V}] \quad \text{and} \quad I(z) = \frac{V^{+}}{Z_0}\left(e^{j\beta z} - \Gamma_L e^{-j\beta z}\right) \quad [\text{A}] \tag{14.93}$$

Similarly, the line impedance of a lossless transmission line is found by setting $\gamma = j\beta$ in **Eq. (14.89)**:

$$\boxed{Z(z) = Z_0 \frac{Z_L + jZ_0 \tan\beta z}{Z_0 + jZ_L \tan\beta z} = Z_0 \frac{Z_L \cos\beta z + jZ_0 \sin\beta z}{Z_0 \cos\beta z + jZ_L \sin\beta z} \quad [\Omega]} \tag{14.94}$$

where $\tanh(j\beta z) = j\tan(\beta z)$ was used. In general, the line impedance is a complex value, as we should expect. The latter expression is also useful in that it indicates explicitly the periodic nature of the line impedance and that the period is directly related to the term βz, which in **Section 14.4** we called the ***electrical length*** of the transmission line. Not surprisingly, the electrical length of the line plays an important role in line behavior.

The generalized reflection coefficient for lossless lines was obtained in **Eq. (14.91)**. Like the line impedance, the reflection coefficient is periodic along the line. This is best seen if the exponential function is written as $e^{-j2\beta z} = \cos 2\beta z - j\sin 2\beta z$. The generalized reflection coefficient now is

$$\boxed{\Gamma(z) = \Gamma_L e^{-j2\beta z} = |\Gamma_L| e^{j(\theta_\Gamma - 2\beta z)} = |\Gamma_L| [\cos(\theta_\Gamma - 2\beta z) + j\sin(\theta_\Gamma - 2\beta z)]} \tag{14.95}$$

Thus, the generalized reflection coefficient for lossless transmission lines can be viewed as having constant amplitude equal to that of $|\Gamma_L|$ but varying in phase along the line as

$$\boxed{\phi_{\Gamma(z)} = \theta_\Gamma - 2\beta z} \tag{14.96}$$

Because of this phase angle, the generalized reflection coefficient has maxima and minima along the line. However, it is more convenient to talk of maxima and minima in voltage or current, or both. Consider **Eq. (14.93)**. Rearranging the terms, we get the voltage on the line as

$$V(z) = V^{+}\left(e^{j\beta z} + \Gamma_L e^{-j\beta z}\right) = V^{+} e^{j\beta z}\left(1 + \Gamma_L e^{-j2\beta z}\right) = V^{+} e^{j\beta z}(1 + \Gamma(z)) \quad [\text{V}] \tag{14.97}$$

Similarly, the current on the line is

$$I(z) = \frac{V^+}{Z_0}\left(e^{j\beta z} - \Gamma_L e^{-j\beta z}\right) = \frac{V^+}{Z_0} e^{j\beta z}\left(1 - \Gamma_L e^{-j2\beta z}\right) = \frac{V^+}{Z_0} e^{j\beta z}(1 - \Gamma(z)) \quad [\mathrm{A}] \tag{14.98}$$

Now, we can discuss the maximum and minimum magnitudes of the voltage. First, we note that the term $e^{j\beta z}$ varies between -1 and $+1$. Thus, its magnitude is 1. Similarly, the generalized reflection coefficient $\Gamma(z)$ varies between $-\ \Gamma(z)$ and $+\ \Gamma(z)$ because the term $e^{-j2\beta z}$ varies between -1 and $+1$. Thus, we can write the maximum and minimum magnitudes of voltage as

$$V_{max} = \left|V^+\right|\left(1 + \left|\Gamma(z)\right|\right) \quad [\mathrm{V}] \tag{14.99}$$

$$V_{min} = \left|V^+\right|\left(1 - \left|\Gamma(z)\right|\right) \quad [\mathrm{V}] \tag{14.100}$$

The same can be done for the current. Following identical steps but starting with **Eq. (14.98)**, we get

$$I_{max} = \frac{V_{max}}{|Z_0|} = \left|\frac{V^+}{Z_0}\right|\left(1 + \left|\Gamma(z)\right|\right) \quad [\mathrm{A}] \tag{14.101}$$

$$I_{min} = \frac{V_{min}}{|Z_0|} = \left|\frac{V^+}{Z_0}\right|\left(1 - \left|\Gamma(z)\right|\right) \quad [\mathrm{A}] \tag{14.102}$$

The ratio between the maximum and minimum voltage (or current) is called the ***standing wave ratio*** (SWR) and is defined as

$$\boxed{\mathrm{SWR} = \frac{V_{max}}{V_{min}} = \frac{I_{max}}{I_{min}} = \frac{1 + \left|\Gamma(z)\right|}{1 - \left|\Gamma(z)\right|} \quad [\text{dimensionless}]} \tag{14.103}$$

The standing wave ratio varies between 1 and ∞. If the reflection coefficient is zero (no reflected waves), the standing wave ratio is 1. If the magnitude of the reflection coefficient is 1, the standing wave ratio is ∞. Thus, a matched load produces no reflected waves and the line should have a standing wave ratio of 1.

Sometimes, the standing wave ratio is known or may be measured. In such cases, the magnitude of the generalized reflection coefficient can be calculated from the standing wave ratio as

$$\boxed{\left|\Gamma(z)\right| = \frac{\mathrm{SWR} - 1}{\mathrm{SWR} + 1}} \tag{14.104}$$

This expression can be substituted in **Eqs. (14.99)** and **(14.100)** to obtain the minimum and maximum voltage on the line in terms of the standing wave ratio:

$$V_{max} = |V^+|(1 + |\Gamma(z)|) = |V^+|\left(1 + \frac{\mathrm{SWR} - 1}{\mathrm{SWR} + 1}\right) = |V^+|\left(\frac{2\mathrm{SWR}}{\mathrm{SWR} + 1}\right) \quad [\mathrm{V}] \tag{14.105}$$

$$V_{min} = |V^+|(1 - |\Gamma(z)|) = |V^+|\left(1 - \frac{\mathrm{SWR} - 1}{\mathrm{SWR} + 1}\right) = |V^+|\left(\frac{2}{\mathrm{SWR} + 1}\right) \quad [\mathrm{V}] \tag{14.106}$$

From the last three equations, and recalling from **Eq. (14.95)** that for lossless lines $|\Gamma_Z| = |\Gamma_L|$, it is apparent that the effect of the standing wave ratio is as follows:

(1) The larger the standing wave ratio, the larger the maximum voltage and the lower the minimum voltage on the line.
(2) If SWR $= 1$, the reflection coefficient is zero. In this case, $V_{max} = V_{min} = |V^+|$. The magnitude of the voltage on the line does not vary. The phase of course varies. This corresponds to a matched load.

(3) If SWR $=\infty$, the magnitude of the reflection coefficient equals 1 ($\Gamma(z)=-1$ or $\Gamma(z)=+1$). In this case, $V_{max}=2|V^+|$ and $V_{min}=0$. We will see shortly that this corresponds to either a short circuit ($\Gamma(z)=-1$) or an open circuit ($\Gamma(z)=+1$). This condition in plane waves was called a complete standing wave.

Now that we have all the tools to calculate the reflection coefficient anywhere on the line as well as the standing wave ratio, we can return to the equations for current and voltage and see how these behave along the line. The basis of calculation are **Eqs. (14.97)** and **(14.98)**. Voltage and current anywhere on the line (including at the load) are

$$\boxed{V(z)=V^+e^{j\beta z}\left(1+\Gamma_L e^{-j2\beta z}\right)=V^+e^{j\beta z}\left(1+|\Gamma_L|e^{j\theta_\Gamma}e^{-j2\beta z}\right)\quad [\mathrm{V}]} \tag{14.107}$$

$$\boxed{I(z)=\frac{V^+}{Z_0}e^{j\beta z}\left(1-\Gamma_L e^{-j2\beta z}\right)=\frac{V^+}{Z_0}e^{j\beta z}\left(1-|\Gamma_L|e^{j\theta_\Gamma}e^{-j2\beta z}\right)\quad [\mathrm{A}]} \tag{14.108}$$

We can also calculate the voltage and current at the load. These are obtained by setting $z=0$:

$$\boxed{V_L=V^+\left(1+|\Gamma_L|e^{j\theta_\Gamma}\right)\quad [\mathrm{V}]}\quad \text{and}\quad \boxed{I_L=\frac{V^+}{Z_0}\left(1-|\Gamma_L|e^{j\theta_\Gamma}\right)\quad [\mathrm{A}]} \tag{14.109}$$

To completely characterize the voltage and current waves, we must find the locations of the minima and maxima on the line. Suppose we plot the voltage and current starting at the load and going toward the generator. For any given load, the load reflection coefficient is known and we can calculate the voltage and current at the load [**Eq. (14.109)**] and the maximum and minimum voltage and current [from **Eqs. (14.99)** through **(14.102)**]. We could, in fact, use **Eqs. (14.107)** and **(14.108)** to plot the voltage and current directly. The only other bit of information needed is the location of minima and maxima in the voltage and current waves. These are found as follows.

From inspection of **Eqs. (14.107)** and **(14.108)**, the minimum in voltage must occur at locations on the line at which the phase $\theta_\Gamma - 2\beta z$ equals $-\pi, -3\pi, -5\pi$, etc. The general condition to be satisfied (taking z to be positive to the left and away from the load) is

$$\theta_\Gamma-2\beta z=-(2n+1)\pi,\quad n=0,1,2,\ldots \tag{14.110}$$

This condition can be verified by direct substitution in **Eq. (14.100)** or **Eq. (14.107)**. On the other hand, the current is maximum at these points because of the negative sign in front of Γ_L in **Eq. (14.102)** or **Eq. (14.108)**. The location of the first minimum in voltage (maximum in current) occurs at

$$\theta_\Gamma-2\beta z_{min}=-\pi\quad\rightarrow\quad z_{min}=\frac{\theta_\Gamma+\pi}{2\beta}\quad [\mathrm{m}] \tag{14.111}$$

The next minimum occurs at

$$\theta_\Gamma-2\beta z=-3\pi\quad\rightarrow\quad z=\frac{\theta_\Gamma+3\pi}{2\beta}=\frac{\theta_\Gamma+\pi}{2\beta}+\frac{\pi}{\beta}\quad [\mathrm{m}] \tag{14.112}$$

From the definition of wavelength, we can also write these relations in terms of the wavelength by using the relation

$$\lambda=\frac{2\pi}{\beta}\quad\rightarrow\quad\frac{\pi}{\beta}=\frac{\lambda}{2}\quad\rightarrow\quad\frac{1}{2\beta}=\frac{\lambda}{4\pi} \tag{14.113}$$

Thus, the conditions for minima are
For the first minimum:

$$\boxed{z_{min}=\frac{\lambda}{4\pi}(\theta_\Gamma+\pi)\quad [\lambda]} \tag{14.114}$$

The unit [λ] shows that the distance is indicated in wavelengths. For any minimum:

$$z_{min} = \frac{\lambda}{4\pi}(\theta_\Gamma + (2n+1)\pi) \quad [\lambda], \quad n = 0,1,2,\ldots \tag{14.115}$$

This has the advantage of being described in terms of increments of $\lambda/2$.

The maxima occur at a distance of $\lambda/4$ on each side of a minimum. We know this must be so since the conditions on the line repeat at increments of $\lambda/2$. Between every two minima there is a maximum. Thus, we can calculate the location of the first voltage maximum by adding $\lambda/4$ in **Eq. (14.115)**. Voltage maxima (current minima) occur at

$$z_{max} = \frac{\lambda}{4\pi}(\theta_\Gamma + (2n+1)\pi) + \frac{\lambda}{4} \quad \rightarrow \quad z_{max} = \frac{\lambda}{4\pi}(\theta_\Gamma + 2n\pi) \quad [\lambda], \quad n = 0,1,2,\cdots \tag{14.116}$$

The complete description of voltage and current on the line is now shown in **Figure 14.20**. Note in particular that the minima are sharper than the maxima. In other words, the voltage or current do not vary sinusoidally. Whenever measurements of standing wave ratio are required, the minima are usually easier to identify. Note also that **Figure 14.20** assumes, arbitrarily, that $V_L > 0$ at the load. This does not have to be so: V_L can be negative or zero.

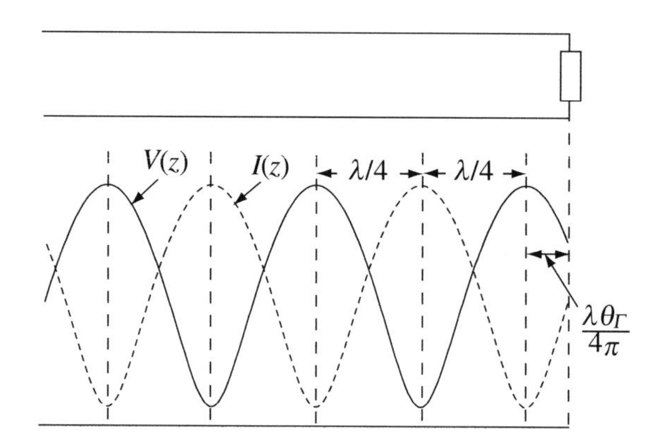

Figure 14.20 Locations of voltage maxima and minima on a transmission line and the relation between voltage and current minima and maxima on the line

The maxima in line impedance occur at locations of voltage maxima (current minima) and minima in line impedance occur at location of voltage minima (current maxima).

From the foregoing discussion, it is clear that voltage and current are highly dependent on the load reflection coefficient and they vary from point to point. From **Eq. (14.94)**, we can also tell that the line impedance varies from point to point. The above relations are general and apply to any load. The only restriction in the above discussion is that the line be lossless.

A number of particular solutions may be obtained for particular loads. These loads are useful because they lead to simple, practical solutions. These are as follows:

(1) Matched load: $Z_L = Z_0$. The load reflection coefficient is zero ($\Gamma_L = 0$).
(2) Short circuited load: $Z_L = 0$. The load reflection coefficient is $\Gamma_L = -1$.
(3) Open circuit load: $Z_L = \infty$. The load reflection coefficient is $\Gamma_L = +1$.
(4) Resistive load: $Z_L = R_L + j0$. The load reflection coefficient is real: $-1 < \Gamma_L < 1$.

These particular types of terminated transmission lines are discussed in the following sections.

Example 14.13 A transmission line with characteristic impedance of 100 Ω and a load of 50 – j50 Ω is connected to a matched generator. The line is very long and the voltage measured at the load is 50 V. Calculate:

(a) The maximum and minimum voltage on the line (magnitude only).
(b) Location of maxima and minima of voltage on the line starting from the load.

Solution: The voltage on the line is best calculated from the standing wave ratio, which, in turn, is calculated from the load reflection coefficient. The location of a voltage maximum is that location at which the impedance is maximum. We first calculate the reflection coefficient and then the standing wave ratio. From these, the minimum and maximum amplitudes are calculated using **Eqs. (14.105)** and **(14.106)**. The location of minima and maxima is calculated from **Eqs. (14.115)** and **(14.116)**.

(a) The reflection coefficient at the load is

$$\Gamma_L = \frac{Z_L - Z_0}{Z_L + Z_0} = \frac{-50 - j50}{150 - j50} = \frac{-1 - j1}{3 - j1} = \frac{-1 - j2}{5} = \frac{1}{\sqrt{5}} e^{-j0.6476\pi}$$

The standing wave ratio is calculated from the magnitude of the load reflection coefficient:

$$\text{SWR} = \frac{1 + |\Gamma(z)|}{1 - |\Gamma(z)|} = \frac{1 + |\Gamma_L|}{1 - |\Gamma_L|} = \frac{1 + 1/|\sqrt{5}|}{1 - 1/|\sqrt{5}|} = \frac{|\sqrt{5}| + 1}{|\sqrt{5}| - 1} = 2.618$$

To calculate the minimum and maximum voltage on the line, we first need to calculate the forward wave amplitude. The total amplitude at the load is known. From **Eq. (14.109)**, we get at the load

$$V_L = V^+(1 + \Gamma_L) \quad \rightarrow \quad V^+ = \frac{V_L}{(1 + \Gamma_L)} = \frac{50}{(1 + (-1 - j2)/5)} = \frac{125}{2 - j1} = 25(2 + j1) \quad [\text{V}]$$

The magnitude of the forward-propagating voltage is

$$|V^+| = |25(2 + j1)| = |25\sqrt{5}| = 55.9 \quad [\text{V}]$$

Thus, the maximum and minimum voltages are

$$V_{max} = |V^+|\left(\frac{2\text{SWR}}{\text{SWR} + 1}\right) = 55.9\left(\frac{2 \times 2.618}{3.618}\right) = 80.9 \quad [\text{V}]$$

$$V_{min} = |V^+|\left(\frac{2}{\text{SWR} + 1}\right) = 55.9\left(\frac{2}{3.618}\right) = 30.9 \quad [\text{V}].$$

(b) The first voltage minimum from the load is calculated from **Eq. (14.114)**:

$$z_{min} = \frac{\lambda}{4\pi}(\theta_\Gamma + \pi) = \frac{\lambda}{4\pi}(-0.6476\pi + \pi) = 0.0881\lambda$$

All other minima are to the left, in increments of $\lambda/2$. Thus, the minima occur at $z = 0.088\lambda, 0.588\lambda, 1.088\lambda, 1.558\lambda, 2.088\lambda, 2.558\lambda$, etc.

The maxima are one-quarter wavelength to the left and right of the above minima. These occur at $z = 0.338\lambda, 0.838\lambda, 1.338\lambda, 1.838\lambda, 2.338\lambda, 2.838\lambda$, etc.

14.7.4 The Lossless, Matched Transmission Line

A matched transmission line is a line on which the load is equal to the characteristic impedance of the line:

$$\boxed{Z_L = Z_0} \tag{14.117}$$

Substitution of this condition in **Eq. (14.83)** results in a zero reflection coefficient at the load: $\Gamma_L = 0$. Thus, the line impedance anywhere on the line is

$$Z(z) = Z_0 \frac{Z_0 + jZ_0 \tan \beta z}{Z_0 + jZ_0 \tan \beta z} = Z_0 \quad [\Omega] \tag{14.118}$$

Therefore, the impedance on the line for a matched load is constant and equal to Z_0.

The other relations on the line are also obtained by substituting $Z_L = Z_0$ and $\Gamma_L = 0$. Thus, the standing wave ratio is SWR $= 1$ anywhere on the line. The voltage and current on the line are

$$V(z) = V^+ e^{j\beta z} \quad [\mathrm{V}] \quad \text{and} \quad I(z) = \frac{V^+}{Z_0} e^{j\beta z} \quad [\mathrm{A}] \tag{14.119}$$

That is, the line voltage and current have only forward-propagating terms, as we expect with a zero reflection coefficient.

In summary, a matched load produces no reflected waves and, therefore, no standing waves on the line. All power on the line is transferred to the load.

14.7.5 The Lossless, Shorted Transmission Line

A shorted transmission line is characterized by $Z_L = 0$. From **Eq. (14.83)**, the reflection coefficient is $\Gamma_L = -1$. For the same reason, SWR $= \infty$. The line impedance is now [see **Eq. (14.94)**]

$$Z(z) = Z_0 \frac{jZ_0 \tan \beta z}{Z_0} = jZ_0 \tan \beta z \quad [\Omega] \tag{14.120}$$

The line impedance of a shorted transmission line is purely imaginary and varies between $-\infty$ and ∞. It has the following properties:

(1) $\Gamma_L = -1$, SWR $= \infty$.

(2) The line impedance is zero at the load and at any value $\beta z = n\pi$, $n = 1,2,\ldots$. In terms of wavelength, the line impedance is zero at $z = n\lambda/2$, $n = 0,1,2,\ldots$ and is infinite at $z = n\lambda/2 + \lambda/4$, $n = 0,1,2,\ldots$.

(3) The line impedance is purely imaginary and alternates between positive and negative values, as shown in **Figure 14.21**. The impedance is positive (inductive) for $n\lambda/2 < z < n\lambda/2 + \lambda/4$ and negative (capacitive) between $n\lambda/2 + \lambda/4 < z < n\lambda/2 + \lambda/2$, $n = 0,1,2,\ldots$ The line impedance changes from $+\infty$ to $-\infty$ at $z = n\lambda/2 + \lambda/4$.

(4) A shorted transmission line behaves as an inductor or a capacitor, depending on the location on the line. A capacitance or an inductance may be designed by simply cutting a line of appropriate length as indicated in **(3)**. In this sense, shorted transmission lines are viewed as circuit elements.

(5) The conditions on a shorted transmission line repeat at intervals of $\lambda/2$; that is, if we add or remove a section of length $\lambda/2$ (or any integer multiple of $\lambda/2$), the conditions on the line do not change.

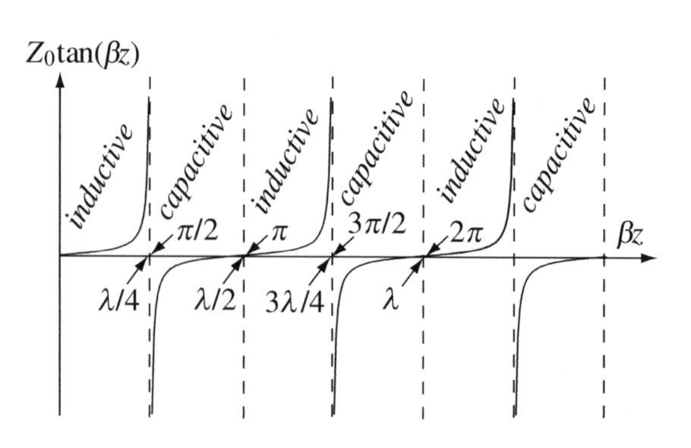

Figure 14.21 Line impedance on a shorted transmission line

The line voltage and line current are [setting $\Gamma_L = -1$ in **Eqs. (14.107)** and **(14.108)**]

$$\boxed{V(z) = V^+ e^{j\beta z}\left(1 - e^{-j2\beta z}\right) \quad [\text{V}]} \quad \text{and} \quad \boxed{I(z) = \frac{V^+}{Z_0} e^{j\beta z}\left(1 + e^{-j2\beta z}\right) \quad [\text{A}]} \tag{14.121}$$

In particular, at the load ($z = 0$), we get

$$V_L = 0 \quad [\text{V}] \quad \text{and} \quad I_L = \frac{2V^+}{Z_0} \quad [\text{A}] \tag{14.122}$$

Thus, whereas the voltage at the load must be zero, the current must be twice the forward-propagating current. This, of course, is a consequence of the fact that there is no transfer of power into the load and the reflected current is equal in magnitude and phase to the forward current.

14.7.6 The Lossless, Open Transmission Line

An open transmission line has an infinite impedance as load. Since $Z_L \to \infty$, the reflection coefficient at the load is $\Gamma_L = +1$. For the same reason, SWR $= \infty$. Substitution of Z_L into the line impedance in **Eq. (14.94)** gives (since $Z_L \gg Z_0$)

$$\boxed{Z(z) = Z_0 \frac{Z_L + jZ_0 \tan\beta z}{Z_0 + jZ_L \tan\beta z} = Z_0 \frac{Z_L}{jZ_L \tan\beta z} = -jZ_0 \cot\beta z \quad [\Omega]} \tag{14.123}$$

This result is very similar to the result for the shorted transmission line. The properties of this line are summarized as follows:

(1) $\Gamma_L = +1$, SWR $= \infty$.
(2) The line impedance is infinite at the load and at any value $\beta z = n\pi$, $n = 1,2,\ldots$ In terms of wavelength, the line impedance is infinite at $z = n\lambda/2$, $n = 0,1,2,\ldots$ The line impedance is zero at $z = n\lambda/2 + \lambda/4$, $n = 0,1,2,\ldots$
(3) The line impedance is purely imaginary and alternates between positive and negative values, as shown in **Figure 14.22**. The impedance is negative (capacitive) for $n\lambda/2 < z < n\lambda/2 + \lambda/4$ and positive (inductive) between $n\lambda/2 + \lambda/4 < z < n\lambda/2 + \lambda/2$, $n = 0,1,2,\ldots$ The line impedance changes from $+\infty$ to $-\infty$ at $z = n\lambda/2$.
(4) An open transmission line behaves as an inductor or a capacitor, depending on the location on the line. A capacitance or an inductance may be designed by simply cutting a line of appropriate length as indicated in **(3)**. Open transmission lines may also be viewed as circuit elements.
(5) The conditions on an open transmission line repeat at intervals of $\lambda/2$; that is, if we add or remove a section of length $\lambda/2$ (or any integer multiple of $\lambda/2$), the conditions on the line are not affected.
(6) The conditions on an open transmission line are identical to those of a shorted transmission line if their length differs by an odd multiple of $\lambda/4$. This can be seen by direct comparison of **Figures 14.22** and **14.21**.

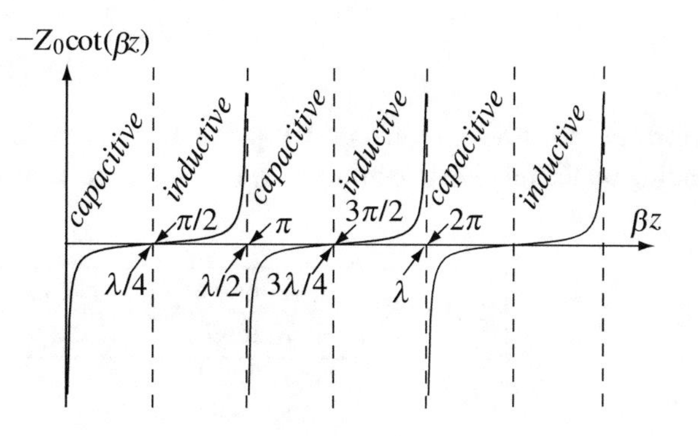

Figure 14.22 Line impedance on an open transmission line

The line voltage and line current on the open transmission line are

$$\boxed{V(z) = V^+ e^{j\beta z}\left(1 + e^{-j2\beta z}\right) \quad [\mathrm{V}]} \quad \text{and} \quad \boxed{I(z) = \frac{V^+}{Z_0} e^{j\beta z}\left(1 - e^{-j2\beta z}\right) \quad [\mathrm{A}]} \tag{14.124}$$

In particular, at the load ($z = 0$), we get

$$V_L = 2V^+ \quad [\mathrm{V}] \quad \text{and} \quad I_L = 0 \tag{14.125}$$

Thus, maximum voltage occurs at the load, whereas maximum current occurs at $\lambda/4$ from the load. Again, there is no transfer of power into the load and the reflected voltage is equal to the forward-propagating voltage (and in the same phase).

Suppose now that we perform an experiment. First, we short a transmission line and obtain the line impedance at a point z. Then, we open the line and obtain the impedance at the same point. The shorted and open line impedances are those given in **Eqs. (14.120)** and **(14.123)**. If we take the product of these two impedances, we get

$$(jZ_0 \tan\beta z)(-jZ_0 \cot\beta z) = Z_0^2 \tag{14.126}$$

Perhaps a bit unexpected result but it gives us yet another way of calculating or measuring the characteristic impedance of a transmission line. The characteristic impedance of any lossless line is given as

$$\boxed{Z_0 = \sqrt{Z_{short} Z_{open}} \quad [\Omega]} \tag{14.127}$$

where Z_{short} is the line impedance with shorted load and Z_{open} is the line impedance with open load.

Exercise 14.6 The characteristic impedance of a transmission line is not known. To determine it, it is suggested to measure the short and open impedances at a point on the line; that is, the load is shorted and the impedance at a point on the line is measured. Then, the load is disconnected (open line) and the impedance at the same point on the line is again measured. These measurements give an open line impedance of $-j50$ and a shorted line impedance of $j75$. Calculate the characteristic impedance of the line.

Answer 61.24 Ω.

Exercise 14.7 Show that the relation in **Eq. (14.127)** holds for real or complex characteristic impedances.

14.7.7 The Lossless, Resistively Loaded Transmission Line

The discussion in **Sections 14.7.1** and **14.7.2** was in terms of a general load, but it applies equally well for a resistive load: $Z_L = R_L + j0$. The reflection coefficient at the load is

$$\boxed{\Gamma_L = \frac{V^-}{V^+} = \frac{R_L - Z_0}{R_L + Z_0}} \tag{14.128}$$

and since Z_0 is also real for lossless lines [see **Eq. (14.49)**], the reflection coefficient is real. It can be either positive or negative depending on the relative magnitudes of R_L and Z_0 and varies between -1 and $+1$. The line impedance is now given as

$$\boxed{Z(z) = Z_0 \frac{R_L + jZ_0 \tan\beta z}{Z_0 + jR_L \tan\beta z} = Z_0 \frac{R_L \cos\beta z + jZ_0 \sin\beta z}{Z_0 \cos\beta z + jR_L \sin\beta z} \quad [\Omega]} \tag{14.129}$$

This impedance is maximum at locations of maximum voltage and minimum at locations of minimum voltage, as described in **Section 14.7.3**. The main difference between a resistive load and a general load is that for a general load, the phase angle of the load reflection coefficient can have any value. On the other hand, for a resistive load, the phase angle can be either zero or $-\pi$. This can be seen from **Eq. (14.128)**. There are two possible situations:

(1) $R_L > Z_0$. In this case, Γ_L is real, positive and we can write

$$\Gamma_L = \frac{R_L - Z_0}{R_L + Z_0} \quad \rightarrow \quad \Gamma_L = |\Gamma_L| e^{j0} \tag{14.130}$$

Now, if we substitute $\theta_\Gamma = 0$ in **Eqs. (14.107)** and **(14.108)**, we obtain the general voltage and current waves on the line:

$$\boxed{V(z) = V^+ e^{j\beta z}\left(1 + \Gamma_L e^{-j2\beta z}\right) \quad [\text{V}]} \tag{14.131}$$

$$\boxed{I(z) = \frac{V^+}{Z_0} e^{j\beta z}\left(1 - \Gamma_L e^{-j2\beta z}\right) \quad [\text{A}]} \tag{14.132}$$

The voltage and current at the load are

$$V_L = V^+(1 + \Gamma_L) \quad [\text{V}] \quad \text{and} \quad I_L = \frac{V^+}{Z_0}(1 - \Gamma_L) \quad [\text{A}] \tag{14.133}$$

The locations of voltage minima are now [see **Eqs. (14.114)** and **(14.115)**]

$$\boxed{z_{min} = (2n + 1)\frac{\lambda}{4} \quad [\lambda], \quad n = 0,1,2,\ldots} \tag{14.134}$$

Thus, the first minimum in voltage occurs at $n = 0$:

$$z_{min} = \frac{\pi}{2\beta} = \frac{\lambda}{4} \quad [\lambda] \tag{14.135}$$

Similarly, the locations of voltage maxima are **[Eq. (14.116)]**

$$\boxed{z_{max} = n\frac{\lambda}{2} \quad [\lambda], \quad n = 0,1,2,\ldots} \tag{14.136}$$

The first voltage maximum is at the load ($z = 0$). The following voltage maxima (current minima) are at increments of $\lambda/2$ from the load. The voltage and current minima and maxima are shown in **Figure 14.23a**.

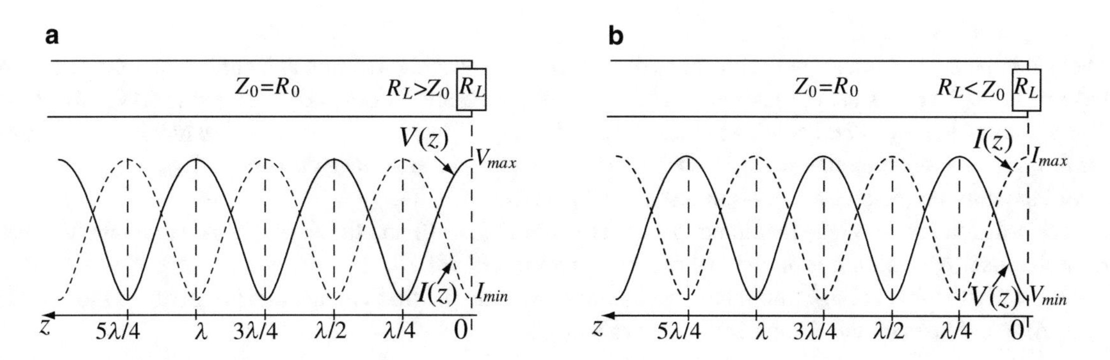

Figure 14.23 (**a**) Voltage and current maxima and minima for $R_L > Z_0$. (**b**) Voltage and current maxima and minima for $R_L < Z_0$

(2) $R_L < Z_0$. In this case, Γ_L is real and negative and we can write

$$\Gamma_L = \frac{R_L - Z_0}{R_L + Z_0} \quad \rightarrow \quad \Gamma_L = -|\Gamma_L| = |\Gamma_L| e^{-j\pi} \tag{14.137}$$

Now, if we substitute $\theta_\Gamma = -\pi$ in **Eqs. (14.107)** and **(14.108)**, we obtain the general voltage and current waves on the line:

$$\boxed{V(z) = V^+ e^{j\beta z}\left(1 + |\Gamma_L| e^{-j\pi} e^{-j2\beta z}\right) \quad [\text{V}]} \tag{14.138}$$

$$\boxed{I(z) = \frac{V^+}{Z_0} e^{j\beta z}\left(1 - |\Gamma_L| e^{-j\pi} e^{-j2\beta z}\right) \quad [\text{A}]} \tag{14.139}$$

The voltage and current at the load are

$$V_L = V^+(1 - |\Gamma_L|) \quad [\text{V}] \quad \text{and} \quad I_L = \frac{V^+}{Z_0}(1 + |\Gamma_L|) \quad [\text{A}] \tag{14.140}$$

The locations of minima in voltage are

$$\boxed{z_{min} = n\frac{\lambda}{2} \quad [\lambda], \quad n = 0,1,2,\ldots} \tag{14.141}$$

Thus, the first minimum in voltage occurs at $z = 0$. Subsequent minima occur at intervals of $\lambda/2$ from the load. The first voltage maximum occurs at $z = \lambda/4$ and the general relation for voltage maxima is

$$\boxed{z_{max} = (2n + 1)\frac{\lambda}{4} \quad [\lambda], \quad n = 0,1,2,\ldots} \tag{14.142}$$

The complete description of voltage and current on the line for $R_L < Z_0$ is shown in **Figure 14.23b**. Note also that the maximum and minimum line impedance are given as

$$\boxed{Z_{max} = Z_0\frac{(1 + |\Gamma_L|)}{(1 - |\Gamma_L|)} = Z_0\text{SWR} \quad [\Omega]} \tag{14.143}$$

$$\boxed{Z_{min} = Z_0\frac{(1 - |\Gamma_L|)}{(1 + |\Gamma_L|)} = \frac{Z_0}{\text{SWR}} \quad [\Omega]} \tag{14.144}$$

The properties of line impedance on a resistively loaded line are

(1) $-1 < \Gamma_L < +1$, $1 < \text{SWR} < \infty$.
(2) The line impedance is maximum at locations of voltage maxima and minimum at locations of voltage minima. These locations are given in **Eqs. (14.134)** and **(14.136)** for $R_L > Z_0$ and in **Eqs. (14.141)** and **(14.142)** for $R_L < Z_0$.
(3) The line impedance can be complex as can be seen from **Eq. (14.129)**, but it is always real at locations of voltage maxima and voltage minima for any lossless line. The impedance at voltage maxima is $Z_{max} = Z_0 * \text{SWR}$, whereas at voltage minima (current maxima), it is $Z_{min} = Z_0/\text{SWR}$.
(4) For $R_L > Z_0$, the first voltage maximum occurs at the load ($z = 0$) and the first voltage minimum at a distance $\lambda/4$ from the load. All conditions on the line repeat at intervals of $\lambda/2$.
(5) For $R_L < Z_0$, the first voltage minimum occurs at the load and the first voltage maximum at a distance $\lambda/4$ from the load. All conditions on the line repeat at intervals of $\lambda/2$.

In effect, the main difference between a general load and a resistive load is the location of the minima and maxima. If the load is such that the magnitude of the reflection coefficient at the load is the same for resistive and arbitrary loads, the voltage and current on the line will be the same in both cases but displaced by the value of z_{min} in **Eq. (14.134)** or **(14.115)**. In other words, if we take an arbitrary load and calculate all circuit parameters, we obtain the standing wave pattern for the line. The line can now be shortened by the magnitude of z_{min} or lengthened by $\lambda/2 - z_{min}$ to obtain an identical circuit but with a resistive loading which has the same reflection coefficient magnitude. We will use this property of transmission lines in the following chapter.

14.8 Power Relations on a General Transmission Line

The power relation on a line can be written directly from the current and voltage on the line. As a rule, when we use the term "power" we mean time-averaged power. The power at a distance z_0 from the load can be calculated by assuming that the load is at $z = 0$ and the input is at $z = z_0$ as shown in **Figure 14.24**. For this condition, the line voltage and current for a general lossy line are given in **Eq. (14.86)**. Setting $z = z_0$ gives the voltage and current as

$$V(z_0) = V^+(e^{\gamma z_0} + \Gamma_L e^{-\gamma z_0}) \quad [\text{V}] \quad \text{and} \quad I(z_0) = \frac{V^+}{Z_0}(e^{\gamma z_0} - \Gamma_L e^{-\gamma z_0}) \quad [\text{A}] \tag{14.145}$$

where Z_0 is the line characteristic impedance given in **Eq. (14.42)** and is, in general, a complex number. Now, the power entering this section of the transmission line is calculated from the current and voltage on the line at this point:

$$\begin{aligned}
P_{z_0} &= \frac{1}{2}\text{Re}\left\{V_{z_0} I^*_{z_0}\right\} = \frac{1}{2}\text{Re}\left\{[V^+(e^{\gamma z_0} + \Gamma_L e^{-\gamma z_0})]\left[\frac{V^+}{Z_0}(e^{\gamma z_0} - \Gamma_L e^{-\gamma z_0})\right]^*\right\} \\
&= \frac{|V^+|^2}{2}\text{Re}\left\{\left(e^{(\alpha+j\beta)z_0} + |\Gamma_L| e^{j\theta_{\Gamma_L}} e^{-(\alpha+j\beta)z_0}\right)\frac{\left(e^{(\alpha-j\beta)z_0} - |\Gamma_L| e^{-j\theta_{\Gamma_L}} e^{-(\alpha-j\beta)z_0}\right)}{Z_0^*}\right\} \\
&= \frac{|V^+|^2}{2|Z_0|}\text{Re}\left\{\left(e^{2\alpha z_0} + |\Gamma_L| e^{j(\theta_{\Gamma_L} - 2\beta z_0)} - |\Gamma_L| e^{-j(\theta_{\Gamma_L} - 2\beta z_0)} - |\Gamma_L|^2 e^{-2\alpha z_0}\right) e^{-j\theta_{Z_0}}\right\} \\
&= \frac{|V^+|^2}{2|Z_0|}\text{Re}\left\{\left(e^{2\alpha z_0} + j2|\Gamma_L|\sin(\theta_{\Gamma_L} - 2\beta z_0) - |\Gamma_L|^2 e^{-2\alpha z_0}\right) e^{-j\theta_{Z_0}}\right\} \\
&= \frac{|V^+|^2}{2|Z_0|}\left(e^{2\alpha z_0} - |\Gamma_L|^2 e^{-2\alpha z_0}\right)\cos(\theta_{Z_0}) \quad [\text{W}]
\end{aligned} \tag{14.146}$$

where θ_{Γ_L} is the phase angle of the reflection coefficient, θ_{Z_0} is the phase angle of the characteristic impedance and θ_{Z_0} is the phase angle at $z = z_0$. To summarize

$$\boxed{P_{z_0} = \frac{|V^+|^2}{2|Z_0|}\left(e^{2\alpha z_0} - |\Gamma_L|^2 e^{-2\alpha z_0}\right)\cos(\theta_{Z_0}) \quad [\text{W}]} \tag{14.147}$$

This relation has two components: The first is forward-propagating toward the load (in the negative z direction according to our convention which defines the load as reference) and the second in the positive z direction (from load to generator). Both are real powers and the total power is the sum of the two.

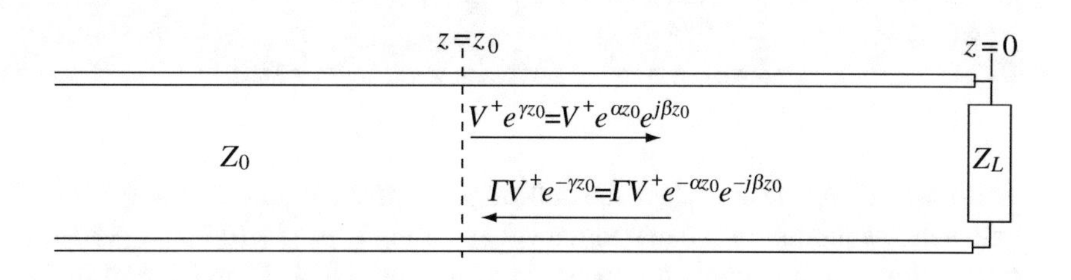

Figure 14.24 Notation used to calculate power relations on the transmission line

The power entering any section of the line may now be evaluated by setting the correct value for z_0. The power at the load may be found by setting $z_0 = 0$:

$$P_{load} = \frac{|V^+|^2}{2|Z_0|}\left(1 - |\Gamma_L|^2\right)\cos(\theta_{Z_0}) \quad [\text{W}] \tag{14.148}$$

The general power relation in **Eq. (14.147)** may be simplified under certain conditions. If there is only a forward-propagating wave, the forward-propagating voltage, current, and power are

$$V^+(z_0) = V^+e^{\gamma z_0} \quad [\text{V}], \quad I^+(z_0) = \frac{V^+}{Z_0}e^{\gamma z_0} \quad [\text{A}], \quad P^+(z_0) = \frac{|V^+|}{2|Z_0|}e^{2\alpha z_0}\cos(\theta_{Z_0}) \quad [\text{W}] \tag{14.149}$$

For the backward propagating wave alone

$$V^-(z_0) = V^+\Gamma_L e^{\gamma z_0} \quad [\text{V}], \quad I^-(z_0) = -\frac{V^+}{Z_0}\Gamma_L e^{\gamma z_0} \quad [\text{A}], \tag{14.150}$$

$$P^-(z_0) = \frac{|V^+|^2|\Gamma_L|^2}{2|Z_0|}e^{-2\alpha z_0}\cos(\theta_{Z_0}) \quad [\text{W}] \tag{14.151}$$

For lossless lines, the attenuation constant is zero and the characteristic impedance is real. The power at any point on the line is therefore

$$\boxed{P_{z_0} = \frac{|V^+|^2}{2Z_0}\left(1 - |\Gamma_L|^2\right) \quad [\text{W}]} \tag{14.152}$$

It is worth mentioning again that this power is positive propagating from generator to load. Note that if the reflection coefficient is zero, all power on the line is transferred to the load, although this does not imply maximum power transfer from generator to load.

The expressions in **Eq. (14.147)** and **Eq. (14.152)** were obtained from the general circuit expression $P = (1/2)\text{Re}\{VI^*\}$. This means that if the voltage on and current in the line are available or can be obtained by any means, they can be used to calculate power without resorting to the equivalent expressions in **Eq. (14.147)** and **Eq. (14.152)**, which require calculation of V^+ and Γ_L.

The instantaneous power on the line is calculated similarly by multiplying the instantaneous voltage by instantaneous current.

Example 14.14 Consider, again, the transmission line in **Example 14.13** but with a resistive load of 50 Ω. The characteristic impedance of the line is 100 Ω. The line is very long and the voltage measured at the load is 50 V. Calculate:

(a) Maximum and minimum voltage on the line (magnitude only).
(b) Location of voltage maxima and minima on the line (starting from the load).
(c) The minimum and maximum impedance on the line. Where do these occur?
(d) Power transmitted to the load.

Solution: First, we calculate the reflection coefficient and standing waves ratio. From the reflection coefficient, we calculate the minimum and maximum voltage using **Eqs. (14.105)** and **(14.106)**. From the standing wave ratio, we calculate the minimum and maximum impedance using **Eqs. (14.143)** and **(14.144)**. The power transmitted to the load is given in **Eq. (14.152)**.

(a) The reflection coefficient at the load is

$$\Gamma_L = \frac{Z_L - Z_0}{Z_L + Z_0} = \frac{50 - 100}{50 + 100} = -\frac{1}{3} = \frac{1}{3}e^{-j\pi} \quad \rightarrow \quad |\Gamma_L| = \frac{1}{3}, \quad \theta_L = -\pi$$

The standing wave ratio is

$$\text{SWR} = \frac{1 + |\Gamma_L|}{1 - |\Gamma_L|} = \frac{3+1}{3-1} = 2$$

To calculate the minimum and maximum voltage on the line, we first need to calculate the forward wave amplitude. The total amplitude at the load is known. From **Eq. (14.109)**, we get at the load

$$V_L = V^+(1 + \Gamma_L) \quad \rightarrow \quad V^+ = \frac{V_L}{(1 + \Gamma_L)} = \frac{50}{(1 - 1/3)} = \frac{50}{2/3} = 75 \quad [\text{V}]$$

Thus, the maximum and minimum voltages are

$$V_{max} = |V^+|\left(\frac{2\text{SWR}}{\text{SWR} + 1}\right) = 75 \times \frac{4}{3} = 100 \quad [\text{V}],$$

$$V_{min} = |V^+|\left(\frac{2}{\text{SWR} + 1}\right) = 75 \times \frac{2}{3} = 50 \quad [\text{V}]$$

Note that the minimum voltage occurs at the load, as required for $R_L < Z_0$ ($V_{min} = V_L = 50$ V).

(b) The first voltage minimum is at the load. The minima therefore occur at $z = 0, 0.5\lambda, 1.0\lambda, 1.5\lambda, 2.0\lambda$, etc. The voltage maxima are one-quarter wavelength to the left of the minima. These occur at: $z = 0.25\lambda, 0.75\lambda, 1.25\lambda, 1.75\lambda, 2.25\lambda$, etc.

(c) The minimum and maximum impedances on the line are

$$Z_{min} = \frac{Z_0}{\text{SWR}} = 50, \quad Z_{max} = Z_0 * \text{SWR} = 200 \quad [\Omega]$$

Minimum impedance occurs at the points of minimum voltage and maximum impedance at the points of maximum voltage.

(d) Power transmitted into the load [from **Eq. (14.152)**] is

$$P_L = \frac{|V^+|^2}{2Z_0}\left[1 - |\Gamma_L|^2\right] = \frac{75^2}{200}\left[1 - \left(\frac{1}{3}\right)^2\right] = 25 \quad [\text{W}]$$

This is about 89% of the power of the incident wave and about 11% of the power is reflected.

Exercise 14.8 Repeat **Example 14.14** but with the load and characteristic impedance interchanged. That is, $Z_L = 100\ \Omega$ and $Z_0 = 50\ \Omega$. The voltage measured at the load is 50 V.

Answer

(a) $\Gamma_L = 1/3$, $\theta_L = 0$, SWR $= 2$, $V^+ = 37.5$ V, $V_{max} = 50$ V (at load), $V_{min} = 25$ V.

(b) Voltage maxima (current minima) at $z = 0, 0.5\lambda, 1.0\lambda, 1.5\lambda, 2.0\lambda$, etc. Voltage minima (current maxima) at $z = 0.25\lambda, 0.75\lambda, 1.25\lambda, 1.75\lambda, 2.25\lambda$, etc.

(c) $Z_{min} = 25\ \Omega$, $Z_{max} = 100\ \Omega$.

(d) $P_L = 12.5$ W.

14.9 Resonant Transmission Line Circuits

We have already described the shorted and open transmission lines. These lines have an impedance which is purely imaginary and can be either positive or negative. Thus, a segment of shorted transmission line of appropriate length will behave as an inductor or as a capacitor. Similarly, an impedance of almost any value may be obtained by appropriate choice of lines and loads. It is, therefore, possible to use these line segments to build particular circuits with given properties. We will take advantage of this aspect of transmission lines in the following chapter, where we make use of shorted, open, and loaded transmission lines to match line to line or load to line.

To demonstrate the use of transmission lines as circuit elements, we discuss here the design of resonant transmission lines. We recall that for a circuit to resonate, it must have the following properties:

(1) A capacitance and inductance must be present.
(2) The capacitance and inductance may be connected in series to form a series resonator **(Figure 14.25a)** or in parallel to form a parallel resonator **(Figure 14.25b)**.
(3) Lossy elements in the form of resistances or conductances may exist in either case **(Figures 14.25c** and **14.25d)**.
(4) At resonance, the impedance of the resonant circuit is real.
(5) As the frequency changes from below to above resonance, the nature of the circuit changes from capacitive to inductive or vice versa.

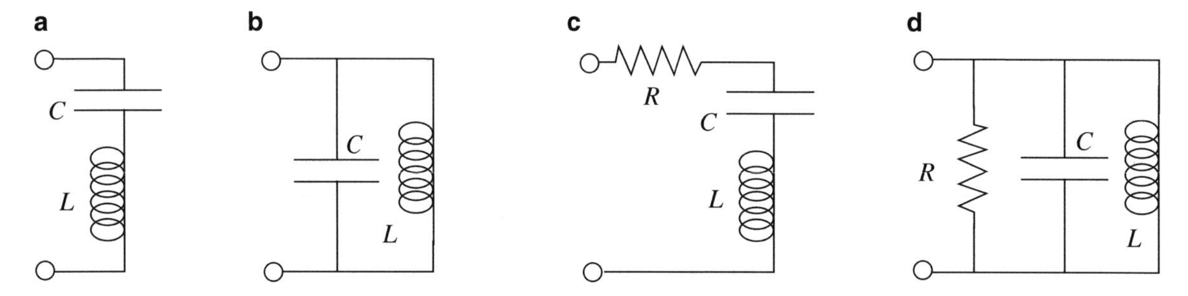

Figure 14.25 Resonating circuits. (**a**) Series resonator. (**b**) Parallel resonator. (**c**) Lossy series resonator. (**d**) Lossy parallel resonator

A parallel resonant circuit is shown in **Figure 14.26** together with a possible transmission line implementation. The lengths d_1 and d_2 must be determined to satisfy the required conditions for resonance. We will use lossless transmission lines for the calculations that follow. Suppose that the left branch in **Figure 14.26b** is made so that it is equivalent to a capacitance C. The input impedance of this segment must be

$$Z_{in1} = jZ_{01}\tan\beta_1 d_1 = \frac{1}{j\omega C} \quad [\Omega] \tag{14.153}$$

where Z_{01} is the characteristic impedance of the shorted line forming this segment and β_1 is the phase constant of the segment.

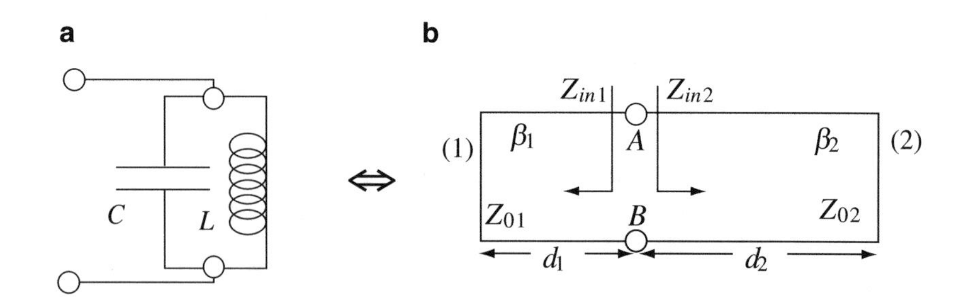

Figure 14.26 (**a**) A parallel resonant circuit. (**b**) The transmission line implementation of the parallel resonant circuit

Line (2), which is also shorted, must behave as an equivalent inductor of inductance L. Its input impedance is

$$Z_{in2} = jZ_{02}\tan\beta_2 d_2 = j\omega L \quad [\Omega] \tag{14.154}$$

At resonance, the impedance of the circuit is ∞ or, alternatively, the admittance of the parallel circuit is zero. Using the latter, we can write

$$\frac{1}{Z_{in1}} + \frac{1}{Z_{in2}} = \frac{1}{jZ_{01}\tan\beta_1 d_1} + \frac{1}{jZ_{02}\tan\beta_2 d_2} = 0 \tag{14.155}$$

Rearranging terms, we get the required condition for resonance:

$$\boxed{Z_{01}\tan\beta_1 d_1 + Z_{02}\tan\beta_2 d_2 = 0} \tag{14.156}$$

or in terms of the frequency itself ($\beta = \omega/v_p = 2\pi f/v_p$), we can write

$$Z_{01}\tan\frac{2\pi f d_1}{v_{p1}} + Z_{02}\tan\frac{2\pi f d_2}{v_{p2}} = 0 \tag{14.157}$$

This is a transcendental equation and we cannot solve it explicitly. However, since β_1, β_2, Z_{01}, and Z_{02} are known from the line parameters, all that remains to be defined are d_1 and d_2. This can be done in two ways: If the frequency is given, then a relation between d_1 and d_2 is obtained. We fix one value and find the second such that it satisfies the relation. Alternatively, we can fix both d_1 and d_2 and find the frequencies at which the resulting circuit resonates. Any method of solving the transcendental equation in **Eqs. (14.156)** or **(14.157)** is acceptable for solution.

Note that we should expect multiple solutions from the periodic nature of the tangent functions. The resonant circuit resonates at an infinite number of discrete frequencies.

In general, shorted transmission lines are preferred for a variety of reasons, including noise, but sometimes, especially when the line is used to measure external conditions, an open line is more practical. For example, resonant coaxial transmission lines are often used to measure water content in snow packs to evaluate runoff levels and water reserves. An open resonant circuit is most useful since it can then simply be pushed into the snow pack for measurement purposes. Similarly, a resonator that measures pollutants in air must be open in one way or another. If coaxial lines are used, the resonator must be made of open segments. If, on the other hand, parallel plates are used, shorted lines may be used because the structure itself is open. Also to be noted is that we assumed lossless lines. Lossy lines do not change the resonance condition but reduce the quality factor of the resonator and as a consequence the resonance is less sharp and impedance at resonance is finite.

The resonant structure discussed here is sometimes called a tapped transmission line resonator. There are other implementations of transmission line resonators including quarter wavelength shorted line resonator (parallel resonance) and half-wavelength open line resonator (parallel resonance). Their analysis is also based on impedance but we will not pursue these.

Example 14.15 It is required to design a transmission line resonator made of a parallel plate transmission line. The dimensions and material properties are shown in **Figure 14.27a**. The length of the line at left is 0.4 m. The length of the line at right is 0.2 m.

(a) Calculate the first two resonant frequencies of the resonator.

(b) Suppose the short on the right-hand side is removed **(Figure 14.27b)**. What are now the two lowest resonant frequencies?

Solution: We will use **Eq. (14.157)**. The intrinsic line impedance and phase velocity are calculated first. Because both line segments are identical in properties, the characteristic impedance is not needed, but we will use it nevertheless for generality. With these, we calculate the expression in **Eq. (14.157)** for a range of frequencies (say from 1 MHz to 3,000 MHz, in increments of 1 MHz) and plot the result. The intersections with the real axis gives the resonant frequencies, provided the transition is smooth.

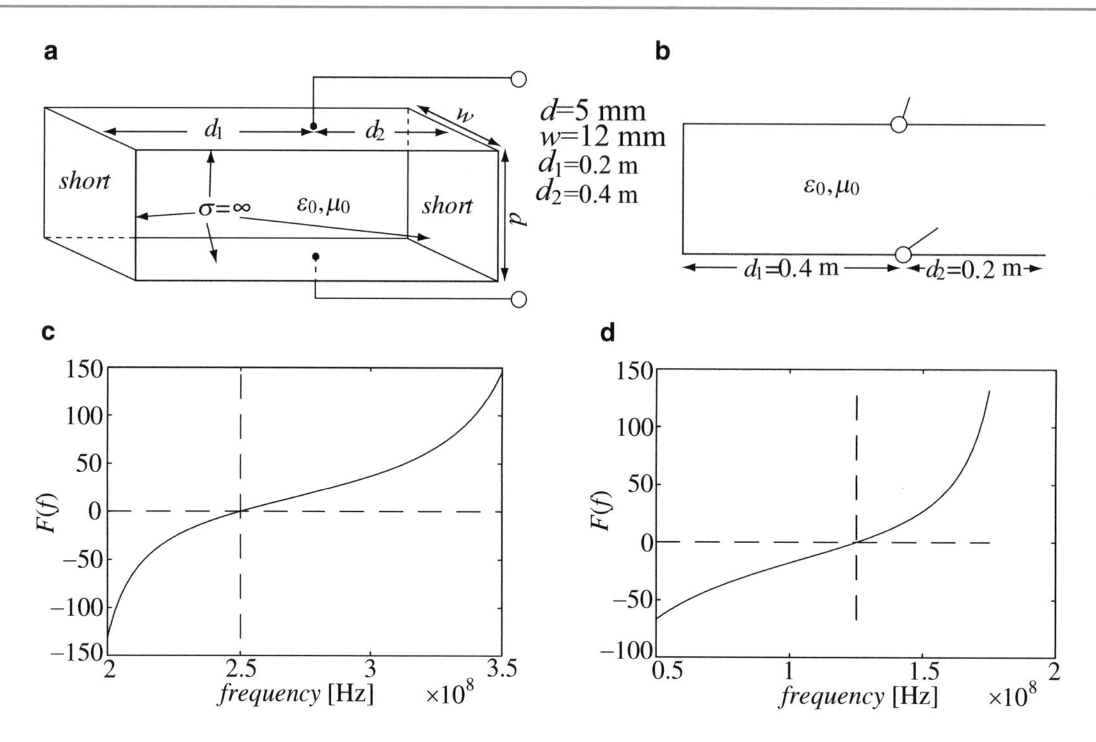

Figure 14.27 (**a**) A transmission line resonator with both ends shorted. (**b**) Transmission line resonator with right side open. (**c**) A resonant frequency for the circuit in (**a**). (**d**) A resonant frequency for the circuit in (**b**)

(**a**) The line is lossless; therefore, the speed of propagation and intrinsic impedance on the line (from **Table 14.1**) is

$$v_p = \frac{1}{\sqrt{\varepsilon_0 \mu_0}} = 3 \times 10^8 \quad [\text{m/s}]$$

$$Z_0 = \sqrt{\frac{L}{C}} = \sqrt{\frac{\mu_0 d/w}{w\varepsilon_0/d}} = \frac{d}{w}\sqrt{\frac{\mu_0}{\varepsilon_0}} = \frac{0.005}{0.12} 377 = 15.7 \quad [\Omega]$$

Both line segments are identical in their properties; therefore, the condition for resonance is

$$F(f) = Z_{01}\tan\frac{2\pi f d_1}{v_p} + Z_{02}\tan\frac{2\pi f d_2}{v_p} = 15.7\tan\frac{2 \times \pi \times f \times 0.4}{3 \times 10^8} + 15.7\tan\frac{2 \times \pi \times f \times 0.2}{3 \times 10^8} = 0$$

To find the resonant frequencies, we plot this expression as a function of frequency and identify the zero crossings. Those crossings which are smooth and change the nature of the circuit (from capacitive to inductive or vice versa) are resonant frequencies. Others, such as when the value of the function tends to $\pm\infty$, are not resonant frequencies. **Figure 14.27c** shows a plot between 200 and 350 MHz with resonance at 250 MHz. A similar plot (not shown) results in a second resonant frequency at 500 MHz.

(**b**) In this case, one end is open. This means that a short circuit exists (in terms of conditions on the line), a distance $\lambda/4$ from the open circuit on either side. Thus, if we take the distance d_1 to be the distance from the shorted end and $d_2 = 0.2 \pm \lambda/4$ the distance from the artificial new short end (i.e., a short end that exists at either side of the open end), we get the condition for resonance:

$$F(f) = \tan\frac{2\pi f d_1}{v_p} + \tan\frac{2\pi f d_2}{v_p} = \tan\frac{2 \times \pi \times f \times 0.4}{3 \times 10^8} + \tan\frac{2 \times \pi \times f \times \left(0.2 - 3 \times 10^8/4f\right)}{3 \times 10^8} = 0$$

where $\lambda/4 = c/4f = 3 \times 10^8/4f$ was used and the shorter section on the right side was chosen.

Again plotting this relation, we get the two resonant frequencies as 125 MHz and 375 MHz. **Figure 14.27d** shows the first of these resonant frequencies. Note that one resonant frequency is lower than for the shorted line, the second higher. These correspond to "lengthening" of the line since in both cases, one-quarter wavelength is larger than (or equal to) the length of the open line ($\lambda/4 = 0.6$ m at 125 MHz, and $\lambda/4 = 0.2$ m at 375 MHz). Higher resonant frequencies may correspond to either "lengthening" or "shortening" of the line since open lines may behave as being shorted on either side of the open.

Exercise 14.9 In **Example 14.15**, calculate the next two resonant frequencies for **(a)** and **(b)**.

Answer **(a)** 750 MHz, 1,000 MHz. **(b)** 625 MHz, 875 MHz.

Exercise 14.10 Consider again **Figure 14.27a**. However, now a dielectric with relative permittivity 4 fills the space between the plates. The ends of the resonator are shorted.

(a) What is the condition for resonance?
(b) Calculate the first three resonant frequencies.

Answer **(a)** $F(f) = \tan(1.6755 \times 10^{-8}f) + \tan(0.8377 \times 10^{-8}f) = 0$. **(b)** 125 MHz, 250 MHz, and 375 MHz.

14.10 Applications

Application: Frequency Domain Reflectometry on Transmission Lines The following procedure may be used to detect conditions on a transmission line such as short circuits, discontinuities, and the like. Suppose a line is shorted at some point and it is desirable to find this location. A matched generator is connected to the line and the line voltage or current is measured at some location between the generator and load, usually close to the generator, as shown in **Figure 14.28**. Now, the frequency of the generator is increased or decreased until a maximum or minimum in voltage (or current) is detected by the measuring instrument. This then represents a maximum or a minimum in the standing wave pattern produced by the forward- and backward-propagating waves along the line. Usually, the minima are used because they are sharper than the peaks. Because the number of minima between the given location and the load is not known, we may assume there are n minima, the distance between each two being one-half wavelength at the given frequency. Now, the frequency is increased until the next minimum appears at the measuring instrument. There are now $n + 1$ minima and, again, the distance between every two minima is $\lambda/2$ at the new, higher frequency. By measuring the two frequencies, the following may be written for the distance to the fault on the line:

$$d = \frac{\lambda_1}{2}n = \frac{\lambda_2}{2}(n+1) \tag{14.158}$$

where d is the distance between the measuring instrument and the short. Because $\lambda f = v_p$, we can write

$$\frac{v_{p_1}}{2f_1}n = \frac{v_{p_2}}{2f_2}(n+1) \tag{14.159}$$

Assuming the phase velocity does not change with frequency $\left(v_{p_1} = v_{p_2}\right)$, the number of minima, n, is

$$n = \frac{\lambda_2}{\lambda_1 - \lambda_2} = \frac{f_1}{f_2 - f_1} \tag{14.160}$$

Substituting for n in **Eq. (14.158)**, the length of the line is

$$d = \frac{\lambda_2 \lambda_1}{2(\lambda_1 - \lambda_2)} = \frac{v_p}{2(f_2 - f_1)} \tag{14.161}$$

If the phase velocity on the line is known (i.e., if the line properties are known), the distance to the discontinuity may be calculated. However, there is a small difficulty here: A shorted or open line will give identical indications, but they will be a distance $\lambda/4$ away. For this reason, the location of the discontinuity may only be found to within $\pm\lambda/4$. Often, this is quite sufficient. In cable TV applications, this means about 1.5 m at the lowest frequency (50 MHz). The higher the frequency used for measurement, the closer the discontinuity may be located, thus the use of high frequencies for location of discontinuities.

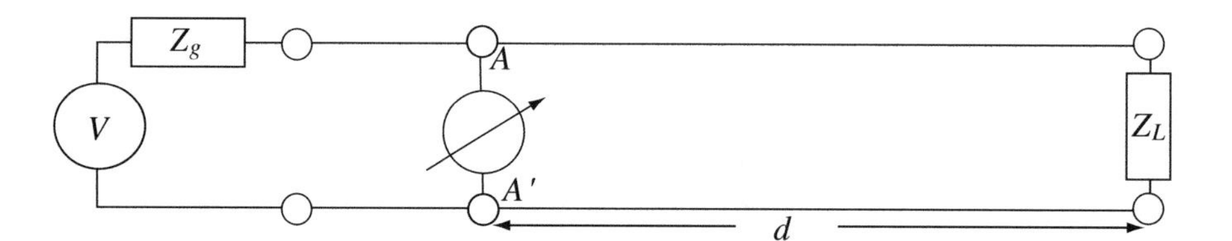

Figure 14.28 Frequency domain reflectometry. The signal is measured at A–A'

Application: Transmission Line Methods of Oil Recovery An interesting proposed application of transmission lines is recovery of oil in oil shale deposits. The idea is rather simple: Vertical wells are drilled at given distances which then lean horizontally into the shale deposits. Cables are introduced and connected to sources in the range of a few MHz. A transverse electromagnetic (TEM) wave propagates in the line, with the shale serving as the lossy dielectric between the lines. Because the loss is rather high and the thermal insulating nature of shale and soil, considerable heat is produced reducing the oil-shale viscosity. The shale is then pumped out by pressurized steam or water. A schematic view of this method is shown in **Figure 14.29**. Multiple transmission lines can heat large areas of the deposits at almost any depth.

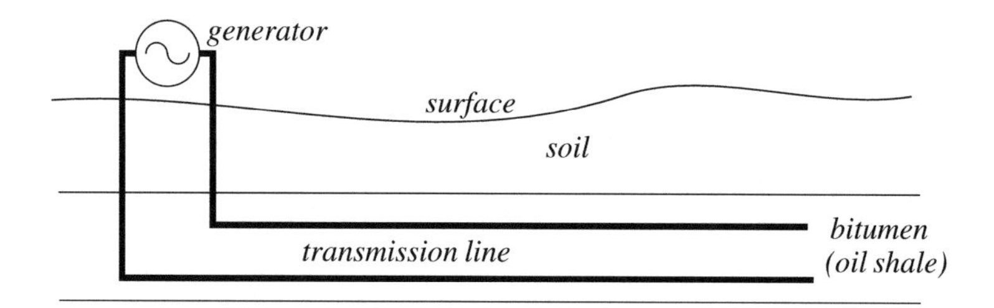

Figure 14.29 Transmission line method of oil recovery from oil shale

Application: Monitoring of Dielectrics and Lossy Dielectrics Resonant transmission lines are often used to monitor or measure the permittivity of materials during production. A simple device designed to monitor the thickness of a known dielectric (such as plastic film, polymer or paper) is shown in **Figure 14.30**. A parallel plate transmission line resonates at a frequency f for nominal thickness. The frequency is adjusted and measured. Any deviation in the material thickness changes the effective permittivity in the resonator and, therefore, changes the resonant frequency of the device. An increase in thickness reduces the frequency, whereas a decrease in thickness increases the frequency. A device of this kind can be easily incorporated in automatic production, with feedback to the appropriate control devices.

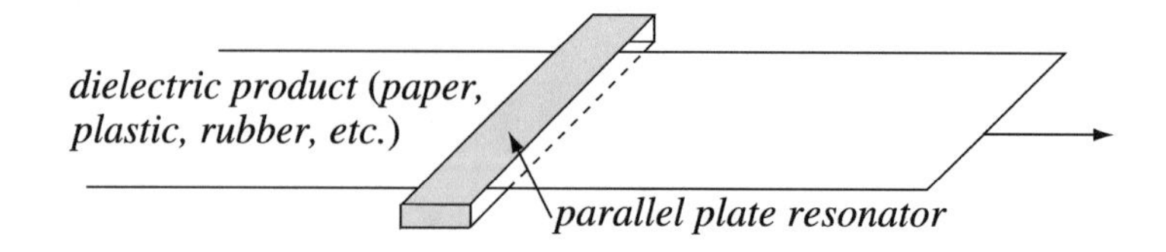

Figure 14.30 A transmission line resonator used to monitor material thickness during production

Similar devices can be used to monitor curing or drying of materials. The method is based on the fact that wet materials have higher permittivities than dry or cured products. As the water or solvent dries, the relative permittivity decreases and the resonant frequency of the device increases. Since resonance is usually quite sharp very sensitive measurements can be made. This method may be used in monitoring of polymers, paper production, drying of grain, or even baked products.

The same method is used to measure water content in snow packs for prediction of runoff and water shortages. A transmission line resonator, usually in the form of a coaxial, open-ended resonator, is calibrated at a given frequency. Then, the device is pushed into the snow until the snow fills the resonator. Now, the frequency is lower because snow has a higher permittivity than air. The lower the frequency, the higher the water content since wetter snow has higher permittivity than dryer snow.

14.11 Summary

The current chapter treats transmission lines in the frequency domain. The behavior of voltages and currents on the line is that of plane waves propagating along the line. Many of the relations in **Chapter 12** find new use here.

Reminders: Impedance: $Z = V/I$ [Ω], admittance: $Y = 1/Z = I/V$ [(1/Ω) = S].

Phase velocity $v_p = \omega/\beta = \lambda f$ [m/s], $\lambda = 2\pi/\beta$ [m]. In vacuum $v_p = 3 \times 10^8$ [m/s]

Transmission Line Parameters Series parameter, L, R are given as resistance/m, and inductance/m. Shunt parameters G, C are given as capacitance/m and conductance/m. Each parameter is calculated for a 1 m length of the transmission line based on principles in **Chapters 4**, **7**, and **9**. **Table 14.1** summarizes the parameters for three common lines.

Transmission line equations are obtained by application of Kirchoff's laws (see **Figure 14.10)** for a section of length dl at an arbitrary location, l on the line. In the frequency domain

$$\frac{dV(l)}{dl} = -I(l)[R + j\omega L] \qquad (14.24) \qquad\qquad \frac{dI(l)}{dl} = -V(l)[G + j\omega C] \qquad (14.27)$$

Wave equations for voltage and current on the line, with reference to $l = 0$ at the generator:

$$\frac{d^2V(l)}{dl^2} - \gamma^2 V(l) = 0 \qquad (14.31) \qquad\qquad \frac{d^2I(l)}{dl^2} - \gamma^2 I(l) = 0 \qquad (14.32)$$

where

$$\gamma = \alpha + j\beta = \sqrt{[G + j\omega C][R + j\omega L]} \qquad (14.33)$$

γ = ***propagation constant***, α = ***attenuation constant***, β = ***phase constant***.

Solutions for voltage and current on the line

$$V(l) = V^+ e^{-\gamma l} + V^- e^{\gamma l} \quad [\text{V}] \qquad (14.34) \qquad\qquad I(l) = I^+ e^{-\gamma l} + I^- e^{\gamma l} \quad [\text{A}] \qquad (14.35)$$

The first term in each solution is the forward (generator to load) propagating wave, and the second is the backward (load to generator) propagating wave.

General properties

$$Z_0 = \frac{V^+}{I^+} = -\frac{V^-}{I^-} = \sqrt{\frac{R + j\omega L}{G + j\omega C}} \quad [\Omega] \qquad (14.45) \qquad\qquad \lambda = \frac{2\pi}{\beta} \quad [\text{m}], \quad v_p = \frac{\omega}{\beta} \quad \left[\frac{\text{m}}{\text{s}}\right] \qquad (14.47)$$

Properties of particular types of lines

	Attenuation constant α [Np/m]	Phase constant β [rad/m]	Z_0 [Ω]	v_p [m/s]	Other relations	Equations
Lossless line	$\alpha = 0$	$\beta = \omega\sqrt{LC}$	$Z_0 = \sqrt{L/C}$	$v_p = 1/\sqrt{LC}$	$R = 0, G = 0,$ $\mu\varepsilon = LC$	**(14.48)–(14.54)**
Long line	Can be any type of line. The defining property is the lack of reflection from the load					
Distortionless line	$\alpha = R\sqrt{C/L}$	$\beta = \omega\sqrt{LC}$	$Z_0 = \sqrt{L/C}$	$v_p = 1/\sqrt{LC}$	$R/L = G/C$	**(14.57)–(14.61)**
Low resistance line	$\gamma = \alpha + j\beta = j\omega\sqrt{LC}\sqrt{1+\dfrac{G}{j\omega C}}$		$Z_0 = \sqrt{\dfrac{j\omega L}{G + j\omega C}}$	$v_p = 1/\sqrt{LC}$	$R = 0$ $\mu\varepsilon = LC$ $\sigma/\varepsilon = G/C$	**(14.62)–(14.65)**

Field approach to transmission line analysis follows the ideas of plane waves in **Chapter 12** and is therefore not summarized here. For a discussion and examples see **Section 14.6**.

Finite transmission lines. By referencing all calculations to the load rather than generator, that is, the load is placed at $z = 0$ and the generator is in the positive z-direction, we get (see **Figure 14.15**)

$$\boxed{V(z) = V^+e^{\gamma z} + V^-e^{-\gamma z} \quad [\text{V}]} \quad \text{and} \quad \boxed{I(z) = \frac{V^+}{Z_0}e^{\gamma z} - \frac{V^-}{Z_0}e^{-\gamma z} \quad [\text{A}]} \tag{14.78}$$

Load reflection coefficient

$$\Gamma_L = \frac{Z_L - Z_0}{Z_L + Z_0} \tag{14.83}$$

Line impedance on a general lossy line at an arbitrary distance z from the load:

$$Z(z) = \frac{V(z)}{I(z)} = Z_0\frac{e^{\gamma z} + \Gamma_L e^{-\gamma z}}{e^{\gamma z} - \Gamma_L e^{-\gamma z}} = Z_0\frac{Z_L + Z_0\tanh(\gamma z)}{Z_0 + Z_L\tanh(\gamma z)} \quad [\Omega] \tag{14.87, 14.89}$$

Generalized reflection coefficient on a general lossy line at an arbitrary distance z from the load:

$$\Gamma(z) = \Gamma_L e^{-2\gamma z} = |\Gamma_L|e^{-2\alpha z}e^{j\theta_\Gamma}e^{-j2\beta z} \tag{14.91}$$

Phase on line (starting from load)

$$\theta_{\Gamma(z)} = \theta_\Gamma - 2\beta z \quad [\text{rad}] \tag{14.92}$$

where θ_Γ is the phase angle of the reflection coefficient.

Lossless line:

$$Z(z) = Z_0\frac{Z_L + jZ_0\tan\beta z}{Z_0 + jZ_L\tan\beta z} = Z_0\frac{Z_L\cos\beta z + jZ_0\sin\beta z}{Z_0\cos\beta z + jZ_L\sin\beta z} \quad [\Omega] \tag{14.94}$$

$$\Gamma(z) = \Gamma_L e^{-j2\beta z} = |\Gamma_L|e^{-j(2\beta z - \theta_\Gamma)} \tag{14.95}$$

$$\theta_{\Gamma(z)} = \theta_\Gamma - 2\beta z \quad [\text{rad}] \tag{14.96}$$

Standing Wave Ratio—SWR

$$\text{SWR} = \frac{V_{max}}{V_{min}} = \frac{I_{max}}{I_{min}} = \frac{1 + |\Gamma_L|}{1 - |\Gamma_L|} = \frac{1 + |\Gamma(z)|}{1 - |\Gamma(z)|} \tag{14.103}$$

Maximum and minimum voltage (magnitude) on the line

$$V_{max} = |V^+|(1 + |\Gamma(z)|) = |V^+|\left(\frac{2\text{SWR}}{\text{SWR}+1}\right) \quad [\text{V}] \tag{14.105}$$

$$V_{min} = |V^+|(1 - |\Gamma(z)|) = |V^+|\left(\frac{2}{\text{SWR}+1}\right) \quad [\text{V}] \tag{14.106}$$

Voltage and current anywhere on the line

$$V(z) = V^+ e^{j\beta z}\left(1 + \Gamma_L e^{-j2\beta z}\right) \quad [\text{V}] \tag{14.107}$$

$$I(z) = \frac{V^+}{Z_0} e^{j\beta z}\left(1 - \Gamma_L e^{-j2\beta z}\right) \quad [\text{A}] \tag{14.108}$$

At the load ($z = 0$):

$$V_L = V^+\left(1 + |\Gamma_L| e^{j\theta_\Gamma}\right) \quad [\text{V}] \quad \text{and} \quad I_L = \frac{V^+}{Z_0}\left(1 - |\Gamma_L| e^{j\theta_\Gamma}\right) \quad [\text{A}] \tag{14.109}$$

Locations of voltage maxima (current minima)

$$z_{max} = \frac{\lambda}{4\pi}(\theta_\Gamma + 2n\pi), \quad n = 0,1,2,\ldots \quad [\text{m}] \tag{14.116}$$

Location of impedance or voltage minima (current maxima): $z_{max} \pm \lambda/4$:

Lossless matched line [Eqs. (14.117) through **(14.119)]**

$$Z_L = Z_0, \quad Z(z) = Z_0 \quad [\Omega], \quad V(z) = V^+ e^{j\beta z} \quad [\text{V}], \quad I(z) = \frac{V^+}{Z_0} e^{j\beta z} \quad [\text{A}]$$

Lossless Shorted and Open Transmission Lines The main relations for shorted and open lines are summarized in the table below **[Eqs. (14.120)** through **(14.125)]**.

	Z_L [Ω]	$Z(z)$ [Ω]	V_L [V]	I_L [A]	$V(z)$ [V]	$I(z)$ [A]
Shorted line	0	$jZ_0 \tan\beta z$	0	$\frac{2V^+}{Z_0}$	$V^+ e^{j\beta z}(1 - e^{-j2\beta z})$	$\frac{V^+}{Z_0} e^{j\beta z}\left(1 + e^{-j2\beta z}\right)$
Open line	∞	$-jZ_0 \cot\beta z$	$2V^+$	0	$V^+ e^{j\beta z}(1 + e^{-j2\beta z})$	$\frac{V^+}{Z_0} e^{j\beta z}\left(1 - e^{-j2\beta z}\right)$

The most significant properties of shorted and open lines are summarized in the table below.

	Γ_L	SWR	$Z(z) = 0$ *at*	$Z(z) = \infty$ *at*	$Z(z) > 0$ (inductive) for	$Z(z) < 0$ (capacitive) for
Shorted line	-1	∞	$z = n\lambda/2$ $n = 0,1,2,\ldots$	$n\lambda/2 + \lambda/4$,	$n\lambda/2 < z < n\lambda/2 + \lambda/4$	$n\lambda/2 + \lambda/4 < z < n\lambda/2 + \lambda/2$
Open line	1	∞	$n\lambda/2 + \lambda/4$, $n = 0,1,2,\ldots$	$z = n\lambda/2$	$n\lambda/2 + \lambda/4 < z < n\lambda/2 + \lambda/2$	$n\lambda/2 < z < n\lambda/2 + \lambda/4$

Notes:

(1) An open line behaves as if it were a shorted line lengthened or shortened by $\lambda/4$.
(2) A shorted line behaves as if it were an open line lengthened or shortened by $\lambda/4$.
(3) All properties and relations on any line repeat at intervals of $\lambda/2$.

Also, on any lossless transmission line

$$Z_0 = \sqrt{Z_{short} Z_{open}} \quad [\Omega] \tag{14.127}$$

Resistively Loaded Lossless Transmission Line The reflection coefficient is

$$\Gamma_L = \frac{R_L - Z_0}{R_L + Z_0} \quad (14.128)$$

The following table summarizes the main relations on resistively loaded line

Load resistance	Γ_L	Voltage on the line [V]	Current in the line [A]	Minimum impedance occurs at: [wavelengths]
$R_L > Z_0$	$\Gamma_L > 0$ $\theta_\Gamma = 0$	$V^+ e^{j\beta z}(1 + \Gamma_L e^{-j2\beta z})$ **Eq. (14.131)**	$\frac{V^+}{Z_0} e^{j\beta z}\left(1 - \Gamma_L e^{-j2\beta z}\right)$ **Eq. (14.132)**	$z_{min} = (2n+1)\frac{\lambda}{4}$ $n = 0,1,2,\ldots$ **Eq. (14.134)**
$R_L < Z_0$	$\Gamma_L < 0$ $\theta_\Gamma = -\pi$	$V^+ e^{j\beta z}(1 + \|\Gamma_L\| e^{-j\pi} e^{-j2\beta z})$ **Eq. (14.138)**	$\frac{V^+}{Z_0} e^{j\beta z}\left(1 - \|\Gamma_L\| e^{-j\pi} e^{-j2\beta z}\right)$ **Eq. (14.139)**	$z_{min} = n\frac{\lambda}{2}$ $n = 0,1,2,\ldots$ **Eq. (14.141)**

Additional Properties

(1) $-1 < \Gamma_L < +1$, $1 < \text{SWR} < \infty$.
(2) The line impedance can be complex [see **Eq. (14.129)**],
(3) Impedance is real at locations of voltage maxima and voltage minima for lossless lines.
(4) The impedance at voltage maxima is $Z_{max} = Z_0 * \text{SWR}$, whereas at voltage minima (current maxima), it is $Z_{min} = Z_0/\text{SWR}$.
(5) For $R_L > Z_0$, the first voltage maximum occurs at the load ($z = 0$), the first voltage minimum at $\lambda/4$ from the load.
(6) For $R_L < Z_0$, the first voltage minimum occurs at the load, the first voltage maximum at $\lambda/4$ from the load.

Power on Transmission Lines Power at any point z_0 on a general, lossy line is

$$P_{z_0} = \frac{|V^+|^2}{2|Z_0|}\left(e^{2\alpha z_0} - |\Gamma_L|^2 e^{-2\alpha z_0}\right)\cos(\theta_{Z_0}) \quad [\text{W}] \quad (14.147)$$

where θ_{Z_0} is the phase angle of the characteristic impedance.

Power at any point z_0 on a lossless line is

$$P_{z_0} = \frac{|V^+|^2}{2Z_0}\left(1 - |\Gamma_L|^2\right) \quad [\text{W}] \quad (14.152)$$

Resonant Transmission Lines (See **Figure 14.26)** The resonant frequency is found by finding the frequencies at which the following function is satisfied:

$$Z_{01}\tan\frac{2\pi f d_1}{v_{p1}} + Z_{02}\tan\frac{2\pi f d_2}{v_{p2}} = 0 \quad (14.157)$$

where d_1, d_2 are the distances to the shorts from the connection point, Z_{01}, Z_{02} are the characteristic impedances of the two lines, and v_{p1}, v_{p2} the phase velocities on the two lines.
Reminder: 1 Np/m = 8.69 dB/m.

Problems

Transmission Line Parameters

14.1 Application: Coaxial Transmission Line. Two coaxial transmission lines, each made of an inner and an outer conducting shell, are given. The radius of the inner shell is $a = 10$ mm and the radius of the outer shell is $b = 20$ mm. The lines are shown in cross section in **Figure 14.31**. Permittivity and permeabilities are given. Calculate the line parameters for each line assuming perfect conductors.

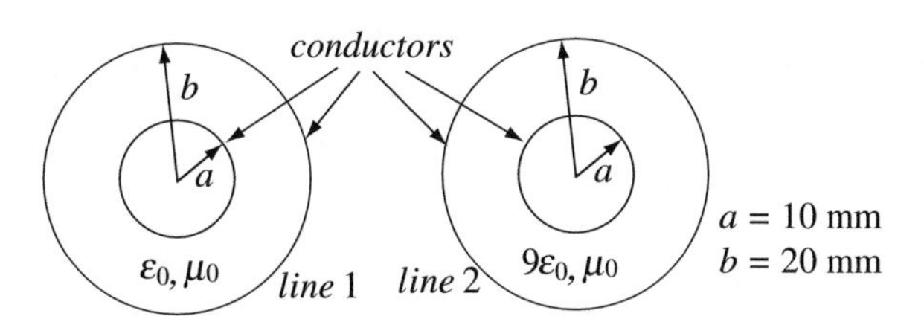

Figure 14.31

14.2 Application: Overhead Transmission Line with Ground Return. Two students live 1 km apart. They decide to connect between them a line so that they may use a private telephone without having to pay for the service. To do so, they need a transmission line and this requires two conductors. Since they do not have enough money, they decide to use a single wire and use the ground as the return conductor, assuming the ground is a good conductor. The wire is 1 mm thick and is strung at a height of 5 m above ground.

(a) Is this a sound approach?
(b) If so, calculate the line parameters for the transmission line made of the single conductor and ground, assuming a perfectly conducting ground.
(c) Suppose the installation was a bit flimsy and during a storm, their wire fell to the ground. The wire is insulated and the insulation is 0.5 mm thick. What are the line parameters now? Use permittivity and permeability of free space for the insulating material.

Long, Lossless Lines

14.3 Application: Parameters of Coaxial Line. The RG-11/U coaxial line has the following properties: $Z_0 = 75\ \Omega$, $v_p = 2c/3$ m/s. Assuming the line to be lossless, calculate its inductance and capacitance per unit length.

14.4 Application: Coaxial Line and Delay on the Line. A coaxial transmission line is made of two circular conductors as shown in **Figure 14.32**. The inner conductor has diameter b and the outer conductor diameter a. The conductors are perfect conductors and are separated by a lossless ferrite (a material with finite permeability, finite permittivity, and zero conductivity). Given: $\varepsilon_0 = 8.854 \times 10^{-12}$ F/m, $\mu_0 = 4\pi \times 10^{-7}$ H/m, $\varepsilon_1 = 9\varepsilon_0$ [F/m], $\mu_1 = 100\mu_0$ [H/m], $a = 8$ mm, $b = 1$ mm.

(a) Calculate the characteristic impedance and the phase velocity on the line.
(b) Two coaxial transmission lines connect two memory banks to a processor. Line A is the same as calculated in part **(a)**. Line B has the same dimensions, but the ferrite is replaced with free space. What must be the minimum execution cycle (maximum frequency) of the processor if the two memory signals must reach the processor within one execution cycle. Assume the length of each line is $d = 100$ mm. The execution cycle is defined by the slower of the signals.

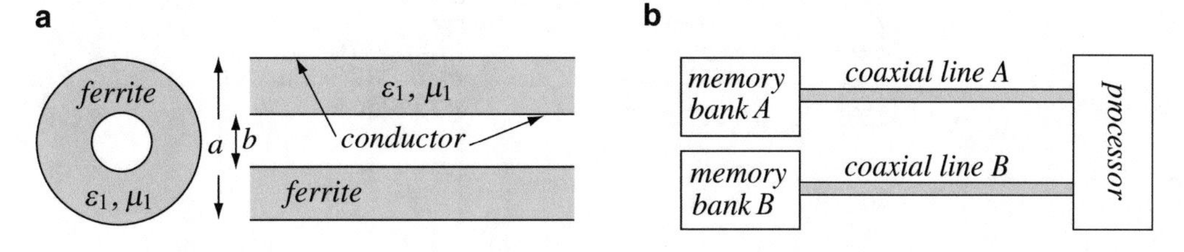

Figure 14.32

14.5 Application: Microstrip Line. A conducting strip with dimensions as shown in **Figure 14.33** is located above a conducting surface. The strip is very long and the conducting surface is infinite. Also, $w \gg d$. The strip and surface serve as a transmission line. Calculate assuming perfect conductors:

(a) Speed of propagation of waves on this line.
(b) Characteristic impedance of the transmission line for $w = 10$ mm, and $d = 0.1$ mm.

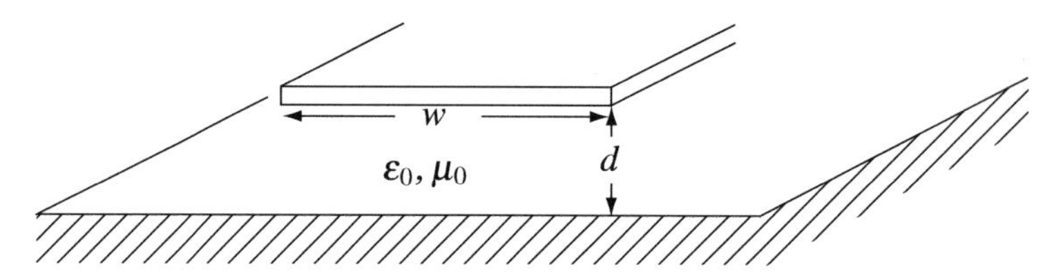

Figure 14.33

The Distortionless Transmission Line

14.6 Application: The Distortionless Line. A coaxial line is made as shown in **Figure 14.34**. Material properties and dimensions are given in the figure. The design calls for a distortionless transmission line at 250 MHz.

(a) Assuming the outer radius must remain constant ($b = 20$ mm), what must be the radius a of the inner conductor for the line to be distortionless? Assume material properties μ_0 [H/m], ε_0 [F/m], $\sigma_c = 5.7 \times 10^7$ [S/m] for copper.
(b) What is the characteristic impedance of the line and what is its attenuation constant for the design in **(a)**?
(c) If reduction of at most 10 dB in amplitude is allowed before an amplifier is required, calculate the distance between each two amplifiers on the line.

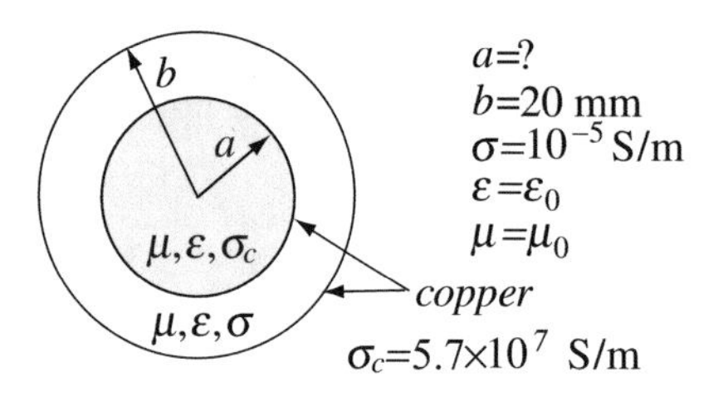

Figure 14.34

14.7 Application: Distortionless Parallel Plate Line. A parallel plate line is made as shown in **Figure 14.35**. Material properties and dimensions are given in the figure. The design calls for a distortionless transmission line operating at 1,000 MHz.

(a) Calculate the required distance d between the two conductors to produce a distortionless line at the given frequency.
(b) What are the characteristic impedance and attenuation constant of the line?
(c) If a reduction of at most 75 dB in amplitude is allowed before an amplifier is required, calculate the distance between each two amplifiers on the line.

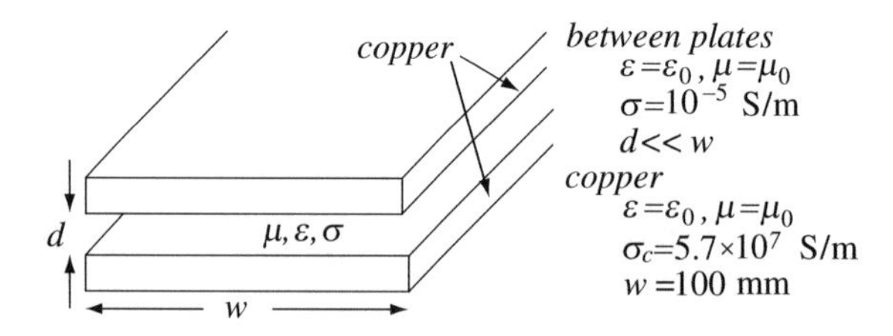

Figure 14.35

The Low-Resistance Transmission Line

14.8 Application: Low-Resistance Power Line. A power line is made of two round, parallel conductors, 20 mm in diameter, separated by 0.8 m, and suspended above ground, in air. Neglect any ground effects. Properties of air are: ε_0 [F/m], μ_0 [H/m], $\sigma = 10^{-5}$ S/m. The line is made of aluminum ($\mu = \mu_0$ [H/m], $\varepsilon = \varepsilon_0$ [F/m], $\sigma_c = 3.5 \times 10^7$ S/m) and operates at 60 Hz. Calculate:

(a) The propagation constant on the line. Show that this may be assumed to be a low-resistance line.
(b) Under the assumption of a low-resistance line, calculate the characteristic impedance of the line and its attenuation and phase constants.

14.9 Application: Properties of Telephone Lines. A two-wire, open-air telephone line is made of round wires, 1 mm in radius, and are supported 200 mm apart on insulating cups.

The line is made of copper, with conductivity of 5.7×10^7 S/m, and the surrounding space is air (with properties of free space). The line is 3 km long, may be assumed to be low loss, and is used to transmit data at 10 kHz.

(a) What is the load impedance if it must equal the characteristic impedance of the line?
(b) If the load requires 0.01 W to operate, how much power must be supplied to the line input?
(c) Suppose you are free to adjust the distance between the lines. Can the line be redesigned to be distortionless? If so, how?
(d) Is the assumption that the line is a low-loss line justified? Explain.

14.10 Properties of Long Lines. A long line has parameters R, L, C, $G = 0$, and operates at 100 kHz. The characteristic impedance is $Z_0 = 300 - j10\ \Omega$ and the phase velocity on the line is $v_p = 2c/3$ m/s. Find:

(a) The values of R, L, and C assuming the line is a low-loss line.
(b) Calculate the propagation constant on the line.
(c) Is the assumption in **(a)** that the line is low loss justified?

The Field Approach to Transmission Lines

14.11 Application: Power Relations on Coaxial Lines. Coaxial lines are usually used for transmission of signals at high frequencies. In some cases however, the lines are also required to carry considerable power at lower frequencies. One example is the cable TV coaxial line. These carry both the high-frequency video signals and 60 Hz power to operate devices on the line (such as amplifiers). Typical requirements are for the lines to carry up to 10 A at about 60 V. Consider the following example:

A coaxial cable with the dimensions shown in **Figure 14.36** has a characteristic impedance of 75 Ω. The line is connected to an AC source at 60 V (peak), 60 Hz. Neglect internal impedance in the source. The line may be considered long and the load is matched to the line. Calculate:

(a) The electric and magnetic field intensities in the dielectric of the line.
(b) Show that application of the Poynting theorem anywhere on the line gives the same time-averaged power as that obtained from the current voltage relations.

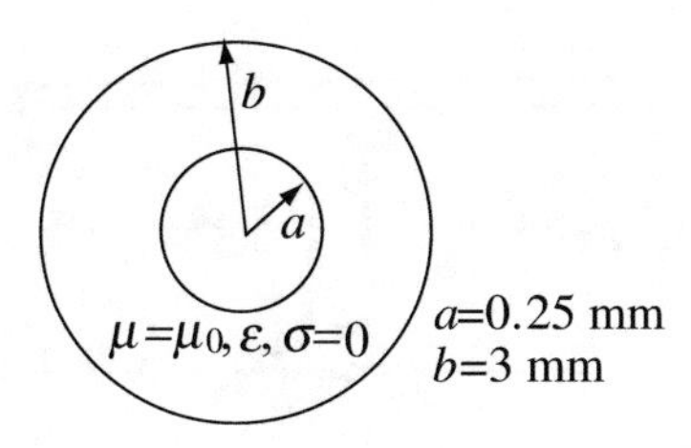

Figure 14.36

14.12 Application: Exposure to Electromagnetic Fields. One recent health concern is with exposure to both low- and high-frequency electromagnetic fields. These are believed by some to cause cancer. An AM station transmitting at 1 MHz decided to minimize exposure of personnel due to power on the transmission line leading from the generator to the antenna. The transmission line is a two-wire exposed (no insulation) line, 10 mm in diameter and separated by 0.2 m. The transmission line is parallel to the ground and leads to an antenna located at some distance from the transmitter, on top of a hill. The station decided that to minimize the risk due to the fields generated by the line, it will not allow the peak magnetic flux density at a distance 1 m from the center of the transmission line to exceed 10^{-6} T.

(a) What is the peak power the station can transmit without violating the maximum magnetic flux density requirement? Assume a lossless line and matched conditions at load and generator. Neglect effects due to the ground.
(b) Suppose the station needs to transmit more power than allowed in **(a)**. To do so, the engineers decide to move the two wires closer together so that now they are 0.1 m apart. How much power can the station transmit now without violating the maximum field requirement?
(c) What are the peak electric field intensities in **(a)** and **(b)** at the location of the peak magnetic flux density?

Finite Transmission Lines

14.13 Application: Voltage and Current on Transmission Lines. The voltage and current at the center of the lossless line in **Figure 14.37** are given as $V_c = 18\angle 72^\circ$ V, $I_c = 0.2\angle -36^\circ$ A. If the wavelength is 1 m, calculate:

(a) The forward- and backward-propagating voltages.
(b) The forward- and backward-propagating currents.
(c) The load impedance.

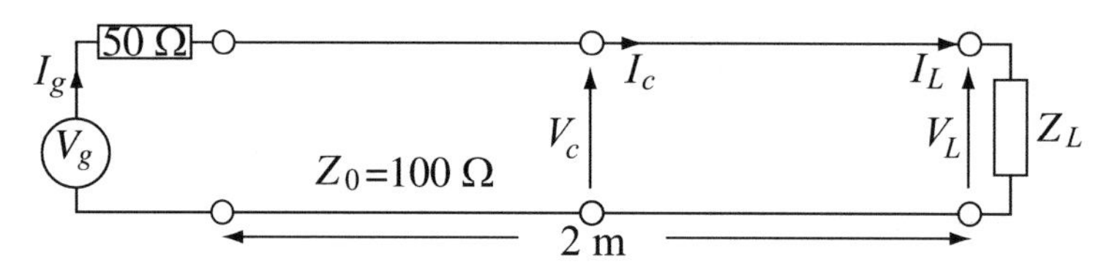

Figure 14.37

14.14 Application: Voltage and Current on Transmission Lines. The voltage and current at the load of the lossless line in **Figure 14.37** are given as $V_L = 50$ V, $I_L = 0.2$ A. If the wavelength is 1 m, calculate:

(a) The voltage and current of the generator.
(b) The power supplied by the generator.

14.15 Multiple Loads on a Line. A lossless transmission line with characteristic impedance of 100 Ω is connected to a generator and three loads as shown in **Figure 14.38** are connected to the line. Calculate:

(a) The power dissipated in each of the loads if the amplitude of the voltage across load (2) is given as 12 V.
(b) The power supplied by the generator to the line.
(c) The power dissipated in the generator itself.

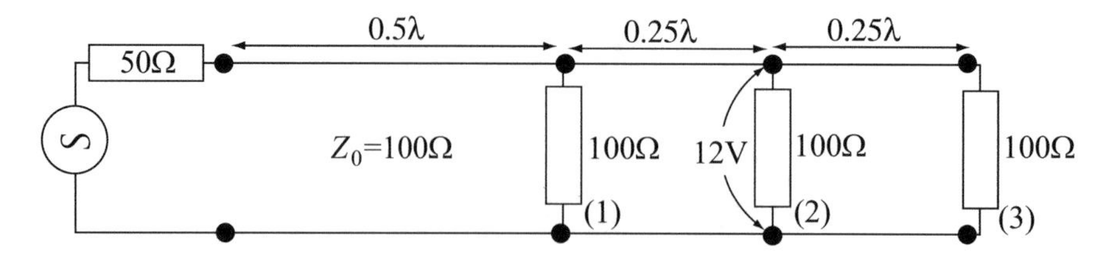

Figure 14.38

14.16 Application: Terminated Line, Matched Power Transfer. A load impedance $Z_L = 50\ \Omega$ is connected to a generator of impedance $Z_G = 50\ \Omega$ through a lossless transmission line with characteristic impedance $Z_0 = 200\ \Omega$.

(a) What must be the length d of the line (in wavelengths) for matched power transfer from generator to load?

(b) If the characteristic impedance of the line is changed to $Z_0 = (200 + j91)\ \Omega$, what is the length of the line (in wavelengths) for matched power transfer between generator and load?

(c) Show that at any value of d other than those found in **(a)** or **(b)**, the load is not matched to the generator.

14.17 Terminated, Mismatched Line. The line in **Figure 14.39** is given. The generator can be connected anywhere on the line. It is required that the generator be matched to the line so that there are no reflections into the generator. Find the closest location d (in wavelengths) to the load at which you can move the generator so that the generator is best matched to the line (i.e., the reflection coefficient at the generator is minimum). Assume the phase constant on the line is known as β_0 [rad/m].

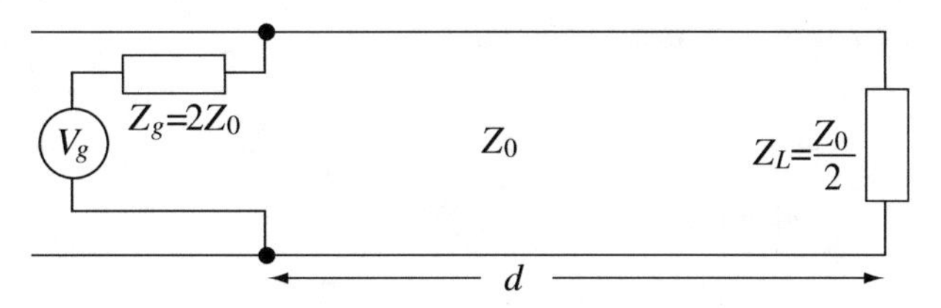

Figure 14.39

Line Impedance, Reflection Coefficient, Etc

14.18 Generalized Reflection Coefficient on a Line. Calculate the generalized reflection coefficient and the line impedance at a distance a wavelengths from the discontinuity shown in **Figure 14.40**. Assume the generator is matched to the line but the load is not. Dimensions are in wavelengths.

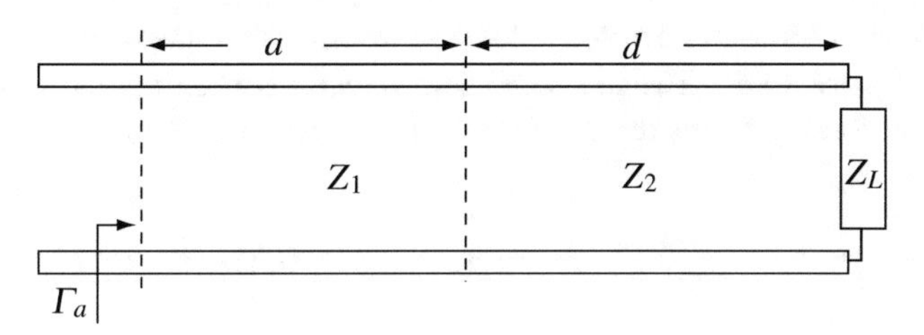

Figure 14.40

14.19 Application: Input Impedance of Striplines. A transmission line is made of two strips separated a distance d [m]. The width of the strips is w [m] and their thickness is b [m]. The strips are made of a conductor with conductivity σ [S/m]. The material between the strips is air (free space). A load Z_L [Ω] is connected at one end of the line (see **Figure 14.41**). Calculate the input impedance of the line at a distance $r = 1.8$ wavelengths from the load. Given: $w \gg d$, $\sigma_c = 1.0 \times 10^7$ S/m, $w = 20$ mm, $d = 0.5$ mm, $b \gg \delta$, $\mu = \mu_0$ [H/m], $\varepsilon = \varepsilon_0$ [F/m], $Z_L = 75\ \Omega, f = 1$ GHz.

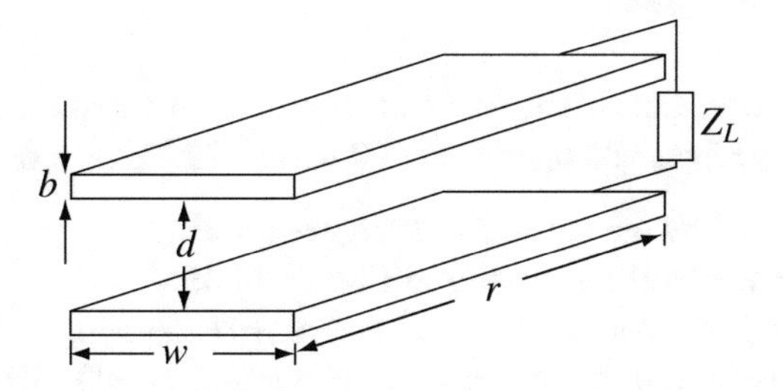

Figure 14.41

14.20 Line Impedance on Line with Multiple Loads. The lossless transmission line in **Figure 14.42** represents a distribution line and a load that has been shorted at its end as shown. Calculate the line impedance at a distance of 0.5λ to the left of the load.

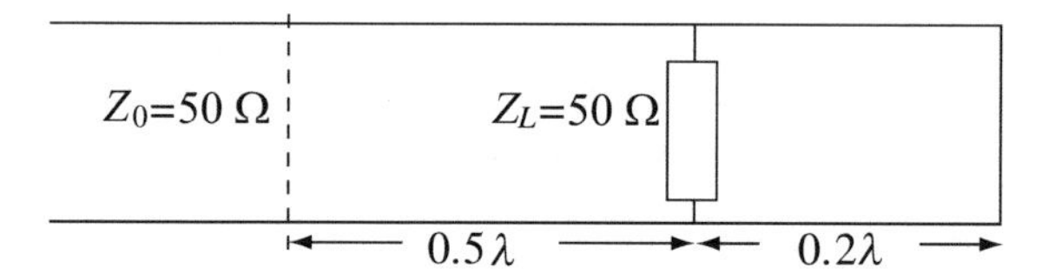

Figure 14.42

14.21 Line Impedance on Line with Multiple Loads. The transmission line in **Figure 14.43** is a distribution line normally operating at matched conditions. A second load at the end of the line has opened inadvertently. Calculate the line impedance at a distance 0.25λ from the load as shown.

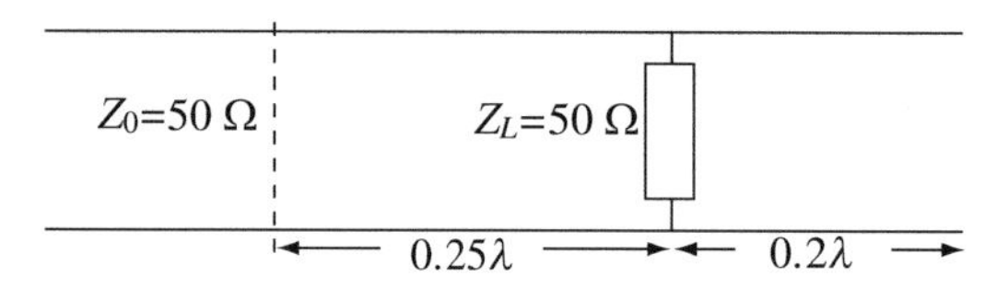

Figure 14.43

14.22 Voltage and Power on Lossless Line. A 75 Ω, lossless transmission line is connected on one side to a generator and the other to a load. The generator has internal impedance of 150 Ω and open circuit voltage amplitude of 24 V. Wavelength is 0.32 m and the line is 125 m long. The load is an antenna with impedance $75 + j42$ Ω. Calculate:

(a) The voltage at the load.
(b) The power supplied to the line by the generator.

14.23 Connection of Transmission Lines in Parallel. A lossless transmission line with characteristic impedance of 50 Ω is connected as shown in **Figure 14.44**. The measured impedance between points A–A' is 100 Ω. Calculate the two impedances Z_{L1} and Z_{L2} if the input impedance of the two lines when disconnected is equal.

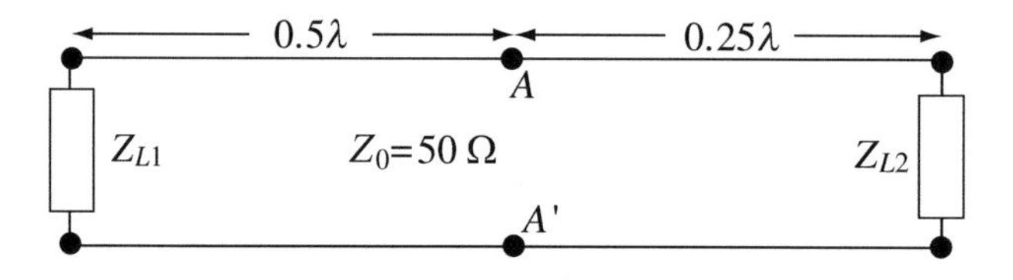

Figure 14.44

Shorted and Open Transmission Lines

14.24 Open and Shorted Transmission Lines. The two configurations in **Figure 14.45** are given. The lines are lossless with air as insulator between the lines and operate at 300 MHz. The voltage of the generator is 12 V rms.

(a) Calculate the current supplied by the generator in **Figure 14.45a**.
(b) Calculate the current supplied by the generator in **Figure 14.45b.**
(c) How much power does the generator supply in **(a)** and in **(b)**? Explain.
(d) What are the answers to **(a)** and **(b)** if the frequency is doubled to 600 MHz?
(e) What are the answers to **(a)** and **(b)** if the frequency is halved to 150 MHz?

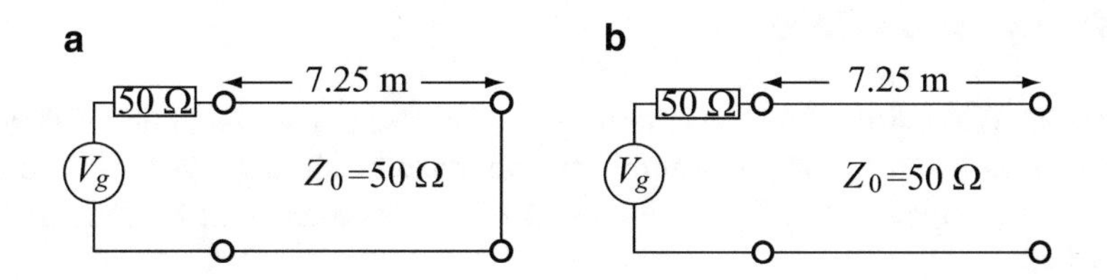

Figure 14.45

14.25 Application: Design of Network Elements. In the design of a transmission line network, it is required to design a reactance of $j100\ \Omega$ as an element in the network. Show how this can be accomplished with:

(a) A lossless, shorted transmission line with characteristic impedance of 50 Ω at a wavelength of 1 m.
(b) A lossless open transmission line with characteristic impedance of 75 Ω at a wavelength of 1 m.
(c) If a capacitive reactance of $-j100\ \Omega$ is needed instead, what are the answers to **(a)** and **(b)**?

14.26 Application: Experimental Evaluation of Line Parameters. A transmission line is 10 km long and operates under matched conditions. An engineer requires the properties of the line; that is, its characteristic impedance, its attenuation constant, and its phase constant. It is not possible to subject the line to testing equipment, but the voltage and currents can be measured at the load and at the generator. These are given as $V = 10$ V, $I = 0.1$ A at the generator and $V = 3 - j2$ V at the load. Find:

(a) The characteristic impedance, attenuation, and phase constants on the line.
(b) The time-averaged power loss on the line.

14.27 Application: Measurements of Line Conditions. A lossless transmission line is connected to a matched load. The characteristic impedance of the line is 50 Ω. An additional load is connected to the line. To determine the additional load, the maximum voltage on the line is measured as 48 V and the minimum voltage on the line is measured as 30 V. From these measurements, calculate:

(a) The additional load connected across the matched load.
(b) What are the maximum and minimum line impedances and where do these occur?

14.28 Line Conditions on Shorted and Open Lines. The standing wave ratio on a lossless transmission line is measured and found to be infinite.

(a) Can you tell from this measurement if the load is a short or an open circuit? Explain.
(b) Suppose you can measure the line impedance directly and you find that at one point, the line impedance is zero. Moving toward the load a very short distance, you find that the line impedance is purely reactive and increases. Can you now tell if the line is shorted or open? If so, how?

14.29 Design with Open and Shorted Transmission Lines. The filter shown in **Figure 14.46** needs to be implemented with transmission line segments so that it may be inserted in a line circuit. Show how this may be done and calculate the line segment lengths in your implementation. The lines are lossless two-wire lines with air as insulator and a characteristic impedance of 75 Ω. Assume that the distance between the two wires in the line segments is very small compared to the wavelength. The frequency of operation is 450 MHz. Note that there is more than one solution possible.

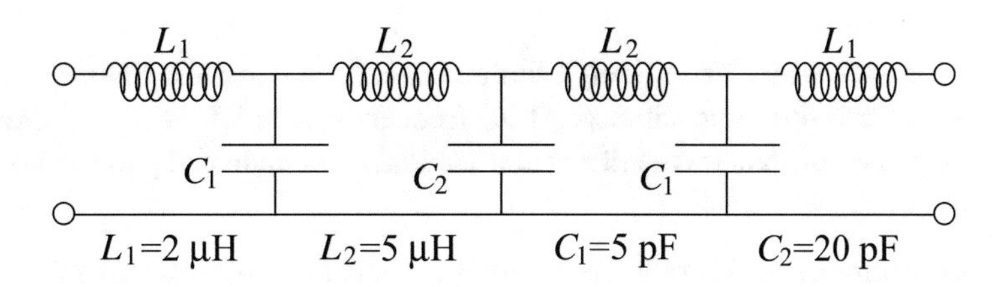

Figure 14.46

Resistive Loads on Transmission Lines

14.30 Application: Standing Wave Ratio Measurements. On a lossless transmission line, SWR $= 5$ and $Z_0 = 50\ \Omega$. To identify the conditions on the line, three measurements are made by sliding a voltmeter along the line. (1) The first maximum is found at a distance 0.25 m from the load. (2) The next maximum is found at 0.75 m from the load. (3) $V_L = 100$ V. Calculate:

(a) The load impedance.
(b) The amplitudes of the forward- and backward-propagating voltage waves.
(c) The amplitudes of the forward- and backward-propagating current waves.

14.31 Application: Standing Wave Ratio and Minima and Maxima on the Line. For a line with characteristic impedance $Z_0 = 50\ \Omega$ and a load $R = 220\ \Omega$, calculate:

(a) The standing wave ratio on the line.
(b) If the voltage at the load is 100 V, find the maximum and minimum voltage on the line.
(c) Where do the voltage maxima and minima occur?

14.32 Application: Power Supplied to Resistively Loaded Line. Two lossless transmission line segments are connected as shown in **Figure 14.47**. Calculate the power supplied by the generator to the line.

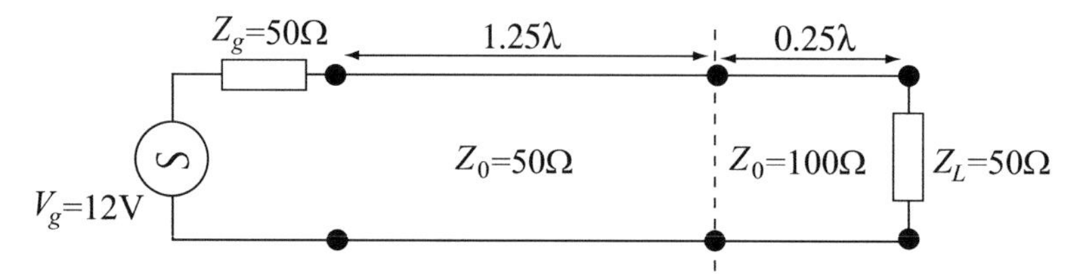

Figure 14.47

14.33 Application: Measurement of Characteristic Impedance. To measure the characteristic impedance of a lossless transmission line it is proposed to measure the voltage at the load and at a distance d [m] from the load. The line is made of two wires in free space (air) and is connected to a 100 Ω load. The voltage at the load is measured as 50 V and at a distance d it is 120 V.

(a) Calculate the characteristic impedance of the line.
(b) If the line operates at a wavelength of 1.8 m, what is the distance d?

14.34 Application: Standing Wave Ratio and Minima and Maxima on the Line. A lossless line has a characteristic impedance $Z_0 = 50\ \Omega$ and a load $R = 25\ \Omega$. Calculate:

(a) The standing wave ratio on the line.
(b) If the voltage at the load is 100 V, find the maximum and minimum voltage on the line.
(c) Where do the voltage maxima and minima occur?

Capacitive and Inductive Loads on Transmission Lines

14.35 Application: Line Impedance on a Capacitively Loaded Line. A long lossless transmission line with characteristic impedance 50 Ω is terminated with a capacitance. At the frequency at which the line operates, the impedance of the load is $-j50\ \Omega$ and the phase constant on the line is 20π [rad/m]. Calculate and plot the line impedance as a function of distance from load.

14.36 Application: Line Impedance on an Inductively Loaded Line. A long lossless transmission line with characteristic impedance 50 Ω is terminated with an inductance. At the frequency at which the line operates, the reactance of the load is 50 Ω and the phase constant is 20π [rad/m]. Calculate and plot the line impedance as a function of distance from load.

14.37 Line Impedance on Line with General Load. A long lossless transmission line with characteristic impedance 50 Ω is terminated with a load as in **Figure 14.48**. The line operates in free space (material between the two lines is air) at 1 MHz. Calculate:

(a) The generalized reflection coefficient on the line.
(b) The standing wave ratio on the line.
(c) The location of the first minimum voltage on the line.
(d) The standing wave ratio as a function of frequency. Plot for frequencies between 0.5 and 1 MHz.

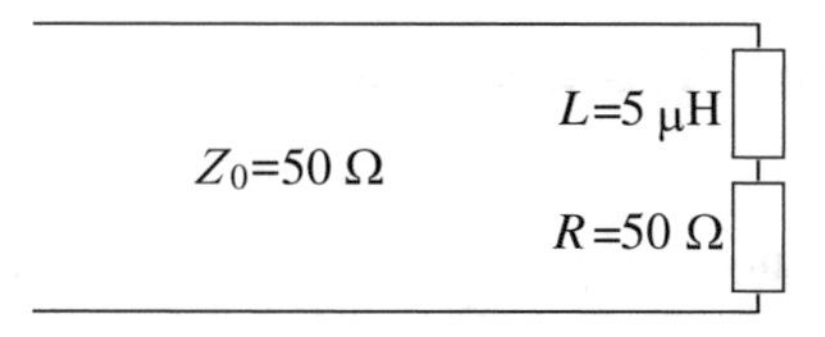

Figure 14.48

14.38 Line Impedance on Line with General Load. A long lossless transmission line with characteristic impedance 50 Ω is terminated with a load as in **Figure 14.49**. The line operates in free space (material between the two lines is air) at 1 MHz. Calculate:

(a) The generalized reflection coefficient on the line.
(b) The standing wave ratio on the line.
(c) The location of the first minimum voltage on the line.
(d) The standing wave ratio as a function of frequency. Plot for frequencies between 0.5 and 1 MHz.

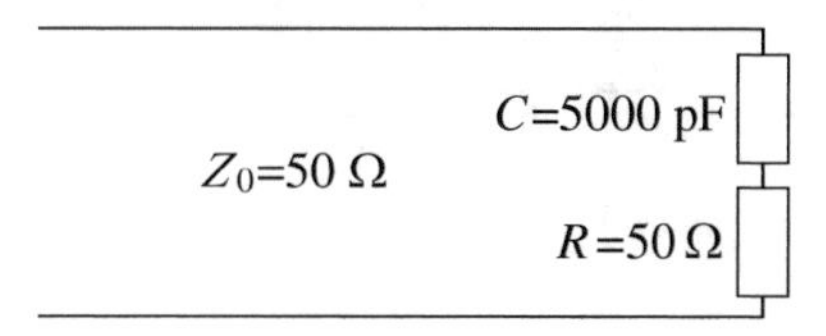

Figure 14.49

Power Relations on Transmission Lines

14.39 Power on Lossless Line. A lossless transmission line is given as shown in **Figure 14.50**. Calculate the power dissipated in the load. The generator is matched to the line and has a voltage of 100 V rms.

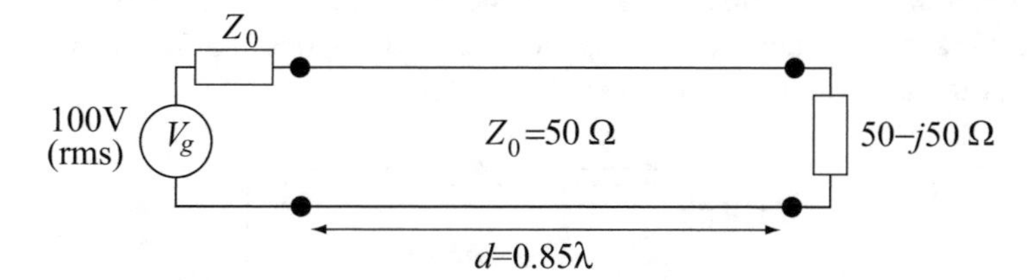

Figure 14.50

14.40 Voltage and Power on a Coaxial Line. A coaxial line is made as follows **(Figure 14.51):** The inner wire diameter is 0.5 mm and the outer one is 6 mm. The material between the conductors is a polymer with relative permittivity 4 and conductivity 10^{-7} S/m. The line is matched at generator and load and is used as a cable TV line at 100 MHz. The peak input voltage to the line is 10 V and the line is 200 m long. Assuming ideal conductors, calculate:

(a) The voltage at the load.
(b) The time averaged power loss on the line.

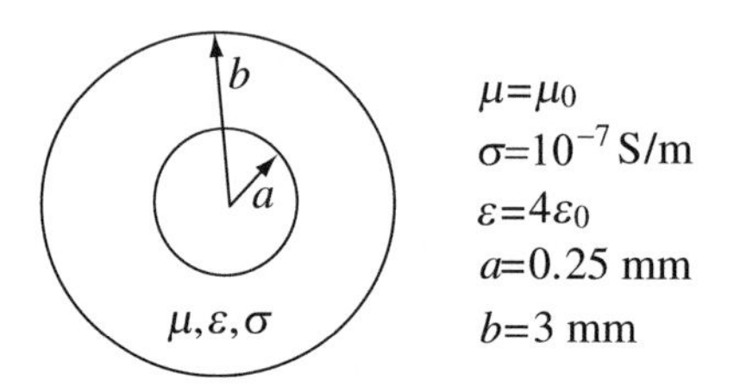

Figure 14.51

14.41 Application: Power Loss on Lossy Transmission Line. A distortionless transmission line has the following parameters:

$$R = 0.2\ [\Omega/\text{m}],\ L = 1\ [\mu\text{H/m}],\ G = 10^{-5}\ [\text{S/m}],\ C = 50\ [\text{pF/m}]$$

The line is matched at the load and is 12.2 wavelengths long at a frequency of 120 MHz. If a power P_0 [W] is supplied at its input terminals, what is the percentage of power lost on the line?

14.42 Power on Lossy Line. In the line shown in **Figure 14.52** it is required that the load dissipates 10 W. Calculate the net power entering the line at location A to produce the required power in the load. Note: net power means the power that propagates towards the load and is the difference between the forward propagating power and the backward propagating power.

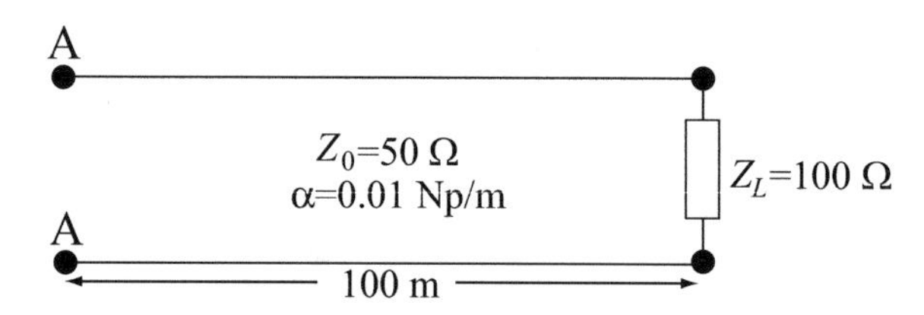

Figure 14.52

14.43 Power Transfer on Lossy Line. A lossy 2-wire transmission line with an attenuation constant $\alpha = 0.1$ Np/m is matched at the generator. That is, $Z_g = Z_0 = 50\ \Omega$. The load is $Z_L = 100\ \Omega$. The medium between the wires is air. The line operates at a wavelength of 0.5 m. The voltage at a distance $d = 3.2$ wavelengths from the load is measured as $V = 100$ V. Calculate the power delivered to the load.

Resonant Transmission Lines

14.44 Application: Transmission Line Resonator. A transmission line resonator is made by connecting a 10 pF capacitor across the input of a 75 Ω lossless line and shorting the other end **(Figure 14.53)**. What must be the length of the line in wavelengths so that the lowest resonant frequency is 100 MHz?

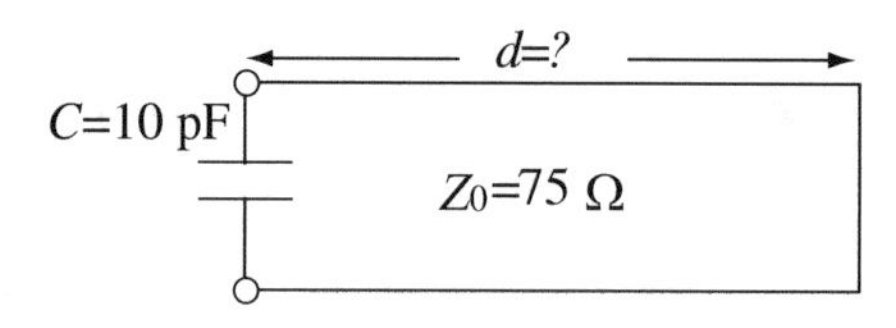

Figure 14.53

14.45 Application: Transmission Line Resonator. A lossless transmission line is cut into a length $b = 0.5$ m and shorted at both ends **(Figure 14.54)**. Properties of the line are $Z_0 = 75\ \Omega$ and $v_p = c/3$ [m/s].

(a) If a connection is made at a point $x = b/2$, what is the impedance of the line at this point?
(b) A connection is now made on the line such that $x = 0.1$ m. Find the first four resonant frequencies for this line.

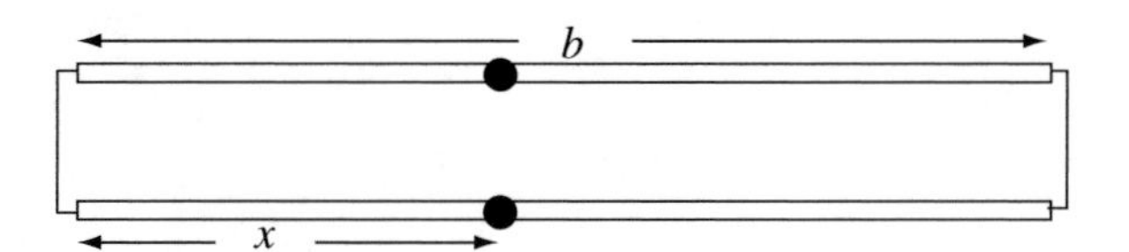

Figure 14.54

14.46 Application: Transmission Line Sensors. A transmission line resonator is made as shown in **Figure 14.55**. The device is used to sense moisture content of dough on a production line as the dough passes through the device, given as a percentage. Dough has a permittivity given by $\varepsilon = \varepsilon_0(1 + 14k)$ [F/m], where k is the percentage of water in the dough. Calculate:

(a) The resonant frequency of the device when empty.
(b) The resonant frequency for 15%, 10%, and 5% moisture in the dough.
(c) Comment on the applicability of this design for moisture measurements.

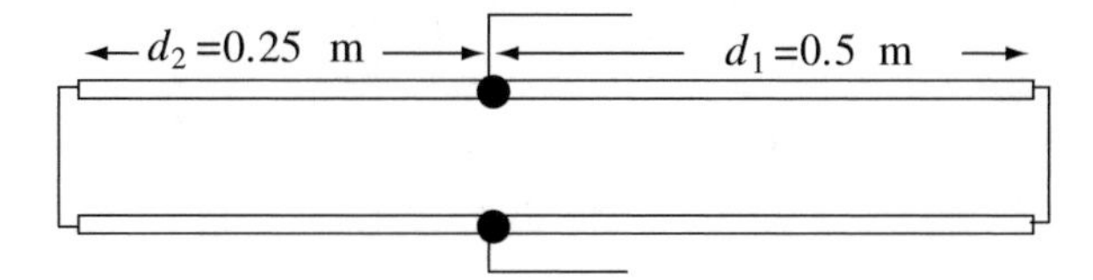

Figure 14.55

14.47 Application: Detection of Faults in Buried Lines. A buried cable has been cut at an unknown point. A signal generator is connected to the cable input and the frequency is swept. The first maximum at the generator occurs at a frequency f_1 [Hz] and the first minimum at a frequency $f_2 > f_1$. What is the distance from the generator to the short if the phase velocity is known and equals v_{p0} [m/s] at both frequencies. The generator is matched to the line.

14.48 Application: Detection of Faults in Buried Lines. A buried cable has water leaking into it, at a location along its length. The water leak manifests itself as an unknown impedance on the cable at the location of the leak. If the same results as in **Problem 14.47** are obtained at the generator:

(a) Is it possible with the data given in **Problem 14.47** to find the exact location of the leak?
(b) If yes, what is that location? If not, what is the minimum section that must be dug to ensure that the leak is found?

The Smith Chart, Impedance Matching, and Transmission Line Circuits

15

Errors using inadequate data are much less than those using no data at all

—attributed to Charles Babbage (1791–1871)
Designer of the "difference machine" –
the first programmable computing machine and
predecessor to modern computers

15.1 Introduction

A look back at much of what we did with transmission lines reveals that the dominant feature in all our calculations is the use of the reflection coefficient. The reflection coefficient was used to find the conditions on the line, to calculate the line impedance, and to calculate the standing wave ratio. Voltage, current, and power were all related to the reflection coefficient. The reflection coefficient, in turn, was defined in terms of the load and line impedances (or any equivalent load impedances such as at a discontinuity). You may also recall, perhaps with some fondness, the complicated calculations which required, in addition to the use of complex variables, the use of trigonometric, harmonic and hyperbolic functions. Thus, the following proposition: Build a graphical chart (or an equivalent computer program) capable of representing the reflection coefficient as well as load impedances in some general fashion and you have a simple method of designing transmission line circuits without the need to perform rather tedious calculations. This has been accomplished in a rather general tool called the Smith chart. The Smith chart is a chart of normalized impedances (or admittances) in the complex reflection coefficient plane. As such, it allows calculations of all parameters related to transmission lines as well as impedances in open space and circuits and networks. Although the Smith chart is rather old, it is a common design tool in electromagnetics. Some measuring instruments such as network analyzers actually use a Smith chart to display conditions on lines and networks. Naturally, any chart can also be implemented in a computer program, and the Smith chart has, but we must first understand how it works before we can use it either on paper or on the screen. The examples provided here are solved using graphical tools and a printed Smith chart, rather than the computer program, to emphasize the techniques and approximations involved although some of the numerical results listed were obtained with a computerized Smith chart (**The_Smith_Chart.m**) available with this text (see **page x**).

The Smith chart is an impedance chart. As such it does not provide for direct calculations of voltages, currents, or power. Nevertheless, it is a useful tool in the calculation of voltages and currents as well as power since it provides important information such as the generalized reflection coefficient, standing wave ratio, and the location of voltage and current maxima and minima. With the information available from the Smith chart, the formulas developed in **Chapter 14** can then be used to obtain the required values or conditions.

N. Ida, *Engineering Electromagnetics*, https://doi.org/10.1007/978-3-030-15557-5_15

15.2 The Smith Chart[1]

To better understand the Smith chart and to gain some insight in its use, we will "build" a Smith chart, gradually, based on the definitions of the reflection coefficient. Then, after all aspects of the chart are understood, we will use the chart in a number of examples to show its utility. In the process, we will also define a number of transmission line circuits for which the Smith chart is commonly used.

Consider the lossless transmission line circuit in **Figure 15.1**. The line impedance Z_0 is real but the load is a complex impedance $Z_L = R_L + jX_L$, where R_L is the load resistance and X_L the load reactance. The reflection coefficient [see **Eqs.** (14.83) and (14.83)] may be written in one of two forms. The first is a rectangular form (i.e., written in complex variables):

$$\Gamma_L = \frac{Z_L - Z_0}{Z_L + Z_0} = \frac{(R_L - Z_0) + jX_L}{(R_L + Z_0) + jX_L} = \Gamma_r + j\Gamma_i \tag{15.1}$$

The reflection coefficient is not modified by normalizing the numerator and denominator by Z_0:

$$\Gamma_L = \frac{(Z_L - Z_0)/Z_0}{(Z_L + Z_0)/Z_0} = \frac{(R_L/Z_0 - 1) + jX_L/Z_0}{(R_L/Z_0 + 1) + jX_L/Z_0} = \frac{(r-1) + jx}{(r+1) + jx} = \Gamma_r + j\Gamma_i \tag{15.2}$$

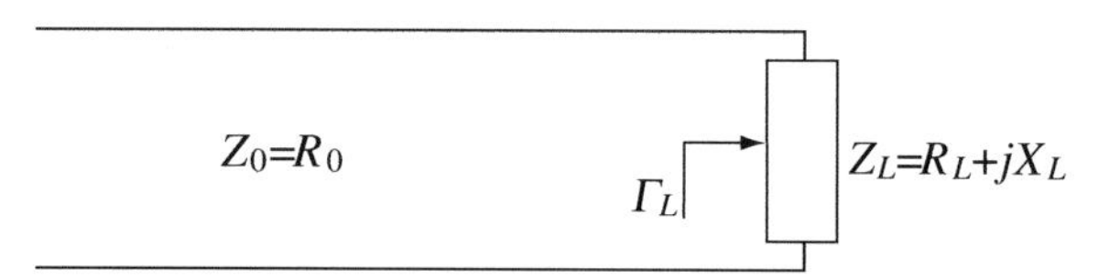

Figure 15.1 A simple lossless transmission line used to introduce the Smith chart

To obtain this result, we substituted $r = R_L/Z_0$ and $x = X_L/Z_0$ as the normalized resistance and reactance. For much of the remainder of this chapter, we will drop the specific notation for load partly to simplify notation but mostly because the magnitude of the reflection coefficient remains constant along lossless lines, therefore, the results we obtain apply equally well for any impedance on the line (see **Figure 15.2**). In the latter case, the generalized reflection coefficient is obtained and this can be written in exactly the same form as **Eq. (15.1)** or **(15.2)** by replacing Z_L with $Z(z)$. **Equation (15.2)** defines a complex plane for the reflection coefficient as shown in **Figure 15.3a**. Any normalized impedance (load impedance or line impedance) is represented by a point on this diagram.

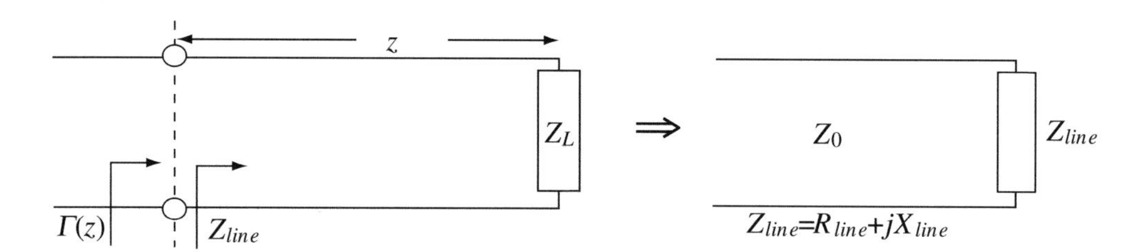

Figure 15.2 Use of an equivalent transmission line to describe the line impedance at a distance z from the load

The second form of the reflection coefficient is the polar form. This may be written as

$$\Gamma_L = |\Gamma|e^{j\theta_\Gamma} = |\Gamma|(\cos\theta_\Gamma + j\sin\theta_\Gamma) \tag{15.3}$$

where θ_Γ is the phase angle of the load reflection coefficient as discussed in **Section 14.7.1**. For a given magnitude of the reflection coefficient, the phase angle defines a point on the circle of radius $|\Gamma_L|$. Thus, since $|\Gamma_L| \leq 1$, only that section of the

[1] The Smith chart was introduced by Phillip H. Smith in January 1939. Smith developed the chart as an aid in calculation and called it a "transmission line calculator." In spite of its age, the chart is as useful as ever as a standard tool in analysis either in its printed form, slide-rule form, or as computer programs and instrument displays.

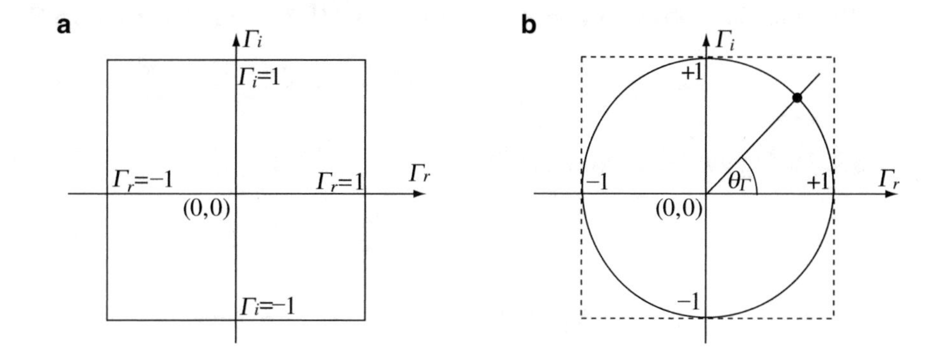

Figure 15.3 The complex plane representation of the reflection coefficient.
(**a**) In rectangular form.
(**b**) In polar form

rectangular diagram enclosed by the circle of radius 1 is used, as shown in **Figure 15.3b**. The polar form is more convenient to use than the rectangular form, but we will, for the moment, retain both.

We now go back to the rectangular representation and calculate the real and imaginary parts of the reflection coefficient in terms of the normalized impedance. The starting point is **Eq. (15.2)**:

$$\Gamma_r + j\Gamma_i = \frac{(r-1)+jx}{(r+1)+jx} \tag{15.4}$$

Cross-multiplying gives

$$(r+1)\Gamma_r - \Gamma_i x + j\Gamma_i(r+1) + jx\Gamma_r = (r-1)+jx \tag{15.5}$$

Separating the real and imaginary parts and rearranging terms, we get two equations:

$$(\Gamma_r - 1)r - \Gamma_i x = -(\Gamma_r + 1) \tag{15.6}$$

$$\Gamma_i r + (\Gamma_r - 1)x = -\Gamma_i \tag{15.7}$$

We now write two equations: one for r and one for x, by first eliminating x and then, separately, r. From **Eq. (15.7)** we write

$$x = -\frac{\Gamma_i(r+1)}{\Gamma_r - 1} \tag{15.8}$$

Substituting this into **Eq. (15.6)** we get

$$(\Gamma_r - 1)r + \frac{\Gamma_i^2(r+1)}{\Gamma_r - 1} = -(\Gamma_r + 1) \tag{15.9}$$

Multiplying both sides by $\Gamma_r - 1$ and rearranging terms, this gives
After rearranging terms, this gives

$$\Gamma_r^2(r+1) - 2\Gamma_r r + \Gamma_i^2(r+1) = 1 - r \tag{15.10}$$

Dividing by the common term $(r + 1)$,

$$\Gamma_r^2 - \frac{2\Gamma_r r}{(r+1)} + \Gamma_i^2 = \frac{1-r}{(r+1)} \tag{15.11}$$

Adding $r^2/(r + 1)^2$ to both sides of the equation and rearranging terms, we get

$$\boxed{\left(\Gamma_r - \frac{r}{r+1}\right)^2 + \Gamma_i^2 = \frac{1}{(r+1)^2}} \tag{15.12}$$

Repeating the process, we now eliminate r in **Eq. (15.7)** by first writing from **Eq. (15.6)**:

$$r = -\frac{(\Gamma_r + 1) - \Gamma_i x}{(\Gamma_r - 1)} \quad (15.13)$$

Substituting this back into **Eq. (15.7)**,

$$\Gamma_i \frac{(\Gamma_r + 1) - \Gamma_i x}{\Gamma_r - 1} + (\Gamma_r - 1)x = -\Gamma_i \quad (15.14)$$

Multiplying both sides of **Eq. (15.14)** by $\Gamma_r - 1$ and rearranging terms we get

$$(\Gamma_r - 1)^2 x + \Gamma_i^2 x - 2\Gamma_i = 0 \quad (15.15)$$

The equation now is divided by x:

$$(\Gamma_r - 1)^2 + \Gamma_i^2 - 2\Gamma_i\left(\frac{1}{x}\right) = 0 \quad (15.16)$$

To bring this into a useful form, we add $1/x^2$ to both sides of the equation and rearrange terms. We get:

$$\boxed{(\Gamma_r - 1)^2 + \left(\Gamma_i - \frac{1}{x}\right)^2 = \left(\frac{1}{x}\right)^2} \quad (15.17)$$

Both **Eqs. (15.12)** and **(15.17)** describe circles in the complex Γ plane.

Equation (15.12) is the equation of a circle, with its center at $\Gamma_r = r/(r + 1)$, $\Gamma_i = 0$ and radius $1/(r + 1)$. The center of the circle is on the real axis and can be anywhere between $\Gamma_r = 0$ for $r = 0$ and $\Gamma_r = 1$ for $r \to \infty$. For example, for $r = 1$, the center of the circle is at $\Gamma_r = 0.5$ and its radius equals 0.5. A number of these circles are drawn in **Figure 15.4a**. The larger the normalized resistance, the smaller the circle. All circles pass through $\Gamma_r = 1, \Gamma_i = 0$. The normalized resistance r can only be positive. Should there ever be a need to describe normalized impedances with negative real part, these must be multiplied by -1 before analysis using the Smith chart can commence.

From **Eq. (15.17)**, we obtain a second set of circles for x. Since x can be positive or negative, the circles are centered at $\Gamma_r = 1, \Gamma_i = 1/x$ for positive values of x and at $\Gamma_r = 1, \Gamma_i = -1/x$ for x negative. These circles are shown in **Figure 15.4b** for a number of values of the normalized reactance x. **Figure 15.5** shows the r and x circles on the Γ plane, truncated at the circle $|\Gamma| = 1$. This is the basic Smith chart. A number of properties of the two sets of circles are immediately apparent:

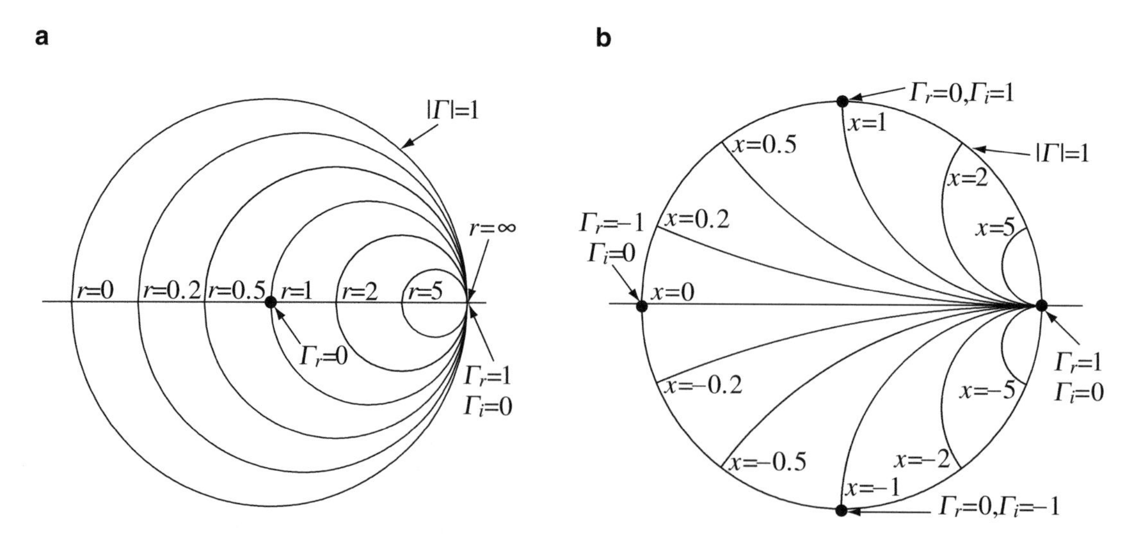

Figure 15.4 The basic components of the Smith chart. (**a**) Circles of constant values of r. (**b**) Circles of constant values of x or $-x$

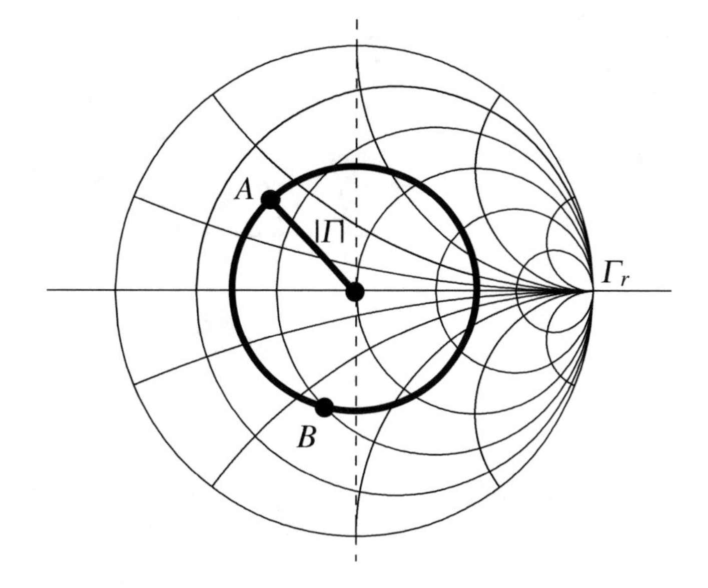

Figure 15.5 The Smith chart. A normalized impedance is a point on the Smith chart defined by the intersection of a circle of constant normalized resistance r and a circle of constant normalized reactance x

(1) The circles are loci of constant r or constant x.
(2) x and r circles are orthogonal to each other.
(3) There is an infinite number of circles for r and for x.
(4) All circles pass through the point $\Gamma_r = 1, \Gamma_i = 0$.
(5) The circles for x and $-x$ are images of each other, reflected about the real axis.
(6) The center of the chart is at $\Gamma_r = 0, \Gamma_i = 0$.
(7) The intersections of the r circles with the real axis, for $r = r_0$ and $r = 1/r_0$, occur at points symmetric about the center of the chart ($\Gamma_r = 0, \Gamma_i = 0$).
(8) The intersections of the x circles with the outer circle ($|\Gamma| = 1$) for $x = x_0$ and $x = -1/x_0$ occur at points diametrically opposite each other.
(9) The intersection of any r circle with any x circle represents a normalized impedance point.
(10) The real part of the normalized impedance, r, can only be positive but x can be negative or positive.

The chart as described above is an impedance chart since we defined all points in terms of normalized impedances. We will see how to use the chart as an admittance chart later.

In addition to the properties of the r and x circles given above, we note the following:

(1) The point $\Gamma_r = 1, \Gamma_i = 0$ (rightmost point in **Figure 15.5**) represents $r = \infty, x = \infty$. This is the impedance of an open transmission line. This point is therefore the ***open circuit point***.
(2) The diametrically opposite point, at $\Gamma_r = -1, \Gamma_i = 0$ represents $r = 0, x = 0$. This is the impedance of a short circuit and is called the ***short circuit point***.
(3) The outer circle represents $|\Gamma| = 1$. The center of the diagram represents $|\Gamma| = 0$. Any circle centered at the center of the diagram ($\Gamma_r = 0, \Gamma_i = 0$) with radius a is a circle on which the magnitude of the reflection coefficient is constant, $|\Gamma| = a$. Moreover, if we take the intersection between any r and x circles, the distance between this point to the center of the diagram is the magnitude of the reflection coefficient for this normalized impedance. A circle drawn through this point represents the generalized reflection coefficient at different locations on the line for this normalized load impedance. The intersection of the reflection coefficient circle with r and x circles represents line impedances at various locations on the line. These aspects of the use of transmission lines are shown in **Figure 15.5**. For example, point A represents a normalized impedance $r_A + jx_A$ and point B represents a normalized impedance $r_B + jx_B$, but the magnitude of the reflection coefficient is the same. This will later be used to calculate the line impedance as well as voltages and currents on the line.
(4) Any point on the chart represents a normalized impedance, say, $z = r + jx$. The admittance of this point is $y = 1/(r + jx) = (r - jx)/(r^2 + x^2)$. The admittance point corresponding to an impedance point lies on the reflection coefficient circle that passes through the impedance point, diametrically opposite the impedance point. Thus, if we mark a normalized impedance as z and draw the reflection coefficient circle through point z, this circle passes through the admittance point $y = 1/z$. The admittance point y is found by passing a line through z and the center of the diagram. These steps are shown in **Figure 15.6a**. These considerations will later be used to calculate admittances instead of impedances.

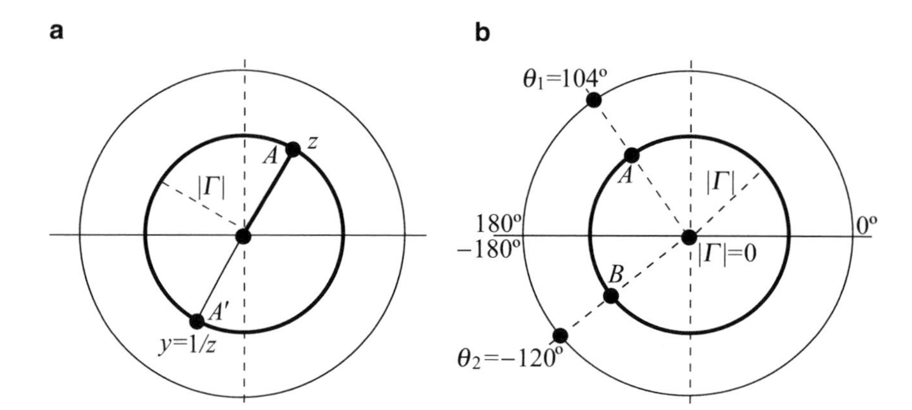

Figure 15.6 (a) Normalized impedance, reflection coefficient, and normalized admittance. (**b**) Indication of phase angle of the reflection coefficient on the Smith chart

The Smith chart also provides for calculation of phase angles and lengths of transmission lines. For this purpose, the Smith chart is equipped with a number of scales, marked on the outer periphery of the diagram. These are defined as follows:

(1) A given impedance corresponds to a point on the chart. If the line connecting the center of the chart with the impedance point is continued until it intersects the outer ($\Gamma = 1$) circle, the location of intersection gives the phase angle of the reflection coefficient in degrees. This is the second set of values given on the circumference of the Smith chart and is shown in **Figure 15.6b**. Note that the open circuit point has zero phase angle ($\Gamma = +1$) and the short circuit point has either a $180°$ or $-180°$ phase angle. The difference is in the sign of the imaginary part of the load impedance (below or above the real axis). Intermediate points will vary in phase depending on the distance from the load. For example, for point A in **Figure 15.6b**, the phase angle of the reflection coefficient is $104°$, whereas for point B it is $-120°$. The phase angle of the transmission coefficient is shown in the first set of values on the circumference of the chart (see **Figure 15.7**).

(2) We recall that the distance between a point of maximum voltage and a point of minimum voltage was found to be $\lambda/4$ in **Section 14.7.3**. In particular, the impedance of a shorted transmission line changes from zero to infinity (or negative infinity) if we move a distance $\lambda/4$ from the short. Thus, the distance between the short circuit and open circuit points is $\lambda/4$. This fact is indicated on the outer circle of the chart, starting at the short circuit point. Since the short (or any other load) can be anywhere on a line, we may wish to move either toward the generator or toward the load to evaluate the line behavior. These two possibilities are indicated with arrows showing the direction toward load and toward generator (**Figure 15.7**). Although the distance is marked from the short circuit point, the distance is always relative: if a point is given at any location on the chart, movement on the chart, a distance $\lambda/4$ represents half the circumference of the chart.

(3) The direction toward the generator is the clockwise direction. If we wish to calculate the line impedance starting from the load, we move in the clockwise direction toward the generator. If, on the other hand, we wish to calculate the line impedance starting from the generator going toward the load or, starting at the load and going away from the generator, we must move in the counterclockwise direction and use the appropriate distance charts (see **Figure 15.7**).

(4) The whole Smith chart encompasses one-half wavelength. This, of course, is due to the fact that all conditions on lines repeat at intervals of $\lambda/2$ regardless of loading or any other effect that may happen on the line. If we need to analyze lines longer that $\lambda/2$, we simply move around the chart as many half-wavelengths as are necessary. Only the remainder length (length beyond any integer numbers of half-wavelengths) needs to be analyzed.

The Smith chart also allows for the calculation of standing wave ratios. The standing wave ratio is calculated from the reflection coefficient as

$$\mathrm{SWR} = \frac{1 + |\Gamma|}{1 - |\Gamma|} \quad [\text{dimensionless}] \tag{15.18}$$

We note that the circle of radius $|\Gamma|$ intersects the positive real axis at $x = 0$. At this point, the normalized impedance is equal to r and the reflection coefficient is given as $\Gamma = (r - 1)/(r + 1)$. Substituting this into the relation for SWR, we get

$$\mathrm{SWR} = r \tag{15.19}$$

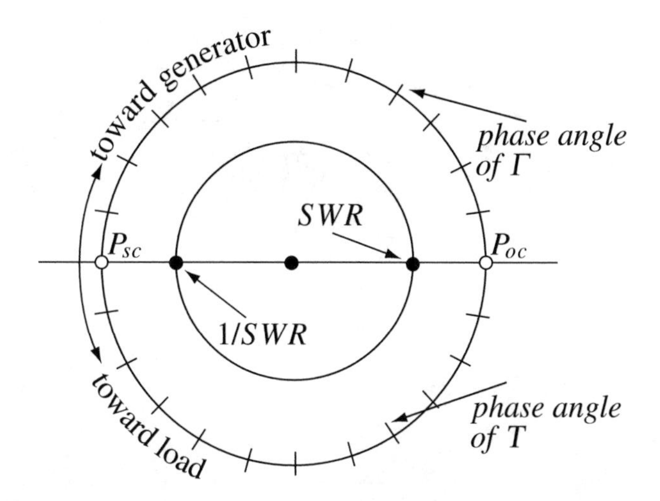

Figure 15.7 Directions on the Smith chart and indication of SWR. The distance between short and open circuit points is $\lambda/4$

Thus, the standing wave ratio equals the value of normalized resistance at the location of intersection of the reflection coefficient circle and the real axis, right of the center of the Smith chart. From property (7) above, the intersection of the reflection coefficient circle with the real axis, left of the center of the chart, is at point $1/r$. This point equals 1/SWR. The two points are shown for the reflection coefficient in **Figure 15.7**.

Now that we discussed the individual parts making up the Smith chart, it is time to put it all together. The result is the Smith chart shown in **Figure 15.8**. You will immediately recognize the r and x circles as well as the scales discussed. There are, however, a number of other scales given at the bottom of the chart as well as a number of indications on the chart itself which we have not discussed. These have to do with losses on the line (which we have neglected) and the use of the chart as an admittance rather than impedance chart (which we will take up later).

Although the chart is relatively simple, it contains considerable information and can be used in many different ways and for purposes other than transmission lines. To see how the chart is used, we will discuss next a number of applications of the Smith chart to design of transmission lines. Because the chart gives numerical data, the examples must also be numerical, but, in general, the equations in the previous chapter can also be used for this purpose. The main difference in the Smith chart solution and the analytic solution is that the Smith chart uses normalized impedances, whereas in analytic calculations, we tend to use the actual values of the impedance. Also, because it is a graphical chart, the results are approximate and depend on our ability to accurately read the values off the chart. The Smith chart is available as a paper chart as well as computer software. The advantage of a software-based Smith chart is that calculations are exact in addition to the ease of analysis and display of results.

Example 15.1 Calculation of Line Conditions **The_Smith_Chart.m**

A long line with characteristic impedance $Z_0 = 50\ \Omega$ operates at 1 GHz. The speed of propagation on the line is c and load impedance is $75 + j100\ \Omega$. Find:

(a) The reflection coefficient at the load.
(b) The reflection coefficient at a distance of 20 m from the load toward the generator.
(c) Input impedance at 20 m from the load.
(d) The standing wave ratio on the line.
(e) Locations of the first voltage maximum and first voltage minimum from the load.

Solution:

(a) **(1)** Normalize the load impedance: $z_l = (75 + j100)/50 = 1.5 + j2$. Enter this on the Smith chart at the intersection of the resistance circle equal to 1.5 and reactance circle equal to 2. This is point P_2 in **Figure 15.9**.

(2) With center at origin (point P_1), draw a circle that passes through point P_2. This circle is the reflection coefficient circle and gives $|\Gamma|$ anywhere on the line. Measure the length of the radius (distance between P_1 and P_2) and divide

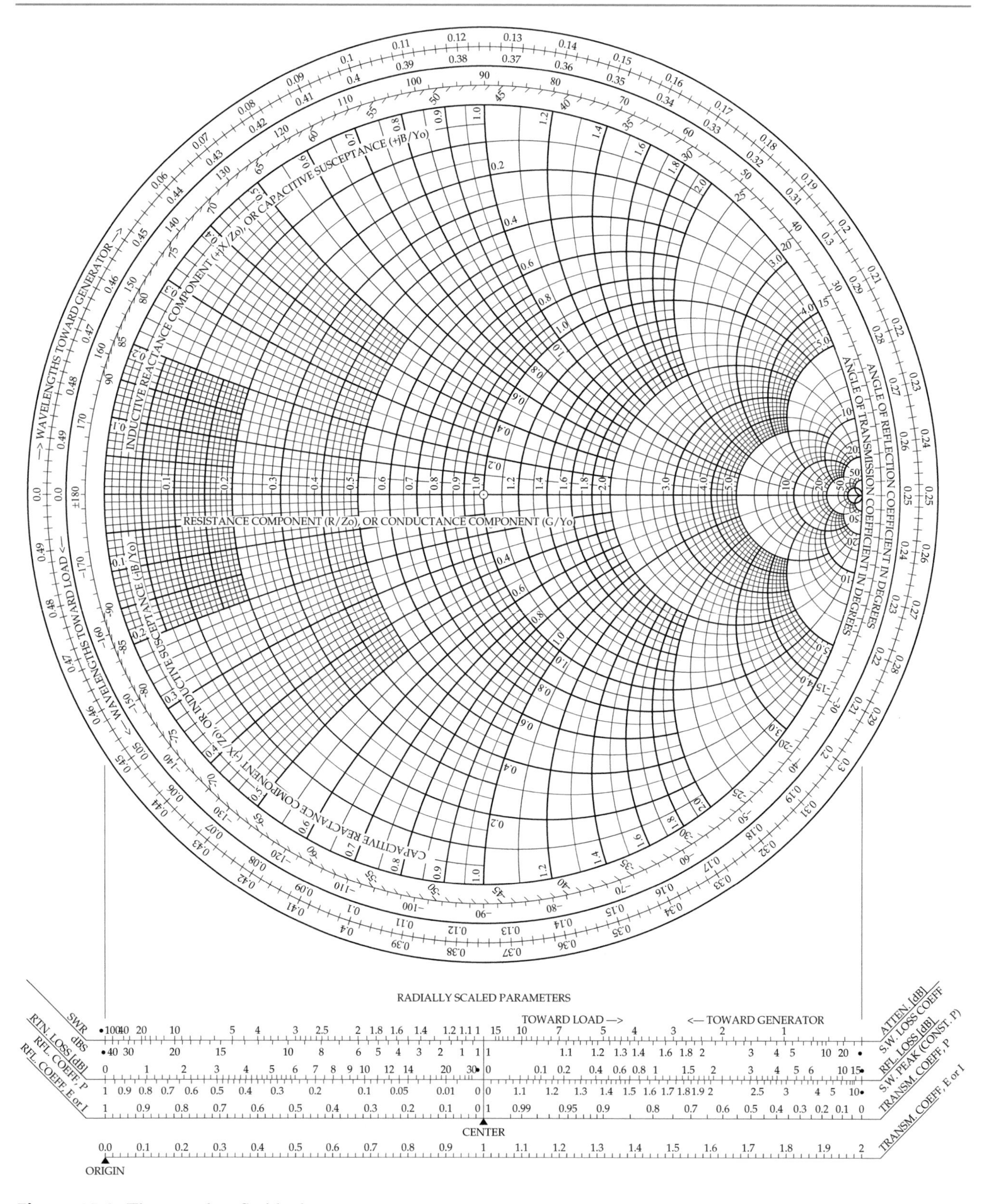

Figure 15.8 The complete Smith chart

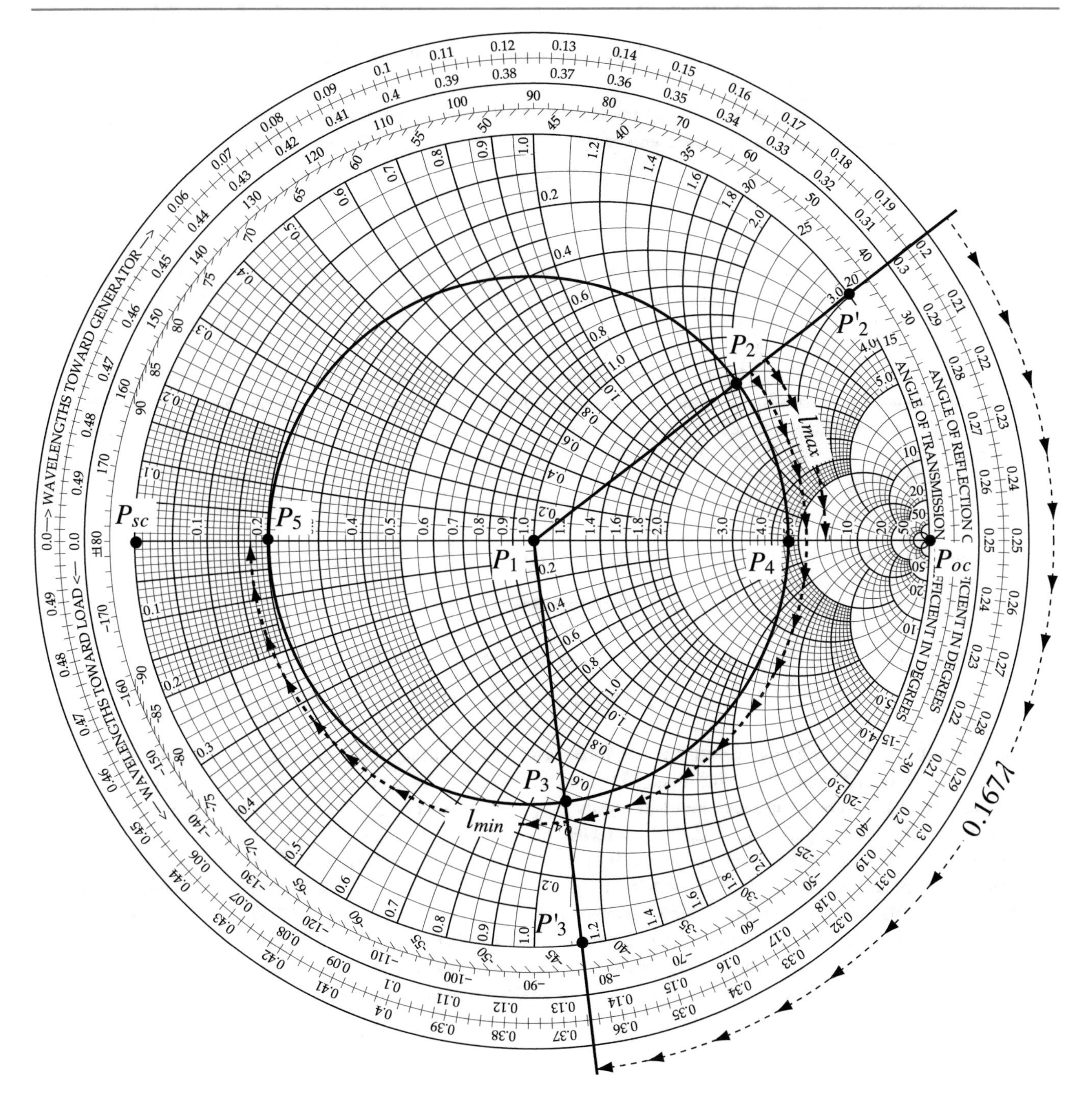

Figure 15.9 The Smith chart for **Example 15.1**

by the radius of the Smith chart's outer circle (distance between P_1 and P_2'). This gives the magnitude of the reflection coefficient. In this case, $|\Gamma| = 0.6439$.

Note: The radius of the Smith chart should be equal to 1, but to facilitate reading, the size differs from one chart to another, thus the need to calculate the magnitude of the reflection coefficient.

(3) Draw a straight line between P_1 and P_2 and extend it to the periphery of the chart to point P_2'. The angle (in degrees, on the periphery) is the phase angle of the reflection coefficient at the load. In this case, it is 37.3°. Alternatively, read the "wavelength toward generator" circle. This is equal to 0.198 at point P_2'. To calculate the angle, subtract this value from the value on the real axis (open circuit point) and multiply by 4π: $(0.25 - 0.198) \times 4\pi = 0.208\pi$ radians or 37.3°. Thus, the answer to **(a)** is

$$\Gamma_L = |\Gamma_L| e^{j\theta r} = 0.6439 e^{j0.208\pi} = 0.6439 \underline{/37.3^\circ}$$

(b) To calculate the reflection coefficient at 20 m from the load, moving toward the generator, we first calculate the wavelength because the chart can only accommodate wavelengths:

$$\lambda = \frac{c}{f} = \frac{3 \times 10^8}{10^9} = 0.3 \quad [\text{m}]$$

Since the circumference of the Smith chart represents 0.5λ (or 0.15 m), the 20 m distance represent (20/0.15) = 133.3334 half-wavelengths. Thus, we move around the reflection coefficient circle toward the generator 133 times, starting at P_2. This puts us exactly where we started (at point P_2). The remainder is one-third of a half-wavelength or $\lambda/6$ (0.167λ).

We now move from point P_2 along the reflection coefficient circle, a distance of 0.167 wavelengths toward the generator to point P_3. Connecting this point with the center of the chart and with the circumference gives the intersection with the reflection coefficient circle at P_3 and with the circumference at P_3'. This point gives the phase angle of the reflection coefficient as $-82.7°$. Thus the reflection coefficient at 20 m from the load is

$$\Gamma = 0.6439 \underline{/-82.7^\circ}.$$

(c) The input impedance 20 m from the load is represented at point P_3. The normalized input impedance is

$$z(l = 20\,\text{m}) = 0.468 - j1.02$$

Multiplying by the characteristic line impedance ($Z_0 = 50\ \Omega$), we get the actual line impedance as

$$Z(l = 20\,\text{m}) = 23.4 - j51.1 \quad [\Omega].$$

(d) The reflection coefficient circle intersects the real axis at point P_4. At this point, $r = 4.62$. This is the standing wave ratio: SWR = 4.62. At point P_5 (on the other side of the reflection coefficient circle) $r = 1/\text{SWR} = 0.217$. At point P_4, the line impedance is real and maximum and equals $Z_{max} = Z_0 \times 4.62 = 230.8\ \Omega$ = At point P_5, the impedance is minimum and real and equals $Z_{min} = Z_0/4.62 = 10.83\ \Omega$.

(e) Location of maximum voltage is on the real axis at the same point where SWR = 4.76 since, at this point, the line impedance is maximum (and real). Thus, moving from point P_2 to the positive real axis, we reach a voltage maximum: the distance is the difference in wavelengths between point P_{oc} and point P_2 or $l_{max} = 0.25\lambda - 0.198\lambda = 0.052\lambda$ from the load. The voltage minimum is a quarter-wavelength away (where 1/SWR = 0.21) at point P_5 or $l_{min} = 0.302\lambda$ from the load. In terms of actual distance the first maximum occurs at a distance of $0.052 \times 0.3 = 0.0156$ m, or 15.6 mm from the load. The first minimum occurs at $0.302 \times 0.3 = 0.0906$ m or 90.6 mm from the load.

15.3 The Smith Chart as an Admittance Chart

We mentioned earlier that the Smith chart may be used as an admittance chart. In **Figure 15.6a**, we showed that for any given normalized impedance, the admittance is found by locating the normalized impedance point $z = r + jx$ on the Smith chart, drawing the reflection coefficient circle, and then drawing a straight line that passes through the impedance point, the center of the chart, and then intersects the reflection coefficient circle, on a point diametrically opposite the impedance point, at point y. This point represents the normalized admittance of the load. Any normalized impedance may be converted into its equivalent admittance using this simple step.

In addition to this, we note that an infinite normalized impedance (open circuit point on the impedance Smith chart) represents infinite admittance on the admittance Smith chart. Similarly, the short circuit point on the impedance Smith chart represents zero admittance on the admittance Smith chart (see **Figure 15.10**).

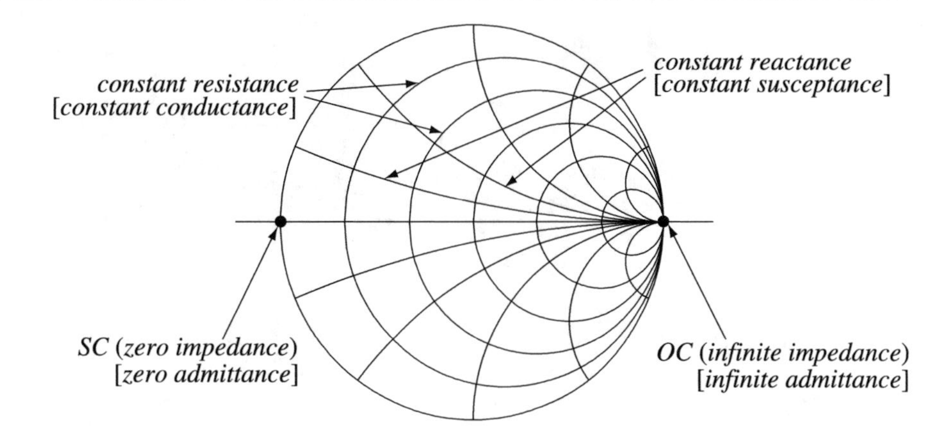

Figure 15.10 Relations between the impedance and admittance Smith charts. Descriptions in *square brackets* are for the admittance chart

The admittance may be written in terms of the impedance at point z as

$$y = \frac{1}{r + jx} = \frac{r}{r^2 + x^2} - j\frac{x}{r^2 + x^2} = g - jb \tag{15.20}$$

Since we use the same chart, the constant resistance circles now become constant conductance circles, and the constant reactance circles become constant susceptance circles. All other aspects of the chart, including phase angles, distances, etc., remain unchanged.

The use of the Smith chart as an admittance chart is shown in **Figure 15.10**, in comparison with the impedance chart.

Example 15.2 Use of the Smith Chart as Admittance Chart

The_Smith_Chart.m

A load, such as an antenna, of impedance $Z_L = 50 - j100\ \Omega$ is connected to a lossless transmission line with characteristic impedance $Z_0 = 100\ \Omega$. The line operates at 300 MHz and the speed of propagation on the line is $0.8c$:

(a) Calculate the input admittance a distance 2.5 m from the load.
(b) Calculate the input impedance a distance 2.5 m from the load.
(c) Suppose the load is shorted accidentally. What is the input admittance at the same point?

Solution: To calculate the input admittance, we first calculate the wavelength on the line. The load is then located on the impedance chart and the admittance is found on the reflection coefficient circle. Then, we move toward the generator a distance 2.5 m (in wavelengths) to find the normalized input admittance. The admittance is found by multiplying the normalized admittance with the characteristic admittance of the line. The input impedance can be found from the input admittance by finding the diametrically opposite point on the reflection coefficient circle.

(a) The normalized load impedance is

$$z_L = \frac{50 - j100}{100} = 0.5 - j1$$

This is marked on the chart as point P_2 in **Figure 15.11**. The reflection coefficient circle is drawn around point P_1, with a radius equal to the distance between P_2 and P_1. The admittance point is P_3. The normalized load admittance is

$$y_L = 0.4 + j0.8$$

The wavelength on the line is $\lambda = 0.8c/f = 2.4 \times 10^8/3 \times 10^8 = 0.8$ m. The given distance represents $2.5/0.8 = 3.125$ wavelengths. To find the input admittance, we move from the load admittance point toward the generator a distance of 0.125λ (the three wavelengths mean simply moving six times around the chart to get to

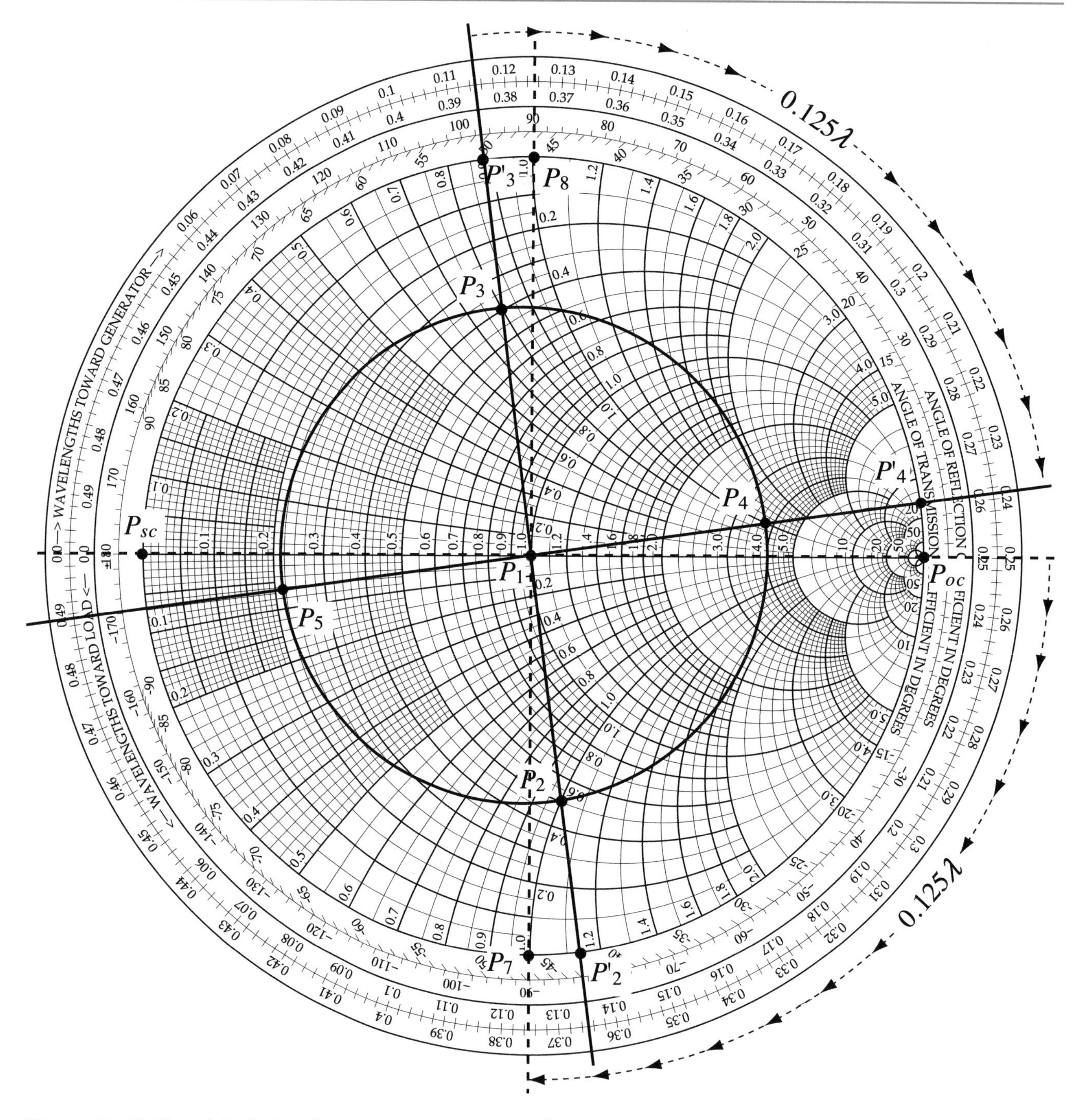

Figure 15.11 Use of the Smith chart as an admittance chart **(Example 15.2)**

the initial point). Moving from point P_3' a distance 0.125λ brings us to point P_4' ($0.114\lambda + 0.125\lambda = 0.239\lambda$). Connecting this point with P_1 intersects the reflection coefficient circle at point P_4. The normalized input line admittance is

$$y_{in} = 4.0 + j1.0$$

The input line admittance is the normalized input line admittance above multiplied by the characteristic line admittance, which equals 0.01:

$$Y_{in} = 0.04 + j0.01 \quad [1/\Omega].$$

(b) The normalized input impedance is found by locating point P_5, which is the diametrically opposite point to P_4, on the reflection coefficient circle. The normalized line impedance at this point is $0.235 - j0.059$. The line impedance is found by multiplying this normalized impedance by the characteristic impedance of the line:

$$Z_{in} = 23.5 - j5.9 \quad [\Omega].$$

(c) If the load is shorted, the load impedance is zero and the line admittance is infinite. This is represented at point P_{oc} on the admittance chart. From here, we move 0.125 wavelengths toward the generator on the outer circle, since for shorted loads, $|\Gamma| = 1$. This point is shown as P_7. The normalized input line admittance is $-j1$. The line admittance is, therefore, $-j0.01$ (line impedance is $j100$, at point P_8).

15.4 Impedance Matching and the Smith Chart

15.4.1 Impedance Matching

When connecting a transmission line to a generator, a load, or another transmission line, the impedances are, in general, mismatched and the result is a reflection coefficient at the load, generator, or discontinuity, which, in turn, generates standing waves on the line. The effect of this reflection was discussed at some length in **Chapter 14**. It is often necessary to match a transmission line to a load or to a generator, for the purpose of eliminating standing waves on the line. Similarly, if a discontinuity exists, such as the connection of an unmatched line section, it is often necessary to eliminate this mismatch before the line can be used.

A transmission line is matched to a load if the load impedance is equal to the characteristic impedance. Similarly, if the line impedance is equal to the generator impedance, the two are matched. To match a load to a line (or a generator for that matter), a matching network is connected between the line and the mismatched generator as in **Figure 15.12a**, or load, as in **Figure 15.12b**, or between two lines with different characteristic impedances as shown in **Figure 15.12c**.

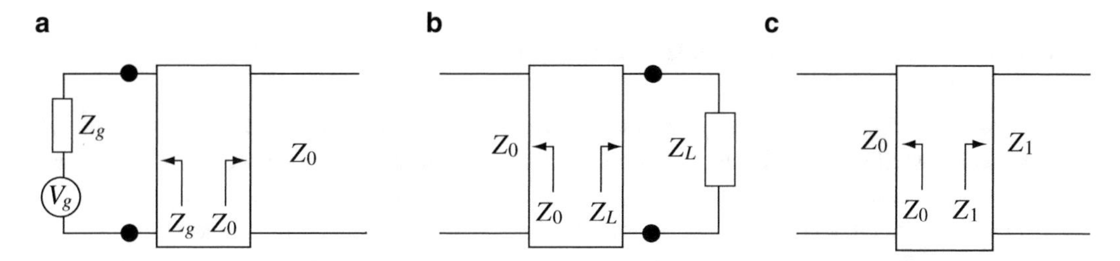

Figure 15.12 Matching networks at (**a**) generator (**b**) load (**c**) between two lines

The location of the matching network depends on the application. If we wish to reduce the standing waves on the line, the matching network should be located as closely as possible to the mismatched impedance. If, however, the line can operate with standing waves, then a more convenient location, at some distance away, can be found. The latter approach is possible since all conditions on the line repeat at intervals of $\lambda/2$. Thus, if a matching network has been designed to be located at a given point on the line, the network can now be moved a distance $n\lambda/2$ (n is an integer) without affecting the line conditions.

The selection of a matching network depends on a number of parameters including frequency, bandwidth, availability of appropriate components and even type of line on which the network is used. At low frequencies, the network may be made of lumped components (capacitors and inductors) or of transformes. At high frequencies there are two types of impedance matching networks that are particularly useful. One is the so-called stub matching, which makes use of properties of shorted (or open) transmission lines. In this type of network, the impedance on the line is altered by connecting shorted or open transmission lines in parallel or in series with the line to adjust the impedance. The second method of impedance matching is based on the properties of transformers. In effect, we build a transmission line impedance transformer which then can match two impedances in a manner similar to that discussed in **Section 10.7.1**.

The following sections discuss these methods and develop the relations required to design matching networks. We use the Smith chart in the design of matching networks for two reasons: First, in many cases, the design is greatly simplified by the use of the Smith chart. Second, and more importantly, the Smith chart is routinely used for this type of application either by itself or as part of instruments such as the network analyzer.

15.4.2 Stub Matching

The idea of stub matching is to connect open- or short-circuited sections of transmission lines, either in parallel or in series with the transmission line as shown in **Figure 15.13**. The impedance of the stub and/or location on the line is chosen such that the combined impedance of line and stubs is equal to the characteristic impedance of the line. The details of design of the stubs for the three methods in **Figure 15.13** are discussed next. Although **Figure 15.13** shows matching at the load, the method of matching at the generator is identical—the generator's internal impedance is viewed as a load for matching purposes since the purpose of matching is to eliminate standing waves for the waves propagating towards the generator.

Consider first the matching network in **Figure 15.13a**. Assuming a characteristic impedance Z_0 (or admittance Y_0) and a line admittance $Y_0 + jB_0$, at a distance d_1 from the load, the two can be matched by adding a stub in parallel, at distance d_1 from the load, such that the admittance of the stub is $-jB_0$. The distance d_1 defines the imaginary part of the line admittance from **Eq.** (14.94). l_1 is then that length of the shorted transmission line stub that cancels the imaginary part of the line admittance at the location of the stub. The choice of l_1 and d_1 is not unique, but any practical combination that satisfies the above conditions may be used.

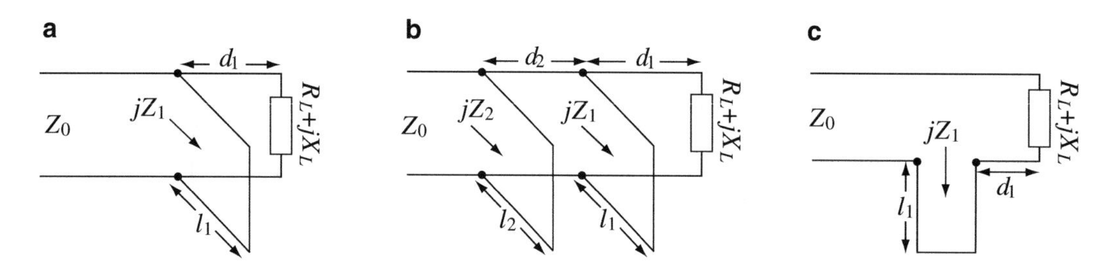

Figure 15.13 (**a**) Single stub matching. (**b**) Double stub matching. (**c**) Series stub matching

Although a single stub may be used to match any load (except for a purely imaginary load) to any line which has real characteristic impedance, sometimes the physical conditions of the line do not allow perfect matching with a single stub because of physical constraints. In such cases, two stubs, at two fixed locations, may be used. This method is similar to the single stub method, but now we must design the lengths l_1 and l_2 whereas d_1 and d_2 are fixed as shown in **Figure 15.13b**.

In the series matching method in **Figure 15.13c**, the idea is the same as in single parallel stub matching: we must choose a stub length l_1 and place it a distance d_1 from the load so that the sum of the line impedance at that point with that of the stub equals Z_0.

To summarize, in the single stub matching method, we choose the length and position of the stub. In the double stub matching method, we choose the lengths of two stubs whereas their positions are fixed and often prescribed by the device being matched.

15.4.2.1 Single Stub Matching

The idea of single stub matching relies on the fact that the line impedance varies along the line and a parallel or series stub changes only the reactive part of the line impedance. To see how this is accomplished, consider a load impedance $Z_L = R_L + jX_L$ connected on a line of characteristic impedance Z_0. For the load to be matched, its impedance must be changed so that $Z'_L = Z_0$. This is done as follows:

(1) Move along the line from the load (**Figure 15.13a**) and find a point at which $Z(z) = Z_0 + jX(z)$. Note that Z_0 does not have to be real, but in most cases, it will be.

(2) At this point (a distance d_1 from the load), connect a shorted or open transmission line of length l_1 such that the term $jX(z)$ cancels. As a result, the line sees a total impedance equal to Z_0 and the new load (which now is the whole line section to the right of the location of the stub) is matched.

These steps are implemented with the use of the Smith chart with the following differences:

(1) The impedance is first normalized to conform with the requirements of the Smith chart.
(2) If the stub is connected in parallel (**Figure 15.13a**), it is easier to work with admittances. Therefore, the normalized load admittance is first located on the chart.
(3) If the stub is connected in series (**Figure 15.13c**), it is easier to work with normalized impedances.

The stubs will be assumed to have the same characteristic impedance as the line, but this is not a necessary condition. The following two examples show the steps and details involved in parallel and series single stub matching.

Example 15.3 Application: Single Parallel Stub Matching at an Antenna The_Smith_Chart.m

An antenna operates at a wavelength of 2 m and is designed with an impedance of 75 Ω. However, because of mistakes in design, the antenna is badly mismatched. The measured impedance after installation is 15 + *j*60 Ω. The antenna is connected to a 75 Ω line as shown in **Figure 15.14**. Calculate:

(a) The required shorted stub and its location on the line to match the antenna to the line. The line and stub have the same characteristic impedance.
(b) The shortest required open circuit stub that will accomplish the same purpose as the short circuit stub in **(a)**.

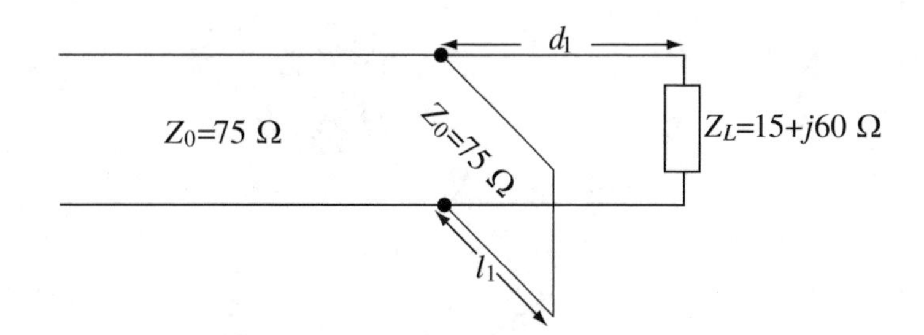

Figure 15.14 Mismatched antenna connected to a line and a parallel stub designed to match the antenna to the line

Solution: First, we find a location on the line at which the real part of the line admittance is equal to the characteristic admittance of the line; that is, find $Z(d_1)$ such that $Y(d_1) = Y_0 + jB(d_1)$. Now, we connect a shorted stub in parallel with the line at this point and of a length such that the imaginary part of the line admittance cancels. The open circuit stub in **(b)** is placed at the same location and its length is that of the short circuit stub $\pm\ \lambda/4$.

(a) In this case, it is simpler to use the Smith chart as an admittance chart. To do so, we first calculate the normalized load impedance:

$$z_L = \frac{15 + j60}{75} = 0.2 + j0.8.$$

(1) We mark this point as P_2 on the Smith chart in **Figure 15.15**, using the chart as an impedance chart. The reflection coefficient circle is now drawn around the center of the chart, with the radius equal to the distance between P_2 and P_1.
(2) To find the load admittance, we draw a straight line from P_2 through P_1 and extend this line to the periphery of the chart. The line intersects the reflection coefficient circle at point P_3. This point is the normalized load admittance:

$$y_L = 0.294 - j1.176.$$

(3) As we move around the reflection coefficient circle, the line admittance changes. To match the load, we must find the location at which the real part of the line admittance equals the characteristic admittance. Since we are working with normalized admittances, this happens when $\text{Re}\{y_L\} = 1$. This happens at the locations at which the reflection coefficient circle intersects the circle $g = 1$. The two possible points are P_4 and P_5. The line admittance at these points is

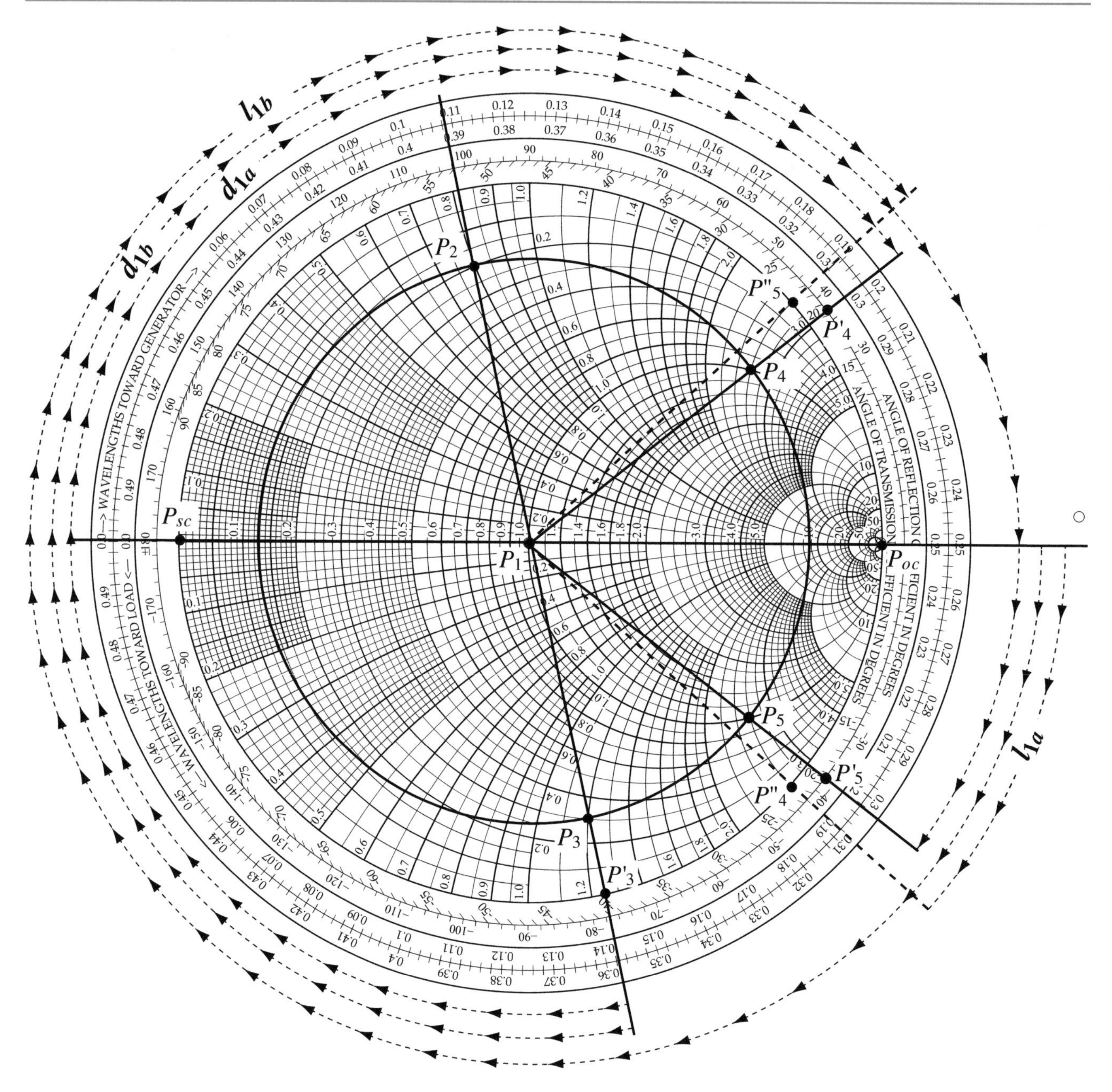

Figure 15.15 Smith chart for **Example 15.3**

At P_4,

$$y_a = 1 + j2.53$$

at P_5,

$$y_b = 1 - j2.53$$

Each one of these points provides one possible solution.

(4) Solution No. 1: Point P_4. The distance d_1 for this solution is the distance traveled from point P_3 to point P_4, on the reflection coefficient circle. The distance in wavelengths is the difference in readings between point P'_3 and P'_4

moving from P_3' to P_4' toward the generator. First, we move a distance of $0.5\lambda - 0.358\lambda = 0.142\lambda$ up to the short circuit point. Then, we move an additional 0.198λ to point P_4'. The total distance is $d_{1a} = 0.142\lambda + 0.198\lambda = 0.34\lambda$.

The normalized line susceptance at this point is 2.53. The stub must, therefore, have a normalized susceptance of -2.53. This point is shown as point P_4''. The length of the stub is the distance from the open circuit point P_{oc} (infinite admittance) to point P_4'' (moving toward the generator). This is $l_{1a} = 0.308\lambda - 0.25\lambda = 0.058\lambda$. Thus, the first possible solution is

$$d_{1a} = 0.337, \quad l_{1a} = 0.058 \quad [\lambda]$$

Since we know the wavelength ($\lambda = 2$ m), we can write the solution in actual lengths:

$$d_{1a} = 0.674, \quad l_{1a} = 0.116 \quad [\mathrm{m}].$$

(5) Solution No. 2. Point P_5. The distance d_1 at this point is the distance between points P_3 and P_5. Again, we move a distance of 0.142λ up to the short circuit point and then a distance of 0.302λ from the short circuit point to point P_5. Thus, $d_{1b} = 0.444\lambda$. The line susceptance at P_5 is -2.53. The stub susceptance must be +2.53. This is marked as point P_5''. The distance from P_{oc} to point P_5'' moving toward the generator, is $l_{1b} = 0.25\lambda + 0.192\lambda = 0.442\lambda$. The second solution is therefore

$$d_{1b} = 0.444\lambda, \quad l_{1b} = 0.442 \quad [\lambda] \quad \text{or} \quad d_{1b} = 0.888, \quad l_{1b} = 0.884 \quad [\mathrm{m}].$$

(b) Because an open line behaves as a shorted line at a distance of $\lambda/4$ from the short, the lines in **(a)** can be replaced by open circuit lines by either shortening the stubs by $\lambda/4$ or lengthening them by $\lambda/4$. Taking in each case the shortest possible stub length (lengthening l_{1a} and shortening l_{1b}), the solutions for open circuit stubs are

$$\begin{array}{ll} d_{1a} = 0.337\,\lambda = 0.674 \quad [\mathrm{m}], & l_{1a} = 0.308\,\lambda = 0.616 \quad [\mathrm{m}] \\ d_{1b} = 0.444\,\lambda = 0.888 \quad [\mathrm{m}], & l_{1b} = 0.192\,\lambda = 0.384 \quad [\mathrm{m}] \end{array}.$$

Exercise 15.1 Suppose that in **Example 15.3**, part **(a)**, it is not physically possible to connect the stub at either location found. The nearest location at which a stub may be connected is 1 m from the load:

(a) What are the solutions for d_1 and l_1?
(b) Are these solutions unique?

Answer 1 m = 0.5λ. The solutions are:

(a)

$$\begin{array}{ll} d_{1a} = (0.337 + 0.5)\,\lambda = 1.674 \quad [\mathrm{m}], & l_{1a} = 0.442\,\lambda = 0.884 \quad [\mathrm{m}] \\ d_{1b} = (0.444 + 0.5)\,\lambda = 1.888 \quad [\mathrm{m}], & l_{1b} = 0.058\,\lambda = 0.116 \quad [\mathrm{m}] \end{array}.$$

(b) No. The addition of any integer number of half-wavelengths to d_1 or l_1 or both is also acceptable solutions.

Example 15.4 Application: Series Stub Matching at an Antenna **The_Smith_Chart.m**

Consider again the transmission line and load in **Example 15.3**. The load has an impedance of 15 + j60 Ω and the line impedance is 75 Ω, as shown in **Figure 15.14**. However, now it is required to match the load using a shorted, series stub similar to that shown in **Figure 15.13c**. Calculate the required length of a series shorted circuit stub and its distance from the load to match the antenna to the line. The line and stub have the same characteristic impedance.

Solution: The solution is similar to that in **Example 15.3**. To match the load, we seek a location d_1 and a stub length l_1 as shown in **Figure 15.13c**. Since the stub's reactance is in series with the line impedance at d_1, the sum of the line impedance and stub reactance must be equal to the line resistance. Therefore, we should now use the Smith chart as an impedance chart. We move a distance d_1 from the load at which location the normalized line impedance is $z_l(d_1) = 1 + jx$. Then, we find a stub length l_1 such that $z_s(l_1) = -jx$. The sum of the two gives the correct match at d_1.

(1) The normalized load impedance is $z_L = 0.2 + j0.8$. This is marked at point P_2 (**Figure 15.16**).

(2) Now, we move toward the generator on the reflection coefficient circle until we intersect the $r = 1$ circle at points P_3 and P_4. Connection of P_1 to P_3 and P_1 to P_4 and extending the lines to the circumference gives points P_3' and P_4'. Each of these is a possible solution.

(3) Solution No. 1: The distance between P_3' and P_2' is the first possible solution for d_1. In this case, we moved a distance $d_{1a} = 0.196\lambda - 0.109\lambda = 0.089\lambda$.

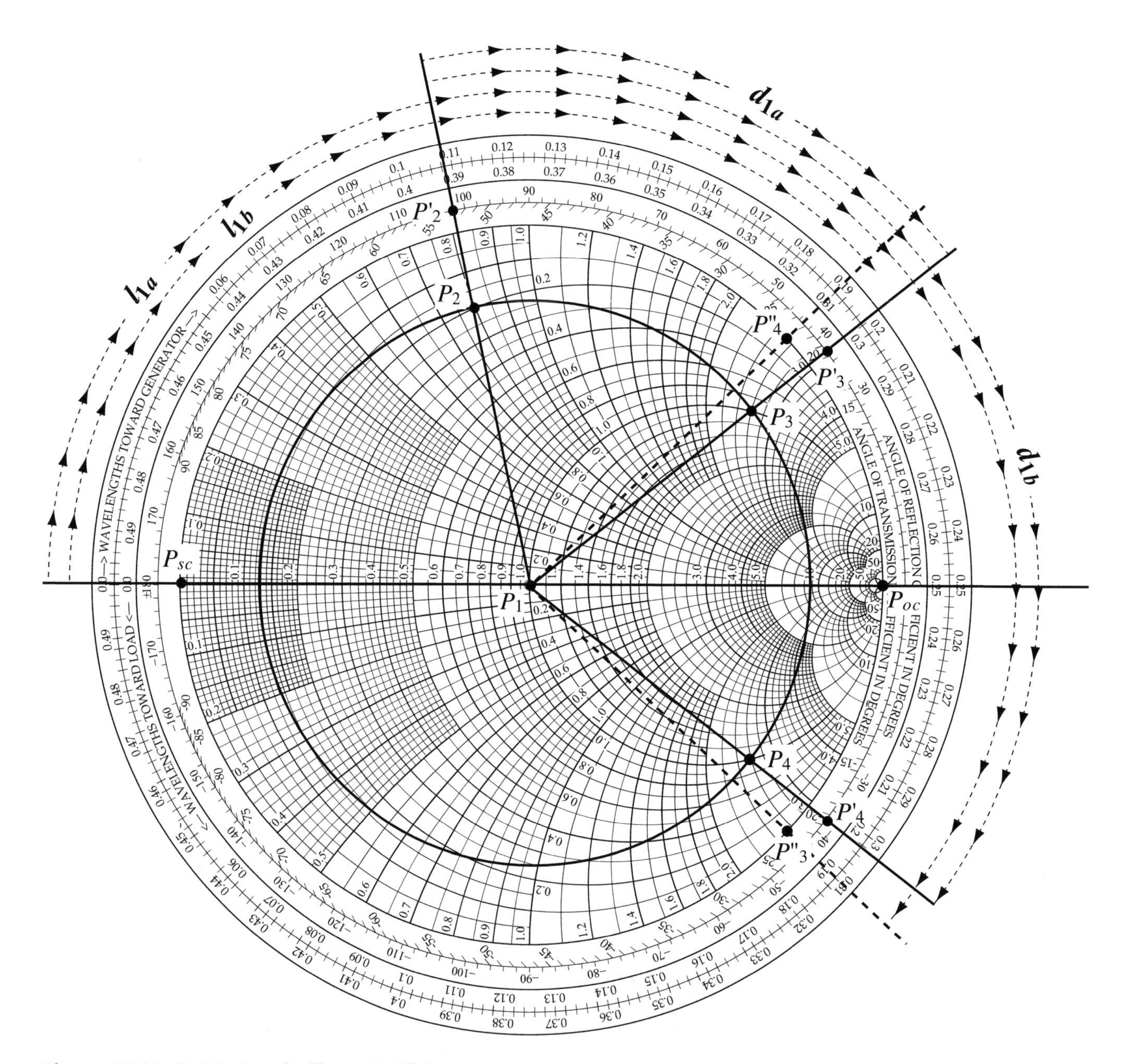

Figure 15.16 Smith chart for **Example 15.4**

The normalized line reactance at point P_3 is $j2.53$. The stub length must be such that its normalized input impedance is $-j2.53$. This required impedance is marked as point P_3''. The distance between the short circuit point P_{sc} and P_3'' moving toward the generator is the stub length necessary. This distance is 0.31λ. Thus, the first solution (with $\lambda = 2$ m) is

$$d_{1a} = 0.087\,\lambda = 0.174 \quad [\text{m}], \quad l_{1a} = 0.31\,\lambda = 0.62 \quad [\text{m}].$$

(4) Solution No. 2: This occurs at point P_4. The distance d_1 now is the distance between point P_4' and P_2' or $d_{1b} = 0.31\lambda - 0.109\lambda = 0.201\,\lambda$.
The normalized line reactance at point P_4 is $-j2.53$. The stub normalized impedance must be $+j2.53$. This impedance is marked at point P_4''. The distance l_{1b} is the distance between the short circuit point to point P_4'' $l_{1b} = 0.190\lambda$. The second solution is therefore

$$d_{1b} = 0.201\,\lambda = 0.402 \quad [\text{m}], \quad l_{1b} = 0.190\,\lambda = 0.38 \quad [\text{m}]$$

Either solution is correct, but perhaps in practical terms, the closest stub to the load (solution no. 1) may be chosen.

15.4.2.2 Double Stub Matching

As mentioned earlier, double stub matching takes a different approach than single stub matching. There are now two stubs at fixed locations d_1 and d_2, as shown in **Figure 15.13b**. Matching is achieved by adjusting the two stub lengths l_1 and l_2. To see how this is accomplished, it is best to look at the process in reverse. Suppose that we have already accomplished matching. From the results for single stub matching, we know that when the load is matched, we must be on a point on the unit circle ($g = 1$). In fact, we know that there will be two points at which matching can be accomplished, but, for clarity, only point P_1 is shown in **Figure 15.17**. The point shown represents the load impedance at a distance $d_1 + d_2$ from the load. Now, we move from P_1 toward the load a distance d_2. For any of the points on the unit circle, this means moving on its reflection coefficient circle. The locus of all points on the unit circle, moved toward the load a distance d_2, is a shifted unit circle, as shown in **Figure 15.17**. This shifted unit circle represents the equivalent load impedance at a distance d_1 from the load (this equivalent load impedance is due to the line impedance and the stub at this point). Point P_1' is the equivalent impedance at the location of stub (2) corresponding to the matched point P_1. Stub (1) only adds a susceptance to the line admittance. Therefore, to get to the load admittance point, we must first remove this susceptance by moving along the circles of constant conductance. This brings us to point P_1'' marked on the chart in **Figure 15.17**. In addition we must move a distance d_1 from P_1' toward the load (not shown on the chart). Note, also, that the difference in susceptance between points P_1' and P_1'' is the susceptance stub (1) must add to the line whereas the susceptance of stub (2) is the imaginary part of the admittance at point P_1.

Of course, when matching a load, we will start with the load impedance, but the above process is more instructive because it explains the need for the shifted unit circle and what the contribution of each stub is. In effect, we may say that the purpose of the first stub (the stub closer to the load) is to modify the line susceptance so that the second stub can then take the line admittance to the unit circle. The following two examples show the steps and the details of double stub matching.

Example 15.5 Double Stub Matching The_Smith_Chart.m

A line with characteristic impedance $Z_0 = 300\ \Omega$ and load impedance $Z_L = 150 + j225\ \Omega$ is given. Design a double stub matching network such that the two stubs are 0.1λ apart, with the first stub at the load as shown in **Figure 15.18**.

Solution: After calculating the normalized load impedance, we draw the reflection coefficient circle and find the normalized load admittance since the two stubs are parallel to the line. Hence the Smith chart is used as an admittance chart. In the single stub case, matching consisted of finding the intersection of the reflection coefficient circle with the $g = 1$ circle. The same principle is used here, but the actual matching is at the second stub from the load (stub (2)) since we want to match the load to the line. Thus, stub (1) is represented by its own unit circle, which is shifted a distance 0.1λ from the $g = 1$ circle toward the load. Now, we start at the load (P_3 in **Figure 15.19**) and move from the admittance point on the constant conductance circle at the load until we intersect the shifted unit circle for stub (1). The intersection points represent the

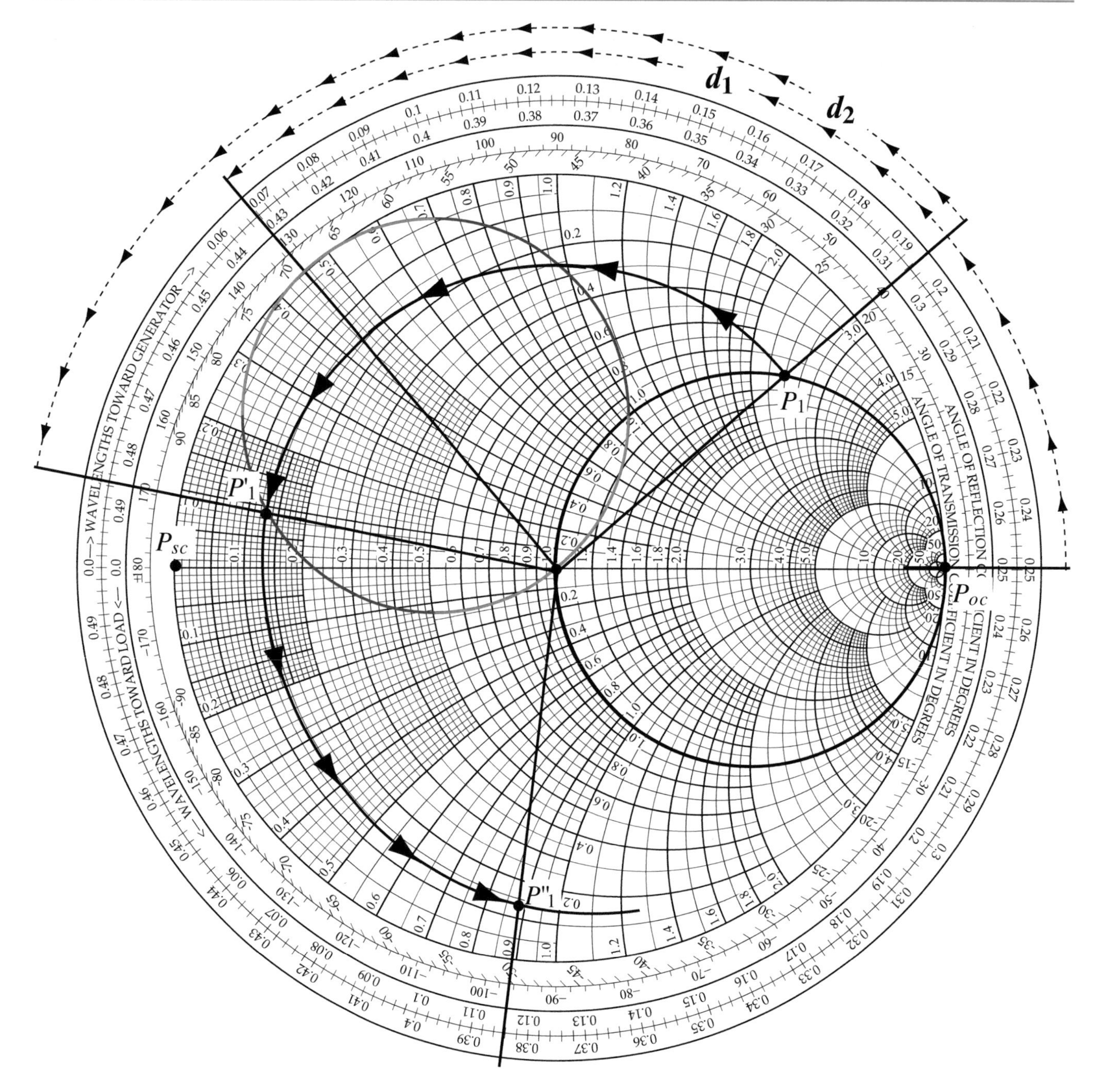

Figure 15.17 Smith chart for **Example 15.4**

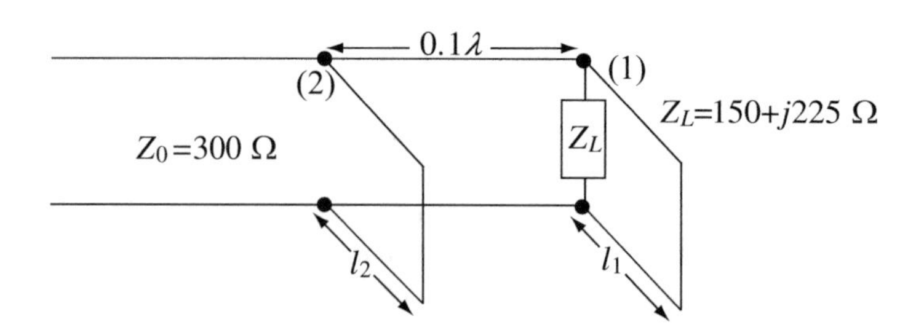

Figure 15.18 A load impedance matched to the line with two stubs

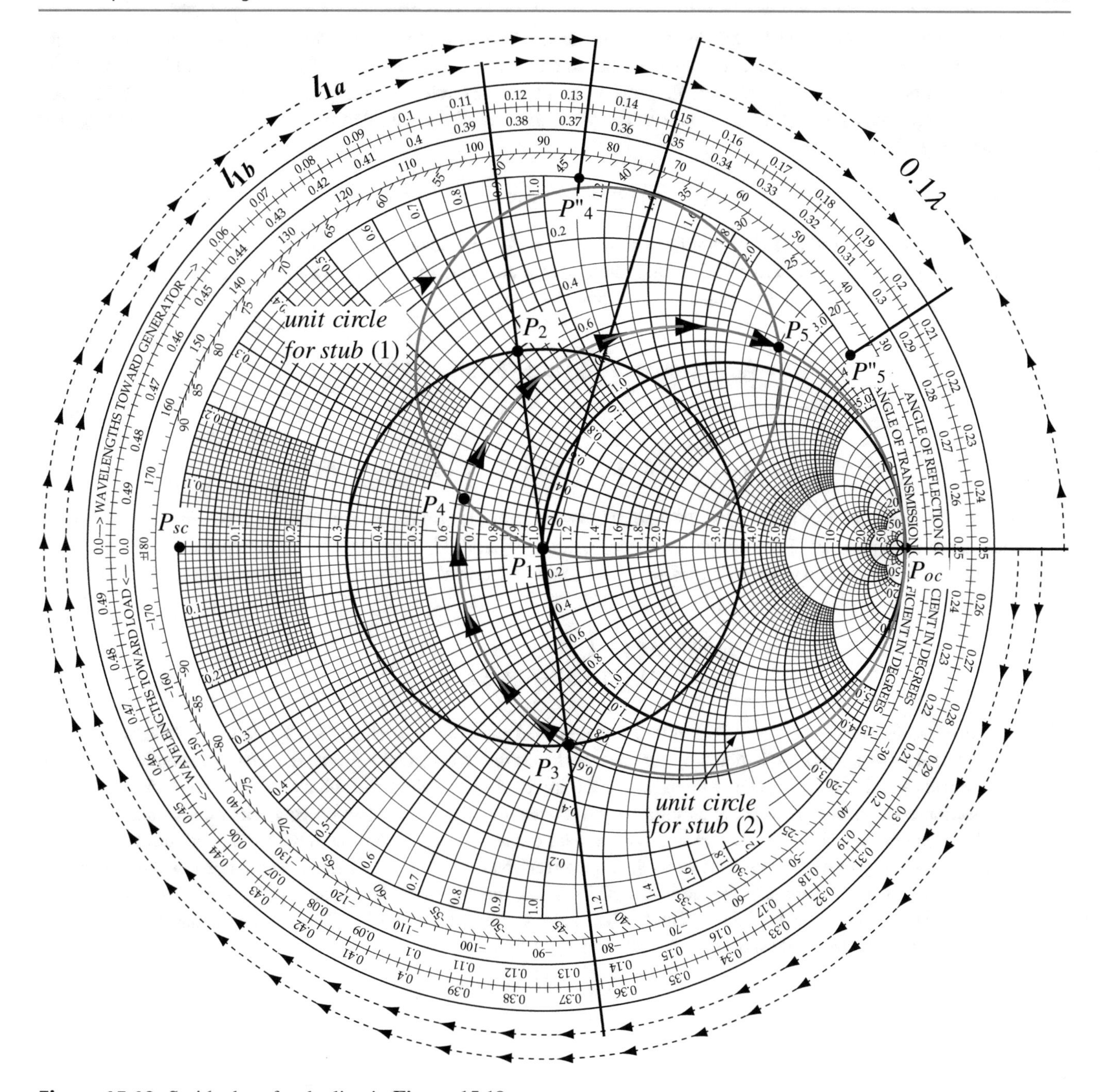

Figure 15.19 Smith chart for the line in **Figure 15.18**

reflection coefficients of the combined load and stub (1). The combined impedance of the load and stub represent a new, modified load with a stub a distance 0.1λ away, toward the generator. This modified line is a load with a single stub; therefore, its treatment is the same as for the single stub matching in that the length of the stub is chosen to cancel the susceptance for each of the two stubs possible at the load. Stubs (1) and (2) refer to the notation used in **Figures 15.13b** and **15.18**, with stub (1) at the load.

(1) The normalized load impedance (without stubs) is $z_L = 0.5 + j0.75$ and is shown at point P_2 in **Figure 15.19**. The normalized load admittance is at point P_3 and is $y_L = 0.615 - j0.923$.

(2) In preparation for the calculation of the stubs, we draw the two unit circles. The unit circle for stub (2) is the $g = 1$ circle of the chart. The unit circle for stub (1) is the same circle, shifted toward the load a distance of 0.1λ, as shown in **Figure 15.19**.

(3) Now, we add the stub at the load. The stub's impedance is purely imaginary. Therefore, it can only change the susceptance of the combined stub and load while the conductance remains the same. To find the combined admittance on the unit circle for stub (1), we move on the constant conductance circle, starting from P_3 (load admittance). This path is shown (gray line) in **Figure 15.19**. The path intersects the shifted unit circle at two points, marked P_4 and P_5. The admittances at P_4 and P_5 are

$$y_{P4} = 0.615 + j0.192, \quad y_{P5} = 0.615 + j2.56.$$

(4) In moving from the load admittance point P_3 to points P_4 and P_5, the change in admittance is only due to the susceptance contributed by stub (1). Subtracting the load admittance from the admittances at points P_4 and P_5 gives the susceptance stub (1) must contribute to the impedance at these points:
At P_4:

$$y_{1a} = y_{P4} - y_L = 0.615 + j0.192 - 0.615 + j0.923 = j1.115$$

At P_5:

$$y_{1b} = y_{P5} - y_L = 0.615 + j2.56 - 0.615 + j0.923 = j3.483$$

These two values are shown at points P_4'' and P_5''. The possible stub lengths are found by moving from the short circuit admittance point (P_{oc}) toward the generator, to points P_4'' and P_5''. For point P_4, the susceptance of the stub must be 1.115. Starting at P_{oc} and moving, in turn, to point P_4'' and P_5'' (toward the generator) gives the two possible lengths for stub (1):

$$\begin{aligned} l_{1a} &= 0.25\lambda + 0.133\lambda = 0.383\ \lambda \quad (\text{at } P_4'') \\ l_{1b} &= 0.25\lambda + 0.205\lambda = 0.455\ \lambda \quad (\text{at } P_5'') \end{aligned}.$$

(5) Now, we consider the admittances at P_4 and P_5 as the new load admittances as shown in **Figure 15.20**. From here on, we treat the problem as a single stub matching for each of these admittances and with the distance between load and stub (2) known and equal to 0.1λ. We start with y_{P4} and use **Figure 15.21**, on which the unit circle has been marked. We draw the reflection coefficient circle for the admittance y_{P4}. As we move on the reflection coefficient circle, starting at P_4, toward the generator, and move 0.1λ, we intersect the unit circle at point P_6. Although we cut the unit circle at another point, symmetrically located about the real axis, this intersection cannot be used since the stub must be a distance 0.1λ from the load. At P_6, the line admittance is $1 + j0.55$. Thus, the stub must have admittance $-j0.55$ so that the line susceptance is canceled at the location of stub (2). The latter is marked as point P_6''. The stub length that will accomplish this is the distance between the short circuit point and P_6''. This is

$$l_{2a} = 0.42\lambda - 0.25\lambda = 0.17\ \lambda$$

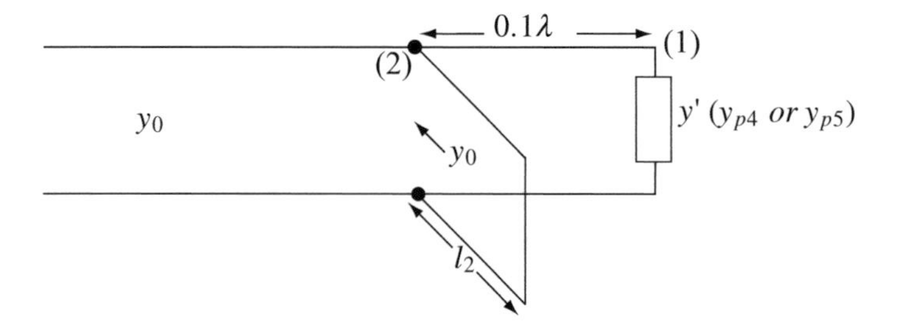

Figure 15.20 The equivalent condition at the location of stub (2) after the load and stub (1) in **Figure 15.18** were taken into account

Similarly, for point P_5, we draw the reflection coefficient circle and move 0.1λ toward the generator, to point P_7. The line admittance at P_7 is $1 - j3.4$. The required stub admittance is $+j3.4$, which is marked as point P_7''. The stub length is the distance between P_{oc} and this point:

$$l_{2b} = 0.25\lambda + 0.203\lambda = 0.453\ \lambda$$

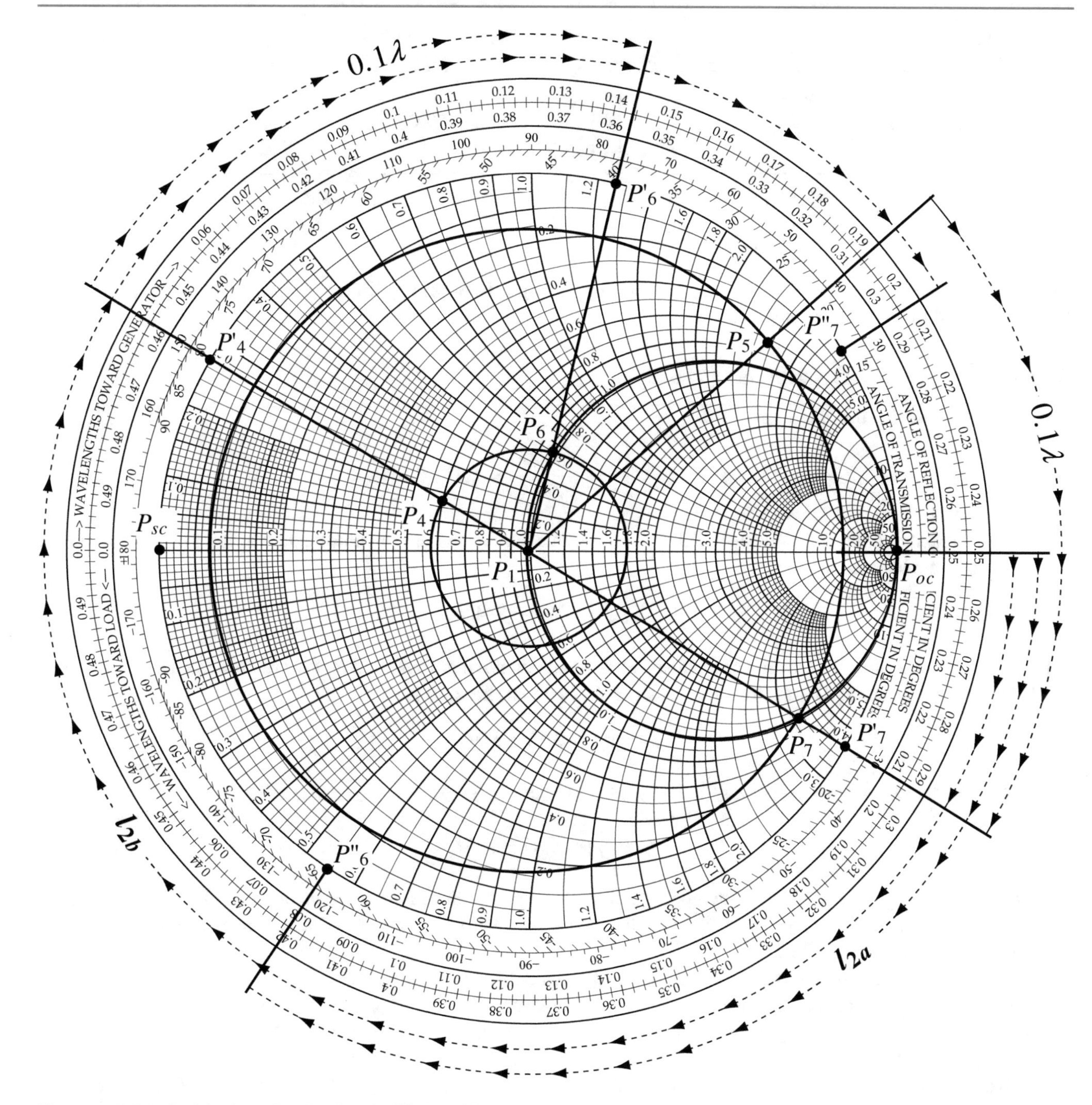

Figure 15.21 Smith chart for the line in **Figure 15.20**

Therefore, the two possible solutions are

$$l_1 = 0.383\,\lambda, \quad l_2 = 0.17\,\lambda \quad \text{or} \quad l_1 = 0.455\,\lambda, \quad l_2 = 0.453\,\lambda.$$

Example 15.6 The_Smith_Chart.m

A transmission line and load are given in **Figure 15.22a**. It is required to calculate the lengths of the stubs so that the load is matched to the line.

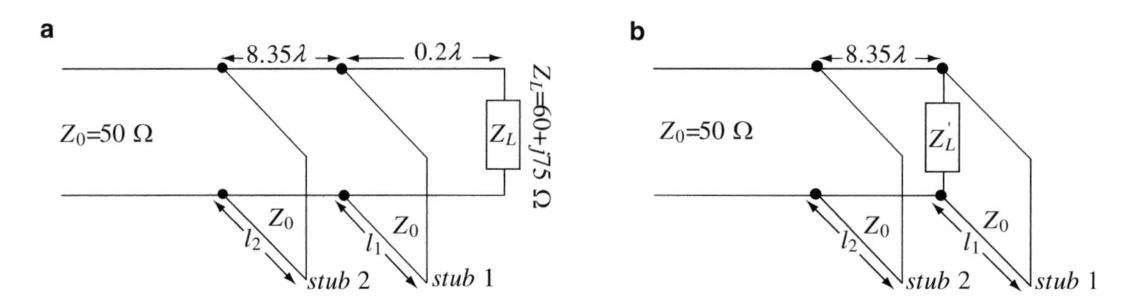

Figure 15.22 (**a**) The double stub matching network for **Example 15.6**. (**b**) Equivalent network after the load impedance has been moved to the location of stub (1)

Solution: The steps in the solution are as follows:

(1) First, we normalize the load impedance:

$$z_L = \frac{Z_L}{Z_0} = \frac{60 + j75}{50} = 1.2 + j1.5$$

This is marked as point P_2 in **Figure 15.23**. The reflection coefficient circle can now be drawn. The point opposite P_2 is P_3. This is the load admittance:

$$y_L = 0.325 - j0.406.$$

(2) The line admittance at the location of the first stub is found by moving from point P_3 toward the generator a distance of 0.2λ. This brings us to point P_4. The admittance at P_4 (without the stub) is

$$y_{p4} = 0.56 + j0.94$$

that is, the normalized line impedance at the location of stub (1), before the stub is added, is

$$z_L^{'} = z_{P5}^{''} = 0.47 - j0.78$$

Note: This is at point P_5, which is the opposite point to point P_4. Since P_4 represents the normalized admittance, P_5 represents the normalized impedance. Now the line and stubs appear as in **Figure 15.22b**. The new load admittance $y_L^{'}$ is marked as P_4 in **Figure 15.24**.

(3) The distance between the two stubs is 8.35λ. Stub (2) is calculated to match the line. Stub (1) must be calculated for a unit circle ($g = 1$) that has been moved toward the load a distance of 8.35λ. **Figure 15.24** shows the actual unit circle and the shifted circle after moving it 8.35λ toward the load (i.e., from stub (1) to stub (2)). Note that this is the same as moving the circle 0.35λ toward the load.

Now, we move on the conductance circle that passes through point P_4 ($g = 0.56$) until the shifted unit circle is intersected at points P_5 and P_6. The $g = 0.56$ circle is shown as a gray line.

(4) The normalized admittances at point P_5 and P_6 are

$$y_{P5} = 0.56 + j0.01, \quad y_{P6} = 0.56 - j1.463$$

To find the length of stub (1), we argue as follows: moving from P_4 to P_5 or P_6, we have changed the imaginary part of admittance only. This change is:
From P_4 to P_5:

$$y_{1a} = y_{P5} - y_{P4} = (0.56 + j0.01) - (0.56 + j0.94) = -j0.95$$

From P_4 to P_6:

$$y_{1b} = y_{P6} - y_{P4} = (0.56 - j1.463) - (0.56 + j0.98) = -j2.43$$

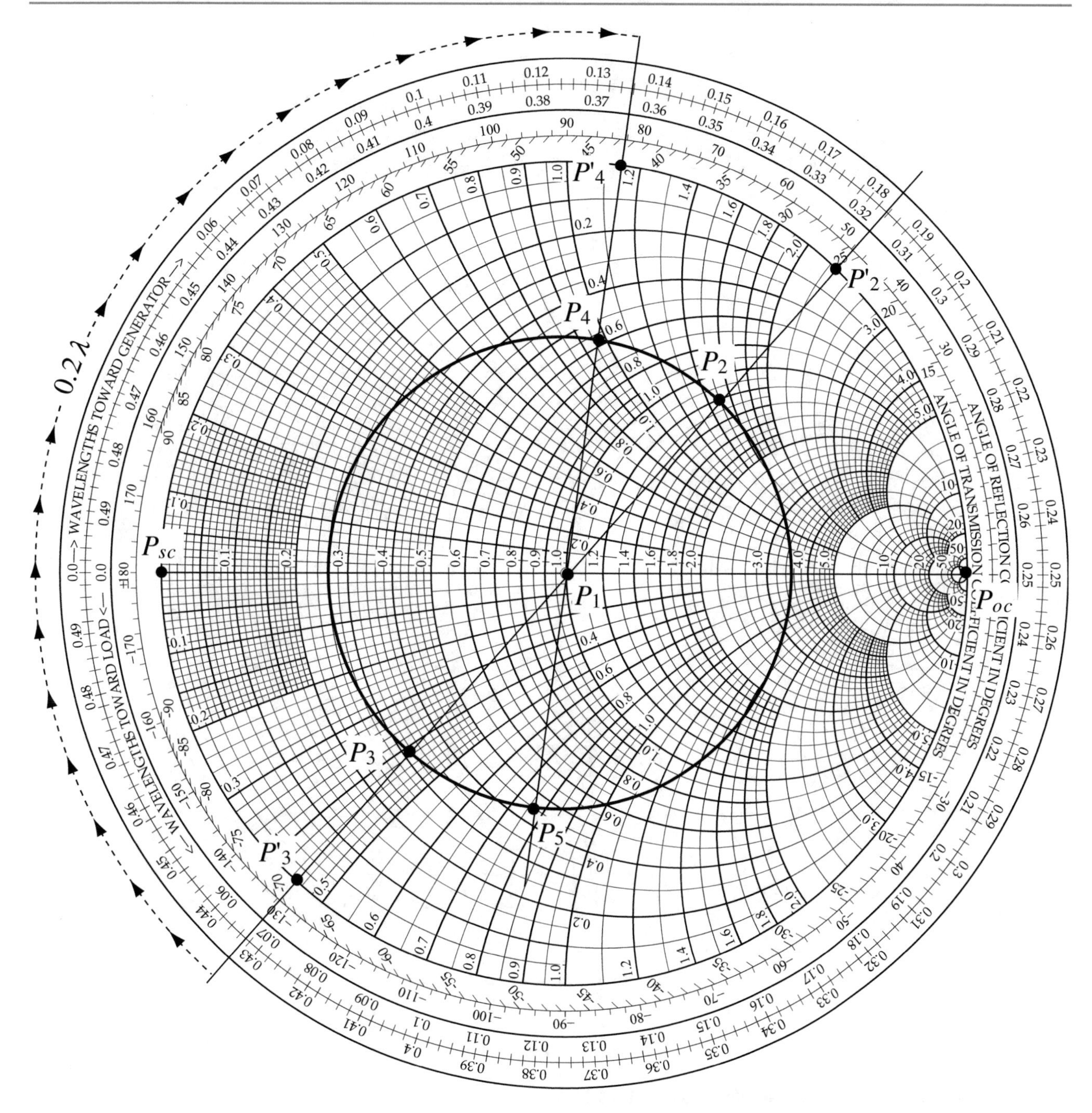

Figure 15.23 Smith chart for the configuration in **Figure 15.22a.** Calculation of the equivalent load impedance at the location of stub (1)

The admittances y_{1a} and y_{1b} are the admittances added to the load by the two possible choices for stub (1). The two admittances required are shown as points P_5'' and P_6''. Thus, the length of stub (1) is the distance between the short circuit point and P_5'' or P_6''. For y_{1a} (P_5), the input stub admittance must be equal to $-j0.95$. We move from the infinite admittance point (point P_{oc} on the chart), toward the generator, on the outer circle of the Smith chart up to point P_5''. The total distance traveled is the length of the stub:

$$l_{1a} = 0.38\lambda - 0.25\lambda = 0.13\,\lambda$$

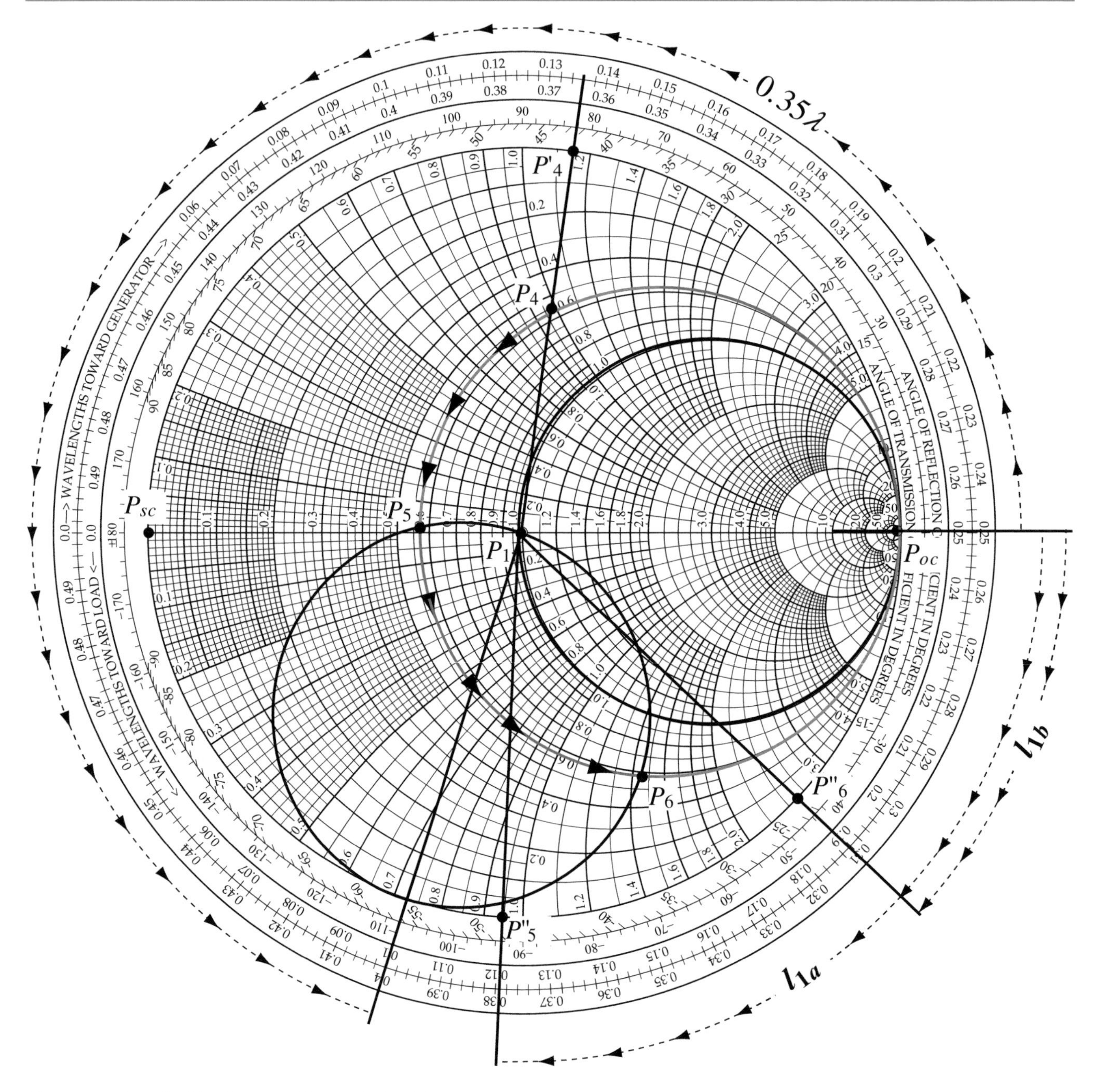

Figure 15.24 Smith chart for **Example 15.6** (continuation)

Similarly, the stub admittance for point P_6 is $y_{1b} = -j2.43$. The stub length is the distance between P_{oc} and P_6''

$$l_{1b} = 0.31\lambda - 0.25\lambda = 0.06\,\lambda.$$

(5) For each one of these solutions, we have an equivalent admittance point: P_5 and P_6. The problem now is that of an equivalent admittance y_{P5} or y_{P6}, and a single stub a distance 8.35λ toward the generator. To avoid confusion, we use the new chart in **Figure 15.25**. Points P_5 and P_6 as well as the unit circle for stub (2) are shown. We now draw the reflection coefficient circles for each of these two admittance points starting with point P_5. From P_5, we move 0.35λ toward the generator. This intersects the unit circle at point P_5''. The line admittance at this point (before connecting stub (2)) is $1 - j0.56$. The admittance of the stub must be $+j0.56$, a value shown at point P_7. The length of stub (2) corresponding to the point is the distance between P_{oc} and P_7, moving toward the generator:

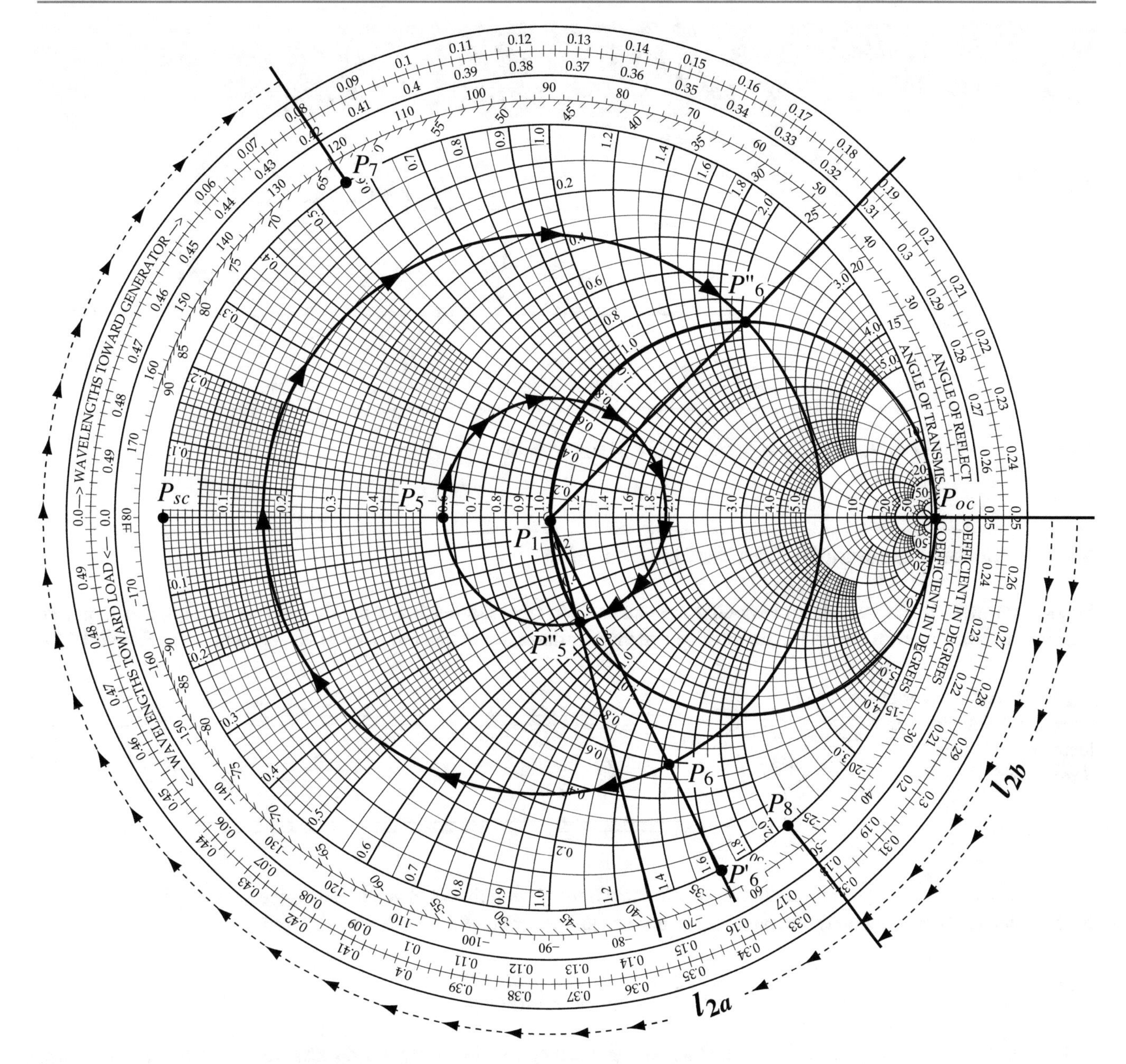

Figure 15.25 Smith chart for **Example 15.6** (continuation)

$$l_{2a} = 0.25\lambda + 0.082 = 0.332\ \lambda$$

Starting with point P_6 and moving 0.35λ toward the generator, we reach point P''_6. The line admittance at this point is 1 + j2.2. The stub admittance must be –j2.2, shown at point P_8. The stub length is therefore

$$l_{2b} = 0.322\lambda - 0.25\lambda = 0.072\ \lambda$$

The two possible solutions are therefore

$$l_{1a} = 0.13\ \lambda, \quad l_{2\text{a}} = 0.332\ \lambda \quad \text{or} \quad l_{1b} = 0.06\ \lambda, \quad l_{2b} = 0.072\ \lambda$$

Exercise 15.2 In **Figure 15.26**, the load impedance is 0.2λ from the first stub (stub (1)) and the distance between the two stubs is 0.1λ. Calculate the lengths of the two stubs to match the load to the line.

Figure 15.26

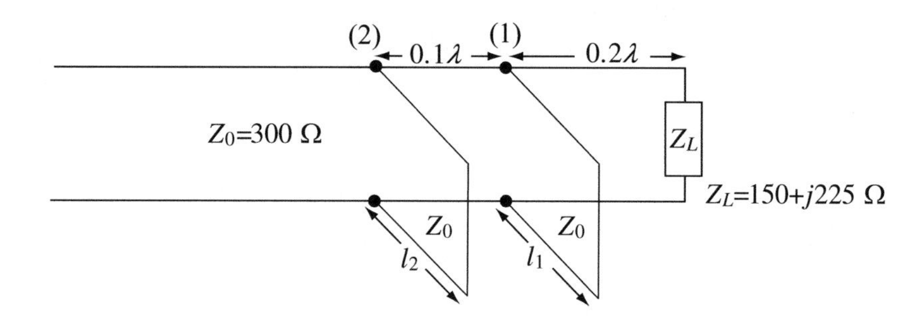

Answer $l_{1a} = 0.257\ \lambda,\ l_{1b} = 0.104\ \lambda$ or $l_{2a} = 0.424\ \lambda,\ l_{2b} = 0.461\ \lambda$.

15.5 Quarter-Wavelength Transformer Matching

Single stub matching is capable of removing a mismatch for any load (except a purely reactive load), but it is not an impedance transformer. If different lines must be matched, a transmission line transformer can be used, as in **Figure 15.27**.

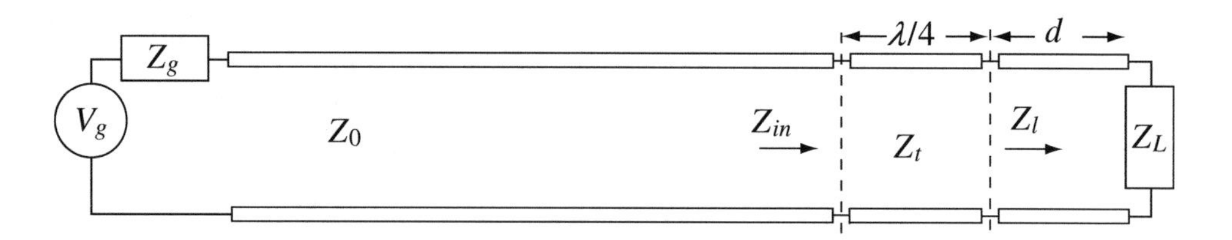

Figure 15.27 A quarter-wavelength transformer located at distance d from load

From **Eq.** (14.94), the line impedance Z_{in} of a lossless transmission line of characteristic impedance Z_0, at a distance z_0, from the load may be viewed as the input impedance of the line section between z_0 and the load:

$$Z_{in} = Z_0 \frac{[Z_L \cos\beta z_0 + jZ_0 \sin\beta z_0]}{[Z_0 \cos\beta z_0 + jZ_L \sin\beta z_0]} \quad [\Omega] \tag{15.21}$$

Now, suppose we chose a transmission line section, with characteristic impedance Z_t, cut it so it is $\lambda/4$ long, and connect it to a load impedance Z_l. Setting $z_0 = \lambda/4$ and $\beta z_0 = \beta\lambda/4 = (2\pi/\lambda)(\lambda/4) = \pi/2$ and replacing Z_L by Z_l and Z_0 by Z_t in **Eq. (15.21)**, we get for the input impedance of the $\lambda/4$ section

$$Z_{in} = Z_t \frac{\left[Z_l \cos\frac{\pi}{2} + jZ_t \sin\frac{\pi}{2}\right]}{\left[Z_t \cos\frac{\pi}{2} + jZ_l \sin\frac{\pi}{2}\right]} = \frac{Z_t^2}{Z_l} \quad [\Omega] \tag{15.22}$$

Referring now to **Figure 15.27**, where Z_l is the line impedance at a distance d from the load, we get the condition for matching using the quarter-wavelength transformer shown:

$$\boxed{Z_t = \sqrt{Z_{in} Z_l} \quad [\Omega]} \tag{15.23}$$

Thus, two different transmission lines or any two impedances may be matched, provided a transformer of proper characteristic impedance Z_t can be found. The quarter-wavelength transformer is normally connected at a point of maximum or minimum voltage since the line impedance is real at that point. The line impedance at a point of minimum voltage is

$$Z_l = \frac{Z_0}{\text{SWR}} \quad [\Omega] \tag{15.24}$$

The location of the minimum voltage on the line for a general load is at a distance [see **Eqs.** (14.114) and (14.115)]

$$d_{min} = \frac{\lambda}{4\pi}(\theta_\Gamma + \pi) + n\frac{\lambda}{2} \quad [\lambda] \tag{15.25}$$

from the load, where n is any integer, including zero. For a resistively loaded line, the location of minimum voltage is either at the load (if $R_L < Z_0$) or at a distance $\lambda/4$ (if $R_L > Z_0$). Thus, the transformer can be located at any of the points in **Eq. (15.25)**. If the characteristic line impedance is Z_0, the characteristic impedance of a transformer located at a point of minimum voltage on the line must be

$$\boxed{Z_t = Z_0\sqrt{\frac{1}{\text{SWR}}} \quad [\Omega]} \tag{15.26}$$

Similarly, if the transformer is located at a point of maximum voltage [by moving it a quarter-wavelength in either direction of any of the points in **Eq. (15.25)**], the characteristic impedance of the transformer for matching is

$$\boxed{Z_t = Z_0\sqrt{\text{SWR}} \quad [\Omega]} \tag{15.27}$$

How can we use the Smith chart to design a quarter-wavelength transformer and, therefore, match two lines or a line and a load? First, we note that two parameters are important in this design. The first is the standing wave ratio SWR. The second is the location of the minimum (or maximum) voltage on the line. For any given load, these are obtainable from the Smith chart. Once the SWR and location of minimum or maximum are found, the transformer impedance is found from **Eq. (15.26)** or **(15.27)**, depending on where the transformer is placed. The following examples discuss the design sequence.

Example 15.7 Application: Matching of Two Different Lines A student has found out that he/she is out of money and cannot pay the cable TV bill. The student decides to cancel the service and go back to the old rooftop antenna. However, the TV input is 75 Ω, whereas the cable coming down from the antenna is 300 Ω. Design a matching network to match the two lines assuming that the antenna is matched to the 300 Ω line and the TV is matched to the 75 Ω line. Where should the matching network be placed?

Solution: A quarter-wavelength transformer can be used, although, because TV reception is in a range of frequencies, the lines will only be matched at the frequency at which the transformer is exactly one-quarter wavelength. The characteristic impedance of the transformer must be

$$Z_t = \sqrt{Z_{in}Z_l} = \sqrt{75 \times 300} = 150 \quad [\Omega]$$

The transformer may be placed anywhere between the antenna and TV because one line is matched to the TV and the second to the antenna and, therefore, the impedance anywhere on each line equals its characteristic impedance. The only important point is that the line between the transformer and the TV must be a 75 Ω line, and between the transformer and the antenna the line must be a 300 Ω line (**Figure 15.28**).

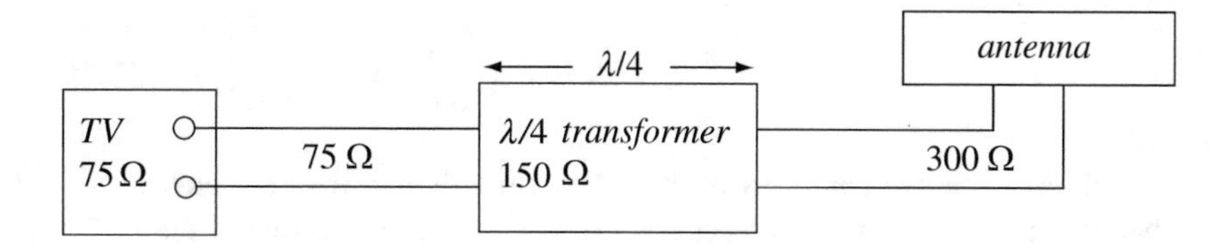

Figure 15.28 A quarter-wavelength transformer used to match two different transmission lines

Example 15.8 Matching a Load to a Line

A load $Z_L = 45 - j60\ \Omega$ is connected to a line with characteristic impedance $Z_0 = 50\ \Omega$. Design a quarter-wavelength transformer to match the load to the line. It is required to connect the transformer as close to the load as possible. Find the required characteristic impedance of the transformer and its location.

Solution: There are two methods to solve this problem. The most obvious is to use **Eqs. (15.21)** through **(15.27)**. The second is to use the Smith chart instead. We will do both, starting with the Smith chart method.

Method A: The Smith Chart

(1) Find the normalized impedance of the load and mark it on an impedance Smith chart. The normalized load impedance is $z_L = (45 - j60)/50 = 0.9 - j1.2$ (point P_2 in **Figure 15.29**).

(2) Find the first extremum in impedance from the load (minimum or maximum). This is done by moving on the reflection coefficient circle, toward the generator, from point P_2, until the real axis of the chart is intersected. This happens at point P_3 at a point of minimum impedance and is a distance of $0.5\lambda - 0.336\lambda = 0.164\lambda$ from the load. At point P_3, the value on the axis is $1/\text{SWR} = 0.3$. Thus, $\text{SWR} = 3.33$.

(3) From **Eq. (15.26)**, the characteristic impedance of the transformer is

$$Z_t = Z_0\sqrt{\frac{1}{\text{SWR}}} = 50\sqrt{0.3} = 27.39 \quad [\Omega].$$

Method B: Direct Calculation First, we find the load reflection coefficient, its magnitude, and its phase angle:

$$\Gamma_L = \frac{Z_L - Z_0}{Z_L + Z_0} = \frac{45 - j60 - 50}{45 - j60 + 50} = \frac{-5 - j60}{95 - j60} = 0.2475 - j0.475$$

$$|\Gamma_L| = \sqrt{(0.2475)^2 + (0.475)^2} = 0.536$$

$$\theta_L = \tan^{-1}\frac{-j0.475}{0.2475} = -62.48° \to \theta_L = -1.09 \quad [\text{rad}]$$

The standing wave ratio is

$$\text{SWR} = \frac{1 + |\Gamma_L|}{1 - |\Gamma_L|} = \frac{1 + 0.536}{1 - 0.536} = 3.310$$

The location of the first minimum from the load [$n = 0$ in **Eq. (15.27)**] is

$$d_{min} = \frac{\lambda}{4\pi}(\theta_L + \pi) = \frac{\lambda}{4\pi}(-1.09 + \pi) = 0.163\ \lambda$$

The transformer's intrinsic impedance is

$$Z_t = Z_0\sqrt{\frac{1}{\text{SWR}}} = 50\sqrt{\frac{1}{3.310}} = 27.48 \quad [\Omega]$$

Note that the two solutions are not identical although they are close. This, of course, is due to the nature of the Smith chart: the precision depends on accuracy of reading the values on the chart. This difficulty is solved with computerized Smith charts since these charts use the actual mathematical relations involved.

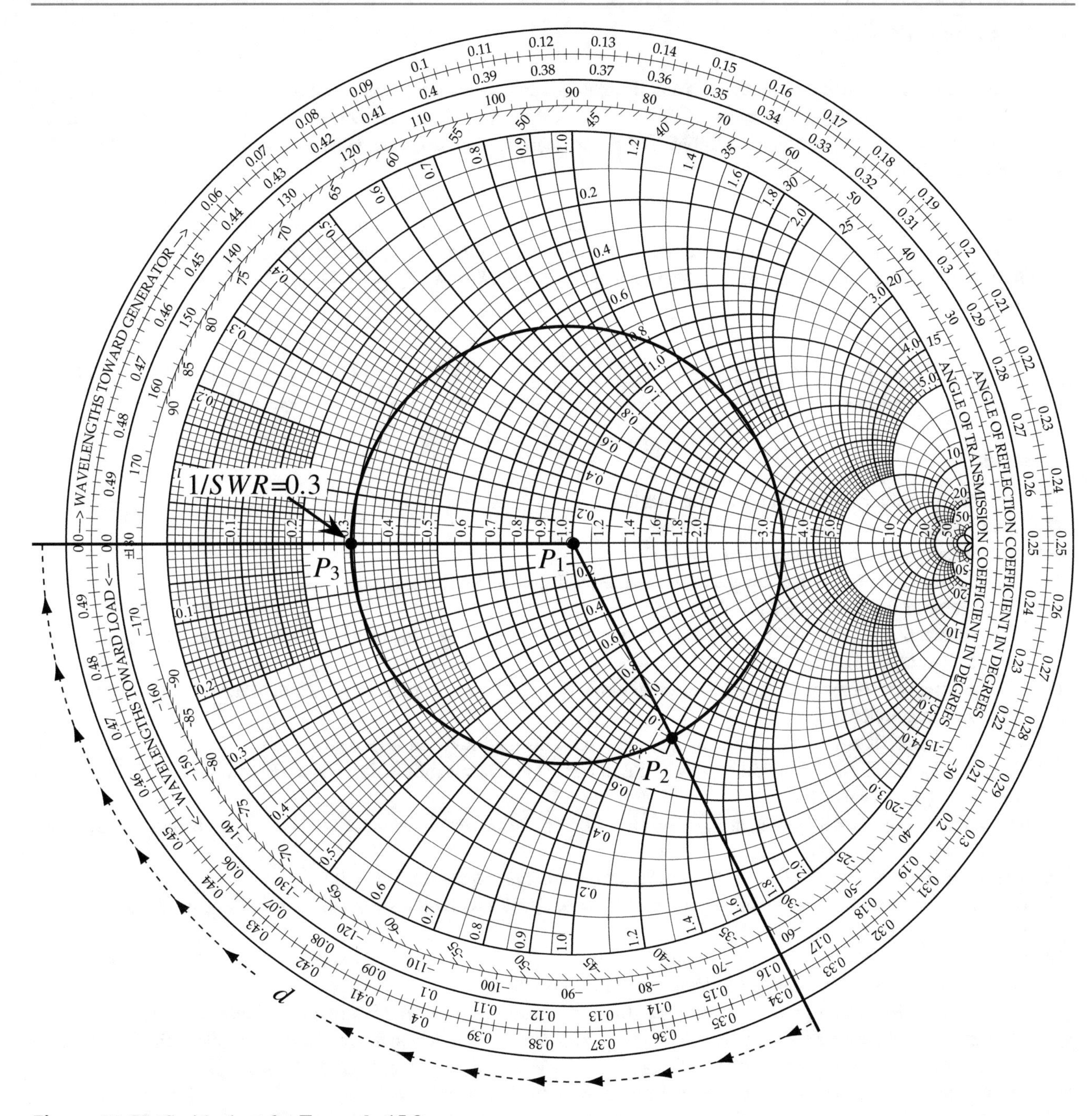

Figure 15.29 Smith chart for **Example 15.8**

Exercise 15.3 Find the location and characteristic impedance of the quarter-wavelength transformer in **Example 15.8** if the transformer is connected at the first voltage maximum.

Answer $d = 0.413\ \lambda,\ Z_t = 90.97\ \Omega.$

15.6 Summary

The Smith chart is a common tool in transmission line calculations and design. It is based on the properties of the load and generalized reflection coefficient. Because of that it allows calculation of impedances, SWR, magnitude and phase of the reflection coefficient, as well as other conditions. The Smith chart does not calculate voltages and currents but can be used as an aid in their calculation.

Smith Chart We assume a lossless line with real characteristic impedance Z_0 (but these are not necessary conditions). Given a load impedance $Z_L = R + jX$ and load reflection coefficient Γ_L the Smith chart defines circles of normalized real and imaginary values, r, x so that the normalized load impedance is $z = (R + jX)/Z_0 = r + jx$ (see **Figure 15.4**). The circles are defined as follows:

$$\left(\Gamma_r - \frac{r}{r+1}\right)^2 + \Gamma_i^2 = \frac{1}{(r+1)^2} \quad (15.12) \qquad (\Gamma_r + 1)^2 + \left(\Gamma_i - \frac{1}{x}\right)^2 = \left(\frac{1}{x}\right)^2 \quad (15.17)$$

Properties

(1) The circles are loci of constant r or constant x.
(2) x and r circles are orthogonal to each other.
(3) All circles pass through the point $\Gamma_r = 1, \Gamma_i = 0$.
(4) The circles for x and $-x$ are images of each other, reflected about the real axis.
(5) The center of the chart is at $\Gamma_r = 0, \Gamma_i = 0$.
(6) The intersections of the r circles with the real axis, for $r = r_0$ and $r = 1/r_0$, occur at points symmetric about the center of the chart ($\Gamma_r = 0, \Gamma_i = 0$).
(7) The intersections of the x circles with the outer circle ($|\Gamma| = 1$) for $x = x_0$ and $x = -1/x_0$ occur at points diametrically opposite each other.
(8) The intersection of any r circle with any x circle represents a normalized impedance point.
(9) The point $\Gamma_r = 1, \Gamma_i = 0$ (rightmost point in **Figure 15.7**) represents infinite impedance ($r = \infty, x = \infty$); hence, it is called the ***open circuit point***.
(10) The diametrically opposite point, at $\Gamma_r = -1, \Gamma_i = 0$, represents zero impedance ($r = 0, x = 0$); hence, it is the ***short circuit point***.
(11) The outer circle represents $|\Gamma| = 1$. The center of the diagram represents $|\Gamma| = 0$.
(12) Any circle centered at the center of the diagram ($\Gamma_r = 0, \Gamma_i = 0$) with radius a is a circle on which the magnitude of the reflection coefficient is constant, $|\Gamma| = a$.
(13) A circle drawn through a point representing a normalized load impedance describes the reflection coefficient at different locations on the line (generalized reflection coefficient).
(14) Any point on the chart represents a normalized impedance, $z = r + jx$. The admittance of this point is $y = (r - jx)/(r^2 + x^2)$. The admittance point corresponding to an impedance point lies on the reflection coefficient circle that passes through the impedance point, diametrically opposite of the impedance point **(Figure 15.6a)**.
(15) Motion toward the generator—clockwise. Toward the load—counterclockwise.
(16) Motion around the chart changes the phase but not the magnitude of the reflection coefficient [**Eq. (14.91)**].
(17) A full circle represents $\lambda/2$.
(18) All distances on the Smith chart are in wavelengths, phases are in degrees.

A common use of the Smith chart is for purposes of impedance matching.

Single Stub Matching Stub matching uses the admittance chart for parallel stubs, impedance chart for series stub. The sequence for parallel stub matching is as follows (see **Figure 15.13**):

(1) A shorted (sometimes open) stub of length l_1, typically of the same characteristic impedance as the line, is placed at a distance d_1 from the load in parallel with the line.
(2) Normalize the load impedance and place the normalized value on the chart. Draw the reflection coefficient circle through that point (P_2).

(3) Find the normalized admittance by drawing a line from P_2 through the center of the chart until it intersects the reflection coefficient circle on the opposite side (P_3).
(4) Identify the points at which the reflection coefficient circle intersects the $r = 1$ circle (P_4 and P_5).
(5) Find the length of the stub, l_1, which when connected in parallel to the line at a distance d_1 from the load cancels the imaginary part of the normalized admittance (susceptance) at the two points in (4). This provides two possible solutions.
(6) The length of the shorted stub is found by starting from the point of infinite admittance on the chart (P_{oc}) and moving clockwise until the desired susceptance is found.
(7) Use of open stubs is possible with the appropriate change in (5) and (6) (see **Example 15.3**).
(8) Series stub matching follows the same process but step (3) is skipped, and all steps are done in terms of impedance rather than admittance (see **Example 15.4**).

Double Stub Matching

(1) In this method, two shorted stubs are placed on the line, at any desired location (typically at the load or close to it). The distance between the two stubs is fixed **(Figure 15.13b)**.
(2) Draw a unit circle, shifted from the $r = 1$ circle toward the load (counterclockwise) a distance in wavelengths equal to the distance between the two stubs (**Figure 15.17**)
(3) Place the normalized load impedance on the chart and draw the reflection coefficient circle.
(4) The normalized load admittance is found diagonally opposite the impedance point.
(5) If the load is not at the stub (i.e., if $d_1 \neq 0$) move along the reflection coefficient circle a distance d_1 to the starting point (see **Example 15.6**).
(6) Move on the constant conductance circle from the load admittance point toward the generator until the shifted unit circle is intersected at two possible points. The difference in susceptance between the two points is due to stub **(1)**.
(7) Find the length l_1 of the stub that will add the necessary susceptance at that point as indicated in **(6)**. There are two possible solutions.
(8) Now consider each of the two points found in **(7)** as a load to the line. Repeat the process for single stub matching for each point to find the two possible solutions for l_2.

Notes

(1) Single stub matching guarantees a match for any line and any load except a purely imaginary load.
(2) Double stub matching does not guarantee a solution for all conditions, but it is often more practical because the matching section can be prefabricated and included with the load (such as an antenna).
(3) Adding any number of half-wavelengths to any stub or to the position of a stub on the line has no effect on the matching conditions.
(4) Matching in transmission lines means the two impedances are equal. It does not mean maximum power transfer, which requires conjugate matching.

$\lambda/4$ Transformer (Figure 15.27)
A section of transmission line, $\lambda/4$ in length loaded with an impedance Z_l, has input impedance:

$$Z_{in} = Z_t^2/Z_l \quad [\Omega] \tag{15.22}$$

We place this section at a distance d from the load so that Z_l at the location of the transformer is real (maximum or minimum voltage point on lossless lines). To ensure matching, select the characteristic impedance of the transformer section, Z_t so that

$$Z_t = \sqrt{Z_{in}Z_l} \quad [\Omega] \tag{15.23}$$

In practical terms, the $\lambda/4$ transformer is placed at the location of voltage maximum or voltage minimum:
At the maximum impedance point

$$Z_t = Z_0\sqrt{SWR} \quad [\Omega] \tag{15.27}$$

At the minimum impedance point

$$Z_t = Z_0/\sqrt{SWR} \quad [\Omega] \tag{15.26}$$

Any number of half-wavelengths may be added to the transformer length or to the location of the transformer without change in the matching conditions.

Problems

General Design Using the Smith Chart

15.1 Application: Line Properties Using the Smith Chart. A long line with characteristic impedance $Z_0 = 100\ \Omega$ operates at 1 GHz. The speed of propagation on the line is c [m/s] and the load impedance is $260 + j180\ \Omega$. Find:

(a) The reflection coefficient at the load.
(b) The reflection coefficient at a distance of 20 m from the load toward the generator.
(c) Standing wave ratio.
(d) Input impedance at 20 m from the load.
(e) Location of the first voltage maximum and first voltage minimum from the load.

15.2 Application: Calculation of Voltage/Current Along Transmission Lines. A transmission line with a characteristic impedance of $100\ \Omega$ and a load of $50 - j50\ \Omega$ is connected to a matched generator. The line is very long and the voltage measured at the load is 50 V. Calculate using the Smith chart:

(a) Maximum voltage on the line (magnitude only).
(b) Minimum voltage on the line (magnitude only).
(c) Location of maxima and minima of voltage on the line (starting from the load).

15.3 Application: Impedance of Composite Line. A transmission line is made of two segments, each 1 m long (**Figure 15.30**). Calculate the input impedance of the combined line using a Smith chart if the speed of propagation on line (1) is 3×10^8 m/s and on line (2) 1×10^8 m/s. The lines operate at 300 MHz.

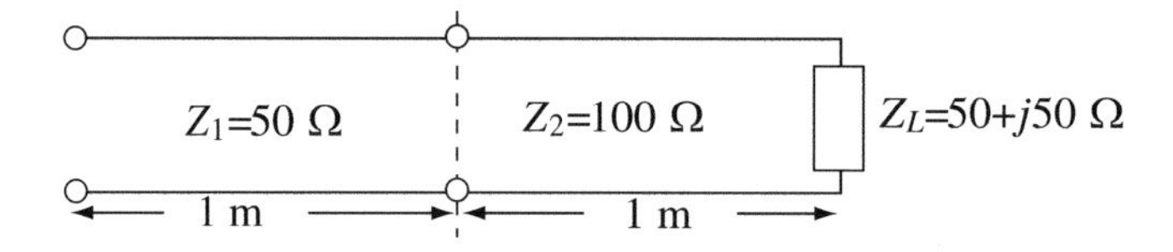

Figure 15.30

15.4 Application: Line Properties. A lossless transmission line has characteristic impedance $Z_0 = 300\ \Omega$, is 6.3 wavelengths long, and is terminated in a load impedance $Z_L = 35 + j25\ \Omega$. Find:

(a) The input impedance on the line.
(b) The standing wave ratio on the main line.
(c) If the load current is 1 A, calculate the input power to the line.

15.5 Application: Line Properties. A lossless transmission line has characteristic impedance $Z_0 = 50\ \Omega$ and its input impedance is $60 - j70\ \Omega$. The line operates at a wavelength of 0.4 m and is 3.85 m long. Calculate:

(a) The load impedance connected to the line.
(b) The location of the voltage minima and maxima on the line, starting from the load.
(c) The reflection coefficient at the load (magnitude and angle) and the standing wave ratio on the line.
(d) The magnitude of the maximum and minimum voltage and current on the line if the load voltage is $22 - j10$ V.

15.6 Application: Design of Transmission Lines. It is required to design a load of $75 - j50\ \Omega$ to simulate a device operating at 100 MHz. It is proposed using a section of a 50 Ω line and connecting to its end a lumped resistance R [Ω]. The line's phase velocity is $c/3$ [m/s].

(a) Calculate the length of line and the required resistance R that will accomplish this.
(b) Is the solution unique? Explain and find all possible solutions if the solution is not unique.

15.7 Application: Line Properties Using the Smith Chart. An unknown load is connected to a 75 Ω lossless transmission line. To find the load, two measurements are performed: (1) The location of the first voltage minimum is found at $0.18\,\lambda$ from the load. (2) The SWR is measured as 2.5. Find using the Smith chart:

(a) The load impedance.
(b) The load reflection coefficient (magnitude and phase angle).

15.8 Application: Evaluation on Line with Unknown Load. A two-wire transmission line with unknown characteristic impedance is 35 m long and connected to an unknown load impedance. Only the input to the line is accessible. To characterize the line and its load, the dimensions of the line are measured as is the input impedance. Each wire is 1.2 mm in diameter and the wires are separated 15 mm apart in air. The input impedance is measured at 915 MHz and found to be $Z_{in} = 350 - j400\ \Omega$ and assuming the line is lossless, find:

(a) The load impedance.
(b) The standing wave ratio on the line.

Stub Matching

15.9 SWR on Line. A general load ($Z_L = Z_0 + jX_L$ [Ω]) is connected on a lossless transmission line as shown in **Figure 15.31**. The shorted section is made of a different line with a different characteristic impedance Z_1 [Ω] and is connected on a lossless transmission line in series with the load.

(a) Assuming the generator is matched, calculate the standing wave ratio on the line.
(b) What must be the length of the shorted line to ensure matching of the load (no reflection). Are there any other conditions that must be satisfied for this to happen?

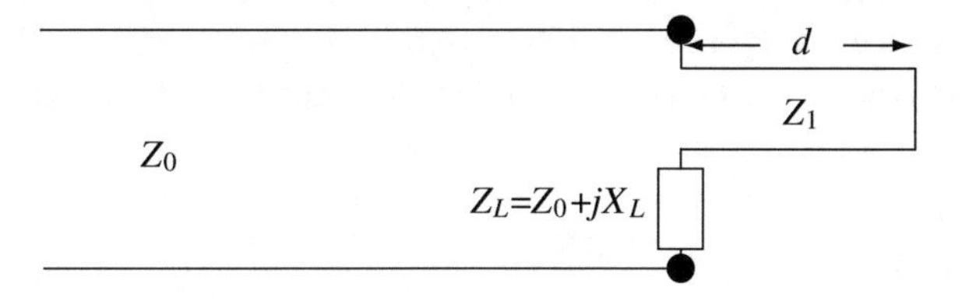

Figure 15.31

15.10 Matching with Shorted/Open Loads. The transmission line network in **Figure 15.32** is given. The shorted transmission line and the open transmission line are part of the network. Show that no stub network will match the two line sections to the main line.

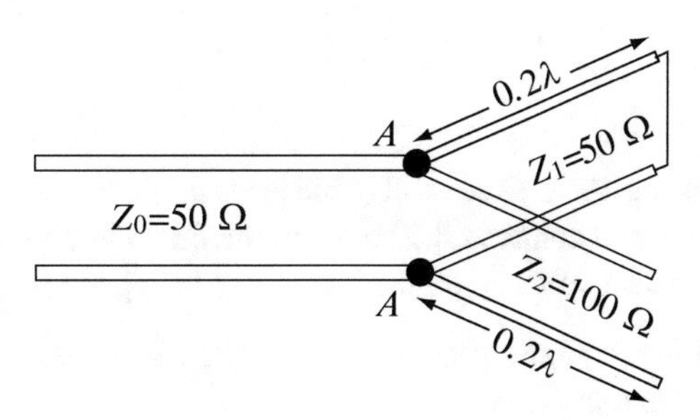

Figure 15.32

15.11 Application: Series Stub Matching. A transmission line of characteristic impedance $Z_0 = 50\ \Omega$ is loaded with an impedance $Z_L = 100 + j80\ \Omega$ (**Figure 15.33**). An open transmission line is connected in series with the line as shown. The open line has the same characteristic impedance as the main line.

(a) Find the length of the open series stub and the location (closest to the load) it should be inserted to match the load to the line.

(b) Since a stub is a reactive component, find the equivalent lumped component (capacitor or inductor) that can replace the stub found in **(a)** if the line operates at 500 MHz.

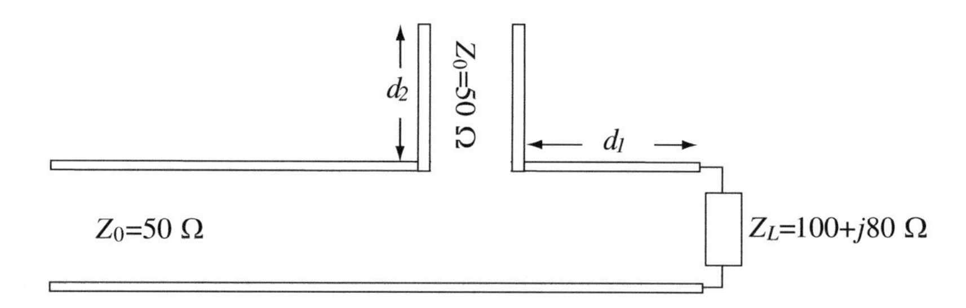

Figure 15.33

15.12 Application: Single Stub Matching. A transmission line is loaded as in **Figure 15.34**. If the wavelength on the line equals 5 m, find a shorted parallel open and its characteristic impedance is the same as that of the line (location and length of stub) placed to the left of points *A–A* to match the load to the line.

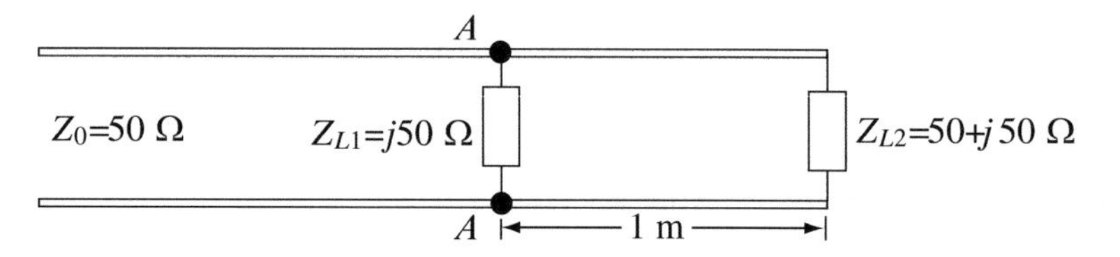

Figure 15.34

15.13 Application: Series Stub Matching. A load is connected to a transmission line as shown in **Figure 15.35**. It is required to match the load to the line (which has a characteristic impedance of 75 Ω). Find the location and length of a stub to match the line. The stub is open and its characteristic impedance is the same as that of the line as shown in **Figure 15.35**.

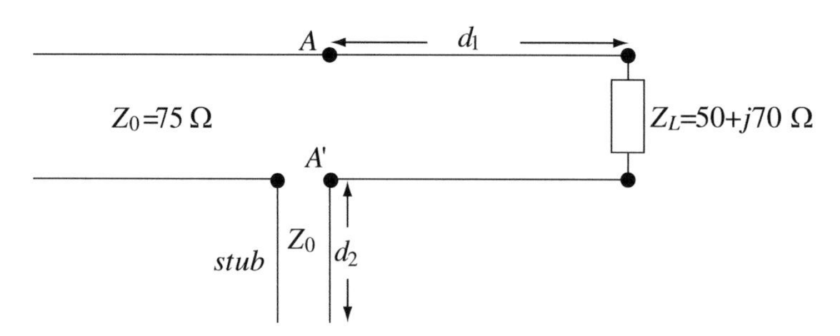

Figure 15.35

15.14 Application: Single Stub Matching. A 75 Ω coaxial cable is used to connect to a TV. The load is matched to the line (**Figure 15.36a**). A second TV must be connected 10 m from the first TV, again with a 12 m section of the same cable (**Figure 15.36b**). Assuming the phase velocity on the line is $c/2$ [m/s], calculate:

(a) The reflection coefficient at the location of connection of the two lines.

(b) The standing wave ratio on the main line.

(c) Design a single stub (its location to the left of the discontinuity and its length) to match the line for TV channel 3 (63 MHz). Use the same line impedance for the stubs.

(d) For the design in **(c)** calculate the reflection coefficient to the left of the stub for channel 2 (57 MHz). What is your conclusion from this calculation as far as stub matching across a range of frequencies?

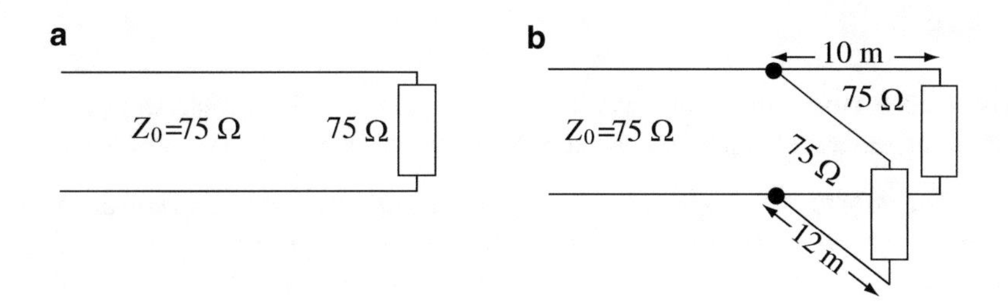

Figure 15.36

15.15 Application: Single Stub Matching of a Generator. Consider the problem of matching a generator with internal impedance of 50 Ω to a line with characteristic impedance of 75 Ω. Find the length (l) of a parallel stub and the distance of the stub (d) from the generator terminals that will achieve the required match (see **Figure 15.37**). Comment on the behavior of the forward and backward waves on the line to the right and to the left of the stub.

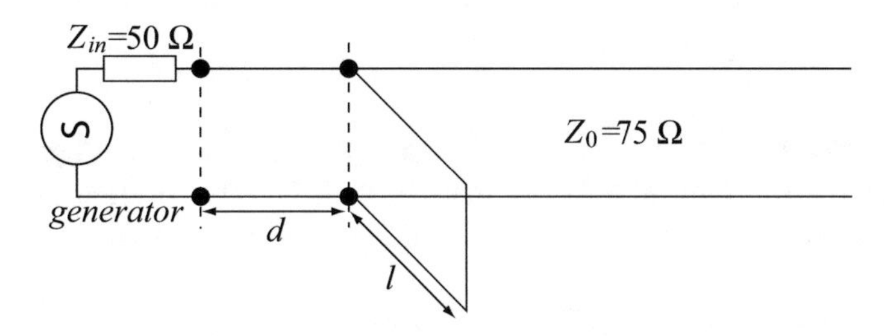

Figure 15.37

15.16 Application: Double Stub Matching. Two stubs are used on a transmission line as shown in **Figure 15.38**. Calculate stub lengths d_1 and d_2 (in wavelengths) to match the load to the line. Is this arrangement of stubs a good arrangement? Why?

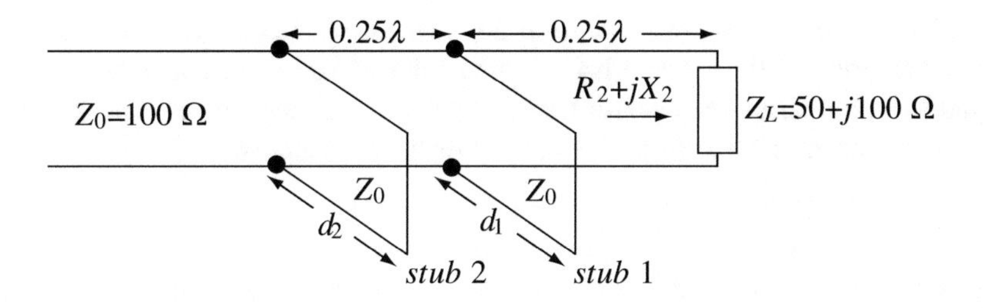

Figure 15.38

15.17 Application: Double Stub Matching. An antenna has an impedance of 68 + j100 Ω. The antenna needs to be connected to a 75 Ω line. Because the antenna goes on a mast, the design engineer decided to fabricate a matching section as shown in **Figure 15.39**. The matching section is then connected to the antenna during installation. At the required frequency, the section is 0.3 λ long and the two stubs are made of 75 Ω line sections. Calculate the lengths of the stubs, if the antenna is connected at $A - A$.

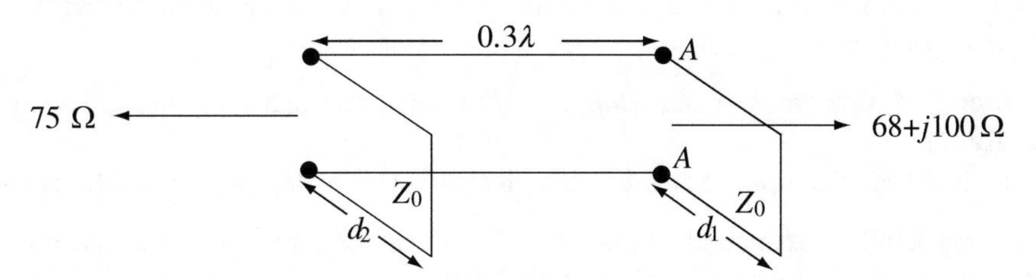

Figure 15.39

Transformer Matching

15.18 Application: $\lambda/4$ Transformer. Show that two lines with any characteristic (real) impedances Z_1 [Ω] and Z_2 [Ω] may be matched with a quarter-wavelength line. What is the characteristic impedance of the matching section?

15.19 Application: $3\lambda/4$ Transformer. A lossless transmission line with characteristic impedance Z_0 [Ω] transfers power to a load Z_L [Ω] (real). To match the line, a matching section is connected as shown in **Figure 15.40**. At what distance d (in wavelengths) from the load must the line be connected (minimum distance) and what must the characteristic impedance of the matching section be?

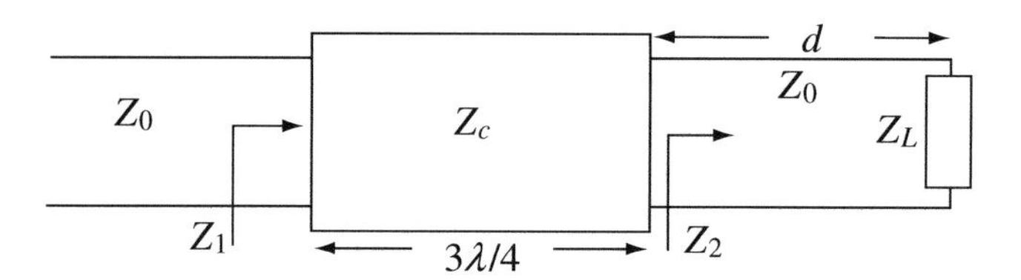

Figure 15.40

15.20 Application: $\lambda/4$ Transformer. A transmission line is given as shown in **Figure 15.41**. If the characteristic impedance of the quarter-wavelength transformer must be real, find the location of the transformer (distance d in the figure, in wavelengths) and the intrinsic impedance of the transformer Z_t [Ω].

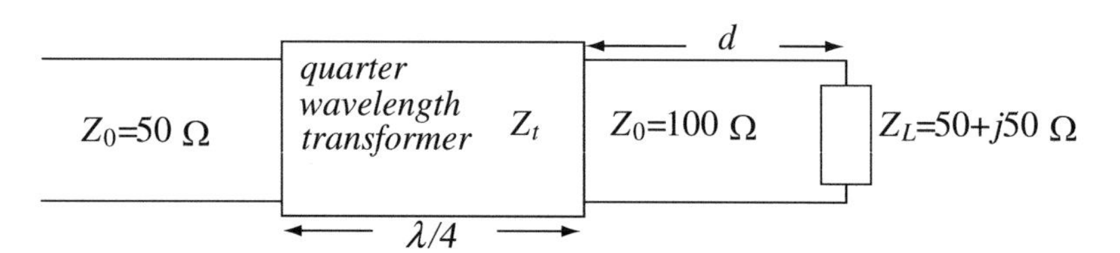

Figure 15.41

15.21 Application: $\lambda/4$ Transformer. A two-wire transmission line has characteristic impedance of 300 Ω and connects to an antenna. The line is long and the antenna has an impedance of 200 Ω and operates at a wavelength of 3.8 m. To match the line and load, a quarter-wavelength transformer is connected on the line, but the location at which the transformer may be connected is 10 m from the antenna or larger. Calculate:

(a) The closest location at which the transformer may be connected.
(b) For the result in **(a)**, the characteristic impedance of the transformer section.
(c) The standing wave ratios on the sections of line between the transformer and antenna and between transformer and generator.

15.22 Application: $\lambda/4$ Transformer Matching at Load and Generator. A 75 Ω transmission line is connected at one end to a generator with internal impedance $Z_{in} = 50 - j75$ Ω and on the other end to a load of impedance $Z_L = 100 + j75$ Ω.

(a) Design a $\lambda/4$ transformer to match the generator to the line; specify the minimum distance from the generator and its characteristic impedance.
(b) Design a $\lambda/4$ transformer to match the load to the line; specify the minimum distance from the load and its characteristic impedance.

15.23 Application: Imperfect Transformer Matching. A $\lambda/4$ transformer can only match exactly at the frequency at which it is exactly $\lambda/4$.
A $\lambda/4$ transformer is designed to operate at 2.4 GHz to match a $24 + j40$ Ω load to a 50 Ω lossless transmission line.

(a) Find the lowest possible characteristic impedance of the transformer and the distance from the load at which it must be located to match the load to the line at 2.4 GHz.
(b) If the frequency changes to 1.9 GHz, calculate the standing wave ratio to the left of the transformer designed in **(a)**.

16 Transients on Transmission Lines

Swift as a shadow, short as any dream,
Brief as the lightning in the collied night,

William Shakespeare, A midsummer night's dream

16.1 Introduction

Chapter 14 discussed the propagation properties of transmission lines with particular emphasis on impedance, the reflection coefficient, and time-harmonic representation. Voltage and current were phasors, and a number of properties such as the speed of propagation, wavelength, and phase and attenuation constants were used as a direct consequence of the time-harmonic nature of the waves. Much of the discussion paralleled that of propagation of plane waves in unbounded domains.

There are, however, important applications in which the single-frequency, time-harmonic representation is not appropriate. For example, when we close a switch on a transmission line connecting the line with the generator, a transient ensues. In effect, we are connecting a step source to the line. Similarly, when disconnecting the line, we should expect a transient. When a power transmission line, which may normally operate under steady-state conditions, is shorted because of a fault or when the load suddenly changes, a transient is again generated. In still other cases, such as in digital communication lines, narrow pulses may be sent at relatively high rates. Similarly, the lines connecting digital circuit components on a board transfer pulses which may be wide or narrow, depending on the application. A number of transient waveforms of this type are shown in **Figure 16.1**. In all of these applications, we cannot use the methods of the previous chapters directly. In fact, many of the basic concepts used in the previous chapters are not properly defined in this new environment. For example, the speed of propagation, wavelength, phase constant, and even impedance are only properly defined in the time-harmonic environment.

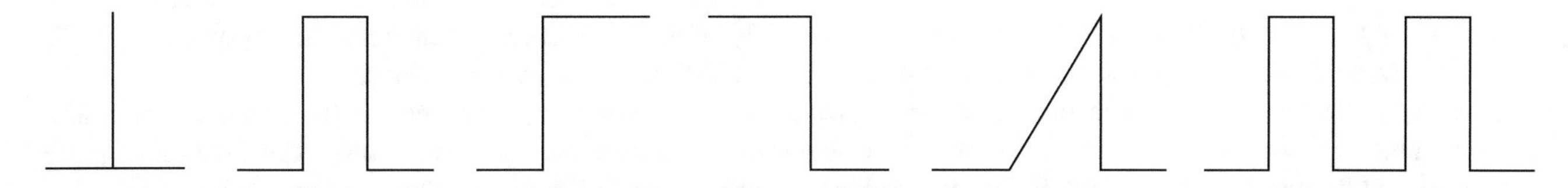

Figure 16.1 Common transients encountered in analog and digital communication lines

The approach adopted here is a very different and fundamental approach. Imagine that we could observe the behavior of the line at all times and at any point we wished. This would give us all the information needed to evaluate the behavior of the line. In effect, we are going to "ride" the various waves that may exist on the line as they propagate. This approach has the great advantage that it is simple and intuitive. It will provide simple solutions to a number of important transmission line applications with few assumptions.

Two types of transients will be discussed here. The first is narrow pulses and the second is the step source. The intermediate case of long pulses will be treated as the superposition of step sources.

N. Ida, *Engineering Electromagnetics*, https://doi.org/10.1007/978-3-030-15557-5_16

16.2 Propagation of Narrow Pulses on Finite, Lossless Transmission Lines

Narrow pulses are common in digital systems but also on communication lines and are characterized by widths which are very small compared with the propagation time along the line. In other words, if a line is of length d [m] and the speed of propagation is v_p [m/s], the time of propagation on the line is $t_p = d/v_p$ [s]. A pulse of width $\Delta t \ll t_p$ is considered a narrow pulse. Note, however, that Δt itself is not necessarily small.

A narrow pulse propagates on a lossless line without distortion since the speed of propagation is independent of frequency. Thus, we can still use the concept of phase velocity even though it was initially defined for time-harmonic waves. The speed of propagation on the line is

$$v_p = \frac{1}{\sqrt{LC}} \quad \left[\frac{\text{m}}{\text{s}}\right] \tag{16.1}$$

where L and C are the inductance and capacitance per unit length of the line, respectively.

Consider first the line in **Figure 16.2**. The load is matched to the line so there will be no reflection from the load. The generator produces a pulse at time $t = 0$. The pulse appears at the input to the line with the following amplitude for voltage and current:

$$V^+ = V_g \frac{Z_0}{Z_0 + Z_g} \quad [\text{V}], \quad I^+ = \frac{V^+}{Z_0} = \frac{V_g}{Z_0 + Z_g} \quad [\text{A}] \tag{16.2}$$

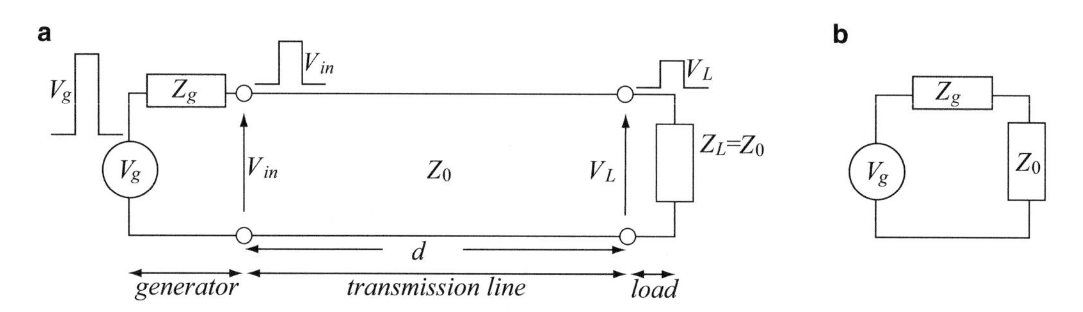

Figure 16.2 (**a**) Propagation of a narrow pulse on a matched line. (**b**) Equivalent circuit at the generator at $t = 0$

This is due to the impedance divider created by the generator's internal impedance and the line impedance. The line current is equal to the forward-propagating voltage divided by the line impedance, which, in this case, equals Z_0 since the pulse has not propagated down the line and the only impedance it sees is the characteristic impedance of the line. This pulse now propagates toward the load, which it reaches after a time $t = d/v_p$. Since the load impedance is equal to the characteristic line impedance, there is no reflection at the load ($\Gamma_L = 0$), and all energy in the forward-propagating pulse is transferred to the load. Nothing more happens on the line unless additional pulses are generated.

Now suppose the line is not matched, as shown in **Figure 16.3**. At time $t = 0$, a pulse appears at the generator terminals. Since nothing happened on the line itself, the generator only sees the characteristic line impedance. Thus, the initial pulse that appears at the generator's terminals is the same as for the matched line in **Eq. (16.2)**. The pulse propagates at the same speed and reaches the load. The pulse is partly transmitted into the load, but because the line and load are not matched, there is a reflection coefficient at the load:

$$\Gamma_L = \frac{V^-}{V^+} = -\frac{I^-}{I^+} = \frac{Z_L - Z_0}{Z_L + Z_0} \tag{16.3}$$

Also, because the sum of the forward and reflected waves must equal the transmitted wave, the transmission coefficient at the load is

$$T_L = 1 + \Gamma_L = 1 + \frac{Z_L - Z_0}{Z_L + Z_0} = \frac{2Z_L}{Z_L + Z_0} \tag{16.4}$$

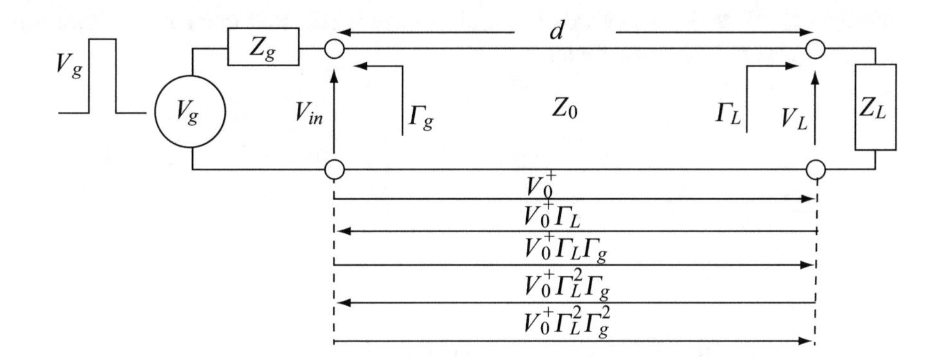

Figure 16.3 Mismatched load and generator. The first few reflected voltages at load and generator are shown

The reflected voltage and current waves are

$$V_1^- = \Gamma_L V^+ = \frac{Z_L - Z_0}{Z_L + Z_0} V^+ = \frac{Z_0}{Z_0 + Z_g}\left(\frac{Z_L - Z_0}{Z_L + Z_0}\right) V_g \quad [\mathrm{V}] \tag{16.5}$$

$$I_1^- = -\Gamma_L I^+ = -\frac{V^+}{Z_0}\left(\frac{Z_L - Z_0}{Z_L + Z_0}\right) \quad [\mathrm{A}] \tag{16.6}$$

The total voltage at the load at time $t = d/v_p$ is the sum of the incoming and reflected waves:

$$V_{L1} = V^+ + V_1^- = V^+(1 + \Gamma_L) \quad [\mathrm{V}] \tag{16.7}$$

where the index 1 indicates that this is the first reflection at the load. Note that although a sum is used, the reflection coefficient can be negative. The current in the load is given from **Eq. (16.3)** as

$$I_{L1} = \frac{V^+}{Z_0}(1 - \Gamma_L) \quad [\mathrm{A}] \tag{16.8}$$

The sum of the forward- and backward-propagating waves only exists for a period equal to the width of the pulse, Δt. After that, only the backward-propagating waves in **Eqs. (16.5)** and **(16.6)** exist on the line. To see how this comes about, the forward-propagating wave and the backward-propagating wave can be viewed as two separate waves propagating in opposite directions, as shown in **Figure 16.4a**. For clarity, we assume that Γ_L is negative, but it may also be positive. After $t_1 = d/v_p$, the pulses add up as shown by the solid lines in **Figure 16.4b**. At a time $t > t_1 + \Delta t$, the only wave on the line is the backward-propagating wave, as shown in **Figure 16.4c**.

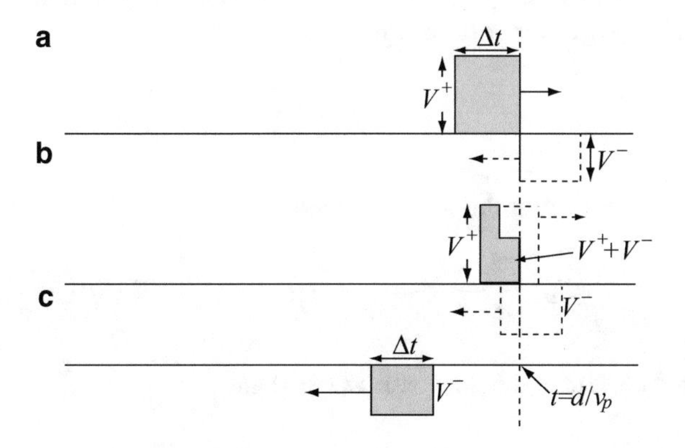

Figure 16.4 Conditions at the load before, during, and after reflection. (**a**) The pulse front reaches the load. (**b**) A reflected wave is generated and propagates toward the generator, partially overlapping the incident pulse. (**c**) After one pulse width, only the backward-propagating pulse is left

The reflected voltage (or current) now travels back and, after an additional time equal to d/v_p, reaches the generator. However, now the generator does not act as a generator but rather like a load Z_g since the source of the reflected wave is at the actual load. As with the load, part of the wave is reflected and part is transmitted into the generator (where it must be dissipated). Thus, the backward-propagating wave is reflected into a new, forward-propagating wave at the generator, with the generator reflection coefficient:

$$\Gamma_g = \frac{V_1^+}{V_1^-} = -\frac{I_1^+}{I_1^-} = \frac{Z_g - Z_0}{Z_g + Z_0} \tag{16.9}$$

The reflected waves at the generator are

$$V_1^+ = \Gamma_g V_1^- = \Gamma_L \Gamma_g V^+ \quad [\text{V}] \quad \text{and} \quad I_1^+ = -\Gamma_g I_1^- = \frac{\Gamma_L \Gamma_g V^+}{Z_0} \quad [\text{A}] \tag{16.10}$$

and the total voltage and current at the generator connections are

$$V_{in1} = V_1^- + V_1^+ = V^+ \Gamma_L \left(1 + \Gamma_g\right) \quad [\text{V}] \quad \text{and} \quad I_{in1} = I_1^- + I_1^+ = -I^+ \Gamma_L \left(1 - \Gamma_g\right) \quad [\text{A}] \tag{16.11}$$

Again, these sums only exist during a time Δt. After that, only the new forward-propagating wave exists. This process repeats itself indefinitely, with each reflection at each end of the line being viewed as a new wave propagating toward the other end. The reflection process is shown schematically in **Figure 16.3** for a few voltage reflections.

If instead of a single pulse, the generator produces a train of pulses, each pulse is reflected as described above. However, both forward-propagating and backward-propagating pulses may meet along the line. When this happens the voltage and current on the line are superposition of the various pulses. Each pulse continues to travel as if it were alone on the line.

Example 16.1 The generator in **Figure 16.5** produces 10 V pulses that are 20 ns wide. Consider a single pulse, produced at $t = 0$. Calculate the voltage and current at the generator and load for all times between zero and 5.5 μs. Assume the line is lossless and speed of propagation on the line is $c/3$ [m/s].

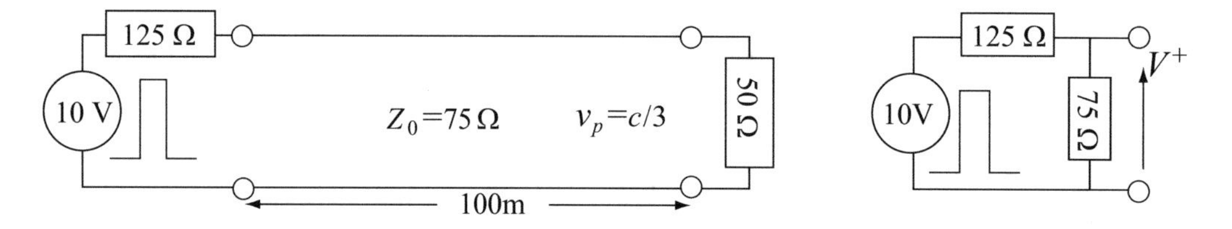

Figure 16.5 A line with mismatched load and generator

Solution: The reflection coefficients at the load (looking into the load) and generator (looking into the generator) are first calculated. Then, we follow the pulse, based on the time of propagation between generator and load. The time it takes the pulse to travel from the generator to load is

$$t = \frac{L}{v_p} = \frac{100}{1 \times 10^8} = 1 \quad [\mu\text{s}]$$

The reflection coefficients at the load and generator are

$$\Gamma_L = \frac{Z_L - Z_0}{Z_L + Z_0} = \frac{50 - 75}{50 + 75} = -0.2, \quad \Gamma_g = \frac{Z_g - Z_0}{Z_g + Z_0} = \frac{125 - 75}{125 + 75} = 0.25$$

The voltage and current at the generator at $t = 0$ are

$$V^+ = V_g \frac{Z_0}{Z_0 + Z_g} = V_g \frac{75}{75 + 125} = 0.375 V_g = 3.75 \quad [\text{V}]$$

$$I^+ = \frac{V_g}{Z_0 + Z_g} = \frac{10}{75 + 125} = 0.05 \quad [\text{A}]$$

These propagate toward the load. After 1 μs, both reach the load. The reflected waves are $V_1^- = \Gamma_L V^+$ and $I_1^- = -\Gamma_L I^+$:

$$V_1^- = \Gamma_L V^+ = -0.2V^+ = -0.75 \quad [\mathrm{V}], \quad I_1^- = -\Gamma_L I^+ = 0.2I^+ = 0.01 \quad [\mathrm{A}]$$

The forward- and backward-propagating waves add up for 20 ns at the load. For these 20 ns, the voltage at the load is $0.8V^+ = 3$ V and the current is $1.2\, I^+ = 0.06$ A. Both reflected waves propagate back to the generator where a second reflection takes place but now with the reflection coefficient of the generator:

$$V_1^+ = \Gamma_g V_1^- = \Gamma_L \Gamma_g V^+ = (-0.2) \times 0.25V^+ = -0.1875 \quad [\mathrm{V}]$$
$$I_1^+ = -\Gamma_g I_1^- = \Gamma_L \Gamma_g I^+ = -0.2 \times 0.25 \times I^+ = -0.0025 \quad [\mathrm{A}]$$

Again, at the generator, the voltage is the sum of the backward- and forward-propagating waves for 20 ns. The process now repeats itself with the new forward-propagating waves. At $t = 3$ μs, we are at the load:

$$V_2^- = \Gamma_L V_1^+ = -0.2V_1^+ = 0.0375 \quad [\mathrm{V}], \quad I_2^- = -\Gamma_L I_1^+ = 0.2I_1^+ = -0.0005 \quad [\mathrm{A}]$$

At $t = 4$ μs, the voltage at the generator is

$$V_2^+ = \Gamma_g V_2^- = 0.009375 \quad [\mathrm{V}], \quad I_2^+ = -\Gamma_g I_2^- = 0.000125 \quad [\mathrm{A}]$$

At $t = 5$ μs, the voltage at the load is

$$V_3^- = V_2^+ \Gamma_L = -0.001875 \quad [\mathrm{V}], \quad I_3^- = -I_2^+ \Gamma_L = 0.000025 \quad [\mathrm{A}]$$

The results are shown in **Figures 16.6a** and **16.6b** for the voltage and current at the generator and load. The sums of the forward and backward waves are shown.

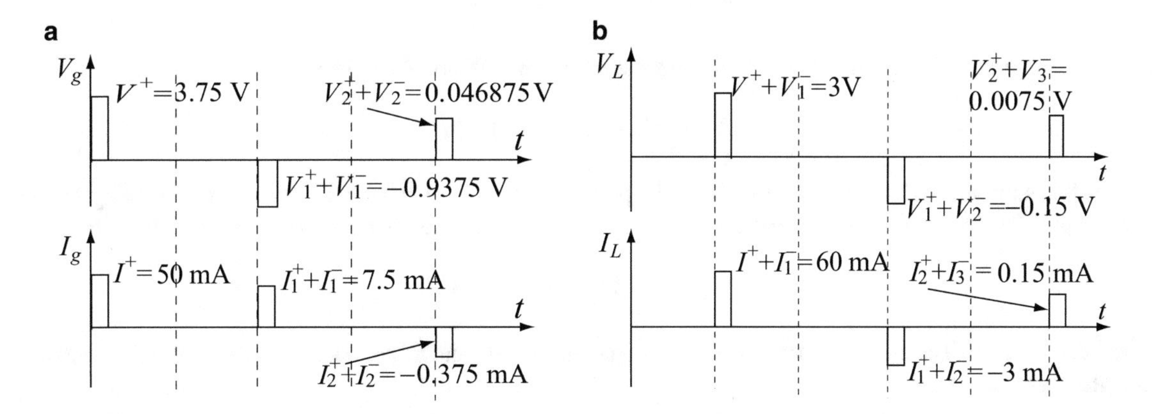

Figure 16.6 (**a**) Voltage and current at the generator in **Figure 16.5**, immediately after the pulses are generated. (**b**) Voltage and current pulses at the load in **Figure 16.5**

16.3 Propagation of Narrow Pulses on Finite, Distortionless Transmission Lines

Although we now assume the line to be lossy, with an attenuation constant α, the line is also assumed to be distortionless (i.e., $R/L = G/C$) so that pulses do not distort. For a single pulse as described in the previous section, all aspects of propagation remain the same, but, in addition, the pulse magnitude is attenuated exponentially as it propagates from generator to load, or load to generator. The problem analyzed here is shown in **Figure 16.7a**.

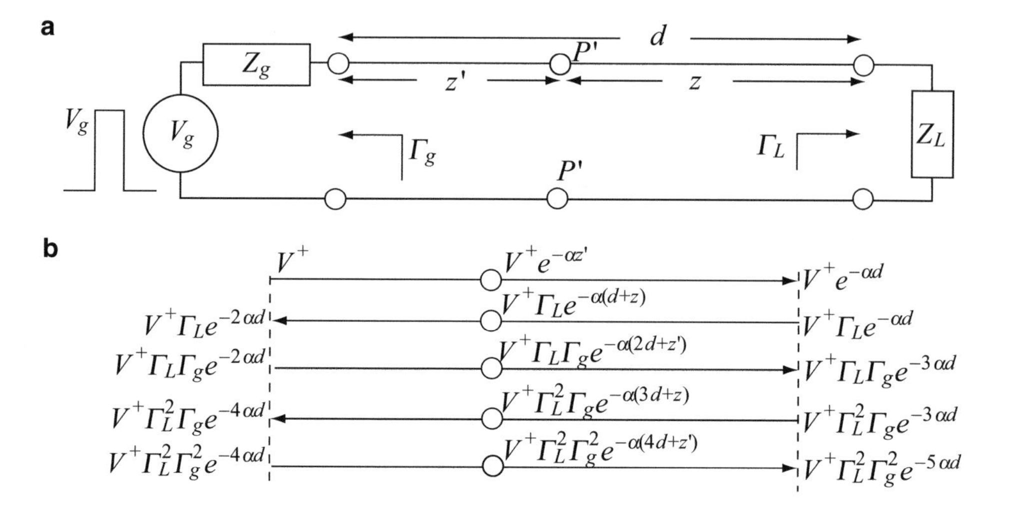

Figure 16.7 (**a**) Distortionless transmission line. (**b**) The voltage waves on the line for a few reflections

With the forward-propagating wave in **Eq. (16.2)**, the wave propagates along the line and is attenuated. For the first wave ($0 < t < d/vp$), the voltage on the line at a point P' is

$$V(z) = V^+e^{-\alpha z'} = V_g \frac{Z_0}{Z_0 + Z_g} e^{-\alpha z'} \quad [\text{V}] \tag{16.12}$$

where z' is the distance from generator to point P' in **Figure 16.7a**. At the load, the forward-propagating wave is

$$V_L^+ = V^+e^{-\alpha d} \quad [\text{V}] \tag{16.13}$$

The reflected wave is

$$V_1^- = \Gamma_L V^+ e^{-\alpha d} \quad [\text{V}] \tag{16.14}$$

At the load, the total voltage is the sum of this and the reflected voltage. This gives

$$V_L = V^+e^{-\alpha d}(1 + \Gamma_L) \quad [\text{V}] \tag{16.15}$$

However, this sum only exists for a time equal to the pulse width Δt. The reflected wave in **Eq. (16.14)** propagates back and is attenuated. The expression for the reflected wave anywhere on the line between load and generator is

$$V_1^-(z) = \Gamma_L V^+ e^{-\alpha d} e^{-\alpha z} \quad [\text{V}] \tag{16.16}$$

This reflected wave reaches the generator and is reflected at the generator unless the generator is matched. At the generator, the first reflection is

$$V_1^-(z = d) = V^+e^{-2\alpha d}\Gamma_L \quad [\text{V}] \tag{16.17}$$

Taking into account the generator reflection coefficient Γ_g, the total voltage at the generator connections is

$$V_{g1} = V^+\Gamma_L e^{-2\alpha d}\left(1 + \Gamma_g\right) \quad [\text{V}] \tag{16.18}$$

This sum also exists for a period Δt. The new forward-propagating wave after the first reflection at the generator is

$$V_1^+(z') = V^+e^{-2\alpha d}e^{-\alpha z'}\Gamma_L\Gamma_g \quad [\text{V}] \tag{16.19}$$

Thus, the attenuation depends on the total distance traveled by the wave, regardless of how many reflections it has undergone. This is shown schematically in **Figure 16.7b**. Note, also, that each pulse is assumed to travel independently of any other pulses on the line. If two pulses meet anywhere on the line, then the voltage and current at that point and time is the superposition of the pulses. This applies particularly to the location of the load and generator, since for any pulse width, the reflected and incident pulses overlap during a time equal to the pulse width. A sum of more than one pulse may exist on the line at other locations if multiple pulses exist on the line and propagate independently.

Example 16.2 Consider, again, **Example 16.1**, but now the line has an attenuation constant $\alpha = 0.002$ Np/m. Draw the voltage and current at the generator for $0 < t < 5.5$ μs.

Solution: From the above discussion, the voltages and currents at any given time are those for the lossless line multiplied by the attenuation from $t = 0$ to the time considered. Thus, from the results in **Example 16.1**, the voltage and current at the generator only exist at times $t = 0$, $t = 2$ μs, and $t = 4$ μs. At $t = 0$, the waves have not propagated. Thus

$$V^+ = 3.75 \quad [\text{V}], \quad I^+ = 0.05 \quad [\text{A}]$$

At time $t = 2$ μs, the waves at the generator are V_1^-, I_1^-, V_1^+ and I_1^+. These are attenuated as if they propagated a distance of 200 m. Thus,

$$\begin{aligned} V_1^- &= -0.75e^{-0.002\times200} = -0.50274 \quad [\text{V}], \\ I_1^- &= 0.01e^{-0.002\times200} = 0.0067 \quad [\text{A}] \\ V_1^+ &= -0.1875e^{-0.002\times200} = -0.1257 \quad [\text{V}], \\ I_1^+ &= -0.0025e^{-0.002\times200} = -0.001676 \quad [\text{A}] \end{aligned}$$

At $t = 4$ μs, at the generator, the total distance traveled by the wave is 400 m. The waves at this time are V_2^-, I_2^-, V_2^+ and I_2^+:

$$\begin{aligned} V_2^- &= 0.0375e^{-0.002\times400} = 0.01685 \quad [\text{V}], \\ I_2^- &= -0.0005e^{-0.002\times400} = -0.0002247 \quad [\text{A}] \\ V_2^+ &= 0.009375e^{-0.002\times400} = 0.0042125 \quad [\text{V}], \\ I_2^+ &= 0.000125e^{-0.002\times400} = 0.00005617 \quad [\text{A}] \end{aligned}$$

The total current and voltage at the generator is the sum of the forward- and backward-propagating waves for the duration of the narrow pulse (20 ns). The resulting voltage and current at the generator are shown in **Figure 16.8a**, which shows the voltage and current on the line at $t = 0$, $t = 2$ μs, $t = 4$ μs, etc. The values shown are the sums of the forward and backward amplitudes.

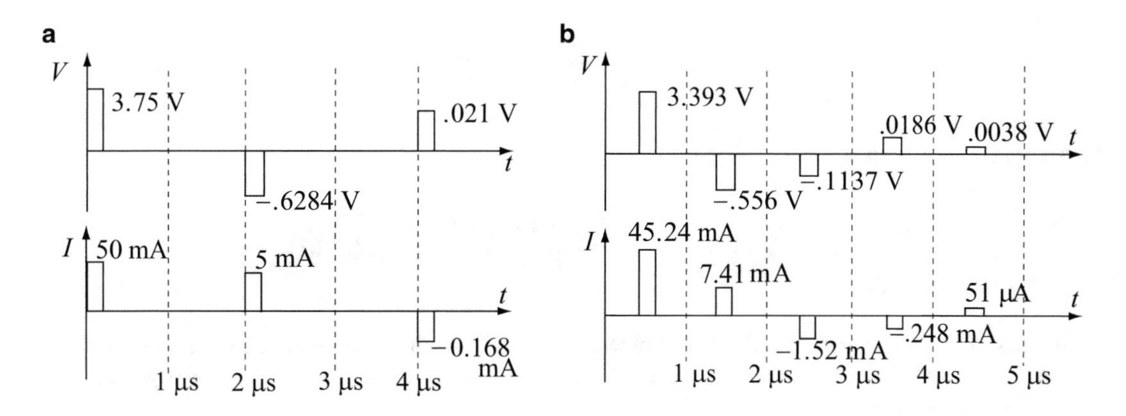

Figure 16.8 (**a**) Voltage and current at the generator in **Example 16.2**. (**b**) Voltage and current in the middle of the line in **Exercise 16.1**

Exercise 16.1 In **Example 16.2**, find the voltage and current in the middle of the transmission line for times $0 < t < 5\ \mu s$.

Answer See **Figure 16.8b**.

Example 16.3 Application: Time Domain Reflectometry Time domain reflectometry (TDR) is a method of testing that relies on reflections from mismatched loads to locate the load. This is very useful in locating short circuits or cuts in inaccessible lines such as underground cables. A pulse is sent on the line and its reflections are recorded on a screen or chart. The distance between every two pulses is twice the time it takes to propagate to the fault. If the speed of propagation on the line is known, the exact location of the fault can be found. From the magnitude, shape, and sign of the signals, it is also possible to evaluate the type of fault (short, low, or high impedance, open) before repair. This can save considerable time and labor, especially if cables are buried.

A lossless underground telephone cable has inductance per unit length of 1 μH/m and capacitance of 25 pF/m. The cable has developed a fault and it is required to locate the fault and identify its nature. The time domain reflectometer reading looks as in **Figure 16.9b**:

(a) Find the distance of the fault from the source.
(b) What kind of fault does the cable have?

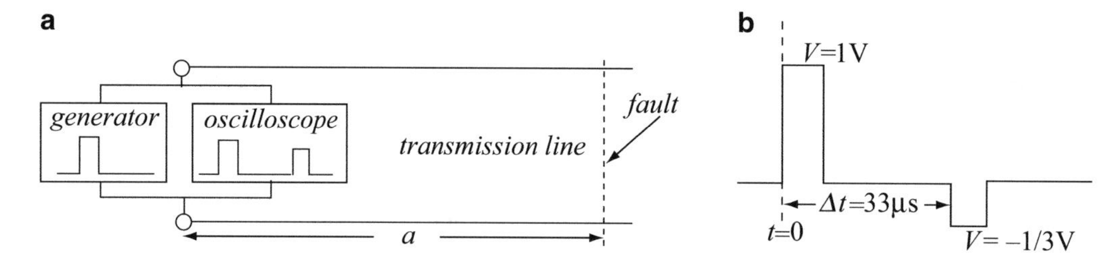

Figure 16.9 (**a**) A time domain reflectometer. (**b**) The signal obtained from the faulty cable

Solution: The distance to the fault is calculated from the time difference between two pulses and the speed of propagation on the line. The type of fault can be identified from the reflection coefficient at the fault:

(a) The speed of propagation on the line is

$$v_p = \frac{1}{\sqrt{LC}} = \frac{1}{\sqrt{1 \times 10^{-6} \times 25 \times 10^{-12}}} = 2 \times 10^8 \quad [\text{m/s}].$$

The distance of the fault is

$$d = \frac{v_p \Delta t}{2} = \frac{2 \times 10^8 \times 3.3 \times 10^{-5}}{2} = 3,300 \quad [\text{m}].$$

(b) Because the first reflection is negative, the impedance at the load is smaller than the line impedance, as can be seen from the formula for the reflection coefficient at the load. The line impedance can be calculated from the inductance and capacitance per unit length:

$$Z_0 = \sqrt{\frac{L}{C}} = \sqrt{\frac{1 \times 10^{-6}}{25 \times 10^{-12}}} = 200 \quad [\Omega]$$

The reflection coefficient is

$$\Gamma_L = \frac{V^-}{V^+} = -\frac{1}{3} = \frac{Z_L - Z_0}{Z_L + Z_0} \quad \rightarrow \quad Z_L = Z_0 \frac{(1-1/3)}{(1+1/3)} = \frac{Z_0}{2} = 100 \quad [\Omega]$$

Thus, the fault is a "partial short," such as may be caused by loss of insulation or water penetration in the cable. The calculation of the fault impedance is only possible if the line is lossless and if the pulses do not distort. In practical applications, the line is never lossless and, therefore, the pulses are distorted. It is much more difficult to classify the fault exactly (although still possible), but the location of the fault is relatively easy to find. Also, step sources are often used and multiple reflection recorded to better analyze the fault.

Exercise 16.2 In **Example 16.3**, suppose that the amplitude of the reflected wave equals 90% of the amplitude of the forward-propagating wave. What is the impedance of the fault if the intrinsic line impedance is $Z_0 = 200\ \Omega$?

Answer $Z_L = 3800\ \Omega$. This is a partially open line.

16.4 Transients on Transmission Lines: Long Pulses

The condition considered here is that of a very long pulse, again, the length being related to the length of the line and speed of propagation. In other words, we assume now that $\Delta t \gg d/v_p$, where Δt is the pulse width, d the length of the line, and v_p the speed of propagation on the line. The main difference between this assumption and the assumption in the previous case is that the pulse can now propagate back and forth from generator to load during the pulse width Δt many times. In particular, a positively going or negatively going step function satisfies this condition. A number of pulses that may be considered here are shown in **Figure 16.10**.

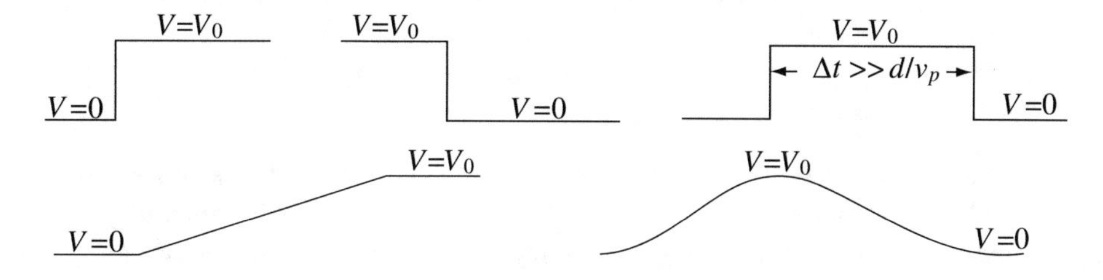

Figure 16.10 Some typical long pulses

Consider the circuit in **Figure 16.11a**. Initially, the switch is open and there is no current on the line. Suppose now the switch is closed at time $t = 0$. Initially, the condition is the same as in the previous case; that is, the disturbance on the line must propagate to the load starting at $t = 0$. The generator "sees" a load equal to Z_0 since no wave has propagated to the load yet. The voltage across the line and the current in the line at $z = 0$ are

$$V^+ = V_g \frac{Z_0}{Z_0 + Z_g} \quad [\mathrm{V}], \quad I^+ = \frac{V_g}{Z_0 + Z_g} \quad [\mathrm{A}] \tag{16.20}$$

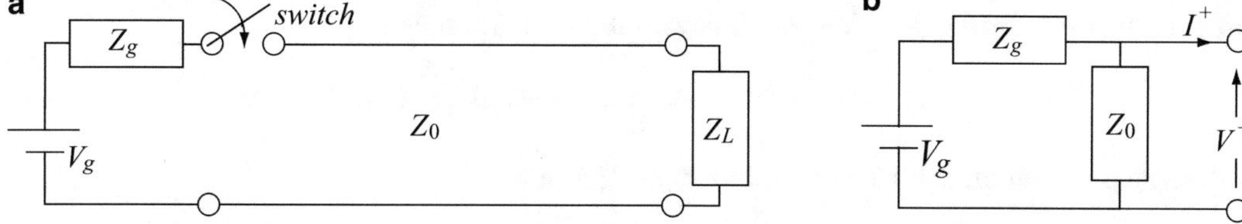

Figure 16.11 (**a**) A step pulse on a line generated by connecting the generator at $t = 0$. (**b**) Calculation of the forward waves V^+ and I^+ at the generator at $t = 0$

The equivalent circuit at $t = 0$ is shown in **Figure 16.11b** and is the same as a lumped parameter circuit. The closing of the switch has created a disturbance on the line: The forward wave V^+ now propagates toward the load at the speed of propagation v_p on the line. For a lossless or distortionless line, this speed is always given by **Eq. (16.1)** and is independent of the frequency content of the pulse. For a line of length d, the time of propagation to reach the end of the line is $\Delta t = d/v_p$. After this time, the forward-propagating wave appears at the load. There are three possible conditions that may occur at the load:

(1) Load impedance equals the characteristic impedance: $Z_L = Z_0$. In this case, the reflection coefficient at the load is zero. There is no reflection at the load and the circuit reaches steady state after a time $t = d/v_p$. The line voltage and line current are shown in **Figure 16.12** for three times.

(2) Load impedance greater than Z_0: $Z_L > Z_0$. In this case, the reflection coefficient is positive and, therefore, the reflected voltage wave is in the same direction as the forward-propagating wave. The reflected current at the load is in the direction opposite the forward current as shown in **Eq. (16.3)**.

(3) Load impedance less than Z_0: $Z_L < Z_0$. In this case, the reflection coefficient is negative ($\Gamma_L < 0$). The reflected voltage wave is opposite in polarity compared to the forward voltage wave, and the current is of the same polarity as the forward current wave.

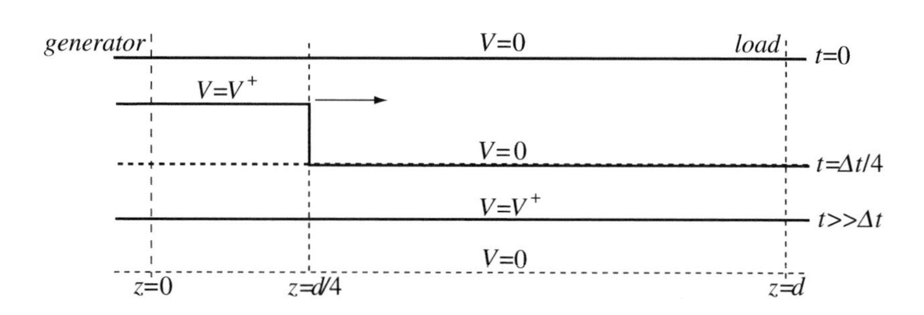

Figure 16.12 Line voltage and current on a line with matched load, at different times and locations

Thus, we can treat cases 2 and 3 in identical fashion using the reflection coefficient, but in actual, numerical calculations, the sign of the reflection coefficient must be taken into account.

After the forward wave reaches the load, it is reflected. We call this the first reflection. The reflected waves are

$$V_1^- = \Gamma_L V^+ \quad [\mathrm{V}], \qquad I_1^- = -\Gamma_L I^+ \quad [\mathrm{A}] \tag{16.21}$$

These two waves propagate back toward the generator as for the narrow pulse, but unlike the narrow pulse situation, the forward-propagating wave still exists on the line (since the pulse is very wide). Thus, the voltage (or current) anywhere on the line is the sum of the forward-propagating wave V^+ and backward-propagating wave V_1^- (I^+ and I_1^- for the current wave). The line voltage and current at any time $\Delta t < t < 2\Delta t$ are

$$V_1 = V^+(1+\Gamma_L) \quad [\mathrm{V}], \quad I_1 = I^+(1-\Gamma_L) \quad [\mathrm{A}], \quad \Delta t < t < 2\Delta t \tag{16.22}$$

After an additional time Δt ($2\Delta t$ from the time the switch was closed), the reflected wave V_1^- reaches the generator. Although the generator has its own voltage, it behaves as a load with load impedance Z_g for the reflected wave. Thus, a reflection coefficient Γ_g exists at the generator, unless $Z_g = Z_0$. For $Z_g \neq Z_0$, the generator reflection coefficient is given in **Eq. (16.9)**. Note that now the forward- and backward-propagating waves have changed roles. This should not be too confusing since the waves reflected from the load propagate backward toward the generator. These waves are reflected at the generator to produce new forward-propagating waves toward the load. These are

$$V_2^+ = \Gamma_g V_1^- = \Gamma_L \Gamma_g V^+ \quad [\mathrm{V}], \quad I_2^+ = -\Gamma_g I^- = \Gamma_L \Gamma_g I^+ \quad [\mathrm{A}] \tag{16.23}$$

The total voltage and current on the line at time $2\Delta t < t < 3\Delta t$ are

$$V_2 = V^+\left(1+\Gamma_L+\Gamma_L\Gamma_g\right) \quad [\mathrm{V}], \quad I_2 = I^+\left(1-\Gamma_L+\Gamma_L\Gamma_g\right) \quad [\mathrm{A}], \quad 2\Delta t < t < 3\Delta t \tag{16.24}$$

After an additional time Δt, the new forward-propagating waves (V_2^+ and I_2^+) reach the load and are reflected again. The new reflected waves, which then propagate backward toward the generator, are

$$V_3^- = \Gamma_L V_2^+ = \Gamma_L^2 \Gamma_g V^+ \quad [\text{V}], \quad I_3^- = -\Gamma_L I_2^- = -\Gamma_L^2 \Gamma_g I^+ \quad [\text{A}] \tag{16.25}$$

and the total line voltage and current are

$$V_3 = V^+\left(1 + \Gamma_L + \Gamma_L \Gamma_g + \Gamma_L^2 \Gamma_g\right) \quad [\text{V}], \quad I_3 = I^+\left(1 - \Gamma_L + \Gamma_L \Gamma_g - \Gamma_L^2 \Gamma_g\right) \quad [\text{A}], \quad 3\Delta t < t < 4\Delta t \tag{16.26}$$

The pattern is now clear: Every reflection adds to (or subtracts from) the previous reflections to produce a total wave. Continuing the pattern, the voltage and current after many reflections may be written as

$$\begin{aligned} V &= V^+\left(1 + \Gamma_L + \Gamma_L \Gamma_g + \Gamma_L^2 \Gamma_g + \Gamma_L^2 \Gamma_g^2 + \Gamma_L^3 \Gamma_g^2 + \ldots\right) \\ &= V^+\left(1 + \Gamma_L \Gamma_g + \Gamma_L^2 \Gamma_g^2 + \Gamma_L^3 \Gamma_g^3 + \ldots\right) + V^+ \Gamma_L \left(1 + \Gamma_L \Gamma_g + \Gamma_L^2 \Gamma_g^2 + \Gamma_L^3 \Gamma_g^3 + \ldots\right) \quad [\text{V}] \end{aligned} \tag{16.27}$$

$$\begin{aligned} I &= I^+\left(1 - \Gamma_L + \Gamma_L \Gamma_g - \Gamma_L^2 \Gamma_g + \Gamma_L^2 \Gamma_g^2 - \Gamma_L^3 \Gamma_g^2 + \ldots\right) \\ &= I^+\left(1 + \Gamma_L \Gamma_g + \Gamma_L^2 \Gamma_g^2 + \Gamma_L^3 \Gamma_g^3 + \ldots\right) - I^+ \Gamma_L \left(1 + \Gamma_L \Gamma_g + \Gamma_L^2 \Gamma_g^2 + \Gamma_L^3 \Gamma_g^3 + \ldots\right) \quad [\text{A}] \end{aligned} \tag{16.28}$$

The term in parentheses is a geometric series (since $|\Gamma_L| < 1$, $|\Gamma_g| < 1$), and for a large number of terms, we get

$$1 + \Gamma_L \Gamma_g + \Gamma_L^2 \Gamma_g^2 + \Gamma_L^3 \Gamma_g^3 + \ldots = \frac{1}{1 - \Gamma_L \Gamma_g}, \quad |\Gamma_L|, |\Gamma_g| < 1 \tag{16.29}$$

Substituting in **Eq. (16.27)**, we get

$$\boxed{V_\infty = V^+ \frac{1}{1 - \Gamma_L \Gamma_g} + V^+ \Gamma_L \frac{1}{1 - \Gamma_L \Gamma_g} = V^+ \frac{1 + \Gamma_L}{1 - \Gamma_L \Gamma_g} \quad [\text{V}]} \tag{16.30}$$

Performing similar operations for I in **Eq. (16.28)**, we get

$$\boxed{I_\infty = I^+ \frac{1 - \Gamma_L}{1 - \Gamma_L \Gamma_g} \quad [\text{A}]} \tag{16.31}$$

where the index indicates an infinite number of reflections (infinite time). This gives the steady-state solution for voltage and current on the line. Substituting for Γ_L and Γ_g from **Eqs. (16.3)** and **(16.9)**, and rearranging terms, we get

$$V_\infty = V^+ \frac{1 + \Gamma_L}{1 - \Gamma_L \Gamma_g} = V^+ \frac{Z_L (Z_0 + Z_g)}{Z_0 (Z_g + Z_L)} \quad [\text{V}] \tag{16.32}$$

Now, substituting for V^+ from **Eq. (16.20)**, we get for the voltage on the line, which is also the voltage on the load at steady state,

$$\boxed{V_\infty = V_g \frac{Z_L}{Z_g + Z_L}} \quad [\text{V}] \tag{16.33}$$

This is the steady-state solution for the circuit, as required. Similarly, for the current in the circuit (load), we get the steady-state solution as

$$\boxed{I_\infty = \frac{V_g}{Z_g + Z_L}} \quad [\text{A}] \tag{16.34}$$

Although the method is simple and intuitive, it is rather lengthy, except for the steady-state solution. However, it is possible to reduce the method into a simple diagram which may be viewed as a tool for keeping track of the various reflections that occur. The diagram is called a reflection diagram (also called a bounce or Bewley diagram) and is shown in **Figures 16.13** through **16.15**. The method consists of the following:

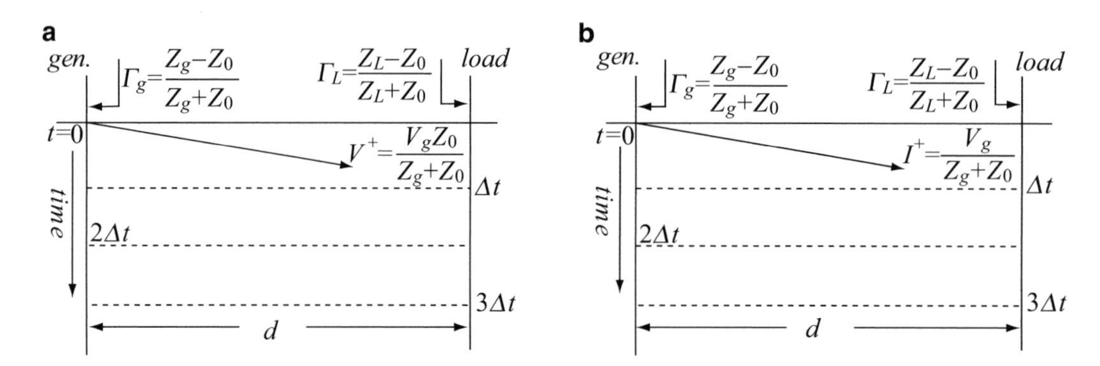

Figure 16.13 Preparatory steps in the reflection diagram. (**a**) Voltage reflection diagram. (**b**) Current reflection diagram

(1) The generator and load are replaced by two perpendicular lines separated a distance d apart. The horizontal distance represents location on the line, and the vertical axis represents time with $t = 0$, usually at the generator. The reflection coefficient at the generator (looking from the line into the generator) is placed on the left vertical line, whereas the reflection coefficient at the load (looking from the line into the load) is placed on the right vertical line. The same applies to the current diagram. These considerations are shown in **Figure 16.13**.

(2) Time is indicated along the lines starting from top to bottom in increments of $2\Delta t$. The left line is marked 0, $2\Delta t$, $4\Delta t$, $6\Delta t$, etc. The right line is marked Δt, $3\Delta t$, $5\Delta t$, $7\Delta t$, etc. This conforms with the above notation and indicates that a wave propagates between generator and load or vice versa in a time $\Delta t = d/v_p$.

(3) The initial voltage and current, at time $t = 0$, are calculated from **Eq. (16.20)**. These are marked at time $t = 0$ on the diagram, pointing toward the load as shown in **Figure 16.13**.

(4) The foregoing steps give the initial or preparatory steps. Now, we allow the initial waves to propagate, and each encounter with a reflection coefficient multiplies the wave by that reflection coefficient **[Eq. (16.21)]** and changes the direction of propagation. **Figure 16.14** shows a few steps in the diagram. All odd-numbered reflections occur at the load; all even-numbered reflections occur at the generator.

(5) To calculate the voltage or current at any point on the line and at any time, we proceed by marking the location at which the values are required. For example, suppose we wish to calculate the line voltage and line current at point z_0 in **Figure 16.14**. A line parallel to the load or generator line is drawn at $z = z_0$. This line shows the voltage or current at any point in time from zero (top) to infinity (bottom). The line $z = z_0$ intersects the reflected voltages and currents at times t_1, t_2, t_3, etc., as shown. The line voltage and current are shown in **Figure 16.15**. Note that in this figure, both Γ_L and Γ_g are assumed to be positive. Thus, the voltage at z_0 increases in diminishing steps. The values of voltage or current remain constant between two reflections, until an additional reflection reaches the same point. Note the alternating signs of the current.

(6) The voltage or current at any given time at a given point between generator and load is calculated by summing up all reflections for all times up to the required time, at the required point. As an example, the voltage and current at time $t = t_0$ at $z = z_0$ in **Figure 16.14** is the sum of the first four reflections and the initial voltage. In this case,

$$V_0 = V^+\left(1 + \Gamma_L + \Gamma_L\Gamma_g + \Gamma_L^2\Gamma_g + \Gamma_L^2\Gamma_g^2\right) \text{ [V]}, \quad I_0 = I^+\left(1 - \Gamma_L + \Gamma_L\Gamma_g - \Gamma_L^2\Gamma_g + \Gamma_L^2\Gamma_g^2\right) \text{ [A]} \qquad (16.35)$$

These values are shown in **Figure 16.15**.

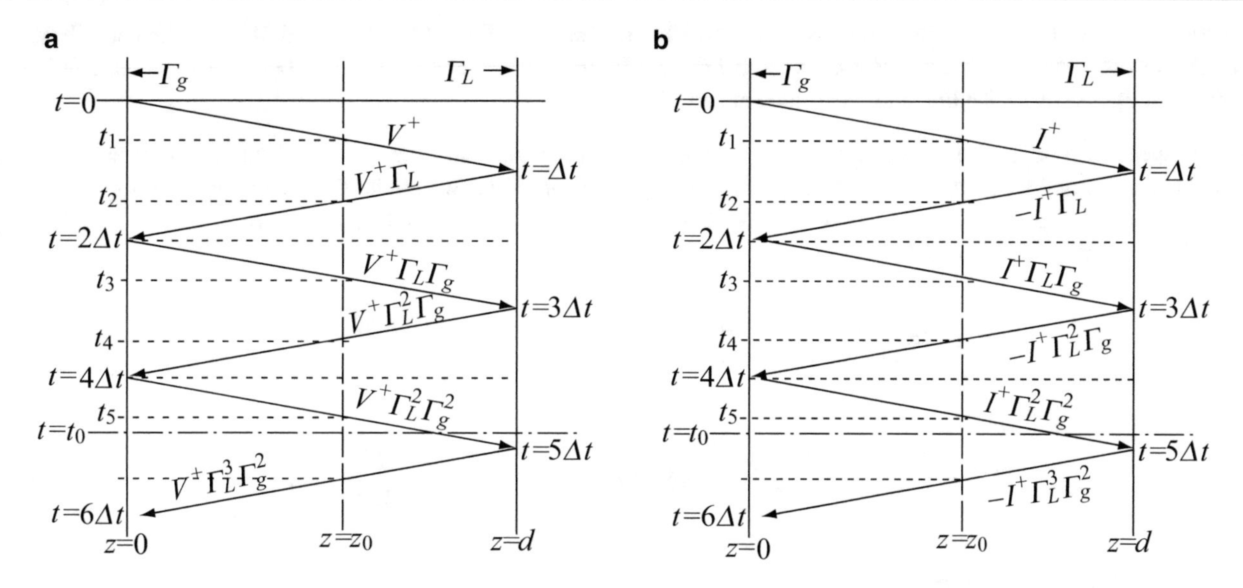

Figure 16.14 (**a**) The voltage reflection diagram for a general transmission line with reflection coefficients Γ_L and Γ_g. (**b**) The current reflection diagram for the conditions in (**a**)

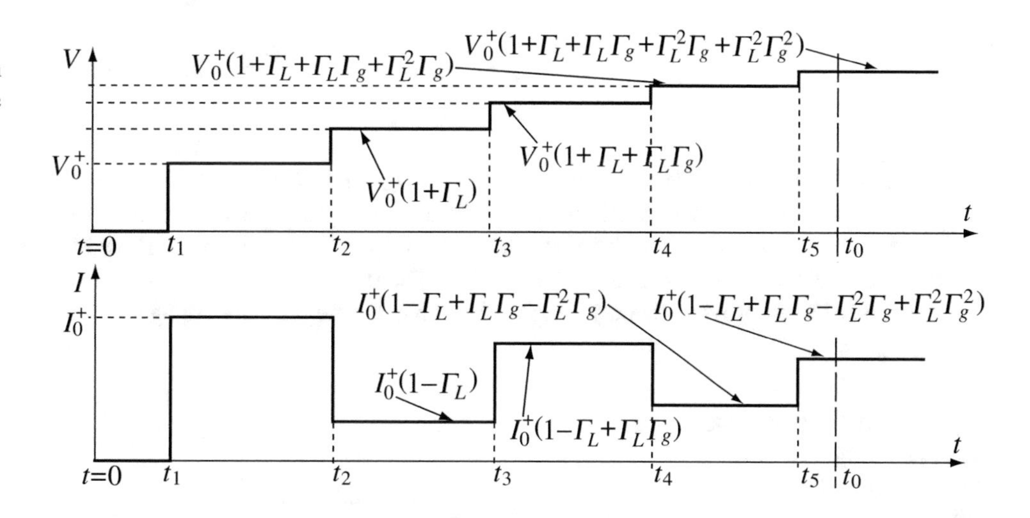

Figure 16.15 Voltage and current on the line at a given location as a function of time

Example 16.4 A lossless transmission line is connected as shown in **Figure 16.16**. The inductance per unit length of the line is 5 μH/m, and the capacitance per unit length is 5 pF/m. The switch is closed at $t = 0$. Calculate:

(a) The steady-state voltage and current on the line.
(b) The voltage at the load as measured by an oscilloscope between $t = 0$ and $t = 3$ μs.
(c) The current midway between generator and load as measured between $t = 0$ and $t = 3$ μs.

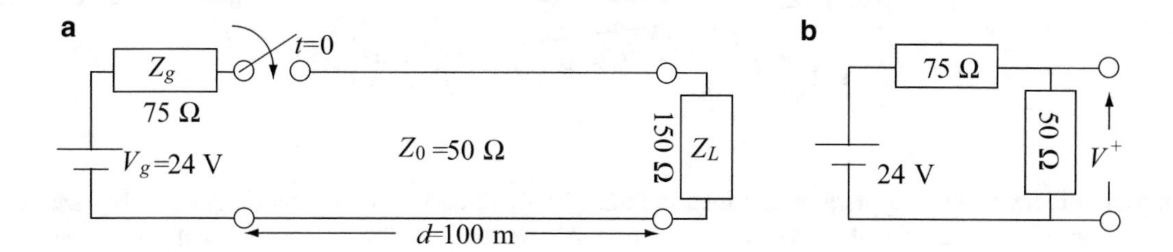

Figure 16.16 A transmission line on which the generator is switched on at $t = 0$

Solution: For steady state, we can either use **Eqs. (16.30)** and **(16.31)** or **Eqs. (16.33)** and **(16.34)**. The former will be used here. As for the transient solution, we use **Eqs. (16.27)** and **(16.28)** with the appropriate number of reflections. The latter is found from the length of the line and speed of propagation:

(a) The speed of propagation on the line is $v_p = 1/\sqrt{LC} = 2 \times 10^8$ m/s. Thus, the time required for propagation between the generator and load is 0.5 μs. To calculate the steady-state solution and to build the reflection diagram, we need the reflection coefficients at the load and generator (looking into the load or generator, respectively) and the initial voltage and current at $t = 0$ (V^+ and I^+). These are

$$\Gamma_L = \frac{Z_L - Z_0}{Z_L + Z_0} = \frac{150 - 50}{150 + 50} = 0.5, \quad \Gamma_g = \frac{Z_g - Z_0}{Z_g + Z_0} = \frac{75 - 50}{75 + 50} = 0.2$$

$$V^+ = \frac{V_0 Z_0}{Z_g + Z_0} = \frac{24 \times 50}{125} = 9.6 \quad [\text{V}], \quad I^+ = \frac{V_0}{Z_g + Z_0} = \frac{24}{125} = 0.192 \quad [\text{A}]$$

The steady-state solution is

$$V_\infty = V^+ \frac{1 + \Gamma_L}{1 - \Gamma_L \Gamma_g} = 9.6 \times \frac{1 + 0.5}{1 - 0.5 \times 0.2} = 16 \quad [\text{V}],$$

$$I_\infty = I^+ \frac{1 - \Gamma_L}{1 - \Gamma_L \Gamma_g} = 0.192 \times \frac{0.5}{0.9} = 0.1067 \quad [\text{A}].$$

(b) The reflection diagram for voltages is now as in **Figure 16.17a**, where the first few reflections are shown. The time $t = 3$ μs is shown as a horizontal line. The voltage at the load is the sum of all values at the load from $t = 0$ to $t = 3$ μs since all remain on the line indefinitely (the pulse is very long). These are shown in **Figure 16.17b**. Note the way the diagram is drawn in comparison to **Figure 16.15**. The steady state in this case is reached quite fast. At $t = 3$ μs, the load voltage is 15.984 V which is only 16 mV lower than the steady state voltage.

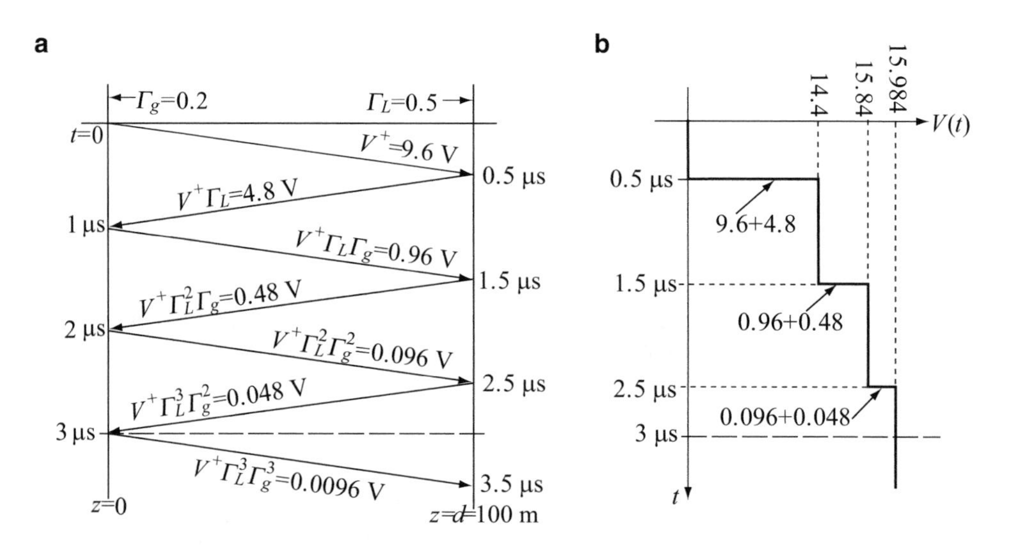

Figure 16.17 (**a**) Voltage reflection diagram for the line in **Figure 16.16**. (**b**) Voltage at the load in **Figure 16.16**

(c) The current midway between generator and load is found from the current reflection diagram in **Figure 16.18a**. The horizontal line at $t = 3$ μs and the vertical line at $z = d/2$ are shown. The plot of current with time is shown in

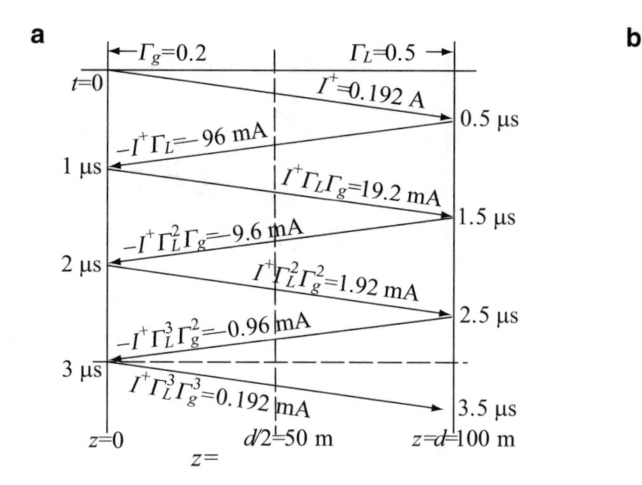

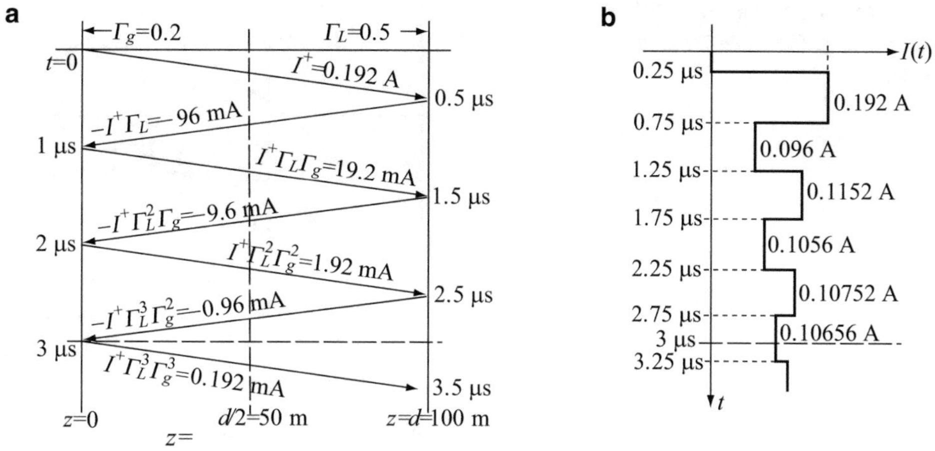

Figure 16.18 (**a**) Current reflection diagram for **Figure 16.16**. (**b**) Current midway between load and generator in **Figure 16.16**

Figure 16.18b. Note that the current is zero for the first 0.25 μs. Then, it remains constant for 0.5 μs until the reflected wave reaches this point again, and so on. The current at $t = 3$ μs is 0.10656 A. The line is almost at steady state.

16.5 Transients on Transmission Lines: Finite-Length Pulses

In the preceding two sections, we discussed the behavior of two types of pulses. One was a very short pulse and the second was very long. If, instead, a finite-width pulse is prescribed, we can use the superposition of solutions we already obtained to calculate the transmission line response to the pulse. A method of obtaining a pulse of width T is shown in **Figure 16.19**. In essence, we create a finite duration pulse as a superposition of two step functions. The first step function is applied at a time $t = 0$ and the second is applied at a time $t + T$. This, of course, is done so that we may use the solutions in the previous section. Each step function is evaluated separately, and the results are added to obtain the pulse response. The additional important point is to displace the second step function by a time T to ensure that the correct pulse width is created. This method can be extended to almost any pulse shape, although the method may be lengthy. For example, a triangular pulse may be approximated by any number of steps. If the steps are small and a large number of steps are used, the pulse may be approximated quite accurately. The approximation for a triangular pulse is shown in **Figure 16.20**, using four steps on the rising edge and four steps on the falling edge. The first four pulses are exactly the same, but the first pulse starts at $t_0 + T/16$ and each subsequent pulse is displaced an additional $T/8$. The net effect is a narrowing of the pulse compared to the actual triangular pulse, but this is of minor concern since we can decrease this narrowing by increasing the number of pulses we use. The last four pulses are the same in magnitude but are negative. The following example shows how this method is applied.

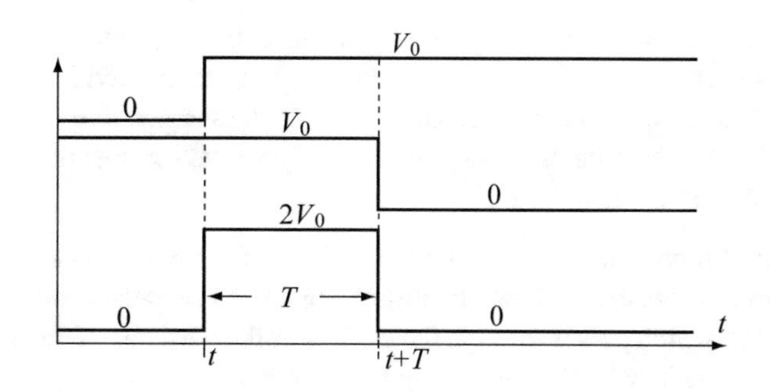

Figure 16.19 The superposition of two shifted step pulses results in a finite duration pulse

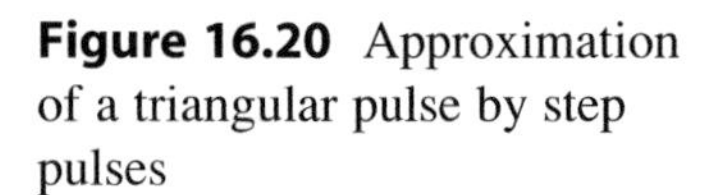

Figure 16.20 Approximation of a triangular pulse by step pulses

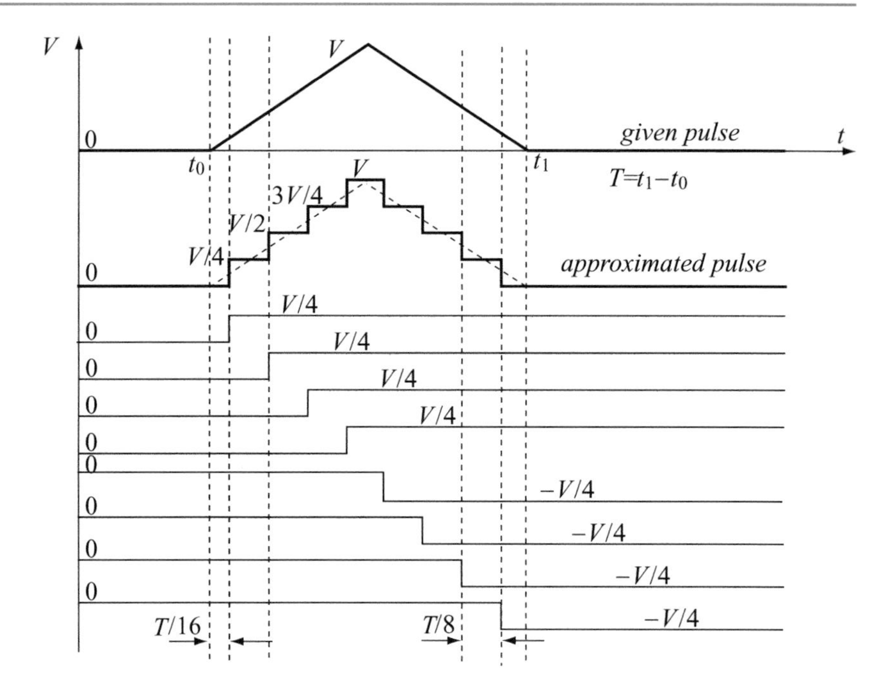

Example 16.5 Transient Due to a Triangular Pulse The transmission line in **Figure 16.21a** is driven with a single triangular pulse as shown. The speed of propagation on the line is 10^8 m/s:

(a) Find the current in the load at all times between $t = 0$ and $t = 50$ μs.
(b) Find the steady-state voltage on the line.

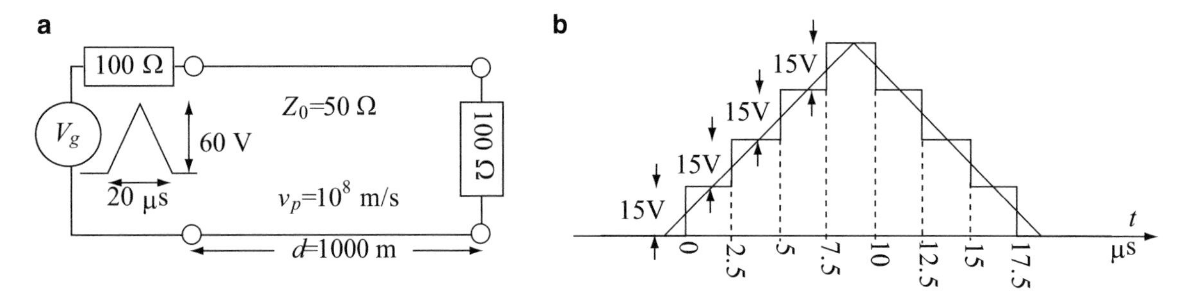

Figure 16.21 (**a**) A transmission line driven by a single triangular pulse. (**b**) Representation of the triangular pulse by step increments/decrements in voltage

Solution: To solve the problem, we can divide the pulse into any number of steps. The larger the number of steps, the better the approximation to the exact solution. Here, we choose to divide the pulse into four steps on each slope, as in **Figure 16.21b**. The solution is then a superposition of four positive steps and four negative steps, each of magnitude 15 V/150 Ω = 0.1 A. The reflection diagram for one positive (or negative) step is shown in **Figure 16.22a**. The reflection coefficients are shown on the diagram:

(a) The solution involves some approximations. The most obvious is the use of the finite number of steps. The second approximation necessary is shown in **Figure 16.21b**. The pulses are chosen such that they approximate the original triangular pulse which passes through the centers of the vertical and horizontal lines forming the pulse. The width of the approximate pulse is only 17.5 μs with each pulse displaced 2.5 μs with respect to the other. Also, the first pulse starts 1.25 μs from the time the true triangular pulse starts, but, in the interest of simplicity, we start the first pulse at $t = 0$.

From the diagram in **Figure 16.22a**, the current in the load is calculated and shown in **Figure 16.22b** for the first step. Note that the first jump occurs at $t = 10$ μs and is equal to $0.1 - 0.1/3 = 0.0667$ A. The second jump at $t = 30$ μs adds $0.1/9 - 0.1/27 = 0.0074$ A. The remaining three pulses are the same, but are displaced to the right by 2.5 μs each. Similarly, the negative pulses are identical in form but negative, and they are also displaced by 2.5 μs each with respect to the previous pulse. If we draw the eight pulses with the proper shift in time, we get the result in **Figure 16.23**. The result is the sum of all eight pulses and is shown at the bottom of the diagram. Note, in particular, the multiple pulses produced by the multiple reflections. These pulses die out with time.

(b) The steady-state voltage on the line is zero. This can be seen from **Figure 16.23**. The steady-state response to each step is identical except for signs. There are four positive responses and four negative responses. Their sum is zero; that is, the pulse is eventually dissipated.

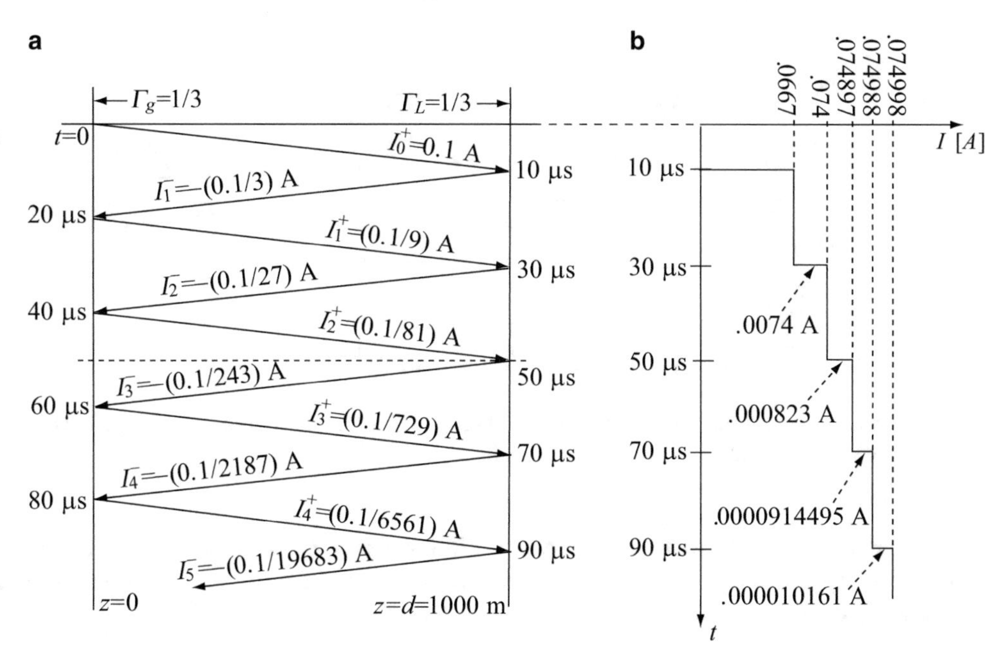

Figure 16.22 (**a**) Current reflection diagram for the first step in **Figure 16.21b**. (**b**) Current at the load due to the first pulse in **Figure 16.21b**

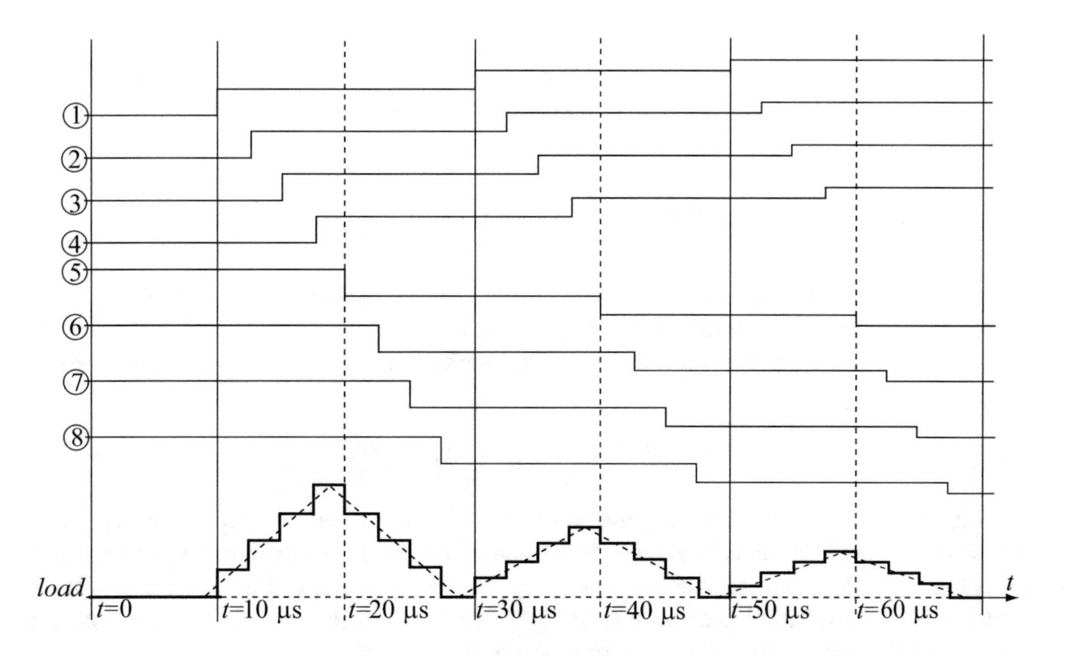

Figure 16.23 Superposition of the responses of the eight pulses that make up the triangular signal at the load

16.6 Reflections from Discontinuities

A discontinuity on a line is any condition that changes the impedance on the line. For example, the connecting point between two lines of different characteristic impedances is a discontinuity that will cause a reflection at the point of discontinuity. Similarly, a matched line on which a load has been connected somewhere on the line becomes a discontinuous line. These two situations are shown in **Figures 16.24a** and **16.24b**. A similar situation is caused by connecting more than one transmission line at the end of a transmission line as shown in **Figure 16.24c**. The introduction of a discontinuity causes both reflection and transmission of waves at the discontinuity as well as at any other location at which there is a mismatch in impedance. To understand the behavior of the transient waves in the presence of a discontinuity, consider **Figure 16.24a**. The waves are found as for the mismatched load in **Section 16.4**, but now we have three locations to deal with: load, generator, and discontinuity. If there is more than one discontinuity, each discontinuity must be treated separately.

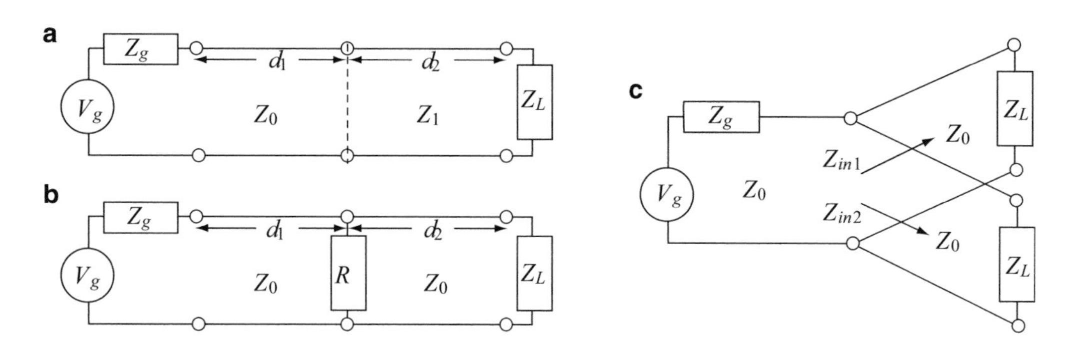

Figure 16.24 Discontinues on transmission lines. (**a**) Due to connection of two lines. (**b**) Due to connection of additional loads on the lines. (**c**) Due to a distribution point

To understand how the waves behave, we will follow the propagation of waves in **Figure 16.24a** and draw the reflection diagram as we go along. For simplicity, we assume that the generator is matched ($Z_g = Z_0$). Therefore, the forward-propagating wave launched by the generator at time $t = 0$ is

$$V_0^+ = \frac{V_g Z_0}{Z_0 + Z_0} = \frac{V_g}{2} \quad [\text{V}] \tag{16.36}$$

This wave propagates at any point on line 1 at a speed of propagation v_{p1}. After a time $\Delta t_1 = d_1/v_{p1}$, the wave reaches the discontinuity. Part of the wave is reflected and part of it is transmitted with the reflection and transmission coefficients Γ_{12} and T_{12}, respectively:

$$\Gamma_{12} = \frac{Z_1 - Z_0}{Z_1 + Z_0}, \quad T_{12} = \frac{2Z_1}{Z_1 + Z_0} \tag{16.37}$$

The reflection coefficient Γ_{12} is the reflection coefficient at the interface between line 1 and line 2, and the transmission coefficient indicates the transmission from line 1 to line 2. These two coefficients are shown in **Figure 16.25**, where the arrows indicate the direction of the waves being reflected and transmitted. The reflected and transmitted voltage waves at d_1 are

$$V_1^- = V_0^+ \Gamma_{12}, \quad V_1^+ = V_0^+ T_{12} \quad [\text{V}] \tag{16.38}$$

The reflected wave V_1^- propagates back to the generator and reaches the generator after a time Δt_1. Since the reflection coefficient at the generator is zero, no additional reflections occur at this point. The wave transmitted across the discontinuity, V_1^+, propagates toward the load at a speed of propagation v_{p2} and reaches the load after an additional time $\Delta t_2 = d_2/v_{p2}$. At the load, the wave is partly reflected and partly transmitted into the load (where it is dissipated or, in the case of an antenna, radiated). The reflection and transmission coefficients at the load are

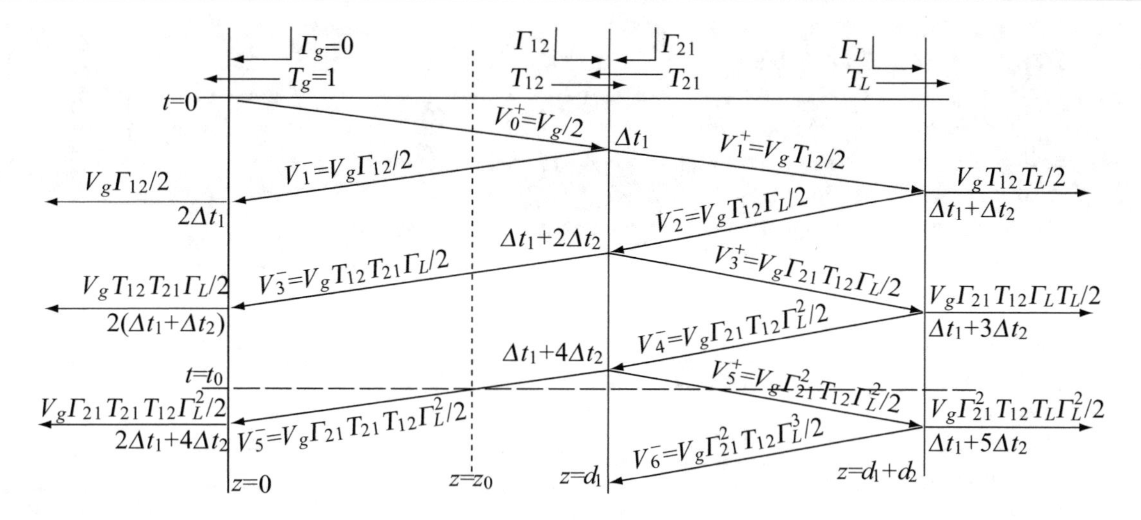

Figure 16.25 Voltage reflection diagram for the line in **Figure 16.24a**, with $Z_g = Z_0$

$$\Gamma_L = \frac{Z_L - Z_1}{Z_L + Z_1}, \quad T_L = \frac{2Z_L}{Z_L + Z_1} \tag{16.39}$$

Thus, the reflected and transmitted waves are

$$V_2^- = V_2^+ \Gamma_L = V_0^+ T_{12} \Gamma_L, \quad V_{L1}^+ = V_0^+ T_{12} T_L \quad [\mathrm{V}] \tag{16.40}$$

V_2^- propagates back toward the discontinuity, which it reaches after an additional time Δt_2. At the discontinuity, there will be a reflected and transmitted wave, but since the wave reaches the discontinuity from line 2, the reflection and transmission coefficients are different. These are denoted Γ_{21} and T_{21}:

$$\Gamma_{21} = \frac{Z_0 - Z_1}{Z_1 + Z_0}, \quad T_{21} = \frac{2Z_0}{Z_1 + Z_0} \tag{16.41}$$

The reflected wave (into line 2) and the transmitted wave (from line 2 into line 1) are

$$V_3^+ = V_2^- \Gamma_{21} = V_0^+ T_{12} \Gamma_L \Gamma_{21}, \quad V_3^- = V_0^+ T_{12} \Gamma_L T_{21} \quad [\mathrm{V}] \tag{16.42}$$

Now, these two waves propagate in opposite directions. V_3^+ propagates toward the load whereas V_3^- propagates toward the generator. The sequence repeats itself indefinitely. A few reflections are shown in **Figure 16.25**, together with the definitions of reflection and transmission coefficients at the various locations.

All other aspects of propagation remain as discussed in **Section 16.4**. Note, in particular, the times at which the waves reach various locations on the line. The main difficulty in treating discontinuities is in keeping track of the increasing number of reflections and transmissions and the associated times. We note also that the reflection and transmission coefficients at the discontinuity depend on the direction of propagation. The following relations hold:

$$\Gamma_{21} = -\Gamma_{12}, \quad T_{21} = 1 - \Gamma_{12} \tag{16.43}$$

and these can be obtained from **Eqs. (16.37)** and **(16.41)**. Once the diagrams are defined, the waves at any location on the line may be found as previously, by finding the intersection of the time line and position line (t_0 and z_0 in **Figure 16.25**) and summing all terms up to that time along the time line. These aspects of calculation are demonstrated in **Example 16.6**. Clearly, an essentially identical process applies to the current diagram.

Example 16.6 Application: Line Patching A segment of a lossless transmission line of finite length $d = 100$ m and characteristic impedance $Z_2 = 75\ \Omega$ is connected between two infinite lossless lines, each with characteristic impedance $Z_1 = Z_3 = 50\ \Omega$ as a temporary fix until the proper line can be obtained, as shown in **Figure 16.26**. A step voltage V_0 arrives at the connection between lines 1 and 2 at $t = 0$ from the left. The speed of propagation on the lines is $v_p = 10^8$ m/s. With the properties given in the figure, calculate the voltage on each line at $t = 5.8$ μs. In lines 1 and 3, calculate the voltage at the discontinuity. In line 2, calculate it midway.

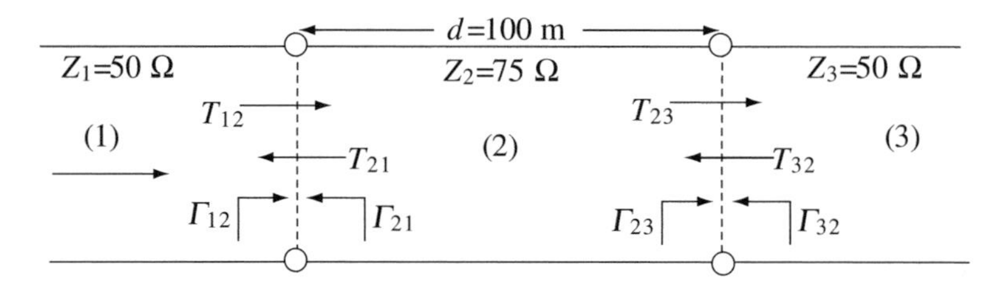

Figure 16.26 A finite transmission line segment connected between two infinite lines. The various reflection and transmission coefficients are shown

Solution: In the two infinite lines, there can be no reflections except at the two connections shown. At the discontinuities there are two reflection coefficients and two transmission coefficients as shown in **Figures 16.26** and **16.27**. The latter figure also shows the first few reflected and transmitted waves at both discontinuities. These are the only waves possible. To find the wave on each line at a given time, the time and position lines are drawn, shown as dashed lines in **Figure 16.27**, and the terms up to the given time and position are summed up.

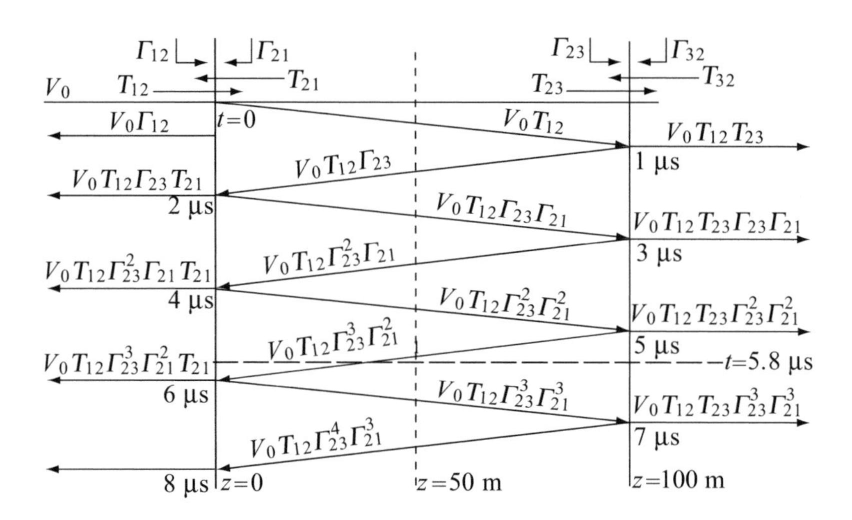

Figure 16.27 Voltage reflection diagram for the line in **Figure 16.26**

On line 1 (immediately to the left of the connection ($z = 0^-$)):

$$V_{t=5.8s} = V_0[1 + \Gamma_{12} + T_{12}T_{21}\Gamma_{23}(1 + \Gamma_{23}\Gamma_{21})] \quad [\text{V}]$$

On line 2 (at the center of the line ($z = 50$ m)):

$$V_{t=5.8s} = V_0T_{12}\left[1 + \Gamma_{23} + \Gamma_{23}\Gamma_{21} + \Gamma_{23}^2\Gamma_{21} + \Gamma_{23}^2\Gamma_{21}^2 + \Gamma_{23}^3\Gamma_{21}^2\right] \quad [\text{V}]$$

On line 3 (immediately to the right of the connection ($z = 100\ \text{m}^+$)):

$$V_{t=5.8s} = V_0T_{23}T_{12}\left[1 + \Gamma_{23}\Gamma_{21} + \Gamma_{23}^2\Gamma_{21}^2)\right] \quad [\text{V}]$$

The various reflection and transmission coefficients needed are

$$\Gamma_{12}=\frac{Z_2-Z_1}{Z_2+Z_1}=\frac{75-50}{75+50}=0.2,\qquad T_{12}=\frac{2Z_2}{Z_2+Z_1}=\frac{150}{125}=1.2,$$

$$\Gamma_{21}=-\Gamma_{12}=-0.2,\qquad T_{21}=\frac{2Z_1}{Z_2+Z_1}=\frac{100}{125}=0.8,$$

$$\Gamma_{23}=\frac{Z_3-Z_2}{Z_3+Z_2}=\frac{50-75}{50+75}=-0.2,\qquad T_{23}=\frac{2Z_3}{Z_3+Z_2}=\frac{100}{125}=0.8,$$

The voltages are as follows:
In line 1, immediately to the left of the discontinuity:

$$V_{t=5.8s}=V_0[1+0.2-1.2\times0.8\times0.2\times(1+0.04)]=1.00032V_0\quad[\mathrm{V}]$$

In line 2, at the center of the line:

$$V_{t=5.8s}=V_0\times1.2\times[1-0.2+0.2\times0.2-0.04\times0.2+0.04\times0.04-0.008\times0.04]=0.999936V_0\quad[\mathrm{V}]$$

In line 3, immediately to the right of the discontinuity:

$$V_{t=5.8s}=V_0\times0.8\times1.2\times[1+0.2\times0.2+0.04\times0.04]=0.999936V_0\quad[\mathrm{V}]$$

Exercise 16.3

(a) Calculate the steady-state voltage on the three lines in **Example 16.6** using the general coefficients.
(b) With the constants found in **Example 16.6**, show that the steady-state voltages are equal to V_0.

Answer

(a)

$$V_1=V_0\left[1+\Gamma_{12}+\frac{\Gamma_{23}T_{12}T_{21}}{1-\Gamma_{23}\Gamma_{21}}\right],\quad V_2=V_0T_{12}\frac{1+\Gamma_{23}}{1-\Gamma_{23}\Gamma_{21}},\quad V_3=V_0\frac{T_{23}T_{12}}{1-\Gamma_{23}\Gamma_{21}}\quad[\mathrm{V}].$$

16.7 Transients on Lines with Reactive Loading

The transient representation in the previous section was based on the concept of reflection and the reflection coefficient. The reflection coefficient is only properly defined if the reflected wave is directly proportional to the forward-propagating wave. In other words, to calculate the reflection coefficient, we assumed that $V^-=\Gamma V^+$. If, however, the reflected wave depends on the forward wave's amplitude in a nonlinear fashion, then the reflection coefficient is not a constant and the method of the previous sections cannot be used. As an example, suppose that a line is terminated with a nonlinear resistor, whose resistance depends on the line voltage as

$$Z_L=R_0\left(1+kV^2\right)\quad[\Omega]\tag{16.44}$$

where V is the total voltage on the load. Assuming the characteristic impedance of the line is $Z_0=R_0$, the reflection coefficient is

$$\Gamma_L=\frac{R_0\left(1+kV^2\right)-R_0}{R_0\left(1+kV^2\right)+R_0}=\frac{kV^2}{kV^2+2}\tag{16.45}$$

This reflection coefficient cannot be used in the relations in **Sections 16.2** through **16.6** because it is not a constant. Thus, we must resort to other means when trying to find the transients on the line. Note that if we had a method of evaluating the voltage in **Eq. (16.45)**, then Γ_L could be evaluated and the methods of the previous section would apply. Thus, the basic method is to calculate the forward-propagating wave and, from this, to calculate the reflected wave without resorting to the use of the reflection coefficient. To see how this is done, we consider two situations: the first deals with capacitive loading and the second with inductive loading.

16.7.1 Capacitive Loading

Consider a transmission line with characteristic impedance $Z_0 = R_0$ connected to a generator with internal impedance $Z_g = R_g$ and a capacitor as a load as shown in **Figure 16.28**.

The calculation starts by calculating the forward-propagating wave, as in **Eq. (16.20)**. The initial voltage and current on the line (immediately after closing the switch) are

$$V^+ = V_g \frac{R_0}{R_0 + R_g}, \quad I^+ = \frac{V_g}{R_0 + R_g} \quad [\mathrm{V}] \tag{16.46}$$

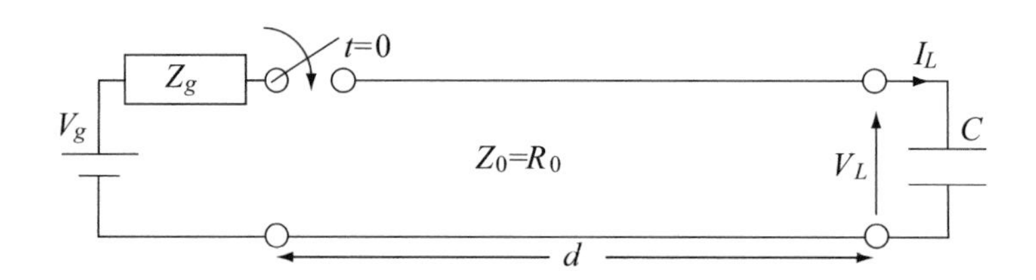

Figure 16.28 A capacitively loaded transmission line

These waves propagate toward the load at a speed v_p defined by the line parameters. At the load, however, the reflected voltage and current must be calculated from the differential equation relating current and voltage for a capacitor, because a reflection coefficient based on impedances cannot be used:

$$i_L(t) = C\frac{d}{dt}(v_L(t)) \quad [\mathrm{A}] \tag{16.47}$$

where $v_L(t)$ is the total voltage at the load. Note also that this voltage is time dependent, whereas V^+ is a constant voltage, and that $i_L(t)$ only exists after a time $t \geq \Delta t$. We can also write at the load the general relations

$$v_L(t) = V^+ + V^-(t) \quad [\mathrm{V}], \quad i_L(t) = \frac{V^+ - V^-(t)}{R_0} \quad [\mathrm{A}] \tag{16.48}$$

Solving for $i_L(t)$,

$$i_L(t) = \frac{2V^+ - v_L(t)}{R_0} \quad [\mathrm{A}] \tag{16.49}$$

Substituting this in **Eq. (16.47)** and rearranging terms gives

$$C\frac{d}{dt}(v_L(t)) + \frac{1}{R_0}v_L(t) - \frac{2V^+}{R_0} = 0 \tag{16.50}$$

Since V^+ is known from **Eq. (16.46)**, we can solve this differential equation for any time $t \geq \Delta t$. The solution gives the voltage at the load:

$$\boxed{v_L(t) = 2V^+\left(1 - e^{-(t-\Delta t)/R_0C}\right) = \frac{2V_gR_0}{R_0 + R_g}\left(1 - e^{-(t-\Delta t)/R_0C}\right) \quad [\text{V}], \quad t \geq \Delta t} \tag{16.51}$$

The current in the load is

$$\boxed{i_L(t) = \frac{2V^+ - v_L(t)}{R_0} = \frac{2V^+e^{-(t-\Delta t)/R_0C}}{R_0} = \frac{2V_ge^{-(t-\Delta t)/R_0C}}{R_0 + R_g} \quad [\text{A}], \quad t \geq \Delta t} \tag{16.52}$$

Now, the reflected voltage and current waves can be calculated from **Eq. (16.48)**:

$$V_1^-(t) = V^+\left(1 - 2e^{-(t-\Delta t)/R_0C}\right) = \frac{V_gR_0}{R_0 + R_g}\left(1 - 2e^{-(t-\Delta t)/R_0C}\right) \quad [\text{V}] \tag{16.53}$$

$$I_1^-(t) = -\frac{V^-(t)}{R_0} = -\frac{V^+}{R_0}\left(1 - 2e^{-(t-\Delta t)/R_0C}\right) = -\frac{V_g}{R_0 + R_g}\left(1 - 2e^{-(t-\Delta t)/R_0C}\right) \quad [\text{A}] \tag{16.54}$$

The total voltage and current at any point on the line are given by the sum of the forward- and backward-propagating waves. The forward, reflected, and total voltages on the line are shown in **Figure 16.29a**. The load voltage and current are shown in **Figure 16.29b**.

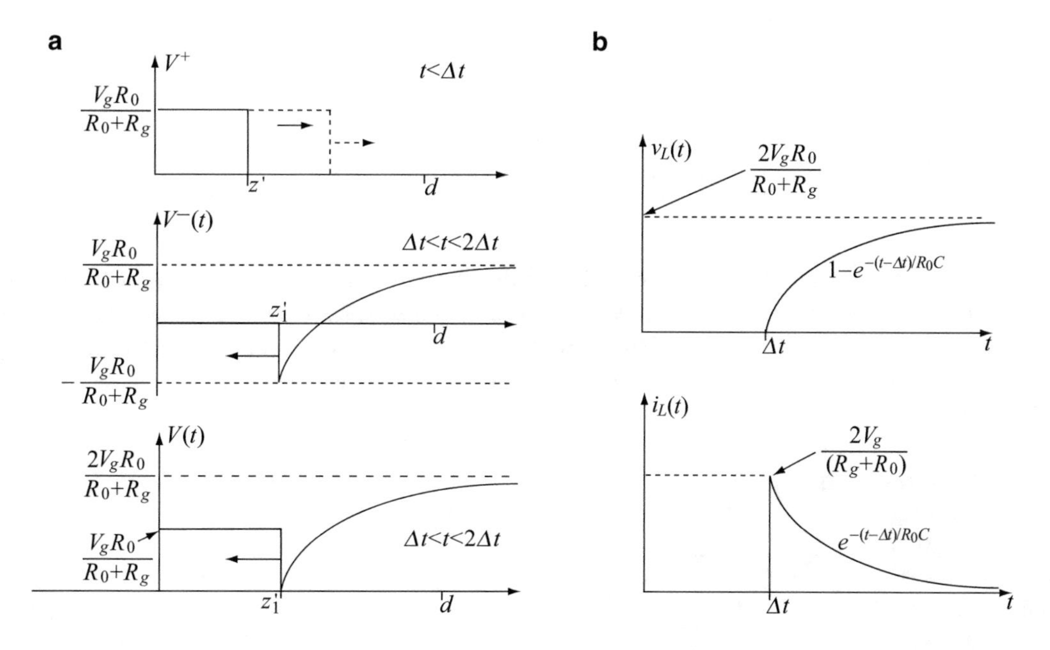

Figure 16.29 (**a**) Forward, reflected, and total voltage on the line in **Figure 16.28**. (**b**) Load voltage and current for the line in **Figure 16.28**

The backward-propagating waves in **Eqs. (16.53)** and **(16.54)** propagate toward the generator. If the generator is matched to the line (i.e., $R_g = R_0$), there will be no reflection at the generator and the solutions in **Eqs. (16.51)** and **(16.52)** apply. If, on the other hand, the generator is not matched, there will be a reflection at the generator as well, but because the generator impedance is resistive, the reflection coefficient Γ_g can be used as in the previous cases. A new forward-propagating wave is obtained which again travels toward the load, and the above steps are repeated. Any number of reflections may be considered in this way, and a steady state is achieved only after a large (infinite) number of reflections have occurred.

16.7.2 Inductive Loading

If an inductor replaces the capacitor in **Figure 16.28**, the treatment is similar except that now the basic equation to deal with is

$$v_L(t) = L\frac{d}{dt}(i_L(t)) \quad [\text{V}] \tag{16.55}$$

All other aspects, including the relations at the load **[Eqs. (16.48)** and **(16.49)]** and the forward-propagating wave, are the same as for the capacitive load.

The differential equation to solve at the load is now

$$L\frac{d}{dt}(i_L(t)) + R_0 i_L(t) - 2V^+ = 0 \tag{16.56}$$

This gives the current at the load as

$$\boxed{i_L(t) = \frac{2V^+}{R_0}\left(1 - e^{-(t-\Delta t)R_0/L}\right) = \frac{2V_g}{R_0 + R_g}\left(1 - e^{-(t-\Delta t)R_0/L}\right) \quad [\text{A}], \quad t \geq \Delta t} \tag{16.57}$$

and the voltage as

$$\boxed{v_L(t) = 2V^+ e^{-(t-\Delta t)R_0/L} = \frac{2V_g R_0}{R_0 + R_g} e^{-(t-\Delta t)R_0/L} \quad [\text{V}], \quad t \geq \Delta t} \tag{16.58}$$

The reflected voltage and current are

$$V_1^-(t) = v_L(t) - V^+ = V^+\left(2e^{-(t-\Delta t)R_0/L} - 1\right) = \frac{V_g R_0}{R_0 + R_g}\left(2e^{-(t-\Delta t)R_0/L} - 1\right) \quad [\text{V}] \tag{16.59}$$

$$I_1^-(t) = -\frac{v_L(t) - V^+}{R_0} = -\frac{V_g}{R_0 + R_g}\left(2e^{-(t-\Delta t)R_0/L} - 1\right) \quad [\text{A}] \tag{16.60}$$

Figure 16.30a shows these relations and their variation on the line and with time, and **Figure 16.30b** shows the voltage and current at the load. As was the case with the capacitive loading in the previous section, if the generator is matched, the results here describe the behavior of the line at all times. If the generator is not matched, the above behavior only applies up to a time $t = 2\Delta t$. At this time, the backward-propagating wave reaches the generator and is reflected, generating a new forward-propagating wave.

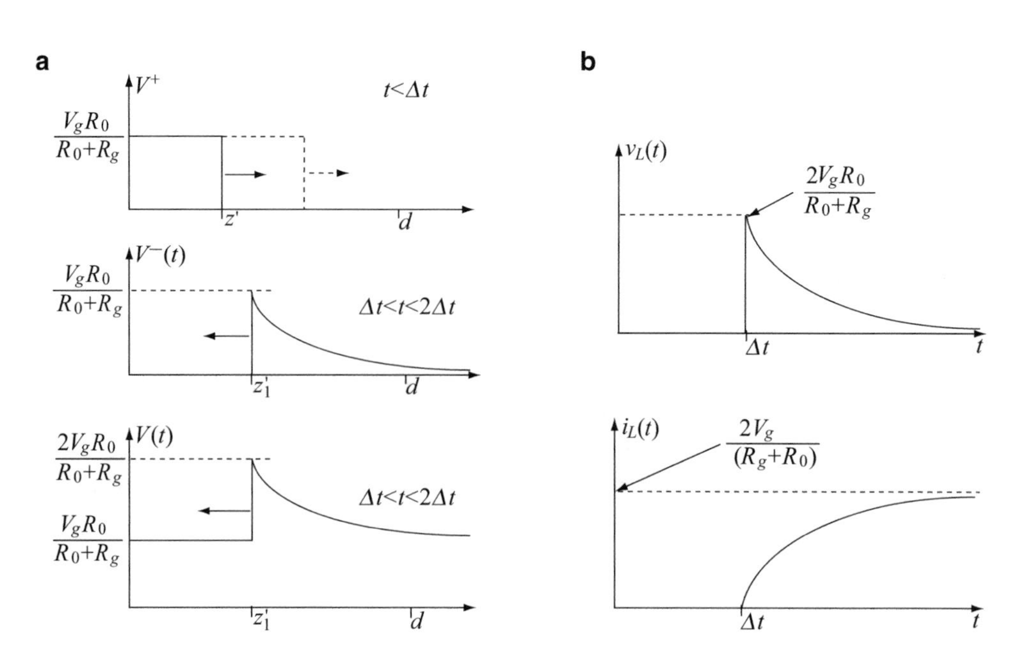

Figure 16.30 (**a**) Forward, reflected, and total voltage on the line in **Figure 16.28** after the capacitance was replaced with an inductance L. (**b**) Load voltage and current on the line for the conditions in (**a**)

Example 16.7 A transmission line with matched generator is 120 m long and terminated by a capacitor as shown in **Figure 16.31**. The characteristic impedance of the line is $Z_0 = 50\ \Omega$, the load capacitance is $C = 100$ pF, and the speed of propagation on the line is c [m/s]. Calculate the voltage at the load for all times.

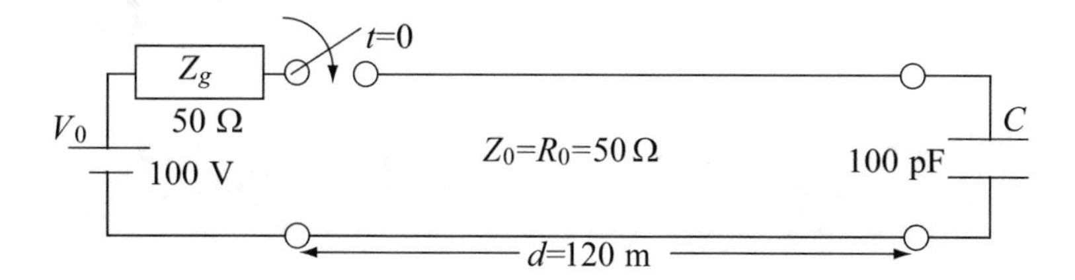

Figure 16.31 A capacitively loaded transmission line with matched generator

Solution: Because the generator is matched ($Z_g = Z_0$), the amplitude of the forward-propagating wave is $V_0/2$. There will be one reflection at the load, and after the backward-propagating wave reaches the generator, there will be no more reflections. Beyond that, the capacitor continues to charge until it reaches steady state. At steady state, the capacitor's voltage equals V_0.

At $t = 0$, the switch is closed and the forward-propagating wave is generated. This travels toward the load at the speed of propagation $v_p = 3 \times 10^8$ m/s. The forward-propagating wave reaches the load at time $\Delta t = d/v_p = 0.4$ μs. During this time, the voltage at the load is zero. The voltage on the line varies from point to point, depending on time, as shown in **Figure 16.32a**. At time $t = \Delta t = 0.4$ μs, the backward-propagating wave is generated. The backward-propagating wave is

$$V^-(t) = V^+\left(1 - 2e^{-(t-\Delta t)/R_0 C}\right) = 50\left(1 - 2e^{-\left(t-4\times 10^{-7}\right)/5\times 10^{-9}}\right) \quad [\text{V}], \quad t \geq \Delta t$$

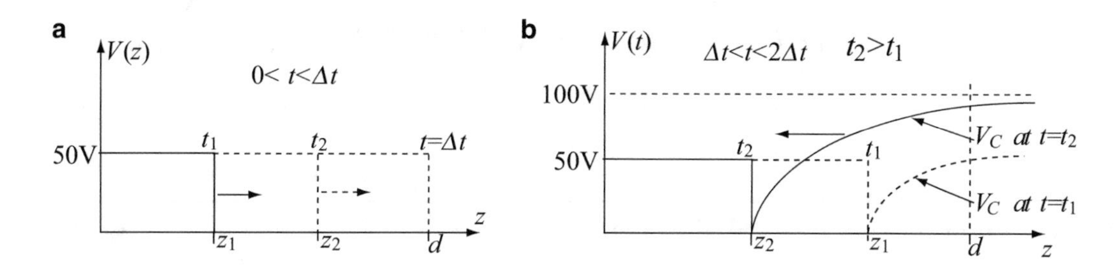

Figure 16.32 (**a**) Propagation of the voltage wave in **Figure 16.31** for $t < \Delta t$. (**b**) The reflected and forward waves for $\Delta t < t < 2\Delta t$

The voltage on the line and load is the sum of the forward- and backward-propagating waves:

$$v(t) = V^+ + V^+\left(1 - 2e^{-(t-\Delta t)/R_0 C}\right) = 100\left(1 - e^{-\left(t-4\times 10^{-7}\right)/5\times 10^{-9}}\right) \quad [\text{V}], \quad t \geq \Delta t$$

This is shown in **Figure 16.32b** for two times, t_1 and t_2, before the backward-propagating wave reaches the generator. The direction of propagation of the waveform is also shown. The capacitor's voltage increases with time until, after considerable time (relative to the time constant) has expired, the capacitor is at a voltage equal to V_g.

After time $t = 2\Delta t = 0.8$ μs, the backward-propagating wave has reached the generator, and since there is no reflection at the generator, the line voltage continues its climb toward steady state as shown in **Figure 16.33a**. The voltage at the load as it varies with time is shown in **Figure 16.33b**. The load voltage is zero between $0 = t < 0.4$ μs. After that it is the sum of the incident and reflected waves and shows steady charging from $v_L = 0$ toward $v_L = V_g$.

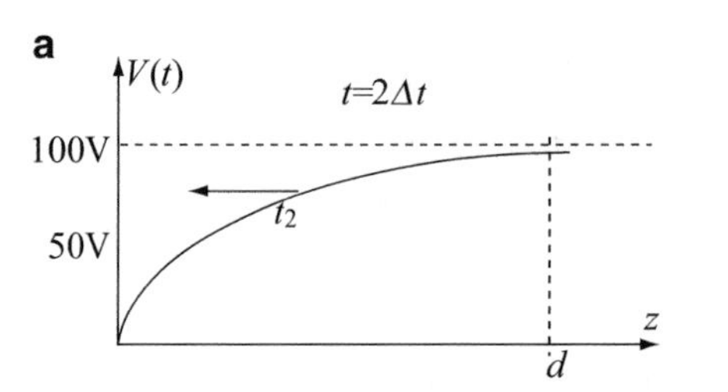

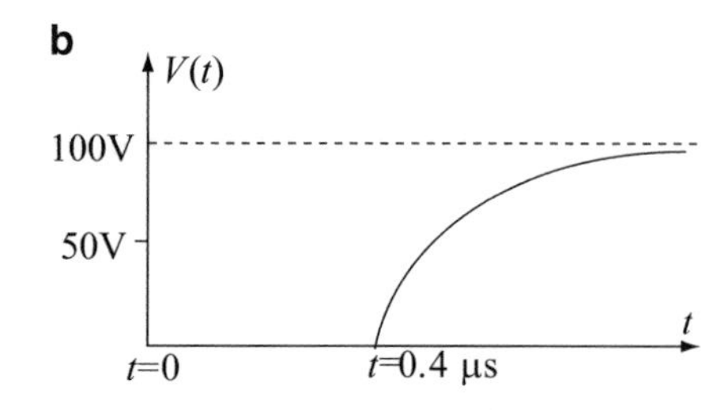

Figure 16.33 (**a**) The total wave at $t = 2\Delta t$ (in **Figure 16.31**). (**b**) Load voltage as a function of time

Exercise 16.4 The line in **Figure 16.31** is given. Find the load current for all times.

Answer

$$i_L(t) = 0, \quad 0 \le t \le 0.4\ \mu\text{s},$$

$$i_L(t) = 2e^{-\left(t-4\times10^{-7}\right)/5\times10^{-9}}\ \ [\text{A}], \quad t \ge 0.4\ \mu\text{s}.$$

Exercise 16.5 The line in **Figure 16.31** is given, but the load is an inductor $L = 100\ \mu\text{H}$. Find the load voltage and current for all times.

Answer

$$v_L(t) = 0, \quad 0 \le t \le 0.4\Delta\text{s}, \quad v_L(t) = 100e^{-\left(t-4\times10^{-7}\right)5\times10^{5}}\ \ [\text{V}], \quad t \ge 0.4\Delta s$$
$$i_L(t) = 0, \quad 0 \le t \le 0.4\Delta s, \quad i_L(t) = 2\left(1 - e^{-\left(t-4\times10^{-7}\right)5\times10^{5}}\right)\ \ [\text{A}], \quad t \ge 0.4\Delta s.$$

16.8 Initial Conditions on Transmission Lines

There is one additional condition that may exist on a line that we have not considered yet. Until now, we assumed that the line was completely neutral before the transient on the line was introduced. This means, for example, that no current or voltage was present anywhere on the line. In the case of the capacitive or inductive loading, this meant that the solution started with the capacitor or inductor discharged. There are, however, a number of situations in which the conditions are different. For example, a transmission line may operate in its steady-state mode when at some time $t = t_0$, a disturbance occurs. A short on a power line is of this type. Another example may be a line, operating at a given steady-state condition, on which the load is changed suddenly. A line which is at some initial voltage and current at the time the disturbance occurs is called an initially charged line. Treatment of transients on this type of line is performed by superposition of the steady-state line conditions and the conditions due to the transient.

Consider an open line on which the voltage is constant and equals V_0, as shown in **Figure 16.34a**. The initial conditions on the initially charged line **(Figure 16.34b)** are

$$V = V_0 \quad [\text{V}], \quad I = 0 \quad [\text{A}] \tag{16.61}$$

When the load resistance is connected at time $t = 0$, the reflection coefficient changes at the load. Initially, the reflection coefficient was 1, but now it changes to a smaller value $\Gamma_L = (R_L - R_0)/R_L + R_0)$ and may be positive or negative. Regardless of the magnitude of the reflection coefficient, a backward-propagating wave is generated, which we denote as

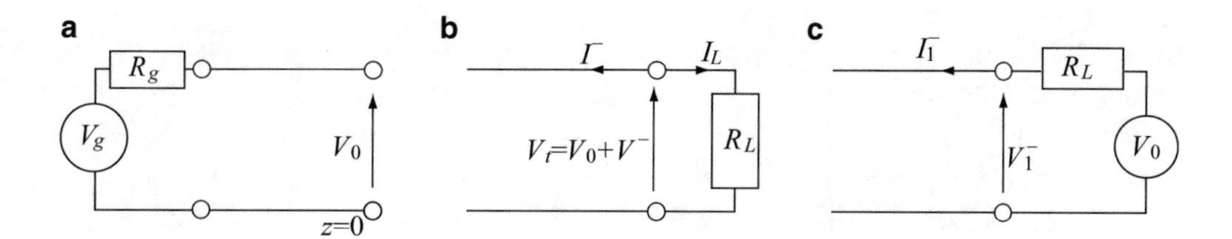

Figure 16.34 (**a**) Open line in steady state. (**b**) A load connected across the line in (**a**). (**c**) Equivalent circuit at the load representing the conditions in (**a**) and (**b**)

V_1^-, indicating that this is the first reflection. The total voltage across the load is the sum of the previously existing condition and the reflected voltage:

$$V_t = V_0 + V_1^- \quad [\mathrm{V}] \tag{16.62}$$

The initial current in the line was zero. Now, however, there must be a current I_1^- reflected from the load. Similarly, from the fact that the current in the line must be continuous, we can write

$$I_1^- = -I_L \quad [\mathrm{A}] \tag{16.63}$$

From the equivalent circuit in **Figure 16.34b**, we have

$$I_L = \frac{V_t}{R_L} = \frac{V_0 + V_1^-}{R_L} = -I_1^- \quad [\mathrm{A}] \tag{16.64}$$

On the other hand, on the line itself, we must have

$$I_1^- = \frac{V_1^-}{Z_0} \quad [\mathrm{A}] \tag{16.65}$$

Thus, we can write

$$V_0 + V_1^- = -I_1^- R_L = -\frac{V_1^- R_L}{Z_0} \quad [\mathrm{V}] \tag{16.66}$$

From this, we obtain the reflected voltage wave as

$$\boxed{V_1^- = -V_0 \frac{Z_0}{Z_0 + R_L} \quad [\mathrm{V}]} \tag{16.67}$$

and from **Eq. (16.65)**, the reflected current wave is

$$\boxed{I_1^- = -\frac{V_0}{Z_0 + R_L} \quad [\mathrm{A}]} \tag{16.68}$$

Now, we can replace the problem by the equivalent circuit at the load as given in **Figure 16.34c**. This equivalent source produces the initial condition for the transient. In other words, this equivalent circuit only exists for the purpose of generating the backward-propagating wave which, in this case, may be viewed as a generator output. Now, we may use the reflection diagram as for any other transient, except that the generator is at the load (the load generates the input signal that causes the transient). To this, we must add the initial conditions on the line. These points are further clarified in **Example 16.8**.

Example 16.8 A high-voltage DC (HVDC) line operates at steady state. The voltage on the line is 10^6 V, and the current is zero (no load). The characteristic line impedance is 200 Ω and the generator impedance is 300 Ω. The line length (distance between generator and load) is 1,000 km. Assume a lossless line and the speed of propagation is 2.5×10^8 m/s. A 300 Ω load is connected on the line at $t = 0$:

(a) Calculate the voltage and current at the load at $t = 10$ ms.
(b) Calculate the new steady-state voltage and current on the line.

Solution: Because the load is connected when the line is at the steady-state voltage, the reflection caused by the connection of the load becomes the generator for the transient. This transient is then superimposed on the initial line voltage (or current). The line after connecting the load is shown in **Figure 16.35a** and the equivalent circuit for the transient shown in **Figure 16.35b**:

(a) First, we calculate the transient voltage and current using the circuit in **Figure 16.35b**. The reflection coefficients at the load and generator are

$$\Gamma_L = \frac{R_L - Z_0}{Z_0 + R_L} = \frac{300 - 200}{300 + 200} = 0.2, \quad \Gamma_g = \frac{R_g - Z_0}{Z_0 + R_g} = \frac{300 - 200}{300 + 200} = 0.2$$

The reflected voltage and current due to connection of the load are given in **Eqs. (16.67)** and **(16.68)**:

$$V_1^- = -V_0\frac{Z_0}{Z_0 + R_L} = -10^6 \times \frac{200}{500} = -0.4 \times 10^6 \quad [\text{V}],$$

$$I_1^- = -\frac{V_0}{Z_0 + R_L} = -\frac{10^6}{500} = -2,000 \quad [\text{A}]$$

The time it takes the current or voltage wave to propagate the length of the line is

$$\Delta t = \frac{L}{v_p} = \frac{10^6}{2.5 \times 10^8} = 0.004 \quad [\text{s}]$$

These now form the basis of two bounce diagrams shown in **Figures 16.36a** and **16.36b**. Note that the reflection-coefficients for the voltage diagram are both positive, whereas for current, we use the negatives of the reflection coefficients as indicated in **Eq. (16.3)** and, more directly, in **Eq. (16.6)**. The propagation starts from the load.
At 10 ms, the waves have bounced once from the generator and once from the load. The transient voltage at the load is

$$V_L(10\text{ms}) = V_1 + V_1\Gamma_g + V_1\Gamma_g\Gamma_L = -0.4 \times 10^6(1 + 0.2 + 0.04) = -0.496 \times 10^6 \quad [\text{V}]$$

To this is added the initial condition on the line of 10^6 V to give the actual load voltage as $10^6 - 0.496 \times 10^6 = 0.504 \times 10^6$ V. In other words, the load voltage has dropped to almost half its initial value. The current in the line is

$$I_{line}(10\text{ ms}) = I_1 - I_1\Gamma_g + I_1\Gamma_g\Gamma_L = -2,000(1 - 0.2 + 0.04) = -1,680 \quad [\text{A}]$$

Since the initial current on the line is zero, the total line current at the load also equals −1680 A. The current in the load is in the opposite direction to the line current, as can be seen in **Figure 16.34**. Thus, the load current is 1680 A.

(b) In the steady state, we can use **Eqs. (16.31)** and **(16.32)**. The steady-state voltage and currents on the line due to the transient only are

$$V_\infty = V^+ \frac{1+\Gamma_L}{1-\Gamma_L\Gamma_g} \quad \rightarrow \quad V_\infty = V_1^- \frac{1+\Gamma_g}{1-\Gamma_L\Gamma_g} = -0.4\times 10^6 \frac{1+0.2}{1-0.04} = -0.5\times 10^6 \quad [\mathrm{V}]$$

$$I_\infty = I^+ \frac{1-\Gamma_L}{1-\Gamma_L\Gamma_g} \quad \rightarrow \quad I_\infty = I_1^- \frac{1-\Gamma_g}{1-\Gamma_L\Gamma_g} = -2{,}000 \frac{1-0.2}{1-0.04} = -1{,}666.67 \quad [\mathrm{A}]$$

As previously, we must add to these the initial values at the load. With these and recalling that the current in the load is opposite the current in the line, we get the steady-state voltage and current of the load as

$$V_L = 0.5\times 10^6 \quad [\mathrm{V}], \quad I_L = 1{,}666.67 \quad [\mathrm{A}].$$

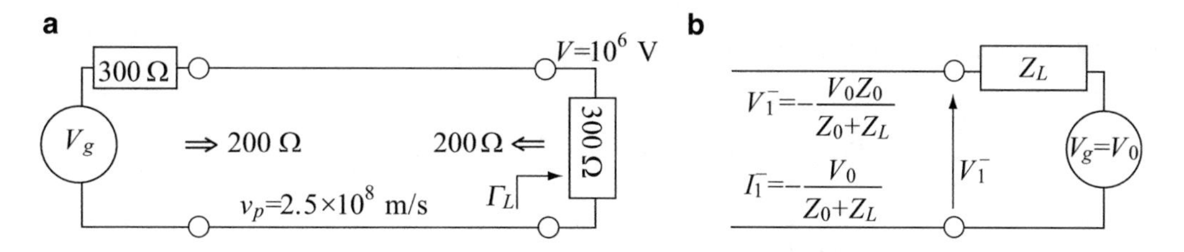

Figure 16.35 (**a**) A load connected across a high-voltage line at steady state. (**b**) The equivalent circuit used to find the transient

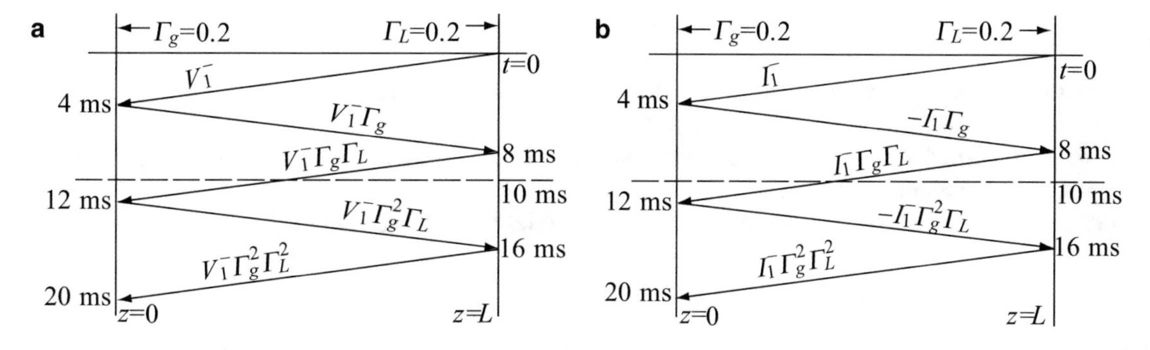

Figure 16.36 (**a**) Voltage reflection diagram for the transient due to **Figure 16.35b**. (**b**) Current reflection diagram for the transient due to **Figure 16.35b**

Note: in power lines the source internal impedance is much smaller that what this example implies, resulting in smaller voltage drops and higher transient currents

16.9 Summary

Following the frequency domain analysis in **Chapters 14** and **15**, this chapter discusses transient analysis and propagation of pulses on transmission lines. Again, the dominant issues are the reflection and transmission coefficients at discontinuities on transmission lines but the analysis is in the time domain.

Narrow Pulses Narrow pulses propagate, attenuate as they propagate, reflect, and transmit at all discontinuities. The forward-propagating waves generated by the generator (such as when closing a switch):

$$V^+ = V_g \frac{Z_0}{Z_0 + Z_g} \quad [\mathrm{V}] \qquad I^+ = \frac{V_g}{Z_0 + Z_g} \quad [\mathrm{A}] \tag{16.2}$$

When the pulse reaches the load (**Figure 16.3**), the first reflection is

$$V_1^- = V^+ \Gamma_L \quad [\mathrm{V}] \quad (16.5) \qquad I_1^- = -\frac{V^+ \Gamma_L}{Z_0} \quad [\mathrm{A}] \tag{16.6}$$

The total voltage and current at load during the length of the pulse after first reflection:

$$V_{L1} = V^+(1 + \Gamma_L) \quad [\mathrm{V}] \quad (16.7) \qquad I_{L1} = \frac{V^+}{Z_0}(1 - \Gamma_L) \quad [\mathrm{A}] \tag{16.8}$$

Back at the generator, the first reflection of the backward-propagating wave:

$$V_1^+ = \Gamma_L \Gamma_g V^+ \quad [\mathrm{V}], \quad I_1^+ = \frac{V^+ \Gamma_L \Gamma_g}{Z_0} \quad [\mathrm{A}] \tag{16.10}$$

The reflection coefficients at the load and generator are

$$\Gamma_L = \frac{Z_L - Z_0}{Z_L + Z_0} \quad (16.3) \qquad \Gamma_g = \frac{Z_g - Z_0}{Z_g + Z_0} \tag{16.9}$$

Notes:

(1) Reflections repeat indefinitely unless the load and/or generator are matched.
(2) The process stops at a matched location (no reflection).
(3) Total voltage or current at a given location during the width of the pulses is the sum of the voltages (currents) at that point (load and generator in particular).
(4) Attenuation (if any) is cumulative—it only depends on the total distance traveled by the pulse.

Step Pulses The step pulse propagates, reflects, and transmits at any discontinuity on the line.

Reflection Diagram A space–time diagram showing the propagation of the wave in space and time:

(1) Time is horizontal, space is vertical (see **Figures 16.13** and **16.14**).
(2) Voltages and currents reflected from all discontinuities are traced through time and space.
(3) The voltage (or current) at any point on the line is the sum of all voltages (or currents) at that location up to that time.

Steady-State Voltages and Currents on Lossless Lines

$$V_\infty = V^+ \frac{1 + \Gamma_L}{1 - \Gamma_L \Gamma_g} = V_g \frac{Z_L}{Z_g + Z_L} \quad [\mathrm{V}] \tag{16.30, 16.33}$$

$$I_\infty = I^+ \frac{1 - \Gamma_L}{1 - \Gamma_L \Gamma_g} = \frac{V_g}{Z_g + Z_L} \quad [\mathrm{A}] \tag{16.31, 16.34}$$

Finite-length pulses

(1) Finite-length pulses are viewed as superposition of positive and negative step pulses (**Figure 16.19**).
(2) Treat the positive going step pulse and the negative going step pulse separately using the reflection diagram and add the results together (see **Example 16.5**).
(3) Can also generate shaped pulses by superposition of pulses of various amplitudes and widths (**Example 16.5**).

Reactive Loads The reflection coefficient is not properly defined—it depends on the amplitude of voltage (or current). Calculate the reflected voltage by solving a differential equation at the reflecting point (for example, at the load) as follows.

For Capacitive Loading

$$i_L(t) = C\frac{d}{dt}(v_L(t)) \quad [\text{A}] \tag{16.47}$$

Given a transmission line with characteristic impedance R_0, internal generator impedance R_g, and a capacitor C as load, the reflected voltages and currents at the load are [see **Eq. (16.46)** for calculation of V^+]

$$V_1^-(t) = V^+\left(1 - 2e^{-(t-\Delta t)/R_0C}\right) = \frac{V_gR_0}{R_0 + R_g}\left(1 - 2e^{-(t-\Delta t)/R_0C}\right) \quad [\text{V}] \tag{16.53}$$

$$I_1^-(t) = \frac{-V^-(t)}{R_0}\left(1 - 2e^{-(t-\Delta t)/R_0C}\right) = -\frac{V_g}{R_0 + R_g}\left(1 - 2e^{-(t-\Delta t)/R_0C}\right) \quad [\text{A}] \tag{16.54}$$

For Inductive Loading

$$v_L(t) = L\frac{d}{dt}(i_L(t)) \quad [\text{V}] \tag{16.55}$$

Given a forward-propagating voltage V^+, the reflected voltage and current at the load are

$$V_1^-(t) = V^+\left(2e^{-(t-\Delta t)R_0/L} - 1\right) = \frac{V_gR_0}{R_0 + R_g}\left(2e^{-(t-\Delta t)R_0/L} - 1\right) \quad [\text{V}] \tag{16.59}$$

$$I_1^-(t) = -\frac{V_g}{R_0 + R_g}\left(2e^{-(t-\Delta t)R_0/L} - 1\right) \quad [\text{A}] \tag{16.60}$$

These then propagate on the line and may reflect again off the generator (unless it is matched).

Initial Conditions on Lines A line at steady state is characterized by a constant voltage V_0 and current I_0. Change in loading then adds reflected voltages and currents which take the line to a new steady state after these generated voltages and currents settle. The reflected voltage and current due to connection of a load, R_L, to an open line with characteristic impedance Z_0 are

$$V_1^- = -V_0\frac{Z_0}{Z_0 + R_L} \quad [\text{V}] \quad (16.67) \qquad I_1^- = -\frac{V_0}{Z_0 + R_L} \quad [\text{A}] \quad (16.68)$$

These now propagate on the line exactly as any step voltage and current and add to the existing conditions on the line. Any discontinuity will create additional reflections until a new steady state is achieved.

Time Domain Reflectometry In this method, often used for testing of line conditions, a narrow pulse or a step pulse is sent on the line and the reflected pulse is received after a time Δt. The distance to the discontinuity that caused the reflection is $d = v\Delta t/2$ where v is the speed of propagation on the line. By measuring time one can identify the location of discontinuity provided the speed of propagation on the line is known. It is also possible, to an extent, to characterize the fault (see **Example 16.3**).

Problems

Propagation of Narrow Pulses on Finite, Lossless, and Lossy Transmission Lines

16.1 Narrow Pulses on Mismatched Line. A generator is matched to a line. A single, narrow pulse is applied to the line. The load equals $2Z_0$ [Ω], where $Z_0 = 50$ Ω is the characteristic impedance of the line. If the pulse is 20 ns wide and the delay on the line (time of propagation to load) is 100 ns, calculate the line voltage and current at the load for $t > 0$ for a generator voltage of 1 V.

16.2 Narrow Pulses on Mismatched Line. A generator with an internal impedance $2Z_0$ [Ω] is connected to a line of characteristic impedance $Z_0 = 50$ Ω. A single, narrow pulse is applied to the line. The load equals $2Z_0$ [Ω]. If the pulse is 20 ns wide and the delay on the line is 100 ns, calculate the line voltage and current for $t > 0$ for a generator voltage of 1 V.

16.3 Application: Transients in Digital Circuits. Two sensors are connected as inputs to an AND gate as shown in **Figure 16.37**. The lines have characteristic impedance of 50 Ω. The inputs to the AND gate and the sensors are matched to the lines. Each sensor generates a single pulse, 50 ns wide at $t = 0$, of open circuit voltage 10 V. Each AND gate has a threshold of 3.25 V (i.e., if both inputs are above this value, the output is 5 V; if one or both are below 3.25 V, the output is zero). One line is 10 m long, the second is 100 m long, and the speed of propagation is $0.1c$ [m/s]:

(a) Calculate the gate output for $t > 0$.
(b) What must be the minimum pulse width for the output to ever be 5 V? What are your conclusions from this result?

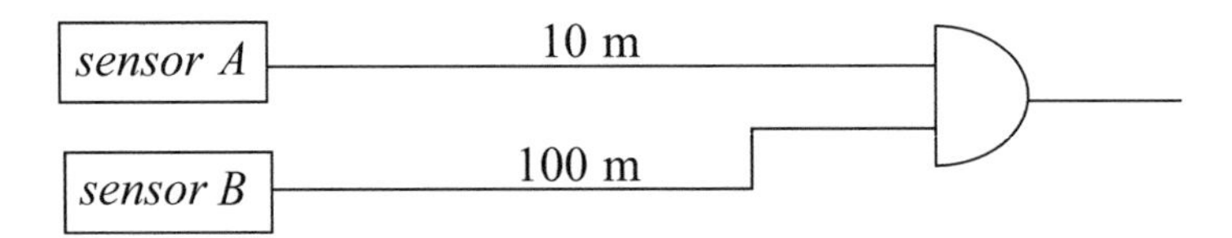

Figure 16.37

16.4 Application: Reflectometry (Narrow Pulses). A lossless cable TV coaxial transmission line is matched to both generator and load. As a routine test, a signal is applied to the input and sent down the line. The distance to the receiver is known to be $d = 1$ km. The speed of propagation on the line is $v_p = c$ [m/s], and the characteristic impedance on the line is $Z_0 = 75$ Ω:

(a) The signal in **Figure 16.38a** is obtained on the oscilloscope screen. If $\Delta t = 0.1$ μs, what happened to the line and at what location?
(b) The signal in **Figure 16.38b** is obtained on the oscilloscope screen. If $\Delta t = 0.2$ μs, what happened on the line and at what location?

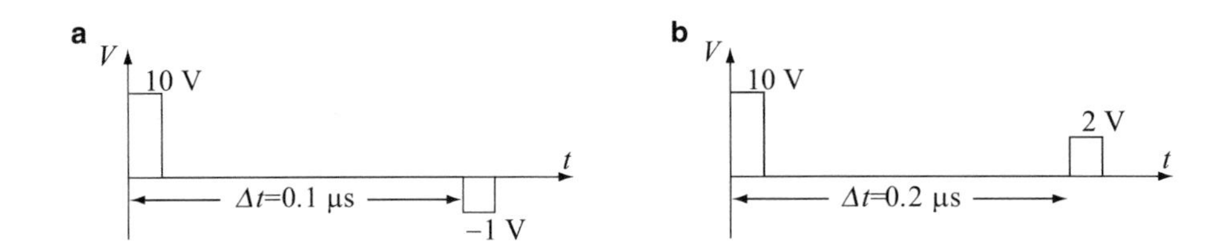

Figure 16.38

16.5 Application: Reflections on Lossy Line. The cable in **Problem 16.4** is given again. However, now the line is considered distortionless, with an attenuation constant of 0.001 Np/m:

(a) The signal in **Figure 16.38a** is obtained on the oscilloscope screen. If $\Delta t = 0.1$ μs, what happened to the line and at what location?

(b) The signal in **Figure 16.38b** is obtained on the oscilloscope screen. If $\Delta t = 0.2$ μs, what happened on the line and at what location?

(c) Compare the location of the fault on the line and magnitude of fault impedance with those for the lossless line in **Problem 16.4**.

Transients on Transmission Lines: Long Pulses

16.6 Transients on an Open Line. A lossless open transmission line is given as shown in **Figure 16.39**. The line is 10 m long, has capacitance of 200 pF/m and inductance of 0.5 μH/m. Calculate the transient voltage at a distance of 5 m from the DC source:

(a) 0.5 μs after closing the switch.
(b) 50 μs after closing the switch.

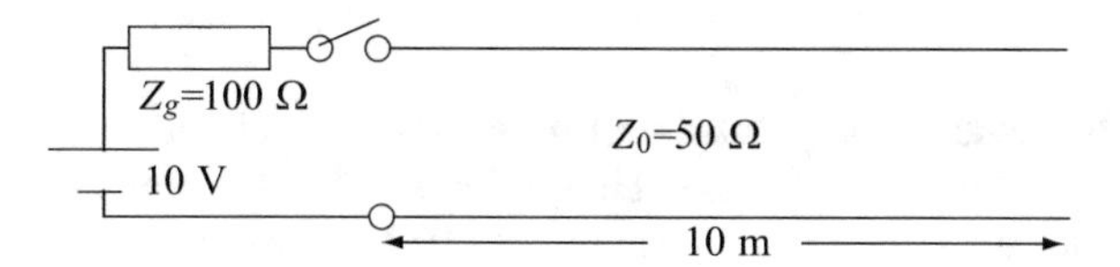

Figure 16.39

16.7 Line Voltage on Long, Loaded Line. A lossless line is very long and the speed of propagation on the line is 10^8 m/s. Assume the ideal DC source has been switched on. The voltage wave reaches the load at time t_0. Calculate the voltage at point $A - A'$ (2 m from the load) for $t > t_0$ and for times $t < t_0$ **(Figure 16.40)**.

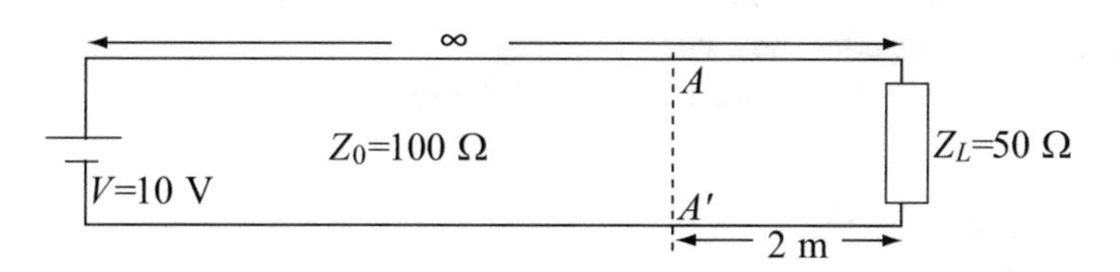

Figure 16.40

16.8 Transient and Steady-State Voltages on Lossless Line. A lossless transmission line of length d [m] is given as in **Figure 16.41**. The transmission line has capacitance per unit length C_0 [F/m] and inductance per unit length L_0 [H/m]. The switch is closed at time $t = 0$. Given: $L_0 = 10$ μH/m, $C_0 = 1{,}000$ pF/m, $d = 1{,}000$ m, $R_g = 100$ Ω, $R_L = 50$ Ω, and $V_0 = 100$ V:

(a) Calculate the steady-state voltage on the line.
(b) Calculate the steady-state current in the line.
(c) How long does it take the voltage to reach steady state at the load?
(d) How long does it take the voltage to reach steady state at the generator?

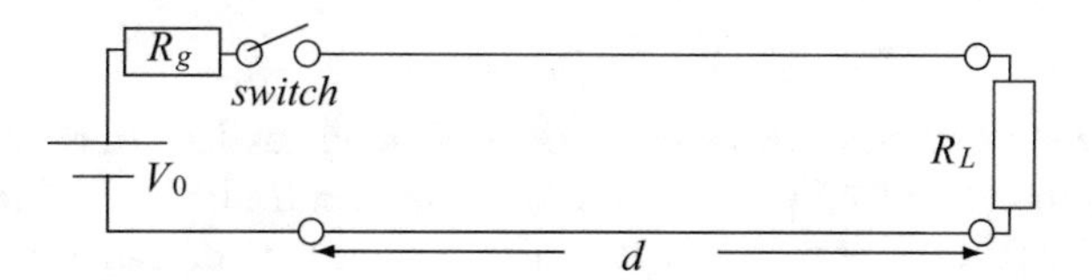

Figure 16.41

Transients on Transmission Lines: Finite-Length Pulses

16.9 Transient Due to a Single Short Pulse. The transmission line in **Figure 16.42** is given. The generator supplies a single pulse as shown. Calculate:

(a) The voltage and current at the generator 10 μs after the pulse began.
(b) The steady-state current and voltage on the line.

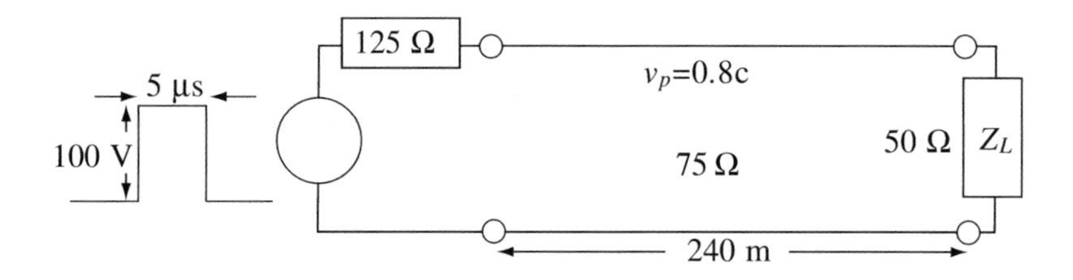

Figure 16.42

16.10 Transient Due to a Single Square Short Pulse on Lossy Line. The circuit in **Figure 16.42** is given. In addition to the data in the figure, the line has an attenuation constant $\alpha = 0.0001$ Np/m. Assume the line is distortionless and calculate the voltage and current at the generator terminals 10.5 μs after the pulse began.

Reflections from Discontinuities

16.11 Reflections from Discontinuities. Three sections of lines are connected as shown in **Figure 16.43**. The propagation time on each section is indicated:

(a) If both the load and generator are matched, calculate the line voltage at g, L, and on both sides of the discontinuities a and b, 45 ns after the switch is closed.
(b) Same as **(a)** but the generator impedance is 50 Ω and the load is matched.

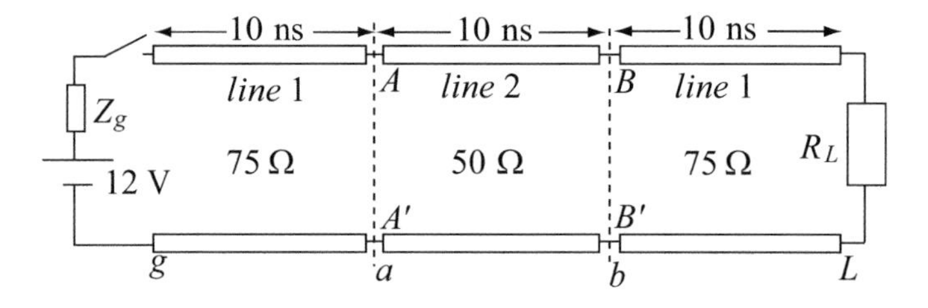

Figure 16.43

16.12 Reflections from Discontinuities. Use the same figure and data as in **Problem 16.11**. The load now is a short circuit. Given a matched generator, calculate the voltage and current at g, L, and on both sides of the discontinuities a and b, 45 ns after the switch is closed.

Reactive Loading

16.13 Application: Capacitively Loaded Transmission Line. A long lossless transmission line with a characteristic impedance of 50 Ω is terminated with a 1 μF capacitor. The length of the line is 100 m and the speed of propagation on the line is $c/3$ [m/s]. At $t = 0$, a 100 V matched generator is switched on. Calculate and plot:

(a) The load voltage and current for $t > 0$.
(b) The line voltage and current at any point on the line for $t > 0$.

16.14 Application: Inductively Loaded Transmission Line. A long lossless transmission line with characteristic impedance of 50 Ω is terminated with a 1 μH inductor. The line is 10 km long and the speed of propagation on the line is $c/3$ [m/s]. At $t = 0$, a 100 V matched generator is switched on:

(a) Calculate and plot the load voltage and current for $t > 0$.
(b) Calculate and plot the line voltage and current at any point on the line for $t > 0$.

Initially Charged Lines

16.15 Application: Initially Charged Line. A 300 m long, lossless transmission line has characteristic impedance of 75 Ω and speed of propagation of $c/3$ [m/s]. The transmission line is matched at the generator and is open ended. The generator's voltage is 100 V. After the line has reached steady state, the generator is disconnected and a resistor $R = 125$ Ω is connected across the open end. Calculate and plot the voltage on and the current in R.

16.16 Application: Initially Charged Line. A 100 m long lossless transmission line has characteristic impedance of 75 Ω and speed of propagation of $0.2c$ [m/s]. The transmission line is matched at the generator and is open ended. The generator's voltage is 100 V. After the line has reached steady state, the generator is disconnected and a resistor $R = 125$ Ω is connected across the open end:

(a) Calculate the voltage on and current in R.
(b) How long does it take for the voltage on R to be below 1 V?

Time Domain Reflectometry

16.17 Application: Time Domain Reflectometry. An underground cable used for transmission of power has developed a fault. The speed of propagation on the line is known and equal to v_p [m/s]. To locate the fault before starting to dig, time domain reflectometry is performed. A 1 V step pulse is applied to the input with matched impedance and the output in **Figure 16.44a** is obtained on the oscilloscope. The characteristic impedance of the cable is $Z_0 = 50$ Ω. Use $v_p = 0.2c$ [m/s] and calculate:

(a) The location of the fault.
(b) Type of fault: calculate the impedance on the line at the fault.

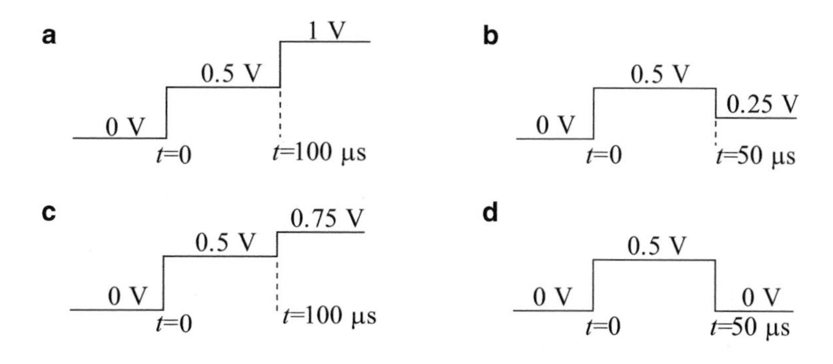

Figure 16.44

16.18 Application: Time Domain Reflectometry. The measurement in **Problem 16.17** is performed on a line and the signal in **Figure 16.44b** is recorded on the time domain reflectometer. Using the data in **Problem 16.17**, calculate:

(a) The location of the fault.
(b) Type of fault: calculate the impedance on the line at the fault.

16.19 Application: Time Domain Reflectometry. The measurement in **Problem 16.17** is performed on a line and the signal in **Figure 16.44c** is recorded on the time domain reflectometer. Using the data in **Problem 16.17**, calculate:

(a) The location of the fault.
(b) Type of fault: find the impedance on the line at the fault.

16.20 Application: Time Domain Reflectometry. The measurement in **Problem 16.17** is performed on a line and the signal in **Figure 16.44d** is recorded on the time domain reflectometer. Using the data in **Problem 16.17**, calculate:

(a) The location of the fault.
(b) Type of fault: find the impedance on the line at the fault.

Waveguides and Resonators

17

He alone is free who lives with free consent under the entire guidance of reason.

Baruch Spinoza (1632–1677),
philosopher

17.1 Introduction

Much of the discussion in the previous three chapters centered around lossless transmission lines. For these lines, the various relations were independent of conductivity of the materials involved. In fact, even for a distortionless transmission line, although an attenuation constant was present, all other basic properties of the line were independent of conductivity. The speed of propagation depended only on material properties of the line (ε and μ), and the phase constant and wavelength were also independent of conductivity of the line. If this is the case, are the conductors necessary? If so, what are their roles in the propagation of energy in the lossless or distortionless transmission line?

Perhaps this question is confusing since in a transmission line, we dealt with line voltage and line current. If there is a current, a conductor must be present. But we also saw that power can be transmitted through space without the need for conductors and, even transmission lines may be analyzed from a field point of view rather than through voltages and currents. What then is the role of conductors in transmission lines? The answer is surprising and simple; the conductors are not necessary in general. What they do is to confine the energy being transmitted to the line itself and guide it along the line. Of course, the conductors may have other effects. If conductivity is low, there may be losses in the conductors, but these are secondary effects.

To convince yourself of this, consider the following "communication system" often used in old ships: A tube connects the bridge of the ship with the quarters below. The tube may bend and may be made of any material. This is an example of a guided transmission system. The tube itself is immaterial, other than to provide the path for propagation. A glass tube may work just as well as a steel or brass tube. The guiding system described here has certain advantages over an open system: the amount of energy required is smaller (no need to shout), the distance over which propagation takes place is longer, and there is less interference from and with external systems (only those listening to the tube can hear the conversation). So much for the obvious and the known. Now, we ask ourselves another question: If the material of the tube is not important, are the shape and size important? What if we tried to use a capillary tube for this purpose, or a tube 2 m in diameter? Are the two going to behave the same way? How about a square rather than a cylindrical tube? What happens if we plug the tube, or insert some material in it? How do we couple energy into and out of the tube? Pretty interesting questions from the design point of view and they all have to do with properties of waves in this environment. The same questions and considerations apply to electromagnetic guiding structures.

We also saw in **Chapter 13** that an electromagnetic wave impinging on a conducting surface at an angle propagates along the surface of the conductor (see, for example, **Section 13.3.1** and **Example 13.7**). We concluded there that the surface of the conductor has a guiding effect on the wave in addition to generating standing waves in the direction perpendicular to the conductor.

In this chapter, we will discuss a very special form of transmission lines: lines or structures that guide waves. In many ways, this is an extension of the results obtained for transmission lines. There are, however, major differences. In most cases, there will only be one apparent conductor rather than two conductors, and in some cases, no conductors at all. The

N. Ida, *Engineering Electromagnetics*, https://doi.org/10.1007/978-3-030-15557-5_17

conductors are explicitly used for guiding the waves and treatment of line behavior will be in terms of field parameters (electric and magnetic field intensities) rather than current and voltage.

17.2 The Concept of a Waveguide

The idea of a waveguide[1] can best be explained using the optic spectrum because we can see the effects involved. Consider, first, two large, parallel mirrors facing each other as shown in **Figure 17.1a**. The light emitted from object A reflects repeatedly off the mirrors and eventually generates an image on the other side of the waveguide or anywhere in the waveguide. The two mirrors may be viewed as a guiding structure because they guide the waves to any location we wish. If the mirrors were flexible (for example by silver coating flexible surfaces), we could create bent waveguides. One particularly useful optical guiding system is the optical fiber shown in **Figure 17.1b**. For simplicity you can view the optical fiber as a long, thin glass rod, coated so that there is total reflection inside the fiber (total reflection also occurs without coating because the fiber's permittivity is higher than that of the surrounding space—see **Section 13.4.4**). This then constitutes a dielectric waveguide and conductors are not necessary. Since optical fibers are flexible, waves may be guided in almost any path. Although most of the discussion here will deal with frequencies much lower than optical frequencies, the picture drawn here is useful as a background.

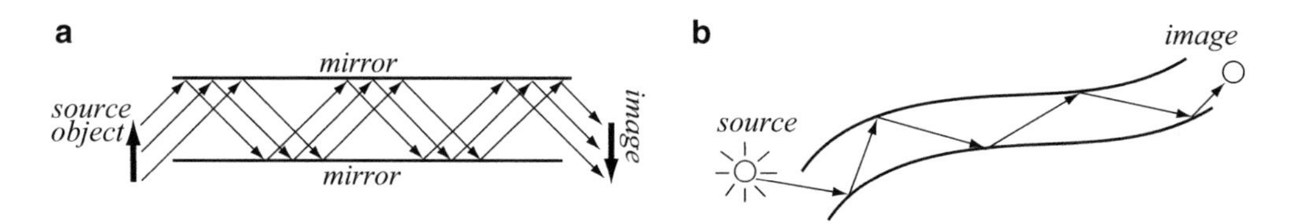

Figure 17.1 (**a**) An "optical waveguide" made of two parallel mirrors. (**b**) An optical fiber: total internal reflections at the interfaces ensure guidance of waves

Before we can look at practical waveguide structures we digress a little and look at the general properties of propagating waves in unbounded domain and in the presence of conducting surfaces since these will become very useful in our study. In particular, we define two specific types of waves, in addition to the plane waves we already saw in **Chapters 12** and **13**. These are the transverse electric (TE) waves and transverse magnetic (TM) waves. The plane waves discussed in **Chapters 12**, **13**, and **14** were transverse electromagnetic (TEM) waves.

17.3 Transverse Electromagnetic, Transverse Electric, and Transverse Magnetic Waves

In **Section 12.7**, we discussed uniform plane waves as they propagate in free space and in dielectrics. One of the most important aspects of propagation was the fact that neither the electric nor the magnetic field intensity had any component in the direction of propagation. Both the electric and magnetic field intensities were perpendicular to each other and to the direction of propagation. For this reason, we called them ***transverse electromagnetic*** (TEM) waves. There are, however, situations in which the electric or magnetic fields have components in the direction of propagation. If a wave has an electric field intensity that is entirely perpendicular to the direction of propagation, but with a component of the magnetic field intensity in the direction of propagation, this wave is called a ***transverse electric*** (TE) wave. Similarly, if the magnetic field

[1] The first mention of a waveguide was in 1894 by Sir Oliver Joseph Lodge (1851–1940). He discovered the effect when he surrounded a spark generator, of the type used by Hertz to demonstrate propagation of waves, with a conducting tube. Three years later, Lord Rayleigh (John William Strutt (1842–1919)) developed much of the theory of guided waves. However, waveguides did not feature in electromagnetics until the early 1930s, when experiments on their properties were conducted at Bell Laboratories, first for propagation in dielectrics (water) and later in air. The main impetus for their development was the then newly developed microwave tubes and work on radar. From then on, waveguides became the basis of microwave work and are used wherever transmission of significant power above about 1 GHz is required. The various optical fibers are also waveguides and their utility in communication is prevalent.

intensity is entirely perpendicular to the direction of propagation and the electric field intensity has a component in the direction of propagation, this is a ***transverse magnetic*** (TM) wave. The differences among the three types of waves are shown in **Figure 17.2**.

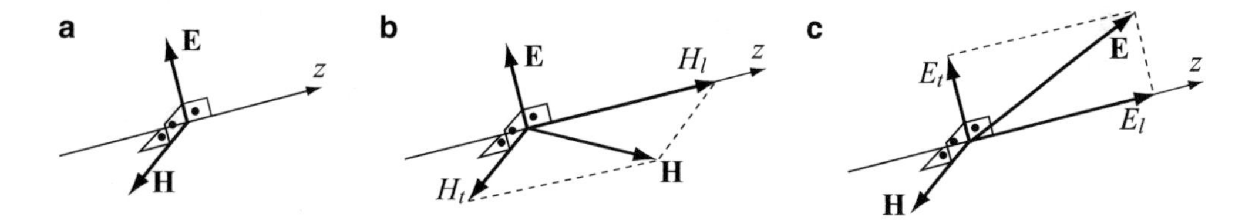

Figure 17.2 Types of waves. (**a**) TEM wave. (**b**) TE wave. (**c**) TM wave

Before proceeding with description of waves in waveguides, we need to define the conditions under which TEM, TM, and TE waves can exist. These conditions are then used to define the possible fields or modes of propagation in waveguides.

The starting point is with Maxwell's two curl equations. Each is expanded into three scalar equations by equating the vector components on both sides, as shown in **Eqs. (17.1)** through **(17.8)**. By doing so, one component of one field (electric or magnetic) is written in terms of the transverse components of the other field. **Table 17.1** summarizes these steps.

Table 17.1 Maxwell's curl equations and their transverse components

$\nabla \times \mathbf{H} = j\omega\varepsilon\mathbf{E}$ (17.1)	$\nabla \times \mathbf{E} = -j\omega\mu\mathbf{H}$ (17.5)
Components of the electric field in terms of the transverse components of the magnetic field	Components of the magnetic field in terms of the transverse components of the electric field
$\frac{\partial H_z}{\partial y} - \frac{\partial H_y}{\partial z} = j\omega\varepsilon E_x$ (17.2)	$\frac{\partial E_z}{\partial y} - \frac{\partial E_y}{\partial z} = -j\omega\mu H_x$ (17.6)
$\frac{\partial H_x}{\partial z} - \frac{\partial H_z}{\partial x} = j\omega\varepsilon E_y$ (17.3)	$\frac{\partial E_x}{\partial z} - \frac{\partial E_z}{\partial x} = -j\omega\mu H_y$ (17.7)
$\frac{\partial H_y}{\partial x} - \frac{\partial H_x}{\partial y} = j\omega\varepsilon E_z$ (17.4)	$\frac{\partial E_y}{\partial x} - \frac{\partial E_x}{\partial y} = -j\omega\mu H_z$ (17.8)

The waves are assumed to propagate in the z direction, with a propagation constant $\gamma = \alpha + j\beta$. Assuming no backward-propagating waves (no reflections), the electric and magnetic field intensities for TEM waves have the following general form:

$$E_x = E_0 e^{-\gamma z} \quad [\mathrm{V/m}], \quad H_y = H_0 e^{-\gamma z} \quad [\mathrm{A/m}] \tag{17.9}$$

where the $e^{j\omega t}$ variation is also implied (phasors). The propagation constant is assumed for the moment to be general, but in most of the discussion that follows, we will use lossless materials. Because all fields vary with the z parameter only, the derivatives with respect to z in **Eqs. (17.2)**, **(17.3)**, **(17.6)**, and **(17.7)** are nonzero:

$$\frac{\partial H_y}{\partial z} = -\gamma H_y, \quad \frac{\partial H_x}{\partial z} = -\gamma H_x, \quad \frac{\partial E_y}{\partial z} = -\gamma E_y, \quad \frac{\partial E_x}{\partial z} = -\gamma E_x \tag{17.10}$$

Important If we were to assume only a backward-propagating wave of the form $E_x = E_0 e^{+\gamma z}$, then all terms in **Eq. (17.10)** would be positive; that is, γ is replaced by $-\gamma$. Similarly, if both a forward- and a backward-propagating wave exist, the derivatives of the total wave are the sum of the derivatives of the forward-propagating waves and those of the backward-propagating waves. At this juncture, we will assume that only forward-propagating waves exist to keep things simple, but if for any reason there are reflections in the system, the backward-propagating wave will have to be added.

$$\frac{\partial H_z}{\partial y} + \gamma H_y = j\omega\varepsilon E_x \tag{17.11}$$

$$-\gamma H_x - \frac{\partial H_z}{\partial x} = j\omega\varepsilon E_y \tag{17.12}$$

$$\frac{\partial H_y}{\partial x} - \frac{\partial H_x}{\partial y} = j\omega\varepsilon E_z \tag{17.13}$$

$$\frac{\partial E_z}{\partial y} + \gamma E_y = -j\omega\mu H_x \tag{17.14}$$

$$-\gamma E_x - \frac{\partial E_z}{\partial x} = -j\omega\mu H_y \tag{17.15}$$

$$\frac{\partial E_y}{\partial x} - \frac{\partial E_x}{\partial y} = -j\omega\mu H_z \tag{17.16}$$

Now, **Eqs. (17.2)** through **(17.4)** and **(17.6)** through **(17.8)** are

Since the nature of the waves is defined by the longitudinal components of the field (components in the direction of propagation), it is necessary to rewrite these equations such that the transverse components of the fields (those perpendicular to the direction of propagation), in this case, E_x, E_y, H_x, and H_y, are written in terms of the longitudinal components, E_z and H_z. As an example, using **Eqs. (17.11)** and **(17.15)**, we can eliminate H_y, and write the component E_x in terms of H_z and E_z. Substitution of H_y from **Eq. (17.15)** in **Eq. (17.11)** gives

$$\frac{\partial H_z}{\partial y} + \frac{\gamma}{j\omega\mu}\left(\gamma E_x + \frac{\partial E_z}{\partial x}\right) = j\omega\varepsilon E_x \tag{17.17}$$

Multiplying both sides by $j\omega\mu$, rearranging terms, and substituting $\omega\sqrt{\mu\varepsilon} = k$, we get

$$E_x = \frac{1}{\gamma^2 + k^2}\left(-\gamma\frac{\partial E_z}{\partial x} - j\omega\mu\frac{\partial H_z}{\partial y}\right) \quad \left[\frac{\text{V}}{\text{m}}\right] \tag{17.18}$$

Repeating the process for the other three transverse components (E_y, H_x, and H_y), we get

$$E_y = \frac{1}{\gamma^2 + k^2}\left(-\gamma\frac{\partial E_z}{\partial y} + j\omega\mu\frac{\partial H_z}{\partial x}\right) \quad \left[\frac{\text{V}}{\text{m}}\right] \tag{17.19}$$

$$H_x = \frac{1}{\gamma^2 + k^2}\left(-\gamma\frac{\partial H_z}{\partial x} + j\omega\varepsilon\frac{\partial E_z}{\partial y}\right) \quad \left[\frac{\text{A}}{\text{m}}\right] \tag{17.20}$$

$$H_y = \frac{1}{\gamma^2 + k^2}\left(-\gamma\frac{\partial H_z}{\partial y} - j\omega\varepsilon\frac{\partial E_z}{\partial x}\right) \quad \left[\frac{\text{A}}{\text{m}}\right] \tag{17.21}$$

These equations are now used to define TEM, TE, and TM propagation of waves by imposing the necessary conditions for each type of propagation. The transverse components for backward-propagating waves may be obtained by replacing γ by $-\gamma$ in **Eqs. (17.18)** through **(17.21)**.

17.3.1 Transverse Electromagnetic Waves

As mentioned earlier, TEM waves require that there be no field in the direction of propagation. The condition for TEM propagation is $E_z = H_z = 0$. Substituting this into **Eqs. (17.18)** through **(17.21)** leads to the requirement that all four transverse components are zero, unless $\gamma^2 + k^2 = 0$. Thus, the constant of propagation for TEM waves must be

$$\gamma^2 = -k^2 \quad \text{or} \quad \gamma = j\omega\sqrt{\mu\varepsilon} \tag{17.22}$$

This is the condition we obtained for the propagation of uniform plane waves in lossless media. Indeed, all the properties we obtained for the propagation of plane waves in unbounded space, including the definition of intrinsic impedance (or wave impedance) and phase velocity, apply here as well. In particular, if we replace the permittivity ε with the complex permittivity $\varepsilon(1 - j\sigma/\omega\varepsilon)$, we obtain the propagation constant for a general lossy medium:

$$\boxed{\gamma_{TEM} = j\omega\sqrt{\mu\varepsilon}\sqrt{1 - j\frac{\sigma}{\omega\varepsilon}}} \tag{17.23}$$

This relation was obtained in **Chapter 12 [Eq.** (12.83)] and was the basis of study of propagation in lossless, low-loss, and high-loss media.

The defining equation for transverse electromagnetic waves (plane waves) was the Helmholtz equation for the electric field:

$$\boxed{\frac{\partial^2 E_x}{\partial z^2} + k^2 E_x = 0} \tag{17.24}$$

In particular, if a plane wave propagates in unbounded space in the z direction and has an electric field intensity in the x direction, then the magnetic field intensity is in the y direction as required by the Poynting theorem:

$$E_x = E_0 e^{-jkz} \quad [\mathrm{V/m}] \quad \text{and} \quad H_y = \frac{E_0}{\eta_{TEM}} e^{-jkz} \quad \left[\frac{\mathrm{A}}{\mathrm{m}}\right] \tag{17.25}$$

The wave impedance in the domain in which the waves propagate was defined as the ratio between the transverse components of the electric and magnetic field intensities:

$$\boxed{\eta_{TEM} = \frac{E_x}{H_y} = \sqrt{\frac{\mu}{\varepsilon}} \quad [\Omega]} \tag{17.26}$$

These properties will be used to contrast transverse electromagnetic waves with transverse electric waves and transverse magnetic waves and to point out the differences. In particular, properties of TE and TM waves are often written in terms of known properties of TEM waves. For example, it is often useful to write the wave impedance of TE and TM waves in terms of the wave impedance of TEM waves.

17.3.2 Transverse Electric (TE) Waves

For TE waves to exist, E_z must be zero; that is, the only field component in the direction of propagation is a magnetic field intensity component H_z. Substituting this condition in **Eqs. (17.18)** through **(17.21)**, we get the transverse components for TE propagation:

$$\boxed{E_x = \frac{-j\omega\mu}{\gamma^2 + k^2}\frac{\partial H_z}{\partial y} \quad \left[\frac{\mathrm{V}}{\mathrm{m}}\right]} \tag{17.27}$$

$$\boxed{E_y = \frac{j\omega\mu}{\gamma^2 + k^2}\frac{\partial H_z}{\partial x} \quad \left[\frac{\mathrm{V}}{\mathrm{m}}\right]} \tag{17.28}$$

$$\boxed{H_x = \frac{-\gamma}{\gamma^2 + k^2}\frac{\partial H_z}{\partial x} \quad \left[\frac{\mathrm{A}}{\mathrm{m}}\right]} \tag{17.29}$$

$$\boxed{H_y = \frac{-\gamma}{\gamma^2 + k^2}\frac{\partial H_z}{\partial y} \quad \left[\frac{\mathrm{A}}{\mathrm{m}}\right]} \tag{17.30}$$

Although not immediately apparent from these relations, it is possible to write these as a wave equation in H_z, which is the only longitudinal component in a TE wave. Doing so allows representation of fields and properties in terms of the solution to the wave equation. Since we have already obtained the wave equations for propagation in free space and in transmission lines, this approach will allow us to build on the existing solutions. We start by taking the derivative with respect to y of **Eq. (17.27)** and the derivative with respect to x of **Eq. (17.28)**:

$$\frac{\partial E_x}{\partial y} = \frac{-j\omega\mu}{\gamma^2 + k^2}\frac{\partial^2 H_z}{\partial y^2} \quad \text{and} \quad \frac{\partial E_y}{\partial x} = \frac{j\omega\mu}{\gamma^2 + k^2}\frac{\partial^2 H_z}{\partial x^2} \tag{17.31}$$

Subtracting the first of these relations from the second and rearranging terms gives

$$\frac{\partial^2 H_z}{\partial x^2} + \frac{\partial^2 H_z}{\partial y^2} + \frac{\gamma^2 + k^2}{j\omega\mu}\left(-\frac{\partial E_y}{\partial x} + \frac{\partial E_x}{\partial y}\right) = 0 \tag{17.32}$$

Now, using **Eq. (17.8)** to eliminate the components of **E**, we get

$$\boxed{\frac{\partial^2 H_z}{\partial x^2} + \frac{\partial^2 H_z}{\partial y^2} + (\gamma^2 + k^2)H_z = 0} \tag{17.33}$$

This is a wave equation in H_z alone and will be taken from now on as the defining equation for TE waves whenever we need to do so.

Comparison of this equation with **Eq. (17.24)** shows that they are of the same form if we replace the term $\gamma^2 + k^2$ by a single term, which is denoted as k_c^2:

$$\boxed{k_c^2 = \gamma^2 + k^2} \tag{17.34}$$

Now, we can write the propagation constant as

$$\boxed{\gamma_{TE}^2 = k_c^2 - k^2 \quad \rightarrow \quad \gamma_{TE} = \sqrt{k_c^2 - k^2} = \sqrt{k_c^2 - \omega^2\mu\varepsilon}} \tag{17.35}$$

that is, the propagation constant for TE waves is not the same as for TEM waves. Whereas the propagation constant for TEM waves in **Eq. (17.22)** is only zero for $\omega = 0$, the propagation constant for TE waves is zero if

$$k_c = \omega\sqrt{\mu\varepsilon} \quad [\text{rad/m}] \tag{17.36}$$

Since a zero propagation constant means no propagation, this condition is quite important in propagation of TE waves. The following conditions may be distinguished:

(1) $k_c^2 = \omega^2\mu\varepsilon$. This is the condition for no propagation. k_c is called the ***cutoff wave number*** or, alternatively, we call the frequency for which this happens the ***cutoff frequency:***

$$\boxed{f_c = \frac{k_c}{2\pi\sqrt{\mu\varepsilon}} \quad [\text{Hz}]} \tag{17.37}$$

and the corresponding wavelength the ***cutoff wavelength.***

(2) $\omega^2\mu\varepsilon < k_c^2$. For these values of k_c, the propagation constant γ_{TE} is real. From the definition of the propagation constant as $\gamma = \alpha + j\beta$, this condition leads to $\gamma = \alpha$; that is, there is no propagation (phase constant is zero), but there is attenuation. This type of attenuated, non-propagating wave is called an ***evanescent wave*** and occurs for $f < f_c$. Note that the attenuation here has nothing to do with losses: It occurs in lossless media as well. If we substitute this condition in **Eq. (17.35)**, we get

$$\gamma_{TE} = \sqrt{k_c^2 - \omega^2\mu\varepsilon} = \sqrt{(2\pi)^2 f^2\mu\varepsilon\left(\frac{f_c^2}{f^2} - 1\right)} = \pm\omega\sqrt{\mu\varepsilon}\sqrt{\frac{f_c^2}{f^2} - 1}, \quad f < f_c \tag{17.38}$$

The attenuation constant for evanescent waves [taking the positive solution in **Eq. (17.38)**] is

$$\boxed{\alpha_e = \omega\sqrt{\mu\varepsilon}\sqrt{\frac{f_c^2}{f^2} - 1} \quad \left[\frac{\text{Np}}{\text{m}}\right]} \tag{17.39}$$

(3) $\omega^2\mu\varepsilon > k_c^2$. Now, the propagation constant in **Eq. (17.35)** is purely imaginary and the wave propagates, without attenuation (lossless media). This occurs for any frequency above the cutoff frequency f_c.

In conclusion, for TE waves to propagate, the frequency of the waves must be above a given cutoff frequency. We will also see that this frequency depends on the conditions under which the wave propagates.

Substituting the condition for propagation in **Eq. (17.35)**, we get

$$\gamma_{TE} = \sqrt{k_c^2 - \omega^2\mu\varepsilon} = \sqrt{(-1)(2\pi)^2 f^2\mu\varepsilon\left(1 - \frac{f_c^2}{f^2}\right)} = \pm j\omega\sqrt{\mu\varepsilon}\sqrt{1 - \frac{f_c^2}{f^2}}, \quad f > f_c \tag{17.40}$$

that is, the phase constant ($\gamma = j\beta$ for lossless media) is

$$\boxed{\beta_{TE} = \omega\sqrt{\mu\varepsilon}\sqrt{1 - \frac{f_c^2}{f^2}} \quad \left[\frac{\text{rad}}{\text{m}}\right], \quad f > f_c} \tag{17.41}$$

where, again, we took only the positive form of the phase constant. The phase constant for TE waves is smaller than that for TEM waves since the term under the square root is smaller than 1 for any frequency $f > f_c$.

By definition, the wavelength is

$$\boxed{\lambda_{TE} = \frac{2\pi}{\beta_{TE}} = \frac{2\pi}{\omega\sqrt{\mu\varepsilon}\sqrt{1 - f_c^2/f^2}} = \frac{\lambda}{\sqrt{1 - f_c^2/f^2}} \quad [\text{m}]} \tag{17.42}$$

where λ is the wavelength for TEM waves in unbounded space. The TE wavelength is larger than that for TEM waves for any given frequency for which the two waves propagate.

From the phase constant, we can calculate the phase velocity, again by definition:

$$\boxed{v_{TE} = \frac{\omega}{\beta_{TE}} = \frac{1}{\sqrt{\mu\varepsilon}\sqrt{1 - f_c^2/f^2}} = \frac{v_p}{\sqrt{1 - f_c^2/f^2}} \quad \left[\frac{\text{m}}{\text{s}}\right]} \tag{17.43}$$

where $v_p = 1/\sqrt{\mu\varepsilon}$ is the phase velocity for TEM waves. Therefore, the phase velocity for TE waves is always larger than the phase velocity for TEM waves.

Although we talked about the propagation and phase constants, we neglected the attenuation constant so far and, together with it, the possibility of losses in the domain in which TE waves propagate. Losses may be easily introduced by starting with the propagation constant in **Eq. (17.35)** and replacing the permittivity ε by the complex permittivity $\varepsilon_c = \varepsilon(1 - j\sigma/\omega\varepsilon)$, where σ is the conductivity of the dielectric and ε is its permittivity. This is identical to what we did in **Section 12.7** for plane (TEM) waves. In addition, we will assume here that losses are small: high-loss TE and TM propagation is of little interest in the context of this chapter. Introducing the complex permittivity in the propagation constant in **Eq. (17.35)**, we get

$$\gamma_{TE} = \sqrt{k_c^2 - \omega^2\mu\varepsilon\left(1 - j\frac{\sigma}{\omega\varepsilon}\right)} = \sqrt{-1\left[\omega^2\mu\varepsilon\left(1 - j\frac{\sigma}{\omega\varepsilon}\right) - k_c^2\right]} = j\sqrt{\left(\omega^2\mu\varepsilon - k_c^2\right) - j\omega\mu\sigma} \tag{17.44}$$

where $j = \sqrt{-1}$ was used to rearrange the expression. **Equation (17.44)** may be written as follows:

$$\gamma_{TE} = j\sqrt{\left(\omega^2\mu\varepsilon - k_c^2\right) - j\omega\mu\sigma} = j\sqrt{\omega^2\mu\varepsilon - k_c^2}\left(1 - \frac{j\omega\mu\sigma}{\omega^2\mu\varepsilon - k_c^2}\right)^{1/2} \tag{17.45}$$

For low losses (σ small), the expression in parentheses may be expanded using the binomial expansion so that the attenuation and propagation constants may be separated. The low-loss condition now requires that the second term in the parentheses in **Eq. (17.45)** be small with respect to 1 (see **Section 12.7.2**):

$$\frac{\omega\mu\sigma}{(\omega^2\mu\varepsilon - k_c^2)} \ll 1 \tag{17.46}$$

Expansion of the term in parentheses in **Eq. (17.45)** using the binomial expansion gives

$$\left(1 - \frac{j\omega\mu\sigma}{\omega^2\mu\varepsilon - k_c^2}\right)^{1/2} = 1 - \frac{1}{2}\left(\frac{j\omega\mu\sigma}{\omega^2\mu\varepsilon - k_c^2}\right) - \frac{1}{8}\left(\frac{j\omega\mu\sigma}{\omega^2\mu\varepsilon - k_c^2}\right)^2 + \frac{1}{16}\left(\frac{j\omega\mu\sigma}{\omega^2\mu\varepsilon - k_c^2}\right)^3 + \cdots \tag{17.47}$$

Retaining only the first two terms in the expansion gives an approximation to the propagation constant in **Eq. (17.45)** as

$$\gamma_{TE} \approx j\sqrt{\omega^2\mu\varepsilon - k_c^2}\left(1 - \frac{1}{2}\left(\frac{j\omega\mu\sigma}{\omega^2\mu\varepsilon - k_c^2}\right)\right) = \frac{\omega\mu\sigma}{2\sqrt{\omega^2\mu\varepsilon - k_c^2}} + j\sqrt{\omega^2\mu\varepsilon - k_c^2} \tag{17.48}$$

The first term on the right-hand side is the attenuation constant and the second is the phase constant. To put these in a form compatible with other expressions in this section, we can write k_c in terms of the cutoff frequency [see **Eq. (17.37)**] as $k_c^2 = (2\pi f_c)^2\mu\varepsilon = \omega_c^2\mu\varepsilon$. Substituting this in **Eq. (17.48)** gives

$$\begin{aligned}\gamma_{TE} = \alpha + j\beta &\approx \frac{\omega\mu\sigma}{2\sqrt{\omega^2\mu\varepsilon - \omega_c^2\mu\varepsilon}} + j\sqrt{\omega^2\mu\varepsilon - \omega_c^2\mu\varepsilon} = \frac{\omega\mu\sigma}{2\sqrt{\mu\varepsilon}\sqrt{\omega^2 - \omega_c^2}} + j\sqrt{\mu\varepsilon}\sqrt{\omega^2 - \omega_c^2} \\ &= \frac{\omega\mu\sigma}{2\omega\sqrt{\mu\varepsilon}\sqrt{1 - (\omega_c/\omega)^2}} + j\omega\sqrt{\mu\varepsilon}\sqrt{1 - \left(\frac{\omega_c}{\omega}\right)^2} = \frac{\sigma}{2}\sqrt{\frac{\mu}{\varepsilon}}\frac{1}{\sqrt{1 - (f_c/f)^2}} + j\omega\sqrt{\mu\varepsilon}\sqrt{1 - (f_c/f)^2}\end{aligned} \tag{17.49}$$

Separating the real and imaginary parts of the propagation constant gives the attenuation and phase constants. The phase constant is identical to that obtained in **Eq. (17.41)** for the lossless case (i.e., under the assumption of low losses and neglecting the higher-order terms in the expansion in **Eq. (17.47)**, the phase constant in the lossy case is the same as for the lossless case). The attenuation constant is

$$\boxed{\alpha_{TE} = \frac{\sigma\eta}{2\sqrt{1 - (f_c/f)^2}} \quad \left[\frac{\text{Np}}{\text{m}}\right]} \tag{17.50}$$

where $\eta = \sqrt{\mu/\varepsilon}$ is the intrinsic impedance of the material in which the waves propagate. This relation only holds above cutoff since the condition we imposed in **Eq. (17.46)**, in effect, requires that $f > f_c$. At cutoff, there is no propagation, whereas below cutoff, the relation in **Eq. (17.39)** must be used.

Finally, we can also calculate the wave impedance by dividing the transverse electric field intensity by the transverse magnetic field intensity [E_x in **Eq. (17.27)** and H_y in **Eq. (17.30)** or E_y in **Eq. (17.28)** and H_x in **Eq. (17.29)**] and imposing the direction of propagation in the positive z direction:

$$\boxed{Z_{TE} = \frac{E_x}{H_y} = -\frac{E_y}{H_x} = \frac{j\omega\mu}{\gamma} \quad [\Omega]} \tag{17.51}$$

Substituting the propagation constant for waves above cutoff from **Eq. (17.40)**, we get

$$\boxed{Z_{TE} = \frac{j\omega\mu}{j\omega\sqrt{\mu\varepsilon}\sqrt{1 - f_c^2/f^2}} = \sqrt{\frac{\mu}{\varepsilon}}\frac{1}{\sqrt{1 - f_c^2/f^2}} = \frac{\eta}{\sqrt{1 - f_c^2/f^2}} \quad [\Omega]} \tag{17.52}$$

The wave impedance for TE waves is frequency dependent and always larger than the wave impedance for TEM waves except at $f \to \infty$, where the two are equal. At cutoff ($f = f_c$), the wave impedance for TE propagation is infinite. This result gives another interpretation of cutoff: infinite wave impedance which, of course, means that for any finite electric field intensity at cutoff, the magnetic field intensity is zero and, therefore, there can be no propagation of power.

17.3.3 Transverse Magnetic (TM) Waves

The condition for existence of TM waves is $H_z = 0$, that is, the only field component in the direction of propagation is an electric field intensity component E_z. Substituting this condition in **Eqs. (17.18)** through **(17.21)** yields the field equations for TM waves since these waves will only have components of the magnetic field transverse to the direction of propagation as required. The steps are identical to those for the TE waves. First, we write the field components by substituting the condition $H_z = 0$ in **Eqs. (17.18)** through **(17.21)**:

$$\boxed{E_x = \frac{-\gamma}{\gamma^2 + k^2}\frac{\partial E_z}{\partial x} \quad \left[\frac{\text{V}}{\text{m}}\right]} \qquad (17.53)$$

$$\boxed{E_y = \frac{-\gamma}{\gamma^2 + k^2}\frac{\partial E_z}{\partial y} \quad \left[\frac{\text{V}}{\text{m}}\right]} \qquad (17.54)$$

$$\boxed{H_x = \frac{j\omega\varepsilon}{\gamma^2 + k^2}\frac{\partial E_z}{\partial y} \quad \left[\frac{\text{A}}{\text{m}}\right]} \qquad (17.55)$$

$$\boxed{H_y = \frac{-j\omega\varepsilon}{\gamma^2 + k^2}\frac{\partial E_z}{\partial x} \quad \left[\frac{\text{A}}{\text{m}}\right]} \qquad (17.56)$$

The wave equation equivalent to these four equations is obtained by taking the derivative of H_x with respect to y in **Eq. (17.55)** and the derivative of H_y with respect to x in **Eq. (17.56)**. Subtracting the second from the first and then using **Eq. (17.4)** to eliminate the components of **H** (see **Exercise 17.1**) gives

$$\boxed{\frac{\partial^2 E_z}{\partial x^2} + \frac{\partial^2 E_z}{\partial y^2} + \left(\gamma^2 + k^2\right)E_z = 0} \qquad (17.57)$$

Comparison of **Eqs. (17.57)** and **(17.33)** reveals that the two equations are identical in form. Therefore, we should expect all relations in **Eqs. (17.34)** through **(17.50)** to remain unchanged since these were obtained from **Eq. (17.33)**. Also the same are the definitions of cutoff, propagation above cutoff, and attenuation below cutoff.

However, the wave impedance is calculated from the transverse components of the electric and magnetic field intensities and these are different, as can be seen from **Eqs.(17.53)** through **(17.56)**. For propagation in the z direction, we take the transverse components as E_x and H_y from **Eqs. (17.53)** and **(17.56)** [or E_y and $-H_x$ from **Eqs. (17.54)** and **(17.55)**] and get the wave impedance for TM waves:

$$Z_{TM} = \frac{E_x}{H_y} = -\frac{E_y}{H_x} = \frac{\gamma}{j\omega\varepsilon} = \frac{j\omega\sqrt{\mu\varepsilon}\sqrt{1 - f_c^2/f^2}}{j\omega\varepsilon} = \sqrt{\frac{\mu}{\varepsilon}}\sqrt{1 - \frac{f_c^2}{f^2}} = \eta\sqrt{1 - \frac{f_c^2}{f^2}} \quad [\Omega] \qquad (17.58)$$

This impedance is lower than the TEM wave impedance η, except at $f \to \infty$, where the two are equal. At cutoff ($f = f_c$), the wave impedance for TM propagation is zero. This means that for any given magnetic field intensity at cutoff, the electric field intensity is zero and, therefore, there can be no propagation of energy.

Table 17.2 summarizes the results we obtained for TE and TM waves in comparison with TEM waves, all above cutoff ($f > f_c$). The definitions of TE and TM waves in **Eqs. (17.27)** through **(17.30)** and **Eqs. (17.53)** through **(17.56)** were in completely general terms. These apply universally since no other conditions (such as boundary or interface conditions) were required. How we use these equations and how we define the interface conditions between different materials and, in particular, for conducting interfaces will define the properties of the waves. In particular, we will look next to the properties of the waves as they propagate in guiding structures which we call waveguides.

Table 17.2 Properties of TEM, TE, and TM waves

	TEM waves	TE waves	TM waves
Cutoff frequency: f_c [Hz]	$f_c = 0$	$f_c = \frac{k_c}{2\pi\sqrt{\mu\varepsilon}}$	$f_c = \frac{k_c}{2\pi\sqrt{\mu\varepsilon}}$
Lossless phase constant: β [rad/m]	$\omega\sqrt{\mu\varepsilon}$	$\omega\sqrt{\mu\varepsilon}\sqrt{1-\frac{f_c^2}{f^2}}$	$\omega\sqrt{\mu\varepsilon}\sqrt{1-\frac{f_c^2}{f^2}}$
Lossless propagation constant: γ	$j\omega\sqrt{\mu\varepsilon}$	$j\omega\sqrt{\mu\varepsilon}\sqrt{1-\frac{f_c^2}{f^2}}$	$j\omega\sqrt{\mu\varepsilon}\sqrt{1-\frac{f_c^2}{f^2}}$
Lossless phase constant: β [rad/m]	$\omega\sqrt{\mu\varepsilon}$	$\omega\sqrt{\mu\varepsilon}\sqrt{1-\frac{f_c^2}{f^2}}$	$\omega\sqrt{\mu\varepsilon}\sqrt{1-\frac{f_c^2}{f^2}}$
Low-loss attenuation constant: α [Np/m]	$\frac{\sigma\eta}{2}$	$\frac{\sigma\eta}{2\sqrt{1-(f_c/f)^2}}$	$\frac{\sigma\eta}{2\sqrt{1-(f_c/f)^2}}$
Low-loss propagation constant: γ	$\frac{\sigma\eta}{2}+j\omega\sqrt{\mu\varepsilon}$	$\frac{\sigma\eta}{2\sqrt{1-(f_c/f)^2}}$ $+j\omega\sqrt{\mu\varepsilon}\sqrt{1-f_c^2/f^2}$	$\frac{\sigma\eta}{2\sqrt{1-(f_c/f)^2}}$ $+j\omega\sqrt{\mu\varepsilon}\sqrt{1-f_c^2/f^2}$
Wavelength: λ [m]	$\frac{1}{f\sqrt{\mu\varepsilon}}$	$\frac{1}{f\sqrt{\mu\varepsilon}\sqrt{1-f_c^2/f^2}}$	$\frac{1}{f\sqrt{\mu\varepsilon}\sqrt{1-f_c^2/f^2}}$
Phase velocity: v_p [m/s]	$\frac{1}{\sqrt{\mu\varepsilon}}$	$\frac{1}{\sqrt{\mu\varepsilon}\sqrt{1-f_c^2/f^2}}$	$\frac{1}{\sqrt{\mu\varepsilon}\sqrt{1-f_c^2/f^2}}$
Wave impedance: Z [Ω]	$\sqrt{\frac{\mu}{\varepsilon}}$	$\sqrt{\frac{\mu}{\varepsilon}}\frac{1}{\sqrt{1-f_c^2/f^2}}$	$\sqrt{\frac{\mu}{\varepsilon}}\sqrt{1-f_c^2/f^2}$

Note: $\eta = \sqrt{\mu/\varepsilon}$ is the no-loss intrinsic impedance of the medium in which the waves propagate

Exercise 17.1 Derive **Eq. (17.57)** from **Eqs. (17.53)** through **(17.56)**.

Example 17.1 An electromagnetic wave, propagating in a guiding structure, is given as follows:

$$\mathbf{E} = \hat{\mathbf{z}} jE_0 e^{-j\beta x} \quad [\text{V/m}] \quad \text{and} \quad \mathbf{H} = -(\hat{\mathbf{y}} jH_0 + \hat{\mathbf{x}} H_0)e^{-j\beta x} \quad [\text{A/m}]$$

(a) The medium in the waveguide is free space, the frequency is $f = 3$ GHz, and its phase constant is $\beta = 12\pi$ [rad/m]. Determine the type of wave.
(b) Find the cutoff frequency, wave impedance, and phase velocity of the wave.
(c) Calculate the time-averaged power density in the structure.
(d) Calculate the magnitude of the electric and magnetic field intensities E_0 and H_0 if the time-averaged power density is uniform in the structure and equals 100 W/m^2.

Solution: The type of wave is determined from the longitudinal component of the wave. After determining the type of wave, its propagation properties are determined from **Table 17.2**:

(a) The wave propagates in the positive x direction and has transverse components in the y and z directions. The longitudinal component, H_x, is a magnetic field component. Therefore, this is a TE wave.

(b) The wave propagates in free space. From **Table 17.2**, we get

$$\beta = \omega\sqrt{\mu_0\varepsilon_0}\sqrt{1-\frac{f_c^2}{f^2}} \quad \left[\frac{\text{rad}}{\text{m}}\right] \quad \rightarrow \quad f_c = \sqrt{f^2 - \frac{\beta^2 f^2}{\omega^2\mu_0\varepsilon_0}} = \sqrt{f^2 - \frac{\beta^2 c^2}{4\pi^2}} \quad [\text{Hz}]$$

With the given values, this is

$$f_c = \sqrt{9\times 10^{18} - \frac{144\times\pi^2\times 9\times 10^{16}}{4\times\pi^2}} = 2.4\times 10^9 \quad [\text{Hz}]$$

Any wave below this frequency will not be propagated in the given structure.

The wave impedance is given as

$$Z_{TE} = \sqrt{\frac{\mu_0}{\varepsilon_0}}\frac{1}{\sqrt{1-f_c^2/f^2}} = \frac{377}{\sqrt{1-\left(\frac{2.4\times 10^9}{3\times 10^9}\right)^2}} = 628 \quad [\Omega]$$

and the phase velocity is

$$v_{TE} = \frac{1}{\sqrt{\mu_0\varepsilon_0}\sqrt{1-f_c^2/f^2}} = \frac{c}{\sqrt{1-f_c^2/f^2}} = \frac{3\times 10^8}{0.6} = 5\times 10^8 \quad [\text{m/s}]$$

The wave impedance is larger than the intrinsic impedance of free space, and the phase velocity is larger than the speed of light. The fact that the phase velocity can be larger than the speed of light will be discussed in the following sections and was also discussed in **Sections 12.7.4** and **13.3.1**.

(c) The time-averaged Poynting vector is

$$\mathcal{P}_{av} = \frac{\text{Re}\{\mathbf{E}\times\mathbf{H}^*\}}{2} = \frac{\text{Re}\left\{\left(\hat{\mathbf{z}}jE_0e^{-j\beta x}\right)\times\left([\hat{\mathbf{y}}jH_0 - \hat{\mathbf{x}}H_0]e^{+j\beta x}\right)\right\}}{2} = \frac{\text{Re}\{\hat{\mathbf{x}}E_0H_0 - \hat{\mathbf{y}}jE_0H_0\}}{2} = \hat{\mathbf{x}}\frac{E_0H_0}{2} \quad \left[\frac{\text{W}}{\text{m}^2}\right]$$

and, as expected, the time-averaged power propagates in the x direction. The transverse component of the power density is imaginary.

(d) The magnitude of the electric and magnetic field intensities may be calculated from the wave impedance, which is equal to the ratio between the transverse electric field intensity and the transverse magnetic field intensity, and the total power per unit area:

$$Z_{TE} = -\frac{E_z}{H_y} = \frac{E_0}{H_0} = 628 \quad [\Omega], \quad \mathcal{P}_{av} = \frac{E_0H_0}{2} = 100 \quad \left[\frac{\text{W}}{\text{m}^2}\right]$$

From these, we get

$$E_0 = 354.4 \quad [\text{V/m}], \quad H_0 = 0.564 \quad [\text{A/m}].$$

17.4 TE Propagation in Parallel Plate Waveguides

Now that we defined TE and TM waves and their general properties, it is time to define the conditions under which TE and TM waves can exist. This will lead to the definition of waveguides and to the properties of waves in waveguides.

To keep in line with a simple explanation of guided waves but also in accordance with the definition of transverse electric and transverse magnetic fields, we consider here the guiding of electromagnetic waves between two parallel conducting surfaces, as shown in **Figure 17.3**. We will discuss TE waves first, followed by TM waves but will try to give a physical feel to both the phenomenon of guided waves and to the properties of these waves. For this reason, we will rely less on the properties defined in the previous sections and more on physical properties of the waveguide.

The parallel plate waveguide is made of two surfaces, which may be viewed as perfect conductors. This separation into two surfaces **(Figure 17.3)** allows analysis of the reflections at each surface separately. This is convenient since we have already discussed many of the properties of reflection at conducting surfaces in **Chapter 13** and should be able to use those results.

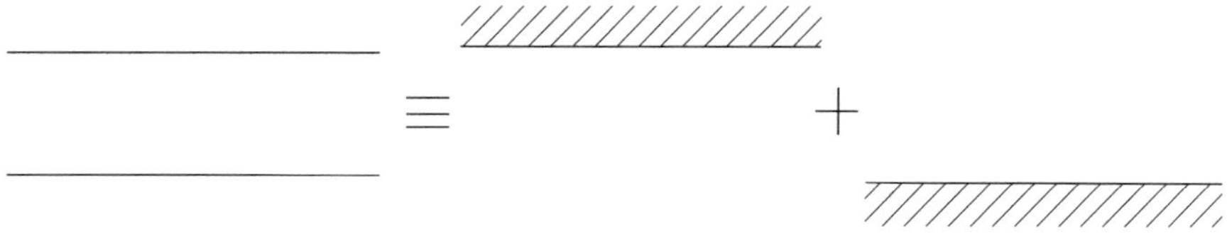

Figure 17.3 Construction of a parallel plate waveguide

$\equiv$ $+$

Suppose a uniform plane wave impinges on the lower surface at an angle of incidence θ_i and the wave is polarized perpendicular to the plane of incidence. The incident and reflected electric and magnetic field intensities are as shown in **Figure 17.4a**. The reflected wave from the lower surface then reflects off the upper surface, as shown in **Figure 17.4b**. The wave propagates between the plates by repeatedly reflecting off the conductors. To calculate the fields and the propagation properties between the plates, we calculate the fields above the lower plate using **Figure 17.4a** and then take into account the effect of the upper plate, without the need to calculate reflections off the upper plate. The incident wave propagates in the direction $\hat{\mathbf{p}}_i = -\hat{\mathbf{x}}\cos\theta_i + \hat{\mathbf{z}}\sin\theta_i$ and the reflected wave is in the direction $\hat{\mathbf{p}}_r = \hat{\mathbf{x}}\cos\theta_i + \hat{\mathbf{z}}\sin\theta_i$. Based on these (see also **Section 13.3.1** and **Exercise 13.6**), we obtain the incident electric field as

$$\mathbf{E}_i(x,z) = \hat{\mathbf{y}}E_i e^{-j\beta(-x\cos\theta_i + z\sin\theta_i)} \quad [\mathrm{V/m}] \tag{17.59}$$

$$\mathbf{H}_i(x,z) = \frac{E_i}{\eta}(-\hat{\mathbf{x}}\sin\theta_i - \hat{\mathbf{z}}\cos\theta_i)e^{-j\beta(-x\cos\theta_i + z\sin\theta_i)} \quad [\mathrm{A/m}] \tag{17.60}$$

where β and η are the phase constant and intrinsic impedance in the medium between the plates. The reflected electric and magnetic field intensities are

$$\mathbf{E}_r(x,z) = -\hat{\mathbf{y}}E_i e^{-j\beta(x\cos\theta_i + z\sin\theta_i)} \quad [\mathrm{V/m}] \tag{17.61}$$

$$\mathbf{H}_r(x,z) = \frac{E_i}{\eta}(\hat{\mathbf{x}}\sin\theta_i - \hat{\mathbf{z}}\cos\theta_i)e^{-j\beta(x\cos\theta_i + z\sin\theta_i)} \quad [\mathrm{A/m}] \tag{17.62}$$

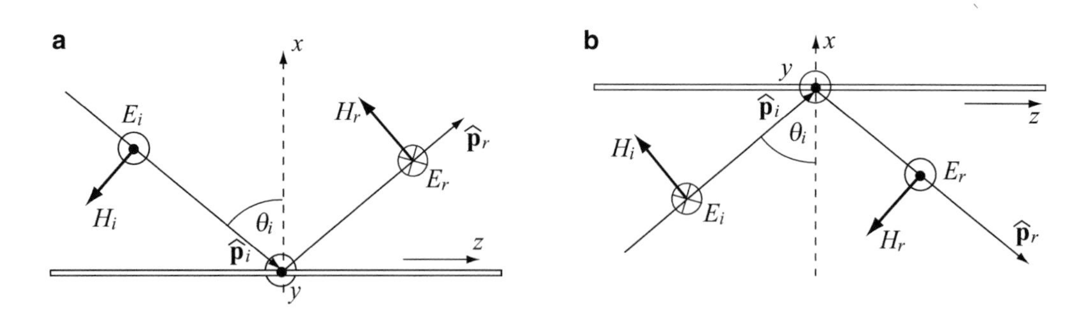

Figure 17.4 (**a**) Reflection of the incident wave off the lower plate in a parallel plate waveguide. (**b**) Reflection off the upper plate

The total electric and magnetic field intensities are (after rearranging terms and using the relation $e^{j\beta x\cos\theta} - e^{-j\beta x\cos\theta} = j2\sin(\beta x\cos\theta)$):

$$\mathbf{E}_1(x,z) = \hat{\mathbf{y}}j2E_i\sin(\beta x\cos\theta_i)e^{-j\beta z\sin\theta_i} \quad [\mathrm{V/m}] \tag{17.63}$$

$$\mathbf{H}_1(x,z) = -2\frac{E_i}{\eta}[\hat{\mathbf{x}}j\sin\theta_i\sin(\beta x\cos\theta_i) + \hat{\mathbf{z}}\cos\theta_i\cos(\beta x\cos\theta_i)]e^{-j\beta z\sin\theta_i} \quad [\mathrm{A/m}] \tag{17.64}$$

This much is almost a direct rewriting of the result in **Section 13.3.1** and has been done in detail in **Exercise 13.6**. We note that the electric field intensity has only a y component whereas the magnetic field intensity has two components: a component in the x direction and a component in the z direction. Therefore, this is a TE wave. Note also that both the incident and reflected waves are TEM waves (both **E** and **H** are perpendicular to the direction of propagation) whereas their sum is a TE wave.

The first task now is to calculate the time-averaged Poynting vector to see how power propagates. The Poynting vector will then tell us the direction of propagation of the wave:

$$\boldsymbol{\mathcal{P}}_{av} = \frac{1}{2}\mathrm{Re}\{\mathbf{E}_1 \times \mathbf{H}_1^*\} \quad [\mathrm{W/m^2}] \tag{17.65}$$

We note the following:

$$\mathbf{H}_1 = -(\hat{\mathbf{z}}H_{1z} + \hat{\mathbf{x}}jH_{1x})e^{-j\beta z\sin\theta_1} \quad \Rightarrow \quad \mathbf{H}_1^* = -(\hat{\mathbf{z}}H_{1z} - \hat{\mathbf{x}}jH_{1x})e^{j\beta z\sin\theta_i} \quad [\mathrm{A/m}] \tag{17.66}$$

From **Eqs. (17.63)** and **(17.64)** and using the relation in **Eq. (17.65)**, we get

$$\boldsymbol{\mathcal{P}}_{av} = \mathrm{Re}\left\{-\hat{\mathbf{x}}j\frac{2E_i^2}{\eta}\sin(2\beta x\cos\theta_i)\cos\theta_i + \hat{\mathbf{z}}\,\frac{2E_i^2}{\eta}\sin^2(\beta x\cos\theta_i)\sin\theta_i\right\} \quad \left[\frac{\mathrm{W}}{\mathrm{m}^2}\right] \tag{17.67}$$

where $\hat{\mathbf{y}} \times (-\hat{\mathbf{x}}) = \hat{\mathbf{z}}$, $\hat{\mathbf{y}} \times (-\hat{\mathbf{z}}) = -\hat{\mathbf{x}}$, $j^2 = -1$, $\sin(\beta\cos\theta_i)\cos(\beta\cos\theta_i) = (1/2)\sin(2\beta\cos\theta_i)$ were used to simplify the expression. The Poynting vector has a real part in the z direction and an imaginary part in the x direction. The real part is

$$\boldsymbol{\mathcal{P}}_{av} = \hat{\mathbf{z}}\,\frac{2E_i^2}{\eta_1}\sin^2(\beta x\cos\theta_i)\sin\theta_i \quad [\mathrm{W/m^2}] \tag{17.68}$$

The conclusion is that the time-averaged power propagates entirely in the z direction. The x component of the Poynting vector is imaginary, and as we saw in **Chapter 13**, this means there is no propagation in this direction. In the x direction, there are only standing waves.

Before continuing and discussing the properties of the waves, it is worth pausing and looking at the physical meaning of the result above. First, from **Figure 17.4a**, we note that the total wave is a superposition of two plane waves (transverse electromagnetic waves). The incident wave, propagates in the direction $\hat{\mathbf{p}}_i = -\hat{\mathbf{x}}\cos\theta_i + \hat{\mathbf{z}}\sin\theta_i$, and the reflected wave, propagates in the direction $\hat{\mathbf{p}}_r = \hat{\mathbf{x}}\cos\theta_i + \hat{\mathbf{z}}\sin\theta_i$. Note, also, that the phase constant of each wave depends on the phase velocity of the wave in the material above the plate and the angle θ_i as if the conductor did not exist. From this, we can draw the following picture: The phase constants of the incident and reflected waves are the same in the direction of propagation of each wave and these depend on the phase velocity in the given material:

$$v_p = \frac{1}{\sqrt{\mu\varepsilon}} \quad \left[\frac{\mathrm{m}}{\mathrm{s}}\right], \quad \lambda = \frac{v_p}{f} \quad [\mathrm{m}], \quad \beta = \frac{2\pi}{\lambda} = \frac{2\pi v_p}{f} = \frac{2\pi}{f\sqrt{\mu\varepsilon}} \quad \left[\frac{\mathrm{rad}}{\mathrm{m}}\right] \tag{17.69}$$

These are exactly the properties we expect from a plane wave propagating in a material with properties ε and μ.

The phase velocities in the guide direction and in the direction transverse to the guide direction may be found from **Figure 17.5**. Consider the front of a plane wave propagating at an angle θ_i to the normal, as wave front A. After some time, the wave front has propagated in the direction of propagation of the wave, at a velocity v_p, and is now at wave front B. The horizontal distance the wave has traveled during the same time is the distance between A and B'. Thus, the wave front has propagated faster horizontally than in the direction of θ_i. From **Figure 17.5**, the horizontal phase velocity is

$$\boxed{v_g = \frac{v_p}{\sin\theta_i} \quad \left[\frac{\text{m}}{\text{s}}\right]} \tag{17.70}$$

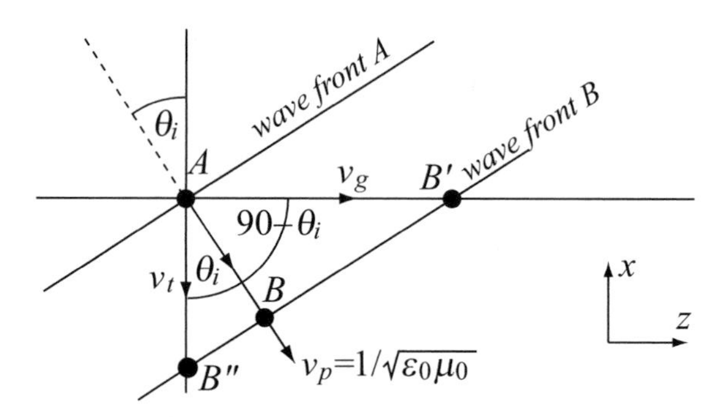

Figure 17.5 Relation between phase velocity of the incident wave and the guide and transverse phase velocities

This is called the ***guide phase velocity***.

Similarly, the vertical distance the wave traveled is the distance between point A and B''. Since this distance is traveled during the same time the wave traveled from point A to B, the phase velocity in the vertical direction is

$$\boxed{v_t = \frac{v_p}{\cos\theta_i} \quad \left[\frac{\text{m}}{\text{s}}\right]} \tag{17.71}$$

The phase velocity in the x direction (perpendicular to the conducting surface) is called the ***transverse phase velocity***.

Note that both the guide and transverse phase velocities are larger than or equal to the phase velocity of the plane wave, v_p, for any angle between zero and $\pi/2$. At zero incidence angle, the guide phase velocity is infinite and the transverse phase velocity is v_p. At an incidence angle equal to $\pi/2$, the transverse phase velocity is infinite and the guide phase velocity is v_p.

This result raises a rather interesting question. Suppose the wave propagates in free space above a conducting surface. In this case, we know the wave propagates at the speed of light ($v_p = c$). This means that the phase velocity is always larger or equal to the speed of light in any direction which does not coincide with the direction of propagation of the plane wave. You probably are distressed by this, but there is really no difficulty, since only the phase moves at this speed. No physical quantity such as power moves at this speed. Perhaps the following example may explain this apparent difficulty: Suppose an ocean wave propagates toward shore at an angle α and a constant speed v as in **Figure 17.6**. Suppose also, that we could mark two points on the wave; one point, B, is the point the wave meets the shore at time t. The second is any point A on the crest of the wave. The wave propagates at a speed v and the time it takes point A to reach the shore (at point A') is $\Delta t = t = d/v$. During the same time, the point the wave meets the shore (point B) has also moved to point A'. Since the distance between B and A' is $d/\sin\alpha$, the speed of propagation of point B is $v = (d/\sin\alpha)/\Delta t = v/\sin\alpha$. For any angle $0 < \alpha < \pi/2$, the speed of point B along the shore is larger than that of point A. If $v = c$, point B travels at speeds larger than the speed of light. However, this is only an apparent speed, since nothing propagates physically from point B to A' (i.e., a surfer cannot surf along the shore at the speed point B moves!).

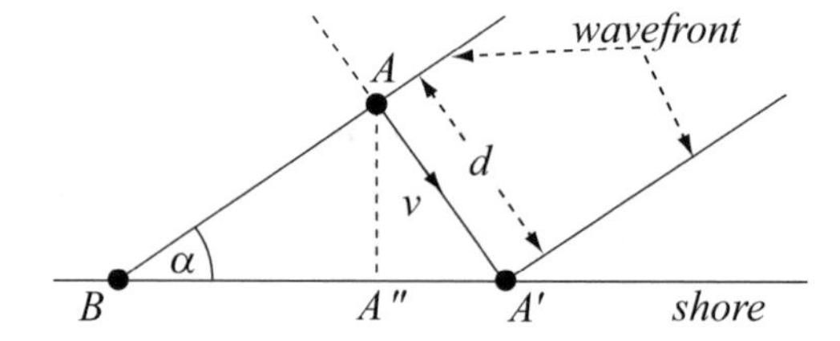

Figure 17.6 Relation between phase velocity and speed of propagation of an ocean wave impinging on the shore

Now that the phase velocities in the guiding and transverse directions are properly understood, we can also define the phase constants and wavelengths in the guiding and transverse directions. These are

$$\beta_t = \frac{\omega}{v_t} = \frac{\omega}{v_p}\cos\theta_i = \beta\cos\theta_i \quad [\text{rad/m}], \quad \lambda_t = \frac{2\pi}{\beta_t} = \frac{2\pi v_p}{\omega\cos\theta_i} = \frac{v_p}{f\cos\theta_i} = \frac{1}{f\sqrt{\mu\varepsilon}\cos\theta_i} \quad [\text{m}] \tag{17.72}$$

$$\beta_g = \frac{\omega}{v_g} = \frac{\omega}{v_p}\sin\theta_i = \beta\sin\theta_i \quad [\text{rad/m}], \quad \lambda_g = \frac{2\pi}{\beta_g} = \frac{2\pi v_p}{\omega\sin\theta_i} = \frac{v_p}{f\sin\theta_i} = \frac{1}{f\sqrt{\mu\varepsilon}\sin\theta_i} \quad [\text{m}] \tag{17.73}$$

The phase constants in the guiding and transverse directions are always smaller than those in the direction of propagation of the oblique wave, whereas the wavelengths are always larger than for the oblique wave.

From the imaginary part of the Poynting vector, we concluded that in addition to propagation (in the z direction), there is also a standing wave in the transverse (x) direction. To see how this standing wave behaves, we return now to the electric field intensity in **Eq. (17.63)**. We note that the field intensity has a sinusoidal variation with respect to x. Thus, the electric field intensity is zero at the conductor's surface and at any other point in space for which

$$\sin(\beta x\cos\theta_i) = 0 \quad \rightarrow \quad \beta x_m\cos\theta_i = m\pi, \quad m = 1, 2, 3, \ldots \tag{17.74}$$

where x_m are the locations of the nodes of the standing wave, a point we also made in **Chapter 13**. While the electric field propagates in the z direction, it changes its amplitude in the x direction, but the nodes of the electric field intensity remain fixed at points x_m such that

$$x_m = \frac{m\pi}{\beta\cos\theta_i} \quad [\lambda], \quad m = 1, 2, 3.... \tag{17.75}$$

Although m can be positive or negative, we will only take absolute values, since the space $x < 0$ in **Figure 17.4a** is assumed to be conducting. x_m can also be written in terms of the transverse wavelength of the wave ($\lambda_t = 2\pi/\beta_t$) or in terms of frequency of the wave ($\lambda f = v_p$). Thus, we can write

$$x_m = \frac{m\lambda}{2\cos\theta_i} = \frac{mv_p}{2f\cos\theta_i} = \frac{m}{2f\sqrt{\mu\varepsilon}\cos\theta_i} = m\frac{\lambda_t}{2} \quad [\lambda], \quad m = 1, 2, 3.... \tag{17.76}$$

where $v_p = 1/\sqrt{\mu\varepsilon}$ is the TEM phase velocity in the material above the conducting plane. Note that x_m is measured in wavelengths.

For now, we note that because the distance between two consecutive nodes of the standing wave pattern is any multiple of half-wavelengths, the distance $x\cos\theta_i$ must be a multiple of $\lambda_t/2$.

Now, suppose that we place the second conducting plane at a point x as given by **Eq. (17.76)**. Since this point is a node in the standing wave pattern, nothing in the standing wave or the propagation properties of the wave would change. There is, however, a very important difference: Waves now must be confined to the space between the two plates. We are free to place the plates at any position x as long as the above condition is satisfied. The geometric configuration described above constitutes a waveguide: the waves propagate in a given direction while they vary spatially in the transverse direction. **Figure 17.7** shows a parallel plate waveguide with a separation $d = m\lambda_t/2$. Note, also, the relation between λ, λ_t, and λ_g for the waveguide.

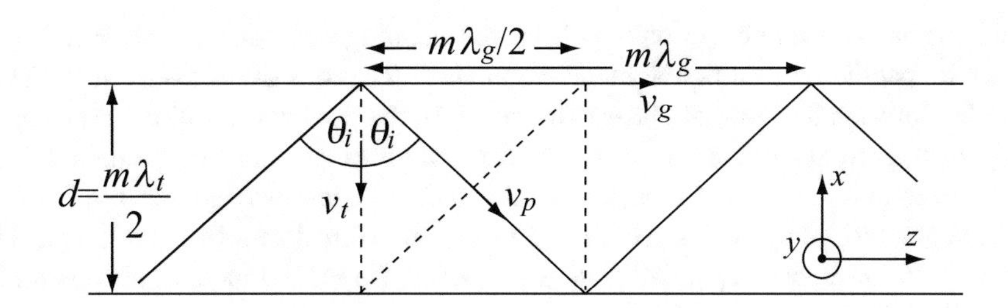

Figure 17.7 A parallel plate waveguide showing the guide and transverse phase velocities and wavelengths

In this case, we placed the conducting plates at very convenient locations: at the nodes of the standing waves. By doing so, we avoided the need to worry about interface conditions since the electric field intensity is zero at the conductors.

Suppose now we start the other way around: Instead of placing the conductors at the distance equivalent to that between two or more nodes of the standing wave, we place them at an arbitrary distance d. What happens to the wave now? We can easily answer this question from known properties: First, the electric field intensity at the conducting surfaces must be zero; that is, the two plates must still be at nodes of the standing wave pattern. Second, neither the wavelength nor, alternatively, the frequency of operation changed because of our choice of location for the plates. Inspection of **Eq. (17.75)** reveals that the only variable possible is the angle of incidence θ_i. The simple answer then is that if we were to place the plates at arbitrary locations, these locations will become nodes in the standing wave pattern (electric field intensity is zero) and only waves with an angle of incidence that satisfy this condition will propagate. It is therefore reasonable to rewrite **Eq. (17.75)** in terms of the angle as

$$\cos\theta_i = \frac{m\lambda}{2d} = \frac{mv_p}{2fd} = \frac{m}{2fd\sqrt{\mu\varepsilon}}, \quad m = 1,2,3\ldots \qquad (17.77)$$

where d is the distance between the plates. From **Eq. (17.77)**, it appears that the larger d, the smaller the value of $\cos\theta_i$ for any given value of m. In the limit, $\cos\theta_i$ approaches zero (θ_i approaches $\pi/2$). The same effect can be observed with regard to frequency: For a given waveguide, the higher the frequency of the wave, the smaller $\cos\theta_i$ and the closer θ_i is to $\pi/2$. On the other hand, as d becomes smaller at a given frequency or as the frequency becomes lower for a given d, the angle θ_i approaches zero. From **Eq. (17.77)**, for any frequency, such that

$$2fd\sqrt{\mu\varepsilon} > m \quad \rightarrow \quad \cos\theta_i < 1, \quad m = 1,2,3,\ldots \qquad (17.78)$$

Thus, $\sin\theta_i > 0$ and the power propagated in the waveguide [based on **Eq. (17.68)**] is real and larger than zero. At

$$2fd\sqrt{\mu\varepsilon} = m \quad \rightarrow \quad \cos\theta_i = 1, \quad m = 1,2,3,\ldots \qquad (17.79)$$

At this frequency, there is no propagation of power in the waveguide since $\sin\theta_i = 0$ and the power in **Eq. (17.68)** is zero. The frequency at which a wave ceases to propagate was defined in **Section 17.3.2** as the cutoff frequency for the wave. From **Eq. (17.79)**,

$$\boxed{f_{cm} = \frac{m}{2d\sqrt{\mu\varepsilon}} \quad [\text{Hz}], \quad m = 1,2,3\ldots} \qquad (17.80)$$

Comparing this with the cutoff frequency, we obtained in **Eq. (17.37)**, the cutoff wave number may be written as

$$\boxed{k_{cm} = \frac{m\pi}{d} \quad \left[\frac{\text{rad}}{\text{m}}\right], \quad m = 1,2,3\ldots} \qquad (17.81)$$

and for each value of m, we have a different cutoff wave number. Similarly, we can write the cutoff wavelength as $\lambda_{cm} = 2\pi/k_{cm}$:

$$\boxed{\lambda_{cm} = \frac{2d}{m} \quad [\text{m}], \quad m = 1,2,3\ldots} \qquad (17.82)$$

Recall that m is the number of half-cycles in the standing wave pattern. We call this a ***mode of propagation*** and the cutoff frequency is specific for a particular mode. For $m = 1$, the mode of propagation is a TE_1 mode. The general mode is called a TE_m mode. Between the parallel plates (**Figure 17.7**), the field in the y direction is uniform. This results in a zero mode (no standing wave pattern) in this direction. Thus we can also call the TE_1 mode a TE_{10} mode or, in general, a TE_{m0} mode. The latter notation is more common, especially since practical waveguides are finite in both transverse directions and are therefore characterized by modes in each of the transverse directions. In a parallel plate waveguide one of the mode indices is always zero. In TE modes in parallel plate waveguides, the second index is always zero, but the first cannot be zero [$m \neq 0$: see **Eq. (17.74)**].

Now that we have the electric and magnetic fields for the parallel plate waveguide and the relations among frequency, dimensions, and modes are known, we can calculate other properties of the waveguide. One important parameter is the ***guide wavelength***, which is discussed next.

Consider a wave at any frequency $f > f_{cm}$ so that the wave propagates between two plates, separated a distance d apart, as in **Figure 17.7**. From **Eq. (17.77)**, the angle of incidence must be

$$\cos\theta_i = \frac{m\lambda}{2d} = \frac{\lambda}{\lambda_{cm}} = \frac{f_{cm}}{f}, \quad m = 1,2,3\ldots \tag{17.83}$$

Note that λ_{cm} differs for each value of m for any given value of d as in **Eq. (17.82)**. Thus, again we have the familiar situation in which the closer the wavelength of the wave to the cutoff wavelength, the closer $\cos\theta_i$ is to 1 and θ_i to zero.

What about the properties of the propagating wave? Since wave properties depend on the angle of incidence, we can write for the wave of frequency $f > f_{cm}$:

$$\sin\theta_i = \sqrt{1 - \cos^2\theta_i} = \sqrt{1 - \frac{\lambda^2}{\lambda_{cm}^2}} = \sqrt{1 - \frac{f_{cm}^2}{f^2}}, \quad m = 1,2,3\ldots \tag{17.84}$$

With this, the guide phase velocity **[Eq. (17.70)]**, the guide wavelength of the wave (λ_g) **[Eq. (17.73)]**, and guide phase constant are

$$\boxed{v_g = \frac{v_p}{\sin\theta_i} = \frac{v_p}{\sqrt{1 - \lambda^2/\lambda_{cm}^2}} = \frac{v_p}{\sqrt{1 - f_{cm}^2/f^2}} \quad \left[\frac{\text{m}}{\text{s}}\right]} \tag{17.85}$$

$$\boxed{\lambda_g = \frac{\lambda}{\sin\theta_i} = \frac{\lambda}{\sqrt{1 - \lambda^2/\lambda_{cm}^2}} = \frac{\lambda}{\sqrt{1 - f_{cm}^2/f^2}} \quad [\text{m}]} \tag{17.86}$$

$$\boxed{\beta_g = \beta\sqrt{1 - \lambda^2/\lambda_{cm}^2} = \beta\sqrt{1 - f_{cm}^2/f^2} \quad \left[\frac{\text{rad}}{\text{m}}\right]} \tag{17.87}$$

With these quantities, we go back to the equations for the electric and magnetic field intensities and write them in terms of the propagation constants. From **Eqs. (17.84)** and **(17.83)**, we write

$$\boxed{\sin\theta_i = \sqrt{1 - \lambda^2/\lambda_{cm}^2}, \quad \cos\theta_i = \frac{m\pi}{\beta d}} \tag{17.88}$$

Also, we will denote the quantity $E_0 = 2E_i$ as the amplitude of the electric field intensity in **Eqs. (17.63)** and **(17.64)** and recall that the term $e^{j\omega t}$ is also present. Substituting these in **Eqs. (17.63)** and **(17.64)** gives the components of the electric and magnetic field intensities. In the time domain, these are

$$\boxed{\begin{aligned}\mathbf{E}_1(x,z,t) &= \hat{\mathbf{y}}jE_0\sin\left(\frac{m\pi x}{d}\right)\cos\left(\omega t - \beta z\sqrt{1 - \lambda^2/\lambda_{cm}^2}\right)\\ &= \hat{\mathbf{y}}E_0\sin\left(\frac{m\pi x}{d}\right)\cos\left(\omega t - \beta_g z + \frac{\pi}{2}\right) = -\hat{\mathbf{y}}E_0\sin\left(\frac{m\pi x}{d}\right)\sin\left(\omega t - \frac{2\pi}{\lambda_g}z\right) \quad \left[\frac{\text{V}}{\text{m}}\right]\end{aligned}} \tag{17.89}$$

where **Eq. (17.87)** was used for β_g together with the relation $\beta_g = 2\pi/\lambda_g$, $j = e^{j\pi/2}$, and $\cos(\alpha + \pi/2) = -\sin\alpha$. Performing identical substitutions for the magnetic field intensity, we get

$$\mathbf{H}_1(x,z,t) = \hat{\mathbf{x}}\frac{E_0}{\eta}\sin\theta_i\sin\left(\frac{m\pi x}{d}\right)\sin\left(\omega t - \frac{2\pi}{\lambda_g}z\right) - \hat{\mathbf{z}}\frac{E_0}{\eta}\cos\theta_i\cos\left(\frac{m\pi x}{d}\right)\cos\left(\omega t - \frac{2\pi}{\lambda_g}z\right) \quad \left[\frac{\text{A}}{\text{m}}\right] \tag{17.90}$$

From **Eqs. (17.86)** and **(17.83)**, we write $\sin\theta_i = \lambda/\lambda_g$ and $\cos\theta_i = \lambda/\lambda_{cm}$. Substituting these into **Eq. (17.90)**,

$$\mathbf{H}_1(x,z,t) = \hat{\mathbf{x}}\frac{E_0}{\eta}\frac{\lambda}{\lambda_g}\sin\left(\frac{m\pi x}{d}\right)\sin\left(\omega t - \frac{2\pi}{\lambda_g}z\right) - \hat{\mathbf{z}}\frac{E_0}{\eta}\frac{\lambda}{\lambda_{cm}}\cos\left(\frac{m\pi x}{d}\right)\cos\left(\omega t - \frac{2\pi}{\lambda_g}z\right) \quad \left[\frac{\text{A}}{\text{m}}\right] \tag{17.91}$$

The electric and magnetic field intensities may also be written in the frequency domain by noting that $\sin(\omega t - 2\pi z/\lambda_g) = \cos(\omega t - 2\pi z/\lambda_g - \pi/2)$, $j = e^{j\pi/2}$, $-j = e^{-j\pi/2}$ and that $\cos(\omega t - 2\pi z/\lambda_g) = \text{Re}\{e^{j\omega t}e^{-j2\pi z/\lambda_g}\}$. Thus, the electric and magnetic field intensities in the frequency domain written with the phasor notation are

$$\mathbf{E}_1(x,z) = \hat{\mathbf{y}}jE_0\sin\left(\frac{m\pi x}{d}\right)e^{-j2\pi z/\lambda_g} \quad \left[\frac{\text{V}}{\text{m}}\right] \tag{17.92}$$

$$\mathbf{H}_1(x,z) = -\hat{\mathbf{x}}j\frac{E_0}{\eta}\frac{\lambda}{\lambda_g}\sin\left(\frac{m\pi x}{d}\right)e^{-j2\pi z/\lambda_g} - \hat{\mathbf{z}}\frac{E_0}{\eta}\frac{\lambda}{\lambda_{cm}}\cos\left(\frac{m\pi x}{d}\right)e^{-j2\pi z/\lambda_g} \quad \left[\frac{\text{A}}{\text{m}}\right] \tag{17.93}$$

The electric and magnetic fields vary in the transverse direction, but the electric field intensity is always zero at the conducting surfaces. The fields in waveguides are usually written as longitudinal and transverse components. Separating **Eqs. (17.92)** and **(17.93)** into their components, we get the following:

Longitudinal component:

$$H_z(x,z) = -\frac{E_0}{\eta}\frac{\lambda}{\lambda_{cm}}\cos\left(\frac{m\pi x}{d}\right)e^{-j2\pi z/\lambda_g} \quad \left[\frac{\text{A}}{\text{m}}\right] \tag{17.94}$$

Transverse components:

$$E_y(x,z) = jE_0\sin\left(\frac{m\pi x}{d}\right)e^{-j2\pi z/\lambda_g} \quad \left[\frac{\text{V}}{\text{m}}\right] \tag{17.95}$$

$$H_x(x,z) = -j\frac{E_0}{\eta}\frac{\lambda}{\lambda_g}\sin\left(\frac{m\pi x}{d}\right)e^{-j2\pi z/\lambda_g} \quad \left[\frac{\text{A}}{\text{m}}\right] \tag{17.96}$$

The power propagated in the wave is calculated using the Poynting vector. The time-averaged power density propagated in the z direction is

$$\mathcal{P}_{av} = -\frac{1}{2}E_y(x,z)H_x^*(x,y) = \frac{E_0^2}{2\eta}\frac{\lambda}{\lambda_g}\sin^2\left(\frac{m\pi x}{d}\right) \quad \left[\frac{\text{W}}{\text{m}^2}\right] \tag{17.97}$$

Finally, since the propagation is in the z direction, we can also define the characteristic impedance of the waveguide by dividing the transverse component of the electric field intensity (E_y) by the transverse component of the magnetic field intensity (H_x):

$$Z_{TE} = \eta\frac{\lambda_g}{\lambda} = \frac{\eta}{\sqrt{1-\lambda^2/\lambda_{cm}^2}} = \frac{\eta}{\sqrt{1-f_{cm}^2/f^2}} \quad [\Omega] \tag{17.98}$$

The wave impedance is, in fact, larger than the impedance for TEM waves. Also, unlike the wave impedance η for TEM waves, the wave impedance for TE waves is frequency dependent. The wave impedance tends to infinity at cutoff ($f = f_{cm}$) and to η as the frequency approaches infinity ($f \gg f_{cm}$) (see also **Section 17.3.1** and **Table 17.2**).

Example 17.2 Application: Use of the Guiding Effects of the Ionosphere to Propagate Low-Frequency Waves The ionosphere is a region of relatively high density of charged particles produced by the solar wind. These particles act as a layer of relatively high conductivity at low frequencies. Thus, the ionosphere, which starts at about 90 km above the surface of the Earth, may be viewed as a conducting spherical surface enclosing the Earth. The surface of the Earth is also a relatively good conductor at low frequencies. These two surfaces create a parallel plate waveguide. Low-frequency waves are reflected back and forth, propagating along the surface of the Earth to long distances. The waves may even encircle the Earth and interact, causing fading of reception in radio receivers. The ionosphere is not fixed in space. It tends to be lower during the day when the supply of charged particles is high, and higher during the night when the Sun is shielded by the Earth.

Assuming the ionosphere and the surface of the Earth to be perfect conductors separated a distance of 90 km, the properties of air to be those of free space and neglecting the curvature of the Earth:

(a) Find the lowest cutoff frequency for a TE wave propagating parallel to the surface of the Earth.
(b) Find and plot the wave impedance for the lowest mode, as a function of frequency.
(c) Find the total electric field intensity at a point after the waves have encircled the globe once. Take the average radius to be 6,400 km. Assume the frequency of the wave is twice the cutoff frequency.

Solution: The propagation of waves is in a parallel plate waveguide (approximately, because the curvature of the Earth is neglected), as shown in **Figure 17.8**. The lowest cutoff frequency is the TE_1 (also called the TE_{10} mode) and it is entirely defined by the distance between the ionosphere and Earth. The total electric field intensity at a point is the sum of the incident field and the same field after it encircles the globe. Since in this example there is no attenuation, only the phase of the electric field intensity has changed:

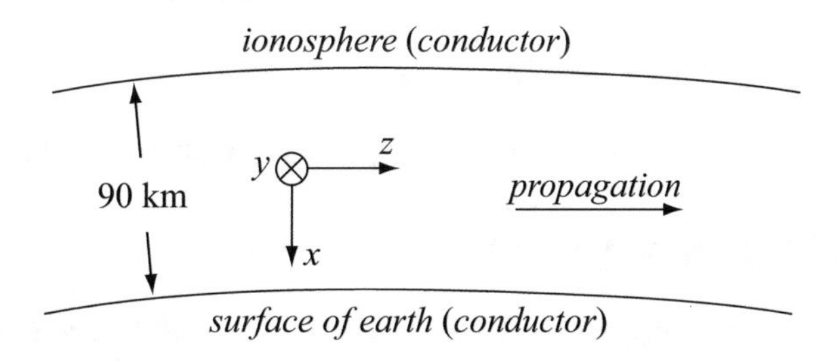

Figure 17.8 The ionosphere and Earth surfaces as a parallel plate waveguide

(a) The lowest mode is the TE_{10} mode. From **Eq. (17.80)**, with $m = 1$, and $d = 90{,}000$ m

$$f_{c10} = \frac{1}{2d\sqrt{\mu_0\varepsilon_0}} = \frac{3\times 10^8}{2\times 90{,}000} = 1{,}667 \quad [\mathrm{H_z}]$$

where $1/\sqrt{\mu_0\varepsilon_0} = 3\times 10^8$ m/s is the speed of light. TE waves below this frequency cannot propagate.

(b) The wave impedance is frequency dependent:

$$Z_{TE} = \frac{\eta_0}{\sqrt{1 - f_{cm}^2/f^2}} = \frac{377}{\sqrt{1-(1{,}667)^2/f^2}} \quad [\Omega]$$

This impedance is infinite at $f = f_c$. As an example at three times f_c, the wave impedance is 399.87 Ω. As frequency increases, the impedance decreases. As the frequency tends to infinity, the wave impedance tends to η_0. The plot of Z_{TE} with frequency is given in **Figure 17.9**.

(c) Assuming the electric field intensity to be as in **Eq. (17.95)**, we get

$$E_y(x,z) = jE_0 \sin\left(\frac{m\pi x}{d}\right)e^{-j2\pi z/\lambda_g} \quad \left[\frac{\mathrm{V}}{\mathrm{m}}\right]$$

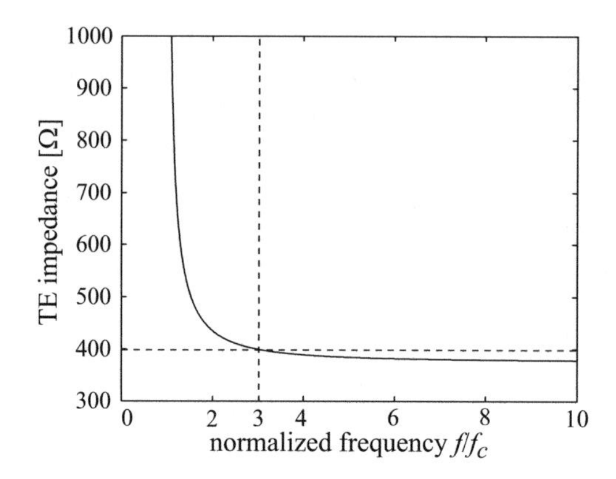

Figure 17.9 Wave impedance for the TE_{10} mode propagating in the ionosphere–Earth waveguide

At the given frequency, the guide wavelength is

$$\lambda_g = \frac{\lambda}{\sqrt{1 - f_{cm}^2/f^2}} = \frac{c}{f\sqrt{1 - f_{cm}^2/f^2}} = \frac{3 \times 10^8}{2 \times 1{,}667\sqrt{1 - (0.5)^2}} = 103{,}902 \quad [\text{m}]$$

Note that this is larger than the wavelength in free space, which is 89,982 m. Substituting λ_g in the expression for the electric field, we get

$$E_y(x, z) = jE_0 \sin\left(\frac{\pi x}{90{,}000}\right) e^{-j2\pi z/103{,}902} \quad [\text{V/m}]$$

The phase constant of the wave is therefore

$$\beta_g = 2\pi/103{,}902 \quad [\text{rad/m}]$$

In encircling the globe, the phase changes by

$$\theta = \beta_g z = 2\pi R\beta_g = 2 \times \pi \times 6{,}400 \times 10^3 \times 2 \times \pi/103{,}902 = 2{,}432 \quad [\text{rad}]$$

Taking the point at $z = 0$ as a reference ($\theta(z = 0) = 0$), the electric field intensity after encircling the globe is

$$E_y(x, 0) = jE_0 \sin\left(\frac{\pi x}{90{,}000}\right) e^{-j2{,}432} \quad [\text{V/m}]$$

This adds to the electric field intensity that already exists at $z = 0$, and the total electric field intensity is

$$E_y(x, 0) = jE_0 \sin\left(\frac{\pi x}{90{,}000}\right) \left(1 + e^{-j2{,}432}\right) \quad [\text{V/m}].$$

17.5 TM Propagation in Parallel Plate Waveguides

Propagation of TM modes can be obtained as for TE modes by starting with a wave impinging obliquely on a surface, but now we will assume parallel polarization; that is, the electric field is parallel to the plane of incidence. Since we have done this in detail for TE waves and since oblique incidence on a conductor for parallel polarization was discussed in **Section 13.3.2**, we will take the electric and magnetic fields obtained in **Exercise 13.8** as the given fields above the conducting surface (in medium (1)). The configuration is shown in **Figure 17.10**. The total electric and magnetic fields are

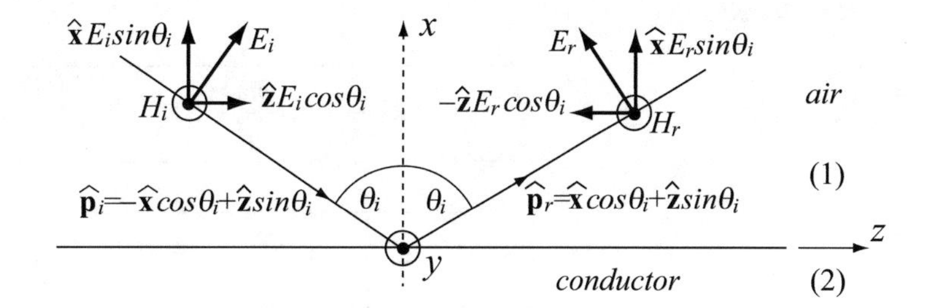

Figure 17.10 One plate of a parallel plate waveguide and the incident and reflected waves for parallel polarization

$$\mathbf{E}_1(x,z) = 2E_i[\hat{\mathbf{x}}\sin\theta_i\cos(\beta x\cos\theta_i) + \hat{\mathbf{z}}j\cos\theta_i\sin(\beta x\cos\theta_i)]e^{-j\beta z\sin\theta_i} \quad [\text{V/m}] \tag{17.99}$$

$$\mathbf{H}_1(x,z) = \hat{\mathbf{y}}2\frac{E_i}{\eta}\cos(\beta x\cos\theta_i)e^{-j\beta z\sin\theta_i} \quad [\text{A/m}] \tag{17.100}$$

The similarity between these equations and **Eqs. (17.63)** and **(17.64)** is easy to see. For the purpose of this discussion, the following properties are important:

(1) From the calculation of the time-averaged Poynting vector (see **Exercise 17.3**), the wave propagates in the positive z direction.
(2) The waves in the x direction are standing waves only.
(3) The magnetic field intensity is in the y direction and is entirely perpendicular to the direction of propagation.
(4) The electric field intensity has components in the x and z directions. Thus, the fields in **Eqs. (17.99)** and **(17.100)** are the fields of a transverse magnetic (TM) wave.
(5) All other relations, including the angle of incidence, the propagation constants, wavelengths, and phase constants remain as defined for TE waves. A comparison between **Eqs. (17.99)** and **(17.64)** shows that the x component of the electric field intensity for TM propagation varies in the same way as the z component of the magnetic field intensity for TE propagation. Similar comparison can be made on the z component of the electric field intensity and the y component of the magnetic field intensity. Thus, we can use the relations obtained for TE waves directly. Since the zeros in the standing wave pattern are the same as for the TE wave, the parallel plate waveguide looks like the waveguide in **Figure 17.7**; that is, the distance between the plates is d and the relations in **Eqs. (17.77)** through **(17.88)** apply here as well. In particular, we use the following:

$$\sin\theta_i = \frac{\lambda}{\lambda_g}, \quad \cos\theta_i = \frac{\lambda}{\lambda_{cm}}, \quad \beta\cos\theta_i = \frac{m\pi}{d}, \quad \beta\sin\theta_i = \beta_g = \frac{2\pi}{\lambda_g} \tag{17.101}$$

Now, we substitute these in **Eqs. (17.99)** and **(17.100)** and obtain

$$\boxed{\mathbf{E}_1(x,z) = \hat{\mathbf{x}}E_0\frac{\lambda}{\lambda_g}\cos\left(\frac{m\pi x}{d}\right)e^{-j2\pi z/\lambda_g} + \hat{\mathbf{z}}jE_0\frac{\lambda}{\lambda_{cm}}\sin\left(\frac{m\pi x}{d}\right)e^{-j2\pi z/\lambda_g} \quad \left[\frac{\text{V}}{\text{m}}\right]} \tag{17.102}$$

$$\boxed{\mathbf{H}_1(x,z) = \hat{\mathbf{y}}\frac{E_0}{\eta}\cos\left(\frac{m\pi x}{d}\right)e^{-j2\pi z/\lambda_g} \quad \left[\frac{\text{A}}{\text{m}}\right]} \tag{17.103}$$

where $E_0 = 2E_i$ is the amplitude of the electric field intensity. Rewriting these as longitudinal and transverse components, we get

Longitudinal component:

$$\boxed{E_z(x,z) = jE_0\frac{\lambda}{\lambda_{cm}}\sin\left(\frac{m\pi x}{d}\right)e^{-j2\pi z/\lambda_g} \quad \left[\frac{\text{V}}{\text{m}}\right]} \tag{17.104}$$

Transverse components:

$$E_x(x,z) = E_0 \frac{\lambda}{\lambda_g} \cos\left(\frac{m\pi x}{d}\right) e^{-j2\pi z/\lambda_g} \quad \left[\frac{\text{V}}{\text{m}}\right] \tag{17.105}$$

$$H_y(x,z) = \frac{E_0}{\eta} \cos\left(\frac{m\pi x}{d}\right) e^{-j2\pi z/\lambda_g} \quad \left[\frac{\text{A}}{\text{m}}\right] \tag{17.106}$$

The time domain expressions are obtained using $\text{Re}\{-je^{-j2\pi z/\lambda g}e^{j\omega t}\} = \sin(\omega t - 2\pi z/\lambda_g)$ and $\text{Re}\{e^{-j2\pi z/\lambda g}e^{j\omega t}\} = \cos(\omega t - 2\pi z/\lambda_g)$:

$$\mathbf{E}_1(x,z,t) = \hat{\mathbf{x}} E_0 \frac{\lambda}{\lambda_g} \cos\left(\frac{m\pi x}{d}\right) \cos\left(\omega t - \frac{2\pi z}{\lambda_g}\right) - \hat{\mathbf{z}} E_0 \frac{\lambda}{\lambda_{cm}} \sin\left(\frac{m\pi x}{d}\right) \sin\left(\omega t - \frac{2\pi z}{\lambda_g}\right) \quad \left[\frac{\text{V}}{\text{m}}\right] \tag{17.107}$$

$$\mathbf{H}_1(x,z,t) = \hat{\mathbf{y}} \frac{E_0}{\eta} \cos\left(\frac{m\pi x}{d}\right) \cos\left(\omega t - \frac{2\pi z}{\lambda_g}\right) \quad \left[\frac{\text{A}}{\text{m}}\right] \tag{17.108}$$

The time-averaged Poynting vector may be calculated from the transverse components in **Eqs. (17.105)** and **(17.106)**. Its magnitude is

$$\mathcal{P}_{av} = \frac{1}{2} \text{Re}\left\{E_x(x,z) H_y^*(x,z)\right\} = \frac{E_0^2 \lambda}{2\eta \lambda_g} \cos^2\left(\frac{m\pi x}{d}\right) \quad \left[\frac{\text{W}}{\text{m}^2}\right] \tag{17.109}$$

We can also calculate the wave impedance of the waveguide as

$$Z_{TM} = \frac{E_x}{H_y} = \eta \frac{\lambda}{\lambda_g} = \eta \sqrt{1 - \lambda^2/\lambda_{cm}^2} = \eta \sqrt{1 - f_{cm}^2/f^2} \quad [\Omega] \tag{17.110}$$

where **Eq. (17.86)** was used to obtain the expressions in terms of the mode cutoff wavelength (λ_{cm}) or mode cutoff frequency (f_{cm}). Note that the wave impedance for TM waves is always smaller than the intrinsic impedance η, whereas at cutoff, $Z_{TM} = 0$ (see also **Section 17.3.3**).

Exercise 17.2 Derive **Eqs. (17.102)** and **(17.103)** for a conducting surface at $x = 0$. Assume the electric and magnetic fields are as in **Figure 17.10**. Show first that given the incident electric field intensity $E_i(x,z)$, the incident magnetic field intensity and reflected electric and magnetic fields must be as shown. Then, evaluate the total fields $E_1(x,z)$ and $H_1(x,z)$ above the conducting surface.

Exercise 17.3 Calculate the time-averaged Poynting vector using the fields in **Eqs. (17.99)** and **(17.100)**. Show that real power propagates in the z direction and reactive power exists in the x direction.

Example 17.3 Application: TM Propagation in Microstrip Waveguides An integrated circuit waveguide is made of a very thin layer of silicon with two strips of aluminum deposited as shown in **Figure 17.11** (hence the name microstrip waveguide). The thickness of the silicon is 0.1 mm and the width of the strips is 2 mm. The structure is used to couple energy between two devices (not shown). TM waves are used in the second TM mode. Assume lossless propagation, with permittivity of silicon equal to $12\varepsilon_0$ [F/m] and permeability equal to μ_0 [H/m]. Also, assume there is no fringing at the edges of the stripline (all energy is contained between the plates):

(a) What is the lowest frequency at which the waveguide can be used and still maintain the second TM mode of propagation?
(b) What is the maximum time-averaged power the waveguide can propagate at 900 GHz without causing breakdown in the silicon? Breakdown in silicon occurs at 32 kV/mm.

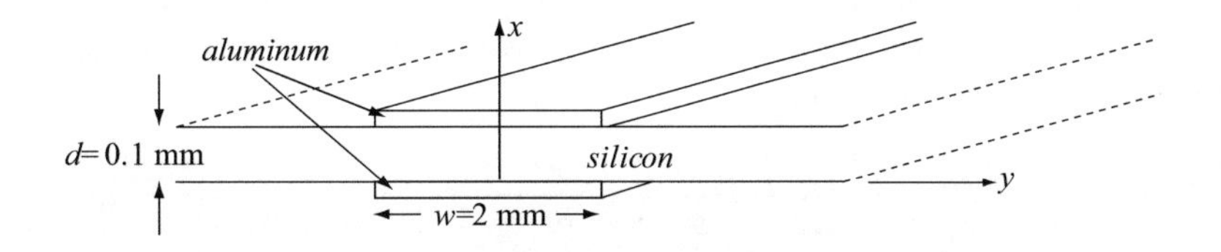

Figure 17.11 Construction and dimensions of a microstrip waveguide

Solution: The cutoff frequency of the second mode is given in **Eq. (17.80)** with $d = 0.1$ mm and $m = 2$. The peak electric field intensity in the waveguide cannot exceed 32,000 V/mm anywhere. Thus, the peak electric field intensity is 3.2×10^7 V/m. The allowed power density is now evaluated from **Eq. (17.109)** using the peak value for E_0. The power density is then integrated over the cross-sectional area of the waveguide to calculate the total power:

(a) The cutoff frequency for the TM_{20} mode is

$$f_{c20} = \frac{2}{2d\sqrt{\mu\varepsilon}} = \frac{2 \times 3 \times 10^8}{0.0002 \times \sqrt{12}} = 866 \quad [\text{GHz}]$$

This waveguide operates well into the millimeter-wave range (see **Section 12.4**).

(b) The time-averaged power density is

$$\mathcal{P}_{av} = \frac{E_0^2 \lambda}{2\eta\lambda_g} \cos^2\left(\frac{m\pi x}{d}\right) \quad \left[\frac{\text{W}}{\text{m}^2}\right]$$

where $\lambda = v_p/f$ is the wavelength, v_p is the phase velocity, $\eta = \sqrt{\mu/\varepsilon}$ is the intrinsic impedance in silicon, and λ_g is the guide wavelength given in **Eq. (17.86)**, all calculated in silicon. These values are

$$v_p = \frac{c}{\sqrt{\varepsilon_r}} = \frac{3 \times 10^8}{\sqrt{12}} = 8.66 \times 10^7 \quad [\text{m/s}]$$

$$\lambda = \frac{v_p}{f} = \frac{8.66 \times 10^7}{9 \times 10^{11}} = 9.622 \times 10^{-5} \quad [\text{m}]$$

$$\eta = \frac{\eta_0}{\sqrt{\varepsilon_r}} = \frac{377}{\sqrt{12}} = 108.83 \quad [\Omega]$$

$$\lambda_g = \frac{\lambda}{\sqrt{1 - (f_c/f)^2}} = \frac{9.622 \times 10^{-5}}{\sqrt{1 - (866/900)^2}} = 3.53 \times 10^{-4} \quad [\text{m}]$$

The time-averaged power density is

$$\mathcal{P}_{av} = \frac{\left(3.2 \times 10^7\right)^2 \times 9.622 \times 10^{-5}}{2 \times 108.83 \times 3.53 \times 10^{-4}} \cos^2\left(\frac{2\pi x}{0.0001}\right) = 1.282 \times 10^{12} \cos^2(20{,}000\pi x) \quad \left[\text{W/m}^2\right]$$

This power density is independent of y but varies with x. Taking a strip parallel to the width of the structure as wdx, multiplying by the power density above, and integrating over x gives the total power flowing through the cross section of the waveguide:

$$P = \int_{x=0}^{x=0.0001} \mathcal{P}_{av} \omega dx = 0.002 \int_{x=0}^{x=0.0001} 1.282 \times 10^{12} \cos^2(20{,}000\pi x) dx$$

$$= 2.564 \times 10^9 \left[\frac{x}{2} + \frac{1}{80{,}000\pi} \sin(2 \times 20{,}000 \times \pi x) \right]_{x=0}^{x=0.0001} = 128.2 \quad [\text{kW}]$$

This is a considerable amount of power for a waveguide this small, but then the electric field intensity is also extremely large. Normally, the maximum electric field intensity is kept well below breakdown. Even so, waveguides can transfer relatively large amounts of power.

Example 17.4 Application: Discontinuities in Waveguides Two parallel plate waveguides are connected as shown in **Figure 17.12**. Both waveguides are 30 mm wide. Assuming TM propagation, calculate:

(a) The lowest frequency that can be propagated if the source is connected on side A of the structure.
(b) The lowest frequency that can be propagated if the source is connected on side B of the structure.
(c) Does it make any difference in (**a**) and (**b**), if the waves are TE waves? Why?
(d) Calculate the wave impedance of the two waveguides for TE and TM waves at a frequency twice the structure's cutoff frequency.
(e) Calculate the reflection and transmission coefficients for waves propagating from A to B and for waves propagating from B to A at the frequency in (**d**) for TE and TM propagation.

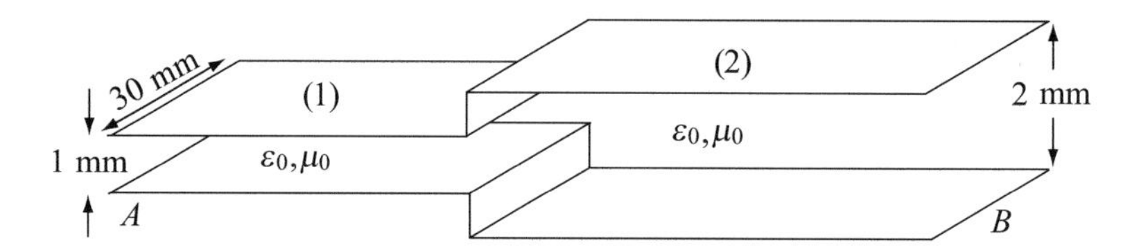

Figure 17.12 Two microstrip waveguides of different spacings connected together

Solution: The cutoff frequency of the combined structure is that below which a wave cannot propagate in the structure even though it may propagate in one of the two waveguides. To solve the problem, we calculate the individual guide's cutoff frequencies and compare to see which of the modes can be propagated in both waveguides:

(a) The cutoff frequencies for TM_{10} mode in the two sections are:
In the small guide:

$$f_{c1}^{(1)} = \frac{1}{2d\sqrt{\mu_0 \varepsilon_0}} = \frac{3 \times 10^8}{2 \times 0.001} = 150 \quad [\text{GHz}]$$

In the large guide:

$$f_{c1}^{(2)} = \frac{1}{2d\sqrt{\mu_0 \varepsilon_0}} = \frac{3 \times 10^8}{2 \times 0.002} = 75 \quad [\text{GHz}]$$

Since the small guide can only propagate above 150 GHz and the large guide above 75 GHz, the combined structure can only propagate above 150 GHz.

(b) The individual cutoff frequencies are the same as in (**a**). The large guide can propagate above 75 GHz, but the small guide cannot propagate between 75 GHz and 150 GHz. The minimum frequency that the structure can propagate is 150 GHz.

(c) No, it makes no difference, since the cutoff frequencies for TE_{m0} and TM_{m0} modes are the same.
Note: This structure is a mismatched structure and causes reflections at the continuity. In practice, this type of connection should be avoided.

(d) The cutoff frequency of each stripline is different. Therefore, their wave impedances must also be different. The cutoff frequency for the composite structure is 150 GHz. Therefore, the wave impedances for TE and TM modes at 300 GHz, for the two striplines, are [from **Eqs. (17.98)** and **(17.110)**]

$$Z_{TM1} = \eta\sqrt{1 - f_{cm1}^2/f^2} = 377\sqrt{1 - (150/300)^2} = 326.49 \quad [\Omega]$$
$$Z_{TM2} = \eta\sqrt{1 - f_{cm2}^2/f^2} = 377\sqrt{1 - (75/300)^2} = 365.03 \quad [\Omega]$$
$$Z_{TE1} = \frac{\eta}{\sqrt{1 - f_{cm1}^2/f^2}} = \frac{377}{\sqrt{1 - (150/300)^2}} = 435.32 \quad [\Omega]$$
$$Z_{TE2} = \frac{\eta}{\sqrt{1 - f_{cm2}^2/f^2}} = \frac{377}{\sqrt{1 - (75/300)^2}} = 389.36 \quad [\Omega]$$

Note: $Z_{TEM} = \eta_0 = 377\ \Omega$ and $Z_{TM} \leq Z_{TEM} \leq Z_{TE}$.

(e) The discontinuity caused by the connection of the two striplines is due to the differences in wave impedances of the two sections. Therefore, the connection may be viewed as the interface between two line sections with different properties. In propagating from A to B:
For TE propagation:

$$\Gamma_{12} = \frac{Z_{TE2} - Z_{TE1}}{Z_{TE2} + Z_{TE1}} = \frac{389.36 - 435.32}{389.36 + 435.32} = -0.0557$$

For TM propagation:

$$\Gamma_{12} = \frac{Z_{TM2} - Z_{TM1}}{Z_{TM2} + Z_{TM1}} = \frac{365.03 - 326.49}{365.03 + 326.49} = 0.0557$$

In propagating from B to A:
For TE propagation:

$$\Gamma_{21} = -\Gamma_{12} = 0.0557$$

For TM propagation:

$$\Gamma_{21} = -\Gamma_{12} = -0.0557.$$

17.6 TEM Waves in Parallel Plate Waveguides

In developing the relations for TE and TM waves in parallel plate waveguides, we relied on the oblique incidence of a TEM wave on the conducting surfaces of the waveguide. This caused reflections at the conducting surfaces, and the sum of the incident and reflected waves produced either a TE or a TM wave, depending on the initial polarization of the TEM wave. We also mentioned that if the incident TEM wave propagates parallel to the surface of the conductors, the conductors do not affect the wave or any of its properties. Therefore, if a TEM wave, such as a plane wave, propagates such that it does not reflect off the conducting surfaces (i.e., if the angle of incidence is $\pi/2$) a TEM rather than a TE or TM wave will propagate in parallel plate waveguides. This possibility was discussed at length in **Chapter 14**, particularly in **Section 14.6**. There is little that needs to be added here except to indicate that TEM waves can indeed exist in parallel plate waveguides and, when they do, the properties in column 1 of **Table 17.1** apply. These are the same properties we used for parallel plate transmission lines and for plane waves in the unbounded domain. However, one point needs to be mentioned again: The cutoff frequency of any TEM wave is zero; TEM waves of any frequency may propagate in parallel plate waveguides, as we have seen for parallel plate transmission lines. Also, unlike TE and TM waves, the phase velocity and wave impedance are independent of frequency (for lossless dielectrics). This also means that the lowest possible mode of propagation is a TEM mode. For any waveguide, the lowest possible mode of propagation is called a ***dominant mode***. In parallel plate waveguides, this is the TEM mode (with cutoff at zero frequency).

TEM waves can only propagate in waveguides made of two conductors. In single, conductor waveguides only TE and/or TM modes may exist.

17.7 Rectangular Waveguides waves.m

In the parallel plate waveguide in the previous sections, the fields only varied in one transverse direction, whereas propagation was along the plates. Because of that, analysis of the fields was simple. True parallel plate waveguides are not practical since all dimensions must be finite. Although structures resembling the parallel plate waveguide can be built (such as the striplines in **Examples 17.3** and **17.4**) and are quite common in microwave integrated circuits, most waveguides are closed structures. A rectangular or cylindrical tube or some other type of enclosed conductor may be used. In the most general sense, the conductor is not a condition of existence of guided waves; only total reflection from a boundary is required. However, to simplify the discussion, we will restrict ourselves to waveguides defined by highly conducting surfaces.

One of the simplest and most common waveguide structures is the rectangular waveguide. You can imagine a rectangular waveguide as the intersection of two pairs of parallel plate waveguides, one lying horizontally and one vertically as shown in **Figure 17.13**. This view has the advantage of defining the waveguide in terms of the parallel plate waveguides we have already discussed. We will not take this approach here since now the structure is two dimensional and the definition of angles of incidence is not as easy to visualize. The actual calculation of fields in the waveguide will be done based on the TE and TM waves described in **Eqs. (17.27)** through **(17.30)** and **Eqs. (17.53)** through **(17.56)**. However, it helps to view the rectangular waveguide as being made of two sets of parallel plates because, then, we can argue that the transverse variation of fields in the rectangular waveguide is a combined variation of the vertical and horizontal plates since both transverse components of the field (if both components exist) are standing waves. The results we obtain here will show these variations.

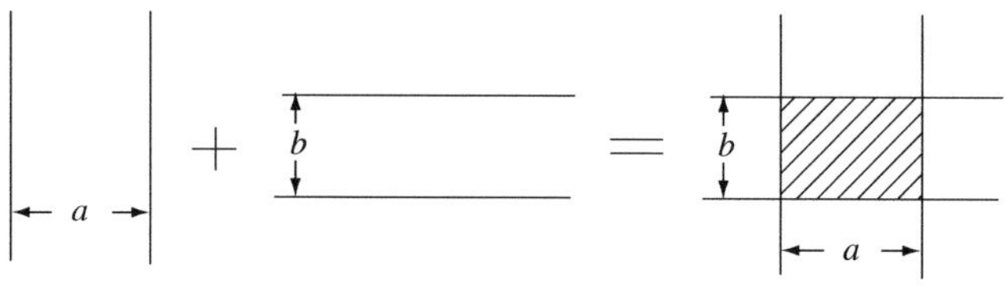

Figure 17.13 A rectangular waveguide (shown in cross section) as a combination of two parallel plate waveguides

A rectangular waveguide is shown in **Figure 17.14**. The dimensions of the waveguide are the internal dimensions and the walls are assumed (for now) to be perfectly conducting. To see how the electric and magnetic field intensities in a waveguide of the type shown in **Figure 17.14** behave, we will solve **Eq. (17.33)** for TE waves or **Eq. (17.57)** for TM waves, subject to boundary conditions on the conducting boundaries. We start with TM waves because the boundary conditions are straightforward.

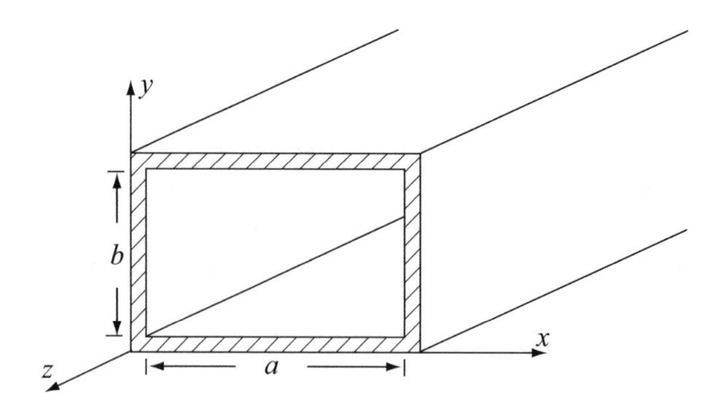

Figure 17.14 Structure and dimensions of a rectangular waveguide

17.7.1 TM Modes in Rectangular Waveguides

The TM modes are defined by the longitudinal (z component) of the electric field intensity through the following equation **[Eq. (17.57)]**:

$$\frac{\partial^2 E_Z}{\partial x^2} + \frac{\partial^2 E_Z}{\partial y^2} + (\gamma^2 + k^2)E_z = 0 \quad (17.111)$$

subject to the condition that the electric field intensity must vanish on the conducting boundaries. These conditions (shown in **Figure 17.15**) are

$$E_z(0, y) = 0 \quad \text{and} \quad E_z(a, y) = 0 \quad (17.112)$$

$$E_z(x, 0) = 0 \quad \text{and} \quad E_z(x, b) = 0 \quad (17.113)$$

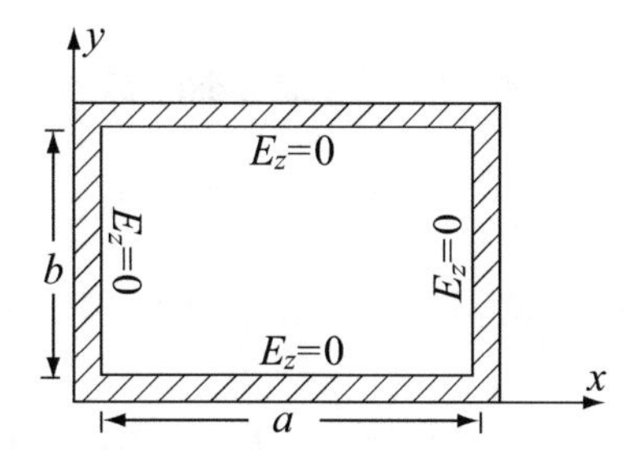

Figure 17.15 Boundary conditions for TM propagation in a rectangular waveguide

Equation (17.111) can be solved directly using separation of variables. This leads to two second-order ordinary differential equations, the solutions to which are sinusoidal functions. The solution process is much the same as for Laplace's equation in **Chapter 5**. To solve this equation using separation of variables, we first use **Eq. (17.34)** to replace the term $\gamma_2 + k_2$ by a single term we denoted k_c^2. Then, we assume a separable solution for E_z in the form

$$E_z(x, y) = X(x)Y(y) \quad (17.114)$$

where $X(x)$ only depends on the x variable and $Y(y)$ only on the y variable. This gives the x and y variations of the field. The z variation is known from the propagation constant. Thus, once we obtain $E_z(x,y)$, we can write

$$E_z(x,y,z) = E_z(x, y)e^{-\gamma z} \quad (17.115)$$

where we assume forward-propagating waves. If backward-propagating waves also exist due to reflections, these must be added, and as required, they must propagate in the negative z direction. Also, for simplicity, we will assume lossless dielectrics in the waveguide ($\gamma = j\beta$) but will retain the general form for the propagation constant to show that, in general, losses exist in the waveguide.

Substituting **Eq. (17.114)** into **Eq. (17.111)** and dividing both sides of the equation by E_z, we get

$$\frac{1}{X(x)}\frac{\partial^2 X(x)}{\partial x^2} + \frac{1}{Y(y)}\frac{\partial^2 Y(y)}{\partial y^2} + k_c^2 = 0 \quad (17.116)$$

For this to be satisfied, the first and second terms must each be equal to a constant which we will take as $-k_x^2$ and $-k_y^2$; that is, the following conditions must be satisfied:

$$\frac{1}{X(x)}\frac{\partial^2 X(x)}{\partial x^2} = -k_x^2 \quad \rightarrow \quad \frac{\partial^2 X(x)}{\partial x^2} + k_x^2 X(x) = 0 \quad (17.117)$$

$$\frac{1}{Y(y)}\frac{\partial^2 Y(y)}{\partial y^2} = -k_y^2 \quad \to \quad \frac{\partial^2 Y(y)}{\partial y^2} + k_y^2 Y(y) = 0 \tag{17.118}$$

The separation constants must also satisfy the following conditions:

$$-k_x^2 - k_y^2 + k_c^2 = 0 \tag{17.119}$$

Equations (17.117) and **(17.118)** have the following solutions (see **Section 5.4.4.1**):

$$X(x) = A_1 \sin k_x x + B_1 \cos k_x x \tag{17.120}$$

$$Y(y) = A_2 \sin k_y y + B_2 \cos k_y y \tag{17.121}$$

Substituting these into **Eq. (17.114)**, we obtain the general solution

$$E_z(x,y) = (A_1 \sin k_x x + B_1 \cos k_x x)\left(A_2 \sin k_y y + B_2 \cos k_y y\right) \tag{17.122}$$

The constants are now evaluated from the boundary conditions in **Eqs. (17.112)** and **(17.113)**:

$$E_z(0,y) = (A_1 \sin k_x 0 + B_1 \cos k_x 0)\left(A_2 \sin k_y y + B_2 \cos k_y y\right) = 0 \quad \to \quad B_1 = 0 \tag{17.123}$$

$$E_z(x,0) = A_1 \sin k_x x\left(A_2 \sin k_y 0 + B_2 \cos k_y 0\right) = 0 \quad \to \quad B_2 = 0 \tag{17.124}$$

$$E_z(a,y) = A \sin(k_x a)\sin\left(k_y y\right) = 0 \quad \to \quad k_x = \frac{m\pi}{a} \tag{17.125}$$

$$E_z(x,b) = A \sin\left(\frac{m\pi}{a}x\right)\sin\left(k_y b\right) = 0 \quad \to \quad k_y = \frac{n\pi}{b} \tag{17.126}$$

The amplitude of the electric field intensity is arbitrary; it does not affect the form of the solution and we will denote it by E_0. Therefore, the solution for the longitudinal component of the electric field intensity is

$$E_z(x,y) = E_0 \sin\left(\frac{m\pi}{a}x\right)\sin\left(\frac{n\pi}{b}y\right) \quad \left[\frac{\text{V}}{\text{m}}\right] \tag{17.127}$$

The general solution also includes the z variation. From **Eq. (17.115)**, we get

$$\boxed{E_z(x,y,z) = E_0 \sin\left(\frac{m\pi}{a}x\right)\sin\left(\frac{n\pi}{b}y\right)e^{-\gamma z} \quad \left[\frac{\text{V}}{\text{m}}\right]} \tag{17.128}$$

Before proceeding with the evaluation of the transverse components, we note the following from **Eq. (17.119)**:

$$\boxed{k_c^2 = k_x^2 + k_y^2 = \left(\frac{m\pi}{a}\right)^2 + \left(\frac{n\pi}{b}\right)^2} \tag{17.129}$$

Also, from **Eq. (17.34)**, we get

$$\boxed{\gamma^2 = \left(\frac{m\pi}{a}\right)^2 + \left(\frac{n\pi}{b}\right)^2 - k^2 = \left(\frac{m\pi}{a}\right)^2 + \left(\frac{n\pi}{b}\right)^2 - \omega^2\mu\varepsilon} \tag{17.130}$$

Now, we can calculate the transverse components of the electric and magnetic fields by substituting the general solution **[Eq. (17.128)]** into **Eqs. (17.53)** through **(17.56)**. This gives

$$E_x(x,y,z) = \frac{-\gamma}{\gamma^2 + k^2} E_0 \frac{m\pi}{a} \cos\left(\frac{m\pi x}{a}\right) \sin\left(\frac{n\pi y}{b}\right) e^{-\gamma z} \quad \left[\frac{\text{V}}{\text{m}}\right] \tag{17.131}$$

$$E_y(x,y,z) = \frac{-\gamma}{\gamma^2 + k^2} E_0 \frac{n\pi}{ba} \sin\left(\frac{m\pi x}{a}\right) \cos\left(\frac{n\pi y}{b}\right) e^{-\gamma z} \quad \left[\frac{\text{V}}{\text{m}}\right] \tag{17.132}$$

$$H_x(x,y,z) = \frac{j\omega\varepsilon}{\gamma^2 + k^2} E_0 \frac{n\pi}{b} \sin\left(\frac{m\pi x}{a}\right) \cos\left(\frac{n\pi y}{b}\right) e^{-\gamma z} \quad \left[\frac{\text{A}}{\text{m}}\right] \tag{17.133}$$

$$H_y(x,y,z) = -\frac{j\omega\varepsilon}{\gamma^2 + k^2} E_0 \frac{m\pi}{a} \cos\left(\frac{m\pi x}{a}\right) \sin\left(\frac{n\pi y}{b}\right) e^{-\gamma z} \quad \left[\frac{\text{A}}{\text{m}}\right] \tag{17.134}$$

These, together with **Eqs. (17.129)** and **(17.130)**, define the transverse fields in the waveguide. We can easily write the time domain form of the equations (see **Exercise 17.4**) by adding the $e^{j\omega t}$ term and writing $e^{j\pi/2}$ for j and $e^{-j\pi/2}$ for $-j$. Also, the longitudinal and transverse components for a backward-propagating wave can be written directly from **Eqs. (17.128)** and **(17.131)** through **(17.134)** by replacing γ by $-\gamma$ wherever these occur (see **Exercise 17.5** and **Problem 17.25**).

Equations (17.128) and **(17.131)** through **(17.134)** are written for the general modes (mn). Since we solved for TM modes, the general mode is a TM_{mn} mode where m, n are any integers, including zero. An infinite number of modes are possible, but usually only the first few modes are used in practice. Also, some modes are not useful. For example, if $m = 0$ and $n = 0$, all components of the field, including the longitudinal component, are zero. This is clearly not a useful mode. Also, if $m = 0$ and $n \neq 0$, or $m \neq 0$ and $n = 0$, all field components become zero as can be seen by substitution in **Eq. (17.128)**. Thus, for TM modes, neither m nor n may be zero. The lowest possible TM mode is the TM_{11} mode. The longitudinal components of the TM_{12} and TM_{22} modes are shown in **Figure 17.16** for a square cross-section waveguide. Similar plots may be obtained for the transverse components of the field.

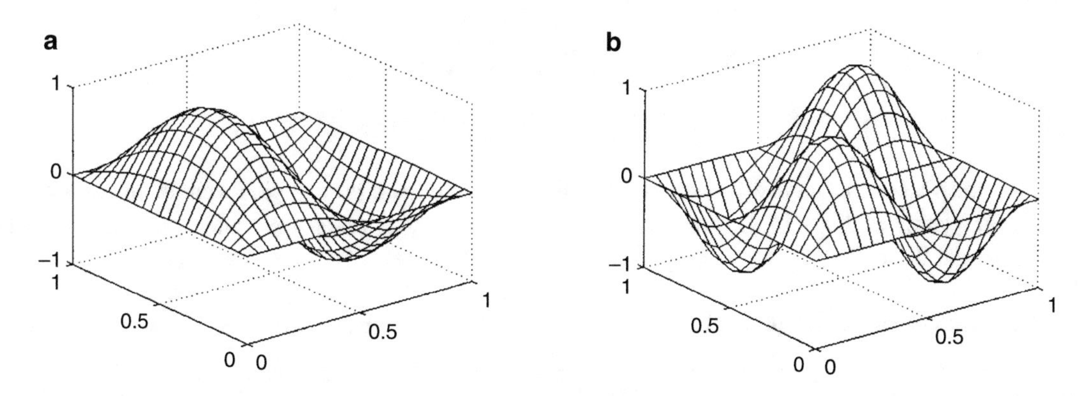

Figure 17.16 The longitudinal electric field intensity distribution in a waveguide. (**a**) For TM_{12} mode. (**b**) For TM_{22} mode

The propagation properties in the waveguide are obtained from **Eqs. (17.129)** and **(17.130)**. The propagation constant for lossless propagation is

$$\boxed{\gamma = j\beta_g = j\sqrt{\omega^2\mu\varepsilon - \left(\frac{m\pi}{a}\right)^2 - \left(\frac{n\pi}{b}\right)^2}} \tag{17.135}$$

where β_g is the guide phase constant:

$$\boxed{\beta_g = \sqrt{\omega^2\mu\varepsilon - \left(\frac{m\pi}{a}\right)^2 - \left(\frac{n\pi}{b}\right)^2} \quad \left[\frac{\text{rad}}{\text{m}}\right]} \tag{17.136}$$

The cutoff wave number is given in **Eq. (17.129)**:

$$k_{cmn} = \sqrt{\left(\frac{m\pi}{a}\right)^2 + \left(\frac{n\pi}{b}\right)^2} \quad \left[\frac{\text{rad}}{\text{m}}\right] \tag{17.137}$$

The cutoff frequency may be obtained from **Eq. (17.137)** by using $k_{cmn} = 2\pi f_{cmn}\sqrt{\mu\varepsilon}$ or by setting the propagation constant in **Eq. (17.130)** to zero:

$$f_{cmn} = \frac{1}{2\sqrt{\mu\varepsilon}}\sqrt{\left(\frac{m}{a}\right)^2 + \left(\frac{n}{b}\right)^2} \quad [\text{Hz}] \tag{17.138}$$

Propagation in the waveguide can only occur above this frequency. As with the parallel plate waveguide, below this frequency there is rapid attenuation of the wave (evanescent wave).

The cutoff wavelength is then

$$\lambda_{cmn} = \frac{v_p}{f_{cmn}} = \frac{1}{\sqrt{(m/2a)^2 + (n/2b)^2}} \quad [\text{m}] \tag{17.139}$$

From **Eq. (17.136)**, we can write

$$\beta_g = \omega\sqrt{\mu\varepsilon}\sqrt{1 - \frac{1}{4\pi^2 f^2 \mu\varepsilon}\left[\left(\frac{m\pi}{a}\right)^2 + \left(\frac{n\pi}{b}\right)^2\right]} \quad \left[\frac{\text{rad}}{\text{m}}\right] \tag{17.140}$$

or, using **Eq. (17.138)**,

$$\beta_g = \beta\sqrt{1 - \frac{f_{cmn}^2}{f^2}} \quad \left[\frac{\text{rad}}{\text{m}}\right] \tag{17.141}$$

where $\beta = \omega\sqrt{\mu\varepsilon}$ is the phase constant in the material filling the waveguide as if propagation were in infinite space. Note that this expression is the same as **Eq. (17.87)** for TM propagation in parallel plate waveguides and **Eq. (17.41)** for TM propagation in general. We called β_g the guide phase constant to distinguish it from β, which is the phase constant in unbounded space.

From β_g, we can also calculate the guide phase velocity v_g and the guide wavelength λ_g as

$$v_g = \frac{\omega}{\beta_g} = \frac{v_p}{\sqrt{1 - \lambda^2/\lambda_{cmn}^2}} = \frac{v_p}{\sqrt{1 - f_{cmn}^2/f^2}} \quad \left[\frac{\text{m}}{\text{s}}\right] \tag{17.142}$$

where $v_p = 1/\sqrt{\mu\varepsilon}$ is the phase velocity in the unbounded space (with properties of the space identical to those in the waveguide), and

$$\lambda_g = \frac{2\pi}{\beta_g} = \frac{\lambda}{\sqrt{1 - \lambda^2/\lambda_{cmn}^2}} = \frac{\lambda}{\sqrt{1 - f_{cmn}^2/f^2}} \quad [\text{m}] \tag{17.143}$$

where $\lambda = 2\pi/\beta$ is the wavelength in unbounded space.

The wave impedance is obtained by taking the ratio between the transverse components of the electric and magnetic field intensities. Since the wave must propagate in the z direction (i.e., the Poynting vector must be in the z direction), we can take either the ratio between E_x and H_y or the negative ratio of E_y and H_x:

$$Z_{TM} = \frac{E_x}{H_y} = -\frac{E_y}{H_x} = \frac{\gamma}{j\omega\varepsilon} \quad [\Omega] \tag{17.144}$$

Or, using **Eq. (17.135)** to replace γ, we can write for lossless propagation

$$\boxed{Z_{TM} = \frac{\beta_g}{\omega\varepsilon} = \sqrt{\frac{\mu}{\varepsilon}}\sqrt{1 - \frac{f_{cmn}^2}{f^2}} = \sqrt{\frac{\mu}{\varepsilon}}\sqrt{1 - \frac{\lambda^2}{\lambda_{cmn}^2}} = \eta\frac{\lambda}{\lambda_g} \quad [\Omega]} \tag{17.145}$$

Also, from **Eqs. (17.130), (17.129),** and **(17.135)**, we have for lossless propagation:

$$k_{cmn}^2 = \gamma^2 + k^2 = \left(\frac{m\pi}{a}\right)^2 + \left(\frac{n\pi}{b}\right)^2 \quad \text{and} \quad \gamma = j\beta_g \tag{17.146}$$

With these, we can write the transverse components in **Eqs. (17.131)** through **(17.134)** in terms of the phase constant β_g and the cutoff wave number k_{cmn} instead of the propagation constant γ. Substituting from **Eq. (17.146)** into **Eqs. (17.131)** through **(17.134)** gives

$$E_x(x,y,z) = \frac{-j\beta_g}{k_{cmn}^2} E_0 \frac{m\pi}{a} \cos\left(\frac{m\pi x}{a}\right) \sin\left(\frac{n\pi y}{b}\right) e^{-j\beta_g z} \quad \left[\frac{\text{V}}{\text{m}}\right] \tag{17.147}$$

$$E_y(x,y,z) = \frac{-j\beta_g}{k_{cmn}^2} E_0 \frac{n\pi}{b} \sin\left(\frac{m\pi x}{a}\right) \cos\left(\frac{n\pi y}{b}\right) e^{-j\beta_g z} \quad \left[\frac{\text{V}}{\text{m}}\right] \tag{17.148}$$

$$H_x(x,y,z) = \frac{j\omega\varepsilon}{k_{cmn}^2} E_0 \frac{n\pi}{b} \sin\left(\frac{m\pi x}{a}\right) \cos\left(\frac{n\pi y}{b}\right) e^{-j\beta_g z} \quad \left[\frac{\text{A}}{\text{m}}\right] \tag{17.149}$$

$$H_y(x,y,z) = \frac{-j\omega\varepsilon}{k_{cmn}^2} E_0 \frac{m\pi}{a} \cos\left(\frac{m\pi x}{a}\right) \sin\left(\frac{n\pi y}{b}\right) e^{-j\beta_g z} \quad \left[\frac{\text{A}}{\text{m}}\right] \tag{17.150}$$

The longitudinal component in **Eq. (17.128)** and the transverse components in **Eqs. (17.131)** through **(17.134)** or in **Eqs. (17.147)** through **(17.150)** may be written in the time domain by multiplying the expressions by $e^{j\omega t}$ and taking the real part of the expression (see **Exercise 17.4** and **Problem 17.24**).

The purpose of a waveguide is to guide waves from a source to a load. Therefore, it must propagate power. To see that this is the case, we calculate the time-averaged power density in the cross section of the waveguide (i.e., for any value of z). This power density must be real and must propagate in the positive z direction. From **Eqs. (17.147)** through **(17.150)**, we see that there are two sets of transverse components of **E** and **H**. Each pair produces average power given by $E_x H_y^*/2$ and $-E_y H_x^*/2$, where the negative sign comes from the fact that $\hat{\mathbf{x}} \times \hat{\mathbf{y}} = \hat{\mathbf{z}}$ and $\hat{\mathbf{y}} \times \hat{\mathbf{x}} = -\hat{\mathbf{z}}$. The total power density is the sum of these two terms:

$$\mathcal{P}_{av}(x,y) = \hat{\mathbf{z}}\frac{1}{2}\text{Re}\left\{E_x(x,y)H_y^*(x,y) - E_y(x,y)H_x^*(x,y)\right\} \quad \left[\frac{\text{W}}{\text{m}^2}\right] \tag{17.151}$$

Substituting for E_x, E_y, H_x, and H_y from **Eqs. (17.147)** through **(17.150)** gives

$$\mathcal{P}_{av}(x,y) = \hat{\mathbf{z}}\frac{\omega\varepsilon\beta_g E_0^2}{2k_{cmn}^4}\left[\left(\frac{m\pi}{a}\right)^2 \cos^2\left(\frac{m\pi x}{a}\right)\sin^2\left(\frac{n\pi y}{b}\right) + \left(\frac{n\pi}{b}\right)^2 \sin^2\left(\frac{m\pi x}{a}\right)\cos^2\left(\frac{n\pi y}{b}\right)\right] \quad \left[\frac{\text{W}}{\text{m}^2}\right] \tag{17.152}$$

The total power in the waveguide cross section is the power density, integrated over the waveguide cross section. Performing the integration over $\mathcal{P}_{av} \cdot d\mathbf{s}$, with $d\mathbf{s} = \hat{\mathbf{z}}dxdy$, gives

$$P = \frac{\omega\varepsilon\beta_g E_0^2}{2k_{cmm}^4}\int_{x=0}^{x=a}\int_{y=0}^{y=b}\left[\left(\frac{m\pi}{a}\right)^2\cos^2\left(\frac{m\pi x}{a}\right)\sin^2\left(\frac{n\pi y}{b}\right)+\left(\frac{n\pi}{b}\right)^2\sin^2\left(\frac{m\pi x}{a}\right)\cos^2\left(\frac{\eta\pi y}{b}\right)\right]dxdy$$

$$= \frac{\omega\varepsilon\beta_g E_0^2\, ab}{k_{cmn}^4\quad 8}\left[\left(\frac{m\pi}{a}\right)^2+\left(\frac{n\pi}{b}\right)^2\right] = \frac{\omega\varepsilon\beta_g E_0^2 ab}{8k_{cmn}^2}\quad [\mathrm{W}] \tag{17.153}$$

where $k_{cmn}^2 = (m\pi/a)^2 + (n\pi/b)^2$. The total power is directly proportional to the cross-sectional area of the waveguide (ab). In any given waveguide, the total power may be increased by using larger fields (electric and magnetic) or increasing the physical dimensions of the waveguide. Also, the power is proportional to frequency and the dielectric constant in the waveguide. Most waveguides use air as the dielectric, but increasing the frequency is feasible up to certain limits, imposed by the circuits used to generate the fields.

Exercise 17.4 Write the time domain expression for the longitudinal component of the electric field intensity for TM propagation in a lossless rectangular waveguide.

Answer $E_z(x,y,z,t) = E_0\sin(m\pi x/a)\sin(n\pi y/b)\cos(\omega t - \beta_g z)$ [V/m].

Exercise 17.5

(a) Find the longitudinal component $E_z(x,y,z)$ for a backward-propagating wave in a general lossy rectangular waveguide. Assume the backward-propagating wave propagates in the negative z direction and the amplitude of the wave is E_0^-.

(b) Find the total longitudinal field in a waveguide if both a forward-propagating wave of amplitude E_0^+ and a backward-propagating wave of amplitude E_0^- exist.

Answer

(a) $E_z^-(x,y,z) = E_0^-\sin(m\pi x/a)\sin(n\pi y/b)e^{\gamma z}$ [V/m].
(b) $E_z(x,y,z) = \sin(m\pi x/a)\sin(n\pi y/b)(E_0^+e^{-\gamma z} + E_0^-e^{\gamma z})$ [V/m].

Note: If the magnitude of the backward-propagating wave equals that of the forward-propagating wave ($|E_0^-| = |E_0^+|$), the term in the last parentheses in **(b)** can be written as sine or cosine functions using the exponential forms. We will use this property in **Section 17.9**.

Example 17.5 A standard rectangular waveguide, designated as EIA WR 75, has internal dimensions $a = 19.05$ mm and $b = 9.525$ mm. The waveguide is air filled and propagates waves at 18 GHz:

(a) Calculate the lowest possible TM mode at which the wave may be excited.
(b) For the mode in **(a)**, calculate the guide wavelength, guide phase constant, guide phase velocity, and the wave impedance for TM propagation (at 18 GHz).
(c) Calculate the maximum time-averaged power transmitted through the waveguide at the mode calculated in **(a)** if the electric field intensity is not to exceed the breakdown level in air (3×10^6 V/m).

Solution: The possible modes are all the modes with cutoff frequencies below 18 GHz. The propagated time-averaged power is given in **Eq. (17.153)**:

(a) The cutoff frequency for the TM modes is given as

$$f_{cmn} = \frac{1}{2\sqrt{\mu_0\varepsilon_0}}\sqrt{\left(\frac{m}{a}\right)^2 + \left(\frac{n}{b}\right)^2} = 1.5 \times 10^8\sqrt{\left(\frac{m}{0.01905}\right)^2 + \left(\frac{n}{0.009525}\right)^2} \quad [\text{Hz}]$$

The possible cutoff frequencies below 18 GHz are:

$$f_{c10} = 1.5 \times 10^8\sqrt{\left(\frac{1}{0.01905}\right)^2 + \left(\frac{0}{0.009525}\right)^2} = 7.874 \quad [\text{GHz}]$$

$$f_{c01} = 1.5 \times 10^8\sqrt{\left(\frac{0}{0.01905}\right)^2 + \left(\frac{1}{0.009525}\right)^2} = 15.748 \quad [\text{GHz}]$$

$$f_{c11} = 1.5 \times 10^8\sqrt{\left(\frac{1}{0.01905}\right)^2 + \left(\frac{1}{0.009525}\right)^2} = 17.60 \quad [\text{GHz}]$$

$$f_{c20} = 1.5 \times 10^8\sqrt{\left(\frac{2}{0.01905}\right)^2 + \left(\frac{0}{0.009525}\right)^2} = 15.75 \quad [\text{GHz}]$$

All other cutoff frequencies are above 18 GHz. Since, in TM modes, m or n cannot be zero, only the third of these, namely, $f_{c11} = 17.60$ GHz, corresponds to a possible TM mode. Thus, the only possible mode is the TM_{11} mode.

(b) At 18 GHz, the waves in the waveguide are TM_{11} from **Eqs. (17.141)** through **(17.145)**, we get

$$\beta_g = \beta\sqrt{1 - \frac{f_{cmn}^2}{f^2}} = \omega\sqrt{\mu_0\varepsilon_0}\sqrt{1 - \frac{f_{cmn}^2}{f^2}} = \frac{2 \times \pi \times 18 \times 10^9}{3 \times 10^8}\sqrt{1 - \left(\frac{17.60}{18}\right)^2} = 79 \quad [\text{rad/m}]$$

$$v_g = \frac{\omega}{\beta_g} = \frac{2 \times \pi \times 18 \times 10^9}{79} = 1.43 \times 10^9 \quad [\text{m/s}]$$

$$\lambda_g = \frac{2\pi}{\beta_g} = \frac{2 \times \pi}{79} = 0.0795 \quad [\text{m}]$$

$$Z_{TM} = \frac{\beta_g}{\omega\varepsilon_0} = \frac{79}{2 \times \pi \times 18 \times 10^9 \times 8.854 \times 10^{-12}} = 78.89 \quad [\Omega].$$

(c) To calculate the power density inside the waveguide, we use the expressions in **Eqs. (17.153)** and **(17.137)** with $m = 1$, $n = 1, a = 0.01905\ m, b = 0.00953\ m, E_0 = 3 \times 10^6$ V/m, $f = 18$ GHz, and $\varepsilon = \varepsilon_0$. Also, we have $\lambda_c(11) = v_p/fc_{11} = 3 \times 10^8/17.6 \times 10^9 = 0.017$ m. With these, we first calculate the cutoff wave number k_{c11} from **Eq. (17.137)**:

$$k_{c11} = \sqrt{\left(\frac{m\pi}{a}\right)^2 + \left(\frac{n\pi}{b}\right)^2} = \sqrt{\left(\frac{\pi}{0.01905}\right)^2 + \left(\frac{\pi}{0.009525}\right)^2} = 368.76 \quad [\text{rad/m}]$$

The maximum total power is

$$P = \frac{\omega\varepsilon\beta_g E_0^2 ab}{8k_c^2} = \frac{2 \times \pi \times 18 \times 10^9 \times 8.854 \times 10^{-12} \times 79 \times 9 \times 10^{12} \times 0.01905 \times 0.009525}{8 \times (368.76)^2} = 118{,}752 \quad [\text{W}]$$

This is almost 119 kW of power for a tube of cross-sectional area of less than 2 cm^2 (181.45 mm^2). In practice however, the fields chosen are much lower than the maximum. Since the power is proportional to the square of the field, the power decreases rapidly.

Exercise 17.6 A rectangular waveguide has dimensions $a = 0.015$ m and $b = 0.0075$ m. Find the first 10 cutoff frequencies for TM modes.

Answer (Frequencies given in GHz, in increasing frequency order)

Mode	TM_{11}	TM_{21}	TM_{31}	TM_{12}	TM_{22}, TM_{41}	TM_{32}	TM_{51}	TM_{13}	TM_{23}, TM_{61}	TM_{52}
f_{cmn}	22.361	28.284	36.056	41.231	44.721	50.0	53.852	60.828	63.246	64.031

Note: Lower modes may have higher frequencies than higher modes, depending on dimensions.

Exercise 17.7 Two standard waveguides are given: EIA WR 3, with dimensions $a = 0.8636$ mm and $b = 0.4318$ mm, operating at 400 GHz, and EIA WR 2300, with dimensions $a = 0.5842$ m and $b = 0.2921$ m operating at 600 MHz. Both waveguides are air filled (these two waveguides represent the smallest and largest standard waveguides):

(a) Find the TM_{11} cutoff frequencies of the two waveguides.
(b) Find the ratio between the total power carried by the two waveguides for a given electric field intensity E_0.

Answer

(a) $f_{c11}(\text{WR 3}) = 388.39$ GHz, $f_{c11}(\text{WR 2300}) = 574.14$ MHz.

(b) $p = \dfrac{P_1}{P_2} = \dfrac{f_1\beta_{g1}a_1b_1k_{c2}^2}{f_2\beta_{g2}a_2b_2k_{c1}^2} = 1.597 \times 10^{-6}$.

(Index 1 stands for the EIA WR 3 waveguide and 2 for the EIA WR 2300 waveguide.)

17.7.2 TE Modes in Rectangular Waveguides

For TE waves to exist, E_z must be zero. The wave equation to solve is now **Eq. (17.33)**:

$$\frac{\partial^2 H_z}{\partial x^2} + \frac{\partial^2 H_z}{\partial y^2} + (\gamma^2 + k^2)H_z = 0 \tag{17.154}$$

subject to the condition that the x and y components of the electric field intensity in **Eqs. (17.27)** and **(17.28)** must vanish on the conducting boundaries. These conditions are shown in **Figure 17.17**:

$$\frac{\partial H_z}{\partial x}(0, y) = 0 \quad \text{and} \quad \frac{\partial H_z}{\partial x}(a, y) = 0 \tag{17.155}$$

$$\frac{\partial H_z}{\partial y}(x, 0) = 0 \quad \text{and} \quad \frac{\partial H_z}{\partial y}(x, b) = 0 \tag{17.156}$$

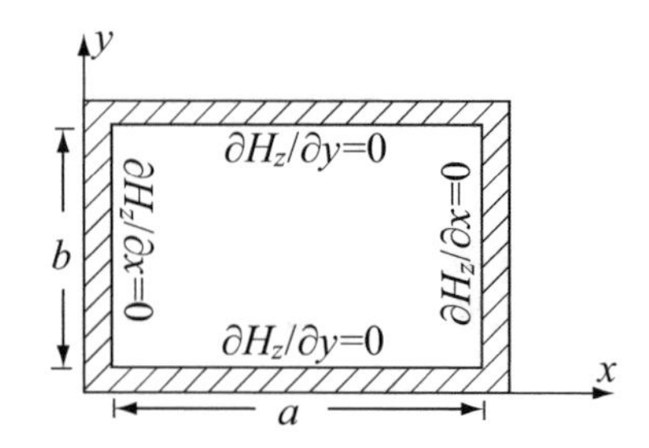

Figure 17.17 Boundary conditions on the magnetic field intensity for TE modes in a rectangular waveguide

Since **Eq. (17.154)** is identical in form to **Eq. (17.111)**, the general solution must also be of the same form as the solution in **Eq. (17.122)**:

$$H_z(x,y) = (A_1\sin k_x x + B_1\cos k_x x)\left(A_2\sin k_y y + B_2\cos k_y y\right) \tag{17.157}$$

To find the constants, we write from the four boundary conditions in **Eqs. (17.155)** and **(17.156)**

$$\left.\frac{\partial H_z(x,y)}{\partial x}\right|_{x=0} = (k_x A_1\cos 0 - k_x B_1\sin 0)\left(A_2\sin k_y y + B_2\cos k_y y\right) = 0 \quad \rightarrow \quad A_1 = 0 \tag{17.158}$$

$$\left.\frac{\partial H_z(x,y)}{\partial y}\right|_{y=0} = B_1\sin k_x x\left(k_y A_2\cos 0 - k_y B_2\sin 0\right) \quad \rightarrow \quad A_2 = 0 \tag{17.159}$$

At this point, the solution is

$$H_z(x,y) = B\cos(k_x x)\cos\left(k_y y\right) \tag{17.160}$$

where $B = B_1 B_2$. From the remaining boundary conditions

$$\left.\frac{\partial H_z(x,y)}{\partial x}\right|_{x=a} = -Bk_x\sin(k_x a)\sin\left(k_y y\right) = 0 \quad \rightarrow \quad k_x = \frac{m\pi}{a}, m = 0, 1, \ldots \tag{17.161}$$

$$\left.\frac{\partial H_z(x,y)}{\partial y}\right|_{y=b} = -B\cos\left(\frac{m\pi x}{a}\right)k_y\sin\left(k_y b\right) = 0 \quad \rightarrow \quad k_y = \frac{n\pi}{b}, n = 0, 1, \ldots \tag{17.162}$$

The solution is obtained by substituting k_x and k_y from **Eqs. (17.161)** and **(17.162)** into **Eq. (17.160)**. Taking the amplitude B in **Eq. (17.160)** as H_0, we get

$$H_z(x,y) = H_0\cos\left(\frac{m\pi x}{a}\right)\cos\left(\frac{n\pi y}{b}\right) \quad \left[\frac{\text{A}}{\text{m}}\right] \tag{17.163}$$

If we now add the variation in z (assuming only a forward-propagating wave), we get the general form of the longitudinal component of the magnetic field:

$$\boxed{H_z(x,y) = H_0\cos\left(\frac{m\pi x}{a}\right)\cos\left(\frac{n\pi y}{b}\right)e^{-\gamma z} \quad \left[\frac{\text{A}}{\text{m}}\right]} \tag{17.164}$$

This is now substituted into the general expressions for TE waves in **Eqs. (17.27)** through **(17.30)** to obtain the transverse components of the electric and magnetic field intensities. Performing the derivatives of H_z with respect to x and y as required in **Eqs. (17.27)** through **(17.30)** gives

$$E_x(x,y,z) = \frac{j\omega\mu}{\gamma^2 + k^2}H_0\frac{n\pi}{b}\cos\left(\frac{m\pi x}{a}\right)\sin\left(\frac{n\pi y}{b}\right)e^{-\gamma z} \quad \left[\frac{\text{V}}{\text{m}}\right] \tag{17.165}$$

$$E_y(x,y,z) = \frac{-j\omega\mu}{\gamma^2 + k^2}H_0\frac{m\pi}{a}\sin\left(\frac{m\pi x}{a}\right)\cos\left(\frac{n\pi y}{b}\right)e^{-\gamma z} \quad \left[\frac{\text{V}}{\text{m}}\right] \tag{17.166}$$

$$H_x(x,y,z) = \frac{\gamma}{\gamma^2 + k^2}H_0\frac{m\pi}{a}\sin\left(\frac{m\pi x}{a}\right)\cos\left(\frac{n\pi y}{b}\right)e^{-\gamma z} \quad \left[\frac{\text{A}}{\text{m}}\right] \tag{17.167}$$

$$H_y(x,y,z) = \frac{\gamma}{\gamma^2 + k^2}H_0\frac{n\pi}{b}\cos\left(\frac{m\pi x}{a}\right)\sin\left(\frac{n\pi y}{b}\right)e^{-\gamma z} \quad \left[\frac{\text{A}}{\text{m}}\right] \tag{17.168}$$

The propagation properties for TE modes in the waveguide are identical to those for TM modes except for the wave impedance. This can be seen either from **Table 17.2** or by direct calculation of the various properties (f_{cmn}, β_g, v_g, etc.). We will not reevaluate these here and simply use the properties in **Eqs. (17.135)** through **(17.143)** and **(17.146)** as given. The wave impedance, however, is different for TE modes and is given by the ratio of the transverse components of the electric and magnetic fields as follows:

$$Z_{TE} = \frac{E_x}{H_y} = -\frac{E_y}{H_x} = \frac{j\omega\mu}{\gamma} \quad [\Omega] \tag{17.169}$$

or, using **Eq. (17.135)** to replace γ by $j\beta_g$, we can write for lossless propagation

$$\boxed{Z_{TE} = \frac{\omega\mu}{\beta_g} = \sqrt{\frac{\mu}{\varepsilon}}\frac{1}{\sqrt{1 - f_{cmn}^2/f^2}} = \sqrt{\frac{\mu}{\varepsilon}}\frac{1}{\sqrt{1 - \lambda^2/\lambda_{cmn}^2}} = \eta\frac{\lambda_g}{\lambda} \quad [\Omega]} \tag{17.170}$$

Using the relations $k_{cmn}^2 = \gamma^2 + k^2$ and $\gamma = j\beta_g$, we can rewrite the transverse components of the electric and magnetic field in **Eqs. (17.165)** through **(17.168)** as

$$E_x(x,y,z) = \frac{j\omega\mu}{k_{cmn}^2}H_0\frac{n\pi}{b}\cos\left(\frac{m\pi x}{a}\right)\sin\left(\frac{n\pi y}{b}\right)e^{-j\beta_g z} \quad \left[\frac{\text{V}}{\text{m}}\right] \tag{17.171}$$

$$E_y(x,y,z) = \frac{-j\omega\mu}{k_{cmn}^2}H_0\frac{m\pi}{b}\sin\left(\frac{m\pi x}{a}\right)\cos\left(\frac{n\pi y}{b}\right)e^{-j\beta_g z} \quad \left[\frac{\text{V}}{\text{m}}\right] \tag{17.172}$$

$$H_x(x,y,z) = \frac{j\beta_g}{k_{cmn}^2}H_0\frac{m\pi}{a}\sin\left(\frac{m\pi x}{a}\right)\cos\left(\frac{n\pi y}{b}\right)e^{-j\beta_g z} \quad \left[\frac{\text{A}}{\text{m}}\right] \tag{17.173}$$

$$H_y(x,y,z) = \frac{j\beta_g}{k_{cmn}^2}H_0\frac{n\pi}{b}\cos\left(\frac{m\pi x}{a}\right)\sin\left(\frac{n\pi y}{b}\right)e^{-j\beta_g z} \quad \left[\frac{\text{A}}{\text{m}}\right] \tag{17.174}$$

TE_{mn} modes are obtained for all possible pairs of the integers m and n, except for $m = 0$ and $n = 0$. Unlike TM modes, in TE modes either m or n can be zero but not both. This indicates that the lowest propagating mode is a TE_{0n} or TE_{m0}, depending on the dimensions a and b of the waveguide. If $a > b$, the lowest cutoff frequency is for a TE_{10} mode. Also to be noted is that TM and TE modes with the same indices have the same cutoff frequency, as can be seen from **Eq. (17.138)**. The longitudinal magnetic field intensity (H_z) distribution in a waveguide with $a = 2b$ for the TE_{11} and TE_{30} modes is shown in **Figure 17.18**.

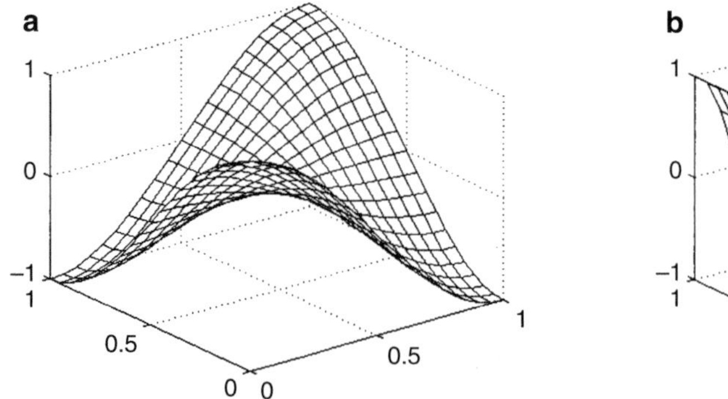

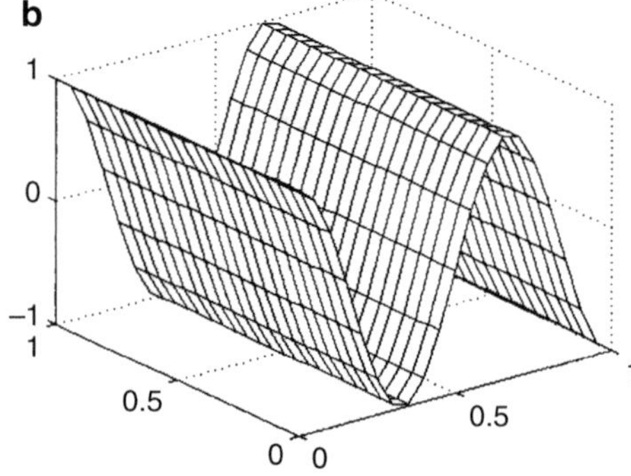

Figure 17.18 The longitudinal magnetic field distribution in a waveguide. (**a**) For the TE_{11} mode. (**b**) For the TE_{30} mode

The power density in a waveguide propagating a TE mode may be calculated using steps identical to those for TM modes, using the Poynting vector for the transverse components of **E** and **H**. Each pair of transverse components produces a time-averaged power density propagating in the z direction given by $E_xH_y^*/2$ and $-\,E_yH_x^*/2$. The total power density is the sum of the two terms as in **Eq. (17.151)**. Substituting for E_x, E_y, H_x, and H_y from **Eqs. (17.171)** through **(17.174)** into **Eq. (17.151)** gives

$$\mathcal{P}_{av}(x,y)=\hat{\mathbf{z}}\frac{\omega\mu\beta_g H_0^2}{2k_{cmn}^4}\left[\left(\frac{n\pi}{b}\right)^2\cos^2\left(\frac{m\pi x}{a}\right)\sin^2\left(\frac{n\pi y}{b}\right)+\left(\frac{m\pi}{a}\right)^2\sin^2\left(\frac{m\pi x}{a}\right)\cos^2\left(\frac{n\pi y}{b}\right)\right]\quad\left[\frac{\mathrm{W}}{\mathrm{m}^2}\right]\tag{17.175}$$

The total power in a rectangular waveguide of dimensions a and b for $m\neq 0$ and $n\neq 0$ is

$$\begin{aligned}P&=\frac{\omega\mu\beta_g H_0^2}{2k_{cmn}^4}\int_{x=0}^{x=a}\int_{y=0}^{y=b}\left[\left(\frac{n\pi}{b}\right)^2\cos^2\left(\frac{m\pi x}{a}\right)\sin^2\left(\frac{n\pi y}{b}\right)+\left(\frac{m\pi}{a}\right)^2\sin^2\left(\frac{m\pi x}{b}\right)\cos^2\left(\frac{m\pi y}{b}\right)\right]dxdy\\&=\frac{\omega\mu\beta_g H_0^2 ab}{8k_{cmn}^2}\quad[\mathrm{W}],\qquad m\neq 0,n\neq 0\end{aligned}\tag{17.176}$$

If $m=0$ or $n=0$, the power is twice as large as can be shown directly from **Eq. (17.176)**. That is:

$$P=\frac{\omega\mu\beta_g H_0^2 ab}{4k_{cmn}^2}\quad[\mathrm{W}],\qquad m=0\ or\ n=0\tag{17.177}$$

In particular, for the TE_{10} mode, we get

$$\boxed{P(\mathrm{TE}_{10})=\frac{\omega\mu\beta_g H_0^2 ab}{4k_{cmn}^2}=\frac{\omega\mu\beta_g H_0^2 ab}{4(\pi/a)^2}=\frac{\omega\mu\beta_g H_0^2 a^3 b}{4\pi^2}\quad[\mathrm{W}]}\tag{17.178}$$

The discussion on waveguides was limited here to rectangular cross-sectional waveguides. Other shapes can be analyzed in a similar fashion. However, since the analysis entails the solution of the wave equation, only simple shapes can be solved exactly. In particular, cylindrical waveguides can be analyzed using steps similar to those presented above, but the solution is in terms of Bessel rather than harmonic functions.

The lowest cutoff frequency mode in any waveguide is called a ***dominant mode***. In rectangular waveguides, this is a TE mode and, usually, the TE_{10} mode. Different modes which have the same cutoff frequency are called ***degenerate modes***.

Example 17.6 Application: The $\mathbf{TE_{10}}$ Mode The TE_{10} mode is the most important mode of propagation in rectangular waveguides. One reason for this is because it is possible to design waveguides with the largest possible mode separation between the TE_{10} and the other possible modes. This gives the largest possible bandwidth for propagation in a waveguide, as well as the lowest cutoff frequency for a given waveguide. Consider the following example.

The EIA WR 34 waveguide has dimensions $a=8.636$ mm and $b=4.318$ mm, is air filled, and may be excited in any mode:

(a) Calculate the lowest possible cutoff frequency.
(b) Calculate the next cutoff frequencies and identify the modes. Decide based on bandwidth (mode separation) which mode is most suitable for general purpose use.
(c) Repeat **(a)** and **(b)** if the waveguide is filled with a perfect dielectric with relative permittivity of 4.

Solution: The cutoff frequencies of the waveguide are calculated from **Eq. (17.138)** for all modes, including TM modes. The only difference between the various modes is that in TM modes, $m=0$ or $n=0$ is not allowed, whereas in TE modes they are. Also, TM_{mn} and TE_{mn} modes are always degenerate (have the same cutoff frequencies). From **Eq. (17.138)**

$$f_{cmn} = 1.5 \times 10^8 \sqrt{\left(\frac{m}{0.008636}\right)^2 + \left(\frac{n}{0.004318}\right)^2} \quad [\text{Hz}]$$

(a) The lowest possible mode must have $m = 1$ and $n = 0$. Since in TM modes, $n = 0$ results in all zero fields, this mode is the TE_{10} mode. Therefore, for any waveguide, with $a > b$, the dominant mode is the TE_{10} mode. The TE_{10} cutoff frequency is

$$f_{c10} = 1.5 \times 10^8 \sqrt{\left(\frac{1}{0.008636}\right)^2 + \left(\frac{0}{0.004318}\right)^2} = 17.369 \quad [\text{GHz}].$$

(b) The remaining modes are calculated similarly. All modes with one index zero are TE modes. All modes that have both indices nonzero are both TE and TM modes. Using the appropriate indices, the remaining (higher) cutoff frequencies are calculated and listed in the first row in **Table 17.3** in ascending order of cutoff frequencies.

The largest mode separation is between the TE_{10} and TE_{01} modes. Although separation between TE_{20} and TE_{02} is larger, the TE_{11}, TM_{11}, TE_{30}, TE_{21}, and TM_{21} modes are also in this range and, therefore, it is not suitable for single-mode operation. The TE_{10} mode can be propagated between 17.369 GHz and 34.738 GHz. Normally, the operation will be about 25% above 17.369 GHz and below 34.738 GHz so as not to be too close to the cutoff frequencies. The recommended range for this waveguide in the TE_{10} mode is 21.7–33 GHz.

(c) If the waveguide is filled with a dielectric, the cutoff frequencies are reduced by a factor of $\sqrt{\varepsilon_r}$. For $\varepsilon_r = 4$, the cutoff frequencies are reduced by a factor of 2. The new cutoff frequencies (in GHz) are shown in the second row in **Table 17.3**, again in ascending order.

Table 17.3 First 10 cutoff frequencies for empty and dielectric filled waveguide in Example 17.6

	TE_{10}	TE_{01}	TE_{11}	TE_{21}	TE_{30}	TE_{31}	TE_{40}	TE_{12}	TE_{22}	TE_{32}
		TE_{20}	TM_{11}	TM_{21}		TM_{31}	TE_{02}	TM_{12}	TE_{41}	TE_{50}
									TM_{22}	TM_{32}
									TM_{41}	
$\varepsilon = \varepsilon_0$	17.369	34.738	38.838	49.127	52.107	62.625	69.476	71.615	77.677	86.845
$\varepsilon = 4\varepsilon_0$	8.685	17.369	19.419	24.564	26.054	31.313	34.738	35.807	38.838	43.423

Example 17.7 Application: The Practical Rectangular Waveguide A practical rectangular waveguide is a tube of any rectangular cross section made of a high-conductivity material. The dimensions of the waveguide are arbitrary but are normally chosen such that the cutoff frequencies of the various modes are not degenerate as much as possible. Every waveguide is designed for a lowest, dominant mode, with the next mode defining the operating range or bandwidth of the waveguide. Normally, the dominant mode is the TE_{10} mode. Thus, for example, the EIA WR 90 waveguide has internal dimensions of 22.86 mm and 10.16 mm. The TE_{10} cutoff frequency is at 6.557 GHz and the normal operating range (recommended) for the TE_{10} mode is 8.2–12.5 GHz (*X*-band). Similarly, the EIA WR 5 waveguide has dimensions of 1.2954 mm and 0.6477 mm (internal) and a cutoff frequency of 115.71 GHz. The recommended operating frequency for the TE_{10} mode is 145–220 GHz. Although waveguides normally operate in the TE_{10} mode, they may also operate in any other mode, including TM modes.

It is required to design a waveguide with lowest cutoff frequency at 10 GHz. Two designs are proposed. One has both dimensions a and b equal. The second is designed such that $a = 2b$:

(a) Find the cutoff frequencies of the first 10 TM and first 10 TE modes in order of ascending frequencies and compare the two waveguides.

(b) What is the dominant mode and which modes are degenerate?

(c) Which waveguide is better suited for use in general purpose applications in terms of mode separation?

Solution: First, we find the dimensions of the two waveguides from the given cutoff frequency. Then, the cutoff frequencies of the remaining modes are found:

(a) The required lowest cutoff frequency is 10 GHz. For the square waveguide we write $b = a$ and substitute in **Eq. (17.138)**:

$$f_{c10} = \frac{1}{2\sqrt{\mu_0\varepsilon_0}}\sqrt{\left(\frac{1}{a}\right)^2 + \left(\frac{0}{a}\right)^2} = \frac{c}{2}\sqrt{\left(\frac{1}{a}\right)^2} \quad \rightarrow \quad a = \frac{c}{2f_{c10}} = \frac{3\times 10^8}{2\times 10^{10}} = 0.015 \quad [\mathrm{m}]$$

The required square waveguide is 15 mm by 15 mm in internal dimensions and the required rectangular waveguide is 15 mm by 7.5 mm in internal dimensions. Substituting these dimensions in the general expression for cutoff frequencies we get for the square waveguide ($b = a = 0.015$ m):

$$f_{cmn} = 1.5\times 10^8\sqrt{\left(\frac{m}{0.015}\right)^2 + \left(\frac{n}{0.015}\right)^2} \quad [\mathrm{Hz}]$$

For the rectangular waveguide ($a = 0.015$ and $b = 0.0075$ m):

$$f_{cmn} = 1.5\times 10^8\sqrt{\left(\frac{m}{0.015}\right)^2 + \left(\frac{n}{0.0075}\right)^2} \quad [\mathrm{Hz}]$$

Substituting the indices for the first ten modes, we get the required TM and TE modes (listed in ascending mode order):

	TM_{11}	TM_{21}	TM_{12}	TM_{22}	TM_{31}	TM_{13}	TM_{32}	TM_{23}	TM_{33}	TM_{41}
$a = b$	14.142	22.360	22.360	28.284	31.622	31.622	36.055	36.055	42.426	41.231
$a = 2b$	22.360	28.284	41.231	44.721	36.055	60.827	50.000	63.245	67.082	44.721

	TE_{10}	TE_{01}	TE_{11}	TE_{20}	TE_{02}	TE_{21}	TE_{12}	TE_{22}	TE_{30}	TE_{03}
$a = b$	10.0	10.0	14.142	20.0	20.0	22.360	22.360	28.284	30.0	30.0
$a = 2b$	10.0	20.0	22.360	20.0	40.0	28.284	41.231	44.721	30.0	60.0

(b) In the square waveguide, the TM_{12} and TM_{21} are degenerate modes as are any of the TM_{mn} and TM_{nm} or TE_{mn} and TE_{nm}. On the other hand, the rectangular waveguide has fewer degenerate modes in the range shown and the cutoff frequencies are much better spaced. For example, in the range between 22.360 and 28.284 GHz, only the TM_{11} mode can propagate. The rectangular waveguide is therefore much better suited for general purpose use than the square waveguide. This is one reason, most standard rectangular waveguides have dimensions which are either exactly a ratio of two to one or very close to this ratio. There are however waveguides with reduced height for special applications. These are typically designed so that $a = 4b$ but other ratios are possible. Most waveguides only propagate the dominant mode and therefore separation of this mode is very important. The dominant TM mode is the TM_{11} mode whereas the dominant TE mode is the TE_{10} mode. Because this is the lowest frequency mode, the dominat mode is the TE_{10} mode.

Example 17.8 A waveguide is given with dimensions $a = 12.954$ mm and $b = 6.477$ mm (EIA WR 51 waveguide). The waveguide is required to propagate at 30 GHz. Suppose we are free to choose any mode with cutoff frequency below 30 GHz. Find the ratio between the powers propagated:

(a) In the TE_{10} and TE_{01} modes.
(b) In the TE_{10} and TM_{11} modes.

Solution: First, we must find the cutoff frequencies, cutoff wavelengths, and the guide propagation constants for the three modes. Then, we use **Eq. (17.177)** to find the power propagated for the TE modes and **Eq. (17.153)** for the TM_{11} mode. The amplitude of the magnetic field intensity is assumed because when calculating the ratio between powers, it cancels out.

(a) The cutoff frequencies for the TE_{10}, TE_{01}, and TM_{11} modes are given in **Eq. (17.138)**. For the waveguide given,

$$f_{c10} = 11.5794\text{ GHz},\quad f_{c01} = 23.1589\text{ GHz},\quad f_{c11} = 25.8924\text{ GHz}$$

All these modes are below 30 GHz and, therefore, appropriate modes for the required wave. The guide propagation constant β_g is

$$\beta_{g10} = \beta\sqrt{1 - \frac{f_{c10}^2}{f^2}} = 2\pi f\sqrt{\mu_0\varepsilon_0}\sqrt{1 - \left(\frac{11.5794}{30}\right)^2} = \frac{2\pi \times 30 \times 10^9}{3 \times 10^8}\sqrt{1 - \left(\frac{11.5794}{30}\right)^2} = 579.63\quad [\text{rad/m}]$$

$$\beta_{g01} = \beta\sqrt{1 - \frac{f_{c01}^2}{f^2}} = 2\pi f\sqrt{\mu_0\varepsilon_0}\sqrt{1 - \left(\frac{23.1589}{30}\right)^2} = \frac{2\pi \times 30 \times 10^9}{3 \times 10^8}\sqrt{1 - \left(\frac{23.1589}{30}\right)^2} = 399.4\quad [\text{rad/m}]$$

$$\beta_{g11} = \beta\sqrt{1 - \frac{f_{c11}^2}{f^2}} = 2\pi f\sqrt{\mu_0\varepsilon_0}\sqrt{1 - \left(\frac{25.8924}{30}\right)^2} = \frac{2\pi \times 30 \times 10^9}{3 \times 10^8}\sqrt{1 - \left(\frac{25.8924}{30}\right)^2} = 317.34\quad [\text{rad/m}]$$

Similarly, the cutoff wave numbers squared are [from **Eq. (17.137)**]

$$k_{c10}^2 = \frac{\pi^2}{a^2} = \frac{\pi^2}{(0.012954)^2} = 5.8815 \times 10^4\quad \left[\text{rad}^2/\text{m}^2\right]$$

$$k_{c01}^2 = \frac{\pi^2}{b^2} = \frac{\pi^2}{(0.006477)^2} = 2.3526 \times 10^5\quad \left[\text{rad}^2/\text{m}^2\right]$$

$$k_{c11}^2 = \frac{\pi^2}{a^2} + \frac{\pi^2}{b^2} = \frac{\pi^2}{(0.012954)^2} + \frac{\pi^2}{(0.006477)^2} = 2.9408 \times 10^5\quad \left[\text{rad}^2/\text{m}^2\right]$$

The ratio between the powers in the TE_{10} and TE_{01} mode is

$$p = \frac{P_{10}}{P_{01}} = \frac{4\omega\mu_0\beta_{g10}H_0^2abk_{c01}^2}{4\omega\mu_0\beta_{g01}H_0^2abk_{c10}^2} = \frac{\beta_{g10}k_{c01}^2}{\beta_{g01}k_{c10}^2} = \frac{579.63 \times 2.3526 \times 10^5}{399.4 \times 5.8815 \times 10^4} = 5.8$$

(b) To calculate the ratio between the powers propagated in the TE_{10} mode and TM_{11} mode, we use **Eq. (17.177)** for the TE mode and **Eq. (17.153)** for the TM mode. Also, from **Eq. (17.144)**, the electric field intensity for the TM wave may be written as

$$E_0 = \frac{\gamma H_0}{j\omega\varepsilon} = \frac{\beta_g H_0}{\omega\varepsilon}\quad \left[\frac{\text{V}}{\text{m}}\right]$$

because $\gamma = j\beta$ (no losses in this case). With this, the ratio between the TE_{10} and TM_{11} modes is

$$p = \frac{P_{10}}{P_{11}} = \frac{8\omega\mu_0\beta_{g10}H_0^2abk_{c11}^2}{4\omega\varepsilon_0\beta_{g11}E_0^2abk_{c10}^2} = \frac{2\omega^2\mu_0\varepsilon_0\beta_{g10}k_{c11}^2}{\beta_{g11}^3k_{c10}^2} = \frac{2\omega^2\beta_{g10}k_{c11}^2}{c^2\beta_{g11}^3k_{c10}^2}$$

$$= \frac{2 \times 4 \times \pi^2 \times 900 \times 10^{18} \times 579.63 \times 2.9408 \times 10^5}{9 \times 10^{16} \times (317.34)^3 \times 5.8815 \times 10^4} = 71.6$$

In either case, the TE_{10} mode carries more power for a given electric or magnetic field intensity, at a given frequency.

Exercise 17.8

(a) Find the longitudinal component $H_z(x,y,z)$ for a backward-propagating TE wave in a rectangular waveguide. Assume lossy propagation in the negative z direction and the amplitude of the wave is H_1.
(b) Find the total longitudinal TE fields in a waveguide if both a forward-propagating wave with amplitude H_0 and a backward-propagating wave with amplitude H_1 exist.

Answer

(a) $H_z^-(x,y,z) = H_1 \cos\left(\frac{m\pi}{a}x\right)\cos\left(\frac{n\pi}{b}y\right)e^{\gamma z} \quad \left[\frac{\text{A}}{\text{m}}\right].$

(b) $H_z(x,y,z) = \cos\left(\frac{m\pi}{a}x\right)\cos\left(\frac{n\pi}{b}y\right)\left(H_0 e^{-\gamma z} + H_1 e^{\gamma z}\right) \quad \left[\frac{\text{A}}{\text{m}}\right].$

Note: If the magnitude of the backward-propagating wave equals that of the forward-propagating wave ($|H_1| = |H_0|$), the term in the last set of parentheses in each component in **(b)** can be written as sine or cosine functions using the exponential forms. This is particularly simple if propagation is in a lossless material ($\gamma = j\beta_g$).

17.7.3 Attenuation and Losses in Rectangular Waveguides

So far in our discussion we have avoided attenuation and losses in waveguides except for the use of the general propagation constant γ in deriving the equations. However, no system can operate without losses. The mechanism for losses in waveguides is the same as in any other transmission line and consists of two parts: (1) losses in the dielectric and (2) losses in the imperfect conductors or wall losses. In addition, below cutoff, the attenuation constant is very high even for perfect dielectrics in the guide and perfectly conducting walls. These losses, the resulting attenuation constants, and their influence on the power relations in the waveguide are discussed next.

17.7.3.1 Dielectric Losses

The medium in waveguides is normally a low-loss dielectric such as air. Therefore, it is safe to assume that the low-loss approximation used in **Sections 17.3.2** and **17.3.3** for TE and TM propagation applies here. The attenuation constant is the same for TM and TE propagation (see **Table 17.2**). Replacing f_c by f_{cmn} to indicate that each mode has a different cutoff frequency and therefore a different attenuation constant and using an index d to indicate that this attenuation constant is due to dielectric losses, we get

$$\alpha_{dTE} = \alpha_{dTM} = \frac{\sigma_d \eta_d}{2\sqrt{1 - f_{cmn}^2/f^2}} \quad \left[\frac{\text{Np}}{\text{m}}\right] \tag{17.179}$$

Normally, the attenuation due to the dielectric for air-filled conducting-wall waveguides is rather small and is normally much smaller than the attenuation caused by losses in the walls of the waveguide. On the other hand, for dielectric waveguides such as optical waveguides, almost all losses are due to dielectric losses.

17.7.3.2 Wall Losses

To calculate the wall losses, we start with the expression for the time-averaged power density in **Eq. (17.151)**, which applies to both TE and TM waves. The only difference is in the expressions for the transverse components. The total power in the waveguide cross section is found by integrating this power density over the cross-sectional area of the waveguide and is given in **Eq. (17.153)** for TM modes and in **Eq. (17.176)** or **Eq. (17.177)** for TE modes.

Now, we assume there are no losses in the dielectric, but there are losses in the wall. The propagation constant is $\gamma = \alpha_\omega + j\beta$. Substituting this in **Eq. (17.151)**, and integrating gives the following expression for the total power in the cross section of the waveguide for TM modes:

$$P = \frac{\omega\varepsilon\beta_g E_0^2 ab}{8k_{cmn}^2} e^{-2\alpha_\omega z} \quad [\mathrm{W}] \tag{17.180}$$

Similarly, for TE modes

$$P = \frac{\omega\mu\beta_g H_0^2 ab}{8k_{cmn}^2} e^{-2\alpha_\omega z}, \quad m \neq 0, n \neq 0 \quad or \quad P = \frac{\omega\mu\beta_g H_0^2 ab}{4k_{cmn}^2} e^{-2\alpha_\omega z}, \quad m = 0 \; or \; n = 0 \quad [\mathrm{W}] \tag{17.181}$$

In general, we can write $P = P_0 e^{-2\alpha_\omega z}$.

In other words, as the wave propagates, starting with some power P_0 (which depends on the mode used), the power is attenuated continuously with the distance z.

We can calculate the attenuation constant as the attenuation per unit length by taking a 1 m length of the waveguide and calculating the power loss in this section. From the Poynting theorem, the rate of decrease in the time-averaged power equals the time-averaged power lost in this section of the waveguide. For $z = 1$ m, we get

$$-\frac{d(P)}{dz} = P_{loss} = 2\alpha_\omega P_0 e^{-2\alpha_\omega 1} \quad \rightarrow \quad \alpha_\omega = \frac{P_{loss}}{2P_0 e^{-2\alpha_\omega}} \quad \left[\frac{\mathrm{Np}}{\mathrm{m}}\right] \tag{17.182}$$

Assuming losses are low, $e^{-2\alpha_\omega} \approx 1$ and we get

$$\boxed{\alpha_\omega \approx \frac{P_{loss}}{2P_{av}} \quad \left[\frac{\mathrm{Np}}{\mathrm{m}}\right]} \tag{17.183}$$

Now, all we have to do is calculate the total power loss in the walls per unit length of the waveguide and the time-averaged power in the waveguide. However, both power loss and time-averaged power in the waveguide are mode dependent. The general method is as follows: We calculate the total time-averaged power density in the waveguide using **Eqs. (17.152)** or **(17.175)**, which depend on the mode. The power in the waveguide may be calculated at any point z, but for simplicity we choose $z = 0$. Power loss occurs only in the wall and is calculated as for any conducting material using the relation $P_{loss} = I^2R/2$. The walls are highly conducting; the only current density in the walls is on and near the surface. Therefore, we assume a surface current density on each of the walls equal to $J_s = |H_t|$, where H_t is the tangential magnetic field intensity at the wall. The total current is the current density, integrated over the width of the wall. The resistance of the wall is the surface resistance. We have already calculated this resistance for lossy parallel plate transmission lines. The surface resistance [see **Eq.** (14.7)] is

$$R_s = \frac{1}{\sigma_c \delta} \quad [\Omega] \tag{17.184}$$

where σ_c is the conductivity of the wall material and $\delta = 1/\sqrt{\pi f \mu \sigma_c}$ its skin depth (see **Section 12.7.3**). The total power loss in the wall is therefore

$$P_{loss} = \frac{R_s}{2}\int_s J_s^2 ds = \frac{R_s}{2}\int_s |H_t|^2 ds \quad [\mathrm{W}] \tag{17.185}$$

where s is the total interior wall surface of the waveguide of length 1 m. This expression looks simple, but the integration must be done on each wall of the waveguide separately, after calculating the tangential component of the magnetic field intensity at the wall. We perform these calculations in **Example 17.9** for the TE_{10} mode. The attenuation constant due to wall losses is therefore

$$\boxed{\alpha_\omega = \frac{R_s}{4P_{av}}\int_s |H_t|^2 ds \quad \left[\frac{\mathrm{Np}}{\mathrm{m}}\right]} \tag{17.186}$$

If both dielectric and wall losses exist, the attenuation constant in the waveguide is the sum of the attenuation due to the dielectric and the attenuation due to the wall losses:

$$\boxed{\alpha = \alpha_d + \alpha_\omega \quad [\text{Np/m}]} \tag{17.187}$$

Attenuation in waveguides depends on a variety of parameters including the mode index, frequency, type of mode, conductivity of walls, and the dielectric in the waveguide. Typical values for waveguides in the microwave region are between about 0.1 dB/m and 0.5 dB/m. On the other hand, optical waveguides have losses that are between 0.5 dB/km to 10 dB/km. The much lower losses in optical waveguides is one of the most important reasons for their widespread use in communication.

17.7.3.3 Attenuation Below Cutoff

Any wave propagating below cutoff will be attenuated because below cutoff, the propagation constant is real. Consider the propagation constant for a mode above cutoff, propagating in a lossless medium ($\alpha_d = 0$), as written in terms of the guide phase constant β_g in **Eq. (17.141)**:

$$\gamma = j\beta_g = j\omega\sqrt{\mu\varepsilon}\sqrt{1 - \frac{f_{cmn}^2}{f^2}}, \quad f > f_{cmn} \tag{17.188}$$

If $f < f_{cmn}$, the term under the square root sign becomes negative and the propagation constant becomes

$$\gamma = -\omega\sqrt{\mu\varepsilon}\sqrt{\frac{f_{cmn}^2}{f^2} - 1}, \quad f < f_{cmn} \tag{17.189}$$

The propagation constant is real and consists of an attenuation constant and zero propagation constant ($\gamma = \alpha + j0$). Taking the positive solution for α (a negative value of α would imply fields increase in magnitude as they propagate which is physically impossible) gives the attenuation below cutoff as

$$\boxed{\alpha_{bc} = \omega\sqrt{\mu\varepsilon}\sqrt{\frac{f_{cmn}^2}{f^2} - 1}, \quad f < f_{cmn} \quad [\text{Np/m}]} \tag{17.190}$$

This attenuation constant is very high and for this reason, we say that waves do not propagate below cutoff. As in parallel plate waveguides, waves below cutoff are evanescent waves.

Example 17.9 An air-filled waveguide has dimensions $a = 28.499$ mm and $b = 12.624$ mm (EIA WR 112 waveguide) and is gold plated ($\sigma_c = 4.7 \times 10^7$ S/m). Conductivity of air is $\sigma_d = 10^{-5}$ S/m and permittivity and permeability of air may be taken as those of free space. The peak magnetic field intensity in the waveguide is 1 A/m and the waveguide operates in the TE_{10} mode at 9 GHz:

(a) Find the surface current densities in the walls of the waveguide.
(b) Find the attenuation constant in the waveguide.
(c) If the waveguide is 100 m long, what must be the total power required from the generator to transfer 1 W to the load? Assume both generator and load are matched.

Solution: In the TE_{10} mode, only the E_y, H_x, and H_z components exist. From these, we calculate the tangential components on each of the walls of the waveguide. The tangential component of the magnetic field intensity is equal in magnitude to the current density in the wall because of the interface conditions. The attenuation constant is the sum of the attenuation constant due to dielectric losses **[Eq. (17.179)]** and the attenuation constant due to wall losses **[Eq. (17.186)]**:

(a) The TE_{10} fields are found by setting $m = 1$, $n = 0$ in **Eqs. (17.164)** through **(17.168)**:

$$H_z(x,y,z) = H_0 \cos\left(\frac{\pi x}{a}\right) e^{-\gamma z} \quad \left[\frac{\text{A}}{\text{m}}\right]$$

$$E_y(x,y,z) = \frac{-j\omega\mu a}{\pi} H_0 \sin\left(\frac{\pi x}{a}\right) e^{-\gamma z} \quad \left[\frac{\text{V}}{\text{m}}\right]$$

$$H_x(x,y,z) = \frac{j\beta_g a}{\pi} H_0 \sin\left(\frac{\pi x}{a}\right) e^{-\gamma z} \quad \left[\frac{\text{A}}{\text{m}}\right]$$

where $\gamma^2 + k^2 = k_c^2 = (\pi/a)^2$ was used to simplify the expressions. Also, in **Eq. (17.167)** we assumed that $\gamma \approx j\beta_g$ (low-loss approximation). All other field components are zero.

At the walls parallel to the y axis, both E_y and H_x are zero. The only tangential component on these walls is H_z. However, on the walls parallel to the x axis, both H_x and H_z are nonzero. At $x = 0$ and $x = a$, the tangential magnetic field intensity is (see, for example, **Figure 17.14**):

$$\mathbf{H}_t(0,y,z) = -\mathbf{H}_t(a,y,z) = \hat{\mathbf{z}} H_0 e^{-\gamma z} \quad [\text{A/m}]$$

At $y = 0$ and $y = b$, the tangential magnetic field intensity is

$$\mathbf{H}_t(x,0,z) = \mathbf{H}_t(x,b,z) = \hat{\mathbf{x}} \frac{j\beta_g a}{\pi} H_0 \sin\left(\frac{\pi x}{a}\right) e^{-\gamma z} + \hat{\mathbf{z}} H_0 \cos\left(\frac{\pi x}{a}\right) e^{-\gamma z} \quad \left[\frac{\text{A}}{\text{m}}\right]$$

To calculate the current on the surfaces, we recall that the relation between current and magnetic field intensity is a curl relation. Therefore, the current density on any surface is $\mathbf{J}_s = \hat{\mathbf{n}} \times \mathbf{H}$ (see **Chapter 11**), where $\hat{\mathbf{n}}$ is the normal unit vector to the surface. Since by definition the normal to a surface points out of the surface (i.e., into the waveguide), the normal to the surface at $x = 0$ is $\hat{\mathbf{n}} = \hat{\mathbf{x}}$, that at $x = a$ is $\hat{\mathbf{n}} = -\hat{\mathbf{x}}$, that on the plate at $y = 0$ is $\hat{\mathbf{n}} = \hat{\mathbf{y}}$, and that on the plate at $y = b$ is $\hat{\mathbf{n}} = -\hat{\mathbf{y}}$. Performing the products $\hat{\mathbf{n}} \times \mathbf{H}$ on the four surfaces, we get

$$\mathbf{J}_s(0,y,z) = -\hat{\mathbf{y}} H_0 e^{-\gamma z}, \quad \mathbf{J}_s(a,y,z) = -\hat{\mathbf{y}} H_0 e^{-\gamma z} \quad [\text{A/m}]$$

$$\mathbf{J}_s(x,0,z) = \hat{\mathbf{y}} \times \left(\hat{\mathbf{x}} \frac{j\beta_g a}{\pi} H_0 \sin\left(\frac{\pi x}{a}\right) e^{-\gamma z} + \hat{\mathbf{z}} H_0 \cos\left(\frac{\pi x}{a}\right) e^{-\gamma z}\right) = \hat{\mathbf{x}} H_0 \cos\left(\frac{\pi x}{a}\right) e^{-\gamma z} - \hat{\mathbf{z}} \frac{j\beta_g a}{\pi} H_0 \sin\left(\frac{\pi x}{a}\right) e^{-\gamma z} \quad \left[\frac{\text{A}}{\text{m}}\right]$$

$$\mathbf{J}_s(x,b,z) = -\hat{\mathbf{y}} \times \left(\hat{\mathbf{x}} \frac{j\beta_g a}{\pi} H_0 \sin\left(\frac{\pi x}{a}\right) e^{-\gamma z} + \hat{\mathbf{z}} H_0 \cos\left(\frac{\pi x}{a}\right) e^{-\gamma z}\right) = -\hat{\mathbf{x}} H_0 \cos\left(\frac{\pi x}{a}\right) e^{-\gamma z} - \hat{\mathbf{z}} \frac{j\beta_g a}{\pi} H_0 \sin\left(\frac{\pi x}{a}\right) e^{-\gamma z} \quad \left[\frac{\text{A}}{\text{m}}\right].$$

(b) To find the attenuation constant, we first calculate the time-averaged power through the cross section of the waveguide. This was calculated in **Eq. (17.178)**. With $k_c^2 = (\pi/a)^2$, we get, at $z = 0$,

$$P_{av} = \frac{\omega\mu_0\beta_g H_0^2 a^3 b}{4\pi^2} \quad [\text{W}]$$

The losses in the wall are now calculated from **Eqs. (17.184)** and **(17.185)**, but first we need to calculate the skin depth δ:

$$\delta = \frac{1}{\sqrt{\pi f \mu_0 \sigma_c}} = \frac{1}{\sqrt{\pi \times 9 \times 10^9 \times 4 \times \pi \times 10^{-7} \times 4.7 \times 10^7}} = 7.7384 \times 10^{-7} \quad [\text{m}]$$

The surface resistance is therefore **[Eq. (17.184)]**

$$R_s = \frac{1}{\sigma_c \delta} = \frac{1}{4.7 \times 10^7 \times 7.7384 \times 10^{-7}} = 2.75 \times 10^{-2} \quad [\Omega]$$

The losses are calculated on each surface separately, but since the current density on each two parallel plates is the same, we only need to calculate one of each and multiply the result by 2. We assume a section of the waveguide, 1 m long, and also assume that $\alpha z \gg 1$ so that $e^{-\alpha z} \approx 1$. At $x = 0$, we have

$$P_{Lav}(x=0) = \frac{I^2 R_s}{2} = \frac{R_s}{2}\int_{y=0}^{y=b} |J_s(0,y,z)|^2 dy = \frac{R_s}{2}\int_{y=0}^{y=b} H_0^2 dy = \frac{R_s H_0^2 b}{2} \quad [\mathrm{W}]$$

At $y = 0$

$$P_{Lav}(y=0) = \frac{R_s}{2}\int_{x=0}^{x=a} \left(|J_{sx}(x,0,z)|^2 + |J_{sz}(x,0,z)|^2\right) dx$$
$$= \frac{R_s}{2}\int_{x=0}^{x=a} \left\{\left[\frac{\beta_g a}{\pi} H_0 \sin\left(\frac{\pi x}{a}\right)\right]^2 + \left[H_0 \cos\left(\frac{\pi x}{a}\right)\right]^2\right\} dx = \frac{R_s H_0^2 a}{4}\left(1 + \frac{\beta_g^2 a^2}{\pi^2}\right) \quad [\mathrm{W}]$$

The total power loss is

$$P_{Lav} = 2P_{Lav}(y=0) + 2P_{Lav}(x=0) = R_s H_0^2\left[b + \frac{a}{2}\left(1 + \frac{\beta_g^2 a^2}{\pi^2}\right)\right] \quad [\mathrm{W}]$$

Dividing this by twice the time-averaged power entering the section at $z = 0$ gives

$$\alpha_\omega = \frac{P_{Lav}}{2P_{av}} = \frac{2\pi^2 R_s}{\omega \mu_0 \beta_g a^3 b}\left[b + \frac{a}{2}\left(1 + \frac{\beta_g^2 a^2}{\pi^2}\right)\right] \quad \left[\frac{\mathrm{Np}}{\mathrm{m}}\right]$$

The attenuation constant due to the dielectric is given in **Eq. (17.178)**

$$\alpha_{dTE_{10}} = \frac{\sigma_d \eta_0}{2\sqrt{1 - f_{c10}^2/f^2}} \quad \left[\frac{\mathrm{Np}}{\mathrm{m}}\right]$$

With the cutoff frequency for the TE_{10} mode equal to $f_{c10} = (c/2)/0.0285 = 5.263$ GHz and $\beta_g = \omega\sqrt{\mu_0 \varepsilon_0}\sqrt{1 - f_{c10}^2/f^2} = 152.9$ rad/m, the total attenuation constant is

$$\alpha = \alpha_\omega + \alpha_d = \frac{2\pi^2 R_s}{\omega \mu_0 \beta_g a^3 b}\left[b + \frac{a}{2}\left(1 + \frac{\beta_g^2 a^2}{\pi^2}\right)\right] + \frac{\sigma_d \eta_0}{2\sqrt{1 - f_{c10}^2/f^2}}$$
$$= \frac{2 \times \pi^2 \times 2.75 \times 10^{-2}}{2 \times \pi \times 9 \times 10^9 \times 4 \times \pi \times 10^{-7} \times 152.9 \times (0.028499)^3 \times 0.012624}$$
$$\times \left[0.012624 + \frac{0.028499}{2}\left(1 + \frac{(152.9)^2 \times (0.028499)^2}{\pi^2}\right)\right]$$
$$+ \frac{10^{-5} \times 377}{2\sqrt{1 - \left(\frac{5.263}{9}\right)^2}} = 0.009282 + 0.002324 = 0.01161 \quad [\mathrm{Np/m}]$$

The attenuation constant is 0.01161 Np/m or $0.01161 \times 8.69 = 0.1$ dB/m, a relatively high attenuation.

(c) For a 100 m long waveguide, the input power required to deliver 1 W to the load is

$$P_{Load} = 1\,\text{W} = P_0 e^{-2\alpha z} = P_0 e^{-200\times 0.01161} \quad \rightarrow \quad P_0 = \frac{1}{e^{-2.322}} = 10.196 \quad [\text{W}]$$

That is, over 90% of the input power is lost (dissipated) in the waveguide itself over the 100 m distance.

Example 17.10 Application: Operation of Waveguides Below Cutoff: Use of Waveguides as High-Pass Filters A waveguide operates at 8 GHz. The cutoff frequency for the mode propagated is 10 GHz. Calculate the attenuation constant below cutoff in the waveguide. Assume an air-filled waveguide.

Solution: From **Eq. (17.190)**:

$$\alpha_{bc} = \omega\sqrt{\mu_0\varepsilon_0}\sqrt{\frac{f_{cmn}^2}{f^2} - 1} = \frac{2\times\pi\times 8\times 10^9}{3\times 10^8}\sqrt{\left(\frac{10}{8}\right)^2 - 1} = 125.66 \quad [\text{Np/m}]$$

To get a better idea of this attenuation, the attenuation in [dB] is $125.66 \times 8.69 = 1{,}092$ dB/m. Thus, waveguides are excellent high-pass filters, blocking any waves below cutoff.

Exercise 17.9 An air-filled waveguide operates below cutoff. The operating frequency is 1 GHz and the cutoff frequency is 1.5 GHz. The power P is supplied to the waveguide at a point $z = 0$. How far does the power propagates before it is attenuated to $10^{-12}P$?

Answer 0.59 m

17.8 Other Waveguides

We discussed in this chapter only waveguides with parallel conducting walls, primarily because of the ease with which concepts could be introduced. However, we also mentioned that the basic requirement is that waves must be totally reflected at the boundaries of the guiding structure. Therefore, any structure that provides this facility may be used as a waveguide. In particular, cylindrical waveguides are very common, but other structures exist. For example, elliptical waveguides are sometimes used because their cross-sectional area remains constant when bent.

Another common waveguide, especially in integrated system is the dielectric waveguide. Since the main requirement of a waveguide is to contain the waves, as long as the waves impinge on dielectric interfaces at angles above the critical angle, guideance of waves within the dielectric is guaranteed (see **Section 13.4.4**). Particularly important applications of dielectric waveguides are in optics including the common optical fiber uses for transmission of data. In some cases, including in optical fibers, a coating is added to prevent leakage of waves below the critical angle.

The analysis of any waveguide is, in principle, the same as that performed for the rectangular waveguides discussed here: the wave equation is solved in a convenient system of coordinates, and then the boundary conditions of the waveguide are satisfied.

Unfortunately, the solution is relatively simple only for rectangular waveguides. Cylindrical waveguides may also be analyzed relatively easily through use of cylindrical coordinates and solution to the Bessel equation. Other, rather complicated-shaped waveguides and modifications of the simple waveguides discussed here exist and are used for specialized applications. Their analysis is often much more involved than the rectangular waveguides presented here and may require approximate analysis, including the use of computational techniques.

Finally it should be mentioned that wave guidance can occur even when no intent of guiding waves is present. A wave impinging on a conducting surface at an angle will be guided by the surface. Similarly, waves can be guided in tunnels, mine galleries, coridors and similar structures.

17.9 Cavity Resonators waves.m

A rectangular cavity resonator is built out of a rectangular waveguide by adding two conducting walls at $z = 0$ and $z = d$, as shown in **Figure 17.19a**. The cavity resonator may be viewed as being made of three parallel plate waveguides, as shown in **Figure 17.19b**. The cavity is a modified waveguide, in which there are standing waves in the z direction as well as in the x and y directions. The main difference between cavities and waveguides is that in cavities, the z direction imposes additional boundary conditions and there is no propagation of waves as in waveguides. The cavity acts as a resonant structure in which there is exchange of energy between the electric and magnetic field at given (resonant) frequencies. This is equivalent to resonant LC circuits in the case of lossless cavities or to RLC circuits in the case of lossy cavities.

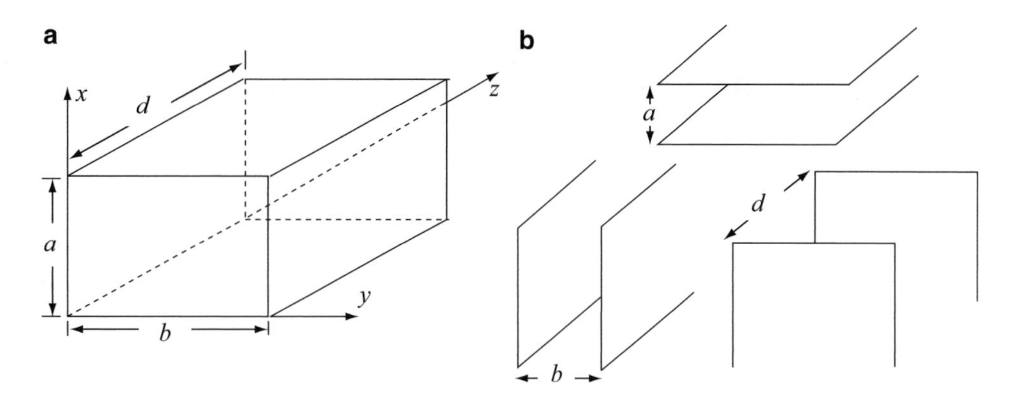

Figure 17.19 (**a**) Structure and dimensions of a rectangular cavity resonator. (**b**) Construction of the cavity resonator as the intersection of three parallel plate waveguides

The analysis of fields in a cavity requires the solution of the full three-dimensional wave equation with the required boundary conditions. The procedure here will be to take the TM and TE waves we have already defined and to modify them to satisfy the additional boundary conditions imposed by the additional conducting walls. However, the TE and TM equations given in **Eqs. (17.33)** and **(17.57)** cannot be used directly. The main reason is that in waveguides, we assumed explicitly that the wave propagates in the z direction and that the transverse directions are the directions perpendicular to the direction of propagation (z direction). In cavities, there is no clear direction we can take as a transverse direction. The approach here is to take the z direction (usually the long dimension of the cavity) as a reference direction, allow the waves to propagate along this direction, and calculate the total waves as the sum of the forward- and backward-propagating waves reflected off the shorting walls. We define the TE and TM waves as:

(1) A TM wave in a cavity resonator is any wave which has no magnetic field component in the z direction of the cavity.
(2) A TE wave in a cavity resonator is any wave which has no electric field component in the z direction of the cavity.

By direct extension of **Eq. (17.57)**, and setting $\gamma^2 = 0$ (no propagation) the TM fields satisfy the following equation:

$$\boxed{\frac{\partial^2 E_z}{\partial x^2} + \frac{\partial^2 E_z}{\partial y^2} + \frac{\partial^2 E_z}{\partial z^2} + k^2 E_z = 0} \tag{17.191}$$

From **Eq. (17.33)**, the TE fields satisfy

$$\boxed{\frac{\partial^2 H_z}{\partial x^2} + \frac{\partial^2 H_z}{\partial y^2} + \frac{\partial^2 H_z}{\partial z^2} + k^2 H_z = 0} \tag{17.192}$$

Comparing these with **Eqs. (17.111)** and **(17.154)**, it is relatively easy to see that the fields E_z and H_z are those in **Eqs. (17.122)** and **(17.157)** multiplied by an additional product due to the z dependence of the field. However, we will find the solutions by formal application of the above ideas to the waveguide fields found in the previous sections.

17.9.1 TM Modes in Cavity Resonators

To find the TM and TE modes in cavity resonators, we can proceed in two ways. One is to start with the wave equations in **Eqs. (17.191)** and **(17.192)** and solve them subject to the boundary conditions on all eight walls of the cavity resonator. This is what we have done for waveguides. Another is to use the waves obtained for the rectangular waveguides and calculate the reflected waves caused by the introduction of the two additional conducting walls. This causes backward-propagating waves, and the required boundary conditions on the conducting walls at $z = 0$ and $z = d$ (see **Figure 17.19**) are used to calculate the reflected waves. Then, the sum of the forward and backward waves gives the correct fields inside the cavity. We will use the first method to evaluate the longitudinal components of the electric and magnetic field intensities, and the second method to evaluate the transverse components for TE and TM modes, to demonstrate the various techniques involved.

For TM modes in a waveguide, we imposed the condition that the tangential components of the electric field intensity must be zero on the conducting walls. These conditions still apply here for the transverse fields. However, the z component of the field is normal to the two walls perpendicular to the z axis. Thus, we cannot impose the zero tangential electric field condition on these boundaries for E_z. On the other hand, since E_z is normal to the surfaces at $z = 0$ and $z = d$, the following conditions apply:

$$\left.\frac{\partial E_z(x,y,z)}{\partial z}\right|_{z=0} = 0, \quad \left.\frac{\partial E_z(x,y,z)}{\partial z}\right|_{z=d} = 0 \tag{17.193}$$

Following the solution process as for the TM modes in waveguides [**Eqs. (17.111)** through **(17.128)**], we write the general solution as

$$E_z(x,y,z) = X(x)Y(y)Z(z) \tag{17.194}$$

Substituting this in **Eq. (17.191)** and dividing by $E(x,y,z)$, we get

$$\frac{1}{X(x)}\frac{\partial^2 X(x)}{\partial x^2} + \frac{1}{Y(y)}\frac{\partial^2 Y(y)}{\partial y^2} + \frac{1}{Z(z)}\frac{\partial^2 Z(z)}{\partial z^2} + k^2 = 0 \tag{17.195}$$

where $k^2 = \omega^2\mu\varepsilon$, as for the waveguide solution. This equation may be separated as follows:

$$\frac{\partial^2 X(x)}{\partial x^2} + k_x^2 X(x) = 0 \tag{17.196}$$

$$\frac{\partial^2 Y(y)}{\partial y^2} + k_y^2 Y(y) = 0 \tag{17.197}$$

$$\frac{\partial^2 Z(z)}{\partial z^2} + k_z^2 Z(z) = 0 \tag{17.198}$$

with the additional condition that

$$-k_x^2 - k_y^2 - k_z^2 + k^2 = 0 \tag{17.199}$$

Note that we have not used here the notation of cutoff wave number k_c as in **Eq. (17.119)** because cutoff has no meaning in cavity resonators; that is, since waves do not propagate, the concept of cutoff does not exist. From the discussion in **Sections 17.7.1** and **17.7.2**, the general solution for E_z is

$$E_z(x,y,z) = (A_1 \sin k_x x + B_1 \cos k_x x)(A_2 \sin k_y y + B_2 \cos k_y y)(A_3 \sin k_z z + B_3 \cos k_z z) \quad [\text{V/m}] \tag{17.200}$$

We already found in **Eqs. (17.123)** through **(17.126)** that

$$B_1 = B_2 = 0 \quad \text{and} \quad k_x = \frac{m\pi}{a}, \quad k_y = \frac{n\pi}{b} \tag{17.201}$$

With these, we have

$$E_z(x,y,z) = A\sin\left(\frac{m\pi x}{a}\right)\sin\left(\frac{n\pi y}{b}\right)(A_3\sin k_z z + B_3\cos k_z z) \quad [\text{V/m}] \tag{17.202}$$

Now, we apply the additional boundary conditions in **Eq. (17.193)**:

$$\left.\frac{\partial E_z(x,y,z)}{\partial z}\right|_{z=0} = A\sin\left(\frac{m\pi x}{a}\right)\sin\left(\frac{n\pi x}{b}\right)k_z(A_3K_z\cos 0 - B_3k_z\sin 0) = 0 \quad \rightarrow \quad A_3 = 0 \tag{17.203}$$

$$\left.\frac{\partial E_z(x,y,z)}{\partial z}\right|_{z=d} = A\sin\left(\frac{m\pi x}{a}\right)\sin\left(\frac{n\pi x}{b}\right)k_z(-\sin k_z d) = 0 \quad \rightarrow \quad k_z = \frac{p\pi}{d} \tag{17.204}$$

where the constant B_3 was absorbed into the general constant A. The solution for E_z is therefore

$$\boxed{E_z(x,y,z) = E_0\sin\left(\frac{m\pi x}{a}\right)\sin\left(\frac{n\pi y}{b}\right)\cos\left(\frac{p\pi z}{d}\right) \quad \left[\frac{\text{V}}{\text{m}}\right]} \tag{17.205}$$

The transverse components in the cavity resonator may be found from those of the waveguide using the following argument: Considering the cavity resonator to be a waveguide in which shorts were introduced at two locations along the length of the waveguide (see **Figure 17.19a**), we can argue that the transverse components of the waves propagating along the guide will be reflected at these shorts, causing, in addition to the forward-propagating wave, a backward-propagating wave. The total electric field intensity, which is the sum of the forward- and backward-propagating waves, must then vanish at the conducting surfaces at $z = 0$ and $z = d$.

For TM waves, we start with **Eqs. (17.131)** through **(17.134)**. To see how this is accomplished we now perform the steps necessary to obtain the x component of the electric field intensity in the cavity resonator. **Equation (17.131)** is a forward-propagating wave in the waveguide:

$$E_x^+(x,y,z) = \frac{-\gamma}{\gamma^2 + k^2}E_0^+\frac{m\pi}{a}\cos\left(\frac{m\pi x}{a}\right)\sin\left(\frac{n\pi x}{b}\right)e^{-\gamma z} \quad \left[\frac{\text{V}}{\text{m}}\right] \tag{17.206}$$

where we added the notation E_0^+ to show explicitly that this is the forward-propagating wave. The backward-propagating wave is obtained by simply replacing z by $-z$ (see **Sections 17.3** and **Exercise 17.5**):

$$E_x^-(x,y,z) = \frac{-\gamma}{\gamma^2 + k^2}E_0^-\frac{m\pi}{a}\cos\left(\frac{m\pi x}{a}\right)\sin\left(\frac{n\pi x}{b}\right)e^{\gamma z} \quad \left[\frac{\text{V}}{\text{m}}\right] \tag{17.207}$$

The total wave is the sum of the forward- and backward-propagating waves:

$$E_x(x,y,z) = E_x^+(x,y,z) + E_x^-(x,y,z) = \frac{-\gamma}{\gamma^2 + k^2}\frac{m\pi}{a}\cos\left(\frac{m\pi x}{a}\right)\sin\left(\frac{n\pi y}{b}\right)\left(E_0^+e^{-\gamma z} + E_0^-e^{\gamma z}\right) \quad \left[\frac{\text{V}}{\text{m}}\right] \tag{17.208}$$

Because the x component of the electric field intensity is zero on the conducting planes at $z = 0$ and $z = d$, these become the boundary conditions from which both the z variation of E_x and the magnitude of the backward-propagating wave E_0^- are found:

$$E_x(x,y,0) = \frac{-\gamma}{\gamma^2 + k^2}\frac{m\pi}{a}\cos\left(\frac{m\pi x}{a}\right)\sin\left(\frac{n\pi y}{b}\right)\left(E_0^+ + E_0^-\right) = 0 \quad \rightarrow \quad E_0^- = -E_0^+ \tag{17.209}$$

$$E_x(x,y,d) = \frac{-\gamma}{\gamma^2 + k^2}\frac{m\pi}{a}\cos\left(\frac{m\pi x}{a}\right)\sin\left(\frac{n\pi y}{b}\right)\left(E_0^+e^{-\gamma d} + E_0^-e^{\gamma d}\right) = 0 \tag{17.210}$$

Writing $E_0^- = -E_0^+$ and $\gamma = jk_z$, **Eq. (17.210)** gives

$$e^{-\gamma d} - e^{\gamma d} = 0 \quad \rightarrow \quad -2\sinh(jk_z d) = 0 \tag{17.211}$$

where the relation $(e^{\gamma d} - e^{-\gamma d})/2 = \sinh(\gamma d) = \sinh(jk_z d)$ was used. Using the relation $\sinh(jk_z d) = j\sin(k_z d)$, we get

$$2\sinh(jk_z d) = 2j\sin(k_z d) = 0 \quad \rightarrow \quad k_z d = p\pi \quad \rightarrow \quad k_z = \frac{p\pi}{d} \tag{17.212}$$

This condition was already obtained in **Eq. (17.204)** for the longitudinal component, but it is worth repeating here to show that the two methods of evaluating the fields are equivalent. Substituting these conditions (i.e., $k_z = p\pi/d$, $\gamma = jk_z$, and $e^{-\gamma z} - e^{\gamma z} = -j2\sin(k_z z) = -j2\sin(p\pi z/d)$ in **Eq. (17.208)** gives

$$E_x(x,y,z) = \frac{-1}{\gamma^2 + k^2}\frac{p\pi}{d}\frac{m\pi}{a}E_0\cos\left(\frac{m\pi x}{a}\right)\sin\left(\frac{n\pi y}{b}\right)\sin\left(\frac{p\pi z}{d}\right) \quad \left[\frac{\text{V}}{\text{m}}\right] \tag{17.213}$$

where $E_0 = 2E_0^+$ is the amplitude of the field. Similar steps for the remaining components of the electric field and magnetic field intensities (see **Exercise 17.5**) lead to the following:

$$E_y(x,y,z) = \frac{-1}{\gamma^2 + k^2}E_0\frac{p\pi}{d}\frac{n\pi}{b}\sin\left(\frac{m\pi x}{a}\right)\cos\left(\frac{n\pi y}{b}\right)\sin\left(\frac{p\pi z}{d}\right) \quad \left[\frac{\text{V}}{\text{m}}\right] \tag{17.214}$$

$$H_x(x,y,z) = \frac{j\omega\varepsilon}{\gamma^2 + k^2}E_0\frac{n\pi}{b}\sin\left(\frac{m\pi x}{a}\right)\cos\left(\frac{n\pi y}{b}\right)\cos\left(\frac{p\pi z}{d}\right) \quad \left[\frac{\text{A}}{\text{m}}\right] \tag{17.215}$$

$$H_y(x,y,z) = \frac{-j\omega\varepsilon}{\gamma^2 + k^2}E_0\frac{m\pi}{a}\cos\left(\frac{m\pi x}{a}\right)\sin\left(\frac{n\pi y}{b}\right)\cos\left(\frac{p\pi z}{d}\right) \quad \left[\frac{\text{A}}{\text{m}}\right] \tag{17.216}$$

From **Eq. (17.199)** we can write the wave number as

$$k^2 = \left(\frac{m\pi}{a}\right)^2 + \left(\frac{n\pi}{b}\right)^2 + \left(\frac{p\pi}{d}\right)^2 = \omega^2\mu\varepsilon \tag{17.217}$$

or the resonant frequency as

$$\boxed{f_{mnp} = \frac{1}{2\sqrt{\mu\varepsilon}}\sqrt{\left(\frac{m}{a}\right)^2 + \left(\frac{n}{b}\right)^2 + \left(\frac{p}{d}\right)^2} \quad [\text{Hz}]} \tag{17.218}$$

where the indices m, n, and p indicate the mode in which the cavity resonates. In resonant cavities, the concept of cutoff is different than in waveguides. Since there is no propagation in a cavity, these are called ***resonant frequencies*** or ***resonant modes*** rather than cutoff frequencies.

Any combination of mode indices m, n, and p results in a resonant frequency of the cavity except for those with $m = 0$ or $n = 0$ [for which the longitudinal component of the field in **Eq. (17.205)** becomes zero]. If m or n or both are zero, all field components become zero. However, p can be zero. The lowest TM resonant mode (assuming $a > b > c$) is TM_{110}.

17.9.2 TE Modes in Cavity Resonators

To find the TE modes in a cavity resonator, we must solve **Eq. (17.192)**, subject to the appropriate boundary conditions on the surfaces of the resonator. For TE modes in a waveguide, we imposed the condition that the normal components of the magnetic field intensity must be zero on the conducting walls. These conditions also apply here for the transverse components of the magnetic field. In the z direction, the magnetic field component, H_z, is normal to the surfaces perpendicular to the z axis. Therefore, the additional condition required for the cavity resonator is for H_z to vanish on the surfaces at $z = 0$ and $z = d$:

$$H_z(x,y,z)\big|_{z=0} = 0, \quad H_z(x,y,z)\big|_{z=d} = 0 \tag{17.219}$$

Following the solution process as for the TM modes in cavity resonators and since **Eq. (17.192)** for TE waves is identical in form to **Eq. (17.191)** for TM waves, the general solution is also identical in form. That is, the magnetic field intensity for the TM waves has the same form as the electric field intensity for the TE waves given in **Eq. (17.200)**:

$$H_z(x,y,z) = (A_1 \sin k_x x + B_1 \cos k_x x)(A_2 \sin k_y y + B_2 \cos k_y y)(A_3 \sin k_z z + B_3 \cos k_z z) \quad [\text{A/m}] \tag{17.220}$$

To obtain TE modes in a cavity resonators, we start with the results obtained for the TE modes in a waveguide in **Eqs. (17.158)** through **(17.162)**:

$$A_1 = A_2 = 0 \quad \text{and} \quad k_x = \frac{m\pi}{a}, \quad k_y = \frac{n\pi}{b} \tag{17.221}$$

With these, the solution is [see **Eq. (17.164)**]

$$H_z(x,y,z) = A\cos\left(\frac{m\pi x}{a}\right)\cos\left(\frac{n\pi y}{b}\right)(A_3 \sin k_z z + B_3 \sin k_z z) \quad [\text{A/m}] \tag{17.222}$$

Now, we apply the boundary conditions in **Eq. (17.219)**:

$$H_z(x,y,z)\big|_{z=0} = A\cos\left(\frac{m\pi x}{a}\right)\cos\left(\frac{n\pi y}{b}\right)(A_3 \sin 0 + B_3 \cos 0) = 0 \quad \rightarrow \quad B_3 = 0 \tag{17.223}$$

and

$$H_z(x,y,z)\big|_{z=d} = A\cos\left(\frac{m\pi x}{a}\right)\cos\left(\frac{n\pi y}{b}\right)\sin k_z d = 0 \quad \rightarrow \quad k_z = \frac{p\pi}{d} \tag{17.224}$$

where the constant A_3 was absorbed into A. The longitudinal component of the magnetic field intensity is therefore

$$H_z(x,y,z) = H_0 \cos\left(\frac{m\pi x}{a}\right)\cos\left(\frac{n\pi y}{b}\right)\sin\left(\frac{p\pi z}{d}\right) \quad \left[\frac{\text{A}}{\text{m}}\right] \tag{17.225}$$

where H_0 is the amplitude of the magnetic field intensity. To obtain the transverse components, we use the same sequence as in the previous section; that is, we write the forward- and backward-propagating TE waves in a shorted waveguide, sum the two waves up, and set the total transverse electric field intensity to zero or the derivatives with respect to z of the total transverse magnetic field intensity to vanish ($\partial H_z(x,y,z)/\partial z = 0$) on the shorting walls at $z = 0$ and $z = d$. This gives (see **Exercises 17.8** and **17.10**)

$$E_x(x,y,z) = \frac{j\omega\mu}{\gamma^2 + k^2} H_0 \frac{n\pi}{b} \cos\left(\frac{m\pi x}{a}\right)\sin\left(\frac{n\pi y}{b}\right)\sin\left(\frac{p\pi z}{d}\right) \quad \left[\frac{\text{V}}{\text{m}}\right] \tag{17.226}$$

$$E_y(x,y,z) = \frac{-j\omega\mu}{\gamma^2 + k^2} H_0 \frac{m\pi}{a} \sin\left(\frac{m\pi x}{a}\right)\cos\left(\frac{n\pi y}{b}\right)\sin\left(\frac{p\pi z}{d}\right) \quad \left[\frac{\text{V}}{\text{m}}\right] \tag{17.227}$$

$$H_x(x,y,z) = -\frac{1}{\gamma^2 + k^2} H_0 \frac{m\pi}{a}\frac{p\pi}{d} \sin\left(\frac{m\pi x}{a}\right)\cos\left(\frac{n\pi y}{b}\right)\cos\left(\frac{p\pi z}{d}\right) \quad \left[\frac{\text{A}}{\text{m}}\right] \tag{17.228}$$

$$H_y(x,y,z) = -\frac{1}{\gamma^2 + k^2} H_0 \frac{n\pi}{b}\frac{p\pi}{d} \cos\left(\frac{m\pi x}{a}\right)\sin\left(\frac{n\pi y}{b}\right)\cos\left(\frac{p\pi z}{d}\right) \quad \left[\frac{\text{A}}{\text{m}}\right] \tag{17.229}$$

From the fields in **Eqs. (17.225)** through **(17.229)**, we see that for TE modes, either m or n can be zero (but not both) while p must be nonzero (otherwise the longitudinal component of the field is zero). For $p = 0$ or for $m = n = 0$, all components of the field are zero. The lowest resonant mode is therefore either the TE_{101} or TE_{011}, depending on the dimensions a, b, and c.

The resonant frequencies for TE modes are the same as for the TM modes:

$$f_{mnp} = \frac{1}{2\pi\sqrt{\mu\varepsilon}}\sqrt{\left(\frac{m}{a}\right)^2 + \left(\frac{n}{b}\right)^2 + \left(\frac{p}{d}\right)^2} \quad [\text{Hz}] \tag{17.230}$$

Some of the modes may have the same resonant frequency even though they are different modes. As an example, for a cubic cavity ($a = b = d$), TE_{011} and TE_{101} have the same frequency. These are called degenerate modes, as in waveguides.

Exercise 17.10 Starting with the TE transverse components in a rectangular waveguide **[Eqs. (17.165)** through **(17.168)]**, derive **Eqs. (17.226)** through **(17.229)** by first writing the backward-propagating waves in the waveguide, summing the forward- and backward-propagating waves, and then applying the appropriate conditions at $z = 0$ and $z = d$.

Example 17.11 A cavity resonator is made in the form of a cubic box, 100 mm on the side. The cavity is air filled and is made of a perfect conductor:

(a) Find the first 15 possible resonant modes of the cavity.
(b) Separate the TE and TM modes.
(c) Which resonant frequency is the dominant mode of the cavity and which modes are degenerate?

Solution: The resonant frequencies of the cavity are calculated from **Eq. (17.230)**. Any combination of the integers m, n, and p may be considered to be a resonant frequency, except, of course, for $m = 0, n = 0, p = 0$. Other combinations may also be inappropriate combinations in the sense that they result in zero fields. We first calculate the first 15 possible resonant frequencies, including those that may not correspond to physical modes and then identify those modes that may exist in the cavity:

(a) Using **Eq. (17.230)**, the possible resonant frequencies are listed in **Table 17.4**.
(b) To identify which of the calculated frequencies correspond to TE and TM modes, we use the properties of the modes:
TE modes: m or n can be zero, while p must be nonzero.
TM modes: m and n must be nonzero, while p can be zero.
This means that the combinations ($00p$), ($0n0$), and ($m00$), where m, n, p are nonzero, are not physical resonant frequencies; that is, the combinations 100, 010, 001, 200, 020, and 002 lead to zero fields and are therefore not resonant modes. The combinations ($mn0$) and (mnp), $m, n, p \neq 0$, are TM modes. These are (110), (111), (210), and (120). The combinations ($0np$), ($m0p$), and (mnp), for $m, n, p \neq 0$, are TE modes. These are (011), (021), (101), (102), (201), and (111). The physical resonant frequencies and their designation are shown in **Table 17.5**.
(c) The dominant mode is the mode with lowest resonant frequency. In this case, there are three modes with lowest frequency. Any of the TM_{110}, TE_{101}, and TE_{011} modes is the dominant mode.

Table 17.4 The first 15 possible modes in a square resonant cavity (frequencies given in GHz)

100	010	001	110	101	011	111	200	020	002	210	201	021	102	120
1.5	1.5	1.5	2.121	2.121	2.121	2.598	3.0	3.0	3.0	3.354	3.354	3.354	3.354	3.354

Table 17.5 Physical modes in a cubic cavity resonator (frequencies given in GHz)

TM_{110}	TE_{101}	TE_{011}	TM_{111}, TE_{111}	TM_{210}	TE_{201}	TE_{021}	TE_{102}	TM_{120}
2.121	2.121	2.121	2.598	3.354	3.354	3.354	3.354	3.354

Modes TM_{110}, TE_{101}, and TE_{011} are degenerate modes as are the TE_{111} and TM_{111} and TE_{210}, TE_{201}, TE_{021}, TE_{102}, and TM_{120} modes. This high-order degeneracy is one reason why cubic cavity resonators are seldom used. Rectangular cavities are usually preferred because they have better mode separation.

17.10 Energy Relations in a Cavity Resonator

Power and energy relations in a cavity are defined by the Poynting theorem. Since there is a certain amount of energy stored in the fields of a cavity, the calculation of this energy is an important aspect of analysis. This is particularly obvious if we recall that in a resonant device, these relations change dramatically at or near resonance. This was true with resonant circuits and is certainly true with resonant cavities. The stored energy and dissipated power in a cavity define the basic qualities of the cavity. A lossless cavity is not practically realizable; therefore, we also define a quantity called quality factor of the cavity, which is a measure of losses in the cavity. A shift in the resonant frequency of the cavity can also be described in terms of energy. These relations can then be used to characterize a cavity and for measurements in the cavity.

To define the energy relations in the cavity, we need to calculate the Poynting vector ($\mathcal{P} = \mathbf{E} \times \mathbf{H}$) in the cavity. From **Eqs. (17.213)** through **(17.216)** or **Eqs. (17.226)** through **(17.229)**, we note that the Poynting vector is purely imaginary; that is, the time-averaged power density in the cavity is zero:

$$\mathcal{P}_{av} = \frac{1}{2}\mathrm{Re}(\mathbf{E} \times \mathbf{H}^*) = 0 \qquad (17.231)$$

This means that there is no real power transferred in or out of the cavity, but there is stored energy in the magnetic and electric fields inside the cavity. From the complex Poynting vector **[Eq.** (12.75)**]**, we have

$$S = j2\omega\int_v \left(\frac{1}{4}\varepsilon\mathbf{E}\cdot\mathbf{E}^* - \frac{1}{4}\mu\mathbf{H}\cdot\mathbf{H}^*\right)dv \quad [\mathrm{W}] \qquad (17.232)$$

The total time-averaged stored electric and magnetic energy in the cavity can now be written as

$$W_0 = \int_v \left(\frac{\varepsilon\mathbf{E}\cdot\mathbf{E}^*}{4} - \frac{\mu\mathbf{H}\cdot\mathbf{H}^*}{4}\right)dv \quad [\mathrm{J}] \qquad (17.233)$$

where **E** and **H** are the fields in the cavity and v the volume of the cavity. This relation is correct at any frequency regardless of resonance.

If the cavity also has wall losses, the time-averaged dissipated power in the cavity walls is

$$P_{loss} = \frac{R_s}{2}\int_s J_s^2 ds = \frac{R_s}{2}\int_s |H_t|^2 ds \quad [\mathrm{W}] \qquad (17.234)$$

where R_s is the surface resistance of the cavity walls, H_t is the tangential magnetic field intensity at the walls surfaces, and s is the internal surface of the cavity walls. This relation is the same as that obtained for waveguides in **Eq. (17.185)**. The calculation of the wall losses is the same as for the waveguide (see **Example 17.9**). In addition, there may also be losses due to the dielectric inside the cavity and these must be added to **Eq. (17.234)**.

17.11 Quality Factor of a Cavity Resonator

The ***quality factor*** of the cavity resonator is defined as the ratio between the stored energy in the cavity and the dissipated power per cycle of the wave:

$$Q = 2\pi\frac{\text{time-averaged stored energy}}{\text{energy loss in one cycle}} = \frac{2\pi W_0}{P_{loss}T} = \frac{\omega_0 W_0}{P_{loss}} \quad [\text{dimensionless}] \qquad (17.235)$$

where T is the period of the wave and ω_0 is the resonant frequency. Since the higher the Q factor, the more selective the cavity is, Q is a measure of the bandwidth of the cavity. It also defines, indirectly, the amount of energy needed to couple into the cavity to maintain an energy balance. Ideal cavities have an infinite quality factor. The calculation of both stored energy W_0 and dissipated power P_{loss} are tedious but straightforward operations [see **Eqs. (17.233)** and **(17.234)**]. The stored energy is calculated by direct integration over the volume of the cavity using either the TE or TM fields and the dissipated power in the walls of the cavity is calculated using the method in **Section 17.7.3** by first finding the current densities in the six walls of the cavity resonator and then integrating **Eq. (17.234)** over the walls.

There are two sources of losses in a cavity. One is the wall loss in conductors, the other is the loss in dielectrics (see **Section 17.7.3**). It is possible to separate the quality factor into a quality factor due to the dielectric, Q_d, and a quality factor due to conductors, Q_c. If this is done, then the quality factor of the cavity may be written as

$$Q = \frac{Q_c Q_d}{Q_c + Q_d} \quad \text{[dimensionless]} \tag{17.236}$$

A small advantage in this separation is that the quality factor due to the dielectric can be calculated in terms of the loss tangent [see **Eq.** (12.80)] or in terms of the complex permittivity [see **Eq.** (12.79)]:

$$Q_d = \frac{\varepsilon'}{\varepsilon''} = \frac{1}{\tan\theta_{loss}} \quad \text{[dimensionless]} \tag{17.237}$$

where ε' is the real part of permittivity and ε'' its imaginary part. In some instances one or the other quality factor may dominate. If, for example, dielectric losses are negligible, Q_d tends to infinity and the quality factor of the cavity is dominated by wall losses.

17.12 Applications

Application: The Slotline A very useful measuring device in waveguide applications is the slotline. The slotline is a section of a waveguide, with a lengthwise slot that allows a probe to measure the electric field intensity in the waveguide as shown in **Figure 17.20a**. The probe can be adjusted with a micrometer over a considerable length and measure the electric field intensity in the waveguide. The basic measurement is that of the maximum and minimum electric fields, indicating the standing wave ratio. However, other measurements may be performed. For example, the slotline may be connected to a waveguide section, and the waveguide caused to reflect some energy back into the slotline, as shown in **Figure 17.20b** by means of a short, an open, or any dielectric material in the cavity. Then, measuring two minima in the standing wave pattern (minima are preferred because they are sharper than maxima), the wavelength and, therefore, the frequency may be measured. The measurement proceeds by identifying one minimum and then moving the probe to the next minimum. The distance between the two minima is always $\lambda/2$. The slotline must be identical in dimensions to the waveguide to which it is connected if reflections due to the connections are to be avoided.

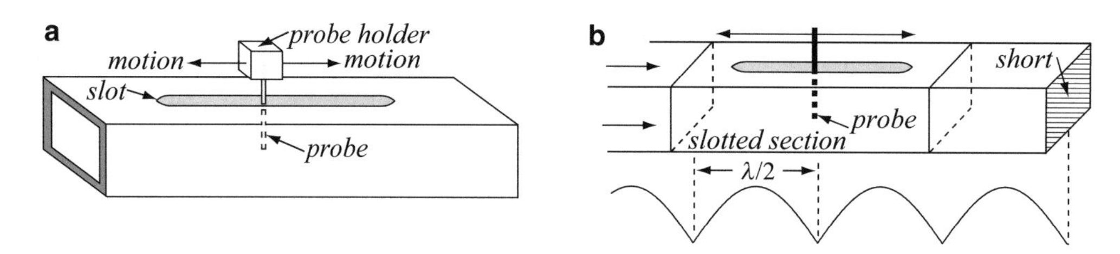

Figure 17.20 (**a**) The slotline. (**b**) Frequency measurement using the slotline

Application: Optical Waveguides Optical waveguides are somewhat different than most other electromagnetic waveguides in that a conductor is not used. The guide is a dielectric guide which relies on total internal reflection for confinement of waves within the waveguide. The most common optical waveguides are made of thin silica fibers (thus, the common name of glass fiber or optical fiber) or plastic. The fiber is normally coated with an opaque dielectric material for protection, even though it is not absolutely necessary for operation. However, the coating or cladding must have lower dielectric constant than the core fiber to ensure total internal reflection. In some cases, the fiber may be coated with metal such as aluminum or nickel. There are a number of types of optical fibers, depending on their constructions. The best optical fibers are the so-called single-mode fibers. These are very thin (1–8 μm) and, as their name implies, allow a single mode to propagate. Propagation is almost entirely parallel to the fiber and because of that, there is little dispersion in the fiber and the attenuation is also low. However, these fibers are difficult to make and are quite expensive. In addition, connection to the fibers is complicated and requires a laser as the source. More common are the multimode fibers and multimode graded-index fibers. The first are about 125–400 μm thick and are made of a uniform material (the index of refraction is constant throughout the cross section of the fiber). This is the simplest fiber but also the worst in terms of dispersion and attenuation. Dispersion in optical fibers refers to the delay in transmission because different modes travel at different speeds depending on the angle of reflection in the fiber. Typical values are 15–30 ns/km meaning that the difference in time of arrival between the slowest/fastest waves is 15–30 ns per km length of the fiber. The second type of fiber has a graded index of refraction which is high at the center and lower toward the outer surface. This reduces dispersion but is more difficult to fabricate.

The optical fiber is commonly used as a waveguide for communication purposes because it has low attenuation, very high bandwidth, no interference from other signals, is thin and lightweight, and is easy to couple energy into. Typical attenuation in fibers is as low as 0.5 dB/km, although some fibers attenuate over 10 dB/km. In addition to optical fibers, optical waveguides can be fabricated in silicon and other semiconducting materials. Optical resonators are also made and can be integrated on silicon chips.

Application: Detection of Materials with Cavity Resonators Cavity resonators with high-quality factors have a very narrow curve (high response) around resonance. Any change in the dielectric constant inside the cavity affects the resonant frequency. This may be utilized to measure properties of materials or to sense the presence of materials. One example of this type of measurement is smoke detection or even detection of explosives. Other applications are drying of materials by sensing moisture content (resonant frequency is lower the higher the amount of water in the cavity) or curing of polymers by sensing the amount of solvent in the vicinity of the drying polymer. The basic application is shown in **Figure 17.21**. It consists of a rectangular cavity, with a few holes that allow penetration of the material (gases) to be detected. The shift in resonant frequency is monitored and any shift indicates the presence of a material with a dielectric constant different than air.

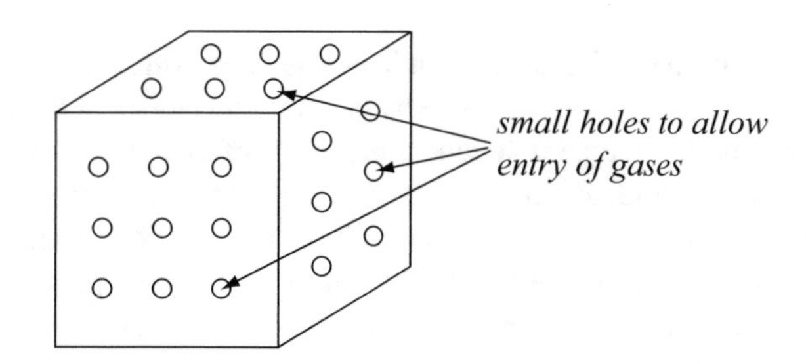

Figure 17.21 A cavity resonator sensor designed for smoke detection

Application: Coupling to Waveguides and Cavities The coupling of electromagnetic energy to waveguides and cavities has not been addressed specifically in the discussion above. If the cavity were ideal, the fields within the cavity would be of infinite amplitude, provided that the necessary modes can be excited. In real cavities, there are always some losses, but these are usually small. The fields are large and the amount of stored energy is also large. However, the small amount of power dissipated has to be compensated for by external sources; otherwise, the cavity would cease to oscillate. This is done by coupling energy into the cavity. A cavity for which the lost energy is exactly balanced is called a critically coupled cavity.

The introduction of energy into the cavity (or a waveguide) can be done in a number of ways. The most obvious of these is to have a source within the cavity that generates the necessary fields. A small loop (**Figure 17.22a)** or a simple probe excitation (**Figure 17.22b)** can be used. A loop generates a magnetic field intensity and this magnetic field intensity excites a mode with magnetic field intensity parallel to that generated by the loop. Different modes can be generated by simply locating the probe or the loop at different locations in the cavity, although, for obvious reasons, these loops or probes must be close to the outer surfaces of the cavity. Similarly, the cavity can be coupled from a waveguide through a small aperture through which a small amount of energy "leaks" into the cavity (**Figure 17.22c)**. In this case, the modes excited are those that have fields parallel to those in the waveguide at the location of the aperture. The three coupling methods in **Figure 17.22** excite different modes.

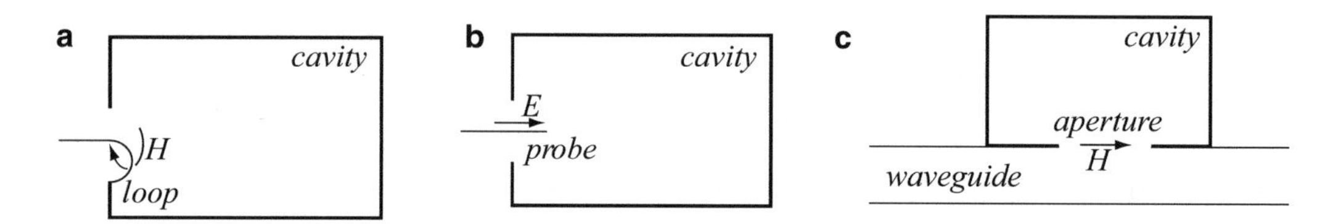

Figure 17.22 (**a**) Coupling to a cavity by a small loop in the cavity. (**b**) Coupling to a cavity by a small probe in the cavity. (**c**) Coupling to a cavity by a small aperture in the wall of the cavity

Application: Frequency Measurement One simple and widely used method for frequency measurement is the tuning of a cavity resonator to resonate at the unknown frequency. Then, by accurate measurements of the cavity dimensions, the frequency may be calculated from **Eq. (17.218)**. In practice, the resonant frequency may be calibrated directly on the cavity. Standard wavemeters are of this type. (They are called wavemeters because often the wavelength is measured rather than the frequency.) Normally, wavemeters are cylindrical cavity resonators as shown in **Figure 17.23**. However, in principle, any cavity resonator may be used. The only requirement is that the modes be separated well and that we either know the mode or the cavity is excited in a known mode.

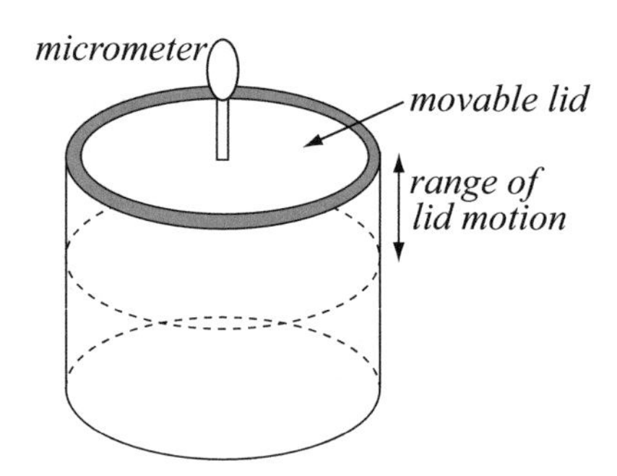

Figure 17.23 A cylindrical cavity wavemeter

Application: Propagation of Microwaves in Tunnels The principles of waveguides apply to propagation of electromagnetic waves in tunnels, mining galleries, subway galleries coridors and the like. In particular, the ideas of cutoff frequencies for propagation into and from tunnels and the dimensions of tunnels and coridors affect the ability of waves to propagate. In addition, in many cases the losses in the walls are high, again affecting the waves. Low frequency waves in general are affected more than high frequencies but the similarity to propagation in waveguides is limited. Tunnel walls have a relatively low conductivity compared to metal waveguide and hence waves penetrate into the walls. Walls are usually rough and sometimes irregular. Tunnels and galleries often have turns that again affect the propagation of electromagnetic waves. As a whole propagation is poor with high attenuation but improves with frequency.

Application: Microwave Ovens The use of microwave ovens is familiar to almost anybody. It has applications in consumer products in the form of microwave ovens in the kitchen with power outputs typically below 1 kW and operating at 2.45 GHz. In industry microwave heating is used in many applications ranging from welding of plastics to freeze drying of foodstuffs to heating and drying applications during production. The frequency typically used is 915 MHz at powers up to about 1 MW, depending on applications. All microwave ovens are essentially an enclosed conducting chamber into which waves are coupled from a source (such as a magnetron) through a waveguide.

Application: Antenna Feeds One of the more common use of waveguides is to feed power into antennas so that the power can be radiated into space. This use is limited of course to the higher frequency range whereas at lower frequencies transmission line (coaxial lines in particular) are a better choice. Waveguides excel in this function when relatively large power levels at high frequencies need to be radiated because the losses are low and the power level possible is high. Waveguides and cavity resonators are also used in conjunction with slot antennas—these are slots cut directly into the walls of the waveguide or cavity resonator. The slots radiate power and their radiation properties depends on dimensions, position on the waveguide or resonator and orientation relative to the fields internal to the structure. Slot antennas in waveguides and cavity resonators can be cut to create arrays of antennas and by so doing generate particular radiation patterns.

17.13 Summary

The theme of this chapter is propagation of waves in waveguides. The starting point is a general description of longitudinal and transverse electric and magnetic field intensities from Maxwell's equations. These, together with the effect of the conducting surfaces, define the properties of the waves. The treatment is separated into ***transverse electromagnetic*** **(TEM)**, ***transverse electric*** **(TE)**, and ***transverse magnetic*** **(TM)** waves.

Assumptions: direction of propagation is z, time-harmonic fields: ($e^{j\omega t}$ time dependency).

General equations—transverse field components in general, lossy media:

$$E_x = \frac{1}{\gamma^2 + k^2}\left(-\gamma\frac{\partial E_z}{\partial x} - j\omega\mu\frac{\partial H_z}{\partial y}\right) \quad \left[\frac{\mathrm{V}}{\mathrm{m}}\right] \tag{17.18}$$

$$E_y = \frac{1}{\gamma^2 + k^2}\left(-\gamma\frac{\partial E_z}{\partial y} + j\omega\mu\frac{\partial H_z}{\partial x}\right) \quad \left[\frac{\mathrm{V}}{\mathrm{m}}\right] \tag{17.19}$$

$$H_x = \frac{1}{\gamma^2 + k^2}\left(-\gamma\frac{\partial H_z}{\partial x} + j\omega\varepsilon\frac{\partial E_z}{\partial y}\right) \quad \left[\frac{\mathrm{A}}{\mathrm{m}}\right] \tag{17.20}$$

$$H_y = \frac{1}{\gamma^2 + k^2}\left(-\gamma\frac{\partial H_z}{\partial y} - j\omega\varepsilon\frac{\partial E_z}{\partial x}\right) \quad \left[\frac{\mathrm{A}}{\mathrm{m}}\right] \tag{17.21}$$

Given $E_z, H_z, \gamma = \alpha + j\beta$, and $k = \omega\sqrt{\mu\varepsilon}$, we can obtain E_x, E_y, H_x, H_y. These fields are used to define TEM, TE, and TM waves by setting conditions on E_z and H_z. **Table 17.2** summarizes the properties of TE, TM, and TEM waves. Some of the properties that show the differences between the three types of waves under lossless conditions are shown in the following table.

	TEM waves	TE waves	TM waves
Conditions on E_z, H_z	$E_z = 0, H_z = 0$	$E_z = 0, H_z \neq 0$	$E_z \neq 0, H_z = 0$
Propagation constant	$\gamma_{TEM} = j\omega\sqrt{\mu\varepsilon}$	$\gamma_{TE} = j\omega\sqrt{\mu\varepsilon}\sqrt{1-(f_c/f)^2}$	$\gamma_{TM} = \gamma_{TE}$
Equations	See note 1	**(17.27)** through **(17.30)**	**(17.53)** through **(17.56)**
Phase velocity [m/s]	$v_p = 1/\sqrt{\mu\varepsilon}$	$v_{TE} = 1/\left(\sqrt{\mu\varepsilon}\sqrt{1-(f_c/f)^2}\right)$	$v_{TM} = v_{TE}$
Wave impedance [Ω]	$\eta = \sqrt{\mu/\varepsilon}$	$Z_{TE} = \eta/\sqrt{1-(f_c/f)^2}$	$Z_{TM} = \eta\sqrt{1-(f_c/f)^2}$

f_c is the cutoff frequency, below which waves do not propagate.

Notes:

(1) TEM waves are plane waves as discussed in **Chapters 12** and **13**.
(2) The equations for TE and TM waves are obtained by substituting the conditions from the first row of the table into **Eqs. (17.18)** through **(17.21)**.
(3) In TE waves there is no longitudinal electric field, whereas in TM waves there is no longitudinal magnetic field.
(4) Phase velocity is given for lossless media.

Properties in lossy media can be easily obtained by modifying γ as follows ($\gamma = \alpha + j\beta$):

$$\gamma_{TE} = \gamma_{TM} = j\sqrt{\left(\omega^2\mu\varepsilon - k_c^2\right) - j\omega\mu\sigma} \tag{17.45}$$

Below cutoff the waves are highly attenuated. The attenuation constant for evanescent waves (below cutoff, at a frequency $f < f_c$) is

$$\alpha_e = \omega\sqrt{\mu\varepsilon}\sqrt{\frac{f_c^2}{f^2} - 1} \quad \left[\frac{\text{Np}}{\text{m}}\right] \tag{17.39}$$

The attenuation below cutoff is very high so that we may safely say that waves do not propagate.

TE and TM Waves in Parallel Plate Waveguides

Mode, m	TE_m	TM_m
Guide phase velocity [m/s]	$v_g = v_p/\sin\theta_i$ (17.70)*	Same as TE
Cutoff frequency [Hz]	$f_{cm} = m/(2d\sqrt{\mu\varepsilon})$ (17.80)	Same as TE
Cutoff wave number [rad/m]	$k_{cm} = m\pi/d, \quad m = 1, 2, 3 \ldots$ (17.81)	Same as TE
Cutoff wavelength [m]	$\lambda_{cm} = 2d/m, \quad m = 1, 2, 3 \ldots$ (17.82)	Same as TE
Guide wavelength [m]	$\lambda_g = \lambda/\sqrt{1 - f_{cm}^2/f^2}$ (17.86)	Same as TE
Guide phase velocity [m/s]	$v_g = v_p/\sqrt{1 - f_{cm}^2/f^2}$ (17.85)*	Same as TE
Guide phase constant [rad/m]	$\beta_g = \beta\sqrt{1 - f_{cm}^2/f^2}$ (17.87)**	Same as TE
Electric field intensity in the waveguide [V/m]	$\mathbf{E}_1(x,z) = \hat{\mathbf{y}}\, jE_0 \sin\left(\frac{m\pi x}{d}\right)e^{-j2\pi z/\lambda_g}$ (17.92)	$\mathbf{E}_1(x,z) = \hat{\mathbf{x}}E_0\frac{\lambda}{\lambda_g}\cos\left(\frac{m\pi x}{d}\right)e^{-j2\pi z/\lambda_g} + \hat{\mathbf{z}}\, jE_0\frac{\lambda}{\lambda_{cm}}\sin\left(\frac{m\pi x}{d}\right)e^{-j2\pi z/\lambda_g}$ (17.102)
Magnetic field intensity in the waveguide [A/m]	$\mathbf{H}_1(x,z) = -\hat{\mathbf{x}}\, j\frac{E_0}{\eta}\frac{\lambda}{\lambda_g}\sin\left(\frac{m\pi x}{d}\right)e^{-j2\pi z/\lambda_g} - \hat{\mathbf{z}}\frac{E_0}{\eta}\frac{\lambda}{\lambda_{cm}}\cos\left(\frac{m\pi x}{d}\right)e^{-j2\pi z/\lambda_g}$ (17.93)***	$\mathbf{H}_1(x,z) = \hat{\mathbf{y}}\frac{E_0}{\eta}\cos\left(\frac{m\pi x}{d}\right)e^{-j2\pi z/\lambda_g}$ (17.103)***
Time-averaged power density in the waveguide [W/m^2]	$\mathcal{P}_{av} = \frac{E_0^2}{2\eta}\frac{\lambda}{\lambda_g}\sin^2\left(\frac{m\pi x}{d}\right)$ (17.97)***	$\mathcal{P}_{av} = \frac{E_0^2}{2\eta}\frac{\lambda}{\lambda}\cos^2\left(\frac{m\pi x}{d}\right)$ (17.109)***
Guide wave impedance [Ω]	$Z_{TE} = \eta\lambda_g/\lambda = \eta/\sqrt{1 - f_{cm}^2/f^2}$ (17.98)***	$Z_{TM} = \eta/\lambda_g = \eta\sqrt{1 - f_{cm}^2/f^2}$ (17.110)***

*$v_p = 1/\sqrt{\mu\varepsilon}$, **$\beta = \omega\sqrt{\mu\varepsilon}$, ***$\eta = \sqrt{\mu/\varepsilon}$

Rectangular Waveguides TM modes, $H_z = 0$. TE modes, $E_z = 0$

Mode, m	$TM_{m,n}$	$TE_{m,n}$
Longitudinal field	$E_z(x,y,z) = E_0\sin\left(\frac{m\pi}{a}x\right)\sin\left(\frac{n\pi}{b}y\right)e^{-\gamma z}$ (17.128)	$H_z(x,y,z) = H_0\cos\left(\frac{m\pi x}{a}\right)\cos\left(\frac{n\pi y}{b}\right)e^{-\gamma z}$ (17.164)
Cutoff frequency [Hz]	$f_{cmn} = \frac{v_p}{2}\sqrt{\left(\frac{m}{a}\right)^2 + \left(\frac{n}{b}\right)^2}$ (17.138)	Same as TM
Cutoff wave number [rad/m]	$k_{cmn} = \sqrt{\left(\frac{m\pi}{a}\right)^2 + \left(\frac{n\pi}{b}\right)^2}$ (17.137)	Same as TM
Cutoff wavelength [m]	$\lambda_{cmn} = \frac{1}{\sqrt{(m/2a)^2 + (n/2b)^2}}$ (17.139)	Same as TM
Guide wavelength [m]	$\lambda_g = \frac{\lambda}{\sqrt{1 - f_{cmn}^2/f^2}}$ (17.143)	Same as TM

(continued)

Mode, m	$\mathrm{TM}_{m,n}$	$\mathrm{TE}_{m,n}$
Guide phase velocity [m/s]	$v_g = \dfrac{v_p}{\sqrt{1 - f_{cmn}^2/f^2}}$ (17.142)*	Same as TM
Guide phase constant [rad/m]	$\beta_g = \beta\sqrt{1 - \dfrac{f_{cmn}^2}{f^2}}$ (17.141)**	Same as TM
Transverse fields	See **Eqs. (17.147)** through **(17.150)**	See **Eqs. (17.165)** through **(17.168)**
Time-averaged power in the waveguide [W]	$P = \dfrac{\omega\varepsilon\beta_g E_0^2 ab}{8k_{cmn}^2}$ (17.153)	$P = \dfrac{\omega\mu\beta_g H_0^2 ab}{8k_{cmn}^2}$ $m, n \neq 0,$ (17.176) $P = \dfrac{\omega\mu\beta_g H_0^2 ab}{4k_{cmn}^2}$ $m = 0$ *or* $n = 0,$ (17.177)
Guide wave impedance [Ω]	$Z_{TM} = \eta\sqrt{1 - \dfrac{f_{cmn}^2}{f^2}} = \eta\dfrac{\lambda}{\lambda_g}$ (17.145)***	$Z_{TE} = \dfrac{\eta}{\sqrt{1 - f_{cmn}^2/f^2}} = \eta\dfrac{\lambda_g}{\lambda}$ (17.170)***
Valid modes	All modes with $m \neq 0, n \neq 0$	All modes except $m = 0, n = 0$

*$v_p = 1/\sqrt{\mu\varepsilon}$, **$\beta = \omega\sqrt{\mu\varepsilon}$, ***$\eta = \sqrt{\mu/\varepsilon}$, ab = cross-sectional area

Note: Range (bandwidth) for each mode is the range between its cutoff frequency and the cutoff frequency of the next, higher mode.

Power propagated in the TE_{10} mode

$$P(\mathrm{TE}_{10}) = \frac{\omega\mu\beta_g H_0^2 a^3 b}{4\pi^2} \quad [\mathrm{W}] \tag{17.178}$$

TE_{10} is the most important and most often used mode.

Losses

In waveguides these are due to wall losses and dielectric losses.

Dielectric attenuation constant:

$$\alpha_{dTE} = \alpha_{dTM} \frac{\sigma_d \eta_d}{2\sqrt{1 - f_{cmn}^2/f^2}} \quad \left[\frac{\mathrm{Np}}{\mathrm{m}}\right] \tag{17.179}$$

Wall attenuation constant:

$$\alpha_w = \frac{P_{loss}}{2P_{av}} \quad \left[\frac{\mathrm{Np}}{\mathrm{m}}\right] \tag{17.183}$$

where P_{loss} is power lost in walls per unit length.

Attenuation below cutoff:

$$\alpha_{bc} = \omega\sqrt{\mu_o \varepsilon_o}\sqrt{\frac{f_{cmn}^2}{f^2} - 1} \quad \left[\frac{\mathrm{Np}}{\mathrm{m}}\right], \quad f < f_{cmn} \tag{17.190}$$

Cavity resonators—treated as shorted waveguides.

Properties: the resonant frequencies (modes) are

$$f_{mnp} = \frac{1}{2\sqrt{\mu\varepsilon}}\sqrt{\left(\frac{m}{a}\right)^2 + \left(\frac{n}{b}\right)^2 + \left(\frac{p}{d}\right)^2} \quad [\mathrm{Hz}] \tag{17.218}$$

TM modes: $m,n \neq 0$, p can be zero. Longitudinal (z-directed) field:

$$E_z(x,y,z) = E_0 \sin\left(\frac{m\pi x}{a}\right)\sin\left(\frac{n\pi y}{b}\right)\cos\left(\frac{p\pi z}{d}\right) \quad \left[\frac{\text{A}}{\text{m}}\right] \tag{17.205}$$

The TM transverse components are obtained from the longitudinal field by adding the conditions imposed by the conducting surfaces that short the guide **(Section 17.9.1)**. These are listed in **Eqs. (17.213)** through **(17.216)**.

TE modes: m *or* n can be zero (but not both), $p \neq 0$

Longitudinal field:

$$H_z(x,y,z) = H_0 \cos\left(\frac{m\pi x}{a}\right)\cos\left(\frac{n\pi y}{b}\right)\sin\left(\frac{p\pi z}{d}\right) \quad \left[\frac{\text{A}}{\text{m}}\right] \tag{17.225}$$

The TE transverse components are obtained from the longitudinal field by adding the conditions imposed by the conducting surfaces that short the guide **(Section 17.9.2)**. These are listed in **Eqs. (17.226)** through **(17.229)**.

Energy and Losses **Stored energy** in the cavity

$$W_0 = \int_v \left(\frac{\varepsilon \mathbf{E}\cdot\mathbf{E}^*}{4} - \frac{\varepsilon \mathbf{H}\cdot\mathbf{H}^*}{4}\right) dv \quad [\text{J}] \tag{17.233}$$

Power loss

$$P_{loss} = \frac{R_s}{2}\int_s J_s^2 ds = \frac{R_s}{2}\int_s |H_t|^2 ds \quad [\text{W}] \tag{17.234}$$

where R_s is the surface resistance **[Eq. (17.184)]**.

Quality factor of the cavity is the ratio of stored energy and power loss per cycle:

$$Q = 2\pi \frac{W_0}{P_{loss}T} = \frac{\omega_0 W_0}{P_{loss}} \quad [\text{dimensionless}] \tag{17.235}$$

Problems

TE, TM, and TEM Propagation in Parallel Plate Waveguides

17.1 Application: TM Modes in Parallel Plate Waveguides. A parallel plate waveguide is made of two strips, $a = 20$ mm wide, separated by $d = 1$ mm and air filled. Neglect edge effects. For an incident electric field intensity of magnitude $E_i = 1$ V, neglect edge effects and calculate at a frequency 20% above the lowest cutoff frequency:

(a) The guide phase velocity, guide wavelength, and wave impedance for TM modes.
(b) The electric field intensity everywhere along the line for TM modes.
(c) The magnetic field intensity along the line for TM modes.
(d) The instantaneous power density in the waveguide for TM modes.

17.2 Application: TE/TM Waves in Parallel Plate Waveguides. A parallel plate waveguide is made of two wide strips, separated by a fiberglass sheet $d = 0.5$ mm thick which has a relative permittivity of 3.5. Neglect any effects due to the edges of the strips (i.e., assume the strips are infinitely wide) and calculate:

(a) The lowest TM mode possible.
(b) The lowest TE mode possible.
(c) If a wave at twice the lowest TE cutoff frequency propagates along the waveguide, calculate the wave impedance for the TE and TM modes and compare with the wave impedance for TEM modes.

17.3 Power Relations in Integrated Microstrip Line. In a microstrip line, the strips are separated a distance 0.8 mm and are 2 mm wide. The material between the strips is Silicon Oxide with a relative permittivity of 3.8. Neglect edge effects. For an incident electric field intensity $E_i = 1$ V/m, calculate:

(a) The total time-averaged power propagated in the lowest TE mode at a frequency 30% above cutoff.
(b) The total time-averaged power propagated in the lowest TM mode at a frequency 30% above cutoff.
(c) Compare the results in **(a)** and **(b)** with the power propagated in the TEM mode at the same corresponding frequencies.

17.4 Propagation in Discontinuous Waveguides. Three parallel plate waveguide sections are connected as shown in **Figure 17.24**. The material between the plates is free space. Assume that the three waveguide sections operate in TE modes only. The source on the left supplies power at all frequencies between 1 MHz and 100 GHz. What is the lowest frequency signal received at the receiver?

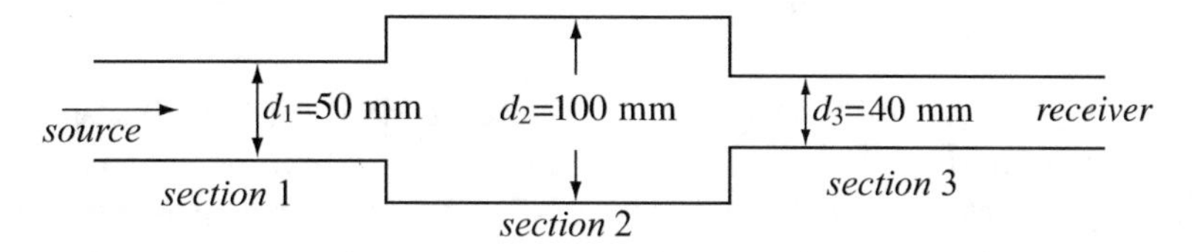

Figure 17.24

17.5 Reflection, Transmission, and SWR in Waveguides. A parallel plate waveguide with dimensions as shown in **Figure 17.25** is very long. A slab of permittivity $\varepsilon_1 = 2.5\varepsilon_0$ [F/m] occupies the right half of the waveguide. Assume TE propagation from left to right, at $f = 2f_c$ [Hz], where f_c is the cutoff frequency of the empty waveguide. Calculate:

(a) The reflection and transmission coefficients at the interface between air and slab.
(b) The standing wave ratio in the waveguide to the left of the interface and to the right of the interface.

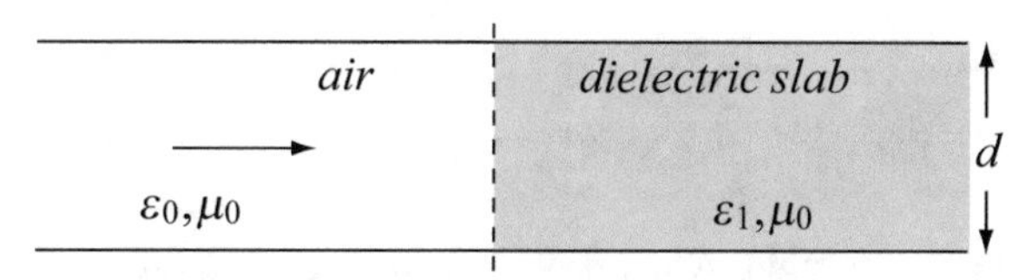

Figure 17.25

17.6 Fields in Shorted Waveguide. A parallel plate waveguide is made of plates 15 mm wide and separated a distance $a = 1$ mm. The space between the plates is air. The waveguide propagates in the lowest TM mode in the positive z-direction at a frequency 10% above cutoff. Assume the magnetic field intensity is directed out of the page in **Figure 17.26** and has amplitude 20 A/m. Now a perfect conductor plate is used to short the two parallel plates (indicated by the dotted line) at $z = 0$. Calculate the electric and magnetic field intensities in the waveguide to the left of the short.

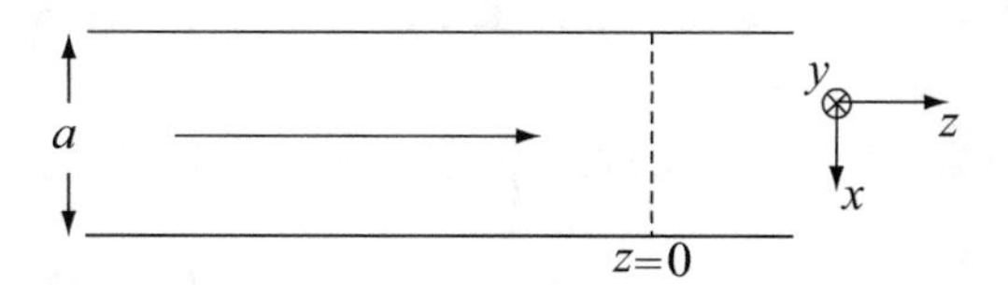

Figure 17.26

17.7 Application: Infrared Detection System. An infrared detection system is made of an optical microstrip line, used to guide infrared waves, and an infrared detector. The system must operate at a wavelength of 1,200 nm and the detector has an impedance of 50 Ω. The microstrip line is made of a thin sheet of glass, with relative permittivity $\varepsilon_r = 1.75$, sandwiched between two conducting sheets as shown in **Figure 17.27**. If the line is matched to the detector, calculate:

(a) The lowest possible mode of propagation that may be used.
(b) The thickness d of the glass sheet that will support the mode calculated in **(a)**.

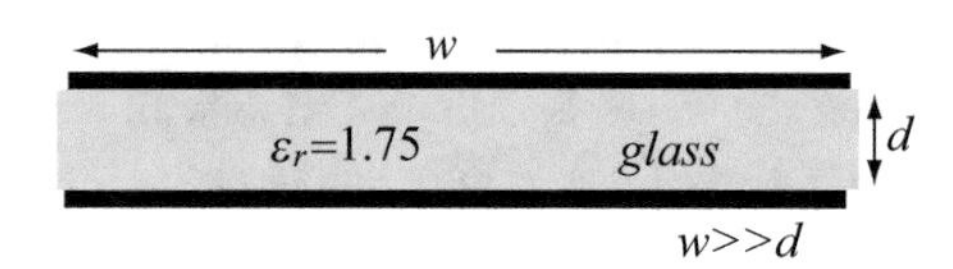

Figure 17.27

17.8 TEM Modes in Parallel Plate Waveguide. A parallel plate waveguide is made with planar conductors as shown in **Figure 17.28b**. Dimensions are $b = 20$ mm, $d = 2$ mm and the space between the plates is Teflon with a relative permittivity of 2.0. The peak magnetic field intensity at $z = 0$ is 3.8 A/m and the propagation is in the positive z direction. Find, assuming perfect conductors for the plates and perfect dielectric for the space between the plates:

(a) The electric and magnetic field intensity for TEM modes as a function of frequency.
(b) The power propagated in the waveguide.

17.9 TE Fields in Parallel Plate Waveguide. A parallel plate waveguide is made with planar conductors as shown in **Figure 17.28a**. Dimensions are $b = 20$ mm, $d = 2$ mm and the space between the plates is Teflon with a relative permittivity of 2.0. The peak electric field intensity at $z = 0$ is 240 V/m and the propagation is in the positive z direction. Find, assuming perfect conductors for the plates and perfect dielectric for the space between the plates:

(a) The electric and magnetic field intensity at the center frequency (middle of the bandwidth) in the lowest TE mode.
(b) The power propagated at the center frequency (middle of the bandwidth) in the lowest TE mode.

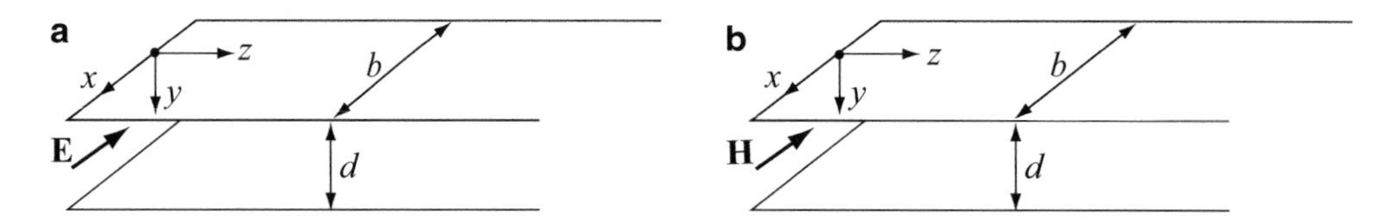

Figure 17.28

17.10 TM Fields in Parallel Plate Waveguide. A parallel plate waveguide is made with planar conductors as shown in **Figure 17.28b**. Dimensions are $b = 20$ mm, $a = 2$ mm and the space between the plates is Teflon with a relative permittivity of 2.0. The peak magnetic field intensity at $z = 0$ is 5 A/m and the propagation is in the positive z direction. Find, assuming perfect conductors for the plates and perfect dielectric for the space between the plates:

(a) The electric and magnetic field intensity at the center frequency (middle of the bandwidth) in the lowest TM mode.
(b) The power propagated at the center frequency (middle of the bandwidth) in the lowest TM mode.

17.11 Application: Dielectric waveguide. A plane electromagnetic wave propagates in a lossless dielectric sheet as shown in **Figure 17.29**. Relative permittivity of the dielectric is 2.25 and relative permeability is 1. The dielectric is 10 mm thick. Calculate:

(a) The smallest incidence angle θ that will still allow propagation in the dielectric without light escaping through the top and bottom surfaces.
(b) The lowest mode (frequency) of waves that can propagate in the dielectric sheet under the conditions in **(a)**.

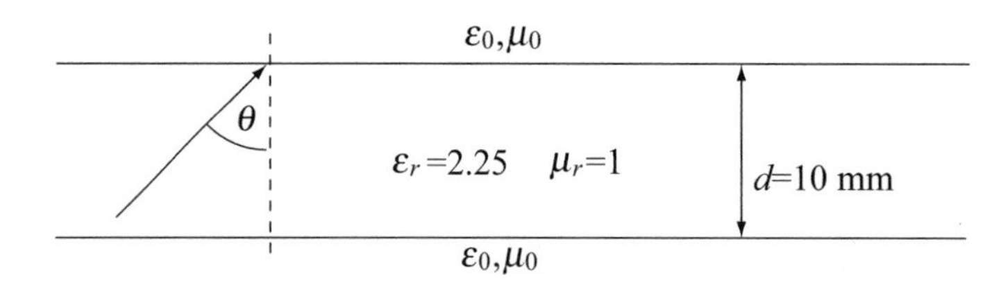

Figure 17.29

TM/TE Modes in Rectangular Waveguides

17.12 Application: Low-Frequency Waveguide–Limitations. An engineer had a bright idea: Why not use rectangular waveguides instead of the coaxial lines used in cable TV? The requirements are as follows: lowest frequency 54 MHz (TV Channel 2), and the waveguide has a ratio of $a = 2b$.

(a) What must be the dimensions of the waveguide to propagate from 54 MHz and up in the TE_{10} mode?
(b) The normal TV range in the VHF band is between 44 MHz and 88 MHz (channels 2 through 6) and from 174 MHz through 216 MHz (Channels 7 through 13) with each channel allocated 6 MHz bandwidth. How many of the TV channels can be propagated in the TE_{10} mode calculated in **(a)?**
(c) Is this a bright idea?

17.13 Application: Mode Separation and Bandwidth. The commercial EIA WR 284 rectangular waveguide has internal dimensions $a = 72.14$ mm and $b = 34.04$ mm. Calculate:

(a) The maximum bandwidth for the TE_{10} mode.
(b) The maximum bandwidth for the TM_{11} mode.
(c) The maximum bandwidth for the TE_{01} mode.

17.14 Application: Modes in Rectangular Waveguide. An EIA WR 112 standard, rectangular waveguide with dimensions $a = 28.499$ mm and $b = 12.624$ mm is used to connect to a radar antenna which operates at a wavelength of 20 mm. Find all propagating modes that can be used at the given wavelength. The waveguide is air filled.

17.15 Application: Fields and Power in Rectangular Waveguide. A rectangular waveguide is used to transmit power from a generator to a radar antenna. The waveguide is an EIA WR 34 waveguide with internal dimensions 8.636 mm and 4.318 mm, operating at 23 GHz in the TE_{10} mode. The power delivered is 50 kW:

(a) Calculate the amplitudes of the electric and magnetic field intensities in the waveguide.
(b) Are these amplitudes acceptable in level? Explain.

17.16 Application: Tunnels as Waveguides. The following communication system is proposed for communication in mine tunnels to avoid the need for cables: the tunnel is used as a waveguide 5 m wide and 2 m high.

(a) What is the lowest frequency that may be used?
(b) If it is desired to propagate a single mode, what is the maximum bandwidth that may be used and still guarantee propagation in the lowest mode?

17.17 Discontinuities in Rectangular Waveguide. A very long rectangular waveguide is filled with two materials as shown in **Figure 17.30**. A TM wave propagates in the waveguide from the left. Calculate:

(a) The lowest frequency (cutoff frequency) that will propagate in the waveguide.
(b) The time it takes the wave to propagate between points A and B.

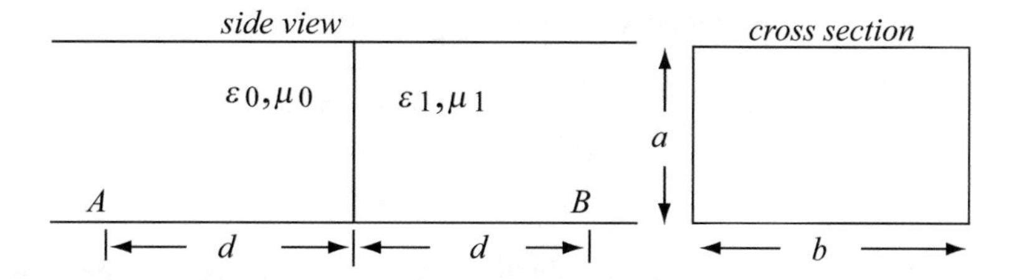

Figure 17.30

17.18 Field Required for Total Power. It is required that a lossless rectangular waveguide carry 100 W (time averaged power) at 4.5 GHz, in the TE_{10} mode. The waveguide is $a = 47.549$ mm wide and $b = 22.149$ mm high (EIA WR 187 waveguide) and is air filled:

(a) Find the longitudinal and transverse components of the electric field intensity.
(b) Find the longitudinal and transverse components of the magnetic field intensity.

17.19 Application: Reduced Height Waveguide. In most waveguides the height equals half the width because this choice improves mode separation (and hence bandwidth of modes). However, in some cases, a different ratio may be chosen for specific purposes. Consider the standard EIA WR 229 waveguide with dimensions $a = 58.166$ mm, $b = 29.083$ mm. The waveguide is also available in what is called the ½ EIA WR 229 with dimensions $a = 58.166$ mm, $b = 14.500$ mm (this is called a half-height waveguide). Calculate:

(**a**) The first 10 lowest modes for the ½ EIA WR 229 waveguide and classify them as TE or TM.
(**b**) Repeat (**a**) for the EIA WR 229 waveguide.
(**c**) Compare the bandwidths of the various modes of the two waveguides.

17.20 Square Cross-section Waveguide. Most rectangular waveguides are made so that $a = 2b$, that is, their height is half their width. Suppose an air-filled waveguide is built so that $a = b = 12$ mm.

(**a**) Calculate the first 8 TE modes and the first 8 TM modes.
(**b**) From (**a**) explain why square cross-section waveguides are not a good idea.

17.21 Application: Power Carried in Lowest Mode. The EIA WR 137 waveguide has dimensions $a = 34.849$ mm, $b = 15.799$ mm. The half-height version is called the ½ EIA WR 137 and has dimensions $a = 34.849$ mm, $b = 7.9$ mm:

(**a**) Find the lowest cutoff frequency and mode for the two waveguides.
(**b**) Calculate the time-averaged power the wave can propagate at a frequency 25% above the cutoff in (**a**), for a given electric field intensity with amplitude $E_0 = 1{,}000$ V/m for the two waveguides.
(**c**) Which waveguide can carry more power for the same field level and why?

17.22 Maximum Power Handling of a Waveguide. A rectangular waveguide has a width to height ratio $a/b = 2.0$ and the ratio between the operating frequency and the cutoff frequency is $f/f_{c10} = 1.25$ at $f = 10$ GHz. What is the maximum time-averaged power that can be transmitted in the waveguide in the TE_{10} mode without exceeding the breakdown electric field intensity of 3×10^6 V/m in air?

17.23 Application: Integrated Optical Waveguide. An integrated optical waveguide is made in the form of a rectangular cross-sectional channel in a silicon substrate, as shown in **Figure 17.31**. The channel is 2 μm wide, 1 μm high, and $\varepsilon = 2\varepsilon_0$ [F/m]:

(**a**) Explain why this structure can function as a rectangular waveguide and outline the conditions necessary for it to operate. **Hint:** Consider the conditions for total reflection in a dielectric.
(**b**) Calculate the lowest frequency that can be propagated. Which mode is it and in what range of the spectrum is this propagation possible?
(**c**) Calculate the peak power that can be propagated at a frequency 25% above the frequency calculated in (**b**) if the peak electric field intensity cannot exceed 1,000 V/mm.

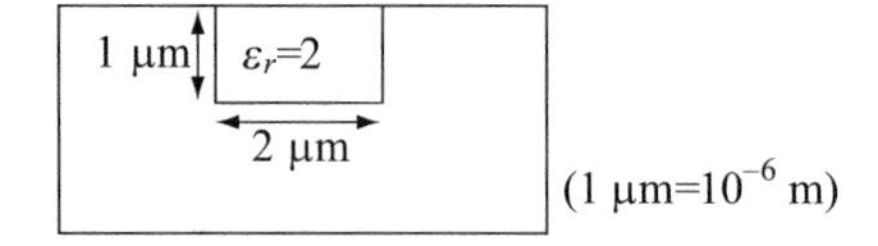

Figure 17.31

17.24 TM Waves in a Waveguide. Write the time domain expressions for the transverse components of the fields for TM propagation in a lossless rectangular waveguide of width a and height b (see **Exercise 17.4**).

17.25 TM Waves in a Waveguide. Find the total wave in a waveguide of width a [m] and height b [m] if a forward-propagating TM wave of amplitude E_0^+ [V/m] and a backward-propagating wave of amplitude E_0^- [V/m] exist.

17.26 TE Waves in a Waveguide. Find the total transverse TE waves in a waveguide of width a [m] and height b [m] if a forward-propagating wave of amplitude H_0 [A/m] and a backward-propagating wave of amplitude H_1 [A/m] exists in the waveguide.

Attenuation and Losses in Rectangular Waveguides

17.27 Dielectric Losses in Waveguides. A rectangular waveguide is filled with a lossy dielectric with relative permittivity $\varepsilon_r = 2$ and conductivity $\sigma_d = 10^{-4}$ S/m. Assuming perfectly conducting walls, find:

(**a**) The attenuation constant in the waveguide at a frequency 1.5 times larger than the lowest cutoff frequency.
(**b**) The percentage of power loss per meter of the waveguide at $f/f_c = 1.75$. Assume the power entering a section of the waveguide is P_0 [W] and calculate the power loss as a percentage of this power.

17.28 Conductor (Wall) Losses. A rectangular waveguide is made of aluminum, which has conductivity of 3.6×10^7 S/m. The walls of the waveguide are thick and the internal dimensions are $a = 38.1$ mm and $b = 25.4$ mm. Assuming the waveguide is empty (free space), calculate:

(**a**) The power loss per meter length in the TE_{01} mode at $f/f_c = 1.5$. Assume the amplitude of the longitudinal magnetic field intensity is 1 A/m.
(**b**) The attenuation constant due to losses in the walls for the conditions in (**a**).

17.29 Application: Waveguide with Dielectric and Wall Losses. The rectangular waveguide in **Problem 17.28** is given again, but now the waveguide is filled with a low-loss dielectric with relative permittivity of 2 and conductivity $\sigma_d = 10^{-4}$ S/m. All other parameters including wall conductivity remain the same. Calculate:

(**a**) The attenuation constant in the waveguide for the TE_{01} mode at $f/f_c = 1.25$.
(**b**) The power loss per meter length in the TE_{01} mode at $f/f_c = 1.25$. Assume the amplitude of the longitudinal magnetic field intensity is 1 A/m.

Cavity Resonators

17.30 Resonant Frequencies in Rectangular Cavity. A rectangular cavity resonator is 60 mm long, 30 mm high, and 40 mm wide and is air filled. Calculate:

(**a**) The TE_{101} resonant frequency.
(**b**) The next three nondegenerate TE resonant modes. Classify the modes.

17.31 Resonant Frequencies in Shorted Waveguide. The EIA WR 284 waveguide is made into a cavity 0.5 m long by shorting the waveguide at two locations. Calculate the first 10 resonant frequencies and classify the modes. The waveguide has dimensions $a = 72.14$ mm and $b = 34.04$ mm.

17.32 Application: Design of a Cavity for Given Resonant Frequencies. A cavity resonator is built from a section of a waveguide with dimensions $a = 47.549$ mm and $b = 22.149$ mm (EIA WR 187) by shorting the waveguide with two conducting plates to create a rectangular cavity of length d [mm]. The cavity is required to resonate at 8 GHz in the TM_{111} mode:

(**a**) Find the length of the shorted section necessary.
(**b**) What is the dominant mode and what is its resonant frequency?

17.33 Application: Microwave Moisture Sensor. A microwave relative humidity (RH) sensor is made as in **Figure 17.21** by drilling holes in a cavity resonator to allow free movement of air. The cavity is 35 mm by 16 mm by 42 mm. The relative permittivity of dry air is 1.0 whereas at saturation humidity (100% RH) it is 1.00215. Assume permittivity increases linearly with relative humidity.

(**a**) Calculate he shift in resonant frequency of the fundamental mode per % RH in the dominant mode.
(**b**) If frequency increments of 1 kHz can be accurately measured, what is the resolution of the sensor in % relative humidity (% RH). Resolution is the smallest increment in % humidity that can be measured.

Antennas and Electromagnetic Radiation

18

Is it a fact—or have I dreamt it—that, by means of electricity, the world of matter has become a great nerve, vibrating thousands of miles in a breathless point of time? Rather, the round globe is a vast head, a brain, instinct with intelligence!

—Nathaniel Hawthorne (1804–1864),
Novelist

18.1 Introduction

After discussing wave propagation, it is time we discuss the sources of the waves. Recall that our whole discussion of waves was based on the solution to the source-free wave equation. Starting with **Chapter 11**, we assumed that a wave was generated in some fashion but did not concern ourselves too much with how the wave was generated. Occasionally, the term "source" or "antenna" was mentioned but only to indicate that the wave must have a source.

However, the sources of waves are extremely important. To transmit power, we must first generate the waves at the proper level and frequencies and, second, we must couple the energy into the appropriate domain. This coupling is done by what we call an antenna. In the following sections, we will discuss the basic principles of antennas starting with the elementary electric and magnetic dipoles. Then, we extend these to other important antennas and discuss the relations between receiving and transmitting antennas, principles of design of antenna systems, as well as some important applications of antennas.

18.2 Electromagnetic Radiation and Radiation Safety

Radiation is the process of emitting energy from a source. If it helps, you may think of the Sun as a source of radiated energy (light, heat, particles) or of a heater in the home as a source of radiated heat. Electromagnetic radiation can be at all frequencies except zero and may take different forms, as can be seen from the electromagnetic spectrum in **Figure 12.8**. We know from experience that radiation at various frequencies is different. At low frequencies, we talk about electromagnetic waves. In the visible domain, the emission is in the form of light. At still higher frequencies, the emission may be ultraviolet or X-ray radiation. Each of these is an electromagnetic wave, but the properties of the wave change with frequency. For example, low-frequency electromagnetic waves are not visible whereas X-rays easily penetrate through our bodies. We also know that X-rays can be damaging to cells and ultraviolet rays are known to harm our eyes and, in some cases, to cause skin cancer. This observation raises more questions than it answers. For example, we may ask: If X-rays are dangerous, why not visible light or microwaves or, indeed, any electromagnetic wave? Or perhaps they are? Even more important is to ask ourselves what makes X-rays dangerous whereas some forms of electromagnetic radiation are not? What, then, is the difference between the various types of radiation? The answer is in the energy associated with the radiation, known as the ***photon energy***, $e = hf$, where h is the Planck constant ($h = 6.63 \times 10^{-34}$ J • s or $h = 4.14 \times 10^{-15}$ eV) and f is the frequency of radiation. This energy, also known as the ***quantum of radiation***, indicates the relative energy in different ranges of

N. Ida, *Engineering Electromagnetics*, https://doi.org/10.1007/978-3-030-15557-5_18

radiation based on frequency. In the visible range (4.2×10^{14} to 7.9 to 10^{14} Hz), the photon energy is between 1.74 and 3.27 eV. In the microwave domain (300 MHz to 300 GHz), the photon energy is between 1.24×10^{-6} and 1.24×10^{-3} eV. In the X-ray range (10^{16} – 10^{21} Hz), the photon energy is between 41.4 eV and 4.14 MeV. In their lowest range, X-rays are at least 10 times more energetic than visible light. For this reason, ultraviolet light, which overlaps part of the X-ray domain, is considered to be harmful. On the other hand, microwave photon energy is at least 4 orders of magnitude lower than the lowest X-ray energy. This distinction between low- and high-energy domains is sometimes made on the basis of the ability of the various emissions to ionize materials through which they pass. Low-frequency (low-energy) radiation is called ***nonionizing*** and includes all frequencies up to the low ultraviolet. High-energy radiation is ***ionizing*** and includes all radiation above the low ultraviolet, including X-rays and γ-rays.

Figure 12.8 shows that we will concern ourselves with radiation in the low-energy domain (nonionizing radiation). This is in contrast with, for example, radioactive radiation, which is ionizing. In this chapter, while using the term radiation, this radiation should be understood as nonionizing radiation.

Example 18.1 Application: Radiation Safety When installing a new radar at an airport, operating at 30 GHz and 50 kW, with a beam diameter of 100 m, there were concerns raised about its safety. The issue was radiation from the radar and its effect on humans. Although what constitutes a safe level of radiation is an unresolved issue, it is common practice to compare microwave radiation to that from the Sun, because both the power density and the photon energy in most microwave applications, including radar, are lower than those of the solar radiation in the visible region. One argument for the safety of microwave radiation is based on this comparison:

(a) Given that the maximum solar power density on Earth is 1,400 W/m^2 and the average frequency in the visible domain is 5×10^{14} Hz, which radiation appears to be more "dangerous"?
(b) Are there any other effects you can think of that might change your view as to safety of either radiation?

Solution: The radar power density may be calculated assuming uniform power density in the beam while the photon energy is $e = hf$:

(a) The radar's beam power density is

$$\mathcal{P}_{av} = \frac{P_r}{A} = \frac{4P_r}{\pi d^2} = \frac{4 \times 50{,}000}{\pi (100)^2} = 6.366 \quad [\mathrm{W/m^2}]$$

which is over 200 times smaller than that of the solar radiation. The photon energy at the radar's frequency is

$$e_r = hf = 4.14 \times 10^{-15} \times 30 \times 10^9 = 1.242 \times 10^{-4} \quad [\mathrm{eV}]$$

The photon energy in the visible range is

$$e_l = hf = 4.14 \times 10^{-15} \times 5 \times 10^{14} = 2.07 \quad [\mathrm{eV}]$$

This is over 16,000 times larger than the photon energy at the radar frequency.
From these considerations, it would appear that the radar radiation is safer than solar radiation.

(b) One effect is the skin depth $\left(\delta = 1/\sqrt{\pi f \mu_0 \sigma}\right)$, which, with conductivity of skin at about 0.1 S/m, gives 9.19 mm penetration for the radar frequency, but only 0.07 mm in the visible-light region. Thus, any induced currents in the visible-light range are limited to a very thin layer of the skin about 70 μm thick (did you ever wonder why your skin blisters when you get sunburned?) whereas the radar waves penetrate 130 times deeper. Also, clothing shields visible light (although less effective in the ultraviolet region and beyond) but does not shield lower-frequency waves like the radar in question.

18.3 Antennas

What is an antenna then? Quite simply, it is any structure that can radiate electromagnetic energy into a medium. The principle is quite simple: An antenna must be supplied with a time-dependent current which, in turn, generates a magnetic and an electric field. When time-dependent electric and magnetic fields exist, power is generated and propagated based on the Poynting theorem. Although this is true in general, it is not immediately obvious why one structure may serve as an antenna whereas others cannot, and, in fact, how an antenna radiates is still not entirely clear.

As an example, consider the output stage of a transmitter shown in **Figure 18.1a**. This antenna is one of the most common electromagnetic structures: that of a straight conductor ("whip" or monopole antenna). How does this structure radiate energy? In fact, at first glance it is not entirely clear how a current may exist in this antenna. Certainly, from a circuit theory point of view, we might conclude that this circuit cannot operate. Another type of antenna is shown in **Figure 18.1b**. A horn antenna is connected to a rectangular waveguide. How does this antenna operate and, most importantly, what is the relation between this and the antenna in **Figure 18.1a**? Surprisingly perhaps, the two antennas in **Figure 18.1** are very similar in operation.

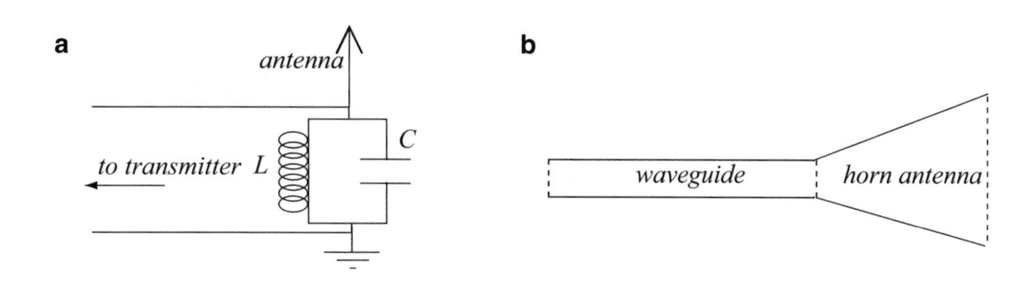

Figure 18.1 Common antennas. (**a**) "Whip" (monopole) antenna and the transmitter's output circuit. (**b**) A microwave horn antenna connected to a waveguide

It is obvious that we have some explaining to do, although, once done, the answer looks surprisingly simple. We will start with the simplest of structures, the elemental dipole radiator, and build on this to define more complex and more practical antennas. There are, in fact, two types of dipoles, as we discussed in **Chapters 4** and **9**. One is an electric dipole, the other a magnetic dipole. Both may be used as radiators and we will discuss both.

18.4 The Electric Dipole

The electric dipole was defined in **Section 3.4.1.3** as two point charges, separated by a very short distance Δl as shown in **Figure 18.2a**. This was the electrostatic definition of the dipole, and because the charges were assumed to be constant, independent of time, all that we could say about this structure is that it produces an electric field but no magnetic field. For a dipole to produce a magnetic field, it must produce a current. For this current to produce a wave, the current must be time dependent. These two conditions can be satisfied if we assume the following structure:

(1) Two point charges are placed a short distance Δl apart. One charge is negative, the other positive.
(2) The point charges are time dependent. We will assume a sinusoidal time dependency, but any other time dependency may be used.
(3) The two point charges are connected through a thin, conducting wire. A time-dependent current can now flow back and forth between the two point charges.

This structure, which we call a ***Hertzian dipole*** or an ***electric dipole***, is shown in **Figure 18.2b**. Note that the total charge at any instant in time must be zero, as required by the law of conservation of charge.

Suppose, now, that we can, by some means, generate time-dependent charges as follows:

$$Q_1 = Q_0 \sin(\omega t) \text{ and } Q_2 = -Q_0 \sin(\omega t) \tag{18.1}$$

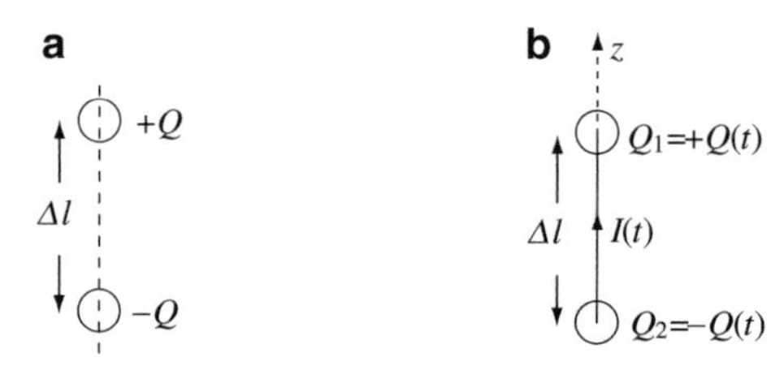

Figure 18.2 (**a**) An electrostatic dipole. (**b**) A Hertzian dipole

With these charges, the current in the wire between charge (2) and (1) is

$$I(t) = \frac{dQ_1}{dt} = -\frac{dQ_2}{dt} = \omega Q_0 \cos(\omega t) = I_0 \cos(\omega t) \quad [\mathrm{A}] \tag{18.2}$$

Since it will prove to be easier to perform calculations in phasor notation, we write the current in the dipole in terms of a phasor I_0:

$$I = I_0 e^{j\omega t} \quad \rightarrow \quad I(t) = \mathrm{Re}\{I_0 e^{j\omega t}\} \quad [\mathrm{A}] \tag{18.3}$$

Similarly, the charges Q_1 and Q_2 may be written as phasors:

$$Q_1 = Q_0 e^{j\omega t}, \quad Q_2 = -Q_0 e^{j\omega t} \quad [\mathrm{C}] \tag{18.4}$$

Now, we can write the current in terms of the charges (dropping $e^{j\omega t}$) as

$$I_0 = j\omega Q_1 = -j\omega Q_2 = j\omega Q_0 \quad [\mathrm{A}] \tag{18.5}$$

The dipole moment, which is also a phasor, may now be calculated using **Figure 18.2b**. With the dipole at the origin and the current in the z direction as shown, the electric dipole moment is

$$\mathbf{p} = \hat{\mathbf{z}} Q_0 \Delta l \quad [\mathrm{C \cdot m}] \tag{18.6}$$

This may be written directly from the result in **Section 3.4.1.3** by replacing the static charge density by the phasor form of **Eq. (18.4)**. Now that the current due to the dipole is known, the magnetic and electric fields produced by the dipole may be calculated. We obtain a solution by calculating the magnetic vector potential of a current segment of length dl', carrying a current I, as was done in **Eq. (8.34)**. There, we obtained the magnetic vector potential due to a current I in a finite segment between two points a and b at a distance $\mathbf{R} - \mathbf{R}'$ from the segment (see **Figure 18.3a)** as

$$\mathbf{A} = \frac{\mu I}{4\pi} \int_a^b \left(\frac{d\mathbf{l}'}{|\mathbf{R} - \mathbf{R}'|} \right) \quad \left[\frac{\mathrm{Wb}}{\mathrm{m}} \right] \tag{18.7}$$

If we reduce this segment to a length $\Delta l'$ and place the current segment at the origin ($\mathbf{R}' = 0$), the magnetic vector potential due to a segment $\Delta l'$ as shown in **Figure 18.3b** is

$$\mathbf{A} = \frac{\mu I \Delta \mathbf{l}'}{4\pi |\mathbf{R}|} = \hat{\mathbf{z}} \frac{\mu I \Delta l'}{4\pi R} \quad \left[\frac{\mathrm{Wb}}{\mathrm{m}} \right] \tag{18.8}$$

where, in effect, we assumed the current to be constant along the dipole, an assumption justified by its very short length (much shorter than a wavelength). This solution may seem simple, but it is incorrect because to define the Biot–Savart law, we used the pre-Maxwell postulates which do not include the displacement currents. This solution may be used for static and slowly varying fields (for which the displacement current is small compared with conduction currents), but does not take into account propagating waves. The net effect of this is to assume that the current I generates fields in the whole of space instantaneously.

So, what can we do? To obtain the correct solution, we can proceed by solving the wave equation we obtained in **Chapter 11** [see **Eq. (11.50)**]. However, the solution to the wave equation is quite complex. Instead of actually solving the wave equation, we rely here on a physical argument that will allow us to obtain the correct solution without the need of actually solving a wave equation.

Suppose that the segment of current generates the magnetic vector potential everywhere in space as in **Eq. (18.8)**. Now, since the solution must also obey the wave equation, the waves generated at the source propagate outward from the source in all directions, at a constant speed, equal to the phase velocity in space, $v_p = 1/\sqrt{\mu\varepsilon}$. If an observer is at a distance R from the source (point P in **Figure 18.3b**), a wave generated at the source will arrive at the observer after a time $t = R/v_p$. Alternatively, the wave that the observer measures at any given time t has been generated by the source a time $t = R/v_p$ earlier. Thus, the wave that arrives at point P is a retarded wave:

$$\mathbf{A}(t) = \hat{\mathbf{z}}\frac{\mu I(t - R/v_p)\Delta l'}{4\pi R} \quad \left[\frac{\text{Wb}}{\text{m}}\right] \tag{18.9}$$

that is, the vector potential measured at point P at a time t has been generated by the current I at time $(t - R/v_p)$. **Equation (18.9)** therefore defines a ***retarded magnetic vector potential***. From this relation, we can also say that the current itself is retarded, since the current is the only time-dependent quantity. Even if the above argument may not seem too rigorous, the solution thus obtained may be substituted in the wave equation to show that it satisfies the wave equation (see **Section 12.3.5**).

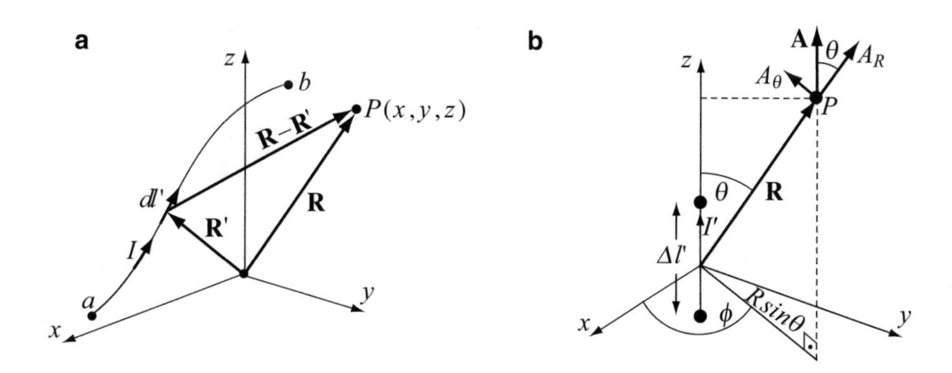

Figure 18.3 (**a**) Use of the Biot–Savart law to calculate the magnetic vector potential of a segment carrying current I. (**b**) The magnetic vector potential of the dipole in spherical coordinates

From **Eqs. (18.2)** and **(18.9)**, the retarded current is

$$I(t - R/v_p) = I_0\cos\left(\omega\left(t - \frac{R}{v_p}\right)\right) = I_0\cos\left(\omega t - \frac{\omega R}{v_p}\right) \quad [\text{A}] \tag{18.10}$$

Since the term ω/v_p is the phase constant β, the retarded magnetic vector potential can be written as

$$\mathbf{A}(R,t) = \hat{\mathbf{z}}\frac{\mu I_0 \Delta l' \cos(\omega t - \beta R)}{4\pi R} \quad \left[\frac{\text{Wb}}{\text{m}}\right] \tag{18.11}$$

or, in phasor form, as

$$\boxed{\mathbf{A}(\text{R}) = \hat{\mathbf{z}}\frac{\mu I_0 \Delta l'}{4\pi R}e^{-j\beta R} \quad \left[\frac{\text{Wb}}{\text{m}}\right]} \tag{18.12}$$

To define the propagation properties and to be able to use much of what we have already defined in previous chapters, we need the electric and magnetic field intensities. These are calculated directly from the magnetic vector potential. Before we do so, we note that since the dipole is approximately a point source, it makes sense to define the fields in a spherical coordinate system centered at the dipole. The magnetic vector potential in **Eq. (18.12)** is written in Cartesian coordinates. Using the transformation between Cartesian and spherical coordinates in **Section 1.5.3**, we get (see **Figure 18.3b**):

$$A_R = A_z\cos\theta = \frac{\mu I_0 \Delta l'}{4\pi R} e^{-j\beta R}\cos\theta, \quad A_\theta = -A_z\sin\theta = -\frac{\mu I_0 \Delta l'}{4\pi R} e^{-j\beta R}\sin\theta, \quad A_\phi = 0 \tag{18.13}$$

Thus, the magnetic vector potential is

$$\mathbf{A} = \hat{\mathbf{R}}\frac{\mu I_0 \Delta l'}{4\pi R} e^{-j\beta R}\cos\theta - \hat{\mathbf{\theta}}\frac{\mu I_0 \Delta l'}{4\pi R} e^{-j\beta R}\sin\theta \quad \left[\frac{\text{Wb}}{\text{m}}\right] \tag{18.14}$$

These can now be used to calculate the magnetic field intensity from $\mathbf{B} = \mu\mathbf{H} = \nabla \times \mathbf{A}$. Because $\mathbf{A}$ has only an R and a θ component, we write

$$\begin{aligned}\mu\mathbf{H} = \nabla \times \mathbf{A} &= \hat{\mathbf{R}}\frac{1}{R\sin\theta}\left(\frac{\partial}{\partial\theta}(A_\phi\sin\theta) - \frac{\partial A_\theta}{\partial\phi}\right) + \hat{\mathbf{\theta}}\frac{1}{R}\left(\frac{1}{\sin\theta}\frac{\partial A_R}{\partial\phi} - \frac{\partial(RA_\phi)}{\partial R}\right) + \hat{\mathbf{\phi}}\frac{1}{R}\left(\frac{\partial(RA_\theta)}{\partial R} - \frac{\partial A_R}{\partial\theta}\right) \\ &= \hat{\mathbf{\phi}}\frac{1}{R}\left(\frac{\partial(RA_\theta)}{\partial R} - \frac{\partial A_R}{\partial\theta}\right) = \hat{\mathbf{\phi}}\frac{1}{R}\left(\frac{\partial}{\partial R}\left[-R\frac{\mu I_0\Delta l'}{4\pi R}e^{-j\beta R}\sin\theta\right] - \frac{\partial}{\partial\theta}\left[\frac{\mu I_0\Delta l'}{4\pi R}e^{-j\beta R}\cos\theta\right]\right)\end{aligned} \tag{18.15}$$

Note that the R and θ components of $\mathbf{H}$ are zero since $A_\phi = 0$ and $\mathbf{A}$ is independent of ϕ. As a consequence the magnetic field intensity must be in planes perpendicular to the current as can also be deduced from the right hand rule. Performing the derivatives and collecting terms,

$$\mathbf{H} = \hat{\mathbf{\phi}}\frac{I_0\Delta l'}{4\pi}e^{-j\beta R}\sin\theta\left(\frac{j\beta}{R} + \frac{1}{R^2}\right) \quad \left[\frac{\text{A}}{\text{m}}\right] \tag{18.16}$$

The components of $\mathbf{E}$ are evaluated from Maxwell's second equation:

$$\begin{aligned}j\omega\varepsilon\mathbf{E} = \nabla \times \mathbf{H} = \hat{\mathbf{R}}\frac{1}{R\sin\theta}\left(\frac{\partial}{\partial\theta}\left(H_\phi\sin\theta\right) - \frac{\partial H_\theta}{\partial\phi}\right) + \hat{\mathbf{\theta}}\frac{1}{R}\left(\frac{1}{\sin\theta}\frac{\partial H_R}{\partial\phi} - \frac{\partial(RH_\phi)}{\partial R}\right) \\ + \hat{\mathbf{\phi}}\frac{1}{R}\left(\frac{\partial(RH_\theta)}{\partial R} - \frac{\partial H_R}{\partial\theta}\right)\end{aligned} \tag{18.17}$$

We first note that $E_\phi = 0$ because $H_\theta = H_R = 0$. Substituting for H_ϕ from **Eq. (18.16)** and performing the derivatives, we get (after dividing both sides by $j\omega\varepsilon$ and rearranging terms)

$$\mathbf{E} = \hat{\mathbf{R}}\frac{\beta}{\omega\varepsilon}\frac{I_0\Delta l'}{2\pi}e^{-j\beta R}\cos\theta\left(\frac{1}{R^2} + \frac{1}{j\beta R^3}\right) + \hat{\mathbf{\theta}}\frac{\beta}{\omega\varepsilon}\frac{I_0\Delta l'}{4\pi}e^{-j\beta R}\sin\theta\left(\frac{j\beta}{R} + \frac{1}{R^2} + \frac{1}{j\beta R^3}\right) \quad \left[\frac{\text{V}}{\text{m}}\right] \tag{18.18}$$

The term $\beta/\omega\varepsilon$ may be replaced with the intrinsic impedance η as follows:

$$\frac{\beta}{\omega\varepsilon} = \frac{1}{v_p\varepsilon} = \frac{\sqrt{\mu\varepsilon}}{\varepsilon} = \sqrt{\frac{\mu}{\varepsilon}} = \eta \quad [\Omega] \tag{18.19}$$

where $v_p = \omega/\beta = 1/\sqrt{\mu\varepsilon}$ for a lossless or low-loss dielectric of permittivity ε and permeability μ.

We now have the electric and magnetic field intensity vectors in **Eqs. (18.16)** and **(18.18)**. Multiplying the numerator and denominator in each equation by the term $(j\beta)^2 = -\beta^2$ and using **Eq. (18.19)**, we obtain the electric and magnetic field intensities of the electric dipole as

$$\mathbf{H} = -\hat{\boldsymbol{\phi}}\,\frac{I_0\Delta l'\beta^2}{4\pi}e^{-j\beta R}\sin\theta\left(\frac{1}{j\beta R}+\frac{1}{(j\beta R)^2}\right)\quad\left[\frac{\mathrm{A}}{\mathrm{m}}\right] \tag{18.20}$$

$$\mathbf{E} = -\hat{\mathbf{R}}\,\frac{\eta I_0\beta^2\Delta l'}{2\pi}e^{-j\beta R}\cos\theta\left(\frac{1}{(j\beta R)^2}+\frac{1}{(j\beta R)^3}\right)-\hat{\boldsymbol{\theta}}\,\frac{\eta I_0\beta^2\Delta l'}{4\pi}e^{-j\beta R}\sin\theta\left(\frac{1}{j\beta R}+\frac{1}{(j\beta R)^2}+\frac{1}{(j\beta R)^3}\right)\quad\left[\frac{\mathrm{V}}{\mathrm{m}}\right] \tag{18.21}$$

These equations give the electric and magnetic field intensities of the Hertzian dipole everywhere in space and for all conditions. These expressions can now be used to calculate anything else we need, such as direction of propagation, power density, total radiated power and the like. However, the expressions are rather complicated, and we will seek to simplify them before applying them to practical antennas.

Because β is constant for a wave of given frequency ($\beta = 2\pi/\lambda$), the only variables that change are the angle θ and distance R. We note that the two fields are composed of terms that diminish as $1/R$, $1/R^2$, and $1/R^3$. Thus, we can define three distinct domains: One is the domain of small values of R, which is called the ***near-field*** domain or the ***Fresnel zone***, and one of large values of R, which is called the ***far-field*** domain or the ***Fraunhofer zone***. A third domain is an intermediate domain where neither assumption holds, called the ***inductive zone***. The inductive zone is of little importance and, therefore, we will not discuss it separately other than to say that it is a transition zone and, if necessary, it may be analyzed using the above general equations.

18.4.1 The Near Field waves.m

The near field is that domain near the source (dipole in this case), for which R is small. Because β is constant, this means that the second term in **Eq. (18.20)** and the last term in the θ and R components in **Eq. (18.21)** are dominant; that is, the following must be satisfied:

$$\frac{1}{\beta R}\ll\frac{1}{(\beta R)^2}\ll\frac{1}{(\beta R)^3}\quad\text{or}\quad\frac{\lambda}{2\pi R}\ll\frac{\lambda^2}{(2\pi R)^2}\ll\frac{\lambda^3}{(2\pi R)^3} \tag{18.22}$$

where $\beta = 2\pi/\lambda$ was used. These conditions reduce to

$$\frac{\lambda}{2\pi R}\gg 1\quad\text{or}\quad R\ll\frac{\lambda}{2\pi} \tag{18.23}$$

In even simpler terms, we require that for a location to qualify as a near-field location, $R \ll \lambda$. Any location that is much closer than one wavelength may be considered to be in the near field. Substituting this condition in **Eq. (18.20)** gives

$$\mathbf{H}\approx -\hat{\boldsymbol{\phi}}\,\frac{I_0\Delta l'\beta^2}{(j\beta R)^2 4\pi}e^{-j\beta R}\sin\theta = \hat{\boldsymbol{\phi}}\,\frac{I_0\Delta l'}{4\pi R^2}e^{-j\beta R}\sin\theta\quad\left[\frac{\mathrm{A}}{\mathrm{m}}\right] \tag{18.24}$$

From the condition in **Eq. (18.23)**, we can also write $\beta R \ll 1$ and, therefore, in the limit, the term $e^{-j\beta R}$ tends to 1. Thus, the magnetic field intensity in the near field may be approximated as

$$\mathbf{H}\approx\hat{\boldsymbol{\phi}}\,\frac{I_0\Delta l'}{4\pi R^2}\sin\theta\quad\left[\frac{\mathrm{A}}{\mathrm{m}}\right] \tag{18.25}$$

Similarly, with the approximations above, the electric field intensity in **Eq. (18.21)** becomes

$$\mathbf{E} \approx \hat{\mathbf{R}}\,\frac{\eta I_0 \Delta l'}{2\pi j\beta R^3}\cos\theta + \hat{\boldsymbol{\theta}}\,\frac{\eta I_0 \Delta l'}{4\pi j\beta R^3}\sin\theta \quad \left[\frac{\mathrm{V}}{\mathrm{m}}\right] \tag{18.26}$$

For a lossless dielectric, $\beta = \omega\sqrt{\mu\varepsilon}$ and $\eta = \sqrt{\mu/\varepsilon}$. Substituting these two relations in **Eq. (18.26)**, we get

$$\boxed{\mathbf{E} \approx \hat{\mathbf{R}}\,\frac{I_0 \Delta l'}{2\pi j\omega\varepsilon R^3}\cos\theta + \hat{\boldsymbol{\theta}}\,\frac{I_0 \Delta l'}{4\pi j\omega\varepsilon R^3}\sin\theta \quad \left[\frac{\mathrm{V}}{\mathrm{m}}\right]} \tag{18.27}$$

Replacing I_0 in **Eq. (18.27)** by $j\omega Q$ from **Eq. (18.5)**, we obtain

$$\mathbf{E} \approx \hat{\mathbf{R}}\,\frac{Q\Delta l'}{2\pi\varepsilon R^3}\cos\theta + \hat{\boldsymbol{\theta}}\,\frac{Q\Delta l'}{4\pi\varepsilon R^3}\sin\theta = \frac{Q\Delta l'}{4\pi\varepsilon R^3}\left(\hat{\mathbf{R}}2\cos\theta + \hat{\boldsymbol{\theta}}\sin\theta\right) \quad \left[\frac{\mathrm{V}}{\mathrm{m}}\right] \tag{18.28}$$

This a rather interesting result since it is identical to the solution for the electrostatic dipole in **Section 3.4.1.3** (**Eq. (3.38)**) and **Example 4.12** if we replace $Q\Delta l'$ with the dipole moment p, with one exception: The charge Q is now a phasor, as shown in **Eq. (18.4)**, and, therefore, the electric field intensity is also a phasor. Thus, we obtain a static-like solution or what we call a ***quasi-static solution.*** In other words, the electric field intensity in the near field of a Hertzian dipole behaves like the electrostatic dipole. For this reason, the near field of the dipole is also called the ***electrostatic field*** (even though it is not static). The same can be said about the magnetic field intensity in **Eq. (18.25)** because the same solution would be obtained for a steady current, but the solution here is for a time-dependent current.

No propagation effects can be seen in either **Eq. (18.25)** or **(18.27)**. The (approximate) solution in the near field shows no wave behavior because the wave effects are small compared to the electrostatic field and these effects were neglected by the approximations used to reach the results in **Eqs. (18.25)** and **(18.27)**.

Thus, in the near field, the dipole does not radiate, as can also be seen from the Poynting vector. Because the magnetic field intensity in **Eq. (18.25)** and the electric field intensity in **Eq. (18.27)** are out of phase, their vector product is imaginary (see **Exercise 18.1**). Therefore, in the near field, there is only storage of energy. Also, since the electric field intensity is much larger than the magnetic field intensity, the stored electric energy is higher than the stored magnetic energy and dominates in the near-field. This means the dipole in the near field is essentially capacitive in nature.

18.4.2 The Far Field waves.m

In the far field (also called the ***radiation field***), R is large compared to the wavelength:

$$\frac{\lambda}{2\pi R} \ll 1 \quad \text{or} \quad R \gg \frac{\lambda}{2\pi} \tag{18.29}$$

To be in the far field, we must be several wavelengths from the antenna (sometimes stated as $\beta R \gg 1$). The terms containing $1/R^2$ and $1/R^3$ in **Eqs. (18.20)** and **(18.21)** can be neglected since they are much smaller than the term containing $1/R$ and we can write the magnetic and electric field intensities in the far field directly from **Eqs. (18.20)** and **(18.21)**:

$$\boxed{\mathbf{H} = \hat{\boldsymbol{\phi}}\,\frac{j\beta I_0 \Delta l'}{4\pi R}e^{-j\beta R}\sin\theta \quad \left[\frac{\mathrm{A}}{\mathrm{m}}\right]} \tag{18.30}$$

$$\boxed{\mathbf{E} = \hat{\boldsymbol{\theta}}\,\frac{j\beta\eta I_0 \Delta l'}{4\pi R}e^{-j\beta R}\sin\theta \quad \left[\frac{\mathrm{V}}{\mathrm{m}}\right]} \tag{18.31}$$

Note: The R component of the electric field intensity vanishes in the far field since it varies at least as $1/R^2$, whereas the θ component of the field varies as $1/R$ and is, therefore, dominant. From these results, we note the following:

(1) The electric and magnetic field intensities in the far field are perpendicular to each other.
(2) The direction of propagation of the wave is in the R direction, as can be seen from the Poynting vector: $\mathcal{P} = \mathbf{E} \times \mathbf{H} = \hat{\boldsymbol{\theta}} \times \hat{\boldsymbol{\phi}} EH = \hat{\mathbf{R}} EH$.
(3) The electric and magnetic field intensities are in phase.
(4) The ratio between the amplitudes of the electric field intensity and magnetic field intensity is equal to η.
(5) The fields in the far-field domain go down as $1/R$, a rate which is much lower than the $1/R^3$ for **E** and $1/R^2$ for **H** in the near field.
(6) The phase changes by 2π radians per wavelength.

In the far field, the wave produced by the Hertzian dipole is, in many ways, similar to a plane wave, although it is not a plane wave. In fact, if we recall that for a plane wave, the magnitude of the field as well as the phase must be constant on a plane, the above fields certainly do not satisfy the first condition (the field intensities are a function of θ). However, the wave approximates a plane wave because at large distances, the spherical surface of radius R approximates a plane.

Example 18.2 The Hertzian Dipole waves.m

A Hertzian dipole operates at 10 GHz in free space. The dipole is 2 mm long and carries a current of 1 A:

(a) What are the approximate near and far-field domains?
(b) Find the magnetic and electric field intensities at $R = 0.1$ mm from the dipole on the center plane of the dipole ($\theta = 90°$).
(c) Find the magnetic and electric field intensities at $R = 100$ m from the dipole on the center plane of the dipole ($\theta = 90°$). Use the general expressions to show which terms may be neglected.

Solution: The near- and far-field zones are calculated from **Eqs. (18.23)** and **(18.29)**, respectively. The magnetic and electric fields for all space are given in **Eqs. (18.20)** and **(18.21)**:

(a) In the near field:

$$\frac{\lambda}{2\pi R} \gg 1 \quad \rightarrow \quad R \ll \frac{\lambda}{2\pi} = \frac{c}{2\pi f} = \frac{3 \times 10^8}{2\pi \times 10 \times 10^9} = 0.00477 \quad [\mathrm{m}]$$

Thus, the near field is defined by $R \ll 0.00477$ m. Taking one-tenth of this, we find the near field to extend up to about 0.5 mm.

In the far field:

$$\frac{\lambda}{2\pi R} \ll 1 \quad \rightarrow \quad R \gg 0.00477 \quad [\mathrm{m}]$$

Taking ten times this distance, we get the far field to be beyond about 0.05 m.

(b) For $R = 0.1$ mm, we are in the near field and could use **Eqs. (18.25)** and **(18.27)**. However, we use the general expressions in **Eqs. (18.20)** and **(18.21)** to demonstrate their use. First, we calculate the phase constant β:

$$\beta = \frac{2\pi f}{v_p} = \frac{2\pi f}{c} = \frac{2\pi \times 10 \times 10^9}{3 \times 10^8} = 209.44 \quad [\mathrm{rad/m}]$$

With this and with $\theta = 90°$, the magnetic field intensity is

$$\mathbf{H} = -\hat{\boldsymbol{\phi}}\frac{I_0\Delta l'\beta^2}{4\pi}e^{-j\beta R}\sin\theta\left(\frac{1}{j\beta R}+\frac{1}{(j\beta R)^2}\right)$$

$$= -\hat{\boldsymbol{\phi}}\frac{1\times 0.002\times(209.44)^2}{4\times\pi}e^{-j209.44\times 0.0001}\left(\frac{1}{j209.44\times 0.0001}+\frac{1}{(j209.44\times 0.0001)^2}\right)$$

$$= \hat{\boldsymbol{\phi}}6.98e^{-j0.0209}(j47.746+2279.7)\quad [\text{A/m}]$$

The electric field intensity at $R = 0.1$ mm and $\theta = 90°$ is

$$\mathbf{E} = \frac{\eta I_0\beta^2\Delta l'}{4\pi}e^{-j\beta R}\left[-\hat{\mathbf{R}}2\cos\theta\left(\frac{1}{(j\beta R)^2}+\frac{1}{(j\beta R)^3}\right)-\hat{\boldsymbol{\theta}}\sin\theta\left(\frac{1}{j\beta R}+\frac{1}{(j\beta R)^2}+\frac{1}{(j\beta R)^3}\right)\right]$$

$$= \frac{1\times 377\times 0.002\times(209.44)^2}{4\times\pi}e^{-j0.0209}\left[-\hat{\boldsymbol{\theta}}\left(\frac{1}{j209.44\times 0.0001}+\frac{1}{(j209.44\times 0.0001)^2}+\frac{1}{(j209.44\times 0.0001)^3}\right)\right]$$

$$= \hat{\boldsymbol{\theta}}2631.97e^{-j0.0209}\left(j47.746+2279.7-j1.088\times 10^5\right)\quad [\text{V/m}]$$

In the magnetic field intensity, the first term in the brackets (far-field term) is negligible, whereas in the electric field intensity, the first and the second terms are negligible. Removing these terms, we obtain the magnetic and electric fields in the near-field zone as

$$\mathbf{E} \approx -\hat{\boldsymbol{\theta}}j2.86\times 10^8 e^{-j0.0209}\quad [\text{V/m}]\quad \text{and}\quad \mathbf{H} \approx \hat{\boldsymbol{\phi}}1.59\times 10^4 e^{-j0.0209}\quad [\text{A/m}]$$

The R component of the electric field intensity is zero because of the $\cos\theta$ term. Also, the electric field intensity and the magnetic field intensity are out of phase, therefore producing no propagating waves. The electric field intensity is over four orders of magnitude larger than the magnetic field intensity. Note also that in the near field, the ratio $|\mathbf{E}|/|\mathbf{H}|$ does not equal η.

(c) For $R = 100$ m, $\theta = 90°$, the R component of the electric field intensity is zero because of the $\cos\theta$ term. The electric and magnetic field intensities are

$$\mathbf{H} = -\hat{\boldsymbol{\phi}}\frac{I_0\Delta l'\beta^2}{4\pi}e^{-j\beta R}\sin\theta\left(\frac{1}{j\beta R}+\frac{1}{(j\beta R)^2}\right) = \hat{\boldsymbol{\phi}}6.98e^{-j2.09\times 10^4}\left(j4.77\times 10^{-5}+2.28\times 10^{-9}\right)\quad [\text{A/m}]$$

$$\mathbf{E} = -\hat{\boldsymbol{\theta}}\frac{\eta I_0\beta^2\Delta l'}{4\pi}e^{j\beta R}\sin\theta\left(\frac{1}{j\beta R}+\frac{1}{(j\beta R)^2}+\frac{1}{(j\beta R)^3}\right)$$

$$= \hat{\boldsymbol{\theta}}2{,}631.97e^{j2.09\times 10^4}\left(j4.77\times 10^{-5}+2.28\times 10^{-9}-j1.09\times 10^{-13}\right)\quad [\text{V/m}]$$

Here, the first term in each field (the term containing R) is dominant. Neglecting the second term in the magnetic field intensity and the second and third terms in the electric field intensity, we get

$$\mathbf{E} = \hat{\boldsymbol{\theta}}j0.1255e^{-j2.09\times 10^4}\quad [\text{V/m}]\quad \text{and}\quad \mathbf{H} = \hat{\boldsymbol{\phi}}j3.33\times 10^{-4}e^{-j2.09\times 10^4}\quad [\text{A/m}]$$

The electric and magnetic field intensities in the far field are in phase, resulting in a real Poynting vector, and the ratio between the two equals η (377 Ω in this case). This is the intrinsic impedance of free space. Thus, in the far field, the waves behave similar to plane waves.

Example 18.3 A short dipole antenna, 10 mm long carries a current $I = 0.1$ A and oscillates at 10 GHz:

(a) Write the time-dependent electric and magnetic field intensities of the dipole at any point in space.
(b) Plot the electric field intensity in the vertical plane (a plane that contains the dipole) for $t = 0$.

Solution:

(a) Adding the term $e^{j\omega t}$ to **Eqs. (18.20)** and **(18.21)** and taking the real part gives the time domain fields:

$$\mathbf{H}(R,t) = -\hat{\boldsymbol{\phi}}\mathrm{Re}\left\{\frac{I_0\Delta l'\beta^2}{4\pi}e^{j(\omega t-\beta R)}\sin\theta\left(\frac{1}{j\beta R}+\frac{1}{(j\beta R)^2}\right)\right\} = -\hat{\boldsymbol{\phi}}\mathrm{Re}\left\{\frac{I_0\Delta l'\beta^2}{4\pi}e^{j(\omega t-\beta R)}\sin\theta\left(-\frac{j}{\beta R}-\frac{1}{(\beta R)^2}\right)\right\}$$

$$= \hat{\boldsymbol{\phi}}\mathrm{Re}\left\{\frac{I_0\Delta l'\beta^2}{4\pi}e^{j(\omega t-\beta R)}\sin\theta\left(\frac{e^{j\pi/2}}{\beta R}+\frac{1}{(\beta R)^2}\right)\right\} = \hat{\boldsymbol{\phi}}\frac{I_0\Delta l'\beta^2}{4\pi}\sin\theta\left(-\frac{\sin(\omega t-\beta R)}{\beta R}+\frac{\cos(\omega t-\beta R)}{(\beta R)^2}\right)\quad\left[\frac{\mathrm{A}}{\mathrm{m}}\right]$$

The electric field intensity may now be written using identical steps: first, we add the $e^{j\omega t}$ term, replace j by $e^{j\pi/2}$, and then take the real part of the expression:

$$\mathbf{E} = \hat{\mathbf{R}}\frac{\eta I_0\Delta l'}{2\pi}\cos\theta\left(\frac{\cos(\omega t-\beta R)}{R^2}+\frac{\sin(\omega t-\beta R)}{\beta}\right)$$
$$+\hat{\boldsymbol{\theta}}\frac{\eta I_0\Delta l'}{4\pi}\sin\theta\left(\frac{\cos(\omega t-\beta R)}{R^2}+\frac{\sin(\omega t-\beta R)}{\beta R^3}-\frac{\beta\sin(\omega t-\beta R)}{R}\right)\quad\left[\frac{\mathrm{V}}{\mathrm{m}}\right].$$

(b) The electric field intensity is drawn in **Figure 18.4a** for $t = 0$. **Figure 18.4b** shows the near field of the dipole. A continuous time representation would show that starting with $t = 0$, the electric field gradually builds up while the pattern propagates away from the dipole. After a quarter period, the electric field intensity in the vicinity of the dipole reaches maximum whereas in the far field, the field lines close on themselves and propagate, expanding radially (run script waves.m by selecting the Dipole option followed by FDTD to see this behavior).

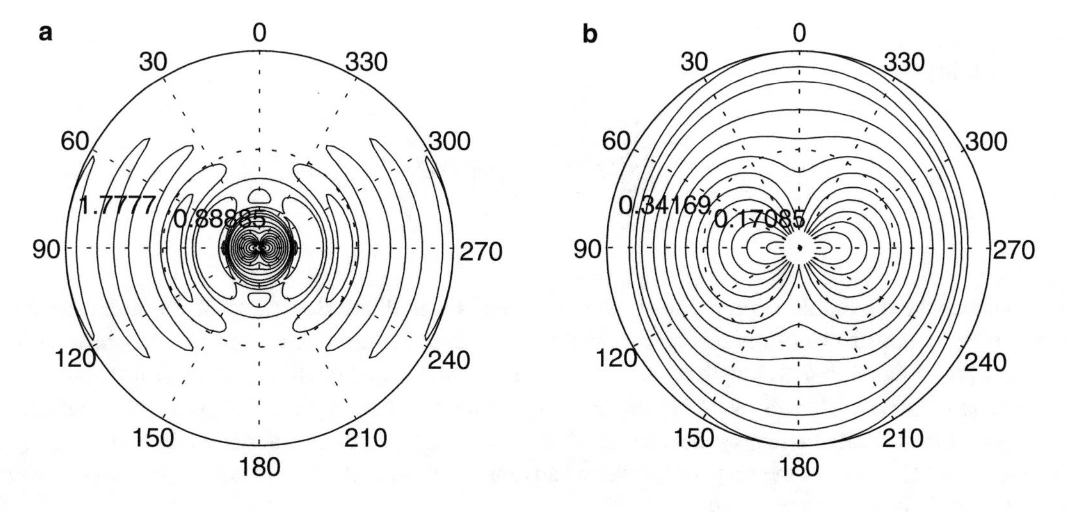

Figure 18.4 (**a**) Plot of the radiation (electric) field of a dipole. (**b**) The near field

Exercise 18.1 Show that:

(a) The time-averaged Poynting vector in the near field of a Hertzian dipole is imaginary.

(b) The time-averaged Poynting vector in the far field of a Hertzian dipole is real and directed in the positive R direction.

18.5 Properties of Antennas

The various properties of antennas are defined next. These include important antenna operation parameters such as the power radiated by the antenna, its efficiency, as well as the concepts of directivity, radiation resistance, radiation patterns, radiation intensity, and gain. To simplify discussion, we use the Hertzian dipole as an example throughout this section. However, the definitions themselves are general and will be used throughout this chapter for other antennas.

18.5.1 Radiated Power

Since we know the electric and magnetic field intensities in the far field, we can calculate the Poynting vector to obtain both the direction of propagation of the wave and the power density in the wave. The time-averaged power density in the far field is obtained from **Eqs. (18.30)** and **(18.31)**:

$$\boxed{\mathcal{P}_{av} = \frac{1}{2}\text{Re}\{\mathbf{E} \times \mathbf{H}^*\} = \hat{\mathbf{R}}\,\frac{\eta I_0^2 \beta^2 (\Delta l')^2}{32\pi^2 R^2}\sin^2\theta \quad \left[\frac{\text{W}}{\text{m}^2}\right]} \tag{18.32}$$

To find the total radiated power, we surround the dipole by a sphere of radius R and calculate the total power traversing the surface of the sphere. Because there are no losses, any sphere will do, as long as it is taken in the far field. The element of area on the sphere is $d\mathbf{s} = \hat{\mathbf{R}}R^2 \sin\theta d\theta d\phi$, and the total radiated power is found by integrating the time-averaged power density in **Eq. (18.32)** over ϕ between 0 and 2π and over θ between 0 and π:

$$P_{rad} = \int_s \mathcal{P}_{av} \cdot d\mathbf{s} = \int_{\phi=0}^{\phi=2\pi} \int_{\theta=0}^{\theta=\pi} \frac{\eta I_0^2 \beta^2 (\Delta l')^2}{32\pi^2 R^2} \sin^2\theta R^2 \sin\theta d\theta d\phi = \frac{\eta I_0^2 \beta^2 (\Delta l')^2}{32\pi^2}\frac{8\pi}{3} = \frac{\eta I_0^2 \beta^2 (\Delta l')^2}{12\pi} \quad [\text{W}] \tag{18.33}$$

With $\beta = 2\pi/\lambda$ [rad/m], we get

$$\boxed{P_{rad} = \frac{\eta I_0^2 \beta^2 (\Delta l')^2}{12\pi} = I_0^2 \frac{\eta\pi}{3}\left(\frac{\Delta l'}{\lambda}\right)^2 \quad [\text{W}]} \tag{18.34}$$

The power radiated by the dipole is proportional to the current squared and the length of the antenna squared. It also depends on the intrinsic impedance of the medium in which the antenna radiates and is directly proportional to frequency squared (inversely proportional to wavelength squared). Thus, a very short dipole will radiate very little power and a longer dipole will radiate more power. We will see in the context of real antennas that, in general, the longer the antenna, the larger the radiated power. From this calculation, we can also see that the true Hertzian dipole, although fundamental, is not the most practical antenna to build because of the very low power it can radiate. Nevertheless there are applications in which Hertzian dipoles are used, exactly because they are short.

18.5.2 Radiation Resistance

An interesting aspect of the radiated power obtained in **Eq. (18.34)** is the form of the equation. Since power is, in general, $P = I^2R$, where R is a resistance, the term multiplying the term I_0^2 must have units of resistance. Because we are using phasors, the power radiated by the source is

$$P_{rad} = I_{rms}^2 R_{rad} \quad [\mathrm{W}] \tag{18.35}$$

where R_{rad} is called the ***radiation resistance*** of the antenna. The root mean squared (rms) value of current is $I_{rms} = I_0/\sqrt{2}$. With this, the radiated power of an antenna can be written as

$$P_{rad} = \frac{I_0^2}{2} R_{rad} \quad [\mathrm{W}] \tag{18.36}$$

Comparing this general relation with the radiated power of the Hertzian dipole in **Eq. (18.34)**, we find the radiation resistance of the dipole as

$$\boxed{R_{rad} = \frac{2\eta\pi}{3}\left(\frac{\Delta l'}{\lambda}\right)^2 \quad [\Omega]} \tag{18.37}$$

In particular, in free space, $\eta_0 = 120\pi$, and we get

$$\boxed{R_{rad} = 80\pi^2\left(\frac{\Delta l'}{\lambda}\right)^2 \quad [\Omega]} \tag{18.38}$$

Radiation resistance is not the ohmic resistance of the antenna but is a characteristic quantity of the dipole described here (and, indeed, of other radiators) and reflects both the antenna structure and dimensions, as well as the environment. It simply indicates the power the dipole can radiate for a given current. Maximization of the radiation resistance means the antenna can radiate more power for any given current.

The power radiated by an antenna of this type is directly proportional to the length of the antenna, $\Delta l'$, and inversely proportional to the wavelength. As a general rule, the longer the antenna, the larger the radiation resistance and the larger the power it can radiate for a given current. Similarly, the higher the frequency, the larger the radiated power (antenna becomes "electrically larger").

Example 18.4 The Short Dipole Antenna A short dipole antenna is 20 mm long and carries a current of 2 A. The dipole radiates at 300 MHz.

(a) Calculate the total power radiated by the dipole in free space.
(b) If it is required to embed the dipole in Teflon ($\varepsilon_r = 2.0$, $\sigma = 0$, $\mu = \mu_0$ [H/m]), what must be the current in the dipole to maintain the same radiated power as in free space?
(c) Calculate the radiation resistance in air and in Teflon and the ratio between the radiation resistance in air and in Teflon.

Solution: First, we must show that this antenna is a Hertzian dipole. Then, the change in radiated power is due to the change in permittivity which affects the intrinsic impedance of the medium. Thus, we calculate the intrinsic impedance in Teflon and substitute in **Eq. (18.34)** to find the current, after calculating the radiated power in free space:

(a) The wavelength in air is

$$\lambda = \frac{c}{f} = \frac{3 \times 10^8}{300 \times 10^6} = 1 \quad [\mathrm{m}]$$

Since the dipole is only 0.02 m long, or 50 times shorter than a wavelength, it is safe to use the Hertzian dipole results for this antenna. The power radiated by the antenna is [from **Eq. (18.34)** with $\eta_0 = 377\ \Omega$]

$$P_{rad} = I_0^2 \frac{\eta_0 \pi}{3} \left(\frac{\Delta l'}{\lambda}\right)^2 = \frac{4 \times 377 \times \pi}{3} \times (0.02)^2 = 0.632 \quad [\mathrm{W}].$$

(b) In Teflon, the intrinsic impedance and wavelength are

$$\eta = \sqrt{\frac{\mu_0}{\varepsilon_0 \varepsilon_r}} = \frac{\eta_0}{\sqrt{\varepsilon_r}} = \frac{377}{\sqrt{2.0}} = 266.6 \quad [\Omega],$$

$$\lambda = \frac{v_p}{f} = \frac{c}{f\sqrt{\varepsilon_r}} = \frac{3 \times 10^8}{300 \times 10^6 \times \sqrt{2.0}} = 0.707 \quad [\mathrm{m}]$$

With these, the radiated power is

$$P_{rad} = I_0^2 \frac{\eta \pi}{3} \left(\frac{\Delta l'}{\lambda}\right)^2 = \frac{4 \times 266.6 \times \pi}{3} \times \left(\frac{0.02}{0.707}\right)^2 = 0.894 \quad [\mathrm{W}]$$

Note that the power transmitted in Teflon is 1.414 times larger because the intrinsic impedance of Teflon is $\sqrt{2} = 1.414$ times lower and the wavelength is 1.414 times shorter (in effect, the antenna has become "longer"). Thus, to maintain the power, we must decrease the current by a factor of $\sqrt{1.414}$. The required current with the antenna embedded in Teflon is $2/\sqrt{1.414} = 1.682$ A.

(c) The general expression of the radiation resistance is **Eq. (18.37)**. The radiation resistance in air is given in **Eq. (18.38)**:

$$R_{air} = 80\pi^2 \left(\frac{\Delta l'}{\lambda_{air}}\right)^2 = 0.316 \quad [\Omega]$$

In Teflon, both the intrinsic impedance and the wavelength are reduced by a factor of $\sqrt{\varepsilon_r} = 1.414$. Thus, the radiation resistance in Teflon is

$$R_{Teflon} = \frac{2\eta_{Teflon}\pi}{3} \left(\frac{\Delta l'}{\lambda_{Teflon}}\right)^2 = \frac{2\left(\eta_0/\sqrt{2.0}\right)\pi}{3} \left(\frac{\Delta l'}{\lambda_{air}/\sqrt{2.25}}\right)^2 = 1.414 R_{air} = 0.447 \quad [\Omega]$$

The ratio between the radiation resistance in air and in Teflon is 1/1.414. The same conclusion may be drawn from (**b**)**:** the increase in radiated power is due to increase in radiation resistance.

18.5.3 Antenna Radiation Patterns

The far-field relations in **Eqs. (18.30)** and **(18.31)** show that the electric and magnetic fields are dependent on the angle θ for any value of R. Thus, we conclude that at least in this case, the electric field intensity, magnetic field intensity, and power densities radiated by the antenna are location dependent. To completely define the characteristics of the antenna, we need to know the field intensity and/or power density at any location in space. Of course, this will vary with distance from the source. The ***antenna radiation pattern*** is defined as the relative strength of the field, at a given distance from the antenna in the far field. The field may be the electric field intensity, magnetic field intensity, or power density, and these may be either absolute values or normalized values. The radiation pattern is then a function of the other coordinates. In most cases, we will use a spherical system of coordinates in which case the radiation pattern for a constant distance R is a function of θ and ϕ. The radiation pattern plot may be given in any convenient system of coordinates. The radiation pattern is an important antenna parameter since it shows the radiation characteristics of the antenna. Furthermore, the radiation patterns are useful in defining other properties of antennas, properties that will be discussed shortly.

18.5.3.1 Planar Antenna Radiation Pattern Plots

The definition of a radiation pattern requires a three-dimensional plot of the field or power density. However, three-dimensional plots are difficult to execute and interpret. It is therefore common to use planar plots. Since a planar plot can be defined on any plane, we will choose those planes that are most useful for our purposes. These are usually the $\phi = 0$ and $\theta = \pi/2$ planes, but others may be useful at times.

Electric Field Antenna Radiation Patterns The electric dipole field in **Eq. (18.31)** is in the θ direction and varies with θ. Thus, if we take a vertical plane, through the dipole, we get an electric field pattern. This plane is called the ***E-plane*** and the pattern is called an ***E-plane pattern.*** The magnitude of the electric field intensity of the dipole is

$$|E| = \left| \frac{j\eta I_0 \beta \Delta l'}{4\pi R} e^{-j\beta R} \sin\theta \right| = \frac{\eta |I_0| \beta \Delta l'}{4\pi R} \left| e^{-j\beta R} \right| |\sin\theta| = \frac{\eta |I_0| \beta \Delta l'}{4\pi R} |\sin\theta| \quad \left[\frac{\mathrm{V}}{\mathrm{m}}\right] \tag{18.39}$$

Thus, at a given value of R, all terms are constant except $\sin\theta$. A plot of **Eq. (18.39)** results in the ***absolute E-field radiation pattern*** for the Hertzian dipole. The ***relative radiation pattern,*** also called a ***normalized pattern,*** is obtained by dividing the field in **Eq. (18.39)** by the amplitude of E. This gives the ***normalized E-plane radiation pattern*** as

$$|f_e(\theta)| = |\sin\theta| \tag{18.40}$$

The normalized and absolute patterns are identical in shape but have different magnitudes. The type of plot shown here is a polar plot, often employed for this purpose.

The plot in **Figure 18.5a** is executed in the z–x (or $\phi = 0$) plane, but this is clearly arbitrary; any other vertical plane that includes the dipole will produce an identical pattern since the pattern only depends on the angle θ. The absolute pattern describes a quantity (in this case, in units of [V/m]) and the normalized pattern describes a shape (no units). If we choose a horizontal plane, an ***H-plane pattern*** is obtained. Any plane may be used, but for standardization, the horizontal plane through the center of the dipole is normally implied. This is obtained by substituting $\theta = \pi/2$ in **Eq. (18.39)**:

$$E = \frac{\eta |I_0| \beta \Delta l'}{4\pi R} \quad \left[\frac{\mathrm{V}}{\mathrm{m}}\right] \tag{18.41}$$

This is a constant value: It does not depend on the angle ϕ which is the angle in the horizontal plane. Thus, this describes a circle of radius E in the horizontal plane, because the electric field intensity of the Hertzian dipole is independent of the angle ϕ. If we normalize the electric field intensity with respect to the maximum value we get the radiation pattern as $f_e(\phi) = 1$. The relative radiation pattern in the H-plane is shown in **Figure 18.5b**, again as a polar plot.

Power Density Radiation Patterns To define the power density pattern, we use the power density as obtained in **Eq. (18.32)**, but first take the magnitude:

$$\mathcal{P}_{av} = \frac{\eta I_0^2 \beta (\Delta l')^2}{32\pi^2 R^2} \sin^2\theta \quad \left[\frac{\mathrm{W}}{\mathrm{m}^2}\right] \tag{18.42}$$

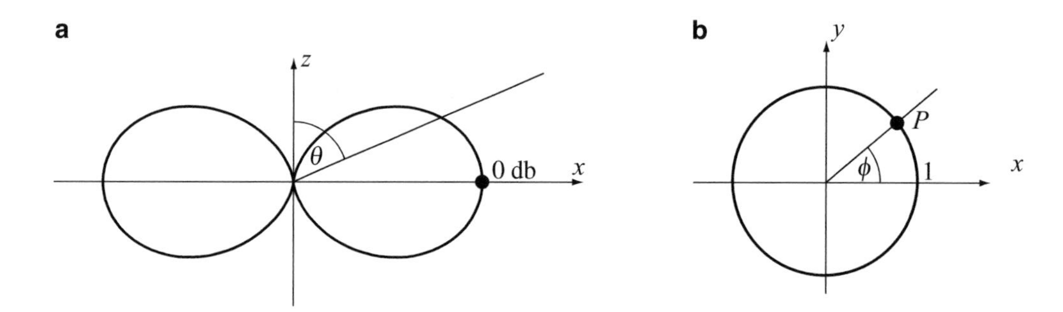

Figure 18.5 (**a**) Normalized electric field radiation pattern in the E-plane. (**b**) Normalized electric field radiation pattern in the H-plane

This may now be plotted in the vertical *(E-plane)* for $\phi = 0$ or in the horizontal plane *(H-plane)* for $\theta = \pi/2$ as was done for the electric field intensity. The normalized power density, or ***power pattern*** f_p, is

$$f_p(\theta) = \sin^2\theta \tag{18.43}$$

Thus, the vertical plane (*E-plane*) plot has the form $\sin^2\theta$, whereas the horizontal plane (*H-plane*) plot is a circle of constant radius [equal to 1 for the normalized plot or to the amplitude in **Eq. (18.42)** for the absolute plot]. The normalized power radiation pattern for the Hertzian dipole is shown in **Figure 18.6a** for the vertical plane. Note, also, that since $\mathcal{P}_{av}$ is proportional to E^2, the normalized power pattern is

$$f_p(\theta) = (f_e(\theta))^2 \tag{18.44}$$

Three-Dimensional Radiation Patterns A three-dimensional radiation pattern may be obtained in two ways: One is to take a large number of vertical plots, each passing through the dipole (much the way you would cut an orange with each cut parallel to the slices). The second method is to perform cuts in the horizontal plane but each for a different value of θ. This can be done for the electric field intensity, magnetic field intensity, or power density. A plot of either method gives the three-dimensional plot in **Figure 18.6b**.

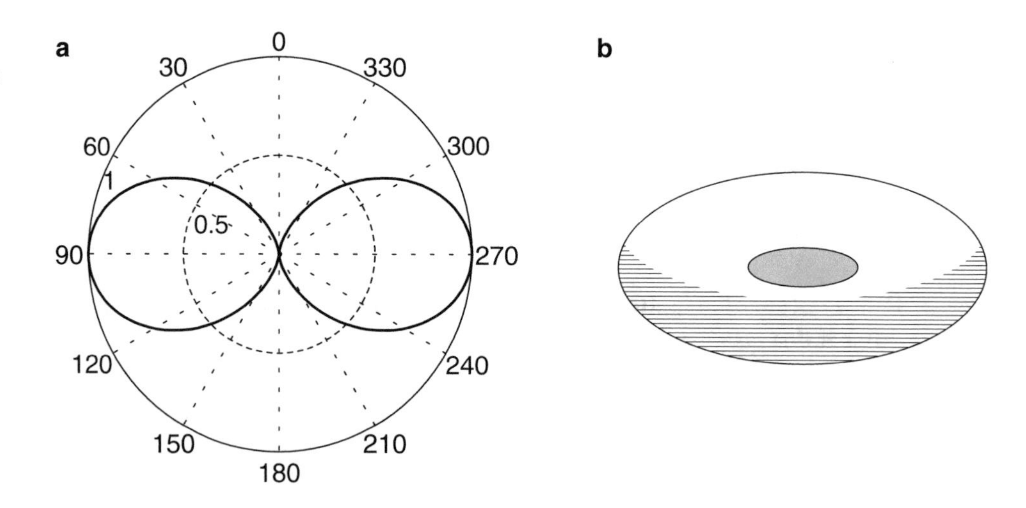

Figure 18.6 (**a**) Normalized power radiation pattern for the Hertzian dipole in the E-plane. (**b**) Three-dimensional normalized power radiation pattern

Notes:

(1) For the Hertzian dipole, the radiation patterns only depend on the angle θ. However, for other antennas, they may also depend on the angle ϕ.

(2) The absolute patterns give the magnitude of the field intensity (or power density) at any location in space, whereas the relative or normalized patterns give the shape of the field intensity (or power density) in space, normalized to 1, and have no units.

(3) The field radiation pattern is the plot of $|f_e(\theta)|$, that is, the absolute value of the normalized field intensity is plotted on a plane or in space. The power radiation pattern is the plot of $f_p(\theta) = (f_e(\theta))^2$.

(4) Magnetic field radiation patterns can be obtained analogous to electric field radiation patterns in both the E-plane and the H-plane. There is however little benefit in doing so since the normalized patterns for the electric and magnetic field intensities are the same whereas the magnitudes are related through the intrinsic impedance (see **Eqs. (18.30)** and **(18.31)**).

18.5.3.2 Rectangular Radiation Pattern Plots

In addition to the plots obtained above, we can also plot the electric field intensity or the power density in a rectangular plot. The plot may be with respect to θ (E-plane) or ϕ (H-plane). This involves simply plotting any of **Eqs. (18.39)** through **(18.43)** as a function of θ or ϕ. For example, allowing θ to vary between $\theta = 0$ and $\theta = \pi$, we obtain the rectangular plot in **Figure 18.7**, showing the E-plane field and power patterns for the Hertzian dipole. It is customary to plot this in terms of the relative values or in dB ($10 \log_{10} |f_p|$ for power plots or $20 \log_{10} |f_e|$ for field plots).

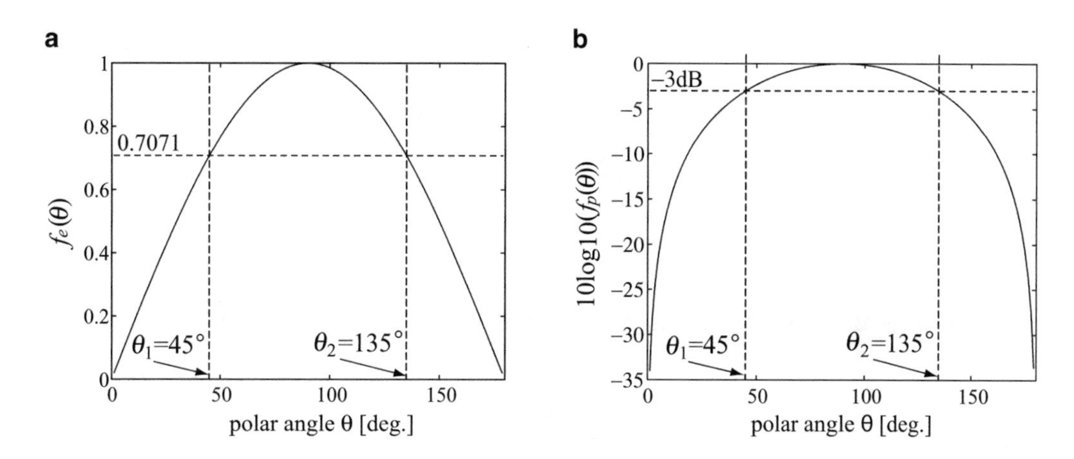

Figure 18.7 Rectangular, normalized radiation patterns for the Hertzian dipole. (**a**) E-plane field pattern. (**b**) E-plane power pattern in dB

Although the plots we obtained here were specifically for the Hertzian dipole, the same method will be applied to any antenna once the far fields are obtained. In most cases, the patterns are more complicated, but the principles and methods remain the same.

18.5.3.3 Beamwidth

In the radiation patterns described above, the plots show the field or power distribution in space on given planes. The beamwidth is defined as the width of the power radiation pattern at the location the beam is 3 db below its maximum value (half-power points). Similarly, in the field radiation pattern, the beamwidth is the width of the pattern at the location where the field is $1/\sqrt{2}$ of its peak. The beamwidth is given as the angle between the two half-power points. The beamwidth is shown in **Figure 18.8** for a polar plot. **Figure 18.7a** shows the beamwidth in rectangular coordinates for the field plot and **Figure 18.7b** for the power plot. The beamwidth is an angle equal to $\Delta\theta = \theta_2 - \theta_1$ (in **Figure 18.7**, the beamwidth is 90°).

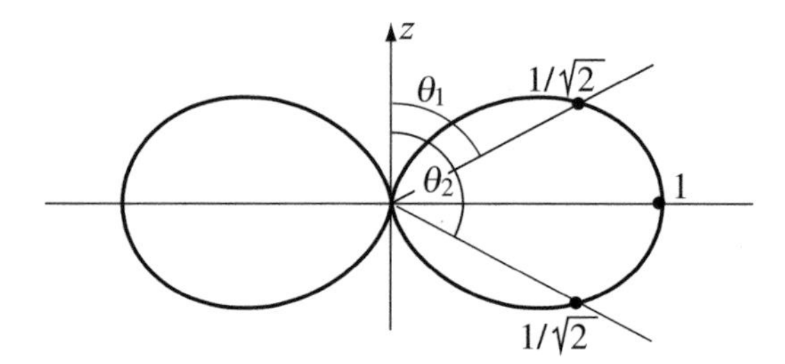

Figure 18.8 Beamwidth shown in an E-plane field plot in polar representation

18.5.4 Radiation Intensity and Average Radiation Intensity

Radiation intensity is defined as the time-averaged power of the antenna per unit solid angle. You will recall that a solid angle is a measure of surface on a sphere in the same way an angle is a measure of arc length. Considering the solid angle in **Figure 18.9**, the element of area is $ds = R^2 \sin\theta d\theta d\phi$. Since $R\sin\theta d\phi$ and $Rd\theta$ are arc lengths (shown in **Figure 18.9**), the area ds can be written as $ds = R^2 d\kappa$, where κ is the solid angle. A unit solid angle represents an area on the surface of the sphere $s = R^2$ [m^2]. The total area of the sphere is $4\pi R^2$. Thus, there are 4π such unit solid angles in a sphere. The unit of solid angle is the radian squared, also called a ***steradian***, and is denoted [sr]. There are 4π steradians in a sphere.

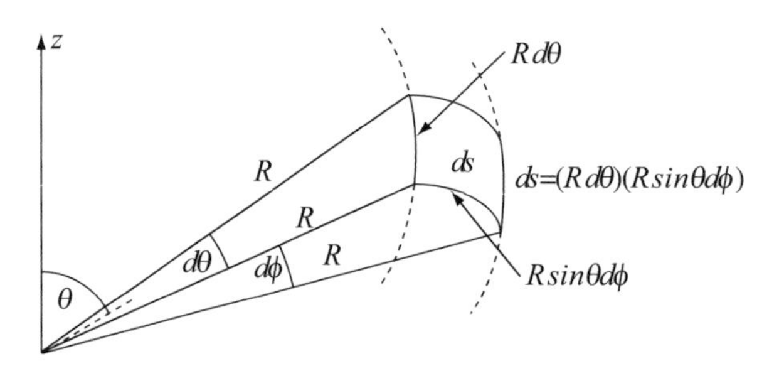

Figure 18.9 The solid angle and the area it represents on the sphere of radius R

With these considerations, the radiated power in **Eq. (18.33)** can be written as

$$P_{rad} = \int_s \mathcal{P}_{av} R^2 \sin\theta d\theta d\phi = \int_s \mathcal{P}_{av} R^2 d\kappa \quad [\mathrm{W}] \tag{18.45}$$

where the integration is over the surface s of the sphere. Since the total radiated power is integrated over 4π unit solid angles, the radiation intensity may be written as

$$\boxed{U(\theta, \phi) = \mathcal{P}_{av} R^2 \quad [\mathrm{W/sr}]} \tag{18.46}$$

The dependency on θ and ϕ was added to indicate that, in general, the radiation intensity depends on both since the magnitude of the time-averaged Poynting vector, $\mathcal{P}_{av}$, depends on both angles.

In the case of the Hertzian dipole, the radiation intensity is [using **Eq. (18.32)** for $\mathcal{P}_{av}$]

$$U(\theta) = \frac{\eta I_0^2 \beta^2 (\Delta l')^2}{32\pi^2} \sin^2\theta = \frac{\eta I_0^2}{8} \left(\frac{\Delta l'}{\lambda}\right)^2 \sin^2\theta \quad \left[\frac{\mathrm{W}}{\mathrm{sr}}\right] \tag{18.47}$$

Note that this quantity is independent of the distance from the dipole and, in this case, only depends on θ.

The radiation intensity is not uniform over a sphere of any radius. We can, however, define an ***average radiation intensity*** by spreading the radiation intensity over the area of the sphere of radius R. This gives

$$U_{av} = \frac{\int_s U(\theta,\phi)d\kappa}{4\pi} = \frac{\int_s \mathcal{P}_{av} R^2 d\kappa}{4\pi} = \frac{P_{rad}}{4\pi} \quad \left[\frac{\text{W}}{\text{sr}}\right] \tag{18.48}$$

The average radiation intensity may be viewed as the radiation intensity of an ***isotropic source*** (an isotropic source is a source that radiates uniformly in all directions) with total radiated power equal to that of the antenna in question. It is, therefore, a convenient quantity to use in computation of antenna parameters and, in particular, when trying to compare antennas with different radiation patterns. For the Hertzian dipole, the radiated power is given in **Eq. (18.34)**. Thus, the averaged radiation intensity of the Hertzian dipole is

$$U_{av} = I_0^2 \frac{\eta}{12}\left(\frac{\Delta l'}{\lambda}\right)^2 \quad \left[\frac{\text{W}}{\text{sr}}\right] \tag{18.49}$$

18.5.5 Antenna Directivity

As we have seen from the radiation patterns in the previous sections, the power density in different directions in space is different. This means that the antenna will necessarily radiate more power in certain directions. As an example, the Hertzian dipole radiates maximum power density in directions perpendicular to the dipole ($\sin^2\theta = 1$), whereas in the direction of the dipole (z axis), the radiated power density is zero. In other words, we may say that the antenna has directive properties. These properties are defined through the directivity and maximum directivity of the antenna.

The ***directivity*** of the antenna is defined as the ratio between the radiation intensity in a given direction (θ,ϕ) and the averaged radiation intensity:

$$D(\theta,\phi) = \frac{U(\theta,\phi)}{U_{av}} = \frac{U(\theta,\phi)}{P_{rad}/4\pi} = \frac{4\pi U(\theta,\phi)}{P_{rad}} \quad [\text{dimensionless}] \tag{18.50}$$

Directivity is a dimensionless quantity and is space dependent as shown by the explicit use of θ and ϕ. Note also that if the antenna radiates uniformly in all directions in space, the average radiation intensity and the radiation intensity are the same and the directivity equals 1. This, of course, indicates the nondirectional properties of the isotropic antenna. For all other antennas, the directivity varies from point to point as expected. When no direction is specified, the direction of maximum directivity is implied.

Maximum directivity of antennas is defined as the ratio of the maximum radiation intensity to the average radiation intensity:

$$D_0 = \frac{U_{max}}{U_{av}} = \frac{U_{max}}{P_{rad}/4\pi} = \frac{4\pi U_{max}}{P_{rad}} \quad [\text{dimensionless}] \tag{18.51}$$

The directivity of the Hertzian dipole is

$$D(\theta,\phi) = \frac{4\pi \dfrac{\eta I_0^2(\Delta l'^2)}{8\lambda^2}\sin^2\theta}{I_0^2 \dfrac{\eta\pi(\Delta l')^2}{3\lambda^2}} = \frac{3}{2}\sin^2\theta \quad [\text{dimensionless}] \tag{18.52}$$

where we used **Eqs. (18.50), (18.47),** and **(18.34)**. The directivity varies from zero (for $\theta = 0$ or $\theta = \pi$) to 1.5 at $\theta = \pi/2$.

The directivity and maximum directivity (denoted as D_0) of the Hertzian dipole are

$$D(\theta,\phi) = \frac{3}{2}\sin^2\theta, \quad D_0 = 1.5 \qquad [\text{dimensionless}] \tag{18.53}$$

18.5.6 Antenna Gain and Radiation Efficiency

The directivity as given in **Eq. (18.50)** defines a ratio between the radiation intensity and averaged radiation intensity, disregarding any losses that may have occurred in the antenna itself. Instead of using the averaged radiation intensity in terms of the radiated power, we can use it in terms of the input power to the antenna. By doing so, the losses in the antenna are included in the calculation. Thus, we define the ***antenna gain*** (also called ***power gain***) as the ratio between the radiation intensity and the average radiation intensity of a perfect isotropic radiator with the same input power:

$$G(\theta,\phi) = \frac{U(\theta,\phi)}{P_{in}/4\pi} = \frac{4\pi U(\theta,\phi)}{P_{in}} \qquad [\text{dimensionless}] \tag{18.54}$$

It is often the practice to use a logarithmic scale for gain. Antenna gain is direction dependent in the same way as directivity. When no direction is given, maximum gain is implied, denoted as G_0.

Note: If there are no losses, $P_{rad} = P_{in}$, antenna gain equals antenna directivity [$G(\theta, \phi) = D(\theta, \phi)$] and maximum antenna gain equals maximum antenna directivity ($G_0 = D_0$).

The ***radiation efficiency*** of the antenna is defined as the ratio between maximum gain and maximum directivity which also equals the ratio between radiated and input power:

$$eff = \frac{G_0}{D_0} = \frac{P_{rad}}{P_{in}} \tag{18.55}$$

Using the idea of radiation resistance, we can write

$$P_{rad} = I^2 R_{rad}/2, \quad P_{in} = I^2(R_{rad} + R_d)/2 \tag{18.56}$$

where I is the peak current in the antenna and R_d is the ohmic resistance of the antenna itself. Thus, we can write

$$eff = \frac{R_{rad}}{R_{rad} + R_d} \tag{18.57}$$

In many cases, the internal resistance of the antenna is relatively small compared to the radiation resistance, and efficiency is close to 100%. In practice, antennas are rather efficient devices. Most of the losses occur due to mismatch between the antenna and the antenna feeding line rather than in the antenna.

Example 18.5 A dipole is 0.1 m long and has an internal resistance of 0.1 Ω. The peak current in the dipole is 1 A. the dipole radiates in air at a wavelength of 5 m. Calculate:

(a) The radiation resistance of the dipole.
(b) The radiated power from the dipole.
(c) The antenna efficiency.
(d) Maximum gain of the antenna.

Solution: The radiation resistance is calculated first, using **Eq. (18.38)**. From this and the current in the dipole, we calculate the radiated power using **Eq. (18.36)**. Antenna efficiency is calculated using the internal and radiation resistances, using **Eq. (18.57)**. The gain of the antenna is calculated using **Eq. (18.54)** after the radiation intensity in **Eq. (18.47)** is calculated:

(a) The radiation resistance of the dipole in air is

$$R_{rad} = 80\pi^2\left(\frac{\Delta l'}{\lambda}\right)^2 = 80\times\pi^2\times\left(\frac{0.1}{5}\right)^2 = 0.3158 \quad [\Omega].$$

(b) The radiated power is

$$P_{rad} = \frac{I^2R_{rad}}{2} = \frac{1\times 0.31588}{2} = 0.158 \quad [\mathrm{W}].$$

(c) The antenna efficiency is

$$eff = \frac{R_{rad}}{R_{rad}+R_d}\times 100 = \frac{0.3158}{0.3158+0.1}\times 100 = 75.94\%.$$

(d) The gain of the antenna is given in **Eq. (18.54)**:

$$G(\theta,\phi) = \frac{4\pi U_{max}(\theta,\phi)}{P_{in}}$$

The radiation intensity for a dipole only depends on θ and is given in **Eq. (18.47)**. The input power P_i equals $P_{in} = P_{rad}/eff = 0.158/0.7594 = 0.208$ W. Thus,

$$U(\theta) = \frac{\eta I_0^2}{8}\left(\frac{\Delta l'}{\lambda}\right)^2\sin^2\theta = \frac{377\times 1^2}{8}\times\left(\frac{0.1}{5}\right)^2\sin^2\theta = 0.01885\sin^2\theta \quad [\mathrm{W/sr}]$$

The maximum gain of the dipole is obtained by setting $\theta = \pi/2$ for maximum radiation intensity:

$$G_0 = \frac{4\pi(U(\theta))_{max}}{P_{in}} = \frac{4\pi\times 0.01885}{0.208} = 1.139$$

On the logarithmic scale the maximum gain is $G_0 = 10\log_{10}(1.139) = 0.565$ dB (above isotropic radiation).

18.6 The Magnetic Dipole

In **Section 9.2.1**, we introduced the ***magnetic dipole*** which, for practical purposes, is merely a small loop carrying a current I. For a steady current, we found that the small loop has a magnetic moment

$$\mathbf{m} = \hat{\mathbf{z}}I\pi a^2 \quad [\mathrm{A\cdot m^2}] \tag{18.58}$$

where a is the loop's radius and the loop is placed at the origin, in the x–y (or r–ϕ) plane. We called this configuration a magnetic dipole because the expression we obtained in **Eq. (9.14)** for the magnetic flux density of the small loop had the same form as the expression for the electric field intensity of the electric dipole.

The magnetic dipole we discuss here is identical to that discussed in magnetostatics with the exception that the current is now a time-dependent current. Again, as with the electric dipole, we will assume the current to be cosinusoidal:

$$I(t) = I_0\cos(\omega t) \quad [\mathrm{A}] \tag{18.59}$$

or in time-harmonic form

$$I = I_0e^{j\omega t} \quad [\mathrm{A}] \tag{18.60}$$

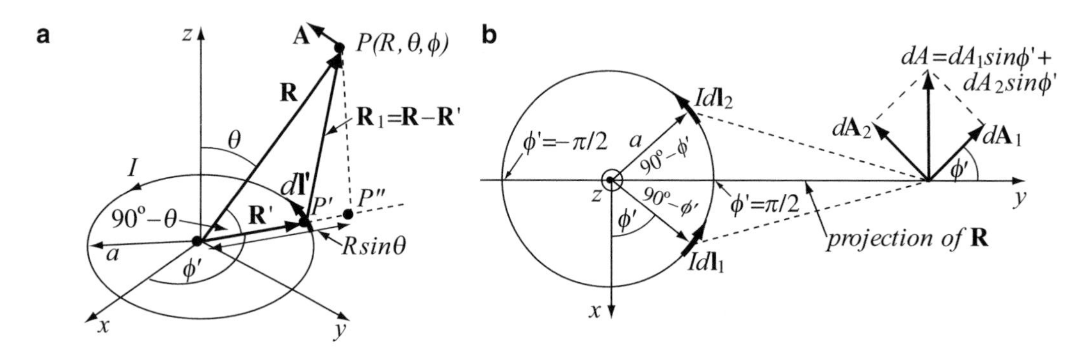

Figure 18.10 (**a**) The magnetic dipole and notation used for calculation. (**b**) Configuration used for integration around the loop

The basic configuration is shown in **Figure 18.10a**. Repeating the basic steps as for the electric dipole, we wish to calculate the electric and magnetic field intensities at a general point in space, P. If we take a segment of current of length dl', we can use the magnetic vector potential due to this segment as in **Eq. (18.12)** by replacing $\Delta l'$ with dl' and $\mathbf{A}$ by $d\mathbf{A}$. Note that now the length of the segment is infinitesimal and in the ϕ direction, whereas in **Eq. (18.12)**, the length $\Delta l'$ was small but not necessarily infinitesimal and in the z direction. We also note that the distance between dl' and P is $\mathbf{R}_1 = \mathbf{R} - \mathbf{R}'$ and $d\mathbf{l}' = \hat{\boldsymbol{\phi}} dl'$. Thus, the magnetic vector potential due to the differential segment dl' is

$$d\mathbf{A} = \hat{\boldsymbol{\phi}} \frac{\mu I_0 dl'}{4\pi R_1} e^{-j\beta R_1} \quad \left[\frac{\text{Wb}}{\text{m}}\right] \tag{18.61}$$

where I_0 is the phasor current. To find the total magnetic vector potential, we integrate this over the circumference of the loop:

$$\mathbf{A} = \hat{\boldsymbol{\phi}} \frac{\mu I_0}{4\pi} \oint_C \frac{dl'}{R_1} e^{-j\beta R_1} \quad \left[\frac{\text{Wb}}{\text{m}}\right] \tag{18.62}$$

where the closed contour is along the loop. The properties of the small loop are now used to simplify this integral so that an analytic integration can be performed. First, we note that R' is small compared to both R and R_1. This is a consequence of the fact that the loop is small but also because we will normally assume R to be large. We will not dwell much on the near-field properties, but, from the development of the previous section, it is safe to assume that in the near field, the results are the same as for a static solution. To simplify the expressions, we now rewrite the magnetic vector potential in **Eq. (18.62)** in terms of R (the distance from the center of the loop to point P) rather than R_1. To do so, we add and subtract R to the exponent in **Eq. (18.62)** as follows:

$$\mathbf{A} = \hat{\boldsymbol{\phi}} \frac{\mu I_0}{4\pi} \oint_C \frac{dl'}{R_1} e^{-j\beta(R_1+R-R)} = \hat{\boldsymbol{\phi}} \frac{\mu I_0}{4\pi} \oint_C \frac{dl'}{R_1} e^{-j\beta R} e^{-j\beta(R_1-R)} \quad \left[\frac{\text{Wb}}{\text{m}}\right] \tag{18.63}$$

The term $e^{-j\beta(R_1-R)}$ may be expanded in a Taylor series as

$$e^{-j\beta(R_1-R)} = 1 - j\beta(R_1 - R) + \frac{[\beta(R_1 - R)]^2}{2} - \cdots \approx 1 - j\beta(R_1 - R) \tag{18.64}$$

where the first two terms in the series were retained based on the assumption that $R - R_1$ is small. Substituting this back into **Eq. (18.63)** gives

$$\mathbf{A} = \hat{\boldsymbol{\phi}} \frac{\mu I_0}{4\pi} \oint_C \frac{dl'}{R_1} e^{-j\beta R}(1 + j\beta R - j\beta R_1) \quad \left[\frac{\text{Wb}}{\text{m}}\right] \tag{18.65}$$

To perform the integration, we use **Figure 18.10b**. The magnetic vector potential at point P is tangential but it is the result of two components generated by segments $d\mathbf{l}_1'$ and $d\mathbf{l}_2'$. The integration can be performed between $\phi' = -\pi/2$ and

$\phi' = +\pi/2$ and the result doubled. You may recall that the same procedure was used to evaluate the magnetostatic dipole in **Section 9.2.1**. Now, since the integration is on the circumference of the loop, we write $dl' = R'd\phi = ad\phi'$. Also, a is constant and can be taken outside the integral sign. Therefore

$$\mathbf{A} = \hat{\boldsymbol{\phi}}\frac{\mu I_0 a e^{-j\beta R}}{4\pi}2\int_{\phi'=-\pi/2}^{\phi'=+\pi/2}\frac{1+j\beta R}{R_1}\sin\phi' d\phi' - \hat{\boldsymbol{\phi}}j\frac{\mu I_0 a e^{-j\beta R}}{4\pi}2\int_{\phi'=-\pi/2}^{\phi'=+\pi/2}\beta\sin\phi' d\phi' \quad \left[\frac{\text{Wb}}{\text{m}}\right] \tag{18.66}$$

Direct evaluation of the second integral shows that it is zero. To integrate the first integral, we must write R_1 in terms of the angle ϕ' since it depends on ϕ'. To do so, we use the following relation:

$$\frac{1}{R_1} = \frac{1}{|\mathbf{R}-\mathbf{R}'|} = \frac{1}{\left(\text{R}^2 + a^2 - 2Ra\sin\theta\sin\phi'\right)^{1/2}} \tag{18.67}$$

where $|\mathbf{R}-\mathbf{R}'| = [(\mathbf{R}-\mathbf{R}')\cdot(\mathbf{R}-\mathbf{R}')]^{1/2}$, $\mathbf{R}\cdot\mathbf{R} = R^2$, $\mathbf{R}'\cdot\mathbf{R}' = a^2$ and $\mathbf{R}\cdot\mathbf{R}' = Ra\sin\theta\sin\phi'$ as can be verified from **Figure 18.10a**. Following the steps in **Eqs. (9.6)** and **(9.7)**, we can write

$$\frac{1}{R_1} \approx \frac{1}{R}\left(1 + \frac{a}{R}\sin\theta\sin\phi'\right) \tag{18.68}$$

Substituting this into the first integral in **Eq. (18.66)**, expanding the terms, and taking constant terms outside the integral sign, we get

$$\mathbf{A} = \hat{\boldsymbol{\phi}}\frac{\mu I_0 a e^{-j\beta R}(1+j\beta R)}{4\pi R}2\int_{\phi'=-\pi/2}^{\phi'=+\pi/2}\sin\phi' d\phi' + \hat{\boldsymbol{\phi}}\frac{\mu I_0 a^2 e^{-j\beta R}(1+j\beta R)\sin\theta}{4\pi R^2}2\int_{\phi'=-\pi/2}^{\phi'=+\pi/2}\sin^2\phi' d\phi' \quad \left[\frac{\text{Wb}}{\text{m}}\right] \tag{18.69}$$

The first integral is zero, as can be shown by direct integration. The second integral is equal to $\pi/2$. Thus, we get the final result for the magnetic vector potential as

$$\mathbf{A} = \hat{\boldsymbol{\phi}}\frac{\mu I_0 \pi a^2 e^{-j\beta R}(1+j\beta R)\sin\theta}{4\pi R^2} \quad \left[\frac{\text{Wb}}{\text{m}}\right] \tag{18.70}$$

The quantity $m = I_0\pi a^2$ is the magnitude of the dipole moment as defined in **Eq. (18.58)**. Therefore

$$\boxed{\mathbf{A} = \hat{\boldsymbol{\phi}}\frac{\mu m}{4\pi R^2}(1+j\beta R)e^{-j\beta R}\sin\theta \quad \left[\frac{\text{Wb}}{\text{m}}\right]} \tag{18.71}$$

The advantage of this form is that it shows that the magnetic dipole discussed here has the same solution as the magnetostatic dipole but multiplied by the term $(1 + j\beta R)e^{-j\beta R}$. The latter term equals 1 at zero frequency. Thus, the solution here is consistent with the magnetostatic solution of **Chapter 9**.

Now that we have a general expression for the magnetic vector potential, we can calculate the magnetic and electric field intensities as for the electric dipole in the previous section. **Equation (18.15)** is used to find the magnetic field intensity, and **Eq. (18.17)** is used to find the electric field intensity. Performing the derivatives in **Eq. (18.15)**, we get the magnetic field intensity as

$$\mathbf{H} = -\hat{\mathbf{R}}\frac{j\omega\mu\beta^2 m}{2\pi\eta}e^{-j\beta R}\cos\theta\left(\frac{1}{(j\beta R)^2} + \frac{1}{(j\beta R)^3}\right) - \hat{\boldsymbol{\theta}}\frac{j\omega\mu\beta^2 m}{4\pi\eta}e^{-j\beta R}\sin\theta\left(\frac{1}{j\beta R} + \frac{1}{(j\beta R)^2} + \frac{1}{(j\beta R)^3}\right) \quad \left[\frac{\text{A}}{\text{m}}\right] \tag{18.72}$$

From **Eq. (18.17)**, the electric field intensity has a ϕ component only:

$$\mathbf{E} = \hat{\boldsymbol{\phi}} \frac{j\omega\mu\beta^2 m}{4\pi} e^{-j\beta R} \sin\theta \left(\frac{1}{j\beta R} + \frac{1}{(j\beta R)^2} \right) \quad \left[\frac{\mathrm{V}}{\mathrm{m}} \right] \tag{18.73}$$

The first thing to note here is that the form (but not the magnitude) of the electric field intensity equation for the magnetic dipole is identical to that of the magnetic field intensity equation for the electric dipole and the form of the magnetic field intensity equation for the magnetic dipole is the same as for the electric field intensity for the electric dipole. To see what the exact relationship between the magnetic and electric dipoles is, we denote by $\mathbf{E}_m$ and $\mathbf{H}_m$ the electric and magnetic field intensities for the magnetic dipole in **Eqs. (18.72)** and **(18.73)** and by $\mathbf{E}_e$ and $\mathbf{H}_e$ the electric and magnetic field intensities for the electric dipole in **Eqs. (18.20)** and **(18.21)**. Then, direct comparison between the two gives

$$\mathbf{E}_m = -\eta \mathbf{H}_e \quad \text{and} \quad \mathbf{H}_m = \frac{\mathbf{E}_e}{\eta} \tag{18.74}$$

if $I_0 \Delta l'$ is replaced by $j\beta m = j\beta I_0 \pi a^2$. This relation is useful and we will use it to derive other properties of the magnetic dipole such as radiated power, efficiency, and others (see **Table 18.1**) from those of the electric dipole. This economizes on formulas and also shows explicitly how similar the two dipoles are.

18.6.1 Near Fields for the Magnetic Dipole

The near-field electric and magnetic field intensities are calculated from the condition $\beta R \ll 1$. Substituting this in **Eq. (18.72)**, we find that the terms containing $(\beta R)^{-3}$ in the magnetic field intensity dominate. Also, $e^{-j\beta R} \approx 1$. With these, the magnetic field intensity is

$$\boxed{\mathbf{H} = \hat{\mathbf{R}} \frac{m}{4\pi R^3} 2\cos\theta + \hat{\boldsymbol{\theta}} \frac{m}{4\pi R^3} \sin\theta = \frac{m}{4\pi R^3} \left(\hat{\mathbf{R}} 2\cos\theta + \hat{\boldsymbol{\theta}} \sin\theta \right) \quad \left[\frac{\mathrm{A}}{\mathrm{m}} \right]} \tag{18.75}$$

This is identical to the magnetic field intensity of a magnetostatic dipole in **Eq. (9.14)** except that the current (and hence the dipole moment **m**) is time dependent.

The near-field electric field intensity is evaluated from **Eq. (18.73)**. Since $\beta R \ll 1$, the second term dominates, $e^{-j\beta R} \approx 1$, and we have

$$\boxed{\mathbf{E} = -\hat{\boldsymbol{\phi}} \frac{j\omega\mu m}{4\pi R^2} \sin\theta \quad \left[\frac{\mathrm{V}}{\mathrm{m}} \right]} \tag{18.76}$$

The electric field intensity is small in comparison to the magnetic field intensity, which varies as $1/R^3$. For this reason the electric field in the near field is not important for the magnetic dipole in the same way that the magnetic field was not important in the near field of the electric dipole. The time-averaged Poynting vector obtained for the near field is zero, showing that in the near field, there is only storage of (mostly magnetic) energy rather than propagation.

18.6.2 Far Fields for the Magnetic Dipole

In the far field, we neglect all terms except those with $1/\beta R$ since now we have $\beta R \gg 1$. **Equations (18.72)** and **(18.73)** become

$$\boxed{\mathbf{E} = \hat{\boldsymbol{\phi}} \frac{\omega\mu\beta m}{4\pi R} e^{-j\beta R} \sin\theta \quad \left[\frac{\mathrm{V}}{\mathrm{m}} \right] \quad \text{and} \quad \mathbf{H} = -\hat{\boldsymbol{\theta}} \frac{\omega\mu\beta m}{4\pi\eta R} e^{-j\beta R} \sin\theta \quad \left[\frac{\mathrm{A}}{\mathrm{m}} \right]} \tag{18.77}$$

Because of the duality of results, the properties of the magnetic dipole and the electric dipole are the same, with the exception of the directions of the fields: These, of course, are perpendicular to one another; that is, the directions of the electric field intensity of the electric dipole and that of the magnetic dipole are at right angles. The time-averaged power density in the far field may be written directly as

$$\mathcal{P}_{av} = \frac{1}{2}\mathrm{Re}\{\mathbf{E}\times\mathbf{H}^*\} = \hat{\mathbf{R}}\frac{\omega^2\mu^2\beta^2 m^2}{32\pi^2\eta R^2}\sin^2\theta \quad \left[\frac{\mathrm{W}}{\mathrm{m}^2}\right] \tag{18.78}$$

The time-averaged Poynting vector in the far field is again in the R direction, indicating outward propagation of energy. Also, by direct comparison with **Eq. (18.32)**, we note that **Eq. (18.78)** may be obtained from **Eq. (18.32)** by replacing $I_0\Delta l'$ with βm.

18.6.3 Properties of the Magnetic Dipole

Now that we have the electric and magnetic fields in the far field, we can calculate the basic properties of the magnetic dipole by starting with the definition of radiated power in **Eq. (18.33)**. Since the steps are identical to those for the electric dipole, we merely list these in **Table 18.1**. The properties of the magnetic dipole may also be obtained from those of the electric dipole in the first column of **Table 18.1** by replacing $I_0\Delta l'$ by βm or $\Delta l'$ by $\beta\pi a^2$ as indicated above.

Note also that although the relative power and field antenna patterns are the same for both antennas, the absolute antenna patterns are not. Also, the E-field pattern for the magnetic dipole corresponds to the H-field pattern for the electric dipole and vice versa.

Table 18.1 Properties of electric and magnetic dipoles as antennas

	Electric dipole	Magnetic dipole
$\mathcal{P}_{av}$ (average power density) $\left[\frac{\mathrm{W}}{\mathrm{m}^2}\right]$	$\hat{\mathbf{R}}\frac{\eta I_0^2\beta^2(\Delta l')^2}{32\pi^2R^2}\sin^2\theta$	$\hat{\mathbf{R}}\frac{m^2\beta^4\eta}{32\pi^2R^2}\sin^2\theta = \hat{\mathbf{R}}\frac{\omega^2\mu^2m^2\beta^2}{32\pi^2\eta R^2}\sin^2\theta$
P_{rad} (radiated power) [W]	$\frac{\eta I_0^2\beta^2(\Delta l')^2}{12\pi}$	$\frac{m^2\beta^4\eta}{12\pi} = \frac{\omega^2\mu^2m^2\beta^2}{12\pi\eta}$
R_{rad} (radiation resistance) [Ω]	$\frac{2\eta\pi}{3}\left(\frac{\Delta l'}{\lambda}\right)^2$	$\frac{\beta^4\pi a^4\eta}{6} = \frac{\omega^2\mu^2\beta^2\pi a^4}{6\eta}$
R_{rad} in air (radiation resistance in air) [Ω]	$80\pi^2\left(\frac{\Delta l'}{\lambda}\right)^2$	$20\pi^2\beta^4a^4 = \frac{\omega^2\mu^2\beta^2a^4}{720}$
$\lvert f_l(\theta)\rvert$ (normalized field radiation pattern)	$\lvert\sin\theta\rvert$	$\lvert\sin\theta\rvert$
$f_p(\theta)$ (normalized power radiation pattern)	$\sin^2\theta$	$\sin^2\theta$
$U(\theta)$ (radiation intensity) $\left[\frac{\mathrm{W}}{\mathrm{sr}}\right]$	$\frac{\eta I_0^2}{8}\left(\frac{\Delta l'}{\lambda}\right)^2\sin^2\theta$	$\frac{\beta^2\eta m^2}{8\lambda^2}\sin^2\theta = \frac{\omega^2\mu^2m^2}{8\eta\lambda^2}\sin^2\theta$
U_{av} (average radiation intensity) $\left[\frac{\mathrm{W}}{\mathrm{sr}}\right]$	$I_0^2\frac{\eta}{12}\left(\frac{\Delta l'}{\lambda}\right)^2$	$\frac{\beta^2\eta m^2}{12\lambda^2}$
$D(\theta)$ (directivity) [dimensionless]	$\frac{3}{2}\sin^2\theta$	$\frac{3}{2}\sin^2\theta$
D_0 (maximum directivity) [dimensionless]	1.5	1.5
eff. (radiation efficiency) [dimensionless]	$\frac{R_{rad}}{R_{rad}+R_d}$	$\frac{R_{rad}}{R_{rad}+R_d}$

Note: $\beta^2\eta^2 = \omega^2\mu^2$, $\beta = 2\pi/\lambda$.

Example 18.6 Application: Microwave Testing of Materials **Figure 18.11** shows a method of testing dielectrics or biological materials for foreign inclusions (such as air inclusions in plastics, foreign objects in meat, water content in snow or fat content in foods). The radiation resistance of the antenna is material dependent, and as the material in its vicinity changes, so does the radiation resistance. The loading of the antenna changes, which is then a measure of the test material properties. A similar method may be used to test for moisture in soils or any other material.

A small loop, 8 mm in diameter, operates at 1 GHz, carries a current of (10 mA), and is used to measure the drying process in freeze-drying of meat. The test material is lossless with a relative permittivity of 38 when fresh (meat is a lossy dielectric, but we will neglect this fact here). When freeze-dried, the meat has no moisture and its relative permittivity is 9. Assume the antenna is inserted deep into the drying product. Calculate:

(a) The range of the radiation resistance of the loop antenna as the product dries.
(b) Suppose the antenna is fed with a constant amplitude current source. What is the range of the radiated power of the antenna?

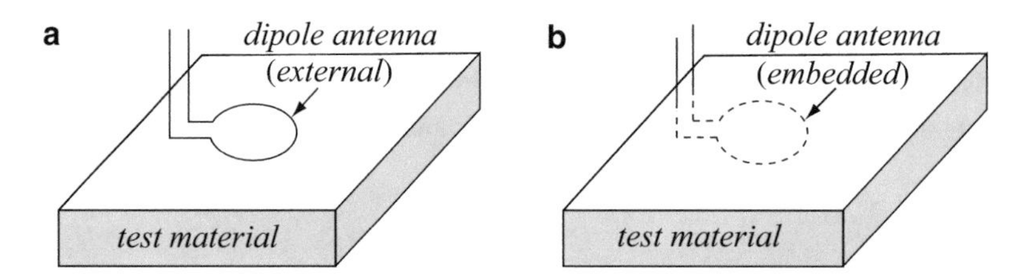

Figure 18.11 Testing of dielectric materials. (**a**) Loop antenna outside the test sample. (**b**) Loop antenna embedded in the test sample

Solution: The radiation resistance depends on the properties of the loop: radius (a), properties of the test material (μ,η,β), and frequency (f). The range in radiated power and radiation resistance is then defined by the change in permitivity of the test material provided all other parameters remain constant:

(a) The radiation resistance and power radiated depend on the intrinsic impedance of the product and its phase constant, which change with the drying process. These are as follows:

$$\beta_{fresh} = \omega\sqrt{\mu_0\varepsilon_{fresh}} = \omega\sqrt{\mu_0\varepsilon_0\varepsilon_r} = \frac{2\times\pi\times 7.5\times 10^9}{3\times 10^8}\sqrt{38} = 968.3 \quad [\mathrm{rad/m}]$$

$$\beta_{dry} = \omega\sqrt{\mu_0\varepsilon_{dry}} = \frac{\omega}{c} = \frac{2\times\pi\times 7.5\times 10^9}{3\times 10^8}\sqrt{9} = 471.24 \quad [\mathrm{rad/m}]$$

$$\eta_{fresh} = \sqrt{\frac{\mu_0}{\varepsilon_0\varepsilon_r}} = \frac{\eta_0}{\sqrt{\varepsilon_r}} = \frac{377}{\sqrt{38}} = 61.16\ [\Omega], \quad \eta_{dry} = \frac{\eta_0}{\sqrt{9}} = 125.67\ [\Omega]$$

The range for the radiation resistance is

$$R_{fresh} = \frac{\omega^2\mu_0^2\beta_{fresh}^2\pi a^4}{6\eta_{fresh}} = \frac{\left(2\times\pi\times 10^9\right)^2\times\left(4\times\pi\times 10^{-7}\right)^2\times(968.3)^2\times\pi\times(0.004)^4}{6\times 61.16} = 128.1 \quad [\Omega]$$

$$R_{dry} = \frac{\omega^2\mu_0^2\beta_{dry}^2\pi a^4}{6\eta_{dry}} = \frac{\left(2\times\pi\times 10^9\right)^2\times\left(4\times\pi\times 10^{-7}\right)^2\times(471.24)^2\times\pi\times(0.004)^4}{6\times 125.67} = 14.87 \quad [\Omega]$$

This is a significant range that is easily monitored.

(**b**) The range of the radiated power is

$$P_{fresh} = \frac{I_0^2 R_{rad}}{2} = \frac{0.01^2 \times 128.1}{2} = 6.4 \quad [\text{mW}],$$

$$P_{dry} = \frac{I_0^2 R_{rad}}{2} = \frac{0.01^2 \times 14.87}{2} = 0.74 \quad [\text{mW}].$$

The power transmitted is reduced from 6.4 mW to 0.74 mW. Since the current is constant, it is sufficient to monitor the voltage (or power) in the transmitter to get an indication of the state of the material. As the material dries, the power coupled into the load the material presents is reduced. In a practical application, the efficiency also changes since the antenna resistance is fixed. The efficiency should also be taken into account because measurements in the transmitter usually monitor the input power rather than radiated power.

18.7 Practical Antennas

The antennas described so far, the Hertzian dipole and the small loop, are small antennas. As such, they can radiate little power and, therefore, are only practical in low power applications. Also, because of the very small sizes required for the analysis to be correct, the results we obtained so far are only exact for infinitesimally small dipoles. For finite-length but small antennas, we may be able to use these results as approximations. However, the Hertzian dipole is useful as a building block of more practical antennas. For example, the Hertzian dipole is based entirely on a current I flowing in a line segment of length Δl. If, instead, we wish to build an antenna of any length and shape, we build it as a stack of elementary dipoles. Each dipole has a current I, which normally varies with location along the antenna and produces a field in the near- and far-field zones. Calculation of the fields of each elemental dipole and summation of all dipole fields provides the field of the antenna. Easier said than done because the current in the antenna must be known everywhere before we start. A constant current cannot, in general, be assumed, as can be seen in **Figure 18.12**. In the Hertzian dipole in **Figure 18.12a**, we assumed that as the charge at one end is depleted, it is replenished at the opposite end, and because the length of the antenna is very short compared with the wavelength, the current is essentially constant. In **Figure 18.12b** and **18.12c** we cannot assume the length of the antenna to be short; therefore, the constant current approximation does not apply.

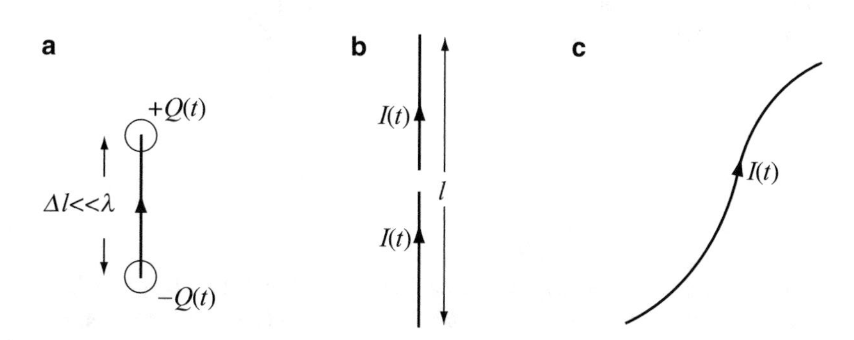

Figure 18.12 (**a**) Short electric dipole (Hertzian dipole). (**b**) Long straight wire antenna. (**c**) Long, arbitrarily shaped wire antenna

To keep things simple and practical, we will discuss in this section only straight element antennas made of very thin wires so that the current distribution in the wire cross section need not be considered. This type of antenna is called a ***linear antenna*** and it may be of two types: One, which we will discuss first, is a two-wire dipole antenna, fed at its center as shown in **Figure 18.13a**. The second is a single wire antenna (monopole) as shown in **Figure 18.13b**. After these two antennas are described, we will also discuss arrays of linear antennas. These antennas are very common and, therefore, it is important to understand their properties.

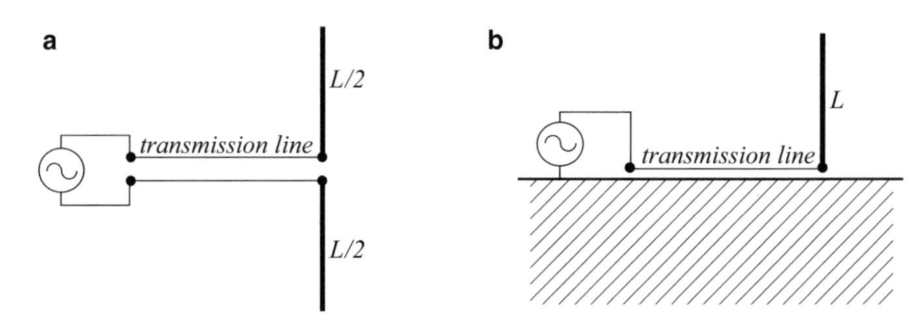

Figure 18.13 (**a**) Dipole antenna. (**b**) Monopole antenna. Both antennas shown with their excitation

18.7.1 Linear Antennas of Arbitrary Length

The antenna considered here is shown in **Figure 18.14a**. It is made of two elements, each $L/2$ in length with a negligible gap between them. The antenna is center-fed at the gap from a source, through a transmission line. To analyze the antenna, we must first define the current distribution in the antenna. Then, we substitute this current into the expressions for the electric and magnetic field intensities of the Hertzian dipole [**Eqs. (18.30)** and **(18.31)**] and integrate the result along the antenna (from $-L/2$ to $+ L/2$).

Regardless of the length of the antenna, the current at the ends of the radiating elements must be zero. Thus, the current for an arbitrarily long dipole antenna of total length L is assumed to be

$$I(z') = I_0 \sin\left[\beta\left(\frac{L}{2} - |z'|\right)\right] \quad [\text{A}] \tag{18.79}$$

This guarantees that the current is zero at $z' = \pm L/2$ and varies sinusoidally along the antenna. At the feed point ($z' = 0$), the current is discontinuous and its amplitude depends on L. As an example, for the half-wavelength antenna shown in **Figure 18.14b**, the current is maximum at the feed point.

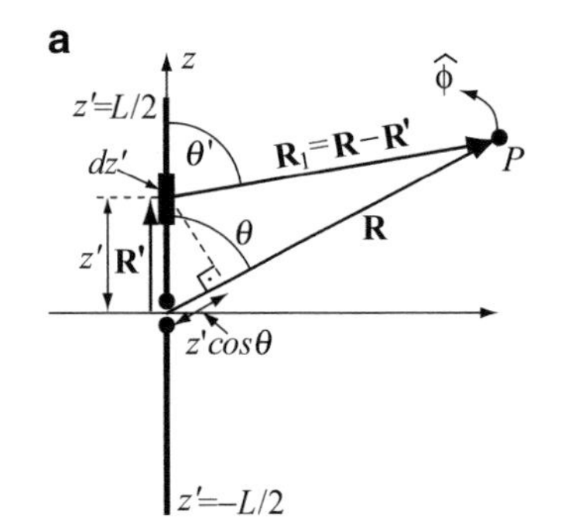

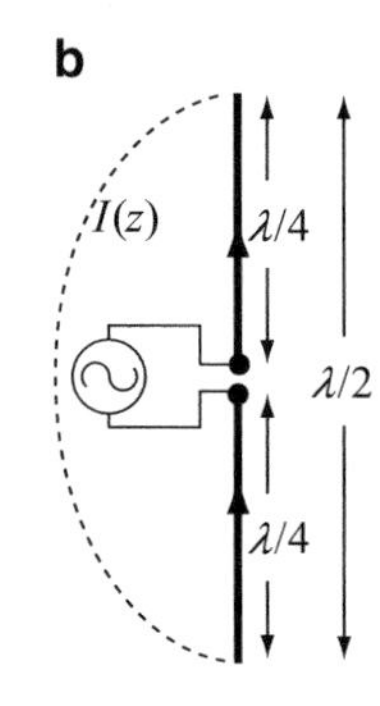

Figure 18.14 The finite-length dipole antenna. (**a**) Notation used for calculation for arbitrary length antennas. (**b**) Half-wavelength dipole and assumed current along the antenna

Now, suppose we place a Hertzian dipole of infinitesimal length dz' at point z' in **Figure 18.14a**. Because the Hertzian dipole is infinitesimal, the current in it is constant and we can use the equations for the Hertzian dipole by setting $\Delta l' = dz'$. The distance between the dipole and point P is R_1 and the angle $\mathbf{R}_1$ makes with the z axis is θ' to distinguish these from the global quantities θ, R, z, which relate to the whole antenna rather than the elemental dipole. Thus, using **Eqs. (18.30)** and **(18.31)**, we get the far-field expressions:

$$d\mathbf{E} = \hat{\boldsymbol{\theta}}' \frac{j\eta I_0 \beta \sin\left[\beta\left(\frac{L}{2} - |z'|\right)\right] dz'}{4\pi R_1} e^{-j\beta R_1} \sin\theta' \tag{18.80}$$

$$d\mathbf{H} = \hat{\boldsymbol{\phi}} \frac{jI_0 \beta \sin\left[\beta\left(\frac{L}{2} - |z'|\right)\right] dz'}{4\pi R_1} e^{-j\beta R_1} \sin\theta' \tag{18.81}$$

To calculate the total fields at point P, we need to integrate these quantities from $z' = -L/2$ to $z' = +L/2$. Before we do so, we introduce some approximations which should simplify our tasks. Since the field is calculated in the far field, $R_1 \gg L$. Because of this, we can also say that $\theta' \approx \theta$. Note that these assumptions, in fact, imply that R and R_1 are parallel to each other, an assumption which can be easily justified for $R_1 \gg L$. As for R_1, we can write

$$R_1 = \sqrt{R^2 + z'^2 - 2Rz'\cos\theta} \approx R - z'\cos\theta \tag{18.82}$$

The term $1/R_1$ may be approximated as $1/R$. However, in the phase, we cannot approximate R_1 by R since β may be large and cause a significant change in phase due to the term $z'\cos\theta$. Thus, for the phase, we retain $R_1 \approx R - z'\cos\theta$. With these approximation, **Eq. (18.80)** becomes

$$\mathbf{E} = \hat{\boldsymbol{\theta}}\frac{j\eta I_0\beta}{4\pi R}e^{-j\beta R}\sin\theta\int_{z'=-L/2}^{z'=L/2}\sin\left[\beta\left(\frac{L}{2} - |z'|\right)\right]e^{j\beta z'\cos\theta}dz' \quad \left[\frac{\text{V}}{\text{m}}\right] \tag{18.83}$$

To evaluate this, we substitute $e^{j\beta z'\cos\theta} = \cos(\beta z'\cos\theta) + j\sin(\beta z'\cos\theta)$ and integrate. Performing the integration in **Eq. (18.83)**, we get

$$\int_{z'=-L/2}^{z'=L/2}\sin\left[\beta\left(\frac{L}{2} - |z'|\right)\right](\cos(\beta z'\cos\theta) + j\sin(\beta z'\cos\theta))dz' = 2\frac{\cos((\beta L/2)\cos\theta) - \cos(\beta L/2)}{\beta\sin^2\theta} \tag{18.84}$$

Substituting this into **Eq. (18.83)** and performing identical steps for the magnetic field, we obtain the far-field electric and magnetic field intensities due to the antenna as

$$\boxed{\mathbf{E} = \hat{\boldsymbol{\theta}}\frac{j\eta I_0}{2\pi R}e^{-j\beta R}\frac{\cos((\beta L/2)\cos\theta) - \cos(\beta L/2)}{\sin\theta} \quad \left[\frac{\text{V}}{\text{m}}\right]} \tag{18.85}$$

$$\boxed{\mathbf{H} = \hat{\boldsymbol{\phi}}\frac{jI_0}{2\pi R}e^{-j\beta R}\frac{\cos((\beta L/2)\cos\theta) - \cos(\beta L/2)}{\sin\theta} \quad \left[\frac{\text{A}}{\text{m}}\right]} \tag{18.86}$$

This is a most remarkable result. It indicates that a linear antenna of any length (provided we are in the far field) produces similar electric and magnetic fields. From a designer point of view, this means considerable flexibility in design.

The fields of linear antennas can therefore be summarized as follows:

(1) For $L = n\lambda/2 = n\pi/\beta$, $n = 1, 3, 5,\ldots$, we can write

$$\mathbf{E} = \hat{\boldsymbol{\theta}}\frac{j\eta I_0}{2\pi R}e^{-j\beta R}\frac{\cos((n\pi/2)\cos\theta)}{\sin\theta} \quad \left[\frac{\text{V}}{\text{m}}\right], \quad \mathbf{H} = \hat{\boldsymbol{\phi}}\frac{jI_0}{2\pi R}e^{-j\beta R}\frac{\cos((n\pi/2)\cos\theta)}{\sin\theta} \quad \left[\frac{\text{A}}{\text{m}}\right] \tag{18.87}$$

(2) For $L = n\lambda = 2n\pi/\beta$, $n = 1, 2, 3,\ldots$,

$$\begin{aligned}\mathbf{E} &= \hat{\boldsymbol{\theta}}\frac{j\eta I_0}{2\pi R}e^{-j\beta R}\frac{\cos((n\pi)\cos\theta) - \cos(n\pi)}{\sin\theta} \quad \left[\frac{\text{V}}{\text{m}}\right], \\ \mathbf{H} &= \hat{\boldsymbol{\phi}}\frac{jI_0}{2\pi R}e^{-j\beta R}\frac{\cos((n\pi)\cos\theta) - \cos(n\pi)}{\sin\theta} \quad \left[\frac{\text{A}}{\text{m}}\right]\end{aligned} \tag{18.88}$$

Although the length L may be arbitrary, it must be relatively large. For antennas shorter than about one-quarter wavelength, the current in the antenna is not sinusoidal but approximately linear. If the fields of such an antenna are required, they must be evaluated separately from the current distribution of the specific antenna (see **Problem 18.8**).

With the above fields, we can calculate the power density from the time-averaged Poynting vector:

$$\boxed{\mathcal{P}_{av} = \frac{1}{2}\mathrm{Re}\{\mathbf{E} \times \mathbf{H}^*\} = \hat{\mathbf{R}}\,\frac{\eta I_0^2}{8\pi^2 R^2}\left(\frac{\cos((\beta L/2)\cos\theta) - \cos(\beta L/2)}{\sin\theta}\right)^2 \quad \left[\frac{\mathrm{W}}{\mathrm{m}^2}\right]} \tag{18.89}$$

Thus, the direction of propagation is outward (in the R direction), and the total power radiated by the antenna through a sphere of radius R can be immediately written as

$$P_{rad} = \int_s \mathcal{P}_{av} \cdot d\mathbf{s} = \int_{\phi=0}^{2\pi}\int_{\theta=0}^{\pi} \mathcal{P}_{av} R^2 \sin\theta d\theta d\phi \quad [\mathrm{W}] \tag{18.90}$$

This gives

$$\boxed{P_{rad} = \frac{\eta I_0^2}{4\pi}\int_{\theta=0}^{\theta=\pi} \frac{[\cos((\beta L/2)\cos\theta) - \cos(\beta L/2)]^2}{\sin\theta} d\theta \quad [\mathrm{W}]} \tag{18.91}$$

where the integral over ϕ contributes 2π. Unfortunately, the integral on θ in **Eq. (18.91)** cannot be evaluated analytically but can be integrated numerically rather easily (see script **Eq_18_92.m**). To aid in the calculation of antenna parameters, the integral in **Eq. (18.91)** is recast in a more convenient form by substituting $\beta = 2\pi/\lambda$ whereas L now is the antenna length in wavelengths:

$$\int_{\theta=0}^{\theta=\pi} \frac{[\cos((L\pi/\lambda)\cos\theta) - \cos(L\pi/\lambda)]^2}{\sin\theta} d\theta \tag{18.92}$$

Eq. (18.92) was evaluated for a range of antenna lengths ranging from 0 to 5λ using script **Eq_18_92.m** and is shown in **Figure 18.15**. A few selected values are listed in **Table 18.2** for easy reference.

The radiation resistance of a dipole of arbitrary length L is [see **Eq. (18.36)**]

$$\boxed{R_{rad} = \frac{\eta}{2\pi}\int_{\theta=0}^{\theta=\pi} \frac{[\cos((L\pi/\lambda)\cos\theta) - \cos(L\pi/\lambda)]^2}{\sin\theta} d\theta \quad [\Omega]} \tag{18.93}$$

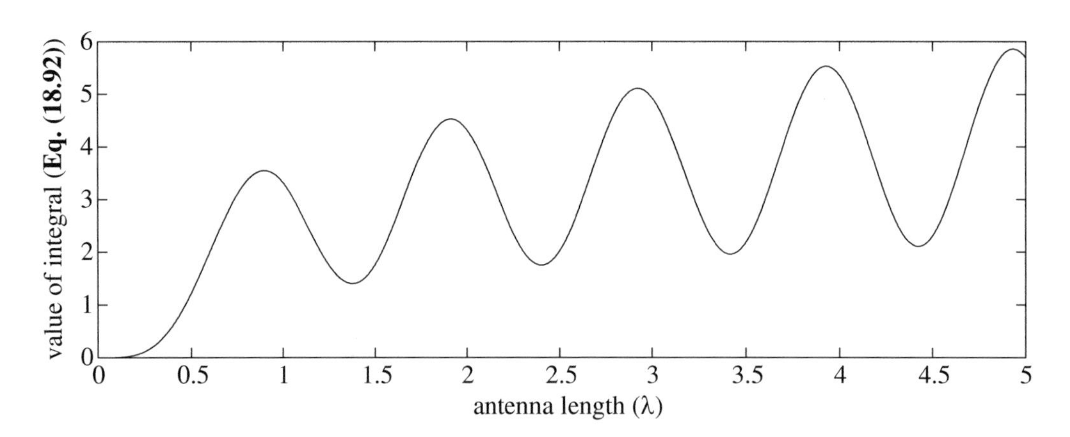

Figure 18.15 The integral in **Eq. (18.92)** as a function of antenna length

Table 18.2 Value of the integral in **Eq. (18.92)** for some antenna lengths **Eq_18_92.m**

Antenna length, L	0.5λ	0.6λ	0.75λ	1λ	1.5λ	2λ	2.5λ	3λ	3.5λ	4λ
Value of **Eq. (18.92)**	1.2188	1.997	3.097	3.318	1.758	4.327	2.013	4.929	2.181	5.358

The directivity of the antenna is calculated from the definition [in **Eq. (18.50)**] as

$$D(\theta) = \frac{U(\theta)}{P_{rad}/4\pi} = \frac{4\pi\mathcal{P}_{av}R^2}{P_{rad}} = \frac{2\left(\dfrac{\cos((L\pi/\lambda)\cos\theta) - \cos(L\pi/\lambda)}{\sin\theta}\right)^2}{\displaystyle\int_{\theta=0}^{\theta=\pi} \frac{(\cos((L\pi/\lambda)\cos\theta) - \cos(L\pi/\lambda))^2}{\sin\theta} d\theta} \tag{18.94}$$

where, again, we make use of the plot in **Figure 18.15** to evaluate the integral in the denominator.

The relative field and power patterns for a general dipole antenna are

$$\boxed{|f_e(\theta)| = \left|\frac{\cos((L\pi/\lambda)\cos\theta) - \cos(L\pi/\lambda)}{\sin\theta}\right|, \quad f_p(\theta) = \left(\frac{\cos((L\pi/\lambda)\cos\theta) - \cos(L\pi/\lambda)}{\sin\theta}\right)^2} \tag{18.95}$$

All other properties of arbitrary length antennas can be found in a similar fashion.

The above calculations assumed that the current along the antenna is distributed in a sinusoidal fashion as given in **Eq. (18.79)** and that the antenna is thin. This is only approximately true. In the practical design of antennas, the current distribution is calculated numerically, but since we wish to obtain the general properties of the antennas in the simplest possible way, there is little benefit in pursuing more accurate expressions for the current in the antenna. Also, we have not taken into account the method of feeding the antenna. Since the two elements of the dipole must be separated physically, there will also be a difference in the actual performance of the antenna due to the gap between the elements. Antenna resistance and matching of the antenna to the feed will also affect performance. Nevertheless, the analysis here is useful for general design. More accurate design can be pursued using numerical calculations.

The relations above may be used to analyze any length antenna, but some antennas are more commonly used. The most common of them is the half-wavelength dipole antenna and the quarter-wavelength monopole antenna. The half-wavelength antenna is described in the following subsection.

18.7.1.1 The Half-Wavelength Dipole Antenna

The half-wavelength dipole antenna is shown in **Figure 18.14b**. The properties of the antenna are obtained by simply substituting $L/2 = \lambda/4$ in the results of the previous section. However, because this particular antenna is very important in practice, the details and final expressions are given below.

First, we note that the current distribution along the antenna is

$$I(z') = I_0 \cos\frac{2\pi z'}{\lambda} \quad [\mathrm{A}] \tag{18.96}$$

This is obtained from **Eq. (18.79)** by setting $L = \lambda/2$ and $\beta = 2\pi/\lambda$. The electric and magnetic field intensities of the half-wavelength dipole are [from **Eqs. (18.85)** and **(18.86)**]:

$$\boxed{\mathbf{E} = \hat{\boldsymbol{\theta}} \frac{j\eta I_0}{2\pi R} e^{-j\beta R} \frac{\cos((\pi/2)\cos\theta)}{\sin\theta} \quad \left[\frac{\mathrm{V}}{\mathrm{m}}\right]} \tag{18.97}$$

$$\boxed{\mathbf{H} = \hat{\boldsymbol{\phi}} \frac{jI_0}{2\pi R} e^{-j\beta R} \frac{\cos((\pi/2)\cos\theta)}{\sin\theta} \quad \left[\frac{\mathrm{A}}{\mathrm{m}}\right]} \tag{18.98}$$

The power density is evaluated using **Eq. (18.89)**:

$$\boxed{\mathcal{P}_{av} = \hat{\mathbf{R}} \frac{\eta I_0^2}{8\pi^2 R^2} \frac{\cos^2((\pi/2)\cos\theta)}{\sin^2\theta} \quad \left[\frac{\mathrm{W}}{\mathrm{m}^2}\right]} \tag{18.99}$$

The radiated power is evaluated using **Eq. (18.91)**, using the value of 1.218 for the value of the integral (first column in **Table 18.2**):

$$\boxed{P_{rad} = \frac{1.218\eta I_0^2}{4\pi} \quad [\mathrm{W}]} \tag{18.100}$$

To find the radiation resistance, we could use **Eq. (18.93)** or simply write $P_{rad} = I_0^2 R_{rad}/2$. Either way,

$$\boxed{R_{rad} = \frac{0.609\eta}{\pi} \quad [\Omega]} \tag{18.101}$$

In particular, in free space, the radiation resistance is (with $\eta = \eta_0 = 120\pi$)

$$R_{rad} = \frac{0.609 \times 120\pi}{\pi} = 73.08 \quad [\Omega] \tag{18.102}$$

The relative field and power radiation patterns are [from **Eq. (18.95)**]

$$\boxed{|f_e(\theta)| = \left|\frac{\cos((\pi/2)\cos\theta)}{\sin\theta}\right|, \quad f_p(\theta) = \frac{\cos^2((\pi/2)\cos\theta)}{\sin^2\theta}} \tag{18.103}$$

The normalized field and power radiation patterns are shown in **Figure 18.16**.
Directivity is found from **Eq. (18.94)**:

$$\boxed{D(\theta) = 1.642\frac{\cos^2((\pi/2)\cos\theta)}{\sin^2\theta}} \tag{18.104}$$

The maximum directivity is obtained by observing that the expression in **Eq. (18.104)** is maximum for $\theta = \pi/2$. Thus, substituting $\theta = \pi/2$ in **Eq. (18.104)**, we get

$$\boxed{D_0 = D(\theta = \pi/2) = 1.642} \tag{18.105}$$

The antenna gain can only be calculated if the antenna resistance R_d is known, using **Eqs. (18.55)** and **(18.56)**, or if the input power is known, using **Eq. (18.54)**.

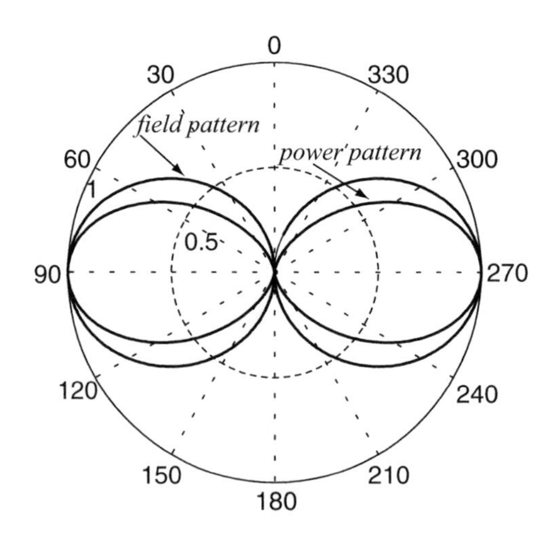

Figure 18.16 Normalized field and power radiation patterns for the half-wavelength antenna

18.7.1.2 Full- and Three-Halves-Wavelength Antennas

As representative of antenna lengths other than half-wavelength, we discuss here, briefly, the full- and three-halves-wavelength antennas. Other antennas may be similarly defined from the general expressions in **Section 18.7.1**:

(1) The ***full wavelength antenna*** is a dipole antenna one wavelength in length. If we substitute $L = \lambda$ in the general equations and use the value 3.318 for the integral in **Eq. (18.92)**, we obtain the properties of the antenna. As an example, the relative E-plane field patterns for the field intensity and power density are

$$|f_e(\theta)| = \left|\frac{\cos(\pi\cos\theta) + 1}{\sin\theta}\right|, \quad f_p(\theta) = \left(\frac{\cos(\pi\cos\theta) + 1}{\sin\theta}\right)^2 \tag{18.106}$$

Similar substitutions in any of **Eqs. (18.85)** through **(18.94)** give the corresponding quantity for the one-wavelength dipole antenna. Note that the relative patterns in **Eq. (18.106)** are not normalized to 1. In fact the maximum value for the field pattern is 2 and that for the power pattern is 4.

(2) The ***three-halves-wavelength antenna.*** This antenna is 1.5 wavelengths long. The value of the integral in **Eq. (18.92)** is 1.758 at $L = 1.5\lambda$. Substituting these in the equations for the arbitrarily long antenna, we obtain the properties of the 1.5λ antenna.

The radiation patterns for the one-wavelength antenna and the 1.5-wavelength antenna are shown in **Figure 18.17a** and **18.17b**. Patterns for any length antenna may be obtained in the same fashion, but it should be noted from **Figure 18.15** and **Table 18.2** that some lengths have "better" properties and are therefore used more often.

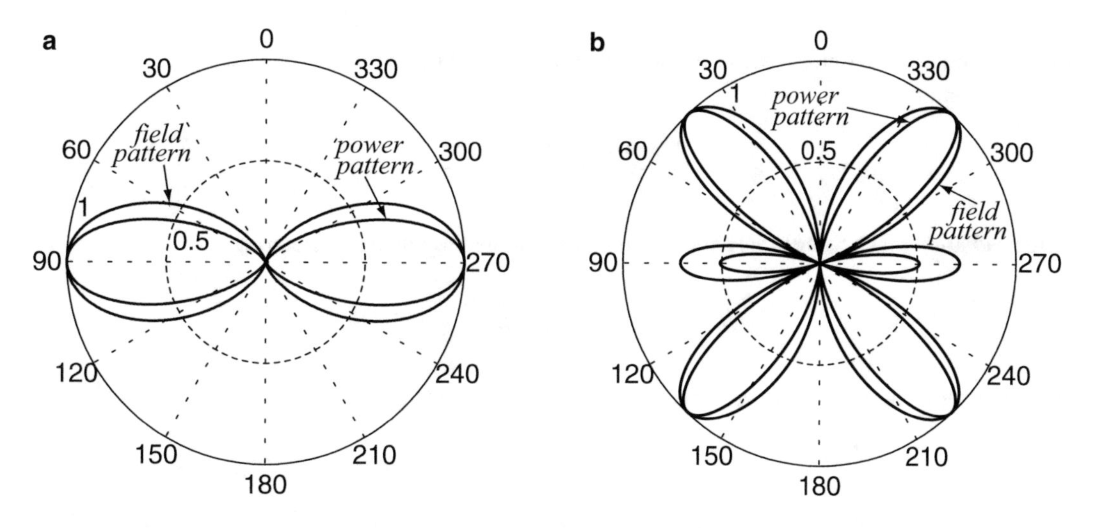

Figure 18.17 Normalized field and power radiation patterns. (**a**) For the 1λ antenna. (**b**) For the 1.5λ dipole antenna

Example 18.7 A radio hobbyist is using the amateur band at 29 MHz. After successfully building a transmitter, he decides to build a half-wavelength, horizontal dipole antenna (see **Figure 18.18a**). The antenna is made of a metal wire such that its resistance at 29 MHz is 5 Ω. The transmitter can supply 100 W (time-averaged power) to the antenna. Neglecting any effects from the ground, calculate:

(a) The required dimensions of the antenna.
(b) The power radiated by the antenna into free space.
(c) The power density at a distance of 100 km from the antenna in the direction of maximum power.
(d) Draw the power radiation pattern of the antenna in a rectangular plot.

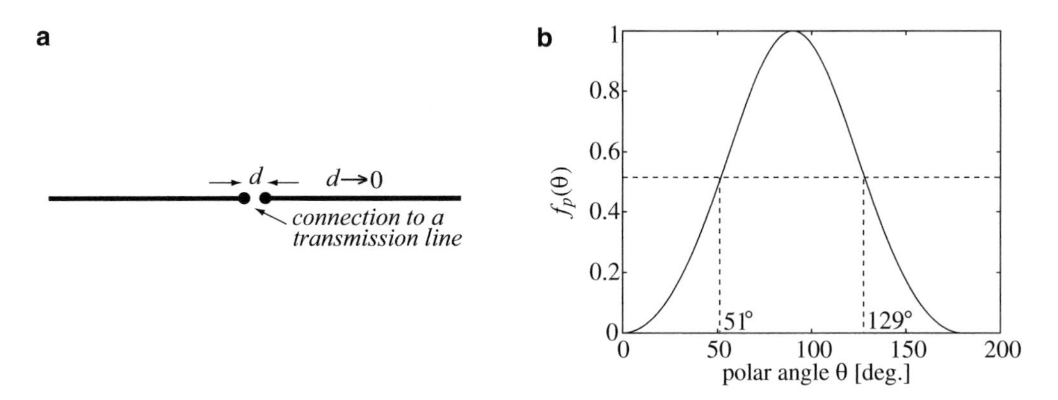

Figure 18.18 (**a**) Horizontal half-wavelength dipole antenna. (**b**) Normalized power pattern: *E*-plane, rectangular plot

Solution: First, we must find the wavelength to define the dimensions of the antenna. The power radiated is calculated from the radiation resistance, the resistance of the antenna, and **Eq. (18.57)**. The power radiation pattern is the same as for any half-wavelength dipole and is given in **Eq. (18.103)**:

(**a**) The wavelength at 29 MHz is

$$\lambda = \frac{c}{f} = \frac{3 \times 10^8}{29 \times 10^6} = 10.345 \quad [\text{m}]$$

Thus, a half-wavelength antenna is 5.172 m long. Each half of the dipole is 2.586 m long.

(**b**) The radiation resistance is

$$R_{rad} = \frac{0.609\eta_0}{\pi} = \frac{0.609 \times 377}{\pi} = 73.08 \quad [\Omega]$$

From the radiation and antenna resistances, we can calculate the antenna efficiency:

$$eff = \frac{R_{rad}}{R_{rad} + R_d} = \frac{73.08}{73.08 + 5} = 0.936$$

Since efficiency is the ratio between radiated and input power, we get

$$eff = \frac{P_{rad}}{P_{in}} = 0.936 \quad \rightarrow \quad P_{rad} = 0.936P_{in} = 0.936 \times 100 = 93.6 \quad [\text{W}]$$

The antenna radiates 93.6 W.

(**c**) The time-averaged power density anywhere in space is given in **Eq. (18.99)**:

$$\mathcal{P}_{av} = \frac{\eta I_0^2 \cos^2((\pi/2)\cos\theta)}{8\pi^2 R^2 \sin^2\theta} \quad \left[\frac{\text{W}}{\text{m}^2}\right]$$

To calculate the power density, we need the current I_0. This is evaluated either from the input power as $P_{in} = I_0^2(R_{rad} + R_d)/2$ or from the radiated power as $P_{in} = I_0^2(R_{rad})/2$. Taking the first, we get

$$I_0^2 = \frac{2P_{in}}{R_{rad} + R_d} = \frac{200}{73.08 + 5} = 2.561 \quad [\text{A}^2]$$

Substituting this, $\eta = \eta_0$, and $R = 100{,}000$ m into **Eq. (18.99)**, we get the time-averaged power density at 100 km from the antenna:

$$\mathcal{P}_{av} = \frac{377 \times 2.561 \times \cos^2((\pi/2)\cos\theta)}{8 \times \pi^2 \times 10^{10} \times \sin^2\theta} = 1.223 \times 10^{-9} \frac{\cos^2((\pi/2)\cos\theta}{\sin^2\theta} \quad \left[\frac{\text{W}}{\text{m}^2}\right].$$

(d) This is plotted in the normalized rectangular plot in **Figure 18.18b**. Note the beamwidth, which is 78° (it was 90° for the Hertzian dipole). Also, because of the orientation of the antenna, maximum radiation is up (or down) relative to the antenna.

Example 18.8 Application: TV Antenna An antenna has been designed as a half-wavelength dipole for use with a TV transmitter at 600 MHz (UHF channel 35) and the transmitter supplies 50 kW to the antenna. The antenna is 6 mm thick and made of aluminum with conductivity $\sigma = 3.5 \times 10^7$ S/m. Calculate:

(a) The radiated power at 600 MHz.
(b) The efficiency of the antenna.

Solution: The antenna at 600 MHz is one-half wavelength. To calculate the radiated power, we must first calculate the current in the antenna. This, in turn, is calculated from the radiation resistance and internal antenna resistance using **Eq. (18.56)**. The radiation resistance of a half-wavelength dipole in free space is given in **Eq. (18.102)**. The internal antenna resistance is calculated by assuming the current to be confined to one skin depth on the surface of the antenna:

(a) First, we must find the length of the antenna. Since it is $\lambda/2$ at 600 MHz, we get

$$\lambda = \frac{c}{f} = \frac{3 \times 10^8}{600 \times 10^6} = 0.5 \quad \rightarrow \quad L = 0.25 \quad [\text{m}]$$

To find the radiated power, we must first find the internal resistance and radiation resistance of the antenna since we only know the input power to the antenna. The internal resistance of the antenna depends only on the resistance of the conductor and this may be calculated from the skin depth of aluminum at the corresponding frequency. We assume the current to be entirely in the skin of the antenna and limited to one skin depth. The skin depth at 600 MHz is

$$\delta = \frac{1}{\sqrt{\pi f \mu_0 \sigma}} = \frac{1}{\sqrt{\pi \times 600 \times 10^6 \times 4 \times \pi \times 10^{-7} \times 3.5 \times 10^7}} = 3.473 \times 10^{-6} \quad [\text{m}]$$

We use the equation for resistance of a hollow cylinder, $R = L/\sigma S$, where L is the length of the cylinder and S is the cross-sectional area of the hollow cylinder. In our case, the cylinder is 6 mm in diameter and one skin depth thick. Thus, the resistance of the antenna is

$$R_d = \frac{L}{S\sigma} = \frac{L}{2\pi r \delta \sigma} = \frac{0.25}{2 \times \pi \times 0.003 \times 3.473 \times 10^{-6} \times 3 \times 10^7} = 0.1091 \quad [\Omega]$$

The radiation resistance of a half-wavelength dipole in free space is $R_{rad} = 73.08\ \Omega$. The current in the antenna at 600 MHz is calculated from **Eq. (18.56)**:

$$P_{in} = \frac{I^2(R_{rad} + R_d)}{2} \quad [\text{W}] \quad \rightarrow \quad I^2 = \frac{2P_{in}}{R_{rad} + R_d} = \frac{10^5}{73.08 + 0.1091} = 1,366.32 \quad [\text{A}^2]$$

The current in the antenna is, therefore, $I = 36.96$ A. Using **Eq. (18.36)**, the radiated power is

$$P_{rad} = \frac{I^2 R_{rad}}{2} = \frac{1{,}366.32 \times 73.08}{2} = 49{,}925.33 \quad [\text{W}].$$

(b) The efficiency of the antenna is given in **Eq. (18.55)**:

$$eff = \frac{P_{rad}}{P_{in}} \times 100 = \frac{49{,}925.33}{50{,}000} \times 100 = 99.85\%$$

That is, only 0.16% of the input power is dissipated in the antenna. This is a very high efficiency for an antenna and is owed to its low internal resistance.

Example 18.9 Application: Shifting of Transmission Frequency A dipole antenna has been designed as a half-wavelength dipole for use with a TV transmitter at 600 MHz and the transmitter supplies 30 A to the antenna. The antenna may be considered to be thin and has an internal resistance of 1 Ω. Because of changes in channel allocation, the TV station must change the frequency to 690 MHz without replacing the antenna. Calculate:

(a) The radiated power, radiation efficiency, and input power supplied by the transmitter at 600 MHz.
(b) The radiated power and radiation efficiency at 690 MHz. Assume the transmitter supplies the same input power to the antenna at both frequencies.
(c) Find the maximum directivity of the antenna at 600 and 690 MHz.

Solution: At 600 MHz, the antenna is one-half wavelength. The wavelength was calculated in **Example 18.8** as 0.5 m. Thus, the antenna is exactly 0.25 m long. Since the antenna radiates in free space, the radiation resistance is 73.08 Ω and the radiated power is calculated using **Eq. (18.100)**. Similarly, the antenna efficiency is calculated from the radiation and internal resistances using **Eq. (18.57)**. Maximum directivity of the antenna at 600MHz is 1.642 [from **Eq. (18.105)**], but at 690 MHz, it must be calculated from the directivity in **Eq. (18.94)**:

(a) The radiated power at 600 MHz is

$$P_{rad} = \frac{I_0^2 R_{rad}}{2} = \frac{30^2 \times 73.08}{2} = 32{,}886 \quad [\text{W}]$$

The radiation efficiency is found from **Eq. (18.57)**:

$$eff = \frac{R_{rad}}{R_{rad} + R_d} = \frac{73.08}{73.08 + 1} = 0.9865$$

The radiation efficiency is 98.65 %. The input power is therefore

$$P_{in} = \frac{P_{rad}}{\eta_{rad}} = \frac{32{,}886}{0.9865} = 33{,}336 \quad [\text{W}].$$

(b) At 690 MHz, the antenna length is 0.575λ since

$$\frac{\lambda_{600}}{\lambda_{690}} = \frac{690}{600} = 1.15 \quad \rightarrow \quad 1.15 \times 0.5\lambda = 0.575\lambda$$

To calculate the radiated power, we must first calculate the radiation resistance, then the radiation efficiency, and then the radiated power from the input power at 600 MHz. Using **Eq. (18.93)** with $L = 0.575\lambda$ we get

$$R_{rad} = \frac{\eta_0}{2\pi} \int_{\theta=0}^{\theta=\pi} \frac{(\cos((0.575\pi)\cos\theta) - \cos(0.575\pi))^2}{\sin\theta} d\theta = \frac{120\pi}{2\pi} \times 1.795 = 107.7 \quad [\Omega]$$

where the value of 1.795 is the value of the integral in **Eq. (18.92)** for $L = 0.575\lambda$. The internal resistance of the antenna remains the same since the physical dimensions of the antenna have not changed (in practice, the resistance increases slightly because of the decrease in skin depth but we neglect this effect). Thus, the radiation efficiency is

$$eff = \frac{R_{rad}}{R_{rad} + R_d} = \frac{107.7}{107.7 + 1} = 0.99$$

This is an efficiency of 99%. The radiated power at 690 MHz for the input power level at 600 MHz is

$$P_{rad} = P_{in}eff = 33,336 \times 0.99 = 33,002 \quad [\mathrm{W}]$$

Thus, the antenna radiates slightly more power at 690 MHz than at 600 MHz since it has a slightly higher efficiency.

(c) Maximum directivity of the antenna is the maximum value of **Eq. (18.104)** for the half-wavelength antenna or of **Eq. (18.94)** for any antenna.

At 600 MHz, the maximum directivity is

$$d_{600} = 1.642$$

At 690 MHz, we must use the value of 1.795 for the integral. Doing so gives

$$d_{690} = \frac{2}{1.795} \times 1.5214 = 1.695$$

where 1.5214 is the maximum value of the power radiation pattern in **Eq. (18.95)** (obtained by substituting $L = 0.575\lambda$ at $\theta = 90°$). The antenna has higher directivity and higher radiation resistance at 690 MHz.

18.7.2 The Monopole Antenna waves.m

The monopole antenna was introduced at the beginning of **Section 18.7** (see **Figure 18.13b**). It consists of an element of some given length, above and perpendicular to a conducting plane. The antenna is fed between one end of the element and the conducting plane. Here, the conducting plane will be called a ground plane, but it may be the metal surface of a car or any other conducting surface. The principle involved is shown in **Figure 18.19a**. Although we call it a monopole antenna, it really is a modified dipole antenna. This can be seen from **Figures 18.19b**. The ground plane, which must be conducting, can be replaced by an image element as shown in **Figure 18.19b**. For example, for a quarter wavelength monopole, the equivalent antenna is a half-wavelength dipole. Therefore, there is no need to analyze monopole antennas of this type separately. We will use the results obtained for the dipole antennas in **Sections 18.4** through **18.7.1** with a few modifications to take into account the existence of the ground plane. Before we do so, we note that from image theory, the following apply:

(1) The length of the image element and the current in it are identical to the length and current of the monopole. The currents in both elements are in the same direction as shown in **Figure 18.19b**. To see that this must be the case, imagine each end of the antenna element to have a charge which produces the current. Because of the opposite nature of image charges, the currents must be in the same direction as shown.

(2) The fields of the dipole exist everywhere in space (above and below the ground plane), but the dipole solution only represents the field of the monopole antenna above the plane. In other words, the fields of the dipole and monopole in **Figure 18.19b** are the same everywhere above the ground plane.

(3) Power is radiated only above the plane. Since the fields of the monopole are identical to the dipole fields in this region, we immediately conclude that the radiated power of a monopole must be half the radiated power of an equivalent dipole (see **Figure 18.19c**).

(4) Because the power radiated is halved, the radiation resistance of the monopole is half the radiation resistance of the equivalent dipole as can be seen from **Eq. (18.35)**.

To summarize these results for the quarter-wavelength monopole, we use the results for the half-wavelength dipole described in **Section 18.7.1.1**. The electric and magnetic field intensities are given in **Eqs. (18.97)** and **(18.98)**, and the time-averaged power density is given in **Eq. (18.99)**.

The radiated power is obtained from **Eq. (18.99)** by integrating over the half-sphere above ground level since the fields below ground do not exist:

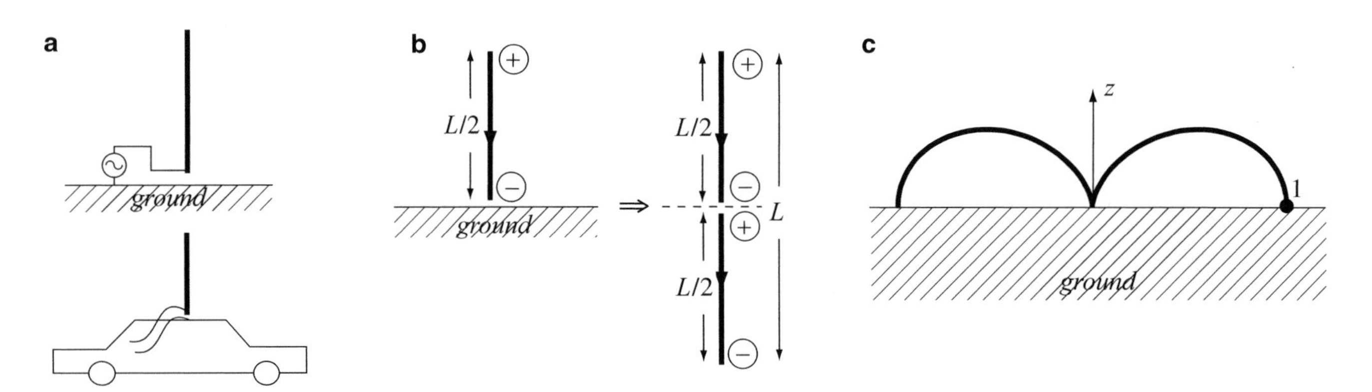

Figure 18.19 The monopole antenna. (**a**) Two monopole antennas showing the ground planes. (**b**) The monopole and the equivalent dipole antenna. (**c**) The radiation pattern of the monopole antenna

$$P_{rad} = \frac{\eta I_0^2}{8\pi^2}\int_{\phi=0}^{\phi=2\pi}\left[\int_{\theta=0}^{\theta=\pi/2}\frac{\cos^2((\pi/2)\cos\theta)}{\sin\theta}d\theta\right] = \frac{1.218\eta I_0^2}{8\pi} \quad \left[\frac{\text{W}}{\text{m}^2}\right] \tag{18.107}$$

The total radiated power is half that of the equivalent dipole. The radiation resistance is

$$R_{rad} = \frac{0.609\eta}{2\pi} \quad [\Omega] \tag{18.108}$$

In free space, the radiation resistance is (with $\eta = 120\pi$)

$$R_{rad} = \frac{0.609 \times 120\pi}{2\pi} = 36.54 \quad \Omega \tag{18.109}$$

and is clearly half the radiation resistance of the equivalent dipole antenna. The radiation pattern of the antenna remains the same as for the dipole, but only that portion of the pattern above $\theta = \pi/2$ exists. The normalized radiation pattern for the electric field and power are obtained from **Eqs. (18.97)** and **(18.99)** as

$$|f_e(\theta)| = \left|\frac{\cos((\pi/2)\cos\theta)}{\sin\theta}\right|, \quad f_p(\theta) = \frac{\cos^2((\pi/2)\cos\theta)}{\sin^2\theta} \quad 0 \le \theta \le \frac{\pi}{2} \tag{18.110}$$

The normalized field radiation pattern is shown in **Figure 18.19c**.

The directivity and maximum directivity remain the same as for the dipole antenna since these are ratios of power:

$$D(\theta) = 1.642\frac{\cos^2((\pi/2)\cos\theta)}{\sin^2\theta}, \quad 0 \le \theta \le \frac{\pi}{2} \tag{18.111}$$

$$D_0 = D(\theta = \pi/2) = 1.642 \tag{18.112}$$

In summary, the monopole behaves like a dipole but radiates only half the power. On the other hand, the power density is the same since it only radiates into the upper half-space. Thus, the monopole is often used in lieu of the dipole antenna because it is a very simple antenna, particularly when a ground plane is readily available, such as on a car, rooftop, and the like. We have discussed here the $\lambda/4$ monopole but the performance of any other monopole can be deduced from the equivalent dipole.

18.8 Antenna Arrays

Now that we can analyze simple linear antennas and know their radiation characteristics, we may wish to ask ourselves the following two questions: First, suppose two or more separate antennas are placed close to each other and operate at the same frequency. What is the net effect of the two antennas radiating? Second, if we wish to obtain a particular radiation pattern, what type of antenna should we use? The first of these questions is encountered when two antennas either happen to be close to each other or when an image antenna is produced either purposely or inadvertently. As an example, suppose two mobile telephones are installed in a car, each with its own (monopole) antenna. Another situation is shown in **Figure 18.20**. A dipole antenna is placed above the conducting surface of a car to transmit or receive, say, TV signals. Because of the conducting surface, an image of the whole antenna is generated below the surface, replacing the effect of the conductor. The effect is the same as having two antennas, each radiating or receiving separately at the same frequency and amplitude; the fields, power, directivity, and all other properties are those of the combined antennas.

Figure 18.20 A horizontal dipole over a ground plane and the equivalent configuration

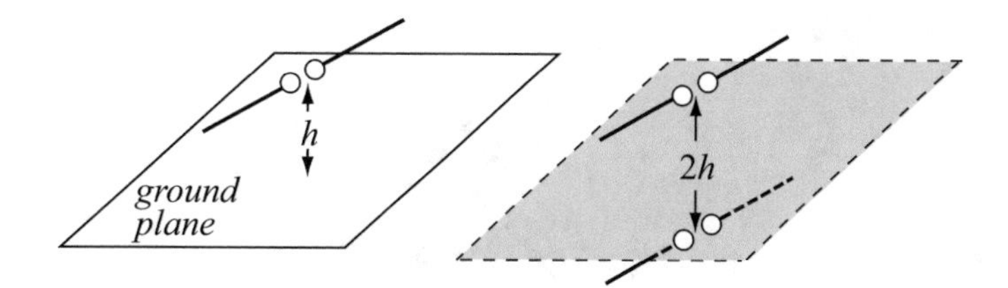

The second question relates to the need to generate specific radiation patterns. For example, a TV station intends to transmit into and around a city. Normally, it would like to transmit uniformly in all horizontal directions. Any energy transmitted upward is wasted. In this case, the vertical dipole may be used since its H-plane pattern is circular as can be seen in **Figure 18.21a**. At any radius, the intensity will be the same and reception will be uniform. Now, if the same TV station is placed on the seashore, transmitting outward toward the sea is again wasteful. The TV station would now like to change the radiation pattern such that none, or very little, of the power it transmits is directed toward the sea. The preferred radiation pattern now would be more like a half-circle, as shown in **Figure 18.21b**. Thus, the question the design engineer must answer is this: how to generate this or any other radiation pattern. The answer is in antenna arrays.

Figure 18.21 Design of a radiation pattern for required coverage. (**a**) Uniform coverage. (**b**) Semicircular coverage

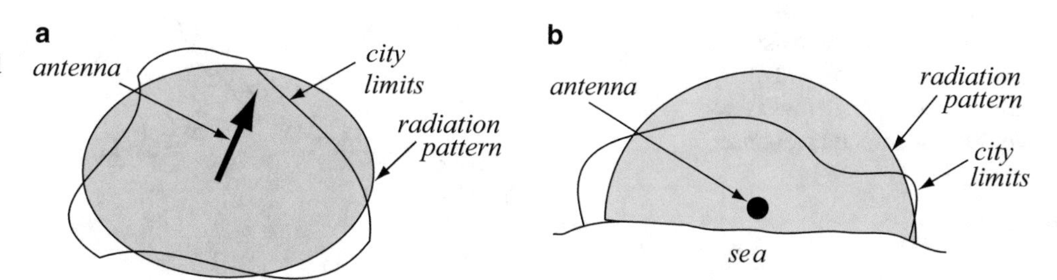

An antenna array is an ensemble of antennas of any type, radiating simultaneously. All antennas may be of the same type or different types and may be of the same size or not, depending on the requirements. In practice, antenna arrays are mostly made of one type of antennas and all antennas are of the same size. However, there is considerable variation in designs. For example, an antenna array may be a linear array made of dipoles, spaced some distance apart as shown in **Figure 18.22a**. Two-dimensional arrays can be viewed as linear arrays of antennas as shown in **Figure 18.22b**. Of course, the dipole shown may be replaced by any other type of antenna. A very large antenna array, made of 27 parabolic (dish) antennas, each 25 m in diameter, exists in New Mexico. The antennas are arranged in a *Y*-shaped pattern, each leg approximately 21 km long. Individual elements can move on tracks so that the array can be reconfigured. The array operates as a radio telescope designed to receive extraterrestrial signals. This array is called the very large array (VLA) and is, currently, the largest array available **(Figure 18.23a)** in terms of number of elements in the array. The signal of the array is synthesized from the signals of the individual antennas and using the Earth's rotation as a means of increasing the effective area (aperture) of the antenna. Another approach, adopted for very large antennas used in astrophysics, is the use of a single antenna or a small number of antennas on a track **(Figure 18.23b)**. The antenna receives a signal and then is moved to a new location and the process is repeated. The signal is then synthesized from the different signals received, taking into account the location of the antenna and the time, phase, propagation constants, and other differences introduced due to the change in location and shifts in time. The latter approach requires considerable computation resources to compensate for the lack of equipment. An array of this type is operational in Calgura, Australia, which operates as a radio telescope. The antennas are 32 m in diameter and move on a rail track. Yet another approach is the use of fixed antennas at different locations. One such array is the Very Long Baseline Array (VLBA). It is a system of 10 radio telescopes located in St. Croix (US Virgin Islands), Hancock (New Hampshire), North Liberty (Iowa), Los Alamos and Pie Town (New Mexico), Kitt Peak (Arizona), Owens Valley (California), Brewster (Washington) and Mauna Kea (Hawaii). The longest baseline it can form is 8611 km. It is controlled from Socorro (New Mexico). The signals from each antenna are recorded and time stamped and then processed to produce what is called baseline interferometry. An array of this type is adaptive and additional antennas can be added, primarily to improve sensitivity.

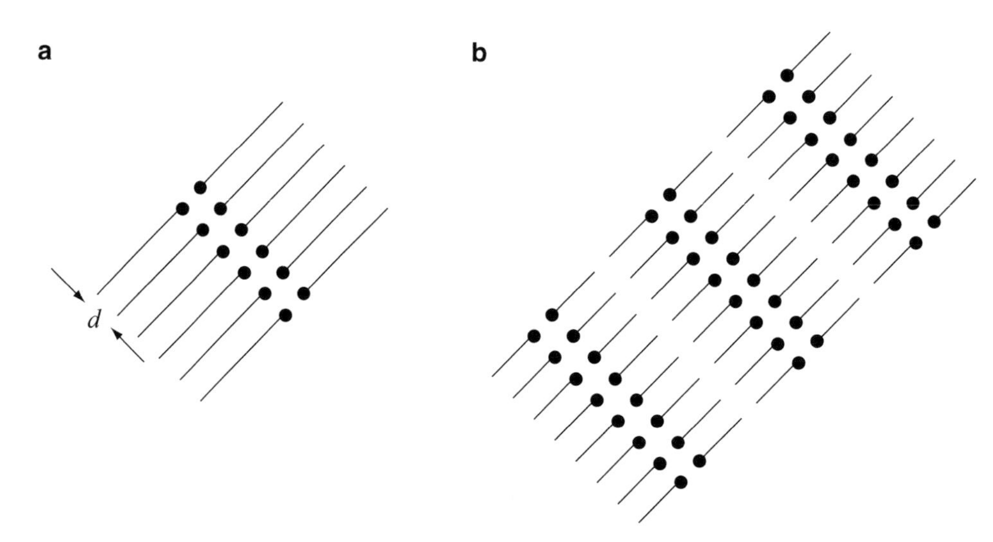

Figure 18.22 (**a**) A linear array of dipole antennas. (**b**) A two-dimensional array of dipole antennas

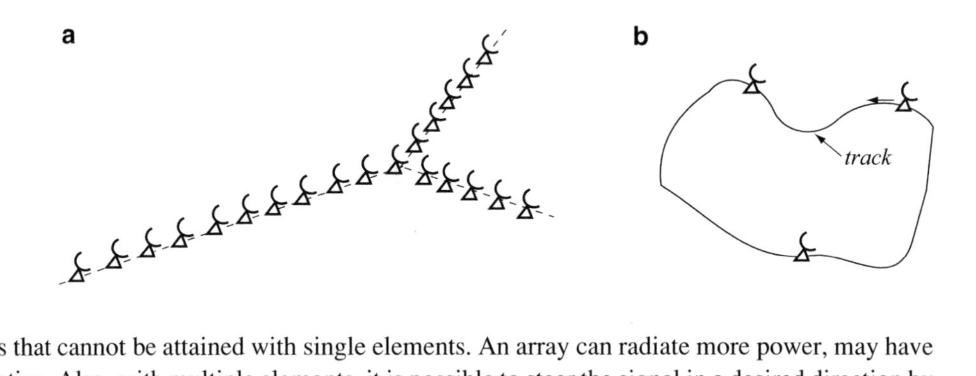

Figure 18.23 (**a**) The Very Large Array (VLA). (**b**) A synthesized array consists of a movable antenna and computers that calculate the equivalent antenna response

Array antennas have properties that cannot be attained with single elements. An array can radiate more power, may have higher gain, and can be more directive. Also, with multiple elements, it is possible to steer the signal in a desired direction by changing the phase of the various elements in the array to produce a desired radiation pattern. This type of array is called a ***phased array***. One particularly useful application is in radars, where scanning is normally done by physically moving the antenna in a pattern. The same can be achieved by an array of fixed antennas and the scanning is accomplished by phasing the elements. There are no moving elements, and scanning can be done in any pattern, often faster than with conventional mechanical scanners.

Another type of modern antenna array is the ***adaptive array***. It consists of a number of antennas, whose properties can be adapted to various, changing conditions such as direction of transmission or required power. These arrays are common in cell phone base stations.

We will discuss antenna arrays here as the superposition of two or more independent antennas, each radiating as if all other antennas did not exist. In this sense, we have already done so in defining the linear antennas in **Section 18.7**, where we used superposition of an infinite number of infinitesimal Hertzian dipoles. The antenna arrays discussed here are different only in that the elements of the antenna arrays are finite in size (and sometimes quite large) and the number of elements in the array is finite and often quite small. We will first discuss the idea of the two-element antenna array as an introduction, followed by the general question of antenna arrays with any number of elements.

18.8.1 The Two-Element Array

As an example of a two-element antenna array, consider **Figure 18.24. Figure 18.24a** shows a dipole antenna (for example, a half-wavelength dipole) above a conducting ground. **Figure 18.24b** shows the equivalent antenna, including the image antenna. This is an array of two parallel antennas. The field above ground is then the combined fields of the two antennas. Below ground, of course, there is no field. Note also that the currents in the two antennas are in opposite directions which means that the antenna cannot be too close to the ground (in terms of wavelengths) or the fields may cancel, at least partially. **Figure 18.25** shows a vertical dipole and the image antenna resulting in a two element collinear array. Again, although the fields will be different than in **Figure 18.24**, the analysis is the same: the field of each antenna is calculated and then summed together.

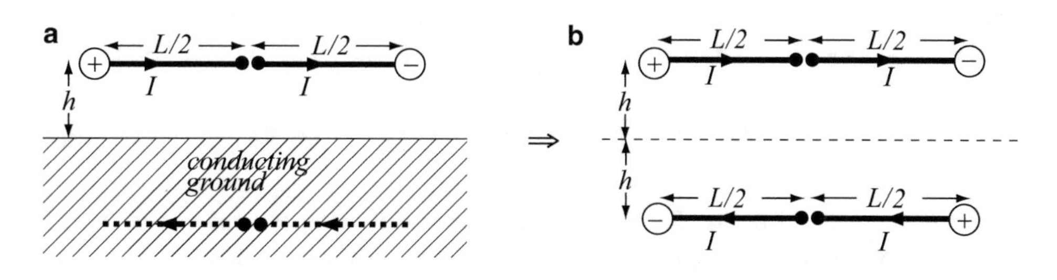

Figure 18.24 Two-element antenna. (**a**) Horizontal dipole above ground. (**b**) The dipole and its image form a two element parallel antennas array

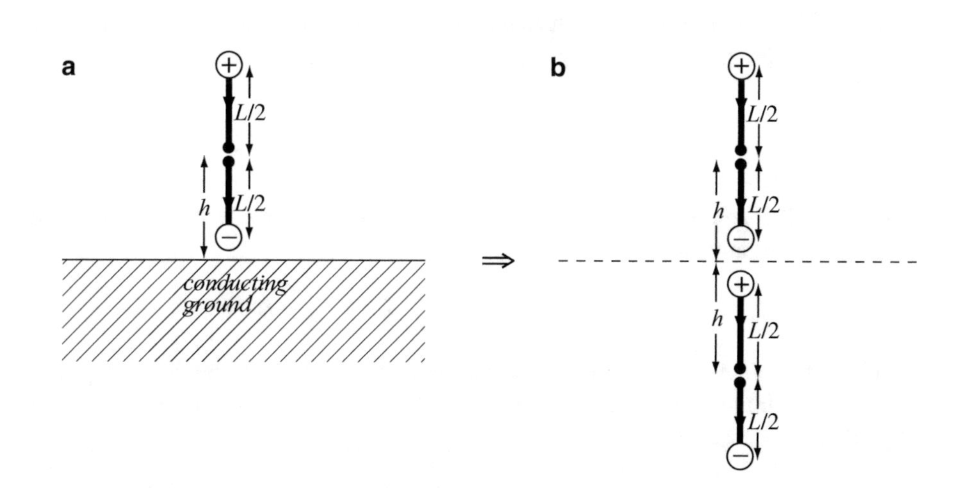

Figure 18.25 (**a**) A vertical dipole above ground. (**b**) The dipole and its image result in a two element collinear array

18.8.1.1 Two Element Parallel Antennas Array

Consider the configuration in **Figure 18.26a**. Two half-wavelength dipole antennas are located on the x axis, separated a distance h apart, and are oriented parallel to the z axis. For simplicity, we will assume the two antennas carry identical currents except for a phase shift between the two currents, which we indicate as φ. The currents in the two antennas are

$$I_1 = I_0 e^{j\omega t}, \quad I_2 = I_0 e^{j\omega t} e^{j\varphi} \quad [\text{A}] \tag{18.113}$$

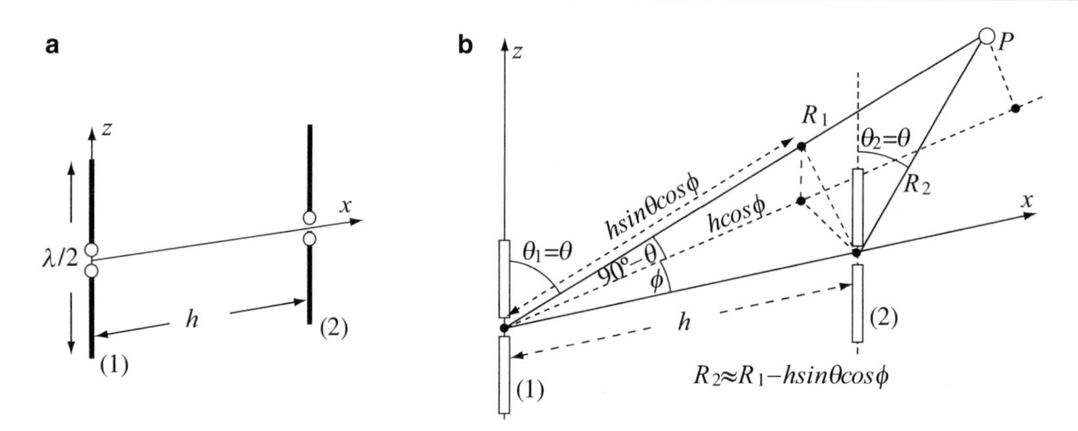

Figure 18.26 (**a**) A two-element array made of two vertical dipoles. (**b**) Relations between the coordinates used for calculation of the radiation field

where we assumed, arbitrarily, that the current in antenna (2) leads by a phase angle φ relative to antenna (1). Now, using **Eq. (18.97)** and **Figure 18.26b**, we can write the electric field intensity of antenna (1) as

$$\mathbf{E}_1 = \hat{\boldsymbol{\theta}}_1 \frac{j\eta I_0}{2\pi R_1} e^{-j\beta R_1} \frac{\cos((\pi/2)\cos\theta_1)}{\sin\theta_1} = \hat{\boldsymbol{\theta}}_1 E_0 \frac{e^{-j\beta R_1}}{R_1} f_1(\theta_1, \phi) \quad \left[\frac{\text{V}}{\text{m}}\right] \tag{18.114}$$

where E_0 is the constant term $j\eta I_0/2\pi$ and $f_1(\theta_1,\phi)$ is the field radiation pattern as defined in **Eq. (18.103)**. Note that the electric field intensity is in the θ direction because the dipoles are parallel to the z axis. Similarly, for antenna (2), we can write

$$\mathbf{E}_2 = \hat{\boldsymbol{\theta}}_2 \frac{j\eta I_0}{2\pi R_2} e^{j\varphi} e^{-j\beta R_2} \frac{\cos((\pi/2)\cos\theta_2)}{\sin\theta_2} = \hat{\boldsymbol{\theta}}_2 E_0 \frac{e^{-j\beta R_2} e^{j\varphi}}{R_2} f_2(\theta_2, \phi) \quad \left[\frac{\text{V}}{\text{m}}\right] \tag{18.115}$$

Because we are in the far field, we can assume that $\theta_1 \approx \theta_2 \approx \theta$, as can be seen from **Figure 18.26b** (for $R \gg h$). Therefore, we can write $f(\theta_1,\phi) \approx f(\theta_2,\phi) \approx f(\theta,\phi)$. Also, we have chosen the center of antenna (1) as the reference point. However, we are free to choose other reference points, including the center of antenna (2) or the middle point between the centers of the two antennas. For an example of the latter choice, see **Example 18.10**.

The total electric field intensity at point P is the sum of the above two fields:

$$\mathbf{E} = \mathbf{E}_1 + \mathbf{E}_2 = \hat{\boldsymbol{\theta}} E_0 f(\theta, \phi) \left(\frac{e^{-j\beta R_1}}{R_1} + \frac{e^{-j\beta R_2} e^{j\varphi}}{R_2} \right) \quad \left[\frac{\text{V}}{\text{m}}\right] \tag{18.116}$$

where ϕ is the azimuthal angle. To simplify the expression, we write R_2 in terms of R_1. From **Figure 18.26b** (assuming $R \gg h$) and for any position in space,

$$R_2 \approx R_1 - h\sin\theta\cos\phi \tag{18.117}$$

The terms in the denominator may be approximated as $R_2 \approx R_1$, but in the exponent, this approximation is not reasonable because the phase constant may be large and we must use the approximations in **Eq. (18.117)**. Substituting these in **Eq. (18.116)**, we get

$$\mathbf{E} = \hat{\boldsymbol{\theta}} \frac{E_0 f(\theta, \phi)}{R_1} e^{-j\beta R_1} \left(1 + e^{j(\beta h\sin\theta\cos\phi + \varphi)} \right) \quad \left[\frac{\text{V}}{\text{m}}\right] \tag{18.118}$$

To simplify notation, we denote

$$\psi = \beta h\sin\theta\cos\phi + \varphi \quad [\text{rad}] \tag{18.119}$$

The term in parentheses in **Eq. (18.118)** now becomes

$$\left(1 + e^{j(\beta h \sin\theta\cos\phi + \varphi)}\right) = \left(1 + e^{j\psi}\right) = e^{j\psi/2}\left(e^{-j\psi/2} + e^{j\psi/2}\right) = e^{j\psi/2} 2\cos\frac{\psi}{2} \quad (18.120)$$

where the function $e^{j\alpha} + e^{-j\alpha} = 2\cos\alpha$ was used. Finally, substituting this into **Eq. (18.118)** gives:

$$\mathbf{E} = \hat{\boldsymbol{\theta}} \frac{E_0 f(\theta, \phi)}{R_1} e^{-j\beta R_1} e^{j\psi/2} 2\cos\frac{\psi}{2} \quad \left[\frac{\text{V}}{\text{m}}\right] \quad (18.121)$$

This may now be viewed as the electric field intensity of the two-element antenna array shown in **Figure 18.27a**. You will recall that we started by discussing the radiation pattern of the antenna. Now, we can write this by noting that the electric field intensity in **Eq. (18.121)** can be written as the product of an amplitude and two antenna patterns:

$$E = \frac{2E_0 e^{-j\beta R_1} e^{j\psi/2}}{R_1} f(\theta, \phi) \cos\frac{\psi}{2} = E_0' f(\theta, \phi) f_n(\theta, \phi, \varphi) \quad \left[\frac{\text{V}}{\text{m}}\right] \quad (18.122)$$

where the pattern $f(\theta,\phi)$ is the individual ***normalized antenna radiation pattern*** or ***element radiation pattern*** and f_n is the ***normalized array factor:***

$$|f_n(\theta, \phi, \varphi)| = \left|\cos\left(\frac{\psi}{2}\right)\right| = \left|\cos\left(\frac{\beta h \sin\theta \cos\phi + \varphi}{2}\right)\right| \quad (18.123)$$

To plot the radiation pattern for the array, we can proceed in two ways. The first is to write the explicit expression

$$E = E_0' \frac{\cos((\pi/2)\cos\theta}{\sin\theta} \cos\left(\frac{\beta h \sin\theta \cos\phi + \varphi}{2}\right) \quad \left[\frac{\text{V}}{\text{m}}\right] \quad (18.124)$$

and then plot the absolute value of this expression for any value of θ, ϕ, and φ. **Equation (18.124)** is the absolute field pattern. Normalized field patterns are obtained by normalizing with respect to E_0'. The plot in **Figure 18.27a** shows the array radiation pattern in the H-plane for $\theta = \pi/2$, ϕ, $\varphi = -\pi/4$, and $h = \lambda/2$. A similar process can be applied to the E-plane pattern.

The second approach is to plot the element radiation pattern and the array factor and calculate the product between the values in the plot. Using the same values as above ($\theta = \pi/2$, ϕ, $\varphi = -\pi/4$, $h = \lambda/2$), we get

$$f(\theta) = 1, \quad |f_n(\phi)| = \left|\cos\left(\frac{\pi}{2}\cos\phi - \frac{\pi}{8}\right)\right| \quad (18.125)$$

Figure 18.27b shows the plot for $|f(\theta = \pi/2)|$ and **Figure 18.27c** shows the plot for $|f_n(\theta = \pi/2, \phi, \varphi = -\pi/8)|$. The product of the two plots gives the third, which is the composite or antenna array radiation pattern. This pattern is maximum at $\phi = \pm 75.6°$ [from **Eq. (18.125)** or from **Figure 18.27c**], which occurs when the point P in **Figure 18.26a** is almost midway between the two antennas (but still in the far field). For this reason, most of the radiation is to the side and the antenna is called a ***broadside antenna***. Other patterns are possible with a two-element antenna depending on the elements themselves (half-wavelength dipole, multiple

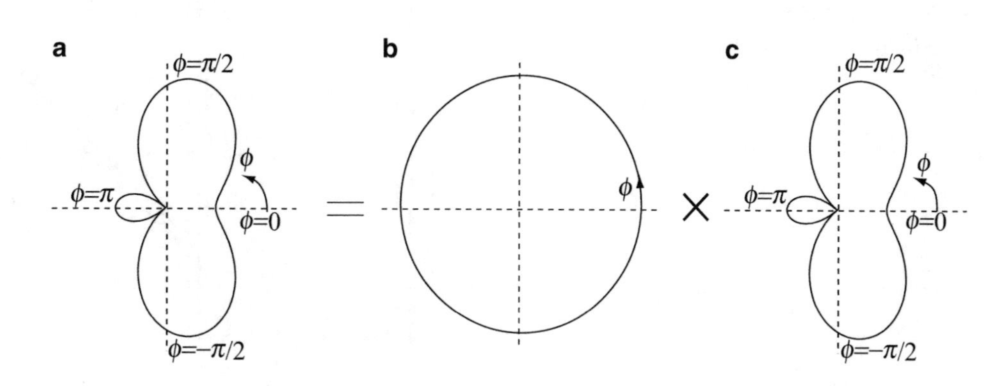

Figure 18.27 (**a**) Normalized radiation pattern of the two-element array in **Figure 18.26a**, as the product of the element radiation pattern (**b**) and the array radiation pattern (**c**)

half-wavelength dipoles, etc.), orientation of the dipoles, the distance between them, and the phase angles between currents. Some of these are explored in the following examples.

Other properties of antennas such as directivity can also be calculated using the relations we used in **Sections 18.5** and **18.7**.

18.8.1.2 Two Element Colinear Antennas Array

Suppose now that we take antenna (2) in **Figure 18.26a** and place it on the z-axis so that the distance between the centers of the two elements is h as shown in **Figure 18.28**. The currents and their phases are as in **Eq. (18.113)**. Therefore, the electric field intensity in the far field is the same as in **Eq. (18.116)**. We can still approximate R_2 in the denominator as $R_2 \approx R_1$ as discussed above. However, in the phase the approximation for R_2 in terms of R_1 is as follows (see **Figure 18.28**):

$$R_2 \approx R_1 - h\cos\theta \tag{18.126}$$

Substituting these approximations in **Eq. (18.116)**, we get

$$\mathbf{E} = \hat{\boldsymbol{\theta}} \frac{E_0 f(\theta, \phi)}{R_1} e^{-j\beta R_1} \left(1 + e^{j(\beta h\cos\theta + \varphi)}\right) \quad \left[\frac{\text{V}}{\text{m}}\right] \tag{18.127}$$

Denoting the term in the exponent as ψ:

$$\psi = \beta h\cos\theta + \varphi \quad [\text{rad}] \tag{18.128}$$

With this **Eq. (18.127)** becomes:

$$\mathbf{E} = \hat{\boldsymbol{\theta}} \frac{E_0 f(\theta, \phi)}{R_1} e^{-j\beta R_1} e^{j\psi/2} 2\cos\frac{\psi}{2} \quad \left[\frac{\text{V}}{\text{m}}\right] \tag{18.129}$$

This is identical inform to **Eq. (18.121)**, the only difference is in the expression for ψ. As a consequence, the difference between the two types of arrays is clearly in the array factor. The normalized array factor now is

$$|f_n(\theta, \phi, \varphi)| = \left|\cos\left(\frac{\psi}{2}\right)\right| = \left|\cos\left(\frac{\beta h\cos\theta + \varphi}{2}\right)\right| \tag{18.130}$$

The array factor then depends on the location and orientation of the elements in the array, the distance between the elements and the progressive phase angle. We will use this property in the following section when we discuss n-element linear arrays.

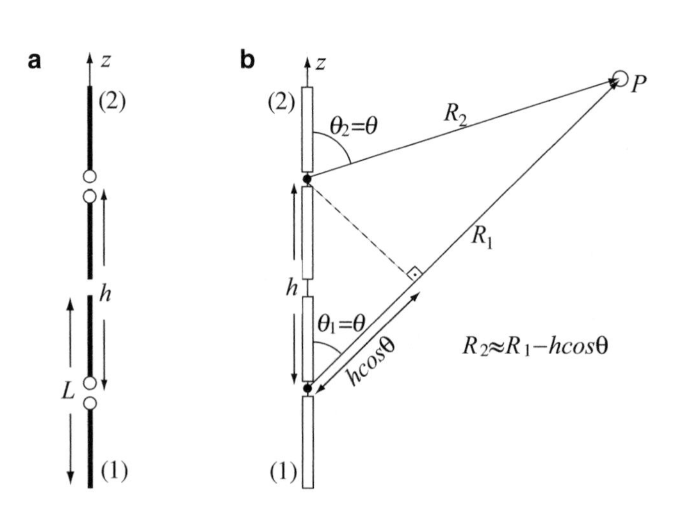

Figure 18.28 A two element co-linear array. (**a**) Geometry. (**b**) Configuration used for calculation of the far field

Example 18.10 A half-wavelength horizontal dipole is placed above a perfectly conducting ground at height h as shown in **Figure 18.29a**. The current in the antenna is $I_0\cos\omega t$ directed as shown:

(a) Find the electric and magnetic field intensities in the far field of the antenna.
(b) Find the normalized power radiation pattern of the antenna.
(c) Show that as the antenna gets closer to the ground, the electric and magnetic field intensities diminish. Show that for $h = 0$ (antenna on the ground), the far fields are zero.

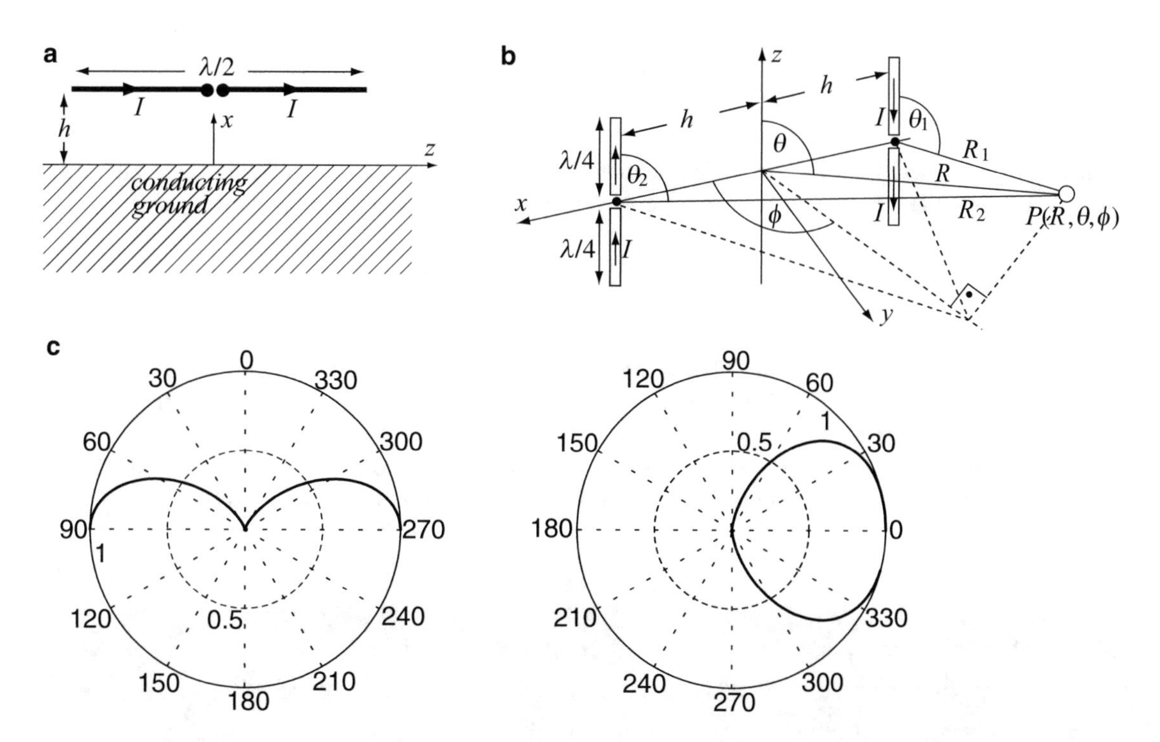

Figure 18.29 (**a**) A horizontal dipole above ground. (**b**) The horizontal dipole and its image, and the dimensions used for calculation of the fields. (**c**) Power radiation pattern of the antenna in the E-plane ($\phi = 0$) and H-plane ($\theta = \pi/2$). Only fields above ground exist

Solution: The dipole produces an image dipole at a depth h below ground. The current in the image antenna is in the opposite direction to that in the dipole. Thus, we have two antennas, at a distance $2h$ with the image antenna at phase π in reference with the dipole. To find the fields, we start with **Eq. (18.116)** but with $\varphi = \pi$ because the current in the image antenna is in the direction opposite that of the dipole. The radiation pattern is then the product of the radiation pattern of the half-wavelength dipole and the array pattern we find in **(a)**:

(a) To calculate the electric field intensity, we use **Figure 18.29b** and assume $\theta_1 \approx \theta_2 \approx \theta$. The reference point is taken at ground level, midway between the elements, and the antennas are placed on the x axis, directed in the z direction so that the relations in **Eqs. (18.114)** through **(18.124)** may be used. This is done for convenience and to show that the reference point may be taken at points other than an antenna. The electric field intensity of the combined antenna and its image is

$$\mathbf{E} = \mathbf{E}_1 + \mathbf{E}_2 = \hat{\boldsymbol{\theta}} E_0 f(\theta)\left(\frac{e^{-j\beta R_1}}{R_1} + \frac{e^{-j\beta R_2}e^{j\pi}}{R_2}\right) = \hat{\boldsymbol{\theta}} E_0 f(\theta)\left(\frac{e^{-j\beta R_1}}{R_1} - \frac{e^{-j\beta R_2}}{R_2}\right) \quad \left[\frac{\text{V}}{\text{m}}\right]$$

where $E_0 = j\eta I_0/2\pi$ and $e^{j\pi} = -1$ were used and $|f(\theta)|$ is the half-wavelength dipole radiation pattern given as

$$|f(\theta)| = \left|\frac{\cos((\pi/2)\cos\theta)}{\sin\theta}\right|$$

Because we are in the far field, the distances R_1 and R_2 may be approximated in terms of the distance R as

$$R_1 \approx R - h\sin\theta\cos\phi, \quad R_2 \approx R + h\sin\theta\cos\phi$$

Substituting these into the expression for the electric field intensity and collecting terms gives

$$\mathbf{E} = \hat{\mathbf{\theta}} E_0 f(\theta)\frac{e^{-j\beta R}}{R}\left(e^{j\beta b\sin\theta\cos\phi} - e^{-j\beta b\sin\theta\cos\phi}\right) = \hat{\mathbf{\theta}} E_0 f(\theta)\frac{e^{-j\beta R}}{R} j2\sin(\beta h\sin\theta\cos\phi) \quad \left[\frac{\text{V}}{\text{m}}\right]$$

Substituting for E_0 and $f(\theta)$, the electric field intensity in the far field is

$$\mathbf{E} = -\hat{\mathbf{\theta}}\frac{\eta I_0 e^{-j\beta R}}{\pi R}\left(\frac{\cos((\pi/2)\cos\theta)}{\sin\theta}\right)(\sin(\beta h\sin\theta\cos\phi)) \quad \left[\frac{\text{V}}{\text{m}}\right].$$

(b) The normalized power radiation pattern of the antenna is the product of the normalized power radiation pattern of the dipole ($f^2(\theta)$) and the array factor squared $f_n^2(\theta,\phi)$. The array normalized power radiation pattern is

$$f_p(\theta,\phi) = \left(\frac{\cos((\pi/2)\cos\theta)}{\sin\theta}\right)^2 (\sin(\beta h\sin\theta\cos\phi))^2$$

This is plotted in two planes in **Figure 18.29c**. The first is for $\phi = 0$. This gives a plot in the plane of the antenna and its image. Note that θ only varies between zero and $\pi/2$ (the pattern only exists above ground). The second is for $\theta = \pi/2$. This gives a plot parallel to the ground.

(c) As h approaches zero, the term $\beta h\sin\theta\cos\phi$ approaches zero. For $h = 0$, the array factor is zero and the electric field intensity (as well as the magnetic field intensity) is zero. The horizontal dipole, therefore, should not be too close to the ground.

Example 18.11 A vertical dipole operates at 600 MHz, carries a current $I = 1$ A, is 1.5 wavelengths long, and radiates in free space. The center of the dipole is 2 m above ground level as shown in **Figure 18.30a**. Calculate:

(a) The electric and magnetic field intensities in the far field.
(b) The array factor and the normalized field and power radiation patterns of the antenna.

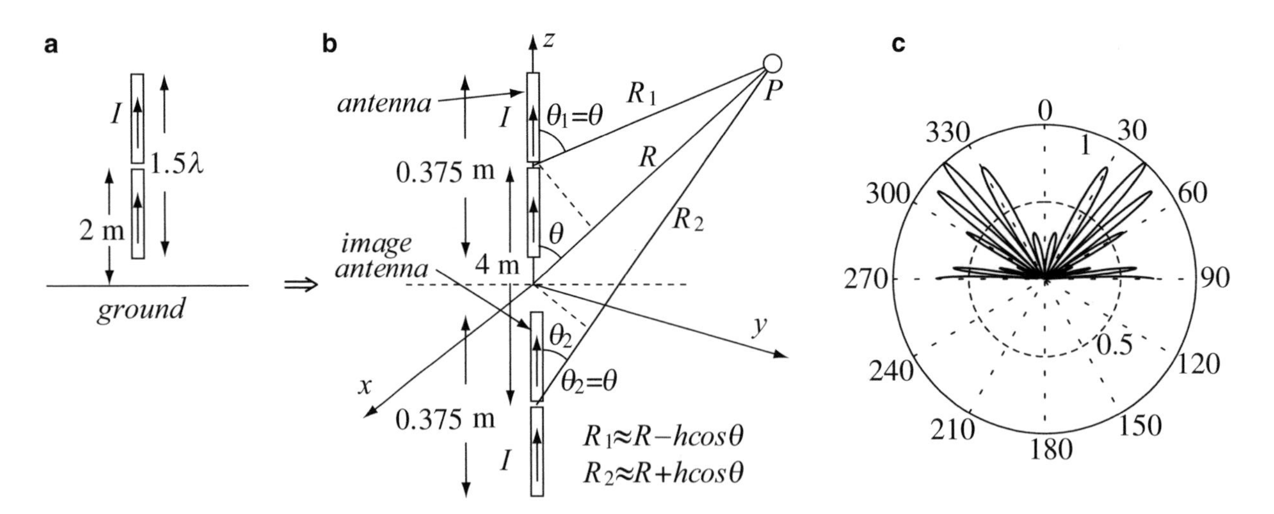

Figure 18.30 (**a**) A vertical dipole above ground. (**b**) The dipole and its image and notation used for calculation of fields. (**c**) Plot of field radiation pattern in the vertical plane (E-plane) above ground

Solution: The wavelength at 600 MHz is 0.5 m. Thus the antenna elements are 0.375 m each. The dimensions of the antenna and its image are as shown in **Figure 18.30b**. The image antenna is also a dipole, identical in dimensions and current to the given antenna, but placed $h = 2$ m below the location of the surface. As with any image method, the ground is removed for the purpose of calculation and the solution only applies above ground. Thus, we can use here the results in **Section 18.8.1.2** with $2h = 4$ m but for a dipole 1.5λ long and using the ground point (the middle between the antenna and its image) as reference. Also, the currents in the two antennas are in phase ($\varphi = 0$):

(a) The electric and magnetic fields of a 1.5λ dipole are given in **Eq. (18.87)**, if we substitute $n = 3$. The electric field intensity is

$$\mathbf{E} = \hat{\boldsymbol{\theta}} \frac{j\eta I}{2\pi R} e^{-j\beta R} \frac{\cos((3\pi/2)\cos\theta)}{\sin\theta} \quad \left[\frac{\text{V}}{\text{m}}\right].$$

Using the notation in **Figure 18.30a**, we can use the result in **Eq. (18.121)** directly if we note the following:

(1) The current in the antenna and its image are in phase. Therefore $\varphi = 0$.
(2) The distance between the antenna and its image is $2h = 4\text{ m} = 8\lambda$.
(3) The normalized element radiation pattern is $|f(\theta)| = |(\cos(3\pi/2)\cos\theta)/\sin\theta|$.
(4) $\beta = 2\pi f\sqrt{\mu_0\varepsilon_0} = 2\pi f/c = 4\pi$ [rad/m]

However, the angle ψ is different. From **Figure 18.30b**, we can write:

$$R_2 \approx R + h\cos\theta, \quad R_1 \approx R - h\cos\theta$$

Substituting these, the intrinsic impedance of free space and the current in the general expression for the electric field intensity, and summing up the fields of the two dipoles, we get for the electric field intensity at a distance R from the center of the array antenna:

$$\mathbf{E} \approx \hat{\boldsymbol{\theta}} \frac{j\eta I\cos((3\pi/2)\cos\theta)}{2\pi\sin\theta}\left(\frac{e^{-j\beta R_2}}{R} + \frac{e^{-j\beta R_1}}{R}\right)$$

$$= \hat{\boldsymbol{\theta}} \frac{j\eta I\cos((3\pi/2)\cos\theta)}{2\pi\sin\theta}\frac{e^{-j\beta R}}{R}\left(e^{-j\beta h\cos\theta} + e^{j\beta h\cos\theta}\right) = \hat{\boldsymbol{\theta}} \frac{j\eta I\cos((3\pi/2)\cos\theta)}{2\pi\sin\theta}\frac{e^{-j\beta R}}{R}(2\cos(\beta h\cos\theta)) \quad \left[\frac{\text{V}}{\text{m}}\right]$$

This gives the electric and magnetic field intensities

$$\mathbf{E} = \hat{\boldsymbol{\theta}} \frac{j120}{R} e^{-j\beta R} \frac{\cos((3\pi/2)\cos\theta)}{\sin\theta}\cos(8\pi\cos\theta) \quad \left[\frac{\text{V}}{\text{m}}\right]$$

$$\mathbf{H} = \hat{\boldsymbol{\phi}} \frac{j}{\pi R} e^{-j\beta R} \frac{\cos((3\pi/2)\cos\theta)}{\sin\theta}\cos(8\pi\cos\theta) \quad \left[\frac{\text{A}}{\text{m}}\right]$$

Note that the time-averaged power transmitted from the antenna propagates in the positive R direction as required and that the antenna and array patterns are independent of the azimuthal angle ϕ.

(b) The array factor is the last term in the electric field intensity equation:

$$|f_n(\theta, \phi, \varphi = 0)| = |\cos(8\pi\cos\theta)|$$

The combined or array radiation pattern is the product of the dipole radiation pattern and the array factor

$$|f_e(\theta, \phi, \varphi = 0)| = \left|\frac{\cos((3\pi/2)\cos\theta)}{\sin\theta}\cos(8\pi\cos\theta)\right|$$

The power radiation pattern of the system made of the two antennas is the field radiation pattern squared:

$$f_p(\theta, \phi, \varphi = 0) = \left(\frac{\cos((3\pi/2)\cos\theta)}{\sin\theta}\cos(8\pi\cos\theta)\right)^2$$

These patterns (which only exist above ground, $0 \le \theta \le \pi/2, 0 \le \phi \le 2\pi$) are shown in **Figure 18.31**. Note the main lobes and the many sidelobes in this pattern.

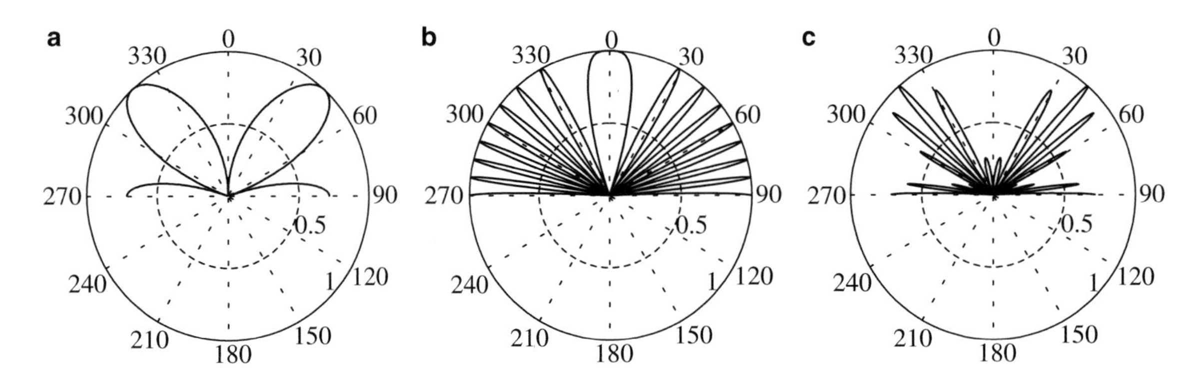

Figure 18.31 (**a**) Element power radiation pattern for a 1.5 λ dipole above ground. (**b**) Array factor. (**c**) Antenna array power radiation pattern

18.8.2 The *n*-Element Linear Array

The two-element array discussed in the previous section may be extended to any number of elements. One method would be to repeat the process in the previous section for an arbitrary number of elements. This has the advantage that the elements may, in principle, be located anywhere and may be of any type. However, to obtain simple useful relations, we restrict ourselves here to a uniform linear array of dipole antennas. In other words, all antennas are identical and uniformly spaced on a straight line. A basic configuration is shown in **Figure 18.32a**, where each element is a half-wavelength dipole (as an example). The number n is arbitrary.

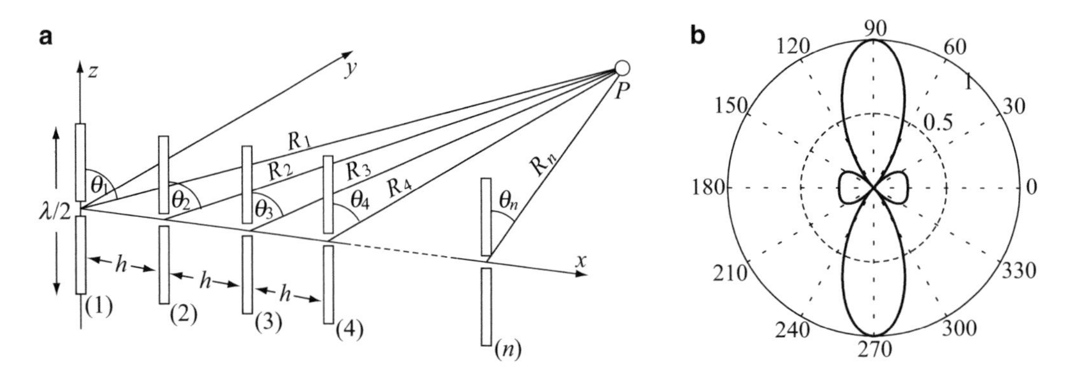

Figure 18.32 (**a**) An n-element uniform linear array of $\lambda/2$ dipoles. (**b**) Plot of the array field radiation pattern for $n = 6$, $\varphi = 0$, $\theta = \pi/2$, and $h = \lambda/4$

The currents in the elements are assumed to be leading each previous element by a constant angle φ. All currents are in the same direction. The proper name for an array with these properties is ***progressive, n-element uniform linear array***.

The currents in individual elements are

$$I_1 = I_0 e^{j\omega t}, \quad I_2 = I_0 e^{j\omega t} e^{j\varphi}, \quad I_k = I_0 e^{j\omega t} e^{j(k-1)\varphi} \tag{18.131}$$

We assume $\theta_1 = \theta_2 = \cdots = \theta_n = \theta$. Also, from this assumption, it is implicit that $R_1, R_2, \ldots, R_n$ are essentially parallel to each other. This approximation is quite good in the far field where $R_k \gg h$. Thus, we may write by extending **Eq. (18.132)**:

$$\begin{aligned} R_2 &\approx R_1 - h\sin\theta\cos\phi, \\ R_3 &\approx R_1 - 2h\sin\theta\cos\phi, \\ R_k &\approx R_1 - (k-1)h\sin\theta\cos\phi \end{aligned} \tag{18.132}$$

We use these approximation in the exponent whereas in the denominator we use $R_2 = R_3 = \ldots = R_k = R_1$. The electric field intensity at point P is now the sum of the electric field intensities of n elements where element k is at a distance R_k from point P. Extending **Eq. (18.118)** as the sum of n electric fields, we get

$$\mathbf{E} = \sum_{k=1}^{n} \mathbf{E}_k = \hat{\boldsymbol{\theta}} \frac{E_0 f(\theta,\phi)}{R_1} e^{-j\beta R_1} \times \left(1 + e^{j\psi} + e^{j2\psi} + \cdots + e^{j(k-1)\psi} + \cdots + e^{j(n-1)\psi}\right) \quad \left[\frac{\text{V}}{\text{m}}\right] \tag{18.133}$$

where

$$\psi = \beta h\sin\theta\cos\phi + \varphi \quad [\text{rad}] \tag{18.134}$$

Performing the summation in the parentheses in **Eq. (18.133)** as a geometric series gives

$$1 + e^{j\psi} + e^{j2\psi} + \cdots + e^{j(k-1)\psi} + \cdots + e^{j(n-1)\psi} = \frac{1 - e^{jn\psi}}{1 - e^{j\psi}} \tag{18.135}$$

To get a more convenient form, we write

$$\frac{1 - e^{jn\psi}}{1 - e^{j\psi}} = \frac{e^{jn\psi/2}\left(e^{-jn\psi/2} - e^{jn\psi/2}\right)}{e^{j\psi/2}\left(e^{-j\psi/2} - e^{j\psi/2}\right)} = e^{j(n-1)\psi/2}\frac{\sin(n\psi/2)}{\sin(\psi/2)} \tag{18.136}$$

The electric field intensity can now be written as

$$\mathbf{E} = \sum_{k=1}^{n} \mathbf{E}_k = \hat{\boldsymbol{\theta}} \frac{E_0 f(\theta,\phi)}{R_1} e^{-j\beta R_1} e^{j(n-1)\psi/2} \frac{\sin(n\psi/2)}{\sin(\psi/2)} \quad \left[\frac{\text{V}}{\text{m}}\right] \tag{18.137}$$

Following the discussion in the previous section, we can now write the radiation pattern as the product of the element radiation pattern and the array factor. The latter is the normalized value of the term $\sin(n\psi/2)/\sin(\psi/2)$. We note that the maximum value of this term is obtained as $\psi \to 0$, and from the L'Hospital rule, we get

$$\lim_{\psi \to 0} \frac{\sin(n\psi/2)}{\sin(\psi/2)} = n\frac{\cos(n\psi/2)}{\cos(\psi/2)} = n \tag{18.138}$$

The expressions in **Eqs. (18.137)** and **(18.138)** are completely general. What changes from one type of array to another is the expression for ψ. For the array in **Figure 18.32a** the *normalized array factor* is

$$|f_n(\theta,\phi,\varphi)| = \frac{1}{n}\left|\frac{\sin(n\psi/2)}{\sin(\psi/2)}\right| = \frac{1}{n}\left|\frac{\sin(n(\beta h\sin\theta\cos\phi + \varphi)/2)}{\sin((\beta h\sin\theta\cos\phi + \varphi)/2)}\right| \tag{18.139}$$

The normalized composite radiation pattern is the product of the normalized array factor and the normalized element radiation pattern in **Eq. (18.103)**.

As with the two-element array, the properties of the n-element linear array depend on the angle ψ, which, in turn, depends on θ, ϕ, and the phase angle φ. Various properties may be obtained for various angles as will be shown in the following examples and in **Problems 18.32** through **18.41**. There are, however, a number of basic properties which are general. A plot of $|f_n(\theta,\phi,\varphi)|$ is given in **Figure 18.32b** for $n = 6$, $\varphi = 0$, $\theta = \pi/2$, and $h = \lambda/4$:

(1) The plot shows that the maximum value occurs at $\phi = \pi/2$ or $\phi = -\pi/2$. This means that the array radiates perpendicular to the array axis. This is called a ***broadside antenna.***
(2) The nulls in the pattern occur at $\sin(n\psi/2) = 0$, which means that $n\psi/2 = \pm k\pi$. These nulls are shown in **Figures 18.32b** and **18.33**, which are equivalent representations.

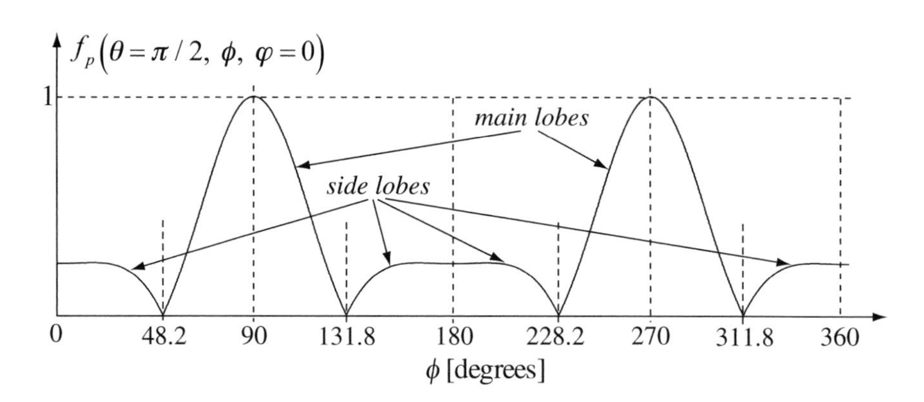

Figure 18.33 Main lobes, sidelobes, and nulls in the array radiation pattern

(3) The ***sidelobes*** occur at $|\sin(n\psi/2)| = 1$, which means that $n\psi/2 = \pm(2k + 1)\pi/2$, where k is called the sidelobe index. $k = 0$ indicates the main lobe, whereas $k \geq 1$ indicates the sidelobes. For example, for $n = 6$, $k = 1, 2, 3$, the sidelobes are at $\psi = 3\pi/6$, $5\pi/6$, and $7\pi/6$, as shown in **Figure 18.33**, together with the main lobes.
(4) An ***end-fire array*** can also be designed by proper choice of the phase angle between the elements. For an end-fire array, the requirement is that maximum radiation occur at $\phi = 0$. This can be satisfied if $\phi = -\beta h$. For these conditions, maximum radiation is parallel to the antenna, hence the name end fire.
(5) Use of other elements of various lengths will produce different results.

The electric and magnetic field intensities of a linear, progressive, uniform n-element antenna array can be viewed as the field of the element of the array multiplied by the array factor. The array factor accounts for the structure of the array (spacing, number of elements, phase progression and element orientation) as can be seen for example, in **Eqs. (18.124)** and **(18.137)**. For convenience we have written the array factor in terms of a quantity ψ (see **Eq. (18.119)**, **Eq. (18.134)** as well as **Example 18.11**). Inspection of **Eqs. (18.128)** and **(18.134)** shows that ψ depends on the orientation of the antennas. Whereas for parallel antennas on the x-axis (see **Figure 18.32**), $\psi = \beta h\sin\theta\cos\phi + \varphi$, for stacked antennas (see **Figure 18.30**), $\psi = \beta h\cos\theta + \varphi$. That means of course, that the expressions for the array factors of the two types of arrays are identical in form but are very different in behavior because of the different expression for ψ.

Example 18.12 A uniform, progressive, linear array of half-wavelength dipoles has five elements, each carrying a 1/2 A current and radiating in free space at 120 MHz. The elements are spaced regularly, at a distance h [m] apart. **Figure 18.34** shows the configuration. The elements are driven with a progressive phase angle φ [rad] as shown. Calculate:

(a) The electric and magnetic field intensities of the array in the far field.
(b) The array factor of the antenna if $h = \lambda$.
(c) The normalized power radiation pattern of the antenna for all elements in phase ($h = \lambda$).
(d) The normalized power radiation pattern for a phase angle of $\varphi = \pi$ ($h = \lambda$).

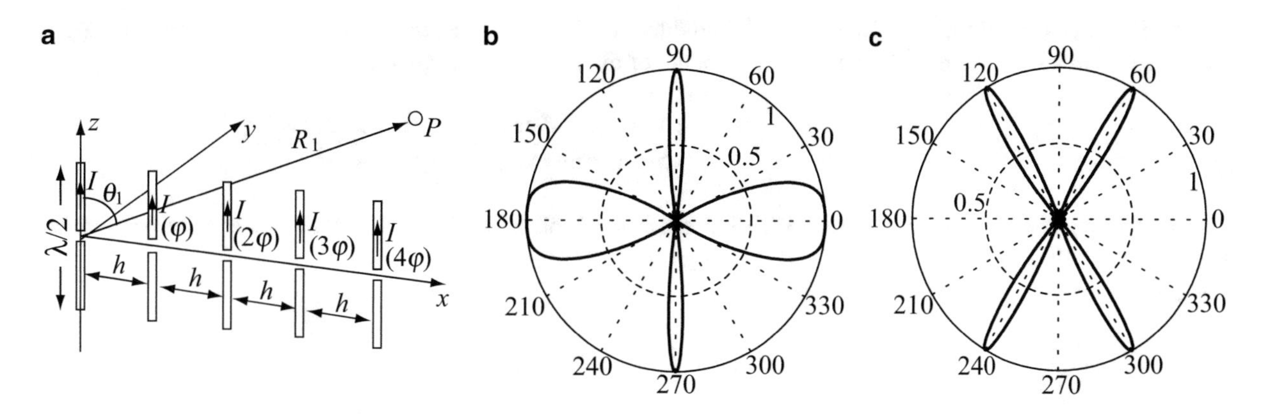

Figure 18.34 (**a**) Five-element uniform array. (**b**) Power radiation pattern for $\varphi = 0$, $\theta = \pi/2$. (**c**) Power radiation pattern for $\varphi = \pi$, $\theta = \pi/2$

Solution: Assuming the antennas are on the x axis, directed in the z direction, the electric field intensity is given in **Eq. (18.137)** and the magnitude of the magnetic field intensity is η_0 times smaller than that of the electric field intensity. The direction of the magnetic field intensity is such that propagation is in the R direction. Antenna No. 1 is taken as a reference antenna, whereas the electric and magnetic field intensities of the individual elements are those in **Eqs. (18.97)** and **(18.98)**:

(**a**) For the array radiating in free space, the intrinsic impedance is $\eta_0 = 120\pi$, and $\beta = \omega/c = 2\pi \times 120 \times 10^6/3 \times 10^8 = 0.8\pi$. With these, and with the given current, the electric field intensity is

$$\mathbf{E} = \hat{\boldsymbol{\theta}}\frac{j30}{R_1}e^{-j0.8\pi R_1}\frac{\cos((\pi/2)\cos\theta)}{\sin\theta}e^{j2\psi}\frac{\sin(5\psi/2)}{\sin(\psi/2)}$$
$$= \hat{\boldsymbol{\theta}}\frac{j30}{R_1}e^{-j0.8\pi R_1}\frac{\cos((\pi/2)\cos\theta)}{\sin\theta}e^{j(0.8\pi h\sin\theta\cos\phi+\varphi)}\frac{\sin(5(0.4\pi h\sin\theta\cos\phi+\varphi/2))}{\sin(0.4\pi h\sin\theta\cos\phi+\varphi/2)}\quad\left[\frac{\mathrm{V}}{\mathrm{m}}\right]$$

where R_1 is the distance from the first half-wavelength dipole, and $f(\theta,\phi)$ was written explicitly.

The magnitude of the magnetic field intensity is $H = E/\eta_0$. The direction of the magnetic field intensity is found from the direction which makes the propagation of the time-averaged power $[\mathcal{P}_{av} = \mathrm{Re}\{(\mathbf{E} \times \mathbf{H}^*)/2\}]$ in the R direction. Thus, **H** must be in the ϕ direction [see **Eq. (18.98)**]:

$$\mathbf{H} = \hat{\boldsymbol{\phi}}\frac{j}{4\pi R_1}e^{-j0.8\pi R_1}\frac{\cos((\pi/2)\cos\theta}{\sin\theta}e^{j(0.8\pi h\sin\theta\cos\phi+\varphi)}\frac{\sin(5(0.4\pi h\sin\theta\cos\phi+\varphi/2))}{\sin(0.4\pi h\sin\theta\cos\phi+\varphi/2)}\quad\left[\frac{\mathrm{A}}{\mathrm{m}}\right].$$

(**b**) The normalized array factor is

$$|f_5(\theta,\phi,\varphi)| = \frac{1}{5}\left|\frac{\sin(5(\beta h\sin\theta\cos\phi+\varphi)/2)}{\sin((\beta h\sin\theta\cos\phi+\varphi)/2)}\right|$$

or substituting $h = \lambda$ and $\lambda = 2\pi/\beta$, we get the normalized array factor as

$$|f_5(\theta,\phi,\varphi)| = \frac{1}{5}\left|\frac{\sin((10\pi\sin\theta\cos\phi+5\varphi)/2)}{\sin((2\pi\sin\theta\cos\phi+\varphi)/2)}\right|.$$

(**c**) If all elements are in phase, $\varphi = 0$, and we get for the array factor:

$$|f_5(\theta,\phi,\varphi)| = \frac{1}{5}\left|\frac{\sin(5\pi\sin\theta\cos\phi)}{\sin(\pi\sin\theta\cos\phi)}\right|$$

The radiation pattern of the array antenna is the product of the element antenna radiation pattern and the array factor. The power radiation pattern of the array is the product of the two patterns squared:

$$f_p(\theta,\phi,0) = [f(\theta)f_5(\theta,\phi,0)]^2 = \left(\frac{1}{5}\frac{\sin(5\pi\sin\theta\cos\phi)}{\sin(\pi\sin\theta\cos\phi)}\right)^2\left(\frac{\cos((\pi/2)\cos\theta)}{\sin\theta}\right)^2$$

This radiation pattern is plotted in **Figure 18.34b** in the $\theta = \pi/2$ plane (ϕ plane). The pattern is end fire, with main lobes parallel to the array (and two sidelobes perpendicular to the array).

(d) For $\varphi = \pi$, we get

$$|f_5(\theta,\phi,\pi)| = \frac{1}{5}\left|\frac{\sin(5\pi\sin\theta\cos\phi + 5\pi/2)}{\sin(\pi\sin\theta\cos\phi + \pi/2)}\right|$$

and

$$|f_p(\theta,\phi,\pi)| = [f(\theta)f_5(\theta,\phi,\pi)]^2 = \left(\frac{1}{5}\frac{\sin(5\pi\sin\theta\cos\phi + 5\pi/2)}{\sin(\pi\sin\theta\cos\phi + \pi/2)}\right)^2\left(\frac{\cos((\pi/2)\cos\theta)}{\sin\theta}\right)^2$$

This radiation pattern is shown in **Figure 18.34c** in the $\theta = \pi/2$ plane (ϕ plane).

18.9 Reciprocity and Receiving Antennas

So far, we discussed only the transmitting properties of antennas. The various properties relate to directivity, radiation pattern, power, and the like, which, at least in principle, imply transmission. We may well ask ourselves, what happens if an antenna is used as a receiving antenna or, perhaps, both as a transmitting and a receiving antenna? For example, when using a cellular telephone, the same antenna is used for transmission and reception. The same applies to many radar systems, walkie-talkies, CB radio and many other transmit-receive systems. What are the receiving antenna properties in relation to the transmitting properties?

At first sight we would be tempted to conclude that they are different. That is, a transmitting antenna acts as a source and generates a wave. This wave has both an electric and a magnetic field which propagate in space. An identical receiving antenna interacts with this wave and produces a current which, although proportional to the incoming wave, is certainly different in magnitude and phase. As a simple illustration, consider **Figure 18.35**. Because the transmitting antenna transmits in a pattern that occupies a portion of space, the receiving antenna can only couple to a small portion of the field produced by the transmitting antenna. This is the same as saying that the light impinging on the lens of a camera is proportional to the size of the lens aperture: The larger the aperture, the larger the amount of light transmitted into the camera. Thus, viewing the aperture as an antenna, only a portion of the light from the source arrives to the camera and only part of it enters the lens.

However, in practice, the properties of an antenna in receiving and transmitting modes are, in fact, the same. This remarkable aspect of antennas is defined by the so-called ***reciprocity theorem***. Consider again the two antennas in **Figure 18.35**. Suppose we first use antenna No. 1 as a transmitter and receive a signal in antenna No. 2. Then, we repeat

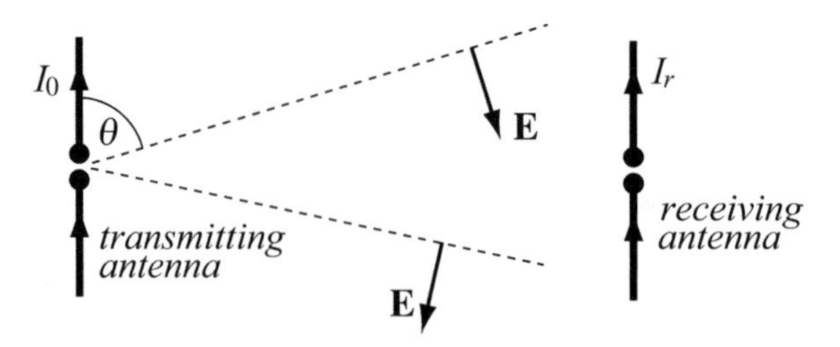

Figure 18.35 Relation between a transmitting and a receiving antenna

the process, using antenna No. 2 as a transmitting antenna and antenna No. 1 as a receiving antenna. The reciprocity theorem then states:

"If a source is applied to antenna No. 1 and a signal is received in antenna No. 2, then, applying the same identical source to antenna No. 2 will produce an identical signal in antenna No. 1."

This theorem (which we do not prove) only applies if the properties of the intervening space between the two antennas are the same in either direction (the space is isotropic). This is normally the case in free space and in many other dielectrics. Although it only applies in isotropic media, the reciprocity theorem is widely used for the analysis of antennas. To show that the receiving properties of the antenna are the same as the transmitting properties, consider a transmitting Hertzian dipole antenna which produces a wave at very large distances. Now, consider a receiving antenna in the form of a Hertzian dipole of length Δl, at some angle θ to the direction of propagation of the wave as shown in **Figure 18.36**. The open circuit voltage along the dipole is

$$V_a = \mathbf{E} \cdot \Delta \mathbf{l} = E\Delta l\cos(90 - \theta) = E\Delta l\sin\theta \quad [\text{V/m}] \tag{18.140}$$

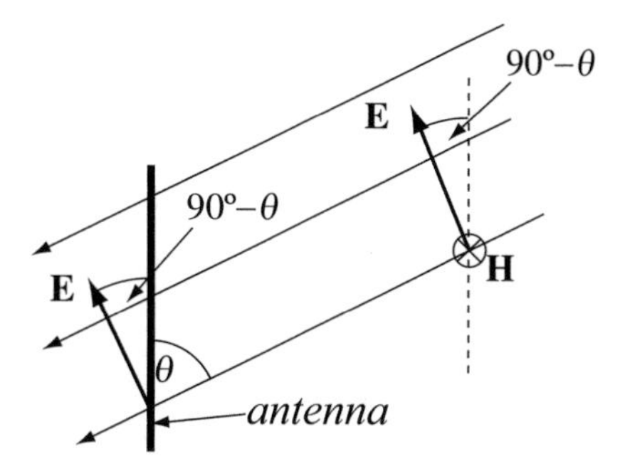

Figure 18.36 A receiving antenna at an angle θ to the direction of propagation of a plane wave

Thus, the voltage in the receiving antenna is proportional to the incident electric field intensity, the length of the antenna, and the angle between the two. The term $\Delta \mathbf{l}$ is called the ***vector effective length*** of the antenna. The ***effective length*** Δl may be defined as the ratio between the magnitude of the open circuit voltage on the antenna and the magnitude of the incident electric field intensity in the direction of the antenna. The current in the receiving antenna depends on the impedance of the receiving antenna, but as long as the impedance is constant, we can use the voltage equally well. From the above relation, we note that if the antenna is aligned in the direction of propagation of the wave, $\theta = 0$ and $\sin\theta = 0$. The most important aspect of this voltage and the current it produces is that it is proportional to $\sin\theta$. If we write the electric field intensity at the dipole as

$$E_d = E\sin\theta \quad [\text{V/m}] \tag{18.141}$$

then we can define a radiation pattern at the receiver that gives the field in terms of the angle θ. We immediately find that the receiving pattern for the dipole is the same as the radiation pattern of the transmitting antenna. Because of this, the directivity of the antenna is also the same as for the transmitting Hertzian dipole.

18.10 Effective Aperture

The next question is how much of the power transmitted by the transmitting antenna is received by the receiving antenna. To understand the process, consider a point source (isotropic or uniformly radiating antenna) that radiates a certain amount of power, say, P [W], uniformly in space. At a distance R [m] from the source (assuming there is no loss or attenuation), the power density is $P/4\pi R^2$ [W/m^2]. Now, suppose an antenna is placed on this sphere and further assume that the receiving antenna has an area A_a [m^2]. The total power incident on the receiving antenna is

$$P_a = \frac{P_{rad}}{4\pi R^2}A_a = \mathcal{P}_{av}A_a \quad [\text{W}] \tag{18.142}$$

where $\mathcal{P}_{av}$ is the time-averaged power density at a distance R from the transmitting antenna. The relationship among power, power density, and antenna surface is shown in **Figure 18.37**. This relationship is derived from simple observation. The same applies to a camera lens: The total amount of light entering the camera depends on the light intensity at the location of the camera and the aperture size. There is only one little difficulty when discussing antennas: If we use a dipole antenna, we can talk about the length of the antenna, but what does the surface of the antenna means? For this reason, and because we know that some power definitely enters an antenna regardless of the shape of the antenna, we define a new quantity which we call the effective antenna area or ***effective aperture*** as

$$A_{ea} = \frac{P_a}{\mathcal{P}_{av}} \quad [\mathrm{m}^2] \tag{18.143}$$

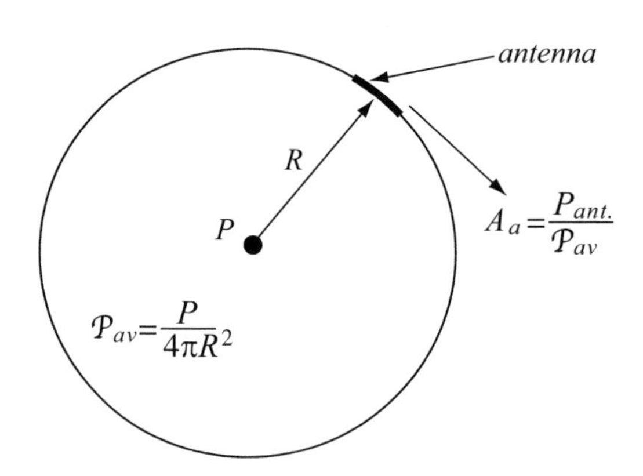

Figure 18.37 Relation between effective area of an antenna and power received

In other words, the effective area of an antenna is the ratio between the total time-averaged power received by the antenna and the time-averaged power density at the location of the antenna.

Now, consider the power relations in the antenna itself. Suppose the antenna develops a voltage as given in **Eq. (18.140)**. Since the antenna is connected to a load (say, a radio receiver), the load can be expressed as an impedance $Z_L = R_L + jX_L$. The antenna itself has an impedance which includes the radiation resistance R_{rad}, the antenna resistance R_d, and, in general, a reactive term we denote here as X_a. Thus, the equivalent circuit of an antenna connected to a load is as shown in **Figure 18.38**. The time-averaged power transferred to any load by a reveiving antenna is

$$P_L = \frac{I^2 R_L}{2} = \frac{1}{2}\left(\frac{V_a}{R_L + R_d + R_{rad}}\right)^2 R_L \quad [\mathrm{W}] \tag{18.144}$$

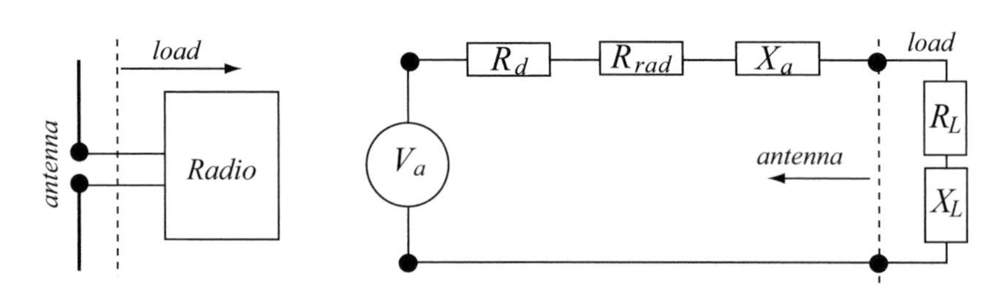

Figure 18.38 Antenna representation as a source and impedance matched to a load

where V_a is the voltage induced in the antenna. First, we note from this relation that maximum antenna efficiency is obtained for $R_d = 0$ (all other impedances being constant). Antennas normally operate at high efficiencies which means that, normally, $R_{rad} \gg R_d$. For the antenna to be matched to the load (conjugate matching), the following must be satisfied:

$$R_{rad} + R_d = R_L \quad \text{and} \quad X_a = -X_L \tag{18.145}$$

Thus, for a matched load, $R_L = R_{rad} + R_d \approx R_{rad}$ and we have

$$P_L = \frac{1}{2}\left(\frac{V_a}{2R_{rad}}\right)^2 R_{rad} = \frac{V_a^2}{8R_{rad}} \quad [\mathrm{W}] \tag{18.146}$$

Suppose we apply this to a Hertzian dipole. Substituting the radiation resistance of the Hertzian dipole from **Eq. (18.37)** and the voltage given in **Eq. (18.140)** into **Eq. (18.146)**, we obtain

$$P_L = \frac{V_a^2}{8R_{rad}} = \frac{E^2(\Delta l)^2 \sin^2\theta}{8R_{rad}} = \frac{3E^2\lambda^2 \sin^2\theta}{16\eta\pi} \quad [\mathrm{W}] \tag{18.147}$$

The time-averaged power density at the antenna under matched conditions is

$$\mathcal{P}_{av} = \frac{E^2}{2\eta} \quad \left[\frac{\mathrm{W}}{\mathrm{m}^2}\right] \tag{18.148}$$

where E is the electric field intensity at the antenna. From **Eq. (18.147)** and **(18.148)**, we obtain the effective aperture or effective area of the Hertzian dipole as

$$A_{ea} = \frac{P_L}{\mathcal{P}_{av}} = \frac{3\lambda^2 \sin^2\theta}{8\pi} \quad [\mathrm{m}^2] \tag{18.149}$$

The maximum voltage induced in a receiving antenna [from **Eq. (18.140)**] is $V_{am} = E\Delta l$. Substituting this voltage and the radiation resistance from **Eq. (18.37)** into **Eq. (18.146)** we obtain the maximum possible power coupled into the load as

$$P_{Lm} = \frac{V_{am}^2}{8R_{rad}} = \frac{E^2(\Delta l)^2}{8(2\eta\pi/3)(\Delta l/\lambda)^2} = \frac{3E^2\lambda^2}{16\eta\pi} \quad [\mathrm{W}] \tag{18.150}$$

Therefore, the maximum effective area of the Hertzian dipole is

$$(A_{ea})_{max} = \frac{P_{Lm}}{\mathcal{P}_{av}} = \frac{3\lambda^2}{8\pi} \quad [\mathrm{m}^2] \tag{18.151}$$

We further observe that for the Hertzian dipole, the directivity is $(3/2)\sin^2\theta$ [from **Eq. (18.52)**]. Thus, we can also write **Eq. (18.149)** as

$$A_{ea} = \frac{3\sin^2\theta}{2}\frac{\lambda^2}{4\pi} = D(\theta)\frac{\lambda^2}{4\pi} \quad [\mathrm{m}^2] \tag{18.152}$$

This is a rather interesting result since it indicates the effective area of the antenna as far as absorbing power, although the result itself is independent of the dimensions of the dipole. However, the effective area of the antenna does depend on the orientation of the antenna as indicated by the directivity in **Eq. (18.152)**. The above result is quite general and applies to any antenna. For other antennas, the directivity may depend on the angle ϕ as well (see **Section 18.5.5**). Thus, we can write for any antenna:

$$\boxed{A_{ea} = D(\theta,\phi)\frac{\lambda^2}{4\pi} \quad [\mathrm{m}^2]} \tag{18.153}$$

The concept of effective area or aperture is useful because it defines the antenna in terms of how much power it can transfer to the load, even if the antenna itself has no "area" in the true sense. It also indicates that the most important aspect of the antenna in the receiving mode is the directivity of the antenna. Note also that the length of the dipole antenna, for which the above relation was defined, does not enter into the relation. The maximum directivity of a Hertzian dipole is 1.5, whereas the maximum directivity for a half-wavelength dipole is 1.642 [see **Eqs. (18.53)** and **(18.105)**]. From these, the maximum effective aperture of a Hertzian dipole and a half-wavelength dipole are

$$\boxed{(A_{ea})_{max} = 1.5\frac{\lambda^2}{4\pi} \quad [\text{m}^2] \quad \text{for Hertzian dipole}} \tag{18.154}$$

$$\boxed{(A_{ea})_{max} = 1.642\frac{\lambda^2}{4\pi} \quad [\text{m}^2] \quad \text{for half-wavelength dipole}} \tag{18.155}$$

Thus, whereas a Hertzian dipole may radiate very little power because its radiation properties depend on its length, the effective area of the dipole is similar to that of a half-wavelength dipole. This means that a very short Hertzian dipole is a rather effective receiving antenna even though it may not be suitable as a transmitting antenna. This property is often used in receivers. For example, a half-wavelength dipole for AM radios would be approximately 150 m long. In practice, a 0.5 m to 1 m "whip" is sufficient for normal reception such as in car radios, making this a very short antenna.

The concept of effective aperture may also be used for two other purposes. First, we can define the ratio between two effective apertures of two different antennas and, second, we can calculate the power available at the receiving antenna from the effective area. Consider, first, two arbitrary antennas A and B. The ratio between their effective apertures is (see **Eq. (18.153)**)

$$\boxed{\frac{A_{eaA}}{A_{eaB}} = \frac{D_A(\theta, \phi)}{D_B(\theta, \phi)}} \tag{18.156}$$

This is a simple and useful relation since the effective aperture of an antenna can then be evaluated from the effective aperture of another known antenna. The ratio applies to any two antennas, regardless of their orientation in space, shape, and size.

Second, we can now go back to the definition of the effective aperture in **Eq. (18.143)** and **Eq. (18.153)** and write

$$P_a = A_{ea}\mathcal{P}_{av} = D_r(\theta, \phi)\frac{\lambda^2}{4\pi}\mathcal{P}_{av} \quad [\text{W}] \tag{18.157}$$

where P_a indicates the power received by the receiving antenna and $D_r(\theta,\phi)$ indicates the directivity of the receiving antenna. The time-averaged power density $\mathcal{P}_{av}$ is generated by the transmitting antenna.

Considering again the time-averaged power density produced by a transmitting antenna, we recall the definition of the directivity as given in **Eq. (18.50)**:

$$D_t(\theta, \phi) = \frac{4\pi\mathcal{P}_{av}R^2}{P_{rad}} \quad \to \quad \mathcal{P}_{av} = \frac{D_t(\theta, \phi)P_{rad}}{4\pi R^2} \quad \left[\frac{\text{W}}{\text{m}^2}\right] \tag{18.158}$$

where $D_t(\theta,\phi)$ indicates the directivity of the transmitting antenna, which, in general, may be different than that of the receiving antenna (i.e., the two antennas may be of different sizes or types). Substituting this into **Eq. (18.157)**, we get the power received as

$$P_a = D_r(\theta, \phi)\frac{\lambda^2}{4\pi}\frac{D_t(\theta, \phi)P_{rad}}{4\pi R^2} = D_r(\theta, \phi)D_t(\theta, \phi)\left(\frac{\lambda}{4\pi R}\right)^2 P_{rad} \quad [\text{W}] \tag{18.159}$$

Where P_a the received power and P_{rad} the transmitted power. The ratio between the two is a well-known relation and may be written as

$$\frac{P_a}{P_{rad}} = \frac{P_{received}}{P_{transmitted}} = \left(D_r(\theta, \phi)\frac{\lambda^2}{4\pi}\right)\left(D_t(\theta, \phi)\frac{\lambda^2}{4\pi}\right)\frac{1}{\lambda^2 R^2} \tag{18.160}$$

From the definition of effective area in **Eq. (18.152)**, the first term in parentheses is the effective area of the receiving antenna and the second that of the transmitting antenna. Therefore, we can write

$$\boxed{\frac{P_{received}}{P_{transmitted}} = \frac{A_{er}A_{et}}{\lambda^2 R^2}} \quad (18.161)$$

This is known as the ***Friis transmission formula*** and is a common design tool. The important aspect of this formula is that it indicates that the ratio of power received by the receiver is directly proportional to the effective areas of both transmitting and receiving antennas and inversely proportional to the wavelength and distance between antennas. It means that if we require a larger power at the receiving antenna, we can either make one or both antennas with larger effective areas (by increasing its directivity), increase the frequency of transmission (decrease the wavelength), or transmit to shorter distances. These simple relations explain why we use very large, highly directive antennas whenever there is a need for transmission to large distances (for example, for deep-space exploration) or when we need to receive signals from relatively weak sources (ground stations for satellites).

Although the discussion here relates to dipoles, the results are applicable to any antenna. The effective aperture of an antenna should not be confused with any physical area, such as the area of a loop.

Example 18.13 Effective Aperture of a Small Loop Calculate the effective aperture of a small loop. The loop is 10 mm in diameter and is used as a receiving antenna at 1.2 GHz. Assume the loop is placed at an arbitrary angle to the direction of propagation of the incoming wave as shown in **Figure 18.39**. Find the maximum effective area of the loop. Compare this with the maximum effective area of the Hertzian dipole.

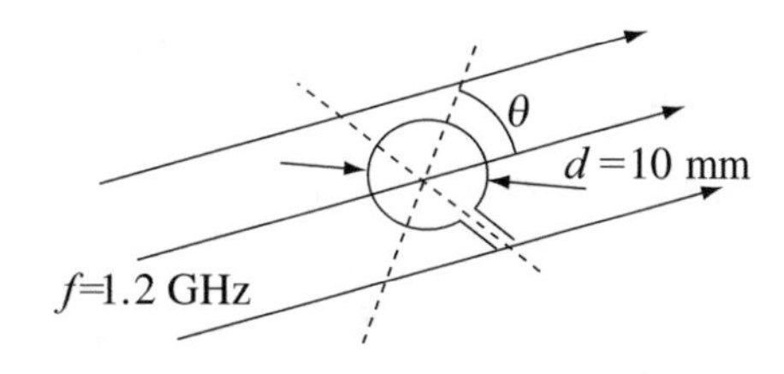

Figure 18.39 A small-loop antenna used as a receiving antenna at 1.2 GHz

Solution: To calculate the effective area or aperture of an antenna, we must calculate the maximum power received by the antenna or use the directivity of the antenna. We will use the former here to demonstrate the process of evaluating the effective area from the fields.

Assuming the electric and magnetic fields at the location of the loop are known, the induced emf in the loop is given as

$$\text{emf} = -j\omega\Phi = -j\omega\mu_0 H\pi a^2$$

where $\mu_0 H = B$ is the magnetic flux density of the received field and Φ is the flux in the loop. Using **Eq. (18.146)** and the radiation resistance of a small loop from **Table 18.1**, we get the maximum power received by the loop as

$$P_{Lm} = \frac{emf^2}{8R_{rad}} = \frac{\omega^2\mu_0^2 H^2 a^4}{160\beta^4 a^4} = \frac{\omega^2\mu_0^2 H^2}{160\beta^4} = \frac{\eta_0^2 H^2}{160\beta^2} \quad [\text{W}]$$

where $\beta = \omega\sqrt{\mu_0\varepsilon_0} = 2\pi/\lambda$, $\eta_0 = \sqrt{\mu_0\varepsilon_0} = 120\pi$ and $\omega^2\mu_0^2/\beta^2 = \eta_0^2$ were used. The time-averaged power density (magnitude) is

$$\mathcal{P}_{av} = \frac{EH}{2} = \frac{E^2}{2\eta_0} = \frac{\eta_0 H^2}{2} \quad \left[\frac{\text{W}}{\text{m}^2}\right]$$

Taking the last relation and using $\beta = 2\pi/\lambda$ and $\eta_0 = 120\pi$ we get the effective area as

$$A_{ea} = \frac{P_{Lm}}{\mathcal{P}_{av}} = \frac{2\eta_0^2 H^2}{160\beta^2\eta_0 H^2} = \frac{\eta_0}{80\beta^2} = \frac{120\pi}{80\beta^2} = 1.5\frac{\pi}{\beta^2} = 1.5\frac{\lambda^2}{4\pi} \quad [\text{m}^2]$$

At the given frequency, the wavelength in air ($\lambda = c/f$) is 0.25 m. The maximum effective aperture of the loop antenna is

$$A_{ea} = \frac{1.5 \times 0.0625}{4 \times \pi} = 0.00746 \quad [\text{m}^2]$$

Note that this is significantly larger than the actual area of the loop, which is only 7.85×10^{-5} m^2. Also, the effective aperture of the loop is identical with the effective aperture of the Hertzian dipole. This conclusion could also have been reached from the fact that a small-loop antenna has the same directivity as a Hertzian dipole. Therefore, the Hertzian dipole can always be interchanged with the small-loop antenna for reception purposes if the loop is also rotated to be perpendicular to the position of the dipole.

Example 18.14 Use of the Friis Formula A half-wavelength dipole is used to transmit 100 W at 80 MHz. A receiving antenna in the form of a Hertzian dipole, $\lambda/50$ m in length, is used to receive, 20 km away. Calculate:

(a) The maximum power received by the Hertzian dipole.
(b) The maximum current (magnitude) in the Hertzian dipole.

Solution: The transmitted and received powers are related through the Friis formula. In this case, the effective areas of the half-wavelength dipole and the Hertzian dipole are calculated from the maximum directivity of the antennas. The current in the antenna is then calculated from the received power and the radiation resistance. Maximum values (power, current) are obtained by using maximum effective areas:

(a) From the Friis formula **[Eq. (18.161)]**, the received power is

$$P_{received} = P_{transmitted} \frac{A_{er} A_{et}}{\lambda^2 R^2} \quad [\text{W}]$$

where A_{et} is the transmitting antenna effective aperture and A_{er} the receiving antenna effective aperture. The receiving antenna maximum aperture is given in **Eq. (18.154)**. The wavelength at 80 MHz is 3.75 m. Thus, the maximum effective aperture of the receiving antenna is

$$A_{er} = \frac{1.5\lambda^2}{4\pi} = \frac{1.5 \times 3.75^2}{4 \times \pi} = 1.6786 \quad [\text{m}^2]$$

The transmitting antenna is a half-wavelength antenna and its maximum aperture is given in **Eq. (18.155)**:

$$A_{et} = \frac{1.642\lambda^2}{4\pi} = \frac{1.642 \times 3.75^2}{4 \times \pi} = 1.8375 \quad [\text{m}^2]$$

The maximum received power is

$$P_{received} = \frac{100 \times 1.6786 \times 1.8375}{3.75^2 \times 20{,}000^2} = 5.48 \times 10^{-8} \quad [\text{W}].$$

(b) The maximum current is calculated from the received power and radiation resistance of the Hertzian dipole in free space [**Eq. (18.38)**], using **Eq. (18.36)** and the reciprocity theorem

$$P_{received} = \frac{I_0^2 R_{rad}}{2} \quad \rightarrow \quad I_0 = \sqrt{\frac{2P_{received}}{R_{rad}}} \quad [\text{A}]$$

Substituting the various values (see **Eq. (18.38)** for the radiation resistance of the Hertzian dipole), the magnitude of the current is

$$I_0 = \sqrt{\frac{2P_{received}\lambda^2}{80\pi^2(\lambda/50)^2}} = \sqrt{\frac{2 \times 5.48 \times 10^{-8} \times 50^2}{80 \times \pi^2}} = 5.89 \times 10^{-4} \quad [\text{A}]$$

18.11 The Radar

Radar[1] stands for Radio Detection And Ranging and is, perhaps, more familiar than any other transmitter–receiver systems even though the details of the system may not be known to most people. Since radars are used for weather prediction, civilian aircraft guidance, as well as military and police work, radar is a household word. However, radar is more than a device to catch speeding drivers and land aircraft. It is extensively used in almost any imaginable application that requires detection of objects or conditions and for measurement of distance, material properties, and the like. For example, measuring the distance between two aircraft or range of aircraft from an airport as well as speed and direction is a common application. Other applications are in collision avoidance, all-weather landing systems, remote sensing of environmental conditions, cloud identification, mapping of planets, detection of buried objects, mapping of ice and snow conditions, and many others, too numerous to mention. It would be hard to find other systems that are as versatile and as useful as radar. It owes its usefulness to its almost trivial principle: that of generating a wave and receiving the reflected wave from any object it encounters. That is not to say that the system is simple. Far from it, but a good analogy to radar is a flashlight: when you turn it on, you can see any object that reflects light.

In this section, we look at the basic radar system as a transmit–receive system, but we will also use the ideas discussed here to define the concept of scattering and scattering cross section. The basic method of radar operation is shown in **Figure 18.40**. There are two modes: One relies on a transmitting antenna and a separate receiving antenna. This is called a bi-static mode and is shown in **Figure 18.41a**. The second mode is an a-static mode, shown in **Figure 18.41b**, and uses the same antenna for transmission and reception. In bi-static mode, antenna A sends a pulse or a continuous wave and the wave is scattered from the object and received back at antenna B. In the a-static mode, the antenna sends a pulse and then is ready to receive a signal. Clearly, a-static operation can only be accomplished with pulses and these must be very narrow so that transmission has been completed by the time the reflected signal arrives.

In either mode of operation, the amount of scattered power at the object (scatterer) depends on a number of properties, including size, shape, and composition of the scatterer. You may think of it as a body that scatters light. The amount of light scattered depends on the shape, size, and reflectivity (material properties) of the scatterer. Now, suppose that the transmitting antenna transmits a power P_{rad}. Part of this power arrives at the scatterer and is reflected back. Only a small part of the power transmitted reaches back to the receiving antenna. We now define one of the most basic of radar properties: the radar or

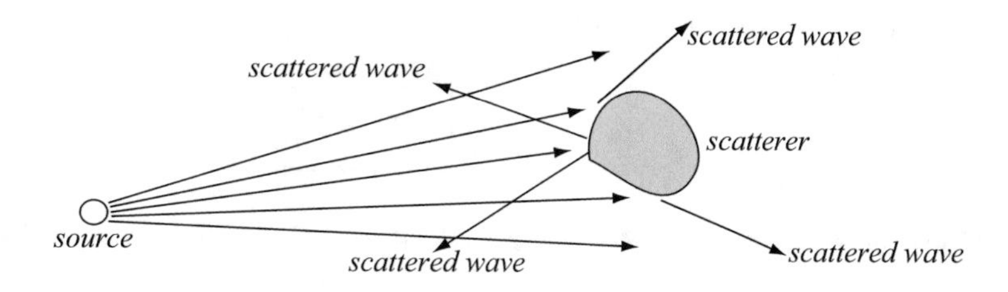

Figure 18.40 Principle of radar operation: scattering of waves from a target

[1] Radar has a long and interesting history. Although it appeared as a military instrument around the beginning of WWII (a radar installation in Hawaii detected the attack planes heading for Pearl Harbor, but the warning was ignored), its development started much earlier. The first reported radar experiment goes back to 1904 when Christian Hulsmeyer used waves from a spark gap (at a wavelength of about 0.5 m) to show azimuthal location capabilities at 3 km, for which he received the first radar patent. He called his device a "telemobiloscope." After that, many more experiments were performed, including experiments by Guglielmo Marconi (1874–1937), and as early as 1925, primitive forms of radar were used for remote sensing of the ionosphere. The development of radar for detection and tracking of aircraft started in 1935 in Britain, and after the development of the magnetron and klystron (both are microwave tubes used for generation of waves at high frequencies), the radar became a reality at the beginning of WWII. From then on, radar became a household name and, today, it is used in many areas from storm prediction and detection to mapping of planets to looking for dinosaurs' bones to collision avoidance.

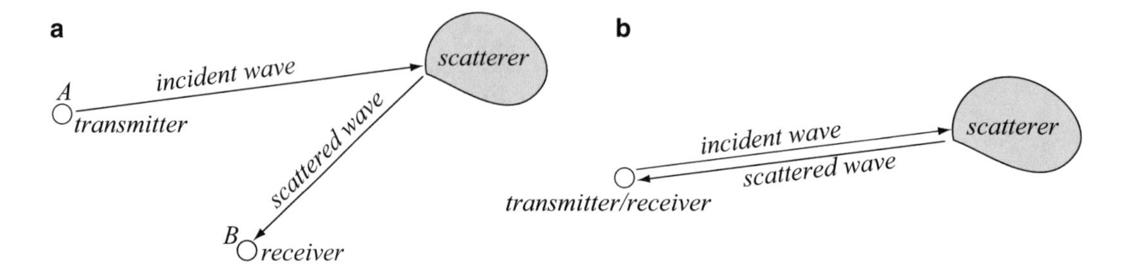

Figure 18.41 (**a**) Bi-static radar. (**b**) A-static

scattering cross section. To define the scattering cross section, consider **Figure 18.42**. In **Figure 18.42a**, the transmitting antenna emits power P_{rad}. How much of this power is reflected by the scatterer? We will assume that the equivalent area of the scatterer is σ and call this area the ***scattering cross section*** of the target. If the time averaged power density at the location of the scatterer is $\mathcal{P}_i$, the total power scattered by the target is σP_i. This scattered power can be viewed as a radiated power, transmitted by the scatterer. This power produces a scattered power density at the receiving antenna, which may be written (assuming uniform distribution of the scattered field since the scattering cannot be, in general, assumed to be directive) as

$$\mathcal{P}_s = \frac{\sigma \mathcal{P}_i}{4\pi R^2} \quad \left[\frac{\text{W}}{\text{m}^2}\right] \tag{18.162}$$

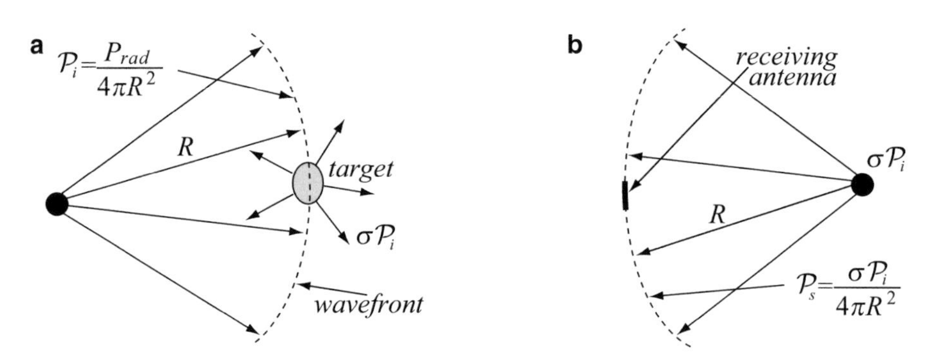

Figure 18.42 (**a**) Power at the target. (**b**) Power scattered by the target, part of which reaches back at the radar antenna

The scattering cross section is therefore

$$\boxed{\sigma = 4\pi R^2 \frac{\mathcal{P}_s}{\mathcal{P}_i} \quad [\text{m}^2]} \tag{18.163}$$

However, the time averaged power density $\mathcal{P}_i$ depends on the radiated power as [see **Eq. (18.158)**]

$$\mathcal{P}_i = \frac{P_{rad}}{4\pi R^2} D_t(\theta, \phi) \quad \left[\frac{\text{W}}{\text{m}^2}\right] \tag{18.164}$$

where $D_t(\theta,\phi)$ is the directivity of the transmitting antenna. Substituting in **Eq. (18.162)**, we get the scattered power density at the receiving antenna:

$$\mathcal{P}_s = \frac{\sigma P_{rad} D_t(\theta, \phi)}{\left(4\pi R^2\right)^2} \quad \left[\frac{\text{W}}{\text{m}^2}\right] \tag{18.165}$$

The total power received by the antenna depends on the scattered power density $\mathcal{P}_s$ and the effective area of the receiving antenna A_{er}. Thus, the received power is

$$P_r = A_{er}\mathcal{P}_s = \frac{A_{er}\sigma P_{rad} D_t(\theta,\phi)}{\left(4\pi R^2\right)^2} \quad [\mathrm{W}] \tag{18.166}$$

Now, using the definition of effective area of the receiving antenna as

$$A_{er} = D_r(\theta,\phi)\frac{\lambda^2}{4\pi} \quad \left[\mathrm{m}^2\right] \tag{18.167}$$

and substituting this in **Eq. (18.166)**, we get

$$P_r = \frac{\sigma P_{rad} D_t(\theta,\phi) D_r(\theta,\phi)\lambda^2}{4\pi\left(4\pi R^2\right)^2} = \frac{\sigma P_{rad} D_t(\theta,\phi) D_r(\theta,\phi)\lambda^2}{(4\pi)^3 R^4} \quad [\mathrm{W}] \tag{18.168}$$

Dividing by P_{rad} we obtain a transmission equation similar to the Friis formula:

$$\frac{P_r}{P_{rad}} = \frac{\sigma D_t(\theta,\phi) D_r(\theta,\phi)\lambda^2}{(4\pi)^3 R^4} \tag{18.169}$$

In the particular case in which the transmitting antenna is used for receiving as well (most radar applications), $D_r(\theta,\phi) = D_t(\theta,\phi) = D(\theta,\phi)$ and we can write

$$\boxed{\frac{P_r}{P_{rad}} = \frac{\sigma\lambda^2}{(4\pi)^3 R^4} D^2(\theta,\phi) = \frac{\sigma A_e^2}{4\pi R^4 \lambda^2}} \tag{18.170}$$

This equation is called the radar equation. It is in many ways similar to the Friis transmission formula, but because the signal traverses the distance between the antenna and scatterer twice, the received power is proportional to $1/R^4$. It is also directly proportional to the scattering (radar) cross section, the frequency of the signal squared, and the square of the directivity of the antenna (or proportional to the directivity of each antenna in bi-static radar). For this reason, radars require considerable power for detection and their range is limited. High-gain antennas are almost always used, and the higher the frequency, the better. Of course, the radar cross section is extremely important since, for a given radar system, it defines detectability of the target. If the scattering cross section is large, detection is relatively easy. If it is small, it is much more difficult. For this reason, many small aircraft are difficult to detect on radars, especially with on-board radars which have limited power and, therefore, limited range. Military aircraft are often designed with reduced radar cross section to avoid detection. The radar cross section is also important in other applications such as mapping and sensing because it defines the limits on sensitivity.

18.11.1 Types of Radar

There are many types of radar in addition to that employing the simple principles above. One particularly useful radar is the Doppler radar, which discerns the frequency shift due to the motion of moving objects. These radars are invaluable in weather prediction and in sensing flying conditions for aircraft. In detecting moving objects against a background, such as in lookdown radars, the Doppler radar is useful where other radars are blind. Doppler radars have been used in other exotic applications such as in the detection of flying insects, measurements in snow and sand storms, and the like.

Another type of radar is the so-called synthetic aperture radar. Since the effective aperture (effective area) of the antenna is important rather than its actual size, anything that will increase the aperture will make the radar more sensitive. In synthetic aperture radar, the aperture is synthesized from the signal such as moving the antenna at a given speed. The area it covers in the time the signals take to propagate to the target and back is roughly the synthetic aperture of the radar. This is particularly useful in space exploration where the effective aperture can be a few kilometers squared, dramatically increasing the sensitivity of the system.

A third type of radar is the ground-penetrating radar. Although the principles of this radar are not different than any other radar, its antennas are quite different because the transmission is at relatively low frequencies. Thus, instead of the familiar parabolic reflector antenna of most radars, the into-the-ground radar, which operates at frequencies as low as 100 MHz, uses dipole antennas. Penetration into the ground is not very deep, but the instrument is relatively small and simple and has been used for a variety of tasks from archaeological digs to the location of buried power lines and inspection of roads and bridges. A variation of this type of radar is a space-borne radar designed to map underground for large features such as aquifers and ancient sites. Flown on the space shuttle and on other spacecraft, this type of radar has been used for general-purpose underground mapping on planet Earth and in space.

There are additional systems that operate on the principles of radar, that is, on reflection of power from targets. An optical equivalent is called LIDAR (LIght Detection And Ranging). It is used primarily for accurate ranging of targets but also for mapping and imaging and in autonomous vehicles.

Example 18.15 Application: Remote Sensing of the Environment A satellite is used to map the surface of a planet using a radar at 60 GHz. The satellite can supply 250 W to the radar antenna at 100% efficiency. The effective area of the antenna is 10 m^2 and the radar encircles the planet at a height of 200 km:

(a) Assuming the receiving circuitry can discern powers as low as 10^{-12} W, what is the smallest feature on the planet that can be detected, in terms of its radar cross section?

(b) In an attempt to improve resolution, the spacecraft is moved into an orbit 100 km high. What is the smallest radar cross section detectable?

(c) How can the resolution be improved further?

Note: The properties and dimensions given here (with the exception of frequency) are close to those of the Magellan spacecraft as it mapped the surface of Venus.

Solution: The radar cross section from **Eq. (18.170)** is

$$\sigma = \frac{P_r 4\pi R^4 \lambda^2}{P_{rad} A_e^2} \quad [\mathrm{m}^2]$$

(a) With a wavelength $\lambda = c/f = 3 \times 10^8/60 \times 10^9 = 0.005$ m, the smallest radar cross section detectable is

$$\sigma = \frac{10^{-12} \times 4 \times \pi \times (200,000)^4 \times 0.005^2}{250 \times (10)^2} = 20.1 \quad [\mathrm{m}^2].$$

(b) At $R = 100$ km, the smallest radar cross section detectable is

$$\sigma = \frac{10^{-12} \times 4 \times \pi \times (100,000)^4 \times 0.005^2}{250 \times (10)^2} = 1.26 \quad [\mathrm{m}^2].$$

(c) The most significant factor is the distance. In addition, the wavelength may be reduced (frequency increased), the power increased, or the antenna may be larger (larger effective aperture). Beyond that, the signal-to-noise ratio may be improved, allowing lower detectable signals. Although the radar cross section is not the actual size of objects, it gives an indication of the size and reflectivity of the objects; that is, smaller, highly reflective objects will appear larger than larger, less reflective objects. This, of course, is not surprising. A similar example is well known: when you fly at high altitudes, you may be able to see the Sun reflecting off a window while not being able to see the house.

18.12 Other Antennas

We have touched on a very small number of antennas here. The variety of antennas in design, size, and implementation is vast. Some antennas are extremely small, such as antennas used for medical applications or for radio tagging of animals. Others may span hundreds of meters such as low-frequency dipoles. Still others may not look like antennas but serve as such. For example, a slot in a waveguide or cavity resonator will radiate into space. These are called aperture antennas, and although they are closely related to dipoles, they do not look like dipoles. Others, like the patch antennas, are no more than a rectangular or circular patch of conducting material, often buried under a layer of paint so that it is difficult to even see them. Still other antennas and antenna arrays have, in addition to the radiating element, guiding and reflecting elements. The parabolic reflector antennas were mentioned earlier. They may be very small, such as the antennas used for digital satellite TV reception (0.45 to 0.9 m in diameter), or extremely large such as antennas used for the deep-space network, which are 70 m in diameter. Other antennas such as waveguide horn antennas come in many varieties, including arrays of horns. Still other antennas may be buried underground or carried on aircraft, may conform to a surface, printed on a printed circuit board, may be made of nanoparticles or may simply be an existing structure such a power cable. The variety is endless and the challenge of designing efficient antennas for mobile equipment is only beginning. Arrays also come in many varieties and not all are uniform arrays. In some cases hundreds or thousands of elements work together as an array, many are used to steer the beam or to produce multiple beams and to adapt to various conditions. Adaptive arrays are particularly useful. For example, cell communication antennas are almost always adaptive arrays but these also find use in various types or radars.

The analysis of these antennas is much more complex than those outlined here for the monopole, dipole, and linear array antennas and, often, the analysis is approximate. In many cases, the analysis must be numerical. For this purpose, the methods discussed in **Chapter 6** and, in particular, the method of moments are often employed. There are of course other methods of analysis that are specific to antennas.

18.13 Applications

Application: The Deep-Space Network The deep-space network is a system of antennas located in the Mojave Desert in California, near Madrid in Spain, and near Canberra in Australia. The system was built by NASA to support its mission in communication with spacecraft and, in particular, with deep-space probes. At each site, there is one 26 m diameter parabolic antenna, at least two 34 m diameter parabolic antenna, and one 70 m diameter parabolic antenna[2] with a central signal processing center which controls the antennas and transmits and receives spacecraft information. This system, which has been in continuous operation since 1966, allowed such remote communication as with the *Pioneer* 10, *Voyager* 1 and *Voyager* 2 spacecraft at distances of billions of kilometers from Earth. The antennas operate at 2.3 GHz and at 8.4 GHz at about 90% efficiency. Similar networks are operated by the space agencies of China, Russia, Japan, India and the European union.

Application: The Large Radio Telescopes A telescope is an instrument designed to receive light and, in some way, to enhance the image received. A radio telescope is a telescope designed to receive lower-frequency electromagnetic waves, usually from extraterrestrial sources, and enhance the signal either through a reflector in the case of a single antenna or by use of synthesized signals in the case of arrays. The large baseline array was mentioned in **Section 18.8**. However, there are other radio telescopes which use a single reflector. These are massive structures. For example, a radio telescope, 100 m in diameter, operates in Effelsberg, Germany. The Arecibo radio telescope in Puerto Rico is 300 m in diameter. The largest of the radio telescopes of this type is the RATAN-600 in the Caucasus Mountains in Russia. The reflector is 576 m in diameter although it does not look as a true reflector. Rather, it is a reflecting ring about 11 m high and 576 m in diameter. The radio telescopes mentioned are fixed structures looking at the sky and are used to detect signals from a variety of electromagnetic sources from pulsars, quasars, radiation from dust clouds, interplanetary matter, other stars and planets, and, yes, also listening for microwave signals from other civilizations, if these are ever detected.

[2] Although parabolic antennas are rather new, it is worth mentioning that in 1888, Heinrich Hertz used a parabolic reflector to focus electromagnetic waves. In his experiments, he used a spark gap which radiated at a wavelength of about 0.6 m. He designed a wooden frame, about 2 m by 1.2 m, and coated it with zinc to obtain the first parabolic reflector.

Application: Automobile Radar—Collision Avoidance The idea of installing a small radar to aid in collision avoidance in cars and, perhaps, as an intelligent autonomous cruise control system is quite old, an idea which could not be developed until recently because of the size and expenses involved. However, Doppler radars of very modest size can be designed, especially since the range required for automobiles is short: perhaps up to 1 km. Experimental radars of this type, at frequencies ranging from about 30 GHz to about 100 GHz, were designed and tested. In car radar, the tendency is toward higher frequencies, primarily because of the antenna size. Collision avoidance radars are a development of the autonomous landing radars in aircraft which also operate at high frequencies. Some cars and trucks come with some form of radar used either for collision avoidance or for cruise control. The push towards autonomous vehicles is likely to increase the use of radar in vehicles including Doppler and Lidar systems.

Application: Microwave Antennas for Therapy An example of the use of very small antennas is in the application of microwaves in medical procedures. One example is the local microwave radiation of tumors. In its simplest form, this consists of a coaxial cable with a small antenna on its end. The antenna need not be too complicated. Exposure of the internal conductor in the cable is enough to create a monopole antenna. Other possibilities are aperture antennas at the end of a waveguide. Microwave hypothermia treatment for cancer is only one aspect of microwave applicators. Another is in the treatment of blocked arteries. In all these applications, a small antenna is used to couple energy into the affected tissue. Most applicators operate at 2.54 GHz (the same frequency used in microwave ovens) at low power and rely on local heating of tumors or blockages in arteries.

Application: Radio Frequency Identification (RFID) of Livestock and Consumer Goods One method often employed in tagging of animals is the implantation of a small microprocessor and an RFID transponder, together with an integrated antenna under the animal's skin. The transponder is activated from outside by a high-frequency source produced by an antenna or a coil. This charges the internal (very small) battery or capacitor and then the transponder sends the required data which may be as simple as an identification code or may contain other information such as age, history, and condition of the animal. The implantable devices are small: about 3 mm long for laboratory animals and somewhat larger for livestock. The antennas are usually small loops or coils and may be integrated within the chip that contains the circuitry. Since the range is short the antennas need not be large and the power required is also rather low. Similar methods are used for inventory control of products for electronic article surveillance on the retail level and for automatic payment of tolls.

18.14 Summary

This present chapter discusses antennas and radiation from antennas. We start with the elemental electric dipole and introduce the idea of retardation of potentials and fields, an idea stemming from the finite speed of propagation of electromagnetic waves. The fields and properties of antennas follow from these simple terms.

Retarded potential is a potential (the magnetic vector potential in this case) produced by a current at time $t–\Delta t$ and measured at a time t at a distance R. Given a sinusoidal current $I_0\cos\omega(t - R/v_p)$, we have in the frequency domain:

$$\mathbf{A}(R) = \hat{\mathbf{z}}\frac{\mu I_0 \Delta l'}{4\pi R}e^{-j\beta R} \quad \left[\frac{\text{Wb}}{\text{m}}\right] \tag{18.12}$$

where $\Delta l'$ is the length of a short wire segment carrying the current I and $\beta = \omega/v_p$. This short segment is called a ***Hertzian dipole*** antenna.

Fields of the Short Dipole—Hertzian Dipole (see Figure 18.3)

$$\mathbf{H} = -\hat{\boldsymbol{\phi}}\frac{I_0 \Delta l' \beta^2}{4\pi}e^{-j\beta R}\sin\theta\left(\frac{1}{j\beta R} + \frac{1}{(j\beta R)^2}\right) \quad \left[\frac{\text{A}}{\text{m}}\right] \tag{18.20}$$

$$\mathbf{E} = -\hat{\mathbf{R}}\frac{\eta I_0 \Delta l' \beta^2}{2\pi} e^{-j\beta R} \cos\theta \left(\frac{1}{(j\beta R)^2} + \frac{1}{(j\beta R)^3} \right) - \hat{\mathbf{\theta}}\frac{\eta I_0 \Delta l' \beta^2}{4\pi} e^{-j\beta R} \sin\theta \left(\frac{1}{j\beta R} + \frac{1}{(j\beta R)^2} + \frac{1}{(j\beta R)^3} \right) \quad \left[\frac{\text{V}}{\text{m}}\right] \tag{18.21}$$

There are three zones that can be defined—***near field, intermediate field,*** and ***far field***. These zones and their main properties are shown in the following table:

	Definition	Approximations	Notes
Near field Also called: ***electrostatic field*** or ***Fresnel zone***	$\beta R \ll 1$ or: $R \ll \lambda/2\pi$	$\mathbf{E} \approx \hat{\mathbf{R}}\frac{I_0 \Delta l'}{j2\pi\omega\varepsilon R^3}\cos\theta + \hat{\mathbf{\theta}}\frac{I_0 \Delta l'}{j4\pi\omega\varepsilon R^3}\sin\theta \quad \left[\frac{\text{V}}{\text{m}}\right]$ (18.27) $\mathbf{H} \approx \hat{\mathbf{\phi}}\frac{I_0 \Delta l'}{4\pi R^2}\sin\theta \quad \left[\frac{\text{A}}{\text{m}}\right]$ (18.25)	**E** is large and **H** small. Terms with $1/(j\beta R)^3$ in **Eqs. (18.20)** and **(18.21)** dominate
Intermediate field or ***inductive zone***		None. Must use **Eqs. (18.20)** and **(18.21)**	The fields in **Eqs. (18.20)** and **(18.21)** are correct in all zones since no approximations are used
Far field Also called: ***radiation field*** or ***Fraunhofer zone***	$\beta R \gg 1$ or: $R \gg \lambda/2\pi$	$\mathbf{H} = \hat{\mathbf{\phi}}\frac{j\beta I_0 \Delta l'}{4\pi R}e^{-j\beta R}\sin\theta \quad \left[\frac{\text{A}}{\text{m}}\right]$ (18.30) $\mathbf{E} = \hat{\mathbf{\theta}}\frac{j\beta\eta I_0 \Delta l'}{4\pi R}e^{-j\beta R}\sin\theta \quad \left[\frac{\text{V}}{\text{m}}\right]$ (18.31)	**E** and **H** are perpendicular to each other and to the direction of propagation Behavior similar to plane waves but in spherical coordinates Terms with $1/(j\beta r)$ in **Eqs. (18.20)** and **(18.21)** dominate

Radiation properties of Hertzian dipoles:

Time-averaged power density

$$\mathcal{P}_{av} = \hat{\mathbf{R}}\frac{\eta I_0^2 \beta^2 (\Delta l')^2}{32\pi^2 R^2}\sin^2\theta \quad \left[\frac{\text{W}}{\text{m}^2}\right] \tag{18.32}$$

Radiated power

$$P_{rad} = \frac{\eta I_0^2 \beta^2 (\Delta l')^2}{12\pi} = I_0^2 \frac{\eta\pi}{3}\left(\frac{\Delta l'}{\lambda}\right)^2 \quad [\text{W}] \tag{18.34}$$

Radiation resistance

$$R_{rad} = \frac{2\eta\pi}{3}\left(\frac{\Delta l'}{\lambda}\right)^2 = 80\pi^2\left(\frac{\Delta l'}{\lambda}\right)^2 \quad [\Omega] \tag{18.37, 18.38}$$

Antenna radiation pattern is the relative (normalized) strength of the electric or magnetic field intensity (field radiation pattern) or its power density (power radiation pattern), in the far field. The pattern is usually given as a plot in polar or rectangular coordinates or as a three-dimensional representation. E-plane plots are radiation patterns in a plane that includes the antenna; H-plane plots are in the plane perpendicular to the antenna.

Beamwidth is the angle between the two half-power density points on the radiation pattern.

Radiation intensity

$$U(\theta, \phi) = \mathcal{P}_{av} R^2 \quad [\text{W/sr}] \tag{18.46}$$

Average radiation intensity

$$U_{av} = P_{rad}/4\pi \quad [\mathrm{W/sr}] \tag{18.48}$$

Antenna directivity

$$D(\theta,\phi) = U(\theta,\phi)/U_{av} = \frac{4\pi U(\theta,\phi)}{P_{rad}} \quad [\text{dimensionless}] \tag{18.50}$$

Maximum directivity

$$D_0 = \frac{4\pi U_{max}}{P_{rad}} \quad [\text{dimensionless}] \tag{18.51}$$

Antenna power gain

$$G(\theta,\phi) = \frac{4\pi U(\theta,\phi)}{P_{in}} \quad [\text{dimensionless}] \tag{18.54}$$

Antenna radiation efficiency

$$eff = \frac{G_0}{D_0} = \frac{P_{rad}}{P_{in}} = \frac{R_{rad}}{R_{rad}+R_d} \quad [\text{dimensionless}] \qquad (18.55, 18.57)$$

Magnetic Dipole (Small-Loop Antenna, radius ≪ λ): See **Figure 18.10**

Near fields

$$\mathbf{H} = \frac{m}{4\pi R^3}\left(\hat{\mathbf{R}}2\cos\theta + \hat{\mathbf{\theta}}\sin\theta\right) \quad \left[\frac{\mathrm{A}}{\mathrm{m}}\right] \tag{18.75}$$

$$\mathbf{E} = -\hat{\mathbf{\phi}}\frac{j\omega\mu m}{4\pi R^2}\sin\theta \quad \left[\frac{\mathrm{V}}{\mathrm{m}}\right] \tag{18.76}$$

Far fields

$$\mathbf{E} = \hat{\mathbf{\phi}}\frac{\omega\mu\beta m}{4\pi R}e^{-j\beta R}\sin\theta \quad \left[\frac{\mathrm{V}}{\mathrm{m}}\right], \quad \mathbf{H} = -\hat{\mathbf{\theta}}\frac{\omega\mu\beta m}{4\pi\eta R}e^{-j\beta R}\sin\theta \quad \left[\frac{\mathrm{A}}{\mathrm{m}}\right] \tag{18.77}$$

where $m = \pi d^2 I$ [$\mathrm{A \cdot m^2}$] is the magnetic dipole moment of the loop with d its radius and I its current.

Table 18.1 summarizes the properties of Hertzian and magnetic dipoles.

Arbitrarily Long Antennas. The properties of arbitrarily long antennas are obtained by viewing them as stacks of Hertzian dipoles. In essence we assume a sinusoidal current along the antenna so that it is zero at its ends, assume a differential of length along the antenna, assign to it the electric and magnetic fields of the Hertzian dipole **[Eqs. (18.30)** and **(18.31)]**, and integrate to find the fields in the far field (see **Figure 18.14**). The table below summarizes the main properties of arbitrarily long antennas together with the particular but important case of a half-wavelength antenna.

	Arbitrarily long antenna (length = L)		$\lambda/2$ antenna ($L = \lambda/2$)	
$I(z')$ [A]	$= I_0\sin\left(\beta\left(\frac{L}{2}-\lvert z'\rvert\right)\right)$	(18.79)	$= I_0\cos\frac{2\pi z'}{\lambda}$	(18.96)
$\mathbf{E}(R,\theta)$ [V/m]	$= \hat{\mathbf{\theta}}\frac{j\eta I_0}{2\pi R}e^{-j\beta R}\frac{\cos((\beta L/2)\cos\theta)-\cos(\beta L/2)}{\sin\theta}$	(18.85)	$= \hat{\mathbf{\theta}}\frac{j\eta I_0}{2\pi R}e^{-j\beta R}\frac{\cos((\pi/2)\cos\theta)}{\sin\theta}$	(18.97)
$\mathbf{H}(R,\theta)$ [A/m]	$= \hat{\mathbf{\phi}}\frac{jI_0}{2\pi R}e^{-j\beta R}\frac{\cos((\beta L/2)\cos\theta)-\cos(\beta L/2)}{\sin\theta}$	(18.86)	$= \hat{\mathbf{\phi}}\frac{jI_0}{2\pi R}e^{-j\beta R}\frac{\cos((\pi/2)\cos\theta)}{\sin\theta}$	(18.98)
$\mathcal{P}_{av}(R,\theta)$ [W/m^2]	$= \hat{\mathbf{R}}\frac{\eta I_0^2}{8\pi^2R^2}\left(\frac{\cos((\beta L/2)\cos\theta)-\cos(\beta L/2)}{\sin\theta}\right)^2$	(18.89)	$= \hat{\mathbf{R}}\frac{\eta I_0^2}{8\pi^2R^2}\left(\frac{\cos((\pi/2)\cos\theta)}{\sin\theta}\right)^2$	(18.99)

(continued)

	Arbitrarily long antenna (length = L)	$\lambda/2$ antenna ($L = \lambda/2$)
P^{rad} [W]	$= \dfrac{\eta I_0^2}{4\pi}\displaystyle\int_{\theta=0}^{\theta=\pi}\frac{(\cos((\beta L/2)\cos\theta) - \cos(\beta L/2))^2}{\sin\theta}d\theta$ (18.91)	$= \dfrac{1.218\eta I_0^2}{4\pi}$ (18.100)
R^{rad} [Ω]	$= \dfrac{\eta}{2\pi}\displaystyle\int_{\theta=0}^{\theta=\pi}\frac{(\cos((\beta L/2)\cos\theta) - \cos(\beta L/2))^2}{\sin\theta}d\theta$ (18.93)	$= \dfrac{0.609\eta}{\pi}$ (18.101)
$D(\theta)$	$= \dfrac{4\pi \mathcal{P}_{av}R^2}{P_{rad}}$ (18.94)	$= 1.642\dfrac{\cos^2((\pi/2)\cos\theta)}{\sin^2\theta}$ (18.104)
Do	= Maximum value of $D(\theta)$	$= D(\theta = \pi/2) = 1.642$ (18.105)
$\lvert f^e(\theta)\rvert$	$= \left\lvert\dfrac{\cos((\beta L/2)\cos\theta) - \cos(\beta L/2)}{\sin\theta}\right\rvert$ (18.95)	$= \left\lvert\dfrac{\cos((\pi/2)\cos\theta)}{\sin\theta}\right\rvert$ (18.103)

Note: The integral in **Eqs. (18.91)** and **(18.93)** is calculated numerically (listed in **Table 18.2** and in **Figure 18.15**)

Monopole Antennas These may be viewed as "half dipoles," that is, a single element perpendicular above a conducting "ground." Its properties are identical to the equivalent dipole except that the radiated power and radiation resistance are half those of the equivalent dipole. Also, the radiation pattern only exists above ground.

Antenna arrays are multiple antennas in a geometric configuration and driven with specific currents, all designed to produce a given radiation pattern.

The fields of antenna arrays are obtained by simple superposition of the fields of individual elements.

Linear Uniform phased array—N elements, equally spaced on an axis carrying currents of identical magnitude and constant phase difference between two neighboring elements. Also called progressive, linear uniform arrays.

Broadside array—The main beam (lobe) radiates sideways relative to the axis of the array.

End-fire array—The main beam radiates along the axis of the array.

Two-element array: (see **Figure 18.26**)

$$\mathbf{E} = \hat{\theta}\frac{E_0 f(\theta,\phi)}{R_1}e^{-j\beta R_1}e^{j\psi/2}2\cos\frac{\psi}{2} \quad (18.121) \qquad \psi = \beta h\sin\theta\cos\phi + \varphi \quad [\text{rad}] \quad (18.119)$$

R_1 is the distance from the first element in the array, E_0 and $f(\theta,\phi)$ are the amplitude and radiation pattern of a single element, h the separation between elements, and φ the phase difference between two consecutive elements. In this case the antennas are parallel to each other and perpendicular to the x axis, directed in the z direction. If the antennas are configured differently (see **Figure 18.30**) the results are different because ψ is different (see **Example 18.11**).

N-element uniform linear array (see **Figure 18.32**):

$$\mathbf{E} = \sum_{k=1}^{n}\mathbf{E}_k = \hat{\boldsymbol{\theta}}\frac{E_0 f(\theta,\phi)}{R_1}e^{-j\beta R_1}e^{j(n-1)\psi/2}\frac{\sin(n\psi/2)}{\sin(\psi/2)} \quad \left[\frac{\text{V}}{\text{m}}\right] \quad (18.137)$$

Normalized array factor of an n-element progressive uniform linear array:

$$|f_n(\theta,\phi,\varphi)| = \frac{1}{n}\left|\frac{\sin(n\psi/2)}{\sin(\psi/2)}\right| = \frac{1}{n}\left|\frac{\sin(n(\beta h\sin\theta\cos\phi + \varphi)/2)}{\sin((\beta h\sin\theta\cos\phi + \varphi)/2)}\right| \quad (18.139)$$

The terms, R_1, ψ, h, and φ are as for the two-element array.

Reciprocity and receiving antennas. Reciprocity theorem (only applies to isotropic media): "*If a source is applied to antenna No. 1 and a signal is received in antenna No. 2, then, applying the same identical source to antenna No. 2 will produce an identical signal in antenna No. 1.*"

Effective length is the ratio between the magnitude of the open circuit voltage at the antenna and the electric field intensity in the direction of the antenna (**Figure 18.36**).

Effective aperture is the ratio of power received by the antenna and the time average power density at the location of the antenna:

$$A_{ea} = \frac{P_a}{\mathcal{P}_{av}} \quad [\mathrm{m}^2] \quad \text{or} \quad A_{ea} = D(\theta, \phi)\frac{\lambda^2}{4\pi} \quad [\mathrm{m}^2] \tag{18.143, 18.153}$$

Maximum effective area:

For Hertzian dipole:

$$(A_{ea})_{max} = 1.5\frac{\lambda^2}{4\pi} \quad [\mathrm{m}^2] \tag{18.154}$$

For $\lambda/2$ dipole:

$$(A_{ea})_{max} = 1.642\frac{\lambda^2}{4\pi} \quad [\mathrm{m}^2] \tag{18.155}$$

Friis transmission formula is a common design tool in antennas and communication. It defines the power received ($P_{rec.}$) in terms of transmitted power ($P_{trans.}$) and properties of the receiving and transmitting antennas (indicated with indices r and t, respectively) in lossless media:

$$\frac{P_{rec.}}{P_{trans.}} = \frac{A_{er}A_{et}}{\lambda^2 R^2} = \left(D_r(\theta, \phi)\frac{\lambda^2}{4\pi}\right)\left(D_t(\theta, \phi)\frac{\lambda^2}{4\pi}\right)\frac{1}{\lambda^2 R^2} \tag{18.160, 18.161}$$

Radar and Radar Cross Section

Radar cross section is the apparent area of a target based on the power it reflects:

$$\sigma = 4\pi R^2 \frac{\mathcal{P}_s}{\mathcal{P}_i} \quad [\mathrm{m}^2] \tag{18.163}$$

where $\mathcal{P}_s$ [W/m^2] is the time averaged power density scattered by the target and $\mathcal{P}_i$ [W/m^2] is the time averaged incident power density at the location of the target.

Radar Equation. This is a modification of the Friis formula for the specific conditions of radar.

$$\frac{P_r}{P_{rad}} = \frac{\sigma A_e^2}{4\pi R^4 \lambda^2} = \frac{\sigma \lambda^2}{(4\pi)^3 R^4} D_t(\theta, \phi), D_r(\theta, \phi) \tag{18.169, 18.170}$$

Problems

Hertzian Dipole

18.1 Near and Far Fields of Antennas. In preparation for a propagation experiment, a researcher places a field measuring instrument (an antenna) at a distance of 150 m from a transmitting antenna, which operates at 30 GHz. Is the instrument in the near or the far field of the transmitting antenna?

18.2 The Hertzian Dipole. A short dipole antenna is 0.02 λ long and carries a current of 2 A at 150 MHz. Calculate (in free space):

(a) The electric and magnetic field intensities in the near field.
(b) The electric and magnetic field intensities in the far field.
(c) The radiation resistance and radiated power of the antenna.
(d) Maximum range in the direction of maximum power density if the time-averaged power density required for reception is 10^{-10} W/m^2.

18.3 Radiated Power. An antenna produces an electric field intensity in the far field:

$$\mathbf{E} = \hat{\boldsymbol{\theta}}\frac{jV_0}{R}e^{-j\beta_0 R}\sin\theta \quad \left[\frac{\text{V}}{\text{m}}\right]$$

where β_0 [rad/m] is the phase constant in free space, R [m] is the distance from the source (antenna), and θ is the angle with respect to the vertical (z) axis. Calculate:

(a) The radiated time-averaged power density.
(b) The total time-averaged power radiated by the antenna.

18.4 Application: The Dipole Antenna. A dipole antenna is 1 m long and is fed with a current of amplitude 2 A. Find the radiated power of the antenna in free space:

(a) At 540 kHz (lowest AM band frequency).
(b) At 1.6 MHz (highest AM band frequency).
(c) To what do you attribute the difference between the radiated powers in **(a)** and **(b)**?

18.5 Power Line as Antenna. Calculate the power radiated by one conductor of a single phase high voltage power line one km long carrying a current of 250 A at 60 Hz. Neglect the effect of the ground and the proximity of other lines. Treat air as free space.

Magnetic Dipole

18.6 Application: Magnetic Dipole (Loop) Antenna. A transmitter uses a Hertzian dipole, 0.02λ long. Because of changes in design, it becomes necessary to replace the Hertzian dipole by an equivalent magnetic dipole (loop) antenna, keeping the current in the loop the same as the current in the dipole:

(a) What is the radius of the loop that will produce the same fields in the far field?
(b) What is the orientation of the loop with respect to the Hertzian dipole?

18.7 The Small-Loop Antenna. A magnetic dipole (loop) antenna is made of 10 turns, closely wound. The turns are of radius a [m]. A current $I_0\cos\omega t$ [A] flows in the antenna:

(a) Find the radiated power of the antenna in free space.
(b) Find an expression for the radiation resistance of the antenna in free space.
(c) Extend the expressions in **(a)** and **(b)** to an n-turn loop antenna.

Linear Antennas of Arbitrary Length

18.8 Application: The Short Dipole Antenna. A short dipole is a dipole that is too long to be considered a Hertzian dipole but too short to be an arbitrarily long antenna. This usually means an antenna between $\lambda/50$ and $\lambda/10$ in. length. Given: In the short dipole, the current is linear that is, $I(z') = I_0(1 - (2/l)|z'|)$ [A], where $l = \lambda/20$ is the total length of the dipole and I_0 [A] is a constant amplitude. Calculate:

(a) The far-field electric field intensity, magnetic field intensity, and power density.
(b) The radiation resistance and radiation pattern of the dipole.

18.9 Arbitrary-Length Dipole. A communication system uses a dipole antenna for transmission. Calculate:

(a) The maximum power density received 5 km away given the following: $I_0 = 2$ A, dipole length is $d = 80$ cm, wavelength is $\lambda = 1$ m, propagation is in free space.
(b) The radiated power of the antenna.
(c) The radiation resistance of the antenna.
(d) Maximum directivity of the antenna.

18.10 Radiation Efficiency. Find the radiation efficiencies of the following antennas. They are made of round copper wire of radius $a = 0.4$ mm and conductivity $\sigma = 5.7\times10^7$ S/m. Permeability of copper is the same as that of free space.

(a) A dipole of length 2 m operating at 1 MHz.
(b) A dipole of length 1.5 m operating at 100 MHz.

The Half-Wave Dipole Antenna

18.11 Application: $\lambda/2$ Dipole. A half-wavelength antenna is made of copper wire with conductivity $\sigma = 5.7 \times 10^7$ S/m and permeability $\mu_0 = 4\pi \times 10^{-7}$ H/m. The antenna is 3 mm thick and 1.5 m long:

(a) Calculate the impedance of the antenna and its efficiency when radiating in free space.
(b) In an attempt to reduce the impedance and increase efficiency, the antenna is made of a copper tube, 12 mm in diameter. What is the efficiency of the antenna? Assume current flows entirely within one skin depth, on the surface of the wire.

18.12 Application: Simple Dipole Antenna. Suppose you need a good antenna in a hurry for a transmitter operating at 27 MHz. A friend suggests the following: Take a two-wire cable, with characteristic impedance of 72 Ω. Split the cable for a length of 2.77 m. Bend the two wires so that they are perpendicular to the cable and on a line as shown in **Figure 18.43**. To support it, the friend suggests tacking it onto a nonconducting wall:

(a) What kind of antenna is this antenna?
(b) If the conductor is 1 mm thick and made of copper ($\sigma = 5.7 \times 10^7$ S/m, $\mu = 4\pi \times 10^{-7}$ H/m), what is the efficiency of the antenna?

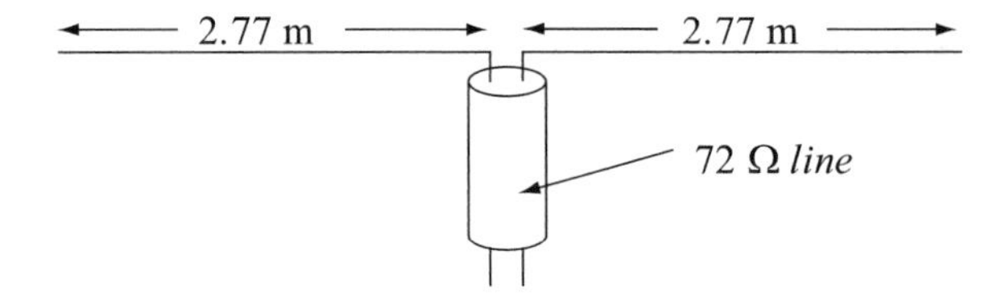

Figure 18.43

18.13 Radiated Power of $\lambda/2$ Dipole. A half-wavelength antenna is given. The maximum magnetic field intensity of the antenna at a distance of 120 m is measured and found to be equal to 1 mA/m (peak value). Radiation is at a wavelength of 1 m.

(a) Assuming free space, calculate the radiated power of the antenna.
(b) Calculate the radiated power of the antenna if the space also has a conductivity of 10^{-6} S/m.

18.14 Properties of $\lambda/2$ Dipole. A half-wavelength dipole is driven at 100 MHz with a current of 10 A (peak). The antenna may be assumed to be at ground level and perpendicular to the ground. Neglect the effect of the ground. An airplane receives transmission from the antenna when it is at 100 km horizontal distance from the antenna and at 10 km elevation above ground. Calculate assuming air has properties of free space:

(a) The time-averaged power density at the airplane.
(b) The antenna gain in the direction of the airplane.

18.15 Fields of $\lambda/2$ Dipole. A half wavelength dipole is directed vertically (z-axis) and radiates a time averaged power density of 28 W in free space at a frequency of 275 MHz. Find the amplitudes of the electric and magnetic field intensities at a distance $R = 100$ m from the center of the antenna, at $\theta = 90°$ and $\phi = 30°$.

Various Length Dipole Antennas

18.16 Radiation Resistance and Standing Wave Ratio. An antenna is connected to a 50 Ω transmission line. The antenna radiates in free space. Calculate the standing wave ratio on the transmission line if:

(a) The antenna is a half-wavelength dipole with zero internal resistance.
(b) The antenna is a half-wavelength dipole with internal resistance of 1.2 Ω.
(c) The antenna is a Hertzian dipole, $\lambda/50$ long and has no internal resistance.

18.17 Various Length Dipoles. Find and plot the current distribution along a dipole antenna of the following length:

(a) $\lambda/2$.
(b) λ.
(c) $3\lambda/2$.
(d) $3\lambda/4$.
(e) 2λ.
(f) $5\lambda/4$.

18.18 Various Length Dipoles. Find and plot the normalized E-plane field radiation pattern of the following dipole antenna lengths:

(a) 0.75λ.
(b) 1λ.
(c) 1.5λ.
(d) 2λ.
(e) 2.5λ.

18.19 Radiation Resistance, Directivity, and Radiated Power. Calculate the radiation resistance, directivity, and radiated power (in free space) of a $3\lambda/2$ dipole carrying a peak current of 0.2 A.

18.20 Radiation Resistance, Directivity, and Radiated Power

(a) Estimate the radiation resistance of a 1.25λ dipole in free space using **Table 18.2** and **Figure 18.15**.
(b) Calculate the exact radiation resistance of the 1.25λ dipole in free space.
(c) Calculate the directivity of the 1.25λ dipole using the estimate in (**a**) and by exact calculation.

The Monopole Antenna

18.21 Application: $\lambda/2$ Monopole. A half-wavelength monopole antenna is used for FM transmission at 100 MHz. The antenna is placed vertically above a conducting surface in free space. Find:

(a) The antenna length.
(b) The radiation resistance of the monopole.
(c) The radiated power of the monopole for a current of amplitude I_0.
(d) The directivity of the monopole at $\theta = \pi/2$.

18.22 Application: Short Monopole. A short monopole is used in a mobile telephone. The limitation is that a telephone of this type should not transmit more than 0.6 W. If the telephone transmits at 76 MHz and the antenna length is 0.1 m:

(a) Calculate the required current in the antenna. Assume a perfect conductor for the antenna.
(b) What is the maximum range of the telephone if the amplitude of the electric field intensity at the receiving antenna should be no lower than 10 mV/m?

18.23 Monopole Antennas: Efficiency. Two quarter-wavelength monopoles are used at the same frequency. Both are made of copper, with conductivity 5.7×10^7 S/m. Both transmit at 1 GHz. One is 1 mm thick, the second is 10 mm thick.

(a) Which antenna has higher efficiency?
(b) What are the efficiencies of the two antennas if both transmit in free space?
(c) Calculate the ratio between the (power) gains of the two antennas.

18.24 Application: The Monopole Antenna. A short monopole antenna is used at 1 MHz for AM transmission. The antenna is 0.5 m long and placed vertically above a conducting surface in air (free space). Find:

(a) The radiation resistance of the monopole.
(b) The radiated power of the monopole for a given current.
(c) The maximum directivity of the monopole.

18.25 Application: λ/4 Monopole. A quarter-wavelength monopole antenna is used at 100 MHz. The antenna is placed vertically above a conducting surface in air (free space). Find:

(a) The physical length of the antenna.
(b) The radiation resistance of the monopole.
(c) The radiated power of the monopole for a given current.
(d) The maximum directivity of the monopole.

Two-Element Image Antennas

18.26 Two-Element Array. A Hertzian dipole antenna of length d is placed horizontally above a perfectly conducting ground at a height h [m]. Assuming a current in the antenna of amplitude I_0 [A]:

(a) Find the electric and magnetic fields in the far field of the antenna.
(b) Find the radiation pattern of the antenna.

18.27 Two-Element Array. A wire antenna is placed perpendicularly above the ground. The wire is 3 m long and is used in conjunction with a transmitter which operates at 100 MHz:

(a) Find the field radiation pattern of the antenna.
(b) For a current $I = 1$ A, find the radiated power of the antenna in free space.
(c) Suppose the wire length is adjusted so that it is exactly 1.5 wavelengths. What are now the field radiation pattern and radiated power? Compare with **(a)** and **(b)**.

18.28 Two-Element Arrays. Two half-wavelength dipole antennas are placed a distance d apart, parallel to each other. Calculate and plot the field radiation patterns for:

(a) $d = 0$, $\varphi = 0°$ (zero separation, in phase)
(b) $d = 0.5\lambda$, $\varphi = 0$.
(c) $d = 0.5\lambda$, $\varphi = \pi$ [rad].
(d) $d = 0.5\lambda$, $\varphi = \pi/2$ [rad].
(e) $d = 0.5\lambda$, $\varphi = \pi/4$ [rad].
(f) $d = 1\lambda$, $\varphi = 0$.
(g) $d = 1\lambda$, $\varphi = \pi$ [rad].
(h) $d = 1\lambda$, $\varphi = \pi/2$ [rad].
(i) $d = 1\lambda$, $\varphi = \pi/4$ [rad].
(j) $d = 1.5\lambda$, $\varphi = 0$.
(k) $d = 1.5\lambda$, $\varphi = \pi$ [rad].
(l) $d = 1.5\lambda$, $\varphi = \pi/2$ [rad].
(m) $d = 1.5\lambda$, $\varphi = \pi/4$ [rad].

18.29 Two-Element Array. Two half-wavelength antennas are spaced 10 wavelengths apart:

(a) If the two antennas are driven in phase, find the radiation pattern of the array.
(b) If the two antennas are driven with a phase difference of $\pi/2$ [rad], find the field radiation pattern of the array.
(c) Describe how the pattern changes as the antennas are moved further apart.

18.30 Magnetic and Hertzian Dipoles. An antenna is built as a combination of a Hertzian dipole of length 0.05λ and a small magnetic dipole (loop) loop of diameter 0.025λ. The dipole is placed at the center of the loop so that the plane of the loop is perpendicular to the dipole, at its center (see **Figure 18.44**). The two antennas are driven at 100 MHz with a current of magnitude $I_0 = 0.1$ A. For convenience place the loop on the x–y plane at $z = 0$ and the dipole on the z axis, symmetrically about the x–y plane in free space:

(a) Find the electric and magnetic intensities in the far field if the currents in the antennas are in phase.
(b) Find the electric and magnetic field intensities in the far field if the current in the loop lags behind the current in the dipole by 180°.

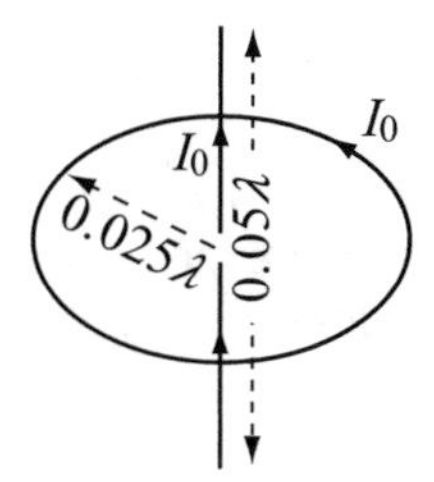

Figure 18.44

18.31 Magnetic and Hertzian Dipoles. In **Problem 18.30** assume the currents in the dipole and in the loop are I_1 and I_2, but they are in phase:

(a) Find the polarization of the wave in the far field.
(b) Suppose the current in the magnetic dipole (loop) and in the dipole are individually adjustable, but their phase is fixed. What are the types of polarization that may be achieved?
(c) Suppose the currents in the loop and Hertzian dipole remain constant in magnitude, but their phases may be changed individually. What are the types of polarization achievable? See **Section 12.8** for the concept of polarization.

The *n*-Element Linear Array

18.32 Application: Linear Antenna Array. An array antenna is made of six identical elements, each a half-wavelength dipole, and is fed with identical currents, all in phase. Spacing of the elements is one-half wavelength. Assume the antennas are placed on the x axis and are parallel to the z axis. Find:

(a) The normalized array field radiation pattern.
(b) The normalized array power radiation pattern.
(c) The direction of maximum radiation.
(d) The directions of the sidelobes.

18.33 Linear Arrays. An antenna array is made of n half-wavelength dipoles, parallel to each other, all placed on the x-axis and directed in the z direction. Find and plot the normalized field array antenna radiation patterns for the following number of elements N, spacing h [m], phase constant $\beta = \pi$ [rad/m], and progressive phase angle φ [rad] of each consecutive element with respect to the previous element starting with zero phase angle for the first element in the array. Plot the patterns for $0 \le \theta \le \pi$ and $\phi = \pi/4$:

(a) $N = 5$, $h = 0.6$ m, $\varphi = -0.6\pi$ [rad].
(b) $N = 6$, $h = 0.6$ m, $\varphi = -\pi$ [rad].
(c) $N = 10$, $h = 0.7$ m, $\varphi = -0.75\pi$ [rad].
(d) $N = 5$, $h = 0.7$ m, $\varphi = \pi$ [rad].

18.34 Five-Element Array. An antenna array is made of five Hertzian dipoles, spaced $\lambda/4$ apart, and driven in phase. The dipoles are parallel to each other. Using the notation in **Figure 18.33**:

(a) Find the array radiation pattern of the antenna.
(b) What is the direction of maximum power?

18.35 Application: 5 Element Array. A three-element array as shown in **Figure 18.45** carries currents $I_1 = I_3 = I_0$ [A] and $I_2 = 2I_0$ [A]. Calculate and plot the antenna radiation pattern, with all elements in phase:

(a) In the plane of the elements.
(b) In the plane perpendicular to the elements.

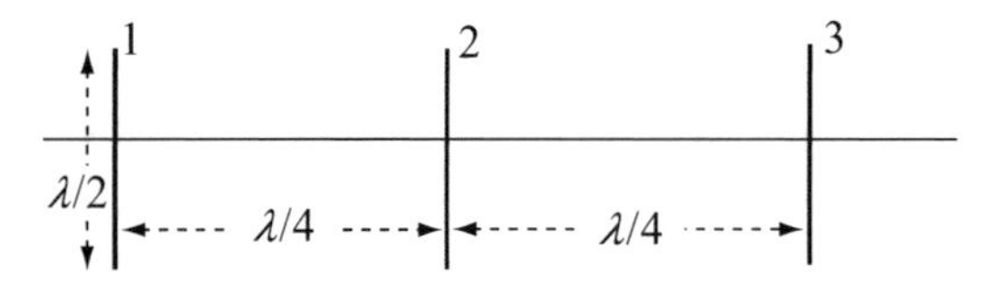

Figure 18.45

18.36 Linear Array of Monopoles. An array is made of n monopoles, all perpendicular to the ground and separated one-half wavelength from each other. The monopoles are 0.25λ long and are all in phase. For orientation purposes place the monopoles on the x-axis and calculate:

(a) The array field radiation pattern and plot it in a plane that includes the monopoles and in a plane perpendicular to the monopoles.
(b) The electric and magnetic field intensities at a general point in space for a given current. All monopoles carry identical currents.

18.37 Linear Magnetic Dipole (Loop) Array. An antenna array is made of n magnetic dipoles (loops), all lying flat on a surface in a line, separated a distance $2d$ apart where $d = 0.02\lambda$ is the diameter of the loop. For orientation purposes place the loops on the x-axis.

(a) Calculate the electric and magnetic field intensities in the far field of the array, given a current I [A] in each loop and a phase difference φ between each two consecutive loops. Assume free space propagation.
(b) Calculate the array factor and the normalized array factor if all elements are in phase.
(c) Explain the relation between the result obtained here and that for a linear array of Hertzian dipoles.

18.38 Application: Co-Linear Array. Five (5) half-wavelength dipole antennas are stacked on a line as in **Figure 18.46** to form a co-linear array. The centers of each two dipoles are a distance λ apart and the antennas are in free space. The currents in all elements have the same magnitude and the same phase. Use the system of coordinates shown and write the expression in terms of the radial distance R from the center of the array and calculate:

(a) The electric and magnetic field intensity of the array in the far field.
(b) The time-averaged power density of the array.
(c) Plot the E-plane field radiation pattern of the array.

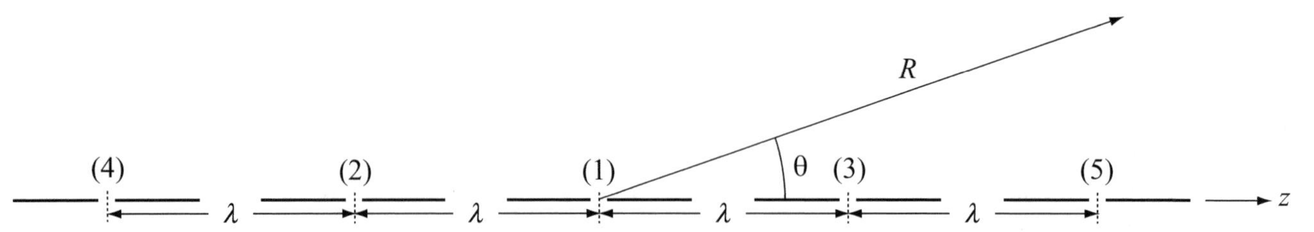

Figure 18.46

18.39 Application: Superposition of Linear Arrays. A Hertzian dipole is placed at a distance $\lambda/2$ in front of a corner conductor to form a reflector antenna as shown in **Figure 18.47**. The dipole is parallel to the corner and symmetrically placed with respect to the conducting surfaces, with its current in the negative z direction. The corner conductor may be assumed to be very large. The dipole operates at a wavelength λ and transmits in free space. Assume there are no losses in the system (i.e., the dipole and the corner conductor are made of perfect conductors). For orientation purposes, use the system of coordinates shown with the origin at the corner. Calculate:

(a) The array factor of the reflector antenna.
(b) The time-averaged power density of the reflector antenna.

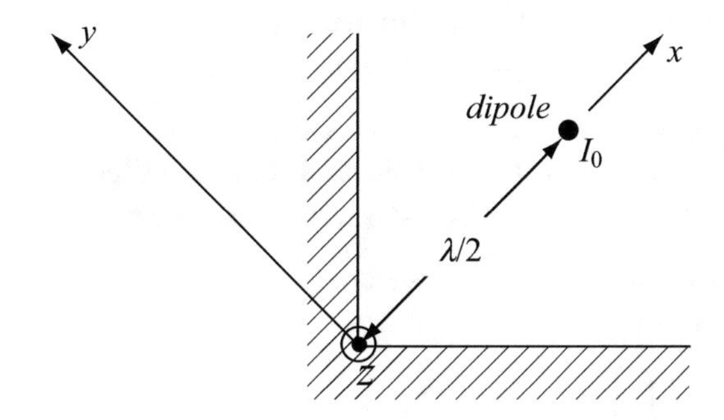

Figure 18.47

18.40 Application: Superposition of Linear Arrays. A seven (7)-element linear array is made of half-wavelength dipoles, all parallel, and the distance between each two elements is $h = \lambda/4$. Taking the middle of the array as reference, calculate the time-averaged power density of the array in the far field given that the magnitude of the current in all elements is I_0 [A] but their directions alternate (see **Figure 18.48**. You may assume the antennas are oriented in the z direction and placed on the x axis, symmetrically about the origin in free space.

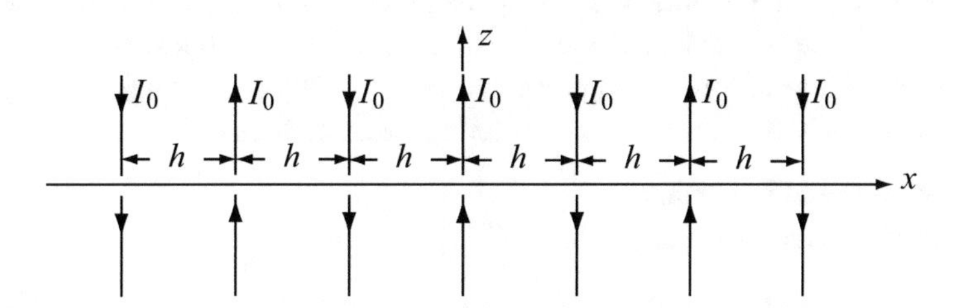

Figure 18.48

18.41 Nonuniform Linear Array. A 6 element array is made of Hertzian dipoles, each $\Delta l = 2$ cm long with their centers lying on the x axis in free space. The spacings are shown in **Figure 18.49**. The amplitude of each element is $I_0 = 0.1$ A. The array radiates at 432 MHz and all elements are in phase.

(a) Find the time averaged power density at the center of the main beam at a distance of 250 m from the center of the array.
(b) Calculate the beamwidth of the main beam.

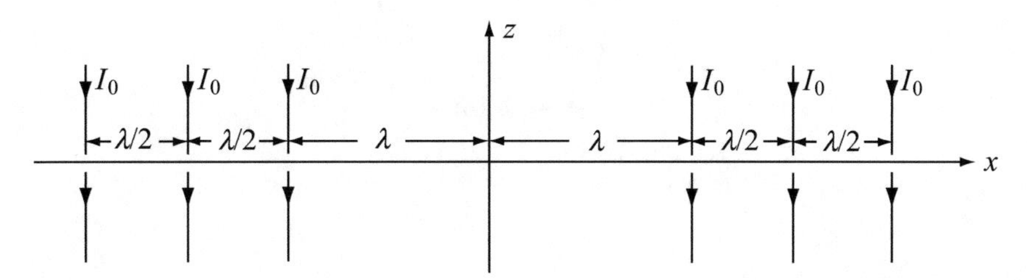

Figure 18.49

Reciprocity and Receiving Antennas

18.42 Effective Area of a Monopole. The receiving antenna in an AM car radio is a 1 m long monopole, vertical above the body of the car. AM reception is between 520 kHz and 1.6 MHz. Calculate the effective area of the antenna at the two extremes of the frequency range, assuming properties of free space.

18.43 Application: Circular Magnetic Dipole (Loop) Receiving Antenna. A small circular loop of radius a [m] is used as a receiving antenna. The antenna is placed at a distance R [m] in the field of an isotropic antenna which radiates P [W] at a frequency f [Hz] in free space:

(a) What is the maximum power received by the loop antenna?
(b) What is the maximum peak current in the loop antenna?

18.44 Application: Receive–Transmit System. Two units communicate with each other at 120 MHz and 10 W, in free space. Unit A uses a half-wavelength dipole antenna placed as shown in **Figure 18.50**. Unit B uses a 1/100 wavelength long dipole antenna and is 10 km from the transmitting antenna as shown:

(a) Find the current in antenna B when unit A transmits.
(b) Find the current in antenna A when unit B transmits.

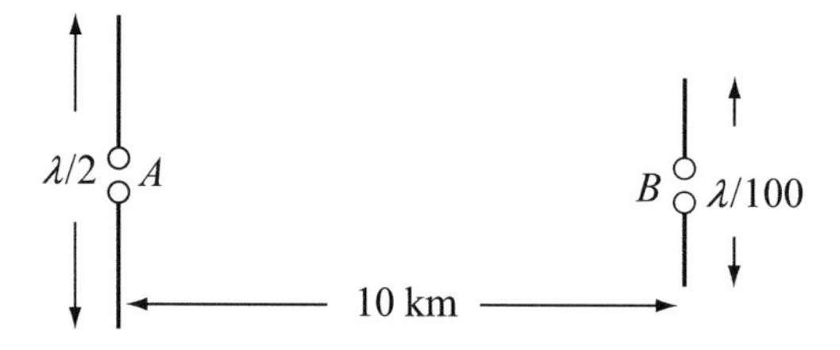

Figure 18.50

18.45 Application: Receive–Transmit System. To evaluate the properties of an antenna, two identical antennas are used in a receive–transmit system. The two antennas are placed a distance $d = 10$ km apart in free space and one antenna transmits 10 W at 300 MHz. The second antenna receives 10 μW. The two antennas are adjusted so that the direction of maximum directivity coincides with the line connecting them. Calculate:

(a) The maximum directivity of the antenna.
(b) The maximum effective area of the antenna.

18.46 Application: Send–Receive System. A half-wavelength dipole antenna is placed vertically (to the surface of the earth) and transmits 1 W (time-averaged power) at a wavelength of 3 m. A second, identical antenna, also placed vertically, is on top of a mountain 1 km high and 1 km away (horizontally) from the first antenna as in **Figure 18.51**. Neglect all effects due to the earth and assume properties of free space for air:

(a) Calculate the current the receiving antenna supplies to the receiver assuming both antennas are matched.
(b) In an attempt to increase this current, both antennas are adjusted by tilting them from the vertical direction. How must they be tilted (sketch or give accurate description) and what is the maximum current obtainable by doing so?

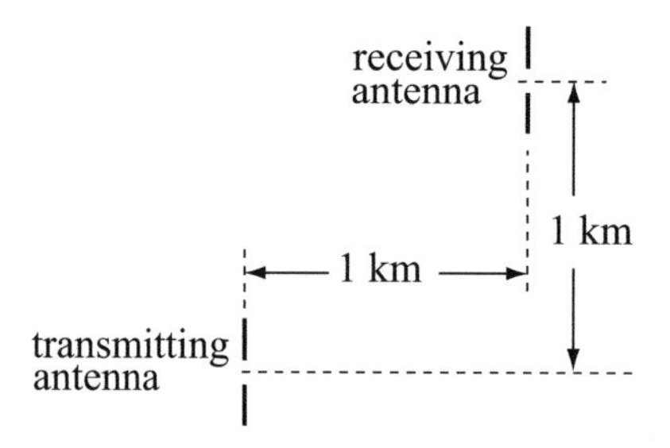

Figure 18.51

Radar

18.47 Application: Radar Cross Section. A radar is used for navigational purposes. A fixed radar station transmits at 3 GHz and 100 kW. If the smallest ocean-going ship has a radar cross section of 200 m^2 and the radar antenna gain is 15 dB, find the effective range of the radar if detection requires a minimum of 1 nW/m^2 at the radar antenna. Neglect losses in the antenna and assume free space properties for air.

18.48 Application: $\lambda/2$ antenna in a Radar. A half wavelength antenna is used as the transmitting antenna of a radar system. A second, identical antenna is used as the receiving antenna. An object with radar cross section of 100 m^2 reflects the wave and this wave is received by the receiving antenna. The transmitting antenna transmits 50 kW at 9 GHz, in free space. To calculate the range of the object, the amplitude of the current at the receiver input is measured as 0.1 μA. Calculate the range of the object if:

(a) The antennas are oriented so that the current measured is maximum.
(b) The two antennas are arbitrarily oriented.

18.49 Application: Radar Cross Section. A small radar is used to measure the radar cross section of a small airplane as part of the certification process. The radar transmits 100 W at 6 GHz. The antenna has a gain of 20 and is placed 100 m from the airplane (in the far field) so that the plane is in the direction of maximum antenna directivity. The received power at the radar's antenna is 100 nW. If propagation is in free space and there are no losses in the antenna, calculate the radar cross section of the airplane.

18.50 Application: Jamming of Radar Signals. A ground-based radar transmits a power P_0 [W] and tries to detect an incoming aircraft at a distance R [m]. The aircraft, trying to evade, transmits a jamming signal toward the radar based on the fact that if the jamming signal is larger than the received radar signal the aircraft cannot be detected. Assume the transmission and reception is lossless, the radar cross section of the aircraft is σ [m^2], the gain of the radar antenna is G_0, and the gain of the aircraft antenna is G_a. Calculate the minimum power P_j [W] the aircraft must transmit to jam the ground-based radar. You may assume the antennas are lossless, directed so that maximum signals are obtained and propagation is in free space.

Answers

Chapter 1

1. **(a)** $A_x = 5, A_y = -3, A_z = -1$. **(b)** $A_x = 1, A_y = 0, A_z = 1$. **(c)** $\sqrt{35}$.
2. **(a)** $\mathbf{v} = \hat{\mathbf{E}}\,35.36 + \hat{\mathbf{N}}35.36$ km/h. **(b)** 84 h, 51 min. **(c)** 4,242.64 km.
3. **(a)** $82^\circ.49'$ from north. **(b)** 220.48 m/s. **(c)** 22,678 s (6 h, 18 min).
4. **(a)** 5.916. **(b)** $\hat{\mathbf{x}}2 + \hat{\mathbf{y}}8 - \hat{\mathbf{z}}3$. **(c)** $\hat{\mathbf{x}}8 - \hat{\mathbf{y}}2 + \hat{\mathbf{z}}$. **(d)** $-\hat{\mathbf{x}}8 + \hat{\mathbf{y}}2 - \hat{\mathbf{z}}$. **(e)** $(-\hat{\mathbf{x}}8 + \hat{\mathbf{y}}2 - \hat{\mathbf{z}})/\sqrt{69}$.
5. **(b)** $\hat{\mathbf{x}}1 + \hat{\mathbf{y}}2 + \hat{\mathbf{z}}8$, $-\hat{\mathbf{x}}1 + \hat{\mathbf{y}}6 + \hat{\mathbf{z}}4$, $\hat{\mathbf{x}}5 - \hat{\mathbf{z}}2$, $\hat{\mathbf{x}}7 - \hat{\mathbf{y}}4 + \hat{\mathbf{z}}2$, $-\hat{\mathbf{x}}1 + \hat{\mathbf{y}}6 + \hat{\mathbf{z}}4$, $-\hat{\mathbf{x}}1 + \hat{\mathbf{y}}6 + \hat{\mathbf{z}}4$.
6. **(a)** 3,066.25 m/s (before), 2,788.47 m/s (after). **(b)** $k = 0.9094$.
7. **(a)** $\hat{\mathbf{v}} = (\hat{\mathbf{x}}150 + \hat{\mathbf{y}}25 - \hat{\mathbf{z}}50)/160.08$ m/s. **(b)** 160.08 m/s.
9. $(-\hat{\mathbf{x}}3 - \hat{\mathbf{y}}4 - \hat{\mathbf{z}}1)/\sqrt{26}$.
10. **(a)** $102^\circ 51'$. **(b)** $\mathbf{v}_{1g} = \hat{\mathbf{x}}3 + \hat{\mathbf{y}}1$ [m/s], $\mathbf{v}_{2g} = -\hat{\mathbf{x}}2 + \hat{\mathbf{y}}3$ [m/s]. **(c)** $105^\circ 31'$.
11. **(a)** $(-\hat{\mathbf{x}} - \hat{\mathbf{y}} - \hat{\mathbf{z}})/\sqrt{3}$. **(b)** $\hat{\mathbf{x}}0.784 - \hat{\mathbf{y}}0.588 + \hat{\mathbf{z}}0.196$. **(c)** $(\hat{\mathbf{x}}a + \hat{\mathbf{y}}b - \hat{\mathbf{z}})/\sqrt{a^2 + b^2 + 1}$.
12. $(\hat{\mathbf{x}}9 + \hat{\mathbf{y}}3 - \hat{\mathbf{z}}6)a/14x$.
16. **(a)** 57.688°. **(b)** $\hat{\mathbf{x}} - \hat{\mathbf{y}}3$.
17. **(a)** $9/\sqrt{17}$ N.
18. **(a)** 27.04.
19. $7x - 7y - 7z = 0$.
20. **(a)** $15y - 2x^2y + 12x^4$. **(b)** $\hat{\mathbf{x}}45x^2 + \hat{\mathbf{y}}18x^3 + \hat{\mathbf{z}}(11xy - 30x^3)$. **(c)** $12x^4 + 15y - 2x^2y$.
25. **(a)** 3.13. **(b)** 0.911. **(c)** 0.738.
26. **(b)** (1,2) and (0,2).
29. **(a)** $P_1(\sqrt{2},45^\circ,1)$, $P_2(\sqrt{2},45^\circ,0)$, $P_3(1,90^\circ,1)$, $P_1(\sqrt{3},54.736^\circ,45^\circ)$, $P_2(\sqrt{2},90^\circ,45^\circ)$, $P_3(\sqrt{2},45^\circ,90^\circ)$. **(b)** $y = 1$. **(c)** $r\sin\phi = 1$. **(d)** $R\sin(\theta)\sin(\phi) = 1$.
30. **(a)** $x^2 + y^2 + z^2 = a^2$. **(b)** $r^2 ++ z^2 = a^2$. **(c)** $R = a$.
31. **(a)** $P(a,\theta,\phi)$. **(b)** $P(a\sin\theta\cos\phi, a\sin\theta\sin\phi, a\cos\theta)$. **(c)** $P(a\sin\theta,\phi,a\cos\theta)$.
32. **(a)** $\hat{\mathbf{r}}(-4.598\cos\phi - 1.5\sin\phi) + \hat{\boldsymbol{\phi}}(4.598\sin\phi - 1.5\cos\phi) + \hat{\mathbf{z}}3$.
    (b) $\hat{\mathbf{R}}(4.598\sin\theta\cos\phi + 1.5\sin\theta\sin\phi - 3\cos\theta) + \hat{\boldsymbol{\theta}}(4.598\cos\theta\cos\phi + 1.5\cos\theta\sin\phi + 3\sin\theta)$
    $+\hat{\boldsymbol{\phi}}(-4.598\sin\phi + 1.5\cos\phi)$
    (c) 5.691.
33. **(a)** 6.124. **(b)** $-\hat{\mathbf{r}}0.5(\cos\phi + \sin\phi) + \hat{\boldsymbol{\phi}}0.5(\sin\phi - \cos\phi) + \hat{\mathbf{z}}0.707$.
    (c) $\hat{\mathbf{R}}(-0.5\sin\theta\cos\phi - 0.5\sin\theta\sin\phi - 0.707\cos\theta) + \hat{\boldsymbol{\theta}}(-0.5\cos\theta\cos\phi - 0.5\cos\theta\sin\phi + 0.707\sin\theta) + \hat{\boldsymbol{\phi}}0.5(\sin\phi - \cos\phi)$
34. $\mathbf{A} = -\hat{\mathbf{R}}4.276 - \hat{\boldsymbol{\theta}}4.229 + \hat{\boldsymbol{\phi}}1.1094$.
35. **(a)** $\mathbf{A} = \hat{\mathbf{x}}(3\cos^2\phi + 2\sqrt{r}\sin\phi) + \hat{\mathbf{y}}(3\cos\phi\sin\phi - 2\sqrt{r}\cos\phi) + \hat{\mathbf{z}}r\phi$. **(b)** $A_R = 3\cos\phi\sin\theta + r\phi\cos\theta$, $A_\theta = 3\cos\phi\cos\theta - r\phi\sin\theta$, $A_\phi = -2\sqrt{r}$.
36. **(a)** $\mathbf{r}_1 = \hat{\mathbf{x}}x_1 + \hat{\mathbf{y}}y_1 + \hat{\mathbf{z}}z_1$. **(b)** $\mathbf{r}_2 = \hat{\mathbf{x}}x_2 + \hat{\mathbf{y}}y_2 + \hat{\mathbf{z}}z_2$. **(c)** $\mathbf{R} = \hat{\mathbf{x}}(x_1 - x_2) + \hat{\mathbf{y}}(y_1 - y_2) + \hat{\mathbf{z}}(z_1 - z_2)$.
37. **(a)** $\mathbf{r}_1 = \hat{\mathbf{z}}3$, $\mathbf{r}_2 = \hat{\mathbf{x}}1.5 + \hat{\mathbf{y}}1.5 + \hat{\mathbf{z}}3\sqrt{2}/2$. **(b)** $\hat{\mathbf{x}}1.5 + \hat{\mathbf{y}}1.5 + \hat{\mathbf{z}}(3\sqrt{2} - 6)/2$.
38. $(A_yB_z - A_zB_y)x + (A_zB_x - A_xB_z)y + (A_xB_y - A_yB_x)z = 0$.

N. Ida, *Engineering Electromagnetics*, https://doi.org/10.1007/978-3-030-15557-5

Chapter 2

1. **(a)** $\tan^{-1}(b/a)$. **(b)** $\tan^{-1}(b/a)$.
3. **(a)** 12. **(b)** 12.
4. π.
5. $\pi^2/12 - \pi(\sqrt{3} - 2)$.
6. 3.
7. $\pi a^2/\sqrt{13}$.
8. $\pi a h^3/6$.
9. **(a)** $7\pi a^4 d/6 + \pi d^3 a^2/12$. **(b)** $4.98a^5$.
10. $10^8 \pi h_0[a^3 + a^2 b + ab^2 + 9b^3]/6$.
11. **(a)** $\hat{\mathbf{z}}8/3$. **(b)** $\hat{\mathbf{x}}16 + \hat{\mathbf{y}}8 + \hat{\mathbf{z}}32/3$.
12. $\hat{\mathbf{x}}50/3 + \hat{\mathbf{y}}75/2$ m/s.
13. $\hat{\mathbf{x}}10/3 + \hat{\mathbf{y}}10$.
14. Zero.
15. **(a)** $|\nabla P| = 2\sqrt{(x-2)^2 + (y-2)^2 + (z+1)^2}$, $\widehat{\nabla P} = [\hat{\mathbf{x}}(x-2) + \hat{\mathbf{y}}(y-2) + \hat{\mathbf{z}}(z+1)]/|\nabla P|$. **(b)** $2(z + 1)$. **(c)** $\sqrt{2}(x + y - 4)$.
16. **(a)** $\hat{\mathbf{r}}\cos^2\phi + \hat{\boldsymbol{\phi}}(-2r\cos\phi\sin\phi + z\cos\phi)/r + \hat{\mathbf{z}}\sin\phi$. **(b)** $\hat{\mathbf{x}}(x^3 + 2xy^2 - xyz)/(x^2+y^2)^{3/2} + \hat{\mathbf{y}}(-x^2y + x^2z)/(x^2+y^2)^{3/2} + \hat{\mathbf{z}}y/(x^2+y^2)^{1/2}$. **(c)** $\hat{\mathbf{R}}(\cos^2\phi\sin\theta + \cos\theta\sin\phi) + \hat{\boldsymbol{\theta}}(\cos^2\phi\cos\theta - \sin\theta\sin\phi) - \hat{\boldsymbol{\phi}}(2\cos\phi\sin\phi - \cos\theta\cos\phi/\sin\theta)$.
17. **(a)** $(\hat{\mathbf{x}}5 + \hat{\mathbf{y}}3 + \hat{\mathbf{z}}1)/\sqrt{35}$. **(b)** $(\hat{\mathbf{x}}4 - \hat{\mathbf{y}}3 + \hat{\mathbf{z}}1)/\sqrt{26}$. **(c)** $(\hat{\mathbf{x}}a + \hat{\mathbf{y}}b - \hat{\mathbf{z}}1)/\sqrt{a^2 + b^2 + 1}$.
18. **(a)** $\hat{\mathbf{z}}1$. **(b)** $(\hat{\mathbf{x}} + \hat{\mathbf{y}}2 - \hat{\mathbf{z}}2)/3$. **(c)** $(\hat{\mathbf{x}} - \hat{\mathbf{y}} - \hat{\mathbf{z}})/\sqrt{3}$.
19. **(a)** $2x$. **(b)** $2z^2/r$. **(c)** $x/\sqrt{x^2+z^2} + y/\sqrt{x^2+y^2}$.
20. 4.
21. 4.
22. 3.6.
25. **(a)** 3. **(b)** 3.
26. **(a)** $8\pi A(e^{-\alpha R_2} - e^{-\alpha R_1})/3$. **(b)** $8\pi A(e^{-\alpha R_2} - e^{-\alpha R_1})/3$.
27. **(a)** $-\hat{\mathbf{x}}2y$. **(b)** $\hat{\boldsymbol{\phi}}(4z + 6r) + \hat{\mathbf{z}}10$. **(c)** $\hat{\mathbf{y}}z/\sqrt{x^2+z^2} + \hat{\mathbf{z}}x/\sqrt{x^2+y^2}$.
28. **(a)** Zero.
29. **(a)** $\hat{\mathbf{x}}150x\sin(\omega t + 50z) + \hat{\mathbf{z}}3\cos(\omega t + 50z)$.
30. **(b)** $\nabla\cdot\mathbf{A} = 0$, $\nabla\cdot\mathbf{B} = 0$.
32. **(a)** $2z$. **(b)** $-\hat{\mathbf{z}}5$. **(c)** -20π.
33. $\pm(ba + ca + bc)/2$.
35. **(a)** $2(y-2)^2(z+1)^2 + 2(x-2)^2(z+1)^2 + 2(x-2)^2(y-2)^2$. **(b)** $12z$. **(c)** $\nabla^2 p = -\dfrac{a}{R^4}\cos\theta\left(1 + \dfrac{2}{R}\right)$.
36. **(a)** $\nabla^2\mathbf{A} = \hat{\mathbf{z}}2$. **(b)** $\nabla^2\mathbf{B} = \hat{\mathbf{r}}(6/r)\cos\phi$. **(c)** $\nabla^2\mathbf{C} = \hat{\mathbf{R}}e^{-j\beta R}\left(\dfrac{\cos 2\theta}{R^4\sin\theta} + \dfrac{j2\beta}{R^3}\sin\theta - \dfrac{\beta^2}{R^2}\sin\theta\right) + \hat{\boldsymbol{\theta}}e^{-j\beta R}\dfrac{2\cos\theta}{R^4}$.
38. **(a)** $\hat{\mathbf{x}}4x + \hat{\mathbf{y}}1$. **(b)** $10x^2 + 4y$. **(c)** 4. **(d)** $\nabla^2\mathbf{R} = 0$. **(e)** $\hat{\mathbf{x}}z - \hat{\mathbf{y}}4xz + \hat{\mathbf{z}}(4xy - x)$.

Chapter 3

1. **(a)** 8.615×10^{13} C.
2. 2.3×10^{12} N.
3. **(a)** 9.57 cm. **(b)** 13.54 cm.
4. 15.4 mm.
5. **(a)** 1.168×10^{-11} C.
6. **(a)** $\mathbf{a} = -\hat{\mathbf{x}}4qQx/\pi m\varepsilon_0\left(4x^2 + d^2\right)^{3/2}\left[\mathrm{m/s^2}\right]$, a_{max} occurs at $x = \pm d/2\sqrt{2}$ [m].

7. **(a)** to $x = 0.414d$ [m] from Q_1.
8. **(b)** 0.1 m to the right of negative charge.
10. **(a)** $\mathbf{E} = \frac{q}{4\pi\varepsilon_0}\left(-\frac{\mathbf{R}}{|\mathbf{R}|^3} + \frac{(\mathbf{R}-\mathbf{d})}{|\mathbf{R}-\mathbf{d}|^3} + \frac{(\mathbf{R}+\mathbf{d})}{|\mathbf{R}+\mathbf{d}|^3}\right)\left[\frac{\mathrm{N}}{\mathrm{C}}\right]$. **(b)** $\mathbf{E} = \frac{3(2q)d^2}{4\pi\varepsilon_0 R^4}\left(\hat{\mathbf{R}}\frac{3\cos^2\theta - 1}{2} + \hat{\boldsymbol{\theta}}\cos\theta\sin\theta\right)\left[\frac{\mathrm{N}}{\mathrm{C}}\right]$.
11. **(a)** ∞. **(b)** $\mathbf{E} = \hat{\mathbf{z}}Q/2\pi\varepsilon_0 a^2$ [N/C]. **(c)** Zero.
12. **(a)** 0.303 m.
13. **(a)** 0.303 m.
15. 3.82×10^{-13} C/h/m^2.
16. **(a)** $0.289L$ or $0.125L$ from centers of sides of the triangle and at infinity.
17. **(a)** 6.131×10^{10} N/C. **(b)** 3.066×10^{10} N/C.
18. **(a)** 9.81×10^{-7} C. **(b)** 9.81 m/s^2. **(c)** 1.53×10^7 s.
19. 8×10^{-16} kg.
20. $\mathbf{E}_1 = \hat{\mathbf{y}}\sqrt{2}\rho_l/2\pi\varepsilon_0 L$ [N/C], $\mathbf{E}_2 = \hat{\mathbf{x}}\rho_l(\sqrt{5}-1)/2\pi\varepsilon_0 L\sqrt{5} + \hat{\mathbf{y}}\rho_l/\pi\varepsilon_0 L\sqrt{5}$ [N/C], $\mathbf{E}_3 = \hat{\mathbf{x}}\rho_l/3\pi\varepsilon_0 L$ [N/C], $\mathbf{E}_4 = 0$.
21. **(a)** $E = \rho_l ah/2\varepsilon_0(h^2 + a^2)^{3/2}$ [N/C].
22. 1.797×10^{-5} N.
23. 1.78×10^{-4} N (attraction).
24. $F = \frac{Q_1 Q_2}{4\pi\varepsilon L_1 L_2}\ln\frac{(a+L_2)(a+L_1)}{a(L_1+a+L_2)}$ [N].
25. $F = \frac{\rho_l^2}{2\pi\varepsilon_0}\ln\left(\frac{a+b}{a}\right)$ [N].
26. **(a)** $\mathbf{E} = \hat{\mathbf{z}}\frac{\rho_0}{2\varepsilon_0}\left[1 - \frac{h}{\sqrt{h^2+a^2}}\right]\left[\frac{\mathrm{N}}{\mathrm{C}}\right]$. **(b)** 112.94 [N/C], at $h = 0$.
27. **(a)** $\mathbf{E} = \hat{\mathbf{z}}\frac{\rho_0 h}{2\varepsilon_0}\left[\sqrt{a^2+h^2} + \frac{h^2}{\sqrt{a^2+h^2}} - 2h\right]\left[\frac{\mathrm{N}}{\mathrm{C}}\right]$.
28. $\mathbf{E} = \hat{\mathbf{z}}\rho_s/2\varepsilon_0$ [N/C].
29. $\mathbf{E} = \hat{\mathbf{z}}\rho_s/2\sqrt{2}\varepsilon_0$ [N/C].
30. $\mathbf{E} = \hat{\mathbf{z}}\frac{\rho_s}{2\varepsilon}\left[\frac{L^2}{a\sqrt{L^2+a^2}} + \frac{\sqrt{L^2+a^2}}{a} - 1\right]\left[\frac{\mathrm{N}}{\mathrm{C}}\right]$.
31. **(a)** $\mathbf{E} = \hat{\mathbf{z}}\frac{\rho_v}{2\varepsilon_0}\left[2z + \sqrt{(L/2-z)^2 + (L/4)^2} - \sqrt{(z+L/2)^2 + (L/4)^2}\right]\left[\frac{\mathrm{N}}{\mathrm{C}}\right]$. **(b)** $\pm L/2$ [m].
32. **(a)** $\mathbf{E}(R) = -\hat{\mathbf{R}}1.44 \times 10^{-9}/R^2$ [N/C]. **(b)** $\mathbf{E}(R) = -\hat{\mathbf{R}}1.44 \times 10^{-9}/R^2$ [N/C]
33. $\mathbf{E} = \hat{\mathbf{R}}\rho_0 a^4/20\varepsilon_0 R^2$ [N/C].
34. **(a)** $8.988 \times 10^{-7}/d^2$ [N], repulsion. **(b)** $1.11 \times 10^{-8}/d^2$ [N], repulsion.
35. **(a)** $\mathbf{D} = \hat{\mathbf{R}}q/4\pi R^2$ [C/m^2]. **(b)** $\mathbf{E} = \hat{\mathbf{R}}q/4\pi\varepsilon_0 R^2$ [N/C] for $R < d_1/2$ and $R > d_2/2$, $\mathbf{E} = \hat{\mathbf{R}}q/4\pi\varepsilon R^2$ [N/C] for $d_1/2 < R < d_2/2$. **(c)** $\Phi = q$ [C]. **(d)** $\Phi = q$ [C]. **(e)** $\Phi = q$ [C].

Chapter 4

1. **(a)** $\rho(x) = 3k\varepsilon x^2$ [C/m^3].
2. **(a)** $4\varepsilon_0 kR$ C/m^3.
4. **(a)** $\mathbf{E} = \hat{\mathbf{R}}3{,}764.78R$ [N/C]. **(b)** $\mathbf{E} = \hat{\mathbf{R}}3.012 \times 10^{-5}/R^2$ [N/C]. **(c)** $\mathbf{E} = \hat{\mathbf{R}}(5.12 \times 10^{-4}/R^2 - 7{,}529.55R)$ [N/C]. **(d)** $\mathbf{E} = -\hat{\mathbf{R}}1.1144 \times 10^{-3}/R^2$ [N/C].
5. Zero.
6. **(a)** $\mathbf{E} = \hat{\mathbf{R}}\frac{Q}{4\pi\varepsilon_0 R^2}\left[\frac{\mathrm{N}}{\mathrm{C}}\right]$, $R < a$, $\mathbf{E} = \hat{\mathbf{R}}\left(\frac{a^2\rho_s}{\varepsilon_0 R^2} + \frac{Q}{4\pi\varepsilon_0 R^2}\right)\left[\frac{\mathrm{N}}{\mathrm{C}}\right]$, $R > a$. **(b)** $\rho_s = -\frac{Q}{4\pi a^2}\left[\frac{\mathrm{C}}{\mathrm{m}^2}\right]$.
7. **(a)** $\mathbf{E} = \hat{\mathbf{z}}\rho_0/2\varepsilon$ [N/C], $z > 0$, $\mathbf{E} = -\hat{\mathbf{z}}\rho_0/2\varepsilon$ [N/C], $z < 0$. **(b)** $\mathbf{E} = \hat{\mathbf{z}}\rho_0/2\varepsilon_0$ [N/C], $z > 0$, $\mathbf{E} = -\hat{\mathbf{z}}\rho_0/2\varepsilon_0$ [N/C], $z < 0$.
8. $\mathbf{E} = -\hat{\mathbf{y}}\rho_v a/\varepsilon_0$ [N/C], $y < -b$. $\mathbf{E} = -\hat{\mathbf{y}}(\rho_s + \rho_v a)/\varepsilon_0$ [N/C], $-b < y < -a$. $\mathbf{E} = -\hat{\mathbf{y}}\rho_v|y|/\varepsilon_0$ [N/C], $-a < y < 0$. $\mathbf{E} = \hat{\mathbf{y}}\rho_v y/\varepsilon_0$ [N/C], $0 < y < a$. $\mathbf{E} = \hat{\mathbf{y}}(-\rho_s + \rho_v a)/\varepsilon_0$ [N/C], $a < y < b$. $\mathbf{E} = \hat{\mathbf{y}}\rho_v a/\varepsilon_0$ [N/C], $y > b$.

9. **(a)** $\mathbf{E}_1 = \hat{\mathbf{x}}\rho_s/2\varepsilon_1 + \hat{\mathbf{y}}\rho_s/2\varepsilon_1$, $\mathbf{E}_2 = -\hat{\mathbf{x}}\rho_s/2\varepsilon_0 + \hat{\mathbf{y}}\rho_s/2\varepsilon_0$, $\mathbf{E}_3 = -\hat{\mathbf{x}}\rho_s/2\varepsilon_2 - \hat{\mathbf{y}}\rho_s/2\varepsilon_2$, $\mathbf{E}_4 = \hat{\mathbf{x}}\rho_s/2\varepsilon_3 - \hat{\mathbf{y}}\rho_s/2\varepsilon_3$ [N/C]. **(b)** $\mathbf{F}_2 = -\hat{\mathbf{x}}q\rho_s/2\varepsilon_0 + \hat{\mathbf{y}}q\rho_s/2\varepsilon_0$ [N] (in second quadrant).

10. $\mathbf{E} = \hat{\mathbf{x}}\rho_0 x^2/2\varepsilon_0$ [N/C], $0 < x < d/2$, $\mathbf{E} = -\hat{\mathbf{x}}\rho_0 x^2/2\varepsilon_0$ [N/C], $-d/2 < x < 0$, $\mathbf{E} = \hat{\mathbf{x}}\rho_0 d^2/8\varepsilon_0$ [N/C], $x \geq d/2$, $\mathbf{E} = -\hat{\mathbf{x}}\rho_0 d^2/8\varepsilon_0$ [N/C], $x \leq -d/2$.

11. **(a)** $\mathbf{E} = \hat{\mathbf{y}}\dfrac{\rho_l y}{2\pi\varepsilon_0(y^2+z^2)} + \hat{\mathbf{z}}\left[\dfrac{\rho_l z}{2\pi\varepsilon_0(y^2+z^2)} + \dfrac{\rho_s}{2\varepsilon_0}\right]\left[\dfrac{\text{N}}{\text{C}}\right]$, $z > z_0$, $\mathbf{E} = \hat{\mathbf{y}}\dfrac{\rho_l y}{2\pi\varepsilon_0(y^2+z^2)} + \hat{\mathbf{z}}\left[\dfrac{\rho_l z}{2\pi\varepsilon_0(y^2+z^2)} - \dfrac{\rho_s}{2\varepsilon_0}\right]\left[\dfrac{\text{N}}{\text{C}}\right]$, $z < z_0$. **(b)** $\mathbf{D}(0,0,1) = \hat{\mathbf{z}}(\rho_l/2\pi \pm \rho_s/2)$ [C/m^2].

12. **(a)** $\mathbf{E} = 0, 0 < r < a, \mathbf{E} = \hat{\mathbf{r}}a\rho_s/\varepsilon r$ [N/C], $a < r < b, \mathbf{E} = \hat{\mathbf{r}}\rho_s(a-b)/\varepsilon_0 r$ [N/C], $r > b$. **(c)** $\rho_l = 2\pi\rho_s(b-a)$ [C/m].

13. **(a)** Zero. **(b)** 4,493.87 N/C.

14. **(a)** $\rho_v = \varepsilon|y|$ [C/m^3]. **(b)** $E = \varepsilon d^2/8\varepsilon_0 \left[\dfrac{\text{N}}{\text{C}}\right]$.

15. **(a)** $\rho_v = 4\varepsilon b$ [C/m^3]. **(b)** $\Phi = \varepsilon b^4\pi$ [C].

16. **(a)** $\rho_{smax} = 26.5$ μC/m^2. **(a)** $\rho_{smax} = 79.5$ μC/m^2.

17. **(a)** $V = 498.15$ V, $E = 0$. **(b)** $V = 481.18$ V, $E = 234.8$ V/m.

18. **(a)** $V(R) = -R^2\rho_v/6\varepsilon_0 + Q/4\pi\varepsilon_0 R + a^2\rho_v/2\varepsilon_0$ [V], $R \leq a$, $V(R) = Q/4\pi\varepsilon_0 R + a^2\rho_v/3\varepsilon_0 R$ [V], $R > a$.

19. **(a)** $\mathbf{E} = -\hat{\mathbf{x}}\rho_s/\varepsilon_0$ [V/m] between plates, $\mathbf{E} = 0$ outside. **(b)** $V = \rho_s x/\varepsilon_0$ [V] between plates. $V = 0$ to the left of the left plate, $V = \rho_s d/\varepsilon_0$ [V] to the right of the right plate.

20. **(a)** $\rho_{sb} = V\varepsilon_0/b\ \ln(b/a)$ [C/m^2]. **(b)** $\rho_{sa} = V\varepsilon_0/a\ln(a/b)$ [C/m^2].

21. $V = \rho_0\left(\sqrt{a^2+d^2} - d\right)/2\varepsilon_0$ [V].

22. **(a)** $\mathbf{E} = \hat{\mathbf{x}}\rho_v x/\varepsilon$ [V/m] for $-a < x < a$ (the $x = 0$ plane is midway between the plates), $\mathbf{E} = \hat{\mathbf{x}}\rho_v a/\varepsilon_0$ [V/m], $x > a$, $\mathbf{E} = -\hat{\mathbf{x}}\rho_v a/\varepsilon_0$ [V/m], $x < -a$. **(b)** $V = \rho_v(a^2 - x^2)/2\varepsilon$ [V], $-a < x < a$, $V = \rho_v a(a - |x|)/\varepsilon_0$ [V], $x < -a$, $x > a$.

23. **(a)** $V(x,y) = \dfrac{Q}{4\pi\varepsilon_0}\left[\dfrac{1}{\sqrt{x^2+y^2}} + \sum_{n=1}^{N}\dfrac{(-1)^n}{\sqrt{(x-na)^2+y^2}} + \sum_{n=1}^{N}\dfrac{(-1)^n}{\sqrt{(x+na)^2+y^2}}\right]$ [V].

(b)
$$\mathbf{E} = \hat{\mathbf{x}}\frac{Q}{4\pi\varepsilon_0}\left[\frac{x}{(x^2+y^2)^{3/2}} + \sum_{n=1}^{N}\frac{(x-na)(-1)^n}{\left((x-na)^2+y^2\right)^{3/2}} + \sum_{n=1}^{N}\frac{(x+na)(-1)^n}{\left((x+na)^2+y^2\right)^{3/2}}\right]$$
$$+\hat{\mathbf{y}}\frac{Q}{4\pi\varepsilon_0}\left[\frac{y}{(x^2+y^2)^{3/2}} + \sum_{n=1}^{N}\frac{y(-1)^n}{\left((x-na)^2+y^2\right)^{3/2}} + \sum_{n=1}^{N}\frac{y(-1)^n}{\left((x+na)^2+y^2\right)^{3/2}}\right]\left[\frac{\text{V}}{\text{m}}\right].$$

24. **(a)** $\rho_v = -6a\varepsilon$ [C/m^3]. **(b)** $\mathbf{E} = -\hat{\mathbf{R}}2aR$ [V/m], $R < b$, $\mathbf{E} = -\hat{\mathbf{R}}2\varepsilon ab^3/\varepsilon_0 R^2$ [V/m], $R > b$.

25. **(a)** $\mathbf{E} = \hat{\mathbf{x}}\dfrac{0.5}{(x-2)^2} + \hat{\mathbf{y}}\dfrac{0.5}{(y-1)^2} - \hat{\mathbf{z}}\dfrac{1}{(z+2)^2}\left[\dfrac{\text{V}}{\text{m}}\right]$. **(c)** $\rho_v = -\varepsilon_0\left(\dfrac{1}{(x-2)^3} + \dfrac{1}{(y-1)^3} - \dfrac{2}{(z+2)^3}\right)\left[\dfrac{\text{C}}{\text{m}^3}\right]$.

26. **(a)** $\mathbf{E}(R,\theta) = \hat{\mathbf{R}}\dfrac{6C}{R^4}\left(\dfrac{3\cos^2\theta - 1}{2}\right) + \hat{\boldsymbol{\theta}}\cos\theta\sin\theta$ [V/m].

27. **(a)** $Vy/2b$ [V] (y is the distance from lower plate). **(b)** $Vy/(2b - 2c)$ [V], $0 < y \leq d$ and $Vd/(2b - 2c) + V(y - (d + 2c))/(2b - 2c)$ [V], $d + 2c < y \leq 2b$.

28. $V = Q/4\pi\varepsilon_0 R$ [V].

29. $\rho_{sc} = \dfrac{V_0}{c^2(1/a\varepsilon_1 - 1/b\varepsilon_1 + 1/b\varepsilon_2 - 1/c\varepsilon_2)}$, $\rho_{sa} = -\dfrac{V_0}{a^2(1/a\varepsilon_1 - 1/b\varepsilon_1 + 1/b\varepsilon_2 - 1/c\varepsilon_2)}\left[\dfrac{\text{C}}{\text{m}^2}\right]$.

30. **(a)** In **Fig. 4.68a**: $\mathbf{E} = -\hat{\mathbf{R}}\dfrac{Q}{4\pi\varepsilon_0 R^2}\left[\dfrac{\text{V}}{\text{m}}\right]$, $V = -\dfrac{Q}{4\pi\varepsilon_0 R} + \dfrac{\rho_{sa}a}{\varepsilon_0} - \dfrac{\rho_{sb}b}{\varepsilon_0}$ [V], $0 < R < a$, $\mathbf{E} = -\hat{\mathbf{R}}\dfrac{Q}{4\pi\varepsilon_0 R^2} + \hat{\mathbf{R}}\dfrac{\rho_{sa}a^2}{\varepsilon_0 R^2}\left[\dfrac{\text{V}}{\text{m}}\right]$, $V = -\dfrac{Q}{4\pi\varepsilon_0 R} + \dfrac{\rho_{sa}a^2}{\varepsilon_0 R} - \dfrac{\rho_{sb}b}{\varepsilon_0}$ [V], $a < R < b$, $\mathbf{E} = -\hat{\mathbf{R}}\dfrac{Q}{4\pi\varepsilon_0 R^2} + \hat{\mathbf{R}}\dfrac{\rho_{sa}a^2}{\varepsilon_0 R^2} - \hat{\mathbf{R}}\dfrac{\rho_{sb}b^2}{\varepsilon_0 R^2}\left[\dfrac{\text{V}}{\text{m}}\right]$, $V = -\dfrac{Q}{4\pi\varepsilon_0 R} + \dfrac{\rho_{sa}a^2}{\varepsilon_0 R} - \dfrac{\rho_{sb}b^2}{\varepsilon_0 R}$ [V], $R > b$; in **Fig. 4.68b**: $\mathbf{E} = -\hat{\mathbf{R}}\dfrac{Q}{4\pi\varepsilon_0 R^2}\left[\dfrac{\text{V}}{\text{m}}\right]$, $V = -\dfrac{Q}{4\pi\varepsilon_0 R} + \dfrac{Q}{4\pi\varepsilon_0 a} - \dfrac{Q}{4\pi\varepsilon_0 b}$ [V], $0 < R < a$, $\mathbf{E} = 0$, $V = -\dfrac{Q}{4\pi\varepsilon_0 b}$ [V], $a < R < b$, $\mathbf{E} = -\hat{\mathbf{R}}\dfrac{Q}{4\pi\varepsilon_0 R^2}\left[\dfrac{\text{V}}{\text{m}}\right]$, $V = -\dfrac{Q}{4\pi\varepsilon_0 R}$ [V], $R > b$. **(c)** $\left|\dfrac{\rho_{sa}}{\rho_{sb}}\right| = \dfrac{b^2}{a^2}$.

31. $\mathbf{P} = \hat{\mathbf{r}}\rho_l(\varepsilon_1 - \varepsilon_0)/2\pi\varepsilon_1 r$ [C/m^2], $a < r < c$, $\mathbf{P} = \hat{\mathbf{r}}\rho_l(\varepsilon_2 - \varepsilon_0)/2\pi\varepsilon_2 r$ [C/m^2], $c < r < b$.

32. **(a)** $E = 120$ kV/m. **(b)** $P = 3.187 \times 10^{-6}$ C/m^2.

33. **(a)** $\mathbf{D} = \hat{\mathbf{x}}\left[\frac{\rho_l x}{2\pi(x^2+y^2)} + \frac{\rho_s}{2}\right] + \hat{\mathbf{y}}\frac{\rho_l y}{2\pi(x^2+y^2)}\left[\frac{\text{C}}{\text{m}^2}\right], x > 1, \mathbf{D} = \hat{\mathbf{x}}\left[\frac{\rho_l x}{2\pi(x^2+y^2)} - \frac{\rho_s}{2}\right] + \hat{\mathbf{y}}\frac{\rho_l y}{2\pi(x^2+y^2)}\left[\frac{\text{C}}{\text{m}^2}\right], x < 1.$
(b) $\mathbf{D} = \hat{\mathbf{x}}6.367 \times 10^{-7}\text{C/m}^2$. **(c)** $\mathbf{P} = \hat{\mathbf{x}}\left[\frac{3\rho_l x}{8\pi(x^2+y^2)} + \frac{3\rho_s}{8}\right] + \hat{\mathbf{y}}\frac{3\rho_l y}{8\pi(x^2+y^2)}\left[\frac{\text{C}}{\text{m}^2}\right], x > 1,$
$\mathbf{P} = \hat{\mathbf{x}}\left[\frac{3\rho_l x}{8\pi(x^2+y^2)} - \frac{3\rho_s}{8}\right] + \hat{\mathbf{y}}\frac{3\rho_l y}{8\pi(x^2+y^2)}\left[\frac{\text{C}}{\text{m}^2}\right], x < 1.$

34. **(a)** $E = 198.76$ V/m in air. **(b)** $V = 15{,}093.75$ V.

35. **(a)** $\rho_s = 26.55\ \mu\text{C/m}^2$. **(b)** $V = 3 \times 10^6 a$ [V]. **(c)** $Q = 1.35 \times 10^{10}$ C.

36. **(a)** $V = 4.577 \times 10^5$ V. **(b)** 5.024×10^5 V.

37. $\mathbf{D}_2 = \hat{\mathbf{x}}5\varepsilon_0 + \hat{\mathbf{y}}6\varepsilon_0$ [C/m^2].

38. **(a)** $\mathbf{E} = \hat{\mathbf{R}}119.86$ V/m. **(b)** $\rho_s = -10.6$ nC/m^2.

39. **(a)** $\mathbf{E}_2 = -\hat{\mathbf{x}}\varepsilon_0 E/\varepsilon_1$, $\mathbf{E}_3 = 0$, $\mathbf{E}_4 = -\hat{\mathbf{x}}\varepsilon_0 E/\varepsilon_1$, $\mathbf{E}_5 = -\hat{\mathbf{x}}E$ [V/m]. **(b)** $V = 2\varepsilon_0 Ed/\varepsilon_1$ [V].

40. **(a)** $E = 21.76$ V/m. **(b)** $E = 50$ V/m.

41. **(a)** $\mathbf{E}_2 = -\hat{\mathbf{y}}E_1\sin\theta - \hat{\mathbf{x}}(\varepsilon_0/\varepsilon_1)E_1\cos\theta$, $\mathbf{E}_3 = -\hat{\mathbf{y}}E_1\sin\theta - \hat{\mathbf{x}}(\varepsilon_0/\varepsilon_2)E_1\cos\theta$, $\mathbf{E}_4 = \mathbf{E}_1 = -\hat{\mathbf{y}}E_1\sin\theta - \hat{\mathbf{x}}E_1\cos\theta$ [V/m].
(b) $V = E_1 a\varepsilon_0\cos\theta((\varepsilon_2 + \varepsilon_1)/\varepsilon_1\varepsilon_2)$ [V].

42. **(a)** $P_{1t}(\varepsilon_1 - \varepsilon_0) = P_{2t}(\varepsilon_2 - \varepsilon_0)$, $\varepsilon_1 P_{1n}/(\varepsilon_1 - \varepsilon_0) = \varepsilon_2 P_{2n}(\varepsilon_2 - \varepsilon_0)$ for $\rho_s = 0$, $P_{1t}(\varepsilon_1 - \varepsilon_0) = P_{2t}(\varepsilon_2 - \varepsilon_0)$,
$\varepsilon_1 P_{1n}/(\varepsilon_1 - \varepsilon_0) - \varepsilon_2 P_{2n}(\varepsilon_2 - \varepsilon_0) = \rho_s$ for $\rho_s \neq 0$. **(b)** $P_{1t}/P_{2t} = (\varepsilon_1 - \varepsilon_0)/(\varepsilon_2 - \varepsilon_0)$, $P_{1n}/P_{2n} = \varepsilon_2(\varepsilon_1 - \varepsilon_0)/\varepsilon_1(\varepsilon_2 - \varepsilon_0)$.

43. $C_{total} = 60\varepsilon_0/(20d_2 + 3d_1)$ [F].

44. $C = \frac{2\pi\varepsilon_0}{\ln(a/b)}\left[\frac{\text{F}}{\text{m}}\right].$

45. $C = 899.6$ pF.

46. **(a)** $x = 0.636$ mm. **(b)** $x = 0.65$ mm.

47. $C = 4\pi\varepsilon_1\varepsilon_2 bca/(\varepsilon_1 a(c - b) + \varepsilon_2 c(b - a))$ [F].

48. $C = 4\pi\varepsilon_0/\ln(b/a)$ [F/m].

49. $W = \frac{Q^2}{4\pi\varepsilon_0}\left[\frac{2}{a} - \frac{1}{a+b} - \frac{2}{b}\right]$ [J].

50. **(a)** $W_s = \frac{11\pi\rho_0^2 a^9}{6{,}300\varepsilon}$ [J]. **(b)** $W_{out} = \frac{\rho_0^2\pi a^9}{200\varepsilon_0}$ [J]. **(c)** $W = \frac{\rho_0^2\pi a^9}{100}\left(\frac{11}{63\varepsilon} + \frac{1}{2\varepsilon_0}\right)$ [J].

51. $W = 2\pi\varepsilon_0 aV^2$ [J].

52. **(a)** $\Delta W = \varepsilon V^2 a/2d(d - a)$. **(b)** $\Delta Q = 2.656 \times 10^{-10}$ C. **(c)** $\Delta W = \frac{\varepsilon_0 V^2}{2}\left(\frac{16d - 12a}{(4d - 3a)^2} - \frac{1}{d}\right)$ [J], $\Delta Q = 1.87 \times 10^{-12}$ C.

53. $\Delta W = \frac{Q^2(\varepsilon_r - 1)}{8\pi\varepsilon_0\varepsilon_r}\left[\frac{1}{a} - \frac{1}{b}\right]$ [J].

54. **(a)** $w = q^2/32\pi^2\varepsilon_0 R^4, 0 < R < a$, $w = (q + 4\pi a^2\rho_0)^2/32\pi^2\varepsilon_0 R^4, a < R < b$, $w = (q + 4\pi a^2\rho_0 - 4\pi b^2\rho_0)^2/32\pi^2\varepsilon_0 R^4, R > b$ [J/m^3].
(b) $W = \frac{(q + 4\pi a^2\rho_0)^2}{8\pi\varepsilon_0}\left[\frac{1}{a} - \frac{1}{b}\right]$ [J].

55. $W = \rho_l^2\ln(b/a)/4\pi\varepsilon$ [J/m].

56. $p = \varepsilon_0 V^2/d^2$ [Pa].

57. $V = (d/b)\sqrt{2(P - k)g/\varepsilon}$ [V].

58. $F = -\varepsilon_0\varepsilon_r V^2 b/2d$ [N] in the direction opposite the displacement.

59. $F = V_0^2 b(\varepsilon - \varepsilon_0)/2d$ [N].

60. **(a)** $P = \rho_0^2/2\varepsilon_0$ [Pa]. **(b)** $\rho_s = 42\ \mu\text{C/m}^2$. **(c)** $P = 4.427$ MPa.

Chapter 5

3. $V(x) = 2{,}500x$ [V].

4. $V(x) = -\rho_0 x^2/2\varepsilon_0 + (V_0/d + \rho_0 d/2\varepsilon_0)x$ [V].

5. $V(x) = -9{,}411.94x^4 + 37.64777x^3 + (50{,}000 - 7.52955 \times 10^{-5})x$ [V].

6. **(a)** $V(r) = -23.083\ln r - 127.45$ V. **(b)** $\mathbf{E} = \hat{\mathbf{r}}23.083/r$ [V/m].

7. (**a**) $V(\phi) = -(4V_0/\pi)\phi + V_0$ [V], $0 < \phi < \pi/4$. (**b**) $V(\phi) = (4V_0/7\pi)\phi - V_0/7$ [V], $\pi/4 < \phi < 2\pi$.

8. (**a**) $V(R) = 25.25/R - 5{,}000$ V. (**b**) $\mathbf{E} = \hat{\mathbf{R}}25.25/R^2$ [V/m].

9. $\rho_v = -6\varepsilon c$ [C/m^3].

10. (**a**) $\rho_s = -1.99 \times 10^{-10}$ C/m^2. (**b**) $\rho_s = -1.424 \times 10^{-10}$ C/m^2.

11. (**b**) $\mathbf{E} = \hat{\mathbf{z}}\dfrac{q}{4\pi\varepsilon_0}\left[\dfrac{4a}{(x^2+y^2+4a^2)^{3/2}} - \dfrac{2a}{(x^2+y^2+a^2)^{3/2}}\right]\left[\dfrac{\text{V}}{\text{m}}\right].$

12. (**a**) $\rho_s = -\rho_l h/\pi(x^2 + h^2)$ [C/m^2].

13. (**a**) $V(2d,y) = \dfrac{\rho_l}{2\pi\varepsilon_0}\ln\dfrac{\sqrt{4d^2+(y+d)^2}}{\sqrt{4d^2+(y-d)^2}}$ [V]. (**b**) $\mathbf{E}(2d,y) = \hat{\mathbf{x}}\dfrac{4d\rho_l}{2\pi\varepsilon_0}\left(\dfrac{1}{4d^2+(y-d)^2} - \dfrac{1}{4d^2+(y+d)^2}\right)\left[\dfrac{\text{V}}{\text{m}}\right].$

14. (**b**) $\rho_s = \pm\dfrac{q\sqrt{2}}{4\pi}\left[\dfrac{b-a}{((x-a)^2+(x-b)^2)^{3/2}} + \dfrac{b+a}{((x-a)^2+(x+b)^2)^{3/2}} - \dfrac{b+a}{((x+a)^2+(x-b)^2)^{3/2}} - \dfrac{b-a}{((x+a)^2+(x+b)^2)^{3/2}}\right]\left[\dfrac{\text{C}}{\text{m}^2}\right],$

where $b = 0.3827d$, $a = 0.9239d$.

15. (**a**)
$$\mathbf{E} = -\hat{\mathbf{x}}\frac{q}{4\pi\varepsilon_0}\sum_{i=1}^{12}\frac{(-1)^{i+1}\left(x - a\cos\left(15^\circ + (i-1)30^\circ\right)\right)}{\left[\left(x - a\cos\left(15^\circ + (i-1)30^\circ\right)\right)^2 + \left(y - a\sin\left(15^\circ + (i-1)30^\circ\right)\right)^2\right]^{3/2}}$$
$$-\hat{\mathbf{y}}\frac{q}{4\pi\varepsilon_0}\sum_{i=1}^{12}\frac{(-1)^{i+1}\left(y - a\sin\left(15^\circ + (i-1)30^\circ\right)\right)}{\left[\left(x - a\cos\left(15^\circ + (i-1)30^\circ\right)\right)^2 + \left(y - a\sin\left(15^\circ + (i-1)30^\circ\right)\right)^2\right]^{3/2}}\left[\frac{\text{V}}{\text{m}}\right]$$

16. (**b**) $V(d/2,0) = 2.2112q/4\pi\varepsilon_0 d$ [V].

17. (**a**)
$$V(x,y) = \frac{q}{4\pi\varepsilon_0}\sum_{i=1,5,9,13,\ldots}^{N}\frac{1}{\sqrt{(x-(i-1)d/2-3d/4)^2+y^2}} - \sum_{i=2,6,10,14,\ldots}^{N}\frac{1}{\sqrt{(x-id/2-d/4)^2+y^2}} -$$
$$\sum_{i=3,7,11,15,\ldots}^{N}\frac{1}{\sqrt{(x+(i-3)d/2+3d/4)^2+y^2}} + \sum_{i=4,8,12,16,\ldots}^{N}\frac{1}{\sqrt{(x+(i-2)d/2+d/4)^2+y^2}}\left[\frac{\text{V}}{\text{m}}\right].$$

21. $E(0,0) = 16.178$ V/m (down).

22.
$$\mathbf{E} = \hat{\mathbf{x}}\frac{\rho_l}{2\pi\varepsilon_0}\left(\frac{(x-2r)}{(x-2r)^2+y^2} - \frac{(x-r/2)}{(x-r/2)^2+y^2} + \frac{(x+r/2)}{(x+r/2)^2+y^2} - \frac{(x+2r)}{(x+2r)^2+y^2}\right)$$
$$+\ \hat{\mathbf{y}}\frac{\rho_l y}{2\pi\varepsilon_0}\left(\frac{1}{(x-2r)^2+y^2} - \frac{1}{(x-r/2)^2+y^2} + \frac{1}{(x+r/2)^2+y^2} - \frac{1}{(x+2r)^2+y^2}\right)\left[\frac{\text{V}}{\text{m}}\right].$$

23. (**a**) $\mathbf{E}(a,0,0) = -\hat{\mathbf{x}}7q/8\pi\varepsilon_0 a^2$ [V/m]. (**b**) Zero.

24. (**b**) $E_R(r,\theta) = \dfrac{q}{4\pi\varepsilon_0 r}\dfrac{(-a^2-2ar)}{\left(r^2+(a+r)^2-2r(a+r)\cos\theta\right)^{3/2}}\left[\dfrac{\text{V}}{\text{m}}\right]$. (**c**) $\rho_s(r,\theta) = -\dfrac{q}{4\pi r}\dfrac{(a^2+2ar)}{\left(r^2+(a+r)^2-2r(a+r)\cos\theta\right)^{3/2}}\left[\dfrac{\text{C}}{\text{m}^2}\right].$

25. (**b**) $\mathbf{E} = \hat{\mathbf{R}}aV_0/R^2$ [V/m], $R \geq a$.

26. (**b**) $\mathbf{E}(r,\theta) = \hat{\mathbf{R}}\left(\dfrac{\rho_s}{\varepsilon_0} - \dfrac{q}{4\pi\varepsilon_0 r}\dfrac{(a^2+2ar)}{\left(r^2+(a+r)^2-2r(a+r)\cos\theta\right)^{3/2}}\right)\left[\dfrac{\text{V}}{\text{m}}\right].$

27. (**a**) $\mathbf{E}(y,z) = \hat{\mathbf{y}}\dfrac{qy}{4\pi\varepsilon_0}\left(\dfrac{1}{R_1^3} - \dfrac{1}{R_2^3} + \dfrac{a}{cR_3^3} - \dfrac{a}{dR_4^3}\right) + \hat{\mathbf{z}}\dfrac{q}{4\pi\varepsilon_0}\left(\dfrac{z-d}{R_1^3} - \dfrac{z+c}{R_2^3} + \dfrac{az+a^3/c}{cR_3^3} - \dfrac{az-a^3/d}{dR_4^3}\right)\left[\dfrac{\text{V}}{\text{m}}\right].$

(**c**) $\mathbf{E}(y,z) = \hat{\mathbf{y}}\dfrac{qy}{4\pi\varepsilon_0}\left(\dfrac{1}{R_1^3} - \dfrac{1}{R_2^3} + \dfrac{c}{aR_3^3} - \dfrac{d}{aR_4^3}\right) + \hat{\mathbf{z}}\dfrac{q}{4\pi\varepsilon_0}\left(\dfrac{z-d}{R_1^3} - \dfrac{z+c}{R_2^3} + \dfrac{cz+a^2}{aR_3^3} - \dfrac{dz-a^2}{aR_4^3}\right)\left[\dfrac{\text{V}}{\text{m}}\right].$

$R_1 = \sqrt{y^2+(z-d)^2}$, $R_2 = \sqrt{y^2+(z+c)^2}$, $R_3 = \sqrt{y^2+(z+a^2/c)^2}$, $R_4 = \sqrt{y^2+(z-a^2/d)^2}$ [m].

28. $V(x,y) = \dfrac{4V_0}{\pi}\displaystyle\sum_{m=1,3,5,\ldots}^{\infty}\dfrac{1}{m}\sin\dfrac{m\pi y}{b}e^{-m\pi x/b}$ [V].

29. **(a)** $V(x,y) = \frac{40}{\pi} \sum_{m=1,3,5,\dots}^{\infty} \frac{1}{m} \sin\frac{m\pi x}{a} \frac{\sinh(m\pi y/a)}{\sinh(m\pi b/a)}$ [V].

30. $V(x,y) = \frac{10}{\sinh\left(\frac{\pi b}{a}\right)} \sinh\left(\frac{\pi y}{a}\right)\sin\left(\frac{\pi x}{a}\right)$ [V].

31. $V(x,y) = \frac{4}{\pi}\left[V_2 \sum_{m=1,3,5,\dots}^{\infty} \frac{1}{m}\sin\frac{m\pi y}{b} \frac{\sinh(m\pi x/b)}{\sinh(m\pi a/b)} + V_1 \sum_{m=1,3,5,\dots}^{\infty} \frac{1}{m}\sin\frac{m\pi y}{b} \frac{\sinh(m\pi(a-x)/b)}{\sinh(m\pi a/b)}\right]$ [V].

32. $V(x,y,z) = \frac{16V_0}{\pi^2} \sum_{m=1,3,5,}^{\infty} \sum_{n=1,3,5,}^{\infty} \left[\frac{\sin(n\pi y/a)}{mn\sinh\left(\pi\sqrt{m^2+n^2}\right)} \sin\frac{m\pi x}{a}\sinh\left(\frac{\pi z}{a}\sqrt{m^2+n^2}\right) + \sin\frac{m\pi z}{a}\sinh\left(\frac{\pi x}{a}\sqrt{m^2+n^2}\right)\right]$ [V].

33. **(a)** $V(r,\phi) = \frac{2V_0}{\pi} \sum_{k=1,3,5,\dots}^{\infty} \frac{1}{k}\left(\frac{r}{a}\right)^k \sin k\phi$ [V], $r < a$, $V(r,\phi) = \frac{2V_0}{\pi} \sum_{k=1,3,5,\dots}^{\infty} \frac{1}{k}\left(\frac{a}{r}\right)^k \sin k\phi$ [V], $r > a$.

(b) $\mathbf{E} = -\frac{2V_0}{a\pi} \sum_{k=1,3,5,\dots} \frac{1}{k}\left(\frac{r}{a}\right)^{k-1}\left[\hat{\mathbf{r}}\sin k\phi + \hat{\boldsymbol{\phi}}\cos k\phi\right]\left[\frac{\text{V}}{\text{m}}\right]$, $r < a$, $\mathbf{E} = \frac{2V_0}{a\pi} \sum_{k=1,3,5,\dots}^{\infty} \frac{1}{k}\left(\frac{a}{r}\right)^{k-1}\left[\hat{\mathbf{r}}\sin k\phi - \hat{\boldsymbol{\phi}}\cos k\phi\right]\left[\frac{\text{V}}{\text{m}}\right]$, $r > a$.

34. **(a)** $V(r,\phi) = \frac{4\times 10^5}{\pi} \sum_{k=1,3,5,\dots}^{\infty} \frac{1}{k}\left(\frac{r}{a}\right)^k \left[\sin k\phi + \sin k\left(\phi + \frac{\pi}{2}\right)\right]$ [V], $r < a$, $V(r,\phi) = \frac{4\times 10^5}{\pi} \sum_{k=1,3,5,\dots}^{\infty}$ $\frac{1}{k}\left(\frac{a}{r}\right)^k \left[\sin k\phi + \sin k\left(\phi + \frac{\pi}{2}\right)\right]$ [V], $r > a$.

(b) $\mathbf{E} = -\frac{4\times 10^5}{a\pi} \sum_{k=1,3,5,\dots}^{\infty} \frac{1}{k}\left(\frac{r}{a}\right)^{k-1}\left[\hat{\mathbf{r}}\sin k\phi + \hat{\boldsymbol{\phi}}\cos k(\phi + \pi/2)\right]\left[\frac{\text{V}}{\text{m}}\right]$, $r < a$, $\mathbf{E} = -\frac{4\times 10^5}{a\pi} \sum_{k=1,3,5,\dots}^{\infty}$ $\frac{1}{k}\left(\frac{a}{r}\right)^{k-1}\left[\hat{\mathbf{r}}\sin k\phi - \hat{\boldsymbol{\phi}}\cos k(\phi + \pi/2)\right]\left[\frac{\text{V}}{\text{m}}\right]$, $r > a$.

Chapter 6

1. **(a)** $V_1 = 0$, $V_2 = 3$ V, $V_3 = 6$ V, $V_4 = 9$ V, $V_5 = 12$ V. **(b)** $V_1 = 0$, $V_2 = 1.5$ V, $V_3 = 3$ V, $V_4 = 4.5$ V, $V_5 = 6$ V, $V_6 = 7.5$ V, $V_7 = 9$ V, $V_8 = 10.5$ V, $V_9 = 12$ V. **(c)** $V(x) = 12x$ [V].
2. **(a)** $V_1 = 0$, $V_2 = 3.09375$ V, $V_3 = 6.125$ V, $V_4 = 9.09375$ V.
3. **(b)** $V(x) = 1.4118 \times 10^4(x - x^2)$ [V], $E = -1.4118 \times 10^4(1 - 2x)$ [V/m].
7. 1.343 to 5.59 pF.
8. **(a)** 2.5116 pF. **(b)** 1.5547 pF.
9. **(a)** ≈ 3 pF.
10. 3.54 pF (8 patches), 3.99 pF (16 patches).
11. **(a)** 4.913×10^{-17} F. **(b)** 4.395×10^{-17} F.
12. **(a)** 2.1943 pF (2×2 subdomains on each plate). **(b)** $V_P = 0$ V, $\mathbf{E}_P = \hat{\mathbf{y}}1.574 - \hat{\mathbf{z}}1.574$ [V/m]. **(c)** 1,024 subdomains.
13. **(a)** 1.096 pF. **(b)** 1.4784 pF, 11×11 division on each plate.
14. **(a)** $N_1 = (x_2 - x)/(x_2 - x_1)$, $N_2 = (x - x_1)/(x_2 - x_1)$. **(b)** $\phi(x) = (x_2 - x)\phi_1/(x_2 - x_1) + (x - x_1)\phi_2/(x_2 - x_1)$.
15. **(c)** $V(x) = -\rho_0 x^4/12\varepsilon + \rho_0 x^3 d/6\varepsilon + 10x/d - \rho_0 x d^3/12\varepsilon$ [V].
16. **(c)** $\mathbf{E} = -\hat{\mathbf{x}}(-\rho_0 x^3/\varepsilon + 10/d + \rho_0 d/2\varepsilon)$ [V/m].
18. **(a)** 1.425 V. **(b)** 3.914 μm.
19. **(a)** 3,260 V. **(b)** 1,500 V.

Chapter 7

1. **(a)** 400,000 C. **(b)** 314 km^2. **(c)** 6,164
2. **(a)** 10.294×10^6 A. **(b)** 205.88×10^6 A.
3. 0.235 mm/s
4. 413 years, 7 months, 8 days

5. Inside beam: $\mathbf{E} = -\hat{\mathbf{r}}7.19 \times 10^7 r$ [V/m], outside beam: $\mathbf{E} = -\hat{\mathbf{r}}17.97/r$ [V/m].

6. **(a)** 1,667 S/m. **(b)** 3.125×10^{21} carriers.

7. **(a)** $\mathbf{E} = -\hat{\mathbf{r}}V/r\ln(b/a)$ [V/m]. **(b)** $\mathbf{J} = -\hat{\mathbf{r}}\sigma V/r\ln(b/a)$ [A/m^2]. **(c)** $I = 2\pi L\sigma V/\ln(b/a)$ [A]. **(d)** $R = \ln(b/a)/2\pi L\sigma$ [Ω].

8. **(a)** 5.1473×10^{-5} Ω. **(b)** –0.198%.

9. **(a)** 1.7243×10^{-5} Ω. **(b)** 1.5936×10^{-5} Ω.

10. $R_{min} = 17.98$ Ω, $R_{max} = 29.3$ Ω.

11. **(a)** $\sigma = 1.273 \times 10^3/V$ [S/m].

12. $\sigma = \dfrac{\sigma_0(b^2 - a^2)}{a^2}\left[\dfrac{(b^2 - a^2)(L - p) + pb^2}{9pb^2 - (L - p)(b^2 - a^2)}\right]\left[\dfrac{\text{S}}{\text{m}}\right].$

13. $V = Il_1/\sigma_1\pi r_1^{\,2} + Il_2/\sigma_2\pi r_2^{\,2} + Il_3/\sigma_3\pi r_3^{\,2}$ [V].

14. **(a)** $R = 2d^2 / \sigma_0^{\,2}b^2$ [Ω]. **(b)** $V = 2d^2I/\sigma_0^{\,2}b^2$ [V].

15. $V_{AB} = 2.944Ja/\sigma$ [V].

16. 15.92 Ω.

17. $\sigma = I_0(a + b)/2\pi abV_0$ [S/m].

18. **(a)** 1.628 mm for series connection, 7.28 mm for parallel connection. **(b)** 33.30 kg for series connection, 666 kg for parallel connection.

19. **(a)** 3.054 Ω. **(b)** 754.42 W.

20. **(a)** $J_{iron} = V\sigma/a$ [A/m^2], $J_{copper} = 5V\sigma/a$ [A/m^2]. **(b)** $P_{iron} = V^2\sigma bc/a$ [W], $P_{copper} = V^2 5\sigma bc/a$ [W].

21. **(a)** $\Delta R = 0.0357$ Ω. **(b)** $\Delta R = 2.589 \times 10^{-4}$ Ω.

22. **(a)** $R = 0.02076$ Ω. **(b)** $R = 8.275 \times 10^{-4}$ Ω.

23. **(a)** 16 A.

24. **(b)** $I = 1.2e^{-100t}$ [C/s].

25. **(a)** $Q_1 = -\varepsilon_1 V\sigma_2\pi b^2/(d_1\sigma_2 + d_2\sigma_1)$, $Q_2 = V\pi b^2(\varepsilon_1\sigma_2 - \varepsilon_2\sigma_1)/(d_1\sigma_2 + d_2\sigma_1)$, $Q_3 = \varepsilon_2 V\sigma_1\pi b^2/(d_1\sigma_2 + d_2\sigma_1)$ [C].
(b) $Q_1 = \varepsilon_1 V\sigma_2\pi b^2/(d_1\sigma_2 + d_2\sigma_1)$, $Q_2 = V\pi b^2(\varepsilon_2\sigma_1 - \varepsilon_1\sigma_2)/(d_1\sigma_2 + d_2\sigma_1)$, $Q_3 = -\varepsilon_2 V\sigma_1\pi b^2/(d_1\sigma_2 + d_2\sigma_1)$ [C].

26. **(a)** $E_{copper} = 0.056$ V/m, $E_{aluminum} = 0.0884$ V/m. **(b)** $w_{copper} = 177.76$ kW/m^3, $w_{aluminum} = 281.45$ kW/m^3.
(c) $\rho_s = -2.88 \times 10^{-13}$ C/m^2.

27. **(a)** $V = 6.324$ mV, $I = 2{,}483.4$ A. **(b)** 1.754×10^7 W/m^3.

Chapter 8

1. **(a)** $B = 3\mu_0 I/2\pi a$ [T], (into the page). **(b)** $\mu_0 I/2\pi\sqrt{3}a$ [T], (into the page)

2. $\mathbf{B}_1 = -\hat{\mathbf{y}}\dfrac{\mu I}{2\pi a}\left(\dfrac{1}{\sqrt{2}} - 1\right)$ [T], $\mathbf{B}_2 = -\hat{\mathbf{y}}\dfrac{\mu I}{4\pi d}$ [T].

3. $2\sqrt{3}/\pi$.

5. $H = \rho_s\omega(b - a)/2$ [A/m], (out of page).

6. $\mathbf{B} = \hat{\mathbf{z}}\dfrac{2\mu_0 Iab}{\pi\sqrt{4h^2 + b^2 + a^2}}\left[\dfrac{1}{(4h^2 + b^2)} + \dfrac{1}{(4h^2 + a^2)}\right]$ [T]

7. $B = \displaystyle\sum_{i=1}^{N}\dfrac{\mu_0 Ia^2 i^2}{2N^2\left(h^2 + i^2(a/N)^2\right)^{3/2}}$ [T], (up).

8. **(b)** $\mathbf{B} = -\hat{\mathbf{x}}58.4$ nT.

9. $\mathbf{B} = \hat{\mathbf{x}}\dfrac{\mu_0 Iz}{2\pi a^2} - \hat{\mathbf{y}}\dfrac{\mu_0 I}{2\pi d}$ [T], $0 < z < a$. $\mathbf{B} = \hat{\mathbf{x}}\dfrac{\mu_0 I}{2\pi z} - \hat{\mathbf{y}}\dfrac{\mu_0 I}{2\pi d}$ [T], $z > a$. $\mathbf{B} = -\hat{\mathbf{x}}\dfrac{\mu_0 Iz}{2\pi a^2} - \hat{\mathbf{y}}\dfrac{\mu_0 I}{2\pi d}$ [T], $-a < z < 0$.
$\mathbf{B} = -\hat{\mathbf{x}}\dfrac{\mu_0 I}{2\pi|z|} - \hat{\mathbf{y}}\dfrac{\mu_0 I}{2\pi d}$ [T], $z < -a$.

10. $\mathbf{B} = \hat{\mathbf{x}}\left[\dfrac{\mu_0 Ja^2 z}{2\left(z^2 + (x - d)^2\right)} - \dfrac{\mu_0 Jb^2(y - d)}{2\left(x^2 + (y - d)^2\right)}\right] + \hat{\mathbf{y}}\dfrac{\mu_0 Jb^2 x}{2\left(x^2 + (y - d)^2\right)} - \hat{\mathbf{z}}\dfrac{\mu_0 Ja^2(x - d)}{2\left(z^2 + (x - d)^2\right)}$ [T].

11. **(a)** $\mathbf{B} = \hat{\mathbf{x}}\dfrac{\mu_0 Ih}{2\pi}\left(\dfrac{1}{(x - a)^2 + h^2} - \dfrac{1}{(x + a)^2 + h^2}\right) + \hat{\mathbf{y}}\dfrac{\mu_0 I}{2\pi}\left(\dfrac{x - a}{(x - a)^2 + b^2} - \dfrac{x + a}{(x + a)^2 + h^2}\right)$ [T].
(b) $\mathbf{B} = -\hat{\mathbf{y}}5.87 \times 10^{-6}$ T.

12. $B = \frac{\mu_0 I}{2\pi}\left[\frac{1}{a}\ln\frac{2a+b}{b} + \frac{1}{c}\ln\frac{2c+b}{b}\right]$ [T], *downward.*
13. **(b)** $-\hat{\mathbf{x}}3.0186 \times 10^{-4} + \hat{\mathbf{y}}3.0186 \times 10^{-4}$ T.
14. $\mathbf{B} = -\hat{\mathbf{y}}\mu_0 NI/2$ [T], $x < 0$. $\mathbf{B} = \hat{\mathbf{y}}\mu_0 NI/2$ [T], $x > 0$.
15. **(a)** $B = 0, 0 < r < r_1$. $\mathbf{B} = \hat{\boldsymbol{\phi}} J\mu_0\left(\pi r^2 - \pi r_1^2\right)/2\pi r$ [T], $r_1 < r < r_2$. $\mathbf{B} = \hat{\boldsymbol{\phi}} J\mu_0\left(\pi r_2^2 - \pi r_1^2\right)/2\pi r$ [T], $r > r_2$.
(b) $\mathbf{B} = \hat{\boldsymbol{\phi}}\mu_0 I/2\pi r$ [T], $0 < r < r_1$. $\mathbf{B} = \hat{\boldsymbol{\phi}}\mu_0\left(I + \left(\pi r^2 - \pi r_1^2\right)J\right)/2\pi r$ [T], $r_1 < r < r_2$. $\mathbf{B} = \hat{\boldsymbol{\phi}}\mu_0\left(I + \left(\pi r_2^2 - \pi r_1^2\right)J\right)/2\pi r$ [T], $r > r_2$. **(c)** $\mathbf{B} = -\hat{\boldsymbol{\phi}}\mu_0 I/2\pi r$ [T], $0 < r < r_1$. $\mathbf{B} = \hat{\boldsymbol{\phi}}\mu_0\left(\left(\pi r^2 - \pi r_1^2\right)J - I\right)/2\pi r$ [T], $r_1 < r < r_2$.
$\mathbf{B} = \hat{\boldsymbol{\phi}}\mu_0\left(\left(\pi r_2^2 - \pi r_1^2\right)J - I\right)/2\pi r$ [T], $r > r_2$.
16. $\mathbf{B} = \hat{\mathbf{z}}\mu_0\omega b\rho_s$ [T], $r \le b$. $\mathbf{B} = 0$, $r > b$.
17. $\mathbf{B} = \hat{\mathbf{x}}\left(-\frac{\mu_0 y_0 J\left(b^2 - a^2\right)}{2\left(x_0^2 + y_0^2\right)} + \frac{\mu_0 y_0 J c^2}{2\left((x_0 - d)^2 + y_0^2\right)}\right) + \hat{\mathbf{y}}\left(\frac{\mu_0 x_0 J\left(b^2 - a^2\right)}{2\left(x_0^2 + y_0^2\right)} - \frac{\mu_0 (x_0 - d) J c^2}{2\left((x_0 - d)^2 + y_0^2\right)}\right)$ [T].
18. **(a)** $\mathbf{B} = -\hat{\boldsymbol{\phi}}\mu_0 I_1/2\pi r$ [T]. **(b)** $\mathbf{B} = \hat{\boldsymbol{\phi}}\mu_0\left(\pi\left(r_2^2 - r_1^2\right)J_1 - I_1\right)/2\pi r$ [T].
19. **(a)** $\mathbf{B} = 0, 0 < r < r_1$. $\mathbf{B} = -\hat{\mathbf{z}}\mu_0 nI$ [T], $r_1 < r < r_2$. $\mathbf{B} = 0$, $r > r_2$. **(b)** $\mathbf{B} = -\hat{\mathbf{z}}2\mu_0 nI$ [T], $0 < r < r_1$.
$\mathbf{B} = -\hat{\mathbf{z}}\mu_0 nI$ [T], $r_1 < r < r_2$. $\mathbf{B} = 0$, $r > r_2$.
20. $H = \frac{J}{2}\left(\frac{b^2}{a} - \frac{c^2}{a-d}\right)\left[\frac{\text{A}}{\text{m}}\right]$.
21. $\mathbf{B}_A = \hat{\mathbf{y}}\frac{\mu_0 I}{2\pi a}\left[\frac{b^2}{2b^2 + 3ab + a^2}\right]$, $\mathbf{B}_B = \hat{\mathbf{x}}\frac{\mu_0 I b^2}{4\pi a\left(4a^2 + b^2\right)}$, $\mathbf{B}_C = 0$, $\mathbf{B}_D = \hat{\mathbf{y}}\frac{\mu_0 I}{2\pi a}\left[\frac{b^2}{2b^2 - 3ab + a^2}\right]$ [T]
22. **(a)** $\mathbf{A}(\mathbf{r}) = \hat{\mathbf{z}}\frac{\mu_0 I}{2\pi}\ln\frac{2L}{a}\left[\frac{\text{Wb}}{\text{m}}\right]$. **(b)** ∞.
23. **(a)** $B_c/B_s = \pi/4\sqrt{2}$. **(b)** Zero.
24. **(a)** $\Phi = \mu_0 Ia(\ln 3)/2\pi$ [Wb]. **(b)** $\Phi = \mu_0 Ia(\ln 3)/2\pi$ [Wb].
25. **(a)** Zero. **(b)** $\Phi = 1.257 \times 10^{-6}$ [Wb].
26. $\psi_{NS} = 7.958 \times 10^3$ A.
27. $H = 1{,}000$ A/m, (down).

Chapter 9

1. **(a)** $m = \pi a^2 I$ [A · m²]. **(c)** $\mathbf{B} = \frac{\mu_0 m}{4\pi R^3}\left(\hat{\mathbf{R}}2\cos\theta + \hat{\boldsymbol{\theta}}\sin\theta\right)$ [T], $m = \pi a^2 I$ [A · m²] for circular loop, $m = a^2 I$ [A · m²] for square loop.
2. **(a)** $\Phi = \mu_0 I\pi a^2 b^2/2h^3$ [Wb]. **(b)** $\Phi = \mu_0 I\pi a^2 b^2/4h^3$ [Wb]. **(c)** $\Phi = \mu_0 I\pi a^2 b^2/8h^3$ [Wb].
3. $I = M/n$ [A].
4. $\mathbf{H} = \mathbf{M}$ [A/m].
5. $\mathbf{B}_2 = \hat{\mathbf{x}}\mu_1 H\cos\alpha_1 - \hat{\mathbf{y}}\mu_2 H\sin\alpha_1$ [T], $\mathbf{B}_3 = \hat{\mathbf{x}}\mu_1 H\cos\alpha_1 - \hat{\mathbf{y}}\mu_3 H\sin\alpha_1$ [T].
6. $89°40'$ to the surface.
7. $B_{2n} = 0.00866$ T, $B_{2t} = 1.0$ T, $\theta_2 = 89°30'$. $B_{3n} = 0.00866$ T, $B_{3t} = 0.25$ T, $\theta_3 = 88°$, $B_{4n} = 0.00866$ T, $B_{4t} = 0.005$ T, $\theta_4 = 30°$.
8. **(a)** $L_{11} = \frac{\mu_1 N^2 c}{2\pi}\ln\frac{d}{b}$ [H]. **(b)** $L_{22} = \mu_0 n^2 \pi a^2$ [H/m].
9. **(a)** $L_{11} = \mu N_1^2 c\ln(b/a)/2\pi$ [H], $L_{22} = \mu N_2^2 c\ln(b/a)/2\pi$ [H], $L_{33} = \mu N_3^2 c\ln(b/a)/2\pi$ [H].
(b) $L_{21} = L_{12} = 0$, $L_{31} = L_{13} = \mu N_1 N_3 c\ln(b/a)/2\pi$ [H], $L_{32} = L_{23} = \mu N_2 N_3 c\ln(b/a)/2\pi$ [H].
(c) $L_1 = \mu N_1 c(N_1 + N_3)\ln(b/a)/2\pi$ [H], $L_2 = \mu N_2 c(N_2 + N_3)\ln(b/a)/2\pi$ [H], $L_3 = \mu N_3 c(N_1 + N_2 + N_3)\ln(b/a)/2\pi$ [H].
10. $L_{11} = \mu_0 n^2 \pi b^2 + \mu_1 n^2 \pi(a^2 - b^2)$ [H/m].
11. $L_{11} = \mu_0 \pi(b^2 - a^2)$ [H/m].
12. **(a)** $L_{12} = \frac{\mu_0 c}{2\pi}\ln\frac{a+b}{c}$ [H]. **(b)** Zero.
13. $L_{12} = L_{21} = \mu_0 n\pi b^2$ [H].

14. **(a)** $L_{12}=2\pi a^2\mu_0 N$ [H]. **(b)** $L_{12}=2\pi a^2\mu_0 N$ [H].

15. **(a)** $L_{ab}=L_{ba}=\mu_0\pi a^2b^2/2h^3$ [H]. **(b)** $L_{ab}=L_{ba}=\mu_0\pi a^2b^2/4h^3$ [H]. **(c)** $L_{ab}=L_{ba}=\mu_0\pi a^2b^2/8h^3$ [H].

16. **(a)** $3L$ [H]. **(b)** $L/3$ [H]. **(c)** $1.5L$ [H]. **(d)** $L_a=3L+6M$ [H], $L_b=(L+2M)/3$ [H], $L_c=1.5L+3M$ [H].

17. $L_{11}=\mu_0\ln(a/b)/2\pi$ [H/m].

18. **(a)** $\mu_0/8\pi$ [H/m]. **(b)** $\dfrac{\mu_0}{2\pi}\left(\dfrac{c^2}{c^2-a^2}\right)^2\ln\dfrac{c}{a}-\dfrac{\mu_0}{2\pi}\left(\dfrac{c^2}{c^2-a^2}\right)+\dfrac{\mu_0}{8\pi}\left(\dfrac{c^2+a^2}{c^2-a^2}\right)\left[\dfrac{\text{H}}{\text{m}}\right]$. **(c)** $\mu_0\ln(a/b)/2\pi$ [H/m].

(d) $L=\dfrac{\mu_0}{8\pi}+\dfrac{\mu_0}{2\pi}\ln\dfrac{a}{b}+\dfrac{\mu_0}{2\pi}\left[\left(\dfrac{c^2}{c^2-a^2}\right)^2\ln\dfrac{c}{a}-\left(\dfrac{c^2}{c^2-a^2}\right)+\dfrac{1}{4}\left(\dfrac{c^2+a^2}{c^2-a^2}\right)\right]\left[\dfrac{\text{H}}{\text{m}}\right]$.

19. **(a)** $\dfrac{\mu_0}{2\pi}\ln\dfrac{(2a-d)^2}{d^2}\left[\dfrac{\text{H}}{\text{m}}\right]$. **(b)** $L_{12}=\dfrac{\mu_0}{2\pi}\ln\dfrac{(2a+2b-d)(2a+2b+d)}{(4a+2b-d)(2b+d)}\left[\dfrac{\text{H}}{\text{m}}\right]$.

20. **(a)** 1.271 μH/m. **(b)** 1.371 μH/m. **(c)** 3.39 nH.

21. **(a)** $\Delta W=-2\mu_0 n_1n_2\pi a^2I^2$ [J/m]. **(b)** $\Delta W=-\left(\mu_0n_2^2\pi b^2+2\mu_0n_1n_2\pi a^2\right)I^2/2$ [J/m].
(c) $\Delta W=-\left(\mu_0n_1^2\pi a^2+2\mu_0n_1n_2\pi a^2\right)I^2/2$ [J/m].

22. $\Delta W=(\mu_1-\mu_0)\dfrac{N^2I^2c}{4\pi}\ln\dfrac{r_2}{r_1}$ [J].

23. $\Delta W=\dfrac{\mu_1N^2I^2(r_2-r_1)c}{2}\left[\dfrac{1}{\pi(r_2+r_1)}-\dfrac{\mu_0}{\mu_0\pi(r_2+r_1)+l_g(\mu_1-\mu_0)}\right]$ [J].

24. **(a)** $W_{total}=\dfrac{\mu N_1^2(b-a)c}{4\pi(b+a)}\left(I_1^2+I_2^2+4I_3^2+2I_1I_2+4I_1I_3+4I_2I_3\right)$ [J].

(b) $W_{min}=0$, $L_{eq}=0$. **(c)** $W_{max}=\dfrac{8\mu N_1^2(b-a)I^2c}{\pi(a+b)}$ [J], $L_{eq}=\dfrac{16\mu N_1^2(b-a)c}{\pi(a+b)}$ [H].

25. $w=\mu_0M^2/2$ [J/m^3].

26. $\Delta W=\dfrac{\mu_0n^2I^2\pi b^2}{2}(1-\mu_r)\left[\dfrac{\text{J}}{\text{m}}\right]$.

27. **(a)** $\Delta W=-\dfrac{\mu_0Nb^2I^2}{d}\left[\dfrac{\text{J}}{\text{m}}\right]$. **(b)** $\Delta W=-\dfrac{\mu_0N\pi a^2I^2}{d}\left[\dfrac{\text{J}}{\text{m}}\right]$.

28. **(a)** $\Phi=\mu Ic\ln(d/b)/2\pi$ [Wb]. **(b)** $\Phi=\dfrac{\mu\mu_0Ic(d-b)}{\mu_0\pi(d+b)+l_g(\mu-\mu_0)}$ [Wb].

29. **(a)** 3.725×10^{-8} [Wb]. **(b)** $H_{min}=0.085$ A/m in central leg, $H_{max}=216.25$ A/m in gap (1).

30. $L_{12}=\dfrac{\pi a^2\mu\mu_0N}{(2\pi r_0-l_g)\mu_0+l_g\mu}\approx\dfrac{\pi a^2\mu\mu_0N}{2\pi r_0\mu_0+l_g\mu}$ [H].

31. $H=\dfrac{\mu_1\mu_2[N_1I_1+N_2I_2]}{\mu_0\mu_2[d+a-2b]+\mu_0\mu_1[d+a-2b]+\mu_1\mu_2l_g}\left[\dfrac{\text{A}}{\text{m}}\right]$.

32. $B=\dfrac{2\mu_0\mu_1\mu_2NI}{\mu_0[\pi(r_1+r_2)-4d](\mu_2+\mu_1)+8\mu_1\mu_2d}$ [T].

33. **(a)** $W=\dfrac{(N_1I_1+N_2I_2)^2\mu_0\mu_1\mu_2bc}{2\left[\mu_0\left(d+a-2b-l_g\right)(\mu_1+\mu_2)+2\mu_1\mu_2l_g\right]}$ [J]. **(b)** $I_2=-N_1I_1/N_2$ [A].

34. $W=\dfrac{\mu_0bc(N_1I_1-N_2I_2)^2}{2(e_1+e_2)}$ [J].

35. $L_{12}=0.8377$ mH.

36. $F_m=0.0005$ N (perpendicular to **B**).

37. **(a)** $\mathbf{F}=\hat{\mathbf{r}}\dfrac{\mu_0I_2I_1c}{2\pi}\left(\dfrac{b-a}{ab}\right)$ [N]. **(b)** $\mathbf{F}=-\hat{\mathbf{r}}\dfrac{\mu_0I_2I_1c}{2\pi}\left(\dfrac{b-a}{ab}\right)$ [N]. **(c)** $\mathbf{F}_w=\hat{\mathbf{r}}\dfrac{\mu_0I_1I_2cb}{2\pi a(b-a)}$ [N].

(d) $\mathbf{F}_w=\hat{\mathbf{r}}\dfrac{2\mu_0I_1I_2c}{\pi b}$ [N]. **(e)** $\mathbf{F}_w\to\infty$.

38. **(a)** $\mathbf{F}_1=\hat{\mathbf{y}}\,9.6\times10^{-6}$ [N], $\mathbf{F}_2=\hat{\mathbf{y}}\,3\times10^{-6}$ [N]. **(b)** $\mathbf{F}_{wire}=-\hat{\mathbf{y}}\,12.6\times10^{-6}$ [N].

39. **(c)** 0.9206 N/m (repulsion).

40. **(c)** $F=4.37\times10^{-7}$ N.

41. **(a)** $241.274I$ [N/A]. **(b)** $I=6.375$ A.

42. $F = -\frac{\mu_0\mu^2 cd}{2}\left[\frac{N_1I_1 - N_2I_2}{\mu_0(a+3b-2g)+2\mu g}\right]^2$ [N].

43. **(a)** Zero. **(b)** Zero.

44. **(a)** Zero. **(b)** $\mathbf{T} = \hat{\mathbf{z}}B_0Ia^2/8$ [N · m].

45. **(a)** Zero. **(b)** Zero. **(c)** $\mathbf{T} = \hat{\mathbf{\phi}}\frac{\mu_0\sqrt{2}\pi I_sIa^2b^2}{8h^3}$ [N · m]. **(d)** $\mathbf{T} = -\hat{\mathbf{\phi}}\frac{\mu_0\pi II_sa^2b^2\sqrt{2}}{8R^3}$ [N · m].

46. **(b)** $\mathbf{T} = \hat{\mathbf{y}}\mu_0I_sI_c\pi a^4/2\pi d^3$ [N · m]. **(c)** $\mathbf{T} = -\hat{\mathbf{y}}\mu_0NI_sI_c\pi a^4/4\pi d^3$ [N · m]. **(d)** $\mathbf{T} = -\hat{\mathbf{y}}\mu_0N^2I_sI_c\pi a^4/4\pi d^3$ [N · m].

Chapter 10

1. **(a)** $I = 3.056$ mA. **(b)** $I(t) = \frac{(v_2+v_1)B_0d}{2r[P-(v_2+v_1)t]}$ [A].

2. **(a)** $v = V/B_0d$ [m/s]. **(c)** $D = V/2Ir$ [m].

3. **(a)** $\mathbf{F} = -\hat{\mathbf{x}}\omega d^3B_0^2\cos(\omega t)\sin(\omega t)/R$ [N]. **(b)** Zero.

4. **(a)** $|\text{emf}| = v_0\mu_0Jr^2/2d$ [V]. **(b)** $|\text{emf}| = v_0\mu_0Jr^2/2d$ [V].

5. $V_{ab} = \frac{B_0vd\sqrt{3}}{2}\cos\omega t$ [V].

6. $\text{emf}_{round}/\text{emf}_{square} = \pi$.

7. **(a)** $emf = (\mu_0H_0b^2)\omega_0\sin\omega_0t\cos\omega_1t + (\mu_0H_0b^2)\omega_1\cos\omega_0t\sin\omega_1t$ [V].
(b) $emf = (\mu_0H_0\pi b^2)\omega_0\sin\omega_0t\cos\omega_0t + (\mu_0H_0\pi b^2)\omega_0\cos\omega_0t\sin\omega_0t$ [V].

8. $emf = -2fN\mu_0SdI_0/(d^2+h^2)$ [V (rms)]

9. **(a)** $V = \frac{\mu_0I_0v}{2\pi}\ln\left(\frac{a+b}{a}\right)$ [V]. **(c)** $V = \frac{\mu_0vI_0\sin\omega t}{2\pi}\ln\left(\frac{a+b}{a}\right)$ [V]. **(d)** $V = \frac{\mu_0I_0v_0\sin\omega t}{2\pi}\ln\left(\frac{a+b}{a}\right)$ [V].

10. $\text{emf} = 2\pi fBcd\sin(2\pi ft)$ [V].

11. **(a)** $\text{emf} = \omega B_0l^2/2$ [V]. **(b)** $\text{emf} = \omega B_0(1-(l+1)e^{-l})$ [V].

12. **(a)** $V = -\omega d^2\mu_0 10^6\cos(\omega t)/8$ [V].

13. $\text{emf}_{AB} = -0.1\pi\sin(800\pi t)$ [V].

14. $T = 0.1$ N · m.

15. **(a)** $V = 11$ V, $I = 13.64$ A. **(b)** $I = 1.364$ A. **(c)** $Z_p = 80.7\ \Omega$, $Z_s = 0.807\ \Omega$, $Z_p/Z_s = 100$.

16. **(a)** $emf = 10.19\sin(200\pi t)$ [mV]. **(b)** $emf = 0.1\sin(200\pi t)$ [mV]

17. **(a)** $N_1/N_2 = 0.024$, $I_1 = 40{,}000$ A, $I_2 = 960$ A. **(b)** $N_1/N_2 = 0.024$. **(c)** $I_2 = 950.4$ A, $\Phi = 0.25$ Wb.

18. **(a)** 17.68 turns. **(b)** 5.3 turns.

19. **(a)** $V_1 = 4.11\sin(314t)$ [V], $V_2 = 0.41\sin(314t)$ [V]. **(b)** $V_3 = 49.32\sin(314t)$ [V].

20. **(a)** $\text{emf}_1 = 2{,}000\pi\cos(2\pi\times10^5t)$ [V], $\text{emf}_2 = 200\pi\cos(2\pi\times10^5t)$ [V]. **(b)** $\text{emf}_1 = 1{,}990\pi\cos(2\pi\times10^5t)$ [V], $\text{emf}_2 = 100\pi\cos(2\pi\times10^5t)$ [V].

21. $\text{emf} = -\frac{\mu_0\omega cI_0}{2\pi}\ln\frac{a+b}{a}\cos\omega t$ [V].

22. **(a)** 156 turns. **(b)** 19.23 W.

23. **(a)** 50,661 turns. **(b)** 25 turns.

24. **(a)** $V = 40$ V. **(b)** $R = 0.02\ \Omega$. **(c)** $P = 20$ kW.

Chapter 11

1. $\mathbf{J}_d = \hat{\mathbf{x}}(0.5/\mu)(\cos100t)\sin5z$ [A/m²].

2. $I_d = 4\pi\varepsilon_0ab\omega V_0\cos\omega t/(b-a)$ [A].

3. **(a)** $\mathbf{J}_d = \hat{\mathbf{r}}\omega\varepsilon_r\varepsilon_0V_0\cos\omega t/r\ln(b/a)$ [A/m²]. **(b)** $I_d = 2\pi\varepsilon_r\varepsilon_0\omega LV_0\cos\omega t/\ln(b/a)$ [A].

5. $f = 7.992$ GHz.

6. $|J_{cond}|/|J_{dissip}| = \sigma/\omega\varepsilon$.

7. $t_c = 35.416 \times 10^{-9}$ s.
8. $I = 494.3$ [mA, rms].
9. 0.8564 mA.
10. **(a)** $\beta = 0.01$ rad/m, $H_0 = 1$ A/m. **(b)** $\beta = 0.02$ rad/m, $H_0 = 2$ A/m.
14. 97.6 μA.
16. **(a)** $\mathbf{E} = \hat{\mathbf{y}}1.997 \times 10^9 e^{j(10^4 t + 10^{-4} z)}$ [V/m]. **(b)** $\mathbf{D} = \hat{\mathbf{y}}0.159 e^{j(10^4 t + 10^{-4} z)}$ [C/m^2], $\mathbf{H} = \hat{\mathbf{x}}1.59 \times 10^7 e^{j(10^4 t + 10^{-4} z)}$ [A/m].
17. $\mathbf{H}(t) = \hat{\mathbf{y}}1{,}250\cos(10^6 t - 50z)$ [A/m], $\mathbf{B}(t) = \hat{\mathbf{y}}5 \times 10^{-4}\pi\cos(10^6 t - 50z)$ [T].
18. $\nabla^2 \mathbf{J}_e - \mu\varepsilon \dfrac{\partial^2 \mathbf{J}_e}{\partial t^2} = \mu\sigma \dfrac{\partial \mathbf{J}_e}{\partial t}$.
20. **(a)** $-\nabla^2 \mathbf{A} + \nabla(\nabla \cdot \mathbf{A}) = \mu \mathbf{J}$. **(b)** $\mathbf{E} = -\partial \mathbf{A}/\partial t$. **(c)** $\nabla^2 \mathbf{A} = -\mu \mathbf{J}$.
21. **(a)** $\mathbf{H} = -\nabla\psi$ [A/m]. **(b)** $-\nabla^2 \mathbf{F} + \nabla(\nabla \cdot \mathbf{F}) = -\varepsilon\mu \dfrac{\partial^2 \mathbf{F}}{\partial t^2} - \varepsilon\mu \dfrac{\partial}{\partial t}(\nabla\psi)$. **(c)** $\nabla \cdot \mathbf{F} = -\mu\varepsilon \dfrac{\partial}{\partial t}(\nabla\psi)$.
22. **(b)** $\nabla^2 \mathbf{F} = \mu \mathbf{J} + \mu\varepsilon \dfrac{\partial^2 \mathbf{F}}{\partial t^2}$.
23. $\mathbf{E} = \omega^2 \mu\varepsilon \boldsymbol{\pi} + \nabla(\nabla \cdot \boldsymbol{\pi})$ [V/m], $\mathbf{H} = j\omega\varepsilon\nabla \times \boldsymbol{\pi}$ [A/m].
24. $\nabla^2 \boldsymbol{\pi}_m + \omega^2 \mu\varepsilon \boldsymbol{\pi}_m = 0$, gauge: $\nabla \cdot \boldsymbol{\pi}_m = -\phi$, ϕ is the magnetic scalar potential.
25. $J_{dmax} = 0.885$ A/m^2.
26. **(a)** $|\mathbf{E}_1| = 80$ V/m, $|\mathbf{D}_1| = 1.42$ nC/m^2, $|\mathbf{E}_2| = 61.18$ V/m, $|\mathbf{D}_2| = 1.62$ nC/m^2. **(b)** $|\mathbf{E}_1| = 80$ V/m, $|\mathbf{D}_1| = 1.42$ nC/m^2, $|\mathbf{E}_2| = 79.66$ V/m, $|\mathbf{D}_2| = 2.11$ nC/m^2.
27. $\mathbf{B}_2 = 4\pi \times 10^{-7}(\hat{\mathbf{x}}24 + \hat{\mathbf{y}}5 - \hat{\mathbf{z}}10)$ [T].
28. $H_{1t} = H_{2t} = 0$.
29. **(a)** $\mathbf{J} = -\hat{\mathbf{x}}318.3 + \hat{\mathbf{y}}159.15$ A/m. **(b)** $\mathbf{B}_2 = \hat{\mathbf{z}}0.015$ T.
30. **(a)** $\mathbf{H} = \hat{\mathbf{z}}10^4$ A/m. **(b)** $\mathbf{J}_s = -\hat{\mathbf{x}}2 \times 10^5 + \hat{\mathbf{y}}10^5$ A/m.
31. $\mathbf{A} \cdot \mathbf{A} = a^2 - b^2 + j2\mathbf{a} \cdot \mathbf{b}$, $\mathbf{A} \cdot \mathbf{A}^* = a^2 + b^2$, $\mathbf{A} \cdot \mathbf{B} = (\mathbf{a} \cdot \mathbf{c} - \mathbf{b} \cdot \mathbf{d}) + j(\mathbf{b} \cdot \mathbf{c} + \mathbf{a} \cdot \mathbf{d})$, $\mathbf{A} \cdot \mathbf{B}^* = (\mathbf{a} \cdot \mathbf{c} + \mathbf{b} \cdot \mathbf{d}) + j(\mathbf{b} \cdot \mathbf{c} - \mathbf{a} \cdot \mathbf{d})$, $\mathbf{A}^* \cdot \mathbf{B} = (\mathbf{a} \cdot \mathbf{c} + \mathbf{b} \cdot \mathbf{d}) + j(\mathbf{a} \cdot \mathbf{d} - \mathbf{b} \cdot \mathbf{c})$, $\mathbf{A} \times \mathbf{A} = 0.$, $\mathbf{A} \times \mathbf{A}^* = -j2(\mathbf{a} \times \mathbf{b})$, $\mathbf{A} \times \mathbf{B} = (\mathbf{a} \times \mathbf{c} - \mathbf{b} \times \mathbf{d}) + j(\mathbf{b} \times \mathbf{c} + \mathbf{a} \times \mathbf{d})$, $\mathbf{A} \times \mathbf{B}^* = (\mathbf{a} \times \mathbf{c} + \mathbf{b} \times \mathbf{d}) + j(\mathbf{b} \times \mathbf{c} - \mathbf{a} \times \mathbf{d})$, $\mathbf{A}^* \times \mathbf{B} = (\mathbf{a} \times \mathbf{c} + \mathbf{d} \times \mathbf{b}) + j(\mathbf{a} \times \mathbf{d} - \mathbf{b} \times \mathbf{c})$, $\mathbf{B} \times \mathbf{A} = (\mathbf{b} \times \mathbf{d} - \mathbf{a} \times \mathbf{c}) - j(\mathbf{b} \times \mathbf{c} - \mathbf{a} \times \mathbf{d})$, $\mathbf{B} \times \mathbf{A}^* = (\mathbf{c} \times \mathbf{a} + \mathbf{d} \times \mathbf{b}) + j(\mathbf{d} \times \mathbf{a} - \mathbf{c} \times \mathbf{b})$, $\mathbf{B}^* \times \mathbf{A} = (\mathbf{c} \times \mathbf{a} + \mathbf{d} \times \mathbf{b}) + j(\mathbf{c} \times \mathbf{b} - \mathbf{d} \times \mathbf{a})$.
32. $\mathbf{H}(z,t) = \hat{\mathbf{y}}5\cos(\omega t - \beta z)$ [A/m].
33. $H(x,y,z) = H_0 \sin\dfrac{m\pi x}{a}\cos\dfrac{n\pi y}{b}[\cos(kz) - j\sin(kz)]$ [A/m], $H(x,y,z) = H_0 \sin\dfrac{m\pi x}{a}\cos\dfrac{n\pi y}{b}\angle -kz$ [A/m], $H(x,y,z) = H_0 \sin\dfrac{m\pi x}{a}\cos\dfrac{n\pi y}{b}e^{-jkz}$ [A/m].
34. $E(z) = E_1\angle -kz + \psi + E_2\angle kz + \psi$ [V/m], $E(z) = E_1 e^{j(-kz+\psi)} + E_2 e^{j(kz+\psi)}$ [V/m].
35. $E(x,z,t) = E_0\cos(\omega t - \beta_0(x\sin\theta_i + z\cos\theta_i))$ [V/m].
36. **(a)** $E_x(z) = E_0 e^{-j(kz-\phi)}$. **(b)** $E_x(z) = j\omega E_0 e^{-j(kz-\phi)}$ [V/m].
37. **(a)** $\mathbf{H} = -\hat{\mathbf{x}}1{,}250\cos(10^6 t - 50z) + \hat{\mathbf{y}}1{,}250\cos(10^6 t - 50z)$ [A/m].
(b) $\mathbf{H} = -\hat{\mathbf{x}}1{,}250 e^{-j50z} + \hat{\mathbf{y}}1{,}250 e^{-j50z}$ [A/m].
38. **(a)** $\mathbf{H}(x,y,z,t) = (\hat{\mathbf{x}}h_x + \hat{\mathbf{y}}h_y + \hat{\mathbf{z}}h_z)\cos(\omega t + \beta z + \phi) - (\hat{\mathbf{x}}g_x + \hat{\mathbf{y}}g_y + \hat{\mathbf{z}}g_z)\sin(\omega t + \beta z + \phi)$ [A/m].
(b) $\mathbf{H}(x,y,z) = |\hat{\mathbf{x}}h_x + \hat{\mathbf{y}}h_y + \hat{\mathbf{z}}h_y|\angle \beta z + \phi + |\hat{\mathbf{x}}g_x + \hat{\mathbf{y}}g_y + \hat{\mathbf{z}}g_z|\angle \beta z + \phi + \pi/2$ [A/m].
39. **(a)** $\mathbf{E}_1(t) = \hat{\mathbf{x}}[20\cos(\omega t + 0.3\pi z) - 20\sin(\omega t + 0.3\pi z)] + \hat{\mathbf{y}}[10\cos(\omega t + 0.3\pi z) + 20\sin(\omega t + 0.3\pi z)]$ [V/m], $\mathbf{E}_2(t) = \hat{\mathbf{x}}[-20\cos(\omega t + 0.3\pi z) - 10\sin(\omega t + 0.3\pi z)] + \hat{\mathbf{y}}[20\cos(\omega t + 0.3\pi z) - 20\sin(\omega t + 0.3\pi z)]$ [V/m].
(b) $\mathbf{E}_1 + \mathbf{E}_2 = \hat{\mathbf{x}}j30 e^{j0.3\pi z} + \hat{\mathbf{y}}30 e^{j0.3\pi z}$ [V/m], $\mathbf{E}(t) = \mathbf{E}_1(t) + \mathbf{E}_2(t) = -\hat{\mathbf{x}}30\sin(\omega t + 0.3\pi z) + \hat{\mathbf{y}}30\cos(\omega t + 0.3\pi z)$ [V/m].
(c) $\mathbf{E} = \mathbf{E}_1 - \mathbf{E}_2 = \hat{\mathbf{x}}(40 + j10)e^{j0.3\pi z} - \hat{\mathbf{y}}(10 + j40)e^{j0.3\pi z}$ [V/m], $\mathbf{E}(t) = \mathbf{E}_1(t) - \mathbf{E}_2(t) = \hat{\mathbf{x}}[40\cos(\omega t + 0.3\pi z) - 10\sin(\omega t + 0.3\pi z)] - \hat{\mathbf{y}}[10\cos(\omega t + 0.3\pi z) - 40\sin(\omega t + 0.3\pi z)]$ [V/m].
(d) $\mathbf{E}_1 \times \mathbf{E}_2 = \hat{\mathbf{z}}300 e^{j(0.6\pi z + \pi/2)}$ [V^2/m^2], $\mathbf{E}_1(t) \times \mathbf{E}_2(t) = \hat{\mathbf{z}}300\cos(2\omega t + 0.6\pi z + \pi/2)$ [V^2/m^2].
(e) $\mathbf{E}_1 \cdot \mathbf{E}_2 = 400 e^{j(0.6\pi z - \pi/2)}$ [V^2/m^2], $\mathbf{E}_1(t) \cdot \mathbf{E}_2(t) = 400\cos(2\omega t + 0.6\pi z - \pi/2)$ [V^2/m^2].

Chapter 12

1. **(a)** $\nabla^2\mathbf{E} - \nabla\left(\frac{\rho}{\varepsilon}\right) - \mu\frac{\partial \mathbf{J}}{\partial t} = 0$. **(b)** $\nabla^2\mathbf{E} - \varepsilon\mu\frac{\partial^2\mathbf{E}}{\partial t^2} = 0$.
2. $\nabla^2\mathbf{B} - \mu\varepsilon\frac{\partial^2\mathbf{B}}{\partial t^2} = 0$.
3. $\nabla^2\mathbf{A} + \omega^2\mu\varepsilon\mathbf{A} = 0$.
4. $\nabla^2\mathbf{D} + \omega^2\mu\varepsilon\mathbf{D} = 0$.
5. **(a)** $\mathbf{E} = \omega^2\mu\varepsilon\mathbf{\Pi}_e - \nabla V$. **(b)** $\nabla^2\mathbf{\Pi}_e + \omega^2\mu\varepsilon\mathbf{\Pi}_e = 0$, Gauge: $\nabla\cdot\mathbf{\Pi}_e = -V$.
6. **(a)** $\mathbf{E} = -\hat{\mathbf{z}}120e^{-j1.592y}$ [V/m], $\mathbf{E} = -\hat{\mathbf{z}}120\cos(4.775\times10^8 t - 1.592y)$ [V/m]. **(b)** $\mathbf{H} = -\hat{\mathbf{x}}0.318e^{-j1.592y}$ [A/m], $\mathbf{H} = -\hat{\mathbf{x}}0.318\cos(4.775\times10^8 t - 1.592y)$ [A/m].
7. **(b)** $v_p = 3\times10^8$ m/s, $f = 10.5$ GHz.
8. $\mathbf{E} = \hat{\mathbf{x}}H_1\eta_0 e^{j\beta y} + \hat{\mathbf{z}}H_0\eta_0 e^{j\beta y}$ [V/m].
9. 663.13 W.
10. **(a)** 68 min, 37 s. **(b)** 34 min, 18 s. **(c)** 9.15 cents.
11. $P_{diss.} = I_0^2/2\pi R^2\sigma$ [W/m].
12. **(a)** $\mathbf{H} = \hat{\boldsymbol{\phi}}\frac{12\pi}{\eta R}e^{-j2\pi R}\sin\theta$ [A/m]. **(b)** $\boldsymbol{\mathcal{P}}_{av} = \hat{\mathbf{R}}\frac{72\pi^2}{\eta R^2}\sin^2\theta$ [W/m^2]. **(c)** $P_{rad} = 15.79$ W.
13. **(a)** $E = 1027$ V/m, $H = 2.72$ A/m. **(b)** $E = 275$ V/m, $H = 0.73$ A/m.
14. **(a)** $\boldsymbol{\mathcal{P}}_c = \hat{\mathbf{r}}j\omega\varepsilon_0 rV_0^2/2d^2$ [W/m^2].
15. 2.4×10^{-9} J.
17. **(b)** $\eta(100\text{ Hz}) = 9.934\times10^{-3} + j9.934\times10^{-3}\,\Omega$, $\eta(100\text{ MHz}) = 9.934 + j9.934\,\Omega$, $\eta(10\text{ GHz}) = 74.47 + j10.92\,\Omega$.
18. $\mathbf{E}(x = 1\text{ m}) = -\hat{\mathbf{z}}3.2\times10^{-16}e^{-j41.3}$ [V/m], $\mathbf{H}(x = 1\text{ m}) = \hat{\mathbf{y}}2.3\times10^{-17}e^{-j42.03}$ [A/m].
19. **(a)** $P_{av} = 2{,}097.1$ W. **(b)** $P_{av} = 0.012$ W.
20. **(b)** $\boldsymbol{\mathcal{P}}(t) = \hat{\mathbf{y}}2(E_0^2/\eta)e^{-2\alpha y}\cos^2(\omega t - \beta y)$ [W/m^2]. **(c)** $|\mathbf{H}| = 0.057$ A/m.
21. 0.266 W.
22. **(a)** $P_1 = 1.0\times10^{-3}$ W, $P_2 = 1.026\times10^{-11}$ W. **(b)** $v_{p1} = 2.268\times10^8$ m/s, $\lambda_1 = 604.75$ nm, $\eta_1 = 284.98\,\Omega$, $v_{p2} = 1.896\times10^8$ m/s, $\lambda_2 = 506$ nm, $\eta_2 = 238.4\,\Omega$. **(c)** $\phi_2 - \phi_1 = 2.04\times10^{10}$ rad.
23. **(a)** $\eta = 362.77\,\Omega$, $\beta = 2.177\times10^{-8}f$ [rad/m], $v_p = 2.887\times10^8$ m/s, -3.77%. **(b)** $\eta_0 = 377\,\Omega$, $\beta_0 = 2.094\times10^{-8}f$ [rad/m], $v_{p0} = 3\times10^8$ m/s.
24. $\alpha = 8.04\times10^{-6}$ Np/m.
25. **(a)** $\eta = 2.63\times10^{-3} + j2.63\times10^{-3}\,\Omega$. **(b)** $\eta = 2.63\times10^{-3} + j2.63\times10^{-3}\,\Omega$.
26. **(a)** 1.11 m. **(b)** 1.09 m. **(c)** 1.71 m.
27. **(a)** $z_1/z_2 = 10^{-4}$. **(b)** $v_{p1}/v_{p2} = 10^{-4}$.
28. $\sigma = 3.36\times10^7$ S/m, $\alpha = 230.26$ Np/m.
29. **(a)** 2.198 m. **(b)** 695.295 m.
30. **(a)** $\delta_{cu} = 0.667$ μm. **(b)** $\delta_{hg} = 5.03$ μm.
31. **(a)** In copper: $\eta_{60\text{Hz}} = 2.04\times10^{-6} + j2.04\times10^{-6}\,\Omega$, $\eta_{10\text{GHz}} = 2.63\times10^{-2} + j2.63\times10^{-2}\,\Omega$. In iron: $\eta_{60\text{Hz}} = 1.54\times10^{-4} + j1.54\times10^{-4}\Omega$, $\eta_{10\text{GHz}} = 0.095 + j0.095\,\Omega$.
33. 0.162 m.
34. $\delta = 2.98\times10^{-9}$ m.
35. **(a)** $f < 90$ MHz. **(b)** $f > 140$ GHz.
36. $J = 62$ μA/m^2.
37. **(a)** 790 A. **(b)** 2317 A. **(c)** 12,566 A.
38. **(a)** $d_{cu} = 2.9123\times10^{-4}$ m, $d_{al} = 3.665\times10^{-4}$ m, $d_{mu} = 3.1\times10^{-6}$ m, $d_{po} = 90.676$ m. **(b)** cost: *aluminum*. weight: *mumetal*. volume: *mumetal*.
39. **(b)** $v_p = \frac{1}{\sqrt{\mu\varepsilon}\sqrt{1-\omega_c^2/\omega^2}}\left[\frac{\text{m}}{\text{s}}\right]$, $v_g = \frac{\sqrt{\omega^2-\omega_c^2}}{\omega\sqrt{\mu\varepsilon}}\left[\frac{\text{m}}{\text{s}}\right]$. **(c)** $v_p = \infty$, $v_g = 0$.

40. **(b)** $v_g = \dfrac{\sqrt{\omega^2\mu\varepsilon - \pi^2/a}}{\omega\mu\varepsilon}\left[\dfrac{\text{m}}{\text{s}}\right]$, $v_p \dfrac{\omega}{\sqrt{\omega^2\mu\varepsilon - \pi^2/a}}\left[\dfrac{\text{m}}{\text{s}}\right]$.

42. **(a)** $v_p = 1.498 \times 10^8$ m/s. **(b)** $v_g = 1.503 \times 10^8$ m/s. **(c)** $v_e = 1.4979 \times 10^8$ m/s.

43. **(a)** $v_g = \dfrac{1}{\sqrt{\mu\varepsilon} - \dfrac{\sigma^2\sqrt{\mu/\varepsilon}}{8\varepsilon\omega^2}}\left[\dfrac{\text{m}}{\text{s}}\right]$.

Chapter 13

1. $\mathbf{E}_r(z) = -\widehat{\mathbf{x}}0.552E_ie^{\,j2.089z}$, $\mathbf{E}_t(z) = \widehat{\mathbf{x}}0.448E_ie^{-5.44\times10^{-8}z}e^{-j7.255z}$ [V/m].

2. **(a)** $E_{1max} = 11.716$ V/m. **(b)** $H_{1max} = 0.031$ A/m.

3. $\dfrac{E_{max}}{E_{min}} = \dfrac{1+ \mid (\eta_2 - \eta_0)/(\eta_2 + \eta_0) \mid}{1- \mid (\eta_2 - \eta_0)/(\eta_2 + \eta_0) \mid}$.

4. **(a)** $\mathcal{P}_{converted} = 329.51$ W/m^2, $eff = 29.42\%$. **(b)** $\mathcal{P}_{converted} = 336$ W/m^2, $eff = 30\%$.

5. **(a)** $E_{glass} = 7.414$ V/m, $H_{glass} = 0.0264$ A/m, **(b)** $P_{glass} = 7.686 \times 10^{-10}$ W.

6. 77.25 pW.

7. $\Delta L/L = 5.9$ nm/m.

9. 731.89 W.

10. $P = 84.9$ W, $W = 509.4$ W • h.

11. 4.93 W.

12. **(a)** 3.25λ. **(b)** 192.8 V/m.

13. $E/H = -\,j\eta_0 \tan \beta z$.

14. **(a)** $\mathbf{J}_s = -\widehat{\mathbf{y}}2E_i/\eta_0$ [A/m].

15. **(a)** $E_{1t} = E_{2t}$, $H_{1t} - H_{2t} = J_s$. **(b)** $\eta_2 = (1+j)\dfrac{1}{\delta_2\sigma_2}$ [Ω], $\delta_2 = \dfrac{1}{\sqrt{\pi f\mu_2\sigma_2}}$ [m].

16. **(a)** $\mathbf{H}_i = 0.265(\widehat{\mathbf{y}}\cos\alpha - \widehat{\mathbf{z}}\sin\alpha)e^{-j2{,}094.4(y\sin\alpha+z\cos\alpha)}$ [A/m].
(b) $\mathbf{E}_r = -\widehat{\mathbf{x}}\,100e^{-j2{,}094.4(y\sin\alpha-z\cos\alpha)}$ [V/m], $\mathbf{H}_r = 0.265(\widehat{\mathbf{y}}\cos\alpha + \widehat{\mathbf{z}}\sin\alpha)e^{-j2{,}094.4(y\sin\alpha+z\cos\alpha)}$ [A/m].
(c) $\mathbf{J}_x = \widehat{\mathbf{x}}\,0.530\cos\alpha e^{-j2{,}094.4y\sin\alpha}$ [A/m].

18. **(a)** $\mathbf{E}_i = 37{,}700(-\widehat{\mathbf{y}}\cos\alpha + \widehat{\mathbf{z}}\sin\alpha)e^{-j2{,}094.4(y\sin\alpha+z\cos\alpha)}$ [V/m].
(b) $\mathbf{E}_r = 37{,}700(\widehat{\mathbf{y}}\cos\alpha + \widehat{\mathbf{z}}\sin\alpha)e^{-j2{,}094.4(y\sin\alpha-z\cos\alpha)}$ [V/m], $\mathbf{H}_r = \widehat{\mathbf{x}}100e^{-j2{,}094.4(y\sin\alpha-z\cos\alpha)}$ [A/m].
(c) $\mathbf{J}_s = -\widehat{\mathbf{y}}\,200\;e^{-j2{,}094.4y\sin\alpha}$ [A/m].

19. **(a)** $\mathbf{E}_1(x,y,z) = \widehat{\mathbf{y}}\,j9{,}794.46\sin(173.2z)e^{-j100y}$ [V/m], $\mathbf{H}_1(x,y,z) = \widehat{\mathbf{x}}\,30\cos 173.2ze^{-j100y}$ [A/m]. **(b)** $E_{peak} = \pm 9{,}794.46$ V/m at $z = n\pi/346.4$ m, $n = 1, 3, 5, 7, \ldots$, $H_{peak} = \pm 30$ A/m at $z = n\pi/173.2$, $n = 0, 1, 2, 3, 4, \ldots$.
(c) $\mathcal{P}_{av}(x,z) = \widehat{\mathbf{y}}\,84{,}825\cos^2(173.2z)$ [W/m^2].

20. **(a)** $v_{px} = c/\sin\theta_i$. **(b)** $v_{px} = 3 \times 10^8$ m/s.

21. **(a)** $\mathbf{J} = \widehat{\mathbf{y}}0.0459$ A/m. **(b)** $\mathbf{J} = \widehat{\mathbf{x}}0.053$ A/m.

23. **(a)** $\mathcal{P}_{av} = \widehat{\mathbf{y}}\dfrac{E_{i1}^2\cos\theta_i}{2\eta_0}\left(1 + \Gamma_\perp^2 + \Gamma_\perp 2\cos(2\beta_0 z\cos\theta_i)\right) + \widehat{\mathbf{z}}\dfrac{E_{i1}^2\sin\theta_i}{2\eta_0}\left(1 - \Gamma_\perp^2\right)\left[\dfrac{\text{W}}{\text{m}^2}\right]$.
(b) $\mathcal{P}_{av} = \dfrac{E_{i1}^2T_\perp^2}{2\eta_2}(\widehat{\mathbf{z}}\cos\theta_t + \widehat{\mathbf{y}}\sin\theta_t)\left[\dfrac{\text{W}}{\text{m}^2}\right]$.

24. **(a)** $\theta_i = 63°26'$, $\theta_t = 46°55'$. **(c)** $\Gamma_{||} = 0.11$, $T_{||} = 0.726$. **(d)** $E_{tx} = 5.93$, $E_{ty} = 5.55$ V/m, $E_{rx} = -1.1$, $E_{ry} = 0.55$ V/m.

25. **(a)** $\phi_{T_\perp} = \tan^{-1}\dfrac{\eta_0\cos\theta_t}{2\sqrt{\dfrac{\omega\mu_2}{2\sigma_2}}\cos\theta_i + \eta_0\cos\theta_t}$. **(b)** $\phi_{\Gamma_\perp} = \tan^{-1}\dfrac{2\eta_0\sqrt{\dfrac{\omega\mu_2}{2\sigma_2}}\cos\theta_i\cos\theta_t}{\dfrac{\omega\mu_2}{\sigma_2}\cos^2\theta_i - \eta_0^2\cos^2\theta_t}$.

27. $\Gamma_\perp = \dfrac{\cos\theta - \sqrt{2.1 - \sin^2\theta}}{\cos\theta + \sqrt{2.1 - \sin^2\theta}}$, $\Gamma_{||} = \dfrac{\sqrt{2.1 - \sin^2\theta} - 2.1\cos\theta}{\sqrt{2.1 - \sin^2\theta} + 2.1\cos\theta}$.

28. **(a)** $\Gamma_{||} = 1$, $T_{||} = 0$, $\Gamma_\perp = -1$, $T_\perp = 0$. **(b)** $\Gamma_{||} = \Gamma_\perp = (\eta_2 - \eta_1)/(\eta_2 + \eta_1)$, $T_{||} = T_\perp = 2\eta_2/(\eta_2 + \eta_1)$.

29. $d = (d_2/2d_1)\sqrt{0.0016 + 3d_1^2}$ [m].

30. **(a)** $\theta_b = 78°28'$. **(b)** $\theta_b = 55°23'$. **(c)** $\theta_b = 55°23'$.

31. $\varepsilon_2 = 3.537\varepsilon_0$.

32. **(a)** $\theta_c = 11°47'$. **(b)** $\theta_c = 41°25'$. **(c)** $\theta_c = 43°38'$.

33. $\varepsilon_r = 2.894$.

34. **(a)** $\Gamma_{||} = -0.12388$, $T_{||} = 2.2478$. **(b)** $\theta_c = 30°$.

35. **(b)** $\theta_c = 76°33'$. **(c)** $\theta_c = 47°20'$. **(d)** $\theta_c = 47°20'$.

37. **(a)** $\eta_1 = \eta_3 = 377\ \Omega$, $\eta_2 = 188.5\ \Omega$.

(b) $$E_1 = E_{i0}\left(e^{-j20\pi z/3} + 3\frac{-1 + e^{-j80\pi d/3}}{9 - e^{-j80\pi d/3}}e^{j20\pi z/3}\right)\left[\frac{\text{V}}{\text{m}}\right],\quad H_1 = \frac{E_1}{377}\left[\frac{\text{A}}{\text{m}}\right],\quad z < 0,$$
$$E_2 = E_{i0}\left(\frac{6e^{-j40\pi z/3} + 2e^{-j80\pi d/3}e^{j40\pi z/3}}{9 - e^{-j80\pi d/3}}\right)\left[\frac{\text{V}}{\text{m}}\right], H_2 = \frac{E_2}{188.5}\left[\frac{\text{A}}{\text{m}}\right],\quad 0 < z < d,$$
$$E_3 = E_{i0}\left(\frac{8e^{-j20\pi d/3}}{9 - e^{-j80\pi d/3}}e^{-j20\pi z/3}\right)\left[\frac{\text{V}}{\text{m}}\right], H_3 = \frac{E_3}{377}\left[\frac{\text{A}}{\text{m}}\right],\quad z > d$$

(c) $E_1(z=0) = 0.8581 - j0.255$ [V/m], $H_1(z=0) = 3.03 \times 10^{-3} + j6.76 \times 10^{-4}$ [A/m],
$E_2(z = 0.005m) = 0.866 - j0.368$ [V/m],
$H_2(z = 0.005m) = 2.68 \times 10^{-3} - j2.85 \times 10^{-4}$ [A/m],
$E_3(z = 0.01m) = 0.914 - j0.281$ [V/m], $H_3(z = 0.01m) = 2.42 \times 10^{-3} - j7.46 \times 10^{-4}$ [A/m].

38. **(a)** $d = 0.075n$ [m], $n = 0, 1, 2 \ldots$.

39. $d = 7.5$ mm.

40. $d = 31.4$ mm.

41. **(a)** $\eta_1 = \eta_3 = 377\ \Omega$, $\eta_2 = 188.37 - j0.42\ \Omega$.

(c) $E_1(z=0) = 0.864 - j0.254$ [V/m], $H_1(z=0) = 3.0 \times 10^{-3} + j6.72 \times 10^{-4}$ [A/m],
$E_2(z = 0.005m) = 0.679 - j0.197$ [V/m],
$H_2(z = 0.005m) = 3.61 \times 10^{-3} - j1.104 \times 10^{-3}$ [A/m],
$E_3(z = 0.01m) = 0.887 - j0.32$ [V/m], $H_3(z = 0.01m) = 2.35 \times 10^{-3} - j8.5 \times 10^{-4}$ [A/m].

42. **(a)** $\mathbf{E}_1 = \hat{\mathbf{x}}[1e^{-j\pi z} + (-0.46693 - j0.12261)e^{j\pi z}]$ [V/m], $z < 0$, $\mathbf{H}_1 = \hat{\mathbf{y}}[1e^{-j\pi z} + (0.46693 + j0.12261)e^{j\pi z}]/377$ [A/m], $z < 0$, $\mathbf{E}_2 = \hat{\mathbf{x}}\left[(0.75552 - j0.020435)e^{-j1.5\pi z} - (0.22246 + j0.10218)e^{j1.5\pi z}\right]$ [V/m], $0 < z < 0.1$ m,
$\mathbf{H}_2 = \hat{\mathbf{y}}[(0.75552 - j0.020435)e^{-j1.5\pi z} + (0.22246 + j0.10218)e^{j1.5\pi z}]/251.34$ [A/m], $0 < z < 0.1$ m,
$\mathbf{E}_3 = \hat{\mathbf{x}}[(0.65441 + j0.055284)e^{-j2\pi z} - (0.16564 + j0.14313)e^{j2\pi z}]$ [V/m], 0.1 m $< z < 0.2$ m,
$\mathbf{H}_3 = \hat{\mathbf{y}}[(0.65441 + j0.055284)e^{-j2\pi z} + (0.16564 + j0.14313)e^{j2\pi z}]/188.5$ [A/m], 0.1 m $< z < 0.2$ m,
$\mathbf{E}_4 = \hat{\mathbf{x}}[(0.74918 + j0.45302)e^{-j\pi z}]$ [V/m], $z > 0.2$ m, $\mathbf{H}_4 = \hat{\mathbf{y}}[(0.74918 - j0.45302)e^{-j\pi z}]/377$ [A/m], $z > 0.2$ m.
(b) $\Gamma_{slab} = -0.46693 - j0.12261$, $T_{slab} = 0.74918 - j0.45302$.

43. **(a)** $\Gamma_{slab} = -0.9226 + j0.3858$. **(b)** $d = 0.0745n$ [m], $n = 0, 1, 2, \ldots$.

Chapter 14

1. $C_1 = 80.26$ pF/m, $L_1 = 0.1386$ μH/m, $C_2 = 722.34$ pF/m, $L_2 = 0.1386$ μH/m.

2. **(b)** $C = 5.6$ pF/m, $L = 1.98$ μH/m. **(c)** $L = 0.2634$ μH/m, $C = 42.24$ pF/m.

3. $C = 66.67$ pF/m, $L = 0.375$ μH/m.

4. **(a)** $Z_0 = 415.9\ \Omega$, $v_p = 1 \times 10^7$m/s. **(b)** 9.97 ns.

5. **(a)** $v_p = c$ [m/s]. **(b)** $Z_0 = 3.77\ \Omega$.

6. **(a)** 14 mm (or 1.05 mm). **(b)** $Z_0 \approx 21.4\ \Omega$ (or $Z_0 \approx 176.65\ \Omega$), $\alpha = 0.00376$ Np/m. **(c)** 307.8 m.

7. **(a)** $d = 0.0117$ m. **(b)** $Z_0 \approx 44.1\ \Omega$, $\alpha = 3.77 \times 10^{-3}$ Np/m. **(c)** $d = 2{,}289$ m.
8. **(b)** $Z_0 = 8.84 + j8.83\ \Omega$, $\alpha = 4.863 \times 10^{-5}$ Np/m, $\beta = 4.864 \times 10^{-5}$ rad/m.
9. **(a)** $Z_L = Z_0 = 635.68 - j19.9\ \Omega$. **(b)** $P = 0.0104$ W.
10. **(a)** $C = 16.67$ pF/m, $L = 1.499\ \mu$H/m, $R = 0.0628\ \Omega$/m. **(b)** $\gamma = 1.047 \times 10^{-4} + j3.14 \times 10^{-3}$.
11. **(a)** $\mathbf{E} = \hat{\mathbf{r}}(24.15/r)e^{-j\beta z}$ [A/m], $\mathbf{H} = \hat{\boldsymbol{\phi}}(0.127/r)e^{-j\beta z}$ [A/m]. **(b)** $P = 24$ W.
12. **(a)** $P = 282$ kW. **(b)** $P = 902.5$ kW. **(c)** $H_{max} = 0.796$ A/m, $E_{max} = 300$ V/m.
13. **(a)** $V^+ = 11.197\ \angle 13^\circ 51'$ V, $V^- = 15.382\ \angle 69^\circ 49'$ V. **(b)** $I^+ = 0.112\ \angle 13^\circ 51'$ A, $I^- = 0.154\ \angle 69^\circ 49'$. **(c)** $Z_L = 90\ \angle 108^\circ\ \Omega$.
14. **(a)** $V_i = 50$ V (rms), $I_i = 0.2$ A (rms). **(b)** 12 W.
15. **(a)** $P_{(1)} = 2.88$ W. $P_{(2)} = 0.72$ W. $P_{(3)} = 0.72$ W. **(b)** 4.32 W. **(c)** 3.24 W.
16. **(a)** $d = n\lambda/2$, $n = 0, 1, 2, \ldots$. **(b)** Same as **(a)**.
17. $d = \lambda/4$.
18. $$\Gamma_a = \frac{Z_2\left(1 + \Gamma_L e^{-j2\beta_2 d}\right) - Z_1\left(1 - \Gamma_L e^{-j2\beta_2 d}\right)}{Z_2\left(1 + \Gamma_L e^{-j2\beta_2 d}\right) + Z_1\left(1 - \Gamma_L e^{-j2\beta_2 d}\right)} e^{-j2\beta_1 a}.$$
19. $1.855 + j2.986$ [Ω].
20. $45.23 + j14.69$ [Ω].
21. $Z_{in} = 50 + j153.9\ \Omega$.
22. **(a)** $8.11 - j5.25$ V. **(b)** 0.474 W.
23. $Z_{L1} = 200\ \Omega$, $Z_{L2} = 12.5\ \Omega$.
24. **(a)** Zero. **(b)** $I = 0.24$ A (rms). **(c)** zero in **Fig. 14.45a**, 2.88 W in **Fig. 14.45b**, dissipated on the internal resistance. **(d)** 0.24 A (rms) in **Fig. 14.45a**, zero in **Fig. 24.45b**. **(e)** $0.12 + j0.12$ A (rms) in **Fig. 14.45b**, $0.12 - j0.12$ A (rms) in **Fig. 14.45a**.
25. **(a)** $d = 0.1762 + 0.5n$, $n = 0, 1, 2, \ldots$ [m]. **(b)** $d = 0.8976 + 0.5n$, $n = 0, 1, 2, \ldots$ [m]. **(c)** $d = 0.8238 + 0.5n$, for **(a)** and $d = 0.1024 + 0.5n$ for **(b)**, $n = 0, 1, 2, \ldots$ [m].
26. **(a)** $Z_0 = 100\ \Omega$, $\alpha = 1.02 \times 10^{-4}$ Np/m, $\beta = 5.88 \times 10^{-5}$ rad/m. **(b)** 0.435 W.
27. **(a)** 83.69 Ω. **(b)** $Z_{min} = 31.25\ \Omega$, $Z_{max} = 80\ \Omega$. Minima at: $z = 0, \lambda/2, \lambda, 1.5\lambda, \ldots$, Maxima at: $z = \lambda/4, 3\lambda/4, 5\lambda/4, \ldots$.
30. **(a)** $Z_L = 10\ \Omega$. **(b)** $V^+ = 300$ V, $V^- = -200$ V. **(c)** $I^+ = 6$ A, $I^- = 4$ A.
31. **(a)** SWR = 4.4. **(b)** $V_{max} = 100$ V, $V_{min} = 22.73$ V. **(c)** V_{max} at $z = n\lambda/2$, $n = 0, 1, 2, \ldots$, V_{min} at $z = (2n + 1)\lambda/4$, $n = 0, 1, 2, \ldots$
32. 0.23 W.
33. **(a)** 240.136 W. **(b)** $(0.45 + n0.9)$ m, $n = 0, 1, 2, 3\ldots$.
34. **(a)** 2.0. **(b)** $V_{min} = 100$ V, $V_{max} = 200$ V. **(c)** V_{min} at $z = n\lambda/2$, $n = 0, 1, 2, \ldots$, V_{max} at $z = (2n + 1)\lambda/4$, $n = 0, 1, 2, \ldots$
35. **(a)** $Z(z) = j50\dfrac{-1 + \tan(20\pi z)}{1 + \tan(20\pi z)}$ [Ω].
36. **(a)** $Z(z) = j50\dfrac{1 + \tan(20\pi z)}{1 - \tan(20\pi z)}$ [Ω].
37. **(a)** $\Gamma(z) = 0.2997e^{-j(0.042z - 1.2664)}$. **(b)** 1.856. **(c)** $z = 105.45$ m. **(d)** $\text{SWR} = \dfrac{2.5\omega^2 + 10^{15} + \left|2.5\omega^2 + j5 \times 10^7\omega\right|}{2.5\omega^2 + 10^{15} - \left|2.5\omega^2 + j5 \times 10^7\omega\right|}$.
38. **(a)** $\Gamma(z) = 0.3033e^{-j(0.042z + 1.2626)}$. **(b)** 1.87. **(c)** $z = 120.07$ m. **(d)** $\text{SWR} = \dfrac{2.5\omega^2 + 10^{13} + \left|10^{13} - j5 \times 10^6\omega\right|}{2.5\omega^2 + 10^{13} - \left|10^{13} - j5 \times 10^6\omega\right|}$.
39. 160 W.
40. **(a)** $V_L = 9.98\ \angle 111.55^\circ$ V. **(b)** 3 mW.
41. 4%.
42. 82.95 W.
43. 96.78 W.
44. $d = 0.1798$ wavelengths.
45. **(b)** $f = 100, 200, 300, 400$ MHz.
46. **(a)** $f = 200$ MHz. **(b)** $f = 113.6, 129.2, 153.4$ MHz.
47. $d = v_{p0}/4(f_2 - f_1)$ [m].
48. **(b)** $d_{max} = v_{p0}(2f_1 - f_2)/4f_1 f_2$ [m].

Chapter 15

1. **(a)** $\Gamma = 0.598\angle 21.6^\circ$. **(b)** $\Gamma = 0.598\angle -98.2^\circ$. **(c)** 3.98. **(d)** $42 - j75\ \Omega$. **(e)** V_{max} at 0.03λ, V_{min} at 0.28λ.
2. **(a)** $V_{max} = 80.5$ V. **(b)** $V_{min} = 30.7$ V. **(c)** Minima: $0.088\lambda + n(0.5\lambda)$, $n = 0, 1, 2, \ldots$, maxima: $0.338\lambda + n(0.5\lambda)$, $n = 0, 1, 2, \ldots$
3. $Z = 48 - j44\ \Omega$.
4. **(a)** $Z_{in} = 210 - j648\ \Omega$. **(b)** SWR $= 9$. **(c)** $P_{in} = 17.5$ W.
5. **(a)** $Z_L = 75 + j75\ \Omega$. **(b)** Minima: $(0.307 + 0.5n)\lambda$, $n = 0, 1, 2, 3....18$. Maxima: $(0.057 + 0.5n)\lambda$, $n = 0, 1, 2, 3....19$. **(c)** $\Gamma_L = 0.412{+}j0.353$, SWR $= 3.372$. **(d)** $|V_{max}| = 25.612$ V, $|V_{min}| = 7.597$ V, $|I_{max}| = 0.512$ A, $|I_{min}| = 0.152$ A.
6. **(a)** $l = 0.058\lambda$, $R = 125\ \Omega$.
7. **(a)** $Z_L = 97.5 - j78\ \Omega$. **(b)** $\Gamma = 0.431\angle -50^\circ$.
8. **(a)** $Z_L = 297 + j355\ \Omega$. **(b)** SWR $= 2.8$.
9. **(a)** $\text{SWR} = \dfrac{\sqrt{4Z_0^2 + (X_L + Z_1\tan(\beta_1 d))^2} + |X_L + Z_1\tan(\beta_1 d)|}{\sqrt{4Z_0^2 + (X_L + Z_1\tan(\beta_1 d))^2} - |X_L + Z_1\tan(\beta_1 d)|}$. **(b)** $d = (\lambda/2\pi)\tan^{-1}(-X_L/Z_1)$.
11. **(a)** $(d_{1a} = 0.462\lambda,\ d_{2a} = 0.102\lambda)$ or $(d_{1b} = 0.12\lambda,\ d_{2b} = 0.397\lambda)$. **(b)** $L = 21.167$ nH for $d_{2b} = 0.397\lambda$, $C = 4.787$ pF for $d_{2a} = 0.102\lambda$.
12. Position: 0.224λ (1.12 m), length: 0.153λ (0.765 m) or: position 0.422λ (2.11 m), length: 0.347λ (1.735 m).
13. $d_{1a} = 0.029\lambda$, $d_{2a} = 0.105\lambda$ or: $d_{1b} = 0.188\lambda$, $d_{2b} = 0.396\lambda$.
14. **(a)** $\Gamma = 1/3\angle -180^\circ$. **(b)** 2.0. **(c)** location: 0.403λ (0.959 m), length: 0.153λ (0.364 m) or: location: 0.098λ (0.233 m), length: 0.347λ (0.826 m). **(d)** $\Gamma_L = 0.22\angle 58^\circ$ or $\Gamma_L = 0.14\angle 110^\circ$.
15. $(d_a = 0.388\lambda,\ l_a = 0.19\lambda)$, $(d_b = 0.11\lambda,\ l_b = 0.313\lambda)$.
16. $d_{21} = 0.375\lambda$, $d_{11} = 0.177\lambda$ or: $d_{22} = 0.125\lambda$, $d_{12} = 0.094\lambda$.
17. $d_{11} = 0.199\lambda$, $d_{21} = 0.081\lambda$ or: $d_{12} = 0.346\lambda$, $d_{22} = 0.384\lambda$.
18. $Z_t = \sqrt{Z_1 Z_2}$ [Ω].
19. $d = \lambda/4$, $Z_c = Z_0\sqrt{(1 - |\Gamma_L|)/(1 + |\Gamma_L|)}$ [Ω], if $Z_L < Z_0$, $Z_c = Z_0\sqrt{(1 + |\Gamma_L|)/(1 - |\Gamma_L|)}$ [Ω] if $Z_L > Z_0$.
20. $d = 0.1611\lambda$, $Z_t = 114\ \Omega$ or: $d = 0.4111\lambda$, $Z_t = 43.87\ \Omega$.
21. **(a)** $d_{min} = 2.75\lambda$. **(b)** $Z_t = 367.42\ \Omega$. **(c)** 1.5 and 1.0.
22. **(a)** $d = 0.12\lambda$, $Z_t = 40.67\ \Omega$. **(b)** $d = 0.068\lambda$, $Z_t = 116.67\ \Omega$.
23. **(a)** $d = 0.382\lambda$, $Z_t = 26.35\ \Omega$. **(b)** SWR $= 5.62$.

Chapter 16

3. **(a)** Zero. **(b)** > 3 μs.
6. **(a)** 9.26 V. **(b)** 10 V.
8. **(a)** 33.33 V. **(b)** 0.667 A. **(c)** 0.1 ms. **(d)** 0.2 ms.
9. **(a)** $V(10\mu s) = -22.32$ mV, $I(10\mu s) = 178.6\ \mu$A.
10. $V(10.5\ \mu s) = -19.37$ mV, $I(10.5\ \mu s) = 155\ \mu$A.
11. **(a)** $V_g = 7.1424$ V, $V_L = 6.912$ V, $V_{a-} = 7.1424$ V, $V_{a+} = 7.1424$ V, $V_{b-} = 7.1885$ V, $V_{b+} = 7.1885$ V. **(b)** $V_g = 7.11$ V, $V_L = 6.912$ V, $V_{a-} = 7.373$ V, $V_{a+} = 7.373$ V, $V_{b-} = 7.465$ V, $V_{b+} = 7.465$ V.
12. $V_g = 5.952$ V, $V_L = 0$ V, $V_{a-} = 5.952$ V, $V_{a+} = 5.952$ V, $V_{b-} = 1.3824$ V, $V_{b+} = 1.659$ V. $I_g = 0.08064$ A, $I_L = 0.1536$ A, $I_{a-} = 0.08064$ A, $I_{a+} = 0.05376$ A, $I_{b-} = 0.11008$ A, $I_{b+} = 0.16589$ A.
13. **(a)** $v_L(t) = 100\left(1 - e^{-(t-10^{-6})/50\times 10^{-6}}\right)$ [V], $t \geq \Delta t$, $i_L(t) = 2e^{-(t-10^{-6})/50\times 10^{-6}}$ [A], $t \geq \Delta t$.
14. **(a)** $v_L(t) = 100e^{-(t-10^{-4})50/1\times 10^{-6}}$ [V], $t \geq \Delta t$, $i_L(t) = 2\left(1 - e^{-(t-10^{-4})50/1\times 10^{-6}}\right)$ [A], $t \geq 10^{-4}$ s. **(b)** $V(t) = 100e^{-(t-10^{-4})50/1\times 10^{-6}}$ [V], $I(t) = 2\left(1 - e^{-(t-10^{-4})50/1\times 10^{-6}}\right)$ [A].

17. **(a)** 3,000 m.
18. **(a)** 1,500 m. **(b)** $Z = 16.7\ \Omega$.
19. **(a)** 3,000 m. **(b)** $Z = 150\ \Omega$.
20. **(a)** 1,500 m.

Chapter 17

1. **(a)** $v_g = 5.43 \times 10^8$ m/s, $\lambda_g = 0.003$ m, $Z_{TM} = 208.4\ \Omega$.
(b) $\mathbf{E}(x,z) = \hat{\mathbf{x}}1.1134\cos(1{,}000\pi x)e^{-j2{,}094.4z} + \hat{\mathbf{z}}j1.67\sin(1{,}000\pi x)e^{-j2{,}094.4z}$ [V/m].
(c) $\mathbf{H}(x,z) = \hat{\mathbf{y}}0.0053\cos(1{,}000\pi x)e^{-j2{,}094.4z}$ [A/m].
(d) $\mathcal{P} = \hat{\mathbf{z}}5.9\cos^2(1{,}000\pi x)e^{-j4{,}148.8z} - \hat{\mathbf{x}}j8.85\sin(1{,}000\pi x)\cos(1{,}000\pi x)e^{-j4{,}148.8z}$ [mW/m²].
2. **(a)** 160 GHz. **(b)** 160 GHz. **(c)** $Z_{TE} = 232.7\ \Omega$, $Z_{TM} = 174.5\ \Omega$, $Z_{TEM} = 201.5\ \Omega$.
3. **(a)** 2.71 nW. **(b)** 2.71 nW. **(c)** $\mathcal{P}_{av_{TEM}} = 8.488$ nW.
4. 3.75 GHz.
5. **(a)** $\Gamma = -0.268$, $T = 0.732$. **(b)** SWR = 1.732 (left), SWR = 1 (right).
6. $\mathbf{E} = -\hat{\mathbf{x}}j6{,}282.9\cos(3{,}141.59x)\sin(1{,}439.78z) - \hat{\mathbf{z}}j13{,}709.2\sin(3{,}141.59x)\sin(1{,}439.78)$ [V/m],
$\mathbf{H} = \hat{\mathbf{y}}40\cos(3{,}141.59x)\cos(1{,}439.78z)$ [A/m].
7. **(a)** TM_1. **(b)** 609.45 nm.
8. **(a)** $\mathbf{E}(z) = -\hat{\mathbf{y}}1{,}013e^{-j4.714\omega z}$ [V/m], $\mathbf{H}(z) = -\hat{\mathbf{x}}3.8e^{-j4.714\times 10^{-9}\omega z}$ [A/m]. **(b)** 77 mW.
9. **(a)** $\mathbf{E}(y,z) = \hat{\mathbf{x}}j240\sin(1{,}570.79y)e^{-j1{,}756.06z}$ [V/m],
$\mathbf{H}(y,z) = -\hat{\mathbf{y}}j0.671\sin(1{,}570.79y)e^{-j1{,}756.06z} - \hat{\mathbf{z}}0.6\cos(1{,}570.79y)e^{-j1{,}756.06z}$ [A/m]. **(b)** 1.6 mW.
10. **(a)** $\mathbf{E}(y,z) = \hat{\mathbf{y}}993.53\cos(1{,}570.79y)e^{-j1{,}756.06z} - \hat{\mathbf{z}}j888.71\sin(1{,}570.79y)e^{-j1{,}756.06z}$ [V/m],
$\mathbf{H}(y,z) = \hat{\mathbf{x}}5\cos(1{,}570.79y)e^{-j1{,}756.06z}$ [A/m]. **(b)** 50.5 mW.
11. **(a)** 41.48°. **(b)** 13.4 GHz.
12. **(a)** $a = 2.778$ m, $b = 1.389$ m. **(b)** 9.
13. **(a)** 2.0793 GHz. **(b)** 1.1865 GHz. **(c)** 465.9 MHz.
14. TE_{10}(5.2633 GHz), TE_{01}(11.882 GHz), $TE_{11} = TM_{11}$(12.996 GHz), TE_{20}(10.527 GHz), $TE_{21} = TM_{21}$(11.587 GHz).
15. **(a)** $E = 1.756 \times 10^6$ V/m, $H = 3{,}053.81$ A/m.
16. **(a)** > 30 MHz. **(b)** 30 MHz.
17. **(a)** $f_{c0} = c\sqrt{\left(\frac{1}{2a}\right)^2 + \left(\frac{1}{2b}\right)^2}$ [Hz]. **(b)** $t = d\left[\sqrt{\mu_0\varepsilon_0}\sqrt{1-\left(\frac{f_{c0}}{f}\right)^2} + \sqrt{\mu_1\varepsilon_1}\sqrt{1-\left(\frac{f_{c1}}{f}\right)^2}\right]$ [s].
18. **(a)** $E_y(x,y,z) = -j14.17\sin(66.07x)e^{-j67.2z}$ [kV/m], $E_z = 0$. **(b)** $H_x(x,y,z) = j26.8\sin(66.07x)e^{-j67.2z}$ [A/m],
$H_z(x,y,z) = 26.8\cos(66.07x)e^{-j67.2z}$ [A/m].
21. **(a)** TE_{10}, 4.304 GHz for both waveguides. **(b)** $P = 219$ mW for EIA WR 137, 109.5 mW for ½ EIA WR 137.
22. 629.4 kW.
23. **(b)** 5.3×10^{13} Hz (infared). **(c)** 1.125 mW.
27. **(a)** $\alpha_{dTE_{10}} = 0.0179$ Np/m. **(b)** 3.2%.
28. **(a)** 2.08 mW/m. **(b)** $\alpha_w = 0.0068$ Np/m.
29. **(a)** $\alpha = 0.0336$ Np/m. **(b)** 3.93 mW/m.
30. **(a)** 4.5 GHz. **(b)** TE_{011}(5.59 GHz), TE_{201}(7.9 GHz), TE_{103}(8.38 GHz).
31. TE_{101}(2.1 GHz), TE_{102}(2.164 GHz), TE_{103}(2.266 GHz), TE_{104}(2.4 GHz), TE_{105}(2.564 GHz), TE_{106}(2.75 GHz), TE_{107}(2.955 GHz), TE_{108}(3.175 GHz), TE_{109}(3.408 GHz), $TE_{1,0,10}$(3.65 GHz).
32. **(a)** $d = 0.0524$ m. **(b)** TE_{101}(4.2598 GHz).
33. **(a)** −59.875 kHz/%RH. **(b)** 0.0167 %RH.

Chapter 18

2. **(a)** $\mathbf{E} \approx \hat{\mathbf{R}}(1.53/jR^3)\cos\theta + \hat{\boldsymbol{\theta}}(0.763/jR^3)\sin\theta$ [V/m], $\mathbf{H} \approx \hat{\boldsymbol{\phi}}(6.37 \times 10^{-3}/R^2)\ \sin\theta$ [A/m].
(b) $\mathbf{E} = \hat{\boldsymbol{\theta}}(\ j7.54/R)\sin\theta e^{-j\pi R}$ [V/m], $\mathbf{H} = \hat{\boldsymbol{\phi}}(\ j0.02/R)\sin\theta e^{-j\pi R}$ [A/m]. **(c)** $R_{rad} = 0.316\ \Omega$, $P_{rad} = 0.632$ W.
(d) 27.5 km.

3. **(a)** $\mathcal{P}_{av} = \hat{\mathbf{R}}(V_0^2/2\eta_0 R^2)\sin^2\theta$ [W/m^2]. **(b)** $P_{rad} = 4\pi V_0^2/3\eta_0$ [W].

4. **(a)** $P_{rad} = 5.12$ mW. **(b)** $P_{rad} = 44.92$ mW.

5. 0.987 W.

6. **(a)** 0.0318λ.

7. **(a)** $P_{rad} = 25\omega^2\mu^2\beta^2\pi a^4 I_0^2/3\eta$ [W]. **(b)** $R_{rad} = 50\omega^2\mu^2\beta^2\pi a^4/3\eta$ [Ω].

8. **(a)** $\mathbf{E} = \hat{\boldsymbol{\theta}}(j\beta\eta I_0 L/8\pi R)\sin\theta e^{-j\beta R}$ [V/m], $\mathbf{H} = \hat{\boldsymbol{\phi}}(j\beta I_0 L/8\pi R)\sin\theta e^{-j\beta R}$ [A/m], $\mathcal{P}_{av} = \hat{\mathbf{R}}(\beta^2\eta I_0^2 L^2/128\pi^2 R^2)\sin^2\theta$ [W/m^2].
(b) $R_{rad} = 20\pi^2(L/\lambda)^2$ [Ω].

9. **(a)** $\mathcal{P}_{av-max} = 2.5$ μW/m^2. **(b)** $P_{rad} = 401.36$ W. **(c)** $R_{rad} = 200.68\ \Omega$. **(d)** $d = 1.957$.

10. **(a)** 14.3%. **(b)** 97.9%.

11. **(a)** $R_{rad} = 73.08\ \Omega$, $eff = 99.43\%$. **(b)** $eff = 99.86\%$.

12. **(b)** $eff = 96.8\%$.

13. **(a)** $P_{rad} = 20.773$ W. **(b)** $P_{rad} = 21.734$ W.

14. **(a)** $\mathcal{P}_{av} = 4.659 \times 10^{-8}$ W/m^2. **(b)** $G = 1.618$.

15. $|\mathbf{E}| = 0.525$ V/m, $|\mathbf{H}| = 1.393 \times 10^{-3}$ A/m.

16. **(a)** SWR = 1.4615. **(b)** SWR = 1.4857. **(c)** SWR = 159.

17. **(a)** $I(z') = I_0\cos(\pi|z'|/L)$ [A]. **(b)** $I(z') = I_0\sin(2\pi|z'|/L)$ [A]. **(c)** $I(z') = -I_0\cos(3\pi|z'|/L)$ [A].
(d) $I(z') = I_0\cos(\pi/4 - 3\pi|z'|/2L)$ [A]. **(e)** $I(z') = -I_0\sin(4\pi|z'|/L)$ [A]. **(f)** $I(z') = -I_0\sin(\pi/4 - 5\pi|z'|/2L)$ [A].

18. **(a)** $|f_e(\theta)| = \left|\left(\cos((3\pi/4)\cos\theta) + 1/\sqrt{2}\right)/\sin\theta\ \right|$. **(b)** $|f_e(\theta)| = |(\cos((\pi)\cos\theta) + 1)/\sin\theta|$.
(c) $|f_e(\theta)| = |\cos((3\pi/2)\cos\theta)/\sin\theta|$. **(d)** $|f_e(\theta)| = |(\cos((2\pi)\cos\theta) - 1)/\sin\theta|$. **(e)** $|f_e(\theta)| = |\cos((5\pi/2)\cos\theta)/\sin\theta|$.

19. $R_{rad} = 105.48\ \Omega$, $D(\theta) = 1.1376(\cos((1.5\pi)\cos\theta))^2$, $P_{rad} = 2.1$ W.

20. **(a)** $R_{rad} = 111\ \Omega$. **(b)** $R_{rad} = 108.162\ \Omega$. **(c)** $D(\theta) \approx 1.08(\cos((1.25\pi)\cos\theta + 0.707))^2$, $D(\theta) = 1.1094(\cos((1.25\pi)\cos\theta + 0.707))^2$.

21. **(a)** 1.5 m. **(b)** $R_{rad} = 99.54\ \Omega$. **(c)** $49.77 I_0^2$ [W]. **(d)** $D = 2.41$.

22. **(a)** $I = 1.088$ A. **(b)** 1.04 km.

23. **(b)** 99.46% and 99.94%. **(c)** 0.995.

24. **(a)** 4.39 mΩ. **(b)** $0.002195 I_0^2$ [W]. **(c)** $D_0 = 1.5$.

25. **(a)** 0.75 m. **(b)** 36.54 Ω. **(c)** $18.27 I_0^2$ [W]. **(d)** 1.642.

26. **(a)** $\mathbf{E} = \hat{\boldsymbol{\theta}}\dfrac{\beta\eta I_0 d e^{-j\beta R}}{2\pi R}\sin\theta\sin(\beta h\sin\theta\cos\phi)$ [V/m], $\mathbf{H} = \hat{\boldsymbol{\phi}}\dfrac{\beta I_0 d e^{-j\beta R}}{2\pi R}\sin\theta\sin(\beta h\sin\theta\cos\phi)$ [A/m].
(b) $|f_e(\theta,\phi)| = |\sin\theta\sin(\beta h\sin\theta\cos\phi)|$, $0 \le \theta \le \pi/2$, $0 \le \phi \le 2\pi$.

27. **(a)** $|f_e(\theta)| = |(\cos(2\pi\cos\theta) - 1)/\sin\theta|$. **(b)** 64.9 W. **(c)** $|f_e(\theta)| = |(\cos(3\pi\cos\theta) + 1)/\sin\theta|$, 73.9 W.

28. **(a)** $|f_a(\theta,\phi,\varphi)| = |\cos((\pi/2)\cos\theta)/\sin\theta|$. **(b)** $|f_a(\theta,\phi,\varphi)| = |\cos((\pi/2)\cos\theta)\cos((\pi/2)\sin\theta\cos\phi)/\sin\theta|$.
(c) $|f_a(\theta,\phi,\varphi)| = |-\cos((\pi/2)\cos\theta)\sin((\pi/2)\sin\theta\cos\phi)/\sin\theta|$. **(d)** $|f_a(\theta,\phi,\varphi)| = |\cos((\pi/2)\cos\theta)\cos((\pi/2)\sin\theta\cos\phi + \pi/4)/\sin\theta|$.
(e) $|f_a(\theta,\phi,\varphi)| = |\cos((\pi/2)\cos\theta)\cos((\pi/2)\sin\theta\cos\phi) + \pi/8)/\sin\theta|$. **(f)** $|f_a(\theta,\phi,\varphi)| = |\cos((\pi/2)\cos\theta)\cos(\pi\sin\theta\cos\phi)/\sin\theta|$.
(g) $|f_a(\theta,\phi,\varphi)| = |-\cos((\pi/2)\cos\theta)\sin((\pi\sin\theta\cos\phi)/\sin\theta|$. **(h)** $|f_a(\theta,\phi,\varphi)| = |\cos((\pi/2)\cos\theta)\cos(\pi\sin\theta\cos\phi + \pi/4)/\sin\theta|$.
(i) $|f_a(\theta,\phi,\varphi)| = |\cos((\pi/2)\cos\theta)\cos(\pi\sin\theta\cos\phi + \pi/8)/\sin\theta|$. **(j)** $|f_a(\theta,\phi,\varphi)| = |\cos((\pi/2)\cos\theta)\cos((3\pi/2)\sin\theta\cos\phi)/\sin\theta|$.
(k) $|f_a(\theta,\phi,\varphi)| = |-\cos((\pi/2)\cos\theta)\sin((3\pi/2)\sin\theta\cos\phi)/\sin\theta|$. **(l)** $|f_a(\theta,\phi,\varphi)| = |\cos((\pi/2)\cos\theta)\cos((3\pi/2)\sin\theta\cos\phi + \pi/4)/\sin\theta|$. **(m)** $|f_a(\theta, \phi,\varphi)| = |\cos((\pi/2)\cos\theta)\cos((3\pi/2)\sin\theta\cos\phi + \pi/8)/\sin\theta|$.

29. **(a)** $|f_a(\theta,\phi,\varphi)| = \left|\dfrac{\cos((\pi/2)\cos\theta)}{\sin\theta}\cos(10\pi\sin\theta\cos\phi)\right|$. **(b)** $|f_a(\theta,\phi,\varphi)| = \left|\dfrac{\cos((\pi/2)\cos\theta)}{\sin\theta}\cos\left(10\pi\sin\theta\cos\phi + \dfrac{\pi}{4}\right)\right|$.

30. **(a)** $\mathbf{E} = \left(\hat{\boldsymbol{\theta}}j0.9425 + \hat{\boldsymbol{\phi}}0.233\right)\dfrac{e^{-j2.094R}}{R}\sin\theta$ [V/m], $\mathbf{H} = \left(\hat{\boldsymbol{\theta}}j2.5 \times 10^{-3} - \hat{\boldsymbol{\phi}}6.168 \times 10^{-4}\right)\dfrac{e^{-j2.094R}}{R}\sin\theta$ [A/m].
(b) $\mathbf{E} = \left(\hat{\boldsymbol{\theta}}j0.9425 - \hat{\boldsymbol{\phi}}0.233\right)\dfrac{e^{-j2.094R}}{R}\sin\theta$ [V/m], $\mathbf{H} = \left(\hat{\boldsymbol{\theta}}j2.5 \times 10^{-3} + \hat{\boldsymbol{\phi}}6.168 \times 10^{-4}\right)\dfrac{e^{-j2.094R}}{R}\sin\theta$ [A/m].

32. **(a)** $|f_e(\theta,\phi,\varphi)| = \dfrac{1}{6}\left|\dfrac{\sin(3\pi\sin\theta\cos\phi)}{\sin((\pi/2)\sin\theta\cos\phi)}\right|\left|\dfrac{\cos((\pi/2)\cos\theta)}{\sin\theta}\right|$. **(b)** $|f_p(\theta,\phi,\varphi)| = \left(\dfrac{1}{6}\left|\dfrac{\sin(3\pi\sin\theta\cos\phi)}{\sin((\pi/2)\sin\theta\cos\phi)}\right|\left|\dfrac{\cos((\pi/2)\cos\theta)}{\sin\theta}\right|\right)^2$.

(c) $\theta = 90^\circ$. **(d)** $\phi = 60^\circ, \phi = 90^\circ, \phi = 120^\circ, \phi = 240^\circ, \phi = 270^\circ$ and $\phi = 300^\circ$.

33. **(a)** $|f_a(\theta,\phi,\varphi)| = \frac{1}{5}\left|\frac{\cos((\pi/2)\cos\theta)}{\sin\theta}\right|\left|\frac{\sin(3\pi(\sin\theta\cos\phi - 1)/2)}{\sin(0.6\pi(\sin\theta\cos\phi - 1)/2)}\right|$.

(b) $|f_a(\theta,\phi,\varphi)| = \frac{1}{6}\left|\frac{\cos((\pi/2)\cos\theta)}{\sin\theta}\right|\left|\frac{\sin(6\pi(0.6\sin\theta\cos\phi - 1)/2)}{\sin(\pi(0.6\sin\theta\cos\phi - 1)/2)}\right|$.

(c) $|f_a(\theta,\phi,\varphi)| = \frac{1}{10}\left|\frac{\cos((\pi/2)\cos\theta)}{\sin\theta}\right|\left|\frac{\sin(10\pi(0.7\sin\theta\cos\phi - 0.75)/2)}{\sin(\pi(0.7\sin\theta\cos\phi - 0.75)/2)}\right|$.

(d) $|f_a(\theta,\phi,\varphi)| = \frac{1}{5}\left|\frac{\cos((\pi/2)\cos\theta)}{\sin\theta}\right|\left|\frac{\sin(5\pi(0.7\sin\theta\cos\phi - 1)/2)}{\sin(\pi(0.7\sin\theta\cos\phi - 1)/2)}\right|$.

34. $|f_a(\theta,\phi,0)| = \frac{1}{5}|\sin\theta|\left|\frac{\sin((5\pi/4)\sin\theta\cos\phi)}{\sin((\pi/4)\sin\theta\cos\phi)}\right|$.

35. **(a)** $|f_a(\theta,\phi)| = \left|\frac{\cos((\pi/2)\cos\theta)}{\sin\theta}\right|\left|\cos^2\left(\frac{\pi\sin\theta\cos\phi}{4}\right)\right|$.

36. **(a)** $|f_a(\theta,\phi)| = \left|\frac{\cos((\pi/2)\cos\theta)}{\sin\theta}\right|\left|\frac{\sin((n\pi/2)\sin\theta\cos\phi)}{n\sin((\pi/2)\sin\theta\cos\phi)}\right|, 0 \le \theta \le \pi/2$.

(b) $\mathbf{E} = \hat{\boldsymbol{\theta}}\frac{j\eta I_0}{2\pi R}e^{-j\beta R}e^{j(n-1)(\pi/2)\sin\theta\cos\phi}\frac{\cos((\pi/2)\cos\theta)}{\sin\theta}\frac{\sin((n\pi/2)\sin\theta\cos\phi)}{\sin((\pi/2)\sin\theta\cos\phi)}\left[\frac{\text{V}}{\text{m}}\right], \mathbf{H} = \hat{\boldsymbol{\phi}}\frac{E_\theta}{\eta}\left[\frac{\text{A}}{\text{m}}\right], 0 \le \theta \le \pi/2$.

37. **(a)** $\mathbf{E} = \hat{\boldsymbol{\phi}}\frac{\omega\mu\beta I\pi d^2/4}{4\pi R}e^{-j\beta R}\sin\theta e^{j(n-1)\psi/2}\frac{\sin(n\psi/2)}{\sin(\psi/2)}\left[\frac{\text{V}}{\text{m}}\right], \mathbf{H} = -\hat{\boldsymbol{\theta}}\frac{E_\phi}{\eta}\left[\frac{\text{A}}{\text{m}}\right], \psi = 0.08\,\pi\sin\theta\cos\phi + \varphi$

(b) $AF(\theta,\phi,\varphi) = \frac{\sin(0.04n\pi\sin\theta\cos\phi)}{\sin(0.04\pi\sin\theta\cos\phi)}, |f_a(\theta,\phi,\varphi)| = \frac{1}{n}\left|\frac{\sin(0.04n\pi\sin\theta\cos\phi)}{\sin(0.04\pi\sin\theta\cos\phi)}\right|$.

38. **(a)** $\mathbf{E} = \hat{\boldsymbol{\theta}}\frac{j\eta I_0}{2\pi}\frac{\cos((\pi/2)\cos\theta)}{\sin\theta}\frac{e^{-j\beta R}}{R}(1 + 2\cos(2\pi\cos\theta) + 2\cos(4\pi\cos\theta))\left[\frac{\text{V}}{\text{m}}\right], \mathbf{H} = \hat{\boldsymbol{\phi}}\frac{E_\theta}{\eta}\left[\frac{\text{A}}{\text{m}}\right]$.

(b) $\mathcal{P}_{av} = \hat{\mathbf{R}}\frac{\eta I_0^2}{8\pi^2R^2}\left(\frac{\cos((\pi/2)\cos\theta)}{\sin\theta}\right)^2[1 + 2\cos(2\pi\cos\theta) + 2\cos(4\pi\cos\theta)]^2\left[\frac{\text{W}}{\text{m}^2}\right]$.

39. **(a)** $AF = 2[(\cos(\pi\sin\theta\sin\phi) - \cos(\pi\sin\theta\cos\phi))]$.

(b) $\mathcal{P}_{av} = \hat{\mathbf{R}}\frac{\beta^2\eta I_0^2(\Delta l)^2}{8\pi^2R^2}\sin^2\theta[\cos(\pi\sin\theta\sin\phi) - \cos(\pi\sin\theta\cos\phi)]^2$ [W/m²].

40. $\mathcal{P}_{av} = \hat{\mathbf{R}}\frac{I_0^2\eta}{2\pi^2R^2}\frac{\cos^2((\pi/2)\cos\theta)}{\sin^2\theta}(2 + \cos(\pi\sin\theta\cos\phi) - \cos((\pi/2)\sin\theta\cos\phi) - \cos((3\pi/2)\sin\theta\cos\phi))^2\ [\text{W/m}^2]$.

41. **(a)** $\mathcal{P}_{av} = \hat{\mathbf{R}}1.762 \times 10^{-7}\sin^2\theta[\cos(2\pi\sin\theta\cos\phi) + \cos(3\pi\sin\theta\cos\phi) + \cos(4\pi\sin\theta\cos\phi)]^2$ [W/m²]. **(b)** 11.378°.

42. At 520 kHz: $A_e = 39{,}726.7\sin^2\theta$ [m²], at 1.6 MHz: $A_e = 4{,}196.5\sin^2\theta$ [m²].

43. **(a)** $P_{recieved|max} = \frac{3c^2P}{32\pi^2f^2R^2}$ [W]. **(b)** $I_{peak} = \frac{3c^3}{8f^3Ra^2\pi^3}\sqrt{\frac{P}{2\pi\eta_0}}$ [A].

44. **(a)** $I = 0.497$ mA. **(b)** $I = 16.33$ μA.

45. **(a)** 125.66. **(b)** 10 m².

46. **(a)** 12.8 μA. **(b)** 32.44 μA (antennas are tilted 45° to the left of the normal).

47. 7,955 m.

48. **(a)** 4,733 m **(b)** $4{,}733\frac{\cos((\pi/2)\cos\theta)}{\sin\theta}$ [m].

49. $\sigma = 198.44$ m².

50. $P_j = \frac{P_0\sigma G_0}{4\pi R^2G_a}$ [W].

Appendix: Summary of Vector Relations and Physical Constants

Gradient, Divergence, Curl, and the Laplacian in Various Coordinates

Cartesian coordinates:

$$\nabla U = \hat{\mathbf{x}}\frac{\partial U}{\partial x} + \hat{\mathbf{y}}\frac{\partial U}{\partial y} + \hat{\mathbf{z}}\frac{\partial U}{\partial z},$$

$$\nabla \cdot \mathbf{A} = \frac{\partial A_x}{\partial x} + \frac{\partial A_y}{\partial y} + \frac{\partial A_z}{\partial z},$$

$$\nabla \times \mathbf{A} = \hat{\mathbf{x}}\left(\frac{\partial A_z}{\partial y} - \frac{\partial A_y}{\partial z}\right) + \hat{\mathbf{y}}\left(\frac{\partial A_x}{\partial z} - \frac{\partial A_z}{\partial x}\right) + \hat{\mathbf{z}}\left(\frac{\partial A_y}{\partial x} - \frac{\partial A_x}{\partial y}\right),$$

$$\nabla^2 U = \frac{\partial^2 U}{\partial x^2} + \frac{\partial^2 U}{\partial y^2} + \frac{\partial^2 U}{\partial z^2}$$

Cylindrical coordinates:

$$\nabla U = \hat{\mathbf{r}}\frac{\partial U}{\partial r} + \hat{\boldsymbol{\phi}}\frac{1}{r}\frac{\partial U}{\partial \phi} + \hat{\mathbf{z}}\frac{\partial U}{\partial z},$$

$$\nabla \cdot \mathbf{A} = \frac{1}{r}\frac{\partial (rA_r)}{\partial r} + \frac{1}{r}\frac{\partial A_\phi}{\partial \phi} + \frac{\partial A_z}{\partial z},$$

$$\nabla \times \mathbf{A} = \hat{\mathbf{r}}\left(\frac{1}{r}\frac{\partial A_z}{\partial \phi} - \frac{\partial A_\phi}{\partial z}\right) + \hat{\boldsymbol{\phi}}\left(\frac{\partial A_r}{\partial z} - \frac{\partial A_z}{\partial r}\right) + \hat{\mathbf{z}}\frac{1}{r}\left(\frac{\partial (rA_\phi)}{\partial r} - \frac{\partial A_r}{\partial \phi}\right),$$

$$\nabla^2 U = \frac{\partial^2 U}{\partial r^2} + \frac{1}{r}\frac{\partial U}{\partial r} + \frac{1}{r^2}\frac{\partial^2 U}{\partial \phi^2} + \frac{\partial^2 U}{\partial z^2}$$

Spherical coordinates:

$$\nabla U = \hat{\mathbf{R}}\frac{\partial U}{\partial R} + \hat{\boldsymbol{\theta}}\frac{1}{R}\frac{\partial U}{\partial \theta} + \hat{\boldsymbol{\phi}}\frac{1}{R\sin\theta}\frac{\partial U}{\partial \phi},$$

$$\nabla \cdot \mathbf{A} = \frac{1}{R^2}\frac{\partial}{\partial R}\left(R^2 A_R\right) + \frac{1}{R\sin\theta}\frac{\partial}{\partial \theta}\left(A_\theta \sin\theta\right) + \frac{1}{R\sin\theta}\frac{\partial A_\phi}{\partial \phi},$$

$$\nabla \times \mathbf{A} = \hat{\mathbf{R}}\frac{1}{R\sin\theta}\left(\frac{\partial (A_\phi \sin\theta)}{\partial \theta} - \frac{\partial A_\theta}{\partial \phi}\right) + \hat{\boldsymbol{\theta}}\frac{1}{R}\left(\frac{1}{\sin\theta}\frac{\partial A_R}{\partial \phi} - \frac{\partial (RA_\phi)}{\partial R}\right) + \hat{\boldsymbol{\phi}}\frac{1}{R}\left(\frac{\partial (RA_\theta)}{\partial R} - \frac{\partial A_R}{\partial \theta}\right),$$

$$\nabla^2 U = \frac{1}{R^2}\frac{\partial}{\partial R}\left(R^2\frac{\partial U}{\partial R}\right) + \frac{1}{R^2\sin\theta}\frac{\partial}{\partial \theta}\left(\sin\theta\frac{\partial U}{\partial \theta}\right) + \frac{1}{R^2\sin^2\theta}\frac{\partial^2 U}{\partial \phi^2}$$

N. Ida, *Engineering Electromagnetics*, https://doi.org/10.1007/978-3-030-15557-5

The divergence theorem:

$$\int_v (\nabla \cdot \mathbf{A}) dv = \oint_s \mathbf{A} \cdot d\mathbf{s}$$

Stokes' theorem:

$$\int_s (\nabla \times \mathbf{A}) \cdot d\mathbf{s} = \oint_L \mathbf{A} \cdot d\mathbf{l}$$

Some useful vector identities:

$$\begin{aligned}
\nabla \times (\nabla U) &= 0, \\
\nabla \cdot (\nabla \times \mathbf{A}) &= 0, \\
\nabla^2 \mathbf{A} &= \nabla(\nabla \cdot \mathbf{A}) - \nabla \times (\nabla \times \mathbf{A}), \\
\nabla(UQ) &= U(\nabla Q) + Q(\nabla U), \\
\nabla \cdot (U\mathbf{A}) &= U(\nabla \cdot \mathbf{A}) + (\nabla U) \cdot \mathbf{A}, \\
\nabla \cdot (\mathbf{A} \times \mathbf{B}) &= -\mathbf{A} \cdot (\nabla \times \mathbf{B}) + (\nabla \times \mathbf{A}) \cdot \mathbf{B}, \\
\nabla \times (U\mathbf{A}) &= U(\nabla \times \mathbf{A}) + (\nabla U) \times \mathbf{A}, \\
\nabla \cdot \nabla U &= \nabla^2 U, \\
\nabla^2 \mathbf{A} &= \hat{\mathbf{x}} \nabla^2 A_x + \hat{\mathbf{y}} \nabla^2 A_y + \hat{\mathbf{z}} \nabla^2 A_z
\end{aligned}$$

Some physical constants:

Charge of the electron (q_e)	$-1.602129 \times 10^{-19}$ C
Rest mass of the electron (m_e)	$9.1093897 \times 10^{-31}$ kg
Speed of light in vacuum (c)	2.997992×10^{8} m/s
Permittivity of free space (ε_0)	8.854187×10^{-12} F/m
Permeability of free space (μ_0)	$4\pi \times 10^{-7}$ H/m
Planck's constant (h)	6.62620×10^{-34} J $\cdot$ s
Intrinsic impedance in free space (η_0)	376.7304 Ω

Index

N. Ida, *Engineering Electromagnetics*, https://doi.org/10.1007/978-3-030-15557-5

N

Printed in the United States
by Baker & Taylor Publisher Services